Thin Layer Chromatography in Drug Analysis

CHROMATOGRAPHIC SCIENCE SERIES

A Series of Textbooks and Reference Books

Editor:
Nelu Grinberg

Founding Editor:
Jack Cazes

1. Dynamics of Chromatography: Principles and Theory, J. Calvin Giddings
2. Gas Chromatographic Analysis of Drugs and Pesticides, Benjamin J. Gudzinowicz
3. Principles of Adsorption Chromatography: The Separation of Nonionic Organic Compounds, Lloyd R. Snyder
4. Multicomponent Chromatography: Theory of Interference, Friedrich Helfferich and Gerhard Klein
5. Quantitative Analysis by Gas Chromatography, Josef Novák
6. High-Speed Liquid Chromatography, Peter M. Rajcsanyi and Elisabeth Rajcsanyi
7. Fundamentals of Integrated GC-MS (in three parts), Benjamin J. Gudzinowicz, Michael J. Gudzinowicz, and Horace F. Martin
8. Liquid Chromatography of Polymers and Related Materials, Jack Cazes
9. GLC and HPLC Determination of Therapeutic Agents (in three parts), Part 1 edited by Kiyoshi Tsuji and Walter Morozowich, Parts 2 and 3 edited by Kiyoshi Tsuji
10. Biological/Biomedical Applications of Liquid Chromatography, edited by Gerald L. Hawk
11. Chromatography in Petroleum Analysis, edited by Klaus H. Altgelt and T. H. Gouw
12. Biological/Biomedical Applications of Liquid Chromatography II, edited by Gerald L. Hawk
13. Liquid Chromatography of Polymers and Related Materials II, edited by Jack Cazes and Xavier Delamare
14. Introduction to Analytical Gas Chromatography: History, Principles, and Practice, John A. Perry
15. Applications of Glass Capillary Gas Chromatography, edited by Walter G. Jennings
16. Steroid Analysis by HPLC: Recent Applications, edited by Marie P. Kautsky
17. Thin-Layer Chromatography: Techniques and Applications, Bernard Fried and Joseph Sherma
18. Biological/Biomedical Applications of Liquid Chromatography III, edited by Gerald L. Hawk
19. Liquid Chromatography of Polymers and Related Materials III, edited by Jack Cazes

Thin Layer Chromatography in Drug Analysis

Edited by
Łukasz Komsta
Monika Waksmundzka-Hajnos
Joseph Sherma

CRC Press
Taylor & Francis Group
Boca Raton London New York

CRC Press is an imprint of the
Taylor & Francis Group, an **informa** business

CRC Press
Taylor & Francis Group
6000 Broken Sound Parkway NW, Suite 300
Boca Raton, FL 33487-2742

First issued in paperback 2020

ISBN 13: 978-0-367-57622-6 (pbk)
ISBN 13: 978-1-4665-0715-9 (hbk)

Library of Congress Cataloging-in-Publication Data

Thin layer chromatography in drug analysis / editors, Lukasz Komsta, Monika Waksmundzka-Hajnos, Joseph Sherma.
    p. ; cm. -- (Chromatographic science series ; 106)
    Includes bibliographical references and index.
    ISBN 978-1-4665-0715-9 (hbk. : alk. paper)
    I. Komsta, Lukasz, editor of compilation. II. Waksmundzka-Hajnos, Monika, editor of compilation. III. Sherma, Joseph, editor of compilation. IV. Series: Chromatographic science ; v. 106. 0069-3936
    [DNLM: 1. Chromatography, Thin Layer--methods--Laboratory Manuals. 2. Pharmaceutical Preparations--analysis--Laboratory Manuals. 3. Drug Evaluation--methods--Laboratory Manuals. 4. Pharmaceutical Preparations--chemistry--Laboratory Manuals. W1 CH943 v.106 2014 / QV 25]

RS420
615.1'9--dc23   2013038207

Visit the Taylor & Francis Web site at
http://www.taylorandfrancis.com

and the CRC Press Web site at
http://www.crcpress.com

This book is dedicated to the memory of Jack Cazes, the founding editor of the Chromatographic Science Series, who passed away on February 16, 2010. In addition to his excellent research and teaching accomplishments in gel permeation (size exclusion) chromatography, high-performance liquid chromatography, countercurrent chromatography (centrifugal partition chromatography), and analytical instrumentation, Jack was undoubtedly one of the most proficient and prolific journal and book editors of all time, with vast expertise in all areas of the separation sciences. For Marcel Dekker Inc. and then CRC Press/Taylor & Francis Group, Jack not only edited and authored books in the Chromatographic Science Series, he also edited the Journal of Liquid Chromatography *(which was renamed the* Journal of Liquid Chromatography & Related Technologies *in 1996), the* Encyclopedia of Chromatography *(three editions), the* Journal of Immunoassay and Immunochemistry, *and the journals* Instrumentation Science & Technology *and* Preparative Biochemistry and Biotechnology. *He also coedited the series Advances in Chromatography with Eli Grushka and Phyllis Brown and wrote Volume I of* Liquid Chromatography of Polymers and Related Materials *as well as edited two other volumes within the Chromatographic Science Series.*

One of the editors of this book (Joseph Sherma) met Jack while on leave from Lafayette College and working at Waters Associates (now Waters Corp.) in the summer of 1977. With Jack's strong and continual encouragement, advice, and support, he had the honor to coauthor or coedit 11 volumes in his Chromatographic Science Series in the period 1982–2011 (Volumes 17, 35, 55, 66, 71, 81, 89, 95, 98, 99, and 102), which had a profound effect on the development of his academic career. Another of this book's editors (Monika Waksmundzka-Hajnos) has coedited two volumes in his series (Volumes 99 and 102).

This tribute by the editors is a heartfelt gesture of appreciation for Jack's many great contributions that have immeasurably enhanced the field of chromatography for us and a vast number of others in the chromatography community throughout the world.

Łukasz Komsta

Monika Waksmundzka-Hajnos

Joseph Sherma

Contents

PART I Theory of Thin Layer Chromatography in Context of Pharmaceutical Analysis

PART II   Planar Chromatography of Particular Drug Groups

Preface

Although in the modern world a drug is usually seen by consumers as simply another commodity, it cannot be fully perceived in this way for several reasons. The main reason is that medical patients, who are the drug consumers, cannot evaluate product quality on their own. This is in contrast to other commodities such as food, for which everyone is able to detect aspects of poor quality such as decomposition or unpleasant taste. In the case of pharmaceutical formulations, the patient is forced to trust manufacturing companies and cannot be held responsible for using bad quality medicine that reaches the market. Therefore, legal regulations on drug quality are very restrictive around the world, and routine checking of drug quality is one of the most important parts of the production cycle.

Many analytical methods can be used in drug quality analysis, including separation techniques such as high-performance column liquid chromatography (HPLC), ultraperformance column liquid chromatography, capillary electrophoresis, and thin layer chromatography (TLC). Although the current trend in official (pharmacopoeial) methods is to rely mostly on HPLC methods with new techniques of detection (mass spectrometry or, at least, diode array), TLC methods still have an important role in drug control. TLC finds broad application in drug quality control, and many analyses are better carried out by using TLC. TLC methods can be seen as alternatives, providing comparative or only slightly poorer results than column HPLC and requiring less complicated or expensive equipment. TLC can be performed under field conditions using small amounts of solvents. Quantitative evaluation can be performed with a densitometer or, as a recent trend, by proper image processing of a videoscan. The main interests in the application of TLC in drug control are noted in resource-limited countries, where a significant number of new applications have been observed. In the case of applications not involving quantitative analysis (e.g., drug identification), TLC can significantly outperform the other methods due to its low cost and simplicity.

The only published book that is known to us to provide extensive coverage of pharmaceutical applications of TLC was written in 1972 by K. Macek, but this book also included, to a large degree, information on the now outdated method of paper chromatography. Although some books (including previous ones in the CRC Press/Taylor & Francis Group Chromatographic Science Series) touch on the subject while also presenting information on many other techniques of drug analysis, there is no up-to-date, complete reference book dedicated to the techniques and applications of pharmaceutical analysis by TLC. This led us to undertake this book project, which is designed to fill the gap present in the market for 40 years.

The aim of this book is to provide up-to-date information on the most important methods in pharmaceutical applications of TLC, that is, analysis of bulk drug material and pharmaceutical formulations, degradation studies, analysis of biological samples, optimization of the separation of drug classes, and lipophilicity estimation. The book is divided into two major parts. Part I is devoted to the general topics related to TLC in the context of drug analysis, while Part II provides a comprehensive overview of a wide spectrum of applications of TLC to separation and analysis of particular drug groups, based on the current literature. Each of the chapters contains an introduction about the structures and medicinal actions of the described substances, and a complete literature review of their TLC analysis.

The book can be treated as a manual, reference book, or teaching source and will hopefully be useful to chromatographers, pharmacists, students, and analytical chemists. It should also prove useful in research and development laboratories in the pharmaceutical industry, as well as in clinical, medical, and forensic laboratories.

We would like to thank Barbara Glunn, senior editor—chemistry, CRC Press/Taylor & Francis Group, for her support of our book proposal and during all aspects of our subsequent editorial work. We also thank the chapter authors for their exceptionally valuable contributions.

Editors

Łukasz Komsta currently serves as assistant professor in the Department of Medicinal Chemistry, Faculty of Pharmacy, Medical University of Lublin, Lublin, Poland. His research interests include the theory and application of liquid chromatography, especially in connection with computational chemistry and chemometrics, as well as the application of chromatographic methods, especially thin layer chromatography (TLC), in the analysis of drugs. To further his chemometric interests, he had three short stints in the Department of Chemometrics of Silesian University, Poland, as a member of the team of Prof. Beata Walczak.

Dr. Komsta is the author or coauthor of more than 60 scientific papers and about 50 scientific conference papers. He was invited as a guest editor of a special section on "Chemometrics in Pharmaceutical Analysis" published in the *Journal of AOAC International* in 2012. Since 2011, he has been a member of the editorial board of the journal *Acta Chromatographica*. He also serves as section editor for the Polish scientific journal *Current Issues in Pharmacy and Medical Sciences*.

Dr. Komsta received an award in 2004 from the Ministry of Health in Poland and received several awards from the rector of the Medical University of Lublin. He has reviewed about 200 papers for journals such as *Acta Chromatographica*, *Analytica Chimica Acta*, the *Journal of AOAC International*, *Central European Journal of Chemistry*, *Chemometrics and Intelligent Laboratory Systems*, the *Journal of Chromatography A and B*, the *Journal of Chromatographic Science*, the *Journal of Liquid Chromatography & Related Technologies*, the *Journal of Pharmaceutical and Biomedical Analysis*, the *Journal of Planar Chromarography—Modern TLC*, the *Journal of Pharmacy and Pharmacology*, and the *Journal of Separation Science*. In 2011, he was awarded the Silver Cross of Merit by the president of the Republic of Poland for his scientific achievements.

Dr. Komsta teaches courses in medicinal chemistry and introductory courses on chemometrics for pharmacy and medical chemistry students. He has also taught courses on pharmaceutical terminology for students of medicine. Over the past ten years, Dr. Komsta has directed the research programs of more than ten MSc pharmacy students. He currently supervises the research of two PhD students and is head of the students' research group in his department.

Monika Waksmundzka-Hajnos graduated in chemistry and received her PhD in analytical chemistry from the Faculty of Chemistry of Maria Curie-Skłodowska University in Lublin, Poland, in 1980. She currently serves as full professor of pharmacy and head of the Department of Inorganic Chemistry at the Faculty of Pharmacy of the Medical University of Lublin. Her research interests include the theory and application of liquid chromatography, taking into consideration the optimization of chromatographic systems for the separation and quantitation of drugs and their degradation products in pharmaceutical preparations and drugs and their metabolites in body fluids, as well as isolation and/or separation of secondary metabolites in extracts of plant tissues. Her scientific work focuses mainly on sample preparation before chromatographic analysis, including extraction from solid plant material to obtain crude plant extracts and purifying them by the classic liquid–liquid extraction (LLE) method or by the solid phase extraction (SPE) method.

Professor Waksmundzka-Hajnos is the author or coauthor of about 150 papers and approximately 300 conference papers. She has published review articles in journals such as the *Journal of Chromatography A*, the *Journal of Chromatography B*, and the *Journal of Liquid Chromatography*. She was the guest editor of a special issue of *Journal AOAC International* and *Medicinal Chemistry*. She is also the author and coauthor of several chapters in known chromatographic textbooks. She has coedited the book titled *Thin Layer Chromatography in Phytochemistry* (published as Vol. 99 in the Chromatographic Science Series by CRC Press/Taylor & Francis Group) with Professor

Kowalska and Professor Sherma. She has also coedited the book titled *HPLC in Phytochemical Analysis* (published as Vol. 102 in the Chromatographic Science Series by CRC Press/Taylor & Francis Group) with Professor Sherma. She has received five awards from the Ministry of Health in Poland and two awards from the Polish Pharmaceutical Society for her scientific achievements.

Dr. Waksmundzka-Hajnos has taught courses in inorganic chemistry to pharmacy and medical chemistry students for more than 35 years. She has also taught courses in instrumental analysis to students of pharmacy. Over the past 20 years, she has directed programs for over 50 MSc pharmacy students involved in the theory and practice of different liquid chromatographic techniques. She has also supervised six PhD students researching separation science.

Since 2008, Dr. Waksmundzka-Hajnos has served as the editor of *Acta Chromatographica*, the quarterly journal from the University of Silesia, Katowice (published by Akademiai Kiado), and has been a member of the editorial board of the *Journal of Planar Chromatography*. Since 2011, she has been a member of the editorial board of *The Scientific World Journal—Analytical Chemistry*.

Joseph Sherma received his BS degree in Chemistry from Upsala College, East Orange, NJ, in 1955 and a PhD degree in analytical chemistry from Rutgers, the State University, New Brunswick, NJ, in 1958 under the supervision of the renowned ion-exchange chromatography expert Wm. Rieman III. Professor Sherma is currently John D. and Francis H. Larkin Professor Emeritus of Chemistry at Lafayette College, Easton, Pennsylvania; he taught courses in analytical chemistry for more than 40 years, was head of the Chemistry Department for 12 years, and continues to supervise research students at Lafayette. During sabbatical leaves and summers, Professor Sherma did research in the laboratories of the eminent chromatographers Dr. Harold Strain, Dr. Gunter Zweig, Professor James Fritz, and Professor Joseph Touchstone.

Professor Sherma has authored, coauthored, edited, or coedited more than 750 publications, including research papers and review articles in approximately 55 different peer reviewed analytical chemistry, chromatography, and biological journals; approximately 30 invited book chapters; and more than 60 books and U.S. government agency manuals in the areas of analytical chemistry and chromatography.

In addition to his research in the techniques and applications of thin layer chromatography (TLC), Professor Sherma has a very productive interdisciplinary research program in the use of analytical chemistry to study biological systems with Bernard Fried, Kreider Professor Emeritus of Biology at Lafayette College, with whom he has written the book *Thin Layer Chromatography* (1st–4th editions) and edited the *Handbook of Thin Layer Chromatography* (1st–3rd editions), both published by Marcel Dekker, Inc., as well as editing *Practical Thin Layer Chromatography* for CRC Press. With Dr. Zweig, Professor Sherma wrote a book titled *Paper Chromatography* for Academic Press and the first two volumes of the *Handbook of Chromatography* series for CRC Press, and coedited 22 more volumes of the chromatography series and 10 volumes of the series *Analytical Methods for Pesticides and Plant Growth Regulators* for Academic Press. After Dr. Zweig's death, Professor Sherma edited five additional volumes of the chromatography handbook series and two volumes in the pesticide series. The pesticide series was completed under the title *Modern Methods of Pesticide Analysis* for CRC Press, with two volumes coedited with Dr. Thomas Cairns. Three books on quantitative TLC and advances in TLC were edited jointly with Professor Touchstone for Wiley-Interscience. Professor Sherma coedited with Professor Teresa Kowalska, *Preparative Layer Chromatography and Thin Layer Chromatography in Chiral Separations and Analysis,* coedited with Professor Kowalska and Professor Monika Waksmundska-Hajnos *Thin Layer Chromatography in Phytochemistry*, and coedited with Professor Waksmundska-Hajnos *High Performance Liquid Chromatography in Phytochemical Analysis* for the CRC/Taylor & Francis Group.

Professor Sherma served for 23 years as editor for residues and trace elements of the *Journal of AOAC International* and is currently that journal's Acquisitions Editor. He has guest edited with Professor Fried 13 annual special issues on TLC of the *Journal of Liquid Chromatography and Related Technologies* and regularly guest edits special sections of issues of the *Journal of*

AOAC International on specific subjects in all areas of analytical chemistry. For 12 years, he also wrote an article on modern analytical instrumentation for each issue of the *Journal of AOAC International*. Professor Sherma has written biennial reviews of planar chromatography for the American Chemical Society journal *Analytical Chemistry* since 1970. He is now on the editorial boards of the *Journal of Planar Chromatography-Modern TLC; Acta Chromatographica; Journal of Environmental Science and Health, Part B; and Journal of Liquid Chromatography & Related Technologies.*

Professor Sherma was recipient of the 1995 ACS Award for Research at an undergraduate institution sponsored by Research Corporation. The first 2009 issue, Volume 12, of the journal *Acta Universitatis Cibiensis, Seria F, Chemia* was dedicated in honor of Professor Sherma's teaching, research, and publication accomplishments in analytical chemistry and chromatography.

Contributors

Danica Agbaba
Faculty of Pharmacy
Department of Pharmaceutical
    Chemistry
University of Belgrade
Belgrade, Serbia

Anna Apola
Faculty of Pharmacy
Department of Inorganic and Analytical
    Chemistry
Jagiellonian University
Kraków, Poland

Rada M. Baošić
Faculty of Chemistry
Department of Analytical Chemistry
University of Belgrade
Belgrade, Serbia

Anna Berecka
Faculty of Pharmacy
Department of Medicinal Chemistry
Medical University of Lublin
Lublin, Poland

Ravi Bhushan
Department of Chemistry
Indian Institute of Technology
Roorkee, India

Mirza Bojić
Faculty of Pharmacy and Biochemistry
Department of Medicinal Chemistry
University of Zagreb
Zagreb, Croatia

Irena Choma
Faculty of Chemistry
Department of Chromatographic
    Methods
University of Maria Curie-Skłodowska
Lublin, Poland

Łukasz Cieśla
Faculty of Pharmacy
Department of Inorganic Pharmacy
Medical University of Lublin
Lublin, Poland

and

Department of Biochemistry and Crop Quality
Institute of Soil Science and Plant
    Cultivation
State Research Institute
Puławy, Poland

Claudia Cimpoiu
Faculty of Chemistry and Chemical
    Engineering
Department of Analytical Chemistry
University of Cluj-Napoca
and
Faculty of Chemistry and Chemical
    Engineering
Babes-Bolyai University
Cluj-Napoca, Romania

Monika Dąbrowska
Faculty of Pharmacy
Department of Inorganic and Analytical
    Chemistry
Jagiellonian University
Kraków, Poland

Željko Debeljak
Department of Clinical Laboratory
    Diagnostics
Clinical Hospital Center Osijek
Osijek, Croatia

Marta de Diego
Faculty of Pharmacy
University of Concepción
Concepción, Chile

Shuchi Dixit
Department of Chemistry
College of Science
Yeungnam University
Gyeongsan, South Korea

Joanna Drozd
Faculty of Pharmacy
Department of Medicinal Chemistry
Medical University of Lublin
Lublin, Poland

Tadeusz H. Dzido
Faculty of Pharmacy
Department of Physical Chemistry
Medical University of Lublin
Lublin, Poland

Radosław J. Ekiert
Faculty of Pharmacy
Department of Inorganic and Analytical
    Chemistry
Jagiellonian University
Kraków, Poland

Carmen Gloria Godoy
Faculty of Pharmacy
University of Concepción
Concepción, Chile

Anna Gumieniczek
Faculty of Pharmacy
Department of Medicinal Chemistry
Medical University of Lublin
Lublin, Poland

Radosław Gwarda
Faculty of Pharmacy
Department of Physical Chemistry
Medical University of Lublin
Lublin, Poland

Anna Hawrył
Faculty of Pharmacy
Department of Inorganic Chemistry
Medical University of Lublin
Lublin, Poland

Mirosław Hawrył
Faculty of Pharmacy
Department of Inorganic Chemistry
Medical University of Lublin
Lublin, Poland

Anamaria Hosu
Faculty of Chemistry and Chemical
    Engineering
Department of Analytical Chemistry
University of Cluj-Napoca
Cluj-Napoca, Romania

Urszula Hubicka
Faculty of Pharmacy
Department of Inorganic and Analytical
    Chemistry
Jagiellonian University
Kraków, Poland

Tadeusz Inglot
Faculty of Pharmacy
Department of Medicinal Chemistry
Medical University of Lublin
Lublin, Poland

Wioleta Jesionek
Faculty of Chemistry
Department of Chromatographic
    Methods
University of Maria Curie-Skłodowska
Lublin, Poland

Grzegorz Jóźwiak
Faculty of Pharmacy
Department of Inorganic Chemistry
Medical University of Lublin
Lublin, Poland

Eliangiringa Kaale
Department of Medicinal Chemistry
and
Pharm R&D Laboratory
School of Pharmacy
Muhimbili University of Health and Allied
    Sciences
Dar es Salaam, Tanzania

Huba Kalász
Department of Pharmacology and
    Pharmacotherapy
Semmelweis University
Budapest, Hungary

Roman Kaliszan
Faculty of Pharmacy
Department of Biopharmacy and
    Pharmacodynamics
Medical University of Gdańsk
Gdańsk, Poland

Łukasz Komsta
Faculty of Pharmacy
Department of Medicinal Chemistry
Medical University of Lublin
Lublin, Poland

Ewelina Kopciał
Faculty of Pharmacy
Department of Physical Chemistry
Medical University of Lublin
Lublin, Poland

Dorota Kowalczuk
Faculty of Pharmacy
Department of Medicinal Chemistry
Medical University of Lublin
Lublin, Poland

Iwona Kowalska
Department of Biochemistry
Institute of Soil Science and Plant
    Cultivation
State Research Institute
Pulawy, Poland

Teresa Kowalska
Institute of Chemistry
University of Silesia
Katowice, Poland

Jan Krzek
Faculty of Pharmacy
Department of Inorganic and Analytical
    Chemistry
Jagiellonian University
Kraków, Poland

Thomas Layloff
Supply Chain Management System
Arlington, Virginia

Michał J. Markuszewski
Faculty of Pharmacy
Department of Biopharmacy and
    Pharmacodynamics
Medical University of Gdańsk
Gdańsk, Poland

Michał Piotr Marszałł
Faculty of Pharmacy
Department of Medicinal Chemistry
Ludwik Rydygier Collegium Medicum in
    Bydgoszcz
Nicolaus Copernicus University
Toruń, Poland

Anna Maślanka
Faculty of Pharmacy
Department of Inorganic and Analytical
    Chemistry
Jagiellonian University
Kraków, Poland

Marica Medić-Šarić
Faculty of Pharmacy and Biochemistry
Department of Medicinal Chemistry
University of Zagreb
Zagreb, Croatia

Sigrid Mennickent
Faculty of Pharmacy
University of Concepción
Concepción, Chile

Dušanka M. Milojković-Opsenica
Faculty of Chemistry
Department of Analytical Chemistry
University of Belgrade
Belgrade, Serbia

Maja M. Natić
Faculty of Chemistry
Department of Analytical Chemistry
University of Belgrade
Belgrade, Serbia

Katarina Nikolić
Faculty of Pharmacy
University of Belgrade
Belgrade, Serbia

Beata Paw
Faculty of Pharmacy
Department of Medicinal Chemistry
Medical University of Lublin
Lublin, Poland

Anna Petruczynik
Faculty of Pharmacy
Department of Inorganic Chemistry
Medical University of Lublin
Lublin, Poland

Rafał Pietraś
Faculty of Pharmacy
Department of Medicinal Chemistry
Medical University of Lublin
Lublin, Poland

Beata Polak
Faculty of Pharmacy
Department of Physical Chemistry
Medical University of Lublin
Lublin, Poland

Arkadiusz Pomykalski
Faculty of Pharmacy
Department of Medicinal Chemistry
Medical University of Lublin
Lublin, Poland

Gordana Popović
Faculty of Pharmacy
Department of General and Inorganic
   Chemistry
University of Belgrade
Belgrade, Serbia

Alina Pyka
Faculty of Pharmacy
Department of Analytical Chemistry
Medical University of Silesia
Sosnowiec, Poland

Fred Rabel
ChromHELP, LLC
Woodbury, New Jersey

Peter Risha
Pharm R&D Laboratory
School of Pharmacy
Muhimbili University of Health and Allied
   Sciences
Dar es Salaam, Tanzania

Mieczysław Sajewicz
Institute of Chemistry
University of Silesia
Katowice, Poland

Joseph Sherma
Department of Chemistry
Lafayette College
Easton, Pennsylvania

Robert Skibiński
Faculty of Pharmacy
Department of Medicinal Chemistry
Medical University of Lublin
Lublin, Poland

Bernd Spangenberg
University of Applied Sciences
Offenburg, Germany

Małgorzata Starek
Faculty of Pharmacy
Department of Inorganic and Analytical
   Chemistry
Jagiellonian University
Kraków, Poland

Mariusz Stolarczyk
Faculty of Pharmacy
Department of General and Inorganic
   Chemistry
Jagiellonian University
Kraków, Poland

Oliwia Szerkus
Faculty of Pharmacy
Department of Biopharmacy and
   Pharmacodynamics
Medical University of Gdańsk
Gdańsk, Poland

Przemysław Talik
Faculty of Pharmacy
Department of Inorganic and Analytical
  Chemistry
Jagiellonian University
Kraków, Poland

Živoslav Lj. Tešić
Faculty of Chemistry
Department of Analytical Chemistry
University of Belgrade
Belgrade, Serbia

Jelena Đ. Trifković
Faculty of Chemistry
Department of Analytical Chemistry
University of Belgrade
Belgrade, Serbia

Mario Vega
Faculty of Pharmacy
University of Concepción
Concepción, Chile

Monika Waksmundzka-Hajnos
Faculty of Pharmacy
Department of Inorganic Chemistry
Medical University of Lublin
Lublin, Poland

Duygu Yeniceli
Faculty of Pharmacy
Department of Analytical Chemistry
Anadolu University
Eskişehir, Turkey

Part I

Theory of Thin Layer Chromatography in Context of Pharmaceutical Analysis

1 Overview of Drug Analysis and Structure of the Book

Łukasz Komsta, Monika Waksmundzka-Hajnos, and Joseph Sherma

CONTENTS

1.1 OVERVIEW OF DRUG ANALYSIS

Pharmaceutical analysis is a very wide topic that includes drug identification, identity confirmation, checking quality of drug formulations, quantitative estimation in various matrices, and decomposition studies. It also covers analysis of xenobiotics, toxins, and pesticides in various sources, including the environment. However, this book focuses only on drug analysis. Drug analysis includes active pharmaceutical ingredients (APIs), excipients (substances that act as the base of tablets or other formulations), decomposition products, impurities remaining from drug synthesis, and metabolites.

Thin layer chromatography (TLC) can be used in a very broad context in drug analysis. It is a separation technique in which the separation process occurs in a uniform planar layer of sorbent placed on a glass or aluminum plate or plastic sheet. The sorbent is called the stationary phase. During analysis, the plate is immersed in the mobile phase, generally a mixture of two to four solvents, and developed vertically or horizontally. The separation process occurs due to various mechanisms such as adsorption (e.g., hydrogen bond interactions), partitioning between the stationary and mobile phases, or ion exchange, depending on the nature of the sorbent. After development, compounds can be detected (visualized) and identified by their natural color or fluorescence, quenching of fluorescence on a layer containing a fluorescent indicator, or by creating colored spots after treating the plate with a chromogenic detection reagent by spraying, dipping, or exposure to vapors. The main reason for continuous high interest in TLC is that it does not require any complicated equipment, and commercially available plates can be developed in almost any laboratory.

The strength of retention is dependent on the structure of a drug, and differences in retention are the main factor responsible for separations. As there is a low chance of obtaining the same retention behavior between two substances in a well-chosen chromatographic system comprising the stationary phase and mobile phase, the identity of the drug can be proven by comparing the retention in several systems with a reference standard. Additionally, the drug in the sample and the reference drug should form spots of the same color after spraying with detection reagents. Although a recent trend in TLC involves mass spectrometry detection as additional proof of the identity of a compound, this approach is used very rarely. If a pure reference standard is available, one can be almost certain about the identity of an unknown drug after examining retention and visualization behavior.

However, there is a question about what "several systems" one should use for identification. Although some "standard TLC systems" were proposed in the literature for a large group of compounds, the continuous evolution of TLC resulted in papers recommending optimal TLC systems for separation and identification of all drug groups. These dedicated systems are described in the second part of this book.

TLC is also a useful tool in purity testing. For almost any drug, one can find in the literature recommended TLC conditions for purity testing (separation of the API and impurities or degradants). The existence of degradants can be proven by seeing additional spots on the plate, and their identity can be confirmed as described earlier if the reference compounds are available.

Besides identification and purity testing, the main application of TLC in drug analysis is drug formulation control. Quantitative analysis in TLC is available by scanning the plate with a densitometer or videoscanner. This enables precise and accurate drug determination in tablets, capsules, solutions, ointments, and many other formulation types. By the use of a selective TLC system, one can separate and detect impurities and drug decomposition, which is essential in drug quality control.

TLC analytical methods have to be validated for selectivity (peak purity), precision, accuracy, linearity, robustness, and ruggedness. The detection and quantification limits also have to be evaluated. In most cases, it is possible to meet all desired validation requirements and achieve results that are comparable to, or negligibly poorer than, results from other much more complicated approaches. The current requirements are set by the *International Conference on Harmonization* (ICH), and most of the methods presented in the literature are designed and validated according to these requirements.

The analysis of biological material by TLC (e.g., examining the drug levels in plasma) is quite rarely reported, due to required complicated extraction and cleanup procedures and some difficulties with insufficiently low detection limits. However, some successful approaches are present in the literature, usually making use of an internal standard, and this book also covers biological analysis.

1.2 ORGANIZATION OF THE BOOK

This book covers all topics important in pharmaceutical TLC applications. The first part contains chapters devoted to the general information on TLC with particular emphasis on topics important in the context of drug analysis. After this introductory chapter, Chapter 2 describes the chemical basis of TLC, how the structure of drugs can affect their retention, and what problems must often be solved during method development. Chapter 3 presents current knowledge about sorbents and layers used in pharmaceutical applications of TLC. Besides the sorbents, selection of the mobile phase composition is important, and mobile phase optimization issues are described in Chapter 4. Many drugs are ionic in nature, and the analysis of these compounds is covered in Chapter 5, both in the context of stationary phase and mobile phase development. Chapter 6 addresses the optimization of mobile phases such as acetone–water for assays of pharmaceutical formulations by addition of an ionic liquid to suppress free silanols in RP-18 layers.

Many drugs are chiral substances, and TLC allows also enantioselective separations and analysis of these compounds. This methodology is often used in pharmaceutical research, and the major aim is testing the chiral purity of the drugs. Chapter 7 is a comprehensive source of current knowledge about TLC chiral separations in the context of drug analysis.

Every TLC plate is developed in some type of chamber. Chapter 8 presents information on the most frequently used chambers as well as techniques and instruments for application of standard and sample solutions to the plate prior to development. Additionally, the newest advances in TLC instrumentation are discussed, including electrochromatography.

One of the major advantages of TLC is a possibility of two-dimensional development, which significantly increases the selectivity and range of separated analytes. Chapter 9 is devoted exclusively to two-dimensional TLC and presents up-to-date information about this technique.

As drug control relies on quantitative analysis, Chapter 10 was included to describe current techniques of quantification of chromatograms by both densitometry and videoscanning. The quantitative results must be properly evaluated statistically, and the correct statistical processing of data is very important in this case because densitometry and videoscanning have nonlinear response; the statistical recommendations are presented in Chapter 11.

Chapter 12 describes the variety of detection and identification methods used in TLC, and this aspect of TLC analysis can be treated as a unique advantage. TLC is the only chromatographic technique that allows the use of detection reagents in order to distinguish analytes based on color forming reactions.

Chapter 13 on lipophilicity explores a topic that is very important for evaluating drugs for their pharmaceutical activity based on TLC, as every drug candidate is evaluated in terms of its lipophilicity properties. Chapter 14 describes the use of TLC for solving one of the main problems in the drug market, the presence of substandard and fake pharmaceutical products. In addition to qualitative and semiquantitative TLC screening methods, the chapter explores efforts to transfer TLC screening methods to HPTLC–densitometry and develop new methods based on this technology, which is more suitable for support of regulatory compliance actions.

The second part of the book provides a comprehensive overview of a very wide spectrum of applications of TLC to separation and analysis of particular drug groups, based on the current literature. These chapters contain introductions about the structures and medicinal actions of the described substances and a complete literature review of their TLC analysis.

2 Chemistry of Drugs and Its Influence on Retention

Oliwia Szerkus and Michał J. Markuszewski

CONTENTS

2.1 INTRODUCTION

The chromatographic separation can be conducted when the stationary phase is in a thin layer. This kind of liquid chromatography is called planar chromatography or thin layer chromatography (TLC). When the stationary phase is a layer of tissue paper, then we mean paper chromatography; and if the stationary phase is distributed as a layer on a glass, aluminum, or plastic plate, we name it thin layer chromatography [1].

An increased interest in this technique was caused by the introduction of the so-called high-performance thin layer chromatography (HPTLC), which apparently improved the resolution capabilities. HPTLC and high-performance liquid chromatography (HPLC) are complementary methods with the a similar mechanism of retention. The principles of liquid chromatography defined for HPLC are also applied for optimization of separation conditions in TLC. As in HPLC, silica-based stationary phases are the most popular in TLC; therefore, the examples described in this chapter concern reversed-phase thin layer chromatography (RPTLC) [2].

In great simplification, the phenomena occurring in the chamber can be summarized as follows. The mobile phase migrates because of the capillary action along the sorbent layer (stationary phase), and depending on the energy effect of interactions between solute and phases, the different degree of retention is exhibited. As a result, the different migration rate is obtained and the substances occur in different places on the plate [1].

Currently, the mechanisms of retention are attempted to be explained in RPTLC. Three competing theories are interpreted in terms of partitioning, solvophobic, and adsorption processes [3].

Retention in the solvophobic theory is described as a function of surface tension and dipole–dipole interactions between the polar groups of an analyte and the mobile phase. Hydrophobic interactions of a compound with the hydro-organic mobile phase are presented as the driving force for

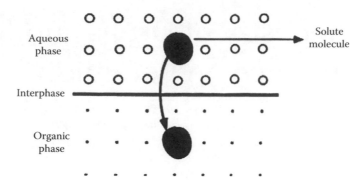

FIGURE 2.1 Partition process between phases: aqueous and organic. (Adapted from *J. Chromatogr. A*, 829, Vailaya, A. and Horvath, Cs., Retention in reversed-phase chromatography: Partition or adsorption?, 1–27, Copyright 1998 from Elsevier.)

retention. The most important shortcoming in this model is the fact that the stationary reversed phase is thought to behave like a passive part of the system [3].

The partitioning theory is explained by the good correlations between the logarithms of octanol-1/water partition coefficients (log $P_{o/w}$). Molecular details of the partition process are illustrated schematically in Figure 2.1. Generally speaking, a solute-sized cavity appears in the organic phase, then a molecule of the solute is moved from the aqueous into the organic phase. Eventually, the cavity is closed and left by the molecule of a compound in the aqueous phase. The application of the partitioning theory is again limited due to the fact that it does not comprehensively explain the shape selectivity and unlikeness of octanol and the organic ligands associated with one end of a substrate [4].

In the third and last theory, the partitioning and adsorption processes are combined together. This theory seems to provide an extensive and satisfying description of the retention mechanism. An important feature of the theory of this mechanism is two-step. At the beginning, a solvent–stationary interphase layer is created. Afterward, the solvent molecules displace among the interphase and the mobile phase [3].

The distribution of a solute between the mobile and stationary phases is characterized by three variables: the chemical structure of the solute molecules, the physicochemical properties of the mobile phase, and the physicochemical properties of the stationary phase. It concerns the distribution at constant temperature. The type of intermolecular interactions that occur between the analyte molecules and the two phases—mobile and stationary—influences the retention of the mentioned analyte [5]. The type of intermolecular interactions is a consequence of chemical parameters of the solute and the phases, and these are dependent on the chemical structures of the solute and the mobile and stationary phases.

Although the reversed-phase liquid chromatography (RPLC) is presently very popular, the mechanisms of retention are not entirely comprehended when taking into account the molecular level. The problem in explanation lies in the complexity of the system and the occurring interactions. The bonded phase is characterized by a composited structure, composed of a porous silica support and the solvation layer [6]. In the interphase region, there are large contributions from the chemical structures of the silica surface, the solvent, and the chemically bonded species. Taking into account all these contributions, varying interactions with solutes of different polarity and geometry in chromatographic separations occur [7]. The 3D structure of a surface layer that is solvated performs like a stationary phase in respect of a change that is dependent on temperature and solvation. The change concerns a conformation of the bonded hydrocarbon moieties [8].

The molecular interactions model describes the distribution between the two phases as a result of the various molecular forces existing between the solute molecules and the molecules of the stationary and mobile phases. On condition that the nature of the intermolecular interactions can be defined, the behavior of a particular solute molecule in a given chromatographic system might be prognosticated [7,8].

When defining the nature of the interactions at the molecular level that rule the chromatographic retention and separation, one has to state that the interactions cannot cause definite chemical alternations of a solute through, for example, oxidation or reduction reactions. The types of intermolecular interactions which have to be taken into consideration are dipole–dipole directional interactions, inductive dipole interactions, dispersion interactions, electron pair donor–electron pair acceptor interactions, as well as solvophobic interactions and hydrogen bonding [5].

Intermolecular forces, which are considered to be nonspecific, are named Van der Waals forces. They take place between closed-shell molecules. The directional, induction, and dispersion forces are included in that group. These forces are said to be "more physical" in their nature, whereas the other group including the electron pair donor–electron pair acceptor forces as well as hydrogen bonding reflect a character that is rather of a "chemical" nature. Directional, specific forces belong to the last group [9].

2.2 TYPES OF INTERMOLECULAR INTERACTIONS

- *Ion–dipole interactions*: From the chromatographic point of view, ion–dipole interactions are important. The electrons that occur in the bond between two diverse atoms in terms of electronegativity are shared unequally. That leads to a lasting dislocation of the electron probability in the direction of the more electronegative atom. The ion-produced dipole that is sited in the electric field will orient itself in order to direct ion-to-ion with the end with opposite charge [8,10]. The ion–dipole potential energy of the interaction, E_{i-d}, is specified as follows [5]:

$$E_{i-d} = \frac{-W^2 Z\mu(\cos\alpha)}{\varepsilon r^2}$$

  where
  Z is the ion charge
  μ is the dipole of the neutral molecule
  r is the distance between the center of the dipole and the ion
  α is the dipole angle related to the line r joining the dipole center and the ion

  The higher the relative electric solvent permittivity (ε), the lower is the energy of attractive ion–dipole interactions. If one added ethanol ($\varepsilon = 27$) to water ($\varepsilon = 81$), the electrostatic interactions are observed to increase. This reliance of intermolecular interactions on electric solvent permittivity gives an explanation to why the solubility of ionic crystals in solvents that are organic is low when compared to water [5,10].

- *Dipole–dipole interactions*: Van der Waals interactions concern long-distance forces, which are formed by the molecule's electric field. For highly polar molecules, the orientation or dipole–dipole interactions are characteristic. Worth mentioning is the fact that with temperature (T) the probability of orientations that are energetically beneficial decreases, and at high values of the temperature all dipole orientations are equally populated. The potential energy is then equal to zero. The temperature reliance of the potential energy in dipole–dipole interactions (E_{d-d}) needs to be specially noted from the chromatographic point of view. So, if the dipole–dipole interactions are of importance for the separation of a particular group of substances at a particular temperature, then their effect at a higher temperature on the separation forces may be unimportant. In fact, different structure parameters may govern retention when measurements are conducted at different temperatures [5,10].

- *Dipole-induced dipole interactions*: A molecule may induce an electric dipole in a close-by molecule. The induced dipole moment always lies in the direction of the inducing dipole. Thus, attraction always takes place between the two molecules, and is independent of the temperature [5] (Figure 2.2).

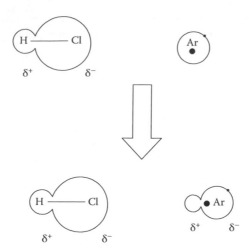

FIGURE 2.2 Dipole-induced dipole attraction. A dipole is induced by a polar molecule of HCl in a non-polar molecule of argon. (Adapted from Kaliszan, R.: *Quantitative Structure—Chromatographic Retention Relationships*. 1987. Copyright John Wiley & Sons.)

- *Instantaneous dipole-induced dipole interactions*: When atoms and molecules are characterized by the lack of a permanent dipole moment, the continuous electron density fluctuations create a dipole moment that is rather negligible. Such a momentary dipole can induce a dipole in nearby molecules and atoms [10].
- *Hydrogen bonding interactions*: If hydrogen atoms that are covalently bonded create a second bond to another atom, that bond is considered as a hydrogen bond. Such formed hydrogen-bonding interactions are of chromatographic importance—they are 10 times stronger than the nonspecific intermolecular interaction forces. The hydrogen bonding conception was initially created to explain high boiling points in liquids that have groups of hydrogen atoms or hydroxyl groups. The following groups may behave as proton donors in hydrogen bond formations: –O–H, Br–H–S–H, Cl–H, F–H, I–H, = N–H [8].
- *Electron pair donor–electron pair acceptor interactions*: Electron pair donor–electron pair acceptor (EPD–EPA) interactions belong to "chemical" intermolecular interactions. Whereas in a typical chemical bond one electron is delivered by one atom to the bond, in EPD–EPA interactions one molecule supplies the pair of electrons and the other molecule offers the vacant molecular orbital [11].
- *Hydrophobic interactions*: Hydrophobic or otherwise called solvophobic interactions are highly important in RP-LC studies. Solvophobic interactions are the composite net effect of the previously described physicochemical interactions [8,10].

2.3 CHROMATOGRAPHIC RETENTION FROM THE POINT OF VIEW OF INTERMOLECULAR INTERACTIONS

When we analyze RP-LC on a chemically bonded hydrocarbon silica stationary phase, interactions occur between the solute, the hydrocarbon bonded to silica, the mobile phase that is adsorbed on the stationary phase system, and additionally, the free silanol groups of the silica support and the eluent's constituents.

A typical feature of RP-LC is an increase of the retention factor logarithm (log k) of a solute when the number of carbon atoms for a homologous series increases. Taking into account the interactions between a hydrocarbonaceous moiety of the stationary phase and the solute, one can identify the dispersive forces as differentiating the homologous sets. The contribution to separation associated with the orientation interactions look as if they were of minor importance as the polarity (dipole moment) of the hydrocarbon part of the compound molecule nearly equals zero. Because the dipole moments within homologous sets are alike, one can conclude that the dipole-induced dipole interactions ought to be similar as well. On condition that the solute polarizability increases, the number of dispersion interactions increases evidently, which in fact reflects their molecular size (bulkiness). When the interactions of homologues amongst a polar eluent and the solute molecules occur, then the orientation as well as inductive forces are tougher than for the solute–nonpolar hydrocarbon of stationary phase interactions. As it was observed, the orientation interactions between all homologues and eluents are congenial by reason of the very similar dipole moments [8].

The dispersive and inductive interactions entail that the homologous compounds attract the mobile phase. The dispersive interactions are predominant among all intermolecular interactions. The increased retention of larger homologues will stand for the "net effect" of Van der Waals interactions.

It has been scientifically confirmed that with increasing amount of organic modifier in the mobile phase, the retention of a solute decreases in RP-LC. As it was proven, the polarizability of the organic/water eluent is higher when relating only to water. When the content of the organic modifier is greater in a binary aqueous eluent, then the dispersive interactions present amongst the solute and the mobile phase decrease the retention.

As it was proved for LC, with the increase of ionic strength of aqueous eluents, the retention increases too. The direct interactions between a solute and the silica matrix in stationary phase will not be influenced by alternations in the ionic strength of the eluent, whereas the Van der Waals interactions will be reduced with growing concentration of ions.

One more observable fact in liquid chromatography that is worth analyzing is the influence of eluent pH on the retention of weak acids and bases on a silica layer of stationary phase. It could be stated that the dispersive interactions of solute ions with each phase do not vary much from interactions for nonionized molecules of the solute. Nevertheless, when we consider ionic solutes, the ion–dipole interactions are leading. This appears significant when we analyze solute ions and polar molecules of the mobile phase and entirely diverse nonpolar hydrocarbon moieties from the stationary one. As concluded, the solutes' retention decreases if their degree of ionization increases. Wholly ionized acids are characterized by lower retention in comparison to the ionized bases of comparable bulkiness (molecular size). Attempting to clarify such observations, free silanols are thought to create exclusion effects (through electrostatic interactions) toward anions, while in the case of cations, retention increases. At higher ionic strength of the eluent, this kind of interactions is diminished [12].

Another typical phenomenon occurring in RP-LC is the decrease of solute retention with temperature. Apparently, when the temperature is increasing, the dielectric constant of water is being reduced notably. Water is usually a major constituent of eluents. If the dispersive interactions are unaltered as a result of temperature modifications, the decreasing retention at upper temperatures is to predict what is caused by the associated decrease of the dielectric constant, and hence the stronger affinity of an analyte to a mobile phase [8].

2.4　QUANTITATIVE STRUCTURE–RETENTION RELATIONSHIPS (QSRR)

QSRR was practically proposed and theoretically described by Kaliszan [5]. QSRR models are regression models employed for the prediction of a relationship amongst analyte retention and its chemical structure. In order to achieve a coherent association, one requires a lot of property parameters. This makes liquid chromatography a valuable technique for studies of quantitative and reproducible retention data obtained for large amount of structurally varied compounds [13].

2.5 FUNDAMENTALS OF QUANTITATIVE STRUCTURE–RETENTION RELATIONSHIPS

In view of ascribing properties to chemical structures, the traditional thermodynamic approach may be improper [7]. This is because a thermodynamic approach is rather of physical nature than the system's chemical nature defined by the QSRR method. It is established that much more data are provided by an extrathermodynamic approach since it is characterized by linear free-energy interactions (LFER). The free-energy changes linked to the chromatographic distribution process are characterized by retention parameters. In view of this, a chromatographic column can act like a "free-energy transducer," translating changes of the chemical potentials of analyzed compounds caused by some differences in the structure. This is performed by a translation of the retention parameters into quantitative differences. Taking into account the assumptions of LFER, any certain chromatographic retention parameter (i.e., log k, R_M, Kováts index) may be pronounced by a set of a compound's structural descriptors:

$$\text{Retention parameter} = f(a_1 x_1, \ldots, a_n x_n)$$

Computer-aided multiple regression analysis is used for the calculations of the coefficient $a_1 - a_n$ for individual n descriptors [14].

2.6 STRUCTURAL DESCRIPTORS OF SOLUTES

1. Structural descriptors of solutes based on LFER: The contributions of the substituents to retention have been discovered to stay constant and additive regarding any particular chromatographic RP system. Hammet constants and Taft steric constants are the most known parameters of substituents, which can be associated with free energy. They are valuable especially when one tries to predict and calculate the differences in physicochemical properties between series of derivatives. They are also found useful in isolation and quantification of more complex electronic effects (e.g., mesomeric, inductive one) [15].
2. Solvatochromic parameters: The linear solvation energy relationship (LSER) is another approach, which was developed as a result of detailed research about the nature of separations in RP-LC. The introduced solvatochromic method is established on a general equation, where solute properties are defined. Properties of an analyte such as molar volume, polarizability, hydrogen bond acidity, as well as hydrogen bond basicity can be denoted [16].
3. Nonempirical structural descriptors: All the structural parameters, which can be calculated with the use of computational chemistry and which are based entirely on the structural compound's formula are reflected by this term. The parameters described in the following belong to this group of structural descriptors [17,18]:
   a. CLOGP—the log P calculated by fragmental methods: Basing on the n-octanol-water partition system, log P is obtained by the fragmental method. In the fragmental method, hydrophobicity is computed from the structural formulae of analyzed compounds [19].
   b. Refractivity: Molar refractivity refers to the ability to undergo inductive and dispersive interactions between the analyte and the constituents of mobile and stationary phases. Refractivity can be computed by summing the atomic and the group or the bond-type increments [19].
   c. Polarizability: Polarizability is affiliated with the "bulkiness" parameters (similarly to the refractivity). The calculation of the analyzed solute's polarizability is established on the basis of the additivity scheme of Miller. Various increments are assigned to various types of atoms [20].
   d. Molar volume: This molecular descriptor is used to define the compound's ability to participate in dispersive and nonspecific interactions. Molar volume is one of the "bulkiness" parameters whose dimension is volume [15].

e. Van der Waals surface area: Evaluation of the Van der Waals surface area can be currently performed with computer-aided graphics and calculation software.

f. Solvent-accessible surface area: The solvent-accessible surface area is the molecular descriptor, which one may calculate by rolling an examined analyte's spherical estimation over the Van der Waals surface.

g. Molecular mechanics and quantum chemistry indices: Structural descriptors supported by theoretical chemistry are commonly used in QSRR analysis. One such obtained structural descriptor is the total energy, E_t. The main feature and advantage of E_t, over the other dispersion interaction parameters, is the fact that E_t distinguishes isomers and conformers.

The dipole moment, μ, is another quantum chemically derived recognizable structural parameter employed to calculate theoretically dielectric constants and in general a chemical compound polarity.

The following two parameters are also assumed to be influential for the retention determination: the quantum chemically calculated energies of the highest occupied molecular orbital (HOMO) and the lowest unoccupied molecular orbital (LUMO) [21].

One has to notice that there exists plenty of other quantum chemical descriptors of solutes; however, they are rarely employed successfully in QSRR studies [8].

h. Structural descriptors derived from molecular formula
   • Molecular shape parameters: For liquid chromatography, the influence of the steric effects on retention is of minor importance in comparison to differences affected by polar and dispersive interactions [13].
   • Graph theoretical indices: The main characteristic of this type of structural descriptors is that they are obtained by counting graph edges or vertices (sometimes both). Calculations are being made in order to get numerical indices, which are valuable in finding information about the properties of analytes [5,8,13].

i. Empirical physicochemical parameters in QSRR analysis: When we look into a particular chromatographic system, the experimental physicochemical parameters as independent variables are found to be useful, especially from the point of view of retention mechanism explanation.

The most significant and commonly applied empirical physicochemical parameter in QSRR studies is the logarithm of the n-octanol-water partition coefficient, log P. There are numerous reports in which log P is a single retention descriptor (mostly in TLC and RP-HPLC). Due to its importance, this hydrophobicity parameter will be described separately [13].

2.7   GENERAL REMARKS ON QSRR DESCRIPTORS

The previously presented structural descriptors are noticed to be statistically the most significant and the most frequently met in QSRR analysis. There are other numerous parameters which can be assigned for an individual compound. Whether various transformations and variations of descriptors are taken into account, the number of descriptors is practically unlimited.

In 2000, Todeschini and Consonni, based on available references, collected and proposed about 1800 molecular descriptors. All descriptors are characterized in detail by means of their physicochemical or topological nature. Now, there is also an appropriate software available for calculation of individual descriptors [22].

All available molecular descriptors calculated from a chemical formula or molecular graph appear appealing and impressive. However, the question arises on relations of these designed descriptors with a real physical or biological property of a compound or they are only its symbolic representation. Although several nonempirical structural descriptors were reported to contribute to numerous multivariable QSRR equations, most often they can be useful for retention prediction in chromatography [13,22].

TABLE 2.1

Basic Structural Descriptors Most Often Employed in QSRR Studies

Molecular Bulkiness-Related Descriptors	**Molecular Polarity-Related (Electronic) Descriptors**
Carbon number	Dipole moments
Molecular mass	Atomic and fragmental electron excess charges
Refractivity	Orbital energies of HOMO and LUMO
Polarizability	Partially charged areas
Van der Waals volume and area	Local dipoles
Solvent-accessible volume and area	Submolecular polarity parameters
Total energy	
Calculated partition coefficient (CLOG P)	
Molecular Geometry-Related (Shape) Descriptors	**Molecular Graph-Derived (Topological) Descriptors**
Length-to-breadth ratio	Molecular connectivity indices
STERIMOL parameters	Kappa indices
Moments of inertia	Information content indices
Shadow area parameters	Topological electronic index
Physicochemical Empirical and Semiempirical Parameters	**Indicator Variables**
	Zero-one indices
Hammett constants	
Hansch constants	**Combined Molecular Shape/Polarity Parameters**
Taft steric constants	Comparative molecular field analysis (CoMFA) parameters
Hydrophobic fragmental parameters	Comparative molecular surface (CoMSA) parameters
Solubility parameters	
Linear solvation energy relationships (LSER) parameters	
Partition coefficients (LOG P)	
Boiling temperatures	
pKa values	

Source: Adapted from Kaliszan, R. and Markuszewski, M.J., *Chem. Anal.*, 48, 373, 2003.

In Table 2.1, the classification of the most often used structural descriptors concerning QSRR is presented.

2.8 QUANTITATIVE RELATIONSHIPS BETWEEN HYDROPHOBICITY OF SOLUTES AND THEIR CHROMATOGRAPHIC RETENTION

It was much talked over the meaning of the word "hydrophobicity." Generally, this term is assumed to express a measure of the tendency of an analyte to favor a nonaqueous over an aqueous environment. Hydrophobicity is also said to measure the tendency of two (and more) molecules of an analyte to aggregate in aqueous solutions [24]. The term "hydrophobicity" describes those parameters which reflect the complex net effect of the identical physicochemical interactions that are used to illustrate the state of all the matter like, for example, orientation, inductive, dispersive, and hydrogen bonding as well as the charge–transfer interactions [13,23].

It is presumed that the driving force for the liquid–liquid partitioning is the hydrophobic effect. Pharmacokinetic processes of drug absorption into the organism's circulation, the drug distribution between compartments, and excretion are assumed to involve a partition model, in this case between lipid membranes and aqueous fluids. All these processes are influenced by drug hydrophobicity. On the hydrophobicity of the drug rely the attraction of the drug headed to nonpolar binding sites of blood proteins [23].

It was reported that the hydrophobicity of solutes depends predominantly on the environment. Nevertheless, when one compares behavior of different solutes in the identical environment, it can be concluded that a quantitative scale which can be gained reveals the variances in ability of an

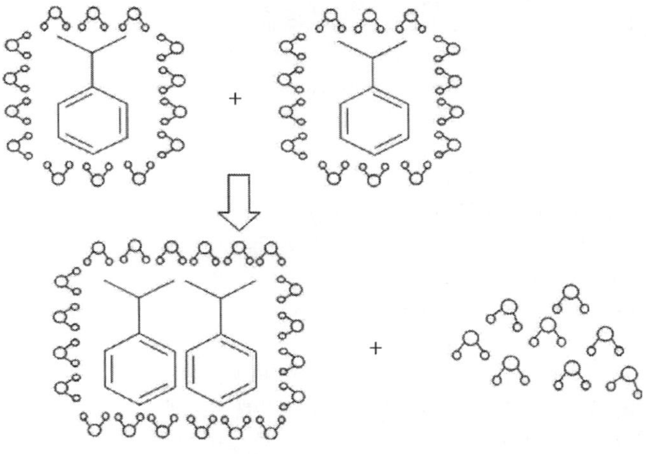

FIGURE 2.3 Illustrative scheme of the formation of hydrophobic interaction between two nonpolar molecules. (Adapted from Nasal A. et al., *Curr. Med. Chem.*, 10, 381–426, 2003.)

individual analyte to participate in hydrophobic interactions. One can construct this kind of scale with the help of distribution studies, in which the compounds are examined between an immiscible solvent polar and nonpolar pair.

The partition coefficients for an individual analyte may be varied provided that it is determined in different organic-water solvent systems. What is interesting is that their logarithms are observed to be related linearly. When the differences of the polarity among organic solvents in the two aqueous systems are larger, then the linear relationship becomes weaker. The number of hydrophobicity parameters is similar to the number of partition systems available [23] (Figure 2.3).

The difference between the expressions "hydrophobicity" and "lipophilicity" needs to be explained because both seem to mean the same feature of objects. According to IUPAC, the hydrophobicity stands for the association of nonpolar molecules in an environment which is an aqueous one. This results from the tendency to the exclusion of nonpolar molecules by water. Lipophilicity describes the affinity between molecules (or moieties) and a lipophilic environment. It can be easily measured through its distribution in a diphase system—a liquid–liquid one (e.g., partition coefficient in n-octanol-water system) or a solid–liquid one (e.g., retention on TLC systems) [13,23].

2.9 THIN LAYER CHROMATOGRAPHY IN LIPOPHILICITY DETERMINATION

TLC is a very good and frequently used chromatographic technique especially with regard to the lipophilicity determination. In this aspect, this method is even more advantageous than the HPLC. For example, the number of eluents which can be chosen for the lipophilicity studies is wider in TLC.

The retention parameter in TLC is the R_M value defined by Bate-Smith and Westall

$$R_M = \log\left(\frac{1}{R_f - 1}\right)$$

where R_f means the ratio of a distance passed by an analyte to that reached by the front of the solvent [24].

Best result in terms of accuracy is gained when R_f values are between values 0.2 and 0.8. With the purpose of obtaining the most reliable R_M values, the mobile phase composition has to be optimized especially when dealing with a series of compounds of numerous retentive properties. To overcome this kind of obstacles, R_M is determined at several compositions of binary eluents and then the linear extrapolation of the relationship between R_M and volume percent of eluent component is performed to obtain a fixed value. That fixed value stands for a zero percent of organic modifier employed in mobile phase. This extrapolation of R_M is represented by R_M^0 value. The R_M^0 is supposed to be the same on a certain stationary phase for a certain analyte, independent of the water-organic system used for the determination of R_M value as well as for the extrapolation to R_M^0 or R_M^W. Such an extrapolation is a comfortable and suitable method for the standardization of the retentive data obtained by chromatography [23–25].

In literature, there were some articles related to application of TLC to lipophilicity determination of xenobiotics published recently.

In order to quantitate the lipophilicity of water-soluble eburnane alkaloids, an RP-TLC method was developed and employed [26]. Vincamine, vinpocetine, and related derivatives are valuable agents of a therapeutic interest, mostly in cardiovascular and cerebral therapy. The parameter commonly used for the lipophilicity quantification is log P, which was employed in these studies as expressed by chromatographic retention parameters. The retention data, R_M values, for analytes were expressed as extrapolated R_M^0 ones, which are related to an eluent with no organic co-solvent. The lipophilicity for all examined compounds was in the range 2.9s–4.8. As it was concluded, the structure–lipophilicity relationships were associated with structural parameters like character and length of side chain, conformational features, and moiety–solvent interactions. Molecular mechanics calculations enabled the interpretation of the results from RP-TLC experiments.

The lipophilicity expressed as R_M^0 and specific hydrophobic surface area of seven 1,2-benzisothiazol-3(2H)-ones was determined by RP-TLC. These compounds are assumed to possess high antifungal and antibacterial activity. The purpose of this study was to find the relationship between the concentration of organic modifiers in the mobile phase and the chromatographic properties of the examined compounds. The second objective was to get to know about the influence of the types of substituents and their position on the change of the lipophilicity in examined compounds [27]. The high correlations between the volume fraction of the organic solvent and the retention parameters were found for every solute distinctly. The calculated R_M values were extrapolated to 0% of organic modifier concentration $\left(R_M^0\right)$ according to the following equation:

$$R_M = R_M^0 + bC$$

where
C means the concentration in volume percent of the organic solvent in the mobile phase
b states for the change in R_M created by shift of organic modifier amount in the mobile phase that is related to the specific hydrophobic surface area of an analyte

As it turned out, the R_M values of the analytes decrease linearly when concentration of organic modifier increases. The correlations between volume fraction of the organic modifier in the mobile phase and the R_M values are found to be linear (correlation coefficients are higher than 0.99). The results are presented graphically in Figure 2.4.

Also, it was attempted to determine the lipophilicity of 14 anti-hypoxia drugs in eluent systems comprising various concentrations of acetonitrile and potassium dihydrogen phosphate. Studies were performed in a RP-TLC system [28]. The significant correlation between lipophilicity and the specific hydrophobic surface of these drugs was reported. It can be concluded that drugs act as homologues. RP-LC proved to be an applicable method for the assessment of molecular lipophilicity.

Some articles regarding the analysis of the other retention parameters in order to predict the retention of the solute are presented below.

Kiridena and Poole describe a novel approach to method development using RP-TLC, which was established on the basis of molecular mechanics calculation using the solvation parameter model [29]:

$$SP = c + mV_X + rR_2 + s\pi_2^H + a\sum\alpha_2^H + b\sum\beta_2^0$$

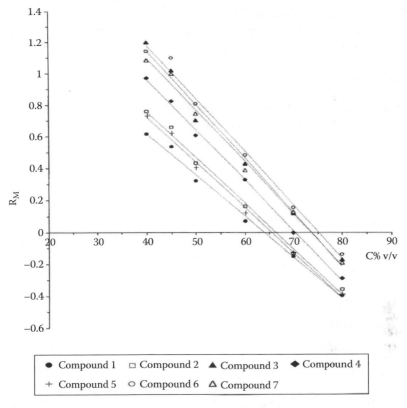

FIGURE 2.4 Relationships between R_M values and volume fraction of the organic modifier (methanol) in the mobile phase for the examined seven compounds. (Adapted from Sławik, T. and Kowalski, C, Lipophilicity of a series of 1,2-benzisothiazol-3(2H)-ones determined by reversed-phase thin-layer chromatography, *J. Chromatogr. A*, 952, 295, 2002.)

where SP means the experimentally observed retention property (here the R_M value). To the solute descriptors belong McGowan's characteristic volume V_X, excess molar refraction R_2, π_2^H which describes the ability of the solute to stabilize a bordering dipole (dipolarity/polarizability), $\sum \alpha_2^H$ and $\sum \beta_2^0$—the solute's effective hydrogen bond acidity and hydrogen bond basicity, respectively. It has to be mentioned that the solute descriptors can be assigned to a great amount of compounds (more than 2000), and there exist also others which are available through computational methods [30]. The system constants from the above equation are characterized by their interactions, corresponding to the solute descriptors. The r constant describes the difference in capacity to interact with solvated molecule's n- or π-electrons, whereas the s constant determines the ability of the system to participate in dipole–dipole and dipole–induced dipole interactions. The a and b constants reflect the differences in hydrogen bond acidity and basicity, respectively, of the solvated sorbent layer and the mobile phase systems. The m constant characterizes the easiness of the solute to create a cavity in both a sorbent layer and mobile phase. The constants in any TLC system are obtainable by employing the multiple linear regression analysis of R_M values of solutes with identified descriptors. System constants are calculated for aqueous binary mobile phase mixtures containing various organic solvents. In the first, the solvents were chosen which were reflected by appropriate values of Kamlet–Taft solvatochromic parameters. As it was reported by the authors, satisfying compatibility among experimental and predicted R_F values for examined compounds (steroids, naphthalene derivatives, phenols) was shown for mobile phase optimization. The solvation parameters can be used as a convenient approach in order to predict the retention in RP-TLC different systems basing on characteristic properties of analytes.

Study of the retention in RP-TLC of some α-adrenergic and imidazoline receptor ligands was conducted by Eric et al. [31]. The retention constant $\left(R_M^0\right)$ was defined for 11 receptor ligands studied. The main aim of the study was to investigate the hydrophobic parameters and their potency in rationalization of drug action. Hydrophobicity is mostly related to following features of the molecules: polarity, molecular size, and the hydrogen bonding. The chemical structures of examined compounds are shown in Figure 2.5.

All molecular mechanics calculations were performed with the use of computer programs (Codessa and Spartan software). The surface area and electrostatic potential and other molecular descriptors such as constitutional, topological, geometrical, and electrostatic descriptors were calculated.

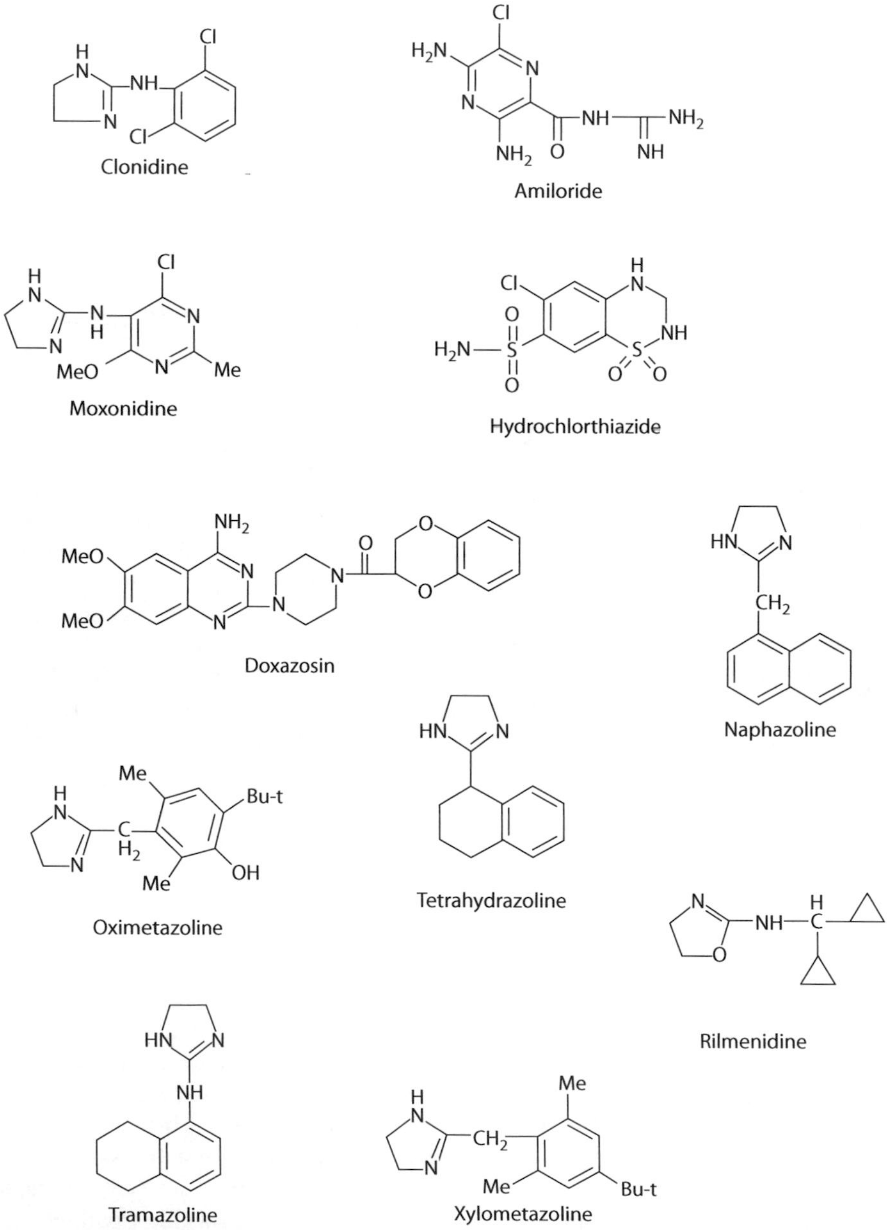

FIGURE 2.5 The chemical structures of investigated compounds. (Adapted from Eric, S. et al., *J. Chromatogr. Sci.*, 45, 1, 2005.)

It was concluded that the retention behavior of studied compounds is mostly subjected to geometrical properties (size and shape) as well as electrostatic properties (e.g., energy of H–N) and hydrogen bonding interactions. Linear correlations between R_M^0 and log P for all 11 analytes were reported.

Komsta [32] worked on a QSRR approach which was based on functional group counts. The new method was used in TLC in order to predict the retention of small organic molecules in different solutes. Proposed method is based on the assumption that the retention of a small molecule might be calculated by means of the sum of the contributions of each of the substituents in the molecule. It was concluded that the presented QSRR models can be useful to determine the retention of candidates for innovative drugs.

Csiktusnádi-Kiss et al. [33] investigated the retention parameters of ditetrazolium salts by RP-TLC in comparison with RP-HPLC. In order to find correlation between the physicochemical parameters of studied compounds, and their retention data, PCA and cluster analysis were employed. The physicochemical parameters which were taken into account in the calculations were polarizability (P), refractivity (ρ), logarithm of the lipophil–hydrophil character (log π), Van der Waals surface (VdW$_{surface}$), Van der Waals volume (VdW$_{volume}$), water-accessible surface (WA$_{surface}$), water-accessible volume (WA$_{volume}$), total energy (E$_{total}$), binding energy (E$_{bind}$), heat of formation (Q$_f$), energy of the highest occupied molecular orbital (eV$_{HOMO}$), energy of the lowest unoccupied molecular orbital (eV$_{LUMO}$), minimum charge on the atoms in the molecule (δ_{min}), maximum charge on the atoms in the molecule (δ_{max}), and dipole moment (ξ). It was observed that most of the calculated physicochemical parameters of molecules had significant loadings in the first PC component. The results were correlated with the clear effect of these parameters on retention. It was established that the shape and the polarity of molecules have predominant influence on their retention, whereas the lipophyl–hydrophyl character was not characterized by such strong effect. It was observed that despite the TLC and HPLC having very similar mechanism of retention, the results obtained for TLC are much different from those for HPLC under circumstances described in this research. The 2D map of PCA loadings is presented in Figure 2.6. The distant location of the cluster characterizing the HPLC parameters points out the difference from TLC in particular parameters of examined compounds.

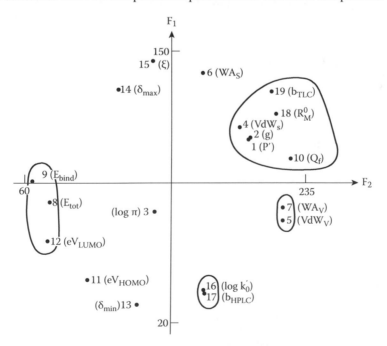

FIGURE 2.6 Two-dimensional nonlinear map of PC loadings. Similarities and dissimilarities among the retention and physicochemical parameters of studied salts. (Adapted from Csiktusnádi-Kiss, G.A. et al., Application of multivariate mathematical-statistical methods to compare reversed-phase thin-layer and liquid chromatographic behaviour of tetrazolium salts in Quantitative Structure-Retention Relationships (QSRR) studies, *Analusis*, 26, 400, 1998.)

One has to notice that the retention of the drugs depends also on the effect of solute ionization. Ionization is dependent on the change of the pH of the mobile phase. To theoretically interpret this phenomenon, pKa values, pH, and ionic strength as well as the mean ionic activity coefficient should be denoted for compositions of the mobile phase [34].

Retention behavior of four groups of drugs was examined in liquid chromatography (LC) systems under different conditions [35]. The studies were applied to peptides and peptide hormones, quinolones, and diuretic drugs. The descriptive model for the behavior of the retention depending on the pH, ionic strength, and composition of the mobile phase was founded. The authors established equations that allow to predict the optimal pH by means of minimum number of experiments. The change of the pH of mobile phase and, as a consequence, the change in the retention of the analyte are linked to the chemical structure of the solute. The tripeptides examined during the study typically have two functional groups. In the acidic range, pKa values are ascribed to the carboxylic acid function, whereas pKa values in the basic range describe the protonated amino groups dissociation. Therefore, peptides act as typical zwitterion-like-compounds. Carboxylic acid C-terminal and amino-N-terminal group are functional groups that are possessed by peptide hormones. As it was earlier observed, typical for peptide drugs is high retention at low pH. With the pH increase, the k value decreases and stabilizes at isoelectric point related to appropriate pH and remains unchangeable. The examined quinolones (e.g., ciprofloxacin, norfloxacin) possesses the carboxylic and the ammonium groups on the piperazine ring, thus they are characterized by two acid–base equilibria within the pH range [35].

2.10 CONCLUSION

TLC is a popular and widely used technique for separation and quantitative and qualitative analysis of a great variety of the compounds. Nonetheless, its great potency in different applications is not identified entirely and its physicochemical principles are still to examine.

TLC, in particular in reversed-phase mode, is a highly valuable system for studying structure–retention relationships, basing on physicochemical structure of the drugs and intermolecular interactions, which results from the fact that during a controllable chromatographic process all conditions stay constant except the structure of an analyte, which remains a variable.

REFERENCES

1. Sherma J. and Fried B. 2003. *Handbook of Thin-Layer Chromatography*. New York: CRC Press.
2. Bączek T. et al. 2005. Behavior of peptides and computer-assisted optimization of peptides separations in a normal-phase thin-layer chromatography system with and without the addition of ionic liquid in the eluent. *Biomedical Chromatography* 19: 1–8.
3. Claessens H.A. 1999. Characterization of stationary phases for reversed-phase liquid chromatography. Column testing, classification and chemical stability. In *Synthesis, Retention Properties and Characterization of Reversed-Phase Stationary Phases*, pp. 7–27. Eindhoven, Technische Universiteit Eindhoven.
4. Vailaya A. and Horvath Cs. 1998. Retention in reversed-phase chromatography: Partition or adsorption? *Journal of Chromatography A* 829: 1–27.
5. Kaliszan R. 1987. *Quantitative Structure—Chromatographic Retention Relationships*. New York: John Wiley & Sons.
6. Rafferty J.L., Zhang L., Siepmann J.I., and Schure M.R. 2007. Retention mechanism in reversed-phase liquid chromatography: A molecular perspective. *Analytical Chemistry* 79: 6551–6558.
7. Szepesy L. 2002. Effect of molecular interactions on retention and selectivity in reversed-phase liquid chromatography. *Journal of Chromatography A* 960: 69–83.
8. Kaliszan R. 1997. *Structure and Retention in Chromatography: A Chemometric Approach*. Amsterdam, the Netherlands: Harwood Academic Publishers.
9. Tomlinson E., Poppe H., and Kraak J.C. 1981. Thermodynamics of functional groups in reversed-phase high performance liquid-solid chromatography. *International Journal of Pharmaceutics* 7: 225–243.

10. Forgacs E. and Cserhati T. 1997. Molecular basis of chromatographic separation. In *Adsorption Phenomena and Molecular Interactions in Chromatography*, pp. 10–13. New York: CRC press Inc.
11. Gutmann V. 1978. *The Donor-Acceptor Approach to Molecular Interactions*. New York: Plenum Press.
12. Knox J.H., Kaliszan R., and Kennedy G.J. 1980. Enthalpic exclusion chromatography. *Faraday Symposia of the Chemical Society* 15: 113–125.
13. Kaliszan R. 2007. QSRR: Quantitative structure-(chromatographic) retention relationships. *Chemical Reviews* 107: 3212–3246.
14. Kaliszan R. and Markuszewski M.J. 2003. Studies on correlation between structure of solutes and their retention. *Chemia Analityczna* 48: 373–395.
15. Charton M. and Motoc I. 1983. *Steric Effects in Drug Design*. Berlin, Germany: Akademie Verlag.
16. Snyder L.R., Carr P.W., and Rutan S.C. 1993. Solvatochromically based solvent-selectivity triangle. *Journal of Chromatography A* 656: 537–547.
17. Kamlet M.J., Abboud J.L.M., and Taft R.W. 1981. Linear solvation energy relationships. 23. A comprehensive collection of the solvatochromic parameters, π, α, and β, and some methods for simplifying the generalized solvatochromic equation. *Journal of Organic Chemistry* 48: 2877–2887.
18. Park J.H., Dallas A.J., Chau P., and Carr P.W. 1994. Study of the hydrogen bond donor acidity of binary aqueous mixtures and their role in reversed-phase liquid chromatography. *Journal of Chromatography A* 677: 757–769.
19. Hansch C. and Leo A. 1979. *Substituent Constants for Correlation Analysis in Chemistry and Biology*. New York: Wiley.
20. Miller K.J. 1990. Additivity methods in molecular polarizability. *Journal of the American Chemical Society* 112: 8533–8542.
21. García-Raso A., Saura-Calixto F., and Raso M.A. 1984. Study of gas chromatographic behaviour of alkenes based on molecular orbital calculations. *Journal of Chromatography A* 302(19): 107–117.
22. Todeschini R. and Consonni V. 2000. *Handbook of Molecular Descriptors*. Weinheim, Germany: Wiley-VCH.
23. Nasal A., Siluk D., and Kaliszan R. 2003. Chromatographic retention parameters in medicinal chemistry and molecular pharmacology. *Current Medicinal Chemistry* 10: 381–426.
24. Bate-Smith E.C. and Westall R.G. 1950. Chromatographic behaviour and chemical structure I. Some naturally occurring phenolic substances. *Biochimica et Biophysica Acta* 4: 427–440.
25. Ben-Naim A. 1980. *Hydrophobic Interactions*. New York: Plenum Press.
26. Mazak K. et al. 2003. Lipophilicity of vinpocetine and related compounds characterized by reversed-phase thin-layer chromatography. *Journal of Chromatography A* 996: 195–203.
27. Sławik T. and Kowalski C. 2002. Lipophilicity of a series of 1,2-benzisothiazol-3(2H)-ones determined by reversed-phase thin-layer chromatography. *Journal of Chromatography A* 952: 295–299.
28. Wallerstein S., Cserhati T., and Fischer J. 1993. Determination of the lipophilicity of some anti-hypoxia drugs: Comparison of TLC and HPLC methods. *Chromatographia* 35: 275–281.
29. Kiridena W. and Poole C.F. 1998. Structure-driven retention model for solvent selection and optimization in reversed-phase thin-layer chromatography. *Journal of Chromatography A* 802: 335–347.
30. Lowery A.H., Cramer C.J., Urban J.J., and Famini G.R.1995. Quantum chemical descriptors for linear solvation energy relationships. *Computers & Chemistry* 19: 209–215.
31. Eric S., Pavlovic M., Popovic G., and Agbaba D. 2005. Study of retention parameters obtained in RP-TLC system and their application on QSAR/QSPR of some alpha adrenergic and imidazoline receptor ligands. *Journal of Chromatographic Science* 45: 1–6.
32. Komsta Ł. 2008. A functional-based approach to the retention in thin layer chromatographic screening systems. *Analytica Chimica Acta* 629: 66–72.
33. Csiktusnádi-Kiss G.A., Forgács E., Markuszewski M.J., and Balogh S. 1998. Application of multivariate mathematical-statistical methods to compare reversed-phase thin-layer and liquid chromatographic behaviour of tetrazolium salts in Quantitative Structure-Retention Relationships (QSRR) studies. *Analusis* 26: 400–406.
34. Snyder L.R., Kirkland J.J., and Glajch J.L. 1997. Practical HPLC method sevelopment. In *Non-Ionic Samples: Reversed- and Normal-Phase HPLC*, pp. 233–288. New York: John Wiley & Sons.
35. Barbosa J., Toro I., Berge's R., Sanz-Nebot V. 2001. Retention behaviour of peptides, quinolones, diuretics and peptide hormones in liquid chromatography. Influence of ionic strength and pH on chromatographic retention. *Journal of Chromatography A* 915: 85–96.

3 Sorbents and Layers Used in Drug Analysis

Fred Rabel and Joseph Sherma

CONTENTS

3.1 INTRODUCTION

Because most chromatographers doing drug analyses today will be using commercially available precoated thin-layer chromatography (TLC) and high-performance TLC (HPTLC) plates, this chapter will only cover the available varieties of these plates. Using precoated plates saves considerable time and gives more reproducible results. A few bulk sorbents are available commercially for preparing homemade plates, but because of the time, cost, and experience needed to produce them reproducibly, these will not be discussed.

3.2 SUPPORTS

Any TLC plate has to be made on a rigid or semirigid support. Initially the most practical support was glass. Glass still is the most widely used support for commercially available layers. It is the usual borosilicate glass found throughout the world, but thinner than most glass used in windows, and is "float" glass (made by floating molten glass on molten metal). This manufacturing technique results in a very flat and even-thickness glass. This glass is more expensive compared to glass formed by rolling but guarantees that the thin layers coated onto it will be reproducible.

Many commercial thin layers are now available on aluminum or polymer (polyterephthalate) sheets. These have the advantages of being lighter in weight and taking up less storage space, and they can easily be cut to smaller sizes. (See Section 3.6.) Because of these advantages, aluminum-backed silica gel plates are designated for use in field kits for semiquantitative drug screening [1,2] (Chapter 14).

Further information on the thickness of the layers appropriate for each support is given in Section 3.5.

3.3 BINDERS

Most of the sorbents used for thin layers would not adhere to the support (glass, aluminum, or polymer) without the inclusion of a binder. The classic binder used was gypsum (G), calcium sulfate hemihydrate, which gives a soft layer plate that needs careful handling to prevent damaging the layer during spotting, developing, or visualizing. This binder is present in a concentration level of about 10% so is not insignificant. These plates are available from Analtech (Newark, DE) and Merck KGaA (Darmstadt, Germany or their North American division, EMD Millipore [Billerica, MA]).

There are TLC plates available with no foreign binder, designated as H, or high purity silica, designated as HR. One of the possible binders used in these plates is a sodium silicate solution (water glass), which when dried forms a version of a polymeric silica for adhering the layer. These, too, are relatively soft, fragile layers.

In the mid-1960s, manufacturers began to use inert polymeric binders at 1%–2% concentration levels. These binders begin as water soluble monomers (like polyvinyl alcohol or polyvinyl pyrrolidone) put into the slurries for plate coating. These monomers cross-link on drying with heat (and even more so when heat activation is used in the protocol). These binders have been chosen to be as inert as possible, and they vary from manufacturer to manufacturer. Therefore, comparative testing should be done to insure compatibility with the protocol being used or developed, including any detection (visualization) techniques since these, too, might be changed by the differences between the plates.

The advantages of these organic polymeric binders were that they resulted in "hard" layers, which were easier to handle through all of the TLC steps, and offered even better reproducibility since these binders were moisture barriers to prevent rapid humidity-related deactivation [3]. A general recommendation when using polymer-bound plates is to reformulate any TLC detection reagent so that the solvent(s) in which they are dissolved includes 5%–10% methanol or ethanol. The inclusion of one of these alcohols allows better penetration into the plate for reaction with any solutes. If only water, some polymer bound plates, particularly any reversed-phase (RP) bonded version, will repel the visualizer solution. The alcohol breaks the surface tension of the solution allowing it to easily penetrate the layer.

Depending on the sorbent, the support, and the expected mobile phases (polar or nonpolar), the composition of the coating slurries might be changed. Thinner layers are placed on the flexible supports, bonded RP plates will be developed with organic solvent/water combinations, and a different percentage of carbon is bonded on each type of RP. Each slurry combination of the sorbent and binder has had to be optimized to give the correct properties seen in the final product.

Many generic pharmaceutical companies use pharmacopeial methods in their laboratories and often think only the classical G or H plates can be used, since these were the only plates available when early TLC methods were written. The European Pharmacopeia (Ph Eur), British Pharmacopeia (BP), U.S. Pharmacopeia (USP), and Deutsches Arzneibuch (DAB; German Pharmacopeia) say that other binders in TLC plates can be used as long as the chromatographic results are the same. Since most of the polymeric binders used today were chosen so as not to give different results from the classical G plates, most could be used once their equivalence was determined.

3.4 FLUORESCENT INDICATORS

As an aid for detection of various classes of compounds, TLC plate manufacturers have included inorganic fluorescent indicators, or phosphors, in their plates, noted as "F" or "UV" followed by the activation wavelength. These indicators are various inorganic salts added in different percentages to the slurry before the supports are coated. If observed in the fluorescent light needed to activate the indicator, the plate will appear blue, white, or green depending on the indicator incorporated. This fluorescence should be an even color across the entire plate. If it appears blotchy or uneven, these plates should not be used for any quantitative analyses since poor results will be obtained. This could mean that the slurry (composed of water, silica gel, binder, and fluorescent indicator) was not stable, and the more dense inorganic indicator settled unevenly in the slurry during the coating or drying operation.

Organic indicators are not used since they might be moved by a mobile phase used for development, which would defeat the purpose of the indicators. Such indicators are often used as visualization solutions applied after development and drying of the TLC plate when their solubility and mobility are of no concern.

Certain compounds with double bonds and/or aromatic rings in their structures, including many drugs, will absorb the fluorescent wavelengths used to activate the indicators. This is referred to as "quenching." Such areas on a TLC plate will appear as dark spots or bands surrounded by the brighter background fluorescence. With standards and a densitometer, it is possible to quantify compounds using fluorescence quenching. A recent innovation in using these indicators is to increase their percentage in the layers so that brighter fluorescence is noted, which is considered more desirable by some users.

The indicators generally incorporated in TLC plates are [4]

- UV254s or F254s—alkali earth wolframate that is acid stable; fluoresces blue white
- UV254 or F254—manganese activated zinc silicate that is not acid stable; fluoresces green

Only strong acids like sulfuric or nitric destroy the structure of the indicator, but not acetic or formic acid.

To get the best results with a plate containing a fluorescent indicator, after developing, the plate should be dried completely so that no quenching is caused by any remaining components of the mobile phase solvents. Acids, like acetic or formic, and amines, like triethylamine, are less volatile and take time to evaporate completely (for safety, perform TLC plate drying in a hood). The speed of drying of the TLC plate can be increased by placing it in a forced air oven for a few minutes at 60°C–80°C. This assumes that the compounds to be detected are stable at these temperatures. In place of an oven, a plate heater with selectable temperature between 25°C and 200°C (programmed and actual temperatures are digitally displayed) can be used (e.g., from CAMAG USA, Wilmington, NC).

When using an F/UV plate, look first inside a viewing box using the 254 nm lamp to check for solute quenching, then with the 366 nm lamp for zones exhibiting native fluorescence. After the results of these two visualization techniques are recorded, any appropriate universal or selective detection reagents can be sprayed onto the plate or applied by dipping.

Most TLC plates sold contain a fluorescent indicator since it is there if required, but it does not interfere or elute when doing any other typical operation, as mentioned previously.

3.5 TLC VERSUS HPTLC

As with any chromatographic technique, smaller particle size sorbents give better separation in a shorter amount of time. Around 1975, HPTLC plate became commercially available. The standard TLC plate is made of silica gel particles whose average size is 15 μm with particle size range of 5–20 μm and layer thickness of 250 μm. HPTLC plates are made of silica gel particles whose average size is 5 μm with a particle size range of 4–7 μm and layer thickness of 100–200 μm. For better layer stability of preparative plates (0.5, 1, and 2 mm thicknesses are available), the average particle size is 25 μm, with a particle size range of 20–40 μm. These properties are summarized in Table 3.1.

Because of the increased resolution and tighter spots or bands, the separation times for HPTLC are reduced considerably, and many separations can be accomplished in 5–10 min using only 5 × 5 or 10 × 10 cm plates. Sample sizes are also reduced to improve the resolution, generally 0.1–0.5 μL for HPTLC compared to 1–5 μL for TLC. Table 3.2 shows a comparison of HPTLC and TLC regarding their separation parameters and detection limits. It is obvious that sample throughput is greatly increased with the use of HPTLC since decreased development times are seen, and more samples can be spotted or streaked across the same area of a plate. Since radial diffusion is less on an HPTLC plate, the more compact spots or bands give lower limits of detection.

Within the past few years, Merck KGaA has also made available HPTLC plates with spherical particles. These are made with 3–5 or 6–8 μm particle size versions of their well-known LiChrospher® Si60. These spherical particle layer plates offer improved efficiency, spot capacity, and detection limits compared to irregular particle layer plates. These are also shown in Table 3.1.

At the request of instrument manufacturers, special HPTLC plates for doing AMD (automated multiple development) with thinner layers are available. Special laser etched TLC and HPTLC plates

TABLE 3.1
Summary of Precoated Plates Available from Various Manufacturers

	TLC	HPTLC	Spherical HPTLC	PLC
Mean particle size (μm)	10–12	5–6	4 or 7	25
Distribution (μm)	5–20	4–8	3–5 or 6–8	20–40
Layer thickness (mm)	250 (glass)	100–200	100–200	0.5–2
	200 (plastic, aluminum)			

Note: Exact details may vary from manufacturer to manufacturer.

TABLE 3.2
Comparison of Typical HPTLC and TLC
Separation Parameters and Detection Limits

	HPTLC	TLC
Typical plate height	12 μm	30 μm
Typical migration distance	3–6 cm	10–15 cm
Typical migration time	3–20 min	20–200 min
Number of samples per plate	<36 (72)	<10
Sample volume	0.1–0.5 μL	1–5 μL
Detection limits—absorption	100–500 pg	1–5 ng
Detection limits—fluorescence	5–10 pg	50–100 pg

for meeting the requirements of "good laboratory practice" (GLP) with lot, batch, and individual numbers are also available. Information on these can be obtained from the manufacturers listed or by visiting their websites given in Table 3.4, the final table in this chapter.

3.6 PLATES WITH PREADSORBENT ZONES, CHANNELS, AND PRESCORING

For the convenience of the user, manufacturers have made a multilayered plate containing an area onto which samples can be more easily placed by manual application. This area is referred to as a preadsorbent or concentration zone. It is composed of a diatomaceous earth (kieselguhr) or wide-pore (50,000 Å) small-surface silica gel. With no adsorptive character, this portion of the layer is designed to only hold the samples placed on it. As mobile phase moves through this zone, it causes migration of the sample and all of its components as an integral solution to concentrate them together as a streak or band ahead of the active silica gel or bonded main layer. When the mobile phase carries the sample onto the active layer, the separation of bands begins with the now concentrated components. This process is shown in Figure 3.1. Preadsorbent plates offer significant advantages for chromatographers who apply samples and standards manually with a micropipet rather than using an automated application instrument. In many studies, bands have been shown to resolve better than spots during mobile phase development.

These multilayer plates can be of immense help in applying biological samples containing drugs in which debris can cause problems if spotted on an active layer. Such debris when placed on the silica gel layer can form an impenetrable area through which the mobile phase has trouble passing, leading to spot or band distortion.

The regular preadsorbent portion of the plate may hold some strongly adsorbed sample components during mobile phase development, and it can also be impregnated to impart different capture characteristics to the plate for on-plate cleanup. This can be done with buffers (an acidic buffer to capture bases once spotted or a basic buffer to capture acids) or with a complexing reagent. These reagents can be dipped or sprayed onto this area and allowed to dry, and then the samples are applied. This allows for different types of sample cleanup on the plate and avoids sample loss that frequently presents problems with sequential off-line techniques. On-plate cleanup can be quite successful with knowledge of components that might be in the sample but would best be removed to simplify the analysis.

Often a preadsorbent zone plate is used for preparative work since larger volumes of sample can more easily be applied to this zone. If used for this application, experimentation is necessary to determine the maximum sample loading and concentration to be placed in this area. This will insure that the final sample passed onto the active layer will not mass overload the plate resulting in poor resolution and tailing. A preparative tapered plate from Analtech containing a preadsorbent zone is described in Section 3.8.

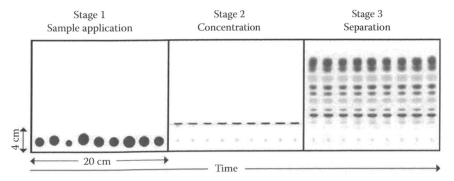

FIGURE 3.1 Stages in the use of a preadsorbent zone TLC plate. (Illustration courtesy of Merck KGaA, Darmstadt, Germany.)

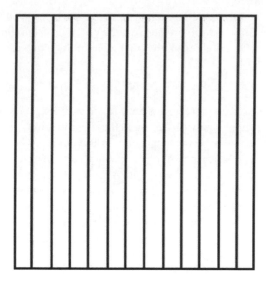

FIGURE 3.2 Channeled TLC plate showing blank areas between sorbent channels.

Channeled TLC plates (Figure 3.2) with small 1 mm sorbent-less spaces between 9 mm wide strips or tracks are available only from Analtech, with or without a preadsorbent area. Each sample is placed onto its individual sorbent track on a channeled plate. Their initial use some years ago was to provide two adjacent lanes, one with a sample and one sample-free, to cancel out any background noise with early densitometers that scanned both channels simultaneously. These plates are preferred by some users to insure sample integrity. If the developing chamber is properly leveled, any migration into another sample's space is unlikely to occur even with unchanneled, standard TLC plates.

Prescored TLC plates are precoated glass plates that the manufacturer has scored with a glass scoring device on the back of the plate. These scores allow the user to easily break the plate into smaller sizes should a larger plate size not be needed. Figure 3.3 shows the appearance of the back of a prescored plate, with the dashed lines indicating the scoring lines. Care and practice are needed in breaking any glass plate into smaller sizes whether they are prescored or are scored

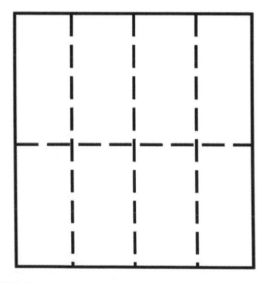

FIGURE 3.3 Prescored TLC plate back showing 20 × 20 cm plate that can be broken into smaller sizes.

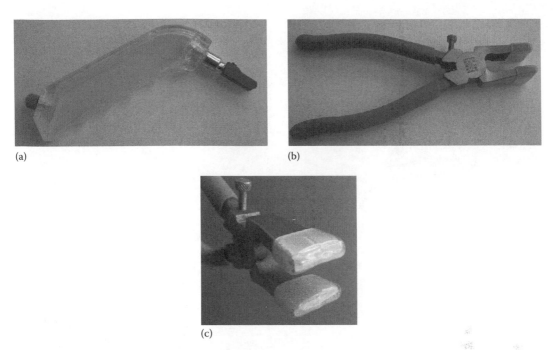

(a)   (b)

(c)

FIGURE 3.4   Glass TLC plate scoring and breaking tools: (a) pistol grip glass scorer that dispenses cutting oil; (b) grozier running pliers to grip and break along a scored line; (c) close-up of grozier pliers showing curved gripping end. (Image courtesy of Fred Rabel.)

by the user. Always wear safety glasses and gloves and use proper scoring and breaking tools as shown in Figure 3.4 that can be found in hobby or craft stores. If not using the special grozier running pliers to break along the score, the manual breaking is a combination of bending and pulling with gloved hands (often cotton gloves will suffice since the more protective leather gloves are too heavy and bulky to allow the handling of the precoated plate). Figure 3.5 shows techniques for breaking the prescored TLC plates. Fortunately, with a little experience, the polymer bound plates with their harder surface can be easily cut without damage to the surface.

Cutting flexible-backed (plastic or aluminum support) TLC plates is much easier, but good technique and proper tools make the task easier as shown in Figure 3.6. Although many different sizes of plates are available, having a box of each size can be costly, so cutting to the size needed can be

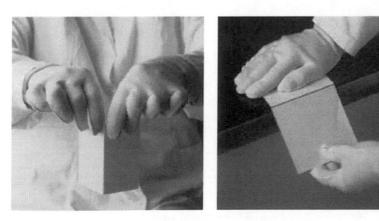

FIGURE 3.5   Techniques for breaking prescored TLC plates. (Photo courtesy of Merck KGaA, Darmstadt, Germany, ChromBook 2012, p. 96.)

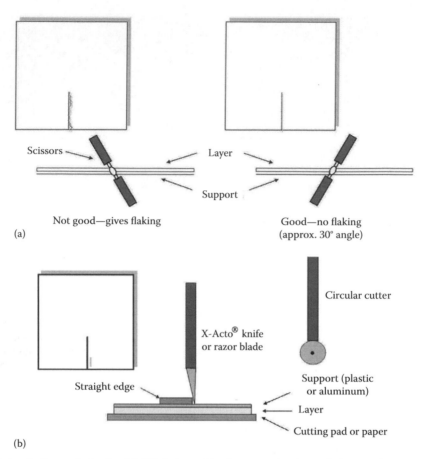

FIGURE 3.6 Cutting tools for flexible TLC plates: (a) scissors need to be angled correctly to avoid flaking of the layer; (b) circular or blade cutting of flexible layer plates. (Image courtesy of Fred Rabel.)

more economical. After cutting any TLC plates to smaller sizes, clean any newly cut edges with a brush or soft cloth to remove loose sorbent. If not cleaned (sometimes referred to as "edging the plate"), then mobile phase can wick onto the layer from the sides causing a distorted or crooked mobile phase front.

3.7 PRETREATMENT OF TLC PLATES

With time, any TLC plates may absorb extraneous compounds from their manufacture, shipping, storage, or packaging materials. If any box or seal has been broken or used over time, then this, too, can lead to compounds being absorbed from the laboratory environment. From the point of view of reproducibility, it is always best to preclean a TLC plate before use. This is accomplished by developing the silica gel (or other) TLC plate to the top with methanol [5]. The plate is then dried in a clean drying (or forced air) oven or in a clean fume hood on a TLC plate heater (CAMAG) for 30 min at 120°C.

An activated silica gel plate will naturally absorb moisture from the relative humidity in the laboratory (ideally controlled to 40%–60% and recorded routinely throughout the day). This aids in producing a TLC plate of lower activity so that less polar mobile phases will cause components to migrate further.

Some protocols may not call for plate activation, but plates still might need to be heated if they have been exposed to high humidity. If sample components are migrating with higher R_f values than normally seen, then this might be the cause.

Other reasons to heat-activate a TLC plate include strengthening the organic binder that might have "softened" in high humidity; heating helps to redry and cross-link it further. This is recommended when applying multiple visualization sprays and when doing on-plate laying of bioautographic detection solutions.

More complete suggestions for initial treatment, prewashing, activation, and conditioning of different types of glass- and foil-backed layers have been published [6].

3.8 PREPARATIVE LAYER CHROMATOGRAPHY

Although Preparative Layer Chromatography (PLC) might not often be used during drug analysis, there may come a time when an unknown zone appears in a sample chromatogram, either arising from a new synthesis, scale-up of a traditional route of a known drug, or drug degradation. Likewise, with the growing discovery of drugs derived from phytochemicals, this can certainly be the case depending on its source. Such instances would dictate scale-up to isolate and identify the unknown(s), perhaps even beginning additional clinical work to show any effects. Thus, this section has been included.

One of the advantages of TLC is that preparative work to produce a few milligrams of material is relatively easy. All that any chromatographer has to do is scale up the analytical work previously done on a standard TLC plate. This can be done by using a number of standard plates, most likely streaked with sample (rather than spotted). However, all silica gel plates and a few bonded phase plates are available with thicker layers on glass to make this process even faster with fewer plates used. These allow for more sample to be applied to obtain the required preparative amounts more rapidly. Some of these preparative plates also come with preconcentration zones to aid in their sample application. One of these is the Analtech tapered plate with a 700 μm thick preadsorbent zone and a silica gel layer progressing in thickness from 700 μm (bottom) to 1700 μm (top), which offers easy sample application as a long band and high sample resolution.

When compared with the standard TLC plates of 0.25 mm layer thickness, preparative layers of 0.5, 1, or 2 mm thickness might be assumed to allow application of 2, 4, and 8 times as much sample, respectively. Theoretically, this is true, but in real life, exceptions are often the rule. Loadability for any scale-up work is always related to the best separation obtainable for the critical pair and the solubility of the compounds in the mobile phase chosen for the separation. Thus, there is no absolute answer to the amounts you can put on a single TLC plate with a thicker layer. This must be determined by trial and error. This is particularly true since the thicker silica gel layers generate more heat of solvation, so the interior of a developing tank will be warmer when developing a thicker-layer plate compared with a thinner one. This affects the solubility and mobility of the compounds being separated, so often higher R_f values are seen on the thicker-layer plate. It is often necessary to re-optimize the mobile phase for a preparative separation to overcome these effects to obtain a better separation with lower R_f values and improved resolution.

Remember, too, that placing more than one standard-thickness silica gel plate in a developing chamber (a technique sometimes used in PLC) will lead to this increased heat effect, resulting in high R_f values. So using more multiple standard TLC plates for preparative work still has its considerations.

Another aid to keep in mind when doing preparative work is the possible use of multiple development with the same or different mobile phase(s). After drying, the plate is re-introduced into the developing chamber for a second (or more) development(s) with fresh mobile phase. Other versions of multiple development using devices to automate this process can be found in the text by Kowalska and Sherma [7] and contact with CAMAG USA.

An instrument called the Cyclograph™ (Analtech) is another approach to PLC where the separation on the precoated plate is driven by centrifugation to speed the separation of a sample into fractions. Precoated silica gel GF rotors are available from 2 to 8 mm in thickness. The possibility of multiple applications of the same sample exists with this type of plate, whether in the prewetted state (with the developing solvent) or its removal and drying before another sample is applied. This device is illustrated in Figure 3.7

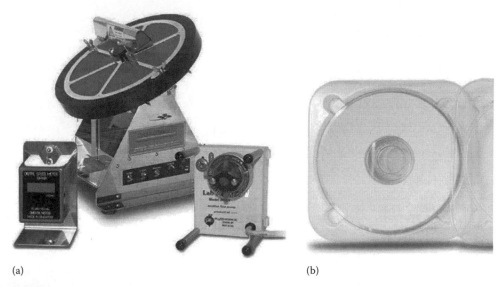

(a)   (b)

FIGURE 3.7   (a) The Cyclograph™ centrifugal PLC instrument; (b) precoated silica gel GF rotor. (Photo courtesy of Analtech, Inc, Wilmington, DE.)

3.9   SORBENTS USED IN TLC

All of the available precoated sorbents will be outlined later. The predominant precoated TLC and HPTLC layers used in most drug work continue to be the classical silica gel 60 (Si60 F_{254}), followed by RP-18/C-18 chemically bonded layers. Even if not used as often, other precoated layers could be of interest for special applications or selectivities in the future, for example, for drug determinations in clinical samples with complex matrices. A general reference is given for most sorbents, although many more references are available in other chapters in this book.

3.9.1   SILICA GEL

The sorbent used the most for TLC work is silica gel. It is an amorphous, porous matrix made by the addition of acid to a sodium silicate solution. With proper control of the hydration process, the silica gel will cross-link to form a hydrogel, which is then washed and dried, yielding a product with pore sizes of approximately 60 Å (the often referenced silica gel 60) and a surface area of about 500 m²/g. Other pore sizes/surface area silica gels can be made by changing the conditions of the hydrolysis, but the 60 Å silica gel has been the standard used in TLC. The active part of any silica gel surface is the Si–O–H groups (silanols). They strongly bind polar groups on any compound through a combination of hydrogen bonding, dipole–dipole interaction, and electrostatic interaction allowing for differentiation and separation. Water also binds strongly with silanols, so the control of moisture in the air and the mobile phase solvents is important for getting reproducible results when using this type of plate. The inherent pH of most silica gels is neutral (pH 7.0) allowing for the separation of neutral, acidic, and basic site compounds; hence, its wide use continues today. There is probably no class of compounds that has not been successfully separated on silica gel layers, so the literature is easily found for any new attempts by users to apply this layer for their application, including drug analysis.

Typical properties of silica gels suitable for TLC are as follows [8]: mean silanol group density ca. 8 μmol/m², mean pore diameter between 40 and 120 Å (4–12 nm), specific pore volume between 0.5 and 1.2 mL/g, and specific surface area S_{BET} between 400 and 800 m²/g.

After the initial manufacturing, the resulting large silica gel particles are crushed to a powder-like consistency. Once screened or air-classified, the various particle size ranges are used for the different plates manufactured. The newer spherical silica gels are synthesized as spheres, but they also need air classification to give the size ranges needed for their HPTLC layers. These sizes were discussed previously and are shown in Table 3.1.

3.9.2 Bonded Phase Silica Gels

Because of the silanols distributed throughout the structure of silica gel, it is relatively easy to bond various organosilanes (often with halide or methoxy leaving groups) to any of these groups that are stereochemically accessible. Once made into TLC plates, these bonded phase silica gels allow many different modes of chromatography to be available for developing new and complementary separations to the classical silica gel plates. With a few exceptions, most of these bonded phases are available only on HPTLC plates to take advantage of their enhanced performance mentioned previously.

3.9.2.1 Reversed-Phase Bonded Silica Gels

Early attempts to accomplish hydrophobic RP-TLC were done by impregnating silica gel or diatomaceous earth layers with paraffin or silicone oils. The development had to be with mobile phases saturated with the same oils to prevent stripping of the impregnated layers. This approach never gained wide acceptance because of its complexity. In the early 1970s, nonpolar phases bonded to silica gel and made into TLC plates came to the market to facilitate RP-TLC.

Today, a number of alkylsiloxane-bonded silica gels are available. These include dimethyl (RP-2 or C-2), octyl (RP-8 or C-8), and octadecyl (RP-18 or C-18.) The bonded CN (propylcyano) can be considered a bonded RP, with only slightly greater polarity. LiChrospher RP-18 layers were used to determine the drug purity of pentoxifylline (blood thinner) [9]. HPTLC C-8 layers were used to determine quetiapine (antipsychotic) in tablets [10].

Most drugs contain polar groups to impart water solubility allowing them to be absorbed into the body. This also makes them ideal candidates for separation in the RP mode, as shown in other chapters in this book. Separation on these bonded RPs is accomplished using combinations of polar organic solvents (methanol, acetonitrile, and tetrahydrofuran are the ones most often used) and water or buffers. Generally, varying the organic solvent changes the selectivity and peak order of the separation. Adjustments are made to the buffer salt concentration and pH to change the peak shape and peak order. As with any chromatographic method development, optimization of these parameters is necessary to get a suitable separation that is written into a protocol.

Depending on the bonding reagents (mono-, di-, or trifunctional organosilanes) used and the conditions of reaction (anhydrous or with moisture added), mono- or polymeric layers can result from bonding to the available silanols. This means that each manufacturer's bonded C-18 (or other bonded phase) will be different regarding the concentration of both bonded phase and residual surface silanols, giving different selectivities, hence different separations.

Alkylsiloxane-bonded layers with a higher percentage of bonded phase are incompatible with highly aqueous mobile phases, and often swelling of the layer is seen making the layer useless for any separation. If the aqueous portion of the mobile phase is substituted with 3% NaCl, this can make these layers work. The alternative is to use a specially formulated TLC plate, "water wettable," often designated with a W in the description. These layers with a lower degree of surface coverage and more residual silanol groups exhibit mixed hydrophilic and hydrophobic character. As such, they can be used for RP-TLC and normal-phase (NP)-TLC with purely organic, aqueous–organic, and purely aqueous mobile phases.

3.9.2.2 Impregnated Reversed-Phase Bonded Silica Gel for Enantiomer Separations

There has been an interest in performing enantiomeric separations by TLC. Only one commercial layer is available under the name Chiralplate (Macherey-Nagel, Düren, Germany) for separation of enantiomers of amino acids and amino alcohols by the mechanism of ligand exchange. These consist of a glass plate coated with C-18 bonded silica gel and impregnated with the Cu(II) complex of (2S, 4R, 2′RS)-N-(2′-hydroxydodecyl)-4-hydroxyproline as a chiral selector. Other more versatile thin layers have not been produced because of the demands of the special bonding reagents that might be required making many costs prohibitive. It is possible to do a considerable number of these separations, however, by putting the chiral selectors in the mobile phase. A book has been published containing information on the theory, materials, and techniques of chiral TLC, with applications to many compound types including drugs [11].

3.9.2.3 Hydrophilic Bonded Layers

A group of polar bonded phase silica gel precoated plates has been available for many years, but they have not found as much use as their silica gel and RP bonded siblings for drug analysis or other applications. These are hydrophilic bonded silica gel containing cyano, amino, or diol groups bonded to silica gel through short-chain nonpolar spacers (a trimethylene chain [-(CH$_2$)$_3$-] in the case of NH$_2$ and CN plates; the diol bonded phase is derived from a glycerol silane); they are wetted by all solvents, including aqueous mobile phases and exhibit multimodal mechanisms. Polarity varies as follows: unbonded silica > diol-silica > amino-silica > cyano-silica > RP materials [12].

3.9.2.3.1 Amino Bonded Layers

The amino bonded layer is the most versatile of all bonded phases. It can be used in NP, RP, or weak basic anion exchange modes by applying the appropriate mobile phase. In NP-TLC, compounds are retained on amino layers by a combination of hydrogen bonding with the amino groups and adsorption on the silanols. The activity is less than silica gel, and the selectivity is different. The separation of amphetamines in 1- and 2D systems has been studied on the amino bonded plates [13].

A limitation of this bonded phase as with others is the problem with highly aqueous mobile phases and layer stability when trying to do RP or ion exchange separations. As mentioned previously, it is possible to use organic/water combinations with buffers or solutions of NaCl to stabilize these layers. Because of this limitation, fewer applications have been reported in these modes.

A special feature of amino precoated layers is that many compounds (e.g., carbohydrates, catecholamines, and fruit acids [14]) can be detected as stable fluorescent zones by simple heating of the plate between 105°C and 220°C (thermochemical activation).

With a few RP separations of polar compounds using HPLC, there can be a problem retaining certain compounds, even when pure water is used. This recently led to the interest and use of hydrophilic interaction chromatography (HILIC) in HPLC. This mode of chromatography is not new, but is simply a version of NP partition chromatography. With this mode, polar sorbents (silica gel or bonded polar phases—amino or diol) are able to separate polar compounds when using a polar mobile phase. A classic example of the HILIC mode is the separation of carbohydrates on a silica gel or bonded amino plate with acetonitrile/water combinations [15]. This mode has certainly already been used for separation of various drugs and should be considered as a possible solution to any new drug separations being developed.

3.9.2.3.2 Diol Bonded Layers

Hydroxyl (C–OH) functional groups are found in the diol bonded TLC plates in addition to the remaining active silanol groups on the silica gel support. The vicinal diol groups (the silane reagent is derived from glycerol) are bonded to silica with a nonpolar alkyl ether spacer group. Diol bonded layers, like other polar bonded phases, can operate in NP- or RP-TLC modes, depending on the mobile phase and solutes. Polar compounds are retained by a combination of hydrogen bond and dipole type interactions in the NP mode. In the RP mode, the retention with polar mobile phases is low but higher than with amino layers. The separation of antihyperlipidemic drugs has been done on diol bonded layer plates [16], and a study of the mixed mechanisms on cyano, amino, and diol layers was reported [17].

3.9.2.3.3 Cyano Bonded Layers

Cyano layers can act as NP or RP sorbents, depending on the characteristics of the mobile phase, with properties similar to a low-capacity silica gel and a short-chain RP bonded layer, respectively. This allows for the possibility of doing 2D TLC on the same plate using appropriate mobile phases with development at right angles to one another. This bonded phase can be used in NP separations when less activity is desired compared to a silica gel layer and will also differ in selectivity. A group of nonsteroidal inflammatory drugs was studied on this layer with comparison to an RP-18 layer [18].

3.9.3 NONSILICA SORBENTS

3.9.3.1 Alumina (Aluminum Oxide)

Alumina in various forms is another polar metal oxide used in thin-layer applications. Its chemical formula is Al_2O_3. Aluminum oxide has a sorbent structure much different from silica gel, being more complex with hydroxyl groups, partial positive and negative charges on its surface, and water sorbed thereon. In the manufacturing process, the aluminum oxide can be made to have a basic surface (pH 9–10), a neutral surface (pH 7–8), or an acidic surface (pH 4–4.5). This allows for a different type of adsorption separation based solely on the surface pH. It has a high density of hydroxyl groups, about 13 $\mu mol/m^2$, so, too, can adsorb moisture readily. It is also activated at 120°C for 10 min before use [19]. Using alumina, silica gel, amino, cyano, diol, and polyamide plates with mixtures of n-hexane and six polar modifiers (acetone, dioxane, diethylamine, ethanol, isopropanol, and tetrahydrofuran) as mobile phases, Skibinski et al. studied the retention behavior of six atypical antipsychotic drugs in NP-TLC [20].

Other properties of the different aluminas available for TLC are shown in Table 3.3. The different pore size and surface area types will also impart different separation characteristics to these adsorbent layers.

3.9.3.2 Cellulose

Cellulose is derived from natural sources and, in its original fibrous form, has a polymerization of beta-glucopyranose units of 400–500 glucose in length. Any fiber can be broken to smaller sizes more suitable for TLC. And the native fiber can also be hydrolyzed to a smaller fiber, with different properties. This latter material is usually referred to as "microcrystalline" (Avicel®) and is about 40–200 glucose units in length. The mechanism of separation is NP partition with water sorbed onto its surface, with possible adsorption effects also taking part. Although the solvent migration in TLC on cellulose is slow and radial diffusion can lead to larger spots or bands, it has a unique selectivity for many compounds.

Cellulose binds very well to glass because of the fiber overlay and hydrogen bonding with the glass surface; thus, cellulose layers are binder free. To prevent cracking when an analytical layer is made, the thickness is usually only 0.1 mm. This layer has been used to separate many different natural products and drugs.

TABLE 3.3
Alumina Specifications for TLC

Type of Aluminum Oxide	Pore Diameter (Å)	Surface Area (m²/g)
60 (Type E)	60	180–200
90	90	100–130
150 (Type T)	150	70

Native cellulose has some chiral recognition properties, but often chiral selective reagents are added to the mobile phase or the layer to enhance such separations. These are shown in the following two examples, one using a cellulose layer, one a silica gel layer. The separation of the antiasthmatic, budesonide, into its stereoisomers was accomplished by impregnating the layer with 1% β-cyclodextrin and developing with a mobile phase of 1% aqueous solution of β-cyclodextrin/methanol (15:1) [21]. The second-generation cephalosporin drug cefuroxime axetil was chirally separated by HPTLC on silica gel layers using a mobile phase composed of 1% aqueous beta-cyclodextrin–methanol (15:1) [22].

3.9.3.3 Other Less Used TLC Plates

In this section, precoated sorbent layer plates that are seldom used for drug analysis are listed. This section is included should the researcher want to know about other available layers that might give a different selectivity for their research or analysis.

3.9.3.3.1 Polyamide

Only one polyamide TLC plate is available today. It is the polyamide 6 or polycaprolactam. The uniqueness of this thin layer is its ability to separate compounds through the formation of ionic, dipole, and electron donor/acceptor interactions as well as hydrogen bonding with the amide and carbonyl groups. Thus, depending on the type of analyte and mobile phase, three separation mechanisms can operate with polyamide: adsorption, partition (NP and RP), and ion exchange. These layers are free of binders and are manufactured with only a 0.1 mm thickness, so care is needed so as not to overload this plate. As mentioned previously, polyamide was included in a drug retention study using NP-TLC [20].

3.9.3.3.2 Ion Exchange Layers

Although in the past micronized ion exchange resins had been placed on supports for thin-layer separations, only two are available today from Macherey-Nagel [Ionex-25 SA-Na (strong acid cation exchanger) and Ionex-25 SB-AC (strong base anion exchanger)] formulated with silica gel to give the mixed layers.

The other ion exchange layers available are derivatized celluloses. Today only polyethyleneimine (PEI) (a strong basic anion exchanger), diethylaminoethyl (DEAE) (a weak anion exchanger), and acetylated (AC) celluloses (an RP type) are available. Others (carboxymethyl [CM], aminoethyl [AE], and phosphorylated [P]) are no longer manufactured. These plates were little used after the development of ion chromatography and ion exchange separations accomplished with HPLC systems.

3.9.3.3.3 Diatomaceous Earth (Kieselguhr)

Kieselguhr/diatomaceous earth is a natural product mined in various parts of the world and is mostly silica gel. It has very high porosity and is completely inactive. When cast into thin layers, it acts as a support for impregnations or for sample application (the preadsorbent zone plates described in Section 3.6). Another chromatographic use for diatomaceous earth is in various-sized columns that are used for "green" liquid–liquid extraction (using minimal amounts of solvent) to replace the often used separatory funnel. The sorbent holds the aqueous sample as the immiscible organic is passed through for extraction. These extraction tubes have brand names of Extrelut (Merck KGaA), CleanElut (Agilent Technologies, Santa Clara, CA), and Cleanelute (UTC, Bristol, PA).

3.9.4 Miscellaneous Layers

3.9.4.1 Impregnated Layers

It is possible to add other chemicals to the slurry when preparing a TLC plate. These might be added to adjust the pH of the layer or to aid in the separation or detection of various compounds. Some of these special layers have been made available by a few manufacturers. Since these special layers are not used routinely, they should be purchased as needed.

Among those commercially available are TLC plates with silver nitrate (added in varying amounts to allow separation of cis/trans isomers and other double bond containing compounds), ammonium sulfate (added to allow self-charring when the plate is developed and then heated), and carbomer (a generic name for synthetic high molecular weight polymer of acrylic acid, added for the analysis of mannitol/sorbitol that might be components in a drug formulation). There are other impregnating reagents, but for compounds other than drugs contact Analtech for further information on these.

If in need of any of these plates, it is possible to dip or spray these chemicals onto a plain silica gel plate (use the W type mentioned previously for better layer stability), dry, activate, and spot. This could be done initially to see how the method works before purchasing the plates commercially. If interested in any of these plates, a call can be made to the manufacturer to inquire about samples for trial use.

Impregnation of silica gel plates with oils was mentioned in Section 3.9.2.1 as the first type of RP layer. Oil-impregnated layers are still being used along with other RP layers in studies of lipophilicity, a property related to a compound's pharmaceutical and biomedical activity, to simulate results obtained using the classic octanol–water shake-flask method. As an example, the lipophilicity of parabens, which have been used as preservatives in pharmaceutical products, was estimated on RP-18, RP-18W, CN, and diol plates as well as silica gel impregnated with paraffin, olive, sunflower, or corn oils, with methanol–water mobile phases [23]. Layers used in lipophilicity studies almost always contain a fluorescent indicator to facilitate analyte detection.

3.9.4.2 Mixed-Phase Layers

One combination of ion exchange resins and silica gel was mentioned previously. Other mixed-phase layers available are combinations of aluminum oxide G/acetylated cellulose, cellulose/silica gel, kieselguhr/silica gel, and Avicel cellulose/DEAE cellulose. References for their use in drug analysis are seldom found today. Consult the manufacturers listed later for availability.

3.9.4.3 Instant TLC Stationary Phases

Although containing "TLC" in their designation, Instant TLC (ITLC) stationary phases are binderless glass microfiber chromatography paper sheets impregnated with silica gel (ITLC-SG) or salicylic acid (ITLC-SA). ITLC is widely used in thin-layer radiochromatography (TLRC; also termed radio-TLC), for example, as specified in the Eur Ph and by manufacturers of nuclear medicine agents (radiopharmaceuticals) for quality control measurement of radiochemical purity. Formerly a product of Gelman Sciences and Pall Corporation, they are now available from Agilent Technologies. An example of the use of ITLC-SG is the quality control analysis of the therapeutic agent 177 Lu-bleomycin with 10 mM pH 4 diethylenetriaminopentaacetic acid and 10% ammonium acetate–methanol (1:1) mobile phases and measurement of radioactive zones with a Bioscan Inc. (Washington, DC) AR-2000 gas-filled counter imaging scanner [24].

3.10 EARLIER SOURCES OF INFORMATION ON LAYERS FOR DRUG TLC

Biennial reviews of TLC have been written continually by Sherma from 1970 to 2010 for the ACS journal *Analytical Chemistry* and the one for 2012 for the *Journal of AOAC International* [25]. Each of these reviews covers the entire field of TLC, including theory, fundamental studies, methods, instruments, and applications for all compound classes, with many references on TLC/HPTLC-densitometry assays of drugs in bulk form and pharmaceutical products and retention and separation studies, lipophilicity determinations, chiral separations, PLC, and TLRC of drugs. Information on layers for these topics are also included in a review of HPTLC drug analysis for the period 1996–2009 [26] and an encyclopedia chapter [27]. Layers for determination of drugs in clinical samples [28] and pharmaceutical analysis [29] were discussed in book chapters. Another book chapter included tabulated migration data for more than 70 drugs in eight standardized TLC systems including four mobile phases on silica gel impregnated with KOH for basic drugs and four mobile

TABLE 3.4

Current Major TLC Plate Manufacturers[a]

North American/European TLC Plate Manufacturers	Web URL
Analtech (Wilmington, DE)	www.ichromatography.com
Dynamic Adsorbents, Inc. (Norcross, GA)	www.dynamicadsorbents.com
Mallinckrodt Baker (Phillipsburg, NJ)	Contact the local distributor
Macherey-Nagel (Düren, Germany)	http://www.mn-net.com
Merck Millipore (Darmstadt, Germany)	www.merckmillipore.de
In North America: EMD Millipore (Billerica, MA)	www.emdmillipore.com
	(click on Chemicals & Reagents)

[a] Other TLC plate manufacturers may be found around the world, but current publications tend to cite these manufacturers' plates. If using other manufacturer's plates, run standards to compare with reference results or protocols to check their suitability.

phases on plain silica gel for acidic and neutral drugs [30]. Retention data for 443 drugs and metabolites were reported for four other standardized silica gel systems by Romano et al. for use in identification by principal components analysis [31].

3.11 SOURCES OF PRECOATED TLC PLATES

Table 3.4 provides a list of the major manufacturers of thin-layer plates in Europe and North America and their websites. Since product availability changes with usage and demand, their latest offerings are best found on their websites or in their latest catalogs. Most of these products are available from various laboratory supply distributors throughout the world. Often these manufacturers have databases of applications on their products that might be of help when developing new TLC methods.

3.12 AVAILABILITY OF PRECOATED TLC PLATES

The TLC precoated plates discussed previously were available when this chapter was written (mid-2012). There are plates that might be found in references but are no longer available. Among these are diphenyl bonded plates, some brands of channeled TLC plates, and a dual-layer plate combining silica gel and bonded RP-18 adjacent to one another. In April 2012 Whatman declared they would no longer be making TLC plates, but in 2013, it was announced their manufacture would be continued by Dynamic Adsorbents (Norcross, GA) (www.dynamicadsorbents.com). If references using discontinued products are found, then alternative plates must be found for the same method or new methods might have to be developed. This is an unfortunate situation but one that might be encountered from time to time. Some manufacturers have advertised replacements for some of these discontinued plates, but as mentioned previously, tests with standards have to be run to assure equivalency.

REFERENCES

1. Global Pharma Health Fund E.V., Minilab TLC screening kit methods, http://www.gphf.org (accessed June 18, 2012).
2. O'Sullivan, C. and Sherma, J., A model procedure for the transfer of TLC pharmaceutical product screening methods designed for use in developing countries to quantitative HPTLC-densitometry methods, *Acta Chromatogr.*, 24, 241–252, 2012.
3. Rabel, F., Sorbents and precoated layers in Thin-Layer Chromatography, in *Handbook of Thin-Layer Chromatography*, 3rd Edn., eds. J. Sherma and B. Fried, Marcel Dekker Inc., New York, 2003, pp. 123–124.

4. Jork, H., Funk, W., Fischer, W., and Wimmer, H., *TLC Reagents & Detection Methods—Physical & Chemical Detection Methods: Fundamentals, Reagents, I*, Vol. 1a, Wiley, New York, 1990, p. 10.

5. Hahn-Deinstrop, E., *Applied Thin-Layer Chromatography, Best Practice and Avoidance of Mistakes*, 2nd Edn., Wiley-VCH, New York, 2007, p. 41.

6. Hahn-Dienstrop, E., TLC plates: Initial treatment, pre-washing, activation, conditioning, *J. Planar Chromatogr. Mod. TLC*, 6, 313–318, 1993.

7. Kowalska, T. and Sherma, J. (Eds.), *Preparative Layer Chromatography*, CRC Press, Taylor & Francis Group, Boca Raton, FL, 2006.

8. Hauck, H.E. and Schulz, M., Sorbents and precoated layers in PLC, in *Preparative Layer Chromatography*, eds. T. Kowalska and J. Sherma, CRC Press, Taylor & Francis Group, Boca Raton, FL, 2006, pp. 41–60.

9. Grozdanovic, O., Antic, D., and Agbaba, D., Development of an HPTLC method for in-process purity testing of pentoxifylline, *J. Sep. Sci.*, 28, 575–580, 2005.

10. Skibiński, R., Komsta, L., and Kosztyla, I., Comparative validation of quetiapine determination in tablets by NP-HPTLC and RP-HPTLC with densitometric and videodensitometric detection, *J. Planar Chromatogr. Mod. TLC*, 21, 289–294, 2008.

11. Sherma, J., Commercial precoated layers for enantiomer separations and analysis, in *Thin Layer Chromatography in Chiral Separations and Analysis*, eds. T. Kowalska and J. Sherma, CRC Press, Taylor & Francis Group, Boca Raton, FL, 2007, pp. 43–64.

12. Lepri, L. and Cincinelli, A., TLC sorbents, in *Encyclopedia of Chromatography*, 2nd Edn., ed. J. Cazes, CRC Press, Boca Raton, FL, 2005, pp. 1645–1649.

13. Fater, Z., Tasi, G., Szabady, S., and Nyiredy, S., Identification of amphetamine derivatives by unidimensional multiple development and two-dimensional HPTLC combined with postchromatographic derivatization, *J. Planar Chromatogr. Mod. TLC*, 11, 225–229, 1998.

14. Klaus, R., Fischer, W., and Hauck, H.E., Application of a thermal in-situ reaction for fluorimetric detection of carbohydrates on NH2-layers, *Chromatographia*, 29, 467–472, 1990.

15. Maloney, M.D., Carbohydrates, in *Handbook of Thin-Layer Chromatography*, 3rd Edn., eds. J. Sherma and B. Fried, Marcel Dekker Inc., New York, Chapter 16.

16. Markowski, W., Czapinska, K.L., Misztal, G., and Komsta, L., Analysis of some fibrate-type antihyperlipidemic drugs by AMD, *J. Planar Chromatogr. Mod. TLC*, 19, 260–266, 2006.

17. Kowalska, T. and Witkowska-Kita, B., A study on the mixed mechanism of solute retention in selected normal phase HPTLC systems, *J. Planar Chromatogr. Mod. TLC*, 9, 92–97, 1996.

18. Sarbu, S. and Todor, S., Determination of lipophilicity of some non-steroidal inflammatory agents and their relationships by using principle component analysis based on thin-layer chromatographic retention data, *J. Chromatogr. A*, 822, 263–269, 1998.

19. Stahl, E. (Ed.), *Thin Layer Chromatography, A Laboratory Handbook*, 2nd Edn., Springer-Verlag, New York, 1969.

20. Skibiński, R., Misztal, G., Komsta, L., and Korólczyk, A., The retention behavior of some atypical antipsychotic drugs in normal-phase TLC, *J. Planar Chromatogr. Mod. TLC*, 19, 73–80, 2006.

21. Krzek, J., Hubicka, M., Dabrowska-Tylka, M., and Leciejewicz-Ziemecka, E., Determination of budesonide R(+) and S(−) isomers in pharmaceuticals by thin-layer chromatography with UV densitometric detection, *Chromatographia*, 56, 759–762, 2002.

22. Dabrowska, M. and Krzek, J., Chiral separation of diastereoisomers of cefuroxime axetil by high performance thin layer chromatography, *J. AOAC Int.*, 93, 771–777, 2010.

23. Casoni, D. and Sarbu, C., The lipophilicity of parabens estimated on reversed phases chemically bonded and oil impregnated plates and calculated using different computation methods, *J. Sep. Sci.*, 32, 2377–2384, 2009.

24. Yousefina, H., Jalilian, A.R., Zolghardi, S., Bahrami-Samani, A., Shirvani-Arani, S., and Gjannadi-Maragheh, M., Preparation and quality control of lutemium-177 bleomycin as a possible therapeutic agent, *Nukleonika*, 55, 285–291, 2010.

25. Sherma, J., Biennial review of planar chromatography: 2009–2011, *J. AOAC Int.*, 25(4), 992–1009, 2012.

26. Sherma, J., Review of HPTLC in drug analysis, *J. AOAC Int.*, 93, 754–764, 2010.

27. Sherma, J., Thin layer chromatography, in *Encyclopedia of Pharmaceutical Technology*, eds. J. Swarbrick and J.C. Boylan, Vol. 15, Marcel Dekker Inc, New York, 1997, pp. 81–206.

28. Jain, R., Thin layer chromatography in clinical chemistry, in *Practical Thin Layer Chromatography—A Multidisciplinary Approach*, eds. B. Fried and J. Sherma, CRC Press, Boca Raton, FL, 1996, pp. 131–152.

29. Dreassi, E., Ceramelli, G., and Corti, P., Thin layer chromatography in pharmaceutical analysis, in *Practical Thin Layer Chromatography–A Multidisciplinary Approach*, eds. B. Fried and J. Sherma, CRC Press, Boca Raton, FL, 1996, pp. 231–247.

30. Ng, L., Pharmaceuticals and Drugs, in *Handbook of Thin Layer Chromatography*, eds. J. Sherma and B. Fried, Marcel Dekker Inc., New York, 1991, pp. 717–755.

31. Romano, G., Caruso, G., Musumarra, G., Pavone, D., and Cruciani, G., Qualitative organic analysis. part 3. identification of drugs and their metabolites by PCA of standardized TLC data, *J. Planar Chromatogr. Mod. TLC*, 7, 1994, 233–241.

4 Optimization of the Mobile Phase Composition

Mirosław Hawrył and Anna Hawrył

CONTENTS

4.1 INTRODUCTION

The main aim of the chromatographic analysis—the separation of analyzed mixture (R_S)—is described by well-known Purnell's equation [1]:

$$R_S = \frac{\sqrt{N}}{4} \cdot \frac{\alpha-1}{\alpha} \cdot \frac{k}{1+k} \tag{4.1}$$

where

N is the number of theoretical plates
α is the chromatographic system selectivity
k is the retention factor

Three parts of this equation show the efficiency, selectivity, and retention, respectively.

The last part of Equation 4.1 presents the fraction of separated substance in the stationary phase, and depending on retention factor value (k), it can receive values from 0 (for the compound not adsorbed by stationary phase) to 1 (for strong retention—big values of k). The resolution factor $R_S = 0$ when $k = 0$ and the analyzed compounds could not be separated, and for $k > 10$ the last segment of Equation 4.1 approaches the maximal value 1. However, the big values of retention factor (k) are not advantageous because of the peak diffusion and the long time of analysis:

$$V_R = V_M(1+k) \tag{4.2}$$

where

V_R is the retention volume
V_M is the dead retention volume

For this reason the optimal range of retention factor is limited for liquid chromatography and is enclosed between 0.1 and 10 for column chromatography and $0.1 < R_F < 0.9$ for thin-layer chromatography (TLC).

In fact for high-performance liquid chromatography (HPLC) of two to three component mixtures, the range of k is more narrow (1–5) unless in the mixture ballast substances (which have small values of k) exist. Then the increase of k of analyzed compounds is needed for complete separation.

Thus, the regulation of k values is one of the fundamental aspects in the optimization of liquid chromatography systems. Nowadays, the assortment of the stationary phases used in TLC is tight: silica, Florisil, alumina, polyamide, and cellulose—polar stationary phases—and silanized silicas (RP, CN, DIOL, and NH_2)—nonpolar stationary phases—so the regulation and optimization of retention in TLC comes down to quantitative and qualitative selection of eluents used in chromatographic process [2–11].

There are three main types of separation mechanisms in TLC:

1. Adsorption
2. Partition
3. Ion exchange

If the sorbent surface is directly involved in the separation, this is known as *adsorption chromatography*. If the stationary phase merely affords a mechanism to immobilize a liquid phase, this is described as *partition chromatography*. The *ion-exchange* process is dependent on the sorbent containing ions that are capable of exchanging with ions of like charge in the sample or mobile phase. The exchange of ions is dependent on the affinity of the support for the various ionic species. The mobile phase acts as an electrolyte solution.

The solvent system used in TLC performs the following main tasks:

1. To dissolve the mixture of substances
2. To transport the substances to be separated across the sorbent layer
3. To give hRf values in the medium range, or as near to this as possible
4. To provide adequate selectivity for the substance mixture to be separated

They should also fulfill the following requirements:

• Adequate purity
• Adequate stability
• Low viscosity
• Linear partition isotherm
• A vapor pressure that is neither very low nor very high
• Toxicity that is as low as possible

Three main factors exert an influence on the result of chromatographic separation in TLC: (1) the polarity of the stationary phase in partition chromatography or its activity in adsorption chromatography, (2) the type and polarity of mobile phase, and (3) the composition of vapors that contact with adsorbent layer. Although all three factors are important, the composition and the type of mobile phase seem to be the most significant and have the most influence on the efficiency of chromatographic separation in TLC.

4.2 ELUOTROPIC SERIES

In adsorption liquid chromatography, some specific intramolecular interactions are of great importance, so the polarity of the mobile phase is a very important factor. The ability of chromatographed compounds to create hydrogen bonds has a special role. In the theory of adsorption on polar adsorbents (like SiO_2 or Al_2O_3), generally one catches on that the adsorption of molecules of chromatographed substances is connected with displacement of solvent molecules from the surface of stationary phase [8,12–14].

The empirical list of solvents from less to most polar in the form of *eluotropic series* [13,15] (Table 4.1) can be interpreted by the competitive character of the substance and solvent adsorption. Molecules of aliphatic hydrocarbons are weakly adsorbed and they are not competitive for the adsorbate molecules, so in this case the strong adsorption (high values of k) of compounds creating hydrogen bonds with hydroxyl groups on the adsorbent surface exists when solvents from the start of eluotropic series are applied. Benzene, the solvent that can form π-complexes, is the stronger eluent, and diethyl ether, acetone, ethanol, and more basic solvents (e.g., diethylamine type) have greater elution strength, respectively.

Besides the basic character of electronegative atom, the steric situation plays a part in the adsorption as well. The most eliminating strength has polar molecules with small areas that are capable of interacting with a large number of hydroxyl groups on the adsorbent surface; from this it follows the order: ethers < ketones < alcohols in eluotropic series. However, in homologous series of solvents, the elution strength decreases when the size of molecules increases, so the elution strength is larger for methyl-alkyl ketones than isomeric dialkyl ketones [16].

The competitive adsorption of solvent molecules is most important, but other factors must be taken into consideration, for example, adsorbate–solvent interactions in the volumetric phase [17–19] as well. Because the competitive character of the sorbent and substance adsorption is dominant, the eluotropic series for a given adsorbent type (e.g., silica) is the universal characteristic

TABLE 4.1
Properties of Solvents for Use in TLC

	(MG/d_{20})	SiO_2	Al_2O_3	A_P
n-Pentane	114.9	0.0	0.0	5.9
n-Octane	163.0	0.0	0.0	7.6
Isopropyl ether	142.0	0.28	0.28	5.1
Toluene	106.5	0.22	0.29	6.8
Benzene	89.2	0.25	0.32	6.0
Ethyl ether	104.4	0.43	0.38	4.5
Methylene chloride	64.4	0.3	0.4	4.1
1,2-Dichloroethane	63.7	0.32	0.47	4.8
MTBE	119.1	0.35	0.48	—
Acetone	73.8	0.5	0.58	4.2
Methyl ethyl ketone	89.3		0.51	4.6
THF	81.2	0.53	0.45	5.0
Acetonitrile	52.7	0.6	0.55	3.1
Ethyl acetate	98.1	0.48	0.6	4.5
Dioxane	85.5	0.6	0.61	6.0
Methanol	40.6	0.7	0.95	8.0
2-Propanol	76.8	0.6	0.82	8.0

Note: Listed are the molar masses in g/mol and solvent strength e^0 and A_P.

of the solvent elution strength. Alumina has a different character of adsorption centers, so the sequence of solvents in eluotropic series is somewhat changed (see Table 4.1).

Nonpolar (RP) adsorbents have quite different structure than polar adsorbents, and interactions of solvents with these adsorbent surfaces are radically different. In this case hydrophobic interactions play a principal role, so nonpolar substances (e.g., PAHs) from water solutions are most strongly adsorbed. The dilution of water by a polar modifier (methanol, acetonitrile, dioxane, tetrahydrofuran, etc.) causes weak adsorption. The eluotropic series for nonpolar (reversed-phase) stationary phases are inversed in comparison with eluotropic series for polar adsorbents.

The number of solvents in eluotropic series can be strongly limited by the application of mixed solvents, for example, by application of n-hexane—acetone mixtures—the whole polarity range can be obtained for continuous regulation of k values.

The elution strength of solvent mixtures that have quite different polarities changes approximately linearly (in logarithmic scale) when the concentration of polar component in the binary mobile phase changes.

The principle of the variation of solvent composition while holding solvent strength constant was first developed by Neher [13,20] for separation of steroids. Figure 4.1 is a representation of the eluotropic series of Neher for application in this way. Six solvents, which will act as solvent S1 (100% concentration, on the left), are arranged vertically, whereas the same solvents, acting as solvent S2 (100% concentration, on the right of the horizontal lines) in a binary solvent eluotropic series, are arranged horizontally. We can obtain a very large number of binary systems, of the same or different strength, by means of this nomogram. The dashed line X determines 12 compositions of binary systems of the same average eluotropic properties. These systems are called equieluotropic systems. The nomogram in Figure 4.1 corresponds to adsorption on silica gel [21].

Saunders [22] obtained another nomogram by using six very common solvents. With the help of the nomogram presented in Figure 4.2, we can achieve binary solvent mixtures of certain strength in the interval 0.0–0.75. In this graph, ε^0 is plotted across the top and in various binary solvent compositions in each of the horizontal lines below it. Each line corresponds to the range 0%–100% by volume of binary solvent composition. Its manner of use is similar to that described for Neher's nomogram.

The application of mixtures with various qualitative compositions fixed on the basis of equieluotropic series guarantees similar ranges of k factor of components in chromatographed mixture. However, individual effects of interactions in the adsorption phase and volume phase cause that k values for individual eluents are not identical, but they demonstrate the diversity that has an influence on the chromatographic system selectivity. So in the case of unsatisfactory selectivity by use of one chromatographic system (after the optimization of mobile phase), the second system should be optimized according to a second vertical line in the equieluotropic table (which has similar elution strength but changed selectivity and gives better separation).

4.3 QUANTITATIVE RELATIONSHIPS OF k VALUES VERSUS ELUENT COMPOSITION

Eluents used in liquid chromatography are usually binary or ternary. There are many papers in which the optimization and theoretical investigations of adsorption using mixed solvents were described [4,6–8,10,14,16–19,23–30]. Theoretical investigations lead to several equations that describe quantitative retention versus eluent composition relationships. Among many theories, Snyder–Soczewiński's model will be described in this chapter as the most realistic and simple model for the prediction of conditions for TLC in normal-phase and reversed-phase systems [8,13,14,16,19,31,32].

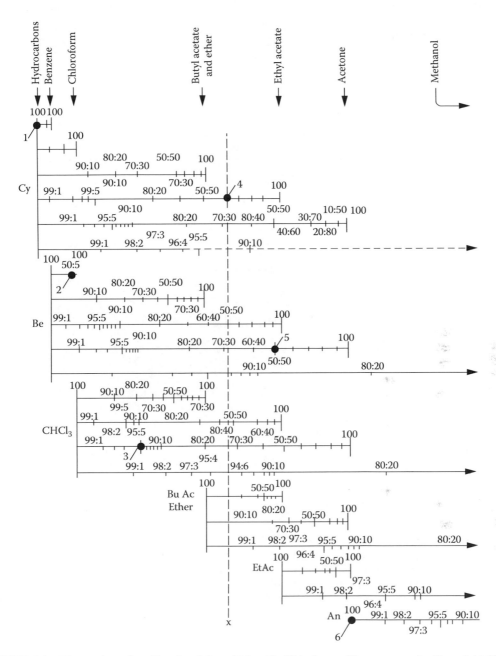

FIGURE 4.1 Eluotropic series. (Reprinted from Neher, R., *Thin-Layer Chromatography*, Copyright 1964, with permission from Elsevier.)

4.3.1 POLAR ADSORBENTS (NORMAL-PHASE CHROMATOGRAPHIC SYSTEMS)

In the adsorption theory on polar adsorbents, the competitive character of interactions between the stationary phase and mobile phase is assumed. The fundamentals of this theory can be introduced as follows:

- The air humidity has great influence on R_F values.
- The system overload leads to form ghost peaks.

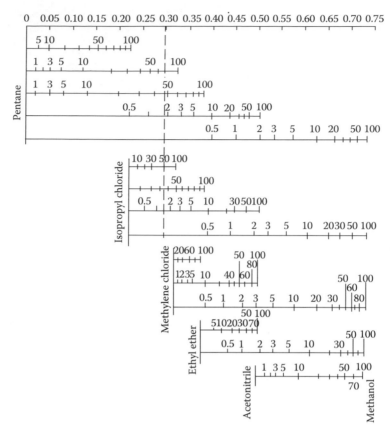

FIGURE 4.2 Eluotropic series. (Reprinted from Saunders, D.L., *Anal. Chem.*, 46, 470, 1974. With permission.)

- Adsorbents of the inorganic oxide type allow the separation according to polarity and the type of functional groups more than the dimension of molecules. Hydrogen bonds play an important role in the separation process.
- The mobile phase contains low or average quantities of polar solvents that have low volatility. The type of developing chamber (especially the volume of gas phase saturating the chamber) has the greater influence on the separation result than in partition chromatography.

Basic Snyder's equation describing the influence of various factors on the R_M value can be presented as [13,33,34]

$$R_M = lg \frac{W_a}{V_P} V_a + \alpha_a (S^0 - \varepsilon^0 A_P) \tag{4.3}$$

where

V_a is the volume of adsorbed solvent monolayer per gram of adsorbent

W_a is the weight of adsorbent layer (g)

V_P is the pore volume freely accessible to the solvent (cm³)

α_a is the energy component of activity parameters, with α_a 1 for the most highly active aluminum oxide

S^0 is the adsorption energy of the sample

ε^0 is the solvent parameter

A_P is the area occupied by a sample molecule (at adsorption center)

Thin-layer (adsorbent) parameters are placed in logarithmic part of Equation 4.3. The elution strength of the solvent is described by the activity parameter. The properties of the sample are described by the S^0 parameter (adsorption energy of the sample) and A_P (the area occupied by a sample molecule on the adsorbent layer). The solvent parameter (ε^0) shows the characteristic of the mobile phase. One molecule of the sample occupies the area A_P and can force out a corresponding number of solvent molecules that were adsorbed on the active centers of the stationary phase. This process releases the adsorption energy S^0. For aliphatic compounds the adsorption energy equals 0.

In the ideal system all previously cited parameters are independent of each other. In practice, however, the adsorption energy for various compounds depends on adsorbent properties, similarly as parameters A_P and ε^0. From a qualitative point of view, Snyder's equation shows that the greater adsorption energy of the sample causes the increase of R_M value of this sample and the decrease of its R_F value. The greater value of the ε^0 parameter causes the increase of R_F values and for low polar compounds that cannot saturate active centers of adsorbent $\varepsilon^0 = 0$ and $S^0 = 0$, and only the first part of Equation 4.3 is actual. When $S^0 \ll A_P\varepsilon^0$ (the mobile phase reduces the interactions with the adsorbent layer), R_M values assume negative values and R_F values approach 1. For $S^0 \gg A_P\varepsilon^0$ (the mobile phase cannot displace the sample from the adsorbent layer), R_M values are positive and R_F values are near 0.

4.3.1.1 Solvent Elution Strength (ε^0)

The elution strength of the solvent in adsorption chromatography (according to Snyder [13,33]) determines the influence of solvent molecules on the adsorption of the molecule of the chromatographed substance. Snyder defines the elution strength of the mobile phase as the relationship of the adsorption energy (E_P) of solvent molecules that are desorbed by the sample and the surface (A_P) initially occupied by mobile phase molecule on the surface of the stationary phase. The parameter ε^0 describes the adsorption energy of the solvent molecule per unit surface area of the adsorbent [13]:

$$\varepsilon^0 = \frac{E_P}{A_P} = \frac{\Delta G^0}{2.3RTA_P} \tag{4.4}$$

where
  E_P is the adsorption energy of the solvent molecule
  A_P is the adsorption area required by a solvent molecule
  ΔG^0 is the standard free energy change

The previously mentioned definition of ε^0 shows that the elution strength not only is the function of the solvent but depends on the properties of the stationary phase as well.

To determine the elution strength of the mobile phase, chromatography, by using the solvent with $\varepsilon^0 = 0$ (e.g., pentane) and adsorbent with activity $\alpha_a = 1$ (most active alumina), is performed to obtain the R_M values. From R_M values parameters A_P and S^0 are calculated according to Equation 4.3:

$$R_M = lg\frac{W_a}{V_P}V_a + \alpha_a S^0 - \alpha_a A_P\varepsilon^0 = C - \alpha_a A_P\varepsilon^0$$

For solvents with various elution strengths on the given adsorbent, the relationship R_M versus elution strength is linear with a sloped intercept C. Assuming that for pentane $\varepsilon^0 = 0$ and calculating $\alpha_a A_P$ and C values, it is easy to determine ε^0 for any solvent. Tabulation of ε^0 values in descending order is known as *eluotropic series* [33] (see Chapter 42).

4.3.1.2 Elution Strength of Binary Mixtures of Solvents

For binary mixtures of solvents, Snyder [33] drew a conclusion about a logarithmic dependence of the elution strength and the composition of the mobile phase.

From Equation 4.3 it results that $R_M = \log k$ (k—retention factor) and for the difference of two values of R_M $R_{M_B} - R_{M_A} = \log(k_B/k_A) = \alpha_a A_P \left(\varepsilon_A^0 - \varepsilon_B^0 \right)$.

For a mixture of two solvents A and B, the analogous equation $R_{M_B} - R_{M_{AB}}$ is true. After the reduction

$$\frac{k_B}{k_{AB}} \approx \frac{n_B}{n_A + n_B} = N_B$$

When $N_B > 0.2$ and $\alpha_a \left(\varepsilon_A^0 - \varepsilon_B^0 \right) > 0.2$, the equation can be reduced to

$$\varepsilon_{AB}^0 = \varepsilon_B^0 + \frac{\log N_B}{\alpha_a A_{PB}} \tag{4.5}$$

Equation 4.5 can be proved experimentally.

Figure 4.3 shows the relationship between the elution strength of the binary mixture (AB) and the percentage of the B component in this mixture. The low percentage of polar solvent in n-pentane, n-octane, or dichloromethane causes a strong increase of the elution strength. In higher concentrations of the polar component, the elution strength gradually achieves the asymptotic value. Figure 4.3 shows the logarithmic relationship between the elution strength of the solvent and the composition of the mixture of eluents.

Geiss [33] described the method of the acceleration of the selection of the optimal mobile phase on the basis of the elution strength idea. He proposed the mixture of 13 various solvents to optimize appropriate elution strength (see Table 4.2). The optimal separation is achieved when the R_F value is 0.33. When the appropriate elution strength is known for this value, the selectivity of separation by combination of individual mixtures with others can be optimized (when the same elution strength is kept).

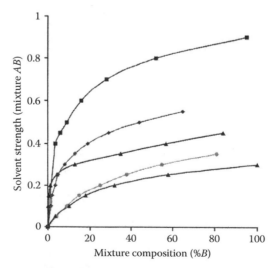

FIGURE 4.3 Solvent strength $e^0 AB$ of binary mixture AB at increasing volume fraction of the polar solvent B. From top to bottom: dichloromethane in octane, diethyl ether in pentane, methyl tert-butyl ether in octane, methyl acetate in pentane, and methanol in dichloromethane. (Reprinted from Geiss, F., *Principles of Adsorption Chromatography*, Dekker, New York, 1987. With permission.)

TABLE 4.2

Solvent Mixtures with Equal Solvent Strengths e^0

ε^0	DCE/ OCT	MTBE/ OCT	CAN/ PEN	MTBE/ DCE	ACN/ DCE	MeOH/ DCE
0.00	0	0	0	0	0	0
0.05	3.5					
0.10	10	0.2	0.3			
0.15	18	0.6	0.6			
0.20	32	1.4	1.1			
0.25	58	4.3	2			
0.30	100	13	3.5	0	0	0
0.35		35	8	30	12	
0.40		37	24	60	30	3.5
0.45		84	52	88	55	6
0.50			88		88	9
0.60			100		100	16
0.70						28
0.80						52
0.90						95

The elution strength of this new mixture can be calculated according to Equation 4.5. Correct results of this method were experimentally confirmed.

Eluotropic series also can be used for the prediction and optimization of the retention in reversed-phase systems taking some restrictions into consideration:

1. In reversed-phase systems, the elution strength increases when the polarity of the solvent decreases.
2. Water is the solvent whose elution strength equals 0.
3. The selection of modifiers is limited to solvents mixing with water.

4.3.1.3 Partition Chromatography Systems

In partition chromatography, the sample is divided between the stationary and mobile phase. The separation is observed when partition coefficients of the sample are different:

$$K = \frac{c_s}{c_m}$$

where
 K is the partition coefficient (repartition coefficient)
 c_s is the substance concentration in the stationary phase
 c_m is the substance concentration in the mobile phase

Thin-layer partition chromatography is characterized by several factors:

- The stationary phase is formed by mobile phase during development. Methanol, acetonitrile, and tetrahydrofuran are the most often used mobile phases. Cellulose, 3-aminopropylsiloxane-bonded silica, and silica are the most often used stationary phases.
- The mobile phase is rich in water and all active centers of stationary phase are deactivated.

- The humidity has practically no influence on the retention (as in adsorption chromatography).
- The thickness of the liquid phase immobilized on the stationary phase must be large enough that the sample can be dissolved in it like in a "real" liquid.
- Systems with water as stationary phase have the influence on the separation of very polar compounds, like sugars, acids, or amino acids. In this case mobile phase butanol–acetic acid–water (4 + 4 + 1 (V/V)) is most often used.
- Separations in which RP-18 phases are used can be discussed as the separation process using lipophilic stationary phases and hydrophilic mobile phases. RP-2 phases do not demonstrate "pure" partition mechanism because the coating of the surface is too small.
- Zone broadening is greater than in adsorption chromatography because the broadening process in the stationary phase is negligible.

Partition chromatography is relatively rarely used compared to adsorption chromatography.

4.3.2 SOLVENT THEORY ACCORDING TO SNYDER: SELECTIVITY TRIANGLE OF SOLVENTS

In reversed-phase HPLC and TLC systems, Snyder's selectivity model is most often applied for partition systems [35,36]. Snyder divided solvents into eight groups according to their interactions with three solutes on the basis of their partition coefficients in the gas–liquid system corrected for differences in molecule values of the solvent, polarizability, and dispersion interactions. Each value was corrected empirically to obtain value 0 of polar partition constants for solvents of the type n-alkanes. For these calculations the database of partition coefficients gas–liquid according to Rohrschneider for 81 solvents dissolved in n-octane, toluene, ethanol, ethyl-methyl ketone, dioxane, and nitromethane was used [36]. Snyder chooses nitromethane, ethanol, and dioxane as a probe of ability of solvents for interactions, respectively: dipole–dipole, hydrogen-bond donor, and hydrogen-bond acceptor [37]. Detailed method of the calculation of corrected R'_M values can be found in the previous paper [38]. Next, Snyder defined the concept of polarity index (P'):

$$P' \equiv \log(K'')_{\text{Ethanol}} + \log(K'')_{\text{Dioxane}} + \log(K'')_{\text{CH}_3\text{NO}_2}$$

Using this relationship, the strength of interactions of solvent and ethanol, dioxane, and nitromethane was calculated, for example,

$$x_e = \frac{\log(K'')_{\text{Ethanol}}}{P'}$$

Thus, the following relationship is true:

$$P' = x_e P' + x_d P' + x_n P' \quad (x_e + x_d + x_n = 1).$$

where
  P' is the polarity scale (solvent strength parameter)
  x_e is the proportion of solvent interacting capacity with ethanol (proton donor character)
  x_d is the proportion of solvent interacting capacity with dioxane (proton acceptor character)
  x_n is the proportion of solvent interaction capacity with nitromethane (dipole character)

The elution strength parameter (ε^0) is only partially comparable with the polarity scale (P'). Approximate equation of the relationship ε^0 versus P' can be shown as

$$\varepsilon^0 \approx 0.1P'$$

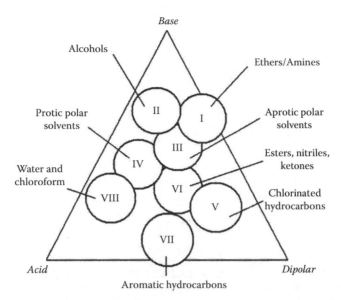

FIGURE 4.4 Snyder's selectivity triangle. (Reprinted from Snyder, L.R., *J. Chromatogr. A*, 92, 223, 1974. With permission.)

Three experimentally obtained selectivity parameters (x_e, x_a, x_n) characterize important features of solvent mixtures. Taking various polar partition constants (selectivity parameters) into consideration for all investigated solvents, Snyder divided them into eight groups (Figure 4.4). Each group contains solvents that have similar characteristic. Strength-adjusting solvents (e.g., *n*-pentane, iso-octane, cyclohexane, and *n*-hexane) are not classified in any group, because their elution strength is 0 and they have no selectivity. Because the sum of selectivity coefficients is always equal 1, the quotient of two factors is sufficient to obtain all three values.

On the basis of Snyder's studies, Nyiredy defined the selectivity factor as [39,40]

$$S_f = \frac{x_e}{x_d}$$

where
  S_f is the selectivity factor
  x_e is the proton acceptor character
  x_d is the proton donator character

When the average value of selectivity factors of a given group is plotted depending on the elution strength, the graph is two lines (Figure 4.5). The solvent clusters I–IV and VIII form one line, and the solvent clusters I, V, and VII form the second line.

Solvent cluster I (ethers and amines) shows the highest selectivity factors, that is, the greatest proton acceptor effects. Solvent clusters VIII and VII show the lowest proton acceptor characteristics. Clusters VIa and VIb lie between the two straight lines. Solvents from both of these clusters are suitable as solvent mediators. They include ethyl acetate, methyl ethyl ketone, cyclohexanone, dioxane, acetone, and acetonitrile.

The selection of the mobile phase and its optimization procedure using the selectivity triangle is presented in Figure 4.6.

The values of selectivity factors are listed in Table 4.3 [32].

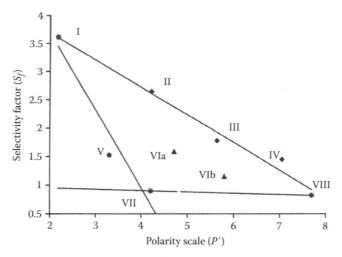

FIGURE 4.5 The selectivity factors (S_f), plotted against the solvent strength (P'). (Reprinted from Nyiredy, Sz., Solvent classification for liquid chromatography, In *Chromatography, Celebrating Michael Tswett's 125th Birthday*, InCom Sonderband, D€usseldorf, Germany, pp. 231–239, 1997; Nyiredy, Sz. et al., *J. Planar Chromatogr.*, 7, 406, 1994. With permission.)

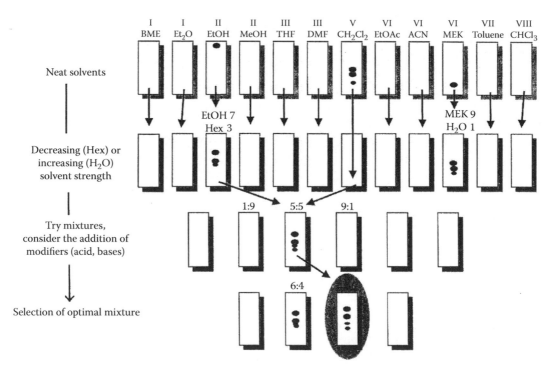

FIGURE 4.6 Selection and optimization of the mobile phase. Stage 1, pure solvents; stage 2, decreasing or increasing solvent strength by mixing two solvents; stage 3, more complex mixing of solvents and addition of acid or base modifiers; stage 4, selection of optimal solvent mixture. BME, butylmethyl ether; THF, tetrahydrofuran; DMF, dimethylformamide; DCM, dichloromethane; EtOAc, ethyl acetate; ACN, acetonitrile; MEK, butanone; CHCl3, chloroform. (Reprinted from Wall, P.E., *Thin-Layer Chromatography. A Modern Practical Approach*, RSC, Cambridge, U.K., 2005. With permission.)

TABLE 4.3
Polarity Scale, Weighting Factors, and Group Membership (with Interpolated Values for Reversed-Phase Chromatography in Brackets)

Solvent	P' Value	x_e	x_d	x_n	Group	γ/η (25°)[a]
n-Hexane	0.1	—	—	—	0	56
n-Pentane	0.1	—	—	—	0	67
Cyclohexane	0.2	—	—	—	0	28
Di-n-butyl ether	2.1	0.44	0.18	0.38	I	91
Di-isopropyl ether	2.4	0.48	0.14	0.38	I	91
Toluene	2.4	0.25	0.28	0.47	VII	48
Triethylamine	1.9 (2.2)	0.56 (0.66)	0.12 (0.08)	0.32 (0.26)	I	52
Methyl-t-butyl ether	2.7	0.49	0.14	0.37	I	72
Diethyl ether	2.8	0.53	0.13	0.34	I	71
Methylene chloride	3.1 (4.3)	0.29 (0.27)	0.18 (0.33)	0.53	V (VII)	62
n-Octanol	3.4	0.56	0.18	0.25	II	3.7
1,1-Dichlorethane	3.5	0.30	0.21	0.49	V	41
2-Propanol	3.9	0.55	0.19	0.27	II	8.7
n-Butanol	3.9	0.56	0.19	0.25	II	8.3
THF	4.0 (4.4)	0.38	0.20	0.42	III	56
1-Propanol	4.0	0.54	0.19	0.27	II	11
t-Butanol	4.1	0.56	0.20	0.24	II	7.3
Chloroform	4.1 (4.3)	0.25 (0.31)	0.41 (0.35)	0.34 (0.34)	VIII	47
Ethanol	4.3 (3.6)	0.52	0.19	0.29	II	19
Ethyl acetate	4.4	0.34	0.23	0.43	VIa	52
Bis-(2-ethoxyethyl)-ether	4.6	0.37	0.21	0.42	VIa	—
Cyclohexanone	4.7	0.36	0.22	0.42	VIa	40[b]
Methyl ethyl ketone	4.7	0.35	0.22	0.43	VIa	57
Dioxane	4.8	0.36	0.24	0.40	VIa	26
Chinolin	5.0	0.41	0.23	0.36	III	26
Acetone	5.1 (3.4)	0.35	0.23	0.42	VIa	74
Methanol	5.1 (3.0)	0.48	0.22	0.31	II	38
Pyridine	5.3	0.41	0.22	0.36	III	39
Methoxy ethanol	5.5	0.38	0.24	0.38	III	18
Benzyl alcohol	5.7	0.40	0.30	0.30	IV	—
Acetonitrile	5.8 (3.1)	0.31	0.27	0.42	VIb	75
Acetic acid	6.0	0.39	0.31	0.30	IV	21
Formic acid	6.0	—	—	—	IV	—
Nitromethane	6.0	0.28	0.31	0.41	VII	—
Methyl formamide	6.0	0.41	0.23	0.36	III	—
DMF	6.4	0.39	0.21	0.40	III	40
Ethylene glycol	6.9	0.43	0.29	0.28	IV	2.3
DMSO	7.2	0.39	0.23	0.39	III	2.4
m-Cresol	7.4	0.38	0.37	0.25	VIII	2.8
Dodecafluoroheptanol	8.8	0.33	0.40	0.27	VIII	—
Formamide	9.6	0.37	0.33	0.30	IV	17
Water	10.2 (0.0)	0.37	0.37	0.25	VIII	73

Italics indicate values according to Reference [46].
[a] γ/η [m/s].
[b] Value of methyl butyl ketone.

4.4 PRISMA MODEL

A further development to this selectivity triangle is a prism. Nyiredy et al. [41–44] proposed such a system, calling it the PRISMA model. This is a 3D model used to correlate the selectivity of the solvent with the solvent strength. Silica gel is taken as the stationary phase and a preliminary solvent selection made according to the previously described Snyder classification using three solvents from the eight solvent groups. This is the first step in the optimization. Step two involves the use of an unusual prism model, the upper part of which is an irregular frustum, and the middle and lower parts are regular triangular prisms as shown in Figure 4.7. The base of the prism represents the modifier (in the case of normal-phase separations, this will be n-hexane with an e^0 value of 0.00). The heights of the prism at each edge (S1, S2, and S3) represent the solvent strength for the neat solvents selected from step one. An isoeluotropic plane (where e^0 values are equivalent) is formed by increasing the solvent strength at the corners of the prism. Points along the edges are therefore combinations of two solvents, points on the sides, three solvent mixtures, and points within the prism, combinations of four solvents. A modifier can be added to each solvent at the start, increasing the overall solvent mixture to five, if required.

In practice, for nonpolar samples the starting point is the center of the triangle on the top face of the main prism (marked A in Figure 4.7). Having run the chromatogram under these conditions, the solvent composition is diluted with hexane until the solutes are in the required *Rf* range. The apexes of a triangle drawn through this plane (B) in the prism give three further solvent compositions that are now tried. These represent the extremes of selectivity for the solvent system. With the information gleaned from these initial chromatograms, selectivity points for further chromatograms can be chosen until the optimum solvent composition has been found. If the best chromatogram does not adequately resolve the analytes, then one or more of the primary solvents will need to be replaced and the optimization procedure repeated. For polar solutes, the face *C* of the upper frustum is used for normal-phase chromatography and the solvent optimized in a similar way [45].

4.5 APPLICATIONS

Table 4.4. shows some applications of the optimization of TLC and HPTLC systems in some drug analyses.

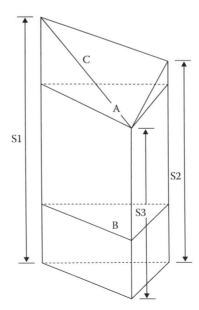

FIGURE 4.7. PRISMA model. (Reprinted with permission from Wall, P.E., *Thin-Layer Chromatography. A Modern Practical Approach*, Cambridge, U.K., RSC, 2005.)

TABLE 4.4

Some Applications of the Optimization of HPTLC and TLC Methods in Drug Analysis

Analyzed Drugs	Optimization Method	References
Forskolin in roots of *Colchicum forskohlii*	TLC method Adsorbent: TLC aluminum plates precoated with silica gel 60F-254 Solvent system: benzene:methanol (9:1; v/v) Densitometric analysis at 545 nm after spraying with anisaldehyde sulfuric acid	[47]
Colchicine in the seeds of *Colchicum autumnale* (meadow saffron)	TLC method Adsorbent: TLC aluminum plates precoated with silica gel 60F-254 Solvent system: chloroform:acetone:diethylamine (5:4:1) Densitometric measurement at 350 nm	[48]
Atenolol, hydrochlorothiazide, amlodipine besylate	HPTLC metod Adsorbent: TLC aluminum plates precoated with silica gel 60 F-254 Solvent system: chloroform:methanol:acetic acid (8:2:0.2; v/v/v). Densitometric analysis at 232 nm	[49]
Levocetirizine hydrochloride nimesulide	HPTLC method Adsorbent: TLC aluminum plates precoated with silica gel 60 F-254 Solvent system: toluene:ethyl acetate:methanol:ammonia (9:1:1:0.5; v/v/v/v) Densitometric analysis at 238 nm	[50]
Losartan potassium metolazone	HPTLC method Adsorbent: TLC aluminum plates precoated with silica gel 60 F-254 Solvent system: toluene:ethyl acetate:methanol:glacial acetic acid (6:4:1:0.1; v/v/v/v) Densitometric analysis at 237nm	[51]
Diclofenac sodium misoprostol	HPTLC method Adsorbent: TLC aluminum plates precoated with silica gel 60 F-254 Solvent sytem: toluene:ethyl acetate:ethanol:glacial acetic acid (8:2:1:0.1; v/v/v/v) Densitometric analysis at 220 nm	[52]
Nifedipine	HPTLC method The extraction solvent: methanol The mobile phase: chloroform:ethyl acetate:cyclohexane (19:2:2; v:v:v). Adsorbent: HPTLC aluminum plates precoated with silica gel 60F-254 Densitometric analysis at 238 nm	[53]
Valsartan	HPTLC method Adsorbent: TLC aluminum plates precoated with silica gel 60 F-254 Solvent system: chloroform:acetonitrile:toluene:glacial acetic acid (1:8:1:0.1; (v/v) (v/v)) UV detection was performed at 254 nm	[54]

(continued)

TABLE 4.4 (continued)
Some Applications of the Optimization of HPTLC and TLC Methods in Drug Analysis

Analyzed Drugs	Optimization Method	References
Levocetirizine	TLC method Adsorbent: TLC aluminum plates precoated with silica gel 60 F-254 Solvent system: ethyl acetate:methanol:ammonia (9:2.5:1.5; v/v/v) Densitometric analysis at 230 nm	[55]
Telmisartan ramipril	HPTLC method Adsorbent: TLC aluminum plates precoated with silica gel 60 F-254 Solvent system: acetone:benzene:ethyl acetate:glacial acetic acid (5:3:2:0.03; v/v/v/v) Densitometric analysis at 274 nm	[56]
Barbiturates benzodiazepines	TLC method Adsorbent: TLC aluminum plates precoated with silica gel 60 F-254 Solvent system: ethyl acetate:methanol:25% ammonia (85:10:5; v/v/v). Visualization of barbiturates is best achieved by the use of a mercuric chloride–diphenylcarbazone reagent	[57]
Closapine	TLC method Adsorbent: TLC aluminum plates precoated with silica gel 60 F-254 Solvent system: toluene:methanol:ethyl acetate:ammonia (8:2:1:0.1; v/v/v/v) Densitometric analysis at 280 nm	[58]
Nelfinavir mesylate	TLC method Adsorbent: TLC aluminum plates precoated with silica gel 60 F-254 Solvent system: toluene:methanol:acetone (7:1.5:1.5; v/v/v) Densitometric analysis at 250 nm	[59]
Atenolol aspirin	HPTLC method Adsorbent: TLC aluminum plates precoated with silica gel 60 F-254 Solvent system: n-butanol:water:acetic acid (8:2:0.2; v/v/v) Densitometric analysis at 235 nm	[60]
Dutasteride	HPTLC method Adsorbent: TLC aluminum plates precoated with silica gel 60 F-254 Solvent system: toluene:methanol:triethylamine (9:2:1; v/v/v) Densitometric analysis at 274 nm	[61]
Rabeprazole sodium aceclofenac	HPTLC method Adsorbent: TLC aluminum plates precoated with silica gel 60 F-254 Solvent system: toluene:ethyl acetate:methanol:acetic acid (6:4:1:0.2; v/v/v/v) Densitometric analysis at 279 nm	[62]

TABLE 4.4 (continued)

Some Applications of the Optimization of HPTLC and TLC Methods in Drug Analysis

Analyzed Drugs	Optimization Method	References
Thiocolchicoside-glafenine, thicolchicoside-floctafenine	TLC method Adsorbent: TLC aluminum plates precoated with silica gel 60 F-254 Solvent system: ethyl acetate:methanol:acetic acid (84:13:3%; v/v/v) Densitometric analysis at 375 nm	[63]
Paracetamol	TLC method Adsorbent: TLC aluminum plates precoated with silica gel 60 F-254 Solvent system: ethyl acetate:benzene:acetic acid (1:1:0.05; v/v/v) Absorbance measurement at 250 nm	[64]
Ecdysteroid alkyl ethers	HPTLC method Adsorbent: HPTLC aluminum plates precoated with silica gel 60 F-254 Solvent system: chloroform:methanol (7:1 or 10:1 or 15:1; v/v) Samples were visualised by UV fluorescence quenching at 254 nm	[65]
Clotrimazole, ketoconazle, fluconazole	TLC method Adsorbent: TLC aluminum plates precoated with silica gel 60 F-254 Solvent system: chloroform:acetone:25%ammonia (7:1:0.1)	[66]
Azaphenothiazines	TLC method Adsorbent: TLC aluminum plates precoated with silica gel 60 F-254 Solvent sytem: I: SiO_2/$CHCl_3$:EtOH (5:1; v/v) or II: Al_2O_3/$CHCl_3$:EtOH (10:0.5; v/v) The resulting spots were observed in daylight, under UV light at 254 and 365 nm	[67]
Diazepam, nitrazepam, flunitrazepam	The TLC plate was developed in an unsaturated TLC tank containing 100 mL of chloroform–acetone (1:1); this mobile phase was allowed to travel 15 cm. After evaporation of solvents, the spots of each lane were scanned at 254 nm	[68]
Tryptophan, serotonin, psychoactive tryptamines	Two types of TLC plates (silica gel and RP-18) and five solvent systems: I: methanol–28% ammonia (100:1.5; v/v) II: acetone–benzene–28% ammonia (20:10:1; v/v) III: methanol–28% ammonia (20:1; v/v) IV: acetonitrile–28% ammonia (20:1; v/v) V: acetone–28% ammonia (20:1; v/v) TLC with fluorescence detection These compounds form fluorophores on the developing plate by heating after spraying with sodium hypochlorite, hydrogen peroxide, or potassium hexacyanoferrate(III)-sodium hydroxide reagent. Fluorescent spots (vivid blue) were observed by irradiation with ultraviolet light at 365 nm	[69]

(continued)

TABLE 4.4 (continued)

Some Applications of the Optimization of HPTLC and TLC Methods in Drug Analysis

Analyzed Drugs	Optimization Method	References
Tizanidine hydrochloride	TLC method Adsorbent: TLC aluminum plates precoated with silica gel 60 F-254 Solvent system: toluene:acetone:ammonia (5:5:0.1; v/v/v)	[70]
Nelfinavir mesylate	TLC method Adsorbent: TLC aluminum plates precoated with silica gel 60 F-254 Solvent system: toluene:methanol:acetone (7:1.5:1.5; v/v/v) Densitometric analysis at 250 nm	[71]
Vinpocetine	TLC method Adsorbent: TLC aluminum plates precoated with silica gel 60 F-254 Densitometric analysis at 279 nm	[72]
Sulfamethoxazole trimethoprim	HPTLC method Adsorbent: TLC aluminum plates precoated with silica gel 60 F-254 Solvent system: toluene:ethylacetate:methanol (50:28.5:21.5; v:v:v) Detection wavelength was 254 nm	[73]
New analogues of amodiaquine	HPTLC method Adsorbent: TLC aluminum plates precoated with silica gel 60 F-254 Solvent system: CH_2Cl_2:MeOH:$NH_3 \cdot H_2O$ (80:20:1) UV light was used as the visualizing agent	[74]

REFERENCES

1. Purnell, J. H. 1960. The correlation of separating power and efficiency of gas-chromatographic columns. *J. Chem. Soc.* 256: 1268–1274.
2. Bieganowska, M. and Soczewiński, E. 1978. Some theoretical aspects of chromatographic investigations in QSAR. In: *Quantitative Structure—Activity Analysis*, eds. N. Franke and P. Oehme, p. 29. Berlin, Germany: Akademie Verlag.
3. Issaq, H. J. 1980. Modification of adsorbent, sample and solvent in TLC. *J. Chromatogr.* 3: 1423–1435.
4. Karger, B. L., LePage, J. N., and Tanaka, N. 1980. Secondary chemical equilibria in HPLC. In: *High Performance Liquid Chromatography—Advances and Perspectives*, ed. Horvath, Cs., Vol. 1, pp. 113–206. New York: Academic Press.
5. Ościk, J. 1979. *Adsorption* (in Polish). Warszawa, Poland: PWN.
6. Slaats, E. H. 1978. Study of the influence of competition and solvent interaction on retention in liquid-solid chromatography by measurement to activity coefficients in the mobile phase. *J. Chromatogr.* 149: 255–270.
7. Slaats, E. H., Markowski, W., Fekete, J., and Poppe, H. 1981. Distribution equilibria of solvent components in reversed-phase liquid chromatographic columns and relationship with the mobile phase volume. *J. Chromatogr.* 207: 299–323.
8. Snyder, L. R. 1974. Role of solvent in liquid-solid chromatography. *Anal. Chem.* 46: 1384–1393.
9. Soczewiński, E. 1968. Prediction and control of zone migration rates in ideal liquid-liquid partition chromatography. *Adv. Chromatogr.* 5: 3–78.
10. Soczewiński, E. 1980. Quantitative retention-eluent composition relationships in liquid chromatography. *J. Liq. Chromatogr.* 3: 1781–1786.

11. Wawrzynowicz, T. and Soczewiński, E. 1979. Solvent demixing effects in continuous thin-layer chromatography and their elimination. *J. Chromatogr.* 169: 191–203.
12. Kiselev, A. V. and Yashin, Y. I. 1969. *Gas-Adsorption Chromatography.* New York: Plenum Press.
13. Snyder, L. R. 1968. *Principles of Adsorption Chromatography.* New York: Marcel Dekker.
14. Snyder, L. R. and Poppe, H. 1980. Mechanism of solute retention in liquid-solid chromatography and the role of the mobile phase in affecting separation: Competition *versus* "sorption". *J. Chromatogr.* 184: 363–413.
15. Stahl, E. 1967. *Dunnschicht Chromatographie.* Berlin, Germany: Springer Verlag.
16. Soczewiński, E. and Gołkiewicz, W. 1971. Application of the law of mass action to thin-layer adsorption chromatography systems of the type electron donor solvent-Silica gel. *Chromatographia* 4: 501–507.
17. Gołkiewicz, W. 1976. A simple molecular model for adsorption chromatography. IX. Comparison of adsorption mechanisms in thin-layer chromatography and high-speed liquid chromatography. *Chromatographia* 9: 113–118.
18. Jaroniec, M. and Ościk-Mendyk, B. 1981. Application of excess adsorption data measured for components of the mobile phase for characterizing chromatographic systems. *J. Chem. Soc. Faraday Trans. 1* 77: 1277–1284.
19. Soczewiński, E. 1977. Solvent composition effects in liquid-solid systems. *J. Chromatogr.* 130: 23–28.
20. Neher, R. 1964. *Thin-Layer Chromatography,* ed. B. G. Martini-Bettolo. Amsterdam, the Netherlands: Elsevier.
21. Gocan, S. 2010. Eluotropic series of solvents for TLC. In: *Encyclopedia of Chromatography,* 3rd edn., ed. J. Cazes, pp. 730–735. Boca Raton, FL: CRC Press.
22. Saunders, D. L. 1974. Solvent selection in adsorption liquid chromatography. *Anal. Chem.* 46: 470–473.
23. Colin, H. and Guiochon, G. 1977. Introduction to reversed-phase high-performance liquid chromatography. *J. Chromatogr.* 141: 289–312.
24. Hara, S., Fujii, Y., Hirasawa, M., and Miyamoto, S. 1978. Systematic design of binary solvent systems for liquid-solid chromatography via retention behaviour of mono- and di-functional steroids on silica gel columns. *J. Chromatogr.* 149: 143–159.
25. Hametsberger, H., Klar, H., and Ricken, H. 1980. Donor-acceptor complex chromatography preparation of a chemically bonded acceptor-ligand and its chromatographic investigation. *Chromatographia* 13: 277–286.
26. Jaroniec, M., Różyło, J. K., and Ościk-Mendyk, B. 1979. Liquid adsorption chromatography with mixed mobile phases: III. Influence of molecular areas of solvents and chromatographed substances on the capacity ratio. *J. Chromatogr.* 179: 237–245.
27. Martire, D. E. and Boehm, R. E. 1980. Molecular theory of liquid adsorption chromatography. *J. Liq. Chromatogr.* 3: 753–774.
28. Ościk, J. and Różyło, J. K. 1971. Possibility for the theoretical calculation of the function between the RM values and the composition of two-component solvent mixtures used as the mobile phase and its application for the calculation of the optimum separation conditions in liquid adsorption chromatography. *Chromatographia* 4: 516–523.
29. Borówko, M. and Jaroniec, M. 1983. Association effects in adsorption from multicomponent solutions on solids and liquid adsorption chromatography. *J. Chem. Soc. Faraday Trans. 1* 79: 363–372.
30. Schoenmakers, P. J., Billiet, A. H., and De Galan, L. 1979. Influence of organic modifiers on the retention behaviour in reversed-phase liquid chromatography and its consequences for gradient elution. *J. Chromatogr.* 185: 179–195.
31. Soczewiński, E. and Jusiak, J. 1981. A simple molecular model of adsorption chromatography XIV. RF or RM? Secondary retention effects in thin-layer chromatography. *Chromatographia* 14: 23–31.
32. Spangenberg, D., Poole, C. F., and Weins, Ch. 2011. *Quantitative Thin-Layer Chromatography: A Practical Survey.* Berlin, Germany: Springer.
33. Geiss, F. 1987. *Principles of Adsorption Chromatography.* New York: Dekker.
34. Frey, H. P. and Zieloff, K. 1993. *Qualitative und Quantitative Dünnschichtchromatographie,* Grundlagen and Praxis. Weinheim, Germany: Chemie.
35. Barwick, V. J. 1997. Strategies for solvent selection—A literature review. *Trends Anal. Chem.* 16: 293–309.
36. Rohrschneider, L. 1973. Solvent characterization by gas–liquid partition coefficients of selected solutes. *Anal. Chem.* 45: 1241–1247.
37. Snyder, L. R. 1978. Classification of solvent properties of common liquids. *J. Chromatogr. Sci.* 16: 223–234.

38. Snyder, L. R. 1974. Classification of solvent properties of common liquids. *J. Chromatogr. A* 92: 223–234.
39. Nyiredy, Sz. 1997. Solvent classification for liquid chromatography. In *Chromatography, Celebrating Michael Tswett's 125th Birthday*, pp. 231–239, Düsseldorf, Germany: InCom Sonderband.
40. Nyiredy, Sz., Fater, Z., and Szabady, B. 1994. Identification in planar chromatography by use of retention data measured using characterized mobile phases. *J. Planar Chromatogr.* 7: 406–409.
41. Nyiredy, Sz., Dallenbach-Toelke, K., and Sticher, O. 1988. The "PRISMA" optimization system in planar chromatography. *J. Planar Chromatogr.* 1: 336–342.
42. Nyiredy, Sz., Erdalmeier, C. A. J., Meier, B., and Sticher, O. 1985. "PRISMA": Ein Modell zur Optimierung der mobilen Phase für die Dünnschichtchromatographie, vorgestellt anhand verschiedener Naturstofftrennungen. *Planta Med.* 51: 241–246.
43. Nyiredy, Sz., Meier, B., Dallenbach-Toelke, K., and Sticher, O. 1986. Optimization of overpressured layer chromatography of polar, naturally occurring compounds by the "PRISMA" model. *J. Chromatogr.* 365: 63–71.
44. Nyiredy, Sz. 2002. Planar chromatographic method development using the PRISMA optimization system and flow charts. *J. Sep. Sci.* 40: 1–11.
45. Wall, P. E. 2005. *Thin-Layer Chromatography. A Modern Practical Approach*. Cambridge, U.K.: RSC.
46. Rutan, S. C., Carr, P. W., Cheong, W. J., Park, J. H., and Snyder, L. R. 1978. Re-evaluation of the solvent triangle and comparison to solvachromic based scales of solvent strength and selectivity. *J. Chromatogr.* 463: 21–37.
47. Ahmad, S., Rizwan, M., Parveen, R., Mujeeb, M., and Aquil, M. 2008. A validated stability-indicating TLC method for determination of forskolin in crude drug and pharmaceutical dosage form. *Chromatographia* 67: 441–447.
48. Bodoki, E., Oprean, R., Vlase, L., Tamas, M., and Sandulescu, R. 2005. Fast determination of colchicine by TLC-densitometry from pharmaceuticals and vegetal extracts. *J. Pharm. Biomed. Anal.* 37: 971–977.
49. Bhusari, V. K. and Dhaneshwar, S. R. 2011. Validated HPTLC method for simultaneous estimation of Atenolol, Hydrochlorothiazide and Amlodipine Besylate in bulk drug and formulation. *Int. J. Anal. Bioanal. Chem.* 1: 70–76.
50. Dhaneshwar, S. R., Rasal, K. S., Bhusari, V. K., Salunkhe, J. V., and Suryan, A. L. 2011. Validated HPTLC method for simultaneous estimation of Levocetirizine Hydrochloride and Nimesulide in formulation. *Der Pharmacia Sinica* 2: 117–124.
51. Dubey, R., Bhusari, V. K., and Dhaneshwar, S. R. 2011. Validated HPTLC method for simultaneous estimation of Losartan potassium and Metolazone in bulk drug and formulation. *Der Pharmacia Lettre* 3: 334–342.
52. Dhaneshwar, S. R. and Bhusari, V. K. 2011. Validated HPTLC method for simultaneous estimation of Diclofenac Sodium and Misoprostol in bulk drug and formulation. *Asian J. Pharm. Biol. Res.* 1: 15–21.
53. Patravale,V. B., Nair,V. B., and Gore, S. P. 2000. High performance thin layer chromatographic determination of nifedipine from bulk drug and from pharmaceuticals. *J. Pharm. Biomed. Anal.* 23: 623–627.
54. Grace, D., Parambi, T., Mathew, M., and Ganesan, V. 2011. Quantitative analysis of valsartan in tablets formulations by High Performance Thin-Layer Chromatography. *J. Appl. Pharm. Sci.* 1: 76–78.
55. Bhusari, V. K., and Dhaneshwar, S. R. 2010. Application of a stability-indicating TLC method for the quantitative determination of Levocetirizine in pharmaceutical dosage forms. *Int. J. Adv. Pharm. Sci.* 1: 387–394.
56. Patel, V. A., Patel, P. G., Chaudhary, B. G., Rajgor, N. B., and Rathi, S. G. 2010. Development and Validation of HPTLC method for the simultaneous estimation of Telmisartan and Ramipril in combined dosage form. *Int. J. Pharm. Biol. Res.* 1: 18–24.
57. Cole, M. D. 2003. The analysis of controlled pharmaceutical drugs—Barbiturates and Benzodiazepines. In *The Analysis of Controlled Substances*, ed. M. D. Cole, pp. 139–160. Chichester, England, U.K.: John Wiley & Sons.
58. Zaheer, Z., Farooqui, M., and Dhaneshwar, S. R. 2010. Stability-indicating High Performance Thin Layer Chromatographic determination of Clozapine in tablet dosage form. *Asian J. Exp. Biol. Sci.* 1: 660–668.
59. Kaul, N., Agrawal, H., Paradkar, A. R., and Mahadik, K. R. 2004. Stability indicating high-performance thin-layer chromatographic determination of nelfinavir mesylate as bulk drug and in pharmaceutical dosage form. *Anal. Chim. Acta* 502: 31–38.
60. Bhusari, V. K. and Dhaneshwar, S. R. 2012. Validated HPTLC method for simultaneous estimation of Atenolol and Aspirin in bulk drug and formulation. *Anal. Chem.* 2012, doi:10.5402/2012/609706.

61. Patel, D. B., Patel, N. J., Patel, S. K., and Patel, P. U. 2011. Validated stability indicating HPTLC method for the determination of Dutasteride in pharmaceutical dosage forms. *Chromatogr. Res. Int.* 2011, doi:10.4061/2011/278923.
62. Bharekar, V. V., Mulla, T. S., Yadav, S. S., Rajput, M. P., and Rao, J. R. 2011. Validated HPTLC method for simultaneous estimation of Rabeprazole Sodium and Aceclofenac in bulk drug and formulation. *Pharm. Glob. (IJCP)* 2: 1–4.
63. El-Ragehy, N. A., Ellaithy, M. M., El-Ghobashy, M. A. 2003. Determination of thiocolchicoside in its binary mixtures (thiocolchicoside-glafenine and thiocolchicoside-floctafenine) by TLC-densitometry. *Il Farmaco* 58: 463–468.
64. Mostafa, N. M. 2010. Stability indicating method for the determination of paracetamol in its pharmaceutical preparations by TLC densitometric method. *J. Saudi Chem. Soc.* 14: 341–344.
65. Lapenna, S. and Dinan, L. 2009. HPLC and TLC characterisation of ecdysteroid alkyl ethers *J. Chromatogr. B* 877: 2996–3002.
66. Abdel-Moety, E. M., Khattab, F. I., Kelani, K. M., and AbouAl-Alamein A. M. 2002. Chromatographic determination of clotrimazole, ketoconazole and fluconazole in pharmaceutical formulations. *Il Farmaco* 57: 931–938.
67. Jeleń, M., Morak-Młodawska, B., and Pluta, K. 2011 Thin-layer chromatographic detection of new aza-phenothiazines. *J. Pharm. Biomed. Anal.* 55: 466–471.
68. Bakavoli, M. and Kaykhaii, M. 2003. Quantitative determination of diazepam, nitrazepam and flunitrazepam in tablets using thin-layer chromatography-densitometry technique. *J. Pharm. Biomed. Anal.* 31: 1185–1189.
69. Kato, N., Kojima, T., Yoshiyagawa, S., Ohta, H., Toriba, A., Nishimura, H., and Hayakawa, K. 2007. Rapid and sensitive determination of tryptophan, serotonin and psychoactive tryptamines by thin-layer chromatography/fluorescence detection. *J. Chromatogr. A* 1145: 229–233.
70. Mahadik, K. R., Paradkar, A. R., Agrawal, H., and Kaul, N. 2003. Stability-indicating HPTLC determination of tizanidine hydrochloride in bulk drug and pharmaceutical formulations. *J. Pharm. Biomed. Anal.* 33: 545–552.
71. Kaul, N., Agrawal, H., Paradkar, A. R., and Mahadik, K. R. 2004. Stability indicating high-performance thin-layer chromatographic determination of nelfinavir mesylate as bulk drug and in pharmaceutical dosage form. *Anal. Chim. Acta* 502: 31–38.
72. Mazáka, K., Vámosa, J., Nemesb, Á., Rácza, A., and Noszála, B. 2003. Lipophilicity of vinpocetine and related compounds characterized by reversed-phase thin-layer chromatography. *J. Chromatogr. A* 996: 195–203.
73. Shewiyoa, D. H., Kaaleb, E., Rishab, P. G., Dejaegherc, B., Smeyers–Verbekec, J., and Vander Heydenc, Y. 2009. Development and validation of a normal-phase high-performance thin layer chromatographic method for the analysis of sulfamethoxazole and trimethoprim in co-trimoxazole tablets. *J. Chromatogr. A* 1216: 7102–7107.
74. Delarue-Cochin, S., Paunescu, E., Maes, L., Mouray, E., Sergheraert, Ch., Grellier, P., and Melnyk, P. 2008. Synthesis and antimalarial activity of new analogues of amodiaquine. *Eur. J. Med. Chem.* 43: 252–260.

5 Chromatographic Analysis of Ionic Drugs

Monika Waksmundzka-Hajnos

CONTENTS

5.1 INTRODUCTION

Organic electrolytes exhibit pharmacological activity because their acidic groups such as carboxylic, phenolic, and sulfoxylic or basic groups such as amine and heterocyclic nitrogen give the possibility of ionic interactions with ionic groups of receptors. The ionic interactions play a fundamental role in drug–receptor bonds.

Aromatic carboxylic acids and their derivatives are widely applied as nonsteroidal anti-inflammatory drugs (NSAID); derivatives of salicylic, phenylacetic, phenylpropionic, and 2-aminobenzoic acids; anesthetic drugs used externally or to spinal anesthesia; derivatives of benzoic, 4-hydroxybenzoic, and 4-aminobenzoic acids; bacteriostatic drugs; 4-aminosalicylic acid (PAS) (bacteriostatic activity on the *Mycobacterium tuberculosis*); and vitamins, 4-aminobenzoic acid (PABA) that is one of the B-group vitamins.

Compounds possessing heterocyclic nitrogen or amine group are included in several groups of drugs; examples are presented in Table 5.1.

TLC (thin layer chromatography) separation of ionic samples tends to be more complicated than separation of neutral compounds [1]. It is due to the fact that these compounds are normally weak acids or bases and are present in solutions both in ionized and unionized forms, especially in solutions with polar solvents. These forms interact differently with the active sites of the adsorbent that leads to problems in their chromatographic analysis.

This chapter describes the retention dependence of ionic samples, in reversed-phase (RP) TLC (RP TLC) systems, on such conditions as pH, buffer type, concentration, polar modifiers, and ion-pair (IP) reagents. Different TLC methods, commonly used in the analysis of ionic compounds, RP and IP TLC, are described. Also the separation of organic electrolytes in commonly used normal-phase systems is discussed.

TABLE 5.1
Selected Basic Drugs

Name of Drug	Formula	Activity
Benzodiazepines		Sedative Anticonvulsant Hypnotic Anesthesia supporting
1,4-Dihydropyridine derivatives		Hypotensive Calcium-channel blockers
4-Aminoquinoline derivatives		Anti-inflammatory Antimalaria
8-Hydroxyquinoline derivatives		Antiseptics
Acridine derivatives		Antiseptics
Quinoline derivatives		Anesthesia supporting
Phenothiazine derivatives		Sedative Spasmolytic Antihistamine
Imidazole derivatives		Antifungal
Imidazoline derivatives		Antihistamine

TABLE 5.1 (continued)
Selected Basic Drugs

Name of Drug	Formula	Activity
Pyridine derivatives	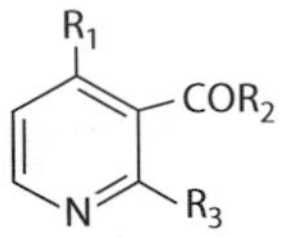	Hypotensive Cardiac
	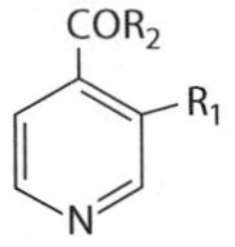	Tuberculostatic
Monobactams	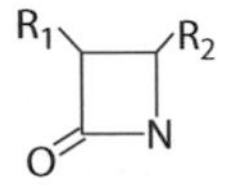	Antibiotics
Piperazine derivatives	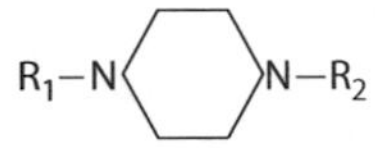	Antidepressant
Morpholine		Analgesic Beta-adrenolytics Anticonvulsant
Piperidine derivatives		Psychotropic Antihistamine

5.2 ACIDIC AND BASIC ANALYTES

Organic electrolytes are soluble in polar solvents giving two forms—ionized and non-ionized. Also in aqueous mobile phases, commonly used in RP TLC systems, given acidic or basic compound exists in ionized and non-ionized forms [2]. The retention factor (k) of a solute is given by the following equations:

For acidic compounds

$$k = k_0(1 - F^-) + k_{-1}F^- \tag{5.1}$$

For basic compounds

$$k = k_0(1 - F^+) + k_1 F^+ \tag{5.2}$$

where
 k_0, k_{-1}, and k_1 are the k values for non-ionized and ionic forms
 F^- and F^+ are the fractions of ionized solute molecules:

$$F^- = \frac{1}{\{1 + ([H^+]/K_a)\}} \tag{5.3}$$

$$F^+ = \frac{1}{\{1 + (K_a/[H^+])\}} \tag{5.4}$$

The data, obtained on the basis of these equations, remain in agreement with experimental values for a wide pH range [3–5].

Computer programs based on Equations 5.1 through 5.4 are able to predict retention and resolution of ionic samples as a function of pH. However predicted retention values are usually more accurate for acidic rather than basic solutes, due to silanol effects, which is more significant for basic compounds.

Changing pH values of the mobile phase strongly influences the retention of ionic compounds on silanized silica, as it can be seen in Figure 5.1 for sulfonamides [6]. The results are illustrated graphically as plots of R_M ($\log k$) versus pH of the phosphate buffer solution in aqueous methanol (MeOH) as the mobile phase (Figure 5.1). As expected, the retention is a sigmoidal function of the hydrogen ion concentration and is typical for sulfonamides. The pH dependence of retention provides preliminary information on the acidic–basic properties of the solutes. The inflection points are at the pH equal to the pK_a of the compounds, where solute retention is very sensitive to the mobile-phase pH. Therefore a small change in pH results in a large change in the R_M value of the solute. At low pH retention is relatively strong and has a maximum value at pH 3–5; at pH 6–7 retention decreases dramatically, and at higher pH values (pH > 7), the retention of the anionic form of sulfonamides is low and independent of pH.

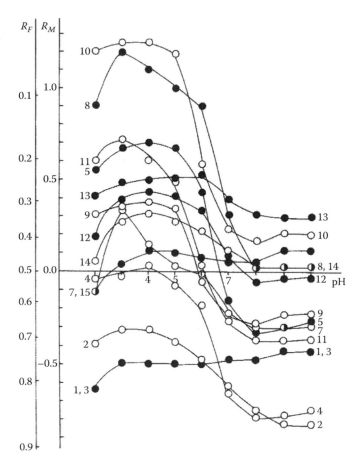

FIGURE 5.1 Plots of R_M values versus pH of the aqueous mobile phase of metanol–0.09 M phosphate buffer (1:3 v/v). Adsorbent silanized silica. Solutes: 1, sulfanilamide; 2, sulfacarbamide; 3, sulfaguanidine; 4, sulfacetamide; 5, sulfamethoxazole; 7, sulfafurazole; 8, sulfathiazole; 9, sulfamethizole; 10, sulfaproxyline; 11, sulfadiazine; 12, sulfamerazine; 13, sulfadimidine; 14, sulfisomidine; 15, sulfadimethoxine. (Reprinted from Bieganowska, M.L. et al., *J. Pharm. Biomed. Anal.*, 11, 241, 1993. With permission.)

By changing the pH values, the retention factors of acids and bases can even change by a factor of 10. As pH increases, RPC retention for an acid decreases and retention for a base increases. In order to maintain an optimal k range for the resulting separation ($1 < k < 20$), it is necessary to combine pH optimization with variation of solvent strength [7,8]. Under such conditions the retention factor of an ionizable compound is a function of pH and volume fraction of organic modifier in mobile phase (φ):

$$k = f([H^+], \varphi) \tag{5.5}$$

Because of that, the retention can be controlled by changing pH and modifier concentration in the aqueous eluent. The retention of ionic compounds depends mostly on processes such as ion pairing with other ions, solvophobic effects of the ionic strength, and co-ion exclusion resulting from ionization of the residual silanol groups on the adsorbent surface [7,9].

5.3 REVERSED-PHASE THIN-LAYER SEPARATION OF IONIC ANALYTES

The optimization in RP separation and the selectivity control of ionic samples can be performed similarly as for nonionic compounds by variation of the solvent strength to obtain satisfactory retention and separation selectivity: the variation of pH, of modifier concentration, or by change of the layer type (C2, C8, C18, diol, cyano) [10]. Figure 5.2 [11] presents the dependence of retention of selected antidepressive basic drugs on C18 layer with changing amount of organic modifier (MeOH). As it is seen, the best separation selectivity was obtained by use of 70% MeOH in mobile phase [11].

5.3.1 PH, BUFFER TYPE, AND CONCENTRATION

Whenever ionic samples are analyzed, the addition of proper buffer is advised. It is important to carefully choose suitable buffer by taking several properties into consideration: buffer capacity, its UV absorbance, solubility, stability, and interactions with the sample and chromatographic systems [5]. The buffer is effective in controlling pH in the range $pK_a \pm 1.5$, when buffer ionization occurs. The greater buffer concentration, the greater its capacity. However higher buffer concentrations may lead to problems in solubility (salting-out effect). Therefore a buffer concentration of 10–50 mM is advised; 25 mM is the best starting point. Commonly used buffers (e.g., phosphate buffer) are more soluble in aqueous MeOH rather than in eluents containing acetonitrile (ACN) or tetrahydrofuran (THF) as organic modifiers.

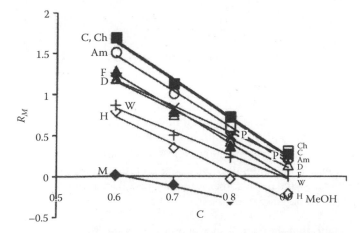

FIGURE 5.2 Plots of R_M versus C (volume fraction) of MeOH in the mobile phase. System: RP-18/MeOH + water + 1% ammonia. Symbols: Am, amizepin; Ch, chlorpromazine; C, clomipramine; D, doxepin; F, flupentixol; H, haloperidol; M, moclobemide; P, perazine; W, wenlafaxine. (Reprinted from Petruczynik, A. et al., *J. Liq. Chromatogr.*, 31, 1913, 2008. With permission.)

RP separations are performed with the use of silica-based sorbents/layers that are stable in the pH range 2–8. Therefore the following buffers are commonly used: phosphate buffer (2.1–3.1 and 6.2–8.2), acetate buffer (3.8–5.8), citrate buffer (2.1–6.4), and carbonate buffer (3.8–4.8).

Whenever a buffer is added into the mobile phase, a proper selection of organic modifier is of crucial importance due to different buffer salt solubilities in water–organic solvent mixtures. Usually aqueous MeOH is recommended as the starting point [1].

Mobile-phase pH may change in case of the volatile buffers used, for example, carbonate buffer, due to loss of CO_2. Some buffers may also change the separation conditions, for example, ion pairing can occur as, for example, in the case of eluents containing trifluoroacetate buffers with cationic samples.

5.3.2 Solvents

Three main solvents are used as organic modifiers in RP TLC: ACN, MeOH, and THF. However other solvents may also be incorporated into the mobile phase. RPC solvent strength varies as water < MeOH < ACN < ethanol < THF < propanol.

In Figure 5.3 [12] R_M values obtained on silanized silica plates are presented as diagram by various modifiers in aqueous eluent in order to compare the strength and selectivity of the four modifiers; their concentration in the mobile phase was kept constant. MeOH–water was the most selective eluent for sulfonamides.

A very important task in RPC method development is the fact that organic modifiers cause the change of the pK_a value [2,13]. In eluents containing greater amounts of organic modifier, the activity coefficients decrease and their influence on pH and pK_a values cannot be neglected. It should be a rule of thumb to adjust the pH of the mobile phase before the addition of organic solvent [10]. Otherwise the obtained results may appear to be irreproducible, as electrode response tends to drift. The problem of the retention factor as the combined function of pH and modifier concentration in the aqueous mobile phase was analyzed by several researchers [9,14–19].

5.3.3 Silanol Blockers and Ion-Suppressing Agents

In case of basic samples, further problems appear due to the interaction of underivatized free silanols with ionic compounds [20]. It appears due to retention by ion-exchange process that involves protonated bases and ionized silanols [5]:

$$BH^+ + SiO^-K^+ \rightarrow K^+ + SiOBH^+$$

$$H^+ + SiO^-K^+ \rightarrow K^+ + SiOH$$

This case leads to increased retention, band tailing, and plate-to-plate irreproducibility. These silanol interactions may be minimized by choice of appropriate experimental conditions [21]. The use of low-pH mobile phase (2.0 < pH < 3.5) minimizes the concentration of ionized silanols due to suppression of their ionization. The use of high pH (>7.0) is also recommended, as the ionization of weak bases is suppressed, thus eliminating ion interactions with acidic silanols. A higher buffer concentration (>10 mM) and proper cations that are strongly held by the silanols (e.g., triethylammonium$^+$, dimethyloctylammonium$^+$) block sample retention. Successful analysis of basic compounds may be also obtained after the incorporation of amines into mobile phase that compete with the analytes for adsorbent silanol sites. The addition of basic silanol-blocking agents causes two effects, depending on the concentration of the blocking molecules:

- At lower concentration they are responsible for blocking free silanol sites; it leads to a decrease in the analyzed base retention (the blockage of the ion-exchange interactions).
- At higher concentrations they cause an increase in base retention because of the suppression of basic compounds' dissociation.

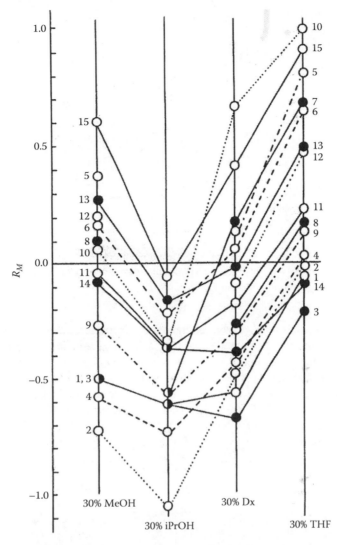

FIGURE 5.3 Graphical representation of the dependence of the R_M values of sulfonamide drugs on silanized silica gel on the nature of the aqueous mobile-phase modifier. Compound symbols as in Figure 5.1. (Reprinted from Bieganowska, M.L. et al., *J. Planar Chromatogr.*, 6, 121, 1993. With permission.)

In a typical ion suppression technique, the ionization of a weak acid or the protonation of a weak base is suppressed by adjusting the pH of the mobile phase. The separation is then achieved on an RP layer using MeOH or ACN and an aqueous buffer solution as the mobile phase. By buffering the mobile phase in the apparent pH 2–5, weakly acidic solutes can be retained on an RP adsorbent. In addition weakly basic substances can also be separated on a similar sorbent by ion suppression if the pH of the mobile phase is maintained in the range 7–8 [22]. In the analysis of acidic compounds, it is also a common procedure to add organic acids, such as formic or acetic acid, into the mobile phase instead of a buffer. It leads to the suppression of dissociation of the analyte, which becomes more lipophilic, thus more strongly retained [23].

RP systems with RP-18 HPTLC plates were also optimized for separation of the basic antidepressive drugs [11]. Because of the basic character of the drugs, they were strongly retained on the layer with asymmetric spots. The use of buffered mobile phase at pH = 3.5 or at pH = 7.8 did

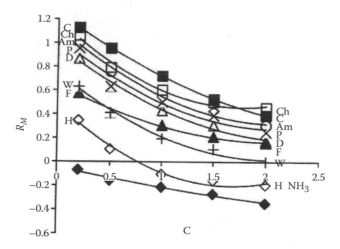

FIGURE 5.4 Plots of R_M versus C% concentration of ammonia in the mobile phase. System: RP-18/70% MeOH + water + NH$_3$. Symbols as in Figure 5.2. (Reprinted from Petruczynik, A. et al., *J. Liq. Chromatogr.*, 31, 1913, 2008. With permission.)

not give satisfactory results—spots were elongated and separation selectivity was insufficient. The use of aqueous ammonia improved separation selectivity and system efficiency. Figure 5.4 [11] shows dependencies of retention versus ammonia concentration. The increase of ammonia concentration from 0.2% to 2% causes a decrease of retention, but the selectivity of the separation is still sufficient and peak profiles are significantly more symmetric and narrow. The best selectivity was obtained with 70% MeOH in water + 1% ammonia. The videoscan of the plate with the standards and mixture of selected drugs separated on RP-18 layer in previously mentioned eluent system is presented in Figure 5.5. It is seen that spots are compact, symmetric, and well separated. The system was used for quantitative analysis of amitriptyline and doxepin in human serum [11].

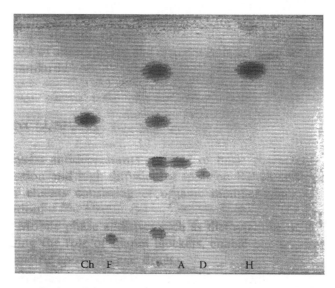

FIGURE 5.5 Videoscan of selected drugs and their mixture on RP-18 layer. Eluent: 70% MeOH + water + 1% of ammonia. Symbols as in Figure 5.2. (Reprinted from Petruczynik, A. et al., *J. Liq. Chromatogr.*, 31, 1913, 2008. With permission.)

5.3.4 Application of RP TLC Systems in Drug Analysis

RP TLC systems were applied in the analysis of basic drugs with the use of RP-8 or RP-18 layers and aqueous eluents containing modifier (MeOH, ACN, THF) of acidic pH that enables suppression of surface silanols' dissociation: for the analysis of feprazone and its metabolites in human plasma by use of the mobile phase acidified with formic acid [24], tetracycline and its impurities by use of the mobile phase acidified with oxalic acid at pH 2 [25,26] or at pH 3 [27], basic drugs (quaternary bases) by use of the mobile phase acidified with hydrochloric acid [28,29], nonselective calcium-channel blockers (prenylamine, lidoflazine, bepril, fendiline) by use of the mobile phase acidified with phosphate buffer at pH 2.06 (see Figure 5.6) [30], timolol and betaxolol by use of the mobile phase acidified with acetic acid [31], amphetamine-type drugs by use of the mobile phase acidified with hydrochloric acid [32], and aminoglycosides by use of the mobile phase acidified with acetic acid [33]. Various layers (RP-8, RP-18, aminopropyl, DIOL, cyanopropyl) and aqueous mobile phases were tested for the separation of basic antipsychotic drugs, and the best results were obtained on RP-8 with buffered dioxane at pH 3.5 [34] (see Figure 5.7). Benzodiazepines were separated in RP systems by use of aqueous mobile phase buffered by acetate buffer at pH 3 [35].

The use of basic additives for the analysis of basic drugs was also reported. As was previously mentioned, they play the role of ion suppressants and/or silanol blockers. Basic drugs were chromatographed in RP systems by use of buffered mobile phase at pH 11 and/or with n-decylamine as aqueous mobile-phase additive [36]. Ranitidine and two analogues were analyzed in RP system with aqueous ammonia as additive [37] (see Figure 5.8). Similar systems with aqueous ammonia were applied in the analysis of benzimidazole derivatives [38]. Aliphatic amines were also applied as mobile-phase additives [39].

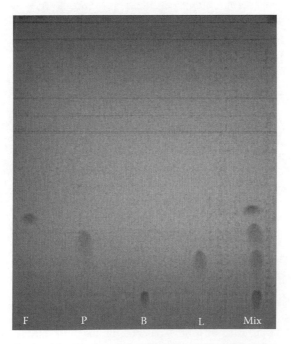

FIGURE 5.6   Videogram obtained by the use of the Desaga A-CCD-300E video camera after chromatography of the drugs on silanized silica gel RP-18 plates with 50% ACN in 0.09 M phosphate buffer, pH 2.06, as mobile phase. Symbols: B, bepridil; F, fendiline; L, lidoflazine; P, prenylamine. (Reprinted from Misztal, G. et al., *J. Planar Chromatogr.*, 16, 433, 2003. With permission.)

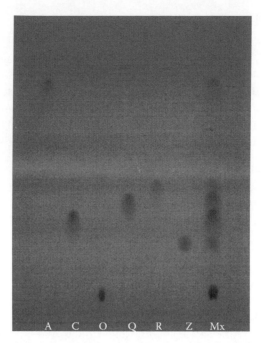

FIGURE 5.7 Chromatogram obtained from separation of amisulpride (A), clozapine (C), olanzapine (O), quetiapine (Q), risperidone (R), and ziprasidone (Z) on an RP-8 plate with dioxane–pH 3.51 phosphate buffer, 40 + 60 (*v/v*), as mobile phase. (Reprinted from Skibiński, R. et al., *J. Planar Chromatogr.*, 20, 75, 2007. With permission.)

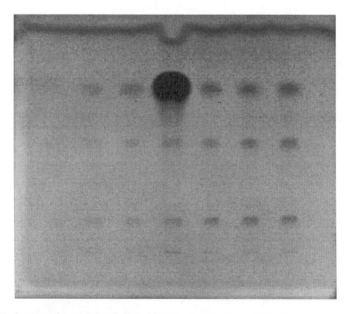

FIGURE 5.8 CCD image of ranitidine hydrochloride and the two related compounds scanned at 254 nm (100 mg of ranitidine hydrochloride spiked with 0.4 mg of related compound A and 0.5 mg of related compound B) in the middle of the plate and six calibration points for the three substances: 0.1, 0.2, and 0.4 mg (A) and 0.6, 0.8, and 1.0 mg (B). System: RP-18W/MeOH—3% ammonia (4;1). (Reprinted from Simonovska, B. et al., *J. Chromatogr. B*, 715, 425, 1998. With permission.)

Acidic drugs are more often separated in RP systems by use of "classic" eluents without any additives. However, sometimes ion-suppressant additive is necessary. For example, the use of hydrochloric acid [24] or acetic acid [40] is reported. Aqueous buffered mobile phase at pH 4.6 for the separation of bile acids was also applied [41].

5.4 ION-PAIR CHROMATOGRAPHY

If a successful separation of a sample containing ionic compounds has not been obtained with the use of RP TLC, IP chromatography provides an important additional selectivity option [5]. IP and RP TLCs share several features. The adsorbent and mobile phase used for the separations are generally similar, differing mainly in the addition of an ion-pairing reagent.

There are several theories explaining the mechanism of adsorption in ion-association systems [42–47]. The IP model assumes an association between the sample ion and oppositely charged ion-pairing reagent in a liquid polar mobile phase before its adsorption on the hydrophobic stationary phase [3,48–50]. According to this theory the retention factor of an ionic solute depends on the kind and concentration of the counterion. The second model of the IPC mechanism is the ion-exchange model [51–55]. It assumes the adsorption of lipophilic counterions on the nonpolar surface of the stationary phase, which then behaves like an ion exchanger. The greater the concentration of the IP reagent and its hydrophobicity, the greater the retention of the chromatographed ionic compounds [56]. Bidlingmeyer et al. proposed an ion interaction model—a model of a double electric layer [57]. According to this theory the dynamic equilibrium of the lipophilic ion in the double electric layer formed on the sorbent surface occurs. The retention of the analyte is caused by the charge of the double layer formed by the ions of the ion-pairing reagent. Stahlberg proposed the electrostatic model, which assumes the ion-pairing reagent is fully ionized in the applied pH range and influences first of all the retention of ionized form of the solute [58].

Except for IP reagent type, concentration, and mobile-phase pH, the retention and selectivity in IP–RP systems can be controlled by a change of type and concentration of the organic modifier in the aqueous mobile phase [10,59,60]. The retention of solutes decreases with the increase in concentration of the organic modifier, and the R_M (log k) values are linear functions of the volume concentration of the modifier in accordance with equation [61]:

$$\log k = R_M = R_{Mw} - b C_{mod} \tag{5.6}$$

where
R_{Mw} is the retention factor for pure water or aqueous buffer solution
b is constant

5.4.1 pH AND ION PAIRING

Mobile-phase pH is an important determinant in IP–RP chromatography [45,62,63]. It should be selected to obtain maximal ionization of solute and ion-pairing molecules to possibly form an IP. For example, for acidic solutes the pH usually used is in the range 7.0–7.5; greater pH values may cause the destruction of silica-based adsorbent [10]. When $pK_a - 2 < pH < pK_a + 2$, the solute molecules exist in ionic and nonionic forms, and the adsorption of both forms (ionic and nonionic) and ion pairing occur.

When pH and IP reagent and concentration are varied simultaneously, considerable control is achievable over both retention range and band spacing. This is a result of the simultaneous retention of the sample by both RP and ion-exchange process.

5.4.2 Ion-Pair Type and Concentration

An IP reagent might cause a very large change in chromatographic properties, enabling the analyte to be moved well out of previously cochromatographic materials present in the sample that do not interact with the IP reagent [1].

In case of acidic compounds, the following cationic ion-pairing reagents have been employed: alkylammonium compounds [64,65], organic amines, and other basic compounds [66–67]. In case of basic compounds, anionic ion-pairing reagents are used, such as sulfonic acids, alkyl sulfonates, or other acids, for example, bis(2-ethylhexyl)ortho-phosphoric acid (HDEHP).

The elongation of alkyl chain of IP reagent leads to increase of R_M (log k) value [68].

In a limited range of concentrations, a linear relationship of R_M (log k) and concentration of IP reagent can be obtained [69,70]. After the surface is saturated by hydrophobic counterions, a further increase of concentration does not lead to significant changes in retention [71,72].

The change of type and concentration of the counterion often causes variations in the selectivity of separation [59,60].

Figure 5.9 [73] presents retention parameters of acidic drugs as a function of percentage concentration of IP reagent (cetrimide) in a mobile phase. The increase of the concentration of cetrimide in a mobile phase causes increasing retention of analytes. Similar results can be observed for basic drugs chromatographed in IP TLC systems with HDEHP as counterion in a mobile phase (see Figure 5.10 [73]).

5.4.3 Application of IP–RP TLC Systems in Drug Analysis

IP TLC is rarely applied in drug analysis. Sometimes the use of IP reagent as stationary-phase additive is reported. It is performed in two ways—plates can be dipped in the solution of IP reagent, and

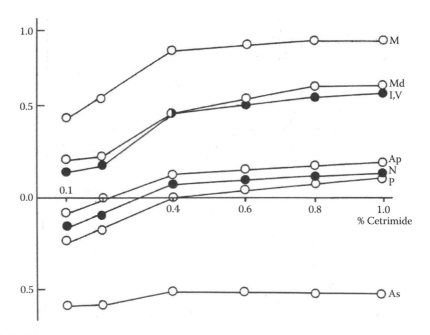

FIGURE 5.9 Plots of R_M versus C% concentration of cetrimide in the mobile phase. System: RP-18/50% MeOH + water + 0.006 M of phosphate buffer at pH 7.38. Symbols: Ap, Apranax; As, aspirin; I, ibuprofen; M, Mefacit; Md, Metindol; N, Nevigramon; P, Profenid; V, Voltaren. (Reprinted from Petruczynik, A., *Thin-Layer Chromatography (TLC) and High Performance Liquid Chromatography (HPLC) of Some Organic Electrolytes in Ion-Pair Systems*, Medical University, 1993. With permission.)

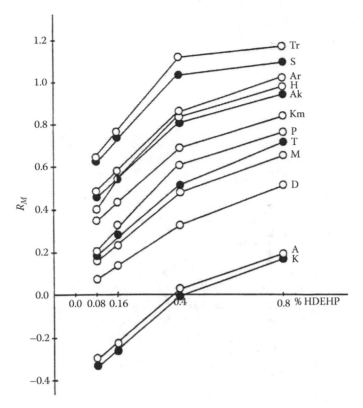

FIGURE 5.10 Plots of R_M versus C% concentration of HDEHP in the mobile phase. System: silanized silica/50% MeOH + water + 0.05 M of phosphate buffer at pH 3.0. Symbols: K, Karion; M, Mydocalm; P, Pridinol; Ar, Artane; Ak, Akineton; Km, Kemadrin; D, Dolargan; H, haloperidol; T, Tremblex; S, Sormodren; Tr, Tramaril; A, atropine. (Reprinted from Petruczynik, A., *Thin-Layer Chromatography (TLC) and High Performance Liquid Chromatography (HPLC) of Some Organic Electrolytes in Ion-Pair Systems*, Medical University, 1993. With permission.)

after solvent evaporation plates remain as covered with IP reagent, or portion of the reagent can be added to the suspension in the preparation process of plates. The examples of IP–RP TLC can be found in literature in separation of basic drugs [74–77] and acidic drugs [78–82]. Figure 5.11 [79] shows videoscan obtained for NSAIDs separated on RP-18 layer by the use of cetyltrimethylammonium bromide (CTMABr) as an aqueous mobile-phase additive.

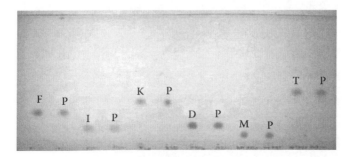

FIGURE 5.11 Separation of the drugs (standards and preparations) on silica gel RP-18 plates by horizontal development with phosphate buffer, pH 5.73–10% CTMABr in MeOH, 3.5 + 6.5 (*v/v*), as mobile phase. F, I, K, D, M, and T denote fenbufen, ibuprofen, ketoprofen, diclofenac sodium, mefenamic acid, and tiaprofenic acid standards. P denotes preparations. (Reprinted from Hopkała, H. and Pomykalski, A., *J. Planar Chromatogr.*, 17, 383, 2004. With permission.)

5.5 NORMAL-PHASE SYSTEMS

Normal-phase (NP) systems are most often used in TLC analysis of ionic drugs. Various layers (silica, alumina, polyamide, cellulose, polar bonded phases) and different combinations of solvents can be applied. Because organic electrolytes are strongly retained on the polar surfaces of adsorbents, the use of polar modifiers and polar or medium polar diluents is often necessary. Silica, polyamide, and alumina plates eluted with chloroform—MeOH mixtures were used in the separation of sulfonamides [12]. Antiparkinsonian basic drugs were separated by the use of silica eluted by MeOH (diisopropyl ether mixtures) and of alumina eluted by methyl ethyl ketone (dichloromethane mixtures) [83]. Sometimes the use of water, as eluent additive, having the highest eluent strength in NP systems, is reported [84–87]. Water molecules deactivate also surface silanols and acidic surface sites of other adsorbents that make the interaction of ionic analytes with the adsorbent surface less strong.

In NP systems the use of ion suppressants is also reported. Antidepressive basic drugs were strongly retained on silica layer, and the use of MeOH—medium polar diluent (dichloromethane or diisopropyl ether) mixtures of high eluent strength—was necessary, but even a 70% solution of MeOH in diisopropyl ether did not give satisfactory results. The obtained spots were wide and asymmetric. Because of this, the use of aqueous ammonia or diethyl amine as mobile-phase additive was decided. Figure 5.12 [11] presents R_M versus ammonia concentration and shows a significant decrease of retention with the increased ammonia concentration. The use of ammonia or diethylamine causes also an increase of system efficiency—spots were compact and symmetric, but simultaneously separation selectivity became worse. The best eluent system for separation of the investigated drugs on silica was 50% MeOH in diisopropyl ether with 0.05M DEA.

The use of basic additive (ammonia or DEA) to the eluent for the separation of basic drugs on silica layers was often reported [32,88–102]. Figure 5.13 presents separation of lamivudine (LVD), stavudine (STV), and nevirapine (NVP) on silica layer developed with nonaqueous mobile phase containing concentrated ammonia as additive [93]. It is seen that peaks are narrow and symmetric. For separation of five basic antiarrhythmic drugs on alumina as well as on silica layers, an eluent containing concentrated ammonia as additive was used (see Figure 5.14 [95]).

The use of acidic additives such as organic acids (acetic acid and formic acid) was also applied in the separation of acidic drugs on silica layers [103,104].

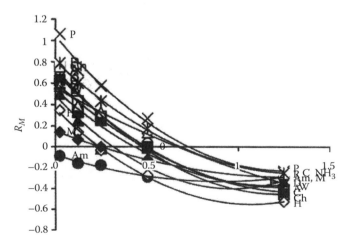

FIGURE 5.12 Plots of R_M versus C% concentration of ammonia in the mobile phase. System: SiO_2/50% MeOH + iPr_2O + NH_3. Symbols as in Figure 5.2. (From Petruczynik, A. et al., *J. Liq. Chromatogr.*, 31, 1913, 2008.)

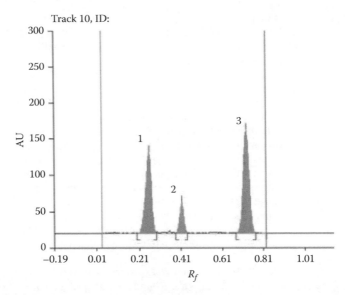

FIGURE 5.13 Chromatogram showing LVD (peak 1), STV (peak 2), and NVP (peak 3) from the solution of spiked tablet matrix on silica layer. Mobile phase: ethyl acetate, MeOH, toluene, and concentrated ammonia (12:6:12:1, *v/v/v/v*). Detection at 254 nm. (Reprinted from Shewiyo, D. et al., *J. Pharm. Biomed. Anal.*, 54, 445, 2011. With permission.)

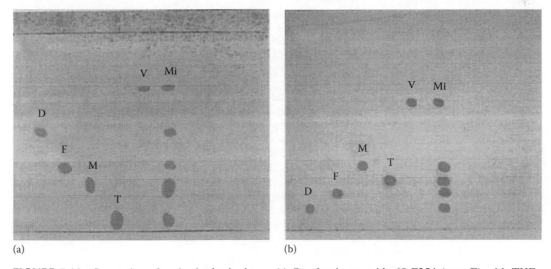

(a)   (b)

FIGURE 5.14 Separation of antiarrhythmic drugs. (a) On aluminum oxide 60 F254 (type E) with THF–hexane–25% ammonia, 5 + 4.8 + 0.2 (v/v), as mobile phase. (b) On silica gel 60 F254 with chloroform–THF–ethanol–25% ammonia, 8.1 + 1.9 + 2 + 0.1 (v/v), as mobile phase. Symbols: D, disopyramide; F, flecainide; M, mexiletine; T, tocainide; V, verapamil; Mi, mixture of the five drugs. (Reprinted from Pietras, R. et al., *J. Planar Chromatogr.*, 17, 213, 2004. With permission.)

REFERENCES

1. Snyder, L.R., Kirkland, J.J., and Glajch, J.L., *Practical HPLC Method Development*, 2nd edn., Wiley, New York, 1997.
2. Roses, M., Canals, I., Allemann, H., Siigur, K., and Bosch, E., Retention of ionizable compounds on HPLC. 2: Effect of pH, ionic strength, and mobile phase composition on the retention of weak acids, *Anal. Chem.*, 68, 4094–4100, 1996.

3. Horvath, C., Melander, W., Molnar, I., and Molnar, P., Enhancement of retention by ion-pair formation in liquid chromatography with non polar stationary phases, *Anal. Chem.*, 49, 2295–2305, 1977.

4. Deming, S.N. and Turoff, M.L.H., Optimization of reverse-phase liquid chromatographic separation of weak organic acids, *Anal. Chem.*, 50, 546–548, 1978.

5. Snyder, L.R., Dolan, J.W., and Lommen, D.C., High-performance liquid chromatographic computer simulation based on a restricted multi-parameter approach: II. Applications, *J. Chromatogr.*, 535, 75–92, 1990.

6. Bieganowska, M.L., Petruczynik, A., and Doraczyńska-Szopa, A., Thin-layer reversed-phase ion-pair chromatography of some sulphonamides, *J. Pharm. Biomed. Anal.*, 11, 241–246, 1993.

7. Lewis, J.A., Lommen, D.C., Raddatz, W.D., Dolan, J.W., Snyder, L.R., and Molnar, I., Computer simulation for the prediction of separation as a function of pH for reversed-phase high-performance liquid chromatography: I. Accuracy of a theory-based model, *J. Chromatogr.*, 592, 183–195, 1992.

8. Lewis, J.A., Dolan, J.W., Snyder, L.R., and Molnar, I., Computer simulation for the prediction of separation as a function of pH for reversed-phase high-performance liquid chromatography: II. Resolution as a function of simultaneous change in pH and solvent strength, *J. Chromatogr.*, 592, 197–208, 1992.

9. Marques, R.M.L. and Schoenmakers, P.J., Modeling retention in reversed-phase liquid chromatography as a function of pH and solvent composition, *J. Chromatogr.*, 592, 157–182, 1992.

10. Waksmundzka-Hajnos, M., Chromatographic separations of aromatic carboxylic acids, *J. Chromatogr. B*, 717, 93–118, 1998.

11. Petruczynik, A., Brończyk, M., Tuzimski, T., and Waksmundzka-Hajnos, M., Analysis of selected antidepressive drugs by High-Performance Thin-Layer Chromatography, *J. Liquid Chromatogr.*, 31, 1913–1924, 2008.

12. Bieganowska, M.L., Doraczyńska-Szopa, A., and Petruczynik, A., The retention behaviour of some sulfonamides on different thin layer plates, *J. Planar Chromatogr.*, 6, 121–128, 1993.

13. Espinoza, S., Bosch, E., and Roses, M., Acid–base constants of neutral bases in acetonitrile–water mixtures, *Anal. Chim. Acta*, 454, 157–166, 2002.

14. Lewis, J.A., Dolan, J.W., Snyder, L.R., and Molnar, I., Computer simulation for the prediction of separation as a function of pH for reversed-phase high-performance liquid chromatography: II Resolution as a function of simultaneous change in pH and solvent strength, *J. Chromatogr.*, 592, 197–208, 1992.

15. Kiel, J.S., Morgan, S.L., and Abramson, R.K., Computer-assisted optimization of a high-performance liquid chromatographic separation for chlorpromazine and thirteen metabolites, *J. Chromatogr.*, 485, 585–596, 1989.

16. Grushka, E., Jang, N.I., and Brown, P.R., Reversed-phase liquid chromatographic retention and selectivity surfaces: II. Deoxyribonucleosides, *J. Chromatogr.*, 485, 617–630, 1989.

17. Lema, M., Otero, J., and Marco, J., Two-parameter mobile phase optimization for the simultaneous high-performance liquid chromatographic determination of dopamine, serotonin and related compounds in micro dissected rat brain nuclei, *J. Chromatogr.*, 547, 113–120, 1991.

18. Haddad, P.R., Drouen, A.C.J.H., Billiet, H.A.H., and de Galan, L., Combined optimization of mobile phase pH and organic modifier content in the separation of some aromatic acids by reversed-phase high-performance liquid chromatography, *J. Chromatogr.*, 282, 71–81, 1983.

19. Otto, M. and Wegscheider, W., Multi factor model for the optimization of selectivity in reversed-phase chromatography, *J. Chromatogr.*, 258, 11–22, 1983.

20. Waksmundzka-Hajnos, M., Retention behavior of heterocyclic bases in liquid chromatographic systems, *Trends Heterocycl. Chem.*, 9, 129–166, 2003.

21. Vervoort, R.J.M., Debets, A.J.J., Claessens, H.A., Cramers, C.A., and de Jong, G.J., Optimisation and characterisation of silica-based reversed-phase liquid chromatographic systems for the analysis of basic pharmaceuticals, *J. Chromatogr. A*, 897, 1–22, 2000.

22. Dash, A.K., Ion-pairing procedures applicable to drug quality control, *Process Control Qual.*, 10, 229–241, 1997.

23. Dzido, T.H. and Polit, D., Retention of aromatic hydrocarbons with polar groups in binary reversed-phase thin-layer chromatography, *J. Planar Chromatogr.*, 14, 80–87, 2001.

24. Spahn, H. and Mutschler, E., Determination of feprazone and one of its metabolites in human plasma after HPLC or TLC separation, *J. Chromatogr.*, 232, 145–153, 1982.

25. Oka, H. and Suzuki, M., Improvement of chemical analysis of antibiotics. VII. Comparison of analytical methods for determination of impurities in tetracycline pharmaceutical preparations, *J. Chromatogr.*, 314, 303–311, 1984.

26. Oka, H., Ikai, Y., Hayakawa, J., Matsuda, K., Harada, K., and Suzuki, M., Improvement of chemical analysis of antibiotics. Part XIX. Determination of tetracycline antibiotics in milk by liquid chromatography and thin-layer chromatography/fast atom bombardment mass spectrometry, *J. AOAC Int.*, 77, 891–895, 1994.

27. Oka, H., Ikai, Y., Kawamura, N., Uno, K., Yamada, M., Harada, K., and Suzuki, M., Improvement of chemical analysis of antibiotics. XII. Simultaneous analysis of seven tetracyclines in honey, *J. Chromatogr.*, 400, 253–261, 1987.

28. Ojanpera, I., Vartiovaara, A., Ruchonen, A., and Vuori, E., Combined use of normal and reversed phase thin-layer chromatography in the screening for basic and quaternary drugs, *J. Liquid Chromatogr.*, 14, 1435–1446, 1991.

29. Ojanpera, I. and Vuori, E., Thin layer chromatographic analysis of basic and quaternary drugs extracted as bis (2-ethylhexyl)phosphate ion-pairs, *J. Liquid Chromatogr.*, 10, 3595–3604, 1987.

30. Misztal, G., Paw, B., Skibiński, R., Komsta, Ł., and Kołodziejczyk, J., Analysis of non-selective calcium-channel blockers by reversed-phase TLC, *J. Planar Chromatogr.*, 16, 433–437, 2003.

31. Hopkała, H., Pomykalski, A., Mroczek, T., and Ostęp, M., Densitometric and videodensitometric TLC determination of timolol and betaxolol in ophthalmic solutions, *J. Planar Chromatogr.*, 16, 280–285, 2003.

32. Ojanpera, I., Lillsunde, P., Vartiovaara, J., and Vuori, E., Screening for amphetamines with a combination of normal and reversed phase thin layer chromatography and visualization with Fast Black K salt, *J. Planar Chromatogr.*, 4, 373–378, 1991.

33. Bhushan, R. and Arora, M., Separation of aminoglycosides by normal- and reversed-phase TLC, *J. Planar Chromatogr.*, 14, 435–438, 2001.

34. Skibiński, R., Komsta, Ł., and Misztal, G., The reversed chase retention behaviour of some atypical antipsychotic drugs, *J. Planar Chromatogr.*, 20, 75–80, 2007.

35. Kastner, P. and Klimes, J., Analysis of benzodiazepines by adsorption and ion-pair TLC, *J. Planar Chromatogr.*, 9, 382–387, 1996.

36. Giaginis, C., Dellis, D., and Tsantili-Kakoulidou, A., Effect of the aqueous component of the mobile phase on RP-TLC retention and its implication on the determination of lipophilicity for a series of structurally diverse drugs, *J. Planar Chromatogr.*, 19, 151–156, 2006.

37. Simonovska, B., Prosek, M., Vovk, I., and Jelen-Zmitek, A., High-performance thin-layer chromatographic separation of ranitidine hydrochloride and two related compounds, *J. Chromatogr. B*, 715, 425–430, 1998.

38. Petrovic, S.M., Acanski, M., Persic-Janic, N.U., and Vlaovuc, D.J., Reversed-phase liquid chromatography of some benzimidazole derivatives, *Chromatographia*, 37, 98–104, 1993.

39. Lambroussi, V., Piperaki, S., and Tsantili-Kakoulidou, A., Formation of inclusion complexes between cyclodextrins as mobile phase additives in RPTLC and fluoxetine, norfluoxetine and promethazine, *J. Planar Chromatogr.*, 12, 124–128, 1999.

40. Baranowska, I. and Kądziołka, A., RPTLC and derivative spectrophotometry for the analysis of selected vitamins, *Acta Chromatogr.*, 6, 61–71, 1996.

41. Pyka, A. and Dołowy, M., Separation of selected bile acids by TLC. VII Separation by reversed partition HPTLC, *J. Liq. Chromatogr.*, 28, 1573–1581, 2005.

42. Crommen, J., Schill, G., Hackzell, L., and Westerlund, D., Indirect detection in liquid chromatography. I. Response models for reversed-phase ion-pairing systems, *Chromatographia*, 24, 252–260, 1987.

43. Inczedy, J. and Szokoli, F., Contributions to the calculation of retention data in ion-pair chromatography, *J. Chromatogr.*, 508, 309–317, 1990.

44. Stahlberg, J., A quantitative evaluation of the electrostatic theory for ion pair chromatography, *Chromatographia*, 24, 820–826, 1987.

45. Jandera, P., Churaček, J., and Taraba, B., Comparison of retention behavior of aromatic sulphonic acids in reversed-phase systems with mobile phases containing ion-pairing ions and in systems with solutions of inorganic salts as the mobile phases, *J. Chromatogr.*, 262, 121–140, 1983.

46. Zou, H., Zhang, Y., and Lu, P., Effect of organic modifier concentration on retention in reversed-phase ion-pair liquid chromatography, *J. Chromatogr.*, 545, 59–71, 1991.

47. Afrashtehfar, S. and Cantwell, F.C., Chromatographic retention mechanism of organic ions on a low-capacity ion exchange adsorbent, *Anal. Chem.*, 54, 2422–2427, 1982.

48. Wittmer, D.P., Nuessle, N.O., and Haney Jr, W.G., Simultaneous analysis of tartrazine and its intermediates by reversed phase liquid chromatography, *Anal. Chem.*, 47, 1422–1423, 1975.

49. Tilly-Melin, A., Askemark, Y., Wahlund, K.G., and Schill, G., Retention behavior of carboxylic acids and their quaternary ammonium ion pairs in reversed phase chromatography with acetonitrile as organic modifier in the mobile phase, *Anal. Chem.*, 51, 976–983, 1979.

50. Horvath, C., Melander, W., and Molnar, I., Solvophobic interactions in liquid chromatography with non polar stationary phases, *J. Chromatogr.*, 125, 129–156, 1976.

51. Kraak, J.C., Jonker, K.M., and Huber, J.F.K., Solvent-generated ion-exchange systems with an ionic surfactants for rapid separations of amino acids, *J. Chromatogr.*, 142, 671–688, 1977.

52. Riley, C.M., Tomlinson, E., and Jefferies, T.M., Functional group behaviour in ion-pair reversed-phase high-performance liquid chromatography using surface-active pairing ions, *J. Chromatogr.*, 185, 197–224, 1979.

53. Crombeen, J.P., Kraak, J.C., and Poppe, H., Reversed-phase systems for the analysis of catecholamines and related compounds by high-performance liquid chromatography, *J. Chromatogr.*, 167, 219–230, 1978.

54. Knox, J.H. and Hartwick, R.A, Mechanism of ion-pair liquid chromatography of amines, neutrals, zwitterions and acids using anionic hetaerons, *J. Chromatogr.*, 204, 3–21, 1981.

55. Goldberg, A.P., Nowakowska, E., Antle, P.E., and Snyder, L.R., Retention-optimization strategy for the high-performance liquid chromatographic ion-pair separation of samples containing basic compounds, *J. Chromatogr.*, 316, 241–260, 1984.

56. Hansen, H. and Helboe, P., High-performance liquid chromatography on dynamically modified silica: V. Influence of nature and concentration of organic modifier in eluents containing cetyltrimethylammonium bromide, *J. Chromatogr.*, 285, 53–61, 1984.

57. Bidlingmeyer, B.A., Deming, S.N., Price, W.P., Sachok, B., and Petrusek, M., Retention mechanism for reversed-phase ion-pair liquid chromatography, *J. Chromatogr.*, 186, 419, 1979.

58. Stahlberg, J., Electrostatic retention model for ion-exchange chromatography, *Anal. Chem.*, 66, 440–449, 1994.

59. Bieganowska, M.L., Petruczynik, A., and Gadzikowska, M., Retention of some organic electrolytes in ion-pair reversed-phase high-performance liquid and reversed-phase high-performance thin-layer chromatographic systems, *J. Chromatogr.*, 520, 403–410, 1990.

60. Bieganowska, M.L. and Petruczynik, A., Retention parameters of some isomeric 2-benzoylbenzoicacids in reversed-phase ion-pair high performance thin-layer and column chromatography, *Chromatographia*, 40, 453–457, 1995.

61. Bieganowska, M.L., Petruczynik, A., and Doraczyńska-Szopa, A., The retention of some organic acids in ion-pair HPLC systems, *J. Liq. Chromatogr.*, 13, 2661–2676, 1990.

62. Bartha, A., Vigh, G., Billiet, H.A.H., and de Galan, L., Effect of the type of ion-pairing reagent in reversed-phase ion-pair chromatography, *Chromatographia*, 20, 587–590, 1985.

63. Knox, J.H. and Jurand, J., Zwitterion-pair chromatography of nucleotides and related species, *J. Chromatogr.*, 203, 85–92, 1981.

64. Tilly-Melin, A., Ljungcrantz, M., and Schill, G., Reversed-phase ion-pair chromatography with an adsorbing stationary phase and a hydrophobic quaternary ammonium ion in the mobile phase: I. Retention studies with tetrabutyl ammonium as cationic component, *J. Chromatogr.*, 185, 225–239, 1979.

65. Grune, T., Siems, W., Gerber, G., Tikhonov, Y.V., Pimenov, A.M., and Toguzov, R.T., Changes of nucleotide patterns in liver, muscle and blood during the growth of Ehrlich ascites cells: Application of the reversed-phase and ion-pair reversed-phase high-performance liquid chromatography with radial compression column, *J. Chromatogr.*, 563, 53–61, 1991.

66. Dimitrova, B. and Budevsky, O., Investigation of an alkyl amine modified and pH-controlled mobile phase for separation of pyrazolone derivatives by high-performance liquid chromatography, *J. Chromatogr.*, 409, 81–89, 1987.

67. Tikhonov, Y.V., Pimenov, A.M., Uzhevko, S.A., and Toguzov, R.T., Ion-pair high-performance liquid chromatography of purine compounds in the small intestinal mucosa of children with coeliac disease, *J. Chromatogr.*, 520, 419–423, 1990.

68. Crommen, J., Reversed-phase ion-pair high-performance liquid chromatography of drugs and related compounds using underivatized silica as the stationary phase, *J. Chromatogr.*, 186, 705–724, 1979.

69. Bidlingmeyer, B.A., Separation of ionic compounds by reversed-phase liquid chromatography: An update of ion- pairing techniques, *J. Chromatogr. Sci.*, 18, 525–539, 1980.

70. Low, K.G.C., Bartha, A., Billiet, H.A.H., and de Galan, L., Systematic procedure for the determination of the nature of the solutes prior to the selection of the mobile phase parameters for optimization of reversed-phase ion-pair chromatographic separations, *J. Chromatogr.*, 478, 21–38, 1989.

71. Deelder, R.S. and Van den Berg, J.H.M., Study on the retention of amines in reversed-phase ion-pair chromatography on bonded phases, *J. Chromatogr.*, 218, 327–339, 1981.

72. Bartha, A., Vigh, G., Billiet, H.A.H., and de Galan, L., Studies in reversed-phase ion-pair chromatography: IV. The role of the chain length of the pairing ion, *J. Chromatogr.*, 303, 29–38, 1984.

73. Petruczynik, A., *Thin-Layer Chromatography (TLC) and High Performance Liquid Chromatography (HPLC) of Some Organic Electrolytes in Ion-Pair Systems*, Medical University, 1993.

74. Ruane, R.J. and Wilson, I.D., Ion-pair reversed-phase thin-layer chromatography of basic drugs using sulphonic acids, *J. Chromatogr.*, 441, 355–360, 1988.

75. Bieganowska, M.L., Petruczynik, A., and Doraczyńska-Szopa, A., Ion pair reversed chase thin layer chromatography of some basic drugs and related pyridine derivatives, *J. Planar Chromatogr.*, 5, 184–191, 1992.

76. Wiater, I., Madej, K., Parczewski, A., and Kala, M., Optimum TLC system for identification of phenothiazines and tri-and tetracycline antidepressants, *Microchim. Acta,* 129, 121–126, 1998.

77. Shalaby, A. and Khalil, H., Reversed-phase ion pair thin layer chromatography of some alkaloids, *J. Liq. Chromatogr.*, 22, 2345–2352, 1999.

78. Jost, W., Hauck, H., and Herbert, H., Reversed-phase thin-layer chromatography of 2-substituted benzoic acids with ammonium compounds as ion-pair reagents, *Chromatographia*, 18, 512–516, 1984.

79. Hopkała, H. and Pomykalski, A., TLC analysis of non-steroidal anti-inflammatory drugs and videodensitometric determination of fenbufen in tablets, *J. Planar Chromatogr.*, 17, 383–387, 2004.

80. Sumina, E.G., Shtykov, S.N., and Dorofeeva, S.V., Ion-pair reversed-phase thin-layer chromatography and high-performance liquid chromatography of benzoic acids, *J. Anal. Chem.*, 57, 210–214, 2002.

81. Kowalczuk, D. and Hopkała, H., Separation of fluoroquinolone antibiotics by TLC on silica gel, cellulose and silanized layers, *J. Planar Chromatogr.*, 19, 216–222, 2006.

82. Wilson, I.D., Ion-pair reversed-phase thin-layer chromatography of organic acids: Reinvestigation of the effect of solvent pH on R_f values, *J. Chromatogr.*, 354, 99–106, 1986.

83. Bieganowska, M.L., Petruczynik, A., and Doraczyńska-Szopa, A., Comparison of the retention behaviour of some basic drugs used mainly against Parkinson's disease in normal- and reversed-phase thin-layer chromatography, *Chem. Anal. (Warsaw)*, 38, 719–731, 1993.

84. Brown, S.M. and Bush, K.L., Direct identification and quantitation of diuretic drugs by fast atom bombardment mass spectrometry following separation by thin-layer chromatography, *J. Planar Chromatogr.*, 4, 189–193, 1991.

85. Sharmila, M. and Sharma, S., Development and validation of an HPTLC method for determination of oseltamivir phosphate in pharmaceutical dosage form, *Indian Drugs*, 47, 68–72, 2010.

86. Saleh, G.A., Mohamed, F.A., El-Shaboury, S.R., and Rageh, A.H., Selective densitometric determination of four alpha-aminocephalosporins using ninhydrin reagent, *J. Chromatogr. Sci.*, 48, 68–75, 2010.

87. Dettwiler, M., Rippstein, S., and Jeger, A., A rapid sensitive determination of carprofen and zomepirac using thin-layer chromatography and gas chromatography-mass spectrometry, *J. Chromatogr.*, 244, 153–158, 1982.

88. Khadry, N.H. and Shaigh, K.M., A new technique for visualizing thin-layer chromatography plates, *Bull. Int. Assoc. Forensic Toxicol.*, 26, 38–41, 1996.

89. Ojanpera, J., Ojansivu, R.L., Nokua, J., and Vuori, E., Comprehensive TLC in 661089 8 toxicology: Comparison of findings in urine and liver, *J. Planar Chromatogr.*, 12, 38–41, 1999.

90. Schmidt, M. and Bracher, F., A convenient TLC method for the identification of local anesthetics, *Pharmazie*, 61, 15–17, 2006.

91. Bhat, L., Bothara, K., and Damle, M., Validated HPTLC method for simultaneous determination of nebivolol hydrochloride and hydrochlorothiazide from tablets, *Indian Drugs*, 45, 948–951, 2008.

92. Rusu, L., Marutoiu, C., Rusu, M., Simionescu, A., Rusu, M., Moldovan, Z., and Barbu, C., HPTLC and MS for separation and identification of some beta-blockers in urine, *Asian J. Chem.*, 22, 4209–4213, 2010.

93. Shewiyo, D., Kaale, E., Ugullum, C., Sigonda, M., Risha, P., Dejanegher, B., Verbeke, J., and Heyden, Y., Development and validation of a normal-phase HPTLC method for the simultaneous analysis of lamivudine, stavudine and nevirapine in fixed dose combination tablets, *J. Pharm. Biomed. Anal.*, 54, 445–450, 2011.

94. Misztal, G. and Skibiński, R., Chromatographic analysis of new antidepressant drugs by normal- and reversed-phase TLC, *J. Planar Chromatogr.*, 14, 300–304, 2001.

95. Pietras, R., Hopkała, H., Kowalczuk, D., and Małysza, A., Normal-phase TLC separation of some antiarrhythmics: Densitometric determination of mexiletine hydrochloride in capsules, *J. Planar Chromatogr.*, 17, 213–217, 2004.

96. Elkady, E.F. and Mahrouse, M.A., Reversed-phase ion-pair HPLC and TLC-densitometric methods for the simultaneous determination of ciprofloxacin hydrochloride and metronidazole in tablets, *Chromatographia*, 73, 297–305, 2011.

97. Mohammad, M.A., Zawilla, N.H., El-Anwar, F.M.B., and El-Moghazy, A.S.M., Stability indicating methods for the determination of norfloxacin in mixture with tinidazole, *Chem. Pharm. Bull.*, 55, 1–6, 2007.

98. Henderson, L., Miller, J.H., and Skellern G.G., Control of impurities in diphenhydramine hydrochloride by an ion-pairing, reverse-phase liquid chromatographic method, *J. Pharm. Pharmacol.*, 53, 323–331, 2001.

99. El Walily, A.F.M., El Gindy, A., and Bedair, M.F., Application of first-derivative UV-spectrophotometry, TLC-densitometry and liquid chromatography for the simultaneous determination of mebeverine hydrochloride and sulpiride, *J. Pharm. Biomed. Anal.*, 21, 535–548, 1999.

100. Bebawy, L.I., Application of TLC-densitometry, first-derivative UV-spectrophotometry and ratio derivative spectrophotometry for the determination of dorzolamide hydrochloride and timolol maleate, *J. Pharm. Biomed. Anal.*, 27, 737–746, 2002.

101. Hassan, S.S.M., Elmosallamy, M.A.F., and Abbas, A.B., LC and TLC determination of cinnarizine in pharmaceutical preparations and serum, *J. Pharm. Biomed. Anal.*, 28, 711–719, 2002.

102. Popa, D.-S., Oprean, R., Curea, E., and Preda, N., TLC-UV densitometric and GC-MSD methods for simultaneous quantification of morphine and codeine in poppy capsules, *J. Pharm. Biomed. Anal.*, 18, 645–650, 1998.

103. Agbaba, D., Grozdanovic, O., Popovic, L., Vladimirov, S., and Zivanov-Stakic, D., HPTLC in the quantitative assay of drugs, *J. Planar Chromatogr.*, 9, 116–119, 1996.

104. Pachaly, P. and Schick, W., Simple thin-layer chromatographic identification of active principles in finished products. Part 2., *Pharm. Ind.*, 55, 259–267, 1993.

6 Ionic Liquid Additives to Mobile Phases

Roman Kaliszan and Michał Piotr Marszałł

CONTENTS

6.1 INTRODUCTION

Although it has some disadvantages, silica is the most widely used material in chromatography, both in high-performance liquid chromatography (HPLC) and thin-layer chromatography (TLC). The undesirable property of silica is caused by its surface acidity due to the free, geminal, or single (isolated) silanols [1]. The adsorption activity and behavior of the different silanols have been discussed and described in the literature. However, the problem from the analytical point of view is that the effects of free silanols on chromatographic retention are difficult to control and are especially problematic as regards the chromatographic behavior of basic analytes. These analytes are inconvenient compounds to separate in liquid chromatography on the silica-based stationary phase due to the interaction between the cationic sites of the compounds and anionic silanols of the silica support [2]. The problem concerns even the highly purified silica supports, which are the most popular in HPLC and TLC [3]. The retention effect of basic solutes on the silica stationary phases is mainly determined by hydrogen bonding, ion pairing, ion exchange, and hydrophobic interactions that produce peak or spot tailing and strong retention.

Numerous mobile-phase manipulations have been tested to suppress undesirable silanol effects in liquid chromatography. The most studied modifiers of mobile phase were primary, secondary, tertiary, and quaternary amine additives [4,5]. Nahum and Horvath recommended them as additives to the eluent employed in reversed-phase chromatography of inorganic substances to suppress the deleterious effect of free silanolic groups in the stationary phase [6]. The class of amine additives involves monoamines, such as dimethyloctylamine (DMOA), ethanolamine, hydroxylamine, triethanolamine, triethylamine (TEA), propylamine, and quaternary tetramethylammonium chloride as well as diamines such as ethylenediamines [7]. The role of the amine additive is to interact with silanol sites and reduce its interaction with basic analytes [1]. In practice, the most effective are tertiary amines that can interact with a silica support through the same mechanism as a basic solute.

The other idea, dynamically modified silica, was proposed by Hansen [8]. The method is based on the mobile phase containing quaternary ammonium salt that allows for the equilibrium of three phases in the chromatographic system: the layer of adsorbed quaternary ammonium ions, micellas obtained when their concentration exceeds the critical micelle concentration (CMC), and dissolved molecules in the mobile phase [1,6]. Although the dynamic modification is easy to perform, the method does not allow gradient elution and hence has significant limitations. Indisputably, DMOA, TEA, and ammonia (NH_4OH) are most often applied as mobile-phase additives in TLC [9]. However, these additives cause slow equilibration of the chromatographic system and are often ineffective in the case of strongly basic analytes [10].

6.2 IONIC LIQUIDS

In the last decade, ionic liquids (ILs) have been used as new potential modifiers of the mobile phase in liquid chromatography with effects similar to those observed for amines [11–17]. ILs are known by many synonyms, such as room-temperature ionic liquids, liquid organic salts, low-temperature molten salts, ambient-temperature molten salts, and also neoteric solvents, meaning new types of solvents [18]. They are organic salts with "dual nature" formed by large, mostly asymmetric organic cations and various anions that are liquid at room temperature (Table 6.1). Their usefulness for analytical chemistry can be due to their favorable physicochemical properties, such as the lack of vapor pressure, good thermal and chemical stability, as well as very good dissolution properties in both organic and inorganic compounds [17]. The physical properties of ILs for their potential application as solvents in liquid chromatography (LC) were described by Poole et al. [19]. The hydrophobic

TABLE 6.1
Most Common Structures of Ionic Liquids Used in Analytical Chemistry (R_1, R_2, R_3, R_4—Alkyl Chains)

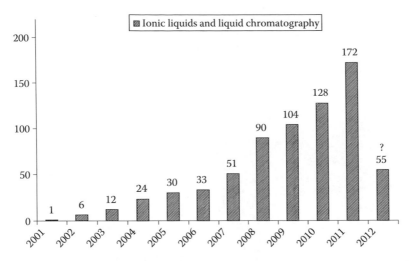

FIGURE 6.1 The number of publications on the subject "ionic liquids and liquid chromatography" identified by Scopus. (From http://www.scopus.com/home.url, dated on June 5, 2012.)

properties of a cation and anion have a significant influence on the solubility of ILs in different chromatographic eluents. The ILs composed of anions such as, BF_4^-, Br^-, $CF_3SO_3^-$, CH_3COO^-, $CH_3SO_4^-$, $C_2H_5SO_4^-$, Cl^-, $(CN)_2N^-$, I^-, NO_3^-, NO_5^-, and SCN^- are water soluble. In turn in liquids composed of PF_6^-, SbF_6^-, and $(CF_3SO_2)_2N^-$ they are water insoluble. The alkylammonium-based ILs in that study had a moderate viscosity, but when mixed with commonly chromatographic solvents (acetonitrile, methanol, water), their viscosity sufficiently decreased to permit their use as mobile phases in LC. Presently, the most commonly used ILs in liquid chromatography based on dialkylimidazolium cations and BF_4^-, Br^-, and Cl^- anions are water stable compounds that dissolve in typical chromatographic solvents [20].

The specific physical and chemical properties of ILs allowed them to be recognized as a novel class of solvents that are applied now in several areas of chemistry. The growing interest in ILs is focused on biocatalytic and organic chemistry as well as electrochemistry. Also, the number of publications presenting original studies on using ILs in liquid chromatography is increasing year by year (Figure 6.1). This fact allows to claim that ILs might be of special importance for the development and improvement of commonly used analytical methods.

6.3 APPLICATION OF IONIC LIQUIDS IN TLC OF BASIC DRUGS

The first study presenting the application of ionic liquid additives to the mobile phases in TLC was reported in 2004 by Kaliszan et al. [11]. The following three ILs based on the dialkylimidazolium cation were subjected to study: 1-ethyl-3-methylimidazolium tetrafluoroborate ([EMIM][BF$_4$]), 1-methyl-3-hexylimidazolium tetrafluoroborate ([HxMIM][BF$_4$]), and 1-hexyl-3-heptyloxymethylimidazolium tetrafluoroborate ([HpOM-HxIM][BF$_4$]). The set of eight basic drugs chlorpromazine, fluphenazine, naphazoline, phenazoline, quinine, thioridazine, tiamenidine, and trifluoropromazine were not moved from the application point either on the normal phase or on octadecylsilica plates using pure acetonitrile as the eluent. The effect of the concentration of the tested ILs on the retardation factor (R_f) of basic drugs is presented in Table 6.2. As evident, the lowest retention was observed when the highest concentration (3% v/v) of ILs was used. Moreover, a more significant advantage on the silanol suppressing potency (lower retention) was observed for octadecylsilica than for bare silica covered plates.

TABLE 6.2

Influence of Addition of Ionic Liquid (1-Ethyl-3-methylimidazolium tetrafluoroborate ([EMIM][BF$_4$]), 1-Methyl-3-hexylimidazolium tetrafluoroborate ([HxMIM][BF$_4$]), and 1-Hexyl-3-heptyloxymethylimidazolium tetrafluoroborate ([HpOM-HxIM][BF$_4$]) on the Retardation Factor, R$_f$, of Tested Drugs with the Use of Octadecylsilica-Covered Plates (RP-18 F$_{254s}$)

Ionic Liquid	Concentration of Ionic Liquid in Acetonitrile% (v/v) as a Mobile Phase						
	0.000	0.125	0.250	0.400	0.500	1.500	3.000
Quinine							
[HpOM-HxIM][BF$_4$]	0.00 (±0.00)	0.05 (±0.01)	0.09 (±0.01)	0.11 (±0.01)	0.15 (±0.01)	0.27 (±0.02)	0.39 (±0.02)
[HxMIM][BF$_4$]	0.00 (±0.00)	0.10 (±0.02)	0.13 (±0.01)	0.17 (±0.01)	0.18 (±0.01)	0.36 (±0.02)	0.50 (±0.03)
[EMIM][BF$_4$]	0.00 (±0.00)	0.12 (±0.01)	0.20 (±0.02)	0.24 (±0.01)	0.29 (±0.01)	0.46 (±0.01)	0.57 (±0.03)
Fluphenazine							
[HpOM-HxIM][BF$_4$]	0.00 (±0.00)	0.04 (±0.01)	0.08 (±0.01)	0.12 (±0.01)	0.15 (±0.01)	0.31 (±0.02)	0.45 (±0.02)
[HxMIM][BF$_4$]	0.00 (±0.00)	0.10 (±0.02)	0.12 (±0.01)	0.15 (±0.02)	0.19 (±0.02)	0.37 (±0.02)	0.54 (±0.03)
[EMIM][BF$_4$]	0.00 (±0.00)	0.11 (±0.01)	0.19 (±0.02)	0.24 (±0.02)	0.31 (±0.02)	0.49 (±0.01)	0.63 (±0.03)
Thioridazine							
[HpOM-HxIM][BF$_4$]	0.00 (±0.00)	0.17 (±0.01)	0.26 (±0.01)	0.34 (±0.02)	0.39 (±0.02)	0.63 (±0.02)	0.75 (±0.02)
[HxMIM][BF$_4$]	0.00 (±0.00)	0.26 (±0.02)	0.39 (±0.01)	0.47 (±0.02)	0.51 (±0.02)	0.73 (±0.01)	0.80 (±0.02)
[EMIM][BF$_4$]	0.00 (±0.00)	0.31 (±0.02)	0.48 (±0.01)	0.58 (±0.01)	0.63 (±0.02)	0.79 (±0.01)	0.84 (±0.01)

Chlorpromazine

[HpOM-HxIM][BF$_4$]	0.00 (±0.00)	0.18 (±0.02)	0.28 (±0.01)	0.36 (±0.02)	0.42 (±0.02)	0.65 (±0.02)	0.77 (±0.03)
[HxMIM][BF$_4$]	0.00 (±0.00)	0.28 (±0.02)	0.41 (±0.01)	0.49 (±0.02)	0.54 (±0.03)	0.76 (±0.02)	0.84 (±0.02)
[EMIM][BF$_4$]	0.00 (±0.00)	0.33 (±0.01)	0.50 (±0.02)	0.61 (±0.01)	0.66 (±0.02)	0.82 (±0.01)	0.88 (±0.01)

Trifluopromazine

[HpOM-HxIM][BF$_4$]	0.00 (±0.00)	0.24 (±0.02)	0.38 (±0.02)	0.49 (±0.03)	0.55 (±0.02)	0.77 (±0.02)	0.85 (±0.03)
[HxMIM][BF$_4$]	0.00 (±0.00)	0.38 (±0.01)	0.52 (±0.01)	0.63 (±0.02)	0.69 (±0.02)	0.83 (±0.01)	0.90 (±0.03)
[EMIM][BF$_4$]	0.00 (±0.00)	0.40 (±0.02)	0.58 (±0.02)	0.69 (±0.01)	0.74 (±0.01)	0.88 (±0.02)	0.91 (±0.02)

Phenazoline

[HpOM-HxIM][BF$_4$]	0.00 (±0.00)	0.35 (±0.02)	0.53 (±0.02)	0.65 (±0.03)	0.71 (±0.02)	0.90 (±0.03)	0.94 (±0.02)
[HxMIM][BF$_4$]	0.00 (±0.00)	0.48 (±0.04)	0.67 (±0.02)	0.77 (±0.03)	0.80 (±0.02)	0.93 (±0.01)	0.95 (±0.02)
[EMIM][BF$_4$]	0.00 (±0.00)	0.52 (±0.01)	0.71 (±0.01)	0.80 (±0.01)	0.84 (±0.01)	0.95 (±0.01)	0.96 (±0.02)

Tiamenidine

[HpOM-HxIM][BF$_4$]	0.00 (±0.00)	0.32 (±0.02)	0.48 (±0.02)	0.62 (±0.04)	0.68 (±0.01)	0.87 (±0.02)	0.91 (±0.02)
[HxMIM][BF$_4$]	0.00 (±0.00)	0.45 (±0.03)	0.63 (±0.02)	0.74 (±0.02)	0.78 (±0.02)	0.92 (±0.01)	0.94 (±0.02)
[EMIM][BF$_4$]	0.00 (±0.00)	0.48 (±0.01)	0.67 (±0.02)	0.77 (±0.01)	0.80 (±0.02)	0.94 (±0.01)	0.96 (±0.02)

Naphazoline

[HpOM-HxIM][BF$_4$]	0.00 (±0.00)	0.29 (±0.02)	0.45 (±0.02)	0.57 (±0.03)	0.63 (±0.01)	0.85 (±0.01)	0.91 (±0.02)
[HxMIM][BF$_4$]	0.00 (±0.00)	0.41 (±0.02)	0.60 (±0.02)	0.71 (±0.03)	0.75 (±0.02)	0.92 (±0.01)	0.93 (±0.02)
[EMIM][BF$_4$]	0.00 (±0.00)	0.44 (±0.02)	0.63 (±0.02)	0.74 (±0.02)	0.78 (±0.01)	0.93 (±0.01)	0.96 (±0.01)

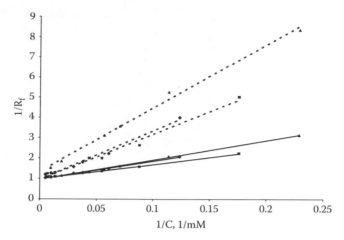

FIGURE 6.2 Plots of reciprocal of retardation factor ($1/R_f$) of tiamenidine on Si 60 F_{254} plates (broken lines) and on octadecylsilica plates (solid lines) with acetonitrile as eluent vs. the reciprocal of concentration ($1/C$) of imidazolium tetrafluoroborate derivatives: [EMIM][BF$_4$] (♦), [HxMIM][BF$_4$] (■) and [HpOM-HxIM][BF$_4$] (▲). (Adapted from *J. Chromatogr. A*, 1030, Kaliszan, R., Marszałł, M.P., Markuszewski, M.J., Baczek, T., and Pernak, J., Suppression of deleterious effects of free silanols in liquid chromatography by imidazolium tetrafluoroborate ionic liquids, 263–271, Copyright 2003, with permission from Elsevier.)

The more effective silanol suppressing potency of the reversed phase was confirmed by comparison of the respective Langmuir plots for three ILs added to the pure acetonitrile eluent (Figure 6.2). Evidently, all tested imidazolium-based ILs are more strongly adsorbed to the octadecylsilica phase (solid lines) than to the silica covered plates (broken lines). That can be due to the hydrophobic interaction of the alkylimidazoilum cation with the alkyl-bonded stationary phase. The strong adsorption of cations may block the direct contact of the basic solute with active silanol sites.

More interesting results from a practical point of view were obtained through a comparison of imidazolium classes of ILs with standard amine additives to eluent (Figure 6.3). The effect of the concentration of standard amino quenchers (DMOA, TEA, and NH$_4$OH) on the retardation factor of tested drugs is much weaker than for the representative IL. The addition of [EMIM][BF$_4$] in an amount of 0.25% (v/v) significantly decreased the retention of test solutes. The saturation of adsorption has been observed when the concentration of IL was reached at 0.5%–1% (v/v). Moreover, the silanol suppressing potency of DMOA, TEA, and NH$_4$OH is much weaker or has a negligible effect, even at a high concentration of the modifiers in the mobile phase.

Comparative study of silanol suppressing activity based on the Langmuir plots of the dependence of $1/R_f$ of a tested analyte (tiamenidine) on the reciprocal of the additive concentration in the eluent provided precious information about the mechanism of the action of ILs and amino quenchers (Figure 6.4). A typical Langmuir adsorption curve resulted in the case of [EMIM][BF4] and TEA. The plots crossed the Y-axis at $1/R_f = 1$, which meant that the test solute would be totally unretained at very high concentrations of the silanol suppressor. The less steep slope of the plot for [EMIM][BF4] than that for TEA proves a more effective adsorption of the former IL compound. The adsorption of DMOA and NH$_4$OH is poor as can be seen in Figure 6.4. This fact disqualifies those substances as potential silanol suppressors.

The effective silanol suppressing potency was also observed in water–acetonitrile eluents of varying compositions but with a 3% (v/v) concentration of [EMIM][BF$_4$], TEA, and NH$_4$OH. In Figure 6.5, the R_M values of eight basic drugs determined on octadecylsilica stationary phases were plotted and compared for tested modifiers of the mobile phase. Because the chromatographic process with the use of DMOA did not produce measurable retention data, the results are not included. In addition, NH$_4$OH gives reliable R_M data, only within a limited range of acetonitrile concentration (60%–100% v/v) in the eluent excluding the tiamenidine. Moreover, the R_M data from reversed-phase TLC systems obtained with the help of an imidazolium tetrafluoroborate additive produce

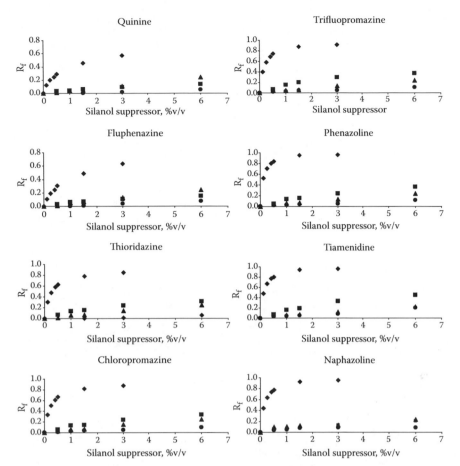

FIGURE 6.3 Thin-layer chromatographic retardation factor, (R_f), on octadecylsilica-covered plates (RP-18 F_{254s}) in relation to volume percent [EMIM][BF$_4$] (◆), TEA (■), DMOA (▲), and NH$_4$OH (●) in pure acetonitrile as the mobile phase. (Adapted from *J. Chromatogr. A*, 1030, Kaliszan, R., Marszałł, M.P., Markuszewski, M.J., Baczek, T., and Pernak, J., Suppression of deleterious effects of free silanols in liquid chromatography by imidazolium tetrafluoroborate ionic liquids, 263–271, Copyright 2003, with permission from Elsevier.)

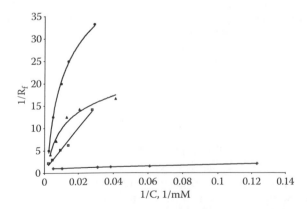

FIGURE 6.4 Plots of reciprocal of retardation factor of tiamenidine on octadecylsilica plates with acetonitrile as eluent versus the reciprocal of concentration of additive in the mobile phase: [EMIM][BF$_4$] (◆), TEA (■), DMOA (▲), and NH$_4$OH (●) . (Adapted from *J. Chromatogr. A*, 1030, Kaliszan, R., Marszałł, M.P., Markuszewski, M.J., Baczek, T., and Pernak, J., Suppression of deleterious effects of free silanols in liquid chromatography by imidazolium tetrafluoroborate ionic liquids, 263–271, Copyright 2003, with permission from Elsevier.)

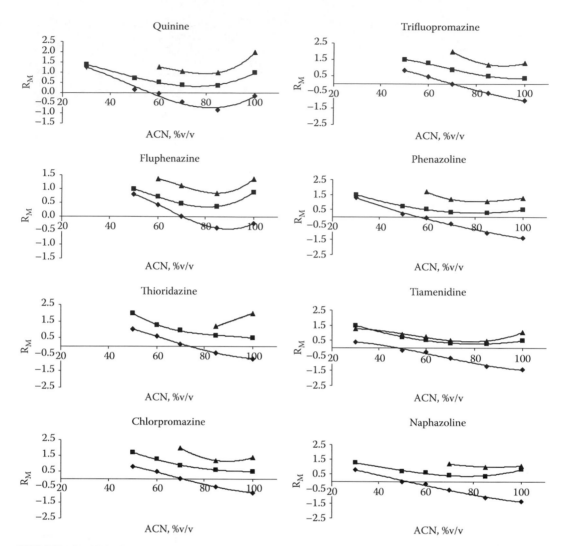

FIGURE 6.5 Thin-layer chromatographic retention parameter, $R_M = \log(1/R_F - 1)$, of basic test drug analytes determined on octadecylsilica-covered plates in relation to the volume percent of acetonitrile in water–acetonitrile eluent. The mobile phases contained 3% (v/v) of [EMIM][BF$_4$] ($\blacklozenge$), TEA ($\blacksquare$) i NH$_4$OH ($\blacktriangle$). (Adapted from *J. Chromatogr. A*, 1030, Kaliszan, R., Marszałł, M.P., Markuszewski, M.J., Baczek, T., and Pernak, J., Suppression of deleterious effects of free silanols in liquid chromatography by imidazolium tetrafluoroborate ionic liquids, 263–271, Copyright 2003, with permission from Elsevier.)

the best fit to the classical linear Snyder–Soczewinski dependence on the organic modifier concentration in the mobile phase. That is due to the complete reduction of uncontrollable attractive interactions of base analytes with free silanols. The reduction is much more effective than that provided by common amine quenchers. The improved linearity of R_M versus percent organic modifier is illustrated in Figure 6.5. It may be used in determinations of liquid chromatographic retention parameters corresponding to 0 percent of organic modifier in the eluent $\left(R_M^0 \text{ or } \log k_w \right)$.

Reliable benefits of the use of ILs as modifiers of the mobile phase are presented in Figure 6.6a through f. Thioridazine, trifluopromazine, phenazoline, naphazoline, tiamenidine, and a mixture of the drugs were spotted on octadecylsilica plates from left to right and chromatographed with the use of a water–acetonitrile 40:60 v/v eluent, either neat or containing 1.5% v/v of various additives. A negligible effect on elution of analytes is evident for the nonmodified mobile phase as well as with NH$_4$OH and DMOA as additives (Figure 6.6a through c). A better but still unsatisfactory resolution

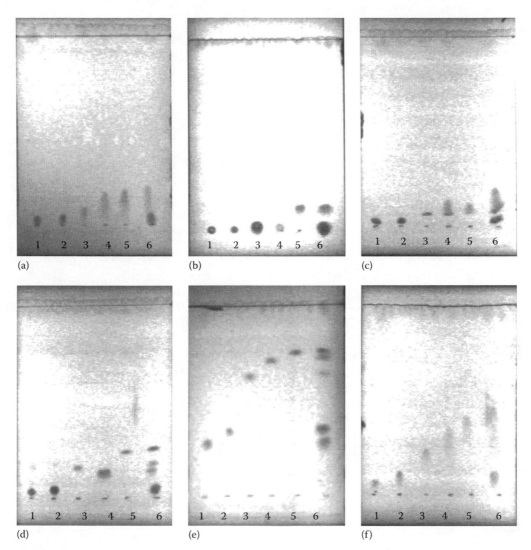

FIGURE 6.6 Chromatograms of thioridazine, trifluoropromazine, phenazoline, naphazoline, tiamenidine, and a mixture of the drugs, as spotted from left to right, on RP-18 F_{254} plates, developed with a water–acetonitrile 40:60 v/v eluent, either pure or containing 1.5% v/v of various additives: (a) no additive; (b) NH_4OH; (c) DMOA; (d) TEA; (e) [EMIM][BF_4]; (f) buffer of pH 2.87. (Adapted with permission from Marszałł, M.P. and Kaliszan, R. Application of ionic liquids in liquid chromatography. *Crit. Rev. Anal. Chem.*, 37, 127, 2007.)

is provided by TEA (Figure 6.6d). An indisputable advantage with the use of ILs is observed in Figure 6.6d. The addition of 1.5% v/v of [EMIM][BF_4] to the neat eluent decreased spot tiling and improved the resolution of basic drugs over a wide range of plate length. Finally, the separation of the components of the mixture of drugs extremely badly separable by liquid chromatography appears to be satisfactory. Because of lower pH (2.87) of 1.5% v/v of [EMIM][BF_4] in the water–acetonitrile 40:60 v/v eluent, the experiment with the use of buffer:acetonitrile 40:60 v/v with the exact same pH as the previous eluent was performed (Figure 6.6e). However, the poor resolution and spot tailing confirm the hypothesis that the separation produced by [EMIM][BF_4] was not due to the pH change caused by the additive.

An additional experiment with the use of acetylosalicylic acid, salicylic acid, phenol, and 2,3-dimetoxytoluene as analytes was performed to check the influence of the tested additives on the chromatographic behavior of acid and neutral analytes. As shown in Figure 6.7, the additive of IL

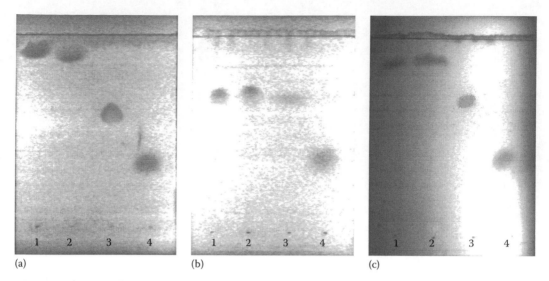

(a)   (b)   (c)

FIGURE 6.7 Chromatograms of acetylosalicylic acid, salicylic acid, phenol, and 2,3-dimethoxytoluene, as spotted from left to right, on RP-18 F_{254} plates, developed with a water–acetonitrile 40:60 v/v eluent either (a) pure, or (b) containing 1.5% (v/v) of [EMIM][BF$_4$], or (c) buffered to pH 2.87.

increased the retention of both acids. This is due to the removal of the repelling interactions of free silanols because of their suppression by the IL ([EMIM][BF$_4$]). Thus, the additive eliminated the exclusion effect of silanols with respect to acidic analytes demonstrated earlier for silica-based stationary phase materials [21]. In Figure 6.7 the chromatogram (c) demonstrates that a stronger retention of acids in the presence of IL 1 (chromatogram (b)) is not due to their decreased ionization at the acidic pH. Chromatogram (c) confirms that a low retention of acids is mostly because of exclusion effects. Comparing chromatograms (a) through (c) in Figure 6.8, one notices that neutral analyte 2,3-dimethoxytoluene retention is not significantly affected by the addition of the IL additive.

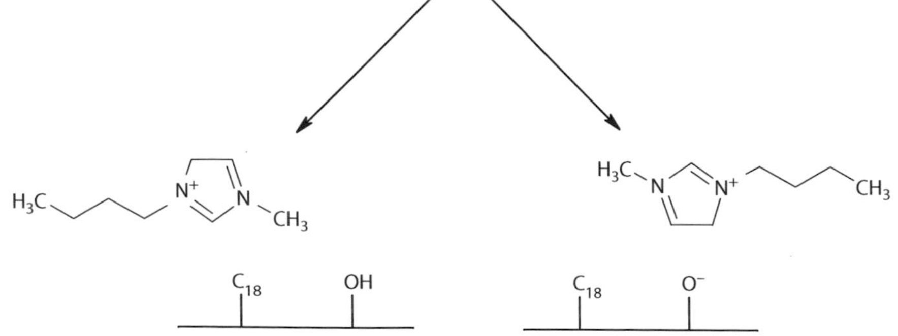

FIGURE 6.8 Scheme of the proposed interaction between ionic liquid and 1-ethyl-3-methylimidazolium cation and octadecylsilica-covered plates.

Based on performed studies, it can be concluded that adding imidazolium tetrafluoroborates to the mobile phases produces liquid chromatographic partition systems universally applicable for basic, acidic, and neutral analytes.

Summarizing the observations from the studies and the literature on the suppression of deleterious effects of free silanols by imidazolium-based ionic liquids, the following conclusions can be drawn. ILs are dual modifiers (with a cationic and anionic character), which means that both cation and anion can be adsorbed on the stationary phase, giving rise to interesting interactions with the anionic free silanols and the cationic basic drugs [22]. However, from our practice the reducing silanol activity is caused by imidazolium cations that are adsorbed on the silica surface (Figure 6.8) [11]. According to the other hypothesis, both anionic and cationic parts of the IL additive can participate in the retention mechanism. The cation of IL coats the surface of the stationary phase on which suppress free silanols and at the same time they become competing agents for the solutes [16,22,23]. Moreover, the chaotropic character of the IL's anions is probably responsible for ion pairing with cationic solutes.

6.4 APPLICATION OF IONIC LIQUIDS IN TLC OF PEPTIDES

The different chromatographic behavior of peptides in TLC was also observed after the addition of ILs to the mobile phase in comparison to the eluent without the modifier, using the normal phase on a silica support stationary phase [24,25]. Regardless of the presence or absence of IL, a nonlinear dependence of the retention coefficient, R_M, of peptides on the volume percentage of acetonitrile in the eluent was found. Generally, R_M increased with increasing concentration of acetonitrile. Depending on the modifier of the mobile phase, R_M can be described well with a quadratic function or with a third-degree polynomial function. On the basis of experimental data, a quadratic model was selected in the case of a mobile phase without IL [24]:

$$R_M = a + bX + cX^2,$$ (6.1)

where
$R_M = \log[(1 - R_f)/R_f]$
a, b, and c are constants for a given analyte and a TLC system
X is the volume fraction of the stronger solvent in the mobile phase (X = %B/100)

A third-degree polynomial model proved to have the best fit for the TLC system with the presence of IL in the mobile phase.

$$R_M = a' + b'X + c'X^2 + d'X^3,$$ (6.2)

where a', b', c', and d' are constants for the respective analyte and the TLC system. Although the computer-assisted simulation of peptide separation did not allow for a satisfactory separation of some peptides, the best resolution (R_S = 0.45) was calculated with 46% (v/v) acetonitrile in the mobile phase (Figure 6.9a). The best resolution of peptides in the TLC system without IL, but worse with comparison to the previous eluent with IL, was achieved at the level of only R_S = 0.075 (Figure 6.9b). Moreover, the performed studies demonstrated that the applied eluent with IL works efficiently in the presence of α-cyano-4-hydroxycinnamic acid (CHCA), conventionally used as a matrix in matrix-assisted laser desorption/ionization mass spectrometry [24]. Hence, the proposed chromatographic system of peptides involving 1-ethyl-3-methylimidazolium tetrafluoroborate IL as a mobile-phase additive may be of special importance in proteomic analysis.

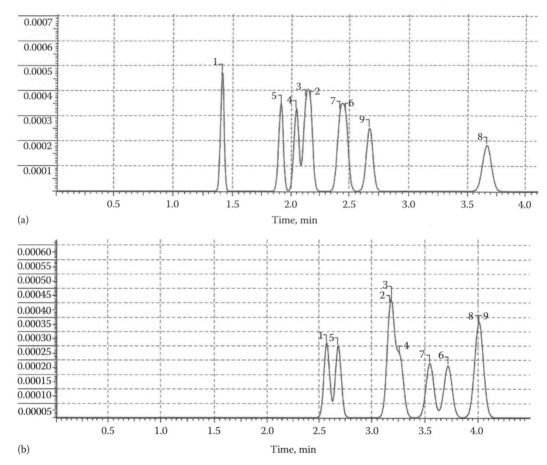

FIGURE 6.9 Simulated TLC chromatogram of a mixture of nine peptides: 1—ETS, 2—KETS, 3—AKETS, 4—VAKETS, 5—TVAKETS, 6—HTVAKETS, 7—WHTVAKETS, 8—HWHTVAKETS and 9—LHWHTVAKETS based on the (a) third-degree polynomial retention model of the mixture of nine peptides with 46% (v/v) acetonitrile in the mobile phase with ionic liquid, (b) quadratic retention model of the mixture of nine peptides with 70% (v/v) of acetonitrile in the mobile phase without ionic liquid. (From Bączek, T., Marszałł, M.P., Kaliszan, R., Walijewski, L., Makowiecka, W., Sparzak, B., Grzonka, Z., Wiśniewska, K., and Juszczyk, P: Behavior of peptides and computer-assisted optimization of peptides separations in a normal-phase thin-layer chromatography system with and without the addition of ionic liquid in the eluent. *Biomedical Chromatography.* 2005. 19. 1–8. Copyright Wiley-VCH Verlag GmbH & Co. KGaA. Adapted with permission.)

REFERENCES

1. Nawrocki, J. 1997. The silanol group and its role in liquid chromatography. *Journal of Chromatography A* 779: 29–71.
2. Berthod, A., Ruiz-Angel, Mt., Carda-Broch, S. 2008. Ionic liquids in separation techniques. *Journal of Chromatography A* 1184: 6–18.
3. Marszałł, M.P., Kaliszan, R. 2007. Application of ionic liquids in liquid chromatography. *Critical Reviews in Analytical Chemistry* 37: 127–140.
4. Kiel, J.S., Morgan, S.L., Abramson, R.K. 1985. Effects of ionic modifiers on peak shape and retention in reversed phase high performance liquid chromatography. *Journal of Chromatography* 320: 313–323.
5. Vervoort, R.J.M., Debets, A.J.J., Debets, A.J.J., Claessens, H.A., Cramers, C.A., de Jong, G.J. 2000. Optimisation and characterisation of silica-based reversed-phase liquid chromatographic systems for the analysis of basic pharmaceuticals. *Journal of Chromatography A* 897: 1–22.
6. Nahum, A., Horvath, C. 1981. Surface silanols in silica-bonded hydrocarbonaceous stationary phases. *Journal of Chromatography* 203: 53–63.

7. Righetti, P.G., Gelfi, C., Verzola, B., Castelletti, L. 2001. The state of the art of dynamic coatings. *Electrophoresis* 22: 603–611.
8. Hansen, S.H. 1981. Column liquid chromatography on dynamically modified silica. *Journal of Chromatography* 209: 203–209.
9. Claessens, H.A. 2001. Trends and progress in the characterization of stationary phases for reversed-phase liquid chromatography. *Trends in Analytical Chemistry* 20: 563–583.
10. Snyder, L.R., Kirkland, J.J., Glajch, J.L. 1977. *Practical HPLC Method Development*, 2nd edn., Wiley, New York, p. 178.
11. Kaliszan, R., Marszałł, M.P., Markuszewski, M.J., Baczek, T., Pernak, J. 2004. Suppression of deleterious effects of free silanols in liquid chromatography by imidazolium tetrafluoroborate ionic liquids. *Journal of Chromatography A* 1030: 263–271.
12. Zhang, W., He, L., Gu, Y.L., Liu, X., Jiang, S. 2003. Effect of ionic liquids as mobile phase additives on retention of catecholamines in reversed-phase high-performance liquid chromatography. *Analytical Letters* 36: 827–838.
13. Xiao, X., Zhao, L., Liu, X., Jiang, S. 2004. Ionic liquids as additives in high performance liquid chromatography: Analysis of amines and the interaction mechanism of ionic liquids. *Analytica Chimica Acta* 519: 207–211.
14. Marszałł, M.P., Baczek, T., Kaliszan, R. 2005. Reduction of silanophilic interactions in liquid chromatography with the use of ionic liquids. *Analytica Chimica Acta* 547: 172–178.
15. Berthod, A., Ruiz-Angel, M.J., Huguet, S. 2005. Nonmolecular solvents in separation methods: Dual nature of room temperature ionic liquids. *Analytical Chemistry* 77: 4071–4080.
16. Ruiz-Angel, M.J., Carda-Broch, S., Berthod, A. 2006. Ionic liquids versus triethylamine as mobile phase additives in the analysis of β-blockers. *Journal of Chromatography A* 1119: 202–208.
17. Marszałł, M.P., Bączek, T., Kaliszan, R. 2006. Evaluation of the silanol-suppressing potency of ionic liquids. *Journal Separation Sciences* 29: 1138–1145.
18. Wilkes, J.S. 2002. A short history of ionic liquids—From molten salts to neoteric. *Green Chemistry* 4: 73–80.
19. Poole, C.F., Kersten, B.R., Ho, S.S.J., Coddens, M.E., Furton, K.G. 1986. Organic salts, liquid at room temperature, as mobile phases in liquid chromatography. *Journal of Chromatography* 352: 407–425.
20. Koel, M. 2000. Physical and chemical properties of ionic liquids based on the dialkylimidaolium cation. *Proceedings of the Estonian Academy of Sciences Chemistry* 49: 145–155.
21. Knox, J.H., Kaliszan, R., Kennedy, G.J. 1980. Enthalpic exclusion chromatography. *Faraday Discussions of Royal Society of Chemistry* 15: 113–125.
22. Fernandez-Navarro, J.J., Garcia-Alvares-Coque, M.C., Ruiz-Angel, M.J. 2011. The role of the dual nature of ionic liquids in the reversed-phase liquid chromatographic separation of basic drugs. *Journal of Chromatography A* 1218: 398–407.
23. Buszewski, B., Studzińska, S. 2008. A review of ionic liquids in chromatographic and electromigration techniques. *Chromatographia* 68: 1–10.
24. Bączek, T., Marszałł, M.P., Kaliszan, R., Walijewski, L., Makowiecka, W., Sparzak, B., Grzonka, Z., Wiśniewska, K., Juszczyk, P. 2005. Behavior of peptides and computer-assisted optimization of peptides separations in a normal-phase thin-layer chromatography system with and without the addition of ionic liquid in the eluent. *Biomedical Chromatography* 19: 1–8.
25. Bączek, T., Sparzak, B. 2005. Ionic liquids as novel solvent additives to separate peptides. *Zeitschrift fur Naturforschung* 61: 827–832.

7 Chromatographic Analysis of Chiral Drugs

Shuchi Dixit and Ravi Bhushan

CONTENTS

7.1 INTRODUCTION

Chirality of pharmaceutical products drew an immense concern of clinicians, pharmaceutical industries, and drug regulatory authorities, at first, in the 1960s, when the administration of the racemic sedative drug *thalidomide* to pregnant women resulted in thousands of seriously crippled babies with underdeveloped limbs. Later, it was observed that the (*R*)-enantiomer of the drug acted as a potential sedative and the (*S*)-counterpart was strongly teratogenic (Muller, 1997). This medical tragedy served as the starting point for massive research efforts to study pharmacological consequences of chiral drug administration and to improvise the technological processes leading to synthesis, analysis, and registration requirements of new enantiomeric drugs.

Enantioselectivity plays an important role not only in pharmacodynamics, involving the interaction of drugs with enzymes and receptors in the target organs, but also in pharmacokinetics involving their absorption, distribution, metabolic conversion, and excretion (ADME). The different pharmacodynamics, pharmacokinetics, and toxicological activities of enantiomers of a drug lead to a variety of effects, that is, one enantiomer may have the desired pharmacological activity, in which case the other enantiomer may be regarded as an impurity; the enantiomers may have an activity that is qualitatively similar but quantitatively different, and the enantiomers may have qualitatively different pharmacological activities (Cayen, 1991). Therefore, the use of a desired enantiomer of a drug is justified clinically (i.e., for more selective pharmacokinetic profile and a lower incidence of interactions with other drugs due to changed metabolic fate) and economically (i.e., for improvement of therapeutic indices reducing overall treatment costs) as well.

International Conference on Harmonization (ICH) and its parties (i.e., European Federation of Pharmaceutical Industries and Associations [EFPIA], U.S. Food and Drug Administration [FDA], Pharmaceutical Research and Manufacturers of America [PhRMA], and Japan Pharmaceutical Manufacturers Association [JPMA]) have produced a comprehensive set of guidelines for registration of new chiral pharmaceuticals (ICH, 1996). These guidelines demand investigations of stereospecific fate of drugs in the body and enantiomeric purity determination of chiral drugs before their introduction into the market and during industrial manufacture. Till date, with "a few" exceptions, chiral synthetic drugs are produced and used as racemic mixture. In general, more than 40% of drugs (over the counter and prescribed) are chiral, and of these 25% are prescribed as pure enantiomers (Cannarsa, 1996). Therefore, the development of efficient analytical methods for enantioseparation has become an essential part of the drug development process.

Chromatography encompasses a diverse but related group of methods that permit separation, isolation, identification, and quantification of components in a mixture. It is widely used for the analysis and estimation of both simple and complex components present in bulk drugs and their formulations. In fact, the widespread acceptance of chromatography for pharmaceutical analysis has made these techniques a natural choice for chiral separation in industry and academia. Thin-layer chromatography (TLC) is a simple, rapid, versatile, sensitive, and inexpensive analytical technique for the separation of components of a mixture. It has been used in virtually all areas of analysis as it provides a relatively high degree of assurance that all possible components of a mixture are separated. Literature survey shows that over a few years, some books have appeared that exclusively deal with applications of TLC for the analysis of various types of compounds (Kowalska and Sherma, 2007; Sherma and Fried, 2003; Sherma and Kowalska, 2007). The role of TLC in control of enantiomeric purity of pharmaceuticals and amino acids has been reviewed by Bhushan and Martens (1997, 2001).

This chapter deals with TLC analysis/separation of certain important enantiomeric drugs. This may be regarded as an exhaustive but not a complete literature source on the subject. Experimental practices experienced in the authors' laboratory have largely been referred to along with the methods described by various authors in literature.

7.2 APPROACHES FOR ENANTIOSEPARATION

Chromatographic enantioseparations may be performed using two basic approaches called "direct" and "indirect."

The separation of a pair of enantiomers without resorting to derivatization prior to separation process is termed as direct approach. The resolution is possible through reversible diastereomeric association between the chromatographic chiral environment and the solute enantiomers. The enantiomers may interact during the course of chromatographic process with (1) a chiral impregnating reagent

(CIR) mixed with/immobilized on the stationary phase, (2) a chiral stationary phase (CSP), or (3) a chiral selector (CS) added to the mobile phase (termed as chiral mobile phase additive [CMPA]).

The resolution of a pair of enantiomers by reacting them with a suitable chiral derivatizing reagent (CDR), that is, the formation of diastereomers followed by their separation in an achiral environment, is considered as an indirect approach. Since diastereomers have different chemical and physical properties, enantioseparation using this approach is considered to be a simpler and easier experimental task in comparison to using direct approach. On the contrary, the loss of identity of the native enantiomers and complex and time-consuming derivatization process can be considered the limitations of this approach.

7.3 VARIOUS STEPS IN TLC ENANTIOSEPARATION

The availability of numerous sorbents, such as silica, cellulose, alumina, polyamides, ion exchangers, and other inorganic and organic sorbents, make TLC considerably versatile in terms of nature of substances that can be separated. Depending upon the requirement/approach of chiral separation, homemade/commercial achiral/chiral TLC plates can be used in normal-/reversed-phase (NP/RP) mode. The homemade TLC plates generally are referred to as the plates prepared in the laboratory by spreading the slurry of neutral silica gel (containing 13% gypsum, iron, and chloride 0.03% each) on glass plates of a particular size. Commercially available silica gel or chemically bound polar diol F_{254} high-performance (HP) TLC plates and hydrophobic octadecyl-, ethyl-, and diphenyl-RPTLC plates, as well as polyamide, are frequently employed as achiral stationary phases. Achiral precoated plates have been commercially produced by Merck (Darmstadt, Germany), Macherey-Nagel (Duren, Germany), Whatman Chemical Division (Clifton, New Jersey), Alltech Associates, Inc. (Deerfield, Illinois), and Fisher Scientific (Fair Lawn, New Jersey). The manufacturers recommend precleaning of plates before the application of samples to remove extraneous materials that can be introduced in commercial plates due to manufacture, shipping, or storage conditions.

Commercial chiral layers are available for the separation of enantiomers by the mechanism of ligand exchange under the name Chiralplate (Macherey-Nagel) or Chir (Merck). Chiralplates consist of a glass plate coated with a 0.25 mm layer of RP-modified (C-18) silica gel coated (not chemically bonded) with $(2S, 4R, 2'RS)$-N-(2'-hydroxydodecyl)-4-hydroxyproline and Cu(II) ions. The chiral phase is bonded to the stationary phase strongly through hydrophobic interactions so that the mobile phases with a high degree of organic composition can be used without bleeding of the copper complex from the support. For optimal separations and reproducible retention factors (R_F) values on Chiralplates, the manufacturer recommends the activation of plates before spotting (15 min at 100°C) and developing in a saturated chamber.

Commercial native fibrous cellulose plates (MN 300; having layer thicknesses of 0.10, 0.25, and 0.50 mm on glass and 0.10 mm on polyester or aluminum), microcrystalline cellulose plates (Avicel MN 400; having 0.10 mm layers on glass or polyester), and acetylated cellulose plates (with 0.1 mm fibrous cellulose layers having 10% or 20% acetyl content on glass or with 10% on polyester sheets) are also available from Macherey-Nagel and certain other firms.

7.3.1 SAMPLE APPLICATION

Aliquots (25 μL) of racemic and pure enantiomers of analytes are applied side by side on TLC plates using Hamilton syringe. The distance of the spots from the edge of the plates and the distance between spots are generally kept 15 and 10 mm, respectively.

7.3.2 DEVELOPMENT OF CHROMATOGRAMS

The chromatograms are developed at specific temperatures in clean and dry rectangular glass chambers previously equilibrated with the mobile phase. The developed chromatograms are dried in oven/fume hood.

7.3.3 DETECTION

Separated compounds are detected on the layers by viewing their natural color, natural fluorescence, or quenching of fluorescence. Substances that cannot be seen in visible or ultraviolet (UV) light are visualized with suitable detection reagents to form colored, fluorescent, or UV-absorbing compounds. Examples include the formation of red to purple zones by reaction of amino acids with ninhydrin and detection of acidic and/or basic analytes with the indicator bromocresol green.

Universal reactions such as iodine absorption are quite unspecific and are valuable for completely locating an unknown sample. The absorption of iodine vapor from its crystals in a closed chamber produces brown spots on a light yellow background with almost all organic compounds. Iodine staining is nondestructive and reversible upon evaporation.

Chromatographic separation and detection takes place separately in TLC, which enables carrying out the analysis at a different time and making full use of detection techniques for the analysis of constituents. An advantage of TLC is the possibility of localization of all constituents on chromatograms, which always appear between the start line and the solvent front, contrary to HPLC, where lack of detection may be observed in case of the retention of a constituent on the column.

7.3.4 RECOVERY OF NATIVE ENANTIOMERS

Pure separated enantiomers could be obtained by cutting the two spots, corresponding to the two *in situ* diastereomers formed between analyte enantiomers and CS, from the TLC plate followed by extraction with suitable solvent.

For example, Bhushan and Gupta reported separation of enantiomers of verapamil on vancomycin-impregnated TLC plates followed by the recovery of native enantiomers (Bhushan and Gupta, 2005). The enantiomers were located by exposure to iodine vapor. Afterward, the spots were marked, and the iodine was allowed to evaporate. The spots were then scraped (from nearly 40 chromatograms) and extracted with ethanol. The combined extracts for each enantiomer were filtered and concentrated *in vacuo*. These were then dried, and the residues were dissolved in methanol. For each extract, optical density was measured at λ_{max}, and the concentration was estimated by the use of calibration plots, while enantiomeric purity was established using a polarimeter. Because vancomycin is insoluble in ethanol, only pure enantiomers of verapamil from the spots went into solution.

7.3.5 QUANTIFICATION

Densitometry provides an excellent tool for determining limit of detection (LOD) and relative amounts of separated enantiomers by *in situ* scanning of the spots under visible absorption, UV absorption, or fluorescence of zones.

Günther and Möller used calibration plots (scan area vs. percent composition of two enantiomers) for quantification of D- and L-*tert*-leucine enantiomers, which were resolved on a Chiralplate (Günther and Möller, 2003). Low microliter volumes of test sample and standard solutions were applied manually with Microcaps (Drummond, Broomall, Pennsylvania) or with a Nanomat II or Linomat IV instrument (Camag), and the zones were scanned using a Shimadzu (Columbia, Maryland) CS 930 or Desaga (Wiesloch, Germany) CD 60 densitometer at 520 nm.

Spell and Stewart separated enantiomers of pindolol on C-8 silica gel followed by quantification using densitometry (Spell and Stewart, 1997). Samples and standards (9 µL) were applied by means of a Linomat IV instrument (Camag) in 5 mm bands, and the separated zones were scanned at 256 nm in the absorbance mode with a slit 4 mm long and 3 mm wide (aliquot scanning method).

External standard methodology was applied to determine the amount of each enantiomer in each band of a drug substance or dosage form:

$$\% \text{Enantiomer} = \frac{\left[(PH_{sam}/PH_{std})(SM)\right]}{2}$$

where

PH$_{sam}$ is the height of the sample scan peak

PH$_{std}$ is the height of the standard peak

SM is the standard mass of the enantiomer in the standard solution

7.4 ENANTIOSEPARATION USING DIRECT APPROACH

The *direct* approach of enantioseparation involves the formation of *in situ/labile* diastereomers between the CS and the enantiomers of analyte. These diastereomers have different thermodynamic stabilities, provided that at least three active points of the CS interact with the corresponding sites of the enantiomers of analyte.

7.4.1 SEPARATION MECHANISMS

There could be considered three main/basic mechanisms for the resolution of enantiomers using direct approach, that is, ion pairing, ligand exchange, and inclusion complexation (guest–host steric interaction). The chemical nature of the CS and the structure of the compounds being separated, along with the influence of the CS on the adsorption/or partition, are together held responsible for the overall results of resolution.

7.4.1.1 Ion Pairing

TLC enantiomeric separations based on diastereomeric ion pairing were published by Bhushan and Ali for the first time (Bhushan and Ali, 1987). The formation of ion pair between the ion of the analyte and the oppositely charged counterion (from the CS) is considered to be responsible for chiral discrimination. An appropriate pH is required to ensure that the analyte and the CS remain in opposite charged forms. For example, L-tartaric acid was employed as a CIR for enantioseparation of (±)-penicillamine (PenA) by Bhushan and Agarwal (2008a). The CS was impregnated with the silica gel slurry used for preparing the thin-layer plates. Resolution studies were carried out on plates prepared in the solution of pH 4, 4.5, 5, 5.5, and 6. The separation of enantiomers was observed at pH 5. On the basis of these findings, it was suggested that the chiral recognition mechanism was based on *in situ* formation of noncovalent bonded diastereomers of type [L-tartaric acid—(+)-PenA$^+$] and [L-tartaric acid—(−)-PenA$^+$]; at pH 5, the CIR (L-tartaric acid) was in the anionic form, and the analyte [(±)-PenA] was in the cationic form. The differences in the physical properties of the diastereomers led to overall separation. The lowering of pH might result in neutral tartaric acid molecules, while at higher pH, the analytes might exist as neutral molecules, providing no sites for ionic interactions.

The use of different CSs suggests that, in general, such reagents cause variation in retention, reduction in tailing, and improvement in selectivity, particularly in the presence of oppositely charged counterions. Large cations are employed for the separation of anions, whereas for the separation of cations, hydrophobic or hydrophilic ion-pair reagents are used.

7.4.1.2 Ligand Exchange

TLC enantiomeric separations based on ligand exchange were published by Günther et al. for the first time (Günther et al., 1984). The basic principle of chiral ligand exchange chromatography

(CLEC) is the reversible coordination of chelating analyte species from the mobile phase into the coordination sphere of a metal ion that is immobilized by complexation with a chelating CS (Davankov, 2004). The formation of mixed ternary metal ion/CS/analyte complex is considered to be responsible for chiral discrimination (Davankov, 2004).

For example, Cu(II)-L-Arg complex was employed for the enantioseparation of atenolol, propranolol, and metoprolol by Bhushan and Gupta (2006). The chiral recognition mechanism was based on *in situ* formation of [Cu (L-Arg) (*S*-propranolol)] and [Cu (L-Arg) (*R*-propranolol)] type diastereomers. Different rates of formation and/or thermodynamic stabilities of these diastereomeric complexes give rise to different retention times of the corresponding solute enantiomers.

7.4.1.3 Inclusion Complexation

The formation of reversible inclusion complexes by fitting of analyte molecules into the cavity of CSs is considered to be responsible for chiral discrimination. Cyclodextrins (CDs), macrocyclic antibiotics (e.g., erythromycin and vancomycin), and polysaccharides (e.g., cellulose and amylose) are the most widely used CSs in this category.

7.4.1.3.1 Cyclodextrins

The characteristics and role of CDs in direct enantioseparation of amino acids and their derivatives, in particular, have been reviewed in literature (Alak and Armstrong, 1986; Armstrong et al., 1994; Martens and Bhushan, 1989). CDs are cyclic oligosaccharides; α-, β-, and γ-CD consist of 6, 7, and 8 glucose units connected via α-1,4 linkages that adopt the shape of a truncated cone with cavity diameters of 0.57, 0.78, and 0.95 nm, respectively. β-CD is the most widely used CD due to easy availability and because its cavity is well suited to a wide variety of drugs. The internal surface of the CD cavity is hydrophobic, originating from the carbon backbone of the sugar moieties, while the upper and lower rim surfaces are hydrophilic due to the presence of hydroxyl groups. Chiral discrimination is achieved by inclusion of the hydrophobic group of the drugs into the hydrophobic cavity of the CD and by differentiated hydrogen bonding interactions between the hydroxyl groups on the CD rims and the functional groups on the analytes. A variety of water-soluble as well as water-insoluble compounds can fit into the hydrophobic cavity of the CD molecule, thereby forming reversible inclusion complex of different stability. If a chiral molecule fits exactly into the cavity with its less polar side, a separation into the enantiomers can be expected. If the guest molecules are small, they are completely enclosed by the CD and cannot be separated. On the other hand, if the molecules are larger than the CD cavity, there may be little or no interaction.

The hydroxyl groups can be modified to other functional moieties to provide a variety of derivatized CDs. The introduced functionalities may modulate the conformational flexibility of the CD, alter the cavity size and access to it, and provide additional supportive binding sites, which lends them distinct enantiorecognition profiles as compared to their underivatized (native) CD counterparts (Bressolle et al., 1996; Mitchell and Armstrong, 2004). It has been indicated that to perform enantioseparation using CDs, there should be at least one aromatic ring in the structure of the analyte to form a tight inclusion with the CD cavity.

There are several advantages of using CDs as CSs, such as stability over the wide range of pH, nontoxicity, resistance to light, and UV transparency within the wavelength range commonly used for chromatographic detections. The interactions of chiral analytes with CDs are quite complex; in addition to inclusion–complexation interactions, many other nonspecific forms of interaction, namely, external absorption to the hydroxyl groups of a CD moiety, interaction with the linkage chain, or binding to free silanol groups can occur in the chromatographic system (Han and Armstrong, 1990). Therefore, the interaction of the analyte molecules with CDs depends on numerous factors, for example, their polarity, hydrophobicity, size, and spatial arrangement (Lepri et al., 1990).The formation of the CD inclusion complexes is strongly affected by pH, temperature, and composition of the mobile phase.

7.4.1.3.2   Macrocyclic Antibiotics

Macrocyclic antibiotics have many useful enantioselective properties of proteins and other polymeric selectors. They contain ionizable groups and hydrophobic and hydrophilic moieties and are somewhat flexible, affecting resolution by variation of the solution environment. They share a common heptapeptide aglycone core with aromatic residues that are bridged to each other forming a basket-like structure with shallow pockets for inclusion complexation. A number of potential interaction sites closely attached to stereogenic centers, including various H-donor/acceptor functionalities, aromatic rings for π–π interaction, and acidic and basic groups, may be involved in electrostatic interactions. Enantioseparation may be possible via π–π complexation, hydrogen bonding, inclusion in a hydrophobic pocket, dipole stacking, steric interactions, or a combination thereof.

Vancomycin is an amphoteric glycopeptide with a molecular weight of 1449. There are three macrocyclic portions and five aromatic rings in the molecule. Vancomycin contains 18 stereogenic centers, 9 hydroxyl groups, 2 amine groups (1 primary and 1 secondary), 7 amide groups, and 2 chlorine atoms (Armstrong and Zhou, 1994; Wilson et al., 1977). Some of the groups may be acidic or basic and are ionizable. Other groups are hydrophobic. These groups are also known to be useful for enantioselective molecular interaction with chiral analytes; their size, shape, and geometric arrangement in vancomycin give it the semirigid openness of a C-shaped aglycone basket, which is an important enantioselective property. Vancomycin has a pI of ca. 7.2 and a pH range of operation of 4–9. It occurs as a mono cation at pH 4 and as a mono anion at pH 9 (Antipas et al., 2000).

Erythromycin is a complex (14-membered ring) antibiotic, characterized by a molecular structure containing a large lactone ring linked with amino sugars through glycosidic bonds. It is a therapeutically useful wide-range antibiotic produced by a strain of *Streptomyces erythreus* and contains one methoxyl, two *N*-methyl, and eight or more (18%) *C*-methyl groups. Of the erythromycins, only erythromycin-A is certified by the US FDA; it is basic in nature (pKa = 8.6) and has a specific rotation $[\alpha]_D^{25} = -78°$ (c = 1.9, ethanol).

7.4.1.3.3   Polysaccharides

Since the introduction of cellulose triacetate as CS for enantioseparation of Tröger's base by Hesse and Hagel, polysaccharides have been serving as excellent CSs in liquid chromatography (Hesse and Hagel, 1973). Cellulose is a linear polymer of D-glucopyranose units connected through 1,4-β linkages in which the pyranose residues assume the energetically favored chain conformation; its chains are arranged on a partially crystalline fiber structure with helical cavities. Amylose is similar to cellulose, but the acetal linkage is in an α configuration. This inversion of configuration at one asymmetric center is accompanied by an important conformational change of the structure. Native cellulose is mostly composed of linear chains, whereas amylose chains tend to adopt a helical conformation.

The enantioselectivity, easy availability from natural sources, and possibility to prepare a wide range of derivatives by substitution and functionalization of the glucose unit hydroxyl groups make cellulose and amylose attractive candidates to prepare CSPs. Though both of them are very poor CSs in their native states, they exhibit enhanced selectivity when derivatized as carbamate or ester derivatives. The broad chiral recognition ability of these derivatives results from the stereoregular structure of the polymer chains that give rise to a supramolecular organization.

In particular, phenylcarbamate and benzoate derivatives of cellulose have been widely used for TLC enantioresolution of chiral drugs. The chiral recognition abilities of these derivatives depend on the type and position of the substituents introduced into the phenyl group. Although a few short-comings were observed to be associated with these derivatives, for example, difficulty in preparing a stable thin layer on glass microslide owing to adhesion and cracking of the layer, the long elution time of mobile phase owing to low polarity of solvents used for separation, limited visualization of the spots due to an intense UV absorption of these derivatives containing strong chromophores, and the expense associated with the plate production, it was found that the addition of an equal proportion of microcrystalline cellulose to the derivative alleviated many of the said problems (Suedee and Heard, 1997). The dominant mechanism for chiral resolution on cellulose-based CSPs is believed to be inclusion, with contributions from hydrogen bonding and dipolar interactions.

7.4.1.4 Factors Affecting Enantioresolution

Enantioresolution by all three mechanisms, as described earlier, takes place via the formation of an *in situ* diastereomeric pair. The formation and mobility of diastereomeric pair require an optimum concentration of CS, pH, and temperature:

* *Concentration of CS*: While lower concentration may lack the efficient level to make an *in situ* diastereomeric pair, a higher concentration may block the capillary action, and thus resolution is decreased.
* *Temperature*: While an optimum temperature provides the desired mobility to the *in situ* diastereomeric ion pair, any change in temperature adversely affects the resolution.
* *pH*: An appropriate pH is required to ensure that the analyte and the CS remain in opposite charged forms. A change in pH changes the ionic states of the CS and the analyte, resulting in noninteraction among the ionizable groups and a failure of enantiomer separation.

7.4.2 ENANTIOSEPARATION USING CHIRAL IMPREGNATING REAGENTS

The incorporation of a suitable reagent with the adsorbent without covalently affecting its inert character, prior to the development of chromatogram, is termed as impregnation. The adsorption characteristics are changed without affecting the inert character of adsorbent. The impregnation of thin-layer material with different suitable CSs provides an inexpensive wider choice of separation conditions for direct enantiomeric resolution; that is, the analyte can be applied to the plates in the form of racemic or scalemic mixture as such, without resorting to the derivatization of the enantiomeric mixture with one or the other chiral reagent that would otherwise involve several experimental stages of synthesis and purification. Direct resolution of enantiomers, by impregnated TLC, has been reviewed for amino acids and their derivatives by Bhushan and Martens (1997).

7.4.2.1 Preparation of Plates Impregnated with Chiral Selectors

7.4.2.1.1 Mixing of CIR with the Silica Gel Prior to Plate Making

This is considered to be the simplest method for impregnation and has been widely and successfully applied in the author's laboratory. A general protocol for the preparation of impregnated TLC plates using this method is described in the following text.

The solution (0.5%) of CIR is prepared in distilled water (100 mL), and the desired pH of the solution is maintained by adding a few drops of aqueous NH_3/acetic acid. Slurry of silica gel G (50 g) is prepared with the solution of CS and spread over glass plates with a Stahl-type applicator to give plates of 20×20 cm and thickness of 0.5 mm. The plates were activated overnight at 60°C.

7.4.2.1.2 Immersion of Plain Plates into Solution of CIR

In this method, TLC plates (either homemade or commercially available) are immersed into an appropriate solution of CIR. The method should be used carefully and slowly so as not to disturb the thin layer; peeling off of the layers has been experienced with the commercially available plates too. General protocols for the preparation of impregnated TLC plates using this method are described in the following text.

Precoated silica gel TLC plates, without fluorescent indicator, were dipped into a large dish for 60 min, containing a solution of the CIR. Varying amounts of CIRs were initially dissolved in 5 mL distilled water, then made up to 100 mL with ethanol. The plates were then removed from the solution and left to dry completely in a fume hood for 24 h. These were then activated in an oven at 100°C for 20 min (Günther and Möller, 2003).

In another approach, commercial TLC plates of silica gel 60 F_{254} were carefully washed, by predevelopment with methanol–water (9:1, v/v), then dried at ambient temperature for 3 h, and these were then impregnated with a solution of CIR (e.g., 3×10^{-2} mol/L solution of L-Arg in methanol) by conventional dipping for 2 s (Sajewicz et al., 2004, 2005). The concentration of the impregnating solution was calculated as that depositing 0.5 g CIR per 50 g of the dry silica gel adsorbent layer.

7.4.2.1.3 Ascending or Descending Development of TLC Plates in CIR Solution

A solution of CIR is allowed to ascend or descend on unmodified homemade/commercial TLC plates in a normal manner of development. The method is least apt to damage the TLC plates. General protocols for the preparation of impregnated TLC plates using this method are described in the following text.

Homemade unmodified TLC plates (10 mm × 5 mm, thickness 0.5 mm) were prepared by spreading a slurry of silica gel G (25 g) in distilled water (50 mL). The plates were activated for 8–10 h at 60°C ± 2°C. These homemade unmodified plates (or commercially available precoated plates) were impregnated by ascending development of these blank plates in solutions (0.5%) of CIR for 15–20 min. The plates were then air-dried and used for enantioseparation (Bhushan and Tanwar, 2009, 2010).

An interesting modification of this approach was employed in the author's laboratory for using ligand exchange reagents (LERs, namely, Cu(II)-L-Pro, Cu(II)-L-Phe, Cu(II)-L-Trp, Cu(II)-L-His, and Cu(II)-N,N-Me$_2$-L-Phe complexes) as CIRs. The plates were impregnated by ascending development in the solutions of respective L-amino acids (100 mL of 2 mM). The plates were dried at 60°C. The plates were developed using the solution of Cu(II) acetate as the mobile phase additive since these were impregnated with L-amino acids only (Bhushan and Tanwar, 2010).

In addition to the methods described earlier, methods like exposing the plain plates to the vapors of the impregnating reagent or spraying the impregnating reagent (or its solution) onto the plain plates have also been employed; spraying provides a less uniform dispersion than by development or immersion. The reagents and the methods used for impregnation are not to be confused with locating or spray reagents, which of course are required for the purpose of identification even on impregnated plates. The addition of CIRs to the mobile phase is excluded from the basic definition of impregnation.

7.4.2.2 Applications

Impregnated TLC plates have been widely used for chiral separation of drugs owing to the ease of impregnation process and availability of a wide variety of CSs (i.e., acids, bases, metal complexes, and amino acids). In addition, only little amounts of CSs are needed to prepare impregnated plates making this technique cost-effective in comparison to CSPs. Chromatographic data for enantiosep-aration of certain drugs by using various CIRs have been compiled in Table 7.1.

7.4.2.2.1 Acids as CSs

Lucic et al. separated enantiomers of (±)-metoprolol tartrate on silica gel 60$_F$ plates using the CS, D-(−)-tartaric acid, both impregnated into the layer and as mobile phase additive (Lucic et al., 2005). No separation was obtained using L-(+)-tartaric acid, whereas the D-(−)-counterpart gave rise to satisfactory results. Chromatography was effected by ascending technique, at 25°C, on 5 × 10 cm precoated plates of silica gel 60 F_{254}s with concentrating zone (Merck), impregnated with 11.6 mM D-(−)-tartaric acid in 70% aqueous ethanol by predeveloping the plates for 90 min in Camag twin-trough TLC chambers, which were saturated with mobile phase. The eluent composition was similar to that of the impregnating solution. Migration distance was 6 cm and development time about 50 min. The visualization was performed with both UV light (254 nm) and iodine vapors.

TABLE 7.1

Chromatographic Data for Enantioseparation of Various Drugs by Using CIRs

Analytes	CIRs	Mobile Phase	hR_F		References
Acids as CIRs					
Metoprolol	D-Tartaric acid	11.6 mM D-tartaric acid in 70% aqueous ethanol	50 (S)	65 (R)	Lucic et al. (2005)
Penicillamine	L-Tartaric acid	$CH_3CN–CH_3OH–H_2O$ (5:1:1, v/v)	30 (−)	61 (+)	Bhushan and Agarwal (2008a)
Penicillamine	(R)-mandelic acid	$CH_3COOAc–CH_3OH–H_2O$ (3:1:1, v/v)	39 (−)	52 (+)	Bhushan and Agarwal (2008a)
Ketamine	L-Tartaric acid	$CH_3CN–CH_3OH–CH_3COOAc$ (7:1:0.9, v/v)	67 (+)	44 (−)	Bhushan and Agarwal, (2008b)
Ketamine	(R)-mandelic acid	$CH_3COOAc–CH_3OH–H_2O$ (3:1:1, v/v)	79 (+)	67 (−)	Bhushan and Agarwal (2008b)
Lisinopril	L-Tartaric acid	$CH_3CN–CH_3OH–H_2O–CH_2Cl_2$ (7:1:1:0.5, v/v)	81 (+)	37 (−)	Bhushan and Agarwal (2008b)
Lisinopril	(R)-mandelic acid	$CH_3COOAc–CH_3OH–H_2O$ (3:1:1, v/v)	27 (+)	17 (−)	Bhushan and Agarwal (2008b)
Atenolol	L-Tartaric acid	$CH_3CN–CH_3OH–CH_2Cl_2$ (3:3:4, v/v)	28 (R)	52 (S)	Bhushan and Tanwar (2008)
Atenolol	(R)-mandelic acid	$CH_3CN–CH_3OH–HOAc$ (5:5:0.5, v/v)	68 (R)	85 (S)	Bhushan and Tanwar (2008)
Propranolol	LS-Tartaric acid	$CH_3CN–CH_3OH–HOAc$ (8:1:0.5, v/v)	60 (R)	86 (S)	Bhushan and Tanwar (2008)
Propranolol	(R)-mandelic acid	$CH_3CN–CH_3OH–HOAc$ (5:5:0.5, v/v)	72 (R)	90 (S)	Bhushan and Tanwar (2008)
Bases as CIRs					
Ibuprofen	(−)-Brucine	$CH_3CN–CH_3OH$ (16:3, v/v) first dimension and $CH_3CN–CH_3OH–H_2O$ (16:3:0.4, v/v) second dimension	93 (+)	86 (−)	Bhushan and Thiong'o (1999)
Flurbiprofen	(−)-Brucine	$CH_3CN–CH_3OH$ (16:3, v/v) first dimension and $CH_3CN–CH_3OH–H_2O$ (16:3:0.4, v/v) second dimension	96 (+)	90 (−)	Bhushan and Thiong'o (1999)
Acidic Amino Acids as CIRs					
Bisoprolol	L-Glutamic acid	$CH_3CN–CH_3OH–CH_2Cl_2–H_2O$ (5:1.5:0.5:1, v/v)	65 (+)	49 (−)	Bhushan and Agarwal (2008c)
Metoprolol	L-Glutamic acid	$CH_3CN–CH_3OH–CH_2Cl_2–H_2O$ (5:1.5:0.5:1, v/v)	54 (+)	44 (−)	Bhushan and Agarwal (2008c)
Carvedilol	L-Glutamic acid	$CH_3CN–CH_3OH–CH_2Cl_2–H_2O$ (5:1.5:0.5:1, v/v)	71 (+)	52 (−)	Bhushan and Agarwal (2008c)
Propranolol	L-Glutamic acid	$CH_3CN–CH_3OH–CH_2Cl_2–H_2O$ (5:1.5:1.5:1, v/v)	53 (+)	43 (−)	Bhushan and Agarwal (2008c)
Salbutamol	L-Glutamic acid	$CH_3CN–CH_3OH–CH_2Cl_2–H_2O$ (5:1.5:0.5:1, v/v)	59 (+)	50 (−)	Bhushan and Agarwal (2008c)
Labetalol	L-Glutamic acid	$CH_3CN–CH_3OH–CH_2Cl_2–H_2O$ (5:1.5:0.5:1, v/v)	69 (+)	59 (−)	Bhushan and Agarwal, (2008c)
Bisoprolol	L-Aspartic acid	$CH_3CN–CH_3OH–CH_2Cl_2–H_2O$ (5:1.5:0.5:1.5, v/v)	60 (+)	44 (−)	Bhushan and Agarwal (2008c)

TABLE 7.1 (continued)
Chromatographic Data for Enantioseparation of Various Drugs by Using CIRs

Analytes	CIRs	Mobile Phase	hR_F		References
Metoprolol	L-Aspartic acid	$CH_3CN–CH_3OH–CH_2Cl_2–H_2O$ (5:1.5:0.5:1.5, v/v)	47 (+)	38 (−)	Bhushan and Agarwal (2008c)
Carvedilol	L-Aspartic acid	$CH_3CN–CH_3OH–CH_2Cl_2–H_2O$ (5:1.5:0.5:1.5, v/v)	61 (+)	50 (−)	Bhushan and Agarwal (2008c)
Propranolol	L-Aspartic acid	$CH_3CN–CH_3OH–CH_2Cl_2–H_2O$ (5:1:1.5:1, v/v)	58 (+)	41 (−)	Bhushan and Agarwal (2008c)
Salbutamol	L-Aspartic acid	$CH_3CN–CH_3OH–H_2O$ (5:3:0.5, v/v)	43 (+)	27 (−)	Bhushan and Agarwal (2008c)
Labetalol	L-Aspartic acid	$CH_3CN–CH_3OH–H_2O–CH_3COOH$ (5:1:0.5:0.7, v/v)	71 (+)	60 (−)	Bhushan and Agarwal (2008c)
Basic Amino Acids as CIRs					
Ibuprofen	L-Arginine	$CH_3CN–CH_3OH–H_2O$ (5:1:1, v/v)	77 (−)	80 (+)	Bhushan and Parshad (1996)
Ibuprofen	L-Arginine	$CH_3CN–CH_3OH–H_2O$ (5:1:1, v/v)	76 (−)	83 (+)	Sajewicz et al. (2004)
Propranolol	L-Arginine	$CH_3CN–CH_3OH$ (16:4, v/v)	12 (−)	23 (+)	Bhushan and Thiong'o (1998)
Metoprolol	L-Arginine	$CH_3CN–CH_3OH$ (15:4, v/v)	15 (−)	39 (+)	Bhushan and Thiong'o (1998)
Atenolol	L-Arginine	$CH_3CN–CH_3OH$ (14:6, v/v)	22 (−)	30 (+)	Bhushan and Thiong'o (1998)
Propranolol	L-Lysine	$CH_3CN–CH_3OH$ (16:2, v/v)	4 (−)	15 (+)	Bhushan and Thiong'o (1998)
Metoprolol	L-Lysine	$CH_3CN–CH_3OH$ (15:5, v/v)	15 (−)	25 (+)	Bhushan and Thiong'o (1998)
Atenolol	L-Lysine	$CH_3CN–CH_3OH$ (16:4, v/v)	3 (−)	10 (+)	Bhushan and Thiong'o (1998)
Cinacalcet	L-Histidine	$CH_3CN–CH_3OH–H_2O$ (4:3.5:1, v/v)	62 (S)	26 (R)	Bhushan and Dubey (2011)
Cinacalcet	L-Arginine	$CH_3CN–CH_3OH–H_2O$ (5:4:1, v/v)	87 (S)	63 (R)	Bhushan and Dubey (2011)
Neutral Amino Acids as CIRs					
Ibuproxam	L-Serine	$CH_3CN–CH_3OH–H_2O$ (16:4:0.5, v/v)	28 (+)	95 (−)	Aboul-Enein et al. (2003a)
Ketoprofen	L-Serine	$CH_3CN–CH_3OH–H_2O$ (16:4:0.5, v/v)	57 (+)	83 (−)	Aboul-Enein et al. (2003a)
Tiaprofenic acid	L-Serine	$CH_3CN–CH_3OH–H_2O$ (16:3:0.5, v/v)	53 (+)	93 (−)	Aboul-Enein et al. (2003a)
Macrocyclic Antibiotics as CIRs					
Verapamil	Vancomycin	$CH_3CN–CH_3OH–H_2O$ (15:2.5:2.5, v/v)	49 (+)	39 (−)	Bhushan and Gupta (2005)
Atenolol	Vancomycin	$CH_3CN–CH_3OH–H_2O–CH_2Cl_2$ (7:1:1:1, v/v), pH 6.98	63 (+)	50 (−)	Bhushan and Agarwal (2010a)
Metoprolol	Vancomycin	$CH_3CN–CH_3OH–H_2O$ (6:1:1, v/v), pH 6.8	48 (+)	25 (−)	Bhushan and Agarwal (2010a)

(continued)

TABLE 7.1 (continued)

Chromatographic Data for Enantioseparation of Various Drugs by Using CIRs

Analytes	CIRs	Mobile Phase	hR$_F$		References
Propranolol	Vancomycin	CH_3CN–CH_3OH–H_2O (15:1:1, v/v), pH 7.2	53 (+)	18 (−)	Bhushan and Agarwal (2010a)
Labetalol	Vancomycin	CH_3CN–CH_3OH–H_2O (15:1:1, v/v), pH 7.2	71 (+)	46 (−)	Bhushan and Agarwal (2010a)
Atenolol	Erythromycin	C_2H_5OH–$CHCl_3$ (2:1, v/v)	56 (R)	81 (S)	Bhushan and Tanwar (2008)
Propranolol	Erythromycin	C_2H_5OH–$CHCl_3$ (2:1, v/v)	62 (R)	91 (S)	Bhushan and Tanwar (2008)

Under these experimental conditions, the hR$_F$ values for (+)-(S) and (−)-(R) enantiomers of metoprolol tartrate were 50 and 65, respectively. The values of α and R$_S$, calculated from the original chromatogram, were 1.86 and 2.00, respectively. With respect to the chiral discrimination mechanism, the authors suggested that monoethyl tartrate was the actual CS adsorbed on silica gel surface that produced the separation of enantiomers of metoprolol tartrate. The best esterification conditions included warming of impregnating solution at 70°C for 15 min, successive cooling, and chromatographic elution as described earlier. According to this hypothesis, after solute application, the diastereomers [(−)-metoprolol-(−)-ethyl tartrate] and [(+)-metoprolol-(−)-ethyl tartrate] could be formed *in situ*.

Bhushan and Agarwal used L-tartaric acid and (R)-mandelic acid as CIRs for enantioresolution of (±)-PenA (Bhushan and Agarwal, 2008a). When L-tartaric acid was used as CIR, R$_F$ values were observed to be 0.30 and 0.61 for (−) and (+) enantiomers, respectively, with solvent system acetonitrile–methanol–water (5:1:1, v/v); these values were found to be 0.39 and 0.52 for (−) and (+) isomer, respectively, with solvent system ethyl acetate–methanol–water (3:1:1, v/v) while using (R)-mandelic acid as impregnating reagent. The best resolution was observed at pH 5 and 16°C ± 2°C with any of the two CSs. The spots were located with iodine vapors. It was observed that the (+)-isomer eluted before the (−)-isomer in both the cases. The detection limit was found to be 0.12 μg for each enantiomer of PenA with L-tartaric acid and 0.11 μg with (R)-mandelic acid. This method can be applied for the detection of L-PenA in D-PenA up to 0.1%.

Later, the same research team used (+)-tartaric acid and (−)-mandelic acid as CIRs for enantioresolution of (±)-ketamine and (±)-lisinopril (Bhushan and Agarwal, 2008b). When (−)-mandelic acid was used as CIR, the use of ethyl acetate–methanol–water (3:1:1, v/v) as solvent system enabled successful resolution of the enantiomers of both compounds. In the case of (+)-tartaric acid, best results were obtained with acetonitrile–methanol–acetic acid (7:1:0.9, v/v) for ketamine; while for lisinopril, the solvent system acetonitrile–methanol–water–dichloromethane (7:1:1:0.5, v/v) was found to be successful. With any of the two CSs, the best resolution for both ketamine and lisinopril was observed at pH 5 and 16°C ± 2°C. It was observed that the (+)-enantiomer eluted before the (−)-counterpart in each case. LODs were 0.25 and 0.27 μg for each enantiomer of ketamine with (+)-tartaric acid and (−)-mandelic acid, respectively, whereas for lisinopril, LODs were 0.14 and 0.16 μg for each enantiomer with (+)-tartaric acid and (−)-mandelic acid, respectively.

Bhushan and Tanwar used TLC plates impregnated with L-tartaric acid or (R)-mandelic acid for enantioseparation of two β-blockers, namely, atenolol and propranolol (Bhushan and Tanwar, 2008). When (R)-mandelic acid was used as CIR, the highest Rs values for both the compounds (i.e., 2.6 for atenolol and 2.8 for propranolol) were observed with acetonitrile–methanol–acetic acid (5:5:0.5, v/v). In the case of the L-tartaric acid, the highest resolution for atenolol (Rs = 3.1) was obtained with acetonitrile–methanol–dichloromethane (3:3:4, v/v); while for propranolol, acetonitrile–methanol–acetic acid (8:1:0.5, v/v) was found to give the highest resolution (Rs = 3.5). The best resolution occurred at 16°C ± 2°C and pH 6.0 on the plates impregnated with L-tartaric acid and at 18°C ± 2°C and pH 6.5 on the plates impregnated with (R)-mandelic acid. It was observed that the (S)-enantiomer eluted before the (R)-enantiomer in each case. L-Tartaric acid and (R)-mandelic acid were able to resolve 2.2 and 2.5 µg of atenolol and propranolol, respectively.

7.4.2.2.2 Bases as CSs

Bhushan and Thiong'o used a suspension of 30 g neutral silica gel G, 60 mL of aqueous ethanol, and 0.1 g (−)-brucine to prepare impregnated thin-layer plates (Bhushan and Thiong'o, 1999). These plates were used for 2D TLC resolution of enantiomers of two nonsteroidal anti-inflammatory drugs (NSAIDs), namely, ibuprofen and flurbiprofen. The chromatograms were developed with aceto-nitrile–methanol (16:3, v/v) for the first dimension for 20 min and acetonitrile–methanol–water (16:3:0.4, v/v) for the second dimension for 20 min, and the spots were located in an iodine chamber. The method was successful in resolving as little as 0.1 µg of the racemates. The resolution of the two racemates was observed only on layers prepared at pH ranging between 6 and 7. Poor or no resolution was obtained for more acidic pH (included between 4 and 5) or more basic pH (included between 9 and 10). It was therefore proposed that enantiomeric separation was due to electrostatic interactions between COO$^-$ of the compounds and ≡N$^+$H of brucine, π–π interactions between phe-nyl moieties, and also hydrogen bonding. These interactions provided *in situ* formation of diastereo-mers on the impregnated plates, and consequently the resolution was observed.

Bhushan and Gupta used (−)-brucine in a different approach for enantioseparation of (±)-ibuprofen (Bhushan and Gupta, 2004). A 10 µL solution of optically pure (−)-brucine (0.12 mmol) mixed with (±)-ibuprofen (0.24 mmol) was applied on an achiral layer of silica gel G. The premixing resulted in the formation of diastereomeric pairs of the type [(+)-ibuprofen-(−)-brucine] and [(−)-ibuprofen-(−)-brucine] without resorting to any covalent linkage. It was the movement of these ionic diastereomers on TLC plates that resulted in separation. Chromatograms were developed at 28°C ± 2°C for 20 min in acetonitrile–methanol (5:1, v/v), and the spots were located with iodine vapors. The best resolu-tion conditions were the mixing of 0.12 mmol of CS with 0.24 mmol of ibuprofen, temperature maintained at 28°C ± 2°C, and pH between 6 and 7. Two round and compact spots were obtained with R_F values of 0.71 and 0.85 for (−)- and (+)-enantiomers, respectively, and the separation factor was observed to be 1.32. The enantiomeric separation of ibuprofen was much better on achiral lay-ers than on impregnated plates (Bhushan and Thiong'o, 1999). The minimum detection limit was found to be 2.45 µg of each enantiomer.

7.4.2.2.3 Metal Complexes as CSs

Bhushan and Gupta reported enantioresolution of three beta-blockers, namely, propranolol, ateno-lol, and metoprolol, using Cu(II)-L-Arg complex as LER (Bhushan and Gupta, 2006). The complex was prepared by mixing Cu(CH$_3$COO)$_2$ (1 mM) and L-Arg (2 mM) in water–methanol (90:10, v/v) and adjusting the final pH to 6–7 by addition of a few drops of aqueous ammonia. Impregnated thin-layer plates (20 cm × 10 cm, thickness 0.5 mm) were prepared by spreading a slurry of silica gel G (50 g) in a solution of Cu(II)-L-Arg complex (100 mL) and activating the plates overnight at 60°C. The eluent composition acetonitrile–methanol–water was used in various ratios for successful sepa-ration for all three racemic drugs under study. Excellent separations were obtained for metoprolol and propranolol, whereas the resolution of atenolol was much more difficult. Chromatograms were developed at 17°C ± 2°C, and the spots were located by placing the plates in an iodine chamber.

Bhushan and Tanwar reported enantioresolution of two β-blocking agents, atenolol and pro-pranolol, and one vasodilator, salbutamol, on commercial precoated NP-TLC plates by using Cu(II) complexes of five L-amino acids (namely, L-Pro, L-Phe, L-Trp, L-His, and N,N-Me$_2$-L-Phe) (Bhushan and Tanwar, 2010). Three different approaches (A–C) were employed. In approach (A), the chromatograms were developed using the solutions of LERs [Cu(II)-L-amino acid(s) complexes] as CMPAs. In approach (B), the plates were impregnated with each of the five LERs by ascending development of the blank plates in their solutions for 15–20 min. The plates were then air-dried and used for enantioseparation. In approach (C), the plain plates were impregnated by ascending development in solution of L-amino acid (100 mL of 2 mM). The plates were dried at 60°C. The plates were developed using a solution of Cu(II) acetate as the mobile phase additive.

Table 7.2 shows different successful combinations of acetonitrile–methanol–water and the hR_F values obtained using them. The photograph of the chromatogram showing the resolution of racemic atenolol, propranolol, and salbutamol using the plate impregnated with N,N-Me$_2$-L-Phe is shown in Figure 7.1. In all the cases, (R)-enantiomer was eluted before (S)-enantiomer. The significant and important aspect lies in working out a common solvent system and in comparing the performance of the three methods for one mobile phase to discuss the issue of involvement of Cu(II) for the best resolution. A comparison of resolution data (in Table 7.3) from the three methods shows that these can be arranged in the decreasing order (C > B > A) of the resolution of enantiomers. In approaches (A) and (B), the replacement of one of the amino acid molecules is taking place in the already existing complex by the enantiomer in the racemic analyte, while in approach (C), the formation of ternary complex is probably taking place *in situ*, during the chromatographic development, as the

---

TABLE 7.2
Different Combinations of MeCN–MeOH–H$_2$O Showing Successful Resolution of Atenolol, Propranolol, and Salbutamol along with hR_F

		Approach (A)			Approach (B)			Approach (C)		
		Solvent	hR_F		Solvent	hR_F		Solvent	hR_F	
Analytes	CSs	Ratio	(R)	(S)	Ratio	(R)	(S)	Ratio	(R)	(S)
Atenolol	L-Phe	1:2:4	36	28	—	—	—	2:2:5	38	23
	L-His	0.5:3:6	38	30	1:5:4	43	32	2:2:4	30	20
	N,N-Me$_2$-L-Phe	1:2:5	28	17	—	—	—	3:4:5	36	16
	L-Pro	1:3:5	37	29	—	—	—	—	—	—
	L-Trp	—	—	—	—	—	—	3:2:5	37	27
Propranolol	L-Phe	1:2:4	36	27	2:5:5	40	27	2:2:5	38	23
	L-His	0.5:3:6	33	25	1:5:8	33	26	2:2:4	32	16
	N,N-Me$_2$-L-Phe	1:2:5	38	26	1:1:1	34	25	3:4:5	34	17
	L-Pro	1:3:5	31	26	—	—	—	—	—	—
	L-Trp	—	—	—	—	—	—	3:2:5	37	26
Salbutamol	L-Phe	1:2:4	34	26	1:5:1	26	18	2:2:5	32	25
	L-His	0.5:3:6	31	25	1:5:5	42	32	2:2:4	31	16
	N,N-Me$_2$-L-Phe	1:3:5	35	23	—	—	—	3:4:5	35	19
	L-Pro	1:3:5	31	26	—	—	—	—	—	—
	L-Trp	—	—	—	—	—	—	3:2:5	38	28

Source: Bhushan, R. and Tanwar, S., *J. Chromatogr. A*, 1217, 1395, 2010.

Approaches A, B, and C, as described in Section 7.4.2.2.3.

Solvent system: MeCN–MeOH–H$_2$O. The volume of water in the ternary mobile phase represents the volume of aq. solution of the corresponding Cu(II)–amino acid complex.

The three racemic β-blockers were also resolved under the following conditions:

1. Approach A, mobile phase CH$_3$COCH$_3$–MeOH–aq. solution of Cu(II) L-Pro complex (5:1:1).
2. Approach B, mobile phase MeCN–MeOH–CH$_2$Cl$_2$ (1:1:1) using the plate impregnated with Cu(II) complex of N,N-Me$_2$-L-Phe.
3. Approach B, mobile phase MeOH–CH$_2$Cl$_2$ (4:6) for the resolution of atenolol only using the plate impregnated with Cu(II) complex of L-Phe.
4. Approach C, mobile phase of MeOH–H$_2$O–aq.Cu(II) 2mM (2:1:10) using the plate impregnated with L-Pro.

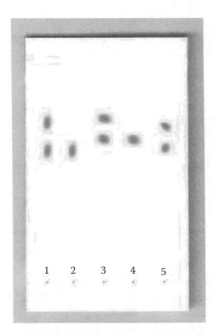

FIGURE 7.1 Photograph of the chromatogram showing resolution of racemic atenolol, propranolol, and salbutamol using the plate impregnated with N,N-Me$_2$-L-Phe (approach C; as described in Section 7.4.2.2.3), run time: 12 min, detection: iodine vapor, solvent front 4 cm. Solvent system: MeCN–MeOH–aq. Cu(II) 2mM (3:4:5), left to right: in track 1, the lower spot is that of (S)-atenolol, and the upper one is that of (R)-atenolol; track 2 is that of pure (S)-atenolol; track 3 lower spot is (S)-propranolol, and the upper one is (R)-propranolol; track 4 is that of pure (S)-propranolol; in track 5, lower spot is that of (S)-salbutamol, and the upper spot is that of (R)-salbutamol. (From Bhushan, R. and Tanwar, S., *J. Chromatogr. A*, 1217, 1395, 2010.)

TABLE 7.3

Comparison of Resolution (R_S) Using Three Approaches with Common Solvent System of Varying Composition (as per Table 7.2)

	(±)-Atl			(±)-Prl			(±)-Sal		
	(A)	(B)	(C)	(A)	(B)	(C)	(A)	(B)	(C)
Amino Acid	R_S			R_S			R_S		
L-Phe	1.7	—	3.0	1.7	2.9	3.0	1.7	1.8	2.4
L-His	2.3	2.5	2.1	1.7	1.6	3.1	1.8	2.3	3.1
N,N-Me$_2$-L-Phe	1.7	—	3.5	2.6	1.4	3.2	2.5	—	4.0

Source: Bhushan, R. and Tanwar, S., *J. Chromatogr. A*, 1217, 1395, 2010.
Approaches A, B, and C, as described in Section 7.4.2.2.3.

Cu(II) is available at the same time to the L-amino acid (present in the impregnated form) and to the analyte molecule spotted on the plate. Therefore, there are differences in resolution and retention. Approach (C) almost in each case outperformed approaches (A) and (B). The best resolution for all the three racemates was obtained at 20°C ± 2°C with any of the CSs used. The LOD was found as 0.18 μg per enantiomer.

The same research group studied enantioseparation of atenolol, propranolol, and salbutamol by using Cu(II)-complexes of L-Pro, L-Phe, L-His, and N,N-Me$_2$-L-Phe as LERs on homemade NP plates (Bhushan and Tanwar, 2009). Of course, one more approach was introduced in this study, that is, approach D: mixing of LER in the silica gel slurry while preparing the plates. However, the best results were obtained using the same approach as was applied for commercial plates (the approach C in the preceding paragraph). It was concluded that approach C of impregnation resulted in better dispersion of the complex on the silica particles (i.e., impregnation) leading to better enantioselectivity as a result of interaction between the analyte and the CS.

7.4.2.2.4 Amino Acids as CSs

7.4.2.2.4.1 Acidic Amino Acids as CSs   Acidic amino acids were used by Bhushan and Ali as CIRs for enantioseparation of racemic alkaloids such as hyoscyamine and colchicines (Bhushan and Ali, 1993). Bhushan and Agarwal reported enantioresolution of six β-blockers, namely, metoprolol, propranolol, carvedilol, bisoprolol, salbutamol, and labetalol using TLC impregnated with L-Glu (0.5%) or L-Asp (0.5%) (Bhushan and Agarwal, 2008c). The chromatograms were developed at 27°C for 20 min. Mobile phases consisting of acetonitrile–methanol–water–dichloromethane and acetonitrile–methanol–water–glacial acetic acid were employed in different proportions. The spots were detected with iodine vapor. All the analytes were baseline resolved with very high R_S values. The detection limits were 0.23, 0.10, 0.27, 0.25, 0.20, and 0.20 μg for each enantiomer of metoprolol, propranolol, carvedilol, bisoprolol, salbutamol, and labetalol, respectively. Bhushan and Arora used L-Asp as CIR for enantioresolution of metoprolol, propranolol, and atenolol (Bhushan and Arora, 2003). The method was successful in detecting 0.26 μg of atenolol and 0.23 μg each of metoprolol and propranolol as racemate.

7.4.2.2.4.2 Basic Amino Acids as CSs   Basic amino acids are the most widely used CIRs and have been successful for a wide variety of acidic and basic compounds. Bhushan and Parshad resolved the enantiomers of ibuprofen on silica gel G layers impregnated with a 0.5% aqueous solution of L-Arg (Bhushan and Parshad, 1996). As the pI of arginine was very high (10.8), trace amounts of acetic acid were added to obtain it in the cationic form. Aliquots (10 μL) of racemic ibuprofen in 70% ethanol (10^{-3} M) were applied to the layer. The visualization was performed by using iodine vapors. The *in situ* formation of diastereomers was verified by spraying 0.2% ninhydrin (in acetone) on the chromatogram, after iodine evaporation. There appeared the characteristic colored spots showing the presence of arginine in both the separated spots, although the whole arginine-impregnated plate reacted with ninhydrin showing a light pink background. An acetonitrile–methanol–water (5:1:1, v/v) mixture was used as eluent, for both 1- and 2D chromatographies. After development at 32°C for 15 min in the first direction and for 20 min in the second direction, calculated hR_F values were 77 and 80 for (−)- and (+)-enantiomers, respectively, with a ΔR_F equal to 0.03.

Kowalska et al. deeply investigated this separation, with an aim to improve the method accuracy and precision, by using standardized and commercially available plates (Sajewicz et al., 2004, 2005). The experimental conditions were adopted as reported by Bhushan and Parshad (1996). The plates were predeveloped in methanol–water (9:1, v/v), dried at room temperature, and coated by dipping for 2 s in a 3 × 10^{-2} M solution of L-Arg in methanol. Visualization was performed by UV light (λ_{max} = 210 nm) by using the Desaga model CD 60 densitometer (Heidelberg, Germany) with Windows compatible ProQuant software. Elution was effected using the mobile phase reported by Bhushan and Parshad (1996), modified by adding acetic acid up to pH 4.8 to have arginine in the cationic form. Both 1- and 2D modes were employed with a migration distance of 15 cm. Using 2D technique, a significant enhancement of the resolution was observed (hR_F (−) = 76 and hR_F (+) = 83). In fact, Δ R_F was 0.07 vs. the value of 0.03 reported previously (Bhushan and Parshad, 1996).

Bhushan and Thiong'o employed silica gel G layers coated with L-Arg (0.5%) or L-Lys (0.5%) for enantioseparation of three β-adrenergic blocking agents, namely propranolol, metoprolol, and atenolol (Bhushan and Thiong'o, 1998). Different binary mixtures of acetonitrile–methanol were used as eluents. Propranolol and metoprolol were resolved at 22°C, while atenolol was resolved at 15°C. High-resolution factors were obtained for the three racemates, particularly for propranolol, on layers impregnated with two basic amino acids. However, taking into account the compactness and dimensions of spots, the best resolutions for the three drugs were achieved employing silica gel layers coated with acidic amino acids (Bhushan and Arora, 2003).

An investigation aimed to assess the best experimental conditions for the resolution of racemic propranolol was carried out by Kowalska and coworkers (Sajewicz et al., 2005) on precoated silica gel 60 F_{254} plates impregnated with L-Arg, maintaining as much as possible the working conditions reported by Bhushan and Thiong'o (1998). In particular, the effect of the eluent pH on chiral resolution was investigated by using an acetonitrile–methanol mixture (15:4, v/v) containing different amounts of ammonia (pH included between 7.75 and 11.0). (S)-(−)-propranolol gave R_F values between 0.01 and 0.04 (the highest at pH 10.9), while the (R)-(+) form showed increasing R_F from 0.11 to 0.18 with the increase of pH. Visualization was performed at 210 nm with densitometric scanning using a Desaga CD 60 densitometer (Heidelberg, Germany). Again, the purpose of this research was to obtain a chiral resolution of this important β-blocker employing standardized commercially available plates and densitometric detection of analytes so as to improve the precision and robustness of the procedure.

Recently, Bhushan and Dubey reported enantioresolution of the calcimimetic drug (R,S)-cinacalcet (Cin) using TLC plates impregnated with L-His and L-Arg (Bhushan and Dubey, 2011). Successful solvent systems were 4:3.5:1(v/v) and 5:4:1(v/v) combinations of acetonitrile–methanol–water for plates impregnated with L-His and L-Arg, respectively. The spots of the two enantiomers were visualized by exposure to iodine vapors. The best resolution was obtained at 18°C ± 2°C with both the impregnating reagents. (S)-Cin eluted before the (R)-Cin in both the cases. The LODs were found to be 0.28 mg for each enantiomer of (R,S)-Cin using L-His and 0.26 mg using L-Arg.

7.4.2.2.4.3 Neutral Amino Acids as CSs   Aboul-Enein et al. investigated the resolution of three profens (namely, ketoprofen, ibuproxam, and tiaprofenic acid) on plates prepared from a slurry of 30 g silica gel G (Fluka, Switzerland) in 70 mL of water–ethanol (97:3, v/v) containing 0.1 g of L-Ser (Aboul-Enein et al., 2003a). The successful solvent systems were acetonitrile–ethanol–water in the ratio of 16:4:0.5 (v/v) for the resolution of ibuproxam and ketoprofen and in the ratio of 16:3:0.5 (v/v) for tiaprofenic acid. The best resolution was observed in the pH range of 6–7 and at 25°C. When an impregnated plate was developed without spotting any of the test compounds, a uniform staining of the entire surface of the plate was visible that indicated that L-serine was immobilized uniformly on the silica gel.

Later, the same team reported enantioseparation of six NSAIDs (namely, flurbiprofen, ibuproxam, benoxaprofen, ketoprofen, pranoprofen, and tiaprofenic acid) on TLC plates impregnated with L-Ser, L-Thr, or 1:1 mixture of L-Ser and L-Thr (Aboul-Enein et al., 2003b). The impregnated plates were prepared by dipping the precoated commercial silica gel plates into a large dish, containing the solution of the selector, for 60 min at ambient temperature. The plates were dried for 24 h in a fume hood and then activated at 100°C for 20 min. Flurbiprofen resolved into its enantiomers only on the plates impregnated with the mixture of L-Ser and L-Thr (0.05 g each) using acetonitrile–methanol–water (16:4:0.5, v/v), when the (S)-(−)-enantiomer eluted first. The best resolution was achieved at 25°C ± 2°C and between pH 6 and 7, and interestingly there was no resolution on the plates impregnated either with L-serine or with L-threonine; however, no explanation was offered for not getting enantioresolution in this condition. TLC plates impregnated with either L-Ser or L-Thr provided separation of ibuproxam, ketoprofen, and pranoprofen into their enantiomers; the separation factors were 1.04, 1.44, and 1.38, respectively, on plates impregnated with L-Ser, while these values were 1.09, 1.52, and 1.56, respectively, on plates impregnated with L-Thr. TLC plates impregnated with both L-Thr and L-Ser were able to resolve the racemic ibuproxam, benoxaprofen, flurbiprofen, ketoprofen, pranoprofen, and tiaprofenic acid.

7.4.2.2.5 Macrocyclic Antibiotics as CSs

Bhushan and Gupta used silica gel G layers impregnated with (−)-vancomycin for enantioresolution of verapamil, a calcium channel blocker (Bhushan and Gupta, 2005). Impregnated TLC plates (20 × 10 cm, thickness 0.5 mm) were prepared by spreading a slurry of silica gel G (30 g) in distilled water (60 mL), containing 0.05 g sterile vancomycin hydrochloride USP. The solvent system acetonitrile–methanol–water (15:2.5:2.5, v/v) was found to be successful on silica gel plates impregnated with vancomycin at 0.34 mM concentration. The two spots were located by exposure to iodine vapors. The hR_F values for (+)- and (−)-enantiomers were 49 and 39, respectively (development time 20 min). Experiments were carried out in the temperature range of 5°C–30°C and pH range of 3–9; it was observed that the best resolution was at 18°C ± 2°C and at pH 6. At this pH, vancomycin exists as cation, whereas the analyte might exist as anion where coulombic interaction could have occurred, favoring enantioseparation. No resolution was observed below pH 6 because protonation of the analyte resulted in neutral molecules. Further, no resolution was observed above pH 6, because at this pH, both vancomycin and the analyte exist as anion. It was therefore proposed that enantiomer separation could be a result of the formation of a hydrophobic pocket (using the C-shaped aglycone basket) and for coulombic interactions between the charged species of the CS and the analyte. The reported technique was successful in detecting as little as 0.147 μg of the racemic mixture (i.e., 0.074 μg of each enantiomer).

Bhushan and Agarwal reported enantioresolution of atenolol, metoprolol, propranolol, and labetalol on vancomycin-impregnated TLC plates (Bhushan and Agarwal, 2010a). The pH of the silica gel slurry in aqueous medium is approximately 7, and aqueous solution of vancomycin (0.56 mM) has a pH of approximately 4. To keep the pH of plates impregnated with vancomycin at approximately 7, the aqueous solution of vancomycin, pH ∼ 4, was first adjusted to pH 5.5 by addition of a few drops of aqueous ammonia; this was then used to make the slurry. The pH of solvent systems found to enable successful resolution of the β-blockers on plates impregnated with vancomycin was very close to 7. The resolution was studied on plates prepared from solutions of pH 4.5, 5.5, and 6.5. The enantiomers were resolved on plates of pH 5.5, but tailing of spots was observed on plates of pH 4.5 and 6.5. The measurement of pH after the development of chromatograms showed that the pH of mobile phases changes when they come into contact with the plates at pH 4.5 or 6.5. Successful resolution of the enantiomers of atenolol, metoprolol, propranolol, and labetalol was achieved by using acetonitrile–methanol–water–dichloromethane (7:1:1:1, v/v), acetonitrile–methanol–water (6:1:1, v/v), acetonitrile–methanol–water–dichloromethane–glacial acetic acid (7:1:1:1:0.5, v/v), and acetonitrile–methanol–water (15:1:1, v/v), respectively. The photograph of chromatogram showing resolution of the enantiomers of (±)-propranolol using vancomycin as CIR has been shown in Figure 7.2. All four beta-blockers separated well at 16°C ± 2°C. The reported method was able to detect small amounts of each enantiomer, that is, 1.3, 1.2, 1.5, and 1.4 μg of atenolol, metoprolol, propranolol, and labetalol, respectively.

Bhushan and Tanwar reported enantioseparation of atenolol and propranolol using TLC impregnated with erythromycin (Bhushan and Tanwar, 2008). Erythromycin (1 g) was dissolved in a minimum volume of chloroform by warming to 40°C. After filtration, the solution was kept at −15°C. The crystals were collected by filtration and dried under vacuum; these were nearly colorless (0.9 g), m.p. 135°C–140°C, and resolidifying and melting at 190°C–193°C. Since erythromycin is sparingly soluble in water, it was first dissolved in 10 mL of ethanol and made up to 50 mL with water; the silica gel slurry was made in this solution to prepare erythromycin-impregnated plates. Impregnated thin-layer plates (10 × 20 cm, thickness 0.5 mm) were prepared by spreading a slurry of silica gel G (25 g) in distilled water (50 mL containing 0.5% CS). A few drops of ammonia were added to the slurry to maintain pH. The solvent system ethanol–chloroform (2:1, v/v) was found to be successful for both the analytes. The best resolution occurred at 18°C ± 2°C and at pH 6.5. The method was successful in resolving as little as 1.5 μg of atenolol and propranolol.

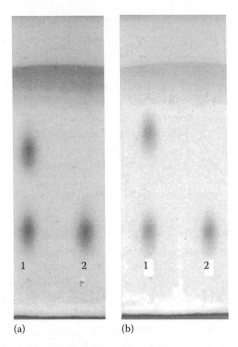

(a)   (b)

FIGURE 7.2 Chromatograms showing resolution of the enantiomers of (±)-propranolol: (a) on plate impregnated with vancomycin; (b) using vancomycin as CMPA. On both plates, track 1 shows the (−) and (+) isomers (lower and upper spots, respectively) resolved from the mixture, whereas track 2 shows the pure (−) isomer. The development time was 10 min, the temperature was 16°C ± 2°C, and detection was with iodine vapor. (From Bhushan, R. and Agarwal, C., *J. Planar Chromatogr.*, 23, 7, 2010a.)

7.4.2.2.6 Miscellaneous CSs

Wall performed the enantioseparation of propranolol, atenolol, and metoprolol on ionically and covalently modified aminopropyl-bonded HPTLC plates with N-(3,5-dinitrobenzoyl)-(R)-(−)-α-phenylglycine (CSP1) and N-(3,5-dinitrobenzoyl)-L-leucine (CSP2) (Wall, 1989). It was observed that the compounds subjected to chiral resolution either on CSP1 or on CSP2 must contain amide, carbonyl, or hydroxyl groups close to the chiral center. Hence, some classes of drugs, such as barbiturates and several benzodiazepines, can be resolved without derivatization, but for the complete enantioseparation, β-blockers required the formation of naphthyl amides. So, the chromatographic resolution of β-blockers was preceded by derivatization with 1-isocyanatonaphthalene. The separated derivatives were visualized as yellow spots on a white background, and no further treatment was necessary. The proposed mechanism involved in chiral discrimination using CSP1 and CSP2 is based on hydrogen bonding, possible dipole stacking, and π–π interactions between the aromatic groups of enantiomers and those in the stationary phases. The separation factors for propranolol, atenolol, and metoprolol were 1.1, 1.3, and 2.2, respectively.

Brunner and Wainer reported enantioseparation of ibuprofen, naproxen, fenoprofen, flurbiprofen, and benoxaprofen in the form of their 3,5-dinitroanilide derivatives on naphthylethyl urea TLC-CSP (Brunner and Wainer, 1989). The CSP was prepared by reacting (R)-(−)-1-(1-naphthylethyl) ethyl isocyanate with a commercially available aminopropyl HPTLC plate. For this purpose, 1 g of isocyanate was dissolved in 100 mL of methylene chloride, and the aminopropyl HPTLC plate was soaked in 20 mL of the solution for 5 min. The plate was removed from the solution and air-dried. It was then washed by immersion in methylene chloride (twice) and air-dried. The amount of isocyanate bound to the plate was calculated from the loss of UV absorbance of the reagent solution, measured at 281 nm; it was found to be 86 mg isocyanate reagent per 10×10 cm^2 plate. The chromatograms were developed in hexane–isopropanol–acetonitrile (20:8:1, v/v), and the detection was carried out at both short (254 nm) and long (360 nm) UV wavelengths. All the analytes were converted to their 3,5-dinitroanilide derivatives, by reacting them first with an acid chloride followed by condensation with nitroaniline, and the derivatives were then chromatographed on the naphthylethyl urea TLC-CSP. In all the cases, (S)-enantiomers were found to be more retained than the (R)-counterparts. Further investigations suggested that π–π interactions between a π-acidic moiety on the solute and the π-basic naphthyl moiety on the CSP are the key aspects in chiral recognition. Thus, the potential solutes for the naphthylethyl urea TLC-CSP that contain an amine or a carboxylic acid moiety should be converted into the corresponding 3,5-dinitrobenzoyl amides or 3,5-dinitroanilides before chromatography.

7.4.3 Enantioseparation Using Chiral Stationary Phases

Various CSPs have been employed for the direct enantioseparation of chiral drugs using planar chromatography (Günther and Möller, 2003; Lepri et al., 2001; Siouffi et al., 2005). However, only a few stationary phases (namely, cellulose and cellulose triacetate) have been commercialized because of the high cost of most of the CSs and the difficulty to detect analytes on stationary phases showing a strong absorbance during exposition to UV lights. In certain laboratories at the international level, commercial plates are preferred to homemade plates as they offer more reproducibility. However, their low versatility either due to limited options of kind or composition of the eluent or due to low enantioselectivity of CSs remains a limiting factor. Therefore, homemade plates are useful to resolve racemic compounds, which otherwise would not be resolved by this technique, to better understand enantioresolution mechanisms of chiral adsorbents in different experimental conditions and, therefore, to stimulate industrial production of cheap and performant chiral precoated plates.

7.4.3.1 Applications

7.4.3.1.1 Polysaccharide-Based CSPs

Suedee and Heard reported the use of various cellulose derivatives (i.e., tris[phenyl]carbamate, tris[2,3-dichlorophenyl] carbamate, tris[2,4-dichlorophenyl] carbamate, tris[2,6-dichlorophenyl] carbamate, tris[2,3-dimethylphenyl] carbamate, tris[3,4-dichlorophenyl] carbamate, tris[3,5 dichlorophenyl] carbamate, and tris[3,5-dimethylphenyl] carbamate) for the enantioseparation of propranolol and bupranolol on noncommercial plates (Suedee and Heard, 1997).

The preparation of the triphenylcarbamates was carried out by the reaction of microcrystalline cellulose with the specific phenyl isocyanate in pyridine at 120°C for 6 h, according to the procedure reported by Okamoto et al. (1984). Each product was washed with methanol to remove the unreacted isocyanate. For the preparation of plates, 300 mg of each derivative (particle size 10–100 μm) and 300 mg of cellulose microcrystalline were mixed with 3 mL water and a small amount of ethanol, as wetting agent, to obtain a suspension that was stratified on 2.6 × 7.6 cm glass microslides. The plates were dried in an oven at 105°C for 5 min to obtain layers having a thickness of about 0.25 mm (Suedee and Heard, 1997).

Ascending development was carried out at room temperature in a saturated 250 mL glass beaker containing 10 mL of mobile phase. The migration distance was 6 cm. Spots were detected by drying the plates at 110°C, spraying with anisaldehyde reagent, and then heating for further 10–15 min. The solutes appeared as yellow-brown spots on a paler yellow-brown background. The ternary phase ethyl acetate–propan-2-ol–water (65:23:12, v/v) and n-hexane–propan-2-ol binary mixtures, in the ratios of 90:10, 80:20, and 70:30 (v/v), were used as eluents. The spots were visualized after the development of chromatograms with anisaldehyde reagent.

(R,S)-propranolol yielded α = 5.5 and Rs nearly unity on stationary phase composed of microcrystalline cellulose: tris(3,5-dimethylphenyl) carbamate of cellulose in 1:1 ratio with eluent n-hexane–propan-2-ol (80:20, v/v). In case of bupranolol, α = 2.53 and Rs value approaching unity were observed on stationary phase composed of microcrystalline cellulose: tris(2,3-dimethylphenyl) carbamate of cellulose in 1:1 ratio using eluent *n*-hexane–propan-2-ol (90:10, v/v) although α-values higher than 4.13 were obtained on the same stationary phase with *n*-hexane–propan-2-ol (80:20, v/v), but the resolution was worse because of the formation of very elongated spots.

Lepri et al. reported the enantioresolution of flurbiprofen (Lepri et al., 1994) and carprofen (Lepri, 1995) on noncommercial microcrystalline cellulose triacetate plates with silica gel 60 G F_{254} as binder. A resolution value of 2.00 and α = 1.44 were observed for flurbiprofen using ethanol–water (40:60, v/v), and for carprofen, Rs = 1.6 and α = 1.23 were observed using propan-2-ol–water (40:60, v/v).

Xu et al. (2002, 2003) reported the enantioseparation of drugs, namely, timolol, ofloxacin, propranolol, atropine, promethazine, atenolol, and nimodipine by using various benzoate derivatives of cellulose (i.e., *tris*-benzoate, *tris*-4 bromo benzoate, *tris*-4 methyl benzoate, *tris*-4 nitro benzoate, *tris*-3,5-dinitro benzoate) as stationary phases.

7.4.3.1.2 Cyclodextrin-Based CSPs

Zhu et al. reported the enantioseparation of several pharmaceuticals (namely, ofloxacin, nimodipine, isoprenaline, carvedilol, and promethazine) on β-CD-bonded CSPs (Zhu et al., 2001). In order to prepare CSP, 10 g of β-CD and 5 g of silica gel H were weighed and dried at 110°C in an oven under vacuum for 12 h. β-CD was dissolved in approximately 120 mL of anhydrous dimethylformamide with a small amount of sodium; the mixture was stirred for 2–3 h at 90°C and filtered. Silica gel H and 2.5 g of KH-560 (3-glycidoxypropyltrimethoxysilane) were added to the filtrate, and the mixture was stirred for 12–18 h at 90°C. The suspension was filtered and washed with DMF, toluene, methanol, water, and again with methanol and then dried in air. To 2 g of obtained β-CD-bonded silica gel, 6 mL of 0.3% aqueous solution of carboxymethyl cellulose sodium was added, and the slurry obtained was spread as a layer of 0.2–0.4 mm thickness on glass plates 2.5 × 7.5 cm in size. The plates were dried in air and activated before use in an oven at 80°C for 1 h. The development was effected at room temperature in a 4 × 4 × 8 cm glass chamber for a distance of 5 cm. The visualization was performed by UV light (365 nm) or iodine vapor.

Compact spots were obtained, because tailing could be minimized by the use of specific eluents and, particularly, of buffer, such as aqueous triethylammonium acetate (aq. TEAA), which increased resolution and efficiency. The disadvantages of this technique were the time spent to prepare homemade plates and stationary phases, which are not commercially available. The highest separation factors for ofloxacin (α = 1.38) and carvedilol (α = 2.28) were observed with methanol–TEAA (pH 1.1) (2:2, v/v). A separation factor of 1.82 was observed for isoprenaline with methanol–TEAA (pH 1.1) (2:1, v/v). The highest α value for nimodipine (2.43) was observed with methanol–acetonitrile–TEAA (pH 1.1) (1.5:0.2:0.1, v/v). The highest separation factor for promethazine (i.e., 2.02) was observed with methanol–acetonitrile–TEAA (pH 1.1) (2:0.2:0.1, v/v).

Wang et al. observed a separation factor of 1.45 for the enantioresolution of DL-naproxen on CD stationary phase with ethyl acetate–chloroform (1:4, v/v) (Wang et al., 2011). Luo et al. (2008) reported resolution of metoprolol on CD stationary phase using methane–dichloromethane (0.87:2.13, v/v). Zhu et al. (2000) reported chiral separation of promethazine, propranolol, carvedilol, nimodipine, ofloxacin, and dobutamine on 3,5-dinitrobenzoyl-substituted CD-bonded CSP. The separation factors were found to be in the range of 1.90–2.03 by using chloroform–acetone–triethylamine (TEA) and acetonitrile–TEAA (pH 4.9)–acetic acid as solvent systems.

7.4.3.1.3 Commercially Available Chiralplates

Mielcarek reported the enantioresolution of felodipine on Chiralplates (Mielcarek, 2001). The CS, (2*S*, 4*R*, 2′*RS*)-*N*-(2′-hydroxydodecyl)-4-hydroxyproline, on the plate and the enantiomers to be separated from diastereomeric mixed chelate complexes with the cupric ion; complexes of different enantiomers have different stabilities, thus achieving TLC separation. Two ternary mobile phases, chloroform–ammonia (25%)–methanol (10:0.02:φ) and acetonitrile–TEA (0.1%)–methanol (5:3:φ), were used as eluents. Sample solution (0.3 μL) at a concentration of 3.039×10^{-4} mol/L was applied on chromatographic plates of 10 × 3 cm size. Chromatograms were developed to a distance of 9 cm in a saturated chamber in dark. The spots were visualized under UV light (300–400 nm). The best separation in mobile phase 1 was obtained at $\varphi = 1.9570 \times 10^{-2}$; the R_F values were 0.6200 for (*R*)-enantiomer and 0.6400 for (*S*)-enantiomer. The best separation in phase 2 was obtained at $\varphi = 1.11 \times 10^{-1}$; the R_F values were 0.3713 for (*R*)-enantiomer and 0.3550 for (*S*)-enantiomer.

Kovacs-Hadady and Kiss reported the chromatographic separation of DL-PenA after derivatization with substituted benzaldehydes and heterocyclic aldehydes (Kovacs-Hadady and Kiss, 1987). For derivatization, 0.1 mmol of PenA was dissolved in 0.1 mL of water, and then 0.5 mL of aldehyde solution of 0.1 mmol/L concentration in methanol was added. After sealing, the mixture was kept at 60°C–65°C for 8 h. Chromatograms were developed at 25°C using methanol–water–acetonitrile (50:50:200, v/v). The hR_F values after the reaction of PenA with 5-nitrofuraldehyde were 72 and 82 for D- and L-PenA, respectively.

7.4.4 ENANTIOSEPARATION USING CHIRAL MOBILE PHASE ADDITIVES

The application of CMPAs to perform direct enantioresolution offers the advantages of flexibility and cost-effectiveness as compared to the corresponding CSPs. When a CS is added to the mobile phase, achiral stationary phases can be used for separation. Chromatographic data for the enantioseparation of various drugs by using CMPAs have been compiled in Table 7.4. The stereoselective separation obtained in a system with a chiral additive in the mobile phase can be due to one or a combination of the following "mechanisms":

- A stereoselective complexation in mobile phase
- Adsorption of the CS to the solid phase
- Formation of the labile diastereomeric complexes with different distribution properties between the mobile and stationary phase

7.4.4.1 Applications

7.4.4.1.1 Ion-Pair Agents as CMPAs

The use of an ion-pair agent as a CS in the mobile phase, for enantioseparation, was introduced to column liquid chromatography by Pettersson and Schill (1981). It was proposed that diastereomeric complexes are formed between the enantiomers and the chiral counterions and therefore the separation can be achieved on an achiral stationary phase. Dissimilar distribution properties of the formed diastereomers enable their discrimination on achiral stationary phases (Duncan et al., 1990).

Gaolan et al. reported the enantioseparation of labarol and bataroc on silica gel GF$_{254}$ plates (2.5 cm × 10 cm) by using ammonium-D-10-camphorsulfonate (CSA) as chiral ion-pair agent, which was added to the mobile phase in the ammonium form (Gaolan et al., 1999). The chromatograms were developed at lower temperature (2°C–4°C) in small glass jars of 250 mL volume. The experiments were carried out in the range of 40%–70% (for methanol) and in the range of 55%–80% (for dichloromethane); the optimum percentage of dichloromethane was reported to be 60% in the mobile phase volume ratio.

Duncan et al. employed CSA and different *N*-carbobenzyloxy (CBZ) amino acid derivatives, namely, *N*-CBZ-glycyl-L-proline (ZGP), *N*-CBZ-isoleucyl-L-proline (ZIP), *N*-CBZ-alanyl-L-proline (ZAP), and *N*-CBZ-proline (ZP) as chiral ion interaction agents on HPTLC diol or HPTLC silica plates for the chiral resolution of adrenergic receptor agonists or antagonists (namely, octopamine, norphenylephrine, isoproterenol, propranolol, pindolol, timolol, and metoprolol) (Duncan, 1990; Duncan et al., 1990). The best resolution values were observed with ZGP. It was also established that the separation could not be achieved on other conventional TLC stationary phases including microcrystalline cellulose, alumina, or ordinary silica gel plates and compared TLC results with HPLC ones (Duncan, 1990). Based on the experimental evidence, the authors emphasized both the low cost and wide versatility of TLC over HPLC. Their results confirmed that TLC can be used as a pilot technique for HPLC as suggested previously (Schilitt and Geiss, 1972).

TABLE 7.4
Chromatographic Data for Enantioseparation of Various Drugs by Using CMPAs

Analytes	CMPAs	Mobile Phase	hR_F		References
Metoprolol	Vancomycin	Acetonitrile–methanol–0.56 mM aqueous vancomycin solution (pH 5.5) 6:1:1 (v/v), pH 6.93	44 (+)	24 (−)	Bhushan and Agarwal (2010a)
Propranolol	Vancomycin	Acetonitrile–methanol–0.56 mM aqueous vancomycin solution (pH 5.5) 15:1:2 (v/v), pH 7.05	43 (+)	17 (−)	Bhushan and Agarwal (2010a)
Labetalol	Vancomycin	Acetonitrile–methanol–0.56 mM aqueous vancomycin solution (pH 5.5)–dichloromethane 9:1:1.5:1 (v/v), pH 6.99	70 (+)	45 (−)	Bhushan and Agarwal (2010a)
Atenolol	Vancomycin	NR	NR	NR	Bhushan and Agarwal (2010a)
Penicillamine	L-Tartaric acid	Acetonitrile–methanol–(0.5% L-tartaric acid in water, pH 5)–glacial acetic acid, 7:1:1.1:0.7, v/v	71 (+)	61 (−)	Bhushan and Agarwal (2008a)
Penicillamine	(R)-Mandelic acid	NR	NR	NR	Bhushan and Agarwal (2008a)
Lisinopril	L-Tartaric acid	Acetonitrile–methanol–(0.5% L-tartaric acid in water, pH 5)–glacial acetic acid, 7:1:1.1:0.7, v/v	25 (+)	17 (−)	Bhushan and Agarwal (2008b)
Lisinopril	(R)-Mandelic acid	NR	NR	NR	Bhushan and Agarwal (2008b)
Ketamine	L-Tartaric acid	NR	NR	NR	Bhushan and Agarwal (2008b)
Ketamine	(R)-Mandelic acid	NR	NR	NR	Bhushan and Agarwal (2008b)

NR, not resolved.

Tivert and Backman used ZGP (mobile phase 5 mM ZGP in DCM) for the enantioseparation of alprenolol, propranolol, and metoprolol and studied the influence of humidity on this procedure (Tivert and Backman, 1989, 1993). They obtained satisfactory and reproducible results for the migration distance and chromatographic performance after keeping chromatographic LiChrosorb Diol plates for about 20 h or longer at high relative humidity. It was established that the humidity of chromatographic system is a critical factor for enantioseparation. This method was successfully applied for the separation of active substances from controlled release tablets containing metoprolol succinate.

The nature of stationary phase is observed to play an important role in deciding the concentration of organic modifier in mobile phase while using CSA or ZGP as CS. The highest concentrations of organic modifiers (30%–50%) are required for conventional silica gel chromatography, somewhat lower (10%–25%) for HPTLC, and the lowest ones (5%–10%) for polar diol-HPTLC. TEA (Duncan et al., 1990) or ethanolamine (Tivert and Backman, 1989, 1993) has been used to provide a higher selectivity. An increase in the concentration of counterions usually causes a decrease in the capacity factor (Duncan, 1990; Tivert and Backman, 1993). This could be attributed to an increased competition between the counterions and the diastereomeric ion pair for adsorption sites on the stationary phase.

The presence of water in a chromatographic system was found to play an important role in the chiral separation of β-blockers while using CSA/N-CBZ-amino acid derivatives as CSs. While Duncan used HPTLC or diol-HPTLC dried plates (Duncan, 1990), Tivert and Backman revealed that a higher relative humidity of stationary phases is required for optimal resolution of enantiomers (Tivert and Backman, 1989). The reason for these quite opposite results is not still fully understood. Temperature was also reported to be a very important factor for chiral discrimination. Huang et al. performed chiral separation of pindolol and propranolol at 5°C in a refrigerator, using homemade silica gel plates and CSA as a chiral ion agent (Huang et al., 1997). However, the enantioseparation of atenolol required temperature of 30°C. This could be explained in terms of a stronger binding of polar amino group of atenolol to the adsorbent comparing with other examined β-blocker racemates.

Lucic et al. (2005) used D-(−)-tartaric acid as a CMPA for the separation of (±)-metoprolol tartrate on silica gel plates preimpregnated with ethanol/water (70:30, v/v) containing D-(−)-tartaric acid as a CS. The experiments were performed with different concentrations of D-(−)-tartaric acid (5.8, 11.6, and 23 mmol/L), and it was revealed that the best resolution was achieved with 11.6 mmol/L D-(−)-tartaric acid in both the mobile phase and the impregnation solution at 25°C ± 2°C. It was proposed that tartaric acid dissolved in excess of ethanol could form monoethyl tartrate, which might play a role of a real CS in this separation system.

Bhushan and Agarwal resolved PenA (Bhushan and Agarwal, 2008a) and lisinopril (Bhushan and Agarwal, 2008b) into their enantiomers by using L-tartaric acid as a CMPA on NP-TLC. L-Tartaric acid (0.5%) was dissolved in double-distilled water, and then the pH was adjusted by adding a few drops of NH_3; this solution was used as a component of the mobile phase for TLC on plain plates. For both the cases, the solvent combination acetonitrile–methanol–(0.5% L-tartaric acid in water, pH 5)–glacial acetic acid (7:1:1.1:0.7, v/v) provided the best resolutions at 16°C ± 2°C. LODs were observed to be 0.12 and 0.14 μg for each enantiomer of PenA and lisinopril, respectively. The use of (−)-mandelic acid as CMPA under conditions identical to those used with (+)-tartaric acid did not enable successful resolution of the enantiomers in both the cases.

Bhushan and Tanwar reported the enantioresolution of atenolol, propranolol, and salbutamol by using Cu(II) complexes of five L-amino acids (namely, L-Pro, L-Phe, L-Trp, L-His, and N,N-Me$_2$-L-Phe) on commercial plates (Bhushan and Tanwar, 2010) and homemade unmodified silica plates (Bhushan and Tanwar, 2009). Table 7.2 shows different successful combinations of acetonitrile–methanol–water and the R_F values obtained using them. The visualization was done with iodine vapors. The best resolution for all the three racemates was obtained at 20°C ± 2°C with any of the CSs used. In all the cases, (R)-enantiomer was eluted before (S)-counterpart. The LODs were found to be 0.18 μg per enantiomer.

7.4.4.1.2 Cyclodextrin as CMPAs

Armstrong et al. were the first to report the application of β-CD as a chiral additive for the separation on RP-TLC plates (Armstrong et al., 1988). Taking into account the requirement for the critical concentration of β-CD in mobile phase for chiral separation, as well as its limited solubility in water (0.017 M at 25°C), saturated solution of urea has been frequently used to increase the solubility of this CD (Armstrong et al., 1988; Lepri et al., 1990).

When chiral separation is performed using RP plates, sodium chloride is generally added to the mobile phase to protect the binding sites from any damage (Armstrong et al., 1988; Lepri et al., 1990). Acetonitrile (LeFevre, 1993; Lepri et al., 1990), methanol (Armstrong et al., 1988; LeFevre, 1993), and pyridine (Huang et al., 1996) represent the frequently used organic modifiers in RP-TLC. The formation of inclusion complex was found to be the strongest in water, becoming weaker upon the addition of organic modifiers, because of competition between organic modifier molecules with the analyte molecules for binding sites within the cavity of CDs. The interaction of organic modifier molecules with the binding sites is much stronger than that of water molecules. However, their interaction with a CD should be weaker than that of the analyte molecules.

Nevertheless, as the organic modifier is always present in great excess, it can still displace the analyte molecules from the binding sites in the CD cavity.

In contrast, under NP conditions, nonpolar solvents, such as hexane or chloroform, predominantly occupy the CD cavity and cannot be easily displaced by the analyte molecules. The mechanism of chiral recognition under NP conditions has not been fully understood (Bereznitski et al., 2001). It has been reported that the temperature may influence the kinetics of complexation between an analyte and CD and therefore can affect TLC enantioseparation. It was also found that the effect of temperature on the enantiomer separation varies considerably with the substituent and its position, but parallel to increasing temperature, a decrease in enantioseparation was mostly observed (Xuan and Lederer, 1994).

Armstrong et al. reported the chiral separation of a variety of racemic compounds using RP-TLC and mobile phases containing highly concentrated solutions of β-CD (Armstrong et al., 1988). A separation factor of 1.07 was obtained with methanol–β-CD (0.262 M) (35:65, v/w) for labetalol. Lambroussi et al. (1999) reported the enantioseparation of fluoxetine, norfluoxetine, and promethazine by using hydroxypropyl (HP)–β-CD (2.5 mM) in methanol buffer (pH 4.5) (55:45; 50:50, v/v), HP–β-CD (2.5 mM) in methanol buffer (pH 6.0) (55:45; 50:50, v/v), and HP–β-CD (2.5 mM) in methanol buffer (pH 6.0) (60:40, v/v), respectively. Taha et al. (2009) reported the enantioseparation of cetirizine on silica gel TLC plates by using acetonitrile–water (17: 3, v/v) containing 1 mM of CS, namely, HP–β-CD, chondroitin sulfate, or vancomycin hydrochloride, as mobile phase. Antic et al. (2011) developed a simple and rapid TLC method using β-CD as a CMPA for direct separation of clopidogrel. The best resolution was achieved on Polygramcel 300 Ac-10% plates using isopropanol–β-CD (0.5 mM) (6:4, v/v) as mobile phase. The spots were detected under UV light and using iodine vapors.

Krzek et al. (2002) reported the enantioseparation of budesonide on cellulose-coated HPTLC plate by using 1% aq. solution of β-CD with methanol as mobile phase. The UV densitometric detection was carried out at 245 nm. Under the conditions established, high sensitivity was achieved, with detection limits of 34.64–43.72 ng, as well as a high recovery of 98.76%–103.60% for the individual enantiomers. Later, Krzek et al. (2005) also developed a TLC densitometric method for the identification and determination of enantiomers of ibuprofen. The chromatographic separation was carried out on RP-TLC plates and β-CD: methanol (15:1, v/v) was used as mobile phase. The UV densitometric detection was carried out at $\lambda_{max} = 222$ nm. The LOD was found to be 1 μg/mg, and high recovery, approximately 99.18%, was obtained for both the enantiomers. The linearity range was reported from 0.01% to 0.30%. The presence of both enantiomers of ibuprofen was observed in all preparations at comparable concentrations from 56.04% to 66.16% for (S)- and from 33.84% to 43.99% for (R)-enantiomer.

7.4.4.1.3 Macrocyclic Antibiotics as CMPAs

Armstrong and Zhou (1994) were the first to describe the application of vancomycin as a CMPA. The racemates of some drugs (e.g., bendroflumethiazide, warfarin, coumachlor, and indoprofen) were successfully resolved employing vancomycin as a chiral additive in the RP mode. The best results were obtained using diphenyl stationary phases and acetonitrile as an organic modifier that produced the most effective separations with the shortest time of development.

Bhushan and Agarwal (2010a) applied vancomycin as a CMPA for the enantioseparation of certain β-blockers, namely, metoprolol, propranolol, labetalol, and atenolol. Vancomycin was dissolved in double-distilled water (0.56 mM), and the pH was adjusted by addition of a few drops of NH$_3$; this solution was used as a component of the mobile phase for TLC on homemade silica plates. Successful resolution of the enantiomers of metoprolol, propranolol, and labetalol was achieved using acetonitrile–methanol–aq. vancomycin (0.56 mM) (pH 5.5) (6:1:1, v/v), acetonitrile–methanol–aq. vancomycin (0.56 mM) (pH 5.5) (15:1:2, v/v), and acetonitrile–methanol–aq. vancomycin (0.56 mM) (pH 5.5)–dichloromethane (9:1:1.5:1, v/v), respectively. The spots were detected by the use of iodine vapor. Chromatographic data have been compiled in Table 7.4. The photograph of chromatogram showing the resolution of the enantiomers of (±)-propranolol using vancomycin as CMPA has been shown in Figure 7.2. The recovery of the enantiomers was in the range of 98%–98.5%. The LODs were 1.3, 1.2, 1.5, and 1.4 µg for each enantiomer of atenolol, metoprolol, propranolol, and labetalol, respectively.

7.5 ENANTIOSEPARATION USING INDIRECT APPROACH

The indirect approach of enantioresolution involves the synthesis of stable covalent diastereomers followed by their separation in an achiral environment.

A few limitations associated with indirect approach are as follows: the derivatization can be tedious and time-consuming; the CDR should be above 99% chirally pure; otherwise, the impurity may result in additional undesired diastereomers; the resulting diastereomeric mixture should have very high chemical and stereochemical stability; and for preparative purposes, an additional synthesis step of cleaving off the CDR from the separated diastereomers is involved, which may introduce impurities or may cause racemization of the just resolved enantiomers.

Nevertheless, once a suitable CDR and derivatization process have been found, the indirect approach offers several advantages, such as the elution order can be predetermined that is important for the determination and control of optical purities; for large-scale resolutions, the achiral media are better handled than, for example, CSPs, and the conditions can be adjusted more easily to obtain the desired resolution (Schulte, 2001). Chromatographic data for enantioseparation of various drugs by using CDRs have been compiled in Table 7.5.

7.5.1 Criteria for Selection of Chiral Derivatizing Reagents

The choice of the CDR has a significant effect on the success of separation, detectability of resulting diastereomer, and accuracy of the method. Therefore, the following criteria are important for choosing the CDR:

- CDR should have high chemical and optical purity.
- CDR should generally be freely soluble in water or water-miscible solvents, such as alcohol and acetonitrile.
- CDR should possess specificity for the target functional group and should quantitatively derivatize the analyte under mild conditions.
- The resulting diastereomers should have good stability.
- The reaction conditions should ensure a complete derivatization reaction and have no chances of racemization or degradation of the analyte.

7.5.2 Marfey's Reagent and Its Chiral Variants as CDRs

Marfey's reagent (MR, 1-fluoro-2,4-dinitrophenyl-5-L-alanine amide, FDNP-L-Ala-NH$_2$, FDAA) can safely be considered as the first chiral variant of Sanger's reagent (2,4-dinitrofluorobenzene, DNFB) as it was prepared by substituting one of the fluorine atoms in difluorodinitrobenzene (DFDNB) with L-Ala-NH$_2$ (Marfey, 1984). The MR takes advantage of the remaining reactive aromatic fluorine that undergoes nucleophilic substitution with the free amino group on L- and D-amino acids (in the mixture), peptide or target molecule, and that of the stereogenic center in its alanine group (the L-form) to create diastereomers. It thus provides a structural feature to replace L-Ala-NH$_2$ by suitable chiral moieties such as amino acids and amines.

The diastereomeric derivatives prepared with MR or its chiral variants have strong absorbance at 340 nm due to highly absorbing dinitrophenyl chromophore. The derivatization with MR and its structural variants provides easier location of diastereomers on TLC plates due

TABLE 7.5

Chromatographic Data for Enantioseparation of Various Drugs by Using CDRs

Analytes	CDRs	Mobile Phase	hR_F		References
Penicillamine(NP)	FDNP-L-Ala-NH$_2$	C$_6$H$_5$OH–H$_2$O (3:1, v/v)	76.9 (D)	61.5 (L)	Bhushan et al. (2007)
Penicillamine (NP)	FDNP-L-Phe-NH$_2$	MeCN/MeOH/H$_2$O (5:1:1, v/v)	67.7 (D)	46.2 (L)	Bhushan et al. (2007)
Penicillamine (NP)	FDNP-L-Val-NH$_2$	EtOAc/MeOH/H$_2$O (3:1:1, v/v)	89.2 (D)	58.5 (L)	Bhushan et al. (2007)
Penicillamine(RP)	FDNP-L-Ala-NH$_2$	MeCN: TEAP buffer (50 mM, pH 5.5) [50:50, v/v]	40.0 (D)	43.1 (L)	Bhushan et al. (2007)
Penicillamine (RP)	FDNP-L-Phe-NH$_2$	MeCN: TEAP buffer (50 mM, pH 5.5) [50:50, v/v]	24.6 (D)	30.7 (L)	Bhushan et al. (2007)
Penicillamine (RP)	FDNP-L-Val-NH$_2$	MeCN: TEAP buffer (50 mM, pH 5.5) [50:50, v/v]	35.4 (D)	44.6 (L)	Bhushan et al. (2007)
Baclofen	FDNP-L-Ala-NH$_2$	MeCN: TEAP buffer (50 mM, pH 5.5) [50:50, v/v]	35.0 (S)	35.0 (R)	Bhushan and Kumar (2008)
Baclofen	FDNP-L-Phe-NH$_2$	MeCN: TEAP buffer (50 mM, pH 5.5) [50:50, v/v]	26.7 (S)	26.7 (R)	Bhushan and Kumar (2008)
Baclofen	FDNP-L-Val-NH$_2$	MeCN: TEAP buffer (50 mM, pH 5.5) [50:50, v/v]	31.7 (S)	31.7 (R)	Bhushan and Kumar (2008)
Baclofen	FDNP-L-Ala-NH$_2$	MeCN: TEAP buffer (50 mM, pH 5.5) [50:50, v/v]	30.1 (S)	30.1 (R)	Bhushan and Kumar (2008)
Baclofen	FDNP-L-Phe-NH$_2$	MeCN: TEAP buffer (50 mM, pH 5.5) [50:50, v/v]	34.1 (S)	34.1 (R)	Bhushan and Kumar (2008)
Metoprolol	FDNP-L-Ala-NH$_2$	MeCN: TEAP buffer (50 mM, pH 5.5) [50:50, v/v]	30.7 (R)	50.8 (S)	Bhushan and Agarwal (2010b)
Metoprolol	FDNP-L-Val-NH$_2$	MeCN: TEAP buffer (50 mM, pH 5.5) [50:50, v/v]	36.0 (R)	54.7 (S)	Bhushan and Agarwal (2010b)
Metoprolol	FDNP-L-Phe-NH$_2$	MeCN: TEAP buffer (50 mM, pH 5.5) [50:50, v/v]	30.7 (R)	48.0 (S)	Bhushan and Agarwal (2010b)
Metoprolol	FDNP-L-Leu-NH$_2$	MeCN: TEAP buffer (50 mM, pH 5.5) [50:50, v/v]	29.3 (R)	48.0 (S)	Bhushan and Agarwal (2010b)
Metoprolol	FDNP-L-Met-NH$_2$	MeCN: TEAP buffer (50 mM, pH 5.5) [50:50, v/v]	40.0 (R)	56.9 (S)	Bhushan and Agarwal (2010b)
Metoprolol	FDNP-D-Phg-NH$_2$	MeCN: TEAP buffer (50 mM, pH 5.5) [50:50, v/v]	32.0 (S)	49.3 (S)	Bhushan and Agarwal (2010b)
Metoprolol	FDNP-L-Pro-NH$_2$	MeCN: TEAP buffer (50 mM, pH 5.5) [50:50, v/v]	NR	NR	Bhushan and Agarwal (2010b)
Carvedilol	FDNP-L-Ala-NH$_2$	MeCN: TEAP buffer (50 mM, pH 5.5) [50:50, v/v]	33.8 (R)	43.1 (S)	Bhushan and Agarwal (2010b)
Carvedilol	FDNP-L-Val-NH$_2$	MeCN: TEAP buffer (50 mM, pH 5.5) [50:50, v/v]	23.1 (R)	36.9 (S)	Bhushan and Agarwal (2010b)
Carvedilol	FDNP-L-Phe-NH$_2$	MeCN: TEAP buffer (50 mM, pH 5.5) [50:50, v/v]	27.7 (R)	33.8 (S)	Bhushan and Agarwal (2010b)
Carvedilol	FDNP-L-Leu-NH$_2$	MeCN: TEAP buffer (50 mM, pH 5.5) [50:50, v/v]	18.5 (R)	32.3 (S)	Bhushan and Agarwal (2010b)
Carvedilol	FDNP-L-Met-NH$_2$	MeCN: TEAP buffer (50 mM, pH 5.5) [50:50, v/v]	26.1 (R)	36.9 (S)	Bhushan and Agarwal (2010b)
Carvedilol	FDNP-D-Phg-NH$_2$	MeCN: TEAP buffer (50 mM, pH 5.5) [50:50, v/v]	NR	NR	Bhushan and Agarwal (2010b)

(continued)

TABLE 7.5 (continued)
Chromatographic Data for Enantioseparation of Various Drugs by Using CDRs

Analytes	CDRs	Mobile Phase	hR$_F$		References
Carvedilol	FDNP-L-Pro-NH$_2$	MeCN: TEAP buffer (50 mM, pH 5.5) [50:50, v/v]	NR	NR	Bhushan and Agarwal (2010b)
Metoprolol	(S)-(+)-BOP-Cl	C$_6$H$_5$CH$_3$/CH$_3$COCH$_3$ (100:10, v/v)	24 (R)	28 (S)	Pflugmann et al. (1987)
Oxprenolol	(S)-(+)-BOP-Cl	C$_6$H$_5$CH$_3$/CH$_3$COCH$_3$ (100:10, v/v)	32 (R)	38 (S)	Pflugmann et al. (1987)
Propranolol	(S)-(+)-BOP-Cl	C$_6$H$_5$CH$_3$/CH$_3$COCH$_3$ (100:10, v/v)	32 (R)	39 (S)	Pflugmann et al. (1987)

NP, normal phase; RP, reversed phase; NR, not resolved.

to their visibility and greater sensitivity for control of enantiomeric purity. At the same time, these chiral derivatives can be easily prepared in a laboratory in micromolar quantities with less expense in comparison to many other CDRs. The literature shows reviews on the application of MR and its variants in structural elucidation of peptides, determination of racemization in peptide synthesis, detection of small quantities of D-amino acids, mechanism of separation, and enantioresolution of a variety of compounds (B'Hymer et al., 2003; Bhushan and Brückner, 2004, 2011; Bhushan and Martens, 2010).

7.5.2.1 Derivatization Process

The representative derivatization procedure of (R,S)-baclofen with FDNP-L-Ala-NH$_2$ is given in the succeeding text (Bhushan and Kumar, 2008). The solution of (R,S)-baclofen (100 μL, 10 mM) was added to FDNP-L-Ala-NH$_2$ (100 μL, 14 mM) prepared in acetone in a plastic tube, and the solution was incubated at 40°C for 1 h with constant stirring. After cooling to room temperature, the reaction was ended by adding HCl (50 μL, 0.2 M). After mixing, the contents were dried in a vacuum desiccator over sodium hydroxide pellets. Acetonitrile (200 μL) was then added to dissolve the diastereomeric derivatives. Since these derivatives are light-sensitive, all procedures were protected from light exposure, and derivatives were kept in the dark at 0°C–4°C (Bhushan and Kumar, 2008).

7.5.2.2 Applications

Bhushan et al. used MR and its two chiral variants (FDNP-L-Phe-NH$_2$ and FDNP-L-Val-NH$_2$) as CDRs in TLC (Bhushan et al., 2007). They reported the enantioseparation of DL-PenA on precoated NP- and RP-TLC plates. The solutions (2 μL, 25 nmol) of diastereomers of DL-PenA, and of pure D- and L-PenA, were spotted on TLC plates. The chromatograms were developed in pre-equilibrated rectangular glass chambers at 25°C. The spots were yellow and visible in ordinary light.

Phenol–water (3:1, v/v) at 25°C in 40 min on NP plates provided the best resolution of the diastereomeric pairs, prepared with all the three CDRs. In RP-TLC, the best resolution was obtained with the combination of acetonitrile–TEAP (triethylamine phosphate) buffer (50 mM, pH 5.5) (50:50, v/v) at 25°C. The hR$_F$ values on both NP- and RP-TLC are given in Table 7.5. The D-derivative was eluted earlier than the L-enantiomer for all the three cases under NP conditions, as expected according to separation mechanism (Brückner and Keller-Hoehl, 1990). The elution order of diastereomers in RP mode was found to be the reverse of NP. Among all the three chiral reagents, the best resolution was of the diastereomers of DL-PenA prepared with FDNP-Val-NH$_2$ under both NP and RP conditions, as indicated by the hR$_F$ values given in Table 7.5.

Bhushan and Kumar (2008) reported the enantioseparation of (R,S)-baclofen on RP-TLC plates using MR and four chiral variants as CDRs (FDNP-L-Phe-NH_2, FDNP-L-Val-NH_2, FDNP-L-Phe-NH_2, and FDNP-L-Val-NH_2). The best separation, for the diastereomers prepared with the four CDRs, was obtained using acetonitrile–TEAP buffer (50 mM, pH 4.0) (50:50, v/v) at 25°C; the diastereomers prepared with FDNP-L-Pro-NH_2 were not successfully resolved. The chromatogram for RP-TLC is shown in Figure 7.2, and hR_F values are given in Table 7.5. It was observed that hR_F values increased with increasing acetonitrile concentration in mobile phase. The reagent FDNP-L-Pro-NH_2 failed to separate the diastereomers of baclofen under the tested chromatographic conditions. The derivative of (R)-enantiomer eluted earlier than that of (S)-counterpart for all the cases under RP conditions as expected according to the separation mechanism (Brückner and Keller-Hoehl, 1990). Among all the chiral reagents, the best resolution was of the diastereomers prepared with FDNP-L-Leu-NH_2, as indicated by the hR_F values given in Table 7.5. The accuracy of the method was determined by investigating the recovery of (S)-enantiomer from the samples of pure (R)-enantiomer. For these studies, samples were prepared by spiking (R)-baclofen with fixed amounts of (S)-baclofen. The results indicate that the method can be applied for the detection of (S)-baclofen in (R)-baclofen up to 0.05%.

Bhushan and Agarwal (2010b) developed a simple and rapid method for the indirect separation of the enantiomers of (R,S)-metoprolol and (R,S)-carvedilol by using RP-TLC. Beta-blockers derivatized with MR and its six structural variants (FDNP-L-Phe-NH_2, FDNP-L-Val-NH_2, FDNP-L-Pro-NH_2, FDNP-L-Leu-NH_2, FDNP-L-Met-NH_2, and FDNP-D-Phg-NH_2) were spotted on precoated plates. The diastereomers were separated most effectively by acetonitrile–TEAP buffer (50 mM, pH 5.5) (50:50, v/v). The time taken for chromatogram development was 7 min; it was 10 min for direct separation of the enantiomers of metoprolol and carvedilol by NP-TLC (Bhushan and Gupta, 2006). The photographs of representative chromatograms showing resolution of the diastereomers of carvedilol are shown in Figure 7.3.

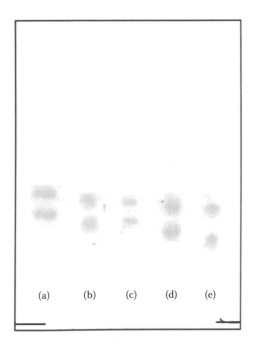

FIGURE 7.3 Photograph of plate showing RP-TLC of diastereomers of carvedilol prepared with (a) FDNP-L-Ala-NH_2, (b) FDNP-L-Val-NH_2, (c) FDNP-L-Met-NH_2, (d) FDNP-L-Phe-NH_2, and (e) FDNP-L-Leu-NH_2. The mobile phase was 50:50 (v/v) TEAP buffer (50 mM, pH 5.5)–acetonitrile, the development distance 6.5 cm, and the temperature 25°C. The upper spot is the (S,S) diastereomer and the lower spot the (R,S)-diastereomer. (From Bhushan, R. and Agarwal, C., *J. Planar Chromatogr.-Modern TLC*, 23, 335, 2010b.)

The diastereomers of metoprolol and carvedilol prepared with FDNP-L-Leu-NH₂ and FDNP-L-Val-NH₂ were better separated than those prepared with the other reagents (Table 7.5). The diastereomers of metoprolol and carvedilol prepared with FDNP-L-Pro-NH₂ did not separate under identical conditions. The diastereomers of (R)-metoprolol migrated further than those of (S)-metoprolol except for those prepared with FDNP-D-Phg-NH₂, because the chiral auxiliary in the CDR is in its D configuration.

7.5.2.3 Effect of Side Chain of Chiral Auxiliary

Bull and Breese (1974) produced a hydrophobicity scale of amino acids by calculating their apparent partial specific volume. Accordingly, the amino acids can be arranged in decreasing order of hydrophobicity as Leu (0.842) > Val (0.777) > Phe (0.756) > Met (0.709) > Ala (0.691) (the values in parentheses are the apparent partial specific volume). In each case, better resolution and increased retention were obtained (Table 7.5) as the hydrophobicity of the amino acid side chain (in the CDR) was increased (Bhushan et al., 2007; Bhushan and Agarwal, 2010b; Bhushan and Kumar, 2008).

7.5.3 (R)-(−)-1-(1-NAPHTHYL)ETHYL ISOCYANATE AS CDR

Gübitz and Mihellyes (1984) employed (R)-(−)-1-(1-naphthyl)ethyl isocyanate (NEIC) as a CDR for the enantioseparation of propranolol, oxprenolol, metoprolol, pindolol, alprenolol, and bunitrolol in the form of their corresponding diastereomeric ureas. For that purpose, to 1.0–50 µmol of the free bases or their salts dissolved in the mixture of dry chloroform and dimethylformamide (8:2, v/v), an equimolar amount of TEA and approximately twofold molar excess of (R)-(−)-NEIC were added. After 20 min reaction time, the excess of the reagent was destroyed by adding diethylamine, and 15 min later, an aliquot of the reaction mixture was applied to the HPTLC plates. No racemization was observed during the reaction.

The chromatograms were developed in benzene–ether–acetone (88:10:5, v/v) over the distance of 5.0 cm and visualized under a UV lamp. It was observed that the retention of (S)-(−)-enantiomers was higher than that of their (R)-(+)-counterparts. The chiral separation factors for the investigated β-blockers ranged from 1.15 to 1.24. It was also established that nanogram amounts of one enantiomer derivative with NEIC were detectable in the presence of a 100-fold excess of the other enantiomer.

7.5.4 TETRA-O-ACETYL-β-GLUCOPYRANOSYLISOTHIOCYANATE AS CDR

Spell and Stewart (1997) used the derivatization reaction of a sugar-based CDR, *tetra-O*-acetyl-β-glucopyranosylisothiocyanate (GITC), with secondary amines of pindolol for the separation of its optical antipodes. A certain volume of each (R)- and (S)-pindolol and (RS)-pindolol was mixed with GITC solution in acetonitrile, and the reaction was carried out for 35 min at 35°C–40°C on a stirrer–hot plate by gentle stirring.

The chromatography was performed on RP thin-layer C₈ F₂₅₄ silica gel plates preconditioned by development in methanol–water (30:70, v/v), air-dried, and activated for 20 min at 120°C. After the application of derivatized (R)- and (S)-pindolol, the chromatogram was developed by ascending chromatography using water–propan-2-ol (70:30, v/v) as the mobile phase. The development distance was about 75 mm and development time 75 min. The separation zones were detected by scanning the plates at 256 nm. The HPTLC method was further evaluated by linearity and accuracy. This approach was found to be applicable for quantitative assessment of each of optical antipodes in a racemic pindolol mixture.

7.5.5 (S)-(+) BENOXAPROFEN CHLORIDE AS CDR

Pflugmann et al. (1987) developed a sensitive method for the determination of optical antipodes of propranolol, oxprenolol, and metoprolol in urine samples after the derivatization with (S)-(+)

benoxaprofen chloride (BOP-Cl). The derivatization was performed overnight at room temperature. The reaction was terminated by adding methanol, and the solution was evaporated to dryness under vacuum. The residue was redissolved in cyclohexane, and a certain volume of the solution was subjected to TLC.

The chromatogram was developed in a glass tank in freshly prepared toluene–acetone (100:10, v/v) and an ammonia-saturated atmosphere, produced by two open 50 mL beakers filled with ammonia (33%) and inserted in the chromatographic tank at room temperature. The separated optical antipodes of all three β-blockers were detected by measuring fluorescence (λ_{ex} = 313 nm, λ_{em} = 365 nm). The R_F values of the examined β-blocker derivatives were 0.24 (R)/0.28 (S), 0.32 (R)/0.38 (S), and 0.32 (R)/0.39 (S) for metoprolol, oxprenolol, and propranolol, respectively. The LODs of all tested β-blockers were ca. 1.6 ng per zone after extraction from urine. The method was evaluated by selectivity in the presence of all other metabolites contained in urine samples, as well as by linearity, accuracy, reproducibility, detection, and quantification limits.

Besides the previously mentioned CDRs, Slegel et al. (1987) reported the enantioseparation of naproxen by using (1R,2R)-(−)-1-(4-nitrophenyl)-2-amino-1,3-propanediol (levobase) and its enantiomer dextrobase as CDRs. Chromatogram was developed using chloroform–ethanol–acetic acid (9:1:0.5, v/v), and the spots were detected under UV (254 nm) light. The R_F values were 0.63 and 0.53 for (S)- and (R)-naproxen, respectively. Rossetti et al. (1986) used (R)-(+)-1-phenylalanine hydrochloride for the chiral separation of ketoprofen, suprofen, and indoprofen with benzene–methanol (93:7, v/v) and chloroform–ethyl acetate (15:1, v/v) mobile phases. The visualization was done under UV light (254 nm). R_F differences were observed in the range of 0.04–0.10, and (S)-enantiomer was found to be eluted earlier than (R)-counterpart in all the cases.

REFERENCES

Aboul-Enein HY, El-Awady MI, and Heard CH (2003a) Enantiomeric resolution of some 2-arylpropionic acids using L-(−)-serine-impregnated silica as stationary phase by thin layer chromatography. *J. Pharm. Biomed. Anal.* 32: 1055–1059.

Aboul-Enein HY, El-Awady MI, and Heard CM (2003b) Thin layer chromatographic resolution of some 2-aryl-propionic acid enantiomers using L-(−)-serine, L-(−)-threonine and a mixture of L-(−)-serine, L-(−)-threonine-impregnated silica gel as stationary phases. *Biomed. Chromatogr.* 17: 325–334.

Alak A and Armstrong DW (1986) Thin layer chromatographic separation of optical, geometrical and structural isomers. *Anal. Chem.* 58: 582–584.

Antic D, Filipic S, Ivkovic B, Nikolic K, and Agbaba D (2011) Direct separation of clopidogrel enantiomers by reversed-phase planar chromatography method using β-cyclodextrin as a chiral mobile phase additive. *Acta Chromatogr.* 23: 235–245.

Antipas AS, Vander Velde DG, Jois SD, Siahaan T, and Stella VJ (2000) Effect of conformation on the rate of deamidation of vancomycin in aqueous solutions. *J. Pharm. Sci.* 89: 742–750.

Armstrong DW, He FY, and Han SM (1988) Planar chromatographic separation of enantiomers and diastereo-isomers with cyclodextrin mobile phase additives. *J. Chromatogr.* 448: 345–354.

Armstrong DW, Tang Y, Chen S, Zhou Y, Bagwill C, and Chen JR (1994) Macrocyclic antibiotics as a new class of chiral selectors for liquid chromatography. *Anal. Chem.* 66: 1473–1484.

Armstrong DW and Zhou Y (1994) Use of a macrocyclic antibiotic as the chiral selector for enantiomeric separations by TLC. *J. Liq. Chromatogr.* 17: 1695–1707.

B'Hymer C, Montes-Bayon M, and Caruso JA (2003) Marfey's reagent: Past, present, and future uses of 1-fluoro-2,4-dinitrophenyl-5-L-alanine amide. *J. Sep. Sci.* 26: 7–19.

Bereznitski Y, Thompson R, O'Neill E, and Grinberg N (2001) Thin layer chromatography—A useful technique for the separation of enantiomers. *J. AOAC Int.* 84: 1242–1251.

Bhushan R and Agarwal C (2008a) Direct enantiomeric TLC resolution of DL-penicillamine using (R)-mandelic acid and L-tartaric acid as chiral impregnating reagents and as chiral mobile phase additive. *Biomed. Chromatogr.* 22: 1237–1242.

Bhushan R and Agarwal C (2008b) Direct TLC resolution of (±)-ketamine and (±)-lisinopril by use of (+)-tartaric acid or (−)-mandelic acid as impregnating reagents or mobile phase additives. Isolation of the enantiomers. *Chromatographia* 68: 1045–1051.

Bhushan R and Agarwal C (2008c) Direct resolution of six beta blockers into their enantiomers on silica plates impregnated with L-Asp and L-Glu. *J. Planar Chromatogr.* 21: 129–134.

Bhushan R and Agarwal C (2010a) Resolution of beta blocker enantiomers by TLC with vancomycin as impregnating agent or as chiral mobile phase additive. *J. Planar Chromatogr.* 23: 7–13.

Bhushan R and Agarwal C (2010b) Liquid chromatographic resolution of the enantiomers of metoprolol and carvedilol in pharmaceutical formulations by use of Marfey's reagent and its variants. *J. Planar Chromatogr. Modern TLC* 23: 335–338.

Bhushan R and Ali I (1987) TLC resolution of enantiomeric mixtures of amino acids. *Chromatographia* 23: 141–142.

Bhushan R and Ali I (1993) Resolution of racemic mixtures of hyoscyamine and colchicine on impregnated silica gel layers. *Chromatographia* 35: 679–680.

Bhushan R and Arora M (2003) Direct enantiomeric resolution of (±)-atenolol, (±)-metoprolol and (±)-propranolol by impregnated TLC using L-aspartic acid as chiral selector. *Biomed. Chromatogr.* 17: 226–230.

Bhushan R and Brückner H (2004) Marfey's reagent for chiral amino acid analysis: A review. *Amino Acids* 27: 231–247.

Bhushan R and Brückner H (2011) Use of Marfey's reagent and analogs for chiral amino acid analysis: Assessment and applications to natural products and biological systems. *J. Chromatogr. B* 879: 3148–3161.

Bhushan R, Brückner H, and Kumar V (2007) Indirect resolution of enantiomers of penicillamine by TLC and HPLC using Marfey's reagent and its variants. *Biomed. Chromatogr.* 21: 1064–1068.

Bhushan R and Dubey R (2011) Indirect reversed-phase high-performance liquid chromatographic and direct thin-layer chromatographic enantioresolution of (*R,S*)-Cinacalcet. *Biomed. Chromatogr.* 25: 674–679.

Bhushan R and Gupta D (2004) Resolution of (±)-ibuprofen using (−)-brucine as a chiral selector by thin layer chromatography. *Biomed. Chromatogr.* 18: 838–840.

Bhushan R and Gupta D (2005) Thin-layer chromatography separation of enantiomers of verapamil using macrocyclic antibiotic as a chiral selector. *Biomed. Chromatogr.* 19: 474–478.

Bhushan R and Gupta D (2006) Ligand-exchange TLC resolution of some racemic beta-adrenergic blocking agents. *J. Planar Chromatogr. Modern TLC* 19: 241–245.

Bhushan R and Kumar V (2008) Indirect resolution of baclofen enantiomers from pharmaceutical dosage form by reversed-phase liquid chromatography after derivatization with Marfey's reagent and its structural variants. *Biomed. Chromatogr.* 22: 906–911.

Bhushan R and Martens J (1997) Direct resolution of enantiomers by impregnated TLC. *Biomed. Chromatogr.* 11: 280–285.

Bhushan R and Martens J (2001) Separation of amino acids, their derivatives and enantiomers by impregnated TLC. *Biomed. Chromatogr.* 15: 155–165.

Bhushan R and Martens J (2010) *Amino Acids: Chromatographic Separation and Enantioresolution.* HNB Publishing, New York.

Bhushan R and Parshad V (1996) Resolution of (±)-ibuprofen using L-arginine impregnated thin layer chromatography. *J. Chromatogr. A* 721: 369–372.

Bhushan R and Tanwar S (2008) Direct TLC resolution of atenolol and propranolol into their enantiomers using three different chiral selectors as impregnating reagents. *Biomed. Chromatogr.* 22: 1028–1034.

Bhushan R and Tanwar S (2009) Direct TLC resolution of the enantiomers of three beta-blockers by ligand exchange with Cu(II)-L-amino acid complex, using four different approaches. *Chromatographia* 70: 1001–1006.

Bhushan R and Tanwar S (2010) Different approaches of impregnation for resolution of enantiomers of atenolol, propranolol and salbutamol using Cu(II)-L-amino acid complexes for ligand exchange on commercial thin layer chromatographic plates. *J. Chromatogr. A* 1217: 1395–1398.

Bhushan R and Thiong'o G (1998) Direct enantioseparation of some beta-adrenergic blocking agents using impregnated thin-layer chromatography. *J. Chromatogr. B* 708: 330–334.

Bhushan R and Thiong'o G (1999) Direct enantiomeric resolution of some 2-arylpropionic acids using (−)-brucine-impregnated thin-layer chromatography. *Biomed. Chromatogr.* 13: 276–278.

Bressolle F, Audran M, Pham TN, and Vallon JJ (1996) Cyclodextrins and enantiomeric separations of drugs by liquid chromatography and capillary electrophoresis: Basic principles and new developments. *J. Chromatogr. B* 687: 303–336.

Brückner H and Keller-Hoehl C (1990) HPLC separation of DL-amino acids derivatized with N^2-(5-fluoro-2,4-dinitrophenyl)-L-amino acid amides. *Chromatographia* 30: 621–629.

Brunner CA and Wainer I (1989) Direct stereochemical resolution of enantiomeric amides via thin layer chromatography on a covalently bonded chiral stationary phase. *J. Chromatogr.* 472: 277–283.

Bull HB and Breese K (1974) Surface tension of amino acid solutions. Hydrophobicity scale of the amino acid residues. *Arch. Biochem. Biophys.* 161: 665–670.

Cannarsa MJ (1996) Single enantiomer drug: New strategies and directions. *Chem. Ind.* 10: 374–378.

Cayen MN (1991) Racemic mixtures and single stereoisomers: Industrial concerns and issues in drug development. *Chirality* 3: 94–98.

Davankov VA (2004) Chiral separation by HPLC using the ligand exchange principle. In *Chiral Separations*, eds. G. Gübitz and M.G. Schmid. Humana Press, Totowa, NJ.

Duncan JD (1990) Chiral separations: A comparison of HPLC and TLC. *J. Liq. Chromatogr.* 13: 2737–2755.

Duncan JD, Armstrong DW, and Stalcup AM (1990) Normal phase TLC separation of enantiomers using chiral ion interaction agents. *J. Liq. Chromatogr.* 13: 1091–1103.

Gaolan L et al. (1999) Enantiomeric separation of aromatic alcohol amino drugs by thin-layer chromatography. *Se Pu* 17: 215.

Gübitz G and Mihellyes S (1984) Optical resolution of β-blocking agents by thin-layer chromatography and high-performance liquid chromatography as diastereomeric (*R*)-(−)-1-(1-naphthyl)ethylureas. *J. Chromatogr.* 314: 462–466.

Günther K, Martens J, and Schickedanz M (1984) Dünnschicht chromatographische enantiomerentrennung mittels ligandenaustausch. *Angew. Chem. Int. Ed. Engl.* 96: 514–515.

Günther K and Möller K (2003) Enantiomer separations. In *Handbook of Thin Layer Chromatography*, eds. J. Sherma and B. Fried, pp. 471–533. 3rd edn., Marcel Dekker, Inc., New York.

Han SM and Armstrong DW (1990) Enantiomeric separation by thin-layer chromatography. *Chem. Anal.* 108: 81–100.

Hesse G and Hagel R (1973) A complete separation of a racemic mixture by elution chromatography on cellulose triacetate. *Chromatographia* 6: 277–280.

Huang MB, Li HK, Li GL, Chang-Tai Y, and Wang LP (1996) Planar chromatographic direct separation of some aromatic amino acids and aromatic amino alcohols into enantiomers using cyclodextrin mobile phase additives. *J. Chromatogr.* 742: 289–294.

Huang MB, Lia GL, Yanga GS, Shia YH, Gaoa JJ, and Liua XD (1997) Enantiomeric separation of aromatic amino alcohol drugs by chiral ion-pair chromatography on a silica gel plate. *J. Liq. Chromatogr. Rel. Technol.* 20: 1507–1514.

ICH-Topic Q2B (1996) Validation of analytical procedures. In: *Proceedings of the International Conference on Harmonization of Technical Requirement for Registration of Pharmaceuticals for Human Use*, Geneva, 1996.

Kovacs-Hadady K and Kiss IT (1987) Attempts for the chromatographic separation of D- and L-penicillamine enantiomers. *Chromatographia* 24: 677–679.

Kowalska T and Sherma J (2007) eds. *Thin Layer Chromatography in Chiral Separations and Analysis*. CRC Press, Boca Raton, FL.

Krzek J, Hubicka U, Dabrowska-Tylka M, and Leciejewicz-Ziemecka E (2002) Determination of budesonide (*R*)-(+) and (*S*)-(−) isomers in pharmaceuticals by thin-layer chromatography with UV densitometric detection. *Chromatographia* 56: 759–762.

Krzek J, Starek M, and Jelonkiewicz D (2005) RP-TLC determination of (*S*)-(+) and (*R*)-(−) ibuprofen in drugs with the application of chiral mobile phase and UV densitometric detection. *Chromatographia* 62: 653–657.

Lambroussi V, Piperaki S, and Tsantili-Kakoulidou A (1999) Formation of inclusion complexes between cyclodextrins as mobile phase additives in RP-TLC, and fluoxetine, norfluoxetine, and promethazine. *J. Planar Chromatogr.* 12: 124–128.

LeFevre JW (1993) Reversed-phase thin-layer chromatographic separations of enantiomers of dansyl-amino acids using β-cyclodextrin as a mobile phase additive. *J. Chromatogr.* 653: 293–302.

Lepri L (1995) Reversed phase planar chromatography of enantiomeric compounds on microcrystalline triacetylcellulose. *J. Planar Chromatogr.* 8: 467–469.

Lepri L, Coas V, Desideri PG, and Checchini L (1990) Separation of optical and structural isomers by planar chromatography with development by β-cyclodextrin solutions. *J. Planar Chromatogr.* 3: 311–316.

Lepri L, Coas V, Desideri PG, and Zocchi A (1994) Reversed phase planar chromatography of enantiomeric compounds on triacetylcellulose. *J. Planar Chromatogr.* 7: 376–381.

Lepri L, Del Bubba M, and Cincinelli A (2001) Chiral separations by TLC. In *Planar Chromatography, A Retrospective View for the Third Millennium*, ed. Sz. Nyiredy, p. 517. Springer Scientific Publishers, Budapest, Hungary.

Lucic B, Radulovic D, Vujic Z, and Agbada D (2005) Direct separation of the enantiomers of (±)-metoprolol tartrate on impregnated TLC plates with D-(−)-tartaric acid as a chiral selector. *J. Planar Chromatogr.* 18: 294–298.

Luo D, Ma L, and Liang B (2008) Direct enantioseparation of metoprolol using thin-layer chromatography. *Huaxue Yanjiu Yu Yingyong* 20: 642–646.

Marfey P (1984) Determination of D-amino acids. II. Use of a bifunctional reagent, 1,5-difluoro-2,4-dinitrobenzene. *Carlsberg Res. Commun.* 49: 591–596.

Martens J and Bhushan R (1989) TLC enantiomeric separation of amino acids. *Int. J. Peptide Protein Res.* 34: 433–444.

Mielcarek J (2001) Normal-phase TLC separation of enantiomers of 1, 4-didydropyridine derivatives. *Drug Dev. Ind. Pharm.* 27: 175–179.

Mitchell CR and Armstrong DW (2004) Cyclodextrin-based chiral stationary phases for liquid chromatography: A twenty-year overview. In *Chiral Separations*, eds. G. Gübitz and M.G. Schmid, Humana Press, Totowa, NJ.

Muller GW (1997) Thalidomide: From tragedy to new drug discovery. *ChemTech* 27: 21–25.

Okamoto Y, Kawashima M, and Hatada K (1984) Useful chiral packing materials for high-performance liquid chromatographic resolution of enantiomers: Phenyl carbamates of polysaccharides coated on silica gel. *J. Am. Chem. Soc.* 106: 5357–5359.

Pettersson C and Schill G (1981) Separation of enantiomeric amines by ion-pair chromatography. *J. Chromatogr.* 204: 179–183.

Pflugmann G, Spahn H, and Mutschler E (1987) Rapid determination of the enantiomers of metoprolol, oxprenolol and propranolol in urine. *J. Chromatogr.* 416: 331–339.

Rossetti V, Lombard A, and Buffa M (1986) The HPTLC resolution of the enantiomers of some 2-arylpropionic acid anti-inflammatory drugs. *J. Pharm. Biomed. Anal.* 4: 673–676.

Sajewicz M, Pietka R, and Kowalska T (2004) Chiral separation of (S)-(+)- and (R)-(−)-ibuprofen by thin layer chromatography—An improved analytical procedure. *J. Planar Chromatogr.* 17: 173–176.

Sajewicz M, Pietka R, and Kowalska T (2005) Chiral separations of ibuprofen and propranolol by TLC—A study of the mechanism and thermodynamics of retention. *J. Liq. Chromatogr. Rel. Tech.* 28: 2499–2513.

Schilitt H and Geiss F (1972) Thin-layer chromatography as a pilot technique for rapid column chromatography. *J. Chromatogr.* 67: 261–276.

Schulte M (2001) Chiral derivatization chromatography. In *Chiral Separation Techniques: A Practical Approach*, ed. G. Subramanian, 2nd edn., Wiley-VCH, Weinheim, Germany.

Sherma J and Fried B (2003) eds. *Handbook of Thin-Layer Chromatography*, 3rd edn. Marcel Dekker, New York.

Sherma J and Kowalska T (2007) eds. *Thin Layer Chromatography in Phytochemistry*. CRC Press, Boca Raton, FL.

Siouffi AM, Piras P, and Roussel C (2005) Some aspects of chiral separations in planar chromatography compared with HPLC. *J. Planar Chromatogr.* 18: 5–13.

Slegel P, Vereczkey-Donath G, Ladanyi L, and Toth-Lauritz M (1987) Enantiomeric separation of chiral carboxylic acids, as their diastereomeric carboxamides, by thin-layer chromatography. *J. Pharm. Biomed. Anal.* 5: 665–673.

Spell JC and Stewart JT (1997) A high-performance thin-layer chromatographic assay of pindolol enantiomers by chemical derivatization. *J. Planar Chromatogr. Modern TLC* 10: 222–224.

Suedee R and Heard CM (1997) Direct resolution of propranolol and bupranolol by thin-layer chromatography using cellulose derivatives as stationary phase. *Chirality* 9: 139–144.

Taha EA, Salama NN, and Wang S (2009) Enantioseparation of cetirizine by chromatographic methods and discrimination by 1H-NMR. *Drug Test. Anal.* 1: 118–124.

Tivert AM and Backman AE (1989) Enantiomeric separation of amino alcohols by TLC using a chiral counter-ion in the mobile phase. *J. Planar Chromatogr.* 2: 472–473.

Tivert AM and Backman AE (1993) Separation of the enantiomers of β-blocking drugs by TLC with a chiral mobile phase additive. *J. Planar Chromatogr.* 6: 216–219.

Wall PE (1989) Preparation and application of HPTLC plates for enantiomer separation. *J. Planar Chromatogr.* 2: 228–232.

Wang L, He Y, Xie D, Wang W, and Deng J (2011) Resolution of DL-naproxen using thin layer chromatography with β-cyclodextrin/SiO$_2$ as chiral stationary phase. *Yingyong Huagong* 40: 1011–1014.

Wilson CD, Gisvold O, and Doerge RF (1977) *Textbook of Organic Medicinal and Pharmaceutical Chemistry*, 7th edn. Lippincott, Singapore.

Xu L. et al. (2002) Zongshan Daxue Xuebao. *Ziran Kexueban Guangzhou* 41: 115–117.

Xu L. et al. (2003) Preparation of cellulose tris(benzoate)s for TLC and their chromatographic properties. *Chinese J. Instrum. Anal.* 22: 1.

Xuan HTK and Lederer M (1994) Adsorption chromatography on cellulose. XI. Chiral separations with aqueous solutions of cyclodextrins as eluents. *J. Chromatogr.* 659: 191–197.

Zhu Q, Yu P, Deng Q, and Zeng L (2001) β-Cyclodextrin-bonded chiral stationary phase for thin-layer chromatographic separation of enantiomers. *J. Planar Chromatogr.* 14: 137–139.

Zhu QH, Deng QY, and Zeng LM (2000) Preparation and evaluation of 3,5-dinitrobenzoyl-substituted β-cyclodextrin bonded stationary phase. *Zhongshan Daxue Xuebao, Ziran Kexueban* 39: 61–65.

8 Chambers, Sample Application, and New Devices in the Chromatography of Drugs

Beata Polak, Radosław Gwarda, Ewelina Kopciał, and Tadeusz H. Dzido

CONTENTS

8.1 INTRODUCTION

There is no single equipment that enables to automatically perform a full TLC/HPTLC (thin-layer chromatography/high-performance thin-layer chromatography) procedure from sample application to evaluation of data so far. Three basic stages of the thin-layer chromatography procedure can be distinguished in contemporary laboratory practice, that is, sample application onto the adsorbent layer of the chromatographic plate, chromatogram development, and chromatogram registration including data processing. One may say that these stages constitute the flexibility of the TLC technique as no other in the laboratory practice. This is an advantage of TLC because each of the stages mentioned can be realized individually in a very simple way or with sophisticated equipment. Therefore, every laboratory all over the world can apply this technique independently within its budget.

It is reasonable to individually discuss the equipment used in TLC and in the present chapter we focus on the first two aspects mentioned, that is, sample application and modern chambers for chromatogram development.

8.2 SAMPLE APPLICATION

The quantity of applied sample onto the adsorbent layer depends on the separation scale (analytical, semipreparative, or preparative) and the type of chromatographic plates (for TLC or HPTLC) as well.

The quantity of components to be separated is correlated with the volume of sample solutions applied onto the chromatographic plates. For TLC plates, it usually varies from 1 to 5 μL, whereas for HPTLC ones it is in the range of 0.1–1 μL. Higher plate efficiency and lower detection limits of systems with HPTLC plates relative to TLC ones are responsible for the difference in sample quantity applied to both the types of chromatographic plates. The starting spot diameter is required to be different dependent on the chromatographic plate quality. It is usually in the range of 2–6 mm for TLC plates and 1.0–1.5 mm for HPTLC ones [1]. However, one should consider that the sample, when applied onto the chromatographic plate, cannot lead to overloading of the separating system. It means that the solute applied onto the chromatographic layer can't exceed the adsorption capacity of the stationary phase. This value for silica is approximately equal to 220 mg/g for total polar materials based on determination with the Langmuir model [2]. The peaks on chromatograms will probably show a tailing effect if the sample quantity exceeds the value of the adsorption capacity.

It is well known that the initial zone size strongly influences the final separation efficiency, especially when the development distance is relatively short. The solute zone can be applied onto the chromatographic plate as a spot, band, or rectangle shape. Spot-shaped application requires less automation and is time-saving. It can be easily performed with a hand-operated microsyringe. However, this operation should be carefully carried out because of its strong contribution to the separation quality of solute components. On the other hand, band-shaped application can be performed with automatic or semiautomatic devices. This application mode is more suitable for quantitative analysis than a spot-shaped one.

Rectangle-shaped starting zones are applied on the chromatographic plate when a large volume or quantity of sample solution is required, especially for preparative separations.

Various shapes of initial sample zones are presented in Figure 8.1.

The solvent type of the sample solution can considerably influence the diameter of the starting zone, especially when direct probe sampling on the chromatographic layer is performed. In such a case, solvents of strong elution strength should be avoided as components of the sample solutions.

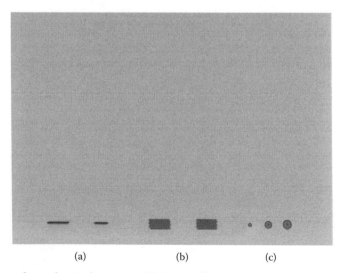

| (a) | (b) | (c) |

FIGURE 8.1 Photo of sample starting zones: (a) bands, (b) rectangles, and (c) spots; sample Eriochrome black T (0.05% w/v solution in methanol), Applicator—ATS-4 (Camag), silica gel plate.

A strong solvent leads to an increase in the starting spot diameter—radial development proceeds during sample application. The solvent volatility is the next solvent property, which is particularly important in respect of sample application. Low volatile solvents elongate their evaporation time from the layer, which can lead to an increase in the width of the starting zone. The solubility of the sample components should be taken into account when the solvent type is considered for the preparation of the sample solution. For a nonpolar solute, a nonpolar solvent is more suitable than a polar one. In addition, the adsorbent type limits the solvent of the sample solution preparation. A sample dissolved in hexane or other nonpolar solvents can be applied onto the silica gel layers, while water or methanol–water sample mixtures are more appropriate for nonpolar adsorbents. However, it should be mentioned that these circumstances cannot be often fulfilled. Then a less suitable solvent must be used for the preparation of the sample solution.

There are some application rules that should be fulfilled regardless of the application method. The flat-ended capillary with a PTFE-coated needle helps to avoid damage of the adsorbent layer. The pipette or syringe needle tip should contact the adsorbent layer at right angle. The whole volume of sample solution ought to be administered at its controlled flow velocity during application. A great flow velocity leads to damage of the layer or sample splashing [3–8].

The main types of sample application are

- Manual (uses contact spot and band application)
- Semiautomatic (uses contact or aerosol spot and band application)
- Fully automatic (uses contact or aerosol spot and band application)
- Nonconventional

A comparison of the characteristics of manual and automatic application modes is presented in Table 8.1.

8.2.1 Manual Sample Application

The manual mode of sample application is still popular in spite of the many advantages of the automatic mode mentioned in Table 8.1. This is mainly because of the considerable costs of the latter. Eppendorf application pipettes, microcapillaries in the holder, or microsyringes can be used as the equipment for this type of sample application. The rule of manual application is the direct spotting of the solute solution onto the adsorbent layer with a suitable tool. Then a solvent is evaporated from the adsorbent layer. This process can be speeded up when a cold or warm air stream flows over the spotted band. The chromatogram development usually follows starting the zone drying process. The main disadvantages of this mode of application are concerned with possible surface damage, nonprecise sample localization on the start line, and problems with automatic zone detection.

TABLE 8.1
Comparison of Application Methods

Manual		Automatic	
Advantage	**Disadvantage**	**Advantage**	**Disadvantage**
Low cost of equipment	High personal qualification	High reproducibility	High cost of equipment
Suitable for qualitative analysis	Heterogonous sample distribution	Indispensable course	
	Chromatographic layer damage by scrapping	Suitable for quantitative analysis	

8.2.2 Semiautomatic Sample Application

In this mode, samples are introduced onto the adsorbent layer at the desired and precise location. This mode usually enables to precisely apply the required volume of the sample solution onto the plate. Minimal or even no damage of the adsorbent layer is another advantage of this application mode. The devices for semiautomatic sample application mode can be simple and sophisticated.

Nanomat 4 (Camag) (Figure 8.2a) is the simplest apparatus [9]. The filled capillary with the sample solution is moved toward the adsorbent layer by manual pressing of the special button in the tool enclosure (shell, housing). The contact of the capillary tip with the adsorbent introduces the dissolved sample with capillary forces onto the chromatographic plate. A solute is applied as a spot. Its quantity depends on the capillary capacity, which varies from 0.5 to 5.0 µL.

Linomat 5 (Camag) is a more complicated device (Figure 8.2b) [10]. This apparatus operates in aerosol mode—the sample solution is sprayed onto the adsorbent layer. However, it requires the external supply of compressed gas (nitrogen or air). The sample solution from the syringe can be sprayed onto the adsorbent layer as spots and/or narrow bands. During the spraying, a solvent of the sample evaporates almost

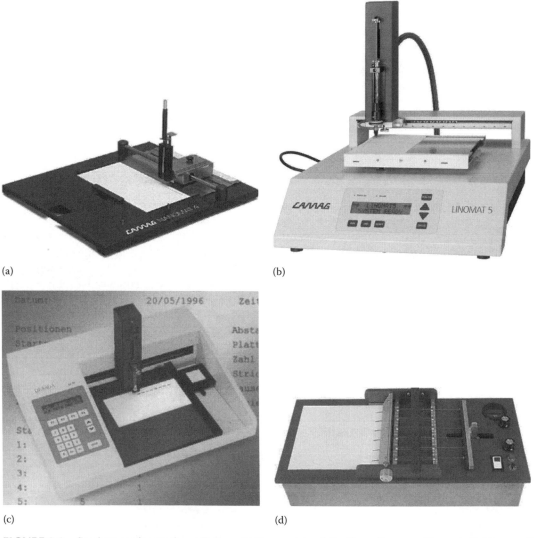

(a)

(b)

(c)

(d)

FIGURE 8.2 Semiautomatic sample applicators: (a) Nanomat 4 and Capillary dispenser (Camag), (b) Linomat 5 (Camag), (c) AS 30 HPTLC Applicator (Sarstedt), and (d) TLC autospotter (Vertical).

entirely, leaving the sample on the adsorbent layer. The operating conditions (e.g., spraying velocity, solute volume, or starting zone shape) might be entered via the keyboard of the device or computer.

The AS 30 HPTLC Applicator, Figure 8.2c, offered by Sarstedt, operates similarly to the previously presented equipment. All parameters for applying up to 30 samples are entered via a keypad [11].

The application of many samples is time-consuming for the devices mentioned previously due to the manual filling of the single microsyringe with the sample solution. The multisyringe system can help to overcome this inconvenience. A few firms such as Analtech, Romer, or Vertical Chromatography offer various models of such equipment (TLC autospotters, Figure 8.2d) [12,13]. The unit can operate up to 18 samples simultaneously. The special needles minimize sample "creep back" and enhance reproducibility of the sample application. The horizontal position of the syringe is another improvement of this equipment. The nozzles of the needles are clamped to the layer surface by the special metal slat. The samples are applied by pressing the syringe pistons by a special device. The apparatus can apply samples at variable rates ranging from 3 min (fastest speed) to 30 min (slowest speed).

Various semiautomatic devices for sample application are presented in Figure 8.2. The main disadvantage of the devices mentioned here is concerned with the manual filling of the capillary or syringe with the sample solution.

8.2.3 Fully Automatic Equipment

AS 30 HPTLC Autosampler (Sarstedt) equipped with the specially designed autosampler indicator (Figure 8.3a) is the device that belongs to this group [11]. This device constitutes the fully automatic system that expands operation possibilities of the AS 30 Applicator mentioned previously. The special software allows entering the method parameters through a dialogue window using a keypad or computer keyboard.

Automatic TLC Sampler 4 (ATS 4, Camag, Figure 8.3b) is a fully automated equipment [14]. Samples can be sprayed onto any adsorbent layer as variously shaped starting zones (spots, bands,

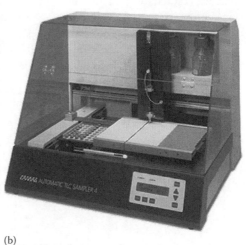

(a)   (b)

FIGURE 8.3 Fully automatic applicators: (a) AS 30 HPTLC Autoapplicator with the specially designed autosampler indicator (Sarstedt), (b) ATS-4 (Camag).

rectangles) and different volumes (from 1 to 500 μL). Syringes of various capacities (10, 25, 100 μL) can be applied to this device. This device enables one to apply sample solutions from different vials at the same zone on the adsorbent layer. The apparatus possesses the predosage system, which generates the reproducible conditions at the needle tip. The heated spray nozzle facilitates the application of large sample volumes diluted in a low-volatile solvent. After the sample application process is finished, the syringe is emptied into a waste recipient. Before the next sample application, the syringe is cleaned up with the rinsing solvent, which is collected into the waste recipient.

8.2.4 NONCONVENTIONAL METHOD OF APPLICATION

It is worth mentioning a solid-phase sample application mode, which is a nonconventional sample application technique. A sample is dissolved in the solvent, and then an inert adsorbent is added to this solution and mixed. After solvent evaporation, the adsorbent with the sample deposited on the surface is introduced into the special channel produced in the adsorbent layer of a chromatographic plate. After this operation, chromatogram development can be run in the conventional mode. This sample application mode is especially suitable for preparative chromatography of nonvolatile samples [15].

8.3 MODES OF CHROMATOGRAM DEVELOPMENT

One may distinguish two basic modes of thin-layer chromatogram development in contemporary laboratory practice: linear and radial developments. The latter can be divided into two modes: circular and anticircular ones. In Figure 8.4, the principle of each development mode is presented. As can be seen in linear development, the mobile phase migrates along a chromatographic plate, and the solvent front is of the same width during the whole separation process. In circular development, the chromatographic plate is fed with eluent solution in its central part, and the solvent front migrates from the center to the periphery of the chromatographic plate. The reverse direction of the solvent front migration features anticircular development. The linear development mode is the most popular in laboratory practice so far. However, the circular mode has its enthusiasts, for example, Prof. Kaiser, who often presents convincing examples of circular development especially devoted to laboratories of low budget [16,17]. Each mode of chromatogram development mentioned here can proceed under conditions of vapor saturation of the TLC chamber atmosphere (conditioning) or without chamber saturation (without conditioning). The former mode of chromatogram development can be performed when the distance between an adsorbent layer and a wall of the developing chamber is greater than 3 mm. If this distance is equal to or smaller than 3 mm, then chromatogram development can be performed under nonsaturated conditions. Conditioning plays an important role when two or more components constitute the mobile phase. When one component stands for the mobile phase, conditioning is not necessary; however, it can be performed in particular cases, for example, when the humidity of the chamber atmosphere and the adsorbent layer has to be adjusted [18].

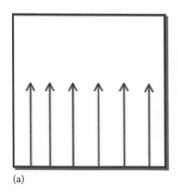
(a)
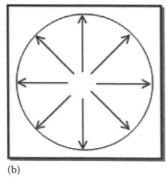
(b)
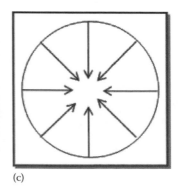
(c)

FIGURE 8.4 Modes of chromatogram development: (a) linear, (b) circular, and (c) anticircular.

8.3.1 CHAMBER TYPES

A few criteria for chamber classification can be distinguished in respect of inner space of TLC chamber, chromatographic plate position in the chamber, and degree of automation.

8.3.1.1 N-Chambers

In respect of chamber space, one can differentiate N-chambers (normal, conventional chambers) and S-chambers (sandwich chambers). The former are characterized by a relatively large volume of gas phase above the adsorbent layer of the chromatographic plate. In the latter the distance between the adsorbent layer and the chamber wall is no larger than 3 mm, as mentioned previously. The N-chambers are probably the most popular chambers in chromatographic practice. They are manufactured as cylindrical or rectangular vessels. A few examples of such chambers are presented in Figure 8.5. Twin-trough cuboid shaped chamber belongs to the same chamber type (Figure 8.6). This chamber is equipped with two troughs located at its bottom. Each trough can be filled up with the same or different solutions, which can be simultaneously and/or consecutively used as conditioning reagent, eluent, and/or humidity control medium [18]. In this way, the twin-trough chamber offers more methodological possibilities of chromatogram development than conventional N-chambers.

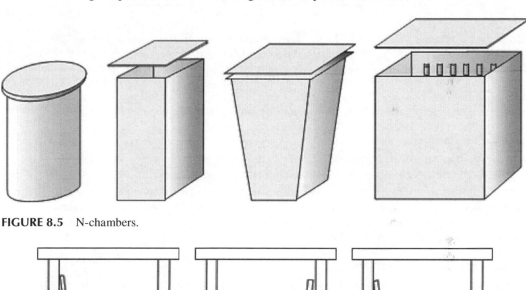

FIGURE 8.5 N-chambers.

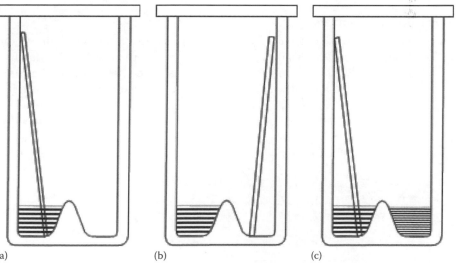

(a)   (b)   (c)

FIGURE 8.6 The twin-trough chamber with variants of chromatogram development: (a) with no saturation with vapors, (b) chromatogram development follows conditioning process, and (c) chromatogram development proceeds under humidity-controlled conditions.

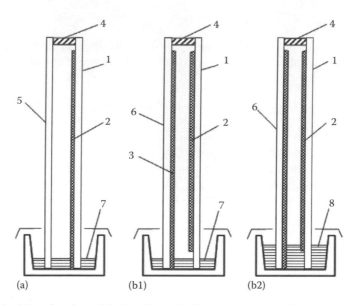

FIGURE 8.7 Principle of action of S-chamber: (a) chromatogram development without conditioning and (b) chromatogram development with conditioning (b1)—wetting of adsorbent layer of the counter chromatographic plate, (b2)—chromatogram development, (1) carrier plate of the chromatographic plate, (2) adsorbent layer, (3) adsorbent layer of the counter chromatographic plate, (4) plug, (5) cover plate, (6) carrier plate of the counter chromatographic plate, (7) solvent as the mobile phase (a), and solvent used for conditioning (b1), (8) solvent used as the mobile phase after conditioning process (b2).

8.3.1.2   S-Chambers

In Figure 8.7 the sandwich chamber is presented [19]. In the contemporary market, Analtech offers a vertical type of this chamber [20]. These chambers are characterized by relatively small solvent consumption because they are mainly applied to chromatogram development without conditioning the chamber space. However, these chambers can be also operated with the conditioning mode. Methodological possibilities of chromatogram development with conditioning and without satura-tion of chamber space are presented in Figure 8.7 [21].

8.3.1.3   Horizontal Chambers

As mentioned previously, the developing chamber can be divided in respect of chromatographic plate location. In the contemporary market, two chamber types can be distinguished in this regard, that is, in horizontal and vertical chambers, the chromatographic plate is located horizontally or vertically, respectively. In N-, S-, and twin-trough chambers, the chromatographic plate is usually positioned almost vertically. So these chambers belong to vertical ones as well. In Figures 8.8 and 8.9, examples of horizontal chambers are presented. First, the horizontal developing chamber (Camag) should be mentioned (Figure 8.8) [22,23]. This chamber is devoted to glass plates and manufactured for two plate dimensions, 10×10 cm and 20×10 cm. In the horizontal DS chambers (Chromdes)

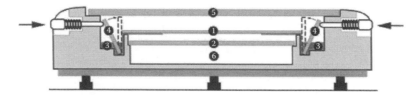

FIGURE 8.8 Horizontal developing chamber: 1—chromatographic plate with layer face down, 2—counter plate (removable), 3—troughs for solution of the mobile phase, 4—glass strip for transfer of the mobile phase by capillary action to the chromatographic plate, 5—cover glass plate. (From www.camag.com)

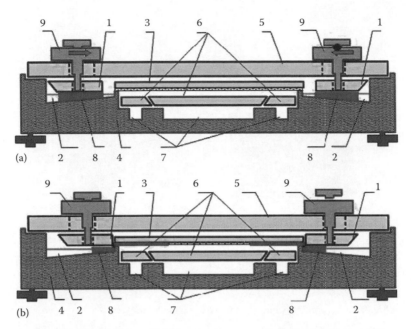

FIGURE 8.9   Horizontal DS-II Chamber: (a) before development, (b) during development; 1—cover plate of the mobile-phase reservoir, 2—mobile-phase reservoir, 3—chromatographic plate with layer face down, 4—body of the chamber, 5—main cover plate, 6—cover plates (removable) of the troughs for vapor saturation, 7—troughs for saturation solvent, 8—mobile phase, 9—mobile-phase distributor/injector. (From www.chromdes.com)

(Figure 8.9), glass- and foil-backed plates can be used [24–26]. These chambers are produced in seven different types for plate dimensions from 5 × 10 cm to 20 × 20 cm. It should be mentioned that the horizontal chambers have been gaining growing laboratory application due to their many methodological possibilities. These methodological possibilities have been described in several studies [24,25]. The main advantage of the horizontal developing chambers and the horizontal DS chambers is that these enable the chromatogram development of a double number of samples on one plate in comparison to that with conventional chambers. This is due to feeding the chromatographic plate with a solvent from its two opposite edges simultaneously; compare Figures 8.8 and 8.9. In one of these figures, Figure 8.9, the horizontal DS chamber is shown during chromatogram development from two opposite edges. Minimal consumption of solvents is another advantage of these chambers. These chambers can operate with conditioning and without conditioning of the chamber atmosphere, so they can be regarded as universal chambers. Some authors claim that the horizontal configuration of the chromatographic plate leads to an increase in the flow velocity of the mobile phase; however, it seems to be of minor relevance to the efficiency of separation in this case. Other horizontal chambers should be also mentioned in this section, that is, H-separating chamber (Sarstedt) [27], which is manufactured for 5 × 5 cm and 10 × 10 cm plates. This chamber enables chromatogram development from one side of the chromatographic plate, and it also features various methodological possibilities [28]. Based on the scientific literature, another horizontal chamber has been reported; however, these have not been launched on the market so far [29–33].

8.3.1.4   Chambers for Circular Development

Professor Kaizer popularizes circular chromatogram development as mentioned here. The chromatographic plate in the chambers for such a chromatogram development is horizontally located, so the chamber belongs to horizontal ones as well. This chamber can be easily produced at home. In Figure 8.10, a simple device for circular development is presented. This chamber is commercially available from Analtech [20]. In circular development, components of sample mixtures showing low retardation factor values can be separated with higher efficiency in comparison to linear development [21].

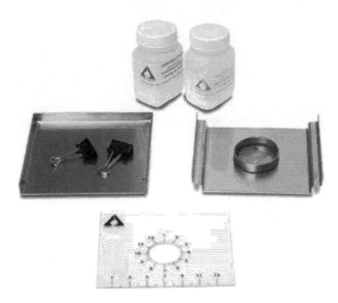

FIGURE 8.10   Circular developing system. (From www.analtech.com)

8.3.1.5   Automatic Developing Chambers

Chromatogram development can be currently performed with a few chamber types, which are characterized by various degrees of automation. A glass rectangular chamber equipped with a solvent front detector is the simplest one. The operator can stop the chromatogram development when the device gives acoustic and/or light signals when the front of the mobile phase approaches the desired migration distance. More sophisticated chambers enable one to perform the development process with no participation of the operator. One example is the chamber named as "Baron DC-Lift mit Frontdetektion" (Figure 8.11) [34]. Chromatogram development in this chamber starts when the chromatographic plate edge is dipped automatically in

FIGURE 8.11   Automatic developing chamber, "Baron DC-Lift mit Frontdetektion." (From www.baron-lab.de)

FIGURE 8.12   Automatic developing chamber, ADC 2. (From www.camag.com)

the mobile-phase solution. The chamber enables conditioning the adsorbent layer with mobile-phase vapors for a selected time before chromatogram development. Automatic removing of the chromatographic plate from the chamber, when the mobile phase reaches the desired migration distance, breaks the chromatogram development. This chamber can be optionally equipped with a humidity control facility.

Camag produces a fully automatic device named as an automatic developing chamber, ADC 2 Figure 8.12 [35]. The equipment uses a twin-trough chamber for chromatogram development. So the device enables one to automatically perform the conventional procedure of chromatogram development with no influence of the operator and environmental conditions. In addition, the chamber can operate under a desired humidity of the atmosphere. All operating conditions (conditioning time, distance of solvent front migration, drying time of the plate, and humidity of the chamber atmosphere) can be introduced via a keypad and/or computer keyboard applying a special software. In this way, the chamber can be especially applied to the routine analysis of various samples under established and reproducible conditions.

Camag also produces the most sophisticated equipment in the contemporary market, the AMD 2 system, which is devoted to automatic chromatogram development under the gradient of the mobile phase, Figure 8.13 [36]. The principle of action of this device is based on the procedure previously worked out by Burger [37]. Chromatogram development can be repeatedly performed. Each successive development follows an evaporation of the mobile phase from the previous development, and in each next development step the solvent front migrates a longer distance and is of lower elution strength. The number of steps can exceed 20. Narrow peaks characterize the final chromatogram, so the number of separated zones can be considerably higher than when conventional development is used. As specified by the manufacturer, the typical bandwidth is equal to about 1 mm. So about 40 components can be completely separated if the development distance reaches 80 mm. All operations with the AMD 2 system can be performed via a computer keyboard using a special software.

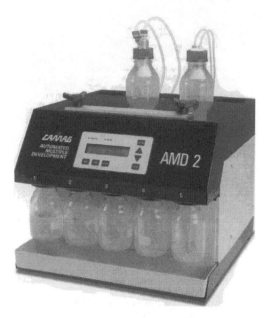

FIGURE 8.13 Equipment for automated multiple development of chromatograms, AMD 2. (From www. camag.com)

8.4 SPECIAL AND NEW DEVICES FOR CHROMATOGRAM DEVELOPMENT

In spite of the many advantages of TLC, some weak points characterize this technique. These make its application quite limited, especially in comparison to high-performance column and capillary techniques. Despite introducing high-performance layers for planar chromatography, its performance still remains worse than that of high-performance liquid chromatography (HPLC), capillary electrophoresis (CE), or capillary electrochromatography (CEC). One of the major problems with reference to TLC is the poor flow characteristic of the mobile phase. The solvent migrates fast only at the beginning of the chromatogram development, and its flow gradually decreases with the progress of the separation process. This results in a prolongation of the separation time and an increase of band broadening [38,39].

To overcome these disadvantages, forced-flow planar techniques were introduced. Overpressured layer chromatography (also called optimum performance laminar chromatography, OPLC) is the technique in which the adsorbent layer of the chromatographic plate is covered with a flexible membrane, which is pressed to it under high external pressure. This gives a closed system, analogous to that in column chromatography. Thanks to this, a solvent mixture can be forced to flow with a high-pressure pump, for example, HPLC pump, through the stationary phase bed. This results in a relatively rapid flow and high performance of the separating system, and enables one to shorten the separation process [40]. A conceptual view of an OPLC chamber is shown in Figure 8.14. There are many modes of chromatogram development with this device—linear, two-dimensional, circular, single- or multilayer, on-line or off-line, etc. [40].

The firm named OPLC-NIT (Budapest, Hungary) manufactures an OPLC device, which is commercially available. The automated OPLC system (Figure 8.15) is equipped with a separation chamber, a liquid delivery system controlled by a microprocessor, and a dedicated software. It gives the possibility of setting various parameters of the separation process (such as volume and flow velocity of the mobile phase, external pressure). Development time is calculated automatically. The device is able to withstand pressure up to 5 MPa and allows one to run isocratic or stepwise gradient separations [40].

OPLC has found many applications, inter alia, in drug analysis. The literature reports its use in the separation of antibiotics (e.g., tetracyclines, penicillines, cephalosporins), benzothiazepines, barbiturates, numerous natural plant compounds (coumarines, flavonoids), and many others [41].

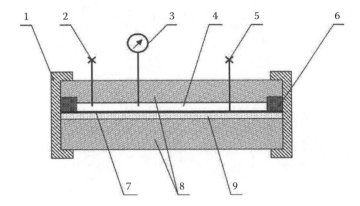

FIGURE 8.14 Conceptual view of OPLC chamber: 1—fastener, 2—gas or water inlet, 3—manometer, 4—space filled with pressured gas or water, 5—mobile phase inlet, 6—rubber, 7—flexible membrane covering chromatographic plate, 8—support blocks, and 9—chromatographic plate. (Adapted with kind permission from Springer Science+Business Media: *Planar Chromatography*, Overpressured-layer chromatography [optimum performance laminar chromatography] [OPLC], 2001, 137, Tyihak, E. and Mincsovics, E.)

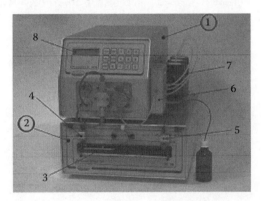

FIGURE 8.15 Automated OPLC instrument: 1—liquid delivery system, 2—separation chamber, 3—chromatographic plate cassette, 4—mobile phase inlet, 5—mobile phase outlet, 6—mobile phase switching valve, 7—mobile phase reservoir, and 8—LCD display. (Reproduced from *J. Chromatogr. A*, 1232, Tyihak, E., Mincsovics, E., and Móricz, A.M., Overpressured layer chromatography: From the pressurized ultramicro chamber to BioArena system, 3–18, Copyright 2012, with permission from Elsevier.)

In planar chromatography, the mobile phase can also be driven into movement by an electric field, which generates electroosmotic flow. This phenomenon is widely used in capillary electrophoresis and electrochromatography. Electroosmotic flow is characterized by a flat profile, in contrast to the laminar profile of mobile-phase flow driven by hydrostatic pressure (as in HPLC or OPLC). This results in higher performance of CE and CEC techniques relative to the HPLC and TLC/HPTLC ones. Additionally, the selectivity of electromigration methods is principally different from that of liquid chromatography. Therefore, electrochromatography and electrophoresis offer additional separation potential [42].

In 1974, Pretorius et al. for the first time applied an electric field to drive the mobile phase into flow in a planar chromatography system [43]. In 1994, Prosek and coworkers introduced the name planar electrochromatography (PEC) for a separation process in an open thin-layer chromatography system [44]. However, Nurok et al. for the first time used a closed planar chromatography system under an electric field. The authors named this technique as pressurized planar electrochromatography (PPEC) [45].

Since then, research on the development of prototypes of PPEC devices and conditions of substance separation by this method has been carried out by three research groups [46–48]. The device presented by Nurok et al. enables one to perform PPEC with a chromatographic plate in a vertical

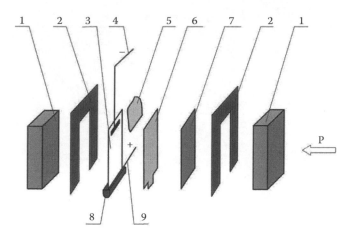

FIGURE 8.16 Expanded scheme of elements of Nurok's et al. PPEC chamber: 1—die blocks, 2—polyacetal frame, 3—chromatographic plate, 4—cathode, 5—paper wick, 6—Teflon foil, 7—ceramic sheet, 8—mobile phase reservoir, and 9—anode. (Adapted from Adapted from *J. Chromatogr. A*, 1218, Dzido, T.H., Płocharz, P.W., Chomicki, A., Hałka-Grysińska, A., and Polak B., Pressurized planar electrochromatography, 2636–2647, Copyright 2011, with permission from Elsevier.)

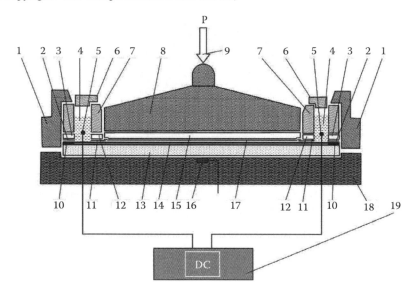

FIGURE 8.17 Longitudinal cross section of the PPEC chamber worked out by our group chamber; 1—body of the chamber, 2—Teflon foils, 3—flexible rubber, 4—electrode compartments filled with mobile phase, 5—electrodes, 6—electrode compartment covers, 7—partitions, 8—chamber cover, 9—external hydraulic press, 10—margins of adsorbent layer, 11—Teflon foil, 12—flexible rubber, 13—chromatographic plate, 14—adsorbent layer, 15—flexible rubber, 16—temperature sensor, 17—Teflon foil, 18—base of the chamber, and 19—high-voltage power supply. (Adapted from Adapted from J. Chromatogr. A, 1218, Dzido, T.H., Płocharz, P.W., Chomicki, A., Hałka-Grysińska, A., and Polak B., Pressurized planar electrochromatography, 2636–2647, Copyright 2011, with permission from Elsevier.)

position, Figure 8.16 [46]. On the other hand, the device worked out by our group operates with a chromatographic plate positioned in a horizontal configuration, Figure 8.17 [47]. The studies on PPEC report that this technique is characterized by a few important advantages: separation performance is comparable to HPLC, separation time is considerably shorter than that of TLC/HPTLC, separation selectivity is different in relation to liquid chromatography and electrophoresis techniques [42].

In spite of the advantages mentioned here. this technique suffers some inconveniences and challenges in relation to others. At first, it is necessary to coat the edges of a chromatographic plate with a special sealant [46,47] (but it is necessary in OPLC as well [40]). The sealant margins obtained have to form a hermetic region, which prevents the mobile phase flows outside the chromatographic plate. This means that in PPEC an additional procedure of chromatographic plate preparation is necessary. Anyway, elaboration of proper routines and their automation may resolve this inconvenience. Chromatographic plates used in PPEC come from the market and are devoted to thin-layer chromatography. These plates do not generate efficient electroosmotic flow of the mobile phase. The next inconvenience of the PPEC prototypes is that an operator has to perform tedious manual operations, which are necessary to run the separation process. In other words, PPEC equipment is still not sufficiently familiar enough for the operator. Anyway, if the devices are further developed and some operating inconveniences are overcome, there is no doubt that they can find wide application in laboratory practice. Future successful introduction of dedicated plates to PPEC can facilitate the commercialization and promotion of this technique too.

It should be mentioned that one study has been published on PPEC application to drug analysis (mixture of acetaminophen, propyphenazone, and caffeine) so far [49]. The technique was successfully used for the separation of other compounds such as dyes [47], amino acids [50], peptides, and oligonucleotides [51]. This makes PPEC a potent method for future application in the field of drug analysis.

REFERENCES

1. Makuch, B. 2004. Chromatografia cienkowarstwowa. In *Chromatografia cieczowa*. M. Kamiński and R. Kartanowicz (eds.), pp. 174–191, Gdańsk, Poland: Politechnika Gdańska, CEEAM.
2. Mijagi, A. and Nakajima, M. 2003. Regeneration of used frying oils using adsorption processing. *J. Am. Oil Chem. Soc.* 80:91–96.
3. Omori, T. 2001. Modern sample application methods. In *Planar Chromatography, A Retrospective Review for the 3rd Millennium*. Sz. Nyiredy (ed.), pp. 120–136, Budapest, Hungary: Springer.
4. Sherma, J. 2003. Basic TLC techniques, materials and apparatus. In *Handbook of Thin-Layer Chromatography*, 3rd edn., revised and expanded. J. Sherma and B. Fried (eds.), pp. 1–21, New York: Marcel Dekker.
5. Morlock, G.E. 2006. Sample application and chromatogram development. In *Preparative Layer Chromatography*. T. Kowalska and J. Sherma (eds.), pp. 100–129, Boca Raton, FL: Taylor & Francis Group.
6. Spandenberg, B., Pole, C.F., and Weins, C. 2011. *Quantitative Thin-Layer Chromatography, A Practical Survey*. Berlin, Germany: Springer-Verlag.
7. Hahn Deinstrop, E. 2007. *Applied Thin-Layer Chromatography. Best Practice and Avoidance of Mistakes*. 2nd revised and enlarged edition. Weinheim, Germany: Willey-VCH.
8. Wall, P.E. 2005. *Thin-Layer Chromatography. A Modern Practical Approach*. Cambridge, U.K.: The Royal Society of Chemistry.
9. Camag. 2005. *Basic Equipment for Modern Thin-Layer Chromatography*. Muttenz, Switzerland: Camag.
10. Camag. 2011. *Linomat 5. Bandwise Sample Application by Spray-On Technique for Planar Chromatography*. Muttenz, Switzerland: Camag.
11. Saerstedt. 2011. http://www.sarstedt.com
12. Romerlabs. 2011. http://www.romerlabs.com
13. Vertichrom. 2011. http://www.vertichrom.com
14. Camag. 2011. *Automatic TLC Sampler 4. Setting New Standards in Planar Chromatography*. Muttenz, Switzerland: Camag.
15. Botz, L., Nyiredy, Sz., and Sticher, O. 1990. A new solid phase sample application method and device for preparative planar chromatography. *J. Planar Chromatogr.* 3: 10–14.
16. Kaiser, R.E. 2005. Is the future of analytical quantitative PLC CIRCULAR. In *Proceedings of the International Symposium on Planar Separations, Planar Chromatography 2005*, Sz. Nyiredy (ed.), pp. 39–41, Siofok, Hungary, May 29–31, 2005.
17. Kaiser, R.E. 2008. *Quantitative Micro Planar Chromatography—MPLC, International Symposium for High Performance Thin-Layer Chromatography*, Helsinki, Stockholm, June 11–13, 2008.
18. Geiss, F. 1988. The role of vapor phase in planar chromatography. *J. Planar Chromatogr.* 1: 102–115.
19. Jänchen, D. 1968. The apparent influence of layer thickness on Rf values of thin-layer chromatograms. *J. Chromatogr.* 33: 195–198.
20. Analtech. 2011. www.analtech.com

21. Geiss, F. 1987. *Fundamentals of Thin Layer Chromatography (Planar Chromatography)*. Heidelberg, Germany: Dr. Alfred Hüthig Verlag.
22. Camag. 2011. http://www.camag.com/v/products/development/horizontal.html
23. Reich, E. 2003. Instrumental thin-layer chromatography (planar chromatography). In *Handbook of Thin-Layer Chromatography*, 3rd edn., revised and expanded. J. Sherma and B. Fried (eds.), pp. 135–152, New York: Marcel Dekker Inc.
24. Dzido, T.H. and Soczewiński, E. 1990. A new modification of horizontal sandwich chamber for thin-layer chromatography. *J. Chromatogr.* 516: 461–466.
25. Dzido, T.H. 2001. Modern TLC chambers. In *Planar Chromatography, A Retrospective View for the Third Millennium*. Sz. Nyiredy (ed.), pp. 68–87, Budapest, Hungary: Springer Scientific Publisher.
26. Chromdes. 2011. www.chromdes.com
27. Saerstedt. 2011. http://www.sarstedt.com/php/main.php
28. Kraus, L. 1996. *Concise Practical Book of Thin-Layer Chromatography*. pp. 79–85, Berlin, Germany: Springer Verlag.
29. Soczewiński, E. 1986. Equilibrium sandwich TLC chamber for continuous development with a glass distributor. In *Planar Chromatography*, R.E. Kaiser (ed.), Vol. 1, pp. 79–118, Heidelberg, New York: Verlagsgruppe Huthig, Jehle Rehm, GmbH.
30. Berezkin, V.G. and Khrebtova, S.S. 2011. The chromatographic process in the S-chamber with the counter plate. *J. Chromatogr. A* 201: 8273–8280.
31. Berezkin, V.G. and Chausov, A.V. 2012. The simple chromatographic chamber and its application in circular and linear TLC. *J. Liq. Chromatogr. Related Technol.* 35: 294–307.
32. Zarzycki, P.K. 2008. Simple horizontal chamber for thermostated micro-thin-layer chromatography. *J. Chromatogr. A.* 1187: 250–259.
33. Wang, Y., Wang, D. et al. 2004. A new instrument for automated multiple development in thin-layer chromatography. *J. Planar Chromatogr.* 17: 290–296.
34. Baron-de. 2011. www.baron-lab.de
35. Camag. 2011. http://www.camag.com/v/products/development/adc2.html
36. Camag. 2011. http://www.camag.com/v/products/development/amd2.html
37. Burger, K. 1984. PC-PMD, Dünnschichtchromatographie mit Gradienten-Eluation im Vergleich zur Säulenflüssigkeitschromatographie. *Fresenius Z Anal. Chem.* 318: 228–233.
38. Nurok, D. 2004. Planar electrochromatography. *J. Chromatogr. A* 1044(1–2): 83–96.
39. Kemsley, J.N. 2009. Modernizing TLC. *Chem. Eng. News* 87(20): 11–18.
40. Tyihak, E., Mincsovics, E. et al. 2012. Overpressured layer chromatography: From the pressurized ultra-micro chamber to BioArena system. *J. Chromatogr. A* 1232: 3–18.
41. Tyihak, E. and Mincsovics, E. 2001. Overpressured-layer chromatography (optimum performance laminar chromatography) (OPLC). In *Planar Chromatography*. Sz. Nyiredy (ed.), pp. 137–176, Budapest, Hungary: Springer Scientific Publisher.
42. Dzido, T.H., Polak, B. et al. 2011. Planar chromatography using electroosmotic flow. In *Encyclopedia of Analytical Chemistry: Applications, Theory and Instrumentation*. R.A. Meyer (ed.), pp. 1–18, Chichester, U.K.: Wiley.
43. Pretorius, V., Hopkins, B.J. et al. 1974. Electro-osmosis: A new concept for high-speed liquid chromatography. *J. Chromatogr. A* 99: 23–30.
44. Pukl, M., Prosek, M. et al. 1994. Planar electrochromatography Part 1. Planar electrochromatography on non-wetted thin-layers. *Chromatographia* 38: 83–87.
45. Nurok, D., Koers, J.M. et al. 2004. Apparatus and initial results for pressurized planar electrochromatography. *Anal. Chem.* 76(6): 1690–1695.
46. Novotny, A.L., Nurok, D. et al. 2006. Results with an apparatus for pressurized planar electrochromatography. *Anal. Chem.* 78(8): 2823–2831.
47. Dzido, T.H., Plocharz, P.W. et al. 2011. Pressurized planar electrochromatography. *J. Chromatogr. A* 1218(19): 2636–2647.
48. Tate, P.A. and Dorsey, J.G. 2006. Linear voltage profiles and flow homogeneity in pressurized planar electrochromatography. *J. Chromatogr. A* 1103(1): 150–157.
49. Hałka-Grysińska, A., Ślązak, P. et al. 2012. Simultaneous determination of acetaminophen, propyphenazone and caffeine in cefalgin preparation by pressurized planar electrochromatography and high performance thin-layer chromatography. *Anal. Methods* 4: 973–982.
50. Chomicki, A., Kloc, K. et al. 2011. Two-dimensional separation of some amino acids by HPTLC and pressurized planar electrochromatography. *J. Planar Chromatogr.* 24(1): 6–9.
51. Woodward, S.D., Urbanova, I. et al. 2010. Separation of peptides and oligonucleotides using a monolithic polymer layer and pressurized planar electrophoresis and electrochromatography. *Anal. Chem.* 82(9): 3445–3448.

9 2D Planar Chromatography

Huba Kalász

CONTENTS

9.1 INTRODUCTION

It is an astonishing sight when a 20 × 20 cm plate has tens or even hundreds of colorful spots. Its qualitative and quantitative evaluation may give useful information on the composition of the active ingredients in certain medications, on the pesticide contamination of soil sample, etc. As high-performance liquid chromatography (HPLC), capillary zone electrophoresis (CZE), and other chromatographic procedures have progressed, so has planar chromatography. Sophisticated methods have been established for the development of the plates, for the quantitative evaluation of the separated spots, and also for the identification of the compounds by online UV–visible spectra, mass spectrometry (MS), and nuclear magnetic resonance spectroscopy, and the available information was further enhanced by 2D planar chromatography (2D-thin-layer chromatography [2D-TLC]).

Multidimensional separations can be carried out in two or more steps. A 2D-TLC is the case when the separation may be easily and inexpensively realized. At the same time, it is the most documented and most frequently used method among the multidimensional separations.

A 2D-TLC represents advancement in planar chromatography. It started with 2D paper chromatography, which included isoelectric focusing combined with electrophoresis. The progress of TLC permitted to substitute the paper stationary phase with a wide choice of sorbent, such as cellulose, silica, alumina, and polyamide. However, the trend of separation science goes in the direction of very specific and very sensitive detections, and only either the MS or the tandem MS fulfills it. Therefore, the future prospects promise the multidimensional liquid (column) chromatography combined with MS to separate a wide choice of the solutes including proteins and peptides of proteomics. This chapter outlines the presence of the 2D separations.

The practice of 2D-TLC is utilizing both capillary-controlled flow and forced flow of the mobile phase. The capillary-controlled flow is characterized by its simplicity. The development is fast enough in the majority of the generally used mobile phases. The procedure is surprisingly simple. The samples are loaded onto the plate with a syringe or using one of the sample application devices. One of the parameters is the shape of application that can be either a circularly shaped spot or a band with a definite length. The plate is developed in a chamber, generally made of all glass with glass lid. The results of TLC depend on the chamber geometry, the saturation of the vapor phase, etc.

After the age of prechromatographic procedures, Tswett was the first to use the method we today call chromatography. He used 1D liquid chromatographic separations of various samples, having the stationary phases packed into columns: liquid in the name of the technique refers to the condition of the mobile phase. Since then, the conditions and parameters of chromatography widened: in addition to a liquid mobile phase also can be either a gas or a supercritical fluid. In addition to packed in a column, the stationary phase can also be distributed on a planar surface. Additional variables can be either the speed or the pressure.

Izmailov and Schraiber were the first who described in 1938 the chromatographic separations on a planar stationary phase representing a thin layer of particles [1]. The most striking features of TLC are as follows:

- The entire chromatographic process can generally be followed visually.
- Separation of the colored spots can be visually followed through the process.
- The method is fast.
- The method can be validated by the use of the adequate instrumentation.
- Using stationary phases having fluorescent additives, the majority of the separated spots can be located under the UV light at 254 nm.
- A wide scale of specific and sensitive color reagents serve to improve the identification of the spots of the sample components.

9.2 HISTORICAL OVERVIEW

The 2D separations may be easily arranged: the plate is developed, dried, turned by 90°, and immersed into the mobile phase for the second dimensional development.

The properly done 2D-TLC essentially improves the separation including both the resolution and the number of separable components. This improvement was recognized relatively early, as Liesegang [2] initiated double development in the so-called capillary analysis in two directions as early as in 1943. The method was called then "Kreuzkapillaranalyse," meaning cross-capillary analysis. In chromatography, Martin's group [3] used first 2D development in paper chromatography.

Their intention originated from the necessity to separate basic and acidic amino acids (diamino monocarboxylic acids and monoamino dicarboxylic acids), which was not possible in column chromatography using silica gel. When the silica stationary phase was substituted with cellulose or, more exactly, with paper, it was easy to carry out 2D developments. Kollidon (a mixture of dimethyl homologues of pyridine) and phenol–water–ammonia mixture were used for the first and second dimensional developments, respectively.

In paper chromatography, the moving forces can be varied: capillarity and electric field were used alternatively. Paper chromatography combined with paper electrophoresis was called "making a fingerprint," indicating the high efficiency of the method [4].

Changing the pH between the first and second dimensional runs is an interesting example of the effective separation by 2D developments. Organic acids were separated using pH 4.5 and 8.9 in the first and second dimensional developments, respectively. Both the acidic and complex characteristics of the organic acids were utilized as the basis of the separation [5].

Poole and Poole published a review on multidimensional chromatography. Only a relatively short part of the paper (10%) was dealing with multidimensional (2D) planar chromatography, and even their experimental chromatograms [6] indicate separations of only either one or two groups arranged along one or two lines. However, real 2D separation has separated spots over the majority of the plate.

The 2D-TLC is the most frequently used technique in 2D developments. The generation of altering differentiation mechanisms in the consecutive developments is possible by the use of the following possibilities:

- Using a heterogeneous stationary phase. One strip represents one of the stationary phases, while the remaining stationary phase represents the other one (here, the second stationary phase may be straight [normal], RP, ion exchanger, etc.).
- Using altering mobile phases in the first and in the second dimensional developments, generating basically different separation mechanisms. These mobile phases may differ from each other in their pH, absence or presence of water content, absence or presence of ion-pairing additives, generating elution or displacement mode of development, etc.
- Carrying out a reaction between the first and second dimensional developments.

9.3 METHODOLOGICAL ASPECTS

9.3.1 STATIONARY PHASES

There are excellent reviews that overview stationary phases used for TLC [7,8]. At the same time, an essential part of TLC is done using silica stationary phase. There are several practical reasons for this, including the relatively low price of plain silica. A 2D separation sometimes means that different stationary phases are used in the consecutive (successive) developments.

Khatib et al. [9] used silica gel stationary phase for the estimation of the presence and quantity of compounds having anti-tracheospasmolytic active compounds in the leaves of *Vitex trifolia* (L.). The mobile phases were chloroform–methanol (9:1) in the first dimensional development followed by ethyl acetate–chloroform–methanol (28:28:44) in the second dimension.

Tuzimski [10] used 2D-TLC on a moderate polarity CN-modified silica gel. Eighteen pesticides were separated on that stationary phase, which works both in normal-phase (NP) and reversed-phase (RP) mode of separations to offer large selectivity differences. A combination of nonaqueous NP mobile phases (tetrahydrofuran or ethyl acetate in n-heptane) and aqueous RPs (a polar solvent [methanol or acetonitrile] in water) was arranged on the consecutive developments.

Tuzimski and Wojtowicz [11] used a stationary phase combination of an NP and an RP on a plate of dual-adsorbent layers containing a narrow zone of silica and adjacent to a wide zone of RP (octadecyl silica), or vice versa.

Tuziminski [12] used 2D-TLC with stationary phase gradients for the separation of the same mixtures of pesticides.

Petruczynik et al. [13] used 2D-TLC on different stationary phases to separate alkaloids or plant extracts (*Chelidonium majus*, *Fumaria officinalis*, or *Glaucium flavum*) using both NP and RP systems. The most selective systems are used for the separation of the alkaloid mixtures by a 2D-TLC with an adsorbent gradient method.

Hawryl and Waksmundzka-Hajnos [14] used cyano-bonded polar stationary phases for the separation of phenolic compounds extracted from *Polygonum hydropiper* L. and *Polygonum cuspidatum* L. Nonaqueous solvents were used in the first direction, and aqueous solvents were used in the second directional development.

Ilić et al. [15] used 2D-TLC on RP-18 silica plates to separate phenols. Optimum efficiency was found when aqueous mobile phase (first dimensional development) was followed by a nonaqueous mobile phase (second dimensional run).

However, there are also chromatographic reasons for this selection. The separation on silica stationary phase (NP chromatography) completes well the separation on an octadecyl-silica phase used in RP separation.

A 2D separation has its own requirements and rules. Usually, the basic requirement is to use the nonaqueous mobile phase in the first dimensional run and the water-containing mobile phase in the second dimensional development. The dominating procedure is adsorption between the water-free mobile phase and the silica or alumina. However, the water from any aqueous mobile phase is retarded by the stationary phase, remaining there even after drying the plate on air.

This phenomenon is utilized in the separation of certain steroids, turning around the two mobile phases: now the water is retained by the stationary phase when using water-containing mobile phase in the first dimension, the water adsorbed on the stationary phase improves the separation during the second dimensional development.

A wide choice of ready-made TLC plates is supplied by a number of manufacturers with sizes varying from 5 × 5 cm to 20 × 20 cm, and certain stationary phases are also supplied in the form of a roll.

9.3.2 CHAMBERS

A 2D-TLC can be carried out in any of the most widely applied glass chambers. The plate is developed in the first dimension, perfectly dried, and developed in the second dimension—after turning it by an angle of 90° relative to the first dimension.

Berezkin and Kormishkina [16] constructed a 2D-TLC system with a semiclosed stationary phase (adsorbent) layer. An ascending version of the setup used a closed stationary phase that considerably decreased (by about 55%) the time for development, when a 5 × 5 cm plate for 2D separation is compared to a single dimensional one using a 10 × 10 cm plate.

Zarzycki and Zarzycka [17] used a small, temperature-controlled horizontal micro-TLC to separate and quantify testosterone, methyltestosterone, testosterone propionate, isobutyrate, phenylpropionate, isocaproate, enanthate, and caprate. The elution distance was 45 mm using a wide variety of methanol–water, acetonitrile–water, methanol–dichloromethane, and acetone–hexane mobile phases. They also studied the effect of various temperatures (from −20°C to +60°C), and the separations were also done in both saturated and unsaturated chambers. Their studies indicated the strong dependence on mobile-phase polarity (RP18W plates may work either as an NP or an RP system).

Dzido et al. [18] used pressurized planar electrochromatography (PPEC). They discussed the theoretical backgrounds, development, examples of separations, constructional details, and the principle of action of the devices (PPEC). The effect of various operating parameters, such as temperature, mobile-phase composition, mobile-phase flow velocity, external pressure, variables (composition of the mobile phase, pressure exerted on the adsorbent layer, mobile-phase flow velocity, temperature of separating system, etc.), as well as its 2D combination with the conventional TLC, were investigated.

White et al. [19] compared their method (multi-one-dimensional thin-layer chromatography [MOD-TLC]) and 2D-TLC in order to reach a high resolution and are liable to quantification of biologically active neutral lipids and phospholipids. These two methods gave well-comparable results for major classes of lipid species, but MOD-TLC is able to analyze several samples at the same time.

Olah et al. [20] compared the separation power of 1D-TLC, 1D automated multiple development (AMD), and 2D-AMD by the separation of four related plant extracts—*Artemisia abrotanum*, *Artemisia absinthium*, *Artemisia vulgaris*, and *Artemisia cina*. The separation in two consecutive directions yielded better results than that in one dimension.

9.3.3 STABILITY CONTROL

Perjési et al. [21] used 2D-TLC to prove that the E/Z isomerization of some cyclic chalcone analogues is a reversible process.

9.3.4 QUANTITATIVE EVALUATION

The quantitative evaluation of 2D-TLC can be a special task. Generally, the whole TLC plate is scanned in linear mode using normal slit shape, such as 0.2 × 8 mm, with as small as 0.5 or 1 mm steps between scans [22]. Special computer programs enable scanning and evaluation. Each scanning has to be processed as a regular densitogram including baseline subtraction, location of the peak, and integration. The computer program is used to synchronize the individual scans mainly based on the data of the highest peak. Finally, the selected scans are reintegrated and the peak heights and areas reported. In addition to the automatic mode, a manual mode is also available. Prosek et al. [22] demonstrated their method by the separation of phenolic compounds from propolis, certain characteristic flavonoids, etc.

Dévényi et al. [23,24] constructed a setup for densitometric evaluation of the whole chromatogram as early as in 1976. The entire chromatogram was irradiated by the light source, and the image was taken by a Vidicon camera.

Drusany et al. [25] evaluated 2D-TLC using a CAMAG TLC scanner connected and controlled by an IBM PC-AT, equipped with td/QTLC software. Aflatoxins were separated, and the standard deviation was between 3.9 and 6.9 for aflatoxin B and aflatoxin G2, respectively. Another approach with the CAMAG TLC scanner was published by Prosek et al. [26].

The results of forced-flow TLC (also called as overpressured-layer chromatography, OPLC) was compared to TLC and HPLC by Vuorela et al. [27]. The composition of the mobile phase was optimized by the use of the PRISMA system. The retention points in TLC were measured at 37 points and 13 points for HPLC and OPLC.

Petrović et al. [28] performed the quantification of the separated spots using a camera following 2D-TLC separation. They used an artificial mixture of 10 pesticides (2-methyl-4-chlorophenoxyacetic acid (MCPA), atrazine, propham, chlorpropham, ofurace, triadimefon, bitertanol A, bitertanol B, tetramethrin, and β-cypermethrin) as substances to be separated on cyano-plates. The 2D-TLC was used, and the developments were done in methanol–water (first dimensional run) and in hexane–diethyl ether (second dimensional development). The separated spots were detected/recorded using a sensitive color charged-coupled device (CCD) camera, while the evaluation was done with CAMAG VideoScan software. The quantification was done with 30 tracks positioned without overlapping.

Sajewicz et al. [29] used TLC-LC-MS to identify phenolic acids and flavonoids from herbal extracts. They developed this novel method to analyze these materials in the extracts of *Salvia lavandulifolia*. Moreover, they compared the performance of TLC-HPLC-MS to that of a 2D-TLC-HPLC-MS. As expected, the 2D mode was found to give better performing compared to the single dimensional method for scouting important constituents of extracts of botanical origin.

Komsta and Szewczyk [30] introduced the concept of kernel density to estimate the performance of 1- and 2D-TLC systems with large retention datasets.

9.3.4.1 Computer Simulation of Continuous Development 2D-TLC

Johnson and Nurok [31] suggested a computer simulation for the optimization of the continuous development of 2D-TLC. They discussed the separation of 30 steroids with NP–NP, NP–RP, and RP–RP 2D-TLC. A good agreement was found between the calculated and experimental results.

Hemmateenejad et al. [32] developed an image-processing system (multivariate image analysis [MIA]-TLC) to monitor the progress of organic chemical reactions, such as alkaline hydrolysis of phenyl benzoate and the reduction of certain benzaldehyde derivatives. Their program converted image data into multidimensional chromatograms.

9.3.4.1.1 Application of 2D-OPLC-DAR

Klebovich et al. [33] used a 2D forced-flow TLC (2D-OPLC) combined with digital autoradiography (DAR) for the detection of metabolites.

9.3.4.1.2 Screening of Metal Ions by 2D-TLC

Horvat et al. [34] used 2D-TLC for screening the metal ions present in honey. The optimal stationary phase was cellulose. Acetonitrile–water–hydrochloric acid and 2-pentanol–acetonitrile–hydrochloric acid were used for the first and second dimensional developments, respectively. For the detection of a mixture of quercetin, dimethylglyoxime and ammonia were used. Various metal ions were traced in their low ppm level in honey, such as Al^{3+}, Cr^{3+}, Fe^{3+}, Mn^{2+}, Pb^{2+}, Ni^{2+}, and Sn^{4+}. The method was used to scout the metal content, and the simplicity of 2D-TLC advocates its use. However, the development using capillary forces was pretty slow, as the first dimensional run took 40 min and the second dimensional development took 150 min. One reason for the slow development and also of the good separation obtained may be the absorbance of water on the microcrystalline cellulose during the second run: it remained there even after air-drying. Unfortunately, the paper did not give enough information on the drying procedure, neither between the first and second dimensional development nor after completing the 2D-TLC.

9.3.5 Calculations

An expression for the spot capacity (n) was given by Guiochon and Siouffi [35]: it gives the number of components that can be separated with a resolution of one:

$$n = 1 + \frac{vN_{max}}{2} \tag{9.1}$$

where N_{max} is the maximal value of the number of theoretical plates. Using standard TLC plates (particle size about 11 μm), about 16–18 components (spots) can be separated with a front distance of 16–16.5 cm. Using HPTLC plates (particle size about 5 μm), about the same number of components can be separated but in 4.5–5.5 cm front distance. The efficiency of TLC and thus the number of separable components can be increased by the optimization of the flow velocity [36]. Recent estimations approximate the separable components using TLC (where the mobile phase is progressed by capillary forces) up to 25, and the use of forced-flow TLC can increase it up to 150 [37].

The shape of a spot is mainly determined by the second dimensional development. This is especially true in the case of displacement 2D-TLC with a relatively high displacer concentration; very concentrated spots are generated with small width in the direction of the second dimensional development. The highly concentrated nature of the spot is definitely advantageous when radiolabeled components are investigated [38,39].

Some authors claim multiplication of the spot capacity from 1D- to 2D-TLC: spot capacities as high as 100 were estimated for 2D-TLC and up to 1500 in 2D overpressured TLC. The displacement effect of the mobile-phase front may essentially influence the spot capacity. The spots are concentrated by the mobile-phase multifronts; however, the number of the fronts limits the number of separable components. The second dimensional development defines the shape of the spots.

An expression for the peak capacity in 1D development was derived by Guiochon and Siouffi [35]. They started from the mathematical expression of Giddings [40], simplified by Grushka [41], assuming that the number of theoretical plates is constant for all solutes with different retention. This approximation is valid with both packed columns and TLC. According to them, the peak capacity for a packed column can be expressed in the following way:

$$n = 1 + \frac{\sqrt{N}}{4} \cdot \ln(1 + k') \qquad (9.2)$$

where
  n is the peak capacity
  N is the plate number
  k' is the retention factor

In another work, Guiochon has shown the column that reaches the maximum of its peak capacity for $k' \approx 6.4$ [then $\ln(1 + k') = 2$]; therefore, Equation 9.1 can be simplified to Equation 9.2:

$$n = 1 + \frac{\sqrt{N}}{2} \qquad (9.3)$$

For TLC, the number of spots (n) with a resolution of unity on a short path of the TLC (dz) is given by Equation 9.3:

$$dn = \frac{dz}{4\sigma} \qquad (9.4)$$

Even considering the diffusion, the mobile-phase flow velocity, and the migration distance,

$$n_M = \frac{\sqrt{L/H}}{2} = \frac{\sqrt{N}}{2} \qquad (9.5)$$

where n_M is the maximum of the spot capacity.

The spot capacity of 2D development is equal to the multiple of the individual developments:

$$n_{M,2D} = n_{M,1} \cdot n_{M,2} \qquad (9.6)$$

Guiochon and Siouffi [35] calculated and also measured the spot capacity of TLC for unidimensional development. The 2D-TLC serves as the separation of multicomponent mixtures. The general requirement is to separate each component from the nearest ones. Special requirement is to separate all components from the others as far as possible. This latter condition can be arranged by the use of optimization. Formulas were formulated by several authors to calculate the possible maximum of distances of the center of the spots [42]. Two relationships were suggested to characterize the separation quality with 2D-TLC:

$$D_A = \sum_{i=1}^{k-1} \sum_{j=i+1}^{k} [(x_i - x_j)^2 + (y_i - y_j)^2] \tag{9.7}$$

$$D_B = \sum_{i=1}^{k-1} \sum_{j=i+1}^{k} \frac{1}{(x_i - x_j)^2 + (y_i - y_j)^2} \tag{9.8}$$

where
   x_i and y_i are the coordinates of the spot
   k is the number of solutes in the mixture
   D_A is the sum of all possible distances between all pairs of spots
   D_B is the inverse of distances detailed in D_A

To prevent D_B from increasing too much nonseparated pairs of spots must be replaced by an arbitrary distance: 3 h R_F. D_A gives an importance to the situation when the plate is covered by a multiplicity of spots. D_A can well characterize the situation when several spots are poorly separated.

Steinbrunner et al. [43] gave formulas for similar reasons, marked as DF and IDF:

$$DF = \sum_{i=1}^{k-1} \sum_{j=i+1}^{k} S_{ij} \tag{9.9}$$

$$IDF = \sum_{i=1}^{k-1} \sum_{j=i+1}^{k} \frac{1}{S_{ij}} \tag{9.10}$$

where S_{ij} is the distance between the centers of the spots marked with i and j. The distance is calculated such as

$$S_{ij} = \sqrt{(x_i - x_j)^2 + (y_i - y_j)^2} \tag{9.11}$$

For efficient separation, DF has to be maximized, and IDF has to be minimized [43]. Nurok et al. [44] introduced the term of planar response function (PRF) for the optimization of separation. The expression is based on the calculation of S_{ij} and S_{SPEC}:

$$PRF = \sum_{i=1}^{k-1} \sum_{j=i+1}^{k} \ln \frac{S_{ij}}{S_{SPEC}} \tag{9.12}$$

where
   S_{ij} is the actual separation in the simulated chromatogram
   S_{SPEC} is the accepted minimum separation

If any separation is better than S_{SPEC}, 10 h R_F is substituted there. Thereby the too good separation has no influence on the further optimization. PRF has to approximate to zero when all spots are separated to the desired extent. If there are unresolved peaks, their component is substituted with an arbitrary 3 h R_F; therefore, they are not influencing the calculation either.

In the comparison of the expressions DF, IDF, and PRF, one can state that the last is more sensitive to poorly separated spot pairs than to the unnecessarily well separated ones. Risley et al. [45] have used PRF to quantify the quality level of 2D-TLC.

De Spiegeleer et al. [46] used to minimize the so-called correlation coefficient:

$$r = \sqrt{\frac{\sum [(x_i - x_c)(y_i - y_c)]^2}{\sqrt{\sum (x_i - x_c)^2 \sum (y_i - y_c)^2}}} \qquad (9.13)$$

The correlation is calculated using the h R_F values from the unidimensional development as the x_i and y_i values and the center of the data x_c/y_c. As more of all spots are distributed over the entire TLC plate, the difference between the selectivity of the two solvents increases. The best separation results with the lowest value of the correlation coefficient [47].

Quantitative evaluation of 2D chromatogram is possible by the use of CCD camera [28]. The video documentation system VideoStore 2 of CAMAG (Muttenz, Switzerland) can be used for this purpose. As the usual scanning wavelength is 254 nm, it is also generally used in the evaluation. The software is adequate for the on-screen manipulation of the image document for positioning the tracks. There are several possibilities to the scanning of a TLC. The major mode is scanning the whole plate along small tracks with a defined width. Another scanning possibility works with wide tracks tilted and bent to include the spots. The software of the CCD camera constructs the proper 3D picture [28]. The repeatability of video-densitometric determination has also been evaluated. The relative standard deviation depends on the focal distance and aperture (F-stop number), and both have to be optimized. If the F-stop number (aperture) is too low, the image will be too bright, while if it is too high, the image was too dark. Values higher or lower of the optimum aperture result in the reduction of the signal-to-noise ratio. The scanning procedure involves the selection of adequate number of tracks, which can be scanned without any overlapping. An increase of the number of tracks will decrease the RSD in the range of 15 through 30 tracks. A further increase of the number of tracks does not decrease the RSD; however, the time needed for data acquisition is increased. Special scanning arrangements can also be set; however, their value depends on the special circumstances.

There are several advantages of using 2D-TLC in monitoring the isolation of natural products or metabolites. The compound to be isolated generally has a characteristic signal; therefore, tracing this compound among the large number of other compounds is possible. Monitoring certain specific characteristics, such as radioactivity or biological activity, may indicate the location of the compound of our interest.

A major possibility is to monitor the metabolic pathway by the help of radiolabeling. For example, if the alkyl (methyl) substituent of a phenylalkylamine was ^{14}C-labeled, a series of metabolites can be easily traced, including the corresponding aldehyde of the alkyl substituent. At the same time, all metabolites containing the aromatic ring may be easily located by their UV absorbance at 254 nm.

9.3.5.1 Spot and Mobile-Phase Front Anomalies in Planar (Thin-Layer) Chromatography

Kalász [48] outlined some anomalies of behavior of spot at 2D planar chromatography. The reproduction of R_F values depends on the fact that the spot remains on the start following its development using the first dimension. His method was the use of 2D-TLC and reversing the order of developments.

9.3.6 Samples Subjected to 2D-TLC

Zakaria et al. [49] give many examples of classes of compounds that were separated using 2D-TLC: lipids (compound belonging to a wide range of compounds, which are soluble in organic solvents and much less soluble in water, such as fatty acids and their esters, steroids, phospholipids, and glycolipids), pigments, alkaloids, amino acids, peptides, proteins, carbohydrates, glycopeptides, nucleic acids (and their constituents), environmental pollutants, pesticides, inorganic ions, and miscellaneous compounds. Zakaria et al. [49] refer to over 200 papers published on these examples. Certain selected and more recent separations refer to the following.

There are several fields where 2D-TLC has been successfully used such as

1. Analysis of multicomponent mixtures, natural and synthetic drugs, and their metabolites
   a. Metabolites of radiolabeled compounds
   b. Analysis of biologically active components from multicomponent mixtures
   c. Analysis of natural products in extracts of plants, organs, tissues, etc.
   d. Drugs with similar chemical structures
2. Stability monitoring

The 2D-TLC can be widely utilized in the study on drug metabolism. The reason is that after development, every component is present on the TLC plate, including the components with the fastest migration or those that do not migrate at all: the fastest migrating components are at the solvent front, and the nonmigrating components are at the start. Both component groups can be easily separated in the second dimensional development.

Szunyogh et al. [50] used a major variation of forced-flow (overpressured) TLC combined with DAR for the detection of radiolabeled metabolites. The parent compound was ^{14}C-deramciclane (phenyl-^{14}C). In addition to several 1D-TLC separations, 2D-TLC demonstrated the real power of planar chromatography. The numbers of distinguishable metabolites were increased and were well observable especially at the solvent front. The comparison can be done with the standards spotted at the side tracks of 2D-TLC. Moreover, the DAR detection locates exactly the spots that can be identified by spectroscopic methods, such as MS and NMR.

Drug abuse is generally detected by gas chromatography/MS of urine samples. The 2D-HPTLC is advised as an alternative method. According to Novakova, "this (2D-HPTLC) method can serve as a method of choice for the detection of low concentration of opiates by laboratories not equipped with GC-MS" [51]. Morphine can be detected as its dansyl derivative on the 2D-HPTLC plate, while the other opiates are visualized by the use of either Dragendorff's or Marquis' reagent. The sensitivity of detection is about 10 ng mL^{-1} urine (or 10 ng mL^{-1} spot) for morphine and 50–100 ng mL^{-1} urine (or 50–100 ng mL^{-1} spot) for other opiates. This detection limit is about 20–100 times lower than using conventional TLC plates, and it is comparable to the limit of GC-MS.

9.3.6.1 Natural Products, Plant Extracts

Another wide area is the isolation of natural products from extracts of plants, animal organs, tissues, and even cells. Compounds with a certain chemical structure may be specifically detected either by the use of color reagents or by their combination. This is the case of ecdysteroids (insect molting hormones with specific biological effects on mammalians), which are monitored using a specific triple detection. Extracts of several plants were analyzed by 2D-TLC [52]. RP-18 silica TLC was used to separate ecdysteroids present in *Silene italica* ssp. *nemoralis*. The separation was carried out using two different organic modifiers, tetrahydrofuran and methanol. While the composition of tetrahydrofuran–water mixture was 45:55, it was 55:45 for methanol–water. The majority of the compounds migrated further in the 45% aqueous tetrahydrofuran than in the 55% of aqueous methanol. Unfortunately, the pair of polypodine B and 20-hydroxyecdysone was not separated in either mobile phase, as it is a generally known deficiency of any mobile phase when RP chromatography is used. However, a multitude of other ecdysteroids is adequately separated on an RP-TLC plate, just as they can be well separated using RP-HPLC. The polypodine B–20-hydroxyecdysone pair can be separated using silica stationary phase with either planar or column technique [4]. Planar chromatography can explain an important question, while silica columns can be used for several runs only, and after it the stationary phase has to be regenerated. The figures give 2D separations on TLC silica, where the only difference between Figures a and b is the order of the use of the mobile phases for development. The stationary phase was TLC silica, and the mobile phases were chloroform–methanol–benzene (25:3:3) and toluene–acetone–96% ethanol–25% aqueous ammonia (100:140:32:9). This latter mobile phase contains 2.85% of water. When the water-free mobile phase was used for the first development and the water-containing mobile phase was used for the second, the spread of the spots was not so expressed. On the other hand, when the water-containing mobile phase was used in the first development, the silica stationary phase adsorbed water, and thereby a partition-type process can take place between chloroform–methanol–benzene (applied as the second mobile phase) and the water-containing silica stationary phase. For TLC, this 3% water is good enough to produce a water-containing stationary phase. However, in the case of HPLC, the supply of a low concentration of water in the mobile phase would continuously saturate the silica packing of the column, after some time, and oversaturation occurs.

The intention of these investigations was to identify the ecdysteroids present in the brown algae *Fucus serratus*. When standards were added to the mixture, 1D development was unable to differentiate adequately; however, using 2D-TLC, the standard 20-hydroxyecdysone was fully separated from the ecdysteroids of *F. serratus*.

Phenolic compounds were separated by Hawryl et al. [53] using 2D-TLC on cyanopropyl-bonded stationary phase. The separation was carried out using polar solvents in a nonpolar basic solvent (e.g., in hexane) as the mobile phase for NP chromatography and polar solvent in water for RP chromatography. The combination of nonaqueous and aqueous mobile phases in the first and second dimensions permitted to combine NP and RP chromatography on the same stationary phase. The 2D-TLC chromatograms demonstrated satisfactory separation of the phenolic compounds of a Flos Sambuci extract when acetone–hexane (6:4) and methanol–water (1:1) were used in the first and second dimensional run, respectively. The results were just as good as when propan-2-ol–n-hexane (4:6) was combined either with tetrahydrofuran–water (1:1) or with 1,4-dioxane–water (1:1).

In the case of the analysis of the phytochemical products from a plant extract, a number of biologically active natural products are present. In situ monitoring of certain products is possible by 2D-TLC separation, with the use of a proper biological test and also by the detection at the adequate wavelength.

Hawryl and Soczewinski [54] published an NP 2D-TLC separation of flavonoids and phenolic acids from *Betula sp.* leaves. In addition to the flavones (acacetin, apigenin, astragalin, hyperoside, kaempferol, myricetin, quercetin, quercitrin, rutin) and flavanones (hesperidin, naringenin, naringin), three phenolic acids (caffeic acid, chlorogenic acid, and ferulic acid) were also present. Although several parameters of all compounds were widely investigated, even the optimal 2D-TLC was unable to make adequate separations. The majority of components were spread over in a curve, more or less separated into groups of four to five partially separated spots. The effective tool for differentiation was the successive use of 2-(diphenylboryoxy)-ethylamine and polyethylene glycol 4000, as the polychromatic spray reagent and scanning the plate at 365 nm.

Fater et al. [55] separated barbiturates on a silica gel stationary phase, supplying the mobile phase by forced-flow TLC. The majority of the 16 barbiturates were located in a zone approximating a straight line. As neither one of the mobile phases contained water, similar retention/differentiation mechanism could be expected in both directions of the development. Muller and Ebel [42] separated antihistamines by TLC on silica gel and cyano-phase.

Smolarz and Waksmundzka-Hajnos [56] improved the results of paper chromatography by the use of 2D-TLC on cellulose. 100 × 100 × 0.1 mm cellulose plates were used, and phenolic acids of plant extracts were separated. The first mobile phase was water-free, consisting of a mixture of methanol–acetonitrile–benzene–acetic acid. This was then followed by sodium formate–formic acid–water (10:1:200, w/v/v) in the other direction. The results depended on both the conditioning of the mobile-phase vapors and also on the composition of the conditioning liquid/vapor.

Effective 2D-TLC may be executed when the proper order of the mobile phases has been found. Plant extracts often contain flavonoids, carbohydrates, and also plant steroids. When the separation of plant steroids has a priority, the use of aqueous mobile phase in the first dimensional run may be preferable. The water becomes physically bound to the silica, and thus the next chromatographic development may act as partition chromatography. This procedure is generally used for steroid analysis [52].

Atanda et al. [57] determined the aflatoxin M1 contamination (in the range of 2.04–4.00 μg) of certain dairy products and also in milk in Ogun State, Nigeria. They used 2D-TLC.

Cieśla and Waksmundzka-Hajnos [58] rendered experimental proof of the usefulness of 2D-TLC of medicinal plants. Methodological aspects were detailed concerning how to perform the best 2D-TLC: either on one adsorbent or on bilayer plates, by the use of a graft TLC as well as using the hyphenated methods. Finally, a description of variable 2D methods was given for the analysis of the most important compounds of plant origin.

Hawryl et al. [59] used 2D-TLC on diol polar-bonded stationary phase in the analysis of antioxidant phenolic compounds from *Eupatorium cannabinum* extracts. The plates were sprayed by 2-(diphenylboryoxy)-ethylamine and PEG4000 or DPPH before photographing.

Ciesla et al. [60] used 2D-TLC of hydrophilic interaction mode to separate highly polar glycosidic compounds, such as iridoids and triterpene saponins, from the flowers of certain Verbascum species. The separations were performed on TLC silica gel, and the mobile phases were ethyl acetate–methanol–water–aqueous ammonia (55:35:9:1) and methanol–ethyl acetate–water–formic acid (10:90:26:22).

Cieśla et al. [61,62] separated a mixture of several structural analogues of coumarins using 2D-TLC. Structural analogues of coumarins were well separated when working on connected thin layers—either silica with RP-18W or CN-silica with silica CN-silica [60]. Both aqueous and nonaqueous mobile phases were used. They also reached a complete separation when using graft TLC. The best separation result was produced when silica and octadecyl-silica layers were used for the first and second dimensional separations, respectively. Furthermore, cyano-silica followed by plain silica (triple development using aqueous 30% AcN (acetonitrile) and also triple development with 35% ethyl acetate in n-heptane) were used as stationary and mobile phases in the first and second dimensions, respectively.

Medić-Šarić et al. [63] separated flavonoids of propolis by 2D-TLC.

9.3.6.2   Lipids

Hafenbradl et al. [64] analyzed lipids isolated from *Methanopyrus kandleri* using 2D-TLC.

Hoischen et al. [65] analyzed the lipid and fatty acid composition (cardiolipin [CL], lysocardiolipin [LCL], phosphatidylethanolamine [PE1 and PE2], lysophosphatidylethanolamine [LPE], phosphatidylinositol mannoside [PIM], phosphatidic acid [PA], dilyso-cardiolipin-phosphatidylinositol [DLCL-PI], and the 13 main fatty acids) of cytoplasmic membranes from *Streptomyces hygroscopicus* using 2D-TLC.

Mougios and Petridou [66] separated individual phospholipids by 2D-TLC. Skeletal muscle phosphatidylcholine, lysophosphatidylcholine, phosphatidylethanolamine, phosphatidylserine, phosphatidylinositol, CL, and sphingomyelin were successfully isolated.

9.3.6.3   Steroids

Nienstedt [67] characterized C19 steroids by 2D-TLC. Brown et al. [68] also used 2D-TLC for the separation of steroids of secretory and neuroendocrine functions.

9.3.6.3.1   Nucleoside Diphosphate Kinase

Dorion et al. [69] scouted autophosphorylation of nucleoside diphosphate kinase (NDPK). Following the incubation of recombinant *Solanum chacoense* cytosolic NDPK with [^{32}P] ATP, 32P-labeled P-Ser was identified in an acid hydrolysate of the protein by 2D-TLC. Gadzikowska et al. [70] analyzed tropane alkaloids from *Datura innoxia* Mill using 2D-TLC.

9.3.6.3.2   Biogenic Amines

Naguib et al. [71] determined eight biogenic amines (histamine, putrescine, cadaverine, tyramine, tryptamine, spermine, spermidine, and β-phenylethylamine) from fish samples. Four of the amines could be well separated in a 1D-TLC, while the other four were not satisfactorily separated. The 2D-TLC using mobile phases as benzene–triethylamine–acetone (10:2:1) and benzene–triethylamine (5:1) resulted in a simple, easy, and inexpensive way for the evaluation of biogenic amine composition.

9.3.6.3.3 Phosphopeptides

Panchagnula et al. [72] separated phosphopeptides by 2D-PEC/TLC. Their analysis from the plates was done by matrix-assisted laser desorption/ionization time-of-flight (MALDI-TOF) MS and also by tandem MS.

Sajewicz et al. [48,73,74] performed enantioseparation of the selected 2-arylpropionic acids by 2D-TLC. One of their chiral selector elements could be the microcrystalline silica gel stationary phase, while the other one could be L-arginine, applied as an impregnating chiral selector.

9.3.6.4 DNA and DNA Damages

Golden et al. [75] investigated how the reactive oxygen species damage DNA to form hydroxylated and other altered DNA products. Adduct detection was measured by labeling with 32P and separated using 2D-TLC.

Petruzelli et al. [76] used 2D-TLC with autoradiography (3–18 h exposition) to detect 32P-labeled normal and N7-methylated dGp. This 2D-TLC method could be used for diagnostic purposes, as various alkylating agents (e.g., tobacco-specific N-nitrosamines) produce methyl-DNA adducts, and also to induce lung tumors.

9.3.6.5 Alkaloids

Gadzikowska et al. [70] analyzed tropane alkaloids from *D. innoxia* Mill using 2D-TLC.

9.3.6.6 Antibiotics

Gallo et al. [77] detected chloramphenicol residues in the feed of animals. Their results were comparable to the analyses done by RP-HPLC.

9.3.6.7 Cations

The 2D-TLC permits the successful separation of inorganic cations on cellulose, using aqueous mobile phases in both directions [77]. 22 spray reagents were used for the specific detection of the ions, and the ionic forms of Tl, Ag, Hg, Sn, Sb, Bi, As, Zn, Cd, Fe, Cu, Co, Pb, Mn, Cr, Ni, Al, Mg, Ca, Sr, Ba, K, Na, and Li were adequately separated [78]. Their mobile phases were n-butanol saturated with 3 M nitric acid–1 M hydrochloric acid (1:1) and methanol–concentrated hydrochloric acid (10:3).

9.3.6.8 2D Elution–Displacement Chromatography

Elution–displacement 2D-TLC can be improved by the use of spacers at the stage of displacement development [79]. The spacers push away the otherwise overlapping zones of the displacement train, and also it proves visibility, if colored spacers are applied. Even colorless displaced compounds can be well observed as they give colorless spots appearing white on the colored background of the spacers.

Bathori developed a special kind of 2D-TLC and sesquidimensional TLC. Her investigations concentrated on phytochemistry, when flavonoids and phytosteroids have to be separated. Sesquidimensional TLC permits group separation in the first dimensional development for all compounds except one; then, in the second dimensional development, all components are separated. Therefore, the main target compounds are spread over on the majority of TLC plate surface but not around the start and front line.

9.3.6.9 Flip-Flop Chromatography

A special method for 2D planar chromatography may be the "flip-flop" development. The method was originally published for column chromatography by Martin and coworkers [80]. The expression "flip-flop" serves to emphasize sudden changes in the direction of development. Various mobile phases were consecutively used, such as heptane, methylene chloride, methanol, and water. The technique of forced-flow TLC seems to be an optimal method for flip-flop chromatography.

The 3D-TLC has been widely applied in the phytochemical analysis where multicomponent mixtures are generally separated. Coumarins were separated using 2- and 3D TLC by Vuerela et al. [27].

9.3.6.10   Methyl β-Cyclodextrin in the Second Dimensional Development

Momose et al. [81] used methyl β-cyclodextrin.

9.3.7   ARGENTATION TLC

Gallo et al. [77] used specially prepared TLC plates for 2D semipreparative TLC of fatty acid methyl esters (FAMEs). One strip of TLC plate was impregnated with urea (first dimensional run), and silver-nitrate impregnation was carried out on the remaining part of the stationary phase. The first dimensional run resulted in separation on the basis of chain structure, while the second dimensional run made a distinction based on the unsaturation (number of double bonds) of FAMEs.

RP and argentation zones of the stationary phase were used to separate phospholipids by 2D-TLC [82].

9.3.8   2D ELUTION–DISPLACEMENT TLC

The displacing effects in chromatography were discovered at the very beginning. Tswett outlined that the stronger adsorbing components are displacing the weaker adsorbing ones. This situation assumes the binding sites, and the binding capacity of the stationary phase is totally utilized by one portion of the solutes, and therefore a competition takes place for it. The essence of competition indicates that the separation is sure toward the direction where the discrimination works. As displacement chromatography highly concentrates the bands (spots), the displacement type of development gives special benefit by concentrating the minor components into sharp and well-defined zones. As the second dimensional development has the dominating role in the formation of the spots' shape, the use of displacement chromatography is preferred at the second dimensional development.

Elution and displacement developments were combined by Kalasz et al. [83,84] in the analysis of ecdysteroids using silica gel stationary phase. The mobile phase for elution chromatography was either dichloromethane–ethanol (85:15) or ethyl acetate–methanol–ammonia (85:10:5); both separated the standard ecdysteroids and the ecdysteroids present in plant extracts. The applied method can be used with the sensitive detection. The displacement run was generated by dichloromethane–2-propanol–dimethylaminopropylamine (80:30:5). The second dimensional displacement run completed the separation and also concentrated the ecdysteroid spots present in the displacement front. One specialty of forced-flow planar displacement chromatography is to decrease the flow velocity of the displacement mobile phase: this may duplicate the displacer front by giving the first and the second front.

In the investigation of metabolic pathways of drugs by 2D-TLC, elution served to differentiate the metabolites, while displacement served both separation and concentration of the spots [85,86]. For various drugs, chloroform–triethanolamine displacer was generally used in the second dimensional run.

An additional feature of 2D-TLC is the absence of zone deformation. In zone deformation, the spot either is tailing or has a U shape, and it is characteristic when a 1D-TLC system is overloaded [87]. However, zone deformation is unusual in 2D-TLC, particularly after the second dimensional elution–displacement development.

9.3.8.1   2D Column Chromatography

Guiochon and coworkers [6] described the construction of 2D column chromatography (2D-CC), which is basically a 2D-TLC system with continuous development. Their system is really a thin-layer plate tightly covered on both sides with arrangements for continuous development after the second dimensional development. A 2D column chromatography is considered as a method combining the advantages of column methods (constant and variable flow velocity, excellent efficiency, good reproducibility, online detection) with the features of 2D-TLC (easy successive developments in two different directions using two different retention mechanisms). Guiochon et al. [6] made calculations to estimate the peak capacity of 2D column chromatography: the peak capacity was found well in excess of 500 and up to several thousands. A 100 × 100 mm column with the thickness of 1 mm can be well packed with 10 μm particles. This planar column could be operated at a reduced velocity of 10, resulting in an estimated reduced peak height of about 2. Considering a moderate diffusion coefficient ($D = 5 \times 10^{-6}$ cm^2 s^{-1}) and a solvent viscosity of 1 cP, the pressure drop may be only 5 atm with a flow rate of 3 mL min^{-1}. The sample spot should be 1 mm or less.

However, certain disadvantages of the 2D-CC have also been found in practice. The removal of the mobile phase used in the first dimensional development is not always easy and simple before the second dimensional run. Specific mobile phases for TLC (such as acetone and phenol) are not proper organic modifiers for 2D-CC. Gonnord and Siouffi [88] solved the problem of ultrafast detection by the use of a diode-array detector. They used continuous detection and monitored the UV absorbance, and their data acquisition system allowed more than 100 readings of the 1024 pixels of the array/s.

It is usual to impregnate the TLC plate before the second dimensional development, but it is unusual to do it in both strips. Rezenka [89] separated FAMEs on semipreparative scale. One part of the TLC plate (a 20 cm × 5 cm strip for the first dimensional separation) was impregnated with urea, while the other part was impregnated with AgNO$_3$. The silica impregnated with urea facilitated the separation according to the structure of the chain (branched or nonbranched chain), while the silver nitrate permitted separation according to the number of the double bonds. Butyl acetate was used as the mobile phase in the urea-impregnated plate. The separation on this plate permits the differentiation of the unsubstituted, monosubstituted, and polysubstituted esters. After the first dimensional development, the plate was dried, turned around 90°, and developed using hexane–diethyl ether–methanol (90:10:1). This ternary mixture disrupts the urea complex and also separates the individual fatty acids. Using Ag-TLC, five fractions are usually well separated, the FAMEs with 0–4 double bonds. If FAMEs with 0–6 double bonds are to be separated, double developments are necessary.

9.3.8.2 Grafted Planar Chromatography

Grafted planar chromatography was described in 1979 by Pandey et al. [90]. It is a multiple system with the same or with different stationary phases for the isolation of compounds of either natural or synthetic origin. Grafted planar chromatography serves mainly preparative purposes; its advantage is the use of two stationary phases and transfer of the spot with or without scraping it, as shown by Székely [91]. Grafted planar chromatography was solely used for preparative purposes, as the transfer of the spots was not really quantitative.

Certain authors have called grafted TLC as adsorbent gradient [51,92–94]. The method well serves the separation of multicomponent plant extracts, as shown by Glensk et al. [92] for the separation of saponins from a ginseng preparation. The applied stationary phases were RP-18W HPTLC plate (from E. Merck) developed in methanol–water (7:3) and silica gel (also from E. Merck) developed in chloroform–methanol–water (70:30:4). The spots were transferred from one plate to the other by the use of methanol, as also detailed in other publications [93,94]. The RP-TLC plate is developed first; then, before the second development, this plate and the silica gel plate are attached firmly in a perpendicular direction. The sample components (here saponins [93]) are transferred by the use of 90% methanol. The second dimensional development is carried out on the silica gel plate using chloroform–methanol–formic acid–water (10:4:1:1). It had been demonstrated that unidimensional TLC was unable to separate plant saponins, and it can be done well by 2D-TLC [47].

9.3.8.3 Section-Formed Development in the 2D-TLC

Lan et al. [95] have developed the TLC plate with one mobile phase in the first dimension but with several phases in the second dimension. After the first dimensional separation was completed, they divided the TLC plate into several zones. The sectioning was done by the use of several holes and a slope distributor. The individual sections were developed with different mobile phases in the second dimension. The reason for the differentiation is that the first dimensional run selects the spots into groups, and each group may require an individually selected mobile phase. This method is the simplest variation of grafted planar chromatography, and the different separations in the second dimension may be more effective than using a single development. At the same time, however, the simplicity of the 2D-TLC can be totally lost when using this more complicated method.

9.3.8.4   2D Multiple Development

An interesting example of 2D-TLC is to improve resolution with double development in the same direction as done by Fater and coworkers [96]. Unidimensional multiple development was a useful tool for the prediction of the effect of multiple development. The separation of compounds migrating with relatively low R_F values is especially increased by the second development in the same direction. The identification of these closely related compounds (in their work, amphetamine derivatives) can be even further completed by the use of multiple detections.

9.3.8.5   Online Coupling of Liquid Chromatography with TLC

A very promising 2D technique is the online connection of HPLC with TLC. It allows the combination of two basically different chromatographic techniques and separation principles. Stan and Schwarzer [97] use RP-HPLC for the first dimensional development and AMD on an NP-TLC plate for the second dimensional run. The interface is computer controlled, and the whole effluent from the microbore HPLC column is transferred in fractions to the TLC plate. Typical applications were pesticide analysis from foodstuffs or environmental samples.

Forced-flow TLC was used as a cleanup method for HPLC separation by Mincsovics et al. [98]. Depending on the chromatographic characteristics of the unwanted portion of the sample, the impurities were either migrating with the solvent front or remaining at the start. The sample components to be analyzed were directly transferred into the HPLC column.

9.3.8.6   Offline Coupling of TLC with Gas Chromatography

Betti et al. [99] described a multidimensional approach to the coupling of TLC with capillary gas chromatography. The components of *Matricaria chamomilla* (L.) were separated by multiple development on silica gel using solvents of multiple polarity. One part of the sample was detected by derivatization, while the other underivatized part was dissolved and transferred for further separation into a GC-MS system.

9.4   DISCUSSION

Features of planar chromatography: The qualitative analysis of certain samples does not involve the loss of the nonmigrating components. Unlike the other chromatographic and electrophoretic methods, TLC and 2D-TLC have full access to evaluate the radioactivity and other characteristics of the spots remaining at the location of the original load. Methods used for TLC (and also for 2D-TLC) can be validated similarly to any other analytical methods. However, the real (validated) quantitative analysis of the sample constituents is rather unusual.

ACKNOWLEDGMENTS

This work was sponsored by OTKA 100155. Advices of late Dr. Leslie S. Ettre, Mr. Zoltán Dmeter, and Mr. János Horváth are appreciated.

REFERENCES

1. Izmailov, N.A. and M.S. Shraiber. 1938. A drop-chromatographic method of analysis and its application to pharmacy (in Russian: Kapelno khromatograficeskij metod analiza I jego primenenije v farmacii). *Farmatsiya*. 3 vols., pp. 1–7. Moscow, Russia: RTS Farmedinfo.
2. Liesegang, R. E. 1943. Kreuz-Kapillaranalyse. *Naturwissenschaften* 31(348).
3. Consden, R., A. H. Gordon, and A. J. Martin. 1944. Qualitative analysis of proteins: A partition chromatographic method using paper. *Biochemical Journal* 38(3): 224–232.

4. Hannig, K. and G. Pascher. 1969. Thin-layer electrophoresis. In *Thin-Layer Chromatography*, ed., E. Stahl, p. 113. Berlin, Germany: Springer.

5. Gross, D. 1959. Two-dimensional high-voltage paper electrophoresis of amino—And other organic acids. *Nature* 184: 1298–1301.

6. Guiochon, G., L. A. Beaver, M. F. Gonnord, A. M. Siouff, and M. Zakaria. 1983. Theoretical investigation of the potentialities of the use of a multidimensional column in chromatography. *Journal of Chromatography* 255: 415–437.

7. Kowalska, T. 2001. Absorbents in thin-layer chromatography. In *Planar Chromatography*, ed., S. Nyiredy, pp. 33–46. Budapest, Hungary: Springer.

8. Gocan, S. 2002. Stationary phases for thin-layer chromatography. *Journal of Chromatographic Science* 40(10): 538–549.

9. Khatib, A., A. C. Hoek, S. Jinap, M. Z. I. Sarker, I. Jaswir, and R. Verpoorte. 2010. Application of two dimensional thin layer chromatography pattern comparison for fingerprinting the active compounds in the leaves of *Vitex trifolia* Llinn possessing anti-tracheospasmolytic activity. *Journal of Liquid Chromatography & Related Technologies* 33(2): 214–224. doi: Pii 918142228. DOI 10.1080/10826070903439416.

10. Tuzimski, T. 2004. Separation of a mixture of eighteen pesticides by two- dimensional thin-layer chromatography on a cyanopropyl-bonded polar stationary phase. *Journal of Planar Chromatography-Modern TLC* 17(5): 328–334.

11. Tuzimski, T. and J. Wojtowicz. 2005. Separation of a mixture of pesticides by 2D-TLC on two-adsorbent-layer Multi-K SC5 plate. *Journal of Liquid Chromatography & Related Technologies* 28(2): 277–287. doi: DOI 10.1081/Jlc.200041336.

12. Tuzimski, T. 2005. Two-dimensional TLC with adsorbent gradients of the type silica-octadecyl silica and silica-cyanopropyl for separation of mixtures of pesticides. *Journal of Planar Chromatography-Modern TLC* 18(105): 349–357.

13. Petruczynik, A., M. Waksmundzka-Hajnos, T. Plech, T. Tuzimski, M. L. Hajnos, G. Jozwiak, M. Gadzikowska, and A. Rompala. 2008. TLC of alkaloids on cyanopropyl bonded stationary phases. Part II. Connection with RP18 and silica plates. *Journal of Chromatographic Science* 46(4): 291–297.

14. Hawryl, M. A. and M. Waksmundzka-Hajnos. 2011. Two-dimensional thin- layer chromatography of selected *Polygonum* sp extracts on polar-bonded stationary phases. *Journal of Chromatography A* 1218(19): 2812–2819. doi: DOI 10.1016/j.chroma.2010.12.020.

15. Ilic, S., M. Natic, D. Dabic, D. Milojkovic-Opsenica, and Z. Tesic. 2011. 2D TLC separation of phenols by use of RP-18 silica plates with aqueous and non-aqueous mobile phases. *Journal of Planar Chromatography-Modern TLC* 24(2): 93–98. doi: DOI 10.1556/Jpc.24.2011.2.1.

16. Berezkin, V. G. and E. V. Kormishkina. 2008. Two-dimensional planar chromatography with a closed adsorbent layer. *Russian Journal of Physical Chemistry A* 82(3): 415–419. doi: DOI 10.1134/S0036024408030163.

17. Zarzycki, P. K. and M. B. Zarzycka. 2008. Application of temperature- controlled micro planar chromatography for separation and quantification of testosterone and its derivatives. *Analytical and Bioanalytical Chemistry* 391(6): 2219–2225. doi: DOI 10.1007/s00216-008-1919-x.

18. Dzido, T. H., P. W. Plocharz, A. Chomicki, A. Halka-Grysinska, and B. Polak. 2011. Pressurized planar electrochromatography. *Journal of Chromatography A* 1218(19): 2636–2647. doi: 10.1016/j.chroma.2011.03.014.

19. White, T., S. Bursten, D. Federighi, R. A. Lewis, and E. Nudelman. 1998. High-resolution separation and quantification of neutral lipid and phospholipid species in mammalian cells and sera by multi-one-dimensional thin-layer chromatography. *Analytical Biochemistry* 258(1): 109–117.

20. Olah, N. K., L. Muresan, G. Cimpan, and S. Gocan. 1998. Normal-phase high-performance thin-layer chromatography and automated multiple development of hydroalcoholic extracts of *Artemisia abrotanum*, *Artemisia absinthium*, *Artemisia vulgaris*, and *Artemisia cina*. *Journal of Planar Chromatography-Modern TLC* 11(5): 361–364.

21. Perjesi, P., M. Takacs, E. Osz, Z. Pinter, J. Vamos, and K. Takacs-Novak. 2005. In-solution and on-plate light-catalyzed E/Z isomerization of cyclic chalcone analogues. Lipophilicity of E- and Z-2-(X-benzylidene)-1- benzosuberones. *Journal of Chromatography Science* 43(6): 289–295.

22. Prosek, M., I. Drusany, and A. Golcwondra. 1991. Quantitative 2-dimensional thin-layer chromatography. *Journal of Chromatography* 553(1–2): 477–487.

23. Devenyi, T. 1976. Quantitative evaluation of thin-layer chromatograms by video-densitometry. I. Determination of lysine in plant materials. *Acta Biochimica et Biophysica; Academiae Scientiarum Hungaricae* 11(1): 1–10.

24. Pongor, S., J. Kovacs, P. Kiss, and T. Devenyi. 1978. Quantitative evaluation of thin-layer ion-exchange chromatograms by video-densitometry III. Determination of D-penicillamine in blood samples dried on filter paper. *Acta Biochimica et Biophysica; Academiae Scientiarum Hungaricae* 13(3): 123–126.
25. Drusany, I., R. Kravanja, and M. Prosek. 1991. Quantitative two-dimensional TLC of aflatoxins. *Journal of Planar Chromatography* 4: 490–492.
26. Prosek, M., M. Pukl, A. Golcwondra, and D. Fercej-Temoljotov. 1989. Quantitative evaluation of thin-layer chromatograms. 9. Special scanning procedures. *Journal of Planar Chromatography* 2: 464–468.
27. Vuorela, P., E. L. Rahko, R. Hiltunen, and H. Vuorela. 1994. Overpressured layer chromatography in comparison with thin-layer and high-performance liquid-chromatography for the determination of coumarins with reference to the composition of the mobile-phase. *Journal of Chromatography A* 670(1–2): 191–198.
28. Petrovic, M., M. Kastelan-Macan, and S. Babic. 1998. Quantitative evaluation of 2D chromatograms with a CCD camera. *Journal of Planar Chromatography-Modern TLC* 11(5): 353–356.
29. Sajewicz, M., D. Staszek, M. Natic, L. Wojtal, M. Waksmundzka-Hajnos, and T. Kowalska. 2011. Tlc-Ms versus Tlc-Lc-Ms fingerprints of herbal extracts. Part Ii. Phenolic acids and flavonoids. *Journal of Liquid Chromatography & Related Technologies* 34(10–11): 864–887. doi: Pii 937705063. DOI 10.1080/10826076.2011.571131.
30. Komsta, L. and K. Szewczyk. 2009. The kernel density estimate as a measure of the performance of one and two-dimensional TLC systems with large retention datasets in the context of their use in fingerprinting. *Acta Chromatographica* 21(1): 13–27. doi: DOI 10.1556/Achrom.21.2009.1.2.
31. Johnson, E. K. and D. Nurok. 1984. Computer-simulation as an aid to optimizing continuous development two-dimensional thin-layer chromatography. *Journal of Chromatography* 302: 135–147.
32. Hemmateenejad, B., M. Akhond, Z. Mohammadpour, and N. Mobaraki. 2012. Quantitative monitoring of the progress of organic reactions using multivariate image analysis-thin layer chromatography (MIA-TLC) method. *Analytical Methods* 4(4): 933–939. doi: DOI 10.1039/C2ay25023c.
33. Klebovich, L., G. Morovjan, L. Hazai, and E. Mincsovics. 2002. Separation and assay of C-14-labeled glyceryl trinitrate and its metabolites by OPLC coupled with on-line or off-line radioactivity detection. *Journal of Planar Chromatography-Modern TLC* 15(6): 404–409.
34. Horvat, A. J. M., Z. Soljic, and M. Debelic. 2002. Qualitative identification of metal ions in honey by two-dimensional thin-layer chromatography. *Journal of Planar Chromatography-Modern TLC* 15(5): 367–370.
35. Guiochon, G. and A. M. Siouffi. 1982. Study of the performances of thin- layer chromatography—Spot capacity in thin-layer chromatography. *Journal of Chromatography* 245(1): 1–20.
36. Kalasz, H., L. S. Ettre, and M. Bathori. 1997. Past accomplishments, present status, and future challenges of thin-layer chromatography. *Lc Gc-Magazine of Separation Science* 15(11): 1044–1050.
37. Poole, C. F. 2001. A contemporary view of the kinetic theory of planar chromatography. In *Planar Chromatography, A Retrospective View for the Third Millennium*, ed., S. Nyiredy, pp. 13–32. Budapest, Hungary: Springer.
38. Ettre, L. S. and H. Kalasz. 2001. The story of thin-layer chromatography. *Lc Gc North America* 19(7): 712+.
39. Kalasz, H., T. Szarvas, A. Szarkane-Bolehovszky, and J. Lengyel. 2002. TLC analysis of formaldehyde produced by metabolic N-demethylation. *Journal of Liquid Chromatography & Related Technologies* 25(10–11): 1589–1598.
40. Giddings, J. C. 1967. Maximum number of components resolvable by gel filtration and other elution chromatographic methods. *Analytical Chemistry* 39: 1027–1028.
41. Grushka, E. 1970. Chromatographic peak capacity and the factors influencing it. *Analytical Chemistry* 42: 1142–1147.
42. Muller, D. and S. Ebel. 1997. Two-dimensional thin-layer chromatographic separation of H-1-antihistamines. *Journal of Planar Chromatography- Modern TLC* 10(6): 420–426.
43. Steinbrunner, J. E., E. K. Johnson, S. Habibigoudarzi, and D. Nurok. 1986. Computer-aided evaluation of continuous development two-dimensional thin layer chromatography. In *Planar Chromatography*, ed., R. E. Kaiser, pp. 239–256. Heidelberg, New York: Huthig.
44. Nurok, D., S. Habibigoudarzi, and R. Kleyle. 1987. Statistical approach to solvent selection as applied to two-dimensional thin-layer chromatography. *Analytical Chemistry* 59(19): 2424–2428.
45. Risley, D. S., R. Kleyle, S. Habibigoudarzi, and D. Nurok. 1990. *Journal of Planar Chromatography* (3): 216–221.
46. De Spiegeleer, B., W. Van den Bossche, P. De Moerloose, and D. Massart. 1987. *Chromatographia* (23): 407–411.

47. WSA Würzburger Skripten zur Analytik. 1992, 1995. Reihe Statistik—Teil 1.3: Lineare Kalibrierfunktionen; Teil 1.5: Spezielle Verfahren der linearen Kalibrierung; Teil 1.7: Nichtlineare Kalibrierfunktionen. Würzburg, Germany: Institut für Pharmazie.

48. Kalasz, H. 2005. Spot and mobile-phase front anomalies in planar (thin-layer) chromatography. *Chromatographia* 62: S57–S62. doi: DOI 10.1365/s10337-005-0605-9.

49. Zakaria, M., M. F. Gonnord, and G. Guiochon. 1983. Applications of two- dimensional thin-layer chromatography. *Journal of Chromatography* 271(2): 127–192.

50. Szunyog, J., E. Mincsovics, I. Hazai, and I. Klebovich. 1998. A new tool in planar chromatography: Combination of OPLC and DAR for fast separation and detection of metabolites in biological samples. *Journal of Planar Chromatography-Modern TLC* 11(1): 25–29.

51. Novakova, E. 2000. The detection of opiates by two-dimensional high- performance thin-layer chromatography. *Journal of Planar Chromatography-Modern TLC* 13(3): 221–225.

52. Bathori, M., G. Blunden, and H. Kalasz. 2000. Two-dimensional thin-layer chromatography of plant ecdysteroids. *Chromatographia* 52(11–12): 815–817.

53. Hawryl, M. A., A. Hawryl, and E. Soczewinski. 2002. Application of normal- and reversed-phase 2D TLC on a cyanopropyl-bonded polar stationary phase for separation of phenolic compounds from the flowers of *Sambucus nigra* L. *Journal of Planar Chromatography-Modern TLC* 15(1): 4–10.

54. Hawryl, M. A. and E. Soczewinski. 2001. Normal phase 2D TLC separation of flavonoids and phenolic acids from *Betula* sp. leaves. *Journal of Planar Chromatography-Modern TLC* 14(6): 415–421.

55. Fater, Z., B. Szabady, and S. Nyiredy. 1995. 2-dimensional overpressured layer chromatographic-separation of barbiturate derivatives. *Journal of Planar Chromatography-Modern TLC* 8(2): 145–147.

56. Smolarz, H. D. and M. Waksmundzkahajnos. 1993. 2-dimensional Tlc of phenolic-acids on cellulose. *Journal of Planar Chromatography-Modern TLC* 6(4): 278–281.

57. Atanda, O., A. Oguntubo, O. Adejumo, J. Ikeorah, and I. Akpan. 2007. Aflatoxin M-1 contamination of milk and ice cream in Abeokuta and Odeda local governments of Ogun State, Nigeria. *Chemosphere* 68(8): 1455–1458. doi: DOI 10.1016/j.chemosphere.2007.03.038.

58. Ciesla, L. and M. Waksmundzka-Hajnos. 2009. Two-dimensional thin-layer chromatography in the analysis of secondary plant metabolites. *Journal of Chromatography A* 1216(7): 1035–1052. doi: DOI 10.1016/j.chroma.2008.12.057.

59. Hawryl, M. A., R. Nowak, M. Waksmundzka-Hajnos, R. Swieboda, and M. Robak. 2012. Two-dimensional thin layer chromatographic separation of phenolic compounds from *Eupatorium cannabinum* extracts and their antioxidant activity. *Medicinal Chemistry* 8(1): 118–131.

60. Ciesla, L., M. Hajnos, and M. Waksmundzka-Hajnos. 2011. Application of hydrophilic interaction TLC systems for separation of highly polar glycosidic compounds from the flowers of selected *Verbascum* Species. *Journal of Planar Chromatography-Modern TLC* 24(4): 295–300. doi: Doi 10.1556/Jpc.24.2011.4.4.

61. Ciesla, L., A. Petruczynik, M. Hajnos, A. Bogucka-Kocka, and M. Waksmundzka-Hajnos. 2008. Two-dimensional thin-layer chromatography of structural analogs. Part II. Method for quantitative analysis of selected coumarins in plant material. *Journal of Planar Chromatography-Modern TLC* 21(6): 447–452. doi: DOI 10.1556/Jpc.21.2008.6.10.

62. Ciesla, L., A. Petruczynik, M. Hajnos, A. Bogucka-Kocka, and M. Waksmundzka-Hajnos. 2008. Two-dimensional thin-layer chromatography of structural analogs. Part I: Graft TLC of selected coumarins. *Journal of Planar Chromatography-Modern TLC* 21(4): 237–241. doi: DOI 10.1556/Jpc.21.2008.4.2.

63. Tuzimski, T. 2004. Separation of a mixture of eighteen pesticides by two- dimensional thin-layer chromatography on a cyanopropyl-bonded polar stationary phase. *Journal of Planar Chromatography-Modern TLC* 17(5): 328–334.

64. Hafenbradl, D., M. Keller, and K. O. Stetter. 1996. Lipid analysis of *Methanopyrus kandleri*. *FEMS Microbiology Letters* 136(2): 199–202.

65. Hoischen, C., K. Gura, C. Luge, and J. Gumpert. 1997. Lipid and fatty acid composition of cytoplasmic membranes from *Streptomyces hygroscopicus* and its stable protoplast-type L form. *Journal of Bacteriology* 179(11): 3430–3436.

66. Mougios, V. and A. Petridou. 2012. Analysis of lipid profiles in skeletal muscles. *Methods in Molecular Biology* 798: 325–355. doi: 10.1007/978-1-61779-343- 1_19.

67. Nienstedt, W. 1985. Characterization of C-19 steroids by two-dimensional thin-layer chromatography. *Journal of Chromatography* 329(1): 171–177.

68. Brown, J. W., A. Carballeira, and L. M. Fishman. 1988. Rapid two-dimensional thin-layer chromatographic system for the separation of multiple steroids of secretory and neuro-endocrine interest. *Journal of Chromatography* 439(2): 441–447.

69. Dorion, S., F. Dumas, and J. Rivoal. 2006. Autophosphorylation of Solanum chacoense cytosolic nucleoside diphosphate kinase on Ser117. *Journal of Experimental Botany* 57(15): 4079–4088. doi: DOI 10.1093/Jxb/Erl175.

70. Gadzikowska, M., A. Petruczynik, M. Waksmundzka-Hajnos, M. Hawryl, and G. Jozwiak. 2005. Two-dimensional planar chromatography of tropane alkaloids from *Datura innoxia* Mill. *Journal of Planar Chromatography-Modern TLC* 18(102): 127–131.

71. Naguib, K., A. M. Ayesh, and A. R. Shalaby. 1995. Studies on the determination of biogenic-amines in foods.1. Development of a Tlc method for the determination of 8 biogenic-amines in Fish. *Journal of Agricultural and Food Chemistry* 43(1): 134–139.

72. Panchagnula, V., A. Mikulskis, L. Song, Y. Wang, M. Wang, T. Knubovets, E. Scrivener et al. 2007. Phosphopeptide analysis by directly coupling two-dimensional planar electrochromatography/thin-layer chromatography with matrix-assisted laser desorption/ionization time-of-flight mass spectrometry. *Journal of Chromatography A* 1155(1): 112–123. doi: DOI 10.1016/j.chroma.2007.04.029.

73. Sajewicz, M., R. Pietka, G. Drabik, E. Namyslo, and T. Kowalska. 2006. On the stereochemically peculiar two-dimensional separation of 2-arylpropionic acids by chiral TLC. *Journal of Planar Chromatography-Modern TLC* 19(110): 273–277.

74. Sajewicz, M., M. Gontarska, A. Dqbrowa, and T. Kowalska. 2007. Use of video densitometry and scanning densitometry to study an impact of silica gel and L-arginine on the retention of ibuprofen and naproxen in TLC systems. *Journal of Liquid Chromatography & Related Technologies* 30(13–16): 2369–2383. doi: DOI 10.1080/10826070701465548.

75. Golden, M. C., S. J. Hahm, R. E. Elessar, S. Saksonov, and J. J. Steinberg. 1998. DNA damage by gliotoxin from *Aspergillus fumigatus*. An occupational and environmental propagule: Adduct detection as measured by P-32 DNA radiolabelling and two-dimensional thin-layer chromatography. *Mycoses* 41(3–4): 97–104.

76. Petruzzelli, S., L. M. Tavanti, A. Celi, and C. Giuntini. 1996. Detection of N-7- methyldeoxyguanosine adducts in human pulmonary alveolar cells. *American Journal of Respiratory Cell and Molecular Biology* 15(2): 216–223.

77. Gallo, P., G. deGrado, and L. Serpe. 1996. Chloramphenicol detection in feed by means of chromatographic techniques. *Italian Journal of Food Science* 8(1): 49–56.

78. Yoshinaga, T., T. Miyazaki, and H. Akei. 1998. Simultaneous identification of mixtures of inorganic cations by two-dimensional thin-layer chromatography. *Journal of Planar Chromatography-Modern TLC* 11(4): 295–299.

79. Kalasz, H. and M. Bathori. 2001. Displacement chromatography and its application using a planar stationary phase. In *Planar Chromatography, A Retrospective View for the Third Milennium*, ed., S. Nyiredy, p. 220. Budapest, Hungary: Springer-Hungarica.

80. Martin, A. J. P., I. Halasz, H. Engelhardt, and P. Sewell. 1979. Flip-Flop chromatography. *Journal of Chromatography* 186: 15–24.

81. Momose, T., M. Mure, T. Iida, J. Goto, and T. Nambara. 1998. Method for the separation of the unconjugates and conjugates of chenodeoxycholic acid and deoxycholic acid by two-dimensional reversed-phase thin-layer chromatography with methyl beta-cyclodextrin. *Journal of Chromatography A* 811(1–2): 171–180.

82. Kennerly, D. A. 1988. Two-dimensional thin-layer chromatographic—Separation of phospholipid molecular-species using plates with both reversed-phase and argentation zones. *Journal of Chromatography* 454: 425–431.

83. Kalasz, H., M. Bathori, L. Kerecsen, and L. Toth. 1993. Displacement thin- layer chromatography of some plant ecdysteroids. *Journal of Planar Chromatography* (6): 38–42.

84. Kalasz, H., M. Bathori, L. S. Ettre, and B. Polyak. 1993. Displacement thin- layer chromatography of some ecdysteroids by forced-flow development. *Journal of Planar Chromatography-Modern TLC* 6(6): 481–486.

85. Kalasz, H. 1983. Carrier displacement chromatography for the identification of deprenyl and its metabolites. *Journal of High Resolution Chromatography & Chromatography Communications* 6(1): 49–50.

86. Kalasz, H., L. Kerecsen, and J. Nagy. 1984. Conditions and parameters dominating displacement thin-layer chromatography. *Journal of Chromatography* 316: 95–104.

87. Bariska, J., T. Csermely, S. Furst, H. Kalasz, and M. Bathori. 2000. Displacement thin-layer chromatography. *Journal of Liquid Chromatography & Related Technologies* 23(4): 531–549.

88. Gonnord, M. F. and A. M. Siouffi. 1990. Ultrafast UV detection for bidimensional chromatography. *Journal of Planar Chromatography* (3): 206–209.

89. Rezanka, T. 1996. Two-dimensional separation of fatty acids by thin-layer chromatography on urea and silver nitrate silica gel plates. *Journal of Chromatography A* 727(1): 147–152.

90. Pandey, R. C., R. Misra, and K. L. Rinehart. 1979. Graft thin-layer chromatography. *Journal of Chromatography* 169: 129–139.
91. Szekely, G. 1969. Trockentransfer dunnsicht-chromatographisch abgetrennter Flecken. *Journal of Chromatography A* 42: 543–544.
92. Glensk, M., M. Czekalska, and W. Cisowski. 2001. Resolution of saponins from a ginseng preparation by 2D TLC with an adsorbent gradient. *Journal of Planar Chromatography-Modern TLC* 14(6): 454–456.
93. Glensk, M. and W. Cisowski. 2000. Two-dimensional TLC with an adsorbent gradient for analysis of saponins in *Silene vulgaris* Garcke. *Journal of Planar Chromatography-Modern TLC* 13(1): 9–11.
94. Matysik, G. 1997. *Problemy Optymalizacji Chromatografii Cienkowarstwowej.* Lublin, Poland: Akademia Nediczna w Lublinie.
95. Lan, M. N., D. Y. Wang, and J. Han. 2002. Improvement of two-dimensional development in TLC. *Journal of Planar Chromatography-Modern TLC* 15(2): 144–146.
96. Fater, Z., G. Tasi, B. Szabady, and S. Nyiredy. 1998. Identification of amphetamine derivatives by uni-dimensional multiple development and two- dimensional HPTLC combined with postchromatographic derivatization. *Journal of Planar Chromatography-Modern TLC* 11(3): 225–229.
97. Stan, H. J. and F. Schwarzer. 1998. On-line coupling of liquid chromatography with thin-layer chromatography. *Journal of Chromatography A* 819(1–2): 35–44.
98. Mincsovics, E., M. Garami, and E. Tyihak. 1991. Direct coupling of OPLC with HPLC: Clean-up and separation. *Journal of Planar Chromatography* 4: 299–303.
99. Betti, A., G. Lodi, N. Fuzzati, S. Coppi, and S. Benedetti. 1991. On the role of planar multiple development in a multidimensional approach to TLC-gc. *Journal of Planar Chromatography* 4: 360–364.

10 Quantitative Detection of Drugs by Densitometry and Video Scanning

Bernd Spangenberg

CONTENTS

10.1 INTRODUCTION

Thin layer chromatography (TLC) and, if modern plates are used, the high-performance thin layer chromatography (HPTLC) are versatile methods for the quantification of pharmaceutically active compounds in pharmaceutical preparations. Both methods work in parallel and allow simultaneous separations of different samples to take place. The big advantage in comparison to other separation methods is that HPTLC and TLC work with disposable plates. Therefore the pretreatment procedure can be often reduced to just a sample dissolving step. This helps to reduce the uncertainty in the assay. In this chapter we will no longer distinguish between TLC and HPTLC. HPTLC is the modern version of TLC and it is strongly recommended to use HPTLC plates for modern separations.

10.2 OBJECTIVES AND GENERAL REMARKS

Quantitative detection of drugs implies quantification within a specified range. Commonly pharmaceutically active compounds (the analytes) are guarantied within a ±5% range of uncertainty. Half of this range is given as production uncertainty, and the other half is what the analysis can claim as their range of uncertainty. The analytical aim for pharmaceutical products in general is therefore to quantify with a confidence interval of better than ±2.5% of the labeled content.

Quantitative chromatography needs a calibration that also extends the measurement uncertainty. Assuming that the measuring variance is roughly the same as we observe as calibration variance, we must reduce the relative standard deviation of the sample measurement from 2.5% to 1.77%. We commonly measure six repetitions. The Student factor for this procedure is 2.571 at a significance level of $\alpha =$ 0.05% (±2σ). To achieve a final confidence interval of better than ±2.5%, we therefore must measure a sample with a standard deviation of better than $2 \times 1.77/2.571 = 1.38$. This is a very ambitious aim [1,2].

10.2.1 MEASUREMENT RANGE

To measure within a ±2.5% range of uncertainty, there is no need to work in a large range of linearity. It is mostly sufficient to measure analyte standards at 80% and 120% of the declared amount. Within this range, the method's selectivity and its precision and accuracy must be determined. This procedure is called "method validation."

10.2.2 SELECTIVITY (SPECIFICITY)

The term selectivity is used to describe a satisfactory separation of a substance from all other accompanying substances. In general, the substances to be quantified in a sample solution are surrounded by the accompanying substances (the so-called matrix). The term selectivity refers to a method that provides responses for a number of chemical compounds, which may or may not be distinguished from each other. The method is said to be selective if the analyte response is distinguished from all other responses.

In a specific measurement, the analyte can be determined in a mixture or in the presence of matrix without interference from any other components. To avoid confusion, the term specificity should not be used in the sense of selectivity. A method is either specific or not. Since there are few methods that provide a response for a single analyte, the term selectivity is usually more appropriate than specificity for a typical chromatographic method.

The weakness of the thin-layer chromatography (TLC) method is its restricted chromatographic resolution. It is scarcely possible to separate more than ten compounds in a single run. In this respect, high-performance liquid chromatography (HPLC) is simply the better choice. In pharmaceutical analysis, we commonly have the aim to separate just a few compounds. For this task, high-performance thin-layer chromatography (HPTLC) is well suited, and HPLC is mostly oversized.

10.2.3 METHOD PRECISION

The term method precision comprises all steps of analysis: sample weighting, dissolving, and all other kinds of sample pretreatments, sample application on plate, chromatographic separation, plate drying, and plate scanning. To measure the method precision, all these steps must be independently performed for each sample. We have to, for example, apply six independently prepared sample solutions to measure the method precision with six measured values. Preparing a single sample solution, applying this solution as a single spot, and separating and replicate scanning the track will simply measure the scanning precision. Replicate applications of a single sample solution and scanning all the different tracks will only measure the precision of sample application and scanning procedure. In the last two cases, the precision of the analytical method was not determined because the pretreatment procedure was not taken into consideration!

HPTLC is well suited to measuring the method precision of a pharmaceutical formulation because the method works in parallel. It is therefore merely necessary to apply the different pre-treated samples on a single plate. The separation can then be performed for all samples under identical separation conditions. This is a very effective way of reducing the separation uncertainty, that is, we can effectively increase the method precision.

10.2.4 TRUENESS (ACCURACY)

Trueness describes the difference between the measured values and the true value for the sample. Often the term "accuracy" is used for this. To check for trueness, the evaluation of a certified reference sample can be used. The method is obviously accurate if it provides a value close to the certified value for the reference sample (within its method precision). The method can also be tested with a second independent measurement process. For HPTLC, quantifications by two independent analytical processes, for example, by determining absorption and fluorescence of a sample zone after normal- or reversed-phase separations or by determining a pure and a stained sample zone, afford an indication of accuracy.

The matrix often has a decisive influence on the accuracy of the analytical result. Therefore, pretreatments for standards and samples should be performed under the same conditions. Mostly, it is too complicated to prepare an artificial matrix for the standard, which contains all interfering substances. In this case, either the internal standard method or the standard addition method is applicable. The internal standard method requires the sample to be mixed with a substance having similar properties to the analyte and performing all pretreatment steps with this modified sample. If a mass selective MS detector is available, the internal standard can be an isotope-labeled analyte, because the internal standard method can compensate for analyte losses during sample preparation. For the standard addition method, various amounts of standard are mixed with the sample. In contrast to the internal standard method, the standard itself is measured in the presence of the matrix, and this method is preferred over the internal standard method for obtaining accurate values. The recovery rate for a reference material is usually expressed as the percent value for the measured results compared with their certified value.

Both methods, the internal standard method and the standard addition method, are applicable to HPTLC. Both methods require a relatively large number of separations, which are commonly time-consuming. The advantage of HPTLC is that the pretreatment procedure can be kept simple by using disposable plates. Another advantage of HPTLC is the parallel separation of different samples on a single plate under the same separation conditions, which decisively reduces analysis time.

10.3 DENSITOMETRY

The most important step in pharmaceutical analysis is the measuring process. The pattern of the separated compounds is transformed into digital data, which can be used for further calculations.

At the present time, there are four differently working systems on the market, which all allow quantifying thin-layer plates: mono-wavelength scanners, diode-array scanners, charge-coupled device (CCD) cameras, and flatbed scanners. The first two are commonly called scanners. The last two classes of items are classified as videodensitometric devices. Mono-wavelength scanners and diode-array scanners allow measuring a spectral distribution directly on a TLC or HPTLC plate. We call these methods densitometry because the optical density of light (the spectra-dependent light intensity) is measured. We therefore measure a densitogram and not a chromatogram. According to M. Tswett (the inventor of chromatography), a chromatogram is the space-separated distribution of compounds in the stationary phase and not its measurement curve [1].

In all these previously mentioned systems, light is used for detecting separated sample spots by illuminating the TLC or HPTLC plate from above with light of known intensity. If the illuminating light shows a higher intensity than the reflected light, a fraction of light must have been absorbed by the sample (the analyte) and/or the layer. Increasing sample amounts will induce decreasing light reflection. We therefore need a transformation algorithm to turn decreasing light intensity into increasing signal value. Ideally, there should be a linear relationship between the transformed measurement data and the analyte amount [2].

To measure a TLC or HPTLC plate, the surface is illuminated by white or monochromatic light of the intensity I_0. The clean plate surface should reflect the light intensity J_0 and the sample spot the light intensity J. Both values are combined in the expression of reflectance R (10.1):

$$R = \frac{J}{J_0} \qquad (10.1)$$

where

    $R(\lambda)$ is the reflectance
    $J(\lambda)$ is the intensity distribution of the analyte zone
    $J_0(\lambda)$ is the intensity distribution of the clean layer

(All intensity abbreviations in the text are used without a wavelength indication. This is done to ease readings although the intensity values are obviously wavelength-dependent.)

Scanners consist of different parts contributing statistically defined noise to the measurement result. The light source is most important, but the detector, signal amplifier, and A/D converter also contribute to the noise. Expression 10.1 is calculated from two error-dependent values (J and J_0), which project their uncertainties into reflectance R. In other words, both measurements contribute to the total error in R.

The larger the value, the better it can be measured. The brighter a clean plate surface reflects the incident light as J_0, the better we will measure J. If we assume that a sample spot reflects nearly no light ($J = 0$), the uncertainty in measuring this sample spot will be unpredictably high. From these facts, we can draw a simple conclusion to keep the measurement uncertainty low: neither J nor J_0 should have a value that is too low.

On the other hand, we measure the amount of light loss, which is calculated as the difference between J and J_0. If this difference is too small, the relative calculation error will also dramatically increase. If we take both effects into account, a minimum in the total measurement error is given for a situation in which both measurement values have the same size; thus, $R = 0.5$. From this result, we can conclude that it is a fairy tale that best results are achieved in measuring at peak maximum [1]. This simple conclusion allows us to predict that the lowest error will be achieved if we measure around an R-value of $R \approx 0.5$. To achieve a minimum uncertainty, the measurement conditions must be prepared in such a way that all R-values will be close to one-half [1].

As stated earlier, there should be a linear relationship between the transformed measurement data and the analyte amount. Mostly, this is fulfilled using Expression 10.2 for the transformation of absorption measurements:

$$KM = \frac{1}{R} - 1 = m \frac{a_m}{1 - a_m} \qquad (10.2)$$

and Expression 10.3 for fluorescence evaluations [1,3]:

$$Fl = R - 1 = m \frac{a_m}{1 - a_m} \qquad (10.3)$$

where

    $R(\lambda)$ is the relative reflectance
    a_m is the total mass absorption coefficient
    m is the analyte mass

Modern TLC and HPTLC scanners can measure absorption and fluorescence. These scanners cover the whole wavelength range from 200 up to 1000 nm. The disadvantages of TLC and HPTLC scanners are their high purchase price and maintenance costs.

10.3.1 Monowavelength Scanner

If we use monochromatic light, the reflected light is directly used as J in Equation 10.1. This is the basic measurement principle of commonly used monowavelength scanners. These systems have different advantages. The first advantage is that this kind of scanner has been on the market for more than 30 years. A second advantage is that they work with low-intensity monochromatic light. Chemical reactions induced by light are almost completely suppressed. Another advantage of such systems is that scanning a single track can be performed within seconds (Figure 10.1).

A disadvantage is that the spectral information of a track is difficult to achieve. Of course, these systems have the ability to measure spectra directly from the plate, but this needs time and that is a real problem. First of all, the track must be measured at different wavelengths to reveal the optimum R-value. That is time-consuming and commonly not performed, but only measuring at R-values around 0.5 will achieve the recommended confidence interval.

For quantifications, it is necessary to check whether the analyte peak is fully separated from any by-products. In other words, the selectivity of the separation must be assured. This can be done by measuring the spectral distribution in the peak. Peak-purity measurements are mostly carried out by spectral measurements at peak start, middle, and end. This cannot be done automatically if you are using a mono-wavelength scanner. Here, the analyst must define the scanning positions for each peak.

The range of linearity for fluorescence measurements is large (but restricted to low concentrations) because all systems use Equation 10.3. In contrast, the range of linearity for absorption measurements is narrow because mono-wavelength scanners do not use Equation 10.2. This is not a real problem in pharmaceutical analysis except for impurity quantifications in pure substances.

In principle, mono-wavelength scanners allow checking selectivity, precision, and trueness. Using them, it is difficult to check selectivity and to perform spectral optimization for achieving high precision.

10.3.2 Diode-Array Scanner

Diode-array scanners use white light for illumination purposes. Usually, the plate is moved below an interface that illuminates the plate at different wavelengths and detects the reflected light. The reflected white light is then dispersed into the spectral colors. In this way, the scanner provides the whole spectrum of the reflected light for each track location. Thus, the reflected light intensity (J) and the light intensity (I_0) are measured at different wavelengths. A contour plot comprises the reflected light intensities at different wavelengths and different track locations. These spectral data can be used to extract the UV–vis absorption and fluorescence spectra of different zones, using Equations 10.2 and 10.3.

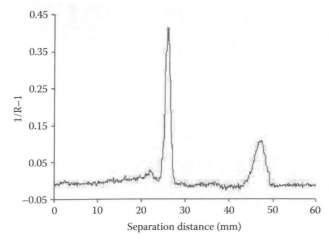

FIGURE 10.1 Scanning result of a monowavelength scanner. Shown is the separation of 500 ng dequalinium chloride in an ointment sample, measured at 238 nm, using methanol and 1 m solution of $NH_4^+COO^-$ in water (17 + 3 V/V) as the mobile phase. The ointment matrix is visible at 47 mm separation distance.

To show the performance of diode-array scanner, a dequalinium chloride containing ointment was analyzed. Dequalinium chloride [decamethylene-bis-(4-aminoquinaldinium)-chloride] is an antibacterial and antifungal agent, active against many Gram-positive and Gram-negative bacteria. It is used pharmaceutically to treat infections of the mouth and the throat in the form of ointments, liquids, or lozenges [4]. The HPTLC quantification of the ointment is easy. The amount of 1.25 g ointment is dissolved in methanol to a final concentration of ca. 4 mg/10 mL dequalinium chloride. This solution is applied in a volume of 1 μL (bandwise, 7 mm) directly on plate. Here, HPTLC plates (10 × 10 cm) with the stationary phase silica gel K60 nano-DC plates (Merck 5629 with a fluorescent dye) are used. The plates were developed in a vertical developing chamber without vapor saturation to a distance of 65 mm from the starting point (immersion line), using methanol and 1 m $NH_4^+COO^-$ in water (17 + 3 V/V) as the mobile phase. Dequalinium chloride shows an R-value of 0.58. The quantification is so simple because the matrix remains on the disposable TLC plate (Figures 10.2 and 10.3).

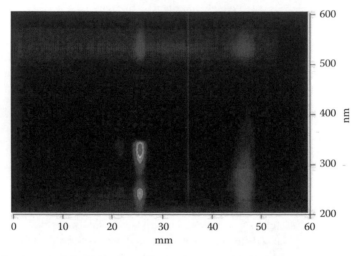

FIGURE 10.2 Contour plot of an ointment absorption evaluated according to Equation 10.2. The signal of dequalinium chloride can be seen at 25 mm separation distance. Sample application was carried out at 3 mm and the mobile phase moved to 50 mm distance. The ointment matrix signal can be seen at 46 mm separation distance. The fluorescence quenching signal of the plate fluorescence can be seen at 550 nm.

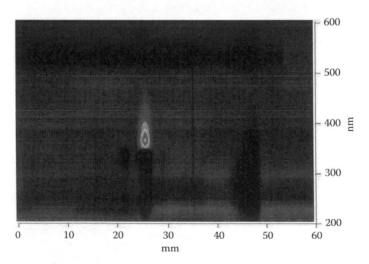

FIGURE 10.3 Fluorescence contour plot of an ointment sample using Equation 10.3 for evaluation. The signal of dequalinium chloride can be seen at 25 mm separation distance. Sample application was done at 3 mm and the mobile phase moved to 50 mm.

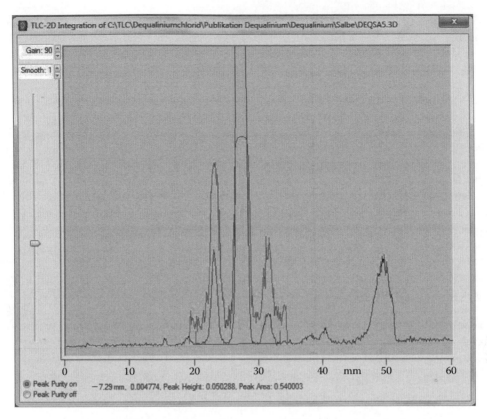

FIGURE 10.4   A separation of dequalinium chloride at a separation distance of 27 mm. The matrix signal is located at 50 mm separation distance. The peak-purity function (brighter line) shows a rectangular shape over the dequalinium peak. The two neighboring peaks show lower peak-purity values, indicating that their spectra differ from the dequalinium spectrum.

The advantage of a minimal pretreatment procedure is the low uncertainty in the final result. This accelerates the measurement. The relative standard deviations of six samples can be dropped below 2% [4].

A contour plot reveals peak purity at first glance. Symmetric signals provide a strong hint for peak purity. Mathematically, peak purity is simply checked by comparing all spectra of the peak with the spectrum in the peak maximum, preferably by using a cross-correlation algorithm [5]. Identical spectra result in the value on using this algorithm. A pure peak consists of identical spectra. This will result in a rectangular peak-purity function with a value near one [5] (Figure 10.4).

10.3.3   Videodensitometric Measurements (Videodensitometry)

Videodensitometric evaluations have become increasingly popular over the last few years. They are performed either with white light (daylight) or with UV light, mostly using wavelengths 254 or 366 nm.

It is the high price of modern scanners that makes image analysis in TLC so interesting [1,6]. Many TLC and HPTLC applications are designed to work in the wavelength range from 400 to 800 nm, using human eyes as detectors. Scanning equipment like CCD cameras or flatbed scanners, working in the visible range, are cheaply available and can be used for plate evaluations [1,6]. The term videodensitometer has also been introduced for such scanning devices.

The advantage of videodensitometric devices is their small size and low weight, which make these camera systems portable [6]. The use of a CCD device provides ultraviolet response and a larger dynamic range than a photomultiplier [6]. The evaluation of 2D separations is easily possible in contrast to slit scanners that are difficult to use for such measurements. The principle of CCD scanning is not restricted to cameras. A flatbed scanner can also be used for plate measuring [6]. The commonly used flatbed scanners illuminate the plate with three different colors (by red, green, and blue LEDs) scanning with a single CCD device. Even fluorescence can be measured if the flatbed scanner is equipped with a UV lamp [7]. The cheapest CCD–TLC evaluation is to use a hand scanner [8].

The disadvantage of a videodensitometer is that spectral information is not available. This makes spectral peak identification and spectral peak-purity tests impossible. Otherwise, most substances show no light absorption or fluorescence in the visible range. To make separation more specific and visible in the visible range, a staining step is recommended. Such staining steps often make spectral identification and peak-purity testing superfluous.

What features should be taken into consideration when buying a CCD camera? Quantitative videodensitometric measuring needs a detector that can linearly digitize light intensities. Double-fold light intensity must result in doubled signal values, which can be checked by changing the measurement time. Therefore, double measurement time must result in doubled measured values. In other words, the gamma factor must equal 1 because all other gamma factors digitize light intensities nonlinearly.

The digital resolution of commonly used cameras is 8 bit. A signal is rendered in $2^8 = 256$ different increments (gray levels), which is not sufficient for quantification purposes because at least 12 bit capacity is necessary for quantifying ($2^{12} = 4096$ increments). CCD cameras with a resolution of 16 bits are much better because such cameras render $2^{16} = 65536$ gray scales. Relatively inexpensive cameras with suitable software that meet these requirements are available for astronomy observations. These cameras produce tagged image file format (TIFF) pictures, because the TIFF supports 16 bit data storing.

A 16 bit system is necessary for quantitative evaluations. A commonly used all day CCD camera is not suitable for quantifying TLC and HPTLC plates. These systems should only be used for documentation purposes. Nevertheless, a 16 bit flatbed scanner suitable for planar chromatographic quantifications has been commercially available since 2010 [9].

10.4 THEORY OF DIFFUSE REFLECTANCE

Impurities of pharmaceutical active compounds can only be accepted at a very low level (e.g., 0.1%). For impurity quantifications, we therefore need a range of linearity of at least three magnitudes of power. This is not possible to achieve by simply using Equation 10.1 for scanning a TLC track. We therefore need a linearization algorithm like the *Beer–Lambert* law, which is valid for clear solutions and widely used in HPLC and UV–vis spectrometry.

The first attempt to solve this problem was made by *Kubelka* and *Munk* in the year 1931 [10,11]. The important point in this theory is that *Kubelka* and *Munk* described that the quotient of absorption and scatter is independent of the layer thickness. Thus, this quotient is a measure of the light absorbing sample mass:

$$KM = \frac{(1-R_\infty)^2}{2R_\infty} = \frac{a}{s} \tag{10.4}$$

where
  R_∞ is the absolute reflectance of an infinitely thick layer
  a is the absorptivity (absorption coefficient)
  s is the scattering coefficient

10.4.1 Is the Quotient of Absorption Coefficient and Scattering Coefficient Linear with Respect to the Analyte Mass?

In planar chromatography, light absorption of the layer is constant, but the sample light absorption depends on the sample mass. The sample distribution within a spot has a distribution of unknown shape.

Layer models profoundly demonstrate that this (unknown) sample distribution is very important for the intensity of the reflection signal. The deeper a sample is located within the layer, the weaker its absorption signal! Quantitative TLC does not need a constant sample concentration within the layer because this cannot be verified experimentally. A constant sample distribution in each sample zone is essential for quantitative TLC. This makes the mobile phase evaporation before scanning important. Uneven drying of the layer results in different sample distributions for the same compound at different tracks [1].

The *Kubelka–Munk* theory is a suitable start to theoretically describe all the different processes that take place when a distributed sample in a TLC layer is illuminated. The *Kubelka–Munk* theory is based on the assumption that half of the scattered light flux is directed forward and half backward. Both light fluxes show the same intensity. According to this theory, scattering in every layer will illuminate the next layer above and below it with half of its nonabsorbed light intensity. The question to be answered in this section is not simple: can it be proved by theory that the quotient of the absorption coefficient and the scattering coefficient is linear with respect to the analyte mass?

In the case of a scattering material like TLC plates, part of the scattered light is emitted as diffuse reflectance J when viewed perpendicularly to the plate surface. In general, the vector I represents the light flux in the direction of the incident light, and the vector J describes the light intensity in the antiparallel direction. If J is the scattered light in the direction of the layer surface and I_{abs} is the abbreviation for the light absorbed by the layer and sample, the incident light intensity I_0 is split into scattered and absorbed light. The abbreviation a stands for the absorption coefficient and s stands for the scattering coefficient:

$$I_0 = J + I_{abs} = sI_0 + aI_0 = I_0(s+a) \tag{10.5}$$

where
$I_0(\lambda)$ is the incident light intensity
$J(\lambda)$ is the diffuse reflectance at the plate surface ($J = sI_0$)
s is the scattering coefficient
a is the absorption coefficient
$I_{abs}(\lambda)$ is the light absorption

In Equation 10.5, the absorbed share of the incident light I_0 is aI_0, and the fraction of scattered light is sI_0. If all incident light is either absorbed or scattered and if there is no other light loss, the sum of both shares must be the original incident light intensity. There is a simple connection between the scattering coefficient and the absorption coefficient [1,2]:

$$s = (1-a) \tag{10.6}$$

This equation explains the observation that opaque TLC plates shine brightly when light from the surface is reflected at a low angle. In this case, light is not penetrating the stationary phase; thus, there is no light absorption; hence, $a = 0$. All the light is scattered from the layer surface.

Regarding the first approximation, the incident light beam with intensity I_0 is scattered in all directions by particles inside the layer, and some radiation may be absorbed by either the sample or the layer itself. The diffuse reflected light J from the top of the plate can be easily measured and provides the desired information regarding how much light was absorbed by the sample. The stationary phase in TLC will absorb light as does the sample. Beside this loss of incident light by absorption, light is also lost by transmission I_T from the back and edges of the layer. We can gather up all losses of light in the absence of sample into the expression $I_{abs,u}$. The expression $I_{abs,s}$ describes the light absorbed by the sample. The scattered light J in the direction of the layer surface is

$$J = I_0 - I_{abs} = I_0 - (I_{abs,u} + I_{abs,s})$$

where
  $J(\lambda)$ is the reflectance
  $I_0(\lambda)$ is the incident light intensity
  $I_{abs}(\lambda)$ is the light absorbed by plate and sample
  $I_{abs,u}(\lambda)$ is the light absorbed by the plate material
  $I_{abs,s}(\lambda)$ is the light absorbed by the sample

With this expression, we assume that these two different types of absorption occur simultaneously. A sample, therefore, "recognizes" the light intensity I_0 of the illumination lamp minus the light intensity $I_{abs,u}$ absorbed by the layer. The sample absorption coefficient is now defined as

$$a_s \equiv \frac{I_{abs,s}}{I_0 - I_{abs,u}} \tag{10.7a}$$

The same is true for the absorption of light by the layer, which is illuminated by I_0 minus the light that is absorbed by the sample. The plate absorption coefficient is therefore defined as

$$a_u \equiv \frac{I_{abs,u}}{I_0 - I_{abs,s}} \tag{10.7b}$$

The replacement of $I_{abs,u}$ from Formula 10.7a in Expression 10.7b and vice versa results in

$$\frac{I_{abs,s}}{I_0} = \frac{a_s - a_s a_u}{1 - a_s a_u} \quad \text{and} \quad \frac{I_{abs,u}}{I_0} = \frac{a_u - a_s a_u}{1 - a_s a_u}$$

The absorption coefficient of the layer and sample then becomes the sum of both factors [1,2,12]:

$$a = \frac{I_{abs,s}}{I_0} + \frac{I_{abs,u}}{I_0} = \frac{a_s + a_u - 2a_s a_u}{1 - a_s a_u} \tag{10.8}$$

This absorption factor describes all absorptions within the layer. The problem is that a is not proportional to the mass of the sample. To prove whether the quotient of a and the scattering factor $(1-a)$ are linear with respect to the sample mass m, Equation 10.8 is substituted into the quotient expression:

$$\frac{a}{1-a} = \frac{a_s + a_u - 2a_s a_u}{(1 - a_s a_u) - (a_s + a_u - 2a_s a_u)} = \frac{a_s(1 - 2a_u) + a_u}{(1 - a_s)(1 - a_u)}$$

The light intensity that is absorbed by the sample mass m depends on the mass absorption coefficient a_m and must be written as

$$I_{abs,s} \equiv m a_m I_0$$

The rest of the light either leaves the layer or is absorbed by the stationary phase:

$$I_{abs,u} = (1-m)a_m I_0$$

It then follows that the sample absorption coefficient a_s becomes

$$a_s = \frac{ma_m}{1 - a_m(1-m)}$$

Replacement of a_s leads to the following expression:

$$\frac{a}{(1-a)} = \frac{ma_m(1-2a_u) + a_u[1 - a_m(1-m)]}{[1 - a_m(1-m) - ma_m](1-a_u)}$$

$$\frac{a}{(1-a)} = \frac{ma_m - ma_m a_u}{(1-a_m)(1-a_u)} + \frac{a_u - a_u a_m}{(1-a_u)(1-a_m)}$$

and we obtain Equation 10.9:

$$\frac{a}{(1-a)} = m\frac{a_m}{(1-a_m)} + \frac{a_u}{(1-a_u)} \tag{10.9}$$

There is a linear connection between the mass m of a compound and its corresponding signal. The intercept in Equation 10.9 only contains the constant a_u and therefore describes the layer absorption [1,2,12].

10.4.2 What Is the Mathematical Connection between Measured Data and the Analyte Mass?

We have just theoretically proved that the quotient a/s is proportional to the mass of the analyte [3,7]. This is the main result of the so-called *Kubelka–Munk* theory, which is often used to describe reflectance on TLC plates [5]. The *Kubelka–Munk* theory therefore describes the quotient $a/(1-a)$ as proportional to the amount of analyte in the TLC layer.

Besides the sample, there is also the TLC layer that absorbs light. The absorption coefficient a comprises both kinds of absorption. The absorption coefficient describes the fraction of incident light aI_0, which is absorbed. The nonabsorbed radiation will be scattered inside the layer, and only scattered light will be emitted as J_1 from the layer surface, which we also call the first virtual layer:

$$J_1 = I_0 - I_{abs} = I_0 - aI_0 = I_0(1-a)$$

Opaque TLC plates shine brightly when light from the surface is reflected at a low angle. This observation from all TLC and HPTLC plates led to the conclusion that remission is influenced not only by the incident light intensity but also by the angle of the incident light. Thus, the extent of reflection is a function of the incident light intensity and the incident light angle. This angle not only changes according to the position of the incident light source placed over the plate; it also changes with the crystal structure of the stationary phase material and its orientation within the layer.

If we assume that every layer will illuminate the next layer above and below with different light intensities, we need to introduce a backscattering factor k. The factor k describes the fraction of the remaining light in a virtual layer, which is scattered in the direction of the layer surface. Logically, the fraction $(1-k)$ describes the remaining light in a virtual layer, which is scattered in the direction of the plate bottom, which is also the direction of the incident light. To make the situation clear, the abbreviation $q = k(1-a)$ represents the fraction of nonabsorbed light toward the layer surface, and $p = (1-k)(1-a)$ stands for the fraction of light in the direction of the incident light. Both expressions together describe how much light was scattered in this process. Therefore, we can conclude the following equation:

$$s = p + q = (1 - a) \tag{10.10}$$

The schematic diagram in Figure 10.5 illustrates the interactions of the incident light (I_0) with at least five virtual layers.

For the light intensity of the first layer $J_1 = J$, the layer model is expressed as

$$\frac{J_1}{I_0} = 2^0 q + \frac{1}{1*2} 2^1 pq^2 + \frac{1*3}{1*2*3} 2^2 p^2 q^3$$

$$+ \frac{1*3*5}{1*2*3*4} 2^3 p^3 q^4 + \frac{1*3*5*7}{1*2*3*4*5} 2^4 p^4 q^5$$

$$+ \frac{1*3*5*7*9}{1*2*3*4*5*6} 2^5 p^5 q^6 + \cdots + \tag{10.11}$$

The sequence of factors (1, 2, 5, 14, 42, 132, 429,...) is well known as the *Catalan numbers (Cns)*. They occur in various problems of combinatorial mathematics. The *Catalan* sequence was first described in the eighteenth century by *Leonhard Euler* (1707–1783). They are named after the

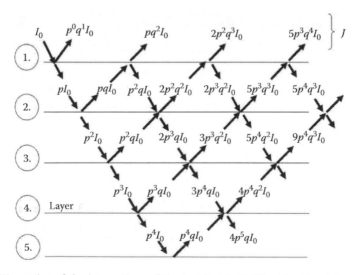

FIGURE 10.5 Illustration of the interactions of the incident light (I_0) with at least five virtual layers. The fraction q of I_0 is reflected to the top, and the fraction p is scattered in the incident light direction. The expression for reflectance from the top five virtual layers is $J = pq^2 I_0 + 2p^2 q^3 I_0 + 5p^3 q^4 I_0 + 14p^4 q^5 I_0 + \cdots +$.

Belgian mathematician *Eugène Charles Catalan* (1814–1894). The nth *Cn* can be calculated directly (for $n \geq 0$) in terms of binomial coefficients [13,14]:

$$C_n = \frac{(2n)!}{(n+1)!n!}$$

Equation 10.11 can be written as the sum of the *Cns* [13,14]:

$$R_0 \equiv \frac{J_1}{I_0} = q + 1pq^2 + 2p^2q^3 + 5p^3q^4 + 14p^4q^5$$

$$+ 42p^5q^6 + 132p^6q^7 + 429p^7q^8 + \cdots + = \sum_{n=0}^{\infty} C_n q(pq)^n$$

Equation 10.12 results if we multiply the function that contains all of the *Cns* by itself (see References [13,14] for details):

$$pR_0^2 + q = R_0 \qquad (10.12)$$

To prove this, we just calculate

$$pR_0^2 + q = q + p\left(\sum_{n=0}^{\infty} C_n q(pq)^n\right)^2$$

$$= q + p\left[\left(q + pq^2 + 2p^2q^3 + \cdots\right)\left(q + pq^2 + 2p^2q^3 + \cdots\right)\right]$$

$$= q + pq^2 + p\left(pq^2q + qpq^2\right) + p\left(q2p^2q^3 + pq^2pq^2 + 2p^2q^3q\right) + \cdots$$

$$= q + pq^2 + 2p^2q^3 + 5p^3q^4 + \cdots = \sum_{n=0}^{\infty} C_n q(pq)^n$$

Equation 10.12 is a quadratic equation that we can solve by using the quadratic formula:

$$R_0^2 - \frac{R_0}{p} + \frac{q}{p} = 0$$

$$\left(R_0 - \frac{1}{2p}\right)^2 = \frac{1}{4p^2} - \frac{q}{p}$$

$$R_0 = \frac{1}{2p} + \sqrt{\frac{1}{4p^2} - \frac{q}{p}}$$

The relative reflectance value R_0, defined as the quotient of J_1 and I_0, converges in Equation 10.13 to an expression containing p and q only:

$$R_0 = \frac{1}{2p} - \frac{1}{2p}\sqrt{1 - \frac{4p^2q}{p}} = \frac{1}{2p}\left(1 - \sqrt{1 - 4pq}\right) \qquad (10.13)$$

This equation must be resolved to the quotient $a/(1-a)$ because only this expression is proportional to the analyte mass. We can rearrange Equation 10.13:

$$1 - 4pq = (1 - 2pR_0)^2 = 1 - 4pR_0 + 4p^2R_0^2$$

When we erase the value 1 on both sides of the equation and if we also divide by 4, we obtain

$$-pq = -pR_0 + p^2R_0^2$$

Dividing by p results in the following equation:

$$-q = -R_0 + pR_0^2 = R_0(pR_0 - 1)$$

After rearranging and by the help of Equation 10.10, we can write

$$pR_0 + \frac{q}{R_0} = 1 = a + p + q$$

$$\frac{q}{R_0} - q + pR_0 - p = a$$

$$q\left(\frac{1}{R_0} - 1\right) + p(R_0 - 1) = a$$

If this equation is resolved to an expression as a function of the total absorption coefficient a and the backscattering factor k, we obtain Equation 10.14, the modified *Kubelka–Munk* equation [10]:

$$KM(R_0, k \geq 0, k \leq 1)$$

$$= k\left(\frac{1}{R_0} - 1\right) + (1-k)(R_0 - 1) = \frac{a}{(1-a)} \tag{10.14}$$

Equations 10.9 and 10.14 demonstrate linearity between the transformed intensity data and the quotient $a/(1-a)$. We obtain direct linearity between the transformed data and the sample mass. With the transformed intensity data according to Equation 10.9, the fraction of light absorbed by the sample can be separated from the light absorbed by the layer. The diffuse reflected light intensity J_0 can be determined at layer positions free of absorbing compounds. If this signal is used in Equation 10.14 instead of the lamp intensity I_0 to calculate the relative reflection R, we assume that the TLC plate does not show any loss of light instead of sample absorption. Thus, the corrected relative reflection R can be written as

$$R = \frac{J}{I_0 - I_{abs,u}} = \frac{J}{J_0} \tag{10.15}$$

As a result, the plate absorption intensity $I_{abs,u}$ becomes zero, but in fact the light intensity of the lamp I_0 is only replaced by J_0. Mathematically, the original light flux of the lamp I_0 is reduced to J_0 by all loss of light at the plate surface, and hence, we assume $I_{abs,u}$ to be zero. In addition, this turns

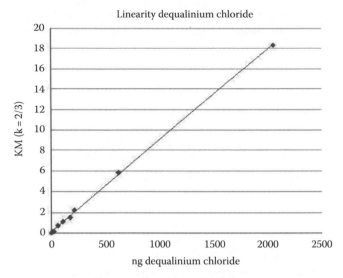

FIGURE 10.6 Linearity over a wide range makes different evaluations possible. For example, the external, internal, and standard addition evaluation methods are all based on calibration functions that need a large range of linearity without intercept. Especially the external standard method simplifies evaluations.

a_u to zero as well. With Equations 10.9, 10.14, and 10.15, we obtain the fundamental expressions (10.16) for quantitative planar chromatography:

$$k\left(\frac{1}{R}-1\right)+(1-k)(R-1) = m\frac{a_m}{(1-a_m)}$$

$$(10.16)$$

$$k\left(\frac{1}{R}-R\right)+(R-1) = m\frac{a_m}{(1-a_m)}$$

where
  $R(\lambda)$ is the relative reflectance ($R = J/J_0$)
  k is the backscattering factor ($k \geq 0$ and $k \leq 1$)
  m is the analyte mass
  a_m is the analyte absorption coefficient

For absorptions in scattering media, Expression 10.16 is the equivalent law to the *Lambert–Beer* law in clear solutions (also known as the *Beer–Lambert–Bouguer* law) (Figure 10.6).

A wide range of linearity helps construction of a direct relationship between the signal area and the area of a low concentration by-product. External standard calculation allows the quantification of by-products at a level of 0.1% analyte concentration if linearity of three magnitudes is achieved.

10.5 THEORY OF FLUORESCENCE IN DIFFUSE REFLECTANCE MATERIAL

The basic idea behind the theory of remission is the observation that opaque TLC plates show a brightly shining light reflected from the surface at a low angle. The extent of reflection is a function of the incident light intensity and the incident light angle. This angle not only changes with the position of the incident light source over the plate but also changes with the crystal structure of the stationary phase material and its orientation within the layer. The backscattering factor k exactly describes this observation and sets the orientation of particles in relation to incident, emitted, and fluorescent light. The value of the backscattering factor k can be larger than 0 and smaller than 1.

For the extreme case of $k = 1$, we must assume that no scattered light illuminates the top of the second virtual layer. In this case, all light is absorbed in the first virtual layer or is scattered from the top of it. No light reaches the second virtual layer. This expression, therefore, transforms all light absorptions into positive values. Therefore, this expression can be beneficially used in trace analysis.

The second extreme case of Equation 10.16 is $k = 0.5$. In this case, Equation 10.16 renders the well-known *Kubelka–Munk* formula (Equation 10.4) with $R = R_\infty$. This equation is suitable for using at high analyte concentrations.

The third extreme case of the extended *Kubelka–Munk* equation is $k = 0$. This is the transformation formula for fluorescence signals. With $k = 0$, no incident light is scattered toward the layer surface. Therefore, light emitted from the plate surface must be fluorescent light. A contour plot evaluated by the fluorescence formula instantly reveals compounds showing fluorescence. The equation for $k = 0$ can be used in trace analysis and for high analyte concentrations, but this is seldom used in quantitative pharmaceutical analysis.

Sample molecules located in light-scattering media exhibit fluorescence J_F if the absorbed light is transformed into fluorescent emission. The extent of this transformation is described by the quantum yield factor q_F. The light intensity absorbed by the sample can be calculated from the light intensity reflected from the clean plate surface (J_0) minus the light intensity (J) reflected from the sample:

$$J_F = q_F(J_0 - J) = q_F J_0 (1 - R) \tag{10.17}$$

where
    $J_F(\lambda)$ is the emitted fluorescence intensity
    $q_F(\lambda)$ is the fluorescence quantum yield factor
    $J_0(\lambda)$ is the light intensity reflected from a clean plate surface
    $J(\lambda)$ is the light intensity reflected from the sample
    $R(\lambda)$ is the relative reflectance ($R = J/J_0$)

To achieve minimal uncertainty in fluorescence measurements, conditions must be prepared so that the fluorescence is as bright as possible. The greater the fluorescence, the lower the uncertainty. The drawback is that high fluorescence signals often run into saturation. In other words, the relation between analyte mass and fluorescence signal becomes nonlinear. This is not a fluorescence property but is related to the nonlinear absorption of the analyte in scattering media. This can be corrected by using the advanced *Kubelka–Munk* theory.

10.5.1 TLC FLUORESCENCE FOR LOW SAMPLE CONCENTRATIONS

In the case of trace analysis (which means we have k-values near $k \sim 1$), we observe linearity according to the extended *Kubelka–Munk* equation (10.16) for the reflectance formula $(1/R)-1$ (10.2). By multiplying Expression 10.17 by $1/R$ and taking (10.10) and (10.16) into account, we observe

$$\frac{J_F}{J_0 R} = q_F \frac{(1-R)}{R} = q_F m \frac{a_m}{s}$$

If R is substituted by $R = s/(a+s)$ (derived from expression $(1-R)/R = a/s$) and if we take into account that $(s+a) = 1$,

$$J_F = m J_0 q_F \frac{a_m}{s} \frac{s}{(a+s)} = m J_0 q_F a_m \tag{10.18}$$

The reflected sample light intensity (J) is the sum of light scattering J_{0F} and fluorescence at the fluorescence wavelength ($J_F = J - J_{0F}$). Equation 10.18 can therefore be rewritten as

$$\frac{J - J_{0F}}{J_{0F}} = R_F - 1 = \frac{J_0}{J_{0F}} m q_F a_m \tag{10.19}$$

where
  $J(\lambda)$ is the scattered light intensity of the sample (at the wavelength of the fluorescence)
  $J_{0F}(\lambda)$ is the intensity of the reference at the wavelength of fluorescence
  $J_0(\lambda)$ is the intensity of the reference at the wavelength of absorption
  m is the sample mass
  $q_F(\lambda)$ is the fluorescence quantum yield factor
  a_m is the mass absorption coefficient

At low analyte concentrations, the fluorescence intensity is directly proportional to the sample amount in the layer. In Equation 10.19, we can see the crucial advantage of fluorescence measurements compared with absorption: the fluorescence signal rises with the increasing intensity of the reference intensity at the wavelength of absorption (i.e., incident lamp intensity J_0) [1].

10.5.2 TLC FLUORESCENCE FOR HIGH ANALYTE CONCENTRATIONS

To measure the fluorescence of a sample at high concentration, the extended *Kubelka–Munk* equation (10.4) with $k \sim 0.5$ describes a linear dependency between fluorescence and sample mass. The sample absorbs light, and the fluorescence emission J_F is described by the quantum yield factor q_F according to Equation 10.17. If this equation is squared and the denominator extended by $2R$ and if the *Kubelka–Munk* equation (10.4) is taken into account, then we will observe the expression (10.20) [15]:

$$\frac{J_F^2}{J_0^2 2R} = q_F^2 \frac{(1-R)^2}{2R} = q_F^2 m \frac{a_m}{s} \tag{10.20}$$

Substituting for R, derived from Expressions 10.10 and 10.13 as well as $p = (1-k)(1-a)$ and $k = 1/2$,

$$R = \frac{1}{s} - \frac{1}{s}\sqrt{1 - s^2} = \frac{1}{1/s + 1/s\sqrt{1 - s^2}} = \frac{s}{1 + \sqrt{1 - s^2}}$$

provides

$$J_F^2 = m J_0^2 q_F^2 \frac{2a_m}{1 + \sqrt{1 - s^2}}$$

If we take into account that with large sample amounts (with strong light absorption) the scattering coefficient is $s \ll 1$, Equation 10.21 results in

$$J_F^2 = m J_0^2 q_F^2 a_m \tag{10.21}$$

where
  $J_F(\lambda)$ is the emitted fluorescence intensity
  m is the sample mass
  $q_F(\lambda)$ is the fluorescence quantum yield factor
  $J_0(\lambda)$ is the light intensity reflected from a clean plate surface
  a_m is the mass absorption coefficient

For large analyte mass, the squared fluorescence intensity is directly proportional to the analyte amount in the layer [1,15]. Taking Equations 10.18 and 10.21 into consideration, a general expression for fluorescence in scattering media can be described as

$$J_F^{1/k} = mJ_0^{(1/k)} q_F^{(1/k)} a_m \tag{10.22}$$

Equation 10.22 extends quantitative TLC and can be used for quantitative pharmaceutical analysis [15].

10.6 CONCLUSION

TLC is a modern separation method that can be beneficially used in pharmaceutical analysis. In comparison to other chromatographic methods like HPLC, the method is faster and less expensive. For example, the precision and accuracy of HPLC and HPTLC were compared using phospholipids as test compounds, and they were found to be comparable. Based upon a consideration of all possible cost sources, HPTLC was shown to be more cost-efficient than HPLC, leading to a 2.5-fold reduction in operating costs [16]. In combination with the modern theory of evaluation, planar chromatographic methods like TLC and HPTLC are well suited for all analytical demands in modern pharmaceutical analysis.

REFERENCES

1. Spangenberg, B., Poole, C. F., and Weins, C. 2010. *Quantitative Thin Layer Chromatography: A Practical Survey*, Springer, Berlin, Germany.
2. Ellison, S. L. R., Rosslein, M., and Williams, A. 2000. Eurachem/CITAC guide quantifying analytical uncertainty, 2nd edn., http://www.eurachem.org/guides/pdf/QUAM2000-1.pdf (accessed July 2, 2013).
3. Spangenberg, B. 2006. Does the Kubelka-Munk theory describe TLC evaluations correctly? *J. Planar Chromatogr.* **19**: 332–341.
4. Hiegel, K. and Spangenberg, B. 2009. A new method for the quantification of Dequalinium-cations in pharmaceutical samples by absorption and fluorescence diode array Thin-Layer Chromatography. *J. Chromatogr. A* **1216**: 5052–5056.
5. Ahrens, B., Spangenberg, B., and Klein, K. F. 2001. TLC-analysis in forensic sciences using a diode-array detector. *Chromatographia* **53**: S438–S441.
6. Broszat, M., Ernst, H., and Spangenberg, B. 2010. A simple method for quantifying triazine herbicides using thin-layer chromatography and a CCD camera. *J. Liq. Chrom. Rel. Technol.* **33**: 948–956.
7. Stroka, J., Peschel, T., Tittelbach, G., Weidner, G., van Otterdijk, R., and Anklam, E. 2001. Modification of an office scanner for the determination of aflatoxins after TLC separation. *J. Planar Chromatogr.* **14**: 109–112.
8. Spangenberg, B., Stehle, S., Ströbele, Ch. 1995. Quantitative DC mit einem Handscanner: Co^{2+}-Bestimmung. *GIT* **39**: 461–464.
9. Milz, B. and Spangenberg, B. A validated quantification of benzocaine in lozenges using TLC and a flatbed scanner. *Chromatographia.* DOI 10.1007/s10337-013-2436-4.
10. Kubelka, P. and Munk, F. 1931. Ein Beitrag zur Optik der Farbanstriche. *Z. Tech. Physik* **11a**: 593–601.
11. Kubelka, P. 1948. New contributions to the optics of intensely light-scattering materials: Part I. *J. Opt. Soc. Am.* **38**: 448–457 and **38**: 1067.
12. Spangenberg, B., Post, P., and Ebel, S. 2002. Fibre optical scanning in TLC by use of a diode-array detector—Linearization models for absorption and fluorescence evaluations. *J. Planar Chromatogr.* **15**: 88–93.
13. Koshy, T. 2008. *Catalan Numbers with Applications*. Oxford University Press, Oxford, U.K., ISBN 0-1953-3454-X.
14. Stanley, R. P. 1999. Enumerative combinatorics, *Cambridge Studies in Advanced Mathematics*, Vol. 2, p. 62, Cambridge University Press, Cambridge, U.K., ISBN 978-0-521-56069-6.
15. Spangenberg, B. and Weyandt-Spangenberg, M. 2004. Quantitative thin-layer chromatography using absorption and fluorescence spectroscopy. *J. Planar Chromatogr.* **17**: 164–168.
16. Renger, B. 1999. Benchmarking HPLC and HPTLC in pharmaceutical analysis. *J. Planar Chromatogr.* **12**: 58–62.

11 Statistical Evaluation and Validation of Quantitative Methods of Drug Analysis

Łukasz Komsta

CONTENT

The validation of an analytical method is defined as checking the correctness of the method, detection of its limits and estimation, if the method is useful to the assumed task, and if it fulfills the initial assumed requirements (Buick et al. 1990, Shah et al. 1991). Only fully validated method preserves correct analytic process and gives the results one may rely on. The validation is not only the statistical processing of the results—it is only a part of full validation (Cardone et al. 1990).

The definition of validation does not contain any requirements (Kollipara et al. 2011). Every analytical method has its own assumptions and requirements, so they can be various. Therefore, the first step of the validation is the estimation of the requirements and the parameters that can characterize them. The validation can be done in the following steps:

1. Setting the aim of the method and its assumptions
2. Experimental design of validation procedure
3. Estimation of instrumental requirements
4. Preparation of reagents and reference samples
5. Performing all comprehensive validation experiments (with possible corrections of the method to improve the results)
6. Analysis of the results, statistical evaluation of parameters, and interpretation
7. Checking if the aforementioned results fulfill the initial assumptions and requirements
8. Writing the procedures for routine analysis
9. Writing report and proposal of revalidation procedures

Actual recommendations can be found in documents publicly available from International Conference of Harmonization (ICH) (at the moment of writing of the chapter, the current version is designed as Q2(R1)).

What should we validate?

1. Selectivity (specificity) is defined by IUPAC as "the extent to which the method can be used to determine particular analytes in mixtures or matrices without interferences from other components of similar behavior." ICH defines it as "the ability to assess unequivocally the analyte in the presence of components which may be expected to be present." In the context of this definition, it is unneeded (and even impossible) to check for interferences of all possible drugs and excipients. We must take into the account only possible degradants, metabolites, and excipients occurring in analyzed formulations. In the case of TLC, there is a need

to prove that the spot of analyzed substance is pure (homogeneous) and not contaminated by any other substance. Additionally, ICH recommends the "lack of blank matrix response" method—developing of blank matrix chromatograms and proving there is no spot in the place where analyte spot occurs.

2. Accuracy is strictly connected with systematic error. It is defined as the difference between the result and true analyte content or concentration. The significance of this difference is most often tested against some reference value (e.g., the declared label content) or the result of another reference method. As the method can be biased by constant or varying systematic error (correlated with the concentration), the accuracy should be estimated at least at three concentration levels. The so-called fortified samples are often used for such estimation, as the content of reference substance is then accurately known.

3. Precision is connected with a random error (Thompson 1998). It is defined as the dispersion of results around some mean value. In most cases, it is presented as relative standard deviation (SD) (coefficient of variation) as a percentage value. The value is often compared with some reference values (known for such instrumentation) or results of alternative reference method. In TLC densitometry, a value in range 1%–5% can be considered as acceptable, whereas in videoscanning, literature often presents larger errors, up to 10%. The precision can be estimated at different levels, to estimate separately three sources of variation: measurement precision (when the same spot is measured several times by densitometer), spotting precision (when the same solution is spotted several times), and overall repeatability (precision of whole analysis in the same conditions, done by the same analyst during the same day).

4. Reproducibility is validated to preserve the possibility of method application in different laboratories, on different instrumentation, and by different analysts. Method that is not validated in this contest can be useless in other laboratories.

5. Sensitivity is most often expressed as limit of detection (LOD) and limit of quantitation (LOQ). The LOD is defined as concentration at which precision rises up to 33.3(3)%. The detection limit is met at precision equal to 10%. These values can be estimated from the calibration curve (by dividing standard error [SE] of residuals by slope at zero, then multiplying by coefficient equal to 3.3 or 10). The other method is to perform a series of analyses at different levels, finding the level with desired precision experimentally. The method used by analysts must be clearly given in the analysis report.

6. Stability is evaluated to ensure that sample does not decompose during analysis or storage, which would affect the results of whole analysis. The most often checked parameters are the following: long-term stability, freeze–thaw stability, and in-process stability.

7. Robustness or ruggedness is "a measure for the susceptibility of a method to small changes, that might occur during routine analysis." In the case of TLC, a small change in mobile phase composition, temperature, or development distance should be tested.

8. Calibration is one of the most extensive parts of validation—it relies on the estimation of real dependence between analyte amount and analytical response and the required error diagnostics.

The results of quantitative estimation of some analyte in a sample are small statistical samples from infinite population characterized by normal distribution (Hibbert 2006, 2007). The statistical evaluation of the results is indeed a statistical inference based on this assumption. Therefore, before statistical evaluation, it is a need to be sure that the results come from normal distributed population. Several tests are designed to check the results, normality; however, the most recommended one (due to small number of results) is Shapiro–Wilk test. It is based on the correlation of sample quantiles with ideal distribution quantiles. If this test detects significant deviation from normality (at 99% significance level), there is a need to inspect the data for possible outliers by Dixon or Grubbs test. Of course, a closer look for possible outliers is needed also in a sample not having deviation from normality. It is also worthy to underline that Dixon and Grubbs tests for outliers are not recursive, so after removing an outlier, one cannot test the remaining data for another possible outlier. Both Dixon and Grubbs tests have some variants for one or several suspected outliers.

If the sample is not deviated from normality and no outliers are detected, the first parameter to compute is the arithmetic mean of the results. The mean of the results is an estimator of true (unknown) mean from the whole infinite population. By dividing it by a reference value, one can obtain the recovery expressed in percents.

The main precision parameter is the SD of the results, which is the square root of the variance. It also can be treated as an estimator of variance and SD in infinite population. The SD has the same unit as analysis results; the variance has its square. In practice, the relative standard deviation (RSD) is most often given as percentage value (SD is divided by mean and multiplied by 100).

The next important parameter is the SE of the mean. It is calculated by dividing the SD by the square root of the number of results. In contrast to SD, it represents not the precision but the accuracy (an error of estimation of population mean by sample mean). It is most often given not directly but converted to confidence interval (an interval in which lies the true population mean by 95% or 99% probability). The confidence interval is calculated from Student's t-distribution, as the sample mean from normal distribution follows this distribution. It is also a good idea to compute a confidence interval for SD or RSD, based on chi-square distribution.

If the reference value (label one or given by reference method) lies inside the confidence interval, the method does not show any significant deviation in its accuracy. In the opposite case, one can assume that the method is biased by a significant systematic error. The computing of confidence interval is equivalent to computing t-test for one mean.

The precision between two methods should be compared by F-Snedecor test for homogeneity of variance. It is based on dividing variances of two samples—larger by smaller. The resulting ratio follows F-Snedecor distribution and the value that is larger than a reference value indicates a significant difference in precision.

The accuracy of two methods should be compared by Student's t-test for two means. This test assumes the homogeneity of variance, so it can be used only in the case of no differences in precision. If the difference in precision is detected, the Wilcoxon test should be used as a nonparametric alternative.

If one considers more than two methods to compare, the comparison can be done by Bartlett test (precision) and ANOVA test (accuracy). These tests allow only to estimate if there are any differences between these methods. The detailed comparison must then be done by multiple comparison tests (Dunnett, Duncan, or Tukey test). Similar to t-test, ANOVA should not be used in the case of difference in precision—the Kruskal–Wallis test is a good nonparametric alternative here.

One of the most comprehensive parts of method validation is the calibration. It is estimating of the dependence between analyte concentration and the analytical response. In most cases, the dependence should be linear (Sayago and Asuero 2004, Sayago et al. 2004); however, it is not the rule. In densitometric quantitative methods, there is often a nonlinear dependence observed, as the analytical response follows not Beer–Lambert linear law but Kubelka–Munk nonlinear dependence (Asnin et al. 2005, Soni et al. 2009).

The calibration is also an estimation of infinite population parameters, done on small sample obtained by analyzing a number of standard solutions. There should be at least six points on the calibration curve and each should be done in several replicates (for review on the dependence, see Laborda et al. 2004). This is the only way to obtain the dependence and characterize it from a statistical point of view. The true population parameters will not be ever known, so the good sample is essential to perform a good estimation.

The calibration response will never fit perfectly the obtained dependence, as it is biased by a measurement uncertainty. Therefore, a response can be treated as a random variable from a certain distribution, most often normal.

The ordinary least squares (OLS) method is the most frequent approach for calibration, both for linear and polynomial curves (Asuero and Bueno 2011, Asuero and Gonzalez 2007).

However, the analyst must be aware of its assumptions to use it correctly (Baumann and Watzig 1997, Betta and Dell'Isola 1996). The most often used coefficient of correlation (r), called Pearson's correlation coefficient, is not a very good measure of the fit and taking into account only this value may result in improper calibration process. One should always check the graph of residuals versus x (to detect outliers, curvilinearity, or heteroscedascity), chronologic residuals (to detect autocorrelation), and quantile–quantile plot (to detect outliers and residual deviation from normality).

The OLS method has the following assumptions:

1. The dependence is indeed linear. Fitting the straight line to nonlinear data causes a significant accuracy error, biasing all the results. As the correlation coefficient is not a measure of linearity (Kirkup and Mulholland 2004, Mulholland and Hibbert 1997), one should also inspect Mandel's test (comparison of linear and quadratic model by ANOVA) or lack-of-fit test (checking by ANOVA the within-concentration and inter-concentration variance of residuals) and see the shape of the residuals. If the nonlinearity is detected, a quadratic polynomial regression must be used (Nagaraja et al. 1999); nonlinear regression is used rarely (Andersen 2008, Tellinghuisen 2000). The cubic and higher-order polynomials would always fit better the calibration data, but their use is justified only if Akaike's information criterion (AIC) value is lower than for quadratic polynomial.

2. The uncertainty of the concentration (x variable) is equal to zero. It is impossible in practice, but calibration samples should be prepared with as small an error as possible. However, new approaches are present in literature, which propose some regression alternatives (Tellinghuisen 2010).

3. The distribution of uncertainty of analytical response is normal. It can be disturbed when data are transformed before calibration. Therefore, the transformation of the data from nonlinear to linear by logarithmic, Box–Cox transform (Box and Cox 1964) is currently in general not recommended, as one can achieve linear data but with bad residual distribution (McLean et al. 1990). The residuals should be always inspected on quantile–quantile plot and by Shapiro–Wilk normality test (Kimanani 1998, Kimanani and Lavigne 1998, Kimanani et al. 1998).

4. Homoscedascity of the error. If the residuals have "trumpet" shape, the error increases with the concentration, which indicates heteroscedascity. It can be confirmed by Bartlett test on residuals. In such cases, the weighted regression (De Levie 1986, Garden 1980, Tellinghuisen 2005, 2007, 2008) should be used and the weights should be optimized to minimize the mean relative error (Jain 2010). Another approach is to assume some variance function (Zeng et al. 2008). In the case of TLC, the heteroscedascity is quite a rare case; however, one must always check for it (Almeida et al. 2002).

5. The impresence of systematic error. Each calibration sample should be prepared from scratch. If one prepares one solution to make a series of calibration samples, the whole calibration is biased by common error of preparing this solution. The presence of systematic errors can be also detected (regardless of curvilinearity) by lack-of-fit test.

6. The lack of error autocorrelation. It can occur with temperature drift, overused lamp in densitometer, and so on. If the autocorrelation of residuals is detected, the calibration has to be repeated. The standard Durbin–Watson test can be used; however, it requires that samples are measured in random order, because one must be sure if it is correlation with time, not with concentration.

7. The impresence of outliers in calibration. They must be detected (Kimanani et al. 1996) and removed or robust regression method must be used. If one is unsure about some outlying observation, a standard diagnostic value, such as studentized residuals or Cook's distances, is useful.

If the proper regression method is chosen, the validation report should contain the significance values of estimators (calculated by t-test). The intercept term should be as insignificant as possible, whereas linear and quadratic terms should be highly significant. The significant intercept term suggests some matrix effect and a closer look at the method is required.

REFERENCES

Almeida, A., M. Castel-Branco, and A. Falcao. 2002. Linear regression for calibration lines revisited: Weighting schemes for bioanalytical methods. *Journal of Chromatography B: Analytical Technologies in the Biomedical and Life Sciences* 774(2): 215–222.

Andersen, J. 2008. Investigation of the performance of the instrument by nonlinear calibration. *Microchimica Acta* 160(1–2): 89–96.

Asnin, L., W. Galinada, G. Gatmar, and G. Guiochon. 2005. Calibration of a detector for nonlinear chromatography. *Journal of Chromatography A* 1076(1–2): 141–147.

Asuero, A. and J. Bueno. 2011. Fitting straight lines with replicated observations by linear regression. IV. Transforming data. *Critical Reviews in Analytical Chemistry* 41(1): 36–69.

Asuero, A. and G. Gonzalez. 2007. Fitting straight lines with replicated observations by linear regression. III. Weighting data. *Critical Reviews in Analytical Chemistry* 37(3): 143–172.

Baumann, K. and H. Watzig. 1997. Regression and calibration for analytical separation techniques. Part I: Design considerations. *Process Control Quality* 10(1–2): 59–73.

Betta, G. and M. Dell'Isola. 1996. Optimum choice of measurement points for sensor calibration. *Measurement: Journal of the International Measurement Confederation* 17(2): 115–125.

Box, G.E.P. and D.R. Cox. 1964. An analysis of transformations. *Journal of the Royal Statistical Society* 26: 211–252.

Buick, A.R., M.V. Doig, S.C. Jeal, G.S. Land, and R.D. McDowall. 1990. Method validation in the bioanalytical laboratory. *Journal of Pharmaceutical and Biomedical Analysis* 8(8–12): 629–637.

Cardone, M.J., S.A. Willavize, and M.E. Lacy. 1990. Method validation revisited: A chemometric approach. *Pharmaceutical Research* 7(2): 154–160.

De Levie, R. 1986. When, why, and how to use weighted least squares. *Journal of Chemical Education* 63(1): 10–15.

Garden, J.S. 1980. Nonconstant variance regression techniques for calibration-curve-based analysis. *Analytical Chemistry* 52(14): 2310–2315.

Hibbert, D. 2006. The uncertainty of a result from a linear calibration. *Analyst* 131(12): 1273–1278.

Hibbert, D. 2007. Systematic errors in analytical measurement results. *Journal of Chromatography A* 1158(1–2): 25–32.

Jain, R. 2010. Comparison of three weighting schemes in weighted regression analysis for use in a chemistry laboratory. *Clinica Chimica Acta* 411(3–4): 270–279.

Kimanani, E. 1998. Bioanalytical calibration curves: Proposal for statistical criteria. *Journal of Pharmaceutical and Biomedical Analysis* 16(6): 1117–1124.

Kimanani, E. and J. Lavigne. 1998. Bioanalytical calibration curves: Variability of optimal powers between and within analytical methods. *Journal of Pharmaceutical and Biomedical Analysis* 16(6): 1107–1115.

Kimanani, E., M. Mihailovici, J. Gaudreault, J. Lavigne, and R. Lalonde. 1996. A statistical procedure to fit and detect outliers in calibration curves. *Pharmaceutical Science* 13(9): 22.

Kimanani, E., M. Mihailovici, J. Lavigne, and J. Gaudreault. 1998. Bioanalytical calibration curves, variability of optimal powers across different analytical techniques and analytes. *Journal of Pharmaceutical Science* 13(9): 22.

Kirkup, L. and M. Mulholland. 2004. Comparison of linear and non-linear equations for univariate calibration. *Journal of Chromatography A* 1029(1–2): 1–11.

Kollipara, S., G. Bende, N. Agarwal, B. Varshney, and J. Paliwal. 2011. International guidelines for bioanalytical method validation: A comparison and discussion on current scenario. *Chromatographia* 73(3–4): 201–217.

Laborda, F., J. Medrano, and J. Castillo. 2004. Influence of the number of calibration points on the quality of results in inductively coupled plasma mass spectrometry. *Journal of Analytical Atomic Spectrometry* 19(11): 1434–1441.

McLean, A., D. Ruggirello, C. Banfield, M. Gonzalez, and M. Bialer. 1990. Application of a variance-stabilizing transformation approach to linear regression of calibration lines. *Journal of Pharmaceutical Science* 79(11): 1005–1008.

Mulholland, M. and D.B. Hibbert. 1997. Linearity and the limitations of least squares calibration. *Journal of Chromatography A* 762(1–2): 73–82.

Nagaraja, N., J. Paliwal, and R. Gupta. 1999. Choosing the calibration model in assay validation. *Journal of Pharmaceutical and Biomedical Analysis* 20(3): 433–438.

Sayago, A. and A. Asuero. 2004. Fitting straight lines with replicated observations by linear regression: Part II. Testing for homogeneity of variances. *Critical Reviews in Analytical Chemistry* 34(3–4): 133–146.

Sayago, A., M. Boccio, and A. Asuero. 2004. Fitting straight lines with replicated observations by linear regression: The least squares postulates. *Critical Reviews in Analytical Chemistry* 34(1): 39–50.

Shah, V.P., K.K. Midha, S. Dighe, I.J. McGilveray, J.P. Skelly, A. Yacobi, T. Layloff et al. 1991. Analytical methods validation: Bioavailability, bioequivalence and pharmacokinetic studies. *European Journal of Drug Metabolism and Pharmacokinetics* 16(4): 249–255.

Soni, T., N. Chotai, P. Patel, L. Hingorani, R. Shah, N. Patel, and T. Gandhi. 2009. Evaluation of an optimum regression model for high-performance thin-layer chromatographic analysis of aceclofenac in plasma. *Journal of Planar Chromatography—Modern TLC* 22(2): 101–107.

Tellinghuisen, J. 2000. A simple, all-purpose nonlinear algorithm for univariate calibration. *Analyst* 125(6): 1045–1048.

Tellinghuisen, J. 2005. Simple algorithms for nonlinear calibration by the classical and standard additions methods. *Analyst* 130(3): 370–378.

Tellinghuisen, J. 2007. Weighted least-squares in calibration: What difference does it make? *Analyst* 132(6): 536–543.

Tellinghuisen, J. 2008. Weighted least squares in calibration: The problem with using "quality coefficients" to select weighting formulas. *Journal of Chromatography B: Analytical Technologies in the Biomedical and Life Sciences* 872(1–2): 162–166.

Tellinghuisen, J. 2010. Least squares in calibration: Dealing with uncertainty in x. *Analyst* 135(8): 1961–1969.

Thompson, M. 1988. Variation of precision with concentration in an analytical system. *The Analyst* 113(10): 1579–1587.

Zeng, Q., E. Zhang, H. Dong, and J. Tellinghuisen. 2008. Weighted least squares in calibration: Estimating data variance functions in high-performance liquid chromatography. *Journal of Chromatography A* 1206(2): 147–152.

12 Detection and Identification in TLC Drug Analysis

Łukasz Cieśla and Iwona Kowalska

CONTENTS

12.1 INTRODUCTION

Thin-layer chromatography (TLC), among other chromatographic and electrophoretic techniques, is an important tool in the analysis of drugs. It is used to quantify active principles or degradation products in pharmaceutical formulations or biological fluids, to detect adulterations, or in the development of new active principles. One of the TLC advantages is the possibility of using different detection

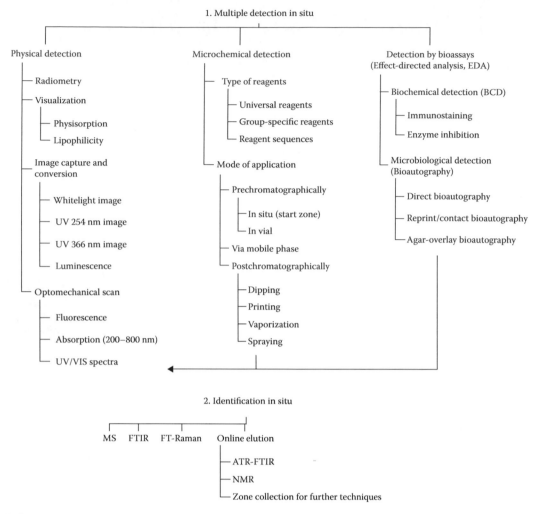

FIGURE 12.1 Detection and identification modes applied in TLC. (From Morlock, G. and Schwack, W., *J. Chromatogr. A*, 1217, 6600, 2010. With permission.)

methods for the separated compounds. Generally, detection techniques used in TLC can be divided into three main categories—physical, chemical, and biological, as presented in Figure 12.1 [1].

TLC detection methods can also be ascribed to one of the two other groups—nondestructive and destructive ones. In the former case the compounds adsorbed on the plate surface remain unchanged after completing the detection step. The majority of physical modes of detection can be classified as nondestructive ones. In case of destructive methods, the detection step leads to changes in chemical and physical properties of a compound.

Direct identification of the separated compounds is not possible with the application of traditional chemical and optical methods. Comparing R_F values and UV–Vis or fluorescence excitation and emission spectra, obtained for the analytes and references, is in many cases insufficient to confirm the identity of a compound. One of the possibilities is scraping off the separated substances, their extraction, purification, and concentration, which precede the use of spectroscopic and spectrometric techniques for structure elucidation (nuclear magnetic resonance [NMR], mass spectrometry [MS]). For direct structural characterization, several coupled techniques (TLC hyphenations) have been proposed, for example, TLC–MS, TLC–infrared (IR), TLC–NMR, TLC–Raman spectroscopy. From the previously mentioned ones, only TLC–MS has gained a meaningful position in routine analysis; several interfaces enabling direct coupling of TLC with MS are commercially available [2].

Most recently, the coupling of different detection methods after a single TLC run has gained much attention, as proved in the review article by Morlock and Schwack [1]. Such an approach, belonging to hypernation techniques, is especially useful in effect-directed analysis (EDA) aimed at detection and identification of compounds with desired activity (e.g., antimicrobials, antioxidants, or enzyme inhibitors).

This chapter gives an overview of different techniques used for detection and identification of compounds separated by means of planar chromatography. There are several fundamental books and book chapters that cover similar subjects, just to mention some of the recent ones by Spangenberg [3], Tyihak et al. [4], Reich and Schibli [5], Botz et al. [6], Cimpan [7], and others. In order not to duplicate information contained within these chapters, the most recent solutions have been presented in more detail, as, for example, the application of TLC coupled with bioautography for detection of enzyme inhibitors and direct antioxidants. Already described solutions have been presented in a more condensed form.

12.2 PHYSICAL METHODS OF DETECTION AND IDENTIFICATION

In most popular physical detection modes, absorption and fluorescence properties of the analyzed compounds are used [8]. The aforementioned physical methods of detection are nondestructive ones; therefore, after completing the detection step, the adsorbed compounds may undergo further analyses. Apart from spectrophotometric techniques, the following coupled methods fall into physical modes of detection—TLC–IR, TLC–Raman spectroscopy, TLC–NMR, and TLC–MS. As already mentioned, only the last one has gained much attention for routine analyses. Contrary to techniques making use of absorption or fluorescence properties, TLC coupled with MS is a destructive technique. Generally in all TLC–MS modes, the compound molecules are first desorbed from the layer and further transported to the MS apparatus [9]. During the analysis, the molecules undergo fragmentation into ions. Frequently TLC–MS is combined with other detection techniques, for example, UV/Vis, fluorescence detection, bioautography, or chemical derivatization [1].

In the subchapters, the most popular physical detection modes used in TLC are described.

12.2.1 TLC with Spectrophotometric Detection (TLC–UV/Vis, TLC–FLD, TLC–DAD)

Spectrophotometric mode of detection is the most popular one in TLC that involves illuminating a TLC plate with light of known intensity [3]. The TLC plate can be inspected by means of a human eye (in the case of compounds absorbing or emitting fluorescent light in visible region) or by means of a densitometer (also referred to as scanner) that scans the plate with single or multiple wavelengths. Depending on the mode used, spectrophotometric detection techniques can be referred to as TLC–UV/Vis, detection performed in absorbance mode at single wavelength; TLC diode-array densitometry (TLC–DAD), detection performed in absorbance mode at multiple wavelength; and TLC–fluorescence detection (TLC–FLD), performed in fluorescence mode.

A densitometer measures the reflected light, and more precisely the difference between the signal from a sample zone and sample-free adsorbent is measured [10]. A compound that absorbs the light causes the reflected light to show lower intensity than the illuminating one [3]. Unfortunately in the case of TLC, apart from light absorption, light scattering appears due to the presence of adsorbent layer and glass support [10]. There is no linear relationship between compound concentration and signal intensity. The calibration curves of peak areas against compound concentration are hyperbolic in shape, according to the Kubelka/Munk theory [11]. For lower compound concentrations, pseudo-linear range can be obtained [10].

In fluorescence measurements, a TLC plate surface is irradiated with a specific excitation wavelength, and a compound produces fluorescent light of longer wavelength than that of the exciting light [10]. Direct detection in fluorescence has several advantages over absorption measurements, namely, greater selectivity, higher sensitivity, linear relationship between fluorescent intensity and compound concentration, and band shape not influencing the signal [10]. In order to enhance fluorescence, the plate may be treated with some reagents, for example, exposure to ammonia vapors.

A common practice in TLC analysis is also the use of plates containing fluorescence indicator, excited by UV light, usually at 254 or 366 nm, resulting in a green or blue background. Compounds absorbing at approx. 254 or 366 nm cause fluorescence quenching, appearing as dark spots on the colored background.

Planar chromatography with diode array scanning is not as popular as HPLC–DAD in routine laboratory analysis. However, it has several advantages over single wavelength scanning: It permits parallel recording of chromatograms and UV–Vis spectra in a wide wavelength range; each compound is quantified at its λ_{max}, ascertaining optimum sensitivity; the compounds' UV–Vis spectra can be compared with library spectra; and peak purity can be assessed [12,13]. Ahrens et al. proved the applicability of fiber-optical scanning in TLC for identification of codeine, propyphenazone, tramadol, flupirtine, lidocaine, and diphenhydramine in urine samples [14].

12.2.2 TLC–MS

TLC hyphenated with mass spectrometric detection is more difficult to perform when compared with GC/MS or HPLC/MS due to adsorption of analyte compounds on the TLC layer [2]. More recently, TLC–MS has gained much attention among analysts and several review articles summarizing technical solutions and applications have been published [1,2,9,15].

TLC hyphenation with MS can be classified into one of the two different modes: (1) indirect and (2) direct analyses, as also indicated in Figure 12.2 [2]. Taking into account the working environment of the ion source, the TLC–MS online detection can be further divided into two subcategories: vacuum-based and ambient MS.

In indirect sampling TLC–MS, several pretreatment steps are needed before introducing the sample to the ion source, as also indicated in Figure 12.2 [2]. Prior to the analysis, the use of nondestructive detection methods is needed in order to localize the spot to be scraped off. For direct TLC–MS, the use of special interfaces is needed to transfer the adsorbed compound to the ion source without any pretreatment.

In indirect TLC–MS, after all the aforementioned pretreatment steps, the sample is introduced into the ion source using a direct inlet system. Such an approach has been used, for example, to analyze diazepam in cream biscuits [16]. After separation on the TLC layer, the plates were scanned

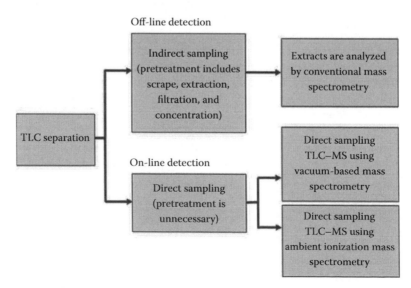

FIGURE 12.2 Types of TLC–MS techniques. (From Cheng, S.C. et al., *J. Chromatogr. A*, 1218, 2700, 2011. With permission.)

at 230 nm and spots of target compounds were scraped from the plate, extracted with ethanol, and analyzed by MS. The analyte molecules were ionized through electron impact ionization (EI).

The following vacuum-based desorption/ionization techniques have been applied in direct sampling TLC–MS: fast atom bombardment (FAB), liquid secondary ion mass spectrometry (LSIMS), laser desorption/ionization (LDI), matrix-assisted laser desorption/ionization (MALDI), and surface-assisted laser desorption/ionization (SALDI) [2]. These techniques are suitable to ionize nonvolatile and thermally labile compounds, but however, are of limited use for analyzing highly volatile compounds. In these techniques, a TLC plate, with separated target compounds, is placed in a vacuum chamber. Due to small dimensions of the chamber, the TLC plate should be cut into smaller pieces to fit it [2].

FAB and LSIMS analyses use fast atoms (Ar) and ions (Cs$^+$) to desorb and ionize the sample molecules, as also seen in Figure 12.3. Prior to the analysis, the TLC plate should be covered with glycerol [2]. The desorbed and ionized compounds are detected usually by quadrupole or time-of-flight (TOF) analyzers [2]. Cheng et al. stress that higher-resolution imaging is possible with the use of LSIMS when compared to FAB, due to better-focused fast ion beam [2]. The use of FAB or LSIMS TLC–MS is restricted to polar and nonvolatile compounds. These techniques are also not suitable for compounds with molecular weights greater than 2000 Da, as they are not desorbed. As far as drug analysis is considered, TLC–MS with FAB has been used, for example, in the analysis of morphine or coccidiostats [2].

Other techniques for direct TLC–MS, using vacuum-based ionization, involve the application of pulsed laser irradiation for desorption and ionization and include LDI, MALDI, and SALDI [2]. In these techniques, after attaching the plate to the sample probe, the target compounds are desorbed and ionized with the use of a pulsed laser beam. In the LDI technique, the laser beam irradiates directly the sample spot, while in MALDI and SALDI, the laser is used in the presence of an additional matrix put on the spot surface. In MALDI, an organic matrix solution is used, while in SALDI, it is a suspension of one of the following substances: carbon powder, Co, TiN, TiO$_2$, and silicon.

The development of atmospheric pressure ionization MS was another step ahead that significantly progressed TLC–MS analyses. In atmospheric pressure techniques, the TLC plate does not have to be placed in a vacuum chamber for ionization. Contrary to vacuum-based ionization techniques, volatile samples can also be analyzed. The following techniques have been applied for ambient ionizing of compounds adsorbed on solid surfaces: laser desorption/atmospheric pressure chemical ionization (LD/APCI), laser ablation inductively coupled plasma (LA-ICP), desorption electrospray ionization (DESI), direct analysis in real time (DART), electrospray laser desorption ionization (ELDI), and laser-induced acoustic desorption/electrospray ionization (LIAD/ESI) [2].

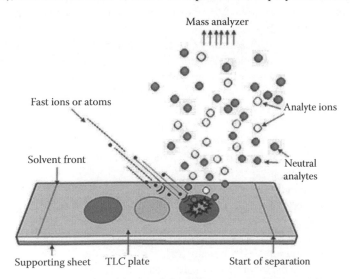

FIGURE 12.3 Schematic representation of FAB and LSIMS analyses of chemical compounds on the surface of a TLC plate. (From Cheng, S.C. et al., *J. Chromatogr. A*, 1218, 2700, 2011. With permission.)

FIGURE 12.4 TLC–MS interface commercially available with CAMAG. (From http://www.camag.com/v/products/tlc-ms/. With permission.)

Most frequently, TLC is interfaced with the following solution-based atmospheric pressure ionization mass spectrometry techniques—ESI/MS or APCI/MS. Several interfaces have been proposed to extract and feed the adsorbed compounds to ESI or APCI sources, and some of them have been commercialized, as the one presented in Figure 12.4 [17].

TLC/MS with ESI ion sources is suitable for the analysis of high- and mid-polar compounds. To desorb and introduce the separated compounds to an ESI mass spectrometer, several sampling methods have been proposed, as one can see in Figure 12.5.

Two different devices have been constructed to extract and deliver the extracted solution to the ESI source [2]. In general the extraction junction for sampling consists of a coaxial tube, which is fixed vertically to the plate surface. It delivers the extraction solvent to the plate through the space between the inner and outer tubes. The extracted compound, in a solution, is pumped to the ESI source through the inner tube [2]. In the other device, two separate capillaries are used—one to deliver the extracting solvent and the other one to transport the extracted solution to the ESI source [2].

The compounds separated and adsorbed on the TLC surface can also be introduced to the ESI source using continuously eluting devices. For such solutions, generation of ESI requires modifications of a TLC plate, as seen in Figure 12.6 [2].

The compounds adsorbed on the surface of a TLC plate can be also desorbed with the use of laser irradiation and further ESI—a technique called electrospray-assisted ELDI [2]. After desorption, neutral molecules of a compound enter the methanol/water ESI plume where they are ionized. Cheng et al. proposed a modification of this technique, called laser-induced acoustic desorption/electrospray ionization mass spectrometry (LIAD–ESI/MS) [18]. In this approach, the rear side of an aluminum TLC plate is irradiated with a pulsed IR laser. To desorb compounds more efficiently, a glass slide is attached to the TLC plate (the rear side) and the gap between them filled with glycerol [18].

In DESI, the TLC surface is impacted with pneumatically charged solvent droplets causing compound desorption and ionization [2].

Apart from ESI, TLC can be interfaced with APCI, using approaches similar to that applied in ESI-based TLC/MS, namely, continuous eluting device, desorption with further ionization, and desorption and ionization directly on the plate [2].

From among TLC/MS analyses coupled with APCI ionization, DART is an interesting solution [2]. In this technique, excited hot inert gas, for example, helium, is used to desorb and ionize the analyzed compounds. One of the greatest advantages of this solution is the fact that DART does not require the use of liquids, which are responsible for the spot's shape distortions in other ambient desorption ionization techniques, as in the case of DESI [15]. TLC–DART–MS analyses have been proved to be

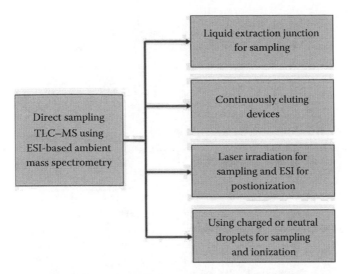

FIGURE 12.5 Examples of direct TLC–MS sampling methods with the use of ESI-based MS. (From Cheng, S.C. et al., *J. Chromatogr. A*, 1218, 2700, 2011. With permission.)

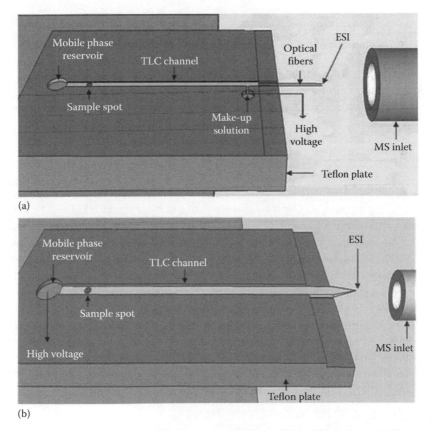

FIGURE 12.6 Examples of continuously eluting devices for direct TLC–MS analyses. (a) Two bound optical fibers are inserted into the C18 gel bed at the other end of the channel. (b) A small aluminum TLC strip with a sharpened end. (From Cheng, S.C. et al., *J. Chromatogr. A*, 1218, 2700, 2011. With permission.)

FIGURE 12.7   An example of the most recently introduced DART ion source model for performing direct TLC–MS analyses. (From Morlock, G.E. and Chernetsova, E.S., *Cent. Eur. J. Chem.*, 10, 703, 2012. With permission.)

suitable for the analysis of different classes of compounds, including drugs, for example, haloperidol, erythromycin, and acetaminophen [15]. The newest setup, presented in Figure 12.7, is commercially available and has been proved to deliver reliable results of qualitative and quantitative analyses [15].

As far as drug analysis is concerned, TLC–MS techniques have been applied for the analysis of the following exemplary compounds: prilates [19]; oseltamivir [20]; proguanil, ibuprofen, sitamaquine, paracetamol, and simvastatin [21]; diclofenac sodium, tramadol, and carbamazepine [22]; and many others.

12.2.3   OTHER PHYSICAL DETECTION AND IDENTIFICATION MODES

Apart from the aforementioned techniques, other physical modes of detection have gained less interest and applications. Some of them were characterized in a review article by Somsen et al. [23].

For identification purposes, TLC–NMR hyphenation has been proposed [24]. The serious drawbacks that cause the infrequent use of TLC–NMR hyphenation are related mainly to high instrument costs as well as sensitivity limitations [24]. Gössi et al. described TLC–NMR analysis for several exemplary compounds of both natural and synthetic origin, for example, amiodarone, rutin, caffeic acid, and chlorogenic acid [24]. The zones corresponding to the analyzed compounds were extracted from the TLC plate with a CAMAG TLC–MS interface [24]. The authors proved the applicability of TLC–NMR for quantitative purposes.

Planar chromatography has been also hyphenated with Fourier transform infrared spectroscopy (TLC–FTIR), enabling identification of the separated compounds [25]. The applicability of this technique has been proved for the drugs of abuse (e.g., amphetamines, members of the ecstasy group, or morphine derivatives) or benzodiazepines [25]. More detailed method description can be found elsewhere [25].

12.3   CHEMICAL METHODS OF DETECTION

The use of TLC for the separation, identification, and quantification of different organic and inorganic substances is often limited by the visualization of the separated and determined substances. In this subchapter, attention is paid to the different chemical reactions applied to obtain colored substrates of the reaction between separated substances and the chosen derivatizing reagents. The subject of chemical derivatization has been widely covered in literature; therefore, the most important issues related to this detection mode are summarized and presented here in a condensed form.

Chemical visualization consists in moistening of the plate with a reagent or a mixture of reagents, which results in developing the colored products of the reaction with the analytes. Sometimes additional plate handling is required, for example, heating, or stabilization of colored spots. If a developed plate is meant for further analysis, it is important to use the substances that will not destroy the remaining spots [26,27].

Detection is simple when the compounds of interest are naturally colored or fluorescent or absorb UV light. However, application of a detection reagent by spraying or dipping is required to produce color or fluorescence for most compounds [28].

Since relatively many samples can be analyzed by TLC in several hours, it is widely used as a simple method for the detection and tentative identification of drugs. For detecting each spot, a reagent solution specially prepared can be sprayed on the plate to detect a compound specifically [29]. Chemical methods of derivatization are complementary to physical, nondestructive techniques. Chemical derivatization is used to enhance analyte detection, to improve quantitation, and to facilitate structure elucidation. The derivatization process can be applied before or after chromatographic development, as prechromatographic and postchromatographic derivatization.

12.3.1 DERIVATIZATION MODES

12.3.1.1 Prechromatographic Derivatization

Prechromatographic derivatization is an easy way of changing the properties of the entire sample or some portions of it. In TLC, derivatization can be performed on the plate immediately after the samples have been applied or even before the application step. The plate can be dipped into a reagent or, more elegantly, the reagent can be sprayed only onto the samples [5].

Prechromatographic derivatization, which introduces a chromophore leading to the formation of strongly absorbing or fluorescent derivatives, increases the selectivity of the separation and the sensitivity of detection and improves the linearity.

The aim of prechromatographic derivatization is usually different from that of postchromatographic derivatization, where the aim is first to detect the substance and then only secondarily to characterize it. In the case of prechromatographic derivatization, the main goal is usually the improvement of separation of substances with similar chromatographic properties. Advantages of prechromatographic derivatization include reduction of the detection limit by introducing chromophoric groups in the structure of the molecule, simple and rapid application of the reagent(s), and increasing linearity in quantitative analysis. Prechromatographic derivatization can be also advantageous when the parent compounds are too volatile for TLC. Prechromatographic derivatization is used to form less volatile derivatives. Such derivatives are easier to separate from other sample constituents; they are characterized with greater stability and more successfully extracted and/or cleaned up, and they are also more sensitively and/or selectively detected. Prechromatographic derivatization may be also advantageous in case other compounds are unstable on the surface of a TLC plate. A disadvantage of prederivatization is that in some cases, the introduction of usually high molecular weight functional groups into the derivative may equalize the chromatographic properties of similar substances and make separation more difficult [28].

The following reaction types may be applied in the prechromatographic derivatization step: hydrolysis, oxidation and reduction, halogenations, nitration and diazotization, hydrazone formation, esterification and etherification, dansylation, etc. [7].

The following drugs can be given as an example of the application of prechromatographic derivatization in routine analysis: cephalosporins, benzodiazepines, analgesic compounds, carboxylic acids, and so on [7,30].

12.3.1.2 Postchromatographic Derivatization

Postchromatographic derivatization is the most popular method used to visualize separated compounds after TLC separation. After the compound separation and drying step, the plate is evenly covered with derivatizing agent. Very often, additional plate handling step is required to complete the visualization procedure, for example, heating. Postchromatographic derivatization in TLC allows the reaction of all standards and samples simultaneously under the same conditions, and the separation of the solutes is not changed by the reaction [28]. Such reactions can be performed with the use of universal reagents or selective ones based on the differences caused by functional groups present in the analyzed molecules. Postchromatographic chemical derivatization for compound detection can be followed by quantification of the separated substances.

The aim of a postchromatographic derivatization is first the detection of the chromatographically separated substances in order to be able to visually evaluate the chromatogram. But equally important is also increasing the selectivity, which is often associated with this and improving the detection sensitivity [31]. One of the greatest advantages of postchromatographic derivatization is the variety of reagents that can be used to visualize the separated compounds.

According to Cimpan, the postchromatographic derivatization can be divided into two subcategories: nondestructive and destructive derivatization [7]. The majority of the applied reagents cause changes in physical and chemical properties of the analyzed compounds; therefore, they are included into the group of destructive agents. Nondestructive techniques involve the application of agents that react with the analyzed compounds in a reversible way.

Undeniably, the possibility of using postchromatographic specific or nonspecific chemical derivatization is a strong point of TLC when compared to other chromatographic techniques.

12.3.2 METHODS OF APPLYING DERIVATIZING AGENTS

12.3.2.1 Dipping

From among all the derivatization procedures, dipping has become the most popular one as it allows obtaining a homogenous reagent layer, which is important for reliable quantitative analysis. What is more, derivatization results depend less on the skills of the operator, when compared with spraying. The procedure is also environmentally friendly.

However, it should always be kept in mind that this method should be used only if the substances separated on the plate are not dissolved by the dipping reagent [7]. Compared with spraying, dipping reagents are usually less concentrated than the corresponding spray solutions, and the solvents should be carefully chosen to avoid the dissolution of the layer, of the separated substances, or of the reaction products. One of the dipping disadvantages is the fact that it is a more expensive way than spraying due to larger volumes of reagent solution that have to be prepared.

In most TLC analyses, dipping is the preferred method and should be used whenever possible. Figure 12.8 presents a commercially available CAMAG dipping device.

The standard procedure, with the use of a dipping device, involves the following steps:

- Charging the tank of an immersion device with enough reagent to ensure complete immersion of the chromatogram
- Placing plate in holder of immersion device, setting conditions according to the method, and pressing "start"
- Letting excess reagent drip off plate, wiping back of plate with paper towel, and removing plate from plate holder
- Drying plate with cold air (vertically in direction of chromatography) or heating it directly in accordance with the method

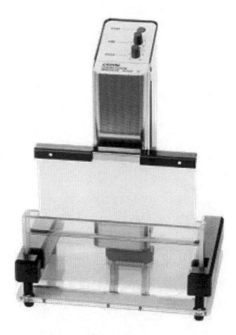

FIGURE 12.8 Chromatogram immersion device commercially available with CAMAG. (From http://www. camag.com/en/tlc_hptlc/products/derivatization/chromatogram_immersion_device.cfm. With permission.)

12.3.2.2 Spraying

During method development, spraying is usually the primary choice when searching for the most suitable reagent. Spraying advantages are as follows: small volumes of reagent are used, it is simple and quick especially when very small plates have to be derivatized, and no expensive equipment is needed. However, a serious drawback relates to the fact that it is difficult to achieve a homogenous and defined derivatization across the plate [5].

Spraying includes the following steps (when performed in a TLC spray cabinet):

- Charging the bottle of the sprayer with up to 50 mL reagent
- Placing plate in spray cabinet or fume hood upright against a filter paper or a paper towel
- Spraying plate with horizontal and vertical motion until it is homogeneously covered with reagent
- Drying plate with cold air or heat it directly according to method

12.3.2.3 Exposure to Vapor

Another way of plate derivatization is exposing the plate to reagent vapors. Usually ammonia vapors are used to induce fluorescence in the case of nonfluorescent compounds [7]. Iodine vapor is also commonly used to derivatize separated compounds. It causes the appearance of yellow and brown spots. The process is reversible. Other reagents that may be used include hydrochloric acid, t-butyl hypochlorite, trifluoroacetic acid, and sulfuryl chloride. To perform derivatization, the plate and reagent are placed in twin-trough chambers and led to stand for several minutes. Horizontal chambers may also be applied; in this case, the plate is placed face down and the reagent is poured into a trough at the bottom of the chamber.

12.3.2.4 Rolling

Most recently, a new concept of application of a derivatizing agent has emerged—rolling. It was developed to detect compounds with antimicrobial properties (Section 12.4.1) after chromatographic separation.

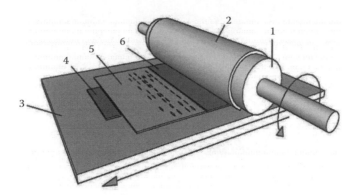

FIGURE 12.9 A rolling device for covering a TLC plate with bacterial strain suspension. (For description of numbers, please refer to the original submission.) (From Baumgartner, V. et al., *J. Chromatogr. A*, 1218, 2692, 2011. With permission.)

Using only simple appliances, a device has been constructed (Figure 12.9) that enabled even covering of a TLC plate with a derivatizing agent. The application of this method caused the peaks of the separated compounds to be narrower with a higher signal-to-noise ratio and without any tailing [32].

12.3.2.5 Derivapress

Derivatization can also be performed with the use of a device called Derivapress [33]. The application of this appliance is particularly useful in pharmaceuticals, cosmetics, chemicals, food, and biological applications. It allows the application of the derivatization reagent on the TLC plate, with optimal conditions of uniformity, safety, and environment preservation. Contrary to spray techniques, the uniformity of the distribution of the reagent by Derivapress allows quantitative post-derivatization TLC analysis.

Apart from the aforementioned techniques, the plates can be derivatized with the use of a reagent incorporated into the stationary or mobile phase.

12.3.2.6 Derivatization Completing Steps

Chemical derivatization is often completed by other steps including plate heating or stabilization of the colored spots/bands. The two principal heating devices are ovens and hot plates (plate heaters). For obtaining reliable results, heating the chromatogram plate with a plate heater specifically designed for this purpose is highly recommended. However, in everyday laboratory practice, the majority of TLC analyses are completed by heating in an oven. An important thing to remember is to place the plate on a smooth surface; otherwise the pattern of the oven grid may become visible. In the case of the use of a plate heater, the following steps are required:

- Turning on plate heater and selecting temperature
- Waiting until the temperature is stable
- Placing plate on plate heater
- After the time specified by the method, removing hot plate from heater [5]

12.3.3 Exemplary General and Specific Derivatizing Agents

The difference between general and specific derivatizing agents is in some sense conventional. A reagent can be regarded as general derivatizing agent when it can be applied for the visualization of a vast group of substances, differing significantly in their structures. Specific agents react only with compounds from one or only few chemical groups. Chosen examples of general and specific derivatizing agents are given in Tables 12.1 and 12.2.

TABLE 12.1
Examples of the Most Popular Universal Reagents

Reagent Name	Preparation, Use	Detection and Effect	Reference
Sulfuric acid	20 mL H_2SO_4 + 180 mL ice-cold methanol. Heat the plate at 100°C for 5 min	General reagent; dark spots of organic compounds (white light and UV light—366 nm)	[3,5]
Vanillin–sulfuric acid	15 g vanillin dissolved in 250 mL ethanol and 2.5 mL concentrated H_2SO_4. Heat the plate at 100°C for 5 min	General reagent; different colors for variety of compounds (white light and UV light—366 nm)	[5]
Iodine vapors	TLC plate inserted to a chamber saturated with iodine vapors	Unsaturated and aromatic compounds; dark brown spots on a yellowish background	[7]
Anisaldehyde–sulfuric acid	0.5 mL anisaldehyde + 10 mL glacial acetic acid + 85 mL methanol + 5 mL concentrated H_2SO_4, in that order. Heat the plate at 100°C for 5–10 min	Universal reagent; different colors for variety of compounds	[5]
Vanillin–glacial acetic acid	0.8 g vanillin + 40 mL glacial acetic acid + 2 mL concentrated H_2SO_4. Heat the plate for 3–5 min at 110°C	General reagent; different colors for variety of compounds (white light and UV light—366 nm)	[5]
Fast blue salt B	0.5 g fast blue salt B dissolved in 100 mL water	Different colors for compounds with phenolic groups	[5]
Iron (III) chloride	1 g iron (III) chloride + 5 mL water diluted to 100 mL with ethanol. The plate may be heated at 110°C for 10 min	Compounds with phenolic groups	[5]
Phosphomolybdic acid	5–20 g phosphomolybdic acid in 100 mL ethanol	Universal reagent, usually used for reducing substances; dark-blue or grey zones on yellowish background	[3,5]
Dinitrophenylhydrazine (DNP)	12 g of 2,4-dinitrophenylhydrazine + 60 mL concentrated H_2SO_4. + 80 mL water in 200 mL of 95% ethanol	Mainly aldehydes and ketones; yellow or orange spots	[7]
Potassium permanganate	1.5 g $KMnO_4$ + 10 g K_2CO_3 + 1.25 mL 10% NaOH in 200 mL water. Heating the plate may be needed for some of the compounds	Compounds with functional groups prone to oxidation; bright yellow or brown spots on a purple background	[7]
Dragendorff's reagent	Solution A: 0.85 g bismuth nitrate + 10 mL glacial acetic acid + 40 mL water under heating. Solution B: 8 g potassium iodide in 30 mL water. Just before spraying solution A and B are mixed with 4 mL acetic acid and 20 mL water	Compounds containing heterocyclic nitrogen; colored zones on a white background	[3,5]
Antimony (III) chloride	8 g antimony chloride in 200 mL chloroform. Heat the plate at 110°C for 5–10 min	Compounds with carbon double bonds (white light and UV light—366 nm)	[3,5]

TABLE 12.2

Examples of Selected Specific Derivatizing Agents

Reagent Name	Preparation	Detection and Effect	Reference
Chloramine–trichloroacetic acid reagent	10 mL 3% aqueous chloramine T solution mix with 40 mL 25% ethanolic trichloroacetic acid. Heat the plate at 100°C for 5–10 min	Cardiac glycosides; UV light—366 nm—blue fluorescent zones	[7]
Coomassie brilliant blue G-250	100 mg G-250 dissolved in 20 mL ethanol + 20 mL phosphoric acid + 160 mL 50% ethanol	Proteins; blue color protein complexes	[7]
Emmer–Engel's reagent	Dissolved 0.5% dipyridyl and 0.2% ferric (III) chloride in methanol	Tocopherols; red spots on a colorless background	[7]
Kedde reagent (3,5-dinitrobenzoic acid KOH-reagent)	Equal amounts of 2% methanolic 3,5-dinitrobenzoic acid and 1 M methanolic KOH	Cardenolides; purple zones of cardiac glycosides	[7]
Marquis' reagent	Dilute 3 mL formaldehyde to 100 mL with concentrated sulfuric acid. Evaluate the plate in vis, immediately after spraying or dipping	Morphine, codeine, thebaine; colored spots on light pink background	[5,7]
Ninhydrin	30 mg ninhydrin in 10 mL n-butanol and mix with 0.3 mL 98% acetic acid. Heat the plate at 120°C for 5–10 min	Amino acids, biogenic amines, cyclic and linear peptides, cyclotides, cyclopeptides. Red or blue zones	[3,5,7]
PDAB-NaNO$_2$ reagent— Ehrlichs, Van Urks reagent	Solution A: 0.1% p-dimethylaminobenzaldehyde (PDAB) in concentrated HCl. Solution B: 0.1% aqueous NaNO$_2$ sol. After spraying the plate with A solution, heat the plate then spray with B one	Blue spots of 2-unsubstittuted indoles	[7]
Fast Black K solution	First spray the plate with 0.5% aqueous Fast Black K solution, after drying spray it with 0.5 M NaOH solution and Fast Black K solution once again	Amphetamines; orange-red and violet spots	[7]
Cacotheline aqueous reagent	1–50 mg cacotheline aqueous solution. Heat the plate at 110°C for about 2 min	Ascorbic acid; brownish or violet spots on a yellow background	[7]
Bratton–Marshall reagent	1% NaNO$_2$ solution in 0.1 M HCl; heat the plate at 100°C for 5 min and spray with 0.2% β-naphtol in 0.1 M NaOH	Sulfonamides; fluorescent zones	[7]
2-Aminodiphenyl solution	1% 2-aminodiphenyl ethanolic solution and sulfuric acid; heat the plate after derivatization	Vitamin B$_6$; fluorescent spots	[7]

Far more examples of specific derivatizing agents can be found in subchapters devoted to the analysis of specific drug groups.

12.4 BIOLOGICAL METHODS OF DETECTION (BIODETECTION)

Planar chromatography coupled with biological detection is usually used to search for active compounds found in complex mixtures. TLC is ideally suited to perform such tests due to its several advantages, namely, flexibility, high throughput, direct access to separated compounds, large amount of spraying agents, and evaporation of eluent solvents that may interfere with test reagents [34,35]. It has been first described by Godall and Levi, who determined the types and amounts of penicillins after resolution on a buffered paper strip [36]. The resolved antibiotics were detected with the use of bacteria strains. Since that time, TLC coupled with biodetection has become an important tool for searching new compounds possessing antibacterial and antifungal properties. Several different approaches have been elaborated for these tests, which already have been summarized in several book chapters and review articles [e.g., 4,6,37,38]. Apart from detection of microbiologically active compounds, TLC can be used to search for selected enzyme inhibitors as well as antioxidants. In the case of antioxidant tests, they are considered as biological; however, detection in such tests is performed with the use of chemical (e.g., relatively stable free radicals: 2,2-diphenyl-1-picrylhydrazyl (DPPH·) or 2,2'-azinobis-(3-ethylbenzothiazoline)-6-sulfonic acid (ABTS·+); β-carotene or crocin) not biological agents (such as bacteria, fungi strains, or isolated enzymes). However, as free radicals play an important role in the pathophysiology of numerous human ailments, these tests have been usually incorporated into the group of biological ones [34,39].

Detection of biologically active compounds can be realized on the surface of TLC plates in two ways: without sample development and after resolution of compounds under optimized chromatographic conditions. In the first case, it is a so-called "dot-blot" test, which resembles a microplate assay. An example of TLC dot-blot test can be seen in Figure 12.10.

After the sample application and drying step, the plate is derivatized with a proper staining reagent. Such an approach can be used as a preliminary test to choose the samples exhibiting the desired activity, which then will be passed to further tests [40]. Alternatively it may be applied to check the activity of individual compounds, isolated from complex matrices or newly synthesized [41,42].

While performing bioautography tests, it is of crucial importance to remember that false-positive results may occur. For example, the possible pitfall in using tetrazolium salts in direct bioautography (DB), for detection of antimicrobials, is the enhanced reduction of these salts by some compounds, for example, trans-resveratrol or genistein [43]. It was observed that some simple aldehydes and amines may produce false-positive results in tests for acetylcholinesterase (AChE) and butyrylcholinesterase (BuChE) inhibitors [44]. In the case of the TLC–DPPH· method for detection of free radical scavengers in volatile samples, white, instead of yellowish zones, may appear due to the hydrophobic character of the applied sample [34]. Therefore, proper tests for false-positives should be performed to eliminate them. It should also be kept in mind that simple benchtop bioassays provide only preliminary results; therefore, overpredictive claims of the analyzed activity should be very cautious [45].

An interesting feature of TLC is the possibility of coupling different modes of detection, physical, chemical, and biological, to screen the samples for the presence of active compounds [46].

In the subsequent sections, the strategies for realization of TLC coupled with biological detection tests are discussed in more detail.

12.4.1 Detection of Antibiotics and Antimicrobial Compounds

Apart from diffusion and dilution methods, thin-layer bioautography is an important tool used to screen complex samples and individual compounds for their antifungal and/or antibacterial properties. Hundreds of papers have been published concerning the use of this methodology since its first application reported by Godall and Levi [36]. Depending on the modes applied, TLC bioautographic techniques can be divided into three main categories: agar diffusion (contact bioautography), DB,

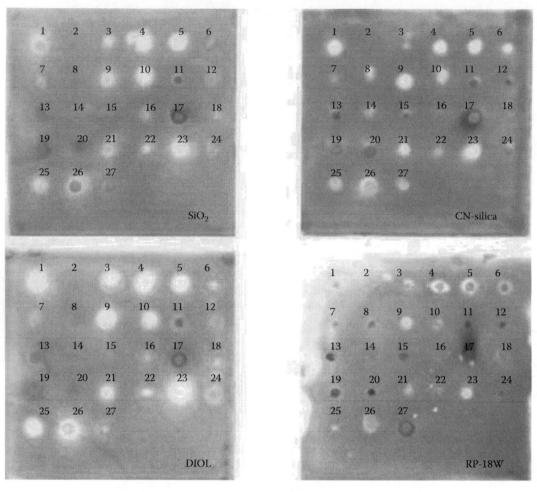

FIGURE 12.10 An example of TLC "dot-blot test" performed for detection of samples exhibiting AChE inhibitory activity. (From Cieśla, Ł. et al., *J. Planar Chromatogr. Mod. TLC*, 25, 225, 2012. With permission.)

and agar overlay (immersion bioautography) [37,39]. The first technique from the previously mentioned ones is currently of limited use. The other two are much more popular, and their selection depends mainly on the microorganism used to investigate antifungal or antibacterial properties.

In contact bioautography, the plate is placed on the agar layer to let the separated compounds diffuse from the adsorbent to the agar plate inoculated with a microorganism [6]. The basic problem is associated with the diffusion process that is different for various compounds. Lipophilic compounds hardly diffuse to the agar plate. This technique has been applied, for example, to determine amphotericin B, which is not a homogenous substance [39]. After separation the TLC plate was put on the surface of the agar plate inoculated with *Candida albicans*. Clear inhibition spots were observed with detection limit 0.8 ng per spot. Ramirez et al. applied contact bioautography for the detection of multiple antibiotic residues (chloramphenicol, ampicillin, benzylpenicillin, dicloxacillin, and erythromycin) in cow's milk [47].

12.4.1.1 TLC–Direct Bioautography

In DB, microorganisms grow directly on the TLC plate. Only those bacteria and fungi can be used whose growth is possible on the plate surface. In the case of fungi, usually spore-producing ones are applied (*Aspergillus, Penicillium, Cladosporium*), while for bacteria, the choice is much wider; however, the following bacteria strains have been used most extensively: *Bacillus subtilis, Escherichia*

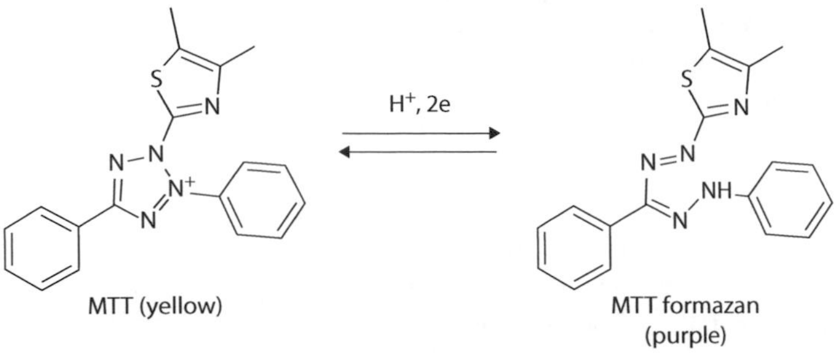

MTT (yellow) $\xrightarrow{\text{H}^+, 2e}$ MTT formazan (purple)

FIGURE 12.11 Reduction of yellow tetrazolium salt to violet formazan by bacterial dehydrogenases, constituting the basis for detection of antimicrobial compounds on the TLC surface. (From Marston, A., *J. Chromatogr. A*, 1218, 2676, 2011. With permission.)

coli, and *Pseudomonas savastanoi pv. phaseolica* [37,48]. In general, after running the plate with the use of a proper solvent system, it should be dried thoroughly. Subsequently, it is sprayed with or dipped into a fungal or bacterial suspension and incubated. Compounds with antifungal properties appear as clear zones against a gray background and no further derivatization is needed. For detection of antibacterial substances, the incubated plate is dipped in a tetrazolium salt solution, for example, methyl-thiazolyl-tetrazolium (MTT bromide). Yellow tetrazolium salts are reduced with bacterial dehydrogenases to bluish formazans (see Figure 12.11). The inhibition zones appear as clear spots against a colored background.

When a MTT solution is used as spraying reagent, the inhibition zones appear clear against a purple background, as in Figure 12.12.

For several years, the Chrom Biodip Antibiotics Test Kit™ from Merck had been in use to perform DB screening, which is now commercially unavailable [37]. It comprised of *B. subtilis* spore suspension, nutrient medium, and MTT staining reagent [37]. To perform the test, first, the suspension of the bacteria should be prepared: the nutrient medium is mixed with 0.5 M TRIS buffer, adjusted to pH 7.2 with 1 M hydrochloric acid, and finally sterilized in an autoclave for 20 min. Subsequently, the sterile medium is inoculated with *B. subtilis* spore suspension and incubated at 35°C. The developed TLC plates are immersed in the suspension and incubated overnight at 28°C. Zones of antibacterial compounds are visualized by spraying with MTT solution. Several modifications of the procedure have been also proposed [37].

Nagy et al. described optimum test conditions to perform TLC–DB with the use of the Gram-positive bacterium, *B. subtilis* [49], and the Gram-negative one, *E. coli* [50]. The authors stressed the importance of using the test bacteria in their logarithmic growth phase on the TLC plate. For *B. subtilis*, approximately 5 h incubation time at 37°C was found optimum, while for *E. coli*, it was 3 h

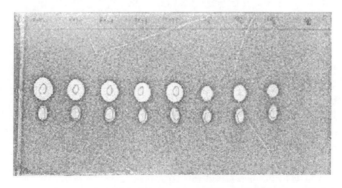

FIGURE 12.12 Example of the results of TLC–DB test to detect enrofloxacin and ciprofloxacin residues in milk. (From Choma, I.M., *J. Planar Chromatogr. Mod. TLC*, 19, 104, 2006. With permission.)

at the same temperature. The same group established also optimum conditions for TLC–DB with *C. albicans* as the test fungus [51].

Tyihak et al. presented an interesting application of TLC–DB for studying biochemical interactions between microorganisms and biologically active compounds, such as antibiotics, mycotoxins, or secondary plant metabolites, the so-called BioArena system [38]. In this respect, the BioArena system can be considered as a further development of DB [38]. In the majority of their studies, the authors have applied overpressured layer chromatography (OPLC) that gave better separation results when compared with conventional TLC [38]. It enables, for example, the investigation of the role of cofactor molecules in the antibiotic effect. The authors have checked the influence of L-arginine, reduced glutathione, and $CuSO_4$ on antimicrobial properties of trans-resveratrol [38]. With the application of BioArena system, the authors have postulated that salicylic acid acts through formaldehyde and its reaction products [4].

Müller et al. proposed a TLC bioautographic technique to detect estrogenic compounds using yeast cells [52]. In this method, yeast cells, containing the human estrogen receptor, were grown directly on high-performance thin-layer chromatographic plates. In the presence of estrogenic substances, the production of β-galactosidase is induced. The enzyme converts the substrate into the product that can easily be detected. In these tests, two different substrates were applied: chlorophenol red β-D-galactopyranoside (CPRG), added directly to the test suspension, and 4-methylumbelliferyl β-D-galactopyranoside (MUG), which was used to spray the incubated plates. β-Galactosidase hydrolyzes CPRG and liberates chlorophenol red, which leads to a color change from yellow to red. Therefore, estrogenic compounds appear as red zones against a yellow background. The authors reported that the contrast between the zones and the background was insufficient to perform densitometric absorbance measurements at 575 nm, λ_{max} of chlorophenol red. In the second case, MUG is cleaved by the enzyme to a fluorescent 4-methylumbelliferone. The measurements are performed at 365 nm excitation wavelength after exposing the plates to ammonia vapors. The emission maximum is observed at approx. 460 nm, at pH values greater than 8 [52]. The application of MUG caused estrogenic compounds that could be detected with better sensitivity when compared to CPRG. 17 β-Estradiol was used as a positive control and limit of detection was 2.75 pg [52].

TLC–DB has been widely applied to detect substances with antibacterial and antifungal properties in plant extracts, pharmaceutical formulations, environmental or food samples, etc. [37]. As far as drugs are considered, TLC–DB has been used to screen different samples for the presence of, for example, antibiotic residues in milk—enrofloxacin and ciprofloxacin [53], flumequine and doxycycline [54], and cefacetrile [55]; amphotericin B [56]; sulfonamides [57]; and still others.

In order to obtain reliable results, several precautions should be taken, namely, the solvents used to develop the plate should not be toxic and should be volatile to be evaporated easily without residues from the plate [58]. The method seems to be difficult to be adjusted for the analysis of thermolabile and volatile compounds as they may be lost during the drying procedure, which usually is performed at higher temperatures, for example, 60°C for 1 h [58].

12.4.1.1.1 TLC–Bioluminescence Screening

TLC coupled with bioluminescence detection was first reported by Eberz et al. in 1996 and it is a new variant of DB [59]. In this technique, after the development and drying, the plate is covered with a suspension of fluorescent bacteria emitting greenish light, which is a product of bacterial respiration [39]. *Vibrio fischeri*, also referred to as *Photobacterium phosphoreum*, is the most extensively used bacterial strain; however, there are reports indicating the application of genetically modified bacteria with incorporated bioluminescence gene, for example, *Acinetobacter* [37] or *Pseudomonas savastanoi pv. maculicola* [60]. A commercially available test, utilizing *V. fischeri* bacteria, was designed by ChromaDex, sold under the name Bioluminex™ [61]. To detect bioluminescence patterns, CAMAG has developed the BioLuminizer™ system [62]. To perform the test, *V. fischeri* suspension should be first prepared using the Bioluminex™ medium. The culture should be grown overnight, and directly before performing the assay, a buffer solution (buffer A from the

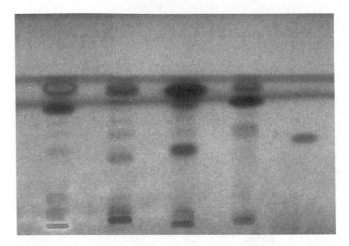

FIGURE 12.13 Example of the results of TLC–DB test with the use of the fluorescent bacteria *V. fischeri*. (From Baumgartner, V. and Dytkiewitz, E., http://www.clubdeccm.com/PDF/20101014/20101014_3-4-5temp. pdf, 2010. With permission.)

Bioluminex™ kit) is added. After sample development, the plate is immersed into luminescent bacteria suspension. Bioluminescence is measured by means of BioLuminizer™ after 2 s exposure and 3 min. incubation time. Substances inhibiting bacterial growth (antibiotics as well as toxic substances) appear as dark spots against a luminescent background, as also seen in Figure 12.13.

The expression of bacterial luminescence is correlated with bacterial cell density, as it is activated after the cell density reaches its critical value [46].

It is claimed that for proper test performance, the plate coating should be polar, as the nonpolar ones are hardly wettable with bacterial suspension [58].

12.4.1.2 Agar-Overlay TLC

In this technique, after the migration of a sample, a TLC plate is covered with a suspension of microorganisms in agar and left to solidify. During incubation, separated compounds diffuse from the adsorbent to the agar layer, where they exert their activity. For better diffusion of analyzed compounds, it is advised to keep the plate with agar overlay, at a low temperature for a few hours before incubation [37]. As in contact bioautography, problems arise from the diffusion process, as lipophilic compounds may remain adsorbed on the TLC plate. Another problem relates to the fact that the agar layer dilutes the antibacterial compounds that reduce the method's sensitivity [49]. Detection of inhibition zones is analogical to the process already described for DB, as usually tetrazolium salt solution is used as a derivatizing agent.

There are also examples of using bacterial strains producing colored colonies, for example, red-colored bacterium *Serratia marcescens* [39]. In order to obtain clearer results, incorporation of phenol red into agar solution is advised [39]. With the application of MTT as staining reagent, the inhibition zones appear as dark red against a blue background [39]. p-Iodonitrotetrazolium violet is reported to be the most sensitive reagent from among tetrazolium salts [39].

Agar-overlay bioautography is a technique suitable also to detect compounds with antifungal properties, for example, active against *C. albicans*, which cannot grow directly on the TLC plate. After migration, inoculation, and incubation, a tetrazolium salt solution is used to detect inhibition zones. Alternatively to tetrazolium salt, it has been proved that inhibition zones may be detected with the use of an esculin solution. Pathogen organisms, *Erwinia carotovora* and *Erwinia herbicola*, hydrolyze esculin to esculetin, which forms brown complexes with ferrum ions [39]. The use of fungi producing dark mycelia, for example, *Valsa ceratosperma*, gives the possibility to detect active compounds without further derivatization [63].

12.4.2 Detection of Enzyme Inhibitors

In several human ailments, the inhibition of some enzymes may bring beneficial effects and become a basis for a successful therapy. For example, in patients suffering from Alzheimer's disease, a substantial drop in acetylcholine brain level is observed. The majority of currently approved drugs for this disease are AChE and BuChE inhibitors.

In all TLC bioautographic techniques aimed at the detection of enzyme inhibitors, the plates after development and drying are sprayed or dipped in enzyme solution. Dipping is usually recommended as it allows more uniform covering of the plate's surface when compared to spraying. Enzyme stability in solution may be a problem while performing the test, as the use of buffers is usually insufficient. One of the approaches aimed at increasing enzyme stability is addition of bovine serum albumin into buffered solution [64,65]. In this case, it is important not to dry the plate thoroughly as it may lead to enzyme inhibition and collapse of the test. The other approach includes enzyme immobilization by gel entrapment in agar [66,67]. After enzyme dissolution, agar is evenly distributed over the TLC layer and left to solidify. This approach resembles the agar-overlay technique used to detect compounds with antimicrobial properties. Incubation of the plates covered with enzyme is the subsequent step. The plates should be incubated in humid atmosphere at 37°C, although other temperatures have also been reported [68]. Humidity may be ascertained by placing the plate face down on a petri dish filled with water [40]. After incubation, the plate should be dipped into the enzyme substrate solution in order to detect the inhibition zones on the TLC plate. The substrate is converted to a product that is subsequently detected with the use of a proper spraying/dipping reagent. Usually the product reacts with the staining reagent to form a colored compound. The substrate does not react with the same agent; therefore, inhibition zones appear as bright zones on the colored background.

12.4.2.1 Detection of Acetyl- and Butyrylcholinesterase Inhibitors

As already mentioned, AChE and BuChE inhibitors constitute the largest group of drugs used in the therapy of Alzheimer's disease, for example, donepezil, rivastigmine, or galantamine. However, the limited amount of drugs as well as their side effects imposes the need to search for new ones. TLC coupled with biodetection plays an important role in screening the samples for the presence of AChE and BuChE inhibitors. The first assay incorporating the use of planar chromatography was elaborated in the mid-1960s by Menn et al. [69,70]. In this assay, an impregnated cellulose plate was sprayed with human plasma (enzyme source) followed by spraying with acetylcholine (substrate) and alkaline bromothymol blue (derivatizing agent) solution. The reaction followed the scheme in Figure 12.14.

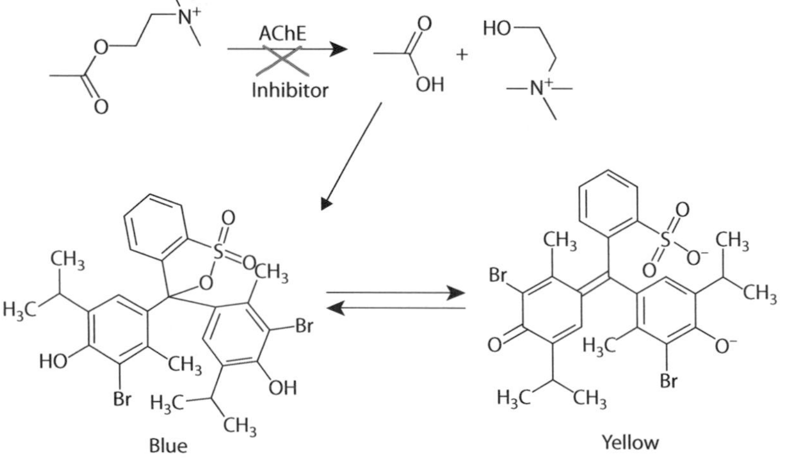

FIGURE 12.14 Schematic representation of reactions used to detect AChE inhibitors on the TLC surface according to the Menn and McBain method. (From Menn, J.J. and McBain, J.B., *Nature*, 209, 1351, 1966.)

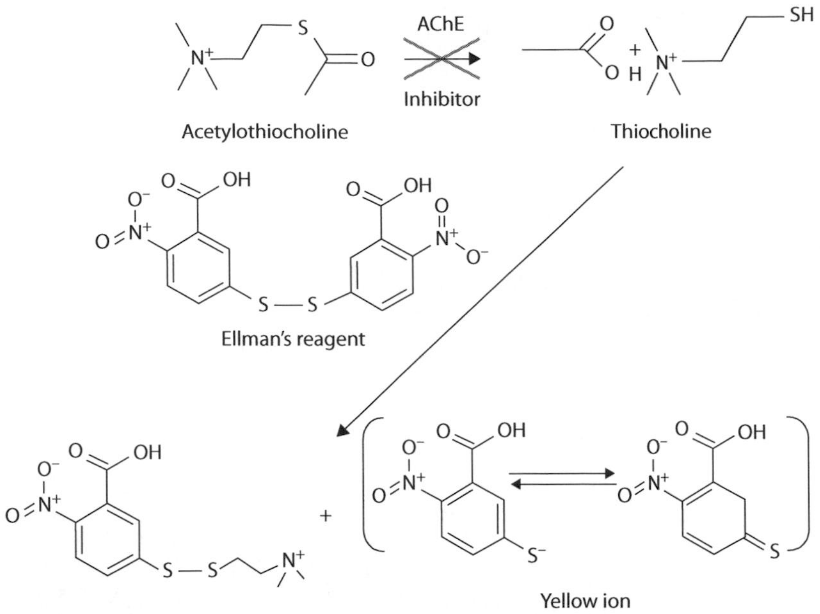

FIGURE 12.15 Schematic representation of reactions used to detect AChE inhibitors on the TLC surface according to a method with Ellman's reagent. (From Marston, A., *J. Chromatogr. A*, 1218, 2676, 2011. With permission.)

The enzyme converts acetylcholine into choline and acetic acid. Under acidic conditions, a blue indicator changes into a yellow form. Inhibition zones appear as blue against the yellow background (please notice that contrary to this assay in the majority of TLC tests, inhibitory zones appear pale against a darker background). The use of human plasma can be considered as the greatest drawback of this assay.

Nowadays, two other TLC-based techniques are in use, utilizing a purified enzyme instead of human plasma. The first one is based on the spectrophotometric method developed by Ellman et al. [71]. The method was adjusted for the performance of TLC test by Kiely et al. [72]. The authors stained the plates just after the application of the samples without their further chromatographic separation (dot-blot test). Rhee et al. were the first to report the application of this method for screening the samples after their separation [64]. Schematic representation of the reactions constituting the test's basis is presented in Figure 12.15.

The plate after development is sprayed with a buffered enzyme solution and incubated. Subsequently, it is covered with acetylthiocholine iodide solution (substrate) and Ellman's reagent (5,5'-dithiobis-(2-nitrobenzoic acid) (DTNB); staining reagent). Thiocholine liberated by the enzyme reacts with DTNB to form a yellow ion. AChE inhibitors appear as white spots on the yellow background. The plates should be scanned at 412 nm, although other wavelengths have also been reported, for example, 405 nm [44]. It is advised to document and evaluate the results up to 15 min after staining as they disappear after 20–30 min [64]. One of the method's drawbacks is only little difference between the white zones, indicating the presence of active compound, and the background [34]. The other one is the possibility of appearing false-positive results [44,73]. To exclude the false-positive results, another plate should be first covered with DTNB solution and subsequently sprayed with a solution obtained after incubation of acetylthiocholine iodide and AChE at 37°C for 20 min [73]. It was observed that the yellow background appeared only on the surface of the silica and aluminum oxide neutral type E [64]. In the case of the following plates, test failure was reported: aluminum oxide plates type T, RP-18, cellulose, and polyamide [64].

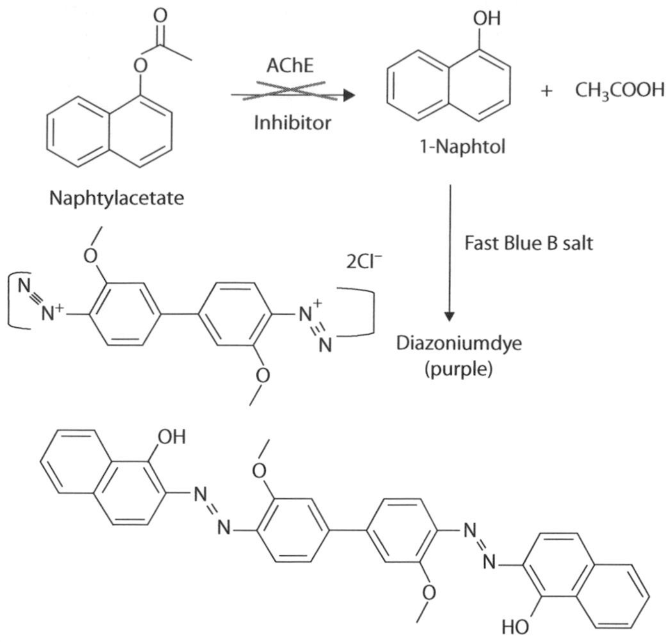

FIGURE 12.16 Schematic representation of reactions used to detect AChE and BuChE inhibitors on the TLC surface according to diazotization method. (From Marston, A., *J. Chromatogr. A*, 1218, 2676, 2011. With permission.)

In another technique, 1-naphthyl acetate is used as enzyme substrate [65]. AChE converts it into 1-naphthol that subsequently reacts with Fast Blue B salt to form a purple diazonium dye. Enzyme inhibitors appear as clear spots on the purple background. A schematic representation of the reactions constituting the test's basis is presented in Figure 12.16.

After covering the plate with enzyme solution, it should be incubated in humid atmosphere at 37°C for 20 min. Subsequently, the plate is derivatized with the use of a mixture prepared ex tempore from 1-naphthyl-acetate ethanolic solution and aqueous Fast Blue B salt solution. The active compounds appear 1–2 min after completing the previously mentioned procedure [65]. The scans should be recorded at 565 nm [73]. The reported detection limits for physostigmine and galantamine were 1 and 10 ng, respectively [65]. The method with diazotization cannot be applied to screen the activity of polyphenolic compounds as they react with Fast Blue B salt and form brownish complexes that may mask inhibitory activity [40]. In such cases, the use of Ellman's method is recommended [74]. Detection limits for known AChE inhibitors seem to be lower with the application of diazotization method when compared to the technique with Ellman's reagent [65]. Differences in the colored background as well as method sensitivity have been observed depending on the type of adsorbent used [40]. The observed divergences have been explained as a probable result of enzyme or test compound interactions with the active sites of TLC layers [40,75]. It was concluded that bioautography tests with AChE and BuChE should preferably be performed on the surface of polar adsorbents (e.g., silica gel or CN silica) [40].

Several modifications of the technique have been published so far. Yang et al. reported changes in the concentrations of the enzyme and reagents, and the time of the reaction that resulted in a remarkable reduction of detection limits [74]. Mroczek proposed to incorporate 2-naphthyl acetate into the elution solvent mixture that results in uniform covering of the TLC plate and in increased sensitivity of the test [76]. The author underlines the fact that with the use of 2-naphthyl acetate, white inhibition zones were stable even for 24 h or more [76].

Cieśla et al. proposed a TLC-based bioautographic technique to screen volatile samples for the presence of AChE inhibitors [34]. Volatile compounds were separated by means of low-temperature TLC. Essential oils of selected *Lamiaceae* plants contained compounds inhibiting AChE activity.

Yang et al. have also proposed a low-cost TLC bioautographic method to screen complex samples for the presence of AChE inhibitors [77]. 4-Methoxyphenyl acetate was applied as an enzyme substrate in that assay. It is converted by AChE into 4-methoxyphenol that subsequently reacts with a solution of potassium hexacyanoferrate(III) ($K_3(FeCN)_6$) and iron chloride hexahydrate ($FeCl_3 \cdot 6H_2O$). The liberated 4-methoxyphenol forms aquamarine blue complexes with iron ions. Inhibition zones appear as yellowish against the colored background [77]. But contrary to the previously described methods, the plate was not directly immersed in solution of chromogenic agent. A filter paper with the same size as the TLC plate was soaked with a mixture of potassium hexacyanoferrate(III) and iron chloride hexahydrate solutions. Next, this soaked paper was placed on the surface of the TLC plate and removed after 1 min. Low consumption of the enzyme is this method's advantage; however, the detection limits of known AChE inhibitors are not as low as in the previously reported techniques. False-positive results may appear as some compounds have been observed to inhibit the reaction of 4-methoxyphenol with chromogenic agent [77].

12.4.2.2 Detection of α- and β-Glucosidase Inhibitors

α-Glucosidase and β-glucosidase hydrolyze poly- and oligosaccharides into monomers as well as bonds between carbohydrate(s) and an aglycone [67]. Their inhibitors may be used in the therapy of type 2 diabetes as they lower postprandial levels of glucose, as currently used drugs, acarbose and miglitol [39]. These enzymes also take part in glucosylation of viral membrane proteins. Therefore, their inhibitors may be potentially applied as antiviral agents [39].

Salazar and Furlan described a method for the detection of β-glucosidase inhibitors in which the enzyme's stability on the TLC plate was ascertained with gel entrapment [67]. After development, the plate was evenly covered with enzyme entrapped in agar. Ferric chloride hexahydrate (staining reagent) was also added to agar solution before solidification. After solidification of staining solution, the plate was incubated at 37°C for 120 min. and immersed in 0.2% (w/v) solution of esculin (enzyme substrate). The enzyme converts esculin (glycoside) into esculetin (aglycone) as presented in Figure 12.17.

Enzyme inhibitors appear as clear zones against the dark brown background, which results from the complex formed by esculetin and ferric ions [67]. The method turned out to be compatible with different stationary phases (silica gel, cellulose, RP-18, CN silica, and diol-silica); however, the detection limit varied depending on the adsorbent used. One of the method's drawbacks is the application of esculin as the enzyme substrate; therefore, it is of limited use in the

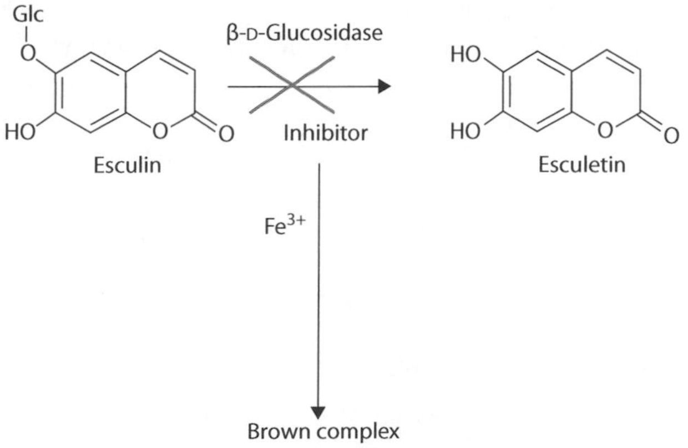

FIGURE 12.17 Schematic representation of reactions used to detect β-glucosidase inhibitors on the TLC surface according to the Salazar et al. method. (From Salazar, M.O. and Furlan, R.L.E., *Phytochem. Anal.*, 18, 209, 2007. With permission.)

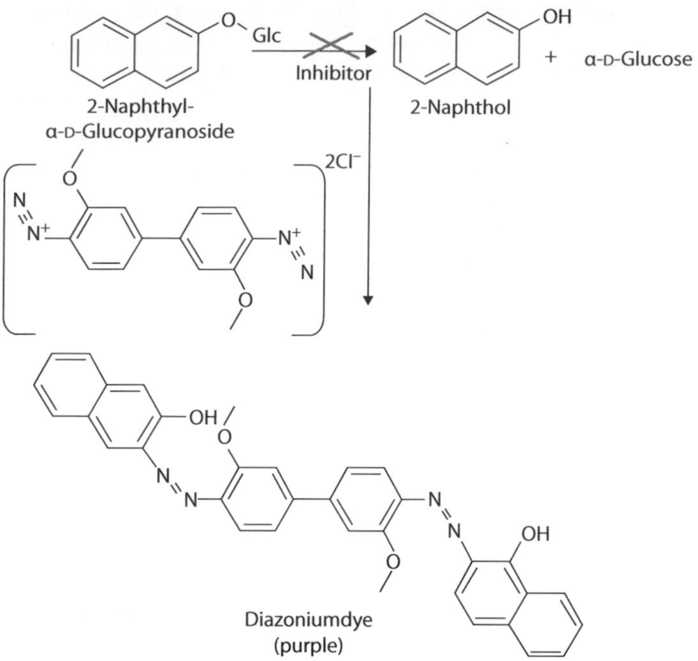

FIGURE 12.18 Schematic representation of reactions used to detect α- and β-glucosidase inhibitors on the TLC surface according to diazotization method. (From Marston, A., *J. Chromatogr. A*, 1218, 2676, 2011. With permission.)

case of samples containing coumarins. Furlan's group applied the same procedure to discover β-glucosidase inhibitors in chemically engineered extracts, prepared through ethanolysis [78] or sulfonylation [79].

The other technique used to detect both α- and β-glucosidase inhibitors is analogous to the TLC assay for the detection of cholinesterase inhibitors [68]. A schematic representation of the reactions constituting the test's basis is presented in Figure 12.18.

After sample migration, TLC plates were sprayed with enzyme solution (α-D-glucosidase or β-D-glucosidase) prepared with the use of sodium acetate buffer and incubated in humid atmosphere. For α-D-glucosidase inhibitors, the plates were incubated at room temperature for 60 min. and for β-D-glucosidase inhibitors at 37°C for 20 min. Subsequently, the plates were covered with a mixture of enzyme substrate (2-naphthyl-α-D-glucopyranoside for α-D-glucosidase inhibitors; 2-naphthyl-β-D-glucopyranoside for β-D-glucosidase inhibitors) and chromogenic agent (Fast Blue B salt) [68]. Enzyme inhibitors appeared as clear spots against a purple background, which resulted from the formation of diazonium dye, analogically to Marston's test for detection of AChE inhibitors [65]. The developed test is sensitive to pH and temperature; therefore, proper control of these parameters is required.

Both groups performed also tests with the use of *o*- and *p*-nitrophenyl glucopyranosides as substrates. However, in both cases, the obtained results were much less satisfying when compared to those observed with the application of esculin or 2-naphthyl-α-D-glucopyranoside.

12.4.2.3 Detection of Other Enzyme Inhibitors

Apart from the previously mentioned techniques, TLC bioautographic methods have been also applied for the detection of inhibitors of the following enzymes: xanthine oxidase (XO) [66], tyrosinase [80], and lipase [81].

XO is an enzyme catalyzing the oxidation of hypoxanthine and xanthine to uric acid [66]. In the process of purine oxidation, the following reactive oxygen species (ROS) are formed: superoxide radical and hydrogen peroxide, whose overproduction may result in elevated oxidative stress. The inhibitors of this enzyme may be potentially used in pathological states caused by XO hyperactivity. Currently, XO inhibitors are used in the therapy of hyperuricemia, for example, allopurinol [66]. Ramallo et al. described a TLC-based method used to detect XO inhibitors [66]. To perform the test, XO was dissolved in agar solution containing chromogenic agent—nitroblue tetrazolium (NBT) salt. This solution was distributed evenly on the TLC plate and left to solidify. Subsequently, the plate was immersed in xanthine solution and incubated at 38°C for 20 min in the dark. Superoxide radicals produced during xanthine oxidation reduce NBT to formazan, the way it is reduced by bacteria reductases in DB. XO inhibitors appear as white spots against a purple background [66]. To differentiate pure XO inhibitors from superoxide radical scavengers, the authors have developed two nonenzymatic tests. In the first one, a riboflavin solution was entrapped in agar instead of XO; in the other one, NADH (reduced nicotinamide adenine dinucleotide) was added instead of the enzyme. In riboflavin test after solidification, the reaction was initiated with the use of 20 W fluorescent lamp. In the case of NADH test, after solidification the plate was immersed in phenazine methosulfate solution at room temperature. In both cases, the spots of superoxide radical scavengers appeared clear against a purple background. Better results were obtained with the use of riboflavin test [66]. Pure XO inhibitors did not produce inhibition zones in both riboflavin and NADH tests.

Tyrosinase catalyzes hydroxylation of monophenols to o-diphenols and subsequent oxidation to o-quinones [80]. Its inhibitors may be potentially used as whitening agents in cosmetics and anti-browning agents in fruits. To detect tyrosinase inhibitors, first, a TLC plate is sprayed with enzyme solution prepared in phosphate buffer (pH 6.8) and subsequently sprayed with L-tyrosine solution [80]. The plate is not incubated after spraying. The inhibitors appear as white spots against a brownish-purple background, which results from formation of colored quinones out of L-tyrosine. After spraying with tyrosinase solution, the plate should not be dried thoroughly before L-tyrosine spraying as it may result in decreased enzyme activity and paler background [80]. The positive results appear on the TLC surface after 3–4 min. Kojic acid, arbutin, glabridin, kojic acid dipalmitate, and standard tyrosinase inhibitors were used as a positive control in that study.

Lipase is an enzyme involved in the digestion of triacylglycerols; its inhibition reduces the absorption of fat; therefore, inhibitors may be potentially applied in body weight reduction [81]. Hassan proposed a TLC bioautographic method to detect lipase inhibitors, which is analogical to Marston's technique used to detect AChE inhibitors. The enzyme was dissolved in tris(hydroxymethyl)aminomethane buffer and was stabilized by adding bovine serum albumin [81]. After sample migration and drying step, the plate was sprayed with enzyme and α-naphthyl-acetate (substrate) and incubated at 37°C for 20 min at humid atmosphere. After incubation, the plate was sprayed with Fast Blue B salt solution (chromogenic agent). Lipase inhibitors appeared as clear spots against the purple background [81]. Orlistat, a well-known lipase inhibitor, could be easily detected with the application of this methodology.

12.4.3 DETECTION OF ANTIOXIDANTS

Results of many scientific researches indicate that ROS may be responsible for the development and progress of some age-related diseases, for example, atherosclerosis, neurodegenerative ailments (Alzheimer's and Parkinson's disease), and cancer [82,83]. Samples of synthetic and natural origin are screened to detect new potent antioxidants (free radical scavengers, metal chelators, oxygen scavengers, UV light absorbers, enzymatic antioxidants). TLC has become an important technique used to detect antioxidants mainly in the samples of plant origin [34,39]. The most popular TLC-based test involves the application of the relatively stable free radical DPPH·; however, other staining reagents [ABTS·+, β-carotene, crocin, or phosphomolybdic acid] have been also reported [34].

12.4.3.1 TLC–DPPH Test

DPPH was proposed for the chemical detection of antioxidants by Blois, who discovered that it reacts quantitatively with, for example, cysteine, ascorbic acid, or tocopherol [84]. Due to the delocalization of its odd electron over the molecule, DPPH does not dimerize and possesses absorption maximum at 517 nm [85]. Upon reduction, the characteristic violet color of DPPH solution changes into yellow caused by the picric acid group.

TLC coupled with DPPH staining was first proposed by Glavind and Holmer who investigated free radical scavenging properties of selected tocopherols in natural products [86]. Since that time, it has been mainly applied to screen complex natural extracts for the presence of free radical scavengers [e.g., 87–98]. However, it has been also used for studying direct antioxidant properties of synthetic compounds, candidates for new drugs [99].

In TLC–DPPH test after sample application and/or development, the plate is dipped or sprayed with DPPH solution. Usually 0.2% (w/v) DPPH methanolic solution is applied for derivatization; however, other concentrations have also been reported [34]. The major problem concerning the use of the TLC–DPPH technique was the lack of one standard approach for performing the test, leading to difficulties in comparing results obtained in different laboratories. Apart from various DPPH concentrations, different stationary phases have been used to perform the assay, for example, silica gel, RP-18W, CN silica, or diol-silica [100]. It has been proved that chemical groups present on the surface of TLC layers influence the observed result of radical–antioxidant reaction [100]. Silica gel (specifically positive adsorbent) strengthened the result of the previously mentioned reaction, while specifically negative, polar-bonded stationary phase (CN silica) weakened it. It is assumed that the enhanced direct antioxidant properties on the surface of specifically positive adsorbents are due to protonation of hydroxyl groups of polyphenolics that favors faster reaction with DPPH. It is recommended to perform the TLC–DPPH assay on nonspecific adsorbents, for example, RP-18, as they do not possess groups able to form hydrogen bonds [100]. The observed results were also influenced by the solvent type used to prepare DPPH staining reagent. In order to observe nonaltered radical–antioxidant reaction, it is advised to dissolve DPPH in nonpolar solvents.

Another problem is associated with the time that elapses between immersion and documentation. Usually it is advised to document the results after 30 min; however, other approaches have also been reported [100]. An interesting alternative may be the results' documentation every 5 min, as they change in time.

For many years, TLC–DPPH assay has been considered as only a preliminary tool to screen samples for the presence of antioxidants. The inability to obtain quantitative data has been considered as one of its main drawbacks. However, the results of recent research indicate that TLC–DPPH test coupled with image processing may be a useful tool to measure quantitatively direct antioxidant properties of a sample [97,98].

TLC–DPPH test has been proved useful to screen volatile samples for the presence of free radical scavengers [40]. In case of lipophilic volatile compounds, false-positive results have been observed [34].

12.4.3.2 Other TLC Autography Tests for Identification of Antioxidants

Other TLC-based tests to screen samples for the presence of antioxidants are less popular than the TLC–DPPH assay. However, in order to properly characterize a compound as an antioxidant, it should be screened with the use of different methods. β-Carotene bleaching tests are sometimes performed to detect antioxidants. To perform the assay, a TLC plate is dipped in chloroformic β-carotene solution and exposed to sunlight or 366 nm UV light. Antioxidant spots remain orange against a pale yellow/white background [39]. The β-Carotene bleaching test may also be applied to screen the compound's ability to stabilize lipid oxidation. First, the plate is dipped in ethanolic solution of linoleic acid and subsequently in chloroformic solution of β-carotene [39]. After exposure to sunlight, antioxidants appear as orange spots against a yellowish background. The greatest drawback of β-carotene assays is only a slight difference between the spots of active compounds and the background [34].

Phosphomolybdenum assay is applied to detect compounds with reducing power toward transition metals [34]. Active compounds appear as blue/violet zones on the chromatogram after heating the plate at 95°C. This test is not a selective one as also other compound classes, apart from antioxidants, may give positive results, for example, fatty acids or steroids [34].

An assay utilizing $ABTS^{·+}$ after TLC separation to detect antioxidants in liquid and semisolid pharmaceutical formulations or in isolated compounds has also been described [101]. ABTS radical cation is generated by the reaction of ABTS with potassium persulfate. Zampini et al. proposed two TLC autographic systems to detect antioxidants with the use of $ABTS^{·+}$—direct spraying of a TLC plate with staining solution (system I) and $ABTS^{·+}$ entrapment in agar (system II) [101]. System II was found to be more sensitive and reproducible than system I. Active compounds appear as colorless or pink zones against a green background [102], as also presented in Figure 12.19.

The plates can be scanned at 734 nm ($ABTS^{·+}$ absorption maximum). Soler-Rivas et al. compared TLC–DPPH$^·$ and TLC–$ABTS^{·+}$ assay, concluding that the latter radical was less stable on the layer [102].

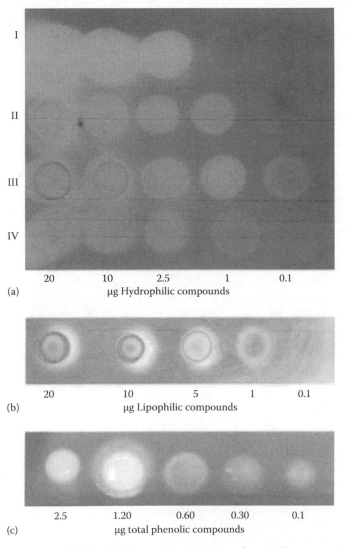

FIGURE 12.19 An example of TLC–$ABTS^{·+}$ test for detection of antioxidant capacity of pharmaceutical formulations. (a) Hydrophilic compounds: I ascorbic acid, II rutin, III quercetin, and IV naringenin. (b) Lipophilic compounds: β-carotene. (c) Liquid pharmaceutical preparation. (From Zampini, I.C. et al., *AAPS Pharm. Sci. Tech.*, 11, 1159, 2010. With permission.)

REFERENCES

1. Morlock, G. and W. Schwack. 2010. Hyphenations in planar chromatography. *J. Chromatogr. A* 1217: 6600–6609.
2. Cheng, S.C., M.Z. Huang, and J. Shiea. 2011. Thin layer chromatography/mass spectrometry. *J. Chromatogr. A* 1218: 2700–2711.
3. Spangenberg, B. 2008. Derivatization, detection (quantification) and identification of compounds online. In *Thin Layer Chromatography in Phytochemistry*, eds., M. Waksmundzka-Hajnos, J. Sherma, and T. Kowalska, pp. 175–191. Boca Raton, FL: CRC Press.
4. Tyihak, E., A. Moricz, and. P.G. Ott. 2008. Biodetection and determination of biological activity of natural compounds. In *Thin Layer Chromatography in Phytochemistry*, eds., M. Waksmundzka-Hajnos, J. Sherma, and T. Kowalska, pp. 193–213. Boca Raton, FL: CRC Press.
5. Reich, E. and A. Schibli. 2007. *High-Performance Thin-Layer Chromatography for the Analysis of Medicinal Plants*, Thieme, Stuttgart, Germany.
6. Botz, L., S. Nagy, and B. Kocsis. 2001. Detection of microbiologically active compounds. In *Planar Chromatography. A Retrospective View for the Third Millennium*, ed., Sz. Nyiredy, pp. 489–516. Budapest, Hungary: Springer Scientific Publisher.
7. Cimpan, G. 2001. Pre- and post-chromatographic derivatization. In *Planar Chromatography. A Retrospective View for the Third Millennium*, ed., Sz. Nyiredy, pp. 410–445. Budapest, Hungary: Springer Scientific Publisher.
8. Skorupa, A. and A. Gierak. 2011. Detection and visualization methods used in thin-layer chromatography. *J. Planar Chromatogr. Mod. TLC* 24: 274–280.
9. Tuzimski, T. 2011. Application of different modes of thin-layer chromatography and mass spectrometry for the separation and detection of large and small biomolecules. *J. Chromatogr. A* 1218: 8799–8812.
10. Dammertz, W. and E. Reich. 2001. Planar chromatography and densitometry. In *Planar Chromatography. A Retrospective View for the Third Millennium*, ed., Sz. Nyiredy, pp. 234–246. Budapest, Hungary: Springer Scientific Publisher.
11. Cieśla, Ł., A. Petruczynik, M. Hajnos, A. Bogucka-Kocka, and M. Waksmundzka-Hajnos. 2008. Two-dimensional thin-layer chromatography of structural analogs. Part II. Method for quantitative analysis of selected coumarins in plant material. *J. Planar Chromatogr. Mod. TLC* 21: 447–452.
12. Tuzimski, T. 2008. Application of SPE-HPLC-DAD and SPE-TLC-DAD to the determination of pesticides in real water samples. *J. Sep. Sci.* 31: 3537–3542.
13. Tuzimski, T. 2011. Basic principles of planar chromatography and its potential for hyphenated techniques. In *High-Performance Thin-Layer Chromatography (HPTLC)*, ed., M.M. Srivastava, pp. 247–310. New York: Springer.
14. Ahrens, B., D. Blankenhorn, and B. Spangenberg. 2002. Advanced fibre optical scanning in thin-layer chromatography for drug identification. *J. Chromatogr. B* 772: 11–18.
15. Morlock, G.E. and E.S. Chernetsova. 2012. Coupling of planar chromatography with direct analysis in real time mass spectrometry. *Cent. Eur. J. Chem.* 10: 703–710.
16. Ghosh, P., M.M.K. Reddy, V.B. Ramteke, and B.S. Rao. 2004. Analysis and quantitation of diazepam in cream biscuits by high-performance thin-layer chromatography and its confirmation by mass spectrometry. *Anal. Chim. Acta* 508: 31–35.
17. http://www.camag.com/v/products/tlc-ms/ (accessed 18 August, 2012).
18. Cheng, S.C., M.Z. Huang, and J. Shiea. 2009. Thin-layer chromatography/laser-induced acoustic desorption/electrospray ionization mass spectrometry. *Anal. Chem.* 81: 9274–9281.
19. Vovk, I, G. Popović, B. Simonovska, A. Albreht, and D. Agbaba. 2011. Ultra-thin-layer chromatography mass spectrometry and thin-layer chromatography mass spectrometry of single peptides of angiotensin-converting enzyme inhibitors. *J. Chromatogr. A* 1218: 3089–3094.
20. Heinig, K., T. Wirz, F. Bucheli, and A. Gajate-Perez. 2011. Determination of oseltamivir (Tamiflu®) and oseltamivir carboxylate in dried blood spots using offline or online extraction. *Bioanalysis* 3: 421–437.
21. Abu-Rabie, P. and N. Spooner. 2009. Direct quantitative bioanalysis of drugs in dried blood spot samples using a thin-layer chromatography mass spectrometer interface. *Anal. Chem.* 81: 10275–10284.
22. Liu,Y.-R., H.-S. Ge, K.-H. Zhao, and L. Yu. 2008. Determination of three chemical components added illegally in a analgesia traditional Chinese medicine preparation by TLC-MS. *Chinese Pharm. J.* 43: 1747–1750.
23. Somsen, G.W., W. Morden, and I.D. Wilson. 1995. Planar chromatography coupled with spectroscopic techniques. *J. Chromatogr. A* 703: 613–665.

24. Gőssi, A., U. Scherer, and G. Schlotterbeck. 2012. Thin-layer chromatography-nuclear magnetic resonance spectroscopy—A versatile tool for pharmaceutical and natural products analysis. *Chimia* 66: 347–349.
25. Rager, I.O.C. and K.A. Kovar. 2001. Planar chromatography and IR. In *Planar Chromatography. A Retrospective View for the Third Millennium*, ed., Sz. Nyiredy, pp. 247–260. Budapest, Hungary: Springer Scientific Publisher.
26. Wagner, H. and S. Bladt. 1996. *Plant Drug Analysis—A Thin Layer Chromatography Atlas*, Springer, Berlin, Germany.
27. Knapp, D.R. 1979. *Handbook of Analytical Derivatization Reactions*, John Wiley & Sons, New York.
28. Sherma, J. and B. Fried. 2003. *Handbook of Thin-Layer Chromatography*, Marcel Dekker, New York.
29. Suzuki, O. and K. Watanabe. 2005. *Drug and Poisons in Humans, A Handbook of Practical Analysis*, Springer, Berlin, Germany.
30. Knauer, A.R., R. Wintersteiger, W. Markl, and H.J. Sametz. 1999. Derivatization of carboxylic acid with fluorescent reagents. *J. Planar Chromatogr. Mod. TLC* 12: 211–214.
31. Jork, H., W. Funk, W. Fischer, and H. Wimmer. 1990. *Thin-Layer Chromatography, Reagents and Detection Methods*, Vol. 1a, VCH, Weinheim, Germany.
32. Baumgartner, V., C. Hohl, and W. Schwack. 2011. Rolling—A new application technique for luminescent bacteria on high-performance thin-layer chromatography plates. *J. Chromatogr. A* 1218: 2692–2699.
33. Vasta, J.D., M. Cicchi, J. Sherma, and B. Fried. 2009. Evaluation of thin-layer chromatography systems for analysis of amino acids in complex mixtures. *Acta Chromatogr.* 21: 29–38.
34. Cieśla, Ł. 2012. Thin-layer chromatography with biodetection in the search for new potential drugs to treat neurodegenerative diseases—State of the art and future perspectives. *Med. Chem.* 8: 102–111.
35. Poole, C.F. 2003. Thin-layer chromatography: Challenges and opportunities. *J. Chromatogr. A* 1000: 963–984.
36. Godall, R.R. and A.A. Levi. 1946. A microchromatographic method for the detection and approximate determination of the different penicillins in a mixture. *Nature* 158: 675–676.
37. Choma, I.M. and E.M. Grzelak. 2011. Bioautography detection in thin-layer chromatography. *J. Chromatogr. A* 1218: 2684–2691.
38. Tyihak, E., E. Mincsovics, and A.M. Moricz. 2012. Overpressured layer chromatography: From the pressurized ultramicro chamber to BioArena system. *J. Chromatogr. A* 1232: 3–18.
39. Marston, A. 2011. Thin-layer chromatography with biological detection in phytochemistry. *J. Chromatogr. A* 1218: 2676–2683.
40. Cieśla, Ł., J. Kryszeń, A. Stochmal, W. Oleszek, and M. Waksmundzka-Hajnos. 2012. Low-temperature thin-layer chromatography bioautographic tests for detection of free radical scavengers and acetylcholinesterase inhibitors in volatile samples. *J. Planar Chromatogr. Mod. TLC* 25: 225–231.
41. Kancheva V.D., L. Saso, P.V. Boranova, A. Khan, M.K. Saroj, M.K. Pandey, S. Malhorta et al. 2010. Structure-activity relationship of dihydroxy-4-methylcoumarins as powerful antioxidants: Correlation between experimental & theoretical data and synergistic effect. *Biochimie* 92: 1089–1100.
42. Kancheva V.D., P.V. Boranova, J.T. Nechev, and I.I. Manolov. 2010. Structure-activity relationships of new 4-hydroxy bis-coumarins as radical scavengers and chain-breaking antioxidants. *Biochimie* 92: 1138–1146.
43. Moricz, A.M., E. Tyihak, and P.G. Ott. 2010. Usefulness of transgenic luminescent bacteria in direct bioautographic investigation of chamomile extracts. *J. Planar Chromatogr. Mod. TLC* 23: 180–183.
44. Rhee, I.K., R.M. van Rijn, and R. Verpoorte. 2003. Qualitative determination of false-positive effects in the acetylcholinesterase assay using thin layer chromatography. *Phytochem. Anal.* 14: 127–131.
45. Houghton, P.J., M.-J. Howes, C.C. Lee, and G. Steventon. 2007. Uses and abuses of in vitro tests in ethnopharmacology: Visualizing an elephant. *J. Ethnopharmacol.* 110: 391–400.
46. Klőppel, A., W. Grasse, F. Brűmmer, and G.E. Morlock. 2008. HPTLC coupled with bioluminescence and mass spectrometry for bioactivity-based analysis of secondary metabolites in marine sponges. *J. Planar Chromatogr. Mod. TLC* 21: 431–436.
47. Ramirez, A., R. Gutierrez, G. Diaz, C. Gonzalez, N. Perez, S. Vega, and M. Noa. 2003. High-performance thin-layer chromatography-bioautography for multiple antibiotic residues in cow's milk. *J. Chromatogr. B* 784: 315–322.
48. Hostettmann, K. and A. Marston. 2002. Twenty years of research into medicinal plants: Results and perspectives. *Phytochem. Rev.* 1: 275–285.
49. Nagy, S., B. Kocsis, T. Kőszegi, and L. Botz. Optimization of conditions for culture of the test bacteria used for direct bioautographic detection. 1. The Gram-positive test bacterium *Bacillus subtilis*. *J. Planar Chromatogr. Mod. TLC* 15: 132–137.

50. Nagy, S., T. Kőszegi, L. Botz, and B. Kocsis. 2003. Optimization of conditions for culture of the test bacteria used for direct bioautographic detection. 2. Gram-negative test bacterium: *Escherichia coli*. *J. Planar Chromatogr. Mod. TLC* 16: 121–126.

51. Nagy, S., B. Kocsis, T. Kőszegi, and L. Botz. 2007. Optimization of growth conditions for test fungus cultures used in direct bioautographic TLC detection. 3. Test fungus: *Candida albicans*. *J. Planar Chromatogr. Mod. TLC* 20: 385–389.

52. Müller, M.B., C. Dausend, and F.H. Frimmel. 2004. A new bioautographic screening method for the detection of estrogenic compounds. *Chromatographia* 60: 207–211.

53. Choma, I.M. 2006. Screening of enrofloxacin and ciprofloxacin residues in milk by HPLC and by TLC with direct bioautography. *J. Planar Chromatogr. Mod. TLC* 19: 104–108.

54. Choma, I., A. Choma, and K. Staszczuk. 2002. Direct bioautography-thin-layer chromatography of flumequine and doxycycline in milk. *J. Planar Chromatogr. Mod. TLC* 15: 187–191.

55. Choma, I.M., C. Kowalski, R. Lodkowski, A. Burmańczuk, and I. Komaniecka. 2008. TLC-DB as an alternative to the HPLC method in the determination of cefacetril residues in cow's milk. *J. Liq. Chromatogr. Relat. Technol.* 31: 1903–1912.

56. Fittler, A., B. Kocsis, Z. Matus, and L. Botz. 2010. A sensitive method for thin-layer chromatographic detection of amphotericin B. *J. Planar Chromatogr. Mod. TLC* 23: 18–22.

57. Williams, L. and O. Bergersen. 2001. Towards an integrated platform for combinatorial library synthesis and screening. *J. Planar Chromatogr. Mod. TLC* 14: 318–321.

58. Baumgartner, V. and E. Dytkiewitz. 2010. HPTLC-bioluminescence-coupling using *Vibrio fischeri*. http://www.clubdeccm.com/PDF/20101014/20101014_3-4-5temp.pdf. Accessed on February 15, 2012.

59. Eberz, G., H.G. Rast, K. Burger, W. Kreiss, and C. Weisemann. 1996. Bioactivity screening by chromatography—Bioluminescence coupling. *Chromatographia* 43: 5–9.

60. Moricz, A.M., Sz. Szarka, P.G. Ott, E.B. Hethelyi, E. Szőke, and E. Tyihak. 2012. Separation and identification of antibacterial chamomile components using OPLC, bioautography and GC-MS. *Med. Chem.* 8: 85–94.

61. Bioluminex, www.bioluminex.com. Accessed on February 20, 2012.

62. BioLuminizer, www.camag.com/v/products/evaluation/bioluminizer.html. Accessed on February 20, 2012.

63. Islam, N., S.A. Parveen, N. Nakazawa, A. Marston, and K. Hostettmann. 2003. Bioautography with the fungus *Valsa ceratosperma* in the search for antimycotic agents. *Pharm. Biol.* 41: 637–640.

64. Rhee, I.K., M. van de Meent, K. Ingkaninan, and R. Verpoorte. 2001. Screening for acetylcholinesterase inhibitors from Amaryllidaceae using silica gel thin-layer chromatography in combination with bioactivity staining. *J. Chromatogr A.* 915: 217–223.

65. Marston, A., J. Kissling, and K. Hostettmann. 2002. A rapid TLC bioautographic method for the detection of acetylcholinesterase and butyrylcholinesterase inhibitors in plants. *Phytochem. Anal.* 13: 51–54.

66. Ramallo, I.A., S.A. Zacchino, and R.L.E. Furlan. 2006. A rapid TLC autographic method for the detection of xanthine oxidase inhibitors and superoxide scavengers. *Phytochem. Anal.* 17: 15–19.

67. Salazar, M.O. and R.L.E. Furlan. 2007. A rapid TLC autographic method for the detection of glucosidase inhibitors. *Phytochem. Anal.* 18: 209–212.

68. Simoes-Pires, C.A., B. Hmicha, A. Marston, and K. Hostettmann. 2009. A TLC bioautographic method for the detection of α- and β-glucosidase inhibitors in plant extracts. *Phytochem. Anal.* 20: 511–515.

69. Menn, J.J., J.B. McBain, and M.J. Dennis. 1964. Detection of naturally occurring cholinesterase inhibitors in several crops by paper chromatography. *Nature* 202: 697–698.

70. Menn, J.J. and J.B. McBain. 1966. Detection of cholinesterase-inhibiting insecticide chemicals and pharmaceutical alkaloids on thin-layer chromatograms. *Nature* 209: 1351–1352.

71. Ellman, G.L., D. Courtney, V. Andies, and R.M. Featherstone. 1961. A new and rapid colorimetric determination of acetylcholinesterase activity. *Biochem. Pharmacol.* 7: 88–95.

72. Kiely, J.S., W.H. Moos, M.R. Pavia, R.D. Schwarz, and G.L. Woodard. 1991. A silica gel plate-based qualitative assay for acetylcholinesterase activity: A mass method to screen for potential inhibitors. *Anal. Biochem.* 196: 439–442.

73. Mroczek, T. and J. Mazurek. 2009. Pressurized liquid extraction and anticholinesterase activity-based thin-layer chromatography with bioautography of *Amaryllidaceae* alkaloids. *Anal. Chim. Acta* 633: 188–196.

74. Yang, Z., X. Zhang, D. Duan, Z. Song, M. Yang, and S. Li. 2009. Modified TLC bioautographic method for screening acetylcholinesterase inhibitors from plant extracts. *J. Sep. Sci.* 32: 3257–3259.

75. Di Giovanni, S., A. Borloz, A. Urbain, A. Marston, K. Hostettmann, P.A. Carrupt, and M. Reist. 2008. In vitro screening assays to identify natural or synthetic acetylcholinesterase inhibitors: Thin layer chromatography versus microplate methods. *Eur. J. Pharm. Sci.* 33: 109–119.

76. Mroczek, T. 2009. Highly efficient, selective and sensitive molecular screening of acetylcholinesterase inhibitors of natural origin by solid-phase extraction-liquid chromatography/electrospray ionization-octopole-orthogonal acceleration time-of-flight- mass spectrometry and novel thin-layer chromatography-based bioautography. *J. Chromatogr. A* 1216: 2519–2528.

77. Yang, Z.D., Z.W. Song, J. Ren, M.J. Yang, and S. Li. 2011. Improved thin-layer chromatography bioautographic assay for the detection of acetylcholinesterase inhibitors in plants. *Phytochem. Anal.* 22: 509–515.

78. Ramallo, I.A., M.G. Sierra, and R.L.E. Furlan. 2012. Discovery of β-glucosidase inhibitors from a chemically engineered extract prepared through ethanolysis. *Med. Chem.* 8: 112–117.

79. Salazar, M.O., O. Micheloni, A.M. Escalante, and R.L.E. Furlan. 2011. Discovery of a β-glucosidase inhibitor from a chemically engineered extract prepared through sulfonylation. *Mol. Divers.* 15: 713–719.

80. Wangthong, S., I. Tonsiripakdee, T. Monhaphol, R. Nonthabenjawan, and S.P. Wanichwecharungruang. 2007. Post TLC developing technique for tyrosinase inhibitor detection. *Biomed. Chromatogr.* 21: 94–100.

81. Hassan, A.M.S. 2012. TLC bioautographic method for detecting lipase inhibitors. *Phytochem. Anal.* 23: 405–407. DOI 10.1002/pca.1372.

82. Harman, D. 1993. Free radical involvement in aging: Pathophysiology and therapeutic implications. *Drugs & Aging* 3: 60–80.

83. Ruffin, IV. M.T. and C.L. Rock. 2001. Do antioxidants still have a role in the prevention of human cancer? *Curr. Oncol. Rep.* 3: 306–313.

84. Blois, M.S. 1958. Antioxidant determinations by the use of a stable free radical. *Nature* 181: 1199–1200.

85. Molyneux, P. 2004. The use of the stable free radical diphenylpicrylhydrazyl (DPPH) for estimating antioxidant activity. *Songklanakarin J. Sci. Technol.* 26: 211–219.

86. Glavind, J. and G. Holmer. 1967. Thin-layer chromatographic determination of antioxidants by the stable free radical α,α′-diphenyl-β-picrylhydrazyl. *J. Am. Oil Chem. Soc.* 44: 539–542.

87. Yrjőnen, T., L. Peiwu, J. Summanen, A. Hopia, and H. Vuorela. 2003. Free radical-scavenging activity of phenolics by reversed-phase TLC. *JAOCS* 80: 9–14.

88. Jaime, L., J.A. Mendiola, M. Herrero, C. Soler-Rivas, S. Santoyo, F.J. Senorans, A. Cifuentes, and E. Ilbanez. 2005. Separation and characterization of antioxidants from Spirulinaplatensis micro alga combining pressurized liquid extraction, TLC, and HPLC-DAD. *J. Sep. Sci.* 28: 2111–2119.

89. Pozharitskaya, O.N., S.A. Ivanova, A.N. Shikov, and V.G. Makarov. 2007. Separation and evaluation of free radical-scavenging activity of phenol components of Emblica officinalis extract by using an HPTLC–DPPH method. *J. Sep. Sci.* 30: 1250–1254.

90. Pozharitskaya, O.N., S.A. Ivanova, A.N. Shikov, V.G. Makarov, and B. Galambosi. 2007. Separation and evaluation of free radical-scavenging activity of phenol components of green, brown, and black leaves of *Bergenia crassifolia* by using HPTLC-DPPH method. *J. Sep. Sci.* 30: 2447–2451.

91. Pozharitskaya, O.N., S.A. Ivanova, A.N. Shikov, and V.G. Makarov. 2008. Separation and free radical-scavenging activity of major curcuminoids of Curcuma longa using HPTLC-DPPH method. *Phytochem. Anal.* 19: 236–243.

92. Bhattarai, H.D., B. Paudel, S.G. Hong, H.K. Lee, and J.H. Yim. 2008. Thin layer chromatography analysis of anti oxidant constituents of lichens from Antarctica. *J. Nat. Med.* 62: 481–484.

93. Zhao, J., J.S. Zhang, B. Yang, G.P. Lv, and S.P. Li. 2010. Free radical scavenging activity and characterization of sesquiterpenoids in four species of Curcuma using a TLC bioautography assay and GC-MS analysis. *Molecules* 15: 7547–7557.

94. Hosu, A., C. Cimpoiu, M. Sandru, and L. Seserman. 2010. Determination of the antioxidant activity of juices by thin-layer chromatography. *J. Planar Chromatogr. Mod. TLC* 23: 14–17.

95. Cieśla, Ł. and M. Waksmundzka-Hajnos. 2010. Application of thin-layer chromatography for the quality control and screening the free radical scavenging activity of selected pharmaceutical preparations containing *Salvia officinalis* extract. *Acta Pol. Pharm. Drug Res.* 67: 481–485.

96. Cieśla, Ł., D. Staszek, M. Hajnos, T. Kowalska, and M. Waksmundzka-Hajnos. 2011. Development of chromatographic and free radical scavenging activity finger prints by thin-layer chromatography for selected Salvia species. *Phytochem. Anal.* 22: 59–65.

97. Olech, M., Ł. Komsta, R. Nowak, Ł. Cieśla, and M. Waksmundzka-Hajnos. 2012. Investigation of anti radical activity of plant material by thin-layer chromatography with image processing. *Food Chem.* 132: 549–553.

98. Cieśla, Ł., I. Kowalska, W. Oleszek, and A. Stochmal. 2013. Free radical scavenging activities of polyphenolic compounds isolated from *Medicago sativa* and *Medicago truncatula* assessed by means of thin-layer chromatography DPPH˙ rapid test. *Phytochem. Anal.* 24: 47–52. doi: 10.1002/pca.2379.

99. Williams, L., R. Sjovik, and M.L. Falck-Pedersen. 2004. ChemScreen: Planar synthesis, separation and screening of antioxidants. *J. Planar Chromatogr. Mod. TLC* 17: 244–249.

100. Cieśla, Ł., J. Kryszeń, A. Stochmal, W. Oleszek, and M. Waksmundzka-Hajnos. 2012. Approach to develop a standardized TLC-DPPH˙ test for assessing free radical scavenging properties of selected phenolic compounds. *J. Pharm. Biomed. Anal.* 70: 126–135.

101. Zampini, I.C., R.M. Ordonez, and M.I. Isla. 2010. Autographic assay for the rapid detection of antioxidant capacity of liquid and semi-solid pharmaceutical formulations using ABTS˙+ immobilized by gel entrapment. *AAPS Pharm. Sci. Tech.* 11: 1159–1163.

102. Soler-Rivas, C., J.C. Espin, and H.J. Wichers. 2000. An easy and fast test to compare total free radical scavenger capacity of foodstuffs. *Phytochem. Anal.* 11: 330–338.

13 TLC Determination of Drug Lipophilicity

Živoslav Lj. Tešić and Dušanka M. Milojković-Opsenica

CONTENTS

13.1 INTRODUCTION

According to popular definition, lipophilicity is a characteristic of a substance that is much more easily soluble in lipids than in water. However, that is a very simplified description of fundamental physicochemical property of compound traditionally used in quantitative structure–activity relationship (QSAR), quantitative structure–retention relationship (QSRR), and quantitative structure–property relationship (QSPR) studies and in drug and pesticide design as well as in toxicology studies either in the pharmaceutical or environmental sciences.

While medicinal chemists prefer the term "lipophilicity," in the chromatographic literature, "hydrophobicity" was often used for expression of the relative tendency of a substance to prefer a nonaqueous over an aqueous environment (Nasal et al. 2003). Indeed, there has been certain confusion in the use of terms "lipophilicity" and "hydrophobicity." Although these two expressions were often considered as synonyms, their meanings are quite different (Pliška et al. 1996). In accordance to International Union of Pure and Applied Chemistry (IUPAC) recommendation (van de Waterbeemd et al. 1997), operational definitions of terms lipophilicity and hydrophobicity are as follows:

- *Hydrophobicity* is the association of nonpolar groups of molecules in an aqueous environment, which arises from the tendency of water to exclude nonpolar molecules.
- *Lipophilicity* represents the affinity of a molecule or a moiety for a lipophilic environment. It is commonly measured by its distribution behavior in a biphasic system, either liquid–liquid (e.g., partition coefficient in 1-octanol–water) or solid–liquid (retention on reversed-phase high-performance liquid chromatography [RP-HPLC] or thin-layer chromatography [TLC] system).

Therefore, hydrophobicity contributes toward lipophilicity, but the difference between these two molecular characteristics is evident.

For the first time, the correlation between partition in olive oil/water system and biological effects of drugs (anesthetics) was reported by Meyer (1899) and Overton (1901) at the end of the nineteenth and the beginning of the twentieth century. After that pioneering work in which it was demonstrated that narcotic potency tended to increase with oil/water partition coefficient, further investigations, such as a correlation between heptane–water partition coefficients of certain drugs and the rate of their penetration through the blood–brain barrier (Gaudette and Brodie 1959), were performed. A comprehensive research on lipophilicity by Hansch and Fujita (1964) and by Leo et al. (1971) resulted in the introduction of the octanol–water system as a reference system for the determination of this property and in development of QSAR analysis, known also as Hansch analysis. The lipophilicity expressed as octanol–water partition coefficient in its logarithmic form, $\log P$, becomes one of the most commonly reported physicochemical characteristics of drugs, pesticides, and other chemicals (Giaginis and Tsantili-Kakoulidou 2008a; Han et al. 2011). The concept of lipophilicity takes an important role in life sciences and the environment. Lipophilicity affects compound solubility (Hill and Young 2010); determines the passive transport of drugs through biological membranes, including gastrointestinal absorption and blood–brain barrier crossing (Henchoz et al. 2009); and represents an important component of drug–receptor binding (Waring 2009). The biodistribution, protein binding, metabolism, and pharmacokinetics of drugs may also be altered by their lipophilicity (Gleeson 2008). In addition, nonspecific toxicity of substances is correlated with their tendency to accumulate in cell membranes and therefore their lipophilicity (Tarcsay et al. 2012). The lipophilicity of various substances is also important in environmental chemistry, as it can be related to bioavailability and bioconcentration of environmentally relevant chemicals in soil, plants, fish, and other animals (Poole and Poole 2003). The distribution of substance between water and soil (or sediment), expressed as $\log K_{o/c}$ (Andrić et al. 2010), is an important factor that indicates the physical movements of pollutants, chemical degradation, and biodegradation activity of a given species in the environment. This parameter, which is of great significance for the environmental risk assessment of organic chemicals, can be estimated based on a compound's lipophilicity.

The determination and expression of lipophilicity for any commercially available chemical are required by the European Union regulations, and appropriate procedures are implemented in the guidelines for testing chemicals (OECD, www.oecd.org).

13.2 EXPRESSION OF LIPOPHILICITY

The partition of solute (A) between two immiscible phases is an equilibrium phenomenon that is governed by Nernst's distribution law (Nernst 1891). For life sciences and environment, the partition system consisting of aqueous and lipid-like organic phase is of special importance. The resulting equilibrium for such partition system can be expressed as

$$A(w) \rightleftarrows A(o) \tag{13.1}$$

where letters in parentheses refer to the aqueous (subscript "w") and organic (subscript "o") phases, respectively. The corresponding equilibrium constant, known as distribution coefficient P (or K), can be expressed as the ratio of activities of A in organic and aqueous phases, respectively:

$$P(\text{or } K) = \frac{(a_A)_o}{(a_A)_w} \tag{13.2}$$

At given temperature, P is constant independent of total concentration of A. If A is neutral species in a sufficiently diluted solution, the distribution coefficient is approximately equal to the ratio of equilibrium molar concentration of A in organic and aqueous phases, respectively:

$$P(\text{or } K) = \frac{[A]_o}{[A]_w} \tag{13.3}$$

Since the measurable P values are ranged over several orders of magnitude, they are usually expressed logarithmically; thus, $\log P$ (or $\log K$) is often used instead of P for practical reasons.

However, when any secondary chemical reaction occurs, that is, if the substance participates in an additional chemical equilibrium such as ionization, complexation, and participation, in either or both phases, the distribution ratio D, defined for the total analytical concentrations, must be used (Nowosielski and Fein 1998):

$$D = \frac{\Sigma\{A\}_o}{\Sigma\{A\}_w} \qquad (13.4)$$

If the molecule is ionizable, the pH of the aqueous phase will influence the concentrations of the ionized and unionized forms of the molecule. Since most medicinal substances are ionizable (Avdeef 2008), their overall partition ratio, which depends on the pH of the aqueous phase and the acidity (pK_a) of the species, should be described as the distribution ratio, D, that is, $\log D$ (Kah and Brown 2008).

The partition coefficient between water and a water-immiscible (lipid-like) organic solvent is generally used for the expression of a substance's lipophilicity. In scope of systematic research on lipophilicity, the saturated biphasic system consisting of 1-octanol and water has been introduced by Hansch and Fujita (1964) and by Leo et al. (1971) as a reference system for modeling of biological partition. In accordance to IUPAC recommendations (Rice et al. 1993), the partition coefficient $P_{o/w}$ or $K_{o/w}$ is a "measure of lipophilicity by determination of the equilibrium distribution between octan-1-ol and water, as used in pharmacological studies and in the assessment of environmental fate and transport of organic chemicals." Although P is the partition coefficient notation generally used in the pharmaceutical and medicinal chemistry literature, in environmental and toxicological sciences, the term $\log K_{o/w}$ is commonly used. As it was mentioned earlier, the lipophilicity of ionizable compounds is expressed by $\log D$ values.

The partition of ionizable compounds in the octanol–water system is commonly accompanied with dissociation in the aqueous phase. The dissociation of a monoprotic acid in solution, written as

$$[HA] + H_2O \rightleftarrows [H^+] + [A^-] \qquad (13.5)$$

is governed by a dissociation constant:

$$K_a = \frac{[H_3O^+][A^-]}{[HA]} \qquad (13.6)$$

The corresponding distribution ratio can be written as

$$D = \frac{[HA]_o + [A^-]_o}{[HA]_w + [A^-]_w} \qquad (13.7)$$

Assuming that only the unionized species are present in the organic phase and the effect of ionic strength of solution is neglected, D of acids can be calculated from the dissociation constant, pK_a and P:

$$D = \frac{P}{1 + 10^{(pH - pK_a)}} \qquad (13.8)$$

However, the pass of ionized forms of substances into organic phase and their crossing through a membrane are experimentally confirmed (Jafvert et al. 1990; van de Waterbeemd 2003). If the octanol–water partition coefficient of ionic species is expressed as

$$P^I = \frac{[A^-]_o}{[A^-]_w} \qquad (13.9)$$

D might be then defined as

$$D = P^N[HA] + P^I[A^-] \tag{13.10}$$

and calculated by the equation

$$D = \frac{P^N + P^I 10^{(pH-pK_a)}}{1 + 10^{(pH-pK_a)}} \tag{13.11}$$

where P^N is the partition coefficient of neutral (unionized) species.

The dependence between distribution coefficient and pH of various ionizable substances is widely investigated and reviewed (Avdeef 1996; Berthod et al. 1999; Berthod and Carda-Broch 2004). It was shown (Berthod et al. 1999) that in case of monoprotic acids for pH $< pK_a - 2$, $\log D$ corresponds to $\log P^N$, while for pH greater than pK_a, $\log D$ linearly decreases with an increasing pH according to the relation:

$$\log D = \log P^N + pK_a - pH \tag{13.12}$$

Based on similar investigations for monoprotic bases, it has been proposed that $\log D$ corresponds to $\log P^N$ for pH $> pK_a + 2$, while for pH smaller than pK_a, $\log D$ linearly decreases according to

$$\log D = \log P^N - pK_a + pH \tag{13.13}$$

In case of amphoteric substances, the partial dissociation constants play a role in distribution processes. The situation becomes complex especially if the difference between mentioned constants is smaller than $4\ pK_a$ units (protonation steps are not well separated from each other), that is, in the case of zwitterionic species. The results of the investigation of partitioning behavior of zwitterionic compounds have also been reported in the literature (Takács-Novák et al. 1995; Avdeef 2001).

13.3 MEASUREMENT OF LIPOPHILICITY

The partition coefficient was first defined in 1872 by Berthelot and Jungfleisch (Berthelot and Jungfleisch 1872), and it was used to correlate the potencies of biologically active substances by both Meyer and Overton in 1899 and in 1901, respectively. From these pioneering works, the numerous methods were developed for the measurement or estimation of the partition coefficient of substances between two phases as a measure of their lipophilicity. From the first comprehensive review of partition coefficients in 1971 (Leo et al. 1971) till now, an enormous number of publications from this field were issued (Poole and Poole 2003; Gocan et al. 2005; Giaginis and Tsantili-Kakoulidou 2008a; Komsta et al. 2010; Trifković et al. 2010; Šegan et al. 2011; Rabtti et al. 2012).

n-Octanol–water is the widely accepted reference system for the determination of lipophilicity.

n-Octanol was chosen because of its superficial similarity to lipids: a long alkyl chain plus a functional group having both hydrogen bond accepting and donating characteristics (Leo et al. 1971). However, the $\log P_{oct}$ lipophilicity scale alone is not sufficient to model membrane permeation, due to major differences in biophysical properties (Bouchard et al. 2002).

Therefore, the partition in other biphase solvent systems is sometimes a better model for the in vivo process. For example, the alkane–water systems (including hexane, cyclohexane, heptane, isooctane, dodecane, or hexadecane) are better models for a transport across the blood–brain barrier (Gocan et al. 2005). Because of similar properties, better dissolving capacity 1,2-dichloroethane has been used to replace alkanes. This system is also well suited for cyclic voltammetry studies (Bouchard et al. 2002). Other investigated systems include chloroform–water and di-n-butylether or propylene glycol dipelargonate (van de Waterbeemd and Manhold 1996).

As it is mentioned earlier, numerous methods are available to measure or estimate the octanol–water partition coefficients. Each of them has advantages and disadvantages, as well as some practical limitations such as unreliability for certain classes of substances. Thus, before determining $P_{o/w}$, it is useful to have preliminary information on the structural formula, dissociation constant, water solubility, hydrolysis, n-octanol solubility, and surface tension of the substance. Before selecting which procedure to use, it is helpful to obtain a preliminary estimation of the $P_{o/w}$ from a suitable calculation method.

The Organization for Economic Co-operation and Development (OECD) Guidelines for the Testing of Chemicals describes the following methods for the determination of octanol–water partition coefficients ($\log P_{o/w}$): Test No. 107, shake-flask method (OECD 1995); Test No. 117, HPLC method (OECD 2004); Test No. 123, slow-stirring method (OECD 2006); and Draft Guideline 122, pH metric method for ionizable substances (OECD 2000).

13.3.1 SHAKE-FLASK METHOD

Traditionally, the octanol–water partition coefficients are determined by the so-called shake-flask method. Since it was adopted in 1981, this method is still the reference for any new procedure of lipophilicity measurement. The principle of the method is shaking of test substance with mixture of mutually saturated octanol and water until equilibrium is reached. Then, the two liquid phases are separated by centrifugation, and the solute concentration is determined in both of them by appropriate analytical methods, which include spectroscopic and separation methods (photometry, GC, and HPLC). The total quantity of substance present in both phases should be calculated and compared with the quantity originally introduced.

In addition to reliability in the applicability range of $\log P_{o/w}$ from −2 to 4, the main advantage of the shake-flask method is that it is a direct method for which reference substances do not need to be employed. On the other hand, the method is tedious and time-consuming and requires relatively large amounts of pure compounds, and it is unsuitable for substances with high lipophilicity ($\log P_{o/w} > 5$) because of insufficient reliability caused by the formation of octanol emulsions in water (Danielsson and Zhang 1996). The formation of stable emulsions makes difficult the complete separation of the octanol and water phases and consequently affects accuracy of measurement of the analyte concentration. This is of special importance in the case of high lipophilic substances where any droplets of octanol in water can contain significant quantities of the investigated compound and can lead to large errors. Also, the measurement of both very low and very high $\log P_{o/w}$ values by shake-flask method requires a disproportionate volume ratio of the aqueous and solvent phases, making it difficult to sample a reasonable amount for the concentration determination and automated sampling almost impossible (Gocan et al. 2005). Last but not least, if the tested substance is ionizable, the method should only be applied to its unionized form using an appropriate buffer with a pH of at least one unit below (protonated form) or above (deprotonated form) the pK_a (OECD 1995).

Several alternative methods have been developed to overcome the difficulties of the shake-flask method. In order to avoid the problem related to the formation of emulsions, the shake-flask method is improved into the stir flask, that is, slow-stirring method in which the two phases are not shaken but stirred in a vessel (Brooke et al. 1986; De Bruijn et al. 1989). The result reported for the usage of this method is shown to be reproducible and in good agreement with data obtained by other techniques. However, with exception of emulsification, all the previously mentioned disadvantages of shake-flask method remain. The slow-stirring method was adopted in 2006 as a suitable experimental approach for the direct determination of $P_{o/w}$ of highly hydrophobic substances (OECD 2006, Test 123).

Several automated approaches have been developed since the shake-flask approach was decided to be the reference method in industry for $\log P/\log D$ measurements (Gocan et al. 2005). Among them, an automated micro shake-flask method described by Valkó (2000) and similar method transferred onto 96-well plate technology and a robotic liquid handler used for sample preparation (Hitzel et al. 2000) should be mentioned.

Finally, as modern variations on the shake-flask method, a flow injection extraction (FIE) was used to determine the octanol–water partition coefficients (Danielsson and Zhang 1994). FIE refers to a continuous extraction system based on flow injection analysis (FIA) and a liquid–liquid segmented flow (Danielsson and Zhang 1996). The main advantages are that the technique is fast, robust, and easy to automate, and it requires small volumes of sample and organic phase. Because of that, it is valuable as a screening method for determining partition coefficients of potential drug substances (Carlsson and Karlberg 2000).

13.3.2 Liquid Chromatographic Methods

Among the indirect methods for the determination of lipophilicity, that is, the octanol–water partition coefficient, the various chromatographic methods are of great importance.

In accordance to IUPAC recommendations from 1993 (Ettre 1993), chromatography is defined as a physical method of separation in which the components to be separated are distributed between two phases, one of which is stationary (stationary phase) while the other (the mobile phase) moves in a definite direction. At given temperature, chromatographic process is determined by intermolecular interactions such as the solute–stationary phase, solute–mobile phase, and mobile–stationary phase. These interactions include van der Waals forces, hydrogen bonding, dipole–dipole interactions, donor–acceptor (charge–transfer) interactions, and solvophobic interactions. Among numerous chromatographic methods, liquid chromatography was of particular interest for lipophilicity assessment because of the similarity between the solute partition in a chromatographic system, consisting of a liquid mobile phase and a liquid-like stationary phase, and the solute partition in a dual liquid phase environment (Gocan et al. 2005). Indeed, the literature is rich in research articles investigating similarities/dissimilarities between octanol–water partitioning and chromatographic retention (Abraham et al. 1994; Tsantili-Kakoulidou 2010). It is generally accepted that chromatography provides an easy, reliable, and accurate way to determine the molecular lipophilicity of compounds based on their retention factors (Valkó 2004). Reversed-phase chromatographic techniques, in particular high-performance liquid chromatography (RP-HPLC) and thin-layer chromatography (RP-TLC), have proved to simulate octanol–water partitioning and are considered as popular alternatives for lipophilicity assessment. The main advantages of these methods in comparison to traditional shake-flask procedure are speed, reproducibility, insensitivity to impurities or degradation products, broader dynamic range, online detection, and reduced sample handling and sample sizes (van de Waterbeemd et al. 1996).

The OECD Guidelines for the Testing of Chemicals, Test 117 (OECD 2004) based on a paper of Eadsforth and Moser (1983), describes a method for the determination of partition coefficients ($\log P_{o/w}$) using RP-HPLC. The principle of method is partitioning of test substance between nonpolar alkyl-modified silica (C8 or C18) and a polar aqueous phase, that is, binary methanol–water mobile phase with more than 25% of water. In particular, under the guidelines, appropriate reference substances with $\log P_{o/w}$ values encompassing the $\log P_{o/w}$ of the test substances are chromatographed in isocratic mode under the same conditions as the test substances. In the following, a calibration graph obtained by correlating the measured retention data, $\log k$, of reference substances with their partition coefficients is used for the determination of the $\log P_{o/w}$ values of the test substances by the relation:

$$\log P_{ow} = a + b \times \log k \tag{13.14}$$

where a and b are linear regression parameters. This equation is determined by linear regression of $\log P_{o/w}$ of reference substance, with $\log P$ values near the ones expected for test substances, against logarithm of their capacity factors. The RP-HPLC method enables an estimation of $\log P_{o/w}$ values in range from 0 to 6 with the reliability of the ±0.5 $\log P_{o/w}$ unit of the shake-flask value.

In some cases, by the modification of the mobile phase (Eadsforth 1986), the applicability range may be expanded to $\log P_{o/w}$ values between 6 and 10. However, the HPLC method is not applicable to strong acids and bases, metal complexes, or surface-active agents. In case of $\log P_{o/w}$ measurement on unionized form of ionizable substances, an appropriate buffer must be used. For the use in environmental hazard classification or in environmental risk assessment, the measurement of $\log P_{o/w}$ values should be performed in the pH range relevant for the natural environment, that is, in the pH range of 5–9 (OECD 2004).

The quality of the obtained $\log P_{o/w}$ values is significantly dependent on reference substance used. Under the guidelines, at least six such substances structurally related to the test substances, preferably, are proposed to be used. Also, a list of reference substances covering $\log P_{o/w}$ values from 0.3 to 6.5 is proposed. This list, containing sixty neutral, basic, as well as acidic substances, may be adapted to a given case (Cichna et al. 1995). In order to improve quality of correlation between $\log P_{o/w}$ and $\log k$, some additional constants taking into account characteristics of column and mobile phase, as well as inductive electronic properties of solutes, were introduced in corresponding relationship (Yamagami et al. 1999).

Doubtlessly, the most widely used chromatographically obtained lipophilicity parameter is $\log k_w$, the extrapolated solute retention factor in pure water (Braumann 1986). The $\log k_w$ is derived from linear dependence of $\log k$ on composition of mobile phase by extrapolation of straight line to pure water, that is, 0% of organic modifier (Du et al. 1998). As every other extrapolated value, $\log k_w$ can yield large errors. This is proved by the fact that the extrapolated $\log k_w$ value, obtained with the same solute and column, may differ significantly depending on the organic modifier used: methanol or acetonitrile (Berthod and Carda-Broch 2004).

In addition, two other reversed-phase retention parameters have been introduced: φ_0 and chromatographic hydrophobicity index (*CHI*) (Valkó and Slegel 1993; Valkó et al. 1997). Both parameters are derived from the solvent strength needed to elute the compound from an RP-HPLC column. The φ_0 index is derived from a series of isocratic measurements as the volume percent of concentration of organic modifier in the mobile phase for which is $\log k = 0$. On the other hand, *CHI* is derived from the retention time in a calibrated generic gradient HPLC experiment and was introduced as a high-throughput method for lipophilicity determination (Valkó et al. 1997).

Apart from RP-HPLC, other separation methods were used for lipophilicity assessment such as immobilized artificial membrane (IAM) chromatography, immobilized liposome chromatography (ILC), micellar liquid chromatography, capillary electrophoresis (EC), micellar electrokinetic capillary chromatography (MEKC), microemulsion electrokinetic capillary chromatography (MEEKC), and other electrokinetic chromatography techniques using liposomes, LEKC, and vesicles, VEKC (Kaliszan et al. 2003; Martel et al. 2008), as well as countercurrent chromatography (CCC) (Berthod and Carda-Broch 2004). CCC is a separation technique that uses a liquid mobile phase with a stationary phase that is also liquid. There is no solid support for the liquid stationary phase. This allows innovative uses of the technique that cannot be conceived with any other chromatographic technique with a solid or a solid support for the stationary phase. Centrifugal fields maintain the two immiscible liquid phases together. The only physicochemical interaction that is responsible for solute retention in a CCC column is liquid–liquid partitioning (Berthod and Carda-Broch 2004).

13.3.3 CALCULATION METHODS

Numerous methods for calculation of $\log P$, that is, prediction of lipophilicity of substances, have been developed and described in the literature (Leo 1993; Mannhold and Ostermann 2008; Tetko and Poda 2008). They may be classified as either substructure-based or property-based methods. The principle of substructure-based calculation is cutting molecules into fragments (so-called fragmental methods) or down to the level of single atom (atom-based methods) and consequently summing the substructure contributions into the final $\log P$ value. On the other hand, the principle of property-based approaches is the usage of various descriptors of the whole molecule to quantify $\log P$. To this class belong methods that utilize 3D structure representation and methods that are based on topological descriptors.

Due to the importance of calculated $\log P_{ow}$ for selecting appropriate experimental method as well as for checking the plausibility of experimentally obtained values, a short introduction to the calculation of octanol–water partition coefficients is accompanied as an annex to OECD Test 117. The suggested methods are based on the theoretical fragmentation of the molecule into suitable substructures with known reliable $\log P_{o/w}$ increments. Log $P_{o/w}$ is obtained by summing the fragment values and the correction terms for intramolecular interactions (OECD 2004; Pomona College, Medicinal Chemistry Project, Claremont, California 91711, USA, Log P Database and Med. Chem. Software [Program CLOGP-3]). On the basis of the fragmental methods, that is, Fujita–Hansch method (Fujita et al. 1964), Rekker method (Rekker 1977), and c–Leo method (Hansch and Leo 1979) as well as their combinations, different software packages are developed. Some of them are KLOGP, KOWWIN, CLOGP, and ACD/LogP. Software packages MOLCAD, TSAR, PROLOGP, ALOGP98, and XLOGP are derived from atom-based calculation methods.

In addition to previously mentioned substructure-based methods, the procedures related to the prediction of $\log P$ on the basis on molecular descriptors have aroused a great interest during the last years. Despite a great variety of developed methods, all of them indicate the importance of molecular size and hydrogen bonding for an accurate prediction of lipophilicity. The agreement between these parameters and those pointed as the most important in the so-called empirical methods such as linear salvation energy relationships (LSERs) has been reported (Tetko and Poda 2008).

The assumption that structure descriptors of substances are correlated with their properties, retention, and biological activity is widely used in quantitative structure–property relationships (QSPR), QSRR, and QSAR studies (Héberger 2007; Šegan et al. 2011).

These studies are unquestionably of great importance in modern chemistry. The concept of QSAR/QSPR is to transform searches for compounds with desired properties into a mathematical model. Once a correlation between structure and activity/property is found, any number of compounds, including those not yet synthesized or examined in in vitro and/or in vivo experiments, can be readily screened on the computer in order to select structures with the properties desired. It is then possible to select the most promising compounds to further synthesis and testing in order to obtain, for example, new drugs or materials (Karelson et al. 1996). Most of the mentioned structure–property relationships reported in the literature were derived by means of the multivariate statistical (chemometrical) methods using molecule property, retention data, or activity as a dependent variable, and various empirical, semiempirical, and nonempirical structural parameters as the independent ones. Various chemometrical methods may be used such as multiple linear regression (MLR), principal component analysis (PCA), hierarchical cluster analysis (HCA), principal component regression (PCR), partial least squares projection of latent structures (PLS), and artificial neural networks (ANNs) (Varmuza and Filzmoser 2009; Trifković et al. 2010). Many papers related to this subject can be found in the available literature covering various research fields including analytical chemistry, chemometrics, bioinformatics, and pharmacy (Kaliszan 2007). For example, in a study performed by Eros et al. (2002), the published methods were critically reviewed and the predictive power of different softwares (CLOGP, KOWWIN, SciLogP/ULTRA) was compared to each other and to their own automatic QS(P)AR program based on the strictest mathematical and statistical rules (Eros et al. 2002). Based on the investigation of a very diverse set of 625 known drugs (98%) and drug-like molecules with experimentally validated $\log P$ values, the authors emphasized the importance of data quality, descriptor calculation, and selection and presented a general, reliable descriptor selection and validation technique for such kind of studies. It was found that the $\log P$ prediction by an MLR, partial least squares regression, and ANN models showed acceptable accuracy for new compounds and therefore it can be used for "in silico screening" and/or planning virtual/combinatorial libraries.

13.4 ESTIMATION OF LIPOPHILICITY BY TLC

Numerous applications of TLC in lipophilicity assessment of pharmaceuticals, drugs, and potential drugs are reported in the literature and reviewed periodically. Apart from publications in specialized journals such as *Journal of Planar Chromatography: Modern TLC*, *Acta Chromatographica*, *Journal of Liquid Chromatography & Related Technologies*, and *Journal of Pharmaceutical and Biomedical Analysis*, biennial reviews in *Analytical Chemistry* titled *Planar Chromatography*, which are published every even year (Sherma 2010), and *Pharmaceuticals and Related Drugs*, published every odd year (Gilpin and Gilpin 2011), are of invaluable importance. Some of the characteristic aspects of assessment of lipophilicity by means of TLC are considered in the succeeding text.

13.4.1 REVERSED-PHASE TLC

The previously mentioned OECD Guidelines for the Testing of Chemicals in scope of Test 117 describes a method for the determination of $\log P_{o/w}$ using RP-HPLC (OECD 2004). However, in several publications, the HPLC method is substituted with TLC, keeping the same principles as in the original test (Rabtti et al. 2012). Indeed, the main difference between TLC and HPLC is a configuration of stationary phase, while the separation mechanisms for these liquid chromatographic techniques are the same. Modern TLC, named high-performance TLC, is an instrumental technique that not only is comparable by reliability but also has some preferences over HPLC. The main advantages of RP-TLC method are as follows:

1. Small amount of sample is needed for estimation.
2. Better precision and accuracy caused by simultaneous analysis of both samples and standards under the same conditions.
3. Low sensitivity to impurities.
4. There is no possibility of interference from previous analysis as fresh stationary phase is used for each analysis.
5. Mobile phase consumption per sample is extremely low.
6. Instrumentation is simple, inexpensive, and easy to handle.
7. Several samples can be separated in parallel on the same plate resulting in a high-throughput and a rapid low-cost analysis.
8. There are wide choices of adsorbents and developing solvents.
9. Greater applicability to compounds with higher lipophilicity (Šegan et al. 2011; Gupta et al. 2012; Shewiyo et al. 2012).

Therefore, HPTLC is rapidly gaining importance in pharmaceutical analysis, biochemistry, and pharmacokinetic studies, which is evidenced by a large number of papers published in different journals (Sherma 2010).

The first chromatographic technique used for the determination of partition coefficients, that is, the estimation of lipophilicity, was just TLC. The commonly used expression of analyte retention in TLC is the retardation factor, R_F:

$$R_F = \frac{x}{f} \tag{13.15}$$

where
 x is the single zone distance from the start point
 f is the solvent distance from the start point to solvent front

The first relation between TLC retention and lipophilicity was appointed by Martin and Synge (Kowalska and Prus 2003):

$$P = \frac{V_M}{V_S}\left(\frac{1}{R_F} - 1\right)$$ (13.16)

where
  P is the partition coefficient, that is, the concentration in the stationary phase divided by the concentration in the mobile phase
  V_M is the volume of the mobile phase
  V_S is the volume of the stationary phase

In addition, a retention constant, R_M, analogous to capacity factor in HPLC, was proposed by Bate-Smith and Westall (1950):

$$R_M = \log\left(\frac{1}{R_F} - 1\right)$$ (13.17)

A substitution of Equation 13.17 into 13.16 and rearrangement give a relation between retention constant and partition coefficient:

$$R_M = \log P + \log r$$ (13.18)

where r is constant for the given system and represents the ratio between volumes of stationary and mobile phases, V_S/V_M.

Similar to Equation 13.14, the solute TLC retention may be correlated with its lipophilicity via the relation:

$$\log P (\text{or} \log D) = a + b R_M$$ (13.19)

A large number of very good linear correlations between the R_M and $\log P$ values for neutral and ionic compounds can be found in the literature (Braumann 1986; Dorsey and Khaledi 1993; Nasal et al. 2003; Poole and Poole 2003).

Lipophilicity estimations by RP-TLC are usually performed on silica gel layers, impregnated with either a nonpolar liquid phase (paraffin, silicon, or vegetable oils) or an alkyl-modified (octyl or octadecyl) and aqueous organic mobile phases. In the case of ionizable substances, the pH and the ionic strength of the eluent will modify the apparent lipophilicity according to the ratio of neutral and ionized species. For example, the influence of mobile phase pH and ionic strength on retention was investigated for phenothiazine drugs (Hulshoff and Perrin 1976) and reviewed by Gocan et al. (2005).

Similar to RP-HPLC methodology, the reliability of $\log P$ values estimated by the described procedure is dependent on properties of the selected reference substances. The calibration with substances structurally and chemically similar to those to be investigated provides sufficiently precise $\log P$ values (Takács-Novak et al. 2001). However, in certain cases, for example, in the early phases of drug research, many structurally unrelated compounds must be characterized often in a relatively short time. At such stage, because of an unavailability of calibration set of structurally similar reference substances, a diverse set of compounds must be used. Valkó et al. (2001) developed a general HPLC method for the estimation of the lipophilicity of chemically diverse compounds. With the same idea, the possibility of the development of a similar RP-TLC method using a chemically unrelated calibration set was studied (Völgyi et al. 2008). A validated method is proposed for parallel estimation of $\log P$ of chemically diverse neutral substances, as well as weak acids and bases. In order to cover a wide range of lipophilicity, two chromatographic systems were used: one for $\log P = 0\text{--}3$ and the other for highly lipophilic substances ($\log P = 3\text{--}6$). Silanized silica gel as a stationary phase combined with water/acetone mobile phases was used. Two sets of chemically stable, well-known, and readily available calibration compounds were selected to cover a wide range of moderate and high lipophilicity. For both systems, very good correlations between $\log P$ determined by shake-flask method and R_M values were obtained with correlation coefficients greater than 0.99. The applicability of the method was confirmed on 20 randomly selected structurally diverse compounds.

Because of difficulties related to reference substances, it seems that extrapolated retention factors corresponding to pure water as mobile phase are more representative lipophilicity indices (Giaginis and Tsantili-Kakoulidou 2008a,b).

Generally, in RP-TLC, the R_M value of the solute depends linearly on the concentration, C, of the organic component in the mobile phase (Soczewinski and Matysik 1968):

$$R_M = R_M^0 + bC \tag{13.20}$$

However, when a wide range of mobile phase compositions is considered, a lesser or greater deviation from linear dependence was observed regardless of organic modifier used (Torres-Lapasió et al. 2000; Gaica et al. 2002). Thus, a quadratic equation of the type

$$R_M = R_M^0 + b_1 C + b_2 C^2 \tag{13.21}$$

describes better the retention of the substances subjected to the separation, depending on mobile phase composition.

Based on the relation between retention and TLC system composition, different lipophilicity indices were derived (Andrić et al. 2010; Sârbu and Briciu 2010; Casoni and Sârbu 2012).

The intercept R_M^0, sometimes denoted as R_{Mw}, represents extrapolated R_M value to 0% v/v of organic modifier in binary water–organic solvent system. It can be considered as an estimation of the solute partitioning between pure water and nonpolar stationary phase, measuring in that way solute lipophilicity. That is the reason why the R_M^0 is sometimes preferably used as quantitative TLC retention descriptor instead of R_M value, which is highly dependent on given chromatographic conditions and reflects the partitioning between a nonpolar stationary phase and the mixture of water and organic modifier as a mobile phase instead of pure water (Trifković et al. 2010).

The parameter b, that is, the slope in Equation 13.20, indicates the dependence of the solute solubility, and therefore retention, on mobile phase composition. The slope is commonly considered as the specific hydrophobic surface area of solute. Indeed, two theories relate these terms. The first one correlates the slope to the number of mobile phase molecules in the solvation sphere of the solute, which are released after the formation of the solute–stationary phase complex (Murakami 1979). In the case of reversed-phase chromatography, it obviously depends on the nonpolar (hydrophobic) area of a molecule. Another approach, developed by Horváth and coworkers, is based on the explanation that the surface tension of the mobile phase changes with its composition, thereby altering the energy of vacancy formation required for the accommodation of solute molecules (Horváth et al. 1976).

The correlation between intercept, R_M^0, and slope, b (Equation 13.20), is usually linear for a series of congeneric, that is, structurally related compounds, and can be expressed by the relation:

$$R_M^0 = a_0 + a_1 b \tag{13.22}$$

The congenericity can be broken by the presence of ionizable groups, which can modify the interactions of the components with the nonpolar stationary phase or polar mobile phase.

In addition, Bieganowska et al. (1995) introduced another hydrophobicity parameter, which is defined as the intercept $\left(R_M^0 \right)$ and the slope (b) ratio:

$$C_0 = -\frac{R_M^0}{b} \tag{13.23}$$

Analogous to parameter φ_0 (Valkó 1987) in HPLC, C_0 could be understood as the concentration of an organic modifier in the mobile phase for which the distribution of the solute between the two phases is equal, that is, $R_M = 0$, $R_F = 0.5$. It could also be interpreted as the hydrophobicity per unit of specific hydrophobic surface area.

The extensive studies of relation between chromatographic behavior and lipophilicity carried out during the last decade showed that retention is much better correlated with lipophilicity parameters if PCR is employed (Sârbu et al. 2001; Onisor et al. 2010). This is understandable if the principles of correlation are taken into account; namely, the principal component (PC) combines all chromatographic data in one single feature, acquiring in this way properties of interpolated quantities, while R_M^0, b, and C_0 are extrapolated values and, therefore, accompanied with higher uncertainty.

All four chromatographic descriptors, that is, R_M^0, b, C_0, and $PC1$, are equally present in the chromatographic literature and are commonly used for lipophilicity assessment of unknown solutes (Andrić et al. 2010).

It is well known that retention in RPLC represents a very complex process governed by a variety of both specific and nonspecific interactions that include van der Waals, electrostatic, and hydrogen bonding interactions (Vailaya 2005). The most important among them are solute–solvent interactions, solute–solute interactions in both mobile and stationary phases, solvent–solvent interactions in the mobile (and possibly on the stationary) phase, and intramolecular and intermolecular interactions of the alkyl chains of the stationary phase among themselves and also with the solute and solvent molecules, respectively. Also, conventional RP silica-based sorbents have about 30%–50% unmodified silanol groups accessible to both solute and solvent—even when coated and end capped (Nikitas and Pappa-Louisi 2009). The complexity of the mentioned molecular interactions that are difficult to describe exactly resulted in two approaches to an explanation of solute separation mechanism in RPLC: the solvophobic theory, advocated by Horváth and coworkers, and the lipophilic theory, advocated by Carr and coworkers (Wick et al. 2004). Although no one of the mentioned approaches can adequately describe complex situation that occurs in actual RP systems, the effect of different factors on retention, and therefore on earlier denoted lipophilicity indices, has widely been studied. Some of these factors will be mentioned here.

As one of three essential elements of chromatographic system, the stationary phase plays a very important role in the chromatographic behavior of various solutes. The large majority of earlier lipophilicity studies by RP-TLC are performed on silica gel layers impregnated with hydrophobic, oily liquids such as octanol (Dzimiri et al. 1987), silicone oil (Biagi et al. 1994a,b), or paraffin oil (Cserháti et al. 1983). Recently, a variety of oil- (paraffin, olive, sunflower, corn, castor, cod liver) and fat- (margarine, butter, pig, sheep, pullet, human) impregnated TLC plates were indirectly evaluated and characterized from the lipophilicity point of view by employing a series of experimental lipophilicity parameters estimated for a representative group of natural sweeteners from retention data (Sârbu and Briciu, 2010). It was found that no significant differences between oil and fat impregnated TLC-silica gel plates and recommend them as an alternative in the field of lipophilicity estimation.

However, the main drawback of described method of TLC estimation of lipophilicity is the interference of supports' original adsorptive characteristics (Van der Giesen and Janssen 1982) as well as the influence of amount and quality of coating substance on measured R_M values (Cserháti et al. 1983).

Since the mid-70s, as an alternative to the stationary phases dynamically coated with a lipophilic substance, the physically stable hydrophobic stationary phases have been used (Kaliszan 2007). Although the interference of silanophilic interactions, attributed to the remaining free silanols, has been recognized for a long time as a serious drawback, for years, different so-called alkyl-bonded silica stationary phases were commonly used in the studies of lipophilicity of substances. Among them, octadecyl-modified silica gel, denoted as C-18 or RP-18, is the preferred material for reversed-phase TLC/HPTLC plates for the assessment of drug lipophilicity. Some studies were performed on less hydrophobic bonded phases, such as cyano-and diol-modified silicas (Poole and Poole 2003), but these gave generally worse results (Komsta et al. 2010).

A difference between lipophilicity parameters $\left(R_M^0 \text{ value} \right)$ obtained on C-18 and CN silica for a series of bis-steroidal tetraoxanes as potential antimalarics was observed (Šegan et al. 2009). Such retention behavior was explained by stronger hydrophobic interactions between the analytes and octadecyl alkyl chains of RP-18 silica in comparison to less hydrophobic propyl chains of CN silica.

Also, Dabić et al. noticed differences in lipophilicity indices obtained in scope of estimation of lipophilicity of some biologically active N-substituted 2-alkylidene-4-oxothiazolidines by means of RP-TLC on C-18 and CN silica (Dabić et al. 2011). Moreover, QSRR study of these retention data was performed by MLR and partial least squares regression in order to select molecular descriptors governing specific chromatographic behavior of the mentioned substances (Dabić et al. 2011). Since the mobile phase (THF/water) was the same in both chromatographic systems, the differences in selected descriptors were attributed to the differences in stationary phases. It was found that three descriptors, hydrogen bond donor (HBD), solubility parameter (SP), and hydration energy (HE), exhibit different influences on R_M^0 in PLS models obtained for CN and C18 stationary phases; namely, compounds with higher SP and HBD values are weaker retained on the nonpolar C18 stationary phase, while HE, which has been considered as the descriptor expressing the hydrophobic effect, has positive influence on the R_M^0.

Finally, regarding some new and better approaches related to stationary phases for lipophilicity determination by means of TLC, such as IAMs, one can say that they are not common mainly due to the high costs of analysis (Komsta et al. 2010).

Similar to RP-HPLC, the most commonly employed organic modifiers for lipophilicity assessment by RP-TLC are methanol and acetonitrile. In case of alkyl-bonded silicas as stationary phases, the former modifier has some advantages; namely, as a protic solvent, methanol can interact with residual silanols, reducing in that way zone tailing. Additionally, during equilibration, molecules of methanol form a monolayer at the surface of the stationary phase, which provides a hydrogen bonding capacity in better agreement with 1-octanol–water partition (Giaginis and Tsantili-Kakoulidou 2008). A great advantage of TLC in comparison to HPLC is the possibility to employ solvents that can strongly absorb in UV, for example, acetone.

The long-time chromatographic work of Biagi and coworkers has been focused on the determination of lipophilicity by means of TLC (Biagi et al. 1994a,b). The mobile phase in these studies was aqueous buffer alone or in different proportions with acetone, acetonitrile, or methanol, while a silica gel G impregnated with silicone oil has been used as stationary phase. On the basis of more than 700 TLC equations, it has been found that from linear relationships between R_M values and the organic solvent concentration in mobile phase, a theoretical R_M value at 0% organic solvent can be calculated even for those substances that do not migrate with an aqueous buffer alone. The validity of the extrapolated method is confirmed by very good correlation between experimental and extrapolated R_M values. Another interesting finding illustrated by means of five classes of substances was that the intercept of the TLC equations is not dependent on the nature of the organic modifier (acetone, acetonitrile, or methanol). In addition, a linear relationship between slopes and intercepts of the TLC equation was shown and could be considered as a basic aspect of the chromatographic determination of lipophilicity of a series of strongly congeneric substances. In scope of the investigation of the effect of the organic modifier on the slope of the TLC equation, a relation between the slope and reciprocal of the solvent strength (ε^0) was shown.

Numerous papers dealing with lipophilicity determination by means of RP-TLC have been published by Cserháti and coworkers. In the framework of one of them, the effect of various organic modifiers on the determination of the hydrophobicity parameters of nonhomologous series of 21 commercial anticancer drugs was investigated (Forgács and Cserháti 1995). For that purpose, Polygram UV 254 plates impregnated by paraffin oil was used as stationary phase in combination with binary water–organic mobile phases. As organic modifier, methanol, ethanol, 1-propanol, 2-propanol, acetonitrile, dioxane, and tetrahydrofuran were used in a wide range of concentrations. The retention data were analyzed by means of various multivariate statistical methods such as the spectral mapping technique, PCA, and cluster analysis. It was found that the solvent strength and selectivity of organic modifiers are strongly correlated to the steric characteristics of the solvent molecules suggesting competition between them and anticancer drugs for the hydrophobic surface of the stationary phase.

The same stationary phase combined with water–methanol mobile phase comprising from 0 to 35 vol% of organic modifier has been used for RP-TLC determination of lipophilicity and specific hydrophobic areas of 18 nonsteroidal anti-inflammatory drugs as well as for subsequent QSAR studies. The results showed that the hydrophobicity parameters for the investigated substances are highly dependent on the presence of acetic acid, sodium acetate, and sodium chloride in the mobile phase (Forgács et al. 1998).

The role of the aqueous component of the mobile phase in the RP-TLC retention behavior and hence in lipophilicity estimation of a series of structurally diverse drugs including 12 basic, 2 neutral, and 1 acetic substances has also been investigated (Giaginis et al. 2006). For that purpose, RP-18 silica thin layer was used as stationary phase. The mobile phases have been composed of methanol and different proportions of phosphate buffer, phosphate-buffered saline, and morpholinepropanesulfonic acid with or without the addition of n-decylamine as masking agent to suppress silanophilic interactions, at pH 7.4. In addition, phosphate buffer at pH 11.0 was employed in order to evaluate the extent of ionization suppression and its influence on retention. It was established that the buffer constituents of the mobile phase play an active role in retention affecting both extrapolated R_{Mw} values and slopes, as well as their interrelationship. The R_{Mw} values were correlated with $\log P$ and $\log D_{7.4}$ values, and it was found that the usage of $\log D$ combined with an ionization correction term leads to an improved correlation.

Basic compounds represent the major fraction in drug-related databases, rendering their lipophilicity assessment as an urgent requirement in drug design (Giaginis and Tsantili-Kakoulidou 2008). The analysis of basic drugs and determination of their lipophilicity by means of RP-TLC are difficult because of the interaction of ionic and nonionic forms of drugs with both ligands from alkyl-modified stationary phases and residual silanols. In order to avoid the mentioned interactions, different mobile phase additives can be used such as buffers at high pH where ionization of bases is suppressed, ammonia that acts as ion suppressant and simultaneously blocker of surface silanols, as well as ion-pair reagents that form associates with the cations of the bases. The two latter approaches have been applied in the determination of the lipophilicity of 12 psychotropic drugs by means of RP-TLC on C-18 silica (Hawryl et al. 2008). The mobile phases consisted of different amounts of water and the organic modifiers—methanol, dioxane, acetonitrile, or tetrahydrofuran. Acetate buffer, dodecyl sulfate, and ammonia solution were used as mobile phase additives, which played the role of ion suppressants, silanol blockers, or reagents for the formation of undissociated ion pairs with the cations of the basic drugs. It was found that the most suitable systems for lipophilicity determination of basic drugs are aqueous mobile phases containing ammonia and dioxane, acetone or acetonitrile, as well as buffered acetonitrile–water solvent system containing sodium dodecyl sulfate as ion-pairing reagent.

Numerous articles related to the estimation of lipophilicity by RP-TLC, correlation of chromatographically obtained lipophilicity parameters with calculated values, as well as correlation of structure and lipophilicity of various substances including drugs and potential drugs are continuously published in various journals.

13.4.2 SALTING-OUT TLC

Salting-out thin-layer chromatography (SOTLC) represents a typical reversed-phase method based on the use of moderately to strongly concentrated solutions of salts as mobile phases with or without any miscible organic solvent. The high polarity of the mentioned mobile phases enables the use of polar adsorbents in RP-TLC, for example, unmodified silica gel. Since the first paper of Lederer (1980) until the present time, this TLC method has been successfully used to study the retention behavior of various organic and inorganic polar substances on thin layers of different adsorbents (Janjić et al. 1997). Due to its high solubility in water and significant salting-out effect, ammonium sulfate was commonly used for preparation of mobile phases. In a few papers, the applicability of SOTLC to the determination of lipophilicity of drugs has been described.

The chromatographic behavior of five myorelaxant drugs (atracurium besylate, vecuronium bromide, pancuronium bromide, rocuronium bromide, and suxamethonium chloride) has been investigated under the SOTLC conditions on cellulose and alumina thin layers (Aleksić et al. 2003). For this purpose, aqueous ammonium sulfate solutions of different concentrations ranging from 0.5 to 4.0 M, in steps of 0.5 M, were used as mobile phases. Generally, a stronger retention was observed on alumina, but it was found that cellulose is a more suitable sorbent for the mentioned separations. It was established that retention of the studied drugs always increased in parallel to increasing salt concentrations in mobile phase. A considerable salting-out effect in the case of cellulose was expressed by linear relationship between the R_M values and the ammonium sulfate content of the mobile phase with very high values of regression coefficients ($r > 0.97$). Regression data of these straight lines were used to assess the lipophilicity parameters R_M^0 and C_0. The lipophilicity parameters determined in this way were correlated with calculated ACDlog P values. In all instances, very good linear relationships were obtained with $r = 0.923$ for R_M^0 and $r = 0.949$ for C_0. Therefore, it was found that the correlation of C_0 was better, indicating that this parameter is a more reliable measure of the lipophilicity of the examined substances than R_M^0 values.

Also, the behavior of five ACE inhibitors (enalapril, quinapril, fosinopril, lisinopril, and cilazapril) and four of their active degradation products has been examined in SOTLC on silica gel, cellulose, and polyacrylonitrile sorbent (PANS) with aqueous ammonium sulfate solutions of different concentrations (0.5–2.5 M) as mobile phases (Odović et al. 2005). It is observed that active metabolites (diacid forms of ACE inhibitors) exhibit weaker retention in comparison to the parent compounds (prodrugs). Also, it was found that the retention of the investigated substances decreased in the following order of stationary phases: silica gel > PANS > cellulose. In all instances, an increase of salt concentrations in the mobile phase was accompanied by an increased retention of the investigated drugs, which can be presented as a linear relationship ($r > 0.98$). The obtained plots were used to determine the lipophilicity parameters R_M^0 and C_0, which were correlated with calculated log P values. Finally, for comparison of SOTLC with conventional RP-TLC, drugs were chromatographed on C-18 plates with binary water–methanol mixtures, containing from 40% to 80% (v/v) organic modifier, as mobile phases. The highest values of R_M^0 were obtained under the conditions used for conventional RP-TLC.

Finally, the chromatographic behavior of five macrolide antibiotics (roxithromycin, midecamycin, erythromycin, azithromycin, and erythromycin ethylsuccinate) has been studied by SOTLC with cellulose as an adsorbent and aqueous ammonium sulfate solutions of different concentrations as mobile phases (Tosti et al. 2005). Lipophilicity was determined from the linear relationships between analyte R_M values and the ammonium sulfate content of the mobile phase. A very good correlation between the slope and the intercept of regression lines ($r = 0.993$) confirmed the suitability of the system investigated for estimation of the lipophilicity. The quality of correlation of chromatographically obtained lipophilicity with calculated log P values clearly revealed that the method is suitable for rapid and simple determination of this important physicochemical parameter for the mentioned group of drugs.

13.4.3 Normal-Phase TLC

As it is already mentioned, the chromatographic determination of lipophilicity is based on the distribution of the analyte between an expressively nonpolar stationary phase (usually octadecyl-modified silica gel) and a polar mobile phase (binary system water–organic solvent, commonly methanol, acetonitrile, or acetone, with relatively high water content). Taking into consideration that under the conditions of a normal-phase chromatography (NPC) the analyte is distributed during the chromatographic procedure between the two phases significantly differing from each other by polarity, it is to be expected that this chromatographic method might be employed for the estimation of relative lipophilicity. There are even several reports in the available literature describing such attempts (Bieganowska et al. 1995; Perišić Janjić et al. 2005a,b; Odović et al. 2009; Poša et al. 2011).

In the first of these papers (Bieganowska et al. 1995), the chromatographic behavior of 15 sulfonamide drugs observed under RP-TLC and NP-TLC on thin layers of polyamide, silica gel, Florisil, and silanized silica gel has been compared. Methanol, ethanol, isopropanol, acetonitrile, ethyl acetate, ethyl methyl ketone, dioxane, and tetrahydrofuran were used as mobile phase modifiers in combination with dichloromethane in NPC and water or phosphate buffer in RPC. The retention behavior of the mentioned substances has been investigated as a function of the mobile phase composition. The linear relationships between R_M values and content of modifier in NP conditions can be expressed by the equation

$$R_M = R_M^0 - m \cdot \log C [\%]$$ (13.24)

while the corresponding relation for RP conditions is

$$R_M = R_M^0 - m \cdot C [\%]$$ (13.25)

With the exception of polyamide, a very good linear correlation was obtained ($r > 0.99$). In addition, linear correlations between intercepts and slopes of TLC equations were found on the basis of which a new hydrophobicity parameter C_0 was defined. The C_0, that is, the ratio of intercept and slope, represents the volume fraction of the organic modifier in the mobile phase for which $R_M = 0$ and can be used as a measure of lipophilicity in just the same way as R_M^0 and parameter φ_0 defined for HPLC.

The chromatographic behavior of seven 16-oximino derivatives of 3β-hydroxy-5-androstene has been investigated by means of NP-TLC on silica gel with three different mobile phases consisting of nonpolar diluent (benzene) and polar modifier, that is, acetonitrile, ethyl acetate, or dioxane (Perišić Janjić et al. 2005a). On the basis of the linear relationships between the retention constants (R_M) and the logarithm of the organic modifier content in the mobile phase, the R_M^0 values were calculated. Good linear correlations between both R_M^0 and C_0 and calculated log P values were established. In addition, by comparison of the relative lipophilicity determined previously by RP-TLC with NP-TLC data, it was found that both NP-TLC and RP-TLC can be used to express lipophilicity but only for the structurally similar substances.

In scope of the systematic investigation of the chromatographic behavior of different primarily biologically active substances, the retention of five ACE inhibitors and their four active metabolites has been investigated by NP-TLC on silica gel using several binary nonaqueous solvent systems, that is, mixtures of ethanol with ethyl methyl ketone, carbon tetrachloride, or toluene (Odović et al. 2009). The linear relationship between the R_M values and composition of mobile phase used was obtained. From regression lines, the hydrophobicity parameters R_M^0 and C_0 were determined analogous to RP-TLC and were correlated with calculated log P values. Based on that, it was concluded that NP-TLC represents the reliable method for the estimation of the lipophilicity of the previously mentioned substances. In addition, the results obtained were correlated with the lipophilicity of the studied ACE inhibitors and their metabolites previously estimated by RP-TLC (Odović et al. 2006). The comparison of these results revealed no significant differences between RP-TLC and NP-TLC with regard to the estimation of lipophilicity.

Finally, a special mode of NPC named hydrophilic interaction chromatography (HILIC), which is an increasingly popular alternative to conventional HPLC for drug analysis, should be mentioned. It offers increased selectivity and sensitivity and improved efficiency when quantifying drugs and related compounds in complex matrices such as biological and environmental samples, pharmaceutical formulations, food, and animal feed (Ares and Bernal 2012). HILIC represents a chromatographic technique where the analytes (highly polar compounds) interact with a hydrophilic stationary phase (similar to NPLC) and are eluted with a relatively hydrophilic binary eluent (water/acetonitrile, water/acetone, or water/methanol—similar to RPLC) in which water is, opposite to RPLC, the stronger eluting member (Radoičić et al. 2009). This mode avoids problems associated with highly aqueous mobile phases used for polar compounds within reversed-phase systems and helps to eliminate the problem associated with low solubility. Although HILIC retention has been exploited to determine the octanol–water partition coefficients of some β-blockers and local anesthetics by means of HPLC on ZIC-pHILIC column (Bard et al. 2009), there is no similar publication dealing with TLC analysis of drugs. As the only exception of that, preliminary results of the investigation of chromatographic behavior of some anesthetics were reported (Radoičić et al. 2009).

ACKNOWLEDGMENTS

The authors are grateful to the Ministry of Education and Science of the Republic of Serbia (Grant No. 172017) for financial support and to Miss Aleksandra Radoičić for technical help.

REFERENCES

Abraham, M.H., H.S. Chandha, and A. Leo. 1994. Hydrogen-bonding. 35. Relationship between high-performance liquid chromatography capacity factors and water-octanol partition-coefficients. *J. Chromatogr. A* 685: 203–211.

Aleksić, M., J. Odović, D. M. Milojković-Opsenica, and Ž. Lj. Tešić. 2003. Salting-out thin-layer chromatography of several myorelaxants. *J. Planar Chromatogr. Mod. TLC* 16: 144–146.

Andrić, F.Lj., J.D. Trifković, A.D. Radoičić, S.B. Šegan, Ž.Lj. Tešić, and D.M. Milojković-Opsenica. 2010. Determination of the soil-water partition coefficients (log KOC) of some mono- and poly-substituted phenols by reversed-phase thin-layer chromatography. *Chemosphere* 81: 299–305.

Ares, A.M. and J. Bernal. 2012. Hydrophilic interaction chromatography in drug analysis. *Cent. Eur. J. Chem.* 10: 534–553.

Avdeef, A. 1996. Assessment of distribution—pH profiles. In *Methods and Principles in Medicinal Chemistry*, eds., V. Pliška, B. Testa, and H. Van de Waterbeemd, pp. 109–139. Weinheim, Germany: VCH.

Avdeef, A. 2001. Physico-chemical profiling (solubility, permeability and charge state). *Curr. Top. Med. Chem.* 1: 277–351.

Avdeef, A. 2008. Drug ionization and physicochemical profiling. In *Molecular Drug Properties Measurement and Prediction*, ed. R. Mannhold, pp. 55–83. Weinheim, Germany: Wiley-VCH.

Bard, B., P.-A. Carrupt, and S. Martel. 2009. Lipophilicity of basic drugs measured by hydrophilic interaction chromatography. *J. Med. Chem.* 52: 3416–3419.

Bate-Smith, E.C. and R.G. Westall. 1950. Chromatographic behavior and chemical structure.1. Some naturally occurring phenolic substances. *Biochim. Biophys. Acta* 4: 427–440.

Berthelot, M. and E. Jungfleisch. 1872. On the laws that operate for the partition of a substance between two solvents. *Ann. Chim. Phys.* 4: 26.

Berthod, A. and S. Carda-Broch. 2004. Determination of liquid–liquid partition coefficients by separation methods. *J. Chromatogr. A* 1037: 3–14.

Berthod, A., S. Carda-Broch, and M.C. Garcia-Alvarez-Coque. 1999. Hydrophobicity of ionizable compounds. A theoretical study and measurements of diuretic octanol–water partition coefficients by countercurrent chromatography. *Anal. Chem.* 71: 879–888.

Biagi, G.L., A.M. Barbaro, and A. Sapone. 1994a. Determination of lipophilicity by means of reversed-phase thin-layer chromatography. I: Basic aspects and relationship between slope and intercept of TLC equations. *J. Chromatogr. A* 662: 341–361.

Biagi, G.L., A.M. Barbaro, A. Sapone, and M. Recanatini. 1994b. Determination of lipophilicity by means of reversed-phase thin-layer chromatography. II. Influence of the organic modifier on the slope of the thin-layer chromatographic equation. *J. Chromatogr. A* 669: 246–253.

Bieganowska, M.L., A. Doraczynska-Szopa, and A. Petruczynik. 1995. The retention behavior of some sulfon-amides on different TLC plates. 2. Comparison of the selectivity of the systems and quantitative determination of hydrophobicity parameters. *J. Planar Chromatogr. Mod. TLC* 8: 122–128.

Bouchard, G., P.-A. Carrupt, B. Testa, V. Gobry, and H.H. Girault. 2002. Lipophilicity and solvation of anionic drugs. *Chem. Eur. J.* 8(15): 3478–3484.

Braumann, T. 1986. Determination of hydrophobic parameters by reversed-phase liquid chromatography: Theory, experimental techniques, and application in studies on quantitative structure-activity relationships. *J. Chromatogr.* 373: 191–225.

Brooke, D.N., A.J. Dobbs, and N. Williams. 1986. Octanol:water partition coefficient (P). Measurement, estimation, and interpretation, particularly for chemicals with P > 105. *Ecotoxicol. Environ. Safe.* 11: 251–260.

Carlsson, K. and B. Karlberg. 2000. Determination of octanol-water partition coefficients using a micro-volume liquid-liquid flow extraction system. *Anal. Chim. Acta* 423: 137–144.

Casoni, D. and C. Sârbu. 2012. Comprehensive evaluation of lipophilicity of biogenic amines and related compounds using different chemically bonded phases and various descriptors. *J. Sep. Sci.* 35(8): 915–921.

Cichna, M., P. Markl, and J.F.K. Huber. 1995. Determination of true octanol-water partition coefficients by means of solvent generated liquid-liquid chromatography. *J. Pharm. Biomed. Anal.* 13: 339–351.

Cserháti, T., Y.M. Darwish, and G. Matolcsy. 1983. Effect of support characteristics on the determination of the lipophilicity of some neutral, acidic and alkaline compounds by reversed-phase thin-layer chromatography. *J. Chromatogr.* 270: 97–104.

Dabić, D., M. Natić, Z. Džambaski et al. 2011. Estimation of lipophilicity of N-substituted 2-alkylidene-4-oxothiazolidines by means of reversed-phase thin-layer chromatography. *J. Liq. Chromatogr. Relat. Technol.* 34: 791–804.

Dabić, D., M. Natić, Z. Džambaski, R. Marković, D. Milojković-Opsenica, and Ž. Tešić. 2011. Quantitative structure–retention relationship of new N-substituted 2-alkylidene-4 oxothiazolidines. *J. Sep. Sci.* 34(18): 2397–2404.

Danielsson, L.-G. and Y.-H. Zhang. 1994. Mechanized determination of n-octanol/water partition constants using liquid-liquid segmented flow extraction. *J. Pharm. Biomed. Anal.* 12: 1475–1481.

Danielsson, L.-G. and Y.-H. Zhang. 1996. Methods for determining n-octanol-water partition constants. *Trends Anal. Chem.* 15: 188–196.

De Bruijn, J., F. Busser, W. Seinen, and J. Hermens. 1989. Determination of octanol/water partition coefficients for hydrophobic organic chemicals with the "slow stirring" method. *Environ. Toxicol. Chem.* 8: 499–512.

Dorsey, J. and M. Khaledi. 1993. Hydrophobicity estimations by reversed-phase liquid chromatography. Implications for biological partitioning processes. *J. Chromatogr. A* 656: 485–499.

Du, C.M., K. Valko, C. Bevan, D. Reynolds, and M.H. Abraham. 1998. Rapid gradient RP-HPLC method for lipophilicity determination: A solvation equation based comparison with isocratic methods. *Anal. Chem.* 70: 4228–4234.

Dzimiri, N., U. Fricke, and W. Klaus. 1987. Influence of derivatization on the lipophilicity and inhibitory actions of cardiac glycosides on myocardial sodium-potassium ATPase. *J. Pharmacol.* 91: 31–38.

Eadsforth, C.V. 1986. Application of reverse H.P.L.C. for the determination of partition coefficient. *Pestic. Sci.* 17: 311–325.

Eadsforth, C.V. and P. Moser. 1983. Assessment of reverse phase chromatographic methods for determining partition coefficients. *Chemosphere* 12: 1459–1475.

Eros, D., I. Kovesdi, L. Orfi, K. Takacs-Novak, G. Acsady, and G. Keri. 2002. Reliability of logP predictions based on calculated molecular descriptors: A critical review. *Curr. Med. Chem.* 9(20): 1819–1829.

Ettre, L.S. 1993. Nomenclature for chromatography. IUPAC Recommendations. *Pure Appl. Chem.* 65(4): 819–872.

Forgács, E. and T. Cserháti. 1995. Effect of various organic modifier on the determination of the hydrophobicity parameters of non-homologous series of anticancer drugs. *J. Chromatogr. A* 697: 59–69.

Forgács, E., T. Cserháti, R. Kaliszan, P. Haber, and A. Nasal. 1998. Reversed-phase thin-layer chromatographic determination of the hydrophobicity parameters of nonsteroidal anti-inflammatory drugs. *J. Planar Chromatogr.* 11: 383–387.

Fujita, T., J. Iwasa, and C. Hansch. 1964. A new substituent constant, π, derived from partition coefficients. *J. Am. Chem. Soc.* 86: 5175–5180.

Gaica, S.B., D.M. Opsenica, B.A. Šolaja, Ž.Lj. Tešić, and D.M. Milojković-Opsenica. 2002. The retention behavior of some cholic acid derivatives on different adsorbents. *J. Planar Chromatogr. Mod. TLC* 15: 299–305.

Gaudette, L.E. and B.B. Brodie. 1959. Relationship between lipid solubility of drugs and their oxidation by liver microsomes. *Biochem. Pharmacol.* 2: 89–96.

Giaginis, C., D. Dellis, and A. Tsantili-Kakoulidou. 2006. Effect of the aqueous component of the mobile phase on RP-TLC retention and its implication in the determination of lipophilicity for a series of structurally diverse drugs. *J. Planar Chromatogr. Mod. TLC* 19: 151–156.

Giaginis, C. and A. Tsantili-Kakoulidou. 2008a. Alternative measures of lipophilicity: From octanol–water partitioning to IAM retention. *J. Pharm. Sci.* 97: 2984–3004.

Giaginis, C. and A. Tsantili-Kakoulidou. 2008b. Current state of the art in HPLC methodology for lipophilicity assessment of basic drugs. A review. *J. Liq. Chromatogr. Relat. Technol.* 31: 79–96.

Gilpin, R.K. and C.S. Gilpin. 2011. Pharmaceuticals and related drugs. *Anal. Chem.* 83(12): 4489–4507.

Gleeson, M.P. 2008. Generation of a set of simple, interpretable ADMET rules of thumb. *J. Med. Chem.* 51: 817–834.

Gocan, S., G. Cimpan, and J. Comer. 2005. Lipophilicity measurements by liquid chromatography. In *Advances in Chromatography*, ed., E. Grushka and N. Grinberg, pp. 79–176. Boca Raton, FL: CRC Press.

Gupta, S., K. Shanker, and S.K. Srivastava. 2012. HPTLC method for the simultaneous determination of four indole alkaloids in *Rauwolfia tetraphylla*: A study of organic/green solvent and continuous/pulse sonication. *J. Pharm. Biomed. Anal.* 66: 33–39.

Han, S.-Y., J.-Q. Qiao, Y.-Y. Zhang et al. 2011. Determination of n-octanol/water partition coefficient for DDT-related compounds by RP-HPLC with a novel dual-point retention time correction. *Chemosphere* 83: 131–136.

Hansch, C. and T. Fujita. 1964. ρ-σ-π analysis. A method for the correlation of biological activity and chemical structure. *J. Am. Chem. Soc.* 86: 1616–1626.

Hansch, C. and A.J. Leo. 1979. *Substituent Constants for Correlation Analysis in Chemistry and Biology*. New York: John Wiley.

Hawryl, A., D. Cichocki, and M. Waksmundzka-Hajnos. 2008. Determination of the lipophilicity of some psychotropic drugs by RP-TLC. *J. Planar Chromatogr.* 21: 343–348.

Héberger, K. 2007. Quantitative structure–(chromatographic) retention relationships. *J. Chromatogr. A* 1158: 273–305.

Henchoz, Y., B. Bard, D. Guillarme, P.A. Carrupt, J.L. Veuthey, and S. Martel. 2009. Analytical tools for the physicochemical profiling of drug candidates to predict absorption/distribution. *Anal. Bioanal. Chem.* 394: 707–729.

Hill, A.P. and R.J. Young. 2010. Getting physical in drug discovery: A contemporary perspective on solubility and hydrophobicity. *Drug Discov. Today* 15(15/16): 648–655.

Hitzel, L., A.P. Watt, and K.L. Locker. 2000. An increased throughput method for the determination of partition coefficients. *Pharm. Res.* 17: 1389–1395.

Horváth, C., W. Melander, and I. Molnar. 1976. Solvophobic interactions in liquid chromatography with non-polar stationary phases. *J. Chromatogr.* 125: 129–156.

Hulshoff, A. and J.H. Perrin. 1976. Chromatographic characterization of phenothiazine drugs by a reversed-phase thin-layer technique. *J. Chromatogr.* 129: 249–262.

Jafvert, C.T., J.C. Westall, E. Grieder, and R.P. Schwarzenbach. 1990. Distribution of hydrophobic ionogenic organic compounds between octanol and water: Organic acids. *Environ. Sci. Technol.* 24: 1795–1803.

Janjić, T.J., V.M. Živković-Radovanović, and M.B. Ćelap. 1997. Planar salting-out chromatography. *J. Serb. Chem. Soc.* 62: 1–17.

Kah, M. and C.D. Brown. 2008. LogD: Lipophilicity for ionizable compounds. *Chemosphere* 72: 1401–1408.

Kaliszan, R. 2007. QSRR: Quantitative structure-(chromatographic) retention relationships. *Chem. Rev.* 107: 3212–3246.

Kaliszan, R., A. Nasal, and M.J. Markuszewski. 2003. New approaches to chromatographic determination of lipophilicity of xenobiotics. *Anal. Bioanal. Chem.* 377: 803–811.

Karelson, M., V.S. Lobanov, and A.R. Katritzky. 1996. Quantum-chemical descriptors in QSAR/QSPR studies. *Chem. Rev.* 96: 1027–1043.

Komsta, Ł., R. Skibiński, A. Berecka, A. Gumieniczek, B. Radkiewicz, and M. Radon. 2010. Revisiting thin-layer chromatography as a lipophilicity determination tool—A comparative study on several techniques with a model solute set. *J. Pharm. Biomed. Anal.* 53: 911–918.

Kowalska, T. and W. Prus. 2003. Theory and mechanism of thin-layer chromatography. In *Encyclopedia of Chromatography*, pp. 821–825. New York: Marcel Dekker.

Lederer, M. 1980. Simple and fast separation of the iodotyrosines by thin-layer chromatography. *J. Chromatogr.* 194: 270–272.

Leo, A., C, Hansch, and D. Elkins. 1971. Partition coefficients and their uses. *Chem. Rev.* 71: 525–616.

Leo, J. 1993. Calculating logPoct from structure. *Chem. Rev.* 93(4): 1281–1306.

Mannhold, R. and C. Ostermann. 2008. Prediction of log *P* with substructure-based methods. In *Molecular Drug Properties Measurement and Prediction*, ed., R. Mannhold, pp. 357–379. Weinheim, Germany: Wiley-VCH.

Martel, S., D. Guillarme, Y. Henchoz et al. 2008. Chromatographic approaches for measuring log P. In *Molecular Drug Properties—Measurement and Prediction*, ed., R. Mannhold, pp. 331–357. Weinheim, Germany: Wiley-VCH.

Meyer, H. 1899. Zur Theorie der Alkoholnarkose I. Welche Eigenschaft der Anaesthetika bedingt ihre narkotische Wirkung? *Archiv f. Experimentelle Pathologie und Pharmakologie.* 42: 109–118.

Murakami, F. 1979. Retention behavior of benzene derivatives on bonded reversed-phase columns. *J. Chromatogr.* 178: 393–399.

Nasal, A., D. Siluk, and R. Kaliszan. 2003. Chromatographic retention parameters in medicinal chemistry and molecular pharmacology. *Curr. Med. Chem.* 10: 381–426.

Nernst, W. 1891. Verteilung eines Stoffes zwischen zwei Lösungsmitteln und zwischen Lösungsmitteln und Dampfraum. *Z. Phys. Chem.* 8: 110–139.

Nikitas, P. and A. Pappa-Louisi. 2009. Retention models for isocratic and gradient elution in reversed-phase liquid chromatography. *J. Chromatogr. A* 1216: 1737–1755.

Nowosielski, B.E. and J.B. Fein. 1998. Experimental study of octanol–water partition coefficients for 2,4,6-trichlorophenol and pentachlorophenol: Derivation of an empirical model of chlorophenol partitioning behaviour. *Appl. Geochem.* 13: 893–904.

Odović, J., M. Aleksić, B. Stojimirović, D. Milojković-Opsenica, and Ž. Tešić. 2009. Normal-phase thin-layer chromatography of some angiotensin converting enzyme (ACE) inhibitors and their metabolites. *J. Serb. Chem. Soc.* 74(6): 677–688.

Odović, J., B. Stojimirović, M. Aleksić, D. Milojković-Opsenica, and Ž. Tešić. 2006. Reversed-phase thin-layer chromatography of some angiotensin converting enzyme (ACE) inhibitors and their active metabolites. *J. Serb. Chem. Soc.* 71(6): 621–628.

Odović, J.V., B.B. Stojimirović, M.B. Aleksić, D.M. Milojković-Opsenica, and Ž.Lj. Tešić. 2005. Examination of the hydrophobicity of ACE inhibitors and their active metabolites by salting-out thin-layer chromatography. *J. Planar Chromatogr. Mod. TLC* 18: 102–107.

OECD Guidelines for the Testing of Chemicals, OECD Organization for Economic Cooperation and Development, www.oecd.org (accessed July 3, 2013).

OECD Guideline for the testing of chemicals, Test No. 107. 1995. Partition coefficient (n-octanol/water): Shake flask method. OECD, Paris. www.oecd.org (accessed July 2, 2013).

OECD Guideline for the testing of chemicals, Proposal for Test No. 122. 2000. Partition coefficient (n-octanol/water), pH-Metric method for ionisable substances. OECD, Paris.www.oecd.org (accessed July 2, 2013).

OECD Guideline for the testing of chemicals, Test No. 117. 2004. Partition coefficient (n-octanol/water), HPLC method. OECD, Paris. www.oecd.org (accessed July 2, 2013).

OECD Guideline for the testing of chemicals, Test No. 123. 2006. Partition coefficient (n-octanol/water): Slow-stirring method. OECD, Paris. www.oecd.org (accessed July 2, 2013).

Onisor, C., M. Poša, S. Kevrešan, K. Kuhajda, and C. Sârbu. 2010. Estimation of chromatographic lipophilicity of bile acids and their derivatives by reversed-phase thin layer chromatography. *J. Sep. Sci.* 33: 3110–3118.

Overton, E. 1901. *Studien über die Narkose, zugleich ein Beitrag zur allgemeinen Pharmakologie.* Jena, Switzerland: Gustav Fischer.

Perišić Janjić, N.U., T.Lj. Đaković-Sekulić, S.Z. Stojanović, and K.M. Penov-Gaši. 2005a. HPTLC chromatography of androstene derivates. Application of normal phase thin-layer chromatographic retention data in QSAR studies. *Steroids* 70: 137–144.

Perišić Janjić, N.U., G.S. Uščumlić, and N. Valentić. 2005b. The retention behavior of some uracil derivatives in normal and reversed-phase chromatography. Lipophilicity of the compounds. *J. Planar Chromatogr.* 18: 92–97.

Pliška, V., B. Testa, and H. van de Waterbeemd. 1996. Lipophilicity: The empirical tool and the fundamental objective. An introduction. In *Lipophilicity in Drug Action and Toxicology*, eds., V. Pliška, B. Testa, and H. van de Waterbeemd, pp. 1–6. Weinheim, Germany: VCH Publishers.

Poole, S.K. and C.F. Poole. 2003. Separation methods for estimating octanol–water partition coefficients. *J. Chromatogr. B* 797: 3–19.

Poša, M., M. Rašeta, and K. Kuhajda. 2011. A contribution to the study of hydrophobicity (lipophilicity) of bile acids with an emphasis on oxo derivatives of 5β-cholanoic acid. *Hem. Ind.* 65(2): 115–121.

Rabtti, El.H.M.A., M.M. Natić, D.M. Milojković-Opsenica et al. 2012. RP TLC-based lipophilicity assessment of some natural and synthetic coumarins. *J. Braz. Chem. Soc.* 23(3): 522–530.

Radoičić, A., H. Majstorović, T. Sabo, Ž. Tešić, and D. Milojković-Opsenica. 2009. Hydrophilic-interaction planar chromatography of some water-soluble Co(III) complexes on different adsorbents. *J. Planar Chromatogr.* 22(4): 249–253.

Radoičić, A., J. Trifković, D.M. Milojković-Opsenica, Ž.Lj. Tešić, D. Vučović, and M. Aleksić. 2009. Hydrophilic interaction planar chromatography of some anesthetics. *Proceedings of the XXXIInd Symposium on "Chromatographic Methods of Investigating the Organic Compounds"*, Katowice-Szczyrk, Poland, p. 61.

Rekker, R.F. 1977. *The Hydrophobic Fragmental Constant. Pharmacochemistry Library*. New York: Elsevier.

Rice, N.M., H.M.N.H. Irving, and M.A. Leonard. 1993. Nomenclature for liquid-liquid distribution (solvent extraction). IUPAC Recommendations. *Pure. Appl. Chem.* 65(11): 2373–2396.

Sârbu, C. and R.D. Briciu. 2010. Lipophilicity of natural sweeteners estimated on various oils and fats impregnated thin-layer chromatography plates. *J. Liq. Chromatogr. Relat. Technol.* 33: 903–921.

Sârbu, C., K. Kuhajda, and S. Kevrešan. 2001. Evaluation of the lipophilicity of bile acids and their derivatives by thin-layer chromatography and principal component analysis. *J. Chromatogr. A* 917: 361–366.

Šegan, S., F. Andrić, A. Radoičić et al. 2011. Correlation between structure, retention and activity of cholic acid derived *cis-trans* isomeric bis-steroidal tetraoxanes. *J. Sep. Sci.* 34(19): 2659–2667.

Šegan, S.B., D.M. Opsenica, B.A. Šolaja, and D.M. Milojković-Opsenica. 2009. Planar chromatography of cholic acid-derived *cis–trans* isomeric bis-steroidal tetraoxanes. *J. Planar Chromatogr.* 22(3): 175–181.

Sherma, J. 2010. Planar chromatography. *Anal. Chem.* 82: 4895–4910.

Shewiyo, D.H., E. Kaale, P.G. Risha, B. Dejaegher, J. Smeyers-Verbeke, and Y. Vander Heyden. 2012. HPTLC methods to assay active ingredients in pharmaceutical formulations: A review of the method development and validation steps. *J. Pharm. Biomed. Anal.* 66: 11–23.

Soczewinski, E. and G. Matysik. 1968. Two types of R_M-composition relationships in liquid-liquid partition chromatography. *J. Chromatogr.* 32: 458–471.

Takács-Novák, K., M. Józan, and G. Szász. 1995. Lipophilicity of amphoteric molecules expressed by the true partition coefficient. *Int. J. Pharm.* 113: 47–55.

Takács-Novak, K., P. Pal, and V. Jozsef. 2001. Determination of logP for biologically active chalcones and cyclic chalcone analogs by RPTLC. *J. Planar Chromatogr. Mod. TLC* 14(1): 42–46.

Tarcsay, Á., K. Nyíri, and Gy.M. Keserű. 2012. Impact of lipophilic efficiency on compound quality. *J. Med. Chem.* 55: 1252–1260.

Tetko, I.V. and G.I. Poda. 2008. Prediction of Log P with property-based methods. In *Molecular Drug Properties—Measurement and Prediction*, ed., R. Mannhold, pp. 381–406. Weinheim, Germany: Wiley-VCH.

Torres-Lapasió, J.R., M. Rosés, E. Bosch, and M.C. Garcia-Alvarez-Coque. 2000. Interpretive optimisation strategy applied to the isocratic separation of phenols by reversed-phase liquid chromatography with acetonitrile-water and methanol-water mobile phases. *J. Chromatogr. A* 886: 31–46.

Tosti, T.B., K. Drljević, D.M. Milojković-Opsenica, and Ž.Lj. Tešić. 2005. Salting-out thin-layer chromatography of some macrolide antibiotics. *J. Planar Chromatogr. Mod. TLC* 18: 415–418.

Trifković, J.Đ., F.Lj. Andrić, P. Ristivojević, D. Andrić, Ž.Lj. Tešić, and D.M. Milojković-Opsenica. 2010. Structure-retention relationship study of arylpiperazines by linear multivariate modeling. *J. Sep. Sci.* 33: 2619–2628.

Tsantili-Kakoulidou, A. 2010. Lipophilicity: Assessment by RP/TLC and HPLC. In *Encyclopedia of Chromatography*, pp. 1400–1407. Boca Raton, FL: CRC Press.

Vailaya, A. 2005. Fundamentals of reversed phase chromatography: Thermodynamic and exothermodynamic treatment. *J. Liq. Chromatogr. Relat. Technol.* 28: 965–1054.

Valkó, K. 1987. The role of chromatography in drug design. *TrAC, Trends Anal. Chem.* 6(8): 214–219.

Valkó, K. 2000. Separation methods in drug synthesis and purification. In *Handbook of Analytical Separations*, pp. 539–542. Amsterdam, the Netherlands: Elsevier.

Valkó, K. 2004. Application of high-performance liquid chromatography based measurements of lipophilicity to model biological distribution. *J. Chromatogr. A* 1037: 299–310.

Valkó, K., C. Bevan, and D. Reynolds. 1997. Chromatographic hydrophobicity index by fast-gradient RP-HPLC: A high-throughput alternative to log P/log D. *Anal. Chem.* 69: 2022–2029.

Valkó, K., C.M. Du, C. Bevan, D.P. Reynolds, and M.H. Abraham. 2001. Rapid method for the estimation of water/octanol partition coefficient (logPoct) from gradient RP-HPLC retention and a hydrogen bond acidity term (sigma alpha2H). *Curr. Med. Chem.* 8: 1137–1146.

Valkó, K. and P. Slegel. 1993. New chromatographic hydrophobicity index (φ_0) based on the slope and the intercept of the log*k'* *versus* organic phase concentration plot. *J. Chromatogr.* 631: 49–61.

van de Waterbeemd, H. 2003. Physico-chemical approaches to drug absorption. In *Drug Bioavailability: Estimation of Solubility, Permeability, Absorption and Bioavailability*, ed., H. van de Waterbeemd, H. Lennernäs, and P. Artursson, pp. 3–20. Weinheim, Germany: Wiley-VCH.

van de Waterbeemd, H., R.E. Carter, G. Grassy et al. 1997. Glossary of terms used in computational drug design. *Pure Appl. Chem.* 69: 1137–1152.

van de Waterbeemd, H., M. Kansy, B. Wagner, and H. Fischer. 1996. Lipophilicity measurement by high performance liquid chromatography (RPHPLC). In *Lipophilicity in Drug Action and Toxicology*, ed., V. Pilška, B. Testa, and H. van de Waterbeemd, pp. 73–87. Weinheim, Germany: VCH Publishers.

van de Waterbeemd, H. and R. Manhold. 1996. Lipophilicity descriptors for structure-property correlation studies: Overview of experimental and theoretical methods and a benchmark of logP calculations. In *Lipophilicity in Drug Action and Toxicology*, ed., V. Pilška, B. Testa, and H. van de Waterbeemd, pp. 401–418. Weinheim, Germany: VCH Publishers.

Van der Giesen, W.F. and L.H.M. Janssen. 1982. Adsorption behaviour of several supports in reversed-phase thin-layer chromatography as demonstrated by the determination of relative partition coefficients of some 4-hydroxycoumarin derivatives. *J. Chromatogr.* 237: 199–213.

Varmuza, K. and P. Filzmoser. 2009. *Introduction to Multivariate Statistical Analysis in Chemometrics*. Boca Raton, FL: CRC Press.

Völgyi, G., K. Deák, J. Vámos, K. Valkó, and K. Takacs-Novák. 2008. RPTLC determination of logP of structurally diverse neutral compounds. *J. Planar Chromatogr.* 21: 143–149.

Waring, M.J. 2009. Defining optimum lipophilicity and molecular weight ranges for drug candidates-Molecular weight dependent lower logD limits based on permeability. *Bioorg. Med. Chem. Lett.* 19: 2844–2851.

Wick, C.D., J.I. Siepmann, and M.R. Schure. 2004. Simulation studies on the effects of mobile-phase modification on partitioning in liquid chromatography. *Anal. Chem.* 76: 2886–2892.

Yamagami, C., K. Kawase, and T. Fujita. 1999. Hydrophobicity parameters determined by reversed-phase liquid chromatography. XIII A new hydrogen-accepting scale of monosubstituted (di)azines for the relationship between retention factor and octanol-water partition coefficient. *Quant. Struct. Act. Relat.* 18: 26–34.

14 Screening of Substandard and Fake Drugs in Underdeveloped Countries by TLC

Eliangiringa Kaale, Peter Risha, Thomas Layloff, and Joseph Sherma

CONTENTS

14.1 INTRODUCTION

The analysis of a pharmaceutical product for the active pharmaceutical ingredient (API) generally requires a chromatographic procedure to separate the API from degradation products and/or other interfering substances. In industrialized economies, these separations are usually effected using high-performance liquid chromatography (HPLC). The HPLC instrumentation involves relatively sophisticated technology, and proper maintenance requires skilled staff with ready access to replacement parts. In addition, HPLC separations are performed on a relatively expensive column used repetitively; the column must be protected from damage and particulate accumulations that could affect its separation characteristics. The HPLC system must also have a reliable and consistent supply of electricity; a power failure would stop the pumping of the mobile phase allowing migration of the analytes in the column, and the thermostat chambers would not be controlled.

Thin-layer chromatography (TLC; also termed planar chromatography) may also be used to separate the API from the other substances in a pharmaceutical product for its determination. In TLC experiments, the separation process is driven by the capillary action of the mobile phase through the layer, so no external pumps or maintenance is required; this also frees the process from the need for continuous electrical power. In addition, the stationary phase in TLC (plate or layer) is used only once, so each experiment uses a new stationary phase. However, the TLC separation power (number of theoretical plates) is much lower than for HPLC. But most pharmaceutical products contain a high percentage of the API and there generally are no more than three or four APIs and/or impurities or degradation products in a chromatogram, so high efficiency frequently is not necessary.

Since the introduction of the Global Pharma Health Fund (GPHF; formerly German Pharma Health Fund) Minilab, a registered trademark, in the late 1990s, there has been a huge expansion in the use of TLC to detect substandard and fake products in developing countries. There now are over 450 Minilabs in use in more than 80 countries. This high rate of adoption is due to the relatively low cost of a complete system, which is provided in a very convenient package with comprehensive manuals and supplies sufficient to perform approximately 1000 assessments; in addition, the Minilab does not require electricity or a laboratory environment to perform the tests. The testing inventory of the Minilab continues to expand, and it can now be used to assess products that contain over 50 APIs to determine if the correct drug is present in approximately the correct amount. As a side benefit, the extensive adoption of the Minilab has created a large cadre of users trained in performing basic planar chromatography; this knowledge base can be used to escalate the quality of the assessments by adopting instrumental high-performance TLC (HPTLC). Because of convenience, low capital investment, low operating costs, minimal training requirements, and the ability to change analytes easily, pharmaceutical product quality assessments using planar chromatography will continue to expand both in developing and developed countries. Accurate quantitative assessments cannot be performed with the Minilab due to variations in the amount of sample manually applied to the plate and the visual estimates of the relative spot amounts that are limited by the user's visual acuity. These limitations can be overcome by shifting the experiment to a laboratory environment, where accurate and precise automatic application devices can be used to replace manual sample application and a densitometer can be used to assess the amount of material on the plate in lieu of visual estimates of amounts.

14.2 HISTORY

The use of TLC procedures for field assessments of drug quality attributes was proposed in a paper by Flinn et al. [1] in which a simple low-cost TLC procedure to assess the quality of a tableted pharmaceutical product was described along with a simple device that could be used to carry out the test in the field. The approach was demonstrated for theophylline, but it could be used to assay the drug content of any tablet and determine its dissolution or disintegration characteristics. The procedure could also be used in the field without the need for any instrumentation. The procedure described the use of well-characterized reference tablets for visual content estimation on the TLC plate, thus eliminating the need for a balance and for disintegration time comparisons. The apparatus described therein was redesigned into a more robust and durable apparatus called the "Speedy TLC Kit." [2].

The GPHF developed this concept into a very convenient test kit called the Minilab. The Minilab is packaged in a sturdy suitcase that contains all the necessary reagents and supplies to perform more than 1000 TLC-based pharmaceutical drug quality assays [3]. In addition to TLC assays, reagents and manuals for simplified color reaction tests are included in a separate package for drug identity color reaction tests that complement the TLC assays. The Minilab also includes a provision for simple disintegration tests intended as a proxy for assessing the dissolution potential of pharmaceutical solid dosage forms. After development, the Minilab system was field-tested in 1997 and 1998 in Kenya, Ghana, Tanzania, and the Philippines. For the field testing, it was deployed in health facilities as stand-alone stations for assessing the quality of pharmaceutical products. The results of the field testing showed that it was an effective tool for verification of identity and approximating the quantity of pharmaceuticals in these settings [4]. Other attributes of the Minilab are that the kit is inexpensive, can be easily transported, and can be deployed in a nonlaboratory environment with only a workbench and running water. Furthermore, the solvents used to perform the assays are of low toxicity and are used in relatively small quantities, thus, no extensive safety or environmental protection measures are required. These attributes make the Minilab a very useful tool for quality assessment in resource-constrained settings where laboratory facilities are not available. However, Minilab application in field testing showed significant among-user variations, so the procedure can only be used as a very useful screening tool to help identify products with potential quality defects; regulatory sanctions based on quantitative differences cannot be supported, but the procedure can be used if no API is present, that is, the absence of a component in the test is unequivocal. Suspect failures in content must be confirmed through further legal standard tests such as those in pharmacopoeias.

The ready applicability of TLC for the convenient assessment of drug quality has resulted in its application in several initiatives including the Minilab (Table 14.1) and the *Rapid Examination Methods Against Counterfeit and Substandard Drugs* by the Japan International Corporation of Welfare Services (JICWELS; Table 14.2). This publication describes a three-tier product quality assessment model for 63 APIs, where the first tier consists of organoleptic tests, the second tier TLC assessments, and the third tier the usual collaborated monograph technologies needed to support litigation, and the *Compendium of Unofficial Methods for Rapid Screening of Pharmaceuticals by Thin-Layer Chromatography* by Kenyon and Layloff (Table 14.3). The Minilab and Kenyon and Layloff TLC methods all use inexpensive, relatively innocuous solvents similar to those found in household paints. The JICWELS methods use a wider array of solvents that are more expensive.

The concept was expanded and densitometry was introduced to quantify the amounts of drugs, and the inventory of applicable products was expanded by Kenyon and his colleagues to over 70 APIs [5]. In addition, a TLC method was developed to detect diethylene or ethylene glycol adulteration in glycerin to help reduce the risk of using this contaminated product in pharmaceutical preparations that could cause illness and deaths [6].

14.3   USE OF MINILAB AS A TOOL FOR SCREENING SUBSTANDARD AND FAKE DRUGS: TANZANIA INITIATIVES

In the early 1990s the Tanzania Pharmacy Board, which was the medicines regulatory authority at the time, had no testing facilities; this placed the country at high risk of fake/substandard medicines in distribution. Although the Pharmacy Board was able in 1997 to establish a medium size central testing laboratory, the capacity to monitor the quality of medicines was very limited, and thus the risk of having fake/substandard drugs in distribution remained high as evidenced by reports on the presence of substandard medicines on the market. For example, in 2001 inspectors confiscated ampoules of expired chloroquine injection fraudulently relabeled as quinine dihydrochloride injection with a new expiry date. Other examples included fake ampicillin capsules that contained only lactose and paracetamol tablets fraudulently relabeled as the antimalarial product sulfamethoxypyrazine/pyrimethamine [7]. In addition to these findings, a survey revealed that about 46% of the samples were not registered [8] and were of unknown quality. Since these examples were likely an indication of more widespread problems, the medicines regulatory authority began investigating the establishment of a cost-effective risk-based quality assurance system that would improve its capacity to monitor the marketplace.

The potential of using TLC as a method for detecting fake/substandard products was highlighted by the World Health Organization (WHO) report on factors contributing to quality failure for pharmaceutical products reported between 1982 and 1999. The three major factors identified in the quality failures were lack of API, incorrect API, and incorrect amount of API in the samples, which accounted for about 93% of all failures [9]. These failures could be reasonably detected with simple TLC techniques, and TLC would have unequivocally identified about 76% of these failures caused by products that contained no API or had an incorrect API. The TLC techniques also were shown to be a useful drug quality screening tool in the successful Minilab field test conducted by the GPHF [4] and the United States Pharmacopeia—Drug Quality Information (USP-DQI) experience on monitoring the quality of antimalarial drugs in the Mekong Valley in South East Asia [10]. Based on these successful applications, the Tanzania medicines regulatory authority adopted a two-tier

TABLE 14.1
Minilab API TLC Testing Inventory and Solvents Used

TLC API testing inventory

Acetylsalicylic acid	Lamivudine
Albendazole	Levofloxacin
Aminophylline	Lumefantrine
Amodiaquine	Mebendazole
Amoxicillin	Mefloquine
Ampicillin	Metronidazole
Artemether	Moxifloxacin
Artesunate	Nevirapine
Atovaquone	Oseltamivir
Azithromycin	Paracetamol
Cefalexin	Phenoxymethylpenicillin
Cefixime	Piperaquine
Cefuroxime	Praziquantel
Chloramphenicol	Prednisolone
Chloroquine	Primaquine
Ciprofloxacin	Proguanil
Cloxacillin	Prothionamide
Cycloserine	Pyrazinamide
Didanosine	Pyrimethamine
Dihydroartemisinin	Quinine
Erythromycin	Rifampicin
Ethambutol	Salbutamol
Ethionamide	Stavudine
Furosemide	Sulfadoxine
Glibenclamide	Sulfamethoxazole
Griseofulvin	Tetracycline
Halofantrine	Trimethoprim
Indinavir	Zidovudine
Isoniazid	

Reagents and solvents used

Acetic acid	Sulfuric acid
Acetone	Toluene
Ammonium hydroxide concentrated 25%–26%	Disodium or ripotassium EDTA
Ethyl acetate	Iodine
Hydrochloric acid, concentrated	Magnesium chloride hexahydrate
Methanol	Ninhydrin

Source: Global Pharma Health Fund Minilab, http://www.gphf.org/web/en/minilab

quality assurance program that linked structured basic inspections with Minilab quality screening tests at the Central Quality Control Laboratory (CQCL). In this program, samples collected during postmarketing surveillance and those taken during inspections of drug consignments at ports of entry were screened for quality using TLC-based Minilab tests. The samples that were deemed to fail the Minilab screening tests were referred to the CQCL for full monograph testing. The overall objective of the program was to improve regulatory capacity and reach, specifically to develop testing capacity that would provide timely screening of the quality of medicines as they enter and circulate in the market.

TABLE 14.2
JICWELS API Testing Inventory and Reagents
and Solvents

Paracetamol	Gentamicin sulfate
Albendazole	Griseofulvin
Allopurinol	Halofantrine hydrochloride
Amikacin sulfate	Hydrocortisone acetate
Amoxicillin	Hydrocortisone succinate
Ampicillin	Ibuprofen
Ascorbic acid	Indomethacin
Aspirin	Kanamycin sulfate
Benzylpenicillin potassium	Lincomycin hydrochloride
Betamethasone valerate	Mefloquine
Bromovalerylurea	Metronidazole
Caffeine	Ornidazole
Calciferol	Pentoxyverine citrate
Cefoperazone sodium	Praziquantel
Ceftazidime	Prednisolone
Chloroquine phosphate	Prednisolone succinate
Chloramphenicol sodium succinate	Pyridoxine hydrochloride
Chlorpheniramine maleate	Pyrimethamine
d-Chlorpheniramine maleate	Quinine sulfate
Clindamycin hydrochloride	Retinol acetate
Cloxacillin sodium	Riboflavin
Codeine phosphate	Rifampicin
Colchicine	Streptomycin sulfate
Cyanocobalamin	Sulfadoxine
Dexamethasone	Testosterone
Dextromethorphan hydrobromide	Testosterone propionate
Diazepam	Tetracycline hydrochloride
Dihydrocodeine phosphate	Thiamine hydrochloride
Diloxanide furoate	Thiamine nitrate
Erythromycin	Tobramycin
Estriol	dl-α-Tocopherol
Ethinylestradiol	

Reagents and developing solvents used

Acetic acid	Ethanol (99.5%)
Acetone	Ethyl acetate
Ammonia solution (28%)	Hydrochloric acid
Butanol	Iodine
Butyl acetate	Methanol
Choroform	Ninhydrin
Diethylamine	Sodium acetate
Dimethylformamide	Sodium hydroxide
Dipotassium hydrogen phosphate	Toluene
Disodium hydrogen phosphate	Water

Source: Japan International Corporation of Welfare Services, (JICWELS), *Rapid Examination Methods against Counterfeit and Substandard Drugs*, 1st edn., 1997; 2nd edn., 2000, Ministry of Health and Welfare, Japan, Tokyo.

TABLE 14.3

TLC API Test Inventory and Solvents Used

TLC API testing inventory

Acetylsalicylic acid	Kanamycin sulfate
Allopurinol	Ketoconazole
Aminophylline	Mebendazole
Amoxicillin	Medroxyprogesterone acetate
Ampicillin	Medroxyprogesterone acetate injection
Atropine	Methyldopa
Betamethasone	Metronidazole
Carbamazepine	Neomycin sulfate
Cephalexin	Nitrofurantoin
Cephradine	Norgestrel
Chloramphenicol	Nystatin
Chloroquine phosphate	Paracetamol
Chlorpheniramine maleate	Penicillin/procaine
Ciprofloxacin HCl	Phenobarbital
Cloxicillin	Phenytoin
Dexamethasone	Praziquantel
Dexamethasone sodium phosphate	Prednisolone
Diazepam	Prednisone
Diethylene Glycol in Glycerin	Promethazine
Digoxin	Pyrimetamine
Diphenhydramine HCl	Quinine sulfate
Ergotamine tartrate	Rifampicin
Erythromycin ethylsuccinate	Salbutamol (albuterol)
Erythromycin estolate	Streptomycin sulfate
Erythromycin & erythromycin stearate	Sulfamethazine
Estradiol cypionate	Sulfamethoxazole
Ethambutol	Testosterone cypionate
Furosemide	Tetracycline
Gentamycin sulfate	Theophylline
Hydrochlorothiazide	Triamcinolone
Hydrocortisone	Triamterene
Ibuprofen	Trifluoperazine HCl
Imipramine HCl	Trimethoprim
Indomethacin	Vitamin A (retinol)
Isoniazid	Warfarin

Reagents and solvents used

Acetic acid	Iodine
Acetone	Methanol
Ammonium hydroxide	Ninhydrin
Distilled water	Potassium iodide
Ethanol (ethyl alcohol)	Sodium hydroxide pellets
Ethyl acetate	Toluene
Hydrochloric acid	

Source: Kenyon, A.S. and Layloff, T.P., *A Compendium of Unofficial Methods for Rapid Screening of Pharmaceuticals by Thin-Layer* Chromatography, Food and Drug Administration (FDA), Division of Testing and Applied Analytical Development, St. Louis, MO, October 1999, http://www.pharmweb.net/pwmirror/library/pharmwebvlib.html

14.3.1 IMPLEMENTATION

The Tanzania pharmaceutical product quality assurance program was developed in three components: the first was training inspectors on Minilab testing techniques, the second was deployment of Minilabs at ports-of-entry and other critical areas in the medicine distribution chain, and the third was the development of a proficiency testing program to help assure the proper implementation of the technology.

A five-day Minilab training curriculum was developed and implemented jointly by the regulatory authority in collaboration with the School of Pharmacy, Muhimbili University of Health and Allied Sciences (MUHAS), Dar es Salaam, Tanzania. The curriculum included theoretical background on TLC along with hands-on Minilab monograph testing procedures. Five training stations equipped with Minilab supplies were established at MUHAS so the materials could be used for inspector training and in addition could be used in the preservice training curriculum for undergraduate pharmacy students to serve as a catalyst for broadening the scope of TLC-based screening in the future.

Standard Operating Procedures (SOPs) that addressed sampling procedures, sample chain of custody, screening procedures, and decision trees on screening results were developed and included as part of the training curriculum to assist in structuring and standardizing the implementation process. After training, the inspectors along with Minilabs were deployed at three major ports-of-entry and seven other critical zonal centers in the supply chain. At the ports-of-entry, screening tests were initially conducted on all imported consignments of antimalarial products; as inspectors gained experience and their competency improved, the testing inventory was expanded to include antibiotic and antiretroviral products. The port-of-entry testing was complemented with postmarketing surveillance testing of batches of the targeted medicines collected during inspections carried out in dispensing outlets and wholesale warehouses. All eight zonal Minilab test centers (Figure 14.1) were linked to the CQCL to follow up on batches that were suspected of being substandard.

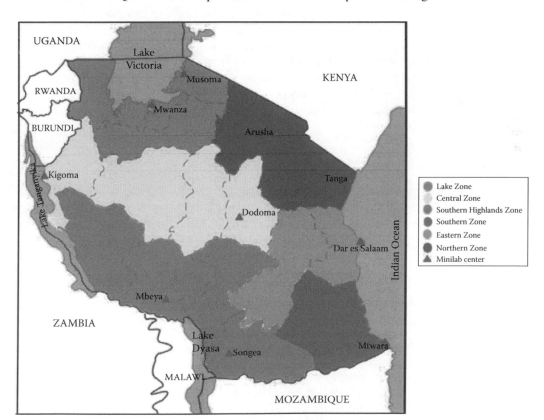

FIGURE 14.1 Map of Tanzania showing zonal distribution of Minilab centers.

14.3.2 USE OF MINILABS AS A TOOL TO SCREEN DRUG QUALITY: LIMITATIONS

Although Minilab testing offers key advantages in terms of affordability, high sample throughput, and ease of deployment and use, it has limitations in that the reliability of the results depends on the technical competency of an individual to process the test samples as well as on the individual's visual acuity to discern differences in spot size/intensity between the reference and test samples. With the understanding of these limitations, the quality assurance program design included proficiency testing as a measure to improve the reliability of test results.

The TLC test protocol as implemented in Minilab has five basic test elements. First is the preparation of the test and reference solutions, which involves placing reference and test tablets into separate containers followed by the addition of solvent to dissolve the analyte from the matrix. Since there are no weighing operations, the sources of error in concentration can arise only from the measurement of the added solvent or incomplete dissolution of the analyte from the dosage form matrix. The next step is to fill by capillary action a volumetric capillary tube to a mark. The individual performing the test must be careful to ensure that the solution fills the capillary to the mark for both the reference and test solutions. The third step is spotting, again by capillary action, the solutions contained in the capillary tubes onto the plate at the designated locations. It is important to conduct the spotting of the reference and test solutions in the same fashion so the spotted materials have similar patterns on the plate. The fourth step is to develop the plate with the appropriate mobile phase to the designated test position. The fifth step is to visually compare the developed reference and test spots and make a decision as to whether the spots sizes and intensities are the same or different. This is a subjective decision based on the visual acuity of the viewer and spot sizes. If the spots were placed on the plate in the same fashion, they should develop in the same fashion, and the comparison is then the relative intensity of the developed spots.

Test elements 1 through 3 are very critical since they involve handling two solutions in the same fashion, which is technique-dependent. The careful performance of these tests must be instilled in the training program, and illustrative failed demonstrations should be included so the trainees can observe the results of not performing them correctly. The development step 4 is least critical since the spots are developed with the same mobile phase on the same plate. Step 5 is again very critical since the decision depends on the visual acuity of the individual viewer and how well the solutions were spotted on the plate, that is, the applied spots should be similar since the same volumetric amounts should have been spotted.

14.3.3 PROFICIENCY TESTING AND RETRAINING

Having identified the potential limitations and their influence on the reliability of results from TLC Minilab-based quality screening, it was important for the quality assurance program to carry out continuous monitoring of the performance of those who performed the tests. Proficiency testing was included as a key component of the program to provide assurance that the Minilab screening tests were being competently performed. A proficiency testing protocol was designed to assess:

1. Inspectors' technique and skill in Minilab screening of batches of targeted medicines
2. Reproducibility of the Minilab screening test results among the field inspectors
3. Inspectors' adherence to SOPs and instructions provided

At program level, the proficiency test results would also provide an insight on the adequacy of the training programs as well as the associated SOPs.

The proficiency test [11] was conducted by providing inspectors with powdered samples having concentrations of 100%, 40%, and 0% using pure starch as a diluent. Each powdered sample was marked to be equivalent to one tablet/capsule. The samples were labeled with the name of the drug and designated as either sample A, B, or C with a random allocation in such a way that samples of different drugs but same drug levels, for example, 40%, were not labeled with the same letter. The inspectors were instructed to add the contents of the labeled sample to a container, add the solvent, shake, and allow settling. They were also asked to similarly prepare a reference tablet. Aliquot portions of the sample and reference solutions were spotted onto the plate, which was then allowed to dry, followed by development using the Minilab procedures. After mobile-phase development and drying, the inspectors were asked to observe the plate under an ultraviolet (UV) lamp and to record whether the sample contained 0%, 40%, or 100% of the amount contained in the reference spot. The 0% result would correspond to no drug or the wrong drug in the sample, the 40% would correspond to a markedly substandard product, and the 100% would correspond to the appropriate amount of drug in the product.

The results of the proficiency testing showed that all the inspectors could correctly identify all drug samples with 0% corresponding to no drug or the wrong drug in the sample and 100% (passing) drug content. However, the majority of the drug samples with lower than the specified drug content were not identified properly by the inspectors; only three of the 36 substandard drug samples were correctly identified. Because of this very significant deficiency, the question arose as to whether the inspectors could discern differences in spot size and/or intensity of drug samples with different drug concentrations or if the training did not adequately address improving the competency of inspectors to identify samples that contained less that the specified amount of drug.

To address the shortcomings observed from the results of this proficiency testing, it was decided to include in the training a session for spotting two different concentrations (40% and 100%) of the standard drugs while emphasizing discerning differences in spot intensity/spot sizes between the different concentrations. The training was also modified to include powdered samples of known drugs, deliberately composed to be "substandard" by diluting with pure starch so the concentration could not be guessed by the trainees. The trainees then were asked to process the sample and judge if it met or did not meet specifications. These two additions improved the skill and confidence of inspectors to identify differences in spot size for samples containing 40% or 100% drug.

Four months after the training was conducted, a second round of proficiency testing was administered. The results of the second round showed marked improvement in the inspectors' performance; five inspectors made the correct inference on all nine samples, while two inspectors failed to identify one out of the nine samples. Two inspectors failed to identify samples containing 50% drug, that is, three out of the nine samples.

The first round proficiency test procedure identified a major problem in the training program, and changes were instituted to include more hands-on work including discerning "substandard" samples. These modifications markedly improved the performance of the inspectors to detect substandard products. The second round proficiency test procedure demonstrated, as did the first round, that all of the inspectors were able to discern the 0% and 100% cases, while two of the inspectors continued to fail to discern the substandard products. The observations from the proficiency testing indicates that one approach to improve reliability on the test results is to include periodic proficiency testing for all participating staff and take measures to ensure that only those shown to be proficient are allowed to continue in the testing program.

14.3.4 Outputs of the Quality Assurance Program and Lessons Learned

The implementation of the Minilab for quality screening provided a high sample throughput. During the first two-year implementation of the quality assurance program, over 1200 samples were tested using the Minilabs. This was almost double the previous testing capacity of the authority. In addition to samples that were Minilab tested, sample flow to and tested by the CQCL greatly increased because of the structured programs. The Minilab implementation contributed to increased regulatory reach and visibility of the authority throughout the country, which served as a deterrent to the entry of substandard medicines into the market [12]. The observations made during the implementation of the program also added to the already available evidence that it can reliably detect only grossly substandard or wrong drug samples. As recommended by the GPHF, it should not be used as an independent testing resource but in conjunction with a full-service quality control laboratory capable of auditing reported substandard results.

14.4 USE OF TLC-BASED MINILABS FOR SUBSTANDARD OR FAKE DRUG DETECTION IN OTHER COUNTRIES

The low cost of deployment combined with its robustness makes the TLC—Minilab a valuable tool for detecting the presence of fake and markedly substandard medicines in the marketplace. Through the efforts of the WHO, GPHF, USP-DQI, and other international organizations, TLC-based Minilab technology has been used as a first line of defense in the fight against fake/substandard drugs in many developing countries, especially those with resource constraints and/or weak quality assurance systems.

14.4.1 EXPERIENCE FROM AFRICAN COUNTRIES

Since 2003, Minilabs have been deployed increasingly in the detection of fake medicines in African countries, particularly those with minimal or no laboratory testing facilities. The USP-DQI supported the regulatory authority of Madagascar, Agence du medicament de Madagascar, in establishing Minilab sentinel sites for regular screening of the quality of medicines, particularly antimalarial drugs in the marketplace [13]. Similar efforts to detect and reduce the prevalence of fake medicines in the market have been undertaken in Senegal with support from USP-DQI [14].

The WHO has also supported the spread of the use of TLC-based Minilab testing in countries with inadequate pharmaceutical product testing capacities. From 2005 to 2008, the WHO supported medicines regulatory authorities in Tanzania and Zambia to acquire and deploy Minilabs as part of their drug quality assurance system [12,15]. The units were deployed as either stand-alone sentinel sites or as part of a risk-based, tiered quality assurance system in which the basic testing is linked to a fully functional laboratory. At the time of Minilab deployment in Zambia, it was the only pharmaceutical products testing resource available in the country [15].

The high throughput capability of TLC was useful in screening large numbers of samples in two large multicountry survey studies conducted in 2007 and 2008 by the WHO and USP-DQI. The two surveys conducted in Uganda, Madagascar, and Senegal were intended to obtain a "snapshot" of the quality of antimalarial medicines, sulfadoxine/pyrimentamine (SP), and artesunate-based combination therapy products (ACTs) in those countries. The Minilabs were used in a two-tiered testing approach to screen the quality of 444 samples of antimalarial drugs collected from the market. On the basis of the Minilab results, about 50% (197) of the processed samples were sent for further laboratory full monograph tests [16]. The WHO also undertook a larger survey of the quality of antimalarial medicines in six sub-Saharan African countries. In the WHO survey, 935 antimalarial drug samples (SP and ACTs) collected from the markets were screened for quality using Minilabs. Based on the screening results, about 30% of samples were selected for follow-up laboratory testing using monograph methods [17].

14.4.2 EXPERIENCE FROM ASIAN COUNTRIES

Due to widespread reports in 2003 of fake artesunate-based antimalarial drugs [18,19] in the countries of the Mekong Valley, the region became one of the focus areas for the implementation of TLC techniques for pharmaceutical quality assessment. The USP-DQI deployed Minilabs to screen antimalarial drugs in 17 field test centers in three countries that did not have laboratory testing facilities [10]. These sentinel site surveys played a key role in identifying fake products in the markets [20].

The use of Minilabs has also expanded into other countries in Central Asia as well where the USP-DQI collaborated with the Medicines Transparency Alliance (MeTA) to improve the capacity of the regulatory authority of Kyrgyzstan to detect fake/substandard drugs in the supply chain. This joint effort offered training on medicine sampling, visual and physical inspection, TLC, and simple disintegration tests at the Kyrgyzstan DRA laboratory in Bishkek. In addition, after the training, Minilabs were deployed as the first-tier testing for monitoring the quality of medicines in the market [21].

14.4.3 ANTICIPATED WIDER USE OF MINILABS AS A DRUG QUALITY SCREENING TOOL

The use of Minilabs for detecting fake drugs in Africa is also increasing. The GPHF estimates that the majority of the more than 470 kits currently in use worldwide are deployed in Africa. Nigeria ranks first with 44 Minilabs, and Tanzania is second with 27 [22]. The GPHF is projecting that by the end of 2012 the 500th Minilab will be sold. In addition, the GPHF continues to expand the testing inventory; starting with 15 drugs in 1997, the inventory will grow to include 57 commonly faked drugs [16]. With this increase in use, the TLC-based assay procedures will gain more recognition and wider acceptance and assume more importance as a simple and effective and affordable means for screening the quality of essential medicines in resource-constrained settings.

14.5 CONVERSION OF TLC METHODS TO HPTLC

TLC and HPTLC are closely related planar chromatography technologies that are increasingly used in pharmaceutical analysis. The advantages and disadvantages of the former technology are described elsewhere [23]. With advancements in the stationary phases and the introduction of automatic sample application devices and densitometers, the techniques can achieve a precision and accuracy comparable to those obtained with HPLC [24].

Until the 1970s, TLC was used extensively for qualitative identification, but that application declined with the development of the HPLC methods [23]. In the mid- to late 1970s, there was a "new beginning" of the TLC applications due to improved methodology along with new instrumentation and automation of operations. Other developments have included new approaches in TLC plate development and the coupling of TLC with spectrometry detection. Automation of operations in TLC has focused on the particular steps with the greatest potential for error. These included application of sample solutions, chromatographic development, detection, and quantitative in situ evaluation.

Recent advances in technology have contributed to a marked improvement in repeatability and reliability of TLC-based testing [24]. Automating the TLC sample application step has greatly improved the repeatability of this process, and thereby of the overall test procedure. In addition, densitometric technology has been developed to measure the intensity of a spot of interest on a plate, which by comparison to standards can be related to drug content. With the aid of software, the complex mathematics needed to calculate the drug content from the amount of reflected light can be easily performed. These two key developments have made TLC-densitometry a reliable method for pharmaceutical analysis. The separation media have also been improved by reducing the particle size and uniformity, leading to the manufacture of HPTLC plates. HPTLC offers all of the advantages of TLC but with improved separation capacity and plate numbers that approach those available by use of conventional HPLC columns.

These developments have increased the acquisition costs for HPTLC versus TLC plates, but the new systems have resulted in improved versatility, throughput, and robustness for the planar technique while retaining the low running and maintenance costs. In addition, where Minilabs have been in use, the existing basic chromatography skills can be adapted with minimal efforts for HPTLC densitometric analysis. This has made HPTLC-densitometry testing a more useful technology for analysis in resource-constrained settings since the acquisition, maintenance, and operation costs generally are less than those for HPLC [24].

HPTLC applications can be developed by identifying an existing TLC method for the drug substance or a compound that is structurally related. In some cases, you may find that a TLC method exists, but it uses noxious or environmentally unfriendly solvents. The method development process should be geared toward "green chromatography" [25,26].

Method development and subsequent validation in TLC and HPTLC are the most critical steps in qualitative and quantitative analysis applications. Nyiredy [27] and Kaale et al. [24] divided this process into rational steps: selection of the stationary phase, saturated vapor phase, suitable solvents, optimization of the mobile phase, selection of the development mode, extensive single laboratory validation (SLV) studies, transfer of the optimized mobile phase to another appropriate laboratory, and selection of other operating parameters.

In order to be able to exploit the cost and convenience advantages of HPTLC-densitometry, there is a need to improve and/or adapt existing methods and develop new methods of analysis. Currently, there are a large number of HPTLC-densitometry studies originating from India and Pakistan that reflect the wide acceptance of this technology there [24]. These additional methods increase the testing inventory and upon qualification will be useful to further expand the use of the technology.

The International Conference on Harmonization of Technical Requirements for Registration of Pharmaceuticals for Human Use (ICH) has published a guidance (Q2R1) for the validation of pharmaceutical analytical methods [28]. Any new method or alteration/improvement of existing methods should be validated in accordance with ICH guidelines to demonstrate the reliability of the obtained results. For the case of revalidation, the extent of needed validation will be determined on a case-by-case basis after performance of risk analysis. As in all methods including HPLC, the following parameters are an integral part of the validation requirements: accuracy; precision (repeatability, intermediate precision, and reproducibility expressed in terms of relative standard deviation [RSD]); specificity; linearity; range; limit of detection; limit of quantification; and robustness.

Method validation is generally categorized in three different levels: extensive SLV, peer verification (PV), and full collaborative study. SLV applies to a specific laboratory, technician, and equipment. For example, staff members and students in our laboratory have developed and extensively validated, in accordance with ICH analytical method validation guidelines for SLV of HPTLC-densitometry, the following assessment methods: metronidazole [29]; quinine [30]; nevirapine [31]; a fixed dose combination of lamivudine and zidovudine [32]; sulfamethoxazole and trimethoprim [33]; an improved method for a triple ARV combination of lamivudine, stavudine, and nevirapine [34]; fluconazole [35]; and a triple lamivudine, tenofovirdisoproxilfumarate, and efavirenz fixed dose combination [36]. To date there have been a number of HPTLC-validated methods published by African scientists; the earliest dates back to 2001 when a method for the simultaneous determination of benazepril hydrochloride and hydrochlorothiazide was developed [37]. In this paper, separation was carried out using Merck HPTLC silica gel 60 F_{254} aluminum backed plates and ethyl acetate–methanol–chloroform (10:3:2) mobile phase. The method performance was compared to HPLC. The same group developed a method for the assay of lisinopril and hydrochlorothiazide in binary mixtures in 2001 [38]. That group has continued to be very active and has published papers on nicergoline in the presence of its hydrolysis-induced degradation product in 2002 [39], rabeprazole in the presence of its degradation products in 2003 [40], zolpidemhemitartrate in 2003 [41], vincamine in the presence of its degradation products in 2005 [42], alfuzosin hydrochloride in 2006 [43], cilostazolin in 2007 [44], and sulpiride and mebeverine hydrochloride in 2010 [45]. The latter seven methods are stability-indicating assays. They have sufficient resolution to allow the determination of the active substance of interest in the presence of potential interfering substances from the formulation matrix and/or degradation products. The results of several of these assessments were compared to results obtained with HPLC, and very good correlations were obtained.

These experiments represent an important milestone in the application of planar chromatography in pharmaceutical analysis because in these cases the use of the HPTLC plates made it possible to resolve the drug substances from impurities and achieve separations and determinations comparable to HPLC. In 2009, a comparison of performance in the determination of oxybutynin hydrochloride and its degradation products using HPTLC-densitometry and HPLC was modeled using a chemometrics approach [46]. Shewiyo et al. in 2011 published a method for quantitative analysis of fluconazole in a finished formulation by HPTLC [35]. The validation of all of these methods has been based on extensive SLV studies. PV and full interlaboratory studies for some other drugs are described elsewhere.

The PV method applies to a limited number of laboratories (two to seven) and is intended to provide information on how a method is interpreted outside of the original laboratory. This stage would involve the originator laboratory performing extensive SLV on a new, existing, or improved analytical method. This is followed by the preparation of method protocols, composite samples, and standard materials that are sent to secondary laboratories to perform the methods. Recently an extensive two-laboratory PV method investigation on the use of HPTLC-densitometry to perform assays of lamivudine-zidovudine, metronidazole, nevirapine, and quinine composite samples was successfully completed [24]. For the four samples analyzed, excellent reproducibility was obtained. In addition, the method used less toxic organic solvents, thus moving toward a greening of chromatography. The full collaborative cross validation study is discussed in Section 14.6.

14.6   SYSTEM ROBUSTNESS

14.6.1   CAMAG COLLABORATIVE STUDY

Substandard and fake finished pharmaceutical products (FPPs) and APIs are becoming increasingly rampant worldwide [47–49]. It is a great concern to government regulatory agencies, pharmaceutical companies, health care providers, and consumers [50,51]. Cost-effective and reliable TLC methods for rapid screening of fake drugs have been reviewed [52]. These are methods that can be carried out in the field by inspectors with limited expertise, and are based on the use of portable kits with standard reference tablets to eliminate weighing. The only drawback is that they are subjective and have low accuracy levels of up to 10% or more. HPTLC has increasingly been shown to be a simple and reliable method for the determination of pharmaceutical products, both in bulk form as well as in single or fixed dose combination dosage forms. The precision levels achieved by the HPTLC procedures with the improved technologies can approach those levels generally attained with HPLC [53]. It should be noted that HPLC procedures may be capable of resolving many more compounds than TLC or HPTLC in a single experiment, but pharmaceutical products generally have a limited number of compounds that need to be separated. Both technologies are necessary, and the choice between them should be based on the consideration of performance needs and costs.

Recently, simple and precise HPTLC densitometric methods for the determination of antibiotic, antiviral, and antimalarial products [29–32] were developed and extensively validated via SLV experiments in the Pharm R&D Lab, MUHAS. These methods were performed in accordance with the ICH Q2R1guidelines for analytical method validation.

To bring the HPTLC technology into wider routine use in analysis, it is necessary to demonstrate that the achievable separation and determination results in one laboratory can be reliably replicated in another. This work presents the results of an interlaboratory collaboration using the above cited methods in the analysis of four composite samples of pharmaceutical products, one of them being a fixed dose combination [24]. Recently Kaale and his coworkers [29–35] did a systematic conversion of the TLC method to HPTLC as described earlier. This work was followed by a two-laboratory cross-validation investigation on the use of HPTLC to perform assays on lamivudine-zidovudine, metronidazole, nevirapine, and quinine composite samples. O'Sullivan and Sherma reported a model procedure for the transfer of four Kenyon and Layloff Compendium methods to HPTLC-densitometry and applied it to the formulations of the analgesics drugs acetominophen, acetylsalicyclic acid, and ibuprofen and the antihistamine drug chlorpheniramine maleate [54]. The procedure involved use of Merck silica gel glass plates with a fluorescent indicator, extraction procedures and mobile phases employing the allowed environmentally friendly solvents (Table 14.3), application of standard and sample zones with a CAMAG Linomat, development of plates in a CAMAG twin trough chamber, and densitometry of the separated zones at 254 nm with a CAMAG Scanner 3 to produce a calibration curve from the standards and interpolate unknown sample weights for calculation of recovery. The methods are suitable for full validation according to ICH guidelines if required for support of regulatory compliance actions. Transfer of other Compendium and Minilab methods is now underway using this same approach with additions of confirmations of drug identity and system selectivity using the peak comparison and peak purity check options, respectively, of the winCATS software, and of accuracy validation using the standard addition method.

To minimize the effects of differences in analysts' techniques, the laboratories in the interlaboratory collaboration conducted the study with automatic sample application devices in conjunction with variable wavelength scanning densitometers to evaluate the plates. The HPTLC procedures used relatively innocuous, inexpensive, and readily available chromatography solvents used in the Kenyon-Layloff [55] or Minilab [56] TLC methods. The use of automatic sample applications in conjunction with variable wavelength scanning densitometry demonstrated an average repeatability or within-laboratory relative standard deviation (RSD) of 1.90%, with 73% less than 2% and 97% at 2.60% or less, and an average reproducibility or among-laboratory RSD of 2.74% [24].

The HPTLC technology with the automatic sample application devices and variable wavelength scanning densitometers provides a pharmaceutical analytical assessment capability similar to that obtained with HPLC [24] but with a simpler and robust chromatography technology. These HPTLC procedures should be amenable for use in enforcing product quality standards.

14.7   FUTURE EXPANSION

For an analytical method to be adopted into official compendia and subsequent application for enforcement of regulatory actions, evidence of its transferability must be demonstrated. This information can only be obtained by an interlaboratory collaborative study. The highest level of validation is a full collaborative study, which requires eight or more laboratories providing acceptable data using the same method protocols, composite samples, and standard materials. A successful collaborative study provides a high level of confidence that the method is reproducible in different laboratories. Collaborative studies are not always practical or possible to manage because they require significant commitments of personnel, equipment, and support resources concurrently in multiple laboratories. It is important that laboratories employ a level of validation that is suitable for the method's intended use; it should ensure that the methodology is accurate, precise, and rugged for the specified analyte and concentration range. Given this challenge and level of complexity in organizing the collaborative studies, there is a need to identify technologies that would not burden participating laboratories in terms of time and financial resources. In analytical testing, HPTLC-densitometry is at the heart of this requirement of keeping costs low as well as time required to provide results similar to those obtained with HPLC. No such study has been reported involving laboratories from Africa. To our present knowledge, only one such study was reported for HPTLC of sucralose involving 14 laboratories from five different countries [57]. The AOAC PV method involving two independent laboratories was also conducted for an HPTLC method for identification of *Echinacea* species and their common adulterants [58]. Another method for determination of the mycotoxin fumonisin B1 was collaboratively studied in 14 laboratories using four duplicate maize meal samples and a spiked sample for determination of recovery [59]. This study also demonstrated that no significant difference was observed between the levels determined by HPLC or TLC.

There is an ongoing effort to organize an interlaboratory collaborative study involving seven laboratories from Africa, one from the United States, and one from Europe.

REFERENCES

1. Flinn, P.E., Juhl, Y.H., and Layloff, T.P., A simple, inexpensive thin-layer chromatography method for the analysis of theophylline tablets, *Bull. WHO*, 67(5), 555–539, 1989.
2. http://www.gphf.org/web/en/minilab (accessed July 2013).
3. http://www.patentgenius.com/patent/5350510.html (accessed July 2013).
4. Jahnke, R.W., Kusters, G, and Fleischer, K., Low—Cost quality assurance of medicines using the GPHF Minilab, *Drug Inf. J.*, 35, 941–945, 2001.
5. http://www.pharmweb.net/pwmirror/library/pharmwebvlib.html (accessed July 2013).
6. Kenyon, A.S., Xiaoye, S., Yan, W., Har, N.W., Prestridge, R., and Sharp, K., Simple, at-site detection of diethylene glycol/ethylene glycol contamination of glycerin and glycerin-based raw materials by thin-layer chromatography, *J. AOAC Int.*, 81(1), 44–50, 1998.
7. Ndomondo–Sigonda, M., *Counteracting Counterfeiting: Strategies for Improving the Integrity of the Medicinal Marketplace in Tanzania*, Presentation at the FIP Congress, Cairo, Egypt, Septemeber 2005, http://www.msh.org/seam/reports/fip-cairo_2005/57_counteracting.pdf (accessed May 2012).
8. Centre for Pharmaceutical Management, Access to essential medicines: Tanzania 2001, Management Sciences for Health, Arlington, VA, 2003.
9. World Health Organization, Unpublished paper of the WHO, Division of Drug Management and Policies: Summary of WHO counterfeit drug database as of April 1999, Geneva, Switzerland, 1999.
10. Smine, A., Malaria sentinel surveillance site assessment report, Mekong Region, USP-DQI, Rockville, MD, April 16–May 17, 2002.
11. Risha, P., Msuya, Z., Ndomondo-Sigonda, M., and Layloff, T., Proficiency testing as a tool to assess the performance of visual TLC quantitation estimates, *J. AOAC Int.*, 89(5), 1300–1304, 2006.
12. Risha, P., Msuya, Z., Clark, M., Johnson, K., Ndomondo-Sigonda, M., and Layloff, T., The use of Minilabs to improve testing capacity of regulatory authorities in resource limited settings: Tanzanian experience, *Health Policy J.*, 87, 217–222, 2008.
13. Hajjou, M., Supervision of minilab training, review of drug quality monitoring results, and preparation for upcoming pharmacovigilance activities, Antananarivo, Madagascar, October 9–17, 2008, Trip Report to USP- DQI, Rockville, MD, http://pdf.usaid.gov/pdf_docs/PDA/CM685.pdf (accessed May 2012).
14. Smine, A., Diouf, K., and Blum, N.L., USP-DQI antimalarial drug quality in senegal 2002, The U.S. agency for international development by the United States pharmacopeia drug quality and information program, United States pharmacopeial convention, Rockville, MD, http://pdf.usaid.gov/pdf_docs/PNACW987.pdf (accessed May 2012).
15. Tran, D. and Risha, P., Quality assurance of medicines in Zambia—An assessment visit of the Zambia pharmaceutical regulatory authority, Trip Report, The U.S. agency for international development by the rational pharmaceutical management plus program, Management Sciences for Health, Arlington, VA, http://pdf.usaid.gov/pdf_docs/PDACJ294.pdf (accessed May 2012).
16. United States pharmacopeia drug quality and information program. 2010, Survey of the quality of selected antimalarial medicines circulating in Madagascar, Senegal, and Uganda: November 2009, The United States Pharmacopeial Convention, Rockville, MD, http://www.usp.org/worldwide/dqi/resources/technicalReports (accessed May 2012).
17. World Health Organization. *Survey of the Quality of Selected Antimalaria Medicines Circulating in Six Countries of Sub-Saharan Africa*, WHO, Geneva, Switzerland, 2011.
18. Lon, C.T., Tsuyuoka, R., Phanouvong, S., Nivanna, N., Socheat, D., Sokhan C., Blum, N., Cristophel, E.M., and Smine, A., Counterfeit and substandard antimalaria drugs in Cambodia, *Trans. Roy. Soc. Trop. Med. Hyg.*, 100, 1019–1024, 2006.
19. Counterfeit and Substandard Drugs in Myanmar and Viet Nam - Report of a Study Carried out in Cooperation with the Governments of Myanmar and Viet Nam. EDM Research Series N0. 029, http://apps.who.int/medicinedocs/en/d/Js2276e/ (accessed July 2013).
20. Phanouvong, P. and Blum, N., Mekong malaria initiative antimalarial drug quality monitoring and evaluation, USP-DQI, Rockville, MD, March 2004, http://www.usp.org/sites/default/files/usp_pdf/EN/dqi/drugqualitymonitoringindicators.pdf (accessed May 2012).
21. Smine, A. and Burimski, K., USP-DQI—MeTA training on basic tests using Minilabs, September 14–18, 2009, Submitted to the U.S. agency for international development by the United States pharmacopeia drug quality and information program. United States Pharmacopeial Convention, Rockville, MD.
22. Global Pharma Health Fund, The GPHF-minilab-protection against counterfeited and substandard pharmaceuticals 2007, www.gphf.org/web/en/minilab/index.htm (accessed May 2012), 2007.

23. Striegel, M.F. and Hill, J., An overview of Thin-Layer Chromatography, in *Thin-Layer Chromatography for Binding Media Analysis*, Striegel, F. and Hill, J. (Eds.), The J. Paul Getty Trust, Malibu, CA, pp. 5–15, 1996.

24. Kaale, E., Risha, P., Reich, E., and Layloff, T.P., An interlaboratory investigation on the use of high-performance thin layer chromatography to perform assays of lamivudine-zidovudine, metronidazole, nevirapine, and quinine composite samples, *J. AOAC Int.*, 93(6), 1836–1843, 2010.

25. Chen, K., Lynen, F., De Beer, M., Hitzel, L., Ferguson, P., Hanna-Brown, M., and Sandra, P., Selectivity optimization in green chromatography by gradient stationary phase optimized selectivity liquid chromatography, *J. Chromatogr. A*, 1217(46), 7222–7230, 2010, Epub September 17, 2010.

26. Anastas, P.T. and Warner, J.C., *Green Chemistry: Theory and Practice*, Oxford University Press, Oxford, UK,135+xi pages, 1998, ISBN 0-19-850234-6.

27. Nyiredy, Sz., Essential guides to method development in thin-layer (planar) chromatography, in *Encyclopedia of Separation Sciences*, Wilson, I.D., Adlard, E.R., Cooke, M., and Poole, C.F. (Eds.), Vol. 10, Academic Press, London, U.K., 2000, pp. 4652–4666.

28. The international conference on harmonization of technical requirements for registration of pharmaceuticals for human use (ICH Q2R1), Geneva, Switzerland, 1994.

29. Hasan, S., Development and validation of an HPTLC densitometric method for quantitative analysis of metronidazole in tablets, http://www.muhas.ac.tz/dmdocuments/HPTLC%20of%20Metronidazole_student%20thesis%20work_2007.pdf (accessed June 2012).

30. Sinda, M., Development and validation of an HPTLC densitometric method for assay of quinine in tablets, http://www.muhas.ac.tz/dmdocuments/HPTLC%20of%20Quinine%20_student%20thesis%20work_2007.pdf (accessed June 2012).

31. Evarist, J., Development and validation of an HPTLC densitometric method for assay of nevirapine in tablets, http://www.muhas.ac.tz/dmdocuments/HPTLC%20of%20Nevirapin_student%20thesis%20work_2007.pdf (accessed June 2012).

32. Ikombola, J., Development and validation of an HPTLC densitometric method for simultaneous assay of lamivudine and zidovudine in tablets, http://www.muhas.ac.tz/dmdocuments/HPTLC%20of%20Zido_Lam%20_student%20thesis%20work_2007.pdf (accessed June 2012).

33. Shewiyo, D.H., Kaale, E., Risha, P.G., Dejaegher, B., Smeyers–Verbeke, J., and Vander Heyden, Y., Development and validation of a normal-phase high-performance thin layer chromatographic method for the analysis of sulfamethoxazole and trimethoprim in co-trimoxazole tablets, *J. Chromatogr. A*, 1216, 7102–7107, 2009.

34. Shewiyo, D.H., Kaale, E., Ugullum, C., Sigonda, M.N., Risha, P.B., Dejaegher, B., Smeyers–Verbeke, J., and Vander Heyden, Y., Development and validation of a normal-phase HPTLC method for the simultaneous analysis of lamivudine, stavudine, and nevirapine in fixed-dose combination tablets, *J. Pharm. Biomed. Anal.*, 54, 445–450, 2011.

35. Shewiyo, D.H., Kaale, E., Risha, G.P., Sillo, H.B., Dejaegher, B., Smeyers-Verbeke, J., and Vander Heyden, Y., Development and validation of a normal-phase HPTLC–densitometric method for the quantitative analysis of fluconazole in tablets, *J. Planar Chromatogr. Mod. TLC*, 24(6), 529–533, 2011.

36. Nyamweru, B.C., Kaale, E., Mugoyela, V., and Chambuso, M., Development and validation of an HPTLC densitometric method for simultaneous analysis of lamivudine, tenofovir, disoproxil fumarate, and efavirenz (LTE) in tablets, http://www.muhas.ac.tz/dmdocuments/HPTLC%20od%20Lam_Tenofovir_efavirenz_student%20thesis%20work%20_2012.pdf (accessed June 2012).

37. El-Gindy, A., Ashour, A., Abdel-Fattah, L., and Shabana, M.M., Application of LC and HPTLC-densitometry for the simultaneous determination of benazepril hydrochloride and hydrochlorothiazide, *J. Pharm. Biomed. Anal.*, 25, 171–179, 2001.

38. El-Gindy, A., Ashour, A., Abdel-Fattah, L., and Shabana, M.M., Spectrophotometric and HPTLC-densitometric determination of lisinopril and hydrochlorothiazide in binary mixtures, *J. Pharm. Biomed. Anal.*, 25, 299–307, 2001.

39. Ahmad, A.K., Kawy, M.A., and Nebsen, M., First derivative ratio spectrophotometric, HPTLC-densitometric, and HPLC determination of nicergoline in presence of its hydrolysis-induced degradation product, *J. Pharm. Biomed. Anal.*, 30, 479–489, 2002.

40. El-Gindy, A., El-Yazby, F., and Maher, M.M., Spectrophotometric and chromatographic determination of rabeprazole in presence of its degradation products, *J. Pharm. Biomed. Anal.*, 31, 229–242, 2003.

41. El Zeany, B.A., Moustafa, A.A., and Farid, N.F., Determination of zolpidemhemitartrate by quantitative HPTLC and LC, *J. Pharm. Biomed. Anal.*, 33, 393–401, 2003.

42. Shehata, M.A., El Sayed, M.A., El Taras, M.F., and El Bardicy, M.G., Stability indicating methods for determination of vincamine in presence of its degradation product, *J. Pharm. Biomed. Anal.*, 38, 72–78, 2005.

43. Salah Fayed, A., Abdel-Aaty Shehata, M., Hassan, N.Y., and El-Weshahy, S.A., Validated HPLC and HPTLC stability indicating methods for determination of alfuzosin hydrochloride in bilk powder and pharmaceutical formulations, *J. Sep. Sci.*, 29, 2716–2724, 2006.
44. Salah Fayed, A., Abdel-Aaty Shehata, M., Hassan, N.Y., and El-Weshahy, S.A., Validated stability indicating methods for determination of cilostrazol in the presence of its degradation products according to ICH guidelines, *J. Pharm. Biomed. Anal.*, 45, 407–416, 2007.
45. Naguib, I.A. and Abdelkawy, M., Development and validation of stability indicating methods for determination of sulpiride and mebeverine hydrochloride according to ICH guidelines, *Eur. J. Chem.*, 45, 3719–3725, 2010.
46. Wagieh, N.E., Hegazy, M.A., Abdelkawy, M., and Abdelaleem, E.A., Quantitative determination of oxybutynin hydrochloride by spectrophotometry, chemometry, and HPTLC in the presence of its degradation product and additives in different pharmaceutical dosage forms, *Talanta*, 80, 2007–2015, 2007, Epub November 10, 2009.
47. Medicines: Spurious/falsely-labeled/falsified/counterfeit (SFFC) medicines, WHO Fact Sheet No. 275, January 2010, http://www.who.int/mediacentre/factsheets/fs275/en (accessed April 2012).
48. WHO General Information on Counterfeit Medicines, http://www.who.int/medicines/services/counterfeit/overview (accessed April 2012).
49. GPHF 2004: Counterfeit Medicines—An Unscrupulous Business, http//www.gphf.org/web/en/minilab/hintergrund_arzneimittelfaelschungen.htm (accessed April 2012).
50. Report of the U.S. Food and Drug Administration, Combating Counterfeit Drugs, 2012, http://www.fda.gov/Drugs/ResourcesForYou/Consumers/BuyingUsingMedicinesSafely/CounterfeitMedicine/default.htm (accessed April 2012).
51. Desingh, A.K., Pharmaceutical counterfeiting, *Analyst*, 130, 271–279, 2005.
52. Sherma, J., Analysis of counterfeit drugs by thin layer chromatography, *Acta Chromatogr.*, 19, 5–20, 2007.
53. Anbazhagan, S., Indumathy, N., Shanmugapandiyan, P., and Krishnan, S.S., Simultaneous quantification of stavudine, lamivudine, and nevirapine by UV spectroscopy, reversed phase HPLC, and HPTLC in tablets, *J. Pharm. Biomed. Anal.*, 30, 801–804, 2005.
54. O'Sullivan, C. and Sherma, J., A model procedure for transfer of TLC pharmaceutical product screening methods designed for use in developing countries to quantitative HPTLC-densitometry methods, *Acta Chromatogr.*, 24(2), 241–256, 2012.
55. Kenyon, A.S. and Layloff, T.P., Screening of pharmaceuticals by thin layer chromatography, PHARM/95.290, World Health Organization, Geneva, Switzerland, 1995.
56. http://www.gphf.org/web/en/minilab/index.htm (accessed April 2012).
57. Stroka, J., Doncheva, I., and Spangenberg, B., Determination of sucralose in soft drinks by high performance thin layer chromatography: Interlaboratory study, *J. AOAC Int.*, 92(4), 1153–1159, 2009.
58. Reich, E., Blatter, A., Jorns, R., Kreuter, M., and Thiekoetter, K., An AOAC peer-verified method for identification of *Echinacea* species by HPTLC, *J. Planar Chromatogr. Mod. TLC*, 15(4), 244–251, 2002.
59. Shephard, G.S. and Sewrama, V., Determination of the mycotoxin fumonisin B1 in maize by reversed phase thin layer chromatography: A collaborative study, *Food Addit. Contam.*, 21(5), 498–505, 2004.

Part II

Planar Chromatography of Particular Drug Groups

15 TLC of Antidepressants and Neuroleptics

Robert Skibiński

CONTENT

Antipsychotic and antidepressant drugs are the most important and the most often used psychotropic drugs. Additionally, antipsychotic drugs are historically the first psychotropic drugs introduced to the market in the early 1950s (chlorpromazine—by Paul Charpentier).

Antipsychotic drugs, which are sometimes called neuroleptics, today are classified into two generations—old (classical) and new (atypical) generation—which is connected with its pharmacological mechanism. In the first case, its activity is connected with the inhibition of dopaminergic system (D_2); in the case of atypical, its activity depends the most on the inhibition of serotoninergic system. In chemical aspect, antipsychotic drugs are classified in the eight main groups: tricyclic neuroleptics (promazine, chlorpromazine, triflupromazine, acepromazine, levomepromazine, cyamemazine, thioridazine, periciazine, perazine, prochlorperazine, trifluoperazine, thioproperazine, perphenazine, fluphenazine, chlorprothixene, clopenthixol, flupenthixol), dibenzoepine neuroleptics (clozapine, olanzapine, quetiapine, clothiapine, loxapine, zotepine, clorotepine), butyrophenone neuroleptics (haloperidol, bromperidol, trifluoperidol, moperone, fluanisone, spiperone, droperidol), diphenylbutylpiperidine neuroleptics (pimozide, penfluridol, fluspirilene), benzamide neuroleptics (sulpiride, remoxipride, raclopride, sultopride, amisulpride, tiapride), indole neuroleptics (ziprasidone, molindone, sertindole), benzisoxazole neuroleptics (risperidone), and quinoline neuroleptics (aripiprazole).

Pharmacological activity of antidepressants is based on the activation of noradrenergic or serotoninergic systems by selective or nonselective reuptake inhibition of these neuromediators or monoaminooxidase (MAO) inhibition. The new generation of antidepressant drugs is also like the new generation of neuroleptics called "atypical" and is characterized by higher selectivity in the activation mechanism than the earlier generation. Nowadays, especially developed atypical antidepressants belong to selective serotonin reuptake inhibitors (SSRIs). Their efficacy is comparable to tricyclic antidepressants, but they are better tolerated and are characterized by a lower risk of causing adverse effects. Antidepressant drugs are usually discussed with the use of chemical–pharmacological classification, and they are categorized in tricyclic antidepressants (imipramine, desipramine, metapramine, opipramol, dibenzepin, amoxapine, amitriptyline, nortriptyline, butriptyline, amineptine, noxyptiline, protriptyline, demexiptiline), anthracene-derivative antidepressants (melitracen, dimetacrine), tetracyclic antidepressants (mianserin, mirtazapine, maprotiline), serotonin–norepinephrine reuptake inhibitor (SNRI) antidepressants (venlafaxine, milnacipran), SSRI antidepressants (sertraline, paroxetine, fluoxetine, fluvoxamine, zimelidine, citalopram, oxaflozane, minaprine), NRI antidepressants (reboxetine, viloxazine), MAO inhibitors (moclobemide, toloxatone, tranylcypromine, selegiline), and the other antidepressants (tianeptine, trazodone, etoperidone).

Thin-layer chromatographic (TLC) method was used in three main aspects in the case of analysis of neuroleptics and antidepressants:

1. Analysis in biological materials
2. Separation and retention behavior study
3. Pharmaceutical analysis

The most popular TLC system was in all cases normal-phase (NP) chromatography with silica gel stationary phase.

In the first case, antipsychotic and antidepressant drugs were separated and identified as well as quantified in various biological fluids. Qualitative analysis was usually performed with the use of densitometric methods in this case. Breyer and Villumsen (1976) performed measurement of plasma levels of perazine, clozapine, imipramine, and amitriptyline and their demethylated metabolites by ultraviolet (UV) reflectance photometry. The procedure included the extraction of alkalinized plasma samples with benzene or toluene and TLC of the extracts on nonfluorescent silica gel plates with the use of various mobile phases, chloroform–isopropanol (20:2) as a first eluent, and then one from four mobile phases was used (in the same dimension) composed with isopropanol–chloroform–ammonium hydroxide–water or isopropanol–chloroform–acetone–2M ammonium hydroxide. 2D chromatography with the use of 1,2-dichloroethane–ethyl acetate–ethanol–acetic acid–water (15:26:12:8:7.5) as a second mobile phase was also performed. The recoveries of compounds added in therapeutic concentrations were between 70% and 98%, and the limits of detection were 5–10 ng/g plasma. Silica gel plates were also used for the determination of chlorpromazine, amitriptyline, nortriptyline, imipramine, desipramine, phenobarbital, and phenytoin in plasma (Fenimore et al. 1978). As a mobile phase, mixtures of hexane–acetone–diethylamine (77:20:3), benzene–acetone–ammonium hydroxide (80:20:0.2), and ethyl acetate–acetic acid–water–acetone–isopropanol (40:5:5:2.5:2.5) were used in this case. Similar tricyclic antidepressants (amitriptyline, protriptyline, imipramine, and nortriptyline) and some additional benzodiazepines were separated and detected in serum on silica gel plates with the use of acetone–dichloromethane–ammonium hydroxide (2:3:0.2) as a mobile phase (Meola et al. 1981).

In the early 1980s, one of the newest antidepressants, citalopram, was determined together with its metabolites in plasma by high-performance thin-layer chromatographic (HPTLC) method (Overo 1981). Samples were extracted with n-hexane containing 1% of triethylamine and developed on silica gel plates with the use of dichloromethane–ethyl acetate–ethanol–acetic acid–water (15:26:12:8:7.5) as a mobile phase and next scanned densitometrically in fluorescence mode.

Chlorpromazine and thioridazine were also quantitatively determined in plasma by the use of silica HPTLC plates and toluene–acetone (50:50) and toluene–acetone–ammonium hydroxide (50:50:2.4) mixtures as a mobile phase (Davis and Harrington 1984). Samples were extracted by pentane and scanned by reflectance densitometry at 365 nm (chlorpromazine) and 375 nm (thioridazine).

In this decade, some psychotropic drugs (fluphenazine, imipramine, levomepromazine, nortriptyline) after liquid–liquid extraction were also separated and identified in urine (Farnke et al. 1988). Cyclohexane–toluene–diethylamine (75:15:10) mixture was used as a mobile phase in this case.

In the 1990s, one of the new-generation antidepressant drugs, fluoxetine, in the presence of three first-generation antidepressants (doxepin, imipramine, and opipramol) was separated and detected in human plasma (Misztal et al. 1997). Samples were extracted by diethyl ether, and TLC analysis was performed on silica gel plates in 15 elution systems (i.e., chloroform–methanol [9:1], methanol–acetone–cyclohexane [8:1:1], or methanol–benzene–ethyl acetate–chloroform [6:2:1:1]).

In this time, TLC method was also used for drug screening in forensic toxicology (Ojanpera et al. 1999). Various drugs including chlordiazepoxide, fluoxetine, mianserin, thioridazine, and trimipramine were identified in urine, blood, and liver by the use of different NP- and reversed-phase (RP)-TLC systems.

In the next decade, TLC method was often used for the analysis of new (second) generation of antidepressants as well as new neuroleptics. In this period, six new antidepressant drugs (fluoxetine, paroxetine, fluvoxamine, citalopram, nefazodone, and moclobemide) were separated in different chromatographic systems on silica gel (NP-TLC) and RP_{18} plates (Misztal and Skibiński 2001). The analyses were performed in horizontal technique, and the detection was made by videoden-sitometric method at 254 nm. The best system (benzene–acetone–ethanol–ammonium hydroxide [9:7:2:1] on silica gel plates) was used for the separation and identification of these drugs in plasma. Oztunc et al. (2002) proposed 7,7,8,8-tetracyanoquinodimethane as a new derivatization reagent for high-performance liquid chromatographic (HPLC) and TLC analysis of some antidepressants (paroxetine, desipramine, fluoxetine, nortriptyline, maprotiline). TLC was performed on silica gel plates by the use of benzene–diethyl ether (1:4) as a mobile phase and on CN HPTLC plates and of hexane–diethyl ether–acetonitrile (4:4:1) and petroleum ether–diethyl ether–acetonitrile–ethyl methyl ketone (5:4:0.5:0.5) as a mobile phase. Developed methods were applied to the identification of these drugs in plasma samples. Kazlauskiene et al. (2003) separated and identified fluoxetine, ami-triptyline, and codeine in sudden poisoning cases by NP-TLC method. Five mobile phases, diethyl acetate–methanol–ammonium hydroxide (85:10:5), buthylacetate–methanol–ammonium hydroxide (85:10:5), cyclohexane–diethyl acetate–diethyl amine (70:15:15), cyclohexane–buthylacetate–diethyl amine (70:15:15), and acetone–dioxane–ammonium hydroxide (30:68:2), were used in this case. Mennickent et al. (2007b) developed HPTLC method for quantitative analysis of clozapine in human serum. Clozapine was extracted with n-hexane–isoamyl alcohol (75:25), and chromatographic sepa-ration was achieved on silica gel plates using a mixture of chloroform–methanol (9:1) as mobile phase. Quantitative analyses were carried out by densitometry at a wavelength of 290 nm.

The retention behavior of some psychotropic drugs (chlorpromazine, clomipramine, doxepin, flupentixol, haloperidol, moclobemide, perazine, risperidone) was also studied in this time by Petruczynik et al. (2008). The investigation was performed on silica, RP_{18}, and CN layers by the use of various eluents. RP_{18} plates eluted with methanol–water–ammonium hydroxide (70:30:1) were used for quantitative analysis of amitriptyline and doxepin in serum after solid-phase extraction (SPE). Phenothiazines were also analyzed in plasma by Blendea et al. (2009). The authors present a simple TLC method to qualitatively and semiquantitatively assay chlorpromazine, thioridazine, and levomepromazine in human plasma spiked with the three substances, following liquid–liquid extraction with dichloromethane. Several mobile phases were tested, and methanol–n-butanol (60:40) with an addition of 0.1 M NaBr was selected as the best system.

In the last years, fluoxetine antidepressant was analyzed in biological materials. Mennickent et al. (2010b) developed HPTLC method for the quantification of fluoxetine in human serum. Analyzed substance was extracted by liquid–liquid extraction method with diethyl ether, and imipramine was used as internal standard. The chromatographic separation was achieved on silica gel plates using a mixture of toluene–acetic acid glacial (4:5) as mobile phase. 4-Dimethylamino-azobenzene-4-sulphonyl chloride was used as derivatization reagent, and densitometric detection was performed at 272 nm in this case. Maślanka et al. (2011) developed TLC densitometric method for the identifica-tion and determination of fluoxetine and as well as haloperidol, promazine, doxepin, clonazepam, alprazolam, and risperidone in plasma and urine. Silica gel plates and two mobile phases, ace-tone–diethylamine–cyclohexane (2.5:2.5:20) and trichloromethane–acetone–ammonium hydroxide (25:25:0.5), were used for the separation of the analyzed drugs. Densitometric detection was per-formed at 254 nm.

The second important aspect was the separation and retention behavior study of these psychotro-pic drugs with the use of different adsorbents and mobile phases. These researches are very useful to the optimization of TLC methods for the identification and detection of these drugs and they are very helpful in QSAR or lipophilicity study.

One of the first papers that concern this kind of research was noticed by Musumarra et al. (1983). They performed principal component analysis (PCA) of R_f data obtained by TLC analysis of 54 drugs including neuroleptics (promazine, chlorpromazine, triflupromazine, trifluoperazine, fluphenazine, clopenthixol, chlorprothixene). Eight different eluent mixtures and silica gel plates were used by these authors. In the same year, Steinbrecher (1983) presented complete retention data for 20 phenothiazine neuroleptics in four different solvent systems: diethyl ether–ethyl acetate–ammonium hydroxide (1:1:saturated), methylene chloride–methanol (50:10), methanol–buthan-1-ol (60:40), and chloroform–isopropanol–ammonium hydroxide (70:30:1) on silica gel plates.

In the early 1990s, Garcia Sanchez et al. (1993) separated various phenothiazines (acepromazine, alimemazine, chlorpromazine, profenamine, promazine, promethazine, propiomazine, thioridazine) and collected its retention data in six different solvent systems. The best separation and fluorescence intensities were achieved in methanol–acetic acid (95:5) system, and qualitative and quantitative analysis was enabled in this case. In the same year, Joseph-Charles and Bertucat (1993) separated zotepine, amoxapine, loxapine, fluperlapine, clotiapine, nitroxazepine, and fluradoline on silica gel plates in six different elution systems. Good results were observed in clohexane–ethanol–n-butanol–ammonium hydroxide (60:20:10:1) and cyclohexane–ethyl acetate–ethanol–n-butanol–25% ammonium hydroxide (70:30:10:20:1) systems.

In the same decade, the lipophilicity study of serotoninergic ligands including mianserin with the use of RP-TLC method was also undertaken (Biagi et al. 1996). The R_M values were measured using RP_{18} plates and acetone, acetonitrile, or methanol as the organic modifier of the mobile phase. Obtained results were well correlated with the calculated log P values.

In the last year, discerning retention behavior study of new-generation neuroleptics was undertaken. Chromatographic behavior in NP-TLC has been investigated for amisulpride, clozapine, olanzapine, quetiapine, risperidone, and ziprasidone (Skibiński et al. 2006). The drugs were separated on silica gel, alumina, NH_2, CN, diol, and polyamide plates with a mixture of n-hexane and six polar modifiers (acetone, dioxane, diethylamine, ethanol, isopropanol, and tetrahydrofuran) as mobile phases. The linearity of relationships between R_M and volume fraction of modifier, the logarithm of the volume fraction, the molar fraction, and the logarithm of the molar fraction were tested. The results usually fitted the Snyder–Soczewiński equation, with r > 0.9. The best separation was achieved on silica gel plates with n-hexane–ethanol–ammonium hydroxide (5:5:0.15) as a mobile phase. Similarly in RP conditions on RP_{18}, RP_8, RP_2, CN, diol, and NH_2 adsorbents, retention behavior of these drugs was also studied (Skibiński et al. 2007b). The mixtures of phosphate buffers and six modifiers (acetone, acetonitrile, dioxane, ethanol, methanol, and tetrahydrofuran) were used as a mobile phase in this case. The best separation was achieved on RP_8 plates with dioxane–phosphate buffer pH 3.5 (4:6) as a mobile phase.

Hawrył et al. (2008) estimated the lipophilicity of some psychotropic drugs (alprazolam, amitriptyline, carbamazepine, chlorpromazine, clomipramine, doxepin, flupentixol, haloperidol, midazolam, moclobemide, perazine, and risperidone) by the use of RP-TLC method. Drugs have been chromatographed on RP_{18} HPTLC plates with mobile phases containing water, an organic modifier (methanol, dioxane, acetone, acetonitrile, or tetrahydrofuran), and ion-pair reagents or ammonia. The relationships between solute retention and modifier concentration were described by the Soczewiński–Wachtmeister and Schoenmaker equations in this case.

The quantitative structure–activity relationship (QSAR) analysis of 20 drugs with affinity for serotonin (5-HT) receptors, including clopenthixol, clozapine, mianserin, mirtazapine, olanzapine, risperidone, tiapride, trazodone, and trifluoperazine, was also carried out with the use of TLC method (Zydek and Brzezińska 2011). TLC was performed on silica gel and RP2 plates impregnated with solutions of aspartic acid, serine, phenylalanine, tryptophan, tyrosine, asparagine, threonine, and their mixtures, with two mobile phases: acetonitrile–methanol–ammonium acetate pH 7.4 (4:4:2) and acetonitrile–methanol–methylene chloride–ammonium acetate pH 7.4 (6:1:1:2). The relationships between chromatographic data and molecular descriptors and biological activity data were found by means of regression analysis. The correlations obtained for the compounds with serotoninergic activity represent their interaction with the proposed biochromatographic models.

The third group of publications concerns the pharmaceutical analysis of discussed drugs, especially determination in pharmaceutical formulations, detection of impurities, and stability study.

One of the first publications concerning the application of TLC method in stability study was a paper describing the photodegradation of moclobemide on silica gel plates (Nakai et al. 1989). The author confirmed that this drug is not stable under UV 254 irradiation and yields degradation to N-oxide derivative. The mixture of methanol–water (3:2) was used as a mobile phase in this study, and the obtained results were confirmed on HPLC system.

In the 1990s, the identification of phenothiazine drugs in pharmaceuticals was undertaken by Revanasiddappa and Ramappa (1995). Triflupromazine, prochlorperazine, and fluphenazine were separated on silica gel plates with the use of chloroform–methanol (9:1), chloroform–ethyl acetate (7:2), and cyclohexane–benzene–diethylamine (50:40:10) solvents and identified in several pharmaceutical formulations.

In the same period, quantitative analysis of sulpiride and its impurities in pharmaceuticals by HPTLC and scanning densitometry was realized (Agbaba et al. 1999). Methylene chloride–methanol–25% ammonium hydroxide (18:2.8:0.4) solvent system was used for separation and quantitative evaluation of chromatograms. The chromatographic plates were first scanned at 240 nm to locate chromatographic zones corresponding to sulpiride and methyl-5-sulfamoyl-2-methoxybenzoate. Next, the second impurity 2-aminomethyl-1-ethylpyrrolidine was derivatized with ninhydrin, and the resulting colored spots were measured at 500 nm.

Sulpiride was also analyzed together with mebeverine in combined pharmaceuticals by TLC densitometry (El Walily et al. 1999). After separation on silica gel plates using ethanol–diethyl ether–triethylamine (70:30:1) as a mobile phase, the chromatographic zones corresponding to the spots of mebeverine and sulpiride were scanned at 262 and 240 nm, respectively.

The other important neuroleptic, haloperidol, was determined densitometrically with its impurity (4,4-bis [4-(p-chlorophenyl)-4-hydroxypiperidino]butyrophenone) in pharmaceuticals (Krzek and Maślanka 2000). Silica gel HPTLC plates and chloroform–methanol–25% ammonium hydroxide (90:9:1) were used for chromatographic separation, and detection has been carried out at 350 nm.

From the early 2000s, the TLC method became much more popular in pharmaceutical analysis. El-Gindy et al. (2002) developed densitometric TLC method for the determination of trifluoperazine and its oxidative degradation products. The separation was carried out on aluminum sheet of silica gel using chloroform–methanol (7:3) as mobile phase. Krzek and Maślanka (2002) developed densitometric HPTLC method for the determination of fluoxetine and its impurities. The compounds were separated on silica HPTLC plates with chloroform–methanol–ammonium hydroxide (45:4.5:0.5) as a mobile phase. UV (260 nm) and visible (VIS) (530 nm), after visualization of the chromatograms with ninhydrin solution, densitometric scanning were performed. Fluoxetine and paroxetine were also determined in pharmaceuticals by the use of densitometric and videodensitometric TLC methods (Skibiński et al. 2003). The drugs were chromatographed on silica gel plates in horizontal chambers with benzene–acetone–ethanol–ammonium hydroxide (9:7:2:1) as mobile phase. Densitometric detection was performed at 218 and 293 nm for fluoxetine and paroxetine, respectively. Videodensitometric detection was performed at 254 nm for both drugs.

Mahadik et al. (2003) estimated trifluoperazine, trihexyphenidyl, and chlorpromazine simultaneously in tablets by HPTLC method. Silica gel on aluminum foil and a mobile phase comprising of toluene–methanol–acetone–ammonium hydroxide (7:1:2:0.1) were used in this case. Densitometric quantification was performed at 213 nm.

Vujic et al. (2003) developed densitometric TLC method for the determination of maprotiline, desipramine, and moclobemide in pharmaceutical dosage forms. After separation on silica gel plates using propanol–ethanol–ammonium hydroxide (8:2:0.3) as the mobile phase, the chromatographic zones corresponding to the spots were scanned at 254 nm. Moclobemide and fluvoxamine were also analyzed in pharmaceuticals by the use of videodensitometric as well as densitometric methods (Skibiński and Misztal 2004). The drugs were chromatographed on silica gel plates in horizontal technique with benzene–acetone–ethanol–ammonium hydroxide (9:7:2:1) as a mobile phase. Densitometric detection and quantification were performed at 249 and 236 nm, respectively, for fluvoxamine and moclobemide. Videodensitometric detection was performed at 254 nm for both drugs.

In this decade, one of the newest second-generation neuroleptics, ziprasidone, was also determined in bulk powder and pharmaceutical formulations by thin-layer densitometric method (El-Sherif et al. 2004). The separation was carried out on aluminum sheet of silica gel using chloroform–methanol–glacial acetic acid (75:5:4.5) as the mobile phase, and densitometric measurement was performed at 247 nm. The developed method was also used for the stability study of ziprasidone. The same team developed also TLC method for the determination of risperidone in the presence of its degradation products in bulk powder and tablets (El-Sherif et al. 2005). The separation was carried out on aluminum sheet of silica gel using acetonitrile–methanol–propanol–triethanolamine (8.5:1.2:0.6:0.2) as the mobile phase and densitometric measurements at 280 nm.

In the same time, TLC methods were still developed for analysis of the first-generation antipsychotic and antidepressant drugs. Maślanka and Krzek (2005) developed densitometric TLC method for the identification and quantitative analysis of haloperidol, amitriptyline, sulpiride, promazine, fluphenazine, doxepin, diazepam, trifluoperazine, clonazepam, and chlorpromazine in pharmaceuticals. The separation was performed on silica gel plates, and chromatograms were developed in various mobile phases (30 phases were tested). Haloperidol and its three metabolites were also analyzed by Ali et al. (2005) on RP$_{18}$ plates. As a mobile phase, methanol containing 0.001% of triethylamine was used in this case. The detection and quantification were performed by exposing TLC plates to iodine vapor and by UV–VIS spectrometry. The method was applied for wastewater samples.

In this period, new-generation psychotropic drugs were analyzed in pharmaceutical formulation. Citalopram was determined in tablets by densitometric and videodensitometric HPTLC methods (Skibiński and Misztal 2005). Chromatography was performed on silica gel plates with a mixture of benzene–acetone–ethanol–ammonium hydroxide (45:40:10:5) as a mobile phase. The detection and quantification were carried out at 226 (densitometry) and 254 nm (videodensitometry). The obtained results were compared with HPLC method, and no statistical differences were observed between all the elaborated methods. Saxena et al. (2006) developed also HPTLC method for the determination of olanzapine in pharmaceuticals. The method employed aluminum plates precoated with silica gel as the stationary phase. The solvent system consisted of toluene–methanol–ethyl acetate–ammonium hydroxide (8:2:1:0.1). The method was applied also as a stability-indicating assay.

In the same year, the analytical methods for the determination of classical neuroleptics and antidepressants in pharmaceuticals were continually developed. Wójciak-Kosior et al. (2006) separated and quantified promethazine, promazine, and thioridazine on silica gel plates. Acetone–methanol–ammonium hydroxide (90:10:2) mixture was used as elution system. Densitometric detection was performed at 247 nm. Maślanka and Krzek (2007) determined either phenothiazine derivatives (chlorpromazine, trifluoperazine, and promazine) or doxepin and its impurities. Densitometric detection was also used in this case. Mennickent et al. (2007a) developed densitometric HPTLC method for quantitative analysis of haloperidol in tablets. Chromatographic separation was achieved on silica gel plates using a mixture of acetone–chloroform–n-butanol–acetic acid glacial–water (5:10:10:2.5:2.5) as a mobile phase. Quantitative analysis was carried out at a wavelength of 254 nm. These authors by the use of the previously described method studied also photostability of haloperidol in injection (Mennickent et al. 2008).

In the next years, new-generation psychotropic drugs were mainly analyzed by TLC method. Simultaneous assay of olanzapine and fluoxetine in tablets by HPTLC method was described by Shah et al. (2007). The separation was achieved on aluminum sheets coated with silica gel using methanol–toluene (4:2) as the mobile phase. The quantitation was achieved by measuring UV absorption at 233 nm by densitometer. The developed method was also used for stability study of these drugs (Shah et al. 2008). Amisulpride was also determined in tablets by HPTLC method (Skibiński et al. 2007a). Two chromatographic systems (NP- and RP-HPTLC) and two instrumental detections were proposed in this case. NP-HPTLC was carried out using HPTLC silica plates and developed with hexane–ethanol–propylamine (5:5:0.1). RP-HPTLC was developed with

RP$_8$ plates, with mobile phase of tetrahydrofuran–phosphate buffer pH 3.50 (4:6). Both analyses were performed in horizontal chambers and scanned with densitometer at 275 nm or videodensitometer at 254 nm. Venkatachalam and Chatterjee (2007) developed stability-indicating HPTLC method of analysis of paroxetine in formulations. The method employed aluminum precoated with silica gel as the stationary phase and butanol–acetic acid–water (8:2:0.5) as a solvent system. Densitometric analysis of was carried out in the absorbance mode at 295 nm. The authors reveal that paroxetine does not undergo degradation with oxidation but gets affected in acidic and alkaline conditions. Venkateswarlu et al. (2007) developed HPTLC densitometry method for the analysis of alprazolam and sertraline in pharmaceutical formulations. The authors used silica gel plates with fluorescent indicators, and mobile phase consisted of carbon tetrachloride–methanol–acetone–ammonium hydroxide (12:3:5:0.1). The detection was performed at 254 nm by using UV absorption densitometry. Gondova et al. (2008) proposed another TLC densitometric method for the determination of citalopram, sertraline, fluoxetine, and fluvoxamine in pharmaceuticals. The analyzed drugs were separated on silica gel plates using a mobile phase composed of acetone–benzene–ammonium hydroxide (50:45:5). Densitometric detection and quantification were carried out in the reflectance mode at 240 nm. Patel et al. (2008) developed stability-indicating HPTLC method for the analysis of moclobemide in bulk drug and in formulations. Aluminum TLC plates precoated with silica gel were used with benzene–methanol–ammonium hydroxide 7:3:0.1 (v/v) as mobile phase and densitometric detection at 238 nm. The authors reveal that moclobemide yields degradation under acidic, basic, and oxidizing conditions. One of the newest neuroleptics, quetiapine, was determined in tablets by NP- and RP-HPTLC methods with densitometric and videodensitometric detection (Skibiński et al. 2008). NP chromatography was developed with silica gel plates and hexane–dioxane–propylamine (1:9:0.4) as mobile phase. RP chromatography was carried out using RP$_8$ plates with tetrahydrofuran–phosphate buffer pH 9.0 (5:5) as mobile phase. Both analyses were performed in horizontal chambers and scanned with a densitometer at 243 nm and a videodensitometer at 254 nm. The NP-HPTLC method was also used for stability-indicating assay in this case. Soliman et al. (2008) developed TLC densitometric method for the determination of sulpiride, tiapride, and veralipride. The analysis was carried out on silica gel plates using developing system: acetonitrile–water–glacial acetic acid (10: 10: 1) and densitometric measurements of the spots at 290 nm. Jagadeeswaran et al. (2009) proposed another HPTLC method for the determination of fluoxetine in pharmaceuticals. Silica gel aluminum foil and acetonitrile–chloroform (1:9) system were used in this case. Maślanka et al. (2009) developed TLC method of the determination of risperidone in tablets in the presence of its degradation products and placebo-derived constituents. Silica gel plates as a stationary phase and n-butanol–acetic acid–water (12:3:5) as a mobile phase were used for separation. Densitometric measurements were done for all constituents at 280 nm. A decrease in stability of risperidone was observed in acidic, basic, and antioxidant solutions. Zaheer et al. (2009) developed stability-indicating HPTLC method for the analysis of clozapine in pharmaceutical dosage form. The method employed aluminum plates precoated with silica gel as the stationary phase and the solvent system comprising of toluene–methanol–ethyl acetate–ammonium hydroxide (8:2:1:0.1). Densitometric analysis of clozapine was carried out in the absorbance mode at 280 nm. The drug undergoes degradation under acidic, basic, photochemical, and thermal degradation conditions.

Patel et al. developed several TLC analytical methods for the analysis of psychotropic drugs in pharmaceuticals in the last years. Alprazolam and sertraline were determined in tablets on HPTLC silica plates with the use of acetone–toluene–ammonium hydroxide (6:3:1) as a mobile phase (Patel et al. 2009a). In similar chromatographic condition, alprazolam was also analyzed together with fluoxetine (Patel et al. 2009b). Fluoxetine was also analyzed by these authors together with olanzapine (Patel and Patel 2009b). The same stationary phase was used in this case, and as a mobile phase, the mixture of acetone–methanol–triethylamine (5:3:0.5) was selected. Another method for the determination of alprazolam together with amitriptyline, trifluoperazine, and risperidone was also proposed by these authors (Patel and Patel 2009a). TLC aluminum sheets of silica and carbon tetrachloride–acetone–triethylamine (8:2:0.3) system were used in this case. Risperidone and olanzapine were one more time analyzed individually on silica gel plates with the use of methanol–ethyl acetate (8:2) mixture (Patel et al. 2010a,b). Patel and Patel (2010) determined either imipramine or chlordiazepoxide in pharmaceutical formulations. Carbon tetrachloride–acetone–triethylamine pH 8.3 (6:3:0.3) system was used for this assay.

In the same time period, Devala et al. (2010) developed HPTLC method for the analysis of risperidone in bulk and pharmaceutical dosage form. Precoated silica gel plates were used as stationary phase, and the separation was carried out using dichloromethane–methanol–ethanol–triethylamine (12:12:6:0.1) as a mobile phase. The densitometry scanning was carried out at 280 nm. Gondova and Petrikova (2010) described TLC densitometry method for the simultaneous separation and determination of the two antidepressants, mirtazapine and mianserin, in commercially available tablets. Silica gel plates and mixture of hexane–isopropanol–ammonium hydroxide (70:25:5) were used as a chromatographic system. Densitometric analysis was carried out in the absorbance mode at 280 nm in this case. Gupta et al. (2010) developed quantitative densitometric HPTLC method for the analysis of sertraline in pharmaceutical preparations. The drug was separated and identified on silica gel plates with chloroform–ethyl acetate–triethylamine (5:3:0.2) as a mobile phase. Densitometric quantification was performed at 279 nm by reflectance scanning. Mennickent et al. (2010a) elaborated instrumental planar chromatographic method for the quantification of fluphenazine in injections. Chromatographic separation was performed on silica gel HPTLC plates, and methanol–water (9:1) was used as an elution system. Densitometric analysis was performed at 306 nm. Stability-indicating HPTLC method for the determination of sulpiride and mebeverine was developed by Naguib and Abdelkawy (2010). The separation was performed on silica gel plates using ethanol–methylene chloride–triethylamine (7:3:0.2) as a mobile phase and scanning of the separated bands at 221 nm.

One of the newest neuroleptics, ziprasidone, was also analyzed by HPTLC method (Skibiński and Komsta 2010). Two separation systems—NP and RP-HPTLC and videodensitometric and densitometric detection—were proposed for this assay. For NP chromatography, silica gel plates and hexane–dioxane–propylamine (1:9:0.4) system were used. For RP chromatography, RP$_8$ plates and tetrahydrofuran–phosphate buffer p 9 (5:5) system were employed. Videodensitometric detection was performed at 254 nm, and densitometric detection was carried out at 243 nm.

Shirvi et al. (2010) described HPTLC densitometric method for quantitative analysis of venlafaxine in tablets. Chromatographic separation was performed on aluminum plates coated with silica gel with toluene–methanol (7:3.5) as mobile phase. Densitometric evaluation was performed at 228 nm. Venlafaxine was also analyzed in pharmaceutical formulations by Ramesh et al. (2011). These authors developed quantitative densitometric HPTLC method for stability-indicating analysis of this drug. Venlafaxine was separated and identified on silica plates with the use of butanol–acetic acid–water (6:2:2) as a mobile phase. Densitometry was performed at 254 nm in this case.

In the last years, stability-indicating methods were especially developed. Rao et al. (2011) elaborated stability-indicating HPTLC method for the quantitative analysis of sertraline in pharmaceuticals. Aluminum foil plates precoated with silica gel and toluene–ethyl acetate–ethanol–ammonium hydroxide (8:2:0.5:0.1) were used as a chromatographic system, and densitometric analysis was performed at 273 nm.

REFERENCES

Agbaba, D., T. Miljkovic, V. Marinkovic, D. Zivanov-Stakic, S. Vladlmirov. 1999. Quantitative analysis of sulpiride and impurities of 2-aminomethyl-1-ethylpyrrolidine and methyl-5-sulphamoyl-2-methoxybenzoate in pharmaceuticals by high-performance thin-layer chromatography and scanning densitometry. *J. AOAC Int.* 82(4): 825–829.

Ali, I., V. Gupta, P. Singh, H. Pant. 2005. RPTLC analysis and haloperidol and its metabolites in wastewater after solid-phase extraction. *J. Planar Chromatogr. Mod. TLC* 18(105): 388–390.

Biagi, G., A. Barbaro, A. Sapone, P. Borea, K. Varani, M. Recanatini. 1996. Study of lipophilic character of serotonergic ligands. *J. Chromatogr. A* 723(1): 135–143.

Blendea, L., D. Balalau, C. Gutu, M. Ilie, D. Baconi. 2009. Qualitative and semiquantitative TLC analysis of certain phenothiazines in human plasma. *Farmacia* 57(5): 542–548.

Breyer, U., K. Villumsen. 1976. Measurement of plasma levels of tricyclic psychoactive drugs and their metabolites by UV reflectance photometry of thin layer chromatograms. *Eur. J. Clin. Pharmacol.* 9(5–6): 457–465.

Davis, C., C. Harrington. 1984. Quantitative determination of chlorpromazine and thioridazine by high-performance thin layer chromatography. *J. Chromatogr. Sci.* 22(2): 71–74.

Devala, R., S. Kathirvel, S. Satyanarayana. 2010. Development and validation of TLC-densitometry method for the estimation of antipsychotic drug in bulk and tablet formulation. *Int. J. ChemTech. Res.* 2(4): 2063–2069.

El-Gindy, A., B. El-Zeany, T. Awad, M. Shabana. 2002. Derivative spectrophotometric, thin layer chromatographic-densitometric and high performance liquid chromatographic determination of trifluoperazine hydrochloride in presence of its hydrogen peroxide induced-degradation product. *J. Pharm. Biomed. Anal.* 27(1–2): 9–18.

El-Sherif, Z., B. El-Zeany, O. El-Houssini. 2005. High performance liquid chromatographic and thin layer densitometric methods for the determination of risperidone in the presence of its degradation products in bulk powder and in tablets. *J. Pharm. Biomed. Anal.* 36(5): 975–981.

El-Sherif, Z., B. El-Zeany, O. El-Houssini, M. Rashed, H. Aboul-Enein. 2004. Stability indicating reversed-phase high-performance liquid chromatographic and thin layer densitometric methods for the determination of ziprasidone in bulk powder and in pharmaceutical formulations. *Biomed. Chromatogr.* 18(3): 143–149.

El Walily, A., A. El Gindy, M. Bedair. 1999. Application of first-derivative UV-spectrophotometry, TLC-densitometry and liquid chromatography for the simultaneous determination of mebeverine hydrochloride and sulpiride. *J. Pharm. Biomed. Anal.* 21(3): 535–548.

Fenimore, D., C. Davis, C. Meyer. 1978. Determination of drugs in plasma by high-performance thin-layer chromatography. *Clin. Chem.* 24(8): 1386–1392.

Franke, J., R. De Zeeuw, J. Wijsbeek. 1988. Impact of urine matrix and isolation procedure on retention behaviour of basic drugs in thin-layer chromatography. *J. Pharm. Biomed. Anal.* 6(4): 415–420.

Garcia Sanchez, F., A. Navas Diaz, M. Fernandez Correa. 1993. Image analysis of photochemically derivatized and charge-coupled device-detected phenothiazines separated by thin-layer chromatography. *J. Chromatogr. A* 655(1): 31–38.

Gondova, T., D. Halamova, K. Spacayova. 2008. Simultaneous analysis of new antidepressants by densitometric thin-layer chromatography. *J. Liq. Chromatogr. Relat. Technol.* 31(16): 2429–2441.

Gondova, T., I. Petrikova. 2010. Determination of new antidepressants in pharmaceuticals by thin-layer chromatography with densitometry. *J. AOAC Int.* 93(3): 778–782.

Gupta, K., M. Tajne, S. Wadodkar. 2010. A validated high-performance thin-layer chromatographic method for quantification of sertraline in tablets. *J. Planar Chromatogr. Mod. TLC* 23(2): 134–136.

Hawryl, A., D. Cichocki, M. Waksmundzka-Hajnos. 2008. Determination of the lipophilicity of some psychotropic drugs by RP-TLC. *J. Planar Chromatogr. Mod. TLC* 21(5): 343–348.

Jagadeeswaran, M., S. Mahibalan, N. Gopal. 2009. Estimation of fluoxetine in capsule dosage form by HPTLC method. *Int. J. Pharmacy Pharm. Sci.* 1(2): 71–73.

Joseph-Charles, J., M. Bertucat. 1993. Separation, identification and assay of fluradoline hydrochloride in the presence of other tricyclic antidepressants or neuroleptics by spectral and thin-layer and liquid chromatographic methods. *Anal. Chim. Acta* 284(1): 45–52.

Kazlauskiene, D., P. Vainauskas, D. Rakauskaite. 2003. Identification of the drugs in the mixture using thin-layer chromatography in sudden poisoning cases. *Medicina (Kaunas)* 39(Suppl 2): 132–136.

Krzek, J., A. Maślanka. 2000. Densitometric determination of impurities in pharmaceuticals. Part VI. Determination of 4,4-bis [4-(p-chlorophenyl)-4-hydroxypiperidino]butyrophenone in haloperidol. *Acta Pol. Pharm. Drug Res.* 57(1): 23–26.

Krzek, J., A. Maślanka. 2002. Densitometric determination of impurities in pharmaceuticals. Part VIII. Simultaneous determination of fluoxetine and (1RS)-3-methylamino-1-phenylpropan-1-ol, N-methyl-3-phenylpropan-1-amine, and C-1 as drug impurities. *J. Planar Chromatogr. Mod. TLC* 15(1): 50–55.

Mahadik, K., H. Aggarwal, N. Kaul. 2003. Simultaneous HPTLC estimation of trifluoperazine HCL, trihexyphenidyl HCL and chlorpromazine HCL in tablet dosage form. *Indian Drugs* 40(6): 340–344.

Maślanka, A., J. Krzek. 2005. Densitometric high performance thin-layer chromatography identification and quantitative analysis of psychotropic drugs. *J. AOAC Int.* 88(1): 70–79.

Maślanka, A., J. Krzek. 2007. Use of TLC with densitometric detection for determination of impurities in chlorpromazine hydrochloride, trifluoperazine dihydrochloride, promazine hydrochloride, and doxepin hydrochloride. *J. Planar Chromatogr. Mod. TLC* 20(6): 463–475.

Maślanka, A., J. Krzek, A. Patrzałek. 2009. Determination of risperidone in tablets in the presence of its degradation products and placebo-derived constituents. *Acta Pol. Pharm. Drug Res.* 66(5): 461–470.

Maślanka, A., J. Krzek, B. Zuromska, M. Stolarczyk. 2011. Identification and determination of compounds belonging to the group of OUN pharmaceutical agents by thin-layer chromatography with densitometric detection in biological material. *Acta Chromatogr.* 23(2): 247–266.

Mennickent, S., J. Contreras, C. Reyes, M. Vega, M. Diego. 2010a. Validated instrumental planar chromatographic method for quantification of fluphenazine hydrochloride in injections. *J. Planar Chromatogr. Mod. TLC* 23(1): 75–78.

Mennickent, S., R. Fierro, M. Vega, M. De Diego, C. Godoy. 2010b. Quantitative determination of fluoxetine in human serum by high performance thin layer chromatography. *J. Sep. Sci.* 33(14): 2206–2210.

Mennickent, S., L. Pino, M. Vega, M. de Diego. 2008. Chemical stability of haloperidol injection by high performance thin-layer chromatography. *J. Sep. Sci.* 31(1): 201–206.

Mennickent, S., L. Pino, M. Vega, C. Godoy, M. de Diego. 2007a. Quantitative determination of haloperidol in tablets by high performance thin-layer chromatography. *J. Sep. Sci.* 30(5): 772–777.

Mennickent, S., A. Sobarzo, M. Vega, C. Godoy, M. de Diego. 2007b. Quantitative determination of clozapine in serum by instrumental planar chromatography. *J. Sep. Sci.* 30(13): 2167–2172.

Meola, J., T. Rosano, T. Swift. 1981. Thin-layer chromatography, with fluorescence detection, of benzodiazepines and tricyclic antidepressants in serum from emergency-room patients. *Clin. Chem.* 27(7): 1254–1255.

Misztal, G., H. Hopkała, T. Sławik. 1997. Chromatographic analysis (TLC) of fluoxetine, doxepine, imipramine and opipramol in human plasma. *Acta Pol. Pharm. Drug Res.* 54(4): 257–259.

Misztal, G., R. Skibiński. 2001. Chromatographic analysis of new antidepressant drugs by normal- and reversed-phase TLC. *J. Planar Chromatogr. Mod. TLC* 14(4): 300–304.

Musumarra, G., G. Scarlata, G. Romano, S. Clementi. 1983. Identification of drugs by principal components analysis of R(f) data obtained by TLC in different eluent systems. *J. Anal. Toxicol.* 7(6): 286–292.

Naguib, I., M. Abdelkawy. 2010. Development and validation of stability indicating HPLC and HPTLC methods for determination of sulpiride and mebeverine hydrochloride in combination. *Eur. J. Med. Chem.* 45(9): 3719–3725.

Nakai, S., T. Kobayashi, T. Ezawa. 1989. Photodecomposition of moclobemide on a silica gel thin-layer chromatographic plate. *J. Chromatogr.* 479(2): 459–463.

Ojanpera, I., R.-L. Ojansivu, J. Nokua, E. Vuori. 1999. Comprehensive TLC drug screening in forensic toxicology: Comparison of findings in urine and liver. *J. Planar Chromatogr. Mod. TLC* 12(1): 38–41.

Overo, K. 1981. Fluorescence assay of citalopram and its metabolites in plasma by scanning densitometry of thin-layer chromatograms. *J. Chromatogr.* 224(3): 526–531.

Oztunc, A., A. Onal, S. Erturk. 2002. 7,7,8,8-Tetracyanoquinodimethane as a new derivatization reagent for high-performance liquid chromatography and thin-layer chromatography: Rapid screening of plasma for some antidepressants. *J. Chromatogr. B Biomed. Anal. Technol. Biomed. Life Sci.* 774(2): 149–155.

Patel, R., A. Patel, M. Patel, M. Shankar, K. Bhatt. 2009a. Estimation of alprazolam and sertraline in pure powder and tablet formulations by high-performance liquid chromatography and high-performance thin-layer chromatography. *Anal. Lett.* 42(11): 1588–1602.

Patel, R., B. Patel, M. Patel, K. Bhatt. 2010b. HPTLC method development and validation for analysis of risperidone in formulations, and in-vitro release study. *Acta Chromatogr.* 22(4): 549–567.

Patel, R., M. Patel, K. Bhari, B. Patel. 2010a. Development and validation of an HPTLC method for determination of olanzapine in formulations. *J. AOAC Int.* 93(3): 811–819.

Patel, R., M. Patel, M. Shankar, K. Bhatt. 2009b. Simultaneous determination of alprazolam and fluoxetine hydrochloride in tablet formulations by high-performance column liquid chromatography and high-performance thin-layer chromatography. *J. AOAC Int.* 92(4): 1082–1088.

Patel, S., R. Keshalkar, M. Patel. 2008. Stability-indicating HPTLC method for analysis of moclobemide, and use of the method to study degradation kinetics. *Chromatographia* 68(9–10): 855–859.

Patel, S., N. Patel. 2009a. TLC determination of amitriptyline HCl, trifluoperazine HCl, risperidone and alprazolam in pharmaceutical products. *Chromatographia* 69(3–4): 393–396.

Patel, S., N. Patel. 2009b. Simultaneous RP-HPLC and HPTLC estimation of fluoxetine hydrochloride and olanzapine in tablet dosage forms. *Indian J. Pharm. Sci.* 71(4): 477–480.

Patel, S., N. Patel. 2010. Simultaneous determination of imipramine hydrochloride and chlordiazepoxide in pharmaceutical preparations by spectrophotometric, rp-hplc, and hptlc methods. *J. AOAC Int.* 93(3): 904–910.

Petruczynik, A., M. Brończyk, T. Tuzimski, M. Waksmundzka-Hajnos. 2008. Analysis of selected anti-depressive drugs by high performance thin-layer chromatography. *J. Liq. Chromatogr. Relat. Technol.* 31(13): 1913–1924.

Ramesh, B., P. Narayana, A. Reddy, P. Devi. 2011. Stability-indicating HPTLC method for analysis of venlafaxine hydrochloride, and use of the method to study degradation kinetics. *J. Planar Chromatogr. Mod. TLC* 24(2): 160–165.

Rao, J., M. Kumar, L. Sathiyanarayanan, S. Yadav, V. Yadav. 2011. Application of a stability-indicating HPTLC method for quantitative analysis of sertraline hydrochloride in pharmaceutical dosage forms. *J. Planar Chromatogr. Mod. TLC* 24(2): 140–144.

Revanasiddappa, H., P. Ramappa. 1995. A new thin-layer chromatographic system for the identification of phenothiazine drugs. *Indian Drugs* 32(2): 73–77.

Saxena, V., Z. Zaheer, M. Farooqui. 2006. High-performance thin layer chromatography determination of olanzapine in pharmaceutical dosage form. *Asian J. Chem.* 18(2): 1212–1222.

Shah, C., N. Shah, B. Suhagia, N. Patel. 2007. Simultaneous assay of olanzapine and fluoxetine in tablets by column high-performance liquid chromatography and high-performance thin-layer chromatography. *J. AOAC Int.* 90(6): 1573–1578.

Shah, C., B. Suhagia, N. Shah, D. Patel, N. Patel. 2008. Stability-indicating simultaneous HPTLC method for olanzapine and fluoxetine in combined tablet dosage form. *Indian J. Pharm. Sci.* 70(2): 251–255.

Shirvi, V., K. Channabasavaraj, G. Kumar, T. Mani. 2010. HPTLC analysis of venlafaxine hydrochloride in the bulk drug and tablets. *J. Planar Chromatogr. Mod. TLC* 23(5): 369–372.

Skibiński, R., Ł. Komsta. 2010. Validation of NP-HPTLC and RP-HPTLC methods with videodensitometric detection for analysis of ziprasidone in pharmaceutical formulations. *J. Planar Chromatogr. Mod. TLC* 23(1): 23–27.

Skibiński, R., Ł. Komsta, H. Hopkała, I. Suchodolska. 2007a. Comparative validation of amisulpride determination in pharmaceuticals by several chromatographic, electrophoretic and spectrophotometric methods. *Anal. Chim. Acta* 590(2): 195–202.

Skibiński R., Ł. Komsta, I. Kosztyła. 2008. Comparative validation of Quetiapine determination in tablets by NP-HPTLC and RP-HPTLC with densitometric and videodensitometric detection. *J. Planar Chromatogr. Mod. TLC* 21: 289–294.

Skibiński, R., Ł. Komsta, G. Misztal. 2007b. The reversed-phase retention behavior of some atypical antipsychotic drugs. *J. Planar Chromatogr. Mod. TLC* 20(1): 75–80.

Skibiński, R., G. Misztal. 2004. Determination of fluvoxamine and moclobemide in tablets by densitometric and videodensitometric TLC. *J. Planar Chromatogr. Mod. TLC* 17(3): 224–228.

Skibiński, R., G. Misztal. 2005. Determination of citalopram in tablets by HPLC, densitometric HPTLC, and videodensitometric HPTLC methods. *J. Liq. Chromatogr. Relat. Technol.* 28(2): 313–324.

Skibiński, R., G. Misztal, Ł. Komsta, A. Korólczyk. 2006. The retention behavior of some atypical antipsychotic drugs in normal-phase TLC. *J. Planar Chromatogr. Mod. TLC* 19(107): 73–80.

Skibiński, R., G. Misztal, M. Kudrzycki. 2003. Determination of fluoxetine and paroxetine in pharmaceutical formulations by densitometric and videodensitometric TLC. *J. Planar Chromatogr. Mod. TLC* 16(1): 19–22.

Soliman, M., N. Mohamed, S. Khalile, H. Ibrahim. 2008. Determination of some antipsychotic drugs through spectrophotometric, TLC-densitometric and HPLC methods. *Egypt. J. Chem.* 51(5): 635–649.

Steinbrecher, K. 1983. Thin-layer chromatographic identification of phenothiazine derivative drugs. *J. Chromatogr.* 260(2): 463–470.

Venkatachalam, A., V. Chatterjee. 2007. Stability-indicating high performance thin layer chromatography determination of Paroxetine hydrochloride in bulk drug and pharmaceutical formulations. *Anal. Chim. Acta* 598(2): 312–317.

Venkateswarlu, K., R. Venisetty, N. Yellu, S. Keshetty, M. Pai. 2007. Development of HPTLC-UV absorption densitometry method for the analysis of alprazolam and sertraline in combination and its application in the evaluation of marketed preparations. *J. Chromatogr. Sci.* 45(8): 537–539.

Vujic, Z., D. Radulovic, B. Lucic, S. Eric, V. Kuntic. 2003. UV-densitometric determination of maprotiline, desipramine and moclobemide in pharmaceutical dosage forms. *Chromatographia* 57(9–10): 687–689.

Wójciak-Kosior, M., A. Skalska, A. Matysik. 2006. Determination of phenothiazine derivatives by high performance thin-layer chromatography combined with densitometry. *J. Pharm. Biomed. Anal.* 41(1): 286–289.

Zaheer, Z., M. Farooqui, S. Dhaneshwar. 2009. Stability-indicating high performance thin layer chromatographic determination of clozapine in tablet dosage form. *J. Pharm. Sci. Res.* 1(4): 158–166.

Zydek, G., E. Brzezińska. 2011. Normal and reversed phase thin layer chromatography data in quantitative structure-activity relationship study of compounds with affinity for serotonin (5-HT) receptors. *J. Chromatogr. B Anal. Technol. Biomed. Life Sci.* 879(20): 1764–1772.

16 TLC of Anxiolytics and Sedatives

Claudia Cimpoiu and Anamaria Hosu

CONTENTS

16.1 INFORMATION ABOUT THE DRUG GROUP

Benzodiazepines are a class of central nervous system (CNS) depressant drugs used as anxiolytic drugs, but also as sedato-hypnotics, anticonvulsants, muscle relaxants, or anesthetic drugs. The benzodiazepine group consists of 1,4-benzodiazepines (diazolobenzodiazepines), 1,5-benzodiazepines, triazolobenzodiazepines, imidazo-1,4-benzodiazepines, and thienodiazepines. This class of drugs has a wide range of both therapeutic potency and elimination half-lives and are the most widely prescribed drugs in the world. Since the discovery of benzodiazepines as anxiolytics in the 1960s, the classical structures of this class of compounds have been widely varied, resulting in benzodiazepine ligands that bind to specific subtypes of the GABAA receptors (Singh et al. 2010).

Barbiturates have been known since 1864 and are classified as CNS depressants as they are a relatively homogeneous group of synthetic drugs used as sedatives, hypnotics, anesthetics, and anticonvulsants in relatively high doses. They have been replaced by the benzodiazepines in many of the developed countries.

There are approximately 40 benzodiazepines and 20 barbiturates approved for clinical use throughout the world (Table 16.1). Often they occur as mixtures with other substances. These drugs may reduce the ability to drive a car or to work at machines, and they may lead to addiction or to severe intoxication (Maurer 1999). There are important reasons for the screening and quantification of these drugs in clinical and forensic toxicology because their abuse is prevalent.

Several presumptive tests are available for benzodiazepines and barbiturates, namely the Zimmerman test for benzodiazepines and the Dille–Koppanyi test for barbiturates, but they have the disadvantage that they do not discriminate between the drugs within the specific class (Cole 2003), so further analysis is required (Cole 2003). The separation of these drugs from each other or from their impurities or other compounds is of special interest due to the large number of compounds with very similar structures and chemical properties. One of the techniques used for this aim is thin-layer chromatography (TLC) together with its alternative high-performance thin-layer chromatography (HPTLC).

TABLE 16.1
The Characteristics of Anxiolytics and Sedatives

Drug	Structural Formula	Chemical Formula	Definition	Molecular Mass
Benzodiazepine				
Alprazolam		$C_{17}H_{13}ClN_4$	8-Chloro-1-methyl-6-phenyl-4*H*-[1,2,4]triazolo[4,3-*a*][1,4]-benzodiazepine	308.8
Bromazepam		$C_{14}H_{10}BrN_3O$	7-Bromo-5-(pyridin-2-yl)-1,3-dihydro-2*H*-1,4-benzodiazepin-2-one	316.2
Brotizolam		$C_{15}H_{10}BrClN_4S$	2-Bromo-4-(2-chlorophenyl)-9-methyl-6*H*-thieno[3,2-f][1,2,4]triazolo[4,3-a][1,4]diazepine	393.7

Camazepam

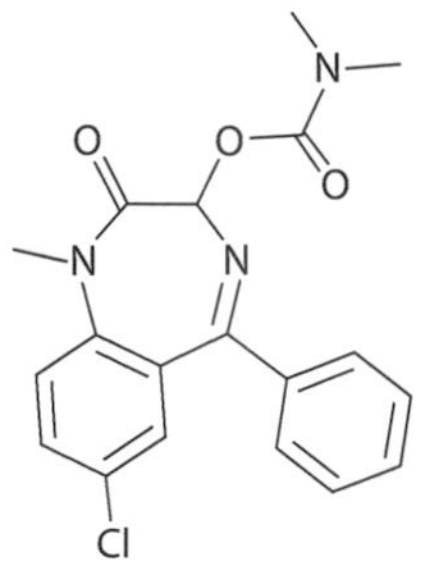

C$_{19}$H$_{18}$ClN$_3$O$_3$

(9-chloro- 2-methyl- 3-oxo- 6-phenyl- 2,5-diazabicyclo [5.4.0] undeca- 5,8,10,12-tetraen- 4-yl) N,N-dimethylcarbamate

371.8

Chlordesmethyldiazepam
(Delorazepam)

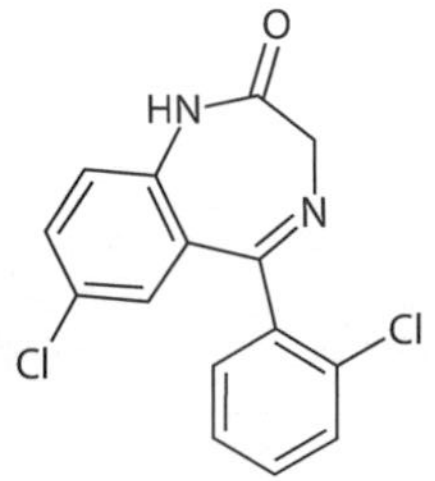

C$_{15}$H$_{10}$Cl$_2$N$_2$O

7-Chloro-5-(2-chlorophenyl)-1,3-dihydro-1, 4-benzodiazepin-2(2H)-one

304.02

Chlordiazepoxide

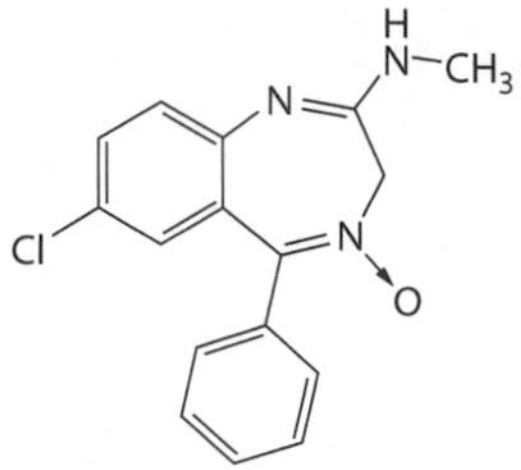

C$_{16}$H$_{14}$ClN$_3$O

7-Chloro-N-methyl-5-phenyl-3H-1,4-benzodiazepin-2-amine 4-oxide

299.8

Clobazam

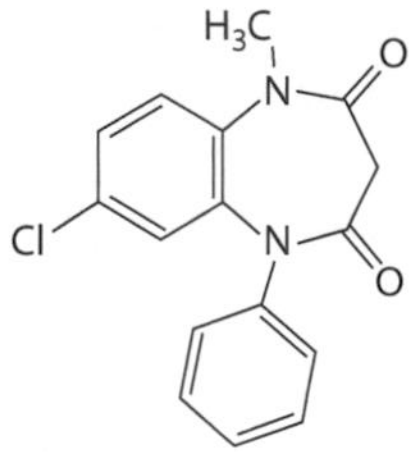

C$_{16}$H$_{13}$ClN$_2$O$_2$

7-Chloro-1-methyl-5-phenyl-1,5-dihydro-3H-1, 5-benzodiazepine-2,4-dione

300.7

(continued)

TABLE 16.1 (continued)
The Characteristics of Anxiolytics and Sedatives

Drug	Structural Formula	Chemical Formula	Definition	Molecular Mass
Clonazepam		$C_{15}H_{10}ClN_3O_3$	5-(2-Chlorophenyl)-7-nitro-1,3-dihydro-2H-1,4-benzodiazepin-2-one	315.7
Clorazepate		$C_{16}H_{11}ClN_2O_3$	7-chloro-2,3-dihydro-2-oxo-5-phenyl- 1H-1,4-benzodiazepine-3-carboxylic acid	314.72
Clotiazepam		$C_{16}H_{15}ClN_2OS$	2-(2-chlorophenyl)-9-ethyl-6-methyl-8-thia-3,6-diazabicyclo[5.3.0]deca-2,9,11-trien-5-one	318.8

Name	Structure	Molecular Formula	Chemical Name	MW
Cloxazolam		$C_{17}H_{14}Cl_2N_2O_2$	13-chloro-2-(2-chlorophenyl)-3-oxa-6,9-diazatricyclo[8.4.0.0^{2,6}] tetradeca-1(10),11,13-trien-8-one	349.2
Diazepam		$C_{16}H_{13}ClN_2O$	7-Chloro-1-methyl-5-phenyl-1,3-dihydro-2H-1,4-benzodiazepin-2-one	284.7
Estazolam		$C_{16}H_{11}ClN_4$	8-Chloro-6-phenyl-4H-1,2,4-triazolo(4,3-a)-1,4-benzodiazepine	294.7
Etizolam		$C_{17}H_{15}ClN_4S$	7-(2-chlorophenyl)- 4-ethyl- 13-methyl- 3-thia- 1,8,11,12-tetraazatricyclo [8.3.0.0^{2,6}] trideca- 2(6),4,7,10,12- pentaene	342.7

(continued)

TABLE 16.1 (continued)
The Characteristics of Anxiolytics and Sedatives

Drug	Structural Formula	Chemical Formula	Definition	Molecular Mass
Fludiazepam		$C_{16}H_{12}ClFN_2$	9-chloro-6-(2-fluorophenyl)- 2-methyl-2,5-diazabicyclo [5.4.0] undeca- 5,8,10,12-tetraen-3-one	302.7
Flumazenil		$C_{15}H_{14}FN_3O_3$	Ethyl 8-fluoro-5-methyl-6-oxo-5,6-dihydro-4H-imidazo[1,5-a] [1,4]benzodiazepine-3-carboxylate	303.3
Flunitrazepam		$C_{16}H_{12}FN_3O_3$	5-(2-Fluorophenyl)-1-methyl-7-nitro-1,3-dihydro-2H-1, 4-benzodiazepin-2-one	313.3

Flurazepam	$C_{21}H_{23}ClFN_3 \cdot HCl$	7-Chloro-1-[2-(diethylamino)ethyl]-5-(2-fluorophenyl)-1,3-dihydro-2H-1,4-benzodiazepin-2-one monohydrochloride	424.3
Halazepam	$C_{17}H_{12}ClF_3N_2O$	7-chloro-5-phenyl-1-(2,2,2-trifluoroethyl)-1,3-dihydro-2H-1,4-benzodiazepin-2-one	352.7
Haloxazolam	$C_{17}H_{14}BrFN_2O_2$	13-bromo-2-(2-fluorophenyl)-3-oxa-6,9-diazatricyclo[8.4.0.0^{2,6}] tetradeca-1(10),11,13-trien-8-one	377.2
Ketazolam	$C_{20}H_{17}ClN_2O_3$	11-chloro-8,12b-dihydro-2,8-dimethyl-12b-phenyl-4H-[1,3] oxazino[3,2-d][1,4]benzodiazepine-4,7(6H)-dione	368.8

(continued)

TABLE 16.1 (continued)
The Characteristics of Anxiolytics and Sedatives

Drug	Structural Formula	Chemical Formula	Definition	Molecular Mass
Loprazolam		$C_{23}H_{21}ClN_6O_3$; CH_4O_3S, H_2O	(Z)-6-(2-chlorophenyl)-2,4-dihydro-2-(4-methylpiperazin-1-ylmethylene)-8-nitroimidazo[1,2-a][1,4]benzodiazepin-1-one methanesulphonate monohydrate	579.1
Lormethazepam		$C_{16}H_{12}Cl_2N_2O_2$	(RS)-7-chloro-5-(2-chlorophenyl)-1,3-dihydro-3-hydroxy-1-methyl-1,4-benzodiazepin-2-one	335.2
Lorazepam		$C_{15}H_{10}Cl_2N_2O_2$	(3RS)-7-Chloro-5-(2-chlorophenyl)-3-hydroxy-1,3-dihydro-2H-1,4-benzodiazepin-2-one	321.2

Medazepam	C$_{16}$H$_{15}$ClN$_2$	9-chloro-2-methyl-6-phenyl-2,5-diazabicyclo[5.4.0] undeca-5,8,10,12-tetraene	270.8
Midazolam	C$_{18}$H$_{13}$ClFN$_3$	8-Chloro-6-(2-fluorophenyl)-1-methyl-4*H*-imidazo[1,5-*a*][1,4] benzodiazepine	325.8
Nitrazepam	C$_{15}$H$_{11}$N$_3$O$_3$	7-nitro-5-phenyl-1,3-dihydro-2*H*-1,4-benzodiazepin-2-one	281.3
Nordazepam	C$_{15}$H$_{11}$ClN$_2$O	7-chloro-5-phenyl-1,3-dihydro-2*H*-1,4-benzodiazepin-2-one	270.7

(continued)

TABLE 16.1 (continued)
The Characteristics of Anxiolytics and Sedatives

Drug	Structural Formula	Chemical Formula	Definition	Molecular Mass
Oxazepam		$C_{15}H_{11}ClN_2O_2$	(3RS)-7-Chloro-3-hydroxy-5-phenyl-1,3-dihydro-2H-1, 4-benzodiazepin-2-one	286.7
Oxazolam		$C_{18}H_{17}ClN_2O_2$	(2R,11bS)-10-chloro-2-methyl-11b-phenyl-2,3,5, 7-tetrahydro-[1,3]oxazolo[3,2-d][1,4] benzodiazepin-6-one	328.8
Phenazepam		$C_{15}N_2H_{10}OBrCl$	7-Bromo-5-(2-chlorophenyl)-1,3-dihydro-2H-1, 4-benzodiazepin-2-one	349.6

Pinazepam	$C_{18}H_{13}ClN_2O$	7-Chloro-5-phenyl-1-prop-2-yn-1-yl-1,3-dihydro-2H-1,4-benzodiazepin-2-one	308.8
Prazepam	$C_{19}H_{17}ClN_2O$	7-Chloro-1-(cyclopropylmethyl)-5-phenyl-1,3-dihydro-2H-1,4-benzodiazepin-2-one	324.8
Quazepam	$C_{17}H_{11}ClF_4N_2S$	7-Chloro-5-(2-fluorophenyl)-1-(2,2,2-trifluoroethyl)-1,3-dihydro-2H-1,4-benzodiazepin-2-thione	386.8
Temazepam	$C_{16}H_{13}ClN_2O_2$	(3RS)-7-Chloro-3-hydroxy-1-methyl-5-phenyl-1,3-dihydro-2H-1,4-benzodiazepin-2-one	300.7

(continued)

TABLE 16.1 (continued)
The Characteristics of Anxiolytics and Sedatives

Drug	Structural Formula	Chemical Formula	Definition	Molecular Mass
Tofisopam		$C_{22}H_{26}N_2O_4$	1-(3,4-Dimethoxyphenyl)-5-ethyl-7,8-dimethoxy-4-methyl-5H-2,3-benzodiazepine	382.5
Triazolam		$C_{17}H_{12}Cl_2N_4$	8-Chloro-6-(2-chlorophenyl)-1-methyl-4H-[1,2,4]triazolo[4,3-a][1,4]benzodiazepine	343.2
5HT1A agonists Buspirone		$C_{21}H_{31}N_5O_2$	8-[4-(4-Pyrimidin-2-ylpiperazin-1-yl)butyl]-8-azaspiro[4.5]decane-7,9-dione	385.5

Gepirone		$C_{19}H_{29}N_5O_2$	4,4-Dimethyl-1-[4-(4-(4-pyrimidin-2-ylpiperazin-1-yl)butyl] piperidine-2,6-dione	359.5
Ipsapirone		$C_{19}H_{23}N_5O_3S$	9,9-Dioxo-8-[4-(4-(4-pyrimidin-2-ylpiperazin-1-yl)butyl]-$9\lambda^6$-thia-8-azabicyclo[4.3.0]nona-1,3,5-trien-7-one	401.5
Zalospirone		$C_{24}H_{29}N_5O_2$	(3aα,4α,4aβ,6aβ,7α,7aα)-hexahydro-2-(4-(4-(2-pyrimidinyl)-1-piperazinyl)butyl)-4,7-etheno-1*H*-cyclobut(f) isoindole-1,3(2*H*)-dione	419.5

(continued)

TABLE 16.1 (continued)
The Characteristics of Anxiolytics and Sedatives

Drug	Structural Formula	Chemical Formula	Definition	Molecular Mass
Other anxiolytics				
Barbiturates				
Allobarbital		$C_{10}H_{12}N_2O_3$	5,5-Diprop-2-enyl-1,3-diazinane-2,4,6-trione	208.2
Amobarbital		$C_{11}H_{18}N_2O_3$	5-Ethyl-5-(3-methylbutyl) pyrimidin-2,4,6(1*H*,3*H*,5*H*)-trione	248.3
Amobarbital sodium		$C_{11}H_{17}N_2NaO_3$	Sodium derivative of 5-ethyl-5-(3-methylbutyl) pyrimidin-2,4,6(1*H*,3*H*,5*H*)-trione	226.3

Name	Structure	Formula	Chemical name	MW
Aprobarbital		$C_{10}H_{14}N_2O_3$	5-Propan-2-yl-5-prop-2-enyl-1,3-diazinane-2,4,6-trione	210.2
Barbital		$C_8H_{12}N_2O_3$	5,5-Diethylpyrimidine-2,4,6(1H,3H,5H)-trione	184.2
Butobarbital		$C_{10}H_{16}N_2O_3$	5-Butyl-5-ethyl-1,3-diazinane-2,4,6-trione	212.3
Cyclobarbital		$C_{12}H_{16}N_2O_3$	5-(1-Cyclohexenyl)-5-ethyl-1,3-diazinane-2,4,6-trione	236.3
Heptobarbital		$C_{11}H_{10}N_2O_3$	5-Methyl-5-phenylpyrimidine-2,4,6(1H,3H,5H)-trione	218.2

(continued)

TABLE 16.1 (continued)
The Characteristics of Anxiolytics and Sedatives

Drug	Structural Formula	Chemical Formula	Definition	Molecular Mass
Hexobarbital		$C_{12}H_{16}N_2O_3$	(5RS)-5-(cyclohex-1-enyl)-1, 5-dimethylpyrimidine-2,4,6(1H,3H,5H)-trione	236.3
Metharbital		$C_9H_{14}N_2O_3$	5,5-Diethyl-1-methylpyrimidine-2,4,6(1H,3H,5H)-trione	192.2
Methohexital		$C_{14}H_{18}N_2O_3$	5-hex-3-yn-2-yl-1-methyl-5-prop-2-enyl-1,3-diazinane-2, 4,6-trione	262.3
Methylphenobarbital (Mephobarbital)		$C_{13}H_{14}N_2O_3$	(5RS)-5-ethyl-1-methyl-5-phenylpyrimidine-2, 4,6(1H,3H,5H)-trione	246.3

Name	Structure	Molecular formula	Chemical name	MW
Pentobarbital		$C_{11}H_{18}N_2O_3$	5-Ethyl-5-[(1RS)-1-methylbutyl]pyrimidine-2,4,6(1H,3H,5H)-trione	226.3
Pentobarbital Sodium		$C_{11}H_{17}N_2NaO_3$	Sodium derivative of 5-ethyl-5-[(1RS)-1-methylbutyl]pyrimidine-2,4,6(1H,3H,5H)-trione	248.3
Phenobarbital		$C_{12}H_{12}N_2O_3$	5-Ethyl-5-phenylpyrimidine-2,4,6(1H,3H,5H)-trione	232.2
Phenobarbital sodium		$C_{12}H_{11}N_2NaO_3$	Sodium derivative of 5-ethyl-5-phenylpyrimidine-2,4,6(1H,3H,5H)-trione	254.2
Primidone		$C_{12}H_{14}N_2O_2$	5-Ethyl-5-phenyldihydropyrimidine-4,6(1H,5H)-dione	218.3

(continued)

TABLE 16.1 (continued)
The Characteristics of Anxiolytics and Sedatives

Drug	Structural Formula	Chemical Formula	Definition	Molecular Mass
Secbutabarbital (Butabarbital)		$C_{10}H_{16}N_2O_3$	5-butan-2-yl-5-ethyl-1,3-diazinane-2,4,6-trione	212.2
Secobarbital		$C_{12}H_{18}N_2O_3$	5-[(2R)-pentan-2-yl]-5-prop-2-enyl-1,3-diazinane-2,4,6-trione	238.3
Thiopental sodium		$C_{11}H_{17}N_2NaO_2S$	Sodium derivative of 5-ethyl-5-[(1RS)-1-methylbutyl]-2-thioxo-2,3-dihydropyrimidine-4,6(1H,5H)-dione	264.3

Vinylbarbital		$C_{11}H_{16}N_2O_3$	5-Ethyl-5-[(1E)-1-methylbut-1-en-1-yl] pyrimidine-2,4,6(1H,3H,5H)-trione	224.3
Vinylbital		$C_{11}H_{16}N_2O_3$	5-(1-Methylbutyl)-5-vinylpyrimidine-2,4,6(1H,3H,5H)-trione	224.3
Other sedatives Zolpidem		$C_{42}H_{48}N_6O_8$	Bis[N,N-dimethyl-2-[6-methyl-2-(4-methylphenyl)imidazo[1,2-a] pyridin-3-yl]acetamide] (2R,3R)-2,3-dihydroxybutanedioate	307.4
Zopiclone		$C_{17}H_{17}ClN_6O_3$	(5RS)-6-(5-Chloropyridin-2-yl)-7-oxo-6,7-dihydro-5H- pyrrolo[3,4-b]pyrazin-5-yl 4-methylpiperazine-1-carboxylate	388.8

TLC/HPTLC is still used for large-scale multiple drug screening programs. Advantages of TLC/HPTLC over other methods are its simplicity, rapidity, versatility, applicability to a large number of samples, cost-efficacy of operation, and ability to screen multiple drugs simultaneously. Moreover, this method detects current use of drugs rather than of the past (Jain 2000). These techniques are widely used in clinical toxicology laboratories for qualitative screening of tissue and fluid samples (Beesley 2010). Toxicological drug screening by TLC/HPTLC is just turning into an instrumental technique, providing new challenges and possibilities for all stages of process (Ojanpera 1992). Furthermore, screening procedures for the detection of toxicologically relevant substances have become of ever-increasing importance due to the rapid development of new substances. Also, TLC/HPTLC is used in pharmacopoeias for the separation of a particular drug from its impurities or related compounds, but the methods described are less suitable for identification purposes.

This chapter intends to present a retrospective view regarding the use of TLC in anxiolytics and sedatives, emphasizing the modern and instrumental aspects without claiming to be exhaustive.

16.2 POSSIBLE DEGRADATION COMPOUNDS AND METABOLITES

Benzodiazepines are relatively unstable substances easily hydrolyzed in acidic solution and also decomposed in UV light. Hydrolysis in acidic solution leads generally to 2-aminobenzophenone derivates, through the split of the N1-C2 bond of the diazepinic ring. TLC of benzophenones obtained by acid hydrolysis of benzodiazepines derivates is widely used for identification purposes, but this method is not specific because different compounds can give the same benzophenone derivate and others (e.g., alprazolam) do not form benzophenones (Schütz 1996). The assessment of the degradation of benzodiazepines can be done after in situ acidic hydrolysis on a chromatographic plate (Hancu et al. 2011). Dilute sulfuric acid (10%) was placed over each spot of benzodiazepines, and then the plate was covered with a glass plate and kept for 15 min in an oven at 120°C. The plate was cooled to room temperature and ammonia solution (25%) was applied on each spot, and the spots were kept at 120°C for 5 min. The plate was then developed with ethyl acetate–methanol–conc. ammonia 17:2:1 v/v/v in chromatographic chambers pre-saturated with the mobile phase for 30 min. Thereafter, the plates were examined first under UV light and then sprayed with Dragendorff reagent.

Metabolism of 1,4-benzodiazepines generally leads to oxazepam and oxazepam glucuronide, usually via nordazepam or to compounds analogous to oxazepam but differing mainly at the 5-aryl position (Drummer 1998). Benzodiazepines with a hydroxy group are directly metabolized to the appropriate glucuronides, and those with a nitro group are reduced to an amino group, which reacts to form glucuronides, as also do the amino groups of 7-aminoflunitrazepam and 7-aminonitrazepam (Drummer 1998). Benzodiazepines with a structure of 1,3-diazole ring at the 1,2 position are rapidly metabolized with the formation of hydroxyl groups in the 3-position of the benzodiazepine ring and on the diazole ring, with subsequent glucuronidation (Drummer 1998).

The barbiturates are also unstable and degraded by hydrolysis. Due to the presence of one (or more) acidic protons, barbiturates can be converted to water-soluble salt forms. Barbiturates are hepatically metabolized (except phenobarbital), and these metabolites are mostly inactive, water-soluble, and excreted in the urine. Only very small amounts of barbiturates are excreted unchanged by the kidney. Oxidation is the most significant metabolic pathway, and the metabolites are readily excreted in the urine or conjugated with gluconuride and excreted in bile. Through secondary hepatic metabolic inactivation, barbiturates lose their affinity for the GABA receptor complex and thus CNS depressant activity.

The degradation products and metabolites of some anxiolytics and sedatives are presented in Table 16.2.

TABLE 16.2
Metabolites and Degradation Products of Some Benzodiazepines and Barbiturates

Compound	Metabolites	Compound	Metabolites
Anxyolitics			
Alprazolam	α-Hydroxyalprazolam	Bromazepam	3-OH ABBP
Brotizolam	α-Hydroxybrotizolam	Camazepam	Oxazepam
	4- Hydroxybrotizolam		Temazepam
Chlordesmethyldiazepam	Lorazepam	Chlordiazepoxide	Desmethyldiazepam
(Delorazepam)			Oxazepam
			Nordiazepam
Clobazam	N-desmethyl-clobazam	Clonazepam	7-Aminoclonazepam
	4'-Hydroxyclobazam		7-Acetaminoclonazepam
			3-Hydroxy clonazepam
Clorazepate	Oxazepam	Clotiazepam	Desmethylclotiazepam
	Nordiazepam		Y-10247
Cloxazolam	Delorazepam	Diazepam	Oxazepam Desmethyldiazepam
			Temazepam
			Nordiazepam
			4'-Hydroxydiazepam
Estazolam	M-II (Estazolam)	Etizolam	8-Hydroxyetizolam
	M-IV (Estazolam)		
Fludiazepam	Desmethylfludiazepam	Flumazenil	[11C]-flumazenil acid
Flunitrazepam	7-Aminoflunitrazepam	Flurazepam	1-Ethanolflurazepam
	7-Acetamidoflunitrazepam		Desalkylflurazepam
	3-Hydroxyflunitrazepam		
Halazepam	Desmethyldiazepam	Haloxazolam	RAZ-609
	Oxazepam		
	3-Hydroxyhalazepam		
Ketazolam	Oxazepam	Loprazolam	Piperazine N-oxide
Lormetazepam	Lorazepam	Lorazepam	Lorazepam glucuronide
Medazepam	Desmethyldiazepam	Midazolam	α-Hydroxymidazolam
	Oxazepam		
	Nordiazepam		
Nitrazepam	Desmethylnitrazepam	Nordazepam	Oxazepam
	7-Aminonitrazepam		
	7-acetamidonitrazepam		
Oxazolam	Desmethyldiazepam	Phenazepam	3-Hydroxy-phenazepam
	Oxazepam		5-Bromo-(2-chlorophenyl)-2-aminobenzophenone 6-Bromo-(2-chlorophenyl) quinazoline-2-one
Pinazepam	Nordiazepam	Prazepam	3-Hydroxyprazepam
			Oxazepam
			Nordiazepam
Quazepam	N-Desalkyl-2-oxoquazepam	Temazepam	Oxazepam
Tofisopam	Desmethyltofisopam	Triazolam	1-Hydroxymethyltriazolam
Buspirone	6-Hydroxybuspirone	Gepirone	1-(2-Pyrimidinyl)-piperazine
Ipsapirone	1-(2-Pyrimidinyl)-piperazine	Zalospirone	1-(2-Pyrimidinyl)-piperazine
Sedatives			
Allobarbital	Hydroxybarbital	Amobarbital	3'-Hydroxyamobarbital
	Allodiol		1-(β-D-glucopyranosyl)amobarbital

(*continued*)

TABLE 16.2 (continued)
Metabolites and Degradation Products of Some Benzodiazepines and Barbiturates

Compound	Metabolites	Compound	Metabolites
Aprobarbital	N-Hydroxyaprobarbitone	Barbital	Barbital N-glucoside
Butobarbital	Phenobarbital 5-Ethyl-5-(1-methyl-2-carboxyethyl) barbituric acid	Cyclobarbital	Cyclohexenoyl-ethylbarbituric acid
Heptobarbital	3'-Hydroxyheptobarbital 3'-Oxoheptobarbital	Hexobarbital	3'-Hydroxyhexobarbital 3'-Oxohexobarbital
Metharbital	Barbital	Methohexital	4'-hydroxymethohexital
Methylphenobarbital	p-Hydroxyphenobarbital Phenobarbital p-Hydroxymethylphenobarbital	Pentobarbital	Ethyl(3-hydroxy-1-methylbutyl) barbituric acid
Phenobarbital	p-Hydroxyphenobarbital 2-Ethyl-2-phenylmalonamide	Primidone	Phenobarbital Phenylethylmalonamide
Secbutabarbital	5-Ethyl-5-(1-methyl-2-carboxyethyl) barbituric acid	Secobarbital	Hydroxysecobarbital Secodiol
Thiopental	Pentobarbital	Zolazepam	1,8-Desmethylzolazepam
Zolpidem	4-[3-(2-N,N-dimethylamino-2-oxoethyl)-6-yl)-6-methylimidazo [1,2-a]pyridin-2-yl]benzoic acid	Zolpiclone	Desmethylzopiclone

16.3 SAMPLE PREPARATION IN CONTEXT OF MATRIX

In the case of pharmaceutical preparations, the extraction is simple and generally implies dissolving in a solvent and then centrifugation or filtration for removing any solid material. Benzodiazepines can easily be extracted into methanol; while barbiturates can be dissolved without difficulty in methanol, even they are in the free acid or in salt forms or in ethyl acetate if they are in the free acid form or are converted to this form. Stability testing of barbiturates in aqueous solutions revealed a lower degradation rate between pH 8.0 and 9.5 than at higher pH-values (Thoma and Struve 1985). For 2,3-benzodiazepines, methanol, chloroform, dioxane, dichloromethane, and ethyl acetate can also be used leading to recoveries well over 60% (Rizzo 2000).

These drugs can be extracted from various biological samples by liquid–liquid extraction (LLE) or solid-phase extraction (SPE) (Gal-Szabo and Perneczki 1991). In the 1960s, LLE was the routine technique, but this required a great volume of samples and solvents and was time-consuming, and the assays were subjected to many interferences (Pippenger 1989). LLE was replaced in many cases by SPE, which was easier, reduced the solvent volumes required, and improved the reproducibility (Gal-Szabo and Perneczki 1991, Edler and Schlüter 1994), but is more expensive. Regardless of the extraction method used, the samples should be spiked with internal standard solution.

The extraction of benzodiazepines from biological samples uses the following solvents: n-butyl chloride (Spangenberg et al. 2005), dichloromethane, hexane–dichloromethane 70:30 v/v, n-butyl chloride–ethyl acetate 1:4 v/v, ethly ether and ethyl acetate (International Drug Control Programme 1997), and n-hexane–ethyl acetate 75:25 v/v (Otsubo et al. 1995). Urine samples should be hydrolyzed before extraction.

Different solvents (hexane, ethyl ether, toluene, n-butyl chloride, and chloroform) were tested for their extraction efficiencies of barbiturates from urine and plasma (Bailey and Kelner 1984). The results indicated that n-butyl chloride, chloroform, and ethyl ether gave adequate recoveries, while hexane and toluene gave relatively poor recoveries. The literature also indicates dichloromethane (Mangin et al. 1987), hexane–ethyl acetate (6:4 v/v) (Mule and Casella 1989), and toluene–ethyl acetate (4:1 v/v) (Lillsunde et al. 1996) for the extraction of barbiturates from urine. It must be mentioned that the hydrolysis step is not required for the analysis of barbiturates in urine, but the pH must be adjusted to acidic conditions by the addition of 1 M sulfuric acid, phosphoric acid, tartaric acid (Baselt 1987), or phosphate buffer (Mangin et al. 1987, Mule and Casella 1989). In the case of barbiturates, analysis from plasma solvents used chloroform ethyl ether and n-butyl chloride, and the pH is adjusted with phosphate buffer (McElwee 1979, Baselt 1987). The barbiturates from blood are extracted using chloroform (Budd and Mathis 1982) or ethyl acetate (Chan and Chan 1984). The extract evaporated to dryness can be purified by partition between hexane and acetonitrile and by removing the hexane layer (Logan and Stafford 1989).

In the case of SPE, various adsorbents were used depending on the type of samples as follows: *benzodiazepines*—urine samples—bonded silica (Casas et al. 1993), diatomaceous earth (Edler and Schlüter 1994), Amberlite XAD-2 Resin (Fujimoto and Wang 1970, Sawada et al. 1976); blood samples—different non-polar bonded phases (Mußhoff and Daldrup 1992, Casas et al. 1993); *barbiturates*—urine and blood sample—bonded silica (Chen et al. 1992, Pocci et al. 1992); and diatomaceous earth (Logan and Stafford 1989, Ferrara et al. 1992).

16.4 CHROMATOGRAPHIC SYSTEMS USED IN SEPARATION AND/OR QUANTITATION OF PARTICULAR DRUGS

Many studies on the TLC or HPTLC of anxiolytic and sedative drugs have been reported in the literature. The chromatographic system used silica gel plates with or without a fluorescence indicator, but in some cases other stationary phases were tested, namely cellulose and polyamide (Sánchez-Moyano et al. 1986, Csehárti and Hauck 1990), RP-18 (Kastner and Klimes 1996, Sumina et al. 2002, Cheng et al. 2009), RP-2 (Żydek and Brzezińska 2011), and aluminum oxide (Paw and Misztal 2000). Although using even CN, NH_2, and diol or other precoated HPTLC plates leads to increase in separating power, the best stationary phase for separation of these drugs is silica gel.

Different mobile phases were tested or, in some cases, even a combination of mobile phases was used to resolve many of these drugs. The mobile-phase systems which have been used for benzodiazepine separation include *n*-propanol–water 70:30 v/v (Haefelfinger 1970); chloroform–methyl acetate–methanol 70:25:5 v/v/v (Steidinger and Schmid 1970); benzene–isopropanol–25% NH_3 v/v/v (Ebel and Schütz 1977); chloroform–methanol–ethylic ether 85:15:10 v/v/v (Howarth and Clegg 1978); dichlorethan–methanol–water 95:5:0.2 v/v/v (Battista et al. 1979); heptane–ethyl acetate–iodine in methanol–etanol–NH_3 5:5:1:0.3 v/v/v/v/v (Šoljić et al. 1977); toluene (Schuetz et al. 1983, Schütz and Fitz 1985); benzene–ether–methanol 15:20:0.7 v/v/v (Li and Wang 1985); chloroform–ethyl acetate–ethanol–NH_3 81:5.5:11:2 v/v/v/v (St-Pierre and Pang 1987); toluene–ethylic ether–methanol 15:20:0.7 v/v/v (De Souza and Hamon 1990); ethyl acetate–toluene–chloroform–acetone 7:5:1:2 v/v/v/v (Gal-Szabo and Perneczki 1991); ethyl acetate–ethanol–NH_3 100:10:3 v/v/v (Rochholz et al. 1994); chloroform–acetone–NH_3 60:30:5 v/v/v (Edler and Schlüter 1994); dichloromethane–acetone 12:1 v/v (Rochholz et al. 1994); upper layer of n-butanol–acetic acid–water 4:1:2 v/v/v (Farina et al. 1996); isopropanol–butanone–NH_3 50:50:1 v/v/v (Farina et al. 1996); isopropanol–acetone–NH_3 70:30:1 v/v/v (Farina et al. 1996); chloroform–ethanol 9:1 v/v (Kamble and Dongre 1997); optimized composition of chloroform–acetone–isopropanol mixture: 84:13:3 v/v/v (Cimpoiu et al. 1997), 86:13:1 v/v/v (Cimpoiu et al. 1998), 80:15:5v/v/v (Cimpoiu and Hodisan 1999) and 73:1:26 v/v/v (Cimpoiu et al. 1999); chloroform–acetone 85:15 v/v (Rochholz et al. 1994), 4:19 v/v (Tames et al. 2000) and 4:1 v/v (Dongre et al. 2000); butanone–toluene 3:2 v/v (Paw and Misztal 2000); chloroform–benzene–ether–tetrahydrofuran–acetone–acetic acid 35:15:16:10:5:3 v/v/v/v/v/v (Saelzer et al. 2001); toluene–1,4-dioxane–triethylamine 12:7:1 v/v/v (Paw and Misztal 2000); cyclohexane–ethyl acetate 1:1 v/v (Paw and Misztal 2000); cyclohexane–toluene–diethylamine 75:15:10 v/v/v (Cole 2003); chloroform–methanol 9:1 v/v (Cheng et al. 2009); and n-hexane–acetone–methanol 16:6:1 (Mali and Garad 2005).

In the case of barbiturates, the mobile-phase systems used in TLC separations are chloroform–acetic acid 500:1 v/v (Roering et al. 1975); isopropanol–chloroform–23% NH_3 45:45:10v/v/v (Thoma and Struve 1985, Maurer 1999); chloroform–ethanol 9:1 v/v (Volford et al. 1996); ethyl acetate–methanol–25% ammonia 85:10:5 v/v/v (Cole 2003); methanol–triethylamine 39:1 v/v (Pushpalatha et al. 2009); acetonitrile–triethylamine 39:1 v/v (Pushpalatha et al. 2009); chloroform–methanol–triethylamine 38:2:1 v/v/v ((Pushpalatha et al. 2009); and acetonitrile–methanol–triethylamine 34:4:1 v/v/v (Pushpalatha et al. 2009).

After development of the plate, the target compounds were detected either in UV light at characteristic wavelengths or in visible light after derivatization with common specific reagents. In the case of benzodiazepines, the principal difficulty is the fact that there is no specific reagent, the most common being 1 M H_2SO_4. After heating, the sprayed plates were visualized under UV light and then were sprayed with acidified potassium iodoplatinate reagent, when purple spots were formed (Cole 2003) or with 10% ammonium chloride solution and then, after 5 min, diazotized with sodium nitrite reagent and treated with N-(1-naphthylethylenediamine) in ethanol when intense violet spots will appear (Dongre et al. 2000). The Bratton-Marshall reagent (N-(1-naphthyl)ethylenediamine in water:acetone-8.7:2 v/v) is the most common and very sensitive detection reagent for benzophenone, the benzodiazepines hydrolysis product, when the substituent is a secondary amino group. The detection of barbiturates is best achieved by derivatization with a freshly prepared mixture of mercuric chloride (0.1 g in 50 mL of ethanol) and diphenylcarbazone (0.1 g in 50 mL of methanol). The barbiturates appear as blue–violet spots on a pink background. One of the major inconveniences is the high toxicity of mercuric compounds, and consequently special safety procedures are needed to protect human health, to avoid potential contamination of the working area, and for waste disposal (Cole 2003). An alternative is the plate exposure to concentrated ammonia vapors and the observation under UV light at 254 nm (Moffat 1986).

In many cases, the visualization reaction gives same color spots for compounds, which make their identification from samples difficult. For this reason, finding a specific reagent for target compounds is tried.

A color reaction of some benzodiazepines is obtained after spraying the TLC plate with 0.1% bromine solution in carbon tetrachloride, and after 5 min spraying with 0.1% o-toluidine (in 0.5% acetic acid) reagent (Kulkarni et al. 1998). The obtained colored spots are due to the oxidation product of o-toluidine. The detection of certain benzodiazepines (nitrazepam, diazepam, medazepam, and chlordiazepoxide) can be done by spraying with stannous chloride in 1:1 hydrochloric acid, heating at 100°C for 10 min, and, after cooling, spraying with 10% ammonium chloride solution and then, after 5 min, with sodium nitrite reagent followed by 0.1% Griess reagent (0.2% naphthylene-diamine dihydrochloride, and 2% sulfanilamide in 5% phosphoricacid) (Patil and Shingare 1993). Another detection method for some benzodiazepines (oxazepam, diazepam, nitrazepam, chlordiazepoxide, and medazepam) consists of spraying with 2% ethanolic dinitrobenzene solution and after drying with 10% ethanolic sodium hydroxide solution, by exposure to HCl vapor, by placing the plate into a cooled TLC chamber with nitrous gases, or by spraying with Dragendorff's reagent (Kastner and Klimes 1996). Some benzodiazepines (alprazolam, diazepam, lorazepam, nitrazepam, and oxazepam) can be detected in the presence of phenobarbital and other compounds (heroin, opium, procaine hydrochloride, mandrax, and caffeine) by spraying with 1% mercuric chloride–1% potassium ferricyanide 1:1 followed by heating at 80°C for about 5 min (Kamble and Dongre 1997). The Bratton-Marshall reagent can also be used for detection of fourteen 1,4-benzodiazepines (diazepam, bromazepam, dipotassium chlorazepate, chlordiazepoxide, clonazepam, flurazepam fluni-trazepam, nitrazepam, medazepam, lormetazepam, oxazepam, prazepam, temazepam, tofisopam; clobazam, midazolam, triazolam, alprazolam, and clotiazepam) after thermal treatment, which gives color reactions with target compounds (Volf 1998). In addition, this reagent can be used for detection of diazepam, desmethyldiazepam, temazepam, and oxazepam after diazotization with $NaNO_2$ and HCl (Schuetz et al. 1983) or after photolytic desalkylation and diazotation (Schütz and Fitz 1985). High sensitivity of specific chromogenic reagent for detection of diazepam (~5 µg/spot) in the presence of oxazepam, nitrazepam, lorazepam, chlordiazepoxide, and flurazepam is obtained. Violet bands were obtained for diazepam, while the other compounds did not react when the plate was sprayed with 5% sodium hydroxide solution followed by 1% *m*-dinitrobenzene in dimethyl sulfoxide (Daundkar et al. 2008). The acidic hydrolysis products of some 1,4-benzodiazepines (haloxazolam, N-desalkyl-2-oxoquazepam, 3-hydroxy-N-desalkyl-2-oxoquazepam, ethyl-loflazepate, cloxazolam, delorazepam, quazepam, N-1-desmethylfludiazepam, 2-oxoquazepam, pinazepam, fludiazepam, nimetazepam, and 3-hydroxy-2-oxoquazepam) separated on HPTLC silica gel plates with chloroform–methanol 90:10 v/v or acetic acid ethyl ester–methanol–conc. NH_3 85:10:5 v/v/v were detected with Bratton-Marshall-reagent after photolytic desalkylation (Schütz et al. 1987). A new technique for visualizing the diazepam together with other drugs involves the TLC plate exposure to vapor of formaldehyde 38%–acetic acid anhydride (4:1) and then sulfuric acid 98%. After evaluation, the plate is placed in a jar saturated with water and again visually evaluated (Khdary and Shaikh 1996). The detection of specific impurities of serotonergic anxiolytics (buspirone, gepirone, ipsapirone, and zalospirone) can be done by spraying with hydroxylamine solution obtaining detection limits of a few nanograms at a signal-to-noise ratio 3:1 (Farina et al. 1996).

Some barbiturates (hexo-, amo-, cyclo-, and phenobarbital) can be detected by the following procedure: spraying with a reagent prepared by suspending 5 g mercury (II)oxide in 100 mL water, 20 mL conc. sulfuric acid, and, after cooling, diluting to 250 mL with water; drying; spraying with a solution obtained from 100 mg of diphenylcarbazone dissolved in 100 mL chloroform; and visualization under UV at 366 nm (Anonymous 1988). Visualization can also use 1% silver acetate followed by diphenylcarbazone spray (Baselt 1987) or exposure to chlorine vapors followed by 2,7-dichlorofluorescein and Dragendorff's reagent (McElwee 1979). Zolpidem and zoplicone can be detected by evaluation in daylight of colored zones obtained after spraying with chloranilic acid reagent (0.5 chloranilic acid in dioxane) (Pushpalatha et al. 2009). A photocatalytic reaction can be used for visualizing the phenobarbital in the presence of 17 other drugs, such as antibiotics, analgesics, anaesthetics, anti-rheumatic, anti-inflammatory, antitussive, broncholytic, spasmolytic, sympathomimetic, and C vitamin. Detection was achieved by spraying with 10 mL each of a solution of 0.25 g titanum dioxide in 0.1 mol/L potassium permanganate (A), 0.25 g titanium dioxide in 1.0 mol/L potassium iodide (B), 0.25 g titanium dioxide in 1.0 mol/L potassium bromide (C), and 0.25 g titanium dioxide in 1.0 mol/L potassium chloride (D). The most sensitive reagent was solution C. After spraying with reagents C and D, the plates were illuminated by use of UV lamps with the radiation at 366 nm for 10 min, sprayed with 0.1 mol/L silver nitrate solution, and illuminated again for 3 min (Makowski et al. 2010). The phenobarbital can also be detected in the presence of diazepam and compounds from other classes (chloral hydrate, saccharin, and anthranilic acid) by evaluation under UV at 366 nm after spraying with 2% sodium hydroxide solution, followed by spraying with 0.5% orcinol solution and heating in an oven at 90°C for 10 min (Mali and Garad 2005).

16.5 USE OF TLC IN PHARMACEUTICAL FORMULATION ANALYSIS

Nowadays, many sedatives and anxiolytics are controlled by specific regulations according to national and international legislation. Most of the pharmaceutical formulations of these drugs are tablets or capsules, although injectable solutions and powders may be used in some cases. Also, in many cases these drugs appear together with other drugs in complex pharmaceutical formulations. The dosage forms and the pharmaceutical formulation must be carefully controlled because the taken doses influence their effect and, on the other hand some of them could be purchased from illicit sources. The unequivocal determination of a drug in pharmaceutical formulations is as important as determination in complex matrices, because the pharmaceutical product quality is directly related to patient health (Bonfilio et al. 2010). For these purposes, many efforts have been made in order to develop a rapid, sensitive, and selective analytical method for the estimation of these drugs in their single or combined dosage forms, including TLC analysis. In Table 16.3 some of the TLC systems used for the analysis of various pharmaceutical preparations are presented.

TABLE 16.3
Chromatographic Conditions for Analysis of Anxiolytics and Sedatives from Different Pharmaceutical Preparations

Drug	Pharmaceutical Preparation	Stationary Phase	Mobile Phase	Detection	References
Clotiazepam	Tablets	TLC silica gel F_{254}	Toluene/methanol 9:1 v/v	Fluorescence	Busch et al. (1984)
Diazepam	Injections	TLC silica gel F_{254}	Benzene/ether/methanol 15:20:0.7 v/v/v	UV–245 nm	Xue and Dou (1985)
Flurazepam hydrochloride	Capsules	TLC silica gel F_{254}	Ether/dichloromethane/diethylamine/triethylamine 90:10:2:1 v/v/v	Detection by UV	Klein and Lau-Cam (1986)
Diazepam	Drugs	TLC silica gel F_{254}	Ethyl acetate and ethyl acetate/methanol 4:1 v/v	Detection by UV	By et al. (1989)
Chlordiazepoxide hydrochloride	Tablets	HPTLC silica gel F_{254} prewashed with ethyl acetate and aq. NH_3/methanol 1:9 v/v	Aq. ammonia/acetonitrile 1:999 v/v	UV–245 nm	White et al. (1991a)
Diazepam	Tablets	HPTLC silica gel F_{254} prewashed with chloroform/ethyl acetate 3:1 v/v and aq. NH_3/methanol 1:9v/v and activated for 15 min at 160°C	Chloroform/ethyl acetate 3:1 v/v	UV–254 nm	White et al. (1991b)
Clordiazepoxide and Amitriptyline	Combined tablets	HPTLC silica gel F_{254}	Ethyl acetate/methanol/diethylamine 190:10:1 v/v/v	UV–245 nm	Shirke et al. (1994)
Diazepam, Nitrazepam and Oxazepam	Ayurvedic preparations	HPTLC silica gel F_{254}	Hexane/methanol/acetone 15:3:2 v/v/v; Benzene/acetone/ethanol 8:1:1 v/v/v and Chloroform/acetone 8:2 v/v	UV–254 nm and Dragendorff's reagent	Kamble et al. (1995)
Diazepam Chlordiazepoxide	Cold drinks Tablets	HPTLC silica gel F_{254} 2-D TLC silica gel F_{254}	Chloroform/acetone 17:3 v/v (1) Ethyl acetate (2) Methanol	UV–230 nm FTIR and UV–230 nm	Sarin et al. (1998) Stahlmann and Kovar (1998)
Diazepam, Nitrazepam, and Flunitrazepam	Tablets	TLC silica gel F_{254}	Chloroform/acetone 9:1 v/v		Bakavoli and Kaykhaii (2003)

Drug	Sample	Stationary phase	Mobile phase	Detection	Reference
Alprazolam and Sertraline	Combined tablets	HPTLC silica gel F$_{254}$	Toluene/ethyl acetate/methanol/acetic acid 90:30:20:3 v/v/v/v	UV–217 nm	Verma and Joshi (2005)
Clonazepam and Escitalopram oxalate	Combined tablets	HPTLC silica gel F$_{254}$	Toluene/ethyl acetate/triethylamine 7:3.5:3 v/v/v	UV–258 nm	Dhavale et al. (2008)
Chlordiazepoxide and Trifluoperazine HCl	Combined dosage form	HPTLC silica gel F$_{254}$	Carbon tetrachloride/acetone/triethylamine 12:6:1 v/v/v	UV–262 nm	Sheladia et al. (2008)
Alprazolam, Amitriptyline HCl, Trifluoperazine HCl, and Risperidone	Combined dosage form	HPTLC silica gel F$_{254}$	Carbon tetrachloride/acetone/triethylamine 80:20:3 v/v/v	UV–250 nm	Patel and Patel (2008)
Oxazepam	Tablets	HPTLC silica gel F$_{254}$	Benzene/ethanol 5:1 v/v	UV–204 nm	Koba et al. (2009)
Alprazolam	Pure powder and tablets	HPTLC silica gel F$_{254}$	Acetone/toluene/ammonia 6:3:1 v/v/v	UV–230 nm	Patel et al. (2009a)
Alprazolam and Fluoxetine hydrochloride	Pure powder and Combined tablets	HPTLC silica gel F$_{254}$	Acetone/toluene/ammonia 6:3.5:0.5 v/v/v	UV–230 nm	Patel et al. (2009b)
Clonazepam and Escitalopram oxalate	Combined tablets	HPTLC silica gel F$_{254}$	Ethanol/toluene/triethylamine 1:3.5:0.1 v/v/v	UV–253 nm	Kakde et al. (2009)
Chlordiazepoxide and Imipramine hydrochloride	Combined tablets	HPTLC silica gel F$_{254}$	Carbon tetrachloride/acetone/triethylamine 6:3:0.3 v/v/v	UV–240 nm	Patel and Patel (2010)
Phenobarbital, Aminopyrine, Caffeine, and Phenacetin	Analgesic tablets	TLC silica gel F$_{254}$	Cyclohexane/chloroform/diethylamine 5:4:2 v/v/v	Detection by UV	Guo (1985)
Phenobarbital	Yinaoningxian tablets	TLC silica gel F$_{254}$	Chloroform/2-propanol/25% NH$_3$ 45:45:10 v/v/v	UV–254 nm	Yue and Gao (1990)
Melatonin		HPTLC silica gel F$_{254}$	Dichlormethane/methanol 95:5 v/v	UV–223 nm	Constantini and Paoli (1998)
Phenobarbital	Dosage form	HPTLC silica gel F$_{254}$ prewashed with methanol and acetone	Dichloromethane-ethyl acetate-formic acid 9.5:0.5:0.1 v/v/v	UV–210 nm	Wojciak-Kosior et al. (2006)
Melatonin	Tablets	HPTLC silica gel F$_{254}$	Toluene/ethyl acetate/formic acid 10:9:1 v/v/v	UV–290 nm	Agarwal et al. (2008)

On the other hand, TLC is recommended, among chromatographic methods, in the pharmacopoeias, mostly for impurity testing, but also for pharmaceutical preparations. The use of this technique is due to the fact that in the case of HPLC time-consuming pretreatment of sample is mandatory before chromatographic analysis. Furthermore, most pharmacopoeias suggest the development and use of methods in which the amount of used reagents and materials is reduced. TLC methods recommended in the monographs of these drugs from some pharmacopoeias (British Pharmacopoeia, European Pharmacopoeia, Japanese Pharmacopoeia, United States Pharmacopoeia, and Romanian Pharmacopoeia) are presented in Table 16.4.

16.6   USE OF TLC IN ANALYSIS OF BIOLOGICAL FLUIDS AND OTHER SAMPLES

Thin-layer chromatography had a significant importance in the identification of substances in biological fluids. TLC applications in the identification of drugs in biological fluids started to appear in the 1960s (Langman and Kapur 2006) and were systematically used for drug screening in urine and other biological materials until the 1980s, especially for forensic purposes due to their simplicity and the possibility to handle rapidly multiple spots in the same plate and screen parent, metabolite, and interfering drugs simultaneously (Yonamine and Cortez Sampaio 2006). Blood, urine, saliva, and hair are possible biological samples used for drug analyses. The levels of drug excretion depend on the drug half-life, which defines the ability of the body to remove a drug. The time required for the body to remove half of the drug absorbed is not a constant, and it varies for a particular drug from person to person, and within a person (Drummer and Odell 2001).

Worldwide, anxiolytic and sedative drugs are frequently used for medical treatment, but they are also potential drugs of abuse leading to dependency. For these reasons, the illicit use of these drugs must be controlled.

Literature data describe the TLC analysis of benzodiazepines, barbiturates, and their metabolites in different biological fluids, such as diazepam and other 1,4-benzodiazepines in serum, in urine, and in rat tissues (Sun 1978, Jain 1993, Zhang et al. 1995); clotiazepam from blood plasma (Busch et al. 1984); nitrazepam and its main metabolites from urine (Inoue and Niwaguchi 1985); flunitrazepam and its metabolites in urine (Van Rooij et al. 1985); phenobarbital from urine (Zhong 1985); brallobarbital and secobarbital (Noir-Falise et al. 1987); primidone and its metabolites in urine (Aboul-Enein and Thiffault 1991); thiopental in rat tissue (Büch et al. 1991); alprazolam in plasma (Zhang et al. 1993a); midazolam in the serum (Okamoto et al. 1993); diazepam, methaqualone, and estazolam in human serum (Zhang et al. 1993b); sedative hypnotics in blood (Zhang et al. 1994); benzodiazepines and zopiclone in human serum (Otsubo et al. 1995); phenobarbital, clonazepam, and clobazam in plasma (Dreassi et al. 1996); sixteen benzodiazepines and zopiclone in serum (Li and Sun 1996); tofisopam and its metabolites from biological samples—post mortem stomach, intestine contents, urine, and feces (Rizzo 2000); flunitrazepam and its major metabolite from urine (Wu et al. 2002); diazepam and its metabolite nordiazepam in blood, urine, liver, kidney, and muscle (Musshoff et al. 2004); 28 benzodiazepines and 5 related compounds in blood (Spangenberg et al. 2005); clonazepam and alprazolam together with haloperidol, fluxetine, promazine, doxepin, and risperidone in plasma and urine (Maslanka et al. 2011).

The diode-array HPTLC method developed by Spangenberg et al. (2005) for forensic drug analysis from blood implies the combination of R_f values and UV spectra into a single fit value, which enables automated identification without manual intervention. The advantage of diode-array scanners is that the spectral information is available directly after scanning of a track. Twenty-eight benzodiazepines and five related compounds were separated on silica gel K60 F_{254} HPTLC plates heated at 110°C for 10 min after prewashing first with methanol and then with CH_2Cl_2–methanol 95:5 v/v. Plates preequilibrated for 10 min were developed in a saturated horizontal development chamber, to a distance of 70 mm, with three developing systems in three runs: CH_2Cl_2–methanol 95:5 v/v;

TABLE 16.4
Chromatographic Condition for Drugs Analysis Recommended by the Pharmacopoeias

Drug	Pharmacopoeia	Chromatographic Conditions
Benzodiazepines		
Alprazolam	British	*Related Substances Test*—**SP:** TLC silica gel GF$_{254}$; **MP:** glacial acetic acid/water/methanol/ethyl acetate-2:15:20:80 v/v/v/v; 5 μL; elution on 12 cm; **D:** UV-254 nm after drying in air
	European	*Related Substances Test*—**SP:** TLC silica gel GF$_{254}$; **MP:** glacial acetic acid/water/methanol/ethyl acetate-2:15:20:80 v/v/v/v; 5 μL; elution on 12 cm; **D:** UV-254 nm after drying in air
	Japanese	*Related Substances Test*—**SP:** silica gel F$_{254}$; **MP:** acetone/hexane/ethyl acetate/ethanol (95%)-4:2:2:1 v/v/v/v; 20 μL; elution on 10 cm; **D:** UV-254 nm after drying in air
	United States	Other analytical methods
Bromazepam	British	Other analytical methods
	European	*Related Substances Test*—**SP:** TLC silica gel GF$_{254}$; **MP:** alcohol/ triethylamine/methylene chloride/light petroleum-5:5:20:70 v/v/v/v; 5 μL; elution on 7.5 cm; **D:** UV-254 nm after drying in air for 20 min *Identification*—**SP:** TLC silica gel GF$_{254}$; **MP:** diethylamine, ether-30:70 v/v; 5 μL; elution on 10 cm; **D:** UV-254 nm after drying in air
	Japanese	*Related Substances Test*—**SP:** silica gel F$_{254}$; **MP:** ethyl acetate/ammonia solution (28%)/ethanol (99.5%)-38:1:1 v/v/v; 20 μL; elution on 12 cm; **D:** UV-254 nm after drying in air
Brotizolam	British	Other analytical methods
	European	Other analytical methods
Camazepam	Not stipulated in any Pharmacopoeias	
Chlordesmethyldiazepam (Delorazepam)	Not stipulated in any Pharmacopoeias	
Chlordiazepoxide	British	*Related Substances Test*—**SP:** silica gel GF$_{254}$; **MP:** chloroform/ methanol/13.5 M ammonia-85:14:1 v/v/v; 2 and 20 μL; elution on 12 cm; **D:** (a) UV-254 nm after drying in air, (b) Spraying with freshly 1% w/v NaNO$_2$ in 1 M HCl, dry in cold air and spraying with 0.4% w/v N-(1-naphthyl)ethylene diamine · 2 HCl in ethanol (96%)
	European	Other analytical methods
	Japanese	*Related Substances Test*—**SP:** silica gel F$_{254}$; **MP:** ethyl acetate/ethanol (99.5%)-19:1 v/v; 25 μL; elution on 12 cm; **D:** (a) UV-254 nm after drying in air, (b) spraying with NaNO$_2$ in 1 M HCl (1 in 100) and N-(1-naphthyl)-N′-diethylethylenediamine oxalate-acetone
	United States	*Related Substances Test*—**SP:** silica gel; **MP:** ethyl acetate; 50 μL; elution on 3/4 of plate; **D:** spraying with 2 N H$_2$SO$_4$, drying at 105°C for 15 min and spraying successively with NaNO$_2$ (1 in 1000), ammonium sulfamate (1 in 200) and N-(1-naphtyl)ethylene diamine dihydrochloride (1 in 1000)
	Romanian	*Related Substances Test*—**SP:** silica gel; **MP:** chloroform/toluene/ methanol-100:40:10 v/v/v; 10 μL; elution on 15 cm; **D:** spraying with potassium tetraiodobismutate (III)
Clobazam	British	Other analytical methods
	European	Other analytical methods

(*continued*)

TABLE 16.4 (continued)
Chromatographic Condition for Drugs Analysis Recommended by the Pharmacopoeias

Drug	Pharmacopoeia	Chromatographic Conditions
Clonazepam	British	*Related Substances Test*—**SP:** silica gel G; **MP:** (a) chloroform/ether-20:80 v/v, (b) ether/nitromethane-10:90 v/v; 50 μL; **D:** heating at 2 kPa at 120° for 3 h, cooling and spraying with 10% w/v $ZnCl_2$ in 0.1 M HCl, dry in air, and expose to nitrous fumes *Identification*—**SP:** Merck silica gel GF_{254}; **MP:** 13.5 M ammonia/n-heptane/nitromethane/ether-2:15:30:60 v/v/v/v; 10 μL; elution on 10 cm; **D:** drying in cold air, spraying with 2 M NaOH, and heating at 120°C for 15 min
	European	Other analytical methods
	Japanese	*Related Substances Test*—**SP:** silica gel F_{254}; **MP:** nitromethane/acetone-10:1 v/v; 10 μL; elution on 12 cm; **D:** UV-254 nm after drying in air
	United States	*Related Substances Test*—**SP:** 0.25 mm silica gel; **MP:** acetone/n-heptane-3:2 v/v; 20 μL; **D:** spraying with 2 M H_2SO_4, drying at 105°C for 15 min and successively spraying with 0.01 sodium nitrite, 9 mM ammonium sulfamate and N-(1-naphthyl)ethylenediamine dihydrochloride TS and drying with air current
Clorazepate dipotassium	United States	Other analytical methods
Clotiazepam	Japanese	*Related Substances Test*—**SP:** silica gel F_{254}; **MP:** chloroform/acetone-5:1 v/v; 10 μL; elution on 10 cm; **D:** UV-254 nm after drying in air
Cloxazolam	Japanese	*Related Substances Test*—**SP:** silica gel F_{254}; MP: toluene/acetone-5:1 v/v; 10 μL; elution on 10 cm; **D:** UV-254 nm after drying in air
Diazepam	British	*Related Substances Test-Oral Solution and Tablets*—**SP:** silica gel GF_{254}; **MP:** ethyl acetate/hexane-1:1 v/v; 25, 20, and 5 μL, respectively; elution on 12 cm; **D:** UV-254 nm after drying in air *Identification-Rectal Solution and Tablets*—**SP:** silica gel; **MP:** methanol/chloroform-1:10 v/v; 10 and 2 μL, respectively; **D:** UV-365 nm after spraying with 10% H_2SO_4 in absolute ethanol and heating at 105° for 10 min
	European	Other analytical methods
	Japanese	*Related Substances Test*—**SP:** silica gel F_{254}; **MP:** ethyl acetate/hexane-1:1 v/v; 5 μL; elution on 12 cm; **D:** UV-254 nm after drying in air
	United States	*Related Substances Test*—**SP:** 0.25 mm silica gel; **MP:** ethyl acetate/n-heptane-1:1 v/v; 10 μL; **D:** UV-254 nm after drying in air
	Romanian	*Related Substances Test*—**SP:** silica gel F_{254}; **MP:** ethyl acetate/n-heptane-50:50 v/v; 10 μL; elution on 12 cm; **D:** (a) UV-254 nm after drying in air, (b) UV-366 nm after spraying with alcohol/nitric acid-10:0.5 and heating at 105° for 10 min
Estazolam	Japanese	*Related Substances Test*—**SP:** silica gel F_{254}; **MP:** hexane/chloroform/vmethanol-5:3:1 v/v/v; 10 μL; elution on 10 cm; **D:** UV-254 nm after drying in air
Etizolam	Not stipulated in any Pharmacopoeias	
Fludiazepam	Japanese	*Related Substances Test*—**SP:** silica gel F_{254}; **MP:** chloroform/ethyl acetate-10:7 v/v; 20 μL; elution on 12 cm; **D:** UV-254 nm after drying in air
Flumazenil	British	Other analytical methods
	European	Other analytical methods

TABLE 16.4 (continued)
Chromatographic Condition for Drugs Analysis Recommended by the Pharmacopoeias

Drug	Pharmacopoeia	Chromatographic Conditions
	United States	*Related Substances Test*—**SP:** 0.25 mm silica gel; **MP:** chloroform/glacial acetic acid/alcohol/water-75:15:7.5:2.5 v/v/v/v; 10 μL; elution on 12 cm; **D:** (a) UV-254 nm after drying in cold air, (b) spraying with ninhydrine and heating at 105°C for 15 min
Flunitrazepam	British	Other analytical methods
	European	Other analytical methods
	Japanese	*Related substances Test*—**SP:** silica gel F$_{254}$; **MP:** 1,2-dichloroethane/diethyl ether/ammonia solution (28%)-200:100:3 v/v/v; 10 μL; elution on12 cm; **D:** UV-254 nm after drying in air
Flurazepam	British	*Related Substances Test*—**SP:** silica gel GF$_{254}$; **MP:** diethylamine/ether-2.5:97.5 v/v; 10 μL; elution on 12 cm; **D:** UV-254 nm after drying in air
	European	Other analytical methods
	Japanese	*Related Substances Test*—**SP:** silica gel F$_{254}$; **MP:** cyclohexane/acetone/ammonia solution (28%)-60:40:1 v/v/v; 10 μL; elution on 12 cm; **D:** UV-254 nm after drying in air
	United States	*Related Substances Test*—**SP:** 0.25 mm silica gel; **MP:** ethyl acetate/ammonium hydroxide-200:1 v/v; 10 μL; elution on 12 cm; **D:** (a) UV-254 nm after drying in air
Flurazepam hydrochloride	Japanese	*Related Substances Test*—**SP:** silica gel F$_{254}$ keep in a chamber filled with ammonia vapor for ~15 min; **MP:** diethyl ether/diethylamine -39:1 v/v; 20 μL; elution on 12 cm; **D:** UV-254 nm after drying in air
Halazepam	Not stipulated in any Pharmacopoeias	
Haloxazolam	Not stipulated in any Pharmacopoeias	
Ketazolam	Not stipulated in any Pharmacopoeias	
Loprazolam Mesilate	British	*Related substances Test*—**(1) SP:** silica gel 60; **MP:** chloroform/methanol/13.5 M ammonia-80:20:2 v/v/v; 10 μL band; **D:** drying in air, heating at 110° for 15 min and spraying with 2% w/v KI, 10% w/v chloroplatinic(IV) acid-50:1 v/v
		(2) SP: silica gel 60 prewashed with methanol and heating at 100°C to 105°C for 1 h; **MP:** chloroform/methanol-80:20 v/v; 10 μL; **D:** drying in air, spraying with 5% w/v TiCl$_3$ in 10% w/v HCL, then spraying with solution of 4-dimethyl-aminocinnamaldehyde (0.4 g) in 6 M HCl, ethanol (96%)-20:100 v/v, and heating at 100°C until spots appear (~10 min)
		Identification-Tablets—**SP:** Merck silica gel 60 F$_{254}$; **MP:** chloroform/methanol-80:20 v/v; 10 μL; **D:** UV-254 nm after drying in air
Lorazepam	British	*Injection and Tablets*—**SP:** Merck silica gel 60 F$_{254}$ prewashed on 17 cm with methanol and heating at 100°C–105°C for 1 h; **MP:** chloroform/methanol-10:1 v/v; 40 μL; **D:** UV-254 nm after drying in air
		Identification-Injection and Tablets—**SP:** Merck silica gel 60; **MP:** toluene; 10 μL; **D:** drying, spraying with freshly 1.25% w/v NaNO$_2$ in 0.5 M HCl, heating at 100°C for 5 min, cooling and spraying with 0.1%w/v N-(1-naphthyl)ethylenediamine dihydrochloride in absolute ethanol
	European	*Related Substances Test*—**SP:** TLC silica gel F$_{254}$ pre-washed with methanol on 17 cm and heating at 100°C–105°C for 1 h; **MP:** methanol/methylene chloride-10:100 v/v; 20 μL; elution on 12 cm; **D:** UV-254 nm after drying in air

(continued)

TABLE 16.4 (continued)
Chromatographic Condition for Drugs Analysis Recommended by the Pharmacopoeias

Drug	Pharmacopoeia	Chromatographic Conditions
	Japanese	*Related Substances Test*—**SP:** silica gel F_{254}; **MP:** chloroform/1, 4-dioxane/acetic acid (100%)-91:5:4 v/v/v; 10 μL; elution on 15 cm; **D:** UV-254 nm after drying in air
	United States	*Related Substances Test*—(**1**) **SP:** 0.25 mm silica gel previously wash with chloroform/ethyl acetate/methanol-2:1:1 v/v/v; **MP:** chloroform/dioxane/ glacial acetic acid-91:5:4 v/v/v; 10 μL; elution until 2–3 cm from the top; **D:** UV-254 nm after drying in air for 30 min. (**2**) **SP:** 0.25 mm silica gel; **MP:** chloroform/dioxane/glacial acetic acid-91:5:4 v/v/v; 50 and 10 μL; elution on minimum 10 cm; **D:** spraying with 2N H_2SO_4, drying at 105°C for 15 min and spraying successively with $NaNO_2$ (1 in 1000), ammonium sulfamate (1 in 200) and N-(1-naphtyl)ethylenediamine dihydrochloride (1 in 1000)
		Related Substances Test-Injection—**SP:** 0.25 mm silica gel; **MP:** chloroform/*n*-heptane/alcohol-10:10:1 v/v/v; 50 and 10 μL; elution on minimum 10 cm; **D:** spraying with freshly $NaNO_2$ in 0.5N HCl (1 in 80), heating at 100°C for 5 min, cooling and spraying with N-(1-naphtyl)ethylenediamine dihydrochloride in alcohol (1 in 1000)
		Identification-Injection—**SP:** 0.25 mm silica gel; **MP:** toluene; 50 and 10 μL; elution on 15 cm; **D:** spraying with freshly $NaNO_2$ in 0.5N HCl (1 in 80), heating at 100°C for 5 min, cooling and spraying with N-(1-naphtyl)ethylenediamine dihydrochloride in alcohol (1 in 1000)
Lormetazepam	British	*Identification-Tablets*—**SP:** silica gel GF_{254}; **MP:** chloroform/ methanol-10:1 v/v; 10 Ml; **D:** UV-254 nm after drying in air
Medazepam	Japanese	*Related substances Test*—**SP:** silica gel F_{254}; **MP:** cyclohexane/acetone/ ammonia solution (28%)-60:40:1 v/v/v; 10 μL; elution on 10 cm; **D:** UV-254 nm after drying in air
Midazolam	British	*Related Substances Test*—**SP:** TLC silica gelF_{254}; **MP:** glacial acetic acid/ H_2O/methanol/ethyl acetate-2:15:20:80 v/v/v/v; 5 μL; elution on 2/3 of plate; **D:** UV-254 nm after drying in air
	European	*Related Substances Test*—**SP:** silica gel GF_{254}; **MP:** glacial acetic acid/ H_2O/methanol/ethyl acetate-2:15:20:80 v/v/v/v; 5 μL; elution on 12 cm; **D:** UV-254 nm after drying in air
Nitrazepam	British	*Related Substances Test*—**SP:** TLC silica gel F_{254}; **MP:** ethyl acetate/ nitromethane-15:85 v/v; 5 μL; elution on 12 cm; **D:** UV-254 nm after drying in air
		Identification-Oral Suspension—**SP:** TLC silica gel GF_{254}; **MP:** ethyl acetate/nitromethane-15:85 v/v; 10 μL; **D:** drying in air and expose to nitrous fumes
		Identification-Tablets—**SP:** TLC silica gel G; **MP:** methanol/ chloroform-1:10 v/v; 10 μL; **D:** UV-365 nm after spraying with 10% v/v H_2SO_4 in absolute ethanol and heating at 105°C for 10 min
	European	*Related Substances Test*—**SP:** silica gel GF_{254}; **MP:** ethyl acetate/ nitromethane-15:85 v/v; 10 μL; elution on 12 cm; **D:** UV-254 nm after drying in air
	Japanese	*Related Substances Test*—**SP:** silica gel F_{254}; **MP:** nitromethane/ethyl acetate-17:3 v/v; 10 μL; elution on 10 cm; **D:** UV-254 nm after drying in air

TABLE 16.4 (continued)
Chromatographic Condition for Drugs Analysis Recommended by the Pharmacopoeias

Drug	Pharmacopoeia	Chromatographic Conditions
	Romanian	*Related Substances Test*—**SP:** silica gel F$_{254}$; **MP:** nitromethane/ethyl acetate-85:15 v/v; 10 µL; elution on 15 cm; **D:** UV-254 nm after drying in air
Nordazepam	Not stipulated in any Pharmacopoeias	
Oxazepam	British	Other analytical methods
	European	*Related Substances Test*—Protected from light!!! - **SP:** silica gel F$_{254}$ pre-washed with methanol on 17 cm and heatinged at 100°C–105°C for 30 min; **MP:** methanol/methylene chloride-10:100 v/v; 20 µL; elution on 15 cm; **D:** UV-254 nm after drying in air
	United States	Other analytical methods
Oxazolam	Japanese	*Related Substances Test*—**SP:** silica gel F$_{254}$; **MP:** toluene/acetone-8:1 v/v; 10 µL; elution on 10 cm; **D:** UV-254 nm after drying in air
Pinazepam	Not stipulated in any Pharmacopoeias	
Prazepam	British	*Related Substances Test*—**SP:** TLC silica gelF$_{254}$; **MP:** ethyl acetate/heptane-50:50 v/v; 5 µL; elution on 10 cm; **D:** UV-254 nm and 366 nm after drying in air
	European	*Related Substances Test*—**SP:** silica gel F$_{254}$; **MP:** ethyl acetate/heptan-50:50 v/v; 5 µL; elution on 10 cm; **D:** UV-254 and 366 nm after drying in air
	Japanese	*Related Substances Test*—**SP:** silica gel F$_{254}$; **MP:** chloroform/acetone-9:1 v/v; 5 µL; elution on 10 cm; **D:** UV-254 nm after drying in air
Quazepam	United States	*Related Substances Test Substance and Tablets*—**SP:** 0.25 mm silica gel F$_{254}$; **MP:** cyclohexane/ethyl acetate/ether-170:40:25 v/v/v; 5 µL; elution on 3/4 of plate; **D:** UV-254 nm after drying in air
Temazepam	British	*Related Substances Test-Oral Solution*—**SP:** Merck silica gel 60 F$_{254}$; **MP:** chloroform/methanol-92.5:7.5 v/v; 10 µL; elution on 12 cm; **D:** UV-365 nm after drying in warm air
		Related Substances Test-Tablets—**SP:** Merck silica gel 60 F$_{254}$; **MP:** methanol/dichlormethane-2:98 v/v; 20 µL; **D:** UV-254 nm after drying in air
		Identification-Oral Solution—**SP:** Merck silica gel 60 F$_{254}$; **MP:** cyclohexane/chloroform/diethylamine-50:40:10 v/v/v; 2 µL; **D:** UV-254 nm after drying in air
		Identification-Tablets—**SP:** Merck silica gel 60 F$_{254}$; **MP:** diethylamine/ether-90:10 v/v; 5 µL; **D:** UV-254 nm after drying in air
	European	Other analytical methods
	United States	*Identification*—**SP:** 0.25 mm silica gel; **MP:** cyclohexane/chloroform/methanol/ammonium hydroxide-50:40:12:1 v/v/v/v; 10 µL; elution on 10 cm; **D:** UV-254 nm after drying in air
		Identification-Capsules—**SP:** 0.25 mm silica gel; **MP:** toluene/dioxane/methanol/ammonium hydroxide-65:30:5:1 v/v/v/v; 20 µL; **D:** UV-254 nm after drying in air
Tofisopam	Japanese	*Related Substances Test*—**SP:** silica gel F$_{254}$; **MP:** ethyl acetate/acetone/methanol/formic acid-24:12:2:1 v/vv/v; 10 µL; elution on 10 cm; **D:** UV-254 nm after drying in air
Triazolam	United States	Other analytical methods

(continued)

TABLE 16.4 (continued)

Chromatographic Condition for Drugs Analysis Recommended by the Pharmacopoeias

Drug	Pharmacopoeia	Chromatographic Conditions
5HT1A agonists		
Buspirone	British	Other analytical methods
	United States	Other analytical methods
Gepirone	Not stipulated in any Pharmacopoeias	
Ipsapirone	Not stipulated in any Pharmacopoeias	
Zalospirone	Not stipulated in any Pharmacopoeias	
Barbiturates		
Allobarbital	Not stipulated in any Pharmacopoeias	
Amobarbital and Amobarbital sodium	British	*Related Substances Test*—**SP:** silica gel GF_{254}; **MP:** lower layer of conc. ammonia/alcohol/chloroform-5:15:80 v/v/v; 20 µL; elution on 15 cm; **D:** (a) UV-254 nm immediately, (b) spraying with diphenylcarbazone mercuric reagent, dryig in air and spraying with freshly alcoholic KOH diluted 1 in 5 with aldehyde-free alcohol and heating at 100°C–105°C for 5 min
		Identification—**SP:** silica gel GF_{254}; **MP:** lower layer of conc. ammonia/alcohol/chloroform-5:15:80 v/v/v; 10 µL; elution on 18 cm; **D:** UV-254 nm immediately
	European	*Related Substances Test*—**SP:** silica gel GF_{254}; **MP:** lower layer of conc. ammonia/alcohol/chloroform-5:15:80 v/v/v; 20 µL; elution on 15 cm; **D:** (a) UV-254 nm immediately, (b) spraying with diphenylcarbazone mercuric reagent, drying in air and spraying with freshly alcoholic KOH diluted 1 in 5 with aldehyde-free alcohol, and heating at 100°C–105°C for 5 min
		Identification—**SP:** silica gel GF_{254}; **MP:** lower layer of conc. ammonia/alcohol/chloroform-5:15:80 v/v/v; 10 µL; elution on 18 cm; **D:** UV-254 nm immediately
Aprobarbital	Not stipulated in any Pharmacopoeias	
Barbital	British	*Related Substances Test*—**SP:** silica gel GF_{254}; **MP:** lower layer of conc. ammonia/alcohol/chloroform-5:15:80 v/v/v; 20 µL; elution on 15 cm; **D:** (a) UV-254 nm, (b) spraying with diphenylcarbazone mercuric reagent, drying in air and spraying with freshly alcoholic KOH diluted 1 in 5 with aldehyde-free alcohol, heating at 100°C–105°C for 5 min
		Identification—**SP:** silica gel GF_{254}; **MP:** lower layer of conc. ammonia/alcohol/chloroform-5:15:80 v/v/v; 10 µL; elution on 18 cm; **D:** UV-254 nm
	European	**SP:** silica gel GF_{254}; **MP:** lower layer of conc. ammonia/alcohol/chloroform-5:15:80 v/v/v; 20 µL; elution on 15 cm; **D:** (a) UV-254 nm, (b) spraying with diphenylcarbazone mercuric reagent, drying in air and spraying with freshly alcoholic KOH diluted 1 in 5 with aldehyde-free alcohol, heating at 100°C–105°C for 5 min
		Identification—**SP:** silica gel GF_{254}; **MP:** lower layer of conc. ammonia/alcohol/chloroform-5:15:80 v/v/v; 10 µL; elution on 18 cm; **D:** UV-254 nm
	Japanese	Other analytical methods
	Romanian	Other analytical methods

TABLE 16.4 (continued)

Chromatographic Condition for Drugs Analysis Recommended by the Pharmacopoeias

Drug	Pharmacopoeia	Chromatographic Conditions
Butobarbital sodium	United States	*Identification*—**SP:** 0.25 mm silica gel; **MP:** acetone/methylene chloride/ methanol/ammonium hydroxide-5:3:1:1 v/v/v/v; 10 μL; elution on 3/4 of plate; **D:** drying in air and spraying with mercurous nitrate hydrate in 0.15N HNO_3 (1 in 100)
Cyclobarbital	Romanian	*Related Substances Test*—**SP:** silica gel G; **MP:** benzene/alcohol/acid acetic-80:10:80 v/v/v; 10 μL; elution on 10 cm; **D:** keep in iodine vapor
Heptobarbital	Not stipulated in any Pharmacopoeias	
Hexobarbital	British	*Related Substances Test*—**SP:** silica gel GF_{254}; **MP:** lower layer of conc. ammonia/alcohol/chloroform-5:15:80 v/v/v; 20 μL; elution on 15 cm; **D:** UV-254 nm
		Identification—**SP:** silica gel GF_{254}; **MP:** lower layer of conc. ammonia/ alcohol/chloroform-5:15:80 v/v/v; 10 μL; elution on 18 cm; **D:** UV-254 nm
	European	*Related Substances Test*—**SP:** silica gel GF_{254}; **MP:** lower layer of conc. ammonia/alcohol/chloroform-5:15:80 v/v/v; 20 μL; elution on 15 cm; **D:** UV-254 nm
		Identification—**SP:** silica gel GF_{254}; **MP:** lower layer of conc. ammonia/ alcohol/chloroform-5:15:80 v/v/v; 10 μL; elution on 18 cm; **D:** UV-254 nm
Metharbital	Not stipulated in any Pharmacopoeias	
Mephobarbital	United States	Other analytical methods
Methohexital	United States	Other analytical methods
Methylphenobarbital	British	*Related Substances Test*—**SP:** silica gel GF_{254}; **MP:** lower layer of conc. ammonia/alcohol/chloroform-5:15:80 v/v/v; 20 μL; elution on 15 cm; **D:** (a) UV-254 nm, (b) spraying with diphenylcarbazone mercuric reagent, drying in air and spraying with freshly alcoholic KOH diluted 1 in 5 with aldehyde-free alcohol, heating at 100°C–105°C for 5 min
		Identification—**SP:** silica gel GF_{254}; **MP:** lower layer of conc. ammonia/ alcohol/chloroform-5:15:80 v/v/v; 10 μL; elution on 18 cm; **D:** UV-254 nm
	European	*Related Substances Test*—**SP:** silica gel GF_{254}; **MP:** lower layer of conc. ammonia/alcohol/chloroform-5:15:80 v/v/v; 20 μL; elution on 15 cm; **D:** (a) UV-254 nm, (b) spraying with diphenylcarbazone mercuric reagent, drying in air and spraying with freshly alcoholic KOH diluted 1 in 5 with aldehyde-free alcohol, heating at 100°C–105°C for 5 min
		Identification—**SP:** silica gel GF_{254}; **MP:** lower layer of conc. ammonia/ alcohol/chloroform-5:15:80 v/v/v; 10 μL; elution on 18 cm; **D:** UV-254 nm
Pentobarbital and Pentobarbital sodium	British	*Related Substances Test*—**SP:** silica gel GF_{254}; **MP:** lower layer of conc. ammonia/alcohol/chloroform-5:15:80 v/v/v; 20 μL; elution on 15 cm; **D:** (a) UV-254 nm, (b) spraying with diphenylcarbazone mercuric reagent, drying in air and spraying with freshly alcoholic KOH diluted 1 in 5 with aldehyde-free alcohol, heating at 100°C–105°C for 5 min
		Identification—**SP:** silica gel GF_{254}; **MP:** lower layer of conc. ammonia/ alcohol/chloroform-5:15:80 v/v/v; 10 μL; elution on 18 cm; **D:** UV-254 nm

(continued)

TABLE 16.4 (continued)
Chromatographic Condition for Drugs Analysis Recommended by the Pharmacopoeias

Drug	Pharmacopoeia	Chromatographic Conditions
	European	*Related Substances Test*—**SP:** silica gel GF$_{254}$; **MP:** lower layer of conc. ammonia/alcohol/chloroform-5:15:80 v/v/v; 20 µL; elution on 15 cm; **D:** (a) UV-254 nm, (b) spraying with diphenylcarbazone mercuric reagent, drying in air and spraying with freshly alcoholic KOH diluted 1 in 5 with aldehyde-free alcohol, heating at 100°C–105°C for 5 min *Identification*—**SP:** silica gel GF$_{254}$; **MP:** lower layer of conc. ammonia/ alcohol/chloroform-5:15:80 v/v/v; 10 µL; elution on 18 cm; **D:** UV-254 nm
	Japanese	Other analytical methods
	United States	*Identification*—**SP:** 0.25 mm silica gel; **MP:** isopropyl alcohol/ammonium hydroxide/chloroform/acetone-9:4:2:2 v/v/v/v; 50 µL; elution on 3/4 of plate; **D:** UV-254 nm after drying in air
Phenobarbital and Phenobarbital sodium	British	*Related substances Test*—**SP:** silica gel GF$_{254}$; **MP:** lower layer of conc. ammonia/alcohol/chloroform-5:15:80 v/v/v; 20 µL; elution on 15 cm; **D:** (a) UV-254 nm, (b) spraying with diphenylcarbazone mercuric reagent, drying in air and spraying with freshly alcoholic KOH diluted 1 in 5 with aldehyde-free alcohol, heating at 100°C–105°C for 5 min *Identification*—**SP:** silica gel GF$_{254}$; **MP:** lower layer of conc. ammonia/ alcohol/chloroform-5:15:80 v/v/v; 10 µL; elution on 18 cm; **D:** UV-254 nm
	European	*Related Substances Test*—**SP:** silica gel GF$_{254}$; **MP:** lower layer of conc. ammonia/alcohol/chloroform-5:15:80 v/v/v; 20 µL; elution on 15 cm; **D:** (a) UV-254 nm, (b) spraying with diphenylcarbazone mercuric reagent, drying in air and spraying with freshly alcoholic KOH diluted 1 in 5 with aldehyde-free alcohol, heating at 100°C–105°C for 5 min *Identification*—**SP:** silica gel GF$_{254}$; **MP:** lower layer of conc. ammonia/ alcohol/chloroform-5:15:80 v/v/v; 10 µL; elution on 18 cm; **D:** UV-254 nm
	Japanese United States	Other analytical methods
	Romanian	Other analytical methods
Primidone		Other analytical methods in all Pharmacopoeias
Secbutabarbital (butabarbital) sodium	United States	*Related Substances Test*—**SP:** 0.25 mm silica gel; **MP:** acetone/methylene chloride/methanol/ammonium hydroxide-5:3:1:1 v/v/v/v; 10 µL; elution on 3/4 of plate; **D:** spraying with mercurous nitrate dehydrate in 0.15N nitric acid (1 in 100)
Secobarbital (quinalbarbitone)	United States	Other analytical methods
Thiopental sodium	British	*Related Substances Test*—**SP:** silica gel GF$_{254}$; **MP:** lower layer of conc. ammonia/alcohol/chloroform-5:15:80 v/v/v; 20 µL; elution on 15 cm; **D:** UV-254 nm *Identification*—**SP:** silica gel GF$_{254}$; **MP:** lower layer of conc. ammonia/ alcohol/chloroform-5:15:80 v/v/v; 10 µL; elution on 18 cm; **D:** UV-254 nm
	European	*Related Substances Test*—**SP:** silica gel GF$_{254}$; **MP:** lower layer of conc. ammonia/alcohol/chloroform-5:15:80 v/v/v; 20 µL; elution on 15 cm; **D:** UV-254 nm *Identification*—**SP:** silica gel GF$_{254}$; **MP:** lower layer of conc. ammonia/ alcohol/chloroform-5:15:80 v/v/v; 10 µL; elution on 18 cm; **D:** UV-254 nm
	Japanese	Other analytical methods
	United States	Other analytical methods

TABLE 16.4 (continued)
Chromatographic Condition for Drugs Analysis Recommended by the Pharmacopoeias

Drug	Pharmacopoeia	Chromatographic Conditions
Vinbarbital	Not stipulated in any Pharmacopoeias	
Vinylbital	Not stipulated in any Pharmacopoeias	
Other sedatives		
Zolazepam	United States	*Related Substances Test*—**SP:** 0.25 mm silica gel; **MP:** toluene/acetone/ammonium hydroxide-75:18:7 v/v/v; 5 μL; elution on 3/4 of plate; **D:** spraying with Dragendorff's reagent
Zolpidem Tartrat	British	*Identification*—**SP:** silica gel GF$_{254}$; **MP:** diethylamine/cyclohexane/ethyl acetate-10:45:45 v/v/v; 5 μL; elution on 12 cm; **D:** UV-254 nm after drying in air
	European	*Identification*—**SP:** silica gel GF$_{254}$; **MP:** diethylamine/cyclohexane/ethyl acetate-10:45:45 v/v/v; 5 μL; elution on 12 cm; **D:** UV-254 nm after drying in air
Zopiclone	British	*Identification*—**SP:** silica gel GF$_{254}$; **MP:** triethylamine/acetone/ethyl acetate-2:50:50 v/v/v; 10 μL; elution on 15 cm; **D:** UV-254 nm after drying in air
	European	*Identification*—**SP:** silica gel GF$_{254}$; **MP:** triethylamine/acetone/ethyl acetate-2:50:50 v/v/v; 10 μL; elution on 15 cm; **D:** UV-254 nm after drying in air

SP, stationary phase; MP, mobile phase; D, detection.

CH_2Cl_2–methanol 95:5 v/v and then ethyl acetate–cyclohexane–25% ammonia solution 500:400:1 v/v/v; and cyclohexane–acetone–methyl *t*-butyl ether 3:2:1 v/v/v. Spectral data from whole track were obtained by acquiring 900 spectra with a spectral resolution of 512 data points for each spectrum in the range 197–612 nm over a distance of 90 mm. Samples of blood were alkalinized with 0.5 M boric acid buffer 2:1 v/v, adjusted to pH 9 with $NaHCO_3$ and then extracted with the same volume of butyl chloride and centrifuged. The organic layer was separated and evaporated to dryness in a stream of argon at 40°C, and the residue was redissolved in 25 μL methanol. The method allows the identification of benzodiazepines from a very small volume of sample on the basis of four independent compound characteristics (three R_f value and spectral data).

An interesting method for labeling of chlorodiazepoxide with ^{131}I and studying its biodistribution on rats was developed by Ünak et al. (2002). The labeled chlordiazepoxide is separated by instant thin-layer chromatography (ITLC) using 10 × 1.5 cellulose-coated plastic sheets with a thickness of 0.1 mm developed in a Sigma ITLC chamber using two different solvent systems: n-butanol–water–acetic acid 4:2:1 v/v/v and isopropano–butanol–0.2 N ammonium hydroxide 2:1:1 v/v/v. After elution, the ITLC sheets were cut into 0.5 cm sections, counted by a Cd(Te) detector equipped with RAD 501 single-channel analyzer, and the ITLC chromatograms and the R_f values were obtained from these counts.

In many cases, the (HP)TLC-proposed methods have advantages over GC/MS, namely less interference by large concentrations of other drugs and low time requirement (Edler and Schlüter 1994), and also lead to obtaining a low limit of detection (LOD) and quantification (LOQ), such as: LOD—2.5 ng/mL for clotiazepam from blood plasma (Busch et al. 1984), LOD—5–10 ng/mL for nitrazepam and its main metabolites from urine (Inoue and Niwaguchi 1985), LOD—0.5 ng/mL for diazepam and its metabolites (2-amino-5-chlorobenzophenone and 5-chloro-2-methyl-aminobenzophenone) from urine (Jain 1993), LOD between 0.1–0.4 μg/mL for 16 benzodiazepines (alprazolam, bromazepam, brotizolam, chlordiazepoxide, clotiazepam, cloxazolam, diazepam, estazolam, etizolam, flunitrazepam, flurazepam, haloxazolam, lormetazepam, medazepam, nitrazepam, triazolam) and zopiclone from serum, LOD—60, 60, 400 ng for diazepam, nitrazepam, estazolam, respectively, from human urine with good precision (1.78%–3.21% within plate and 2.75%–4.05% inter plate) (Li and Sun 1996), LOD—5 ng/mL and LOQ—15 ng/mL for 7-aminoflunitrazepam, the major metabolite of flunitrazepam, from urine (Wu et al. 2002), LOD from 0.009 to 0.260 µg for selected psychotropic drugs containing diazepam, clonazepam, haloperidol, amitriptyline, sulpiride, promazine, fluphenazine, doxepin, trifluoperazine, and chlorpromazine (Maslanka and Krzek 2005), LOD—20 µg/L (40 ng/band) and LOQ—40 µg/L (80 ng/band) for the new ultra-shortacting thiazolodiazepine (ethyl 8-oxo-5,6,7,8-tetrahydrothiazolo[3,2-a][1,3]diazepin-3-carboxylate–HIE-124) from human plasma (Abourashed et al. 2009), LOD < 500 ng/mL for phenobarbital from urine (Zhong 1985). The sensitivity increased about ten fold over the conventional method (LOD = 0.1–0.4 µg/mL) obtained for benzodiazepines and zopiclone in human serum accurately identified by means of the values of $R_f \times 100$ and the spot color in three systems (Otsubo et al. 1995). Improvement of LOD and LOQ up to the picomole range can be achieved by hyphenating the TLC with mass spectrometry (MS), such as: TLC/fast atom bombardment (FAB)MS in the determination midazolam in the serum of intoxication patient (Okamoto et al. 1993) and ultrathin-layer chromatography (UTLC)/atmospheric pressure matrix-assisted laser desorption/ionization (AP-MALDI) MS for midazolam, diazepam, lorazepam, oxazepam, N-desalkyl-flurazepam, triazolam, and nitrazepam bioanalysis (Salo et al. 2007). Moreover, by hyphenating the TLC separation with MS, the results are improved and the confirmation of the identity of abused drugs is without doubt (Tames et al. 2000, Cheng et al. 2009).

A computer program (SPOT CHEK) that assists in matching the data from a particular chromatogram with those obtained for known drugs recovered from serum, urine, or solid dosage forms (e.g., tablets, pills) and mystery powders was developed (Siek et al. 1997). The logic of the SPOT CHEK technique is toward discovery of drug classes (e.g., benzodiazepines, barbiturates), functional groups (primary, secondary, and tertiary amines), and conjugation (as in aromatic structures) by fluorescamine, ferric chloride/perchloric acid/nitric acid, Dragendorff, Marquis, Mandelin, and iodinated Dragendorff reagents, vapor from chlorine or hydrochloric acid, and UV absorbance. Detection limits of 5–200 ng per sample spot were obtained for drugs in the database.

The literature also presents TLC methods used for analysis of such drugs in food matrices with possible application in forensic screening. A recent study investigated the possible addition of drugs in food and the adulteration of hypnotic sedatives (alprazolam, bromazepam, diazepam, flurazepam, flunitrazepam, methaqualone, nitrazepam, oxazolam, barbital, phenobarbital, secobarbital sodium, and triazolam) in food products that claim to have medicinal effects and/or are assigned to a specific category of drugs (Tsai et al. 2004). The TLC method involves the compound separation on Merck Kieselgel 60F$_{254}$ plates eluted with three developing solvent systems: chloroform–ethanol 9:1 v/v; ethyl acetate–ether 4:1 v/v; and chloroform–acetone 4:1 v/v. Three detection reagents were used, which give different color spots and allow a certain detection of compounds: Dragendorff's spray reagent—orange spots for all compounds; Zwikker's reagent—pink spots for barbital, phenobarbital, and secobarbital sodium; and Marquis reagent—different color spots for all compounds. All compounds are also visible in UV light at 254 nm. The R_f values of compounds were in the range of 0.17–0.92. The identification of compounds after TLC separation was done using UV spectra. The spot was scratched, dissolved in methanol, centrifuged, and filtrated, and the filtrate was examined with a UV spectrophotometer. The combination of R_f values and UV spectra furnishes much more information making compound identification unquestionable. The results are in good agreement with those obtained by GC/MS and LC/MS.

A rapid and reliable method for qualitative and quantitative analysis of cream biscuits or similar food matrix spiked with diazepam was developed by Ghosh et al. (2004). The method is based on HPTLC separation of diazepam from samples and medazepam as internal standard (IS) on silica gel 60F$_{254}$ plates prewashed with methanol and dried at 105°C for 1 h, developed in a paper-lined twin-trough chamber, and saturated for 20 min with chloroform–acetone 85:15 v/v as mobile phase.

The densitograms and spectra were obtained by UV-Vis scanning at 230 nm. The corresponding spots were scraped from the plate, extracted with ethanol, and analyzed by MS operating at 70 eV using a direct inlet system (electron impact ionization, EI). The LOD and LOQ of diazepam in the samples were 7.5 ng/6 mm band and 28 ng/6 mm band. The proposed method could be used in enforcement laboratories for routine analysis of diazepam from a similar food matrix.

Another TLC method with sensitive detection for determination of diazepam and phenobarbital together with saccharin from toddy samples, a fermented juice from a coconut, brab, date, or any kind of palm tree containing no more than 5% alcohol, was developed and tested (Mali et al. 2005). The toddy samples were separately spiked with each of the adulterants diazepam, Phenobarbital, and saccharin and left for one day. Then, the acidified solutions (1 M HCl, pH = 3) were extracted separately twice with diethyl ether. Finally, the solvent was evaporated at room temperature, and residues were dissolved separately in ethanol. The chromatographic separation was done on 0.25 mm silica gel TLC plates activated by heating in an oven at 100°C for 1 h before use and stored in a desiccator. The plates were developed in a presaturated glass chamber by the ascending technique, to a distance of ~10 cm with three mobile phases: chloroform–acetic acid 9:1 v/v, n-hexane–acetone–methanol 16:6:1 v/v/v, and n-hexane–acetone–butanol 24:16:1 v/v/v from the point of application. The plate was dried in air and then was placed for ca. 5 min in a chamber containing chlorine gas (prepared 10 min earlier). After excess chlorine removal, the plate was uniformly sprayed with o-tolidine, then with 1% phosphomolybdic acid, when intense blue spots appear on a white background. The color of the spots, which became faint after approximately 1 h, can be stabilized by spraying with 1% phosphoric acid solution. The limits of detection for diazepam, phenobarbital, and saccharin with this reagent are 0.5 µg/spot (8 µg/cm^2), 0.3 µg/spot (7 µg/cm^2), and 0.1 µg/spot (9 µg/cm^2), respectively.

In conclusion, the obtained results showed that HPTLC, LC, and GC techniques are comparable for the determination of such compounds in the requested working range to be analyzed in raw materials and in clinical and toxicological screening.

REFERENCES

Aboul-Enein, H.Y. and Thiffault, C.H. 1991. Identification of some primidone urinary metabolites in dyslexic patient by thin-layer chromatography. *Anal. Lett.* 24: 209–216.

Abourashed, E.A., Hefnawy, M.M., and El-Subbagh, H.I. 2009. HPTLC analysis of a new ultra-shortacting thiazolodiazepine hypnotic (HIE-124) in spiked human plasma. *J. Planar Chromatogr. Mod. TLC* 22: 183–186.

Agarwal, S., Gonsalves, H., and Khar, R. 2008. HPTLC method for the analysis of melatonin in bulk and pharmaceutical formulations. *Asian J. Chem.* 20: 2531–2538.

Anonymous. 1988. Thin-layer chromatographic detection methods. *Merck Spect. (Darmstadt)* 2: 61.

Bailey, D.N. and Kelner, M. 1984. Extraction of acidic drugs from water and plasma: Study of recovery with five different solvents. *J. Anal. Toxicol.* 8: 26–28.

Bakavoli, M. and Kaykhaii, M. 2003. Quantitative determination of diazepam, nitrazepam and flunitrazepam in tablets using thin-layer chromatography-densitometry technique. *J. Pharm. Biomed. Anal.* 31: 1185–1189.

Baselt, R.C. 1987. *Analytical Procedures for Therapeutic Drug Monitoring and Emergency Toxicology*, 2nd edn., Littleton, MA: PSG Publishing Co. Inc.

Battista, H.J., Udermann, H., Henning, G., and Vycudilik, W. 1979. Detection of benzodiazepines in forensic chemistry. *Beitr. Gerichtl. Med.* 37: 5–28.

Beesley, T. 2010. Evolution of chromatography: One scientist's 51-year journey. *LC/GC North Am.* 28: 960–971.

Bonfilio, R., De Araújo, M.B., and Nunes Salgado, H.R. 2010. Recent applications of analytical techniques for quantitative pharmaceutical analysis: A review. *WSEAS Trans. Biol. Biomed.* 7: 316–338.

British Pharmacopoeia, 2009. London, U.K.: The Stationery Office.

Büch, U., Altmayer, P., Isenberg, J.C., and Büch, H.P. 1991. Increase of thiopental concentration in tissues of the rat due to an anestesie with halothane. *Arzneim.-Forsch./Drug Res.* 41: 363–366.

Budd, R.D. and Mathis, D.F. 1982. GLC screening and confirmation of barbiturates in postmortem blood apecimens. *J. Anal. Toxicol.* 6: 317–320.

Busch, M., Ritter, W., and Moehrle, H. 1984. Dünnschicht-densitometrische Bestimmung von Clotiazepam in Tabletten und Blutplasma. *Arzneim. Forsch.* 35: 547–551.

By, A., Ethier, J.C., Lauriault, G. et al. 1989. Traditional oriental medicine. I. Blank pearl: Identification and chromatographic determination of some undeclared medicinal ingredients. *J. Chromatogr.* 469: 406–411.

Casas, M., Berrueta, L.A., Gallo, B., and Vicente, F. 1993. Solid phase extraction of 1,4benzodiazepines from biological fluids. *J. Pharm. Biomed. Anal.* 11: 277–284.

Chan, E.M. and Chan, S.C. 1984. Screening for acidic and neutral drugs by high performance liquid chromatography in postmortem blood. *J. Anal. Toxicol.* 8: 173–176.

Chen, X.-H., Franke, J.-P., Wijsbe, J., and De Zeeuw, R.A. 1992. Isolation of acidic, neutral, and basic drugs from whold blood using a single mixed-mode solid-phase extraction column. *J. Anal. Toxicol.* 16: 351–355.

Cheng, S.-C., Huang, M.-Z., and Shiea, J. 2009. Thin-layer chromatography/laser-induced acoustic desorption/electrospray ionization mass spectrometry. *Anal. Chem.* 81: 9274–9281.

Cimpoiu, C. and Hodisan, T. 1999. Application of numerical taxonomy techniques to the choice of optimum mobile phase in high-performance thin-layer chromatography (HPTLC). *J. Pharm. Biomed. Anal.* 21: 895–900.

Cimpoiu, C., Hodisan, T., and Nascu, H. 1997. Comparative study of mobile phase optimization for the separation of some 1,4-benzodiazepines. *J. Planar Chromatogr. Mod. TLC* 10: 195–199.

Cimpoiu, C., Jantschi, L., and Hodisan, T. 1998. A new method for mobile phase optimization in high-performance thin-layer chromatography (HPTLC). *J. Planar Chromatogr. Mod. TLC* 11: 191–194.

Cimpoiu, C., Jantschi, L., and Hodisan, T. 1999. A new mathematical model for the optimization of the mobile phase composition in HPTLC and the comparison with other models. *J. Liq. Chromatogr. Relat. Technol.* 22: 1429–1441.

Cole, M.D. 2003. *The Analysis of Controlled Substances*, Hoboken, NJ: John Wiley & Sons, Ltd.

Constantini, A. and Paoli, F. 1998. Melatonin: Quantitative analysis in pharmaceutical oral dosage forms using thin-layer chromatography (TLC) densitometry. *Il Farmaco* 53: 443–447.

Csehárti, T. and Hauck, H.E. 1990. Retention characteristics of CN, NH_2 and diol precoated high-performance thin-layer chromatographic plates in the adsorption and reversed-phase separation of some benzodiazepine derivatives. *J. Chromatogr. A* 514: 45–55.

Daundkar, B.B., Malve, M.K., and Krishnamurthy. 2008. A specific chromogenic reagent for detection of diazepam among other benzodiazepines from biological and nonbiological samples after HPTLC. *J. Planar Chromatogr. Mod. TLC* 21: 249–250.

De Souza, A. and Hamon, M. 1990. Degradation of diazepam in aqueous solution: Simulation essay by action of diluted hydrogen peroxide. *Ann. Pharm. Fr.* 48: 7–12.

Dhavale, N., Gandhi, S., Sabnis, S., and Bothara, K. 2008. Simultaneous HPTLC determination of escitalopram oxalate and clonazepam in combined tablets. *Chromatographia* 67: 487–490.

Dongre, V.G., Kamble, V.W., and Janrao, D.M. 2000. High-performance thin-layer chromatographic detection of 1,4-benzodiazepines. *J. Planar Chromatogr. Mod. TLC* 13: 468–470.

Dreassi, E., Corbini, G., Corti, P., Ulivelli, M., and Rocchi, R. 1996. Quantitative analysis of lamotrigine in plasma and tablets by planar chromatography and UV spectrophotometry. *J. AOAC Int.* 79: 1277–1280.

Drummer, O.H. 1998. Methods for the measurement of benzodiazepines in biological samples. *J. Chromatogr. B Biomed. Sci. Appl.* 713: 201–225.

Drummer, O.H. and Odell, M. 2001. *The Forensic Pharmacology of Drugs of Abuse*, London, U.K.: Arnold.

Ebel, S. and Schütz, H. 1977. Determination of clonazepam and its main metabolites. *Arzneim.-Forsch.* 27: 325–328.

Edler, M. and Schlüter, R. 1994. Dünnschichtchromatographischer Nachweis von Flunitrazepam (RohypnolR) im Urin. *T+K* 61: 74–79.

European Pharmacopoeia, 6th edn., 2007. Strasbourg, France: Council of Europe.

Farina, A., Doldo, A., Cotichini, V., and Rajevic, M. 1996. Assay and purity control of new serotonergic anxiolytics by HPTLC and scanning densitometry. *J. Planar Chromatogr. Mod. TLC* 9: 185–188.

Ferrara, S.D., Tedeschi, L., Frison, G., and Castagna, F. 1992. Solid-phase extraction and HPLC-UV confirmation of drugs of abuse in urine. *J. Anal. Toxicol.* 16: 217–222.

Fujimoto, J.M. and Wang, R.I.H. 1970. A method of identifying narcotic analgesics in human urine after therapeutic doses. *Toxicol. Appl. Pharmacol.* 16: 186–190.

Gal-Szabo, Z.S. and Perneczki, S. 1991. Comparison of solid phase and liquid-liquid extraction of diazepam and nitrazepam. *Acta Pharm. Hung.* 61: 239–245.

Ghosh, P., Krishna Reddya, M.M., Ramteke, V.B., and Sashidhar Rao, B. 2004. Analysis and quantitation of diazepam in cream biscuits by high-performance thin-layer chromatography and its confirmation by mass spectrometry. *Anal. Chim. Acta* 508: 31–35.

Guo, D. 1985. Analysis of analgesic tablets by thin-layer chromatography. *Chinese J. Pharm. Anal.* 5: 306–307.

Haefelfinger, P. 1970. Empfindlicher nachweis von substanzen mit primaren aminogruppen auf dunnschicht-tplatten. *J. Chromatogr.* 48: 184–187.

Hancu, G., Fulop, E., Rusu, A., Mircia, E., and Gyeresi, A. 2011. Thin layer chromatographic separation of benzodiazepines derivatives. *Anal. Univ. Bucuresti* 20: 181–188.

Howarth, A.T. and Clegg, G. 1978. Simultaneous detection and quantitation of drugs commonly involved in selfadministered overdoses. *Clin. Chem.* 24: 804–807.

Inoue, T. and Niwaguchi, T. 1985. Determination of nitrazepam and its main metabolites in urine by thin-layer chromatography and direct densitometry. *J. Chromatogr.* 339: 163–169.

International Drug Control Programme. 1997. *Recommended Methods for the Detection and Assay of Barbiturates and Benzodiazepines in Biological Specimens—Manual for Use by National Laboratories*, New York: United Nations.

Jain, R. 1993. Simplified method for simultaneous determination of diazepam and its metabolites in urine by thin-layer chromatography and direct densitometry. *J. Chromatogr.* 615: 365–368.

Jain, R. 2000. Utility of thin layer chromatography for detection of opioids and benzodiazepines in a clinical setting. *Addict. Behav.* 25: 451–454.

Japanese Pharmacopoeia XVI, 2011. Tokyo, Japan: The Minister of Health, Labour and Welfare.

Kakde, R., Satone, D., and Bawane, N. 2009. HPTLC method for simultaneous analysis of escitalopram oxalate and clonazepam in pharmaceutical preparations. *J. Planar Chromatogr. Mod. TLC* 22: 417–420.

Kamble, V.W. and Dongre, V.G. 1997. TLC detection and identification of heroin (diacetylmorphine) in street samples: Part II. *J. Planar Chromatogr. Mod. TLC* 10: 384–386.

Kamble, V.W., Garad, M.B., and Padalikar, S.V. 1995. Identification and estimation of diazepam in Ayurvedic preparations. *J. Planar Chromatogr. Mod. TLC* 8: 143–144.

Kastner, P. and Klimes, J. 1996. Analysis of benzodiazepines by adsorption and ion-pair RP TLC. *J. Planar Chromatogr. Mod. TLC* 9: 382–387.

Khdary, N.H. and Shaikh, K.M. 1996. A new technique for visualizing thin layer chromatography plates. *Bull. TIAFT* 26: 38–41.

Klein, C.M. and Lau-Cam, C.A. 1986. Simple thin-layer chromatographic method for the investigation of six related compounds in flurazepam hydrochloride and its capsules. *J. Chromatogr.* 350: 273–278.

Koba, M., Koba, K., and Bączek, T. 2009. Determination of Oxazepam in pharmaceutical formulation by HPTLC UV-densitometric and UV-derivative spectrophotometry methods. *Anal. Lett.* 42: 1831–1843.

Kulkarni, R.R., Patil, V.B., Bhoi, A.G., and Knandode, S. 1998. A new spray reagent for the detection of phentoin by thin layer chromatography. *J. Planar Chromatogr. Mod. TLC* 11: 309–310.

Langman, L.J. and Kapur, B.M. 2006. Toxicology: Then and now. *Clin. Biochem.* 39: 498–510.

Li, B. and Wang, S. 1985. Determination of diazepam by TLC and scanning densitometry. *Chinese J. Pharm. Anal.* 5: 176–177.

Li, T. and Sun, Y. 1996. Simultaneous determination of diazepam, nitrazepam, estazolam in human urine by thin-layer chromatography. *Chinese J. Hosp. Pharm.* 16: 418–419.

Lillsunde, P., Michelson, L., Forsstrom, T. et al. 1996. Comprehensive drug screening in blood for detecting abused drugs or drugs potentially hazardous for traffic safety. *Forensic Sci. Int.* 77: 191–210.

Logan, B.K. and Stafford, D.T. 1989. Liquid/solid extraction on diatomaceous earth for drug analysis in post-mortem blood. *J. Forensic Sci.* 34: 553–564.

Makowski, A., Adamek, E., and Baran, W. 2010. Use of photocatalytic reactions to visualize drugs in TLC. *J. Planar Chromatogr. Mod. TLC* 23: 84–86.

Mali, B.D. and Garad, M.V. 2005. Thin-layer chromatographic detection of chloral hydrate in an alcoholic beverage. *J. Planar Chromatogr. Mod. TLC* 18: 397–399.

Mali, B.D., Rathod, D.S., and Garad, M.V. 2005. Thin-layer chromatographic determination of diazepam, phenobarbitone, and saccharin in toddy sample. *J. Planar Chromatogr. Mod. TLC* 18: 330–332.

Mangin, P., Lugnier, A.A., and Chaumont, A.J. 1987. A polyvalent method using HPLC for screening and quantification of 12 common barbiturates in various biological materials. *J. Anal. Toxicol.* 11: 27–30.

Maslanka, A. and Krzek, J. 2005. Densitometric high performance thin-layer chromatography identification and quantitative analysis of psychotropic drugs. *J. AOAC Int.* 88: 70–79.

Maslanka, A., Krzek, J., Zuromska, B., and Stolarczyk, M. 2011. Identification and determination of compounds belonging to the group of OUN pharmaceutical agents by thin-layer chromatography with densitometric detection in biological material. *Acta Chromatogr.* 23: 247–266.

Maurer, H.H. 1999. Systematic toxicological analysis procedures for acidic drugs and/or metabolites relevant to clinical and forensic toxicology and/or doping control. *J. Chromatogr. B* 733: 3–25.

McElwee, D.J. 1979. A new method for the detection of sedative drugs on thin-layer chromatograms using chlorine vapors and 2,7-dichlorofluorescein. *J. Anal. Toxicol.* 3: 266–268.

Moffat, A.C. 1986. *Clarke's Isolation and Identification of Drugs*, 2nd edn., London, U.K.: The Pharmaceutical Press.

Mußhoff, F. and Daldrup, T. 1992. A rapid solid phase extraction and HPLC/DAD procedure for simultaneous determination and quantitation of different benzodiazepines in serum, blood and postmortem blood. *Int. J. Legal Med.* 105: 105–109.

Mule, S.J. and Casella, G.A. 1989. Confirmation and quantitation of barbiturates in human urine by gas chromatography/mass spectrometry. *J. Anal. Toxicol.* 13: 13–16.

Musshoff, F., Padosch, S., Steinborn, S., and Madea, B. 2004. Fatal blood and tissue concentrations of more than 200 drugs. *Forensic Sci. Int.* 142: 161–210.

Noir-Falise, A., Dodinval, P., Quiriny, J., and Schreber, J. 1987. Death through injection of barbiturates. *Forensic Sci. Int.* 35: 141–144.

Ojanpera, I. 1992. Toxicological drug screening by thin-layer chromatography. *Trends Anal. Chem.* 11: 222–229.

Okamoto, M., Kakamu, H., Oka, H., and Ikai, Y. 1993. Evaluation of thin-layer chromatography—Fast atom bombardment mass spectrometry: Application to the determination of midazolam intoxication by use of 3-glycidoxypropyltreated thin-layer chromatographic plates. *Chromatographia* 36: 293–296.

Otsubo, K., Seto, H., Futagami, K., and Oishi, R. 1995. Rapid and sensitive detection of benzodiazepines and zopiclone in serum using high-performance thin-layer chromatography. *J. Chromatogr. B* 669: 408–412.

Patel, R.B., Patel, A.B., and Patel, M.R. 2009a. Estimation of alprazolam and sertraline in pure powder and tablet formulations by high-performance liquid chromatography and high-performance thin-layer chromatography. *Anal. Lett.* 42: 1588–1602.

Patel, R.B., Patel, M.R., Shankar, M.B., and Bhatt, K.K. 2009b. Simultaneous determination of alprazolam and fluoxetine hydrochloride in tablet formulations by high-performance column liquid chromatography and high-performance thin-layer chromatography. *J. AOAC Int.* 92: 1082–1088.

Patel, S. and Patel, N. 2008. HPTLC estimation of amitriptyline HCl, trifluoperazine HCl, risperidone and alprazolam in pharmaceutical products using single mobile phase. Paper presented at *60th Indian Pharmaceutical Congress*, New Delhi, India.

Patel, S.K. and Patel, N.J. 2010. Simultaneous determination of imipramine hydrochloride and chlordiazepoxide in pharmaceutical preparations by spectrophotometric, RP-HPLC, and HPTLC methods. *J. AOAC Int.* 93: 904–910.

Patil, V.B. and Shingare, M.S. 1993. Thin layer chromatographic detection of certain benzodiazepines. *J. Planar Chromatogr. Mod. TLC* 6: 497–498.

Paw, B. and Misztal, G. 2000. Chromatographic analysis (TLC) of zopiclone and benzodiazepines. *J. Planar Chromatogr. Mod. TLC* 13: 195–198.

Pippenger, C.E. 1989. Therapeutic drug monitoring in the 1990s. *Clin. Chem.* 35: 1348–1351.

Pocci, R., Dixit, V., and Dixit, V.M. 1992. Solid phase extraction and GC/MS confirmation of barbiturates from human urine. *J. Anal. Toxicol.* 16: 45–47.

Pushpalatha, P., Sarin, R.K., Rao, M.A., and Baggi, T.R.R. 2009. A new thin-layer chromatographic method for analysis of zolpidem and zopiclone. *J. Planar Chromatogr. Mod. TLC* 22: 449–451.

Rizzo, M. 2000. Chromatographic separation of 2,3-benzodiazepines. *J. Chromatogr. B* 474: 203–216.

Rochholz, G., Ahrens, B., and Schütz, H. 1994. Modified screening procedure with fluorescence detection for flunitrazepam and its metabolites via acridine derivatives. *Arzneim.-Forsch./Drug Res.* 44: 469–471.

Roering, D.L., Lewand, D.L., Mueller, M.A., and Wang, R.I.H. 1975. Comparison of radioimmunoassay with thin-layer chromatographic and gas-liquid chromatographic methods of barbiturates detection in human urine. *Clin. Chem.* 21: 672–675.

Romanian Pharmacopoeia X, 2005. Bucuresti, Romania: Medicala.

Saelzer, R., Godoy, G., Vega, M., De Diego, M., Godoy, R., and Rios, G. 2001. Instrumental planar chromatographic determination of benzodiazepines: Comparison with liquid chromatography and gas chromatography. *J. AOAC Int.* 84: 1287–1295.

Salo, P.K., Vilmunen, S., Salomies, H., Ketola, R.A., and Kostiainen, R. 2007. Two-dimensional ultra-thin-layer chromatography and atmospheric pressure matrix-assisted laser desorption/ionization mass spectrometry in bioanalysis. *Anal. Chem.* 79: 2101–2108.

Sánchez-Moyano, E., Herráez, M., and Plá-Delfina, J.M. 1986. A contribution to the pharmaceutical analysis of benzodiazepines. *Pharm. Acta Helv.* 61: 167–176.

Sarin, R.K., Sharma, G.P., Varshney, K.M., and Rasool, S.N. 1998. Determination of diazepam in cold drinks by high-performance thin-layer chromatography. *J. Chromatogr. A* 822: 332–335.

Sawada, H., Hara, A., Asano, S., and Matsumoto, Y. 1976. Isolation and identification of benzodiazepine drugs and their metabolites in urine by use of Amberlite XAD-2 resin and thin-layer chromatography. *Clin. Chem.* 22: 1596–1603.

Schütz, H. 1996. *Dunnschichtchromatograpische Suchanalyse fur 1,4-Benzodiazepine in Harn, Blut und Mageninhalt. Mitteilung VI der Denatskommision fur Klinisch-toxikologische Analytik*, Stuttgart, Germany: VCH Verlagsgessellschaft mbH.

Schütz, H., Borchert, A., Koch, E.M., Schneider, W.R., and Schölermann, K. 1987. Dünnschichtchromatographischer Nachweis von Benzodiazepinen unter Berücksichtigung neuerer tetrazyklischer Substanzen. *J. Clin. Chem. Clin. Biochem.* 25: 628–629.

Schütz, H. and Fitz, H. 1985. Screening des neuen Anxiolyticums Hlazepam. *Fresenius Z. Anal. Chem.* 321: 359–362.

Schuetz, H., Fitz, H., and Suphachearabhan, S. 1983. Screening und Nachweis von Ketazolam und Oxazolam. *Arzneim. Forsch.* 33: 507–512.

Sheladia, S., Patel, S., Patel, N., Patel, S., and Patel, T. 2008. Simultaneous estimation of trifluoperazine HCl and chlordiazepoxide by HPTLC method in pharmaceutical formulation. Paper presented at *60th Indian Pharmaceutical Congress*, New Delhi, India.

Shirke, P., Patel, M.D., Tirodkar, V.B., Tamhane, V., and Sethi, P.D. 1994. Estimation of amitriptyline hydrochloride and chlordiazepoxide in combined dosage form by HPTLC. *East. Pharm.* 37: 179–180.

Siek, T.J., Stradling, C.W., McCain, M.W., and Mehary, T.C. 1997. Computer-aided identifications of thin-layer chromatographic patterns in broad-spectrum drug screening. *Clin. Chem.* 43: 619–623.

Singh, H., Sattayasai, J., Lattmann, P., Boonprakob, Y., and Lattmann, E. 2010. Antidepressant/anxiolytic and anti-nociceptive effects of novel 2-substituted 1,4-Benzodiazepine-2-ones. *Sci. Pharm.* 78: 155–169.

Šoljić, Z., Grba, V., and Bešić, J. 1977. Separation of some compounds from the 1,4-benzodiazepine groupe by thin-layer chromatography. *Chromatographia* 10: 751–754.

Spangenberg, B., Seigel, A., Kempf, J., and Weinmann, W. 2005. Forensic drug analysis by means of diodearray HPTLC using R_F and UV library search. *J. Planar Chromatogr. Mod. TLC* 18: 336–343.

Stahlmann, S. and Kovar, K.-A. 1998. Analysis of impurities by high-performance thin-layer chromatography with Fourier transform infrared spectroscopy and UV absorbance detection in situ measurement: Chlordiazepoxide in bulk powder and its tablets. *J. Chromatogr.* 813: 145–152.

Steidinger, J. and Schmid, E. 1970. Studies on the metabolism of oxazepam. *Arzneim.-Forsch.* 20: 1232–1235.

St-Pierre, M.V. and Pang, K.S. 1987. Determination of diazepam and its metabolites by high-performance liquid chromatography and thin-layer chromatography. *J. Chromatogr.* 421: 291–307.

Sumina, E.G., Shtykov, S.N., and Dorofeeva, S.V. 2002. Ion-pair reversed-phase thin-layer chromatography and high-performance liquid chromatography of Benzoic Acids. *J. Anal. Chem.* 57: 210–214.

Sun, S.R. 1978. Fluorescence-TLC densitometric determination of diazepam and other 1,4-benzodiazepines in serum. *J. Pharm. Sci.* 67: 1413–1415.

Tames, F., Watson, I.D., Morden, W.E., and Wilson, I.D. 2000. Analysis of benzodiazepines by thin-layer chromatography and tandem mass spectrometry. *J. Planar Chromatogr. Mod. TLC* 13: 432–436.

Thoma, K. and Struve, M. 1985. Dtabilität von Barbitursäurederivaten in wässrigen Lösunge. *Dtsch. Apoth. Ztg.* 125: 2062–2068.

Tsai, F.-I., Wei, N.-T., Wu, C.-Y., Chen, Y.-P., and Wen, K.-C. 2004. Adulteration of drugs in food—Glucocorticoids, anorexics and hypnotic-sedatives (I). *J. Food Drug Anal.* 12: 84–96.

Ünak, P., Yurt, F., and Zümrüt Biber, F. 2002. Labeling of chlorodiazepoxide with ^{131}I and biodistribution studies on rats. *J. Radioanal. Nucl. Chem.* 251: 253–256.

United States Pharmacopeia 30-NF25, e-Book. 2007. Rockville, MD: United States Pharmacopeial Convention.

Van Rooij, H.H., Fakiera, A., Verrijk, R., and Soudijn, W. 1985. The identification of flunitrazepam and its metabolites in urine samples. *Anal. Chim. Acta* 170: 153–158.

Verma, J. and Joshi, A. 2005. Simultaneous estimation of alprazolam and sertraline in tablet dosage form by HPTLC. *Indian Drugs* 42: 805–807.

Volf, K. 1998. Detection of 1,4-benzodiazepines and benzophenones with a secondary amino group by thermal conversion and diazotization and spraying with Bratton-Marshall reagent. *J. Planar Chromatogr. Mod. TLC* 11: 132–136.

Volford, O., Takacs, M., and Vamos, J. 1996. Ion exchange interaction on silica gel in thin-layer chromatography. Part IV. On plate investigations by UV spectroscopy. *Acta Pharm. Hung.* 66: 133–140.

White, D.J., Stewart, J.T., and Honigberg, I.L. 1991a. Quantitative analysis of chlordiazepoxide hydrochloride and related compounds in drug substance and tablet dosage form by HPTLC and scanning densitometry. *J. Planar Chromatogr. Mod. TLC* 4: 330–332.

White, D.J., Stewart, J.T., and Honigberg, I.L. 1991b. Quantitative analysis of diazepam and related compounds in drug substance and tablet dosage form by HPTLC and scanning densitometry. *J. Planar Chromatogr. Mod. TLC* 4: 413–415.

Wojciak-Kosior, M., Skalska, A., Matysik, G., and Kryska, M. 2006. Quantitative analysis of phenobarbital in dosage form by thin-layer chromatography combined with densitometry. *J. AOAC Int.* 89: 995–998.

Wu, Y., Tan, J., and Xia, Y. 2002. Determination of in urine by thin-layer chromatography. *Chinese J. Chromatogr.* 20: 182–184.

Xue, S. and Dou, X. 1985. Quality control of diazepam injection. *Chinese J. Pharm. Anal.* 5: 98–99.

Yonamine, M. and Cortez Sampaio, M. 2006. A high-performance thin-layer chromatographic technique to screen cocaine in urine samples. *Leg. Med.* 8: 184–187.

Yue, Q. and Gao, B. 1990. Determination of phenobarbital in Yinaoningxian tablets by thin-layer chromatography/ultraviolet spectrophotometry. *Chinese Anal. Chem.* 18: 161–163.

Zhang, A., Bu, Zh., Zhang, Y., Wang, Q., Li, W., and Qi, L. 1993a. Determination of alprazolam in plasma by thin-layer chromatography. *Chinese J. Pharm.* 28: 233–234.

Zhang, Y., Geng, Y., and Wu, Y. 1994. Determination of sedative hypnotics in blood by thinlayer chromatography. *Chinese J. Hosp. Pharm.* 14: 3–5.

Zhang, Y., Ren, J., and Xing, L. 1993b. Determination of diazepam, methaqualone and estazolam in human serum by thin-layer chromatography. *Chinese J. Pharm.* 28: 550–552.

Zhang, Sh., Xu, Q., and Hu, W. 1995. Determination of diazepam in rat tissues by thin-layer chromatography and UV spectrophotometry. *J. Chinese Pharm. Univ.* 26: 20–22.

Zhong, H. 1985. Separation of urinary hypnotic sedatives by thin-layer chromatography. *Chinese J. Pharm. Anal.* 16: 362–363.

Żydek, G. and Brzezińska, E. 2011. Normal and reversed phase thin layer chromatography data in quantitative structure–activity relationship study of compounds with affinity for serotonin (5-HT) receptors. *J. Chromatogr. B* 879: 1764–1772.

17 TLC of Morphine Analogs

Grzegorz Jóźwiak

CONTENTS

17.1 MORPHINE AND ITS ANALOGS

The name morphine derives from Morpheus, the Greek god of dreams. The use of a wide variety of opium alkaloids, obtained from the dried sap of *Papaver somniferum*, has been reported in several cultures. The potential of opium was known and used in ancient Greece, Rome, as well as Arab cultures. Morphine, which is the benzylisoquinoline alkaloid, occurs in significant amounts in opium. First isolated by Friedrich Sertürner in 1804, morphine has been successfully used as a powerful narcotic painkiller, sedative, and hypnotic [1]. Apart from morphine, contemporary medicine also uses its semisynthetic/synthetic analogs. Morphine and its analogs, both depicting a strong biological activity, require various methods to control their production and use.

The derivatives of morphine can be divided into the following groups:

1. Agonists
   - Natural narcotic substances that appear in opium, for example, morphine or codeine
   - Semisynthetic analogs such as hydromorphone, oxymorphone, or oxycodone
   - Synthetic compounds such as levorphanol, methadone, sufentanil, fentanyl, or remifentanil
2. Mixed agonists–antagonists: Nalbuphine and pentazocine display an agonist activity at some receptors, whereas their antagonist activity is observed at the other receptors. Apart from nalbuphine and pentazocine, there are also the so-called partial agonists such as butorphanol and buprenorphine.
3. Narcotic antagonists such as naloxone do not display an agonist activity at any of the opiate receptors. Antagonists block the opiate receptor and inhibit the pharmacological activity of the agonist, which proves helpful in the therapy and treatment of addicted patients.

Morphine is a principal alkaloid obtained from the opium poppy (*P. somniferum*). Because of its natural source, morphine and morphine derivatives are referred to as opiates. The prototypic narcotic analgesic is (–)-morphine (Figure 17.1); the remaining narcotic analgesics will be classified on the basis of their structural derivation from morphine [2].

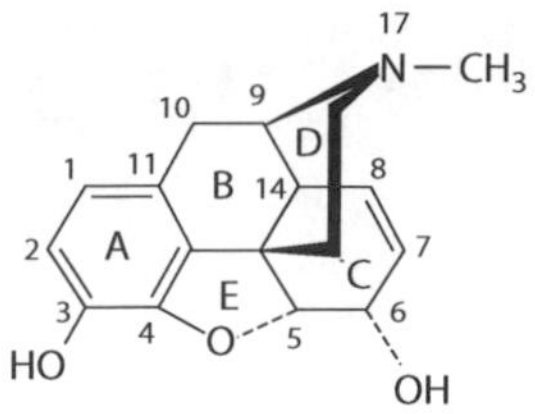

FIGURE 17.1   (−)-Morphine.

17.2   TLC AND FREQUENTLY APPLIED TLC SYSTEMS

Thin-layer chromatography (TLC) and high-performance thin-layer chromatography (HPTLC) are useful techniques thoroughly applied in the industrial, toxicological, and forensic analysis of morphine and its analogs. They are also used in the qualitative and quantitative analysis of opium—the natural source of morphine and its derivatives. In Asian countries, where opium is still used in simple pharmaceutical preparations such as tinctures, TLC and HPTLC are also frequently used in local pharmaceutical industry. As far as toxicological and forensic analyses are concerned, the previously mentioned planar techniques prove useful especially in cases where rapid qualitative and quantitative determination of substances or their metabolites comprised in partially cleaned or even raw biological samples is required for the analysis and where other methods, such as HPLC, might be more time-consuming during sample preparation process. Moreover, a quick result of the analysis is also required for the qualitative analysis of specific samples (as in the case of requisitioned drugs from illegal sources) for their rapid detoxification and further judicial decisions. Qualitative and quantitative analysis by TLC and HPTLC, supported with densitometry, finds its widespread application as a successful method in the analysis of opiates. The chemical nature of these compounds requires the use of systems with polar stationary phases, especially silica gel, as the solid phase. Tombesi et al. applied alumina as the stationary phase with different mixtures of eluents for the separation of picrates of morphine alkaloids (i.e., morphine and codeine) [3]. Szumilo and Flieger used silica gel impregnated with the solution of the various metal salts with mixture of nonaqueous solvents as the mobile phase to separate morphine and codeine [4]. Stationary phase modified with bismuth salts was also used for the chromatographic analysis of dextropropoxyphene [5].

The use of RP-TLC systems is not as popular as NP-TLC. Nevertheless, RP-TLC systems are also applied in the analysis of morphine and its analogs [6,7]. The literature also proposes the use of diol as the chemically modified stationary phase in the earlier application [8].

The mixtures of solvents of medium polarity are used as the mobile phase in the development of plates. The complicated matrix, which usually contains a large number of pollutants or metabolites, causes the complexity of the mobile phase. In view of the fact that the majority of opiate compounds have a nitrogen atom in the molecule (weak nitrogenous bases), the solution of either ammonia or amines is often used as the mobile phase modifier. They play either a role of ion suppressing or a role of blocker of surface free residue silanols. The solution of ammonia is most frequently used because of the possibility of its simple evaporation from the developed chromatographic plates. This is important because of the possibility of subsequent reaction with the reagent used to visualize the separated substances.

The analysis of complex mixtures also requires special chromatogram developing techniques, such as 2D-TLC, automated multiple development (AMD), and overpressured layer chromatography (OPLC). 2D-TLC is the easiest technique that does not need any complex instrumentation [6,9], whereas such methods as AMD and OPLC need special equipment. Both AMD and OPLC have been used to develop the chromatograms in the analysis of opium alkaloids [10].

17.3 METHODS OF DETERMINATION

The method of detection used in TLC is as important as the selection of optimized chromatographic system. For this reason, the use of standards of chromatographed substances to compare the retention values for opiate identification is a good solution. In case of the determination of colorless substances, such as morphine and its analogs, the use of UV light for the detection might also prove helpful [11,12].

17.3.1 DERIVATIZING REAGENTS

In the opiate analysis, there is the need to use specific reagents that react only with determined substances and change their properties (such as color or specific wavelength absorbance), which allows to distinguish those substances from the others. The substances being the result of such reactions are localized in visible or UV light [13], but they are not so typical to be clearly identified as other compounds of the opiate and opiate-related groups (e.g., derivatives or metabolites) and may give a similar response.

Dragendorff's reagent is very popular in the detection of nitrogenous bases (orange spots and the yellow background) [3,9,14]. Marquis reagent [9] is also popular for the identification of alkaloids; however, it contains a concentrated sulfuric acid that may give false signals from carbonized compounds (e.g., sugars). Opiates are also examined as DANS derivatives (with the reaction before and after chromatographic separation) [9,15] or as picrates [3].

The examples of reagents applied in the visualization of morphine and its derivatives are presented in Table 17.1.

17.3.2 DENSITOMETRY

Densitometry was originally applied as a useful method for electrophoresis. Nowadays, the technique is mainly used in visualization and evaluation of TLC chromatograms. The evaluation of reflected light enables to draw a chromatogram/densitogram for each track, thus permitting qualitative (retention values) and quantitative analysis. Currently used devices allow to measure the spectra of substances in spots to confirm their identity. Before the densitometric scanning, the chromatographic plate is usually sprayed with a suitable reagent to obtain the derivatives, which manifests with the adequate wavelength. Using scanner with diode array detector (DAD) is the newest method that gives more information than classical densitometry. The UV spectra of separated substances taken from an HPTLC silica plate were used to identify codeine, tramadol, flupirtine, and lidocaine [16]. Densitometry as the detection method finds its application in clinical, toxicological, and forensic analytical analysis. The examples of the application of densitometry to the analysis of opiates are presented in Table 17.2.

Moreover, the classical spectrophotometric method is used to determine drugs eluted after development. Quantitative analysis at 258 nm for dextropropoxyphene (with paracetamol, oxyphenbutazone, and ibuprofen) was performed [17].

17.4 INDUSTRY/CLINICAL/TOXICOLOGY/FORENSIC APPLICATION

TLC has been used in quantitative and qualitative analysis of opium, which is still used as the source of morphine and its derivatives. To examine opium alkaloids, the following systems are used: silica/benzene + chloroform + acetone + MeOH + diethylamine (12:3:3:1:1), spraying with 0.5 M solution of sulfuric acid, absorbance at 285 nm [13], and silica/AcOEt + MeOH + NH_3 (85:10:5), detected under UV light, 254 nm, with the quantification with densitometry at 280 nm [18] or at 288 nm [19].

The quantitative analysis of morphine and codeine on seeds and shells of *P. somniferum* [20] was performed using simple TLC system: silica/cyclohexane + ethylenediamine (4:1), with densitometry

TABLE 17.1

Examples of Reagents Applied in the Visualization of Morphine and Its Derivatives

lp.	Reagent	Compound
1.	1% Aq. ferric chloride, next 1% acidified ethanolic 2,2-dipirydyl, heating at 100°C for 10 min	Heroin, morphine, codeine [27]
2.	Iodoplatinate reagent and methanolic iodoplatinate reagent	Morphine [4,6,24], codeine [4,24], methadone, norpropoxyphene [24], 6-β-naltrexol, naltrexone [46]
3.	5% Phosphomolybdic acid and heating at 120°C for 5 min (for RP-18 layers)	Morphine [6]
4.	Dragendorff's reagent	Codeine [3,4,9,30,31], morphine [3,4,25], iso-, pseudo-, and allopseudo codeine [30]
5.	Silver acetate 1% in water	Morphine, codeine, methadone, norpropoxyphene [24]
6.	Mercuric sulfate solution	Morphine, codeine, methadone, norpropoxyphene [24]
7.	Diphenylcarbazone 0.01% in acetone/water	Morphine, codeine, methadone, norpropoxyphene [24]
8.	0.5% Sulfuric acid	Morphine, codeine, methadone, norpropoxyphene [24]
9.	Iodine–potassium iodide reagent	Morphine, codeine, methadone, norpropoxyphene [24]
10.	Iodine vapors	Morphine [3,11], codeine [3], 10-oxo-morphine [11]
11.	1% Mercuric chloride–1% potassium ferricyanide 1:1, heating at 80°C for 5 min	Heroin, remain opiates [34]
12.	Marquis reagent	Codeine, dihydrocodeine, hydrocodone, hydromorphone [9]
13.	Fast blue B salt	Codeine, dihydrocodeine, hydrocodone, hydromorphone [9]
14.	Fast black K salt	Codeine, dihydrocodeine, hydrocodone, hydromorphone [9]
15.	Potassium iodobismuthate reagent	Methyl codeine, codeine [32], nalorphine [40]
16.	Vapor of formaldehyde 38%, acetic acid anhydride (4:1), and then sulfuric acid 98%	Morphine, heroin, codeine [28]
17.	2% Sodium nitrate in 1.5 M hydrochloric acid	6-β-Naltrexol, naltrexone [46]

at 320 nm. The content of morphine on simple pharmaceutical preparation as camphor tincture was examined on silica/AcOEt + MeOH + NH_3 (43:25:2), with the quantification by densitometry [12].

TLC method was used to investigate morphine both as the substance and as the component occurring in pharmaceutical preparations. Morphine hydrochloride of pharmacopoeial grade and 10-oxo-morphine (decomposition product of morphine) in two systems on silica layer was examined: chloroform + hexane + MeOH (65:25:10) and chloroform + acetone + diethylamine (5:4:1), detection with iodine vapors and UV 366 nm [11]. The example of the application of reversed phase systems can be found both in 2D-TLC fingerprinting and in confirmation procedures (more complicated procedures) performed on a special plate with two layers (silica and RP-18) of morphine with methanol + 0.5 M NaCl (65:35), left for 20 min at the temperature of 65°C; AcOEt + MeOH + water + 28% NH_3 (170:27:2:1); and lubricating oil with MeCN + water + acetic acid (90:10:1) and

TABLE 17.2

Examples of the Application of Densitometry to the Analysis of Morphine and Its Derivatives

lp.	Compound	Parameters of Densitometric Measurement
1.	Morphine	Quantification by densitometry at 280 nm [18]
		Quantification by densitometry at 278 nm [19]
		Detection by densitometry at 230 nm [20]
		Quantification by densitometry at 288 nm [21]
		Quantification by densitometry at 439 nm [22]
		Quantification by densitometry at 287 nm [23]
2.	Codeine	Quantitation by densitometry between 198 and 610 nm (the UV spectra of the separated substances are used for the identification) [16]
		Detection by densitometry at 230 nm [20]
		Quantification by densitometry at 288 nm [21]
3.	Codeine, methyl codeine	Quantification by densitometry at λ_S 278 nm and λ_R 400 nm or λ_R 500 nm and λ_S 750 nm after using a potassium iodobismuthate reagent [32]
4.	Tramadol	Quantitation by densitometry between 198 and 610 nm (the UV spectra of the separated substances are used for the identification) [16]
		Quantification by densitometry at 254 nm [35,37]
		Quantification by densitometry 270 nm [36]
		Quantification by densitometry 273 nm [39]
5.	Nalorphine	Quantification by densitometry at 515 nm after using a potassium iodobismuthate reagent [40]
6.	Dextromethorphan	Quantification by densitometry at 223 nm [43]
		Detection by densitometry at 257 nm [44]
		Quantification by densitometry at 225 nm [45]
7.	Methadone	Quantification by densitometry at 254 nm [8]

hexane + chloroform (8:20). For the detection of morphine, methanolic iodoplatinate reagent was used [6]. The determination of morphine and codeine phosphate on silica/AcOEt + MeOH + NH$_3$ (85:10:5) with the detection at 254 nm and the quantification with densitometry at 288 nm [21] was also reported. Morphine as the component of pharmaceutical preparation—Platycodon [22]—was chromatographed in the same system [21], with the densitometry at 439 nm.

TLC and its special techniques such as AMD and OPLC were applied to examine the opium extract and opiates, for example, morphine, codeine, thebaine, papaverine, noscapine, ethylmorphine, pholcodine, heroin, buprenorphine, dextromethorphan, and methadone (as reference substances). TLC was performed on silica with toluene + acetone + EtOH + 28% NH$_3$ (45:45:7:1), OPLC on aluminum oxide with ethyl acetate as the mobile phase, and AMD with methanol, acetone, ethyl acetate, ethyl acetate + dichloromethane, and dichloromethane. OPLC proved to be the fastest method of separation, whereas AMD depicted the best resolution (no diffusion of spots, elution without oxidation products). That is why AMD is considered to be a very interesting method for further application [10].

The application of TLC in opiate analysis plays a fundamental role in toxicological and forensic experiments. The properties of morphine and its analogs generate the black market and the need of forensic analysis. The concentration of opiates in biological fluids needs to be controlled because their high level in organism may cause addiction. To control the concentration of morphine chloride in plasma, the following system is often used: silica/AcOEt + MeOH + NH$_3$ (86:10:5), followed by the quantification with densitometry at 287 nm [23]. To find and quantify morphine, codeine, and methadone in urine, four different systems on silica are applied: AcOEt + cyclohexane + MeOH + H$_2$O (70:15:2:0.5), AcOEt + cyclohexane + MeOH + NH$_3$ (56:40:0.8:0.4), AcOEt + cyclohexane + MeOH + NH$_3$ (70:15:2.8:0.5), and AcOEt + cyclohexane/NH$_3$ 50:40:01. Different reagents, 1% silver acetate in water, mercuric sulfate solution, 0.01% diphenylcarbazone in mixture of acetone and water, 0.5% sulfuric acid, iodoplatinate, and iodine–potassium reagent [24], were used for the detection. The rapid detection of morphine in urine of opiate addicts is obtained in the system, silica/AcOEt + MeOH + NH$_3$ (80:15:5), with the detection using Dragendorff's reagent [25]. The combination of 2D HPTLC with sensitive and/or specific detection gives a suitable and reliable method for the detection of low concentrations of morphine and structurally related opiates in urine with the detection limit of 20–100 times lower than TLC. Two systems with two eluents on silica, AcOEt + MeOH + NH$_3$ (17:2:1) and MeOH + NH$_3$ (99:1) and the second toluene + diethylamine + MeOH AcOEt (60:10:10:1) and acetone + water + NH$_3$ (20:20:1), were used for the analysis of opiates, with the following visualization: Dragendorff's reagent, Marquis reagent, fast blue B salt, fast black K, and morphine as dansyl derivative. The previously mentioned method allows to determine not only morphine but also codeine, dihydrocodeine, hydrocodone, and hydromorphone [9]. The quantitative HPTLC of dansyl derivatives of morphine and monoacetylmorphine was performed on silica/MeOH + NH$_3$ 25% (99:1) where the plate was dried in cold air and dipped in hexane solution of paraffin and again dried in cold air and quantified with fluorimetry (densitometry) at 366 nm [26].

Morphine, heroin, codeine, and other drugs obtained as street samples from different illegal sources were detected and identified on silica, chloroform/EtOH 9:1. After development, the plates were sprayed for the visualization with 1% aqueous solution of FeCl$_3$ and then with 1% acidified ethanolic 2,2-dipyridyl solution and heated at 100°C for 10 min [27]. The TLC of morphine, heroin, codeine, and other drugs was performed on silica with three different developing mixtures: AcOEt + chloroform + MeOH + NH$_3$ (100:60:24:3), AcOEt + MeOH + NH$_3$ (17:2:1), and benzene + acetone + EtOH + NH$_3$ (45:45:7:3), respectively. The visualization by exposure to vapor of formaldehyde in concentration of 38% with acetic acid anhydride (4:1) and then with sulfuric acid in concentration of 98% was performed. After visual evaluation, the plate was placed in a jar saturated with water and again visually evaluated [28].

To identify codeine with TLC and morphine and codeine with RP-TLC, a combined dual system of R$_F$ values and UV spectra library was performed on silica/toluene + acetone + 94% EtOH + 25% NH$_3$ (45:45:7:3) and on the reversed phase—RP18/MeOH + water + 37% HCl (50:50:1). The densitometry at 220 nm and the measurement of spectra at the range of 190–400 nm were made and compared with library records [7]. Skalican et al. examined different systems on silica and alumina and 68 alkaline mobile phases as the means to identify morphine, codeine, and other opiate-related drugs as ionic associates formed with bromoxylenol blue, cresol red, and eriochromecyanine-R [29].

Codeine is available as sulfate and phosphate salt and also as a free base, all being present in various pharmaceutical forms. The drug is used as an analgesic and antitussive but less powerful than morphine. To analyze codeine and its isomers after N-demethylation in postreaction mixture, two mobile phases on silica, benzene + MeOH (6:2 or 1:1) and chloroform + acetone + diethylamine (5:4:1), with Dragendorff's reagent for visualization were used [30]. The determination of codeine phosphate in tablets was performed using silica with EtOH + NH$_3$–water in various proportions. The quantification with densitometry at 510 nm after spraying with Dragendorff's reagent [31] was carried out. The TLC of codeine and methyl codeine developed on silica with toluene + AcOEt + EtOH + NH$_3$ (45:45:7:4) was performed, with the detection using UV light or with spraying using potassium iodobismuthate reagent. The quantification by densitometry at 278 and 400 nm or 500 and 750 nm after spraying with reagent [32] was carried out. The same systems and techniques as in the case of morphine analysis described earlier [7,9,10,16,20,24,27–29] were used for TLC analysis of codeine.

Diacetylmorphine (heroin), displaying a strong activity and being an addictive substance, is usually prescribed as a strong painkiller for patients in agony. However, the general necessity to determine chromatographically heroin samples is more frequently applied in forensic investigation. Simultaneous detection of impurities that occur in heroin possessed illegally might be performed with two consecutive mobile phases (multiple development technique) and proves to be a new simple, sensitive, and specific HPTLC procedure. For heroin, morphine, methaqualone, papaverine, narcotine, caffeine, and paracetamol, HPTLC was performed on silica with benzene + MeCN + MeOH (8:1:1) and developed to 5 cm. After drying, the second development to 8 cm with cyclohexane + toluene + diethylamine (75:15:9) with visualization under UV 254 nm [33] was performed. Also, the detection and identification of heroin and opium in street samples using TLC method were carried out in the system: silica/chloroform + EtOH (9:1). After the development, the plate was sprayed with 1% mercuric chloride and 1% potassium ferricyanide and heated at 80°C for 5 min. As the result, blue spots for heroin and opiates were obtained [34]. Other systems used for heroin determination are also described earlier [10,27].

Tramadol, the popular synthetic opiate drug, is used as a strong painkiller. For its determination, it is TLC and more often HPTLC that is used in pharmaceutical research as well as in quality control. The TLC of tramadol and paracetamol mixture was developed on silica, with chloroform + EtOH (7:3), with densitometric evaluation at 264 nm. The method was linear over a concentration range of 2.5–32.5 pg/band for tramadol and 10–50 pg/band for paracetamol, and it did not interfere with excipients from the dosage form [35]. Tramadol and paracetamol in pharmaceutical dosage form were developed on silica with chloroform + MeOH + glacial acetic acid (90:20:1) with densitometric quantification at 270 nm. The method was linear in the range of 500–2000 ng/band for both compounds. The recovery was 98.9%–99.7% [36]. On reversed phases, TLC and HPTLC estimation of tramadol hydrochloride and paracetamol in combination on silica gel with AcOEt + toluene + NH_3 (60:40:1) was performed. The absorbance measurement was carried out at 254 nm. The method was linear in the range of 0.1–0.5 μg/mL and 0.9–4.5 μg/mL for tramadol and paracetamol, respectively. The method was suitable for routine analysis [37]. HPTLC of tramadol and chlorzoxazone (as the internal standard) in pharmaceutical preparations on silica gel with AcOEt/MeOH/ NH_3 (7:1 and 1 drop of ammonia) was performed. The quantitative determination with absorbance measurement at 275 nm was done [38]. Tramadol hydrochloride and chlorzoxazone were either examined on silica HPTLC plates, prewashed with methanol, or previously saturated with AcOEt + toluene + NH_3 (35:15:1) (twin through chamber was used). The densitometric quantitative determination at 273 nm [39] was performed. The toxicological identification of codeine, tramadol, flupirtine, and lidocaine with TLC and HPTLC on silica gel with AcOEt/MeOH/NH_3, 17:2:1 was carried out. The quantitation with DAD densitometry in range of 198 and 610 nm was performed. The UV spectra of the separated substances are used for the identification [16].

To determine nalorphine hydrochloride and other three drugs in injection, TLC on silica with two mobile phases, i-PrOH + MeOH + H_2O + acetic acid (7:3:3:0.5) and CH_2Cl_2 + acetone + MeOH (8:2:0.6), was used. The quantification by densitometry at 515 nm after spraying with potassium iodobismuthate reagent was done [40]. For buprenorphine hydrochloride that usually occurs in pharmaceutical preparations, HPTLC on silica with cyclohexane + toluene + diethylamine + MeOH (100:30:33:10) proved to be a simple, reliable, sensitive, and fast method. The quantification by densitometry was performed, and the recovery at 98.8% and the detection limit at 16 ng were observed [41].

TLC of different analgesics in combined dosage forms (with the content of dextropropoxyphene) was performed on silica in phosphate buffer (pH 6.0) with AcOEt + chloroform + MeOH + NH_3 (110:80:10:1). The determination of separated drugs after the elution with spectrophotometry at 242 nm for paracetamol, 258 nm for dextropropoxyphene, and 265 nm for oxyphenbutazone and ibuprofen [17] was done. Dextropropoxyphene hydrochloride, paracetamol, and dicyclomine hydrochloride in pharmaceutical formulations were determined by quantitative TLC. The TLC of analgesics was performed on silica (1 mm) with AcOEt + chloroform + MeOH + water (75:15:5:3:2).

For the quantification (after the elution with methanol—1 M HCl 99:1), the spectrophotometric method was used, with the parameters at 242 nm for paracetamol, 258 nm for dextropropoxyphene HCl, and 620 nm for dicyclomine HCl, after previous derivatization with cobalt chloride reagent [42]. The determination of dextropropoxyphene as performance-enhancing drug among 14 others required a specific stationary phase, that is, bismuth silicate gel (prepared from 75 mL bismuth nitrate gel with 14 g silica gel powder). To perform chromatography, 21 organic, aqueous, and organic–aqueous mobile phases were investigated. The drugs were detected with iodine vapors [5].

Dextromethorphan, the most popular antitussive drug, used in different pharmaceutical formulations was determined using TLC and HPTLC. As the mixture with doxylamine succinate in syrups, dextromethorphan hydrobromide was densitometrically quantified at 223 nm. The TLC of dextromethorphan hydrobromide was previously performed on silica with MeOH + 25% NH_3 (14:1) [43]. HPTLC system on silica gel with CH_2Cl_2 + acetone + MeOH + triethylamine (70:40:50:2) with densitometry at 257 nm was used for simultaneous determination of pseudoephedrine sulfate, azatadine maleate, and dextromethorphan hydrobromide [44]. The determination of dextromethorphan hydrobromide in pharmaceutical forms such as caplets, gelcaps, and tablets with HPTLC was performed on silica gel, prewashed by chromatography with CH_2Cl_2 + MeOH (1:1) and with AcOEt + MeOH + NH_3 (17:1:2) with previous chamber saturation. For quantitative determination, the densitometry with measurement at 225 nm [45] was used.

The determination of methadone (D + L) and L-polamidon in urine with HPTLC was used in the system with diol layers and toluene + hexane + diethylamine (3:6:1) as the mobile phase. The substances were detected after derivatization with modified Dragendorff's reagent and quantified by densitometry at 254 nm [8]. The TLC on silica with AcOEt/MeOH/NH_3 (170:20:4:6) was used for quantitative determination of 6-β-naltrexol and naltrexone. The detection by spraying with 2% sodium nitrate in 1.5 M hydrochloric acid or with iodoplatinate reagent was carried out [46].

REFERENCES

1. Aragon-Poce, F., Martinez-Fernandez, E., Marquez-Espinos, C., Perez, A., Mora, R., Torres, L.M. 2002. History of opium. *Int. Cong. Ser.* 1242:19–21.
2. DeRuiter, J. 2000. Narcotic analgesics: Morphine and "peripherally modified" morphine analogs. *Princ. Drug Action* 2(Fall):1–9.
3. Tombesi, O.L., Maldoni, B.E., Bartolome, E.R., Haurie, H.M., Faraoni, M.B. 1994. Purification of alkaloids by thin-layer chromatographic decomposition of their picrates. *J. Planar Chromatogr.* 7:77–79.
4. Szumilo, H., Flieger, J. 1999. Application of differently modified silica gel in the TLC analysis of alkaloids. *J. Planar Chromatogr.* 12:466–470.
5. Hassankhani-Majd, Z., Ghoulipour, V., Husain, S.W. 2006. Chromatographic behaviour of performance-enhancing drugs on thin layers of bismuth silicate ion exchanger. *Acta Chromatogr.* 16:173–180.
6. Beesley, T.E., Heilweil, E. 1982. Two phase, two dimensional TLC for fingerprinting and confirmation procedures. *J. Liq. Chromatogr.* 5:1555–1566.
7. Ojanpera, I., Nokua, J., Vuori, E., Sunila, P., Sippola, E. 1997. Novel for combined dual-system RF and UV library search software: Application to forensic drug analysis by TLC and RPTLC. *J. Planar Chromatogr.* 10:281–285.
8. Koch, A. 1997. Quantitation of methadone (D+L) and L-polamidon in soft drinks. *Proceedings of the 9th International Symposium on Instrumental Planar Chromatography*, Interlaken, Switzerland, April 9–11, pp. 155–159.
9. Novakova, E. 2000. The detection of opiates by two-dimensional high-performance thin-layer chromatography. *J. Planar Chromatogr.* 13:221–225.
10. Pothier, J., Galand, N., Viel, C. 2001. Separation of opiates and derivatives—Analgesics, antitussives and narcotic compounds by different techniques of planar chromatography (TLC, OPLC, AMD). *Proceedings of the International Symposium on Planar Separations, Planar Chromatography*, Lillafüred, Hungary, pp. 321–326.
11. Szigeti, J., Mezey, G., Bulyaki, M. 1984. Investigation of morphine hydrochloride of pharmacopoeial grade. *Acta Pharm. Hung.* 54:58–63.

12. Wang, L., Feng, J., Bi, S. 1985. Rapid TLC densitometric assay of morphine in compound camphor tincture. *Chin. J. Pharm. Anal.* 5:109–110.
13. Ling, L. 1985. TLC-UV spectrophotometric determination of morphine in opium tincture. *Chin. J. Pharm. Anal.* 5:229–230.
14. Huo, X., He, W., Luo, Y. 1992. Rapid detection of morphine in urine of opiate addicts by thin-layer chromatography. *Chin. J. Acad. Mil. Med. Sci.* 16:300–302.
15. Jeger, A.N., Briellmann, T.A. 1994. Quantitative applications of TLC in forensic science. *J. Planar Chromatogr.* 7:157–159.
16. Spangenberg, B., Ahrens, B. 2003. Identification of substance in diode array thin-layer chromatography. (Substanzidentifikation in der Dioden-Array DUnnschichtchromatographie). *GIT Fachz. Lab.* 47(6):658–660.
17. Parimoo, P., Bharathi, A., Shajahan, M. 1994. Estimation of oxyphenbutazone and ibuprofen in presence of paracetamol and dextropropoxyphene in dosage form by quantitative thin-layer chromatography. *Indian Drugs* 31(4):139–143.
18. Gu, Y., Li, L., Lin, K. 1993. Determination of morphine in opium by thin-layer chromatography. *Chin. J. Pharm. Anal.* 13:236–237.
19. Li, W., Zhang, H., Chen, J., Jiang, J. 1994. Determination of morphine in pulverized opium by thin-layer chromatography. *Chin. J. Anal. Lab.* 13:92–93.
20. Li, Zh., Zhao, X., Su, Y., Chen, Y. 1990. Quantitative determination of major alkaloids in the seeds and shells of *Papaver somniferum* L. by thin-layer chromatography. *Chin. J. Chromatogr.* 8:388–389.
21. Liu, J., Yu, Q., Zhang, G. 1995. Quality control of Chinese medicine, Zhisouhuatanconji by thin-layer chromatographic determination of morphine and codeine phosphate. *J. Chin. Trad. Patent Med.* 17:10–12.
22. Zhang, Z., Jiang, Y. 1989. Dual-wavelength TLC determination of morphine in compound Platycodon tablets. *Chin. J. Pharm. Ind.* 20:28–30.
23. Zhou, J., Yuan, Y. 1991. Determination of morphine chloride in plasma by thin-layer chromatography. *Chin. J. Pharm. Anal.* 11:293–294.
24. Kaistha, K.K., Tadrus, R. 1983. Improved cost effective thin-layer detection techniques for routine surveillance of commonly abused drugs in drug abuse urine screening and proficiency testing programs with built-in quality assurance. *J. Chromatogr.* 267:109–116.
25. Huo, X., He, W., Luo, Y. 1992. Rapid detection of morphine in urine of opiate addicts by thin-layer chromatography. *Chin. J. Acad. Mil. Med. Sci.* 16:300–302.
26. Funk, W., Donnevert, G., Patzsch, K., Kaferstein, H., Schutz, H. 1990. Quantitative HPTLC of acetylmorphine. *Merck Spectr.* 1990: 34–37.
27. Dongre, V.G., Kambie, V.W. 2003. HPTLC detection and identification of heroin (diacetylmorphine) in forensic samples. Part III. *J. Planar Chromatogr.* 16:456–460.
28. Khdary, N.H., Shaikh, K.M. 1996. A new technique for visualizing thin layer chromatography plates. *Bull. Int. Assoc. Forensic Toxicol. (TIAFT)* 26(3):38–41.
29. Skalican, Z., Halamek, E., Kobliha, Z. 1997. Study of the potential of thin-layer chromatographic identification of psychotropic drugs in field analysis. *J. Planar Chromatogr.* 10:208–216.
30. Hosztafi, S., Makleit, S., Miskolczi, Z. 1983. N-demethylation of morphine alkaloids, II. Codeine isomers. *Acta Chim.* 114:63–68.
31. Dong, Y., Han, B., Zhang, Z.H., Wang, X. 1985. Determination of codeine phosphate in tablets by thin-layer chromatography. *J. Hebei Med. Coll.* 6:78–80.
32. Zhu, W., Chen, W. 1989. Determination of methyl codeine in codeine phosphate by thin-layer chromatography. *Chin. J. Pharm. Anal.* 9:37–38.
33. Krishnamurtthy, R., Srivastava, A.K. 1997. Simultaneous detection of adulterants and coextractants in illicit heroin by HPTLC with two successive mobile phases. *J. Planar Chromatogr.* 10:388–390.
34. Kamble, V.W., Dongre, V.G. 1997. TLC detection and identification of heroin (diacetylmorphine) in street samples: Part II. *J. Planar Chromatogr.* 10:384–386.
35. Solomon, W., Anand, P., Shukla, R., Sivakumar, R., Venkatnarayanan, R. 2010. Application of TLC-densitometric method for simultaneous estimation of tramadol HCl and paracetamol in pharmaceutical dosage forms. *Int. J. ChemTech Res.* 2(2):1188–1193.
36. Roosewelt, C., Harihrishnan, N., Gunasekaran, V., Chandrasekaran, S., Haribaskar, V., Prathap, B. 2010. Simultaneous estimation and validation of tramadol and paracetamol by HPTLC in pure and pharmaceutical dosage form. *Asian J. Chem.* 22(2):850–854.
37. Gandhimathi, M., Ravi, T.K. 2008. RP-HPTLC and HPTLC estimation of tramadol hydrochloride and paracetamol in combination. *Asian J. Chem.* 20(6):4940–4942.

38. Meyyanathan, S., Kumar, P., Suresh, B. 2003. Analysis of tramadol in pharmaceutical preparations by high performance thin layer chromatography. *J. Sep. Sci.* 26:1359–1362.
39. Gandhimathi, M., Ravi, T.K. 2008 Simultaneous densitometric analysis of tramadol hydrochloride and chlorzoxazone by high-performance thin-layer chromatography. *J. Planar Chromatogr.* 21:305–307.
40. Li, Y., Xie, J. 1988. Determination of nalorphine hydrochloride and other three drugs in injections by thin-layer chromatography. *Chin. J. Pharm. Anal.* 8:170–172.
41. Chandrashekhar, T.G., Rao, P.S.N., Sneth, D., Vyas, S.K., Dutt, C. 1994. Determination of buprenorphine hydrochloride by HPTLC: A reliable assay method for pharmaceutical preparations. *J. Planar Chromatogr.* 7:249–250.
42. Parimoo, P., Mounisswamy, M., Bharathi, A., Lakshmi, N. 1994. Determination of paracetamol, dextropropoxyphene HCl and dicyclomine HCl in pharmaceutical formulations by quantitative thin-layer chromatography. *Indian Drugs* 31(5):211–214.
43. Indryanto, G. 1996. Simultaneous densitometric determination of dextromethorphan hydrobromide and doxylamine succinate in syrups and its validation. *J. Planar Chromatogr.* 9:282–285.
44. Sodhi, R.A., Chawla, J.L., Sane, R.T. 1997. Simultaneous determination of pseudoephedrine sulphate, azatadine maleate and dextromethorphan hydrobromide by high-performance thin-layer chromatography. *Indian Drugs* 34(8):433–436.
45. Di Gregorio, D., Harnett, H., Sherma, J. 1999. Quantification of dextromethorphan hydrobromide and clemastine fumarate in pharmaceutical caplets, gelcaps and tablets by HPTLC with ultraviolet absorption densitometry. *Acta Chromatogr.* 9:72–78.
46. Verebey, K., Alarzi, J., Lehrer, M., Mule, S.J. 1986. Determination of 6-beta-naltrexol and naltrexone by bonded-phase adsorption thin-layer chromatography. *J. Chromatogr.* 378:261–266.

18 TLC of Nonopioid Analgesics, Anti-Inflammatics, and Antimigraine Drugs

Arkadiusz Pomykalski

CONTENT

Pain has accompanied man for ages and in the 1970s was defined in many ways, for example, such as unpleasant feelings causing a specific defense mechanism and characteristic emotional state. In the late 1990s, pain has been defined as a complex phenomenon of unpleasant sensation of different type and intensity, showing the damage or danger of damaging tissues as well as accompanying emotional reactions and many vegetative symptoms. According to the International Association for the Study of Pain (IASP), pain is an unpleasant sensational and emotional feeling, accompanying the existence or the danger of damaging tissue or connected with such damage. Humankind has been fighting with pain for ages. Despite the great development in medicine, pain is still present in our lives, and the fight with it is still one of the priorities of our healthcare. There are different methods of killing pain such as pharmacological, surgical, psychotherapeutic, and physiotherapeutic. An effective pain cure is very essential not only for individuals—directly concerned people—but for the whole society since pain that is not cured, lasts for a long time, and chronic pain can lead to serious mental and physical disorders. Even though pain is an unpleasant sensation, it is an alarm of a changed ill organism. The pain border is very individual, and it changes under the influence of external and internal factors (Makulska-Nowak 2004).

Pain is caused by the irritation of sensory receptors and is present along with many acute and chronic diseases. Pain is felt because of endogenous pain modulators, prostaglandin and substance P, which intensify it, but endorphins and encephalin put a stop to it.

A defense reaction of the organism is an inflammatory condition, which is an indirect body response. It is created as a result of stimulus: chemical, physical, and biological. Inflammation can be caused by infection, mechanical injuries (micro injuries), and immunological reactions as a response to foreign antigens or their degeneration changes in the osteoarticular system, local tissue irritation by microcrystal compounds (putting aside gout salt in articulations), as well as developing cancerous changes. The clinical picture of the inflammation process in the tissues and organs may be significantly different depending on the cause of the inflammation. The difference concerns mostly the duration of inflammatory condition, pathomorphology changes, symptom intensification, and consequences. Despite these differences, inflammation pathomechanism development is very similar. In the beginning, if the inflammation process is of acute type, usually, later it can become chronic.

Inflammatory reaction starts in an acute phase by pathogenic factors, which set off a number of defense mechanisms both local and systemic, which often have feedback type. Under their influence, effector cells become degranulated (mastocytes, epithelium cells, thrombocytes, and others). These cells, during degranulation, release from their granules preformed mediators, such as histamine, serotonin, bradykinin, chemotactic substances acting on neurofile and eozynofile, as well as substances that reduce the effects of the inflammatory response, (α_1-antitrypsin, α_2-macroglobulin, plasminogen activator blockers, and others). These mediators affect the secondary effector cells, flowing into the inflammatory reaction sites—monocytes, segmented neutrophils and eosinophils, and lymphocytes. Secondary effector cells are capable of phagocytosis, the destruction of microorganisms, their toxins, and breakdown products of cells. Cytokines produced by monocytes and macrophage play important roles in the inflammatory process. These are the interleukins IL-1 and IL-6, interferon α and β, and tumor necrosis factor—TNF. Their function is to transmit to the cells and to other organs information about the inflammatory process. At the site of inflammatory reaction, lymphocytes T and B appear. The first ones are responsible for the cytotoxic reaction of the cell, and the latter condition the humoral response. In the case of longer inflammation, production changes happen with tissue overgrowth. At the site of inflammation, fibroblasts occur, cell proliferation happens, and collagen synthesis and synthesis of glycosaminoglycans occur, which constitute bulk of the newly formed tissue. During the inflammatory process, backward and degenerative changes may happen.

The inflammatory process has been accompanied by pain arising from the accumulation of cAMP in the tissue and peripheral sensitization of pain receptors. In this process, prostaglandins play an important role and lower the threshold of excitability of pain receptors, increasing their susceptibility to histamine, serotonin, or bradykinin pain activity.

Prostaglandins belong to a wider group of eicosanoids, which include prostacyclin, thromboxanes, and leukotrienes. Eicosanoids derived from arachidonic acid (5, 8, 11, 14-eikozatetraenic acid) consist of 20 carbon atoms. Prostaglandin biosynthesis occurs in the microsomes with the participation of numerous enzymes. Among them, the key role is played by phospholipase A_2 and cyclooxygenase, also known as PGH synthase, a bifunctional enzyme with cyclooxygenase and hydroperoxydase. The various prostaglandins are formed from peroxide with the participation of the respective synthases (Bartoszuk and Zarzycki 2004).

In 1989, it was found that there are at least two isoforms of cyclooxygenase: COX-1 and COX-2 encoded by genes located on different chromosomes (Rosenstock et al. 2001). COX-1 is a constitutive enzyme present in physiological conditions and is stimulated by hormones and by growth factors. This protein is produced by most cells in the body, which regulates normal metabolic processes, and products of its activity are involved in the protection of cytoprotective gastrointestinal mucosa of the stomach and duodenum in particular through increased secretion of sodium bicarbonate, while stopping the secretion of hydrochloric acid (PGE_2 and PGI_2), inhibiting the formation and development of atherosclerotic plaques within the blood vessels (PGI_2), to maintain proper flow in vascular renal excretory function (PGE_2 and PGI_2) and the intensity of platelet aggregation caused by thromboxane (TXA_2). An important discovery of recent years was to show that this enzyme participates in the early stages of inflammation, particularly in the initial period lasting 1 h.

Inducible isoform COX-2 is activated in pathological states, and products of its activity are essentially identical to those emerging under the influence of COX-1 and are responsible for the inflammation process and the liberation of mediators of inflammation and pain sensations. Under physiological conditions, in most tissues COX-2 is present in very low concentrations or not detectable, and in high concentrations is present in the inflammatory environment. However, note that the izoenzyme COX-2 is also a constitutive enzyme and occurs in significant concentrations in the following tissues and organs:

- The cerebral cortex and hippocampus, which affects the central nervous system
- Women's reproductive organs, affecting the course of the ovulatory cycle and the implantation of a fertilized egg
- The kidneys, where it influences the hydroelectrolyte balance
- In the ciliary body epithelial cells, which reveal the presence of both isoenzymes

Recent studies indicate the presence of COX-3 isoenzyme. This isoform is attributed to the possibility of synthesis of prostaglandins involved in the process of quenching the inflammatory response.

The chemical structure of spatial enzymes COX-1 and COX-2 is well known. Both enzymes have a similar weight and arachidonic area acid binding activity. However, COX-2 has higher activity in both area and extent of access of structured substrates. The important difference is that they are coded by different genes located on different chromosomes.

The second way of changing arachidonic acid takes place with the enzyme 5-lipoxygenase, which is located in the lungs, leukocytes, and platelets. With the participation of this enzyme leukotrienes are produced, which are stronger than histamin and shrink bronchial mucosa. With the biosynthesis of leukotrienes Samuelson won the Nobel Prize in 1980.

Drugs that block the synthesis of arachidonic acid metabolites under the influence of cyclo-oxygenase include nonsteroidal anti-inflammatory drugs (NSAIDs). These drugs do not modify the course of rheumatic diseases, but they inhibit inflammation and analgesic, allowing one to keep mobility of joints affected by the disease and carry out rehabilitation. The origins of the introduction of NSAIDs for the treatment is widely believed to be in salicylates. The sodium salt of salicylic acid produced from the glycoside salicin was introduced into medical practice in 1875 and gained huge success. In 1899, acetylsalicylic acid was introduced and called the aspirin form—Spiraea, from which salicylic acid is also obtained. For over 100 years, aspirin has been used as anti-inflammatory, antipyretic, and more rarely as painkiller or urikozuric. Currently, the drug is used in so-called cardiac doses (75–150 mg) as an antiplatelet for prevention of recurrence of partial ischemic attacks, ischemic stroke, as well as in reducing mortality due to myocardial infarction and the frequency of relapses. The dominance of aspirin persisted since the early 60s. In 1962, ibupro-fen was introduced, and in 1963—indomethacin. These drugs have given rise to other NSAIDs. The mechanism of action of these drugs, however, remained unclear, despite many speculations and hypotheses. In 1971, a team under professor J. Vane demonstrated their inhibitory effect on cyclooxygenase activity. He got the Nobel prize in 1983 for this discovery. It turned out that this mechanism is common to many drugs, which were called nonsteroidal anti-inflammatory drugs, in contrast to steroidal inflammatory drugs that inhibit the production of reaction products of both cyclooxygenase and lipoxygenase.

All drugs in this group show an acidic quality, and the pKa values are in the range 3–5. Acidic prop-erties largely determine their ability to bind to proteins. In the NSAIDs, the molecule has hydrophilic moiety (carboxyl group or enolate) and is lipophilic (aromatic ring, halogen atoms), but these groups cannot be in the same range. Introducing an additional center of the molecule lipophilic NSAIDs increases the strength of their actions, it but also affects some pharmacokinetic parameters.

Assuming that the inhibition of COX-2 leads to favorable clinical features of NSAIDs and deter-mines their effectiveness in rheumatic diseases, the degenerative-inflammatory quality, inhibiting the COX-1, determines the unfavorable, adverse effects of NSAIDs. Older NSAIDs showed a greater affinity for COX-1, and so they were affected by greater side effects. No wonder there is a race to obtain the compounds that selectively inhibit the activity of possible COX-2. NSAIDs were divided, taking into account the ratio of COX-2/COX-1 inhibitory concentrations, and the IC_{50} value is less than when the more selective the drug is tested on COX-2 (Griswold and Adams 1996, Laneuville et al. 1994, Vane and Batting 1996, Warner et al. 1999). Given this value, NSAIDs are divided into three generations:

- Traditional drugs that more strongly inhibit COX-1 than COX-2 belong to the first genera-tion. These are the salicylates, acid derivatives: indoloacetic, phenylacetic, phenylpropi-onic, anthranilic, and enol acid derivatives.
- Preferentially acting on COX-2—etodolac, meloxicam, nabumetone, and nimesulide belong to the second generation. These drugs were introduced into medical practice in the years 1994–1995.
- Drugs that act selectively on the so-called COX-2. Coxibes—celecoxib, rofecoxib, etoricoxib, valdecoxib, parecoxib, and lumiracoxib belong to third generation. These drugs were introduced in 1999.

In the case of first-generation drugs, IC_{50} values of COX-2/COX-1 relations vary in a wide range, but definitely the worst show they are for piroxicam, aspirin, and tolmentin (250, 160, and 175, respectively), with a relatively favorable ratio for ibuprofen, diclofenac, and naproxen (15, 0.7, 0.6, respectively).

The incidence of gastrointestinal lesions is especially serious: ulcers, bleeding, perforation of ending deaths, by using first-generation NSAIDs, are smaller in the case of ibuprofen and very high for piroxicam. More rarely, second-generation drugs preferentially induce side effects in the gastrointestinal tract than the first-generation drugs. In fact, the evaluation of their efficiency and safety is difficult because they have not managed to gain popularity, as the third generation drugs have already appeared, selectively acting on COX-2 and potentially safer. In 1999, they were first introduced into medical drugs, selectively acting on COX-2. Activity of COX-2 inhibiting concentrations are 200-fold lower than those that inhibit COX-1. The first recorded coxib are celecoxib and rofecoxib; both agents were in the top 20 best-selling drugs in 2000 and therefore were recorded shortly following coxib: valdecoxib, parecoxib, and etoricoxib, which are a drug precursor of valdecoxib. Introduction of coxib was to begin a golden era in rheumatology; these drugs should be free of side effects typical of traditional NSAIDs. However, as a result of expanded research on long-term use it has been established that coxib has an adverse effect on the cardiovascular system—increased thromboembolic events; gastrointestinal—nausea, diarrhea, ulcers, bleeding, and perforation (Langman et al. 1999); cardiovascular—hypertension, palpitation, heart failure, heart attacks; and effects on cardiac centers—hallucinations, depression, confusion, which disappear after cessation of treatment (Breckenridge et al. 2000). Both celecoxib and rofecoxib cause renal impairment (Whelton 2001). Swelling of vascular, bronchospasm, and exacerbation of asthma are rare side effects. It was also found in animal studies that they delayed wound healing of bone fractures. In addition, NSAIDs are responsible for infertility in women, which occurs after the use of rofecoxib, but also after diclofenac, and disappears after several months following discontinuation of therapy (Norman 2001). Furthermore, the presence of glaucoma in the classical NSAIDs is not described, but with the use of selective COX-2 inhibitors, glaucoma is a contraindication of these drugs, because both isoforms are constitutive enzymes in epithelial cells of the ciliary body. The data do not indicate that coxibs meet expectations placed on them; maybe the absolute selectivity in inhibiting COX-2 may not be the final goal. New directions of research on the development of NSAIDs are associated with nitric oxide-releasing substances and the substances that inhibit both cyclooxygenase and 5-lipoxygenase. Both directions of research have theoretical grounds; interactions between systems, NO synthases, and cyclooxygenases have long been known (Vane and Batting 1996). Many authors believe that the inhibition of COX-2 is a privileged direction of change that is second leukotriene formation under the influence of 5-lipoxygenase. It is now known that leukotrienes damage gastrointestinal mucosa, especially in poor production of prostaglandins, and they may cause the occurrence of "asthma aspirin derivative." In medicine, there are 5-lipoxygenase inhibitors (zileuton) and angiotensin receptor cysteine-leukotriene (montelukast, zafirlukast), which are successfully used to treat asthma but do not show action in combating inflammatory processes. Administration of naproxen and zileuton together in the fight against experimentally induced arthritis in mice showed that these drugs are highly inflammatory, in the absence of the effectiveness of these drugs used alone (Nickerson-Nutter and Medvedeff 1996), and so began the search for compounds that would impede the activity of both COX and 5—lipoxygenase. Quite a large group of such compounds are in preclinical development.

Frequently, pain is the only troublesome symptom of the disease and constitutes a warning signal, which is why analgesics should be considered as acting only symptomatically. Dangers of improper use are combined with the fact that these drugs are the most widely used group of drugs. It is assumed that the process of self-medication of NSAIDs should not exceed 10 days. The so-called character of ceiling analgesic effect, namely the lack of increase in activity over a specific dose for each drug, is also important. This increases the probability of potential adverse effects. Also, do not associate two or more NSAIDs, because synergism is uncertain, but most likely increases side effects. Due to the lack of anti-inflammatory effect, paracetamol does not belong to NSAIDs and its association with this class of drugs is indicated. In addition, it is important that it does not increase the adverse effects of NSAIDs in the gastrointestinal tract. Ternary connections are permitted such as classic NSAIDs—paracetamol—opioid (Tatarkiewicz 2004).

It is worth noting that one of the strongest painkillers is water for injection, injected subcutaneously. The development and progress of pharmacology led to the discontinuation of this method, formerly used for the elimination of cancer pain. Water must be free of minerals, since its effect is due to irritation of nerve endings resulting from the osmotic pressure difference of water and tissues. The analgesic effect lasts for several hours (Martensson et al. 2000). Water injection is also used in low back pain during labor due to its analgesic effect (Martensson and Wallin 1999) and in chronic neck and shoulder pain after certain injuries (Byrn et al. 1993). Extremely important is the fact that the injection of water causes fewer side effects than most other analgesics.

A common complaint among people is migraine, which is defined as repeated, usually unilateral, pulsating headache. Migraine usually lasts from 4 to 72 h and is characterized by varying degrees of severity and frequency of occurrence, the pain made worse under the influence of emotion or physical exertion. There is photophobia, excessive sensitivity to sound (fonofobia) and odors. There are also nausea and vomiting. Sometimes before a migraine episode, there may be a so-called aura, which occurs in 10% of patients with migraine (Herold 2008, Lewis et al.), in the form of paresthesia, visual field defects, the appearance of scotoma, paralysis, and aphasia. Studies in twins suggest that genetic influences are 60%–65% (Gervil et al. 1999, Young et al. 1992). Presumably, migraine is associated with a genetic predisposition to hypersensitive neuro vascular reactions. The substrate may be a malfunction of ion channel receptors of the cerebral cortex, neurons regulating cerebral blood flow and blood platelets, or macrophages, which chemically stimulate the intracranial perivascular nerve fibers (Glaubic-Łątka et al. 2004). It seems that the expanding cortical depression (cortical spreading depression)—a wave propagating—reduced bioelectrical activity of the cerebral cortex and is the dominant mechanism of migraine with aura (Domitrz 2007, Lauritzen 1994). According to another hypothesis, the so-called vascular responsible for the aura phase of cerebral vasospasm, which passes at a later stage in their overextension of the increased permeability of the walls, leading to pain attack. In migraine attacks, NSAIDs are also used.

The use of nonsteroidal anti-inflammatory drugs (NSAIDs) is nowadays widespread. It seems therefore advisable to develop new analytical methods for their qualitative and quantitative analysis.

Thin-layer chromatography (TLC) and the method of high-performance thin-layer chromatography (HPTLC) were used in the analysis of nonsteroidal anti-inflammatory drugs. The TLC method was used for chromatographic separation of study medications in stock solutions (Suedee et al. 2001), pharmaceutical preparations (Fartushnii et al. 1983, Glisovic et al. 1989, Sun et al. 1994), and biological preparations (Stanislavchuk et al. 1989, Schumacher et al. 1980). It was used also to determine the metabolic products of diclofenac sodium (Stanislavchuk et al. 1989). TLC has been used in a simple phase (Fartushnii et al. 1983, Glisovic et al. 1989, Stanislavchuk et al. 1989, Suedee et al. 2001); the solid phase was silica gel (Fartushnii et al. 1983, Glisovic et al. 1989, Schumacher et al. 1980), and the mobile phase mixtures of eluents were as follows:

- Chloroform + methanol (9 + 1 v/v) (Glisovic et al. 1989)
- Benzene + propan-2-ol (9 + 1 v/v) (Fartushnii et al. 1983)
- Benzene + methanlol + acetone (7 + 2 + 3 v/v) (Fartushnii et al. 1983)
- n-Hexane + ethyl acetate + acetic acid anhydrous (139 + 60 + 1 v/v) (Stanislavchuk et al. 1989)
- Dichloromethane + methanol + tetrahydrofuran (170 + 30 + 1 v/v) (Schumacher et al. 1980)
- Ethyl acetate + chloroform + methanol (53 + 40 + 7 v/v) (Parimoo et al. 1994)

The densitometric method was used for identification and determination of diclofenac sodium, using a silica gel solid phase and the mobile phase cyclohexane + chloroform + methanol (12 + 6 + 1 v/v) (Krzek and Starek 2002).

Identification of diclofenac sodium suppositories and stability testing were performed using a solid-phase silufol and hexane + chloroform + acetone + anhydrous acetic acid (60 + 60 + 30 + 1 v/v) as the mobile phase (Budukova et al. 1989).

Ketoprofen was determined in pharmaceutical formulations and spiked urine using micellar thin-layer chromatography and quantitative analysis applied to HPTLC (Mohammad et al. 2010).

Naproxen and ibuprofen tablets were analyzed by HPTLC, using chromatography plates coated with silica gel, and mobile phase: ethyl acetate + anhydrous acetic acid (19 + 1 v/v) (Lippstone and Sherma 1995). Using this method, applying silica gel 60 F_{254}, an analysis of enantiomers of keto-profen, indoprofen, and suprofenu was conducted. Two mobile phases have been used: benzene + methanol (93 + 7 v/v) and chloroform + ethyl acetate (15 + 1 v/v). For better separation of indopro-fen, a mobile-phase chloroform + ethyl acetate (3 + 1 v/v) has been used (Rossetti et al. 1986).

Stability analysis of piroxicam was also carried out by HPTLC using silica gel 60 F_{254} as the solid phase and toluene + acetic acid (8 + 2 v/v) as mobile phase (Puthli and Vavia 2000). HPTLC method was used for separation of indoprofen, ketoprofen, and suprofen enantiomers (Rosetti et al. 1986) and the analysis of ibuprofen, mefenamic acid, acetaminophen, and naproxen in pharmaceutical prepara-tions (Argekar and Sawant 1999, Lippstone and Sherma 1995, Shirke et al. 1993). Contents of ibu-profen were determined by densitometry at a wavelength $\lambda = 265$ nm (Shirke et al. 1993), mefenemic acid and paracetamol, using a wavelength $\lambda = 263$ nm (Argekar and Sawant 1999). An indication of paracetamol and diclofenac sodium with silica gel 60 F_{254} HPTLC and the mobile phase chloroform + methanol + ammonia 25% (100 + 25 + 1 v/v), developing chromatograph over a distance of 7 cm. Densitometric determination was carried out at a wavelength $\lambda = 286$ nm (Shinde et al. 1994).

HPTLC method indicated the degradation products of piroxicam, using silica gel 60 F_{254} and the mobile phase: ethyl acetate + toluene + buthylamine (2 + 2 + 1 v/v) or toluene + 95° ethanol + anhydrous acetic acid (80 + 12 + 5 v/v) and detection at a wavelength $\lambda = 296$ nm (Tomankova and Sabartova 1989). By TLC (Starek et al. 2009) and HPTLC, stability of piroxicam was investigated (Puthli and Vavia 2000).

Quantitative analysis was performed on chlorzoxazone, paracetamol, and diclofenac sodium, using gas chromatography, high-performance liquid chromatography, and high-performance thin-layer chromatography (Chawla et al. 1996).

Lipophilicity study of selected nonsteroidal anti-inflammatory drugs was performed using the method of thin-layer chromatography on reversed-phase using as a mobile phase the mixture of water and methanol containing 50 mM HOAc, NaOAc, or sodium chloride (Forgáts et al. 1998).

Lipophilicity study was also conducted on the following anti-inflammatory drugs: aspirin, diclofenac, ibuprofen, indomethacin, ketoprofen, naproxen, niflumoic acid, phenylbutazone, piroxi-cam, and tenoxicam using HPTLC plates coated with RP-18W/UV$_{254}$, nano-sil CN/UV$_{254}$, or nano-durasil-20/UV$_{254}$ coated with paraffin oil. The mixture of methanol and water was used as the mobile phase (Sârbu and Todor 1998).

Arylpropionic acid derivative in biological material was studied using HPLC, HPTLC, LC, GC/MS, CE, and optical methods [45].

Thin-layer chromatography with densitometric detection was used for enantioseparation and oscillatory transenantiomerization of S, R-(±)-ketoprofen (Sajewicz et al. 2007).

Aspirin (Panchal et al. 2009), diclofenac sodium, paracetamol (Dighe et al. 2006), ethamsylate, mefenamic acid (Jaiswal et al. 2005), flurbiprofen (Kilinc and Aydin 2009), etoricoxib (Maheshwari et al. 2007), celecoxib (Sane et al. 2004), celecoxib, etoricoxib, and vakdecoxib (Starek and Rejdych 2009) were extracted by methanol from pharmaceutical formulations. Aspirin, salicylic acid, and sulfosalicylic acid were extracted by ethanol from tablets (Panahi et al. 2010). Piroxicam, meloxicam, tenoxicam, and isoxicam were extracted by acetone from pharmaceutical formulations (Starek 2011). Nimesulide from plasma was extracted with dichloromethane (Miljković et al. 2003).

Chromatography TLC and HPTLC were comparatively less used in the analysis of nonsteroidal anti-inflammatory drugs, while the great interest was aroused by high-performance liquid chroma-tography (HPLC). Table 18.1 shows the TLC methods used for the analysis of NSAIDs.

TABLE 18.1

TLC Methods Used for the Analysis of NSAIDs

Compound	Stationary Phase	Mobile Phase (v/v)	Detection	Reference
Eighteen NSAIDs; hydrophobicity parameters	RP-TLC F_{254S}	Water + methanol containing 50 mM acetic acid, sodium acetate, or sodium chloride	UV	Forgáts et al. (1998)
Lipophilicity of phenolic drugs	RP-8 F_{254s}, RP-18 F_{254s}	Methanol + water in different volume proportions	UV	Pyka and Gurak (2007)
Fenbufen, ibuprofen, ketoprofen, diclofenac sodium, mefenamic acid and tiaprofenic acid; substance and pharmaceutical preparations	NP-TLC silica gel 60 F_{254} RP-18 F_{254}	Chloroform + methanol + ammonia 25% (6.7 + 2.5 + 0.8) pH 5.73 phosphate buffer + 10% CTMA-Br in methanol (3.5 + 6.5)	254 nm	Hopkała and Pomykalski (2004)
Pranoprofen, loxoprofen sodium, ketoprofen, naproxen, fenoprofen calcium, flurbiprofen, alminoprofen, ibuprofen, mefenamic acid, diclofenac sodium, aluminum flufenamate, tolfenamic acid, acemetacin, indomethacin, aspirin and phenylbutazone; substance	HPTLC silica gel 60 F_{254}	n-Butyl ether + n-hexane + acetic acid (20 + 4 + 1)	Spraying the plates with 0.1% dichlorophenolindophenol ethanol reagent followed by heating in oven at 100°C for 10 min. Phenylbutazone could be detected with the Liebermann reagent	Shinozuka et al. (1996)
Aceclofenac, paracetamol; bulk drug and in tablets	HPTLC silica gel 60 F_{254}	Toluene + ethyl acetate + methanol + acetic acid (6 + 4 + 0.8 + 0.4)	256 nm	Bharekar et al. (2011)
Aspirin; capsules	HPTLC silica gel F_{254}	Methanol + benzene + ethyl acetate + glacial acetic acid (0.36 + 5.6 + 4.0 + 0.04)	210 nm	Panchal et al. (2009)
Aspirin; pharmaceutical preparations	HPTLC silica gel G 60 F_{254} impregnated with 2% (w/v) boric acid in ethanol	Chloroform + methanol + ammonia + water (120 + 75 + 2 + 6)	254 nm	Panahi et al. (2010)

(continued)

TABLE 18.1 (continued)
TLC Methods Used for the Analysis of NSAIDs

Compound	Stationary Phase	Mobile Phase (v/v)	Detection	Reference
Aspirin, acetaminophen; substance	RP-18W HPTLC F_{254S}	Planar chromatography(TLC) and pressurized planar electrochromatography (PPEC) 15% acetonitrile in pH 3.8 buffer; 60% acetonitrile in pH 3.8 buffer, the polarization potential 2.0 kV; 30% acetonitrile in pH 3.0 buffer, the polarization potential 1.5 kV	UV	Hałka et al. (2010)
Diclofenac; pharmaceutical preparations	TLC silica gel F_{254}	Cyclohexane + chloroform + methanol (12 + 6 + 1)	248 nm	Krzek and Starek (2002)
Diclofenac sodium; serum	HPTLC silica gel 60 F_{254}	Diclofenac sodium was extracted with ethyl acetate from serum samples Toluene + acetone + glacial acetic acid (80 + 30 + 1)	280 nm	Lala et al. (2002)
Diclofenac; tablets	HPTLC silica gel 60 F_{254}	Chloroform + methanol + ammonia (10 + 25 + 0.25)	277 nm	Shah et al. (2003)
Diclofenac sodium; pharmaceutical preparations and bulk drug powder	HPTLC silica gel 60 F_{254}	Toluene + ethyl acetate + methanol + formic acid (5.0 + 4.0 + 1.0 + 0.01)	260 nm	Dighe et al. (2006)
Diclofenac; gel	HPTLC silica gel 60 F_{254}	Toluene + ethyl acetate + glacial acetic acid (6 + 4 + 0.02)	283 nm	Panchal et al. (2008)
Diclofenac, Ibuprofen; substance	Cyanopropyl bonded plates	CH_2Cl_2 + methanol + cyclohexane (95 + 5 + 40)	254 nm	Seigel et al. (2011)
Dipyrone; pure form, pharmaceutical preparations	HPTLC silica gel F_{254}	Water + methanol (95 + 5)	260 nm	Aburjai et al. (2000)
Etodolac; pharmaceutical preparations	HPTLC silica gel 60 F_{254}	n-Hexane + ethyl acetate + glacial acetic acid (6 + 2 + 0.4)	282 nm	Sane et al. (1998)
Flurbiprofen; plasma	HPTLC silica gel 60 F_{254}	The drug was extracted with Hexane + diethyl ether (80 + 20)	247 nm	Dhavse et al. (1997)

TABLE 18.1 (continued)
TLC Methods Used for the Analysis of NSAIDs

Compound	Stationary Phase	Mobile Phase (v/v)	Detection	Reference
		Mobile phase: n-hexane + ethyl acetate + glacial acetic acid (60 + 30 + 10)		
Flurbiprofen; tablets	HPTLC silica gel 60 F_{254}	Chloroform + acetone + xylene (5.0 + 2.0 + 1.0)	247 nm?	Kilinc and Aydin (2009)
Ibuprofen; lipophilicity	RP-2 F_{254}, RP-8 F_{254s}, RP-18 F_{254s}	Methanol + water in different volume	UV	Pyka (2009)
Ibuprofen; pharmaceutical preparations	RP-TLC	β-Cyclodextrin + methanol (15 + 1)	222 nm	Krzek et al. (2005)
Ibuprofen; tablets	HPTLC silica gel 60 F_{254}	t. butanol + ethyl acetate + glacial acetic acid + water (7 + 4 + 2 +2)	254 nm	Chitlange et al. (2008)
Ibuprofen; pharmaceutical preparations	NP-TLC silica gel 60 F_{254}	n-Hexane + ethyl acetate + acetic acid (15 + 5 + 0.7)	200 nm	Pyka and Bocheńska (2010)
	RP-18 F_{254}	Methanol + water (9 + 1)	224 nm	
Ibuprofen; tablets	HPTLC silica gel 60 F_{254}	Propanol + ethyl acetate + ammonia + water (4 + 3 + 2 + 1)	254 nm	Sam Solomon et al. (2010)
Ibuprofen; pharmaceutical preparations	TLC silica gel 60 F_{254}	Toluene + ethyl acetate + glacial acetic acid (17 + 13 + 1)	UV	Starek and Krzek (2010)
Ketorolak tromethamine; substance	HPTLC silica gel 60 F_{254}	Chloroform + ethyl acetate + glacial acetic acid (3 + 8 + 0.1)	323 nm	Devarajan et al. (2000)
Ketorolac tromethamine; human plasma	HPTLC silica gel 60 F_{254}	n-Butanol + chloroform + acetic acid + ammonium hydroxide + water (9 + 3 + 5 + 1 + 2)	323 nm	López-Bojórquez et al. (2008)
Mefenamic acid; tablets	HPTLC silica gel 60 F_{254}	Chloroform + methanol + acetic acid (10 + 8 + 0.2)	300 nm	Jaiswal et al. (2005)
Mefenamic acid; tablets	HPTLC silica gel 60 F_{254}	Methanol + toluene + triethylamine (1 + 7.5 + 0.2)	241 nm	Maliye et al. (2006)

(continued)

TABLE 18.1 (continued)
TLC Methods Used for the Analysis of NSAIDs

Compound	Stationary Phase	Mobile Phase (v/v)	Detection	Reference
Lornoxicam, meloxicam, piroxicam and tiaprofenic acid; substance and pharmaceutical preparations	NP-TLC silica gel 60 F_{254}	Toluene + acetic acid 99.5% + methanol (11 + 1 + 0.5) Chloroform + methanol + ammonia 25% (6.6 + 2.75 + 0.65)	254 nm	Hopkała and Pomykalski (2003)
	RP-TLC RP-18 F_{254S}	pH 5.54 phosphate buffer + 10% CTMA-Br in methanol (2 + 8)		
Oxicams: piroxicam, meloxicam, tenoxicam and isoxicam; substance	TLC silica gel 60 F_{254}	Ethyl acetate + ethanol + toluene (6 + 3 + 1) + 2 drops of ammonia 25%	360 nm	Starek (2011)
Meloxicam; bulk drug and pharmaceutical preparations	HPTLC silica gel 60 F_{254}	Ethyl acetate + cyclohexane + glacial acetic acid (6.5 +3.5 + 0.02%)	353 nm	Desai and Amin (2008)
Piroxicam; potential degradation products	HPTLC silica gel with fluorescence indicator	Ethyl acetate + toluene + diethylamine (10 + 10 + 5) Toluene + absolute ethanol + glacial acetic acid (8 + 1.2 + 0.5)	296 nm	Tománková and Šabartová (1989)
Piroxicam; substance	HPTLC silica gel 60 F_{254}	Toluene + acetic acid (8 + 2)	360 nm	Puthli and Vavia (2000)
Piroxicam; degradation products in drugs	TLC silica gel 60 F_{254}	Ethyl acetate + toluene + butylamine (2 + 2 + 1)	360 nm	Starek et al. (2009)
Nabumetone, paracetamol; combined tablet dosage form	HPTLC silica gel 60 F_{254}	Toluene + 2-propanol + acetic acid (8 + 2 + 0.1)	236 nm	Gandhi et al. (2011)
Naproxen, metabolite; urine	TLC silica gel F_{254}	Chloroform + methanol (85 + 15)	232 nm	Abdel-Moety et al. (1988)
Naproxen; rat serum	HPTLC silica gel F_{254}	Toluene + ethyl acetate + acetic acid (82 + 15 + 3)	260 nm	Guermouche et al. (2000)
Naproxen, naproxen methyl ester, its impurity; pharmaceutical preparations	TLC silica gel F_{254}	Cyclohexane + chloroform + methanol (12 + 6 + 1)	223 nm	Krzek and Starek (2004)
Nimesulide; pharmaceutical preparations	HPTLC silica gel 60 F_{254}	Cyclohexane + ethylacetate (60 + 40)	295 nm	Patravale et al. (2001)
Nimesulide; plasma	TLC silica gel 60 F_{254}	Toluene + acetone (100 + 10)	310 nm	Miljković et al. (2003)

TABLE 18.1 (continued)
TLC Methods Used for the Analysis of NSAIDs

Compound	Stationary Phase	Mobile Phase (v/v)	Detection	Reference
Paracetamol (acetaminophen) and mefenamic acid; tablets	HPTLC silica gel 60 F_{254}	Toluene + acetone + methanol (8 + 1 + 1)	263 nm	Argekar and Sawant (1999)
Paracetamol (acetaminophen); pharmaceutical preparations	RP-HPTLC-W18 F_{254S}	Methanol + glacial acetic acid + water (25 + 4.3 + 70.7)	254 nm	Soponar et al. (2009)
Paracetamol; bulk and pharmaceutical preparations	HPTLC silica gel 60 F_{254}	n-Hexane + ethyl acetate + glacial acetic acid (5 + 5 + 0.2)	223 nm	Chitlange and Soni (2010)
Paracetamol, diclofenac potassium; bulk drug, tablets	HPTLC silica gel 60 F_{254}	Toluene + acetone + methanol + formic acid (5 + 2 + 2 + 0.01)	274 nm	Khatal et al. (2010)
Paracetamol, lornoxicam; tablets	HPTLC silica gel 60 F_{254}	Ethyl acetate + methanol + toluene + glacial acetic acid (7 + 2.5 + 1 + 0.5)	270 nm	Patel and Patel (2010)
Paracetamol, dexketoprofen trometamol; tablets	HPTLC silica gel G 60 F_{254}	Toluene + ethyl acetate + acetic acid (6 + 4 + 0.2)	256 nm	Rao et al. (2011)
Paracetamol; pharmaceutical preparations	TLC with a fluorescence plate reader	n-Hexane + ethyl acetate + ethanol (2.5 + 1.5 + 0.4)	254 nm	Tavallali et al. (2011)
Celecoxib; in the presence of their degradation products	TLC silica gel 60 F_{254}	Cyclohexane + dichloromethane + diethylamine (50 + 40 + 10)	253 nm	Bebawy et al. (2002)
Celecoxib; pharmaceutical preparations	HPTLC silica gel 60 F_{254}	n-Hexane + ethyl acetate (60 + 40)	262 nm	Sane et al. (2004)
Celecoxib, etoricoxib, valdecoxib; pharmaceutical preparations	TLC silica gel 60 F_{254}	Chloroform + acetone + toluene (12 + 5 + 2)	254 nm and 290 nm	Starek and Rejdych (2009)
Etoricoxib; pharmaceutical preparations	HPTLC silica gel F_{254}	Chloroform + methanol + toluene (4 + 2 + 4)	289 nm	Shah et al. (2006)
Etoricoxib; in the bulk drug and in pharmaceutical preparations	HPTLC silica gel 60 F_{254}	Toluene + 1.4 dioxane + methanol (8.5 + 1.0 + 0.5)	235 nm	Maheshwari et al. (2007)
Etoricoxib; tablets	HPTLC silica gel 60 F_{254}	Ethyl acetate + methanol (8 + 2)	290 nm	Rajmane et al. (2010)

(continued)

TABLE 18.1 (continued)
TLC Methods Used for the Analysis of NSAIDs

Compound	Stationary Phase	Mobile Phase (v/v)	Detection	Reference
Rofecoxib; pharmaceutical preparations	HPTLC silica gel 60 F_{254}	Toluene + methanol + acetone (7.5 + 2.5 + 1.0)	311 nm	Kaul et al. (2005)
Rofecoxib; tablets	HPTLC silica gel G 60 F_{254}	n-Butyl acetate + formic acid + chloroform (6 + 4 + 2)	315 nm	Ravi et al. (2006)
Rofecoxib; pure and tablets	HPTLC silica gel 60 F_{254}	Acetone + methanol (1 + 1)	UV	Roosewelt et al. (2007)
Valdecoxib; bulk drug and pharmaceutical preparations	HPTLC silica gel 60 F_{254}	Toluene + ethyl acetate (5 + 5)	262 nm	Baviskar et al. (2007)

TABLE 18.2
TLC Methods Used for the Analysis of Antimigraine Drugs

Compound	Stationary Phase	Mobile Phase (v/v)	Detection	Reference
QSAR	NP-TLC silica gel 60 F_{254}	Acetonitrile + methanol + buffer (40 + 40 +20)	UV	Żydek and Brzezińska (2011)
	RP-TLC RP-2 silica gel 60 F_{254} (silanized)	Acetonitrile + methanol + methylene chloride + buffer (60 + 10 + 10 + 20)		
Almotriptan malate; tablets	HPTLC silica gel 60 F_{254}	Butanol + acetic acid + water (3 + 1 + 1)	300 nm	Suneetha and Syamasundar (2010)
Rizatriptan benzoate; in bulk and tablets	HPTLC silica gel 60 F_{254}	Dichloromethane + acetone + acetic acid (3 + 2 +0.2)	230 nm	Syama Sundar and Suneetha (2010)
Sumatriptan succinate; in the presence of its degradation products	HPTLC silica gel 60 F_{254}	Cyclohexane + dichloromethane + diethylamine (50 + 40 + 10)	228 nm	Bebawy et al. (2003)
Sumatriptan succinate; pharmaceutical preparations	HPTLC silica gel 60 F_{254}	Chloroform + methanol + ethyl acetate + ammonia (7.2 + 1 + 1.8 + 0.2)	247 nm	Tipre and Vavia (1999)
Sumatriptan; tablets	HPTLC silica gel 60 F_{254}	Methanol + water + glacial acetic acid (4 + 8 +0.1)	230 nm	Shah et al. (2008)

In summary, it can be stated that the thin-layer chromatography and high-performance thin-layer chromatography method are quite often used in the analysis of nonopioid analgesics, nonsteroidal anti-inflammatory and antimigraine drugs. Collected literature data testify to the fact that they were conducted both in normal and reversed phase using nonpolar and polar adsorbents as well as using the detection at selected wavelengths or dyeing reagents.

Sesquiterpenes and sesquiterpene lactone are used in the treatment of migraine. Sesquiterpenes were determined using various analytical methods such as GC, C-NMR, and HPLC, and in the analysis sesquiterpenes of lactones used the following methods: HPLC, GC, TLC, and OPLC (Merfort 2002). Stability-indicating methods of sumatriptan succinate have been studied using a densitometric determination at the wavelength $\lambda = 228$ nm. The studies used a mobile-phase composition: cyclohexane + dichloromethane + diethylamine (50 + 40 + 10 v/v) (Bebawy et al. 2003). QSAR analysis was used in the studies of serotoninergic drugs, using a solid-phase silica gel 60 F_{254} impregnated with solutions of amino acid analogues and their mixtures (Żydek and Brzezińska 2010). To mark 29 adulterations in lipid-regulating, antihypertensive, antitussive, and antiasthmatic herbal remedies, TLC was used as the solid-phase silica gel GF_{254} and a mobile phase of isopropyl alcohol + ethyl acetate + ammonia 25% (15 + 25 + 1.5 v/v) and detection at a wavelength $\lambda = 254$ nm (Wang et al. 2010). Indirect reversed-phase high-performance liquid chromatographic and direct thin-layer chromatographic enantioresolution were used for the study of (R, S)-Cinacalcet (Bhushan and Dubey 2011). Table 18.2 shows the TLC methods used for the analysis of anti-migraine drugs.

REFERENCES

Abdel-Moety, E.M., Al-Obaid, A.M., Jado, A.I., and E.A. Lotfi. 1988. Coupling of TLC and UV-measurement for quantification of naproxen and its main metabolite in urine. *European Journal of Drug Metabolism and Pharmacokinetics* 13:267–271.

Aburjai, T., Amro, B.I., Aiedeh, K., Abuirjele, M., and S. Al-Khalil. 2000. Second derivative ultraviolet spectrophotometry and HPTLC for the simultaneous determination of vitamin C and dipyrone. *Pharmazie* 55:751–754.

Argekar, A.P. and J.G. Sawant. 1999. Simultaneous determination of paracetamol and mefenamic acid in tablets by HPTLC. *Journal of Planar Chromatography—Modern TLC* 12:361–364.

Bartoszuk, M.A. and P.K. Zarzycki. 2004. Współczesne poglądy na metabolizm prostaglandyn oraz ich znaczenie w medycynie. *Farmacja Polska* 60:683–686.

Baviskar, D.T., Jagdale, S.C., Girase, N.O., Deshpande, A.Y., and D.K. Jain. 2007. Determination of valdecoxib from its bulk drug and pharmaceutical preparations by HPTLC. *Indian Drugs* 44:734–737.

Bebawy, L.I., Moustafa, A.A., and N.F. Abo-Talib. 2002. Stability-indicating methods for the determination of doxazosin mezylate and celecoxib. *Journal of Pharmaceutical and Biomedical Analysis* 27:779–793.

Bebawy, L.I., Moustafa, A.A., and N.F. Abo-Talib. 2003. Stability-indicating methods for the determination of sumatriptan succinate. *Journal of Pharmaceutical and Biomedical Analysis* 32:1123–1133.

Bharekar, V.V., Mulla, T.S., Rajput, M.P., Yadav, S.S., and J.R. Rao. 2011. Validated HPTLC method for simultaneous estimation of Rabeprazole sodium, paracetamol and aceclofenac in bulk drug and formulation. *Der Pharma Chemica* 2011:171–179.

Bhushan, R. and R. Dubey. 2011. Indirect reversed-phase high-performance liquid chromatographic and direct thin-layer chromatographic enantioresolution of (R,S)-Cinacalcet. *Biomedical Chromatography* 25:674–679.

Budukova, L.A., Kondratèva, T.S., Uribe-Echevarria, V.D., Arzamastsev, L.P., and V.I. Volchenok. 1989. Composition, preparation, analysis and stability studies of orthofen [diclofenac sodium] suppositories. *Farmatsiya (Moscow)* 38:16–20.

Byrn, C., Olsson, I., Falkheden, L., Lindh, M., Hosterey, U., Fogelberg, M., Linder, L.-E., and O. Bunketorp. 1993. Subcutaneous sterile water injections for chronic nec and shoulder pain following whiplash injuries. *Lancet* 341:449–452.

Chawla, J.L., Sodhi, R.A., and R.T. Sane. 1996. Simultaneous determination of chlorzoxazone, paracetamol and diclofenac sodium by different chromatographic techniques. *Indian Drugs* 33:171–178.

Chitlange, S., Sakarkar, D., Wankhede, S., and S. Wadodkar. 2008. High performance thin layer chromatographic method for simultaneous estimation of ibuprofen and pseudoephedrine hydrochloride. *Indian Journal of Pharmaceutical Sciences* 70:398–400.

Chitlange, S.S. and R. Soni. 2010. Development of stability indicating HPTLC method for simultaneous estimation of paracetamol and dexibuprofen. *Journal of Chemical and Pharmaceutical Sciences* 3:1–4.

Breckenridge, A.M., Kendall, M., Raine, J.M., Waller, P.C., Arlett, P., and L. Henderson. 2000. In fokus: Rofecoxib (Vioxx). *Current Problems in Pharmacovigilance* 26:13–14.

Desai, N. and P. Amin. 2008. Stability indicating HPTLC determination of meloxicam. *Indian Journal of Pharmaceutical Sciences* 70:644–647.

Devarajan, P.V., Gore, S.P., and S.V. Chavan. 2000. HPTLC determination of ketorolac tromethamine. *Journal of Pharmaceutical and Biomedical Analysis* 22:679–683.

Dhavse, V.V., Parmar, D.V., and P.V. Devarajan. 1997. High-performance thin-layer chromatographic determination of flurbiprofen in plasma. *Journal of Chromatography B: Biomedical Applications* 694:449–453.

Dighe, V.V., Sane, R.T., Menon, S.N., Tambe, H.N., Pillai, S., and V.N. Gokarn. 2006. Simultaneous determination of diclofenac sodium and paracetamol in a pharmaceutical preparation and in bulk drug powder by high-performance thin-layer chromatography. *Journal of Planar Chromatography—Modern TLC* 19:443–448.

Domitrz, I. 2007. Współczesne poglądy na patogenezę aury migrenowej. *Neurologia Polska* 41:70–75.

Fartushnii, A.F., Muzhanovskii, E.B., and A.I. Sedov. 1983. Identification and determination of voltaren [diclofenac sodium]. *Farm. Zhurnal* 1:52.

Forgáts, E., Cserháti, T., Kaliszan, R., Haber, P., and A. Nasal. 1998. Reversed-phase thin-layer chromatographic determination of the hydrophobicity parameters of nonsteroidal anti-inflammatory drugs. *Journal of Planar Chromatography – Modern TLC* 11:383–387.

Gandhi, S.V., Ranher, S.S., Deshpande, P.B., and D.K. Shah. 2011. Simultaneous HPTLC determination of nabumetone and paracetamol in combined tablet dosage form. *Journal of the Brazilian Chemical Society* 22:1068–1072.

Gervil, M., Ulrich, V., Kaprio, J., Olesen, J., and M.B. Russell. 1999. The relative role of genetic and environmental factors in migraine without aura. *Neurology* 53:995–999.

Glaubic-Łątka, M., Łatka, D., Bury, W., and K. Pierzchała. 2004. Współczesne poglądy na patofizjologię migreny. *Neurobiologia i Neurochirurgia Polska* 38:307–315.

Glisovic, L., Agbaba, D., Popovic, R., and I. Zgradic. 1989. Densitometric determination of ketoprofen in synovial fluid. *Pharmazie* 44:298–299.

Griswold, D.E. and J.L. Adams. 1996. Constitutive cyclooxygenase (COX–1) and inducible cyclooxygenase (COX–2): Rationale for selective inhibition and progress to date. *Medicinal Research Reviews* 16:181–206.

Guermouche, M.-H., Atik, N., and H. Chader. 2000. Assay of naproxen in rat serum by high-performance thin-layer chromatography/densitometry. *Journal of AOAC International* 83:1489–1492.

Hałka, A., Płocharz, P., Torbicz, A., and T. Dzido. 2010. Reversed-phase pressurized planar electrochromatography and planar chromatography of acetylsalicylic acid, caffeine, and acetaminophen. *Journal of Planar Chromatography—Modern TLC* 23:420–425.

Herold, G. 2008. *Medycyna Wewnętrzna*. Warszawa PZWL.

Hopkała, H. and A. Pomykalski. 2003. TLC analysis of inhibitors of cyclooxygenase and videodensitometric determination of meloxicam and tiaprofenic acid. *Journal of Planar Chromatography—Modern TLC* 16:107–111.

Hopkała, H. and A. Pomykalski. 2004. TLC analysis of non-steroidal anti-inflammatory drugs and videodensitometric determination of fenbufen in tablets. *Journal of Planar Chromatography—Modern TLC* 17:383–387.

Jaiswal, Y.S., Talele, G.S., and S.J. Surana. 2005. Quantitative analysis of ethamsylate and mefenamic acid in tablets by use of planar chromatography. *Journal of Planar Chromatography—Modern TLC* 18:460–464.

Kaul, N., Dhaneshwar, S.R., Agrawal, H., Kakad, A., and B. Patil. 2005. Application of HPLC and HPTLC for the simultaneous determination of tizanidine and rofecoxib in pharmaceutical dosage form. *Journal of Pharmaceutical and Biomedical Analysis* 37:27–38.

Khatal, L.D., Kamble, A.Y., Mahadik, M.V., and S.R. Dhaneshwar. 2010. Validated HPTLC method for simultaneous quantitation of paracetamol, diclofenac potassium, and famotidine in tablet formulation. *Journal of AOAC International* 93:765–770.

Kilinc, E. and F. Aydin. 2009. Stability-indicating HPTLC analysis of flurbiprofen in pharmaceutical dosage forms. *Journal of Planar Chromatography—Modern TLC* 22:349–354.

Krzek, J. and M. Starek. 2002. Densitometric determination of diclofenac, 1-(2,6-dichlorophenyl)indolin-2-one and indolin-2-one in pharmaceutical preparations and model solutions. *Journal of Pharmaceutical and Biomedical Analysis* 28:227–243.

Krzek, J. and M. Starek. 2004. Densitometric determination of naproxen, and of naproxen methyl ester, its impurity, in pharmaceutical preparations. *Journal of Planar Chromatography—Modern TLC* 17:137–142.

Krzek, J., Starek, M., and D. Jelonkiewicz. 2005. RP-TLC determination of S(+) and R(−) ibuprofen in drugs with the application of chiral mobile phase and UV densitometric detection. *Chromatographia* 62:653–657.

Lala, L.G., D'Mello, P.M., and S.R. Naik. 2002. HPTLC determination of diclofenac sodium from serum. *Journal of Pharmaceutical and Biomedical Analysis* 29:539–544.

Laneuville, O., Breuer, D.K., and D.L. Dewitt et al. 1994. Differential inhibition of human prostaglandin endoperoxide H synthases-1 and -2 by nonsteroidal anti-inflammatory drugs. *Journal of Pharmacology and Experimental Therapeutics* 271:927–934.

Langman, M.J., Jensen, D.M., Watson, D.J. et al. 1999. Adverse upper gastrointestinal effects of rofecoxib compared with NSAIDs. *JAMA* 282:1929–1933.

Lauritzen, M. 1994. Pathophysiology of the migraine aura. The spreading depression theory. *Brain: A Journal of Neurology* 117:199–210.

Lewis, P.R. and T.A. Pedley (Eds.). 2010. Primary and secondary headaches. In: *Merritt's Neurology (Neurology (Merritt's))*. Lippincott Williams & Wilkins, Philadelphia, PA, p. 152.

Lippstone, M.B. and J. Sherma. 1995. Analysis of tablets containing naproxen and ibuprofen by HPTLC with ultra-violet absorption densitometry. *Journal of Planar Chromatography—Modern TLC* 8:427–429.

López-Bojórquez, E., Castañeda-Hernández, G., González-De La Parra, M., and S. Namur. 2008. Development and validation of a high-performance thin-layer chromatographic method, with densitometry, for quantitative analysis of ketorolac tromethamine in human plasma. *Journal of AOAC International* 91:1191–1195.

Maheshwari, G., Subramanian, G.S., Karthik, A., Ranjithkumar, A., Musmade, P., Ginjupalli, K., and N. Udupa. 2007. High-performance thin-layer chromatographic determination of etoricoxib in the bulk drug and in pharmaceutical dosage form. *Journal of Planar Chromatography—Modern TLC* 20:335–339.

Makulska-Nowak, H.E. 2004. Walka z bólem. *Farmacja Polska* 60:987–992.

Maliye, A.N., Walode, S.G., Kasture, A.V., and S.G. Wadodkar. 2006. Simultaneous estimation of mefenamic acid and drotaverine hydrochloride in tablets by high performance thin layer chromatography. *Asian Journal of Chemistry* 18:667–672.

Martensson, L., Nyberg, K., and G. Wallin, 2000. Subcutaneous versus intracutaneous injections of sterile water for labour analgesia: A comparison of perceived pain during administration. *BJOG* 107:1248–1251.

Martensson, L. and G. Wallin. 1999. Labour pain treated with cutaneous injections of sterile water: A randomized controlled trial. *British Journal of Obstetrics & Gynaecology* 106:633–637.

Merfort, I. 2002. Review of the analytical techniques for sesquiterpenes and sesquiterpene lactones. *Journal of Chromatography A* 967:115–130.

Miljković, B., Brzaković, B., Kovacević, I., Agbaba, D., and M. Pokrajac. 2003. Quantitative analysis of nimesulide in plasma by thin-layer chromatography: Application to pharmacokinetic studies in man. *Journal of Planar Chromatography—Modern TLC* 16:211–213.

Mohammad, A., Sharma, S., and S.A. Bhawani. 2010. Identification of ketoprofen in drug formulation and spiked urine samples by micellar thin layer chromatography and its quantitative estimation by high performance liquid chromatography. *International Journal of PharmTech Research* 2:89–96.

Nickerson-Nutter, C.L. and E.D. Medvedeff. 1996. The effect of leukotriene synthesis inhibitors in models of acute and chronic inflammation. *Arthritis & Rheumatism* 39:515–521.

Norman, R.J. 2001. Reproductive consequences of COX-2 inhibition. *Lancet* 358:1287–1288.

Panahi, H.A., Rahimi, A., Moniri, E., Izadi, A., and M.M. Parvin. 2010. HPTLC separation and quantitative analysis of aspirin, salicylic acid, and sulfosalicylic acid. *Journal of Planar Chromatography—Modern TLC* 23:137–140.

Panchal, H.J., Rathod, I.S., and S.A. Shah. 2008. Development of validated HPTLC method for quantitation of diclofenac in diclofenac gels. *Indian Drugs* 45:301–306.

Panchal, H.J., Suhagia, B.N., and N.J. Patel. 2009. Simultaneous HPTLC analysis of atorvastatin calcium, ramipril, and aspirin in a capsule dosage form. *Journal of Planar Chromatography—Modern TLC* 22:265–271.

Parimoo, I., Bharathi, A., and M. Shajahan. 1994. Estimation of oxyphenbutazone and ibuprofen in presence of paracetamol and dextropropoxyphene in dosage forms by quantitative thin layer chromatography. *Indian Drugs* 31:139–143.

Patel, D.J. and V.P. Patel. 2010. Simultaneous determination of paracetamol and lornoxicam in tablets by thin layerchromatography combined with densitometry. *International Journal of ChemTech Research* 2:1929–1932.

Patravale, V.B., D'Souza, S., and Y. Narkar. 2001. HPTLC determination of nimesulide from pharmaceutical dosage forms. *Journal of Pharmaceutical and Biomedical Analysis* 25:685–688.

Puthli, S.P. and P.R. Vavia. 2000. Stability indicating HPTLC determination of piroxicam. *Journal of Pharmaceutical and Biomedical Analysis* 22:673–677.

Pyka, A. 2009. Lipophilicity investigations of ibuprofen. *Journal of Liquid Chromatography and Related Technologies* 32:723–731.

Pyka, A. and P. Bocheńska. 2010. Comparison of NP-TLC and RP-TLC with densitometry to quantitative analysis of ibuprofen in pharmaceutical preparations. *Journal of Liquid Chromatography and Related Technologies* 33:825–836.

Pyka, A. and D. Gurak. 2007. Use of RP-TLC and theoretical computational methods to compare the lipophilicity of phenolic drugs. *Journal of Planar Chromatography—Modern TLC* 20:373–380.

Rajmane, V.S., Gandhi, S.V., Patil, U.P., and M.R. Sengar. 2010. High-performance thin-layer chromatographic determination of etoricoxib and thiocolchicoside in combined tablet dosage form. *Journal of AOAC International* 93:783–786.

Rao, J.R., Mulla, T.S., Bharekar, V.V., Yadav, S.S., and M.P. Rajput. 2011. Simultaneous HPTLC determination of paracetamol and dexketoprofen trometamol in pharmaceutical dosage form. *Der Pharma Chemica* 3:32–38.

Ravi, T., Gandhimathi, M., Sireesha, K., and S. Jacob. 2006. HPTLC method for the simultaneous estimation of tizanidine and rofecoxib in tablets. *Indian Journal of Pharmaceutical Science* 68:234–236.

Roosewelt, C., Harikrishnan, N., Muthuprasanna, P., Shanmugapandiyan, P., and V. Gunasekaran. 2007. Validated high performance thin layer chromatography method for simultaneous estimation of rofecoxib and tizanidine hydrochloride in pure and tablet dosage forms. *Asian Journal of Chemistry* 19:4286–4290.

Rosenstock, M., Danon, A., Rubin, M., and G. Rimon. 2001. Prostaglandin H synthase-2 inhibitors interfere with prostaglandin H synthase-1 inhibition by nonsteroidal anti-inflammatory drugs. *European Journal of Pharmacology* 412:101–108.

Rossetti, V., Lombard, A., and M. Buffa. 1986. HPTLC resolution of the enantiomers of some 2–arylpropionic acid anti-inflammatory drugs. *Journal of Pharmaceutical and Biomedical Analysis* 4:673–676.

Sajewicz, M., Gontarska, M., Wróbel, M., and T. Kowalska. 2007. Enantioseparation and oscillatory transenantiomerization of S,R-(±)-ketoprofen, as investigated by means of thin layer chromatography with densitometric detection. *Journal of Liquid Chromatography and Related Technologies* 30:2193–2208.

Sam Solomon, W.D., Kumar, R.A., Vijai Anand, P.R., Sivakumar, R., and R. Venkatnarayanan. 2010. Derivatized HPTLC method for simultaneous estimation of glucosamine and ibuprofen in tablets. *Journal of Pharmaceutical Research and Health Care* 2:156–162.

Sane, R.T., Francis, M., and A.R. Khatri. 1998. High-performance thin-layer chromatographic determination of etodolac in pharmaceutical preparations. *Journal of Planar Chromatography—Modern TLC* 11:211–213.

Sane, R.T., Pandit, S., and S. Khedkar. 2004. High-performance thin-layer chromatographic determination of celecoxib in its dosage form. *Journal of Planar Chromatography—Modern TLC* 17:61–64.

Sârbu, C. and S. Todor. 1998. Determination of lipophilicity of some non-steroidal anti-inflammatory agents and their relationships by using principal component analysis based on thin-layer chromatographic retention data. *Journal of Chromatography A* 822:263–269.

Schumacher, A., Geissler, H.E., and E. Mutschler. 1980. Quantitative determination of plasma levels of diclofenac sodium by absorption measurement with direct evaluation of thin-layer chromatograms. *Biomedical Applications* 181:512–515.

Seigel, A., Schröck, A., Hauser, R., and B. Spangenberg. 2011. Sensitive quantification of diclofenac ond ibuprofen using thin layer chromatography coupled with a vibrio fisheri bioluminescence assay. *Journal of Liquid Chromatography and Related Technologies* 34:817–828.

Shah, C.R., Suhagia, B.N., Shah, N.J., and R.R. Shah. 2008. Development and validation of a HPTLC method for the estimation of sumatriptan in tablet dosage forms. *Indian Journal of Pharmaceutical Sciences* 70:831–834.

Shah, H.J., Rathod, I.S., Shah, S.A., Savale, S.S., and C.J. Shishoo. 2003. Sensitive HPTLC method for monitoring dissolution profiles of diclofenac from different tablets containing combined diclofenac and acetaminophen. *Journal of Planar Chromatography—Modern TLC* 16:36–44.

Shah, N.J., Shah, S.J., Patel, D.M., and N.M. Patel. 2006. Development and validation of HPTLC method for the estimation of etoricoxib. *Indian Journal of Pharmaceutical Science* 68:788–789.

Shinde, V.M., Tendolkar, N.M., and B.S. Desai. 1994. Simultaneous determination of paracetamol and diclofenac sodium in pharmaceutical preparations by quantitative TLC. *Journal of Planar Chromatography* 7:50–53.

Shinozuka, T., Terada, M., Ogamo, A., Nakajima, R., Takei, S., Murai, T., Wakasugi, C., and J. Yanagida. 1996. Data on high-performance thin-layer chromatography of analgesic and antipyretic drugs. *Japanese Journal of Forensic Toxicology* 14:246–252.

Shirke, P.P., Patel, M.K., Tamhane, V.A., Tirodkar, V.B., and P.D. Sethi. 1993. Simultaneous estimation of ibuprofen and paracetamol in combined dosage formulation by high-performance thin-layer chromatography. *Indian Drugs* 30:653–654.

Soponar, F., Mot, A.C., and C. Sârbu. 2009. Quantitative evaluation of paracetamol and caffeine from pharmaceutical preparations using image analysis and RP-TLC. *Chromatographia* 69:151–155.

Stanislavchuk, N.A., Pentyuk, A.A., and N.B. Lutsyuk. 1989. Methods for the determination of voltaren [diclofenac sodium] in biological materials. *Khimiko Farmatsevticheskii Zhurnal* 23:1131–1133.

Starek, M. 2011. Separation and determination of four oxicams in pharmaceutical formulations by thin-layer chromatographic-densitometric method. *Journal of Planar Chromatography—Modern TLC* 24:367–372.

Starek, M. and J. Krzek. 2010. TLC chromatographic-densitometric assay of ibuprofen and its impurities. *Journal of Chromatographic Science* 48:825–829.

Starek, M., Krzek, J., Tarsa, M., and M. Zylewski. 2009. Determination of piroxicam and degradation products in drugs by TLC. *Chromatographia* 69:351–356.

Starek, M. and M. Rejdych. 2009. Densitometric analysis of celecoxib, etoricoxib and valdecoxib in pharmaceutical preparations. *Journal of Planar Chromatography—Modern TLC* 22:399–403.

Suedee, R., Srichana, T., Saelim, J., and T. Thavonpibulbut. 2001. Thin-layer chromatographic separation of chiral drugs on molecularly imprinted chiral stationary phases. *Journal of Planar Chromatography—Modern TLC* 14:194–198.

Sun, S.W., Fabre, H., and H. Maillols. 1994. Test procedure validation for the TLC assay of a degradation product in a pharmaceutical formulation. *Journal of Liquid Chromatography* 17:2495–2509.

Suneetha, A. and B. Syamasundar. 2010. Development and validation of HPTLC method for the estimation of almotriptan malate in tablet dosage form. *Indian Journal of Pharmaceutical Sciences* 72:629–632.

Syama Sundar, B. and A. Suneetha. 2010. Development and validation of HPTLC method for the estimation of rizatriptan benzoate in bulk and tablets. *Indian Journal of Pharmaceutical Sciences* 72:798–801.

Tatarkiewicz, J. 2004. Farmakoterapia bólu: wybrane zagadnienia. *Farmacja Polska* 60:1001.

Tavallali, H., Zareiyan, S.F., and M. Naghian. 2011. An efficient and simultaneous analysis of caffeine and paracetamol in pharmaceutical formulations using TLC with a fluorescence plater reader. *Journal of AOAC International* 94:1094–1099.

Tipre, D.N. and P.R. Vavia. 1999. Estimation of Sumatriptan succinate in pharmaceutical dosage form by spectrophotometric and HPTLC method. *Indian Drugs* 36:501–505.

Tomankova, H. and J. Sabartova. 1989. Determination of potential degradation products of piroxicam by HPTLC densitometry and HPLC. *Chromatographia* 28:197–202.

Vane, J.R. and R.M. Betting (Eds.). 1996. Overview—Mechanisms of action of anti-inflammatory drugs. In: *Improved Non-Steroidal Anti-Inflammatory Drugs. COX-2 Enzyme Inhibitors.* Kluwer Academic Press Publisher, Dordrecht, the Netherlands.

Wang, T.-S., Tong, Y., Zheng, J., Zhang, Z., Gao, Q., and B.-Q. Che. 2010. Detection of 29 adulterations in lipid-regulating, antihypertensive, antitussive and antiasthmatic herbal remedies by TLC. *Chinese Pharmaceutical Journal* 45:857–861.

Warner, T.D., Giuliano, F., Vojnovic et al. 1999. Nonsteroid drug selectivities for cyclo-oxygenase-1 rather than cyclo-oxygenase-2 are associated with human gastrointestinal toxicity: A full in vitro analysis. *Proceedings of the National Academy of Sciences of the United States of America* 96:7563–7568.

Whelton, A. 2001. Renal aspects of treatment with conventional non-steroidal anti-inflammatory drugs versus cyclooxygenase-2-specific inhibitors. *American Journal of Medicine* 110 (3 Suppl. 1):33S–42S.

Young, C.A., Humphrey, P.R., Ghadiali, E.J., Klapper, P.E., and G.M. Cleator. 1992. Short-term memory impairment in an alert patient as a presentation of herpes simplex encephalitis. *Neurology* 42:260–261.

Żydek, G. and E. Brzezińska. 2010. Application of QSAR analysis in the studies of serotoninergic drugs. *Farmaceutyczny Przegląd Naukowy* 7:34–42.

Żydek, G. and E. Brzezińska. 2011. Normal and reversed phase thin layer chromatography data in quantitative structure-activity relationship study of compounds with affinity for serotonin (5-HT) receptors. *Journal of Chromatography B: Analytical Technologies in the Biomedical and Life Sciences* 879:1764–1772.

19 TLC of Ergot Alkaloid Derivatives

Anna Petruczynik

CONTENTS

19.1 INFORMATION ABOUT THE DRUG GROUP

The ergot alkaloids are a complex family of mycotoxins derived from prenylated tryptophan in several species of fungi. They are well known for their historical role in human toxicoses. The ergot alkaloids have a high biological activity and a broad spectrum of pharmacological effects; hence, they are of considerable importance to medicine. They have adrenoblocking, antiserotonin, and dopaminomimetic properties. Ergot alkaloids have a therapeutic effect on some forms of migraine, postpartum hemorrhages, and mastopathy, and a sedative effect on the central nervous system. Several semisynthetic compounds based on ergotamine such as lisuride, methysergide, and nicergoline have been used in medicine.

Dihydroergotamine is a semisynthetic ergot derivative that has a complex mechanism of therapeutic action. It is a partial antagonist on noradrenergic, dopaminergic, and serotoninergic receptors. It causes a construction of the cerebral and peripheral arteries through a stimulation of the serotoninergic receptor 5-HT$_{1D}$. Dihydroergotamine has been used for years to treat migraine. Dihydroergotamine also has a venoconstructive and hypertensive effect.

Dihydroergocristine exercises a double agonistic/antagonistic activity on dopaminergic and adrenergic receptors; it also shows a noncompetitive antagonistic effect on serotonin (5-HT) receptors. The central effects of dihydroergocristine depend on the initial cerebrovascular resistance. Dihydroergocristine increases the cerebral blood flow and the oxygen consumption of the brain. It protects the brain against the metabolic effects of ischemia by acting at cellular level. Dihydroergocristine exercises a vasoregulating amphoteric action that depends on the initial tonus.

Lisuride is a semisynthetic ergot derivative that was first used clinically as an antimigraine agent. Low doses of lisuride are mediated by stimulation of 5-HT$_{1A}$ receptors. Although lisuride has high affinity for D$_2$ dopamine receptors, it produces unusual effects relative to other dopamine agonists. The drug is used usually in combination with levodopa to treat Parkinson's disease. It reduces symptoms such as rigidity of movement, muscle rigidity, and the trembling when trying to move.

Methysergide, a D-lysergic acid derivative and 5-HT receptor antagonist, is used in the prophylactic treatment of migraine. A metabolite of methysergide, methylergometrine, is used in obstetrics for its uterotonic effect.

Pizotifen is an H$_1$ receptor antagonist used for prophylaxis of migraine.

Nicergoline has a broad spectrum of action: As an alpha(1)-adrenoreceptor antagonist, it induces vasodilation and increases arterial blood flow; it enhances cholinergic and catecholaminergic neurotransmitter function; it inhibits platelet aggregation; it promotes metabolic activity, resulting in increased utilization of oxygen and glucose; and it has neurotrophic and antioxidant properties.

19.2 SAMPLE PREPARATION

Dihydroergotamine was determined by thin-layer chromatography (TLC) in human plasma (Riedel et al. 1982). Before quantitative analysis the alkaloid was extracted twice by dichloromethane, then the organic phase was removed under nitrogen, and the residue resolved in ethanol and applied on silica TLC plate.

Methysergide was extracted from sodium chloride-saturated biological fluid into n-heptane at a basic pH. Solvent was then evaporated at 45°C, and dry residue was redissolved in one of the solvents applied and spotted on TLC plates (Bianchine et al. 1967).

Anderson et al. have analyzed dihydroergotamine produced by the strain of *Claviceps purpurea* (Anderson et al. 1979). Before TLC analysis the culture was filtered with suction and the liquor was passed through a cation-exchange column. The alkaloids were eluted with 5% ammonia and the eluate was concentrated to dryness on the rotary evaporator.

Tablets (Mosegor) containing pizotifen before TLC analysis were transferred into a volumetric flask, shaken with the mobile phase for 25 min using ultrasonic bath, and filtered, and aliquot of the filtrate was diluted to 10 mL with the mobile phase (Abounassif et al. 2005).

For high-performance thin layer chromatography (HPTLC)-densitometric method for determination of nicergoline, a standard solution of the substance in the concentration range 5–45 µg mL^{-1} was prepared in ethanol (Ahmad et al. 2002).

Rossi et al. described the identification of metabolites formed in vitro after incubation of oxetorone with rat liver microsomes (Rossi et al. 1978). The incubation mixture was adjusted to pH 9 with 1 M sodium hydroxide and evaporated to dryness in a water bath at 60°C under nitrogen. The residue was dissolved in methanol.

19.3 CHROMATOGRAPHIC SYSTEMS USED IN THE SEPARATION OF DRUGS

TLC is a technique that has been used successfully in the separation and determination of ergot alkaloid derivatives. TLC analysis of these compounds was most often performed in normal-phase (NP) system. Most TLC procedures for separating ergot alkaloid derivatives use a silica gel, often with fluorescent agent, as stationary phase. Usually polar eluents containing strongly polar modifier (methanol, ethanol) and medium polar diluent such as chloroform, ethyl acetate, and dichloromethane are used for separation of these drugs. Because ergot alkaloid derivatives are bases, often strongly basic eluents have been used. The use of basic additives to the eluent such as ammonia, or rarely amines, for example, diethylamine, is often reported.

Reversed-phase (RP) system was rarely applied in analysis of the group of drugs. The RP plates were used for these purposes with eluents containing organic modifiers and aqueous buffers.

Dihydroergotamine from *Claviceps purpurea* was analyzed on silica gel layer in different eluent systems containing chloroform–methanol (4:1), ethyl acetate–methanol (19:1), or ethyl acetate–dimethylformamide–ethanol (13:1.9:0.1) (Anderson et al. 1979).

Prosek et al. determined dihydroergocristine, dihydroergocornine, and dihydroergocryptine on cellulose plates (Prošek et al. 1976). The eluent contains mixture of ethyl acetate–n-heptane–diethylamine (4:6:0.2). The spots on chromatograms were determined by fluorodensitometric method.

Dihydroergocristine was analyzed by TLC on Silufol plates and chloroform–ethanol–25% NH$_3$ solution (45:5:0.2) as the mobile phase (Zvonkova et al. 1998).

Lisuride was purified by TLC on silica plate developed with eluent containing chloroform–methanol–ammonia (19:1:0.1) (Filer et al. 2006). The UV-visualized and scraped band was eluted with ethanol and cochromatographed on TLC in the same system. TLC was also applied to purification of iodolisuride—derivative used to study human striatal D2 dopamine receptor (Farouk et al. 2011). Purification was performed on silica gel plates by the use of chloroform–methanol (95:3).

Quantitative determination and stability investigations of lisuride hydrogen maleate in pharmaceutical preparations using TLC were described by Amin (1987). Analysis was performed on silica gel plates with mixture of chloroform and methanol (17:3) as mobile phase. The spots were detected spectrophotometrically.

Abounassif et al. analyzed the fraction extracted from tablets containing pizotifen on silica gel plates developed with chloroform–methanol–ammonia (8:2:0.2) (Abounassif et al. 2005).

Żydek and Brzezińska carried out quantitative structure–activity relationship (QSAR) analysis of 20 drugs (inter alia pizotifen) with affinity for 5-HT receptor using TLC (Żydek and Brzezińska 2011). The analysis of pizotifen was performed in NP and RP systems on silica gel and RP2 plates, respectively. A mixture of acetonitrile–methanol–acetate buffer at pH 7.4 (40:40:20) or acetonitrile–methanol–methylene chloride–acetate buffer at pH 7.4 (60:10:10:20) was used as eluents.

Banno et al. described a method for quantitative determination of nicergoline by combination of TLC and secondary ion mass spectrometry (TLC/SIMS) (Banno et al. 1991). The analysis was performed on aluminum TLC plates using an eluent containing a mixture of dichloromethane–acetone–water (100:10:1). After development, the portion of the plate with the nicergoline and internal standard spots was cut off at the TLC plate and was attached to the SIMS holder. The method allowed the determination of the drug down to a 10 ng level.

Ahmad et al. developed tree methods for the determination of nicergoline in the presence of its hydrolysis-induced degradation product (Ahmad et al. 2002). One of the methods was based on HPTLC separation of nicergoline and degradation products. The analysis was performed on silica gel plates using methanol–ethyl acetate–glacial acetic acid (5:7:3). After development they were scanned at 287 nm. The method was successfully applied to the analysis of nicergoline in Sermion tablets.

For identification of nicergoline on TLC silica gel plates, iodic acid was applied. The spots of nicergoline were a light yellow color after drying (Gavazzutti et al. 1983).

Methysergide was analyzed on silica gel or aluminum oxide plates with different eluents containing a mixture of chloroform and acetone, benzene, cyclohexane, methanol, or ethanol. Eluents containing a mixture of chloroform and benzene, with the addition of acetic acid, or a mixture of chloroform and cyclohexane, with the addition of diethylamine, were applied. A mixture of chloroform and diethylamine was also used successfully as mobile phase. Spots were localized in UV light, eluted, and quantitated spectrophotofluorometrically.

Oxetorone and its metabolites were analyzed on silica gel plates with mobile phase containing benzene–acetone–ammonia (50:50:1) (Rossi et al. 1978). These substances were detected under UV light at 254 and 365 nm.

HPTLC silica gel plates were used for identification of different drugs; among others were dihydroergotamine, methysergide, and nicergoline (Musumarra et al. 1985). Four eluent systems containing ethyl acetate–methanol–ammonia (85:10:15), cyclohexane–toluene–diethylamine (65:25:10), ethyl acetate–chloroform (50:50), and acetone were applied for principal component analysis (PCA). For eluent containing only acetone, the plates were dipped in 0.1 M potassium hydroxide methanolic solution and dried before application of the drugs.

19.4 USE OF TLC IN CLINICAL ANALYSIS

TLC was rarely applied in clinical analysis of ergot alkaloid derivatives. The dihydroergotamine and its metabolites (determined in plasma and urine) were purified on silica gel plates using a mixture of dichloromethane, methanol, and water (80:18:2 v/v) (Maurer and Frick 1984).

In order to determine nicergoline pharmacokinetics after oral administration to humans, Ezan et al. developed the radioimmunoassay method (Ezan et al. 2001). Metabolites were iodinated, and after 1 min incubation the mixture was analyzed by TLC on silica gel plates with a mixture of chloroform–methanol (80:20) as mobile phase.

REFERENCES

Abounassif, M.A., El-Obeid, H.A., Gadkariem, E.A. 2005. Stability studies on some benzocycloheptane antihistaminic agents. *J. Pharm. Biomed. Anal.* 36: 1011–1018.

Ahmad, A.K.S., Kawy, M.A., Nebsen, M. 2002. First derivative ratio spectrophotometric, HPTLC-densitometric, and HPLC determination of nicergoline in presence of its hydrolysis-induced degradation product. *J. Pharm. Biomed. Anal.* 30: 479–489.

Amin, M. 1987. Quantitative determination and stability of lisuride hydrogen maleate in pharmaceutical preparations using thin-layer chromatography. *Analyst* 112: 1663–1665.

Anderson, J.A., Kim, I.-S., Lehtonen, P., Floss, H.G. 1979. Conversion of dihydrolysergic acid to dihydroergotamine in an ergotamine-producing strain of *Claviceps purpurea*. *J. Nat. Prod.* 42: 271–273.

Banno, K., Matsuoka, M., Takahashi, R. 1991. Quantitative analysis by thin-layer chromatography with secondary ion mass spectrometry. *Chromatographia* 32: 179–181.

Bianchine, J.R., Niec, A., Macaraeg, P.V.J. 1967. Thin-layer chromatographic separation and detection of methysergide and methergine. *J. Chromatogr.* 31: 255–257.

Ezan, E., Delestre, L., Legendre, S., Riviere, R., Doignon, J.-L., Grognet, J.-M. 2001. Immunoassays for the detection of nicergoline and its metabolites in human plasma. *J. Pharm. Biomed. Anal.* 25: 123–130.

Farouk, N., Abdel-Aziz, H.M., Ayoub, S. 2011. Separation of [125]IBZM and [125]ILIS using polyacrylamide-acrylic acid resin. *J. Radioanal. Nucl. Chem.* 290: 587–593.

Filer, C.N., Hainley, C., Nuget, R.P. 2006. Tritiation of dopaminergic ligands (−)-lisuride and (+/−)-nomifensine. *J. Radioanal. Nucl. Chem.* 267: 345–348.

Gavazzutti, G., Gagliardi, L., Amato, A., Profili, M., Zagarese, V., Tonelli, D., Gattavecchia, E. 1983. Colour reactions of iodic acid as reagent for identifying drugs by thin-layer chromatography. *J. Chromatogr.* 268: 528–534.

Maurer, G., Frick, W. 1984. Elucidation of the structure and receptor binding studies of the major primary, metabolite of dihydroergotamine in man. *Eur. J. Clin. Pharmacol.* 26: 463–470.

Musumarra, G., Scarlata, G., Cirma, G., Romano, G., Palazzo, S., Clementi, S., Giulirtti, G., 1985. Qualitative organic analysis. I. Identification of drugs by principal components analysis of standardized thin-layer chromatographic data in four eluent systems. *J. Chromatogr.* 350: 151–168.

Prošek, M., Kučan, E., Katić, M., Bano, M. 1976. Quantitative fluorodensitometric determination of ergot alkaloids. II. Determination of hydrogenated ergot alkaloids of the ergotoxine group. *Chromatographia* 9: 325–327.

Riedel, E., Kreutz, G., Hermsdorf, D. 1982. Quantitative thin-layer chromatographic determination of dihydroergot alkaloids. *J. Chromatogr. B* 229: 417–423.

Rossi, E., de Pascale, A., Negrini, P., Zanol, M., Frigerio, A. 1978. Oxetorone metabolism in vitro. *J. Chromatogr.* 152: 228–233.

Zvonkova, E.N., Sheichenko, V.I., Lapa, G.B., Monakhova, T.E., Anufrieva, V.V., Bykov, V.A. 1998. Solubility of dihydroergocristine mesylate. *Pharm. Chem. J.* 32: 567–568.

Żydek, G., Brzezińska, E. 2011. Normal and reversed phase thin-layer chromatography data in quantitative structure-activity relationship study of compounds with affinity for serotonin (5HT) receptors. *J. Chromatogr. B* 879: 1764–1772.

20 Thin-Layer Chromatography of Anesthetics

Maja M. Natić

CONTENTS

20.1 INTRODUCTION

Anesthetics are the most heterogeneous class in pharmacology. They include a variety of drugs that differ in chemical structure. Modern anesthetics can be divided into two categories: general and local anesthetics. According to the route of administration, anesthetics can be divided into inhalation anesthetics, intravenous anesthetics, and local anesthetics.

General anesthetics produce controlled, reversible depression of the functional activities of the central nervous system, which causes the loss of sensation and consciousness. Volatile general anesthetics are administered by inhalation and are further subdivided into gases (e.g., ethyl chloride, nitrous oxide) and liquids (e.g., diethyl ether, halothane, chloroform, and trichloroethylene). Nonvolatile anesthetics are administered by the intravenous route.

Local anesthetics produce insensitivity in a limited area by blocking the generation and conduction of nerve impulses. The drugs are bound to specific receptors located inside the sodium channels of cell membranes, and they thus block the sodium ion permeability. They are applied locally or injected to produce the loss of sensation in the target area.

The purpose of this chapter is to present a summary of thin-layer chromatographic (TLC) systems suitable for the analysis, separation, and qualitative and quantitative determination of some of the anesthetic drugs that are currently in use. TLC has a resolution that is adequate for identifying many drugs from the anesthetic group, and it has been shown to be a useful screening technique in toxicology laboratory. In comparison with conventional TLC, high-performance thin-layer chromatography (HPTLC) is more suitable for qualitative and quantitative assays, and it provides better precision and accuracy.

The focus will be on the analysis of local anesthetics. Apart from them, some selected nonvolatile intravenous general anesthetics will be studied. As far as inhalational anesthetics are concerned, chromatographic methods such as gas-liquid chromatography (GLC) or high-performance liquid chromatography (HPLC) are superior for the analysis (Heusler 1985, Baniceru et al. 2011). The volatility of these agents restricts the analytical possibilities of their estimation, so TLC and HPLC techniques are applicable only in a very limited extent. The structures of the members of the group of general anesthetics are shown in Figure 20.1, while the structures of local anesthetics are shown in Figure 20.3.

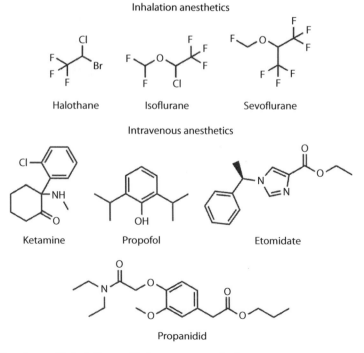

Inhalation anesthetics

Halothane   Isoflurane   Sevoflurane

Intravenous anesthetics

Ketamine   Propofol   Etomidate

Propanidid

FIGURE 20.1 Structures of inhalation and intravenous anesthetics.

This chapter covers the most recent papers, but it also includes a large number of previously published ones (older literature). A thorough survey of the literature was undertaken for that purpose. TLC systems and conditions used in the investigation of either one drug in particular or several drugs simultaneously are presented. Publications on the chromatographic conditions used in pharmaceutical formulation analysis, degradant impurities, and possible metabolites were considered when available. TLC conditions are discussed both in the text and in the tables. Names of drugs analyzed have been abbreviated, and the list of abbreviations is presented following the tables in alphabetical order, to serve as a quick reference.

Attention is also devoted to the summary of various developed methods used for the isolation and sample preparation of a particular matrix, both pharmaceutical formulations and biological material.

Apart from routine therapeutic monitoring, the analysis of anesthetics is often performed in cases of accidental overdose. Many publications have dealt with the analysis of abused drugs and drugs of unknown origin available on the illicit market. For this reason, this chapter closes with a survey of some of the most important papers dealing with TLC in toxicology and drug abuse monitoring.

20.2 SCREENING, IDENTIFICATION, AND QUANTIFICATION OF ANESTHETICS USING TLC

In the 1970s and 1980s, a lot of effort was put into the standardization of TLC systems for the identification of drugs and other substances. TLC appeared especially useful for the identification of unknown toxic substances. Several papers on the separation of a wide range of neutral, basic, and acidic drugs were published. TLC procedures were standardized and comprehensive, multi-laboratory databases for approximately 6000 compounds were established (Zeeuw et al. 1992).

Comprehensive databases used for the identification of drugs by TLC include R_F values mainly for local anesthetics. In a small number of cases, the members of the group of intravenous agents can also be found in such comprehensive studies, while local anesthetics are much more numerous.

TLC systems were selected based on their discriminating power to identify basic drugs (Fike 1966, Moffat 1975, Stead et al. 1982). Davidow et al. (1968) described an efficient screening procedure for the detection of drug abuse. Using a single solvent system, acidic, neutral, and basic drugs were extracted simultaneously from urine and separated by means of TLC.

Extensive work in this area was carried out by Stead et al. (1982), who reported the R_F values of almost 800 basic, neutral, and acidic drugs in eight carefully standardized TLC systems and ordered the drugs according to their increasing R_F values in each eluent in order to facilitate the identification of unknown samples. The hR_F values were corrected graphically using reference compounds in each system. In this way, the chances of correct identification by allowing comparison of R_F values of an unknown compound using different systems were improved. The previously mentioned study included hR_F data for ketamine (KET), barbiturates (namely, etomidate [ETD] and propanidid [PPD]), and numerous local anesthetics, including procaine (PRC), cocaine (COC), lidocaine (LID), and bupivacaine (BUP). The data are presented in Table 20.1.

Brzezinka et al. (1999) described a method for off-line coupling of TLC and electron-impact ionization mass spectrometry (EIMS) that is well suited for routine forensic and toxicological investigation of a large number of samples. The advantages and drawbacks of this approach are discussed. Several TLC systems for 493 compounds, including numerous local anesthetics (COC, PRC, cinchocaine [CIN], benzocaine [BEN], tetracaine [TET], mepivacaine [MEP], LID, and BUP) and intravenous anesthetic ETD, were described.

A chromatographic scheme consisting of one normal-phase TLC system, one reversed-phase thin-layer chromatography (RPTLC) system, and sequential analyte detection through four stages of characteristic color reactions was described in a paper by Harper et al. (1989). Eighty-one basic and neutral drugs were analyzed with this scheme, and three of these drugs were anesthetics (KET, LID,

TABLE 20.1
TLC Conditions for Determination of Anesthetic Drugs

Compound (Matrix)	Stationary Phase	Mobile Phase	Development Mode; Extraction; Remarks	Detection; Visualizing Agent	S_1	S_2	S_3	S_4	S_5	Reference
							hR_F			
BEN	Silica gel	S_1—benzene–acetone–ammonia (80:20:1, v/v)	—	—	—					Messerschmidt (1971)
BEN	Silica gel	S_1—benzene–dioxane–acetic acid (90:75:8, v/v)	—	—	—					Wan et al. (1972)
LID	Silica gel GF$_{254}$	S_1—chloroform–ether–methanol–conc. ammonia hydroxide (15:20:5:1, v/v) S_2—ethyl acetate–n-propanol–conc. ammonia hydroxide (40:30:3, v/v) S_3—methanol–conc. ammonia hydroxide (100:1.5, v/v) S_4—alcohol USP–acetic acid–water (60:30:10, v/v)	—	UV	80	82	84	60		Masoud (1976)
CIN (injectable solution and plasma serum)	HPTLC silica gel 60 F$_{254}$	S_1—ether–benzene–cyclohexane–diethylamine (20:12.5:10:3.5, v/v)	Chamber saturation, 15 min	—	—					Gübitz and Wintersteiger (1978)
PRC (serum)	Silica gel G dipped in 0.1 M methanolic KOH and dried	S_1—acetone	—	—	30					Andrey and Moffat (1979)
COC (tablets, ampoules, urine)	HPTLC silica gel G60 F$_{254}$	S_1—toluene–methanol–ammonia conc. (50:50:1, v/v) S_2—Isopropanol–n-heptane–ammonia conc. (50:50:1, v/v)	Camag twin trough chambers; Extraction from tablets and ampoules by dichloromethane–2-propanol (85:15)	254 and 365 nm, densitometric	75	54				Gübitz and Wintersteiger (1980)

Compound	Stationary phase	Solvent systems	Treatment	Detection	S1	S2	S3	S4	Reference
PRC	Silica gel	S1—isopropyl ether–acetone–diethylamine (8:1.5:0.5, v/v)	Extraction of urine (pH 9) on Extrelut columns by dichloromethane–2-propanol (85:15)	—	—	—			Panić (1981)
BEN		S2—benzene–chloroform–methanol–tetrahydrofuran (5:3:1:0.5, v/v)							
BEN	Silica gel GF$_{254}$	S1—toluene–ethanol (95:5, v/v)	—	Ehrlich's reagent, UV	—	—			Ali and Steinbach (1981)
KET	Silica gel dipped in or sprayed with 0.1 M KOH and dried	S1—methanol–conc. ammonia solution (100:1.5, v/v)	Saturated chambers	—	63	37	63	64	Stead et al. (1982)
ETD		S2—cyclohexane–toluene–diethylamine (75:15:10, v/v)			67	26	71	52	
PPD		S3—chloroform–methanol (90:10, v/v)			66	20	70	55	
AMY		S4—acetone			73	60	67	63	
BEN					67	6	57	66	
BUP					76	9	68	69	
BUT					71	9	30	64	
CIN					63	25	34	35	
COC					65	47	47	54	
LID					70	35	73	29	
MEP					65	27	62	48	
PRI					77	29	64	60	
PRC					54	6	31	30	
CIN	Silica gel G	S1—acetone–toluene–methanol–conc. ammonia (30:50:5:1, v/v)	—	UV	—	—	—	—	Padmanabhan (1983)
		S2—glacial acetic acid–ethyl acetate–conc. hydrochloric acid–water (35:55:5:5, v/v)							
		S3—chloroform–methanol–ammonium hydroxide–water (80:20:1:1, v/v)							
		S4—chloroform–acetone–diethylamine (5:4:1, v/v)							

(continued)

TABLE 20.1 (continued)
TLC Conditions for Determination of Anesthetic Drugs

Compound (Matrix)	Stationary Phase	Mobile Phase	Development Mode; Extraction; Remarks	Detection; Visualizing Agent	hR_F S₁	S₂	S₃	S₄	S₅	Reference
		S₅—chloroform–diethylamine (9:1, v/v)								
		S₆—methanol–ammonium hydroxide (100:1.5, v/v)								
		S₇—n-butanol–acetic acid–water (5:3:2, v/v)								
		S₈—chloroform–methanol (9:1, v/v)								
		S₉—dioxane–water (9:1, v/v)								
		S₁₀—dioxane–water–chloroform (8:1:1, v/v)								
		S₁₁—dioxane–water–toluene (8:1:1, v/v)								
		S₁₂—acetone–benzene–methanol–conc. ammonium hydroxide (30:50:5:1, v/v)								
		S₁₃—chloroform–methanol–water (80:20:2, v/v)								
LID	Silica gel G60 F₂₅₄ (untreated) Silica gel G60 F₂₅₄ (treated with 0.1 M KHSO₄)	S₁—KBr (0.01 M) in methanol S₂—KBr (0.01 M) in methanol	—	UV	84	54				Sundholom (1983)
PRC	HPTLC silica gel 60 F₂₅₄	S₁—chloroform–methanol (80:10, v/v) S₂—cyclohexane–benzene–diethylamine (75:15:10, v/v)	Unsaturated twin trough chambers	—	23	8				De Spiegeleer et al. (1987)
CIN					4	1				
TET					37	22				
AMY					51	58				
PYR					46	31				

Compound	Sorbent	Mobile phase	Sample prep / technique	Detection	hRf	Reference
COC					38, 42	
LID					60, 35	
BUT					37, 11	
ISO					80, 10	
OXY					40, 28	
MEP					47, 36	
FOM (urine, feces)	Silica gel GF$_{254}$	S$_1$—benzene–ethanol (70:30, v/v); S$_2$—n-pentane–acetone (70:30, v/v); S$_3$—benzene–methanol–acetic acid (70:20:10, v/v); S$_4$—benzene–methanol–diethylamine (70:15:15, v/v)	Extraction with chloroform	254 and 366 nm Dragendorff's, Gibbs, or potassium iodoplatinate reagents	69, 32, 72, 88	Oelschlager et al. (1975)
PRC	Silica gel	S$_1$—methanol–25% ammonia solution (200:3, v/v)	—	288 nm, densitometric	55	Melent'eva et al. (1984)
LID (serum)	HPTLC silica gel 60 F$_{254}$	S$_1$—benzene–ethyl acetate–methanol (4:4:1, v/v)	Twin trough chamber with saturation; Extraction with benzene	220 nm, densitometric	34	Lee et al. (1978)
PRC	Silica gel	S$_1$—benzene–ethanol (9:1, v/v)	—	—	—	Sarsunova et al. (1985)
LID	TLC silica gel 60 F$_{254}$	S$_1$—methanol–chloroform (1:9, v/v)	—	254 nm	56	Allgire et al. (1985)
LID	TLC silica gel F$_{254}$	S$_1$—dioxane–xylene–toluene–isopropanol–15 M ammonia (1:2:1:4:2, v/v)	—	—	—	Jarzebinski and Ciszewska-Jedrasik (1986)
LID (ointments, suppositories, gels)	TLC silica gel GF$_{254}$	S$_1$—diisopropyl ether–acetone–diethyl amine (85:10:5, v/v)	Extraction with absolute ethanol	254 nm, densitometric	—	Živanović et al. (1988)
LID	HPTLC silica gel 60 F$_{254}$ with a concentrating zone	S$_1$—chloroform–methanol–ammonia (90:10:0.5, v/v)	Extraction with chloroform–2-propanol (95:5)	254 and 365 nm Dragendorff's reagents	81	Tanaka et al. (1989)
COC					76	
PRC					46	
CIN (plasma)					6	
CIN	TLC silica gel	S$_1$—ethanol–ethyl acetate (50:50, v/v)	—	266 nm, densitometric	—	Li and Lubman (1989)

(continued)

TABLE 20.1 (continued)
TLC Conditions for Determination of Anesthetic Drugs

Compound (Matrix)	Stationary Phase	Mobile Phase	Development Mode; Extraction; Remarks	Detection; Visualizing Agent	hRF					Reference
					S_1	S_2	S_3	S_4	S_5	
LID COC PRC ART (urine)	HPTLC silica gel 60 F_{254}	S_1—ethyl acetate–methanol–25% ammonia (85:10:5, v/v)	Extraction of urine (buffered with ammonia, pH 9) with ether–ethyl acetate (1:9)	254 nm	81 79 72 75					Daldrup et al. (1989)
COC KET LID	TLC TOXI-GRAM blank A	S_1—95.2% ethyl acetate, 3.2% methanol, and 1.6% water, 20 µL of concentrated ammonium hydroxide solution	—	Dragendorff's reagent	80 83 85					Harper et al. (1989)
COC KET LID	RPTLC TOXI-GRAM C8	S_1—48% acetonitrile, 48% water, 2% isopropanol, and 2% ethyl acetate with 100 µL of concentrated ammonium hydroxide solution	—	Dragendorff's reagent	58 64 60					Harper et al. (1989)
LID COC PRC	TLC silica gel 60 F_{254} TOXI-GRAM A glass fiber	S_1—ethyl acetate–methanol–25% ammonia (85:10:5, v/v) S_2—methanol S_3—methanol–n-butanol (60:40, v/v) containing 0.1 M sodium bromide S_4—cyclohexane–toluene–diethyl amine (75:15:10, v/v) S_5—ethyl acetate–methanol–water (87:3:1.5, v/v) containing 5 mL/L of 30% ammonia solution	—	366 nm, densitometric	80 77 70	70 35 33	69 30 42	35 47 6	88 79 64	De Zeeuw et al. (1990)
LID PRC BUP	Silica gel	S_1—cyclohexane–benzene–ethylenediamine (15:3:2, v/v)	—	223 nm, densitometric	—					Xu et al. (1991)

Compound	Stationary phase	Mobile phase	Conditions	Detection	Recovery		Reference
BEN	Silica gel 60 F$_{254}$	S$_1$—ethyl acetate–methanol–ammonia–water (43.5:0.5:0.5:1.5, v/v); S$_2$—methanol–ammonia (50:0.5, v/v)	Urine adjusted to pH 8–9 with Na$_3$HPO$_4$, elution with dichloromethane–isopropanol (90:10) from ChemElut extraction tube	Dragendorff's reagent and iodoplatinate	—	88	Lillsunde and Korte (1991)
BUP					89	—	
COC					86	—	
LID					88	90	
PRI					90	92	
KET (urine)					81	84	
PRC	Silica gel G	S$_1$—toluene–anhydrous acetic acid–acetone–methanol (14:1:1:4, v/v)		284 nm, densitometric	—		Zhu and Hu (1992)
LID (tablets)	Silica gel	S$_1$—chloroform–ether–methanol–25% ammonia (77:15.3:7.6:0.1, v/v)		254 and 366 nm, densitometric	—		Nagy et al. (1992)
LID (tablets)	TLC silica gel	S$_1$—ethanol–benzene–dioxane–ammonia (2:20:16:3, v/v)	Ascending technique	—	—		Witek and Przyborowski (1996)
LID	TLC silica gel GF$_{254}$	S$_1$—diisopropyl ether–acetone–diethyl amine (85:10:5, v/v)		254 nm, densitometric	—		Živanović et al. (1996)
LID	TLC silica gel	S$_1$—toluene–acetone–94% ethanol–25% ammonia (45:45:7:3, v/v)		220 nm, densitometric	—		Ojanperä et al. (1997)
COC	RP18	S$_2$—methanol–water–37% hydrochloric acid (50:50:1, v/v)			—		
CIN (ointment)	Silica gel 60 F$_{254}$	S$_1$—toluene–acetone–acetic acid–ethanol mobile phase (40:30:5:2, v/v)	Chamber saturation, 15 min; Extraction with methanol	240 nm, densitometric	—		Morlock and Charegaonkar (1998)
PRC	TLC silica gel	S$_1$—acetic acid–dibutyl ether–hexane (1:20:4, v/v)	Horizontal chamber	254 nm	—		Galais et al. (2000)
LID	TLC silica gel 60 F$_{254}$	S$_1$—ethyl acetate–methanol–ammonia (4:1:0.4, v/v)	Chamber saturation, 30 min	262 nm, densitometric	75		Devarajan et al. (2000)
PHE (drug delivery system)					23		

(continued)

TABLE 20.1 (continued)
TLC Conditions for Determination of Anesthetic Drugs

Compound (Matrix)	Stationary Phase	Mobile Phase	Development Mode; Extraction; Remarks	Detection; Visualizing Agent	hR_F					Reference
					S_1	S_2	S_3	S_4	S_5	
LID COC (urine)	—	S_1—chloroform–methanol (9:1, v/v) S_2—hydrochloride–methanol–28% aqueous ammonia (100:1.5, v/v)	Urine was adjusted to pH 11 with aqueous ammonia, extracted with diethyl ether, evaporated to dryness, and redissolved in methanol	—	—	—				Kato and Ogamo (2001)
CIN (ointment)	HPTLC silica gel	S_1—toluene–acetone–methanol–25% ammonia (25:15:2.5:0.5, v/v)	Filter paper-lined twin trough chamber; Extraction with methanol	312 and 366 nm, densitometric	—					Essig and Kovar (2001)
LID	HPTLC silica gel 60 F_{254}	S_1—ethyl acetate–methanol–25% ammonia (85:10:5, v/v)	Saturation	230 nm	—					Stroka et al. (2002)
CIN (suppositories)	HPTLC silica gel 60 F_{254}	S_1—n-butanol–toluene–ethanol–water–100% acetic acid (10:8:7:4:1, v/v)	Extraction with methanol	313 nm, densitometric	—					Jehle et al. (2004)
LID PRC BEN ART TET PRI BUP PRA	TLC silica gel 60 F_{254}	S_1—ethyl acetate–methanol–32% ammonia (48:11.5, v/v)	—	254 nm	84 57 92 78 38 78 78 8					Schmidt and Bracher (2006)
CIN	TLC silica gel	S_1—methanol–tetrahydrofuran–acetic acid (45:5:0.5, v/v)	Chamber saturation, ascending	327 nm, densitometric	55					Mohammad et al. (2007)

Compound	Stationary phase	Mobile phase	Technique	Detection	hRf	Reference
BEN (gel)	HPTLC silica gel 60 F$_{254}$	S$_1$—hexane–ether (80:20, v/v)	Vapor-equilibrated chamber, 25 min Extraction with methanol	280 nm, densitometric	45	Bhawara (2008)
LID	TLC silica gel	S$_1$—ethyl acetate–methanol–28% ammonia (85:10:5, v/v) S$_2$—cyclohexane–toluene–diethyl amine (65:25:10, v/v)	2D TLC	—	— — —	Kurita et al. (2008)
	HPTLC silica gel	S$_3$—methanol S$_4$—ethyl acetate–methanol–25% ammonia (85:10:5, v/v)	2D HPTLC			
LID PRI	HPTLC silica gel	S$_1$—methanol–*n*-butanol–water–toluene–glacial acetic acid (2:3:1:2:0.1, v/v)		220 and 237 nm, densitometric	— —	Riad et al. (2008)
MEP ROP BUP PIP EPI	Silica gel GF$_{254}$	S$_1$—1-butanol–glacial acetic acid–water (12:3:5, v/v) S$_2$—2-propanol–glacial acetic acid–water (12:3:5, v/v) S$_3$—ethanol–glacial acetic acid–water (12:3:5, v/v) S$_4$—methanol–glacial acetic acid–water (12:3:5, v/v)	Chamber saturation, 2 h	Potassium permanganate solution and Dragendorff's reagent	45 51 68 49 59 72 52 60 72 43 50 65 47 56 70	Nemák et al. (2008)
COC	Silica gel G	S$_1$—ammonia solution–methanol (1.5:100, v/v)	Chamber saturation, 1 h	—	60	Clarke (1978)
COC	Silica gel F 1500 LS 254 activated at 110°C	S$_1$—methanol–aqueous ammonia (100:1.5, v/v) S$_2$—ethanol–chloroform (1:1, v/v)	—	230 nm	7 31	Bertulli et al. (1978)
COC	Silica gel, activated	S$_1$—ethyl acetate–*n*-propanol–28% ammonium hydroxide (40:30:3, v/v)	—	Acidic iodoplatinate	79 73	Brown et al. (1973)
COC	Silica gel G	S$_2$—ethyl acetate–methanol (17:2, v/v) and 20 mL of 50% ammonia v/v) in a beaker		Dragendoff then iodoplatinate reagent		

(continued)

TABLE 20.1 (continued)
TLC Conditions for Determination of Anesthetic Drugs

Compound (Matrix)	Stationary Phase	Mobile Phase	Development Mode; Extraction; Remarks	Detection; Visualizing Agent	hR_F					Reference
					S_1	S_2	S_3	S_4	S_5	
BNZ	Silica gel G	S_1—chloroform–methanol–ammonium hydroxide (100:20:1, v/v)	Chamber saturation; Extraction with 20% ethanol in chloroform	Dragendorff's reagent and 20% sulfuric acid	20					Wallace et al. (1975)
COC (urine)					87					
COC	TLC silica gel G	S_1—hexane–chloroform–diethylamine (80:10:10, v/v)	Extraction with chloroform–isopropanol	Acidified iodoplatinate	74	52	59			Budd et al. (1980)
NOR		S_2—hexane–acetone–diethylamine (70:30:1, v/v)			47	28	32			
PRP (urine)		S_3—hexane–chloroform–diethylamine (85:10:5, v/v)			88	64	75			
PPF	HPTLC silica gel 60 F$_{254}$	S_1—toluene	Horizontal chamber	276 nm, densitometric	70					Salomies et al. (1995)
PPD	TLC silica gel G	S_1—methanol–concentrated ammonia (100:1.5, v/v)	—	0.2% solution of chloranil in acetonitrile, heating at 105°C	73					Taha and Abd El-Kader (1979)

Compound	Adsorbent	Solvent system	Detection	R_f	Reference
(±)-KET	Silica gel G impregnated with (−)-mandelic acid	S$_1$—ethyl acetate–methanol–water (3:1:1, v/v) S$_2$—acetonitrile–methanol–acetic acid (7:1:0.9, v/v)	Iodine vapor	67 44	Bhushan and Agarwal (2008)
	Silica gel G impregnated with (+)-tartaric acid	S$_3$—ethyl acetate–methanol–water (3:1:1, v/v) S$_4$—acetonitrile–methanol–acetic acid (7:1:0.9, v/v)	—	79 67	
KET	Silica gel 60 (Merck)	S$_1$—methylene chloride–n-butanol–aqueous ammonia (85:15:0.2, v/v)	Potassium iodoplatinate	82 93	Cone et al. (1979)
	Silica gel (Quanta Gram)	S$_2$—methylene chloride–n-butanol–aqueous ammonia (85:15:0.2, v/v)		92	
	Glass fiber plates impregnated with silicic acid (Gelman ITLC-SA)	S$_3$—ethyl acetate–methanol–aqueous ammonia–water (29:1:0.25:0.5, v/v)	—	80	
	Silica gel 60	S$_4$—ethyl acetate–methanol–dimethylamine (40% aqueous solution) (90:10:1.6, v/v) S$_5$—ethyl acetate–methanol–diethylamine (90:10:1.6, v/v)		75	

Abbreviations: AMY, Amylocaine; ART, Articaine; BEN, Benzocaine; BNZ, Benzoylecgonine; BUP, Bupivacaine; BUT, Butacaine; CIN, Cinchocaine; COC, Cocaine; EPI, *N*-Ethylpipecoloxylidide; ETD, Etomidate; FOM, Fomocaine; ISO, Isocaine; KET, Ketamine; LID, Lidocaine; MEP, Mepivacaine; NOR, Norpropoxyphene; OXY, Oxybuprocaine; PHE, Phenylephrine hydrochloride; PIP, Pipecoloxylidide; PPD, Propanidide; PPF, Propofol; PRA, Procainamide; PRC, Procaine; PRI, Prilocaine; PRP, Propoxyphene; PYR, Pyrrocaine; ROP, Ropivacaine; TET, Tetracaine.

and COC; see Table 20.1). Musumarra et al. (1983) reported the retention factors of 54 drugs in 8 eluent mixtures on silica gel TLC plates. Principal component analysis (PCA) of these data provided a significant two-component model, with the eluent mixtures clustered in three groups. The PCA model that used only three eluents (one for each group) allowed unambiguous identification of the drugs. In another paper, Musumarra et al. (1984) reported the PCA of the R_F values of 55 basic and neutral drugs in 40 solvent mixtures with the purpose of selecting the minimum number of eluent systems that have the maximum information content. The examined drugs belonged to various classes of compounds (tranquilizers, analgesics, natural and synthetic opiates, alkaloids, antihistamines, anesthetics, etc.) and differed in their structural and biological properties. Four eluent mixtures—ethyl acetate–methanol–30% ammonia (85:10:5, v/v), cyclohexane–toluene–diethylamine (65:25:10, v/v), ethyl acetate–chloroform (1:1, v/v), and acetone with the plate dipped in potassium hydroxide solution—provided a significant two-component PC model that can be used for the identification of unknown samples. The corresponding hR_F values for several local anesthetics and KET obtained by using 40 different solvent mixtures are presented in the paper. These hR_F values are presented in a separate table (Table 20.2) for the purpose of this chapter.

There is not a universally accepted TLC procedure for screening, identification, and quantification of anesthetics. The literature shows that numerous TLC procedures are available, and most of them are summarized in this section. These methods have been developed in order to measure one drug (and its metabolites) or to detect or confirm the presence of several drugs using one assay.

Heusler (1985) provided a survey of the methods of quantitative analysis of anesthetics, with special attention being paid to practical applications. A summary of chromatographic (e.g., gas chromatography [GC], HPLC) and non-chromatographic (e.g., radioimmunoassay [RIA], enzyme-multiplied immunoassay) analytical techniques suitable for quantitative analysis of the most popular inhalational (halothane, methoxyflurane, enflurane, isoflurane, and nitrous oxide), intravenous (barbiturate, benzodiazepines, ETD, althesin, morphine, fentanyl, alfentanil, sufentanil, droperidol, and KET), general, and amide-type local (LID, MEP, etidocaine, and BUP) anesthetic agents and some of their metabolites in biological material is given. In the case of inhalational anesthetics, attention was given to pollution measurement and breath-to-breath monitoring. Recently, gas and liquid chromatographic methods for local anesthetics and/or their metabolites in biological samples have been reviewed (Baniceru et al. 2011).

20.3 GENERAL ANESTHESIA: INHALATION AND INTRAVENOUS AGENTS

20.3.1 INHALATIONAL ANESTHETICS

A number of inhalational anesthetics have been introduced into clinical practice and some of them are listed in Figure 20.1. Except for ethyl vinyl ether, all agents introduced after 1950 contain fluorine. Agents such as ether, chloroform, and trichloroethylene, which were once used, have been replaced by new fluorinated anesthetics. The inhalational anesthetic agents that are currently used are halothane, enflurane, isoflurane, sevoflurane, and methoxyflurane. These agents are volatile organic liquids.

The limited number of publications dealing with their analysis by utilizing TLC is expected considering their volatility. Only a few procedures were established for the identification of their nonvolatile metabolites in samples (urine and liver).

20.3.1.1 Halothane (2-Bromo-2-Chloro-1,1,1-Trifluoroethane)

Halothane is a polyhalogenated hydrocarbon used as an inhalation anesthetic. Two products of halothane biotransformation are trifluoroacetic acid (TFA) and trifluoroethanol. Two investigations

TABLE 20.2
Mobile Phase Composition and hR_F Values

Mobile Phase Composition	hR_F				
	BUP	COC	LID	PRC	KET
[a]Toluene–acetone–ethanol–30% ammonia (45:45:7:3, v/v)	83	81	77	64	17
[a]Ethyl acetate–benzene–methanol–30% ammonia (60:35:6:2.5, v/v)	84	82	79	60	79
[a]Benzene–dioxane–ethanol–30% ammonia (50:40:7.5:2.5, v/v)	84	81	80	70	80
[a]Methanol–30% ammonia (100:1.5, v/v)	80	71	73	65	76
[a]Benzene–isopropanol–methanol–30% ammonia (70:30:20:5, v/v)	86	87	86	82	86
[a]Ethyl acetate–methanol–30% ammonia (85:10:5, v/v)	83	82	80	73	79
[a]Acetone–7.5% ammonia (90:10, v/v)	90	89	89	85	89
[a]Cyclohexane–toluene–diethylamine (65:25:10, v/v)	45	46	35	8	41
[a]Cyclohexane–toluene–diethylamine (75:15:10, v/v)	41	41	30	5	33
[a]Cyclohexane–benzene–methanol–diethylamine (70:20:10:5, v/v)	32	38	28	16	32
[a]Chloroform–acetone–diethylamine (50:40:10, v/v)	84	81	84	66	81
[a]Cyclohexane–chloroform–diethylamine (50:40:10, v/v)	74	72	73	24	66
[a]Benzene–ethyl acetate–diethylamine (50:40:10, v/v)	81	80	77	54	16
[a]Xylene–methyl ethyl ketone–methanol–diethylamine (40:40:6:2, v/v)	69	52	66	37	61
[a]Diethyl ether–diethylamine (95:5, v/v)	73	72	64	50	66
[a]Ethyl acetate–chloroform (50:50, v/v)	22	6	25	1	21
[b]Ethyl acetate–chloroform (50:50, v/v)	56	24	54	11	37
[a]Butanol–methanol (40:60, v/v)	69	30	68	29	66
[b]Butanol–methanol (40:60, v/v)	87	57	84	53	79
[a]Chloroform–methanol (90:10, v/v)	65	42	71	23	65
[b]Chloroform–methanol (90:10, v/v)	71	60	74	40	64
[a]Acetone	65	35	67	25	74
[b]Acetone	85	73	83	57	80
[c]Acetone	76	71	73	70	72
[a]Benzene–acetonitrile (70:30, v/v)	25	5	29	1	37
[b]Benzene–acetonitrile (70:30, v/v)	64	34	56	7	56
[a]Benzene–tetrahydrofuran (80:20, v/v)	20	8	16	2	25
[b]Benzene–tetrahydrofuran (80:20, v/v)	48	30	37	9	33
[a]Chloroform–ethyl acetate–methanol (40:40:20, v/v)	64	29	68	20	64
[b]Chloroform–ethyl acetate–methanol (40:40:20, v/v)	81	67	79	55	74
[a]Chloroform–n-hexane–methanol (65:75:10, v/v)	56	34	60	16	53
[b]Chloroform–n-hexane–methanol (65:25:10, v/v)	77	63	76	40	66
[a]Dichloromethane–methanol (95:5, v/v)	45	21	53	7	45
[b]Dichloromethane–methanol (95:5, v/v)	68	44	67	21	55
[a]Chloroform–methanol (75:25, v/v)	85	67	87	41	84
[b]Chloroform–methanol (75:25, v/v)	87	79	87	70	85
[a]Acetic acid–ethanol–water (30:60:10, v/v)	74	40	52	48	67
[a]Ethyl acetate–dimethylformamide–ethanol (86.5:12.5:1, v/v)	49	49	49	33	49
[a]Methanol–acetone–triethanolamine (50:40:1.5, v/v)	62	44	62	54	63
[a]Chloroform–acetone–methanol–triethylamine (30:40:10:20, v/v)	88	86	83	68	83

Source: Musumarra, G. et al., *J. Chromatogr.*, 295, 31, 1984.
Stationary phase: HPTLC silica gel 60 F$_{254}$ plates; chamber saturation 30 min.
[a] Not treated.
[b] Dipped in 0.1 M KOH methanolic solution and dried.
[c] After application of the drugs, the plate was kept for 30 min in a tank saturated with 30% ammonia solution and then transferred into the elution tank.

were designed to detect these two compounds in the urine of men after administration of ^{14}C-halothane. A halothane metabolite, TFA, was determined in urine samples (Stier 1964, Blake et al. 1972). After intravenous injection of the labeled halothane, radioactivity in urine was characterized by liquid–liquid extraction (LLE) and TLC on silica gel plates developed with ammonia–isopropyl alcohol (1:4, v/v). Urine was acidified to pH < 1 with sulfuric acid, extracted with diethyl ether, and ether-portion-back extracted into concentrated ammonium hydroxide. TFA was the only detectable radioactive metabolite of halothane excreted in the urine of human volunteers with $R_F = 0.77$ (Blake et al. 1972).

20.3.1.2 Isoflurane (2-Chloro-2-(Difluoromethoxy)-1,1,1-Trifluoroethane)

TFA was also determined as isoflurane metabolite in urine samples. The metabolism of isoflurane was studied in rats and men by Hitt et al. (1974). In order to examine the excretion of nonvolatile urinary metabolites of isoflurane, the collected urine was extracted with chloroform–methanol (1:1, v/v), and the aqueous phase was chromatographed on silica gel with ethanol–chloroform–ammonia (5:2:1, v/v). After drying, the plates were sprayed with bromothymol blue. This study indicated that metabolism of isoflurane into both ionic fluoride and nonionic fluoride metabolites occurs. The authors proposed some possible pathways of isoflurane metabolism into TFA.

20.3.1.3 Sevoflurane (1,1,1,3,3,3-Hexafluoro-2-(Fluoromethoxy)propane)

Sevoflurane is a nonflammable general anesthetic administered by inhalation of vaporized liquid. A paper dealing with the body distribution of sevoflurane in a sevoflurane-induced death was published by Burrows et al. (2004). Serum, urine, and gastric contents from the deceased were screened for numerous drugs and metabolites using a combination of TLC, colorimetric, and immunoassay techniques.

20.3.2 INTRAVENOUS ANESTHETICS

20.3.2.1 Propofol (2,6-Diisopropylphenol)

Propofol is a liquid general anesthetic drug for short-term infusion narcosis. Chemically, it is 2,6-diisopropylphenol. It is used for the induction and maintenance of anesthesia during surgical procedures. Clinically relevant concentrations of propofol (3–8 μg/mL; 20–50 μM) were also reported to have anticancer activities (Siddiqui et al. 2005). The drug is only slightly soluble in water and is administered as an oil-in-water emulsion.

Salomies et al. (1995) developed an HPTLC method for the determination of propofol and for monitoring the effect of different infusion containers on the stability of propofol in 5% glucose. The stability of propofol in polypropylene-lined infusion bags was evaluated and compared with the stability of propofol in glass bottles and PVC bags. The authors proposed a simple, rapid, and stability-indicating HPTLC method for determining the concentrations of propofol in samples. The repeatability of the application showed the relative standard deviation (RSD) values to be less than 3%. RSD measuring the repeatability of the whole method was 1.5%, 1.8%, and 1.8% for samples with concentrations of 1.0, 1.5, and 2.0 mg/mL, respectively. The accuracy of the method was good since the RSD was 2.9%. The reliability of the method was tested by comparing the results with those obtained using the HPLC method.

TLC with petroleum ether–ethyl acetate (92:8, v/v) as the solvent was performed on the product of the chemical synthesis of propofol–docosahexaenoic acid conjugate in a study by Siddiqui et al. (2005), who described the synthesis, purification, characterization, and evaluation of two novel anticancer conjugates. Kumpulainena et al. (2008) synthesized ethyl dioxy phosphate prodrug of propofol. All reactions described here were monitored by TLC using aluminum sheets precoated with Merck silica gel 60 F_{254}. The samples were visualized by ultraviolet (UV) light.

20.3.2.2 Etomidate (Ethyl 3-[(1R)-1-Phenylethyl]imidazole-5-Carboxylate)

ETD is a short-acting imidazole hypnotic that is commonly used to induce anesthesia. Analysis of pharmaceutical formulations and determination of ETD in biological fluids are mainly performed using some techniques that are more sensitive than TLC. Only one TLC system, the one reported by Chang and Martin (1983), was found by searching the literature, and it is also used for its major degradation product, ETD acid. The solvent system of chloroform–methanol–ammonium hydroxide (100:100:2, v/v) was used to induce the migration of the compound, and shortwave UV was used to detect ETD and the degradation product. The R_F was 0.74 for ETD. If present, the major degradation product, ETD acid, has an R_F of 0.45.

20.3.2.3 Propanidid (Propyl {4-[2-(Diethylamino)-2-Oxoethoxy]-3-Methoxyphenyl}acetate)

PPD is a nonbarbiturate general anesthetic. PPD is used as a short-acting general anesthetic. It also possesses local anesthetic activity.

Görög et al. (1997) proposed a general scheme for the rational use of chromatographic, spectroscopic, and hyphenated techniques in drug impurity profiling studies. The authors gave several examples, one of them being PPD. The procedure of impurity profiling began with detecting the impurities on thin-layer chromatograms, GC, and analytical and preparative HPLC.

Taha and Abd El-Kader (1979) developed a selective method for detecting N-ethyl drugs. This was achieved by utilizing chloranil as the detection reagent, which selectively oxidizes then condenses with the two-carbon chain of the tertiary N-ethyl moiety, yielding blue aminovinylquinone derivatives. PPD and other N-alkyl analogs were found not to interfere, forming brown color. The chromatograms were observed in daylight. Results for PPD are presented in Table 20.1.

Zawisza and Przyborowski (1992) identified PPD and ETD using TLC on silica gel. The drugs were extracted from blood plasma with ether. Ether–acetone (3:1, v/v) was used as the solvent system for PPD and dioxane–acetic acid was used for ETD. The detection limit of 1 μg/cm^3 was obtained for PPD, but not for ETD.

20.3.2.4 Ketamine ((RS)-2-(2-Chlorophenyl)-2-(Methylamino)cyclohexanone)

KET is a cyclohexanone derivative, the pharmacological actions of which are quite different from those of other described anesthetics. At therapeutic doses, KET induces dissociative anesthesia and is used as a preanesthetic.

Sass and Fusari (1977) reviewed chromatographic conditions for determination of KET. Tritium-labeled metabolites and intact KET hydrochloride were separated on silica gel GF using chloroform–ethyl acetate–methanol–ammonium hydroxide (60:35:5:1, v/v). The separation of unresolved metabolites was accomplished on aluminum oxide HF using chloroform–cyclohexane–diethylamine (60:40:2, v/v). Chloroform–cyclohexane–ethyl acetate–ammonia (25:50:25:5, v/v) was used to separate KET ($R_F = 0.58$) and the N-dealkylated metabolite ($R_F = 0.41$) using LQ6D plate. They were all visualized by exposure to iodine. Concentrated ammonium hydroxide–methanol (1.5:100, v/v) was also used to develop samples on activated silica gel G. KET at $R_F = 0.72$ was made visible with acidified iodoplatinate spray.

KET is commercially marketed as a racemic mixture. There are a few reports on resolution of the enantiomers of KET. These include the use of chiral stationary phases for direct resolution of the enantiomers by GC and liquid chromatography (LC), supercritical and subcritical fluid chromatography, and capillary electrophoresis. Bhushan and Agarwal (2008) reported a direct TLC resolution of (±)-lisinopril and (±)-KET by using (−)-mandelic acid and (+)-tartaric acid as chiral impregnating reagents. hR_F values of the enantiomers of KET and the mobile phase that enabled the successful resolution are shown in Table 20.1.

TLC can be successfully used to identify KET when it is present in higher concentrations (Ondra et al. 2006). Urine was extracted with diethyl ether under alkaline (pH = 10) conditions. After the extraction, the fractions were examined for the presence of tropane alkaloids and KET. The TLC procedure was performed using Dragendorff's solution and 1% iodine solution in chloroform as detection reagents.

Cone et al. (1979) described the application of TLC and GLC methods for the separation and identification of the phencyclidine precursors, metabolites, and analogs. The results for KET are presented in Table 20.1.

20.4 LOCAL ANESTHETICS

All local anesthetic drugs except COC are synthetic. Traditionally, there have been two main groups available for use—the esters and the amides (Figure 20.2). All local anesthetics have some similar structural features with three main structural parts—aromatic ring, intermediate chain, and the amino group (Figure 20.3). The hydrophilic region is a secondary or tertiary amine, and the hydrophobic region is the aromatic residue. In most cases the linkage between the two includes an ester or amide bond. The type of bond greatly determines the persistence of the drug.

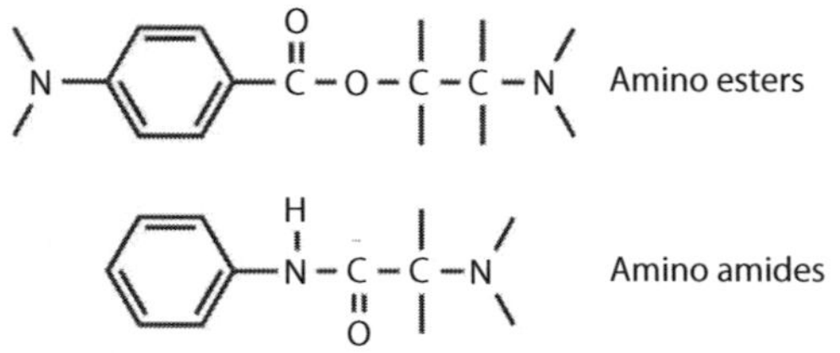

FIGURE 20.2 Basic structure of ester- and amide-type local anesthetics.

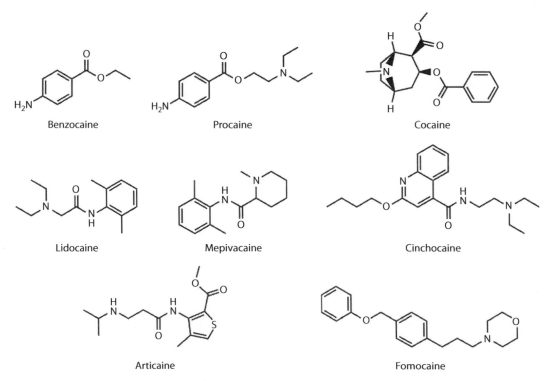

FIGURE 20.3 Chemical structures of local anesthetics.

Local anesthetic agents are widely used in surgery, dentistry, and ophthalmology to block the transmission of impulses in peripheral nerve endings. The ester agents include BEN, COC, PRC, amethocaine, TET, and chloroprocaine, while amides include LID, prilocaine (PRI), MEP, and BUP. There are important practical differences between these two groups of local anesthetic agents. Esters are relatively unstable in solution and are rapidly hydrolyzed in the body. The amide form of the drug is more stable and resistant to hydrolysis. In clinical practice esters have largely been superseded by amides.

Local anesthetic drugs are weak bases with a pK between 7.7 and 9.3. Practically all free-base forms of the drugs are liquids. For this reason, most of these drugs are used as salts (chloride, sulfate, etc.) that are water soluble, odorless, and crystalline solids.

It is possible to identify unknown drugs due to the large amount of R_F data available in the literature and the ability to perform a chemical reaction using a wide spectrum of different reagents in situ. A lot of different methods for the detection of local anesthetics have been described, ranging from nonselective to highly selective ones, and they are presented here.

In most of the papers presented here, the object of study was a single anesthetic. However, there are some papers in which a large number of local anesthetics were analyzed at the same time. Chromatographic conditions applied in these papers and the relevant data are described in the following text and/or presented in Table 20.1 (De Spiegeleer et al. 1987, Daldrup et al. 1989, Tanaka et al. 1989, De Zeeuw et al. 1990, Lillsunde and Korte 1991, Schmidt and Bracher 2006).

De Spiegeleer et al. (1987) demonstrated the applicability and usefulness of 2D HPTLC for the separation of 14 local anesthetics. The choice of the two systems in 2D was based on the absolute values of the correlation matrix elements. The results of the 1D experiment are shown in Table 20.1. The authors proposed 1D HPTLC as the method of quantitative determination by reflectance scanning by UV at the optimum wavelength. Xu et al. (1991) reported a densitometric method for determination of PRC, LID, and BUP in a mixture. Anesthetic mixture was dissolved in absolute ethanol and detected by TLC scanning at 223 nm. The linear calibrations were at 1–8, 2.5–20, and 2.5–20 µg, respectively. Tanaka et al. (1989) developed a simple TLC screening procedure for some basic, neutral, and acidic drugs using horse plasma. COC, CIN, LID, and PRC were identified among the investigated drugs by an HPTLC plate and by spraying successively with Dragendorff's detection reagents.

A convenient method for the identification and distinction between seven local anesthetics (BEN, PRC, TET, LID, PRI, BUP, and articaine [ART]) and the related antiarrhythmic drug, procainamide (PRA), in which TLC and a combination of two detection reagents (cobalt(II) thiocyanate solution and Ehrlich's reagent) were used, was proposed by Schmidt and Bracher (2006). Distinction between PRI and ART was not achieved with the TLC systems used. The authors proposed a color reaction with copper(II) sulfate in a test tube of the two anesthetics as the appropriate way of distinguishing between them. Dissolving the drugs in dilute hydrochloric acid, followed by addition of copper(II) sulfate solution and a slight excess of sodium hydroxide solution gave a clear reddish-brown solution in the case of PRI and its hydrochloride, whereas ART gave a brownish-green solution. LID hydrochloride gave a deep-blue solution, while the other drugs gave light-blue solutions, clear or opalescent. This color reaction could not be used for the detection of the spots on TLC plates.

The previously mentioned paper and similar ones that deal with different color reagents for detection of local anesthetics on TLC plates are of particular importance. Nonselective methods, such as quenching of UV light on fluorescence plates, iodine vapor or iodine spray reagents, and concentrated sulfuric acid, are used for detection of all kinds of organic compounds in TLC.

The reactivity of iodic acid with many classes of organic compounds of pharmaceutical interest, including some local anesthetics, was examined in order to establish if it can be used for detecting and identifying drugs separated by TLC (Cavazzutti et al. 1983). No positive reaction was observed for COC and LID, while PRC gave a positive reaction with iodic acid. Blue color was observed after heating at 120°C for 2 min.

Specific reagents and methods by which local anesthetics can be selectively detected are of particular importance. Selective and specific alkaloid reagents are various modifications of Dragendorff's reagent and potassium iodoplatinate. Both reagents react with tertiary and quaternary nitrogen atoms. Apart from the previously mentioned publication by Schmidt and Bracher (2006), several other reports dealing with different coloring reagents for detection of local anesthetics on TLC plates were published (Baker and Gough 1979, O'Neal et al. 2000, Kato and Ogamo 2001, Deakin 2003, Makowski et al. 2010).

Kato and Ogamo (2001) reported citric acid-acetic anhydride reagent as the color reagent selective for tertiary amines in solution, which can be used for the detection of abused tertiary amino drugs on the TLC plate (Table 20.1). The plate is pretreated by a brief immersion in phosphoric acid-acetone solution to suppress coloration. After suppressing, the plate is sprayed with a reagent and heated at 100°C, causing tertiary amines to turn reddish purple. COC and LID were studied among other tertiary amino drugs. Citric acid acetic anhydride reagent was found to be superior to Dragendorff's reagent in the analysis of real urine samples because tertiary amines in urine samples could be detected without interference by components usually found in urine. LLE method was utilized for the extraction of the stimulant in urine.

For some drugs, a compilation of TLC data has been elaborated and stored in computer-based information systems. The possibility of using color reactions in combination with R_F values for the identification of unknown substances by means of computerized retrieval from large databases is also available for local anesthetics. De Zeeuw et al. (1990) developed a system that makes color reactions on TLC plates amenable to computer handling. The system was based on a series of four color reactions carried out in sequence on the same spot and numeric encoding of the observed color by means of a color reference chart. Ojanperä et al. (1997) introduced novel software for the simultaneous processing of qualitative data from two parallel TLC analyses based on the comparison of libraries of corrected hR_F values and in situ UV spectra. The applied TLC conditions that were used for analysis of COC on silica and LID on RP-18 silica are listed in Table 20.1.

20.4.1 Sample Preparation

20.4.1.1 Pharmaceutical Formulations

Over the years, different methods have been developed in order to determine anesthetics in biological materials. Local anesthetics are basic drugs and are therefore extracted from aqueous alkaline samples into organic solvents. In general, alkaline pH is used with chloroform-based solvents. The pH of the extraction should be carefully controlled, and the extraction procedure should be as rapid as possible as some of the ester-type local anesthetics are prone to hydrolysis.

Various solvents are used for the preparation, depending on the type of pharmaceutical formulation that is being analyzed. Most of the extraction conditions and described procedures are rather old, but similar conditions are used nowadays as well. Only a short overview of the publications is given at this point. The results of chromatographic studies that were obtained in the previously mentioned papers are described in detail in this chapter and are presented systematically in Tables 20.1 and 20.3.

Aqueous solutions can be applied directly (Reichelt 1955), or the solution can be made alkaline with sodium hydroxide and the drug then extracted with ether (Tatscuzawa et al. 1968). Oil solutions, suppositories, or ointment are dissolved in light petroleum and then extracted with 1 M hydrochloric acid. The hydrochloric acid extract is made alkaline and reextracted with chloroform.

TABLE 20.3
TLC Conditions for Determination of Some Anesthetic Drugs in Forensic Toxicology and Doping Control

Compound (Matrix)	Stationary Phase	Mobile Phase	Development Mode; Extraction	Detection; Visualizing Agent	hR_F							Reference
					S_1	S_2	S_3	S_4	S_5	S_6	S_7	
HER	TLC silica gel	S_1—ethyl acetate–*n*-propanol–28% ammonium hydroxide solution (40:30:3, v/v)	Extraction with 95% ethanol	Iodoplatinate	45							Brown et al. (1973)
TET				*p*-dimethylaminobenzaldehyde	55							
PRC					70							
COC					79							
LID					87							
BUT					89							
BEN					89							
HOL					93							
PRC	Silica gel G	$S_1 = (S_{1a} + S_{1b})$	Urine buffered to pH 9.5 passed through XAD-2 resin; elution of drugs with 1,2-dichloroethane–ethyl acetate (4:6)	350 nm	60	72	92					Bussey and Backer (1974)
BEN (urine)		S_{1a}—ethyl acetate–cyclohexane–methanol–water–ammonium hydroxide (70:15:8:0.5:2, v/v) S_{1b}—ethyl acetate–methanol–water (80:15:5, v/v) S_2—methanol–ammonium hydroxide (100:1.5, v/v) S_3—ethyl acetate–methanol–ammonium hydroxide (80:10:5, v/v)			90	86	94					

(continued)

TABLE 20.3 (continued)

TLC Conditions for Determination of Some Anesthetic Drugs in Forensic Toxicology and Doping Control

Compound (Matrix)	Stationary Phase	Mobile Phase	Development Mode; Extraction	Detection; Visualizing Agent	hR_F							Reference
					S_1	S_2	S_3	S_4	S_5	S_6	S_7	
BNZ	Silica gel	S_1—ethyl acetate–cyclohexane–methanol–ammonium hydroxide (70:15:10:5, v/v)	Urine was acidified with HCl to pH 2, extracted with ether; aqueous layer was buffered with ammonium hydroxide to pH 9, extracted with chloroform–isopropyl alcohol (3:1) to obtain COC and its metabolites	Visualizing reagents in the following sequence: Dragendorff's reagent, iodoplatinate reagent, Van Urk reagent	4	26	3	9	2	28	33	Rafla and Epstein (1979)
COC		S_2—100% methanol			74	52	43	62	53	61	88	
CET		S_3—ethyl acetate–methanol (80:20, v/v)			74	50	45	64	58	60	88	
BEN		S_4—chloroform–methanol (80:20, v/v)			78	76	69	69	15	72	88	
LID		S_5—cyclohexane–benzene–diethylamine (65:20:15, v/v)			67	71	64	75	50	75	90	
PRC		S_6—chloroform–methanol (50:50, v/v)			68	52	36	44	17	54	83	
TET (urine)		$S_7 = S_1 + S_6$			70	49	29	50	35	58	85	
LID	Silica gel 60 F_{254}	S_1—ethyl acetate–methanol–conc. ammonia (85:10:5, v/v)	Tanks were equilibrated 30 min	UV irradiation 254 and 366 nm and iodoplatinate reagent	76	85						Baker and Gough (1979)
COC		S_2—methanol–2 N ammonium hydroxide–1 N ammonium nitrate (27:2:1, v/v)			63	83						
PRC					57	80						
BEN					79	81						
CIN					60	83						
MEP					71	75						
BUP					79	83						
AMY					70	87						
PRI					74	85						
PIP					51	86						
TPC					30	86						
TET					56	79						
PMT					63	83						
BUT					63	84						
CCM					51	83						

Compound	Sorbent	Mobile phase	Sample preparation	Detection	R_f values					Reference
COC (tablets, ampoules, urine)	HPTLC silica gel G60 F$_{254}$	S$_1$—toluene–methanol–ammonia conc. (50:50:1, v/v); S$_2$—isopropanol–n-heptane–ammonia conc. (50:50:1, v/v)	Extraction from tablets and ampoules by dichloromethane–2-propanol (85:15); Urine (pH 9) extraction on Extrelut columns by dichloromethane/2-propanol (85:15)	254 and 365 nm, densitometric	75	54				Gübitz and Wintersteiger (1980)
CET, COC (urine)	Silica gel	S$_1$—hexane–toluene–diethylamine (65:20:5, v/v)	SPE of urine buffered to pH 9.3	Iodoplatinate reagent	—	—				Bailey (1994)
BNZ	Silica gel 60 F$_{254}$	S$_1$—ethyl acetate–methanol–ammonia (60:30:6, v/v); S$_2$—chloroform–methanol–ammonia (100:20:1, v/v); S$_3$—chloroform–methanol–ammonia–water (70:30:1:0.5, v/v); S$_4$—methanol–chloroform (4:1, v/v); S$_5$—chloroform–methanol (4:1, v/v)	—	Dragendorff's reagent with sulfuric acid	12	8	8	16	7	Kiszka and Madro (2002)
COC					81	80	85	30	40	
CET					84	80	86	31	40	
LID (tissues, urine)					83	81	89	60	64	
COC	HPTLC silica gel 60 F$_{254}$	S$_1$—ethyl acetate–cyclohexane–ammonia (25:10:0.1, v/v)	Chamber (horizontal) saturation; urine buffered to pH 6 with phosphate buffer, elution from column with methylene chloride–2-propanol–ammonium hydroxide (80:20:2 v/v); Diazomethane added to convert BNZ to COC	Dragendorff's reagent	44					Yonamine and Sampario (2006)
LID					63					
PRC					24					

(continued)

TABLE 20.3 (continued)

TLC Conditions for Determination of Some Anesthetic Drugs in Forensic Toxicology and Doping Control

Compound (Matrix)	Stationary Phase	Mobile Phase	Development Mode; Extraction	Detection; Visualizing Agent	hR_F							Reference
					S_1	S_2	S_3	S_4	S_5	S_6	S_7	
CET COC BNZ (urine)	HPTLC silica gel 60 F$_{254}$	S$_1$—hexane–toluene–diethylamine (65:20:5, v/v)	Horizontal chamber with saturation SPE, LiChrolut TSC elution with dichloromethane–2-propanol–25% ammonium hydroxide (80:20:2)	234 nm, densitometric	44 36 2							Antonilli et al. (2001)
COC PRC BEN TET HER	Eastman chromatogram	S$_1$—40 mL Chloroform, 10 mL ethyl acetate and 10 drops conc. NH$_4$OH	—	—	84 45 67 11 34							Comer and Comer (1967)
LID COC	TLC silica gel 60 F$_{254}$	S$_1$—chloroform–methanol–conc. ammonia (90:10:1, v/v) S$_2$—ethyl acetate–methanol–water–conc. ammonia (85:10:3:1, v/v)	—	—	90 90	89 87						Jukofsky et al. (1980)
COC BNZ LID	TLC silica gel 60 F$_{254}$	S$_1$—chloroform–methanol (90:10, v/v) S$_2$—cyclohexane–toluene–diethylamine (75:15:10, v/v)	Chamber saturation, 1 h	UV Dragendorff's reagent	47 1 71	47 0 35						Ensing and De Zeeuw (1991)
LID COC	TLC silica gel 60 F$_{254}$	S$_1$—methyl chloride–methanol (9:1, v/v)	—	Dragendorff's and iodoplatinate reagent	—							Bernardo et al. (2003)
LID COC PRC	Alumina 60	S$_1$—cyclohexane–ethanol (9.5:0.5, v/v)	—	Cobalt thiocyanate reagent with glycerol impregnated directly on the TLC plate	61 73 25							Haddoub et al. (2011)

Compound	Stationary phase	Mobile phase	Chamber	Detection	hRf values	References
LID COC PRC BEN	Silica gel 60 GF$_{254}$	S$_1$—methanol–chloroform–acetic acid (20:75:5, v/v) S$_2$—acetone	Horizontal chamber	UV	47 83 15 45 36 36 89 92	Sabino et al. (2011)
LID (ecstasy tablets)	TLC silica gel GF$_{254}$	S$_1$—chloroform–methanol (50:50, v/v) S$_2$—chloroform–methanol–acetic acid (20:75:5, v/v) S$_3$—chloroform–ammonia solution (98:2, v/v) S$_4$—isopropanol–ammonia solution (95:5, v/v)	Horizontal chamber; Tablets were pulverized, dissolved in methanol, and centrifuged; the upper layer was analyzed by TLC	UV	— — —	Sabino et al. (2010)
LID (ecstasy tablets)	TLC silica gel GF$_{254}$	S$_1$—chloroform–methanol (50:50, v/v) S$_2$—chloroform–methanol–acetic acid (20:75:5, v/v) S$_3$—chloroform–ammonia solution (98:2, v/v) S$_4$—isopropanol–ammonia solution (95:5, v/v)	Horizontal chamber; Tablets were pulverized, dissolved in methanol, and centrifuged; the upper layer was analyzed by TLC	UV	— — —	Sabino et al. (2010)
LID PRC BEN MEP TET	Silica gel 60 GF$_{254}$ plates	S$_1$—chloroform–methanol (9:1, v/v) S$_2$—toluene–dioxane–acetic acid (90:25:5, v/v) S$_3$—chloroform–ethyl acetate–methanol–propionic acid (65:15:15:7, v/v)	—	Dragendorff's reagent 254 nm 366 nm	90 72 25 89 65 88 77 38 17 17 96 32 17 60 92 93 93 90 54 88 95 67 11 11 65 69 82 55 16 16 28 91	Salvadori et al. (1988)

(continued)

TABLE 20.3 (continued)
TLC Conditions for Determination of Some Anesthetic Drugs in Forensic Toxicology and Doping Control

Compound (Matrix)	Stationary Phase	Mobile Phase	Development Mode; Extraction	Detection; Visualizing Agent	S_1	S_2	S_3	S_4	S_5	S_6	S_7	Reference
		S_4—ethyl acetate–methanol–ammonia (85:10:5, v/v)										
		S_5—chloroform–methanol–propionic aid (72:18:10, v/v)										
		S_6—dichloromethane–dioxane (2:1, v/v) saturated with water										
		S_7—chloroform–methanol (4:1, v/v)										
COC (saliva, plasma, urine)	Silica gel GF_{254}	S_1—hexane–acetone–diethylamine (6:3:1, v/v)	Extraction with $CHCl_3$	—	86							Debackere and Laruelle (1968)
BNZ (urine)	—	S_1—chloroform–methanol–ammonia–water (70:30:1:0.5, v/v)	Urine extracts were reconstituted in chloroform–isopropanol	Potassium triiodide	23							Lewis (1980)

KET NKT DHN (urine)	Silica gel 60 F$_{254}$	S$_1$—chloroform–methanol–propionic acid (72:18:10 v/v)	Urine mixed with carbonate buffer and extracted with dichloromethane–isopropanol (3:1) Enzymatic hydrolysis with 3-glucuronidase	Dragendorff's reagent	54 64 72	Sams and Pizzo (1987)
KET (ecstasy tablet)	Silica gel 60 F$_{254}$	S$_1$—methanol–concentrated ammonia (98.5:1.5, v/v) S$_2$—ethyl acetate–methanol–concentrated ammonia (85:10:5, v/v) S$_3$—cyclohexane–toluene–diethylamine (75:15:10, v/v)	Extraction with methanol	—	—	Khajeamiri et al. (2011)

Abbreviations: CET, Cocaethylene; DHN, 5,6-Dehydronorketamine; HER, Heroin; HOL, Holocaine; PPC, Piperocaine; TPC, Tropacocaine; PMT, Proxymetacaine; CCM, Cyclomethycaine; NKT, Norketamine.

Jehle et al. (2004) prepared a suspension of suppositories in methanol to remove fat. Hydrocortisone and CIN were extracted with methanol from lanolin ointments. Morlock and Charegaonkar (1998) and Essig and Kovar (2001) extracted CIN hydrochloride from a commercial ointment with methanol to obtain a clear supernatant. Ointments can be dissolved in benzene (Maggiorelli 1960).

Tablets can be extracted directly with methanol (Reichelt 1955, Sabino et al. 2010) or with dichloromethane–2-propanol (Gübitz and Wintersteiger 1980). Similar solvents are useful for the extraction from ampoules.

20.4.1.2 Biological Material

Drug monitoring in the field of anesthetics may involve quantitation in microgram (10^{-6} g or ppm) or nanogram (10^{-9} g or ppb) per milliliter concentration range. These concentrations are present in biological fluids or tissues.

The most common sample preparation methods used for the extraction of local anesthetics from biological samples generally belong to two types: LLE and liquid–solid extraction (mainly solid-phase extraction [SPE]). Most of the reports presented in this chapter favor LLE. Table 20.1 shows examples of the procedures applied to different types of biological samples and to a smaller number of formulations of different type.

Several papers focused on the methods for the isolation of the assay preparations from biological fluids. An article in Russian by Stoliarov et al. (2009) described analytical procedures for the detection of certain local anesthetics, including LID, trimecaine, ART, and anilocaine in blood and urine.

Daldrup (1989) presented a technique for the extraction of free and conjugated drugs and related compounds from small quantities of urine. Urine samples buffered with ammonia (pH = 9) were extracted with ether–ethyl acetate (1:9, v/v) and after evaporation to dryness dissolved in methanol to obtain a mixture of conjugated bases. The detection of the substances was performed by HPTLC silica gel 60 F_{254} using different reagents, such as ninhydrin, Dragendorff's, iodoplatinate, and ferric chloride solutions. Among local anesthetics, ART, COC, LID, and PRC were studied (Table 20.1).

The potential differences in TLC behavior between pure drugs and extracted drugs were investigated by Bogusz et al. (1985). Drug-free samples of autopsy blood and liver homogenate were spiked with the appropriate mixture of drugs used as a correction standard. Spiked samples were extracted with chloroform at pH = 9.0; organic solvent was collected, dried, evaporated, and reconstituted with methanol. Mixtures of pure drugs in methanol were also used. The results showed that erroneous TLC retention data could be avoided by using reference standards spiked in the biological matrix being examined.

There have been several reports on the application of liquid–solid extraction procedures suitable for local anesthetics. Two of these procedures use EXtrelut, a diatomaceous earth absorbent, a porous material that acts as a support for the aqueous phase (Matsubara et al. 1984, Stewart et al. 1984). In both methods the sample (plasma and urine) was pH-adjusted with sodium hydroxide and then applied to the Extrelut column. Apart from the previously mentioned absorbent, Stewart et al. (1984) used bonded-phase extraction cartridges. Silica, cyano, C-18 phase, and XAD-2 were among the examined SPE adsorbents.

Single- and two-step extraction and thin-layer detection procedures for benzoylecgonine (BNZ), a COC metabolite, alone or in combination with a wide variety of commonly abused drugs in urine-screening programs were presented by Kaistha and Tadrus (1977). The described procedures involved the use of ion-exchange resin-loaded paper. In the two-step extraction method, a wide variety of abused drugs are extracted by the first step, and the BNZ left in the aqueous buffer phase is extracted in the second step.

Lillsunde and Korte (1991) described a simple and sensitive SPE method for the detection of a broad spectrum of drugs. Chem Elut extraction tubes were used for the isolation of drugs from urine.

About 300 substances, including all potentially abused drugs and their metabolites, were covered. Urine was adjusted to pH = 8–9 with solid disodium phosphate and poured into a ChemElut extraction tube. Drugs were eluted with dichloromethane–isopropanol (90:10, v/v). The extract was evaporated and the residue was dissolved in ethanol. Compounds were screened by TLC and confirmed by gas chromatography–mass spectrometry (GC–MS). The detection limit of the local anesthetics was 5 μg. The plates were examined under UV light (366 and 254 nm) and then sprayed with Dragendorff's reagents and iodoplatinate solution.

Nowadays, the most frequently used RP sorbent for extraction of local anesthetics from biological samples (e.g., plasma, serum, urine) is C-18 silica. In a few recently published papers, SPE was used in order to extract and concentrate analytes from the liquid matrix. At acidic or neutral pH, local anesthetics exist largely in their ionized form, in which they are poorly retained by this lipophilic sorbent. Consequently, in order to secure a more complete retention, the sample must be adjusted to alkaline pH. In most cases, mixed-mode SPE cartridges are applied for COC and its metabolites. The mixed mode (i.e., C18, C8, or C4 with cation exchange-functionalized silica) reduces matrix effects, which causes ion suppression. In comparison with RP SPE cartridges, the mixed mode is stable over a wide range of pH and gives high-purity drug extracts and a better signal-to-noise ratio (Janicka et al. 2010).

20.4.2 Application of TLC in Analysis of Individual Local Anesthetics

20.4.2.1 Lidocaine (2-(Diethylamino)-N-(2,6-Dimethylphenyl)acetamide)

LID (lidocaine, Xylocaine) is widely used as a local anesthetic in dentistry. Apart from COC, LID has been the most closely examined.

LID was identified on silica gel TLC plates by either short-wavelength UV light or by a positive test with acidified iodoplatinate sprays (Masoud 1976, Sundholom 1983). The chromatographic conditions and the solvent systems that were used in these two papers are presented in Table 20.1.

Furthermore, the enantiomer ratio of D- to L-epinephrine in 70 LID–epinephrine samples of various dosage forms and concentrations was successfully analyzed by TLC (Allgire et al. 1985). Epinephrine was isolated via Sep-Pak cartridges while the retained LID was eluted with methanol and examined by TLC (Table 20.1).

LID is useful for controlling ventricular arrhythmias, particularly in patients with acute myocardial infarction, and therefore, it is commonly used as an antiarrhythmic agent. Several antiarrhythmic drugs, including LID, were simultaneously determined in serum by HPTLC (Lee et al. 1978). Plates were developed with solvents of different polarity. LID was scanned at 220 nm. Within the quality control of multidrug pharmaceuticals, Nagy et al. (1992) presented the results obtained for the identification of LID in antiarrhythmic tablets using TLC and spectrophotometric determination. Witek and Przyborowski (1996) described simple and sensitive conditions for separation and identification of diltiazem in tablets by a TLC method on silica gel in the presence of five other antiarrhythmic drugs, including LID. The chromatographic conditions of all three determinations are presented in Table 20.1.

Apart from conventional TLC, HPTLC has also been used for the quantitative analyses of LID in pharmaceutical preparations and samples that contain other local anesthetic drugs (Waraszkiewicz et al. 1981, Noggle and Clare 1983, Duschi and Hackett 1985).

Jarzebinski and Ciszewska-Jedrasik (1986) reported the application of densitometry in the determination and separation of active substances in dosage forms containing naphazoline and diphenhydramine, boric acid, antazoline, sulfathiazole, LID, or methylene blue (Table 20.1). In order to study the degradation product, 2,6-dimethylaniline in LID–hydrochloride injection TLC on silica gel G plate, was used with acetic acid–methanol–benzene (0.4:0.8:4.5, v/v) as solvent and p-dimethylaminobenzaldehyde or ninhydrin as color reagent (Zheng 1982).

Živanović et al. (1988) published the results obtained for a TLC UV-densitometric method for the analysis of LID in ointments, suppositories, and gels (Table 20.1). The active substances were extracted from pharmaceutical dosage forms with absolute ethanol. The ethanolic extracts were applied to the chromatographic plates. LID was separated from other components in the pharmaceutical dosage forms without preliminary extraction. Recoveries of LID from pharmaceutical formulations were in the range of 99.2%–100.7%. The RSDs were less than 3.0% for all the formulations.

The same mobile phase was used to develop an HPLC method for determination of LID hydrochloride in liquid dosage forms (Živanović et al. 1996). The results obtained by RP HPLC and TLC–UV densitometry were compared. The linear response was achieved up to 10 μg/mL (HPLC) and 8 mg/mL (TLC). The recoveries were in the ranges of 99.6%–100.2% and 99.2%–100.7% for HPLC and TLC, respectively. The RSD of the peak areas was 1.71% for HPLC and 0.55% for TLC, with recoveries in the range of 99.6%–100.2% for HPLC and 99.2%–100.7% for TLC.

Devarajan et al. (2000) established a simple, rapid, specific, and selective technique for the quantitative determination of LID and phenylephrine hydrochloride (PHE), the agent that produces prolongation in anesthesia, individually and in combination both as a bulk drug and from pharmaceutical preparations. The chromatographic conditions are shown in Table 20.1. The densitometric determination of LID and PHE was carried out at 262 and 291 nm, respectively. The calibration curves of LID and PHE were linear in the range of 8–18 and 4–9 mg, respectively. The detection limits of LIG and PHE were found to be 4 and 2 mg, respectively. The method was validated regarding the system precision, method precision, recoveries, and intra- and inter-day variation. RSD of less than 2% suggested system suitability and precision of the developed method.

Four simple, selective, and accurate methods were adopted for quantitative determination of LID and PRI in the presence of their major degradation products (Riad et al. 2008). LID and PRI were detected at 220 and 237 nm over concentration ranges of 3–15 and 0.4–10 μg/spot with mean percentage recoveries 99.51 ± 0.53 and 99.97 ± 0.76 for LID and PRI, respectively. The proposed methods were successfully applied in the analysis of the dosage forms.

Stroka et al. (2002) reported TLC densitometric analysis of LID, and the chromatographic conditions are given in Table 20.1. Kurita et al. (2008) established a simple and rapid method for screening for drugs in health food using 2D TLC. LID was determined using four developing systems. Two systems were run as 2D HPTLC (S_1 and S_2), while S_3 and S_4 were run as conventional 2D TLC. The chromatographic conditions used in these two papers are also shown in Table 20.1.

20.4.2.2 Articaine ((RS)-Methyl 4-Methyl-3-(2-Propylaminopropanoylamino) thiophene-2-Carboxylate)

ART hydrochloride, classified as a local amide anesthetic, contains a thiophene instead of a benzene ring. ART hydrochloride is commonly used in clinical dentistry and is marketed as a racemic mixture. There is not much literature on ART identification and quantitative determination using TLC systems (Daldrup et al. 1989, Schmidt and Bracher 2006). Chromatographic conditions and retardation factors for ART are given in Table 20.1 together with several other local anesthetic drugs.

20.4.2.3 Benzocaine (Ethyl 4-Aminobenzoate)

BEN is an active ingredient in many over-the-counter anesthetic ointments. BEN sprays are used in medical practice to locally numb mucous membranes of the mouth and the throat for minor surgical procedures or when a tube must be inserted into the stomach or airways. The use of sprays is known to be occasionally associated with methemoglobinemia. However, some cases of methemoglobinemia are also due to the incorrect use of BEN sprays (e.g., longer duration or more frequent sprays than recommended).

A survey of older literature shows that BEN can be separated using silica gel thin-layer plates and different solvent mixtures (Sunshine and Fike 1964, Messerschmidt 1971, Wan et al. 1972, Ali and Steinbach 1981) or using cellulose powder impregnated with oleyl alcohol and aqueous buffer solutions as mobile phases (Büchi and Fresen 1966). The mobile phases used in these reports are presented in Table 20.1.

Röder et al. (1969) used mixtures of different homogeneous azeotropic solvents for the TLC separation of BEN from other local anesthetics. Chloroform–methanol (80:10, v/v) and ethanol, the mobile phases, and silica gel plates treated with 0.1 M sodium hydroxide were used for the separation of BEN from other local anesthetics (Higuchi 1967). Detection of BEN can be performed by quenching in short-wavelength UV light, after spraying with Ehrlich's reagent, which gives yellow spots, and diazotization followed by coupling with α-naphthol (Wagner and Zimmer 1955) or with iodide solution (Sarsunova 1963). BEN was also determined after TLC separation by reacting it with the dansyl chloride reagent, and the subsequent determination of fluorescent intensity of adducts was performed using TLC–fluorimetry (Messerschmidt 1971).

In more recent literature, a simple, sensitive, and reproducible HPTLC method for the determination of BEN in mucopain gel by HPTLC densitometry was published by Bhawara (2008) (Table 20.1). The active constituent was extracted with methanol. The limit of detection of BEN was found to be 10 µg/mL.

Apart from BEN itself, papers in which BEN is analyzed together with other local anesthetics can often be found in the literature. Such examples are given in the following text and presented in Table 20.1.

20.4.2.4 Procaine (2-(Diethylamino)ethyl 4-Aminobenzoate)

PRC (generic name: novocaine) is the ester of *p*-aminobenzoic acid and diethylaminoethanol. It is one of the oldest ester-type local anesthetic agents used. It is primarily used to reduce the pain caused by intramuscular injection of penicillin. It is also used in dentistry. PRC is used less frequently nowadays since there are more effective alternatives such as LID.

Al-Badr and Tayel (1999) reviewed several TLC methods that had been used for the analysis of PRC up to that date. Selected examples of the applied TLC conditions and the corresponding literature are presented in Table 20.1 (Andrey and Moffat 1979, Melent'eva et al. 1984, Sarsunova et al. 1985, Zhu and Hu 1992).

Maulding and Michaelis (1975) reported results of kinetic investigations of PRC and quantitation of its stability. The hydrolysis of PRC at ambient temperature was monitored spectrally and by TLC. PRC is hydrolyzed enzymatically at a very high rate with the formation of *p*-aminobenzoic acid and diethylaminoethanol. The TLC method turned out to be useful for the identification of *p*-aminobenzoic acid as the decomposition product of PRC. Buesing and Grigat (1988) reported a convenient method for the identification of *p*-aminobenzoic acid based on its separation from PRC followed by diazotization in the aqueous phase and coupling with thymol to give the azo dye. Galais et al. (2000) reported the assay of *p*-aminobenzoic in cardioplegia solution drug used in heart surgery (CP1, Fabiani's solution). CP1 contains PRC that is unstable at pH > 7 and can be hydrolyzed yielding *p*-aminobenzoic acid and diethylaminoethanol. PRC was assayed by UV spectrophotometry at 290 nm. Under these conditions (Table 20.1), PRC remained at the origin of the plate and *p*-aminobenzoic acid (the product of hydrolysis) migrated ca 12 mm.

Quantitative determination of adrenaline hydrochloride, ephedrine hydrochloride, and PRC hydrochloride in ointments was performed using TLC–UV densitometry by scanning at 254 nm (Radulović et al. 1990). The RSD and recovery were 2.77–3.26 and 98.0%–98.75%, respectively.

PRC is a common component of antibiotic mixtures. Penicillin G/PRC is a combination of benzylpenicillin and PRC. Streptomycin, penicillin G, and PRC/penicillin G were separated by TLC with acetone–methanol–0.1 M sodium chloride (2:2:1, v/v) solvent system and detected quantitatively (Shahjahan et al. 1972). Separation zones were observed using 0.4% aqueous potassium permanganate.

Methods of analysis and determination of pharmaceuticals in wastewater samples from pharmaceutical industry are reported in the literature (Mutavdžić et al. 2006, Babić et al. 2007). Mutavdžić et al. (2006) reported HPTLC of enrofloxazine, norfloxazine, oxytetracycline, trimethoprim, sulfamethazine, sulfadiazine, and penicillin G/PRC on the cyano phase with 0.05 M oxalic acid–methanol (81:19, v/v). The SPE–TLC determination was validated for linearity, precision, quantification, and detection limit. Babić et al. (2007) reported HPTLC of seven pharmaceuticals, enrofloxazine, oxytetracycline, trimethoprim, sulfamethazine, sulfadiazine, sulfaguanidine, and penicillin G/PRC in production wastewater (obtained by SPE on hydrophilic–lipophilic balance cartridges with methanol) on the cyano phase with 0.05 M oxalic acid–methanol (81:19, v/v). Quantification was performed by videodensitometry at 254 and 366 nm.

TLC separation and spectrophotometric determination of PRC and BEN in various samples were performed (Bussey and Backer 1974, Panić 1981, Wollmann and Patrunky 1982). BEN and PRC were separated from drug mixtures by TLC on silica gel using two solvents (Panić 1981). The spots were localized under UV light, eluted by methanol, and quantified by spectrophotometry after treatment with Ehrlich's reagent. The isothermal stability of bisacodyl, BEN, and PRC hydrochloride was investigated by TLC and DTA (Wollmann and Patrunky 1982). Ehrlich's reagent and Dragendorff's reagent were used in TLC for the detection of BEN and PRC hydrochloride, respectively. The detection limits ranged from 0.04 to 1 µg. No decomposition product (>1%) could be detected in BEN and PRC hydrochloride. Kovar et al. (1981) used charge-transfer complexes formed in a reaction with polynitrobenzenes for quantitative determination of PRC in mixtures with caffeine and for identifying PRC in the presence of other local anesthetics and other bases on TLC.

20.4.2.5 Cinchocaine (2-Butoxy-N-[2-(Diethylamino)ethyl]quinoline-4-Carboxamide)

CIN (also known as dibucaine) is a local amide-type anesthetic, which is suitable for surface or spinal anesthesia. It is used in various formulations such as creams, ointments, suppositories, and injections. It is the active ingredient in some topical hemorrhoid creams, and it is also a component of the veterinary drugs used for euthanasia of horses and cattle. It is sold under the brand names Cincain, Nupercainal, Nupercaine, and Sovcaine.

A number of TLC systems were developed for the identification of the CIN and for the determination of the related compounds (Padmanabhan 1983), and they are listed in Table 20.1 (Systems S_1–S_{13}). System 1 can be employed to control the impurities present from the synthesis of the drug. System 3 was proposed for the estimation of transformation products in formulations, while System 2 was used to determine the content of 3-chlorodibucaine.

CIN was detected directly from the silica gel TLC plate using the pulsed laser desorption/volatilization method with subsequent detection using laser-induced resonant by R2PI at 266 nm (Li and Lubman 1989). Gübitz and Wintersteiger (1978) reported an HPTLC system for the quantification of CIN in injectable solutions and in plasma serum samples.

Several papers dealing with determination of CIN in ointment were published. Morlock and Charegaonkar (1998) reported a rapid and economic HPTLC procedure for determining hydrocortisone and CIN (Table 20.1). Hydrocortisone and CIN were extracted with methanol from lanolin ointment and separated on silica gel plates. The intermediate precision (n = 3) was found to be at 0.53% (peak height) and 1.7% (peak area), respectively, for hydrocortisone and at 1.6% (peak height) and 3.3% (peak area), respectively, for CIN. Essig and Kovar (2001) described HPTLC method for the quantification of CIN hydrochloride (dibucaine) in commercial Faktu ointment. Classical slit scanning densitometry was compared with videodensitometry by evaluating the plates by line scanning and then by image analysis. CIN hydrochloride was extracted from the commercial ointment with methanol and filtered through a 0.45 µg membrane filter to obtain clear supernatant.

CIN hydrochloride spots were detected by videodensitometry at 366 nm and by scanning densitometry at 312 nm. The authors proposed videodensitometry as a fast alternative for quantifying thin-layer chromatograms in CIN analysis.

Jehle et al. (2004) validated an HPTLC method for the determination of the content of the CIN hydrochloride in suppositories, which are used to treat hemorrhoids (proctologic). When preparing the samples, the suspension of suppositories in methanol was vigorously shaken in order to remove fat. Precision was determined to be 2% and recovery 101%.

LC and TLC–densitometry methods were optimized for simultaneous determination of acediasulfone and CIN (Mohammad et al. 2007). In the LC method, the separation and quantitation of the two drugs was achieved on a Zorbax C8 column using a mobile phase composed of methanol–phosphate buffer, pH = 2.5 (66:34, v/v), with UV detection at 300 and 327 nm for acediasulfone and CIN, respectively. Both methods proved to be specific and accurate for the analysis of these drugs in laboratory-prepared mixtures and dosage forms. HPTLC method showed linearity over concentration range 2–9 µg/spot for CIN with limit of quantitation (LOQ) 0.05 ng/spot and limit of detection (LOD) 0.015 ng/spot. A mobile phase composition and chromatographic conditions for the TLC–densitometry are presented in Table 20.1.

20.4.2.6 Fomocaine (4-[3-[4-(Phenoxymethyl)phenyl]propyl]morpholine)

Fomocaine (FOM) is a basic ether-type local anesthetic with good surface anesthesia and low toxicity. FOM is used as a topical anesthetic in dermatological practice. Modifications of the FOM molecule have been pursued for many years, for example, to improve its physicochemical properties and to find some possible new (systemic) applications, for example, in the treatment of migraine. It has also been successfully tested as an antiarrhythmic agent.

Oelschlager et al. (1975) qualitatively investigated the metabolism of FOM after oral administration to rats and dogs. Samples of urine and feces were collected from animals treated with FOM for 24 h. After extraction with chloroform, concentrated extracts were applied to the precoated TLC plates. Feces were homogenized in 10 volumes of phosphate buffer (pH = 7.4) and treated in the same way as the urine samples. Development was carried out using the four different solvent systems presented in Table 20.1. It was found that the rats and dogs had excreted some unchanged FOM in the urine together with *p*-hydroxyfomocaine (free and conjugated), fomocaine-*N*-oxide, *p*-hydroxy-fomocaine-*N*-oxide and *p*-(*y*-morpholinopropyl) benzoic acid.

Toxicity of FOM, five *N*-free FOM metabolites and two chiral FOM derivatives in rats was compared with PRC (Fleck et al. 2001). The results of pharmacological and toxicological testing of the enantiomers of two chiral FOM alkyl morpholine derivatives in comparison with their in vitro interactions on drug metabolism in rats were also published (Lupp et al. 2006).

20.4.2.7 Mepivacaine ((*RS*)-*N*-(2,6-Dimethylphenyl)-1-Methyl-Piperidine-2-Carboxamide)

MEP is chemically and pharmacologically related to amide-type local anesthetics. It contains an amide linkage between the aromatic nucleus and the amino group. MEP is marketed under various trade names (including Carbocaine and Polocaine).

MEP is supplied as the hydrochloride salt of racemate. Although pure enantiomers have greater pharmacological effects, the racemic forms are used in most commercial formulations. Studies on the anesthetic activities and toxicity of individual enantiomers of BUP and MEP generally indicate that *S*-enantiomers are longer-acting and less toxic than the *R*-enantiomers.

Nemák et al. (2008) studied the resolution of enantiomers of each member of the group of five pipecoloxylidides (PIP) (MEP, PIP, *N*-ethylpipecoloxylidide [EPI], ropivacaine [ROP], and BUP). The authors developed an efficient TLC method for studying the diastereomeric salt-forming behavior of the compounds and for testing the chemical purity of the diastereomeric salts. TLC of a racemic mixture of diastereomeric salts on silica gel with 1-butanol–glacial acetic acid–water (12:3:5, v/v). TLC was also performed using 2-propanol, ethanol, and methanol instead of 1-butanol.

Santos et al. (2004) performed a fast screening of anesthetics levobupivacaine and MEP by TLC followed by direct on-spot matrix-assisted laser desorption/ionization time-of-flight mass spectrometry (MALDI TOF MS) identification. The technique is fast and sensitive, requires little sample preparation and manipulation, and is therefore suitable for fast screening with TLC separation and MS identification of low molecular weight compounds.

20.4.2.8 Cocaine (Methyl (1*R*,2*R*,3*S*,5*S*)-3-(Benzoyloxy)-8-Methyl-8-Azabicyclo[3.2.1]octane-2-Carboxylate)

COC is a natural agent, an alkaloid obtained from the leaves of *Erythroxylon coca*. Chemically, it is methyl BNZ. COC is rapidly biotransformed by the body into a major metabolite, BNZ, and a minor metabolite, ecgonine. For the detection of BNZ, the metabolite should be converted into the parent compound, which gives a lower LOD and good separation from interferences present in biological material.

Most papers in which COC is analyzed deal with toxicological screening and analysis of drugs on the illicit market. Therefore, these papers will be presented in the next section that deals with TLC in toxicology and drug abuse monitoring. An overview of the literature dealing with current chromatographic systems and the conditions of determination of COC and its major metabolite, BNZ, will be given in this section.

COC can be determined qualitatively by using TLC performed on silica gel. A number of different solvent systems are available for this. The compounds can be examined visually under UV light (254 and 360 nm) and with general visualization reagents, such as 1% potassium permanganate in water, acidified potassium iodoplatinate, or Dragendorff's reagent.

Muhtadi and Al-Badr (1986) pointed out numerous TLC systems for the analysis of COC hydrochloride. For the purpose of this chapter, the selected systems are presented in Table 20.1 (Brown et al. 1973, Bertulli et al. 1978, Clarke 1978, Jukofsky et al. 1980).

Janicka et al. (2010) reviewed analytical methods that were proposed for the verification of COC and its metabolites in biological matrices (e.g., blood, hair, saliva, plasma, urine, internal organs, and meconium) during the past two decades. Different methods of sample preparation and extraction techniques are described, as well as a variety of analytical techniques that enabled identification and quantitative determination of drugs of abuse, screening, and confirmatory techniques. Approaches using TLC are discussed, particularly the possibility of simultaneous determination of parent, metabolite, and interfering drugs in urine samples, as well as the identification of markers of combined consumption of COC and alcohol.

Wallace et al. (1975) reported 20% ethanol in chloroform (v/v) as an effective solvent for the extraction of both unchanged COC and BNZ in urine. A developing solvent (Table 20.1) was very effective for the separation of BNZ from urine contaminants. Dragendorff's reagent followed by a light spraying of 20% aqueous sulfuric acid and a subsequent brief exposure to iodine vapors provided the optimum sensitivity for the detection of both COC and BNZ.

The ^{125}I-RIA for BNZ in urine was evaluated by comparison with GLC and TLC and enzyme-multiplied immunoassay technique (EMIT) (Mule et al. 1977). Two hundred human urine samples were analyzed. GLC was considered the reference technique, that is, all analytical results were compared to GLC. Full agreement between GLC and RIA was observed in 95.5% of the analyses, between GLC and TLC in 84.0% of the analyses and between GLC and EMIT in 86.0% of the analyses. Mueller et al. (1977) proposed a TLC method that can be used for detection of BNZ in human urine in a rapid screening program of abused drugs. Amberlite XAD-2 resin was utilized for the extraction of BNZ from urine, and two separate solvent systems were used for the identification of BNZ. The first part of the procedure utilizes two TLC solvent systems to identify a variety of drugs, including BNZ. The second part is specific for BNZ and it can be used as a confirmation method. The use of two chromatographic systems significantly reduces the possibility of interference. The procedure is sensitive to 3–4 µg/mL urine.

Budd et al. (1980) developed a TLC procedure for the analysis of urine samples for BNZ and norpropoxyphene (NOR), the major metabolites of COC and propoxyphene (PRP) respectively (Table 20.1). Urine is made slightly acidic and methylated with dimethyl sulfate to convert BNZ back into much more readily extractable COC. The urine is then made basic and extracted with chloroform/isopropanol. The organic layer containing COC and NOR is separated and evaporated to dryness. The residue is reconstituted and spotted on a TLC plate. The mobile phase used, hexane–chloroform–diethylamine (80:10:10, v/v), separates COC and NOR without interference from other drugs, metabolites, and urinary substances. The two substances are visualized with acidified iodo-platinate spray and can be detected at levels of 2.0 μg/mL for BNZ and 1.0 μg/mL for NOR. The mixture of hexane–acetone–diethylamine was proposed as the best solvent system for COC analysis.

Funk and Droeschel (1991) developed an HPTLC method for quantitative determination of COC and its major metabolites: ecgonine methyl ester, ecgonine, and BNZ. The mobile phase was opti-mized according to the PRISMA model. After development, two different postchromatographic derivatization methods were tested (modified Dragendorff's reagent gave red chromatographic zones on the yellow background). Quantitative analysis could be performed after heating the devel-oped HPTLC plates to 260°C for a few minutes. This yielded pale blue, fluorescent derivatives, which could be quantitated at levels greater than 50 ng/spot. Wössner and Kovar (1996) reported a simple and sensitive 2D TLC procedure for detecting COC and its metabolite BNZ on silica using acetic acid–methanol–hexane mixture in the first dimension and methanol–dichloromethane in the second dimension.

20.5 CLINICAL AND FORENSIC TOXICOLOGY AND DOPING CONTROL

The identification of toxic compounds in biological material is of utmost importance in clinical toxicology and in forensic institutes. The samples that are most commonly submitted for toxicologi-cal analysis used to detect local anesthetics are of biological origin (blood, urine, or plasma). Urine, hair, and saliva are generally used for screening assays. The main advantage of using the screen-ing methods applied to urine and saliva samples is direct measurement with little or no sample preparation.

The use of local anesthetic substances is not entirely without a risk to the patient, and a long-term application of some anesthetic drugs requires monitoring. Cases of deaths associated with the administration of local anesthetics are not rare.

Two fatal cases of intoxication that were the result of the use of LID and MEP were the subject of the paper published by Sunshine and Fike (1964). Large doses of local anesthetic were administered prior to cosmetic surgery. The qualitative identification of these 2 and 17 other local anesthetics was performed using TLC silica gel G plates prepared in 0.1 M sodium hydroxide, with cyclohex-ane–benzene–diethylamine (75:15:10, v/v). Christie (1976) published a review of the literature on fatal consequences of local anesthesia. Furthermore, qualitative analysis of tissue distribution of LID in the resultant extracts after fatal accidental injection was performed by TLC (Poklis et al. 1984). LID was extracted from blood and tissue (brain, heart, kidney, lung, spleen) homogenates at pH = 9.5 (carbonate buffer) with hexane–isoamyl alcohol (98:2, v/v), which was back extracted in 0.5 N sulfuric acid. The separated acid solution was then made basic with solid carbonate buffer and extracted with fresh hexane–isoamyl alcohol (98:2, v/v). The organic layer was separated and evaporated to dryness under nitrogen at room temperature. LID was indicated by iodoplatinate posi-tive spots of $R_F = 0.76$ in the TLC development system by Davidow et al. 1968.

Some anesthetic compounds are subject to abuse and are available on the illicit market. Emergency drug analysis is required in suspected drug-related deaths by accidental overdose or in cases when a drug or drugs of unknown origin have been taken. Apart from psychoactive com-ponents, drug tablets contain diluents and adulterants. While diluents are compounds without any significant pharmacological properties, intentionally added to a street drug to increase its volume, adulterants have similar pharmacological, sensory, and physicochemical properties as the drug.

Analysis of illicit drug tablets is usually carried out using chromatographic methods. Many standardized solvent systems have been developed for toxicologically relevant substances. A review of the application of chromatography to the analysis of drugs of abuse was published by Gough and Baker (1983) and covered the period up to 1980. TLC and HPTLC are used in such investigations for presumptive identification of a drug and some other components of the sample. Forensic relevance of measuring anesthetics in different biological matrices and a review of the literature on chromatographic methodologies available for drug monitoring are presented further in this chapter.

COC has a long history of use as a stimulant, and it is one of the most commonly abused street drugs. Therefore, determination of COC in biological material is of great importance in clinical and forensic toxicology.

Qualitative identification of COC and its metabolites in urine samples is generally carried out by an immunoassay technique followed by a GC-MS confirmation. Several studies have demonstrated the applicability of TLC and HPTLC to the analysis of COC and its metabolite BNZ and the transesterification product cocaethylene (CET) in urine samples containing illicit drug.

CET is an important marker of combined consumption of COC and alcohol. Rafla and Epstein (1979) reported a method that can be used to confirm the presence of COC and its major metabolites, CET and ethylecgonine, in urine samples positive for ethyl alcohol. Six TLC solvent systems were proposed for separating COC and other abused drugs, such as morphine and methadone, from their metabolites (Table 20.3). BEN, LID, PRC, and TET were identified among adulterants. GC–MS was used to establish the presence of ecgonine, BNZ, methylecgonine, ethylecgonine, COC, and CET. Bailey (1994) developed a TLC method for the detection of CET and COC in human urine samples of patients who had ingested COC with ethanol. Statistically significant correlations between the intensity of the urine TLC detection and the measured urine concentrations of CET and COC were noted. Both CET and COC were visualized with an iodoplatinate spray. The chromatographic conditions are shown in Table 20.3. The presented TLC method showed low sensitivity, with LOD values over 1 μg/mL, even when methylating BNZ back into the parent drug was performed.

The usefulness of TLC for the identification of COC and BNZ in urine and tissues was evaluated by Kiszka and Madro (2002). They studied the possibility of separating COC from BNZ, as well as of separating these two compounds from the "biological background." Thirty developing systems were examined, four of which (see Table 20.3) were assessed as optimal for the analysis of extracts from the liver and the kidney due to good separation of the investigated xenobiotics from the "biological background." As far as the analysis of urine extracts was concerned, the first three systems (ethyl acetate–methanol–ammonia (60:30:6, v/v), chloroform–methanol–ammonia (100:20:1, v/v), and methanol–chloroform (4:1, v/v)), and the mixture chloroform–methanol (4:1, v/v) turned out to be optimal for separation. The optimal system of staining chromatograms was a combination of Dragendorff's reagent with a solution of sulfuric acid.

More recently, Yonamine and Sampario (2006) proposed HPTLC on silica gel for the detection of COC in urine samples of drug addicts. Urine samples were submitted to SPE prior to methylation (to convert BNZ to COC) with diazomethane. The limit of detection was 100 ng of COC. The method was shown to be useful for detecting COC in biological matrices and for discriminating it from interfering substances (local anesthetic, caffeine, and nicotine). The presented mobile phase (Table 20.3) could not be used to separate COC from its transesterification product, CET, which had the same hR_F value as COC. Antonilli et al. (2001) studied the potential of HPTLC to simultaneously detect CET, COC, and BNZ in 16 urine specimens of drug addicts, previously tested positive for BNZ at immunoenzymatic screening (Table 20.3). HPTLC limit of quantitation was 1.0 μg/mL for the three compounds. The RSD ranged from 1.03% to 12.60% and 1.56% to 16.6% for intra- and inter-day HPTLC analysis, respectively. Accuracy and precision, as well as detection and quantitation limits of the method, were compared with HPLC. HPTLC was shown to be suitable for detection of the three analytes only for samples with high concentrations.

Local anesthetics are often added to COC and other illicit drugs before they are put on the illegal market. The anesthetics that are the most frequently used as COC adulterants are LID, BEN, PRC, and TET. Some other local anesthetics are also used, although less often. Many authors have described procedures that are suitable for screening illicit COC and heroin (HER) samples, both pure and adulterated with local anesthetics, for the presence or absence of these compounds. A procedure useful for rapid screening of illicit COC and HER samples and tentative identification of local anesthetics used as adulterants was published by Brown et al. (1973) (Table 20.3). The results showed that a single extraction with 95% ethanol and a TLC evaluation of the ethanolic extract were suitable for the extraction, separation, and tentative identification of COC, HER, and selected local anesthetics. Baker and Gough (1979) described the application of field tests and chromatography in the detection of COC and 14 other local anesthetics, which were used to adulterate COC. Initial screening of the samples by field tests, followed by concurrent TLC and GC, enabled a rapid identification of these compounds. The detection of compounds separated by TLC was performed using two different methods (Table 20.3). The first method involved the inspection of the dried plate under UV irradiation (254 and 366 nm), whereas the second one involved spraying with an iodoplatinate reagent.

Jukofsky et al. (1980) reported a simple procedure for semiquantitative determination of COC–LID mixtures known as "rock-COC." The presence of LID was confirmed and it was semiquantitatively measured by the differences in form and color reaction between the two drugs. The results of the TLC analysis are given in Table 20.3. Gübitz and Wintersteiger (1980) combined HPTLC with specific fluorescent reagents for screening for a number of drugs of abuse. The method was applied to pure substances, some preparations, and urine (Table 20.3). The extraction from tablets and ampoules was carried out from aqueous solution using dichloromethane–2-propanol mixture, following pH adjustment in carbonate buffer pH = 9. The authors used Extrelut columns for extraction of urine samples adjusted with carbonate buffer to pH = 9. The column was eluted with dichloromethane–2-propanol.

Della Casa and Martone (1986) described a suitable HPTLC method with direct UV-densitometric measurements for the separation and quantitation of HER and COC in street samples. The solvent system for COC samples was chloroform–methanol (50:50, v/v). The RSDs were 8.76 and 9.14 for HER and COC samples, respectively. The HPTLC method was proven to be sensitive, precise, and accurate for quantitative determination of drugs in street samples. Furthermore, the method was proposed as useful in forensic chemistry for showing a fingerprint of each sample. By means of TLC, COC, BNZ, and LID were determined and several isomers of truxillines were observed, isolated, and identified in illicit COC (Ensing and De Zeeuw 1991). The conditions used are presented in Table 20.3. Much earlier, Comer and Comer (1967) reviewed a number of papers describing the use of TLC for the separation of different kinds of drugs. They listed 16 different TLC systems for separating COC from HER, other local anesthetics, and a number of analgesics–antipyretics.

Two hundred and nine samples of street COC were screened using TLC (Bernardo et al. 2003) (Table 20.3). Confirmation and quantification of active adulterants and COC were performed by gas chromatography/flame ionization detector (GC/FID). Apart from LID, PRI was another local anesthetic found, although in a smaller number of samples than LID. Haddoub et al. (2011) published a new test for the identification of COC (in its acid or base form), LID and PRC on precoated aluminum oxide with coloring identification involving a cobalt(II) isothiocyanate complex stabilized on alumina by the presence of glycerol. hR_F values and TLC conditions are shown in Table 20.3.

Sabino et al. (2011) described the use of TLC coupled with easy ambient sonic-spray ionization mass spectrometry (TLC EASI MS) for rapid and certain analysis of COC and crack COC. Fifteen COC samples were analyzed and all of them had positive TLC EASI MS results for COC, but other drugs and adulterants, such as LID, caffeine, BEN, lactose, BNZ, and ecgonidine, were also detected. The investigated drug samples were pulverized and white powder samples of COC were homogenized and each sample was dissolved in methanol. After centrifugation, the upper layer was transferred to a glass vial and analyzed by TLC EASI MS. The limit of detection found for COC on TLC plates was 2 μg. See Table 20.3 for the chromatographic conditions.

Ecstasy, which can be found in various forms, is a common street name for illicit tablets. Apart from the main active component, 3,4-methylenedioxymethamphetamine, ecstasy tablets contain amphetamine, but many other psychoactive components were discovered in the seized tablets. Recently, Sabino et al. (2010) have reported the combination TLC EASI MS as a relatively simple and powerful screening tool for analysis of ecstasy tablets (Table 20.3). The ecstasy tablets were pulverized and partially dissolved in methanol. After centrifugation, the upper layer was transferred to a glass vial and analyzed by TLC. Separation of seven common ecstasy drugs (including LID) was attained. EASI MS was then performed directly on the surface of each TLC spot for MS characterization.

Kochana et al. (2005) tested the usefulness of several multicomponent eluents for TLC screening and identification of active components of ecstasy tablets. The solid drug samples were dissolved in methanol and phosphate buffer. The influence of individual additive (both adulterants and diluents) on TLC chromatograms was tested. The optimum eluent was found to be chloroform–dioxane–methanol–ammonia–acetonitrile (3.5:15:2:1.5:15, v/v) on silica gel 60 F_{254} in a horizontal developing chamber. It was found that most of the studied additives influenced the separation of the drugs and that the influence was stronger if drugs were dissolved in methanol. PRC extinguished fluorescence of the TLC plate at 254 nm, but the spots did not interfere with spots of the tested drugs.

KET, among other ingredients, was found in ecstasy tablets in the study by Khajeamiri et al. (2011). TLC that used three solvent systems was carried out (Table 20.3). GS/MS and LC MS were used for quantitative determination.

The presence of PRC and/or its metabolites in urine can be presumptive of HER, COC, or other street drugs usage. A positive PRC test could also be due to the use of PRC penicillin (a salt of penicillin with PRC used for gradual release of penicillin in the body). Reports on qualitative identification of PRC in HER powders on silica gel plates can be found in the literature (Bussey and Backer 1974, Roesener 1982, Kamble and Dongre 1997, Dongre and Kamble 2003).

Bussey and Backer (1974) described a TLC procedure for distinguishing several drugs that may be encountered in urine-screening programs (Table 20.3). Amphetamine, mescaline, 4-methyl-2,5-dimethoxyamphetamine, 3,4-methylenedioxyamphetamine, chlorphentermine, phenylpropanolamine, β-phenylethylamine, BEN, and PRC were separated from each other. From buffered urine samples (pH = 9.5), drugs were separated on XAD-2 column, and elution was performed with 1,2-dichloroethane:ethyl acetate mixture. The resulting residue is spotted on a silica gel plate. Different developing solvents were studied, and results of multiple developments in two solvents are shown in Table 20.3.

Dongre and Kamble (2003) proposed a new chromogenic spray for detection and identification of HER in forensic samples after HPTLC separation on silica plates with chloroform–ethanol (9:1, v/v). The red complex was formed on reaction of HER with the ferric chloride–2,2′-dipyridyl mixture. PRC and other adulterants that usually occur in forensic HER samples did not give a positive reaction with this reagent.

In the most recent paper, Kuwayama et al. (2012) reported a method applicable to rapid drug screening and precise identification of toxic substances in the cases of poisoning and postmortem examinations. The authors examined a simple preparation and highly sensitive analysis of drugs in samples such as urine, plasma, and organs using TLC coupled with matrix-assisted laser desorption/ionization mass spectrometry (TLC MALDI MS). The developing solvent was acetone–28%-aqueous ammonium (99:1, v/v). All the psychotropic compounds tested were detected at levels of 0.05–5 ng (LOD for KET 0.5 [ng/spot]). The type of a TLC plate and the layer thickness of plate did not affect detection sensitivity.

Unlawful misuse of drugs in greyhound and horse racing is another type of abuse where local anesthetic drugs find wide application. UV spectrophotometry, TLC, HPLC, and GC were proposed for the identification and subsequent confirmation of 46 drugs related to the doping of thoroughbred horses (Salvadori et al. 1988). Seven TLC systems were used to evaluate silica gel plates. The results presented in this chapter provide the choice of the most suitable techniques for the identification and confirmation of each individual drug. System S_5 was proposed as an alternative system for local anesthetics (Table 20.3).

Debackere and Laruelle (1968) published a TLC method that could be used to detect 12 alkaloids with a sensitivity of 2–5 μg. The procedure was applied to the isolation and detection of alkaloids from biological materials after their injection and passage through the bodies of horses. Samples of saliva, plasma, and urine were examined at different times after injection. Brucine, caffeine, COC, codeine, HER, lobeline, morphine, nikethamide, papaverine, quinine, strychnine, and sparteine were extracted from aqueous solution. The chromatographic conditions are given in Table 20.3. Lewis (1980) reported a qualitative screening procedure for the detection of BNZ in the urine of greyhounds. Following the administration of COC hydrochloride to greyhounds, no unchanged COC was detected in the urine. BNZ was detected in the urine of greyhounds 1–8 h after COC hydrochloride administration using TLC (Table 20.3).

A study was undertaken to identify the metabolites of KET in the urine of adult horses and to evaluate methods for detecting and confirming KET administration (Sams and Pizzo 1987). The urine samples were extracted using two different procedures, one being basic extraction with dichloromethane–isopropanol mixture and the other one being enzymatic hydrolysis. KET and its two metabolites, norketamine (NKT) and 5,6-dehydronorketamine (DHN) were detected by TLC analysis of both extracts (Table 20.3).

ACKNOWLEDGMENTS

The author is grateful to the Ministry of Science of the Republic of Serbia (Grant No. 172017) for financial support and to Miss Dragana Č. Dabić for technical assistance.

REFERENCES

Al-Badr, A. A. and M. M. Tayel. 1999. Procaine hydrochloride. *Anal. Prof. Drug Sub. Excip.* 26:359–458.
Ali, S. L. and D. Steinbach. 1981. Benzocaine in formulations for mouth and throat therapy. Comparative studies on content uniformity. *Pharmz. Ztg.* 126(8):1549–1552.
Allgire, J. F., E. C. Juenge, C. P. Damo, G. M. Sullivan, and R. D. Kirchhoefer. 1985. High-performance liquid chromatographic determination of d-/l-epinephrine enantiomer ratio in lidocaine-epinephrine local anesthetics. *J. Chromatogr.* 325:249–254.
Andrey, R. E. and A. C. Moffat. 1979. A compilation of analytical data for the identification of lysergide and its analogues in illicit preparations. *J. Forensic Sci. Soc.* 19(4):253–282.
Antonilli, L., C. Suriano, M. C. Grassi, and P. Nencini. 2001. Analysis of cocaethylene, benzoylecgonine, and cocaine in human urine by high-performance thin-layer chromatography with ultraviolet detection: A comparison with HPLC. *J. Chromatogr. B Biomed. Sci. App.* 751(1):19–27.
Babić, S., D. Mutavdžić, D. Ašperger, A. J. M. Horvat, and M. Kaštelan-Macan. 2007. Determination of veterinary pharmaceuticals in production wastewater by HPTLC-videodensitometry. *Chromatographia* 65:105–110.
Bailey, D. N. 1994. Thin-layer chromatographic detection of cocaethylene in human urine. *Am. J. Clin. Pathol.* 101(3):342–345.
Baker, P. B. and T. A. Gough. 1979. The rapid determination of cocaine and other local anesthetics using field tests and chromatography. *J. Forensic Sci.* 24(4):847–855.
Baniceru, M., C. V. Manda, and S. M. Popescu. 2011. Chromatographic analysis of local anesthetics in biological samples. *J. Pharm. Biomed. Anal.* 54:1–12.
Bernardo, N. P., M. E. Pereira Bastos Siqueira, M. J. Nunes de Paiva, and P. P. Maia. 2003. Caffeine and other adulterants in seizures of street cocaine in Brazil. *Int. J. Drug Policy* 14:331–334.
Bertulli, G., L. Mosca, and G. Pedroni. 1978. Rapid method for the identification and quantitative determination of cocaine, heroin and morphine in illicit drugs. *Boll. Chim. Farm.* 117(3):170–175.
Bhawara, H. S. 2008. Determination of benzocaine by HPTLC-densitometry. *Indian J. Criminol. Criminal.* 29(3):247.

Bhushan, R. and C. Agarwal. 2008. Direct TLC resolution of (±)-ketamine and (±)-lisinopril by use of (+)-tartaric acid or (2)-mandelic acid as impregnating reagents or mobile phase additives: Isolation of the enantiomers. *Chromatographia* 68:1045–1051.

Blake, D. A., J. Q. Barry, and H. F. Cascobi. 1972. Qualitative analysis of halothane metabolites in man. *Anesthesiology* 36(2):152–154.

Bogusz, M., J. Gierz, R. A. De Zeeuw, and J. P. Franke. 1985. Influence of the biological matrix on retention behavior in thin-layer chromatography: Evidence of systematic differences between pure and extracted drugs. *J. Chromatogr. B Biomed. Sci. App.* 342(1):241–244.

Brown, J. K., R. H. Schingler, M. C. Chaubal, and M. H. Malone. 1973. A rapid screening procedure for some "street-drugs" by thin-layer chromatography: II. Cocaine, heroin, local anesthetics and mixtures. *J. Chromatogr.* 87:211–214.

Brzezinka, H., P. Dallakian, and H. Budzikiewicz. 1999. Thin-layer chromatography and mass spectrometry for screening of biological samples for drugs and metabolites. *J. Planar Chromatogr.—Modern TLC* 12(2):96–108.

Büchi, J. and J. A. Fresen. 1966. Thin-layer chromatographic investigations of homologous and analogous series of local anesthetics. *Pharm. Acta Helv.* 41(10):551–574.

Budd, R. D., D. F. Mathis, and F. C. Yang. 1980. TLC analysis of urine for benzoylecgonine and norpropoxyphene. *Clin. Toxicol.* 16(1):1–5.

Buesing, G. and H. Grigat. 1988. The azo dye from 4-aminobenzoic acid and thymol. Proof of 4-aminobenzoic acid as a decomposition product of procaine. *Arch. Pharm.* 321(7):433.

Burrows, D. L., A. Nicolaides, G. C. Stephens, and K. E. Ferslew. 2004. The distribution of sevoflurane in a sevoflurane induced death. *J. Forensic Sci.* 49(2):1–4.

Bussey, R. J. and R. C. Backer. 1974. Thin-layer chromatographic differentiation of amphetamine from other primary-amine drugs in urine. *Clin. Chem.* 20(2):302–304.

Cavazzutti, G., L. Gagliardi, and A. Amato et al. 1983. Colour reactions of iodic acid as reagent for identifying drugs by thin-layer chromatography. *J. Chromatogr.* 268:528–534.

Chang, Z. L. and J. B. Martin. 1983. *Anal. Prof. Drug Sub.* 12:191–214.

Christie, J. L. 1976. Fatal consequences of local anesthesia: Report of five cases and a review of the literature. *J. Forensic Sci.* 21(3):671–679.

Clarke, E. G. C. 1978. *Isolation and Identification of Drugs*. The pharmaceutical press, London, U.K., vol. 1, p. 267.

Comer, J. P. and J. Comer. 1967. TLC of cocaine, procaine and other compounds of forensic interest. *J. Pharm. Sci.* 56:413.

Cone, E. J., W. D. Darwin, D. Yousefnejad, and W. F. Buchwald. 1979. Separation and identification of phencyclidine precursors, metabolites and analogs by gas and thin-layer chromatography and chemical ionization mass spectrometry. *J. Chromatogr.* 177:149–153.

Daldrup, T., A. Rickert, and Z. Fresenius. 1989. Drug screening in urine by TLC with special regard to reagents with low toxicity. *Anal. Chem.* 334(4):349–353.

Davidow, B., N. L. Petri, and B. Quame. 1968. A thin-layer chromatographic procedure for detecting drugs of abuse. *Am. J. Clin. Pathol.* 50(6):714–719.

De Spiegeleer, B., W. Van den Bossche, P. De Moerloose, and D. Massart. 1987. A strategy for two-dimensional, high-performance thin-layer chromatography, applied to local anesthetics. *Chromatographia* 23(6):407–411.

De Zeeuw, R., J. Franke, F. Degel, G. Machbert, H. Schütz, and J. Wijs-beek. 1992. *Thin-Layer Chromatographic R_F-Values of Toxicologically Relevant Substances on Standardized Systems*. Report XVII of the DFG commission for clinical-toxicological analysis. Special issue of the TIAFT Bulletin; VCH, Weinheim, New York Basel Cambridge.

De Zeeuw, R. A., D. T. Witte, and J. P. Franke. 1990. Identification power of thin-layer chromatographic colour reactions and integration of colour codes in a database for computerized identification in systematic toxicological analysis. *J. Chromatogr.* 500:661–671.

Deakin, A. L. 2003. A study of acids used for the acidified cobalt thiocyanate test for cocaine base. *Microgram J.* 1:40–43.

Debackere, M. and L. Laruelle. 1968. Isolation, detection, and identification of some alkaloids or alkaloid-like substances in biological specimens from horses, with special reference to doping. *J. Chromatogr.* 35(2):234–247.

Della Casa, E. and G. Martone. 1986. A quantitative densitometric determination of heroin and cocaine samples by high-performance thin-layer chromatography. *Forensic Sci. Int.* 32(2):117–120.

Devarajan, P. V., M. H. Adani, and A. S. Gandhi. 2000. Simultaneous determination of lignocaine hydrochloride and phenylephrine hydrochloride by HPTLC. *J. Pharm. Biomed. Anal.* 22:685–690.

Dongre, V. G. and V. W. Kamble. 2003. HPTLC detection and identification of heroin (diacetylmorphine) in forensic samples. Part III. *J. Planar Chromatogr.* 16:458–460.

Duschi, J. L. and P. L. Hackett. 1985. Simultaneous determination of lidocaine, mexiletine, disopyramide, and quinidine in plasma by high-performance liquid chromatography. *J. Anal. Toxicol.* 9(2):67–70.

Ensing, J. G. and R. A. De Zeeuw. 1991. Detection, isolation, and identification of truxillines in illicit cocaine by means of thin-layer chromatography and mass spectrometry. *J. Forensic Sci.* 36(5):1299–1311.

Essig, S. and K. A. Kovar. 2001. Fluorimetric determination of cinchocaine in a pharmaceutical drug by scanning and video densitometry. *Chromatographia* 53:321–322.

Fike, W. W. 1966. Structure-R_F correlations in the thin layer chromatography of some basic drugs. *Anal. Chem.* 38(12):1697–1702.

Fleck, C., M. Kämena, and L. Tschritter et al. 2001. Local anaesthetic effectivity and toxicity of fomocaine, five N-free fomocaine metabolites and two chiralic fomocaine derivatives in rats compared with procaine. *Arzneimittelforschung* 51(6):451–488.

Funk, W. and S. Droeschel. 1991. Qualitative and quantitative HPTLC determination of cocaine, ecgonine, ecgonine methyl ester and benzoylecgonine. *J. Planar Chromatogr.—Modern TLC* 4:123–126.

Galais, Ph., C. Dauphin, D. Pradeau, and A. Chevallier. 2000. Assay of para aminobenzoic acid formed by hydrolysis of procaine in CP1B solution. *Chromatographia* 52:115–119.

Görög, S., M. Babják, and G. Balogh et al. 1997. Drug impurity profiling strategies. *Talanta* 44(9):1517–1526.

Gough, T. A. and P. B. Baker. 1983. Identification of major drugs of abuse using chromatography: An update. *J. Chromatogr. Sci.* 21:145–153.

Gübitz, G. and R. Wintersteiger. 1978. Rapid determination of local anesthetics by high-performance thin-layer chromatography. *Sci. Pharm.* 46(4):275–280.

Gübitz, G. and R. Wintersteiger. 1980. Identification of drugs of abuse by high-performance thin-layer chromatography. *J. Anal. Toxicol.* 4(3):141–144.

Haddoub, R., D. Ferry, P. Marsal, and O. Siri. 2011. Cobalt thiocyanate reagent revisited for cocaine identification on TLC. *New J. Chem.* 35(7):1351–1354.

Harper, J. D., P. A. Martel, and C. M. O'Donnell. 1989. Evaluation of a multiple-variable thin-layer and reversed-phase thin-layer chromatographic scheme for the identification of basic and neutral drugs in an emergency toxicology setting. *J. Anal. Toxicol.* 13(1):31–36.

Heusler, H. 1985. Quantitative analysis of common anaesthetic agents. *J. Chromatogr.* 340:273–319.

Higuchi, W. I. 1967. Diffusional models useful in biopharmaceutics: Drug release rate processes. *J. Pharm. Sci.* 56(3):315–324.

Hitt B. A., R. I. Mazze, M. J. Cousins, H. N. Edmunds, G. A. Barr, and J. R. Trudell. 1974. Metabolism of isoflurane in Fischer 344 rats and man. *Anesthesiology* 40(1):62–67.

Janicka, M., A. Wasik, and J. Namieśnik. 2010. Analytical procedures for determination of cocaine and its metabolites in biological samples. *TRAC—Trends Anal. Chem.* 29(3):209–224.

Jarzebinski, J. and M. Ciszewska-Jedrasik. 1986. Application of densitometry in determination of active substances in drugs. XII. Determination of substances in compound preparations containing naphazoline nitrate. *Acta Pol. Pharm.* 43(3):264–269.

Jehle, H., I. Mesaros, S. Siggert, B. Wochner, D. Kleiber, and N. Sinner. 2004. Content uniformity test of cinchocaine hydrochloride. *CBS TLC* 92:1–3.

Jukofsky, D., K. Verebey, and S. J. Mule. 1980. Qualitative differentiation between cocaine, lidocaine and cocaine–lidocaine mixtures ("rock-cocaine") using thin-layer chromatography. *J. Chromatogr.* 198:534–535.

Kaistha K. K. and R. Tadrus. 1977. Single- and two-step extraction and thin-layer detection procedures for benzoylecgonine (cocaine metabolite) alone or in combination with a wide variety of commonly abused drugs in urine screening programs. *J. Chromatogr.* 135:385–393.

Kamble, V. W. and V. G. Dongre. 1997. TLC detection and identification of heroin (diacetylmorphine) in street samples: Part II. *J. Planar Chromatogr.* 10:384–386.

Kato, N. and A. Ogamo. 2001. A TLC visualization reagent for dimethylamphetamine and other abused tertiary amines. *Sci. Justice* 41(4):239–244.

Khajeamiri, A. R., F. Kobarfard, R. Ahmadkhaniha, and G. Mostashari. 2011. Profiling of ecstasy tablets seized in Iran. *Iran. J. Pharm. Res.* 10(2):211–220.

Kiszka, M. and R. Madro. 2002. The usefulness of the thin-layer chromatography method in the identification of cocaine and its metabolite benzoylecgonine in autopsy material. *Z Zagadnień Nauk Sądowych* 51:7–28.

Kochana, J., A. Zakrzewska, A. Parczewski, and J. Wilamowski. 2005. TLC screening method for identification of active components of "ecstasy" tablets: Influence of diluents and adulterants. *J. Liq. Chromatogr. Related Technol.* 28(18):2875–2886.

Kovar, K. A., W. Mayer, and H. Auterhoff. 1981. Molecular complexes and radicals of procaine. *Arch. Pharm.* 314(5):447–458.

Kumpulainena, H., T. Järvinena, and A. Mannilaa et al. 2008. Synthesis, in vitro and in vivo characterization of novel ethyl dioxy phosphate prodrug of propofol. *Eur. J. Pharm. Sci.* 34:110–117.

Kurita, H., K. Mizuno, and K. Kuromi et al. 2008. Screening test of drugs in health foods using 2-dimensional TLC. *Yakugaku Zasshi* 128(3):487–493.

Kuwayama, K., K. Tsujikawa, H. Miyaguchi, T. Kanamori, Y. T. Iwata, and H. Inoue. 2012. Rapid, simple, and highly sensitive analysis of drugs in biological samples using thin-layer chromatography coupled with matrix-assisted laser desorption/ionization mass spectrometry. *Anal. Bioanal. Chem.* 402:1257–1267.

Lee, K. Y., D. Nurok, A. Zlatkis, and A. Karmen. 1978. Simultaneous determination of antiarrhythmia drugs by high-performance thin-layer chromatography. *J. Chromatogr. A* 158:403–410.

Lewis, J. H. 1980. Qualitative screening procedure for the detection of benzoylecgonine in the urine of greyhounds. *J. Chromatogr.* 196(2):337–341.

Li, L. and D. M. Lubman. 1989. Resonant two-photon ionization spectroscopic analysis of thin-layer chromatography using pulsed laser desorption/volatilization into supersonic jet expansions. *Anal. Chem.* 61(17):1911–1915.

Lillsunde, P. and T. Korte. 1991. Comprehensive drug screening in urine using solid-phase extraction and combined TLC and GC/MS identification. *J. Anal. Toxicol.* 15(2):71–81.

Lupp, A., J. Wange, H. Oelschläger, and C. Fleck. 2006. Pharmacological and toxicological testing of the enantiomers of two chiral fomocaine alkylmorpholine derivatives in comparison to their in vitro interactions on drug metabolism in rats. *Arzneimittelforschung* 56:369–376.

Maggiorelli, E. 1960. Identification and determination of lidocaine: Its differentiation from acetanilide, procaine, benzocaine, amylocaine. *Boll. Chim. Farm.* 99:8–14.

Makowski, A., E. Adamek, and W. Baran. 2010. Use of photocatalytic reactions to visualize drugs in TLC. *J. Planar Chromatogr.—Modern TLC* 23(1):84–86.

Masoud, A. N. 1976. Systematic identification of drugs of abuse II: TLC. *J. Pharm. Sci.* 65(11):1585–1589.

Matsubara, K., C. Maseda, and Y. Fukui. 1984. Quantitation of cocaine, benzoylecgonine and ecgonine methyl ester by GC-CI-SIM after extrelut extraction. *Forensic Sci. Int.* 26:181–192.

Maulding, H. V. and A. F. Michaelis. 1975. Practical kinetics II: Quantitation of procaine stability by TLC. *J. Pharm. Sci.* 64(2):275–278.

Melent'eva, G. A., A. F. Solodova, and G. M. Rodionova. 1984. Analysis of chloramphenicol and novocaine in a drug mixture. *Farmatsiya* 33:69–71.

Messerschmidt, W. 1971. Qualitative and quantitative thin layer chromatography of p-aminobenzoate esters as 5-(dimethylamino)naphthalene-1-sulfonic acid amides. *Deutsche Apotheker Zeitung* 111:597–599.

Moffat, A. C. 1975. The standardization of thin-layer chromatographic systems for the identification of basic drugs. *J. Chromatogr.* 110:341–347.

Mohammad, M. A. A., N. H. Zawilla, F. M. El-Anwar, and S. M. El-Moghazy Aly. 2007. Column and thin-layer chromatographic methods for the simultaneous determination of acediasulfone in the presence of cinchocaine, and cefuroxime in the presence of its hydrolytic degradation products. *J. AOAC Int.* 90:405–413.

Morlock, G. and D. Charegaonkar. 1998. Determination of hydrocortisone and cinchocaine in lanolin ointments. *GIT Fachz. Lab.* 2:96–98.

Mueller, M. A., S. M. Adams, D. L. Lewand, and R. I. H. Wang. 1977. Detection of benzoylecgonine in human urine. *J. Chromatogr.* 144(1):101–107.

Muhtadi, F. J. and A. Al-Badr. 1986. Cocaine hydrochloride. *Anal. Prof. Drug Sub.* 15:151–231.

Mule, S. J., D. Jukofsky, M. Kogan, A. De Pace, and K. Verebey. 1977. Evaluation of the radioimmunoassay for benzoylecgonine (a cocaine metabolite) in human urine. *Clin. Chem.* 23(5):796–801.

Musumarra, G., G. Scarlata, and G. Cirma et al. 1984. Application of principal components analysis to the evaluation and selection of eluent systems for the thin-layer chromatography of basic and neutral drugs. *J. Chromatogr.* 295:31–47.

Musumarra, G., G. Scarlata, G. Romano, and S. Clementi. 1983. Identification of drugs by principal components analysis of R_F data obtained by TLC in different eluent systems. *J. Anal. Toxicol.* 7(6):286–292.

Mutavdžić, D., S. Babić, D. Asperger, A. J. M. Horvat, and M. Kastelan-Macan. 2006. Comparison of different solid-phase extraction materials for sample preparation in the analysis of veterinary drugs in water samples. *J. Planar Chromatogr.* 19:454–462.

Nagy, A., J. Bartha, and L. Nagy. 1992. Determination of drugs in multidrug pharmaceuticals by thin-layer chromatography. *Gyogyszereszet* 36(5):279–282.

Nemák, K., E. Fogassy, A. Bényei, and I. Hermecz. 2008. The role of TLC in investigation of diastereomeric salt formation by a group of pipecoloxylidides. *J. Planar Chromatogr.* 21:125–128.

Noggle, T. F. and R. C. Clare. 1983. Liquid chromatographic analysis of samples containing cocaine, local anesthetics, and other amines. *J. Assoc. Off. Anal. Chem.* 66(1):151–157.

Oelschlager, H. A. H., D. J. Temple, and C. F. Temple. 1975. The metabolism of the local anaesthetic fomocaine (1-Morpholino-3-[p-phenoxymethylphenyl]propane) in the rat and dog after oral application. *Xenobiotica* 5(5):309–323.

Ojanperä, I., J. Nokua, E. Vuori, P. Sunila, and E. Sippola. 1997. Novel for combined dual-system R_F and UV library search software: Application to forensic drug analysis by TLC and RPTLC. *J. Planar Chromatogr.—Modern TLC* 10:281–285.

Ondra, P., K. Zedníková, and I. Válka. 2006. Detection and determination of abused hallucinogens in biological material. *Neuro Endocrinol. Lett.* 27:125–129.

O'Neal, C. L., D. J. Crouch, and A. A. Fatah. 2000. Validation of twelve chemical spot tests for the detection of drugs of abuse. *Forensic Sci. Int.* 109:189–201.

Padmanabhan, G. R. 1983. Dibucaine and dibucaine hydrochloride. *Anal. Prof. Drug Sub.* 12:105–134.

Panić, D. 1981. Chromatographic separation and photometric determination of procaine and benzocaine in different pharmaceuticals. *Arhiv za Farmaciju* 31:325–331.

Poklis, A., M. A. Mackell, and E. F. Tucker. 1984. Tissue distribution of lidocaine after fatal accidental injection. *J. Forensic Sci.* 29(4):1229–1236.

Radulović, D., Lj. Živanović, S. Antonijević, and L. Glisović. 1990. UV-densitometric determination of mixture of adrenaline hydrochloride, ephedrine hydrochloride and procaine hydrochloride in ointment. *Arhiv za Farmaciju* 40(4):115–120.

Rafla, F. K. and R. L. Epstein. 1979. Identification of cocaine and its metabolites in human urine in the presence of ethyl alcohol. *J. Anal. Toxicol.* 3(2):59–63.

Reichelt, J. 1955. Paper chromatography of local anesthetics. *Cesk Farm.* 4(6):297–301.

Riad, S. M., M. A. A. Mohammad, and A. O. Mohamed. 2008. Stability-indicating methods for the determination of lidocaine and prilocaine in presence of their degradation products. *Bull. Fac. Pharm.* 46(2):35–48.

Röder, E., E. Mutschler, and H. Rochelmeyer. 1969. Use of homogenous azeotropic mixtures in thin-layer chromatography. 4. Separation of some local anesthetics. *Pharm. Acta Helv.* 44:644–646.

Roesener, H. U. 1982. Rapid detection of extenders in heroin powders. *Lebensmittelchem. Gerichtl. Chem.* 36(4):89–90.

Sabino, B. D., W. Romão, and M. L. Sodré et al. 2011. Analysis of cocaine and crack cocaine via thin-layer chromatography coupled to easy ambient sonic-spray ionization mass spectrometry. *Am. J. Anal. Chem.* 2:658–664.

Sabino, B. D., M. L. Sodré, and E. A. Alves et al. 2010. Analysis of street Ecstasy tablets by thin layer chromatography coupled to easy ambient sonic-spray ionization mass spectrometry. *Br. J. Anal. Chem.* 1:6–12.

Salomies, H., P. Lautala, and M. Toppila. 1995. High-performance thin-layer chromatographic method to determine sorption of propofol to infusion containers. *J. Chromatogr. A* 697:597–601.

Salvadori, M. C., M. E. Velletri, M. M. A. Camargo, and A. C. P. Araujo. 1988. Identification of doping agents by chromatographic techniques and UV spectrophotometry. *Analyst* 113(8):1189–1195.

Sams, R. and P. Pizzo. 1987. Detection and identification of ketamine and its metabolites in horse urine. *J. Anal. Toxicol.* 11(2):59–62.

Santos, L. S., R. Haddad, N. F. Hoeehr, R. A. Pilli, and M. N. Eberlin. 2004. Fast screening of low molecular weight compounds by thin-layer chromatography and "on-spot" MALDI-TOF mass spectrometry. *Anal. Chem.* 76(7):2144–2147.

Sarsunova, M. 1963. Thin-layer chromatography without a binding agent in drug analysis. 4. The separation of commonly used local anesthetics. *Pharmazie* 18:748–750.

Sarsunova, M., Z. Perina, and K. Kisonova. 1985. Stability of injection solutions containing local anesthetics. *Farm. Obz.* 54(2):87–97.

Sass, W. C. and S. A. Fusari. 1977. Ketamine. *Anal. Prof. Drug Sub.* 6:298–322.

Schmidt, M. and F. Bracher. 2006. A convenient TLC method for the identification of local anesthetics. *Pharmazie* 61(1):15–17.

Shahjahan, M., B. K. Dutta, and A. Rashid. 1972. Separation and determination of antibiotic mixtures by TLC. *Bangladesh Pharm. J.* 1(1):15–17.

Siddiqui, R. A., M. Zerouga, M. Wu, A. Castillo, K. Harvey, G. P. Zaloga, and W. Stillwell. 2005. Anticancer properties of propofol-docosahexaenoate and propofol-eicosapentaenoate on breast cancer cells. *Breast Cancer Res.* 7(5):645–654.

Stead, A. H., R. Gill, T. Wright, J. P. Gibbs, and A. C. Moffat. 1982. Standardised thin-layer chromatographic systems for the identification of drugs and poisons. *Analyst* 107:1106–1168.

Stewart, J. T., T. S. Reeves, and I. L. Honigberg. 1984. A comparison of solid-phase extraction techniques for assay of drugs in aqueous and human plasma samples. *Anal. Lett.* 11:1811–1826.

Stier, A. 1964. Trifluoroacetic acid as a metabolite of halothane. *Biochem. Pharmacol.* 13(11):1544.

Stoliarov, E. E., I. N. Karpenko, and T. L. Malkova. 2009. Detection of certain local anesthetics in biological fluids by chemo-toxicological analysis. *Sud. Med. Ekspert.* 52(3):24–27.

Stroka, J., B. Spangenberg, and E. Anklam. 2002. New approaches in TLC-densitometry. *J. Liq. Chromatogr. Related Technol.* 25:1497–1513.

Sundholom, E. G. 1983. More economical use of high-performance thin-layer plates for chromatographic screening of illicit drug samples *J. Chromatogr.* 265:285–291.

Sunshine, I. and W. W. Fike. 1964. Value of thin-layer chromatography in two fatal cases of intoxication due to lidocaine and mepivacaine. *New Engl. J. Med.* 271:487–490.

Taha, A. M. and M. A. Abd El-Kader. 1979. Selective detection of tertiary N-ethyl drugs on thin-layer chromatograms. *J. Chromatogr.* 177(2):405–408.

Tanaka, T., S. Aramaki, and A. Momose. 1989. Thin-layer chromatographic screening procedure for some drugs in horse plasma. *J. Chromatogr.* 496(2):407–415.

Tatscuzawa, M., S. Hashiba, and A. Okawara. 1968. Determination of dibucaine and β-(diethylamino)ethyl p-(butylamino)benzoate (T-caine) in pharmaceutical preparations. *Bunseki Kagaku* 17(9):1116–1118.

Wagner, G. and V. Zimmer. 1955. Paper chromatography and paper iontophoresis of various local anesthetics of the procaine series and their decomposition products. *Pharm. Acta Helv.* 30:385–407.

Wallace, J. E., H. E. Hamilton, H. Schwertner, D. E. King, J. L. McNay, and K. Blum. 1975. Thin-layer chromatographic analysis of cocaine and benzoylecgonine in urine. *J. Chromatogr.* 114(2):433–441.

Wan, S. H., B. V. Lehmann, and S. Riegelman. 1972. Renal contribution to overall metabolism of drugs III: Metabolism of p-aminobenzoic acid. *J. Pharm. Sci.* 61(8):1288–1292.

Waraszkiewicz, E., A. Milan, and R. DiRubio. 1981. Stability-indicating high-performance liquid chromatographic analysis of lidocaine hydrochloride and lidocaine hydrochloride with epinephrine injectable solutions. *J. Pharm. Sci.* 70(11):1215–1218.

Witek, A. and L. Przyborowski. 1996. Chromatographic (TLC) analysis of diltiazem in pharmaceutical form in presence of other antiarrhythmics. *Acta Pol. Pharm.* 53(1):9–12.

Wollmann, H. and M. Patrunky. 1982. Stability testing of some drugs containing ester groups: Bisacodyl, benzocaine and procaine hydrochloride. Part 12: Stability of drugs and preparations containing the drugs. *Zbl. Pharm., Pharmakother. Lab.* 121(9):857–866.

Wössner, A. and K. A. Kovar. 1996. Influence and determination of doping drugs. *Deutsche Apotheker Zeitung* 136:17–26.

Xu, P., L. Zheng, and Ch. Hai. 1991. Quantitative determination of mixtures of three local anesthetic drugs by thin-layer chromatography. *Chin. J. Pharm. Ind.* 22:218–219.

Yonamine, M. and M. C. Sampario. 2006. A high-performance thin-layer chromatographic technique to screen cocaine in urine samples. *Leg. Med.* 8(3):184–187.

Zawisza, P. and L. Przyborowski. 1992. Propanidid and etomidate identification in blood by thin-layer chromatography. *Acta Pol. Pharm.* 49:15–17.

Zheng, H. 1982. TLC test of lidocaine hydrochloride preparations. *Chin. J. Pharm. Anal.* 2(2):114–115.

Zhu, Z. and G. Hu. 1992. HPTLC scanner for determination of PABA in procaine hydrochloride injection. *Chin. Pharm. J.* 27(4):221–222.

Živanović, Lj., S. Agatonović-Kustrin, M. Vasiljević, and I. Nemcova. 1996. Comparison of high-performance and thin-layer chromatographic methods for the assay of lidocaine. *J. Pharm. Biomed. Anal.* 14:1229–1232.

Živanović, Lj., D. Živanov-Stakić, and D. Radulović. 1988. Determination of lidocaine in pharmaceutical preparations using thin-layer chromatographic densitometry. *J. Pharm. Biomed. Anal.* 6:809–812.

21 TLC of Psychostimulants

Joanna Drozd

CONTENTS

21.1 PURINES

Among the drugs that stimulate the central nervous system are the analeptic, psychotonic (stimulant and psychostimulant), and nootropic (psychoenergizing) substances. This division is conventional, because many of them have multiactivity. Analeptics mainly stimulate respiratory and vasomotor centers, but some of them also affect the intellectual functions of the brain.

Caffeine and derivatives as central nervous system stimulants belong to a loosely defined group of drugs that tend to increase behavioral alertness, agitation, or excitation. They work by a variety of mechanisms, but usually not by direct excitation of neurons. They work as phosphodiesterase inhibitors, which inhibit or antagonize the biosynthesis or actions of phosphodiesterases and compounds that bind to and block the stimulation of purinergic P1 receptors. Theophylline and theobromine mainly are bronchodilator agents that cause an increase in the expansion of a bronchus or bronchial tube. Aminophylline has a strengthening effect on the heart or can increase cardiac output. It may be cardiac glycosides or sympathomimetics; they are used after myocardial infarction, cardiac surgical procedures, in shock, or in congestive heart failure (heart failure). Pentoxyphylline is a platelet aggregation inhibitor that antagonizes or impairs any mechanism leading to blood platelet aggregation, whether during the phases of activation and shape change or following the dense-granule release reaction and stimulation of the prostaglandin–thromboxane system. As a radiation-protective agent, it is used to protect against ionizing radiation. It is usually used in radiation therapy but has been considered for other, for example, military, purposes and in vasodilator agents used to cause dilation of the blood vessels. Doxophylline works as an antitussive agent that suppresses cough. It acts centrally on the medullary cough center, is also used in the treatment of cough, and acts locally. As a bronchodilator agent, it causes an increase in the expansion of a bronchus or bronchial tube.

Purines are rapidly distributed into body tissues, readily crossing the placenta and blood–brain barrier.

Allopurinol is the drug that is chemically similar to naturally occurring metabolites, but differs enough to interfere with normal metabolic pathways. It is the enzyme inhibitors—compounds or agents that combine with an enzyme in such a manner as to prevent the normal substrate–enzyme combination and the catalytic reaction. Free radical scavengers are substances that influence the course of a chemical reaction by ready combination with free radicals. Among other effects, this combining activity protects pancreatic islets against damage by cytokines and prevents myocardial and pulmonary perfusion injuries. Gout suppressants that increase uric acid excretion by the kidney (uricosuric agents) decrease uric acid production (antihyperuricemics) or alleviate the pain and inflammation of acute attacks of gout.

Allopurinol and oxypurinol, despite the chemical structure of both compounds, do not belong to a group of psychostimulants.

Chemical structure of purines is shown in Table 21.1.

TABLE 21.1

Chemical Structure of Purines

1,3,7-Trimethyl-3,7-dihydro-1H-purine-2,6-dione
Caffeine

1,3-Dimethyl-3,7-dihydro-1H-purine-2,6-dione
Theophylline

3,7-Dimethyl-3,7-dihydro-1H-purine-2,6-dione
Theobromine

1,3-Dimethyl-3,7-dihydro-1H-purine-2,6-dione-ethane-
 1,2-diamine (2:1)
Aminophylline

7-(2,3-Dihydroxypropyl)-1,3-dimethyl-3,7-dihydro-
 1H-purine-2,6-dione
Diprophylline

3,7-Dimethyl-1-(5-oxohexyl)-3,7-dihydro-1H-purine-2,6-
 dione
Pentoxyphylline

7-(2-Hydroxypropyl)-1,3-dimethyl-3,7-dihydro-1H-
 purine-2,6-dione
Proxyphylline

7-(2-Hydroxyethyl)-1,3-dimethyl-3,7-dihydro-1H-purine-
 2,6-dione
Etophylline

(1,3-Dimethyl-2,6-dioxo-1,2,3,6-tetrahydro-7H-
 purin-7-yl)acetic acid
Acephylline

7-(1,3-Dioxolan-2-ylmethyl)-1,3-dimethyl-3,7-dihydro-1H-
 purine-2,6-dione
Doxophylline

TABLE 21.1 (continued)
Chemical Structure of Purines

7-[2-(Diethylamino)
ethyl]-1,3-dimethyl-3,7-dihydro-1H-purine-
2,6-dione
Etamiphylline

7-{2-[bis(2-Hydroxyethyl)amino]
ethyl}-1,3-dimethyl-3,7-dihydro-1H-purine-2,6-dione

Vephylline

7-[2-Hydroxy-3-[4-(3-phenylsulfanylpropyl)
piperazin-1-yl]propyl]-1,
3-dimethylpurine-2,6-dione
Taziphylline

8-Benzyl-7-[2-[ethyl(2-hydroxyethyl)amino]ethyl]-1,3-
dimethylpurine-2, 6-dione
Bamiphylline

1,2-Dihydropyrazolo[3,4-d]pyrimidin-4-one
Allopurinol

1,2-Dihydropyrazolo[3,4-d]pyrimidine-4,6-dione
Oxypurinol

   The first studies on the use of thin-layer chromatography in the analysis of purines were published in the 1960s. Some possibilities for use of thin-layer chromatography without binding agents in drug analysis and qualitative detection of purines and barbiturates was described in 1963 (Sarsunova and Schwarz 1963). Rink and Gehl used thin-layer chromatographic separation of some purine derivatives, including uric acid (Rink and Gehl 1966). A thin-layer chromatographic method was used for separation of caffeine, theobromin, and theophylline in 1965 (Szász et al. 1965). TLC as a method for detection of caffeine, theophylline, theobromine, and some of their derivatives was described in 1966 (Paulus et al. 1966). A sensitive method for the quantitative detection of the purine bases caffeine, theobromine, and theophylline was described in 1968. The method was based on thin-layer chromatography and had the added advantage that it did not require any expensive apparatus (Senanayake and Wijesekera 1968).

Investigations on factors influencing behavior in thin-layer chromatography were described by Kraus and Dumont (1970). The influence of the pH value in thin-layer chromatography of caffeine, theophylline, and theobromine on uniform layers was described. The paper starts with a review of work published on the thin-layer chromatography of caffeine, theophylline, and theobromine with regard to the different chromatographic conditions. The separation results obtained by the different methods are briefly discussed. The second part reports systematic investigations on the influence of the pH value of the stationary phase after climatization with solutions of ammonia or acids and investigations on the influence of the pH value of the mobile phase (Kraus and Dumont 1970). The relationship between the chromatographic behavior of caffeine, theophylline, and theobromine and the pH value of the silica gel layer was investigated by the pH gradient technique. For the development of the chromatograms, use was made of several neutral organic mobile phases described in earlier publications by other authors (Dumont and Kraus 1970). The determination of accompanying active compounds such as adenosine, caffeine, strychnine, and theophylline in pharmaceutical dosage forms containing nikethamide by thin-layer chromatography was described in 1971. Individual qualitative examination and quantitative assay of the eluted drugs were performed by ultraviolet spectrophotometry (Carmichael 1971). The internal standard method for quantitative thin-layer chromatography was described and used for determination of caffeine with acetophenetidin as the internal standard. Since there is no necessity to calculate a calibration factor, a large number of analyses can be done on the same TLC plate. The reproducibility of the method is about 1.4% (approximate relative standard deviation) (Ebel and Herold 1975a). The use of silver halide-impregnated supports for high-pressure liquid chromatographic (HPLC) and thin-layer chromatographic (TLC) analysis of drug substances has been studied. Successful separations of xanthines and mixtures containing barbiturates, xanthines, ergot alkaloids, and tropane alkaloids have been achieved with isocratic conditions or by using simple gradients. The Ag supports show a similar behavior as many chemically bonded stationary phases, such as modification of adsorption sites on silica gel to give lower retention but better specificity, and rapid reconditioning in connection with gradient elution (3–5 min). Reasonable cost, simplicity of preparation and usage, and good chemical and physical stability render the silver halide phases applicable in routine analysis and quite competitive with many commercially available chemically bonded stationary phases (Aigner et al. 1976). Pavlik has described identification of caffeine using the TLC method (Pavlik 1973). Separation of amino-phenazone, phenprobamate, caffeine, and ascorbic acid using thin-layer chromatography was described by Kubiak et al. (1974). Identification caffeine and other drugs using TLC method was provided by Allen (1974).

TLC methods for the determination of phenacetin or paracetamol, acetylsalicylic acid, and caffeine in pharmaceutical analysis using calibration curves and the method with an internal standard were compared. For routine analysis as many determinations as possible should be done on the same TLC plate. The accuracy of the two methods as relative standard deviations was 1.5%–3% (internal standard method, six analyses per TLC-plate) or 1.8%–3% (calibration curves, concentration levels within lower and higher tolerance, four analyses per plate) (Ebel and Herold 1975b). A TLC method for the determination of phenacetin and caffeine in pharmaceutical preparations using a densitometer in line with an integrator and electronic desk calculator was described by the same author. The calibration curves of each compound were prepared where the higher and lower tolerance concentration levels were defined. The accuracy of the method is 1.51% (phenacetin) and 2.56% (caffeine) for two experiments carried out on each plate; 1.50% (phenacetin) and 2.64% (caffeine) for four experiments for each plate (relative standard deviation, n = 18 and n = 36, respectively) (Ebel and Herold 1975c). The methylxanthines can be determined in foods and biological systems by the chromatographic methods of TLC, GC, and HPLC. Ultraviolet spectroscopy following a separation procedure was also used (Hurst et al. 1984). Aspirin, phenacetin, and caffeine in two commercial analgesic tablets were determined by scanning of fluorescence-quenched zones after separation on high-performance TLC plates. Assays for the components ranged from 100% to 88% of label values, and agreement between duplicate samples was better than 5%. Use of phenacetin as an internal standard for determination of aspirin and caffeine was not advantageous. The analysis serves well as an introduction to quantitative TLC in undergraduate laboratory courses (Sherma et al. 1985).

Cesko-Slovenska Farmacie described TLC (HPTLC)—spectrodensitometry in pharmaceutical analysis for spectrodensitometric simultaneous determination of paracetamol, caffeine, guaiacol-glycerinether (Tomankova and Vasatova 1988a), and acetylsalicylic acid in analgesic-antipyretic pharmaceuticals and simultaneous determination of propyphenazone, paracetamol, and caffeine in the preparation in progress valetol tablets (Tomankova and Vasatova 1988b).

A traditional Chinese medicine, which was supplied as pills by a temple and used as Ch'i tonic, was obtained for testing illegal adulteration with chemical drugs. In the qualitative analysis, two-dimensional TLC, with chloroform:ethanol (9:1, v/v) and ethyl acetate:methanol:ammonia water (8:1:1, v/v) as the solvent systems, established the presence of acetaminophen, piroxicam, hydrochlorothiazide, caffeine, ethoxybenzamide, chlorzoxazone, and nicotinamide in the pills. In quantitative analysis, a reversed-phase HPLC method has been established for simultaneous determination of the seven chemical drugs including caffeine (Tseng et al. 1996). A traditional Chinese medicine, which was brought by a consumer for treating heart disease and uricosuria, was obtained and tested for illegal adulteration with chemical drugs. In the qualitative analysis, two-dimensional TLC, with chloroform:ethanol (9:1, v/v) and ethyl acetate:methanol:ammonia water (8:1:1, v/v) as the solvent systems, established the presence of caffeine, indomethacin, ethoxy-benzamide, chlorzoxazone, and diazepam in the pills. For quantitative analysis, a reversed-phase HPLC method was used for simultaneous determination of the five chemical drugs including caffeine (Tseng et al. 1997).

By the use of a combination of silica gel 60 and a weak infrared-active, reflection-enhancing material, the quality of DRIFT (diffuse reflectance infrared Fourier transform spectroscopy) spectra was improved and the signal-to-noise ratio of adsorbed substances increased. Of the reflection-enhancers tested, magnesium tungstate proved the most suitable; the optimum proportion was 50%. For this type of layer, the signal-to-noise ratios for caffeine, paracetamol, and phenazone were increased by factors of 3.4, 3.1, and 2.3, respectively, compared with those obtained on pure silica gel 60. Use of material of small mean particle-diameter and narrow particle-size distribution provided an almost optimum diffuse reflecting surface. Addition of strongly reflecting metallic powders to the silica gel had the effect of increasing diffuse Fresnel reflection in particular; this did not, however, contain the desired spectral information. The same effect was also achieved with reflecting sorbent supports. In contrast with aluminum-coated glass plates, normal glass plates absorb most of the IR radiation, a small proportion only being reflected. When aluminum-coated glass plates were used as supports, small fluctuations in layer thickness resulted in artifact peaks in the chromatograms. Layer thickness sensitivity was less for normal glass plates. A layer thickness of 100 μm for a 50:50 (m/m) mixture of silica gel 60 and magnesium tungstate on a normal glass plate proved to be the most suitable combination. The chromatographic, analytical, and spectroscopic performance of the optimized sorbent layer was investigated using separation numbers, detection limits, and the quality of the DRIFT spectra of selected substances. Addition of 50% magnesium tungstate resulted in sorbent layers with separation performance practically identical with that of silica gel 60 F(254s), even though the retention properties were different. By use of this optimized sorbent layer, detection limits were reduced by a factor of 3.7 for caffeine and 2.3 for both paracetamol and phenazone.

In addition, the evaluable IR range was extended by approximately 100 wavenumbers to approximately 1270 cm^{-1} in the "fingerprint" range. This enabled the detection of two additional caffeine bands, which, if a conventional silica gel layer had been used, would have been superimposed by the wide interference bands occurring at approximately 1340 cm^{-1} (Bauer et al. 1998).

Simultaneous determination of naproxen with diflunisal (mixture I), paracetamol with chlorzoxazone (mixture II), and chlorphenoxamine hydrochloride with 8-chlorotheophylline and caffeine (mixture III) in multicomponent mixtures was conducted by a thin-layer chromatography densitometric method. The mobile-phase ethyl acetate:methanol:ammonia 25% (85:15:5, v/v) was used for the separation of the components of mixtures (I) and (II). Efficient separation of the components of mixture (III) was attained using ethyl acetate as mobile phase with R_f values of 0.12, 0.62, and 0.42 for chlorphenoxamine hydrochloride, 8-chlorotheophylline, and caffeine, respectively. Linearity ranges, mean recoveries, and relative standard deviations in calibration graphs of the proposed method were calculated. The method has been successfully applied to pharmaceutical formulations, sugar-coated tablets, capsules, and suppositories. The results obtained were statistically compared with those obtained by applying the reported alternate methods (Bebawy and El-Kousy 1999).

In the therapy of pain of weaker genesis, frequently used drugs usually represent a mix of analgoantipyretics of different chemical structures, mostly derivatives of salicylic acid, pyrazolone, and p-aminophenol as well as derivatives of propionic and acetylsalicylic acid. For the determination of these drugs, different chromatographic methods have been applied, mostly HPLC, due to the lower polarity (pyrazolone derivatives) and thermolability, as well as nonvolatility of compounds investigated. The TLC method, considering advantages that include simplicity, reasonable sensitivity, rapidity, excellent resolving power, and low cost, has been successfully explored for the determination of analgoantipyretic compounds. The aim of this work was to develop a simple and rapid HPTLC method for the determination of acetylsalicylic acid, paracetamol, caffeine, and phenobarbitone in dosage form. The determination of analgoantipyretics was performed on precoated HPTLC silica gel plates (10 × 20 cm^2) by development in the mobile-phase dichlormethane–ethyl acetate–cyclohexane–isopropanol–0.1 M HCl–formic acid (9:8:3:1.5:0.2:0.2, v/v/v/v/v/v). Migration distances (68.6 + 0.2 mm, 54.1 + 0.1 mm, 36.4 + 0.14 mm, and 85.9 + 0.11 mm for acetylsalicylic acid, paracetamol, caffeine, and phenobarbitone, respectively) with low RSD values (0.13%–0.39%) showed a satisfactory reproductivity of the chromatographic system. A TLC scanner was used for direct evaluation of the chromatograms in the reflectance/absorbance mode. Established calibration curves (r > 0.999), precision (0.3%–1.02%) and detection limits, as well as recovery values (96.51%–98.1%) were validated and found to be satisfactory. The method was found to be reproducible and convenient for the quantitative analysis of compounds investigated in their dosage forms (Franeta et al. 2001).

Application of citric acid/acetic anhydride reagent (CAR), a color reagent selective for tertiary amines in solution, improved the detection of abused tertiary amino drugs on the TLC plate. The plate was pretreated by a brief immersion in phosphoric acid/acetone solution to suppress coloration. After suppressing, the plate was sprayed with CAR and heated at 100°C, causing tertiary amines to turn red purple within 3 min. The sensitivity of this new CAR method was 2.5–15 times greater than that of conventional detection with Dragendorff reagent for some of the tertiary amines dimethylamphetamine, methylephedrine, levomepromazine, chlorpromazine, caffeine, theophylline, theobromine, and nicotine. This present method provided rapid TLC detection of abused tertiary amino drugs such as phenethylamine, phenothiazine, xanthine derivative, nicotine, and narcotics (Kato and Ogamo 2001).

A quantitative method using silica gel HPTLC plates with a fluorescent indicator, automated sample application, and UV absorption densitometry has been developed for the determination of caffeine in pharmaceutical preparations designed to promote alertness. Tablet, coated tablet, and coated caplet products containing caffeine as the active ingredient were analyzed to test the applicability of the new method. Precision was evaluated by replicate analysis of samples and accuracy by analysis of spiked blank samples containing inactive ingredients in common with the caffeine products. The amount of caffeine in the preparations analyzed ranged from 101% to 121% of label values. Precision ranged from 1.8% to 2.3% relative standard deviation, and the errors from three spiked blank analyses averaged 0.983% compared with fortification levels. The limit of detection was 200 ng caffeine (Ruddy and Sherma 2002).

A quantitative method using silica gel HPTLC plates with fluorescent indicator, automated sample application, and UV absorption densitometry has been developed for the determination of caffeine and acetaminophen in pharmaceutical preparations. Multicomponent analgesic tablets containing caffeine, acetaminophen, and acetylsalicylic acid as the active ingredients were analyzed to test the applicability of the new method. Precision for the caffeine analysis was evaluated by replicate analysis of samples and accuracy by analysis of two spiked blank samples containing inactive ingredients in common with the multicomponent analgesic tablets. The amount of caffeine in the tablets analyzed ranged from 96% to 115% of the label value. Precision was 1.19% relative standard deviation, and the errors from the two spiked blank analyses averaged 1.90% compared with fortification levels. The limit of detection was 0.200 µg of caffeine (Sullivan and Sherma 2003).

Sigh et al. described a study undertaken to develop thin layers of silica gel G impregnated with transition metal ions for the separation, identification, and estimation of purines. The influence of transition metal ions and eluting solvents on chromatographic behavior has been studied. The method was applied for qualitative analysis of purines in the mixtures and quantitative analysis of purine bases in the mixture as well as in pharmaceutical formulations. The results were compared statistically with those obtained by official methods. The method is simple, reproducible, and accurate within 1.3% ± 0.6% (Singh et al. 2003).

Desorption electrospray ionization (DESI) was demonstrated as a means to couple thin-layer chromatography (TLC) with mass spectrometry. The experimental setup and its optimization are described. Development lanes were scanned by moving the TLC plate under computer control while directing the stationary DESI emitter charged droplet plume at the TLC plate surface. Mass spectral data were recorded in either selected reaction monitoring mode or in full scan ion trap mode using a hybrid triple quadrupole linear ion trap mass spectrometer. Fundamentals and practical applications of the technique were demonstrated in positive ion mode using selected reaction monitoring detection of rhodamine dyes separated on hydrophobic reversed-phase C8 plates and reversed-phase C2 plates, in negative ion full scan mode using a selection of FD&C dyes separated on a wettable reversed-phase C18 plate, and in positive ion full scan mode using a mixture of aspirin, acetaminophen, and caffeine from an over-the-counter pain medication separated on a normal-phase silica gel plate (Van Berkel et al. 2005).

The ChromImage flatbed scanner densitometer with Galaxie-TLC software has been used for quantification of silica gel high-performance thin-layer chromatography (HPTLC). The visible mode was evaluated by determination of the recovery of a rhodamine B standard dye from a four-dye mixture and by determination of the precision of replicate analysis. Determinations of caffeine in multicomponent analgesic tablets and a cola beverage were performed in the fluorescence-quenching mode. Previously published methods for the tablet and beverage analyses were modified by using HPTLC plates with a brilliant ultraviolet indicator, and analysis of the beverage was further modified by automated application of standard and sample bands by use of a Linomat. Accuracy, precision, linearity, and sensitivity of these analyses with the ChromImage are reported (Halkina and Sherma 2006).

A new high-performance thin-layer chromatography/electrospray ionization mass spectrometry (HPTLC/ESI-MS) method for the quantification of caffeine in pharmaceutical and energy drink samples was developed using stable isotope dilution analysis. After sample preparation, the samples and caffeine standard were applied on silica gel 60 F_{254} HPTLC plates and over-spotted with caffeine-d^3 used for correction of the plunger positioning. After chromatography, densitometric detection was performed by UV absorption at 274 nm. The bands were then eluted by means of a plunger-based extractor into the ESI interface of a single-quadrupole mass spectrometer. For quantification by MS, the $[M^+H]^+$ ions of caffeine and caffeine-d^3 were recorded in the positive ion single ion monitoring (SIM) mode at 195 and 198 m/z, respectively. The calibration showed a linear regression with a determination coefficient (R^2) of 0.9998. The repeatability (RSD, n = 6) in matrix was $\leq \pm 3.75\%$. The intermediate precisions (RSD, n = 2) for two samples of different brand names were determined three times and ranged between RSD $\pm 0.68\%$ and $\pm 2.64\%$ (sample 1) and between $\pm 3.44\%$ and $\pm 8.60\%$ (sample 2). The method's accuracy was evaluated by comparing the results obtained by HPTLC/SIDA-ESI-MS with those from the validated HPTLC/UV method. The results for pharmaceutical and energy drink samples were (ng per band) 99.82 $\pm$ 3.75 and 338.09 $\pm$ 4.87 by HPTLC/SIDA-ESI-MS and 104.74 $\pm$ 1.51 and 334.86 $\pm$ 5.63 by HPTLC/UV. According to the F-test (homogeneity of variances) and the t-test (comparison of means), the two methods show no significant difference. The detection and quantification limits were 75 and 250 μg L^{-1} (0.75 and 2.5 ng per band), respectively, which were a factor of 13 lower than those established for HPTLC/UV. The positioning error (RSD $\pm 6\%$) was calculated by comparing HPTLC/SIDA-ESI-MS with HPTLC/ESI-MS. However, using SIDA the positioning error was nullified. HPTLC/SIDA-ESI-MS was demonstrated to be a highly reliable method for the quantification of compounds by planar chromatography coupled online with mass spectrometry (Aranda and Morlock 2007a).

A new high-throughput method was developed to quantify caffeine, ergotamine, and metamizol in a solid pharmaceutical formulation. After dissolution, the compounds were separated on silica gel 60 F_{254} high-performance thin-layer chromatography (HPTLC) plates with ethyl acetate–methanol–ammonia (90:15:1, v/v/v) as the mobile phase. Detection was performed by UV absorption at 274 nm for caffeine and metamizol, and by fluorescence at 313/>340 nm for ergotamine. Calibrations were linear or polynomial with determination coefficients (R^2) $\geq$ 0.9986. Recoveries of the three compounds were between 95% and 102% at three different concentration levels. Repeatability (relative standard deviation [RSD]) of all substances in the matrix was between $\pm 0.9\%$ and $\pm 1.7\%$. Intermediate precision (RSD) of the three compounds range from $\pm 2.0\%$ to $\pm 3.1\%$. Mass confirmation was performed by a single quadrupole mass spectrometry in positive electrospray ionization full scan mode for caffeine and ergotamine and in negative mode for metamizol. The results proved that this method was a simple and reliable alternative to routine analysis (Aranda and Morlock 2007b).

Criminalistic investigation often has as a task the identification of the compounds from confiscated materials (powders, tablets, etc.). Sometimes the speed and flexibility of the method chosen for the purpose are crucial. A simple, yet robust, analytical method for the identification of organic compounds from mixtures is thin-layer chromatography (TLC). This paper presents a TLC method for the separation and identification of caffeine, codeine, and phenobarbital from mixtures, using silicagel F_{254} precoated glass plates as a stationary phase. Best results for the separation were obtained using the solvent mixture ethyl acetate:methanol:ammonia (85:10:5). Two components out of three unknown confiscated drug samples could be identified in these conditions (Gheorghe et al. 2008).

A reversed-phase high-performance thin-layer chromatographic method combined with image analysis was developed and validated for simultaneous quantitative evaluation of paracetamol and caffeine in pharmaceutical preparations. RP-HPTLC-W18 chromatographic plates were used as the stationary phase and methanol:glacial acetic acid:water (25:4.3:70.7; v/v/v) as the mobile phase. The detection of the spots and the image documentation were carried out under 254 nm UV radiation. Quantitative evaluation of the studied compounds was performed by Sorbfil TLC software. The proposed analytical method was characterized by good linearity, sustained by the correlation coefficient of 0.9974 for paracetamol and 0.9982 for caffeine. The limit of detection and limit of quantitation were found to be 0.100 and 0.191 μg per spot, for paracetamol, as well as 0.040 and 0.076 μg per spot, for caffeine. The results of the recovery studies were in the range of 99.56%–106.84%, and the repeatability of the method was shown to be excellent (RSD < 1.88%) (Soponar et al. 2009).

The direct analysis of pharmaceutical formulations and active ingredients from nonbonded reversed-phase thin-layer chromatography (RP-TLC) plates by desorption electrospray ionization (DESI) combined with ion mobility mass spectrometry (IM-MS) was reported. The analysis of formulations containing analgesic (paracetamol), decongestant (ephedrine), opiate (codeine), and stimulant (caffeine) active pharmaceutical ingredients was described, with and without chromatographic development to separate the active ingredients from the excipient formulation. Selectivity was enhanced by combining ion mobility and mass spectrometry to characterize the desorbed gas-phase analyte ions on the basis of mass-to-charge ratio (m/z) and gas-phase ion mobility (drift time). The solvent composition of the DESI spray using a step gradient was varied to optimize the desorption of active pharmaceutical ingredients from the RP-TLC plates. The combined RP-TLC/ DESI-IM-MS approach had potential as a rapid and selective technique for pharmaceutical analysis by orthogonal gas-phase electrophoretic and mass-to-charge separation (Harry et al. 2009).

To establish TLC methods for the fast detection of lipid-regulating agents, antihypertensive agents, and antitussive and antiasthmatic agents in herbal remedies simultaneously, the sample was extracted with methanol and separated by TLC plates precoated with silica gel GF_{254} as stationary phase. Lipid regulating agents were eluted with the mobile phase consisting of n-hexane–ethyl acetate–diethyl ether–glacial acetic acid (10:8:2:1), determined at UV 254 nm and stained with 5% vanillin sulfuric acid solution. Antihypertensive agents were eluted with the mobile phase consisting of ethyl acetate–methanol–diethylamine (40:0.5:3) and examined with UV 254 nm and UV 365 nm. Antiasthmatic agents were separated with the mobile phase consisting of isopropyl alcohol–ethyl acetate–ammonia (15:25:1.5) and examined at UV 254 nm. Theophylline, diprophylline, terbutaline, oxyphylline, caffeine, doxofylline, and many other drugs were detected, respectively. The methods are fast, sensitive, and easy to operate and are fit to be applied for routine test of illegal adulteration (Wang et al. 2010). The use of pressurized planar electrochromatography (PPEC) and planar chromatography (TLC) for reversed-phase separation of a mixture of acetylsalicylic acid, caffeine, and acetaminophen was investigated. The mixture was separated on C18 plates; the mobile phase was prepared from acetonitrile (ACN), buffer, and bidistilled water. The effects of operating conditions such as mobile-phase composition, type of the stationary phase, and mobile-phase buffer pH on migration distance, separation selectivity, and separation time in TLC and PPEC were compared. The results showed that pressurized planar electrochromatography of these drugs is characterized by faster separation, better performance, and different separation selectivity. In conclusion, PPEC is a very promising mode for future application in pharmaceutical analysis (Hałka et al. 2010). A simple, rapid, and efficient method using TLC with a fluorescence plate reader has been described for simultaneous determination of caffeine and paracetamol. Determination was carried out using the fluorescence-quenching action of caffeine and paracetamol on a TLC plate with a fluorescent indicator at $\lambda = 254$ nm in the linear ranges of 0.2–1.9 and 0.03–1.5 $\mu g\ L^{-1}$, respectively. Separation of caffeine and paracetamol was performed on the TLC plate, and the best results were obtained using the optimized mobile-phase n-hexane–ethyl acetate–ethanol (2.5:1.5:0.4, v/v). Some important parameters, such as solvent type and ratio of the mobile phase, the presence of other components, and instrumental parameters, were studied. Caffeine and paracetamol detection limits were 0.025 and 0.032 $\mu g\ L^{-1}$, and RSD values for 0.6 $\mu g\ L^{-1}$ caffeine and 0.06 $\mu g\ L^{-1}$ paracetamol (n = 5) were 1.93% and 2.06%, respectively. Using this technique, some pharmaceuticals containing caffeine and paracetamol were analyzed with satisfactory results (Tavallali et al. 2011).

The thin-layer chromatography method was successfully applied to the determination of xanthines in biological materials. The methodology for the unequivocal identification of caffeine and 13 possible metabolites (mono, di-, and tri-N-methylated xanthine and uric acid derivatives) based on TLC, UV, and mass spectrometry has been developed. Upon administration of caffeine-3H to the rat, 64%–67% of the radioactivity was recovered in the urine over a period of 24 h. The chloroform–methanol (9:1) extract of the urine accounted for about 37% of the administered radioactivity. Water-soluble metabolites constitute approximately 30% of the injected caffeine-3H. With the aid of preparative TLC, 8.8% of unchanged caffeine and the following metabolites were isolated from a chloroform–methanol extract of urine: theophylline (1.2%), theobromine (5.1%), paraxanthine (8.8%), and trace amounts of 1,3,7-trimethyluric acid and 3-methyluric acid. Two unidentified metabolites (metabolite A, 11.4% and metabolite B, 1.3%) have also been isolated (Khanna et al. 1972). The urinary metabolites of [2 14C] and [1 Me 14C] caffeine administered orally have been separated by two-dimensional thin-layer chromatography from whole urine and localized by autoradiography. The scanning of chromatograms gives quantitative results showing the importance of the metabolic pathway leading to trimethyldihydrouric acid. The following three new metabolites were identified in rat's urine: 1,7 dimethyluric, N-methylurea, and NN'- dimethylurea. The excretion kinetics of each metabolite shows that caffeine, theophylline, trimethylallantoin, and an unknown derivative that is either an isomer or a precursor of trimethylallantoin were excreted later than the other metabolites as a whole (Arnaud 1976).

A thin-layer chromatographic method is described for the quantitation of caffeine and its dimethylxanthine metabolites, theophylline, theobromine, and paraxanthine. The method was used to evaluate the effect of polycyclic aromatic hydrocarbons (PCH), polychlorinated biphenyls (Aroclor 1254), or phenobarbital on the pharmacokinetics of caffeine and its dimethylxanthine metabolites. Oral administration of benzo[a]pyrene (BP) or Aroclor 1254 to rats for 3 days markedly increased the plasma clearance of caffeine and its dimethylxanthine metabolites. Similar results on the plasma clearance of caffeine were obtained with benzanthracene, dibenzanthracene, chrysene, or pyrene. Although the elimination of caffeine from plasma was increased in rats treated with phenobarbital for 3 days, it was less effective in this respect than the PCH or Aroclor 1254. In addition, phenobarbital did not significantly affect the rate of elimination of the dimethylxanthine metabolites from rat plasma following an intravenous dose of caffeine. Following the intravenous administration of caffeine to rats pretreated with Aroclor 1254 or BP, there was a marked increase in the appearance of theophylline, theobromine, and paraxanthine in plasma. The time to achieve peak plasma levels of these metabolites was reduced from 6 to 7 h in control rats to 1 h in Aroclor 1254-treated rats and to less than 3 h in rats treated with BP. Moreover, the plasma elimination of the dimethylxanthine metabolites formed from caffeine was greatly accelerated after pretreatment with BP or Aroclor 1254. A dose–response study with BP indicated that as little as 1.0 mg kg^{-1} of BP administered orally for 3 days markedly increased the plasma clearance of caffeine; however, pretreatment with BP did not affect the absolute bioavailability of caffeine. The area under the caffeine plasma curve after oral administration was identical to the area when the same dose was administered intravenously. These studies indicate that the plasma clearance of caffeine is markedly increased in rats pretreated with phenobarbital, PCH, or Aroclor 1254 and suggest that the metabolism and pharmacology of caffeine may be considerably altered in human subjects exposed to these substances (Welch et al. 1977).

The described TLC and GLC methods are considerably more sensitive than the previously described GLC assay, which required a 5 mL blood sample to achieve a reported sensitivity of 0.25 µg mL^{-1}. Both techniques are therefore suitable for use in studies of caffeine pharmacokinetics (Bradbrook et al. 1979). The metabolism of caffeine was investigated in liver slices of young and adult rats. Liver slices from adult rats metabolized caffeine at an initial rate of 48.31 ± 3.71 nmol (g liver)$^{-1}$ h^{-1} to four main metabolite fractions. By a combination of thin-layer radiochromatography and high-performance liquid chromatography, theophylline, paraxanthine, and 1,3,7-trimethyldihydrouric acid were identified as caffeine metabolites. The apparent Vmax of the overall reaction was 83.30 nmol caffeine metabolites formed (g liver)$^{-1}$ h^{-1}. Theophylline competitively inhibited caffeine metabolism (the apparent K[m] was 19.20 µM in the absence of theophylline, the apparent K[i] was 36.50 µM in the presence of theophylline [100 µM]). SKF 525-A inhibited caffeine metabolism; the formation of all of the metabolite fractions was inhibited to a similar extent. Allopurinol (100 µM) had no effect. The specific activity of the enzyme system was extremely low when liver slices of 2-day-old-rats were used [1.46 ± 0.08 nmol caffeine metabolites formed (g liver)$^{-1}$ h^{-1}]: the reaction velocity increased gradually with increasing age and reached a peak [52.26 ± 1.41 nmol caffeine metabolites formed (g liver)$^{-1}$ h^{-1}] at 30 days of age. Changes in the formation of the four metabolite fractions with age followed the pattern of the overall caffeine metabolism. These results demonstrate that the liver of the newborn rat has an extremely limited capacity to metabolize caffeine in vitro and are consistent with the proposed involvement of the liver microsomal cytochromes P-450 monooxygenase system in the metabolism of caffeine. N-Demethylation is the main pathway of in vitro caffeine metabolism in the rat liver at all ages (Warszawski et al. 1981).

Caffeine was given intravenously to three groups of six sheep each in doses of 25, 2.5, and 1.8 mg kg^{-1} body weight, respectively. Thin-layer chromatography was used for the analysis of caffeine and its metabolites. The results showed: 25 mg kg^{-1} body weight induced increase in motility and slight tremor during the first 3 h after injection; an increase in urine volume was observed directly after application of 25 mg kg^{-1}, with return to normal within 2 days; all doses of caffeine resulted in comparable plasma concentration curves with a maximum after half an hour, followed by a gradual decrease; the plasma half-life of caffeine is related to the dose and was found to be 3.6 and 7.8 h after a dose of 1.8 and 25 mg kg^{-1} body weight, respectively; the percentage of excreted unchanged caffeine in urine in the period from 0 to 48 h was not related to the dose and was 7.12% ± 1.42%; 7.9% ± 1.9%, and 6.89% ± 1.48% for doses of 25, 2.5, and 1.8 mg, respectively; the smallest amount of caffeine excreted in feces was recorded for the highest doses (1.76 ± 0.29; 4.76 ± 1.52, and 3.75 ± 1.43 for doses of 25, 2.5, and 1.8 mg, respectively), because excretion in urine is increased by high caffeine dosage; and in addition to the excretion of unchanged caffeine in urine, 1,7-dimethylxanthine, 7-methyluric acid, theobromine, theophylline, and 7-methylxanthine were quantitatively estimated up to 48 h (Aly 1981).

Thin-layer chromatography-tandem mass spectrometry (TLC-MS-MS) allowed the detection and confirmation of caffeine and nicotine in human urine and of butorphanol, beclamethasone, and clenbuterol in equine urine. In most cases of trace analysis of labile compounds, the drugs could not be identified unless they were developed on a TLC plate, scraped from the plate, and the TLC scrape eluted with a suitable organic solvent prior to MS-MS. Usually a sample prepared in this way still had several components in it, but was sufficiently cleaned up to allow collision-induced dissociation (CID) experiments to unequivocally identify the drug. In contrast, trace levels of labile drugs could not be identified by CID experiments either directly from the raw urine extracts or by thermally desorbing them from the TLC scrape (Henion et al. 1983).

The metabolic disposition of [1-Me14C] caffeine has been studied and compared in three male rodent species: the rat, the mouse, and the Chinese hamster. No interspecies differences appeared in urinary and fecal excretion of radioactivity. However, 1-methyl demethylation was significantly more important in the rat with 20.6% ± 0.8% of the dose recovered as 14CO$_2$ compared with the Chinese hamster, 16.1% ± 2%, and the mouse, 13.9% ± 0.9%. HPLC and TLC analysis of 1-methyl-labeled metabolites showed that the rat exhibits a significantly higher urinary excretion of the four trimethyl derivatives: caffeine, 1,3,7-trimethyluric acid, trimethylallantoin, and 6-amino-5-[N-formylmethylamino]-1,3-dimethyluracil (40.8% of total urine radioactivity) when compared with the Chinese hamster (21.1%) and the mouse (19.7%). Compared with man (6%), these rodents have a greater ability to excrete caffeine without any demethylation. The rat was also characterized by a higher excretion of theophylline, while the Chinese hamster excreted more paraxanthine, 1-methylxanthine, and the uracil derivative of paraxanthine. In the mouse, in addition to 1-methylxanthine and 1-methyluric acid, higher amounts of 1,3- and 1,7-dimethyluric acid were found. The mouse was particularly characterized by the presence of an unknown polar metabolite amounting to 22% ± 3% of urine radioactivity. This metabolite must be produced from paraxanthine because its quantitative formation was inversely related to the excretion of paraxanthine and its metabolites. The observations that this metabolite is neither 5-acetylamino-6-amino-3-methyluracil nor 5-acetylamino-6-formylamino-3-methyluracil reported in humans demonstrate that both quantitative and qualitative interspecies differences occur for caffeine metabolism. The identification of this metabolite and studies on pharmacokinetics and tissue exposure to caffeine and its metabolites must be initiated in order to explain interstrain differences in toxicity response reported in mice (Arnaud 1985).

Caffeine metabolites in urine from premature infants were analyzed by TLC and HPLC. Caffeine, dimethyluric acids, mono- and dimethyl-xanthines, and, for the first time, a uracil derivative (6-amino-5-[N-methylformylamino]-1,3-dimethyluracil) were identical (Gorodischer et al. 1986a). A fast and simple method for the simultaneous quantification of paracetamol, caffeine, phenobarbital, and propyphenazone in plasma using diffuse reflectance measurements on HPTLC-plates was described. The concentration range was from 0.5 to 2.5 μg μL^{-1} plasma for paracetamol, caffeine, and propyphenazone, and from 0.25 to 1.0 μg μL^{-1} plasma for phenobarbital. The precision was 0.2 μg μL^{-1} (Kosmeas and Clerc 1989).

Determination of xanthines by high-performance liquid chromatography and thin-layer chromatography in horse urine after ingestion of guarana powder was performed in 1994. The seeds of Guaraná are rich in xanthines and are used for the preparation of guaraná powder, which is commonly given to horses as a "tonic" in Brazil. In this chapter, the xanthine content of guaraná powder was determined, in addition to its clearance time in horses. Thin-layer chromatography was used as a screening procedure, and high-performance liquid chromatography was performed to quantify the drugs in both the powder and urine samples. The guaraná powder was found to contain 2.16, 1.10, and 36.78 mg g^{-1} of theobromine (TB), theophylline (TP), and caffeine (CF), respectively, and in urine it was possible to detect TB and TP up to 13 days and CF up to 9 days after the administration of guaraná powder (Salvadori et al. 1994). A simple high-performance thin-layer chromatographic (HPTLC) method was described for the identification and semiquantitative determination of morphine, caffeine, and paracetamol in urine. The results obtained were further confirmed and quantified by high-performance liquid chromatography (HPLC). All the urine samples were first screened by use of the enzyme-multiplied immunoassay technique (EMIT), and positive samples were then hydrolyzed, extracted, and analyzed by HPTLC and HPLC. HPTLC was performed on silica gel 60 F$_{254}$ plates with ethyl acetate–methanol–ammonia, 17:2:1 (v/v) as mobile phase. The developed plates were viewed in a UV cabinet at $\lambda = 254$ nm and photographed. The concentrations of these drugs found in the urine of addicts by use of this method were in the ranges 0.1–4 μg mL^{-1} for morphine and caffeine and 0.1–6 μg mL^{-1} for paracetamol. HPTLC detection limits were 0.5 μg for morphine and 0.2 μg for caffeine and paracetamol. HPLC analysis of these drugs was performed on a LiChrospher 100 RP-18 column with potassium dihydrogen phosphate (0.05 M) buffer–acetonitrile, 80:20 (v/v) as mobile phase, methyldopa as internal standard, and UV detection at $\lambda = 280$ nm. Calibration curves were linear, and the HPLC detection limits were 20 ng mL^{-1} for morphine, 5 ng mL^{-1} for caffeine, and 10 ng mL^{-1} for paracetamol. Detection of the drugs in the urine confirmed the abuse of brown sugar seized from the addicts (Krishnamurthy et al. 2000). A quantitative method using thin-layer chromatography plates with concentrating zone and ultraviolet absorption densitometry was described for caffeine determination in human saliva and urine. The applicability of the caffeine method was tested for saliva and urine from male individuals. The changes of salivary caffeine, urinary caffeine concentration, and of urinary caffeine excretion correspond with results of previous investigations using gas chromatography, high-performance liquid chromatography, radio-immunoassay, or enzyme immunoassay methods. The thin-layer chromatography/densitometry method described was easy to perform, reproducible, and accurate. Thus, it was suitable for routine analysis of caffeine in human saliva or urine, as it was an effective alternative to other, more expensive, and more time-consuming chromatographic or immunological methods (Fenske 2007).

Rapid and precise identification of toxic substances is necessary for the urgent diagnosis and treatment of poisoning cases and for establishing the cause of death in postmortem examinations. However, identification of compounds in biological samples using gas chromatography and liquid chromatography coupled with mass spectrometry entails time-consuming and labor-intensive sample preparations. In this study a simple preparation was examined and a highly sensitive analysis was made of drugs in biological samples such as urine, plasma, and organs using thin-layer chromatography coupled with matrix-assisted laser desorption/ionization mass spectrometry (TLC/MALDI/MS). When the urine containing 3,4-methylenedioxymethamphetamine (MDMA) without sample dilution was spotted on a thin-layer chromatography (TLC) plate and was analyzed by

TLC/MALDI/MS, the detection limit of the MDMA spot was 0.05 ng per spot. The value was the same as that in aqueous solution spotted on a stainless steel plate. The 11 psychotropic compounds, including caffeine, tested on a TLC plate were detected at levels of 0.05–5 ng, and the type (layer thickness and fluorescence) of TLC plate did not affect detection sensitivity. In addition, when rat liver homogenate obtained after MDMA administration (10 mg kg^{-1}) was spotted on a TLC plate, MDMA and its main metabolites were identified using TLC/MALDI/MS, and the spots on a TLC plate were visualized by MALDI/imaging MS. The total analytical time from spotting of intact biological samples to the output of analytical results was within 30 min. TLC/MALDI/MS enabled rapid, simple, and highly sensitive analysis of drugs from intact biological samples and crude extracts. Accordingly, this method could be applied to rapid drug screening and precise identification of toxic substances in poisoning cases and postmortem examinations (Kuwayama et al. 2012).

Fourier transform infrared photoacoustic spectroscopy (FTIR-PAS) is used to analyze substances deposited on thin-layer chromatographic silica gel plates. In spectral regions of high silica gel absorption, little structural information is derived. Regions of lesser absorption provide infrared spectral information of the adsorbate and interactions between adsorbed species and the silica gel substrate. Detection limits of 1 µg were reported for photoacoustic analysis of caffeine deposited on silica gel. The total time required for sample preparation and photoacoustic analysis was less than 5 min per sample (White 1985).

Laser desorption/laser ionization time-of-flight mass spectra of caffeine, theophylline, theobromine, and xanthine were reported. These mass spectra were compared with the published spectra obtained using electron impact ionization. Mass spectra of caffeine and theophylline obtained by IR laser desorption from thin-layer chromatography plates were also described. The laser desorption of materials from thin-layer chromatography plates was discussed (Rogers et al. 1993).

Complete validation of an HPTLC method for quantitative determination of caffeine in Coca-Cola has been performed according to ICH guidelines, taking into the consideration the special features of the method. The validation parameters were: selectivity, stability of the analyte, linearity, precision (repeatability and intermediate precision), and robustness. The plates were quantified with a slit-scanning densitometer (Camag TLC Scanner LT) and with an image-analyzing system (Camag Video Documentation System). The same plates were used in both validation procedures to reveal possible differences between the different modes of quantification (Vovk et al. 1997).

Alkaloids extracted from green tea were separated by high-speed counter-current chromatography. A series of experiments have been performed to investigate the effects of different solvent system. A system of $CHCl_3–CH_3OH–NaH_2PO_4$ (23 mmol L^{-1}) (4:3:2) was selected, in which the upper phase was used as the stationary phase, and the lower phase as mobile phase. When acidity of solvent system is pH 5.6, three chemical components were very efficiently isolated by one injection of 50 mg sample mixture. Analyzing the eluted fractions by TLC, it was determined that one is caffeine, and the other is theophylline. In comparing the separation results by high-speed counter-current chromatography with those by TLC, the advantages of this method were verified. This technology for the separation of crude mixture of plant components should find wide applications (Yuan et al. 1998).

Callus and root suspensions from *Camellia sinensis* have been established to produce and accumulate caffeine and theobromine as secondary metabolites. Leaf fragments from a mature greenhouse tea plant, grown on MS medium supplemented with IAA (5.7 µM) produced roots, while leaf explants of the same plant cultured on MS medium supplemented with 2,4-D (4.5 µM) and BA (0.45 µM) gave rise to the formation of friable callus. Both callus and roots when transferred to MS liquid medium supplemented with 2,4-D (4.5 µM) and BA (0.45 µM) produced caffeine and theobromine, which were detected by TLC, UV, and GC (Shervington et al. 1998).

TLC-MS, more specifically TLC-SPE-APCI-MS, has been evaluated for determination of caffeine in soft drinks. Caffeine standards (between 1.0 and 20 µg per spot) and samples of different soft drinks were applied to TLC plates, and the components were separated and quantified by densitometry with image analysis. After the in situ evaluation, spots were marked and the stationary phase containing the compounds was scraped from the plates, transferred to SPE cartridges, and eluted by use of solid-phase extraction (SPE) equipment. The substance was diluted 100-fold during the elution procedure; the eluted samples (10 μL) were injected into an LCQ-MSn spectrometer equipped with an atmospheric-pressure chemical ionization (APCI) interface. The TLC-SPE-APCI-MS analytical procedure was validated, and the limit of detection, limit of quantitation, precision, recovery, and linearity were evaluated. MS and MS-MS spectra of caffeine were obtained by direct injection of standards into the spectrometer. Results from validation showed the proposed off-line TLC-MS combination to be sensitive and repeatable. Acceptable MS and MS-MS spectra were obtained from 20 ng caffeine (Prošek et al. 2000). High-performance thin-layer chromatography is a frequently used separation technique that works well for quantification of caffeine and quinine in beverages. Competing separation techniques, for example, high-performance liquid chromatography or gas chromatography, are not suitable for sugar-containing samples, because these methods need special pretreatment by the analyst. In HPTLC, however, it is possible to separate "dirty" samples without time-consuming pretreatment, because disposable HPTLC plates are used. A convenient method for quantification of caffeine and quinine in beverages, without sample pretreatment, is presented later. The basic theory of in situ quantification in HPTLC by use of remitted light is introduced and discussed. Several linearization models are discussed. A homemade diode-array scanner has been used for quantification; this, for the first time, enables simultaneous measurements at different wavelengths. The new scanner also enables fluorescence evaluation without further equipment. Simultaneous recording at different wavelengths improves the accuracy and reliability of HPTLC analysis. These aspects result in substantial improvement of in situ quantitative densitometric analysis and enable quantification of compounds in beverages (Spangenberg et al. 2002).

The development of ultrathin-layer silica gel plates with a monolithic structure opens up a new dimension in thin-layer chromatography. The very small layer thickness of approximately 10 pmol and the absence of any kind of binder in combination with the framework of this stationary phase leads to new and improved properties of these ultrathin-layer chromatographic (UTLC) silica-gel plates compared with conventional TLC and high-performance TLC (HPTLC) precoated layers. First of all, the advantages of the UTLC plates are the very short migration distances and, in combination with this, the short development times as well as the very low consumption of solvents as the mobile phase in connection with high sensitivity. The separations of amino acids, pesticides, pharmaceutically active ingredients, phenols, and plasticizers effectively demonstrate the possibilities of the new ultrathin-layer silica-gel plates. Furthermore, a comparison of UTLC, HPTLC, and TLC concerning retention behavior, efficiency, detection limits, migration times, and solvent consumption is performed effectively by the separation of caffeine and paracetamol (Hauck and Schulz 2002).

Cocaine sold as a "street drug" in white powder usually contains several adulterants (including caffeine) and diluents. The purpose of this research was to quantify cocaine content and to check the presence of adulterants in illicit samples seized by the Narcotic and Toxic Substances Regional Department of Alfenas and Varginha districts, Brazil, in the year 2001. The identification of cocaine and adulterants in 209 samples, was screened by thin-layer chromatography (TLC) and confirmed/quantified by gas chromatography/flame ionization detector (GC/FID). The pharmacologically inactive adulterants (starch, carbonates/bicarbonates, and sugars) were analyzed by qualitative tests or TLC. 88.9% of the samples analyzed gave positive results for cocaine in concentrations between 4.3% and 87.1% of the powder. Active compounds detected were caffeine (50.2% of the samples, in concentrations ranging from 2.8% to 63.3% of the powder), lidocaine (65% of the samples, concentrations ranging 0.5%–92%), and prilocaine (11% of the samples, concentrations ranging 1.4%–20.7%). Carbonate/bicarbonate showed positive results in 41.2%, starch in 51.2%, and sugars in 9.6% of the samples. The presence of other substances in cocaine street samples is worrying because they could modify or intensify signs and symptoms of the intoxication, acute or fatal, due to drug (Bernardo et al. 2003). Thin-layer chromatography and high-performance TLC commercially precoated silica gel plates, with enhanced brightness ultraviolet (UV)-indicator, were compared with comparable plates formulated with standard indicators. Caffeine, acetaminophen, and salicylamide were used as the model test compounds. Results showed that the increased UV-indicator plates had visually brighter backgrounds; however, the limits of detection (LOD) did not improve. Differences among the plates in terms of efficiency, resolution, and sensitivity and linearity of quantification with slit-scanning densitometry are also reported (Sullivan and Sherma 2004).

The caffeine content of selected herbal products and energy drinks available in the Saudi market was determined by HPTLC-UV densitometric analysis. Precoated HPTLC silica gel plates (20 cm × 10 cm) were used for the analysis. The solvent system consisted of ethyl acetate–methanol (85:15, v/v), and caffeine was detected at 275 nm. The developed method was validated for specificity, repeatability (C.V. < 5%), recovery (98.90 ± 3.46), and accuracy (99.84 ± 2.87). The levels of caffeine were 4.76%–13.29% (w/w) and 0.011%–0.032% (w/v), for the herbal products and the energy drinks respectively (Abourashed and Mossa 2004).

Seeds of guaraná (*Paullinia cupana* H.B.K. var. *sorbilis* [Mart.] Ducke), obtained from different localities, were tested in regard to physicochemical aspects (loss on drying, determination of extracts, dry residue) and their total tannin and methylxanthine contents. TLC chromatographic analysis showed a R_f = 0.43 for caffeine and R_f = 0.72 and 0.71 for catechin and epicatechin, respectively. Analysis of the semipurified fraction by HPLC showed retention times for catechin, epicatechin, and caffeine of 6.17, 8.85, and 11.91 min, respectively. The quality and similarity of the chemical components of the raw material from different sources were confirmed (Antonelli-Ushirobira et al. 2004).

Thin-layer chromatographic separation with densitometric detection has been used for quantitative determination of caffeine in different tea beverages prepared from green, oolong, pu-erh, and black tea. Chromatographic separation on 60 F_{254} silica gel layers was performed with chloroform–ethyl acetate–formic acid, 5:4:1, (v/v/v) as a mobile phase. The spots were developed densitometrically at λ = 280 nm. The method was validated in terms of selectivity, stability of the analyte, linearity, precision, and limits of detection and determination. The level of caffeine in different manufactured teas decreased in the order: black tea > oolong $\cong$ pu-erh > green tea. Under the established experimental conditions, repeatable and accurate results were obtained. The method was simple, reliable, economical, and convenient for isolation and quantification of caffeine in tea beverages (Buhl et al. 2006).

A fully automated interface to couple high-performance thin-layer chromatography with mass spectrometry (MS) is described. This universal hands-free interface connects intact normal-phase plates to any liquid chromatography/mass spectrometry (LC/MS) system without any adjustments or modifications to the mass spectrometer. The interface extracts the complete substance band with its depth profile and thus allows detections in the picogram per band range. The high performance of the automated interface was evaluated through caffeine quantification in real samples: energy drinks and pharmaceutical tablets, without internal standard. Following chromatographic separation on silica gel 60 F_{254} HPTLC plates, caffeine bands were eluted from the plate by means of the automated interface to the electrospray ionization (ESI) source of a triple-quadrupole mass spectrometer. Since in full scan mode only the protonated molecule $[M^+H]^+$ was observed, caffeine quantification was performed using the selected-ion monitoring (SIM) mode at m/z 195. The validation showed highly reliable results for the linear range (R^2 = 0.9973), repeatability (RSD = 5.6%, n = 6) and intermediate precision (RSD = 1.5%, n = 3). Regarding accuracy, the results obtained by HPTLC/MS were not statistically different (F-test, t-test) from those obtained by validated HPTLC/UV methods. Hence, this interface proved to be one of the most reliable and universal interfaces for HPTLC/MS (Luftmann et al. 2007).

Methods of standardization and quality control of homeopathic matrix tinctures of Coffea arabica and Coffea tosta obtained using green and roasted coffee beans are described. They include the organoleptic characteristics, qualitative reactions, TLC, quantitative estimation of caffeine and total phenolcarboxylic acids (calculated as caffeic acid), relative density, and percentage of dry residue.

The period of expiration on storage is established at 2 years. These methods are included in drafts of pharmacopoeial monographs for homeopathic matrix tinctures of Coffea arabica and Coffea tosta (Kopyt'ko 2008).

A simple, precise, and accurate high-performance thin-layer chromatographic (HPTLC) method has been established and validated for screening and quantitative estimation of caffeine in different extracts of tea samples (*Camellia sinensis*). Separation was performed on silica gel 60 F_{254} HPTLC plates with ethyl acetate:methanol in the proportion of 27:3 (v/v), as a mobile phase. The determination was carried out in the UV region using the densitometric remission-absorbance mode at 274 nm. Maximum recovery of caffeine was achieved when extracted with 5% diethyl amine in DM water (v/v). The maximum concentration of caffeine in tea samples was found to be 2.145%, dry weight basis. Caffeine response was found to be linear over the range of 2–14 g per zone. Limits of detection and quantitation were found to be 40 and 120 ng per spot, respectively. The HPTLC method was validated in terms of precision, accuracy, sensitivity, and robustness. Some rare parameters for the HPTLC method like calculation of flow constant (k) and plate efficiency (N) are included specially (Misra et al. 2009).

Researchers have paid particular attention to biologically active ingredients, especially alkaloids and polyphenols in food and beverages due to their positive effects on human health. Cimpoiu et al. developed a simple, fast, and economical method for simultaneous quantitative determination of methylxanthine compounds based on TLC combined with image analysis. To obtain certain results, both extraction and chromatographic separation were optimized. The optimum extraction conditions were maceration in ethanol–water 8:2, v/v. The chromatographic separations were done on the silica gel F_{254} TLC plates developed with chloroform–dichloromethane–isopropanol, 4:2:1 v/v/v. Detection was performed under UV lamp at 254 nm and the evaluation of the chromatographic plate was based on digital processing of chromatographic images. The developed TLC method was validated for parameters such as specificity, linearity and range, LOD and LOQ, precision, robustness, and accuracy. This method was then applied for determination of caffeine, theobromine, and theophylline in different types of tea, commercially available. Moreover, the content of methylxanthines detected and determined in commercial tea samples can be used as chemical marker in quality control (Cimpoiu et al. 2010).

Yalçindag described using the TLC method for identification and quantitative determination of theophylline derivatives (Yalçindag 1970). Separation and detection of products of decomposition of cyclovegantine, ethionamide, and euphylline by means of thin-layer chromatography were used to study the stability of drugs in lyophilized suppository bases described Cieszyński (1971). For determination theophylline concentration, thin-layer chromatography was used by Schmidtova et al. (1987). Purine derivative drugs, aminophylline, theophylline, xantinolnicotinate, 6-mercaptopurine, 6-thyoguanine, azathioprine, 1-methylxanthine, 3-methylxanthine, 1-methyluric acid, 1,3-dimethyluric acid, 6-thioxanthine, 2-amino-6-methylmercaptopurine, were separated by thin-layer chromatography on rice starch and cellulose by three different solvent systems. Conditions for quantitative fluorodensitometric determination of purine derivatives drugs were investigated (Popović and Perišić-Janjić 1988).

This chapter described the determination and quantification of impurities and synthetic by-products in theophylline by videodensitometry and classical TLC scanning densitometry, in each instance by the use of equipment from two different manufacturers. The performance of the two scanners did not differ significantly in terms of limit of detection (approximately 24 ng for Scanner I and 23 ng for Scanner II) and precision (coefficients of variation 3.2% and 4.1%, respectively). Detection by videodensitometry is possible for concentrations larger than 25 ng per spot; quantification requires approximately 100 ng per spot (depending on the types of impurity present). The European Pharmacopoeia usually limits impurities in pharmaceutical drugs to 0.1%. These limits could not be quantified by videodensitometry at its current state of development. The regression curves obtained by videodensitometry show average coefficients of variation of approximately 17.4% for videodensitometer I and 5.9% for videodensitometer II. The use of videodensitometry for impurity testing is nevertheless useful in special cases, where higher deviations are considered acceptable for assays, and when higher concentrations are available for quantitation of the main component (Essig and Kovar 1999).

A simple, low-cost thin-layer chromatography procedure to estimate the quality of simple pharmaceuticals in tablet form is described together with easily built equipment to carry out the test in the field. The approach is demonstrated for theophylline, but can be used to assay the drug content of any tablet or to determine its dissolution or disintegration characteristics. A TLC aluminum plate has been precoated with a 0.2 mm layer of silica gel 60 F_{254}. Developing solvent consisted of glass–distilled chloroform and acetone (1:1, v/v). A two-component agents: A—potassium iodide in 95% ethanol and dissolve of iodine in this solution; B—concentrated hydrochloric acid with water (1:3) and 95% ethanol add up to 100 mL and C—mix solution A and solution B to obtain the visualizing agent. The procedure can be used in the field without the need for any instrumentation (Flinn et al. 1989).

The content and dissolution rate of theophylline, diprophylline, and proxyphylline from a sustained release formulation were determined by UV in situ densitometry. After separation, the chromatographic zones corresponding to the spots of theophylline, diprophylline, and proxyphylline on the high-performance thin-layer chromatographic plates were scanned in reflectance/absorbance mode at 275 nm. Quantification was performed with a second-degree polynomial function over the range 40–200 ng for theophylline and 60–300 ng for diprophylline and proxyphylline. Percentages of dissolved theophylline, diprophylline, and proxyphylline were monitored over 1, 3, and 6 h. The method was found to be simple, accurate, reliable, timesaving (up to 18 samples can be determined simultaneously), and low cost (Agbaba et al. 1992).

Using the HPTLC analysis of theophylline in an effervescent tablet as an example, a procedure for the validation of analytical procedures in pharmaceutical analysis has been described (Renger et al. 1995).

Simple, rapid, accurate, precise, reliable, and economical thin-layer chromatographic and spectrophotometric methods have been proposed for the resolution and determination of salbutamol sulfate (SS), ambroxol hydrochloride (AH), and theophylline (THE) in pure and pharmaceutical formulations, respectively. The developed methods show best results in terms of resolution, linearity, accuracy, precision, limit of detection (LOD), and limit of quantification (LOQ) for standard laboratory mixtures of pure drugs and marketed formulations. The R_f value for SS was found to be 0.25, for AH R_f value was found to be 0.89, and for THE R_f value was found to be 0.72. The range for SS, AH, and THE were found to be 1–35, 5–35, and 6–60 µg mL^{-1}, respectively. The values of LOD were 0.316, 0.22, and 0.29 µg mL^{-1}, and the values LOQ were 0.957, 0.74, and 0.967 µg mL^{-1} for SS, AH, and THE, respectively. The precision values were less than 2% in terms of % relative standard deviation for the developed method. The common excipients and additives did not interfere in their determinations (Dave Hiral et al. 2010).

A simple, precise, accurate, and rapid high-performance thin-layer chromatographic method has been developed and validated for the estimation of etophylline and theophylline in the combined dosage form. The stationary phase used was precoated silica gel 60 F_{254}. The mobile phase used was a mixture of toluene:isopropyl alcohol:acetic acid (12:12:1 v/v/v). The detection of spots was carried out at 261 nm. The method was validated in terms of linearity, accuracy, precision, and specificity. The calibration curve was found to be linear between 200 and 400 ng for etophylline and 60–80 ng for theophylline with a regression coefficient of 0.9997 and 0.9994. The proposed method can be successfully used to determine the drug content of marketed formulation (Venkatesh et al. 2011).

A simple spectrodensitometric method for the direct determination of theophylline was developed from measurement of the absorbance of the compound on silica gel layers irradiated at 275 nm. Quantities as low as 0.010 µg can be detected, and a linear relationship was obtained between peak area and the amount of the drug in the spots from 0.025 to 0.200 µg. The recovery over the usual range of plasma concentration (2.5–20 µg mL^{-1}) was 95%–107%. The method is sufficiently sensitive and specific for clinical purposes, and the time for the assay is about 2 h. Caffeine, frequently present in human plasma, was well separated from theophylline at all concentration levels as were several other drugs commonly used in respiratory problems (Wesley-Hadzija and Mattocks 1975). Thin-layer chromatography was used for micro method for the determination of caffeine and theophylline allowing direct application of biological fluids to thin-layer chromatography plates (Riechert 1978). A quantitative thin-layer chromatographic procedure for theophylline has been evaluated. An internal standard, 3-isobutyl-1-methylxanthine, a compound similar in properties to those of theophylline was added to the specimen prior to extraction to eliminate the need for accurate measurements of volume during extraction or analysis. This method did not show interference from other xanthines and from a number of drugs commonly prescribed with theophylline. It had an acceptable correlation with a reference gas-liquid chromatographic procedure. This procedure was capable of handling batches of six or more samples faster than the serial processing of either high-performance liquid chromatography or gas–liquid chromatography (Gupta et al. 1978).

Chromatographic characteristics of urinary metabolites of theophylline were studied by two-dimensional thin-layer chromatography, high-performance liquid chromatography, and gas chromatography-mass spectrometry. Quantitative data for the urinary metabolites of theophylline in asthmatic children were given. It was shown that 1,3-dimethyluric acid is the predominant excretory product. In addition, smaller amounts of 1-methyluric acid, 3-methyl-xanthine and unchanged theophylline were found. Excretory patterns after theophylline ingestion before and during the administration of allopurinol in asthma patients and in rats suggest the existence of three metabolic pathways of theophylline. The administration of this drug to a patient with xanthine oxidase deficiency resulted in the excretion of 1-methyluric acid in addition to 1,3-dimethyluric acid, 3-methyl-xanthine, 1-methylxanthine and unchanged theophylline. It was concluded that in man the oxidation of theophylline was not catalyzed by xanthine oxidase (Van Gennip et al. 1979).

The direct application of 20 μL of serum to thin-layer chromatograms was a rapid and sensitive method for the determination of theophylline. Chromatograms were developed on LK6DF silica gel TLC plates and the mobile-phase chloroform:isooctane:methanol (70:20:10) on distance 7.5 cm and they were scanning on the densitometer by fluorescence quenching at 270 nm. Theophylline had an R_f of 0.35 while theobromine and caffeine showed R_f values of 0.29 and 0.46, respectively. Ten common drugs were shown not to interfere. A comparison with the enzyme immunoassay values obtained in an independent laboratory showed the utility of the method (Heilweil and Touchstone 1981).

Pokrajac and Varagic used spectrodensitometric method for determination of theophylline in plasma (Pokrajac and Varagic 1983). TLC determination of theophylline in plasma was developed in 1985 (Li et al. 1985). Theophylline and caffeine in blood serum by thin-layer chromatography-densitometry were determined in Japan in 1985 (Kawamoto and Yamane 1985). The pharmacokinetics and bioavailability of theophylline from a commercial oral elixir of theophylline, a rectal suppository of aminophylline, and a rectal enema of theophylline monoethanolamine was compared in six normal subjects. Using a complete crossover design, the fasted subjects received a single dose of each dosage form. Blood and saliva samples were collected at frequent time intervals for 24 h, and the plasma assayed for theophylline by a specific thin-layer chromatography densitometric method. No statistically significant differences existed among the three dosage forms with respect to C(max) and AUC corrected for the elimination rate constant and the dose (mg kg^{-1}). However, t(max) was significantly larger for the suppository. While the rate of absorption was significantly lower for the suppository, no differences in the extent of absorption existed among the three dosage forms. A one-compartment open model with apparent first-order absorption adequately described the plasma concentration-time data for the elixir and enema, whereas the suppository data were best fitted by a one-compartment open model with apparent zero-order absorption and a lag time. A rate-limiting, concentration-independent release of drug from the base most likely accounts for the slow absorption of theophylline from the suppository. While the saliva:plasma ratio remained fairly constant for most of the study period, the large variability found during the absorption phase following drug administration limits the usefulness of this parameter as a monitor of theophylline plasma concentrations (Cole and Kunka 1984).

The metabolism of theophylline was studied in liver slices of young and adult rats. Theophylline and six metabolite fractions were recognized in adult liver by thin-layer radiochromatography and high-performance liquid chromatography: 1-methyluric acid; 1-methylxanthine; 1,3-dimethyluric acid and/or 3-methylxanthine; caffeine; a uracil derivative and two unknown polar compounds. Preincubation with caffeine or theobromine inhibited theophylline metabolism. Allopurinol decreased the formation of three metabolite fractions but markedly increased the production of 1-methylxanthine. SKF 525-A inhibited the overall metabolism of theophylline. The specific activity of the enzyme system was 3.2 ± 0.4 nmol (g liver)$^{-1}$ h^{-1} in the 4–5-day old rat and increased to a peak of 25.7 ± 1.7 in the 28-day-old; values for K(m) and V(max) in the 7- and 28-day-olds were 132.1 and 67.5 μM, and 23.9 and 52.1 nmol (g liver)$^{-1}$ h^{-1}, respectively. Theophylline and the same six metabolites were identified in young and adult rats, but the development pattern was not uniform. Peak age-related activity and involvement of mixed-function oxidase system are features that are common to theophylline and caffeine metabolism. Xanthine oxidase played a role in theophylline metabolism. Formation of caffeine from theophylline was not dependent on a lack of activity of other pathways (Gorodischer et al. 1986b). A thin-layer chromatographic screening procedure for some basic, neutral, and acidic drugs including theophylline was developed using 3 mL of horse plasma. Chloroform–2-propanol (95:5, v/v) was used as the extraction solvent. The drugs were identified by a high-performance thin-layer chromatographic plate and spraying successively with some detection reagents. In this study, the extraction recovery rates and the detection limits were determined at the same time (Tanaka et al. 1989).

Percutaneous absorption of theophylline in human skin from five sources was examined by use of a flow-through in vitro diffusion system. The metabolites and unchanged drug were estimated by thin-layer chromatography. Correlation was evident in the percentage of the applied dose that diffused through the five skin samples (range $2.8\% \pm 0.5\%$–$7.7\% \pm 0.8\%$); however, the percentage of applied dose absorbed varied between different skin samples (range $3.6\% \pm 0.9\%$–$33.4\% \pm 2.4\%$). Between $0.2\% \pm 0.1\%$–$4.6\% \pm 0.2\%$ of the doses applied were metabolized, and over 60% of the total metabolites formed diffused through the skin. The uptake and metabolism of theophylline by microsomes obtained from four of the human skin samples were measured. All preparations showed detectable activities for the metabolism of theophylline. Microsomal preparations from skin sources A, B, and D, and B, C, and E biotransformed theophylline to 1,3,7-trimethyluric acid and 1,3-dimethyluric acid, respectively. The activities of microsomes from skin samples C and E on the drug produced the pharmacologically active metabolite 3-methylxanthine. The specific activities of the microsomes from skin sources A–E for the formation of 1,3-dimethyluric acid and 3-methylxanthine varied fivefold. However, the variation in specific activities of the microsomes for the formation of 1,3,7-trimethyluric acid was twofold (range 2.8 ± 0.1–6.2 ± 0.5 pmol min^{-1} [mg protein]$^{-1}$). These metabolic data may be of value in the development of transdermal theophylline systems. The results indicate that a high level of absorption enhancement will be required before transdermal theophylline preparations could produce therapeutic plasma concentrations (Ademola et al. 1992).

A high-performance thin-layer chromatographic method for the determination of theophylline in plasma has been developed. Plasma samples were prepared by the internal reference method. Denaturation was carried out through cold centrifugation. Separation of the drug from the complex matrix was accomplished within a short 50 mm migration distance and 25 min migration time on silica gel 60 F_{254} layers using toluene–isopropanol–acetic acid (80:20:5) as the mobile phase. The analyte was quantified by absorbance/reflectance densitometry using peak-area ratio analysis. Validation parameters were investigated considering the special features of the method (Jamshidi et al. 1999).

A high-performance thin-layer chromatographic method for quantification of theophylline from plasma is described. The calibration curves of theophylline in methanol and in plasma were linear in the range 20–100 ng. The correlation coefficients were 0.9971 ± 0.0011 and 0.9955 ± 0.0003 for standard curves in methanol and in plasma, respectively. The limit of quantitation of theophylline in human plasma (assay sensitivity) was 20 ng and no interference from endogenous compounds was observed. The recovery of theophylline from human plasma using the described assay procedure was 89%. The mean relative standard deviations for intra- and inter-day analyses were 1.67% and 2.34% for 50 ng and 2.25% and 3.14% for 75 ng theophylline concentration, respectively. The method was utilized to monitor plasma concentration of theophylline post-administration of sustained release tablets in human patient volunteers (Devarajan et al. 1999a).

A sustained-release formulation of theophylline with an innovative release mechanism was evaluated in adult asthmatics. The pharmacodynamics and pharmacokinetic behavior of this formulation was compared with a market formulation (Theobid). The formulations, each containing 200 mg of anhydrous theophylline, were evaluated in six male subjects, 40–55 years of age, 151–169 cm in height, 41–60 kg in weight, who were nonsmokers with moderate chronic obstructive pulmonary disease (COPD); the study was a randomized, single-dose, open, complete crossover study with an interval of 1 week. Written consent was obtained from the patients prior to the trial. Plasma samples were obtained at 0, 1, 2, 4, 6, 8, 10, and 12 h postadministration. Pulmonary functions were simultaneously recorded using an Erich Jaeger spirometer. Plasma theophylline assays were performed using high-performance thin-layer chromatography. Individual bioavailability parameters were obtained using the S-Inv computer program. Pharmacodynamic–pharmacokinetic correlation was studied using the Excel 95 version 7.0 Regression Statistics program. The test formulation (innovator) was found to be comparable with the marketed product with respect to tmax, t1/2 and Kel ($p < 0.05$). A significant difference in the means of Cmax and AUC0–12 between the innovator and the market formulation indicated a superior extent of absorption from the innovator formulation. A good pharmacodynamic–pharmacokinetic correlation was observed when plasma theophylline concentration was compared with forced expiratory volume (Devarajan et al. 1999b).

A simple assay method for theophylline in plasma using thin-layer chromatography was developed in 2002. The method involved extraction of the drug and internal standard (acetaminophen) by chloroform–isopropanol (75:25) followed by separation on TLC silica plates using a mixture of acetic acid–isopropanol–toluene (1:12:6), as the eluting solvent. Both peak height ratios and peak are ratios showed high correlation coefficient ($r > 0.98$, $p < 0.001$). However authors used peak heights for the determinations. Within-day and between-day coefficients of variation were less than 4.4% and 7.8%, respectively. The assay proved inexpensive, accurate, and reproducible with a limit of detection of 100 ng mL^{-1} that makes it suitable for bioavailability studies (Mirfazaelian et al. 2002).

The determination of the lipophilicity of xanthines by reversed-phase liquid chromatography was provided in 2004 by Gondová et al. (2004).

A method using sodium fluoride-preserved post-mortem blood samples has been established to enable evaluation of theophylline levels in the blood at the time of death. The simple, rapid, and accurate high-performance thin-layer chromatography method with UV-densitometric detection at 277 nm was validated for analysis of theophylline in postmortem blood. Theophylline was extracted at pH 8.5 with chloroform–isopropanol 8:2 from post-mortem blood after acid hydrolysis. Recovery ranged from 89.1% to 93.4% at a concentration of 10 µg mL^{-1} in the pH range 8.3–8.6. An average analytical recovery of 89.9% was achieved at pH 8.5 with a relative standard deviation of 2.2%. Chromatographic separation was performed on precoated silica gel 60 F_{254} plates, with chloroform–methanol 9:1 as mobile phase. Polynomial regression of the data points (0.5–20 µg mL^{-1}) for blood theophylline resulted in a calibration plot with a regression coefficient (r^2) = 0.998 and the detection limit was 0.5 µg mL^{-1} (S/N = 3). Intra-day and inter-day repeatability ranged from 0.5% to 0.8% and from 0.5% to 1.3%, respectively, for three different concentrations in the range 0.5–10 µg mL^{-1}. There was no evidence of degradation of theophylline either in the methanolic solution or during chromatography. The method is highly selective, because retardation factor (RF), peak resolution (RS), and peak purity were reproducible. The proposed method enables simple and rapid separation from post-mortem blood (completed in 1 h after extraction), uses less solvent, and does not require an extensive clean-up procedure. The high sensitivity (LOD 0.5 µg mL^{-1}) is comparable with that of other chromatographic techniques (Sanganalmath et al. 2009).

Thin-layer chromatography (TLC) and high-performance liquid chromatography (HPLC) methods have been developed for the determination of the xanthine drug, pentoxifylline, and three of its metabolites (a secondary alcohol and two carboxylic acids) in microbial extracts. The methods require initial extraction of acidified media with dichloromethane–2-propanol (4:1). Extracts were submitted to TLC development on silica gel G layers using three solvents and HPLC development on an C18 column using methanol–phosphoric acid (0.02 M, pH 5) (3:7) as mobile phase. All systems provide good separations of the drug and its metabolites. Quantitative analyses of pentoxifylline and its metabolites by HPLC were accurate and precise. The HPLC method was applied to studies of the metabolism of pentoxifylline by two microorganisms (Smith et al. 1983).

An analytical method to study the bioavailability of newly developed pentoxifylline sustain-release tablets has been developed and assessed in experiments on dogs. For the isolation of pentoxifylline and its metabolites from serum solid–liquid extraction was applied by involving the internal standard probe. HPTLC plates with a preconcentration zone were used for separation of the analyzed substances, using chloroform–methanol (95:5, v/v). Quantification was by densitometric detection. The detector response was linear in the concentration range investigated for pentoxifylline: $0.02-1.5$ μg mL^{-1} of serum (Bauerová et al. 1991).

This chapter presents the applicability of thin-layer chromatographic methods with a subsequent densitometric or video densitometric quantitation for determination of residues in controlling pharmaceutical equipment cleanliness. Analytical methods were developed for monitoring residues of pentoxifylline at 10 mg/M2 and mebendazol at 1 mg/M2 on stainless steel surfaces. Simulated samples were prepared by addition of a calculated amount of pharmaceutical (as a solution) on a 35×35 cm stainless steel surface. After evaporation of solvent, the residues were wiped with wetted cotton. The cotton was extracted with dichloromethane–methanol (1 + 1). Filtered extract was concentrated by vacuum evaporation and an aliquot applied to the plate, where standards were also applied. In the narrow concentration range near the acceptable residue limits, linear calibration curve could be obtained for both substances. The mean recovery (n = 4) obtained by densitometric quantitation was 93.4% for pentoxifylline and 85.6% for mebendazol, with coefficients of variation of 3.5% and 8.3%, respectively. Results of video densitometric quantitation did not differ significantly. However, data acquisition and evaluation are faster compared with densitometry and allow better archiving possibilities as required by the regulatory authorities. Both quantitation modes can be applied to routine control of pharmaceutical equipment cleanliness (Vovk and Simonovska 2001).

A HPTLC method for the separation and identification of pentoxifylline and related substances, impurities of reaction partners, and side-reaction products has been developed using different mobile and stationary phases was described. For quantitative assay of possible by-products as impurities, LiChrospher RP-18 F$_{254}$s chromatoplates, acetone–chloroform–toluene–dioxane (2:2:1:1 v/v) as a mobile phase, and detection at 275 nm were employed. Linearity (r ≥ 0.997), recovery (86.5%–115.5%), and determination limit (0.1%–0.6%) were evaluated and found to be satisfactory. This method enables monitoring of the synthesis, as well as purity control of pentoxifylline-containing raw materials and pharmaceuticals (Grozdanovic et al. 2005).

Doxophylline, a new antibronchospastic drug, being more active and less toxic than aminophyline, was detected by high-performance thin-layer chromatography. The pharmacokinetics of doxophylline have been characterized in rats, whose serum concentration were monitored after 100, 200, 400 mg kg^{-1} oral dose. The drug was found to conform to a one-compartment model and can be bio-transformed quickly in rats. The Cmax, AUC and CL/F appeared to be dose-dependent. T1/2 (Ke) was 1.17 ± 0.13 h after the 100 mg kg^{-1} dose, 2.54 ± 0.60 h after the 200 mg kg^{-1} dose and 3.75 ± 0.92 h after the 400 mg kg^{-1} dose. The doxophylline concentration in tissues decreased rapidly. Total excretion of the drug in urine, bile, and feces was 5.2% of the dose. Plasma protein binding was about 25% (Guo et al. 1997). A simple, sensitive, selective, precise, and stability-indicating high-performance thin-layer chromatographic method for analysis of doxofylline as the bulk drug and in formulations has been developed and validated. The method used aluminum plates coated with silica gel 60 F$_{254}$ as stationary phase and toluene–methanol 8:2 (v/v) as mobile phase, followed by densitometric measurement at 254 nm. The RF value of doxofylline was 4.3. The drug was subjected to acidic, alkaline, oxidative, and photolytic stress to establish a validated stability-indicating HPTLC method. The method was validated in accordance with ICH guidelines; there was no chromatographic interference from tablet excipients. The drug was found to be stable under wet and dry heat conditions. Because the method could effectively separate the drug from its degradation products it can be regarded as stability-indicating (Patre et al. 2009).

Methods for the quantitative TLC chromatographic separation and spectrophotometric analysis are reported for allopurinol [4-hydroxypyrazolo(3,4,-d) pyrimidine] in the presence of its alkaline decomposition products. Stability data are presented using rate constants obtained from Arrhenius-type plots; t90% for an unbuffered sodium allopurinol solution is approximately 150 days at 25°C. The formate salt of 3-amino-4-pyrazolecarboxamide accounts for essentially all the degradation under pharmaceutically usable conditions (Gressel and Gallelli 1968).

Simultaneous removal of indomethacine, papaverine, and allopurinol from aqueous solution by using submerged aquatic plant *Nasturtium officinale* biomass in high-performance thin-layer chromatography was studied. Optimum biosorption conditions were determined as a function of contact time, pH, removal capacity of the amount of biomass, and initial dry concentration. Experiments were performed in batch conditions. Concentrations of the drugs in the remaining solutions were simultaneously analyzed by HPTLC. Langmuir and Freundlich models were applied to describe the biosorption isotherm of the drugs by aquatic plant Nasturtium officinale biomass. According to the results, optimum parameters were found as 2.0 g biomass, pH: 5.0 and 60 min contact time. Obtained from plots of Langmuir and Freundlich adsorption models, the highest drug uptakes were calculated from Langmuir isotherm and found to be 43.10, 39.68, and 38.61 mg g^{-1} for indomethacine, papaverine, and allopurinol, respectively (Okumus et al. 2010).

21.2 AMPHETAMINE AND DERIVATIVES

Amphetamine, also known as desoxynorephedrine, fenopromin, mydrial, protioamphetamine, 1-phenyl-2-aminopropane, 1-methyl-2-phenylethylamine, or beta-aminopropylbenzene, is a powerful central nervous system stimulant and sympathomimetic. Amphetamine has multiple mechanisms of action including blocking uptake of adrenergics and dopamine, stimulation of release of monamines, and inhibiting monoamine oxidase. As adrenergic agent—that acts on adrenergic receptors or affects the life cycle of adrenergic transmitters, included here are adrenergic agonist and antagonist that affect the synthesis, storage, uptake, metabolism, or release of adrenergic transmitters. Adrenergic uptake inhibitors block the transport of adrenergic transmitters into axon terminals or into storage vesicles within terminals, and dopamine uptake inhibitors aredrugs that block the transport of dopamine into axon terminals or into storage vesicles within terminals. These drugs also block the transport of serotonin. Amphetamine is also a drug of abuse and a psychotomimetic. Sympathomimetics drugs mimic the effects of stimulating postganglionic adrenergic sympathetic nerves. Amphetamine is a central nervous system stimulant—a loosely defined group of drugs that tend to increase behavioral alertness, agitation, or excitation. The l-form has less central nervous system activity but stronger cardiovascular effects. The d-form is dextroamphetamine (Table 21.2).

Amphetamine belongs to the phenetylamine class of compounds, while the analogue is a subgroup of the substituted phenetylamine class. Typical reaction is substitution by methyl and sometimes ethyl groups at the amine and phenyl sites. The most common illegally consumed amphetamine analogues in the world are methamphetamine, 3,4-methylenedioxyamphetamine, 3,4 methylenedioxymethamphetamine, 2,5-dimethoksy-4-methylamphetamine, and 3,4 methylenedioksy-N-ethylamphetamine. These drugs are substances that have potent stimulating effects on the central nervous system. These effects are classified into three main classes: emphatogenic, hallucinogenic, and psychanaleptic. Their pharmacological activities differ depending on the enantiomeric composition. All these substance are chiral compounds, and it is well known that the S- and R-enantiomers have different pharmacological effects. The S-enantiomers exhibit grater pharmacological potency that the R-enantiomer

TABLE 21.2
Chemical Structure of Amphetamine and Derivatives

1-Phenylpropan-2-amine
Amphetamine (AM)

N-Methyl-1-phenylpropan-2-amine
Methamphetamine (MAM)

4-[2-(Methylamino)propyl]phenol
p-Hydroxymethamphetamine (pOHMA)

1-(1,3-Benzodioxol-5-yl)propan-2-amine
3,4-Methylenedioksyamphetamine (MDA)

4-(2-Aminopropyl)-2-methoxyphenol
4-Hydroxy-3-methoxy-amphetamine (pOHAM)

1-(1,3-Benzodioxol-5-yl)-N-methylpropan-2-amine
3,4-Methylenedioxymethamphetamine (MDMA)

2-Methoxy-4-[2-(methylamino)propyl]phenol
4-Hydroxy-3-methoxymethamphetamine

1-(1,3-Benzodioxol-5-yl)-N-ethylpropan-2-amine
3,4-Methylenedioxy-N-ethylamphetamine (MDEA)

4-[2-(Ethylamino)propyl]-2-methoxyphenol
4-Hydroxy-3-methoxy-N-ethylamphetamine

N-ethyl-1-phenylpropan-2-amine
Ethylamphetamine EA

1-(1,3-Benzodioxol-5-yl)-N-hydroxypropan-2-amine
N-Hydroxy-methylenedioxyamphetamine

2-(Methylamino)-1-phenylpropan-1-ol
Ephedrine/pseudoephedrine (EP/PEP)

1-(2,5-Dimethoxy-4-methylphenyl)propan-2-amine
2,5-Dimethoksy-4-methylamphetamine (DOM)

and are about five times as active as R-enantiomers. The enatiomeric determination and separation of various amphetamine compounds can be carried out by a variety of analytical techniques including gas chromatography, high-performance liquid chromatography, thin-layer chromatography, capillary electrophoresis, and nuclear magnetic resonance spectroscopy. For TLC analysis, as stationary phases for chiral separations of amphetamine and related drugs, molecular imprinted polymers, monomers (methacrylic acid or itaconic acid), synthetic polymers imprinted quinine ($CaSO_4$—as a binder), silica gel plates impregnated with optically-pure L-tartaric acid and L-histidine can be used. As a mobile phase, methanol, acetonitrile, or a different combination of mobile phases can be used (i.e., MeCN–MeOH–H_2O) were used (Płotka et al. 2011).

A rapid and precise identification of toxic substances is necessary for urgent diagnosis and treatment of poisoning cases and for establishing the cause of death in postmortem examinations. Identification of compounds in biological samples using gas chromatography and liquid chromatography coupled with mass spectrometry entails time-consuming and labor-intensive sample preparations. Thin-layer chromatography is a widely used conventional method for separation analysis and still used in parallel with highly sensitive instrumental analyses. Combination of TLC with MALDI/MS complements the disadvantages of each technique. A simple preparation and highly sensitive analysis of drugs in biological samples such as urine, plasma, and organs using thin-layer chromatography coupled with matrix-assisted laser desorption/ionization mass spectrometry (TLC/MALDI/MS) was examined. The sample (human urine, human plasma, and rat liver homogenate) was developed with acetone–28% aqueous ammonium (99:1, v/v) in a 50 mL screw-top glass vial. The time required to develop a distance of 5 cm on the condition was approximately 10 min. α-Cyano-4-hydroxycinnamic acid 20 mg mL^{-1} in 5% trifluoroacetic acid–MeCN (3:7, v/v) was sprayed on the plate. The image of the plate was scanned, and the plate was introduced into the MALD/MS instrument. To observe the spots compounds using conventional spot detection method, a sample was spotted at three points on TLC plates with fluorescence, which was irradiated with UV light at 254 nm and then sprayed with Dragendorff reagent. When the urine containing 3,4-methylenedioxymethamphetamine (MDMA) without sample dilution was spotted on a thin-layer chromatography (TLC) plate and was analyzed by TLC/MALDI/MS, the detection limit of the MDMA spot was 0.05 ng per spot. The value was the same as that in an aqueous solution spotted on a stainless steel plate. All the psychotropic compounds tested (MDMA, 4-hydroxy-3-methoxymethamphetamine, 3,4-methylenedioxyamphetamine, methamphetamine, p-hydroxymethamphetamine, amphetamine, ketamine, caffeine, chlorpromazine, triazolam, and morphine) on a TLC plate were detected at levels of 0.05–5 ng, and the type (layer thickness and fluorescence) of TLC plate did not affect detection sensitivity. In addition, when a rat liver homogenate obtained after MDMA administration (10 mg kg^{-1}) was spotted on a TLC plate, MDMA and its main metabolites were identified using TLC/MALDI/MS, and the spots on a TLC plate were visualized by MALDI/imaging MS. The total analytical time from spotting of intact biological samples to the output of analytical results was within 30 min. TLC/MALDI/MS enabled rapid, simple, and highly sensitive analysis of drugs from intact biological samples and crude extracts. Accordingly, this method could be applied to rapid drug screening and precise identification of toxic substances in poisoning cases and postmortem examinations (Kuwayama et al. 2012).

HPLC and TLC methods were developed for the separation and detection of some amphetamine analogs: methamphetamine; 3,4-methylenedioxymethamphetamine ("ecstasy"); and 3,4-methylenedioxy-N-ethylamphetamine in spiked plasma samples. The methods are based on purple chromogens formed by displacement reaction of these secondary aliphatic amine-bearing drugs with 7,7,8,8-tetracyanoquinodimethane at 80°C for 25 min. For HPLC, both normal phase (silica gel) and RP (C18) columns were used. With the former, good detection limits in plasma were obtained with a 6 min run: 70, 100, and 500 ng mL^{-1} for MDMA, MA, and MDEA, respectively. For TLC, hexane–chloroform (1:9) and benzene–diethyl ether–petroleum ether (40°–60°)–acetonitrile–ethyl methyl ketone (2:3.5:3.5:0.5:0.5) were used as mobile phases for silica gel 60 TLC and cyano-bonded silica gel HPTLC plates, respectively. The former offered more sensitive results than the latter.

Influence of evaporation steps on recovery and interferences for the HPLC and TLC methods were investigated. The developed methods are selective, simple, and easily applicable (Oztunç et al. 2010).

3,4-Methylenedioxymethamphetamine (MDMA) is the major ingredient of ecstasy illicit pills. It is a hallucinogen, central nervous system stimulant, and serotonergic neurotoxin that strongly releases serotonin from serotonergic nerves terminals. Moreover, it releases norepinephrine and dopamine from nerves terminal, but to a lesser extent than serotonin. Poisoning and even death from abusing MDMA-containing ecstasy illicit pills among abusers is usual. Thus, quantitative determination of MDMA content of ecstasy illicit pills in illicit drug bazaars must be done regularly to find the most high-dose ecstasy illicit pills and removing them from illicit drug bazaar. In the present study, MDMA contents of 13 most abundant ecstasy illicit pills were determined by quantitative thin-layer chromatography. Two procedures for quantitative determination of MDMA contents of ecstasy illicit pills by TLC were used: densitometric and so-called "scraping off" methods. The former was done in a reflection mode at 285 nm, and the latter was done by absorbance measurement of eluted scraped off spots. The limit of detection (LOD), considering signal-to-noise ratio (S/N) of 2, and limit of quantification (LOQ), regarding S/N of 10, of densitometric and scraping off methods were 0.40, 1.20, and 6.87, 20.63 µg, respectively. Repeatabilities (within-laboratory error) of densitometric and scraping off methods were 0.5% and 3.6%, respectively. The results showed that the ecstasy illicit pills contained 24–124.5 and 23.9–122.2 mg MDMA by densitometric and scraping off methods, respectively (Shetab Boushehri et al. 2009).

The study takes advantage of the presently available effective physicochemical methods (isolation, crystallization, determination of melting point, TLC, GLC, and UV spectrophotometry) for an objective and reliable qualitative and quantitative analysis of frequently abused drugs. The authors determined the conditions for qualitative and quantitative analysis of active components of the secured evidence materials containing amphetamine sulfate, methylamphetamine hydrochloride, 3,4-metylenedioxymethamphetamine hydrochloride, as well as delta (9)-tetrahydrocannabinol (delta(9)-THC) as an active component of cannabis (marihuana, hashish). Isolation of amphetamines was performed using Kiesselgel 60 F_{254} Merck plates and mobile-phase $CH_3COOC_2H_5$–MeOH–NH_4OH (8.5:1.0:0.5), applied on the plate 0.1 mL 1% ethanolic extract solutions of the tested compounds. After developing and drying, the chromatogram was identified by UV $\lambda = 254$ nm and next sprayed with 1% aqueous solution Fast Black K. The usefulness of physicochemical tests of evidence materials for opinionating purposes was subject to a detailed forensic toxicological interpretation (Wachowiak and Strach 2006).

A highly sensitive analytical method for the detection of p-hydroxymethamphetamine (pOHMA) in urine is presented. The proposed method combines liquid–liquid extraction with acetonitrile and solid-phase extraction by thin-layer chromatography with oxidation, using potassium hexacyanoferrate (III) and sodium hydroxide to detect the fluorophor of pOHMA. The TLC silica gel 60, nonfluorescence plates as stationary phase, and isopropanol–28% ammonia solution (95:5, v/v) as developing solvent were used. pOHMA, which was spotted onto TLC plate, could be detected as a fluorescent spot of vivid blue under UV irradiation when the reagent of 1 M NaOH–20 mM potassium hexacyanoferrate (III) (1:1, v/v) was sprayed upon the TLC plate. In addition, upon spraying a 50% methanol solution onto the TLC plate after heating, the fluorescence intensity was increased. The detection limit for the spot of pOHMA was 10 ng (n = 3), and R_f value was 0.26. The analysis of pOHMA in forensic samples was successfully performed, without interference from endogenous fluorophors, yielding concentrations in the appropriate range for methamphetamine abusers (Kato et al. 2005).

The abuse of the designer amphetamines such as 3,4-methylenedioxymethamphetamine is increasing throughout the world. They have become popular drugs at all night techno dance parties, and their detection is an important issue. The objective of the presented study was to identify an unknown compound detected by thin-layer chromatography in the urine of an illicit drug abuser. The compound was isolated by TLC and analyzed by gas chromatography-mass spectrometry (GC-MS) in electron ionization (EI) and positive ion chemical ionization (PICI) mode to elucidate its chemical structure. Based on EI-MS and PICI-MS mass spectral data, the unknown compound was indicated to be a structure similar to MDMA, substituted by a single chlorine atom—a chlorinated MDMA (Cl-MDMA). To confirm the Cl-MDMA structure, the unknown compound was silylated, trifluoroacetylated, acetylated, heptafluorobutyrylated, and analyzed by GC-MS. The position of the chlorine atom cannot be assigned exactly from the mass spectral data presented here; however, we believe that the unknown compound could be 6-Cl-MDMA (Maresova et al. 2005).

With the increasing number of intoxications with some preparations, at the same time there is a lack of literature data about chemical-toxicological research of psychostimulator mixtures. The aim of the study was to analyze the possibility of identification of four preparations in the mixture. The thin-layer chromatography method was used. The most acceptable solvent systems were determined: ethylacetate–methanol–conc. ammonia hydroxide (40:4:2:25) and dioxane–benzenum–conc. ammonia hydroxide–acetone (28:14:2:25) (Ivanauskaite et al. 2003).

Application of citric acid/acetic anhydride reagent (CAR), a color reagent selective for tertiary amines in solution, improves detection of abused tertiary amino drugs on the TLC plate. TLC silica gel F_{254} plates and isopropanol–28% aqueous ammonia (95:5, v/v) as developing solvent for stimulants were used. The plate was pretreated by a brief immersion in phosphoric acid/acetone solution to suppress coloration. After suppressing, the plate is sprayed with CAR and heated at 100°C, causing tertiary amines to turn red purple within 3 min. The sensitivity of this new CAR method is 2.5–15 times greater than that of conventional detection with Dragendorff reagent for some of the tertiary amines dimethylamphetamine, methylephedrine, purines, and other drugs. This present method provides rapid TLC detection of abused tertiary amino drugs including narcotics (Kato and Ogamo 2001).

Methods for the analysis of narcotics belonging to amphetamine methylene dioxy derivatives (MDD) were reviewed. The characteristics of these agents, their metabolism, and methods used for their detection and identification (TLC, GC, HPLC, GC/MS) were described. Methods for their extraction from biological objects (human urine and hair) were described. Efficacy of MDMA and MDEA from the urine by different extractants were assessed. The data demonstrate different potentialities for detection and identification of amphetamine MDD, including those in biological specimens (human urine and hairs), by numerous chromatographic methods (Veselovskaia et al. 1999).

The consumption of 3,4-methylenedioxy-N-ethylamphetamine (MDEA), an analogue of ecstasy, can be detected by direct in situ HPTLC-FTIR measurement of the main metabolite N-ethyl-4-hydroxy-3-methoxyamphetamine (HME). HME can, like the other important metabolite 3,4-methylenedioxyamphetamine (MDA) and unchanged MDEA, be determined quantitatively in urine by HPTLC-UV after two-step automatic development. Sample preparation for determination of HME (with enzymatic hydrolysis) and of MDEA and MDA (both without enzymatic hydrolysis) was carried out separately. HPTLC silica gel 60 F_{254} plates with a layer thickness of 200 μm were employed. The development was carried out in two steps each of 15 min with toluene–acetone–ethanol–ammonia 25% (45:45:7:3) followed by acetone–methanol (50:50) in the AMD apparatus. Detection was carried out at $\lambda = 283$ nm. The identification of MDEA and HME was carried out by recording the HPTLC-FTIR spectra maxima of the chromatograms and comparing the spectra obtained with the contents of an HPTLC-FTIR spectrum library. The quantitation was carried out in UV fluorescence mode. The internal standard MDMA and MDE were measured at $\lambda = 283$ nm, HME at $\lambda = 278$ nm and MDA by fluorescence after postchromatographic derivatization with fluorescamine dipping reagent after excitation at $\lambda = 365$ nm using a 450 nm cut-off filter. The results have been compared with those obtained using an HPLC method. The differences were not generally significant. Small deviations were attributable to the different sample preparation methods necessary. The working range for the HPTLC method was between 0.1 and 8.2 μg mL^{-1} and for the HPLC method between 0.2 and 60.0 μg mL^{-1}. The method standard deviations were 2.66%–4.91% (HPTLC) and 0.48%–3.67% (HPLC) (Pisternick et al. 1997).

Klimes and Pilarová (1996) investigated the chromatographic conditions for the TLC analysis of metamphetamine on silica gel and the lipophilic stationary phase RP V 18 to qualitatively analyze the drug in urine samples. Attention was also paid to chemical detection. Five detection reagents were tested, out of which Fast Black K salt, yielding orange–red spots with a detection limit of 1 µg, proved to be the best. In the analysis of metamphetamine in the samples of model urine, the best results were achieved on Kieselgel. Using the developing system ethyl acetate–ethanol–conc. solution of ammonia (36:2:2), a complete separation of the metamphetamine spot from the spots of ballast from the biological matrix was achieved (Klimes and Pilarová 1996).

Recently, there have been claims among drug users that some herbal drinks interfere with urinalysis for drugs of abuse and yield false positive results. Proof of such claims has yet to be shown. Screening for drugs of abuse is usually carried out using fluorescence polarization immunoassay (FPIA) or thin-layer chromatography (TLC). Fifty herbal samples that are considered among the most purchased herbs in the consumer market were used to investigate such claims. The drug groups that were tested for included amphetamines, opiates, barbiturates, cocaine metabolite, methadone, and their analogs. The herbs were analyzed at different concentrations (0.1, 1, 3, and 5 g [100 mL]$^{-1}$ of distilled water) using TLC and FPIA to determine if any interfere with urinalysis for drugs of abuse and yield false positive results. For the FPIA test, the sample infusions were analyzed directly using the automated ADX analyzer (Abbott Laboratory). For TLC, infusions of the herbs were added to a solid-phase extraction column (pH 9.25), then extracted with a methylene chloride–isopropanol solvent system. At this pH, neutral, basic, and acidic drugs of abuse are extractable. The developed chromatographic plates were sprayed sequentially with several reagents. None of the herbs in the concentration ranges screened showed any interference with TLC or FPIA, indicating the invalidity of such claims (Winek et al. 1993).

Methamphetamine and amphetamine were screened, and their levels were determined using the TOXI-LAB® thin-layer chromatography system and gas chromatography-mass spectrometry, respectively, in the blood, urine, and stomach contents from 211 emergency medical care and 417 autopsy cases. MAM and AM were detected in five emergency medical cases, and the blood MA and AMP concentrations ranged from 0.697 to 0.041 µmol (100 g)$^{-1}$ and from 0.0944 to 0.0003 µmol (100 g)$^{-1}$, respectively. MAM and AM were detected in 19 autopsy cases, in which blood MAM and AM concentration ranged from 14.3 to 0.123 µmol (100 g)$^{-1}$ and from 0.256 to 0.0017 µmol (100 g)$^{-1}$, respectively. The autopsy cases included five cases of sudden death with blood MAM concentration of less than 3 µmol (100 g)$^{-1}$. MAM and AM screening and determination in emergency medical care and autopsy cases provide useful information and are indispensable in clarifying the dimensions of MAM abuse in Japan (Takayasu et al. 1995).

Using the TOXI-LAB drug detection system, emergency toxicological screening was performed in autopsy cases and emergency cares. In 280 autopsy cases (male 182 cases 65%, female 98 cases 35%), drug positive cases were 28 cases of male (15%) and 24 cases of female (24%). The age groups that showed higher rate of drug positive cases were 10–40s in male (approximately 20%) and 20s in female (67%). In the 238 cases of emergency care (male 129 cases 54%, female 104 cases 44%, unknown 5 cases 2%), drugs were positive in 29 cases of male (22%) and 32 cases of female (30%). The age groups that showed relatively higher rate of drug positive cases were 40s in male (64%), 20s (71%) and 30s (89%) in female. Forty-four different kinds of drugs were detected in TOXI-LAB positive cases, in which the psychotropic drugs and the sedative-hypnotic drugs amounted to approximately 70%. Methamphetamine and amphetamine, which were the main abused drugs showing a socially important problem, were detected in total 15 cases. TOXI-LAB was based on thin-layer chromatography (TLC); however, from the extraction to development, coloration and detection have been accelerated (about 50 min) and simplified. In order to perform the forensic toxicological practice in Japan, it becomes more useful that TOXI-LAB is used in autopsy cases and emergency cares, if the drugs, which have caused poisoning cares in Japan, are added to TOXI-LAB. The present study describes the advantage and problem of TOXI-LAB drug detection system through demonstrating the practical cases of autopsy and emergency cares (Nishigami et al. 1993).

Several diazonium salts and thin-layer chromatographic (TLC) solvent systems were evaluated for the characterization of amphetamines and other basic and nonbasic drugs. With the Fast Black

K (FBK) salt, it was possible to discriminate between primary amines (e.g., amphetamines) and secondary amines (e.g., methylamphetamines) by their different colored derivatives. A TLC solvent system consisting of methanol:acetone:ammonia (100:24:1.6) was found to give reproducible R_f data for these drugs. The optimum reaction conditions were determined for the derivatization procedure by isolation and identification of the products. These conditions provided minimum detection levels of 50 ng for amphetamine and methylamphetamine (Munro and White 1995).

The analytical method for 3,4-methylenedioxy-N-ethylamphetamine, one of the hallucinogenic phenethylamine derivatives newly controlled in Japan, was investigated in several aspects of forensic chemical analysis. The standard MDEA was synthesized from piperonylmethylketone. Researchers analyzed MDEA, its related compounds (3,4-methylenedioxymethamphetamine, 3,4-methylene-dioxyamphetamine and ethylamphetamine [EA]), and a seized sample containing MDEA, metham-phetamine, and caffeine using color test, thin-layer chromatography (TLC), infrared spectrometry (IR), gas chromatography (GC) and gas chromatography mass spectrometry (GC-MS). Under the analytical conditions described here, they succeeded in complete separation and identification of these drugs. It may also be possible to apply our method for the intake of drug abuse such as MDEA (Ohshita et al. 1995).

Identification and purities of N-ethyl methylenedioxyamphetamine, N-hydroxy methylene-dioxyamphetamine, mecloqualone, 4-methylaminorex, phendimetrazine, and phenmetrazine were described. The reference standards of all these substances were chemically prepared from commercial chemicals. Their purities determined by HPLC were more than 99.8%. The standard spectra and chromatograms of the standards such as TLC, UV, IR, HPLC, GC/MS, and NMR were measured. For the identification of these six drugs in forensic laboratory, their mass fragmentation and NMR spectra were discussed (Shimamine et al. 1993).

A rapid, low-cost method for detecting opioids, cocaine, and amphetamine in the urine of drug abusers was present in 1990. Rapid solid-phase extraction of 2 mL of urine using octadecylsilane cartridges (Bond Elut C18) concentrates compounds of interest (including the polar metabolite of cocaine, benzoylecgonine), in a small volume of methanol, which is easily dried down. Four micro-liters of aliquots of samples and standards were spotted on to 5 cm^2 polyester-backed silica gel plates and developed (8–10 min) in a Desaga H-Chamber. R_f values of all relevant compounds were presented for two mobile phases using an iodoplatinate spray. The sensitivity of the method was not the same for all drugs but was always at least 1 µg mL^{-1} urine. A number of investigations can be performed on a single extract from 2 mL of urine including confirmatory tests using different solvent systems or sprays (Wolff et al. 1990).

Five extraction methods were examined for the analysis of methamphetamine and its major metab-olites in tissue samples. The extraction methods studied were an acetone extraction method, an ethanol extraction method, an ammonium sulfate method, dialysis, and a direct solvent extraction. Acetone, ethanol, and dialysis methods showed no interference from endogenous components using thin-layer chromatography and gas chromatography and gave satisfactory recovery of methamphetamine, amphetamine, and p-hydroxymethamphetamine when added to rabbit liver. These methods, however, proved time-consuming. The ammonium sulfate method and direct solvent extraction method were simple and more rapid, but recovery of the polar metabolite was poor (Inoue and Suzuki 1986).

Rasmussen and Knutsen (1985) reviewed analytical techniques for the detection and identification of amphetamines and amphetamine-like substances in nonbiological samples. It shows the wide range of methods available, from simple testing procedures to the use of the most powerful instruments available in analytical chemistry. The following techniques are discussed: color tests, microcrystal tests, ultraviolet spectrophotometry, fluorescence spectrometry, infra-red and Raman spectrophotometry, thin-layer chromatography, gas chromatography, and high-performance liquid chromatography (Rasmussen and Knutsen 1985).

The excretion of amphepramone, amphetamine, ephedrine, and prolintane was estimated qualitatively from saliva and urine using thin-layer chromatography (TLC) 1, 2, 4, 8, 12, and 24 h after a single intake of therapeutic doses by healthy volunteers. The purpose was to compare the suitability of these two biological fluids for doping tests, and it was therefore important that either the drug or its metabolites should be found in all subjects. Ephedrine was the only one to be detected regularly in saliva already 1 h after drug intake, and it was found there up to 8 h, while in urine it was found up to 24 h. In saliva, prolintane or its main metabolite did not appear during 24 h, whereas it or its metabolite were consistently found in urine 4–12 h after drug intake. Neither amphepramone nor its metabolites were detected at any time in saliva, but this drug was found, either in unchanged or in metabolized form, in urine 1–24 h after intake. The same was true for amphetamine, except at 1 h in urine. The findings did not clearly correlate with the pH of either saliva or urine samples. It is concluded that in TLC screening saliva is inferior to urine in doping tests (Vapaatalo et al. 1984).

The paper in 1981 reviewed the most common synthesis of amphetamines that may be found in illicit traffic. Emphasis was laid on the detection, isolation, and identification of impurities in illicit amphetamines through gas chromatography and thin-layer chromatography. The latter method was also used for isolation purposes. Impurities were identified by mass spectral and NMR spectroscopic methods and the relevant data are presented (Sinnema and Verweij 1981).

A rapid, inexpensive, and simple screening procedure for the detection of amphetamine abuse was developed for use by laboratories without sophisticated equipment. A small volume of extract from a pH-adjusted urine specimen was used to spot a high-resolution micro TLC plate. The developed TLC plate was sprayed with a solution of fluorescamine in dried acetone. When viewed under ultraviolet illumination, amphetamines and other compounds with a primary amino group complexed with fluorescamine appear as greenish or bluish-white fluorescent spots. Secondary or tertiary amines did not react with fluorescamine. About 20 min was required to perform the procedure; the lower limit of detectability was approximately 100 ng mL^{-1} urine (Decker and Thompson 1978).

The isolation and identification of two pyrimidines, five pyridines, and one pyridone as impurities in illicit amphetamines prepared by the Leuckart synthesis are reported. Isolation was achieved by repeated thin-layer chromatography with various solvent mixtures, while identification was done by both high and low-resolution mass spectrometry and 1H and 13C NMR spectroscopy. Some chromatographic data were reported, and a quantitative analysis of a reaction mixture and an illicit amphetamine was given (van der Ark et al. 1978).

21.3 NOOTROPICS

There are two psychostimulant groups of nootropics that stand out:

- The right nootropics drugs to normalize the metabolism of neurons through reducing the oxygen demand or preventing the uncoupling of phosphorylation in the mitochondria
- Improve blood flow, which contributes to better supply to the brain with oxygen

In the chemical character, nootropics are pyrrolidin or aminoethanol derivatives (Table 21.3).

In biological samples after oral doses of piracetam this drug was detectable in serum as well as in urine after formation of the red Fe(III)–hydroxamate complex by thin-layer densitometry. Detection in serum requires extraction with a mixture of dichloromethane/methanol, while urine can be used directly after dilution. The limit of quantification in urine is 100 μg mL^{-1} and in serum 4.0 μg mL^{-1}. This method was validated by HPLC. The coefficient of correlation was 0.9999% for determination in urine and 0.9986% for determination in serum (Bockhard et al. 1997).

A simple and reproducible TLC method has been described for the separation of piracetam from its manufacturing and degradation impurities. All the analytes were separated from each other on both conventional and HPTLC Kieselgel 60 F$_{254}$ plates with a quaternary mobile phase of pentyl acetate–ethyl acetate–ethanol–glacial acetic acid (10:10:9:1, v:v:v:v). Detection of piracetam and its impurities was accomplished using Gibb's reagent–ammonia vapor or dual-wavelength UV scanning densitometry (sample: 210 nm; reference: 230 nm). The methods are thought to be suitable for the routine control of the four impurities because all substances exhibited similar detector responses (Ovalles et al. 2000).

TABLE 21.3
Chemical Structure of Nootropics

Pyrrolidin derivatives

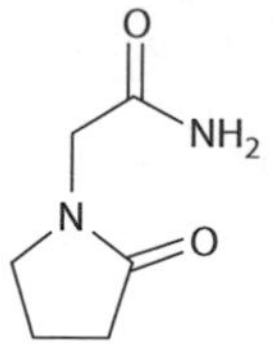

2-(2-Oxopyrrolidin-1-yl) acetamide
Piracetam

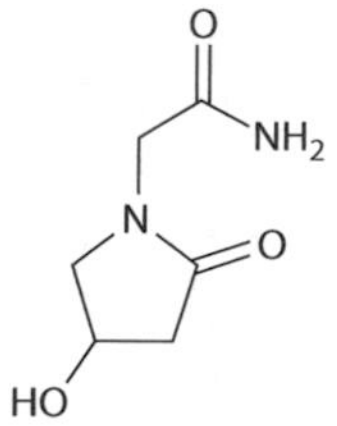

2-(4-Hydroxy-2-oxopyrrolidin-1-yl) acetamide
Oxiracetam

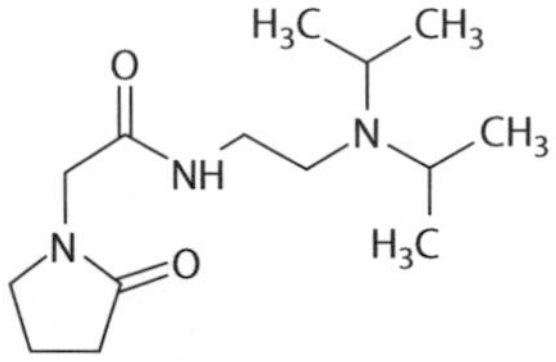

N-[2-(diisopropylamino)ethyl]-2-(2-oxopyrrolidin-1-yl)
    acetamide
Pramiracetam

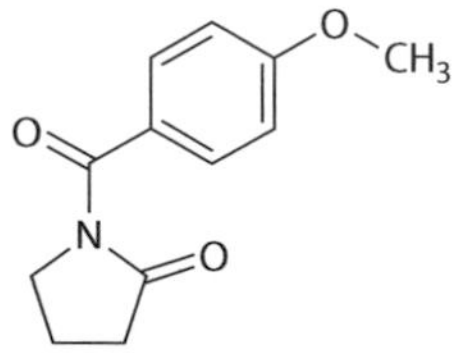

1-(4-Methoxybenzoyl)pyrrolidin-2-one

Aniracetam

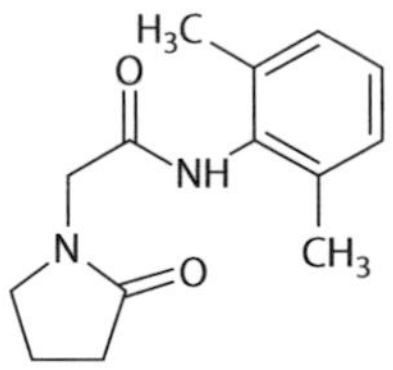

N-(2,6-dimethylphenyl)-2-(2-oxopyrrolidin-1-yl) acetamide

Nefiracetam

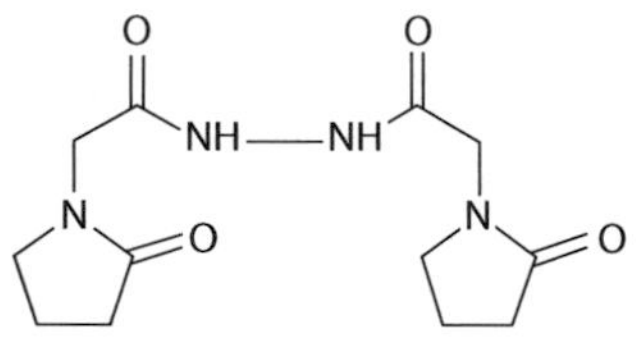

2-(2-Oxopyrrolidin-1-yl)-*N*'-[(2-oxopyrrolidin-1-yl)
    acetyl]acetohydrazide
Dupracetam

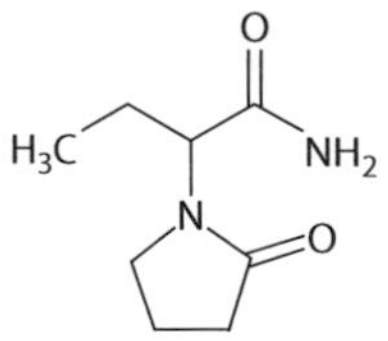

2-(2-Oxopyrrolidin-1-yl)butanamide
Etiracetam

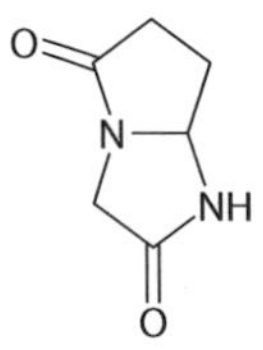

Dihydro-1H-pyrrolo[1,2-a]imidazole-2,5(3H,6H)-dione
Dimiracetam

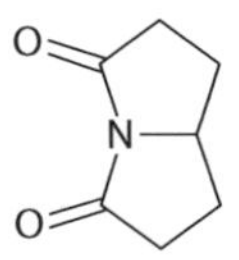

Dihydro-1H-pyrrolizine-3,5(2H,6H)-dione
Rolziracetam

TABLE 21.3 (continued)
Chemical Structure of Nootropics

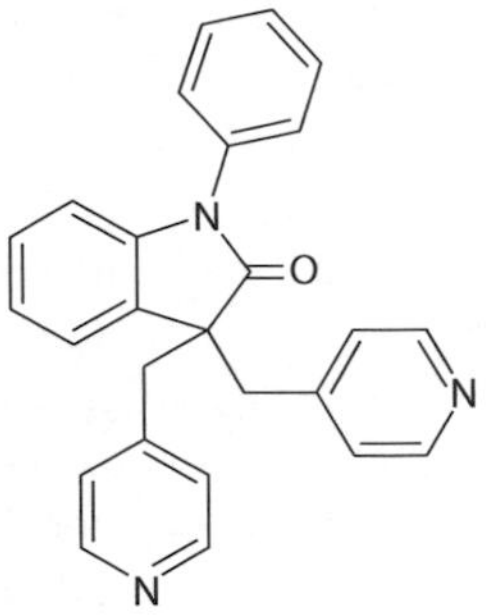

1-Phenyl-3,3-bis-(pyridin-4-ylmethyl)-1,3-dihydro-2H-
indol-2-one

Linopirdine

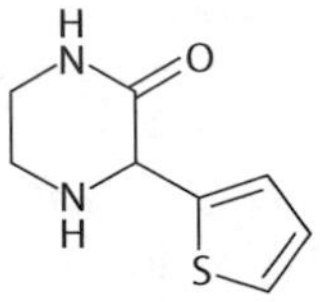

3-(2-Thienyl)piperazin-2-one

Tenilsetam

Aminoethanol derivatives

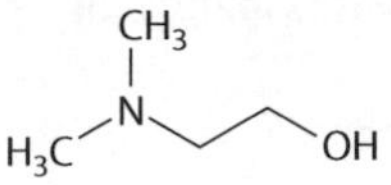

2-(Dimethylamino)ethanol

Deanol

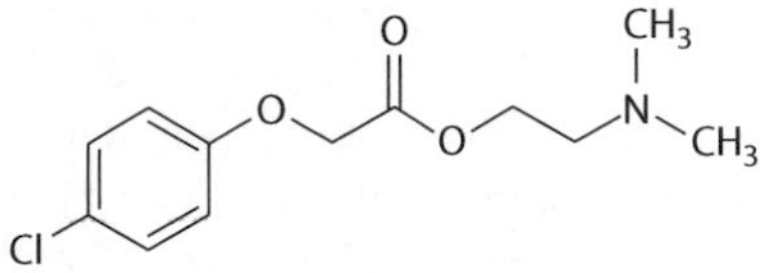

2-(Dimethylamino)ethyl (4-chlorophenoxy) acetate

Meclophenoxate

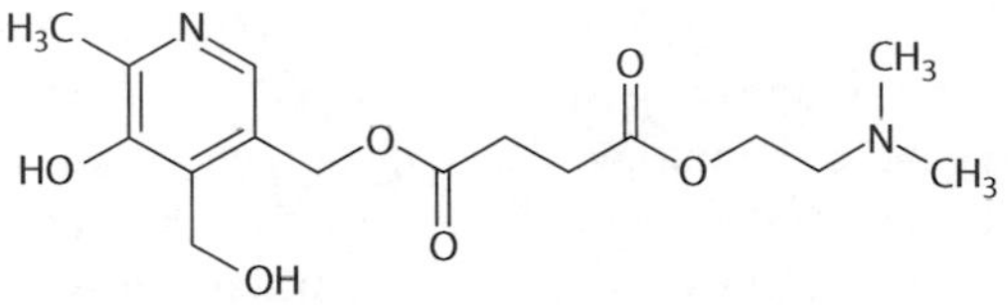

2-(Dimethylamino)ethyl [5-hydroxy-4-(hydroxymethyl)-6-
methylpyridin-3-yl]methyl succinate

Pirisudanol

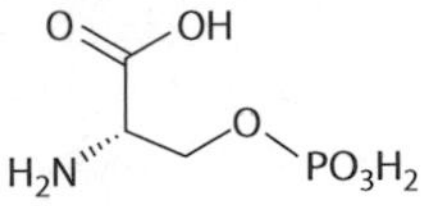

(2S)-2-Amino-3-(phosphonooxy)propanoic acid

Dexfosfoserine

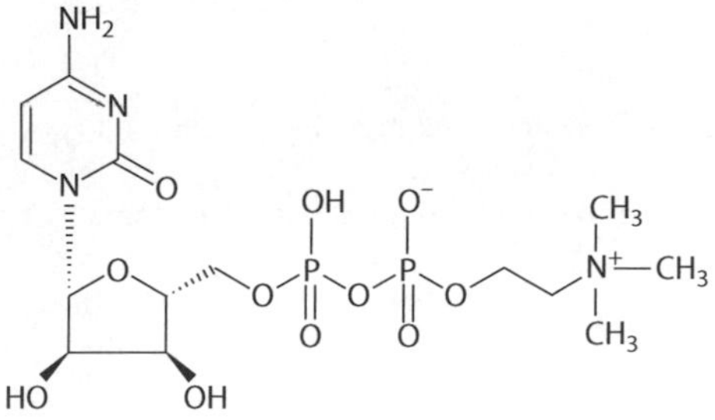

[[(2R,3S,4R,5R)-5-(4-amino-2-oxopyrimidin-1-yl)-3,
4-dihydroxyoxolan-2-yl]methoxy-hydroxyphosphoryl]
2-(trimethylazaniumyl)ethyl phosphate

Citicoline

An analysis of alcoholic extracts so-called ecstasy tablets and their hydrolysates obtained at various pH was carried out by thin-layer chromatography in a screening system, UV spectrophotometry, HPLC-DAD and mass spectrometry. The preparation offered, in the form of powders or tablets, contain mainly amphetamine and its derivatives in pure form or with admixture of medicines, mainly analgesics (paracetamol, salicylic acid) or sugar (saccharose). Preliminary screening quantitative examination were performed, which did not indicate the presence of amphetamine, its dioxy-derivatives or methamphetamine ephedrine, morfine, and cocaine. Methanol extract of the tablets and standard solutions were examined by means of thin-layer chromatography. The analyses were performed on Merck aluminum plates coated with silica gel G with the addition of fluorescein. The chromatograms were preliminary eluted in chloroform–acetone (90:10) and methanol–ammonia (99:1) system. The chromatographic zones were revealed by observation of the plates in UV light at 254 nm, and use of the following tests kit: Dragendorff, Bratton-Marshal, Marquis, Mandelin, aqueous solution of $FeCl_3$, solution of $HgCl_2$ and chlorobenzydine reagent. In a further study performed using the TLC method, a developing system was used composed of benzene–dioxane–acetic acid (90:25:4) and the chlorobenzydine tests as developing reagent. The results of color reactions, as well as the values of separation coefficient, suggested the presence of among others piracetam in the analyzed tablets. This comprehensive analysis allowed to detect four active components in these tablets: piracetam, caffeine, pemoline, and salicylic acid. Such a quantitative composition makes this patent drug a strong stimulant (Kulikowska et al. 2002).

Etiracetam, also known as Etiracetamum, alpha-ethyl-2-oxo-1-pyrrolidineacetamide, alpha-ethyl-2-oxo-2-(2-oxopyrrolidin-1-yl)butanamide, or 1-pyrrolidineacetamide, is an enantiomer (2S)-2-(2-oxopyrrolidin-1-yl)butanamide known as Levetiracetam—an anticonvulsant used to prevent seizures or reduce their severity and nootropic agents—drugs used to specifically facilitate learning or memory, particularly to prevent the cognitive deficits associated with dementias. These drugs act by a variety of mechanisms. While no potent nootropic drugs have yet been accepted for general use, several are being actively investigated. Levetiracetam TLC analysis is discussed in the group of anticonvulsive drugs.

The available literature on the use of analytical method thin-layer chromatography does not include the following medicinal substances: oxiracetam, pramiracetam, aniracetam, linopirdyne, tenilsetam, nefiracetam, dupracetam, dimiracetam, rolziracetam, deanol, meclophenoxate, pirisudanol, citicoline, and dexphosphoserine.

REFERENCES

Abourashed, E. A., J. S. Mossa. 2004. HPTLC determination of caffeine in stimulant herbal products and power drinks. *Journal of Pharmaceutical and Biomedical Analysis* 36:617–620.

Ademola, J. I., R. C. Wester, H. I. Maibach. 1992. Cutaneous metabolism of theophylline by the human skin. *Journal of Investigative Dermatology* 98:10–14.

Agbaba, D., D. Zivanov-Stakić, N. Vukićević. 1992. Dissolution assay of theophylline, diprophylline and proxyphylline from a sustained release dosage form by high performance thin layer chromatography. *Biomedical Chromatography* 6:141–142.

Aigner, R., H. Spitzy, R. W. Frei. 1976. Separation of drug substances by modern liquid chromatography on silver impregnated silica gels. *Journal of Chromatographic Science* 14:381–385.

Allen, L. 1974. Quantitative determination of carisoprodol, phenacetin, and caffeine in tablets by near IR spectrometry and their identification by TLC. *Journal of Pharmaceutical Science* 63:912–963.

Aly, Z. H. 1981. Studies on caffeine and theobromine in sheep. II. Caffeine. *Journal of Veterinary Medicine, Series A* 28:701–710.

Antonelli-Ushirobira, T. M., E. Yamaguti, L. M. Uhemura, J. C. Palazzo De Mello. 2004. Quality control of samples of Paullinia cupana H.B.K. var. sorbilis (Mart.) Ducke. *Acta Farmaceutica Bonaerense* 23:383–386.

Aranda, M., G. Morlock. 2007a. Simultaneous determination of caffeine, ergotamine, and metamizol in solid pharmaceutical formulation by HPTLC-UV-FLD with mass confirmation by online HPTLC-ESI-MS. *Journal of Chromatographic Science* 45:251–255.

Aranda, M., G. Morlock. 2007b. New method for caffeine quantification by planar chromatography coupled with electrospray ionization mass spectrometry using stable isotope dilution analysis. *Rapid Communications in Mass Spectrometry* 21:1297–1300.

Arnaud, M. J. 1976. Identification, kinetic and quantitative study of [2–14C] and [1 Me-14C]caffeine metabolites in rat's urine by chromatographic separations. *Biochemical Medicine* 16:67–76.

Arnaud, M. J. 1985. Comparative metabolic disposition of [1-Me14C] caffeine in rats, mice, and Chinese hamsters. *Drug Metabolism and Disposition* 13:471–478.

Bauer, G. K., A. M. Pfeifer, H. E. Hauck, K. A. Kovar. 1998. Development of an optimized sorbent for direct HPTLC-FTIR on-line coupling. *Journal of Planar Chromatography—Modern TLC* 11:84–89.

Bauerová, K., L. Soltés, Z. Kállay, K. Schmidtová. 1991. Determination of pentoxifylline in serum by high-performance thin-layer chromatography. *Journal of Pharmaceutical and Biomedical Analysis* 9:247–250.

Bebawy, L. I., N. M. El-Kousy. 1999. Simultaneous determination of some multicomponent dosage forms by quantitative thin layer chromatography densitometric method. *Journal of Pharmaceutical and Biomedical Analysis* 20:663–670.

Bernardo, N. P., M. E. P. B. Siqueira, M. J. N. De Paiva, P. P. Maia. 2003. Caffeine and other adulterants in seizures of street cocaine in Brazil. *International Journal of Drug Policy* 14:331–334.

Bockhard, H., H. Oelschläger, R. Pooth. 1997. Fast thin-layer densitometric determination of the nootropic piracetam in biological material. Pharmazie 52:357–361.

Bradbrook, I. D., C. A. James, P. J. Morrison, H. J. Rogers. 1979. Comparison of thin-layer and gas chromatographic assays for caffeine in plasma. *Journal of Chromatography* 163:118–122.

Buhl, F., S. Anikiel, U. Hachuła. 2006. Densitometric determination of caffeine in tea beverages after TLC separation. *Chemia Analityczna* 51:603–611.

Carmichael, W. M. 1971. The determination of nikethamide and other compounds in pharmaceutical dosage forms by thin-layer chromatography. *The Analyst* 96:716–720.

Cieszyński, T. 1971. Studies on stability of drugs in lyophilized suppository bases. 1. Separation and detection of products of decomposition of cyclovegantine, ethionamide and euphylline by means of thin-layer chromatography. *Acta Poloniae Pharmaceutica—Drug Research* 28:59–69.

Cimpoiu, C., A. Hosu, L. Seserman, M. Sandru, V. Miclaus. 2010. Simultaneous determination of methylxanthines in different types of tea by a newly developed and validated TLC method. *Journal of Separation Science* 33:3794–3799.

Cole, M. L., R. L. Kunka. 1984. Pharmacokinetics and bioavailability of theophylline following enema and suppository administration in man. *Biopharmaceutics and Drug Disposition* 5:229–240.

Dave Hiral, N., C. Mashru Rajeshree, K. Patel Alpesh. 2010. Thin layer chromatographic method for the determination of ternary mixture containing Salbutamol sulphate, Ambroxol hydrochloride and Theophylline. *International Journal of Pharmaceutical Sciences* 2:390–394.

Decker, W. J., J. D. Thompson. 1978. Rapid detection of amphetamine in urine by micro thin-layer chromatography and fluorescence. *Clinical Toxicology* 13:545–549.

Devarajan, P. V., P. N. Sule, D. V. Parmar. 1999a. Comparative pharmacodynamic–pharmacokinetic correlation of oral sustained-release theophylline formulation in adult asthmatics. *Drug Development and Industrial Pharmacy* 25:529–534.

Devarajan, P. V., P. N. Sule, D. V. Parmar. 1999b. High-performance thin-layer chromatographic determination of theophylline in plasma. *Journal of Chromatography B: Biomedical Sciences and Applications* 736:289–293.

Dumont, E., L. Kraus. 1970. Untersuchungen von faktoren, die das dünnschichtchromatographische verhalten der stoffe beeinflussen. 2. Mitt. dünnschichtchromatographie von coffein, theophyllin und theobromin auf pH-gradient-schichten. *Journal of Chromatography A* 48:106–112.

Ebel, S., G. Herold. 1975a. Quantitative thin layer chromatography using an internal standard. Part I. Determination of caffeine by in situ reflectance measurement. *Chromatographia* 8:35–37.

Ebel, S., G. Herold. 1975b. Quantitative evaluation of thin-layer chromatograms: Multicomponent analysis from calibration curves and by internal standard. *Chromatographia* 8:569–572.

Ebel, S., G. Herold. 1975c. Quantitative evaluation of thin-layer chromatograms: On-line coupling with programmable electronic desk calculators—Part 4. Two-component analysis. *Fresenius' Zeitschrift für Analytische Chemie* 273:7–9.

Essig, S., K. A. Kovar. 1999. Impurity test of theophylline: Comparison of classical slit scanner with videodensitometry. *Journal of Planar Chromatography—Modern TLC* 12:63–65.

Fenske, M. 2007. Caffeine determination in human saliva and urine by TLC and ultraviolet absorption densitometry. *Chromatographia* 65:233–238.

Flinn, P. E., Y. H. Juhl, T. P. Layloff. 1989. A simple, inexpensive thin-layer chromatography method for the analysis of theophylline tablets. *Bulletin of the World Health Organization* 67:555–559.

Franeta, J. T., D. D. Agbaba, S. M. Eric, S. P. Pavkov, S. D. Vladimirov, M. B. Aleksic. 2001. Quantitative analysis of analgoantipyretics in dosage form using planar chromatography. *Journal of Pharmaceutical and Biomedical Analysis* 24:1169–1173.

Gheorghe, M., D. Bălălău, M. Ilie, D. L. Baconi, A. M. Ciobanu. 2008. Qualitative analysis of confiscated illegal drugs by thin-layer chromatography. *Farmacia* 56:541–546.

Gondová, T., M. Vincová, K. Flórián. 2004. Determination of the lipophilicity of xanthines by reversed-phase liquid chromatography. *Journal of Planar Chromatography—Modern TLC* 17:156–158.

Gorodischer, R., A. Yaari, M. Margalith et al. 1986a. Changes in theophylline metabolism during postnatal development in rat liver slices. *Biochemical Pharmacology* 35:3077–3081.

Gorodischer, R., E. Zmora, Z. Ben-Zvi et al. 1986b. Urinary metabolites of caffeine in the premature infant. *European Journal of Clinical Pharmacology* 31:497–499.

Gressel, P. D., J. F. Gallelli. 1968. Quantitative analysis and alkaline stability studies of allopurinol. *Journal of Pharmaceutical Sciences* 57:335–338.

Grozdanovic, O., D. Antic, D. Agbaba. 2005. Development of a HPTLC method for in-process purity testing of pentoxifylline. *Journal of Separation Science* 28:575–580.

Guo, H., X. Y. Xu, L. Y. He, J. H. Huang. 1997. Preliminary study on the absorption, distribution and excretion of doxophylline in rats. *Yaoxue Xuebao* 32:81–84.

Gupta, R. N., F. Eng, M. Stefanec. 1978. Estimation of theophylline in plasma by thin-layer chromatography. *Clinical Biochemistry* 11:12–15.

Hałka, A., P. Płocharz, A. Torbicz, T. Dzido. 2010. Reversed-phase pressurized planar electrochromatography and planar chromatography of acetylsalicylic acid, caffeine, and acetaminophen. *Journal of Planar Chromatography—Modern TLC* 23:420–425.

Halkina, T., J. Sherma. 2006. Use of the ChromImage flatbed scanner for quantification of high-performance thin layer chromatograms in the visible and fluorescence-quenching modes. *Acta Chromatographica* 17:250–260.

Harry, E. L., J. C. Reynolds, A. W. T. Bristow, I. D. Wilson, C. S. Creaser. 2009. Direct analysis of pharmaceutical formulations from non-bonded reversed-phase thin-layer chromatography plates by desorption electrospray ionisation ion mobility mass spectrometry. *Rapid Communications in Mass Spectrometry* 23:2597–2604.

Hauck, H. E., M. Schulz. 2002. Ultrathin-layer chromatography. *Journal of Chromatographic Science* 40:550–552.

Heilweil, E., J. C. Touchstone. 1981. Theophylline analysis by direct application of serum to thin layer chromatograms. *Journal of Chromatographic Science* 19:594–597.

Henion, J., G. A. Maylin, B. A. Thomson. 1983. Determination of drugs in biological samples by thin-layer chromatography-tandem mass spectrometry. *Journal of Chromatography* 271:107–124.

Hurst, W. J., R. A. Martin, S. M. Tarka Jr. 1984. Analytical methods for quantitation of methylxanthines. *Progress in Clinical and Biological Research* 158:17–18.

Inoue, T., S. Suzuki. 1986. Comparison of extraction methods for methamphetamine and its metabolites in tissue. *Journal of Forensic Sciences* 31:1102–1107.

Ivanauskaite, K., R. Marksiene, G. Kiliuviene, D. Kazlauskiene. 2003. The chemical-toxicological research of psychostimulator mixture by thin-layer chromatography. *Medicina (Kaunas)* 39:113–116.

Jamshidi, A., M. Adjvadi, S. Shahmiri, A. Masoumi, S. W. Husain, M. Mahmoodian. 1999. A new high performance thin-layer chromatography (HPTLC) method for determination of theophylline in plasma. *Journal of Liquid Chromatography & Related Technologies* 22:1579–1587.

Kato, N., H. Kubo, H. Homma. 2005. Fluorescence analysis of p-hydroxymethamphetamine in urine by thin-layer chromatography. *Analytical Sciences* 21(9):1117–1119.

Kato, N., A. Ogamo. 2001. A TLC visualisation reagent for dimethylamphetamine and other abused tertiary amines. *Science & Justice-Journal of the Forensic Science Society* 41(4):239–244.

Kawamoto, H., T. Yamane. 1985. Determination of thephylline and caffeine in blood serum by thin layer chromatography-densitometry. Rinsho byori. *The Japanese Journal of Clinical Pathology* 33:217–221.

Khanna, K. L., G. S. Rao, H. H. Cornish. 1972. Metabolism of caffeine-3H in the rat. *Toxicology and Applied Pharmacology* 23:720–723.

Klimes, J., P. Pilarová. 1996. Thin-layer chromatography analysis of methamphetamine in urine samples. *Ceska a Slovenska Farmacia* 45(6):279–283.

Kopyt'ko, Y. F. 2008. Standardization of homeopathic Coffea arabica (Coffea cruda) and Coffea tosta matrix tinctures. *Pharmaceutical Chemistry Journal* 42:647–649.

Kosmeas, N., J. T. Clerc. 1989. A quick DC method for simultaneous quantitative determination of paracetamol, caffeine, phenobarbital and propyphenazone in plasma. *Pharmaceutica Acta Helvetiae* 64:2–7.

Kraus, L.J., E. Dumont. 1970. Factors influencing behavior in thin-layer chromatography. I. The influence of the pH value in thin-layer chromatography of caffeine theophylline and theobromine on uniform layers. *Journal of Chromatography A* 48:96–105.

Krishnamurthy, R., M. K. Malve, B. M. Shinde. 2000. Simultaneous determination of morphine, caffeine, and paracetamol in the urine of addicts by HPTLC and HPLC. *Journal of Planar Chromatography—Modern TLC* 13:171–175.

Kubiak, Z., J. Porebski, T. Stozek. 1974. TLC separation of aminophenazone, phenprobamate, caffeine and ascorbic acid. *Polish Journal of Pharmacology & Pharmacy* 26:581–584.

Kulikowska, J., R. Celiński, A. Soja, H. Sybirska. 2002. Identification study of tablets of unknown composition originating from illicit drug sale. Z Zagadnień Nauk Sądowych 49:99–113.

Kuwayama, K., K. Tsujikawa, H. Miyaguchi, T. Kanamori, Y. T. Iwata, H. Inoue. 2012. Rapid, simple, and highly sensitive analysis of drugs in biological samples using thin-layer chromatography coupled with matrix-assisted laser desorption/ionization mass spectrometry. *Analytical and Bioanalytical Chemistry* 402:1257–1267.

Li, B., Y. L.Tong, B. T. Zou. 1985. TLC determination of theophylline in plasma. *Acta Pharmaceutica Sinica* 20:398–400.

Luftmann, H., M. Aranda, G. E. Morlock. 2007. Automated interface for hyphenation of planar chromatography with mass spectrometry. *Rapid Communications in Mass Spectrometry* 21:3772–3776.

Maresova, V., J. Hampl, Z. Chundela, F. Zrcek, M. Polasek, J. Chadt. 2005. The identification of a chlorinated MDMA. *Journal of Analytical Toxicology* 29(5):353–358.

Mirfazaelian, A., M. Goudarzi, M. Tabatabaiefar, M. Mahmoudian. 2002. A quantitative thin layer chromatography method for determination of theophylline in plasma. *Journal of Pharmacy & Pharmaceutical Sciences* 5:131–134.

Misra, H., D. Mehta, B. K. Mehta, M. Soni, D. C. Jain. 2009. Study of extraction and HPTLC–UV method for estimation of caffeine in marketed tea (*Camellia sinensis*) granules. *International Journal of Green Pharmacy* 3:47–51.

Munro, C. H., P. C. White. 1995. Evaluation of diazonium salts as visualization reagents for the thin layer chromatographic characterization of amphetamines. *Science & Justice—Journal of the Forensic Science Society* 35(1):37–44.

Nishigami, J., T. Ohshima, T. Takayasu, T. Kondo, Z. Lin, T. Nagano. 1993. Forensic toxicological application of TOXI-LAB screening for biological specimens in autopsy cases and emergency cares. *Nihon Hoigaku Zasshi* 47:372–379.

Ohshita, T., A. Yamaguchi, K. Harafuji. 1995. Analytical method for N-ethyl-3,4-methylenedioxyamphetamine. *Japanese Journal of Toxicology and Environmental Health* 41(1):77–84.

Okumus, V., E. Oral, D. Basaran, A. Onay. 2010. Simultaneous removal of indomethacine, papaverine and allopurinol from aqueous solution by using submerged aquatic plant Nasturtium officinale. *Asian Journal of Chemistry* 22:2081–2089.

Ovalles, J. F., J. N. Tettey, J. H. Miller, G. G. Skellern. 2000. Determination of piracetam and its impurities by TLC. *Journal of Pharmaceutical and Biomedical Analysis* 23:757–761.

Oztunç, A., A. Onal, S. E. Toker. 2010. Detection of methamphetamine, methylenedioxymethamphetamine, and 3,4-methylenedioxy-N-ethylamphetamine in spiked plasma by HPLC and TLC. *Journal of AOAC International* 93:556–561.

Patre, N. G., L. Sathiyanarayanan, M. V. Mahadik, S. R. Dhaneshwar. 2009. A validated, stability-indicating HPTLC method for analysis of doxofylline. *Journal of Planar Chromatography—Modern TLC* 22:345–348.

Paulus, W., S. Goenechea, G. Wienert. 1966. Detection of caffeine, theophylline, theobromine and some of their derivatives by thin layer chromatography. *Archives of Toxicology* 21:362–366.

Pavlik, J. W. 1973. TLC detection of caffeine in commercial products. *Journal of Chemical Education* 50:134–138.

Pisternick, W., K. A. Kovar, H. Ensslin. 1997. High-performance thin-layer chromatographic determination of N-ethyl-3,4-methylenedioxyamphetamine and its major metabolites in urine and comparison with high-performance liquid chromatography. *Journal of Chromatography B: Biomedical Sciences and Applications* 688(1):63–69.

Płotka, J. M., C. Morrison, M. Biziuk. 2011. Common methods for the chiral determination of amphetamine and related compound I. Gas, liquid and thin layer chromatography. *Trends in Analytical Chemistry* 10(7):1139–1158.

Pokrajac, M., V. M. Varagic. 1983. Spectrodensitometric determination of theophylline in plasma. *Acta Pharmaceutica Jugoslavica* 33:23–27.

Popović, M., N. Perišić-Janjić. 1988. Separation and fluorodensitometric determination of some purine derivative drugs by thin-layer chromatography on starch and cellulose. *Chromatographia* 26:244–246.

Prošek, M., A. Golc-Wondra, I. Vovk, S. Andrenšek. 2000. Quantification of caffeine by off-line TLC-MS. *Journal of Planar Chromatography—Modern TLC* 13:452–456.

Rasmussen, K. E., P. Knutsen. 1985. Techniques for the detection and identification of amphetamines and amphetamine-like substances. *Bulletin on Narcotics* 37:95–112.

Renger, B., H. Jehle, M. Fischer, W. Funk. 1995. Validation of analytical procedures in pharmaceutical analytical chemistry: HPTLC assay of theophylline in an effervescent tablet. *Journal of Planar Chromatography—Modern TLC* 8:269–278.

Riechert, M. 1978. Micro method for the determination of caffeine and theophylline allowing direct application of biological fluids to thin layer chromatography plates. *Journal of Chromatography* 146:175–180.

Rink, M., A. Gehl. 1966. Thin-layer chromatographic separation of some purine derivatives, including uric acid. *Journal of Chromatography* 21:143–145.

Rogers, K., J. Milnes, J. Gormally. 1993. The laser desorption/laser ionization mass spectra of some methylated xanthines and the laser desorption of caffeine and theophylline from thin layer chromatography plates. *International Journal of Mass Spectrometry and Ion Processes* 123:125–131.

Ruddy, D., J. Sherma. 2002. Analysis of the caffeine in alertness tablets and caplets by high-performance thin-layer chromatography with ultraviolet absorption densitometry of fluorescence-quenched zones. *Acta Chromatographica* 12:143–150.

Salvadori, M. C., E. M. Rieser, L. M. Ribeiro Neto, E. S. Nascimento. 1994. Determination of xanthines by high-performance liquid chromatography and thin-layer chromatography in horse urine after ingestion of Guaraná powder. *Analyst* 119:2701–2703.

Sanganalmath, P. U., K. M. Sujatha, S. M. Bhargavi, V. G. Nayak, B. M. Mohan. 2009. Simple, accurate and rapid HPTLC method for analysis of theophylline in post-mortem blood and validation of the method. *Journal of Planar Chromatography—Modern TLC* 22:29–33.

Sarsunova, M., V. Schwarz. 1963. Some possibilities for use of thin layer chromatography without binding agents in drug analysis. 3. Qualitative detection of purines and barbiturates used in prescriptions. *Pharmazie* 18:207–209.

Schmidtova, K., E. Timkova, R. Dzurik. 1987. Theophylline concentration determination by thin layer chromatography. *Biochemia Clinica Bohemoslovaca* 16:497–501.

Senanayake, U. M., R. O. B. Wijesekera. 1968. A rapid micro-method for the separation, identification and estimation of the purine bases: Caffeine, theobromine and theophylline. *Journal of Chromatography A* 32:75–86.

Sherma, J., S. Stellmacher, T. J. White. 1985. Analysis of analgesic tablets by quantitative high-performance reversed phase TLC. *Journal of Liquid Chromatography* 8:2961–2967.

Shervington, A., L. A. Shervington, F. Afifi, M. A. El-Omari. 1998. Caffeine and theobromine formation by tissue cultures of *Camellia sinensis*. *Phytochemistry* 47:1535–1536.

Shetab Boushehri, S. V., M. Tamimi, A. Kebriaeezadeh. 2009. Quantitative determination of 3,4-methylenedioxymethamphetamine by thin-layer chromatography in ecstasy illicit pills in Tehran. *Toxicology Mechanisms and Methods* 19:565–569.

Shimamine, M., K. Takahashi, Y. Nakahara. 1993. Studies on the identification of psychotropic substances. IX. Preparation and various analytical data of reference standard of new psychotropic substances, N-Ethyl methylenedioxyamphetamine, N-Hydroxy methylenedioxyamphetamine, Mecloqualone, 4-Methylaminorex, Phendimetrazine and Phenmetrazine. *Bulletin of the National Institute of Hygienic Sciences* 111:66–74.

Singh, D. K., B. Srivastava, A. Sahu. 2003. Thin layer chromatography of purines on silica gel G impregnated with transition metal ions; assay of caffeine and theophylline in pharmaceutical formulations. *Journal of the Chinese Chemical Society* 50:1031–1036.

Sinnema, A., A. M. Verweij. 1981. Impurities in illicit amphetamine: Review. *Bulletin on Narcotics* 33:37–54.

Smith, R. V., S. K. Yang, P. J. Davis, M. T. Bauza. 1983. Determination of pentoxifylline and its major metabolites in microbial extracts by thin-layer and high-performance liquid chromatography. *Journal of Chromatography* 23:281–287.

Soponar, F., A. C. Moț, C. Sârbu. 2009. Quantitative evaluation of paracetamol and caffeine from pharmaceutical preparations using image analysis and RP-TLC. *Chromatographia* 69:151–155.

Spangenberg, B., P. Post, S. Ebel. 2002. Fiber optical scanning in TLC by use of a diode-array detector—Linearization models for absorption and fluorescence evaluation. *Journal of Planar Chromatography—Modern TLC* 15:88–93.

Sullivan, C., J. Sherma. 2003. Development and validation of an HPTLC-densitometry method for assay of caffeine and acetaminophen in multicomponent extra strength analgesic tablets. *Journal of Liquid Chromatography & Related Technologies* 26:3453–3462.

Sullivan, C., J. Sherma. 2004. Comparative evaluation of TLC and HPTLC plates containing standard and enhanced UV indicators for efficiency, resolution, detection, and densitometric quantification using fluorescence quenching. *Journal of Liquid Chromatography & Related Technologies* 27:1993–2002.

Szász, G., M. Szászné Zacskó, V. Polánkay. 1965. A new method for thin-layer chromatographic separation of caffeine, theobromin and theophylline. *Acta Pharmaceutica Hungarica* 35:207–212.

Takayasu, T., T. Ohshima, J. Nishigami, T. Kondo, T. Nagano. 1995. Screening and determination of methamphetamine and amphetamine in the blood, urine and stomach contents in emergency medical care and autopsy cases. *Journal of Clinical Forensic Medicine* 2(1):25–33.

Tanaka, T., S. Aramaki, A. Momose. 1989. Thin-layer chromatographic screening procedure for some drugs in horse plasma. *Journal of Chromatography B: Biomedical Sciences and Applications* 496:407–415.

Tavallali, H., S. F. Zareiyan, M. Naghian. 2011. An efficient and simultaneous analysis of caffeine and paracetamol in pharmaceutical formulations using TLC with a fluorescence plate reader. *Journal of AOAC International* 94:1094–1099.

Tomankova, H., M. Vasatova. 1988a. TLC (HPTLC)-spectrodensitometry in pharmaceutical analysis. I. TLC-spectrodensitometric simultaneous determination of paracetamol, caffeine, guaiacolglycerinether and acetylsalicylic acid in analgesic-antipyretic pharmaceuticals. *Cesko-Slovenska Farmacie* 37:255–259.

Tomankova, H., M. Vasatova. 1988b. TLC (HPTLC)-spectrodensitometry in pharmaceutical analysis. II. TLC-spectrodensitometric simultaneous determination of propyphenazone, paracetamol and caffeine in the preparation in progress valetol tables. *Cesko-Slovenska Farmacie* 37:291–294.

Tseng, M. C., M. J. Tsai, K. C. Wen. 1996. Quantitative analysis of acetaminophen, ethoxybenzamide, piroxicam, hydrochlorothiazide, caffeine, chlorzoxazone and nicotinamide illegally adulterated in Chinese medicinal pills. *Journal of Food and Drug Analysis* 4:54–56.

Tseng, M. C., M. J. Tsai, K. C. Wen. 1997. Quantitative analysis of caffeine, ethoxybenzamide, chlorzoxazone, diazepam and indomethacine illegally adulterated in Chinese medical pills. *Journal of Food and Drug Analysis* 5:78–80.

Van Berkel, G. J., M. J. Ford, M. A. Deibel. 2005. Thin-layer chromatography and mass spectrometry coupled using desorption electrospray ionization. *Analytical Chemistry* 77:1207–1215.

van der Ark, A. M., A. M. Verweij, A. Sinnema. 1978. Weakly basic impurities in illicit amphetamine. *Journal of Forensic Sciences* 23(4):693–700.

Van Gennip, A. H., J. Grift., E. J. Van Bree-Blom et al. 1979. Urinary excretion of methylated purines in man and in the rat after the administration of theophylline. *Journal of Chromatography* 163:351–362.

Vapaatalo, H., S. Kärkäinen, K. E. Senius. 1984. Comparison of saliva and urine samples in thin-layer chromatographic detection of central nervous stimulants. *International Journal of Clinical Pharmacology Research* 4:5–8.

Venkatesh, V., A. E. Prabahar, P. V. Suresh, C. U. Maheswari, N. R. Rao. 2011. HPTLC method for the simultaneous estimation of etophylline and theophylline in tablet dosage form. *Asian Journal of Chemistry* 23:309–311.

Veselovskaia, N. V., E. A. Simonov, V. I. Sorokin et al. 1999. The analysis of the methylenedioxy derivatives of amphetamine. *Sudebno Meditsinskaia Ekspertiza* 42:23–30.

Vovk, I., A. Golc-Wondra, M. Prošek. 1997. Validation of an HPTLC method for determination of caffeine. *Journal of Planar Chromatography—Modern TLC* 10:416–419.

Vovk, I., B. Simonovska. 2001. TLC determination of mebendazol and pentoxifylline as residues on pharmaceutical equipment surfaces. *Journal of AOAC International* 84:1258–1264.

Wachowiak, R., B. Strach. 2006. Analysis of active components of evidence materials secured in the cases of drugs abuse associated with amphetamines and cannabis products. *Archiwum Medycyny Sadowej i Kryminologii* 56:251–257.

Wang, T. S., Y. Tong, J. Zheng, Z. Zhang, Q. Gao, B. Q. Che. 2010. Detection of 29 adulterations in lipid-regulating, antihypertensive, antitussive and antiasthmatic herbal remedies by TLC. *Chinese Pharmaceutical Journal* 45:857–861.

Warszawski, D., Z. Ben-Zvi, R. Gorodischer. 1981. Caffeine metabolism in liver slices during postnatal development in the rat. *Biochemical Pharmacology* 30:3145–3150.

Welch, R. M., S. Y. Hsu, R. L. DeAngelis. 1977. Effect of Aroclor 1254, phenobarbital, and polycyclic aromatic hydrocarbons on the plasma clearance of caffeine in the rat. *Clinical Pharmacology and Therapeutics* 22:791–798.

Wesley-Hadzija, B., A. M. Mattocks. 1975. Specific thin-layer chromatographic method for the determination of theophylline in plasma in the presence of some other drugs. *Journal of Chromatography* 115:501–505.

White, R. L. 1985. Analysis of thin-layer chromatographic adsorbates by Fourier transform infrared photo-acoustic spectroscopy. *Analytical Chemistry* 57:1819–1822.

Winek, C. L., E. O. Elzein, W. W. Wahba, J. A. Feldman. 1993. Interference of herbal drinks with urinalysis for drugs of abuse. *Analytical Toxicology* 17:246–247.

Wolff, K., M. J. Sanderson, A. W. Hay. 1990. A rapid horizontal TLC method for detecting drugs of abuse. *Annals of Clinical Biochemistry* 27:482–488.

Yalçindag, O. N. 1970. Identification and quantitative determination of theophylline derivatives. 3. *Journal de Pharmacie de Belgique* 25:423–429.

Yuan, L., R. Fu, T. Zhang, J. Deng, X. Li. 1998. Separation of alkaloids in tea by high-speed counter-current chromatography. *Se pu = Chinese Journal of Chromatography/Zhongguo hua xue hui* 16:361–362.

22 TLC of Antiepileptics

Beata Paw

CONTENTS

22.1 INTRODUCTION

Epilepsy is one of the most frequent neurological disorders, both in children and adults. About 0.5%–1% of the general population suffer from epilepsy, which means that about 50 million people in the world are affected. Epilepsy treatment is based on pharmacological therapy. Antiepileptic drugs (AEDs) have different mechanisms of action. The most important in inhibiting seizures are the compounds that block sodium and calcium channels and enhance the GABA-ergic neurotransmission. The antiepileptic drugs presented in this chapter belong to different classes including urea derivatives, iminostilbenes, imides, sulfonamides, GABA analogues, and some miscellaneous drugs (Figure 22.1). The aim of this review is the presentation of published work on the TLC analysis of AEDs in bulk, pharmaceutical formulations, and in biological material.

22.2 UREA DERIVATIVES

22.2.1 PHENYTOIN AND MEPHENYTOIN

Phenytoin (PHT; 5,5-diphenylimidazolidine-2,4-dione) is one of the most widely used drugs for the treatment of epilepsy. It affects Na^+, K^+, and Ca^{2+} ion conductance, membrane potentials, and amino acid concentrations. It blocks repetitive firing of neurons and maintains Na^+ channels in the inactivated state.

Mephenytoin, 5-ethyl-3-methyl-5-phenylimidazolidine-2,4-dione, is an anticonvulsant drug (trade name Mesantoin) used in the treatment of epilepsy when less toxic anticonvulsants are ineffective. Its pharmacological effects are similar to those of phenytoin.

Bress et al. described the procedure for separating phenytoin and barbiturates by reversed-phase TLC method. A spotted TLC plate was dipped into the mixture of mineral oil–petroleum ether (1:10)

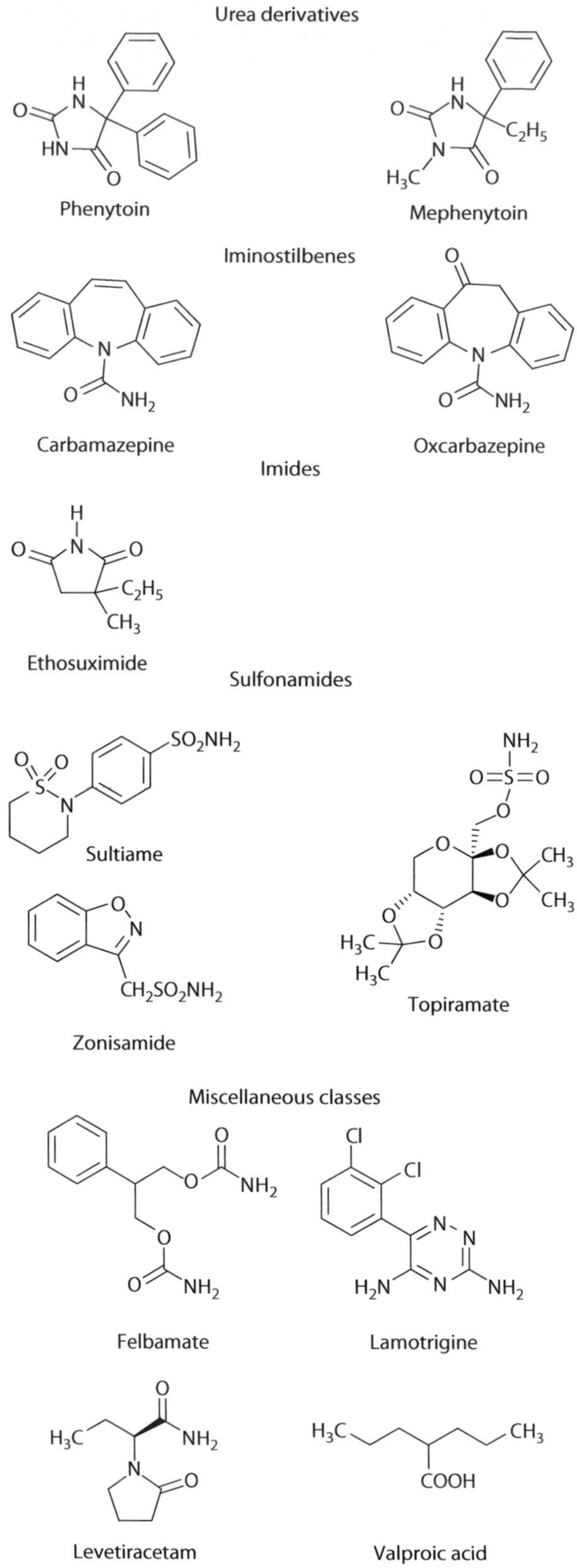

FIGURE 22.1 Chemical structures of antiepileptic drugs.

and developed to a distance of 15 cm by a mobile phase consisting of water–methanol–ammonium hydroxide (80:20:2). Diphenylcarbazone and mercuric sulfate were used as the visualization reagents. The RF value for phenytoin was 29 (Bress et al. 1980).

The TLC method for detection and determination of benzophenone in phenytoin formulations (tablets, capsules, suspension) and drug substances is presented. Separation was provided with TLC silica gel 60F plates as a stationary phase, and mixture of toluene–methanol–ethyl acetate–acetic acid–chloroform (80:20:15:10:5) or ethyl acetate–methanol–ammonium hydroxide (85:10:5) as a mobile phase. The determination was performed at 248 nm. The method was recommended for production monitoring (Matsui and Smith 1975).

Aboul-Enein and Serignese used the TLC method for determination of phenytoin in pharmaceutical formulations (capsules and injectables). Recovery of phenytoin from capsules and injectables was 98.9% ± 0.46% and 100.52% ± 0.25%, respectively. The method was also developed for the identification of PHT and its metabolites: p-hydroxy phenytoin and phenytoin dihydrodiol in the urine of epileptic adult patients (Aboul-Enein and Serignese 1994).

The semiquantitative method for determination of the anticonvulsant drugs in blood and urine by thin-layer chromatography was described. Utilization of SilicAR TLC-7GF plates with benzene:acetone (4:1), chloroform:acetone (9:1), and carbon tetrachloride:acetone (7:3) provides separation of phenytoin, mesantoin, phenobarbital, mysoline, phenylethylmalondiamid, and nirvanol. The substances were visualized by UV irradiation. The liquid–liquid extraction with chloroform was applied for the sample preparation. Recoveries were 80%–105%. The lower limit of detection was approximately 1 µg for phenytoin and mesantoin. The best separations of drugs in this procedure were generally obtained by developing unactivated plates in the unsaturated chambers (Pippenger et al. 1969).

Vedsö et al. reported the TLC method for determination of phenytoin in serum in the presence of barbiturates, sulthiame, and ethosuximide. The extracts of 1 mL serum were applied to the silica gel plates with a dispensing syringe. After development, the plates were dipped into the staining reagents. Under routine conditions, 0.5–40 µg phenytoin could be quantified (Vedsö et al. 1969).

Fenimore et al. developed the high-performance thin-layer chromatography method for determination of phenytoin and phenobarbital in plasma using liquid–liquid extraction. Diethyl ether as the extraction solvent and metharbital as the internal standard were used. Analysis was performed on HPTLC silica gel 60 plates with ethyl acetate–ammonium hydroxide (97:3, v/v) as mobile phase. Measurement was performed at 230 nm. Phenytoin was determined over a range of 0–50 mg/L (Fenimore et al. 1978).

The TLC method was used for determination of phenytoin in blood (Simon et al. 1971, Reszka and Lewandowska 1976), and in serum (Kohsaka 1974), as well as for analysis of phenytoin and its hydroxy metabolites (Rao and McLennon 1977), and to detect the metabolites of phenytoin in the human placental tissue (Henneberg and Serowka 1984).

Shimada et al. developed a simple and rapid method for determination of (S)-mephenytoin 4-hydroxylase activity by human liver microsomal cytochrome P-450. The 4-hydroxylated metabolite of mephenytoin was separated by thin-layer chromatography and quantified (Shimada et al. 1985).

22.3 IMINOSTILBENES

The iminostilbene class of antiepileptic drugs is constituted by the derivatives of carbamazepine.

22.3.1 CARBAMAZEPINE

Carbamazepine (CBZ), 5H-dibenz [b,f]azepine-5-carboxamide, is a drug used in the treatment of focal epilepsies of any type with or without secondary generalized tonic-clonic seizures. It blocks the sodium channel-dependent potential. This drug is extensively metabolized in the liver to an active metabolite carbamazepine epoxide, which possesses anticonvulsant properties similar to those of carbamazepine (Ambrósio et al. 2002).

Liu et al. used a TLC to find and separate the chemical components added illegally to the traditional Chinese medicine preparations. Then the components were detected by MS, and the MS spectra were compared with the standard substance. Carbamazepine, diclofenac sodium, and tramadol in analgesia traditional Chinese medicine preparation were detected by TLC-MS quickly and accurately (Liu et al. 2008).

Patel et al. described the HPTLC method for quantitative determination of carbamazepine as a bulk drug in tablets and in-house developed mucoadhesive microemulsion formulations and solutions. Carbamazepine was chromatographed on silica gel 60 F254 plates using ethyl acetate–toluene–methanol (5:4:1, v/v) as mobile phase. Densitometric assay was performed at 285 nm. RF value for carbamazepine was 0.47. Calibration plots were linear in the range 100–600 ng/spot; $r^2 = 0.9995$. The LOD and LOQ were 16.7 and 50.44 ng/spot, respectively. Total recoveries of carbamazepine from the laboratory-prepared mixtures ranged from 99.4% to 101.9%. The intra- and interday precision expressed as RSD were less than 5%. Recovery of carbamazepine from Mazetol tablets was 99.35% (Patel et al. 2011).

Hundt and Clark presented the TLC method for simultaneously determining serum levels of carbamazepine and two of its major metabolites, carbamazepine-10,11-epoxide and 10,11-dihydroxy-carbamazepine in serum. Serum (1 μL) was spotted directly onto the thin-layer plate. The separated spots were converted into fluorescing compounds by exposing the plates to hydrogen chloride gas for 5 min and then to strong ultraviolet radiation from a mercury lamp for 20 min. The fluorescence was measured quantitatively using a spectrofluorimeter equipped with a thin-layer chromatogram scanning attachment (Hundt and Clark 1975).

In the TLC method described by Li, after exposure to hydrogen chloride gas the plates were heated for 10 min at 160°C. The fluorescence was measured quantitatively, using a CS-920 TLC scanner equipped with a fluorescence attachment. The calibration plot was linear in the range 4–16 μg/mL. The LOD was about 2 μg/mL. The CV was 3.25%–10.0%. The recovery of carbamazepine from serum was about 99% (Bing-Yang 1986).

The TLC method for determining carbamazepine in blood (Faber and Man In 'T Vel 1974) as well as rapid and accurate determination of the level of carbamazepine in serum by ultraviolet reflectance photometry on thin layer chromatograms (Breyer 1975) were described.

The method of determining the level of carbamazepine in the earlobe blood by a CS-930 dual-wavelength TLC scanner was reported. The substance was extracted from plasma by organic solvent and was chromatographed on a TLC plate horizontally. Detection was performed at 280 nm. Recovery of carbamazepine from plasma was 97.15% (Li et al. 1993).

Mennickent et al. developed the HPTLC method for quantitative analysis of carbamazepine in saliva. It involves the derivatization of carbamazepine with 60% perchloric acid in ethanol/water (1:1, v/v) and densitometric detection at 366 nm. Chromatographic separation was carried out on silica gel F254 HPTLC plates, previously washed with methanol and activated at 130°C for 20 min, using a mixture of ethyl acetate–toluene–methanol (5:4:1, v/v) as mobile phase. The calibration plot was linear in the range 0.50–15.0 ng/spot, with a regression coefficient of 0.999. The intraassay variation (repeatability) was between 5.1% and 7.4%, and the interassay (reproducibility) was between 5.6% and 7.4%. The LOD was 0.18 ng, and the LOQ was 0.54 ng. Recovery of carbamazepine from saliva was 109.8%. The method allows separation of carbamazepine from its main metabolites, carbamazepine-dihydroxide and carbamazepine-epoxide (Mennickent et al. 2003).

The chromatographic conditions given by Mennickent et al. (2003) were reported for determination of carbamazepine in human serum using liquid–liquid extraction with dichloromethane. The method was linear in the range of 3 and 20 ng/μL; r = 0.998. The intra- and interassay precisions, expressed as the RSD, were in the range of 0.41%–1.24% (n = 3) and 2.17%–3.17% (n = 9), respectively.

The LOD was 0.19 ng, and the LOQ was 0.57 ng. Accuracy, calculated as percentage recovery, was between 98.98% and 101.96%, with the RSD not higher than 1.52%. The method was selective for carbamazepine and its major metabolite (carbamazepine-epoxide). RF values were 0.55 and 0.69, for the metabolite and the analyte, respectively (Mennickent et al. 2009).

Sajewicz presented the TLC densitometric method for detection of carbamazepine and other drugs in the river water in South Poland. After the solid-phase extraction with octadecyl C18 columns, the drugs were chromatographed on the RP-18 F254 TLC plates with acetonitrile–water–acetic acid (5:5:2, v/v) as mobile phase. Calibration plots were constructed in the range 4–10 µg/µL. Densitometric detection was performed at λ = 220 nm. The RF value for carbamazepine was 0.56 (Sajewicz 2005).

The RP-TLC method was used for determination of the lipophilicity of some psychotropic drugs, for example, carbamazepine. Drugs were chromatographed on RP-18 F254 HPTLC plates with mobile phases containing water, an organic modifier (methanol, dioxane, acetone, acetonitrile, or tetrahydrofuran), and ion-pair reagents or ammonia in horizontal Teflon chambers with eluent distributor. RF was measured for different concentrations of organic modifier. The final location of the spots was determined by use of ultraviolet (UV) light 254 nm (Hawryl et al. 2008).

22.3.2 OXCARBAZEPINE

Oxcarbazepine (OXC), 10,11-dihydro-10-oxo-5H-dibenz [b,f]azepine-5-carboxamide, a 10-keto analog of carbamazepine, is used in mono- and adjunctive therapy in patients with simple or complex partial or generalized seizures. After oral administration, oxcarbazepine is rapidly metabolized to its pharmacologically active 10-monohydroxy metabolite. It acts by blocking the voltage-sensitive sodium channels and also reduces the voltage-activated calcium currents in striatal and cortical neurons, which contributes to the inhibition of repetitive neuronal firing and the reduction of synaptic impulsive activity (Johannessen and Tomson 2006, Bialer et al. 2007).

Paw elaborated rapid and accurate thin-layer chromatography (TLC) methods with densitometric and videoscanning detection for the determination of oxcarbazepine in pharmaceutical preparations (Trileptal and Apydan). The silica gel 60 F254 TLC plates and mixture: ethyl acetate–chloroform–methanol (75:20:5, v/v) were used as the stationary and mobile phases, respectively. Chromatograms were developed to a distance of 9 cm in the horizontal chambers. Densitometric assay was performed at 220 nm and videoscanning quantitation at 254 nm. Calibration plots were constructed in the range 0.2–4.0 µg/spot and were correlated with good correlation coefficients (r220 = 0.9990; r254 = 0.9919). In the densitometric assay, the intra- and interday precisions, as RSD, were in the range 0.56%–1.38% and 0.74%–1.48%, respectively; the LOD and LOQ were 0.02 and 0.05 µg/spot, respectively. In the videodensitometric procedure they were in the range 0.98%–2.36% and 1.26%–3.10%, respectively; the LOD and LOQ were 0.1 and 0.2 µg/spot, respectively. Recovery of oxcarbazepine measured by use of densitometry was 100.74% (RSD 0.76%) from Apydan tablets and 100.45% (RSD 0.74%) from Trileptal tablets. Videodensitometry resulted in recovery of 100.17% (RSD 0.32%) and 100.37% (RSD 0.71%), respectively. The active substance was extracted from tablets with methanol. The *F-Snedecor* test and *t*-test for two means showed there was no significant difference between the precision and accuracy of the methods (Paw 2007).

One HPTLC method was reported for the determination of oxcarbazepine in human plasma. The procedure involved extraction of oxcarbazepine from plasma by protein precipitation with acetonitrile and quantification using the silica gel 60 F254 plates, the mobile phase containing ethyl acetate–toluene–methanol (7:2:1, v/v), densitometric detection at 254 nm, and chlorzoxazone as the internal standard. The calibration plot was linear in the range 10–3000 ng/band. The LOQ was 10 ng. The RSD of accuracy and precision was less than 10% at the LOQ level. The absolute recovery was more than 75%. Processed plasma was stable for just 5 h under ambient conditions and 23 h under refrigeration (Gandhimathi and Ravi 2008).

22.4 SULFONAMIDES

22.4.1 ZONISAMIDE

Zonisamide (ZNS; 1,2-benzisoxazole-3-methanesulfonamide) is effective against partial seizures and various generalized seizure types. Its mode of action is based on blocking the voltage-dependent sodium and T-type calcium channels. It also increases dopaminergic and serotoninergic neurotransmission and weakly inhibits carbonic anhydrase (Johannessen et al. 2003, Pollard and French 2006).

Thin layer chromatography-matrix-assisted-secondary-ion mass spectrometry (TLC/SIMS) was applied for identification of metabolites of zonisamide in rat urine. Metabolites were extracted from 48 h-urine and subjected to TLC/SIMS. Three metabolites, *N*-acetyl-3-sulfamoylmethyl-1, 2-benzisoxazole, 1,2-benzisoxazol-3-carboxylic acid, and the sulfate of (2-hydroxy benzoyl)-methanesulfonamide, were identified (Yoshida et al. 1989).

Zonisamide was determined together with lamotrigine and levetiracetam in human plasma using high-performance thin layer chromatography (Antonilli et al. 2011) (study described in Section 22.6).

22.4.2 TOPIRAMATE

Topiramate (TPM), 2,3:4,5-*bis-O*-(1-methylethylidene)-β-D-fructopyranose sulfamate, is a sugar derivative with no UV or fluorescent chromophore. TPM is used in many types of seizures. Topiramate has multiple modes of action, including modulation of voltage-dependent sodium channels, enhancement of GABA activity, and inhibition of kainite and AMPA receptors. The drug is also a weak inhibitor of carbonic anhydrase (Angenhagen et al. 2003).

Salama et al. used the TLC method for separation of topiramate from its degradants (acid, base, and thermal). Planar chromatographic separation was performed on precoated silica gel F254 TLC plates using a mixture of chloroform–dichloromethane–acetic acid (4:4:2, v/v) as mobile phase. The chromatograms were developed up to 8 cm in the usual ascending way, air dried, and visualized by dipping in potassium permanganate solution. For acid degradants, three spots were detected with RF values zero, 0.29, and 0.51. For the base degradants, two spots appeared with RF 0.51 and zero, while for dry heat, one spot was with RF zero and another one had the RF value similar to that of the drug. The RF value of topiramate was 0.65 (Salama et al. 2010).

Koba et al. described the HPTLC method for determination of topiramate in pharmaceutical formulations. Analysis was performed on silica gel HPTLC plates in horizontal chambers with benzene–ethanol (5:2, v/v) as mobile phase and densitometric detection at 340 nm after topiramate visualization with the use of a chemical reagent. The plates were sprayed using a reagent composed of 30 mg/mL phenol in a mixture of ethanol–sulfuric acid 95:5, v/v, because topiramate is a UV-transparent compound. The RF value for topiramate was 0.61. Quantification was achieved in the range of 0.25–4.0 μg/spot and with adequate precision (RSD 4.16%) and recovery (104.47%) using a nonlinear calibration curve. The active substance was extracted from tablets with methanol. The excipients in tablets did not interfere with chromatographic separation and densitometric scanning at 340 nm (Koba et al. 2012).

22.5 MISCELLANEOUS CLASSES

22.5.1 FELBAMATE

Felbamate (FBM), 2-phenyl-1,3-propanediol dicarbamate, is an antiepileptic drug, similar in structure to meprobamate. It is recommended for treatment of partial seizures with or without secondary generalization in adults, and for Lennox-Gastaut syndrome in children. Because of the risk of aplastic anemia and hepatotoxicity, felbamate is not recommended as first-line treatment (Bialer et al. 2007).

Paw et al. described a simple, rapid, and accurate thin-layer chromatography method with densitometric detection for a quantitative analysis of felbamate in tablets. Chromatography was performed on silica gel 60 F254 TLC plates. Chromatograms were developed to a distance of 9 cm in horizontal Teflon DS chambers (Chromdes, Lublin, Poland) with acetone–chloroform–acetic acid (59:40:1, v/v) as mobile phase. Densitometric assay was performed at 205 nm by means of a Desaga (Heidelberg, Germany) CD 60 densitometer controlled by Desaga ProQuant software. The calibration plot was constructed in the range 6.0–60.0 µg/spot with good correlation coefficient (r = 0.9990). The intraday and interday precisions of the method were estimated by performing five determinations of small (6.0 µg/spot), medium (36.0 µg/spot), and large (60.0 µg/spot) amounts of felbamate. The intraday precisions for felbamate expressed as RSD were 0.78% and 0.81% for the lowest and the highest concentrations, respectively. The respective values for the interday precision were 0.84% and 1.08%. Accuracy of the method was assessed on the basis of determination of felbamate in the laboratory-prepared mixtures at three levels of addition (50%, 100%, and 150% of the drug concentration in tablets). For felbamate, the recovery results ranged from 100.17% to 100.46% for the lowest and the highest concentrations of the drug, with the RSD values ranging from 0.84% to 0.45%. The detection limit (LOD) and quantification limit (LOQ) for felbamate were 2.0 and 4.0 µg/spot. The densitometric method was successfully applied for the determination of felbamate in pharmaceutical preparation. The active substance was extracted from tablets with methanol. The recovery from pharmaceutical product was 100.05% (RSD = 0.36%). Tablet excipients did not interfere with the chromatography (Paw et al. 2009).

Cornford et al. studied the transport of felbamate by the blood-brain barrier in mice, rats, and rabbits. FBM was retained in mouse brain. No FBM metabolites were detected in the brain 5 min after administration. Mouse brain extracts were analyzed by TLC on silica gel using a mobile-phase acetone–hexane (70:30, v/v). A single [14C]-labeled FBM peak was detected—RF = 0.504 (Cornford et al. 1992).

22.5.2 LAMOTRIGINE

Lamotrigine (LTG), 3,5-diamino-6-(2,3-dichlorophenyl)-1,2,4-triazine, is a broad-spectrum antiepileptic agent that is indicated for the adjunctive and monotherapy treatment of partial seizures, both with and without secondary generalization. Lamotrigine inhibits the voltage-sensitive sodium channels, stabilizing neuronal membranes and modulating presynaptic transmitter release of excitatory amino acids such as glutamate and aspartate. Lamotrigine also inhibits the high voltage-activated calcium channels (Bialer et al. 2007). Lamotrigine is a lipophilic weak base, and it is well absorbed after oral administration.

Dreassi et al. developed a method using planar chromatography for determining lamotrigine in tablets and human plasma. LTG was extracted with acetonitrile in the presence of sodium carbonate. 3,5-Diamino-6-(2-methoxyphenyl)-1,2,4-triazine was used as internal standard. The LOD was 0.27 µg/mL plasma, and the recovery from human plasma fortified with various concentrations of LTG was 91.3% ± 3.4% (Dreassi et al. 1996).

Youssef and Taha developed and validated a TLC densitometric method for the determination of lamotrigine in the presence of its main impurity, 2,3-dichlorobenzoic acid. The separation was carried out on silica gel F254 plates using ethyl acetate–methanol–ammonia 35% (17:2:1, v/v) as a mobile phase. The RF values of the drug and its impurity were 0.75 ± 0.01 and 0.23 ± 0.01, respectively. The plates were visualized at 254 nm and scanned at 275 nm by a densitometer. The linear calibration curve in the concentration range 0.5–10 µg/spot showed a correlation coefficient of 0.9984. The intraday and interday relative standard deviation values for the standard solutions were 1.30% and 1.75%, respectively. Accuracy (mean ± RSD) was 99.99% ± 1.33%. The method was applied for the determination of lamotrigine in Lamictal tablets. Total recovery was 97.98% ± 0.51% (mean ± SD) (Youssef and Taha 2007).

Wyszomirska and Czerwińska presented a TLC-densitometric method for assay of lamotrigine in human milk using liquid–liquid extraction. LTG was extracted with methylene chloride from the alkaline solution. Analysis was performed on silica gel GF254 plates with hexane–acetone (1:1, v/v) as a mobile phase. Calibration plot was linear in the range 2–10 µg/mL. Lamotrigine concentration in the human milk was determined at 217 nm. Mean lamotrigine recovery was 94.72%; RSD = 2.81%. Limit of lamotrigine detectability was 0.2 µg (Wyszomirska and Czerwińska 1999).

Two HPTLC methods with densitometric detection and liquid–liquid extraction with ethyl acetate were reported for the determination of lamotrigine in serum (Patil and Bodhankar 2005b, Mennickent et al. 2011). The authors used the silica gel F254 plates as a stationary phase. Patil and Bodhankar applied a mixture of toluene–acetone–ammonia (7:3:0.5, v/v) as a mobile phase and densitometric detection at 312 nm. RF value for lamotrigine was 0.54. The plates were developed to a distance of 80 mm. The calibration curve was constructed in the range 20–300 ng/spot and was correlated with correlation coefficient r = 0.9983. The LOD and LOQ were 6.4 and 10.2 ng, respectively. Accuracy was reported in the range of 92.06%–97.12%. Intraday and interday precisions, as relative standard deviation, were in the range 2.21%–3.70% (n = 7) and 1.16%–3.13% (n = 5), respectively. The method was applied for determination of serum lamotrigine levels in epileptic patients and in the pharmacokinetic study of lamotrigine administered orally to rabbits (Patil and Bodhankar 2005b). Mennickent et al. used the mixture: ethyl acetate–methanol–32% aqueous ammonia (17:2:1, v/v) as a mobile phase, with densitometric detection at 280 nm and chloramphenicol as an internal standard. The method was linear in the range 0.6–300 ng/band. The correlation coefficient was 0.998. The migration distance was 8 cm and development time 15 min. Intra- and interassay precisions, as RSD, were in the range 0.53%–2.91% (n = 3) and 1.58%–2.98% (n = 9), respectively. The limits of detection and quantification were 0.016 and 0.042 ng, respectively. Accuracy, calculated as percentage recovery, was between 94.09% and 101.30%, with RSD no higher than 3.52%. This method has lower LOD and LOQ values and higher recovery than the HPTLC one described by Patil and Bodhankar (Mennickent et al. 2011).

Using lamotrigine as an internal standard, HPTLC silica gel 60 F254 plates as a stationary phase, and water–methanol–*n*-butanol–ammonia solution (5:5:5:0.4, v/v) as a mobile phase, analysis of levofloxacin in pharmaceutical formulations was developed. The RF value of lamotrigine was 0.89 (Meyyanathan et al. 2004).

22.5.3 LEVETIRACETAM

Levetiracetam (LVT), (*S*)-2-(2-oxopyrrolidin-1-yl)-butanamide, is used as adjunctive therapy in the treatment of partial seizures with or without secondary generalization. LVT binds to the synaptic vesicle protein, SV2A, which is believed to impede nerve conduction across synapses (Johannessen and Tomson 2006).

The TLC method was used for determination of levetiracetam together with other drugs in tablets (Reddy and Devi 2007) and human plasma (Antonilli et al. 2011) (study described in Section 22.6).

22.5.4 VALPROIC ACID

Valproic acid (VPA; 2-propylpentanoic acid) and its sodium salt are widely used in the treatment of a variety of seizure types as major AEDs.

Thin-layer chromatography was applied for qualitative and quantitative assays of metabolites of valproic acid in rabbits', rats', and dogs' urine. Urines were collected and hydrolyzed in acidic medium. Sodium valproate and its metabolites were extracted, and these crude extracts were purified. The final extracts were used for assay by TLC method (Ferrandes and Eymard 1977).

The thin-layer chromatographic (TLC) method was described for the determination of valproic acid in plasma. The use of high-performance (HPTLC) plates gave detection limits (4.87 µg/mL) for derivatives of valproic acid, and the reproducibility on the same or different plates was good. Comparison with high-performance liquid chromatography showed a similar performance of plate and column (Corti et al. 1990).

22.6 SIMULTANEOUS DETERMINATION OF ANTIEPILEPTIC DRUGS DISCUSSED IN THIS CHAPTER

Patil and Bodhankar developed and validated the HPTLC method for simultaneous estimation of phenytoin sodium, phenobarbitone sodium, and carbamazepine in tablet dosage forms. A silica gel G60 F254 TLC plate and mixture: acetone–toluene (100:40, v/v) were used as a stationary and mobile phase respectively. Detection and quantification were performed by densitometry at 217 nm (Patil and Bodhankar 2005).

The chromatographic conditions given by Patil and Bodhankar (2005) were reported for simultaneous determination of carbamazepine, phenytoin, and phenobarbitone in human serum. The method was linear in the range of 100–2000 ng, r = 0.998. RF values for carbamazepine, phenytoin, and phenobarbitone were 0.20, 0.41, and 0.49, respectively. The limit of quantification was 30 ng/spot for carbamazepine and 80 ng/spot for phenytoin. The accuracy was found in the range of 88.5%–98.12% and %CV in the range of 1.14–3.87 (Patil and Bodhankar 2005a).

Breyer and Villumsen described the TLC method for the measurement of levels of phenytoin, hexobarbital, phenobarbital, and cyclobarbital in blood, serum, and/or plasma. The lowest concentrations measured were 0.3–0.7 µg/mL. For phenytoin determinations, 5-(p-methylphenyl)-5-phenylhydantoin was used as internal standard. The method has been applied to clinico-pharmacological assays (Breyer and Villumsen 1975).

Wad et al. presented the TLC method and its modification for the simultaneous determination of carbamazepine, carbamazepine-10,11-epoxide, phenytoin, mephenytoin, phenobarbital, and primidone in serum. Liquid–liquid extraction with toluene was applied for the sample preparation. The drugs were chromatographed on TLC plates with chloroform–acetone (87:13, v/v) as a mobile phase. Detection was performed at 215 nm. The authors measured the concentration of carbamazepine and its epoxide in serum from patients treated with carbamazepine and other anticonvulsants. The mean carbamazepine concentration in serum was 4.2 mg/L and the epoxide metabolite 0.9 mg/L (Wadd and Rosenmund 1978, 1978a, Wad et al. 1977, 1980).

A method for the detection of carbamazepine, phenytoin, and other drugs in blood on thin-layer chromatograms using chlorine vapors and 2,7 dichlorofluorescein was described. Blood specimens were extracted with chloroform, the solvent was evaporated, and the residue was spotted, half on each of two chromatographic plates. One plate was developed in ethyl ether–acetone (75:25, v/v), the other one in chloroform–acetone (88:12, v/v). Detection limits for pure substances were from 0.5 to 1.5 µg depending on the drug (McElwee 1979).

Davis and Fenimore presented rapid microanalysis of phenytoin, carbamazepine, phenobarbital, and primidone by high-performance thin-layer chromatography (HPTLC). This procedure involved a single extraction of 50 µL plasma sample. Quantification was performed by densitometry. The coefficient of variation was less than 4%. The extraction efficiency was approximately 95%. The minimum detectable amount of pure drug standards was 5 ng or less for all four anticonvulsants (Davis and Fenimore 1981).

Siek et al. developed a systematic thin-layer chromatographic (TLC) technique for detecting and identifying drugs (e.g., carbamazepine, phenytoin) and drug metabolites on 10-cm-long silica-gel plates with organic binder (fluorescent indicator). A computer program, SPOT CHEK, was used. For visualization and detection of tested drugs, various chemical reagents and UV light of 254 nm were applied. Detection limit for carbamazepine was 5 ng/spot. Drugs were extracted from serum and urine with a solution of dichloromethane/isopropanol (19/1 by vol.) (Siek et al. 1997).

Reddy and Devi described the HPTLC method with densitometric detection for the quantitation of levetiracetam and oxcarbazepine in tablets. The drugs were chromatographed on silica gel 60

F254 HPTLC plates with toluene–acetone–methanol (6:2:2, v/v) as a mobile phase. Before use the plates were washed with methanol and activated at 110°C for 5 min. The RF values for levetiracetam and oxcarbazepine were 0.45 and 0.55, respectively. Densitometric scanning was performed at 200 and 261 nm. The calibration plots were linear in the ranges 100–600 ng/band for oxcarbazepine and 5–10 µg/band for levetiracetam. The intra- and interday precisions for oxcarbazepine, expressed as the RSD, were in the range of 1.02%–1.63% and 1.13%–1.75%, respectively. The respective values for levetiracetam were in the range of 0.72%–0.93% and 0.95%–1.34%. Recovery was 99.30% (RSD 1.64%) for oxcarbazepine and 99.05% (RSD 1.51%) for levetiracetam. The drugs were extracted from tablets with methanol. The LOD and LOQ were 23.72 and 78.99 ng (per band) for oxcarbazepine and 223.64 and 745.46 ng (per band) for levetiracetam, respectively. The drugs were also subjected to photodegradation studies; oxcarbazepine was degraded in 9 h and levetiracetam was not degraded even after a long period (Reddy and Devi 2007).

Antonilli et al. developed the HPTLC method for quantitative determination of lamotrigine, zonisamide, and levetiracetam in human plasma using silica gel 60 F254 plates, as a stationary phase, and ethylacetate–methanol–ammonia (91:10:15, v/v) as a mobile phase. Detection was carried out by densitometry at a wavelength of 312, 240, and 210 nm for lamotrigine, zonisamide, and levetiracetam, respectively. RF values for LVT, LGT, and ZNS were 0.26, 0.45, and 0.71, respectively. Calibration curves were linear over a range of 0–200 ng for LTG and ZNS and 0–400 ng for LVT. The quantification limits were 3.69, 3.75, and 6.85 µg/mL for LTG, ZNS, and LVT, respectively. A fast chemical deproteinization with acetonitrile was applied for the sample preparation. Intra and interassay precisions provided relative standard deviations lower than 10% for all three analytes (Antonilli et al. 2011).

REFERENCES

Aboul-Enein, H. Y., V. Serignese. 1994. Thin layer chromatographic (TLC) determination of phenytoin in pharmaceutical formulations and identification of its hydroxylated urinary metabolites. *Analytical Letters* 27:723–729.
Ambrósio, A. F., P. Soares-da-Silva, C. M. Carvalho, A. P. Carvalho. 2002. Mechanisms of action of carbamazepine and its derivatives, oxcarbazepine, BIA 2-093, and BIA 2-024. *Neurochemical Research* 27:121–130.
Angenhagen, M., E. Ben-Menachem, L. Ronnback, E. Hansson. 2003. Novel mechanisms of action of three antiepileptic drugs, vigabatrin, tiagabine, and topiramate. *Neurochemical Research* 28:333–340.
Antonilli, L., V. Brusadin, F. Filipponi, R. Guglielmi, P. Nencini. 2011. Development and validation of an analytical method based on high performance thin layer chromatography for the simultaneous determination of lamotrigine, zonisamide and levetiracetam in human plasma. *Journal of Pharmaceutical and Biomedical Analysis* 56:763–770.
Bialer, M., S. I. Johannessen, H. J. Kupferberg, R. H. Levy, E. Perucca, T. Tomson. 2007. Progress report on new antiepileptic drugs: A summary of the *Eighth Eilat Conference* (EILAT VIII). *Epilepsy Research* 73:1–52.
Bing-Yang, L.I. 1986. Determination of carbamazepine in serum by thin-layer fluorescence scanner. *Acta Pharmaceutica Sinica* 21:633–635.
Bress, W., K. Ziminski, W. Long, T. Manning, L. Lukash. 1980. Separation of barbiturates using reverse-phase thin-layer chromatography. *Clinical Toxicology* 16:219–221.
Breyer, U. 1975. Rapid and accurate determination of the level of carbamazepine in serum by ultraviolet reflectance photometry on thin layer chromatograms. *Journal of Chromatography* 108:370–374.
Breyer, U., D. Villumsen. 1975. Thin-layer chromatographic determination of barbiturates and phenytoin in serum and blood. *Journal of Chromatography A* 115:493–500.
Cornford, E. M., D. Young, J. W. Paxton, R. D. Sofia. 1992. Blood-brain barrier penetration of felbamate. *Epilepsia* 33:944–954.
Corti, P., A. Cenni, G. Corbini, E. Dreassi, C. Murratzu, A. M. Caricchia. 1990. Thin-layer chromatography and densitometry in drug assay: Comparison of methods for monitoring valproic acid in plasma. *Journal of Pharmaceutical and Biomedical Analysis* 8:431–436.
Davis, C. M., D. C. Fenimore. 1981. Rapid microanalysis of anticonvulsants by high-performance thin-layer chromatography. *Journal of Chromatography* 222:265–270.

Dreassi, E., G. Corbini, P. Corti, M. Ulivelli, R. Rocchi. 1996. Quantitative analysis of lamotrigine in plasma and tablets by planar chromatography and comparison with liquid chromatography and UV spectrophotometry. *Journal of AOAC International* 79:1277–1280.

Faber, D. B., W. A. Man In 'T Vel. 1974. A thin-layer chromatographic method for determining carbamazepine in blood. *Journal of Chromatography A* 93:238–242.

Fenimore, D. C., C. M. Davis, C. J. Meyer. 1978. Determination of drugs in plasma by high-performance thin-layer chromatography. *Clinical Chemistry* 24:1386–1392.

Ferrandes, B., P. Eymard. 1977. Metabolism of valproate sodium in rabbit, rat, dog, and man. *Epilepsia* 18:169–182.

Gandhimathi, M., T. K. Ravi. 2008. Rapid HPTLC analysis of oxcarbazepine in human plasma. *Journal of Planar Chromatography* 20:451–456.

Hawryl, A., D. Cichocki, M. Waksmundzka-Hajnos. 2008. Determination of the lipophilicity of some psychotropic drugs by RP-TLC. *Journal of Planar Chromatography* 21:343–348.

Henneberg, M., E. Serowka. 1984. Phenytoin metabolites in human placental tissue. *Archives of Toxicology* 55:311.

Hundt, H. K. L., E. C. Clark. 1975. Thin-layer chromatographic method for determining carbamazepine and two of its metabolites in serum. *Journal of Chromatography A* 107:149–154.

Johannessen, B. D., D. Berry, M. Bialer, G. Kramer, T. Tomson, P. Patsalos. 2003. Therapeutic drug monitoring of the newer antiepileptic drugs. *Therapeutic Drug Monitoring* 25:347–363.

Johannessen, S. I., T. Tomson. 2006. Pharmacokinetic variability of newer antiepileptic drugs. *Clinical Pharmacokinetics* 45(11):1061–1075.

Koba, M., M. Marszałł, W. Sroka, M. Tarczykowska, A. Buciński. 2012. Application of HPTLC and LC-MS methods for determination of topiramate in pharmaceutical formulations. *Current Pharmaceutical Analysis* 8:44–48.

Kohsaka, M. 1974. Determination of phenobarbital and phenytoin in serum by thin layer chromatography. *No To Hattatsu* 6:441–443.

Li, Y. Z. J., Z. M. Li. 1993. Determing the level of carbamazepine in earlobe plasma by TLCS. *Chinese Journal of Hospital Pharmacy* 6:244–246.

Liu, Y. R., H. S. Ge, K. H. Zhao, L. Yu. 2008. Determination of three chemical components added illegally in a analgesia traditional Chinese medicine preparation by TLC-MS. *Chinese Pharmaceutical Journal* 43:1747–1750.

Matsui, F. F., S. J. Smith. 1975. Detection and determination of benzophenone present in diphenylhydantoin sodium formulations and drug substance. *Fresenius Zeitschrift für Analytische Chemie* 275:365–367.

McElwee, D. J. 1979. A new method for the detection of sedative drugs on thin-layer chromatograms using chlorine vapors and 2,7-dichlorofluorescein. *Journal of Analytical Toxicology* 3:266–268.

Mennickent, S., R. Fierro, M. Vega, M. de Diego, C. G. Godoy. 2009. Instrumental planar chromatographic method for determination of carbamazepine in human serum. *Journal of Separation Science* 32:1454–1458.

Mennickent, S., R. Fierro, M. Vega, M. de Diego, C. G. Godoy. 2011. Quantification of lamotrigine in human serum by high-performance thin-layer chromatography. *Journal of Planar Chromatography* 24:222–226.

Mennickent, S., M. Vega, C. G. Godoy. 2003. Development and validation of a method using instrumental planar chromatography for quantitative analysis of carbamazepine in saliva. *Journal of Chilean Chemical Society* 48:71–73.

Meyyanathan, S. N., G. V. S. Ramasarma, B. Suresh. 2004. Analysis of levofloxacin in pharmaceutical preparations by high performance thin layer chromatography. *Journal of Separation Science* 27:1698–1700.

Patel, R. B., M. R. Patel, K. K. Bhatt, B. G. Patel. 2011. Development and validation of HPTLC method for estimation of carbamazepine in formulations and its in vitro release study. *Chromatography Research International* 2011:8, Article ID 684369, doi:10.4061/2011/684369.

Patil, K. M., S. L. Bodhankar. 2005. Validated high performance thin layer chromatography method for simultaneous estimation of phenytoin sodium, phenobarbitone sodium and carbamazepine in tablet dosage forms. *Indian Journal of Pharmaceutical Sciences* 67:351–355.

Patil, K. M., S. L. Bodhankar. 2005a. High-performance thin-layer chromatography method for therapeutic drug monitoring of antiepileptic drugs in serum. *Indian Drugs* 42:665–670.

Patil, K. M., S. L. Bodhankar. 2005b. High-performance thin-layer chromatographic determination of lamotrigine in serum. *Journal of Chromatography B* 823:152–157.

Paw, B. 2007. Application of videodensitometric and classical densitometric thin-layer chromatography for quantification of oxcarbazepine in pure and pharmaceutical preparations. *JPCCR* 1:158–160.

Paw, B., J. Matysiak, D. Kowalczuk. 2009. Quantitative analysis of felbamate in pharmaceutical preparation by thin-layer chromatography with densitometric UV detection. *Annales UMCS Sectio DDD* 22:139–143.

Pippenger, C. E., J. E. Scott, H. W. Gillen. 1969. Thin-layer chromatography of anticonvulsant drugs. *Clinical Chemistry* 15:255–260.

Pollard, J. R., J. French. 2006. Antiepileptic drugs in development. *The Lancet Neurology* 5:1064–1067.

Rao, G. S., D. A. McLennon. 1977. Thin layer chromatographic analysis of phenytoin and its hydroxyl metabolites. *Journal of Chromatography* 137:231–233.

Reddy, T. S., P. S. Devi. 2007. Validation of high-performance thin-layer chromatographic method with densitometric detection for quantitative analysis of two anticonvulsants in tablets. *Journal of Planar Chromatography* 20:452–456.

Reszka, I., I. Lewandowska. 1976. Assays of microgram levels of diphenylhydantoin in blood serum. *Acta Poloniae Pharmaceutica—Drug Research* 33:365–371.

Sajewicz, M. 2005. Use of densitometric TLC for detection of selected drugs present in river water in south Poland. *Journal of Planar Chromatography* 18:108–111.

Salama, N. N., A. O. Mohamed, E. A. Taha. 2010. Development and validation of spectrofluorometric, spectrophotometric and thin layer chromatography stability indicating method for analysis of topiramate. *International Journal of Pharmacy & Technology* 2:1299–1314.

Shimada, T., J. P. Shea, F. P. Guengerich. 1985. A convenient assay for mephenytoin 4-hydroxylase activity of human liver microsomal cytochrome P-450. *Analytical Biochemistry* 147:174–179.

Siek, T. J., C. W. Stradling, M. W. McCain, T. C. Mehary. 1997. Computer-aided identifications of thin-layer chromatographic patterns in broad-spectrum drug screening. *Clinical Chemistry* 43:619–626.

Simon, G. E., P. I. Jatlow, H. T. Seligson, D. Seligson. 1971. Measurement of 5,5-diphenylhydantoin in blood using thin-layer chromatography. *American Journal of Clinical Pathology* 55:145–151.

Vedsö, S., C. Rud, J. F. Place. 1969. Determination of phenytoin in serum in the presence of barbiturates, sulthiame and ethosuximid by thin-layer chromatography. *Scandinavian Journal of Clinical and Laboratory Investigation* 23:175–180.

Wad, N., E. Hanifl, H. Rosenmund. 1977. Rapid thin-layer chromatographic method for the simultaneous determination of carbamazepine, diphenylhydantoin, mephenytoin, phenobarbital and primidone in serum. *Journal of Chromatography B* 143:89–93.

Wad, N., H. Rosenmund. 1978. Rapid quantitative method for the simultaneous determination of carbamazepine, carbamazepine-10,11-epoxide, diphenylhydantoin, mephenytoin, phenobarbital and primidone in serum by thin layer chromatography. *Journal of Chromatography* 146:167–168.

Wad, N., H. Rosenmund. 1978a. Rapid quantitative method for the simultaneous determination of carbamazepine, carbamazepine-10,11-epoxide, diphenylhydantoin, mephenytoin, phenobarbital and primidone in serum by thin layer chromatography: Improvement of the buffer system. *Journal of Chromatography* 146:361–362.

Wad, N., E. Weidkuhn, H. Rosenmund. 1980. A quantitative method for the simultaneous determination of carbamazepine, mephenytoin, phenylethylmalonamide, phenobarbital, phenytoin and primidone in serum by thin layer chromatography: Additional comments to the analysis of carbamazepine-10,11-epoxide. *Journal of Chromatography* 183:387–388.

Wyszomirska, E., K. Czerwińska. 1999. Identification and assay of lamotrigine in human milk with gas chromatography and densitometry. *Acta Poloniae Pharmaceutica—Drug Research* 56:101–105.

Yoshida, K., K. Matsumoto, A. Kagemoto, S. Arakawa, H. Miyazaki. 1989. Identification of rat urinary metabolites of zonisamide by TLC/SIMS. *Yakubutsu Dotai* 4:403–409.

Youssef, N. F., E. A. Taha. 2007. Development and validation of spectrophotometric, TLC and HPLC methods for the determination of lamotrigine in presence of its impurity. *Chemical and Pharmaceutical Bulletin* 55:541–545.

23 TLC of Alzheimer's Disease Medicines

Łukasz Cieśla

CONTENTS

23.1 INTRODUCTION

Alzheimer's disease (AD) is one of the most popular causes of dementia affecting millions of people all over the world. It was first described at the beginning of the twentieth century by Alois Alzheimer and named after him. Currently it is an incurable disease that eventually leads to death. There are several hypotheses aimed at the explanation of AD development; however, the exact causes are still unknown. Changes in the brains of AD-demented patients as well as neuropathological hallmarks have been well recognized and described [1]. The following pathological structures are observed in neurons of patients suffering from AD: extracellular β-amyloid-containing plaques and intraneuronal neurofibrillary tangles (NFTs) formed by hyperphosphorylated tau proteins. Formation of these structures may lead to neurodegeneration, synaptic instability, neuronal loss, neurotransmitter dysfunction, and finally the development of full symptoms of AD dementia [2]. There are also other hypotheses aiming at the description of neurodegeneration mechanism, for example, oxidative stress, mitochondrial dysfunction, or proinflammatory responses. However, the initiator of the pathological processes still remains unidentified. Degeneration of cholinergic neurons has been correlated with density of amyloid plaques [3]. Some approaches have been proposed to prevent or delay the cognitive symptoms, such as mental and physical exercises or balanced diet. More studies are required to find if there is any correlation between these factors and the development of AD symptoms.

23.2 PHARMACOLOGICAL TREATMENT OF AD SYMPTOMS

The amount of approved drugs for the treatment of the symptoms of AD is limited. The main reason is the fact that the therapy is hypothesis driven, not causal. Five Food and Drug Administration (FDA) (United States) approved drugs are currently used for the treatment of AD symptoms. Four of them are acetylcholinesterase (AChE) inhibitors: donepezil, galantamine, rivastigmine, and tacrine. It has been observed that in patients suffering from AD, the amount of acetylcholine decreases mainly in the areas responsible for memory and cognition, for example, limbic and association cortices (cholinergic hypothesis of cognitive dysfunction) [4]. An interesting fact is that in AD patients, butyrylcholinesterase (BChE) becomes the predominant cholinesterase in the brain [4]. Therefore, the search for BChE inhibitors is equally important as looking for the new AChE inhibitors. Currently used drugs have several common side effects, which include nausea or vomiting. Muscle cramps, bradycardia, or decreased appetite appears less frequently in patients treated with AChE inhibitors. Donepezil is the only AChE inhibitor approved for use in the advanced stage of AD, contrary to others that are applied in mild and moderate AD. Other substances have been also identified as AChE inhibitors, just to mention huperzine A, lycoramine, sanguinine, protopine, palmatine, berberine, sanguinarine, terpenoids, and many others [5]. As far as terpenoids are considered, it has been shown that they are also potent free radical scavengers, which can be beneficial in the treatment of neurodegenerative ailments, for example, AD [5,6]. Terpenoids seem to have a great potential for the treatment of AD symptoms. They are low-molecular lipophilic compounds that can easily penetrate the blood–brain barrier [7]. The side effects from the gastrointestinal tract can be excluded in the future, as terpenoids may be delivered in transdermal systems or could be inhaled and absorbed through the lungs [5,7]. The results of memory tests obtained by demented people also improved after regular administration of standardized *Ginkgo biloba* extract (EGb 761) [8]. Most recently, two substances have undergone clinical trials for their possible use in the treatment of AD symptoms, namely, latrepirdine and huperzine A [2]. Apart from inhibiting AChE and BChE, latrepirdine also inhibits the activation of NMDA receptors as well as acts as a neuroprotector [2]. Unfortunately, latrepirdine turned out to be ineffective in several clinical trials. Huperzine A, apart from being a selective AChE inhibitor, also reduces glutamate-induced cytotoxicity [2].

Apart from the four previously mentioned approved drugs, memantine NMDA receptor antagonist has been approved for the treatment of Alzheimer's-type dementia. Overstimulation of NMDA receptors leads to a process called excitotoxicity, which finally leads to cell death. Such overstimulation can be responsible for neurodegeneration observed in AD patients.

Most recently, researches have focused also on searching for substances altering the amyloid cascade and include the following compounds: secretase modulators, antiaggregants, and immunotherapies [2].

23.3 ELEMENTS OF CHEMISTRY AND PHARMACOKINETICS OF APPROVED DRUGS

23.3.1 AChE Inhibitors

As already mentioned, AChE inhibitors constitute the largest group of compounds used as drugs in the treatment of AD symptoms. The inhibitors increase the amount of acetylcholine in the brain exerting positive effects on memory and cognition in AD patients.

23.3.1.1 Donepezil

The IUPAC name of donepezil is as follows: 2-[(1-benzylpiperidin-4-yl)methyl]-5,6-dimethoxy-2, 3-dihydro-1H-inden-1-one. It is a lead compound of a new class of AChE inhibitors with N-benzylpiperidine and indanone moieties [9]. Its log P value is around 3.60 [10]. Donepezil is used as hydrochloride in the form of tablets. The bulk drug substance is a white crystalline powder with a bitter taste. It is freely soluble in chloroform, soluble in water and in glacial acetic acid, slightly soluble in ethanol and in acetonitrile, and practically insoluble in ethyl acetate and in n-hexane [11]. Donepezil has only weak inhibitory effect on BChE. Donepezil is well absorbed with a relative oral bioavailability of 100% and reaches peak plasma concentrations in 3–4 h [10]. Donepezil's main metabolite, 6-O-desmethyl donepezil, has been reported to inhibit AChE to the same extent as donepezil in vitro [10].

23.3.1.2   Galantamine

Galantamine is used for the treatment of AD symptoms as hydrobromide in the form of tablets, extended-release capsules, and solution. It is a benzazepine derivative and its full chemical name is as follows: (4aS,6R,8aS)-4a,5,9,10,11,12-hexahydro-3-methoxy-11-methyl-6H-benzofuro[3a,3,2-ef] [2]benzazepin-6-ol. Its log P value is around 1.80 [10]. Galantamine is well absorbed from the gastrointestinal system with high bioavailability and linear pharmacokinetics. Apart from inhibiting AChE, galantamine can allosterically modulate nicotinic acetylcholine receptors [2]. It has been found that no metabolites of galantamine exert additional inhibition of AChE activity [12].

23.3.1.3   Rivastigmine

Rivastigmine is a semisynthetic derivative of physostigmine. Its IUPAC name is as follows: 3-[(1S)-1-(dimethylamino)ethyl]phenyl N-ethyl-N-methylcarbamate. Its log P value is 2.30 [10]. It can be administered orally as rivastigmine tartrate or transdermally. The (S) enantiomer of rivastigmine is pharmacologically more active than the (R) enantiomer [13]. It inhibits both AChE and BChE. Rivastigmine is well absorbed in the gastrointestinal tract and is characterized with a short kinetic half-life [9]. This compound inhibits AChE by carbamylating serine residue in the enzyme's active center [9]. Rivastigmine is rapidly metabolized by cholinesterase-mediated hydrolysis [10].

23.3.1.4   Tacrine

The IUPAC name for tacrine is 1,2,3,4-tetrahydroacridine-9-amine. It was the first reverse and nonselective AChE inhibitor to be approved by the FDA for the treatment of AD symptoms. Tacrine possesses a flat configuration. The pKa value for this compound is 9.85, while its log P is around 2.70 [14]. It has a good intestinal permeability; however, it is characterized with poor oral bioavailability due to extensive first-pass metabolism [14]. Tacrine is rapidly absorbed. Absolute bioavailability of tacrine is approximately 17% [10]. One of the greatest disadvantages of the drug is the high rate of patients suffering from side effects. It has been also recognized as a hepatotoxic agent. Due to the aforementioned facts, tacrine has been largely abandoned as a symptomatic medication in AD patients. Its hydroxylated metabolites may also exert AChE inhibitory activity, as, for example, is assumed for 1-hydroxytacrine [14].

23.3.2   NMDA ANTAGONIST

There is only one reversible uncompetitive NMDA antagonist approved for the treatment of AD symptoms—memantine.

23.3.2.1   Memantine

The IUPAC name for memantine is 3,5-dimethyladamantan-1-amine. It is a free base, which is highly basic (pKa 10.42) and lipophilic (log P 3.28). This compound is present in solutions mainly in the form of cations. Memantine is administered orally in the form of water-soluble hydrochloride salt [15]. Memantine hydrochloride is commercially available as tablets and as an oral solution [3]. It is well absorbed after oral administration and is excreted mainly in the urine in the unchanged form as well as polar metabolites. Peak plasma concentrations are reached in 3–7 h [10]. Food has no effect on the absorption of memantine. Memantine is excreted largely unchanged [10]. When combined with donepezil, it provides significant improvement of the patients' cognitive functions when compared to results obtained in subjects treated with donepezil only [16].

23.4   CHROMATOGRAPHIC SYSTEMS

There are no complex formulations, containing more than one substance, approved for the treatment of AD. Therefore, there are no thin-layer chromatography (TLC) methods aimed at separation and identification of these substances, as they do not appear in such combinations. TLC systems described in this chapter concern methods that have been applied to screen complex mixtures for the presence of new or identification of already known compounds, which could be used in managing the symptoms of Alzheimer's-type dementia.

As already mentioned, the majority of currently approved drugs are AChE inhibitors; therefore, this chapter mainly focuses on TLC systems that can be applied in the analysis of these compounds. TLC plates have been used for the analysis of AChE inhibitors in two ways. First, they were applied to screen the activity of several compounds simultaneously without prior separation, so-called dot-blot test (Figure 23.1).

Rhee et al. stress that when it is not necessary to develop the plate, the TLC test for AChE inhibition should be performed on aluminum oxide rather than on silica gel [17,18]. This test can be considered as only a preliminary method to fish out active compounds or mixtures and their subsequent analysis by other techniques. In the other approach, the plate is first developed and then derivatized and documented. An example of a TLC plate that has been developed with the optimized eluent and subsequently derivatized can be seen in Figure 23.2.

In the majority of cases, silica TLC or HPTLC plates have been used in the analysis of AChE inhibitors [6,19,20]. Usually, mixtures of chloroform and methanol in different proportions have been used as eluents [5]. Some exemplary TLC systems have been provided in Table 23.1.

One of the problems that may be encountered in the analysis of basic compounds is poor system efficiency [33]. It may be overcome by incorporating some additives into the mobile phase, for example,

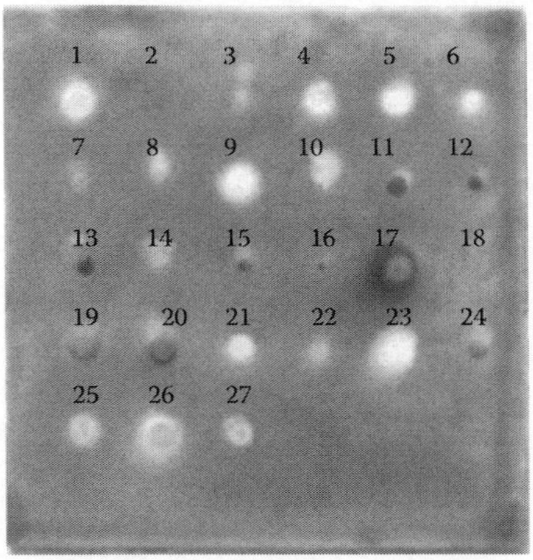

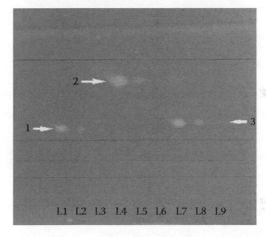

FIGURE 23.1 An example of TLC dot-blot test applied to screen the samples for the presence of AChE inhibitors. (For experimental data, see Cieśla, Ł. et al., *J. Planar Chromatogr. Modern TLC*, 25, 225, 2012. With permission.)

FIGURE 23.2 An example of TLC silica gel plate used for the analysis of AChE inhibitors. Spots L1–L3, galantamine; L4–L6, harmine; and L7–L9, harmaline. (For experimental data, see Zheng, X.Y. et al., *J. Planar Chromatogr. Modern TLC*, 24, 470, 2011. With permission.)

aqueous ammonia or aliphatic amines. However, some of these substances turned out to produce false-positive AChE inhibitory activity [32]. Therefore, the use of aqueous ammonia is rather recommended. If a compound strongly adsorbs on the surface of a silica gel, eluents containing greater amount of water may be advised (buffered aqueous–organic mobile phases) [34]. Such systems used to be termed in some publications as "pseudo-reversed systems" [35,36]. Under these conditions, the more polar compounds are strongly retained on the water-enriched stationary phase and are characterized with lower R_F values [37,38]. The retention mechanism seems to be the same as described by Alpert, who coined the name hydrophilic interaction liquid chromatography (HILIC) [39]. Separation of some basic compounds can be improved after conditioning silica TLC plates with mobile phase vapors containing aqueous ammonia or aliphatic amines, as proved for isoquinoline and tropane alkaloids [40,41].

Apart from standard silica gel adsorbent, other plates can also be used, for example, cyano-silica, diol silica, or nonspecific RP-18 W plates [6]. It has been observed that the results of enzyme–substrate reactions depend on the kind of adsorbent applied in the study [6,17,32]. AChE inhibitory tests performed for terpenes revealed the most distinct difference between the spots of inhibitors and the background on the surface of CN-silica plates. The weakest activity has been observed on the surface of nonspecific

TABLE 23.1

Exemplary TLC Systems Used in the Analysis of Compounds with Potential to Treat AD Symptoms

No.	Adsorbent + Eluent	Analyzed Compounds	Detection	Ref.
1	Silica gel, chloroform–methanol (8:2, v/v)	Galantamine, physostigmine, lycorine, tazettine	UV λ = 254 nm; Dragendorff's reagent; bioassay (Ellman's method)	[17]
2	Silica gel, chloroform–methanol (10:1, v/v)	Galantamine, physostigmine	Bioassay (Ellman's method)	[18]
3	Silica gel, chloroform–methanol–25% aqueous ammonia (9:0.5:0.5, v/v/v)	Galantamine, lycorine, hippeastrine	UV λ = 207 nm; Bioassay (Ellman's method and Marston's method modified by Mroczek)	[19]
4	Silica gel and aluminum oxide, chloroform–methanol–25% aqueous ammonia (8:1:1, v/v/v)	Galantamine, 1,2-dihydrogalantamine, hippeastrine, lycorine	(Marston's method modified by Mroczek)	[20]
5	Silica gel, chloroform–methanol (8:2, v/v)	Donepezil	UV λ = 254 nm	[21]
6	Silica gel, butanol–water–glacial acetic acid (5:4:1, v/v/v)	Donepezil	UV λ = 260 nm	[22]
7	Silica gel, chloroform–methanol–25% ammonia (16:64:0.1, v/v/v)	Donepezil	UV λ = 315 nm	[23]
8	Silica gel, methanol–butanol–water–ammonia (5:4::0.01 v/v/v)	Rivastigmine	UV λ = 263 nm	[24]
9	Silica gel, ethyl acetate–methanol–ammonia (10:1.5:0.5, v/v/v)	Galantamine, harmine, harmaline	Bioassay (Marston's method)	[25]
10	Silica gel, chloroform–methanol (4:6, v/v)	Rivastigmine	UV λ = 210 nm	[26]
11	Silica gel, chloroform–methanol (7:1, v/v); acetone–methanol (4:5, v/v)	Huperzine A*, physostigmine, berberine	Bioassay (Marston's method); bioassay (Yang et al. method)	[27] [28]
12	Silica gel, ethyl acetate–methanol–ammonia (3:1:0.1, v/v/v); silica gel, n-hexane–ethyl acetate–methanol–ammonia (3:3:1:0.1, v/v/v/v)	Galantamine, epigalantamine, sanguinine, epinorgalantamine, protopine, corydaline, bulbocapnine, stylopine, deshydrocorydaline	UV λ = 254 nm; Dragendorff's reagent; bioassay (Ellman's method)	[29]
13	Silica gel, chloroform–methanol–water (65:35:5, v/v/v)	Galantamine, physostigmine	Bioassay (Marston's method)	[30]
14	Silica gel, chloroform–ethyl acetate–methanol (90:7:3, v/v/v); silica gel, ethyl acetate–methanol–water (100:13.5:10, v/v/v)	Physostigmine, chelidonine, harmaline, harmine	Dragendorff's reagent; bioassay (Marston's method)	[31]
15	Silica gel, hexane–ethyl acetate (1:1, v/v)	Galantamine, tacrine, donepezil, huperzine A*, physostigmine, set of 138 natural and synthetic compounds	Bioassay (Marston's method)	[32]

The names of approved drugs are underlined. Huperzine A*—Drug approved in China.

RP-18 W plates [6]. The differences have been explained by the interaction of analyzed substances or AChE with the surface of chromatographic plates, which results in altered affinity of the enzyme for the compounds [6,32]. Di Giovanni et al. underline that the probable reason for the differences in results obtained with microplate and TLC assays can result from a change of conformation of the enzyme for the inhibitors [32]. Adsorbent's thickness may also influence the results of TLC-AChE inhibitory assay [17].

23.5 DETECTION

Apart from memantine, the rest of the drugs approved for the treatment of AD symptoms possess chromophores and therefore may be detected and quantified under the UV light. For example, galantamine was quantified at 207 nm [19,20] and rivastigmine at 210 nm [26]. Other exemplary alkaloids possessing AChE inhibitory activity were quantified at the following wavelengths: 290 nm, lycorine, and 313 nm, hippeastrine [20]. Rhee et al. screened plant extracts for the presence of AChE inhibitors at 254 nm [17]. Detection of alkaloids, used as AD drugs, may be enhanced with the use of Dragendorff's reagent [42]. Details related to its preparation may be found in Section 12.3.3. Memantine hydrochloride derivatization has been proposed with 9-fluorenylmethyl chloroformate (FMOC) and borate buffer solution. However, the conditions have been elaborated only for high-performance liquid chromatography (HPLC) analysis [43].

As the majority of approved drugs used for the treatment of AD are AChE inhibitors, the detection methods have been focused mainly on searching for specific derivatizing agents to detect compounds with such properties. These specific derivatizing agents have been mostly applied to pinpoint active compounds present in complex matrices (e.g., plant extracts). A detailed description of detecting AChE inhibitors can be found in Section 12.4.2.1. Generally, after TLC separation and drying, the plate is covered with a solution containing enzyme. In 1960s, human plasma was used as an enzyme source [44,45]; nowadays, AChE is commercially available in pure form. It is AChE from *Electrophorus electricus*; BChE from horse serum can also be purchased [5,6,17–20,27,28,30,32,44–48]. Some recent reports indicate that apart from AChE, traces of other peptides may be found in the commercially available product [49]. To ensure enzyme stability in the solution, proper buffers should be applied. Bovine serum albumin (BSA) is also used for this purpose, when a lyophilized *E. electricus* AChE is applied to prepare the solution. In a method proposed by Menn et al. [44,45], the addition of BSA is not needed, as the enzyme is stabilized by proteins naturally present in human plasma. After covering the plate with enzyme solution, it is sprayed with or dipped into the substrate solution and incubated at 37°C in humid atmosphere. The following substances have been used as AChE and BChE substrates: (a) acetylcholine (Menn et al. method) [44,45], (b) acetylthiocholine (Ellman's method) [50], (c) 1-naphthyl acetate (Marston's method) [30], (d) 2-naphthyl acetate (Marston's method modified by Mroczek) [19,20], and (e) 4-methoxyphenyl acetate (Yang et al. method) [28]. During the incubation step, the substrates are converted to products that are detected in subsequent step. Accordingly to the previously enumerated substrates, the enzyme converts them into (a) choline and acetic acid, (b) thiocholine and acetic acid, (c) 1-naphthol and acetic acid, (d) 2-naphthol and acetic acid, and (e) 4-methoxyphenol and acetic acid. Subsequently, the plate is covered with derivatizing agent to detect spots/bands of active compounds. The inhibitors, if present in the sample, inhibit the reaction catalyzed by the enzyme, causing the lack of proper product in the inhibition zones. Therefore, those zones appear as differently colored, when compared to the background, where the reaction catalyzed by the enzyme was completed. The following derivatizing agents are used to detect the inhibitors: (a) bromothymol blue (Menn et al. method; inhibitory zones appear as blue against a yellow background) [44,45], (b) 5,5′-dithiobis-(2-nitrobenzoic acid) (Ellman's method; inhibitory zones appear as white spots against a yellow background) [17], (c, d) Fast Blue B salt (Martson's method and Mroczek's modification; inhibitory zones appear as pale spots against a purple background) [19,20,30], and (e) potassium ferricyanide and iron chloride hexahydrate (Yang et al. method; inhibitory zones appear as yellowish against an aquamarine-blue background) [28]. Two of the aforementioned techniques are the most frequently applied in routine analysis: Marston's and Ellman's technique. TLC coupled with bioassay detection is said to be more sensitive when compared with UV or Dragendorff's reagent detection [17]. It has been also found that the minimal detectable amount of AChE inhibitors is lesser in case of TLC test when compared to the microplate assay [17]. Due to the possibility of obtaining false-positive results, verifying procedures have also been elaborated [18–20,28]. For more details related to the previously mentioned techniques, please refer to Section 12.4.2.1.

23.6 QUANTITATIVE ANALYSIS

As already stated in this chapter, there are no formulations containing more than one compound used for the treatment of AD symptoms. Therefore, HPTLC-based quantitative methods focus on single-compound formulations. Personal communication has been presented during the *International Symposium for HPTLC* on simultaneous quantitative determination of memantine and donepezil in synthetic mixture [51]. The compounds were separated on silica gel plates using the following eluent: chloroform–cyclohexane–methanol (2:2:1, v/v/v) [51].

Jagadeeswaran et al. proposed a method to quantify donepezil hydrochloride in the bulk drug and tablets [21]. The analysis was performed on the surface of the silica gel with the use of the chloroform–methanol mixture (see Table 23.1). The plate was scanned at 254 nm. Calibration curves were linear in the range of 200–1000 ng spot^{-1}. The limit of detection (LOD) and limit of quantification (LOQ) were found to be 120 and 165 ng spot^{-1}, respectively. Ali et al. quantified donepezil on the silica gel plates after development with a mixture of butanol, water, and acetic acid (Table 23.1) [22]. Retardation factor of donepezil, under such conditions, was 0.54 ± 0.03. The plate was scanned at 260 nm. The application of this chromatographic system was also suitable for the separation of donepezil and its forced degradation products. Linearity was observed in the range 50–1000 ng spot^{-1}, with the correlation coefficient r^2 = 0.9979. LOD and LOQ values were as follows: 15.40 and 50.90 ng spot^{-1}. Abbas et al. proposed a method to separate donepezil hydrochloride from its oxidative degradate [23]. The analyzed samples were chromatographed on silica gel plates with eluent: chloroform–methanol–25% ammonia (16:64:0.1, v/v/v). The obtained R$_F$ value for donepezil was 0.78 and it was quantified at its λ_{max}—315 nm [23].

Karthik et al. quantified rivastigmine in the bulk drug and in a capsule formulation [26]. Rivastigmine and its forced degradation products were separated on silica gel plates giving a compact spot of rivastigmine at R$_F$ 0.53 ± 0.02. Details related to the applied chromatographic system are given in Table 23.1. The analyzed compounds were quantified in absorbance mode at 210 nm. The authors claim to have obtained linear response between rivastigmine concentration and area under the peak over the range 200–1600 ng spot^{-1} with the correlation coefficient 0.9916 ± 0.008. The limits of detection and quantification were 30 and 100 ng spot^{-1}, respectively. The method was validated and turned out to be reliable for the purpose of rivastigmine determination. Salem et al. managed to separate rivastigmine from its degradation products and subsequently determine the amount of rivastigmine in pharmaceutical dosage form [24]. Details related to the applied procedure are given in Table 23.1.

REFERENCES

1. Alves, L., A.S.A. Correia, R. Miguel, P. Alegria, and P. Bugalho. 2012. Alzheimer's disease: A clinical practice-oriented review. *Frontiers in Neurology* 3: 63, doi: 10.3389/fneur.2012.00063.
2. Tayeb, H.O., H.D. Yang, B.H. Price, and F.I. Tarazi. 2012. Pharmacotherapies for Alzheimer's disease: Beyond cholinesterase inhibitors. *Pharmacology and Therapeutics* 134: 8–25.
3. DailyMed, Current medication information. http://dailymed.nlm.nih.gov/dailymed/lookup.cfm?setid= 03b10fd1-4da2-4e38-b720-d6aeb563697a#nlm34089-3 (accessed September 20, 2012).
4. Bonesi, M., F. Menichini, R. Tundis, M.R. Loizzo, F. Conforti, N.G. Passalacqua, G.A. Statti, and F. Menichini. 2010. Acetylcholinesterase and butyrylcholinesterase inhibitory activity of *Pinus* species essential oils and their constituents. *Journal of Enzyme Inhibition and Medicinal Chemistry* 25: 622–628.
5. Cieśla, Ł. 2012. Thin-layer chromatography with biodetection in the search for new potential drugs to treat neurodegenerative diseases—State of the art and future perspectives. *Medicinal Chemistry* 8: 102–111.

6. Cieśla, Ł., J. Kryszeń, A. Stochmal, W. Oleszek, and M. Waksmundzka-Hajnos. 2012. Low-temperature thin-layer chromatography bioautographic tests for detection of free radical scavengers and acetylcholinesterase inhibitors in volatile samples. *Journal of Planar Chromatography—Modern TLC* 25: 225–231.

7. Mukherjee, P.K., V. Kumar, M. Mal, and P. Houghton. 2007. Acetylcholinesterase inhibitors from plants. *Phytomedicine* 4: 289–300.

8. Bastianetto, S. and R. Quirion. 2002. Natural extracts as possible protective agents of brain aging. *Neurobiology of Aging* 23: 891–897.

9. Sugimoto, H., Y. Yamanishi, Y. Iimura, and Y. Kawakami. 2000. Donepezil hydrochloride (E2020) and other acetylcholinesterase inhibitors. *Current Medicinal Chemistry* 7: 303–339.

10. Drug Bank. Open Data Drug & Drug Target Database. http://www.drugbank.ca/drugs/DB00843 (accessed September 15, 2012).

11. Center for drug evaluation and research, http://www.accessdata.fda.gov/drugsatfda_docs/nda/2010/022568Orig1s000ChemR.pdf (accessed September 15, 2012).

12. Bickel, U., T. Thomsen, W. Weber, J.P. Fischer, R. Bachus, M. Nitz, and H. Kewitz. 1991. Pharmacokinetics of galanthamine in humans and corresponding cholinesterase inhibition. *Clinical Pharmacology & Therapeutics* 50: 420–428.

13. Srinivasu, M.K., B.M. Rao, B.S.S. Reddy, P.R. Kumar, K.B. Chandrasekhar, and P.K. Mohakhud. 2005. A validated chiral liquid chromatographic method for the enantiomeric separation of Rivastigmine hydrogen tartarate, a cholinesterase inhibitor. *Journal of Pharmaceutical and Biomedical Analysis* 38: 320–325.

14. Qian, S., S.K. Wo, and Z. Zuo. 2012. Pharmacokinetics and brain dispositions of tacrine and its major bioactive monohydroxylated metabolites in rats. *Journal of Pharmaceutical and Biomedical Analysis* 61: 57–63.

15. PubChem, http://pubchem.ncbi.nlm.nih.gov/ (accessed September 20, 2012).

16. Periclou, A., D. Ventura, N. Rao, and W. Abramowitz. 2006. Pharmacokinetic study of memantine in healthy and renally impaired subjects. *Clinical Pharmacology & Therapeutics* 79: 134–143.

17. Rhee, I.K., M. van de Meent, K. Ingkaninan, and R. Verpoorte. 2001. Screening for acetylcholinesterase inhibitors from Amaryllidaceae using silica gel thin-layer chromatography in combination with bioactivity staining. *Journal of Chromatography A* 915: 217–223.

18. Rhee, I.K., R.M. van Rijn, and R. Verpoorte. 2003. Qualitative determination of false-positive effects in the acetylcholinesterase assay using thin layer chromatography. *Phytochemical Analysis* 14: 127–131.

19. Mroczek, T. and J. Mazurek. 2009. Pressurized liquid extraction and anticholinesterase activity-based thin-layer chromatography with bioautography of *Amaryllidaceae* alkaloids. *Analytica Chimica Acta* 633: 188–196.

20. Mroczek, T. 2009. Highly efficient, selective and sensitive molecular screening of acetylcholinesterase inhibitors of natural origin by solid-phase extraction-liquid chromatography/electrospray ionization-octopole-orthogonal acceleration time-of-flight-mass spectrometry and novel thin-layer chromatography-based bioautography. *Journal of Chromatography A* 1216: 2519–2528.

21. Jagadeeswaran, M., N. Gopal, M. Gandhimathi, R. Rajavel, M. Ganesh, and T. Sivakumar. 2011. A validated HPTLC method for the estimation of donepezil HCL in bulk and its tablet dosage form. *Eurasian Journal of Analytical Chemistry* 6: 40–45.

22. Ali, J., M. Ali, S. Ahmad, and M. Bhavna. 2009. Stability-indicating HPTLC method for determination of donepezil hydrochloride in bulk drug and pharmaceutical dosage form. *Chemia Analityczna (Warsaw)* 54: 1501–1516.

23. Abbas, S.S., Y.M. Fayez, and L.E.S. Abdel Fatah. 2006. Stability indicating methods for determination of donepezil hydrochloride according to ICH guidelines. *Chemical and Pharmaceutical Bulletin* 54: 1447–1450.

24. Salem, M.Y., A.M. El-Kosasy, M.G. El-Bardicy, and M.K. AbdEl-Rahman. 2010. Spectrophotometric and spectrodensitometric methods for the determination of rivastigmine hydrogen tartrate in presence of its degradation product. *Drug Testing & Analysis* 2: 225–233.

25. Zheng, X.Y., L. Zhang, X.M. Cheng, Z.J. Zhang, C.H. Wang, and Z.T. Wang. 2011. Identification of acetylcholinesterase inhibitors from seeds of plants of genus *Peganum* by thin-layer-chromatography–bioautography. *Journal of Planar Chromatography—Modern TLC* 24: 470–474.

26. Karthik, A., G.S. Subramanian, P. Musmade, A. Ranjithkumar, M. Surulivelrajan, and N. Udupa. 2007. Stability-indicating HPTLC determination of rivastigmine in the bulk drug and in pharmaceutical dosage forms. *Journal of Planar Chromatography* 20: 457–461.

27. Yang, Z., X. Zhang, D. Duan, Z. Song, M. Yang, and S. Li. 2009. Modified TLC bioautographic method for screening acetylcholinesterase inhibitors from plant extracts. *Journal of Separation Science* 32: 3257–3259.

28. Yang, Z.D., Z.W. Song, J. Ren, M.J. Yang, and S. Li. 2011. Improved thin-layer chromatography bioautographic assay for the detection of acetylcholinesterase inhibitors in plants. *Phytochemical Analysis* 22: 509–515.

29. Berkov, S., J. Bastida, M. Nikolova, F. Viladomat, and C. Codina. 2008. Rapid TLC/GC-MS identification of acetylcholinesterase inhibitors in alkaloid extracts. *Phytochemical Analysis* 19: 411–419.
30. Marston, A., J. Kissling, and K. Hostettmann. 2002. A rapid TLC bioautographic method for the detection of acetylcholinesterase and butyrylcholinesterase inhibitors in plants. *Phytochemical Analysis* 13: 51–54.
31. Adhami, H.R., H. Farsam, and K. Krenn. 2011. Screening of medicinal plants from Iranian traditional medicine for acetylcholinesterase inhibition. *Phytotherapy Research* 25: 1148–1152.
32. Di Giovanni, S., A. Borloz, A. Urbain, A. Marston, K. Hostettmann, P.A. Carrupt, and M. Reist. 2008. In vitro screening assays to identify natural or synthetic acetylcholinesterase inhibitors: Thin layer chromatography versus microplate methods. *European Journal of Pharmaceutical Sciences* 33: 109–119.
33. Waksmundzka-Hajnos, M. and Ł. Cieśla. 2010. Separation of ionic analytes: Reversed-phase, ion-pair, ion-exchange, and ion-exclusion HPLC. In *High Performance Liquid Chromatography in Phytochemical Analysis*, M. Waksmundzka-Hajnos and J. Sherma, eds., pp. 195–210. Boca Raton, FL: CRC Press.
34. Waksmundzka-Hajnos, M. and A. Petruczynik. 2008. TLC of isoquinoline alkaloids. In *Thin Layer Chromatography in Phytochemistry*, M. Waksmundzka-Hajnos, J. Sherma, and T. Kowalska, eds., pp. 641–684. Boca Raton, FL: CRC Press.
35. Gołkiewicz, W., M. Gadzikowska, J. Kuczyński, and L. Jusiak. 1993. Micropreparative chromatography of some quaternary alkaloids from the roots of *Chelidonium majus* L. *Journal of Planar Chromatography* 6: 382–385.
36. Gołkiewicz, W. and M. Gadzikowska. 1981. Isolation of some quaternary alkaloids from the extract roots of *Chelidonium majus* L. by column and thin-layer chromatography. *Chromatographia* 50: 52–56.
37. Cieśla, Ł., M. Hajnos, and M. Waksmundzka-Hajnos. 2011. Application of hydrophilic interaction TLC systems for separation of highly polar glycosidic compounds from the flowers of selected *Verbascum* species. *Journal of Planar Chromatography* 24: 295–300.
38. Cieśla, Ł., I. Kowalska, W. Oleszek, and A. Stochmal. 2012. Free radical scavenging activities of polyphenolic compounds isolated from *Medicago sativa* and *Medicago truncatula* assessed by means of thin-layer chromatography DPPH˙ rapid test. *Phytochemical Analysis* 24(1): 47–52, doi: 10.1002/pca.2379.
39. Alpert, A.J. 1990. Hydrophilic-interaction chromatography for the separation of peptides, nucleic acids and other polar compounds. *Journal of Chromatography A* 499: 177–196.
40. Petruczynik, A., K. Śliwka, and M. Waksmundzka-Hajnos. 2010. Effect of the vapour phase on the separation of isoquinoline alkaloids by thin-layer chromatography. *Acta Chromatographica* 22: 391–404.
41. Gadzikowska, M., G.W. Jóźwiak, and M. Waksmundzka-Hajnos. 2010. Effect of the vapour phase on the TLC separation of tropane alkaloids. *Acta Chromatographica* 22: 515–525.
42. Reich, E. and A. Schibli. 2007. *High-Performance Thin-Layer Chromatography for the Analysis of Medicinal Plants*. Stuttgart, Germany: Thieme Publishers.
43. Narola, B., A.S. Singh, P.R. Santhakumar, and T.G. Chandrashekhar. 2010. A validated stability-indicating reverse phase HPLC assay method for the determination of memantine hydrochloride drug substance with UV-detection using precolumn derivatization technique. *Analytical Chemistry Insights* 5: 37–45.
44. Menn, J.J., J.B. McBain, and M.J. Dennis. 1964. Detection of naturally occurring cholinesterase inhibitors in several crops by paper chromatography. *Nature* 202: 697–698.
45. Menn, J.J. and J.B. McBain. 1966. Detection of cholinesterase-inhibiting insecticide chemicals and pharmaceutical alkaloids on thin-layer chromatograms. *Nature* 209: 1351–1352.
46. Choma, I.M. and E.M. Grzelak. 2011. Bioautography detection in thin-layer chromatography. *Journal of Chromatography A* 1218: 2684–2691.
47. Marston, A. 2011. Thin-layer chromatography with biological detection in phytochemistry. *Journal of Chromatography A* 1218: 2676–2683.
48. Kiely, J.S., W.H. Moos, M.R. Pavia, R.D. Schwarz, and G.L. Woodard. 1991. A silica gel plate-based qualitative assay for acetylcholinesterase activity: A mass method to screen for potential inhibitors. *Analytical Biochemistry* 196: 439–442.
49. Drączkowski, P., P. Halczuk, D. Matosiuk, and K. Jóźwiak. 2012. Influence of different inhibitors of herbal origin on acetylcholinesterase kinetics monitored by isothermal titration calorimetry. Paper presented at the *2nd International Conference and Workshop, Plant – The Source of Research Material*, Lublin, Poland.
50. Ellman, G.L., D. Courtney, V. Andies, and R.M. Featherstone. 1961. A new and rapid colorimetric determination of acetylcholinesterase activity. *Biochemical Pharmacology* 7: 88–95.
51. Gubandru, M., D. Balalau, C.M. Gutu, and M. Illie. 2011. Simultaneous quantitative evaluation of memantine and donepezil by HPTLC. *International Symposium for High-Performance Thin-Layer Chromatography*, Basel, Switzerland.

24 TLC of Antiparkinsonians

Sigrid Mennickent, Marta de Diego,
Mario Vega, and Carmen Gloria Godoy

CONTENTS

459

24.1   INFORMATION ABOUT THE DRUG GROUP

The primary deficit in Parkinson's disease (PD) is the loss of the neurons in the substantia nigra pars compacta that provide dopaminergic innervations to the striatum. This parallels the loss of neurons from the substantia nigra, suggesting that replacement of dopamine could restore function. These fundamental observations led to an extensive investigative effort to understand the metabolism and actions of dopamine and to learn how a deficit in dopamine gives rise to the clinical features of PD.

The drugs used for this pathology are levodopa (L-DOPA)/carbidopa, benserazide, pergolide, bromocriptine, lisuride, terguride, cabergoline, pramipexole, selegiline, rasagiline, amantadine, memantine, bupidine, apomorphine, pramipexole, and ropinirole. All of them are dopaminergic drugs, but some anticholinergic drugs are also used in PD, as trihexyphenidyl and benztropine. Moreover, some catechol-O-methyltransferase (COMT) inhibitor drugs had been used, as entacapone and tolcapone [1–5].

24.1.1   L-DOPA

L-DOPA, the levorotatory isomer of dihydroxyphenylalanine, a natural amino acid, is the immediate precursor of the neurotransmitter dopamine. The actions of L-DOPA are mainly those of dopamine. Unlike dopamine, L-DOPA can readily enter the central nervous system and is used in the treatment of conditions, such as PD, that are associated with depletion of dopamine in the brain [1–4]. L-DOPA is considered by many clinicians the drug of choice in the management of idiopathic parkinsonian syndrome [2].

L-DOPA is rapidly decarboxylated in the human body, so that very little unchanged drug is available to cross the blood–brain barrier for central conversion into dopamine. Moreover, dopamine release into the circulation by peripheral conversion of L-DOPA produces gastrointestinal effects. Consequently, L-DOPA is usually given together with a peripheral DOPA-decarboxylase inhibitor such as carbidopa or benserazide to increase the proportion of L-DOPA that can enter the brain and to reduce its adverse effects.

L-DOPA therapy can have a dramatic effect on all of the signs and symptoms of PD. Early in the treatment, L-DOPA improves the degree of tremor, rigidity, and bradykinesia almost completely. With time, the action of the drug is lost, and the patient's motor state may fluctuate dramatically with each dose of L-DOPA. Increasing the dose and the frequency of administration can improve this situation, but high doses can produce dyskinesias [1–3].

24.1.2   DOPAMINE RECEPTOR AGONISTS

These drugs are an alternative to L-DOPA and they have some advantages over L-DOPA: Their activities do not depend on the functional capacities of the nigrostriatal neurons and thus might be more effective than L-DOPA in late PD, they are more selective in their actions (unlike L-DOPA, which leads to activation of all dopamine receptor types throughout the brain, agonists may exhibit relative selectivity for different subtypes of dopamine receptors), they have longer duration of action than that of L-DOPA, and they have the potential to modify the course of the disease by reducing endogenous release of dopamine as well as the need for exogenous L-DOPA.

Two dopamine agonists, bromocriptine and pergolide, currently are available in the United States for treatment of PD. Both are ergot derivatives, and their actions and adverse effects are similar to each other and to those of L-DOPA. Other dopamine receptor agonists are apomorphine, pramipexole, and ropinirole [1–3].

24.1.3   SELEGILINE

At low to moderate doses (10 mg/day or less), selegiline is a selective inhibitor of monoamine oxidase-B (MAO-B), leading to irreversible inhibition of the enzyme. This enzyme is responsible for the majority of oxidative metabolism of dopamine in the striatum. Doses of selegiline higher than 10 mg daily can produce inhibition of MAO-A and should be avoided. Other MAO-B inhibitor is the rasagiline [1–3].

24.1.4 AMANTADINE

Amantadine is an antiviral and a drug with antiparkinsonian actions. The mechanism of amantadine is not clear, maybe it might alter dopamine release or reuptake, and anticholinergic properties also may contribute to its action. In any case, their affectivity is modest and this drug is used as initial therapy of mild PD. Its adverse effects are dizziness, lethargy, and sleep disturbance, as well as nausea and vomiting. The effects depend on doses [1–3].

24.1.5 MUSCARINIC RECEPTOR ANTAGONISTS

They were widely used for the treatment of PD before the discovery of L-DOPA. It seems that they act within the neostriatum, through the receptors that normally mediate the response to the intrinsic cholinergic innervations of this structure. The most used has been the trihexyphenidyl. All of these drugs have a moderate action in PD, which can be useful in the treatment of early disease or an as adjunct to dopamimetic therapy. The adverse effects of these drugs are a result of their anticholinergic properties, mostly sedation and mental confusion, especially in elderly people [1–3].

24.1.6 CATECHOL-O-METHYLTRANSFERASE INHIBITORS

Entacapone and tolcapone are reversible inhibitors of COMT. These drugs are nitrocatechols and are chemically and pharmacologically unrelated to other currently available antiparkinsonian agents. Entacapone, unlike tolcapone, has not been associated with hepatotoxicity (e.g., drug-induced hepatitis, fatal liver failure) [1–3].

24.2 POSSIBLE DEGRADANT IMPURITIES

24.2.1 L-DOPA

In the presence of moisture, oxidation of L-DOPA by atmospheric oxygen causes rapid discoloration. Solid-state L-DOPA showed marked discoloration, which increased with heating time, after storage at 105°C for 24 h. In alkaline solution, oxidation of L-DOPA results in the formation of melanin and related intermediates [6].

24.2.2 BROMOCRIPTINE

As a nonhydrogenated ergot alkaloid, bromocriptine is relatively sensitive to autoxidation both in solid and in dissolved states [7].

24.2.3 PERGOLIDE

Pergolide is known to be a photosensitive drug substance, and all the necessary precautions are taken during the manufacturing of products, for example, opaque blister packaging. The major photodegradation products are pergolide sulfoxide and pergolide sulfone, which are also the main impurities of the bulk drug substance [8].

24.2.4 SELEGILINE

In a study performed by Lester Chafetz, Manisha P. Desai, and Ludmilla Sukonik, a solution of selegiline hydrochloride reference standard, which contained no detectable impurities at the time of its preparation, was found by high-performance liquid chromatography (HPLC) to contain a trace of a compound at the locus of methamphetamine when analyzed after 1 year. Heating selegiline solutions at pH 7 and 105°C produced methamphetamine as the major product at a rate that closely followed the first-order rate equation. Using only these data and worst-case assumptions, rate constants were estimated at various temperatures; the activation energy was estimated to be about 25 kcal, and the stability-indicating validity of the assay used was reaffirmed. Selegiline undergoes degradation at a negligibly slow rate [9].

24.2.5 AMANTADINE

In a study performed by Hassan F. Askal, Alaa S. Khedr, Ibrahim A. Darwish, and Ramadan M. Mahmoud, the compatibility of amantadine with the excipients used was studied. The stress testing results revealed that amantadine was compatible with the combined excipients, whereas no more degradation products were observed when the stress-testing experiments were carried out on amantadine-containing capsules [10].

24.2.6 TRIHEXYPHENIDYL

Some degradant impurities found were 1-phenyl-2-propenone, 3-piperidinopropiophe-none, and 3-aminopropiophenone [11].

24.2.7 ENTACAPONE

Some impurities found in a study were methoxy entacapone (2E)-2-cyano-3-(3-methoxy-4-hydroxy-5-nitrophenyl)-N,N-diethyl-2-propenamide; vanillin entacapone (2E)-2-cyano-3-(3-methoxy-4-hydroxyphenyl)-N,N-diethyl-2-propenamide; and dihydroxy entacapone (2E)-2-cyano-3-(3,4-dihydroxyphenyl)-N,N-diethyl-2-propenamide [12].

24.3 SAMPLE PREPARATION IN CONTEXT OF MATRIX

Sample preparation depends on the employed matrix. Sample preparation has an impact in nearly all the later assay steps, and hence, it is critical for both qualitative and quantitative analyses. It usually includes the isolation and/or preconcentration of compounds of interest from various matrices, making the analytes more suitable for separation and detection. Indeed, sample preparation is a key factor in determining the success of analysis from complex matrices such as biological samples and typically takes 80% of the total analysis time. Due to their complexity and protein content, biological samples are usually not directly compatible with chromatographic analysis of the analytes of interest. Therefore, a given set of more or less complex steps is required for sample preparation. Different strategies can be applied, ranging from a simple dilution of the sample to more complex pretreatment processes. Conventionally, sample preparation has been performed by means of protein precipitation (PP), liquid–liquid extraction (LLE), or solid-phase extraction (SPE); however, many modern approaches such as restricted access material (RAM), molecularly imprinted polymers (MIP), solid-phase microextraction (SPME), liquid–liquid microextraction (LLME), and microextraction by packed sorbent (MEPS) have also been currently used for sample preparation in the process of development of bioanalytical methods. PP is considered to be the simplest and fastest sample preparation technique; however, a considerable amount of interferences still remain in the extract—the supernatant. On the other hand, the extracts obtained from LLE and SPE are cleaner than that of PP, but LLE and SPE procedures are frequently more complex and costly [13].

By other way, accurate analytical data are critical in the pharmaceutical industry. During drug development, this information is used to evaluate and select formulations for use in toxicology and clinical studies, to assess manufacturing processes, and to assess the suitability and stability of clinical supplies. For marketed products, analytical data are used to assess the suitability and stability of the commercial product. Development and use of robust analytical methods is critical for the ability to generate accurate analytical data. Sample preparation is an integral part of the analytical method. Approximately two-thirds of the time spent testing and analyzing samples is spent on the sample preparation portion of the method. In addition, issues related to sample preparation accounted for one-third of the errors generated while performing an analytical method. A number of challenges exist in the sample preparation/extraction of pharmaceutical dosage forms for potency and purity analysis as well as isolation and identification of impurities and degradants. These challenges increase for complex dosage forms such as some controlled-release formulations and other challenging formulations such as suspensions, ointments, and transdermal patches. Challenges in developing rapid and rugged sample preparation methods include complete dispersion of the dosage form to facilitate extraction and solubilization of the analytes of interest, dealing with extracted interfering components, and addressing drug–excipient interactions. A number of factors must be considered and addressed in each of these areas [14].

Some examples founded in literature for PD drugs stability studies are the following.

24.3.1 DEVELOPMENT AND VALIDATION OF AN HPTLC-DENSITOMETRIC METHOD FOR DETERMINATION OF L-DOPA IN SEEDS OF *MUCUNA PRURIENS* AND ITS DOSAGE FORM

Isolation of L-DOPA from the seeds: The seeds of *Mucuna pruriens* were dried in shade and powdered in a mechanical grinder. Then it was passed through sieve no. 20. The seed powder was macerated with 1% aqueous acetic acid containing 0.1% sodium sulfide for 6 h at 60°C. It was filtered and concentrated to half volume by rotary vacuum evaporator. Then the concentrate was freezed to get the crystals of L-DOPA. The crystals of L-DOPA were recrystallized to get off-white-colored pure L-DOPA. The isolated L-DOPA was qualitatively characterized by FTIR and LC–MS [15].

24.3.2 HPTLC METHOD DEVELOPMENT AND VALIDATION FOR THE SIMULTANEOUS ESTIMATION OF DIOSGENIN AND L-DOPA IN MARKETED FORMULATION

Sample preparation: An aphrodisiac polyherbal formulation was taken for quantitative estimation of biomarkers like diosgenin and L-DOPA. Twenty capsules were weighed and the total weight was determined. The capsule granules are subjected to fine powder with the help of mortar and pestle. A 1000 mg of the aforementioned powder was refluxed with methanol, and the final volume was made to 10 mL to get a concentration of 100 mg/mL and used in the study of diosgenin, and for L-DOPA, 10 mg powder dissolved in 5 mL phosphoric acid and diluted up to 10 mL. Standard (STD) solution was prepared by dissolving 10 mg of STD diosgenin in 10 mL methanol and 10 mg STD l-DOPA in 5 mL phosphoric acid. These solutions were next diluted with methanol up to 10 mL and ultrasonicated for 15 min, then used as a stock solution [16].

24.3.3 SIMULTANEOUS RP-HPTLC METHOD FOR DETERMINATION OF L-DOPA, CARBIDOPA, AND ENTACAPONE IN COMBINED TABLET DOSAGE FORM

Analysis of marketed tablet dosage forms: Twenty tablets were weighed, and an amount of tablet powder equivalent to 25 mg of carbidopa was accurately weighed and transferred to a 100 mL volumetric flask. The carbidopa standard (100 mg) was added to the aforementioned tablet powder and dissolved in a 75 mL mixture of methanol–0.05 N HCl (1:1, v/v). The solution was sonicated for 15 min and then diluted to final volume with methanol–0.05 N HCl (1:1, v/v). It was centrifuged for 15 min at 2500 rpm and then filtered by using Whatman filter no. 41. The aforementioned solution was suitably diluted further with methanol to get a solution containing 50 µg/mL of L-DOPA, 62.5 µg/mL of carbidopa, and 100 µg/mL of entacapone. The plate was activated, and 13 µL of the sample solution was spotted and analyzed [17].

24.3.4 QUANTITATIVE DETERMINATION OF L-DOPA IN TABLETS BY HPTLC

Sample preparation: Pharmaceutical preparations were tablets, nominally containing 100 mg of L-DOPA and some nonspecific excipients. They were processed as follows: 20 tablets were weighed and ground into fine powder, and an accurately weighed portion equivalent to 40 mg of L-DOPA was diluted to 100 mL with water–methanol (7:3, v/v). The solution was centrifuged and the supernatant was used. Appropriate dilutions were done for validation of the method [18].

24.4 CHROMATOGRAPHIC SYSTEMS USED IN THE SEPARATION AND/OR QUANTIFICATION OF PARTICULAR DRUGS

Scientific literature reports several chromatographic methods for the determination of L-DOPA in biological fluids and in pharmaceutical preparations, such as HPLC, mostly in biological fluids [19,20–30] and TLC [13–16]. The official method for the quantitative determination of L-DOPA in tablets is by HPLC [31]. Nevertheless, HPLC and other mentioned nonchromatographic techniques have often suffered from diverse disadvantages with regard to cost or selectivity, complex sample preparation procedures, and long analysis time.

High-performance thin-layer chromatography (HPTLC) is a technique carried out within a short period of time, requires few mobile phases, and allows for the analysis of a large number of samples simultaneously. The ability of HPTLC to analyze many samples in parallel has the advantage over other techniques because separation of 10 or 20 samples takes the same time as the separation of one sample, with very good precision, accuracy, and sensitivity.

24.5 USE OF TLC IN PHARMACEUTICAL FORMULATION ANALYSIS

Some HPTLC methods for the quantitative determination of antiparkinsonians are described in this section.

24.5.1 DEVELOPMENT AND VALIDATION OF AN HPTLC-DENSITOMETRIC METHOD FOR DETERMINATION OF L-DOPA IN SEEDS OF *MUCUNA PRURIENS* AND ITS DOSAGE FORM

A simple, sensitive, and selective HPTLC-densitometric method was developed and validated for the determination of L-DOPA in seeds of *M. pruriens* and its herbal dosage form. Analysis of L-DOPA was carried out on HPTLC plates precoated with silica gel $60F_{254}$ as stationary phase. Linear ascending development of the plate was done in a presaturated twin trough glass chamber. The mobile phase consisted of n-butanol–glacial acetic acid–water (5:1:4, v/v/v) at room temperature (25°C ± 2°C). CAMAG TLC scanner III was used for spectrodensitometric scanning and analysis was done in absorption mode at 280 nm. The system gave compact spot for L-DOPA (R_f value of 0.39 ± 0.04). The polynomial regression analysis data for the calibration plots showed correlation coefficient r = 0.999 in concentration range 100–1000 ng/spot with respect to peak area. According to the International Conference on Harmonization (ICH) guidelines, the method was validated for precision, recovery, robustness, and ruggedness. The limits of detection and quantification were determined. The statistical data analysis showed that the method is reproducible and selective for estimation of L-DOPA [15].

24.5.2 HPTLC Method Development and Validation for the Simultaneous Estimation of Diosgenin and l-DOPA in Marketed Formulation

The stationary phase silica gel G60F254 was selected for separation, and the sample was developed using a mixture of toluene–ethyl acetate–formic acid–glacial acetic acid (GAA) in the ratio 2:1:1:0.75 v/v as mobile phase. Quantification was carried out at 194 nm for diosgenin and 280 nm for l-DOPA using absorbance reflectance mode. The R_f value of l-DOPA and diosgenin was found to be 0.27 + 0.2 and 0.61 + 0.2, respectively. Linearity was found to be in the concentration range of 100–700 ng/spot of l-DOPA and 600–1800 for diosgenin. The correlation coefficient values were 0.9954 and 0.9934, respectively.

The results of analysis were validated in terms of accuracy and precision. The limit of detection (LOD) was found to be 1.03 and 5.69 ng for l-DOPA and diosgenin, respectively. The limit of quantification (LOQ) was found to be 3.14 and 17.25 ng/spot for l-DOPA and diosgenin, respectively. The content uniformity test was carried out as per the United States Pharmacopoeia (USP) specification. The proposed HPTLC method provides a faster and effective quantitative control for routine analysis of l-DOPA and diosgenin [16].

24.5.3 Simultaneous RP-HPTLC Method for Determination of l-DOPA, Carbidopa, and Entacapone in Combined Tablet Dosage Form

A simple, rapid, specific, and accurate reverse-phase HPTLC (RP-HPTLC) method was developed and validated for simultaneous quantification of l-DOPA, carbidopa, and entacapone in their combined dosage form. Due to the structural similarity between l-DOPA and carbidopa and the vast difference in their polarity with that of entacapone, it is very challenging to carry out the simultaneous estimation of all three drugs together. In the developed method, chromatography was performed on HPTLC plates with precoated silica gel 60 RP-18 F_{254} using acetonitrile–n-butanol–water–triethylamine (0.5:9.5:1:0.001, v/v/v/v), pH adjusted to 3.6 with o-phosphoric acid, as the mobile phase. Densitometric evaluation was performed at 282 nm. The R_F values were 0.46, 0.64, and 0.87 for l-DOPA, carbidopa, and entacapone, respectively. The polynomial regression data for the calibration plots showed good linear relationship in the concentration range 300–1500 ng/spot for l-DOPA, 200–1000 ng/spot for carbidopa, and 200–2000 ng/spot for entacapone. The suitability of this HPTLC method for quantitative determination of drugs was proved by validation in accordance with the requirements of the ICH guidelines (Q2B). The application of the developed method was evaluated to determine the amounts of each drug in their marketed combined tablet dosage forms with label claims of 100 mg l-DOPA, 25 mg carbidopa, and 200 mg entacapone [17].

24.5.4 Quantitative Determination of l-DOPA in Tablets by HPTLC

A densitometric HPTLC method was developed and validated for quantitative analysis of l-DOPA in tablets. Chromatographic separation was achieved on precoated silica gel F 254 HPTLC plates using a mixture of acetone–chloroform–n-butanol–GAA–water (60:40:40:40:35 v/v/v/v/v) as mobile phase. Quantitative analysis was carried out at a wavelength of 497 nm. The method was linear between 100 and 500 ng/μL, with a correlation coefficient of 0.999. The intraassay variation was between 0.26% and 0.65%, and the interassay variation between 0.52% and 2.04%. The detection limit was 1.12 ng/μL, and the quantification limit was 3.29 ng/μL. The accuracy ranged from 100.40% to 101.09%, with a Coefficient of variation (CV) not higher than 1.40%. The method was successfully applied to quantify l-DOPA in real pharmaceutical samples, including the comparison with HPLC measurements. The method was fast, specific, with a good precision, and accurate for the quantitative determination of l-DOPA in tablets [18].

24.5.5 Determination of Impurities in l-DOPA and Carbidopa by HPLC with Electrochemical Detection

The detection of 3-(3,4,6-trihydroxyphenyl)-alanine and 3-methoxytyrosine in l-DOPA and methyldopa and 3-O-methylcarbidopa in carbidopa by HPLC with electrochemical detection is described. An octyl-bonded reversed-phase column is employed with a buffered aqueous methanol mobile phase containing an "ion-pairing" reagent. All components are well resolved and sensitivity detected by amperometric oxidation at a glassy carbon electrode maintained at +0.90 V versus Ag/AgCl. The impurities can be qualitatively characterized for identification purposes by hydrodynamic voltammetry, in which the peak height is observed for a range of oxidation voltages. The analysis of l-DOPA–carbidopa combination tablets is discussed [19].

REFERENCES

1. Sweetman, S. 2003. *Martindale, Guía Completa de Consulta Farmacoterapéutica*. Barcelona, Spain: Pharma Editores S.L.
2. Mc Evoy, G. 2006. *AHFS Drug Information*. Bethesda, MD: American Society of Health-System Pharmacists.
3. Hardman, J. and Limbird, L. 2003. *Las Bases Farmacológicas de la Terapéutica*. México: McGraw-Hill.
4. Delgado, J. and Remers, W. 1998. *Wilson and Gisvold's Textbook of Organic Medicinal and Pharmaceutical Chemistry*. Philadelphia, PA: Lippincott-Raven.
5. Brody, T., Larner, J., and Minneman, K. 1998. *Human Pharmacology: Molecular to Clinical*. St. Louis, MO: Mosby-Year Book.
6. Lund, W. 1994. *The Pharmaceutical Codex*. London, U.K.: The Pharmaceutical Press.
7. Florey, K. 1979. *Analytical Profiles of Drug Substances*, Vol. 8. New York: Academic Press.
8. Sprankle, D.J., Jensen, E.C., and Britain, H. 1992. *Analytical Profiles of Drug Substances and Excipients, Pergolide Mesylate*, Vol. 21. New York: Academic Press.
9. Chafetz, L., Desai, M., and Sukonik, L. 1994. Trace decomposition of selegiline. Use of worst-case kinetic for a stable drug. *J. Pharm. Sci.* 83(9): 1250–1252.
10. Hassan, A., Khedr, A., Darwish, I., and Mahmoud, R. 2008. Quantitative thin-layer chromatographic method for determination of amantadine hydrochloride. *Int. J. Biomed. Sci.* 4(2): 155–160.
11. Poirier, M.A., Curran, N.M., McErlane, K.M., and Lovering, E.G. 1979. Impurities in drugs III: Trihexyphenidyl. *J. Pharm. Sci.* 68(9): 1124–1127.
12. Shetty, S.K., Surendranath, K.V., Radakrishananand, P., Satish, J., Jogul, J., and Tripathi, U.M. 2009. Stress degradation behavior of Entacapone and development of LC-stability indicating related substances and assay method. *Chromatographia* 10: 1365.
13. Gonçalves, D., Alves, G., Soares-da-Silva, P., and Falcão, A. 2012. Bioanalytical chromatographic methods for the determination of catechol-O-methyltransferase inhibitors in rodents and human samples: A review. *Anal. Chim. Acta* 710: 17–32.
14. Nickerson, B. 2011. *Sample Preparation of Pharmaceutical Dosage Forms: Challenges and Strategies for Sample Preparation and Extraction*. New York: Springer.
15. Behera, A., Gowri Sankar, D., and Chandra Si, S. 2010. Development and validation of an HPTLC—Densitometric method for determination of levodopa in seeds of *Mucuna pruriens* and its dosage form. *Eurasian J. Anal. Chem.* 5(2): 126–136.
16. Kshirsagar, V.B., Deokate, U.A., Bharkad, V.B., and Khadabadi, S.S. 2008. HPTLC method development and validation for the simultaneous estimation of diosgenin and levodopa in marketed formulation. *Asian J. Res. Chem.* 1(1): 36–39.
17. Gandhi, D. and Mehta, P. 2011. Simultaneous RP-HPTLC method for determination of levodopa, carbidopa, and entacapone in combined tablet dosage form. *JPC* 24(3): 236–241.
18. Mennickent, S., Nail, M., Vega, M., and de Diego, M. 2007. Quantitative determination of L-DOPA in tablets by high performance thin layer chromatography. *J. Sep. Sci.* 30(12): 1893–1898.
19. Rihbany, L.A. and. Delaney, M.F. 1982. Determination of impurities in levodopa and carbidopa by high performance liquid chromatography with electro-chemical detection. *J. Chromatogr. A* 248(1): 125–133.
20. Fernández, M., Fernández, T., García, B., Gutiérrez, J., and Iraizoz, A. 2001. Estabilidad acelerada de tabletas de levodopa-carbidopa (250–25 mg). *Rev. Cubana Farm.* 35 (2): 1–5.

21. Karimi, M., Carl, J.L., Loftin, S., and Perlmutter, J.S. 2006. Modified high performance liquid chromatography with electrochemical detection method for plasma measurement of levodopa, 3-*O*-methyldopa, dopamine carbidopa and 3,4-dihydroxyphenyl acetic acid. *J. Chromatogr. B* 19(1–2): 120–123.

22. Saxer, C., Nilna, M., Nakashima, A., Nagae, Y., and Masude, N. 2004. Simultaneous determination of levodopa, and 3-*O*-methyldopa in human plasma by liquid chromatography with electrochemical detection. *J. Chromatogr. B* 802(2): 299–305.

23. Tolokán , A., Klebovich, I., Valgo, K., and Horvai, G. 1997. Automated determination of levodopa and carbidopa in plasma by high-performance liquid chromatography-electrochemical detection using an on-line flow injection analysis sample pretreatment unit. *J. Chromatogr. B* 698: 201–207.

24. Nyholm, D., Lennernans, H., Gomes-Trolin, C., and Aquilonious, S.-M. 2002. Levodopa pharmacokinetics and motor performance during activities of daily living in patients with Parkinson's disease on individual drug combinations. *Clin. Neuropharmacol.* 25(2): 89–96.

25. Chaná, P., Kunstmann, C., Reyes-Parada, M., and Sáez-Briones, P. 2004. Delayed early morning turn "ON" in response to a single dose of levodopa in advanced Parkinson's disease: Pharmacokinetics should be considered. *J. Neurol. Neurosurg. Psychiatr.* 75: 1782–1783.

26. Crevoisier, C., Monreal, A., Metzger, B., and Nilsen, T. 2003. Comparative single-and multiple-dose of levodopa and 3-*O*-methyldopa following a new dual-release and conventional slow-release formulation of levodopa and benserazide in healthy volunteers. *Eur. Neurol.* 49: 39–44.

27. Sagar, K.A. and Smyth, M.R. 2000. Bioavailability studies of oral dosage forms containing levodopa and carbidopa using column-switching chromatography following by electrochemical detection. *Analyst* 125: 439–445.

28. Blandini, F., Matignoni, E., Pachetti, C., Desideri, S., Rivellini, D., and Nappi, G.1997. Simultaneous determination of L-dopa and 3-*O*-methyldopa in human platelets and plasma using high performance liquid chromatography with electrochemical detection. *J. Chromatogr. B* 700 (1–2): 278–282.

29. Dethy, S., Laue, M.A., Van Blercom, N., Damhaut, P., Goldman, S., and Hildebrand, J. 1997. Microdialysis-HPLC for plasma levodopa and metabolites monitoring in parkinsonian patients. *Clin. Chem.* 43: 740–744.

30. Wang, J. and Fang, Y. 2006. Determination, purity assessment and chiral separation of levodopa methyl ester in bulk and formulation pharmaceuticals. *Biomed. Chromatogr.* 20(9): 904–910.

31. *The United States Pharmacopeia.* 2005. Rockville, MD: U.S. Pharmacopeial Convention Inc.

25 Thin-Layer Chromatography of Cardiac Drugs

Rafał Pietraś

CONTENTS

25.1 INTRODUCTION

Lifestyle diseases are the cause of increasing mortality among middle-aged, which constitute the largest group of professionals in the modern society. Cardiovascular diseases, especially cardiac failure, have become synonymous with chronic illnesses plaguing social groups in developed countries. The most prevalent cardiovascular disease is hypertension, which affects nearly half the population. Moreover, the ischemic heart disease, the heart failure, and the arrhythmias of cardiac origin are also frequently mentioned. Pathophysiology of these phenomena has a complex character, and the most important reasons are included: impaired myocardial contractility, lipid disorders, congenital or acquired defects, and an unhealthy lifestyle.

Medicinal products used in the treatment of cardiovascular diseases are characterized by considerable structural diversity. A different construction of particular groups of medicinal substances allows effective therapy based on the mechanism of direct cardiovascular effects and indirect effects on the autonomic and central nervous system. The unusual multiplicity of actions of these drugs results in the presence of many dangerous side effects, especially the very high toxicity.

This chapter presents the application of thin-layer chromatography (TLC) analysis of selected drugs used to treat cardiovascular disease. The present group of medicinal substances show a considerable variety of chemical and multi-therapeutic activity. Therefore, after consideration, the following groups of cardiovascular drugs were selected: cardiac glycosides, antiarrhythmics that block the sodium and potassium channels, and nitrates.

25.2 CARDIAC GLYCOSIDES

The use of cardiac glycosides in medicine was first described in 1785 by William Withering. Cardiac glycosides currently used in therapy is a group of substances obtained from the foxglove plants (*Digitalis purpurea*, *Digitalis lanata*) and sea onion (*Scilla maritima*). Chemically, these compounds are composed of two major parts: a steroid structure with an unsaturated five- or six-membered lactone ring (aglycone) and covalently bound sugar residue (genin). Due to the structure of the lactone ring, glycosides can be divided into two groups: cardenolides with a five-membered butenolide ring and bufadienolides with a six-membered α-pyrone ring.

Sugar residue contains a chain of monosaccharides (1–4 monomers in the form of deoxyhexoses with the exception of D-glucose) connected to the third ring of the aglycone. For the appropriate

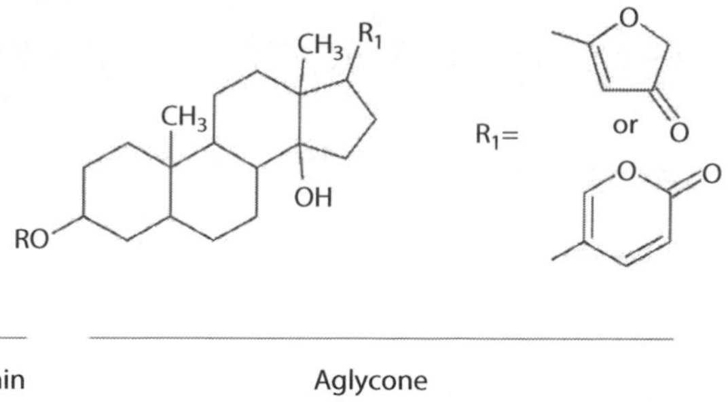

FIGURE 25.1   General structure of cardiac glycosides.

biological activity, *cis*-configuration of the cyclopentanoperhydrophenanthrene system is required. Moreover, the presence of hydroxyl group in 14 β-position and lactone ring in 17 β-position is necessary for the activity of these drugs. Furthermore, the cardiac influence of glycosides is characterized by the specific configuration of their carbohydrate chains linked to the hydroxyl group in 3 β-position. The strongest activity is suitable for glycosides without a single monosaccharide that is usually glucose. The aglycone presence determines the effect on the cardiac myocytes, while sugar residue is responsible for the physical and chemical properties such as solubility in water, blood protein binding, and the rate of absorption from the gastrointestinal tract. General structure of cardiac glycosides is shown in Figure 25.1.

Cardiac glycosides have a direct effect on heart muscle cells, both the myocytes and those in the heart conduction system. The result of their activity is to increase the force of contraction and tension of myocardial fibers. It is also observed an inhibitory effect on cardiac conduction system. Moreover, in a resting phase, glycosides increase the excitability of the ventricular muscles in association with depolarization of the Purkinje fibers.

The mechanism of action of cardiac glycosides is associated with the inhibition of adenosine triphosphatase cell membranes and activated sodium and potassium ions (Na–K ATPase). Inhibition of this enzyme increases the concentration of calcium ions inside the cells of the heart and potentiates the inward calcium current.

On account of a large group of substances belonging to this group, the following section was limited to the cardiac glycosides that are currently used in treatment. Within the group of cardiac glycosides, the following compounds were selected: digoxin, digitoxin, lanatoside, deslanoside, methyldigoxin, strophantoside, scillaren, and proscillaridin.

Chromatographic analysis of the cardiac glycosides is generally difficult due to their complex nature. The chromatographic behavior of cardiac glycosides is determined by the number of functional groups presented both in aglycone and genin moieties. Therefore, only an appropriate selection of analytical conditions provides their adaptability to constantly varying analytical challenges. Because of their physical properties, cardiac glycosides present in plants as complex mixtures are difficult to purify and extract. The glycosides are isolated from plant material with a dilute methanol (Horák 1970), ethanol, or complex mixtures of organic solvents such as chloroform, ethyl acetate, acetone, or hexane. Compounds of this group of drugs were dissolved in polar organic solvents or in their mixtures with water. For chromatography, the solutions were applied automatically onto plates mostly in amounts of 20 μg/spot. The best separations were obtained with the systems consisting of ethyl acetate, benzene, isopropanol, toluene, methyl ethyl ketone, *n*-butanol, acetonitrile, methanol, and acetic acid. The most common materials used for separation of cardiac glycosides are silica gel and octadecyl silane-bonded silica. For the detection of cardiac glycosides, absorbance in UV light (Bethke and Frei 1976, Balbaa et al. 1979), fluorescence in long-wave UV light, and quenching of the fluorescence in short-wave UV light are commonly used. Of the chemical reagents, *m*-dinitronaphthalene, xanthydrol reagent, and strong acids (phosphoric, hydrochloric, trichloroacetic) are the most commonly used. The cardiac glycosides were determined by means of specific TLC procedures in pharmaceutical formulations (Dow et al. 1971, Ponder and Stewart 1994), injections (Mancini and Wakimoto Hanai 1989), human serum, and urine or animal tissues (Marzo and Ghirardi 1974). Moreover, TLC has been used for analysis of digoxin in the presence of its degradation products (Khafagy et al. 1974, Mazei and Pap 1985).

Baranowska et al. (2009) have developed normal- (NP) and reversed-phase (RP) TLC methods for the separation and determination of L-arginine, its metabolites, and selected drugs (digoxine) in model solutions and spiked human urine samples. NP-TLC and RP-TLC methods have been used to study the retention behavior of these compounds. The analyzed drugs were separated in saturated (1 h with mobile phase) glass chambers on silica gel plates with methanol–50% acetic acid (3:1 v/v) and acetonitrile–water (2:3 v/v) and on RP-18 with 5% acetic acid–methanol–acetonitrile (50:35:15 v/v) as mobile phases. Chromatography was performed on silica gel 60 GF_{254} or octadecylsilane RP-18 F_{254s} glass plates (20 × 20 cm). Standard solutions (10 and 20 µL) were spotted by use of a microsyringe. The plates dried at room temperature were visualized by spraying with a 1% ethanolic solution of ninhydrin or in an iodine-vapor chamber. Human urine spiked with appropriate amounts of analyzed compounds was directly applied to TLC plates after specific urine preparation procedure. Other chromatographic procedures have been developed for quantitative determination of cardiac glycosides in *D. lanata* (Horvath 1982, Ikeda et al. 1996) or *D. purpurea* (Fujii et al. 1990) leaves. RP-TLC method was developed for determination of digoxin, digitoxin, lanatoside, and deslanatoside. For extraction from leaf material, a mixture of methanol and water (1:1 v/v) was used. The obtained extract was filtered and evaporated after sonication. The resulting residue was dissolved in a mixture of methanol and water, and then 2 mL of chloroform–acetone–acetic acid (7:3:0.05 v/v) was added. The solution was applied to the Sep-Pak silica cartridge and 30 mL of chloroform–acetone–acetic acid (7:3:0.05 v/v) and 40 mL of chloroform–methanol–water–acetic acid (9:1:0.8:0.05 v/v) were passed through the cartridge. The collected fraction was evaporated to dryness and the residue was dissolved in 2 mL of methanol–water–acetic acid (3.8:6.2:0.02 v/v) and loaded on the Sep-Pak C_{18} cartridge. After elution with methanol–water–acetic acid (6.8:3.2:0.02 v/v), the glycosides were concentrated and analyzed by TLC. Sample solutions were spotted (4 µL) with Microcaps micropipettes and the plates were developed in pre-equilibrated (10 min with mobile phase) glass chambers. The TLC was performed on KC_{18} plates (5 × 20 cm, Whatman) and scanned by CS-920 TLC scanner (Shimadzu) in reflection–absorption mode at λ = 225 nm. Two mobile phases were used in proposed study, acetonitrile–metanol–0.5 M NaCl (12:7:9) and (1:1:1) for secondary and primary glycosides, respectively. Under elaborated condition, the chromatographed compounds migrated as well-shaped spots with R_F values of 0.26, 0.31, 0.44, and 0.6 for digoxin, digitoxin, lanatoside, and deslanatoside, respectively. Calibration plots were constructed by plotting peak area against concentration of analyzed compounds in the range 10–400 µg/spot with a correlation coefficient 0.9946 ± 0.28.

One interesting example of the use of TLC in the analysis of glycosides is a method of identification of the presence of digoxin and digitoxin in serum samples after treatment of plantain (Dasgupta et al. 2006). Dry content of capsule or leaf was extracted with mixtures of methanol or ethanol and water (6:4 v/v) and added to drug-free serum, and apparent digoxin was measured. In the RP-TLC experiment, extracts of dry capsule, leaf, and liquid plantain extract were analyzed for presence of cardiac glycosides by chromatographic procedure elaborated by Ikeda et al. (1996). TLC method was also applied for determination of digitoxin, its metabolite, and digoxin in human serum. Faber et al. (1997) have proposed specific TLC assay for therapy control and toxicology. Digitoxin was isolated from serum by means of a single extraction with chloroform. Fluorescence of the spots was generated by treatment with hydrogen chloride under UV irradiation. High-performance TLC (HPTLC) method was developed by Ponder and Stewart (1994) for the determination of digoxin, its related compounds, and gitoxin in pharmaceutical formulation. Separation was done on a C18 (wettable) glass plate with water–methanol–ethyl acetate (50:48:2 v/v) as mobile phase. The analytes were determined densitometrically using absorbance and fluorescence for digoxin and related compounds, respectively. Dependence of peak height on concentration was recorded and the amount in tablets was calculated by use of linear regression equation (calibration range 320–480 ng/spot, with correlation coefficient 0.9990). The limit of detection (LOD) and limit of quantification (LOQ) values obtained for digoxin were 8 and 64 ng, respectively. Matysik (1994) has described gradient TLC method for analysis of extracts from *D. lanata* and *D. purpurea* that contained digoxin, digitoxin, and lanatosides a–c. The extracts were prepared according to the method in the Polish Pharmacopoeia IV using 100 mL of 50% aqueous methanol solution. After extraction, dry residue was dissolved in 1 mL of methanol–chloroform (1:1) mixture and spotted (10 μL) on silica gel 60 (10 × 10 cm) plates. Chromatograms were developed to a distance of 9 cm in sandwich DS chambers adapted to stepwise gradient elution (increasing concentration of methanol 5%, 10%, 20%, 30%). The mobile phase was composed of ethyl acetate and methanol or ethyl acetate and isopropanol. After drying, the chromatograms were sprayed with 3% aqueous solution of chloramine in a mixture of trichloroacetic acid and ethanol or scanned with Shimadzu CS-930 densitometer at 365 nm. Matysik et al. (1992) have proposed a computer-aided method for optimization of gradient mobile phase composition for TLC of cardiac glycosides extracts. The glycosides obtained from the Digitalis species were separated by stepwise gradient using HPTLC silica gel 60 plates (10 × 10 cm) with mixtures of toluene–ethyl acetate and methanol–ethyl acetate as mobile phases. The values predicted by the computer program written in GW-Basic and experimental R_F values revealed good agreement in the upper part of chromatogram ($R_F > 0.5$). For lower spots, the experimental values were lower by ca. 0.15 R_F unit. However, the differences between these values did not affect the separation on the proposed chromatographic system. Cohnen et al. (1978) have applied TLC for determination of experimental partition coefficients of 48 cardenolides. R_m values of these substances were measured in different TLC systems and related to P values obtained in octanol–water shake-flask method. The best linear fit to the data was obtained using octanol-impregnated silica gel plates and methanol–water mobile phase. Szeleczky (1979) has investigated TLC behavior of cardenolides (digoxin, gitoxin, and lanatoside c) and their derivatives in the presence of boric acid. Introduction of boric acid to the mobile phase reduced the mobility of studied cardenolides containing diol units in their carbohydrates moiety. Formation of the boric acid derivatives resulted in an improvement of the separation and detection of the cardenolides containing diol units. Einig (1976) has described a chromatographic method applied for investigation on stability of proscillaridin and its methyl esters in artificial gastric and intestinal fluids.

Moreover, TLC was used in identification and determination of scillaren (Verbiscar et al. 1986, Dias et al. 2000) and proscillaridin (Ferre et al. 1980) in plant material or plasma.

25.3 ANTIARRHYTHMICS

Antiarrhythmics are a chemically and pharmacodynamically diverse group of medicines designed as substances that normalize both improper impulse generation and improper impulse conduction-dependent cardiac rhythm disturbances. Abnormalities of rate or regularity in the physiological activation of atria and ventricles are reduced by blocking the sodium (class I), potassium (class III), and calcium (class IV) channels, and/or the adrenergic receptors (class II). Due to the fact that some groups of antiarrhythmic agents (beta blockers or calcium channel blockers) are described elsewhere, the following section was limited to the drugs classified as Classes I and III of The Vaughan Williams classification. The class I and III agents are used to suppress a wide range of atrial (class IA and IC) and ventricular (class I and III) arrhythmias. This chapter discusses TLC application in the analysis of ajmaline, bretylium, disopyramide, flecainide, mexiletine, prajmaline, procainamide, propafenone, tocainide, and amiodarone. Due to the wide availability of papers on the analysis of quinidine in the pure form or after extraction from plant material (Verpoorte et al. 1980, Baerheim Svendsen and Verpoorte 1984, Popl et al. 1990, Bogusz and Smith 2000, McCalley 2002), the TLC analysis of *Cinchona* alkaloids has been skipped in this chapter.

TLC is widely applied for the analysis of antiarrhythmic drugs. A variety of specific procedures is available for separation and quantification of these compounds in pharmaceutical formulations and biological samples. Moreover, TLC and HPTLC methods have been used for analysis of antiarrhythmics in the presence of their metabolites (Drayer et. al 1974, Dombrowski et al. 1975, Gupta 1979, Simona and Grandjean 1981, Bonicamp and Pryor 1985), degradation products (Kopelent-Frank and Schimper 1999), stereoisomers (Sosa et al. 1994), and form chromatographic estimation of lipophilicity (Dross et al. 1993). Some authors describe the use of TLC methods for studies of bioavailability (Kark et al. 1983) and metabolism (Hege et al. 1986, McQuinn et al. 1984) of selected antiarrhythmic drugs. From the chemical point of view, the selected antiarrhythmic drugs form a relatively homogeneous group of aliphatic and aromatic amines and amides. Additionally, some compounds contain an amide moiety and dofetilide is distinguished by the presence of a sulfonamide functional group. For application onto plates, the substances are usually dissolved in methanol. Ethanol, acetone, and less polar solvents were used occasionally. The nonpolar solvents, cyclohexane, dichloromethane, chloroform, or hexane, were mostly used in the case of dissolution of dried products formed in applied procedures for compound derivatization. Many solvents, above all low aliphatic alcohols, have usually been used to direct extraction of some compounds from fresh (Habib and Court 1973) or dry plant material (Court and Timmins 1975, Le Xuan et al. 1980). The best TLC systems elaborated for analysis of antiarrhythmic drugs are shown in Table 25.1. A literature survey reveals that selected

TABLE 25.1

Most Efficient TLC Systems Recommended for Analysis of Selected Antiarrhythmic Drugs

Chromatographic System	hR_F					
	A[a]	D[b]	F[c]	M[d]	P[e]	T[f]
Dichloromethane–methanol–ammonia (17:2:1 v/v)/silica gel 60 G	N/A[g]	N/A[g]	N/A[g]	N/A[g]	64	N/A[g]
Ethyl acetate–acetone–ammonia (5:4:1 v/v)/silica gel 60 G	N/A[g]	N/A[g]	N/A[g]	N/A[g]	79	N/A[g]
Acetone–methanol–ammonia (17:2:1 v/v)/silica gel 60 G	N/A[g]	73	65	N/A[g]	N/A[g]	N/A[g]
Ethyl acetate–acetone–ammonia (5:4:1 v/v)/silica gel 60 G	N/A[g]	61	46	N/A[g]	N/A[g]	N/A[g]
Acetone–chloroform–ammonia (30:35:2 v/v)/silica gel 60 G	N/A[g]	N/A[g]	28	N/A[g]	N/A[g]	N/A[g]
Tetrahydrofuran–citrate buffer pH 4.45 (3:7 v/v)/RP8 or RP18	N/A[g]	34	7.8 or 11	25 or 28	N/A[g]	41 or 42
Tetrahydrofuran–hexane–ammonia (5:4.8:0.2 v/v)/aluminum oxide 60	N/A[g]	50	32	23	N/A[g]	5
Chloroform–tetrahydrofuran–ethanol (8.1:1.9:2:0.1 v/v)/silica gel 60	N/A[g]	10	16	28	N/A[g]	22
Chloroform–methanol–ammonia (35:5:05 v/v)/silica gel 60 G	90	N/A[g]	N/A[g]	N/A[g]	N/A[g]	N/A[g]
Ethyl acetate–acetone–water (20:10:0.5 v/v)/silica gel 60 G	25	N/A[g]	N/A[g]	N/A[g]	N/A[g]	N/A[g]

[a] Amiodarone.
[b] Disopyramide.
[c] Flecainide.
[d] Mexiletine.
[e] Propafenone.
[f] Tocainide.
[g] The compounds analyzed in other systems.

antiarrhythmics were also extracted with cold water. Silica gel was mostly selected by authors for separation and determination of this group of compounds. The chromatograms were frequently developed using organic solvents without pH modifications. NP-TLC separation of antiarrhythmics was obtained with the systems consisting of ethyl acetate, benzene, toluene, chloroform, acetone, dioxane, and methanol. Detection of substances of this group of drugs does not as a rule cause major troubles. Most antiarrhythmics quench UV light in short-wave spectrum (Hartmann and Schnabel 1975), and the use of layers impregnated with fluorescence indicators is commonly found in the analysis of these drugs. The most frequently used detection technique in the analysis of antiarrhythmic agents is densitometry. Quantitative and qualitative measurements of separated compounds were done in UV–Vis (Duez et al. 1986) or fluorescence mode (Gupta 1979). Practically only a few detection reagents are applied for chemical detection. For detection of these compounds, Dragendorff's and Benedict's reagents were used. The reagents like solution of potassium permanganate acidified with concentrated sulfuric acid, iodine–potassium iodide solution, Sonnenschein's reagent, or iodine vapor are less frequently used. Among the pharmaceutical formulation most frequently considered are tablets and capsules, which were finely powdered and extracted with methanol. For extraction of the analyzed antiarrhythmic drugs from biological metrices, generally liquid–liquid extraction was used (Gupta 1987).

NP-HPTLC method was developed by Gupta et al. (2006) for separation and identification of the components (ajmaline, reserpine, and ajmalicine) of extracts of *Rauvolfia serpentina* roots. Air-dried and powdered root samples were extracted (3 × 15 mL) with hexane, chloroform, methanol, and water. The obtained extracts were filtered, concentrated under reduced pressure by rotary evaporation, dissolved in chloroform (1 mL), and applied automatically to the plates by means of CAMAG sample applicator. Chromatography was performed on silica gel 60 F_{254} plates (10 × 20 cm) with mixtures of benzene–acetone (86:14 v/v) and toluene–methanol (76:24 v/v) as mobile phases. The plates were stained with Dragendorff's reagent and vanillin solution in H_2SO_4–ethanol (5:95 v/v) and then scanned with CAMAG TLC scanner in absorbance–reflectance mode, at $\lambda = 520$ nm. 2D spectrographic image analysis of obtained chromatograms was also used for profiling the sample under investigation. HPTLC of the *R. serpentina* extracts showed the presence of additional major spots. The limits of detection obtained for ajmaline, reserpine, and ajmalicine were 45, 70, and 50 μg, respectively. Pelander et al. (2003) have described a dual-plate overpressured thin-layer chromatography (OPLC) method for the screening of 47 basic drugs (amiodarone, disopyramide, lidocaine, mexiletine, flecainide, procainamide, propafenone, quinidine, tocainide,) in urine samples. Urine sample (2 mL) extraction was carried out by IST mixed-mode solid-phase extraction (SPE) (C8 and cation exchange) using HCX 130 mg cartridges. SPE was performed according to the manufacturer's application note. The chromatography was performed on aluminum sheets coated with silica gel F_{254} of 5 μm particle size. The compositions of mobile phases trichloroethylene–methyl ethyl ketone–*n*-butanol–acetic acid–water and butyl acetate–ethanol (96.1%)–tripropylamine–water were 17:8:25:6:4 v/v and 85:9.25:5:0.75 v/v, respectively. The external pressure was 50 bar, flow rate 450 μL/min, volume of rapid delivery 300 μL, and mobile phase volume 5500 μL. TLC procedure was also used for the analysis of some benzofuran derivatives (amiodarone) in human plasma (Gumieniczek and Przyborowski 1994). The chromatography was performed on silica gel 60 G, GF_{254} (10 × 20 cm), and HPTLC RP-18 F_{254S} (10 × 10 cm), plates using both ascending and horizontal techniques. The methanolic solution of examined compounds was spotted on plates (10 μL for TLC and 5 μL for HPTLC) by using micropipette. In order to develop appropriate chromatographic condition for separation and detection of drugs, 24 TLC and 14 HPTLC original solvent systems were examined. For extraction of amiodarone and other analyzed compounds from human plasma, liquid–liquid procedure was performed with diisopropyl ether and phosphate buffer pH 4.8 after protein precipitation with acetonitrile. The best separation was achieved using the following mobile phases: methyl ethyl ketone–ammonia (40:0.5 v/v) and benzene–chloroform–diethyl ether–methanol (30:2:2:2 v/v). For detection of all drugs in amount of 0.2 μg for amiodarone and 0.5 μg for other compounds, Dragendorff's reagent was used. Stevens and Moffat (1974) have developed a rapid NP-TLC screening procedure for quaternary ammonium compounds (bretylium) in fluids and tissues. The analyzed compounds were extracted from urine, blood, and samples in the form of its ion pair complex with bromothymol blue. Detection of the drugs was performed colorimetrically and enzymically using two TLC systems. Moreover, the stability of examined substances was studied at various pH values. Stability-indicating method has been developed for the determination of disopyramide phosphate in capsules by Salem et al. (2006). For the separation and determination of disopyramide in the presence of its degradation products, the silica gel 60 F_{254} TLC plates and ethyl acetate–chloroform–ammonium hydroxide (85:10:5 v/v) were used. The TLC method dependent on the quantitative densitometric determination of this drug in concentration range of 0.25–2.5 μg/spot (mean accuracy of 100.3% ± 1.1%) was found to be specific for disopyramide in presence up to 90% of its degradation products. NP-TLC method has been established for separation of five antiarrhythmic drugs—disopyramide, flecainide, mexiletine, tocainide, and verapamil—and determination of mexiletine in capsules (Pietraś et al. 2004, 2010). The analysis was performed in horizontal chambers on aluminum oxide 60 F_{254} and silica gel 60 F_{254} TLC plates. The best separation was achieved using tetrahydrofuran–hexane–25% ammonia (5:4.8:0.2 v/v) and chloroform–tetrahydrofuran–ethanol–25% ammonia (8.1:1.9:2:0.1 v/v) on the alumina and silica plates, respectively. The substances were identified by use of different reagents and under UV irradiation at λ = 254 nm (LOD was found to be 0.5 μg/spot for all the analyzed compounds). Limits of detection obtained with Dragendorff's reagent were found to be 0.2, 0.5, 1.0, 1.0, and 0.5 μg/spot for disopyramide, flecainide, mexiletine, tocainide, and verapamil, respectively. The most sensitive reagent was Marquis reagent for mexiletine (0.05 μg/spot). Quantification of mexiletine in capsules was performed densitometrically at λ = 210 nm. A good correlation coefficient (r = 0.9974) was obtained for the calibration plot constructed in the concentration range 20–45 μg/spot. The active substance was extracted from the formulation with methanol (recovery 97.01% ± 2.39%). The relative standard deviation (RSD) expressing the precision of the proposed method was 5.23%. Pietraś and Kowalczuk (2010) have developed another TLC procedure for analysis of previously described antiarrhythmic drugs and quantification of flecainide acetate in selected tablets. The chromatographic behavior of these compounds has been studied on TLC plates coated with octylsilane and octadecylsilane silica gel plates with organic–aqueous mobile phases containing citrate or acetate buffers at different pH. The best separations were achieved on both RP-8 and RP-18 with tetrahydrofuran–citrate buffer pH 4.45 (3:7 v/v) as mobile phase. To determine the usefulness of elaborated systems for analysis, flecainide was identified and quantified by UV densitometry at two wavelengths, 225 and 310 nm. Linear relationships were obtained between peak height and peak area and amount in the range 6–12 μg/spot (r = 0.9990). The method then successfully applied to the analysis of flecainide in pharmaceutical preparation, with satisfactory precision (RSD 1.14%–5.93%) and accuracy (96.19%–103.59%). Pietraś et al. (2004, 2010) have also developed two validated HPTLC methods for densitometric determination of mexiletine in pharmaceutical formulation. Analyses were performed in horizontal chambers on RP C18F254s and NP amino (NH2) HPTLC precoated plates with the mobile phases tetrahydrofuran–citrate buffer pH 4.45 (3:7 v/v) and chloroform–tetrahydrofuran–hexane–ethylamine (3:2:0.1 v/v), respectively. The plates were developed for a distance of 40 mm in both cases. Densitometric measurements were achieved in the UV mode at 217 nm based on peak areas with semilinear calibration curves (r^2 = 0.97) in the concentration range 0.5–8.0 mg/spot for the NH2 and C18 HPTLC methods. The elaborated chromatographic methods were validated in accordance with *International Conference on Harmonization* guidelines in terms of linearity, accuracy (99.64% for NH2 and 99.53% for C18), precision (intraday RSD 1.16% and 2.71%, respectively), sensitivity (LOD 0.1 mg/spot for both systems), and specificity. Paw and Przyborowski (1995) and Schmidt and Bracher (2006) have developed NP-TLC (silica gel plates) method for identification of flecainide and lorcainide in blood. The methanolic solutions were spotted on plates by means of microsyringe (corresponding to a drugs amount from 0.05 to 10 μg/spot). The plates were developed to a distance of 10 cm in the chambers previously saturated with mobile phase for 1 h, using ascending technique. Liquid–liquid extraction with hexane (flecainide) or heptane–isoamyl alcohol (lorcainide) was performed on human blood serum after alkalization with sodium hydroxide. Twenty-three chromatographic systems were examined in regard to their separation efficiency and development time. A good separation was achieved using acetone–chloroform–ammonia (3:3.5:2 v/v) and benzene–dioxane–ammonia (13:6:1 v/v) as mobile phases. The most sensitive reagent applied for chemical detection of flecainide was 15% solution of $FeCl_3$ in 5% HCl (0.2 µg/spot). The lowest flecainide concentration in human blood that may be analyzed was 0.5 µg/1 mL of blood plasma. TLC has been applied for identification of diltiazem in the presence of five other antiarrhythmics: disopyramide, flecainide, lidocaine, lorcainide, and procainamide (Witek and Przyborowski 1996, 1997). NP-TLC was performed on precoated silica gel 60 and 60 F_{254} plates in classical pre-saturated (mobile phase, 1 h) chambers using ascending technique. The volumes of methanolic solutions equivalent to 5 µg of examined compounds were spotted on plates using microsyringe. The 34 newly proposed chromatographic systems were examined. The best separation was achieved by using ethanol–benzene–dioxane–ammonia (2:20:16:3 v/v) as mobile phase. The most sensitive reagent, which has been applied for detection for proposed study, was Dragendorff's reagent (0.1 µg/spot). The same authors have developed NP-TLC procedure for simultaneous analysis of β-blocking drugs and propafenone in plasma. The chromatographic procedure was performed on precoated silica gel 60 G and 60 GF_{254} plates saturated with mobile phase (1 h) classical TLC chambers using ascending technique. For extraction of the analyzed compounds from plasma, liquid–liquid extraction was used with dichloromethane after alkalization (0.1 M NaOH). The 30 solvent systems were examined and the best separation was obtained using ethyl acetate–acetone–ammonia (5:4:1 v/v). The drug detection was achieved by several dyeing reagents previously described (Witek et al. 1998, 1999). To improve sensitivity of elaborated procedures, Witek et al. (1998, 1999) have developed the TLC method for analysis of propafenone and some β-blocks with conversion to their dabsyl derivatives. The derivatization procedure was as follows: to 1 mL of methanolic solutions of compound in screw-cap vial, 3 mL of dabsylchloride solution was added. Then, 0.5 mL of $NaHCO_3$ (0.5 M) solution was pipetted to vial to adjust the pH of reaction mixture to the optimal value of about 9.0. After mixing, the samples were incubated at 60°C for 10 min and shaken during the 2nd and 7th min of heating. Then, the vials were cooled in the water bath for 5 min to stop the reaction. After acetone was evaporated under the stream of cold air, the samples were extracted twice with 2.5 mL portions of cyclohexane. The organic layers were collected and applied to TLC analysis. The first chromatographic procedure uses precoated silica gel 60 G and aluminum oxide plates and n-hexane–chloroform–acetone (2:5:1 v/v) as mobile phase, which was chosen in regard to its best separative properties for analysis of examined dabsyl derivatives on silica gel plates. The visual detection limits obtained for propafenone dabsyl derivative were 2 ng and 1 ng/spot for silica gel and aluminum oxide plates, respectively. In the second procedure, densitometry at $\lambda = 490$ nm was applied for quantification of this drugs as dabsyl derivatives in pharmaceutical formulations. A good correlation coefficient (r = 0.995) for the calibration plot was constructed in the concentration range 0.04–1.0 mg/band. The obtained LOD and LOQ values were 0.015 and 0.04 mg/band, respectively. Gupta (1979) has described TLC method for the fluorescence determination of disopyramide and its mono-N-dealkylated metabolite in plasma. Simple liquid–liquid extraction procedure with benzene after alkaline was applied to plasma samples (200 µL). The chromatography was performed on silica gel plates with ethyl acetate–methanol–ammonia (40:30:0.5 v/v) as mobile phase. After development, the plates were dipped in 20% solution of sulfuric acid in methanol and scanned at fluorescence mode. Giaginis and Tsantili-Kakoulidou (2009) used the RP-TLC method for lipophilicity estimation of 26 selected basic (mexiletine) and neutral drugs. The effect of buffer constituents and pH on the compound's retention factors has been investigated in different chromatographic condition. The chromatographic study was performed on RP-18 F_{254S} plates (20 × 20 cm) and methanol was used as organic modifier. R_m values obtained under different mobile phase conditions were evaluated for their performance as lipophilicity indices by comparison with octanol–water partition or distribution coefficients taken from previous studies. Ojanpera et al. (1992, 1997) have described a chromatographic method enabling differentiation of primary, secondary, and tertiary aliphatic and aromatic amines with Fast Black K salt. NP-TLC was also used for the detection of procainamide in the presence of some beta-adrenergic blocking drugs and their metabolites. Lee et al. (1978) have applied densitometric detection in their TLC method for the determination of antiarrhythmics in serum. Lidocaine and phenytoin were scanned at 220 nm and procainamide, propranolol, and quinidine at 290 nm.

25.4 NITRATES

Nitrates are described as polyol esters of nitric acid, which possess the characteristic C–O–N sequence and convert to NO in vascular smooth muscle. These drugs produce vasodilation, which reduces blood pressure and volume within the heart. The consequence of nitrate action significantly reduces the cardiac workload and myocyte oxygen demand. The mechanism of action is related to activation of the guanylate cyclase, which increases the synthesis of cGMP inhibitory contractile activity of smooth muscle in blood vessel walls. Nitrates are used to relieve anginal pain caused by abnormal cardiac function in myocardial ischemia. Within the nitrates group, TLC procedures applied to the analysis of glycerol trinitrate, isosorbide dinitrate, and pentaerythritol tetranitrate are discussed. Moreover, TLC of molsidomine is also included.

Overpressure liquid chromatography technique coupled with several radioactive detection methods was applied for the analysis of 14c-labeled glyceryl trinitrate and its metabolites (Klebovich et al. 2002). The proposed method is rapid, economic, and effective as a separation system for application in metabolism research. After separation, identification and quantification of 14c-labeled glyceryl trinitrate, glyceryl trinitrate and its metabolites 1,2- and 1,3-glyceryl dinitrate, and 1- and 2-glyceryl mononitrate were performed by means of digital autoradiography, flow-cell solid scintillation radioactivity detection, and the phosphor-imaging technique. The liquid–liquid extraction technique was used for isolation of analytes from rat plasma. OPLC separation of glyceryl trinitrate and its metabolites was performed on silica gel-coated TLC plates with acetonitrile and dibutyl ether mixtures as mobile phases of stepwise gradient elution. TLC has been also used for the determination of small amounts of nitrate esters such as trinitroglycerin in wastewater samples (Chandler et al. 1974). TLC has been recommended by Azcona et al. (1991) for assay of 5-isosorbide. The elaborated TLC procedure is simple, fast, and reliable for the isosorbide determination in isosorbide dinitrate and isosorbide-5-mononitrate dosage forms. Carlson and Thompson (1986) studied the stability of isosorbide dinitrate and nitroglycerin. Glycerol 1- and 2-nitrate, glycerol 1,2- and 1,3-dinitrate, and isosorbide 2- and 5-nitrate were identified as degradation products. Separation was done using an original planar chromatographic system with silica gel as the stationary phase. Agbaba et al. (1996) developed a simple, fast, and reliable procedure for determination of molsidomine in the presence of ketoprofen and pitofenone in commercial formulation. Separation of analyzed drugs extracted from tablets, injections, suppositories, and oral suspensions was performed on HPTLC plates. The separated compounds were determined by densitometry. The proposed method was validated for accuracy, reproducibility, selectivity, linearity, and detection limits. The HPTLC method was applied for simultaneous determination of pentaerythritol and 2,4,6-trinitrotoluene (Nejad-Darzi et al. 2012). For separation and determination, the mixture of petroleum ether and acetone (2:1 v/v) and silica gel layers (60 F_{254}) was used as chromatographic system. The ultraviolet spectra of these materials were recorded (CAMAG TLC-scaner3), and partial least-squares regression was applied for calibration and quantification of these compounds from artificial samples. Another HPTLC procedure for separation and determination of pentaerythritol and TNT was proposed by Chaloosi et al. (2009) The compounds were separated on silica gel 60 F_{254} plates using a double separation system consisting of a mixture of petroleum ether and acetone (5.7:1 v/v) and trichloroethylene and acetone (4:1 v/v) as mobile phases. Samples and standard solutions were applied automatically. The densitometric technique was applied for the identification and determination of analyzed compounds using detection at 215 and 250 nm for pentaerythritol and 2,4,6-trinitrotoluene, respectively. The linearity of elaborated method was in the range 0.835–83.5 µg/spot for pentaerythritol and 0.045–1.44 µg/spot 2,4,6-trinitrotoluene. Another procedure applied for the analysis of selected nitrates was TLC method combined with photometry for identification and determination of impurities in pentaerythritol tetranitrate (Yasuda 1970). The analyzed compounds were separated by 1D TLC plates on silica gel G–Zn–sodium sulfanilate plates with an acetone and benzene mixture as mobile phase. Conversion to red color diazo dye with *N,N*-dimethyl-1-naphthylamine–acetic acid reagent has been recommended for the detection of pentaerythritol and its impurities.

REFERENCES

Agbaba, D., Grozdanovic, O., Popovic, L., Vladimirov, S., and Zivanov-Stakic, D. 1996. HPTLC in the quantitative assay of drugs. *Journal of Planar Chromatography—Modern TLC* 9: 116–119.

Azcona, T., Martin-Gonzalez, A., Zamorano, P., Pascual, C., Grau, C., and Garcia de Mirasierra, M. 1991. New methods for the assay of 5-isosorbide mononitrate and its validation. *Journal of Pharmaceutical and Biomedical Analysis* 9: 725–729.

Baerheim Svendsen, A. and R. Verpoorte. 1984. *Chromatography of the Cinchona Alkaloids. Part A: Thin-Layer Chromatography*. Amsterdam, the Netherlands: Elsevier.

Balbaa, S.I., Khafagy, S.M., Khayyal, S.E., and Girgis, A.N. 1979. TLC-spectrophotometric assay of the main glycosides of red squill, a specific rodenticide. *Journal of Neutral Products* 42: 522–524.

Baranowska, I., Markowski, P., Wilczek, A., Szostek, M., and Stadniczuk, M. 2009. Norma land reversed-phase thin-layer chromatography in the analysis of L-arginine, its metabolism, and selected drugs. *Journal of Planar Chromatography—Modern TLC* 22: 89–96.

Bethke, H. and R.W. Frei. 1976. Thin layer chromatography/densitometry with transferable calibration factors. *Analytical Chemistry* 48: 50–54.

Bogusz, M. and R.M. Smith. 2000. *Handbook of Analytical Separations*. Amsterdam, the Netherlands: Elsevier Science B.V.

Bonicamp, J.M. and Pryor, L. 1985. Detection of some beta adrenergic blocking drugs and their metabolites in urine by thin layer chromatography. *Journal of Analytical Toxicology* 9: 180–182.

Carlson, M. and Thompson, R.D. 1986. Thin-layer chromatography of isosorbide dinitrate, nitroglycerin and their degradation products. *Journal of Chromatography* 368: 472–475.

Chaloosi, M., Nejad-Darzi, S.K.H., and Ghoulipour, V. 2009. Separation and determination of PETN and TNT by HPTLC. *Propellants, Explosives, Pyrotechnics* 34: 50–52.

Chandler, C.D., Gibson, G.R., and Bolleter, W.T. 1974. Liquid chromatographic determination of nitroglycerin products in waste waters. *Journal of Chromatography* 100: 185–188.

Cohnen, E., Flasch, H., Heinz, N., and Hempelmann, F.W. 1978. Partition coefficients and R(m)-values of cardenolides. *Arzneimittel-Forschung/Drug Research* 28: 2179–2182.

Court, W.E. and Timmins, P. 1975. The thin layer chromatographic behaviour of some *Rauvolfia* alkaloids on silica gel layers. *Planta Medica* 27: 319–329.

Dasgupta, A., Davis, B., and Wells, A. 2006. Effect of plantain on therapeutic drug monitoring of digoxin and thirteen other common drugs. *Annals of Clinical Biochemistry* 43: 223–225.

Dias, C., Borralho Graca, J.A., and Lurdes Goncalves, M. 2000. *Scilla maderensis*, TLC screening and positive inotropic effect of bulb extracts. *Journal of Ethnopharmacology* 71: 487–492.

Dombrowski, L.J., Crain, A.V.R., Browning, R.S., and Pratt, E.L. 1975. Determination of 17 monochloroacetylajmaline and its metabolite in plasma by TLC. *Journal of Pharmaceutical Sciences* 64: 643–645.

Dow, M.L., Kirchhoefer, R.D., and Brower, J.F. 1971. Rapid identification and estimation of gitoxin in digitoxin and digoxin tablets by TLC. *Journal of Pharmaceutical Sciences* 60: 298–299.

Drayer, D.E., Reidenberg, M.M., and Sevy, R.W. 1974. N acetyl procainamide: an active metabolite of procainamide. *Proceedings of the Society for Experimental Biology and Medicine* 146: 358–363.

Dross, K., Sonntag, C., and Mannhold, R. 1993. On the precise estimation of R(M) values in reversed-phase thin-layer chromatography including aspects of pH dependence. *Journal of Chromatography* 639: 287–294.

Duez, P., Chamart, S., and Vanhaelen, M. 1986. Comparison between high-performance thin-layer chromatography-densitometry and high-performance liquid chromatography for the determination of ajmaline, reserpine and rescinnamine in *Rauvolfia vomitoria* root bark. *Journal of Chromatography* 356: 334–340.

Einig, H. 1976. Investigation on the stability of proscillaridine, proscillaridine 3′ methyl ether and proscillaridine 4′ methyl ether in artificial gastric and intestinal fluids. *Arzneimittel-Forschung/Drug Research* 26: 1276–1279.

Faber, D.B., De Kok, A., and Brinkman, U.A.Th. 1997. Thin-layer chromatographic method for the determination of digoxin in human serum. *Journal of Chromatography B* 143: 95–103.

Ferrer, M., Basarte, J.C., and Fernandez, M. 1980. Densitometric methods for determination of proscillaridin, scillaren A and scilliroside. *Farmaco, Edizione Pratica* 35: 32–46.

Fujii, Y., Ikeda, Y., and Yamazaki, M. 1990. Quantitative determination of digitalis glycosides in *Digitalis purpurea* leaves by reversed-phase thin-layer chromatography. *Journal of Liquid Chromatography* 13: 1909–1919.

Giaginis, C. and Tsantili-Kakoulidou, A. 2009. RPTLC retention indices of basic and neutral drugs as surrogates of octanol-water distribution coefficients. Effect of buffer constituents and pH. *Journal of Planar Chromatography—Modern TLC* 22: 217–224.

Gumieniczek, A. and Przyborowski, L. 1994. Thin-layer chromatographic analysis of some benzofuran derivatives in human plasma. *Acta Poloniae Pharmaceutica—Drug Research* 51: 429–432.

Gupta, M.M., Srivastava, A., Tripathi, A.K., Misra, H., and Verma, R.K. 2006. Use of HPTLC, HPLC, and densitometry for quantitative separation of indole alkaloids from *Rauvolfia serpentina* roots. *Journal of Planar Chromatography—Modern TLC* 19: 282–287.

Gupta, R.N. 1979. Fluorescence photometric determination of disopyramide and mono-a/-dealkylated disopyramide in plasma after separation by thin-layer chromatography. *Analytical Chemistry* 51: 455–458.

Gupta, R.N. 1987. Fluorescence photometric quantitation of procainamide and N-acetylprocainamide in plasma after separation by thin-layer chromatography. *Analytical Chemistry* 50: 197–199.

Habib, M.S. and Court, W.E. 1973. The estimation of *Rauwolfia* alkaloids by quantitative thin layer chromatography. *Canadian Journal of Pharmaceutical Sciences* 8: 81–83.

Hartmann, V. and Schnabel, G. 1975. Determination of reserpine and rescinnamine in alkaloid extracts and pharmaceutical preparations by direct quantitative photometry on thin layer chromatograms. *Pharmaceutische Industrie* 37: 451–455.

Hege, H.G., Lietz, H., and Weymann, J. 1986. Studies on the metabolism of propafenone. 3rd Comm.: Isolation of the conjugated metabolites in the dog and identification using fast atom bombardment mass spectrometry. *Arzneimttel-Forschung/Drug Research* 36: 467–474.

Horák, P. 1970. Determination of the content of lanatosides in dry and fresh leaves of *Digitalis lanata* EHRH. *Cesko-Slovenska Farmacie* 19: 213–217.

Horvath, P. 1982. Quantitative analysis of natural drugs II. Quantitative TLC determination of digoxin and other glycosides in pretreated *Digitalis lanata* extracts. *Acta Pharmaceutica Hungarica* 52: 133–140.

Ikeda, Y., Fujii, Y., Umemura, M., Hatakeyama, T., Morita, M., and Yamakazi, M. 1996. Quantitative determination of cardiac glycosides in *Digitalis lanata* leaves by reversed-phase thin-layer chromatography. *Journal of Chromatography A* 746: 255–260.

Kark, B., Sistovaris, N., and Keller, A. 1983. Thin-layer chromatographic determination of procainamide and N-acetylprocainamide in human serum and urine at single-dose levels. *Journal of Chromatography—Biomedical Applications* 277: 261–272.

Khafagy, S.M., Girgis, A.N., and Roefael, N. 1974. Study of the stability of primary glycosides of digitalis in certain solid dosage forms. *Journal of Drug Research* 6: 55–73.

Klebovich, I., Morovján, G., Hazai, I., and Mincsovics, E. 2002. Separation and assay of 14c-labeled glyceryl trinitrate and its metabolites by OPLC coupled with on-line of off-line radioactivity detection. *Journal of Planar Chromatography—Modern TLC* 15: 404–409.

Kopelent-Frank, H. and Schimper, A. 1999. HPTLC-based stability assay for the determination of amiodarone in intravenous admixtures. *Pharmazie* 54: 542–544.

Le Xuan, P., Munier, R.L., and Meunier, S. 1980. Two-dimensional separations and behavior of *Rauvolfia, Corynanthe* and *Pseudocinchona* alkaloids on unmodified silica gel. *Chromatographia* 13: 693–697.

Lee, K.Y., Nurok, D., Zlatkis, A., and Karmen, A. 1978. Simultaneous determination of antiarrhythmia drugs by high-performance thin-layer chromatography. *Journal of Chromatography* 158: 403–410.

Mancini, M.A.D. and Wakimoto Hanai, L. 1989. Chromatographic identifications of cardiac glycosides in tablets and injections. *Revista de Ciencias Farmaceuticas* 11: 171–180.

Marzo, A. and Ghirardi, P. 1974. Subcellular distribution of K strophanthoside (3H) in isolated guinea pig hearts. *Biochem Pharmacol.* 23: 2817–2824.

Matysik, G. 1994. Gradient thin-layer chromatography of extracts from *Digitalis lanata* and *Digitalis purpurea*. *Chromatographia* 38: 109–113.

Matysik, G., Markowski, W., Soczewiński, E., and Polak, B. 1992. Computer-aided optimization of stepwise gradient profiles in thin-layer chromatography. *Chromatographia* 34: 303–307.

Mazei, J. and J. Pap. 1985. A simple reversed phase TLC method for stability-indicating assay of digoxin. *Acta Pharmaceutica Hungarica* 55: 134–137.

McCalley, D.V. 2002. Analysis of the *Cinchona* alkaloids by high-performance liquid chromatography and other separation techniques. *Journal of Chromatography A* 967: 1–19.

McQuinn, R.L., Quarforth, G.T., and Johnson, J.D. 1984. Biotransformation and elimination of 14c-flecainide acetate in humans. *Drug Metabolism and Disposition* 12: 414–420.

Nejad-Darzi, S.K.H., Chaloosi, M., and Gholamian, F. 2012. Simultaneous determination of pentaerythritol tetranitrate and 2,4,6-trinitrotoluene by high performance thin layer chromatography and partial least squares regression (PLSR) method. *Propellants, Explosives, Pyrotechnics* 35: 66–71.

Ojanpera, I., Nokua, J., Vuori, E., Sunila, P., and Sippola, E. 1997. Combined dual-system R(F) and UV library search software: Application to forensic drug analysis by TLC and RPTLC. *Journal of Planar Chromatography—Modern TLC* 10: 281–285.

Ojanpera, I., Wahala, K., and Hase, T.A. 1992. Characterization of amines by fast black K salt in thin-layer chromatography. *Analyst* 117: 1559–1565.

Paw, B. and Przyborowski, L. 1995. Identification of flecainide and lorcainide in blood by means of TLC. *Acta Poloniae Pharmaceutica—Drug Research* 52: 5–7.

Pelander, A., Ojanpera, I., Sistonen, J., Rasanen, I., and Vuori, E. 2003. Screening for basic drugs in 2-ml urine samples by dual-plate overpressured layer chromatography and comparison with gas chromatography-mass spectrometry. *Journal of Analytical Toxicology* 27: 226–232.

Pietraś, R., Hopkała, H., Kowalczuk, D., and Malysza, A. 2004. Normal-phase TLC separation of some anti-arrhythmics. Densitometric determination of mexiletine hydrochloride in capsules. *Journal of Planar Chromatography—Modern TLC* 17: 213–217.

Pietraś, R. and Kowalczuk, D. 2010. RP-TLC separation of antiarrhythmic drugs. Densitometric analysis of flecainide in tablets. *Journal of Planar Chromatography—Modern TLC* 23: 65–69.

Pietraś, R., Skibiński, R., Komsta, Ł., Kowalczuk, D., and Panecka, E. 2010. Validated HPTLC methods for quantification of mexiletine hydrochloride in a pharmaceutical formulation. *Journal of AOAC International* 93: 820–824.

Ponder, G.W. and Stewart, J.T. 1994. High-performance thin-layer chromatographic determination of digoxin and related compounds, digoxigenin bisdigitoxoside and gitoxin, in digoxin drug substance and tablets. *Journal of Chromatography A* 659: 177–183.

Popl, M., Fahnrick, J., and Tatar, V. 1990. *Chromatographic Analysis of Alkaloids.* Marcel-Decker, New York.

Salem, M.Y., Ramadan, N.K., Moustafa, A.A., and El-Bardicy, M.G. 2006. Stability-indicating methods for the determination of disopyramide phosphate. *Journal of AOAC International* 89: 976–986.

Schmidt, M. and Bracher, F. 2006. A convenient TLC method for the identification of local anesthetics. *Pharmazie* 61: 15–17.

Simona, M.G. and Grandjean, E.M. 1981. Simple high-performance thin-layer chromatography method for the determination of disopyramide and its mono-N-dealkylated metabolite in serum. *Journal of Chromatography* 224: 532–538.

Sosa, M.E., Valdes, J.R., and Martinez, J.A. 1994. Determination of ajmaline stereoisomers by combined high-performance liquid and thin-layer chromatography. *Journal of Chromatography A* 662: 251–254.

Stevens, H.M. and Moffat, A.C. 1974. A rapid screening procedure for quaternary ammonium compounds in fluids and tissues with special reference to suxamethonium (succinylcholine). *Journal of the Forensic Sciences Society* 14: 141–148.

Szeleczky, Z. 1979. Thin-layer chromatography of cardenolides in the presence of boric acid. *Journal of Chromatography* 178: 453–458.

Verbiscar, A.J., Petel, J., Banigan, T.F., and Schatz, R.A. 1986. Scilliroside and other scilla compounds in red squill. *Journal of Agricultural and Food Chemistry* 34: 973–979.

Verpoorte, R., Mulder-Krieger, Th., Troost, J.J., and Svendsen, A.B. 1980. Thin-layer chromatographic separation of cinchona alkaloids. *Journal of Chromatography A* 184: 79–96.

Witek, A. and Przyborowski, L. 1996. Chromatographic (TLC) analysis of diltiazem in pharmaceutical form in presence of other antiarrhythmics. *Acta Poloniae Pharmaceutica—Drug Research* 53: 9–12.

Witek, A. and Przyborowski, L. 1997. Thin-layer chromatographic analysis of some β-blocking drugs in plasma. *Acta Poloniae Pharmaceutica—Drug Research* 54: 183–186.

Witek, A., Hopkała, H., and Matysik, G. 1999. TLC-densitometric determination of bisoprolol, labetalol and propafenone, as dabsyl derivatives, in pharmaceutical preparations. *Chromatographia* 50: 41–44.

Witek, A., Hopkała, H., and Przyborowski, L. 1998. Chromatographic separation of bisoprolol, labetalol and propafenone in the form of dabsyl derivatives. *Chemia Analityczna* 43: 817–822.

Yasuda, S.K. 1970. Identification and determination of impurities in pentaerythritol tetranitrate. *Journal of Chromatography A* 51: 253–260.

26 TLC of Antihypertensive and Antihypotensive Drugs

Danica Agbaba and Katarina Nikolić

CONTENTS

26.1 INTRODUCTION

In humans, blood volume and blood pressure are regulated by various physiological systems, mainly the components of the central and peripheral nervous systems, the hormonal systems, kidneys, the peripheral vascular network, etc. Certain abnormalities or diseases of these systems can create a hypertensive state in humans. An enhanced adrenergic activity is a principal contributor to the primary or essential hypertension, although the etiology of this type is not yet completely resolved (Mutschler and Derendorf 1995). Considering the complexity of all factors influencing hypertension in humans, different chemical/pharmacological classes of drugs are used to treat them, namely, angiotensin converting enzyme (ACE) inhibitors; angiotensin receptor antagonists (ATs); calcium channel blockers; vasodilators (arterial, and arterial and venous); and adrenergics (centrally active sympatholytics, ganglion blockers, adrenergic neuron blockers, alpha adrenergic blockers, and others). In parallel with the high blood pressure, most often, the presence of certain other cardiovascular diseases occurs, which requires combination of certain drugs from the aforementioned groups with an additional therapy with diuretics and antilipidemics, and hence, apart from the monocomponent formulations, different multicomponent dosage formulations are launched nowadays as fixed combination of drugs to treat hypertension (Williams 2008).

The states of hypotonia (the low blood pressure) require to be treated, if necessary, with the selected adrenergic/sympathomimetic drugs.

This chapter will cover the reference data concerning application of thin-layer chromatography/ high-performance TLC (TLC/HPTLC), with the aim to assess the pharmaceutical quality of bulk drugs, the mono- or multi-component dosage formulations, and the stability and purity of antihypertensive drugs according to the requirements of the official authorities; to follow the progress in qualitative organic analysis/synthesis and for identification of drugs and their metabolites; as tools for screening toxicological potential of compounds; to investigate the retention behavior and lipophilicity of several structurally different groups of antihypertensives by use of chemometrics; to investigate quantitative structure activity/properties relationship (QSAR and QSPR); and hyphenation in ultra-TLC (UPTLC) and HPTLC, in order to enhance the potential of pharmaceutical analysis on the different thin-layer beds (sorbents).

26.1.1 CHEMISTRY AND PHARMACOLOGY OF ACE INHIBITORS

The angiotensin-converting enzyme (ACE) inhibitors (familiar as *prils*) are widely used as the first-line drugs in the treatment of essential hypertension. Their action is based on the inhibition of ACE, blocking hereby the conversion of angiotensin I to angiotensin II, a potent vasoconstrictor. ACE inhibitors act either as di- or tripeptide substrate analogs having stereochemistry that is consistent with L-amino acids present in the natural substrates. Captopril (CAP) and fosinopril (FOS) are the lone representatives of their respective chemical subclassification, while the majority of the inhibitors contain the dicarboxylate functionality. In order to be absorbed, most of them are used as more lipophilic ethyl ester. These esters are the *prodrugs*, and the active forms of these drugs (called *prilates*) are polar. *Prils* differ among themselves mainly in pharmacokinetic properties, bioavailability, lipid solubility, and the first-pass metabolism (Harrold 2008). ACE inhibitors used in therapy of hypertension are **CAP** (S)-1-(3-Mercapto-2-methyl-1-oxopropyl)-L-proline, **enalapril** (ENA) (S)-1-[N-[1-(Ethoxycarbonyl)-3-phenylpropyl]-L-alanyl]-L-proline, **lisinopril** (LIS) (S)-1-[N2-(1-Carboxy-3-phenylpropyl)-L-lysyl]-L-proline, **perindopril** (PER) (2 S, 3a S, 7a S)-1-[(S)-N-[(S)-1-carboxybutyl]alanyl]hexahydro-2-indolinecarboxylic acid 1-ethyl ester, **ramipril** (RAM) [2 S,3a S,6a S]-1-[(2 S)-2-[[(1 S)-(Ethoxycarbonyl)-3-phenylpropyl]amino]-1-oxopropyl] octahydrocyclopenta[b]pyrrole-2-carboxylic acid, **quinapril** (QUI) (S)-2-[(S)-N-[(S)-1-carboxy-3-phenylpropyl]alanyl]-1,2,3,4-tetrahydro-3-isoquinolinecarboxylic acid 1-ethyl ester, **Benazepril** (BEN) (3 S)-1-(carboxymethyl)-[[(1 S)-1-(ethoxycarbonyl)-3-phenylpropyl] amino]-2,3,4,5-tetrahydro-1 H-[1]benzazepin-2-one, **cilazapril** (CIL) (1 S, 9 S)-9-[[(S)-1-carboxy-3-phenylpropyl]

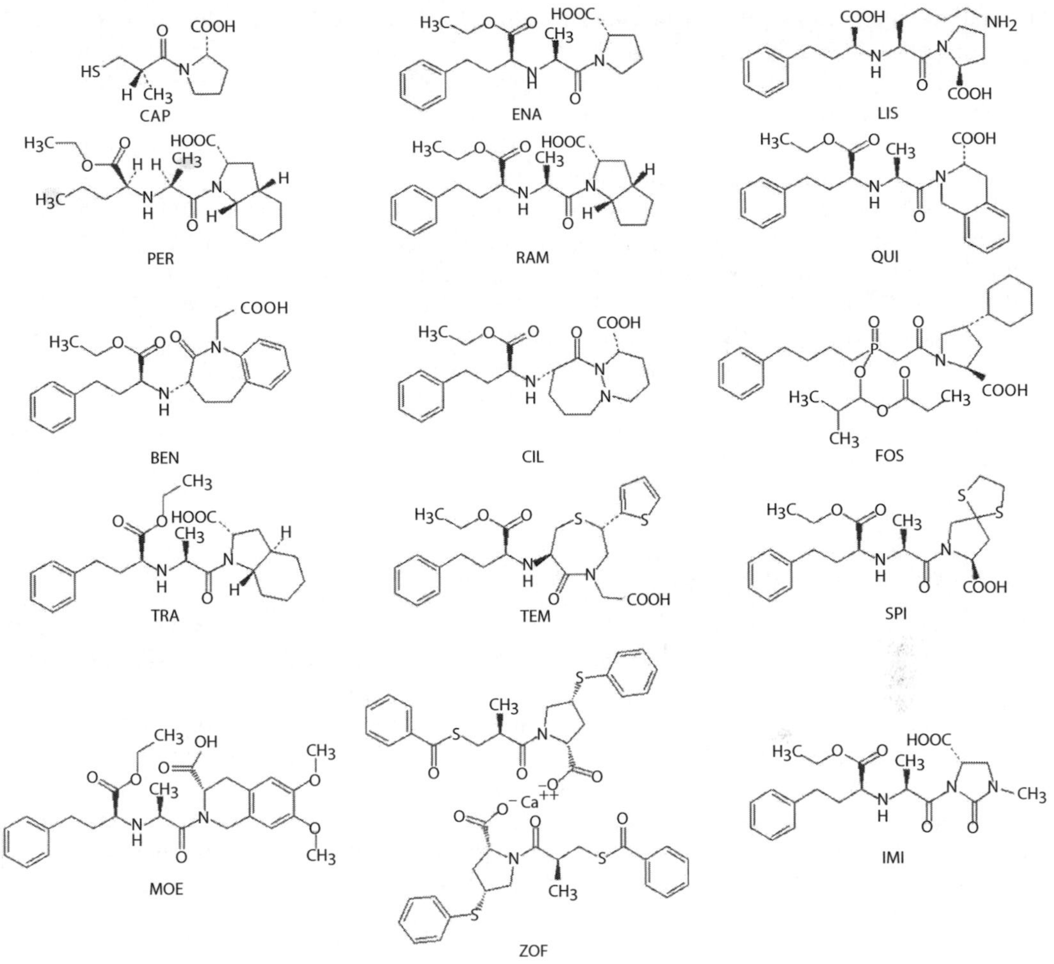

FIGURE 26.1 Chemical structures of ACE inhibitors.

amino]octahydro-10-oxo-6 H-pyridazino[1,2-a][1,2]diazepine-1-carboxylic acid 9-ethyl ester, **FOS** (4 S)-4-cyclohexyl-1-[[[(RS)-1-hydroxy-2-methylpropoxy](4-phenylbutyl)phosphinyl] acetyl]-L-proline propionate, **trandolapril** (TRA) (3a R,7a S)-1-[N-[1(S)-(ethoxycarbonyl)-3-phenylpropyl]-(S)-alanyl]octahydroindole-2(S)-carboxylic acid, **temocapril** (TEM) (+)-(2 S,6 R)-6-[[(1 S)-1-carboxy-3-phenylpropyl] amino]tetrahydro-5-oxo-2-(2-thienyl)-1,4-thiaze-pine-4(5 H)-acetic acid, 6-ethyl ester, **spirapril** (8 S)-7[(S)-N-[(S)-1-carboxy-3-phenyl propyl] alanyl]-1,4-dithia-7-azaspiro [4.4] nonane-8-carboxylic acid 1-ethyl ester, **moexipril** (MOE) [3 S -[2[R *(R *)],3 R *]]-2-[2-[[1-(Ethoxycarbonyl)-3-phenyl propyl]amino]-1-oxopropyl]-1,2,3,4-tetrahydro-6,7-dimethoxy-3-isoquinoline carboxylic acid, and **zofenopril calcium** (ZOF) (4S)-N-[3-(Benzoylsulfanyl)-2(S)-methylpropionyl]-4-(phenylsulfanyl)-L-proline calcium salt (Harrold 2008, Cutler 2011a). The chemical structures of the corresponding *prils* are shown in Figure 26.1.

26.1.2 IN SITU DETERMINATION OF ACE INHIBITORS IN THE PRESENCE OF OTHER DRUGS

Binary mixtures of CAP and indapamide (Mousa et al. 2006), LIS and amlodipine (AML) (Gopani et al. 2011), PER and AML (Borole et al. 2011) or losartan (LOS) (Lakshmi et al. 2010), RAM and telmisartan (TEL) (Potale et al. 2010) or atorvastatin (Panchal and Suhagia 2010),

QUI and hydrochlothiazide (Kowalczuk et al. 2003), BEN and AML (Meyyanathan and Suresh 2005) or hydrochlorothiazide (El-Gindy et al. 2001, Hassib et al. 2000), and TRA and vera-pamil (VER) (Kowalczuk 2005); ENA and hydrochlorothiazide (Stolarczyk et al. 2008); and ternary mixture of RAM, aspirin, and atorvaststin (Panchal et al. 2009) have been determined in situ. Along with an in situ approach, certain ACE inhibitors were simultaneously determined by high-performance liquid chromatography (HPLC) (El-Gindy et al. 2001, Hassib et al. 2000, Panchal and Suhagia 2010) and derivative spectrophotometry (Mousa et al. 2006). The details of chromatographic conditions and analytical results are shown in Tables 26.1 and 26.2.

Potale et al. 2010 describe a stability-indicating method for RAM exposing it to dry heat, oxidative conditions, photolysis, ultraviolet (UV) and the cool white fluorescent light, and to hydrolytic conditions under the different pH values. Major degradation (48%–56%) was observed both for the dry oxidative conditions and for the oxidation running under the reflux and with alkaline hydrolysis. Kowalczuk et al. 2003 tested the lowest detectable amounts for QUI using seven different detection reagents and they found out that the detection by the Folin–Ciocolater reagent and the Forest reagent is more sensitive by the factor of 2.5 than the detection in UV light at 254 nm.

26.1.3 In Situ Determination of Single ACE Inhibitors

TRA was identified and assayed in the bulk and dosage pharmaceutical forms (Kotaiah et al. 2010, Sreekanth et al. 2010). The details of chromatographic conditions and analytical results are shown in Tables 26.1 and 26.2.

Prils are stereoselective drugs, and only one stereoisomer exerts an activity by inhibiting the ACE enzyme. The S(+) enantiomer of LIS is active. Bhushan and Agarwal (2008) developed a method for direct resolution of (±) LIS using (+) tartaric acid or (−) mandelic acid as impreg-nating reagents, or as mobile phase additives. Successful separation of the LIS enantiomers was performed using (−) mandelic acid or (+) tartaric acid as chiral impregnating reagents and ethyl acetate–methanol–water 3:1:1 (v/v) and acetonitrile–methanol–water–dichloromethane 7:1:1:0.5 (v/v) as mobile phases, respectively. When (+) tartaric acid was used as a mobile phase additive, the mobile phase acetonitrile–methanol–(+) tartaric acid (0.5% in water, pH 5)–glacial acetic acid (7:1:1:1:0.7, v/v) enabled a successful resolution of the LIS enantiomers. One studied an effect of temperature, pH, and the amount of chiral selector on the resolution. The best resolution was obtained at 16°C, pH 5, and using the 0.5% impregnating reagent. The resolved spots were detected with the iodine vapor. LODs were equal to 0.14 and 0.16 µg for each enantiomer with (+) tartaric acid (both conditions) and (−) mandelic acid, respectively. Linearity was evaluated on the basis of the correlation between the absorbance and the amount of enantiomers (2.1×10^{-4} to 4.7×10^{-4}) recovered from the plates, and the correlation coeffi-cients were > 0.99. The mean precision values (RSD) were within the range of 0.28%–0.31% for LIS with (+)-tartaric acid and (−) mandelic acid, respectively. The enantiomer separation was achieved within 10 min at 16°C, using the readily available and the low-cost chiral selectors. The mechanism of the enantiomer resolution was discussed.

Mohammad et al. (2009) applied micellar TLC for the identification and quantification of LIS in the bulk drug, the dosage form, and the spiked urine samples. The silica gel H layer as station-ary phase and the 4% aqueous N-cetyl-N,N,N–trimethylammonium bromide as mobile phase were used for the separation of LIS. After the development, the spots of the standards and the LIS sam-ples were scraped off from the plate and quantitatively eluted with dimethylformamide (DMF). Then the ninhydrin solution (2% in DMF) was added as a chromogenic reagent, and the entity was kept at 25°C for 90 min. Absorbance was measured at 595 nm, and linearity was observed in the concentration range of 10–150 µg/mL. The recoveries of LIS (pure, formulated, and urine spiked) were within the range 93%–100.2%, and the LOD value was 5.5 µg/mL.

TABLE 26.1

Chromatographic Separation Conditions Used for ACE Inhibitors

Drug	Plates	Stationary Phase Pretreatment; Mobile Phase Composition	Development Mode	Detection (nm)	Rf	References
CAP	TLC silica gel F_{254}	Chloroform–glacial acetic acid, 6.5:1 v/v	—	212	—	Mousa et al. (2006)
LIS	TLC silica gel $60F_{254}$ aluminum–backed	Methanol, activation at 110°C, 5 min; methanol–ethyl acetate–ammonium sulphate (0.2%), 3:6:4 v/v/	Chamber saturation 30 min, 25°C; ascending	218	0.3	Gopani et al. (2011)
PER	HPTLC silica gel $60F_{254}$	Prewashing; dichloromethane–methanol–glacial acetic acid, 8.5:1.5:0.1 v/v/v	—	238	0.9	Borole et al. (2011)
PER	TLC silica gel $60F_{254}$	Toluene–acetonitrile–formic acid, 5:5:0.3 v/v/v		215	0.3	Lakshmi et al. (2010)
RAM	TLC silica gel $60F_{254}$	Methanol–chloroform, 1:6 v/v	Ascending	210	0.4	Potale et al. (2010)
RAM	Silica gel $60F_{254}$, aluminum-backed	Methanol, activation at 110°C, 5 min; methanol–benzen–glacial acetic acid, 19.6:80.0:0.4 v/v	Chamber saturation 20 min, 25°C; ascending	210	0.2	Panchal and Suhagia (2010)
QUI	HPTLC silica gel $60F_{254}$	Ethyl acetate–acetone–acetic acid, 8 + 2 + 0.5 v/v	Horizontal	210	0.5	Kowalczuk et al. (2003)
QUI	RP-18 TLC F_{254}	Methanol–0.07 M phosphate buffer, pH 2.5, 6:4	Horizontal	210	0.2	Kowalczuk et al. (2003)
BEN	HPTLC silica gel $60F_{254}$, aluminum-backed	Methanol; ethyl acetate:methanol:ammonia (10%) 8.5:2:1 v/v	Ascending	240	0.5	Meyyanathan and Suresh (2005)
BEN	HPTLC silica gel $60F_{254}$, aluminum-backed	Ethyl acetate–methanol–chloroform, 10:3:2 v/v	Chamber saturation ascending	238	0.2	El-Gindy et al. (2001)
BEN	TLC silica gel $60F_{254}$	Ethyl acetate:methanol:ammonia, 85:20:10 v/v/	—	240	0.3	Hassib et al. (2000)
TRA	TLC silica gel $60F_{254}$	Ethyl acetate–ethanol–acetic acid, 8:2:0.5 v/v	Horizontal	215	0.7	Kowalczuk (2005)
ENA	HPTLC silica gel $60F_{254}$, aluminum-backed	Butanol-1–glacial acetic acid–water, 12:3:5 v/v/	Ascending	208	0.6	Stolarczyk et al. (2008)
RAM	HPTLC silica gel $60F_{254}$, aluminum-backed	Methanol, activation at 110°C, 5 min; methanol–benzene–ethyl acetate–glacial acetic acid, 0.36:5.6:4:0.04 v/v	Chamber saturation 30 min, 25°C; ascending	210	0.06	Panchal et al. (2009)
TRA	HPTLC silica gel $60F_{254}$, aluminum-backed	Chloroform–methanol–acetic acid, 8:1.5:0.5 v/v	Ascending	212	0.5	Sreekanth et al. (2010)
TRA	TLC silica gel $60F_{254}$	Chloroform–methanol–acetic acid, 8:1.5:0.5 v/v/v	Ascending	212	0.5	Kotaiah et al. (2010)

TABLE 26.2

Analytical Parameters of the Methods Used for Separation of ACE Inhibitors

Drug	Calibration Mode/Range	LOD Unit/Spot	LOQ Unit/Spot	Precision (RSD%)	Recovery (%)	Matrix	References
CAP	Linear, 6.6–33 µg/spot	—	—	—	—	—	Mousa et al. (2006)
LIS	Linear, 2.0–10.0 µg/spot	80 ng	242 ng	1.1–1.2	98.9–100.0	Tablets	Gopani et al. (2011)
PER	Linear, 6.0–18.0 µg/spot	179 ng	544 ng	—	—	Tablets	Borole et al. (2011)
RAM	Linear, 0.4–2 µg/spot	90 ng	300 ng	Mean 1.8	98.0–102.0	Bulk drug and tablets	Potale et al. (2010)
RAM	Linear, 0.05–0.5 µg/spot	2.3 ng	7.05 ng	1.2–1.4	98.8–100.7	Capsules	Panchal and Suhagia (2010)
QUI	Polynomial, 2–12 µg/spot	380 ng	—	3.5–5.0	96.0–104.0	Tablets	Kowalczuk et al. (2003)
QUI	Polynomial, 2–12 µg/spot	460 ng	—	2.6–4.0	96.0–104.0	Tablets	Kowalczuk et al. (2003)
BEN	Linear, 0.2–2 µg/spot	200 ng	600 ng		99.7–100.2	Tablets	Meyyanathan and Suresh (2005)
BEN	Polynomial, 2–20 µg/spot	—	—	0.75–0.81	99–101.1	Tablets	El-Gindy et al. (2001)
BEN	Linear, 0.7–9.6 µg/spot	120 ng	400 ng	—	98.9–101.1	Tablets	Hassib et al. (2000)
TRA	Linear, 0.5–15 µg/spot	1.25 µg	3.75 µg	0.35–1.2	Mean 103.4	Capsules	Kowalczuk (2005)
ENA	Linear, 0.3–5 µg/spot	170 ng	520 ng	0.4–1.2	97.8–105.0	Tablets	Stolarczyk et al. (2008)
RAM	Linear, 50–300 ng	2.9 ng	8.8 ng	1.1–1.5	99–100.1	Capsules	Panchal et al. (2009)
TRA	Linear 25–150 ng/spot	18 ng	54 ng	0.65–1.26	99.2–99.7	Tablets	Sreekanth et al. (2010)
TRA	Linear, 25–150 ng/spot	10 ng	24 ng	—	—	Bulk drug and dosage forms	Kotaiah et al. (2010)

26.1.4 IN SITU IDENTIFICATION AND DETERMINATION OF ACE INHIBITORS IN MIXTURES

Chromatographic conditions for an in situ identification of CAP and ENA was investigated, using different stationary phases (HPTLC F_{254}, GF_{254}, RP-18F_{254}, and HPTLC RP-8F_{254}) and different mobile phases. For the selected mobile phases, the detection limits for both compounds have been established (visualization both in UV light at 254 nm and with iodine vapors). Densitometric determination was performed after the separation of CAP and ENA on the HPTLC$_{F254}$ plates with chloroform–methanol–glacial acetic acid (30:5:1, v/v). The obtained Rf values were 0.7 and 0.6, respectively. After the development, the separated zones of CAP and ENA were scanned at 212 and 210 nm, respectively. The linearity was tested over the range of 5–50 µg for both inhibitors. The TLC determination results obtained for CAP and ENA from the bulk drug and the tablets were compared with those derived from the gas chromatographic (GC) method, originally developed and presented in the same paper. It was concluded that the densitometric method is simpler, faster, and less expensive than the reference HPLC method (Czerwińska et al. 2001).

Chromatographic conditions for the separation of 13 ACE inhibitors (Figure 26.1 all except for PER and TEM) were investigated, using six different mobile phases and HPTLC F_{254} plates. The detection limits for all tested compounds were established at $UV_{254\,nm}$ and exposing the chromatograms to iodine vapors. The obtained values are in the range 0.025–3 µg/spot. Densitometric measurement was carried out at the maximum wavelength for each individual substance, and the regression equations were established. The proposed method was applied to determine the ACE inhibitors from the tablets, and the results were compared with those obtained with the routine HPLC method. The elaborated method is considered simpler, faster, and less expensive than HPLC (Wyszomirska et al. 2010).

The HPTLC method coupled with densitometric analysis was developed for quantification of BEN and CIL, both pure and in the respective commercial dosage forms. Active substances were extracted from the tablets with methanol (mean recovery 102%) and chromatographed on the silica gel 60 F_{254} HPTLC plates in the horizontal chambers with ethyl acetate–acetone–acetic acid–water, 8:2:0.5:0.5 (v/v) as mobile phase. Chromatographic separation of these ACE inhibitors was followed by the UV densitometric quantification at 215 nm. The calibration plots were constructed in the range 0.4–2.0 µg/µL for BEN (2.0–10.0 µg/spot) and 0.5–1.5 µg/µL for CIL (4.0–12.0 µg/spot), with good correlation coefficients (r = 0.990). The LOD values for BEN and CIL were 0.2 and 0.4 µg/spot, respectively, while the LOQ values were 0.6 and 1.2 µg/spot, respectively. The method was used for determination of BEN and CIL in pharmaceutical preparations with satisfactory precision (1.4% < RSD < 5.6%). The developed mobile phase is suitable for the separation and screening of the BEN, CIL, PER, QUI, and TRA mixture (Kowalczuk et al. 2004). Gumieniczek and Przyborowski (1997) used TLC to analyze BEN, CAP, CIL, and ENA present simultaneously in the sample. Normal- and reversed-phase (RP) systems were examined with respect to their separation efficiency and development time. Mobile phases were prepared separately for each development run, and migration distances were measured. Since the ascending technique was not sufficiently effective for the separation of the examined drugs, horizontal development was applied. On the normal-phase TLC plates, the best separation was achieved using the following mobile phases: isopropanol–acetone–water–acetic acid (5:4:1:0.1, v/v) and acetone–butanol–chloroform–acetic-acid (8:1:1:0.1, v/v). In the case of the reverse-phase analysis, greter selectivity was achieved on the RP-18 plates by increasing the acetonitrile content in mobile phase to 50 vol%. It was found out that the greatest sensitivity of detection (5 µg) was with BEN, CAP, and CIL, and with ENA, it was 20 or 50 µg in the normal-phase and the RP systems, respectively. CAP was detected by the reaction with the iodine solution only, and the remaining compounds could be detected in the UV light also. The established method is simple, rapid, and selective for the routine analysis of the examined ACE inhibitors in tablets.

Active ingredients from the binary formulations, RAM and AML, or ELA and AML, were separated using TLC and chloroform–methanol (6:1, v/v) as mobile phase and also the open-column LC system. Detection was carried out in UV light at 210 nm and using iodine vapors. Solvent conditions from TLC were transferred to the open-column chromatographic system. Quantitative determination was carried out using TLC and the open-column chromatography supplemented with UV spectrophotometry. Recovery was in the range of 82%–93% (Bhushan et al. 2006).

26.1.5 In Situ Identification and Determination of CAP from Biological Fluids

The in vivo metabolism of CAP ^{14}C in rats and dogs was investigated by thin-layer radiochromatography after extraction of this compound from urine samples. The samples were spotted on to the silica-gel-coated chromatographic plates and two different mobile phase systems were developed—benzene–acetic acid (3:1, v/v) and n-butanol–acetic acid–water (4:1:1, v/v)—in order to resolve the metabolites of different polarity. The developed plates were dried and then for 2 weeks brought in the contact with the x-ray-sensitive films, in order to obtain autoradiograms.

The radioactive spots were identified by comparing their Rf values with those of the authentic samples, then scraped off into the counting vials, and supplemented with 10 mL aliquot of the toluene–ethanol liquid scintillator. Radioactivity was measured in liquid scintillation spectrophotometer. In the experiments, in order to enable the chromatographic comparison of the partially purified metabolites with the authentic samples, three additional mobile phases were used. This method enabled simultaneous monitoring of CAP and its five metabolites (Ikeda et al. 1981).

The in vitro biotransformation of CAP, uniformly labeled with ^{14}C in the blood of rats, dogs, and humans, was followed by thin-layer radiochromatography. After incubation of CAP[14]C and extraction, samples were subjected to TLC using the Analtech silica gel GF-coated thin-layer chromatographic plates with chloroform–ethyl acetate–glacial acetic acid (4:5:3 v/v/) as mobile phase. For the generation of the zonal profiles of the extracted radioactive samples, all reference standards of the metabolites were applied to each plate. After the development, the plates with reference standards were visualized by exposing the plates to iodine vapors. Silica gel on each plate containing the samples was divided into the six zones, and the radioactivity of each zone was determined by the liquid scintillation counting. The percentage radioactivity contribution of each zone was calculated against the total radioactivity of each plate equal to 100%. This method enables monitoring of CAP, its metabolites, the disulfide dimer CAP–CAP, CAP-L–cysteine, and CAP–glutathione (Wong et al. 1981).

26.1.6 In Situ Identification, Retention Behavior, and Lipophilicity of ACE Inhibitors

The chromatographic behavior of selected *prils*, ENA, QUI, FOS, CIL, and LIS, and their corresponding *prilates* was investigated using normal-phase TLC and several nonaqueous mono- and two-component mobile phases. It was established that the retention order of the examined substances obtained with alcohols as mobile phases is in agreement with their elution strength as well as polarity, that is, the less polar the solvent, the stronger the retention. The results obtained in the course of examination of the *prils* and *prilates* with use of the two-component solvents demonstrated a decrease of the Rf values, that is, an increased retention of the examined substances in parallel with the increasing concentration of the less polar solvent in mobile phase, which is in agreement with the normal-phase chromatographic mode. Differentiation in chromatographic behavior of the selected *prilates* (including LIS), due to their specific interactions with the silica gel, is much stronger than that of the corresponding *prils*, which contain one carboxylic group only within their respective molecules. Based on the observed correlations of the chromatographically determined hydrophobicity parameters, R_M^o and C_o, and on the computer-assisted calculated log P values, it was established that this method is a reliable tool for the estimation of lipophilicity of the examined drugs (Odović et al. 2009). Aleksić et al. (2001a) examined the chromatographic behavior of CAP, ENA, LIS, QUI, RAM, and CIL on silica gel thin layers and polyacrylonitrile sorbents (PANS). Optional conditions for the separation of these *prils* were investigated, using different multi-component mobile phases. The hRf value order for the examined components was the same on the silica and PANS when the same solvent system was used, but the retention on PANS was weaker than on the silica gel. The retention mechanisms of the investigated compounds were discussed in detail. Aleksić et al. 2001b examined the chromatographic behavior of the same *prils* as those mentioned in the previous paper under the conditions of the salting-out TLC on silica gel, cellulose, and PANS used as sorbents. Different concentrations of the aqueous ammonium sulphate were used as solvents, and it was observed that the hRf values decreased with the increasing salt concentration in the eluent system used. Linear relationship between the corresponding R_M values and the contents of ammonium sulphate in the solvents was observed. The retention mechanisms of the investigated compounds were discussed in detail, and a conclusion was made that this simple and inexpensive chromatographic system can be used as a screening method for the identification of the selected ACE inhibitors.

Separation of LIS, CIL, CAP, QUI, and RAM on the thin layer of aminoplast (carbimide-formaldehyde polymer) and the corresponding retention mechanism were investigated. On optimization of the chromatographic systems, the separation was achieved with use of benzene–cyclohexane–methyl ethyl ketone (15 + 10 + 15, v/v) as mobile phase. It was revealed that normal-phase chromatography occurs on the thin layers of aminoplast. The retention mechanisms of the investigated compounds were discussed in detail (Perišić-Janjić and Agbaba 2002).

Normal- and RP-TLC in the analysis of L-arginine, its metabolites, and the select drugs (including CAP) were investigated. A variety of mobile phases were evaluated and finally, methanol–50% acetic acid (3:1, v/v) on silica gel and 5% acetic acid–methanol–acetonitrile (50:35:15, v/v) on RP-18 were selected for the separation of L-arginine and its metabolites, and for the separation of drugs, acetonitrile–water (2:3, v/v) on silica gel was chosen. The effect of the polar mobile phase modifiers on selectivity was assessed (Baranowska et al. 2009). The inclusion complexes of LIS, FOS, and ZOF with β-cyclodextrin (BCD) and those of FOS and ZOF with hydroxypropyl-β-cyclodextin and randomly methylated BCD were prepared, using the kneading method, in the 1:1 molar ratio. Evaluation of TLC results revealed that the inclusion complexes show the lower hRf values when compared with the two components of the inclusion complex proving that there is an interaction between an active substance and a corresponding cyclodextrin due to the formation of the inclusion complexes. Formation of these complexes is used to improve physical–chemical characteristics of pharmaceutical substances, in order to enhance certain bio-pharmaceutical properties. The details of the chromatographic conditions, preparation of the inclusion complexes, and the retention properties of the examined substances are given in paper (Sbârcea et al. 2010a,b).

A simple and reliable RP-TLC method was used to determine lipophilicity of CAP, DEC, ENA, LIS, and MOE. The silica gel plates were saturated with paraffin using continuous development for 10 h with the 10% paraffin solution in hexane. The four variable volumetric ratios of acetonitrile–water–ammonium hydroxide mixture were used as mobile phases. The R_M values of the five ACE inhibitors followed a linear relationship against the acetonitrile content in the mobile phase. Values characterizing lipophilicity of the five drugs determined by means of RP-TLC (R_M^0, a) are given in paper (Csermely et al. 2008).

26.1.7 Classical and Novel Methods of ACE Detection in TLC

In spite of a considerable advance recently achieved in chromatography, TLC still remains a method widely in use for a simple, easy, and fast control of the reaction progress in the course of the drug synthesis. The reaction progress in the course of synthesis of the optically active *trans* 4-cyclohexyl-L-proline as an intermediate product in the preparation of FOS was controlled by means of TLC performed on the silica gel 60F layers using various different ratios of the mobile phase components, to finally select the chloroform–methanol (12:1, 10:1, v/v) and the ethyl acetate-hexane (2:1, 1:1, 1:2, v/v) mixtures as the best performing mobile phases. Compounds on the chromatograms were detected in the iodine vapors, or by using the 0.2% ninhydrin solution in ethanol, and finally visualizing the spots in UV light (366/254 nm) was performed (Kalisz et al. 2005). Wood and Steiner (2011) proposed the TLC separation prior to the Direct Analysis in Real Time (DART) detection as a time efficient and cost-saving technique for the forensic drug analysis community. LIS was detected in the presence of hydrochlorothazide and after being separated the drug spot was scraped off from the chromatogram, eluted from the stationary phase, and introduced to the DART™ ion source, which proved a quick and simple approach. In such a way, a comparative analysis with the primary standard can be avoided, especially if it is not easily available, very expensive, or with the short duration period.

Separation of the structurally related LIS, CIL, RAM, and QUI, and their corresponding *prilates* with use of the silica gel 60 TLC plates was compared with that with use of the monolithic

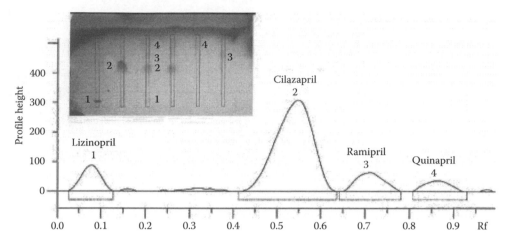

FIGURE 26.2 UTLC image and video densitogram of prils. (From *J. Chromatogr. A.*, 1218, Vovk, I., Popović, G., Simonovska, B., Albreht, A., and Agbaba, D., Ultra-thin-layer chromatography mass spectrometry and thin-layer chromatography mass spectrometry of single peptides of angiotensin-converting enzyme inhibitors, 3089–3094, Copyright 2011, with permission from Elsevier.)

ultra-thin-layer chromatographic (UTLC) plates. The plates were developed in the modified horizontal chromatographic chamber using ethyl acetate–acetone–acetic acid–water (4:1:0.25:0.5, v/v). Detection of the separated compounds was carried out densitometrically at 220 nm and also by the image analysis, after an exposure to the iodine vapors. It was observed that the monolithic layers were more efficient for the separation of the structurally similar polar compounds, such as *prilates*, than the conventional silica gel layers (Figure 26.2). Identification of the compounds was carried out by ES-mass spectrometry (MS), after an on-line extraction from the UTLC and TLC plates by means of the Camag TLC–MS interface (Vovk et al. 2011).

26.2 CHEMISTRY AND PHARMACOLOGY OF ANGIOTENSIN II RECEPTOR ATs

Angiotensin II receptor ATs, also known as **angiotensin receptor blockers** (ARBs), AT_1-**receptor ATs**, or *sartans*, are a group of pharmaceuticals, which modulate the renin–angiotensin–aldosterone system. Their main applications are in the treatment of hypertension (high blood pressure), diabetic nephropathy (kidney damage due to diabetes), and congestive heart failure. The AT_1 receptor ATs currently used are: **LOS**, 2-butyl-4-chloro-1-[[2′-(1H-tetrazol-5-yl)[1,1′-biphenyl]-4-yl]methyl]-1H-imidazole-5-methanol, **valsartan** (VAL) N-(1-oxopentyl)-N-[[2′-(1H-tetrazol-5-yl)[1,1′-biphenyl]-4-yl]methyl]-L-valine, **eprosartan** (EPR) (E)-α-[[2-butyl-1-[(4-carboxyphenyl) methyl]-1H-imidazol-5-yl]methylene]-2-thiophene propanoic acid, **candesartan** (CAN) -2-ethoxy-1-[[2′-(1H-tetrazol-5-yl)[1,1′-biphenyl]-4-yl]methyl]-1H-benzimidazole-7-carboxylic acid, CAN **cilexetil** (CANC)1 [[(cyclohexyloxy)carbonyl]oxy]ethyl 2-ethoxy-1-[[2′-(1H-tetrazol-5-yl)[1,1′-biphenyl]-4-yl]methyl]-1H-benzimidazole-7-carboxylate, **TEL** 4′-[(1,4′-dimethyl-2′-propyl[2,6′-bi-1H-benzimidazol]-1′-yl)methyl][1,1′-biphenyl]-2-carboxylic acid, **irbesartan** (IRB) 2-butyl-3-[[2′-(1H-tetrazol-5-yl)[1,1′-biphenyl]-4-yl]methyl]-1,3-diazaspiro[4.4]non-1-en-4-one, **olmesartan** (OLM) 4-(1-hydroxy-1-methylethyl)-1-[[2′-(1H-tetrazol-5-yl)[1,1′-biphenyl]-4-yl]methyl]-1H-imidazole-5-carboxylic acid, and **OLM medoxomil** (OLMM) 4-(1-hydroxy-1-methylethyl)-2-propyl-1-[[2′-(1H-tetrazol-5-yl)[1,1′-biphenyl]-4-yl]methyl]-1H-imidazole-5-carboxylic acid (5-methyl-2-oxo-1,3-dioxol-4-yl)methyl ester. The chemical structures of *sartans* are shown in Figure 26.3. These drugs are biphenyl methyl derivatives that possess an acidic moiety, which can interact with the respective position at the AT_1 receptor.

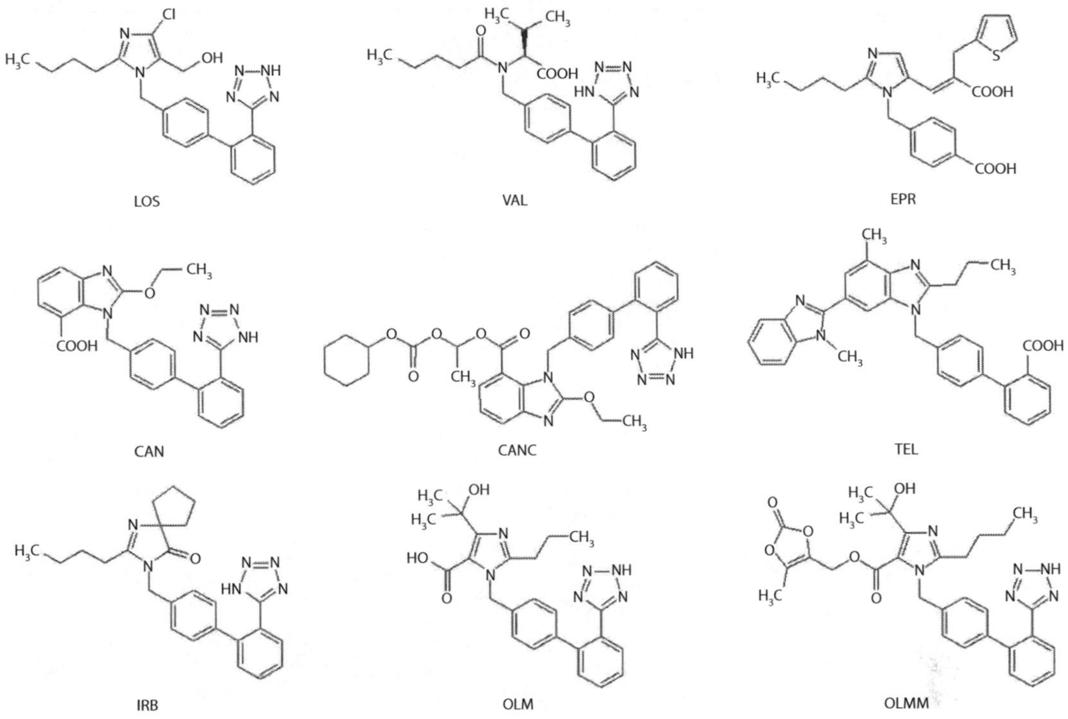

FIGURE 26.3 Chemical structures of ARBs.

The tetrazol ring present in LOS, VAL, IRB, and CAN characterizes with $pKa \approx 6$ and at physiological pH, it is 90% ionized. The carboxylic acid moiety present in VAL, CAN, TEL, and EPR characterizes with the pKa values in the range from 3 to 4 and it is also primarily ionized. All these compounds have low bioavailability due to their poor solubility in lipids. CANC and OLMM give an active form in vivo, and VAL gives active metabolites after hydrolysis or oxidation, respectively. They are administered either alone or in combination with the other antihypertensive drugs, such as ACE inhibitors, diuretics, beta blockers, or calcium channel blockers (Harrold 2008, Cutler 2011a).

26.2.1 In Situ Determination of Single or Mixed ARBs

LOS, VAL, TEL, and OLM present as single component in pharmaceutical formulations (McCarthy et al. 1998; Moussa et al. 2010; Parambi et al. 2010, 2011; Shrivastava et al. 2009; Vekariya et al. 2010; Yadav and Srinivasan 2010) or in plasma (Tambe et al. 2010) were separated and/or in situ assayed from various different excipients present in the dosage units (Parambi et al. 2010, 2011, Yadav and Srinivasan 2010) or separated from their respective impurities also (McCarthy et al. 1998, Moussa et al. 2010, Shrivastava et al. 2009, Vekariya et al. 2010). Along with an in situ approach, certain ARB inhibitors were simultaneously determined by means of the HPLC (McCarthy et al. 1998, Moussa et al. 2010, Tambe et al. 2010, Yadav and Srinivasan 2010). Combination of CAN and LOS was assayed by an in situ classical technique and by videoscanning (Gumieniczek et al. 2011). The details of the chromatographic conditions and the analytical results for the examined ARBs are shown in Tables 26.3 and 26.4.

The stability indicating TLC method was developed for LOS, VAL, TEL, and OLMM (McCarthy et al. 1998, Moussa et al. 2010, Shrivastava et al. 2009, Vekariya et al. 2010), respectively. McCarthy et al. (1998) developed the RP-TLC method for the simultaneous determination of LOS and the two diastereoisomeric degradates assigned as dimer E and F (apparently formed

TABLE 26.3

Chromatographic Separation Conditions Used for ARBs

Drug	Plates	Stationary Phase Pretreatment; Mobile Phase Composition	Development Mode	Detection (nm)	Rf	References
LOS	RP-18 HPTLC slica gelF$_{254}$	Acetonitrile–methanol–acetic acid (0.1%), 35:25:40 v/v	Ascending-AMD	254	0.8	McCarthy et al. (1998)
VAL	HPTLC silica gel 60 F$_{254}$	Chloroform–acetonitrile–toluene–glacial acetic acid, 1:8:1:0.1 v/v	Chamber saturation 20 min ascending	254	0.65	Parambi et al. (2011)
VAL	HPTLC silica gel 60F$_{254}$, aluminum-backed	Methanol, activation at 110°C, 15 min; toluene–ethyl acetate–methanol–formic acid, 60:20:20:1 v/v	Chamber saturation 20 min; ascending	250	0.4	Shrivastava et al. (2009)
TEL	TLC silica gel 60F$_{254}$, aluminum-backed	Ethyl–acetate–dichloromethane–methanlol, 6:2:1 v/v	—	295	0.7	Vekariya et al. (2010)
OLMM	TLC silica gel F$_{254}$	Chloroform–acetonitrile–toluene–acetic acid, 1:8:1:0.1	Chamber saturation 30 min; ascending	301	0.7	Parambi et al. (2010)
OLMM	TLC silica gel F$_{254}$	Methanol, activation at 105°C, 5 min; chloroform–acetone–methanol, 7:2:1 v/v	Chamber saturation 30 min; ascending	256	0.4	Yadav and Srinivasan (2010)
OLMM	Nano silica gel 60F$_{254}$ aluminum-backed	Chloroform–methanol–formic acid, 8:1.5:0.5 v/v	Chamber saturation 60 min; ascending	260	0.7	Moussa et al. (2010)
OLM	HPTLC silica gel 60F$_{254}$	Ethyl acetate–methanol–acetic acid, 8:2:0.05 v/v	Ascending	269	0.3	Tambe et al. (2010)
CANC	TLC silica gel F$_{254}$	Oven dried, 105°C,15 min; 1.4-dioxan–hexane–formic acid (99%) 5:5:0.1 v/v	Horizontal	258 254	0.5	Gumieniczek et al. (2011)
LOS	TLC silica gel F$_{254}$	Oven dried, 105°C,15 min; 1.4-dioxan–hexane–formic acid (99%), 5:5:0.1 v/v	Horizontal	243 254	0.35	Gumieniczek et al. (2011)
LOS	TLC silica gel 60F$_{254}$	Toluene–acetonitrile–formic acid, 5:5:0.3 v/v/v		215	0.5	Lakshmi et al. (2010)
TEL	TLC silica gel 60F$_{254}$	Methanol–chloroform, 1:6 v/v	Ascending	210	0.7	Potale et al. (2010)
VAL	HPTLC silica gel 60F$_{254}$, aluminum-backed	Ethyl acetate–tetrahydrofuran–acetic acid, 8:2:0.5 v/v	Ascending	252	0.9	Stolarczyk et al. (2008)
CAN	HPTLC silica gel 60F$_{254}$, aluminum-backed	Ethyl acetate–tetrahydrofuran–acetic acid, 8:2:0.5 v/v	Ascending	252	0.9	Stolarczyk et al. (2008)

TABLE 26.3 (continued)
Chromatographic Separation Conditions Used for ARBs

Drug	Plates	Stationary Phase Pretreatment; Mobile Phase Composition	Development Mode	Detection (nm)	Rf	References
VAL	TLC silica gel 60F$_{254}$, aluminum-backed	Methanol, activation at 110°C, 5 min; toluene:methanol:acetic acid, 7:3:0.1 v/v	Chamber saturation 30 min; ascending	244	0.5	Dhaneshwar et al. (2009)
VAL	TLC silica gel 60F$_{254}$	Mobile phase, activation at 100°C, 15 min; ethyl acetate–methanol–ammonim hydroxide, 55:45:5 v/v	Chamber saturation 15 min; ascending	237	0.7	Ramadan et al. (2010)

in the course of the stability studies). When developing the normal phase TLC method, the peak tailing was observed, which was overcome, and an improved sensitivity was recorded on the C18 reverse stationary phase. As the pKa values for LOS are 2.36 and 5.56, mobile phases containing a buffer at pH < 2.3 allow avoiding the peak splitting and the peak broadening. Details on the chromatographic working conditions are given in Table 26.3. Under the same chromatographic conditions, the Rf values of the dimeric degradates E and F were found as equal to 0.5 and 0.3, respectively. The LOQ and LOQ values for both dimers were found as equal to 0.05% and 0.1%, respectively, as assessed against the LOS content. Linear relationships between the peak areas and the concentrations of dimers E and F ranging from 0.1% to 2% also referred to the LOS levels. The results of the TLC determination were confirmed by means of HPLC. Shrivastava et al. (2009) developed a stability indicating HPTLC method for the analysis of the bulk drug. VAL when subjected to the acidic, basic, and oxidative (H$_2$O$_2$) degradation underwent a significant degradation, but when subjected to the dry and wet heat, or to photodegradation, it remained stable. The developed method proved to be stability indicating, because it allowed separation of VAL from its degradation products. Moussa et al. 2010 developed the HPTLC and HPLC approaches as the suitable stability indicating methods for an assay of OLMM in the presence of its acid- and base-induced degradation products. It was observed that the acidic and the alkaline degradants were the nearly similar molecules, having the same t$_R$ values in HPLC, and the same Rf values in TLC. On the TLC plates, it was observed that the acid-induced degradants were separated as the two closely neighboring spots, having the same UV and infrared (IR) spectra, hence a suggestion was made that they might be the enantiomeric degradates. Gumieniczek et al. (2011) compared the classical densitometry and the videoscanning method for the analysis of CANC and LOS present in pharmaceuticals. The authors performed a pair-wise comparison of precision in the analysis of tablets using the *Snedecor F* test and the accuracy in the analysis of tablets using the *Student t* and *Wilcoxon* test. The results obtained for the analysis of CANC showed that there were no significant differences between densitometry and videscanning with regard to the accuracy and precision. For LOS, the accuracy of videoscanning was similar to that of densitometry (*Wilcoxon* test), but videoscanning was much more precise than the classical densitometry (*Snedecor F* test). For the established working conditions, videoscanning was much more robust than densitometry with the two drugs considered (*Snedecor F* and *Student t* tests). Despite certain statistically significant differences, the obtained results showed that for the two ARBs and the different pharmaceutical purposes, both methods are suitable for quantitative determination. Tambe et al. (2010) developed and compared two extraction procedures, liquid–liquid extraction (LLE) and solid-phase extraction (SPE), for OLM contained in human plasma. Determination of OLM was performed by means HPTLC

TABLE 26.4

Analytical Parameters of the Methods Used for Separation of ARBs

Drug	Calibration Mode/Range	LOD Unit/ Spot	LOQ Unit/ Spot	Precision (RSD%)	Recovery (%)	Matrix	References
LOS	Linear, 6–15 µg/spot	—	—	Mean 1.2	99.5–100.5	Tablets	McCarthy et al. (1998)
VAL	Linear, 50–500 ng/spot	5 ng	16 ng	0.4–1.4	Mean 99.4	Tablets	Parambi et al. (2011)
VAL	Linear, 0.2–1.6 µg/spot	25 ng	150 ng	0.9–1.5	99.3–101.1	Bulk drug	Shrivastava et al. (2009)
TEL	Linear, 0.3–1.8 µg/spot	19.3 ng	58.5 ng	—	—	Tablets	Vekariya et al. (2010)
OLMM	Linear, 50–500 ng/spot	4.8 ng	16 ng	1.3–1.5	99.3–101.3	Tablets	Parambi et al. (2010)
OLMM	Linear, 200–800 ng/spot	39 ng	128 ng	0.2–0.6	99.9–100.2	Tablets	Yadav and Srinivasan (2010)
OLMM	Linear, 0.05–1 mg/mL	—	—	Mean ± 1.1	Mean 99.8	Bulk drug and tablets	Moussa et al. (2010)
OLM	Linear, 80–600 ng/spot	—	8 ng	0.7–3	80.1–79.6	Plasma	Tambe et al. (2010)
CANC	Linear, 0.2–1.4 µg/spot	—	—	2.5–3.7	103.8–104.9	Tablets	Gumieniczek et al. (2011)
LOS	Linear, 0.2–1.4 µg/spot	—	—	2.5–4.6	100.9–105.4	Tablets	Gumieniczek et al. (2011)
TEL	Linear, 1–5 µg/spot	121 ng	400 ng	Mean 1.85	98.0–102.0	Bulk drug and tablets	Potale et al. (2010)
VAL	Linear, 0.7–6.1 µg/spot	639 ng	1940 ng	Mean 1.14	92.1–97.7	Tablets	Stolarczyk et al. (2008)
CAN	Linear, 0.2–1.8 µg/spot	245 ng	750 ng	Mean 0.8	92.1–94.7	Tablets	Stolarczyk et al. (2008)
VAL	Linear, 1.6–9.6 µg/spot	50 ng	100 ng	0.1–0.4	98.5–100.1	Tablets	Dhaneshwar et al. (2009)
VAL	Linear, 2–12 µg/spot	—	—	Mean 0.5	Mean 100.6	Tablets	Ramadan et al. (2010)
OLMM	Linear, 0.8–5.6 µg/spot	200 ng	400 ng	0.6–1.28	99.8–100.4	Tablets	Desai et al. (2010)
OLMM	Linear, 0.4–4 µg/spot	50 ng	100 ng	0.1%–0.4%	97%–99%	Tablets	Kamble et.al. (2010)
TEL	Polynomial, 1.2–7.2 µg/spot	149 ng	453 ng	0.2–0.5	99–101	Tablet	Vekariya et al. (2009)
LOS	Linear, 0.4–1.2 µg/spot	—	—	0.6–2.7	96.9–102.8	Tablets	Shah et al. (2001)
VAL	Linear, 0.8–5.6 µg/spot	114 ng	387 ng	0.2–0.4	99.5–100.2	Tablets	Kadam and Bari (2007)
IRB	Linear, 0.1–0.7 µg/spot	30 ng	100 ng	0.6–0.8	99.7–101.8	Tablets	Shah et al. (2007a)
IRB	Linear, 0.1–0.6 µg/spot	22 ng	74 ng	0.3–0.45	99.9–100.2	Tablets	Shanmugasundaram and Velraj (2011)

TABLE 26.4 (continued)
Analytical Parameters of the Methods Used for Separation of ARBs

Drug	Calibration Mode/Range	LOD Unit/ Spot	LOQ Unit/ Spot	Precision (RSD%)	Recovery (%)	Matrix	References
TEL	Linear, 0.5–4.5 μg/spot	—	—	Mean 0.2	99.0–100.1	Tablets	Bebawy et al. (2005)
CAN	Linear, 65–325 μg/mL	9 μg	30 μg	Mean 1.4	99.4–100.1	Tablets	Mehta and Morge (2008)

and RP-HPLC. Generally, it was established that both extraction procedures gave satisfactory and repeatable recovery values, but LLE gave higher recovery, as compared with SPE (90.1% vs. 79.6%). Kumar et al. (2009) used TLC (among the other instrumental techniques) to characterize several newly developed VAL formulations, including solid dispersion in skimmed milk, designed with aim to improve the poor solubility of VAL and to enhance its biopharmaceutical properties.

26.2.2 In Situ Determination of ARBs in Presence of Other Drugs

Binary mixtures of *sartans* and *prils* (Lakshmi et al. 2010, Potale et al. 2010), *sartans* and calcium blocker, AML (Desai et al. 2010, Dhaneshwar et al. 2009, Kamble et al. 2010, Ramadan et al. 2010, Vekariya et al. 2009), *sartans*, and diuretics (hydrochlorothiazide) (Bari and Rote 2009; Bebawy et al. 2005; Kadam and Bari 2007; Kadukar et al. 2009; Mehta and Morge 2008; Moussa et al. 2011; Shah et al. 2001, 2007a,b; Shanmugasundaram and Velraj 2011); ternary mixture of *sartans*, beta blocker (atenolol), and diuretics (hydrochlorothiazide) (Sathe and Bari 2007) have been determined in situ. The chromatographic and analytical details concerning determination of LOS and TEL contained in the binary mixtures with *prils* (Lakshmi et al. 2010, Potale et al. 2010), and VAL and CAN individually contained in the binary mixtures with hydrochlorothiazide (Stolarczyk et al. 2008) are shown in Tables 26.3.and 26.4.

Potale et al. (2010), described a stability indicating method for the determination of TEL exposing it to dry heat, oxidative conditions, photolysis, UV light, the cool white fluorescent light, and to hydrolytic conditions under the different pH values. Major degradation (85%) was observed for the oxidative conditions, lesser degradation (8.9%) for the alkaline condition, and the negligible one only for the dry oxidative conditions.

26.2.2.1 In Situ Determination of ARBs in Presence of AML

VAL (Dhaneshwar et al. 2009, Ramadan et al. 2010), OLMM (Desai et al. 2010, Kamble et al. 2010), and TEL (Vekariya et al. 2009) were separated and densitometrically assayed. Details regarding the chromatographic working conditions and the analytical separation results valid for VAL are presented in Tables 26.3 and 26.4. Details regarding the chromatographic working conditions valid for the separation of OLMM and TEL are shown in Table 26.5 and the analytical data are given in Table 26.4.

Dhaneshwar et al. (2009) reported on an optimized and fully validated TLC method for the simultaneous determination of VAL and AML as the fixed labeled drugs (16:1 ratio) in the tablet form. Ramadan et al. (2010), developed the validated TLC and HPLC methods for the simultaneous determination of VAL in the presence of AML from the mixtures prepared in laboratory and containing different ratios of the two drugs (from 16:1 to 6.5). Plasma samples spiked with VAL and AML were analyzed by means of the originally developed and validated HPLC method. Apart from TLC, OLMM and AML were also assayed by means HPLC (Kamble et al. 2010).

TABLE 26.5

Chromatographic Separation Conditions Used for ARBs

Drug	Plates	Stationary Phase Pretreatment; Mobile Phase Composition	Development Mode	Detection (nm)	Rf	References
OLMM	TLC silica gel 60F$_{254}$, aluminum-backed	Chloroform:methanol:toluene: acetic acid, 8:1:1:0.1 v/v	Ascending	254	0.45	Desai et al. (2010)
OLMM	TLC silica gel 60F$_{254}$, aluminum-backed	Methanol, activation at 110°C, 5 min; n-butanol–acetic acid–water, 5:1:0.1 v/v	Chamber saturation 30 min; ascending	254	0.7	Kamble et al. (2010)
TEL	TLC silica gel 60F$_{254}$, aluminum-backed	Tetrahydrofuran:dichloroethane:methanol:ammonia solution, 3:1:0.5:0.2	Chamber saturation 30 min; ascending	326	0.2	Vekariya et al. (2009)
LOS	TLC silica gel 60F$_{254}$, aluminum-backed	Mobile phase washed; chloroform–methanol–acetone–formic acid, 7.5:1.5:0.5:0.03 v/v	Chamber saturation 45 min; ascending	254	0.6	Shah et al. (2001)
VAL	HPTLC silica gel 60F$_{254}$, aluminum-backed	Methanol; chloroform–ethyl acetate–acetic acid 5:5:0.2 v/v	Chamber saturation 30 min; ascending	248	0.4	Kadam and Bari (2007)
IRB	TLC silica gel 60F$_{254}$, aluminum-backed	Acetonitrile:chloroform:glacial acetic acid, 7:3:0.1 v/v	Chamber saturation 30 min; ascending	260	0.6	Shah et al. (2007a)
IRB	TLC silica gel 60F$_{254}$, aluminum-backed	Methanol, activation at 50°C, 5 min; acetonitrile-ethyl acetate, 8:2 v/v	Chamber saturation; ascending	260	0.3	Shanmugasundaram and Velraj (2011)
TEL	TLC silica gel 60F$_{254}$, aluminum-backed	n-butanol–ammonia (25%), 8:2 v/v	Ascending	295	0.6	Bebawy et al. (2005)
CAN	HPTLC silica gel 60F$_{254}$, aluminum-backed	Ethyl acetate–chloroform–acetone–methanol, 3:3:3:0.5 v/v	Chamber saturation 10 min; ascending	280	0.3	Mehta and Morge (2008)
OLMM	TLC silica gel 60F$_{254}$, aluminum-backed	Acetonitrile–chloroform–glacial acetic acid, 7:2:0.5 v/v	Chamber saturation 30 min; ascending	254	0.6	Shah et al. (2007a,b)
OLMM	TLC silica gel 60F$_{254}$, aluminum-backed	Methanol, activation at 120°C, 20 min; acetonitrile–ethyl acetate–glacial acid, 7:3:0.4	Chamber saturation 20 min; ascending	254	0.4	Bari and Rote (2009)

TABLE 26.5 (continued)
Chromatographic Separation Conditions Used for ARBs

Drug	Plates	Stationary Phase Pretreatment; Mobile Phase Composition	Development Mode	Detection (nm)	Rf	References
OLMM	HPTLC silica gel 60F$_{254}$, aluminum-backed	Methanol, activation at 120°C, 20 min; chloroform–methanol–toluene, 6:4:5 v/v	Chamber saturation 20 min; ascending	258	0.6	Kadukar et al. (2009)
OLMM	HPTLC nano silica gel 60F$_{254}$ aluminum-backed	Chloroform–methanol–formic acid, 8:1:5:0.5 v/v	Chamber saturation 60 min; ascending	260	0.7	Moussa et al. (2011)
LOS	TLC silica gel 60F$_{254}$, aluminum-backed	Methanol, activation at 50°C, 5 min; toluene–methanol–triethylamine, 6.5:4:0.5 v/v	Chamber saturation 30 min; ascending	274	0.6	Sathe and Bari (2007)

26.2.2.2 In Situ Determination of ARBs in Presence of Diuretics (Hydrochlorothiazide) and Beta Blockers (Atenolol)

In situ simultaneous determination of the fixed binary combinations of ARBs and hydrochlorothiazide (Bari and Rote 2009; Bebawy et al. 2005; Kadam and Bari 2007; Kadukar et al. 2009; Mehta and Morge 2008; Moussa et al. 2011; Shah et al. 2001, 2007a,b; Shanmugasundaram and Velraj 2011), and the ternary combination of ARBs, hydrochlorothiazide, and atenolol (Sathe and Bari 2007) contained in pharmaceuticals was developed and described in 11 papers for the following drugs: LOS (Sathe and Bari 2007, Shah et al. 2001), VAL (Kadam and Bari 2007), IRB (Shah et al. 2007a, Shanmugasundaram and Velraj 2011), TELM (Bebawy et al. 2005), CANC (Mehta and Morge 2008), and OLMM (Bari and Rote 2009, Kadukar et al. 2009, Moussa et al. 2011, Shah et al. 2007b,). At the same time, some of the fixed mixtures were analyzed by the other analytical techniques, such as first-derivative spectrophotometry (Bebawy et al. 2005, Shah et al. 2001), ratio derivative spectrophotometry (Bebawy et al. 2005), spectrofluorimetry (Bebawy et al. 2005), and HPLC (Bari and Rote 2009). Details about the applied chromatographic working conditions for the separation of LOS, VAL, IRB, TELM, CANC, and OLMM are shown in Table 26.5. The details valid for the analytical data of LOS (Shah et al. 2001), VAL, IRB, TELM, and CANC are shown in Table 26.4, and those valid for the separation of OMLM and LOS (Sathe and Bari 2007) are given in Table 26.6.

Apart from densitometry used for the determination of LOS and hydrochlorothiazide, Shah et al. (2001) additionally developed the first-derivative spectrophotometric method, and both methods were optimized and fully validated. The assay results obtained using both methods and valid for the two drugs in the combined dosage forms were compared by applying the 2-tailed *Student's t* test. There was no significant difference revealed in the content of LOS and hydrochlorothiazide determined by the two methods.

Bebawy et al. (2005) applied the four optimized and validated methods (which were the first-derivative and the ratio derivative spectrophotometry, spectrofluorimetry, and densitometry) for the simultaneous determination of TEL and hydrochlorothiazide from the dosage formulations. Spectrofluorimetric method was found as more sensitive than the remaining ones, so it can be applied to the determination of TEL in plasma.

TABLE 26.6

Analytical Parameters of the Methods Used for Separation of ARBs and Calcium Channel Blockers

Drug	Calibration Mode/Range	LOD Unit/Spot	LOQ Unit/Spot	Precision (RSD%)	Recovery (%)	Matrix	References
OLMM	Linear, 0.5–75 µg/spot	170 ng	150 ng	1.3–1.5	99–100.1	Tablets	Shah et al. (2007b)
OLMM	Linear, 0.2–0.6 µg/spot	20 ng	60 ng	—	98.7–101.3	Tablets	Bari and Rote (2009)
OLMM	Linear, 0.1–0.6 µg/spot	20 ng	60 ng	Mean 0.37	99.9–100.8	Tablets	Kadukar et al. (2009)
OLMM	Linear, 0.5–10 µg/spot	138 ng	460 ng	Mean 1.2	99.3–101.3	Bulk drug and tablets	Moussa et al. (2011)
LOS	Linear, 1–5 µg/spot	188 ng	570 ng	0.5–1.1	99.3–100.2	Bulk drug and tablets	Sathe and Bari (2007)
NIF	Linear, 0.18–0,72 µg/spot	20 ng	40 ng	0.5–0.6	98.1–100.9	Tablets and capsules	Patravale et al. (2000)
NIF	Polynomial, 0.5–3 µg/spot	50 ng	100 ng	Mean 3.5	Mean 100.4	Tablets and plasma	Kowalczuk et al. (2006)
NIT	Polynomial, 0.5–3 µg/spot	25 ng	75 ng	Mean 1.9	Mean 97	Tablets	Kowalczuk (2006)
NIT	Linear, 0.1–1 µg/spot	—	100 ng	1.9–3.9	96.5–100.6	Bulk drug	Tipre and Vavia (2001)
NIS	Linear, 0.4–2.4 µg/spot	5 ng	13 ng	1.1–2.1	99.4–100	Bulk drug and tablets	Gupta et al. (2010)
NIC	Linear, 0.1–0.4 µg/spot	—	—	Mean 0.5	98–101	Capsules	A-Elghany et al. (1997)
AML	Linear, 0.2–1.2 µg/mL	0.2 ng		Mean 1.6	Mean 100.1	Tablets	Chandrashekhar et al. (1994)
ISR	Linear, 0.04–0.36 µg/spot	—	—	0.6–1.5	99.5–101	Tablets	A-Elghany et al. (1996)
NIS	Polynomial, 0.015–0.15 µg/spot	—	—	2.5–3.1	Mean 101.2	Bulk drug and tablets	Agbaba et al. (2004)
AML	Linear, 0.002–0.1 µg/spot	—	—	4–5.5	94–102	Plasma	Pandya et al. (1995)
LAC	Linear, 0.01–0.08 µg/spot	5 ng	10	Mean 1.0	88–102	Urine	Kharat et al. (2002)
NIF	Linear, 2.3–7.6 µg/mL	—	—	0.7–1.1	99–100	Tablets	Ramteke et al. (2010)
NIT	Linear, 1.6–4.0 µg/mL	46 ng	140 ng	—	97–99	Tablets	Argekar and Sawant (1999)
AML	Linear, 2.0–10.0 µg/spot	1 ng	3.2 ng	0.9–1.1	98.1–99.7	Tablets	Gopani et al. (2011)
AML	Linear, 0.5–2.5 µg/spot	10 ng	32 ng	—	—	Tablets	Borole et al. (2011)

TABLE 26.6 (continued)

Analytical Parameters of the Methods Used for Separation of ARBs and Calcium Channel Blockers

Drug	Calibration Mode/Range	LOD Unit/Spot	LOQ Unit/Spot	Precision (RSD%)	Recovery (%)	Matrix	References
AML	Linear, 0.2–0.8 µg/spot	20 ng	80 ng	—	99.7–100.2	Tablets	Meyyanathanand and Suresh (2005)
AML	Linear, 0.1–0.6 µg/spot	50 ng	100 ng	1.5–1.8	98.6–100.4	Tablets	Dhaneshwar et al. (2009)
AML	Linear, 0.5–4 µg/spot	—	—	Mean 0.6	Mean 99.8.6	Tablets	Ramadan et al. (2010)
AML	Linear, 0.2–1.4 µg/spot	80 ng	150 ng	0.8–1.3	99.7–100.6	Tablets	Desai et al. (2010)
AML	Linear, 0.1–1 µg/spot	50 ng	100 ng	0.6–1.5%	97%–99%	Tablets	Kamble et al. (2010)
AML	Polynomial, 0.4–1.4 µg/spot	53 ng	160.8 ng	0.4–0.7	100.7–101.5	Tablets	Vekariya et al. (2009)
AML	Linear, 0.5–2 µg/spot	103 ng	312 ng	Mean 1.8	98.5–99.1	Tablets	Dangi et al. (2010)

Two methods, HPTLC and RP-LC, were also optimized and validated for the determination of OLMM and hydrochlorothiazide in the combined dosage forms and found as suitable for the routine quality control of these two drugs (Bari and Rote 2009).

Moussa et al. (2011) described the TLC densitometric method, which can be used for the simultaneous assay of OLMM and hydrochlorothiazide in the presence of the OLMM degradation products, which appeared as a result of the acid- and base-catalyzed stress stability experiments. The advantage of the method was easy performance, reproducibility, and the lack of a complicated pretreatment prior to the analysis.

Simultaneous determination of LOS, atenolol, and hydrochlorothazide (Sathe and Bari 2007) was performed by means of TLC. The chromatogram of the separated substances is shown in Figure 26.4.

26.2.3 In Situ Separation and Retention Behavior of ARBs Inhibitors

Retention behavior of LOS, VAL, EPR, CAN, and TELM has been investigated using the normal-phase (Inglot et al. 2007) and the (Inglot et al. 2009) RP-TLC mode.

Normal-phase chromatography was tested on the silica gel $60F_{254}$, aluminum oxide $60F_{254}$ (normal type E), NH_2F_{254}, CNF_{254}, and diol F_{254} stationary phases. After sample application, the chromatograms were developed in the horizontal chamber, and the spots were detected by means of the UV fluorescence quenching at 254 nm and by video densitometry. To study the retention mechanism with these ARBs on different stationary phases, binary mixtures of the six modifiers were used, and the linearity of the *Snyder–Soczewinski* relationship was examined. The best separations were obtained on the diol plates with hexane–dioxane–formic acid, 3:7:0.1 (v/v) as mobile phase. For the investigated substances, the following detection limits were found: using video densitometry, 200 ng/spot for VAL, EPR, and CAN, 100 ng/spot for TEL, and 500 ng/spot for LOS; using classical densitometry, 200 ng/spot for VAL, EPR, and CAN and 100 ng/spot for TEL and LOS (Inglot et al. 2007).

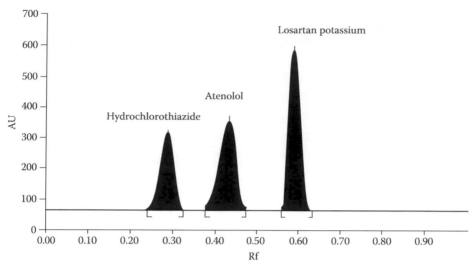

FIGURE 26.4 Typical chromatogram obtained from losartan potassium ($R_F = 0.60$), atenolol ($R_F = 0.43$), and hydrochlorothiazide ($R_F = 0.29$). Detection was at $\lambda = 274$ nm and the mobile phase was toluene–methanol–triethylamine 6.5:4:0.5 v/v. (From Sathe, S.R. and Bari, S.B., *Acta Chromatogr.*, 19, 270, 2007. With permission from Akadémiai Kiadó.)

The retention behavior of the same ARBs was investigated in the RP systems with the $RP8F_{254}$ and $RP18F_{254}$ stationary phases. After sample application, the chromatograms were developed in the horizontal chamber, and the spots were detected by the UV fluorescence quenching at 254 nm and by videodensitometry. To study the mechanism of retention with these ARBs on different stationary phases, mobile phases comprising the 3:7 (v/v) mixtures of phosphate buffer at different (2–8) pH values and the eight polar modifiers were tested. Although all examined *sartans* have ionizable groups, the change of pH exerted no significant effect on their retention. The best spot/peak shapes were achieved at pH 5. Similar drug retention patterns were observed for one and the same mobile phase and the RP8 and RP18 stationary phases. For a number of the binary mobile phases, satisfactory correlation between the R_M values and the mobile phase composition was obtained. Separation of the five investigated *sartans* was also achieved using the dimethyl sulfoxide–phosphate buffer at pH 5 (8:2, v/v) as mobile phase. Depending on the detection system used, the following detection limits were found for the investigated substances: using videodensitometry, 200 ng/spot for VAL, EPR, and CAN, 100 ng/spot for LOS, and 500 ng/spot for TEL; using classical densitometry, 100 ng/spot for VAL, LOS, and CAN and 200 ng/spot for TEL and EPR (Inglot et al. 2009).

26.3 CALCIUM CHANNEL BLOCKERS

Calcium ion plays an important role in the regulation of many cellular processes, such as synaptic transmission and muscular contraction. The role of calcium in these cellular functions is of a second messenger, for example, regulating enzymes and ion channels. The pharmacological class of drugs, known as calcium channel blockers, exerts these effects through interaction with the potential-dependent channel. To date, six functional subclasses (or types) of the potential-dependent calcium channels have been identified, that is, the T, L, N, P, Q, and the R type. L-type is located in skeletal, cardiac, and smooth muscles, causing contraction of the muscle cells. Calcium blockers act on the L-type channel only to produce their pharmacological effect. Specific calcium channel blockers interact with the specific binding sites of the calcium channel protein, leading to the selective inhibition of calcium influx through the cell membranes or affect the release and binding of calcium in the intracellular pools. Depending on interactions of the different calcium blockers, they are used to treat hypertension, angina pectoris, and certain specific types of arrhythmias (Mutschler and Derendorf 1995).

Calcium channel blockers have diverse chemical structures, and they can be grouped according to one of the following chemical classifications (Figure 26.5), each of them producing a distinct pharmacological profile: 1,4-dihydropyridines (1,4-DHPs; e.g., nifedipine (NIF) and others); benzothiazepines (e.g., diltiazem [DIL]); phenylalkylamines (e.g., VER, gallopamil, prenylamine [PRE], terodiline [TER], and fendiline [FEN]); diaminopropanol ethers (e.g., bepridil [BEP]); piperazine (e.g., cinnarizine [CIN], flunarizine, and lidoflazine [LID]); and a non-recognized group classified as perhexiline.

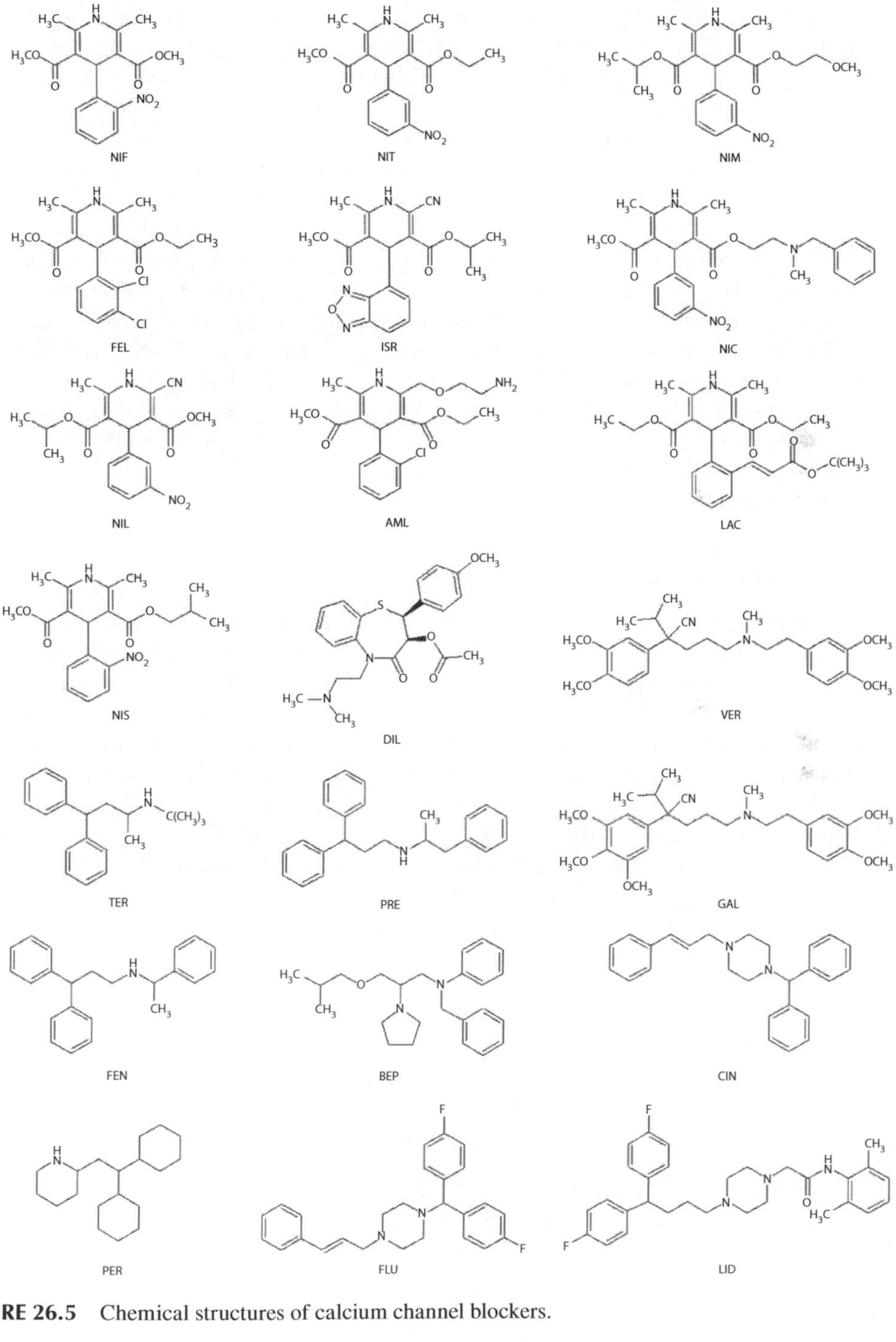

FIGURE 26.5 Chemical structures of calcium channel blockers.

The nomenclature of the currently used calcium channel blockers is as follows **NIF**, 1,4-dihydro-2,6-dimethyl-4-(2-nitrophenyl)-3,5-pyridinedicarboxylic acid dimethyl ester, **nimodipine** (NIM), 1,4-dihydro-2,6-dimethyl-4-(3-nitrophenyl)-3,5-pyridinedicarboxylic acid 2-methoxyethyl 1-methylethyl ester, **nitrendipine** (NIT), 1,4-dihydro-2,6-dimethyl-4-(3-nitrophenyl)-3,5-pyridinedicarboxylic acid ethylmethyl ester, **felodipine** (FEL), 4-(2,3-dichlorophenyl)-1,4-dihydro-2,6-dimethyl-3,5-pyridinedicarboxylic acid ethylmethyl ester, **nilvadipine** (NIL), 2-cyano-1,4-dihydro-6-methyl-4-(3-nitrophenyl)-3,5-pyridinedicarboxylic acid 3-methyl 5-(1-methylethyl) ester, **isradipine** (ISR), 4-(4-benzofurazanyl)-1,4-dihydro-2,6-dimethyl-3,5-pyridinedicarboxylic acid methyl 1-methylethyl ester, **nicardipine** (NIC), 1,4-dihydro-2,6-dimethyl-4-(3-nitrophenyl)-3,5-pyridinedicarboxylic acid methyl 2-[methyl(phenyl methyl)amino]ethyl ester, **nisoldipine** (NIS), 1,4-dihydro-2,6-dimethyl-4-(2-nitrophenyl)-3,5-pyridinedicarboxylic acid methyl 2-methylpropyl ester, AML, 2-[(2-aminoethoxy)methyl]-4-(2-chlorophenyl)-1,4-dihydro-6-methyl-3,5-pyridinedicarboxylic acid 3-ethyl 5-methyl ester, **lacidipine** (LAC), (E)-4-[2-[3-(1,1-dimethylethoxy)-3-oxo-1-propenyl]phenyl]-1,4-dihydro-2,6-dimethyl-3,5-pyridinedicarboxylic acid diethyl ester, **DIL**, (2S-cis)-3-(acetyloxy)-5-[2-(dimethylamino)ethyl]-2,3-dihydro-2-(4-methoxyphenyl)-1,5-benzothiazepin-4(5H)-one, **VER**, α-[3-[[2-(3,4-dimethoxyphenyl) ethyl]methylamino]propyl]-3,4-dimethoxy-α-(1-methylethyl)benzeneacetonitrile, **gallopamil** (GAL), α-[3-[[2-(3,4-dimethoxyphenyl) ethyl]methylamino]propyl]-3,4,5-trimethoxy-α-(1-methylethyl)benzeneacetonitrile, **PRE**, N-(1-methyl-2-phenylethyl)-γ-phenylbenzenepropanamine, **TER**, N-(1,1-dimethylethyl)-α-methyl-γ-phenyl- benzenepropanamine, **FEN**, γ-phenyl-N-(1-phenylethyl)benzenepropanamine, **BEP**, β-[(2-methylpropoxy) methyl]-N-phenyl-N-(phenylmethyl)-1-pyrrolidine-ethanamine, **CIN**, 1-(diphenylmethyl)-4-(3-phenyl-2-propenyl)piperazine, **flunarizine** (FLU), (E)-1-[bis(4-fluorophenyl)methyl]-4-(3-phenyl-2-propenyl)piperazine, **LID**, 4-[4,4-bis(4-fluorophenyl)butyl]-N-(2,6-dimethylphenyl)-1-piperazine-acetamide, and **perhexiline** (PER), 2-(2,2-dicyclohexylethyl)piperidine.

26.3.1 CHEMISTRY AND PROPERTIES OF 1,4-DHPs

The 1,4-DHPs ring is essential for the drugs' activity; its substitution at the N_1 position decreases or eliminates this activity. Phenyl (present in all except ISR) or heteroaryl (ISR) substitution at the C_4 position is important for the size and the position rather than for electronic effects. *Ortho* or *meta* substituted compounds possess an optimal activity. The importance of the *ortho* and *meta* substituents is to provide a sufficient bulk to *lock* the conformation of the 1,4-DHPs such that the C_4 aromatic ring is perpendicular to the 1,4-DHP ring. This conformation has been suggested as essential for the activity of 1,4-DHPs. Ester groups at the C_3 and C_5 positions optimize the activity. With an exception of AML, all other 1,4-DHPs have methyl groups at C2 and C6. An enhanced potency of AML versus NIF suggests that 1,4-DHPs can tolerate larger substituents at this position and that an enhanced activity can be obtained by altering these groups. The first and the last launched 1,4-DHPs are NIF and AML, and they basically differ in the elimination half-life, 2–5 versus 36 h, respectively (Cutler 2011b, Harrold 2008).

Photoreactivity is the most avoidable property of 1,4-DHPs, which can induce molecular changes able to decrease the therapeutic effect and even exert certain toxic effects upon administration. Photo-induced changes of 1,4-DHPs involve oxidation of the dihydropyridine ring to the pyridine ring and reduction of the aromatic nitro group to the nitroso group. Dehydrogenated nitro or nitroso degradants of the corresponding 1,4-DHPs are either the main product of photodegradation running in vitro, or the main metabolites observed in vivo (Roth and Fenner 2000).

26.3.1.1 In Situ Determination of Single 1,4-DHPs

In situ investigations of NIF, NIT, NIS NIC, AML, ISR, and LAC either as single drugs or in the presence of the degradation products in the bulk and the dosage formulations, as well as in biological samples, have been reported (A-Elghany 1997, A-Elghany et al. 1996, Agbaba et al. 2004, Chandrashekhar et al. 1994, Gupta et al. 2010, Kharat et al. 2002, Kowalczuk 2006,

Kowalczuk et al. 2006, Pandya et al. 1995, Patravale et al. 2000, Tipre and Vavia 2001). Simple densitometric methods were developed and validated for determination of NIS and AML in the absence of impurities (Chandrashekhar et al. 1994, Gupta et al. 2010); NIF, NIT, NIC, and ISR, in the presence of impurities originating from stress stability studies (A-Elghany et al. 1996, 1997; Kowalczuk 2006; Kowalczuk et al. 2006; Patravale et al. 2000; Tipre and Vavia 2001); NIS and impurities originating from the synthetic pathways and the degradants (Agbaba et al. 2004) in pharmaceuticals. Kinetics of NIT degradation was measured, followed by the isolation and structure elucidation of the degradation product (Tipre and Vavia 2001). The developed method for the NIF assay in pharmaceuticals was extended to the plasma samples (Kowalczuk et al. 2006). AML and LAC were assayed using the developed HPTLC methods in the plasma and urine samples (Kharat et al. 2002, Pandya et al. 1995). The oxidative degradation study of NIT was followed with use of the developed HPTLC, HPLC, and spectrophotometric methods. Details regarding the chromatographic working conditions valid for the separation of the aforementioned calcium channel blockers are shown in Table 26.7, and the analytical data are given in Table 26.6.

Patravale et al. (2000) developed a method for the in situ determination of NIF, which can be used as stability indicating. After exposing the standard NIF solution for 2 h to the diffused sunlight and using the established chromatographic procedure, an additional spot was observed eluted earlier than NIF and corresponding to its nitrosopyridine analog (the degradation product). This spot was absent from the three different dosage formulations tested.

Kowalczuk et al. (2006) developed a method for the determination of NIF in dosage formulations and for its identification in the plasma samples, in presence of the other calcium channel blockers, NIC, NIT, NIM, and NIT. At the same time, using this method, the authors managed to show that photodegradation (sunlight and UV irradiation) and oxidation (3% hydrogen peroxide) almost entirely (~99%) decomposed the NIF samples. Chemical degradation of NIF under the acidic and alkaline conditions (~90% and ~85%, respectively) was also performed, but its yields were lower than that of photodegradation. The method for the determination of NIT developed by Kowalczuk (2006) can also be used for its identification in the presence of induced degradation products. Pure drug and tablet extract subjected to acidic and alkaline hydrolysis, oxidation, UV degradation, and photodegradation indicated high susceptibility of NIT to these factors.

The oxidative degradation study of NIT by using stability-indicating and originally developed HPLC, HPTLC, and spectrophotometric methods was reported by Tipre and Vavia 2001. At 100°C, drug degradation was found faster in the acidic (0.1 M HCl) medium than in the alkaline (0.1 M NaOH) one. Photodegradation rates were studied with special emphasis laid on the effect of solvents, such as methanol, chloroform, dichloromethane, acetone, and ethyl acetate. In all examined cases, degradation of NIT followed the first-order kinetics, and its major route was established as oxidation, with the degradation product dehydronitrendipine, as confirmed with use of UV, IR, and ^{1}H NMR spectroscopy. Compared with the other developed HPLC and spectrophotometric methods, HPTLC was found sufficiently sensitive and robust.

The degree of NIC degradation, when exposed to acid, base, peroxide, and the short-wavelength UV light, has been investigated using different molarities, solvent strengths, temperatures, and exposure durations. It came out that NIC was degraded in 20%, when exposed to the 254 nm UV light for 4 h and more than in 70%, when the exposure lasted for 24 h (A-Elghany 1997). HPTLC method for the assay of ISR in the presence of acid, base, and hydrogen peroxide was developed and validated (A-Elghany et al. 1996).

Agbaba et al. (2004) established the TLC method for monitoring of the reactant and the reaction side products formed in the process of synthesis and the UV and daylight degradation of NIS as a bulk drug and in pharmaceuticals. The densitogram of the separated substances is shown in Figure 26.6. The proposed method was fully validated and it can be applied for simultaneous determination of NIS and the corresponding impurities, 2-isopropyl-2-(2-nitrobenzylidene) acetoacetate (II-reaction partner), NIF (IV-side reaction product), and the nitrosophenylpiridine analog of NIS

TABLE 26.7

Chromatographic Separation Conditions Used for Calcium Channel Blockers

Drug	Plates	Stationary Phase Pretreatment; Mobile Phase Composition	Development Mode	Detection (nm)	Rf	References
NIF	HPTLC silica gel 60F$_{254}$, aluminum-backed	Chloroform–ethyl acetate–cyclohexane, 19:2:2 v/v	Ascending	238	0.3	Patravale et al. (2000)
NIF	HPTLC silica gel 60F$_{254}$	n-hexane–ethyl acetate–acetone, 6:3:2 v/v	Horizontal	335	0.45	Kowalczuk et al. (2006)
NIT	HPTLC silica gel 60F$_{254}$	n-hexane–ethyl acetate–acetone, 6:3:2 v/v	Horizontal	335	0.6	Kowalczuk (2006)
NIT	HPTLC silica gel 60F$_{254}$	Ethyl acetate–chloroform, 1:9 v/v	Ascending	254	0.7	Tipre and Vavia (2001)
NIS	TLC silica gel 60F$_{254}$, aluminum-backed	Cyclohexane–ethyl acetate–toluene, 3:3:4 v/v	Chamber saturation 20 min; ascending	320	0.4	Gupta et al. (2010)
NIC	HPTLC silica gel 60F$_{254}$	Methanol, activation at 120°C, 30 min; n-hexane–ethyl acetate–ethanol–ammonia (25%) 30:10:5:1 v/v	Chamber saturation 10 min; ascending	280	0.3	A-Elghany et al. (1997)
AML	HPTLC silica gel 60F$_{254}$	Methanol; chloroform–acetic acid–toluene–methanol, 80:10:10:10 v/v	Chamber saturation 60 min; ascending	366	—	Chandrashekhar et al. (1994)
ISR	HPTLC silica gel 60F$_{254}$	Methanol, activation at 120°C, 30 min; chloroform–aqueous ammonia (25%) 500:10:1 v/v	Chamber saturation 10 min; ascending	325	0.4	A-Elghany et al. (1996)
NIS	HPTLC Li chrospher Si60F$_{254}$	Cyclohexane–ethyl acetate–toluene, 7.5:7.5:10 v/v	Chamber saturation 15 min; ascending	280	0.35	Agbaba et al. (2004)
AML	TLC silica gel 60F$_{254}$, aluminum-backed	Chloroform–methanol–acetic acid, 15:2.5:0.4 v/v	Chamber saturation; ascending	365	0.3	Pandya et al. (1995)
LAC	TLC silica gel 60F$_{254}$, aluminum-backed	Toluene–ethyl acetate 6.5:3.5 v/v	Chamber saturation 10 min; ascending	287	0.45	Kharat et al. (2002)
NIF	Kieselghur GF254TLC, aluminum-backed	Cyclohexane–methanol–ethyl acetate–ammonia v/v, 5:1:3:0.5	Ascending	230	0.7	Ramteke et al. (2010)
NIT	HPTLC silica gel 60F$_{254}$	Choroform–methanol–toluene–ammonia (25%), 2:2.5:5.5:0.1 v/v	Ascending	233	0.5	Argekar and Sawant (1999)
AML	TLC silica gel 60F$_{254}$, aluminum-backed	Methanol, activation at 110°C, 5 min; methanol–ethyl acetate–ammonium sulphate (0.2%) 3:6:4 v/v/	Chamber saturation 30 min, 25°C; ascending	218	0.7	Gopani et al. (2011)

TABLE 26.7 (continued)

Chromatographic Separation Conditions Used for Calcium Channel Blockers

Drug	Plates	Stationary Phase Pretreatment; Mobile Phase Composition	Development Mode	Detection (nm)	Rf	References
AML	HPTLC silica gel 60F$_{254}$	Prewashing; dichloromethane–methanol–glacial acetic acid 8.5:1.5:0.1 v/v/v	—	238	0.2	Borole et al. (2011)
AML	HPTLC silica gel 60F$_{254}$, aluminum-backed	Methanol; ethyl acetate:methanol:ammonia (10%) 8.5:2:1 v/v	Ascending	362	0.6	Meyyanathan and Suresh (2005)

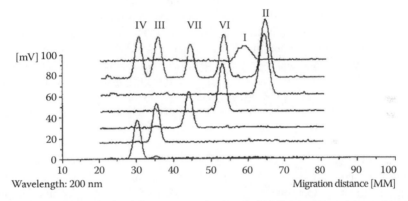

FIGURE 26.6 Densitograms obtained from a test mixture of nisoldipine (III) and impurities I, II, IV, VI, and VII. (With kind permission from Springer Science + Buisness Media: *Chromatographia* Determination of nisoldipine and its impurities in pharmaceuticals, 60, 2004, 223, Agbaba, D., Vučićević, K., and Marinković, V.)

(VI-photo degradant). The obtained LODs and LOQs for impurities II, IV, and VI were 0.06%, 0.04%, and 0.008%; and 0.2%, 0.15%, and 0.03%, respectively. Densitometry with very sensitive fluorimetric detection was applied to the assessment of AML in the tablets. The minimum detectable limit for the proposed method was 0.2 ng (Chandrashekhar et al. 1994).

The same detection mode was used for the HPTLC determination of the total (both free and bonded) AML concentration in plasma when the drug was dispensed to healthy volunteers in the therapeutic dose of 10 mg. This method was reported as economical and faster than the previously published methods, that is, on a single plate at least 10–12 samples can be analyzed in 5–6 h. The method was used to obtain comparative pharmacokinetic information about the drug in healthy volunteers for the two market tablet formulations (Pandya et al. 1995). Kharat et al. (2002) proposed a method for the routine TLC analysis of LAC in pharmaceuticals and for the pharmacokinetic studies in human urine samples. Densitometry was performed after the SPE from the urine samples. The low LOQ values of 8 ng/spot, extraction recoveries of 95% at the lower (10 ng) and the higher (60 ng) concentrations, and the intra- or inter-day RSD values lower than 1% enable reliable application of this method for quantification of LAC in the urine samples.

26.3.1.2 In Situ Determination of 1,4-DHPs in Presence of Other Drugs

NIF and NIT were investigated in situ, in the presence of beta blockers (atenolol) (Argekar and Sawant 1999, Ramteke et al. 2010). As the third-generation 1,4-DHP drug, AML was widely investigated in situ, in the fixed pharmaceutical dosage formulations, that is, in the presence of *prils* (Borole et al. 2011, Gopani et al. 2011, Meyyanathan and Suresh 2005), *sartans* (Desai et al. 2010, Dhaneshwar et al. 2009, Kamble et al. 2010, Ramadan et al. 2010, Vekariya et al. 2009), beta blockers (nebivolol, atenolol, and metoprolol) (Argekar and Powar 2000, Dangi et al. 2010, Kakde and Bawane 2009), and *statins* (atorvastatine) (Patel et al. 2011). Details regarding the chromatographic working conditions valid for the separation of the aforementioned calcium channel blockers are shown in Tables 26.7 and 26.8, and the analytical data are given in Tables 26.6 and 26.9.

26.3.1.3   In Situ Identification of 1,4-DHPs

Marciniec et al. (2002) investigated the effect of gamma and beta radiation in the doses between 10 and 100 kGy on the physicochemical properties of the four 1,4-DHP derivatives, NIF, NIT, FEL, and NIT, in solid state. Irradiated and nonirradiated substances, all in solid state, were examined, simultaneously using organoleptic, chromatographic (TLC and HPLC), spectroscopic (UV, IR, and EPR), microbiological (sterility test), and the other (DSC) methods. The purity profiles of the irradiated and the nonirradiated substances were tested by means of TLC. After having the samples applied to the silica gel $60F_{254}$ plates, the chromatograms were developed in the benzene–methanol (6:1, v/v) mobile phase. Spots of the tested 1,4-DHPs were detected in UV light at 254 nm. The results revealed that the small doses of the ionizing radiation (10–25 kGy) showed sufficient sterilizing properties, yet neither altered the physicochemical properties nor decomposed the 1,4-DHPs tested. The resistance of the derivatives tested toward the gamma radiation increased in the order FEL < NIM < NIT < NIF, whereas that toward the beta radiation in the order FEL < NIM < NIF < NIT. It was concluded that small doses of gamma radiation (10–15 kGy) could be used for sterilization of the tested 1,4-DHPs in order to get them microbiologically pure.

Marciniec and Ogrodowczyk (2006) investigated the effect of temperature and air humidity on the stability of seven 1,4-DHP derivatives (NIF, NIS, NIT, NIM, NIC, FEL, and AML) in the solid state by accelerated testing. Identification of 1,4-DHPs, and their thermodegradation products and the reference standards was made by TLC, UV spectroscopy, and the reaction with $KMnO_4$, and quantification was made spectrophotometrically. The TLC analysis was performed on the silica gel-coated Kiselgel 60F254 plates, using benzene–methanol as mobile phase. The chromatograms were assessed at 254 nm. Thermodegradation of the examined derivatives was not taking place in dry air, whereas in humid air, it occurred as the first-order reaction, with the similar rates for all derivatives. A comparative analysis of the obtained results suggests that thermal stability of the compounds of interest depends on their chemical structure and in particular, on position of the nitro group. The derivatives with the nitro group in *meta* position (NIC, NIT, and NIM) are much more resistant to the temperature effect than those with the nitro group in *ortho* position (NIS and NIF). The main products of thermodegradation were the corresponding nitroso derivatives, formed as a result of aromatization of the dihydropyridine ring, accomplished by elimination of the water molecule. Compounds devoid of this group (FEL and ALM) are more thermally stable than NIS and NIF, but their stabilities differ from those of the other derivatives. It was also observed that the selected 1,4-DHPs in the solid state were highly stable when stored in the dark.

Mielcarek and Daczkowska (1999) studied photochemical decomposition of ISR and its inclusion complexes with methyl-BCD by means of HPTLC, RP-HPLC, and UV spectroscopy. Identification of the complexes and the photodegradation products was performed on the silica gel 60 F_{254} plates, using hexane–acetone–ethyl acetate (4:2.5:2.2, v/v) as mobile phase in the ascending chromatography mode. The chromatograms were detected at 254 nm. The inclusion complexes of ISR with methyl-BCD were found twice more photo stable than ISR alone.

Marciniec and Brzeska (1999) investigated the condition for the differential analysis with five 1,4-DHP derivatives (NIF, NIT, NIS, NIM, and NIC), using TLS, DSC, UV, and IR spectroscopy. The TLC studies were performed on silica gel $60HF_{254}$, using different single or mixed solvents. Successful separation of all five compounds was obtained in the following solvents/solvent systems:

TABLE 26.8

Chromatographic Separation Conditions Used for Calcium Channel Blockers

Drug	Plates	Stationary Phase Pretreatment; Mobile Phase Composition	Development Mode	Detection (nm)	Rf	References
AML	TLC silica gel 60F$_{254}$, aluminum-backed	Methanol, activation at 110°C, 5 min; toluene:methanol:acetic acid, 7:3:0.1 v/v	Chamber saturation 30 min; ascending	244	0.4	Dhaneshwar et al. (2009)
AML	TLC silica gel 60F$_{254}$	Mobile phase, activation at 100°C, 15 min; ethyl acetate–methanol–ammonim hydroxide, 55:45:5	Chamber saturation 15 min; ascending	237	0.3	Ramadan et al. (2010)
AML	TLC silica gel 60F$_{254}$, aluminum-backed	Chloroform:methanol:toluene: acetic acid, 8:1:1:0.1 v/v	Ascending	254	0.15	Desai et al. (2010)
AML	TLC silica gel 60F$_{254}$, aluminum-backed	Methanol, activation at 110°C, 5 min; n-butanol–acetic acid–water, 5:1:0.1 v/v	Chamber saturation 30 min; ascending	254	0.3	Kamble et al. (2010)
AML	TLC silica gel 60F$_{254}$, aluminum-backed	Tetrahydrofuran:dichloroetha ne:methanol:ammonia solution, 3:1:0.5:0.2	Chamber saturation 30 min; ascending	326	0.4	Vekariya et al. (2009)
AML	TLC silica gel 60F$_{254}$, aluminum-backed	Methanol, activation at 110°C, 5 min; ethyl acetate–methanol–ammonia v/v, 8.5:1:1	Chamber saturation 30 min; ascending	240	0.4	Dangi et al. (2010)
AML	HPTLC silica gel 60F$_{254}$, aluminum-backed	Methylene chloride– methanol–ammonia (25%), 8.8:1.3:0.1 v/v	Ascending	230	0.3	Argekar and Powar (2000)
AML	HPTLC silica gel 60F$_{254}$, aluminum-backed	Methanol, activation at 115°C, 30 min; methanol–ethyl acetate–water–toluene–ammonia(25%), 1.5:5:0.3:3:0.3 v/v	Chamber saturation 20 min; ascending	236	0.4	Kakde and Bawane (2009)
AML	TLC silica gel 60F$_{254}$, aluminum-backed	Methylene chloride 1:1, activation at 70°C, 20 min; chloroform–methanol–acetic acid, 85:10:5 v/v	Chamber saturation 5 min; ascending	247	0.3	Patel et al. (2011)
DIL	TLC silica gel 60F$_{254}$, aluminum-backed	Methanol air dried; ethyl acetate–methanol–ammonia (25%), 80:10:10 v/v	Ascending	238	0.5	Devarajan and Dhavse (1998)
DIL	TLC silica gel 60F$_{254}$,	Chloroform–cyclohexane–toluene–diethylamine, 3:2:2:0.3	Ascending	240	39	Agbaba et al. (1997)
VER	TLC silica gel 60F$_{254}$	Ethyl acetate–ethanol–acetic acid, 8:2:0.5 v/v	Horizontal	215	0.3	Kowalczuk (2005)
CIN	TLC silica gel 60F$_{254}$	Methanol–dichloromethane–formic acid, 1:9:0.05 v/v	Ascending	276	0.5	Argekar and Powar (1999)

(continued)

TABLE 26.8 (continued)
Chromatographic Separation Conditions Used for Calcium Channel Blockers

Drug	Plates	Stationary Phase Pretreatment; Mobile Phase Composition	Development Mode	Detection (nm)	Rf	References
CIN	TLC silica gel 60F$_{254}$	Benzene–methanl–formic acid, 80:17:3 v/v	Ascending	250	0.4	Hassan et al. (2002)
CIN	TLC silica gel 60F$_{254}$	Chloroform–n-butanol–ethanol–methanol–acetic acid, 8:3:3:1:1 v/v (twice developed)	Ascending	237	0.4	Sugihara et al. (1984)

diethyl ether, diisopropyl ether, ethyl acetate-chloroform (3:1, v/v), and ethyl acetate-cyclohexane (6:5, v/v). The chromatograms were assessed at 254 nm. The chromatographic system with diisopropyl ether as mobile phase proved particularly recommendable, due to the short development time, and the compact and well-resolved spots. Marciniec et al. (1992) used TLC for the separation of NIF from 10 different marketed pharmaceutical formulations, followed by the elution of NIF and further subjecting it to UV spectrophotometric determination. Separation of NIF was performed on the Kieselgel 60F$_{254}$ pre-coated TLC plates, using benzene–methanol (6:1, v/v) as mobile phase. Quantitative analysis was performed by the originally developed TLC-UV and GC-MS methods. This work revealed that in none of the preparations, the nitro derivative as the NIF decomposition product was found, yet in each preparation, the presence of the nitroso derivative was reported.

Mielcarek (2001) proposed the TLC-based method for the separation of the FEL enantiomers and the racemic mixtures of FEL, NIL, and ISR. The enantioseparation was investigated on Chiralplates (incorporating L-proline and Cu^{2+}), using two mobile phases, containing methanol in different volume proportions (φ) as an organic modifier (phase 1, chloroform–ammonia (25%)–methanol, 10:0.02:φ; phase 2, acetonitrile–triethylamine (0.1%)–methanol, 5:3:φ). The results revealed that the processes taking place in the examined ligand exchange chromatographic systems can be described by Snyder–Soczewinski equation.

Haddad et al. (2008) described an easy ambient sonic-spray ionization-MS (EASI-MS) combined with TLC for the fast screening of the two samples: the drug tablets containing AML and the beta blocker, propranolol. This combined technique was found beneficial for the multidrug tablets (to avoid ion suppression), or for the forensic fingerprinting analysis of the counterfeit tablets, to characterize the known and unknown impurities contained therein.

Hemmateenejad et al. (2010) developed a multivariate image analysis TLC (MIA-TLC) for the simultaneous determination of co-eluting components. An imaging system composed of a dark cabinet, a digital camera, and a MIA program was prepared to record images of the TLC plates after the development of a multi-component solution. The devised program was able to produce the two-dimensional (2D) and three-dimensional chromatograms, which were then used as inputs for the partial least squares as an efficient multivariate method. Simultaneous determination of NIF and its degradation product designed as P-NIF was performed by EASI-MS. Two binary solvent systems were used for the separation of NIF and P-NIF: (1) chloroform–ethyl acetate (40:10, v/v), which fully resolved the two compounds followed by the determination thereof, using the conventional univariate calibration model, and (2) chloroform–ethyl acetate (10:40, v/v), unable to completely resolve the spots. The overlapping spots were resolved by means of multivariate calibration. The 3D chromatograms of a typical NIF and P-NIF mixture after the elution are shown in Figure 26.7.

The experiments aimed for multivariate calibration consume less toxic mobile phase and need a shorter time of analysis. Works in this area are still being developed.

TABLE 26.9

Analytical Parameters of the Methods Used for Separation of Calcium Channel Blockers, Alpha 1 Adrenoceptor Antagonists, Ganglion Blockers, and Alpha 1 Agonists

Drug	Calibration Mode/Range	LOD Unit/Spot	LOQ Unit/Spot	Precision (RSD%)	Recovery (%)	Matrix	Reference
AML	Linear, 0.1–0.5 μg/spot	20 ng	60 ng	0.9–1.7	—	Tablets	Argekar and Powar (2000)
AML	Linear, 18–28 μg/mL	10 ng	20 ng	Mean ± 0.24	98.9–99.6	Tablets	Kakde and Bawane (2009)
AML	Linear, 0.1–0.8 μg/spot	60 ng	100 ng	Mean 1.8	99.8–102.4	Tablets	Patel et al. (2011)
DIL	Linear, 0.04–0.4 μg/spot	20 ng	40 ng	1.5–2.1	95.0–99.0	Bulk drug and tablets	Devarajan and Dhavse (1998)
DIL	Polynomial, 1–6 μg/spot	—	—	1.7–1.9	99.7–101.2	Bulk drug and tablets	Agbaba et al. (1997)
VER	Linear, 0.5–15 μg/spot	150 ng	500 μg	0.5–3.1	Mean 97.1%	Capsules	Kowalczuk (2005)
CIN	Linear, 0.06–1 μg/spot	29 ng	87 ng	—	97.0–99.7	Tablets	Argekar and Powar (1999)
CIN	Linear, 4–30 μg/mL	32 ng	—	Mean 0.6	Mean 98.6	Tablets, serum	Hassan et al. (2002)
TER	Linear, 0.2–2.2 μg/spot	—	—	—	Mean 98.6	Tablets	El Bayoumi et al. (2009)
DOX	Linear, 0.1–0.7 μg/spot	—	—	Mean 0.7	95%–99%	Tablets	Sane et al. (2002)
RES	Linear, 0.2–1.6 μg/spot	6.3 ng	19	Mean 0.52	99.3%–99.7%	Tablets	Deshmukh et al. (2011)
RES	Linear, 0.1–0.6 μg/spot	30 ng	100 ng	—	Mean 97.0	Tablets	Argekar et al. (1996)
PHE	Polynomial, 0.2–6 mg/mL	100 ng	—	Mean 2.7	96–102.5	Tablets	El Sadek et al. (1990)
PHE	Linear, 1–4 μg/spot	—	—	—	98.8–101.2	Tablets	Greshock and Sherma (1997)
PHE	Linear, 4–9 μg/spot	2 μg	—	1–1.3	99.7–100.4	Solution, drug delivery system	Devarajan et al. (2000)

26.3.2 CHEMISTRY AND PHARMACOLOGY OF BENZOTHIAZEPINES AND PHENYLALKYLAMINES

DIL is the only 1.5-benzotiazepine derivative clinically in use. Very few SAR studies are available for DIL. However, it is known that the *cis* arrangement of the acetyl ester and the substituted phenyl ring are responsible for its activity. DIL is a weak base (pKa = 7.7) and at physiological pH, it is protonated. DIL undergoes the first-pass metabolism, and one of its main metabolites is deacetyldiltiazem (DAD), which retains ca. 20%–50% of the DIL activity. VER and galopamil are derived from 2-isopropylvaleronitrile. They differ with the number of the methoxy groups (ether) on the aromatic ring. VER is a stronger base than DIL, (pKa = 8.9), and at physiological pH, it is

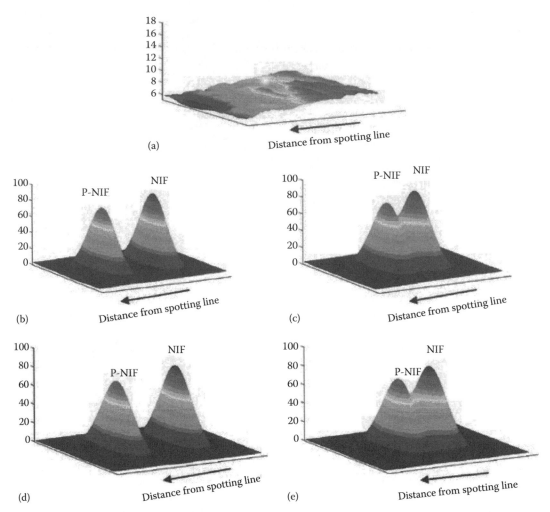

FIGURE 26.7 Three-dimensional chromatograms of the background (a) and mixtures of NIF and P-NIF before subtraction of background (b) and (c), and after subtraction of background (d) and (e). The levels of analytes in the mixture are 3.6 and 1.4 mg for NIF and P-NIF, respectively, and mobile phase in (b) and (d) is chloroform/ethyl acetate (40:10) with 10.0 min TLC development and in (c) and (e) is chloroform/ethyl acetate (10:40) with 2.0 min TLC development. (From Hemmateenejad, B., Mobaraki, N., Shakerizadeh-Shirazi, F., and Miri, R, Multivariate image analysis-thin layer chromatography (MIA-TLC) for simultaneous determination of coeluting components. *Analyst* 135, 1747, 2010. Reproduced by permission from The Royal Society of Chemistry.)

protonated. *Ortho* substitution with the methoxy group on the aromatic ring decreases its activity. Hence, the *meta* methoxy derivative of VER, that is, galopamil, is more liphophilic, slower metabolized, and has the pharmacological properties similar to VER. The S(−) enantiomers of VER and the other phenylalkylamines are more potent than the R(+) enantiomers. VER undergoes stereoselective metabolism, and the S(−) isomer undergoes a more extensive first-pass hepatic metabolism than does the less-active R(+) isomer (Cutler 2011b, Harrold 2008).

26.3.2.1 In Situ Identification and Determination of DIL, VER, and CIN

DIL (Agbaba et al. 1997, Devarajan and Dhavse 1998), VER (Kowalczuk 2005), and CIN (Argekar and Powar 1999, Hassan et al. 2002) were determined in situ from the bulk drug or pharmaceutical preparations. DIL and its metabolites were determined or identified from biological samples (plasma and urine) (Sugihara et al. 1984) or in the presence of several anti-arrhythmic drugs (Witek and Przyborowski 1996). VER was identified as a single component (Salo et al. 2005) or as one belonging to a huge set of the tested drugs (Brzezinka et al. 1999, Ojanperä 1999a). Details regarding the chromatographic working conditions valid for the separation of the aforementioned calcium channel blockers are shown in Table 26.8, and the analytical data are given in Table 26.9.

The in situ developed assay of DIL (Devarajan and Dhavse 1998) contained in pharmaceuticals could be considered as stability indicating. A single spot observed to elute faster than DIL was assumed to be the more polar DAD obtained by hydrolysis of the labile acetyl group of DIL.

Agbaba et al. (1997) developed a fully validated TLC method for the simultaneous determination of the DAD residues present as the main DIL degradant in pharmaceuticals. The polynomial regression function for DAD was in the range of 5–30 ng, the LOD and LOQ values were established as 0.2 and 0.7 ng/spot, respectively (equivalent to the impurity levels 0.02% and 0.07%), the mean recovery value of 100.1% and RSD (%) for precision equal to 1.5%, confirm the capacity of the approach as an in situ purity testing method.

For the binary mixture of CIN and domperidone present in pharmaceuticals, the in situ assay developed and reported by Argekar and Powar (1999) was validated and applied to domperidone as well.

Hassan et al. (2002) presented two methods (TLC and LC) for the determination of CIN in pharmaceuticals, in the presence of its photodegradation products and the related substances, and in the presence of its metabolites in serum. The validated method was used for the assessment of drug purity, stability, bioavailability, bioequivalency, and for the determination of the tablet dissolution rate. Four CIN-related substances and six degradation products were isolated and identified by means of IR spectroscopy and MS.

The human urinary metabolites of DIL were analyzed by means of TLC and GC-MS. An unchanged DIL and its metabolite (N-monodemethyldiltiazem) were determined in human plasma and urine by TLC densitometry. The lower limits of this DIL determination method were ca. 1 ng/mL in plasma and 50 ng/mL in urine (Sugihara et al. 1984).

The TLC analysis of DIL in the presence of the other anti-arrhythmics (flecainide, procainamide, lorcainide, lidocaine, and diisopyramide) was performed on the silica gel 60GF$_{254}$ and 60G pre-coated plates using 34 different mobile phases. The best separation of the tested drugs was achieved using ethanol–benzene–dioxane–ammonia (2:20:16:3, v/v) as the mobile phase. It was confirmed that all six investigated drugs can be detected using any of the 10 detection modes presented in the chapter. The greatest detection sensitivity with DIL (100 ng/spot) was obtained using the Dragendorff reagent or the acidified iodoplatinate reagent (Witek and Przyborowski 1996).

The analysis of small molecules by ultra TLC-atmospheric pressure matrix-assisted laser desorption/ionization MS (UTLC-AP-MALDI-MS) has been reported for several compounds, VER being one of them. Performance of AP-MALDI-MS and the vacuum MALDI-MS1 was compared in the analysis of VER directly from the UTLC plates. LOD values of the tested compound for the different chromatographic modes (in situ or after the elution) and using the different plates (UTLC/HPTLC and UTLC/HPTLC-AP) and the vacuum MALDI-MS were presented. The lowest LOQ values for VER using UTLC and the AP-MALDI-MS in situ mode were obtained (0.5 pmol) (Salo et al. 2005).

The method was reported (Brzezinka et al. 1999) of the off-line coupling TLC with the electron-impact ionization MS, which is well suited for the routine forensic and toxicological investigations of a large number (493) of samples, among them the already mentioned antihypertensive drugs (NIF, VER, etc.), and also those later to be mentioned (clonidine [CLO], phentolamine [PEN], etc.).

A comprehensive TLC drug screening in forensic toxicology, using a broad spectrum of the TLC and RP-TLC procedures for the acidic, basic, amphoteric, and quaternary drugs (including the ion-pair extraction of the hydrophilic drugs in the cationic form) was applied to the urine and liver samples from the 618 medical cases, and 104 different drugs were found in the course of this study. By plotting the RP-TLC Rf values of the drugs against the percentage of occasions when the drug was more readily observed in liver than in urine, the correlation value of −0.58 was obtained. VER was found distributed exclusively in liver, its active metabolite norverapamil occurred more readily in urine than in liver, and DIL was equally distributed in liver and urine (Ojanperä et al. 1999a).

An analytical scheme composed of one NP-TLC and one RP-TLC system, and the sequential analyte detection through the four stages of the color reactions using the commercially available Toxi-Lab set was described. Eighty-one basic drugs (including VER) were analyzed with use of this scheme, and 74 drugs were uniquely characterized with 85% confidence (Harper et al. 1989).

26.3.2.2 In Situ Analysis of Nonselective Calcium Channel Blockers, PRE, FEN, LID, and BEP

Gumieniczek et al. (1995) investigated an impact of different absorbents (silica gel and silica gel RP-18) and the mobile phase compositions on the retention of PRE and dilazep. Concerning the analysis on the NP-TLC plates, the best separation was achieved using the ascending development mode and the following mobile phases: ethyl acetate–methanol–ammonia (45:5:1, v/v), diisopropyl ether–methanol (1:1, v/v), and methanol–benzene–acetic acid (40:5:1, v/v). The chromatograms were detected in UV light at 254 nm and using the modified Dragendorff reagent. The greatest sensitivity of detection with PRE (equal to 0.5 µg/spot) was obtained using the Dragendorff reagent. When it comes to the analysis on the RP-18 plates, the best separation was achieved using the following mobile phases: acetonitrile–phosphate buffer, pH 2.3 (9:1, v/v), acetonitrile–phosphate buffer, pH 2.3 (7:3, v/v), and tetrahydrofuran–phosphate buffer, pH 2.3 (9:1, v/v). The mobile phase acetonitrile–phosphate buffer, pH 2.3 (9:1 v/v) was further used for the determination of PRE and dilazep by means of the HPLC method.

Misztal et al. (2002) investigated the separation efficiency with PRE, FE, LID, and BEP on silica gel 60F254, alumina 60F254, and Florisil with several mobile phases being the mixtures of n-hexane with several polar modifiers. Chromatographic plates were developed in the horizontal chamber and visualized in UV light at 254 nm, using the videoscanning method and different visualizing reagents. Detection was possible at 254 nm and after spraying the plates with the iodine reagent, and the detected amounts were 125 ng for PRE, LID, and FEN and 250 ng for BEP. Separation of all substances of interest was obtained on silica with 40% 1,4-dioxane in n-hexane as mobile phase and on alumina with 20% tetrahydrofuran in hexane as mobile phase, and only the latter system resulted in well-defined oval zones. Misztal et al. (2003) investigated the efficiency and the selectivity of separation of PRE, FEN, LID, and BEP on the silanized silica gel RP8 and RP18 plates also. Optimization of retention of these compounds was achieved by altering the pH and concentration of the organic modifier in the aqueous mobile phases. Substances were separated in the horizontal chamber, and the drugs were detected by video scanning and in UV light at 254 nm. Optimal separation of these drugs on RP8 and RP18 was obtained with 50% acetonitrile in 0.09 M phosphate buffer, pH 2.06, and with 50% ethanol in 0.09 M phosphate buffer, pH 2.06 as mobile phases, respectively.

26.4 DRUGS AFFECTING ADRENERGIC NERVOUS SYSTEM AND OTHER ANTIHYPERTENSIVES

The sympathetic adrenergic nervous system plays a major role in the regulation of arterial pressure. Sympatholytic drugs can block the sympathetic adrenergic system at three different levels. First are the peripheral sympatholytic drugs, such as alpha adrenoceptor ATs and beta adrenoceptor ATs, which block the influence of norepinephrine on the effector organ (the heart or the blood vessels). In therapeutic use are the following selective alpha 1 adrenoceptor ATs, known as *osins*: **prazosin** (PRA), 1-(4-amino-6,7-dimethoxy-2-quinazolinyl)-4-(2-furanylcarbonyl)piperazine, **terazosin** (TER), 1-(4-amino-6,7-dimethoxy-2-quinazolinyl)-4-[(tetrahydro-2-furanyl)carbonyl]piperazine, **doxazosin** (DOX), 1-(4-amino-6,7-dimethoxy-2-quinazolinyl)-4-[(2,3-dihydro-1,4-benzodioxin-2-yl)carbonyl]piperazine, etc. Besides *osins*, **urapidil** 6-[[3-[4-(2-methoxyphenyl)-1-piperazinyl]propyl] amino]-1,3-dimethyl-2,4(1 H, 3 H)-pyrimidinedione is a sympatholytic antihypertensive drug, acting as an α1-adrenoceptor AT and as a 5-HT$_{1A}$ receptor agonist. The nonselective alpha adrenoceptor ATs are the following ones: **phenoxybenzamine (PHEN)**, N-(2-chloroethyl)- N- (1-methyl-2-phenoxyethyl) benzenemethanamine and **PEN**, 3-[[(4,5-dihydro-1 H-imidazol-2-yl)methyl](4-methylphenyl) amino]phenol. The beta adrenoceptor ATs are present in a separate chapter of this book.

The second group of the centrally acting sympatholytics consists of these drugs that block sympathetic activity binding to and activating alpha 2 adrenoceptors, which results in the reduction of systemic vascular resistance and in decreasing arterial pressure. To this group belong **CLO**, N-(2,6-dichlorophenyl)-4,5-dihydro-1H-imidazol-2-amine, **moxonidine** (MOX), 4-chloro-N-(4,5-dihydro-1H-imidazol-2-yl)-6-methoxy-2-methyl-5-pyrimidinamine, **rilmenidine** (RIL), N-(dicyclopropylmethyl)-4,5-dihydro-2-oxazolamine, **guanoxan**, ([(2,3-dihydro-1,4-benzodioxin-2-yl) methyl]guanidine, etc.

The third group consists of the ganglionic blockers that block impulse transmission at the sympathetic ganglia. An adrenergic neuron blocking drug, **reserpine** (RES), 3,4,5-trimethoxybenzoylmethylreserpate, is one of several indole alkaloids, isolated from the root of *Rauwolfia serpentina*, which has been used in India for the centuries now, both as a remedy against snake bites and as a sedative. RES was the first effective antihypertonic drug introduced to Western medicine, but it has been largely replaced in clinical use with agents exerting fewer side effects, such as the ACE inhibitors and others. Synthetic drugs used as ganglion blockers are **guanethidine**, (hexahydro-1(2H)-azocinyl)ethyl]guanidine, **debrisoquine** (DEB), 3,4-dihydro-2(1H)-isoquinolinecarboximidamide, and other structurally related drugs (Cutler 2011b, Griffith 2008, Roth and Fenner 2000, Williams 2008). Chemical structures of these groups of drugs are shown in Figure 26.8.

Relaxation of smooth muscles either in arterial or venous blood vessels causes their dilatation leading to a reduction in systemic vascular resistance, which leads to a fall in arterial blood pressure. Directly acting vasodilators include **hydralazine** (HYD), 1-hydrazinophthalazine, **dihydralazine** (DIH) 1,4-dihydrazinophthalazine, and the potassium channel openers, **minoxidil**, 6-piperidino-2,4-diaminopyrimidine 3-oxide, **diazoxide** (DIA), 3-methyl-7-chloro-1,2,4-benzothiadiazine 1,1-dioxide, and others (Cutler 2011b, Roth and Fenner 2000, Williams 2008). Chemical structures of these groups of drugs are shown in Figure 26.8.

26.4.1 In Situ Separation, Identification, and/or Determination of Alpha 1 Adrenoceptor ATs, Centrally Acting Sympatholytics, and Ganglion Blockers in Dosage Forms

In situ identification and determination of TER (El Bayoumi et al. 2009), DOX (Sane et al. 2002), and RES (Argekar et al. 1996, Deshmukh et al. 2011) was carried out in dosage forms and adulterated dietary supplements. Details of the chromatographic conditions and analytical data are shown in Table 26.9 and 26.10, respectively.

The method for the in situ determination of DOX was established, using TER as an internal standard (Sane et al. 2002). Simultaneous determination of RES and arjunolic acid in tablets, single arjunolic acid in tablets, and the *arjuna* extract was performed by the validated TLC method (Deshmukh et al. 2011). Argekar et al. 1996 developed the TLC method with florescence detection, which enables separation and determination (the fluorescence mode) of RES in the presence of its photo-induced degradates from the tablets. The results obtained by the TLC method for RES were compared with HPLC results. Mikami et al. (2002) reported an easily available, simultaneous identification/determination procedure for PEN and sidenafil adulterated in dietary supplements, by using a combination of three different analytical methods, TLC, LC/UV, and LC/MS. TLC was used for fast and brief identifications of both drugs, using silica gel plates 60F$_{254}$, mobile phases chloroform–ammonia (28)–methanol (70:5:3, v/v) and chloroform–diethylamine–methanol (15:3:2, v/v), and the detection was carried out at 254 nm.

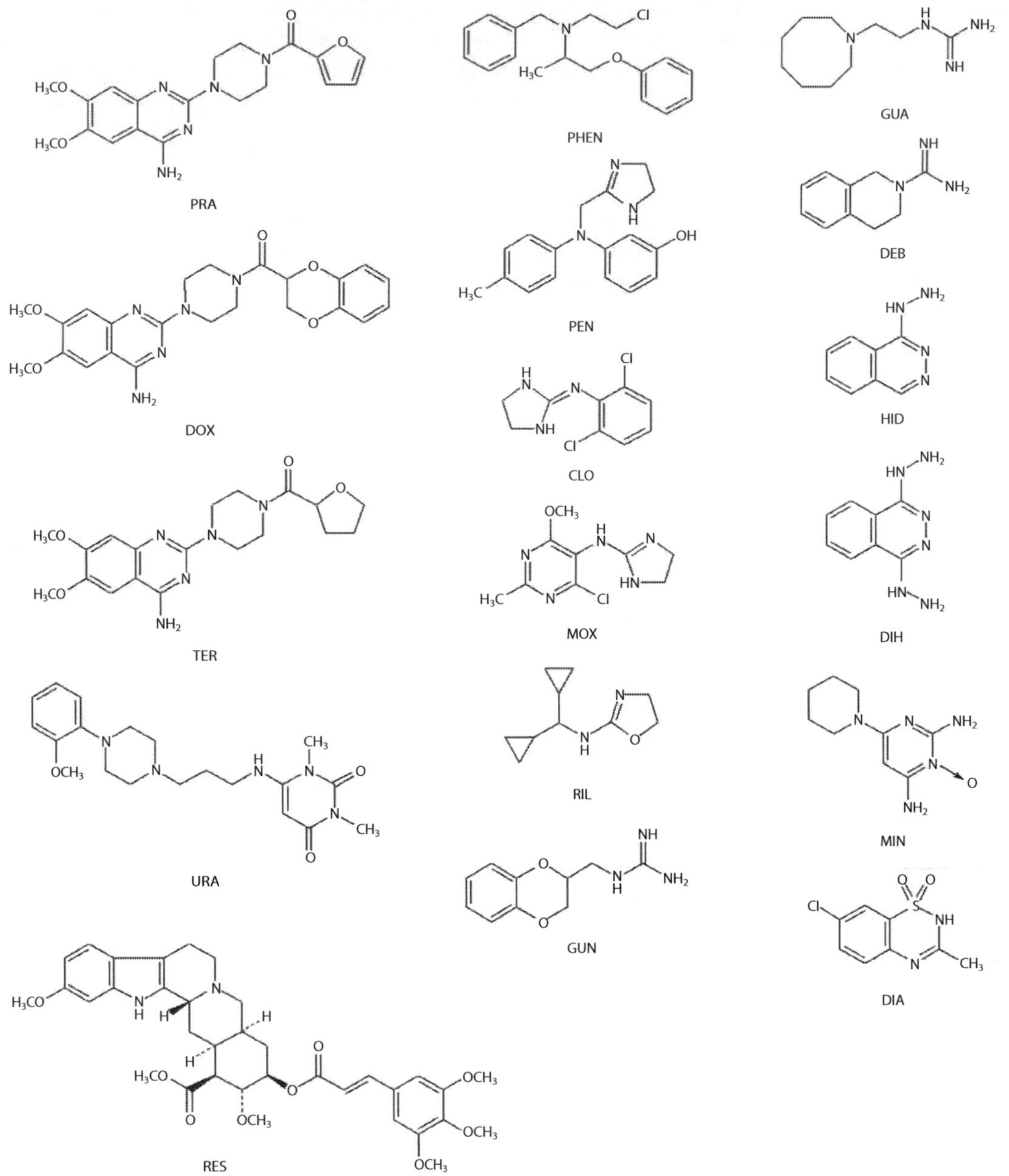

FIGURE 26.8 Chemical structures of alpha 1 adrenoceptor antagonists, centrally acting as sympatholytics, ganglion blockers, and vasodilators.

26.4.2 In Situ Separation, Identification, and/or Determination of Ganglion Blockers in Plants and Biological Fluids

Complete separation of the components belonging to various chemical classes of alkaloids, among them RES, was obtained by overpressured layer chromatography (OPLC) on alumina plates with a single solvent only, that is, ethyl acetate, as mobile phase. This method was successfully applied to the plant extracts and to determination of the hRf values of 81 samples of natural alkaloids or their derivatives (Pothier et al. 1991).

TABLE 26.10

Chromatographic Separation Conditions Used for Alpha 1 Adrenoceptor Antagonist, Ganglion Blockers, and Alpha 1 Agonist

Drug	Plates	Stationary Phase Pretreatment; Mobile Phase Composition	Development Mode	Detection (nm)	Rf	Reference
TER	TLC silica gel 60F$_{254}$	Ethyl acetate–methanol–ammonis, 9:1:0.01 v/v	Ascending	246	—	El Bayoumi et al. (2009)
DOX	HPTLC silica gel 60F$_{254}$, aluminum-backed	Methanol, activation at 110°C, 120 min; Ethyl acetate–methanol, 9:1 v/v	Chamber saturation 30 min; ascending	277	0.65	Sane et al. (2002)
RES	TLC silica gel 60F$_{254}$, aluminum-backed	Methanol, activation at 110°C, 5 min;toluene–ethyl acetate–diethyl amine–acetic acid, 6.5:5:1.5:0.5 v/v	Chamber saturation 30 min; ascending	254	0.4	Deshmukh et al. (2011)
RES	HPTLC silica gel 60F$_{254}$	Ethyl acetate–cycloxane–diethylamine, 21:9:0.1 v/v	Ascending	366	0.4	Argekar et al. (1996)
PHE	HPTLC silica gel 60F$_{254}$	Methylene chloride–ethyl acetate–ethanol–formic acid, 3.5:2:4:0.5 v/v)	Ascending	274	0.2	El Sadek et al. (1990)
PHE	HPTLC silica gel 60F$_{254}$	Dichloromethane–methanol, 1:1, activation air dried; 1-butanol–water–acetic acid glac. 7:2:1 v/v	Ascending	277	0.45	Greshock and Sherma (1997)
PHE	TLC silica gel 60F$_{254}$	Ethyl acetate–methnol–ammonia 4:1:0.4 v/v	Chamber saturation 30 min; ascending	291	0.2	Devarajan et al. (2000)

HPTLC was used for normal-phase separation of the components of the hexane, chloroform, methanol, and water extracts of the *R. serpentina* root. Computerized densitometry was used for the 2D spectrographic image analysis of HPTLC plates. RP-HPLC was also used for the separation of these extracts. This investigation revealed the presence of three indole alkaloid markers: ajmaline, ajmalicine, and RES. Use of chloroform resulted in the most efficient extraction, while de-fatting with hexane might result in the loss of alkaloids (Gupta et al. 2006).

Direct densitometric method for quantification of RES and ajmaline in the whole *Rauwolfia vomitoria* roots and in the bark thereof, upon the extraction and separation, was reported. On separation, the fluorescence measurement mode was applied; the calibration curve was found as linear up to 100 ng/spot for RES and 500 ng for ajmaline, respectively (Katič et al. 1980). Panda et al. (2010) reported a fully validated TLC method for the separation and determination of RES from the different herbal parts of the *Rauvolfia* species, using the silica gel 60F254 pre-coated plates and toluene–ethyl acetate–diethylamine (7:2:1, v/v) as mobile phase. The LOD and LOQ values valid for RES and established for the proposed method were 42 ng and 124 ng, respectively.

A simple, sensitive, and rapid method was reported for the qualitative and quantitative assessment of the fractional composition of the carbonyl functionalities in natural products. Several extracts, some of them containing RES, were directly analyzed by applying a simple colorimetric procedure, by videodensitometry and LC, using *o*-dianisidine as a chromogenic reagent (Abou-Shoer 2008).

Ishii et al. (2001) reported a radio-TLC method for the determination of the ^{14}C-labeled metabolite of 14-hydroxy DEB in the liver microsomes of human, hamster, and rat plasma, using the chloroform–methanol (100:1, v/v) mobile phase and Molecular Dynamics PhosporImager SF. The percent conversion was determined by calculating the ratio of the 14-hydroxy DEB radioactivity to the radioactivity of the entire TLC channel, using the ImageQuant software. Okuyama et al. (1997) developed a TLC-autoradioluminography method for the detection of the DEB metabolites and of those of several other drugs in the rat liver biopsy sample, in a relatively short time span and at a low concentration, similar to those in vivo. Upon application of the reaction mixture on to the silica gel layer, mixture chromatography was performed using ethyl acetate–toluene–water–formic acid (6:2:2:1.5, v/v) as mobile phase. The resultant TLC plate was contacted with an imaging plate, and the image of the radioactivity distribution in the non-metabolized DEF and in its metabolite fractions was analyzed, using a bio-imaging analyzer.

26.4.3 Other Classical and Newly Developed Options in TLC Screening

TLC combined with the diode laser induced desorption/atmospheric pressure chemical ionization was developed. RES was used as a model compound for this investigation. The use of a graphite suspension and the decoupling of desorption and ionization for the first time allowed the diode lasers to be used in TLC/MS (Peng et al. 2004).

The retention parameters obtained in the RP-TLC system for the 11 adrenergic and imidazoline receptor ligands (among them DOX, CLO, MOX, and RIL) and the application thereof to QSAR/QSPR were investigated. The hydrophobicity parameter, R_M^o, plotted vs. log P for all the investigated compounds showed a satisfactory linear correlation, using multilinear regression analysis (Erić et al. 2007).

Toxicological drug screening by OPLC was developed for the screening of the toxicologically relevant basic drugs in a forensic and clinical context. Eighty-two toxicologically important drugs and metabolites (including antihypertensive drugs, such as PRA, DIL, and VER and its metabolite, norverapamil) were tested. The OPLC1 liquid system was trichloroethylene–methyl ethyl ketone–n-butanol–acetic acid–water (17 + 8 + 25 + 6 + 4, v/v), and the OPLC2 liquid system was butyl acetate–ethanol–tripropylamine–water (85:9.25:5:0.75, v/v), with chamber presaturation. Both systems were tested with use of the high-performance silica gel plates. Combination of the two systems was demonstrated to be feasible in drug screening of the autopsy urine samples, utilizing the automated identification hRfc/UV library search, with the combined dual-system reporting (Ojanperä et al. 1999b).

In mid-seventies, the procedure of TLC separation and identification of antihypertensive drugs, RES, HYD and DIA, and 17 thiazide diuretics, using several mobile phases and a variety of possible detection reagents was reported (Stohs and Scratchley 1975).

A simple TLC method for monitoring of the antihypertensive drug compliance in urine of the patients subjected to administration of beta blockers and HID was reported. Separation of the tested compounds was performed on polygram Sil N-HR UV$_{254}$ plates, using the two solvent systems, ethyl acetate–methanol (45:5, v/v) and ethyl acetate–methanol–ammonia (45:5:5, v/v). Detection was performed, exposing the chromatograms to UV illumination at 254 nm and to the derivatizing agents (Jack et al. 1980).

Komsta et al. (2010a) presented a comparative study on several approaches to determination of lipophilicity by means of TLC: a single TLC run, extrapolation of retention, principal component analysis of the retention data matrix, Parallel Factor Analysis (PARAFAC) on a three-way array, and a PLS regression. Each technique was applied to 35 simple-molecule model solutes including DIH, and using nine concentrations of each of the six modifiers, acetonitrile, acetone, dioxane, propan-2-ol, methanol, and tetrahydrofurane. Moreover, Komsta et al. (2010b) investigated in situ the retention of 35 model compounds (including DIH) with the 10 screening mobile phases on the 6 NP and 7 RP adsorbens. The factors formed two cubes with the defined dimensions, which enabled the three-way analysis by PARAFAC. The one-component PARAFAC model was in both cases found better than the two-component model. These results showed that the major variability of the retention factor R_F can be modeled as the product of the three factors related to the solute the mobile phase and the absorbent. Modeling of R_F was found substantially better than using the k or R_M values. Komsta et al. (2011a) reported the trilinear multiplicative modeling of the thin-layer chromatographic retention of 35 compounds (including DIH) as a function of solute, organic modifier, and its concentartions. Chemometric characterisation of 35 simple model compounds (including DIH) using the salting-out chromatography was investigated on silica gel (Komsta et al. 2011b) and cellulose (Komsta et al. 2012). Twelve inorganic salts dissolved in high concentrations (0.5–4 mol/L) in water were investigated. The authors found out that PCA allowed a discovery of certain hidden trends in the salting-out retention dataset of model compounds. It was observed that on the silica adsorbens, the choice of a salt is an impotrant parameter. The highest salting-out effect was observed in the case of calcium chloride, magnesium chloride, and ammonium sulfate, which confirms the importance of the diameter of an ion in the salting-out chromatography.

26.5 ENDOGENOUS NEUROTRANSMITTERS AND THEIR SYNTHETIC ANALOGUES

Norephinephrine (NE, noradrenaline, (R)-4-(2-amino-1-hydroxyethyl)-1,2-benzenediol) and **epinephrine** (E, adrenaline, (R)-4-[1-hydroxy-2-(methylamino)ethyl]-1,2-benzenediol), neurotransmitter and neurohormon, respectively, are known as catecholamines. Pharmacological effects of NE and E are similar, yet not identical, due to different activities exerted toward the alpha and beta adrenergic receptors. NE causes general vasoconstriction with an exception of the coronary arteries and hence, it elevates the systolic and diastolic blood pressure. In the case of E alone and in the high, nonphysiological doses, its activity directed toward the alpha receptor dominates, causing an increase of the blood pressure. NE and E have a limited clinical application, due to nonselective nature of their action. Because of a poor oral bioavailability and rapid metabolism of these two compounds, and of a short duration of their action lasting for 1–2 min only (even when administered by infusion), they cannot be dispensed (used) orally. Like with most phenols, the catechol functional groups in catecholamines are highly susceptible to easy oxidation. E and NE each possess a chiral carbon atom in their respective chemical structures, and due to that, each can form a pair of enantiomers. The enantiomer with (R) configuration is biosynthesized by the body and possesses biological activity. The (R) configuration of many other adrenergic agents also contributes to their high affinity to the corresponding adrenoceptors. NE and E are used to treat hypotensive crises and to stimulate heart in the cardiac arrest. In order to improve selectivity, the in vivo and in vitro stability and bioavailability of the synthetic group of the selective alpha1 agonists has been established and those therapeutically used are **phenylephrine** (PHE), (R)-3-hydroxy-α-[(methylamino)methyl] benzenemethanol, **etilefrine**, α-[(ethylamino)methyl]-3-hydroxy benzenemethanol and others. PHE is used in the treatment of a severe hypotension resulting from a shock or drug administration. It also has a widespread use as a nonprescription drug acting as a nasal decongestant in both, an oral and topical preparation (Griffith 2008, Lin 2011, Roth and Fenner 2000). There are not so many marketed drugs available to treat hypotension, and we are going to mention only these analyzed by means of TLC. The chemical structures of these drugs are given in Figure 26.9.

26.5.1 In Situ Separation, Identification, and/or Determination of Catecholamines and Their Analogues in Bulk, Dosage Formulation, and Biological Fluids

Sleckman and Sherma (1983) are reported to be the first ones who completed separation of the three underivatized catecholamines, NE, E, and dopamine, by means of TLC. The separation was performed on the RP diphenyl stationary phase and using methanol–0.063 M NaCl (6:4, v/v) as mobile phase. Detection of the separated zones was performed by spraying the chromatogram with

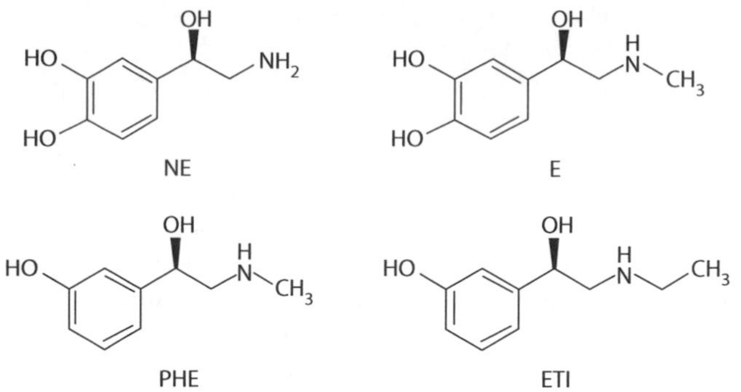

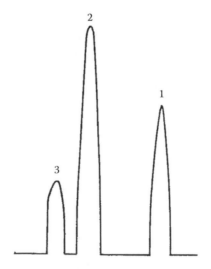

FIGURE 26.9   Chemical structures of catecholamines and structural analogues.

ninhydrin, or the fluoroborate reagent. The minimum amount for visual detection was found to be from 100 ng/spot. The chromatogram of the three separated compounds is shown in Figure 26.10.

The group of 30 phenylethylamine derivatives (including NE and E), phenolic acids, and glycols on different stationary phases impregnated with the anionic and cationic detergents was investigated. Chromatographic separation of 19 phenylethylamines and 11 phenolic acids and glycols on the OPTI-UPC$_{12}$ and Sil 18–50 plates was investigated, using aqueous solutions at different pH values and with the 20% methanol content. The presence of potassium chloride in aqueous eluents accounts for the compactness of the spots. The retention mechanism of the examined compounds on the different stationary phases was discussed (Lepri et al. 1985).

Thirteen new visualizing reagents have been used for the detection of 13 phenolic drugs (including NE) after the separation on silica gel. The mobile phase consisting of acetic acid-n-butanol-water (1:4:1, v/v) was used for the separation of a mixture of five structurally similar compounds, including NE. Alkacidometric and redoximetric indicators were used as detection agents. For each

FIGURE 26.10   Separation of a mixture containing norepinephrine (peak 1; 500 ng), dopamine (peak 2; 500 ng), and epinephrine (peak 3; 1 µg) on a diphenyl bonded TLC plate developed with methanol–0.063 M NaCl (6:4). The spots were visualized with ninhydrin and scanned using a Kontes Chromaflex densitometer and baseline correctior with an attenuation setting of × 50, a scan rate of 6 cm/min, and a recorded chart speed of 5 in/min. (Sleckman, B.P. and Sherma, J: Separation and quantification of epinephrine, norepinephrine, and dopamine by chromatography on diphenyl thin layers. *J. High Res. Chromatogr.* 1983. 6. 156–157. Copyright Wiley-VCH Verlag GmbH & Co. KGaA. With permission.)

compound, the lowest detection levels were established; and with NE, the lowest detection level was found with use of bromothymol blue as equal to 600 ng (Pyka et al. 2002). Synthesis and the application of 4-dipropylaminodiazabenzene-4'-isothiocyanate for the derivatization of eight biogenic amines (including NE and E), followed by the separation thereof in 15 different solvent systems was reported. The thiocarbamoyl derivatives of all tested compounds happen to be less polar than the respective original compounds, which improves the selectivity of chromatographic separation and elevates the detection sensitivity of these compounds to the sub-nanomolar amounts (Pyra et al. 2000).

Direct enantioseparation of E, norephedrine, pseudoephedrine, and ephedrine via TLC and using the molecularly imprinted polymers (MIPs) was done by Suedee et al. 1999a. Two monomeric molecules of methacrylic acid and itaconic acid were used as MIP templates. The mobile phase system comprising either methanol or acetonitrile was used and the effect of the acetic acid content in the mobile phase was also investigated. The best resolution of E was obtained on the plates based on the (−) norephedrine MIP, using itaconic acid as a functional monomer with methanol ($\alpha = 3$) and 1% acetic acid in acetonitrile ($\alpha = 2.5$) as the respective mobile phases. (−) E was retarded on this stationary phase more, than (+) E. Suedee et al. (1999b) extended their previous work, imprinting polymers with the (+) pseudoephedrine, (+) norephedrine, and (+) ephedrine templates, using the thermal polymerization method. Resolution of the E enantiomers with the spots tailing was observed on the plates based on (+) pseudoephedrine as a MIP template, and using 10% acetic acid in acetonitrile ($\alpha = 1.9$) as mobile phase.

Alemany et al. (1996) reported on the TLC method for quantitative determination of NE, E, and dopamine in the rat plasma. After deproteinization, catecholamines were absorbed on acid-alumina and acetylated. The acetyl derivatives were extracted, using a C18 mini-column, resolved on the HPTLC plates and quantified by fluorescence densitometry at 415 nm, using isoprenalin as an internal standard. The chromatograms were developed in chloroform–acetone–ethyl acetate–methanol (34:16:1:0.25, v/v) as mobile phase. The fluorescence was generated by spraying the plates with ethylenediamine-methanol-potassium ferricyanide 1.5% (10:30:10, v/v), incubating them for 5 min at 60°C and leaving at room temperature in the darkness for 15 more minutes to complete the development of fluorescence. Less than 0.5 ng of the amine derivative could be detected on the chromatographic plate. Separation of E, NE, and dopamine from the rat plasma is shown in Figure 26.11.

A series of four organic amines with diverse structures (including E) were directly analyzed by the positive ion TLC, ToF-SIMS. It was observed that the direct analysis of the amines on the normal phase silica gel was facilitated by the gel acidity, that is, by the proton donation of the surface silanol groups (Parent et al. 2006).

The analysis of biogenic amines, E, NE, dopamine, and serotonin, the metabolites thereof as well as alkaloids, caffeine, theobromine, nicotine, and conitine, was carried out on silica gel, the chemically bonded amino and diol plates, and on the RP 18 adsorbent, employing the adsorption and partition TLC. E, NE, dopamine, and serotonin, and also their metabolites, were analyzed by means of RP partition chromatography on the RP 18 layers with the acetate buffer—organic modifier (methanol, acetonitrile, or THF) binary mobile phases, and the effect of the mobile phase modifier on the selectivity separation was studied (Baranowska and Zydron 2000). The possibility of using molecular descriptors to predict both the retention behavior and the retention mechanism of E and NE and certain metabolites thereof in TLC was examined by Baranowska and Zydron (2003). The quantitative structure-retention relationships analysis was completed by a simultaneous solution of a set of linear equations. Basic correlations were found between the retention and the molecular descriptors, calculated by means of a semi-empirical quantum chemistry method. The obtained function provided a molecular-level insight into the mechanism of chromatographic retention. The experimental retention data for the tested compounds additionally served as a basis for cluster analysis of this TLC system.

El Sadek et al. (1990) reported on the spectrodensitometric method for the simultaneous determination of two analgesic mixtures containing PHE, paracetamol, ascorbic acid, and caffeine.

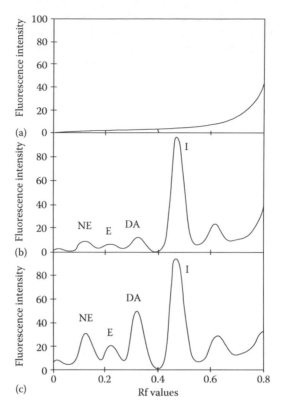

FIGURE 26.11 Fluorescence densitometry chromatograms from: (a) Aqueous blank extract; (b) Aqueous standard extract containing 2.5 ng (equivalent to 5 ng/mL) of each NE, E, and DA; (c) Rat plasma extract. I: lsoprenaline (internal standard). (Alemany, G., Akaârir, M., Rosselló, C., and Gamundi, A.: Thin-layer chromatographic determination of catecholamines in rat plasma. *Biomed. Chromatogr.* 1996. 10. 225–227. Copyright Wiley-VCH Verlag GmbH & Co. KGaA. With permission.)

Greshock and Sherma (1997) reported on the densitometric method for determination of the decongestant and antihistamine drugs from the six different sample compositions containing PHE, chlorpheniramine, brompheniramine, phenylpropanolamine, and guaifenesin. The developed method was found accurate, precise, more selective than spectrophotometry, less complex than HPLC and GC, and applicable to the analysis of a variety of the prescription and non-prescription pharmaceutical dosage forms.

PHE was determined by densitometry in presence of lignocaine from solutions and drug delivery systems (Devarajan et al. 2000). The chromatographic and analytical data provided in the reports given in (El Sadek et al. 1990, Devarajan et al. 2000, Greshock and Sherma 1997) are shown in Tables 26.9 and 26.10.

REFERENCES

Abou-Shoer, M. 2008. Evaluation of carbonyl compounds in natural products by o-dianisidine. *Chromatographia* 68:447–451.

A-Elghany, M.F., Elkawy, M.A., Zeany, B.E., and Stewart, J.T. 1996. High performance thin layer chromatography quantitation of isradipine in the presence of its degradation products. *J. Planar Chromatogr.* 9:290–292.

A-Elghany, M.F., Elkawy, M.A., Zeany, B.E., and Stewart, J.T. 1997. Determination of nicardipine hydrochloride in drug substance and in a capsule dosage form. *J. Planar Chromatogr.* 10:61–63.

Agbaba, D., Solomun, LJ., and Zivanov-Stakic, D. 1997. Simultaneous HPTLC determination of diltiazem and its impurity desacetyldiltiazem in raw material and in dosage forms. *J. Planar Chromatogr.* 10:303–304.

Agbaba, D., Vučićević, K., and Marinković, V. 2004. Determination of nisoldipine and its impurities in pharmaceuticals. *Chromatographia* 60:223–227.

Aleksić, M., Agbaba, D., Milojković-Opsenica, D.M., and Tešić, Ž.LJ. 2001a. Salting-out thin-layer chromatography of some angiotensin converting enzyme inhibitors on different sorbents. *Chromatographia* 53:442–444.

Aleksić, M.B., Agbaba, D., Baošić, R.M., Milojković-Opsenica, D.M., and Tešić, Ž.Lj. 2001b. Thin-layer chromatography of several antihypertensive drugs from the group of angiotensin converting enzyme inhibitors. *J. Serb. Chem. Soc.* 66:39–44.

Alemany, G., Akaârir, M., Rosselló, C., and Gamundi, A. 1996. Thin-layer chromatographic determination of catecholamines in rat plasma. *Biomed. Chromatogr.* 10:225–227.

Argekar, A.P. and Powar, S.G. 1999. Simultaneous HPTLC determination of cinnarizine and domperidone maleate in formulations. *J. Planar Chromatogr.* 12:272–274.

Argekar, A.P. and Powar, S.G. 2000. Simultaneous determination of atenolol and amlodipine in tablets by high-performance thin-layer chromatography. *J. Pharm. Biomed. Anal.* 21:1137–1142.

Argekar, A.P., Raj, S.V. and Kapadia, S.U. 1996. Quantitative determination of reserpine from rauwolfia serpentina tablets by high performance thin layer chromatography. *J. Planar Chromatogr.* 9:148–151.

Argekar, A.P. and Sawant, J.G. 1999. Simultaneous determination of atenolol and nitrendipine in pharmaceutical dosage forms by HPTLC. *J. Liq. Chromatogr. Related Technol.* 22:1571–1578.

Baranowska, I., Marakowski, P., Wilczek, A., Szostek, M., and Stadniczuk, M. 2009. Normal and reversed-phase thin-layer chromatography in the analysis of L-arginine, its metabolites, and selected drugs. *J. Planar Chromatogr.* 22:89–96.

Baranowska, I. and Zydron, M. 2000. Analysis of biogenic amines, alkaloids and their derivatives by TLC and HPLC. *J. Planar Chromatogr.* 13:301–306.

Baranowska, I. and Zydron, M. 2003. Quantitative structure-retention relationships (QSRR) of biogenic amine neurotransmitters and their metabolites on RP-18 plates in thin-layer chromatography. *J. Planar Chromatogr.* 16:102–106.

Bari, P.D. and Rote, A.R. 2009. RP-LC and HPTLC methods for the determination of olmesartan medoxomil and hydrochlorothiazide in combined tablet dosage forms. *Chromatographia* 69:1469–1472. doi: 10.1365/s10337–009–1094-z.

Bebawy, L.I., Abbas, S.S., Fattah, L.A., and Refaat, H.H. 2005. Application of first-derivative, ratio derivative spectrophotometry, TLC-densitometry and spectrofluorimetry for the simultaneous determination of telmisartan and hydrochlorothiazide in pharmaceutical dosage forms and plasma. *Il Farmaco* 60:859–867.

Bhushan, R. and Agarwal, C. 2008. Direct TLC resolution of (±)-ketamine and (±)-lisinopril by use of (+)-tartaric acid or (−)-mandelic acid as impregnating reagents or mobile phase additives. Isolation of the enantiomers. *Chromatographia* 68:1045–1051.

Bhushan, R., Gupta, D., and Singh, S.K. 2006. Liquid chromatographic separation and UV determination of certain antihypertensive agents. *Biomed. Chromatogr.* 20:217–224.

Borole, T.C., Gandhi, S.P., Ladke, A.V., and Damle, M.C. 2011. Validated HPTLC method for determination of amlodipine besylate and perindopril erbumine in bulk drug and in pharmaceutical dosage forms. *Pharma Rev.* 5:113–117.

Brzezinka, H., Dallakian, P., and Budzikiewich, H. 1999. Thin-layer chromatography and mass spectrometry for screening of biological samples for drugs and metabolites. *J. Planar Chromatogr.* 12:96–108.

Chandrashekhar, T.G., Rao, P.S.N., Smrita, K., Vyas, S.K., and Dutt, C. 1994. Analysis of amlodipine besylate by HPTLC with fluorimetric detection: A sensitive method for assay of tablets. *J. Planar Chromatogr.* 7:458–460.

Csermely, T., Kalász, H., Deák, K., Hasan, M.Y., Darvas, F., and Petroianu, G. 2008. Lipophilicity determination of some ACE inhibitors by TLC. *J. Liq. Chromatogr. Related Technol.* 31:2019–2034.

Cutler, S.J. 2011a. Drugs acting on the renal system. In: *Wilson and Gisvold'S Textboook of Organic Medicinal and Pharmaceutical Chemistry*, ed. J.M. Beale, Jr. and J.H. Block, Wolters Kluwer/Lippincott Williams & Wilkins, Philadelphia, PA, pp. 607–616.

Cutler, S.J. 2011b. Cardiovascular agents. In: *Wilson and Gisvold'S Textboook of Organic Medicinal and Pharmaceutical Chemistry*, ed. J.M. Beale, Jr. and J.H. Block, Wolters Kluwer/Lippincott Williams & Wilkins, Philadelphia, PA, pp. 617–665.

Czerwińska, K., Wyszomirska, E., and Kaniewska, T. 2001. Identification and determination of selected medicines reducing hypertention by densitometric and gas chromatographic methods. *Acta Pol. Pharm.—Drug Res.* 58:331–338.

Dangi, M., Chaudhari, D., Sinker, M., Racha, V., and Damle, M.C. 2010. Stability indicating HPTLC method for estimation of nebivolol hydrochloride and amlodipine besylate in combination. *Eurasian J. Anal. Chem.* 5:161–169.

Desai, D.J., More, A.S., Chabukswar, A.R., Kuchekar, B.S., Jagdale, S.C., and Lokhande, P.D. 2010. Validated HPTLC method for simultaneous quantitation of olmesartan medoximal and amlodipine besylate in bulk drug and formulation. *Der. Pharma. Chem.* 2:135–141.

Deshmukh, T.A., Chaudhari, A.B., and Patil, V.R. 2011. Development and validation of HPTLC method for simultaneous determination of reserpine and arjunolic acid in tensowert tablet. *Der. Pharma. Lett.* 3:43:50.

Devarajan, P.V., Adani, M.H., and Gandhi, A.S. 2000. Simultaneous determination of lignocaine hydrochloride and phenylephrine hydrochloride by HPTLC. *J. Pharm. Biomed. Anal.* 22:685–690.

Devarajan P.V. and Dhavse, V.V. 1998. High-performance thin-layer chromatographic determination of diltiazem hydrochloride as bulk drug and in pharmaceutical preparations. *J. Chromatogr. B.* 706:362–366.

Dhaneshwar, S.R., Patre, N.G., and Mahadik, M.V. 2009. Validated TLC method for simultaneous quantitation of amlodipine besylate and valsartan in bulk drug and formulation. *Chromatographia* 69:157–161.

El Bayoumi, A. El. A., Metwally, F.H., Badawey, A.M., and Lami, N.T. 2009. Stability indicating methods for the determination of terazosin hydrochloride dihydrate in the presence of its degradation product. *Bull. Fac. Pharm. Cairo Univ.* 47:49–59.

El-Gindy, A., Ashour, A., Abdel-Fattah, L., and Shabana, M.M. 2001. Application of LC and HPTLC-densitometry for the simultaneous determination of benazepril hydrochloride and hydrochlorothiazide. *J. Pharm. Biomed. Anal.* 25:171–179.

El Sadek, M., El Shanawany, A., Aboul Khier, A., and Rücker, G. 1990. Determination of the components of analgesic mixtures using high-performance thin-layer chromatography. *Analyst* 115:1181–1184.

Erić, S., Pavlović, M., Popović, G., and Agbaba, D. 2007. Study of retention parameters obtained in RP-TLC system and their application on QSAR/QSPR of some alpha adrenergic and imidazoline receptor ligands. *J. Chromatogr. Sci.* 45:140–144.

Gopani, K.H., Havele, S.S., and Dhaneshwar, S.R. 2011. Application of high performance thin layer chromatography densitometry for the simultaneous determination of amlodipine besilate and lisinopril in bulk drug and tablet formulation. *LJPT* 3:2253–2267.

Greshock, T. and Sherma, J. 1997. Analysis of decongestant and antihistamine pharmaceutical tablets and capsules by HPTLC with ultraviolet absorption densitometry. *J. Planar Chromatogr.* 10:460–463.

Griffith, R.K. 2008. Adrenergic receptors and drugs affecting adrenergic neurotransmission. In: *Foye's Principles of Medicinal Chemistry*, ed. T.L. Lemke and D.A. Williams, Wolters Kluwer/Lippincott Williams & Wilkins, Philadelphia, PA, pp. 392–416.

Gumieniczek, A., Inglot, T., and Kończak, A. 2011. Classical densitometry and videoscanning in a new validated method for analysis of candesartan and losartan in pharmaceuticals. *J. Planar Chromatogr.* 24:99–104.

Gumieniczek, A. and Przyborowski, L. 1997. Thin-layer chromatographic analysis of some ACE inhibitors in tablets. *Acta. Pol. Pharm. Drug Res.* 54:13–16.

Gumieniczek, A., Przyborowski, L., and Plizga, J. 1995. Chromatographic analysis (TLC and HPLC) of dilazep and prenylamine. *Acta Pol. Pharm.* 52:451–454.

Gupta, A., Gaud, R.S., and Ganga, S. 2010. Liquid chromatographic method for determination of nisoldipine from pharmaceutical samples. *E-J. Chem.* 7:751–756.

Gupta, M.M., Srivastava, A., Tripathi, A.K., Misra, H., and Verma, R.K. 2006. Use of HPTLC, HPLC, and densitometry for qualitative separation of indole alkaloids from rauvolfia serpentina roots. *J. Planar Chromatogr.* 19:282–287.

Haddad, R., Milagre, H.M.S., Catharino, R.R., and Eberlin, M.N. 2008. Easy ambient sonic-spray ionisation mass spectrometry combined with thin-layer chromatography. *Anal. Chem.* 80:2744–2750.

Harper, J.D., Martel, P.A., and O'Donnell, C.M. 1989. Evaluation of a multiple-variable thin-layer and reversed-phase thin-layer chromatographic scheme for the identification of basic and neutral drugs in an emergency toxicology setting. *J. Anal. Toxicol.* 13:31–36.

Harrold, M. 2008. Angiotensin-converting enzyme inhibitors, antagonists and calcium blockers. In: *Foye's Principles of Medicinal Chemistry*, ed. T.L. Lemke and D.A. Williams, pp. 738–768. Wolters Kluwer/ Lippincott Williams & Wilkins, Philadelphia, PA.

Hassan, S.S.M., Elmosallamy, M.A.F., and Abbas, A.B. 2002. LC and TLC determination of cinnarizine in pharmaceutical preparations and serum. *J. Pharm. Biomed. Anal.* 28:711–719.

Hassib, S.T., El-Sherif, Z.A., El-Bagary, R.I., and Youssef, N.F. 2000. Reversed-phase high performance liquid chromatographic and thin layer chromatographic methods for the simultaneous determination of benazepril hydrochloride and hydrochlorothiazide in cibadrex tablets. *Anal. Lett.* 33:3225–3237.

Hemmateenejad, B., Mobaraki, N., Shakerizadeh-Shirazi, F., and Miri, R. 2010. Multivariate image analysis-thin layer chromatography (MIA-TLC) for simultaneous determination of co-eluting components. *Analyst* 135:1747–1758.

Ikeda, T., Komai, T., Kawai, K. and Shindo, H. 1981. Urinary metabolites of 1-(3-mercapto-2-D-methyl-1-oxopropyl)-L-proline (SQ-14225), a new antihypertensive agent, in rats and dogs. *Chem. Pharm. Bull.* 29:1416–1422.

Inglot, T.W., Dąbrowska, K., and Gumieniczek, A. 2009. The reversed-phase retention behavior of some angiotensin-II receptor antagonists. *J. Planar Chromatogr.* 22:145–155.

Inglot, T.W., Dąbrowska, K., and Misztal, G. 2007. The normal-phase retention behavior of some angiotensin-II receptor antagonists. *J. Planar Chromatogr.* 20:293–301.

Ishii, M., Xu, B.Q., Ding, L.R., Fischer, N.E., and Inaba, T. 2001. Interaction of plasma proteins with cytochromes P450 mediated metabolic reactions: inhibition by human serum albumin and α-globulins of the debrisoquine 4-hydroxylation (CYP2D) in liver microsomes of human, hamster and rat. *Toxicol. Lett.* 119:219–225.

Jack, D.B., Dean, S., Kendall, M.J., and Laugher, S. 1980. Detection of some antihypertensive drugs and their metabolites in urine by thin-layer chromatography. *J. Chromatogr.* 196:189–192.

Kadam, B.R. and Bari, S.B. 2007. Quantitative analysis of valsartan and hydrochlorothiazide in tablets by high performance thin-layer chromatography with ultraviolet absorption densitometry. *Acta Chromatogr.* 18:260–269.

Kadukar, S.S., Gandhi, S.V., Ranjane, P.N., and Ranher S.S. 2009. HPTLC analysis of olmesartan medoxomil and hydrochlorothiazide in combination tablet dosage forms. *J. Planar Chromatogr.* 22:425–428.

Kakde, R. and Bawane, N. 2009. High-performance thin-layer chromatographic method for simultaneous analysis of metoprolol succinate and amlodipine besylate in pharmaceutical preparations. *J. Planar Chromatogr.* 22:115–119.

Kalisz, D., Dąbrowski, Z., Kąkol, B., Bełdowicz, M., Obukowicz, B., and Kamiński, J. 2005. Synthesis of optically active trans 4-cyclohexyl-L-proline as an intermediate product in the preparation of fosinopril. *Acta Pol. Pharm. Drug Res.* 62:121–126.

Kamble, A.Y., Mahadik, M.V., Khatal, L.D., and Dhaneshwar, S.R. 2010. Validated HPLC and HPTLC method for simultaneous quantitation of amlodipine besylate and olmesartan medoxomil in bulk drug and formulation. *Anal. Lett.* 43:251–258.

Katič, M., Kučan, E., Prošek M., and Bano, M. 1980. Quantitative densitometric determination of reserpine and ajmaline in rauwolfia vomitoria by HPTLC. *J. High Res. Chromatogr.* 3:149–150.

Kharat, V.R., Verma, K.K., and Dhake, J.D. 2002. Determination of lacidipine from urine by HPTLC using off-line SPE. *J. Pharm. Biomed. Anal.* 28:789–793.

Komsta, Ł., Radoń, M., Radkiewicz, B., and Skibiński, R. 2011a. Trilinear multiplicative modeling of thin layer chromatography retention as a function of solute, organic modifier and its concentration. *J. Sep. Sci.* 34:59–63.

Komsta, Ł., Skibiński, R., Berecka, A., Gumieniczek, A., Radikiewicz, B., and Radoń, M. 2010a. Revisiting thin-layer chromatography as a lipophilicity determination tool—A comparative study on several techniques with a model solute set. *J. Pharm. Biomed. Anal.* 53:911–918.

Komsta, Ł., Skibiński, R., and Bojarczuk, A. 2012. Chemometric characterization of model compounds retention in salting-out thin layer chromatography on cellulose. *J. of Liq. Chromatogr. Related Technol* 35:1298–1305.

Komsta, Ł., Skibiński, R., Gumieniczek, A., and Wojnar, A. 2010b. Multi-way analysis of retention of model compounds in thin-layer chromatography. *Acta Chromatogr.* 22:27–36.

Komsta, Ł., Skibiński, R., and Radoń, M. 2011b. Chemometric characterization of model compounds retention in salting-out thin layer chromatography on silica. *J. Liq. Chromatogr. Related Technol.* 34:776–784.

Kotaiah, M.R., Ganesh, B., Chandra Sekhar, K.B., Rasheed, S.H., Venkateswarlu, Y., and Dhandapani, B. 2010. HPTLC method development and validation for the estimation of trandolapril in bulk and its formulations. *Asian J. Res. Chem.* 3:158–160.

Kowalczuk, D. 2005. Simultaneous high-performance thin-layer chromatography densitometric assay of trandolapril and verapamil in the combination preparation. *J. AOAC Int.* 88:1525–1529.

Kowalczuk, D. 2006. Determination of nitrendipine in tablets by HPTLC-densitometry. *J. Planar Chromatogr.* 19:135–138.

Kowalczuk, D., Hopkala, H., and Pietraś, R. 2003. Simultaneous densitometric determination of quinapril and hydrochlorothiazide in the combination tablets. *J. Planar Chromatogr.* 16:196–200.

Kowalczuk, D., Pietraś, R., and Hopkała, H. 2004. Development and validation of an HPTLC-densitometric method for determination of ACE inhibitors. *Chromatographia* 60:245–249.

Kowalczuk, D., Wawrzycka, M.B., and Maj, A.H. 2006. Application of an HPTLC densitometric method for quantification and identification of nifedipine. *J. Liq. Chromatogr. Related Technol.* 29:2863–2873.

Kumar, K.V., Arunkumar, N., Verma, P.R.P., and Rani, C. 2009. Preparation and in vitro characterization of valsartan solid dispersions using skimmed milk powder as carrier. *Int. J. Pharm. Tech. Res.* 1:431–437.

Lakshmi, K.S., Sivasubramanian, L., and Pandey, A.J. 2010. A validated HPTLC method for simultaneous determination of losartan and perindipril in tablets. *Res. J. Pharm. Tech.* 3:559–561.

Lepri, L., Desideri, P.G., and Heimler, D. 1985. High-performance thin-layer chromatography of phenylethylamines and phenolic acids on silanized silica and on ammonium tungstophosphate. *J. Chromatogr.* 347:303–309.

Liu, S. 2011. Adrenergic agents, In: *Wilson and Gisvold'S Textboook of Organic Medicinal and Pharmaceutical Chemistry*, eds. J.M. Beale, Jr. and J.H. Block, Wolters Kluwer/Lippincott Williams & Wilkins, Philadelphia, PA, pp. 519–557.

Marciniec, B. and Brzeska, A. 1999. Differentiating analysis of some 1,4-dihydropyridine derivatives. *Chem. Anal. (Warsaw)* 44:849–855.

Marciniec, B., Jaroszkiewicz, E., and Ogrodowczyk, M. 2002. The effect of ionizing radiation on some derivatives of 1,4-dihydropyridine in the solid state. *Int. J. Pharm.* 233:207–215.

Marciniec, B., Kujawa, E., and Ogrodowczyk, M. 1992. Evaluation of nifedipine preparations by chromatographic-spectrophotometric methods. *Pharmazie* 47:502–504.

Marciniec, B. and Ogrodowczyk, M. 2006. Thermal stability of 1,4-dihydropyridine derivatives in solid state. *Acta Pol. Pharm. Drug Res.* 63:477–484.

McCarthy, K.E., Wang, Q., Tsai, E.W., Gilbert, R.E., Ip, D.P., and Brooks, M.A. 1998. Determination of losartan and its degradates in COZAAR® tablets by reversed-phase high-performance thin layer chromatography. *J. Pharm. Biomed. Anal.* 17:671–677.

Mehta, B.H. and Morge, S.B. 2008. HPTLC-densitometric analysis of candesartan cilexetil and hydrochlorothiazide in tablets. *J. Planar Chromatogr.* 21:173–176.

Meyyanathan, S.N. and Suresh, B. 2005. HPTLC method for the simultaneous determination of amlodipine and benazepril in their formulations. *J Chromatogr. Sci.* 43:73–75.

Mielcarek, J. 2001. Normal-phase TLC separation of enantiomers of 1,4-dihydropyridine derivatives. *Drug Dev. Ind. Pharm.* 27:175–179.

Mielcarek, J. and Daczkowska, E. 1999. Photodegradation of inclusion complexes of isradipine with methyl-β-cyclodextrin. *J. Pharm. Biomed. Anal.* 21:393–398.

Mikami, E., Onho, T., and Matsumoto, H. 2002. Simultaneous identification/determination system for phentolamine and sildenafil as adulterants in soft drinks advertising roborant nutrition. *Forensic Sci. Int.* 130:140–146.

Misztal, G., Paw, B., Skibiński, R., Komsta, Ł., and Iwaniak, K. 2002. Analysis of some nonselective calcium-channel blockers in normal-phase chromatographic systems. *J. Planar Chromatogr.* 15:458–462.

Misztal, G., Paw, B., Skibiński, R., Komsta, Ł., and Kołodziejczyk, J. 2003. Analysis of non-selective calcium-channel blockers by reversed-phase TLC. *J. Planar Chromatogr.* 16:433–437.

Mohammad, A., Sharma, S., and Bhawani, S.A. 2009. Identification and quantification of lisinopril from pure, formulated and urine samples by micellar thin layer chromatography. *Int. J. Pharm. Tech. Res.* 1:264–272.

Mousa, B.A., Abadi, A.H., Abou-Youssef, H.E., and Mahrouse, M.A. 2006. Simultaneous determination of indapamide and captopril in binary mixture by derivative spectrophotometry and TLC densitometry. *Bull. Fac. Pharm. Cairo Univ.* 44:63–76.

Moussa, B., Mohamed, M., and Youssef, N. 2010. Acid-alkali degradation study on olmesartan medoxomil and development of validated stability-indicating chromatographic methods. *J. Chil. Chem. Soc.* 55:199–202.

Moussa, B., Mohamed, M., and Youssef, N. 2011. Simultaneous densitometric TLC analysis of olmesartan medoxomil and hydrochlorothiazide in the tablet dosage form. *J. Planar Chromatogr.* 24:35–39.

Mutschler, E. and Derendorf, H. 1995. *Drug Actions Basic Principles and Therapeutic Aspects.* CRS Press, Boca Raton, FL.

Odović, J., Aleksić, M., Stojimirović, B., Milojković-Opsenica, D., and Tešić, Ž. 2009. Normal-phase thin-layer chromatography of some angiotensin converting enzyme (ACE) inhibitors and their metabolites. *J. Serb. Chem. Soc.* 74:677–688.

Ojanperä, I., Goebel, K., and Vuori, E. 1999. Toxicological drug screening by overpressured layer chromatography. *J. Liq. Chromatogr. Related Technol.* 22:161–171.

Ojanperä, I., Ojansivu, R.L., Nokua, J., and Vuori, E. 1999a. Comprehensive TLC drug screening in forensic toxicology: Comparison of findings in urine and liver. *J. Planar Chromatogr.* 12:38–41.

Okuyama, M., Inoue, C., Aijima, K., Nakamura, Y., Kaburagi, T., and Shigematsu, A. 1997. in vitro method for assessing hepatic drug metabolism. *Biol. Pharm. Bull.* 20:1–5.

Panchal, H.J. and Suhagia, B.N. 2010. Simultaneous determination of atorvastatin calcium and ramipril in capsule dosage forms by high-performance liquid chromatography and high-performance thin layer chromatography. *J. AOAC Int.* 93:1450–1457.

Panchal, H.J., Suhagia, B.N., and Patel, N.J. 2009. Simultaneous HPTLC analysis of atorvastatin calcium, ramipril, and aspirin in a capsule dosage form. *J. Planar Chromatogr.* 22:265–271.

Panda, S.K., Pattanayak, P., Oraon, A., and Parhi, P.K. 2010. Quantitative estimation of reserpine in different parts of R. serpentina and R. tetraphylla by using HPTLC. *Der Pharmacia Lettre* 2:363–370.

Pandya, K.K., Satia, M., Gandhi, T.P., Modi, I.A., Modi, R.I., and Chakravarthy, B.K. 1995. Detection and determination of total amlodipine by high-performance thin-layer chromatography: A useful technique for pharmacokinetic studies. *J. Chromatogr. B Biomed. Appl.* 667:315–320.

Parambi, D.G.T., Mathew, M., and Ganesan, V. 2011. Quantitative analysis of valsartan in tablets formulations by high performance thin-layer chromatography. *J. Appl. Pharm. Sci.* 1:76–78.

Parambi, D.G.T., Mathew, S.M., Ganesan, V., Jose, A., and Revikumar, K.G. 2010. A validated HPTLC determination of an angiotensin receptor blocker olmesartan medoxomil from tablet dosage form. *Int. J. Pharm. Sci. Rev. Res.* 4:36–39.

Parent, A.A., Anderson, T.M., Michaelis, D.J., Jiang, G., Savage, P.B., and Linford, M.R. 2006. Direct ToF-SIMS analysis of organic halides and amines on TLC plates. *Appl. Surf. Sci.* 252:6746–6749.

Patel, V.B., Sahu, R., and Patel, B.M. 2011. Simultaneous determination of amlodipine besylate and atorvastatin calcium in pharmaceutical tablet formulation by high performance thin layer chromatographic method. *Int. J. Chem. Tech. Res.* 3:695–698.

Patravale, V.B., Nair, V.B., and Gore, S.P. 2000. High performance thin layer chromatographic determination of nifedipine from bulk drug and from pharmaceuticals. *J. Pharm. Biomed. Anal.* 23:623–627.

Peng, S., Ahlmann, N., Kunze, K., et al. 2004. Thin-layer chromatography combined with diode laser desorption/atmospheric pressure chemical ionization mass spectrometry. *Rapid Commun. Mass Spectrom.* 18:1803–1808.

Perišić-Janjić, N.U. and Agbaba, D. 2002. The retention behavior of antihypertensive drugs of the type angiotensin-converting enzyme inhibitors on thin layers of aminoplast. *J. Planar Chromatogr.* 15:210–213.

Potale, L.V., Damle, M.C., Khodke, A.S., and Bothara, K.G. 2010. A validated stability indicating HPTLC method for simultaneous estimation of ramipril and telmisartan. *Int. J. Pharm. Sci. Rev. Res.* 2:35–39.

Pothier, J., Galand, N., and Viel, C. 1991. Separation of alkaloids in plant extracts by overpressured layer chromatography with ethyl acetate as mobile phase. *J. Planar Chromatogr.* 4:392–396.

Pyka, A., Gurak, D., and Bober, K. 2002. New visualizing reagents for selected phenolic drugs investigated by thin layer chromatography. *J. Liq. Chromatogr. Related Technol.* 25:1483–1495.

Pyra, E., Wawrzycki, S., and Modzelewska-Banachiewicz, B. 2000. Synthesis and application of 4-dipropylaminodiazabenzene-4′-isothiocyanate to the derivatization and chromatographic separation of biogenic amines. *Chromatographia* 51:S313–S315.

Ramadan, N.K., Mohamed, H.M., and Moustafa, A.A. 2010. Rapid and highly sensitive HPLC and TLC methods for quantitation of amlodipine besilate and valsartan in bulk powder and in pharmaceutical dosage forms and in human plasma. *Anal. Lett.* 43:570–581.

Ramteke, M., Kasture, A., and Dighade, N. 2010. Development of high prformance thin layer chromatographic method for simultaneous estimation of atenolol and nifedipine in combined dosage form. *Asian J. Chem.* 22:5951–5955.

Roth, H.J. and Fenner, H. 2000. *Arzneistoffe Strktur-Bioreaktivität-Wirkungsbezogene Eigenschaften.* Detscher Apotheker Verlag Sttutgart, Germany.

Salo, P.K., Salomies, H., Harju, K., et al. 2005. Analysis of small molecules by ultra thin-layer chromatography-atmospheric pressure matrix-assisted laser desorption/ionization mass spectrometry. *J. Am. Soc. Mass. Spectrom.* 16:906–915.

Sane, R.T., Francis, M., Hijli, P.S., Pawar, S., and Pathak, A.R. 2002. Determination of doxazosin in its pharmaceutical formulation by high-performance thin-layer chromatography. *J. Planar Chromatogr.* 15:34–37.

Sathe, S.R. and Bari, S.B. 2007. Simultaneous analysis of losartan potassium, atenolol, and hydrochlorothiazide in bulk and in tablets by high-performance thin-layer chromatography with UV absorption densitometry. *Acta Chromatogr.* 19:270–278.

Sbârcea, L., Udrescu, L., Dragan, L., Trandafirescu, C., Soica, C., and Bojita, M. 2010a. Thin-layer chromatographic studies of some angiotensin converting enzyme inhibitors and their inclusion complexes with β-cyclodextrin. *Studia Universitatis "Vasile Goldiş", Seria Ştiinţele Vieţii* 20:35–38.

Sbârcea, L., Udrescu, L., Drăgan, L., Trandafirescu, C., Szabadai, Z., and Bojiţă, M. 2010b. Thin-layer chromatography analysis for cyclodextrins inclusion complexes of fosinopril and zofenopril. *Farmacia* 58:478–484.

Shah, N.J., Suhagia, B.N., Shah, R.R., and Patel, N.M. 2007a. Development and validation of a HPTLC method for the simultaneous estimation of irbesartan and hydrochlorothiazide in tablet dosage form. *Indian J. Pharm. Sci.* 69:240–243.

Shah, N.J., Suhagia, B.N., Shah, R.R., and Patel, N.M. 2007b. Development and validation of a simultaneous HPTLC method for the estimation of olmesartan medoxomil and hydrochlorothiazide in tablet dosage form. *Indian J. Pharm. Sci.* 69:834–836.

Shah, S.A., Rathod, I.S., Suhagia, B.N., Savale, S.S., and Patel, J.B. 2001. Simultaneous determination of losartan and hydrochlorothiazide in combined dosage forms by first-derivative spectroscopy and high-performance thin-layer chromatography. *J. AOAC Int.* 84:1715–1723.

Shanmugasundaram, R.P. and Velraj, M. 2011. Validated HPTLC method for simultaneous estimation of irbesartan and hydrochlorothiazide in a tablet dosage form. *Der. Pharma. Chem.* 3:310–317.

Shrivastava, A.R., Barhate, C.R., and Kapadia, C.J. 2009. Stress degradation studies on valsartan using validated stability-indicating high-performance thin-layer chromatography. *J. Planar Chromatogr.* 22:411–416.

Sleckman, B.P. and Sherma, J. 1983. Separation and quantification of epinephrine, norepinephrine, and dopamine by chromatography on diphenyl thin layers. *J. High Res. Chromatogr.* 6:156–157.

Sreekanth, N., Awen, B.Z., and Rao, ChB. 2010. HPTLC method development and validation of trandolapril in bulk and pharmaceutical dosage forms. *J. Adv. Pharm. Tech. Res.* 1:172–179.

Stohs, S.J. and Scratchley, G.A. 1975. Separation of thiazide diuretics and antihypertensive drugs by thin-layer chromatography. *J. Chromatogr.* 114:329–333.

Stolarczyk, M., Anna, M., and Krzek, J. 2008. Chromatographic and densitometric analysis of hydrochlorothiazide, valsartan, kandesartan, and enalapril in selected complex hypotensive drugs. *J. Liq. Chromatogr. Related Technol.* 31:1892–1902.

Suedee, R., Songkram, C., Petmoreekul, A., Sangkunakup, S., Sankasa, S., and Kongyarit, N. 1999a. Direct enantioseparation of adrenergic drugs via thin-layer chromatography using molecularly imprinted polymers. *J. Pharm. Biomed. Anal.* 19:519–527.

Suedee, R., Srichana, T., Saelim, J., and Thavornpibulbut, T. 1999b. Chiral determination of various adrenergic drugs by thin-layer chromatography using molecularly imprinted chiral stationary phases prepared with α-agonists. *Analyst* 124:1003–1009.

Sugihara, J., Sugawara, Y., Ando, H., Harigaya, S., Etoh, A., and Kohno, K. 1984. Studies on the metabolism of diltiazem in man. *J. Pharm. Dyn.* 7:24–32.

Tambe, S.R., Shinde, R.H., Gupta, L.R., Pareek, V., and Bhalerao, S.B. 2010. Development of LLE and SPE procedures and its applications for determination of olmesartan in human plasma using RP-HPLC and HPTLC. *J. Liq. Chromatogr. Related Technol.* 33:423–430.

Tipre, D.N. and Vavia, P.R. 2001. Oxidative degradation study of nitrendipine using stability indicating, HPLC, HPTLC and spectrophotometric method. *J. Pharm. Biomed. Anal.* 24:705–714.

Vekariya, N.R., Patel, G.F., Bhatt, H.S., Patel, M.B., Dholakiya, R.B., and Ramani, G.K. 2009. Application of TLC-densitometry method for simultaneous estimation of telmisartan and amlodipine besylate in pharmaceutical dosage form. *Int. J. Pharm. Tech. Res.* 1:1644–1649.

Vekariya, N.R., Patel, G.F., and Dholakiya, R.B. 2010. Stability—Indicating HPTLC determination of telmisartan in bulk and tablets. *Res. J. Pharm. Tech.* 3:900–904.

Vovk, I., Popović, G., Simonovska, B., Albreht, A., and Agbaba, D. 2011. Ultra-thin-layer chromatography mass spectrometry and thin-layer chromatography mass spectrometry of single peptides of angiotensin-converting enzyme inhibitors. *J. Chromatogr. A.* 1218:3089–3094.

Williams, D.A. 2008. Central and peripheral sympatholytics and vasodilators. In: *Foye's Principles of Medicinal Chemistry*, ed. T.L. Lemke and D.A. Williams, Wolters Kluwer/Lippincott Williams & Wilkins, Philadelphia, PA, 769–796.

Witek, A. and Przyborowski, L. 1996. Chromatographic (TLC) analysis of diltiazem in pharmaceutical form in presence of other antiarrhythmics. *Acta Pol. Pharm. Drug Res.* 53:9–12.

Wong, K.K., Lan, S., and Migdalof, B.H. 1981. In vitro biotransformations of [14C] captopril in the blood of rats, dogs and humans. *Biochem. Pharm.* 30:2643–2650.

Wood, J.L. and Steiner, R.R. 2011. Purification of pharmaceutical preparations using thin-layer chromatography to obtain mass spectra with direct analysis in real time and accurate mass spectrometry. *Drug Test. Anal.* 3:345–351.

Wyszomirska, E., Czerwińska, K., and Mazurek, A.P. 2010. Identification and determination of antihypertonics from the group of angiotensin—Convertase inhibitors by densitometric method in comparition with HPLC method. *Acta Pol. Pharm. Drug Res.* 67:137–143.

Yadav, N. and Srinivasan, B.P. 2010. A validated reverse phase HPLC and HPTLC method for estimation of olmesartan medoxomil in pharmaceutical dosage form. *Der. Pharma. Chem.* 2:103–112.

27 TLC of Beta-Blockers and Beta-Agonists

Anna Gumieniczek and Anna Berecka

CONTENTS

27.1 THIN-LAYER CHROMATOGRAPHY OF BETA-BLOCKERS AND BETA-AGONISTS

A review of thin-layer chromatography (TLC) methods for determination of the drugs that influence beta adrenergic receptor (beta-blockers and beta-agonists) is presented. TLC methods are described for determination in pharmaceuticals and in biological material. Different pretreatment modes, including liquid–liquid extraction, solid-phase extraction, or direct spotting onto TLC plates, are shown. For quantitative methods, sensitivity, linearity, and specificity of the presented methods are compared. The stability data concerning some of the mentioned drugs are also described. Suitable methods for separation of these drugs as well as for separation of their enantiomers are also shown.

27.2 BETA-BLOCKERS

Beta-blockers are the drugs that bind to beta-adrenoreceptors and thereby block the binding of natural ligands like norepinephrine and epinephrine. This inhibits normal sympathetic effects that act through these receptors. The first generation of beta-blockers are nonselective, meaning that they block both beta 1 (β_1) and beta 2 (β_2) adrenoreceptors. This group includes alprenolol, bupranolol, carazolol, carvedilol, labetalol, nadolol, oxprenolol, pindolol, practolol, propranolol, sotalol, and timolol (Figure 27.1). Some of these drugs also block alpha-adrenoreceptors, for example, carvedilol and labetalol. The second generation of beta-blockers are more cardio-selective in

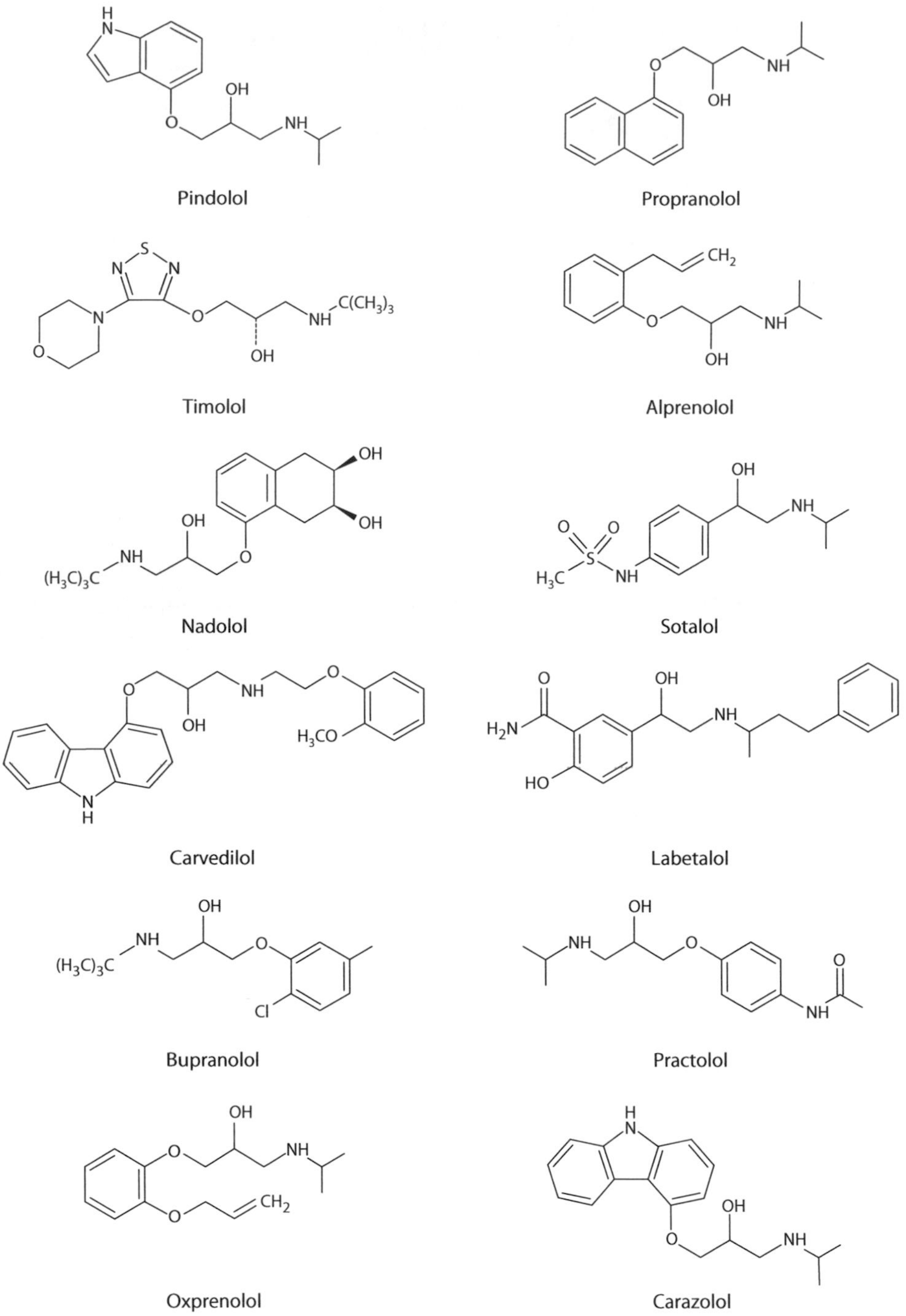

FIGURE 27.1   The chemical structures of nonselective beta-blockers.

that they are more selective for β_1-adrenoceptors. This group includes acebutolol, atenolol, betaxolol, bisoprolol, celiprolol, esmolol, metoprolol, nebivolol, and talinolol (Figure 27.2) (Mehvar and Brocks 2001).

Beta-blockers are used for treating hypertension, angina, myocardial infarction, arrhythmias, and heart failure. For therapy of glaucoma, some of them (especially timolol and betaxolol) are also used in ocular formulations. They can be used alone or in combination with a wide range of drugs

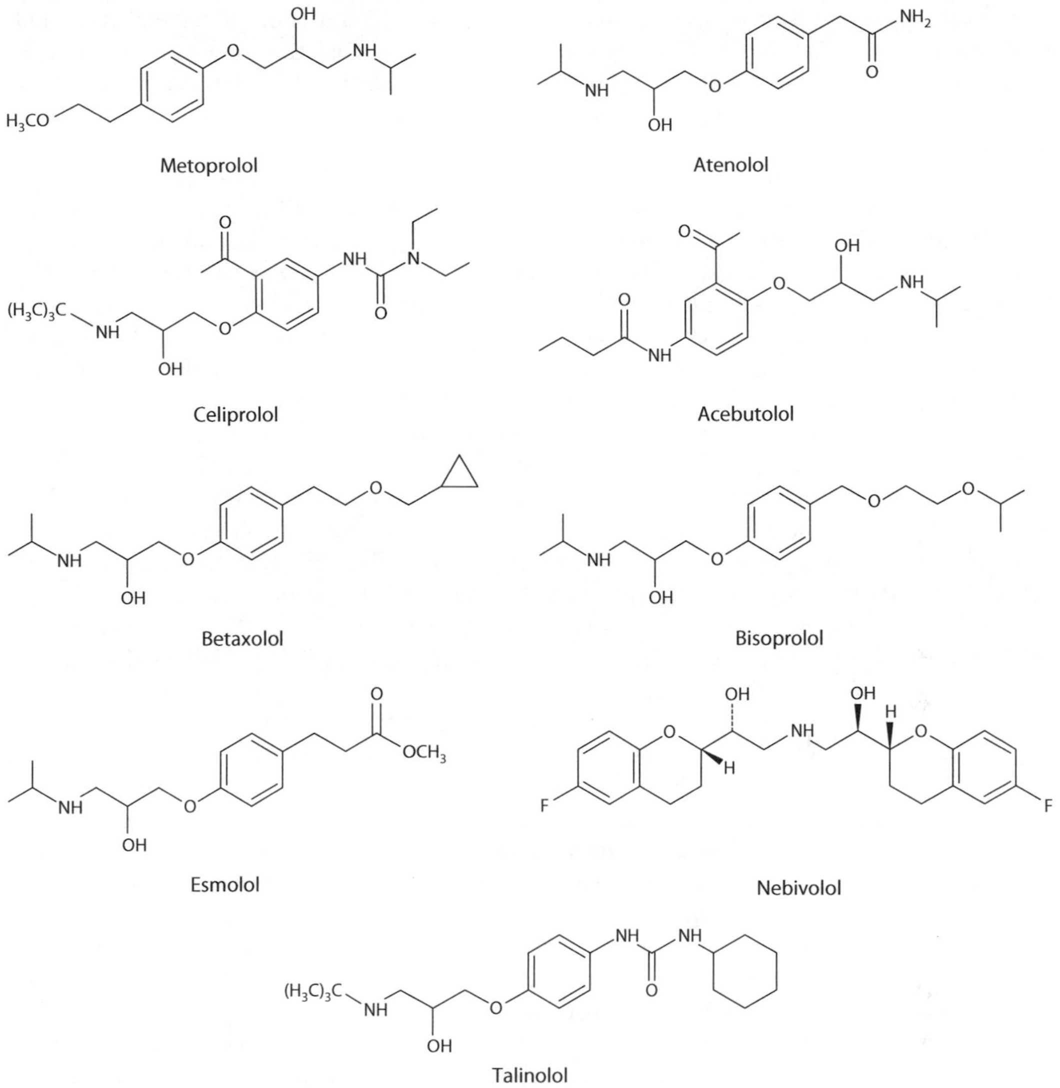

FIGURE 27.2   The chemical structures of selective beta-blockers.

affecting cardiovascular system and glaucoma. The idea behind combining two different drugs is that each drug has a different mechanism of action and thus may help tackle different mechanisms involved in causing the pathological condition. In this way, greater effects may be achieved than with single drug therapy (Haeusler 1990).

27.2.1   TLC ASSAY OF BETA-BLOCKERS IN BULK DRUG AND IN ONE COMPONENT PHARMACEUTICAL FORMULATIONS

For high-performance thin-layer chromatography (HPTLC) assay of carvedilol, silica gel 60F$_{254}$ and the mobile phase consisting of ethyl acetate–toluene–methanol (1:4:3.5, v/v/v) were used. The detection was carried out at 242 nm. The R$_f$ value was 0.65 ± 0.02. The method was validated in terms of linearity, accuracy, and precision. The linearity curves were found to be linear over the range 50–300 ng/spot. The limit of detection (LOD) and limit of quantification (LOQ) were found to be 10 and 35 ng/spot, respectively (Patel et al. 2006b).

A similar method was developed on silica gel $60GF_{254}$ using a mixture of toluene–methanol–ethyl acetate–ammonia 25% (8:2:1:0.2, v/v/v/v) as mobile phase. Quantification was carried out by the use of densitometer at 254 nm. This HPTLC system was quantitatively evaluated in terms of stability, precision, repeatability, specificity, accuracy, and calibration providing the utility in the analysis of its tablet dosage form (Badgujar et al. 2005).

A new simple HPTLC method with a different detection mode was developed for determination of nadolol and pindolol in tablets. The stationary phase was silica gel $60F_{254}$, and the mobile phase was ethyl acetate–methanol–glacial acetic acid (49:49:2, v/v/v). Detection and quantification were done densitometrically at 270 nm and by video scanning at 254 nm. In a densitometric procedure, the linearity range was 0.2–1.2 µg/10 µL for nadolol and pindolol. In video scanning, respective linearity ranges were 2.0–12.0 and 0.2–1.2 µg/10 µL. In densitometric procedure, the RSD obtained for the standard solutions ranged from 1.14% to 2.80% and from 0.74% to 1.85% for nadolol and pindolol, respectively. For the video scanning, the RSD values ranged from 0.68% to 2.36% and from 0.79% to 3.20% for nadolol and pindolol, respectively (Gumieniczek et al. 2002).

A simple, accurate, and precise HPTLC method was developed for estimation of propranolol as bulk drug and in tablet formulations. The assay was performed on silica gel $60F_{254}$ plates by ascending movement of solvent system for 70 mm in a chamber previously saturated with vapors for 20 min. The solvent system consisted of propan-2-ol–ethyl acetate–ammonia 25% (1:8.5:0.5, v/v/v). The spots so developed were densitometrically scanned at 290 nm. The linearity of the method was found to be within the concentration range of 200–2000 ng/spot. The validation parameters were in accordance with the requirements of ICH guidelines (Bhavar and Chatpalliwar 2008).

The HPTLC method for analysis of timolol was reported by Kulkarni and Amin (2000). The mobile phase was ethyl acetate–methanol–propan–2-ol–ammonia 25% (80:20:2:1, v/v/v/v). The calibration curve was linear in the range of 100–600 ng. Densitometric analysis was carried out at 294 nm. The mean (±RSD) value of correlation coefficient was 0.996 ± 0.081. The system precision and the method were excellent with an RSD of 2.8% and 1.0%, respectively. The LOD and LOQ were 10 ng and 40 ng, respectively, while the mean percent recovery was found to be 98.6. Timolol was degraded by exposing to heat, acid, and base conditions. The degraded products were found to be well separated from the pure drug with significantly different R_f values suggesting a stability indicating analysis. The method was utilized to analyze timolol from conventional eye drops and from solid polymeric ocular inserts and oral preparations.

The TLC assay of acebutolol was based on separation of the drug from its acid-induced degradation product followed by densitometric measurement at 230 nm. The separation was carried out on silica gel $60F_{254}$ using ethanol–glacial acetic acid (4:1, v/v) as mobile phase. Using the aforementioned TLC system, the R_f values of the compounds were found to be 0.52 and 0.25 for the drug and degradation products, respectively. Second-order polynomial equation was used for the regression line. The calibration graph was constructed in the range 0.5–10 µg/spot. The LOD of acebutolol in the proposed method was found to be 0.32 µg/mL (El-Gindy et al. 2001).

The next HPTLC method was developed for quantification of esmolol using the mobile phase chloroform–methanol–acetic acid (7:2:0.2, v/v/v). The detection was carried out at 225 nm, while the mean R_f value was found to be 0.53 (Patel et al. 2006a).

Nebivolol from the respective formulations was determined on silica gel $60F_{254}$ HPTLC plates with toluene–ethyl acetate–methanol–formic acid (8:6:4:1, v/v/v/v) as mobile phase. The plates were developed to a distance of 8 cm. Densitometric quantification was performed at 285 nm by reflectance scanning. The calibration plot for nebivolol standard was linear with r = 0.9991. The LOD and LOQ were found to be 18.65 and 62.18 ng/band, respectively. The method was selective and specific with potential application in pharmaceutical analysis (Reddy and Devi 2007). In the second HPTLC method concerning nebivolol, silica gel $60F_{254}$ and the mobile phase consisting of ethyl acetate–toluene–methanol–ammonia 25% (1:6:2:0.1, v/v/v/v) were used. The detection was carried out at 282 nm, and the R_f value was found to be 0.33 ± 0.02. The method was validated in terms of linearity, accuracy, and precision over the range 100–600 ng/spot. The LOD and LOQ were found to be 30 and 100 ng/spot, respectively (Patel et al. 2007).

Next, HPTLC method for determination of nebivolol was developed using silica gel $60F_{254}$, and the solvent system consisted of toluene–methanol–triethylamine (3.8:1.2:0.2, v/v/v). Densitometry was carried out in the absorbance mode at 281 nm. The system was found to give compact spot for nebivolol with R_f value of ca. 0.33. The linear regression analysis data for the calibration plots showed good relationship with $r^2 = 0.9994$ in the concentration range 500–3000 ng/spot. The LOD and LOQ were 63.10 ng/spot and 191.23 ng/spot, respectively. Nebivolol was subjected to acid and alkali hydrolysis, oxidation, thermal degradation, and photodegradation. All the peaks of degradation products were well resolved from the standard drug with significantly different R_f values (Shirkhedkar et al. 2010).

A simple, rapid, reliable, and accurate HPTLC method was developed for quantitative determination of metoprolol in bulk and tablets. Silica gel $60F_{254}$ as stationary phase prewashed with methanol and the mixture toluene–methanol–triethylamine (3:0.5:0.3, v/v/v) as mobile phase were used. The spots were scanned at 274 nm. The R_f value of metoprolol was ca. 0.40. Calibration curves were linear in the range 5,000–10,000 ng per spot. The LOD and LOQ were found to be 431.22 ng/spot and 1306.74 ng/spot, respectively (Sathe et al. 2008). The similar work was done for determination of metoprolol from tablets and ampoules. After separation on silica gel GF_{254} plates using acetone–methanol–triethylamine (2:1:0.1, v/v/v) for tablets and acetone–triethylamine (2.5:0.5, v/v) for ampoules, the chromatographic zones corresponding to the spots of metoprolol were scanned at 275 nm. The calibration function was established in the ranges 1–28 µg for tablets and 1–9 µg for ampoules. The method was precise and reproducible with recovery values of 99.1%–99.4% (Vujic et al. 1997).

HPTLC method was also performed for determination of atenolol, acebutolol, bisoprolol, and propranolol using chloroform–methanol–ammonia 25% (15:7:0.2, v/v/v) as mobile phase. ultraviolet (UV) densitometric measurements were performed at the wavelength of maximum absorption of these drugs. Pharmaceutical preparations from a variety of manufacturers were analyzed. The LOD and LOQ ranged from 30 to 400 ng and recovery from 97.14% to 102.18% (Krzek and Kwiecień 2005).

For identification and determination of bisoprolol and labetalol in pharmaceuticals, TLC method after the derivatization with dabsyl chloride was also described (Witek et al. 1999).

27.2.2 TLC Assay of Beta-Blockers in Two or More Component Pharmaceutical Combinations

For the combined therapy of cardiovascular disorders, cardio-selective beta-blockers like atenolol, nebivolol, and metoprolol are frequently used in combination with diuretics and other antihypertensive agents. Such formulations are now available in one tablet or capsule. The combination of atenolol and indapamide is available as tablet dosage forms in the ratio of 20:1. HPTLC method was developed for analysis of the aforementioned formulation. Atenolol and indapamide were separated on silica gel $60GF_{254}$ using a mixture of toluene–ethanol–acetone–acetic acid (7:2.5:3:0.3, v/v/v/v) as mobile phase. Quantification was carried out at 266 nm in the absorbance mode. The R_f values of atenolol and indapamide were found to be 0.21 and 0.74, respectively. The recovery was found to be 99.42% or 100.51% and 99.07% or 98.65% for atenolol and indapamide by height and by area, respectively (Gupta et al. 2007).

A new HPTLC method was developed for simultaneous determination of different mixtures, atenolol and chlorthalidone (I), atenolol, chlorthalidone and amiloride (II), atenolol, hydrochlorothiazide, and amiloride (III) in bulk powders and in pharmaceutical dosage forms. The adequate mobile phase was dioxane–ethyl acetate–acetonitrile–propan-1-ol (10:7:5.5:3, v/v/v/v). Detection was carried out densitometrically at 283, 266, 226, and 362 for atenolol, chlorthalidone, hydrochlorothiazide, and amiloride, respectively. Calibration curves of the drugs were linear in the range

1–100 µg/mL with correlation coefficients not less than 0.9996. The percentage recoveries ranged from 98.3 ± 1.42 to 100.8 ± 0.79 (Salem 2004).

In the next work, two drugs that are administered in combination to provide greater therapeutic effects, atenolol and chlorthalidone, were selectively determined in the presence of their degradation products, using silica gel plates and chloroform–methanol–ethyl acetate–ammonia 25% (75:28:2:1.6, v/v/v/v) as developing system. The suggested method was used for determination of the studied drugs at 227 nm in their pharmaceutical formulations, and the results were statistically compared to the reported reversed phase high-performance liquid chromatography (RP-HPLC) method (Abdelwahab 2010).

One of the most frequently applied pharmaceutical combination is atenolol with amlodipine or some other dihydropyridine derivative. For simultaneous determination of atenolol and amlodipine in tablets, a HPTLC method was developed using methylene chloride–methanol–ammonia 25% (8.8:1.3:0.1, v/v/v) as mobile phase and $60F_{254}$ silica gel plates. Detection was carried out at 230 nm. The R_f values were ca. 0.33 and 0.75, respectively. Calibration curves were linear in the range 10–500 µg/mL for the both drugs (Argekar and Powar 2000).

A similar work was done by Ilango et al. (2000). The mobile phase consisted of toluene, ethanol, acetone, and ammonia 25% (6.5:0.7:4.0:0.4, v/v/v/v), while the stationary phase was silica gel F_{254}. The R_f value of atenolol and amlodipine were found to be ca. 0.16 and 0.54.

An accurate and precise HPTLC method for simultaneous estimation of atenolol and nifedipine in their combined dosage form was also developed. The study employed kieselguhr $60GF_{254}$ and a mobile phase comprising cyclohexane–methanol–ethyl acetate–ammonia 25% (5:1.5:3:0.5, v/v/v/v). The detection was carried out at 230 nm. The linear detector response for atenolol was observed between 5.7 and 18.9 µg/mL, while for nifedipine it was 2.3 and 7.0 µg/mL. The mean percentage results of recovery were 99.76 for atenolol and 100.04 for nifedipine (Ramteke et al. 2010).

For simultaneous determination of atenolol and nitrendipine in pharmaceutical dosage forms, silica gel 60F HPTLC plates and the mobile phase consisting of chloroform–methanol–toluene–ammonia 25% (2:2.5:5.5:0.1, v/v/v/v) were used. Detection and quantification were done densitometrically at 233 nm. The linearity ranges were 4–10 µg/spot and 1.6–4.0 µg/spot, and the percentage recoveries were 101.10 and 98.43 for atenolol and nitrendipine, respectively (Argekar and Sawant 1999).

In the next paper, chromatographic separation of atenolol and lercanidipine was achieved on silica gel $60F_{254}$ with toluene–methanol–triethylamine (3.5:1.5:0.1, v/v/v) as mobile phase. Detection was performed densitometrically at 275 nm. The R_f values of atenolol and lercanidipine were ca. 0.24 and 0.68, respectively. The reliability of the method was assessed by evaluation of linearity (2,000–12,000 ng/band for atenolol and 400–2,400 ng/band for lercanidipine) and accuracy (98.94% ± 0.30% for atenolol and 99.75% ± 0.69% for lercanidipine) (Deore et al. 2008).

A simple HPTLC method was also developed for the simultaneous determination of atenolol and losartan. The method used silica gel $60F_{254}$ as stationary phase (prewashed with methanol) and toluene–methanol–triethylamine (6:4:0.5, v/v/v) as mobile phase. Detection and quantitation was performed densitometrically at 230 nm. The R_f values of atenolol and losartan were ca. 0.45 and 0.67, respectively. Calibration curves were linear over the ranges 1000–4000 ng/spot for the both drugs. The LOD and LOQ for atenolol and losartan were found to be 211.37 and 640.52, and 207.55 and 628.95 ng/spot with average recovery 99.82% and 99.86%, respectively (Sathe and Barl 2009).

The similar HPTLC method was elaborated for separation and quantitative analysis of atenolol, losartan, and hydrochlorothiazide in pharmaceutical formulations. After extraction with methanol, sample and standard solutions were applied to prewashed silica gel plates and developed with toluene–methanol–triethylamine (6.5:4:0.5, v/v/v) as mobile phase. Zones were scanned densitometrically at 274 nm. The R_f values of atenolol, losartan, and hydrochlorothiazide were 0.43, 0.60, and 0.29, respectively. Calibration plots were linear in the ranges 1000–5000 ng/band for atenolol and losartan and 250–1250 ng/band for hydrochlorothiazide, while the correlation coefficients (r) were 0.9993, 0.9994, and 0.9994, respectively (Sathe and Bari 2007).

In the literature, the validated HPTLC method for simultaneous analysis of metoprolol and amlodipine in pharmaceutical preparations was also found. Separation was achieved on silica gel $60F_{254}$ with methanol–ethyl acetate–water–toluene–ammonia 25% (1.5:5.0:0.3:3.0:0.3, v/v/v/v/v) as mobile phase. Densitometric quantification was performed at 236 nm by reflectance scanning. The R_f values of metoprolol and amlodipine were ca. 0.31 and 0.43, respectively. The linearity of the method was investigated in the range 180–280 µg/mL for metoprolol and 18–28 µg/mL for amlodipine. The method was validated for precision, accuracy, specificity, and ruggedness (Kakde and Bawane 2009).

Two HPTLC methods were elaborated for analysis of metoprolol and atorvastatin in capsules. Both the drugs were separated on silica gel $60F_{254}$ with toluene–methanol–ethyl acetate–glacial acetic acid (7:1.5:1:0.5, v/v/v/v) (Wankhede et al. 2011) or with chloroform–methanol–glacial acetic acid (9:1.5:0.2, v/v/v) (Patole et al. 2011) as mobile phases. Densitometric analysis of the drugs was performed at 276 nm (Wankhede et al. 2011) or at 220 nm (Patole et al. 2011). Good linear relationships between response and concentration of metoprolol were obtained over the range 1,000–6,000 ng/band (Wankhede et al. 2011) or 500–25,000 ng/spot (Patole et al. 2011). In the method of Wankhede et al. (2011), metoprolol and atorvastatin were also subjected to acid, base, peroxide, heat, and UV-induced degradation study.

Nebivolol is another beta-blocker frequently used in combined pharmaceutical combinations. As all beta-blockers, it is usually combined with diuretics or antihypertensive agents. Three HPTLC methods were found for simultaneous determination of nebivolol with hydrochlorothiazide (Bhat et al. 2008, Damle et al. 2010, Dhandapani et al. 2010). In the first method, optimum separation was achieved on silica gel $60F_{254}$ plates with ethyl acetate–methanol–acetic acid (6.5:1:0.5, v/v/v) as mobile phase. Detection and quantification were performed at 280 and 270 nm for nebivolol and hydrochlorothiazide, respectively. The drugs were resolved with R_f values of 0.46 ± 0.02 and 0.78 ± 0.02 for nebivolol and hydrochlorothiazide, respectively. The drugs were subjected to hydrolysis under acidic, basic, and neutral conditions, oxidation, heat, and photolysis. However, degradation products only occurred when the drugs were subjected to oxidative stress. The degradation products resulting from these stress conditions did not interfere with the drug peaks (Damle et al. 2010). In the second HPTLC method, chromatograms were developed using a mobile phase of ethyl acetate–methanol–ammonia 25% (8.5:1:0.5, v/v/v) on silica gel $60F_{254}$ plates. The drugs were quantified by densitometric absorbance mode at 285 nm. The R_f values of nebivolol and hydrochlorothiazide were 0.41 and 0.21, respectively. The recovery studies of 98.88%–102.41%, RSD of not more than 0.8% and correlation coefficient (r) of 0.9954–0.9999, were obtained (Dhandapani et al. 2010). In the third work, the mobile phase consisted of ethyl acetate, methanol, ammonia 25% (8.5:1:0.5, v/v/v), and wavelength of detection was 280 nm. The developed method was validated as per ICH guidelines (Bhat et al. 2008).

A simple and rapid HPTLC method was elaborated for simultaneous determination of propranolol and hydrochlorothiazide. The method was carried out by using benzene–methanol–ethyl acetate–ammonia 25% (8:2:1:0.2, v/v/v/v) as mobile phase. The quantification was done by densitometry at 280 nm. Trimethoprim was used as internal standard. Linear relationship was obtained within the concentration range of 40–200 ng for propranolol and 20–100 ng for hydrochlorothiazide (Suedee and Heard 1997).

Timolol is a beta-blocker frequently used in the treatment of glaucoma-related elevated eye pressure. For this treatment, some combined ocular pharmaceutical formulations also exist in the market. In the analytical literature concerning such formulations, one paper about the simultaneous determination of timolol and dorzolamide by HPTLC method was found (Bebawy 2002). The method used silica gel GF_{254} plates, methanol–ammonia 25% (100:1.5, v/v) as mobile phase and densitometry at 297 for timolol and 253 nm for dorzolamide. The calibration function for timolol was established in the range of 0.5–4.5 µg.

27.2.3 TLC FOR CHIRAL SEPARATION OF BETA-BLOCKERS

Each of the available beta-blockers has one or more chiral centers in its structure, and in all cases, at least one of the chiral carbon atoms residing in the alkyl side chain is directly attached to a hydroxyl group. Most of the beta-blockers with one chiral center (e.g., propranolol, metoprolol, atenolol, pindolol, and acebutolol) are marketed as a racemate consisting of two enantiomers. Labetalol, which has two chiral centers, is marked as a racemate consisting of four isomers. As for nadolol, this drug has three chiral centers in its structure. However, the two ring hydroxyl groups are in the *cis* orientation allowing only for four isomers (Mehvar and Brocks 2001).

The enantiomers of beta-blockers possess markedly different pharmacodynamics, and in some cases, pharmacokinetics. Most beta-blockers depend on (*S*)-enantiomers for the disease therapies. In general, the (*S*)-enantiomers are more potent than the diastomers in 10–500 folds. Some methods were developed for synthesis of single enantiomeric beta-blockers (e.g., for timolol), but separation of their racemates was not carried out easily. Liquid chromatography, capillary electrophoresis, super- and subcritical fluid chromatography, and capillary electrochromatography were used in separation of these drugs in clinical and pharmaceutical analysis. TLC might not be able to compete with HPLC regarding separation efficiency; however, it shows several advantages. TLC is a very simple, inexpensive, rapid, and flexible technique where many samples can be processed parallel on one plate, and very selective detection can be carried out by using spray reagents (Agustain et al. 2010, Gubitz and Schmid 2001).

Chromatography can be used to separate enantiomers either directly with chiral stationary phases (CSPs) or chiral mobile phase additives (chiral selectors) or indirectly with chiral derivatization reagents. Each of these techniques has advantages or disadvantages, and they all are present in the literature concerning beta-blockers.

Cellulose triphenylcarbamate derivatives were used as CSPs for resolution of the enantiomers of propranolol and bupranolol. The derivatives examined were (1) cellulose trisphenylacarbamate, (2) cellulose tris(2,3-dichlorophenyl carbamate), (3) cellulose tris(2,4-dichlorophenyl carbamate), (4) cellulose tris(2,6- dichlorophenyl carbamate), (5) cellulose tris(2,3-dimethylphenyl carbamate), (6) cellulose tris(3,4-dichlorophenyl carbamate), (7) cellulose tris(3,5-dichlorophenyl carbamate), (8) and cellulose tris(3,5-dimethylphenyl carbamate). The best resolution of propranolol racemate was obtained on CSP8 in a mobile phase hexane:propan-2-ol (80:20, v/v). The best resolution of bupranolol racemate was obtained on CSP5 using the same mobile phase. These results demonstrated the potential of cellulose triphenylcarbamates as CSPs in TLC and indicated that this was a useful method for direct, simple, and rapid (within 30 min) resolution of respective racemates. Physical aspects such as problems in cracking of the CSP, adhesion to plate, and interference of spot detection due to triphenylcarbamate chromophores were also discussed (Suedee and Heard 1997).

A variety of racemic compounds including labetalol were resolved using RP-TLC with mobile phases containing highly concentrated solutions of β-cyclodextrin (β-CD). It was possible to resolve some racemates that could not be separated on similar β-CD-bonded phase LC columns. In cases of racemates that could be resolved by either approach, it was found that the retention order was exactly opposite for these two methods. Enantiomeric resolution was highly dependent on the mobile phase composition. In particular, the type and amount of organic modifier as well as the concentration of β-CD affected the observed resolution. Possible reasons for such as chromatographic behavior were discussed (Armstrong et al. 1988).

Between very effective chiral selectors are macrocyclic antibiotics (Gubitz and Schmid 2001).

Resolution of the enantiomers of racemic atenolol, metoprolol, propranolol, and labetalol was achieved on silica gel plates using vancomycin as chiral-impregnating reagent or as chiral mobile phase additive. With vancomycin as impregnating agent, successful resolution of the enantiomers of atenolol, metoprolol, propranolol, and labetalol was achieved by use of the mobile phases: acetonitrile–methanol–water–dichloromethane (7:1:1:1, v/v/v/v), acetonitrile–methanol–water (6:1:1, v/v/v), acetonitrile–methanol–water–dichloromethane–glacial acetic acid (7:1:1:1:0.5, v/v/v/v/v), and acetonitrile-methanol-water (15:1:1, v/v/v). With vancomycin as mobile phase additive, successful resolution of the enantiomers of metoprolol, propranolol, and labetalol was achieved by use of the mobile phases: acetonitrile–methanol–0.56 mM aqueous vancomycin (pH 5.5) (6:1:1, v/v/v), acetonitrile–methanol–0.56 mM aqueous vancomycin (pH 5.5) (15:1:2, v/v/v), and acetonitrile–methanol–0.56 mM aqueous vancomycin (pH 5.5)–dichloromethane (9:1:1.5:1, v/v/v/v). The spots were detected by use of iodine vapor. The LODs were 1.3, 1.2, 1.5, and 1.4 µg for each enantiomer of atenolol, metoprolol, propranolol, and labetalol, respectively (Bhushan and Agarwal 2010).

Another way to perform chiral separation is the use of a chiral ion-pairing reagent as mobile phase additive.

Enantioselectivity of various beta-blockers, that is, propranolol, acebutolol, practolol, and nadolol, was studied by normal-phase ion-pair planar chromatography employing diol-, cyano-, and aminosilica layers, and ammonium-D-10-camphorsulfonate as the chiral counterion in the mobile phase dichloromethane-propan-2-ol. The improved enantioselectivity was observed after simultaneous change of such parameters as temperature of chromatographic system and polarity of mobile or stationary phases. Thermodynamic parameters characteristic of chiral recognition phenomena were estimated from the calculated Van't Hoff plots. Formation of multiple associates between molecules of the chiral counterion and enantiomers was postulated on results of supporting molecular modeling calculations. The influence of the experimental conditions on the detection of each enantiomer during densitometric quantitation was also estimated (Posyniak et al. 1996).

The enantiomeric separation of bisoprolol into its enantiomers was achieved by TLC and HPTLC silica gel plates using optically pure (+)-10-camphorsulphonic acid as a chiral selector in mobile phase and triethylamine–methanol–pentan-1-ol (0.14:9.9:0.18, v/v/v) as solvent system. Spots were located in a UV chamber. The detection limit was 8 µg for TLC and 50 ng for HPTLC for both the isomers. The effect of concentration of chiral selector on separation was studied, and satisfactory results were obtained followed by frequent resolution of the enantiomers. The procedure was applied successfully to resolve commercially available formulation of bisoprolol (Patel et al. 2011).

Several aromatic amino alcohols, including beta-blockers, were resolved on diol and/or silica gel plates using a mobile phase containing ammonium-D-10-camphorsulfonate or N-benzoxycarbonyl-glycyl-L-proline. A comparison was made between various chiral counterions/chiral mobile phase additives (Duncan et al. 1990).

Atenolol and propranolol were resolved into their enantiomers by adopting different modes of loading/impregnating Cu(II) complexes of L-proline. L-phenylalanine, L-histidine, L-N,N-dimethyl-L-phenylalanine, and L-tryptophan on commercial normal phase plates. The three different approaches were using the Cu(II)-L-amino acid complex as chiral mobile phase additive, ascending development in the solution of Cu complex, and using a solution of Cu(II) acetate as mobile phase with TLC plates impregnated in the solutions of amino acids. The spots were located using iodine vapor. The detection limit was 0.18 µg for each enantiomer (Bhushan and Tanwar 2009, 2010).

Resolution of atenolol, metoprolol, and propranolol into their enantiomers was also achieved using silica gel plates impregnated with L-aspartic acid as chiral selector. Different combinations of acetonitrile–methanol–water as mobile phase were found to be successful in resolving the enantiomers. The spots were detected with iodine vapor, and the LODs were found to be 0.26 µg for atenolol and 0.23 µg for metoprolol and propranolol as racemate (Bhushan and Arora 2003). A similar work was done on silica-gel plates impregnated with optically pure L-lysine (0.5%) and L-arginine (0.5%). In all cases, different combinations of acetonitrile–methanol solvent systems were found to be successful in resolving these compounds (Bhushan and Thuku Thiongo 1998).

Resolution of metoprolol, propranolol, carvedilol, bisoprolol, and labetalol was also achieved on silica gel plates impregnated with optically pure L-glutamate and L-aspartate. Acetonitrile–methanol–water–dichloromethane and acetonitrile–methanol–water–glacial acetic acid as mobile phases in different proportions enabled successful separation. The spots were detected with iodine vapor. The detection limits were 0.23, 0.1, 0.27, 0.25, and 0.2 µg for each enantiomer of metoprolol, propranolol, carvedilol, bisoprolol, and labetalol (Bhushan and Agarwal 2008).

Direct separation of metoprolol enantiomers was achieved using silica gel plates previously impregnated with the mobile phase (ethanol–water, 70:30, v/v) containing D-(-)-tartaric acid as chiral selector. The results of experiments with different concentrations of D-(-)-tartaric acid (5.8, 11.6, and 23 mM) revealed that the best resolution was achieved with 11.6 mM D-(–)-tartaric acid in both the mobile phase and the impregnation solution. Spot visualization on chromatograms was performed by use of a UV lamp at 254 nm or an iodine vapor-saturated chamber (Lučić et al. 2005).

Direct resolution of racemic atenolol and propranolol into their enantiomers was achieved on silica gel plates impregnated with optically pure L-tartaric acid, (R)-mandelic acid, and (–)-erythromycin as chiral selectors. Different solvent systems were worked out to resolve the enantiomers. The spots were detected using iodine vapor. The TLC method was validated for linearity, LOD, and LOQ. The influence of pH, temperature, and concentration of chiral selector was studied (Bhushan and Tanwar 2008).

The next approach in chiral separation is the use of molecularly imprinted polymers (MIP). This approach is based on polymerizing a monomer with a cross-linking agent in the presence of a chiral template molecule. After removing the template molecule, a chiral imprinted cavity remains, which shows high stereo-selectivity to the template molecule or closely related compounds (Gubitz and Schmid 2001).

The MIPs of S-timolol were prepared as CSPs. The resolution of the enantiomers of some cardiovascular drugs including propranolol, atenolol, timolol, and nadolol were investigated on these CSP. A mobile-phase system, which consisted of either methanol or acetonitrile, was used, and the effects of acetic acid content in the mobile phase were also investigated. The best resolution was achieved for enantioseparation of propranolol, timolol, and atenolol using methacrylic acid as a functional monomer using acetonitrile containing 5% acetic acid or methanol containing 1% acetic acid as mobile phases (Aboul-Enein et al. 2002).

Several MIPs were also prepared using R-(+)-propranolol, R-(+)- or S-(–)-atenolol, and three different functional monomers. Finally, the MIP of R-(+)-propranolol with 5% acetic acid in acetonitrile as mobile phase resolved the racemate of propranolol into two spots (Suedee et al. 2001).

In the study of Bhushan and Tanwar (2008), (R,S)-atenolol was derivatized with Marfey's reagent (1-fluoro-2,4-dinitrophenyl-5-L-alanine amide) and its four structural variants that reacted quantitatively with amino groups of atenolol. The derivatization reactions were carried out under conventional and microwave heating and compared. The resulting diastereomers were separated on RP-TLC with detection at 340 nm. It was observed that (R)-isomer was eluted before (S). The conditions of derivatization and chromatographic separation were optimized. The method was validated for linearity, repeatability, LOD, and LOQ (Bhushan and Tanwar 2008).

Some other papers concerning these topics exist in the literature, but their abstracts are not available for the authors (Spell and Stewart 1997, Šubert and Šlais 2001).

27.2.4 TLC FOR IDENTIFICATION AND PURITY/STABILITY TESTS OF BETA-BLOCKERS

Celiprolol and talinolol were separated on silica gel and aluminum oxide plates by ascending and horizontal development using various solvent systems. The best separation was achieved on silica gel by ascending technique using methanol–diethyl ether–ammonia 25% (24:20:1, v/v/v) as mobile phase. The substances were identified by UV irradiation at 254 nm or by reactions with a variety of reagents. The minimal detection limits (0.1 µg of each substance) were reached by using UV detection or acidified 15% solution of $FeCl_3$ followed by 15% solution of KI. The elaborated TLC method was employed for detection of celiprolol and talinolol in tablets (Witek and Przyborowski 1996).

In the next HPTLC method, timolol and betaxolol were separated and determined using silica gel $60F_{254}$, silanized C8 silica gel, and aluminum oxide $60F_{254}$ plates with ascending and horizontal developments. The most suitable conditions for separation of these drugs were silica gel with ethyl acetate–methanol–ammonia 25% (80:18:2, v/v/v) as mobile phase. The substances were visualized by UV irradiation at 254 nm or by spraying with suitable reagents. Quantification of timolol and betaxolol were performed densitometrically at 300 and 218 nm or by video scanning at 254 nm (Hopkała et al. 2003).

In the study of Marciniec et al. (2010), alprenolol in a solid state was radiated with a high-energy electron beam (9.96 MeV) at doses from 25 to 400 kGy and analyzed by HPTLC using the mobile phase consisting of methanol–ammonia 25% (99:1, v/v). Densitometric analysis was carried out directly from chromatograms at 270 nm. The applied method was validated and characterized by good precision (RSD = 3.95%) and sufficient accuracy (99.99%). Chromatograms recorded for samples irradiated at the doses of 25 kGy were unchanged; but at higher doses (100–400 kGy), additional peaks corresponding to the radio degradation products appeared ($R_f = 0.24$ and $R_f = 0.40$).

TLC method was also used for stability study of atenolol, acebutolol, and propranolol in acidic environment by Krzek et al. (2006). Hydrolysis was carried out in hydrochloric acid at concentrations of 0.1, 0.5, and 1 M for 2 h at 40°C, 60°C, and 90°C. The degradation processes that occurred in drugs were described with different kinetic parameters. This study demonstrated that the stability of chosen beta-blockers increased with their lipophilicity, that is, propranolol > acebutolol > atenolol.

In the literature concerning beta-blockers, there are also works in which TLC method was used for the purity study of two radio-labeled metoprolol derivatives, that is, ^{125}I-metoprolol and ^{99m}Tc-metoprolol (Ibrahim et al. 2010, Amin et al. 2009), as well as of different synthetic derivatives of oxprenolol, that is, caproil, 2-chlorobenzoil, isobutyryl, capryl, acetyl, pivaloyl, butyryl, and valeryl esthers (Nogowska 1996a,b, 1998, 1999).

27.2.5 TLC in Biological Assays of Beta-Blockers

Atenolol, celiprolol, metoprolol, propranolol, and talinolol were extracted from plasma and separated on silica gel by ascending technique. Isolation of the drugs was performed by using liquid–liquid extraction after alkalization of blood plasma. Suitable conditions for separation were established using ethyl acetate–acetone–ammonia 25% (5:4:1, v/v/v) as mobile phase. The substances were identified by UV irradiation at 254 nm or by reactions with a variety of reagents. The visual limits of determination, after reaction with indicators, were between 100–300 ng. Celiprolol and talinolol are the most sensitively detectable from this group of substances (Witek and Przyborowski 1997).

Using HPTLC coupled with mass spectrometry, labetalol and metoprolol were separated and detected in urine. Separation was performed on silica gel 60 with toluene–ethyl acetate–acetone–ammonia 25% (10:20:20:1, v/v/v/v) as mobile phase. Identification of the drugs was performed in UV light and by mass spectrometry. Urine sample (100 mL) was treated with ammonia up to pH = 9 and then extracted with two 50 mL ethyl acetate portions. The organic phase was separated and dried on anhydrous sodium sulfate and evaporated at 60°C under vacuum. The obtained residue was dissolved in 0.2 mL of methanol. The spots on the chromatographic plate were scraped, extracted with methanol, and analyzed with a double focalization mass spectrometer. The obtained spectra were analyzed and compared with the spots from the data library for metoprolol and labetalol. Quantitative measurements were performed by densitometry at 265 nm. The LOD value for labetalol was 0.01 µg/spot and for metoprolol 0.1 µg/spot (Rusu et al. 2010).

A rapid and sensitive HPTLC method was developed for measurement of celiprolol in human plasma. After simple extraction procedure, a known amount of the extract was spotted on silica gel $60F_{254}$ plates. The average recovery of celiprolol (added to plasma in the range of 20–200 ng/mL) was 72.06%, and the lowest amount of celiprolol that could be detected was 10 ng/mL. Pharmacokinetic parameters of two marketed preparations were also determined after oral administration to 12 healthy human volunteers (Savale et al. 2001).

A densitometric assay of propranolol in plasma was developed based on measurement of the absorbance of the drug on silica gel plates irradiated at 288 nm. Quantities as low as 0.010 µg were detected, while a linear relationship was obtained in the range 0.010–0.400 µg. The percent recovery from plasma spiked with known amounts of the drug was 90.0–102.05. This procedure was used to determine propranolol in the plasma of patients receiving therapeutic doses of the drug (Shinde et al. 1994).

A simple and rapid method was also described, which allowed the analysis of carazolol in tissue of pig. In this method, octadecyl solid-phase column that selectively absorbed the drug and significantly improved sample cleanup procedure was used. The compound was eluted from the column with acidic acetonitrile. Then, TLC analysis was performed for detection and qualitative determination. The LOD value was 10 ng/g in tissue. The recoveries of carazolol from spiked samples were above 80%. This procedure was suitable for the routine residue analysis (Posyniak et al. 1996a).

In the literature, the work in which the possibility to increase the sensitivity of TLC method for oxprenolol detection was also found. The authors proposed the derivatization with 1-ethoxy-4-(dichloro-s-triazinyl)naphthalene (Schaefer and Mutschler 1979).

27.3 BETA-AGONISTS

In the past, beta agonists were utilized for many therapeutic purposes. Nowadays, they are considered as the first-line medications in the treatment of airway narrowing, asthma, and obstructive pulmonary disease. The first bronchodilator epinephrine was a nonselective alpha (α) and beta (β) agonist. Today, it is often used in combination with local anesthetic agents to prolong the duration of anesthetic action. However, a major concern with using vasopressors like epinephrine as well as norepinephrine is their effects on systemic arterial pressure. They are therefore used in the treatment of various shock syndromes and in emergency situations related to bronchial asthma (Westfall and Westfall 2011). The next compound ephedrine originally is an alkaloid from the herb Ephedra. Because of its two chiral centers, ephedrine has four isomers: (+)-ephedrine, (−)-ephedrine, (+)-pseudoephedrine, and (−)-pseudoephedrine. They all act directly on β_1 and β_2 receptors and indirectly on α_1 receptors by causing noradrenaline release. They cause a rise in blood pressure and heart rate and some bronchodilation. They also stimulate the central nervous system to a greater extent than epinephrine, but their central action is less potent than that of amphetamine. Ephedrine and pseudoephedrine are commonly found in several over-the-counter nasal decongestants as well as in antiasthmatic drugs (Westfall and Westfall 2011).

The next step in the field of adrenergic receptor pharmacology was the development of beta selective (β_1 and β_2) receptor agonists void of α adrenergic activity. In turn, selective β_2 agonists with reduced cardiovascular effects were introduced into the therapy of different airway diseases (salbutamol, salmeterol, clenbuterol, and terbutaline). The group of beta-agonist contains also fenoterol that is a direct-acting sympathomimetic agent with predominantly β_2 adrenergic action. It is used as a bronchodilator in the treatment of asthma, in prevention of exercise-induced bronchospasm, and in the management of premature labor (Westfall and Westfall 2011). The chemical structures of the mentioned drugs are given in Figure 27.3.

27.3.1 TLC Methods for Different Types of Detection and Separation of Beta-Agonists

In the study of Wardas et al. (2000), 16 visualizing agents, 13 of which were a group of alkalimetric indicators, were used for the detection of selected drugs, including epinephrine and fenoterol. Visualizing effects for these drugs were investigated after their TLC separation on silica gel, mixture silica gel/kieselguhr, and polyamide. The best separations and the most positive visualizing effects were obtained on silica gel and the least on polyamide. On silica gel, the most profitable detectability of epinephrine and fenoterol were equal 100 ng with the application of basic solution of bromocresol green and brilliant cresol blue as visualizing agents.

The possibility of detection of ephedrine between other active components of "ecstasy" by TLC method was also described. For sample dissolution, methanol and phosphate buffer were used. The usefulness of several multicomponent eluents for TLC was tested. The optimum composition

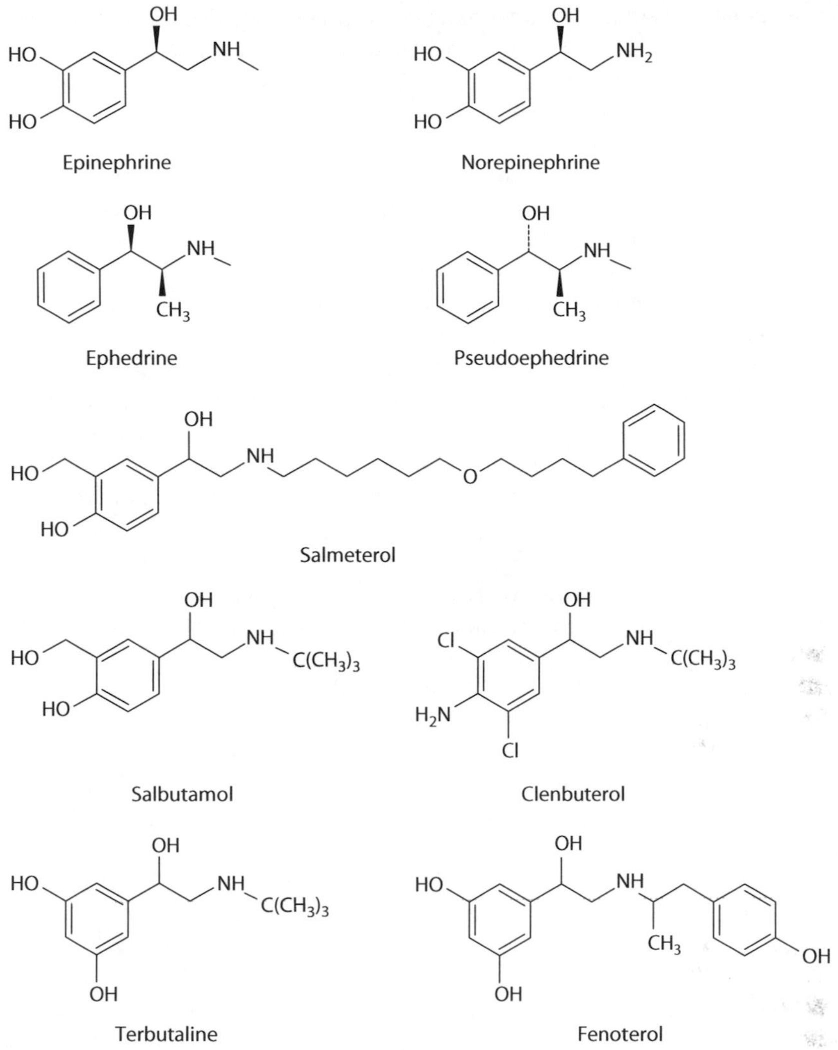

FIGURE 27.3 The chemical structures of beta-agonists.

of eluent was found as chloroform–dioxane–methanol–ammonia 25%–acetonitrile (3.5:15:2:1.5:15, v/v/v/v/v) (Kochana et al. 2005).

In the next paper, fast detection of some antiasthmatic agents like terbutaline and clenbuterol in herbal remedies were described. The samples were extracted with methanol and separated by TLC plates with silica gel GF_{254} as stationary phase with the mobile phase consisting of propan-2-ol–ethyl acetate–ammonia 25% (15:25:1.5, v/v/v) and examined at 254 nm. The method was fit to be applied for the routine test of different illegal adulteration (Wang et al. 2010).

27.3.2 TLC METHODS FOR QUANTITATIVE MEASUREMENTS OF BETA-AGONISTS IN HERBS AND PHARMACEUTICALS

HPTLC method was used for chemical standardization of several herbal species containing ephedrine. The authors used HPTLC silica gel plates and the mobile phases containing toluene–diethyl acetate–diethylamine (7:2:1, v/v/v) and toluene–chloroform–ethanol (13:30:7, v/v/v). Common and distinguishing bands were observed for all species at 366 nm. In the next step, quantitative measurements were done at 200 nm (Khatoon et al. 2005).

A simple quantitative method for determination of pseudoephedrine was developed on silica gel 60F$_{254}$ HPTLC plates. Stability studies were also performed. It was clearly shown that pseudoephedrine band was distinctly separated from degradation products (Lalla et al. 1997).

Five HPTLC methods were described for pseudoephedrine in different combined formulations. The first method was developed for simultaneous estimation of pseudoephedrine and ibuprofen in tablets. Silica gel 60F$_{254}$ was used as stationary phase and tert-butanol–ethyl acetate–glacial acetic acid–water (7:4:2:2, v/v/v/v) as mobile phase. Quantitative measurements were done at 254 nm. Percent recovery was found in the range of 100.91% and 98.27% for pseudoephedrine and ibuprofen, respectively (Chitlange et al. 2008). The second method was done for simultaneous determination of pseudoephedrine and loratadine. This method employed silica gel 60F$_{254}$ plates and the mobile phase comprising n-hexane–dichloromethane–triethylamine (5.5:4.0:0.5, v/v/v). Diltiazem was used as internal standard. Detection was carried out at 235 nm. Linear detector response for pseudoephedrine was observed in the range 12.0–90 µg/spot. The recovery pseudoephedrine was found to be 100.06% (Sane et al. 2001). A simple and precise HPTLC method was also developed for the simultaneous determination of pseudoephedrine, azatadine, and paracetamol. This method employed silica gel 60F$_{254}$ plates and the mobile phase comprising dichloromethane–acetone–methanol–triethylamine (7:4:0.5:0.2, v/v/v/v). Quantification was done by densitometry at 257 nm (Chawla et al. 1996).

In the later works, TLC methods for determination of pseudoephedrine together with guaifenesin (Lippstone et al. 1996) or with paracetamol and chlorpheniramine (Jagadeesh Babu et al. 1999) were described.

A few TLC methods exist in the literature concerning salbutamol. This drug was determined alone or in different combined formulations. Between others, TLC plastic plates with silica gel were used as stationary phase. The solvent system was then acetonitrile–methanol–ammonia 25% (10:85:5, v/v/v). The compact spots of salbutamol and related impurities, namely isopropyl salbutamol, desoxy salbutamol base, salbutamol ketone hydrochloride, and 5-formyl saligenin salbutamol, were obtained. Densitometry was done at 254 nm. The calibration plots exhibited good linear relationship (r = 0.9996) over a concentration range of 5–25 mg/mL of salbutamol. The RSD values for intraday and inter-day analysis were found to be 4.86% and 1.23%, respectively (Aboul-Enein and Abu-Zaid 2001).

A simple, rapid, and precise method was developed for simultaneous determination of salbutamol and bromhexine. These compounds were separated on silica gel 60F$_{254}$ plates with methanol–chloroform–triethylamine (5.5:4.5:0.05, v/v/v) as mobile phase. Detection was performed densitometrically at 276 nm. Calibration curves were linear in the concentration range of 20–580 ng/µL for both compounds. These assays were done with RSD values 1.4% and 1.1% for one brand and 1.62% and 1.41% for another brand (Argekar and Powar 1998).

An accurate and precise TLC method was developed for determination of salbutamol and fenoterol in their pure forms and in pharmaceutical preparations. The method was elaborated on silica gel plates in ethyl acetate–methanol–ammonia 25% (26:4:1, v/v/v) as mobile phase. The spots were measured by densitometry at 280 nm. The linearity ranges were 5–40 and 5–50 µg/spot, respectively. Mean recovery was found to be 101.12% and 100.20% for salbutamol and fenoterol, respectively (Soliman et al. 2009).

A stability-indicating HPTLC method was developed for determination of terbutaline as bulk drug and in pharmaceutical formulations (submicronized dry powder inhalers). Separation was achieved on silica gel 60F$_{254}$ using chloroform–methanol (9.0:1.0, v/v) as mobile phase. Densitometric analysis was carried out at 366 nm. The compact spots of terbutaline appeared at R$_f$ = 0.34 ± 0.02. For the proposed procedure, linearity (r^2 = 0.9956 ± 0.0015), LOQ (28.35 ng/spot), LOD (9.41 ng/spot), recovery (97.06%–99.56%), and precision (≤1.86) were found. Terbutaline was subjected to acid and alkali hydrolyses, oxidation, and photo-degradation treatments. The degraded products, however, were well separated from the pure drug (Faiyazuddin et al. 2010).

An HPTLC method was developed and validated for simultaneous identification and quantification of basic and acidic moieties of salmeterol xinafoate salt. The drug was split into two spots, salmeterol base and xinafoic acid at R_f values ca. 0.48 and 0.36, respectively. Silica gel $60F_{254}$ plates using ethyl acetate–methanol–ammonia 33% (8:1.5:0.5, v/v/v) were used. The spots were located by fluorescence quenching, while quantification of salmeterol base was carried out at 300 nm and xinafoic acid at 250 nm. The calibration curves were linear in the range of 1–6 µg/spot for salmeterol base and 0.5–4 µg/spot for xinafoic acid with good correlation coefficients (0.9964 and 0.9986). The proposed method was found to be reproducible for quantitative analysis of salmeterol xinafoate in drug substance, inhaled pharmaceutical dosage forms, as well as spiked human urine (Ahmed and Youssef 2011).

Next HPTLC method was developed for simultaneous determination of salmeterol xinafoate and fluticasone propionate. The study was performed on silica gel using n-hexane–ethyl acetate–acetic acid (5:10:0.2, v/v/v) as mobile phase. A TLC scanner was set at 250 nm in the reflectance/absorbance mode. Determination coefficients of calibration curves were found to be 0.9977 and 0.9936 in the ranges 100–1000 and 200–2000 ng/band for salmeterol and fluticasone, respectively. This method had the potential to determine these drugs simultaneously from dosage forms without any interference (Kasaye et al. 2010).

The next method was developed for fast detection of terbutaline and clenbuterol in herbal remedies. The samples were extracted with methanol, separated on silica gel GF_{254} with the mobile phase consisting of propan-2-ol–ethyl acetate–ammonia 25% (15:25:1.5, v/v/v), and examined at 254 nm (Wang et al. 2010).

In the literature, the method of analysis of formulations containing ephedrine, paracetamol, codeine, and caffeine was also elaborated using RP-TLC method and desorption electrospray ionization combined with ion mobility mass spectrometry (Harry et al. 2009).

27.3.3 TLC for Chiral Separation of Beta-Agonists

Similar to beta-blockers, beta-agonists have also one or more chiral centers in their structures, and chromatography can be used to separate their enantiomers either directly with CSPs or chiral mobile-phase additives (chiral selectors) or indirectly with chiral derivatization reagents.

In the study of Bazylak (1998), separation of various catecholamines, including epinephrine and norepinephrine as well as beta adrenergic drugs, that is, propranolol, acebutolol, practolol, clenbuterol, labetalol, and nadolol, was studied by normal-phase ion-pair chromatography employing diol-, cyano-, and aminosilica layers and ammonium-D-10-camphorsulfonate as chiral counterion in dichloromethane–propan-2-ol as mobile phase. From the calculated Van't Hoff plots, some thermodynamic parameters characteristic of chiral recognition phenomena were estimated. Formation of multiple associates between molecules of the chiral counterion and solute enantiomers were postulated using molecular modeling calculations. Additionally, the influence of the experimental conditions on detection of some enantiomers was estimated by densitometry.

Also, a TLC method based on MIPs as CSPs was applied to determine enantiomers of various adrenergic drugs including α- and β-agonists as well as some β-blockers. Three polymers imprinted with (+)-ephedrine, (+)-pseudoephedrine, and (+)-norephedrine were prepared by thermal polymerization with methacrylic acid and ethylene glycol dimethacrylate at 40°C for 16 h and coated on a glass support as thin layers. Then, enantiomeric determination of the mentioned adrenergic drugs was carried out. Adrenergic drugs, structurally related to print molecules, were completely resolved into two spots. For α- and β-agonists, acetonitrile/acetic acid (95:5 or 90:10, v/v) was used as mobile phase, and detection was effected with ninhydrin reagent or acidified $KMnO_4$ solution. For β-antagonists, dichloromethane–acetic acid (19:1 or 93:7, v/v) was used as mobile phase, and detection was effected with anisaldehyde reagent (Suedee et al. 1999b).

In a similar study, for resolution of the enantiomers of pseudoephedrine, ephedrine, norephedrine, and epinephrine, two monomers, that is, methacrylic acid and itaconic acid were employed as functional monomers. Mobile-phase system of either methanol or acetonitrile was used, and the effects of acetic acid content in the mobile phase were investigated. The best resolution was achieved for enantioseparation of norephedrine based on MIP of (–)-norephedrine using itaconic acid as functional monomer in the mobile phase containing 1% acetic acid in methanol (Suedee et al. 1999a).

The use of synthetic polymers imprinted with quinine as CSPs was also described. The stereo-selectivity of MIP was investigated for resolution both of diastereomers, including quinine–quinidine and cinchonine–cinchonidine, and of enantiomers, including pseudoephedrine, ephedrine, norephedrine, and epinephrine. The authors found that during the imprinting process, some residual print molecules were retained by the polymer, and then, the absorption of the background and of test substances on the TLC plate at 366 nm was different, which was advantageous for their detection (Suedee et al. 1998).

In the next paper, direct resolution of (±)-ephedrine and atropine into their enantiomers was achieved on silica gel plates impregnated with optically pure L-tartaric acid and L-histidine as chiral selectors. The mobile phases enabling successful resolution were different combinations of acetonitrile–methanol–water. The spots were detected with iodine vapors, and the LODs were 2 and 6 µg, respectively, in terms of the racemate. The effects of concentration of the impregnating reagent, temperature, and pH on resolution were also studied (Bhushan et al. 2001).

27.3.4 TLC for Biological Determinations of Beta-Agonists

A TLC method for quantitative determination of epinephrine, norepinephrine, and dopamine in rat plasma was used. After deproteinization, catecholamines were absorbed on acid-alumina and acetylated. The acetyl derivatives were extracted using a C18 minicolumn, resolved on HPTLC plates and quantitated by fluorescence densitometry at 415 nm using isoprenaline as internal standard (Alemany et al. 1996).

TLC procedure was also applied for rapid determination of tyrosine hydroxylase activity, an indicator of chronic stress in adrenal glands of rats and large animals. Preparation of tissue samples was adapted for rats and pigs. The activity of the enzyme was expressed as the rate of the tyrosine hydroxylation to 3,4-dihydroxyphenylalanine (DOPA) using tritium-labeled tyrosine. The subsequent separation of the radioactive DOPA from the substrate (tyrosine) was accomplished by TLC on silica gel plates and butan-1-ol–acetic acid–water (4:1:1, v/v/v). The scraped zones containing DOPA were estimated by radiometric methods. An advantage of this procedure is its simplicity, reliability, and convenience for the routine assays (Chobotská et al. 1998).

Next, a TLC method was used to study the formation of adrenochrome from epinephrine by enzyme myeloperoxidase via free radicals. The light absorption peaks at 220, 302, and 485 nm and the R_f value on TLC of this products were consistent with those of standard adrenochrome. By the authors, the formation of adrenochrome by the MPO system in vivo might role in the body's self-defense system (Onishi and Odaiima 1997).

In the study of Ersoz and Erlacin (1975), detection of ephedrine in urine probes by TLC was carried out, and then quantitative determination by the gravimetric method was proposed.

In the next paper, terbutaline was assayed in biological samples by densitometry at 366 nm on silica gel $60F_{254}$ as stationary phase and chloroform–methanol (9.0:1.0, v/v) as mobile phase. Terbutaline was well resolved at R_f equal ca. 0.34. In all matrices, the calibration curve appeared linear ($r^2 \geq 0.9943$) in the tested range of 100–1000 ng/spot with LOQ of 18.35 ng/spot. Drug recovery from biological fluids was $\geq 95.92\%$. The method was successfully used to carry out the pharmacokinetic studies of terbutaline from novel drug delivery systems (Faiyazuddin et al. 2011).

In the literature, a rapid method for identification of salbutamol in liver and urine was also described. Salbutamol was extracted from the liver samples with an acid solution, purified on SPE columns, and eluted with methanol, while urine samples were directly applied on SPE columns (DeGroof et al. 1991).

The next method involved optimization of SPE method and TLC densitometric assay for salbutamol and clenbuterol from tissue and urine samples with LOD of 1 ng/g tissue. This method was used on a routine scale for the residue control in samples from meat-producing animals (Posyniak et al. 1996b).

In the study of Degroodt et al. (1989), clenbuterol was liberated from the tissues by an enzymatic digestion, purified on SPE columns using alkaline conditions, and extracted with 0.01 M HCl. Chromatography was performed on HPTLC silica gel plates, and then clenbuterol was visualized by means of the modified Ehrlich's reagent. Since this method was as sensitive as HPLC, it was used to exclude false-positive results. The LOD and LOQ were 0.25 µg/L and 0.5 µg/kg, respectively. The recovery in urine varied from 85% to 90% and in animal tissues from 70% to 74%.

The next method was developed for the analysis of clenbuterol residues in liver with LOD of 0.5 µg/kg. The recovery varied from 55% to 60%. After extraction, a cleanup procedure with C-18 columns was performed. The compound was eluted with methanol and quantified by HPTLC method (Degroodt et al. 1991).

The multistep analytical procedure, routinely applied in for detection of clenbuterol in bovine matrices, was extended to the analysis of brombuterol, a new clenbuterol-like compound. The urine samples were tested by enzyme-linked immunosorbent assay (ELISA) and then the positive samples were analyzed by TLC with the Bratton-Marshall color reaction. The TLC spots corresponding to the suspected compounds were scraped off the plates, collected, and extracted separately with methanol. The β-agonists in the extracts were detected by HPLC (Sangiorgi and Curatolo 1997).

For clenbuterol determination in animal tissues and urine, TLC tandem mass spectrometry was also used (Henion et al. 1983).

REFERENCES

Abdelwahab, N.S. 2010. Determination of atenolol, chlorthalidone and their degradation products by TLC-densitometric and chemometric methods with application of model updating. *Analytical Methods* 2:1994–2001.

Aboul-Enein, H.Y. and S. Abu-Zaid. 2001. Two-dimensional TLC method for identification and quantitative analysis of salbutamol and related impurities in pharmaceutical tablet formulation. *Analytical Letters* 34:2099–2110.

Aboul-Enein, H.Y., M.I. El-Awady, and C.M. Heard. 2002. Direct enantiomeric resolution of some cardiovascular agents using synthetic polymers imprinted with (-)-S-timolol as chiral stationary phase by thin layer chromatography. *Pharmazie* 57:169–171.

Agustain, J., A.H. Kamaruddin, and S. Bhatia. 2010. Single enantiomeric beta-blockers-the existing technologies. *Process Biochemistry* 45:1587–1604.

Ahmed, H.A. and N.F. Youssef. 2011. Validated HPTLC method of salmeterol xinafoate determination in inhaled pharmaceutical product and spiked human urine. *Journal of Planar Chromatography—Modern TLC* 24:423–427.

Alemany, G., M. Akaârir, C. Rosselló, and A. Gamundi. 1996. Thin-layer chromatographic determination of catecholamines in rat plasma. *Biomedical Chromatography* 10:225–227.

Amin, A.M., K.M. El-Azony, and I.T. Ibrahim. 2009. Application of 99Mo/99mTc alumina generator in the labeling of metoprolol for diagnostic purposes. *Journal of Labeled Compounds and Radiopharmaceuticals* 52:467–472.

Argekar, A.P. and S.G. Powar. 1998. Simultaneous determination of salbutamol sulfate and bromhexine hydrochloride in formulations by quantitative thin-layer chromatography. *Journal of Planar Chromatography—Modern TLC* 11:254–257.

Argekar, A.P. and S.G. Powar. 2000. Simultaneous determination of atenolol and amlodipine in tablets by high-performance thin-layer chromatography. *Journal of Pharmaceutical and Biomedical Analysis* 21:1137–1142.

Argekar, A.P. and J.G. Sawant. 1999. Simultaneous determination of atenolol and nitrendipine in pharmaceutical dosage forms by HPTLC. *Journal of Liquid Chromatography and Related Technologies* 22:1571–1578.

Armstrong, D.W., F.Y. He, and S.M. Han. 1988. Planar chromatographic separation of enantiomers and diastereoisomers with cyclodextrin mobile phase additives. *Journal of Chromatography* 448:345–354.

Badgujar, V.B., P.S. Jain, G.S. Talele, and S.J. Surana. 2005. HPTLC method for estimation of carvedilol from tablet formulation. *Indian Drugs* 42:511–515.

Bazylak, G. 1998. Stereoselective analysis of catecholamines and adrenoceptors in normal-phase ion-pair planar chromatography systems with bonded polar silica layers. *Journal de Pharmacie de Belgique* 53:157.

Bebawy, L.I. 2002. Application of TLC-densitometry, first-derivative UV-spectrophotometry and ratio derivative spectrophotometry for the determination of dorzolamide hydrochloride and timolol maleate. *Journal of Pharmaceutical and Biomedical Analysis* 27:737–746.

Bhat, L.R., K.G. Bothara, and M.C. Damle. 2008. Validated HPTLC method for simultaneous determination of nebivolol hydrochloride and hydrochlorothiazide from tablets *Indian Drugs* 45:948–951.

Bhavar, G. and V. Chatpalliwar. 2008. Quantitative analysis of propranolol hydrochloride by high performance thin layer chromatography. *Indian Journal of Pharmaceutical Sciences* 70:395–398.

Bhushan, R. and C. Agarwal. 2008. Direct resolution of six beta blockers into their enantiomers on silica plates impregnated with L-Asp and L-Glu. *Journal of Planar Chromatography—Modern TLC* 21:129–134.

Bhushan, R. and C. Agarwal. 2010. Resolution of beta blocker enantiomers by TLC with vancomycin as impregnating agent or as chiral mobile phase additive. *Journal of Planar Chromatography—Modern TLC* 23:7–13.

Bhushan, R. and M. Arora. 2003. Direct enantiometric resolution of (±)-atenolol, (±)-metropolol, and (±)-propranolol by impregnant TLC using L-aspartic acid as chiral selector. *Biomedical Chromatography* 17:226–230.

Bhushan, R., J. Martens, and M. Arora. 2001. Direct resolution of (±)-ephedrine and atropine into their enantiomers by impregnated TLC. *Biomedical Chromatography* 15:151–154.

Bhushan, R. and S. Tanwar. 2008. Direct TLC resolution of atenolol and propranolol into their enantiomers using three different chirals electors as impregnating reagents. *Biomedical Chromatography* 22:1028–1034.

Bhushan, R. and S. Tanwar. 2009. Direct TLC resolution of the enantiomers of three β-blockers by ligand exchange with Cu(II)-L-amino acid complex, using four different approaches. *Chromatographia* 70:1001–1006.

Bhushan, R. and S. Tanwar. 2010. Different approaches of impregnation for resolution of enantiomers of atenolol, propranolol and salbutamol using Cu(II)-L-amino acid complexes for ligand exchange on commercial thin layer chromatographic plates. *Journal of Chromatography A* 1217:1395–1398.

Bhushan, R. and G. Thuku Thiongo. 1998. Direct enantioseparation of some β-adrenergic blocking agents using impregnated thin-layer chromatography. *Journal of Chromatography B: Biomedical Applications* 708:330–334.

Chawla, J., R.A. Sodhi, and R.T. Sane. 1996. Simultaneous HPTLC determination of azatadine maleate and pseudoephedrine HCl with sequential determination of paracetamol. *Indian Drugs* 33:208–212.

Chitlange, S., D. Sakarkar, S. Wankhede, and S. Wadodkar. 2008. High performance thin layer chromatographic method for simultaneous estimation of ibuprofen and pseudoephedrine hydrochloride. *Indian Journal of Pharmaceutical Sciences* 70:398–400.

Chobotská, K., M. Arnold, P. Werner, and V. Pliška. 1998. A rapid assay for tyrosine hydroxylase activity, an indicator of chronic stress in laboratory and domestic animals. *Biological Chemistry* 379:59–63.

Damle, M., K. Topagi, and K. Bothara. 2010. Development and validation of a stability-indicating HPTLC method for analysis of nebivolol hydrochloride and hydrochlorothiazide in the bulk material and in pharmaceutical dosage forms. *Acta Chromatographica* 22:433–443.

Degroodt, J.M, B.W. deBukanski, H. Beernaert, and D. Courtheyn. 1989. Clenbuterol residue analysis by HPLC-HPTLC in urine and animal tissues. *Zeitschrift für Lebensmittel-Untersuchung und Forschung* 189:128–131.

Degroodt, J.M., B.W. deBukanski, J. DeGroof, and H. Beernaert. 1991. Cimaterol and clenbuterol residue analysis by HPLC-HPTLC in liver. *Cinliver Zeitschriftfür Lebensmittel-Untersuchungund-Forschung* 192:430–432.

DeGroof, J., J.M. Degroodt, B.W. deBukanski, and H. Beernaert. 1991. Salbutamol identification in liver and urine by high-performance thin-layer chromatography and densitometry. *Zeitschrift für Lebensmittel-Untersuchungund-Forschung* 193:126–129.

Deore, P.V., A.A. Shirkhedkar, and S.J. Surana. 2008. Simultaneous TLC-densitometric analysis of atenolol and lercanidipine hydrochloride in tablets *Acta Chromatographica* 20:463–473.

Dhandapani, B., N. Thirumoorthy, and D. Jose Prakash. 2010. Development and validation for the simultaneous quantification of nebivolol hydrochloride and hydrochlorothiazide by UV spectroscopy, RP-HPLC and HPTLC in tablets. *E-Journal of Chemistry* 7:341–348.

Duncan, J.D., D.W. Armstrong, and A.M. Stalcup. 1990. Normal phase TLC separation of enantiomers using chiral ion interaction agents. *Journal of Liquid Chromatography* 13:1091–1103.

El-Gindy, A., A. Ashour, L. Abdel-Fattah, and M.M. Shabana. 2001. First derivative spectrophotometric, TLC-densitometric, and HPLC determination of acebutolol HCl in presence of its acid-induced degradation product. *Journal of Pharmaceutical and Biomedical Analysis* 24:527–534.

Ersoz, B. and S. Erlacin. 1975. The detection of ephedrine in human urine by thin layer chromatography. *Aegean Medical Journal* 4:123–131.

Faiyazuddin, M., S. Ahmad, Z. Iqbal et al. 2010. Stability indicating HPTLC method for determination of terbutaline sulfate in bulk and from submicronized dry powder inhalers. *Analytical Sciences* 26:467–472.

Faiyazuddin, M., A. Rauf, N. Ahmad et al. 2011. A validated HPTLC method for determination of terbutaline sulfate in biological samples: Application to pharmacokinetic study. *Saudi Pharmaceutical Journal* 19:185–191.

Gubitz, G. and M.G. Schmid. 2001. Chiral separation by chromatographic and electromigration techniques. A review. *Biopharmaceutics & Drug Disposition* 22:291–336.

Gumieniczek, A., H. Hopkała, and A. Berecka. 2002. Densitometric and video densitometric determination of nadolol and pindolol in tablets by quantitative HPTLC. *Journal of Liquid Chromatography and Related Technologies* 25:1401–1408.

Gupta, K.R., S.B. Wankhede, M.R. Tajne, and S.G. Wadodkar. 2007. High performance thin layer chromatographic estimation of atenolol and indapamide from pharmaceutical dosage form. *Asian Journal of Chemistry* 19:4183–4187.

Haeusler, G. 1990. Pharmacology of beta-blockers: Classical aspects and recent developments. *Journal of Cardiovascular Pharmacology* 16:S1–S9.

Harry, E.L., J.C. Reynolds, A.W.T. Bristow, I.D. Wilson, and C.S. Creaser. 2009. Direct analysis of pharmaceutical formulations from non-bonded reversed-phase thin-layer chromatography plates by desorption electro spray ionisation ion mobility mass spectrometry. *Rapid Communications in Mass Spectrometry* 23:2597–2604.

Henion, J., G.A. Maylin, and B.A. Thomson. 1983. Determination of drugs in biological samples by thin-layer chromatography-tandem mass spectrometry. *Journal of Chromatography* 271:107–124.

Hopkała, H., A. Pomykalski, T. Mroczek, and M. Ostep. 2003. Densitometric and videodensitometric TLC determination of timolol and betaxolol inophthalmic solutions. *Journal of Planar Chromatography—Modern TLC* 16:280–285.

Ibrahim, I.T., A.M. Amin, and K.M. El-Azony. 2010. Preparation of radioiodo-metoprolol and its biological evaluation as a possible cardiac imaging agent. *Radiochemistry* 52:212–216.

Ilango, K., P.B.S. Kumar, and K.S. Lakshmi. 2000. Simple and rapid high performance thin layer chromatographic estimation of amlodipine and atenolol from pharmaceutical dosages. *Indian Drugs* 37:497–499.

Jagadeesh Babu, R., A. Bharathi, C.N.V.H.B. Gupta, and D. Mamatha. 1999. Determination of paracetamol, pseudoephedrine hydrochloride and chlorpheniramine maleate in dosage forms by quantitative thin layer chromatography. *Asian Journal of Chemistry* 11:703–706.

Kakde, R. and N. Bawane. 2009. High-performance thin-layer chromatographic method for simultaneous analysis of metoprolol succinate and amlodipine besylate in pharmaceutical preparations. *Journal of Planar Chromatography—Modern TLC* 22:115–119.

Kasaye, L., A. Hymete, and A.M.I. Mohamed. 2010. HPTLC-densitometric method for simultaneous determination of salmeterol xinafoate and fluticasone propionate in dry powder inhalers. *Saudi Pharmaceutical Journal* 18:153–159.

Khatoon, S., M. Srivastava, A.K.S. Rawat, and S. Mehrota. 2005. HPTLC method for chemical standardization of Sida species and estimation of the alkaloid ephedrine. *Journal of Planar Chromatography—Modern TLC* 18:364–367.

Kochana, J., A. Zakrzewska, A. Parczewski, and J. Wilamowski. 2005. TLC screening method for identification of active components of "ecstasy" tablets. Influence of diluents and adulterants. *Journal of Liquid Chromatography and Related Technologies* 28:2875–2886.

Krzek, J. and A. Kwiecień. 2005. Application of densitometry for determination of beta-adrenergic-blocking agents in pharmaceutical preparations. *Journal of Planar Chromatography—Modern TLC* 18:308–313.

Krzek, J., A. Kwiecień, and M. Zylewski. 2006. Stability of atenolol, acebutolol and propranolol in acidic environment depending on its diversified polarity. *Pharmaceutical Development and Technology* 11:409–416.

Kulkarni, S.P. and P.D. Amin. 2000. Stability indicating HPTLC determination of timolol maleate as bulk drug and in pharmaceutical preparations. *Journal of Pharmaceutical and Biomedical Analysis* 23:983–987.

Lalla, U.K., S.U. Bhat, S.H. Fesharaki, N.R. Sandu, and N.R. Shastry. 1997. A quantitative high performance thin layer chromatographic method for estimation of pseudoephedrine hydrochloride—Application to stability studies. *Indian Drugs* 34:197–202.

Lippstone, M.B., E.K. Grath, and J. Sherma. 1996. Analysis of decongestant pharmaceutical tablets containing pseudoephedrine hydrochloride and guaifenesin by HPTLC with ultraviolet absorption densitometry. *Journal of Planar Chromatography—Modern TLC* 9:456–458.

Lučić, B., D. Radulović, Z. Vujić, and D. Agbaba. 2005. Direct separation of the enantiomers of (±)-metoprolol tartrate on impregnated TLC plates with D-(-)-tartaric acid as a chiral selector. *Journal of Planar Chromatography—Modern TLC* 18:294–299.

Marciniec, B., M. Ogrodowczyk, and A. Kwiecień. 2010. Effect of radiation sterilization on alprenolol in the solid state studied by high-performance thin-layer chromatography. *Journal of AOAC International* 93:792–797.

Mehvar, R. and D. Brocks. 2001. Stereospecific pharmacokinetics and pharmacodynamics of beta-adrenergic blockers in humans. *Journal of Pharmacology and Pharmaceutical Sciences* 4:185–200.

Nogowska, M. 1996a. Physico-chemical properties and hydrolysis of oxprenolol butyryl and valeryl esters. *Acta Poloniae Pharmaceutica—Drug Research* 53:99–105.

Nogowska, M. 1996b. Physico-chemical properties and hydrolysis of oxprenolol acetyl and pivaloyl esters. *Acta Poloniae Pharmaceutica—Drug Research* 53:349–355.

Nogowska, M. 1998. Physico chemical properties and hydrolysis of oxprenolol isobutyryl and capryl esters. *Acta Poloniae Pharmaceutica—Drug Research* 55:105–110.

Nogowska, M. 1999. Physico-chemical properties and hydrolysis of oxprenolol caproil and 2-chlorbenzoilesters. *Acta Poloniae Pharmaceutica—Drug Research* 56:195–199.

Onishi, M. and T. Odaiima. 1997. Formation of adrenochrome from epinephrine by myeloperoxidase via a free radical. *FASEB Journal* 11:A892.

Patel, D.R., R.C. Mashru, and M.M. Patel. 2011. Enantio separation of bisoprolol fumarate by TLC and HPTLC using (+)-10-camphor sulphonic acid as a chiral selector. *International Journal of Pharmacy and Technology* 3:1593–1602.

Patel, L., B. Suhagia, and P. Shah. 2007. RP-HPLC and HPTLC methods for the estimation of nebivolol hydrochloride in tablet dosage form. *Indian Journal of Pharmaceutical Sciences* 69:594–596.

Patel, L.J., B.N. Suhagia, and P.B. Shah. 2006a. RP-HPLC and HPTLC methods for the estimation of esmolol hydrochloride in bulk drug and pharmaceutical formulations. *Indian Drugs* 43:591–593.

Patel, L.J., B.N. Suhagia, P.B. Shah, and R.R. Shah. 2006b. RP-HPLC and HPTLC methods for the estimation of carvedilol in bulk drug and pharmaceutical formulations. *Indian Journal of Pharmaceutical Sciences* 68:790–793.

Patole, S.M., A.S. Khodke, L.V. Potale, and M.C. Damle. 2011. A validated densitometric method for analysis of atorvastatin calcium and metoprolol tartarate as bulk drugs and in combined capsule dosage forms. *Journal of Young Pharmacists* 3:55–59.

Posyniak, A., J. Niedzielska, S. Semeniuk, and J. Zmudzki. 1996a. Determination of β-agonists in biological samples by thin layer and liquid chromatographic methods. *Polish Journal of Environmental Studies* 5:33–36.

Posyniak, A., S. Semeniuk, J. Niedzielska, and J. Zmudzki. 1996b. Detecting and determining carazolol residues in biological samples. *Polish Journal of Environmental Studies* 5:37–39.

Ramteke, M., A. Kasture, and N. Dighade. 2010. Development of high performance thin layer chromatographic method for simultaneous estimation of atenolol and nefidipine in combined dosage form. *Asian Journal of Chemistry* 22:5951–5955.

Reddy, T.S. and P.S. Devi. 2007. Validation of a high-performance thin-layer chromatographic method, with densitometric detection, for quantitative analysis of nebivolol hydrochloride in tablet formulations. *Journal of Planar Chromatography—Modern TLC* 20:149–152.

Rusu, L.D., C. Marutoiu, M.L. Rusu et al. 2010. HPTLC and MS for separation and identification of some β-blockers in urine. *Asian Journal of Chemistry* 22:4209–4213.

Salem, H. 2004. High-performance thin-layer chromatography for the determination of certain antihypertensive mixtures. *Scientia Pharmaceutica* 72:157–174.

Sane, R.T., F. Mary, K. Sachin, P. Sagar, and M. Atul. 2001. Simultaneous HPTLC determination of pseudoephedrine sulphate and loratadine from their combined dosage form. *Indian Drugs* 38:436–438.

Sangiorgi, E. and M. Curatolo. 1997. Application of a sequential analytical procedure for the detection of the β-agonist brombuterol in bovine urine samples. *Journal of Chromatography B: Biomedical Applications* 693:468–478.

Sathe, S.R. and S.B. Bari. 2007. Simultaneous analysis of losartan potassium, atenolol, and hydrochlorothiazide in bulk and in tablets by high-performance thin-layer chromatography with UV absorption densitometry. *Acta Chromatographica* 19:270–278.

Sathe, S.R. and S.B. Bari. 2009. Quantitative analysis of losartan potassium and atenolol by high performance thin layer chromatography. *Indian Drugs* 46:78–81.

Sathe, S.R., S.B. Bari, and S.J. Surana. 2008. Development of HPTLC method for the estimation of metoprolol succinate in bulk and in tablet dosage form Indian. *Journal of Pharmaceutical Education and Research* 42:32–35.

Savale, H.S., K.K. Pandya, T.P. Gandhi, I.A. Modi, R.I. Modi, and M.C. Satia. 2001. Plasma analysis of celiprolol by HPTLC: A useful technique for pharmacokinetic studies. *Journal of AOAC International* 84:1252–1257.

Schaefer, M. and E. Mutschler. 1979. Fluorimetric determination of oxprenolol in plasma by direct evaluation of thin-layer chromatograms. *Journal of Chromatography* 164:247–252.

Shinde, V.M., B.S. Desai, and N.M. Tendolkar. 1994. Simultaneous determination of propranolol hydrochloride and hydrochlorothiazide in tablets by quantitative. *TLC Indian Drugs* 31:192–196.

Shirkhedkar, A.A., P.M. Bugdane, and S.J. Surana. 2010. Stability-indicating TLC-densitometric determination of nebivolol hydrochloride in bulk and pharmaceutical dosage form. *Journal of Chromatographic Science* 48:109–113.

Soliman, M.H., M. Soltan, S.M. Khalile, and E.A. Moety. 2009. Spectrophotometric and chromatographic determination of fenoterol, salbutamol sulfate and theophylline. *Egyptian Journal of Chemistry* 52:91–103.

Spell, J.C. and J.T. Stewart. 1997. A high-performance thin-layer chromatographic assay of pindolol enantiomers by chemical derivatization. *Journal of Planar Chromatography—Modern TLC* 10:222–224.

Šubert, J. and K. Šlais. 2001. Progress in the separation of enantiomers of chiral drugs by TLC without their prior derivatization. *Pharmazie* 56:355–360.

Suedee, R. and C.M. Heard. 1997. Direct resolution of propranolol and bupranolol by thin-layer chromatography using cellulose derivatives as stationary phase. *Chirality* 9:139–144.

Suedee, R., C. Songkram, A. Petmoreekul, S. Sangkunakup, S. Sankasa, and N. Kongyarit. 1998. Thin-layer chromatography using synthetic polymers imprinted with quinine as chiral stationary phase. *Journal of Planar Chromatography—Modern TLC* 11:272–276.

Suedee, R., C. Songkram, A. Petmoreekul, S. Sangkunakup, S. Sankasa, and N. Kongyarit. 1999. Direct enantioseparation of adrenergic drugs via thin-layer chromatography using molecularly imprinted polymers. *Journal of Pharmaceutical and Biomedical Analysis* 19:519–527.

Suedee, R., T. Srichana, J. Saelim, and T. Thavornpibulbut. 1999. Chiral determination of various adrenergic drugs by thin-layer chromatography using molecularly imprinted chiral stationary phases prepared with α-agonists. *Analyst* 124:1003–1009.

Suedee, R., T. Srichana, J. Saelim, and T. Thavonpibulbut. 2001. Thin-layer chromatographic separation of chiral drugs on molecularly imprinted chiral stationary phases. *Journal of Planar Chromatography—Modern TLC* 14:194–198.

Vujic, Z., D. Radulovic, and D. Agbaba. 1997. Densitometric determination of metoprolol tartrate in pharmaceutical dosage forms. *Journal of Pharmaceutical and Biomedical Analysis* 15:581–585.

Wang, T.S., Y. Tong, J. Zheng, Z. Zhang, Q. Gao, and B.Q. Che. 2010. Detection of 29 adulterations in lipid-regulating, antihypertensive, antitussive and antiasthmatic herbal remedies by TLC. *Chinese Pharmaceutical Journal* 45:857–861.

Wankhede, S.B., N.R. Dixit, and S.S. Chitlange. 2011. Stability indicating HPTLC method for quantitative determination of atorvastatin calcium and metoprolol succinate in capsules. *Der Pharmacia Lettre* 3:1–7.

Wardas, W., I. Lipska, and K. Bober. 2000. TLC fractionation and visualization of selected phenolic compounds applied as drugs. *Acta Poloniae Pharmaceutica—Drug Research* 57:15–21.

Westfall, T.C. and D.P. Westfall. 2011. Adrenergic agonists and antagonists, in: *Goodman & Gilman's the Pharmacological Basis of Therapeutics*, 12th edn., (ed. L.L. Brunton, B.A. Chabner, and B.C. Knollmann), McGraw Hill Medical, New York, pp. 277–333.

Witek, A., H. Hopkała, and G. Matysik. 1999. TLC-densitometric determination of bisoprolol, labetalol and propafenone, as dabsyl derivatives, in pharmaceutical preparations. *Chromatographia* 50:41–44.

Witek, A. and L. Przyborowski. 1996. Identification of celiprolol and talinololin substance and tablets by means of TLC. *Acta Poloniae Pharmaceutica—Drug Research* 53:399–402.

Witek, A. and L. Przyborowski. 1997. Thin-layer chromatographic analysis of some β-blocking drugs in plasma. *Acta Poloniae Pharmaceutica—Drug Research* 54:183–186.

28 TLC of Antithrombotics

Marica Medić-Šarić, Mirza Bojić, and Željko Debeljak

CONTENTS

The prevalence of cardiovascular diseases (e.g., hypertension, coronary heart disease, heart failure, and stroke) in the United States in 2010 was 36.9% resulting in $273 billion in direct and $172 billion in indirect costs (Heidenreich et al. 2011). Antithrombotics are one of the major groups of drugs for the prevention of cardiovascular diseases that act against thrombus (clot) formation. If they repress primary clot formation, usually in the phase of platelet aggregation, they are referred to as antiplatelet or antiaggregatory drugs. On the other hand, anticoagulants act on secondary hemostasis—coagulation cascade. The last group of drugs, fibrinolytics–thrombolytics act on fibrin–thrombus disintegration (Cox 2008).

As seen from list of antithrombotics on the list of World Health Organization (shown in Table 28.1), this group consists of different chemical entities: proteins (enzymes (*-ase*), monoclonal antibodies (*-mab*)) and peptides (*-tide*, hirudin derivatives *-irudin*), and polysaccharides (heparin and derivatives (*-parin*)). Based on the mechanism of action, if they act as factors of coagulation cascade, they have steam *-cog*, inhibitors have ending *-cogin*, specific inhibitors have endings *-gatran* (direct thrombin inhibitor), and *-xaban* (direct factor Xa inhibitor), and *-parinux* (indirect Xa inhibitor) (WHO 2009).

Heparin is a glycosaminoglycan—mostly negatively charged biomolecule due to sulfate groups (Williams 2010). Proteins and peptides are charged depending on the amino acids side chains. Thus, these compounds are usually separated based on the influence of the electric field—electrophoresis. However, when glycosaminoglycans are enzymatically digested, oligosaccharides are formed that can be analyzed by thin-layer chromatography (TLC) (Chai, Rosankiewicz, and Lawson 1995; Qiu, Tanikawa, Akiyama, Toida, Koshiishi, and Imanari 1993). Fibrinolytics are enzymes, thus they are characterized as protein structures that are usually analyzed by techniques other than TLC. One of the most commonly used antithrombotics is acetylsalicylic acid in daily dose of 100 mg. As a non-steroidal anti-inflammatory drug (NSAID), it is used in dosage forms of 300 or 500 mg (Cox 2008), thus TLC of acetylsalicylic acid is described in Chapter 18.

In this chapter, we will focus on classical drugs and use of TLC for analysis of pure substances, pharmaceutical formulations, study of pharmacokinetics (Absorption, Distribution, Metabolism, and Extraction (ADME)), pharmacodynamics (mode of action), and toxicity of antithrombotics.

28.1 TLC OF ANTITHROMBOTICS AS ACTIVE SUBSTANCES IN THE PHARMACEUTICAL DOSAGE FORMS

TLC has been used for the analysis of active pharmaceutical ingredient (API), pharmaceutical dosage form, and stability (stress) testing (Table 28.2). The most common application of TLC in pharmacopoeia is testing of related substances, for example, warfarin sodium and picotamide monohydrate.

TABLE 28.1

Anatomy Therapeutic Chemical (ATC) Classification of Antithrombotics

ATC Classification Group	Drugs
Vitamin K antagonists	Dicoumarol, phenindione, warfarin, phenprocoumon, acenocoumarol, ethyl biscoumacetate, clorindione, diphenadione, tioclomarol, and fluindione
Heparin group	Heparin, antithrombin III, dalteparin, enoxaparin, nadroparin, parnaparin, reviparin, danaparoid, tinzaparin, sulodexide, and bemiparin
Platelet aggregation inhibitors excl. heparin	Ditazole, cloricromen, picotamide, clopidogrel, ticlopidine, acetylsalicylic acid, dipyridamole, carbasalate calcium, epoprostenol, indobufen, iloprost, abciximab, aloxiprin, eptifibatide, tirofiban, triflusal, beraprost, treprostinil, prasugrel, cilostazol, and ticagrelor
Enzymes	Streptokinase, alteplase, anistreplase, urokinase, fibrinolysin, brinase, reteplase, saruplase, ancrod, drotrecogin alfa (activated), tenecteplase, and protein C
Direct thrombin inhibitors	Desirudin, lepirudin, argatroban, melagatran, ximelagatran, bivalirudin, and dabigatran etexilate
Other antithrombotic agents	Defibrotide, dermatan sulfate, fondaparinux, and rivaroxaban

Medicinal herbal drugs that contain coumarins undergo reflux extraction in methanol and are usually developed in normal stationary phase. Detection is performed under ultraviolet (UV) light with or without potassium hydroxide for florescence monitoring (Wagner and Blat 2009).

Agrawal, Kaul, Paradkar, and Mahadik (2003) developed stability-indicating high-performance thin layer chromatography (HPTLC) method for clopidogrel. The method was validated: linearity 200–1000 ng/spot, limit of detection (LOD) 40 ng/spot, limit of quantification (LOQ) 120 ng/spot, accuracy 100.42% ± 1.57%, repeatability of application 1.41%, repeatability of measurement 0.25%, inter-day precision 0.46%, and intra-day precision 0.002%; to ensure repeatability, plates prewashing and chamber saturation was performed.

TLC is not the only technique that can provide identification of degradation products without corresponding standard used. Authors conclude that "The lower R_F values of acidic and base degraded components indicated that they were less polar than analyte itself." As stationary phase (silica gel) is polar, this conclusion is wrong—lower R_F indicates more polar compound, which was confirmed in a later study by Zaazaa, Abbas, Abdelkawy, and Abdelrahman (2009) using infrared spectroscopy and mass spectrometry. Spot $R_F = 0.22$ corresponds to clopidogrel carboxylic acid (XlogP = 1.1) and $R_F = 0.3$ to clopidogrel (XlogP = 3.8).

Zaazaa et al. (2009) developed the spectrodensitometric method for the separation of clopidogrel from degradation products and determination of clopidogrel in pharmaceutical forms. Clopidogrel is an ester that undergoes hydrolysis when analyzed in a stress test induced by acid or base (Figure 28.1). To follow the rate of hydrolysis, mobile phase hexane–methanol–ethyl acetate 8.7:1:0.3 was used. Although clopidogrel has $\lambda_{max} = 220$ nm, due to higher sensitivity and noise reduction, $\lambda = 248$ nm was selected for determination. Validation parameters: linearity 0.6–3 µg/band, LOD = 0.04 µg/band, LOQ = 0.4 µg/band, accuracy 99.97% ± 1.16%, intermediate precision 1.62%, and recovery 100.46% ± 1.43%. Statistical analysis showed no difference when comparing the results of TLC analysis and official HPLC USP method.

In this chapter, you can notice the common disadvantage of expressing results as mass per band. If you want to compare LOD and LOQ to spectrophotometry analysis, LOD of 0.04 µg/band seams lower than 0.8–1.2 µg/mL of different spectrophotometric analysis. Taking into account that 10 µL was applied per band, you come to the concentration of LOD = 0.04 µg/10 µL = 4 µg/mL. Recalculated linearity range would be 60–300 µg/mL. If we go one step back, concentration range of clopidogrel solutions was 6–30 µg/10 mL = 0.6–3 µg/mL. We believe this is unintentional mistake, although sometimes this can be the result of "speeding up" the experiment by applying different volumes and omitting the working solutions preparation.

Shimizu, Osumi, Niimi, and Nakagawa (1985) conducted stability tests of cilostazol in acidic and basic solutions, solid form, long-term stability, stability to heat, humidity, and light. TLC showed that cilostazol is stable in all of the previously mentioned stress tests.

TABLE 28.2

Sample Preparation and Chromatographic Systems for Active Substance and Pharmaceutical Dosage from Analysis

Drug	Sample Preparation	Reference Substance	Stationary Phase	Mobile Phase	Detection	Reference
Clopidogrel	Dissolution in methanol	Clopidogrel ($R_F = 0.30$)	Aluminum plates with silica gel 60 F_{254}	Carbon tetrachloride–chloroform–acetone, 6:6:0.15	Spectrodensitometric ($\lambda = 230$ nm)	Agrawal et al. (2003)
Clopidogrel	Dissolution in methanol	Clopidogrel ($R_F = 0.646$)	Silica gel 60 F_{254}	Hexane–methanol–ethyl acetate, 8.7:1:0.3	Spectrodensitometric ($\lambda = 248$ nm)	Zaazaa et al. (2009)
Cilostazol	Liquid–liquid purification with chloroform, reconstitution in chloroform–methanol	Cilostazol	Kiesel gel 60 F_{254}	Acidic system: chloroform–n-propanol–acetic acid, 400:60:1; basic system: chloroform–methanol–NH_4OH 35:6:1; neutral system: chloroform–methanol, 8:1	UV light ($\lambda = 254$)	Shimizu et al. (1985)
Coumarins (medicinal herbal drugs)	Extraction/dissolution in methanol	Natural occurring coumarins	Silica gel 60 F_{254}	Aglycones: toluene–ether 1:1 (saturated with 10% acetic acid); glycosides: ethyl acetate–formic acid–acetic acid (glac.)–water 100:11:11:26	UV light ($\lambda = 254$), $\lambda = 365$ after spraying with 5%–10% ethanolic KOH, natural products-polyethylene glycol reagent	Wagner and Blat (2009)
Picotamide monohydrate	Dissolution in methanol	Picotamide, picotamide impurity A (4-methoxybenzene-1,3-dicarboxylic acid)	Silica gel F_{254}	Acetic acid (glac.)–water–methanol–butanol, 0.8:1:2.5:8	UV light ($\lambda = 254$)	Ph. Eur. (2010)
Warfarin sodium	Dissolution in acetone	Warfarin, acenocoumarol	Silica gel GF_{254}	Acetic acid (glac.)–chloroform–cyclohexane, 20:50:50	UV light ($\lambda = 254$)	Ph. Eur. (2010)

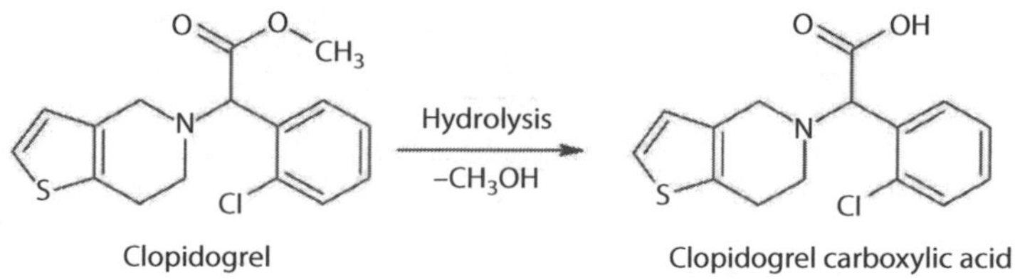

FIGURE 28.1 Degradation of clopidogrel in presence of acid or base.

28.2 TLC STUDY OF ADME OF ANTITHROMBOTICS

TLC was successfully used for metabolism analysis of anticoagulants. In this case, quite often compounds were marked with radioactive isotope, ensuring specificity of the technique used. Sample preparation and chromatography systems used for TLC study of metabolism are shown in Table 28.3.

Barker, Hermodson, and Link (1970) analyzed the metabolism of warfarin, the most often used anticoagulant, in rats (Figure 28.2). Most common metabolites are hydroxylated derivatives at the positions 6, 7, 8, and 4′. Urine and feces profile of metabolites are similar, although major metabolite in feces is 7-hydroxywarfarin. Reduction generates an unstable metabolite that undergoes cyclization forming tetracycle-metabolite. All metabolites lose activity except 4′-hydroxywarfarin. For the analysis of glucuronide conjugates, paper chromatography was used with or without addition of β-glucuronidase.

TLC assay of warfarin was also applied on human samples of plasma and stool (Lewis et al. 1970). As some patients also used NSAID-like acetylsalicylic acid and phenylbutazon, no interactions were observed.

Welling et al. (1970) analyzed pharmacokinetic profile of warfarin. TLC was used for separation and UV spectrophotometry for determination of warfarin. Interestingly, no hydroxy-metabolites were found in plasma. This was explained by rapid elimination and strong binding to albumins of monohydroxywarfarins.

Pohl et al. (1975) synthesized possible monohydroxylated metabolites of phenprocoumon that were used in later studies of phenprocoumon metabolism. To separate these metabolites, two-dimensional (2D) TLC was used. Combination of mobile phase chloroform–acetic acid 100:1 and *tert*-butanol–benzene–NH₄OH–water 45:20:9:3 was used. It was observed that the behavior of monohydroxylated phenprocoumon on TLC plate can be a unique property of individual compound. Contrary to other derivatives, 7-hydroxyphenprocoumon is unstable in mobile phase toluene–acetic acid 9:1, and yellow degradation products appear some time after development. 5-hydroxyphenprocoumon shows light blue and 6-hydroxyphenprocoumon green–blue fluorescence under $\lambda = 254$ nm. Intensity of fluorescence under UV light increases after exposure to ammonium but stays unchanged for 2-hydroxyphenprocoumon.

The same group (Haddock et al. 1975) conducted research of metabolism and elimination of phenprocoumon in rats combining TLC, radiography, and mass spectrometry. Compared to warfarin, major phenprocoumon metabolite in urine was 6-hydroxy vs. 7-hydroxywarfarin in feces and urine.

Metabolic fate of phenprocoumon was studied in humans using TLC, HPLC, and GC-MS (Toon et al. 1985). As with warfarin, samples were first cleaned in basic media and then extracted with acidified water. TLC showed adequate separation, after pretreatment with methanol for band narrowing and fatty material removal with benzene for fecal sample. Forty percent of the drug is eliminated unchanged and 60% in forms of hydroxy metabolites (Figure 28.3). As with warfarin, stereoselectivity is observed. For this purpose, S-2-^{13}C-phenprocoumon and racemic 2-^{14}C-phenprocoumon were used. Hydroxylation at the positions 4′ and 7′ is a preferred reaction for S-phenprocoumon, while no substrate stereoselectivity is observed for 6-hydroxyphenprocoumon. Substrate stereoselectivity was also observed in phase II reactions of biotransformation. However, as a combination of glucuronidase and sulfatase was used in the experiment, further conclusions could not be made.

TABLE 28.3

Sample Preparation and Chromatographic Systems for TLC Study of Absorption, Distribution, Metabolism, and Extraction (ADME) of Antithrombotics

Drug	Sample Preparation	Reference Substance	Stationary Phase	Mobile Phase	Detection	Reference
Acenocoumarol	Plasma and urine: liquid–liquid purification and extraction with ethyl acetate–benzene 20:80; evaporation and reconstitution in chloroform	^{14}C-acenocoumarol, ^{3}H-acethyl-acenocoumarol—synthesized, identified by elemental analysis and mass spectrometry	Silica gel F$_{254}$	Petroleum ether–acetone 140:60; benzene–ethyl acetate 140:60	UV light ($\lambda = 254$ nm), radiography after spot extraction	Le Roux and Richard (1977)
Dermatan sulfate (humans)	Urine: washing with cetylpyridinium chloride, ethanol, and saline with centrifugation; pronase digestion	Dermatan sulfate, chondroitin sulfate	Silica gel plates	n-propanol–NH$_4$OH (25%)–acetone 31.2:36.8:22.0	Extraction: 1 mm with water; lyophilization; electrophoresis	Qiu et al. (1993)
Phenprocoumon (humans)	Urine and fecal (with or without glucuronidase–sulfatase pretreatment) samples: liquid–liquid extraction with ether and ethyl acetate	Deuterated internal standards: phenprocoumon, 4', 6, 7 and 8-hydroxyphenprocoumon	Aluminum plates with silica gel 60 F$_{254}$	(A) twice chloroform–acetic acid 100:1; (B) *tert*-butyl alcohol–benzene–NH$_4$OH–water 40:20:9:3	Spot extraction and radiography, HPLC, GC	Toon et al. (1985)
Phenprocoumon (rats)	Dried feces dissolved in methanol, evaporated, liquid–liquid extraction with ether—cleaning and isolation; urine lyophilized, reconstituted in aqueous H$_3$PO$_4$, liquid–liquid extraction with ether—cleaning and isolation; bile—similar to urine with or without glusulase	^{3}H-phenprocoumon, identity of metabolites confirmed with chemical ionization mass spectrometry	Fluorescent silica gel plates	Toluene–acetic acid 9:1; chloroform–acetic acid 100:1; *tert*-butanol–benzene–NH$_4$OH–water 45:20:9:3; chloroform–ethyl acetate–acetic acid 100:50:1; toluene–ethyl formate–formic acid 10:5:1	UV light ($\lambda = 254$ nm), radiography	Haddock et al. (1975)
Phenprocoumon (synthesis of metabolites)	Synthesis	Synthesized, identity confirmed with UV and chemical ionization mass spectrometry	Fluorescent silica gel plates	Toluene–acetic acid 9:1; Chloroform–acetic acid 100:1; *tert*-butanol–benzene–NH$_4$OH–water 45:20:9:3; Chloroform–ethyl acetate–acetic acid 100:50:1; toluene–ethyl formate–formic acid 10:5:1	UV light ($\lambda = 254$ nm) after exposing to NH$_3$ vapors	Pohl et al. (1975)

(continued)

TABLE 28.3 (continued)
Sample Preparation and Chromatographic Systems for TLC Study of Absorption, Distribution, Metabolism, and Extraction (ADME) of Antithrombotics

Drug	Sample Preparation	Reference Substance	Stationary Phase	Mobile Phase	Detection	Reference
Ticlopidine (rats)	Urine and bile: enzymatic or acid hydrolysis, unhydrolyzed samples followed by liquid–liquid extraction with dichlormethan and ether	Ticlopidine, ticlopidine-N-oxide, tetrahydrothienopyridine, 2-chlorohippuric acid, 2-chlorobenzylalcohol, 2-chlorbenzaldehyd, and 2-chlorobenzoic acid	Silica gel 60 F_{254}	(A) benzene–methanol 7:3; (B) chloroform–methanol–acetic acid 7:3:0.2	UV-light, spraying: iodoplatinate, ninhydrin, SO_2 in methanol, bromocresol	Tuong et al. (1981)
Warfarin (humans)	Acidification of with HCl (ensures neutral form of weak acid phenol groups) and subsequent liquid–liquid extraction with ethylene dichloride, centrifugation, evaporation, and reconstitution in acetone	Warfarin ($R_F = 0.50$), 7-hydroxywarfarin, and warfarin alcohols	Silica gel G	Ethylene dichloride–acetone 9:1; pretreatment of stool sample; benzene for fat elimination	Ammonia vapor–short wave UV light; fluorimetric determination	Lewis et al. (1970)
Warfarin (humans)	Liquid–liquid chromatography with 1,2-ethylendichloride	Warfarin, 4′-, 6- and 7-hydroxywarfarin	Silica gel GF_{254}	1,2-Ethylendichloride–acetone 9:1	UV light ($\lambda = 254$), extraction–UV-Vis spectrophotometry	Welling et al. (1970)
Warfarin (rats)	Acidification of with HCl (ensures neutral form of weak acid phenol groups) and subsequent liquid–liquid extraction with ether; feces samples were dried and extracted in a Soxhlet extractor with ether	4-^{14}C-warfarin, warfarin ($R_F = 0.80$), 7-hydroxywarfarin ($R_F = 0.33$), 8-hydroxywarfarin ($R_F = 0.50$), 4′-hydroxywarfarin ($R_F = 0.65$), 6-hydroxywarfarin ($R_F = 0.67$), and 2,3-dihydro-2-methyl-4-phenyl-5-oxo-γ-pyrano(3,2-c) (1) benzopyran ($R_F = 0.89$)	Silica gel G	Cyclohexane–ethyl formate–formic acid 100:200:1 and terc-butanol–benzene–NH_4OH (conc.)–water 45:20:9:3	Liquid scintillation spectrometer, planched counter; non-radioactive standards–diazotized p-nitroaniline, 20% sodium carbonate solution	Barker et al. (1970)

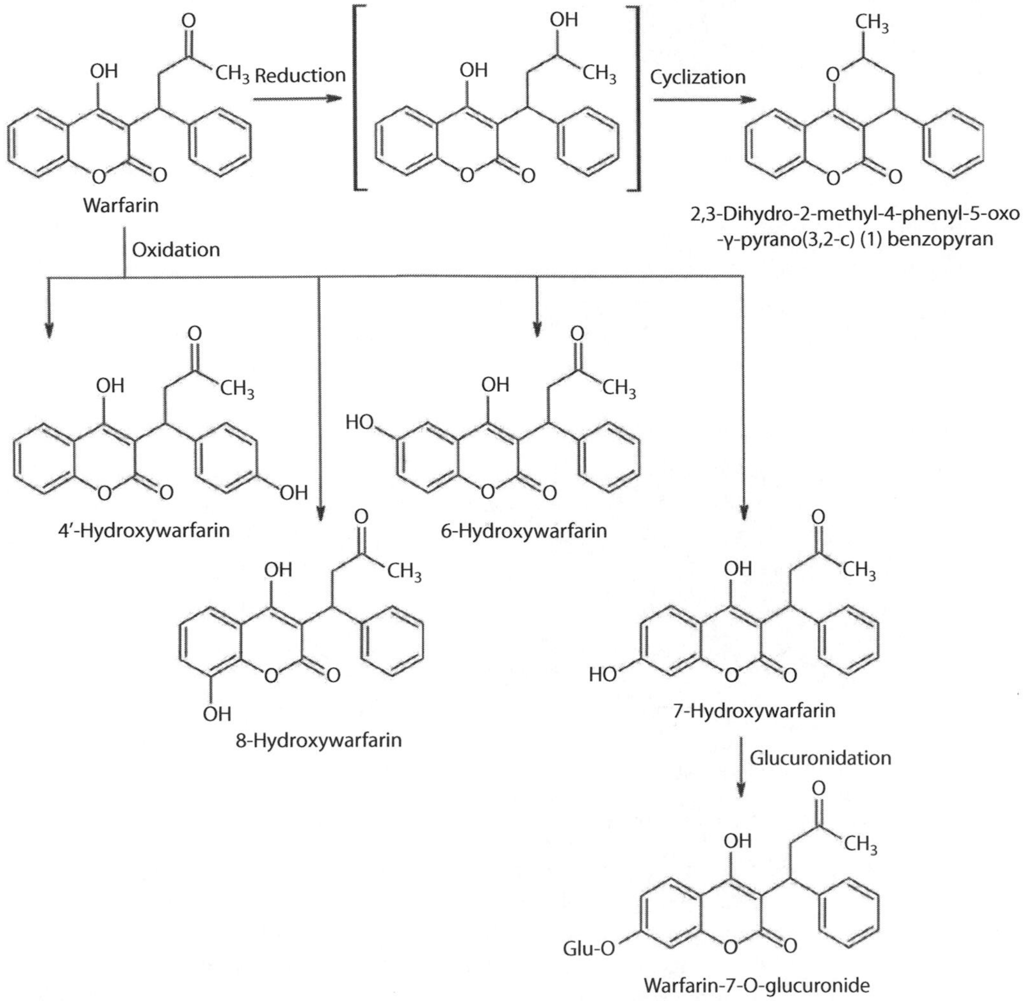

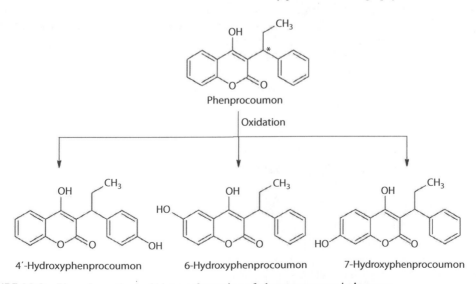

FIGURE 28.2   Metabolic fate of Warfarin in rats obtained by planar chromatography.

FIGURE 28.3   Phase I reactions of biotransformation of phenprocoumon in humans.

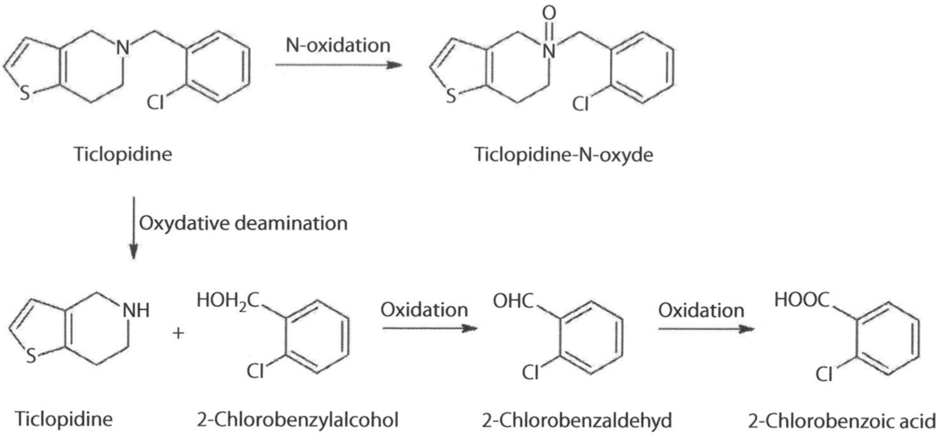

FIGURE 28.4 Metabolism of ticlopidine in rats. Ticlopidine-N-oxyde was confirmed by reduction with SO_2 to Ticlopidine. Unidentified metabolite (T-M) that converts to ticlopidine has been omitted.

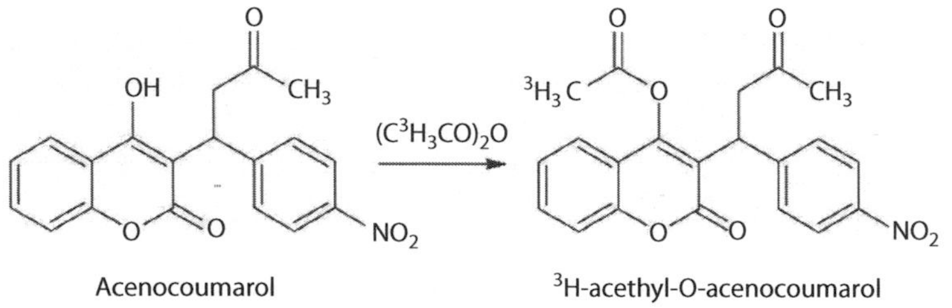

FIGURE 28.5 Acetylation of acenocoumarol.

Tuong et al. (1981) used TLC in combination with gas chromatography (GC) for separation, identification, and separation of ticlopidine metabolites (Figure 28.4). In this case, different spraying reagents were used to speculate about the structure of the metabolites (Table 28.3).

Le Roux and Richard (1977) used derivatization to increase specificity of TLC analysis. As expected, acenocoumarol (Figure 28.5) is a weak acid and is extractable to organic solvents when pH of aqueous media is lower than 7. If extraction is performed at neutral pH, greater volume of organic solvents has to be used (ratio 1:5). Speculated and confirmed metabolites of acenocoumarol include monohydroxylated derivatives at position 6 and 7, alcohol (after reduction of keto group), and amino (after reduction of nitro group) derivative that can be acetylated. To separate acenocoumarol from its metabolites, derivatization with ^{3}H-acetic anhydride was applied. This increased sensitivity to a concentration of 8 nm/ng of acenocoumarol.

Dermatan sulfate is a polysaccharide indicating that size exclusion chromatography would be ideal for separation. However, better results were obtained when preparative TLC was used in combination with electrophoresis after enzymatic digestion. The identity of metabolites was determined by HPLC (Qiu et al. 1993).

28.3 TLC STUDY OF PHARMACODYNAMICS OF ANTITHROMBOTICS

An interesting approach to understanding the possible mechanism of action of antithrombotics represents TLC analysis of arachidonic acid metabolism. It involves several important targets for possible antiplatelet effect: cyclooxygenase and thromboxane A_2 (TxA$_2$) synthase (Shantsila et al. 2008). Based on eicosanoids production, it can be speculated on the possible targets of the analyzed substances. Structure-activity relationship of glycosaminoglycan with possible targets could be explained by TLC analysis of oligosaccharide fragments (Chai et al. 1995). Chromatography systems used for TLC study of pharmacodynamics of antithrombotics are shown in Table 28.4.

TABLE 28.4
Sample Preparation and Chromatographic Systems for TLC Study of Pharmacodynamics of Antithrombotics

Drug	Sample Preparation	Reference Substance	Stationary Phase	Mobile Phase	Detection	Reference
Dipyridamole	Effluent collected after nonrecirculating perfusion, liquid–liquid extraction with ethyl acetate—with or without acidification	Metabolite extract pH 7.4: PGA$_2$, ricinoleic acid, sodium arachidonate; metabolite extract pH 3.5: 6-keto-PGF$_{1\alpha}$, PGF$_{2\alpha}$, PGE$_2$, TxB$_2$, PGD$_2$, 15-keto-PGE$_2$ PGA$_2$, ricinoleic acid, and sodium arachidonate	Aluminum sheets with Kieselgel 60 (0.2 mm)	Metabolite extract pH 7.4: petroleum ether–diethyl ether–acetic acid 50:50:1; metabolite extract pH 3.5: first direction ethyl acetate–iso-octane–acetic acid–water 110:50:20:100, second direction twice ethyl acetate–acetic acid 99:1	Anisaldehyde and radiography	Uotila (1982)
Metabolism of arachidonic acid	Drug addition to platelets prelabeled with ^{14}C-arachidonic acid, incubation, platelet induction; extraction ethyl acetate–petrol ether, evaporation, dissolution in chloroform–methanol	1-^{14}C-arachidonic acid, ^{3}H-labeled eicosanoids: PGF$_{2\alpha}$, PGE$_2$, PGD$_2$, PXB$_2$, and 12-HETE	Silica gel 60 F$_{254}$	Chloroform–methanol–acetic acid–water 87:8:4:1	Radiography	Hornberger and Patscheke (1989)
SAR of disaccharides with possible targets	Ten disaccharides conjugated with amino-lipid DHPE conjugates dissolved in chloroform–methanol–water 25:25:8	Ten disaccharides conjugated with amino-lipid DHPE	Aluminum-backed HPTLC plates	Chloroform–methanol–water 60:35:8	UV light after spraying with primulin reagent (immersion for quantification); LSI-MS	Chai et al. (1995)

Dipyridamole is an antiaggregatory drug that relies mainly on vasodilatation effect. Although it has a long history of use, many mechanisms of antiaggregatory activity have been described; still, they are not completely understood (Cox 2008). The influence of dipyridamol on arachidonic acid has been studied in guinea pig isolated lungs by 2D TLC (Uotila 1982). Although the only difference between treated samples with dipyridamol and control was increased concentration of TxA_2, the effect could not be extrapolated to platelets (Viinikka et al. 1983).

Hornberger and Patscheke (1989) provided research tool for TLC assessment of eicosanoids metabolism. When stimulated by platelet-aggregation agonists (collagen, thrombin, ADP, and ristocetin), arachidonic acid metabolism is induced. To assess whether the new substance has antithrombotic potential by blocking formation of TxA_2, TLC in combination with radioimmunoassay was used. This could be useful for research of cyclooxygenase and TxA_2 synthase inhibitors. However, only inhibition of later enzyme accumulates prostaglandin H_2, which also binds to TxA_2 receptors producing proaggregatory effect on platelets. Due to short half-life, TxA_2 was rapidly converted to TxB_2 that was measured in this assay.

Chai et al. (1995) used TLC for the separation of neoglycolipids produced by reductive amination with dihexadecyl phosphatidylethanolamine (DHPE) for the purpose of finding glycosaminoglycan oligosaccharide chains responsible for interaction with proteins. HPTLC was used in conjugation with liquid secondary-ion mass spectrometry and GC-MS. Seven heparins and three chondroitin sulfates disaccharides, which represent most common products of glycosaminoglycan lysis, were conjugated with amino–lipid DHPE. The yields of conjugation were assessed by fluorescence intensity. There were clear zones of conjugates reflecting sulfatation of disaccharides (zero, mono, di, or tri). The clear advantage of this method is the analysis of 10–50 pmol of sugar (equivalent of 100–500 pmol of neoglycolipid).

The method was further used to analyze cleavage of uronic acid residue from heparin disaccharides and tetrasaccharide (oxymercuration reaction) producing unmodified glycosaminoglycan fragments that can further be tested for biological activity.

28.4 TLC STUDY OF TOXICITY OF ANTITHROMBOTICS

Coumarins are used as anticoagulants, especially as rodenticides, thus their analysis is of interest for forensic toxicology (Table 28.5). Lau-Cam and Chu-Fang (1972) performed mobile phase optimization. Of 11 single solvents, 24 binary solvent systems, 9 ternary solvent systems, 3 basic solvent systems, and 22 acidic solvent systems with different ratios of solvents, 13 mobile phases were found to be most suitable. Silica gel H ($SiO_2 + Al_2O_3$) showed shorter time of development compared with silica gel G ($SiO_2 + CaSO_4$). The unique distinction of individual coumarins was achieved with 2,6-dichloroquinonchloroimid by monitoring the color of the spot immediately, after 10 min, and 1 h.

HPTLC was used for separation, identification, and determination of anticoagulant rodenticides in liver and serum samples (Berny et al. 1995). The method was shown to be specific and usable on real samples of animal intoxication. Validation parameters tested: recovery, 85%–95%; repeatability and reproducibility, <5%; LOD = 0.2 µg/mL; LOQ = 0.5 µg/mL; and linearity 0.5–2.0 µg/mL.

28.5 TLC STUDY OF NEW ANTITHROMBOTIC AGENTS

The major problem of heparin is its parenteral application, for example, low-molecular-weight heparins (LMWH) are given as subcutaneous injections. Efforts have been made to prepare dendrimeric nanoparticles that can entrap LMWH. The toxic potency was reduced and pharmacokinetic profile advanced by pegylation. TLC was used to monitor synthesis and incorporation of LMWH in nanoparticles (Bai and Ahsan 2009) (Table 28.6).

TLC has been commonly used for reaction monitoring and product purity testing, for example, synthesis of new conformationally constrained pentasaccharides as molecular probes for the

TABLE 28.5

Sample Preparation and Chromatographic Systems for TLC Study of Toxicity of Antithrombotics

Drug	Sample Preparation	Reference Substance	Stationary Phase	Mobile Phase	Detection	Reference
Coumarins	Standards dissolved in chloroform; API extracted from dosage forms with methylene dichloride–methanol 9:1	Acenocoumarol, ethyl biscoumacetate, dicumarol, warfarin, phenprocoumon, coumarin	Silica gel GF$_{254}$ and HF$_{254}$	Ether–acetic acid 99:1; ether–chloroform–acetone 60:30:1; ether–ethyl acetate–acetic acid 80:30:1; ether–benzene–acetone–acetic acid 60:50:10:1; ether–benzene–acetone 60:30:10; ether–benzene–acetic acid 60:40:1; toluene–acetone–acetic acid 50:1:1; toluene–ethyl acetate–acetic acid 20:5:1; toluene–p–dioxane–acetic acid 50:10:1.5; toluene–ether–acetic acid 50:7:1; toluene–$tert$-butyl acetate 80:20; toluene–$tert$-butyl acetate–acetic acid 80:20:1; benzene–ethyl acetate–acetic acid 70:30:0.5	UV light (λ = 254 and 365 nm), iodine vapor, 2,6-dichloroquinonchloroimid, anisaldehyde-sulfuric acid, and diazotized sulfanilic acid	Lau-Cam and Chu-Fong (1972)
Rodenticides	Extraction with acetone, centrifugation, diethylether–protein precipitation; evaporation and solubilization in methanol	Brodifacoum (R_F = 0.210), bromadiolone (R_F = 0.482), difenacoum (R_F = 0.232), coumatetralyl (R_F = 0.600), coumachlor (R_F = 0.632), difethialone (R_F = 0.158), difenacoum (R_F = 0.304), and warfarin (R_F = 0.717)	HPTLC RP$_{18}$ (particle diameter 2–10 µm)	Methanol–phosphoric acid (4.72 M) 9:1	Camag TLC3 scanner (λ = 286 nm), R_F in situ spectra (λ = 220–380 nm)	Berny et al. (1995)

TABLE 28.6

Sample Preparation and Chromatographic Systems for TLC of LMWH Dendrimers

Drug	Sample Preparation	Reference Substance	Stationary Phase	Mobile Phase	Detection	Reference
Dendrimers of LMW heparins (synthesis)	Dissolution in methanol	Synthesis monitoring with reactants: mPEG-2000, 4-nitrophenyl chloroformate, 4-nitrophenyl carbonate mPEG, G3 PAMAM dendrimer, and mPEG-dendrimer	Silica gel 60 F_{254}	Chloroform– methanol 2:1; methanol–water 2:1	Iodine vapor	Bai and Ahsan (2009)

investigation of biological activity of heparin (Sisu et al. 2003), synthesis of biphenyl antithrombotics—selective inhibitors of tissue factor FVIIa complex (Kotian et al. 2009), synthesis of 2-substituted benzopyrano[4,3-d]pyrimidin-4-cycloamines and 4-amino/cycloamino-benzopyrano[4,3-d]pyrimidin-5-ones as potential antiplatelet and antithrombotic agents (Bruno et al. 2006), and synthesis of 1-C-(5-thio-D-xylopyranosyl) derivatives as potential orally active venous antithrombotics (Mignon et al. 2003).

HPLC and LC-MS have suppressed TLC, or at least it seems, as the number of papers with TLC as mono-technique diminishes. Although TLC is an "old" technique, it can still produce results in nano and pico size. In this chapter, we reviewed simple and rapid TLC chromatographic systems that can be used in different areas of research: API and dosage form analysis, ADMET, mechanism of action, etc. TLC stays an elegant technique for in-house analysis, irreplaceable for synthesis monitoring.

REFERENCES

Agrawal, H., Kaul, N., Paradkar, A. R., and K. R. Mahadik. 2003. Stability indicating HPTLC determination of clopidogrel bisulphate as bulk drug and in pharmaceutical dosage form. *Talanta* 61:581–589.

Bai, S. and F. Ahsan. 2009. Synthesis and evaluation of pegylated dendrimeric nanocarrier for pulmonary delivery of low molecular weight heparin. *Pharm Res* 26:539–548.

Barker, W. M., Hermodson, M. A., and K. P. Link. 1970. The metabolism of 4-C14-warfarin sodium by the rat. *J Pharmacol Exp Ther* 171:307–313.

Berny, P. J., Buronfosse, T., and G. Lorgue. 1995. Anticoagulant poisoning in animals: A simple new high-performance thin-layer chromatographic (HPTLC) method for the simultaneous determination of eight anticoagulant rodenticides in liver samples. *J Anal Toxicol* 19:576–580.

Bruno, O., Brullo, C., S. Schenone et al. 2006. Synthesis, antiplatelet and antithrombotic activities of new 2-substituted benzopyrano[4,3-d]pyrimidin-4-cycloamines and 4-amino/cycloamino-benzopyrano[4,3-d]pyrimidin-5-ones. *Bioorg Med Chem* 14:121–130.

Chai, W., Rosankiewicz, J. R., and A. M. Lawson. 1995. TLC-LSIMS of neoglycolipids of glycosaminoglycan disaccharides and of oxymercuration cleavage products of heparin fragments that contain unsaturated uronic acid. *Carbohydr Res* 269:111–124.

Cox, D. 2008. Platelet pharmacology. In *Platelets in Hematologic and Cardiovascular Disorders*, eds. P. Gresele, V. Fuster, J. A. López, C. P. Page, and J. Vermylen, pp. 341–66. Cambridge, U.K.: Cambridge University Press.

European Pharmacopoeia 7th edition (Ph. Eur.). 2010. Strasbourg Cedex: EDQM Council of Europe.

Haddock, R. E., Trager, W. F., and L. R. Pohl. 1975. Biotransformation of phenprocoumon in the rat. *J Med Chem* 18:519–523.

Heidenreich, P. A., Trogdon, J. G., O. A. Khavjou et al. 2011. Forecasting the future of cardiovascular disease in the United States: A policy statement from the American Heart Association. *Circulation* 123:933–944.

Hornberger, W. and H. Patscheke. 1989. Transient concentrations and agonist potency of PGH2 in platelet activation by endogenous arachidonate. *Eicosanoids* 2:241–248.

Kotian, P. L., Krishnan, R., S. Rowland et al. 2009. Design, parallel synthesis, and crystal structures of biphenyl antithrombotics as selective inhibitors of tissue factor FVIIa complex. Part 1: Exploration of S2 pocket pharmacophores. *Bioorg Med Chem* 17:3934–3958.

Lau-Cam, C. A. and I. Chu-Fong. 1972. TLC of coumarin anticoagulants. *J Pharm Sci* 61:1303–1306.

Le Roux, Y. and J. Richard. 1977. Determination of acenocoumarol in plasma and urine by double radioisotope derivative analysis. *J Pharm Sci* 66:997–1000.

Lewis, R. J., Ilnicki, L. P., and M. Carlstrom. 1970. The assay of warfarin in plasma or stool. *Biochem Med* 4:376–382.

Mignon, L., Goichot, C., P. Ratel et al. 2003. New 1-C-(5-thio-D-xylopyranosyl) derivatives as potential orally active venous antithrombotics. *Carbohydr Res* 338:1271–1282.

Pohl, L. R., Haddock, R., Garland, W. A., and W. F. Trager. 1975. Synthesis and thin-layer chromatographic ultraviolet, and mass spectral properties of the anticoagulant phenprocoumon and its monohydroxylated derivatives. *J Med Chem* 18:513–519.

Qiu, G., Tanikawa, M., Akiyama, H., Toida, T., Koshiishi, I., and T. Imanari. 1993. Separation and characterization of dermatan sulfate in normal human urine. *Biol Pharm Bull* 16:340–342.

Shantsila, E., Watson, T., and G. Y. H. Lip. 2008. Laboratory investigation of platelets. In *Platelets in Hematologic and Cardiovascular Disorders*, eds. P. Gresele, V. Fuster, J. A. López, C. P. Page, and J. Vermylen, pp. 124–46. Cambridge, U.K.: Cambridge University Press.

Shimizu, T., Osumi, T., Niimi, K., and K. Nakagawa. 1985. Physico-chemical properties and stability of cilostazol. *Arzneimittelforschung* 35:1117–1123.

Sisu, E., Tripathy, S., J. M. Mallet et al. 2003. Synthesis of new conformationally constrained pentasaccharides as molecular probes to investigate the biological activity of heparin. *Biochimie* 85:91–99.

Toon, S., Heimark, L. D., Trager, W. F., and R. A. O'Reilly. 1985. Metabolic fate of phenprocoumon in humans. *J Pharm Sci* 74:1037–1040.

Tuong, A., Bouyssou, A., Paret, J., and T. G. Cuong. 1981. Metabolism of ticlopidine in rats: Identification and quantitative determination of some its metabolites in plasma, urine and bile. *Eur J Drug Metab Pharmacokinet* 6:91–98.

Uotila, P. 1982. Arachidonate metabolism is changed by dipyridamole in guinea pig isolated lungs. *Acta Pharmacol Toxicol (Copenh)* 50:374–378.

Viinikka, L., Salokannel, J., and O. Ylikorkala. 1983. Effect of prolonged treatment with acetylsalicylic acid and dipyridamole on platelet thromboxane A2 production in atherosclerotic subjects. *Prostaglandins Leukot Med* 11:45–50.

Wagner, H. and S. Bladt. 2009. *Plant Drug Analysis: A Thin Layer Chromatography Atlas*, 2nd ed. Heidelberg, Germany: Springer.

Welling, P. G., Lee, K. P., Khanna, U., and J. G. Wagner. 1970. Comparison of plasma concentrations of warfarin measured by both simple extraction and TLC methods. *J Pharm Sci* 59:1621–1625.

WHO. 2009. *The use of stems in the selection of International Nonproprietary Names (INN) for pharmaceutical substances*. Geneva, Switzerland: World Health Organization.

Williams, L. 2010. Thrombophilia. In *Clinical Laboratory Hematology*, 2nd ed., ed. S. B. McKenzie and J. L. Williams, pp. 889–922. New York: Pearson.

Zaazaa, H. E., Abbas, S. S., Abdelkawy, M., and M. M. Abdelrahman. 2009. Spectrophotometric and spectrodensitometric determination of clopidogrel bisulfate with kinetic study of its alkaline degradation. *Talanta* 78:874–884.

29 TLC of Antihyperlipidemics

Łukasz Komsta

CONTENTS

One of the most important achievements of modern medicine is a significant reduction in mortality caused by circulatory system diseases, arising from atherogenesis. A significant increase in the occurrence of such diseases (as a consequence of hypertriglyceridemia and hypercholesterolemia) is undoubtedly caused by changes in diet and mode of life, resulting from the life changes inducted by modern civilization.

Increasing lack of treatment possibilities of such conditions exclusively by changing the nutrition style and increasing physical activity led to significant widespread of antihyperlipidemic agents. The two most important groups of them are fibrates and statins. The other drugs (such as resins) have now negligible application and their thin-layer chromatography (TLC) was not reported, so they will not be described in this chapter.

Fibrates were introduced into treatment in 1970s and are still successfully used worldwide (Chapman 2006). However, the mechanism of their action was completely unknown until the 1990s.

The current knowledge (Fazio and Linton 2004) connects the fibrate action with activation of PPAR-γ receptors, which modulates expression of some particular genes.

From chemical point of view, fibrates are derivatives of fibric (2-methyl-2-phenoxypropionic) acid, and the Ph–O–C(–CH)(–CH)–COOH moiety is believed to be responsible for the action. Historically, first drug from this group was clofibric (4-chlorofibric) acid. It was used in the free form and as the ethyl ester—clofibrate. Other esters: etofibrate, simfibrate, pirifibrate, ronifibrate, or theofibrate undergo hydrolysis inside the body, creating free clofibric acid (so they are prodrugs). In later years, different para-substituents, improving pharmacokinetic parameters and lowering side effects, were introduced into base structure. Bezafibrate contains a N-ethylbenzamide substituent, whereas ciprofibrate possesses dichlorocyclopropyl group. 4-Chlorobenzoyl substituent with isopropyl esterification can be seen in fenofibrate, which undergoes hydrolysis in vivo to active fenofibric acid. The last is directly manufactured and also used in several countries.

Statins are a group of drugs, lowering cholesterol by inhibiting its synthesis. They are of natural origin and are found in microorganisms competing with the other ones needing sterols to stay alive. The first statin, which went into treatment—mevastatin—was isolated from cultures of *Penicillium brevicompactum* in 1976. In 1978, lovastatin was isolated from *Aspergillus terreus* and *Monascus ruber*. The next ones were: simvastatin (semisynthetic lovastatin derivative) and pravastatin (metabolite of mevastatin). In later years, entirely synthetic drugs were created: atorvastatin, fluvastatin, and cerivastatin (the last one was withdrawn from the market after suspicions of bad side effects).

Statins differ in their chemical structure. The oldest of them—lovastatin, pravastatin, and mevastatin—are the esters of 2-methylbutyric acid, whereas in simvastatin, the acid part comes from 2,2-dimethylbutyric acid. The basic structure is 1,2,3,7,8,8 a-hexahydronaphthalene, connected with the acid ester bond in position 1. In position 8, these drugs possess dihydroxyheptanoic acid structure, creating through internal lactone loop the structure of 2-[(2R, 4R)—tetrahydro-4-hydroxy-6-oxo-2H-pyran-2-yl]-ethyl (form-lactone).

In atorvastatin, the partially hydrogenated naphthalene structure is replaced by the 2-(4-fluorophenyl)-5-(1-methylethyl)-3-phenyl-4-[(phenylamino)-carbonyl]-1H-pyrrole, coupled with the acid in position 1 (through nitrogen). Cerivastatin and fluvastatin possess double bond in the acid fragment (dihydroxy-6-heptenoic acid).

Lovastatin and simvastatin, due to their lactone form, belong to the prodrugs requiring activation in liver. Other statins are hydroxy acids, which are directly active without any transformation.

29.1 DETECTION OF FIBRATES AND STATINS

In the twentieth century, there was weak interest in chromatography of fibrates and statins. The oldest found reference (Assem de Juarez and Garber 1969) describes qualitative analysis of clofibrate and xanthinol nicotinate in pharmaceutical preparations. The authors used pure chloroform and a mixture of chloroform and acetic acid (95:5) on silica gel. Next, two main conjugates of clofibric acid together with free form were separated by TLC (Faed and McQueen 1978). This method was applied to characterize the metabolism of clofibric acid, and percentage proportions in urine were studied.

Next, a combined method (Daldrup et al. 1981) of comprehensive identification of medicinal substances by TLC, GC, and high-performance liquid chromatography was reported. The bezafibrate and etofibrate are among detected drugs in this work, and their identification was performed by measuring the RF value relative to bupranolol. Various drugs containing chlorine, bromine, and iodine have also been studied by TLC (Heinisch et al. 1981), including clofibrate, etofibrate, and clofibric acid.

Clofibrate and etofibrate were chromatographed with benzene–ethyl acetate (24:6) on silica gel by measuring the RF relative to vanillin. In the second system with eluent-containing benzene–chloroform (75:50), clofibrate was also analyzed, reporting RF relative to thymol. Etofibrate was also chromatographed using pure ethyl acetate as eluent (relative to vanillin). Clofibric acid was chromatographed with benzene–acetone–formic acid (70:29:1) and ethyl acetate–isopropanol–water (65:24:11) (again relative to vanillin).

In the beginning of 1990s, a work on screening of 500 substances essential in terms of toxicology was described (Schuetz et al. 1990). There was clofibrate among the identified substances. TLC was carried out on silica gel with a mixture of ethyl acetate–methanol–25% ammonia (85:10:5) as the mobile phase.

Recently, a TLC method for simultaneous fast detection of lipid-regulating agents, antihypertensive agents, and antitussive and antiasthmatic agents in herbal remedies was described (Wang et al. 2010). They describe sample extraction with methanol and separation on silica TLC plates. For lipid-regulating agents (pravastatin sodium, rosuvastatin calcium, atorvastatin calcium, fluvastatin sodium, bezafibrate, simvastatin, lovastatin, and fenofibrate are detected among other drugs), they proposed a mobile phase consisting of n-hexane-ethyl acetate-diethyl ether-glacial acetic acid (10:8:2:1).

29.2 RETENTION INVESTIGATION

The significant interest in chromatography of fibrates began in twenty-first century. The first comprehensive study (Misztal and Komsta 2003) presented TLC data of fibrates in normal-phase TLC systems. The substances investigated were: bezafibrate, ciprofibrate, clofibrate, clofibric acid, fenofibrate, and gemfibrozil. The drugs were separated on silica gel, CN, and DIOL plates with various mobile phases, containing n-hexane and five polar solvents: acetone, dioxane, ethyl methyl ketone, ethyl acetate, and tetrahydrofuran. In each case, the RF values (and corresponding RM values) were presented. The optimum mobile phases were also investigated on alumina, NH$_2$, and polyamide phases for comparison. The RM values were regressed against volume fraction, molar fraction, and logarithm of molar fraction of modifier, obtaining linear relationships and coefficients of resulting equations. It was observed, that separation of all the drugs was achieved only on DIOL plates with mobile phases containing 20%–30% of each modifier in n-hexane, and with hexane-acetone (90:10) on CN plates. The recommended system for normal phase fibrated separation is to use DIOL plates with tetrahydrofuran–hexane (20:80) as mobile phase.

The main difficulty, which appeared during this study, was to separate clofibric acid from ciprofibrate. They differ only with one substituent, which does not change polarity in significant manner. No system on silica gel was found being able to achieve separation of these compounds. The second aspect worthy to mention is that two of investigated substances (fenofibrate and clofibrate) are esters, so their RF values are visibly higher and the spots are always well-defined. In the case of free acids, strong tailing was observed on silica gel, which forced the authors to use acetic acid addition to mobile phase.

The same substances were next investigated by means of reversed-phase TLC (Misztal et al. 2004). Investigation was done on RP18 plates by use of phosphate buffer mixed with six water-miscible solvents—acetone, acetonitrile, dioxane, isopropanol, methanol, and tetrahydrofuran. Optimum modifier concentrations were also investigated on RP8 and CN plates for comparison.

In analogous manner, the obtained data were fitted to linear equations between RM and modifier volume fraction, molar fraction, and the logarithm of the molar fraction. In the case of reversed-phase TLC, the main difficulty was to separate bezafibrate and clofibric acid. The best recommended system providing full separation was RP18 plates with 70% dioxane in pH 7.60 phosphate buffer, with increased development distance to 20 cm.

The experimentally obtained coefficients of Snyder–Soczewiński linear equations allow chromatographer to perform mathematical simulations of development to search for better separation conditions. The data obtained in these two papers were next used (Markowski et al. 2006) for optimization of fibrates separation by automatic multiple development. The drugs were separated by multiple development on DIOL plates with tetrahydrofuran-hexane as mobile phase. The main purpose of this investigation was to prove that results of multiple development analysis, predicted by computer simulation, were in agreement with experimental ones. The retention data of fibrates was also used to illustrate the usefulness of two new chromatographic response functions, used to estimate the equal-spreading of separated spots (Komsta et al. 2007a).

The analogous studies in normal-phase and reversed-phase systems was done for five statin-like drugs—atorvastatin, cerivastatin, fluvastatin, lovastatin, and simvastatin. In the case of normal-phase systems, the silica, diol, and CN layers were investigated with mobile phases analogous to these of fibrates (Komsta et al. 2007d). Similarly to previous studies, data was fitted to linear Snyder–Soczewiński models. The correlation between regression coefficients within and between models were also deeply investigated, estimating the coincidence between regression coefficients and the similarity between modifiers. The recommended system to separate these statins is silica gel with hexane–tetrahydrofuran (60:40) as mobile phase. The analogous study was performed also on RP18 plates (Komsta et al. 2007c). As the statins are highly lipophilic, optimal separations were achieved with high modifier concentrations. The optimal mobile phase was methanol–phosphate buffer pH 7.6 (80:20). The most difficult task on RP18 adsorbent was to separate lovastatin and simvastatin, as they differ only by one methyl group.

In the case of both statins and fibrates, no significant changes of retention was observed with change of pH of mobile phase, possibly due to low solubility of these drugs in water.

29.3 QUANTITATIVE DETERMINATION

Both fibrates and statins are very slightly soluble in water and freely soluble in methanol (up to 10 mg/ml). Therefore, no special extraction procedure is needed—the grinded tablets can be ultrasonicated with methanol and filtered.

Bezafibrate, ciprofibrate, fenofibrate, and gemfibrozil were quantified by TLC-densitometry and TLC-video scanning in tablets. The work describing the first two of them (Misztal and Komsta 2005) presents quantitation of bezafibrate in Benzamidine tablets and ciprofibrate in Lipanor capsules. The optimal mobile phase was selected from previous investigations (DIOL plates with hexane–tetrahydro-furan, 8:2). Detection was performed at $\lambda = 227$ nm (densitometry) and 254 nm (video scanning). Due to small visibility of the spots, calibration plots were constructed in the relatively high range 5–30 µg/spot for both drugs. The calibration data were tested using several regression models, and the optimum models were selected (quadratic for videoscanning and nonlinear $y = ax^m + b$ for densitometry. Additional validation tests were performed: The linearity of the method was tested by spotting different amounts of extracted solution (15–30 mg). Statistical evaluation showed there was no significant difference between the precision and accuracy of the methods. Analogous method is presented for fenofibrate and gemfibro-zil (in Fenoratio and Gemfibral tablets, respectively) (Komsta and Misztal 2005).

Densitometric and video densitometric methods for determination of lovastatin and simvas-tatin in commercially available pharmaceuticals have been also described (Komsta et al. 2007b). Analysis was performed using HPTLC Si F254 plates and hexane-methyl ethyl ketone (55:45) mobile phase with densitometric detection at 230 nm and video scanning at 254 nm. Calibration plots were constructed in the range 4–16 µg per spot for both drugs. Quadratic regression was cho-sen. Densitometric method was much more precise than the video densitometric one, but no signifi-cant differences in accuracy between both procedures were observed. Rosuvastatin and fenofibrate was also recently assayed by Devika et al. (2011).

29.4   OTHER APPLICATIONS

Bezafibrate was used (Korany et al. 2006) as a sample substance in chemometric investigation of den-sitometric peak processing. The densitometric data were subjected to derivative treatment followed by convolution of the resulting derivative curves using discrete Fourier functions. It was found to be ben-eficial in eliminating the interference due to background noise. Nicotinic acid derivatives (also used for lowering blood lipids concentration) were quantitatively assayed by Pyka and Klimczok (2007).

REFERENCES

Assem de Juarez, E.M., C. Garber 1969. Separation of clofibrate [ethyl p-chlorophenoxyisobutyrate] and xan-thinol nicotinate in pharmaceutical compositions. *Proanalisis* 2(5): 125–126.
Chapman, M.J. 2006. Fibrates: Therapeutic review. *British Journal of Diabetes & Vascular Disease* 6(1): 11–21.
Daldrup, T., F. Susanto, P. Michalke 1981. Combination of TLC, GLC (OV 1 and OV 17) and HPLC (RP 18) for a rapid detection of drugs, intoxicants and related compounds. *Fresenius' Journal of Analytical Chemistry* 308(5): 413–427.
Devika, G., M. Sudhakar, J. Venkateshwara Rao 2011. A new improved RP-HPLC method for simultaneous estimation of rosuvastatin calcium and Fenofibrate in tablets. *International Journal of Pharmacy and Pharmaceutical Sciences* 3(Suppl. 4): 311–315.
Faed, E., E. McQueen 1978. Separation of two conjugates of clofibric acid (CPIB) found in the urine of subjects taking clofibrate. *Clinical and Experimental Pharmacology and Physiology* 5(2): 195–198.
Fazio, S., M.F. Linton 2004, The role of fibrates in managing hyperlipidemia: Mechanisms of action and clini-cal efficacy. *Current Atherosclerosis Reports* 6(2): 148–157.
Heinisch, G., H. Matous, W. Rank, R. Wunderlich 1981. Differentiation of drugs. 4. Drugs containing chlorine, bromine or iodine as a heteroelement and extractable with ether from acidic aqueous solutions. *Scientia Pharmaceutica* 49(4): 472–482.
Komsta, Ł., W. Markowski, G. Misztal 2007. A proposal for new RF equal-spread criteria with stable distribu-tion as a random variable. *Journal of Planar Chromatography—Modern TLC* 20(1): 27–37.
Komsta, Ł., G. Misztal 2005. Determination of fenofibrate and gemfibrozil in pharmaceuticals by densi-tometric and videodensitometric thin-layer chromatography. *Journal of AOAC International* 88(5): 1517–1524.
Komsta, Ł., R. Skibiński, H. Hopkała, M. Winiarczyk-Serwacka 2007. Comparative validation of densito-metric and videodensitometric determination of lovastatin and simvastatin in pharmaceuticals. *Chemia Analityczna* 52(5): 771–780.

Komsta, Ł., R. Skibiński, A. Iwańczyk, H. Hopkała 2007. Separation of statin-type antihyperlipidemic drugs by reversed-phase TLC. *Journal of Planar Chromatography—Modern TLC* 20 (3): 235–237.

Komsta, Ł., R. Skibiński, A. Iwańczyk, G. Misztal 2007. Retention data for some statin-type antihyperlipidemic drugs in normal-phase TLC. *Journal of Planar Chromatography—Modern TLC* 20 (2): 107–115.

Korany, M., I. Hewala, K. Abdel-Hai 2006. Non-parametric linear regression of discrete Fourier transform convoluted densitometric peak responses. *Journal of Pharmaceutical and Biomedical Analysis* 40 (5): 1048–1056.

Markowski, W., K. Czapińska, G. Misztal, Ł. Komsta 2006. Analysis of some fibrate-type antihyperlipidemic drugs by AMD. *Journal of Planar Chromatography—Modern TLC* 19 (110): 260–266.

Misztal, G., Ł. Komsta 2003. The retention behavior in normal-phase chromatographic systems of some fibrate-type antihyperlipidemic drugs. *Journal of Planar Chromatography—Modern TLC* 16 (5): 351–358.

Misztal, G., Ł. Komsta 2005. Determination of bezafibrate and ciprofibrate in pharmaceutical formulations by densitometric and videodensitometric TLC. *Journal of Planar Chromatography—Modern TLC* 18 (3): 188–193.

Misztal, G., Ł. Komsta, D. Chichecka 2004. Reversed-phase chromatographic retention behavior of some fibrate-type antihyperlipidemic drugs. *Journal of Planar Chromatography—Modern TLC* 17 (2): 123–127.

Pyka, A., W. Klimczok 2007. Application of densitometry for the evaluation of the separation effect of nicotinic acid derivatives. Part III. Nicotinic acid and its derivatives. *Journal of Liquid Chromatography and Related Technologies* 30 (20): 3107–3118.

Schuetz, H., A. Pielmeyer, G. Weiler 1990. Thin-layer chromatographic screening of 500 toxicologically relevant substances by corrected Rf values (Rfc values) in two systems. *Aerztl. Lab.* 36 (5): 113–123.

Wang, T.-S., Y. Tong, J. Zheng, Z. Zhang, Q. Gao, B.-Q. Che 2010. Detection of 29 adulterations in lipid-regulating, antihypertensive, antitussive and antiasthmatic herbal remedies by TLC. *Chinese Pharmaceutical Journal* 45 (11): 857–861.

30 TLC of Spasmolytics

Przemysław Talik and Jan Krzek

CONTENTS

30.1 INTRODUCTION

Spasmolytics are the group of drugs commonly classified as anticholinergics or antiparasympathetics (parasympatholytics). More precisely, however, they are termed as antimuscarinic agents since they antagonize the muscarine-like actions of acetylcholine and other choline esters. The nerve fibers of the parasympathetic system are responsible for the involuntary movements of smooth muscles present in the gastrointestinal tract, urinary tract, lungs, etc. Pharmacologically, spasmolytics are divided into three groups: tropane alkaloids and their semisynthetic analogues, anticholinergic agents with different structure, and musculotropic spasmolytics, which are selective inhibitors of phosphodiesterase 4 and have no anticholinergic effects.

30.2 TROPANE ALKALOIDS AND THEIR SEMISYNTHETIC ANALOGUES

30.2.1 ATROPINE AND HYOSCYAMINE

Atropine

IUPAC name: (1R,3S,5S)-8-methyl-8-azabicyclo[3.2.1]octan-3-yl 3-hydroxy-2-phenylpropanoate.

Chemical formula: $C_{17}H_{23}NO_3$.

Chemical form: sulfate.

Basic properties: highly soluble in water (2200 mg/L); logP 1.8, pK_a 9.43.

Stationary phase: 20 × 20 silica gel $60F_{254}$ plates of 0.25 mm thickness (Merc) were pretreated by dipping them into a methanolic solution of potassium hydroxide (0.1 M) and then allowing them to air-dry; aliquots of 5 µL containing 10 µg of drug and four reference compounds with known R_f values (see subsequent sections) of the same concentration were applied.

Mobile phase: (A) methanol:ammonia 100:1.5 (reference compounds: diazepam, chlorprothixene, codeine, and atropine), (B) Cyclohexane:toluene:diethylamine 75:15:10 (reference compounds: dipipanone, pethidine, desipramine, and codeine), (C) chloroform:methanol 9:1 (reference compounds: meclozine, caffeine, dipipanone, and desipramine), (D) acetone (reference compounds: meclizine, mepivacaine, procaine, and amitriptyline).

Detection: all drugs were located wherever possible by their response to short- (254 nm) and long-wavelength (350 nm) ultraviolet (UV) light. Where no absorption or fluorescence was seen, drugs were detected using one or more spray reagents.

Under established conditions, the $R_f \times 100$ values (A) 18, (B) 6, (C) 3, and (D) 1 were obtained [1].

30.2.1.1   Determination of Atropine and Other Alkaloids from *Datura inoxia*

Sample preparation: exactly weighed samples of *Datura inoxia* were extracted by percolation of a 1% aqueous solution of acetic acid. The extract was evaporated to fixed volume under reduced pressure at temperatures below 60°C, alkalized with 25% aqueous ammonia, and extracted five times with chloroform; chloroform fractions were evaporated to dryness at 40°C and dissolved in methanol. The solutions were transferred to volumetric flasks and diluted to 10 mL.

Stationary phase: normal-phase thin-layer chromatography (TLC) was performed on 10 cm × 10 cm silica gel F_{254} (TLC and high performance TLC [HPTLC]) and HPTLC RP 18 WF_{254S} plates (Merck).

Two-dimensional (2D) TLC separations were performed on 20 cm × 20 cm Multi-K SC5 TLC plates (Whatman), containing a 3 cm × 20 cm strip of silica and a 17 cm × 20 cm layer of C_{18}.

Samples (2 µL) of 2.5% (w/v) solutions of standards in methanol were spotted on the adsorbent layer. Plates were conditioned for 15 min in mobile phase vapor to eliminate the demixing effect.

Mobile phase: HPTLC plates were developed with methanol:acetone:diethylamine 5:4.8:0.2 (v:v:v:v), or methanol:acetone:aqueous ammonia 5:4.5:0.5 (v:v:v:v).

2D TLC: the plates were developed over the whole distance with $MeOH:Me_2CO:$diethylamine 50:48:2 (v:v:v:v), as multicomponent mobile phase. After complete evaporation of the mobile phase from the layer, spots were transferred to the C_{18} layer by use of pure methanol. The silica strip was then cut off, and the plate was developed in the second direction with methanol:aqueous phosphate buffer (pH 3.4) 1:3 (v:v), containing 0.001 M di-(2-ethylhexyl) phosphoric acid.

Detection: HPTLC—densitometer at 520 nm after spraying the chromatograms with Dragendorff's reagent. 2D TLC—the alkaloids atropine, homatropine, scopolamine, scopolamine N-oxide, and tropine were identified in the plant extract by derivatization with Dragendorff's reagent. Tropic acid was identified by UV illumination.

Under established conditions, complete separation of the alkaloids present in *D. inoxia Mill* was done. The error of estimation RSD was <2%, indicating the possibility of application of the method for routine analysis [2].

30.2.1.2   Quantitative and Qualitative Analysis of Tropane Alkaloids from *D. inoxia* by TLC (i.e., Atropine, Homatropine, ʟ-Hyoscyamine, Scopolamine, Scopolamine N-Oxide, Tropine, and Tropic Acid)

Stationary phase: silica gel and RP-18.

Mobile phase: methanol:acetone:diethylamine 25:24:1 and methanol:acetone:NH_3 10:3:1.

Detection: densitometry after spraying with Dragendorff's reagent at 520 nm [3].

Hyoscyamine is a tropane alkaloid. Similar to atropine, it is a secondary metabolite found in certain plants of the *Solanaceae* family. It is the levorotary isomer of atropine.

30.2.1.3   Direct Resolution of (±)-Ephedrine and Atropine into Their Enantiomers by Impregnated TLC

Stationary phase: homemade silica gel plates (20 cm × 20 cm × 0.5 mm); a 50 g slurry of silica gel G (Merc) in double-distilled water (100 mL) containing optically pure ʟ-histidine or ʟ-tartaric acid 0.5% (pH 7–8), activated overnight at 60°C. The solutions of (±)-ephedrine and atropine and their respective optically pure isomers (10^{-3} M) were prepared in 70% ethanol and were applied side by side to the plates at a 10 µL level.

Mobile phase: acetonitrile:methanol:water 11.3:5.4:0.6 (v:v:v).

Chromatograms were developed in a paper-lined glass chamber preequilibrated with the solvent system for 20–25 min.

Detection: iodine vapor.

Under described conditions, the detection limits were 2 and 6 µg, respectively, for (±)-ephedrine and atropine [4].

Other methods are mentioned earlier.

30.2.2 APOATROPINE

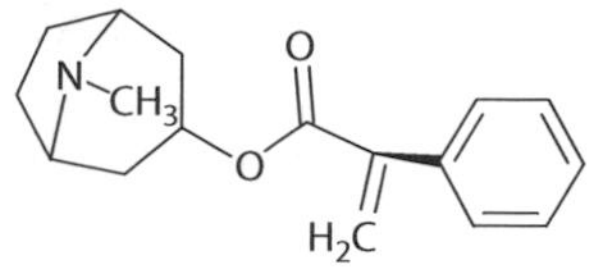

Apoatropine

IUPAC name: (8-methyl-8-azabicyclo[3.2.1]octan-3-yl) 2-phenylprop-2-enoate.

Chemical formula: $C_{17}H_{21}NO_2$.

Chemical form: (hydrochloride salt) soluble in chloroform.

Stationary phase: same as atropine.

Mobile phase: same as atropine.

Detection: same as atropine.

Under established conditions, the $R_f \times 100$ values (A) 22, (B) 18, (C) 7 and (D) 2 were obtained [1].

30.2.3 SCOPOLAMINE

Scopolamine, also known as levo-duboisine or hyoscine is another secondary metabolite of plants from *Solanaceae* family, structurally and pharmacologically similar to atropine.

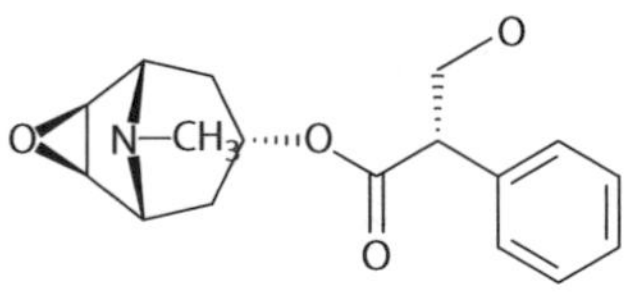

Scopolamine

IUPAC name: (1R,2R,4S,5S,7S)-9-methyl-3-oxa-9-azatricyclo[3.3.1.0²,⁴]nonan-7-yl (2S)-3-hydroxy -2-phenylpropanoate.

Chemical formula: $C_{17}H_{21}NO_4$.

Basic properties: soluble in water 700 g/L, 1 g/20 mL alcohol, slightly soluble in chloroform, insoluble in ether; logP 0.8, pK_a 7.55–7.81.

Stationary phase: same as atropine.

Mobile phase: same as atropine.

Detection: same as atropine.

Under established conditions, the $R_f \times 100$ values (A) 1, (B) 6, (C) 39, and (D) 15 were obtained [1].
Other methods are mentioned earlier.

30.2.4 HOMATROPINE METHYL BROMIDE

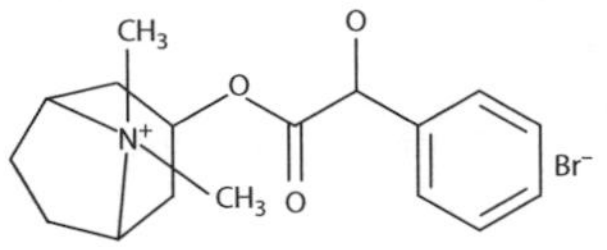

Homatropine methyl bromide

IUPAC name: 3-[(2-hydroxy-2-phenylacetyl)oxy]-8,8-dimethyl-8-azabicyclo[3.2.1]octan-8-ium bromide.

Chemical formula: $C_{17}H_{24}BrNO_3$.

Chemical form: methyl bromide.

Basic properties: soluble in water (50 mg/mL); logP 3.421.

Stationary phase: same as atropine.

Mobile phase: same as atropine.

Detection: same as atropine.

Under established conditions, the $R_f \times 100$ values (A) 1, (B) 1, (C) 1, and (D) 1 were obtained [1]. Other methods are mentioned earlier.

30.3   ANTICHOLINERGIC AGENTS WITH DIFFERENT STRUCTURE

30.3.1 SOLIFENACIN

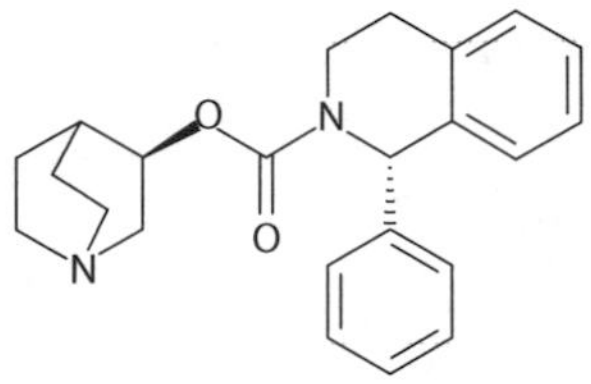

Solifenacin

IUPAC name: butanedioic   acid   (3R)-1-azabicyclo[2.2.2]octan-3-yl   (1S)-1-phenyl-1,2,3,
4-tetrahydroisoquinoline-2-carboxylate.

Chemical formula: $C_{27}H_{32}N_2O_6$.

Chemical form: succinate salt.

Basic properties: highly lipophilic; the octanol:water distribution at pH 7.0 is 50:1; succinate salt is equally soluble in water (at room temperature) and methanol (0.1–1.0 g/mL) and freely soluble in glacial acetic acid and dimethyl sulfoxide (DMSO); pK_a 8.5.

30.3.1.1   Simultaneous Estimation of Alfuzosin (α_1 Receptor Antagonist) and Solifenacin in Pharmaceutical Dosage Form

Stationary phase: HPTLC aluminum plates precoated with silica gel $60F_{254}$ (10 cm × 10 cm) (Merc); standard stock solutions (0.5 mg/mL) of alfuzosin (ALF) and solifenacin were prepared in methanol as solvent. Solutions of 2 μL were applied under nitrogen ambient.

Mobile phase: methanol:ethyl acetate 7:3 (v/v).

TABLE 30.1
Results of Degradation Studies

Stress Condition	Drugs	Time (h)	Mass Balance (%Assay of Recovered + %Impurities + %Degradents)	R_f Values of Degradation Products
Acid hydrolysis	ALF	24	99.88	0.01, 0.17, and 0.89
(0.1 M HCl)	SOL	24	100.20	0.02, 0.23, and 0.37
Alkali hydrolysis	ALF	24	99.96	0.01, 0.06, and 0.80
(0.1 N NaOH)	SOL	24	99.01	0.16, 0.20, and 0.97
Oxidation (3% H_2O_2)	ALF	24	100.11	0.17, 0.63, 0.65, 0.80, and 0.86
	SOL	24	98.99	0.13, 0.71, and 0.81

Detection: densitometric scanning was done in absorbance mode at 254 and 220 nm for ALF and SOL, respectively; the slit dimensions: 5 mm × 0.45 mm, the scanning speed was 20 mm/s, and the data resolution at 100 μm/step.

Under established conditions, the R_f values were 0.71 ± 0.03 and 0.32 ± 0.02 for ALF and SOL, respectively. The drug response was linear over the concentration range between 500 and 2500 ng/band both for ALF and SOL; the limit of detection (LOD) and limit of quantification (LOQ) values were 5.44, 16.5 and 4.35, 13.2 ng/band for ALF and SOL, respectively. The precision: less than 2%.

Forced degradation studies: The amount of drug recovered after degradation studies and the R_f of the degradation products are given in Table 30.1 [5].

30.3.2 TOLTERODINE

Tolterodine

IUPAC name: 2-[(1R)-3-[bis(1-methylethyl)amino]-1-phenylpropyl]-4-methylphenol.

Chemical formula: $C_{22}H_{31}NO$.

Chemical form: tartrate salt.

Basic properties of tartrate salt: solubility in water 12 mg/mL, soluble in methanol, slightly soluble in ethanol, and practically insoluble in toluene. The partition coefficient between n-octanol and water is 1.83 at pH 7.3.; pK_a 9.87.

30.3.2.1 Estimation of Tolterodine in Terol Tablets (2 mg)

Sample preparation: the tablets were powdered, and equivalent to 10 mg tolterodine was weighed and dissolved in methanol. The solution was sonicated for 10 min and filtered. The filtrate was further diluted to a concentration of 100 μg/mL (working stock), and 10 μL was spotted on to the plate followed by chromatographic analysis.

Stationary phase: aluminum plates precoated silica gel 60 F_{254} TLC plates (20 cm × 10 cm) of 0.20 mm thickness (Merc); the assay amount of tolterodine was 1000 ng/band (10 μL).

Mobile phase: toluene:methanol:aqueous ammonia 5:5:0.02 (v:v:v).

Detection: in absorbance mode at 284 nm, the slit dimensions are 5.0 mm × 0.45 mm and the scanning speed: 5 mm/s.

Under described conditions, the R_f of 0.40 ± 0.02 was obtained. The drug response was linear over the concentration range between 200 and 1800 ng/band with LOD 100.0 ng/band, LOQ 200 ng/spot (in compliance with the original text) and RSD below 2% [6].

30.3.2.2 Estimation of Tolterodine in Unknown Drug

Sample preparation: accurately weighed quantity of tablet powder equivalent to 50 mg of the drug was transferred to 10 mL volumetric flask. Eight milliliters of methanol was added and shaken for 10 min, and the volume was adjusted up to the mark with methanol and then filtered through Whatman filter paper no. 40. This solution was used as the sample solution.

Stationary phase: precoated silica gel GF_{254} TLC plates (10 cm × 10 cm) (Merc); 5 mg/mL solution was used as a working standard solution.

Mobile phase: acetonitrile:water:formic acid (50:50:3) (v:v:v).

Detection: scan mode: absorbance/reflectance; wavelength of scanning 281 nm, the slit dimensions 3.0 mm × 0.45 mm.

Under described conditions, the drug response was linear over the concentration range between 10 and 30 μg/mL with LOD 21 ng, LOQ 53 ng (in compliance with the original text), and RSD below 0.1%. The R_f value was not mentioned [7].

30.3.3 PITOFENONE

Pitofenone R = COOCH$_3$
Pitofenone acid R = H

Pitofenone

IUPAC name: methyl 2-[4-(2-piperidinoethoxy)benzoyl]benzoate.

Chemical formula: $C_{22}H_{25}NO_4$.

Chemical form: hydrochloride.

Basic properties of hydrochloride salt: soluble in DMSO and methanol.

Stationary phase: Kieselgel 60 F_{254} TLC plates (Merck). A 20 g/L aliquot of the reaction mixture was spotted.

Mobile phase: chloroform:methanol 4:1.

Detection: UV light (254 nm).

Under described conditions, the R_f values for pitofenone and its metabolite, pitofenone acid, were 0.7 and 0.2, respectively [8].

30.3.4 Dicyclomine

Dicyclomine

IUPAC name: 2-(diethylamino)ethyl 1-cyclohexylcyclohexane-1-carboxylate.

Chemical formula: $C_{19}H_{35}NO_2$.

Chemical form: hydrochloride.

Basic properties of hydrochloride salt: soluble in water 50 mg/mL, logP 5.5.

Stationary phase: same as atropine.

Mobile phase: same as atropine.

Detection: same as atropine.

Under established conditions, the $R_f \times 100$ values (A) 68, (B) 67, (C) 64, and (D) 54 were obtained [1].

30.3.4.1 Determination of Dicyclomine HCl (DCH) in Bulk Drug and Injection Formulation (Cyclopam—Containing 10 mg of DCH)

Stationary phase: precoated silica gel 60 F_{254} TLC plates (10 cm × 10 cm) were prewashed with methanol and activated at 80°C for 5 min prior to the sample application; the final stock solution obtained was 1000 ng/μL. The solution was prepared as per formulation in injection dosage form and stored at 2°C–8°C protected from light.

Analysis of Cyclopam injection dosage form: 1 mL of solution from vial was taken and diluted to 10 mL with methanol, sonicated it for 45 min, and filtered through Whatman filter paper 41.

Mobile phase: toluene:acetone:methanol:concentrated ammonia 5:2:1:0.02 (v/v/v/v).

Detection: densitometric analysis was carried out in the absorbance mode at 523 nm after the plate was sprayed; the slit dimension: 5.0 mm × 0.45 mm and a scanning speed: 20 mm/s.

Preparation of derivatizing agent: 6 g of potassium thiocyanate, 5 g of cobalt chloride, and 3.4 g of sodium acetate were dissolved in water; 2.5 mL of 1N HCl was added and volume was made up to 25 mL with water. Twenty milliliters of this solution was further diluted to 50 mL with methanol, filtered, and stored at room temperature.

Under established conditions, the R_f 0.71 was found. The drug response was linear over the concentration range between 800 ng/spot and 4000 ng/spot with LOD 250 ng/spot, LOQ 800 ng/spot, and RSD 1.17%. The method was robust with the specificity of 0.9983. Statistical analysis proved that the method is suitable for the analysis of dicyclomine HCl (DCH) in pharmaceutical injection dosage form and may be extended to the determination of DCH in plasma and other biological fluids [9].

30.3.4.2 Simultaneous Estimation of Diclofenac Potassium (DCL) (A Nonsteroidal Anti-Inflammatory Agent with Antipyretic and Analgesic Actions) and DCH in a Tablet Formulation (Cataspa), Containing 50 mg DCL and 20 mg DCH

Stationary phase: precoated silica gel 60 F_{254} TLC plates (10 cm × 10 cm) of 0.20 mm thickness were prewashed with methanol and activated at 80°C for 5 min prior to the sample application;

concentration 2000 ng/spot for DCH, 5000 ng/spot for diclofenac potassium (DCL), and the mixed standard stock solution (20 mg/mL DCH, and 50 mg/mL DCL) was spotted onto HPTLC plate (the solution was prepared as per formulation in pharmaceutical tablet). The stock solution was stored at 2°C –8°C protected from light.

Analysis of Cataspa formulation: the weight of the tablet triturate equivalent to 50 mg DCL and 20 mg DCH was transferred into a 25 mL volumetric flask containing 20 mL methanol. The solution was sonicated for 45 min and diluted to 25 mL with methanol. The resulting solution was centrifuged at 3000 rpm for 5 min. Then aforementioned filtered solution gave a concentration of 2000 and 800 μg/mL for DCL and DCH, respectively, and 1 mL of this solution (4000 and 1600 ng/spot for DCL and DCH, respectively) was applied to a HPTLC plate.

Mobile phase: same as mentioned earlier: toluene:acetone:methanol:concentrated ammonia 5:2:1:0.02 (v:v:v:v).

Detection: densitometric analysis was carried out in the absorbance mode at 215 nm for DCL, then the plate was sprayed and scanned again at 523 nm for DCH. Slit dimension: 5.0 mm × 0.45 mm and scanning speed: 20 mm/s.

Under described conditions, the very good separation (R_f 0.51 ± 0.02 for DCL and 0.72 ± 0.03 for DCH) was obtained. The drug response was linear over the concentration range between 1800–9000 ng/spot with LOD 40.0 ng/spot, LOQ 100 ng/spot, and RSD 1.3% for DCL, and 1300–6500 ng/spot with LOD 250 ng/spot, LOQ 800 ng/spot, and RSD 0.95% for DCH. The method was robust with the specificity of 0.999 for both compounds. Statistical analysis proved that the method is suitable for the analysis of DCL and DCH as a bulk drug and in pharmaceutical formulation without any interference from the excipients [10].

30.3.4.3 Different Mobile Phase Used for Simultaneous Estimation of DCL and DCH in the Same Tablet Formulation as Mentioned Earlier (Cataspa)

Stationary phase: precoated silica gel 60 F_{254} TLC plates (20 cm × 10 cm) (Merc) were prewashed with methanol and activated at 105°C for 20 min prior to the sample application; the final working standard 4 μL solution of concentration 200 ng/μL for DCH and 100 ng/μL for DCL was spotted onto HPTLC plate under a stream of nitrogen.

Analysis of Cataspa formulation: the powdered equivalent to 20 mg DCH and 50 mg DCL were accurately weighed and dissolved in 10 mL methanol. The solution was centrifuged for 15 min at 600 rpm then filtered through Whatman filter paper no. 41. One milliliter of this solution was further diluted to 10 mL with methanol to furnish solution of concentration 200 ng/μL for DCH determination. Two milliliters of the second dilution was further diluted to 10 mL so the concentration was the same as that of final standard solutions, that is, 100 ng/μL of DCL, 4 μL of standard, and sample solutions were applied to the plate, which was then developed and scanned.

Mobile phase: toluene:methanol:acetic acid 8:2:0.1 (v/v/v).

Detection: the separated spots were stained with iodine vapors, and densitometric scanning was performed at 410 nm in reflectance mode; slit dimension: 6.0 mm × 0.45 mm.

Under described conditions, the very good separation (R_f 0.41 ± 0.01 for DCL and 0.23 ± 0.01 for DCH) was obtained. The drug response was linear over the concentration range between 0.2–1.6 μg/spot for DCL and 0.4–2.4 μg/spot for DCH. The % assay (Mean ± SD) was found to be 99.25 ± 0. 79 and 99.46 ± 0.58 for DCH and DCL, respectively. Statistical analysis proved that the method is suitable for the analysis of DCL and DCH as a bulk drug and in pharmaceutical formulations [11].

30.3.4.4 Simultaneous Estimation of Ranitidine (a Non-Imidazole Blocker of H₂ Receptors) and DCH in a Tablet Formulation (Ranidic) Containing 150 mg of Ranitidine and 10 mg of DCH

Stationary phase: precoated silica gel 60 F_{254} TLC plates (20 cm × 20 cm) (Merc) were prewashed with methanol and activated at 105°C for 20 min prior to the sample application; the final working standard 4 μL solution of concentration 100 ng/μL for DCH and 150 ng/μL for ranitidine hydrochloride were spotted onto HPTLC plate under a stream of nitrogen.

Analysis of ranidic formulation: the powdered equivalent to 10 mg DCH (150 mg ranitidine hydrochloride) was accurately weighed and dissolved in 10 mL methanol. The solution was centrifuged for 15 min at 600 rpm then filtered through Whatman filter paper no. 41. One milliliter of this solution was further diluted to 10 mL with methanol to furnish solution of concentration 100 ng/μL for DCH determination. One milliliter of the second dilution was further diluted to 10 mL so the concentration was the same as that of final standard solutions, that is, 150 ng/μL of ranitidine hydrochloride; 4 μL of standard and sample solutions were applied to the plate, which was then developed and scanned.

Mobile phase: methanol:water:acetic acid 8:2:0.1 (v/v/v).

Detection: the separated spots were stained with iodine vapors and densitometric scanning was performed at 410 nm in reflectance mode; slit dimension: 6.0 mm × 0.45 mm.

Under described conditions, the very good separation (R_f 0.27 ± 0.01 for ranitidine hydrochloride and 0.67 ± 0.01 for DCH) was obtained. The drug response was linear over the concentration range between 0.150–0.9 μg/spot for ranitidine hydrochloride and 0.4–2.4 μg/spot for DCH. The % assay (mean ± SD) was found to be 99.1 ± 0.847 and 98.80 ± 0.546 for DCH and ranitidine hydrochloride, respectively. Statistical analysis proved that the method is suitable for the analysis of DCL and DCH as a bulk drug and in pharmaceutical formulations [12].

30.3.4.5 Simultaneous Estimation of Nimesulide (A COX-2 Selective, Nonsteroidal Anti-Inflammatory Drug with Analgesic and Antipyretic Properties) and DCH in a Tablet Formulation (Nimek-Spas) Containing 100 mg of Nimesulide and 10 mg of DCH

Stationary phase: silica gel 60 F_{254} TLC plates (20 cm × 10 cm) of 0.25 mm thickness (Merc) were prewashed with methanol and activated at 110°C for 5 min prior to the sample application; standard stock solutions of concentration 1000 μg/mL of Nimesulide and 1000 μg/mL of DCH were prepared separately using methanol, then from the standard stock solution, the mixed standard solution was prepared using methanol to contain 100 μg/mL of Nimesulide and 10 μg/mL of DCH. The stock solution was stored at 2°C–8°C protected from light.

Analysis of nimek-spas formulation: the weight of the tablet triturate, equivalent to 100 mg of nimesulide and 10 mg DCH, was transferred into a 50 mL volumetric flask containing 30–35 mL of methanol, sonicated for 30 min and diluted to 50 mL with methanol. The resulting solution was centrifuged at 3000 rpm for 5 min, and the drug content of the supernatant was determined (2000 and 200 μg/mL for Nimesulide and DCH, respectively). Then, 5 mL of the earlier filtered solution was diluted to produce a concentration of 1000 and 100 μg/mL for nimesulide and DCH, respectively, and 4 μL of this solution (4000 and 400 ng/spot for nimesulide and DCH, respectively) was applied to a HPTLC plate, which was developed in optimized mobile phase.

Mobile phase: toluene:acetone:methanol:concentrated ammonia 7:1.2:1:0.05 (v/v/v/v); approximately 20 min for complete development of the TLC plate.

Detection: after the TLC plate was developed in mobile phase, the derivatizing agent was poured on the plate and dried. Blue spots against light pink background were scanned at 345 nm within 20 min as later background starts getting darker; slit dimension: 5.0 mm × 0.45 mm and a scanning speed: 10 mm/s.

Preparation of derivatizing agent: 6.06 g potassium thiocyanate, 5 g cobalt chloride, and 3.4 g sodium acetate were dissolved in sufficient water, 2.5 mL 1 N HCl was added and the volume was made up to 25 mL with water. Then, 20 mL of this solution was further diluted to 50 mL with methanol, filtered, and stored at room temperature.

Under described conditions, the very good separation (R_f 0.53 ± 0.02 for nimesulide and 0.65 ± 0.02 for DCH) was obtained. The drug response was linear over the concentration range 200–450 ng/spot for nimesulide and 80–180 ng/spot for DCH. The other validation parameters were: precision (intra-day RSD 1.109%–1.73% and inter-day RSD 0.49%–1.878% for nimesulide and intra-day RSD 0.868%–1.44% and inter-day RSD 0.064%–1.07% for DCH) and accuracy (100.10% ± 1.16% for nimesulide and 99.19% ± 0.50% for DCH.

Statistical analysis proved that the method is suitable for the analysis of nimesulide and DCH as a bulk drug and in pharmaceutical formulation without any interference from the excipients [13].

30.4 MUSCULOTROPIC SPASMOLYTICS

30.4.1 ISOQUINOLINE DERIVATIVES AND RELATED ALKALOIDS

30.4.1.1 Papaverine

Papaverine

IUPAC name: 1-[(3,4-dimethoxyphenyl)methyl]-6,7-dimethoxyisoquinoline.

Chemical formula: $C_{20}H_{21}NO_4$.

Basic properties: soluble in water 25 mg/mL; logP 3.

Stationary phase: same as atropine.

Mobile phase: same as atropine.

Detection: same as atropine.

Under established conditions, the $R_f \times 100$ values (A) 61, (B) 8, (C) 65, and (D) 45 were obtained [1].

Chromatographic separations were performed using the automated multiple development (AMD) system. AMD is an instrumental technique of planar chromatography, which uses an eluent gradient starting from the most polar to the least polar.

Stationary phase: HPTLC silica gel F_{254} (10 cm × 20 cm) on glass, layer thickness 0.1 mm (Merc) and HPTLC silica gel F_{254} (10 cm × 20 cm) on glass, layer thickness 0.2 mm (Merc).

Sample preparation: all standard solutions were prepared by dissolving 10 mg of reference substance in 1 mL methanol. The opium extract was obtained by shaking 2 g of raw opium powdered (180) according to *European Pharmacopoeia* (5.0) with 20 mL hydrochloride acid (0.1 M) during 5 min. After filtration, the acidic solution was alkalized by 28% ammonia until pH 10 and extracted three times with 10 mL of dichloromethane. The filtrate was evaporated to dryness, and the residue solved in 2 mL methanol. Solution of 4 μL were applied; the other samples from 3 to 7 μL.

Mobile phase: universal gradient used was: methanol 100, methanol:dichloromethane 50:50, dichloromethane 100, and hexane 100.

Detection: derivatization by iodoplatinate of potassium.
Under established conditions, the R_f value of papaverine (PAP) was 0.52 [14].

30.4.1.1.1 Chemical Analysis of the Reference Substances and Plant Extracts

Sample preparation: the extracts were prepared from fresh petals of *Papaver rhoeas* L.: 30 g of fresh petal powder was added to 300 mL of water, or 10% ethanol or 30% ethanol (WIW) and were left to macerate at room temperature for 12 h. A second maceration was then performed at 35°C for 12 h. After filtration, ethanol was evaporated at low pressure at 35°C, and the extract was freeze-dried.

Extraction of the alkaloids: freeze-dried powder of 5 g was dissolved in 100 mL of methanol. The preparation was heated for 1 h. After cooling, the preparation was filtered, and the solution was evaporated. A 150 mL of 10% sulfuric acid was added to the residue under agitation for 15 min. The preparation was filtered once more, and the pH was adjusted by adding Na_2CO_3 solution at 10%. The filtrate was then extracted three times in 10 mL of chloroform. The chloroform phase was recuperated, then filtered and covered with anhydrous sodium sulfate. The liquid was then evaporated away. The alkaloid residue was dissolved in 1 mL of methanol. The solution was then used for the TLC.

Solutions to be analyzed: the reference substances were narceine, morphine, codeine, thebaine, PAP, and narcotine. Alkaloid extract: 1 mg of powder in 1 mL of methanol.

Stationary phase: silica plate 60 F_{254} (Merc).

Mobile phase: toluene:acetone:ethanol:ammonia (25%) 45:45:7:3 (v:v:v:v).

Detection: observation of fluorescence under UV 365 nm; Dragendorff's reagent (formula of Munier and Macheboeuf [15]).

Under established conditions, the R_f value of PAP was 0.55 [16].
    The effect of stress conditions on the concentrations of secondary metabolites were examined.

Sample preparation: leaf samples of 5 g were placed in 60 mL methanol, extracted (turbo-extractor Polytron), filtered through a filter paper (Siltrak), and concentrated under vacuum (40°C). The dry matter was dissolved in 5 mL methanol, evaporated again, and finally dissolved in 1 mL of the same solvent. Aliquots of 0.5–2.5 μg were spotted for TLC; for HPTLC, the 0.005–1 μg samples were used.

Stationary phase: TLC: 20 cm × 10 cm silica gel 60 F_{254} plates.

HPTLC: 20 cm × 10 cm and 10 cm × 10 cm silica gel 60 plates.

Mobile phase: toluene:acetone:ethanol:aqueous ammonia (25%) 40:40:6:2 (v:v:v:v).

Detection: Dragendorff's reagent with $NaNO_2$ at 520 nm; scanner: 4 mm × 5 mm slit, 30 nm monochromator bandwidth, and tungsten lamp. For UV-mode detection: Naturstoff reagent A (diphenylboric acid 2-amino ethyl ester) performed at 310 nm with a 4 mm × 5 mm slit, 30 nm monochromator bandwidth, and the deuterium lamp [17].

30.4.1.1.2　Qualitative and Quantitative Determination of PAP Either as Raw
　　　　　　Material or in Certain Dosage Forms; Poppy Capsules and Opium

Sample preparation:

- Raw material: 100 mg of PAP was dissolved in 100 mL volumetric flask with methanol tablets. Powdered equivalent to about 100 mg of PAP was dissolved in 100 mL volumetric flask with methanol.
- Ampoules and drops: 10 ampoules were dissolved in 100 mL volumetric flask with methanol; aliquot equivalent to about 100 mg of PAP was transferred to another 100 mL volumetric flask and again completed to the mark with methanol.
- Suppositories: equivalent to 100 mg of PAP was extracted in 100 mL beaker with methanol on water bath for 1 h and filtered into 100 mL volumetric flask; the process was repeated several times with 10 mL of methanol and completed to volume with methanol.
- Crude opium: 500 mg of crude opium with methanol (3 mL × 5 mL) at 50°C; after filtration and evaporation, the residue was dissolved in methanol (3 mL × 1 mL) and filtered to 5 mL volumetric flask and completed to the mark with methanol. One milliliter of the resulted solution was transferred to a 25 mL volumetric flask for procedure 1 and 5 mL for procedure 2. The flasks were adjusted with methanol.
- Poppy capsules: 25 g of the air-dried powdered unripe capsules were extracted with methanol (3 mL × 50 mL) at 50°C; after filtration and evaporation, the residue was triturated with1 N sulfuric acid (3 mL × 2 mL), shaken with chloroform (3 mL × 5 mL), rendered alkaline with ammonia TS, reextracted with chloroform (5 mL × 5 mL), and washed with water (2 mL × 5 mL) and dried. Finally, after filtration and evaporation, the residue was dissolved in chloroform and transferred to a 5 mL volumetric flask for procedure 1 and 1 mL volumetric flask for procedure 2. The flasks were adjusted with methanol.

Stationary phase: silica gel G precoated plate 60 F_{254} 0.25 mm thick (Merc) activated in 110°C for 1 h; aliquots of 1–5 µL of sample preparations were transferred.

Mobile phase: toluene:acetone:ethyl acetate:methanol 70:20:10 (v:v:v).

Detection: spectrodensitometer at 238 nm.

The results were precise and reproducible. The mean recovery of determination was 99.2 ± 1.09. The R_f values were not posted [18].

Sample preparation: for alkaloids and quaternary ammonium salts (methadone, heroin, buprenorphine, opium extract, noscapine, PAP, thebaine, codeine, and morphine), powdered plant material, 1 g/10 mL, was mixed thoroughly with ethanol for approximately 4 h, filtered through a No. 2 glass frit, and the volume was adjusted to 10 mL.

Over pressured layer chromatography (OPLC) technique: the plates used for the alkaloids were aluminum oxide 60 F_{254} type E 20 cm × 20 cm on glass (Merck). The plates were impregnated with Impress No. II; eluent: ethyl acetate 100, external pressure 15 bar, starting mobile phase 7 bar, and flow rate 0.40 mL/min; the separations were checked by visual observation under UV illumination at 365 nm and after spraying with either Dragendorff's or iodoplatinate reagent. Densitograms were recorded at 540 nm after visualization with Dragendorff's reagent.

OPLC semi-preparative mode: the plates were aluminum oxide 60 F_{254} 20 cm × 20 cm, thickness 1.5 mm (Merck), and impregnated with Impress No. II; mobile phases: hexane, ethyl acetate 50:50 (v/v) for fractions 1–8, ethyl acetate 100 for fractions 9–12, and ethyl acetate saturated with 28% ammonia for fraction 13; water cushion pressure 12 bar, and flow rate 0.6 mL/min.

Automated multiple development system (AMD) technique: for opium alkaloids, analysis plates were silica gel 60 F$_{254}$ 10 cm × 20 cm on glass (Merck). The elution gradient was methanol 100, methanol–acetone 50:50 v/v, acetone 100, ethyl acetate 100, ethyl acetate–dichloromethane 50:50 v/v, and dichloromethane 100; 25 development steps; reagent: potassium iodoplatinate.

Under established conditions, the R$_f$ × 100 values of PAP were TLC 68, OPLC 82, and AMD 73. AMD allows to work on very small sample quantities and obtain sharper separations owing to the absence of diffusion in the adsorbent. OPLC presents a great rapidity of analysis on an important number of samples and allows making chromatographic semi-preparative separations allowing pure products obtention by direct elution. Thus, hR$_f$ values reproducible, and each compound is eluted at a defined time [19].

30.4.1.2 Drotaverine

Drotaverine

IUPAC name: (1Z)-1-[(3,4-diethoxyphenyl)methylidene]-6,7-diethoxy-1,2,3,4-tetrahydroisoquinoline.

Chemical formula: C$_{24}$H$_{31}$NO$_4$.

Chemical form: hydrochloride.

Basic properties: moderately soluble in water, soluble in alcohol (96%), easily soluble in chloroform.

30.4.1.2.1 *Simultaneous Determination of Nifuroxazide and Drotaverine HCl in Synthetic Mixtures and a Commercial Preparation (Drotazide Capsules, Labeled to Contain 200 mg Nifuroxazide and 40 mg Drotaverine HCl per Capsule)*

Sample preparation: solution of drotaverine HCl (DRT) (0.4 mg/mL) and nifuroxazide (NIF) (1.2 mg/mL) were separately prepared in 50 mL volumetric flasks using 2 mL DMSO for solubility and volume made up with methanol. Serial dilutions were prepared in two series of 5 mL volumetric flasks containing 0.02–0.4 mg/mL of DRT, and 0.06–1.2 mg/mL of NIF in methanol and were used for the preparation of calibration curves of TLC method.

   Laboratory prepared mixtures were prepared in different ratios of DRT to NIF mainly (1:5) as that of capsule using the same solvent as in the method and containing 0.02–0.4 mg/mL of DRT and 0.06–1.2 mg/mL of NIF.

Assay of capsules: an amount of the powdered capsules equivalent to 10 mg of DRT and 50 mg of NIF was transferred into 50 mL volumetric flask, treated with 2 mL DMSO, and made to volume with methanol. The solutions were filtered, and suitable aliquots of the filtrates were used to quantify the two drugs as described under application of the methods.

Stationary phase: TLC plates precoated with silica gel 60, 10 cm × 10 cm, 0.25 mm thickness, and fluorescent at 254 nm (Merck).

Mobile phase: ethyl acetate:methanol:ammonia 33% 10:1:0.1 (v:v:v).

Detection: the spots of the NIF and DRT were scanned at 287 and 308 nm, respectively, under photo mode (reflection); scan mode: zigzag and swing width: 10 mm.

The R_f values of DRT and NIF were 0.75 ± 0.01 and 0.50 ± 0.01, respectively. The drug response was linear over the concentration range between 0.2 and 4 µg/spot for DRT and 0.6 and 12 µg/spot for NIF [20].

30.4.1.2.2 Simultaneous Determination of Paracetamol and DRT Drugs in Synthetic Mixtures and a Commercial Preparation (Anaspasm Tablets, Labeled to Contain 325 mg Paracetamol and 60 mg DRT per Tablet)

Sample preparation: an amount of the powder equivalent to 50 mg of paracetamol and 10 mg DRT was extracted with methanol by shaking for 10 min, filtered into 50 mL volumetric flask, and the volume was completed with methanol. Then, 10 µL of each solution was applied to TLC plate.

Stock solutions: Paracetamol (1.2 mg/mL) and DRT (0.4 mg/mL) in methanol.

Stationary phase: TLC plates precoated with silica gel 60, 10 cm × 10 cm, 0.25 mm thickness, and fluorescent at 254 nm (Merck).

Mobile phase: ethyl acetate:methanol:ammonia 100:1:5 (v:v:v).

Detection: the spots of the paracetamol and DRT were scanned at 249 and 308 nm, respectively, under photo mode (reflection); scan mode: zigzag and swing width: 10 mm.

The R_f values of paracetamol and DRT were 0.49 ± 0.37 and 0.72 ± 0.01, respectively. The drug response was linear over the concentration range 60–1200 µg/mL for paracetamol and 20–400 µg/mL for DRT; the LOD and LOQ values were 0.26, 0.86 and 0.12, 0.41 for paracetamol and DRT, respectively [21].

30.4.1.2.3 Different Mobile Phase Used for Simultaneous Determination of NIF and DRT in Drotazide (see preceding text)

Sample preparation: 50 mg of both NIF and DRT were accurately weighed and transferred to a 100 mL volumetric flask; 75 mL of absolute ethanol was added, shaked, and diluted to volume with absolute ethanol to prepare a 500 µg/mL stock solution. Then, an accurately measured 20 mL volume of stock solution was transferred into 100 mL volumetric flask and diluted to volume with absolute ethanol to prepare 100 µg/mL working solution.

Stationary phase: TLC plates 20 cm × 20 cm coated with silica gel 60 (Merc).

Mobile phase: chloroform:acetone:methanol:glacial acetic acid 6:3:0.9:0.1 (v:v:v:v).

Detection: 365 nm.

Over the concentration range of 0.2–1 µg/band, the drug response was linear, and LOD values were 0.05 µg/band and 0.1 µg/band for NIF and DRT, respectively. The LOQ of 0.2 µg/band was obtained for both drugs [22].

30.4.1.2.4 Simultaneous Determination of a Ternary Mixture Containing DRT, Caffeine, and Paracetamol and Pharmaceutical Preparation (Petro Tablets Containing 40 mg DRT, 400 mg Paracetamol, and 60 mg Caffeine)

Sample preparation: stock standard solutions: 100 mg of DRT, 150 mg of caffeine, and 1000 mg of paracetamol were transferred into separate 100 mL volumetric flasks, and 50 mL of methanol was added, shaked for a few minutes, and diluted to volume with the same solvent.

Aliquot portions equivalent to 10–100 µg of DRT, 15–150 µg of caffeine, and 25–150 µg of paracetamol from their stock solutions (1, 1.5, and 10 mg/mL, respectively) were transferred into separate series of 10 mL volumetric flasks and diluted to volume with methanol. Then, 10 µL of each series of DRT, caffeine, and paracetamol was applied to a TLC plate.

Application to a pharmaceutical preparation (Petro tablets): an accurately weighed amount of pulverized powder equivalent to 100 mg of DRT, 150 mg of caffeine, and 1000 mg of paracetamol was extracted by sonication with methanol (3 mL × 25 mL), filtered to a 100 mL volumetric flask, and diluted to the line with the same solvent (1 mg/mL DRT, 1.5 mg/mL caffeine, and 10 mg/mL paracetamol). The same procedures were completed on the solution as described earlier.

Stationary phase: TLC plates precoated with silica gel GF_{254} 20 cm × 20 cm 0.25 cm thickness (Merc).

Mobile phase: ethyl acetate:chloroform:methanol 16:3:1 (v:v:v).

Detection: the spots were scanned at 281, 272, and 248 nm for DRT, caffeine, and paracetamol, respectively.

Under described conditions, R_f values of 0.26, 0.5, and 0.68 for DRT, caffeine, and paracetamol, respectively, were obtained. The drug response was linear over the concentration range 2–40 µg/mL for DRT, 6–48 µg/mL for caffeine, and 4–40 µg/mL for paracetamol; the LOD and LOQ values were 0.92, 1.06, 1.10 µg/mL and 1.57, 3.54, 3.68 µg/mL for DRT, caffeine, and paracetamol, respectively [23].

30.4.2 ESTERS OF AMINOALCOHOLS AND AROMATIC ACIDS

30.4.2.1 Mebeverine

Mebeverine

IUPAC name: (RS)-4-(ethyl[1-(4-methoxyphenyl)propan-2-yl]amino)butyl 3,4-dimethoxybenzoate.

Chemical formula: $C_{25}H_{35}NO_5$.

Chemical form: hydrochloride.

Basic properties: very soluble in water, freely soluble in ethanol (96%), and practically insoluble in ether.

Stationary phase: the same as atropine.

Mobile phase: the same as atropine.

Detection: the same as atropine.

Under established conditions, the R_f × 100 values (A) 63, (B) 40, (C) 53, and (D) 49 were obtained [1].

30.4.2.1.1 *Simultaneous Estimation of Sulpiride (SU) (an Antagonist of the Dopamine D_2 Receptors) and Mebeverine Hydrochloride (MB) in Tablets of Colona, Containing 100 mg MB and 25 mg SU per tablet*

Sample preparation: portions of the powder equivalent to about 100 mg of mebeverine hydrochloride (MB) were weighed accurately, transferred to 100 mL volumetric flasks using methanol, and completed to volume with the same solvent. For the derivative and TLC procedures, the suspensions were filtered through a methanolic wetted filter paper and then further diluted, either with 0.1 M HCl or methanol, to suit the calibration graphs for the derivative measurements or the TLC procedure, respectively.

Stationary phase: aluminum plates precoated silica gel GF_{254} TLC plates (20 cm × 10 cm) of 0.25 mm thickness (Merc); 10 μL test solution and 10 μL of the corresponding standard solution were applied.

Mobile phase: ethanol:diethyl ether:triethylamine 70:30:1 (v:v:v). Addition of triethylamine in 1% concentration to the aforementioned system was essential to prevent tailing and to move the drug spots upward.

Detection: scan mode: the zigzag; wavelengths of scanning: 262 and 240 nm for MB and sulpiride (SU), respectively.

Under established conditions, R_f values of the investigated drugs 0.45 and 0.29 for MB and SU, were obtained, respectively. The drug response was linear over the concentration range between 4 and 12 μg/mL, LOD 0.18 and relative sensitivity 0.053 for MB and 2–7 μg/mL, LOD 0.05 and relative sensitivity 0.417 for SU [24].

30.4.2.1.2 Different Mobile Phase for the Same Dosage Form of SU and MB (Colona Tablets)

Stationary phase: silica gel HPTLC F_{254} plates; 200 μg/mL working solution for each compound was prepared in absolute ethanol.

Mobile phase: absolute ethanol:methylene chloride:triethyl amine 7:3:0.2 (v:v:v).

Detection: scan of the separated bands at 221 nm.

Under established conditions, R_f values of the investigated drugs 0.617 ± 0.01 and 0.423 ± 0.01 for MB and SU, respectively, were obtained. The drug response was linear over the concentration range 50–60 μg/mL, LOD 0.04 μg/band, LOQ 0.2 μg/band for MB and 5–40 μg/mL, LOD 0.02 μg/band, and LOQ 0.3 μg/band for SU. RSD% was from 1.273 to 2.568 [25].

30.4.2.2 Oxybutynin

Oxybutynin

IUPAC name: 4-(diethylamino)but-2-yn-1-yl 2-cyclohexyl-2-hydroxy-2-phenylacetate.

Chemical formula: $C_{22}H_{31}NO_3$.

Chemical form: hydrochloride.

Basic properties: (concerns hydrochloride salt) freely soluble in water and alcohol, very soluble in ethyl alcohol and in chloroform, soluble in acetone, and very slightly soluble in hexane.

Determination of oxybutynin (OX) in pharmaceutical formulations: Uripan tablets (5 mg), detronin tablets (5 mg), Uripan syrup (100 mg per 100 mL), and detronin syrup (100 mg of oxybutynin, 100 mg methylparaben, and 20 mg propylparaben per 100 mL).

Sample preparation: an accurate weight of the mixed sample was transferred into a beaker, and 50 mL methanol was added with continuous magnetic stirring for about 10 min. The solution was filtered into a 100 mL volumetric flask, and the volume was completed with methanol.

An accurate volume of the syrup was transferred, measured, and diluted to 100 mL methanol, filtered.

Aliquots equivalent to 2–14 mg OX were accurately transferred from its stock standard solution (1 mg/mL) into a series of 10 mL volumetric flasks dissolved in and diluted to the volume with methanol.

Stationary phase: precoated silica gel aluminum plates 60 F_{254}, ALLUGRAM®SIL G/UV 254 (Machenary-Nagel) 0.2 mm coating thickness (20 cm × 10 cm) were prewashed with methanol and dried at 60°C for 15 min prior to sample application. Then, 10 µL of each solution was applied on the HPTLC plates.

Mobile phase: chloroform:methanol:ammonia solution:triethylamine 100:3:0.5:0.2 (v/v/v/v).

Detection: wavelengths of scanning 220 nm.

Under described conditions, the standard OX R_f was 0.59 and methyl and propyl parabens R_f 0.28. The drug response was linear over the concentration range 2–14 µg/band with LOD 0.56 µg/band, LOQ 1.85 µg/spot, and RSD below 1.5% [26].

30.4.3 MUSCULOTROPIC SPASMOLYTICS WITH DIFFERENT STRUCTURES

30.4.3.1 Naftidrofuryl

Naftidrofuryl

IUPAC name: (*RS*)-2-(diethylamino)ethyl 3-(1-naphthyl)-2-(tetrahydrofuran-2-ylmethyl)propanoate.

Chemical formula: $C_{24}H_{33}NO_3$.

Chemical form: oxolate.

Basic properties: (concerns oxolate salt) freely soluble in water, freely soluble or soluble in alcohol, and slightly or sparingly soluble in acetone.

30.4.3.1.1 Quantification of Naftidrofuryl in Stomach Contents

Sample preparation: 3 mL of the homogenized stomach contents was diluted with phosphate buffer (pH = 6, 18 mL) and applied to the HCX 300 mg solid-phase extraction column. With the flow rate 1.5 mL/min, the column was washed with water (1 mL) and 0.01 M acetic acid (0.5 mL) and dried for 4 min under vacuum. Then, methanol (50 µL) was applied and again dried for 1 min. Next, the column was centrifuged to completely remove solvents. Using acetone:chloroform (1:1 v/v, 4 mL), the fraction of neutral and acidic compounds (fraction 1) was eluted, collected, and taken to dryness. Finally, the fraction of basic and amphoteric compounds (fraction 2) was eluted from the column with dichloro-methane:isopropanol:ammonium hydroxide (8:2:0.5 v/v, 2 mL) and taken to dryness. By qualitative HPTLC with identification of the compounds by their UV remission spectra, it was found that naftidrofuryl was detectable both in fractions 1 and 2.

Fraction 1: 10 μL of solution 1 was diluted with 1190 μL of methanol. The dilution (0.5 and 1 μL) was applied to a TLC plate, alternating with naftidrofuryl calibrators of 100, 200, 400, 600, and 800 ng.

Fraction 2: 10 μL of a solution of fraction 2 in 150 μL of methanol was diluted with 590 μL of methanol. Then, 0.5 and 1 μL of the dilution were applied to a TLC plate; calibrators were as described for fraction 1.

Stationary phase: TLC plates silica gel F_{254} 20 cm × 10 cm (Merck); samples of 1 and 2 μL of the dilution were applied.

Mobile phase: methanol:ammonia 99:1 (v/v).

Detection: UV detector, wavelength of 283 nm.

Under described procedure, the R_f value of 0.52–0.58 was obtained [27].

30.4.3.2 Bencyclane

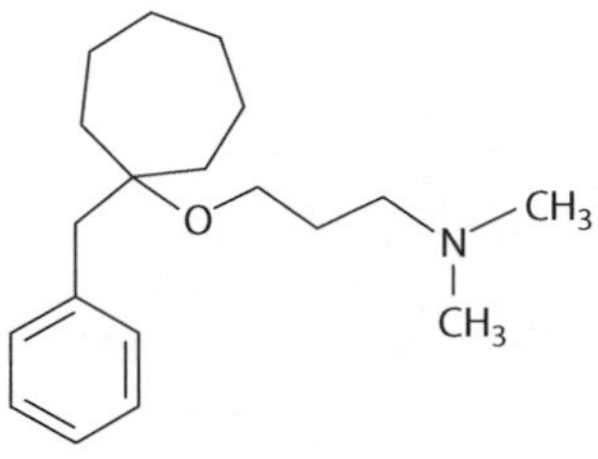

Bencyclane

IUPAC name: 3-[(1-benzylcycloheptyl)oxy]-*N,N*-dimethylpropan-1-amine.

Chemical formula: $C_{19}H_{31}NO$.

Chemical form: fumarate.

Stationary phase: precoated plates of silica gel G (Merc) 0.2 mm thick; 15 μL aliquot of an 1 mg/mL aqueous solution of bencyclamine was applied to the plate.

Mobile phase: cyclohexane:isopropanol:ammonium hydroxide (28%) (80:20:1 (v:v:v).

Detection: the spots were detected by exposure to iodine vapor [28].

30.4.3.3 Cyclandelate

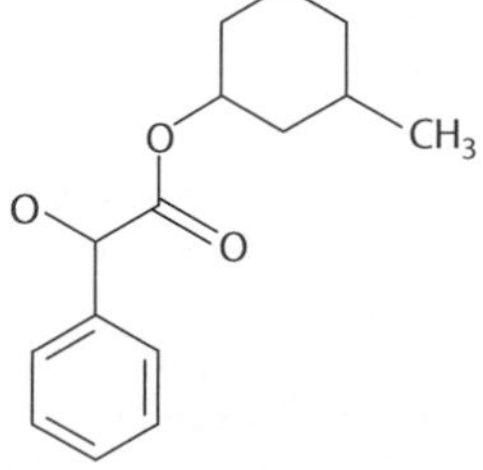

Cyclandelate

IUPAC name: 3,3,5-trimethylcyclohexyl 2-hydroxy-2-phenylacetate.

Chemical formula: $C_{17}H_{24}O_3$.

Chemical form: mandelate.

Basic properties: practically insoluble in water; soluble 1 in about 1 of ethanol; very soluble in chloroform and ether; and freely soluble in acetonitrile, ethyl acetate, DMF, and toluene.

Stationary phase: 20 × 20 silica gel 60F$_{254}$ plates of 0.25 mm thickness (Merc); aliquots of 5 μL containing 10 μg of drug and four reference compounds with known R$_f$ values (see subsequent text) of the same concentration was applied.

Mobile phase: (A) chloroform: acetone 4:1 (reference compounds: methohexitone, quinalbarbitone, clonazepam, and paracetamol), (B) ethyl acetate:methanol:ammonia 85:10:5 (reference compounds: prazepam, temazepam, hydrochlorothiazide, and sulphadimidine), (C) ethyl acetate (reference compounds: quinalbarbitone, salicylamide, phenacetin, and sulphathiazole), and (D) chloroform:methanol 9:1 (reference compounds: prazepam, phenacetin, sulphafurazole, and hydroflumethiazide).

Detection: same as atropine.

Under established conditions, the R$_f$ × 100 values (A) 74, (B) 81, (C) 71, and (D) 73 were obtained [1].

REFERENCES

1. Stead, A.H., Gill, R., Wright, T., Gibbs, J.P., Moffat, A.C. 1982 Standardized thin-layer chromatographic systems for the identification of drugs and poisons. *Analyst*. 107(1279):1106–1168.
2. Gadzikowska, M., Petruczynik, A., Waksmundzka-Hajnos, M. 2005 Two-dimensional planar chromatography of tropane alkaloids from *Datura inoxia* Mill. *J. Planar Chromatogr*. 18:127–131.
3. Fecka, I., Kowalczyk, A., Cisowski, W. 2003 Proc. Intern. Symp. Planar Separations, *Plan. Chrom.*, Sz. Nyiredy, Ed., (Res. Inst. Med. Plants, Budapest, Hungary), 223–230.
4. Bhushan, R., Martens, J., Arora, M. 2001 Direct resolution of (±)-ephedrine and atropine into their enantiomers by impregnated TLC. *Biomed Chromatogr*. 15(3):151–154.
5. Wankhede, S.B., Somani, K., Chitlange, S.S. 2011 Stability indicating normal phase HPTLC method for estimation of alfuzosin and solifenacin in pharmaceutical dosage form. *Int. J. ChemTech Res*. 3(4):2003–2010.
6. Mishra, A., Gowdra, V.S., Arumugam, K., Hussen, S.S., Bhat, K., Udupa, N. 2011 Stability-indicating HPTLC method for analysis of tolterodine in the bulk drug. *J. Plan. Chrom*. 24(2):150–153.
7. Shaiba, M., Maheswari, R., Chakraborty, R., Sai Praveen, P., Jagathi, V. 2011 High performance thin layer chromatographic estimation of tolterodine tartarate. *RJPBCS* 2(1):6–11.
8. Bal-Tembe, S., Bhedi, D.N., Mishra, A.K., Rajagopalan, R., Ghate, A.V., Subbarayan, P., Punekar, N.S., Kulkarni, A.V. 1997 HL 752: A potent and long-acting antispasmodic agent *Bioorg. Med. Chem*. 5(7):1381–1387.
9. Keer, A.R., Havele, S.S., Gopani, K.H., Dhaneshwar, S.R. 2011 Application of high performance thin layer chromatography-densitometry for the determination of dicyclomine HCL in bulk drug and injection formulation. *Der Pharma Chem*. 3(1):549–556.
10. Dhaneshwar, S.R., Keer, A.R., Havele, S.S., Gopani, K.H. 2011 Validated HPTLC method for simultaneous estimation of diclofenac potassium and dicyclomine hydrochloride in tablet formulation. *RJPBCS*. 2(4):314–324.
11. Potawale, S.E., Nanda, R.K., Bhagwat, V.V., Hamane, S.C., Deshmukh, R.S., Puttamsetti, K. 2011 Development and validation of a HPTLC method for simultaneous densitometric analysis of diclofenac potassium and dicyclomine hydrochloride as the bulk drugs and in the tablet dosage form. *J. Pharm. Res*. 4(9):3116–3118.
12. Nanda, R.K., Potawale, S.E., Bhagwat, V.V., Deshmukh, R.S., Deshpande, P.B. 2010 Development and validation of a HPTLC method for simultaneous densitometric analysis of Ranitidine hydrochloride and dicyclomine hydrochloride as the bulk drugs and in the tablet dosage form. *J. Pharm. Res*. 3(8):1997–1999.
13. Dhaneshwar, S.R., Suryan, A.L., Bhusari, V.K., Rasal, K.S. 2011 Validated HPTLC method for nimesulide and dicyclomine hydrochloride in formulation *J. Pharm. Res*.4(7):2288–2290.
14. Pothier, J., Galand, N. 2005 Automated multiple development thin-layer chromatography for separation of opiate alkaloids and derivatives *J. Chrom. A*. 1080:186–191.

15. Munier, R., Macheboeuf, M. 1951 Paper partition microchromatography of alkaloids and of various biological nitrogenous bases. III. Examples of the separation of various alkaloids by the acid solvent phase technic (atropine, cocaine, nicotine, sparteine, strychnine and corynanthine families) *Bull. Soc. Chim. Biol.* 33(7):846–856.

16. Soulimani, R., Younos, C., Jarmouni-Idrissi, S., Bousta, D., Khalouki, F., Laila, A. 2001 Behavioral and pharmaco-toxicological study of *Papaver rhoeas* L. in mice. *J Ethnopharmacol.* 74:265–274.

17. Szabó, B., Lakatos, A., Kõszegi, T., Kátay, G., Botz, L. 2005 Thin-layer chromatography-densitometry and liquid chromatography analysis of alkaloids in leaves of Papaver somniferum under stress conditions. *J AOAC Int.* 88(5):1571–1577.

18. Salama, O.M., Walash, M.I. 1991 Densitometric determination of papaverine in opium, poppy capsules and certain pharmaceutical dosage forms *Anal. Lett.* 24(1):69–82.

19. Galand, N., Pothier, J., Dollet, J., Viel, C. 2002 OPLC and AMD, recent techniques of planar chromatography: Their interest for separation and characterization of extractive and synthetic compounds *Fitoterapia* 73:121–134.

20. Ayad, M.M., Youssef, N.F., Abdellatif, H.E., Soliman, S.M. 2006 A comparative study on various spectrometries with thin layer chromatography for simultaneous analysis of drotaverine and nifuroxazide in capsules. *Chem. Pharm. Bull.* 54(6):807–813.

21. Abdellatef, H.E., Ayad, M.M., Soliman, S.M., and Youssef, N.F. 2007 Spectrophotometric and spectrodensitometric determination of paracetamol and drotaverine HCl in combination. *Spectrochim Acta A.* 66:1147–1151.

22. Metwally, F.H., Abdelkawy, M., Naguib, I.A. 2006 Determination of nifuroxazide and drotaverine hydrochloride in pharmaceutical preparations by three independent analytical methods. *J AOAC Int.* 89(1):78–87.

23. Metwally, F.H., El-Saharty, Y.S., Refaat, M., El-Khateeb, S.Z. 2007 Application of derivative, derivative ratio, and multivariate spectral analysis and thin-layer chomatography-densitometry for determination of a ternary mixture containing drotaverine hydrochloride, caffeine, and paracetamol. *J AOAC Int.* 90(2):391–404.

24. El Walily, A.F.M., El Gindy, A., Bedair, M.F. 1999. Application of first-derivative UV-spectrophotometry, TLC-densitometry and liquid chromatography for the simultaneous determination of mebeverine hydrochloride and sulpiride. *J. Pharm. Biomed. Anal.* 21:535–548.

25. Naguib, I.A., Abdelkawy, M. 2010. Development and validation of stability indicating HPLC and HPTLC methods for determination of sulpiride and mebeverine hydrochloride in combination. *Eur. J. Med. Chem.* 45:3719–3725.

26. Wagieha, N.E., Hegazyb, M.A., Abdelkawyb, M., Abdelaleem, E.A. 2010 Quantitative determination of oxybutynin hydrochloride by spectrophotometry, chemometry and HPTLC in presence of its degradation product and additives in different pharmaceutical dosage forms. *Talanta* 80:2007–2015.

27. Koller, M.F., Schmid, M., Iten, P.X., Vonlanthen, B., Bär, W. 2009 Fatal intoxication with naftidrofuryl. *Leg Med.* 11(5):229–233.

28. Fujioka, K., Kurosaki, Y., Sato, S., Noguchi, T., Nogushi, T., Yamahira, Y. 1983 Biopharmaceutical study of inclusion complexes. I. Pharmaceutical advantages of cyclodextrin complexes of bencyclane fumarate. *Chem. Pharm. Bull.* 31(7):2416–2423.

31 TLC of Mucolytic, Antitussive, and Antiasthmatic Drugs

Radosław J. Ekiert and Jan Krzek

CONTENTS

31.1 INTRODUCTION

Mucolytic, antitussive, and antiasthmatic drugs do not form a homogeneous division of chemically related pharmaceuticals. They even do not exert the same pharmacological action. Only bromhexine and ambroxol are almost identical. In a description of particular compounds, information includes—if only such data are available—possible degradant impurities, sample preparation in the context of matrix (pharmaceutical formulation, biological samples), chromatographic systems used in the separation, and/or quantitation of particular drugs.

There are no relevant publications for many known pharmaceuticals, which should be assigned to the group of drugs under consideration: sulfogaiacol, pentoxiverin, butamirate, nedocromil, mesna, zileuton, zafirlukast, ablukast, and pranlukast. Drugs that exert among others mucolytic, antitussive, or antiasthmatic pharmacological effects not described in this section, for example, codeine, could be found in other chapters.

Thin-layer chromatography (TLC), especially as high-performance thin-layer chromatography with densitometric detection, is suitable and widely used for analysis of almost all pharmaceuticals [1,2]. It is indicated for analysis both in drug formulations and in clinical practice as well. This technique has undeniable advantages: it is simple and economic; enables simultaneous analysis of many samples; does not require complicated sample preparation; has acceptable precision, accuracy, and resolution; and gives the opportunity to color spots in characteristic reactions [3,4]. Nevertheless, TLC is consistently superseded by high-performance liquid chromatography (HPLC) because it is characterized by better resolution and lower detection limit and is simply more convenient. It is a reason why the number of publications describing analytical procedures elaborated for mucolytic, antitussive, and antiasthmatic drugs are limited.

31.2 DRUG ANALYSIS

31.2.1 ACETHYLCYSTEINE

Chemical name: 2-acetamido-3-sulfanylpropanoic acid [5]. The substance is easily soluble in water and ethanol. It is a mucolytic medicine and breaks disulfide bonds in mucus, making it easier to cough up. The compound is metabolized to cystein, N,N-diacetylocystein, cystidine, and taurin [6].

Investigation of interactions between acethylcysteine and different pharmaceutical excipients, microcrystalline cellulose, sodium carboxymethylcellulose, silicon dioxide, PVP, corn starch, saccharose, lactose, and magnesium stearate was performed. Physical mixtures were formed. Different degradation products, whose structures were not elucidated, were determined using the TLC procedure [7].

Stationary phase: cellulose plate

Mobile phase: n-butanol/acetic acid/water (40:10:10, v/v/v)

Detection: in visible light after spraying with reagent ninhydrin in butan-1-ol

Acethylcysteine derivatized with the fluorobenzoxadiazole reagents SBD-F and ABD-F was analyzed by high-performance thin-layer chromatography. Effects of several types of luminescence enhancers were investigated. Acethylcysteine solution was prepared in the water environment, with sodium borate buffer and the addition of disodium EDTA [8].

Stationary phase: silica gel 60 plates without fluorescence indicator

Mobile phase: isopropyl ether/methanol/water/acetic acid (9:8:2:1, v/v/v/v)

Detection: UV lamp at 366 nm, quantitative analysis by means of fluorodensitometry

31.2.2 AMBROXOL

Chemical name: *trans*-4-(2-Amino-3,5-dibrombenzylamino)-cyclohexanol [5]. The substance in the form of hydrochloride is soluble in methanol and slightly soluble in water. It is an effective expectorant drug, secretolytic and secretomotoric agent, used in the treatment of respiratory diseases, acts like bromhexine, but is stronger. It is excreted in urine, mostly as unchanged compound, partly as dibromoanthranilic acid [6].

Identification and determination of ambroxol hydrochloride in bulk drug and pharmaceutical dosage form (tablet) was done. The elaborated procedure is simple, selective, and precise. The compact spot of ambroxol hydrochloride can be observed at R_F value of 0.53. The samples were applied on the plate in the form of 6 mm wide bands, accounting for 8 mm spaces. The development time was approximately 70 min. The linear range was within 100–1000 ng/spot. All solutions, for substance in bulk and in the preparation, were prepared in ethanol [9].

Stationary phase: HPTLC aluminum plates with silica gel 60 F_{254}

Mobile phase: methanol/triethylamine (4:6, v/v)

Detection: densitometric analysis at absorbance of 254 nm (Camag)

The next section is devoted to the characterization of bromhexine and ambroxol in equine urine. The urine samples were prepared by an enzyme hydrolysis (beta-glucuronidase) in saturated phosphate buffer. After several stages of the preparing procedure, finally the sample was resuspended in dichloromethane. The elucidated metabolites in horses were hydroxy-ambroxol and methylenedioxy-ambroxol [10].

Stationary phase: HPTLC plates

Mobile phase: chloroform/methanol/propionic acid (72:18:10)

Detection: visualized by modified Ehrlich's reagent or Dragendorff/sodium nitrate (III) overspray, scanner Camag

Ambroxol metabolism in humans and determination in clinical material were assessed. Some metabolites in human urine and plasma were elucidated after i.v. and p.o. Administration: 6,8-dibromo-3-(trans-4-hydroxycyclohexyl)-1,2,3,4-tetrahydro-quinazoline and 3,5-dibromo-anthranilic acid. The quantification of ambroxol was achieved by the radiochemical derivatization with ^{14}C-labeled formaldehyde. The substance was dissolved in water. The blood samples were treated with heparin. The blood plasma and the urine undergo a complicated procedure including diethyl ether extraction and β-glucuronidase addition [11].

Stationary phase: silica gel 60 F_{254}

Mobile phases: benzene/isopropanol/ammonia conc. (70:25:1, v/v/v), ethyl acetate/chloroform/formic acid (50:50:10, v/v/v)

Detection: after radiochemical derivatization radioactive zones on glass plate with Roentgen film

The thin-layer chromatographic method was used for the resolution of the ternary mixture containing salbutamol sulfate, ambroxol hydrochloride, and theophylline. The developed method is simple, rapid, accurate, precise, reliable, and economical. The obtained R_F coefficient was found to be 0.89. The range covered 5–35 μg/mL. The development time equaled 30 min. Methanol served as an analytical solvent [12].

Stationary phase: silica gel GF_{254} prepared by spreading slurry on plate and activation in oven at 110°C

Mobile phase: methanol/n-hexane (2.1:0.9)

Detection: after placing the plate in the iodine chamber, bands were scraped out and underwent absorbance measurement, for ambroxol hydrochloride at 244.2 nm

The simultaneous determination of levofloxacin hemihydrate and ambroxol hydrochloride was carried out. Both substances were analyzed in bulk and in tablets. The elaborated procedure has many advantages: simplicity, sensitivity, rapidity, and low cost. The R_F value was 0.7. The range was defined as 600–1000 ng/spot. The samples were spotted in the form of bands of width 6 mm. The length of chromatogram run was approximately 70 mm. The tablets were extracted using methanol [13].

Stationary phase: aluminum sheet of silica gel 60 F_{254}

Mobile phase: chloroform/methanol/toluene/ammonia (10:6:3:0.8, v/v/v/v)

Detection: quantification was carried out densitometrically at 245 nm (Camag)

Mohammad and Zawilla have accomplished simultaneous analysis of ambroxol hydrochloride and doxycycline hyclate in a binary mixture [14]. The linear range was 1–10 µg/spot, which is equivalent to 100–1000 µg/mL. The substances were extracted from a capsule content with methanol.

Stationary phase: silica gel

Mobile phase: ethyl acetate/ethanol/glacial acetic acid/water (9:4:0.5:1, v/v/v)

Detection: densitometry, scanned 254

Lakshmi and Niraimathi have determined cetirizine and ambroxol in tablets [15]. The presented procedure is simple, fast, and precise. The retardation factor value lay between 0.4 and 0.5.

Stationary phase: silica gel 60 F_{254}

Mobile phase: methanol/0.067 M potassium dihydrogen phosphate (35:65, v/v)

Detection: scanned at 231 nm

31.2.3 BROMHEXINE

Chemical name: 2,4-dibromo-6-{[cyclohexyl(methyl)amino]methyl}aniline [5]. The substance in the form of hydrochloride is hardly soluble both in water and organic solvents. The expectorant, mucolytic, increases the production of thin mucus and synthesis of surfactant. In the metabolic path, after *N*-demethylation and C-hydroxylation, an active metabolite, namely ambroxol, is formed [6].

Bromhexine is stable in a water solution even despite applying hydrolysis conditions. Very small amounts of three degradation products were isolated (TLC) and their structures were elucidated (MS, UV, IR). Different R_F values were received, depending on the mobile phase used [16].

Stationary phase: silica gel F_{254}

Mobile phases: *n*-hexane/ethyl acetate/methanol–methyl acetate (50:10:15:15), ethyl acetate/trichloromethane/formic acid (50:50:10), *n*-hexane/ethyl acetate/acetic acid (50:50:2.5), trichloromethane/*n*-hexane/methanol/acetic acid (40:40:10:10), benzene/methyl acetate/methanol/6 M ammonia (50:30:20:1.5), cyclohexane/hexane/methanol/methyl acetate/acetic acid (40:40:10:10:10), benzene/methyl acetate/methanol/acetic acid (50:20:30:10).

Detection: UV detection (254 nm) or in visible light after spraying with potassium tetraiodobismuth (III)

The species differences in metabolism and excretion of radiolabeled bromhexine in mice, rats, rabbits, dogs, and man were described. Considerable species differences were observed. The rabbit metabolic pattern is most similar to man and the rat's is the least similar. The analyses of ^{14}C-bromhexine in urine and faces were preformed. The authors reported 3,5-dibromoanthranilic acid as a metabolite. Bromhexine was dissolved in distilled water [17].

Stationary phase: silica gel plates F_{254} with x-ray film

Mobile phase: ethyl acetate/chloroform/formic acid (10:10:1, v/v/v)

Detection: in UV at 254 nm

The simultaneous determination of salbutamol sulfate and bromhexine hydrochloride in liquid and solid (tablet) pharmaceutical formulations was done. The procedure is simple, rapid, accurate, and precise. Bromhexine hydrochloride underwent the assay at a concentration of 1.98 mg/5 mL [18].

Stationary phase: silica gel plates F_{254}

Mobile phase: methanol/chloroform/triethylamine (5.5:4.5:0.05, v/v/v)

Detection: densitometrically at 276 nm

The HPTLC method of determination of amoxicillin trihydrate and bromhexine hydrochloride in oral capsule was developed and validated. Bromhexine spot could be observed at the R_F of approximately 0.74. The authors declare that the method is simple, accurate, precise, and rapid. The method's range was confirmed as 200–1000 ng/band. Both bromhexine and amoxicillin were dissolved in methanol. The analyzed pharmaceutical preparation was extracted with methanol as well [19].

Stationary phase: silica gel 60 F_{254}

Mobile phase: butyl acetate/glacial acetic acid/methanol/water (5:2.5:2.5:1, v/v/v/v)

Detection: densitometric detection 260 nm (Camag)

Sumarlik and Indrayanto [20] determined bromhexine hydrochloride in pharmaceuticals: syrup and tablets. The migration distance in all experiments was 8 cm, and the development time was about 90 min. The R_F of bromhexine hydrochloride equaled 0.63. The procedure seems to be simple, rapid, selective, precise, and accurate. The extraction of the analyte was carried out using a mixture of acetone and water (2:1).

Stationary phase: silica gel plates 60 F_{254}

Mobile phase: *n*-butanol/glacial acetic acid/water (26:7.5:7.5)

Detection: densitometric absorbance–reflectance detection 325 nm

The determination of ternary mixture containing salbutamol sulfate, bromhexine hydrochloride, and etofylline by spectrophotometry with prior TLC resolution procedure was described. The R_F value was about 0.91, and the development time 30 min. The range was determined as 4–40 µg/mL, and the stock methanol solution had a concentration of 1 mg/mL. According to the authors, the procedure is simple, rapid, accurate, and precise [21].

Stationary phase: silica gel GF_{254} prepared and activated before analysis

Mobile phase: methanol/*n*-hexane (2:1).

Detection: in UV (254 nm) and after reaction with iodine

As mentioned earlier, characterization of bromhexine and ambroxol in equine urine was done. The urine samples were prepared by an enzyme hydrolysis (beta-glucuronidase) in saturated phosphate buffer. After several stages of preparing the procedure, finally the sample was resuspended in dichloromethane. The development distance was 6 cm. Some metabolites were elucidated in horses: hydroxy-bromhexine, desbromo-bromhexine, deshexy-bromhexine, desmethyl-bromhexine, bromhexadienone, ambroxol, hydroxy-ambroxol, methylene-dioxy-ambroxol [10].

Stationary phase: HPTLC plates

Mobile phase: chloroform/methanol/propionic acid (72:18:10)

Detection: spots visualized by modified Ehrlich's reagent (0.1% *p*-dimethylaminocinnamaldehyde) or Dragendorff/sodium nitrate (III) overspray scanner Camag

31.2.4 CARBOCYSTEINE

Chemical name: (*R*)-2-amino-3-(carboxymethylsulfanyl)propanoic acid [5]. The substance is insoluble in water and ethanol but soluble in diluted acids and hydroxides. The mucolytic reduces the viscosity of sputum and is used in diseases of the respiratory tract [6].

Carbocysteine metabolism is described in at least three publications [22–24]. It is excreted by the urine untransferred and as *S*-oxide, *S*-methyl-L-cysteine, *S*-methyl-L-cysteine *S*-oxide and *S*-(carboxymethylthio)-L-cysteine. Some specific parameters include [22] silica gel 60 F_{254} TLC glass plates with chloroplatinate visualization and scanning densitometry at 550 nm (Desaga and Schimadzu densitometers).

31.2.5 CROMOGLICATE DISODIUM

Chemical name: 5,5′-(2-hydroxypropane-1,3-diyl)*bis*(oxy)*bis*(4-oxo-4*H*-chromene-2-carboxylic acid) disodium salt [5]. The substance is soluble in water and practically insoluble in organic solvents. Cromoglicate is an antiasthmatic drug, which inhibits release of inflammation mediators and acts symptomatically [6].

Disodium cromoglycate was analyzed in plasma, urine, and mouth of humans after inhalation of medicine, and in urine after oral administration. The human volunteers had inhaled the substance from capsules inserted in a spinhaler. The molecule is fluorescent in aqueous solutions, with a fluorescent maximum at 450 nm. In order to quantify disodium cromoglycate in urine, the compound was heated in alkaline solution in order to obtain *bis-o*-hydroxyacetophenone, which forms a complex with diazotized *p*-nitroaniline, convenient to analyze at 490 nm [25].

Stationary phase: silica gel plate

Mobile phase: propanol/ammonia/water (100:10:20, v/v/v)

Detection: fluorescence under UV lamp (366 nm)

Disodium cromoglycate was determined in ophthalmic solutions, gels, and capsules by TLC-densitometry. The method is rapid, simple, free from interferences by adjuvants and gives precise and accurate results. The standard solution and the sample solutions were prepared by dissolving in water to obtain the concentration of 1 mg/mL [26].

Stationary phase: silica gel GF_{254}

Mobile phase: methanol/water/ethyl acetate (15:45:40, v/v/v)

Detection: densitometrically, scan at 254 nm

The aim of another work was to determine radiolabeled (^{3}H) disodium cromoglycate to assess distribution and metabolism of cromoglycate in rats. The compound was administered by two routes: p.o. and i.v. Determinations of the agent were done in urine, plasma, feces, bile, lung, liver, brain, kidney, and spleen. Preparation of the samples before analysis depended on type of material under investigation. The aqueous solution of disodium cromoglycate was analyzed. TLC was used not only to determine the compound but also to purify the radiolabeled drug [27].

Stationary phase: silica gel plates

Mobile phases: butanol/acetic acid/water (12:5:3, v/v/v), isopropanol/ammonia/water (20:1:2, v/v/v), chloroform/acetic acid/methanol (9:9:2, v/v/v).

Detection: chromatograms were scanned for radioactivity and fluorescence under UV light

31.2.6 EPRAZINONE

Chemical name: 3-[4-(2-ethoxy-2-phenyl-ethyl)piperazin-1-yl]-2-methyl-1-phenyl-propan-1-one [5]. The substance as hydrochloride is sparingly soluble in water and slightly soluble or insoluble in organic reagents. Eprazinone executes the central antitussive effect, and it is non-narcotic amd mucolytic [6].

Metabolism of eprazinone in humans after oral application was described. In the separation of active compound from its major metabolite (2,2′-phenyletoxy-*N*-ethylpiperazine), a hydrolysis product wasobtained. The total amount of degradants in urine was five. Two routes of the degradation has been identified: hydrolysis and cleavage of α-binding to the nitrogen atom with subsequent dealkylation. The urine was acidified with HCl and extracted with chloroform, then it underwent hydrolysis. After preparation of the methanolic solution, it was used in chromatography [28].

Stationary phase: Merck Plates HF$_{254}$

Mobile phases: methanol/ammonia (100:1), cyclohexane/diethylamine (9:1), chloroform/acetone (85:15), benzene/polyvinylalcohol/diethylamine (7:2:1), benzene/ethanol/ammonia (80:20:1)

Detection: UV and Vis after spraying with Dragendorff reagent. Moreover eprazinone and its major metabolite were sprayed with bromocresol green and potassium iodide

31.2.7 ERDOSTEINE

Chemical name: 2-[(2-oxothiolan-3-yl)carbamoylmethylsulfanyl]acetic acid [5]. The substance is slightly soluble in methanol, ethanol, acetone, and water. Erdosteine is a mucolytic thiol derivative used in the treatment of bronchitis [6].

Analysis of erdosteine was carried out in pharmaceutical formulation (capsule). No interferences with excipients were found. The method could effectively separate the drug from its degradation products, but the structure of the degradation products was not established. The band of erdosteine in later mentioned conditions can be observed at the R$_F$ value equal to 0.45. Amounts of 30–1000 ng of erdosteine was applied in the form of methanolic solution to the plates. The bands were 6 mm wide and 6 mm apart. The development distance was 9 cm, and the development time approximately half an hour. Development was done in a twin-trough glass chamber that was previously well saturated [29].

Stationary phase: silica gel 60 F$_{254}$ prewashed with methanol and activated at 60°C for 5 min

Mobile phase: toluene/methanol/acetone/ammonia (3.5:3.5:2.5:0.05, v/v/v/v)

Detection: densitometric detection at wavelength 254 nm (deuterium lamp)

31.2.8 FOMINOBEN

Chemical name: *N*-[3-chloro-2-[[methyl-(2-morpholin-4-yl-2-oxoethyl)amino]methyl]phenyl] benzamide [5]. The substance is soluble in acid, slightly soluble in ethanol, slightly soluble in chloroform, and insoluble in water. Fominoben is a non-narcotic, centrally acting cough suppressant [6].

The metabolism of ^{14}C-fominoben hydrochloride was investigated in human urine and plasma. Many different metabolism products were elucidated. The biotransformation can be characterized by four main pathways: cleavage reactions, hydroxylations, cyclizations, and conjugations. The urine was desiccated, dissolved in methanol, and then acetate buffer and β-glucuronidase/arylsulfatase were added. The methanol extracts underwent chromatographic analysis [30].

Stationary phase: silica gel 60 F_{254}

Mobile phases: ethyl acetate/methyl acetate/ethanol/water (50:20:20:10), benzene/ethyl acetate/methanol/conc. ammonia (50:35:15:5), benzene/ethyl acetate/methanol/conc. ammonia (70:15:15:1), methyl acetate/methanol/water/conc. ammonia (65:20:10:5), *t*-butanol/ethanol/water/conc. ammonia (20:65:10:5), *n*-butanol/water/acetic acid (90:50:10), chloroform/ethanol/acetic acid (80:20:5), benzene/*n*-propanol/conc. ammonia (80:20:1), tetrahydrofuran/methanol/water/conc. ammonia (70:20:5:5)

Detection: fluorescence detector, color diazo reaction according to Bratton–Marshall, autoradiography

Excretion and metabolism of ^{14}C-fominoben was examined in urine and feces of four species: human, rabbit, dog, and rat. The percent of radiolabeled atoms was determined. The urine was extracted in alkaline environment by using acetate ester or ether and afterwards underwent enzymatic hydrolysis using β-glucuronidase/arylsulfatase. The feces were extracted with methanol with ammonia addition. None of all species excreted fominoben unchanged. The first step of metabolism seems to be the detachment of benzoic acid [31].

Stationary phase: silica gel

Mobile phases: *n*-butanol/water/acetic acid (90:50:10), chloroform/ethanol/acetic acid (80:20:5), chloroform/ethanol/acetic acid (80:15:5)

Detection: autoradiography, spraying with *N*-[naphthyl(1)]-ethylenediamine (1% in ethanol) and Vis detection

Study showed that noleptan is almost completely metabolized in human organism. The research was done on ^{14}C-fominoben. Approximately 10 metabolites were separated chromatographically. The principal product is formed by detachment of benzoic acid. The blood was alkaline with NaOH, reacted with hydrogen peroxide and "Digestin (Merck)." The urine underwent reaction with "Diotol" [32].

Stationary phase: silica gel 60 F_{254}

Mobile phases: *n*-butanol/water/acetic acid (90:50:10), chloroform/ethanol/acetic acid (80:20:5), ethyl acetate/methyl acetate/ethanol/water (50:20:20:10), *n*-butanol/water/acetic acid (80:20:20), benzene/ethyl acetate/methanol/conc. ammonia (50:35:15:5)

Detection: fluoresce measurement at 254 nm, color reaction according to Bratton and Marshall and autoradiography

31.2.9 GUAIFENESIN

Chemical name: (*RS*)-3-(2-methoxyphenoxy)propane-1,2-diol [5]. The substance is hardly soluble in water; it could be dissolved in ethanol. Guaifenesin is a synthetic expectorant. It acts as secretolytic agent by irritation of intestinal mucous membrane, which causes impulsive secretion of mucus by bronchi. Guaifenesin underwent biotransformation to β-(2-methoxyphenoxy)-lactic acid [6].

Identification of guaifenesin in OTC drug for sleeping combined with determination by HPLC is described. Ten centimeter plates were used. The R_F value equaled 0.12 [33].

Stationary phase: silica gel 60 F_{254}

Mobile phase: chloroform/acetone (8:2)

Detection: in Vis after spraying with formaldehyde in conc. sulfuric acid (violet spot). This color reaction segregates guaifenesin from hypnotics that do not react. Guaifenesin does not give positive reaction with mercury (I) nitrate or chloro-*o*-toluidine in contrast to hypnotics

Guaifenesin can be identified and determined beside methocarbamol, acetylosalicylic acid, and adjuvant substances from tablet. Guaifenesin is an impurity of methocarbamol and occurs in its preparations. In developed conditions, the R_F equaled 0.46. The 3% methanolic solution of the preparation examined and the 0.06% methanolic solution of the standard substance were applied on the chromatographic plate. Spots received in chromatographic analysis can be revealed after reaction with concentrated sulfuric acid and 40% formaldehyde. Guaifenesin identity was moreover confirmed by the UV and IR spectrums [34,35].

Stationary phase: silica gel 60

Mobile phase: ethyl acetate/methanol/25% ammonia (17:2:1)

Detection: densitometry, 273 nm

Densitometric determination of propyphenazone, paracetamol, guaiacol glyceryl ether, coffeine, and acethylsalicylic acid in marketed analgesic–antipyretic preparations was presented by Tománková and Vašatová [36]. The reference standard was dissolved, and the preparations were extracted with methanol.

Stationary phase: silica gel 60 F_{254}

Mobile phase: ethyl acetate/methanol/acetic acid (8:1:1), ethyl acetate/methanol/acetic acid (7:2:1), ethyl acetate/benzene/methanol/acetic acid (5:4:1:0.1)

Detection: scanned at 276 nm (D2 lamp, scanner Camag)

Simple determination of guaiacol glyceryl ether in fluid and solid form was elaborated. The distance of development was 12 cm. The R_F equaled 0.35. Chloroform served as an extractant [37].

Stationary phase: silica gel G

Mobile phase: ethyl acetate/benzene/acetic acid (74.5:25.0:0.5)

Detection: spectrophotometric detection after spraying with iodine

31.2.10 MONTELUKAST

Chemical name: (*S,E*)-2-(1-((1-(3-(2-(7-chloroquinolin-2-yl)vinyl)phenyl)-3-(2-(2-hydroxypropan-2-yl)phenyl)propylthio)methyl)cyclopropyl)acetic acid [5]. The sodium salt of substance is freely soluble in ethanol, methanol, and water. Montelukast is an antiasthmatic drug, leukotriene CysLT$_1$ receptor antagonist, reduces bronchospasms and inflammation [6].

Determination of montelukast sodium was performed in combined tablet also containing levocetirizine dihydrochloride. The simple, rapid, precise HPTLC method and first-derivative spectrophotometry were utilized for simultaneous determination in tablets. The linearity was obeyed in the range of 400–1200 ng/spot [38].

Stationary phase: silica gel 60 F$_{254}$

Mobile phase: ethyl acetate/methanol/triethylamine (5:5:0.02, v/v/v)

Detection: detection was carried out at 240 nm

The HPTLC determination of montelukast sodium in bulk drug and in pharmaceutical preparation in tablet form was described by Sane et al. [39]. Plates were developed to a distance of 90 mm. The linearity working range was 0.8–10.0 μg. The tablet after powdering was extracted using methanol. This solvent was also used to dissolve bulk substance.

Stationary phase: silica gel 60 F$_{254}$ washed in methanol and dried in an oven at 105°C for 2 h

Mobile phase: toluene/ethyl acetate/glacial acetic acid (6.0:3.4:0.1, v/v/v)

Detection: densitometry 344 nm (Camag scanner)

31.2.11 OXELADIN

Chemical structure: 2-(2-diethylaminoethoxy)ethyl 2-ethyl-2-phenyl-butanoate [5]. The substance is practically insoluble in water; soluble in ethanol, acetone, toluene, and ether. In pharmacopoeial the form of hydrogencitrate is freely soluble in water and slightly to very slightly in organic solvents. The compound exerts antitussive effect, has no narcotic properties, is much weaker than narcotic drugs, indicated in all types of cough, and acts as secretolytic [6].

Oxeladin citrate was determined spectrophotometrically in complex preparations, and TLC served as separation technique. Division of many mixtures was performed including the presence of chloramphenicol, caffeine, etofylline, guaifenesin, paracetamol, disodium dibunate, phenazon, proxyphylline, and theophylline. Water extraction was applied [40].

Stationary phase: silica gel 60 F_{254}, activation with methanol

Mobile phase: methanol/conc. ammonia (100:1.5), chloroform/ethanol (9:1)

Detection: UV light 254 nm

REFERENCES

1. Sherma, J., Modern thin-layer chromatography, *J. AOAC Int.*, 91, 1142, 2008.
2. Sherma, J., Review of HPTLC in drug analysis: 1996–2009, *J. AOAC Int.*, 93, 754, 2010.
3. Kalász, H. and Báthori, M., Pharmaceutical applications of TLC, *LCGC Europe*, May 1, 2001.
4. Ferenczi-Fodor, K., Végh, Z., and Renger, B., Thin-layer chromatography in testing the purity of pharmaceuticals, *Trends Anal. Chem.*, 25, 778, 2006.
5. http://pubchem.ncbi.nlm.nih.gov
6. Zejc, A. and Gorczyca, M. (red.), *Chemistry of Drugs* (in Polish), PZWL, Warszawa, Poland, 2002.
7. Kerč, J. et al., Compatibility study between acetylcysteine and some commonly used tablet excipients, *J. Pharm. Pharmacol.*, 44, 515, 1992.
8. Lin Ling, B. et al., Use of enhancers in the HPTLC fluorescence analysis of thiols, *J. Pharm. Biomed. Anal.*, 7, 1671, 1989.
9. Jain, P.S., Stability-indicating HPTLC determination of ambroxol hydrochloride in bulk drug and pharmaceutical dosage form, *J. Chromatogr. Sci.*, 48, 45, 2010.
10. Uboh, C.E. et al., Characterization of bromhexine and ambroxol in equine urine: Effect of furosemide on identification and confirmation, *J. Pharm. Biomed. Anal.*, 9, 33, 1991.
11. Jauch, R. et al., Ambroxol, Untersuchungen zum Stoffwechsel beim Menschen und zum quantitativen Nachweis in biologischen Proben, *Arzneim.-Forsch./Drug Res.*, 28(I/5a), 904, 1978.
12. Dave Hiral, N., Mashru Rajeshree, C., and Patel Alpesh, K., Thin layer chromatographic method for the determination of ternary mixture containing salbutamol sulphate, ambroxol hydrochloride and theophylline, *Int. J. Pharm. Sci.*, 2, 390, 2010.
13. Agrawal, O.D., Shirkhedkar, A.A., and Surana, S.J., Simultaneous determination of levofloxacin hemihydrate and ambroxol hydrochloride in tablets by thin-layer chromatography combined with densitometry, *J. Anal. Chem.*, 65, 418, 2010.
14. Mohammad, M.A-A. and Zawilla, N.H., Thin-layer and column-chromatographic methods for simultaneous analysis of ambroxol hydrochloride and doxycycline hyclate in binary mixture, *J. Planar Chromatogr.*, 22, 201, 2009.
15. Lakshmi, K. and Niraimathi, V., High performance thin layer chromatographic determination of cetirizine and ambroxol in tablets, *Indian Drugs*, 40, 227, 2003.
16. Göber, B., Lisowski, H., and Franke, P., Zur Stabilität von Bromhexin und zur Struktur seiner Abbauprodukte, *Pharmazie*, 43, 23, 1988.

17. Kopitar, Z. et al., Species differences in metabolism and excretion of bromhexine in mice, rats, rabbits, dogs and man, *Eur. J. Pharmacol.*, 21, 6, 1973.
18. Argekar, A.P. and Powar, S.G., Simultaneous determination of salbutamol sulfate and bromhexine hydrochloride in formulations by quantitative thin-layer chromatography, *J. Planar Chromatogr.*, 11, 254, 1998.
19. Dhoka, M.V., Gawande, V.T., and Joshi, P.P., HPTLC determination of amoxicillin trihydrate and bromhexine hydrochloride in oral solid dosage forms, *J. Pharm. Sci. Res.*, 2, 477, 2010.
20. Sumarlik, E. and Indrayanto, G., TLC densitometric determination of bromhexine hydrochloride in pharmaceuticals and its validation, *J. Liq. Chromatogr. Related Technol.*, 27, 2047, 2004.
21. Dave, H.N., Mashru, R.C., and Patel, A.K., Thin layer chromatography method for the determination of ternary mixture containing salbutamol sulphate, bromhexine hydrochloride and etofylline, *J. Pharm. Sci. Res.*, 2, 143, 2010.
22. Steventon, G.B., A methodological and metabolite identification study of the metabolism of *S*-carboxymethyl-L-cysteine in man, *Chromatographia*, 48, 561, 1998.
23. Gregory, W.L. et al., Re-evaluation of the metabolism of carbocisteine in a British white population, *Pharmacogenetics*, 3, 270, 1993.
24. Steventon, G.B., Diurnal variation in the metabolism of *S*-carboxymethyl-L-cysteine in humans, *Drug Metab. Dispos.*, 27, 1092, 1999.
25. Moss, G.F. et al., Plasma levels and urinary excretion of disodium cromoglycate after inhalation by human volunteers, *Toxicol. Appl. Pharmacol.*, 20, 147, 1971.
26. Kocić-Pešić, V. et al., Determination of sodium cromoglycate in pharmaceutical dosage forms using TLC-densitometry, *Farmaco*, 47, 1563, 1992.
27. Moss, G.F. et al., Distribution and metabolism of disodium cromoglycate in rats, *Toxicol. Appl. Pharmacol.*, 17, 691, 1970.
28. Toffel-Nadolny, von P. and Gielsdorf, W., Zum Metabolismus von Eprazinon, *Arzneim.-Forsch./Drug Res.*, 31, 719, 1981.
29. Mhaske, D.V. and Dhaneshwar, S.R., High-performance thin-layer chromatographic method for determination of erdosteine in pharmaceutical dosage forms, *Acta Chromatogr.*, 19, 170, 2007.
30. Zimmer, A., Krüger, G., and Prox, A., Untersuchungen zum Metabolismus von Fominoben-HCl beim Menschen. *Arzneim.-Forsch./Drug Res.*, 28, 688, 1978.
31. Zimmer, A., Ausscheidung und Metabolitenmuster von Fominoben in Urin und Faeces von Mensch, Kaninchen, Hund und Ratte, *Arzneim.-Forsch./Drug Res.*, 23, 317, 1973.
32. Zimmer, A., Pharmakokinetik und Metabolismus von Fominoben (Noleptan®) beim Menschen, *Arzneim.-Forsch./Drug Res.*, 23, 1798, 1973.
33. Maier, R.D., Zum Nachweis von Guaiphenesin, einem Inhaltsstoff einiger Rezepfreier Schlafmittel, *Arch. Toxicol.*, 45, 123, 1980.
34. Krzek, J., Densitometric determination of impurities in drugs. Part III. Determination of guaifenesin in methocarbamol preparations, *Acta Pol. Pharm.-Drug Res.*, 55, 99, 1998.
35. Krzek, J. and Starek, M., The effect of acetylsalicylic acid on the stability of methocarbamol in complex drugs, *Acta Pol. Pharm.-Drug Res.*, 56, 369, 1999.
36. Tománková, H. and Vašatová, M., Densitometrische Bestimmung von Propyphenazon, Paracetamol, Guajacolglycerinether, Coffein und Acethylsalicylsäure in analgetisch-antipyretischen Präparaten auf Dünnschichtchromatogrammen, *Pharmazie*, 44, 197, 1989.
37. Karting, Th. and Still, F., Eine einfache Bestimmung von Guajakolglycerinäther in flüssigen und fasten Arzneiformen, *Sci. Pharm.*, 43, 6, 1975.
38. Rote, A.R. and Niphade, V.S., Determination of montelukast sodium and levocetirizine dihydrochloride in combined tablet dosage form by HPTLC and first-derivative spectrophotometry, *J. Liq. Chromatogr. Related Technol.*, 34, 155, 2011.
39. Sane, R.T. et al., HPTLC determination of montelukast sodium in bulk drug and in pharmaceutical preparations, *J. Planar Chromatogr.*, 17, 75, 2004.
40. Knopp, C. and Korsatko, W., Bestimmung von Oxeladincitrat In Kombinationspräparaten, *Pharmazie*, 37, 76, 1982.

32 TLC of Neuromuscular Blockers

Dušanka M. Milojković-Opsenica and Maja M. Natić

CONTENTS

32.1 INTRODUCTION

Skeletal muscle relaxants are a heterogeneous group of drugs, which affect skeletal muscle function. Skeletal muscle relaxants can be classified into two different therapeutic groups: (1) neuromuscular blockers, peripheral-acting drugs and (2) spasmolytics, central-acting drugs.

Neuromuscular blocking drugs are the peripheral type and interfere with transmission at the neuromuscular end plate and have no central nervous system (CNS) activity. These compounds are being used primarily for muscle relaxation upon endotracheal intubation and/or general anesthesia for surgical operation.

Spasmolytics are central-acting muscle relaxants and are used for treatments of painful muscle contracture, caused by locomotorial disorders, and for relaxation of muscle stiffness caused by psychotic tension or by neurosis.

32.2 NEUROMUSCULAR BLOCKERS

Nicotinic antagonists are chemical compounds that bind to cholinergic nicotinic receptors but have no efficacy. All therapeutically useful nicotinic antagonists are competitive antagonists; in other words, the effects are reversible with acetylcholine. There are two subclasses of nicotinic antagonists—skeletal neuromuscular blocking agents and ganglionic blocking agents—classified according to the two populations of nicotinic receptors. This section emphasizes nicotinic antagonists used clinically as neuromuscular blocking agents.

A *bis*-quaternary ammonium compound having two quaternary ammonium salts separated by 10–12 carbon atoms (similar to the distance between the nitrogen atoms in tubocurarine, the first known neuromuscular blocking drug) is a requirement for neuromuscular blocking activity. The rationale for this structural requirement was that nicotinic receptors possessed two anionic-binding sites, both of which had to be occupied for a neuromuscular blocking effect. It is important to observe that the current transmembrane model for the nicotinic receptor protein has two anionic sites in the extracellular domain (Craig and Stitzel 1997, Lemke and Williams 2008).

Some of the new *bis*-quaternary ammonium agents produce depolarization of the postjunctional membrane at the neuromuscular junction before causing blockade; other compounds, such as tubocurarine, do not produce this depolarization. Thus, the structural features of the remainder of the molecule determined whether the nicotinic antagonist is a depolarizing or a nondepolarizing neuromuscular blocker (Flood 2005, Toner and Flynn 1994). Table 32.1 summarizes neuromuscular blocking agents according to the polarization effect and a class of chemical compounds to which they belong. The structures of processed agents are presented in Figure 32.1.

Quaternary nitrogen muscle relaxants are difficult to extract and to analyze by conventional analytical methods because of the lack of chromophore, insufficient stability, and also because of the presence of a permanent positive charge. A few procedures were reported using high-performance liquid chromatography (HPLC) with ultraviolet (UV) (García et al. 2008, Zecevic et al. 2002), fluorescence (Bederbck et al. 1996), coulometric electrode array detection (Błazewicz et al. 2008), or electrospray ionization mass spectrometry (Ariffin et al. 2006, Cirimele et al. 2003), and capillary electrophoresis (Wedig 2002).

With its advantages of simplicity, economy, easy operation, and the need for only small amounts of solvent, thin-layer chromatography (TLC) is used widely in various fields to separate or purify mixtures of chemical compounds (Kaale 2011). TLC is not used too extensively in the analysis of neuromuscular agents, which is understandable considering the previously mentioned facts, that is, the chemical characteristics of *bis*-quaternary ammonium compounds. Investigated literature, which become from all relevant biomedical databases, covered TLC methods developed in the past. Most of the published papers date from the 1980s and 1990s, and only a few papers dealing with hydrophobicity of neuromuscular agents are from the first decade of the twenty-first century. Chromatographic parameters, which comprise preparation of standard solution, selected stationary and mobile phase, development mode, detection, retention parameter, and appropriate reference, are presented in Table 32.2.

A series of papers dealing with ion-pair adsorption TLC of quaternary ammonium compounds have been published by Giebelmann during several years. The basic concept of the separation system was ion-pair adsorption chromatography of quaternary amines on a silica layer in which the basic drug cations migrated as uncharged ion pairs using inorganic anions as ion-pair formers. Suxamethonium, succinic acid monocholine ester, and choline were separated using ion-pair adsorption TLC on silica gel with acetone–NaCl–HCl (1:0.5:0.5, v/v) as the mobile phase (Giebelmann 1981a). The separation was based on the chain length and charge number of the quaternary ammonium ions. It was concluded that the higher retention of the aprotic ammonium ions in TLC indicate more hydrophilic mobile phase and the higher charge number.

Later, the same author evaluated the retention equation for silica gel (Giebelmann 1981b). Namely, R_F values of the quaternary ammonium ions (during ion-pair adsorption TLC) reached saturation values with increasing concentration of the counterions (halogens). The retention (R_M) of the quaternary ammonium ions on silica gel was described as a function of the solubility parameters of the organic solvent (δ_i), the charge number of the measuring ions (nQ), the counterion concentration (CA), and the counterion radius by the following equation: $R_M = K\delta \cdot \delta_i + nQ \cdot (\log CA^{-1} + Kr \cdot rA^{-1}) + KR$, where rA = counterion radius and KR = rest retention. The residual retentions of some quaternary ammonium ions were described. Few years after, Giebelmann pronounced simple approximation relation for R_F values (Giebelmann 1984). The R_F values of quaternary ammonium ions in ion-pairing chromatography on silica gel with acetone—counterion solutions ($\leq$0.5M Cl$^-$, Br$^-$, ClO$_4^-$, or NO$_3^-$)—were

TABLE 32.1

Neuromuscular Blocking Agents

Polarization Effect	Chemical Class	Generic Name	Chemical Name
Specific depolarizing agents	Choline-based blocking agents	Succinylcholine (SUC) (suxamethonium)	2,2′-[(1,4-dioxobutane-1,4-diyl)bis(oxy)]bis(N,N,Ntrimethylethanaminium)
Specific nondepolarizing agents	Steroid-based blocking agents	Pancuronium (PAN)	[[2S,3S,5S,8R,9S,10S,13S,14S,16S,17R)-17-acetyloxy-10,13-dimethyl-2,16-bis(1-methyl-3,4,5,6-tetrahydro-2H-pyridin-1-yl)-2,3,4,5,6,7,8,9,11,12,14,15,16,17-tetradecahydro-1H-cyclopenta[a]phenanthren-3-yl]
		Vecuronium (VEC)	[[2S,3S,5S,8R,9S,10S,13S,14S,16S,17S)-17-acetyloxy-10,13-dimethyl-16-(1-methyl-3,4,5,6-tetrahydro-2H-pyridin-1-yl)-2-(1-piperidyl)-2,3,4,5,6,7,8,9,11,12,14,15,16,17-tetradecahydro-1H-cyclopenta[a]phenanthren-3-yl]
		Rocuronium (ROC)	1-((2S,3S,5S,8R,9S,10S,13S,14S,16S,17R)-17-acetoxy-3-hydroxy-10,13-dimethyl-2-morpholinohexadecahydro-1H-cyclopenta[a]phenanthren-16-yl)-1-allylpyrrolidinium
		Pipecuronium (PIP)	(2β,3α,5α,16β,17β)-3,17-bis(acetyloxy)-2,16-bis(4,4-dimethylpiperazin-4-ium-1-yl)androstane
	Tetrahydroisoquinoline-based blocking agents	Atracurium (ATR)	2,2′-{1,5-Pentanediylbis[oxy(3-oxo-3,1-propanediyl)]}bis[1-(3,4-dimethoxybenzyl)-6,7-dimethoxy-2-methyl-1,2,3,4-tetrahydroisoquinolinium]dibenzenesulphonate
		Mivacurium (MIV)	(1R,1′R)-2,2′-[[(4E)-1,8-dioxooct-4-ene-1,8-diyl]bis(oxypropane-3,1-diyl)]bis[6,7-dimethoxy-2-methyl-1-(3,4,5-trimethoxybenzyl)-1,2,3,4-tetrahydroisoquinolinium]
		Doxacurium (DOX)	bis[3-[6,7,8-trimethoxy-2-methyl-1-[(3,4,5-trimethoxyphenyl)methyl]-3,4-dihydro-1H-isoquinolin-2-yl] propyl] butanedioate
	Alkaloid-based blocking agents	d-Tubocurarine (TUB)	6,6′-dimethoxy-2,2,2′,2′-tetramethyltubocuraran-2,2′-diium-7′,12′-diol
		Metocurine (MET) (dimethyltubocurarinium)	(1S,16R)-9,10,21,25-tetramethoxy-15,15,30,30-tetramethyl-7,23-dioxa-15,30-diazaheptacyclo[22.6.2.23,6.18,12.118,22.027,31.016,34]hexatriaconta-3,5,8(34),9,11,18(33),19,21,24,26,31,35-dodecaene-15,30-diium
		Alcuronium (ALC)	4,4′-Didemethyl-4,4′-di-propenyltoxiferin

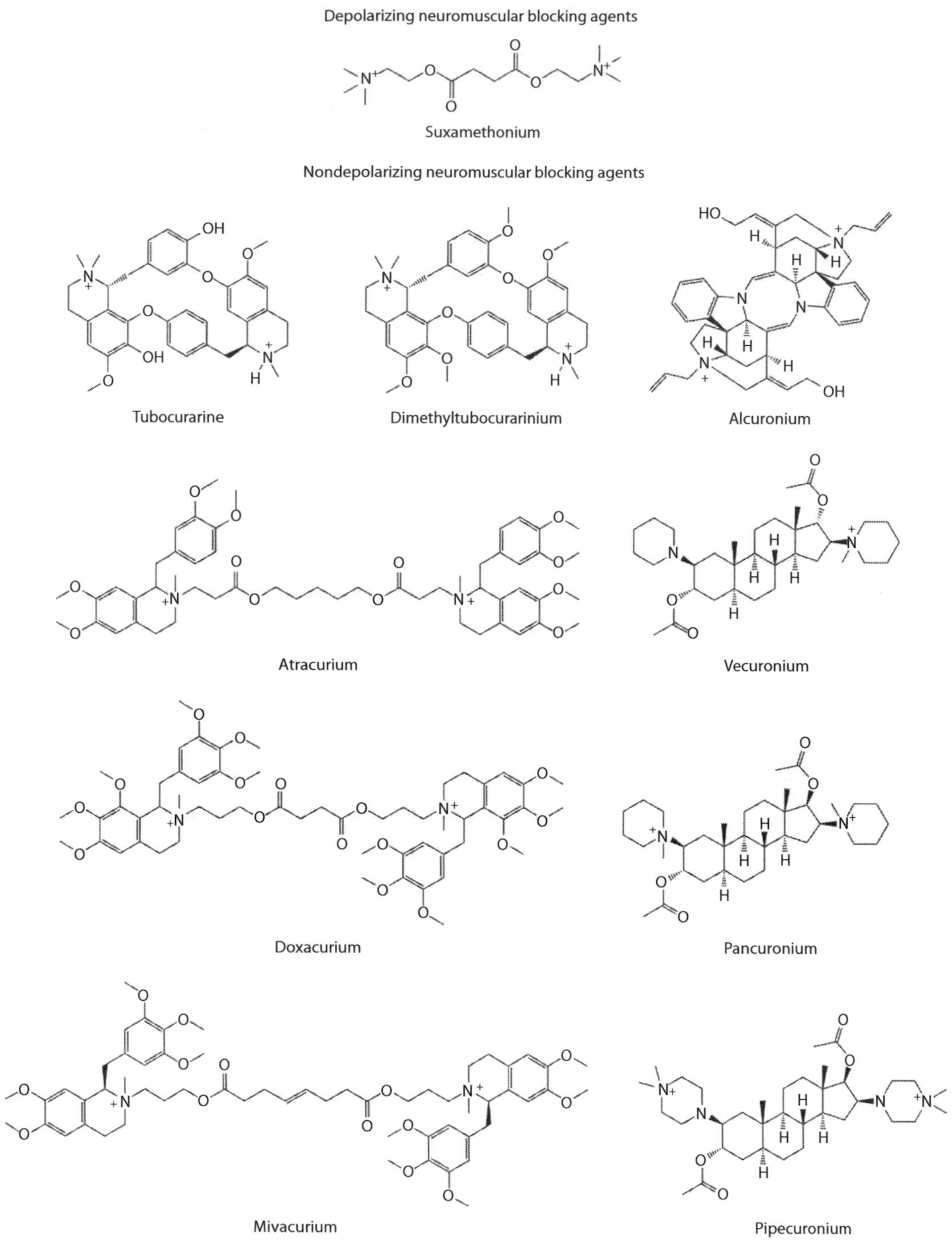

Depolarizing neuromuscular blocking agents

Suxamethonium

Nondepolarizing neuromuscular blocking agents

Tubocurarine

Dimethyltubocurarinium

Alcuronium

Atracurium

Vecuronium

Doxacurium

Pancuronium

Mivacurium

Pipecuronium

FIGURE 32.1 Structures of depolarizing and nondepolarizing neuromuscular blockers.

approximately proportional to the square root of the concentration of the counterions. For *bis* quaternary ammonium ions, the R_F values increased essentially linearly with counterion concentration.
The resulting regression line increased with counterion radius. Finally, the same author published a
three-parameter-system with silica gel (Giebelmann 1986). Ion-pair TLC of quaternary ammonium
compounds on silica gel with monovalent inorganic anions like Cl^-, Br^-, NO_3^-, I^-, and ClO_4^- in the

TABLE 32.2

TLC Conditions Used for Determination of Peripheral-Acting Muscle Relaxants

Drug	Standard Solution	Stationary Phase; Pretreatment	Mobile-Phase Composition	Detection	Retention Factor	Reference
TUB	Solution of tubocurarine chloride in methanol	Silica gel	Chloroform–methanol–12.5% trichloroacetic acid (1:1:1, v/v)	0.1M potassium ferricyanide–0.1M ferric chloride (1:1, v/v); Dragendorff's reagent		Clarke and Raja (1982)
	Solution of tubocurarine chloride in methanol	Silica gel	S_1—0.1 M Hydrochloric acid–acetonitrile (1:1, v/v) S_2—Methanol–tetrahydrofuran–5% formic acid (7:7:6, v/v) S_3—Methanol–chloroform–acetic acid (5:4:1, v/v)	Dragendorff's reagent; iodoplatinate reagent	S_1 $R_F = 0.59$ S_2 $R_F = 0.52$ S_3 $R_F = 0.20$	Nishikawa and Tsuchihashi (2005)
VEC	Aqueous solution (approx. 2 μg/μL) of the vecuronium bromide	Cellulose Alumina	Aqueous solutions of ammonium sulfate at concentrations ranging from 0.5 M to 4.0 M	Iodine vapor	$R_M^0 = -1.56$ $C_0 = -1.74$	Aleksić et al. (2003)
	Solution of vecuronium bromide in methanol	Silica gel	S_1—0.1 M Hydrochloric acid–acetonitrile (1:1, v/v) S_2—Methanol–tetrahydrofuran–5% formic acid (7:7:6, v/v) S_3—Methanol–chloroform–acetic acid (5:4:1, v/v)	Dragendorff's reagent; Iodoplatinate reagent	S_1 $R_F = 0.51$ S_2 $R_F = 0.47$ S_3 $R_F = 0.27$	Nishikawa and Tsuchihashi (2005)
	Aqueous solution of the vecuronium bromide	Silica gel	S_1—0%–100% (v/v) water in ACN, in steps of 10% (v/v) S_2—0%–100% (v/v) water in methanol, in steps of 10% (v/v) S_3—0%–100% (v/v) water in dioxane, in steps of 10% (v/v) S_4—0–100% (v/v) water in THF, in steps of 10% (v/v)	Iodine vapor	—	Radoičić et al. (2009b)
		Aluminum oxide, neutral	0–100% (v/v) water in ACN, in steps of 10% (v/v)			

(continued)

TABLE 32.2 (continued)
TLC Conditions Used for Determination of Peripheral-Acting Muscle Relaxants

Drug	Standard Solution	Stationary Phase; Pretreatment	Mobile-Phase Composition	Detection	Retention Factor	Reference
PAN	—	Silica gel	S_1—Buthanol–pyridine–acetic acid–20% ammonium chloride (60:40:12:48, v/v) S_2—Ammonium acetate–methanol–25% ammonium (4:100:3, v/v)	Iodine vapor	—	Klys et al. (2000)
	—	Pre-coated plates (silica gel); Activation at 110°C, 1 h	n-Butanol–pyridine–acetic acid–20% aqueous solution of ammonium chloride (60:40:12:48, v/v)	Iodine vapor	Pancuronium bromide $R_F = 0.27$ 17β-monoacetate $R_F = 0.31$ 3α-monoacetate $R_F = 0.35$ 3,17-dihydroxy-derivative $R_F = 0.41$	Kinget and Michoel (1976)
	Aqueous solution (approx. 2 μg/μL) of the pancuronium bromide	Cellulose Alumina	Aqueous solutions of ammonium sulfate at concentrations ranging from 0.5 M to 4.0 M	Iodine vapor	$R_M^0 = -1.47$ $C_0 = -2.03$	Aleksić et al. (2003)
	Aqueous solution of the pancuronium bromide	Silica gel	S_1—0%–100% (v/v) water in ACN, in steps of 10% (v/v) S_2—0%–100% (v/v) water in methanol, in steps of 10% (v/v) S_3—0%–100% (v/v) water in dioxane, in steps of 10% (v/v) S_4—0%–100% (v/v) water in THF, in steps of 10% (v/v)	Iodine vapor	—	Radoičić et al. (2009a,b)
		Aluminum oxide, neutral	0%–100% (v/v) water in ACN, in steps of 10% (v/v)			

	Sample	Stationary phase	Mobile phase	Detection	R_F / R_M values	Reference
	Solution of pancuronium bromide in methanol	Silica gel	S_1—0.1 M Hydrochloric acid–acetonitrile (1:1, v/v) S_2—Methanol–tetrahydrofuran–5% formic acid (7:7:6, v/v) S_3—Methanol–chloroform–acetic acid (5:4:1, v/v)	Dragendorff's reagent; Iodoplatinate reagent	$S_1\ R_F = 0.47$ $S_2\ R_F = 0.38$ $S_3\ R_F = 0.10$	Nishikawa and Tsuchihashi (2005)
PIP	Solution of pipecuronium bromide in methanol	HPTLC silica gel	Methanol–acetonitrile–concentrated ammonia solution (514:386:100) containing 5×10^{-3} mol/dm³ ammonium chloride and 8×10^{-3} mol/dm³ ammonium carbonate	Dragendorff's reagent; Scanned in reflectance/absorbance mode at 525 nm	$R_F \sim 0.20$	Gazdag et al. (1985)
ROC	Aqueous solution (approx. 2 µg/µL) of the rocuronium bromide	Cellulose Alumina	Aqueous solutions of ammonium sulfate at concentrations ranging from 0.5 M to 4.0 M	Iodine vapor	$R_M^0 = -1.69$ $C_0 = -2.45$	Aleksić et al. (2003)
	Aqueous solution of the rocuronium bromide	Silica gel Aluminum oxide, neutral	S_1—0%–100% (v/v) water in ACN, in steps of 10% (v/v) S_2—0%–100% (v/v) water in methanol, in steps of 10% (v/v) S_3—0%–100% (v/v) water in dioxane, in steps of 10% (v/v) S_4—0%–100% (v/v) water in THF, in steps of 10% (v/v) 0%–100% (v/v) water in ACN, in steps of 10% (v/v)	Iodine vapor	—	Radoičić et al. (2009a,b)
ATR	Aqueous solution (approx. 2 µg/µL) of the atracurium besilate	Cellulose Alumina	Aqueous solutions of ammonium sulfate at concentrations ranging from 0.5 M to 4.0 M	Iodine vapor	$R_M^0 = -0.97$ $C_0 = -1.17$	Aleksić et al. (2003)

(continued)

TABLE 32.2 (continued)
TLC Conditions Used for Determination of Peripheral-Acting Muscle Relaxants

Drug	Standard Solution	Stationary Phase; Pretreatment	Mobile-Phase Composition	Detection	Retention Factor	Reference
SUX	Aqueous solution (approx. 2 μg/μL) of the suxamethonium chloride	Cellulose Alumina	Aqueous solutions of ammonium sulfate at concentrations ranging from 0.5 M to 4.0 M	Iodine vapor	$R_M^0 = -1.92$ $C_0 = -5.03$	Aleksić et al. (2003)
	Solution of suxamethonium iodide in methanol	Sorbfil plates (PTSKh-II-A; TU 26-11-17- 89). Samples are applied on the plate under cold air flow in a place protected from direct sunlight immediately after the preparation of solutions	Monosubstituted potassium phosphate solution	Dragendorff's reagent	Suxamethonium iodide $R_F = 0.11$ Succinylmonocholine iodide $R_F = 0.17$ Choline iodide $R_F = 0.36$	Arzamastsev et al. (1999)
	Solution of suxamethonium chloride in methanol	Silica gel	S₁—0.1 M Hydrochloric acid–acetonitrile (1:1, v/v) S₂—Methanol–tetrahydrofuran–5% formic acid (7:7:6, v/v) S₃—Methanol–chloroform–acetic acid (5:4:1, v/v)	Dragendorff's reagent; iodoplatinate reagent	$S_1 \ R_F = 0.25$ $S_2 \ R_F = 0.15$ $S_3 \ R_F = 0.02$	Nishikawa and Tsuchihashi (2005)

mobile phase depends on the charge number of the substituent present in the compounds, the concentration (up to saturation value), diameter of the counterions, and the solubility parameter of the organic solvents in the mobile phase. The effect of the counterions and their concentrations on the retention of the quaternary ammonium compounds on silica gel was determinate.

Later, Gazdag et al. (1985) developed TLC methods for the separation and determination of pipecuronium bromide and related steroids. The basic concept of the separation system was ion-pair adsorption chromatography of primary, secondary, tertiary, and quaternary amines on a silica layer. The polarity of the compounds depends on the number of quaternary amino groups and the presence or absence of ester groups. According to this, authors examined the influence of different factors on separation, such as effect of ammonium chloride and ammonium carbonate in the mobile phase, effect of salt concentration in the mobile phase, effect of water concentration in the mobile phase, effect of nature, and concentration of organic solvents.

The neuromuscular blocking agents are susceptible for the hydrolysis. Several papers engaged in developing TLC methods for separation of active drugs from their hydrolysis products. Kinget and Michoel (1976) studied the degradation of pancuronium bromide in order to elucidate the mechanism and the influence of hydrolysis on the preparation and the stability of parenteral solutions of the drug. Hydrolysis occurs at the two acetoxy linkages; and beside $3\alpha,17\beta$-diacetate (pancuronium bromide), three possible degradation products could be expected: 3α-ol-17β-acetate, 3α-acetate-17β-ol, and $3\alpha,17\beta$-diol. Selective determination of the different compounds cannot be achieved by methods based on reactions involving the acetate groups (hydroxamic acid method) or the quaternary ammonium groups (acid dye method), so that preliminary separation is carried out by TLC. By previously reported methods (Kersten et al. 1973, Tanaka et al. 1974), the 17β- from the 3α-monoacetate cannot be separated sufficiently for individual determinations. Kersten et al. (1973) allowed the chromatogram to run for 18–22 h, achieving this by lengthening the layer with a filter paper. Only semi-quantitative results were obtained following visual comparison. Tanaka et al. (1974) used the same chromatographic system, but developed the chromatogram in an open system for 16 h and evaluated the relative proportions of the different compounds by densitometry after spraying the layer with Dragendorff's reagent. The two monoacetates were determined together, and the results were expressed as "monoacetyl derivatives". Kinget and Michoel (1976) also failed to improve the separation by changing the solvent system or the adsorbent, so they paid further attention to the effects of different working conditions. They concluded that better separation was achieved by using laboratory-made plates, unsaturated conditions, and enlarged plates. The final result was complete separation of pancuronium bromide from its hydrolysis products.

Kovacs and Takacsi Nagy (1983) separated pipecuronium bromide from its hydrolysis products, 17β-hydroxy monoester and dihydroxy derivate by TLC on aluminum oxide using sec-BuOH–pyridine–acetic acid–20% NH_4Cl (75:50:15:60, v/v) mobile phase. Later, Lauko et al. (1992) determined 21 synthetic intermediates of pipecuronium bromide by fused-silica capillary gas chromatography, TLC, and potentiometric titration. Eleven solvent systems are given for TLC determination.

There are several described methods for determination of impurities of neuromuscular blocking agents. Arzamastsev et al. (1999) described the method for determining host impurities of suxamethonium iodide (choline iodide, succinylmonocholine iodide, and unreacted residue of succinic acid esters), the presence of which may be caused by violation of the drug production technology. Development of the procedure was preceded by investigation of the possibilities offered by various sorbents (Silufol, Kieselgel 60 [Merck], and Sorbfil) and various mobile phases. It was established that the optimum TLC systems are Sorbfil plates (PTSKh-II-A; TU 26-11-17-89) eluted with a saturated monosubstituted potassium phosphate solution. The resulting TLC chromatogram displayed the spots of ditilin, succinylmonocholine iodide, and choline iodide. Taking into account the ability of suxamethonium iodide to readily decompose on illumination, the application of samples and the chromatographic procedure should be performed in a place protected from light. The total time of sample application onto the plate must not exceed 25 min. If the samples are applied for a longer time and/or in the light, the TLC chromatograms exhibit additional spots on the level of choline iodide and succinylmonocholine iodide.

Several years after their first work on chromatography of *bis*-quaternary amino steroids, Gazdag et al. (1992) analyzed pipecuronium bromide and structurally related impurities in pharmaceuticals by four methods: HPLC, TLC (using the same chromatographic system as it has been previously described [Gazdag et al. 1985]), spectrophotometry, and potentiometric microtitration.

Clarke and Raja (1982) described the TLC system for evaluation of *d*-tubocurarine produced from curare resin, a concentrated aqueous extract of the plant pulp. Besides the mentioned muscle relaxant, curare resin contains a large number of other bisbenzylisoquinoline alkaloids, the principal ones (*d*-chondrocurine, *l*-curine, *l*-curarine, and isochondrodendrine) being potential impurities in the refined drug. The proposed TLC method gave compact spots and excellent resolution of the components of curare resin. Some correlations between the chromatographic mobility of the five alkaloids and the degree of quaternization of the two nitrogen centers were also observed. The diquartenary curarine migrated more slowly than the ditertiary curine, while tubocurarine with a quaternary and a tertiary nitrogen center had an intermediate R_F value.

Planar chromatography has been widely used to estimate the hydrophobicity of active drugs. Aleksić et al. (2003) studied the chromatographic behavior of five myorelaxant drugs (atracurium besylate, pancuronium bromide, rocuronium bromide, vecuronium bromide, and suxamethonium chloride) under the conditions of so-called salting-out TLC (SOTLC). This typically reversed-phase method enables separation of the compounds, even on polar adsorbents, by use of moderately to highly concentrated solutions of ammonium sulfate as mobile phases. Taking into account the chemical structures of the substances examined, cellulose and alumina were chosen as adsorbents for the study. Aqueous ammonium sulfate solutions of different concentration (0.5–4.0 mol dm^{-3}) were used as mobile phases. It was established that hR_F values always decreased in parallel to increasing salt concentrations in mobile phase. Taking into account the composition of the mobile phases used and the considerable salting-out effect established for all the compounds investigated, it was assumed that separation of the myorelaxants is based on their nonspecific hydrophobic interactions with the sorbents. These interactions were more pronounced when cellulose, a much less polar adsorbent than alumina, was used. In case of cellulose as adsorbent, a linear relationship (r > 0.97) was observed between the R_M values and the ammonium sulfate content in the mobile phase. Regression data of the plots obtained were used to determine the lipophilicity parameters R_M^0 (values extrapolated to pure water) and C_0 (derived from ratio of intercept and slope of mentioned plots). Lipophilicity determined in this way has been correlated with calculated ACD/log *P* values.

Hydrophilic interaction liquid chromatography (HILIC) has recently been introduced as a highly efficient chromatographic technique for the separation of a wide range of polar solutes (Hemström and Irgum 2006). In the HILIC mode, an aqueous–organic mobile phase combined with a polar stationary phase was used to provide normal-phase retention behavior. HILIC is often considered as a normal-phase separation in a reversed-phase mode (Radoičić et al. 2009a, Shweshein et al. 2012). This technique is especially suitable for the separation of polar low-molecular-weight compounds such as hydrophilic amino acids, di- and tripeptides, and organic acids, which are often not sufficiently retained in RPLC mode. The chromatographic behavior of seven anesthetics (pancuronium, vecuronium, rocuronium, lidocaine, procaine, etomidate, and fentanyl) was investigated under HILIC conditions on thin layers of silica-gel and alumina, using simple mixtures of water and organic solvent (methanol or acetonitrile) as mobile phase (Radoičić et al. 2009b). Considering the effect of the nature of mobile phase, it was noticed that increasing the amount of the water in mobile phase, relative to the organic component, results in conversion of separation mechanism. The results obtained showed that in a wide range of water content in a mobile phase, chromatographic behavior of the compounds investigated is determinate by hydrophilic interactions in chromatographic system.

In poisoning incidents with the muscle relaxants, the injection solution is occasionally left on the spot. In such a case, TLC is a simple and rapid method for identification. Nishikawa and Tsuchihashi (2005) in the chapter "Muscle relaxants" of the book *Drugs and Poisons in Humans* described a TLC method for determination of tubocurarine, suxamethonium, pancuronium, and vecuronium.

Gajdzinska and Wedrychowski (1980) described a suicide by tubocurarine injection. Liquid found in a syringe near the body of a nurse and tissue samples was extracted in acid-Et$_2$O and alkaline-CHCl$_3$ systems and chromatographed. TLC with 1:1 (v/v) acetone–1 N HCl as solvent system showed the presence of tubocurarine in the syringe fluid. The results were confirmed by paper chromatography using MeOH–acetic acid–H$_2$O (3:6:1, v/v) and BuOH–AcOH–H$_2$O (4:1:5, v/v) solvent systems, followed by elution with 0.1 N H$_2$SO$_4$ and spectrophotometry at 230–320 nm, which gave a maximum at 255 nm. Briglia et al. (1990) described attempted murder with pancuronium. Namely, urine specimens were obtained from the victim shortly after the suspected assaults, and samples were initially tested fluorometrically using Rose Bengal dye as a pairing agent. Both samples were presumptively positive for pancuronium. Confirmation of these results was achieved by pairing the drug with potassium iodide, extracting the complex, and submitting the extract to TLC. Klys et al. (2000) reported a case of suicide by intravenous injection of pancuronium. Pancuronium bromide was identified in the blood and brain by means of ion-pair extraction followed by TLC, spectrophotometry, and electrospray ionization/tandem mass spectrometry.

32.3 SPASMOLYTICS

Spasmolytics, also known as "centrally acting" muscle relaxants, are used to alleviate musculoskeletal pain and spasms and to reduce spasticity in a variety of neurological conditions. In this section, TLC methods are presented for the selected compounds belonging to this group of agents. Their structures are presented in the Figure 32.2.

According to the available information, chromatographic parameters, which comprise selected stationary and mobile phase, retardation factors, and belonging reference, are presented in Table 32.3. The majority of the literature that has been recently published evaluates tizanidine (TIZ). Otherwise, the older literature on the application of TLC in analysis of the drug group is reviewed.

Dantrolene (DAN) is a hydantoin derivative also discussed in this section (see Figure 32.3) that is structurally and pharmacologically different from other skeletal muscle relaxants. It has no significant central effects, but has a direct action on muscle, probably by interfering with the release of calcium from sarcoplasmic reticulum.

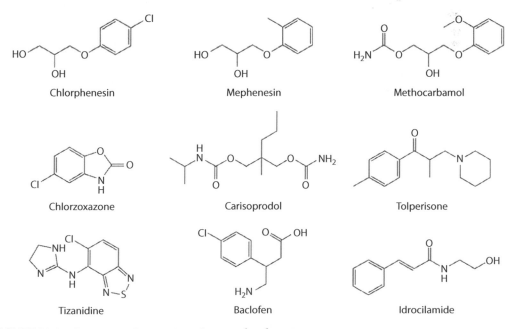

FIGURE 32.2 Structures of centrally acting muscle relaxants.

TABLE 32.3

TLC Conditions Used for Determination of Centrally Acting Muscle Relaxants

Compound	Stationary Phase	Mobile Phase	hR_F	Reference
CHP	Silica gel 60 F_{254}	Ethyl acetate–diisopropyl ether–96% ethanol–25% ammonia (55:30:10:2, v/v)	30	De Kruijf et al. (1987)
	Al_2O_3		14	
CHZ	Silica gel HF_{254}	Chloroform–methanol (19:1, v/v)	32	Ullah et al. (1970)
		Chloroform–methanol (9:1, v/v)	58	
		Chloroform–methanol (4:1, v/v)	87	
		Chloroform–methanol (7:3, v/v)	100	
		Chloroform–methanol (3:2, v/v)	100	
CHZ	Silica gel 60 F_{254}	Toluene–ethyl acetate–85% formic acid (50:45:5, v/v)	69	Egli and Tanner (1979)
		Toluene–isopropanol–concentrated ammonium hydroxide (70:29:1, v/v)	68	
		Toluene–dioxane–methanol–ammonium hydroxide (20:50:20:10, v/v)	55	
CHZ	Silica gel 60 F_{254}	Toluene–acetone–2N acetic acid (30:65:5, v/v)	83	Egli and Keller (1984)
		Toluene–isopropanol–ethyl acetate–2N acetic acid (10:35:35:20, v/v)	95	
	Bonded C18	Methanol–0.5% phosphoric acid–3% sodium chloride (60:10:30, v/v)	32	
		Isopropanol–0.5% phosphoric acid–3% sodium chloride (40:10:50, v/v)	35	
		Isopropanol–methanol–0.5% phosphoric acid–3% sodium chloride (23:23:10:44, v/v)	33	
		Tetrahydrofuran–methanol–0.5% phosphoric acid–3% sodium chloride (28:28:10:34, v/v)	32	
		Methanol–2N ammonia–3% sodium chloride (60:10:30, v/v)	60	
		Isopropanol–2N ammonia–3% NaCl (40:10:50, v/v)	47	
		Isopropanol–methanol–2N ammonia–3% sodium chloride (23:23:10:44, v/v)	43	
		Tetrahydrofuran–methanol–2N ammonia–3% sodium chloride (28:28:10:34, v/v)	55	
		Acetonitrile–3% sodium chloride (50:50, v/v)	46	
		Methanol–3% sodium chloride (60:40, v/v)	32	
		Acetone–3% sodium chloride (50:50, v/v)	25	
		Tetrahydrofuran–3% sodium chloride (43:57, v/v)	18	
		Isopropanol–3% sodium chloride (40:60, v/v)	35	
		Methanol–isopropanol–tetrahydrofuran–3% sodium chloride (17:17:17:49, v/v)	36	
CHZ (ACP)	Silica gel GF_{254}	Ethyl acetate–benzene–acetic acid (1:1:0.05, v/v)	75 (43)	El Bayoumi et al. (1999)
CHZ	Silica gel HF_{254}	Chloroform–methanol (9:1, v/v)	58	Ullah et al. (1970)

Compound	Stationary phase	Mobile phase	R_f	Reference
CHZ (HCZ)	TLC silica gel 60 F$_{254}$	Acetone–hexane (70:30, v/v)	88 (91)	Zerilli et al. (1996)
		Acetone–hexane (60:40, v/v)	84 (89)	
		Acetone–hexane (50:50, v/v)	47 (64)	
		Acetone–hexane (45:55, v/v)	55 (75)	
CHZ	TLC silica gel G 60 F$_{254}$	Ethyl acetate–methanol–25% ammonia (85:15:5, v/v)	32	Bebawy and El-Klousy (1999)
CHZ	HPTLC silica gel 60 GF$_{254}$	Ethyl acetate–toluene–ammonia (7:3:0.2, v/v)	70	Gandhimathi and Ravi (2008)
MET	Silica gel 60 GF$_{254}$	Chloroform–methanol (9:1, v/v)	46	Salvadori et al. (1988)
		Toluene–dioxane–acetic acid (90:25:5, v/v)	18	
		Chloroform–ethyl acetate–methanol–propionic acid (65:15:15:7, v/v)	80	
		Ethyl acetate–methanol–ammonia (85:10:5, v/v)	75	
		Chloroform–methanol–dpropionic acid (72:18:10, v/v)	93	
		Dichloromethane–dioxane (2:1, v/v) saturated with water	60	
		Chloroform–methanol (4:1, v/v)	84	
MET	Silica gel 60 GF$_{254}$	Methanol–toluene (2:8, v/v)	40	Manwar et al. (2007)
MET	Silica gel 60 GF$_{254}$	Ethyl acetate–acetone–triethylamine–formic acid (62:35:6:0.3, v/v)	78	Ali et al. (2012)
IBU			14	
DCL			12	
GUA			54	
MET	Silica gel 60 F$_{254}$	Ethyl acetate–methanol–ammonia–water (43:3:0.5:1.5, v/v)	57	Lillsunde and Korte (1991)
		Methanol–ammonia (50:0.5, v/v)	72	
MET	Silica gel G treated with 0.1M KOH in methanol	Methanol–strong ammonia solution (100:1.5, v/v)	70	Moffat (1986)
		Chloroform–acetone (4:1, v/v)	7	
		Ethyl acetate–methanol–strong ammonia solution (85:10:5, v/v)	49	
		Ethyl acetate	23	
MET	Toxi-Grams Blank	Ethyl acetate–methanol–water (95.2:3.2:1.6, v/v)	67	Harper et al. (1989)
MET	Toxi-Grams C8	Acetonitrile–water–isopropanol–ethyl acetate (48:48:2:2, v/v)	86	
MET	Silica gel plates	Chloroform–acetone (4:1, v/v)	9	Owen et al. (1978)
		Acetic acid–toluene–ether–methanol (18:120:20:1, v/v)	9	
		Ethyl acetate–methanol–ammonia (85:10:5, v/v)	40	
		Ethyl acetate	33	
		Dioxane-toluene–ammonia (20:75:5, v/v)	2	
		Chloroform–methanol (9:1, v/v)	36	

(continued)

TABLE 32.3 (continued)
TLC Conditions Used for Determination of Centrally Acting Muscle Relaxants

Compound	Stationary Phase	Mobile Phase	hR_F	Reference
MET	Silica gel	Acetic acid–carbon tetrachloride–chloroform–water (100:60:90:50, v/v)	22	Daldrup et al. (1981)
	Kieselguhr G impregnated with formamide	Benzene–chloroform (30:120, v/v)	31	
CAR	Silica gel	Ethyl acetate–methanol–conc. ammonium hydroxide (85:10:1, v/v)	36	Broich et al. (1971)
MET			10	
CAR	Silica gel	Chloroform–acetone (4:1, v/v)	34	Owen et al. (1978)
		Acetic acid–toluene–ether–methanol (18:120:20:1, v/v)	35	
		Isopropanol–chloroform–ammonia (45:45:10, v/v)	71	
		Ethyl acetate–methanol–ammonia (85:10:5, v/v)	63	
		Ethyl acetate–dioxane–toluen–ammonia (20:75: 5, v/v)	18	
		Chloroform–methanol (9:1, v/v)	57	
TIZ	TLC silica gel 60F$_{254}$	Toluene–acetone–ammonia (5:5:0.1, v/v)	32	Mahadik et al. (2003)
TOL (LID)	HPTLC silica gel G60	Methanol	61 (91)	Liawruangrath and Liawruangrath (1999)
		Methanol–acetone–butanol (50:45:5, v/v)	50 (86)	
TOL	Silica gel 60 GF$_{254}$	Toluene–ethylacetate–ethanol (6:1.5:2.5, v/v)	40	Patel et al. (2012)
ETO			58	

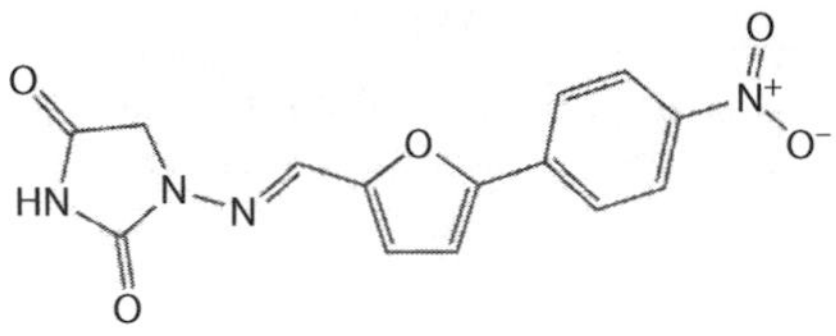

FIGURE 32.3 Structure of peripherally acting muscle relaxant dantrolene.

32.3.1 CHLORPHENESIN

Chlorphenesin (CHP) is 3-(4-chlorophenoxy)propan-1,2-diol, a centrally acting muscle relaxant used to treat muscle pain and spasm. In cosmetics, it is used as a preservative because of its antifungal and antibacterial properties.

Buhler (1964) utilized paper, thin-layer, and gas chromatographic techniques to investigate the excretion of CHP carbamate-3H in rats and the metabolism of the nonradioactive drug in humans. TLC of neutral metabolites in this study was carried out on both silica gel GF and alumina GF plates using ethyl acetate–cyclohexane and chloroform–methanol, respectively, as solvent systems. The same author used paper and TLC to elaborate the nature of the CHP carbamate conjugates excreted by both rats and humans (Buhler 1965). TLC of the methyl acetyl glucuronide was carried out on silica gel GF in 1.5% methanol in chloroform and alumina GF in a 0.75% methanol in chloroform solvent system. The glucuronide esters were detected by their UV absorption and by their reaction with 1% p-dimethylaminobenzaldehyde in 4M hydrochloric acid or 50% sulfuric acid.

Schmahl and Hieke (1980) reported a TLC on silica gel 60 F_{254} with toluene–acetone (4:1, v/v) or ethyl acetate–methanol–10% NH_4OH (65:30:5, v/v) to separate 38 preservatives used in cosmetics. The compounds were identified by color reactions with 12 spray reagent systems.

In order to identify the preservatives in cosmetic products, De Kruijf et al. (1987) developed a screening procedure based on a combination of TLC and HPLC. From a total of 88 preservatives that were tested, 74 were characterized, including CHP. The method consisted of extraction of acidified cosmetics with methanol and separation of the extracts by TLC on Al_2O_3 and silica gel 60 F_{254} plates (Table 32.3). Visualization of the preservatives on the plates using six different detection reagents was done. The following detection reagents were used: bromocresol green spray reagent, Gibbs' spray reagent, Dragendorff's reagent, Millon's reagent, Purpald spray reagent, and ammoniacal silver nitrate spray reagent. The preservatives in 14 commercial cosmetic products were tentatively identified by the procedure described. In general, this method allows routine detection of preservatives in cosmetics in an approximate concentration of 0.1% (w/w).

32.3.2 METHOCARBAMOL

Methocarbamol (MET) is 2-hydroxy-3-(2-methoxyphenoxy)propyl carbamate, a carbamate derivate of guaifenesin (GUA), starting material in MET synthesis. GUA is also considered as the main degradation product of MET, especially in alkaline medium. Structurally, it is related to mephenesin (MEP).

As spinal muscles relaxing drug, MET is often applied in the form of complex pharmaceuticals. That is why a lot of papers are related to the use of chromatographic methods in the analysis of pharmaceutical formulations. The proposed methods can be used for routine quality control analysis of the studied drugs in bulk powders and in pharmaceutical formulations.

Krzek and Starek (1998) described a TLC method for simultaneous determination of MET and acetylsalicylic acid in pharmaceutical formulations. The commercial tablets Mespefam were prepared by extraction with methanol. Silica gel 60 plates were used with developing solvent of methyl acetate–methanol–25% ammonia (17:2:1, v/v) and UV densitometry. Quantitative determination of each component was done at 225 and 272 nm.

Manwar et al. (2007) developed a simple, sensitive, and precise high-performance TLC (HPTLC) method for the estimation of MET in its tablet formulation, *Robinax*. In this method, standard solution and sample solutions of MET were applied on precoated TLC plates and developed using methanol–toluene mixture (Table 32.3). The plates were scanned and quantified at 276 nm using TLC scanner. The method was reported to be simple, sensitive, accurate, specific, and precise and can be used for routine quality control testing of marketed formulations.

MET is also formulated with either ibuprofen (IBU) or diclofenac potassium (DCL), which are nonsteroidal drugs with anti-inflammatory, antipyretic, and analgesic properties. Ali et al. (2012) established sensitive, selective, and accurate analytical methods for the determination of MET and GUA in ternary mixtures with IBU and DCL (Table 32.3). Methanol was used for the extraction of tablets. Linear ascending development was performed in a chromatographic tank previously saturated with solvent. The developed plates were scanned at 222 nm, for the mixture containing MET, IBU, and GUA, and at 278 nm, for the second mixture containing MET, DCL, and GUA. Linearity of the proposed methods was evaluated, and it was in the range of 2–12, 4–20, and 2–10 µg per band for MET, IBU, and GUA, respectively, and 2–12, 0.2–2.2, and 2–10 µg per band for MET, DCL, and GUA, respectively. In the case of MET, accuracy was 100.05 ± 1.55, while the intra day and interday relative standard deviation (%RSD) were 1.23 and 1.45, respectively.

Among other results that are encountered in the literature, most reports are drug screening methods, and MET is one of many compounds included.

TLC was proposed by Salvadori et al. (1988) for the identification and subsequent confirmation of 46 drugs related to doping of thoroughbred horses, including MET. Seven TLC systems were used to evaluate silica gel plates. The results are presented in Table 32.3. After development, the plates were observed under UV light at 254 nm and sprayed with Ehrlich and Mandelin reagent. The combined detection procedure is versatile, sensitive, reliable, and recommended for doping control analysis.

MET, among other basic drugs used illegally in horse racing, was the subject of investigation conducted by Woods et al. (1985). They have investigated the ability of phenylbutazone and oxyphenbutazone to mask or interfere with the detection reagents by Whatman HP-KF HPTLC silica gel plates. Urine samples were extracted with dichloromethane and a few drops of concentrated ammonium hydroxide. Plates were sprayed with phenothiazine and thereafter with Dragendorff's reagent. MET (R_F = 0.37) was not masked with phenylbutazone and oxyphenbutazone in system consisting of chloroform–methanol (9:1, v/v).

Lillsunde and Korte (1991) described a simple and sensitive solid-phase extraction method for the detection of a broad spectrum of drugs (about 300 substances including MET were analyzed in urine samples). Spots were visualized with furfuraldehyde and sulfuric acid or with 1% Fast Black K salt and NaOH. Chromatographic conditions are presented in Table 32.3.

An overview of related papers was given by Alessi-Severini et al. (1994). TLC conditions, hR_F values, and the corresponding literature are presented in Table 32.3 (Daldrup et al. 1981, Harper et al. 1989, Moffat 1986, Owen et al. 1978).

32.3.3 Chlorzoxazone

Chlorzoxazone (CHZ) (5-chloro-3H-benzooxazol-2-one) is an active muscle relaxant with sedative properties. It contains a benzoxazolone ring system, which is highly unstable due to presence of both lactam and lactone functional groups in the fused ring system. Both groups are subject to alkaline-induced hydrolysis. It is sold as *Muscol* or *Parafon Forte*, a combination of CHZ and acetaminophen (*Paracetamol*).

Stewart and Janicki (1987) reviewed literature relating to the application of TLC in CHZ determinations. Ullah et al. (1970) developed a TLC procedure for the quantitative determination of heat-labile CHZ formulations in the presence of decomposition products. Numerous chromatographic systems resulted from the research published by Egli and Tanner (1979) and Egli and Keller (1984). TLC conditions are presented in Table 32.3.

El Bayoumi et al. (1999) described a quantitative thin-layer procedure for estimating CHZ in bulk powders and in dosage forms each in the presence of its degradation product. 2-Amino-4-chlorophenol (ACP) was found to be the alkaline-induced degradation product of CHZ. The method consisted of dissolving the drug in methanol and then spotting this solution on a thin-layer of silica gel GF_{254} (Table 32.3). Plates were scanned densitometrically at 282 nm for CHZ and at 285 nm for 2-ACP. Quantitation was achieved by comparing the areas under the peaks obtained from scanning the thin-layer chromatographic plates in a densitometer.

Several papers were published dealing with determination of formulations of CHZ with other drugs using chromatographic methods. TLC with UV scanning densitometry was applied to resolve CHZ binary mixture with paracetamol in pharmaceutical dosage forms and in biological fluids (Bebawy and El-Klousy 1999). The chromatographic chamber was equilibrated for 1 h prior to use and developed by ascending technique. Components of mixture were separated with R_F values of 0.77 for paracetamol and 0.32 for CHZ. Linearity ranges, mean recoveries, and %RSDs in calibration graphs of the proposed method were calculated. The method has been successfully applied to pharmaceutical formulations, sugar-coated tablets, capsules, and suppositories. Sodhi et al. (1996) studied CHZ and IBU by the use of GC, HPLC, and HPTLC and reported comparable results obtained by the different techniques. In another study, Chawla et al. (1996) reported results of simultaneous determination of CHZ and diclofenac sodium using HPTLC.

CHZ was shown to be a useful in vitro or in vivo probe for the determination of Cytochrome P450 2E1 (CYP2E1) activity, which metabolizes CHZ mainly to a single metabolite, 6-hydroxy CHZ (HCZ). Zerilli et al. (1996) proposed a method for in vitro determination of CYP2E1 using TLC and labeled $[2-^{14}C]CHZ$. The method was performed in liver and kidney microsomes from control and ethanol- or acetone-induced rats. TLC silica gel 60 F_{254} plate was developed in various acetone–hexane mixtures. The corresponding hR_F values of CHZ and HCZ for various solvent ratios are shown in Table 32.3.

CHZ is also marketed in combination with some nonsteroidal anti-inflammatory drugs (NSAIDs). Sane and Gadgil (2002) reported a method for simultaneous determination of paracetamol, CHZ, and nimesulide (which is an NSAID). The tablets were dissolved in methanol. Chromatography was performed on precoated silica gel 60 F_{254} HPTLC plates and developed in a twin-trough chamber previously equilibrated with mobile phase for 10 min. The mobile phase was chloroform–methanol–toluene (6:1:1.5, v/v). Plates were evaluated by densitometry at 270 nm. The plot was linear in the concentration ranges 0.1956–0.5868 mg mL^{-1} paracetamol, 0.2251–0.6754 mg mL^{-1} CHZ, and 0.0607–0.1822 mg mL^{-1} nimesulide. The mean recovery, 98%–102%, is within acceptable limits, indicating the method is accurate.

Abdel-Azeem et al. (2009) presented results on the design and synthesis of different mutual ester prodrugs of CHZ and some NSAIDs. The objective of the work was to minimize gastrointestinal side effects of NSAIDs and to improve pharmacokinetic properties of both CHZ and NSAIDs while maintaining the useful anti-inflammatory and skeletal muscle relaxation activities. The purity of the synthesized prodrugs was ascertained by TLC carried out on pre-coated silica gel 60 F_{254} plates. Spots were detected under UV light.

Gandhimathi and Ravi (2008) reported method for simultaneous densitometric analysis of tramadol hydrochloride and CHZ in the pharmaceutical preparation using HPTLC. For that purpose, tablets were finely powdered and extracted with methanol. Plates were saturated with mobile phase vapor by equilibration for 10 min then developed with ethyl acetate–toluene–ammonia mixture in a glass twin-trough chamber (Table 32.3). Evaluation of developed plates was performed densitometrically at 273 nm. R_F value for tramadol and CHZ were 0.19 and 0.70, respectively. The average RSD was 1.72% for tramadol hydrochloride and 1.94% for CHZ, indicating the precision of quantification was acceptable. The average accuracy was 98.72% for tramadol hydrochloride and 102.46% for CHZ. Minimum detection limits were 30 ng per spot for tramadol hydrochloride and 100 ng per spot for CHZ. Quantitation limits were 100 and 700 ng per spot for tramadol and CHZ, respectively.

32.3.4 CARISOPRODOL

Carisoprodol (CAR) is (RS)-2-{[(aminocarbonyl)oxy]methyl}-2-methylpentyl isopropyl carbamate. It is a skeletal muscle relaxant and also has analgesic action. This agent is metabolized primarily in the liver, with multiple metabolites including meprobamate.

CAR was among the "neutral" drugs investigated by Sunshine et al. (1963) in a study of the utility of TLC in the diagnosis of poisoning. Silica gel G was the absorbent, cyclohexane the mobile phase, and detection was accomplished utilizing chlorinestarch iodide and furfural-concentrated hydrochloric acid. The former reagent yielded a fading blue color with each of the carbamates. Overspraying with furfural and then with concentrated hydrochloric acid after the plate dried resulted in the characteristic black colors for the carbamates.

Broich et al. (1971) described a rapid TLC method suitable for the detection of drug abuse. Drugs, including centrally acting muscle relaxants, were detected using a separate series of chromophoric spray reagents for the acid and basic drugs. Owen et al. (1978) reported numerous TLC systems for the screening of neutral drugs. Among the tested compounds was also CAR. Table 32.3 shows hR_F values as a result of different mobile phases used.

32.3.5 TOLPERISONE

Tolperisone (TOL) is 2-methyl-1-(4-methylphenyl)-3-(1-piperidyl)propan-1-one. Clinically used TOL is a racemic mixture. Very few assay methods of TOL were reported. Liawruangrath and Liawruangrath (1999) described an HPTLC method for the determination of TOL hydrochloride in pharmaceutical preparations. The TLC separations of TOL and lidocaine hydrochloride were carried out using HPTLC silica gel plates and spectrophotometric detection at $\lambda = 260$ nm (Table 32.3). The limit of detection (LOD) was 20.12 ng/spot drug in methanol. When methanol–acetone–butanol (50:45:5, v/v) was used as mobile phase, the detection limit was 4.12 ng/spot.

Patel et al. (2012) proposed an HPTLC method for the simultaneous analysis of TOL and etodolac (ETO) in combined fixed-dose tablet formulation. Liner ascending development was performed in twin-through glass chamber previously equilibrated with mobile phase vapor for 30 min. Densitometric scanning was performed at 260 nm (Table 32.3). The linearity of the response to TOL and ETO was assessed in the concentration ranges 75–450 and 200–1200 ng/band, respectively. The precision of the method, determined as intraday precision (%RSD) was 1.27–1.85 for TOL, while the inter-day precision (%RSD) was 1.78–2.02. The accuracy of the method was determined as multiple-level recovery studies (98.4%–99.3%).

The most recent publication (Hubicka et al. 2012) is related to TLC densitometric method for determination of TOL hydrochloride and its impurities, such as piperidine hydrochloride and 4-methylpropiophenone that are used during synthesis as reagents. These substances may also appear during degradation of TOL. Silica gel TLC F_{254} plates were used as stationary phase and cyclohexane–1,4 dioxane–isopropanol–ethanol–glacial acetic acid (16:0.5:1:4:0.6, v/v) and n-butanol–isopropanol–water–glacial acetic acid (10:7:8:2, v/v) as mobile phases. Densitometric measurements were done at 260 nm for TOL hydrochloride and 4-methylpropiophenone and at 570 nm for piperidine hydrochloride. A broad range of linearity was reported (µg per band) 0.06–1.50, 0.09–0.40, and 0.04–0.25 for TOL, 4-methylpropiophenone, and piperidine hydrochloride, respectively. The method was sensitive with LOD (µg per band) 0.02, 0.03, and 0.02, for TOL, 4-methylpropiophenone, and piperidine hydrochloride, respectively, and LOQ (µg per band) 0.06, 0.09, and 0.04, for TOL, 4-methylpropiophenone, and piperidine hydrochloride, respectively. Precision of the method was good with RSD values of 2.16% for TOL and 3.60% for studied impurities. Average recovery for tested compounds at three concentrations was 99.67% for TOL, 90.73% for 4-methylpropiophenone, and 84.62% for piperidine hydrochloride. The developed method may be used for purity evaluation of TOL as a bulk drug and in pharmaceutical preparations as an alternative to pharmacopeial methods.

32.3.6 Mephenesin

MEP is 3-(2-methylphenoxy)propane-1,2-diol. There are very few reports on analytical methods for analysis of MEP. Fishbein and Zielinski (1967) published review of the literature with reference to the application of chromatographic analysis of miscellaneous carbamates, including CHP carbamate, MEP carbamate, and CAR.

Thomas and Dryon (1964) described a TLC analysis of some psychotropic drugs, which was carried out on silica gel G with solvent systems consisting of benzene–methanol (90:10, v/v) and acetone–methanol–ammonia (50:50:1, v/v). Detection was accomplished with 2.5% *p*-dimethylaminobenzaldehyde in alcoholic sulfuric acid. Among drugs that were chromatographed in the aforementioned procedure were MEP and CAR.

32.3.7 Tizanidine

TIZ is an imidazoline derivative, 5-chloro-N-(4,5-dihydro-1H-imidazol-2-yl)benzo[c][1,2,5]thiadiazol-4-amine. TIZ is a central α-2 adrenergic agonist and centrally active myotonolytic skeletal muscle relaxant approved for treating adult spasticity. Its chemical structure is not related to other muscle relaxants.

Analytical procedures for the estimation of TIZ in bulk and its dosage form are mostly described in recent literature. Mahadik et al. (2003) developed and validated a simple, selective, precise, and stability-indicating HPTLC method for analysis of TIZ hydrochloride, both as a bulk drug and in formulations. TIZ hydrochloride was subjected to acid and alkali hydrolysis, oxidation, and photodegradation. Furthermore, the degraded product was well separated from the pure drug. Densitometric analysis was carried out in the absorbance mode at 315 nm. The linear regression analysis data for the calibration plots showed good linear relationship with $r^2 = 0.9922$ in the concentration range 300–1000 ng/spot. The method was validated for precision, recovery, and robustness. The limits of detection and quantitation were 88 and 265 ng/spot, respectively. It was found that the drug does not undergo degradation under acidic and basic conditions, while it is susceptible to oxidation. The samples degraded with hydrogen peroxide showed additional peak at R_F value of 0.12. As the method could effectively separate the drug from its degradation product, it can be employed as a stability-indicating one.

Shanmugasundaram et al. (2009) reported a simple, precise, accurate, and rapid HPTLC method for the simultaneous estimation of DCL and TIZ hydrochloride in tablet dosage forms. Precoated silica gel plate 60 F_{254} and the mobile phase solvent system consisting of toluene–isopropanol–ammonia (30:20:2.5, v/v) were used. The drugs were scanned at 280 nm. The R_F value of DCL and TIZ were 0.62–0.69 and 0.85–0.92, respectively. The linearity was in the range of 3000–7000 mg/mL and 120–260 mg/mL for DCL and TIZ, respectively. The proposed method can be successfully used to determine the drug contents of marketed formulation.

Sivasubramanian and Bell (2009) presented a method for the simultaneous determination of TIZ and valdecoxib in binary mixture. The method was based on HPTLC separation of the two drugs followed by densitometric measurements of their spots at 254 nm. The separation was carried out using methanol–acetone–toluene (4:4:2, v/v) as mobile phase on HPTLC aluminum sheets of silica gel 60 F_{254}. The linearity was in the range of 10–100 and 100–1000 ng/spot for TIZ and valdecoxib, respectively. The proposed method was successively applied to pharmaceutical formulation. No chromatography interference from the tablet excipients was found. The method was validated in terms of precision, robustness, recovery, and limits of detection and quantitation.

TIZ and rofecoxib (ROF) are available in combined tablet dosage form, where both the drugs act synergistically in relieving the muscle spasms. Several papers have been reported for the estimation of TIZ and ROF from their formulations.

Kaul et al. (2005) described two methods for the simultaneous determination of TIZ and ROF in binary mixtures. The first method was based on HPTLC separation followed by densitometric measurements at 311 nm. The separation was carried out on HPTLC aluminum sheets of silica gel 60 F_{254} using toluene–methanol–acetone (7.5:2.5:1.0, v/v) as mobile phase. The linear regression was used in the range of 10–100 and 100–1500 ng/spot for TIZ and ROF, respectively. The second method was based on HPLC separation of the two drugs on the reversed phase C18 column using mobile phase consisting of phosphate buffer pH 5.5 and methanol (45:55, v/v), with UV detection at 235 nm. Both methods were successively applied to pharmaceutical formulation. The analysis of variance and Student's t-test were applied to correlate the results of TIZ and ROF determination in dosage form by HPTLC and HPLC methods.

Ravi et al. (2006) validated a method for determination of TIZ and ROF in tablet formulation. Tablets were extracted with methanol. Precoated silica gel G 60 F_{254} TLC plates were developed after chamber saturation (20 min) and scanned at 315 nm. Solvent mixture n-butyl acetate–formic acid–chloroform (6:4:2 v/v), as mobile phase, gave R_F values of 0.41 ± 0.03 for TIZ and 0.69 ± 0.05 for ROF. The dynamic linearity range was 2–10 mg/spot for TIZ and 16–80 mg/spot for ROF. The method was validated for precision, accuracy, and reproducibility. The LOD for TIZ and ROF was found to be 200 ng and 20 ng, respectively, and the limit of quantification was found to be 2 and 0.4 mg, respectively. The %RSD was found to be less than 2 for both interday and intraday assay precision.

Another paper dealing with simultaneous estimation of TIZ and ROF in tablet formulation using HPTLC method was developed by Roosewelt et al. (2007). Silica gel 60 F_{254} TLC plates were developed using a mixture of acetone–methanol (1:1, v/v). The method was validated in terms of linearity, accuracy, precision, repeatability, and specificity proving that this method is effective for the simultaneous estimation of the drug content in pure and tablet dosage form.

Pawar et al. (2009) published results of a rapid HPTLC method for the simultaneous determination of TIZ and ROF in dosage form. The HPTLC method was developed and validated for simultaneous determination of ROF and TIZ in tablet dosage form. The separation was achieved using HPTLC silica gel 60 F_{254} plates and mobile phase comprising toluene–ethyl acetate–methanol–triethylamine (6:3:0.5:0.1, v/v), with chamber saturation of 15 min. The plate was scanned and quantified at 235 nm. The linearity of ROF and TIZ were in the range of 3.75–11.25 μg/spot and 0.30 to 0.90 μg/spot, respectively. The LOD for ROF and TIZ was found to be 45.00 and 30.00 ng/spot, respectively. The LOQ for ROF and TIZ was found to be 135.00 ng/spot and 90.00 ng/spot, respectively. The percentage assay was found in the range of 99.58%–103.21% for ROF and 98.73%–101.55% for TIZ, whereas recovery was found between 99.97% and 100.43% for ROF and 100.00%–101.00% for TIZ by standard addition method.

Rao and Varma (2009) developed a simple and rapid HPTLC method and validated for the estimation of ROF and TIZ hydrochloride simultaneously in combined pharmaceutical dosage forms. The stationary phase used was precoated silica gel 60 F_{254}. The mobile phase used was a mixture of methanol–acetone in the ratio of 1:1 (v/v). The detection of spots was carried out at 254 nm. The method was validated in terms of linearity, accuracy, precision, specificity, and reproducibility. The calibration curve was found to be linear in the range of 2.204–3.016 μg/spot, for ROF, and 0.180–0.260 μg/spot, for TIZ hydrochloride. The LOD and LOQ for TIZ hydrochloride were 0.052 and 0.16 μg/spot, respectively. The proposed method can be successfully used to determine the drug content of marketed formulation.

An overpressured layer chromatography (OPLC) method was evaluated for broad-scale screening of basic drugs, and among them was TIZ, in 5 g autopsy liver samples using two parallel OPLC systems (Pelander et al. 2007). Sample preparation included enzymic digestion with trypsin and liquid–liquid extraction with butyl chloride. Chromatographic separation was performed as dual-plate analysis, with trichloroethylene–methylethylketone–n-butanol–acetic acid–water (17:8:25:6:4, v/v) and butyl acetate–ethanol (96.1%)–tripropylamine–water (85:9.25:5:0.75, v/v) as mobile phases. Identification was based on automated comparison of corrected R_F values and in situ UV spectra with library values by dedicated software. The identification limit was determined for 25 basic drugs in liver ranging from 0.5 to 10 mg/kg. The OPLC method proved to be well suited for routine screening analysis of basic drugs in post-mortem samples.

32.3.8 BACLOFEN

Baclofen (BAC), (RS)-4-amino-3-(4-chlorophenyl)butanoic acid. As it is the *p*-chlorophenyl derivative of γ-aminobutyric acid (GABA), BAC achieves its therapeutic effect by being a GABA$_B$ agonist and includes sedation and confusion among its side effects.

The following TLC systems were proposed for chromatographing BAC (Ahuja 1985): toluene–ethanol–ammonium hydroxide (50:40:7, v/v), chloroform–ethyl acetate (7:3, v/v), n-butanol–acetic acid–water (4:1:1, v/v). Glutarimide, the precursor in BAC synthesis, and 4-(4-chlorophenyl)-2-pyrrolidone, the major decomposition product of BAC, can be resolved with all of these systems. However, the system n-butanol–acetic acid–water provides the best resolution for these compounds, and degradation compounds formed at trace levels in the preliminary kinetic studies. Herdeis and Hubmann (1992) presented a stereoselective synthesis of R-BAC. The reaction was monitored by TLC using silica gel 60 F$_{254}$ plates. Reaction compounds were visualized by iodine vapor.

Krauss et al. (1988) reported a method for extraction of BAC from human plasma and urine. Liquid–solid extraction followed by ion-pair extraction was used. A fluorescence derivatization was performed with benoxaprofen chloride followed by silica gel TLC and quantification with fluorescence at 313 and 365 nm. The procedure is applicable for the determination of the fluoro analog of BAC. The lower limits of detection were 10 ng/mL plasma and 20 ng/0.1 mL urine. The procedure was used to study the pharmacokinetics of oral BAC in volunteers.

Baum and Schuster (1998) described the preparation of BAC solutions for intrathecal administration. The sterilizability of the solutions was tested by TLC and HPLC. Other decomposition products were not detectable.

Balerio and Rubio (2002) used TLC to determine blood and brain concentrations of (−)3H-BAC. Ascending chromatograms were run with 2% acetic acid solution on silica gel GF$_{254}$ plates. The spots were visualized using iodine vapor.

32.3.9 IDROCILAMIDE

Idrocilamide (IDR) is N-(2-hydroxyethyl)-3-phenyl-2-propenamide. Two sensitive and selective chromatographic methods have been developed and validated for analysis of IDR in the presence of its degradation products (Salem et al. 2010). Forced degradation studies were performed using HCl, NaOH, and 3% H$_2$O$_2$. The first method was based on TLC separation of the intact drug from its degradation products, followed by densitometric measurement. The second method was based on isocratic reversed-phase HPLC separation of the drug from its degradation products on a C18 column.

32.3.10 DANTROLENE

DAN is 1-{[5-(4-nitrophenyl)-2-furyl]methylideneamino}imidazolidine-2,4-dione (Figure 32.3). It occurs as sodium salt. DAN is unusual among the muscle-relaxant drugs as it works directly on skeletal muscle and inhibits muscle contraction at the level of the nerve-muscle connection. Pharmacologic action is usually attributed to general depression of the CNS, but may involve blockage of nerve impulses that cause increased muscle tone and contraction. DAN inhibits the release of calcium in skeletal muscle cells and thereby decreases the strength of muscle contraction. It is used to relieve spasticity in neurologic disorders (e.g., multiple sclerosis and spinal cord injury) and to prevent or treat malignant hyperthermia, a condition occasionally triggered by anesthetic agents and in which intense muscle spasm is a predominant feature.

Our findings indicate that there is little literature data concerning the use of TLC in DAN analysis. Several methods relating to the application of TLC for identification and quantification of DAN sodium can be found in Moffat (1986). Reported methods comprise silica gel G plates and several mobile phase mixtures consisting of chloroform–acetone (4:1, v/v), ethyl acetate–methanol–strong ammonia solution (85:10:5, v/v), and ethyl acetate. Corresponding R_F values in three mobile phases were 0.19, 0.09, and 0.36, respectively.

Another paper was published by Kuroiwa et al. (1985), who examined the possible metabolism of DAN sodium in fresh stool. The degradation of DAN and the appearance of a metabolite was followed, and the metabolite was identified as 7-aminodantrolene by HPLC and TLC.

ACKNOWLEDGMENTS

The authors are grateful to the Ministry of Science of the Republic of Serbia (Grant No. 172017) for financial support and to Uroš M. Gašić for technical assistance.

REFERENCES

Abdel-Azeem, A. Z., A. A. Abdel-Hafez, G. S. El-Karamany, and H. H. Farag. 2009. Chlorzoxazone esters of some non-steroidal anti-inflammatory (NSAI) carboxylic acids as mutual prodrugs: Design, synthesis, pharmacological investigations and docking studies. *Bioorgan. Med. Chem.* 17:3665–3670.

Ahuja, S. 1985. Baclofen. *Anal. Prof. Drug Sub.* 14:527–548.

Aleksić, M., J. Odović, D. Milojković-Opsenica, and Ž. Tešić. 2003. Salting-out thin-layer chromatography of several myorelaxants. *J. Planar Chromatogr.* 16:144–146.

Alessi-Severini, S., F. Jamali, R. T. Coutts, and F. M. Pasutto. 1994. Methocarbamol. *Anal. Prof. Drug Sub. Excip.* 23:371–398.

Ali, N. W., M. A. Hegazy, M. Abdelkawy, and E. A. Abdelaleem. 2012. Simultaneous determination of methocarbamol and its related substance (guaifenesin) in two ternary mixtures with ibuprofen and diclofenac potassium by HPTLC spectrodensitometric method. *J. Planar Chromatogr. Modern TLC* 25(2):150–155.

Ariffin, M. M. and R. A. Anderson. 2006. LC/MS/MS analysis of quaternary ammonium drugs and herbicides in whole blood. *J. Chromatogr. B* 842:91–97.

Arzamastsev, A. P., T. Yu. Luttseva, N. P. Sadchikova, E. P. Gernikova, and O. A. Vaganova. 1999. Evaluation of the quality of ditilin (suxamethonium iodide) preparations with respect to host impurities. *Pharm. Chem. J.* 33:616–618.

Balerio, G. N. and M. Rubio. 2002. Pharmacokinetic-pharmacodynamic modeling of the antinociceptive effect of baclofen in mice. *Eur. J. Drug Metab. Pharmacokinet.* 27(3):163–169.

Baum, S. and F. Schuster. 1988. Preparation and testing of baclofen solutions. *Pharmazeutische Zeitung* 133(17):28–30.

Bebawy, L. I. and N. M. El-Kousy. 1999. Simultaneous determination of some multicomponent dosage forms by quantitative thin-layer chromatography densitometric method. *J. Pharm. Biomed. Anal.* 20:663–670.

Bederbck, W., G. Aydinciouglou, C. Diefenbach, and M. Theisohn. 1996. Stereoselective high-performance liquid chromatographic assay with fluorometric detection of the three isomers of mivacurium and their cis- and trans-alcohol and ester metabolites in human plasma. *J. Chromatogr. B* 685:315–322.

Błazewicz, A., Z. Fijałek, and K. Samsel. 2008. Determination of pipecuronium bromide and its impurities in pharmaceutical preparation by high-performance liquid chromatography with coulometric electrode array detection. *J. Chromatogr. A* 1201:191–195.

Briglia, E. J., P. L. Davis, M. Katz, and L. A. Dal Cortivo. 1990. Attempted murder with pancuronium. *J. Forensic Sci.* 35(6):1468–1476.

Broich, J. R., D. B. Hoffman, S. Andrayauskas, L. Galante, and C. J. Umberger. 1971. An improved method for rapid, large-scale thin-layer chromatographic urine screening for drugs of abuse. *J. Chromatogr.* 60:95–101.

Buhler, D. R. 1964. The metabolism of chlorphenesin carbamate. *J. Pharmacol. Exptl. Therap.* 145:232–241.

Buhler, D. R. 1965. Characterization of the glucuronide conjugate of chlorphenesin carbamate from the rat and from man. *Biochem. Pharmacol.* 14:371–373.

Chawla, J. L., R. A. Sodhi, and R. T. Sane. 1996. Simultaneous determination of chlorzoxazone, paracetamol and diclofenac sodium by different chromatographic techniques. *Indian Drugs* 33(4):171–178.

Cirimele, V., M. Villaina, G. Pépin, B. Ludes, and P. Kintz. 2003. Screening procedure for eight quaternary nitrogen muscle relaxants in blood by high-performance liquid chromatography–electrospray ionization mass spectrometry. *J. Chromatogr. B* 789:107–113.

Clarke, C. and R. B. Raja. 1982. Improved thin-layer chromatographic system for evaluation of (+)-tubocurarine chloride and commercial curare. *J. Chromatogr.* 244(1):174–176.

Craig, C. R. and R. E. Stitzel. 1997. *Modern Pharmacology with Clinical Applications.* Philadelphia, PA: Lippincott Williams & Wilkins.

Daldrup, T., F. Susanto, and P. Michalke. 1981. Kombination von DC, GC (OV1 und OV17) und HPLC (RP18) zur schnellen erkennung von arzneimitteln, rauschmitteln und verwandten verbindungen. *Fresen J. Anal. Chem.* 308(5):413–427.

De Kruijf, N., M. A. H. Rijk, L. A. Pranoto-Soetardhi, and A. Schouten. 1987. Determination of preservatives in cosmetic products. I. Thin-layer chromatographic procedure for the identification of preservatives in cosmetic products. *J. Chromatogr.* 410(2):395–411.

Egli, R. A. and S. Keller. 1984. Comparison of silica gel and reversed-phase thin-layer chromatography and liquid chromatography in the testing of drugs. *J. Chromatogr.* 29:249–256.

Egli, R. A. and S. Tanner. 1979. Universelle, halogenfreie DC-Fließmittel für Arzneisubstanzen. *Fresen J. Anal. Chem.* 295(5):398–401.

El Bayoumi, A. A., S. M. Amer, N. M. Moustafa, and M. S. Tawakkol. 1999. Spectrodensitometric determination of clorazepate dipotassium, primidone and chlorzoxazone each in presence of its degradation product. *J. Pharmaceut. Biomed.* 20:727–735.

Fishbein, L. and W. L. Zielinski. 1967. Chromatography of carbamates. *Chromatog. Rev.* 9:37–101.

Flood, P. 2005. The importance of myorelaxants in anesthesia. *Curr. Opin. Pharm.* 5:322–327.

Gajdzinska, H. and J. Wedrychowski. 1980. Suicide by tubocurarine injection. *Archiwum Medycyny Sadowej i Kryminologii* 30(3):229–231.

Gandhimathi, M. and T. K. Ravi. 2008. Simultaneous densitometric analysis of tramadol hydrochloride and chlorzoxazone by high-performance thin-layer chromatography. *J. Planar Chromatogr. Modern TLC* 21(4):305–307.

Garcia, P. L., F. P. Gomes, M. I. Rocha, M. Santoro, and E. R. M. Kedor-Hackmann. 2008. Validation of an HPLC analytical method for determination of pancuronium bromide in pharmaceutical injections. *Analytical Letters* 41:1895–908.

Gazdag, M., G. Szepesi, K. Mihalyfi et al. 1992. Analysis of steroids. Part 45. Analytical investigation of pipecuronium bromide (Arduan). *Acta Pharm. Hung.* 62(3):88–96.

Gazdag, M., G. Szepesi, K. Varsanyi-Riedl, Z. Vfigh, and Zs. Pap-Sziklay. 1985. Chromatography of *bis*-quaternary amino steroids, i. separation on silica by thin-layer and high-performance liquid chromatography, *J. Chromatogr.* 328:219–287.

Giebelmann, R. 1981a. Ion-pair adsorption thin-layer chromatography of quaternary ammonium compounds. Part 3. Dependence of the retention on the carbon chain length of the measured ion. *Pharmazie* 36(9):649–650.

Giebelmann, R. 1981b. Ion pair adsorption thin-layer chromatography of quaternary ammonium ions. Part 5. Retention equation for silica gel. *Pharmazie* 36(11):786–787.

Giebelmann, R. 1984. Ion-pair-adsorption thin-layer chromatography of quaternary ammonium ions. Part 13: Simple approximation relation for Rf values. *Pharmazie* 39(7):471–473.

Giebelmann, R. 1986. Ion-pair adsorption thin-layer chromatography of quaternary ammonium ions. Part 17: 3-Parameter-system with silica gel. *Pharmazie* 41(2):106–107.

Harper, J. D., P. A. Martel, and C. M. O'Donnell. 1989. Evaluation of a multiple-variable thin-layer and reversed-phase thin-layer chromatographic scheme for the identification of basic and neutral drugs in an emergency toxicology setting. *J. Anal. Toxicol.* 13(1):31–36.

Hemström, P. and K. Irgum. 2006. Hydrophilic interaction chromatography. *J. Sep. Sci.* 29:1784–1821.

Herdeis, C. and H. P. Hubmann. 1992. Synthesis of homochiral R-baclofen from S-glutamic acid. *Tetrahedron–Asymmetr* 3(9):1213–1221.

Hubicka, U., J. Krzek, and B. Zuromska-Witek. 2012. TLC-densitometric determination of tolperisone and its impurities 4-methylpropiophenone and piperidine in pharmaceutical preparations. *J. Liq. Chromatogr. R. T.* Available online. DOI: 10.1080/10826076.2012.675862.

Kaale, E., P. Risha, and T. Layloff. 2011. TLC for pharmaceutical analysis in resource limited countries. *J. Chromatogr. A* 1218:2732–2736.

Kaul, N., S. R. Dhaneshwar, H. Agrawal, A. Kakad, and B. Patil. 2005. Application of HPLC and HPTLC for the simultaneous determination of tizanidine and rofecoxib in pharmaceutical dosage form. *J. Pharmaceut. Biomed.* 37(1):27–38.

Kersten, U. W., D. K. F. Meijer, and S. Agoston. 1973. Fluorimetric and chromatographic determination of pancuronium bromide and its metabolites in biological materials. *Clin. Chim. Acta* 44:59–66.

Kinget, R. and A. Michoel. 1976. Thin-layer chromatography of pancuronium bromide and its hydrolysis products. *J. Chromatogr.* 120:234–238.

Klys, M., J. Blalka, and B. Bujak-Gizycka. 2000. A case of suicide by intravenous injection of pancuronium. *Legal Medicine* 2(2):93–100.

Kovacs, P. and G. Takacsi Nagy. 1983. Kinetics of pipecuronium bromide hydrolysis. I. Separation of pipecuronium bromide and its hydrolysis products and their determination by thin-layer chromatography and densitometry. *Acta Pharm. Hung.* 53(2):56–63.

Krauss, D., H. Spahn, and E. Mutschler. 1988. Quantification of baclofen and its fluoro analog in plasma and urine after fluorescent derivatization with benoxaprofen chloride and thin-layer chromatographic separation. *Arzneimittel-Forschung* 38(10):1533–1536.

Krzek, J., M. Starek, and A. Kwiecien. 1998. Simultaneous determination of guaiamar carbamate and acetylsalicylic acid by a chromatographic-densitometric method. *Acta Pol. Pharm.* 55(6):429–434.

Kuroiwa M., N. Inotsume, R. Iwaoku, and M. Nakano. 1985. Reduction of dantrolene by enteric bacteria. *Yakugaku Zasshi* 105(8):770–774.

Lauko, A., P. Horvath, F. Trischler, and S. Gorog. 1992. Analysis of steroids. Part 44. Analytical investigation of the intermediates of the synthesis of pipecuronium bromide (Arduan). *Acta Pharm. Hung.* 62(3):82–87.

Lemke, T. L. and D. A. Williams. 2008. *Foye's Principles of Medicinal Chemistry.* Philadelphia, PA: Lippincott Williams & Wilkins.

Liawruangrath, S. and B. Liawruangrath. 1999. High performance thin-layer chromatographic determination of tolperisone hydrochloride. *J. Pharmaceut. Biomed.* 20(1–2):401–404.

Lillsunde, P. and T. Korte. 1991. Comprehensive drug screening in urine using solid-phase extraction and combined TLC and GC/MS identification. *J. Anal. Toxicol.* 15(2):71–81.

Mahadik, K. R., A. R. Paradkar, H. Agrawal, and N. Kaul. 2003. Stability-indicating HPTLC determination of tizanidine hydrochloride in bulk drug and pharmaceutical formulations. *J. Pharmaceut. Biomed.* 33(4):545–552.

Manwar, J. V., A. G. Nerkar, N. S. Bhajipale, and S. S. Laddha. 2007. Estimation of methocarbamol in tablets by HPTLC. *Int. J. Chem. Sci.* 5(1):333–339.

Moffat, A. C. 1986. *Clarke's Isolation and Identification of Drugs in Pharmaceuticals, Body Fluids, and Postmortem Material.* 2nd edn., pp. 752–753. London, U.K.: The Pharmaceutical Press.

Nishikawa, M. and H. Tsuchihashi. 2005. Muscle relaxants. In: *Drugs and Poisons in Humans. A Handbook of Practical Analysis*, eds. O. Suzuki, and K. Watanabe, pp. 359–367. New York: Springer-Verlag.

Owen, P., A. Pendlebury, and A. C. Moffat. 1978. Choice of thin-layer chromatographic systems for the routine screening for neutral drugs during toxicological analyses. *J. Chromatogr.* 161:187–93.

Patel M. J., A. N. Patel, C. N. Patel, and R. Badmanaban. 2012. A simple and sensitive HPTLC method for simultaneous analysis of tolperisone hydrochloride and etodolac in combined fixed-dose oral solid formulation. *J. Planar Chromatogr. Modern TLC* 25(1):85–88.

Pawar, U. D., A. V. Sulebhavikar, A. V. Naik, S. G. Pingale, and K. V. Mangaonkar. 2009. Simultaneous determination of rofecoxib and tizanidine by HPTLC. *E-J. Chem.* 6(1):295–302. http://www.e-journals.in/PDF/V6N1/295–302.pdf.

Pelander, A., D. Backstroem, and I. Ojanperae. 2007. Qualitative screening for basic drugs in autopsy liver samples by dual-plate overpressured layer chromatography. *J. Chromatogr. B* 857(2):337–340.

Radoičić, A., H. Majstorović, T. Sabo, Ž. Tešić, and D. Milojković-Opsenica. 2009a. Hydrophilic-interaction planar chromatography of some water-soluble Co(III) complexes on different adsorbents. *J. Planar Chromatogr.* 22(4):249–253.

Radoičić, A., J. Trifković, D. Milojković-Opsenica, Ž. Tešić, D. Vučović, and M. Aleksić. 2009b. Hydrophilic interaction planar chromatography of some anaesthetics. Paper presented at the *XXXIInd symposium, Chromatographic Methods of Investigating the Organic Compounds*, Katowice. Book of Abstracts, p. 61.

Rao, A. L. and D. P. Varma. 2009. Development and validation of a HPTLC method for the simultaneous estimation of rofecoxib and tizanidine hydrochloride in tablet dosage form. *Int. J. Chem. Sci.* 7(2):986–992.

Ravi, T. K., M. Gandhimathi, K. R. Sireesha, and S. Jacob. 2006. HPTLC method for the simultaneous estimation of tizanidine and rofecoxib in tablets. *Indian J. Pharm. Sci.* 68(2):234–236.

Roosewelt, C., N. Harikrishnan, P. Muthuprasanna, P. Shanmugapandiyan, and V. Gunasekaran. 2007. Validated high performance thin-layer chromatography method for simultaneous estimation of rofecoxib and tizanidine hydrochloride in pure and tablet dosage forms. *Asian J. Chem.* 19(6):4286–4290.

Salem, M. Y., N. N. El Din Salama, L. M. Abd El-Halim, and L. El-Sayed Abdel Fattah. 2010. Use of validated stability-indicating chromatographic methods for quantitative analysis of idrocilamide in a pharmaceutical formulation. *Acta Chromatogr.* 22(4):569–579.

Salvadori, M. C., M. E. Velletri, M. M. A. Camargo, and A. C. P. Araujo. 1988. Identification of doping agents by chromatographic techniques and UV spectrophotometry. *Analyst* 113(8):1189–1195.

Sane R. T. and M. Gadgil. 2002. Simultaneous determination of paracetamol, chlorzoxazone, and nimesulide by HPTLC. *J. Planar Chromatogr. Modern TLC* 15:76–78.

Schmahl, H. J. and E. Hieke. 1980. Separation and identification of some antimicrobials used also in cosmetic products by means of thin-layer chromatography. *Fresen. J. Anal. Chem.* 304(5):398–404.

Shanmugasundaram P., R. K. Raj, S. S. Phanindra, and M. V. Aanandhi. 2009. Validated HPTLC method for the simultaneous determination of diclofenac potassium and tizanidine hydrochloride in tablet dosage form. *Indian J. Chem. A* 8(2):176–178.

Shweshein, K. S. A. M., A. Radoičić, F. Andrić, Ž. Tešić, and D. Milojković-Opsenica. Hydrophilic interaction planar chromatography of geometrical isomers of some Co(III) complexes. *J. Liq. Chromatogr. Related Technol.* 35:1289–1297.

Sivasubramanian, L. D. and K. E. Bell. 2009. HPTLC for the simultaneous determination of tizanidine and valdecoxib in pharmaceutical dosage form. *J. Pharm. Res.* 2(2):189–195.

Sodhi, R. A., J. L. Chawla, and R. T. Sane. 1996. Simultaneous determination of paracetamol, ibuprofen and chlorzoxazone by HPLC, HPTLC and GC methods. *Indian Drugs* 35:280–285.

Stewart, J. T. and C.A. Janicki. 1987. Chlorzoxazone. *Anal. Prof. Drug Sub. Excip.* 16:119–144.

Sunshine, I. 1963. Use of thin-layer chromatography in the diagnosis of poisoning. *Am. J. Clin. Pathol.* 40:576–582.

Tanaka, K., M. Hioki, and H. Shindo. 1974. Determination of pancuronium bromide and its metabolites in human urine by dye-extraction method. Relation between the extractability and structure of quaternary ammonium ions. *Chem. Pharm. Bull.* 22:2599–2606.

Thomas, J. J. and L. Dryon. 1964. Identification of psychotropic drugs by thin-layer chromatography. *J. Pharm. Belg.* 19:481–504.

Toner, C. C. and P. J. Flynn. 1994. New neuromuscular blocking drugs. *Baillibre's Clinical Anaesthesiology* 8(2):441–460.

Ullah, I., D. E. Cadwallader, and I. L. Honigberg. 1970. Determination of degradation kinetics of chlorzoxazone by thin-layer chromatography. *J. Chromatogr.* 46:211–216.

Wedig, M., N. Novatchev, T. Worch, S. Laug, and U. Holzgrabe. 2002. Evaluation of the impurity profile of alcuronium by means of capillary electrophoresis. J Pharm. Biomed. 28:983–990.

Woods, W. E., S. Chay, T. Houston, J. W. Blake, and T. Tobin. 1985. Effects of phenylbutazone and oxyphenbutazone on basic drug detection in high performance thin-layer chromatographic systems. *J. Vet. Pharmacol. Ther.* 8(2):181–189.

Zecevic, M., Lj. Zivanovic, and A. Stojkovic. 2002. Validation of a high- performance liquid chromatography method for the determination of pancuronium in Pavulon injections. *J. Chromatogr. A* 949:61–64.

Zerilli, A., D. Lucas, F. Berthou, L. G. Bardou, and J.-F. Menez. 1996. Determination of cytochrome P450 2E1 activity in microsomes by thin-layer chromatography using [2-^{14}C]chlorzoxazone. *J. Chromatogr. B* 677:156–160.

33 TLC of Antiulcers

Tadeusz Inglot

CONTENT

The primary class of drugs used for gastric acid suppression are the proton pump inhibitors (PPIs)—omeprazole, lansoprazole, pantoprazole, and rabeprazole. The H2-receptor blocking agents—cimetidine, famotidine, nizatidine, and ranitidine—have been used for this purpose, but are now more widely used for maintenance therapy after treatment with the PPIs.

The PPIs block the secretion of gastric acid by gastric parietal cells. The extent of inhibition of acid secretion is dose related. In some cases, gastric acid secretion is completely blocked for over 24 h on a single dose. In addition to their role in treatment of gastric ulcers, the PPIs are used to treat syndromes of excessive acid secretion (Zollinger–Ellison syndrome) and gastroesophageal reflux disease.

Sucralfate (carafate), a substituted sugar molecule with no nutritional value, does not inhibit gastric acid, but rather, reacts with existing stomach acid to form a thick coating that covers the surface of an ulcer, protecting the open area from further damage. A secondary effect is to act as an inhibitor of the digestive enzyme pepsin. Sucralfate does not bind to normal stomach lining. The drug has been used for prevention of stress ulcers, the type seen in patients exposed to physical stress such as burns and surgery. It has no systemic effects.

Next group of antiulcer drugs are M1 receptor antagonists—pirenzepine and telenzepine. Muscarinic blockers inhibit gastric acid secretion at doses that have little effect on heart rate. Although they can reduce gastric motility and the secretion of gastric acid, antisecretory doses produce pronounced side effects, such as dry mouth, loss of visual accommodation, photophobia, and difficulty in urination. As a consequence, patient compliance in the long-term management of symptoms of acid peptic disease with these drugs is poor (Brunton et al. 2005).

In the literature on thin-layer chromatography (TLC) methods used to determine antiulcer drugs, there are papers concerning only the PPI group. All proposed methods for determination of PPI are based on the use of normal phase with silica gel as a stationary phase and densitometric detection. The vast majority of papers describe the methods of analysis of dosage forms or pure substances. Only one paper deals with determination of PPI in biological material. Basic parameters of the methods are shown in Table 33.1, for simple substances, and in Table 33.2, for PPI in combination with other drugs.

Pandya et al. (1997) developed a method for determination of lansoprazole in human plasma and investigated the pharmacokinetics of the drug. Calibration curve was prepared for the range of 10–250 ng/mL. Limit of quantification (LOQ) was determined at 20 ng/mL. The method is validated for accuracy, precision, and linearity. Kinetic studies revealed the highest concentrations of lansoprazole in the blood 2–3 h after administration. It was found that this method can be applied successfully to control patients receiving a daily dose of 30 mg lansoprazole. The method was also tested in terms of translatability to the high-performance liquid chromatography (HPLC) method.

Hegazy et al. (2011) proposed a method for determination of pantoprazole in combination with mosapride in a pharmaceutical form. They conducted an analysis of drug stability after acid hydrolysis. In addition, they proposed an HPLC method. The method allowed satisfactory TLC separation

TABLE 33.1

Basic Parameters of the Chromatographic Systems for TLC Methods on Single Substances

Substances	Form/Method	Mobile Phase	Detection (nm)	Literature
Lansoprazole	Human plasma	Chloroform–methanol (15:1, v/v)	286	Pandya et al. (1997)
Lansoprazole	Stability indicating	Chloroform–methanol–n-hexane (75:25:60 v/v/v)	285	El-Sherif et al. (2005)
Omeprazole	Tablets	Methanol–water (2:1, v/v)	302	Ray and De (1994)
Omeprazole	Stability indicating	Chloroform–methanol (9 + 1, v/v)	302	Jha et al. (2010)
Pantoprazole sodium	Tablets	Methanol–water–ammonium acetate (4 + 1 + 0,5 v/v)	290	Gosavi et al. (2006)
Rabeprazole	Degradation	Acetone–toluene–methanol (9:9:0.6 v/v)	284	El-Gindy et al. (2003)
Rabeprazole sodium	Degradation	Isopropyl alcohol–30% ammonia (80 + 2, v/v)	284	Osman and Osman (2009)
Tenatoprazole	Stability indicating	Toluene–ethyl acetate–methanol (6 + 4 + 1, v/v/v)	306	Dhaneshwar et al. (2009)

of pantoprazole and mosapride and four degradation products of these substances. The performed validation showed good linearity, accuracy, and precision of the method. The limit of detection (LOD) for pantoprazole was 0.11 µg/mL and the LOQ was 0.34 µg/mL.

El-Sherif et al. (2005) analyzed the stability of lansoprazole by TLC, derivative spectrophotometry, and fluorimetry. Developed TLC method allows the separation of lansoprazole and six degradation products. Authors recorded a total destruction of lansoprazole under the influence of 0.1 N HCl after 120 min. The method is optimized for accuracy, precision, and repeatability.

Jha et al. (2010) developed a method for determination of omeprazole in pharmaceutical formulation and analyzed the stability of this substance. The method was validated according to ICH guidelines in terms of accuracy, precision, linearity, specificity, and robustness. Degradation of omeprazole was carried out under the action of acid, bases, 30% H_2O_2, temperature, photochemical, and ultraviolet (UV) factors. The method allowed the separation of eight degradation products of omeprazole.

El-Gindy et al. (2003) proposed three methods for determination of rabeprazole—high-performance thin layer chromatography (HPTLC), HPLC, and UV–VIS. The work also included the study of stability of rabeprazole under the influence of acidic, oxidative, photolytic factors, and kinetics of decomposition of rabeprazole influenced by these factors. Developed HPTLC method allowed the complete separation of rabeprazole and two of its degradation products. The method was validated according to ICH guidelines.

Osman and Osman (2009) developed a TLC, HPLC, and spectrofluorimetric methods for the determination of rabeprazole in pharmaceutical preparations. Stability tests of rabeprazole were also performed under the influence of acid and oxidizing agents. The proposed TLC method allows the observation of nine degradation products of rabeprazole after using the acidic factor. Two major degradation products were subjected to IR and MS analysis. The decomposition pathways of rabeprazole have also been proposed.

Dhaneshwar et al. (2009) described a method capable of determining tenatoprazole in the pharmaceutical formulation and stability studies. Validation of methods confirmed the linearity, precision, specificity, accuracy, and robustness. The LOD and LOQ were 50 ng/mL and 100 ng/mL, respectively. Stability analysis was conducted examining the effect of acid and alkali hydrolysis, oxidation, and photodegradation on stability of tenatoprazole. The method allows separation of four tenatoprazole degradation products.

TABLE 33.2
Basic Parameters of the Chromatographic Systems for TLC Methods on PPI in Conjunction with Other Drugs

Substances	Form/Method	Mobile Phase	Detection (nm)	Literature
Omeprazole cisapride	Capsules	Propanol–toluene (1:3 v/v)	276	Godse et al. (2009)
Omeprazole domperidone	Capsules	Ethyl acetate–methanol–benzene (40:20:40 v/v)	295	Patel et al. (2007a)
Omeprazole rabeprazole ondansetron	Tablets	Dichloromethane–methanol (9:1 v/v)	309	Raval et al. (2008)
Omeprazole pantoprazole impurities	Tablets	Chloroform–2-propanol–25% ammonia–acetonitrile (10.8 + 1.2 + 0.3 + 4 v/v)	298	Agbaba et al. (2004)
Esomeprazole domperidone	Tablets	Chloroform–acetonitrile–ammonia (5:10:0.25 v/v)	222	Roosewelt et al. (2007)
Pantoprazole cinitapride	Capsules	Ethyl acetate–methanol (9:1 v/v)	278	Patel et al. (2011)
Pantoprazole mosapride	Stability indicating	Ethyl acetate–methanol–toluene (4:1:2 v/v)	276	Hegazy et al. (2011)
Pantoprazole domperidone	Capsules	Acetone–toluene–methanol–glacial acetic acid (2:8:2:0.1 v/v)	298	Kakde et al. (2006)
Pantoprazole sodium domperidone	Tablets	Methanol–water–ammonium acetate (4 + 1 + 0.5 v/v)	286	Gosavi et al. 2006)
Pantoprazole domperidone	Capsules	Ethyl acetate–methanol (60 + 40 v/v)	287	Patel et al. (2007b)
Rabeprazole itopride hydrochloride	Capsules	n-butanol–toluene–ammonia (8.5:0.5:1 v/v)	288	Suganthi et al. (2008)
Rabeprazole sodium itopride hydrochloride	Capsules	Toluene–chloroform–methanol–25% ammonia (25:30:10:1 v/v)	225	Mageswari et al. (2007)
Rabeprazole sodium domperidone	Capsules	Ethyle acetate–toluene–methanol (4.8:3.7:1.5 v/v)	284	Patel et al. (2007c)
Rabeprazole domperidone	Capsules	Ethyl acetate–methanol–benzene–acetonitrile (30:20:30:20 v/v)	287	Patel et al. (2008b)
Rabeprazole paracetamol aceclofenac	Tablets	Ethyl acetate–methanol–glacial acetic acid (9:1:0.1 v/v)	275	Mallikarjuna Rao and Gowri (2011)
Rabeprazole mosapride	Tablets	Ethyl acetate–methanol–benzene (2:0.5:2.5, v/v)	276	Patel et al. (2008a)
Rabeprazole sodium aceclofenac	Capsules	Toluene–ethyl acetate–methanol–acetic acid (6:4:1:0.2 v/v)	279	Bharekar et al. (2011)
Dexrabeprazole domperidone	Stability indicating	Acetone–toluene–methanol (4.5:4.5:0,5 v/v)	285	Chitlange et al. (2010)

Agbaba et al. (2004) developed the method for separation of omeprazole, pantoprazole, and their main impurities omeprazole sulfone and N-methylpantoprazole. Developed conditions allow for good separation of drugs and their impurities. The method was validated and can be used for routine quality control of the pharmaceutical preparations.

There are two papers in the literature on antiulcer drugs describing the method of determination of omeprazole (Ray and De 1994) and pantoprazole sodium (Gosavi et al. 2006a) in pharmaceutical preparations. The proposed methods are simple, cheap, and fast and can be used for routine control of the active substance in pharmaceutical preparations.

A large part of papers on the determination of drugs from the group of PPI applies to the determination of multicomponent preparations. Omeprazole was determined in conjunction with cisapride (Godse et al. 2009), domperidone (Patel et al. 2007), rabeprazole, and ondansetron (Raval et al. 2008); esomeprazol with domperidone (Roosewelt et al. 2007); pantoprazole with cinitapride (Patel et al. 2011) and domperidone (Gosavi et al. 2006b, Kakde et al. 2006, Patel et al. 2007); rabeprazole in combination with itopride hydrochloride (Suganthi et al. 2008, Mageswari et al. 2007), domperidone (Patel et al. 2007, 2008), mosapride (Patel et al. 2008), aceclofenac (Bharekar et al. 2011), and aceclofenac and paracetamol (Mallikarjuna Rao and Gowri 2011); and dexrabeprazole with domperidone (Chitlange et al. 2010). Described methods are characterized by simplicity, speed, and low cost implementation. They have been validated and can be used for routine determination of these substances in complex preparations.

REFERENCES

Agbaba, D., Novovic, D., Karljiković-Rajić, K. et al. 2004. Densitometric determination of omeprazole, pantoprazole, and their impurities in pharmaceuticals. *Journal of Planar Chromatography—Modern TLC* 17(3): 169–172.

Bharekar, V.V., Mulla, T.S., Rajput, M.P. et al. 2011. Validated HPTLC method for simultaneous estimation of rabeprazole sodium, paracetamol and aceclofenac in bulk drug and formulation. *Der Pharma Chemica* 3(4): 171–179.

Brunton, L., Lazo J., Parker K. 2005. *Goodman & Gilman's The Pharmacological Basis of Therapeutics* (11 ed.). McGraw-Hill, New York.

Chitlange, S.S., Mulla, A.I., Pawbake, G.R. et al. 2010. Stability-indicating TLC-densitometric method for estimation of dexrabeprazole and domperidone in pharmaceutical dosage form. *Preparative Biochemistry and Biotechnology* 40(4): 337–346.

Dhaneshwar, S.R., Bhusari, V., Mahadik, M.V. et al. 2009. Application of a stability-indicating thin-layer chromatographic method to the determination of tenatoprazole in pharmaceutical dosage forms. *Journal of AOAC International* 92(2): 387–393.

El-Gindy, A., El-Yazby, F., Maher, M.M. 2003. Spectrophotometric and chromatographic determination of rabeprazole in presence of its degradation products. *Journal of Pharmaceutical and Biomedical Analysis* 31(2): 229–242.

El-Sherif, Z.A., Mohamed, A.O., El-Bardeicy, M.G. et al. 2005. Stability-indicating methods for the determination of lansoprazole. *Spectroscopy Letters* 38(1): 77–93.

Godse, V.P., Bhosale, A.V., Gowekar, N.M. et al. 2009. Simultaneous HPTLC estimation of omeprazole and cisapride in pharmaceutical dosage form. *Asian Journal of Chemistry* 21(2): 1239–1243.

Gosavi, S.A., Shirkhedkar, A.A., Jaiswal, Y.S. et al. 2006a. A simple and sensitive HPTLC method for quantitative analysis of pantoprazole sodium sesquihydrate in tablets. *Journal of Planar Chromatography—Modern TLC* 19(109): 228–232.

Gosavi, S.A., Shirkhedkar, A.A., Jaiswal, Y.S. et al. 2006b. Quantitative planar chromatographic analysis of pantoprazole sodium sesquihydrate and domperidone in tablets. *Journal of Planar Chromatography—Modern TLC* 19(110): 302–306.

Hegazy, M.A., Yehia, A.M., Mostafa, A.A. 2011. Stability-indicating chromatographic methods for simultaneous determination of mosapride and pantoprazole in pharmaceutical dosage form and plasma samples. *Chromatographia* 74(11–12): 839–845.

Jha, P., Parveen, R., Khan, S.A. et al. 2010. Stability-indicating high-performance thin-layer chromatographic method for quantitative determination of omeprazole in capsule dosage form. *Journal of AOAC International* 93(3): 787–791.

Kakde, R.B., Gedam, S.N., Kasture, A.V. 2006. Spectrophotometric and HPTLC method for simultaneous estimation of pantoprazole and domperidone in their pharmaceutical preparations. *Asian Journal of Chemistry* 18(2): 1347–1351.

Mageswari, S.D., Surendra, K., Maheswari, R. et al. 2007. HPTLC method for simultaneous estimation of rabeprazole sodium and itopride hydrochloride in capsule and bulk drug. *Asian Journal of Chemistry* 19(7): 5634–5638.

Mallikarjuna Rao, N., Gowri, S.D. 2011. Development and validation of stability indicating HPTLC method for simultaneous estimation of paracetamol, aceclofenac and rabeprazole in combined tablet dosage formulation. *International Journal of PharmTech Research* 3(2): 909–918.

Osman, A., Osman, M. 2009. Spectrofluorometry, thin layer chromatography, and column high-performance liquid chromatography determination of rabeprazole sodium in the presence of its acidic and oxidized degradation products. *Journal of AOAC International* 92(5): 1373–1381.

Pandya, K.K., Mody, V.D., Satia, M.C. et al. 1997. High-performance thin-layer chromatographic method for the detection and determination of lansoprazole in human plasma and its use in pharmacokinetic studies. *Journal of Chromatography B: Biomedical Applications* 693(1): 199–204.

Patel, B., Patel, M., Patel, J. et al. 2007a. Simultaneous determination of omeprazole and domperidone in capsules by RP-HPLC and densitometric HPTLC. *Journal of Liquid Chromatography and Related Technologies* 30(12): 1749–1762.

Patel, B.H., Suhagia, B.N., Patel, M.M. et al. 2008a. High-performance liquid chromatography and thin-layer chromatography for the simultaneous quantitation of rabeprazole and mosapride in pharmaceutical products. *Journal of Chromatographic Science* 46(1): 10–14.

Patel, B.H., Suhagia, B.N., Patel, M.M. et al. 2008b. HPTLC determination of rabeprazole and domperidone in capsules and its validation. *Journal of Chromatographic Science* 46(4): 304–307.

Patel, B.H., Suhagia, B.N., Patel, M.M. et al. 2007b. Simultaneous estimation of pantoprazole and domperidone in pure powder and a pharmaceutical formulation by high-perfomance liquid chromatography and high-performance thin-layer chromatography methods. *Journal of AOAC International* 90(1): 142–146.

Patel, G.H., Prajapati, S.T., Patel, C.N. 2011. HPTLC method development and validation for simultaneous determination of cinitapride and pantoprazole in capsule dosage form. *Journal of Pharmacy and Technology* 4(9): 1428–1431.

Patel, R.D., Bhatt, K.K., Nakarani, N.V. et al. 2007c. Simultaneous HPTLC estimation of rabeprazole sodium and domperidone in capsule dosage form. *Indian Drugs* 44(5): 337–341.

Raval, P., Puranik, M., Wadher, S. et al. 2008. A validated HPTLC method for determination of ondansetron in combination with omeprazole or rabeprazole in solid dosage form. *Indian Journal of Pharmaceutical Sciences* 70(3): 386–390.

Ray, S., De, P.K. 1994. HPTLC and TLC method for rapid quantification and identification of omeprazole. *Indian Drugs* 31(11): 543–547.

Roosewelt, C., Magesh, A.R., Sheeja Rekha, A.C. et al. 2007. Simultaneous estimation and validation of esomeprazole and domperidone by HPTLC in pure and pharmaceutical dosage forms. *Asian Journal of Chemistry* 19(4): 2955–2960.

Suganthi, A., John, S., Ravi, T. 2008. Simultaneous HPTLC determination of rabeprazole and itopride hydrochloride from their combined dosage form. *Indian Journal of Pharmaceutical Sciences* 70(3): 366–368.

34 TLC of Antiemetic Drugs

Łukasz Komsta

CONTENTS

Antiemetics are drugs acting against vomiting and nausea. They are typically used to treat motion sickness. They are also applied to suppress side effects of analgesics, general anesthetics, and anticancer agents.

Antiemetics can be divided into three general groups:

- Antagonists of H1 receptor (oldest group), such as diphenhydramine, cyclizine, or meclozine. They are described in Chapter 40.
- Antagonists of D2 receptor, such as thioproperazine, thiethylperazine, metopimazine, metoclopramide, cisapride, domperidone, and alizapride.
- Antagonists of 5HT3 receptor, such as ondansetron, tropisetron, granisetron, and dolasetron.
- Antagonists of NK1 receptor, such as nabilone and aprepitant.

Although the interest of thin-layer chromatography (TLC) analysis of this group of drugs started in the 1970s, many of these drugs are still not described in context of TLC analysis. This chapter will present the current state of research in this field.

34.1 ANALYSIS OF METOCLOPRAMIDE

Huizing et al. (1979) analyzed quantitatively metoclopramide and analogue clebopride with their metabolic products. They also tested extraction from biological tissues and fluids. Various phases were used on silica gel chromatographic plates. Visualization was carried out by diazotisation, followed by coupling with N-(1-naphtyl)ethylenediammonium dichloride (on the plate). The method was validated against variations of various experimental conditions and it has proven to be satisfactory for routine purpose.

Musumarra et al. (1983) reported comprehensive TLC study of 54 drugs in eight eluent mixtures and carried out chemometric principal component analysis of these results. Although the main aim of the study was clustering of eluent mixtures (the analysis showed that they cluster into three groups), there are metoclopramide among analyzed substances. This work is one of first chemometric insights in TLC, as it proved the abilities to restrict the range of searched substances to a few candidates and to help in identification of the drug.

Biagi et al. (1996) performed a comprehensive study of lipophilicity of serotonergic antagonists, including metoclopramide. They used acetone, acetonitrile, and methanol as modifiers on RP18 plates, obtaining good correlation between reference lipophilicities and chromatographic results.

Another study focused on commonly prescribed drugs with presence of pi-electrons (Onah 1999). Author analyzed 11 such substances, including metoclopramide. The substances were visualized by color reaction with chloranil, and the stability of obtained colors was at least 5 min.

A recent study by Żydek and Brzezińska (2012) presents stepwise discriminant analysis of NP-TLC data of 33 drugs, including metoclopramide and tropisetron. The NP-TLC plates were impregnated with a solution of aspartic acid to investigate molecular interaction models and developed using three modifiers. The drugs were then successfully divided by discriminant analysis into three groups of activity (three receptors as targets) based on retention data in these systems.

34.2 ANALYSIS OF DOMPERIDONE

A high-performance TLC method for the simultaneous determination of cinnarizine and domperidone maleate was reported by Argekar and Powar (1999). They used silica gel plates and performed analysis with mobile phase methanol: dichloromethane: formic acid, 1:9:0.05 (v/v), detecting the analyte at 276 nm. It is worthy to mention low RSD (1.1% and 1.02%) achieved by this method.

Bagade et al. (2005) reported a method for simultaneous estimation of cinnarizine and domperidone. They used silica gel and mobile phase methanol:toluene:ethyl acetate:glacial acetic acid (2:9:0.5:0.5). Detection wavelength was 216 nm. Substances were showing Rf values: 0.61 (cinnarizine) and 0.16 (domperidone). The calibration curve response was investigated in ranges 5–14 µg (cinnarizine) and 4–11 µg (domperidone) per spot.

A very similar method was described parallelly by Vinodhini et al. (2005). They used the same plates with different mobile phase: toluene:ethyl acetate:methanol (70:5:25). Detection was carried out in different ultraviolet (UV) range—271 nm in absorbance mode. The Rf values were then 0.85 (cinnarizine) and 0.4 (domperidone).

Gosavi et al. (2006) presented simultaneous determination of pantoprazole sodium and domperidone in tablets. They used silica gel plates previously washed with methanol with mobile phase methanol–water–ammonium acetate (4:1:0.5). Scanning was performed at 286 nm. Linearity was observed in ranges 2–10 µg per spot (pantoprazole) and 0.5–2.5 µg per spot (domperidone), and the whole method was validated in accordance with ICH guidelines.

A similar method was also reported by Patel et al. (2007) with comparison with high-performance liquid chromatography (HPLC) method. They used for separation silica gel and mobile phase ethyl acetate–methanol (60:40). Substances were detected at 287 nm.

Domperidone was also assayed together with lansoprazole (Susheel et al. 2007). Authors elaborated the method that allows the determination of 100–500 ng/spot of lansoprazole and 100–500 ng/spot of domperidone. They used mobile phase n-butanol:glacial acetic acid:water (9.3:0.25:0.5), with densitometric detection at 288 nm. In the used system, Rf values of lansoprazole and domperidone were 0.78 and 0.21, respectively.

The team of Patel et al. (2008) described next assay of domperidone in combination with pantoprazole. They also used silica gel and ethyl acetate–methanol–benzene–acetonitrile (30:20:30:20 v/v) as mobile phase. Quantitation was performed with UV detection at 287 nm, and linearity was examined over a concentration range 400–1200 ng/spot and 600–1800 ng/spot, respectively.

Another method (Chitlange et al. 2010) presents dexrabeprazole (pure chiral form) together with domperidone. The mobile phase was acetone:toluene:methanol (4.5:4.5:0.5) on silica with detection at 285 nm. Authors obtained Rf values 0.49 and 0.24, respectively. The linearity range was 50–350 ng/spot and 100–700 mg/spot, respectively. Famotidine was also assayed together with domperidone in a two-component tablet dosage form by Deshpande et al. (2010). The authors used wavelengths equal to 267 nm and 285 nm to get linearity in the range of 10–60 mg/mL. These drugs can also be assayed together by another method given by Pawar et al. (2010).

Domperidone was also determined together with paracetamol (Yadav et al. 2009). Authors used silica gel plates with acetone:toluene:methanol (4:4:2) as mobile phase, on 8 cm distance. Two wavelengths were used for assay: 285 and 248 nm for domperidone and paracetamol, respectively.

The authors performed a comprehensive study on the influence of band size, chamber saturation time, and so on to obtain optimal conditions. The drugs were satisfactorily resolved with Rf values 0.52 and 0.74, respectively.

Sathiyanarayanan et al. (2011) described analysis of famotidine together with domperidone. Separation was performed on silica gel using mobile phase ethyl acetate:methanol:water (8.0:1.5:0.3). Plates were scanned at 288 nm. The Rf values were found to be 0.42 and 0.67, respectively. The percentage average recovery was found to be 98.88% and 98.26% for famotidine and domperidone, respectively.

34.3 ANALYSIS OF ONDANSETRON AND ANALOGUES

Raval et al. (2008) described ondansetron determination, in dosage combination with omeprazole and rabeprazole. The stationary phase used by them was also silica gel, and mixture of dichloromethane:methanol (9:1) was applied as a mobile phase. Detection of spots was carried out at 309 nm (ondansetron with omeprazole) and 294 nm (ondansetron with rabeprazole). The Rf values were found to be 0.42 for ondansetron, 0.54 for rabeprazole, and 0.51 for rabeprazole. The linearity was observed in range 0.1–0.5 µg per spot for all three drugs.

Another method for simultaneous ondansetron and rabeprazole determination was then described by Gandhi et al. (2009). The authors used mobile phase toluene:acetone:methanol (4.5:4.5:0.5) with UV detection at 285 nm. Rf values for both substances were 0.53 and 0.32, respectively. Linearity was obeyed in the concentration range of 50–800 ng per spot.

Tropisetron, together with its acidic degradation products, was assayed in tablets by Abdel-Fattah et al. (2010). Authors compared TLC method with HPLC and derivative spectrometry. The degradation products were also identified by IR, NMR, and MS techniques. TLC method described in this article is based on densitometric measurement at 285 nm, on silica gel using methanol–glacial acetic acid (22:3) mobile phase.

A method for granisetron determination was described by Lakshmana Prabu et al. (2010). The chromatographic separation was performed on silica gel as the stationary phase with chloroform:methanol (80:20) as mobile phase. Rf value was 0.45. Densitometric analysis was carried out at 301 nm. The range 400–1600 ng/spot showed satisfactory linearity.

Tropisetron is also described by Żydek and Brzezińska (2012) (see metoclopramide section).

34.4 OTHER DRUGS

Westgate and Sherma (2001) described a quantitative method for determination of meclizine hydrochloride in tablets. They used silica gel plates and UV densitometry. They achieved satisfactory recoveries (97.0%–110%) and precision (RSD 1.58% and 1.26%). The error of a standard addition analysis was also estimated and found to be 0.506%.

Metopimazine was used as internal standard during determination of sulpiride and mebeverine hydrochloride in presence of their reported impurities and hydrolytic degradants (Naguib and Abdelkawy 2010).

REFERENCES

Abdel-Fattah, L.S., El-Sherif, Z.A., Kilani, K.M., and El-Haddad, D.A. 2010. HPLC, TLC, and first-derivative spectrophotometry stability-indicating methods for the determination of tropisetron in the presence of its acid degradates. *Journal of AOAC International* 93(4): 1180–1191.

Argekar, A.P. and Powar, S.G. 1999. Simultaneous HPTLC determination of cinnarizine and domperidone maleate in formulations. *Journal of Planar Chromatography—Modern TLC* 12(4): 272–274.

Bagade, S.B., Walode, S.G., Charde, M.S., Tajne, M.R., and Kasture, A.V. 2005. Simultaneous HPTLC estimation of cinnarizine and domperidone in their combined dose tablet. *Asian Journal of Chemistry* 17(2): 1116–1126.

Biagi, G.L., Barbaro, A.M., Sapone, A., Borea, P.A., Varani, K., and Recanatini, M. 1996. Study of lipophilic character of serotonergic ligands. *Journal of Chromatography A* 723(1): 135–143.

Chitlange, S.S., Mulla, A.I., Pawbake, G.R., and Wankhede, S.B. 2010. Stability-indicating TLC-densitometric method for estimation of dexrabeprazole and domperidone in pharmaceutical dosage form. *Preparative Biochemistry and Biotechnology* 40(4): 337–346.

Deshpande, P., Gandhi, S., Bhavnani, V., Bandewar, R., Dhiware, A., and Diwale, V. 2010. High performance thin layer chromatographic determination of famotidine and domperidone in combined tablet dosage form. *Research Journal of Pharmaceutical, Biological and Chemical Sciences* 1(4): 354–359.

Gandhi, S.V., Khan, S.I., Jadhav, R.T., Jadhav, S.S., and Jadhav, G.A. 2009. High-performance thin-layer chromatographic determination of rabeprazole sodium and domperidone in combined dosage form. *Journal of AOAC International* 92(4): 1064–1067.

Gosavi, S.A., Shirkhedkar, A.A., Jaiswal, Y.S., and Surana, S.J. 2006. Quantitative planar chromatographic analysis of pantoprazole sodium sesquihydrate and domperidone in tablets. *Journal of Planar Chromatography—Modern TLC* 19(110): 302–306.

Huizing, G., Beckett, A.H., and Segura, J. 1979. Rapid thin-layer chromatographic photodensitometric method for the determination of metoclopramide and clebopride in the presence of some of their metabolic products. *Journal of Chromatography* 172: 227–237.

Lakshmana Prabu, S., Selvamani, P., and Latha, S. 2010. HPTLC method for quantitative determination of granisetron hydrochloride in bulk drug and in tablets. *Latin American Journal of Pharmacy* 29(8): 1455–1458.

Musumarra, G., Scarlata, G., Romano, G., and Clementi, S. 1983. Identification of drugs by principal components analysis of R(f) data obtained by TLC in different eluent systems. *Journal of Analytical Toxicology* 7(6): 286–292.

Naguib, I.A. and Abdelkawy, M. 2010. Development and validation of stability indicating HPLC and HPTLC methods for determination of sulpiride and mebeverine hydrochloride in combination. *European Journal of Medicinal Chemistry* 45(9): 3719–3725.

Onah, J.O. 1999. Thin-layer chromatographic detection of some common drugs by Ď€-acceptor complexation. *Acta Pharmaceutica* 49(3): 217–220.

Patel, B.H., Suhagia, B.N., Patel, M.M., and Patel, J.R. 2007. Simultaneous estimation of pantoprazole and domperidone in pure powder and a pharmaceutical formulation by high-perfomance liquid chromatography and high-performance thin-layer chromatography methods. *Journal of AOAC International* 90(1): 142–146.

Patel, B.H., Suhagia, B.N., Patel, M.M., and Patel, J.R. 2008. HPTLC determination of rabeprazole and domperidone in capsules and its validation. *Journal of Chromatographic Science* 46(4): 304–307.

Pawar, S.M., Patil, B.S., and Patil, R.Y. 2010. Validated HPTLC method for simultaneous quantitation of famotidine and domperidone in bulk drug and formulation. *International Journal of Advances in Pharmaceutical Sciences* 1(1): 54–59.

Raval, P., Puranik, M., Wadher, S., and Yeole, P. 2008. A validated HPTLC method for determination of ondansetron in combination with omeprazole or rabeprazole in solid dosage form. *Indian Journal of Pharmaceutical Sciences* 70(3): 386–390.

Sathiyanarayanan, L., Kulkarni, P.V., Nikam, A.R., and Mahadik, K.R. 2011. Rapid densitometric method for simultaneous analysis of famotidine and domperidone in commercial formulations using HPTLC. *Der Pharma Chemica* 3(1): 134–143.

Susheel, J., Lekha, M., and Ravi, T. 2007. High performance thin layer chromatographic estimation of lansoprazole and domperidone in tablets. *Indian Journal of Pharmaceutical Sciences* 69(5): 684–686.

Vinodhini, C., Kalidoss, A.S., and Vaidhyalingam, V. 2005. Simultaneous estimation of cinnarizine and domperidone by high performance thin layer chromatography in tablets. *Indian Drugs* 42(9): 600–603.

Westgate, E. and Sherma, J. 2001. Analysis of the active ingredient, meclizine, in motion sickness tablets by high performance thin layer chromatography with densitometric measurement of fluorescence quenching. *Journal of Liquid Chromatography and Related Technologies* 24(18): 2873–2878.

Yadav, A., Singh, R., Mathur, S., Saini, P., and Singh, G. 2009. A simple and sensitive HPTLC method for simultaneous analysis of domperidone and paracetamol in tablet dosage forms. *Journal of Planar Chromatography—Modern TLC* 22(6): 421–424.

Żydek, G. and Brzezinska, E. 2012. NP TLC data in structure-activity relationship study of selected compounds with activity on dopaminergic, serotoninergic, and muscarinic receptors. *Journal of Liquid Chromatography and Related Technologies* 35(6): 834–853.

35 TLC of Steroids and Analogs

Rada M. Baošić

CONTENTS

35.1   INTRODUCTION

Steroid drugs and their (semi-) synthetic analogues are among the most important groups in drug therapy. Steroid compounds possess the skeleton of cyclopenta[*a*]perhydrophenanthrene or a skeleton derived from one or more bond scissions or ring expansions or contractions. Methyl groups are normally present at C_{10} and C_{13} (angular methyl groups). An alkyl side chain may also be present at C_{17}. The steroid rings are lettered A, B, C, and D, and the 17 ring carbons are numbered as shown in Figure 35.1. The two angular methyl groups are numbered C_{18} and C_{19}. Steroids consist of an essentially lipophilic (or hydrophobic, nonpolar) cyclopentanoperhydrophenanthrene nucleus modified on the periphery of the nucleus or on the side chain by the addition of hydrophilic groups. In addition to steroids, which are widely distributed in nature, many thousands of them have been synthesized in the pharmaceutical and chemical laboratories. Despite the similarities in chemical structures and stereochemistry, each class of steroids demonstrates unique and distinctively different biological activities. Steroids (e.g., androgens, antiandrogens, estrogens, progestagens, anabolics, corticosteroids, 5α-reductase inhibitors, and aromatase inhibitors) have major role as hormones, controlling metabolism, salt balance, and

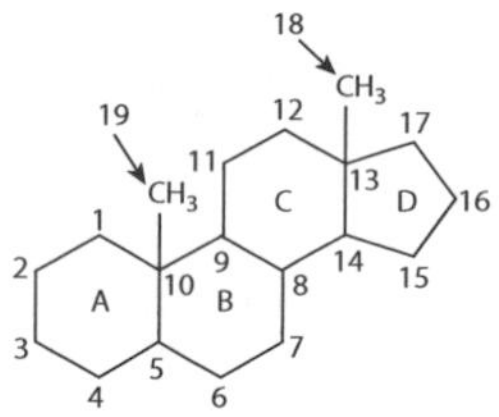

FIGURE 35.1 Basic steroid structure and numbering system.

the development and function of sexual organs as well as other biological differences between the sexes. Steroids, naturally occurring or synthetic, are also used for the treatment of various diseases such as allergic reactions, arthritis, some malignancies, and diseases resulting from hormone deficiencies or abnormal production. In addition, synthetic steroids (e.g., mifepristone [MIF]) that mimic the action of progesterone (PRO) are widely used as oral contraceptive agents. Steroids and their metabolites are analyzed by thin-layer chromatography (TLC) in a variety of samples such as biological samples or plants and pharmaceutical formulations. Many samples can be analyzed simultaneously and quickly at relatively low cost. Multiple separation techniques and detection procedures can be applied, and the detection limits are often in the low nanogram range with accurate and quantitative densitometric methods.

During metabolism, steroids generally become more hydrophilic by reduction, further hydroxylation, and esterification (conjugation) with glucuronic or sulfuric acid. Bile acids (containing a C_{24} carboxylic acid group) may be linked through a peptide bond to glycine or taurine. Despite the addition of these polar groups, the essential nonpolarity of the steroids means that they are all, to varying degrees, soluble in organic solvents and can thus be extracted from aqueous media by a solvent or solvent mixture of suitable polarity (Makin and Gower 2010).

Many procedures used for the quality control and quality assurance of steroids are based on classical methods of analysis. TLC is the method of choice in the field of steroid analysis, especially when many simultaneous analyses have to be carried out. In fact, hundreds of analyses can be performed in a short time and with small demands on equipment and space (Mulja and Gunawan 2010). Steroids usually occur in nature in low concentration and in association with other closely related steroids. This makes their isolation, identification, and specific determination virtually impossible without the use of efficient separation methods. TLC has been used for the analysis of natural and synthetic steroids in various environmental materials (Bhawani et al. 2010).

The initial question to consider is whether it is possible to assay the steroid directly in the medium without prior treatment; and today, this would appear to be a major objective in the development of tests for routine use in clinical laboratories for diagnostic purposes since such procedures are simple and avoid the need for extraction. The difficulty with TLC is the need to identify the areas on the plate, which correspond to the steroid of interest. Steroids, which are ultraviolet (UV) absorbing, can be visualized with the use of UV light. Adsorptive material that contains a fluorescent compound, which enhances the UV absorbance of the steroid of interest, is available. Steroids which do not absorb in the UV may have to be visualized, by spraying of the plate, to identify the position of standards, which have been run together with the samples of interest (Makin and Gower 2010).

35.2 CHOLESTEROL

When discussing steroids and steroid hormones, it is important to mention cholesterol (CHO). **CHO**, *(3β)-cholest-5-en-3-ol*, is known as a sterol, which is a natural product from the steroid nucleus. A structure of the CHO is shown in Figure 35.2. CHO is very important in the

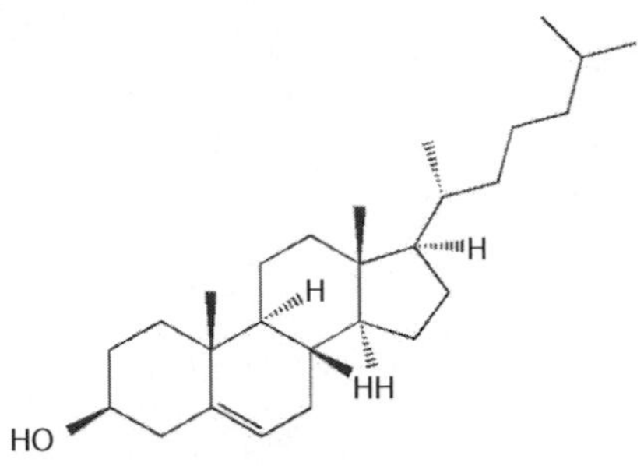

Cholesterol (CHO)

FIGURE 35.2   Structure of cholesterol.

production of steroid hormones; in fact, it is the precursor for biosynthesis of bile acids, steroid hormones, and provitamin D. CHO is incorporated into the cell membrane by lipoproteins, where it plays a role in regulation of membrane fluidity.

Zarzycki et al. (1999) published studies about retention and separation of CHO and bile acids using thermostated TLC. Chromatographic experiments were performed on wettable with water $RP_{18}W$ high-performance TLC (HPTLC) plates at the temperatures 5°C, 10°C, 20°C, 30°C, 40°C, 50°C, and 60°C with methanol–water mobile phases (0%–100% methanol). Spots were visualized by spraying the plates with a 1% solution of phosphomolybdic acid in 2-propanol and then heating at 120°C for 5–10 min. Nielsen (1990) developed the three-step one-dimensional TLC method for separation of neutral lipids including the CHO. Applied TLC plates were silica gels without binder (silica gel 60 HR, silica gel 60 H, and silica gel) and with calcium sulfate as binder (silica gel G as well as a commercial TLC plates silica gel 60). Three different solvents systems were applied: diethyl ether–benzene–ethanol–triethylamine (40:50:2:1 v/v), diethyl ether–hexane–triethylamine (10:90:1 v/v), and diethyl ether–hexane–acetic acid (75:25:2 v/v). The designed three-step TLC system works equally well with different brands of silica gel without binder. In contrast, it does not work with commercial plates and plates prepared with silica gel G. The system employs very short periods of drying without heating between chromatographic developments, and consequently the risk of autoxidation of unsaturated lipids during the chromatographic analysis is small.

Hung and Harris (1988) developed four solvent systems for more effective one-stage separation of CHO and its low-molecular-weight esters from their mixtures by one-dimensional TLC. The stationary phase applied is Silica gel G F_{254} as well as the following mobile phases:

1. Hexane–chloroform–ethyl ether–methanol–acetic acid, 92:1.5:6:0.5:0.2 v/v
2. Hexane–chloroform–ethyl ether–methanol–acetic acid–1-propanol, 92:1.5:6:0.2:0.2:0.3 v/v
3. Hexane–chloroform–ethyl ether–acetic acid–1-propanol, 92:1.5:6:0.2:0.5 v/v
4. Hexane–chloroform–ethyl ether–formic acid–acetic acid–propionic acid, 92:1.5:6:1.5:0.8:0.2 v/v

The spots were detected by spraying with a reagent consisting of 30 g of cupric acetate in 80 mL of reagent-grade phosphoric acid diluted to 1 L and then heating in an oven at 130°C for 5 min. Kovacs et al. (1986) developed a one-dimensional TLC technique for separation of mixture of phospholipids, neutral lipids, and CHO esters subfractions on the same precoated silica gel plate for 85 min by two successive developments in the same direction. Solvent systems used were chloroform–methanol–water (65:25:4 v/v) and *n*-hexane–dimethylketone (100:1 v/v). This method was used to analyze lipids in blood serum of humans.

In Table 35.1 are listed chromatographic and analytical parameters for separation and determination of CHO in different types of matrix.

TABLE 35.1
Analytical Parameters of the Methods for Cholesterol

Matrix	Stationary Phase	Mobile Phase	Detection	Calibration Mode/Range	LOD Unit/Spot	LOQ Unit/Spot	RSD%	Reference
Egg	Silica gel 60 F_{254}	Toluene–acetone–glacial acetic acid, 6:1.3:0.1 v/v	Densitometry at 215 nm (TLC scanner)	Linear, 5–25 µg/spot	1 µg/spot	3 µg/spot	0.57%–86%	Mallikarjuna et al. (2011)
Chicken-breast tissue	HPTLC silica gel	Petroleum ether–diethyl ether–acetic acid, 170:30:2 v/v	Spray reagent: 5% (m/v) phosphomolybdic acid in ethanol (10 s); heating for 10 min at 110°C	Linear, 20–200 ng/spot	10 ng/spot	20 ng/spot	—	Jazbec et al. (2009)
—	Diol F_{254}	Chloroform	Densitometric detection with and without sulfuric acid	—	—	—	—	Pyka (2009)
Degrad. products of egg phosphate dylcholine	RP_{18} silica gel	Butanol–methanol–water–acetic acid, 40:40:20:4 v/v	1% 4-methoxybenzaldehyde in 98% sulfuric acid–acetic acid (96%–98%)–ethanol–water, 2:10:60:30, v/v	Linear, 5–40 µg/spot	—	—	<5%	Gabriels et al. (2002)
Animal tissues	TLC silica gel	Petroleum ether (b.p. 40–0°)–diethyl ether–acetic acid, 80:20:3 v/v	Charring with manganese chloride–sulfuric acid at 110°	—	20 ng/spot	—	—	Thanh et al. (2000)
Biological samples	TLC silica gel	First heptane–diethyl ether–acetic acid, 70:20:4 v/v Second heptane	Molybdatophosphoric acid staining and heating at 120°C for 7 min.	Linear, 10–125 ng/spot	5 ng	—	9%–21%	Asmis et al. (1997)
Serum	TLC silica gel impregn. sodium carboxymethyl cellulose	Petroleum ether–ethyl acetate–glacial acetic acid, 80:20:1 v/v	Spray reagent: 1 g of vanillin in 100 mL of sulfuric acid; densitometry at 550 nm	Linear, 80–700 ng/spot	40 ng/spot	—	—	Li (1990)

35.3 ANDROGENS

The addition of a hydrogen atom at position 5 and an angular methyl group at positions 18 and 19, in relation to the structure of steroids (Figure 35.1), establishes the basic chemical framework for androgenic activity. Androgens are steroid hormones that are secreted primarily by the testis, and **Testosterone (TES)**, *(8R,9S,10R,13S,14S,17S)-17-hydroxy-10,13-dimethyl-1,2,6,7,8,9,11,12,14,15,16,17-dodecahydrocyclopenta[a]phenanthren-3-one*, is the principal androgen secreted. Its primary function is to regulate the differentiation and secretory function of male sex accessory organs. By one of the definitions, androgen is any hormone with TES-like actions. Androgens also possess protein anabolic activity that is manifested in skeletal muscle, bone, and kidneys. As a class, androgens are reasonably safe drugs, having limited and relatively predictable side effects (Schwartz and Miler 1997).

The natural steroid hormone TES belongs to the C_{19} androgen group. Rapid intrahepatic degradation yields androsterone among other metabolites (17-ketosteroids) that are eliminated as conjugates in the urine. Because of rapid hepatic metabolism, TES is unsuitable for oral use. TES, administered orally, is rapidly absorbed, but it is largely converted to inactive metabolites, and only about one-sixth is available in active form. In order to be sufficiently active when given by mouth, TES derivatives are alkylated at the 17 position. This modification reduces the liver's ability to break down these compounds before they reach the systemic circulation. For oral use, **methyltestosterone (MTT)**, *(8R,9S,10R,13S,14S,17S)-17-hydroxy-10,13,17-trimethyl-2,6,7,8,9,11,12,14,15,16-decahydro-1H-cyclopenta[a]phenanthren-3-one* or **fluoxymesterone (FLU)**, *(8S,9R,10S, 11S,13S,14S, 17S)-9-fluoro-11,17-dihydroxy-10,13,17-trimethyl-1,2,6,7,8,11,12,14,15,16-decahydro cyclopenta[a]phenanthren-3-one* is applied. **Danazol (DAN)**, *(1S,2R,13R,14S,17R,18S)-17-ethynyl-2,18-dimethyl-7-oxa-6-azapentacyclo [11.7.0.0^{2,10}.0^{4,8}.0^{14,18}]icosa-4(8),5,9-trien-17-ol*, is a synthetic steroid drug, a 2,3-isoxazol derivative of 17α-ethynyl TES (ethisterone [ETS]) that has weak virilizing and protein anabolic properties. The structures of TES, MTT, DAN, and FLU are shown in Figure 35.3.

MTT and FLU are androgenic steroids used primarily for androgen replacement. Androgen replacement regimens for treating male hypogonadism include long-acting intramuscular injections (e.g., TES enanthate and TES cypionate) and oral preparations (e.g., MTT and FLU). Due to its increased metabolic stability, 17α MTT is effective by the oral route; but because of the hepatotoxicity of C_{17}-alkylated androgens (cholestasis and tumors), its use should be avoided (Lullmann et al. 2000).

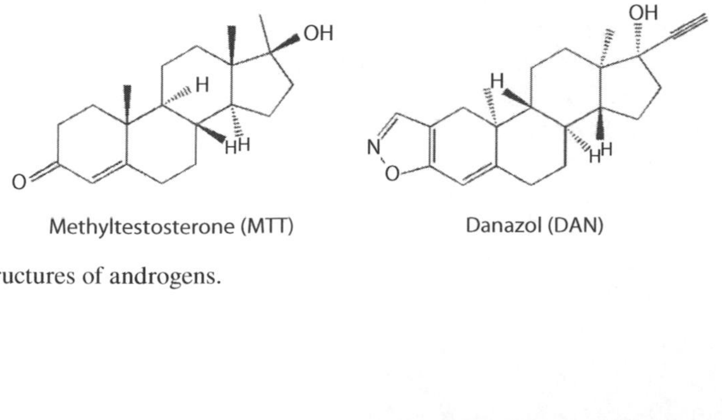

Testosterone (TES)

Fluoxymesterone (FLU)

Methyltestosterone (MTT)

Danazol (DAN)

FIGURE 35.3 Structures of androgens.

Preparation of sample for determination of these androgens is very simple. For example, sample of urine is prepared for determination of TES and its derivates by TLC as follows. After adjustment of pH of the urine to pH 6.5, add glucuronidase, phosphate bluffer, and a few drops of chloroform and mix well. Incubate the mixture for 24 h, then adjust to pH 1 with 6% sulfuric acid, and saturate with 5 g of sodium chloride. Shake the solution with 15 mL of ethyl acetate for 5 min. After centrifuging, discard the urine layer and keep the ethyl acetate layer for another 24 h at 37°C. Wash the ethyl acetate hire successively with 3 mL of concentrated sodium carbonate and 2 mL of water. After separation, transfer 10 mL of the ethyl acetate extract to a tube and evaporate to dryness (Yamaguchi 1982). Thus, the prepared sample is investigated by TLC under conditions listed in Table 35.2.

In Table 35.2 are listed chromatographic parameters for separation of TES, MTT, DAN, and FLU.

35.4 ANABOLICS

Anabolics are synthetic androgen analogues, that is, TES derivatives (e.g., metandienone [MET], oxymesterone [OXY], methenolone, trenbolone [TRE], nandrolone [NAN] [19-nortestosterone], norethandrolone [NOR], and stanozolol [STA]) that are used in debilitated patients, and misused by athletes, because of their protein anabolic effect. Androgens, such as TES, possess both androgenic and anabolic activities. The TES molecule has a number of positions, which can be modified by addition/removal of double bonds, reduction of the keto group, and substitution using heteroatoms, halogens, or addition of functional groups such as hydrocarbon chains and heterocyclic rings. With structural modifications, the anabolic effects of androgens can be enhanced but, even so, these cannot be divorced entirely from their androgenic effects.

Hence, a more accurate term for anabolic steroids is anabolic–androgenic steroids, but for simplicity, the shorter term is used within this chapter (Kicman et al. 2010). Anabolic steroids are widely used as growth-promoting agents. They act via stimulation of androgen receptors and, thus, also display androgenic actions (virilization in females and suppression of spermatogenesis). Anabolic steroids were initially used in medicine to treat hypogonadism, a condition in which testes produce abnormally low TES levels. Bodybuilders and weightlifters first used anabolic steroids in 1930s to increase skeletal muscle mass. Anabolic effects are as follows: (1) muscles become stronger and bulkier and (2) bones become heavier. Anabolic steroids can be absorbed from the gastrointestinal tract, but many compounds undergo such extensive first-pass metabolism in the liver that they are inactive. This steroid has widely been used illegally as a growth-promoting drug and thus they are banned in most sports competitions. The low residue levels in the tissues of the treated animals and the lack of suitable multi-residue methods hamper the efficient control of the use of anabolics. TLC continues to be an important method for simultaneous and quick qualitative and quantitative analysis. In Figure 35.4 are given the structures of some anabolics that are mentioned in this section.

MET, *(8R,9S,10R,13S,14S,17S)-17-hydroxy-10,13,17-trimethyl-7,8,9,11,12,14,15,16-octahydro-6H-cyclopenta[a]phenanthren-3-one,* is a derivative of TES, exhibiting strong anabolic and moderate androgenic properties. The primary urinary metabolites are detectable for up to 4 days, and a recently discovered hydroxymethyl metabolite is found in urine for up to 19 days. **OXY**, *(8R,9S,10R,13S,14S, 17S)-4,17-dihydroxy-10,13,17-trimethyl-2,6,7,8,9,11,12,14, 15,16-decahydro-1H-cyclopenta[a]phenanthren-3-one*, is a potent synthetic derivative of the anabolic steroid 4-hydroxytestosterone. **Metenolone**, *(5S,8R,9S,10S,13S,14S,17S)-17-hydroxy-1,10,13-trimethyl-4,5,6,7,8,9,11,12,14,15,16,17-dodecahydrocyclopenta[a] phenanthren-3-one*, is an anabolic steroid with weak androgenic (TES or androsterone-like) properties. **STA**, *(1S,3aS,3bR,5aS,10aS,10bS,12aS)-1,10a,12a-trimethyl-1,2,3,3a,3b,4, 5,5a,6,7,10,10a,10b,11,12,12a-exadecahydrocyclopenta[5,6]naphtho[1,2-f]indazol-1-ol*, is a synthetic anabolic steroid derived from dihydrotestosterone (DHT). Removal of the 19-methyl group in TES led to the development of the19-nor steroids such as NAN. **NAN**, *8R,9S,10R,13S,14S,17S)-17-hydroxy-13-methyl-2,6,7,8,9,10,11,12,14,15,16,17-dodecahydro-1H-cyclopenta[a]*

TABLE 35.2

Chromatographic Parameters for Determination of Androgens

Comp.[a]	Stationary Phase	Mobile Phase Composition	Detection	Remarks	Reference
TES MTT	TLC silica gel 60 F_{254}	First run 80% methanol–water v/v Second and third runs 20% acetone-*n*–hexane v/v	Plate dipped in the solution of 10% phosphomolybdic acid in methanol; heated for 10 min at 100°C	Two-dimensional and multistep development modes	Zarzycki (2008)
TES	TLC silica gel	Impregnated with $AgNO_3$ Toluene–acetone–chloroform 8:2:5 v/v	Sprayed with a molybdate solution (10% (w/v) ammonium molybdate in 10% (v/v) sulfuric acid) and slowly heated	Ascending conditioned by dipping into a solution of 4.06% (w/v) of $AgNO_3$. The plates were air dried (few min) and heated at 80°C/1 h	Godin et al. (1999)
TES	Ultrathin silica gel	ethyl acetate-cyclohexane 3:2 v/v	254 nm TLC scanner II	Ascending	Hauck and Schulcz (2003)
TES	TLC silica gel 60 F_{254}	Light petroleum–diethyl ether–acetic acid, 48:50:2 v/v Dichloromethane–methanol–water, 225:15:1.5 v/v	UV light or $SbCl_3$ spray	Two-dimensional	Brown et al. (1988)
TES	Diol F_{254}	Chloroform	Densitometric detection with and without sulfuric acid	Ascending	Pyka (2009)
TES	TLC silica gel 60 F_{254}	Isopropyl alcohol–tetrahydrofuran–hexane, 5:15:80 v/v	Spraying with 50% methanol:sulfuric acid and heating at 110°/5 min	Ascending	Shaikh et al. (1979)
TES	TLC silica gel 60 F_{254}	Ethyl acetate–1,1,2-trichlorotrifluoroethane	—	Continuous development	Tecklenburg et al. (1983)
TES	TLC silica gel	First benzene–diethyl ether 9:1 v/v (run twice) Second benzene–methanol 9:1 v/v	By heating with Allen reagent and by spraying with iodine in light petroleum	Two-dimensional	Bicknell and Gower (1971)
TES	Silica gel G	Hexane–ethyl acetate or cyclohexane–ethyl acetate, 1:2 v/v	Spraying with 50% H_2SO_4 in EtOH or MeOH, followed by heating (110°C) for 15 min	Activation plates 110°/30 min ascending techniques	Socic and Belic (1968)
TES	Silica gel 60 F_{254}	Cyclohexane–ether 8:2 v/v	—	Ascending	Bican-Fister (1966)
TES	Ultrathin monolithic silica gel	Ethyl acetate–cyclohexane, 3:2 v/v	Video documentation; plates were evaluated by use of a Camag TLC scanner II	Application volume is 10 nL, in methanol–chloroform, 4:1	Hauck et al. (2001)

(continued)

TABLE 35.2 (continued)

Chromatographic Parameters for Determination of Androgens

Comp.[a]	Stationary Phase	Mobile Phase Composition	Detection	Remarks	Reference
TES MTT	Silica-, octadecyl silica- or alumina	Binary mixts. such as methanol–water, acetonitrile–water, methanol–dichloromethane or acetone–hexane 0–100% v/v	Fluorescence	Small thermostated horizontal chamber unit (temp. ranging from −20 to +60°C); one- and two-dimensional developing modes	Zarzycki and Zarzycka (2008)
TES	HPTLC	—	—	Programmed multiple development technique	Matyska et al. (1991)
FLU	HPTLC silica gel 60 F_{254}	Chloroform–acetone	254 nm; after spraying with an ethanolic soln. of p-toluenesulfonic acid	Horizontal	Lekic et al. (2007)
TES	TLC silica gel 60 F_{254}	Chloroform–methanol–water, 9:1:0.1 v/v Chloroform–acetone, 9:1 v/v Benzene–acetone–methanol, 5:5:2 v/v	254 nm; Radioactive compounds were located by autoradiography or by direct scanning	Development—1 h in each solvent system	Barbieri et al. (1972)
DAN	Impregnated by cholesterol	Methanol–water	—	—	Farkas et al. (2003)
TES	TLC silica gel	Chloroform–acetone, 9:1 v/v Followed by dichloromethane–ethyl acetate–95% ethanol, 7:2:0.5 v/v	254 nm Iodine vapor	Separation of TES metabolites	Agrawal et al. (1995)
TES	TLC silica gel	Ethyl acetate–benzene, 1:1 v/v	Water bath at 40°C; spraying by enzyme reagent; dens. at 500 nm	Determination of urinary 17β-hydroxysteroids	Yamaguchi (1982)
TES	HPTLC silica gel	Benzene–hexane–ethanol	Spraying with 0.005 M 1,4-dihydrazinophthalazine in sulfuric acid–ethanol, 1:1; dens. at 400 nm	LOD 0.125 μg% plasma TES LOD 1 μg/1000 mL 24 h urinary TES	Agbaba et al. (1991)

[a] See Figure 35.3.

phenanthren-3-one, is an anabolic steroid (a muscle-building chemical), which occurs naturally in the human body, but only in small amounts. It was apparently found to be as myotrophic as TES but with greatly reduced androgenic activity. **NOR**, *(8R,9S,10R,13S,14S,17S)-17-ethyl-17-hydroxy-13-methyl-1,2,6,7,8,9,10,11,12,14,15,16-dodecahydrocyclopenta[a] phenanthren-3-one*, represents a moderate anabolic steroid with moderate androgenic properties. It is widely used for medical purpose for treating ulcer, anorexia nervosa, and severe burns. Being a 19-nor steroid, NOR can stop natural production of TES.

Verbake (1979) developed routine procedures for detection of various anabolic residues in tissues or urine contaminated at levels as low as 0.5–10 ppb. The tissues (meat, liver, and kidney) were homogenized in the presence of sodium acetate buffer (0.04 M, pH 5.2). After addition of

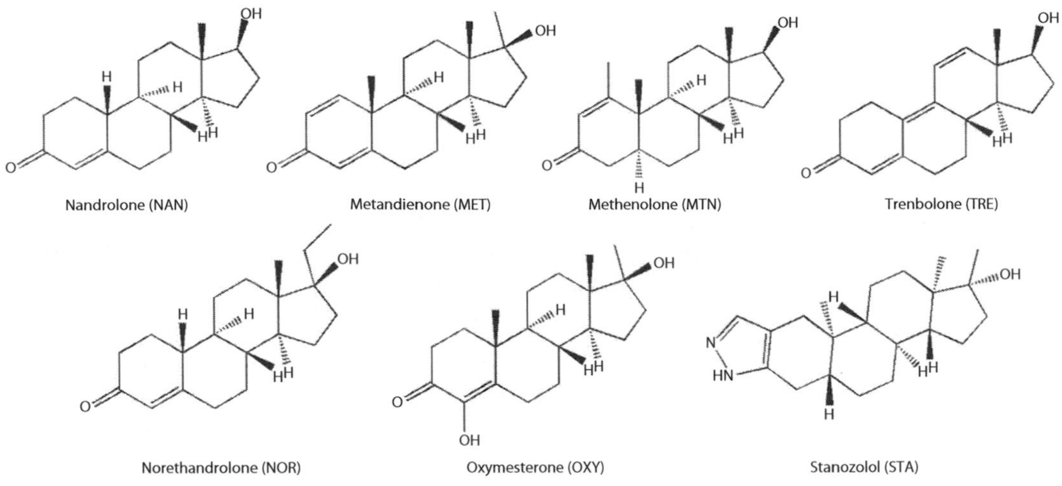

FIGURE 35.4 Structures of anabolics.

glucuronidase–sulfatase, hormone conjugates were hydrolyzed overnight at 37°C. The incubation mixture was then homogenized in the presence of methanol and centrifuged. The methanol phase was then extracted with dichloromethane. These phases were collected and evaporated to dryness. The steroids were eluted by passing methanol through the Amberlite XAD-2 column. Urine was allowed to percolate through an Amberlite column. The conjugated and free steroids were eluted with methanol. The obtained extracts were analyzed by two-dimensional chromatography on pre-coated silica gel 60 plates. Development was carried out in non-saturated tanks. Chromatographic development was carried out using of chloroform–ethanol–benzene (36:1:4 v/v). The plate was air-dried, and the starting point of the sample was over spotted with 5–10 ng of the steroids presumed to be present. The plate was then run in the second direction using the appropriate solvent. The plates were air dried, and the fluorescence reaction was induced by spraying with 5% sulfuric acid in acetic anhydride or in ethanol. The plate was viewed under UV light (366 nm) and then incubated at 95°C during 12 min. The fluorescence was observed under transillumination at 366 nm.

Frequently, in recent years, anabolics are injected into animals as highly concentrated mixtures (the so-called hormone cocktails), which usually stay locally at the site of injection from where they are distributed by a slow diffusion process. The analysis of these injection sites by an HPTLC method following a simple and unselective extraction yields are described by Daleseleire et al. (1994). The injection site was cut and placed in a plastic bag. After adding 5 mL of methanol, the bag contents were blended for 5 min. The methanolic extract was centrifuged. The supernatant was evaporated under a stream of nitrogen. The residue was dissolved in methanol. Separations were carried out on HPTLC silica gel 60 plates. Development was carried out in one direction in a twin-through chamber with chloroform–acetone (90:10 v/v). After drying the plate, the spots on the opposite side were developed in mixtures of cyclohexane–ethyl acetate–methanol (58.5:39:5.5 v/v). After the second elution, the plate was dried under a cool air stream and sprayed with 10% sulfuric acid in methanol, then heated for 10 min at 95° and examined in daylight and under UV light at 366 nm. In these determinations, sample preparation has a great impact in the context of matrix. Meat or kidney fat was cut into small pieces, weighed into a polypropylene flask, and sodium acetate buffer (0.04 mol/L, pH 5.2) was added. The fat samples are melted on a water bath at 70°C for 20 min, 50 mL of methanol is added, and the mixture is homogenized again. The supernatant is filtered over silanized glass wool in a separating funnel. The methanolic supernatant is extracted once with n-hexane, and the hexane phase is discarded. Anabolics are then extracted with diethyl ether. The combined ether phases are washed, once with carbonate buffer (pH 10.25) and twice with water and then evaporated to dryness. The residue is transferred with methanol and concentrated with a

TABLE 35.3

Chromatographic Parameters for Determination of Anabolics

Comp.[a]	Stationary Phase	Stationary Phase Pretreatment; Mobile Phase Composition	Detection	Remarks	Reference
TRE MET STA	TLC silica gel 60 F$_{254}$	Chloroform–methanol, 92:8 v/v or chloroform–acetone, 90:10 v/v	Mixtures of 2,4-dihydroxybenz aldehyde, sulfuric acid, and acetic acid as spray reagent	No significant interferences were found	Huetos et al. (1998)
MET	HPTLC RP$_{18}$W	Plate is washed up by dipping in methanol for 1 min, drying in the air and activating at 110°C for 10 min; acetonitrile and buffer solution	TLC diode array scanner	Horizontally developing DS chamber	Płocharz et al. (2010)
TRE NAN MET	Silica gel	First run cyclohexane–ethyl acetate–ethanol, 60:40:2.5 v/v Second run chloroform–hexane–acetone, 50:40:10 v/v	The plate is dipped into a 5% ethanolic sol. of sulfuric acid (30s); 95°C (10 min.); the spots are identified by viewing by transillumination at 366 nm	Double one-dimensional elution on the same plate	Smets et al. (1993)
NAN TRE	HPTLC plates silica gel 60 F$_{254}$	Chloroform–acetone	254 nm; spraying with an ethanolic soln. of p-toluenesulfonic acid	Horizontal elution	Lekic et al. (2007)
TRE	Chromatoplate	Chloroform–ethanol, 95:5 v/v	UV light (254 nm)	—	Boursier and Chafey (1988)
TRE	Silica gel plates	Chloroform–ethyl acetate, 2:1 v/v	Exposure to HCl vapor; fluorescence at 498 nm (excitation at 365 nm)	—	Oehrle et al. (1975)
STA	silica gel	Ethyl acetate–n-hexane, 1:1 v/v	254 nm; spraying with 25% sulfuric acid in ethanol and heating for 10 min. at 20°C	—	Schludi et al. (2000)
MET	T sealant solution was prepared by mixing components Sarsil Wor Sarsil H50 with hardener 100:4 w/w	Plates were placed in the oven at 105°C –110°C for 45 min to polymerize the sealant then left in a desiccator and used for experiments within 1 day; acetonitrile and buffer solution	—	Pressurized planar electrochromatography, PPEC chamber	Płocharz et al. (2010)

[a] See Figure 35.4.

vacuum evaporator and chromatographed (Smets et al. 1993). The chromatographic parameters for determination of extracted anabolics are listed in Table 35.3. Limit of quantification (LOQ) of MET in tablets, suppositories, solutions for injection, or blood plasma is 8–14 pmol (Wintersteiger and Gamse 1982).

TRE, *(8S,13S,14S,17S) -17-hydroxy-13-methyl-2,6,7,8,14,15,16,17-octahydro-1H-cyclopenta[a] phenanthren-3-one*, is one of the most important xenobiotic anabolic steroids often used as growth promotors for fattening veal calves and cattle. To screen the illegal use of this compound, a method for detecting residues in bovine urine is required. Immunoaffinity chromatography is a very powerful technique when small amounts of a compound have to be isolated from a complex matrix. Detection limits of TRE in urine are 8 ppb, and recoveries are 74.7% (Boursier and Chafey 1988).

In Table 35.3 are listed chromatographic parameters for determination of some anabolics.

35.5 ANTIANDROGENS

By definition, antiandrogens are substances that prevent or depress the action of male hormones in their target organs. Potential sites of action include gonadotropin suppression, inhibition of androgen synthesis, and androgen receptor blockade. Potential clinical uses of antiandrogens include suppression of androgen excess and treatment of androgen-dependent tumors (Craig and Stitzel 1997). Flutamide (FLT), nilutamide (NIL), and bicalutamide (BIC) are nonsteroidal antiandrogen drugs, while cyproterone (CYP) acetate is a steroidal medication, and spironolactone (SPI) is a synthetic steroid that acts as diuretic and used as an antiandrogen. Their structures are shown in Figure 35.5.

CYP, *6-chloro-1β,2β-dihydro-17-hydroxy-3'H-cyclopropa[1,2] pregna-4,6-diene-3,20-dione*, is a synthetic derivative of 17-hydroxyprogesterone (HYP) and acts as an androgenreceptor antagonist with weak protestation and glucocorticoid activity. CYP acts as competitive antagonist of TES. It is available in a topical form in Europe for the treatment of hirsutism. Originally developed as a progestagen, CYP acetate was found to have antiandrogenic properties. Now, it has major clinical applications.

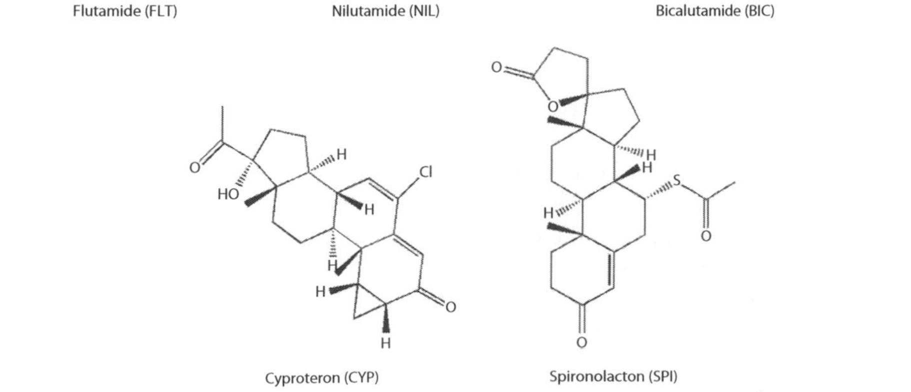

Flutamide (FLT)   Nilutamide (NIL)   Bicalutamide (BIC)

Cyproteron (CYP)   Spironolacton (SPI)

FIGURE 35.5 Structures of antiandrogens.

CYP acts by competition for the androgen receptor, blocking androgen synthesis, and suppressing any compensatory rise in gonadotropins (Wood and Gower 2010). CYP is practically insoluble in water, very soluble in dichloromethane and acetone, soluble in methanol, and sparingly soluble in ethanol. **FLT**, *2-methyl-N-[4-nitro-3-(trifluoromethyl)phenyl]-propanamide*, is a non-steroidal androgen receptor antagonist that inhibits androgen binding to its nuclear receptor. FLT may eventually be used for the treatment of hirsutism and male-pattern baldness in women if a topical preparation is developed. **NIL**, *5,5-dimethyl-3-[4-nitro-3-(trifluoro methyl phenyl]-imidazolidine-2,4-dione*, is an antiandrogen medication used in the treatment of advanced stage prostate cancer. NIL blocks the androgenreceptor, preventing its interaction with TES. **BIC**, *N-[4-cyano-3-(trifluoromethyl)phenyl]-3-[(4-fluorophenyl)sulfonyl]-2-hydroxy-2-methylpropanamide*, is an oral nonsteroidal antiandrogen used in the treatment of prostate cancer and hirsutism. **SPI**, *S-[(7R,8R,9S,10R,13S,14S,17R)-10,13-dimethyl-3,5'-dioxospiro [2,6,7,8,9,11,12,14,15,16-decahydro-1H-cyclopenta[a]phenanthrene-17,2'-oxolane]-7-yl] ethanethioate*, is a synthetic steroid that acts as a competitive antagonist of aldosterone (ALD). It is a synthetic 17-lactone drug in a class of pharmaceuticals called potassium-sparing diuretics, used primarily to treat heart failure, ascites in patients with liver disease, low-renin hypertension, hypokalemia, secondary hyperaldosteronism (such as occurs with hepatic cirrhosis), and is used as an antiandrogen. On its own, SPI is only a weak diuretic because its effects target the distal nephron (collecting tubule), where urine volume can only be slightly modified; but it can be combined with other diuretics to increase efficacy.

Identification and determination of CYP acetate in the presence of ethinyl estradiol (ESD) in pharmaceutical dosage forms (tablets) were performed on LiChrospher silica gel 60 F_{254} HPTLC plates with cyclohexane–ethyl acetate (60:40 v/v) as mobile phase. Densitometry of CYP acetate was performed at 284 nm and of ethinyl ESD (ETH) at 220 nm. The ranges validated were 250–4000 ng (Novakovic et al. 1990; Pavic et al. 2003).

Quantification of BIC in the bulk drug and in a liposomal formulation containing CHO and lecithin and other surfactants were performed on aluminum foil–backed silica gel 60 F_{254} plates with toluene–ethyl acetate (45:55 v/v) as mobile phase in twin-trough chamber saturated for 30 min. Leflunomide (*5-methyl-N-[4-(trifluoromethyl) phenyl]-isoxazole-4-carboxamide*) ($R_F = 0.85$) was used as internal standard. Densitometric detection of BIC ($R_F = 0.45$) was performed in absorbance mode at 273 nm. The method is specific, selective, and free from matrix interferences at the R_F of BIC and the internal standard. The limits of detection (LOD) and LOQ were 50 and 200 ng per band, respectively (Subramanian et al. 2009).

Hegazy et al. (2011) determined SPI in mixture with hydrochlorothiazide. The separation was achieved using silica gel 60 F_{254} TLC plates and ethyl acetate–chloroform–formic acid–triethyl amine (7:3:0.1:0.1 v/v) as a developing system. Sharma et al. (2010) determined SPI in bulk drug and in tablet dosage in mixture with torsemide by TLC using silica gel 50 F_{254}. The mobile phase used was a mixture of ethyl acetate–acetone–acetic acid (10.5:4:1.5 v/v). The detection of spot was carried out at 269.0 nm. LOD and LOQ of torsemide and SPI were found to be 120 and 178 ng/spot, respectively. The linearity range was found to be 360–850 ng/spot. Gaikwad et al. (2010) performed the separation of these compounds using the same sorbent and n-hexane–ethyl acetate–methanol–glacial acetic acid (7:3:1.5:0.5 v/v) with UV detection at 263 nm. The major metabolite of SPI is canrenone. Van der Merwe et al. (1979) has developed a sensitive and highly specific thin-layer spectrofluorimetric method for the simultaneous determination of SPI and canrenone in human serum. The serum was extracted with diethyl ether following acid hydrolysis and centrifuged, then sodium hydroxide was added to the organic phase, extracted, and centrifuged again, and the organic phase was evaporated to dryness. The residue was dissolved in chloroform and spotted on the thin-layer chromatographic plate and quantitated by spectrofluorometry at 370 nm. Chromatographic parameters are: TLC silica gel as sorbent and carbon tetrachloride–ethyl acetate (2:3 v/v) as mobile phase. The method had a sensitivity of 2 and 1 ng/mL for SPI and canrenone, respectively, and a recovery of 95% for both (RSD was ≤6.4% and ≤5.1%, respectively).

35.6 5α-REDUCTASE INHIBITORS

Use of 5α-reductase inhibitors results in increased levels of TES and decreased levels of DHT. They block the action of 5-α-reductase enzymes, which convert TES into DHT, which has greater affinity for androgen receptors. Drugs in this class are **finasteride (FIN)**, *N-(1,1-dimethylethyl)-3-oxo-(5α,17β)-4-aza-androst-1-ene-17-carboxamide*, and **dutasteride (DUT)**, *(5α, 17β)-N-[2,5 bis(trifluoro methyl phenyl]-3-oxo-4-aza-androst-1-ene-17-carboxamide*. Their structures are shown in Figure 35.6.

FIN is 5-α-reductase inhibitor that blocks the conversion of TES to DHT in target tissues. Since DHT is the major intracellular androgen in the prostate, FIN is effective in suppressing DHT stimulation of prostatic growth and secretory function without markedly affecting libido. It is approved for the treatment of benign prostatic hyperplasia.

During the analysis of pharmaceutical formulations, it is very important to make simultaneous determination of tamsulosin hydrochloride (*[(−)-(R)-5-[2-[[2-(-ethoxy phenoxy)ethyl]amino] propyl]-2-methoxybenzenesulfonamide]*) and FIN. Chromatographic separation was performed on silica gel 60 F_{254} using toluene–*n*-propanol–triethylamine (3.0:1.5:0.2 v/v) as mobile phase. Detection was carried out densitometrically at 260 nm. The R_F value of tamsulosin hydrochloride and FIN were 0.32 and 0.54, respectively. The reliability of this method was assessed by evaluation of linearity, which was found to be 200–1200 ng/spot for tamsulosin hydrochloride and 1000–6000 ng/spot for FIN (Bari et al. 2011; Patel and Patel 2010a).

Analysis of FIN in pharmaceutical preparations using loratadine (*Ethyl 4-(8-chloro-5,6-dihydro-11H-benzo[5,6]cyclohepta[1,2-b]pyridine-11-ylidene)-1-piperidine-carboxylate*) as an internal standard was performed on silica gel 60 F_{254} (prewashed with methanol) with chloroform–ethyl acetate (6:4 v/v) as mobile phase. Detection and quantification were performed densitometrically at 228 nm. The linear range of the analysis was 0.2–2.0 µg, and the percentage recovery was 101.8% (Meyyanathan et al. 2001).

For the determination of DUT in pharmaceutical dosage forms, the TLC method employed silica gel G60 F_{254} as stationary phase. The solvent system consisted of mixtures of toluene–methanol–triethylamine (9:2:1 v/v). This solvent system was found to give compact spots with Rf value 0.71 ± 0.01. Densitometric determination was carried out in the absorbance mode at 274 nm. Linear regression analysis showed good linearity with respect to peak area in the concentration range of 200–3000 ng per spot, LOD is 10 ng/spot and LOQ is 50 ng/spot (Patel et al. 2011). This determination can also be done on the same sorbent using mixtures of acetonitrile–acetic acid with UV detection at 210 nm. Calibration curves were linear from 50 to 500 µg/mL. The accuracy of this method ranged 99.17%–99.94% (Kamat et al. 2008).

The simultaneous determination of alfuzosin hydrochloride (ALF) and DUT in a pharmaceutical dosage form was performed on silica gel 60 F_{254} with toluene–methanol–dichloromethane–triethylamine (6:1:1:0.6 v/v) as mobile phase. Densitometric evaluation of the separated zones was

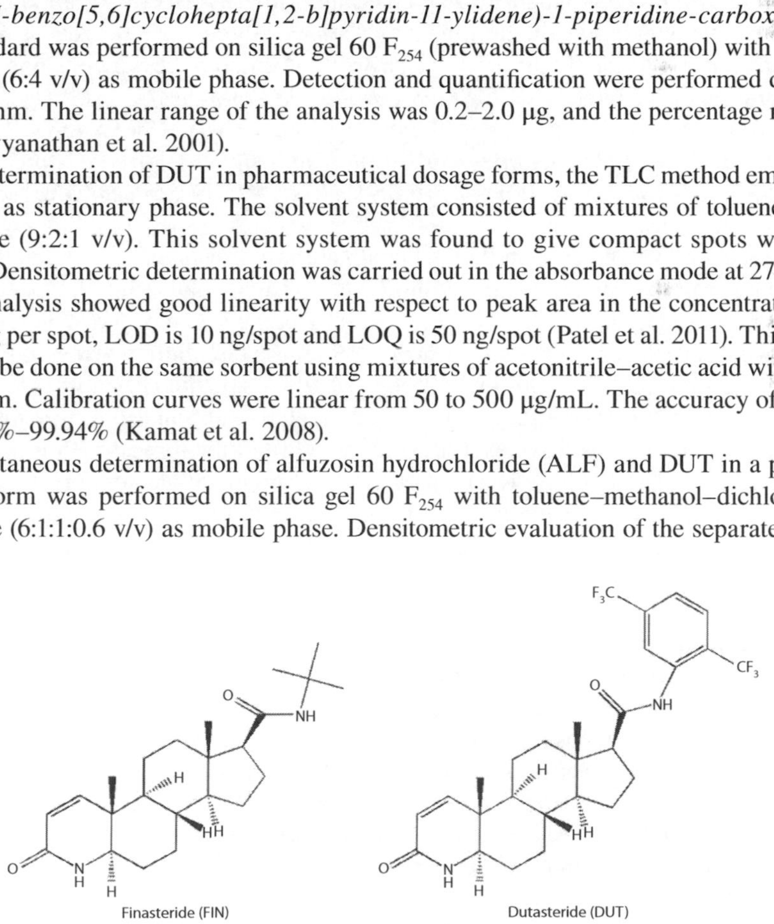

Finasteride (FIN)   Dutasteride (DUT)

FIGURE 35.6 Structures of 5α-reductase inhibitors.

performed at 247 nm. Compact spots were obtained for ALF (R_F 0.46 ± 0.03) and DUT (R_F 0.65 ± 0.03). This method revealed good linearity over the concentration ranges 300–600 ng per band for ALF and 500–1000 ng per band for DUT (Deshmukh et al. 2011). When the determination of DUT in the pharmaceuticals dosage forms was performed in the presence of tamsulosin hydrochloride, the same sorbent was applied, and toluene–methanol–triethylamine (18:3:2, v/v) was applied as mobile phase. Quantification was achieved by UV detection at 280 nm over the concentration range 200–2000 ng per band for both. The recovery was 99.7% and 100.1% for tamsulosin hydrochloride and DUT, respectively (Patel et al. 2010b).

DUT was subjected to acid and alkali hydrolysis, oxidation, photodegradation, dry heat, and wet-heat treatment. Because of that, the drug undergoes degradation under acidic, basic conditions, photolytic, oxidative, and upon wet- and dry-heat treatment. The degraded products were well separated from the pure drug. The stability-indicating method for the determination of DUT both as a bulk drug and as pharmaceutical tablets employed TLC on aluminum plates precoated with silica gel 60 F_{254}, and the mobile phase consisted of mixture of acetonitrile–methanol–dichloromethane (2.0:1.0:2.0 v/v) ($R_F = 0.64$). Densitometric analysis was carried out in the absorbance mode at 244 nm with respect to peak area in the concentration range of 100–600 ng/band, LOD 7.54 ng/band and 22.85 ng/band (Choudhari and Nikalje 2009).

35.7 ESTROGENS

Estrogens originate from the adrenal cortex and gonads and primarily affect maturation and function of secondary sex organs (female sexual determination). The naturally occurring estrogens are C_{18} steroids that contain an aromatic A ring (Figure 35.1) with a hydroxyl group at the third position. Biologically important natural estrogens include **ESD**, *(8R,9S,13S,14S,17S)-13-methyl-6,7,8,9,11,12,14,15,16,17-decahydrocyclopenta [a]phenanthrene-3,17-diol*, **estrone (ESN)**, *(8R,9S,13S,14S)-3-hydroxy-13-methyl-7,8,9,11,12,14,15,16-octahydro-6H-cyclopenta[a]phenanthren-17-one* and **estriol (ESL)**, *(8R,9S,13S,14S,16R,17R)-13-methyl-6,7,8,9,11,12,14,15,16,17-decahydrocyclopenta [a]phenanthrene-3,16,17-triol*.

ESD-17β, the most potent estrogen that is found naturally in women, represents 10%–20% of the circulating estrogen. ESD is practically insoluble in water, soluble in ethanol (1 part in 28), chloroform (1 part in 435), diethyl ether (1 part in 150), acetone, and dioxane. One-tenth of ESN is biologically active as ESD and accounts for 60%–80% of the circulating estrogen. ESN is practically insoluble in water (0.003 g/100 mL at 25°C); soluble in ethanol (1 in 250), chloroform (1 in 110 at 15°C), acetone (1 in 50 at 50°C), dioxane, and vegetable oils; and slightly soluble in diethyl ether and solutions of alkali hydroxides. ESL is the weakest of the three, and it is synthesized by the placenta and is excreted at high levels in the urine of pregnant women, practically insoluble in water, sparingly soluble in ethanol, soluble in acetone, chloroform, dioxane, diethyl ether, and vegetable oils. Both 17β-ESD and ESN are converted by 16α-hydroxylase to yield ESL, which is found in the urine as the glucuronide conjugate. One method to increase the oral bioavailability of ESD is to prevent metabolic oxidation of the ESD C_{17} hydroxyl group to ESN. This is readily accomplished via alkylation of the C_{17} position with a chemically inert alkyne group (e.g., ETH). **ETH**, *(8R,9S,13S,14S,17R)-17-ethynyl-13-methyl-7,8,9,11,12,14,15,16-octahydro-6H-cyclopenta [a]phenanthrene-3,17-diol*, is 15–20 more potent than ESD when orally administrated. This synthetic analogue is several hundred-fold more potent than ESD. In addition, widely applied synthetic estrogen is **epimestrol (EPI)**, *(8R,9S,13S,14S,16R,17S)-3-methoxy-13-methyl-6,7,8,9,11,12,14,15,16,17-decahydrocyclopenta[a]phenanthrene-16,17-diol*. Besides them, are very important and nonsteroid estrogens such as diethylstilbestrol (DES), dienestrol (DIE), chlorotrianisene (CHL), clomifen, and cyclofenil (CYC). Structures of estrogens mentioned earlier are given in Figure 35.7.

Thin-layer chromatographic method is described for the detection of individual natural estrogens in human pregnancy urine. After acidic hydrolysis, the urine was extracted with ether, washed

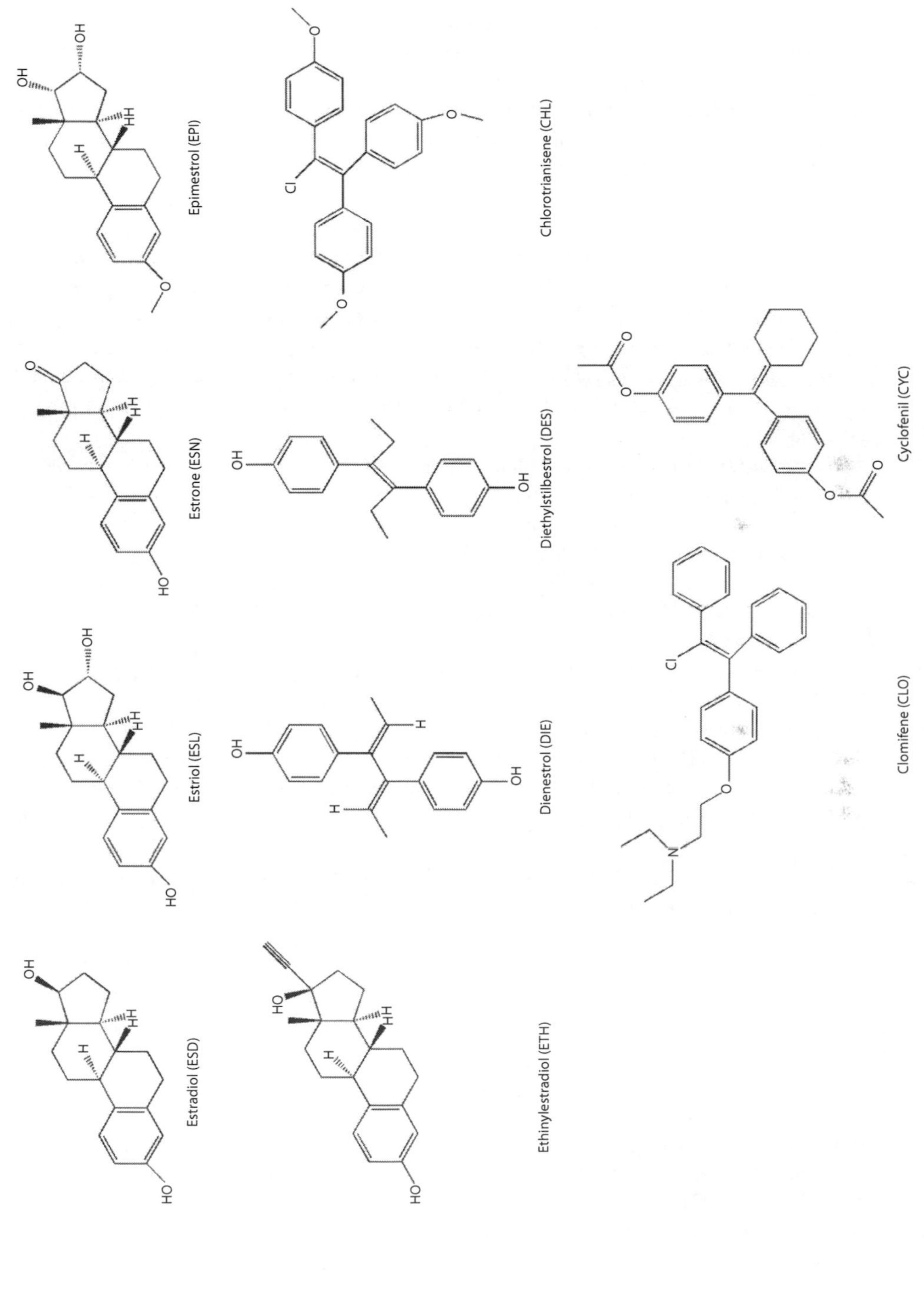

FIGURE 35.7 Structures of estrogens.

with sodium bicarbonate, and the extract chromatographed on silica gel G layer with the solvent system benzene–ethanol (9:1 v/v). Quantification was carried out using Ittrich's color reaction in the presence of adsorbent. The lower limit of the measurement was 0.2 pg of the urinary estrogens. Recovery is 1–2 µg/5 mL (Feher et al. 1967).

For determination of estrogens in meat and liver, one of the proposed procedures in literatures (Wortberg et al. 1978) is as follows: 50 g amount of meat was minced, mixed with 100 mL of ethanol, and extracted by shaking for 1 h. After centrifugation, the extraction procedure was repeated twice. The combined ethanolic extracts were evaporated. The residue was treated with dichloromethane and, after addition of water, the organic phase was separated. Two further extractions with dichloromethane were carried out. The combined dichloromethane extracts were washed three times with 10% sodium carbonate solution (pH 10.5) and evaporated under vacuum. The residue, dissolved in tetrahydrofuran, was separated from lipids by gel chromatography. The fraction was eluted, collected, and concentrated under vacuum. The residue was dissolved in benzene and washed twice with saturated sodium chloride solution. Steroidal estrogens were extracted from benzene solution with 0.1% sodium hydroxide solution. The aqueous phase, collected with water in a separating funnel, was acidified with concentrated hydrochloric acid and re-extracted with benzene. After washing the benzene phase with water to neutrality, it was dried with anhydrous sodium sulfate and evaporated to dryness. The residue was dissolved in ethanol, re-evaporated, and the final residue dissolved ethanol. After that, authors performed chromatographic determination.

A simplified method is described for the identification and determination of urinary estrogens in sheep under conditions whereby the urinary pigments are reduced in vivo by changing the diet from green to dry fodder. The method involves mild hydrolysis, extraction and purification, and quantification. The recoveries were: 68.9%, 65.4%, and 67.2% for the three estrogens, respectively. The sensitivity was 6.5 µg ESN/L. The mean urinary output of total estrogen on the day of estrus was 300.45 µg/L, whereas on the day of parturition, it was 450.60 µg/L (Baksai-Horvath et al. 1977).

Furthermore, TLC is a method of choice for determination of estrogens in the presence of anabolics (such as TRE) in calf urine because its use is regulated or prohibited in certain countries. The compounds were extracted from urine with ether after enzymic hydrolysis, purified on a silica G 60 column, submitted to HPTLC or chromatoplates with a chloroform–ethanol (95:5 v/v) mobile phase, detected under UV light, and identified by comparison with standards. Detection limits for 17α-ESD and 17α-TRE in urine were 2 and 8 ppb, respectively, and recoveries were 46% and 74.7%, respectively (Boursier and Chafey 1988).

A rapid, selective, and precise stability-indicating HPTLC method was developed and validated for the detection of ESD in bulk and pharmaceutical dosage forms. The method employed TLC aluminum plates precoated with silica gel 60 F_{254} as the stationary phase. The solvent system consisted of chloroform–dimethylketone–isopropanol–acetic acid (9:1:0.4:0.1 v/v). Spectrodensitometric scanning-integration was performed on a Camag system using a wavelength of 286 nm. The polynomial regression data for the calibration plots exhibited good linear relationship over a concentration range of 1–8 µg. The recovery data reveals that the RSD for intra-day and inter-day analysis was 1.27% and 1.75%, respectively (Kotiyan and Vavia 2000).

Picogram amounts of ESD (≥40 pg) can be isolated by TLC on silica gel F_{254} plates by development with benzene–ethyl acetate (3:1 v/v) and eluting with 30% methanol in dichloromethane. The separation procedure results in blank values in the endpoint detection by a radioimmunoassay system of <10 pg (Doerr 1971).

TLC, with spectrophotometric detection, can be used for determination of ETH in oral contraceptives (together with levonorgestrel [LEV]). Development can be carried out on both silica gel 60 F_{254} and HPTLC silica gel 60 F_{254}. The spots of ETH could be measured by diffuse reflection only after the plate had been sprayed with 5% methanolic solution of sulfuric acid and after heating for 10 min at 120° (530 nm). Using the fluorescence method, the spots were sprayed with 0.5% methanolic solution of sulfuric acid and heated too. Fluorescence appeared and measured at 560 nm (excited at 485 nm). The RSD for the diffuse reflection and fluorescence methods were 0.44%–2.6% and 3.4%–5.2%, respectively (Amin and Hassenbach 1979).

Lisboa (1966) gave an overview of the separation and characterization of ESD, ESL, ESN, and EPI by ascending chromatography on silica gel G. The estrogens were characterized in several systems, by different color reactions and derivatization formations.

35.7.1 Nonsteroidal Estrogens

DES, *4-[(E)-4-(4-hydroxyphenyl)hex-3-en-3-yl]phenol*, a synthetic nonsteroidal estrogen that was historically widely used to prevent potential miscarriages by stimulating the synthesis of estrogen and PRO in the placenta. It is a white, odorless, crystalline powder at room temperature. It is practically insoluble in water and soluble in alcohol, ether, chloroform, fatty oils, dilute hydroxides, acetone, dioxane, ethyl acetate, methanol, and vegetable oils. Besides, **CHL**, *1-[1-chloro-2,2-bis(4-methoxyphenyl)ethenyl]-4-methoxybenzene*, is a nonsteroidal synthetic estrogen that was formerly used for the treatment of menopause, deficiencies in ovary function, and prostate cancers. **DIE**, *4-[(2E,4E)-4-(4-hydroxyphenyl)hexa-2,4-dien-3-yl]phenol*, is a nonsteroidal estrogen structurally related to stilbestrol.

Determination of DES in plasma and tissues isolated with alumina and ion-exchange membrane columns in tandem were described by Medina and Nagay (1993). The use of an ion-exchange membrane reduced the analysis time by 25%. TLC plates were immersed horizontally in methanol for 10 min, and activated at 85°C for 30 min. It was developed with methylene chloride–methanol–2-propanol (97:1:2 v/v) with channeled TLC plates with preconcentration zones. This developing mode zones increased the minimum detectability from 100 to 25 ppb. Visualization with iodine-starch or by spraying with 0.05% aqueous solution of diazonium dyes and exposure to ammonia vapor for 30 s and heating at 80°C for 2 min. Minimum detectabilities for DES were 25 ppb (LOD 12.5 ng) in fortified tissue extracts, but lower signals (lighter TLC bands) resulted when tissues were fortified prior to extraction and purification.

After enzyme hydrolysis, DES and ESD were extracted from urine with ether, purified on a silica G 60 column, submitted to HPTLC on chromatoplates with a chloroform–methanol (95:5 v/v) mobile phase, detected under UV light, and identified by comparison with referent spots. Detection limits for DES and 17α-ESD in urine were 8 and 2 ppb, and recoveries were 60% and 46%, respectively (Boursier and Chafey 1988). Quantitative determination of DES was performed by TLC procedures for the analysis of DES at the 0.01% level in a water-dispersible suppository base. Following a preliminary separation on an alumina column, the DES is isolated on a silica gel chromatoplate. A color is produced via the Folin–Ciocalteu reagent and evaluated densitometrically directly on the chromatoplate. The standard deviation is 3.4% (Jones et al. 1968). For separation of DES from naturally phenolic substances in urine, Schuller (1967) recommended two extraction steps with ethanol and chloroform.

Clomifene (CLO), *2-[4-[(Z)-2-chloro-1,2-diphenylethenyl]phenoxy]-N,N-diethylethanamine,* is a nonsteroidal selective estrogen receptor modulator similar to tamoxifen (TAM). It increases production of gonadotropins by inhibiting negative feedback on the hypothalamus. This synthetic estrogen is a mixture of two geometric isomers, enclomiphene (*E*-clomifene) and zuclomiphene (*Z*-clomifene). Clomiphene is a triphenylethylene derivative distantly related to DES. **CYC**, *[4-[(4-acetoxyphenyl)-cyclohexylidene-methyl] phenyl]acetate*, is a nonsteroidal selective estrogen receptor modulator. It is very similar to clomiphene in its actions. Sometimes, it is used by anabolic steroid users.

Chromatographic parameters for separation and determination of estrogens are given in Table 35.4.

TABLE 35.4

Chromatographic Parameters for Separation and Determination of Estrogens

Comp.[a]	Stationary Phase	Stationary Phase Pretreatment; Mobile Phase Composition	Detection	Reference
ESD ESN ESL ETH	TLC silica gel 60 F_{254}	Chloroform–glacial acetic acid, 10:1 v/v	Spray reagent (methanol/ sulfuric acid); the plate was heated at 110°C for 10 min; densitometric scanning	Biswara and Jakovljevic (1969)
ESD ESN ESL	TLC silica gel LQDF	Precond. for 30 min at 32% humidity; Chloroform–ethyl acetate, 90:10; 80:20; 50:50 v/v; Chloroform–methanol 98:2 v/v Benzene–ethyl acetate, 60:20 v/v	Iodine vapor	Ruh (1976)
ESD ESN ESL	TLC silica gel H	chloroform-acetic acid 85:15, v/v	—	Lars (1970)
ESD ESN	Silica gel G F_{254}	Benzene–ethyl acetate, 2:1; 3:1; 5:1 v/v Chloroform–diethyl ether, 5:2 v/v Hexane–chloroform, 4:1 v/v Hexane–dichloromethane, 1:2 v/v Hexane–methanol, 4:1 v/v Hexane–ethyl acetate, 5:2; 1:1 v/v	Spots were visualized under UV light, then the plates were sprayed with 50% aqueous sulfuric acid and heated at 70°C for 30 min; if necessary, they were again viewed under UV light	Rajkowski and Broadhead (1974)
ESD ESN ESL	Silica gel 60 HPTLC plates	Twice with diethyl ether–cyclohexane, 80:20 v/v	Densitometric, 520 nm	Wortberg et al. (1978)
ESD ESN ESL	TLC silica gel	2,2,4_Trimethylpentane– ethyl acetate, 4:1 Cyclohexane–acetone, 6:1 v/v Cyclohexane–ethyl acetate, 6:1 v/v	Sprayed with a mixture of ethanol and sulfuric acid (1:1) and heated at 110°C for 20 min	Renwick et al. (1983)
ESD ESN ESL	1. HPTLC silica gel 60 F_{254} 2. HPTLC NH_2 F_{254}	Chloroform–methanol, 5–90% v/v Cyclohexane–methanol, 5%–90% v/v n-hexane–ethanol, 5–90% v/v	254 nm	Grassini-Strazza and Nicoletti (1985)
ESD	Glass plates precoated with Diol F_{254}	Chloroform	Densitometric detection with and without sulfuric acid as visualizing reagents	Pyka (2009)
ETH	HPTLC silica gel	Hexane–chloroform– methanol 1.0:3.0:0.25 v/v	Densitometric, 225 nm; linearity 40–160 ng/spot	Fakhari et al. (2006)
ETH	TLC silica gel 60 F_{254}	Cyclohexane–ether–ethyl acetate, 50:50; 85:15 v/v	254 nm; iodine vapor	Ramic et al. (2006)
ESD ETH	TLC impregnated by cholesterol	Methanol–water	—	Farkas et al. (2003)
EPI	TLC silica gel	Benzene–ethanol, 85:15 v/v	—	Falkay et al. (1973)

TABLE 35.4 (continued)

Chromatographic Parameters for Separation and Determination of Estrogens

Comp.[a]	Stationary Phase	Stationary Phase Pretreatment; Mobile Phase Composition	Detection	Reference
DIE ESD ETH DES	TLC silica gel	Chloroform–methanol, 92:8 v/v Chloroform–acetone, 90:10 v/v	Mixtures of 2,4-dihydroxybenzaldehyde, sulfuric acid, and acetic acid as spray reagent	Huetos et al. (1998)
DIE	TLC silica gel	Cyclohexane–ethyl acetate, 75:25 v/v	Dimethyl sulfoxide as a spray reagent	Agrawal et al. (1984)
ESD ESL ESN	RP-HPTLC	Acetonitrile–methanol Acetonitrile–water Methanol–water	Spraying (mixture of 10 g copper sulfate and 5 mL o-phosphoric acid (86%) dissolved in 95 mL methanol)	Lamparczyk et al. (1990)
ESD DES	TLC silica gel	Cyclohexane–ethyl acetate, 1:1 v/v Benzene–ethyl acetate, 7:1 v/v	Spraying with 20% sulfuric acid in ethanol; detection limit 1–2 ng	Garcia et al. (1991)
ESD	TLC silica gel	Methylene chloride–methanol:2 propanol, 97:1:2 v/v	0.05% aqueous solution of Fast Corinth V and exposing to NH_3 vapor for 30 s	Medina and Schartz (1992)
ETH	TLC silica gel	Cyclohexane–ethyl acetate, 6:4 v/v	Densitometry at 284 nm; LOD 30 ng; linear range 120–1200 ng	Novakovic et al. (1990)
ETH	TLC silica gel	Toluene–ethyl acetate, 7:3 v/v	Spraying with 20% sulfuric acid, heating at 110°C/5 min; 360 nm	Pachaly (1999)
ESD	RP_{18} W	Acetonitrile–water Methanol–water	Spraying with sulfuric acid–methanol 1:9 (v/v) and heating at 120°C (15 min)	Pyka and Babuska (2006)
ESD ESL ESN	HPTLC NH_2-bonded silica	Chloroform–1-propanol– formic acid, 50:10:5 v/v	Heating at 150°C for 3–4 min	Klaus et al. (1994)
ETH	TLC silica gel	Chloroform–ethyl acetate, 8:2 v/v	Densitometry at 280 nm	Molnar et al. (1982)
ETH	TLC silica gel	Toluene–ethyl acetate, 2:1 v/v	Spraying with 1% cerium(IV) sulfate in 10% aqueous sulfuric acid and heating at 110°C/10 min	Brooks et al. (1993)
ETH	HPTLC on spherical silica gel	Cyclohexane–ethyl acetate, 3:2 v/v	Densitometry at 220 nm	Pavic et al. (2003)
DES	Silica gel G	Hexane–diethyl etar– dichloromethane, 4:3:2 v/v	254 nm	Schuller (1967)
ESL ESD DES ETH SHL	Silica gel G	Chloroform:methanol, 98:2 v/v	—	Lee et al. (1973)
ESD	Silica gel 60 F_{254}	Chloroform–acetone– isopropyl alcohol–glacial acetic acid, 9:1:0.4:0.1 v/v	Densitometry at 286 nm. (In pharmaceutical formulation)	Kotiyan and Vavia (2000)

[a] See Figure 35.7.

35.8 AROMATASE INHIBITORS

Aromatase inhibitors are a class of hormone drugs. They block the conversion of androgens to estrogens and, therefore, have the therapeutic potential to control reproductive functions and aid in the treatment of estrogen-dependent cancers, such as breast cancer. These steroidal agents compete with androstenedione for the active site of the aromatase enzyme. The structure–activity relationships for steroidal aromatase inhibitors indicate that the best agents are substrate analogues, with only small structural changes to the A ring and at C_{19} permitted. Analogues that contain aryl functionalities at the 7α position have enhanced affinity for the enzyme. In addition, 4-hydroxy-androstenedione, several androsta-1,4-diene-3,17-diones, and 10β-propynylester-4-ene-3,17-dione act as enzyme-activated irreversible inhibitors in vitro. In Figure 35.8 are shown structural formulas of aromatase inhibitors.

Formestane (FOR), *8R,9S,10R,13S,14S)-4-hydroxy-10,13-dimethyl-2,6,7,8,9,11,12,14,15,16-decahydro-1H-cyclopenta[a]phenanthrene-3,17-dione*, is an antineoplastic aromatase inhibitor, known as a irreversible inhibitor because it permanently binds to the aromatase enzyme. FOR converts to the active androgen 4-hydroxytestosterone, which has about half of the anabolic potency and about 25% of the androgenic potency as TES. Recommended chromatographic parameters for determination of FOR is: silica gel 60 F_{254} as stationary phase and mixtures of chloroform–methanol–water (6:4:1 v/v) as well as chloroform–methanol (19:1 v/v) as mobile phases (Goss 1986). **Testolactone (TST)**, *(4aS,4bR,10aR,10bS,12aS)-10a,12a-dimethyl-3,4,4a,5,6,10a,10b,11,12,12a-decahydro-2H-naphtho[2,1-f]chromene-2,8(4bH)-dione*, is a synthetic antineoplastic agent that is structurally distinct from the androgen steroid nucleus in possessing a six-membered lactone ring in place of the usual five-membered carbocyclic D-ring. Despite some similarity to TES, TST has no in vivo

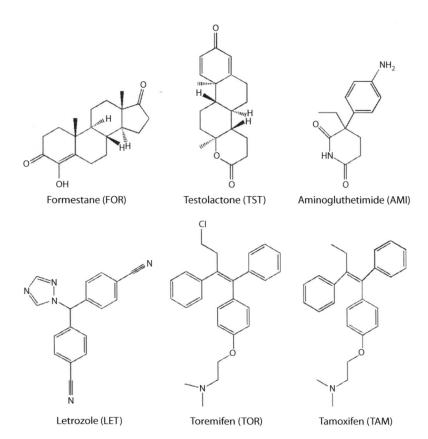

FIGURE 35.8   Structures of aromatase inhibitors.

androgenic effect. No other hormonal effects have been reported in clinical studies in patients receiving TST. TST is a white to off-white, odorless crystalline powder. The following thin-layer chromatographic systems have been reported for separation of TST: (1) Silica gel H F_{254} with a solvent system composed of either ethyl acetate–hexane (5:5 v/v) or chloroform–methanol (97:3 v/v). The bands obtained were located under UV light or by spraying the plate with water (Kusner and Garrett 1971); (2) silica gel F_{254} with a mixture of chloroform–acetone (94:6 v/v) and detection by UV lamp or sulfuric acid spray (Miler 1972); and (3) mixtures of benzene–ethyl acetate (1:1, v/v) as well as (1:2 v/v) as mobile phases on silica gel and detection by spraying with 50% sulfuric acid (Belic et al. 1970).

35.8.1 Nonsteroidal Aromatase Inhibitors

Nonsteroidal aromatase inhibitors inhibit the synthesis of estrogen via reversible competition for the aromatase enzyme. **Aminoglutethimide (AMI)**, *(RS)-3-(4-aminophenyl)-3-ethyl-piperidine-2,6-dione,* is an antisteroid drug marketed under the trade name Cytadren. It blocks the production of steroids derived from CHO and is clinically used in the treatment of Cushing's syndrome and metastatic breast cancer. It is also used by bodybuilders (Gross et al. 2007).

It was initially developed as an anticonvulsant for the treatment of epilepsy, but was subsequently withdrawn because of its inhibitory effects on adrenal function. During drug development, it is important to be able to isolate the enantiomers in order to assess which one is responsible for the potency, the toxicity, and for the side effects. Separation of AMI, acetylaminoglutethimide, and dansylaminoglutethimide was performed by chiral TLC using chiral mobile phase additives selectors, including native cyclodextrins (CDs). Spot visualization was achieved using a fixed wavelength of 254 nm. A mobile phase, consisting of 0.05 M native β-CD–methanol (65:35 v/v) and 30% w/v (carboxymethyl-β-CD)–methanol (65:35 v/v), was used successfully (Aboul-Enein et al. 2000). Furthermore, separation can be done using alumina plate as stationary phase, mixtures of chloroform–benzene (1:1 v/v), and hydroxylamine as visualization reagent (Davies and Nicholls 1965).

Letrozole (LET), *4,4'-(1,2,4-triazol-1-ylmethyl)dibenzonitrile,* is an oral nonsteroidal aromatase inhibitor for the treatment of hormonally responsive breast cancer after surgery. It is a nonsteroidal aromatase inhibitor that dramatically reduces serum levels of ESD, ESN, and ESN sulfate in postmenopausal women by blocking the conversion of adrenal androgens, androstenedione, and TES to ESN and ESD. LET is active when orally administrated and is excreted primarily in the urine. TLC of LET were performed on silica gel using ethyl acetate–acetone (2:1 v/v) and traces of ammonia as mobile phase and UV detection at 254 nm (Gu et al. 2001).

TAM, *Z-2-[4-(1,2-diphenylbut-1-enyl)phenoxy]-N,N-dimethylethanamine,* is an antagonist of the estrogen receptor in breast tissue via its active metabolite, hydroxytamoxifen. In other tissues such as the endometrium, it behaves as an agonist, and thus may be characterized as a mixed agonist/antagonist. Determination of TAM was performed by single-step isocratic HPTLC on plates coated with silica gel 60 F_{254} with densitometric quantitation at 258 nm. Before chromatography, the layers were cleaned by predevelopment to the top with chloroform–methanol (1:1 v/v), then dried in a fume hood and heated on a TLC plate heater at 80°C for 10 min to completly remove the cleaning solvent. Plates were developed with toluene–methanol–glacial acetic acid (57:38:5 v/v) as mobile phase. Linear range was 51.5–309.0 ng per band. The LOD and quantitation were 25 ng and 51.5 ng, respectively, per band. Relative standard deviations were less than 3.0% and 3.1%, respectively, and recovery was 96.6%–103.2% (Jamshidi et al. 2009).

Toremifene, *2-{4-[(1Z)-4-chloro-1,2-diphenyl-but-1-en-1-yl]phenoxy}-N,N-dimethylethanamine,* is a nonsteroidal agent that has demonstrated potent antiestrogenic properties in animal test systems. It is a chlorinated TAM analogue. The antiestrogenic effects may be related to its ability to compete with estrogen for binding sites in target tissues such as breast.

35.9 PROGESTAGENS AND ORAL CONTRACEPTIVE

Steroids in this group include the naturally occurring C_{21} steroids, PRO (*4-pregnene-3,20-dione*) and its metabolites, and synthetic steroids. The group of progestagen hormones includes PRO (corpus luteum hormone) and a number of synthetic preparations with a structure close to that of PRO. The structures of observed progestagens in this section are shown in Figure 35.9.

The progestagen hormones were determined in drug preparations by TLC on Silufol UV-254 plates eluted with ethyl acetate–chloroform (30:70 v/v) mixtures. The spots were detected by the 254 nm fluorescence quenching or by spraying with a 10% solution of molybdophosphoric acid in acetone. The detection limit was 0.1%. This technique is convenient for the identification of steroids as well as detection of impurities (Maslov et al. 1998).

PRO, *pregn-4-ene-3,20-dione*, is the most important naturally occurring progestagen involved in the female menstrual cycle, pregnancy, and embryogenesis of humans and other species. It is excreted as pregnanediol or as a pregnanediol conjugate. Like all steroid hormones, it is hydrophobic. PRO is practically insoluble in water, soluble in ethanol (1 in 8), arachis oil (1 in 60), chloroform (1 in <1), diethyl ether (1 in 16), ethyl oleate (1 in 60), and light petroleum (1 in 100), soluble in acetone, dioxane, and concentrated sulfuric acid, and sparingly soluble in vegetable oils. PRO has chromatographic properties similar to its 6-, 11- and 16-dehydro derivatives on silica gel layers. The separation of PRO from 6- and 11-dehydro-PROs (but not from 16-dehydro-PRO) can only be achieved on silica gel layers after impregnation with silver nitrate. Solvent systems applied were (1) cyclohexane–ethyl acetate (50:50 v/v); (2) hexane–ethyl acetate (75:25 v/v); and (3) benzene–ethyl acetate (50:50 v/v) (Lisboa 1966).

HYP, *17-Hydroxypregn-4-ene-3,20-dione*, is a C_{21} steroid hormone produced during the synthesis of glucocorticoids and sex steroids. It is a natural progestagen, and in pregnancy increases in the third trimester primarily due to fetal adrenal production. As a hormone, 17-OH PRO also interacts with the PRO receptor. It is white or creamy white odorless crystalline powder. Sallam et al. (1969) applied TLC on silica gel with different solvents systems for separation of mixture of C_{21}, C_{19}, and C_{19} steroids among which is HYP. The following solvent systems were used:

1. Cyclohexane–chloroform–acetic acid (80:10:10 v/v)
2. Toluene–petroleum ether (40–60°)–methanol (40:40:20 v/v)
3. Cyclohexane–acetone–chloroform (75:25:20 v/v)
4. Ethylene chloride–acetone (80:20 v/v)
5. Benzene–ethyl acetate–acetone (80:20:20 v/v)
6. Chloroform–cyclohexane–isopropanol (50:100:20 v/v)
7. Chloroform–ethyl acetate (80:20 v/v)
8. Chloroform–acetic acid (90:10 v/v)

The mixtures of 4-chlorosulphonic acid–acetic acid (3:1 v/v) were used for detection of spots (after spraying the plates were heated at 130°C for 5 min). HYP can be successfully separated by application of magnesium silicate as stationary phase and different solvent systems such as benzene–ethanol (98:2 v/v), chloroform, chloroform–ethanol (98:2 v/v), benzene–dioxane (2:1 v/v), and ether–ethanol (98:2 v/v). Furthermore, chromatography can be done on thin layer of alumina with mixtures of benzene–acetone (4:1 v/v). Detection systems for the both stationary phases were: UV light, sulfuric acid (dark brown), sulfuric acid–acetic acid (yellow red), and vanillin–sulfuric acid (dark yellow) (Hara and Mibe 1967; Schwarz 1967).

Medrogestone (MED), *6,17-dimethylpregna-4,6-diene-3,20-dione,* is a synthetic drug with similar effects as PRO, involved in menstrual cycle and pregnancy. It is a derivative of pregna-4,6-diene structurally related to the progestagen chlormadinone (CLM) and the androgen antagonist CYP. MED possesses a lipophilic group at position 6, which is found in other synthetic steroid

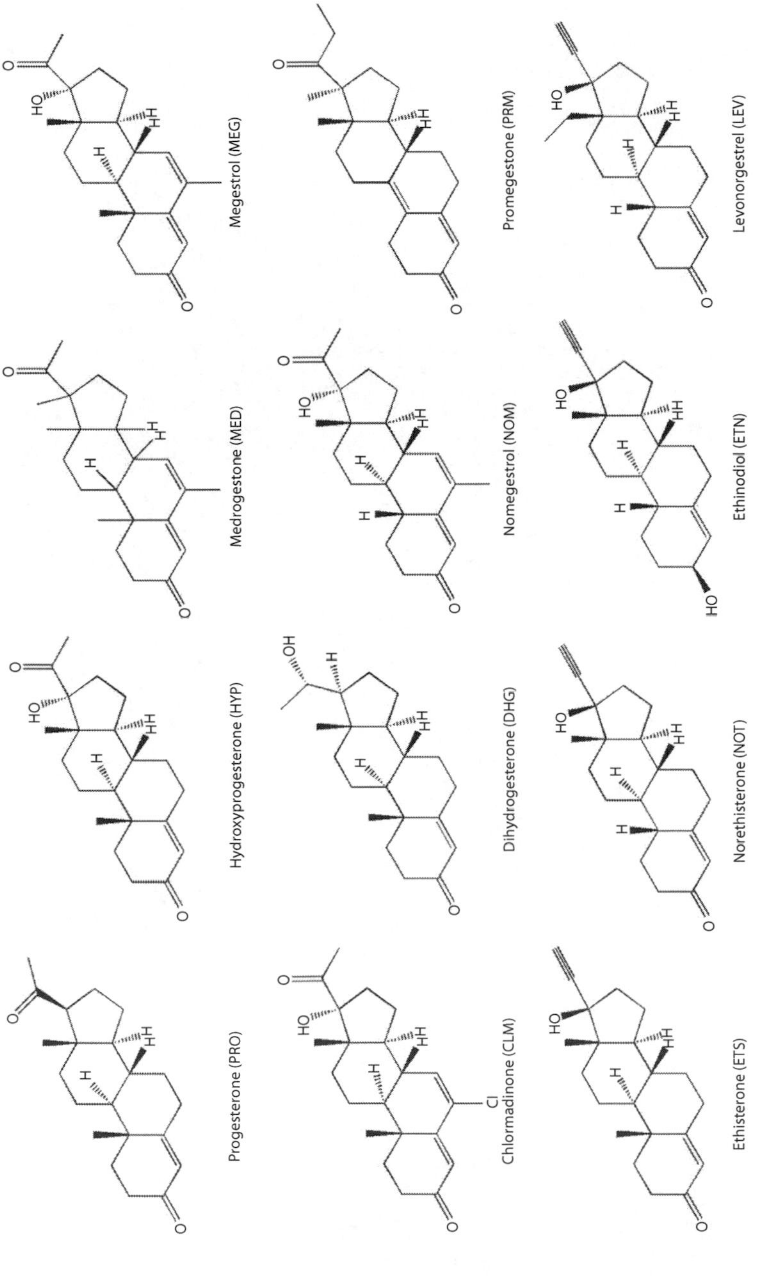

(continued)

FIGURE 35.9 Structures of progestagens.

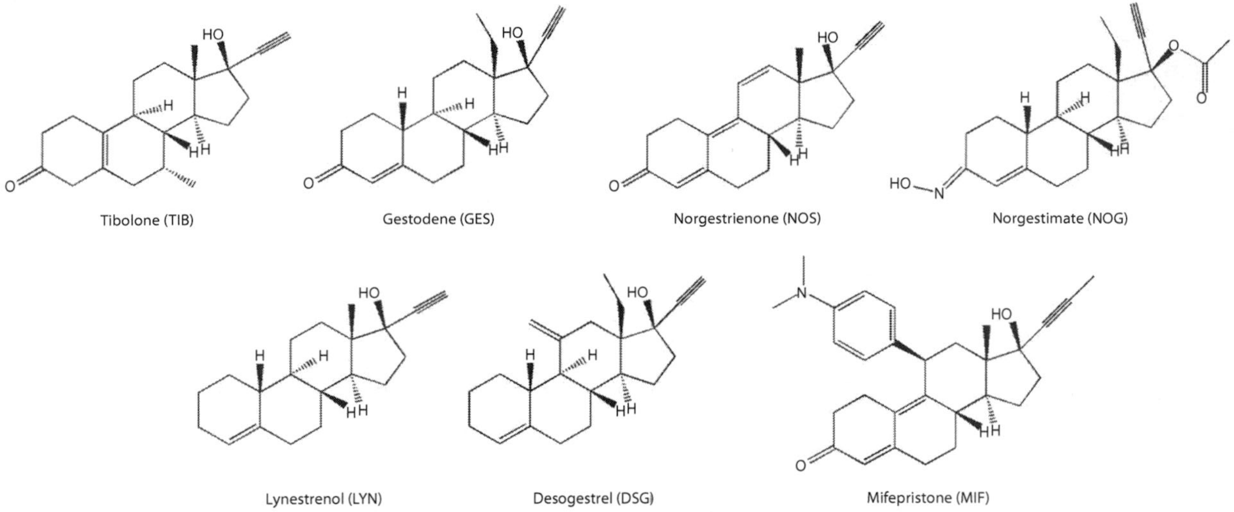

FIGURE 35.9 (continued)   Structures of progestagens.

hormones. MED itself cannot be excreted. The substance is hydroxylized and glucuronidized in the liver and the resulting metabolites are eliminated via urine and feces.

Megestrol (MEG), *17-acetyl-17-hydroxy-6,10,13-trimethyl-2,8,9,11,12,14,15,16- octahydro-1H-cyclopenta[a]phenanthren-3-one,* is synthetic hormone, PRO derivative. Alkyl chain additions to the C_{17} position increase the biological half-life of these compounds. Furthermore, modifications at positions C_6 and C_7 increase their progestational activity. These compounds are metabolized in the same manner as PRO and are excreted in the urine. MEG is practically insoluble in water (2 µg/mL at 37°C), very soluble in chloroform, soluble in acetone, slightly soluble in diethyl ether and fixed oils, and sparingly soluble in ethanol.

CLM, *(9S,14S,17R)-17-acetyl-6-chloro-17-hydroxy-10,13-dimethyl-2,8,9,11,12,14,15,16-octahydro-1H-cyclopenta[a]phenanthren-3-one,* is an orally active synthetic progestagen that shows high affinity and activity at the PRO receptor. It has an antiestrogenic effect and, in contrast to natural PRO, shows moderate antiandrogenic properties. CLM is insoluble in water, very soluble in chloroform, soluble in acetonitrile, and slightly soluble in ethanol and diethyl ether.

Dihydrogesterone (dydrogesterone), **(DHG)**, *17-acetyl-10,13-dimethyl-1,2,8,9,11,12,14, 15,16,17-decahydrocyclopenta[a]phenanthren-3-one,* is a potent, orally active progestagen indicated in a wide variety of gynecological conditions, although similar in molecular structure and pharmacological effects to endogenous PRO. DGS and its metabolites are excreted predominantly by urine. About 85% of an oral dose is excreted within 24 h. DHG is practically insoluble in water, soluble in acetone, chloroform (1 in 2), ethanol (1 in 40), and diethyl ether (1 in 200) slightly soluble in fixed oils and sparingly soluble in methanol.

Chromatographic parameters for separation and qualitative and quantitative analysis of these drugs are listed in Table 35.5.

35.9.1. ORAL CONTRACEPTIVES

Oral contraceptives are among the most effective forms of birth control. They are a class of synthetic steroid hormones that suppress the release of follicle-stimulating hormone (FSH) and luteinizing hormone (LH) from the anterior lobe of the pituitary gland. FSH and LH are called gonadotropic hormones and they stimulate the release of PRO and estrogen from the ovaries, which are responsible for modulating the menstrual cycle. Oral contraceptives are marketed in the form of coated or uncoated tablets containing both a progestagen and estrogen. The progestagen component with pronounced contraceptive effect are norethisterone (NOT), ETS, trengestone, lynestrenol (LYN), ethynodiol, gestonorone, nomegestrol (NOM), promegestone (PRM), desogestrel (DSG), LEV, tibolone (TIB), gestodene (GES), norgestrienone, and MIF (RU 486). They are usually present in milligram amounts, while the estrogen is present (ethinyl estradiol or mestranol) in microgram quantities.

LEV, *(−)-13α-ethyl-17α-hydroxy-18,19-dinor-17α -pregn-4-en-20-yn-3-one,* is a hormonally active levorotatory enantiomer of the racemic mixture norgestrel. Combination of LEV and ethinylestradiol is used in pregnancy prevention in humans. In commonly used low dosage oral contraceptives ethinylestradiol is present at a very low dosage level (0.03–1.0 mg per tablet) in combination with LEV, which is present at a level of 5–30 times that of the ethinylestradiol.

ETS (17-ethynyl TES), *(8R,9S,10R,13S,14S,17R)-17-ethynyl-17-hydroxy-10,13-dimethyl-2,6,7,8,9,11,12,14,15,16-decahydro-1H-cyclopenta[a]phenanthren-3-one,* is a substance with more progestational than androgenic activity. **NOT**, *(8R,9S,10R,13S,14S,17R)-17-ethynyl-17-hydroxy-13-methyl-1,2,6,7,8,9,10,11,12, 14,15,16-dodecahydrocyclopenta[a]phenanthren-3-one*, is widely used as a contraceptive agent, but it is also prescribed for the treatment of endometrial and breast cancer. NOT is practically insoluble in water, slightly to sparingly soluble in ethanol, slightly soluble in diethyl ether, and soluble in chloroform and dioxane. Among TLC and overpressured layer chromatography is

TABLE 35.5

Chromatographic Parameters for Separation and Determination of Progestagens

Comp.[a]	Plate	Stationary Phase Pretreatment; Mobile Phase Composition	Detection	Remarks	Reference
PRO	TLC silica gel 60 F$_{254}$	Benzene–ethyl acetate, 5:1 v/v	Sprayed with a 50% v/v aqueous solution of sulfuric acid and heated to 80°C/45 min	Mixture of progesterone and estrogens	Jambu et al. (2010)
PRO	TLC silica gel	Chloroform–ethanol–benzene, 36:1:4 v/v	Spraying with 5% sulfuric acid in acetic anhydride or ethanol; scanning at 366 nm; then incubated at 95°C/12 min; fluorescence was observed under transillum. at 366 nm	Mixture of progesterone and anabolics	Verbake (1979)
PRO	UTLC silica gel	ethyl acetate–cyclohexane, 3:2 v/v	Video documentation and plates were evaluated by use TLC scanner	Hydrocortisone, progesterone, and testosterone	Hauck et al. (2001)
PRO	Silica gel	Petrol ether–ethyl acetate, 6:4 v/v Chloroform–ethyl acetate–ethanol, 90:10:1 v/v Chloroform–phenol, 9:1 v/v	50% phosphoric acid, heating at 105°C for 15 min	OPLC/mixture of progesterone, cholesterol, corticosterone, aldosterone, cortisol, and cortisone/ qualitative	Szucs et al. (1984)
PRO	Silica gel	Chloroform–ethyl acetate 1:50 v/v	2,4-Dinitrophenylhydrazine	—	Eczely (1985)
PRO	RP-HPTLC	Acetonitrile–methanol acetonitrile–water Methanol–water	Spraying with a mixture of 10 g copper sulfate and 5 mL o-phosphoric acid (86%) dissolved in 95 mL methanol	Mixture with cholesterol, estradiol, estrone, and estriol/quality	Lamparczyk et al. (1990)
PRO	Silica gel	Gradient consisting of methanol–ethyl acetate–chloroform–methylene chloride (First inverse gradient program) Methanol–chloroform (second)	254 nm	Mixture with testosterone, hydrocortisone/ quantification by densitometry	Matyska et al. (1991)
PRO	HPTLC silica gel	Toluene–2-propanol, 9:1 v/v	Visual inspection at 254 nm; quantitative determination in reflectance mode at 252 nm	AMD chamber LOQ 25 ng/zone LOD 5 ng/zone	Jashidi (2004)
PRO	Silica gel H F$_{254}$ hand made	Cyclohexane–ether 8:2 v/v	254 nm	Mixture with testosterone	Bican-Fister (1966)

TABLE 35.5 (continued)

Chromatographic Parameters for Separation and Determination of Progestagens

Comp.[a]	Plate	Stationary Phase Pretreatment; Mobile Phase Composition	Detection	Remarks	Reference
HYP	Silica gel	First dichloromethane–methanol–water, 225:15:1.5 v/v Second light petroleum–diethyl ether–acetic acid, 48:50:2 v/v	UV light on 254 nm or spraying by SbCl$_3$	Two dimensional; total running time is <8h: mixture with estrogens and corticosteroids	Brown et al. (1988)
HYP	Silica gel	Chloroform–methanol, 10:1 v/v	Rubeanic acid as spray reagent; densitometry	The plates were heated after spraying at 110°, and the color responses were observed	Garcia (1985)
MEG CLM PRO	Silica gel	Chloroform–methanol, 92:8 v/v chloroform–acetone, 90:10 v/v	Mixtures of 2,4-dihydroxybenzaldehyde, sulfuric acid, and acetic acid as spray reagent	Mixture with corticosteroids and estrogens	Huetos et al. (1998)
DHF	Silica gel 60 F$_{254}$	Chloroform-ethyl acetate, 4:1, v/v	254 nm	—	Chetrite et al. (2004)
MEG DHG	Silica gel G F$_{254}$	Chloroform–methanol, 9:1, v/v followed by benzene–methanol, 95:5 v/v Benzene–acetone 8:2 v/v, followed by methylene chloride–methanol–water, 150:9:0.5 v/v	254 nm; Spraying with conc. sulfuric acid and heated on 100°C for 30 min	Two dimensional; mixture of gestagens	Frank et al. (1986)
PRM	Silica gel G F$_{254}$	Chloroform–thylacetate, 4:1 v/v ethyl acetate–methanol–ammonium hydroxide, 75:25:2	UV detection	—	Chetrite et al. (1998)

[a] See Figure 35.9.

a suitable method for separation of all the expected impurities of NOT both in the bulk drug substance and in tablet form (Bagocsi et al. 2003). The chromatographic parameters are shown in Table 35.6. **PRM**, *(8S,13S,14S,17S)-13,17-dimethyl-17-propanoyl-1,2,6,7,8,11,12,14,15,16-decahydrocyclopenta[a] phenanthren-3-one,* and **RU 486** or **MIF**, *11β-[p-(Dimethylamino)phenyl]-17β-hydroxy-17-(1-propynyl)estra-4,9-dien-3-one,* are antiprogestagens, which sensitize the myometrium to prostaglandin-induced contractions, softens and dilates the cervix. Its remains the most well-studied antiprogestin of clinical importance to date. **NOM**, *17α-hydroxynor-PRO derivate (17α-acetoxy-6-methyl-19-nor-4,6-pregnadiene-3,20-dione),* is a potent and useful synthetic progestagen for the treatment of menopausal complaints and is under current development for oral contraception. **Etynodiol (ETN)**, *(3S,8R,9S,10R, 13S,14S,17R)-17-ethynyl-13-methyl-2,3,6,7,8,9,10,11,12,14,15,16-dodecahydro-1H-cyclopenta[a]*

TABLE 35.6

Chromatographic Parameters for Separation of Oral Contraceptive Steroids

Comp.[a]	Stationary Phase	Stationary Phase Pretreatment; Mobile Phase Composition	Detection	Remarks	Reference
LEV	TLC silica gel 60 F_{254}	Hexane–chloroform–methanol, 1:3:0.25 v/v	Densitometric scanning (Camag TLC scanner 3) in the reflectance mode at 225 nm	Plates were prewashed by methanol; activated at 80°C/5 min	Fakhari et al. (2006)
LEV	TLC and HPTLC silica gel 60 F_{254}	Cyclohexane–diethyl ether–acetic acid, 50:50:1 v/v	1. Diffuse reflection method at 248 nm 2. Fluorescence method—0.5% methanolic sulfuric acid as spray reagent and heated at 120°C/10 min (excit. at 365 nm/meas. 485 nm)	diffuse reflection method RSD 0.44%–2.6%; LOL 1–5 µg fluorescence method RSD 3.4–5.2%; LOL 50–250 ng	Amin and Hassenbach (1979)
ETN	Silica gel G	Ethyl acetate–cyclohexane, 1:1 v/v or 3:7 v/v	Spraying with a saturated solution of antimony trichloride in chloroform	Mixture with 38 19-nor-steroids	Golab and Layne (1962)
ETN	Silica gel G F	Chloroform–methanol, 90:10 v/v Methylene chloride–methanol–water, 150:9:0.5 v/v	Spraying with concentrated sulfuric acid; heating at 100°C/38 min; short-wave UV light	—	Simard and Lodge (1970)
ETN LEV DES	Silica gel	Cyclohexane–butyl acetate–chloroform, 5:1:4 v/v for LEV Toluene–ethyl acetate–chloroform, 86:7:7 v/v for ETN and DES	—	One- and two-dimensional technique; mixture with estrogens	Ferenczi-Fodor et al. (1999)
NOM	Silica 60 F_{254}	Chloroform–ethyl acetate, 4:1 v/v	254 nm	—	Chetrite et al. (2005)
ETS	Silica gel	Ethyl acetate as strong solvent and different weak solvents	Spaying with 10% sulfuric acid; heating at 110°C/10 min; visualization by fluorescence mode	Weak solvents: chloroform, bromobenzene, decalin, and cyclohexane	Nurok et al. (1993)
ETS	Silica gel G	Cyclohexane–ethyl acetate, 75:5 v/v	Spraying with dimethyl sulfoxide	Mixt. with cholesterol and estrogens	Agarwal and Nwaiwu (1984)
NOT	Silica gel	Chloroform–acetone, 9:1, v/v Chloroform–methanol, 95:5 v/v	Spraying by 10% ethanolic sulfuric acid and heating at 120°C/2 min; 366 nm	Documentation of the chromatograms by VideoStore 2	Bagocsi et al. (2003)

TABLE 35.6 (continued)

Chromatographic Parameters for Separation of Oral Contraceptive Steroids

Comp.[a]	Stationary Phase	Stationary Phase Pretreatment; Mobile Phase Composition	Detection	Remarks	Reference
NOT	Sealed silica gel for OPLC	Continuous development with n-hexane and butyl acetate-chloroform 85:15 v/v	Spraying by 10% ethanolic sulfuric acid and heating at 120°C/2 min; 366 nm	Flow rate 400 μL/min	N/A
LEV DES	Silica gel G F	Benzene–ether, 1:1 v/v 2× Chloroform–ethyl acetate, 85:15 v/v 2× chloroform–ether, 9:1 v/v 2×	UV light, staining with tungstophosphoric acid, and when using a radioactive substrate, with a radio thin-layer scanner	—	Groh et al. (1997)
LEV DES	Alumina	Benzene–ethanol, 98:2 v/v 2×	N/A	—	N/A
TIB	TLC silica gel 60 F$_{254}$	First toluene–cyclohexane, 50:50 v/v Second dichloromethane–diethyl ether, 80:20 v/v	254 nm	Metabolite of tibolone	Blom et al. (2006)
TIB	TLC silica gel	Toluene–methanol, 9:1 v/v	—	—	Raobaikady et al. (2006)
NOG	HPTLC NH$_2$ F$_{254}$ or Silica	n-hexane–methyl ethyl ketone–diethylamine, 8:1:1 v/v	254 nm	—	Ferenczi-Fodor et al. (1987)
LYN NOS	Silica gel	Benzene–acetone, 92.5:7.5; 70:30; 80:20 v/v Chloroform–methanol, 95:5 v/v Chloroform–ethyl acetate, 95:5 v/v	Radioimmunoassay	Direct labeling with I^{125}	Pala and Benagian (1976)

[a] See Figure 35.9.

phenanthrene-3,17-diol, has similar biophysiological effects as PRO. ETN is very slightly soluble to practically insoluble in water, soluble in ethanol, freely to very soluble in chloroform, and freely soluble in diethyl ether.

TIB, *17-Hydroxy-7α-methyl-19-nor-17α-pregn-5(10)-en-20-yn-3-one*, is a 7α-methyl derivative of the synthetic progestagen, norethynodrel, and is used as a hormone replacement therapy for postmenopausal women. TIB was rapidly metabolized in rat uterine and vaginal tissue fragments yielding mainly 3α-OH-TIB, its 3α-reduced estrogenic derivative. **LYN**, *(8R,9R,10R,13S, 14S,17S)-17-ethynyl-13-methyl-2,3,6,7,8,9,10,11, 12,14,15,16-dodecahydro-1H-cyclopenta [a]phenanthren-17-ol*, has a strong progestational effect on the uterine endometrium, inhibits secretion of gonadotropin, suppresses maturation of follicles in the ovaries and ovulation, and reduces menstrual bleeding. This synthetic gestagen is associated with minimal estrogenic, androgenic, and anabolic effects. **GES**, *(8R,9S,10R,13S,14S,17R)-13-Ethyl-17-ethynyl-17-hydroxy-1,2,6,7,8,9,10,11,12,14-decahydrocyclopenta[a]phenanthren-3-one*, is androgenically neutral, meaning that contraceptive pills containing GES do not exhibit androgenic side effects (e.g., acne, hirsutism, and weight gain) often associated with second-generation contraceptive pills, such as those containing LEV. GES does not occur naturally.

DSG, *13-ethyl-17-ethynyl- 11-methylidene- 1,2,3,6,7,8,9,10,12,13,14,15, 16,17-tetradeca hydrocyclopenta[a] phenanthren-17-ol*, **norgestimate (NOG)**, *[(3E,8R,9S, 10R,13S,14S,17R)-13-ethyl-17-ethynyl-3-hydroxyimino-1,2,6,7,8,9,10,11,12,14,15,16-dodecahydrocyclopenta[a] phenanthren-17-yl] acetate*, and **norgestrienone (NOS)**, *(17a)-17-Hydroxy-19-norpregna-4,9,11-trien-20-yn-3-one*, were developed in an attempt to produce agents with more selective progestational activity that would improve cycle control and minimize metabolic changes and adverse events while effectively preventing pregnancy. DSG is practically insoluble in water, slightly soluble in ethanol and ethyl acetate, and sparingly soluble in *n*-hexane.

Bebawy et al. (2000) developed a precise, accurate, and selective method for the determination of GES in the mixture with CYP acetate (androgen) and ETH. After TLC separation on silica gel G F_{254} using dichloromethane–methanol–water (95:5:0.2 v/v) as mobile phase, the chromatograms were scanned at 247 and 281 nm. R_F of GES, CYP acetate, and ETH is 0.64, 0.74, and 0.46, respectively. A linear correlation was obtained in the concentration range of 0.2–1.6 and 0.1–0.7 μg/μL using Hamilton syringe 10 μL with mean percentage recoveries of 100.72% and 99.82% and RSD of 1.90% and 1.75% for GES and CYP acetate, respectively.

In the Table 35.6 are listed chromatographic parameters for separation and determination of oral contraceptive steroids.

35.10 CORTICOSTEROIDS

Corticosteroids are a group of C_{21} steroids produced by the adrenal cortex. The following features characterize them structurally: (1) a double bond at C_4 and an oxo group at C_3 (they therefore belong to the class of active steroid hormones known as 4-en-3-ones); (2) a side chain at C_{17} consisting of a C_{21} hydroxyl group with an oxo group at C_{20}; (3) the presence or absence of a hydroxyl group at C_{17}, thus giving rise to the classification of 17-oxygenated corticosteroids, such as cortisol or 11-deoxycortisol, and 17-deoxy-corticosteroids, such as 11-deoxycorticosterone, corticosterone, and ALD, often confusingly referred to, on the basis of their prime physiological activity, as gluco- and mineralo-corticoids; and (4) a hydroxyl or oxo group, which may or may not be present at C_{11} (Fraser et al. 2010).

35.10.1 GLUCOCORTICOIDS

The glucocorticoids are a class of hormones so called because they are primarily responsible for modulating the metabolism of carbohydrates. Glucocorticoids, such as cortisol, exert effects on a wide range of metabolic events in tissues as diverse as those in liver and skeletal muscle, as well as in adipose and lymphatic tissues. The primary effects are those that occur within 4 h of administration of the hormones, and they include increased glycogen synthesis, increased gluconeogenesis, increased amino-acid uptake, increased hepatic RNA and protein synthesis, and promotion of lipid mobilization.

The naturally occurring and synthetic glucocorticoids are known to have important effects on the inflammatory response and on fibroblast synthesis of collagen and glycosaminoglycans. The principal need to measure cortisol relates to human diseases characterized by the deficiency of adrenal steroid secretion (e.g., Addison's disease) or conversely overproduction of glucocorticoids (e.g., Cushing's syndrome). In both cases, reliable measures of cortisol concentration are required.

Hydrocortisone (cortisol) (HYC), *(8S,9S,10R,11S,13S,14S,17R)-11,17-dihydroxy-17-(2-hydroxyacetyl)-10,13-dimethyl-2,6,7,8,9,11,12,14,15,16-decahydro-1H-cyclopenta[a]   phenanthren-3-one*, is quantitatively the major biologically important product of the adrenal cortex and is found in virtually all body tissues. It is released in response to stress and a low level of blood glucocorticoids. Its primary functions are to increase blood sugar through glycogenolysis, suppress the immune system, and aid in fat, protein, and carbohydrate metabolism. Commercially, it is available as the unchanged hormone and as hydrocortisone acetate, hydrocortisone cypionate, hydrocortisone sodium phosphate, hydrocortisone butyrate, hydrocortisone valerate, and hydrocortisone sodium succinate. Preparation of sample for identification and quantification of HYC in pharmaceutical formulation is very simple. Tablets were ground and powders obtained were shaken with absolute ethanol. After shaking, the solution was filtered through a medium-density filter (Pyka et al. 2011). Chromatographic parameters for investigation of this solution are given in Table 35.7. One of the possibilities is TLC after fluorescence derivatization of hydrocortisone with isonicotinic acid hydrazide (Fenske 1998, 2000).

Cortisone (COR), *(8S,9S,10R,13S,14S,17R)-17-hydroxy-17-(2-hydroxyacetyl)-10,13-dimethyl-1,2,6,7,8,9,12,14,15,16-decahydrocyclopenta[a]phenanthrene-3,11-dione*, is one of the main hormones released by the adrenal gland in response to stress. In chemical structure, it is a corticosteroid closely related to corticosterone. It is used to treat a variety of ailments and can be administered intravenously, orally, intra-articularly, or transcutaneously. In addition to the metabolic effects of the glucocorticoids, these hormones are immunosuppressive and anti-inflammatory.

Prednisone (PRS), *(8S,9S,10R,13S,14S,17R)-17-hydroxy-17-(2-hydroxyacetyl)-10,13-dimethyl-6,7,8,9,12,14,15,16-octahydrocyclopenta[a]phenanthrene-3,11-dione*, is the most commonly prescribed synthetic corticosteroid for arthritis. PRS is four to five times more potent than as cortisol. Some topical pharmaceutical preparations contain **mometasone (MOM)**, *(8S,9R,10S,11S,13S,14S,16R,17S)-9-chloro-17-(2-chloroacetyl)-11,17-dihydroxy-10,13,16-trimethyl-6,7,8,11,12,14,15,16-octahydrocyclopenta[a] phenanthren-3-one*. MOM furoate is a high potent chlorinated glucocorticoid with a favorable ratio between local and systemic side effects (Teng et al. 2001). **Fluticasone** propionate, *S-(fluoromethyl)(6S,8S,9R,10S,11S,13S,14S,16R,17R)-6,9-difluoro-11,17-dihydroxy-10,13,16-trimethyl-3-oxo-6,7,8,11,12,14,15,16-octahydrocyclopenta[a] phenanthrene-17-carbothioate*, is a neutral, highly potent trifluorinated corticosteroid based on the androstane nucleus. It is effective in treatments of asthma and allergic rhinitis because of its antiinflammatory activity (Laugher et al. 1999). Furthermore, a synthetic glucocorticosteroids are: **prednisolone**, *(8S,9S,10R,11S,13S,14S,17R)-11,17-dihydroxy-17-(2-hydroxyacetyl)-10,13-dimethyl-7,8,9,11,12,14,15,16-octahydro-6H-cyclopenta[a] phenanthren-3-one*, **triamcinolone**, *(8S,9R,10S,11S,13S,14S,16R,17S)-9-fluoro-11,16,17-trihydroxy-17-(2-hydroxyacetyl)-10,13-dimethyl-6,7,8,11,12,14,15,16-octahydrocyclopenta[a] phenanthren-3-one*, **dexamethasone (DXM)**, *(8S,9R,10S,11S,13S,14S,16R,17R)-9-fluoro-11,17-dihydroxy-17-(2-hydroxyacetyl)-10,13,16-trimethyl-6,7,8,11,12,14,15,16-octahydrocyclopenta[a] phenanthren-3-one*, **budesonide**, *16,17-(butylidenebis(oxy))-11,21-dihydroxy-*, *(11-β,16-α)-pregna-1,4-diene-3,20-dione*, **beclomethasone**, *(8S,9R,10S,11S,13S,14S,16S,17R)-9-chloro-11,17-dihydroxy-17-(2-hydroxyacetyl)-10,13,16-trimethyl-6,7,8,11,12,14,15,16-octahydro cyclopenta[a]phenanthren-3-one*, **fluocinolone**, *(6S,8S,9R,10S,11S,13S,14S,16R, 17S)-6,9-difluoro-11,16,17-trihydroxy-17-(2-hydroxyacetyl)-10,13-dimethyl-6,7,8,11,12,14,15,16-octahydrocyclopenta[a]phenanthren-3-one*, **amcinonide**, *2-[(1S,2S,4R,8S,9S,11S,12R,13S)-12'-fluoro-11'-hydroxy-9',13'-dimethyl-16'-oxo-5',7'-dioxaspiro[cyclopentane-1,6'-entacyclo[10.8.0.0^{2,9}.0^{4,8}.0^{13,18}]icosane]-14',17'-dien-8'-yl]-2-oxoethyl acetate*, **clobetasol (CLB)**, *(8S,9R,10S,11S,13S,14S,16S,17R)-17-(2-chloroacetyl)-9-fluoro-11,17-dihydroxy-10,13,16-trimethyl-6,7,8,11,12,14,15,16-octahydrocyclopenta[a]phenanthren-3-one*, **methylprednisolone**, *(6S,8S,9S,10R,11S,13S,14S,17R)-11,17-dihydroxy-17-(2-hydroxyacetyl)-6,10,13-trimethyl-7,8,9,11,12,14,15,16-octahydro-6H-cyclopenta[a] phenanthren-3-one*, **fluorometholone**,

TABLE 35.7

Chromatographic Parameters for Separation of Corticosteroids

Comp.[a]	Stationary Phase	Stationary Phase Pretreatment; Mobile Phase Composition	Detection	Remarks	Reference
ALD HYC COR	Silica gel	Dichloromethane–methanol–water, 225:15:1.5, v/v and light petroleum–diethyl ether–acetic acid, 48:50:2 v/v as the second solv.	UV light on 254 nm or spraying by SbCl₃	Two dimensional; total running time is <8 h:mixture with estrogens and progestagens	Brown et al. (1988)
ALD	TLC silica gel 60 F₂₅₄	Benzene–acetone–water, 3:2:0.025 v/v ethyl acetate–acetone–water, 6:1:0.025 v/v	—	Determ. by radioimmunoassay; recovery 73%	Hilfenhaus (1977)
ALD HYC COR	TLC silica gel F₂₅₄	8% Ethanol (96%) in chloroform	UV light	Ascending technique for 1 h at 38.5°C	Bruinvels (1963)
DXM	Polivinyl pyrrolidone plate	Dichloromethane–acetic acid, 94:6 v/v	Spray reagent: 100 mL of a 0.4%/(w/v) of isonicotinic acid hydrazide in methanol with 0.5 mL conc. hydrochloride acid	—	Dalmau et al. (1973)
PRE COR DXM HYC FLD	Kieselgel 60 HPTLC	Chloroform–methanol, 92:8 v/v	Spray reagent: equal amounts of solutions A (1% resorcylaldehyde in acetic acid) and B (10% sulfuric acid in acetic acid); heating at 95°/10 min; 366 nm	Mixtures of corticosteroids	Vanoosthuyze et. al (1993)
ALD COR HYC PRE PRS	Silica gel	Ethyl acetate Isopropanol–n-butyl ether, 3:7 v/v Toluene–ethyl acetate–ethanol, 60:35:5 v/v Toluene–β-ethoxyethyl acetate, 2:3 v/v	254 nm	Quantitative TLC of corticosteroids down to the 0.001 µg	Few and Forward (1968)
HYC	TLC silica gel 60 F₂₅₄	Acetone–n-hexane–glacial acetic acid, 5 mL:5 mL:1 drop	After development, the plates were dried for 20 h at room temperature and scaned at 250 nm (Camag TLC scanner)	Determin. in pharmaceuticals. LOD 0.35 µg/spot, LOQ 1.10 µg/spot	Pyka et al. (2011)

TABLE 35.7 (continued)
Chromatographic Parameters for Separation of Corticosteroids

Comp.[a]	Stationary Phase	Stationary Phase Pretreatment; Mobile Phase Composition	Detection	Remarks	Reference
HYC PRE	HPTLC RP18W	Plate was dipped in methanol for 1 min, acetonitrile and buffer solut.	Horizontal developing DS chamber (type DS-II-5 × 10	TLC 2010 Diode Array Scanner	Płocharz et al. (2010)
HYC PRE	Plate with the sealant solution	At 105°C–110°C/5 min to polymerize the sealant; acetonitrile and buffer solut.	Pressurized planar electrochromatography; PPEC chamber	The sealant solution: mixture Sarsil Wor Sarsil H50 with hardener 100:4 (w/w)	N/A
HYC CHO	Glass plates Diol F$_{254}$	Chloroform	Densitometric detection with and without sulfuric acid as visualizing reagents	Mixtures with estrogens	Pyka (2009)
PRE HYC COR DXM	Silufol UV 254	Distilled water to which 3, 5, 8.3, 12, 15, 17, or 20 mM of sodium dodecyl sulfate (SDS) was added	254 and 365 nm	—	Kartsova and Strel'nikova (2007)
HYC	UTLC monolithic silica gel	Ethyl acetate–cyclohexane, 3:2 v/v	TLC scanner II at 254 nm	Migration time: 345 s Migration distance: 2 cm	Hauck and Schulz (2003); Hauck et al. (2001)
COR DXM	TLC silica gel F$_{254}$	Acetone–toluene, 1:1 v/v	254 nm	Determination of free cortisol in the urine and feces of guinea pigs; recovery 98%	Fenske (1999)
COR HYC	TLC aluminum sheets silica Gel 60 F$_{254}$	Acetone–toluene 1:1 v/v Acetone–chloroform–toluene, 1:1:1 v/v	Plates dipped into isonicotinyl hydrazones; 60 min later into paraffin:chloroform (1:2, v/v); fluorescence (exc. 370 nm, emiss. 460 nm)	Ascending chromatography; matrix: urine	Fenske (2000)
DXM COR HYC	TLC silica gel 60F$_{254}$	Chloroform–methanol, 92:8 v/v Chloroform–acetone, 90:10 v/v	Mixtures of 2,4-dihydroxybenzaldehyde, sulfuric acid, and acetic acid as spray reagent	No significant interferences were found	Huetos et al. (1998)

(continued)

TABLE 35.7 (continued)
Chromatographic Parameters for Separation of Corticosteroids

Comp.[a]	Stationary Phase	Stationary Phase Pretreatment; Mobile Phase Composition	Detection	Remarks	Reference
HYC	Silica gel	First inverse gradient program methanol–ethyl acetate–chloroform–methylene chloride	254 nm	Programmed multiple development of analysis of mixture of steroids	Matyska et al. (1991)
		Second inverse gradient program Methanol–chloroform			
DXM TRI PRE HYC	TLC silica gel	Chloroform–methanol–water, 180:15:1 v/v	240 nm	—	Das et al. (1985)
PRE COR HYC	TLC Silica gel 60 F$_{254}$	Acetone–toluene, 50:50 v/v	254 nm	Linearity 0.5–3000 ng/sample Recovery 78%	Lewbart and Elverson (1982)
COR HYC PRE PRS DXM	Silica gel impregnate using HCONH$_2$	Chloroform	Sulfuric acid color reaction	—	Vanderhaeghe and Hoebus (1976)
TRI DXM PRE FCN FLD HYD CLB FMT COR MPR AMC MOM DFC	Silica gel 60 F$_{254}$	Ethyl acetate–*n*-hexane, 40:60 v/v Ethyl acetate–*n*-hexane, 60:40 v/v Ethyl acetate–*n*-hexane, 80:20 v/v	The plates were air dried at room temperature and observed under UV light; each set was sprayed with detection reagents: 20% solution of sulfuric acid in ethanol, and a mixture of equal volumes of 0.1% tetrazolium blue in methanol and of 16% NaOH in methanol	All standards dissolved in chloroform–methanol, 1:1, v/v; recovery 88%–98%	Gagliardi et al. (2002)
COR PRE MPR DXM TRI BMT HYC FCR FMT	HPTLC Silica gel 60	Chloroform–methanol, 92:8 v/v	The plate was dried under a cool air stream and sprayed with equal amounts of 1% resorcyl aldehyde in acetic acid and 10% sulfuric acid in acetic acid; after spraying, the plate was heated for 10 min at 95°C in an oven and examined in daylight and at 366 nm	18 mobile phases were tested for separation of mixture of corticosteroids; LOD 750 ng/spot	Vanoosthuyze et al. (1993)

TABLE 35.7 (continued)
Chromatographic Parameters for Separation of Corticosteroids

Comp.[a]	Stationary Phase	Stationary Phase Pretreatment; Mobile Phase Composition	Detection	Remarks	Reference
DXM HYD BMT CLB	TLC silica gel G	*n*-Hexane as the first developing solvent (to wash out cream or ointment base) and chloroform–ethyl acetate 1:1 v/v	254 nm	Identification of corticosteroids and their esters in creams and ointments	Datta and Das (1994)
DXM PRE MPR DSM FMT FCN	TLC silica gel	Chloroform–methanol 24:1 v/v	254 nm by HPTLC scanner	—	Zivanovic et al. (1987)
PRE HYC DXM FCN	3-Cyanopropyl trichlorosilane treated silica gel plates	Propanol–water Methanol–water Ethanol–water	Sprayed with 10% sulfuric acid and heated at 100°C for 10 min; the corticoids appeared as fluorescent violet spots on a dark white background	Matrix: ointment recovery 95%–100%	Okamoto and Ohta (1986)
PRE FCN	TLC Silica gel	Chloroform–dimethyl ketone, 70:30 v/v	Iodine vapors	Recovery 96.0% ± 2.0%	Choulis (1968)
BUD	Cellulose	1% aqueous solution of β-cyclodextrin with methanol	Densitometry at 245 nm	LOD 34.64–43.72 ng, recovery 98.76%–103.60%	Krzek et al. (2002)
DXM	Silica gel 60 and F_{254}	Cyclohexane–ethyl acetate, 2:3 v/v	Densitometry 245 nm	—	Krzek et al. (2005)
DXM	HPTLC silica gel/spherical particles	Ethyl acetate–acetonitrile–diethylamine–water, 3:3:1:1 v/v	TLC scanner set at 240 nm in reflectance/absorbance	RSD 1.67%–2.77% recovery 95.95–100.69%	Agbaba et al. (2001)
DXM	HPTLC silica gel 60	First methylene chloride–diethyl ether–methanol–water, 77:15:8:1.2 v/v Second chloroform–ethyl acetate–water, 10:90:1	Spraying with 2,4-dihydroxybenzaldehyde; heating in an oven for 15 min at 90°C	Matrix: feed samples; two dimensionally; Recovery 62%, LOD 50 ng/g,	Huetos et al. (1999)

(continued)

TABLE 35.7 (continued)
Chromatographic Parameters for Separation of Corticosteroids

Comp.[a]	Stationary Phase	Stationary Phase Pretreatment; Mobile Phase Composition	Detection	Remarks	Reference
MOM	HPTLC silica gel 60 F_{254}	Dichloroethane–diethyl ether–ammonia–methanol–ethyl acetate, 6:3:0.2:1.75:3.5	Densitometry at 254 nm	Analysis of drugs in pharm. form.; linearity 100–300 ng/band; RSD 100.3% ± 1.2%	Kulkarni et al. (2010)
MOM	TLC silica gel 60 F_{254}	Dichloromethane–diethyl ether, 3:1 v/v	Densitometry at 260 nm (TLC scanner II)	Analysis of drugs in pharm. form.; linearity 85–400 ng/spot; recovery 100.3% ± 0.8%	Wulandari et al. (2003)
FTC	HPTLC silica gel	n-Hexane–ethyl acetate–acetic acid, 5:10:0.2 v/v	TLC scanner set at 250 nm in reflectance/absorbance	Mixture with antiasthmatic drugs	Kasaye et al. (2010)
HYC PRE PRS MPR	TLC silica gel 60 F_{254}	Cyclohexane–ether, 50:50 v/v Cyclohexane–ether, 85:15 v/v Ethyl acetate	254 nm	—	Ramic et al. (2006)
FNS HYC	TLC silica gel G 60 F_{254}	Chloroform–methanol, 85:15 v/v	Sprayed with iodic acid solution	Investigation of sensitivity of iodic acid reagent	Cavazzutti et al. (1983)

[a] See Figure 35.10.

(6S,8S,9R,10S,11S,13S,14S,17R)-17-acetyl-9-fluoro-11,17-dihydroxy-6,10,13-trimethyl-6,7,8,11,12,14,15,16-octahydrocyclopenta [a]phenanthren-3-one, **desoxymethasone**, *(8S,9R,10S,11S,13S,14S,16R,17S)-9-fluoro-11-hydroxy-17-(2-hydroxyacetyl)-10,13,16-trimethyl-7,8,11,12,14,15,16,17-octahydro-6H-cyclopenta[a]phenanthren-3-one*, **diflucortolone**, *(6S,8S,9R,10S, 11S,13S,14S,16R,17S)-6,9-difluoro-11-hydroxy-17-(2-hydroxyacetyl)-10,13,16-trimethyl-7,8,11,12,14,15,16,17-octahydro-6H-cyclopenta[a]phenanthren-3-one*, and **flunisolide**, *(1S,2S,4R,8S,9S,11S,12S,13R,19S)-19-fluoro-11-hydroxy-8-(2-hydroxyacetyl)-6,6,9,13-tetramethyl-5,7-dioxapentacyclo[10.8.0.0^{2,9}.0^{4,8}.0^{13,18}]icosa-14,17-dien-16-one*. Structures of corticosteroids are shown in Figure 35.10. Chromatographic parameters for separation and determination of these drugs are listed in Tabele 35.7. Advances in analysis of xenobiotic glucocorticoids in biological matrices have been driven largely by the need to control their use in sport, and also because of the need to monitor their use as growth promoters in livestock.

35.10.2 MINERALOCORTICOIDS

Mineralocorticoids (MCs) are steroid hormones with the ability to stimulate sodium (Na$^+$) reabsorption in the distal nephron as well as in the large intestine and salivary glands. They play a key role in controlling salt and water balance. As the name of this class of hormones implies, the MCs control the excretion of electrolytes. This only occurs primarily through actions on the kidneys but also in the colon and sweat glands. The major circulating MC is ALD.

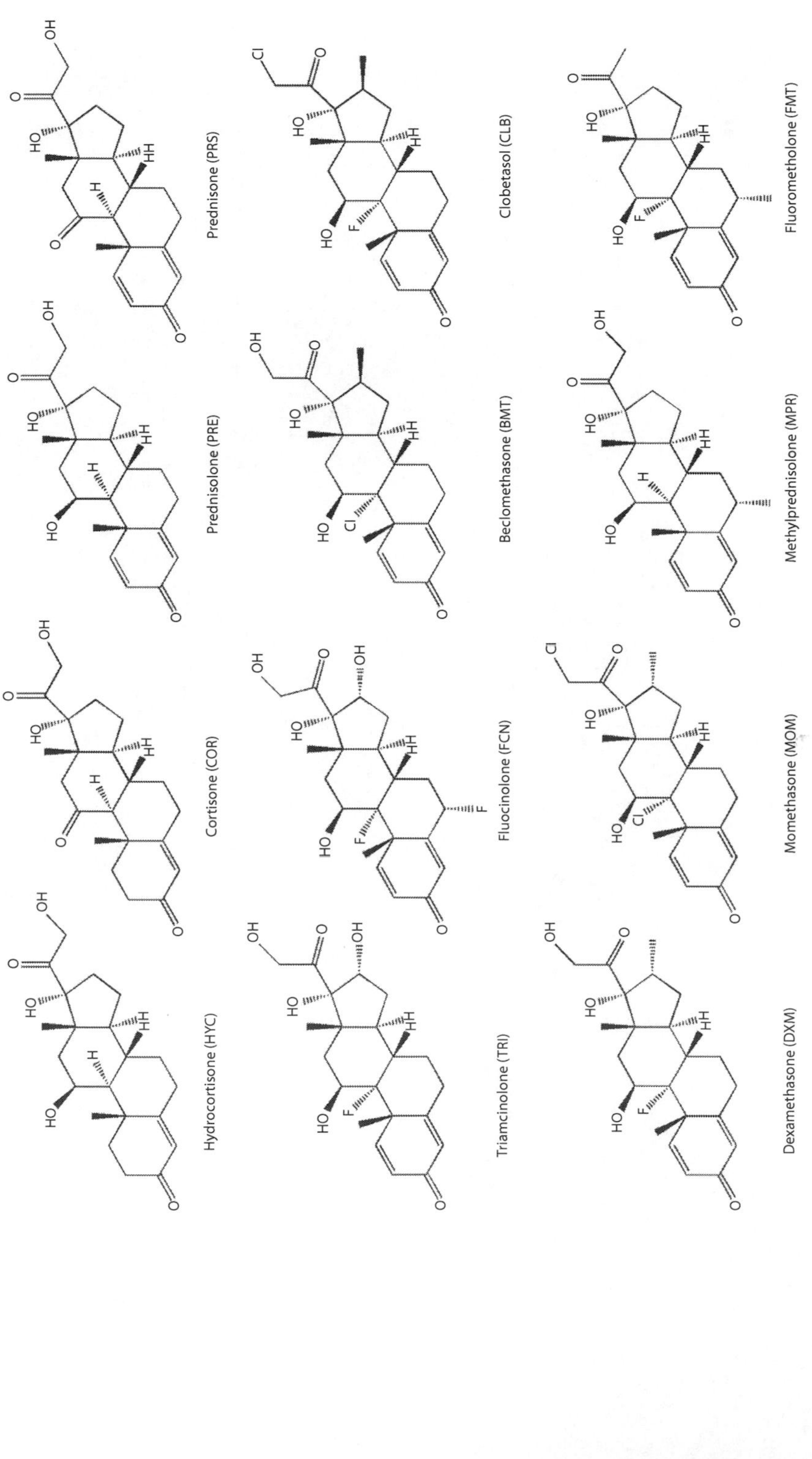

(continued)

FIGURE 35.10 Structures of corticosteroids.

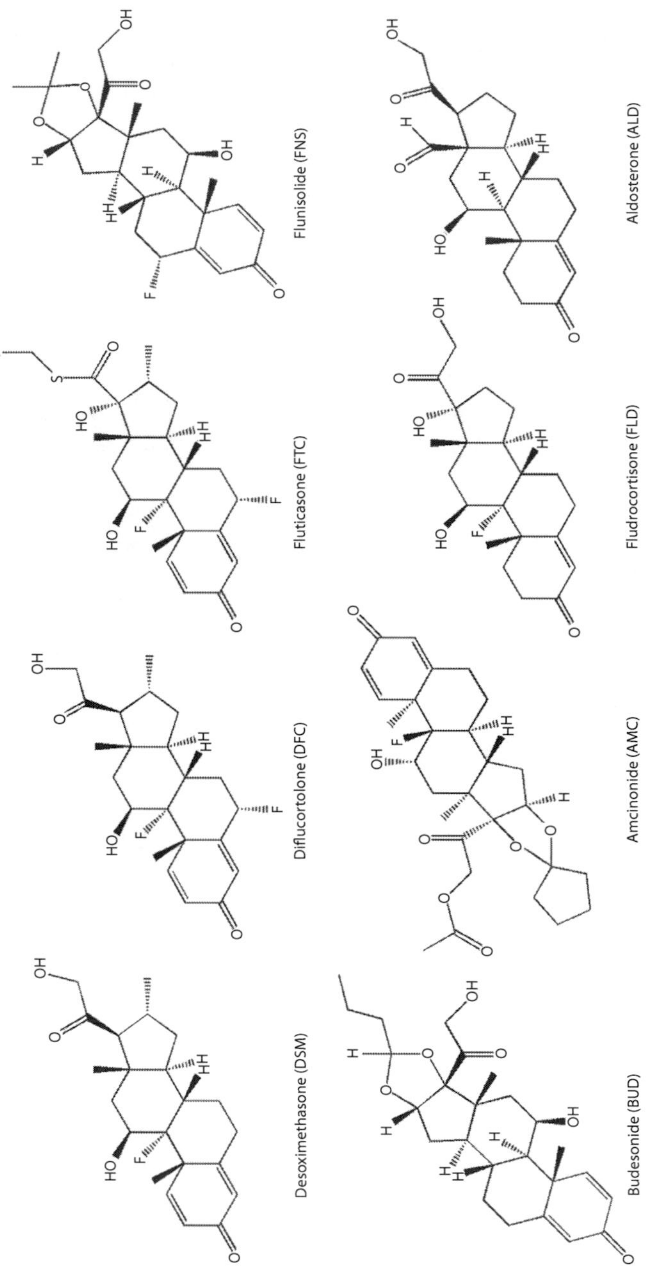

FIGURE 35.10 (continued)   Structures of corticosteroids.

ALD, *(8S,9S,10R,11S,13R,14S,17S)-11-hydroxy-17-(2-hydroxyacetyl)-10-methyl-3-oxo-1,2,6,7,8,9,11,12,14,15,16,17-dodecahydrocyclopenta[a]phenanthrene-13-carbaldehyde*, is much more potent than 11-deoxycorticosterone and is probably the most important in normal subjects. Whether 11-deoxycorticosterone contributes significantly in normal subjects is debatable but in a number of rare inherited diseases, it assumes a dominant role. ALD is a product of the zona glomerulosa, which is under the control of the renin-angiotensin system. In contrast, the origin of almost all circulating 11-deoxycorticosterone is the zona fasciculata, and it is under adrenocorticotropic hormone (ACTH) control. The principle effect of ALD is to enhance sodium reabsorption in the cortical collecting duct of the kidneys.

Synthetic corticosteroids are sometimes referred to as xenobiotic steroids, but both these terms encompass a large group, which have been synthesized by pharmaceutical firms in attempts to emphasize the physiological features associated with various steroid structures in order to use various formulations for clinical treatment. The structures of these steroids are based on cortisol with modifications: an extra double bond at C_1, methylation at C_6, C_9, C_{16}, halogenation (chlorine of fluorine) at C_6 and/or C_9, alkylation at C_{17}, C_{21}, and isomerization of the side chain.

A method for assaying urinary ALD based on successive column and TLC and a final assay with tetrazolium blue was described by Cavina et al. (1969). The urine is acidified and extracted with chloroform; the extract is purified by column chromatography and then on a thin-layer of purified silica gel G F_{254}, first with the chloroform–methanol–water (90:10:0.8 v/v) system and then by a two-dimensional procedure with ethyl acetate–methanol–water (85:15:1 v/v) and acetone–benzene–acetic acid (30:70:0.5 v/v) continuous development for about 4 h. The zones of the extracts containing the ALD were visualized under UV light. Before the two-dimensional chromatography, the zones are scraped and transferred into small column and eluted with chloroform–methanol (1:1 v/v) system. The extracts are evaporated under low pressure at 40°C and concentrated in the tip of the flask to be subjected to a second TLC. Recovery was 70.7%–82.4% and LOQ 2.2 µg/zone.

Fludrocortisone, *(8S,9R,10S,11S,13S,14S,17R)-9-fluoro-11,17-dihydroxy-17-(2-hydroxyacetyl)-10,13-dimethyl-1,2,6,7,8,11,12,14,15,16-decahydrocyclopenta[a] phenanthren-3-one*, is a synthetic corticosteroid, chemically related to ALD. It is used primarily to replace the missing hormone ALD in various forms of adrenalin sufficiency such as Addison's disease and the classic salt wasting. Therefore, it helps keep balance of water and minerals in the body.

35.11 SITOSTEROL

Sitosterol(SIT), *17-(5-Ethyl-6-methylheptan-2-yl)-10,13-dimethyl-2,3,4,7,8,9,11,12,14,15,16,17-dodecahydro-1H-cyclopenta[a]phenanthren-3-ol*, is one of several phytosterols (plant sterols) with chemical structures similar to that of CHO. Structural formula of SIT is shown in Figure 35.11. SITs are white, waxy powders with a characteristic odor. They are hydrophobic and soluble in alcohols. In recent years, there a large number of papers related to the isolation, identification, and determination of SITs, individually or in mixtures in extracts of various plants (Bhagat and Kulkarni 2010;

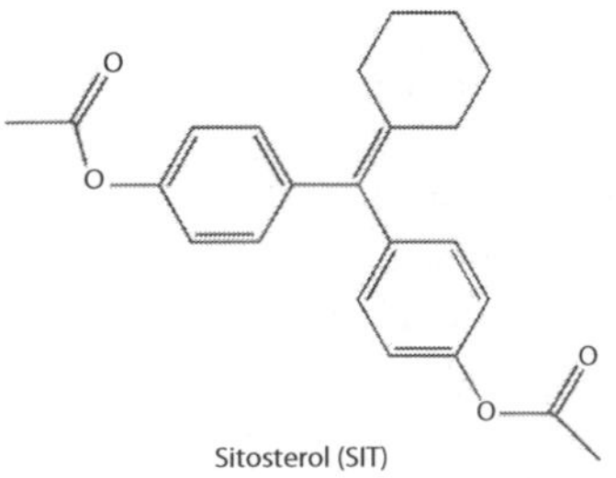

Sitosterol (SIT)

FIGURE 35.11 Structural formula of sitosterol.

Jirge et al. 2010; Misar et al. 2010; Rathee et al. 2010; Shah and Shailajan 2009; Shailajan and Menon 2011; Shamshul and Mangaonkar 2010).

Khan et al. (2012) developed a sensitive, selective, and robust densitometric HPTLC method for determination of SIT in *Betula utilis* stem bark. Dried and powdered *B. utilis* stem bark was extracted with dichloromethane for 4 h, filtered, and extraction was repeated three times, using the same solvent. The combined filtrates were concentrated. The dry residue left after dichloromethane extraction was further extracted with dichloromethane–methanol (1:1 v/v). Then, it was extracted with methanol to obtain the methanol extract for chromatography (extraction recovery is 95%–99%). The TLC was carried out on silica gel 60 plates, using *n*-hexane–ethyl acetate (8:2 v/v) as the mobile phase. The HPTLC densitometry (TLC scanner) was performed at 500 nm wavelength after the post chromatographic derivatization with ceric ammonium sulfate reagent (Linear calibration mode, 100–600 ng/band; LOD 60 ng/band; and LOQ 90 ng/band).

Dharmender et al. (2010) developed the TLC densitometric method for quantification of SIT from *Bergenia ciliata (Haw.) Sternb. forma ligulata Yeo (Pasanbheda)*. SIT was quantified from petroleum ether extract using the solvent system of toluene-methanol (9:1 v/v) on TLC silica gel plate. The linearity range was 80–480 ng/spot, RSD 0.85%, and recovery 99.92%.

Simultaneous determination of SIT and lupeol was performed on silica gel 60 F_{254} HPTLC plate, with toluene–methanol (8:1 v/v), as mobile phase. After development, plates were treated with Lieberman–Burchard reagent (Edward and Morris, 1969), and detection and quantification was performed by densitometry at 366 nm in fluorescence mode. The method was validated in terms of its linearity, LOD, and LOQ and precision following standard protocols. SIT and lupeol were found to be linear in the range of 10–40 µg/mL for both (Rane et al. 2011).

Shah and Shailajan (2010) gave a simple, sensitive, and accurate HPTLC method for quantitation of SIT from the stem bark of *Symplocos racemosa Roxb.* collected from different regions. Separation was performed on aluminum HPTLC plates coated with silica gel 60 F_{254} with toluene–methanol, as mobile phase. After development, the plates were treated with 10% methanolic sulfuric acid reagent. Detection and quantification were performed by densitometry at 366 nm.

Misar et al. (2010) dissolved petroleum ether extract in methanol, and the quantification of analytes was carried out using mobile phase toluene–methanol (9:1 v/v) on precoated aluminum silica gel plates. Densitometric detections were carried out after derivatization with anisaldehyde sulfuric acid reagent, and plates were scanned at 525 nm by absorption/reflection mode (the linearity range was 100–500 g). Martelanc et al. (2009) performed TLC on HPTLC silica gel 60 and C_{18} RP plates. Silica gel plates were prewashed with developing solvent methanol–chloroform (1:1 v/v) and C_{18} RP plates with acetone in a saturated twin-trough chamber and then dried with Camag TLC plate heater at 110°C for 30 min. The separation on the silica gel plate was obtained in 15 min developing solvent *n*-hexane–ethyl acetate (5:1 v/v). C_{18} RP plates were developed in 17 min using acetone–acetonitrile (5:1, v/v) or ethyl acetate–acetonitrile (3:2 v/v). After developing and drying, the plates were dipped for 2 s in the anisaldehyde detection reagent using Camag immersion device II, dried in a stream of warm air, and heated at 110°C with a TLC plate heater for 2 min (silica gel plate) or 30 s (RP_{18} plates).

Jarusiewicz et al. (2005) separated the mixture of sterols (CHO, cholestanol, SIT, stigmasterol, ergosterol, campesterol, desmosterol, and brassicasterol) using reversed phase, multimodal, and argentation TLC. Tested layers included C_{18}, $C_{18}W$, NH_2, CN, diol, C_2, C_8, and Ph-bonded layers; hydrocarbon-impregnated layers; and commercially prepared silica gel precoated with 10% silver nitrate. Optimal system for sterol separations is C_{18} with acetonitrile–chloroform (40:35 v/v) or petroleum ether–acetonitrile–methanol (2:4:4 v/v).

A large number of plants produced secondary metabolites such as steroids. HPTLC fingerprinting is a method of choice for detection and quantification of these metabolites. Yamunadevi et al. (2011) revealed 30 different types of steroids in the different parts of *Aerva lanata* by observing HPTLC profile of methanolic extract. HPTLC fingerprinting was performed on silica gel 60 F_{254}. The plate was pre-washed by methanol and activated at 60° for 5 min. Chloroform–acetone (8:2 v/v) was employed as mobile phase, and the chromatographic development was carried out for two times with the same mobile phase to get good resolution of phytochemical content. The plate was sprayed with anisaldehyde sulfuric acid reagent and dried at 100°C for 3 min and then scanned at 366 nm (photo documentation).

35.12 SAMPLE PREPARATION IN CONTEXT OF MATRIX

Throughout the sample preparation procedure, it is essential to recognize the necessity of utilizing methods that satisfy statistical sampling and analysis requirements. The purpose of sample preparation is to isolate analytes from a matrix. Sample preparation can range from a simple, in which a portion of the sample is dissolved in a solvent for subsequent introduction into an instrument, to complex acid–base–neutral sequential extractions. For qualitative analysis only, simple preparations and cleanups are adequate but if quantitative analysis is needed, rigorous techniques, quantitative extractions, and standardizations are required.

Samples in which qualitative and quantitative determination of steroids are carried out by TLC are different types of tablets, tissues, biological fluids, ointments, and topical creams. In determining a synthetic cotico steroids in ointment by TLC, Okamoto and Ohta (1986) prepared the samples as follows: a 1 g amount of ointment was blended four times using 30 mL of methanol, warmed in a hot bath at 50°C for 5 min, and then cooled in the icebox for 30 min at 5°C. The organic phase was dried and made up to volume with methanol in a 30 mL volumetric flask.

Fenske (2000) determined cortisol and COR in human morning and overnight urine by TLC. Steroids were extracted with 80 mL dichloromethane (recovery: 93.3% for cortisol and 94.2% for COR). Since simple dichloromethane extraction yielded poorly cleaned-up extracts, the organic phase was washed with 5 mL 0.2M NaOH and 5 mL distilled water. The aqueous supernatant was removed by aspiration, and the solvent was allowed to evaporate overnight. Samples were then reconstituted with 2.0 mL ethanol, with careful rinsing of the glass walls. Slightly modified procedure is applied in accordance with the type of setting (determination of ALD by TLC). Volume of 0.05–4 mL of plasma or urine was mixed with known amounts of internal standard for determination of individual recovery. After an equilibration period of 12 h, the samples were extracted with 20 mL of methylene chloride by magnetic stirring for 15 min. The aqueous layer was discarded and the filtered organic extract was evaporated. The dry extracts were dissolved in acetone and transferred to TLC plate (Hilfenhaus 1977).

Preparation of food samples for TLC determination of steroids is significantly different from the aforementioned. Huetos et al. (1999) performed sample preparation for determination of DXM in food samples as follows: 5.0 g of DXM-free feed were spiked with 2 mg of DXM and allowed to stand for 30 min. A 10 mL volume of methylene chloride–hexane (3:1 v/v) was added, thoroughly mixed, extracted for 15 min in an ultrasonic bath, and centrifuged for 15 min at 3000 g. The supernatant was filtered and transferred to a silica cartridge, which was washed with 4 mL of methylene chloride and eluted twice with 5 mL of methylene chloride–methanol (9:1 v/v). The eluate was evaporated to dryness under a stream of nitrogen at 45°C, and the dry extract resuspended in 3 mL of methanol and 9 mL of water. This mixture was applied to a previously prepared C_{18} cartridge, which was rinsed with 4 mL of water, dried with nitrogen, washed with 4 mL of mixtures of hexane–methylene chloride (4:1 v/v), and eluted with mixture of hexane–ethyl acetate (1:1 v/v).

If different steroids and their analogues are determined in pharmaceutical formulations, sample preparation is simpler due to the mode of production tablet. Pyka et al. (2011) in determining of hydrocortisone in pharmaceutical formulations, ground 10 tablets for 5 min. The obtained powders were shaken with absolute ethanol (10 mL) for 5 min. After shaking, the solution was filtered through a medium-density filter. From this solution, a solution with a concentration of active substance equal to 2.05 mg/mL was prepared. This solution (5 mL) was used for the TLC-densitometric analysis and quantitative determination of hydrocortisone in certain pharmaceutical preparations. Simard and Lodge (1970) used methanol, while Bagocsi et al. (2003) used chloroform for extraction in the same procedure. Preparation of samples of topical cream for determination of some steroids by TLC is performed as described by Kulkarni et al. (2010) as well as Wulandari et al. (2003): 1 g of cream was weighed and transferred to a 100 mL volumetric flask and 30 mL of methanol was added, the solution was warmed for 5–10 min, then ultrasonicated for 20 min, and volume was made up to the mark with methanol. The solution was filtered using Whatman paper No. 41. The choice of solvent in the procedures of preparation of samples, among other things, depends on the chromatographic system, which will be applied for separation and quantification of the steroids.

35.13 DETECTION

The detection methods were developed to give the highest sensitivity and specificity in the qualitative and quantitative determination of individual compounds. TLC using staining reagents is a superior method for analyzing organic compounds without chromophores. It is fast, versatile, and sometimes the only viable method (Bajaj and Ahuja 1979; Johnsson et al. 2007). Steroids which do not absorb in the UV may have to be visualized, by spraying part of the plate, to identify the position of standards, which have been run together with the samples of interest. Spray reagents, which are mentioned earlier (Tables 35.1 through 35.7), were not discussed in this section.

There is a wide variety of methods of visualizing steroids on TLC plates. They usually involve spraying and/or heating with a variety of reagents, which may or may not be specific for particular types of steroids, and these have been admirably summarized by Edwards (1969). If standards have not been run, a narrow side strip of the TLC plate can be removed and the steroids located. Fenske (2008) described a TLC method for plasma cortisol, which involved dipping the plate into isonicotinic acid hydrazide and quantifying the cortisol by fluorimetry using a scanner. A further advantage of TLC is that radioactive steroids can be identified by placing the plate in contact with an x-ray film, producing an autoradiogram (Dalla et al. 2004; Jang et al. 2007).

Different spray reagents were developed in order to increase sensitivity and selectivity in the detection of steroids, which are similar in structure. Verbake (1979) proposes a very sensitive method for detection of anabolics in urine and in tissues by sulfuric acid–induced fluorescence (observed under transillumination at 366 nm) with the LOD of 1–10 ng. Rubeanic acid was used as a spray reagent for the detection of up to 0.1 μg of various types of steroids on precoated silica gel plates. The color produced after spraying was observed at room temperature and under shortwavelength UV (Garcia 1985). Yamaguchi et al. (1980) described a TLC-enzyme solution spray technique for the determination of the excretion pattern of 3α-hydroxysteroids in patients with some adrenogenital syndrome. This reagent is very selective. Phenolic hydroxyl, 3β-hydroxy, and 3-keto groups did not react with this enzyme.

Sulfosalicylic and picrylsulfonic acids can be used as spray reagents for detection of steroids and triterpenoids with detection limit of 0.3 μg of steroids and 0.4 μg of triterpenoids (Ghosh and Thakur 1983) as well as mixtures of carbazole and sulfuric acid for detecting ≥0.8 μg triterpenoids and 0.2–6 μg steroids on silica gel G (Ghosh and Thakur 1982a) and dimethyl sulfoxide (Agrawal and Nwaiwu 1984). Folin–Ciocalteu procedure for phenols can be applied for detection of steroids. The blue spot was evaluated densitometrically directly on the chromatoplate (Jones et al. 1968). Furthermore, the metal salts can be applied as spray reagent: boron trifluoride etherate (Ghosh and Thakur 1982b), arsenic(III) chloride in mixture with acetic acid (Kohli 1975), and silicon(IV) chloride (Segura and Navarro 1981).

Vasta and Sherma (2008) compared the quantitative and qualitative results for three methods for application of derivatization reagent phosphomolybdic acid (manual spraying, manual dipping, and use of the "derivapress"). Their results showed that dipping is the best method of reagent application for achieving quantitative data, and spraying is the best method for qualitative purposes.

ACKNOWLEDGMENTS

The authors are grateful to the Ministry of Science of the Republic of Serbia (Grant No. 172017) for financial support and to Nikola Stevanovic for technical assistance.

REFERENCES

Aboul-Enein, H.Y., M.I. El-Awady, and C.M. Heard. 2000. Enantiomeric separation of aminogluthathimide, acetyl aminogluthathimide, and dansyl aminogluthathimide by TLC with β-cyclodextrin and derivatives as mobile phase additives. *J. Liq. Chromatogr. Related Technol.*, 23(17):2715–2726.

Agbaba, D., S. Ivanovic, and D. Zivanov-Stakic. 1991. Fluorodensitometric determination of plasma and urinary testosterone. *J. Planar. Chromatogr.*, 4:267–269.

Agbaba, D., Z. Milojevic, S. Eric, M. Aleksic, G. Markovic, and M. Solujic. 2001. The high-performance thin-layer chromatographic determination of dexamethasone and xylometazoline in nasal drops containing different preservatives. *J. Planar Chromatogr.*, 14(5):322–325.

Agrawal, S.P. and J. Nwaiwu. 1984. Dimethyl sulfoxide as a spray reagent for the detection of triterpenoids and some steroids on thin-layers plates. *J. Chromatogr.*, 295(2):537–542.

Agrawal, A.K., N.A. Pampori, and B.H. Shapiro. 1995. Thin-layer chromatographic separation of regioselective and stereospecific androgen metabolites. *Anal. Biochem.*, 224:455–457.

Amin, M. and M. Hassenbach. 1979. Direct quantitative thin-layer chromatographic determination of levonorgestrel and ethinylestradiol in oral contraceptives by diffuse reflection and fluorescence methods. *Analyst*, 104(1238):404–411.

Asmis, R., E. Buehler, J. Jelk, G. Jennifer, and K. Fred. 1997. Concurrent quantification of cellular cholesterol, cholesteryl esters and triglycerides in small biological samples. Reevaluation of thin-layer chromatography using laser densitometry. *J. Chromatogr. B. Biomed. Sci. Appl.*, 691(1):59–66.

Bagocsi, B., G. Rippel, M. Mezei, Z. Vegh, and K. Ferenczi-Fodor. 2003. OPLC, a method between TLC and HPLC, for purity testing of norethisterone bulk drug substance and tablet. *J. Planar Chromatogr.*, 16(5):359–362.

Bajaj, K.L. and K.L. Ahuja. 1979. General spray reagent for the detection of steroids on thin-layer plates, *J. Chromatogr.*, 172:417–419.

Baksai-Horvath, E. and A.M. Osman. 1977. Determination of urinary estrogen by thin-layer chromatography in sheep under in vivo condition of urinary chromogen reduction. *Zbl. Vet. Med. A.*, 24(2):122–127.

Barbieri, U., A. Massaglia, M. Zannino, and U. Rosa. 1972. Thin-layer chromatography of steroid derivatives for radioimmunoassay. *J. Chromatogr.*, 69(1):151–155.

Bari, S.B., P.S. Jain, A.R. Bakshi, and S. Surana. 2011. HPTLC method validation for simultaneous determination of tamsulosin hydrochloride and finasteride in bulk and pharmaceutical dosage form. *J. Anal. Bioanal. Tech.*, 2(2):100–119.

Bebawy, L.I., M.A. Azza, and R.H. Heba. 2001. Different methods for the determination of gestodene, and cyproterone acetate in raw material and dosage forms. *J. Pharm. Biomed.*, 25:425–436.

Belic, J., E. Pertot, and H. Socic. 1970. Metabolism of progesterone by Fusarium oxysporum. *J. Steroid Biochem*, 1:105–110.

Bhagat, R.B. and D.K. Kulkarni. 2010. Quantification of β-sitosterol from three Jatropha species by high performance thin-layer chromatography. *Asian J. Chem.*, 22(10):8117–8120.

Bhawani, S.A., O. Sulaiman, R. Hashim, and M.N. Mohamad. 2010. Thin-layer chromatographic analysis of steroids: A review. *Trop. J. Pharm. Res.*, 9(3):301–313.

Bican-Fister, T. 1966. Quantitative separation and estimation of steroid mixtures by thin-layer chromatography. II. Determination of progesterone and estradiol benzoate and of progesterone, testosterone propionate, and estradiol benzoate in mixtures *J. Chromatogr.*, 22(2):465–468.

Bicknell, D.C. and D.B. Gower. 1971. Separation of C19–16-unsaturated steroids from C21 and other C19-steroids by two-dimensional thin-layer chromatography. *J. Chromatogr.*, 61(2):358–360.

Biswara, R.H. and I.M. Jakovljevic. 1969. The separation of same estrogens by thin-layer chromatography. *J. Chromatogr.* 41:136–138.

Blom, M.J., M.G. Wassink, M.E. De Gooyer, A.G.H. Ederveen, H.J. Kloosterboer, J. Lange, J.G.D. Lambert, and H.J. Th. Goos. 2006. Metabolism of tibolone and its metabolites in uterine and vaginal tissue of rat and human origin. *J. Steroid Biochem.*, 101(1):42–49.

Boursier, B. and C. Chafey. 1988. Determination of DES, 17α-estradiol, and 17α-trenbolone in the same urine sample of calf by high-performance thin-layer chromatography. *Analysis*, 16(4):249–252.

Brooks, C.J.W., M.I. Walash, M. Rizk, N.A. Zakhari, S.S. Toubar, and D.G. Watson. 1993. Assay of certain oral contraceptive formulations by gas chromatography-mass spectrometry-selected ion monitoring. *Acta Pharm. Hung.*, 63:19–27.

Brown, J.W., A. Carballeira, and L.M. Fishman. 1988. Rapid two-dimensional thin-layer chromatographic system for the separation of multiple steroids of secretory and neuroendocrine interest. *J. Chromatogr.*, 439(2):441–447.

Bruinvels, J. 1963. A simple method for separation and determination of aldosterone, hydrocortisone, and corticosterone. *Experimentia*, 19(10):551–552.

Cavazzutti, G., L. Gagliardi, A. Amato, M. Profili, V. Zagarese, D. Tonelli, and E. Gattavecchia. 1983. Colour reactions of iodic acid as reagent for identifying drugs by thin layer Chromatography. *J. Chromatogr.*, 268:528–534.

Cavina, G., G. Giocoli, and D. Sardini. 1969. Thin-layer chromatographic separation of urinary aldosterone for determination with the tetrazolium blue reaction. *Steroids*, 14(3):315–325.

Chetrite, G., E. Le Nestour, and J.R. Pasqualini. 1998. Human estrogen sulfotransferase (hEST1) activities and its mRNA in various breast cancer cell lines. Effect of the progestin, promegestone (R-5020). *J. Steroid Biochem. Mol. Biol.*, 66:295–302.

Chetrite, G.S., T.H. Hubert, J-C. Philippe, and J.R. Pasqualini. 2004. Dydrogesterone (Duphaston) and its 20-dihydro-derivative as selective estrogen enzyme modulators in human breast cancer cell lines. Effect on sulfatase and on 17β-hydroxysteroid dehydrogenase (17β-HSD) activity. *Anticancer Res.*, 24:1433–1438.

Chetrite, G.S., J-L. Thomas, J. Shields-Botella, J. Cortes-Prieto, J-C. Philippe, and J.R. Pasqualini. 2005. Control of sulfatase activity by nomegestrol acetate in normal and cancerous human breast tissues. *Anticancer Res.*, 25:2827–2830.

Choudhari, V.P. and A.P. Nikalje. 2009. Stability-indicating TLC method for the determination of dutasteride in pharmaceutical dosage forms. *Chromatographia*, 70:309–313.

Choulis, N.H. 1968. Separation and quantitative determination of prednisone, fluocinolone acetonide, and paramethasone acetate using thin-layer chromatography. *Can. J. Pharm. Sci.*, 3(3)76–80.

Craig, C. and R. Stitzel. 1997. *Modern Pharmacology with Clinical Applications*, Boston, MA: Little, Brown & Company.

Daleseleire, E., K. Vanoosthuyze, and C. Van Peteghem. 1994. Application of high-performance thin-layer chromatography and gas chromatography-mass spectrometry to the detection of new anabolic steroids used as growth promoters in cattle fattening. *J. Chromatogr.*, 674:247–253.

Dalla V.L., V. Toffolo, S. Vianello, P. Belvedere, and L. Colombo. 2004. Expression of cytochrome P450c17 and other steroid-converting enzymes in the rat kidney throughout the life-span. *J. Steroid Biochem. Mol. Biol.*, 91:49–58.

Dalmau, J.M., J.M. Pla-Delfina, and A. del Pozo Ojeda. 1973. Separation and determination of the synthetic epimeric corticosteroids dexamethasone and betamethasone in mixtures by thin-layer chromatography on insoluble polyvinylpyrrolidone. *J. Chromatogr.*, 78:165–171.

Das, B., K. Datta, and S.K. Das. 1985. Thin-layer chromatographic method for the quality control of dexamethasone and betamethasone tablets. *J. Liq. Chromatogr.*, 8(16):3009–3016.

Datta, K. and S.K. Das. 1994. Identification and quantitation of corticosteroids and their esters in pharmaceutical preparations of creams and ointments by thin-layer chromatography and densitometry. *J. AOAC Int.*, 77:435–438.

Davies, D. and P.J. Nicholls. 1965. R_f values of some glutarimides. *J. Chromatogr.*, 17:416–419.

Deshmukh, S.S., V.V. Musale, V.K. Bhusari, and S.R. Dhaneshwar. 2011. Validated HPTLC method for simultaneous analysis of alfuzosin hydrochloride and dutasteride in a pharmaceutical dosage form. *J. Planar Chromatogr.*, 24(3):218–221.

Dharmender, R., M. Thanki, R. Agrawal, and S. Anandjiwala. 2010. Simultaneous quantification of bergenin, (+)-catechin, gallicin and gallic acid; and quantification of β-sitosterol using HPTLC from *Bergenia ciliata* (Haw.) Sternb. forma ligulata Yeo (Pasanbheda). *Pharm. Anal. Acta* 1(1):104–107.

Doerr, P. 1971. Thin-layer chromatography and elution of picogram amounts of estradiol. *J. Chromatogr.*, 59(2):452–456.

Eczely, P. 1985. The role of thyroid and adrenal cortical hormones in the modulation of the gonadal function in birds. *Acta Biol. Hung.*, 36:45–70.

Edward, K. and G. Morris.1969. Serum cholesterol assay using a stable Liebermann–Burchard reagent. *Clinical Chem.*, 15(12):1171–1179.

Edwards R.W.H. 1969. Steroids. Methods for the detection of biochemical compounds on paper and thin-layer chromatograms. In: *Data for Biochemical Research*, eds. Dawson, R.M.C., Elliott, D.C., Eliott, W.H., and Jones, K.M. pp. 567–568. Clarendon, Oxford, U.K.

Fakhari, A.R., A.R. Khorrami, and M. Shamsipur. 2006. Stability-indicating high-performance thin-layer chromatographic determination of levonorgestrel and ethinyloestradiol in bulk drug and in low-dosage oral contraceptives. *Anal. Chim. Acta*, 572(2):237–242.

Falkay, G., J. Morvay, and M. Sas. 1973. Clinicochemical analysis of 3-methoxy-17-epiestriol. *Acta Pharm. Hung.*, 43:1–8.

Farkas, O., E. Gere-Paszti, and E. Forgacs. 2003. Study of the interaction of structurally similar bioactive compounds by thin-layer chromatography. *J. Chromatogr. Sci.,* 41:169–172.

Feher, K.G. and M. Csillag. 1967. Determination of estriol, estrone, and estradiol in pregnancy urine by thin-layer chromatography. *Clin. Chim. Acta,* 15(2):343–346.

Fenske, M. 1998. Determination of cortisol in plasma and urine by thin-layer chromatography and fluorescence derivatization with isonicotinic acid hydrazide. *Chromatographia,* 47(11/12):695–700.

Fenske, M. 1999. Rapid and specific determination of free cortisol in guinea pig urine and faeces by thin-layer chromatography-competitive protein-binding assay. *Chromatographia,* 50:428–432.

Fenske, M. 2000. Determination of cortisol and cortisone in human morning and overnight urine by thin-layer chromatography and fluorescence derivatization with isonicotinic acid hydrazide. *Chromatographia,* 52(11/12):810–814.

Fenske, M. 2008. Determination of cortisol in human plasma by thin-layer chromatography and fluorescence derivatization with isonicotinic acid hydrazide. *J. Chromatogr. Sci.,* 46:1–3.

Ferenczi-Fodor, K., I. Kovacs, and G. Szepesi. 1987. Separation and determination of steroid isomers on amino-bonded silica by conventional and overpressurized thin-layer chromatography. *J. Chromatogr.,* 392:464–469.

Ferenczi-Fodor, K., A. Lauko, A. Wiskidenszky, Z. Vegh, and K. Ujszaszy. 1999. Chromatographic and spectroscopic investigation of irreversible adsorption in conventional TLC and HPTLC and in OPLC. *J. Planar Chromatogr.,* 12:30–37.

Few, J.D. and T.J. Forward. 1968. The quantitative thin layer chromatography, of corticosteroids. *J. Chromatogr.,* 36:63–73.

Frank, H., G. Heinisch, and F. Tanzer. 1986. Differentiation of drug substances. Part 6. Some selected keto steroids. *Pharmazie,* 41(7):488–489.

Fraser, R., D.B. Gower, J.W. Honour, M.C. Ingram, A.T. Kicman, L.J.M. Hugh, and P.M. Stewart. 2010. Analysis of corticosteroids. In: *Steroid Analysis,* ed. H.L.J. Makin, and D.B. Gower, 743–836. Springer, the Netherlands.

Gabriels, M., F. Camu, and J. Plaizier-Vercammen. 2002. Improved thin-layer chromatographic method for the separation of cholesterol, egg phosphatidylcholine, and their degradation products. *J. AOAC Int.,* 85(6):1273–1287.

Gagliardi, L., D. De Orsi, M.R. Del Giudice, F. Gatta, R. Porra, P. Chimenti, and D. Tonelli. 2002. Development of a tandem thin-layer chromatography–high-performance liquid chromatography method for the identification and determination of corticosteroids in cosmetic products. *Anal. Chim. Acta,* 457(2):187–198.

Gaikwad, N.V., P.B. Deshpande, S.V. Gandhi, and K.K. Khandagale. 2010. High Performance thin-layer chromatographic determination of spirinolactone and torsemide in combined tablet dosage form. *Res. J. Pharm. Tech.* 3(4):1106–1108.

Garcia, C.P., 1985. Rubeanic acid: A general spray reagent for the detection of steroids on thin-layer plates. *J. Chromatogr.,* 350(2):468–470.

Garcia, G., R. Saelzer, and M. Vega. 1991. Screening of meat for residues of hormonal anabolic growth promoters. *J. Planar Chromatogr.,* 4:223–225.

Ghosh, P. and S. Thakur. 1982a. Spray reagents for the detection of steroids and triterpenoids on thin-layer plates. *J. Chromatogr.,* 258:258–261.

Ghosh, P. and S. Thakur. 1982b. Boron trifluoride etherate as a spray reagent for the detection of steroids and triterpenoids by TLC. *Fresen. Z. Anal. Chem.,* 313(2):144–147.

Ghosh, P. and S. Thakur. 1983. Spray reagents for the detection of steroids and triterpenoids on thin-layer plates. *J. Chromatogr.,* 240(2):515–517.

Godin, C., D. Poirier, C.H. Blomquist, and Y. Tremblay. 1999. Separation by thin-layer chromatography of the most common androgen-derived C19 steroids formed by mammalian cells. *Steroids,* 64:767–769.

Golab, T. and D.S. Layne. 1962. The separation of 19-nor-steroids by thin-layer chromatography on silica gel. *J. Chromatogr.,* 9:312–330.

Goss, P.E. 1986. Metabolism of the aromatase inhibitor 4-hydroxy and rostenedione in vivo. Identification of the glucuronideas a major urinary metabolite in patients and biliary metabolite in the rat. *J. Steroid Biochem.,* 24(2):619–622.

Grassini-Strazza, G. and I. Nicoletti. 1985. High-performance thin-layer chromatography on amino-bonded silica gel: Application to barbiturates and steroids. *J. Chromatogr.,* 322(1):149–158.

Groh, H., R. Schon, M. Ritzau, H. Kasch, K. Undisz, and G. Hobe. 1997. Preparation of 3-ketodesogestrel metabolites by microbial transformation and chemical synthesis. *Steroids,* 62(5):437–443.

Gross, B.A., S.A. Mindea, A.J. Pick, J.P. Chandler, and H.H. Batjer. 2007. Diagnostic approach to Cushing disease. *Neurosurg. Focus*, 23(3):125–129.

Gu, P., Y. Li, Y. Wu, and S. Jiang. 2001. Analysis of lefrozole and its tablets by thin-layer chromatography. *Chinese J. Pharm.*, 32(7):317–311.

Hara, S. and K. Mibe. 1967. Systematic analysis of steroids. VII. Thin-layer chromatography of steroidal pharmaceuticals. *Chem. Pharm. Bull.*, 15:1036–1040.

Hauck H.E., O. Bund, W. Fischer, and M. Schulz. 2001. Ultra-thin layer chromatography (UTLC)–A new dimension in thin-layer chromatography. *J. Planar Chromatogr.*, 14:234–236.

Hauck, H.E. and M. Schulcz. 2003. Ultra thin-layer chromatography. *Chromatographia*, 57:S313– S315.

Hegazy, M.A., F.H. Metwaly, M. Abdelkawy, and N.S. Abdelwahab. 2011. Validated chromatographic methods for determination of hydrochlorothiazide and spironolactone in pharmaceutical formulation in presence of impurities and degradants. *J. Chromatogr. Sci.*, 49(2):129–135.

Hilfenhaus, M. 1977. Evaluation of radioimmunoassay for aldosterone in urine and plasma of rats. *J. Steroid Biochem.*, 8:847–851.

Huetos, O., M. Ramos, M. Martin de Pozuelo, T.B. Reuvers, and M. San Andres. 1999. Determination of dexamethasone in feed by TLC and HPLC. *Analyst*, 124(11):1583–1587.

Huetos, O., T. Reuvers, and J.J. Sanchez. 1998. Comparative study of the thin-layer chromatographic detection of different corticosteroids. *J. Planar Chromatogr.*, 11(4):305–308.

Hung, G.W.C. and A.Z. Harris. 1988. The development of more sensitive solvent systems for separation of cholesterol and low-molecular-weight cholesteryl esters by thin-layer chromatography. *Microchem. J.*, 37(2):174–180.

Jambu, S.C., S. Vilvan, A. Kumar, and M.G. Tyagi. 2010. Chromatographic and spectrophotometric evaluation of progesterone and estrogen. *Rec. Res. Sci. Technol*, 2(6)22–28.

Jamshidi, A. and S. Sharifi. 2009. HPTLC analysis of tamoxifen citrate in drug-release media during development of an in-situ-cross-linking delivery system. *J. Planar Chromatogr.*, 2(3):187–189.

Jang S., Y. Lee, S.L. Hwang, M.H. Lee, S.J. Park, I.H. Lee, S. Kang, S.S. Roh, Y.J. Seo, J.K. Park, J.H. Lee, and C.D. Kim. 2007. Establishment of type II 5alpha-reductase over-expressing cell line as an inhibitor screening model. *J. Steroid Biochem. Mol. Biol.*, 107:245–252.

Jarusiewicz, J., J. Sherma, and B. Fried. 2005. Separation of sterols by reversed phase and argentation thin-layer chromatography. Their identification in snail bodies. *J. Liq. Chromatogr., Related Technol.*, 28(16):2607–2617.

Jashidi, A. 2004. A convenient and high throughput HPTLC method for determination of progesterone in release media of silicon-based controlled-release drug-delivery systems. *J. Planar Chromatogr.*, 17:229–232.

Jazbec, P., A. Smidovnik, M. Puklavec, M. Krizman, J. Sribar, L. Milivojevic, and M. Prosek. 2009. HPTLC and HPLC-MS quantification of coenzyme Q10 and cholesterol in fractionated chicken-breast tissue. *J. Planar Chromatogr.*, 22(6):395–398.

Jirge, S., P. Tatke, and S.Y. Gabhe. 2010. Marker based standardization of commercial formulations and extracts containing beta-sitosterol-D-glucoside using HPTLC. *Int. J. Res. Ayur. Pharm.*, 1(2):616–623.

Johnsson, R., G. Traff, M. Sunden, and U. Ellervik. 2007. Evaluation of quantitative thin layer chromatography using staining reagents. *J. Chromatogr.*, 1164:298–305.

Jones, L.N., M. Seidman, and C.B. Southworth. 1968. Quantitative determination of diethylstilbestrol by thin-layer chromatography. *J. Pharm. Sci.*, 57(4):646–649.

Kamat, S.S., V.T. Vele, V.C. Choudhari, and S.S. Prabhune. 2008. Determination of dutasteride from its bulk drug and pharmaceutical preparations by high-performance thin layer chromatography. *Asian J. Chem.*, 20(7):5514–5518.

Kartsova, L.A. and E.G. Strel'nikova. 2007. Separation of exogenous and endogenous steroid hormones by micellar high-performance thin-layer chromatography. *J. Anal. Chem.*, 62(9):872–874.

Kasaye, L., A. Hymete, and A-M.I. Mohamed. 2010. HPTLC-densitometric method for simultaneous determination of salmeterol xinafoate and fluticasone propionate in dry powder inhalers. *Saudi Pharm. J.*, 18(3):153–159.

Khan, I., P.L. Sangwan, J.K. Dhar, and S. Koul. 2012. Simultaneous quantification of five marker compounds of *Betula utilis* stem bark using a validated high-performance thin-layer chromatography method. *J. Sep. Sci.*, 35(3):392–399.

Kicman, A.T., E. Houghton, and D.B. Gower. 2010. Anabolic steroids: Metabolism, doping and detection in human and equestrian sports. In: *Steroid Analysis*, eds. H.L.J. Makin and D.B. Gower, pp. 743–836. Springer, the Netherlands.

Klaus, R., W. Fischer, and H.E. Hauck. 1994. Analysis and chromatographic separation of some steroid hormones on NH_2 layers. *Chromatographia*, 39:97–102.

Kohli, J.C. 1975. Novel spray reagent for steroids on thin-layer plates. *An Chim.*, 10(3):145–147.

Kotiyan, P.N. and P.R. Vavia. 2000. Stability indicating HPTLC method for the estimation of estradiol. *J. Pharm. Biomed. Anal.*, 22(4):667–671.

Kovacs, L., A. Zalka, R. Dobo, and J. Pucsok. 1986. One-dimensional thin-layer chromatographic separation of lipids into fourteen fractions by two successive developments on the same plate. *J. Chromatogr. Biomed. App.*, 382:308–313.

Krzek, J., U. Hubicka, M. Dabrowska-Tylka, and E. Leciejewicz-Ziemecka. 2002. Determination of budesonide R-(+) and S-(−) isomers in pharmaceuticals by thin-layer chromatography with UV densitometric detection. *Chromatographia*, 56:759–762.

Krzek, J., A. Maslanka, and P. Lipner. 2005. Identification and quantitation of polymyxin B, framycetin, and dexametasone in an ointment by using thin-layer chromatography with densitometry. *J. AOAC Int.*, 88(5):1549–1554.

Kulkarni, A.A., R.K. Nanda, M.N. Ranjane, and P. Ranjane. 2010. Simultaneous estimation of nadifloxacin and mometasone furoate in topical cream by HPTLC method. *Pharma Chem.*, 2(3):25–30.

Kusner, E.J. and R.D. Garrett. 1971. The transformation of selected steroids by Cylindrocarpon radicicola. A new biosynthetic pathway for estrololactone. *Steroids*, 17:521–529.

Lamparczyk, H., R.J. Ochocka, P. Zarzycki, and J.P. Zielinski. 1990. Separation of steroids by reversed-phase HPTLC using various binary mobile phases. *J. Planar Chromatogr.*, 3:34–37.

Lars, T. 1970. Separation of some synthetic estrogens from natural estrogens by thin-layer chromatography. *J. Chromatogr.*, 48(3):560–562.

Laugher, L., T.G. Noctor, A. Barrow, J.M. Oxford, and T. Phillips. 1999. An improved method for the determination of fluticasone propionate in human plasma. *J. Pharm. Biomed. Anal.*, 21:749–758.

Lee, D.K., J.C. Young, Y. Tamura, D.C. Patterson, C.E. Bird, and A.F. Clark. 1973. In vitro effects of estrogens on the 4-reduction of testosterone by rat prostate and liver preparations. *Can. J. Biochem.* 51(6):735–740.

Lekic, M., F. Korac, M. Sober, and A. Marjanovic. 2007. Planar chromatography of steroid hormones and anabolics. *Acta Chim. Slov.*, 54(1):88–91.

Lewbart, M.L. and R.A. Elverson. 1982. Determination of urinary free cortisol and cortisone by sequential thin layer and high-performance liquid chromatography. *J. Steroid Biochem.*, 17(2):185–190.

Li, K. 1990. Simple and rapid thin-layer chromatographic method for quantitative measurement of free cholesterol in serum. *J. Chromatogr. Biomed.*, 532(2):449–452.

Lisboa, B.P. 1966. Thin-layer chromatography of oestrogen on kieselgel G. *Clin. Chim. Acta*, 13(2):179–199.

Lullmann, H., K. Mohr, A. Ziegler, and D. Bieger. 2000. *Color Atlas of Pharmacology.*Thieme Sttutgart, New York.

Makin, H.L.J. and D.B. Gower. 2010. *Steroid Analysis.* Springer, the Netherlands.

Mallikarjuna, R.N., J. Bagyalakshmi, and T.K. Ravi. 2011. Development and validation of an analytical method for the estimation of standard cholesterol and egg cholesterol by high performance thin-layer chromatography (HPTLC). *Res. J. Pharm. Tech.*, 4:1155–1159.

Martelanc, M., I. Vovk, and B. Simonovska. 2009. Separation and identification of some common isomeric plant triterpenoids by thin-layer chromatography and high-performance liquid chromatography. *J. Chromatogr.*, 1216(38):6662–6670.

Maslov, L.G., N.S. Evtushenko, A.N. Shchavlinskii, and A.I. Luttseva. 1998. Methods of control and standardization of drugs containing progestogen hormones. *Pharm. Chem. J.*, 32(4):45–52.

Matyska, M., A.M. Siouffi, and E. Soczewinski. 1991. Programmed multiple development (PMD) analysis of steroids by planar chromatography with a new modification of the horizontal sandwich chamber. *J. Planar Chromatogr.*, 4:255–257.

Medina, M.B. and N. Nagay. 1993. Improved thin-layer chromatographic detection of diethylstilbestrol and zeranol in plasma and tissues isolated with alumina and ion-exchange membrane columns in tandem. *J. Chromatogr. Biomed. Appl.*, 614(2):315–323.

Medina, M.B., D.P. Schartz. 1992. Thin-layer chromatographic detection of zeranol and tissue extracts with Fast Corinth. V. *J. Chromatogr.*, 581:119–128.

Meyyanathan, S.N., G.V.S. Ramasarma, and B. Suresh. 2001. Analysis of finasteride in pharmaceutical preparations by high performance thin-layer chromatography. *J. Planar Chromatogr.*, 14(3):188–190.

Miler, T.L. 1972. The inhibition of microbial steroid D-ring lactonization by high levels of progesterone. *Biochim. Biopys. Acta,* 270:167–180.

Misar, A., A.M. Mujumdar, A. Ruikar, and N.R. Deshpande. 2010. Quantification of β-sitosterol from barks of 3 Acacia species by HPTLC. *J. Pharm. Res.*, 3(11):2595–2596.

Molnar, J., M. Gazdag, and G. Szepesi. 1982. Purification of biological samples by HPLC for the simultaneous determination of anabolics by HPTLC. *Pharmazie*, 37:836–838.

Mulja, M. and I. Gunawan. 2010. Steroids: TLC analysis. In: *Encyclopedia of Chromatography*, ed. J. Cazes, 3, pp. 2259–2262. Taylor & Francis Group, New York.

Nielsen, H. 1990. Three-step one-dimensional thin-layer chromatographic separation of neutral lipids. *J. Chromatogr.*, 498(2):423–427.

Novakovic, J., I. Nemcova, and M. Vasatova. 1990. Quantitative HPTLC determination of syproterone acetate and ethynyl estradiol in pharmaceuticals. *J. Planar Chromatogr.*, 3:407–409.

Nurok, D., R.M. Kleyle, K.B. Lipkowitz, S.S. Myers, and M.L. Kearns. 1993. Dependence of retention and separation quality in planar chromatography on properties of the mobile phase. *Anal. Chem.*, 65(24):3701–3707.

Oehrle, K.L., K. Vogt, and B. Hoffmann. 1975. Determination of trenbolone and trenbolone acetate by thin-layer chromatography in combination with a fluorescence colour reaction. *J. Chromatogr.*, 114 (1):244–246.

Okamoto, M. and M. Ohta. 1986. Behaviour of synthetic corticoids in ointment on 3-cyanopropyltrichlorosi-lane in high-performance thin-layer chromatography. *J. Chromatogr.*, 369(2):403–407.

Pachaly, P. 1999. TLC of ethinyl estradiol. *Dtsch. Apoth. Ztg.*, 139:835–836.

Pala, A. and G. Benagian. 1976. A direct radioiodination technique for the radioimmunoassay of 17α-ethynyl, 17β-hydroxy-4-estren-3-one. *J. Steroid Biochem.*, 7:491–496.

Patel, D.B. and N.J. Patel. 2010. Validated RP-HPLC and TLC methods for simultaneous estimation of tamsulosin hydrochloride and finasteride in combined dosage forms. *Acta Pharm.*, 60(2):197–205.

Patel, D.B. and N.J. Patel. 2010b. Validated reversed-phase high-performance liquid chromatographic and high-performance thin-layer chromatographic methods for simultaneous analysis of tamsulosin hydrochloride and dutasteride in pharmaceutical dosage forms. *Acta Chromatogr.*, 22(3):419–431.

Patel, D.B., N.J. Patel, S.K. Patel, and P.U. Patel. 2011. Validated stability indicating HPTLC method for the determination of dutasteride in pharmaceutical dosage forms. *Chromatogr. Res. Int.*, 2011:1–5.

Pavic, K., O. Cudina, D. Agbaba, and S. Vladimirov. 2003. Quantitative analysis of cyproterone acetate and ethinylestradiol in tablets by use of planar chromatography. *J. Planar Chromatogr.*, 16(1):45–47.

Płocharz, P., A. Klimek-Turek, and Dz. H. Tadeusz. 2010. Pressurized planar electrochromatography, high-performance thin-layer chromatography and high-performance liquid chromatography-comparison of performance. *J. Chromatogr.*, 1217:4868–4872.

Pyka, A. 2009. Spectrodensitometry application to analytical identification of estradiol, hydrocortisone, testosterone and cholesterol on Diol plates. *J. Liq. Chromatogr. Related Technol.*, 32(8):1084–1095.

Pyka, A. and M. Babuska. 2006. Lipophilicity of selected steroids compounds. I. Investigations on RP-18W stationary phase by RP-HPTLC. *J. Liq. Chromatogr. Related Technol.*, 29:1891–1903.

Pyka, A., M. Babuska-Roczniak, and P. Bochenska. 2011. Determination of hydrocortisone in pharmaceutical drug by tlc with densitometric detection in UV. *J. Liq. Chromatogr. Related Technol.*, 34:753–769.

Rajkowski, K.M. and G.D. Broadhead. 1974. Thin-layer chromatography of estrogens, their derivatives, and cholesterol. *J. Chromatogr.*, 89(2):374–379.

Ramic, A., M. Medic-Saric, S. Turina, and I. Jasprica. 2006. TLC detection of chemical interactions of vitamins A and D with drugs. *J. Planar Chromatogr.*, 19(1):27–31.

Rane, N., V.K. Vaidya, V. Kekare, W. Shah, and P. Champanerkar. 2011. HPTLC method for simultaneous quantification of two biologically active compounds in the bark powder of Artocarpus lakoocha Roxb. Plant. *Anal. Chem. Indian J.*, 10(7):449–452.

Raobaikady, B., M.F.C. Parsons, M.J. Reed, and A. Purohit. 2006. Lack of aromatisation of the 3-keto-4-ene metabolite of tibolone to an estrogenic derivative. *Steroids*, 71(7):639–646.

Rathee, S., O.P. Mogla, P. Rathee, and D. Rathee. 2010. Quantification of β-sitosterol using HPTLC from Capparis decidua (Forsk.) Edgew. *Pharma Chem.*, 2(4):86–92.

Renwick, A.G.C., S.M. Pound, and D.J. O'Shannessy. 1983. Thin-layer chromatographic systems for the separation of some estrogen acetates. *J. Chromatogr.*, 256:375–377.

Ruh, T.S. 1976. Simultaneous separation of estrogens and androgens using thin-layer chromatography. *J. Chromatogr.*, 121:82–84.

Sallam, L.A.R., A-M.H. El-Refai, and I.A. El-Kady. 1969. Thin-layer chromatography of some C_{21}, C_{19} and C_{18} steroids. *J. Gen. Appl. Microbiol.*, 15:309–315.

Schludi, H., E. Wolferseder, and K. Zeitler. 2000. Drug falsification, *Dtsch. Apoth. Ztg.*, 140:63–70.

Schuller, P.L. 1967. The detection of diethylstilboestrol (DES) in urine by thin-layer chromatography. *J. Chromatogr.*, 31(1):237–240.

Schwartz, F.L. and R.J. Miler. 1997. Androgens, antiandrogens, and anabolic steroids. In: *Modern Pharmacology with Clinical Applications*, eds. C. Craig, and R. Stitzel, 730–732.Little, Brown & Company, Boston, MA.

Schwarz, V. 1967. Methods for the separation of natural products. *Pharmazie*, 18:122–125.

Segura, R. and X. Navarro. 1981. Use of metal salts as fluorescence-inducing reagents in thin-layer chromatography. *J. Chromatogr.*, 217:329–340.

Shah, S. and S. Shailajan. 2009. Quantitation of β-sitosterol from *Woodfordia fruticosa* (Linn.) Kurz and a polyherbal formulation used for treating female reproductive disorders. *Anal. Chem. Indian J.*, 8(1):82–86.

Shah, S. and S. Shailajan. 2010. High performance thin-layer chromatographic quantification of β-sitosterol from stem bark of *Symplocos racemosa* Roxb. and it's formulation. *J. Herb. Med. Toxicol.*, 4(2):133–139.

Shailajan, S. and S. Menon. 2011. Polymarker based standardization of an ayurvedic formulation, Lavangadi vati using high performance thin-layer chromatography. *J. Pharm. Res.*, 4(2):467–470.

Shaikh, B., M.R. Hallmark, H.J. Issaq, N.H. Risser, and J.C. Kawalek. 1979. Use of high pressure liquid chromatography and thin-layer chromatography for the separation and detection of testosterone and its metabolites from in vitro incubation mixtures. *J. Liq. Chromatogr.*, 2(7):943–956.

Shamshul, S.S. and K.V. Mangaonkar. 2010. High-performance thin-layer chromatographic method for quantification of β-sitosterol from Tridax procumbens. *Anal. Chem. Indian J.*, 9(2):252–255.

Sharma, M.C., S. Sharma, D.V. Kohli, and A.D. Sharma. 2010. Validated TLC densitometric method for the quantification of Torsemide and Spirinolactone in bulk drug and in tablet dosage form. *Pharma Chem.*, 2(1):121–126.

Simard, M.B. and B.A. Lodge. 1970. Thin-layer chromatographic identification of estrogens and progestagens in oral contraceptive. *J. Chromatogr.*, 51:517–524.

Smets, F., Ch. Vanhoenackere, and G. Pottie. 1993. Influence of matrix and applied method on the detection of anabolic residues in biological samples. *Anal. Chim. Acta*, 275:147–162.

Socic, H. and I. Belic. 1968. Separation and identification of steroids produced by fermentative oxidation of progesterone. *Fresen. Z. Anal. Chem.*, 243:291–294.

Subramanian, G.S., A. Karthik, A. Baliga, P. Musmade, and S. Kini. 2009. High-performance thin-layer chromatographic analysis of bicalutamide in bulk drug and liposomes. *J. Planar Chromatogr.*, 22(4):273–276.

Szucs, V., B. Bonsonk, B. Polyak, and L. Boross. 1984. Study of the chromatographic behavior of some steroids by OPLC, *Proc. Inter. Symposium on TLC with special Emphasis on OPLC*, Szeged, 90.

Tecklenburg, R.E., B.L. Maidak, and D. Nurok. 1983. Separation of steroid mixtures by time-optimized thin-layer chromatography. *J. High Res. Chromatogr.*, 6(11):627–628.

Teng, X.W., K. Foe, K.F. Brown, D.J. Cutler, and N.M. Davies. High-performance liquid chromatographic analysis of mometasone furoate and its decomposition products: Application to in vitro degradation studies. 2001. *J. Pharm. Biomed. Anal.*, 26:313–331.

Thanh, N.T.K., G. Stevenson, D. Obatomi, and P. Bach. 2000. Determination of lipids in animal tissues by high-performance thin-layer chromatography with densitometry. *J. Planar Chromatogr.*, 13(5):375–381.

Vanderhaeghe, H. and J. Hoebus. 1976. Identification of steroid hormones. I. Corticosteroids. *J. Pharm. Bel.*, 31(1):25–37.

Van der Merwe, P.J., D.G. Mueller, and E.C. Clark. 1979. Quantitation of spironolactone and its metabolite, canrenone, in human serum by thin-layer spectrofluorimetry. *J. Chromatogr.*, 171:519–521.

Vanoosthuyze, K.E., L.S.G. Van Poucke, A.C.A. Deloof and C.H. Van Peteghem. 1993. Development of a high-performance thin-layer chromatographic method for the multi-screening analysis of corticosteroids. *Anal. Chem. Acta*, 215:177–182.

Vasta, J.D. and J. Sherma. 2008. Comparison of spraying, dipping, and the derivapress for postchromatic derivatization with phosphomolybdic acid in the detection and quantification of neutral lipids by high-performance thin-layer chromatography. *Acta Chromatogr.*, 20:15–23.

Verbake, R. 1979. Sensitive multi-residue method for detection of anabolics in urine and in tissues of slaughtered animals. *J. Chromatogr.*, 177:69–84.

Wintersteiger, R. and E. Gamse. 1982. Quantitation of keto steroids in pharmaceutical formulations and plasma in the picomole range. *Anal. Chem. Symp. Series*, 10:453–456.

Wood, P.J. and D.B. Gower. 2010. Analysis of progestagens. In: *Steroid Analysis*, eds. H.L.J. Makin, and D.B. Gower, pp. 559–603. Springer, the Netherlands.

Wortberg, B., R. Woller, and T. Chulamorakot. 1978. Detection of estrogen-like compounds by thin-layer chromatography. *J. Chromatogr.*, 156(1):205–210.

Wulandari, L., T.K. Sia, and G. Indrayanto. 2003. TLC densitometric determination of mometasone furoate in topical preparations: Validation. *J. Liq. Chromatogr. Related Technol.*, 26:109–117.

Yamaguchi, Y. 1982. Enzymatic determination of urinary 17β-hydroxysteroids on thin-layer chromatograms, *J. Chromatogr.*, 228:317–320.

Yamaguchi, Y., H. Chozo, and K. Miyai. 1980. Enzymic color development of 3α-hydroxysteroids on thin-layers chromatograms for determination of excretion pattern of 3α-hydroxysteroids in patients with some adrenogenital syndrome. *J. Chromatogr. Biomed. Appl.*, 182:430–434.

Yamunadevi, M., E.G. Wesely, and M. Johnson. 2011. Chromatographic finger print analysis of steroids in *Aerva lanata L* by HPTLC technique. *Asia-Pac. J. Trop. Biomed.*, 1(6):428–433.

Zarzycki, P.K. 2008. Simple horizontal chamber for thermostated micro-thin-layer chromatography. *J. Chromatogr.*, 1187:250–259.

Zarzycki, P.K., M. Wierzbowska, and H. Lamparczyk. 1999. Retention and separation studies of cholesterol and bile acids using thermostated thin-layer chromatography. *J. Chromatogr.*, 857(1/2):255–262.

Zarzycki, P.K. and M.B. Zarzycka. 2008. Application of temperature-controlled micro planar chromatography for separation and quantification of testosterone and its derivatives. *Anal. Bioanal. Chem.*, 391(6):2219–2225.

Zivanovic, L.J., D. Zivanov-Stakic, and D. Radulovic. 1987. UV-densitometric determination of corticosteroids. *Arch. Pharm.*, 320(11):1183–1185.

36 TLC of Drugs Used in Obesity and Sexual Dysfunction Treatment

Łukasz Komsta

CONTENTS

It could seem to be strange to present two unlike groups of drugs in one chapter. However, both groups presented in this chapter have one common property: They are often added as illegal additions to herbal preparations used as aphrodisiac or slimming aids. (Miller and Strip, 2007; Talati et al. 2011). Therefore, they are often detected and determined together by various analytical methods, mainly in drugs formulations; one study (Yang et al. 2009) also touches the problem of postmortem TLC analysis in forensic toxicology.

36.1 DETECTION

TLC was used as a detection tool of sildenafil citrate (Viagra) and methyltestosterone, before quantitative analysis by High-performance liquid chromatography (HPLC) (Ku et al. 2002). Separation was carried out on Inertsil ODS-2 reversed-phase column using gradient elution. Calibration curve was constructed in the range 80–800 ug/mL. The RSD of sildenafil was 0.54% (intraday) and 3.56% (interday).

A similar approach (Mikami et al. 2002) was published for phentolamine and sildenafil to detect and quantitate them in adulterants. A combination of TLC, LC/MS, and HPLC/DAD was used, where TLC was only an identification step. The sample solution for TLC was applied to silica gel plates and developed with two solvents: chloroform–ammonia solution–methanol (70:5:3) and chloroform–diethylamine–methanol (15:3:2). The HPLC analysis was performed on a C18 column using water–methanol–acetonitrile–triethylamine (580:250:170:1) mobile phase. This procedure was applied to commercial soft drinks, where both substances were identified and determined.

Cai et al. (2010) developed a set of TLC methods for detection of counterfeits in dietary supplements. This method is able to detect sildenafil, hongdenafil, homosildenafil, hydroxyhomosildenafil, vardenafil, pseudovardenafil, tadalafil, and aminotadalafil with comparable power to HPLC-PDA-MS. The method was used for screening of 36 commercial suspicious dietary supplements.

36.2 RETENTION INVESTIGATION

Baranowska et al. (2009) investigated retention behavior of L-arginine, its metabolites and several selected drugs, including sildenafil (dexamethasone, prednisolone, furosemide, vancomycin, amikacin, fluconazole, digoxin, captopril, dipyrone, metoprolol, and sildenafil). They were chromatographed in solutions and in also in spiked urine. A variety of mobile-phase systems were evaluated, which allowed a discussion on selectivity of the polar modifiers in mobile phases.

Nicoletti (2011) noticed the presence of thiosildenafil in herbal products and developed the HPTLC method for its detection. The analysis was then followed by isolation and analysis of spectroscopic data. HPTLC allowed detection of this counterfeiting without complicated instrumentation.

An up-to-date review of all strategies of detection of these drugs in dietary supplements (discussing also TLC) was given by Singh et al. (2009).

36.3 QUANTITATION

The first HPTLC method for determination of sildenafil citrate in commercial products (Abourashed et al. 2005) describes separation on silica gel, with chloroform–methanol–diethylamine, 90:10:1 (v/v), as mobile phase, and the analyte spots were scanned at 305 nm. Recovery of sildenafil citrate was 100.6% and 98.2%. Four pharmaceutical products were analyzed and three herbal preparations, in which sildenafil were also detected.

Similar method (Reddy et al. 2006) has been described on silica gel HPTLC plates, using mobile phase toluene–acetone–methanol (6:2:2). The plates were developed vertically in saturated chambers, and densitometry was carried out at 312 nm. Calibration range was set to 100–600 ng per spot. Around 3% of relative standard deviation (RSD) value was observed.

A simple and rapid densitometric method (Tampubolon et al. 2006) has been developed for determination of tadalafil citrate, also in pharmaceuticals. Silica gel TLC plates were used, and chromatograms were developed with n-hexane–ethyl acetate–methanol (8:6:2). Quantitative evaluation was performed at 285 nm.

REFERENCES

Abadi, A., D. Abouel-Ella, N. Ahmed, B. Gary, J. Thaiparambil, H. Tinsley, A. Keeton, G. Piazza 2009. Synthesis of novel tadalafil analogues and their evaluation as phosphodiesterase inhibitors and anticancer agents. *Arzneimittel-Forschung/Drug Research* 59(8): 415–421.

Abourashed, E., M. Abdel-Kader, A.-A. Habib 2005. HPTLC determination of sildenafil in pharmaceutical products and aphrodisiac herbal preparations. *Journal of Planar Chromatography—Modern TLC* 18(105): 372–376.

Baranowska, I., P. Markowski, A. Wilczek, M. Szostek, M. Stadniczuk 2009. Normal and reversed-phase thin-layer chromatography in the analysis of l-arginine, its metabolites, and selected drugs. *Journal of Planar Chromatography—Modern TLC* 22(2): 89–96.

Cai, Y., T.-G. Cai, Y. Shi, X.-L. Cheng, L.-Y. Ma, S.-C. Ma, R.-C. Lin, W. Feng 2010. Simultaneous determination of eight PDE5-IS potentially adulterated in herbal dietary supplements with TLC and HPLC-PDA-MS methods. *Journal of Liquid Chromatography and Related Technologies* 33(13): 1287–1306.

Dowling, S., J. Cox, R. Cenedella 2009. Inhibition of fatty acid synthase by orlistat accelerates gastric tumor cell apoptosis in culture and increases survival rates in gastric tumor bearing mice in vivo. *Lipids* 44(6): 489–498.

Hasegawa, T., M. Saijo, T. Ishii, T. Nagata, Y. Haishima, N. Kawahara, Y. Goda 2008. Structural elucidation of a tadalafil analogue found in a dietary supplement. *Journal of the Food Hygienic Society of Japan* 49(4): 311–315.

Hsu, F.-L., C.-H. Chen, C.-H. Yuan, J. Shiea 2003. Interfaces to connect thin-layer chromatography with electrospray ionization mass spectrometry. *Analytical Chemistry* 75(10): 2493–2498.

Kremer, L., C. De Chastellier, G. Dobson, K. Gibson, P. Bifani, S. Balor, J.-P. Gorvel, C. Locht, D. Minnikin, G. Besra 2005. Identification and structural characterization of an unusual mycobacterial monomeromycolyl-diacylglycerol. *Molecular Microbiology* 57(4): 1113–1126.

Ku, Y.-R., Y.-C. Liu, J.-H. Lin 2002. High-performance liquid chromatographic analysis of sildenafil citrate and methyltestosterone adulterants in a herbal medicine. *Chinese Pharmaceutical Journal* 54(4): 307–312.

Lin, M.-C., Y.-C. Liu, J.-H. Lin 2006. Identification of a sildenafil analogue adulterated in two herbal food supplements. *Journal of Food and Drug Analysis* 14(3): 260–264.

Lin, M.-C., Y.-C. Liu, Y.-L. Lin, J.-H. Lin 2008. Isolation and identification of a novel sildenafil analogue adulterated in dietary supplements. *Journal of Food and Drug Analysis* 16(4): 15–20.

Lin, M.-C., Y.-C. Liu, Y.-L. Lin, J.-H. Lin 2009. Identification of a tadalafil analogue adulterated in a dietary supplement. *Journal of Food and Drug Analysis* 17(6): 451–458.

Mikami, E., T. Ohno, H. Matsumoto 2002. Simultaneous identification/determination system for phentolamine and sildenafil as adulterants in soft drinks advertising roborant nutrition. *Forensic Science International* 130(2–3): 140–146.

Miller, G., R. Stripp 2007. A study of western pharmaceuticals contained within samples of Chinese herbal/patent medicines collected from New York City's Chinatown. *Legal Medicine* 9(5): 258–264.

Nicoletti, M. 2011. Identification of thiosildenafil in a health supplement. *Natural Product Communications* 6(7): 1003–1004.

Reddy, T., A. Reddy, P. Devi 2006. Quantitative determination of sildenafil citrate in herbal medicinal formulations by high-performance thin-layer chromatography. *Journal of Planar Chromatography—Modern TLC* 19(112): 427–431.

Reepmeyer, J., D. D'avignon 2009. Use of a hydrolytic procedure and spectrometric methods in the structure elucidation of a thiocarbonyl analogue of sildenafil detected as an adulterant in an over-the-counter herbal aphrodisiac. *Journal of AOAC International* 92(5): 1336–1342.

Singh, S., B. Prasad, A. Savaliya, R. Shah, V. Gohil, A. Kaur 2009. Strategies for characterizing sildenafil, vardenafil, tadalafil and their analogues in herbal dietary supplements, and detecting counterfeit products containing these drugs. *TrAC—Trends in Analytical Chemistry* 28(1): 13–28.

Talati, R., S. Parikh, Y. Agrawal 2011. Pharmaceutical counterfeiting and analytical authentication. *Current Pharmaceutical Analysis* 7(1): 54–61.

Tampubolon, H., E. Sumarlik, S. Saputra, S. Cholifah, W. Kartinasari, G. Indrayanto 2006. Densitometric determination of tadalafil citrate in tablets: Validation of the method. *Journal of Liquid Chromatography and Related Technologies* 29(18): 2753–2765.

Yang, W., S. Lee, Y. Choi, H. Chung 2009. Importance of sildenafil analysis for drug screening of postmortem specimens: Demonstration of five autopsy cases involving sildenafil. *Forensic Toxicology* 27(2): 107–109.

37 TLC of Prostaglandins

Mariusz Stolarczyk, Anna Apola, and Jan Krzek

CONTENTS

37.1 INTRODUCTION

Prostaglandins represent a group of compounds present in all cells and body fluids. Physiologically, prostaglandins are synthesized from arachidonic acid (eicosa-5,8,11,14-tetraenoic acid), a component of cell membrane phospholipids. Arachidonic acid is released under an influence of enzyme phospholipase A_2. Displacement of phospholipase A_2 from cell cytoplasm to phospholipid membranes, and thus release of arachidonic acid, may be caused by numerous factors, for example, antigen–antibody reactions, physical stimulation (change in some ions concentration, heat, or cold), inflammatory reaction, or hormonal factors, for example, angiotensin II or adrenaline. Prostaglandins are formed from arachidonic acid as a result of an activity of cyclooxygenases (COX-1 and COX-2) enzymes.

The history of prostaglandins discovery reaches the year 1935, when Swedish physiologist von Euler [1] and Goldblatt [2] separated independently highly active lipids from prostate (*glandula prostatae*), which caused uterine contraction in vitro.

The study of Bergström allowed to determine the chemical structure of prostaglandins as derivatives of 20-carbon, 4-unsaturated acid (arachidonic acid) [3].

Currently, 16 kinds of prostaglandins are known. In order to systematize the nomenclature, prostaglandins are numbered with subsequent letters of the alphabet, for example, PGA, PGB, and PGH, and also with numbers, which determine the number of unsaturated bonds in a chain, for example, PGA_2 and PGE_2. Greek α or β index in prostaglandins designations, for example, PGF_α or $PGF_{2\alpha}$ determine the stereochemistry of the particle [4–6].

Prostaglandins demonstrate much differentiated activity on humans, the most important include the following:

- PGD_2—synthesized in mast cells, inhibits aggregation of blood platelets and leukocytes, decreases proliferation of T cells and lymphocytes migration, causes blood vessels extension, and increases cyclic adenosine monophosphate (cAMP) production.
- PGE_1—causes blood vessels extension and inhibits blood platelets aggregation.
- PGE_2—synthesized in kidneys, spleen, and heart; increases blood vessels extension and cAMP production, enhances the effect of bradykinin and histamine activity, induces uterine contractions and blood platelets aggregation, maintains an open passage from fetal ductus arteriosus, and decreases proliferation of T cells and lymphocytes migration.
- $PGF_{2\alpha}$—synthesized in kidneys, spleen, and heart; increases smooth muscles contraction of vessels and gills.

- PGH$_2$—is a precursor of thromboxane (TXB$_2$), induces aggregation of blood platelets and vessels contraction.
- PGI$_2$—synthesized in heart and endothelial cells of the vessels; inhibits aggregation of blood platelets and leukocytes in pulmonary vessels, decreases proliferation of T cells and lymphocytes migration, causes blood vessels extension, and cAMP production.

37.2 CHEMICAL STRUCTURE AND PHARMACOLOGICAL ACTIVITY OF PROSTAGLANDINS ANALOGS

Due to the rich and multidirectional activity of prostaglandins, numerous analogs have been synthesized, which are commonly used in numerous disease therapies. The most often used prostaglandins analogs include the following:

- ALPROSTADIL—analog of prostaglandin E$_1$. Extends blood vessels. Used in thromboangiitis obliterans, patent ductus arteriosus, and in erection disturbances. Applied by infusion (arteriovenously or intravenously) in a dose of 20–40 μg (Figure 37.1).
- GEMEPROST—analog of prostaglandin E$_1$. Expands uterine cervix. Applied *per rectum* in doses of 1–5 mg in gynecological treatments and pharmacological abortion (Figure 37.2).
- MISOPROSTOL—analog of prostaglandin E$_1$. Decreases hydrochloric acid release, demonstrates antiulcer activity, especially recommended with an application of nonsteroid anti-inflammatory drugs. Dose 0.4–0.8 mg. Applied rarely since it causes an increase in intestines peristalsis, pains, and diarrhea (Figure 37.3).
- SULPROSTONE—analog of prostaglandin E$_1$. Used in obstetrics for parturition leading, abortion inducing, in postpartum atonia, and bleeding from uterus, maximum 0.5–1 mg (Figure 37.4).

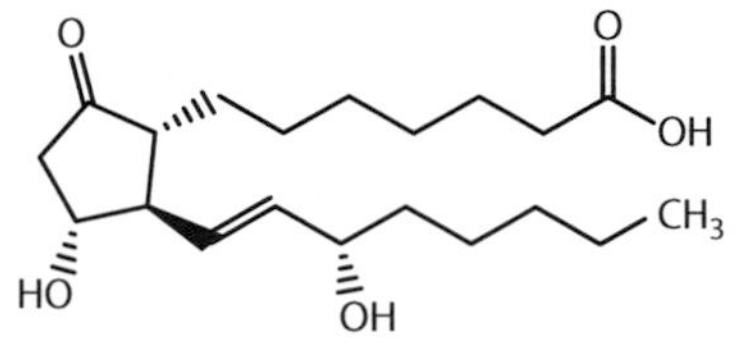

FIGURE 37.1   7-[(1R,2R,3R)-3-hydroxy-2-[(3S)-3-hydroxyoct-1-en-1-yl]-5-oxocyclopentyl]heptanoic acid.

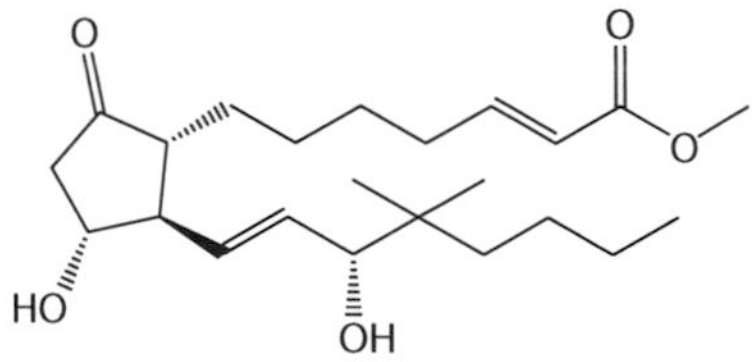

FIGURE 37.2   Methyl (2*E*,11α,13*E*,15*R*)-11,15-dihydroxy-16,16-dimethyl-9-oxoprosta-2,13-dien-1-oate.

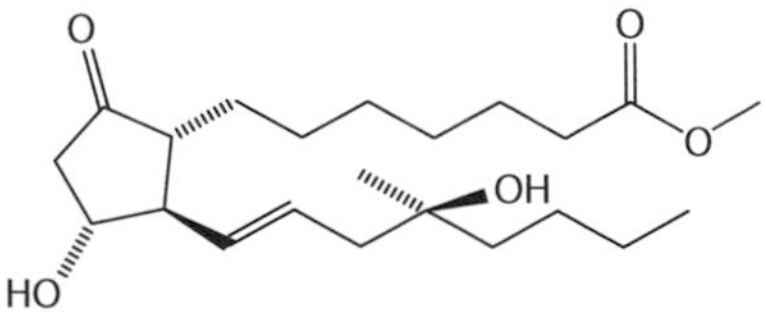

FIGURE 37.3   Methyl 7-((1*R*,2*R*,3*R*)-3-hydroxy-2-((*S*,*E*)-4-hydroxy-4-methyloct-1-enyl)-5-oxocyclopentyl) heptanoate.

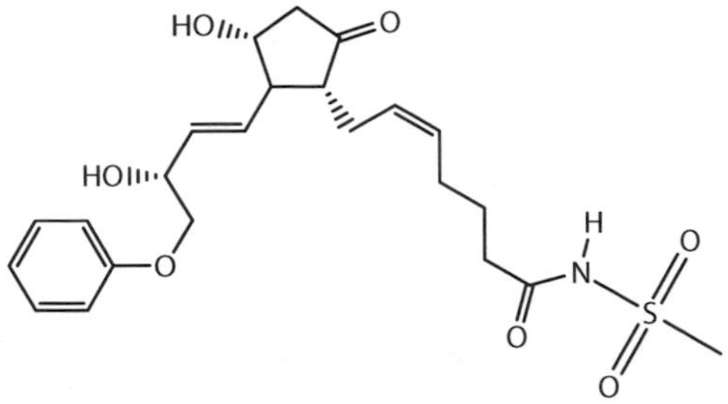

FIGURE 37.4 (Z)-7-[(1R,2R,3R)-3-hydroxy-2-[(E,3R)-3-hydroxy-4-phenoxybut-1-enyl]-5-oxocyclopentyl]-N-methylsulfonylhept-5-enamide.

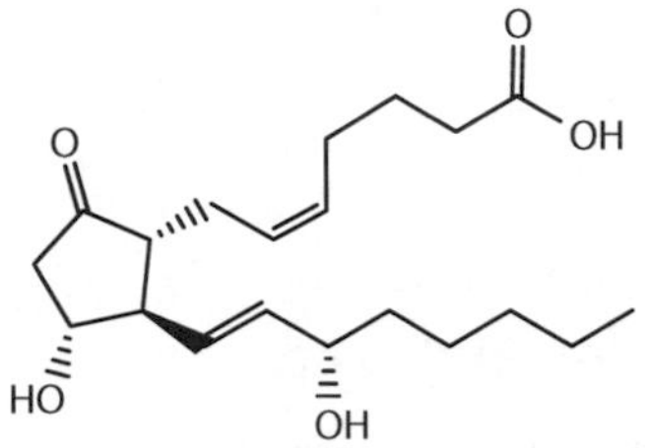

FIGURE 37.5 (5Z)-7-[(1R,2R,3R)-3-hydroxy-2-[(1E,3S)-3-hydroxyoct-1-en-1-yl]-5-oxocyclopentyl]hept-5-enoic acid.

- DINOPROSTONE—analog of prostaglandin E_2. Used in pregnancy leading in doses of 1–3 mg/at parturition, 3–6 mg at abortion (Figure 37.5).
- DINOPROST—analog of prostaglandin $F_{2\alpha}$. Used in atonia and uterus bleeding after parturition. The dose is matched individually, intravenously from 0.25 µg (Figure 37.6).
- LATANOPROST—analog of prostaglandin $F_{2\alpha}$. Decreases intraocular pressure. Used in eye drops in glaucoma treatment, drops 0.005% (Figure 37.7).
- EPOPROSTENOL—analog of prostaglandin I_2. Extends blood vessels, inhibits blood platelets aggregation, and used in atherosclerosis (especially of limbs vessels). Used intravenously in doses of 10 ng/kg and in dialyzed patients, 5 ng/kg (Figure 37.8).

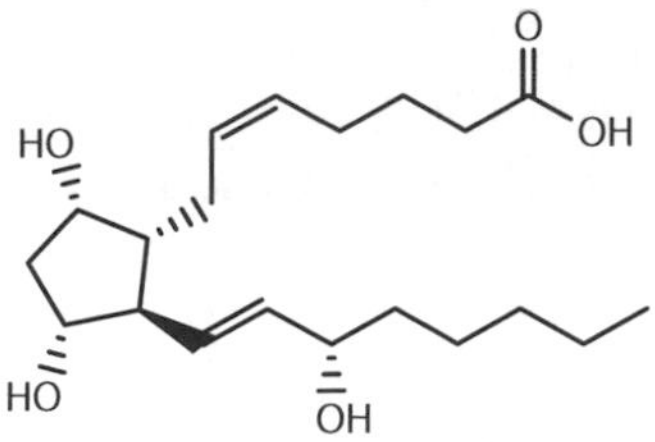

FIGURE 37.6 (5Z)-7-[(1R,2R,3R,5S)-3,5-dihydroxy-2-[(1E,3S)-3-hydroxyoct-1-en-1-yl]cyclopentyl]hept-5-enoic acid; 2-amino-2-(hydroxymethyl)propane-1,3-diol.

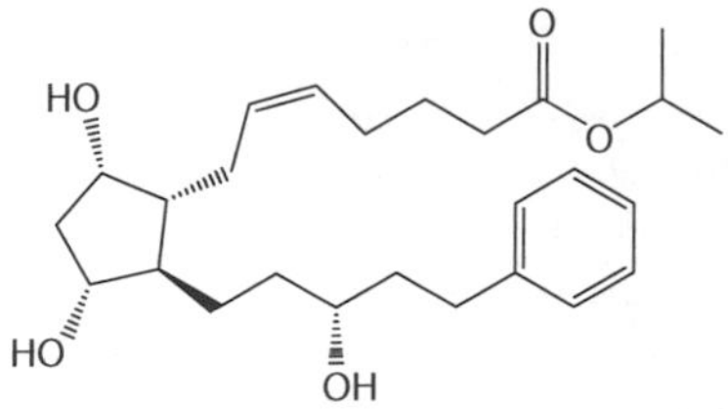

FIGURE 37.7 Isopropyl (Z)-7-[(1R,2R,3R,5S)-3,5-dihydroxy-2-[(3R)3-hydroxy-5-phenylpentyl]-cyclopentyl]hept-5-enoate.

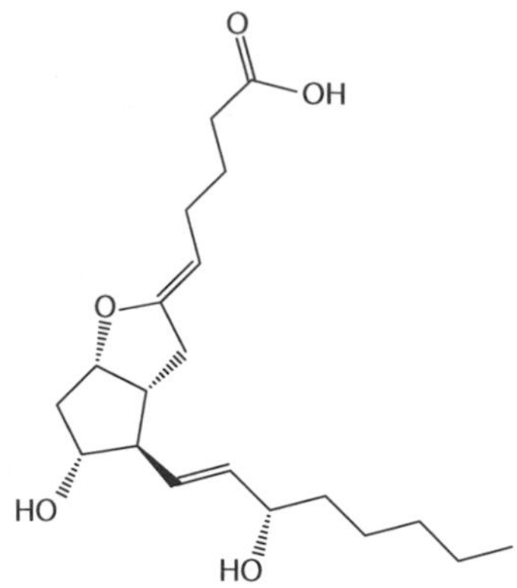

FIGURE 37.8 (Z)-5-[(4R,5R)-5-hydroxy-4-((S,E)-3-hydroxyoct-1-enyl)hexahydro-2H-cyclopenta[b]furan-2-ylidene]pentanoic acid.

37.3 CHROMATOGRAPHIC ANALYSIS

Thin-layer chromatography (TLC) and high-performance TLC methods were used for determination of prostaglandins and their metabolites in body fluids, and rarely in pharmaceutical preparations. Due to an application of various samples of biological origin for the examination, there is a need for their suitable preparation.

37.3.1 Prostaglandins Isolation from Biological Material and Drugs Form

Examination of prostaglandins content concerns biological material in a predominant number of publications. Due to the nature of these substances, their physicochemical properties, and small concentration in the examined samples, the preparation of the sample plays a significant role.

Prostaglandins isolation from human seminal plasma was described by Hamberg and Samuelson [7]. Preliminary purification involved centrifugation of the ethanol solution of semen sample. Next, a portion of ethanol was added to supernatant, and was centrifuged again. Joined supernatants were filtered, and volume was lowered by solvent excess distilling off. After acidification with hydrochloric acid up to pH 3, the sample was extracted with ether. After distilling off ether, the remaining part was rinsed with ethanol–water (2:1) mixture. The final volume of the sample was reduced by solvent excess evaporation.

A similar procedure was applied in the extraction of prostaglandins from group F [8]. A different scheme of sample preparation was used for prostaglandins determination in human and rat skin [9]. An application of suitable extraction technique allowed to determine prostaglandin (PGE) in amounts of 760–2140 ng per gram wet skin (human) and 625–839 ng per gram wet skin (albino rat skin) (Figure 37.9).

Another isolation method was elaborated in the case of an analysis of prostaglandins content from Kupffer cells [10]. Prostaglandins were extracted from the cells at pH 2.4 using ethyl acetate. Extraction was carried out in three stages (supernatant: ethyl acetate): 1:1 in the first stage, 1:2 in the second one, and 1:3 in the third stage (v/v). Joined extracts were distilled off up to the volume of 50 µL. Solid-phase extraction with an application of SEP-PAK 18C-cartriges apparatus (Waters by Millipore) was used in the control experiment.

Solid-phase extraction was also used for isolation of PGE_2 and $PGF_{2\alpha}$ and their metabolites from urine, plasma, or placenta cells homogenate [11]. The extraction was led on Supelclean LC-18 SPE columns. One milliliter of the sample was mixed with 1 mL of ice cold methanol and stored at a temperature of −20°C for 2 h. Next, the supernatant was centrifuged (10,000 × g at 4°C for 15 min).

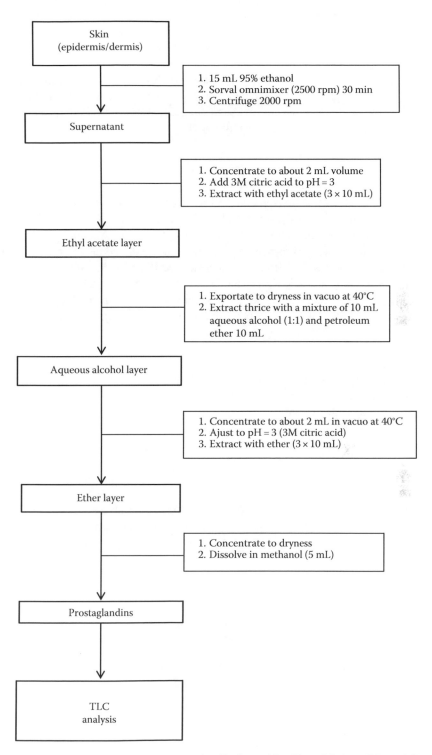

FIGURE 37.9 The scheme of isolation of prostaglandin from skin. (From Mathur, G.P. and Gandhi, V.M., *J. Invest. Dermatol.*, 58(5), 291, 1972.)

The samples were diluted up to 40 mL with water, and next, 0.4 mL of 90% (v/v) formic acid was added. Such prepared mixture was spread on SPE columns and filtered under lowered pressure. Next, the columns were rinsed with methanol–water mixture (4 mL) and dried for 30 min under lowered pressure. Prostaglandins were eluted with 4 × 100% methanol. Methanol was evaporated. Dry residue was dissolved in 100 μL of ethanol.

Determination of prostaglandins in pharmaceutical preparations was preceded with an extraction from fragmented drug form using ethanol–water mixture 7:3 [12,13]. In these cases, the extraction was automated. Heating, cooling, and filtering were controlled in an apparatus for extraction (W. Krannich KG, Gottingen, Republic of Germany).

37.3.2 Separation Conditions

Green and Samuelson [14] separated prostaglandins (free acids forms and methyl esters) from groups E and F on the plates with silica gel (Table 37.1). Glass silica gel G^2 plates were sprinkled with alcohol silver nitrate solution. In order to activate the plates, they were heated at a temperature of 110°–115° for 30 min. Material was spread in an amount of 10–100 μg using Hamilton syringe. After development, the plates were dried at a temperature of 100°C and sprinkled with 10% methanol solution of phosphomolybdic acid (PMA) and dried again at a temperature of 120°C for 15 min. Blue spots coming from prostaglandins were obtained on yellow–green background. The plates with silver nitrate (V) addition were developed using sulfuric acid (VI).

Mobile phases:

M1: Benzene–dioxane, 5:4
M2: Ethyl acetate–methanol–water, 8:2:5
M3: Ethyl acetate–methanol–water, 6:2.5:10
A1: Benzene–dioxane–acetic acid, 10:20:1
A2: Ethyl acetate–acetic acid–methanol–2,2,4-trimethylpentane–water, 110:30:35:10:100

TABLE 37.1
R_f Values of Methyl Esters and Free Acids

| Prostaglandin | R_f of Methyl Esters | | | R_f of Free Acids | |
| | Mobile Phase | | | Mobile Phase | |
	M1[a]	M2	M3	A1[a]	A2
PGE$_1$	0.58	0.65	0.62	0.62	0.80
PGE$_2$	0.57	0.57	0.49	0.62	0.70
PGE$_3$	0.58	0.29	0.20	0.63	0.35
PGE$_{1\alpha}$	0.38	0.47		0.46	0.64
PGE$_{1\beta}$	0.25	0.43		0.35	0.58
PGE$_{2\alpha}$	0.37	0.35		0.47	0.49
PGE$_{2\beta}$	0.26	0.33		0.36	0.48
PGE$_{3\alpha}$	0.38	0.18		0.47	0.23
PGE$_{3\beta}$	0.26	0.18		0.36	0.23
PGE$_{1-278}$			0.90		
PGE$_{2-278}$			0.84		
PGE$_{3-278}$			0.50		

[a] Plate with silver nitrate (V).

Andersen proposed the method of separation of prostaglandins from groups F, E, A, and B and their stereoisomers [15]. The stationary phase was neutral silica plates, alumina plates, and acidic silica plates with fluorescence factor.

The examined samples and reference substances were dissolved in acetone, dioxane, and tetrahydrofuran. The chromatograms were developed on a height of 7–8 cm in the case of neutral silica plates, alumina plates, and 10 cm in the case of acidic silica plates.

The spots were visualized

- With ultraviolet (UV) radiation—PGB and PGA

or after plates sprinkling

- With 1% $SbCl_5$ solution in CCl_4-CH_2Cl_2 mixture (5:1)—PGF, PGE_1, PGA, and PGB
- Vanilin-H_3PO_4-ethanol spray
- Three percent cupric acetate in 15% phosphoric acid spray—characteristic colors of PGA, PGB, PGE, and PGF (after plate warming up to temperature of 120°C)

Mobile phase:

P11: ethyl acetate–hexane–water–methanol–acetic acid, 4:2:2:1:1
C1: chloroform–tetrahydrofuran–acetic acid, 10:2:1
N1: hexane–tetrahydrofuran–methylene dichloride, 1:1:1
H1: hexane–methylene dichloride–tetrahydrofuran–acetic acid, 6:2:2:1
H2: hexane–methylene dichloride–tetrahydrofuran–acetic acid, 30:10:3:3
H4: hexane–methylene dichloride–tetrahydrofuran–acetic acid, 10:10:10:1
D1: benzene–dioxane, 3:2
D2: benzene–dioxane–acetic acid, 40:10:1
D3: benzene–dioxane–acetic acid, 20:10:1
D4: benzene–dioxane–acetic acid, 20:20:1
F1: ethyl acetate–formic acid, 100:1
F4: ethyl acetate–formic acid, 400:5
F5: ethyl acetate–ethanol–acetic acid, 100:1:1
F6: ethyl acetate–acetone–acetic acid, 90:10:1
F7: cyclohexane–ethyl acetate–acetic acid, 60:40:2

The R_f values for the examined prostaglandins and their epimers with stationary and mobile phases are presented in Table 37.2.

Prostaglandins PGE_2 and $PGF_{2\alpha}$ may be separated and determined using TLC chromatography method. One of the methods of visualization of spots coming from examined substances is an application of PMA. Visualization procedures based on PMA are commonly used for determination of bioactive compounds like lipids, saponins, terpenes, and sterols [16–18]. Determination of PGE_2 and $PGF_{2\alpha}$ [18] was conducted on two kinds of plates, K60WF254S (silica gel) and RP-18W (octadecyl silica). Chromatograms were developed in a vertical chamber saturated previously with mobile phase. Working prostaglandin solutions of a concentration of 1 mg mL^{-1} in ethanol were manually spread in an amount of 1 μL using Hamilton syringe. Mobile phases was a mixture of methanol–dichloromethane 1:9 (v/v) and 100% acetonitrile. After separation, the samples were dried at room temperature and sprinkled with a solution of PMA (10% in methanol), heated in convection furnace at a temperature range from 40°C to 140°C for a period from 2 to 40 min. The plates were cooled to room temperature and scanned immediately. It was observed that optimum effect of staining of prostaglandins PGE_2 or $PGF_{2\alpha}$ by PMA is obtained by heating the plates at a temperature from 80°C to 100°C for K60WF254S plates (silica gel), or below 80°C for RP-18W plates (octadecyl silica) for 20–30 min. It was noted, that decrease in a temperature of plates heating with concurrent elongation of its time results in lowering

TABLE 37.2
R_f Value, Stationary Phase, and Mobile Phase

	Neutral Silica													Acidic Silica								Al_2O_3
	P11 (1×)	C1 (2×)	H1 (2×)	H1 (4×)	H2 (5×)	D1 (1×)	D1 (2×)	D2 (3×)	D3 (2×)	D4 (1×)	F1 (4×)	F1 (4×)	N1 (4×)	F4 (2×)	F4 (1×)	F4 (1×)	F5 (1×)	F6 (1×)	F7 (2×)	D3 (1×)	H4 (1×)	H2 (2×)
$PGF_{1\beta}$	0.18	0.5							0.17	0.23				0.15		0.12	0.1	0.14	0.03	0.15	0.08	
$PGF_{1\alpha}$	0.22	0.12							0.26	0.3				0.25		0.23	0.18	0.24	0.07	0.25	0.17	
PGE_1	0.34	0.25	0.24	0.38	0.15			0.15	0.42	0.37	0.32	0.57		0.37	0.24	0.37	0.32	0.39	0.1	0.38	0.27	0.39
11-epi-PGE_1	0.35	0.29	0.31	0.48	0.17			0.21				0.66		0.47	0.29	0.49	0.45	0.54	0.11	0.52	0.39	
15-epi-PGE_1	0.4	0.37	0.36	0.56	0.21			0.28	0.56	0.46	0.48	0.76		0.57	0.36	0.58	0.54	0.62	0.15	0.58	0.48	0.42
11,15-epi-PGE_1	0.42	0.4	0.41	0.59	0.22			0.29				0.76		0.56	0.34	0.59		0.64	0.19	0.6	0.5	
PGA_1	0.57	0.78	0.69	0.82	0.5			0.69	0.78	0.57	0.62			0.76	0.5	0.79	0.55	0.79	0.62	0.75	0.81	0.62
15-epi-PGA_1	0.59	0.8	0.72	0.84	0.57			0.73	0.8	0.58				0.77	0.51	0.83	0.77	0.86	0.68	0.81	0.87	
PGB_1	0.56	0.79	0.67	0.81	0.5				0.77					0.74		0.79	0.81	0.78	0.6	0.76	0.85	
CH_3-PGE_1	0.47	0.38	0.34	0.49	0.2	0.27	0.42			0.45			0.38		0.3		0.76					
CH_3-11-epi-PGE_1				0.57			0.51						0.51									
CH_3-15-epi-PGE_1				0.63			0.58						0.64									
CH_3-11,15-epi-PGE_1				0.65			0.57						0.63									
CH_3-PGA_1	0.64	0.84	0.79		0.69	0.58				0.65	0.65				0.67							0.82
CH_3-PGB_1	0.63	0.85	0.79			0.58					0.76											

Note: 1–4 multiple number of chromatogram development.

of signal coming from the background and increase in signal coming from prostaglandins. This is especially observed in the case of an application of plates with C-18 for separation.

Similar procedure was used for determination of PGE_2 and $PGF_{2\alpha}$ except their inactive metabolites, that is, 15-keto-PGE_2 and 13,14-dihydro-15-keto-PGE_2, and 15-keto-$PGF_{2\alpha}$ and 13,14-dihydro-15-keto-$PGF_{2\alpha}$ [19].

Pestel et al. proposed determination of prostaglandins in biological material with Kupffer cells derived from rat's liver [10]. The study included compounds present in these cells, that is, PGD_2, PGE_2, TXB_2, and $PGF_{2\alpha}$. The preparation of the sample has been described previously. Separation was conducted on silica plates, and the mobile phase was a mixture of ethylacetate–water–isooctane–acetate (110:100:50:20 v/v). The obtained R_f coefficients, both for single reference prostaglandins solutions and for extract obtained from Kupffer cells prove well separation of the examined compounds. R_f values were: PGD_2 $R_f = 0.32$; PGE_2 $R_f = 0.21$; TXB_2 $R_f = 0.17$; $PGF_{2\alpha}$ $R_f = 0.12$— prostaglandins, and extracts obtained from Kupffer cells, PGD_2 $R_f = 0.33$; PGE_2 $R_f = 0.20$; TXB_2 $R_f =$ below detection limit; $PGF_{2\alpha}$ $R_f = 0.13$.

TLC method was also used for prostaglandins isolation from the cells of *Plexaura homomalla* species [20]. Lipid extract obtained from *P. homomalla* was spread on extraction columns (Mallinckrodt Silicar CC-4, 100–200 mesh) and eluted with ethyl acetate solution in benzene. Anhydrous fraction was spread on plates with silica gel (E. Merck). Mobile phase was a mixture of ethylacetate–benzene–formic acid (25:75:1, v/v/v). Visualization of spots on chromatogram was noted after the plate was sprinkled with a solution of PMA and heated at a temperature of 120°C.

Prostaglandins analogs are a basic component of antiglaucoma drugs. Xalatan [21], Travatan [22], and Lumigan [23].

Products of Bimatoprost hydrolysis (analog of prostaglandin $F_{2\alpha}$) were determined using TLC method [24]. Hydrolysis was performed in the presence of cornea extract in phosphate buffer of pH = 7.4 at a temperature of 37°C for 6 and 23 h. Reaction mixture was acidified to pH = 3 and extracted with a mixture of ethyl acetate:hexane. The samples were spread in an amount of 5 µL of organic phase on silica gel G-60 plates (Analtech), and chromatogram was developed in mobile phase composed of ethyl acetate–methanol–acetic acid, 95:5:1. Spots from prostaglandin and hydrolysis product were detected after plates were sprinkled with sulfuric acid. The experiment demonstrated that the product of hydrolysis is present both in the samples incubated for 6 and 23 h, which was confirmed by an application of HPLC method as a reference method.

Determination of PGE_2 and $PGF_{2\alpha}$ and their metabolites in urine, plasma, or placenta cells homogenate was conducted on Silica gel 60 F_{254} plates. Chromatograms were developed horizontally (DS.-L-Chamber, CHROMDES, Lublin, Poland) at a room temperature. Mobile phase was chloroform–methanol mixture (85:15, v/v). After separation, the spots were visualized by sprinkling the plates with 10% PMA in ethanol (w/v) and heating at a temperature of 130°C for 10 min. The determined prostaglandins are visualized as blue stains on yellow background.

Prostaglandins from group E_2 and $F_{2\alpha}$ in human skin after exposure to UV radiation at range of 290–320 nm were examined using TLC method [25].

Prostaglandins samples obtained from exposed cells and methyl esters of prostaglandins were spread on neutral silica gel plates. Plates spread and non-spread with 3% $Ag(NO_3)$ solution were used. The developing system was a mixture of ethyl acetate–acetone–glacial acetic acid (180:20:1). R_f coefficients corresponding to particular prostaglandins are presented in Table 37.3.

Amin [12,13] used TLC method for examination of prostaglandins from subgroup E_2, A_2, and B_2 in pharmaceutical preparations. Separation was led on silica gel 60 F_{254} plates. The examined solutions were spread in an amount of 1.2–5.0 µL. Mobile phase were the mixtures: diethyl ether–methanol–chloroform (65:15:20) at a distance of 2×15 cm, chloroform–methanol (95:5) at a distance of 3×20 cm, and chloroform–acetic acid ester–methanol (80:5:15) at a distance of 2×15 cm. After separation, the chromatograms were dried with hot air stream for 5 min. UV detection was conducted after the plates were sprinkled with 1% methanolic potassium hydroxide solution and drying for 15 min at a temperature of 120°C.

TABLE 37.3
R_f Value of Identified Substances

Compound	R_f Plate without $Ag(NO_3)$	Plate with $Ag(NO_3)$
PGE_1	0.27	0.27
PGE_2	0.28	0.17
$PGF_{1\alpha}$	0.13	0.12
$PGF_{2\alpha}$	0.12	0.05
Arachidonic acid	0.85	0.32
Oleic acid	0.88	0.77
15-keto-$PGF_{2\alpha}$	0.39	0.19
15-keto-13,14 dihydro-$PGF_{2\alpha}$	0.42	0.22
15-keto-13,14 dihydro-PGE_2	0.73	0.47
15-keto-PGE_2	0.72	0.56
PGD_1	0.61	0.37
PGD_2	0.59	0.20
PGA_2	0.75	0.50

37.4 CONCLUSIONS

TLC method is used for prostaglandins determination, especially in biological material, both in body liquids and plant tissue homogenates. It is also used for prostaglandins determination in various forms of drugs. An application of extraction method, stationary and mobile phase, and detection method depends on the matrix and its possible influence on the compounds analyzed. Determination of prostaglandins in biological material requires sample purification from matrix compounds, which causes the need for use of complex extraction process. Determination of prostaglandins in pharmaceutical preparations was preceded with a completely automated extraction process, with an application of suitable laboratory equipment and extraction reagent. Visualization of the spots coming from examined substances was performed with an application of various developing reagents and densitometry as the detection method.

REFERENCES

1. Von Euler U.S. 1935. Über die spezifische blutdrucksenkende Substanz des menschlichen prostata- und Samenblasensekrets. *Wien Klin Wochenschr* 14:1182–1183.
2. Goldblatt M.W. 1935. Properites of human seminal plasma. *The Journal of Physiology* 84:208–218.
3. Bergström S., R. Ryhage, B. Samuelsson, J. Sjövall. 1963. Prostaglandins and related factors: 15. The structures of prostaglandin E_1, $F_{1\alpha}$, and $F_{1\beta}$. *The Journal of Biological Chemistry* 238:3555–3564.
4. Goodwin, G. M. 2010. *Prostaglandins: Biochemistry, Functions, Types and Roles.* New York: Nova Science Publishers.
5. Pace-Asciak C., E. Granstrom. 1983. *Prostaglandins and Related Substances, Vol. 5 (New Comprehensive Biochemistry).* Amsterdam, The Netherlands: Elsevier Science Publishers B.V.
6. Horrobin D. 1978. *Prostaglandins: Physiology, Pharmacology & Clinical Significance.* Montreal, Quebec, Canada: Eden Press.
7. Hamberg M., B. Samuelsson, 1966. Prostaglandins in human seminal plasma. *Journal of Biological Chemistry* 241:257–263.
8. Srivastava K.C. 1977. Extraction of prostaglandins F1α and F2α from human seminal plasma. *Fresenius Zeitschrift fur Analytische Chemie Labor und Betriebsverfahren* 285(1):35–38.
9. Mathur G.P., V.M. Gandhi. 1972. Am Prostaglandin in human and albino rat skin. *Journal of Investigative Dermatology* 58(5):291–295.

10. Pestel S., K. Jungermann, H.L. Schieferdecker. 2005. Re-evaluation of thin layer chromatography as an alternative method for the quantification of prostaglandins from rat Kupffer cells. *Prostaglandins and Other Lipid Mediators* 75(1–4):123–139.

11. Welsh T.N., S. Hubbard, C.M. Mitchell, S. Mesiano, P.K. Zarzycki, T. Zakar. 2007. Optimization of a solid phase extraction procedure for prostaglandin E2, F2α and their tissue metabolites. *Prostaglandins and Other Lipid Mediators* 83(4):304–310.

12. Amin M. 1987. Quantitative thin-layer chromatorgraphic determination of some prostaglandin-derivatives of the subgroups E2, A2 and B2. *Fresenius Zeitschrift fur Analytische Chemie* 329(5):600–602.

13. Amin M. 1989. Stability studies of a prostaglandin derivative of the subgroup E2 in pharmaceutical preparations. *Fresenius Zeitschrift fur Analytische Chemie* 333(2):134–139.

14. Green K., B. Samuelsson. 1964. Thin-layer chromatography of prostaglandins. *Journal of Lipid Research* 5:117–120.

15. Andersen, N.H. 1969. Preperative thin-layer and column chromatography of prostaglandins. *Journal of Lipid Research* 10:316–319.

16. Zarzycki P.K., M.A. Bartoszuk, A.I. Radziwon. 2006. Optimization of TLC detection by phosphomolybdic acid staining for robust quantification of cholesterol and bile acids. *Journal of Planar Chromatography—Modern TLC* 19(107):52–57.

17. Zarzycki P.K., M. Baran, E. Włodarczyk, M.A. Bartoszuk. 2007. Improved detection of ergosterol, stigmasterol, and selected steroids on silica coated TLC plates using phosphomolybdic acid staining. *Journal of Liquid Chromatography and Related Technologies* 30(17):2629–2634.

18. Zarzycki P.K., M.A Bartoszuk. 2008. Improved TLC detection of prostaglandins by post-run derivatization with phosphomolybdic acid. *Journal of Planar Chromatography—Modern TLC* 21(5):387–390.

19. Welsh T., T. Zakar, S. Mesiano, P.K. Zarzycki. 2003. Separation of bioactive prostaglandins and their metabolites by reversed-phase thin-layer chromatography. *Journal of Planar Chromatography—Modern TLC* 16(2):95–101.

20. Light R.J., B. Samuelsson. 1972. Identification of prostaglandins in the gorgonian, Plexaura homomalla. *European Journal of Biochemistry* 28(2):232–240.

21. Camras C.B., A. Alm, P. Watson, J. Stjernschantz, P. Aasved, P. Jangard, H. Lund-Andersen. 1996. Latanoprost, a prostaglandin analog, for glaucoma therapy: Efficacy and safety after 1 year of treatment in 198 patients. *Ophthalmology* 103(11):1916–1924.

22. Stjernschantz J., B. Resul. 1992. Phenyl substituted prostaglandin analogs for glaucoma treatment. *Drugs of the Future* 17(8):691–704.

23. Cantor L.B. 2001. Bimatoprost: A member of a new class of agents, the prostamides, for glaucoma management. *Expert Opinion on Investigational Drugs* 10(4):721–731.

24. Maxey K.M., J.L. Johnson, J. La Brecque. 2002. The hydrolysis of bimatoprost in corneal tissue generates a potent prostanoid FP receptor agonist. *Survey of Ophthalmology* 47(4 Suppl. 1):34–40.

25. Kobza Black A., M.W. Greaves, C.N. Hensby, N.A. Plummer. 1978. Increased prostaglandins E2 and F2(α) in human skin at 6 and 24 h after ultraviolet B irradiation (290–320 nm). *British Journal of Clinical Pharmacology* 5(5):431–436.

38 TLC of Diuretics

Urszula Hubicka, Anna Maślanka, and Jan Krzek

CONTENTS

38.1 INTRODUCTION

In some diseases, collecting of too much water in the body occurs leading to edema in tissues. These symptoms are often connected with heart failure, renal failure, and liver failure. Collecting of too much water in the organism is the effect of sodium ions and chloride ions retention. Diuretics stimulate the kidneys to increase the secretion of urine to rid the body of excess of sodium and water.

The mechanism of action of diuretics is diversified in this group and depends on their structure. Excess diuresis may be achieved by increasing osmotic pressure, by increasing the acidity or alkalinity in the blood, by inhibiting reabsorption of sodium ions, chloride ions, and water or by increasing glomerular filtration.

When assessing the suitability of diuretics, not only the potency of action is taken into account but also their toxicity.

It is important that the increased renal excretion does not affect the electrolyte balance in the body and does not result in acidification or alkalization.

Diuretics may be divided into the following groups:

- Thiazides and thiazide-like duretics
- Anhydrase inhibitors
- Loop diuretics
- Cyclic amidines
- Antagonists of aldosterone
- Uricosurics

38.2 THIAZIDE AND THIAZIDE-LIKE DIURETICS

Thiazide diuretics consist of two distinct groups: those containing benzothiadiazine ring, such a hydrochlorothiazide (HCT) and chlorothiazide, referred to as thiazide diuretics and those that lack this heterocyclic structure but contain an unsubstituted sulfonamide group. The latter are called thiazide-like diuretics; they include metolazone, xipamide, clopamide, and indapamide [1].

Thiazide diuretics act on the distal convoluted tubule and inhibit the sodium-chloride symporter leading to retention of water in the urine, as water normally follows penetrating solutes. Frequent urination is due to the increased loss of water that has not been retained in the body as a result of a concomitant relationship with sodium loss from the convoluted tubule. The short-term antihypertensive action is based on the fact that thiazides decrease preload, decreasing blood pressure. On the other hand, the long-term effect is due to a vasodilator effect that decreases blood pressure by decreasing resistance. Especially at higher doses, administration of some of the thiazides results in some degree of carbonic anhydrase inhibition [1–3]. Moreover, indapamide has the ability to inhibit aggregation of platelets and prolong smooth muscle cell proliferation in "in vitro" culture. It is suggested that this ability may inhibit adverse remodeling of blood vessels and heart muscle caused by long-term hypertension [3].

Thiazide diuretics are commonly combined with other antihypertensive agents in the treatment of moderate to severe hypertension. The usual combination is with β-blockers. However, preparations consisting of low-dose thiazides and angiotensin converting enzyme (ACE) inhibitors, angiotensin receptor blockers, and potassium sparing agents are widely used.

38.2.1 HYDROCHLOROTHIAZIDE

HCT, 6-chloro-3, 4-dihydro-2H-1,2, 4-benzothiadiazine-7-sulfonamide1,1-dioxide, is one of the oldest and widely used thiazide diuretics (Figure 38.1).

38.2.1.1 HCT and β-Blockers

High-performance thin-layer chromatography (HPTLC) methods for the estimation of nebivolol (NBV) and HCT have been developed [4–7].

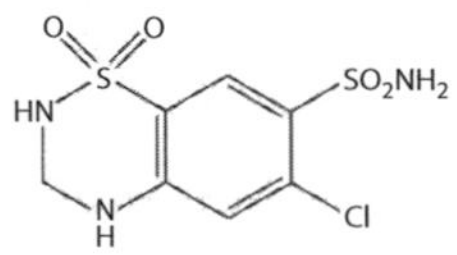

FIGURE 38.1 Hydrochlorothiazide.

Standard solution:

- Methanolic solutions in the concentration range 0.25–1.25 µg/mL for HCT and 0.1–0.5 µg/mL for NBV [5]
- Methanolic solutions 400 µg/mL for HCT and 1000 µg/mL for NBV [7]

Sample preparation:

- An accurately weighed tablet powder, equivalent to 12.5 mg HTC and 5.0 mg NBV, was transferred to a 100 mL volumetric flask. Extraction of the analyte was carried out by using methanol [5].
- An accurately weighed tablet powder, equivalent to 100 mg NBV and 40 mg HCT, was transferred to a 100 mL volumetric flask. Extraction of the analyte was carried out by using methanol [7].

Samples were applied using Linomat 5 sample applicator. The developing solvent mixture was run up to 80 mm in chamber previously saturated for 10 min [5] or 30 min [7]. Stationary and mobile phases used for the separation HCT and NBV are presented in Table 38.1. After development, the plate was dried at 50°C in an oven for about 5 min. Linearity was observed in the concentration range of 0.25–1.25 µg/mL for HTC and 0.1–0.5 µg/mL for NBV [5] or 1.0–5.0 µg/spot for NBV and 0.40–2.0 µg/spot for HTC [7].

38.2.1.2   HCT and ACE Inhibitors

HPTLC-densitometric method was described for the simultaneous determination of benazepril HCl and HCT in binary mixture [8].

Standard solution: methanolic solutions in the concentration range 0.2–2.0 mg/mL for benazepril HCl and 0.25–2.5 mg/mL for HCT.

Sample preparation: 20 tablets were weighed and finely powdered. A portion of the powder equivalent to about 50 mg of benazepril HCl and 62.5 mg HCT was weighed accurately, dissolved in, and diluted to 50 mL with methanol.

Stationary phase: HPTLC aluminum plates of silica gel 60 F_{254} (Merck).

Mobile phase: ethyl acetate–methanol–chloroform (10:3:2, v/v/v).

TABLE 38.1

Chromatographic Conditions for Simultaneous Determination of Nebivolol and Hydrochlorothiazide

Stationary Phase	Mobile Phase	Detection	R_F Value	Reference
Aluminum-backed silica gel 60 F_{254} TLC plates prewashed with methanol	Ethyl acetate–methanol–ammonia (8.5:1:0.5, v/v/v)	Densitometrically at 280 nm or 281 nm	$Rf_{HCT} \approx 0.21$ $Rf_{NBV} \approx 0.41$	[4–5]
Aluminum-backed silica gel 60 F_{254} TLC plates	Ethyl acetate–methanol–acetic acid (6.5:1:0.5, v/v/v)	Densitometrically at 280 nm for NBV and 270 nm for HCT	$Rf_{HBV} \approx 0.46$ $Rf_{HCT} \approx 0.78$	[6]
Aluminum-backed silica gel 60 F_{254} TLC plates, (20 cm × 10 cm, Merck), prewashed with methanol	1,4-Dioxane–toluene–triethylamine (5:3:0.1, v/v/v)	Densitometrically at 281 nm	$Rf_{HCT} \approx 0.43$ $Rf_{NBV} \approx 0.74$	[7]

For quantification, 10 µL of test and 10 µL of different concentrations of the standard solution within the quantitation range were applied. The plate was developed up to the top (over a distance of 8 cm) in the usual ascending way. After elution, the plate was air dried and scanned at 238 nm for benazepril HCl and 275 nm for HCT by using Shimadzu dual wavelength flying spot densitometer model CS-9000.

The R_F values of benazepril HCl and HCT were 0.22 and 0.60, respectively. The calibration graphs were constructed in the range of 2–20 µg/spot for benazepril HCl and 2.5–25 µg/spot for HCT. The calibration curves obtained were nonlinear for both compounds and were expressed by means of second-order polynomial functions.

A HPTLC method for the estimation of captopril (CPP) and HCT has been developed [9].

Stationary phase: TLC aluminum plates of silica gel 60 F_{254}.

Mobile phase: methanol–toluene–ethyl acetate–glacial acetic acid (1:6:3:0.5, v/v/v/v).

Detection: densitometrically at 219 nm.
The R_F values were 0.38 and 0.57 for HCT and CPP, respectively. The calibration curve response was observed between 4 and 14 µg for both drugs.

HPTLC methods for the estimation of enalapril maleate (ELP) and HCT have been developed [10–12].

Standard solution:

- Methanolic solutions in the concentration range 0.02–0.38 mg/mL for HCT and 0.0625–1.0 mg/mL for ELP [11].
- Methanolic solutions 100 µg/mL for ELP and 250 µg/mL for HCT, respectively [12].

Sample preparation:

- Twenty tablets were weighed accurately and finely powdered. Extraction of the analyte was carried out by using methanol. Required dilutions were made to get 0.20 mg/mL for ELP and 0.625 and 0.125 mg/mL for HCT, respectively [11].
- Twenty tablets were weighed accurately, finely powdered, and powder equivalent to 5 mg of ELP and 12.5 mg of HCT were weighed accurately, and extraction of the analyte was carried out by using methanol. Required dilutions were made to get 100 µg/mL for ELP and 250 µg/mL for HCT, respectively [12].

Samples were applied using Linomat 5 sample applicator. The developing solvent mixture was run up to 95 mm [11] or 80 mm [12] in chromatographic chamber previously saturated for 20 min [12]. Stationary and mobile phases used for the separation ENL and HTC are presented in Table 38.2.
Calibration curves were linear in the range:

- 1–100 µg/mL with correlation coefficients not less than 0.9996 [10]
- 0.312–5.00 µg/band for ENL and 0.078–1.25 µg/band for HCT, respectively [11]
- 100–500 ng/mL for ENL and 250–1250 ng/mL for HCT [12]

HPTLC-densitometric method was presented for the simultaneous determination of lisinopril (LNP) and HCT in pharmaceutical tablets [13].

Standard solutions: methanolic solutions in the concentration range 0.25–2.5 mg/mL for HCT and 0.4–2.0 mg/mL for LNP.

Sample preparation: 20 tablets were weighed and finely powdered. A portion of the powder equivalent to 40 mg of LNP and 25 mg of HCT was weighed accurately, dissolved and diluted to 50 mL with methanol. The sample solution was filtered.

TABLE 38.2

Chromatographic Conditions for Simultaneous Determination of Enalapril and Hydrochlorothiazide

Stationary Phase	Mobile Phase	Detection	R_F Value	Reference
Aluminum-backed silica gel 60 F_{254} TLC plates (0.2 mm thickness, Merck)	Ethyl acetate–chloroform–methanol–acetic acid (11:8:7.5:1.5: v/v/v/v)	Densitometrically at 257 nm for ELP and 226 nm for HCT	—	[10]
Aluminum-backed silica gel 60 F_{254} HPTLC plates (10 × 10 cm, Merck)	Butane-1-ol–glacial acetic acid–water (12:3:5, v/v/v)	Densitometrically at 274 nm for ELP and 208 nm for HCT	$Rf_{ELP} \approx 0.61$ $Rf_{HCT} \approx 0.84$	[11]
Aluminum-backed silica gel 60 F_{254} HPTLC plates, (20 × 10 cm, Merck)	Ethanol–toluene (7:3, v/v)	Densitometrically at 211 nm	$Rf_{ELP} \approx 0.52$ $Rf_{HCT} \approx 0.83$	[12]

Stationary phase: HPTLC aluminum plates of silica gel 60 F_{254} (20 × 10 cm, Merck).

Mobile phase: chloroform–ethyl acetate–acetic acid (10:3:2, v/v/v).

For detection and quantification, 10 µL of test and 10 µL of different concentrations of the standard solution were applied. The plate was developed up to the top (over a distance of 8 cm) in the usual ascending way. After elution, the plate was air dried and scanned using Shimadzu dual wavelength flying spot densitometer model CS-9000 at 210 and 275 nm for LNP and HCT, respectively. The R_F of LNP and HCT were 0.31 and 0.88, respectively. The linear and second-order polynomial were used for the regression equation of LNP and HCT, respectively. Calibration curves were estimated in the concentration range of 4–20 µg/spot for LNP and 2.5–25 µg/spot for HCT.

HPTLC-densitometric methods were presented for the simultaneous determination of quinapril (QNP) and HCT in pharmaceutical tablets [14–16].

Standard solution:

- Methanolic solution 100 µg/mL for QNP and 125 µg/mL for HCT [15]
- Methanolic mixtures in the concentration range 50–300 µg/mL for HCT and 80–480 µg/mL for QNP [16]

Sample preparation:

- Twenty tablets were weighed and finely powdered. A portion of the powder equivalent to 10 mg of QNP and 12.5 mg of HCT was transferred into 100 mL volumetric flask. Extraction of the analyte was carried out by using methanol [15].
- Twenty tablets were weighed and finely powdered. A portion of the powder equivalent to 30 mg of QNP and 18.75 mg of HCT was transferred into 25 mL volumetric flask. Extraction of the analyte was carried out by using methanol [16].

Chromatograms were development to a distance of

- 70 mm in twin-trough chamber saturated for 15 min with vapors of mobile phase [15]
- 50 mm in unsaturated horizontal DS chamber [16]

After development, the plates were dried at room temperature. Chromatographic condition used for the separation and determination of QNP and HCT are presented in Table 38.3.

TABLE 38.3

Chromatographic Conditions for Simultaneous Determination of Quinapril and Hydrochlorothiazide

Stationary Phase	Mobile Phase	Detection	R_f Value	Reference
Aluminum-backed silica gel 60 F_{254} HPTLC plates (0.2 mm thickness, Merck)	Toluene–ethyl acetate–glacial acetic acid (1:6:0.5, v/v/v/)	Densitometrically using UV detector in absorbance mode	—	[14]
Aluminum-backed silica gel 60 F_{254} TLC plates (20 × 10 cm, Merck)	Ethyl acetate–acetone–acetic acid (6.5:3:0.5, v/v/v)	Densitometrically at 208 nm	$Rf_{QNP} \approx 0.51$ $Rf_{HCT} \approx 0.76$	[15]
Aluminum-backed silica gel 60 F_{254} HPTLC plates, (20 × 10 cm, Merck), prewashed with methanol	Ethyl acetate–acetone–acetic acid (8:2:0.5, v/v/v)	Densitometrically at 210 nm	$Rf_{QNP} \approx 0.51$ $Rf_{HCT} \approx 0.81$	[16]
Octadecilsilane (RP-18) TLC plates, (20 × 10 cm, Merck)	Methanol–0.07 M phosphate buffer, pH 2.5 (6:4, v/v)		$Rf_{QNP} \approx 0.21$ $Rf_{HCT} \approx 0.78$	

Linearity:

- Linearity was observed in the concentration range of 0.5–3.5 µg/spot for HTC and 0.4–2.8 µg/spot for QNL [15].
- Calibration curves were estimated in the concentration range of 2–12 µg/band for QNP and 1.25–7.5 µg/band for HCT. The calibration curves obtained were nonlinear for both compounds and were expressed by means of second-order polynomial functions [16].

38.2.1.3   HCT and Angiotensin-II Receptor Antagonists

TLC method was described for the simultaneous determination of HCT and candesartan cilexetil (CAN) in combined dosage forms [11,14,17,18].

Standard solution:

- Methanolic solutions in the concentration range 0.02–0.38 mg/mL for HCT and 0.045–0.723 mg/mL for CAN [11].
- Methanolic solutions in the concentration range 0.05–0.25 mg/mL for HCT and 0.065–0.325 mg/mL for CAN [17].

Sample preparation:

- Twenty tablets were weighed accurately and finely powdered. Extraction of the analyte was carried out by using methanol. Required dilutions were made to get 0.16 mg/mL for CAN and 0.625 and 0.125 mg/mL for HCT, respectively [11].
- Ten tablets were weighed and finely powdered. A portion of the powder, equivalent to 16 mg of CAN and 12.5 mg of HCT, was transferred into 20 mL volumetric flask. Extraction of the analyte was carried out by using methanol [17].

Samples were applied using Linomat 5 sample applicator. The developing solvent mixture was run up to 95 mm [11] or 80 mm [17] in chromatographic chamber previously saturated for 10 min [17]. Stationary and mobile phases used for the separation and determination of CAN and HCT are presented in Table 38.4.

TABLE 38.4

Chromatographic Conditions for Simultaneous Determination of Candesartan and Hydrochlorothiazide

Stationary Phase	Mobile Phase	Detection	R_f Value	Reference
Aluminum-backed silica gel 60 F_{254} HPTLC plates (10 × 10 cm, Merck)	Ethyl acetate–tetrahydrofuran–acetic acid (8:2:0.5, v/v/v)	Densitometrically at 252 nm for CAN and 208 nm for HCT	$Rf_{HCT} \approx 0.74$ $Rf_{CAN} \approx 0.89$	[11]
Aluminum-backed silica gel 60 F_{254} HPTLC plates (0.2 mm thickness, Merck)	Toluene–ethyl acetate–glacial acetic acid (2:5:0.1, v/v/v/)	Densitometrically using UV detector in absorbance mode	—	[14]
Aluminum-backed silica gel 60 F_{254} HPTLC plates, (20 × 10 cm, Merck)	Ethyl acetate–chloroform–acetone–methanol (3:3:3:0.5, v/v/v/v)	Densitometrically at 280 nm	$Rf_{CAN} \approx 0.27$ $Rf_{HCT} \approx 0.45$	[17]
Aluminum-backed silica gel 60 F_{254} HPTLC plates, (Merck)	Chloroform–methanol (8:2, v/v)	Densitometrically at 270 nm	—	[18]

Calibration curves were linear in the following ranges:

- 0.226–1.810 µg/band for CAN and 0.100–1.600 µg/band for HCT, respectively [11]
- 65–325 µg/mL for CAN and 50.6–253 µg/mL for HCT, respectively [17]

HPTLC-densitometric method was described for simultaneous determination of eprosartan and HCT in binary mixture [19].

Standard solution: A combined methanolic stock solution containing 12 mg/mL eprosartan and 0.5 mg/mL HCT was prepared.

Sample preparation: 20 tablets were weighed and finely powdered. Powder equivalent to approximately 600 mg eprosartan and 25 mg HCT was weighed accurately, and transferred to a 50 mL volumetric flask. Extraction of the analyte was carried out by using methanol.

Stationary phase: HPTLC aluminum plates of silica gel 60 F_{254} (Merck).

Mobil phase: benzene–methanol–formic acid (7:3:0.1, v/v/v).

Before use, the plates were washed with methanol and stored in a desiccator. Solutions of eprosartan and HCT were applied to the plate by means of Desaga AS 30 Win sample applicator. The plate was developed to a distance of 45 mm in a flat-bottomed twin-trough chamber previously saturated for 30 min with the mobile phase. After elution, the plate was air dried and scanned at 272 nm with Desaga CD 60 densitometer. The R_F values of eprosartan and HCT were 0.76 and 0.57, respectively. The linear range was 4.8–43.2 µg/spot for eprosartan and 0.15–1.35 µg/spot for HCT.

HPTLC methods have been developed and validated for the simultaneous estimation of irbesartan (IRB) and HCT in combined dosage forms [20–23]. Stability of IRB and HCT was studied under dry heat, oxidative, photolytic, and hydrolytic (in different pH values) conditions [21].

Standard solutions:

- Methanolic solutions in the concentration 100 µg/mL of each drug were prepared [20,21].
- Methanolic solutions in the concentration 1 mg/mL of each drug were prepared [22].

TABLE 38.5

Chromatographic Conditions for Simultaneous Determination of Irbesartan and Hydrochlorothiazide

Stationary Phase	Mobile Phase	Detection	R_F Value	Reference
Aluminum-backed silica gel 60 F_{254} TLC plates (10 × 10 cm, Merck)	Acetonitrile–chloroform–glacial acetic acid (7:3:0.1, v/v/v)	Densitometrically at 260 nm	$Rf_{IRB} \approx 0.60$ $Rf_{HCT} \approx 0.70$	[20]
Aluminum-backed silica gel 60 F_{254} TLC plates (250 mm thickness, 10 × 10 cm, Merck)	Acetonitrile–chloroform (5:6, v/v/v)	Densitometrically at 270 nm	$Rf_{IRB} \approx 0.27$ $Rf_{HCT} \approx 0.45$	[21]
Aluminum-backed silica gel 60 F_{254} TLC plates, (10 × 10 cm, Merck)	Acetonitrile–ethyl acetate (8:2, v/v).	Densitometrically at 260 nm	$Rf_{IRB} \approx 0.27$ $Rf_{HCT} \approx 0.87$	[22]
Aluminum-backed silica gel 60 GF_{254} TLC plates	Acetone–chloroform–ethyl acetate–methanol (3:3:3:0.5, v/v/v/v)	Densitometrically at 250 nm	$Rf_{IRB} \approx 0.27$ $Rf_{HCT} \approx 0.37$	[23]

Sample preparation:

- Twenty tablets were weighed accurately and ground to fine powder. The powder equivalent to 25 mg of IRB and 2.083 mg of HCT was transferred to 250 mL volumetric flask. Extraction of the analyte was carried out by using methanol [20].
- Twenty tablets were weighed accurately and ground to fine powder. The powder equivalent to 10 mg of IRB and 0.83 mg of HCT was transferred to 10 mL volumetric flask. Extraction of the analyte was carried out by using methanol. One milliliter of this solution was again diluted to 10 mL using methanol to get 100 µg/mL of IRB and 8.33 µg/mL of HCT [22].

Samples were applied using Linomat 5 sample applicator. The developing solvent mixture was run up to 72–75 mm [20, 22] or 90 mm [21] in twin-trough chamber previously saturated for 30 min [20]. Stationary and mobile phases used for the separation and determination of IRB and HTC are presented in Table 38.5.

The calibration curve was found to be linear between

- 100–700 ng/spot for IRB and 100–350 ng/spot for HCT [20]
- 200–1000 ng/spot for IRB and 200–600 ng/spot for HCT [21]
- 100–600 ng/spot for IRB and 50–250 ng/spot for HCT [22]
- 300.4–1802.4 µg for IRB and 25.0–150.4 µg for HCT [23]

Both of the drugs were not degraded under dry heat and photolytic conditions but showed degradation under hydrolytic condition [21].

HPTLC methods have been developed and validated for the simultaneous determination of losartan (LST) and HCT in combined dosage forms [24].

Standard solutions: methanolic solutions in the concentration 100 µg/mL for LST and 25 µg/mL for HCT were prepared.

Sample preparation: 20 tablets were weighed and finely powdered. The powder equivalent to 25 mg of LST or 6.25 mg of HCT was transferred to 25 mL volumetric flask. Extraction of the analyte was carried out by using methanol. An appropriate volume of this solution was diluted with methanol to obtain a solution containing 100 µg/mL of LST or 25 µg/mL of HCT.

Stationary phase: TLC aluminum plates precoated with silica gel G60 F_{254} (10 × 10 cm, Merck).

Mobil phase: chloroform–methanol–acetone–formic acid (7.5:1.5:0.5:0.03, v/v/v/v).

Before use, the plate was washed with the mobile phase. Next, the plate was developed to a distance of 35 mm in a twin-trough chamber previously saturated for 45 min with the mobile phase. After elution, the plate was dried under infrared lamp and scanned at 254 nm with TLC scanner 3. The R_F values were 0.61 and 0.41 for LST and HCT, respectively. The calibration curve response was observed between 0.4–1.2 µg/spot for LST and 0.1–0.3 µg/spot for HCT.

TLC method was described for simultaneous determination of HCT and olmesartan medoxomil (OLM) in tablet dosage form [25–28].

Standard solution:

- Methanolic solutions of each drug with a concentration of 100 µg/mL [25]
- Methanolic solutions of each drug with a concentration of 1 mg/mL [26]
- Methanolic solutions of each drug with a concentration of 25 ng/µL [27]
- Methanolic solutions of each drug with a concentration of 0.625 mg/mL for HCT
- 1 mg/mL for OLM [28]

Sample preparation:

- Twenty tablets were weighed accurately and ground to fine powder. The powder equivalent to 25 mg of drugs was transferred to a 250 mL flask. Extraction of the analyte was carried out by using methanol. Required dilutions were made to get 100 µg/mL of drugs [25].
- Ten tablets were weighed and powdered. The powder equivalent to 62.5 mg HCT and 100 mg OLM was weighed and transferred to a 100 mL volumetric flask. Extraction of the analyte was carried out by using methanol. A final concentration of 12.5 µg/mL of HCT and 20 µg/mL of OLM were prepared [26].
- Twenty tablets were weighed accurately and ground to fine powder. A quantity of powder equivalent to 12.5 mg of HCT and 20 mg of OLM was weighed and transferred to a 50 mL volumetric flak containing 25 mL methanol and sonicated for 5 min. Required dilutions were made to get 12.5 ng/µL for HCT and 20 ng/µL for OLM [27].
- An amount of powder equivalent to 62.5 mg of HCT and 100 mg of OLM was weighed and transferred to a 100 mL volumetric flask, 80 mL methanol was added and was sonicated for 30 min. A final concentration of 0.655 mg/mL of HCT and 1 mg/mL of OLM was prepared [28].

Samples were applied to the plates using sample applicator. The developing solvent mixture was run up to 72–70 mm [25,26], 90 mm [27], or 170 mm [28] in chromatographic chamber previously saturated with mobile phase for 30 min [25], 20 min [26], 15 min [27] or 1 h [28]. Stationary and mobile phases used for the separation and determination of OLM and HTC are presented in Table 38.6. Calibration curves were linear in the following range:

- 100–600 ng/spot for HCT and 500–750 ng/spot for OLM [25]
- 125–375 ng/spot for HCT and 200–600 ng/spot for OLM [26]
- 50–300 ng/spot for HCT and 100–600 ng/spot for OLM [27]
- 0.05–1.00 mg/mL for the both drugs [28]

HPTLC method was described for simultaneous determination of HCT and telmisartan (TEL) in combined dosage forms [29–31].

TABLE 38.6

Chromatographic Conditions for Simultaneous Determination of Olmesartan and Hydrochlorothiazide

Stationary Phase	Mobile Phase	Detection	R_F Value	Reference
Aluminum-backed silica gel 60 F_{254} TLC plates (10 × 10 cm, Merck)	Acetonitrile–chloroform–glacial acetic acid (7:2:0.5 v/v/v/)	Densitometrically at 254 nm	$Rf_{HCT} \approx 0.68$ $Rf_{OLM} \approx 0.58$	[25]
Aluminum-backed silica gel 60 F_{254} HPTLC plates (20 × 10 cm, Merck)	Acetonitrile–ethyl acetate–glacial acid (7:3:0.4 v/v/v)	Densitometrically at 254 nm	$Rf_{HCT} \approx 0.64$ $Rf_{OLM} \approx 0.44$	[26]
Aluminum-backed silica gel 60 F_{254} HPTLC plates (20 × 10 cm, Merck	Chloroform–methanol–toluene (6:4:5 v/v/v)	Densitometrically at 258 nm	$Rf_{HCT} \approx 0.40$ $Rf_{OLM} \approx 0.58$	[27]
Aluminum-backed silica gel 60 F_{254} HPTLC plates (20 × 20 cm, Macherey-Nagel)	Chloroform–methanol–formic acid (8:1.5:0.5 v/v/v)	Densitometrically at 272 nm for HCT and 260 nm for OLM	$Rf_{HCT} \approx 0.33$ $Rf_{OLM} \approx 0.70$	[28]

Standard solution:

- Solutions were prepared by dissolving each of TEL and HCT in methanol to obtain concentration of 1.00 mg/mL or 25.00 µg/mL for each drug [29].
- Solution was prepared by dissolving of TEL and HCT in solvent mixture (chloroform + methanol 1:1) to obtain a concentration of 2.00 mg/mL for TEL and 0.625 mg/mL for HCT [30].
- Solution was prepared by dissolving each of TEL and HCT in chloroform to obtain concentration of 100.0 µg/mL for each drug [31].

Sample preparation:

- Ten tablets were weighed and powdered. The powder equivalent to 12.5 mg HCT and 80.0 mg TEL was weighed and transferred to a 100 mL volumetric flask. Extraction of the analyte was carried out by using methanol. The solution was diluted with methanol [29].
- Twenty tablets were weighed and powdered. The powder equivalent to 40 mg of TEL and 12.5 mg of HCT was transferred to 20 mL volumetric flask. Extraction of the analyte was carried out by using the solvent mixture (chloroform + methanol 1:1) [30].
- Twenty tablets were weighed and powdered. The powder equivalent to 25 mg of TEL and HCT was transferred to a 250 mL volumetric flask. Extraction of the analyte was carried out by using chloroform. Required dilutions were made to get a concentration of 100 µg/mL [31].

Before use, plates were washed with methanol. Activation of plates was done in oven at 50°C for 5 min. Next, the plate was developed to a distance of 72 mm in a twin-trough chamber previously saturated for 30 min with the mobile phase (Table 38.7) [31].

Calibration curves were linear in the following range:

- 0.5–4.50 µg/spot for both drugs [29]
- 1.6–2.4 mg/mL for TEL and 0.5–0.75 mg/mL for HTC [30]
- 250–500 ng/spot for TEL and 200–700 ng/spot for HCT [31]

HPTLC methods were described for simultaneous determination of HCT and valsartan (VAL) in combined dosage forms [11,32,33].

TABLE 38.7

Chromatographic Conditions for Simultaneous Determination of Telmisartan and Hydrochlorothiazide

Stationary Phase	Mobile Phase	Detection	R_f Value	Reference
Aluminum TLC plates precoated with silica gel 60, F_{254} (20 × 20 cm Merck)	Butanol–ammonia 25% (8:2, v/v)	Densitometrically at 225 nm for HCT and 295 nm for TEL	$Rf_{TEL} \approx 0.64$ $Rf_{HCT} \approx 0.45$	[29]
Aluminum-backed silica gel 60 F_{254} TLC plates prewashed with methanol	Ethyl acetate–chloroform–methanol (10:3:1 v/v/v)	Densitometrically at 270 nm	—	[30]
Aluminum-backed TLC plates of silica gel 60 F_{254} (10 × 10 cm, Merck)	Chloroform–methanol–toluene (2:5:5, v/v/v)	Densitometrically at 272 nm	$Rf_{TEL} \approx 0.53$ $Rf_{HCT} \approx 0.68$	[31]

Standard solution:

- Solution containing 800 ng/μL VAL and 125 ng/μL HCT was prepared [32]
- Methanolic solutions in the concentration range 0.02–0.38 mg/mL for HCT and 0.077–3.31 mg/mL for VAL [11]
- Methanolic solutions of each drug with a concentration of 100 μg/mL [33]

Sample preparation:

- Twenty tablets were weighed and finely powdered. The powder equivalent to approximately 80 mg VAL and 12.5 mg HCT was weighed accurately, dissolved in ethanol, and diluted to 10 mL with the same solvent. The sample solution was then filtered [32].
- Twenty tablets were weighed accurately and finely powdered. Extraction of the analyte was carried out by using methanol. Required dilutions were made to get 4.0 mg/mL for VAL and 0.625 mg/mL and 0.125 mg/mL for HCT, respectively [11]
- Twenty tablets were weighed and finely powdered. The powder equivalent to approximately 25 mg VAL and HCT was transferred to a 250 mL volumetric flask. Extraction of the analyte was carried out by using methanol. Required dilutions were made to get 100.0 μg/mL for VAL and HCT [33]

Samples were applied to the plates by means of a Linomat 5 applicator. Ascending development of the plates was performed at 25°C ± 2°C in a twin-trough TLC chamber previously saturated with the mobile phase for 30 min (Table 38.8) [32,33]. The development distance was 70–72 mm [32,33] or 95 mm [11]. After development, the plates were dried for 5 min in an oven at 50°C [32].

Calibration curves were linear in the following range:

- 800–5600 ng/spot for VAL and 125–875 ng/spot for HCT [32]
- 0.385–6.150 μg for VAL and 0.100–1.600 μg for HCT [11]
- 300–800 ng/spot for VAL and 100–600 ng/spot for HCT [33]

38.2.1.4 HCT and Aldosterone Antagonist

HPTLC method has been developed and validated for determination of spironolactone and HCT in their mixtures and in presence of their impurities and degradation products [34].

Chromatographic conditions of the analysis are described in Section 38.6.1.

TABLE 38.8

Chromatographic Conditions for Simultaneous Determination of Hydrochlorothiazide and Valsartan

Stationary Phase	Mobile Phase	Detection	R_F Value	Reference
HPTLC plates precoated with silica gel 60, F_{254} (10 × 20 cm, Merck)	Chloroform–ethyl acetate–acetic acid, (5:5:0.2, v/v/v)	Densitometrically at 248 nm	$Rf_{VAL} \approx 0.27$ $Rf_{HCT} \approx 0.56$	[32]
HPTLC plates precoated with silica gel 60 F_{254} (10 × 10 cm, Merck,)	ethyl acetate–tetrahydrofuran–acetic acid (8:2:0.5, v/v/v)	Densitometrically at 252 nm for VAL and 208 nm for HCT 3	$Rf_{VAL} \approx 0.88$ $Rf_{HCT} \approx 0.74$	[11]
TLC plates precoated with silica gel 60 F_{254} (10 × 10 cm, Merck), prewashed with methanol	Chloroform–methanol–toluene–glacial acetic acid (6:2:1:0.1, v/v/v/v)	Densitometrically at 260 nm	$Rf_{VAL} \approx 0.36$ $Rf_{HCT} \approx 0.63$	[33]

HPTLC method has been developed and validated for determination of HCT, triamterene, furosemide, and spironolactone in their mixtures and in complex drugs [35].

Chromatographic conditions of the analysis are described in Section 38.6.1.

38.2.1.5   Determination of HCT in Multicomponent Mixtures

HPTLC method has been developed and validated for estimation of VAL, HCT, and amlodipine besylate (AML) in combined dosage form [36,37].

Stationary phase: aluminum-backed silica gel 60 F_{254} TLC plates [36,37].

Mobile phase:

- Ethyl acetate–methanol–10% ammonia (8.5:2:1, v/v/v) [36]
- Chloroform–glacial acetic acid–n-butyl acetate (8:4:2, v/v/v) [37]

The R_F values were 0.82, 0.34, and 0.26 for HCT, VAL, and AML, respectively. The calibration curve response was observed in the range 100–170 ng/spot for VAL and HCT and 50–400 ng/spot for AML [36].

HPTLC method for separation and quantitative analysis of LST potassium, atenolol (ATL), and HCT in bulk and in pharmaceutical formulations has been established and validated [38].

Standard solution: A solution containing 1 mg/mL LST and ATL and 0.25 mg/mL HCT, was prepared.

Sample preparation: 20 tablets were weighed and powdered. An amount of powder equivalent to 50 mg LST, 50 mg ATL, and 12.5 mg HCT was transferred to a 50 mL volumetric flask. Extraction of the analyte was carried out by using methanol.

Stationary phase: HPTLC plates coated with 0.2 mm layers of silica gel 60 F_{254} (10 × 20 cm, Merck).

Mobile phase: toluene–methanol–triethylamine (6.5:4:0.5, v/v/v).

Samples were applied as 6 mm bands by means of an automatic sample applicator. Ascending development of the plate, to a migration distance of 70 mm, was performed at 25°C ± 2°C, in a twin-trough chamber previously saturated with mobile phase vapor for 30 min. Densitometric scanning at 274 nm was then performed with a TLC scanner. The R_F values of LST, ATL, and HCT were 0.60, 0.43, and 0.29, respectively. Calibration plots were linear in the range 1000–5000 ng/band for LST potassium and ATL and 250–1250 ng/band for HCT.

38.2.1.6 Determination HCT in Biological Material

TLC screening method has been developed for pre- and post-race detection of HCT in equine plasma and urine of horses [39].

Sample preparation: 10 mL of urine or 5 mL of plasma were added to teflon-lined screw cap glass test tube. The pH was adjusted to 4.5–5.0 with 3N HCl.

The plasma was then treated with 1 g $(NH_4)_2SO_4$ and 5 mL ethyl acetate was added, but only 5 mL of ethyl acetate to the urine was added.

The mixture was mixed and then centrifuged for 5 min and the upper ethyl acetate layer was transferred into a test tube. One milliliter 0.1 M $NaHCO_3$ was added only to the urine sample, mixed, and centrifuged to separate the layers. The ethyl acetate was evaporated to dryness.

The residue of urine sample was dissolved in 25 μL of ethyl acetate and spotted on a TLC plate.

The residue of plasma sample was dissolved in 100 μL of absolute ethanol. One half mL hexane was added to the solution and mixed. The solution of plasma extract was then added to the top of micro chromatography column and eluted successively with 2 mL each of hexane and dichloromethane, which was discarded. A final elution with ethyl acetate removed thiazide diuretics from the column. The ethyl acetate eluant was filtered and evaporated to dryness. The residue was dissolved in 25 μL of ethyl acetate and spotted on a TLC plate.

Stationary phase: TLC plates (5 × 10 cm, Merck) precoated with 0.2 mm layers of silica gel 60 F_{254}.

Mobile phase: ethyl acetate–methanol–ammonium hydroxide (85:10:5, v/v/v).

The plates were developed for a distance of 5.0 cm. After drying, the plates were observed under shortwave ultraviolet (UV) light. R_F value for HCT was 0.45, and limit of detection was 50 ng/mL.

38.2.2 CHLORTHALIDONE

Chlorthalidone (2-chloro-5-(1-hydroxy-3-oxo-2H-isoindol-1-yl)benzenesulfonamide) is a diuretic with actions and uses similar to those of the thiazide diuretics, even though it does not contain a thiazide ring system (Figure 38.2). It is used for hypertension and edema, including those associated with heart failure [40,41].

A TLC densitometric method has been developed for the selective determination of ATL and chlorthalidone along with their hydrolytic degradation products (Figure 38.3) [42].

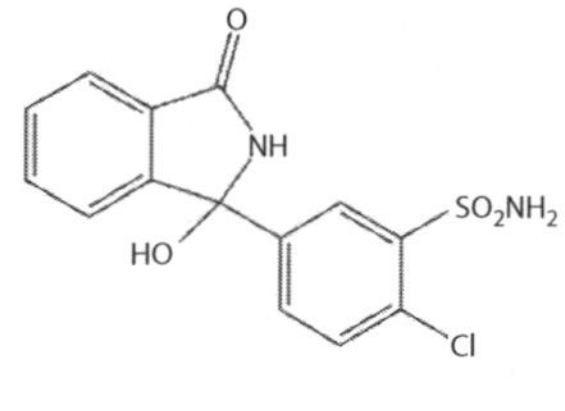

FIGURE 38.2 Chlorthalidone.

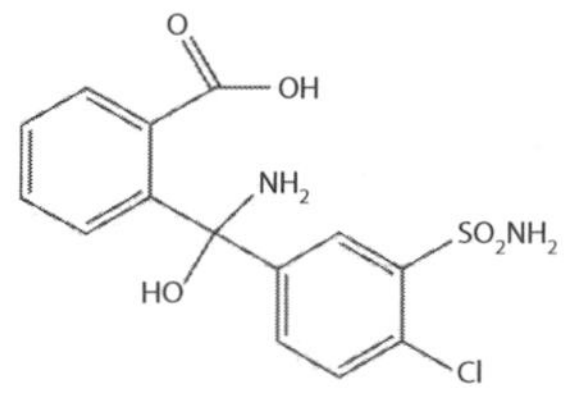

FIGURE 38.3 Chlorthalidone degradation product.

Standard solutions: working standard solutions of ATL and chlorthalidone and working solutions of ATL and chlorthalidone degradation products (CLT Deg) were prepared in methanol in the concentration of 0.1 mg/mL.

Sample preparation: 20 tablets were separately weighed and powdered. An accurately weighted portions equivalent to 100 mg of ATL (and the corresponding amounts of chlorthalidone) were separately transferred into three 100 mL flasks. Extraction of the analytes was carried out by using methanol. Then, 1.0 mL of the solution was diluted to 10.0 mL with methanol.

Preparation of CLT Deg: accurately weighed amount of pure chlorthalidone equivalent to 0.5 mg was refluxed with 50 mL 0.1 N NaOH for 5 h, cooled, and then the pH was adjusted to 2 pH units for complete precipitation of the produced degradation product using 0.1 N HCl. The formed CLT Deg was filtered, washed with distilled water, and then dried at 70°C.

Stationary phase: TLC plates (20 × 10 cm, Merck) coated with silica gel 60 F_{254}.

Mobile phase: chloroform–methanol–ethyl acetate–ammonia solution (75:28:2:1.6, v/v/v/v).

The samples were applied to TLC plates as 6 mm bands by means of a Linomat 5 sample applicator. The plates were developed in chromatographic tank previously saturated for 1 h with the mobile phase by ascending chromatography for a distance of 80 mm. The developed plates were air dried and scanned at 227 nm.

Under developed conditions, ATL, chlorthalidone, and its degradation products were completely separated with R_F values of 0.49, 0.69, and 0.04, 0.17, respectively. Linearity range for chlorthalidone and its degradation product was 0.2–2 μg/band and 0.2–1.8 μg/band, respectively.

HPTLC method has been developed for simultaneous determination of mixtures ATL and chlorthalidone (mix. I) and chlorthalidone and amiloride hydrochloride (mix. II) in bulk powders and in pharmaceutical dosage forms [10].

Stationary phase: HPTLC plates (0.25 mm thickness) precoated with 60 GF_{254} silica gel on aluminum.

Mobile phase:

- Dioxane–acetonitrile–1-propanol–hexane (30:18:23:1; v/v/v/v) for mixture I
- Dioxane–acetonitrile–1-propanol–tetrahydrofuran (20:13:4:15, v/v/v/v) for mixture II

Detection was carried out densitometrically using UV detector at 283, 266, and 362 nm for ATL, chlorthalidone, and amiloride hydrochloride, respectively.

38.2.3 CLOPAMIDE

Clopamide (4-chloro-N-[(2R,6S)-2,6-dimethylpiperidin-1-yl]-3-sulfamoylbenzamide) belongs to drugs with single-ring saluretic group (Figure 38.4).

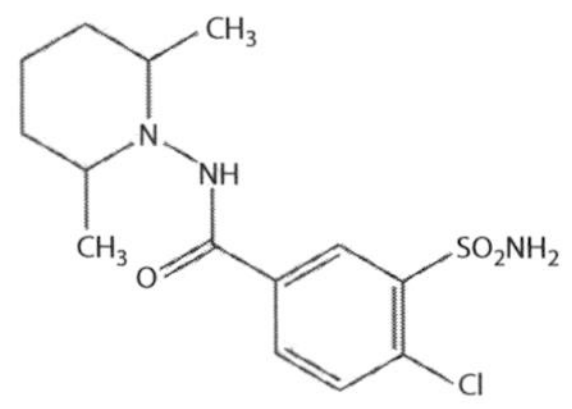

FIGURE 38.4 Clopamide.

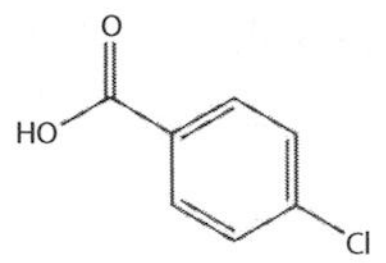

FIGURE 38.5   4-Chlorobenzoic acid.

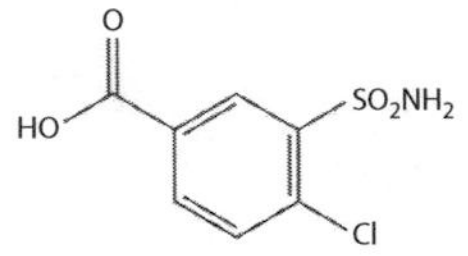

FIGURE 38.6   4-Chloro-3-sulfamoylbenzoic acid.

Chromatographic-densitometric method for determination of clopamide and impurities (4-chloro-benzoic and 4-chloro-3-sulfamoylbenzoic acids) in tablets was developed (Figures 38.5 and 38.6) [43].

Standard solution: methanolic solutions were prepared at concentrations, clopamide (10, 0.15, and 0.1 mg/mL), 4-chlorobenzoic, and 4-chloro-3-sulfamoylbenzoic acids (0.10 and 0.06 mg/mL).

Sample preparation: 20 tablets were weighed and powdered. An accurately weighted portion equivalent to 40 mg of clopamide was transferred into a 10 mL flask. Extraction of the analyte was carried out by using methanol. The obtained clopamide solution with 4.0 mg/mL was used for purity determination. For determination of clopamide content in tablets, the solution was diluted to a final concentration of 0.1 mg/mL.

Mobile phase: n-butanol–2-propanol–water–methylene chloride–methanol (10:7:2:5:3, v/v/v/v/v).

Stationary phase: TLC plates coated with silica gel 60 F_{254} on aluminum 10.8 × 12.5 cm, (Merck).

The samples were applied to TLC plates as 8 mm bands by means of a Linomat 5 sample applicator. Chromatograms were developed in room temperature as far as 11.0 cm from start in a glass chromatographic chamber saturated with vapors of mobile phase for 15 min.

Scanning and recording appropriate spots was done by using TLC Scanner at 235 nm. Under developed conditions, clopamide and its impurities, 4-chlorobenzoic acid and 4-chloro-3-sulfamoylbenzoic acid, were completely separated with the R_F values 0.87, 0.72, and 0.51, respectively. Linearity was observed in the concentration range of 1.6–4.8 µg/spot for clopamide and 0.12–0.72 µg/spot for 4-chlorobenzoic acid and 4-chloro-3-sulfamoylbenzoic acid.

38.2.4   INDAPAMIDE

Indapamide ((RS)-4-chloro-3-sulfamoyl-N-(2-methyl-2,3-dihydro-1*H*-indol-1-yl) benzamide) is a nonthiazide indole derivative of chlorosulfonamide (Figure 38.7) [44].

TLC densitometric method was described for determination of indapamide in the presence of degradation product and related substance [45].

Solutions: 0.6 mg/mL methanolic solution of indapamide and 0.0002% (w/v) methanolic solution of 2-methylnitrosoindoline.

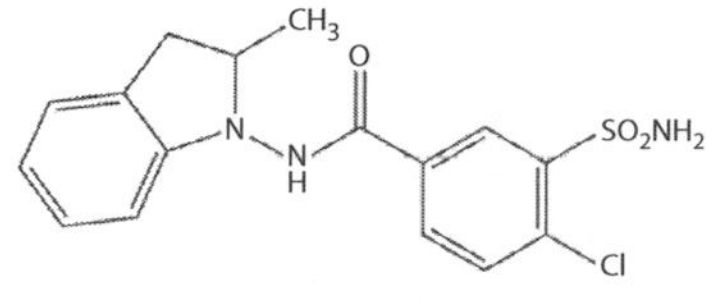

FIGURE 38.7   Indapamide.

Sample preparation: 20 tablets were weighed, decoated, and powdered. An accurate weight of indapamide equivalent to 30 mg was placed in a 50 mL flask. Extraction of the analyte was carried out by using methanol.

Degradation study: 2 mg/mL indapamide solution in 0.1 M NaOH was prepared. The solution was left under direct sunlight and induced pressure inside the well-closed tube for 2 days.

Stationary phase: silica gel 60 F_{254}, with 0.25 mm thickness TLC plates (20 × 20 cm, Merck).

Mobile phase: toluene–ethyl acetate–glacial acetic acid (69:30:1, v/v/v).

A 10 µL aliquot of each solution was applied to TLC plate by using micropipette. The plates were developed in a chromatographic tank previously saturated for 1 h with the mobile phase by ascending chromatography for a distance of 15 cm and dried at room temperature.

The spots were detected under a UV lamp at 254 nm and scanned by using dual wavelength flying spot scanning densitometer (Shimadzu) at 242 nm.

Under developed conditions, indapamide, its degradation product, and related substance (2-methylnitrosoindoline) were completely separated with R_F values of 0.23, 0.02, and 0.86 ± 0.01, respectively. Linearity range for indapamide was 0.6–6 µg/spot.

Indapamide and ATL in combination are available as tablet dosage forms in the ratio of 20:1. HPTLC method was developed for analysis of aforementioned formulation [46].

Standard solution: methanolic solutions containing 0.8 mg/mL of ATL and 0.04 mg/mL of indapamide.

Sample preparation: 20 tablets were accurately weighed and crushed to fine powder. Accurately weighed quantity of tablet powder equivalent to 40 mg of ATL was transferred to a 25 mL volumetric flask. To it, 15 mL of methanol was added and shaken for 10 min, and volume was adjusted up to the mark with methanol and then filtered.

Stationary phase: silica gel 60 G F_{254} TLC plates (10 × 10 cm, Merck).

Mobile phase: toluene–ethanol–acetone–ammonia (7:2.5:3:0.3 v/v/v/v).

The chamber saturation time employed was 10 min, and the plates were developed using ascending technique to a distance of 7 cm. Quantification was carried out by the use of densitometer in absorbance mode at 266 nm. The R_F value of ATL and indapamide was found to be 0.21 and 0.74, respectively. Linearity range for indapamide was 0.2–0.6 µg per spot.

HPTLC method was developed and validated for simultaneous determination of ATL and indapamide from bulk and formulations [47].

Standard solution: 1 mg/mL methanolic solution of indapamide and ATL.

Sample preparation: 20 tablets of the pharmaceutical formulation (containing 50 mg ATL and 2.5 mg indapamide) were assayed. They were crushed to a fine powder, and an amount of the powder corresponding to approximately 50 mg ATL and 2.5 mg indapamide was weighed in a 25 mL volumetric flask. Extraction of the analyte was carried out by using methanol.

Stationary phase: aluminum foil plates coated with 0.2 mm layers of silica gel 60 F_{254} (Merck).

Mobile phase: toluene–ethyl acetate–methanol–ammonia (5:3:3:0.1, v/v/v/v).

Before use, plates were prewashed with methanol then dried and activated. Samples were applied to the plates, as 6 mm bands, by means of a Linomat 5 sample applicator. Plates were developed in a twin-trough chamber previously saturated with mobile phase vapor for 20 min at room temperature (25°C ± 2°C). The development distance was approximately 80 mm.

Detection was performed densitometrically at 229 nm. The R_F of ATL and indapamide were 0.27 and 0.71, respectively. Linearity was observed in the concentration range of 100–600 ng/spot for indapamide.

HPTLC method for determination of perindopril erbumine and indapamide in bulk drug and tablet dosage form was described [48].

Standard solution: 500 µg/mL methanolic solution of perindopril erbumine and 100 µg/mL methanolic solution of indapamide.

Sample preparation: 10 tablets were accurately weighed and powdered. From the powdered mixture, certain amount (equivalent to 10 mg of perindopril erbumine and 4 mg of indapamide) of the powder was accurately weighed and transferred to 10 mL volumetric flask. Extraction of the analyte was carried out by using methanol. The filtered solution was then diluted with methanol to get the final concentration of 400 and 125 µg/mL of perindopril erbumine and indapamide, respectively. Two microliters of this solution was applied on TLC plate for the assay of indapamide.

Stationary phase: aluminum plates precoated with silica gel 60 F_{254}, (10 cm × 10 cm, Merck).

Mobile phase: dichloromethane–methanol–glacial acetic acid (9.5:0.5:0.1 v/v/v).

Samples were applied on the plate as a band with 4 mm width using sample applicator Linomat 5. Plates were developed in a twin-trough chamber previously saturated with mobile-phase vapor for 15 min at room temperature. The development distance was approximately 90 mm. Detection was performed densitometrically at 215 nm. The R_F of perindopril erbumine and indapamide were 0.30 and 0.50, respectively. Linearity was observed in the concentration range of 100–600 ng/spot for indapamide.

HPTLC method was described for simultaneous analysis of AML and indapamide in pharmaceutical formulations [49].

Standard solution: 0.5 mg/mL methanolic solution of AML and 0.15 mg/mL methanolic solution of indapamide.

Sample preparation: 20 tablets were accurately weighed and powdered. Tablet powder equivalent to about 5 mg AML and 1.5 mg indapamide was transferred to 25 mL volumetric flasks. Extraction of the analytes was carried out by using methanol. Then, 1.0 mL of the solution was diluted to 10.0 mL with methanol.

Stationary phase: aluminum HPTLC plates precoated with 250 µm thickness of silica gel (Merck).

Mobile phase: methanol–ethyl acetate–toluene–ammonia (2.5:3.5:4:0.2, v/v/v/v).

The plates were prewashed by methanol and activated at 110°C for 25 min prior to chromatography. Samples were applied as 6 mm wide bands by means of Linomat 5 sample applicator. The plates were then conditioned for 20 min in a presaturated twin-trough glass chamber with the mobile phase in one trough and plates in the other trough. The plates were then placed in the mobile phase and ascending development was performed to a distance of 70 mm from the point of application at ambient temperature.

Detection was performed densitometrically at 242 nm. The R_F of AML and indapamide were 0.58 and 0.76, respectively. Linearity was observed in the concentration range of 150–900 ng/spot for indapamide.

38.2.5 METOLAZONE

Metolazone is chemically 7-chloro-2-methyl-3-(2-methylphenyl)-4-oxo-1,2-dihydroquinazoline-6-sulfonamide (Figure 38.8).

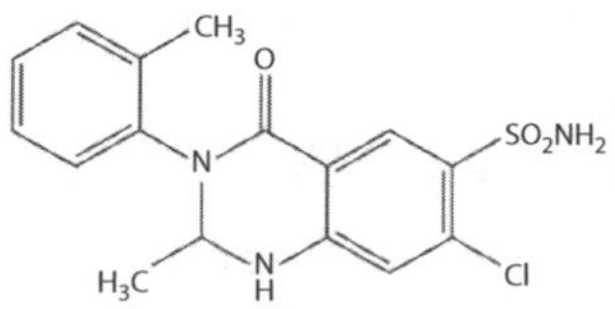

FIGURE 38.8 Metolazone.

HPTLC method was described for simultaneous analysis of LST potassium and metolazone in pharmaceutical formulations [50].

Standard solutions: 500 µg/mL methanolic solution of LST potassium and 50 µg/mL methanolic solution of metolazone.

Sample preparation: 20 tablets were weighed, their mean weight determined, and finely powdered. The weight of the tablet triturate equivalent to 25 mg LST potassium and 2.5 mg metolazone was transferred into a 50 mL volumetric flask. Extraction of the analyte was carried out by using methanol. Then 2 µL of the solution (500 µg/mL for LST potassium and 50 µg/mL for metolazone) was applied on TLC plate.

Stationary phase: HPTLC plates precoated with silica gel 60 F_{254} (Merck).

Mobile phase: toluene–ethyl acetate–methanol–glacial acetic acid (6:4:1:0.1 v/v/v/v).

The samples were spotted in the form of bands of width 6 mm using a Linomat 5 sample applicator. The plates were prewashed with methanol and activated at 110°C for 5 min prior to chromatography. Plates were developed in a twin-trough glass chamber previously saturated with mobile phase vapor for 40 min at room temperature. The development distance was approximately 8 cm. Following the development, the plates were dried in a stream of air with the help of an air dryer in a wooden chamber with adequate ventilation.

Densitometric scanning was performed using a TLC scanner in the reflectance absorbance mode at 237 nm. The R_F values of LST potassium and metolazone were 0.33 and 0.46, respectively. Linearity was observed in the concentration range of 120–420 ng/spot for metolazone.

HPTLC method was described for simultaneous analysis of ramipril and metolazone in pharmaceutical formulations [51].

Standard solutions: 500 µg/mL methanolic solutions of ramipril and metolazone.

Sample preparation: 20 tablets were weighed, their mean weight determined, and finely powdered. The weight of the tablet triturate equivalent to 2.5 mg ramipril and 2.5 mg metolazone was transferred into a 50 mL volumetric flask. Extraction of the analyte was carried out by using methanol. Then, 20 µL of the solution (50 µg/mL for ramipril and 50 µg/mL for Metolazone) was applied on TLC plate.

Stationary phase: HPTLC plates precoated with silica gel 60 F_{254} (Merck).

Mobile phase: toluene–ethyl acetate–methanol–glacial acetic acid (4:4:1:0.2 v/v/v/v).

The samples were spotted in the form of bands of width 6 mm using a Linomat 5 sample applicator. The plates were prewashed with methanol and activated at 110°C for 5 min prior to chromatography. Plates were developed in a twin trough glass chamber previously saturated with mobile-phase vapor for 40 min at room temperature. The development distance was approximately 8 cm. Following the development, the TLC plates were dried in a stream of air with the help of an air dryer in a wooden chamber with adequate ventilation.

Densitometric scanning was performed using a TLC scanner at 223 nm. The R_F values of ramipril and metolazone were 0.33 and 0.59, respectively. Linearity was observed in the concentration range of 100–350 ng/spot for metolazone.

38.2.6 XIPAMIDE

TLC method was described for simultaneous analysis of xipamide and triamterene in pharmaceutical formulations (Figure 38.9) [52].

Sample preparation: 20 tablets (containing 30 mg triamterene and 10 mg xipamide) were weighed, their mean weight determined, and finely powdered. Tablet powder equivalent to about 75 mg of

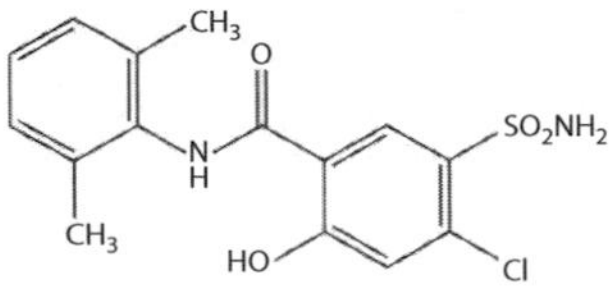

FIGURE 38.9 Xipamide.

triamterene and 15 mg of xipamide were transferred to a 100 mL volumetric flask. Extraction of the analyte was carried out by using methanol.

Stationary phase: TLC plates precoated with silica gel 60 F_{254} (20 × 10).

Mobile phase: chloroform–methanol–33% ammonia solution (8:2:0.2, v/v/v).

The samples (10 μL) were spotted in the form of bands of width 6 mm using a Linomat 5 sample applicator. Plates were developed in a glass chamber previously saturated with mobile phase vapor for 1 h at room temperature. The development distance was approximately 8 cm. Densitometric scanning was performed using a TLC scanner at 254 nm. The R_F values of xipamide and triamterene were 0.22 and 0.64, respectively.

38.2.7 SEPARATION OF OTHER THIAZIDES

TLC method was used to separate and identify thiazide diuretics and other antihypertensive agents [53].

Sample preparation and chromatographic conditions of the analysis are described in Section 38.3.1.

Under the proposed conditions, the following thiazide drugs were examined: benzthiazide, bendroflumethiazide, cyclothiazide, chlorothiazide, chlorthalidone, hydroflumethiazide, HCT, methyclothiazide, metolazone, polythiazide, and trichlormethiazide.

By using any one of the four mobile phases reported, it was possible to separate most of the drugs.

38.3 CARBONIC ANHYDRASE INHIBITORS

Drugs of this group inhibit carbonic anhydrase, the enzyme taking part in the reaction of water with carbon dioxide producing carbonic acid in kidneys that dissociates into hydrogen ion and hydrocarbonate ion.

38.3.1 ACETAZOLAMIDE

Acetazolamide N-[5-(Aminosulfonylo)-1,3,4-tiadiazol-2-ilo]acetamid inhibits carbonic anhydrase in renal proximal tubule by the influence of nitrogen from sulfonamide group on zinc cation of enzyme prosthetic group (Figure 38.10).

TLC method was used to separate and identify 20 of thiazide diuretics and other antihypertensive agents [53].

Sample preparation: the solutions of the drugs were prepared in methylene chloride–methanol (3:2) at the concentration of 2 mg/mL.

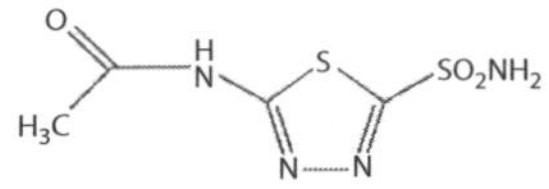

FIGURE 38.10 Acetazolamide.

Stationary phase: 30 g of silica gel G was shaken with 65 mL of water for 60–90 s, and the slurry was spread over 20 × 20 cm glass plates to a thickness of 250 μm. Coated plates were air-dried at room temperature and activated for 1 h at 110°C.

Mobile phase: the analysis was made using four mobile phases:

    Phase 1: methyl ethyl ketone–n-hexane (1:1, v/v)
    Phase 2: methyl ethyl ketone–n-hexane (2:1, v/v)
    Phase 3: methyl ethyl ketone–n-hexane (3:2, v/v)
    Phase 4: chloroform–acetone–triethanolamine (50:50:1.5, v/v/v)

Aliquots of 10–20 μL were applied to the plates. The plates were developed once to the top of the plate in a solvent system and allowed to dry before spraying with one of the detecting systems.

Detection: To visualize the spots of acetazolamide, staining reagents were used as follows:

A. 2 g p-dimethylaminobenzaldehyde was dissolved in 75 mL of 80% acetone and 25 mL of concentrated ammonium

B. 2 g potassium iodide was dissolved in 100 mL of water with 4 mL of 5% platinum chloride solution added

C. To 100 mL of water, 0.50 mL of 0.6 M HNO_3 with 1 g $Hg(NO_3)_2$ was added

D. The Bratton–Marshall reagents were sprayed in the following order: (1) 1% sodium nitrite in 1% sulfuric acid, freshly prepared; (2) 5% ammonium sulfamate, stored in a refrigerator; and (3) 2% N-(1-naphtyl)-ethylene diamine dihydrochloride in 95% ethanol, stored in a refrigerator. Hydrolysis was accomplished by spraying the developed plates with 10 M HCl and heating for 10 min at 100°C.

The R_F values of acetazolamide were found to be $R_F \sim 0.18$ after the development in mobile phase 1, $R_F \sim 0.42$ in phase 2, $R_F \sim 0.40$ in phase 3, and $R_F \sim 0.40$ in phase 4. Acetazolamide gives yellow–orange spots when reagent A is used, reagents B and C give white spots in chromatograms, and reagent D gives purple spots.

TLC method was used for the determination of sulfonamides: sulfacetamide, sulfathiazole, sulfadicarbamide, sulfacarbamide, sulfadimethoxine, sulfadimidine, acetazolamide, sulfaguanidine, sulfafurazole, and sulfanilamide. The influence of metal ions on their separation was studied [54].

Sample preparation: 0.5% methanolic or 1:1 methanol–acetone solutions of drugs were prepared.

Stationary phase: Two types of plates were used for research:

1. Glass plates coated with 0.25 mm layers of plain polyamide
2. Glass plates coated with 0.25 mm layers of plain polyamide impregnated with metal salts: Zn(II), Co(II), Ni(II), Mn(II), Cu(II), Cr(III), and Fe(III). The plates were activated at 80°C for 15 min before development.

Mobile phase: The analysis was made using two mobile phases:

    Phase 1: ethanol–acetone (10:90, v/v)
    Phase 2: ethanol–acetone (5:95, v/v)

The 2 μL aliquots of solution were applied on the plates and developed in a horizontal chamber to a distance of 15 cm.

Detection: the spots were detected under UV illumination (λ = 254 nm), scanned at λ = 254 nm, or visualized by spraying the plates with 4-aminobenzaldehyde reagent.

The retardation factor of acetazolamide after developing in mobile phase 1 on chromatographic plates not impregnated with metal ions was R_F ~ 0.60. Covering the stationary phase with solution of Mn(II) ions does not affect the separation (R_F ~ 0.59), while the impregnation of chromatographic plates with solutions of Zn(II) and Cu(II) ions increases retardation factors' values as follows: R_F ~ 76 in the presence of Zn(II) and R_F ~ 0.67 in the presence of Cu(II). The retardation factors' values of acetazolamide were lower after the development in mobile phase 2 than in mobile phase 1: R_F ~ 0.13 (chromatographic plates not impregnated with metal ions), R_F ~ 0.24 in the presence of Zn(II) ions, R_F ~ 0.21 after impregnation with Ni(II) ions, and R_F ~ 0.41 after impregnation of stationary phase with $FeCl_3$ solution.

HPTLC method was applied for determination of acetazolamide tablets and to study its stability [55].

Sample preparation: the powdered tablet mass containing 2.5 mg of acetazolamide was transferred into a 25 mL volumetric flask containing 15 mL of methanol, sonicated for 30 min, and the volume was made up to 25 mL solution. The sample stock solution was centrifuged and the supernatant was filtered.

Stationary phase: HPTLC plates kieselgel 60 F_{254}. The plates were washed with methanol before use.

Mobile phase: toluene–acetone–methanol (6:2:2, v/v/v).

The 3 μL of the solution was applied to the plates and developed to a distance of 8.5 cm. The chromatographic chamber was saturated with the mobile phase for 30 min.

Detection: densitometric scanning was performed in the absorbance-reflectance mode at λ = 254 nm.

Under the described conditions, the concentration of acetazolamide in tablets and in the presence of degradation products in acidic solutions, basic solutions, and in the presence of oxidizing agent H_2O_2 and sunlight was determined. The retardation factor of acetazolamide was R_F ~ 0.53. Linearity of the method was determined in the concentration range of the acetazolamide: 50–400 ng/spot.

38.3.2 Dorzolamide

Dorzolamide, (4S,6S)-2-ethylamino-4-methyl-5,5-dioxo-5λ^6,7-dithiabicyclo[4.3.0]nona-8,10-diene-8-sulfonamide, was determined next to the timolol in ophthalmic solutions with densitometric method (Figure 38.11) [56].

Sample preparation: 5 mL of ophthalmic solution was transferred into a 25 mL volumetric flask and diluted with methanol.

Stationary phase: TLC plates silica gel GF_{254}.

Mobile phase: methanol–ammonia 25% (100:1.5, v/v)

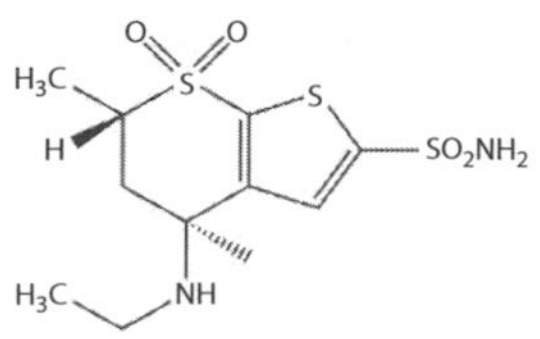

FIGURE 38.11   Dorzolamide.

Aliquots of 5 µL of solution were applied to the plates. The plates were developed by ascending migration over 16 cm. The chromatographic chamber was saturated with the mobile phase for 30 min.

Detection: the chromatograms were visualized under UV lamp at $\lambda = 254$ nm and were scanned at $\lambda = 253$ nm.

The retardation factor for dorzolamide under the described chromatographic conditions was $R_F \sim 0.67$ and $R_F \sim 0.56$ for timolol. The calibration function was established in the ranges of 2–18 µg for dorzolamide and 0.5–4.5 µg for timolol.

38.4 LOOP DIURETICS

Diuretics of this group act mainly by inhibiting sodium reabsorption in the nephron at the thick ascending limb of Henle's loop.

38.4.1 Furosemide

4-C hloro-2-(furan-2-ylmethylamino)- 5-sulfamoylbenzoic acid (Figure 38.12).

TLC method was used to separate and identify 20 of thiazide diuretics and other antihypertensive agents [53].

Sample preparation: the preparation procedure is described in Section 38.3.1.

Stationary phase: preparation of the stationary phase is described in Section 38.3.1.

Mobile phase: the mobile phases and the assay are described in Section 38.3.1.

Detection: To visualize the spots of furosemide except for the reagents described in Section 38.3.1, the following reagents were used:

  E. To 0.5 mL of anisaldehyde, 1.0 mL of sulfuric acid and 50 mL of acetic acid were added
  F. 1% solution of potassium manganate (VII) in water
  G. 0.1% solution of diphenylcarbazone in chloroform
  H. spray reagent D without hydrolysis

The R_F values of furosemide were found: $R_F \sim 0.30$ after the development in the mobile phase 1, $R_F \sim 0.57$ in phase 2, $R_F \sim 0.50$ in phase 3, and $R_F \sim 0.10$ in phase 4.

Furosemide after visualization with reagent A and C gives yellow spots, reagents B and F give white spots in chromatograms, reagent D gives dark red color of spots, reagent E gives brown spots, reagent G gives purple spots, and reagent H gives pink spots of furosemide.

Furosemide and its metabolites were determined by TLC method in biological fluids [57].

Preparation of urine samples: 0.4 mL of phosphate buffer (pH = 2) and 5 mL of diethyl ether were added to 1 mL of urine; after 10 min of shaking, the sample was centrifuged. Four milliliters of the ether phase was transferred to a rotating pear-shaped bottle and evaporated to dryness at reduced pressure at 26°C. Then, 0.5 mL of methanol was added.

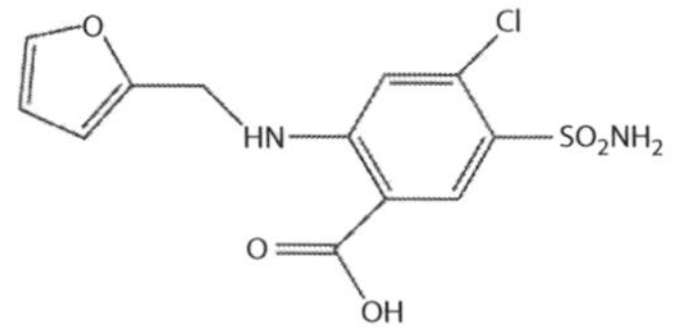

FIGURE 38.12 Furosemide.

Preparation of plasma samples: amounts of 100 μL of 1 M NaOH and 5 mL of chloroform were added to a 2 mL sample. After shaking and centrifuging, 1.5 mL of the supernatant was transferred to another tube and 50 μL of concentrated HCl and 5 mL of diethyl ether were added. Four milliliters of the ether was evaporated and dissolved in 0.5 mL of methanol.

Stationary phase: TLC plates silica gel 60 F_{254}. Before use, the plates were activated for 20 min at 110°C.

Mobile phase: chloroform–methanol–acetic acid (98:6:5, v/v/v).

One hundred microliters of the solutions were applied to the plates, and the chromatograms were developed to a distance of 10 cm.

Detection: scraped silica gel containing furosemide was shaken with 2 mL of phosphate buffer (pH 2). After shaking and centrifuging, the supernatant was analyzed on the spectrofluorometer at $\lambda = 279$ nm and $\lambda = 416$ nm.

Under the described conditions, the R_F value of furosemide was $R_F \sim 0.51$.

TLC method was used to separate and identify furosemide, metoprolol, vancomycin, fluconazole, and sildenafil in the urine [58].

Sample preparation: human urine (0.6 mL) spiked with appropriate amounts of drug standard solution (1 mg/mL) was transferred to volumetric flasks (1 mL), diluted to volume with water, and shaken for 1 min.

Stationary phase: TLC glass plates silica gel 60 GF_{254}.

Mobile phase: acetonitrile–water (2:3, v/v).

Twenty microliters of sample were applied to TLC plates and developed in a chromatographic chamber to a distance of 16 cm. Before analysis, the chamber was saturated with mobile phase vapor for 1 h.

Detection: drugs were visualized in a chamber saturated with iodine-vapor where they became visible as yellow–brown spots on a white background.

Under the described conditions, the R_F value of furosemide was $R_F \sim 0.96$.

A chromatographic-densitometric method has been developed for the simultaneous determination of HCT, triamterene, furosemide, and spironolactone in pharmaceutical preparations [35].

Sample preparation: methanol extracts of drugs were prepared at concentrations: furosemide 0.09 and 0.14 mg/mL, spironolactone 0.25 mg/mL, triamterene 0.26 mg/mL, and HCT 0.13 and 0.26 mg/mL.

Stationary phase: TLC plates silica gel 60 F_{254}.

Mobile phase: hexane-ethyl acetate–methanol–water–acetic acid (8.4:8:30.4:0.2, v/v/v/v/v).

The aliquots of 5 μL of the examined solutions were applied to the plates and developed to a distance of 9.5 cm in a chromatographic chamber. The chromatographic chamber was previously saturated with the mobile phase for 30 min.

Detection: the chromatograms were dried at room temperature and scanned at $\lambda = 264$ nm.

The R_F values obtained were: $R_F \sim 0.58$ for furosemide, $R_F \sim 0.22$ for triamterene, $R_F \sim 0.71$ for spironolactone and $R_F \sim 0.46$ for HCT. The linearity of the method was determined in a concentration range: 0.03–0.53 mg/mL for furosemide, 0.01–0.35 mg/mL for triamterene, 0.02–0.28 mg/mL for spironolactone and 0.03–0.52 mg/mL for HCT.

38.4.2 BUMETANIDE

3-Butylamino-4-phenoxy-5-sulfamoyl-benzoic acid (Figure 38.13).

HPTLC method has been described for the analysis of bumetanide in the bulk drug and tablet dosage form [59].

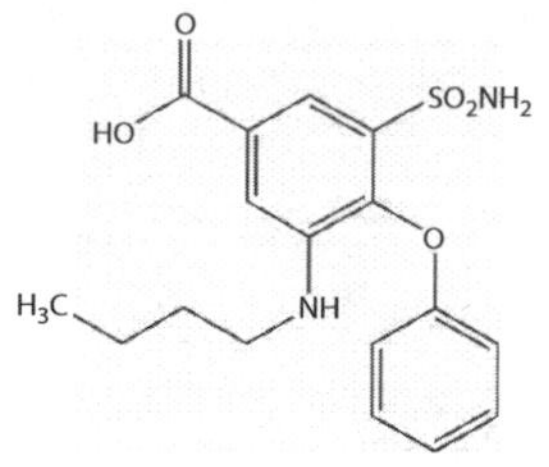

FIGURE 38.13   Bumetanide.

Stationary phase: TLC plates silica gel 60 F_{254}.

Mobile phase: toluene–ethyl acetate–formic acid (7:3.5:0.5, v/v/v).

Detection: the chromatograms were scanned densitometrically at $\lambda = 335$ nm.

The described method was used for the stability studies of bumetanide in acidic solution, alkaline solution, under the influence of the oxidant, and under the influence of electromagnetic radiation and temperature. The R_F value obtained for bumetanide was $R_F \sim 0.96$. Linear regression analysis revealed a good correlation between peak area and concentration in the range 100–800 ng/spot.

38.4.3   PIRETANIDE

3-(Aminosulfonyl)-4-phenoxy-5-pyrrolidin-1-ylbenzoic acid (Figure 38.14).

TLC method for the determination of piretanide in pharmaceutical preparations has been described [60].

Sample preparation: powdered tablet mass containing 50 mg of piretanide was shaken for 1 h with 40 mL of methanol. The suspension was filtered and diluted with methanol to 50 mL.

Stationary phase: TLC plates silica gel 60 F_{254}.

Mobile phase: propane-2-ol–ammonia 33% (8:2, v/v).

TLC plates were developed in a chromatographic chamber, saturated with the mobile phase for 10 min. The migration distance of the mobile phase was 8 cm.

Detection: the chromatograms were dried at room temperature. The spots were observed under UV lamp at $\lambda = 254$ nm and then scanned densitometrically at $\lambda = 255$ nm.

Under the described conditions, piretanide was determined in pharmaceutical preparations and in the presence of its degradation products obtained in alkaline solution. The R_F value for piretanide was $R_F \sim 0.54$. Calibration graph is linear in the concentration range 0.5–10 µg/spot.

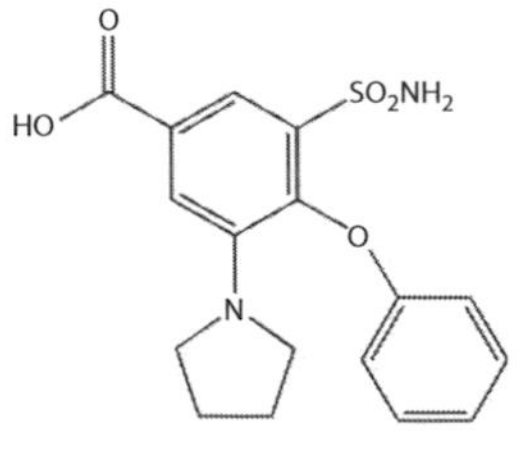

FIGURE 38.14   Piretanide.

38.5 CYCLIC AMIDINES

Amidines block sodium channels in the cell membrane of nephron causing the inhibition of the influx of sodium ions to the final part of urethra. They also reduce the exchange of sodium ions to potassium ions and hydrogen ions, reducing the excretion of potassium and magnesium ions. Their diuretic effect is weak.

38.5.1 AMILORIDE

3,5-Diamino-6-chloro-*N*-(diaminomethylene)pyrazine-2-carboxamide (Figure 38.15).

A HPTLC method for the determination of amiloride next to the HCT, ATL and chlorthalidone in pharmaceutical preparations has been developed [10].

Sample preparation: tested drugs were dissolved in methanol.

Stationary phase: HPTLC plates silica gel 60 GF_{254}.

Mobile phase: for the determination of mixtures containing amiloride, three mobile phases were used:

Phase 1: ethyl acetate–chloroform–propan-1-ol–ammonia 25% (12:9:1:0.2, v/v/v/v)
Phase 2: dioxane–acetonitrile–propan-1-ol–tetrahydrofuran (20:13:4:15, v/v/v/v)
Phase 3: dioxane–ethyl acetate–acetonitrile–propran-1-ol (10:7:5.5:3, v/v/v/v)

Detection: the chromatograms were scanned densitometrically at $\lambda = 362$ nm.

Amiloride and HCT were separated and determined using mobile phase 1. For the separation of amiloride, ATL, and chlorthalidone, the mobile phase 2 was used; and for the simultaneous determination of amiloride, ATL, and HCT mobile phase 3 was used. Calibration curves were linear in the range 1–100 µg/mL.

38.5.2 TRIAMTERENE

6-Phenylpteridine-2,4,7-triamine (Figure 38.16).

TLC method was used for the separation and identification of thiazide diuretics and other antihypertensive agents [53].

Sample preparation: the preparation procedure is described in Section 38.3.1.

Stationary phase: preparation of the stationary phase is described in Section 38.3.1.

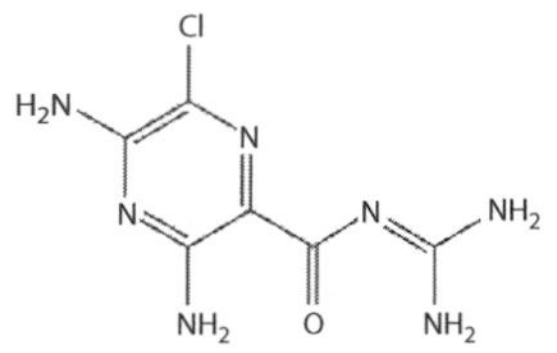

FIGURE 38.15 Amiloride.

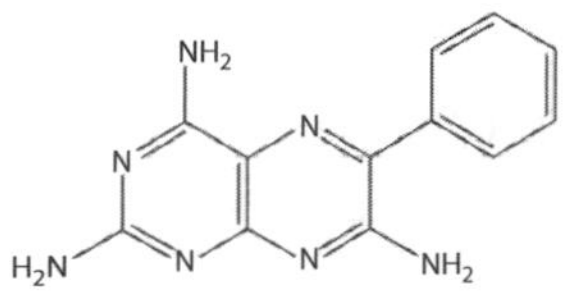

FIGURE 38.16 Triamterene.

Mobile phase: The mobile phases and the assay are described in Section 38.3.1.

Detection: To visualize the spots of triamterene, Dragendorff's reagent was used except for the reagents described in Section 38.4.1:

I. Solution 1—1.7 g of basic bismuth nitrate was dissolved in 100 mL of 20% acetic acid.
II. Solution 2—40 g of potassium iodide was dissolved in 100 mL of water. Immediately before use, mix 5 mL of solution 1, 5 mL of solution 2, 20 mL of acetic acid, and 70 mL of water.

Retardation factors of triamterene were as follows: $R_F \sim 0.12$ after the development in mobile phase 1, $R_F \sim 0.19$ in mobile phase 2, $R_F \sim 0.21$ in mobile phase 3, and $R_F \sim 0.36$ in mobile phase 4. Triamterene gives yellow spots after visualization with reagent C and D, reagents E and H give pale-blue spots in chromatograms, reagent B gives brown spots, reagent A gives yellow–orange spots, reagent G gives purple spots, reagent F gives white spots, and reagent I gives pale-yellow spots.

Chromatographic-densitometric method has been described for the analysis of drugs in urine and liver [61].

Preparation of urine samples: urine samples (5 mL) were incubated for 2 h at 56°C with 50 μL of β-glucuronidase with *Escherichia coli* K12.

Preparation of liver samples: liver samples (5 g) were cut into pieces and digested for 2 h shaking in a water bath at 57°C with 8 mL of phosphate buffer (pH 7.5) and 1 mL of trypsin solution (5 g/mL). After filtration, the sample was extracted with dichloromethane containing 0.01 M bis (2-ethylhexyl) phosphoric acid.

Stationary phase 1: TLC plates silica gel 60 F_{254}.

Stationary phase 2: TLC plates RP-18 F_{254}.

Mobile phase 1: toluene–acetone–ethanol–conc. ammonia (45:45:7:3, v/v/v/v).

Mobile phase 2: methanol–water–conc. hydrochloric acid (50:50:1, v/v/v).
The plates were developed to a distance of 7 cm.

Detection: the chromatograms were scanned in absorbance mode at λ = 220 nm.
Screening of about 300 drugs in urine in toxicological aspect using TLC method has been described [62].

Sample preparation: 20 mL of urine was adjusted to pH 8–9. After 3 min, the drugs were eluted with 2 × 15 mL of dichloromethane–propan-2-ol (90:10). A drop of 1% HCl in ethanol was added and the sample was evaporated. The evaporation residue was dissolved in w 150 μL of ethanol.

Stationary phase: TLC plates silica gel 60 F_{254}.

Mobile phase: ethyl acetate–methanol–ammonia–water (43:5:0.5:1.5, v/v/v/v).

Detection: Chromatograms were observed under UV lamp and then visualized:

A. Dragendorff's reagent and platinum iodide
B. H_2SO_4-$FeCl_3$.

Retardation factor of triamterene under the described conditions was $R_F \sim 0.29$.
Triamterene gives brown spots after visualization with reagent A, and reagent B gives blue spots.
Chromatographic-densitometric method has been developed for simultaneous determination of HCT, triamterene, furosemide, and spironolactone in pharmaceutical preparations [35].

Sample preparation and chromatographic conditions of the analysis are described in Section 38.4.1.

TLC method was described for simultaneous analysis of xipamide and triamterene in pharmaceutical formulations [52].

Sample preparation and chromatographic conditions of the analysis are described in Section 38.2.6.

38.6 ALDOSTERONE ANTAGONISTS

Aldosterone binds to the cytoplasmic receptor. It stimulates reabsorption of sodium ions, chloride ions, and water, increasing excretion of potassium ions. Aldosterone antagonists block receptors in distal convoluted tubule and collecting duct inhibiting sodium channels.

38.6.1 SPIRONOLACTONE

7α-Acetylthio-3-oxo-17α-pregn-4-ene-21,17-carbolactone (Figure 38.17).

Chromatographic-densitometric method has been developed for simultaneous determination of HCT, triamterene, furosemide, and spironolactone in pharmaceutical preparations [35].

Sample preparation and chromatographic conditions of the analysis are described in Section 38.4.1.

TLC densitometric method was used for determination of spironolactone and HCT in pharmaceutical preparations and also in the presence of their impurities and degradation products [34].

Sample preparation: substances were dissolved in methanol (0.1 mg/mL).

Stationary phase: TLC plates silica gel 60 F_{254}.

Mobile phase: ethyl acetate–chloroform–formic acid–triethylamine (7:3:0.1:0.1, v/v/v/v).

The plates were prewashed with methanol and activated at 100°C for 5 min. Chromatograms were developed in a chromatographic chamber, previously saturated with mobile phase for 1 h to a distance of 8 cm.

Detection: the chromatograms were scanned at $\lambda = 235$ nm.

Under the described conditions, satisfactory separation of examined drugs and degradation products was obtained.

Retardation factor of spironolactone was $R_F \sim 0.74$ and HCT $R_F \sim 0.34$. Linear regressions were obtained in the range of 50–5.0 μg/mL for spironolactone and 50–4.0 μg/mL for HCT.

TLC densitometric method has been developed for the quantification of spironolactone and torasemide in pharmaceutical preparations [63].

Sample preparation: A powder equivalent to 20 mg of drugs was transferred to a flask and extracted with methanol (2 × 50 mL) by sonication. The extracts were filtered and combined.

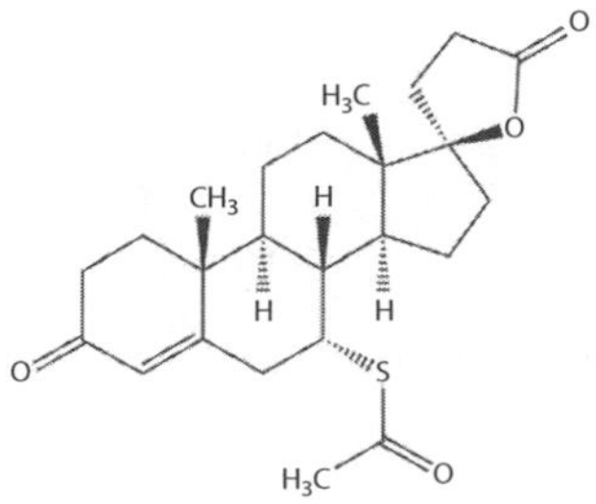

FIGURE 38.17   Spironolactone.

Stationary phase: TLC plates silica gel 60 F_{254}.

Mobile phase: ethyl acetate–acetone–acetic acid (10.5:4:1.5, v/v/v/v).

The plates were prewashed with methanol and activated at 95°C for 30 min. Chromatograms were developed in a chromatographic chamber, previously saturated with mobile phase for 25 min to a distance of 3 cm.

Detection: the chromatograms were scanned at $\lambda = 247$ nm.

Under the described conditions, retardation factor of spironolactone was $R_F \sim 0.19$ and torasemide $R_F \sim 0.33$. The linearity range for spironolactone and torasemide was found to be 360–850 ng/spot.

REFERENCES

1. Friedman, P.A. and W.O. Berndt. Diuretic drugs. pp. 239–255 in: Craig, C.R. and R.E. Stitzel. *Modern Pharmacology in Clinical Application*, 6th edn. Lippincott Williams & Wilkins, New York, 2003.
2. Shah, S.U., Anjum, S., and W.A. Littler. 2004. Use of diuretics in cardiovascular disease: (2) hypertension. *Postgrad Med J* 80: 271–276.
3. Sassard, J., Bataillard, A., and H. McIntyre. 2005. An overview of the pharmacology and clinical efficacy of indapamide sustained release. *Fund Clin Pharmacol* 19: 637–645.
4. Bath, L.R, Bothara, K.G., and M.C. Damle. 2008. A validated HPTLC method for simultaneous determination of nebivolol hydrochloride and hydrochlorothiazide from tablets. *Indian Drugs* 45(12): 948–951.
5. Dhandapani, B., Thirumoorthy, N., and D.J. Prakash. 2010. Development and validation for the simultaneous quantification of nebivolol hydrochloride and hydrochlorothiazide by UV spectroscopy, RP-HPLC and HPTLC in tablets. *E-J Chem* 7: 341–348.
6. Damle, M.C., Topagi, K.S., and K.G. Bothara. 2010. Development and validation of a stability-indicating HPTLC method for analysis of Nebivolol hydrochloride and hydrochlorothiazide in the bulk material and in pharmaceutical dosage forms. *Acta Chromatographica* 22: 433–443.
7. Kumbhar, S.T., Chougule, G.K., Tegeli, V.S., Gajeli, G.B., Thorat, Y.S, and U.S. Shivsharan. 2011. A validated HPTLC method for simultaneous quantification of nebivolol and hydrochlorothiazide in bulk and tablet formulation. *Int J Pharm Sci Drug Res* 3(1): 62–66.
8. El-Gindy, A.E., Ashour, A., Fattah, L.A, and M.M. Shabana. 2001. Application of LC and HPTLC-densitometry for the simultaneous determination of benazepril hydrochloride and hydrochlorothiazide. *J Pharm Biomed Anal* 25: 171–179.
9. Walode, S.G., Charde, M.S., Tajne, M.R., and A.V. Kasture. 2005. Development of HPTLC method for simultaneous estimation of captopril and hydrochlorothiazide in combined dosage form. *Indian Drugs* 42(6): 340–344.
10. Salem, H. 2004. High-performance thin-layer chromatography for the determination of certain antihypertensive mixtures. *Scientia Pharmaceutica* 72(2): 157–174.
11. Stolarczyk, M., Maślanka, A., and J. Krzek. 2008. Chromatographic and densitometric analysis of hydrochlorothiazide, walsartan, kandesartan, and enalapril in selected complex hypotensive drugs. *J Liq Chromatogr Relat Technol* 31: 1892–1902.
12. Manish, K., Rajdeep, G., Vishesh, A., and R. Amit. 2011. High performance thin layer chromatographic determination of enalapril maleate, hydrochlorothiazide in pharmaceutical dosage form. *Int J Pharm Tech Res* 3(3): 1454–1458.
13. El-Gindy, A., Ashour, A., Abdel-Fattah, L., and M.M. Shabana. 2001. Spectrophotometric and HPTLC-densitometric determination of lisinopril and hydrochlorothiazide in binary mixtures. *J Pharm Biomed Anal* 25: 923–931.
14. Gandhimathi, M. and T.K. Ravi. 2009. HPTLC method for the estimation of hydrochlorothiazide from its combined dosage forms. *Indian Drugs* 46(2): 150–153.
15. Bhavar, G.B., Chatpalliwar, V.A., Patil, D.D., and S.J. Surana. 2008. Validated HPTLC method for simultaneous determination of quinapril hydrochloride and hydrochlorothiazide in tablets dosage form. *Indian J Pharm Sci* 70(4): 529–531.
16. Kowalczuk, D., Hopkała, H., and R. Pietraś. 2003. Simultaneous densitometric determination of quinapril and hydrochlorothiazide in the combination tablets. *J Planar Chromatogr—Modern TLC* 16: 96–200.

17. Mehta, B.H. and S.B. Morge. 2008. HPTLC-densitometric analysis of candesartan cilexetil and hydro-chlorothiazide in tablets. *J Planar Chromatogr—Modern TLC* 21: 173–176.
18. Youssef, R.M., Maher, H.M., Hassan, E.M., El-Kimary, E.I., and M.A. Barary. 2010. Development and validation of HPTLC and spectrophotometric methods for simultaneous determination of candesartan cilexetil and hydrochlorothiazide in pharmaceutical preparation. *Int J Appl Chem* 6(2): 233–246.
19. Patel, H.U., Suhagia, B.N., and C.N. Patel. 2009. Simultaneous analysis of eprosartan and hydrochloro-thiazide in tablets by high-performance thin-layer chromatography with ultraviolet absorption densitom-etry. *Acta Chromatographica* 21: 319–326.
20. Shah, N.J., Suhagia, B.N., Shah, R.R., and N.M. Patel. 2007. Development and validation of a HPTLC method for the simultaneous estimation of irbesartan and hydrochlorthiazide in a tablet dosage form. *Indian J Pharm Sci* 69(2): 240–243.
21. Khodke, A.S., Potale, L.V., Damle, M.C., and K.G. Bothara. 2010. A validated stability indicating HPTLC method for simultaneous estimation of irbesartan and hydrochlorthiazide. *Pharm Methods* 1(1): 39–43.
22. Rosangluaia, B., Shanmugasundaram, P., and M. Velraj. 2011. Validated HPTLC method for simulta-neous estimation of irbesartan and hydrochlorthiazide in a tablet dosage form. *Der Pharma Chemica* 3(5):310–317.
23. Mehta, B.H. and S.B. Morge. 2010. Simultaneous determination of irbesartan and hydrochlorothiazide by HPTLC method. *Indian Drugs* 47(2): 71–74.
24. Shah, S.S., Rathod, S., Suhagia, B.N., Savale, S.S., and J.B. Patel. 2001. Simultaneous determination of losartan and hydrochlorothiazide in combined dosage forms by first-derivative spectroscopy and high-performance thin-layer chromatography. *J AOAC Int* 84(6): 1715–1723.
25. Shah, N., Suhagia, B., Shah, R., and N. Patel. 2007. Development and validation of a simultaneous HPTLC method for the estimation of olmesartan medoxomil and hydrochlorothiazide in tablet dosage form. *Ind J Pharm Sci* 69(6): 834–836.
26. Bari, P.D. and A.R. Rote. 2009. RP-LC and HPTLC methods for the determination of olmesartan medoxomil and hydrochlorothiazide in combined tablet dosage forms. *Chromatographia* 69(11–12): 1469–1472.
27. Kadukar, S., Gandhi, S., Ranjane, P., and S. Ranher. 2009. HPTLC analysis of olmesartan medoxomil and hydrochlorothiazide in combination tablet dosage forms. *J Planar Chromatogr—Modern TLC* 22(6): 425–428.
28. Moussa, B., Mohamed, M., and N. Youssef. 2011. Simultaneous densitometric TLC analysis of olmesar-tan medoxomil and hydrochlorothiazide in the tablet dosage form. *J Planar Chromatogr—Modern TLC* 24(1): 35–39.
29. Bebawy, L.I., Abbas, S.S., Fattah, L.A., and H.H. Refaat. 2005. Application of first-derivative, ratio derivative spectrophotometry, TLC-densitometry and spectrofluorimetry for the simultaneous determina-tion of telmisartan and hydrochlorothiazide in pharmaceutical dosage forms and plasma. *Il Farmaco* 60: 859–867.
30. Maheswari, R., Mageswari, S.D., Surendra, K., Gunasekaran, V., and P. Shanmugasundaram. 2007. Simultaneous estimation of telmisartan and hydrochlorothiazide in tablet dosage form by HPTLC method. *Asian J Chem* 19(7): 5582–5586.
31. Shah, N.J., Suhagia, B.N., Shah, R.R., and P.B. Shah. 2007. Development and validation of a HPTLC method for simultaneous estimation of telmisartan and hydrochlorothiazide in tablet dosage form. *Ind J Pharm Sci* 69(2): 202–205.
32. Kadam, B.R. and S.B. Bari. 2007. Quantitative analysis of valsartan and hydrochlorothiazide in tab-lets by high performance thin-layer chromatography with ultraviolet absorption densitometry. *Acta Chromatographica* 18: 260–269.
33. Shah, N.J., Suhagia, B.N., Shah, R.R., and N.M. Patel. 2009. HPTLC method for the simultaneous esti-mation of valsartan and hydrochlorothiazide in tablet dosage form. *Indian J Pharm Sci* 71(1): 72–74.
34. Hegazy, M.A., Metwaly, F.H., Abdelkawy, M., and N.S. Abdelwahab. 2011. Validated chromatographic methods for determination of Hydrochlorothiazide and Spironolactone in pharmaceutical formulation in presence of impurities and degradants. *J Chromatogr Sci* 49(2): 129–135.
35. Maślanka, A., Krzek, J., and M. Stolarczyk. 2009. Simultaneous analysis of hydrochlorothiazide, tri-amterene, furosemide, and spironolactone by densitometric TLC. *J Planar Chromatogr—Modern TLC* 22(6): 405–410.

36. Galande, V.R., Baheti, K.G., and M.H. Dehghan. 2011. Development and validation of RP-HPLC and HPTLC method for the estimation of valsartan, hydrochlorothiazide and amlodipine besylate in combined tablet dosage form. *Indian Drugs* 48(4): 49–56.
37. Varghese, S.J. and T.K. Ravi. 2011. Quantitative simultaneous determination of amlodipine, valsartan, and hydrochlorothiazide in "Exforge HCT" tablets using high-performance liquid chromatography and high-performance thin-layer chromatography. *J Liq Chromatogr Rel Tech* 34(12): 981–994.
38. Sathe, S.R. and S.B. Bari. 2007. Simultaneous analysis of losartan potassium, atenolol, and hydrochlorothiazide in bulk and in tablets by high-performance thin-layer chromatography with UV absorption densitometry. *Acta Chromatographica* 19: 270–278.
39. Henion, J.D. and G.A. Maylin. 1980. Qualitative and quantitative analysis of hydrochlorothiazide in equine plasma and urine by high-performance liquid chromatography. *J Anal Toxicol* 4:185–191.
40. Budavcari, S. 2002. *"The Merck Index," An Encyclopedia of Chemicals, Drugs and Biologicals*, 13th edn. Merck & Co. Inc., Whitehouse Station, NJ.
41. Martindale—Extra Pharmacopoeia. 2005. *The Complete Drug References*, 34th edn. The Pharmaceutical Press, London, U.K.
42. Abdelwahab, N.S. 2010. Determination of atenolol, chlorthalidone and their degradation products by TLC-densitometric and chemometric methods with application of model updating. *Anal. Methods* 2: 1994–2001.
43. Hubicka, U., Krzek, J., and M. Stankiewicz. 2009. Chromatographic-densitometric method for determination of clopamide and 4-chlorobenzoic, and 4-chloro-3-sulfamoylbenzoic acids in tablets. *Current Pharm Anal* 5: 408–415.
44. British Pharmacopeia. 2001. H.M. Stationary Office, London, U.K. Vol. 1, p. 760.
45. Youssef, N.F. 2003. Spectrophotometric, spectrofluorimetric and densitometric methods for determination of indapamide. *J AOAC Int* 86(5): 935–940.
46. Gupta, K.R., Wankhede, S.B., Tajne, M.R., and S.G. Wadodkar. 2007. High performance thin layer chromatographic estimation of atenolol and indapamide from pharmaceutical dosage form. *Asian J Chem* 19(6): 4183–4187.
47. Yadav, S.S. and J.R. Rao. 2011. Simultaneous HPTLC analysis of atenolol and indapamide in tablet formulation. *Pharmacie Globale, Int J Comp Pharm* 2(9): 1–4.
48. Dewani, M., Bothara, K., Madgulkar, A., and M. Damle. 2011. Simultaneous estimation of perindopril erbumine and indapamide in bulk drug and tablet dosage form by HPTLC. *Pharmacie Globale, Int J Comp Pharm* 2(1): 1–4.
49. Patel, D.B., Mehta, F.A., and K.K. Bhatt. 2012. Simultaneous estimation of amlodipine besylate and indapamide in pharmaceutical formulation by thin-layer chromatographic-densitometric method. *Novel Sci Int J Pharm Sci* 1(2): 74–82.
50. Dubey, R., Bhusari, K.V., and S.R. Dhaneshwar. 2011. Validated HPTLC method for simultaneous estimation of losartan potassium and metolazone in bulk drug and formulation. *Der Pharmacia Lettre* 3(2): 334–342.
51. Wayadande, J.A., Dubey, R., Bhusari, V.K., and S.R. Dhaneshwar. 2011. Validated HPTLC method for simultaneous estimation of ramipril and metolazone in bulk drug and formulation. *Der Pharmacia Sinica* 2(4): 286–294.
52. Wagieh, N.E., Abbas, S.S., Abdelkawy, M., and M.M. Abdelrahman. 2010. Spectrophotometric and spectrodensitometric determination of triamterene and xipamide in pure form and in pharmaceutical formulation. *Drug Test Anal* 2(3): 113–121.
53. Stohs, S.J. and G.A. Scratchley. 1975. Separation of thiazidediuretics and antihypertensive drugs by thin layer chromatography. *J Chromatogr* 114(2): 329–333.
54. Szumilo, H. and J. Flieger. 1996. Chromatographic behavior of sulfonamides on polyamide impregnated with various metal salts. *J Planar Chromatogr* 9(6): 462–466.
55. Ramesh, B., Srimannaraya, P., Reddy, A.S., and P. Sitadevi. 2011. Densitometric evaluation of stability-indicating HPTLC method for the analysis of acetazolamide and a study of degradation kinetics. *J Pharm Res* 4(2): 429–433.
56. Bebawy, L.I. 2002. Application of TLC-densitometry, first-derivative UV-spectrophotometry and ratio derivative spectrophotometry for the determination of dorzolamide hydrochloride and timolol maleate. *J Pharm Biomed Anal* 27(5): 737–746.
57. Mikkelsen, E. and F. Andreasen. 1977. Simultaneous determination of furosemide and two of its possible metabolites in biological fields. *Acta Pharmacol Toxico* 41(3): 254–262.

58. Baranowska, I., Markowski, P.,Wilczek, A., Szostek, M., and M. Stadniczuk. 2009. Normal and reversed-phase thin-layer chromatography in the analysis of l-arginine, its metabolites, and selected drugs. *J Planar Chromatogr—Modern TLC* 22(2): 89–96.

59. Kumar, M., Rao, J.R., Yadav, S.S., and S.L. Vikas. 2010. Development and validation of a stability-indicating HPTLC method for analysis of bumetanide in the bulk. *J Pharm Tech* 3(1): 239.

60. Youssef, N.F. 2005. Stability-indicating methods for the determination of piretanide in presence of the alkaline induced degradates. *J Pharm Biomed Anal* 39(5): 871–876.

61. Ojanperä, I., Ojansivu, R.-L., Nokua, J., and E. Vuori. 1999. Comprehensive TLC drug screening in forensic toxicology: Comparison of findings in urine and liver. *J Planar Chromatogr—Modern TLC* 12(1): 38–41.

62. Lillsunde, P. and T. Korte. 1991. Comprehensive drug screening in urine using solid-phase extraction and combined TLC and GC/MS identification. *J Anal Toxicol* 15: 71–81.

63. Sharma, M.C., Sharma, S., Kohli, D.V., and A.D. Sharma. 2010. Validated TLC densitometric method for the quantification of torsemide and spironolactone in bulk drug and in tablet dosage form. *Der Pharma Chemica* 2(1): 121–126.

39 TLC of Antidiabetics

Anna Gumieniczek and Anna Berecka

CONTENTS

39.1 TLC METHODS IN ANALYSIS OF ORAL ANTIDIABETIC DRUGS

A review of TLC methods for the detection and determination of oral antidiabetic drugs—chlorpropamide, tolbutamide, acetohexamide, glibenclamide (glyburide), gliclazide, glimepiride, and glipizide from sulfonylureas; nateglinide and repaglinide from glinides; metformin from biguanides; and ciglitazone, darglitazone, englitazone, pioglitazone, and rosiglitazone from glitazones—is presented. This chapter describes the methods applied for the analysis of these important drugs in bulk, pharmaceuticals (one or more component), new formulations, adulterated health food, and in different biological materials. For quantitative methods, sensitivity, linearity, and precision of the presented methods are shown. The TLC methods applied for the determination of potential impurities of the drug substances are also included, and they are described as stability-indicating TLC methods. In addition, suitable methods for the separation of these drugs in different matrices as well as for studying their lipophilicity are also presented.

39.1.1 ORAL ANTIDIABETICS, THEIR CHEMISTRY, AND MECHANISMS OF ACTION

Sulfonylureas (Figure 39.1) and metformin are the oral agents most commonly used for the treatment of type 2 diabetes mellitus. They are really available and affordable and have convenient dosing. Sulfonylureas act by binding to K_{ATP} channels (SUR1) locating on a membrane of beta cells of the islets of Langerhans. The binding of the drug to SUR1 is followed by depolarization of the cell membrane and extracellular releasing of insulin (hypoglycemic effect). The main side effect of sulfonylureas is hypoglycemia, which can be more severe and prolonged than that produced by insulin. They are also associated with weight gain. Because of it, the first generation of sulfonylureas (acetohexamide, chlorpropamide, and tolbutamide) is rarely ever used these days. The second generation includes glimepiride, glyburide (glibenclamide), glipizide, and gliclazide. They are widely used as they are much safer and more potent than the first-generation drugs (Nyenwe et al. 2011).

Metformin from biguanides (Figure 39.2) improves islet cell responsiveness to a glucose load and improves peripheral glucose utilization. It also reduces hepatic gluconeogenesis by the inhibition of key enzymes in this pathway (antihyperglycemic effect). Metformin may also have long-term benefits of weight reduction and cardiovascular protection in diabetic patients (Nyenwe et al. 2011).

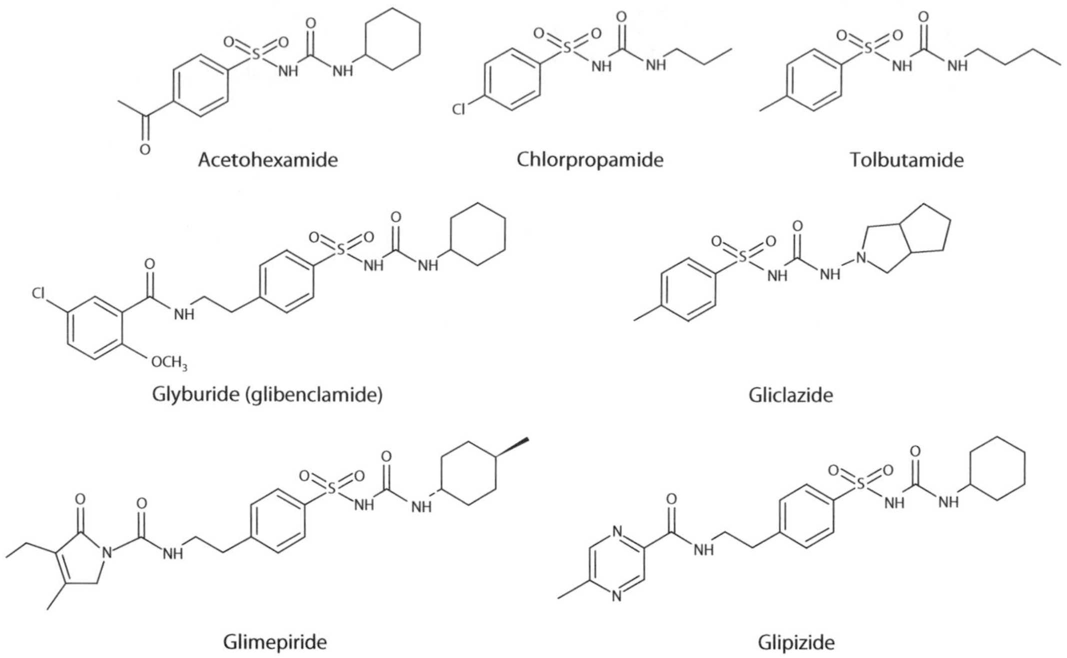

FIGURE 39.1   The chemical structures of sulfonylureas.

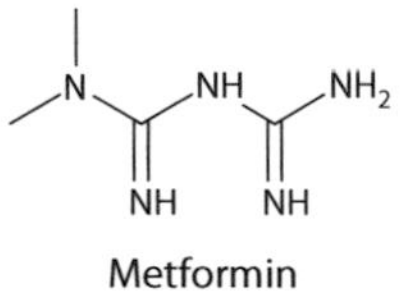

Metformin

FIGURE 39.2   The chemical structure of metformin.

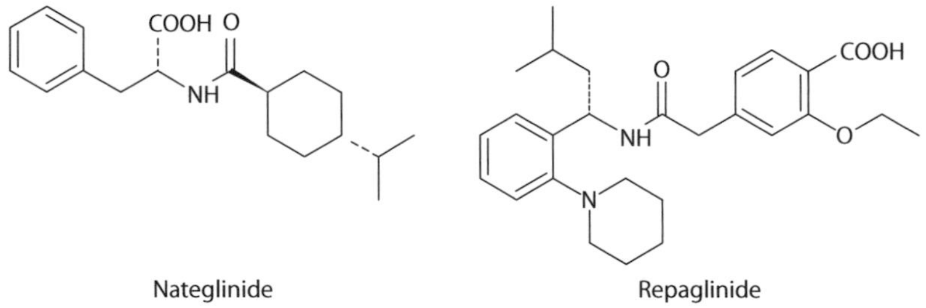

FIGURE 39.3   The chemical structures of nateglinide and repaglinide.

Glinides (nateglinide and repaglinide) (Figure 39.3) act in a mechanism similar to that of sulfo-nylureas by binding to a different part of the sulfonylurea receptor. They have shorter serum half-lives and therefore a lower risk of hypoglycemia than sulfonylureas, but they must be administered immediately before each meal. These drugs are ideally suited for combination with metformin (Nyenwe et al. 2011).

In recent years, much progress has been achieved in the discovery and development of per-oxisome proliferator-activated receptor gamma (PPARγ) modulators, including pioglitazone and rosiglitazone as well as ciglitazone, darglitazone, and englitazone. Chemically, they all are deriva-tives of thiazolidin-2,4-dione so they are called thiazolidinediones or TZDs (Figure 39.4). Their mechanism of action is different from that of the established antidiabetic drugs, including regula-tion of the expression of genes involved in lipid and carbohydrate metabolism in targeted tissues.

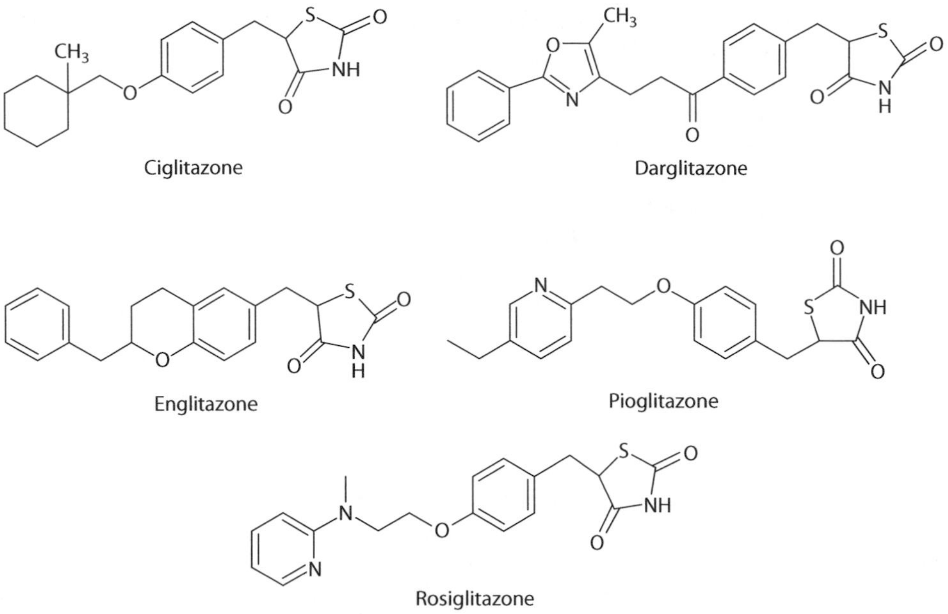

FIGURE 39.4 The chemical structures of TZDs.

TZDs have also been shown to preserve or even improve beta cell secretory function in diabetic patients. However, there are reports associating TZDs with an increased incidence of cardiovascular events. So far, pioglitazone and rosiglitazone have been approved for monotherapy and for combination with sulfonylureas or with metformin (Nyenwe et al. 2011).

39.1.2 QUANTITATIVE DETERMINATION OF ORAL ANTIDIABETICS IN PHARMACEUTICALS

In the papers reviewed in the following, TLC method was used as an effective alternative to other analytical tools, especially to HPLC. Most of the presented quantitative methods were validated in accordance with the requirements of ICH or FDA guidelines. In many cases, their selectivity, linearity, precision, and accuracy were sufficient for their routine use in pharmaceutical analysis. Some of the presented methods are described as stability indicating because their specificity toward stress degradation products was confirmed.

39.1.2.1 TLC for Determination of One-Component Formulations

From this chapter, it is clearly shown that TLC with densitometric detection and quantification was successively applied for the determination of gliclazide (El Kousy 1998), glipizide (El Kousy 1998; Jain and Saraf 2008), glibenclamide (El Kousy 1998), repaglinide (Gumieniczek et al. 2005; Jiladia and Pandya 2009), nateglinide (Kale and Kakde 2011b), pioglitazone (Manoj et al. 2004; Shirkhedkar and Surana 2009), and rosiglitazone (Sane et al. 2002; Gumieniczek et al. 2003; Rao et al. 2003; Walode et al. 2010).

The most commonly used sulfonylurea drugs, that is, gliclazide, glipizide, and glibenclamide, were determined using a stability-indicating densitometric method (El Kousy 1998). The degradation products were prepared by acid hydrolysis (5 M HCl) of the intact drugs. Chromatography was carried out using silica gel 60F$_{254}$ plates and different mobile phases, followed by scanning of the developed chromatograms at 230, 276, and 300 nm for gliclazide, glipizide, and glibenclamide, respectively. Complete resolution within the stress degradation products was obtained using chloroform–methanol (9:1, v/v) for gliclazide and glipizide and chloroform–cyclohexane–glacial acetic acid–ethanol (9:9:1:1, v/v/v/v) for glibenclamide. However, the optimal distance for good separation was equal to 16 cm. Mixtures of the investigated drugs and their degradation products were prepared and analyzed using the proposed methods, with recovery in the range of 100.4%–101.0% and RSD in the range of 0.6%–0.7%.

The TLC stability-indicating method was also elaborated for the determination of gliclazide and glipizide in tablets (Gumieniczek and Berecka 2010) using RP18 plates with 60% acetonitrile in pH 2.3 phosphate buffer as mobile phase. Compact spots were obtained for gliclazide (R_f ca. 0.38) and glipizide (R_f ca. 0.51). Detection and quantification were performed by classical densitometry at 215 nm, which was the wavelength of maximum absorption of both drugs. Calibration plots were constructed in the range of 0.8–1.8 µg/10 µL for both and were linear with good correlation coefficients (r = 0.998 for gliclazide, r = 0.993 for glipizide). The limit of detection (LOD) and limit of quantification (LOQ) were 50 and 200 ng/spot for gliclazide and 60 and 300 ng/spot for glipizide. The samples were also treated with acid (0.1 M HCl), base (0.1 M NaOH), hydrogen peroxide (3%), ultraviolet (UV) light (254 nm), and high temperature (60°C). Alkaline, oxidative, and UV treatment at room temperature did not result in significant degradation. However, samples treated with acid and/or high temperature showed well-separated spots of gliclazide or glipizide and additional peaks at different R_f values.

Glipizide was also determined on silica gel 60GF$_{254}$ HPTLC plates using ethyl acetate, formic acid, and dichloromethane (1:1:2, v/v/v) as mobile phase, while the plates were scanned at 275.5 nm. The linearity range was 200–800 ng/spot (Jain and Saraf 2008).

A relatively new drug, nateglinide, was estimated on silica gel 60F$_{254}$ HPTLC plates with hexane, methanol, and propan-2-ol in the proportion of 7.5:1.5:1 (v/v/v) as mobile phase. Densitometric quantification was performed at 210 nm, while the mean R_f value was ca. 0.56. The calibration curve was linear in the range of 300–1000 ng/band. The LOD and LOQ were 78.95 and 236.84 ng/band, respectively (Kale and Kakde 2011b).

Also, repaglinide was chromatographed on silica gel 60F$_{254}$ HPTLC plates using chloroform, methanol, and ammonia 25% (4.5:0.8:0.05, v/v/v) as mobile phase (Jiladia and Pandya 2009). The drug scanning at 288 nm showed the R_f value of ca. 0.55. The method was validated in terms of linearity (400–2400 ng/spot), precision (intraday variation 0.7%–2.6%, interday variation 0.8%–3.2%), accuracy (97.0%–99.0%), and specificity. The LOD and LOQ for repaglinide were found to be 50 and 300 ng/spot, respectively.

In the next paper (Gumieniczek et al. 2005), repaglinide analysis was performed on RP8 TLC plates with a mixture of acetonitrile–pH of 6.0 phosphate buffer (60:40, v/v) as mobile phase. The detection and quantification were performed by classical densitometry at the wavelength of maximum absorption of repaglinide 225 nm. A calibration plot was constructed in the range of 0.6–3.6 µg/10 µL, while LOQ and LOD were 0.27 and 0.08 µg/10 µL, respectively. Instrumental precision, established at three concentrations of the drug, ranged from 3.92% to 0.97% for the lowest and highest concentrations of repaglinide. The recovery from the model mixtures, at three levels of addition, ranged from 103.1% to 102.5% for the lowest and highest levels. The effect of pH, temperature, and UV light on degradation of repaglinide was also investigated. Degradation study did not reveal any additional peaks that could interfere with the elution of repaglinide. The acid degradation products were shown, but they were well separated from the pure drug with substantially different R_f values.

Three quantitative TLC methods were elaborated for the determination of pioglitazone (Gumieniczek et al. 2004; Manoj et al. 2004; Shirkhedkar and Surana 2009). A simple, rapid, and stability-indicating HPTLC method was developed and validated for the quantitative determination of pioglitazone in tablets. The analysis was performed using the horizontal technique with CNF$_{254}$ plates, 1,4-dioxane-phosphate buffer of pH 4.4 (5:5, v/v), and classical densitometry at the wavelength of maximum absorption of pioglitazone 266 nm. A calibration plot was constructed in the range of 0.4–2.4 µg/10 µL. The precision of the proposed chromatographic method expressed as mean RSD was 4.99% and 2.57%, respectively, for the lowest and the highest calibration levels.

Recovery from the fortified samples ranged from 98.1% to 103.3% (Gumieniczek et al. 2004). Using TLC plates coated with RP18F$_{254}$ and the mixture, acetone–water–acetic acid 4:1:0.1 (v/v/v) as mobile phase, the compact band for pioglitazone with R$_f$ value of ca. 0.68 was obtained in the study of Shirkhedkar and Surana (2009). UV densitometric detection was performed at 225 nm. The method was linear over the concentration range of 500–3000 ng/band, while the LOD and LOQ were 46.62 and 141.30 ng, respectively. Pioglitazone was subjected to acid and alkaline hydrolysis, oxidation, and photochemical and thermal degradation. The drug was degraded under all these conditions, but the most significant changes were observed in acidic conditions (four additive peaks with R$_f$ in the range of 0.27–0.48 were observed). Initially, acetone and water in different ratios were tried, but significant tailing was observed. To overcome this problem, acetic acid was added. Finally, the pure drug and all additive peaks were sufficiently resolved. The peak shape of pioglitazone and reproducibility were improved by previous plate development with methanol followed by drying and presaturation of the TLC chamber for 30 min. In the third paper, separation was carried out on silica gel 60F$_{254}$ HPTLC plate using a mixture of toluene, methanol, acetone, and ammonia 10% in the ratio of 80:40:20:1 (v/v/v/v) as mobile phase. Quantification was carried out by the use of a the densitometer at the absorbance/reflectance mode at 254 nm. The R$_f$ value of pioglitazone lied between 0.49 and 0.55. The linear dynamic response was found to be 2.0–4.0 μg/spot (Manoj et al. 2004).

Also, for rosiglitazone, quantitative TLC methods were found in the literature (Sane et al. 2002; Gumieniczek et al. 2003; Rao et al. 2003; Walode et al. 2010). The first method used silica gel 60F$_{254}$ HPTLC plates as stationary phase and ethyl acetate–toluene–methanol, 45:55:1 (v/v/v) as mobile phase. Prewashing and subsequent drying of plates for 10 min at 110°C as well as equilibrating of TLC chamber for 10 min were applied. After the developing detection was performed at 242 nm, pioglitazone was used as internal standard. The response was found to be linearly dependent on the amount of rosiglitazone between 100 and 1200 ng. Interesting studies were also done for the robustness testing. The authors checked the effects of temperature, humidity, light, and the method of application on chromatographic separation and the spot shapes (Sane et al. 2002). The second method was performed on HPTLC silica gel 60GF$_{254}$ using mobile phase consisting of methanol, toluene, chloroform, and triethylamine (1:8:0.5:0.5, v/v/v/v) when the detection was carried out in the absorbance mode at 264 nm. The linear regression data showed good linear relationship in the concentration range of 1.0–7.0 μg/μL. The mean recovery of drug was carried out by the standard addition method and was found to be 100.2%. Caffeine was used as internal standard. Stability-indicating capability of the proposed method was investigated by applying samples to different stress conditions to access the presence of components that may be expected to be present such as impurities and degradation product. The sample solution was allowed to be stored for 24 h under different stress conditions like 0.1 M HCl (acid), 0.1 M NaOH (alkali), 3% of H$_2$O$_2$ (oxidation), at 60°C (heat), and in UV cabinet at 265 nm (UV light). However, there was no degradation of the sample under the aforementioned stress conditions. Also, the excipients present in the formulation did not interfere with the assay (Walode et al. 2010). The third analysis (Gumieniczek et al. 2003) was performed on silica gel 60F$_{254}$ HPTLC plates in horizontal chambers with chloroform–ethyl acetate–ammonia 25% (5:5:0.1, v/v/v) as mobile phase. Detection and quantification were performed by classical densitometry at 240 and 254 nm. Calibration plots were constructed in the range of 0.2–1.0 μg/10 μL. The precision of the proposed method, expressed as mean RSD, was 3.58% and 2.76% for 240 nm and 8.23% and 6.56% for 254 nm, for the lowest and the highest calibration levels, respectively. The mean recoveries from the fortified samples ranged from 89.5% to 99.4% at 240 nm and from 89.1% to 100.9% at 254 nm. It was concluded that the results obtained at 240 nm were better (the maximum of absorbance of rosiglitazone).

Densitometric determination of rosiglitazone was also performed on silica gel 60GF$_{254}$ HPLC plates using a mixture of toluene–methanol–acetone–ammonia 25% (80:10:20:1.0, v/v/v/v) as mobile phase. The quantification was carried out by the use of a densitometer at the absorbance mode at 241 nm (Rao et al. 2003).

39.1.2.2 TLC for Two or More Component Formulations

The two processes that result in type 2 diabetes mellitus are insulin resistance in the liver and peripheral tissues and dysfunction of the pancreatic β cells. The combination of two or more drugs with different mechanisms of action addresses these both processes. Such combinations could significantly improve the therapy of diabetes so they are now available in the same pill or capsule. Therefore, adequate analytical methods including rapid and simple TLC have been developed for quantitative determination of different oral antidiabetics in many combinations.

In the study of Gayatri et al. (2003), the combination of TZD–rosiglitazone and sulfonylurea–gliclazide was studied using silica gel 60F$_{254}$ HPTLC plates and a mixture of toluene–ethyl acetate–methanol (85:5:10, v/v/v) as mobile phase. The detection of the spots was carried out at 225 nm in the absorbance mode. The retardation factors of gliclazide and rosiglitazone were found to be 0.36 and 0.47, respectively. The linearity range was found to be 1–3 μg/10 μL for gliclazide and 0.05–0.15 μg/10 μL for rosiglitazone.

The combination containing pioglitazone and glimepiride was studied by Patel et al. (2006). The separation was achieved on silica gel 60F$_{254}$ HPTLC plates using toluene–ethyl acetate–methanol (50:45:5, v/v/v) as mobile phase. Quantification was achieved with UV detection at 230 nm over the concentration range of 200–700 and 1500–5250 ng/spot with mean recovery of 98.4% ± 0.68% and 98.8% ± 1.14% for glimepiride and pioglitazone, respectively. The R$_f$ values for glimepiride and pioglitazone were found to be 0.49 and 0.61, respectively. The method was validated in terms of accuracy and precision.

Densitometric determination of pioglitazone and glimepiride was also performed on silica gel 60F$_{254}$, while the solvent system consisted of chloroform, toluene, glacial acetic acid, and ethanol (4.5:4.5:1:1, v/v/v/v). Densitometric evaluation of the separated zones was performed at 228 nm and 268 nm. The two drugs were satisfactorily resolved with R$_f$ values 0.40 and 0.65 for pioglitazone and glimepiride, respectively. The accuracy and reliability of the method were assessed by the evaluation of linearity in the range of 3–15 μg/spot for pioglitazone and 0.1–3 μg/spot for glimepiride, precision (RSD < 1.2% for pioglitazone and glimepiride), accuracy (99.94% for pioglitazone and 100.74% for glimepiride), and specificity, in accordance with ICH guidelines (Rezk et al. 2011).

Recently, formulations containing pioglitazone, glimepiride, and the third component, metformin, have occurred in the market. Such formulation was determined by Kale and Kakde (2011a). Pioglitazone, metformin, and glimepiride were separated and determined on silica gel 60F$_{254}$ HPTLC plates with acetonitrile, methanol, propan-1-ol, and ammonium acetate in the proportion of 7:2:1:1 (v/v/v/v) as mobile phase. Densitometric quantification was performed at 240 nm. Well-resolved bands were obtained with R$_f$ values 0.83, 0.21, and 0.89 for pioglitazone, metformin, and glimepiride, respectively. The method was validated for precision, accuracy, specificity, and robustness. The calibration curve was found to be linear in the concentration range of 0.3–1.2, 10–40, and 0.04–0.16 μg/band with correlation coefficients of 0.995, 0.996, and 0.998 for pioglitazone, metformin, and glimepiride, respectively. The LOD was found to be 56.57, 60.85, and 11.99 ng/band for pioglitazone, metformin, and glimepiride, while the LOQ was found to be 171.43, 69.26, and 36.33 ng/band. The specificity of the method was ascertained by exposing the sample to different stress conditions such as acidic (0.1 M HCl), alkaline (0.1 M NaOH), oxidizing (3% H$_2$O$_2$), heat (60°C), and UV radiations for 24 h and then analyzing them by the proposed method. The estimation of pioglitazone and metformin showed remarkable degradation (nearly 60%) in alkali, while other conditions did not affect the estimation. Glimepiride did not show any degradation with acid, alkali, H$_2$O$_2$, UV light, and heat exposure. Using the earlier method, all additive peaks of pioglitazone and metformin were sufficiently separated. Therefore, the presented HPTLC method was proved to be sufficiently selective and stability indicating.

A simple, precise, and accurate HPTLC method was also developed for simultaneous estimation of metformin with sulfonylureas like gliclazide (combination I) or glimepiride (combination II) present in two component dosage forms (Havele and Dhaneshwar 2011). This method employed silica gel 60F$_{254}$ HPTLC plates and ammonium sulfate (0.25%), methanol, and ethyl acetate in the ratio of 10:2.5:2.5 (v/v/v) with former saturation for 30 min at room temperature. This method allowed quantitation over the range of 100–500 ng/mL for gliclazide and 1,000–5,000 ng/mL for metformin in combination I and 300–500 ng/mL for glimepiride and 150,000–250,000 ng/mL for metformin in combination II.

In one other report, simultaneous determination of glibenclamide (glyburide) and metformin was performed (Ghassempour et al. 2006) using silica gel $60F_{254}$ HPTLC plates as stationary phase and water–methanol–ammonium sulfate (2:1:0.5, v/v/v) as mobile phase. This system gave a good resolution for metformin (R_f value of 0.43 ± 0.01) and glibenclamide (R_f value of 0.64 ± 0.02). The determination was by densitometry in the absorbance mode at 237 nm. The linear regression data for the calibration plot showed a good relationship with r = 0.9958 and 0.9998 for metformin and glibenclamide, respectively. The LOD and LOQ were 25.24 and 84.12 ng/spot, respectively, for metformin and 12.26 and 40.86 ng/spot, respectively, for glibenclamide.

Simultaneous determination of metformin and glipizide was also elaborated using silica gel $60F_{254}$ and water–methanol–0.5% (v/v) ammonium sulfate (6:3:1.5, v/v/v) as mobile phase. This system gave a good resolution for metformin (R_f value of 0.22) and glipizide (R_f value of 0.85). The detection was done at 236 nm. The linear regression data for the calibration plot showed a good relationship with r = 0.9962 and 0.9930 for metformin and glipizide, respectively. The LODs and LOQs were 991.30 and 3003.95 ng/band for metformin and 9.57 and 29.01 ng/band for glipizide, respectively (Modi and Patel 2012).

TLC method was also used for quantitative determination of nateglinide in the presence of metformin (Thomas et al. 2011) where silica gel HPTLC plates and the mixture chloroform–ethyl acetate–acetic acid (4:6:0.1, v/v/v) as mobile phase were selected as optimal conditions. In developing studies, different mobile phase systems like toluene–methanol, chloroform–methanol, chloroform–diethyl ether–ethyl acetate, and chloroform–ethyl acetate–acetic acid at different concentrations were tried. The use of acetic acid was found to be necessary for the elution of metformin from the plate due to the fact that it is a weak basic drug, thereby interacting with the unreacted silanol groups on silica stationary phases. However, increasing the ratio of acetic acid was found to be deleterious and affected the peak shape of nateglinide.

A TLC scanner set at 216 nm was used for direct evaluation of the chromatograms in the reflectance/absorbance mode. The correlation coefficients of calibration curves were found to be 0.996 and 0.995 in the concentration range of 200–2400 and 500–3000 ng/band for nateglinide and metformin, respectively. The method had an accuracy of 99.7% for nateglinide and 100.1% for metformin. Nateglinide and metformin were also subjected to acid (0.001 M HCl), base (0.001 M NaOH), oxidation (1.5% H_2O_2), wet (40°C for 30 min), heat (50°C for 24 h), and photodegradation studies. Both drugs were found to be degraded, and many additive peaks occurred. However, all degradation products obtained during the study were well resolved from the pure drugs with significantly different R_f values.

39.1.3 TLC FOR ORAL ANTIDIABETICS ASSAYS IN BIOLOGICAL MATERIAL

In the literature, a radio thin-layer chromatographic method is described for in vitro measurement of tolbutamide methylhydroxylation as an alternative tool to the commonly used HPLC assay (Ludwig et al. 1998). After the incubation of [^{14}C]tolbutamide with human liver microsomes, the supernatants were directly spotted onto standard silica gel TLC plates and developed in a horizontal chamber using a solvent system consisting of toluene–acetone–formic acid (60:39:1, v/v/v). Dried TLC plates were exposed to a phosphor imager plate and quantified. By the authors, the described method provided a valuable tool for the determination of tolbutamide hydroxylation activity in human liver microsomes.

Also, in vitro test using rat liver samples and TLC autoradioluminography were presented to rapidly assess tolbutamide metabolizing activities of individual patients (Okuyama et al. 1997).

Using this method, the authors could detect drug metabolites in rat liver biopsy samples in a relatively short time and at low concentrations similar to those in vivo. As per the authors, the presented method was expected to be useful for assessing the drug metabolizing activities of patients.

39.1.4 TLC Methods in Oral Antidiabetic Drug Technology

In the studies of Mutalik and Udupa (2004, 2005), HPTLC method was used for the analysis of transdermal patches containing glibenclamide, prepared with different polymers. The silica gel 60GF$_{254}$ plates and mobile phase consisting of ethyl acetate and ammonia 25% (10:0.1, v/v) were applied. The volumes of 5 mL of each solution containing drug alone or drug and different ratios of polymers were spotted onto the plates. The plates were dried in the stream of warm air for 5 min and scanned. It was stated that the R$_f$ value of glibenclamide (ca. 0.90) was not changed in the presence of polymers.

Different analytical methods, including TLC, were used for the characterization of solid dispersion of gliclazide (Hosmani and Thorat 2011). Respective formulations were prepared by solvent evaporation and melt dispersion techniques using poloxamer as hydrophilic carrier. TLC was used to identify any possibility of degradation during the preparation and to optimize melting temperature for melt dispersion batches.

39.1.5 TLC Methods for Separation of Oral Antidiabetics

Some reports from the literature concern pure separation study of different groups of antidiabetic compounds or their chromatographic behavior study.

The normal- and reversed-phase chromatographic behaviors of seven oral antidiabetic drugs—chlorpropamide, tolbutamide, glibenclamide, metformin, pioglitazone, rosiglitazone, and repaglinide—were studied in normal- and reversed-phase systems using the horizontal technique (Gumieniczek et al. 2003). For normal-phase systems, silica gel and alumina layers with mixtures of chloroform, diethyl ether, and ethyl acetate were used. For better resolution, ammonia or acetic acid was added to mobile phases. Generally, the silica gel plates were more suitable than alumina plates. The best separation of six of the seven target drugs was achieved with chloroform–ethyl acetate–acetic acid (5:5:0.1, v/v/v). By the use of this mobile phase, pioglitazone was separated from rosiglitazone and tolbutamide from chlorpropamide. For reversed-phase systems, RP18 plates and mixtures of acetonitrile or propan-2-ol with phosphate buffers at different pH were applied. The mobile phase containing 70% acetonitrile in phosphate buffer of pH 4.4 was best for the resolution of all seven target drugs. Generally, R$_f$ values on RP18 plates enabled better separation than on silica gel, especially for metformin.

The chromatographic behavior of pioglitazone, rosiglitazone, and repaglinide on CNF$_{254}$ plates was also investigated (Gumieniczek et al. 2004). The mobile phases comprising 1,4-dioxane with phosphate buffers at different pH were used, and the influence of pH on separation of these drugs was examined. The horizontal technique was used for separation, and the detection was done at 254 nm. The concentration of the buffer in the mobile phase varied between 20% and 80%, while the pH was in the range of 2.8–7.9. The pH value strongly affected the retention of pioglitazone when the content of organic modifier was 20%. For two other drugs, rosiglitazone and repaglinide, and for other mobile phases, this effect was less apparent. The best separation of the three target drugs was achieved using 50% 1,4-dioxane at any pH studied, 40% at pH 2.8, and 60% at pH 4.4. The mobile phase containing 40% phosphate buffer of pH 6.4 gave the best separation of pioglitazone and rosiglitazone.

The same three drugs were separated on RP8 adsorbent with mobile phases containing acetonitrile, methanol, and propan-2-ol in phosphate buffers at pH 2.4, 4.4, 6.0, and 7.9. The plates were developed in horizontal chambers and visualized by UV at 254 nm. Generally, the retention of the drugs decreased as the modifier concentration increased up to 60%–80%. Finally, several chromatographic systems were proposed for effective identification, separation, and quantification of these three drugs (Berecka et al. 2005).

As far as concerns more TZDs, five of them—ciglitazone, darglitazone, englitazone, pioglitazone, and rosiglitazone—were chromatographed on RP18 TLC plates with binary mobile phases containing water and the organic modifiers, acetone, 1,4-dioxane, or methanol in horizontal chambers. Linear relationships were obtained between the R_M values of the compounds and the concentration of organic modifier in the mobile phase. These R_M values enabled the calculation of R_{M0} values by extrapolation. The study was significantly improved by the use of standardization procedure. Calibration equations were obtained for nine standards of known lipophilicity in the range of 0.83–6.04. From these equations, the partition coefficients (experimental logP) were calculated for the mentioned TZDs (Gumieniczek et al. 2007).

A simple and sensitive method for separation and detection of five antidiabetic drugs extracted from respective tablets (metformin, pioglitazone, rosiglitazone, glibenclamide, and gliclazide) was also elaborated. The plates coated with silica gel were activated overnight at 60°C in an oven. After developing, the detection of the drugs was done by the use of iodine vapor. The mobile phases containing butan-1-ol-acetic acid–water–methanol (12:4:1:2, v/v/v/v) and toluene–ethyl acetate–methanol (8.5:1:8.5, v/v/v) were found to enable successful separation of the mixtures, metformin with pioglitazone, metformin with rosiglitazone, metformin with gliclazide, and metformin with glibenclamide (Bhushan et al. 2006).

Screening and quantitative analysis for six sulfonylureas (tolbutamide, acetohexamide, chlorpropamide, gliclazide, glibenclamide, and glimepiride) in adulterated health food were also developed (Kumasaka et al. 2005). Good separation was achieved with a mixture of n-butyl acetate containing 0.4% formic acid as solvent. UV irradiation at 254 nm was used to detect the drugs. Specificity was obtained with Dragendorff's test solution, 10% phosphomolybdic acid in methanol and 30% sulfuric acid in methanol. Representative R_f values of tolbutamide, acetohexamide, chlorpropamide, gliclazide, glibenclamide, and glimepiride were 0.78, 0.58, 0.69, 0.48, 0.40, and 0.35, respectively. The presented method was applied to 11 types of adulterated health food such as tablets, tea bags, and capsule products.

Similar work was done for 11 oral hypoglycemic agents, using different analytical methods including TLC. In normal-phase and reversed-phase TLC, the conditions to separate and detect 10 ingredients (except nateglinide) were developed (Date et al. 2009).

Also, two other papers on chromatographic behavior of chlorpropamide, tolbutamide, metformin, and glipizide were found; however, the abstracts of them and what is more in the whole papers are not available in the literature (Agarwal et al. 1973; Puncuh et al. 1995).

REFERENCES

Agarwal, S. P., M. I. Walash, and M. I. Blake. 1973. Identification of oral hypoglycaemic and diuretic drugs by TLC using metal ions. *Indian Journal of Pharmacy* 35:181–183.

Berecka, A., A. Gumieniczek, and H. Hopkała. 2005. Retention behavior of the newest oral anti-diabetic drugs in reversed-phase chromatography. *Journal of Planar Chromatography—Modern TLC* 18:61–66.

Bhushan, R., D. Gupta, and A. Jain. 2006. TLC supplemented by UV spectrophotometry compared with HPLC for separation and determination of some antidiabetic drugs in pharmaceutical preparations. *Journal of Planar Chromatography—Modern TLC* 19:288–296.

Date, H., M. Terauchi, M. Sugimura, A. Toyota, and T. Matsuo. 2009. Systematic analysis of oral hypoglycemic agents in health foods. *Yakugaku Zasshi* 129:163–172.

El Kousy, N. M. 1998. Stability-indicating densitometric determination of some antidiabetic drugs in dosage forms, using TLC. *Mikrochimica Acta* 128:65–68.

Gayatri, S., A. Shantha, and V. Vaidyalingam. 2003. Simultaneous HPTLC determination of gliclazide and rosiglitazone in tablets. *Indian Journal of Pharmaceutical Sciences* 65:663–665.

Ghassempour, A., M. Ahmadi, S. N. Ebrahimi, and H. Y. Aboul-Enein. 2006. Simultaneous determination of metformin and glyburide in tablets by HPTLC. *Chromatographia* 64:101–104.

Gumieniczek, A. and A. Berecka. 2010. Quantitative analysis of gliclazide and glipizide in tablets by a new validated and stability-indicating RPTLC method. *Journal of Planar Chromatography—Modern TLC* 23:129–133.

Gumieniczek, A., A. Berecka, and H. Hopkała. 2005. Quantitative analysis of repaglinide in tablets by reversed-phase thin-layer chromatography with densitometric UV detection. *Journal of Planar Chromatography—Modern TLC* 18:155–159.

Gumieniczek, A., A. Berecka, H. Hopkała, and T. Mroczek. 2003. Rapid HPTLC determination of rosiglitazone in pharmaceutical formulations. *Journal of Liquid Chromatography and Related Technologies* 26:3307–3314.

Gumieniczek, A., A. Berecka, D. Matosiuk, and H. Hopkała. 2007. Standardized reversed-phase thin-layer chromatographic study of the lipophilicity of five anti-diabetic thiazolidinediones. *Journal of Planar Chromatography—Modern TLC* 20:261–265.

Gumieniczek, A., H. Hopkała, and A. Berecka. 2004. Reversed-phase thin-layer chromatography of three new oral antidiabetics and densitometric determination of pioglitazone. *Journal of Liquid Chromatography and Related Technologies* 27:2057–2070.

Gumieniczek, A., H. Hopkała, A. Berecka, and D. Kowalczuk. 2003. Normal and reversed-phase thin-layer chromatography of seven oral anti-diabetic agents. *Journal of Planar Chromatography—Modern TLC* 16:271–275.

Havele, S. S. and S. R. Dhaneshwar. 2011. Simultaneous determination of metformin hydrochloride in its multi component dosage forms with sulfonylureas like gliclazide and glimepiride using HPTLC. *Journal of Liquid Chromatography and Related Technologies* 34:966–980.

Hosmani, A. H. and Y. S. Thorat. 2011. Optimization and pharmacodynamic evaluation of solid dispersion of gliclazide for dissolution rate enhancement. *Latin American Journal of Pharmacy* 30:1590–1595.

Jain, S. and S. Saraf. 2008. Development and validation of high performance thin layer chromatographic (HPTLC) technique for quantification of glipizide in tablet dosage forms. *Biosciences Biotechnology Research Asia* 5:425–428.

Jiladia, M. A. and S. S. Pandya. 2009. Estimation of repaglinide in bulk and tablet dosage forms by HPTLC method. *International Journal of Pharmacy and Pharmaceutical Sciences* 1(Suppl. 1):141–144.

Kale, D. and R. Kakde. 2011a. Simultaneous determination of pioglitazone, metformin, and glimepiride in pharmaceutical preparations using HPTLC method. *Journal of Planar Chromatography—Modern TLC* 24:331–336.

Kale, D. L. and R. B. Kakde. 2011b. HPTLC estimation of nateglinide in bulk drug and tablet dosage form. *Asian Journal of Chemistry* 23:4351–4354.

Kumasaka, K., T. Kojima, H. Honda, and K. Doi. 2005. Screening and quantitative analysis for sulfonylurea-type oral antidiabetic agents in adulterated health food using thin-layer chromatography and high-performance liquid chromatography. *Journal of Health Science* 51:453–460.

Ludwig, E., H. Wolfinger, and T. Ebner. 1998. Assessment of microsomal tolbutamide hydroxylation by a simple thin-layer chromatography radio activity assay. *Journal of Chromatography B: Biomedical Applications* 707:347–350.

Manoj, K., P. Muthusamy, and S. Anbazhagan. 2004. HPTLC method for estimation of pioglitazone hydrochloride from tablet formulation. *Indian Drugs* 41:354–357.

Modi, D. K. and B. H. Patel. 2012. Simultaneous determination of metformin hydrochloride and glipizide in tablet formulation by HPTLC. *Journal of Liquid Chromatography and Related Technologies* 35:28–39.

Mutalik, S. and N. Udupa. 2004. Glibenclamide transdermal patches: Physicochemical, pharmacodynamic and pharmacokinetic evaluations. *Journal of Pharmaceutical Sciences* 93:1577–1594.

Mutalik, S. and N. Udupa. 2005. Formulation development, in vitro and in vivo evaluation of membrane controlled transdermal systems of glibenclamide. *Journal of Pharmacy and Pharmaceutical Sciences* 8:26–38.

Nyenwe, E. A., T. W. Jerkins, G. E. Umpierrez, and A. E. Kitabachi. 2011. Management of type 2 diabetes: Evolving strategies for the treatment of patients with type 2 diabetes. *Metabolism Clinical and Experimental* 60:1–23.

Okuyama, M., C. Inoue, K. Aijima, Y. Nakamura, T. Kaburagi, and A. Shigematsu. 1997. In vitro method for assessing hepatic drug metabolism. *Biological and Pharmaceutical Bulletin* 20:1–5.

Patel, J. R., B. N. Suhagia, and M. M. Patel. 2006. Simultaneous estimation of glimepiride and pioglitazone in bulk and in pharmaceutical formulation by HPTLC method. *Asian Journal of Chemistry* 18:2873–2878.

Puncuh, A., E. Morosini-Berus, and B. Kotar-Jordan. 1995. Solid-state interaction studies of glipizide with excipients using thermal analysis and thin layer chromatography (TLC). *Farmacevtski Vestnik* 46:285–286.

Rao, J. R., S. S. Kadam, K. R. Mahadik, and A. K. Aggarwal. 2003. HPTLC method for estimation of rosigli-tazone maleate from tablet formulation. *Indian Drugs* 40:393–396.

Rezk, M. R., S. M. Riad, G. Y. Mahmoud, and A.-A. E. B. A. Aleem. 2011. Simultaneous determination of pioglitazone and glimepiride in their pharmaceutical formulations. *Der Pharma Chemica* 3:176–184.

Sane, R. T., M. Francis, A. Moghe, S. Khedkhar, and A. Anerao. 2002. High-performance thin-layer chromato-graphic determination of rosiglitazone in its dosage form. *Journal of Planar Chromatography—Modern TLC* 15:192–195.

Shirkhedkar, A. and S. Surana. 2009. Application of a stability-indicating densitometric RP-TLC method for analysis of pioglitazone hydrochloride in the bulk material and in pharmaceutical formulations. *Journal of Planar Chromatography—Modern TLC* 22:191–196.

Thomas, A. B., S. D. Patil, R. K. Nanda, L. P. Kothapalli, S. S. Bhosle, and A. D. Deshpande. 2011. Stability-indicating HPTLC method for simultaneous determination of nateglinide and metformin hydrochloride in pharmaceutical dosage form. *Saudi Pharmaceutical Journal* 19:221–231.

Walode, S. G., H. K. Chaudhari, M. S. Saraswat, A. V. Kasture, and S. G. Wadodkar. 2010. Validated high performance thin layer chromatographic determination and content uniformity test for rosiglitazone in tablets. *Indian Journal of Pharmaceutical Sciences* 72:249–252.

40 TLC of Antihistamines

Danica Agbaba and Gordana Popović

CONTENTS

40.1 INTRODUCTION

Histamine exhibits a wide variety of both physiological and pathological functions in different tissues and cells. Actions of histamine that are of interest both from a pharmacological and therapeutic point of view include (a) its important, but limited, role as chemical mediator of hypersensitivity and allergic inflammatory reactions, (b) its major role in the regulation of gastric acid secretion, and (c) its emerging role as a neurotransmitter in the central nervous system (CNS). The term "antihistamine" historically has referred to drugs that block the actions of histamine at the H_1-receptors rather than the other histamine receptor subtypes. H_1-Antihistamines act as inverse agonists that combine with and stabilize an inactive form of the H_1-receptor, shifting the equilibrium toward an inactive state. Based on pharmacological profiles, the H_1-antihistamines are now commonly subdivided into the two broad groups: the first-generation (or classical) antihistamines and the second-generation (or "nonsedating") antihistamines. The first-generation H_1-antihistamines are useful and effective in the treatment of allergic responses (e.g., hay fever, rhinitis, urticaria, and food allergy). These agents also exert effects at cholinergic, adrenergic, dopaminergic, and serotonergic receptors. Many of the first-generation antihistamines readily penetrate the blood–brain barrier because of their lipophilicity and maintain significant CNS concentrations, because they are not the substrates for the P-glycoprotein efflux pump that is expressed on endothelial cells of the CNS vasculature. Blockade of central H_1-receptors results in sedation, drowsiness, and decreased cognitive ability. The primary objective of antihistamine drug development over the past several decades has centered on developing new drugs with higher selectivity for the H_1-receptors than those already existing and the lack of undesirable CNS and cardiovascular actions. These efforts led to the introduction of the second-generation antihistamines, which exert little antagonist activity on other neurotransmitter receptors, including muscarinic receptors, and the cardiac ion channels at therapeutic concentrations (Beale and Block 2011).

This chapter presents the results of works in which thin-layer chromatography (TLC) is applied in order to identify and/or quantify single antihistamines or those in the presence of other antihistamines, to develop the methods for the analysis of antihistamines in the mono- and multicomponent formulations, to test stability of antihistamines, to couple TLC with the other analytical techniques for rapid separation and characterization of drugs, to develop screening systems for identification of antihistamines in toxicological analysis, and to investigate structure–activity relationship (SAR) of antihistamines.

The H_1-antihistamine drugs used in therapy are as follows:

Antazoline (**ANT**), 4,5-dihydro-N-phenyl-N-(phenyl methyl)-1H-imidazole-2-methanamine; 2-(N-benzylanilino methyl)-2-imidazoline; **astemizole** (**AST**), 1-[(4-fluorophenyl)methyl]-N-[1-[2-(4-methoxyphenyl)ethyl]-4-piperidinyl]-1H-benzimidazol-2-amine; **azatadine** (**AZA**), 6,11-dihydro-11-(1-methyl-4-piperidinylidene)-5H benzo[5,6]cyclohepta[1,2-b]pyridine; **bromodiphenhydramine** (**BDI**), 2-[(4-bromophenyl)phenylmethoxy]-N,N-dimethylethanamine; **brompheniramine** (**BPH**), γ-(4-bromophenyl)-N,N-dimethyl-2-pyridinepropanamine; **carbinoxamine** (**CAR**), 2-[(4-chlorophenyl)-2-pyridinylmethoxy]-N,N-dimethylethanamine; **cetirizine** (**CET**), [2-4-[(4-chlorophenyl)phenylmethyl]-1-piperazinyl]ethoxy]acetic acid; **chlorcyclizine** (**CCY**), 1-[(4-chlorophenyl)phenylmethyl]-4-methylpiperazine; **chloropyramine** (**CPY**), N-[(4-chlorophenyl)methyl]-N',N'-dimethyl-N-2-pyridinyl-1,2-ethanediamine; **chlorpheniramine** (**CPH**), γ-(4-chlorophenyl)-N,N-dimethyl-2-pyridinepropanamine; **chlorphenoxamine** (**CPO**), 2-[1-(4-chlorophenyl)-1-phenylethoxy]-N,N-dimethylethanamine; **cinnarizine** (**CIN**), 1-(diphenylmethyl)-4-(3-phenyl-2-propenyl)piperazine; **clemastine** (**CLA**), [R-(R*,R*)]-2-[2-[1-(4-chlorophenyl)-1-phenylethoxy]ethyl]-1-methylpyrrolidine; **clemizole** (**CLI**), 1-[(4-chlorophenyl)methyl]-2-(1-pyrrolidinylmethyl)-1H-benzimidazole; **cyproheptadine** (**CYP**), 4-(5H-dibenzo[a,d]cyclohepten-5-ylidene)-1-methylpiperidine; **deptropine** (**DEP**), endo-3-[(10,11-dihydro-5H-dibenzo[a,d]cyclohepten-5-yl)oxy]-8-methyl-8-azabicyclo[3.2.1]octane; **desloratadine** (**DES**), 8-chloro-6,11-dihydro-11-(4-piperdinylidene)-5H-benzo[5,6]cyclohepta[1,2-b] pyridine; **dimethindene** (**DIM**), N,N-dimethyl-3-[1-(2-pyridinyl)ethyl]-1H-indene-2-ethanamine; **diphenhydramine** (**DHY**), 2-diphenylmethoxy-N,N-dimethylethanamine; **doxylamine** (**DOX**), N,N-dimethyl-2-[1-phenyl-1-(2-pyridinyl)ethoxy]ethanamine; **ebastine** (**EBA**), 1-[4-(1,1-dimethylethyl)phenyl]-4-[4-(diphenylmethoxy)-1-piperidinyl]-1-butanone; **embramine** (**EMB**), 2-[1-(4-bromophenyl)-1-phenylethoxy]-N,N-dimethylethanamine; **fexofenadine** (**FEX**), α,α-dimethyl-4-[1-hydroxy-4-[4-(hydroxydiphenylmethyl)-1-piperidinyl]butyl]benzeneacetic acid; **hydroxyzine** (**HYD**), 2-[2-[4-[(4-chlorophenyl)phenylmethyl]-1-piperazinyl]ethoxy]ethanol; **isothipendyl** (**ISO**), $N,N,$α-trimethyl-10H-pyrido[3,2-b][1,4]benzothiazine-10-ethanamine; **levocabastine** (**LEV**), [3S-[1(cis),3α,4β]]-1-[4-cyano-4-(4-fluorophenyl)cyclohexyl]-3-methyl-4-phenyl-4-piperidine carboxylic acid; **loratadine** (**LOR**), 4-(8-chloro-5,6-dihydro-11H-benzo[5,6] cyclohepta[1,2-b]pyridin-11-ylidene)-1-piperidinecarboxylic acid ethyl ester; **mebhydroline** (**MEB**), 2,3,4,5-tetrahydro-2-methyl-5-(phenylmethyl)-1H-pyrido[4,3-b] indole; **mequitazine** (**MEQ**), 10-(1-azabicyclo[2.2.2]oct-3-ylmethyl)-10H-phenothiazine; **methapyrilene** (**MET**), N,N-dimethyl-N'-2-pyridinyl-N'-(2-thienylmethyl)-1,2-ethanediamine; **orphenadrine** (**ORP**), N,N-dimethyl-2-[(2-methylphenyl)phenylmethoxy]ethanamine; **phenindamine** (**PDA**), 2,3,4,9-tetrahydro-2-methyl-9-phenyl-1H-indeno[2,1-c]pyridine; **pheniramine** (**PIR**), N,N-dimethyl-γ-phenyl-2-pyridine propanamine; **phenyltoloxamine** (**PTO**), N,N-dimethyl-2-[2-(phenylmethyl) phenoxy]ethanamine; **promethazine** (**PRO**), $N,N,$α-trimethyl-10H-phenothiazine-10-ethanamine; **pyrilamine** (**PYR**), N-[(4-methoxyphenyl)methyl]-N',N'-dimethyl-N-2-pyridinyl-1,2-ethanediamine; **pyrrobutamine** (**PYB**), 1-[4-(4-chlorophenyl)-3-phenyl-2-butenyl]pyrrolidine; 1-[γ-p-(chlorobenzyl)cinnamyl]pyrrolidine; **terfenadine** (**TER**), α-[4-(1,1-dimethylethyl)phenyl]-4-(hydroxydiphenylmethyl)-1-piperidinebutanol; **thenyldiamine** (**THE**), N,N-dimethyl-N'-2-pyridinyl-N'-(3-thienylmethyl)-1,2-ethanediamine; **thonzylamine** (**THO**), N-[(4-methoxy

phenyl)methyl]-*N'*,*N'*-dimethyl-*N*-2-pyrimidinyl-1,2-ethanediamine monohydrochloride; **tolpropamine** (**TOL**),
N,*N*,4-trimethyl-γ-phenylbenzenepropanamine; **tripelennamine** (**TPE**), *N*,*N*-dimethyl-*N'*-(phenylmethyl)-*N'*-
2-pyridinyl-1,2-ethanediamine; **triprolidine** (**TPR**), (E)-2-[1-(4-methylphenyl)-3-(1-pyrrolidinyl)-1-propenyl]
pyridine; **tritoqualine** (**TTO**), 7-amino-4,5,6-triethoxy-3-(5,6,7,8-tetrahydro-4-methoxy-6-methyl-1,3-
dioxolo[4,5-g]isoquinolin-5-yl)-1(3*H*)-isobenzofuranone.

Chemical structures of the corresponding antihistamine drugs are shown in Figure 40.1.

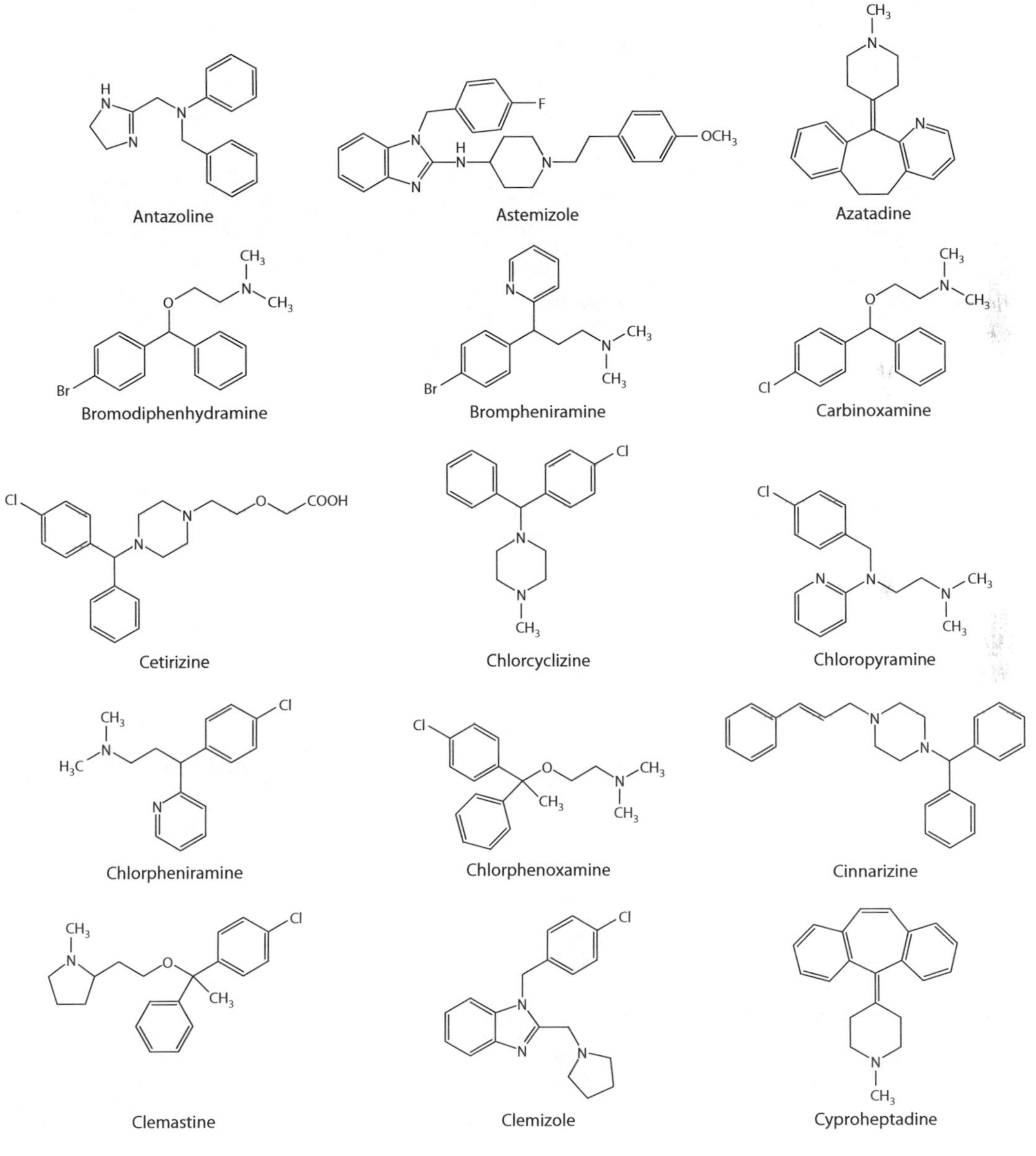

FIGURE 40.1   Chemical structures of antihistamines.

(continued)

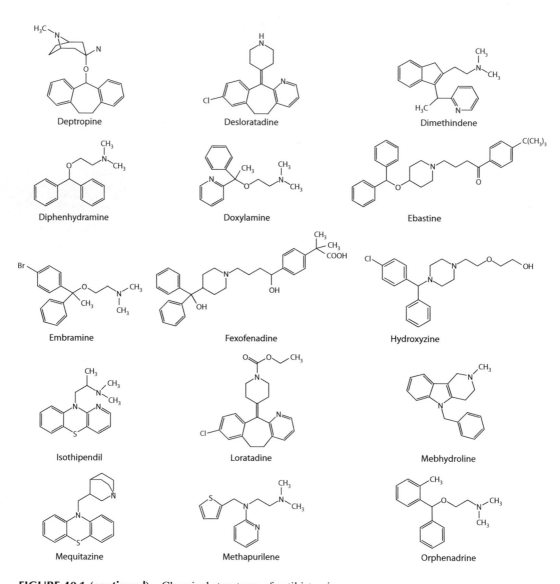

FIGURE 40.1 (continued) Chemical structures of antihistamines.

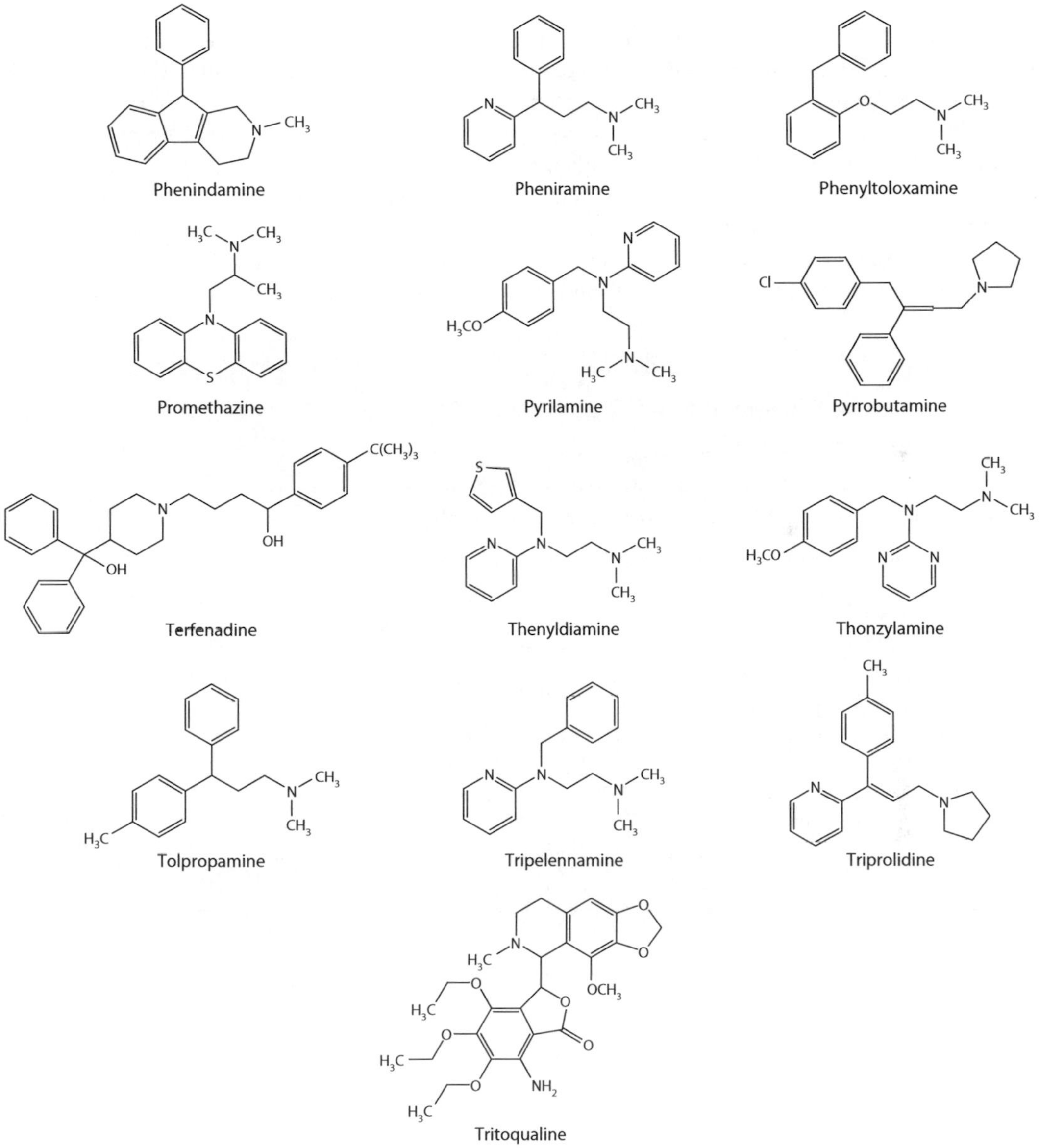

FIGURE 40.1 (continued) Chemical structures of antihistamines.

40.2 DETERMINATION OF SINGLE ANTIHISTAMINES

EBA (Ashok et al. 2003) was determined in pharmaceutical dosage forms, and **AST** (Mangalan et al. 1991), **CET** (Pandya et al. 1996), and **TPR** (DeAngelis et al. 1977) were determined in biological samples. Details of chromatographic conditions and the analytical results are shown in Tables 40.1 and 40.2. **CET** is currently marketed as a racemate, which consists of equal amounts of (R)-levocetirizine and (S)-dextrocetirizine. The H_1-antagonist activity of **CET** is primarily due to (R)-levocetirizine. Taha applied three methods for chiral discrimination of **CET** in the bulk drugs and the drug product using TLC, high-performance liquid chromatography (HPLC), and proton nuclear magnetic resonance spectroscopy (^{1}H-NMR) (Taha et al. 2009). They developed a TLC method based on the enantioseparation of **CET** on the TLC silica gel F_{254}–coated aluminum-backed plates, using different chiral selectors as mobile phase additives. The mobile phase enabling successful resolution was acetonitrile–water 17:3 (v/v) containing 1 mM of the chiral selector (hydroxypropyl-β-cyclodextrin, chondroitin sulfate, or vancomycin hydrochloride). The ascending development was performed, and chromatograms were visualized in the ultraviolet

TABLE 40.1
Chromatographic Separation Conditions Used for Antihistamine Drugs EBA, AST, CET, TPR, and PRO

Drug	Plates	Stationary Phase Pretreatment; Mobile Phase Composition	Development Mode	Detection (nm)	R_f	Reference
EBA	TLC silica gel $60F_{254}$ Aluminum-backed	Prewashing with methanol Methanol–n-hexane 10:1 (v/v)	Ascending	265	0.47	Ashok et al. (2003)
AST	TLC silica gel $60F_{254}$	n-Butanol–acetic acid–water 12:5:2 (v/v/v)	Chamber saturation 2 h; ascending	280	0.45	Mangalan et al. (1991)
CET	TLC silica gel $60F_{254}$	Chloroform–methanol 17:3 (v/v)	Ascending	232	0.31	Pandya et al. (1996)
TPR	TLC silica gel	Methanol–ammonia–chloroform 10:1:89 (v/v/v/v)	Ascending	405		DeAngelis et al. (1977)

TABLE 40.2
Analytical Parameters of the Methods Used for Separation of Antihistamines EBA, AST, CET, and TPR

Drug	Calibration Mode/ Range	LOD[a] Unit/Spot	LOQ[b] Unit/Spot	Precision (RSD%)	Recovery (%)	Matrix	Reference
EBA	Linear, 4–9 µg/spot	100 ng	300 ng		96.0	Tablets	Ashok et al. (2003)
AST	Linear, 0.1–1 µg/spot			2.07–12.37	86	Plasma	Mangalan et al. (1991)
CET	Linear, 0.05–5 µg/spot	50 ng		3.94–5.26	98.1–102.24	Plasma	Pandya et al. (1996)
TPR	Linear, 1–5 ng/spot		0.4 ng	14	83	Plasma	DeAngelis et al. (1977)

[a] Limit of detection.
[b] Limit of quantification.

(UV) light (254 nm) or by exposure to iodine vapors. The diminishing order of enantioselectivity was hydroxypropyl-β-cyclodextrin > chondroitin sulfate > vancomycin hydrochloride. This can be attributed to the size and geometry of **CET** relative to the chiral selectors. The R_f values were 0.46, 0.51, and 0.55 for levocetirizine and 0.76, 0.68, and 0.69 for dextrocetirizine, for the three selected chiral additives, respectively. The detection limit was found equal to 25 ng/spot for each enantiomer after the detection in iodine vapors.

40.3 SIMULTANEOUS IDENTIFICATION AND DETERMINATION OF ANTIHISTAMINES

Selectivity of the separation of **CET**, **ANT**, **DOX**, and **CPY** has been investigated on the silica gel $60F_{254}$ and the aluminum oxide $60F_{254}$ neutral plates with a variety of mobile phases in horizontal TLC chambers (Misztal and Paw 2001). Table 40.3 lists the R_f values for these mobile phases, which enabled the best separation results. The plates were visualized by illumination at 254 nm and by the reaction with different reagents (Dragendorff, potassium iodoplatinate, Benedicta, iodic, ferric(III) chloride with iodine). The highest detection sensitivity (0.06 µg, **CET**; 0.1 µg, **CPY**; 0.2 µg, **ANT**; and 0.2 µg, **DOX**) was obtained by spraying the plates with the potassium iodoplatinate reagent.

Transition metal ions are known to form complexes with molecules containing the electron-pair-donating atoms (e.g., nitrogen), which are present in antihistamines. Complex formation influences the chromatographic behavior of antihistamines on the plates impregnated with a metal ion (Bhushan et al. 1989, Srivastava and Reena 1982, Bhushan and Joshi 1996). **DHY**, **BDI**, **CPH**, and **CAR** were separated on the silica gel plates impregnated with various different metal salts (zinc acetate, manganese acetate, cadmium acetate, zinc sulfate, and zinc chloride) (Srivastava and Reena 1982). The best results (R_f: 0.77, **DHY**; 0.70, **BDI**; 0.47, **CPH**; and 0.30, **CAR**) were obtained on the zinc acetate (1%)–impregnated plates by using ethanol–dimethyl formamide–ammonia 16:5:0.1 (v/v/v) as a solvent system and employing Dragendorff's reagent for visualization. Separation of **CYP**, **MEB**, **PIR**, **ISO**, and **PRO** was performed on the TLC plates precoated with a mixture of silica gel G and the varying amounts of metal salts (Bhushan et al. 1989). The best resolution was obtained on the zinc acetate (1%)–impregnated plate with the solvent system benzene–dimethyl formamide–acetic acid 3:1:0.7 (v/v/v). Spots were visualized by spraying the plates with Dragendorff's reagent (R_f: 0.20, **CYP**; 0.02, **MEB**; 0.26, **PIR**; 0.35, **ISO**; and 0.64, **PRO**). The LODs of 200 ng/spot, **CYP**; 1000 ng/spot,

TABLE 40.3

R_f Values of Antihistamine Drugs CET, ANT, DOX, and CPY on Silica Gel and Aluminum Oxide Plates with Different Mobile Phases 1–4[a]

	Silica Gel $60F_{254}$	Aluminum Oxide $60F_{254}$		
Drug	**1**	**2**	**3**	**4**
CET	0	0	0	0
ANT	0.70	0.22	0.18	0.27
DOX	0.24	0.58	0.80	0.83
CPY	0.40	0.80	0.94	0.95

Source: Misztal, G. and Paw, B., *J. Planar Chromatogr.*, 14, 430, 2001.

[a] 1, benzene–1,4-dioxane–25% aqueous ammonia 10:9:1 (v/v/v); 2, chloroform–ethyl acetate 1:1 (v/v); 3, ethyl acetate; 4, butan-2-one–toluene 7:3 (v/v).

MEB; 1020 ng/spot, **PIR**; 300 ng/spot, **ISO**; and 30 ng/spot, **PRO** were obtained. Bushan and Joshi have separated **TPR**, **MEB**, **CYP**, **PRO**, **PIR**, and **DHY** on the plain silica gel plates (silica gel G with 13% calcium sulfate as a binder) and on those impregnated with metal ions Mn(II), Fe(II), Ni(II), and Cu(II) (Bhushan and Joshi 1996). Separations were performed at 35°C and the spots were detected with iodine vapors. Complete separation of all examined antihistamines was achieved with the solvent systems butanol–acetic acid–water 4:1:0.5 (v/v/v) on the plain plates and with butanol–acetic acid–water (4:1:1 v/v/v) on the plates impregnated with Mn(II).

The TLC method for the detection of 11 drugs (including **DHY**, **CIN**, and **PRO**) commonly having an amino group in the side chain of the molecule or/and basic nitrogen in a ring system was developed (Lang 1990). Separations were performed on the silica gel 60–coated aluminum-backed plates with two mobile phases: (A) 3% ammonia–methanol 7:18 (v/v) and (B) 20% ammonia–isopropanol–ethyl acetate 6:15:79 (v/v/v). The established R_f values (visualization with cobalt(II) thiocyanate and in the UV light, 366 nm) were 0.72 (**DHY**), 0.92 (**CIN**), and 0.67 (**PRO**) with mobile phase A and 0.76 (**DHY**) and 0.73 (**PRO**) with mobile phase B. The developed method permits identification of the examined drugs in biological fluids (urine or the stomach content).

The 2D TLC was applied for the separation of **CET**, **CCY**, **CIN**, and **HYD** (Muller and Ebel 1997). The computer-simulated 2D chromatograms and the optimization functions were used for the method development. Separation was performed on the three stationary phase types (the silica gel $60F_{254}$–coated aluminum-backed TLC plates, reversed-phase (RP) 18W high-performance thin layer chromatography (HPTLC) plates, and CN F_{254} HPTLC plates). The best separation was achieved on silica gel using ethyl acetate–1-propanol–water–ammonia (26%) 5:3:1:0.1 (v/v/v/v) for the first development and cyclohexane–triethylamine 7:3 (v/v) for the second development. The spots were examined in the UV light, at 254 nm.

40.4 DETERMINATION OF ANTIHISTAMINES IN THE PRESENCE OF OTHER SUBSTANCES

40.4.1 DETERMINATION OF ANTIHISTAMINES IN MULTICOMPONENT PHARMACEUTICALS

Binary mixtures of pseudoephedrine with **TPR** (Baseski and Sherma 2000), **BPH** (Greshock and Sherma 1997), **DHY** (Muller and Sherma 1999), and **DOX** with dextromethorphan hydrobromide (Indrayanto 1996) and ternary mixtures of **CPH**, phenylephrine, and methscopolamine nitrate (Greshock and Sherma 1997); **CPH**, phenylephrine, and paracetamol (Tuszynska et al. 1994); and **CPH**, phenylpropanolamine hydrochloride, and paracetamol (Tuszynska et al. 1994) were analyzed. Details of chromatographic conditions and the analytical results are shown in Tables 40.4 and 40.5.

TLC was used to separate, identify, and quantify **CPH** and **PIR**, when present in pharmaceutical preparations (tablets, capsules, injections, pediatric drops, eye- and eardrops, and syrups) with multiple components (caffeine, ephedrine, guaifenesin, phenylephrine, ibuprofen, paracetamol, diclofenac, and codeine) (Subramaniyan and Das 2004). Separation was performed on the silica gel F_{254}–coated aluminum-backed plates with cyclohexane–chloroform–methanol–diethylamine 4.5:4:0.5:1 (v/v/v/v/) as mobile phase. Ascending chromatography was applied and the plates were visualized at 260 nm. **CPH** (R_f 0.663) and **PIR** (R_f 0.663) were satisfactorily separated from the other drugs present in the formulations. Linear calibration curves were obtained within the range of 0.25–8 µg/spot for **CPH** and **PIR**. Recoveries of **CPH** and **PIR** were 100.09% ± 0.77% and 100.09% ± 0.87%, respectively.

40.4.2 DETERMINATION OF ANTIHISTAMINES IN THE PRESENCE OF PRESERVATIVES

TLC was applied for the determination of **LOR** in the presence of the preservatives, methylparaben and propylparaben (Indrayanto et al. 1999, Popovic et al. 2007), and sodium benzoate (Popovic et al. 2007). **LOR** was determined in the pharmaceutical dosage forms (tablets and syrup) on the silica gel F_{254}–coated HPTLC and TLC plates with chloroform–ethyl acetate–acetone 5:7:7 (v/v/v) as mobile phase, by the ascending development (Indrayanto et al. 1999). In syrup, determination of **LOR** was performed in the

TABLE 40.4

Chromatographic Separation Conditions Used for Antihistamine Drugs TPR, BPH, DOX, DHY, and CPH in Multicomponent Pharmaceuticals

Drug	Plates	Stationary Phase Pretreatment; Mobile Phase Composition	Development Mode	Detection (nm)	R_f	Reference
TPR	HPTLC GLP[a] silica gel 60F$_{254}$	Prewashing with dichloromethane–methanol 1:1; ethyl acetate–methanol–ammonia 17:2:1 (v/v/v)	Ascending	267	0.38–0.48	Baseski and Sherma (2000)
BPH	HPTLC silica gel 60F$_{254}$	Prewashing with dichloromethane–methanol 1:1; 1-butanol–water–glacial acetic acid 7:2:1 (v/v/v)	Ascending	265	0.30	Greshock and Sherma (1997)
DOX	TLC silica gel 60F$_{254}$	Methanol–25% ammonia 14:1 (v/v)	Ascending	223	0.70	Indrayanto (1996)
DHY	Silica gel 60F$_{254}$ GLP	Prewashing with dichloromethane–methanol 1:1; ethyl acetate–methanol–ammonia 17:2:1 (v/v/v)	Ascending chamber saturation 15 min	260	0.50–0.60	Muller and Sherma (1999)
CPH	HPTLC silica gel 60F$_{254}$	Prewashing with dichloromethane–methanol 1:1; 1-butanol–water–glacial acetic acid, 7:2:1 (v/v/v)	Ascending	264	0.30	Greshock and Sherma (1997)
CPH	TLC silica gel F$_{254}$	Ethyl acetate–methanol–6% acetic acid 5:3:2 (v/v/v)	Not specified	264		Tuszynska et al. (1994)
CPH	TLC silica gel 60F$_{254}$	Ethyl acetate–methanol–conc. acetic acid 5:3:2 (v/v/v)	Not specified	264		Tuszynska et al. (1994)

[a] Good laboratory practice.

TABLE 40.5

Analytical Parameters of the Methods Used for Separation of Antihistamine Drugs TPR, BPH, DOX, DHY, and CPH in Multicomponent Pharmaceuticals

Drug	Calibration Mode/Range	LOD Unit/Spot	LOQ Unit/Spot	Precision (RSD%)	Recovery (%)	Matrix	Reference
TPR	Polynomial, 0.200–0.800 µg/spot			1.5	98.5	Tablets	Baseski and Sherma (2000)
BPH	Linear, 0.420–1.68 µg/spot			0–4.9	99.4–100.6	Tablets	Greshock and Sherma (1997)
DOX	Linear, 1.5–7.4 µg/spot	140 ng	462 ng	1.81	98.56	Syrup	Indrayanto (1996)
DHY	Linear, 2–8 µg/spot			1.7–1.9	99.19–100	Tablets, capsules	Muller and Sherma (1999)
CPH	Linear, 0.400–1.60 µg/spot			0–4.9	99.08–100.9	Tablets	Greshock and Sherma (1997)
CPH	Linear, 2–6 µg/spot			0.63	100	Tablets	Tuszynska et al. (1994)
CPH	Linear, 1–10 µg/spot			2.15	94.5	Tablets	Tuszynska et al. (1994)

presence of two preservatives, methylparaben and propylparaben. Separation on the HPTLC and TLC plates resulted in almost identical densitograms. **LOR** (R_f 0.51) was well separated from the preservatives (R_f 0.62; preservatives appeared as one peak). Quantitative evaluation of **LOR** was performed at 250 nm. The method was validated for linearity (0.28–0.80 μg/spot), LOD (0.029 μg/spot), LOQ (0.087 μg/spot), accuracy (recovery 98.3%–101.2%), and precision (RSD 1.8%–1.9%).

Simultaneous determination of **LOR** and the preservatives in **LOR**–sodium benzoate and **LOR**–methylparaben–propylparaben mixtures was performed on the silica gel 60 F_{254}–coated aluminum-backed plates (Popovic et al. 2007). The choice of *n*-butyl acetate–carbon tetrachloride–acetic acid–acetonitrile 3:6:0.2:3 (v/v/v/v) as mobile phase enabled satisfactory resolution of **LOR** (R_f 0.47) and sodium benzoate (R_f 0.60) (Figure 40.2). The best resolution of **LOR** (R_f 0.69), methylparaben (R_f 0.30), and propylparaben (R_f 0.38) was achieved with the solvent system ethyl acetate–*n*-hexane–methanol–ammonia–diethylamine 1:4:0.8:0.4:2 (v/v/v/v/v) (Figure 40.3). Ascending chromatography was applied, and the zones were scanned at 240 nm (**LOR**–sodium benzoate) and 275 nm

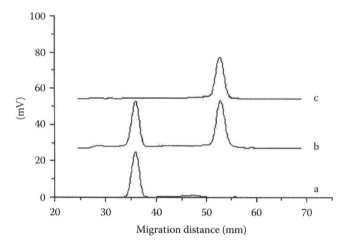

FIGURE 40.2 Densitograms of (a) LOR, (b) Claritin® syrup, and (c) sodium benzoate. Mobile phase: *n*-butyl acetate–carbon tetrachloride–acetic acid–acetonitrile, 3:6:0.2:3 v/v/v/v. Wavelength 240 nm. (Reprinted with permission from Kowalska, T., *Acta Chromatogr.*, 19(19), 161, 2007.)

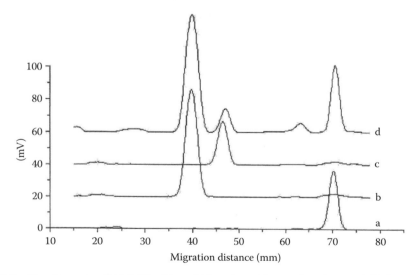

FIGURE 40.3 Densitograms of (a) LOR, (b) methylparaben, (c) propylparaben, and (d) Loratadin® syrup. Mobile phase: ethyl acetate–*n*-hexane–methanol–ammonia–diethylamine, 1:4:0.8:0.4:2 v/v/v/v. Wavelength 275 nm. (Reprinted with permission from Kowalska, T., *Acta Chromatogr.*, 19(19), 161, 2007.)

(**LOR**– methylparaben–propylparaben). Linear calibration curves were obtained within the range of 0.3–0.7 µg/band of LOR. For the mixture of **LOR**–sodium benzoate, the mean recovery of **LOR** was 100.2% (RSD, 1.98%), and LOD, 30 ng/band, and LOQ, 100 ng/band, were obtained. For the mixture of **LOR**–methylparaben–propylparaben, the analogical values were as follows: recovery, 100.8% (RSD, 2.04%); LOD, 10 ng/band; and LOQ, 40 ng/band. The method was used for the simultaneous determination of **LOR** and the preservatives in the commercial medicinal syrups.

40.5 TLC STABILITY STUDIES OF ANTIHISTAMINES

Stability testing provides evidence that the quality of drug substance or drug product changes with time under the influence of various environmental conditions. The International Conference on Harmonization (ICH) guideline requires that stress testing be carried out to elucidate the inherent stability characteristics of an active substance in pharmaceutical preparation. The light testing should be an integral part of the stress testing. The progress in this field has been aided by the advances in procedures available for the separation and identification of the mixture components.

The TLC methods were developed for quantification of **DES** (Sumarlik et al. 2005), **CIN** (Hassan et al. 2002), **FEX** (Bhalekar et al. 2010), and **CET** (Makhija and Vavia 2001) in the presence of their degradation products and/or the related substances and/or the metabolites. Details of chromatographic conditions and the analytical results are summarized in Tables 40.6 and 40.7. To ensure the selectivity of the method for the determination of **DES** (Sumarlik et al. 2005), the forced degradation studies with use of HCl, NaOH, and H_2O_2 were carried out with the laboratory-made tablets. Selectivity of the method was proved by identification and the purity checks of the analyte spots. **CIN** (Hassan et al. 2002) was determined in the presence of the related substances and the acid degradation products. Four **CIN**-related substances (R_f 0.45, 0.23, 0.18, and 0.13), four water-insoluble acid degradation products (R_f 0.87, 0.82, 0.46, and 0.13), and two water-soluble acid degradation products (R_f 0.26 and 0.13) were isolated and identified by the infrared and mass spectrometry (MS). The effect of UV light on the photostability of **FEX** in its solid state was monitored by the HPTLC analysis (Bhalekar et al. 2010). Concentration of **FEX** was reduced to 57.04% after an exposure to 1800 W h/m^2 UV light. The photodegradation products were not detected upon the multiwavelength scanning of the plate in the range from 200 to 400 nm, although the peak area for the drug diminished. The stability-indicating HPLC method was developed for the simultaneous estimation of **CET** and pseudoephedrine (Makhija and Vavia 2001). The drugs were subjected to

TABLE 40.6
Chromatographic Separation Conditions Used for Antihistamine Drugs DES, CIN, FEX, and CET in Stability Studies

Drug	Plates	Stationary Phase Pretreatment; Mobile Phase Composition	Development Mode	Detection (nm)	R_f	Reference
DES	TLC silica gel 60F$_{254}$ aluminum-backed	Ethyl acetate–n-butanol–25% aqueous ammonia–methanol 21:5:4:5 (v/v/v/v)	Ascending	279	0.62	Sumarlik et al. (2005)
CIN	Whatman K6 F silica gel	Benzene–methanol–formic acid 80:17:3 (v/v/v)	Ascending	250	0.4	Hassan et al. (2002)
FEX	TLC silica gel 60F$_{254}$ aluminum-backed	Ethyl acetate–methanol–ammonia (30%) 6:3.5:0.5 (v/v/v)	Ascending	220	0.43	Bhalekar et al. (2010)
CET	TLC silica gel 60F$_{254}$ aluminum-backed	Ethyl acetate–methanol–ammonia (20%) 7:1.5:1 (v/v/v)	Ascending	240	0.38	Makhija and Vavia (2001)

TABLE 40.7

Analytical Parameters of the Methods Used for Separation of Antihistamines DES, CIN, FEX, and CET in Stability Studies

Drug	Calibration Mode/Range	LOD Unit/ Spot (ng)	LOQ Unit/ Spot (ng)	Precision (RSD%)	Recovery (%)	Matrix	Reference
DES	Linear, 1.5–5 µg/spot	96.8	290	0.45–1.58	100.3	Tablets	Sumarlik et al. (2005)
CIN	Linear, 0.08–0.6 µg/spot	16	48	1.3	98.6	Tablets, blood serum	Hassan et al. (2002)
FEX	Linear, 0.2–1 µg/spot	24.91	75.50	<2	100.1	Tablets	Bhalekar et al. (2010)
CET	Linear, 0.2–1.2 µg/spot	0.5	0.8	1.21–1.52	99.7–101.23	Tablets	Makhija and Vavia (2001)

forced degradation under the acidic condition (1 M HCl), base condition (1 M NaOH), and oxidation (H_2O_2) by heating at 70°C for 2 h. The chromatogram of the acid- and base-degraded samples showed the spots of the pure drugs only (R_f 0.38, **CET**; 0.69, pseudoephedrine), while the samples degraded with hydrogen peroxide showed the additional peaks at R_f 0.28 and 0.75 for the **CET** and pseudoephedrine degradation products, respectively.

A TLC/second-derivative spectrophotometric procedure has been developed for low levels of PRO sulfoxide (photooxidation **PRO** product) in the **PRO** raw material and in formulations (syrup, tablet, and injection) (Fadiran and Davidson 1988). Separation was performed on the silica gel $60F_{254}$–coated plastic-backed TLC plates. The best resolution was obtained by using cyclohexane–acetone–diethylamine 8:1:1 (v/v/v) as mobile phase, and the obtained R_f values were 0.09 and 0.42 for PRO sulfoxide and **PRO**, respectively. Spots were located by exposing the plates to UV light (254 nm). Two methods of eluting PRO sulfoxide from the chromatographic plate were investigated. The first method involved scraping off an area of silica gel containing the sulfoxide spot and the second one involved cutting off an area of the plastic-backed plate with the sulfoxide spot. After extraction, quantification of sulfoxide was performed by second-derivative spectrophotometry.

In addition to the HPLC stability study of LOR, TLC was also applied to the fractionation of the UV-degraded drug (Abounassif et al. 2005). Utilizing the silica gel $60F_{254}$ sheets (developed with chloroform–methanol–ammonia 8:2:0.2 v/v/v and visualized in the UV light at 254 and 366 nm), three photodegradation products were revealed more polar than **LOR**, that is, with the respective R_f values lower than the parent compound. Two out of these three photodegradates were fluorescent, with the most polar one demonstrating higher fluorescence.

TLC was used to evaluate the purity of the two **FEX** photodegradation products prior to their characterization by NMR and MS (Breier et al. 2008). TLC analysis was performed on the silica gel $60F_{254}$–coated aluminum sheets with chloroform–methanol 8.5:1.5 (v/v) as mobile phase.

40.6 TLC-COUPLED TECHNIQUES FOR SEPARATION AND CHARACTERIZATION OF ANTIHISTAMINES

The electrospray-assisted laser desorption/ionization mass spectrometry (ELDI-MS) was used to characterize active ingredients (**CPH**, DL-methylephedrine, caffeine, ethoxybenzamide, noscapine, and acetaminophen) in the anticold tablets directly from the TLC plates (Lin et al. 2007). Separation was performed in the normal-phase (NP) mode on the silica gel–coated TLC plates with ethyl acetate–acetic acid–dichloromethane 98:1:1 (v/v/v) as mobile phase. Figure 40.4a displays the photograph of the TLC plate taken after separation of these components. Figure 40.4b and c present their respective mass spectra. The innately high spatial resolution and scanning capability of

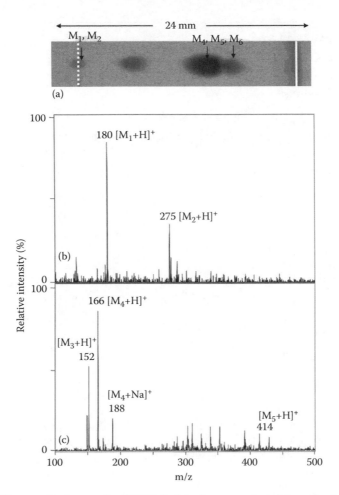

FIGURE 40.4 (a) Photographic image of an NP TLC plate after separation of the active ingredients extracted from an anticold tablet. (b and c) Background-subtracted, averaged positive-ion mass spectra recorded at locations in the chromatogram corresponding to the respective spot positions on the TLC plate: (b) methylephedrine hydrochloride (M_1) and CPH maleate (M_2), (c) acetaminophen (M_3), ethoxybenzamide (M_4), and noscapine (M_5). (Adapted with permission from Lin, S.Y., Huang, M.Z., Chang, H.C., and Shiea, J., Using electrospray-assisted laser desorption/ionization mass spectrometry to characterize organic compounds separated on thin-layer chromatography plates, *Anal. Chem.*, 79, 8789–8795, 2007. Copyright 2012 American Chemical Society.)

the laser beam makes a direct coupling of TLC with ELDI a useful analytical technique for rapid separation and characterization of organic compounds.

Preparative TLC was applied to the separation of N-nitroso derivative of TPE formed in the rats after oral administration of drug. Ethyl acetate–methanol 1:1 (v/v) was used for the development of chromatograms on the silica gel F_{254} layer (Rao et al. 1975). The band corresponding to N-nitroso derivative (R_f 0.7) was scraped off and used for further mass spectrometric identification.

40.7 TLC IN SCREENING OF ANTIHISTAMINES

A comprehensive drug identification system using TLC for preliminary screening and GC/MS for confirmation was developed (Lillsunde and Korte 1991). This system is useful for the screening of the samples in misuse, impaired driving, poisoning, and the other forensic cases. About 300 drugs, including antihistamine drugs (**ANT, BPH, CCY, CIN, CYP, DIM, DHY, HYD, PIR, PRO, TPE,** and **TPR**), can be simultaneously detected with this method. The TLC separations were performed

TABLE 40.8
TLC Data of Antihistamine Obtained with Mobile Phases 1 and 2[a] and Visualized with UV and Different Spray Reagents[b]

			Spray Reagent			Mobile Phase R_f	
Drug	A	B	C/UV/H$_2$SO$_4$/HNO$_3$	D	E	1	2
ANT	bl	br/br	−/bk/−/r	o		0.20	0.24
BPH		br/br				0.30	0.35
CCY		br/br				0.53	
CIN		br/br				0.94	
CYP		y/y				0.47	
DIM		br/br			v	0.35	0.45
DHY		br/br	−/−/y/−			0.55	
HYD		br/bl				0.50	
PIR		br/br		r		0.25	0.45
PRO	r	br/br	r/bk/gn/−	br		0.47	0.51
TPE		br/br				0.40	
TPR		br/br	−/bl/−/−			0.32	

Source: Lillsunde, P. and Korte, T., *J. Anal. Toxicol.*, 15, 71, 1991.

Color codes are: bl = blue, br = brown, bk = black, o = orange, y = yellow, v = violet, r = red, gn = green.

[a] 1, ethyl acetate–methanol–water–ammonia 43:5:1.5:0.5 (v/v/v/v); 2, methanol–ammonia 50:0.5 (v/v).

[b] A, furfuraldehyde–H$_2$SO$_4$; B, Dragendorff's reagent–iodoplatinate; C, H$_2$SO$_4$–FeCl$_3$; D, Fast Black K Salt–NaOH; E, H$_2$SO$_4$–NaNO$_2$–H$_2$NSO$_3$NH$_4$–N-(1-naphthyl)ethylenediamine.

on the aluminum-backed silica gel 60 and 60F$_{254}$ plates and with various mobile phases. The most effective separations of antihistamines were achieved with ethyl acetate–methanol–water–ammonia 43:5:1.5:0.5 (v/v/v/v) and methanol–ammonia 50:0.5 (v/v). The plates were first examined in UV light (366 and 254 nm) and then sprayed with different reagents. The best results for antihistamines were obtained with UV, furfuraldehyde–H$_2$SO$_4$, Dragendorff's reagent, iodoplatinate, H$_2$SO$_4$–FeCl$_3$, concentrated H$_2$SO$_4$, concentrated HNO$_3$, Fast Black K Salt–NaOH, and H$_2$SO$_4$–NaNO$_2$–H$_2$NSO$_3$NH$_4$–N-(1-naphthyl)ethylenediamine. The R$_f$ values and colors developed with different visualization spray reagents are presented in Table 40.8.

An off-line thin layer chromatographic/electron impact mass spectrometry (TLC-EIMS) approach for the screening of biological samples, for example, urine, tissues (lung, liver, kidney), and the stomach content, for the drugs and their metabolites was proposed (Brzezinka et al. 1999). A library of the R$_f$ values and the corresponding EI mass spectra for 493 compounds was developed for the routine toxicological and forensic applications. The silica gel F$_{254}$–precoated TLC plates and six mobile phases were used for the determination of the R$_f$ values. The plates were pretreated with a solution of methanol–1.5% ammonia and reactivated for 2 h at 150°C. The best separations of the antihistamine drugs were achieved with methanol–25% ammonia 100:1.5 (v/v), cyclohexane–toluene–diethylamine 75:15:10 (v/v/v), and chloroform–methanol 90:10 (v/v) as mobile phases. Experimentally determined hR$_f$ values were converted to the corrected values (hR$_f$c) by the graphical standardization procedure (Stead et al. 1982). The hR$_f$c values of 15 antihistamine drugs are listed in Table 40.9.

Based on 8 years of analytical experience, Stead et al. (1982) proposed the standardized TLC systems and a large bank of the hR$_f$ values for identification of drugs and poisons. The 794 drugs chosen

TABLE 40.9

hR$_f$c[a] Measured Using Mobile Phases 1–3[b] for Antihistamine Drugs

Drug	Mobile Phase hR$_f$c		
	1	2	3
CPO	53	47	36
DOX	48	41	10
ORP	55	48	33
DHY	55	45	33
TPE	55	44	27
THE	53	42	25
MET	52	43	26
PRO	50	37	35
CLA	46	48	25
PIR	45	35	13
CIN	76	51	78
TPR	51	39	20
PYB	54	55	37
CLI	78	31	69
PDA	63	45	57

Source: Brzezinka, H. et al., *J. Planar Chromatogr.*, 12, 96, 1999.

[a] R$_f$c × 100.

[b] 1, methanol–25% ammonia 100:1.5 (v/v); 2, cyclohexane–toluene–diethylamine 75:15:10 (v/v/v); 3, chloroform–methanol 90:10 (v/v).

for this study were those that are the most frequently prescribed, abused, or involved in the accidental, suicidal, or homicidal poisonings. The reaction to the selected spray reagents and the use of up to four systems were applied to provide the greatest discrimination between possible compounds. Although the systems used in this study were selected for the general screening purposes, they can also be used for the separation and identification of drugs belonging to discrete pharmacological or chemical groups, for example, antihistamines, local anesthetics, CNS stimulants, sulfonamides, and benzo-diazepines. Eight standardized solvent systems were used (four for the basic drugs and four for the acidic and neutral drugs) for the separation of drugs on the silica gel 60F$_{254}$–precoated plates. The list of hR$_f$ values for 29 antihistaminic drugs obtained with the four mobile phases, (1) methanol–ammonia 100:1.5 (v/v), (2) cyclohexane–toluene–diethylamine 75:15:10 (v/v/v), (3) chloroform–methanol 9:1 (v/v), and (4) acetone, is shown in Table 40.10. The measurement of the R$_f$ values is prone to systematic errors, due to the difficulties in reproducing the working conditions of the TLC systems on a day-to-day basis and from one laboratory to another. To compensate for this irreproducibility, the reference compounds run at the same time as the test drugs can be used to convert practically obtained hR$_f$ values to the corrected hR$_f$c values by a graphical method.

Principal component analysis (PCA) has a great potential for the identification of drugs on the basis of their R$_f$ values in different eluents. PCA has significant advantages over the statistical methods based on the information provided by single systems, due to the fact that it enables direct measure of each system's properties in combination with other systems and indicates both the minimum set of eluents needed and reliable statistical criteria for their selection.

TABLE 40.10
hR$_f$^{a} Values for Antihistamine
Drugs in Four TLC Systems

	Mobile Phase^b hR$_f$			
Drug	1	2	3	4
ANT	37	7	7	3
BDI	54	44	43	13
BPH	45	33	16	6
CCY	57	42	46	14
CPH	45	33	18	2
CPO	53	47	36	17
CIN	46	51	78	69
CLA	46	48	25	9
CYP	51	45	44	13
DEP	13	24	4	1
DIM	42	36	13	6
DHY	55	45	33	15
DOX	48	41	10	9
EMB	54	50	32	17
ISO	52	41	30	14
MEB	57	28	45	20
MEQ	10	6	6	0
MET	52	43	26	15
ORP	55	48	33	16
PDA	63	45	57	21
PIR	45	35	13	3
PTO	53	39	48	15
PRO	50	37	35	17
PYB	54	55	37	18
THE	53	42	25	12
THO	55	38	28	14
TOL	51	52	32	15
TPE	55	44	27	15
TPR	51	39	20	6

Source: Stead, A.H. et al., *Analyst*, 107, 1106, 1982.

^{a} R$_f$ × 100.

^{b} 1, methanol–ammonia 100:1.5 (v/v); 2, cyclohexane–toluene–diethylamine 75:15:10 (v/v/v); 3, chloroform–methanol 9:1 (v/v); 4, acetone.

PCA was applied (Musumarra et al. 1984) to the earlier reported R$_f$ data for 596 basic and neutral drugs in the four different eluent mixtures (Stead et al. 1982). The following conclusions were drawn: (A) The number of the variables can be reduced to the principal component scores (the θ_1 and θ_2 values), which characterize the component in a 2D space. This possibility represents great advantage for identification of the unknowns over the earlier described approach based on the four R$_f$ values. (B) The PCA model allows for a drastic reduction of the inquiry range to the few candidates only, although an ambiguous identification could not be achieved. (C) The eluents methanol–ammonia 100:1.5 (v/v), chloroform–methanol 9:1 (v/v), and acetone provide an approximately identical information, while the eluent cyclohexane–toluene–diethylamine 75:15:10 (v/v/v) provides a distinct information. A significant improvement of this method could be obtained by choosing the appropriate eluent mixtures, each providing an independent information.

PCA was applied to the identification of 805 drugs and their metabolites (including 23 antihistamine drugs) in biological samples on the basis of the respective R_f values obtained in the four eluent systems on the silica gel F_{254}–precoated HPTLC plates (Musumarra et al. 1985, Romano et al. 1994). The eluent composition was (1) ethyl acetate–methanol–30% ammonia 85:10:15 (v/v/v), (2) cyclohexane–toluene–diethylamine 65:25:10 (v/v/v), (3) ethyl acetate–chloroform 50:50 (v/v), and (4) acetone. Before use of eluent 4, the chromatographic plates were dipped in potassium hydroxide solution. The experimentally determined hR_f values were converted to corrected values ($hR_f c$). The list of the $hR_f c$ values for 23 antihistaminic drugs is given in Table 40.11.

TABLE 40.11

$hR_f c^a$ Values for Antihistamine Drugs in Four TLC Systems

	Mobile Phase[b] $hR_f c$			
Drug	1	2	3	4
AST	75	3	0	49
BPH	66	42	0	7
CPH	63	41	0	7
CIN	91	61	41	83
CLA	79	56	0	13
CYP	79	51	1	22
DIM	63	41	0	8
DHY	76	51	0	27
DOX	65	45	0	8
HYD	62	13	0	48
ISO	76	49	0	25
MEB	77	37	1	33
MEQ	35	12	0	4
ORP	79	53	2	29
PDA	81	52	4	4
PIR	61	41	0	6
PTO	77	43	4	38
PRO	74	44	2	28
PYR	70	42	0	30
TER	83	19	1	58
THE	76	47	0	24
TPE	73	52	0	34
TTO	88	35	67	83

Source: Romano, G. et al., *J. Planar Chromatogr.*, 7, 233, 1994.

[a] $R_f c \times 100$.

[b] 1, ethyl acetate–methanol–30% ammonia 85:10:15 (v/v/v); 2, cyclohexane–toluene–diethylamine 65:25:10 (v/v/v); 3, ethyl acetate–chloroform 50:50 (v/v); 4, acetone.

TABLE 40.12

Individual hR_fc^a Values, L_{95}^b Values of Antihistamine Drugs

Drug	RPTLC hR_fc	RPTLC L_{95}	TLC hR_fc	TLC L_{95}	RPTLC/TLC L_{95}
AST	19	15	40	12	3
CIN	15	12	78	4	1
CLA	4	8	30	14	2
DHY	20	14	39	11	3
HYD	20	14	41	12	3
ORP	13	15	45	10	2
PIR	76	6	18	9	1
PRO	11	17	41	12	3
TPR	64	8	25	13	4

Source: Ojanpera, I. et al., *J. Liq. Chromatogr.*, 14, 1435, 1991.

[a] $R_fc \times 100$.
[b] List length value with 95% cumulative probability.

A combined use of the normal and RP TLC was evaluated in drug screening by the mean list length (MLL) method (Ojanpera et al. 1991). An RP system (the RP-18 F_{254}s plates and methanol–water–conc. HCl 50:50:1, v/v/v) was shown as an effective complementary counterpart to the main medium-polar NP system (silica gel 60F$_{254}$ and toluene–acetone–ethanol–ammonia 45:45:7:3, v/v/v/v). The plates were viewed in the UV light, at 254 and 366 nm. Table 40.12 shows the R_fc and L values (list length) for nine antihistaminic drugs (out of the 141 drugs). The L value for a given drug is the number of other drugs, in addition to the considered drug itself that would qualify for identification with a certain cumulative probability level. The MLL value is obtained by averaging the individual L values (Schepers et al. 1983). The shorter the MLL, the better is the given chromatographic system.

The impact of an elevated temperature (33°C–38°C) and a high relative humidity (80%–100%) on applicability of the TLC systems in analytical toxicology for the screening of 64 drugs (including **DHY, HDY, PIR**, and **PRO**) was studied during a 6-month climatologic cycle in Jakarta, Indonesia (De Zeeuw et al. 1994). In general, the R_f values (as measured from the TLC plates) were considerably affected in comparison with those obtained in the moderate climate. Under tropical conditions, most substances characterized with higher R_f values than in the moderate climate, although certain exceptions might also occur. Tropical conditions exerted a negative effect on the reproducibility of the R_f values also. When the graphical standardization procedure was, however, applied to the measured R_f values (making use of the reference mixture of standard drugs developed on each chromatographic plate), the accuracy and reproducibilites of the resulting R_fc values were dramatically improved, and thus, the corrected data were found as compatible with those in the already existing TLC databases developed under moderate climatic conditions. These observations remain in conformity with those based on the results from earlier studies carried out in a relatively dry tropical climate.

40.8 TLC TESTS IN PHARMACOPOEIAL REGULATIONS OF ANTIHISTAMINES

USP31 and EP7.04 pharmacopoeias regulate application of TLC to identification, testing for the related substance, or for purity of antihistamines. Tables 40.13 and 40.14 present the details of these tests.

TABLE 40.13

EP7.04 Pharmacopoeial Requirements for Chromatographic Identification and Testing of Antihistamines and/or Related Substances

Drug	Form of Antihistamine	Stationary Phase	Mobile Phase Composition	Detection; Spraying Reagent	Test
TER	TER	TLC silica gel F_{254} plate	Methanol–methylene chloride 10:90 v/v	254 nm	Identification
PIR	PIR maleate	TLC silica gel F_{254} plate	Water–anhydrous formic acid–methanol–diisopropyl ether 3:7:20:70 v/v/v/v	254 nm	Identification
HYD	HYD hydrochloride	TLC silica gel G plate	Concentrated ammonia–ethanol 96%–toluene 1:24:75 v/v/v	Potassium iodobismuthate solution	Identification
DEP	DEP citrate	Silica gel with a fluorescent indicator	Concentrated ammonia–butanol 8:92 v/v	254 nm; Potassium iodobismuthate solution; sodium nitrite solution; iodine vapors	Related substances
CLA	CLA fumarate	TLC silica gel G plate	Water–anhydrous formic acid–diisopropyl ether 5:25:70 v/v/v	Potassium permanganate solution 16 g/L	Identification
CLA	CLA fumarate	TLC silica gel G plate	Concentrated ammonia–methanol–tetrahydrofuran 1:20:80 v/v/v	Potassium iodobismuthate solution–dilute acetic acid 1:10 v/v; dilute hydrogen peroxide solution	Related substances
CIN	CIN	TLC octadecylsilyl silica gel F_{254} plate	Sodium chloride 58.4 g/L–methanol–acetone 20:30:50 v/v/v	254 nm	Identification
CCY	CCY hydrochloride	Silica gel	Concentrated ammonia–methanol–methylene chloride 2:13:85 v/v/v	Iodine vapor	Related substances
CET	CET dihydrochloride	TLC silica gel GF_{254} plate	Ammonia–methanol–methylene chloride 1:10:90 v/v/v	254 nm	Identification
BPH	BPH maleate	TLC silica gel F_{254} plate	Water–anhydrous formic acid–methanol–diisopropyl ether 3:7:20:70 v/v/v/v	254 nm	Identification

40.9 SAR AND QSAR OF ANTIHISTAMINES

An SAR study of the H_1-, H_2-, and H_3-antihistamine activity was carried out, and chromatographic data were obtained for the selected antihistamine drugs. The SAR studies have been performed using the chromatographic data obtained in the experiments with the NP silica gel $60F_{254}$ and the RP2 $60F_{254}$ (silanized silica gel) layers impregnated with a solution of aspartic acid and its analogue, propionic acid. Aspartic acid is a joint element of the simulated environment interaction between the various antihistamine compounds on the one hand and the biological goal (e.g., the H_1-, H_2-, and H_3-receptors) on the other. Chromatography was carried out using two eluents, acetonitrile–methanol–ammonium acetate buffer pH 7.4 (4:4:2 v/v/v) and acetonitrile–methanol–methylene chloride–ammonium acetate buffer pH 7.4 (6:1:1:2 v/v/v/v). The four proposed

TABLE 40.14
USP31 Pharmacopoeial Requirements for the Chromatographic Identification and Testing of Related Substances and Purity of Antihistamines

Drug	Form of Antihistamine; Preparation	Stationary Phase	Mobile Phase Composition	Detection; Spraying Reagent	Test
ANT	ANT phosphate	Silica gel	Ethyl acetate–methanol–diethylamine 17:2:1 v/v/v	Short-wavelength UV light	Chromatographic purity
AZA	AZA maleate	Silica gel	Toluene–isopropyl alcohol–diethylamine 10:10:1 v/v/v	Short-wavelength UV light	Chromatographic purity
BDI	BDI hydrochloride and codeine phosphate oral solution	Silica gel	Alcohol–ammonium hydroxide 49:1 v/v	Short-wavelength UV light	Identification
CLA	CLA fumarate; CLA fumarate tablets	Silica gel	Diisopropyl ether–formic acid–water 70:25:5 v/v/v	0.1 M Potassium permanganate	Identification
HYD	HYD pamoate	Silica gel	0.1 M hydrochloric acid	Potassium iodoplatinate	Identification
PIR	Naphazoline hydrochloride and PIR maleate ophthalmic solution	Silica gel	Methanol–water–acetic acid 8:1:1 v/v/	Ninhydrin TS	Identification
PRO	PRO hydrochloride	Silica gel	Ethyl acetate–acetone–alcohol–ammonium hydroxide 90:45:2:1 v/v/v/v	Short-wavelength UV light	Related substances
PYR	PYR maleate	Silica gel	Ethyl acetate–diethylamine–n-hexane–methanol 93:7:1:1 v/v/v/v	Short-wavelength UV light	Related substances
TPR	TPR hydrochloride	Silica gel	Chloroform–diethylamine 95:5 v/v	Long- and short-wavelength UV light	Chromatographic purity

discriminant models based on biochromatographic studies proved an efficient tool in the SAR analysis aiming to preliminarily predict the direction of compound activity with the histamine receptors (Brzezinska and Koska 2006).

A quantitative structure–activity relationship (QSAR) between the H_1-antihistamine activity and the TLC data has been studied for the recently synthesized thiazole and benzothiazole derivatives with an antihistamine activity. TLC was performed in the NP mode on the precoated silica gel 60F$_{254}$ plates (Brzezinska et al. 2003a, Brzezinska and Koska 2004) and in the RP mode on the precoated RP2 60F$_{254}$ plates (Brzezinska et al. 2003b, Brzezinska and Koska 2003), impregnated with solutions of selected amino acids. The two developing solvents used were acetonitrile–methanol–ammonium acetate buffer pH 7.4 (4:4:2 v/v/v) and acetonitrile–methanol–methylene chloride–ammonium acetate buffer pH 7.4 (6:1:1:2 v/v/v/v). Using regression analysis, the relationships between chromatographic and biological activity data were found. Correlations obtained in the experiment performed in the NP mode are more significant than those obtained in the RP mode, due to the optimal fitting of the chromatographic system conditions to the solutes lipophilicity in the former mode. Some of the obtained correlation equations can be used to predict pharmacological activity of the new drug candidates.

ACKNOWLEDGMENT

This work was supported by the Ministry for Education and Science of the Republic of Serbia, contract #172033.

REFERENCES

Abounassif, M.A., El-Obeid, H.A., and Gadkariem, E.A. 2005. Stability studies on some benzocycloheptane antihistaminic agents. *J. Pharm. Biomed. Anal.* 36:1011–1018.

Ashok, P., Meyyanathan, S.N., Pandilla, B., and Suresh, B. 2003. Analysis of ebastine in pharmaceutical preparations by high-performance thin-layer chromatography. *J. Planar Chromatogr.* 16:167–169.

Baseski, H.M. and Sherma, J. 2000. Quantification of triprolidine hydrochloride and methscopolamine nitrate in pharmaceutical tablets by HPTLC with ultraviolet absorption densitometry. *J. Planar Chromatogr.* 13:16–19.

Beale, J.M. and Block, J.H. 2011. *Wilson and Gisvold's Textbook of Organic Medicinal and Pharmaceutical Chemistry.* Philadelphia, PA: Lippincott Williams & Wilkins.

Bhalekar, M.R., Shete, T.K., Damle, M.C., and Madgulkar, A.R. 2010. Solid state photodegradation study of fexofenadine hydrochloride. *Anal. Lett.* 43:406–416.

Bhushan, R. and Joshi, S. 1996. TLC separation of antihistamines on silica gel plates impregnated with transition metal ions. *J. Planar Chromatogr.* 9:70–72.

Bhushan, R., Reena, Chauhan, R.S. 1989. Separation of some antihistamines on impregnated TLC silica plates. *Biomed. Chromatogr.* 3:46–47.

Breier, A.R., Nudelman, N.S., Steppe, M., and Schapoval, E.E.S. 2008. Isolation and structure elucidation of photodegradation products of fexofenadine. *J. Pharm. Biomed. Anal.* 46:250–257

Brzezinska, E. and Koska, G. 2003. TLC data in QSAR assay of thiazole and benzothiazole derivatives with H1-antihistamine activity. Part 1. *J. Planar Chromatogr.* 16:451–457.

Brzezinska, E. and Koska, G. 2004. TLC data in QSAR assay of thiazole and benzothiazole derivatives with H1-antihistamine activity. Part 2. *J. Planar Chromatogr.* 17:40–45.

Brzezinska, E. and Koska, G. 2006. A structure-activity relationship study of compounds with antihistamine activity. *Biomed. Chromatogr.* 20:1004–1016.

Brzezinska, E., Koska, G., and Klimczak, A. 2003a. Application of thin-layer chromatographic data in quantitative structure-activity relationship assay of thiazole and benzothiazole derivatives with H1-antihistamine activity. II. *J. Chromatogr. A* 1007:157–164.

Brzezinska, E., Koska, G., and Walczynski, K. 2003b. Application of thin-layer chromatographic data in quantitative structure–activity relationship assay of thiazole and benzothiazole derivatives with H1-antihistamine activity. I. *J. Chromatogr. A* 1007:145–155.

Brzezinka, H., Dallakian, P., and Budzikiewicz, H. 1999. Thin-layer chromatography and mass spectrometry for screening of biological samples for drugs and metabolites. *J. Planar Chromatogr.* 12:96–108.

DeAngelis, R.L., Kearney, M.F., and Welch, R.M. 1977. Determination of triprolidine in human plasma by quantitative TLC. *J. Pharm. Sci.* 66:841–843.

De Zeeuw, R.A., Franke, J.P., van Halema, M., Schaapman, S., Logawab, E., and Siregar, C.J.P. 1994. Thin-layer chromatography under tropical conditions: Impact of high temperatures and high humidities on screening systems for analytical toxicology. *J. Chromatogr. A* 664:263–270.

Fadiran, E.O. and Davidson, A.G. 1988. Determination of low levels of phenothiazine sulfoxides in phenothiazine drug substances and formulations by thin-layer chromatography-second-derivative spectrofluorimetry. *J. Chromatogr.* 442:363–370.

Greshock, T. and Sherma, J. 1997. Analysis of decongestant and antihistamine pharmaceutical tablets and capsules by HPTLC with ultraviolet absorption densitometry. *J. Planar Chromatogr.* 10:460–463.

Hassan, S.S.M., Elmosallamy, M.A.F., and Abbas, A.B. 2002. LC and TLC determination of cinnarizine in pharmaceutical preparations and serum. *J. Pharm. Biomed. Anal.* 28:711–719.

Indrayanto, G. 1996. Simultaneous densitometric determination of dextromethorphan hydrobromide and doxylamine succinate in syrups and its validation. *J. Planar Chromatogr.* 9:282–285.

Indrayanto, G., Darmawan, L., Widjaja, S., and Noorrizka, G. 1999. Densitometric determination of loratadine in pharmaceutical preparations, and validation of the method. *J. Planar Chromatogr.* 12:261–264.

Lang, K.L. 1990. Some observations on the determination of organic bases with cobalt(II) thiocyanate for the identification of synthetic pharmacological products. *Microchem. J.* 41:191–195.

Lillsunde, P. and Korte, T. 1991. Comprehensive drug screening in urine using solid-phase extraction and combined TLC and GC/MS identification. *J. Anal. Toxicol.* 15:71–81.

Lin, S.Y., Huang, M.Z., Chang, H.C., and Shiea, J. 2007. Using electrospray-assisted laser desorption/ionization mass spectrometry to characterize organic compounds separated on thin-layer chromatography plates. *Anal. Chem.* 79:8789–8795.

Makhija, S.N. and Vavia, P.R. 2001. Stability indicating HPTLC method for the simultaneous determination of pseudoephedrine and cetirizine in pharmaceutical formulations. *J. Pharm. Biomed. Anal.* 25:663–667

Mangalan, S., Patel R.B., Gandhi, T.P., and Chakravarthy, B.K. 1991. Detection and determination of free and plasma protein-bound astemizole by thin-layer chromatography: A useful technique for bioavailability studies. *J. Chromatogr.* 567:498–503.

Misztal, G. and Paw, B. 2001. Analysis of some antihistamine drugs in normal-phase chromatographic systems. *J. Planar Chromatogr.* 14:430–434.

Muller, D. and Ebel, S. 1997. Two-dimensional thin-layer chromatographic separation of H_1-antihistamines. *J. Planar Chromatogr.* 10:420–426.

Muller, E.E. and Sherma, J. 1999. Quantitative HPTLC determination of diphenhydramine hydrochloride in tablets, gelcap, and capsule antihistamine pharmaceuticals. *J. Liq. Chromatogr. Relat. Technol.* 22:153–159.

Musumarra, G., Scarlata, G., Cirma, G., Romano, G., Palazzo, S., Clementi, S., and Giulietti, G. 1985. Qualitative organic analysis. I. Identification of drugs by principal components analysis of standardized thin-layer chromatographic data in four eluent systems. *J. Chromatogr.* 350:151–168.

Musumarra, G., Scarlata, G., Romano, G., Clementi, S., and Wold, S. 1984. Application of principal components analysis to TLC data for 596 basic and neutral drugs in four eluent systems. *J. Chromatogr. Sci.* 22:538–547.

Ojanpera, I., Vartiovaara, J., Ruohonen, A., and Vuori, E. 1991. Combined use of normal and reversed-phase thin-layer chromatography in the screening for basic and quaternary drugs. *J. Liq. Chromatogr.* 14:1435–1446.

Pandya, K.K., Bangaru, R.A., Gandhi, T.P., Modi, I.A., Modi, R.I., and Chakravarthy, B.K. 1996. High-performance thin-layer chromatography for the determination of cetirizine in human plasma and its use in pharmacokinetic studies. *J. Pharm. Pharmacol.* 48:510–513.

Popovic, G., Cakar, M., and Agbaba, D. 2007. Simultaneous determination of loratadine and preservatives in syrups by thin-layer chromatography. *Acta Chromatogr.* 19:161–169.

Rao, S. G., Krishna, G., and Gillette, J.R. 1975. Drug nitrite interactions: The lack of toxicity of a N-nitroso derivative of tripelennamine formed in the rat. *Toxicol. Appl. Pharm.* 34:264–270.

Romano, G., Caruso, G., Musumarra, G., Pavone, D., and Cruciani, G. 1994. Qualitative organic analysis. Part 3. Identification of drugs and their metabolites by PCA of standardized TLC data. *J. Planar Chromatogr.* 7:233–241.

Schepers, P.G.A.M., Franke, J.P., and de Zeeuw, R.A. 1983. System evaluation and substance identification in systematic toxicological analysis by the mean list length approach. *J. Anal. Toxicol.* 7:272–278.

Srivastava, S.P. and Reena. 1982. Chromatographic separation of some antihistamines on metal salt-impregnated silica gel thin-layer plates. *Anal. Lett.* 15:451–457.

Stead, A.H., Gill, R., Wright, T., Gibbs, J.P., and Moffat, A.C. 1982. Standardised thin-layer chromatographic systems for the identification of drugs and poisons. *Analyst* 107:1106–1168.

Subramaniyan, S.P. and Das, S.K. 2004. Rapid identification and quantification of chlorpheniramine maleate or pheniramine maleate in pharmaceutical preparations by thin-layer chromatography-densitometry. *J. AOAC Int.* 87:1319–1322.

Sumarlik, E., Tampubolon, H.B., Yuwono, M., and Indrayanto, G. 2005. Densitometric determination of desloratadine in tablets, and validation of the method. *J. Planar Chromatogr.* 18:19–22.

Taha, E.A., Salama, N.N., and Wang, S. 2009. Enantioseparation of cetirizine by chromatographic methods and discrimination by 1H-NMR. *Drug Test. Anal.* 1:118–124.

Tuszynska, E., Podolska, M., Kwiatkowska-Puchniarz, B., and Kaniewska, T. 1994. New methods for determination of active compounds present in multicomponent antihistaminic pharmaceuticals. *Acta Pol. Pharm.* 51:317–323.

41 TLC of Vitamins Including Nicotinic Acid Derivatives

Alina Pyka

CONTENTS

Vitamins are defined as biologically active organic compounds, controlling agents that are essential for an organism's normal health and growth, not synthesized within the organism, available in the diet in small amounts, and carried in the circulatory system in low concentrations to act on target organs or tissues. Vitamins are classified according to their solubility in water and in fats. Hydrophilic vitamins are the vitamins of the groups, B_1 (thiamine), B_2 (riboflavin), B_3 (PP, niacin, nicotinamide, nicotinic acid), B_5 (panthenol and pantothenic acid), B_6 (pyridoxal), B_7 (biotin), B_9 (folic acid), B_{12} (cobalamin), and C (ascorbic acid). Lipophilic vitamins are the vitamins of the groups A, D, E, and K. In vitamin chromatography, the following problems should be solved: identification and the determination of vitamins in pharmaceutical preparations and identification and determination of vitamins and related substances in natural materials and foodstuffs. The isolation of the vitamins, their metabolites, and related substances from natural material is the most difficult task (De Leenheer et al. 2000; Eitenmiller et al. 2008).

Vitamins are the object of wide investigations because of their biological proprieties. High-performance liquid chromatography (HPLC), thin-layer chromatography (TLC), and gas chromatography (GC) are the principal techniques in a range of qualitative and quantitative investigations of hydrophilic and lipophilic vitamins. The problem of analysis of hydrophilic and lipophilic vitamins with the help of liquid chromatography (TLC and HPLC) is continually developed and is the object of many scientific publications (De Leenheer et al. 2000; Pyka 2003, 2009a; Watanabe and Miyamoto 2003, 2009; Cimpoiu and Hosu 2007; Eitenmiller et al. 2008; Hossu et al. 2009).

Generally, one should ascertain that TLC determined the application for the investigation of hydrophilic and lipophilic vitamins in range: purification of samples, qualitative detection, quantitative determination, as well as use of visualizing agents. The presentation of selected works with TLC, which described the analytical separation of hydrophilic and lipophilic vitamins, is the aim of this chapter.

41.1 HYDROPHILIC VITAMINS

41.1.1 VITAMIN B$_1$ (THIAMINE)

The sensitive spectrophotometric method for determination of vitamin B$_1$ in drugs (Vitaminum B-complex, and Multivitaminum tablets, Polfa—Poland) after chromatographic separation on silica gel using methanol–water (9:1, v/v) mobile phase was developed by Hachuła (1997). Beer's law was obeyed for 0.5–1.5 µg·mL^{-1} vitamin B$_1$ ($\varepsilon = 125,000$). Variation coefficients of the determination of vitamin B$_1$ were 0.79%–1.85% and 0.97%–1.57% for Vitaminum B-complex, and Multivitaminum tablets, respectively. No detection limit was given. Mohammad and Zehra (2008) separated vitamin B$_1$ from riboflavin, nicotinic acid, calcium D-pantothenate, pyridoxine hydrochloride, cyanocobalamin, ascorbic acid, and folic acid. TLC analysis was performed on silica gel 60F$_{254}$ plates using dioxane–water (1:1, v/v) mobile phase. This mobile phase enabled specific separation of vitamin B$_1$ from riboflavin, nicotinic acid, calcium D-pantothenate, pyridoxine hydrochloride, cyanocobalamin, and ascorbic acid (R$_F$ were equal to 0.03, 0.95, 0.93, 0.93, 0.84, 0.98, 0.98, and 0.50, respectively.). The detection limit for vitamin B$_1$ was 0.09 µg·spot^{-1}. This method for the identification of vitamin B$_1$ in pharmaceutical preparation was also investigated. The relative standard deviation for vitamin B$_1$ was 14.99%. Funk and Derr (1990) used HPTLC for the characterization and quantitative determination of thiamine hydrochloride in pharmaceutical product. Fifty grams of pharmaceutical product contained: 2.5 g L(+)-lactic acid; 0.04 g thiamine hydrochloride, 0.04 g riboflavin-5' phosphate sodium, 0.07 g nicotinamide, 7.5 g Crataegus, 0.83 g Quassia amara, and 0.08 g Lycopodium. Two milliliters of these pharmaceutical preparations was dissolved in methanol and filled to 20 mL. About 250 µL of this solution was analyzed by HPTLC. HPTLC analysis was performed on silica gel 60 plates using two mobile phases: methanol–ammonia solution (25%)–glacial acetic acid (8:1:1, v/v/v) and methanol–ammonia solution (25%)–glacial acetic acid–chloroform (9:1:0.5:0.5, v/v/v/v). After chromatographic separation of thiamine hydrochloride from all remaining compounds, vitamin B$_1$ was derivatized by two various postchromatographic methods using *tert*-butyl hydrochloride or potassium hexacyanoferrate(III)–sodium hydroxide as reagents. The R$_F$ value of thiamine was 0.40–0.45. The limit of detection of thiamine was equal to 3 ng and 500 pg using *tert*-butyl hydrochloride and potassium hexacyanoferrate(III)–sodium hydroxide as reagents, respectively.

The stability of thiamine hydrochloride in a liquid product was performed on silica gel using butanol–glacial acetic acid–water (20:10:70, v/v/v) (Schlemmer and Kammeri 1973).

Thiamine was detected on thin-layer using *tert*-butyl hypochlorite reagent with the detection limit equal to 3 ng spot^{-1} (Jork et al. 1989), potassium hexacyanoferrate (III)-sodium hydroxide-reagent with the detection limit equal to 500 pg spot^{-1} (Merck 1980; Jork et al. 1989), Reindel–Hoppr reagent (chlorine–*o*-tolidine–potassium iodine reagent), chlorine–*o*-toluidine reagent (Jork et al. 1994), and dipicrylamine (Merck 1980).

41.1.2 VITAMIN B$_2$ (RIBOFLAVIN)

Treadwell et al. (1968) studied the products of the photolysis of riboflavins. Flavin solutions were photolyzed in modified Thunberg tubes made either of Pyrex glass or silica. The source of light was a mercury lamp. TLC analyses were performed on silica gels H or G and using three mobile phases:

n-butanol–acetic acid–water (4:1:5, v/v), *n*-amyl alcohol–acetic acid–water (3:1:3, v/v), acetic acid–2-butanone–methanol–benzene (5:5:20:70, v/v). The flavins on the chromatograms were detected in the light of a long wavelength ultraviolet. In the first system, riboflavin ($R_F = 0.37$), lumichrome ($R_F = 0.68$), and formylmethylflavin ($R_F = 0.60$) were separated well. In the second mobile phase, the R_F value of riboflavin increased to about 0.20. The third mobile phase gives a rapid separation. A total of 29 compounds, including riboflavin, were separated by two-dimensional TLC. Riboflavin and its photodegradation products were separated on silica gel $60F_{254}$ using chloroform–methanol–formic acid (70:20:5, v/v) mobile phase (Granzow et al. 1995).

A simple and very rapid method (Etournaud and Aubort 1991) was proposed for the analysis of riboflavin and riboflavin phosphate in foods. The sample of food was extracted using an acetone-water mixture. Next, the solution was purified by solid-phase concentration on a C_{18} column. TLC analysis was performed on silica gel 60 without indicator plates and using mobile phases: *n*-butanol–ethanol (96%)–water, 20:10:20 (R_F values were equal to 0.6 and 0.4 for riboflavin and riboflavin phosphate, respectively); isobutanol–pyridine–acetic acid (96%)–water, 33:33:1:33 (R_F values were equal to 0.8 and 0.5 for riboflavin and riboflavin phosphate, respectively). Riboflavin and riboflavin phosphate were identified by TLC, and their identities can be confirmed using RP-HPLC. Preparative TLC on silica gel 60 or cellulose to confirm the presence of riboflavin and its derivatives in food was used by Gliszczyńska-Świgło and Koziołowa (2000). Details of the extract preparation were described by Gliszczyńska-Świgło and Koziołowa (2000). For TLC, the following 11 chromatographic systems were used:

1. *n*-Butanol–glacial acetic acid–water (2:1:1, v/v/v), silica gel
2. *n*-Butanol–glacial acetic acid–water (5:2:3, v/v/v), silica gel or cellulose
3. Chloroform–methanol–ethyl acetate (5:5:2, v/v/v), silica gel
4. *n*-Butanol–benzyl alcohol–glacial acetic acid (8:4:3, v/v/v), silica gel
5. Collidine–water (3:1, v/v), cellulose
6. 5% $Na_2HPO_4 \cdot 12H_2O$, silica gel
7. *n*-Butanol–formic acid–water–diethyl ether (77:10:13:15, v/v/v/v), silica gel
8. *n*-Butanol–ethanol–water (10:3:7, v/v/v), silica gel
9. Isoamyl alcohol–ethyl methyl ketone–glacial acetic acid–water (40:40:7:13, v/v/v/v), silica gel
10. *n*-Butanol–isopropanol–water–glacial acetic acid (30:50:10:2, v/v/v/v), silica gel
11. Ethyl methyl ketone–acetic acid–methanol (3:1:1, v/v/v), silica gel

The unknown flavin derivatives were separated from other ones using preparative TLC.

Riboflavin was also identified by TLC on silica gel in plain yogurt (Gliszczyńska and Koziołowa 1999).

Riboflavin was detected on thin layer using Reindel–Hoppr reagent (chlorine–*o*-tolidine–potassium iodine reagent), and chlorine–*o*-toluidine reagent (Jork et al. 1994).

41.1.3 VITAMIN B_3 (NIACIN, NICOTINAMIDE, NICOTINIC ACID) AND NICOTINIC ACID DERIVATIVES

Pyka and Klimczok (2009) investigated the stability of methyl nicotinate on silica gel $60F_{254}$ (using acetone + *n*-hexane mobile phase) in a temperature of 120°C, in ethanolic solution stored in the temperature of 8°C, as well as in aqueous and ethanolic solutions stored in ordinary- and quartz flasks heated at a temperature of 40°C and exposed to UV radiation ($\lambda = 254$ nm). Methyl nicotinate on silica gel was heated at a temperature of 120°C by 1–7 h; after chromatographic separation, the substance being a product of chemical changes remains at the start of the chromatogram. Methyl nicotinate in ethanolic solution undergoes chemical changes during 365 days storage in the temperature of 8°C; the substance being a product of its chemical changes had the R_F value equal to 0.40.

Methyl nicotinate in aqueous and ethanolic solutions stored in ordinary and methyl nicotinate in aqueous solutions stored in quartz flasks, heated at a temperature of 40°C, and exposed to UV radiation (λ = 254 nm) during 200 h undergoes no chemical changes. However, the methyl nicotinate in ethanolic solution stored in quartz flask underwent chemical changes during its exposure to UV radiation (λ = 254 nm) from 5 to 200 h; the substance with R_F value equal to 0.26 was the main product of chemical changes of methyl nicotinate (Pyka and Klimczok 2009). Parys and Pyka (2010) also investigated the chemical stability of nicotinic acid and its esters, namely: methyl nicotinate, ethyl nicotinate, isopropyl nicotinate, butyl nicotinate, hexyl nicotinate, and benzyl nicotinate. These compounds were heated for 1–7 h at 120°C on silica gel 60F$_{254}$. The plates were developed with the following mobile phases: methanol–benzene at the volume composition of 50:50 for nicotinic acid; and acetone–n-hexane at the volume composition of 40:60 for methyl nicotinate, ethyl nicotinate, isopropyl nicotinate, butyl nicotinate, hexyl nicotinate, and benzyl nicotinate. It was stated that the most stable was nicotinic acid from all the compounds analyzed during the 1–7 h of heating. However, the most stable were isopropyl nicotinate and hexyl nicotinate from all the examined esters. The most unstable were ethyl nicotinate, and methyl nicotinate. TLC on silica gel plates (Silufol UV$_{254}$) was also used in the study of retention behavior and stability of mixed-ligand zinc carboxylate complex bound nicotinic acids (Orinak et al. 1998).

Pyka and Klimczok (2007a–c, 2008) investigated the separation of nicotinic acid and its derivatives using NP-TLC and RP-TLC. The investigations concerning the evaluation of chromatographic separations were considered in three groups, namely, first group: nicotinic acid, methyl nicotinate, ethyl nicotinate, isopropyl nicotinate, butyl nicotinate, hexyl nicotinate, benzyl nicotinate; second group: nicotinic acid, nicotinamide, N-methylnicotinamide, N,N-diethylnicotinamide; and third group: nicotinic acid, nicotinamide, 3-pyridinecarbaldehyde, 3-pyridinecarbonitrile, 3-pyridylmethanol, methyl 3-pyridyl ketone. The separation factors ΔR_F, R_F^α, and selectivity α were calculated from the R_F values. The comparison and characteristic of chromatographic bands of the examined compounds were presented on the basis of calculated resolutions. The best separation of the first group compounds was obtained by RP-HPTLC technique on RP18WF$_{254}$ plates and by use of dioxane–water in a volume composition of 50:50 (Pyka and Klimczok 2007b). However, butyl nicotinate from benzyl nicotinate cannot be separated by RP-HPTLC technique. Good separations of isopropyl nicotinate from ethyl nicotinate as well as ethyl nicotinate from methyl nicotinate were also not obtained by RP-HPTLC technique. NP-TLC in the system of a neutral aluminum oxide 60F$_{254}$ and the acetone–n-hexane mobile phase at a volume composition of 20:80 provided the optimum conditions for the complete separation of butyl nicotinate from benzyl nicotinate, methyl nicotinate from ethyl nicotinate, as well as ethyl nicotinate from isopropyl (Pyka and Klimczok 2007b). However, adsorption NP-TLC in the system of a neutral aluminum oxide 60F$_{254}$ and acetone + n-hexane mobile phase in a volume composition of 50:50 provided the optimum conditions for the separation of all the studied compounds of second groups (Pyka and Klimczok 2007a). The best separation of the third group compounds was obtained by RP-HPTLC technique on RP18WF$_{254}$ plates and by use of dioxane–water in a volume composition of 20:80 (Pyka and Klimczok 2007c). However, 3-pyridinecarbaldehyde from 3-pryridinecarbonitrile cannot be separated by RP-HPTLC technique. NP-TLC in the system of a silica gel 60F$_{254}$ and the acetone–n-hexane mobile phase in a volume composition of 80:20 provided the optimum conditions for the complete separation of 3-pyridinecarbaldehyde from 3-pyridinecarbonitrile (Pyka and Klimczok 2007c). Nicotinic acid and its derivatives were also separated (Pyka and Klimczok 2008) using NP-TLC on a mixture of silica gel 60 and Kieselguhr F$_{254}$ nonimpregnated and impregnated with 2.5% and 5% aqueous solutions of CuSO$_4$ and using the mixture of an acetone–n-hexane in different volume compositions was used as mobile phase. It was stated that the impregnation of the mixture of silica gel and kieselguhr with 2.5% and 5% aqueous solutions of CuSO$_4$ causes a decrease in the R_F values of first, second, and third groups of the compounds in relation to the R_F values of these substances obtained on a nonimpregnated plate. Namely, impregnation with water solution of copper (II) sulfate (VI) of the mixture of silica gel and kieselguhr worsens the separation of second and third groups of the investigated compounds. Improvement of the chromatographic separations is observed only in the case of nicotinic acid esters (Pyka and Klimczok 2008).

Nicotinamide (Ropte and Zieloft 1985) was determined in seven deutsche multivitamin preparations (ampoules, tablets, drops). Chromatographic analyses were performed on silica gel $60F_{254}$ using acetone–chloroform–n-butanol-concentrated ammonia solution (30:30:40:5 and 30:30:40:2, respectively) mobile phase. Next chromatograms were densitometric scanned at 262 nm. Sherma and Ervin (1986) determined niacin and niacinamide in vitamin preparations (capsules and tablets). Capsules were emptied, and an amount equivalent to 100 mg of niacin or niacinamide according to the label value was accurately weighed and transferred to a 100 mL volumetric flask. About 50 mL of absolute ethanol was added, and the content was shaken for 4 min. The solution was diluted to volume with ethanol and the mixing was repeated. Tablets were ground and powder equivalent (100 mg) of vitamin was accurately weighed and treated as described earlier for capsules. Capsule and tablet solutions had theoretical concentrations of 1.00 µg mL^{-1}. TLC was performed on silica gel $60F_{254}$ using benzene–methanol–acetone–glacial acetic acid (7:20:5:5, v/v/v/v) as mobile phase. Densitometric analysis was performed at 254 nm. The R_F values in applied chromatographic conditions were equal to 0.61 and 0.43 for niacin and niacinamide, respectively. The linearity ranges for niacin and niacinamide were between 3 and 10 µg. The accuracy of the TLC method was confirmed by analysis of samples fortified with standard niacinamide. The recoveries for niacinamide in vitamin products were from 96% to 104% and for niacinamide added to products 99% and 102%. The recoveries for niacin in vitamin products were from 0% to 74% and for niacin added to products 99% and 102%. Hubicka et al. (2008) elaborated the TLC-densitometric method for determination of N-(hydroxymethyl)nicotinamide in the pharmaceutical preparation Cholamid and stability evaluation in NaOH solutions of different concentrations. TLC analysis was performed on silica gel F_{254} plates with a mixture of chloroform and ethanol (2:3, v/v) as mobile phase. However, densitometric analysis was performed at 260 nm. This method was suitable for quantitative analysis of pharmaceutical products and kinetic analysis.

Nicotinamide and nicotinic acid were detected using 1-chloro-2,4-dintrobenzene-sodium hydroxide–ammonia vapor (Merck 1980; Jork et al. 1994) (the detection limits were equal to 200 ng nicotinamide and nicotinic acid per spot), Reindel–Hoppr reagent (chlorine–o-tolidine–potassium iodine reagent), and potassium hydroxide (Jork et al. 1994). Spectrodensitograms of nicotinic acid and its derivatives on a mixture silica gel 60 and kieselguhr plates impregnated with 2.5% and 5% aqueous solutions of $CuSO_4$ and by use of Rhodamine B as visualizing reagent are different from the spectrodensitograms obtained on the plates without the use of a visualizing reagent. Spectra of these compounds on nonimpregnated plates and impregnated with aqueous solution of $CuSO_4$ and without use of a visualizing reagent have two or three absorption bands. Nicotinic acid and its derivatives on the plates impregnated with aqueous solution of $CuSO_4$ and by use of Rhodamine B as visualizing reagent have one additional absorption band at $\lambda \approx 560$ nm. This fact has analytical significance in the identification of the nicotinic acid and its derivative (Pyka and Klimczok 2008).

41.1.4 VITAMIN B₅ (PANTHENOL AND PANTOTHENIC ACID)

Nag and Das (1992) described the TLC-densitometric method for the identification and quantitation of panthenol and pantothenic acid in pharmaceutical preparations containing other vitamins, amino acids, syrups, and enzymes. The vitamin B₅ was extracted with ethanol (for tablets and capsules) or benzyl alcohol (liquid oral preparations). TLC analysis was performed on silica gel using two mobile phases: first chloroform, and second isopropanol–water (85:15, v/v). For benzyl alcohol extracts, develop plate first with mobile phase to full height. Dry briefly in air (10 min) and develop again in second mobile phase to full height. For ethanol extracts, omit first development. After development, the dried plates were sprayed with ninhydrin reagent. Densitometric analysis was performed at 490 nm. The linear relationship was a concentration range of 0.5–8 µg for both vitamins (panthenol and pantothenic acid). The average recoveries were 99.8% for panthenol and 100.2% for pantothenic acid.

Panthenol and pantothenic acid were detected on thin-layer using Reindel–Hoppr reagent (chlorine–*o*-tolidine–potassium iodine reagent) (Jork et al. 1994), and 2,5-dimethoxytetrahydrofuran-4-(dimethylmino)-benzaldehyd reagent with detection limit equal to 500 ng spot^{-1} of panthenol (Jork et al. 1989).

41.1.5   VITAMIN B$_6$ (PYRIDOXAL)

Ahrens and Korytnyk (1969) separated pyridoxol and its derivatives by TLC and thin-layer electrophoresis. R_F values on silica gel HF$_{254}$ and using 0.2% NH$_4$OH in water (1/139, v/v concentrated NH$_4$OH/H$_2$O) were equal to 0.62, 0.68, 0.12, 0.54, 0.91, 0.91, 0.95, 0.95, and 0.86 for pyridoxol, pyridoxal, pyridoxamine, pyridoxal ethyl acetal, 4-pyridixic acid, 4-pyridixic acid lactone, pyridoxol phosphate, and pyridoxal phosphate, respectively.

Argekar and Sawant (1999) simultaneously determined the vitamin B$_6$ and doxylamine succinate in tablets by HPTLC. Tablets were weighed, powdered, and an amount of the powdered sample equivalent to 50 mg of vitamin B$_6$ and 50 mg of doxylamine succinate was taken in a 50 mL volumetric flask. Methanol (30 mL) was added to the flask, sonicated next for 10 min, and diluted to the mark with methanol. This solution was next filtered. HPTLC analysis was performed on silica gel 60F$_{254}$ using acetone–chloroform–methanol–25% ammonia solution (7:1.5:0.3:1.2, v/v/v/v) as mobile phase. The R_F values were 0.28 and 0.86 for vitamin B$_6$ and doxylamine succinate. The wavelength of 269 nm was selected for densitometic analysis. The linearity range was 0.5–2.0 µg · spot^{-1} for both vitamin B$_6$ and doxylamine succinate. The recoveries were 99.30%–103.00% and 97.70%–101.00% for vitamin B$_6$ and doxylamine succinate. The stability study was also presented. The degradation products of vitamin B$_6$ and doxylamine succinate, with respect to heat, alkaline and acid conditions, were not observed. In stress condition (strong oxidizing condition), both vitamin B$_6$ and doxylamine succinate were degraded. This HPTLC method is simple, precise, accurate, rapid, and can be used for the routine quality control analysis of these drugs.

Tazoe et al. (2000) investigated the biosynthesis of vitamin B$_6$ in *Rhizobium*. The amount of vitamin B$_6$ was quantified by the turbidity method with *S. carlsbergensis* ATCC 9080. TLC analysis was performed on silica gel 60 plates using chloroform–methanol (3:1, v/v) mobile phase. Vitamin B$_6$ was detectable by bioautogram of TLC using the microorganism as an indicator strain.

Pyridoxal (vitamin B$_6$) was detected on thin layer using diphenyl-2-ylamin–sulfuric acid reagent with the detection limit equal 10 ng pyridoxal per spot (Jork et al. 1989), Gibbs'-reagent (2,6-dibromoquinone-4-chloroimide reagent) (Jork et al. 1989), Reindel–Hoppr reagent (chlorine–*o*-tolidine–potassium iodine reagent) (Jork et al. 1994), chlorine–*o*-toluidine reagent (Jork et al. 1989), and 1-chloro-2,4-dinitrobenzene–indicator reagent (Merck 1980).

41.1.6   VITAMIN B$_7$ (BIOTIN)

Zempleni et al. (1997) identified the biotin sulfone, bisnorbiotin metyl ketone, and tetranorbiotin–*L*-sulfoxide in human urine. Identification and determination of biotin by HPLC was confirmed by TLC. TLC analysis was performed on microcellulose plates using two different mobile phases:

1. *n*-Butanol–acetic acid–water, 4:1:1 v/v/v (R_F values were equal to 0.49, 0.78, 0.57, 0.74, 0.29, and 0.22 for the biotin sulfone, bisnorbiotin metyl ketone, tetranorbiotin, tetranorbiotin metyl ketone, tetranorbiotin–*D*-sulfoxide, and tetranorbiotin–*L*-sulfoxide, respectively). Three unknown biotin metabolites were identified as biotin sulfone ($R_F = 0.49$), bisnorbiotin metyl ketone ($R_F = 0.78$), and tetranorbiotin–*L*-sulfoxide ($R_F = 0.22$).
2. *n*-Butanol (R_F values were equal to 0.17, 0.29, 0.21, 0.05, and 0.01 for the biotin sulfone, bisnorbiotin metyl ketone, tetranorbiotin, tetranorbiotin–*D*-sulfoxide and tetranorbiotin–*L*-sulfoxide, respectively). Three unknown biotin metabolites were identified as biotin sulfone ($R_F = 0.17$), bisnorbiotin metyl ketone ($R_F = 0.29$), and tetranorbiotin–*L*-sulfoxide ($R_F = 0.01$).

Zempleni and Mock (1999) used TLC for purification and identification of biotin and its metabolites in different body fluids.

p-Dimethylaminocinnamaldehyde was used as visualizing reagent (Zempleni et al. 1997). 4-(Dimethylamino)-cinnamaldehyde–hydrochloric acid reagent was also used for the detection of bition in thin layer (Jork et al. 1989).

41.1.7 VITAMIN B$_9$ (FOLIC ACID)

Jamil Akhtar et al. (2003) applied TLC for the identification of photoproducts of folic acid and its degradation products in aqueous solution. TLC was performed on silica gel GF$_{254}$ plates using two mobile phases: ethanol–ammonia (13.5 M)–1-propanol (60:20:20, v/v), and acetic acid–acetone–methanol–benzene (5:5:20:70, v/v). The pterin-6-carboxylic acid and *p*-aminobenzoyl-L-glutamic acid were identified as degradation products of folic acid. Blair and Dransfield (1969) also applied TLC for analysis of radioactive folic acid. N-(4-aminobenzoyl-)-L-glutamic acid and 4-aminobenzoic acid were also identified as the degradation products of folic acid (Krzek and Kwiecień 1999). Tablets with folic acid were powered and next 150 mg folic acid accurately weighed and 25 mL methanol were added. This mix was shaken and filtered. Chromatograms were developed using mobile phase: *n*-propanol–ammonia hydroxide–25% ethanol, 2:2:1, v/v/v or toluene–methanol–glacial acetic acid–acetone, 14:4:1:1, v/v/v/v. The first mobile phase is recommended to evaluate folic acid by some pharmacopeial monographs (Ph. Eur, BP, FP). Substances were detected on chromatograms using the Ehrlich reagent. Folic acid, N-(4-aminobenzoyl-)-L-glutamic acid, and 4-aminobenzoic acid give yellow spots with the Ehrlich reagent. In the first mobile phase, the R$_F$ values were equal to 0.52, 0.60, and 0.80, respectively, for folic acid, N-(4-aminobenzoyl-)-L-glutamic acid, and 4-aminobenzoic acid. In the second mobile phase, the R$_F$ values were equal to 0.06, 0.20, and 0.85, respectively for folic acid, N-(4-aminobenzoyl-)-L-glutamic acid, and 4-aminobenzoic acid. The folic acid preparations showed two spots on chromatograms that were derived from folic acid and N-(4-aminobenzoyl-)-L-glutamic acid. In folic acid preparations, no spot deriving from 4-aminobenzoic acid was noted. The chromatographic separation time was about 30 min using second mobile phase, whereas the same time under the pharmacopeial mobile phase (first mobile phase) is about 3 h. These conditions can be applied to evaluate the purity of the folic acid preparations. The determination of the N-(4-aminobenzoyl-)-L-glutamic acid content in investigated series of folic acid tablets with the second mobile phase applied ranges between 0.70% and 0.85% w/w for measurement in UV, and from 0.73% to 0.95% w/w in the visible light; however, with the first mobile phase used from 0.82% to 1.53% w/w and from 0.84% to 1.4% w/w, respectively (Krzek and Kwiecień 1999). Dongre and Bagul (2003) proposed a new HPTLC-densitometric method for identification and quantification of folic acid from pharmaceutical preparations. The contents of 10 *Fesovite* capsules were weighed accurately and transferred quantitatively into a 200 mL standard volumetric flask containing 100 mL diluent (aqueous ammonia–methanol, 9:1, v/v). The contents of 10 *Fe-fol* capsules were weighed accurately and transferred quantitatively into a 100 mL standard volumetric flask containing 50 mL diluent. The contents of 10 *Folvite* tablets were weighed accurately and transferred quantitatively into a 1000 mL standard volumetric flask containing 500 mL diluent. Next, these mixtures were shaken mechanically for 30 min, sonicated for 5 min, and diluted to volume with the same diluent to a concentration of 0.05 mg·mL^{-1}. These solutions were next filtered through Whatman #41 filter paper. TLC was performed on silica gel 60F$_{254}$ HPTLC plates using a mobile phase ipa–ethyl acetate–aqueous ammonia, 4:2:2:2 (v/v/v/v). The R$_F$ value for the folic acid was equal to 0.47. The typical wavelength λ = 280 nm was chosen for quantification. Linear relationship between the peak area and amount of folic acid was in the range 25–70 µg, and it was used for direct determination of the amount of folic acid in *Fesovite* and *Fe-fol* capsules, as well as in *Folvite* tablets. The low *RSD* values are indicative of the accuracy and precision of the method.

Folic acid and its derivatives were detected on thin-layer in UV light (Dongre and Bagul 2003; Jamil Akhtar et al. 2003), using Ehrlich reagent (1 g of 4-dimethylbenzoic aldehyde was dissolved in a mixture of 25 mL hydrochloric acid [concentration 36%] and 75 mL ethyl alcohol [concentration 95%]) (Krzek and Kwiecień 1999) and using Bratton-Marshall reagent (Jork et al. 1989). The detection limit of folic acid after the detection with Bratton-Marshall reagent was equal to 200 ng (Jork et al. 1989).

41.1.8 Vitamin B_{12} (Cobalamin)

Watanabe and Miyamoto (2010) in a review article described the purification and characterization of vitamin B_{12} and related compounds by TLC.

TLC was used for evaluation of photolysis of aqueous cyanocobalamin solution in the presence of vitamins B and C (Ansari et al. 2004). Aqueous cyanocobalamin solutions (pH 1–7) were photolysed in the presence of individual B (thiamine, riboflavin, nicotinamide, and pyridoxine) and C vitamins with visible light. The photolysed solutions of cyanocobalamin were subjected to TLC using silica gel G, silica gel G_{254}, or Whatman CC 41 cellulose powder and the following mobile phases:

1. n-Butanol–acetic acid–0.066 M KH_2PO_4–methanol (36:18:36:10, v/v/v/v)
2. Methanol–water (95:5, v/v)
3. Pyridine–2-butanol–water (66:17:17, v/v/v)
4. Chloroform–methanol–ammonia solution (30%) (70:30:1, v/v/v)
5. n-Butanol–acetic acid–water (40:10:50, v/v/v)
6. n-Butanol–n-propanol–acetic acid–water (50:30:2:18, v/v/v/v)
7. Distilled water
8. Acetic acid–acetone–methanol–benzene (5:5:20:70, v/v/v/v)
9. Ethanol–10% acetic acid (90:10, v/v)

After development, the chromatographic spots were detected visually at 254 and 366 nm or on spraying with 3% aqueous phenylhydrazine solution. Hydroxocobalamin (mobile phases 1 and 2) was found to be the only phytoproduct of cyanocobalamin at pH 1–7, or in the presence of individual B/C vitamins. Thiamine hydrochloride (mobile phases 3 and 4) showed trace amounts of 4-methyl-5-(β-hydroxyethyl)thiazole and 2-methyl-4-amino-5-hydroxymethylpyrimidine. Riboflavin (mobile phases 5 and 6) showed six photodegradation products, namely, formylmethylflavin, lumichrome, lumiflavin, carboxymethylflavin, and two unknown products. Ascorbic acid (mobile phases 8 and 9) was degraded to dehydroascorbic acid. Nicotinamide and pyridoxine hydrochloride did not show any photoproducts. Ahmad et al. (2003) investigated the effect of nicotinamide on the photolysis of cyanocobalamin in aqueous solution. TLC was performed on silica gel GF_{254} plates using two mobile phases: 1-butanol–acetic acid–0.066 M potassium dihydrogen phosphate–methanol (36:18:36:10, v/v), and methanol–water (95:5, v/v). The spots were detected in UV light.

Winkler and Hachuła (1997) determined vitamin B_{12} in ampoule pharmaceuticals (Vitaminum B_{12} from *Polfa—Poland* and Neurobion from *Cascan-Germany*). TLC was performed on plastic-backed silica gel 60 plates using methanol–water (95:5, v/v) as a mobile phase. The R_F value of vitamin B_{12} was 0.68 under chromatographic conditions, and vitamin B_{12} was separated well from vitamins B_1 and B_6 and benzyl alcohol, which occur in Neurobion preparation. Vitamin B_{12} after development was dried, and silica gel containing spots with vitamin B_{12} was scraped from the surface of the plate, extracted, and centrifuged to separate the silica gel. The extract was then acidified and transferred to a 25 mL flask containing borate buffer, Triton X100, cetylpyridine chloride, and phenylfluorone. Absorptivity was measured at 635 nm. The described process of analysis makes it possible to obtain results comparable with declared values.

Vitamin B_{12} was purified from the lyophilized purple laver (*Porphyra yezoensis*) (Watanabe et al. 2000). TLC was performed on silica gel 60 plates using two mobile phases:

1. 1-Butanol–2-propanol–water, 10:7:10 v/v/v (R_F values were equal to 0.18, 0.18, 0.15, 0.16, and 0.16 for B_{12} compound from purple laver, CN-B_{12}, benzimidazolylcyanocobamide, 5-hydroxybenzimidazolylcyanocobamide, pseudovitamin B_{12})
2. 2-Propanol–NH_4OH (28%)–water, 7:1:2 v/v/v (R_F values were equal to 0.59, 0.59, 0.55, 0.47, and 0.46 for B_{12} compound from purple laver, CN-B_{12}, benzimidazolylcyanocobamide, 5-hydroxybenzimidazolylcyanocobamide, pseudovitamin B_{12})

The pink-colored spot on TLC plates was dried, extracted with methanol solution, evaporated to dryness under reduced pressure, and dissolved in 50 µL of distilled water. The solution was next purified by HPLC. The B_{12} concentration was determined by the B_{12} chemiluminescence analyzer (Watanabe et al. 2000). TLC was used for purification of vitamin B_{12} compounds that occur in foods (Watanabe et al. 1998). The purified B_{12} degradation product gave a single red spot on silica gel 60 after TLC separation. The R_F values of purified B_{12} degradation product were equal to 0.01, 0.14, 0.09, 0.04, 0.14, and 0.60 using different mobile phases: *n*-butanol–isopropanol–water (10:7:10), 2-butanol–acetic acid (99:1), 2-butanol–acetic acid–water (127:1:50), 2-butanol–NH_4OH (28%)–water (50:7:18), 2-propanol–NH_4OH (28%)–water (7:1:2), respectively. Vitamin B_{12} in the microwavetreated foods was derived from the conversion of B_{12} to some inactive B_{12} degradation products, namely OH-B_{12}, which predominates in food. OH-B_{12} was treated by microwave heating for 6 min; next it was analyzed on silica gel 60 plates using *n*-butanol–isopropanol–water (10:7:10, v/v/v) mobile phase. The treated OH-B_{12} was separated into three red spots, namely with R_F value of 0.03 (identical R_F of intact OH-B_{12}) and other degradation products with R_F values equal to 0.16 and 0.27. These investigations indicate that the conversion of vitamin B_{12} to the inactive vitamin B_{12} degradation products occurs in foods during microwave heating (Watanabe et al. 1998). Guy et al. (1998) used TLC and RPLC for evaluation of coupling of cobalamin (vitamin B_{12}) to antisense oligonucleotides. The different cobalamin derivatives were analyzed by TLC on silica gel 60 using the buffer containing 71.5% of 2-butanol (v/v), 0.5% acetonitrile (v/v), and 28% water (v/v) as mobile phase. R_F values of carboxycobalamin, aminocobalamin, 1-ethyl-3-3-diaminopropyl)-carbodimide-cobalamin, and biotinylated cobalamin were equal 1, 0.3, 0.2, and 0.8, respectively. R_F was determined using carboxycobalamin as marker reference ($R_F = 1$). TLC was also used for identification of vitamin B_{12} in boiled and dried Japanese Anchovy (*Engraulis japonicus*) (Nishioka et al. 2009) as well as in the short-necked clam (*Ruditapes philippinarum*) extract (Ueta et al. 2010).

The spot of vitamin B_{12} on chromatogram is red in color (Watanabe et al. 1998, 2000).

41.1.9 VITAMIN C (ASCORBIC ACID)

Roomi and Tsao (1998) separated isomers of ascorbic acid (L-ascorbic acid, D-ascorbic acid, D-isoascorbic acid) and their oxidation product (dehydroascorbic acid). Separations were performed by normal phase TLC (on silica gel; cellulose; or metaphosphoric acid [MPA] plates that were prepared by uniformly spraying a 2% MPA solution on silica or cellulose plates), by reversed phase-TLC (silica gel, cellulose, borate plates were uniformly impregnated with silicone oil), and using two mobile phases: acetonitrile–acetone–water–acetic acid (80:5:15:2) and acetonitrile–butylnitrile–water–acetic acid (66:33:15:2). Three ascorbic acid isomers were separated on NP-TLC and RP-TLC cellulose plates and using first mobile phase. But three dehydroascorbic acid isomers were not separated. The separation of all six isomers was performed on NPTLC cellulose plates using NP-TLC cellulose-MPA plates; the R_F values of three stereoisomer ascorbic acids were 0.42, 0.40, and 0.50, and the R_F values of three dehydroascorbic acids were 0.55, 0.52, and 0.47. Saari et al. (1967) investigated the oxidative degradation of ascorbic acid. TLC was used for ascorbic acid separations. TLC analysis was performed on silica gel SG-41 and using acetonitrile–butyronitrile–water (63:33:2, v/v/v) mobile phase. Next, radioautographs were prepared by exposing the developed TLC plates to x-ray film. Alternatively, substances were visualized by Tollen's reagent (0.1% solution of 2,6-dichlorophenolindophenol in ethanol). 2,3-Diketogulonic acid (R_F = 0.06), ascorbic acid (R_F = 0.26), dehydroascorbic acid (R_F = 0.37), oxalic acid (R_F = 0.69), and two unknown substances (R_F = 0.15 and R_F = 0.48) occurred on chromatogram.

Frgacić and Kniewald (1974) applied ascorbic acid as an antioxidant in the TLC of corticosteroids. TLC was performed on silica gel G and on silica gel $60F_{254}$. The plates were sprayed with a saturated solution of ascorbic acid in absolute ethanol. After spotting, the samples on plates were first developed in *n*-heptane to eliminate impurities, and then corticosterones were separated using acetone–benzene (5:9, v/v) mobile phase. After separation, corticosterones on nonimpregnated plates on chromatograms were three spots, namely: corticosterone (R_F = 0.51), dihydrocorticosterone (R_F = 0.57), and 5α-pregnan-3β,11β,21-triol-20-one (R_F = 0.44). After separation, corticosterone impregnated with ascorbic acid plate occurred in one spot.

The scientific literature (Abbaspour et al. 2008) also described a simple and selective sensor for the determination of ascorbic acid in vitamin C tablets based on paptode design on TLC plate. This procedure was based on the reduction reaction of iron (III) with ascorbic acid and the formation of a red complex with 2,2′-dipyridyl on TLC plate. The linear range in the applied condition was from 20 to 200 ppm, and the limit of detection (LOD) was 1 ppm. Aburjai et al. (2000) applied UV spectrophotometry and HPTLC for the simultaneous determination of vitamin C and dipyrone in different pharmaceutical preparations (tablets and ampoules). TLC analysis was performed on silica gel HPTLC plates using water–methanol (95:5, v/v) as mobile phase. UV-densitometric analysis was performed at 260 nm. The R_F values were equal to 0.92 and 0.65 for vitamin C and dipyrone, respectively. The tablet powder equivalent to 200 mg of vitamin C and 200 mg of dipyrone was transferred into a volumetric flask, mixed with 100 mL methanol, and sonicated for 15 min. Next, the solution was filtered, and 250 μL was used for TLC analysis. About 50 μL of the content of an ampoule was transferred to a 10 mL volumetric flask and diluted to volume with methanol, and from this solution 700 μL was used for TLC analysis. The linear range for ascorbic acid was 1.5 ± 35, and 2 ± 10 μg · mL⁻¹ determined by TLC and spectrophotometry, respectively. Buhl et al. (2005) determined the ascorbic acid in "Rutinoscorbic" tablets (Glaxo Welcome, Poznan, Poland), "Vitaral" tablets (Jelfa, Jelenia Góra, Poland), and fresh green pepper juice, after chromatographic separation. "Rutinoscorbic" tablet and "Vitaral" tablet were ground and extracted with water. Freshly prepared pepper juice was centrifuged, to remove suspended material. These prepared samples were spotted on silica gel $60F_{254}$ plates. The glacial acetic acid–acetone–methanol–benzene (3:1:4:14, v/v/v/v) was the best mobile phase for the separation of L-ascorbic acid from other components of Rutinoscorbic tablets. Butanol–formic acid–water (20:1:0.3, v/v/v) was the best mobile phase for the separation of L-ascorbic acid from other components of aqueous "Vitaral" tablets. Water-glacial acetic acid–ethyl methyl ketone–ethyl acetate (1:2:2:5, v/v/v/v) was used for the separation of L-ascorbic acid from other components of green pepper juice sampler. Ascorbic acid was detected by UV light at 254 nm. The chromatographic spots of ascorbic acid were scraped and extracted with water. The analyte was next oxided to dehydroascorbic acid by the iodate (V) anion, and spectrophotometric analysis was performed at λ = 588 nm. These results indicated that the elaborated method was simple, precise, and sensitive for quantification of L-ascorbic acid in pharmaceutical and food products. Ascorbic acid was determined in pharmaceutical preparation also containing paracetamol, caffeine, and phenylephrine (El-Sadek et al. 1990). HPTLC analysis was performed on silica gel $60F_{254}$ using two mobile phases, first: methylene chloride–ethyl acetate–ethanol–formic acid (3.5:2:4:0.5, v/v/v/v), and second: methylene chloride–ethyl acetate–ethanol (5:5:1, v/v/v). These conditions permit the separation of ascorbic acid, paracetamol, caffeine, and phenylephrine, and their degradation products, namely dehydroascorbic acid and p-aminophenol. The powered *Rhino C* tablets were extracted with 80% ethanol. The R_F values using the first mobile phase were equal to 0.53, 0.22, 0.87, 0.69, 0.96, and 0.69 for ascorbic acid, phenylephrine, paracetamol, caffeine, p-aminophenol, and coloring matter in *Rhino C* tablets. The R_F values using the second mobile phase were equal to 0.00, 0.00, 0.67, 0.52, 0.80, and 0.06 for ascorbic acid, phenylephrine, paracetamol, caffeine, *p*-aminophenol, and coloring matter in *Rhino C* tablets. The limit of determination was equal to 100 ng · spot^{-1} for ascorbic acid.

Beljaars et al. (1974) elaborated a TLC with the densitometric method for the quantitative analysis of ascorbic acid in buttermilk sample. This procedure was based on oxidation of ascorbic acid to dehydroascorbic acid, followed by reaction with 2,4-dinitrophenylhydrazine to form the dinitrosazone, which was separated on silica gel G plates using chloroform-ethyl acetate-acetic acid (60:35:5, v/v/v) mobile phase. Densitometric analysis was performed at 525 nm. The analyzed substances had red-brown spots on chromatograms. The calibration plot was linear in the range 0.08–1.00 μg ascorbic acid per spots. The results obtained by TLC-densitometric method were compared with those reported for spectrodensitometric, titrimetric, and potentiometric procedures. The TLC-densitometric method was less accurate. Hosu et al. (2010) elaborated the TLC method for the determination of vitamin C after reaction with 2,2-diphenyl-1-picrylhydrazyl (DPPH). Methanolic solution of vitamin C of different concentration was mixed with methanolic solution of DPPH. After 30 min, 10 mL of each reaction mixture was applied on silica gel 60 plates. In the second step, the elaborated procedure was used for each juice (0.075 mL juice + 0.050 mL DPPH solution). The proposed method was used for the determination of ascorbic acid in juices and was compared with results obtained by spectrophotometric method. It can be used as a simple, rapid, and inexpensive alternative to know spectrophotometric and spectroscopic methods. Ascorbigen is a natural indole derivative of L-ascorbic acid (Katay et al. 2004). Ascorbigen was extracted from plant material (juice) using methanol, and methanolic solution was centrifuged. Next, methanol was removed under reduced pressure. The residual aqueous solution was extracted three times with ethyl acetate. The extracts were dried over Na_2SO_4 and filtered. Next, the solvent was evaporated, and the residue was dissolved in methanol. This methanolic extracts were used for OPLC and MALDI-MS analysis. Separation of ascorbigen was performed by means of two-step OPLC development. The first step was realized using n-hexane and served for elimination of the total wetness front. The second step was realized using chloroform–methanol (9:1, v/v) and served for the separation. Prochazka reagent (reaction with formaldehyde) was used as the visualizing reagent. Densitometric analysis was performed at $\lambda = 460$ nm. The value of the putative ascorbigen was equal to 0.47. The results indicated that the elaborated OPLC method can be used for the separation and quantification of ascorbigen in extracts of *Brassica* vegetables.

Vitamin C was detected on a thin layer using 2,2'-dipiridin–iron (III) chloride–reagent (Jork et al. 1989), Tillmanns reagent (2,6-dichlorophenolindophenol reagent) (Merck 1980; Jork et al. 1989), molybdophosphoric acid, 1,2-phenylendiamin-trichloroacetic acid reagent, Mandelins reagent (ammonium vanadate (V)-sulfuric acid–reagent) (Jork et al. 1989), Cacotheline reagent (Merck 1980; Jork et al. 1994), selenium dioxide reagent (Jork et al. 1994), 4-methoxy-2-nitroaniline (Merck 1980), and titanium(III) chloride-hydrochloric reagent (Jork et al. 1994). The detection limits for both ascorbic acid and dehydroascorbic acid after detection with titanium(III) chloride–hydrochloric reagent were equal to 50 ng substance per spot (Jork et al. 1994). Dare and Ajibola (2007) applied benzyl and benzoin as general spray reagents for visualization of organic compounds on thin-layer chromatograms. The spray reagent of benzyl or benzoin was freshly prepared by dissolving 1.0 g of each active reagent in a solution consisting of 9.0 mL ethanol, 0.5 mL concentrated H_2SO_4, and 3 drops of acetic acid. Vitamin C was chromatographed on silica gel plate using chloroform–acetone (6:1, v/v) mobile phase. The R_F value of vitamin C was equal to 0.62. The color of ascorbic acid spots with benzyl and benzoin was light brown and light brown and purple, respectively. Detection limits of vitamin C with benzoil, benzoin, iodine vapor, UV light, and $KMnO_4$ (10% water) were equal to 0.5, 0.05, 2.0, 1.0, 0.5 μg spot^{-1}, respectively. Vitamin C was also detected by photocatalytic reaction [plate was sprayed with 10 mL KBr (1.0 mol · L^{-1}) with 0.25 g TiO_2, and after 10 min the chromatogram was then sprayed with 0.1 mol · L^{-1} AgNO$_3$]; detection limit was equal 0.5 μg · spot^{-1} (Makowski et al. 2010).

41.1.10 Mixture of Hydrophilic Vitamins

Thiamine (B_1), niacin (nicotinic acid, after derivatization procedure with potassium ferricyanide), and riboflavin (B_2) were separated on HPTLC silica gel plates using methanol–water (70:30, v/v) as mobile phase (Diaz et al. 1993). The R_F values for thiamine derivative, riboflavin, and niacin derivative were 0.73, 0.86, and 0.91, respectively. The densitometric measurements of thiamine, riboflavin, and niacin were performed at λ_{ex} = 390 nm, λ_{em} = 440 nm; λ_{ex} = 464 nm, λ_{em} = 520 nm; and λ_{ex} = 501 nm, λ_{em} = 523 nm, respectively. The detection limit values were 7.4, 3.1, and 2.6 ng for thiamine, riboflavin, and niacin, respectively. Panahi et al. (2011) investigated B_1, B_2, B_6, and B_{12} vitamins. The HPTLC analysis was performed on silica gel 60F$_{254}$ using thirty mobile phases. The ethanol–chloroform–acetonitrile–toluene–ammonia–water (7:4:4.5:0.5:1:1, v/v) was selected as the optimum mobile phase. In these conditions, the R_F values were equal to 0.36, 0.60, 0.85, and 0.46 for B_1, B_2, B_6, and B_{12} vitamins, respectively. The linearity ranges were 200–2500, 10–1100, 20–500, and 20–1000 ng·spot^{-1} for B_1, B_2, B_6, and B_{12} vitamins, respectively. The limits of quantification (LOQ) were equal to 42.52, 12.72, 30.09, and 3.45 ng for B_1, B_2, B_6, and B_{12} vitamins, respectively. LOD were equal to 141.72, 42.41, 100.31, and 11.50 ng for B_1, B_2, B_6, and B_{12} vitamins, respectively. Panahi et al. (2008) also separated B_1 and B_{12} vitamins on silica gel 60F$_{254}$ HPTLC plates impregnated with boric acid. The separation was successfully done using methanol–water–ammonia (7:3:1, v/v/v). The R_F values on nonimpregnated plate were equal to 0.91 and 0.97 for B_1, and B_{12} vitamins, respectively. The R_F values on impregnated plate were equal to 0.71 and 0.97 for B_1, and B_{12} vitamins, respectively. Mohammad and Laeeq (2010) elaborated TLC procedures for the analysis and separation of a variety of organic compounds, including B-complex vitamins. Vitamins (thiamine, pyridoxine, and riboflavin) were successfully resolved on silica gel HPTLC plate using water-in-iol-microemulsion containing *N*-cetyl-*N,N,N*-trimethylammonium bromide (CTAB) as mobile phase. Vitamins were visualized in UV light. The R_F values of thiamine, riboflavin, and pyridoxine were equal to 0.08, 0.27, and 0.40, respectively. These chromatographic conditions were used for the analysis of vitamins in Becosules capsules (Pfizer) and Becozine syrup (Dr. Reddy's). Ponder et al. (2004) separated B_1, B_2, B_3, B_6, B_{12}, C vitamins, and folic acid using 14 mobile phases and commercially available plates precoated silica gel and chemically bonded silica gel for TLC and HPTLC. The best separations of individual and mixed vitamin standards were achieved on silica gel plates with *n*-butanol–chloroform–acetic acid–ammonia–water (7:4:5:1:1, v/v/v/v/v), benzene–methanol–acetone–acetic acid (70:20:5:5, v/v/v/v), and chloroform–ethanol–acetone–ammonia (2:2:2:1, v/v/v/v). Separation of hydrophilic vitamins from samples of *Helisoma trivolvis* snails was evaluated using the previously mentioned mobile phases. Using *n*-butanol–chloroform–acetic acid–ammonia–water (7:4:5:1:1, v/v/v/v/v) mobile phase, the four spots with R_F values 0.60, 0.66, 0.67, and 0.69 were stated on chromatograms. The plate at 0.67 was conformed to be vitamin B_2 by its yellow fluorescence. Using benzene–methanol–acetone–acetic acid (70:20:5:5, v/v/v/v) mobile phase, the spot at 0.20 was identified also as vitamin B_2. The spot at R_F 0.17 or 0.12 could have been vitamin C. Also using chloroform–ethanol–acetone–ammonia (2:2:2:1, v/v/v/v) mobile phase, the spot at 0.30 was identified also as vitamin B_2. Bhushan and Arora (2002) separated vitamin B complex and folic acid by HPLC, RP-TLC, and NP-TLC. RP-TLC was performed on RP18W/UV254 plates using 25% methanol–borate buffer (5 mM, pH 7.0)–acetonitrile (14:1:2, v/v/v) and 25% methanol–borate buffer (5 mM, pH 7.0)–acetonitrile (14:2:2, v/v/v) as mobile phases. The results indicate that mobile phase 25% methanol–borate buffer (5 mM, pH 7.0)–acetonitrile (14:1:2, v/v/v) gave a good separation in reversed phase TLC for vitamins; the R_F values were equal to 0.11, 0.16, 0.26, 0.20, and 0.30 for B_1, B_2, B_6, B_{12} vitamins, and folic acid, respectively. An increase in the amount of borate buffer caused poor separation of B_2 and B_{12} vitamins. NP-TLC was performed on silica gel plates using butanol–chloroform–acetic acid–ammonia–water (7:4:5:1:1, v/v/v/v/v), ethanol–chloroform–acetic acid–ammonia–water (7:4:5:1:1, v/v/v/v/v), and propanol–chloroform–acetic acid–ammonia–water (7:4:5:1:1, v/v/v/v/v) as mobile phases. The results indicate that mobile phase butanol–chloroform–acetic acid–ammonia–water (7:4:5:1:1, v/v/v/v/v) gave a good separation in normal phase TLC for vitamins; the R_F values were equal to 0.28, 0.59, 0.66, 0.42, and 0.12 for B_1, B_2, B_6, B_{12} vitamins, and folic acid, respectively. The replacement of butanol with ethanol or propanol in the same volume ratio was found to give an almost satisfactory separation. Bhushan and Parshad (1994, 1999) also separated vitamin B complex and folic acid on silica gel G plates impregnated in aqueous solutions of different metal ions Mn^{2+}, Fe^{2+}, Co^{2+}, Ni^{2+}, Cu^{2+}, Cd^{2+}, Zn^{2+}, or Hg^{2+} (0.1%, 0.2%, 0.3%, 0.4% of each ion) and using new solvent systems:

1. n-Propanol–n-butanol–water–ammonia (7:5:1:2, v/v/v/v)
2. n-Propanol–n-butanol–water–ammonia (7:5:1:1.5, v/v/v/v)
3. n-Propanol–n-butanol–water–ammonia (7:5:0.75:2, v/v/v/v)
4. Chloroform–n-butanol–acetic acid–ammonia (4:7:5:1, v/v/v/v)
5. Chloroform–n-butanol–water–acetic acid–ammonia (3:5:0.5:5:0.5, v/v/v/v/v)
6. Benzene–butylacetate–n-propanol–acetic acid ammonia (1:4:1:5:1, v/v/v/v/v)
7. Carbon tetrachloride–butylacetate–propionic acid–ammonia (3:7:9:3, v/v/v/v)
8. Carbon tetrachloride–butylacetate–methanol–ammonia (1.5:4.5:7:0.5, v/v/v/v)
9. Carbon tetrachloride–butylacetate–propionic acid–methanol–water (2:3:1:0.5:3, v/v/v/v/v)

The iodine vapors were used as the visualizing reagents for vitamins B complex and folic acid.

Cimpoiu et al. (2005) separated and identified water-soluble vitamins (B_1, B_2, B_3, B_5, B_6, B_9, B_{12}, and C) by TLC and Raman spectroscopy. TLC analysis was performed on silica gel 60F_{254} using one development and programmed multiple development by methanol–benzene in different volume compositions. The detection was performed under UV light at 254 nm, except for vitamin B_5 for which spraying with ninhydrin reagent was used. The best results of separations were obtained using programmed multiple development. The Raman spectra of vitamins were recorded both in the pure solid state and on the chromatographic plate. These methods can be used for monitoring the water-soluble vitamins content of real samples. Postaire et al. (1991) determined the water-soluble vitamins in multivitamin solutions by OPLC with photodensitometric detection. Separations of water-soluble vitamins (B_1, B_2, B_6, B_{12}, C, folic acid, nicotinamide, calcium pantothenate, biotin) were compared after their analysis using TLC, HPTLC, and OPLC on silica gel plates and using n-butanol–pyridine–water (50:35:15, v/v/v) as mobile phase. Results indicated that the best separations were obtained using OPLC. Photodensitometric detection was performed at 254 nm for vitamins B_1, B_2, and nicotinamide, at 290 nm for folic acid and vitamin B_6, at 530 nm for biotin and vitamin B_{12} after dimethylaminocinnamaldehyde reagent spray, and at 375 nm for calcium pantothenate after ninhydrin reagent spray. This method of determination of water-soluble vitamins in multivitamin solutions was fast, accurate, specific, and can be used for routine quality control. Anionic–nonionic and cationic–nonionic surfactants coupled micellar TLC were applied for the evaluation of effect on simultaneous separations of water-soluble vitamins (B_1, B_2, B_6, B_{12}, and folic acid) (Mohammad et al. 2009; Mohammad and Zehra 2010). The best chromatographic system was silica gel impregnated with 0.01% sodium dodecyl sulfate as stationary phase and 0.1% aqueous solution Cween 80 as mobile phase. The vitamins were detected by iodine vapors. The R_F values were equal to 0.05, 0.20, 0.42, 0.70, and 0.99 for vitamins B_1, B_{12}, B_2, B_6, and folic acid, respectively. The LODs were equal to 0.05, 0.5, 0.05, 0.08, and 1 µg spot^{-1} for vitamins B_1, B_2, B_6, B_{12}, and folic acid, respectively. Kawanabe (1985) separated the mixture of water-soluble vitamins (thiamine, riboflavin, pyridoxine·HCl, and nicotinamide) on Wakogel–microcrystaline cellulose (5:2) plates using chloroform–ethanol–acetic acid (100:50:1, v/v/v) mobile phase. The R_F values of thiamine, riboflavin, pyridoxine·HCl, and nicotinamide were equal to 0.00, 0.36, 0.47, and 0.57, respectively. Detection limits in UV light for these compounds were equal to 5.0, 1.0, 5.0, and 5.0 µg, respectively. The complexes of retinol–palmitic acid and retinol–HAc were separated on Wakogel–microcrystaline cellulose (5:2) plates using benzene mobile phase. Retinol complexes were detected by $SbCl_3$. R_F values of these compounds were equal to 0.84 and 0.54, respectively. Mohammad and Zehra (2009) simultaneously separated and identified cyanocobalamin, thiamine, and ascorbic acid on polyoxyethylene sorbitan monooleate–impregnated silica gel plates using double-distilled water as mobile phase. The R_F values of vitamins B_{12}, B_1, and C were equal to 0.60, 0.20, and 0.97, respectively. Detection limits of vitamins B_{12}, B_1, and C were equal to 0.5, 0.5, and 1.0 µg spot^{-1}, respectively. These chromatographic conditions were applied to identification of vitamins B_{12}, B_1, and C in drug samples, namely Polybion, Alamin Forte, Becozyme Forte, Nutrisan, Supradyn, Riconia, Astynin Forte, and Basiton.

Perisic-Janjić et al. (1995) determined vitamins B_1, B_2, B_6, B_{12}, and PP in *Plibex* (Pliva–Zagreb) and *Beviplex* (Galenika–Zemun) lyophilized ampoules. Contents of each lyophilized ampule were dissolved in 2 mL of distilled water. Chromatographic analysis was performed using a newly synthesized carbamide formaldehyde polymer–aminoplast stationary phase and mobile phase: *n*-butanol–methanol–benzene–water (20:10:10:8, v/v/v/v). Chromatographic plates were densitometric scanned at 254 and 366 nm. The results obtained by Perisic-Janjić et al. (1995) indicate that this procedure can successfully be applied for quantitative and qualitative determination of B-complex vitamins, except vitamin B_{12}, which is present only in small quantities in lyophilized ampules. A huge error of the determination of vitamin B_2 was stated, because this vitamin is highly unstable. Bican-Fister and Drazin (1973) quantitatively determined the water-soluble vitamins in multicomponent pharmaceutical preparations. Thiamine hydrochloride, riboflavin, pyridoxine hydrochloride, nicotinamide, and *p*-aminobenzoic acid were determined in tablets and granules. The amount of the tablet powder or granules containing approximately 4 mg of thiamine hydrochloride, 5 mg of riboflavin, 2 mg of pyridoxine hydrochloride, 25 mg of nicotinamide, and 20 mg of *p*-aminobenzoic acid was dissolved in a 10 mL calibrated flask in 50% methanol, with heating on a waterbath for 10 min. After cooling, the solution was diluted with the solvent up to the mark. TLC was performed on silica gel HF_{254} plates using glacial acetic acid–acetone–methanol–benzene as mobile phase. The previously mentioned vitamins were determined after extraction from the silica gel plates by UV spectrophotometry (for thiamine hydrochloride, pyridoxine hydrochloride, and nicotinamide) and by colorimetry after diazotation (for riboflavin, and *p*-aminobenzoic acid). The results obtained were in good agreement with manufacturers' specifications. Thiamine hydrochloride (B_1), pyridoxine hydrochloride (B_6), and cyanocobalamine (B_{12}) in pharmaceutical preparations were determined by spectrophotometry and TLC-densitometry (Elzanfaly et al. 2010). Tablets were weighed and powdered. The powder equivalent to one tablet was accurately weighed and dissolved in 100 mL water in a measuring flask. Next, the solutions were filtered and 1 mL was transferred to a 100 mL measuring flask and completed to volume with water. TLC analysis were performed on silica gel $60F_{254}$ using a mobile phase of chloroform–ethanol–water–acetic acid (2:8:2:0.5, v/v/v/v). After separation, plates were air-dried and sprayed with 1% EDTA solution and dried at 110°C. The plates were scanned at 242, 291, and 360 nm for vitamin B_1, B_6, and B_{12}, respectively. TLC-densitometric method was applied successfully to simultaneous determination of the three vitamins in the range 0.1–1.5, 0.5–3.5, and 0.1–1.5 µg·spot^{-1} for B_1, B_6, and B_{12}, respectively and in their pharmaceutical preparations. The recoveries were found to be 99.66 ± 1.79, 98.89 ± 1.22, and 99.89 ± 1.71 for B_1, B_6, and B_{12}, respectively.

Bauer-Petrovska and Petrushevska-Tozi (2000) analyzed water-soluble vitamins in "Kombucha"—a curative liquor. TLC analysis was performed on silica gel 60 using water as mobile phase. The plates after development were visually examined in UV light at 254 and 366 nm. The water-soluble vitamins in the Kombucha drink were identified by comparing their R_F values with the standard substances. The R_F values were equal to 0.21, 0.73, 0.34, and 0.96 for B_1, B_6, B_{12}, and C vitamins, respectively. Aranda and Morlock (2006) determined riboflabin (B_2), pyridoxine (B_6), nicotinamide (PP, B_3), caffeine, and taurine in energy drink by HPTLC-multiple detection with confirmation by electrospray ionization mass spectrometry. The samples were degassed for 20 min in an ultrasonic bath. For quantification of B_2, B_6, PP, and caffeine, the plates were transferred to an amber sampler vial and directly used for chromatography. HPTLC was performed on silica gel $60F_{254}$ and using chloroform–ethanol–acetic acid–acetone–water (54:27:10:2:2, v/v/v/v/v) as mobile phase. On the plate, the optimal wavelengths obtained by spectra recording were similar, that is, 272, 296, 261, and 275 nm for riboflavin, pyridoxine, nicotinamide, and caffeine, respectively. Repeatabilities were determined for the same sample on the same plate, showing RSD of 0.9% for riboflavin, 0.8% for pyridoxine, and 1.1% for nicotinamide. The intermediate precisions were from the lowest to the highest level spiked: 7.4%, 4.5%, and 3.6% for riboflavin; 2.8%, 6.3%, and 6.2% for nicotnamide; and 0.5%, 2.0%, and 4.0% for pyridoxine. The LOD and the LOQ were 1 and 2 ng $\cdot$ spot^{-1} for riboflavin, 31 and 62 ng $\cdot$ spot^{-1} for nicotinamide, and 23 and 46 ng $\cdot$ spot^{-1} for pyridoxine, respectively. Ten different energy drinks were analyzed, and the values found in the respective samples for each compound were compared to the claimed target values. For identification by HPTLC/MS, all compounds were extracted online from the plate and after electrospray ionization (ESI), recorded in the positive and negative mode.

41.2　LIPOPHILIC VITAMINS

41.2.1　Vitamin A

Kahan (1967) separated on silica gel TLC plates using a mixture of cyclohexane–ethanol (97:3, v/v), vitamin A_1 alcohol, and *retro*-vitamin A_1 alcohol (R_F values were equal to 0.10 and 0.09, respectively) from pair vitamin A_1 acetate/*retro*-vitamin A_1 acetate (R_F values were equal to 0.48 and 0.45, respectively) and vitamin A_1 palmitate (R_F value 0.78) and anhydrovitamin A_1 ($R_F = 0.87$). Geometric isomers of retinol, retinal, retinal oxime, and retinyl ester using silica gel plates with cyclohexane–toluene–ethyl acetate (5:3:2, v/v) mobile phase were separated by Groenendijk et al. (1980). The R_F values were equal to 0.21, 0.23, 0.28, 0.28 for all-*trans*-, 9-*cis*, 11-*cis*, 13-*cis*-retinol, respectively. The R_F values were equal to 0.46, 0.50, 0.53, 0.55 for all-*trans*-, 9-*cis*, 11-*cis*, 13-*cis*-retinal, respectively. The R_F values were equal to 0.45, 0.40, 0.47, 0.39 for *syn* form of all-*trans*-, 9-*cis*, 11-*cis*, 13-*cis*-retinal oxime, respectively. R_F values were equal to 0.21, 0.23, 0.27, 0.33 for *anti* form of all-*trans*-, 9-*cis*, 11-*cis*, 13-*cis*-retinal oxime, respectively. The R_F values were equal 0.70 for all-*trans*-, and 11-*cis* retinyl ester. Under these conditions, the retinyl ester isomers as well as 11-*cis*, and 13-*cis*-retinol cannot be separate. However, 11-*cis* and 13-*cis*-retinol (R_F value 0.28 and 0.23, respectively) were separated with hexane–diethyl ether (1:1, v/v) as mobile phase. Retinol isomers were converted into 2,4-dinitrophenylhydrazones, and next separated on silica gel G using a mixture of petroleum benzine–chloroform–ethyl acetate (30:3:1, v/v) by Dobrucki (1979). R_F values were equal to 0.32, 0.39, and 0.16 for 2,4-dinitrophenylhydrazones of all-*trans*-, 9-*cis*, and 13-*cis*-retinals, respectively. TLC technique with densitometric detection on Silufol plates using ethyl ether–hexane (1:1, v/v) mobile phase was applied for determination of retinol acetate (Kolomoets and Bidnichenko 1992). Analyte concentration was determined by peak area. The relative error of the method was ±3.15% (Kolomoets and Bidnichenko 1992). Parizkova and Blattna (1980) separated fourteen oxidation of retinyl acetate on preparative silica gel HR with a mixture of hexane–diethyl ether (95:5 to 10:90, v/v, depending on the polarities of the substances to be separated) as mobile phase. Sliwiok et al. (1990) separated the vitamin A derivatives on RP-2 F_{254} and Kieselguhr F_{254} (impregnated with 10% paraffin oil in cyclohexane) with methanol–water (95 + 5, v/v). The best separation was obtained on Kieselguhr F_{254} (impregnated with 10% paraffin oil; the R_M values were equal to −1.59, −0.84, −0.22, and 1.18 for all-*trans* retinoic acid, all-*trans* retinal, vitamin A acetate, and vitamin A palmitate, respectively. Vitamin A compounds (Fung et al. 1978) dissolved in chloroform–methanol (1:1, v/v) containing 50 μg butylated hydroxytoluene (BHT) were spotted on silica gel 60F$_{254}$ plates and developed using acetone–light petroleum (18:82, v/v) mobile phase. The R_F values were equal to 0.71, 0.54, 0.40, 0.26, and 0.21 for retinol palmitate, retinol acetate, retinal, retinol, and retinoic acid, respectively. Each spot was scraped from the plate and extracted. The quantitative determination of vitamin A compounds was performed using the Gilford spectrophotometer.

　　Kawanabe (1985) applied thin-layer stick chromatography (TLSC is an advanced version of TLC in a cylindrical form) as the identification test method for drugs contained in preparations in the pharmacopoeia of Japan. Vitamin A palmitate (retinyl palmitate) and vitamin A acetate (retinyl acetate) were separated on a mixture of silica gel (Wako FM-BO, Wako, Japan) and microcrystalline cellulose (Abricel SF) was the benzene mobile phase. The R_F values were equal to 0.84 and 0.51 for retinyl palmitate and retinyl acetate, respectively. De Paolis (1983) described two separate methods of all-*trans*- and 13-*cis*-retinoic acids, one for gel sample, and one for cream sample. Methanolic extract of the gel formulation was analyzed directly on silica gel HPTLC plates using diethyl ether–cyclohexane–acetone–glacial acetic acid (40:60:2:1, v/v) mobile phase. This method gives fast and completed resolution of the two isomers with an LOD equal to 20 ng for both isomers. Methanolic extracts of the cream samples contain interfering excipients, which were required for a precleaning prior to chromatography. Next, the analysis was performed on a NP-TLC and RP-TLC using two-dimensional TLC plates. The cream excipients from the isomers of retinoic acid were separated on C_{18} plate (RP-TLC) using the mixture of ethanol and distilled water (80:20, v/v) as mobile phase. Next, NP-TLC on silica gel plate resolved both isomers: all-*trans*-retinoic acid with R_F value equal to 0.34, and 13-*cis*-retinoic acid with R_F value equal to 0.39 (De Paolis 1983).

α- and β-carotenes are the precursors of vitamin A. Keefer and Johnson (1972) separated retinol, retinal, β-carotene, retinyl acetate and α-carotene on magnesium hydroxide with carbon disulfide. The R_F values were equal to 0.03, 0.15, 0.29, 0.39, and 0.43, respectively (Keefer and Johnson 1972). Daurade-Le-Vagueresse and Bounias (1991) separated β-carotene, phaeophytin a, chlorophyll a and b, lutein, violaxanthin, and neoxanthin extracted from Barley leaves using on HPTLC, CN-coated plates using chloroform–hexane–methanol (5:14:1, v/v). The R_F values were equal to 0.83, 0.51, 0.41, 0.31, 0.19, 0.18, 0.13 for β-carotene, phaeophytin a, chlorophyll a and b, lutein, violaxanthin, and neoxanthin, respectively. Hodisan et al. (1997) determined the carotenoid composition, including β-carotene, of fruits of *Rosa canina* by TLC with densitometric analyses and by HPLC. TLC analysis were performed on silica gel with 15% v/v acetone in petroleum ether. The R_F values were equal to 0.96, 0.90, 0.62, 0.53, and 0.32 for β-carotene, lycopene, rubixanthin, β-chryptoxanthin, and zeaxanthin, respectively. Details about TLC analysis of carotenoids in plant and animal samples were presented in review article (Zeb and Murkovic 2010). Arthur et al. (2006) determined β-carotene in *Biomphalaria glabrata* and *H. trivolvis* by HPTLC.

Retinol was also separated by TLC in foods and next determined by the fluorimetric method (Gerstenberg 1985).

Periquet et al. (1985) determined retinyl palmitate (vitamin A palmitate) in homogenates and sub-cellular fractions of rat liver by TLC and HPLC. Thin-layer chromatography was performed on silica gel using a mixture of petroleum ether–isopropyl ether–acetic acid–water (180:20:2:5, v/v) as mobile phase. Next, the plates were examined under UV light to detect the fluorescent retinyl palmitate. The TLC results were confirmed by HPLC and spectrophotometry techniques.

Carotenoids can be detected in ultraviolet (UV) light and also in visible light; the LOD is 0.01 μg of the carotenoids. Retinal (0.02–0.03 μg) can be detected after spraying of the rhodanine. Many vitamin A compounds fluoresce yellow-green in light of 365 nm wavelength with detection limit (LOD) equal to 0.05 μg. Vitamin A compounds can be detected with antimony(III) chloride (Carr-Price reagent) (Kahan 1967; Jork et al. 1989) and antimony(V)-chlorides (Kawanabe 1985; Kolomoets and Bidnichenko 1992), with concentrated sulfuric acid (De Paolis 1983; Jork et al. 1989), with molybdophosphoric acid, and with potassium dichromate in sulfuric acid (LOD = 0.1–0.3 μg) (Bolliger and König 1969). LOD of retinol isomers converted to 2,4-dinitrophenylhydrazones of retinals is 1 μg (Dobrucki 1979). The use of bromophenol blue for the detection of vitamin A (3 μg) after NP-TLC was obtained (Wardas and Pyka 1995).

41.2.2 Vitamin D

There are various physiological forms known as vitamins D, namely vitamin D_2 (calciferol, ergo-calciferol), vitamin D_3 (cholecalciferol), phosphate esters of D_2, D_3, 25-hydroxycholecalciferol, 1,25-dihydroxycholecalciferol, and 5,25-dihydroxycholecalciferol. D_2 and D_3 vitamins are 9,10 sec-osteroids, which differ structurally in the degree of saturation of an isoprenoid side chain.

The oxidative degradation of ergocalciferol has been known for over 50 years. The degradation products of crystalline ergocalciferol were investigated by Steward et al. (1984). Numerous acidic and neutral oxidation products were formed, resulting in the complete destruction of the triene functionality. The neutral products were separated by preparative TLC.

Norman and DeLuca (1963) separated vitamin D from tachysterol and ergosterol on silica gel plates using chloroform as mobile phase; the R_F values were equal to 0.44, 0.64, and 0.25, respectively. Kocjan and Śliwiok (1994) separated vitamin D_2 (ergocalciferol), and vitamin D_3 (cholecalciferol) using RP-TLC on plates precoated with kieselguhr F_{254} impregnated with 10% paraffin oil in benzene and developed with binary mixtures of methanol–water (9.5:0.5, v/v; R_F = 0.56 and 0.48, respectively) and acetonitrile–water [(9.5:0.5 v/v; R_F = 0.50 and 0.41, respectively); and (9:1, v/v; R_F = 0.38 and 0.29, respectively). Pinelli et al. (1969) used TLC for the separation of vitamin D from cholesterol. TLC analysis was performed on silica gel G, Silica gel HF_{254}, and aluminum oxide G using 11 different mobile phases. The best separation of vitamin D from cholesterol was obtained on silica gel HF_{254} and using ethylene dichloride–benzene–acetone (90:90:20, v/v/v) mobile phase. The R_F values were equal to 0.26 and 0.40 for cholesterol and vitamin D, respectively. Hirayama and Inoue (1967) used cylindrical TLC for the determination of vitamin D in the presence of vitamin A. Vitamin A alcohol in vitamin D_2 preparations was converted to retinene on MnO_2 column. The vitamin D_2 in petroleum ether solution was pipeted on the thin-layer cylinder, and the vitamin D_2 was isolated using the mixed solvent of acetone and petroleum ether (1:4, v/v). The sample was applied to TLC under a stream of N_2 gas, and the chromatogram was developed with chloroform. These results show that with N_2 gas was used, decomposition of vitamin D_2 did not occur. Chen (1965) analyzed vitamins D_2 and D_3 on silica gel using dichloromethane mobile phase. The R_F values were equal to 0.32.

TLC separations of vitamin D hydroxymetabolites on Kieselgel $60F_{254}$ foils produced by E. Merck using chloroform–ethanol–water (183:16:1, v/v/v) mobile phase were presented by Justova and Starka (1981). The R_F values were equal to 0.69, 0.35, 0.56, 0.20, 0.41, and 0.30 for vitamin D_3 and their hydroxymetabolites: 1-OH-D_3, 25-OH-D_3, 1,25-$(OH)_2$-D_3, 24,25-$(OH)_2$-D_3, and 25,26-$(OH)_2$-D_3, respectively. In biological samples, the contents of the hydroxymetabolites of vitamin D_3 were at the nanogram level. These chromatographic conditions were applied by Esparza et al. (1982) to investigate of vitamin D_3 hydroxylated metabolites in *Solanum malacoxylon*. Also, Justova et al. (1985) applied identical conditions as was described previously for identification and determination of 1,25-dihydroxyvitamin D_3 in biological material. These conditions were also applied to determine [^{3}H]1,25-dihydroxycholecalciferol in plasma (Justova et al. 1984). Thierry-Palmer and Gray (1983) developed a solvent system for separating the mono-, di-, and trihydroxylated metabolites of vitamin D_3 by HPTLC and compared these results with the separation of these compounds obtained in conventional TLC. The obtained results indicated that the efficiency of separation by conventional TLC is very similar to that by HPTLC.

Sirec et al. (1978) described a TLC method for the determination of ergocalciferol in multivitamin tablets. Analysis of the extracted solution of lipophilic vitamins was performed on silica gel HF_{254} using cyclohexane–ethyl acetate (3:1, v/v). The vitamin spots were located in UV light at λ = 254 nm, and the spots were directly scanned at λ = 270 nm. It was stated that the results were reliable with samples containing between 0.05 and 3.0 μg of ergocalciferol. The results were compared with the determination of ergocalciferol after TLC separation with the spectrocolorimetric antimony trichloride–acetylchloride method. Results obtained by both methods in multivitamin tablets were statistically compatible.

The vitamin D was extracted from lipid fraction of milk powder and next separated by NP-TLC using cyclohexane–diethyl ether (1 + 1, v/v) mobile phase and determined by densitometry under UV (Du 1983). Similar investigations on whole milk were conducted by Grace and Bernhard (1984). The preparative TLC and the mixture of hexane–isopropanol (85 + 15, v/v) mobile phase were used for the separation of vitamin D from methyl esters. Final analysis was realized by RP-HPLC with methanol–water (97 + 3, v/v) as mobile phase. Blagojevic and Laban-Bozic (1981) separated ergocalciferol and cholecalciferol from retinyl-acetate or palmitate and tocopheryl-acetate using TLC on silica gel HF$_{254}$. After TLC separation, these vitamins were determined by spectrophotometry. This method was used for the determination of ergocalciferol in the preparations *Jecoderm* ointment and *Vidaylin-M* syrup and of cholecalciferol in the preparation *Oligovit* coated tables. Demchenko et al. (2011) elaborated and validated a selective HP-TLC method for quantification of vitamin D$_3$ for analysis of fish oils, lipid food supplements, and drugs. The oil sample was prepared by alkaline hydrolysis. Vitamin D$_3$ was quantitatively eluted with ether–petroleum ether (8:92, v/v) and the solvent was removed by evaporation at low temperature, and the residue was dissolved in 1 mL of chloroform and analyzed on HPTLC silica gel 60F$_{254}$ plates using chloroform–ether (9:1, v/v) as mobile phase. Densitometric analysis was performed at 280 nm without visualizing reagents. This method was applicable for the estimation of the vitamin D$_3$ content in fish oil from liver of cod (ProBio Nutraceuticals, Norway), fish oil Amber drop (Ekko Plus, Russia), Biafishenol, fish oil from salmon (Del'Rios, Russia), fish oil Polien (Polyaris, Russia), Children's Fish (Real Caps, Russia), Cod liver oil, Walleye Pollack liver oil, and Blue whiting liver oil. Vitamin D$_2$ was indentified and isolated in shiitake mushroom (*Lentius edodes*) by TLC using Wako-gel B5FM silica gel plates and benzene–acetone (95:5, v/v) as mobile phase as well as HPLC and identified by thermospray-interface mass spectrometry (Takamura et al. 1991). This procedure prevents the decomposition of vitamin by heat, which was a common problem in the GC-MS of vitamin D$_2$. Saden-Krehula and Tajic (1987) investigated unconjugated vitamin D and its metabolites in the pollen of *Pinus nigra* Ar. and *Pinus sylvestris* L. by TLC. Skliar et al. (1992) used TLC for purification of extract, including vitamin D$_3$ and its derivatives, from *Solanum malacoxylon*. Analysis of an extract of cod liver oils was presented by Das (1994). The sample, for example, feeding-stuff, fish-liver oils, or feed component, was saponified with the addition of an internal standard before extraction with light petroleum and cleanup of the extract by TLC. The ppb levels of vitamin D$_3$ were measured using densitometric analysis in the sample extract.

The unknown spots on the chromatographic plate in the extraction of urine sample were observed by Cheng et al. (1987). GC-MS analysis indicated that the unknown compounds were ergocalciferol metabolites from a vitamin D supplement.

Spots of vitamin D derivatives can be detected in short-wavelength UV light (LOD = 0.025–0.5 µg). The mono- and dihydroxycholecalciferols were visualized under UV light at 254 nm or by spraying with a solution of anisaldehyde in sulfuric acid (Justova and Starka 1981). Vitamins D$_2$ and D$_3$ can be detected with antimony (III) chloride (Carr-Price reagent) (Jork et al. 1989), antimony (V) chloride (LOD = 0.025–0.3 µg), with concentrated sulfuric acid (LOD = 30 µg), with tungstophosphoric acid LOD = 0.2 µg), with molybdophosphoric acid (LOD = 0.3 µg), with trichloroacetic (Jork et al. 1989), and trifluoroacetic acids (LOD = 0.1–0.2 µg) (Strohecker and Henning 1966). Vitamins D$_2$ and D$_3$ separated by NP-TLC were visualized with 0.005% aqueous new fuhsine solution (Kocjan and Śliwiok 1994). Vitamin D$_3$ was also detected with sodium hydroxide (LOD = 1–8 µg), with ferrum (III) chloride (LOD = 8 µg), with iodoplatinate (LOD = 1–8 µg) (Opong-Mensah and Porter 1988); Jork et al. 1994), and with phosphomolybdate (LOD 8 µg) (Opong-Mensah and Porter 1988). Vitamin D$_2$ was also detected by a mixture of vanillin and phosphoric acid (Skorupa and Gierak 2011). The bromocresol green and bromothymol blue as well as helasol green were also used for detection of D$_2$ and D$_3$ vitamins after NP-TLC (LOD = 5 µg) and RP-TLC (LOD = 50 µg), respectively. Fluorescent dyes: Auramine O, Rhodamine 6G, brilliant yellow 6G, acridine yellow, acridine orange, acridine red, brilliant acid, yellow 8G, fluorescein, primuline, safranin, and pyronin were also used as visualizing reagents for vitamins D$_2$ and D$_3$ (Chen 1965).

41.2.3 VITAMIN E (TOCOPHEROL, TOCOPHEROL ACETATE)

In nature, vitamin E occurs in eight different forms (α-, β-, γ-, and δ-tocopherols and α-, β-, γ-, and δ-tocotrienols) with varying biological activities. Of these eight compounds, α-tocopherol is reported to have the highest biological activity (Hosomi et al. 1997).

Sliwiok and Kocjan (1992) separated α-, β-, γ-, and δ-tocopherols on kieselguhr G plates impregnated with a 10% solution paraffin oil in hexane and using a mixture of methanol–water (9.5:0.5, v/v) mobile phase. The R_F values were equal to 0.26, 0.48, 0.50, and 0.65, respectively. The separations of α-, β-, γ-, and δ-tocopherols were also performed on RP-C18-HPTLC using methanol–water, ethanol–water, and n-propanol–water mobile phases in different volume compositions (Pyka and Niestrój 2001; Pyka and Sliwiok 2001). Sliwiok et al. (1993) separated enantiomers of DL-α-tocopherol on Chiralplates (Machery-Nagel, Germany) using 2-propanol–water–methanol (17:2:1, v/v) as mobile phase. The R_F values were equal to 0.72 and 0.62. Tyrpień et al. (2003) separated α-, β-, γ-, δ-tocopherols and α-tocopherol acetate on the C_{30} plate using methanol as mobile phase.

Hachuła and Buhl (1991) described the analytical method of control of α-tocopherol in "Vitaminum E" capsules (Polfa, Poland) and soybean oil. Determination of α-tocopherol (after extraction of samples) was performed by TLC on silica gel plates with benzene–ethanol (99:1) as mobile phase. The results obtained by the authors indicate that they were in good agreement with those expected, and the precision of the measurements was satisfactory. Pyka et al. (2011) elaborated the conditions for quantitative determination of tocopherol acetate in oral fluid *Vitaminum E* by a TLC-densitometric method with regard to obligatory validation (ICH Harmonised Tripartite Guideline 2005; Ferenczi-Fodor et al. 2010). The conditions for separation of tocopherol acetate from its related substance, namely tocopherol, were elaborated. The analysis was performed on silica gel $60F_{254}$ using chloroform + cyclohexane (11:9, v/v) mobile phase. These conditions resulted in optimum migration of tocopherol acetate ($R_F = 0.47 \pm 0.02$) and resolution of the drug from its related substance, namely tocopherol ($R_F = 0.38 \pm 0.02$) (Figure 41.1). Fundamental wavelengths of tocopherol acetate and tocopherol occur at 202 and 272 nm, respectively (Figure 41.2). Densitogram of 6.00 μg tocopherol acetate standard at 202 nm is presented in Figure 41.3. Typical densitograms of 6.00 μg tocopherol acetate coming from a pharmaceutical preparation is presented in Figure 41.4. Two peaks are observed on this densitogram, namely tocopherol acetate (TA), and assistant substance occurs in oral fluid *Vitaminum E* (S). The comparison of spectrodensitogram of tocopherol acetate standard with spectrodensitogram of tocopherol acetate from *Vitaminum E* sample and spectrodensitogram of assistant substance occurs in *Vitaminum E* analyzed by NP-TLC as presented in Figure 41.5. It was observed that excipients present in the formulation did not interfere with the tocopherol acetate peak. The proposed NP-TLC-densitometric method was validated by specificity, range, linearity, accuracy, precision, detection limit, quantitative limit (Table 41.1, Figure 41.6), and robustness (Figure 41.7). These results (Tables 41.1 and 41.2) confirmed statistically that the NP-TLC-densitometric method can be used as a substitute method in relation to GC with flame ionization method recommended by United States, British, and Polish Pharmacopeias for the analysis of tocopherol acetate in pharmaceutical preparations (Pyka et al. 2011).

α- and γ-tocopherols and plastochromanol-8 were determined in linseed oil by Olejnik et al. (1997) using TLC (silica gel; chloroform as first mobile phase, and after reapplication, hexane-diethyl ether (19:1, v/v) as second mobile phase) for purification of unsaponifiable fraction of linseed oil. El-Mallah et al. (1990) used TLC to estimate the deodorizer distillate of soybean oil as a source for tocopherols, which can be used instead of synthetic antioxidants as food preservatives. Analytical and preparative TLC were carried out on silica gel $60GF_{254}$ (E. Merck) with n-hexane–benzene–diethyl ether (40:40:20, v/v) mobile phase. TLC was also used for separation and determination of tocopherol isomers in wheat germ, cottonseed and soybeans (Tatsumi and Izumitani 1981), and other food (Semenova and Kuznetsov 1984; Surai 1988). Hodisan et al. (2008) elaborated the condition for identification and quantification of tocopherols in vegetable oils by TLC. α-, γ-, and δ-tocopherols were determined in samples of vegetable oils, namely sunflower, olive, corn, soy, and almond oils. SPE of tocopherols from vegetable oils was performed after saponification of the oils. The saponification was realized using a solution of 50% NaOH in water. The tocopherols retained on the SPE cartridge were eluted with the elution solvent (hexane–ethyl acetate, 85:15, v/v). Tocopherol content was realized by NPTLC on silica gel 60 using chloroform as mobile phase. After drying of plates, the substances on chromatograms were detected with a

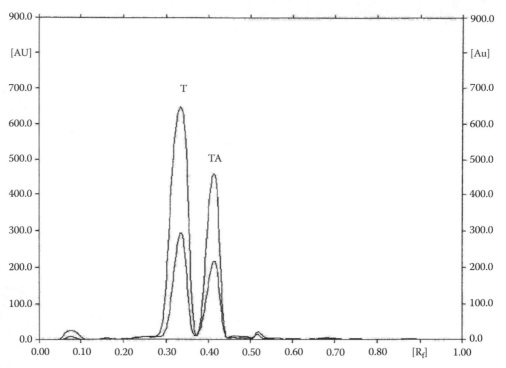

FIGURE 41.1 Densitogram obtained from tocopherol acetate standard (TA) spiked with related substance, namely tocopherol (T) analyzed by NP-TLC (silica gel 60 F$_{254}$; mobile phase: chloroform + cyclohexane, 11:9, v/v) and scanned at 202 and 272 nm. (Reprinted from Pyka, A. et al., *Liq. Chromatogr. Relat. Technol.*, 34, 2548, 2011. With permission.)

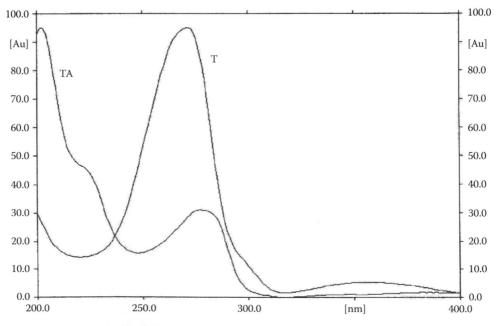

FIGURE 41.2 Spectrodensitograms of tocopherol acetate standard (TA), and tocopherol (T) analyzed by NP-TLC (silica gel 60 F$_{254}$; mobile phase: chloroform + cyclohexane, 11:9, v/v). (Reprinted from Pyka, A. et al., *Liq. Chromatogr. Relat. Technol.*, 34, 2548, 2011. With permission.)

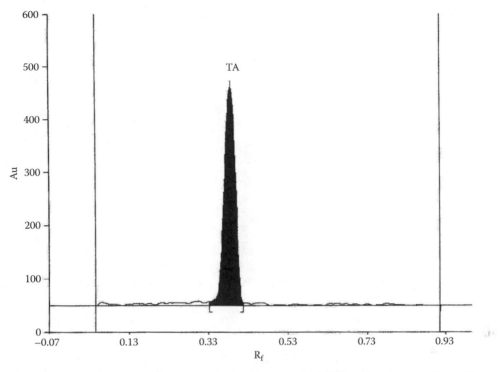

FIGURE 41.3 Densitograms of 6.00 μg tocopherol acetate standard (TA) analyzed by NP-TLC (silica gel 60 F_{254}; mobile phase: chloroform + cyclohexane, 11:9, v/v) and scanned at 202 nm. (Reprinted from Pyka, A. et al., *Liq. Chromatogr. Relat. Technol.*, 34, 2548, 2011. With permission.)

mixture of bipyridyl in methanol (0.5%) and ferric chloride in methanol (0.2%) in equal volume. Pink-red spots were obtained on a white background for all the tocopherol isomers investigated. The order of the elution of tocopherols was δ-tocopherol < γ-tocopherol < α-tocopherol. It was stated that the amounts of tocopherols depend on the type of oil. γ-Tocopherol was determined in sesame oils (Shahidi et al. 1997). Oils were saponified by alkaline hydrolyze. Unsaponificables were extracted with diethyl ether. Next, ether extracts were washed with water and then dried over anhydrous sodium sulfate. Ether was removed, and the residues were dissolved and kept in 5 mL of mixture of chloroform–diethyl ether (4:1, v/v) at –20°C for further TLC analysis. TLC analysis was performed on silica gel using a chloroform–hexane–methanol (60:30:2, v/v/v) mobile phase. The substances were visualized by spraying with 10% phosphomolybdic acid in ethanol–diethyl ether (1:1, v/v) and heating at 110°C for 5 min. γ-Tocopherol was separated from desmethylsterols, mono-methylsterols, sesamin, seamolin, and three unknown substances. Next, quantitative analysis of tocopherols was performed by NP-HPLC. Rybakova et al. (2008) elaborated the condition of TLC with densitometry for the identification and determination of α-tocopherol in vegetable oils and oil extracts. Analysis was performed on silica gel using chloroform as mobile phase. α-Tocopherol was detected using concentrated nitric acid, which gives a reddish-orange spot. The detection limit using this visualizing reagent was 3×10^{-6} g. The content of α-tocopherol in oil extracts was equal between 0.03% and 0.450%. Askinazi et al. (1990) also described a modified TLC method for determining the isomeric composition of tocopherol. Separation of tocopherol isomers from the nonsaponifiable fraction of oils and margarine, which was dissolved in 2 mL of benzene and ana-lyzed on Silufol UV-254 plates using chloroform as mobile phase. Results obtained by TLC method were compared with those obtained by the GLC method. As in the case of the GLC method, the TLC method can be recommended to analyze the tocopherol isomers in fats, vegetable oils, and drugs based on them (Askinazi et al. 1990). Ghoulipour et al. (2010) reported the chromatographic

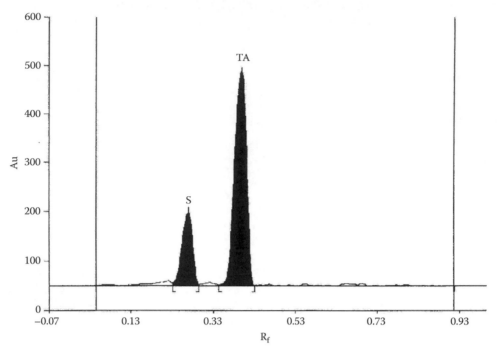

FIGURE 41.4 Densitogram of *Vitaminum E* sample with 6.00 µg tocopherol acetate declared by manufacturer analyzed by NP-TLC (silica gel 60F$_{254}$; mobile phase: chloroform + cyclohexane, 11:9, v/v; where: TA—tocopherol acetate, S—assistant substance occurs in *Vitaminum E*) and scanned at 202 nm. (Reprinted from Pyka, A. et al., *Liq. Chromatogr. Relat. Technol.*, 34, 2548, 2011. With permission.)

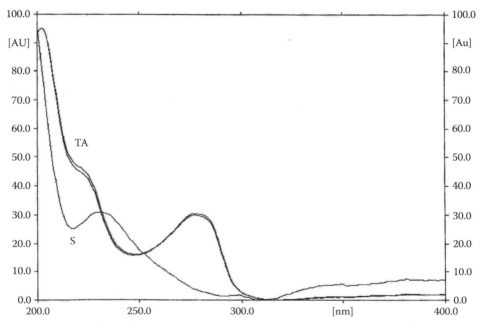

FIGURE 41.5 Comparison of spectrodensitogram of tocopherol acetate standard with spectrodensitogram of tocopherol acetate from *Vitaminum E* sample and spectrodensitogram of assistant substance occurs in *Vitaminum E* analyzed by NP-TLC (silica gel 60 F$_{254}$; mobile phase: chloroform + cyclohexane, 11:9, v/v; where: TA—tocopherol acetate, S—assistant substance occurs in *Vitaminum E*). (Reprinted from Pyka, A. et al., *Liq. Chromatogr. Relat. Technol.*, 34, 2548, 2011. With permission.)

TABLE 41.1

Method-Validation Data for the Quantitative Determination of Tocopherol Acetate by NP-TLC with Densitometry[a]

Method Characteristic	Results
Specificity	Specific
Range (μg spot^{-1})	2.00–8.00
Linearity (μg spot^{-1})	A = 2066.4($\pm$142.1)x + 1371.2($\pm$26.4)
	n = 7 r = 0.999 F = 2001
Accuracy	
For 80% tocopherol acetate added (n = 3)	R = 101.3%; CV = 0.99%
For 100% tocopherol acetate added (n = 3)	R = 99.8%; CV = 0.78%
For 120% tocopherol acetate added (n = 3)	R = 101.5%; CV = 1.15%
Detection limit (DL) (μg spot^{-1})	0.05
Quantitation limit (QL) (μg spot^{-1})	0.15
Precision (CV, [%])	
Repeatability	
For 3.00 μg spot^{-1} (n = 3)	1.28
For 5.00 μg spot^{-1} (n = 3)	1.18
For 7.00 μg spot^{-1} (n = 3)	1.65
Intermediate	
For 3.00 μg spot^{-1} (n = 3)	1.89
For 5.00 μg spot^{-1} (n = 3)	1.42
For 7.00 μg spot^{-1} (n = 3)	1.95
Robustness (CV, %)	Robust

Source: Data from Pyka, A. et al., *Liq. Chromatogr. Relat. Technol.*, 34, 2548, 2011.

[a] A—peak area [AU], x—amount [μg spot^{-1}] of drug analyzed, r—correlation coefficient, R—recovery [%], CV—coefficient of variation [%].

behavior of 30 food additives, including vitamin E, on thin layers of titanium (IV) silicate ion exchanger using several aqueous, organic, and mixed mobile phases. Rapid separation of vitamin E additive from many other food additives, and quaternary and ternary separation were presented. The fats obtained from the seed of cotton plants grown in different regions for separating α-, β-, γ-, and δ-tocopherols were investigated by Bapcum (1984) using TLC (silica gel H$_{254}$ and with light petroleum–isopropyl ether–acetone–ethyl ether–anhydride acetic acid on volume composition 160:30:10:2:1). The quantity of tocopherols was determined by phothometric methods. TLC on silica gel 60F$_{254}$ plates and chloroform–isooctan (1:1, v/v) mobile phase were also used for quantitative determination of α-tocopherol in foods, feeding stuffs, and vitamin concentrate (Manz et al. 1979). Kivcak and Akay (2005) determined the α-tocopherol in *Pistacia lentiscus*, *P. lentiscus* var. *chia*, and *Pistacia terebinthus* using TLC with densitometry and colorimetry. A sample of leaves was accurately weighed, air-dried, powdered, extracted using *n*-hexane, next filtered, and distilled. TLC analysis was performed on silica gel 60F$_{254}$ using cyclohexane–diethyl ether (4:1, v/v) mobile phase. The developed plates were air-dried, then oven-dried (15 min at 100°C), and sprayed with 10% CuSO$_4$–phosphoric acid followed by charring at 190°C (10 min) and next densitometric scanned. Kivcak and Mert (2001) also determined α-tocopherol in *Arbutus unedo* by TLC-densitometry, and the obtained results were compared with those obtained by colorimetry.

Leray et al. (1997) analyzed tocopherols, cholesterol, and phospholipids in the same minute samples of human platelets and on human cultured endothelial cells. NP-TLC on Whatman LK5 silica gel plates impregnated with boric acid and chloroform–ethanol–water–triethylamine (35:30:7:35, v/v)

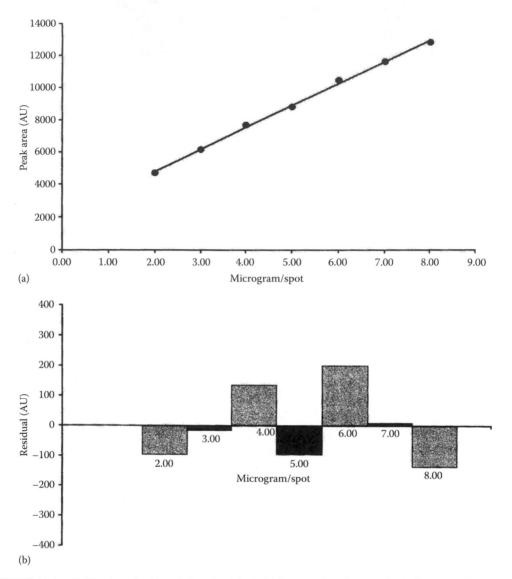

FIGURE 41.6 Calibration plot (a) and plot of residuals (b) for tocopherol acetate in the linear working range. (Reprinted from Pyka, A. et al., *Liq. Chromatogr. Relat. Technol.*, 34, 2548, 2011. With permission.)

containing 0.10 g/L of ascorbic acid and 0.15 g/L of butylated hydroxytoluene as mobile phase were used for the separation of phospholipids, cholesterol, and tocopherols in total lipid chloroform extracts. The visualization was performed by UV after a primuline spray. R_F values of cholesterol and tocopherols were equal to 0.86–0.89. Next, α-, γ-, and δ-tocopherols, tocopherol acetate, and cholesterol were determined by HPLC (Leray et al. 1997). TLC was also used for the estimation of the antioxidant activity of vitamin E and antioxidant activity of different biological samples containing the α-tocopherol (Alary et al. 1982; Cavin et al. 1998; Davino et al. 1998). Preparative TLC was also used for the purification of the extracts of the biological samples. The different antioxidants containing the α-tocopherol can be qualitative evaluated in extract by TLC separation (Gertz and Herrmann 1983). Silica gel plate and cyclohexane–dioxane–acetic acid (80:15:5, v/v) mobile phase were used for the separation of α-tocopherol (R_F = 0.51) from five other antioxidants (Alary et al. 1982). However, on RP-18 plate using methanol–water–acetic acid (82:16:2, v/v) α-tocopherol had R_F = 0, and it was separated from ten other antioxidants (Alary et al. 1982). Lovelady (1973) reported a TLC-GLC

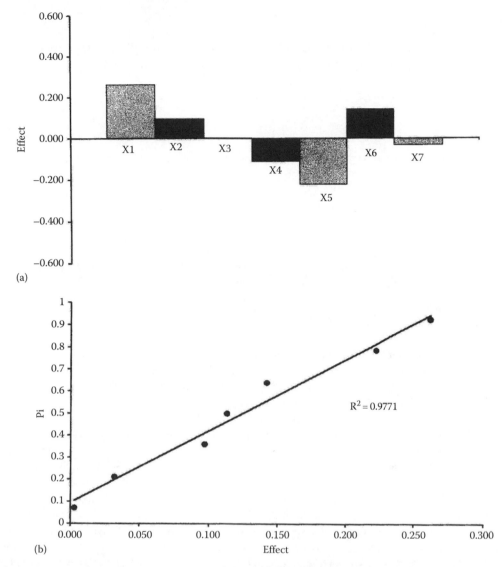

FIGURE 41.7 Robustness test: the effects of factors (a), and half-normal probability plot of effects (b). (Reprinted from Pyka, A. et al., *Liq. Chromatogr. Relat. Technol.*, 34, 2548, 2011. With permission.)

system to separate the individual tocopherols for qualitative and quantitative determination in blood plasma and red blood cells. TLC was performed on silica gel G plates using the following solvent system: cyclohexane–n-hexane–isopropyl ether–ammonium hydroxide (conc.) (40:40:20:2, v/v/v/v). The individual tocopherols were identified by their R_F values. Quantitative analysis of tocopherols (α-, β-, γ-, and δ-) was performed by GLC. Bieri and Prival (1965) described a rapid, TLC procedure that makes possible the separation of α- from β-, γ-, and δ-tocopherols. TLC was performed on silica gel G using a benzene mobile phase. With standards α-, β-, γ-, and δ-tocopherols, three distinct spots were optained representing α, β + γ, and δ-tocopherols. Cholesterol, retinol, δ-tocopherol, β + γ-tocopherols, vitamin K_1, and β-carotene were also separated in these chromatographic conditions. α-Tocopheryl quinone, ubiquinone-50, ubichromenol, α-tocopherol, and martius dimer can also be separated in these chromatographic conditions. The described procedure was used for the determination of vitamin E in serum. Determination of α-tocopherol in serum by one- and two-dimensional TLC, colorimetric, and GLC methods were compared.

TABLE 41.2

Statistical Data Concerning Results of Quantitative Determination of Tocopherol Acetate in 1 mL of Oral Fluid *Vitaminum E* Investigated by NP-TLC with Densitometry

	Analysis Performed by NP-TLC Technique on Plates (# 1.05554)
Number of analysis	10
The label claim of tocopherol acetate in oral fluid *Vitaminum E* (µg)	300
Average amount of tocopherol acetate (µg)	314
Minimum amount of tocopherol acetate (µg)	308
Maximum amount of tocopherol acetate (µg)	320
Variance	11.6
Standard deviation (SD)	3.4
Coefficient of variation (CV, %)	1.13%
Confidence interval of arithmetic mean with confidence level equal 95%	$\mu = 314.0 \pm 2.5$
Amount of tocopherol acetate (%) in relation to the label claim	104.7%

Source: Data from Pyka, A. et al., *Liq. Chromatogr. Relat. Technol.*, 34, 2548, 2011.

Vitamin E compounds can be detected (about 20 µg) as dark spots in UV light. Nonspecific visualization procedures for tocopherols and tocotrienols are based on spraying with sulfuric acid, molybdophosphoric acid (Seher 1959; Bapcum 1984; Jork et al. 1989), antimony(V) chloride, dipyridyl-iron reagent (Jork et al. 1989; Pyka and Sliwiok 2001; Skorupa and Gierak 2011), bathophenanthroline (Tatsumi and Izumitani 1981; Desai 1984), phosphomolibdic acid (Shahidi et al. 1997), concentrated nitric acid (Rybakova et al. 2008), and the Emery-Engel reagent (Askinazi et al. 1990). α-Tocopherol can be also visualized by a coupled redox-complexation reaction with iron (III), phenanthroline, and bromophenol blue (Hachuła and Buhl 1991). Wardas and Pyka (1985) applied 11 visualizing reagents in the form of 13 visualizing systems for detection of E vitamin in adsorption and partition of TLC. Aniline blue, alkaline blue, and bromothymol blue were also effective for detection of E vitamin (1 µg) after NP-TLC. By using β-carotene as a TLC spray reagent, tocopherols (10 µg) appeared as yellow spots against a white background (Silva et al. 2001). Pyka (2009b) investigated six dyes as new visualizing reagents, namely gentian violet, methylene violet, methylene blue, methyl green, malachite green, and Janus blue for the detection of (±)-α-tocopherol and (+)-α-tocopherol acetate on silica gel 60 using toluene as a mobile phase. Rhodamine B and 2,2′-bipyridine–iron (III) chloride reagent were used as the comparative visualizing reagents. Among all studied new visualizing reagents, methylene violet and methyl green were the best to detect (±)-α-tocopherol. However, the best detection way of (+)-α-tocopherol acetate was the densitometric method without using a visualizing reagent. Whereas among all studied new visualizing reagents, gentian violet, methyl green, and Janus blue were the best for detection of (+)-α-tocopherol acetate. These visualizing reagents had similar detection properties of (+)-α-tocopherol acetate in relation to Rhodamine B (Pyka 2009b).

41.2.4 VITAMIN K

There are different physiological forms known as vitamins K, namely vitamin K_1 (phylloquinone, phytonadione), and vitamin K_2 (farnoquinone). There are active analogs and related compounds known as vitamins K, namely menadiol diphosphate, menadione (vitamin K_3), menadione bisulfite, phthiocol, synkayvite, menadiol (vitamin K_4), menaquinone-n (MK-n), ubiquinone (Q-n), and plastoquinone (PQ-n) (Kutsky 1973).

Lichtenthaler (1984) separated vitamins K_1 (phylloquinone), $K_{2(20)}$ (menaquinone-4), $K_{2(45)}$ (menaquinone-9), $K_{2(50)}$ (menaquinone-10), and K_3 (menadion) by NP-TLC on silica gel using different mobile phases. The best separations were obtained using the petroleum hydrocarbon–ethyl ether (5:1, v/v) mobile phase; the R_F values for these vitamins were equal to 0.63, 0.58, 0.64, 0.66, and 0.34, respectively. The previously mentioned vitamin K were also separated by RP-TLC on silica gel impregnated with 5% liquid paraffin in petroleum hydrocarbon using acetone–water mobile phase in different volume compositions. The best separations were obtained using acetone–water (19:1, v/v) mobile phase; the R_F values for these vitamins were equal to 0.59, 0.77, 0.34, 0.26, and 0.90, respectively. Lichtenthaler (1984) also separated vitamins K_1, $K_{2(20)}$, and K_3 by argentation TLC on silica gel impregnated with $AgNO_3$ and activated at 100°C for 1 h. The best separations were obtained using mobile phases: hexane–ethyl acetate–isopropyl ether (2:1:1, v/v/v), petroleum hydrocarbon–chloroform–acetone (50:10:24, v/v/v). For example, using second mobile phase the R_F values were equal to 0.76, 0.62, and 0.56 for vitamins K_1, $K_{2(20)}$, and K_3, respectively. In argentation TLC vitamins K_1, K_2, and K_3 were also separated (Lichtenthaler et al. 1982). Pyka and Gurak (2009) analyzed phylloquinone (vitamin K_1) using the $RP8F_{254s}$, and $RP18F_{254s}$ (E. Merck) plates and methanol as a mobile phase. The obtained densitograms show five and six chromatographic spots on $RP18F_{254s}$ and $RP8F_{254s}$ plates, respectively. Vitamin K_1 had $R_F = 0.19$ and $\lambda_{max} = 270$ nm on $RP18F_{254s}$ plate; however, the remaining densitometric bands of existing compounds besides vitamin K_1 had the R_F values equal to 0.08, 0.36, 0.40, and 0.72. Vitamin K_1 had $R_F = 0.45$ and $\lambda_{max} = 254$ nm on $RP8F_{254s}$ plate; however, the impurities had the R_F values equal to 0.27, 0.54, 0.59, 0.62, and 0.75, respectively. Rittich et al. (1977) reported the separation of naphthoquinones and lipophilic vitamins on silica gel and on silica gel impregnated with polyethylene glycol (PEG) 200. Using system silufol/benzene, the R_F values were equal to 0.10, 0.28, 0.61, 0.30, 0.13, 0.08, 0.20, and 0.09 for 1,2-napthoquinone, 1,4-napthoquinone, phylloquinone, menadione, menadiol diacetate, calciferol, retinyl acetate, and cholesterol, respectively. Using system silufol/benzene–ethyl methyl ketone (3:1, v/v), the R_F values were equal to 0.48, 0.61, 0.80, 0.65, 0.59, 0.53, 0.52, and 0.46 for 1,2-napthoquinone, 1,4-napthoquinone, phylloquinone, menadione, menadiol diacetate, calciferol, retinyl acetate, and cholesterol, respectively. Using system silica gel impregnated with PEG/benzene–light petroleum (1:1, v/v), the R_F values were equal to 0.25, 0.54, 0.75, 0.59, 0.41, 0.30, 0.47, and 0.28 for 1,2-napthoquinone, 1,4-napthoquinone, phylloquinone, menadione, menadiol diacetate, calciferol, retinyl acetate, and cholesterol, respectively.

The sensitive spectrophotometric method for the determination of vitamin K_4 in drugs (Styptobion, E. Merck) after chromatographic separation on silica gel using benzene–acetone (9:1, v/v) mobile phase was developed by Hachuła (1997). Beer's law was obeyed for 0.3–1 $\mu g \cdot mL^{-1}$ vitamin K_4 ($\varepsilon = 196\,000$). Pharmaceutical formulas (tablets, coated tablets, and injection solution) containing vitamin K_1 (phylloquinone), vitamin K_3 (menaphthone), or vitamin K_4 (acetomenaphthone) were also analyzed by TLC-spectrophotometry methods (Marciniec and Stachowicz 1989). Silica gel HF_{254} (E. Merck) and benzene–ethyl acetate (97:3, v/v) mobile phase for K_1, and cyclohexane–chloroform–methanol–acetic acid (2:15:3:1, v/v) mobile phase for K_3, and benzene–acetone (9:1, v/v) mobile phase for K_4 were used. The spots were visualized in UV light at 254 nm. The visualized spots were scraped off; extracted with ethanol (for K_1), water (for K_3), or methanol (for K_4); and next the vitamins were determined at 251, 234, or 225 nm, respectively, by using suitable calibration graphs (Marciniec and Stachowicz 1989).

Sakano et al. (1986) extracted vitamins K from human and animal (monkey, dog, guinea pig, rat, mouse, rabbit, and chicken) feces and purified by TLC on silica gel $60F_{254}$ using the petroleum ether–diethyl ether mixture (85:15, v/v) mobile phase. In this chromatographic system, the R_F values of vitamin K standards were in the range of 0.53 (MK-4) to 0.61 (MK-13). Next, vitamins K were measured by RP-HPLC with fluorometric detection. The vitamin K in human liver was measured by Usui et al. (1989) by gradient elution HPLC, after preparative TLC, using platinum-black catalyst reduction and fluorimetric detection. Preparative NP-TLC was performed on silica gel 60 F_{254} using petroleum ether and diethyl ether (85:15, v/v) mobile phase. In addition, menaphthone (vitamin K_3) can be determined by GC after preparative TLC using benzene or methanol–benzene (1:2 or 1:4, v/v) mobile phases (Haiduc et al. 1988). Vitamin K was determined in cattle and liver samples by Madden and Stahr (1993). RP-TLC plate with indicator and methylene chloride–methanol (7:3, v/v) mobile phase were selected for direct method of detection and indication of vitamin K. The R_F value of vitamin K was equal to 0.75. GC, and densitometry can be used to quantitate vitamin K present in bovine liver. Mass spectroscopy can be used to confirm vitamin K present in the extracts. Hirauchi et al. (1989) elaborated NP-TLC method on silica gel 60 F_{254} for the separation of vitamin K in tissue homogenate extracts. Separated vitamins were extracted from the silica gel and next determined by HPLC. This method was applied in the analysis of liver, spleen, kidney, heart, and muscle. Recoveries were 73.5%–91.8%, and detection limits were in the $pg \cdot g^{-1}$ or $pg \cdot L^{-1}$ range. The phylloquinone (K_1) and menaquinone-4 (MK-4) in plasma and liver were determined by Sakamoto et al. (1996). The NP-TLC on silica gel (E. Merck) and 85% petroleum ether–15% ethyl ether mobile phase were used for purification. However, final separation was performed by HPLC technique. This method is useful for K vitamin studies on rats, which require micro and multisampling methods. Rutherford et al. (1991) isolated menaquinone-7 from Pseudomonas N.C.I.B. 10590 and identified by RP-TLC and GC-MS analysis. Ubiquinones extracted from 24 strains of *Legionella pneumophila* and from 44 strains of other *Legionella* species were also analyzed by RP-TLC on $KC_{18}F$ TLC plates (Whatman) using a mixture acetone–water (19:1, v/v) mobile phase. This method was reproducible, both for qualitative and semiquantitative determination of ubiquinone profile, and provided information to aid in the identification of species of *Legionella* (Mitchell and Fallon 1990). Ubiquinones were also analyzed in the rat tapeworm *Hymenolepis diminuta*; TLC was used for isolation and HPLC for determination (Johnson and Cain 1985; Proksa and Slavikova 1990). Menaquinone was identified by TLC in a purple bacterium (Schoepp-Cothenet et al. 2009).

A simple and highly sensitive method of quantitative determination of the vitamin K amount in *Milfoil plant* was elaborated by Mazulin and Kaloshina (1997). The presence of vitamin K was confirmed using TLC. TLC analysis was performed on Silufol UV254 plates (Kavalier, Czech Republic) using two mobile phases: cyclohexane–diethyl ether (4:1, v/v) and hexane–diethyl ether–acetic acid (9:1:0.1, v/v). Next, a portion of extracted, filtered, coagulated, cooled, and filtered sample was used for spectrophotometric determination at 265 nm. Vitamin K_3 (menadione) was extracted from food products (butter, margarine, yogurt, beef, pork, chicken, cheese, chicken egg, ordinary liquid milk) and next was purified on a Sep-Pak silica cartridge and/or a Sep-Pak silica cartridge followed by TLC, and then measured by HPLC on an ODS-UH column with 45% dioxane saturated with argon and containing 0.2% $NaClO_4$. The detection limit for vitamin K_3 was equal to 50 $pg \cdot g^{-1}$ or $pg \cdot mL^{-1}$ in foods (Hirauchi et al. 1990).

All lipoquinones can be detected in daylight and, with high sensitivity, in UV light. Vitamin K compounds can be detected with iodine vapor, with concentrated sulfuric acid followed by heating (LOD = 3 μg), with molybdophosphoric acid (LOD = 0.5 μg) (Strohecker and Henning 1966; Bolliger and König 1969; De Leenheer and Lambert 1996; Mazulin and Kaloshina 1997), and with potassium hexaiodoplatinate reagent (Jork et al. 1994). Vitamin K was also detected in long- and short-wave UV light; under short-wave UV light, vitamin K showed an intense purple color (Madden and Stahr 1993). Vitamin K_1 was also detected with antimony (III) chloride (LOD = 8 μg), with ferrum (III) chloride (LOD = 8 μg), with sulfuric acid (LOD = 8 μg), with iodoplatinate (LOD = 1–8 μg), with phosphomolybdate (LOD = 8 μg) (Opong-Mensah and Porter 1988), with 0.2% aniline-naphthalenesulfonic acid in methanol and 0.05% rhodamine B in ethanol (UV light), with 0.05% rhodamine 6G in acetone (UV light), and with Dragendorff's reagent (Lichtenthaler 1984). Detection limits of naphthoquinones with perchloric acid were equal to 3, 2, 1, and 0.4 μg spot^{-1} of 1,4-napthoquinone, phylloquinone, menadione, and menadiol diacetate (Rittich et al. 1977).

41.2.5 Mixture of Lipophilic Vitamins

Fat-soluble vitamins A, D_2, and E in multivitamin products were qualitatively and semiquantitatively analyzed by TLC (Hossu et al. 2006). TLC analyses were performed on silica gel 60F_{254} and two different mobile phases: hexane–ether (9:1, v/v) and benzene–chloroform (1:1, v/v). Spots of vitamins were visualized in UV light at 254 nm. Using the first mobile phase, the R_F values were equal to 0.463 and 0.065 for vitamins A, and D_2, respectively. Using the second mobile phase, the R_F values were equal to 0.186 and 0.703 for vitamins D_2, and E, respectively. The semiquantitative determination of vitamin in applied conditions given here resulted in a two-order polynominal regression curve in the field 5–85 µg; as for vitamins D_2 and E, linear regression in the fields: 0.5–10.5 µg for vitamin D_2 and 5–80 µg for vitamin E was received. Lipophilic vitamins A, D_2, and E (α-tocopherol) (R_F values 0.34, 0.44, and 0.62, respectively) were separated on silica gel plates using a mixture of benzene–chloroform–acetone (88.5:8.8:2.7, v/v) as mobile phase (Ranny 1987). Determination of vitamin D_2 and K_1 in the presence of rutin added as stabilizer was described by Bączyk et al. (1981). TLC was performed on silica gel H using chloroform as mobile phase. The R_F values were 0.60 for vitamin K_1, 0.34 for vitamin D_2 and 0 for rutin. Quantitative determination of vitamins D_2 and K_1 was conducted by spectrophotometric analysis. Opong-Mensah and Porter (1988) separated the diphacinone, pindone, valone, warfarin, bromadiolone, vitamins K_1 and D_3 on silica gel plates, and three mobile phases were tested. None of the mobile phases used alone could separate all seven compounds. However, vitamins K_1 and D_3 were satisfactory separated using dichloromethane–methanol–acetic acid (45:4:1, vv) mobile phase (R_F values were equal to 0.75 and 0.59, respectively), and using chloroform–methanol (97:3, v/v) mobile phase (R_F values were equal to 0.92 and 0.68, respectively). Ramic et al. (2006) elaborated the TLC method to investigate possible chemical interactions of vitamins A and D with frequently used therapeutics (estrogens, progestins, corticosteroids, nonsteroidal anti-inflammatory drugs, and other drugs).

Borodina et al. (2007) determined α-tocopherol and vitamin D_2 (ergocalciferol) in bur marigold oily extract and grape seed oil. The samples were prepared by alkaline hydrolysis of an oil according to GOST State Standard 30417 and the independent recovery of vitamins by extraction with organic solvents. The samples of the extracts were separated on silica gel (Sorbfil) plates using n-octane–diethyl ether (7:1, v/v) and tetrachloromethane–diethyl ether (4:1, v/v) mobile phases for α-tocopherol and ergocalciferol, respectively. Nitric acid and sulfuric acid were used as visualizing reagents for α-tocopherol and ergocalciferol, respectively. The detection limit limits of α-tocopherol and ergocalciferol were equal to 2.50 and 0.08 $mg \cdot mL^{-1}$, respectively. The contents of α-tocopherol and ergocalciferol in bur marigold oily extract were equal to 40 ± 2 mg per 100 g and 63 ± 2 mg per 100 mg, respectively. The content of ergocalciferol in grape seed oil was equal to 15 ± 1 mg per 100 g. The determinations of α-tocopherol and ergocalciferol were compared by TLC and HPLC procedures. The resolving power, sensitivity, and performance HPLC were higher than those for TLC. However, HPLC also offers more difficulties in selecting a solvent and a separation medium. Ersoy and Duden (1980) determined the fat-soluble vitamins D-α-tocopherol and K_1 (phyllochinone), as well as β-carotene in spinach by reflectance photometry after chromatography of the nonsaponified raw extracts on HPTLC silica gel plates using benzene or petroleum ether–benzene (6:1, v/v) as mobiles phases. Dinnen and Ebisuzaki (1997) used TLC for purification and determination of lapachol and vitamin K in pau d'arco. Vostokov and Kartaszov (2006) chromatographed vitamins A, D, and E on silufol UV-254 plates using n-hexane–acetone (4:1, v/v) mobile phase. After separation, and visualization on UV light, chromatographic zones were scraped and eluted with chloroform and filtered to flasks (25 mL). The 25% solution of $SbCl_3$ in chloroform was added to flask. Next, quantitative determination of vitamin A was performed by spectrophotometric method (λ = 620 nm). Analogical analysis was performed for vitamins D and E, after reaction with 2,2-dipirydyl. Spectrophotometric analysis was performed at 490 and 520 nm for vitamin D and E, respectively. The elaborated method was used for quantitative analysis of vitamins A, D, and E in fodder bioraw (Vostokov and Kartaszov 2006).

Retinol and α-tocopherol in plasma were determined by Chavan and Khatri (1992). A sample was mixed with methanol and then extracted with heptane containing α-tocopheryl acetate (internal standard). NP-TLC was performed on silica gel F_{254} HPTLC plate and chloroform–cyclohexane (11:9, v/v) mobile phase and evaluated densitometrically with a Camag TLC Scanner II or by diffuse-reflectance absorbance at 290 nm.

41.3   MIXTURE OF HYDROPHILIC AND LIPOPHILIC VITAMINS

Baranowska and Kądziołka (1996) developed a chromatographic system for the RP-TLC separation of water- and fat-soluble vitamins. Water-soluble vitamins (C, PP, B_1, B_6, and rutin) were prepared in methanol; however, fat-soluble vitamins (E, E-acetate, D_3, A-acetate) were prepared in ethanol. RP-TLC analysis was performed on RP-18 plates using water–methanol (5:4, v/v) and water–acetic acid (7:1, v/v) for water-soluble vitamins and acetonitrile–benzene–chloroform (10:10:1, v/v/v) for fat-soluble vitamins. Rutin was detected with 25% lead(II)acetate. Remaining water-soluble vitamins, except vitamin B_6, were detected with a solution of hexaiodoplatinate (IV). Fat-soluble vitamins were detected with a 10% solution of antimony chloride. The R_F values of water-soluble vitamins on RP-18 plates and using water–methanol (5:4, v/v) were equal to 0.94, 0.76, 0.71, 0, and 0.28 for vitamin C, nicotinic acid, nicotinamide, vitamin B_1, and rutin, respectively. The R_F values of water-soluble vitamins on RP-18 plates and using water–acetic acid (7:1, v/v) were equal to 0.91, 0.57, 0.40, 0.68, and 0 for vitamin C, nicotinic acid, nicotinamide, vitamin B_1, and rutin, respectively. The R_F values of fat-soluble vitamins on RP-18 plates and using acetonitrile–benzene–chloroform (10:10:1, v/v/v) were equal to 0.80, 0.80, 0.62, and 0.86 for vitamins E, E-acetate, D_3, and A-acetate, respectively. However, the determination of vitamins B_1, B_6, and A-acetate in mixtures with other vitamins were performed by spectrophotometry.

Thielemann (1981a,b) separated vitamins A, D_2, and E from vitamins B_1, B_2, B_6, C, nicotinamide, and panthenol, which occur in multivitamins Summavit® (Jenapharm) and Turigeran® (Jenapharm). Lipophilic vitamins were separated on silica gel with benzene–petroleum ether–acetic acid (35 + 65 + 1, v/v). R_F values were equal to 0.71, 0.18, and 0.07 for vitamins A, D_2, and E, respectively. R_F values were equal to 0.07, 0.30, 0.24, 0.35, 0.86, and 0.58 for vitamins B_1, nicotinamide, B_2, B_6, C, panthenol, respectively. This method can be used for pharmaceutical experiments and control investigations. Kartsova and Koroleva (2007) simultaneously determined the water- and fat-soluble vitamins in pharmaceutical preparations by HPTLC-densitometric using aqueous micellar mobile phase. The water-soluble vitamins B was analyzed in NP-TLC on Silufol plates and high-performance Silufol plates using acetonitrile–water, methanol–water, and ammonia–water mobile phases in different volume compositions. The full separation of water–soluble vitamins B (B_1, B_2, B_6, and B_{12}) was enabled by the use of high-performance Silufol plates and an ammonium aqueous solution (0.045%) as well as methanol–water (10:90, v/v) as mobile phases. The interaction between vitamin B_1 and silanol hydroxyl groups of the adsorbent could not be reduced, which is significant for quantitative analysis. Water–soluble vitamins were also determined on unmodified plates and on the plates impregnated with methanol solution of sodium dodecyl sulfate (SDS) in a concentration of 0.005 M and using a 0.020 M aqueous micellar solution of SDS as mobile phase. All these conditions enabled to reduce the interaction between vitamin B_1 and silanol hydroxyl groups of silica gel. Vitamins A and D were determined using high-performance Sorbfil plates and different mobile phases. The separation of vitamins A and D was obtained using benzene, benzene–ethyl acetate (20:1, v/v), benzene–ethyl acetate (10:1, v/v), and hexane–ethyl acetate (20:1, v/v) as mobile phases. However, the simultaneous determination of water-soluble vitamins (B_1, B_2, B_6, and B_{12}) and vitamins A and D was performed on HP Sorbfil plates using two mobile phases. Benzene was the first mobile phase, and next the plates were developed using an aqueous micellar solution of SDS as mobile phase as the second eluent. The detection limit in UV light was equal to 0.50, 0.50, 1.00, and 0.30 μg for B_1, B_2, B_6, and B_{12} vitamins, respectively. The LOD in visible light was equal to 0.25 and 0.20 μg for vitamins B_2 and B_{12}, respectively. These chromatographic conditions were also used in the determination of vitamins B_1 and B_6 in pharmaceutical preparations (*Aerovit* and *Komplivit*).

41.4   GENERAL CONCLUSION

This chapter has discussed some parts of investigations that concern the analysis of hydrophilic and hydrophobic vitamins and vitamin PP derivatives by TLC technique. The determinations were performed by TLC technique for standard substances and vitamins in pharmaceutical preparations, foods, and biological samples. All presented materials have indicated the significance and the application of TLC in the analysis of these substances. The studies quoted in this chapter have indicated a necessary modification in the range of analysis of the discussed substances. The elaborated TLC methods for the determination of vitamins must be validated according to the ICH Harmonised Tripartite Guideline.

REFERENCES

Abbaspour, A., Khajehzadeh, A., and A. Noori. 2008. A simple and selective sensor for the determination of ascorbic acid in vitamin C tablets based on paptode. *Anal. Sci.* 24: 721–725.

Aburjai, T., Amro, B.I., Aiedeh, K., Abuirjeie, M., and S. Al-Khalil. 2000. Second derivative ultraviolet spectrophotometry and HPTLC for the simultaneous determination of vitamin C and dipyrone. *Pharmazie* 55: 751–754.

Ahmad, I., Ansari, I., and T. Ismail. 2003. Effect of nicotinamide on the photolysis of cyanocobalamin in aqueous solution. *J. Pharm. Biomed. Anal.* 31: 369–374.

Ahrens, H. and W. Korytnyk. 1969. Pyridoxine chemistry: XXI. Thin-layer chromatography and thin-layer electrophoresis of compounds in the vitamin B_6 group. *Anal. Biochem.* 30: 413–420.

Alary, J., Grosset, C., and A. Coeur. 1982. Identification de molecules antioxydantes par nanochromatographie sur couche mince. *Ann. Pharm. Fr.* 40: 301–309.

Ansari, I.A., Vaid, F.H.M., and I. Ahmad. 2004. Chromatographic study of photolysis of aqueous cyanocobalamin solution in presence of vitamins B and C. *Pak. J. Pharm. Sci.* 17: 19–24.

Aranda, M. and G. Morlock. 2006. Simultaneous determination of riboflavin, pyridoxine, nicotinamide, caffeine and taurine in energy drinks by planar chromatography-multiple detection with confirmation by electrospray ionization mass spectrometry. *J. Chromatogr. A* 1131: 253–260.

Argekar, A.P. and J.G. Sawant. 1999. Simultaneous determination of pyridoxine hydrochloride and doxylamine succinate in tablets by HPTLC. *J. Liq. Chromatogr. Relat. Technol.* 22: 2051–2060.

Arthur, B., Fried, B., and J. Sherma. 2006. Effects of estivation on lutein and β-carotene concentrations in *Biomphalaria glabrata* (NMRI Strain) and *Helisoma trivolis* (Colorado Strain) snails as determined by quantitative high performance reversed phase thin layer chromatography. *J. Liq. Chromatogr. Relat. Technol.* 29: 2559–2565.

Askinazi, A.I., Shelaeva, E.A., Sokolova, L.A., Radchenko, L.M., and V.F. Tsepalov. 1990. TLC assay of tocopherol isomers. *Pharm. Chem. J.* 24: 313–315.

Bączyk, S., Duczmal, L., Sobisz, I., and K. Swidzińska. 1981. Spectrophotometric determination of vitamins D_2 and K_1 in presence of rutin. *Mikrochim. Acta* 2: 151–154.

Bapcum, A. 1984. Separation and quantitative determination of tocopherols in cottonseed oil. *Chim. Acta Turc.* 12: 298–304.

Baranowska, I. and A. Kądziołka. 1996. RPTLC and derivative spectrophotometry for the analysis of selected vitamins. *Acta Chromatogr.* 6: 61–71.

Bauer-Petrovska, B. and L. Petrushevska-Tozi. 2000. Mineral and water soluble vitamin content in the Kombucha drink. *Int. J. Food Sci. Technol.* 35: 201–205.

Beljaars, P.R., Horrock, W.V.S., and T.M.M. Rondags. 1974. Assay of L(+)-ascorbic acid in buttermilk by densitometric transmittance measurement of the dehydroascorbic acid. *J. Assoc. Off. Anal. Chem.* 57: 65–69.

Bhushan, R. and M. Arora. 2002. Separation of vitamin B complex and folic acid by HPLC, normal and reversed phase-TLC. *Natl. Acad. Sci. Lett.* 25: 159–167.

Bhushan, R. and V. Parshad. 1994. Separation of vitamin B complex and folic acid using TLC plates impregnated with some transition metal ions. *Biomed. Chromatogr.* 8: 196–198.

Bhushan, R. and V. Parshad. 1999. Improved separation of vitamin B complex and folic acid using some new solvent systems and impregnated TLC. *J. Liq. Chromatogr. Relat. Technol.* 22: 1607–1623.

Bican-Fister, T. and V. Drazin. 1973. Quantitative analysis of water-soluble vitamins in multicomponent pharmaceutical forms: Determination of tablets and granules. *J. Chromatogr. A* 77: 389–395.

Bieri, J.G. and E.L. Prival. 1965. Serum vitamin E determined by thin-layer chromatography. *Proc. Soc. Exp. Biol. Med.* 120: 554–557.

Blagojevic, Z.O. and O. Laban-Bozic. 1981. Separation and determination of ergocalciferol and cholecalciferol in the presence of other fat soluble vitamins in pharmaceutical preparations. *Arch. Farm* 31: 121–131.

Blair, J.A. and E. Dransfield. 1969. Anomalous behavior of radioactive folic acid on thin layer chromatography. *J. Chromatogr. A* 45: 476–477.

Bolliger, R. and A. König. 1969. Vitamins, including carotenoids, chlorophylls and biologically active quinones. In *Thin-Layer Chromatography, A Laboratory Handbook*. E. Stahl, Ed., pp. 259–311. Berlin, Germany: Springer-Verlag.

Borodina, E.V., Kitaeva, T.A., Safonova, E.F., Selemenev, V.F., and A.A. Nazarova. 2007. Determination of α-tocopherol and ergocalciferol by thin-layer chromatography. *J. Anal. Chem.* 62: 1064–1068.

Buhl, F., Szpikowska-Sroka, B., and M. Gałkowska. 2005. Determination of L-ascorbic acid after chromatographic separation. *J. Planar Chromatogr.—Mod. TLC* 18: 368–371.

Cavin, A., Potterat, O., Wolfender, J.L., Hostettmann, K., and W. Dyatmyko. 1998. Use of on-flow LC/¹H NMR for the study of an antioxidant fraction from *Orophea enneandra* and isolation of a polyacetylene, lignans, and a tocopherol derivative. *J. Nat. Prod.* 61: 1497–1501.

Chavan, J.D. and J.M. Khatri. 1992. Simultaneous determination of retinol and α-tocopherol in human plasma by HPTLC. *J. Planar Chromatogr.—Mod. TLC* 5: 280–282.

Chen, P.S. 1965. Fluorescent dyes and thin layer chromatography applied to detection of vitamin D and related sterols in Tuna liver oil. *Anal. Chem.* 37: 301–302.

Cheng, M.H., Huang, W.Y., and A.I. Lipsey. 1987. Detection of Bromocriptine-like substances in urine of an infant on soy formula. *Clin. Chem.* 33: 414–415.

Cimpoiu, C., Casoni, D., Hosu, A., Miclaus, V., Hodisan, T., and G. Damian. 2005. Separation and identification of eight hydrophilic vitamins using a new TLC method and Raman spectroscopy. *J. Liq. Chromatogr. Relat. Technol.* 28: 2551–2559.

Cimpoiu, C. and A. Hosu. 2007. Thin layer chromatography for the analysis of vitamins and their derivatives. *J. Liq. Chromatogr. Relat. Technol.* 30: 701–728.

Dare, E.O. and A.S. Ajibola. 2007. Benzil and benzoin: General spray reagents for the visualization of organic compounds on thin layer chromatograms. *Chromatographia* 66: 823–825.

Das, B. 1994. Quantitative HPLC determination of vitamin D analogs. *J. Planar Chromatogr.—Mod. TLC* 7: 162–164.

Daurade-Le-Vagueresse, M.H. and M. Bounias. 1991. Separation, purification, spectral properties and stability of photosynthetic pigments on CN-coated HPTLC plates. *Chromatographia* 31: 5–10.

Davino, S.C., Barros, S., Barros, S.B.M., Silva, D.H.S., and M. Yoshida. 1998. Antioxidant activity of *Iryanthera sagotiana* leaves. *Fitoterapia* LXIX: 185–186.

De Leenheer, A.P. and W.E. Lambert. 1996. Lipophilic vitamins. In *Handbook of Thin-Layer Chromatography. Chromatographic Science Series*, Vol. 71, J. Sherma and B. Fried, Eds., pp. 1055–1077. New York: Marcel Dekker, Inc.

De Leenheer, A.P., Lambert, W.E., and J.F. Van Bocxlaer. 2000. *Modern Chromatographic Analysis of Vitamins.* New York: Marcel Dekker.

Demchenko, D.V., Pozharitskaya, O.N., Shikov, A.N., and V.G. Makorov. 2011. Validated HPTLC method for quantification of vitamin D_3 in fish oil. *J. Planar Chromatogr.—Mod. TLC* 24: 487–490.

De Paolis, A.M. 1983. Determination of all-*trans*- and 13-*cis*-retinoic acids by two-phase, two-dimensional thin-layer chromatography in creams and by high-performance thin-layer chromatography in gel formulations. *J. Chromatogr.* 258: 314–319.

Desai, I.D. 1984. Vitamin E analysis methods for animal tissues. *Methods Enzymol.* 105: 138–139.

Diaz, A.N., Paniagua, A.G., and F.G. Sanchez. 1993. Thin layer chromatography and fibre-optic fluorimetric quantitation of thiamine, riboflavin and niacin. *J. Chromatogr. A* 655: 39–43.

Dinnen, R.D. and K. Ebisuzaki. 1997. The search for novel anticancer agents: A differentiation-based assay and analysis of a folklore product. *Anticancer Res.* 17: 1027–1034.

Dobrucki, R. 1979. Chromatograficzny rozdział izomerów witaminy A. *Acta Pol. Pharm.* 36: 217–219. (in Polish.)

Dongre, V.G. and V.A. Bagul. 2003. New HPTLC method for identification and quantification of folic acid from pharmaceutical dosage preparations. *J. Planar Chromatogr.—Mod. TLC* 16: 112–116.

Du, M. 1983. Thin layer chromatographic method for the determination of vitamin D in fortified milk powder. *Yingyang Yuebao* 5: 381–387; *Chem. Abstr.* 100: 84310v, 1984.

Eitenmiller, RR., Ye, L., and W.O. Landen, Jr. 2008. *Vitamin Analysis for the Health and Food Sciences.* Boca Raton, FL: CRC Press, Taylor & Francis Group.

El-Mallah, M.H., El-Shami, S.M., and F.A. Zaher. 1990. Studies on deodorization distillates of soya-bean oil as potential sources of natural tocopherols. *Seifen, Oele, Fette, Wachse* 116: 199–201.

El-Sadek, M., EI-Shanawany, A., and A. Aboul Khier. 1990. Determination of the components of analgesic mixture using high-performance thin-layer chromatography. *Analyst* 115: 1181–1184.

Elzanfaly, E.S., Nebsen, M., and N.K. Ramadan. 2010. Development and validation of PCR, PLS, and TLC densitometric methods for the simultaneous determination of vitamins B_1, B_6 and B_{12} in pharmaceutical formulations. *Pak. J. Phram. Sci.* 23: 409–415.

Ersoy, L. and R. Duden. 1980. Remissionsphotometrische bestimmung von α-tocopherol und vitamin K_1 sowie von β-carotin in spinatextrakten. *Lebensm. Wiss. Technol.* 13: 198–201.

Esparza, M.S., Vega, M., and R.L. Boland. 1982. Synthesis and composition of vitamin D-3 metabolites in *Solanum malacoxylon*. *Biochim. Biophys. Acta* 719: 633–640.

Etournaud, A. and J.D. Aubort. 1991. Recherche des colorants riboflavin et phosphate de riboflavin dans les denrées alimentaires. *Trav. Chim. Aliment. Hyg.* 82: 152–158.

Ferenczi-Fodor, K., Renger, B., and Z. Végh. 2010. The frustrated reviewer—Recurrent failures in manuscripts describing validation of quantitative TLC/HPTLC procedures for analysis of pharmaceuticals. *J. Planar Chromatogr.—Mod TLC* 23: 173–179.

Frgacić, S. and Z. Kniewald. 1974. Ascorbic acid as an antioxidant in thin layer chromatography of corticosteroids. *J. Chromatogr. A* 94: 291–293.

Fung, Y.K., Rahwan, R.G, and R.A. Sams. 1978. Separation of vitamin A compounds by thin-layer chromatography. *J. Chromatogr. A* 147: 528–531.

Funk, W. and P. Derr. 1990. Characterization and quantitative HPTLC determination of vitamin B_1 (thiamine hydrochloride) in a pharmaceutical product. *J. Planar Chromatogr.—Mod. TLC* 3: 149–152.

Gerstenberg, H. 1985. Quantitative bestimmung von retinol (Vitamin A) in lebensmitteln mit hilfe der fluorimetric. *Lebensmittelchem. Gerichtl. Chem.* 39: 1–2.

Gertz, C. and K. Herrmann. 1983. Identifizierung und Bestimmung antioxidativ wirkender Zusatzstoffe in Lebensmitten. *Z. Lebensm. Unters Forsch.* 177: 186–188.

Ghoulipour, V., Amini, S., Haghshenas, A., and S. Wagif-Husain. 2010. Chromatographic behavior of food additives on thin layers of titanium (IV) silicate ion-exchanger. *J. Planar Chromatogr.—Mod. TLC* 23: 250–254.

Gliszczyńska, A. and A. Koziołowa. 1999. Chromatographic identification of a new flavin derivative in plain yogurt. *J. Agric. Food Chem.* 47: 3197–3201.

Gliszczyńska-Świgło, A. and A. Koziołowa. 2000. Chromatographic determination of riboflavin and its derivatives in food. *J. Chromatogr. A* 881: 285–297.

Grace, M.L. and R.A. Bernhard. 1984. Measuring vitamins A and D in milk. *J. Dairy. Sci.* 67: 1646–1654.

Granzow, C., Kopun, M., and T. Krober. 1995. Ribofalvin-mediated photosensitization of *Vinca* alkaloids distorts drug sensitivity. *Cancer Res.* 55: 4837–4843.

Groenendijk, G.W.T., Jansen, P.A., Bonting, S.L., and F.J. Daemen. 1980. Analysis of geometrically isomeric vitamin A compounds. *Methods Enzymol.* 67F: 203–206.

Guy, M., Olszewski, A., Monhoven, N., Namour, F., Gueant, J.L., and F. Plenat. 1998. Evaluation of coupling of cobalamin to antisense oligonucleotides by thin-layer and reversed-phase liquid chromatography. *J. Chromatogr. B* 706: 149–156.

Hachuła, U. 1997. Determination of vitamin K_4 and B_1 in pharmaceutical preparations after chromatographic separation. *J. Planar Chromatogr.—Mod. TLC* 10: 131–132.

Hachuła, U. and F. Buhl. 1991. Determination of alpha-tocopherol in capsules and soya-bean oil after chromatographic separation. *J. Planar Chromatogr.—Mod. TLC* 4: 416.

Haiduc, I., Crisan, C., Gocan, S., and T. Hodisan. 1988. Determinea vitaminei K_3 prin metode cromatografice si spectrofotometrice. *Rev. Chim. (Bucharest)* 39: 623–624.

Hirauchi, K., Notsumoto, S., Nagaoka, T., Fujimoto, K., and Y. Suzuki. 1990. Measurement of vitamin K_3 in foods by high-performance liquid chromatography with fluorometric detection. *Vitamins* 64: 183–186.

Hirauchi, K., Sakano, T., S. Notsumoto et al. 1989. Measurement of K vitamins in animal tissues by high-performance liquid chromatography with fluorimetric detection. *J. Chromatogr. B* 497: 131–137.

Hirayama, K. and K. Inoue. 1967. Determination of vitamin D in the presence of vitamin A by cylindrical thin-layer chromatography. *J. Pharm. Sci.* 56: 444–449.

Hodisan, T., Casoni, D., Beldean-Galea, M.S., and C. Cimpoiu. 2008. Identification and quantification of tocopherols in vegetable oils by thin-layer chromatography. *J. Planar Chromatogr.—Mod. TLC* 21: 213–215.

Hodisan, T., Socaciu, C., Ropan, I., and G. Neamtu. 1997. Carotenoid composition of *Rosa canina* fruits determined by thin layer chromatography and high-performance liquid chromatography. *J. Pharm. Biomed. Anal.* 16: 521–528.

Hosomi, A., Arita, M., Sato, Y. et al. 1997. Affinity for α-tocopherol transfer protein as a determinant oft he biological activities of vitamin E analogs. *FEBS Lett.* 409: 105–108.

Hossu, A.M., Maria, M.F., Radulescu, C., Ilie, M., and V. Mageanu. 2009. TLC application on separation and quantification of fat-soluble vitamins. *Rom. Biotechnol. Lett.* 14: 4615–4619.

Hossu, C., Radulescu, M., Balalau, D., and V. Mageanu. 2006. Qualitative and semiquantitative TLC analysis of vitamins A, D and E. *Rev. Chim.* 57: 1188–1190.

Hosu, A., Cimpoiu, C., Sandru, M., and L. Seserman. 2010. Determination of the antioxidant activity of juices by thin-layer chromatography. *J. Planar Chromatogr.—Mod. TLC* 23: 14–17.

Hubicka, U., Krzek, J., and J. Łuka. 2008. Thin-layer chromatography-densitometric measurements for determination of N-(hydroxymethyl)nicotinamide in tablets and stability evaluation in solutions. *J. AOAC Int.* 91: 1186–1190.

ICH Harmonised Tripartite Guideline. (2005) *Validation of Analytical Procedures: Text and Methodology.* Q2(R1) Geneva, Switzerland. http://www.ich.org (accessed November 2005).

Jamil Akhtar, M., Ataullah Khan, M., and I. Ahmad. 2003. Identification of photoproducts of folic acid and its degradation pathways in aqueous solution. *J. Pharm. Biomed. Anal.* 31: 579–588.

Johnson, W.J. and G.D. Cain. 1985. Biosynthesis of polyisoprenoid lipids in the rat tapeworm *Hymenolepis Diminuta. Comp. Biochem. Physiol. B* 82: 487–495.

Jork, H., Funk, W., Fischer, W., and H. Wimmer. 1989. *Dünnschicht-Chromatographie, Reagenzien und Nachweismethoden, Vol. 1a, Physicalische ind Chemische Nachweismethoden: Grundlagen, Reagenzien I.* Weinheim, Germany: VCH.

Jork, H., Funk, W., Fischer, W., and H. Wimmer. 1994. *Thin-Layer Chromatography: Reagents and Detection Methods, Vol. 1b, Physical and Chemical Detection Methods: Activation Reactions, Reagents Sequences, Reagents II.* Weinheim, Germany: VCH.

Justova, V. and L. Starka. 1981. Separation of functional hydroxymetabolites of vitamin D_3 by thin-layer chromatography. *J. Chromatogr.* 209: 337–340.

Justova, V., Wildtova, Z., and V. Pacovsky. 1984. Determination of 1,25-dihydroxyvitamin D_3 in plasma using thin-layer chromatography and modified competitive protein binding assay. *J. Chromatogr.* 290: 107–112.

Justova, V., Wildtova, Z., and V. Rehak. 1985. Chromatographie na tenke vrstve jako rovnocenna analoga vysoucinne papalinove chromatografie. *Chem. Listy* 79: 1103–1107.

Kahan, J. 1967. Thin-layer chromatography of vitamin a metabolites in human serum and liver tissue. *J. Chromatogr.* 30: 506–513.

Kartsova, L.A. and O.A. Koroleva. 2007. Simultaneous determination of water and fat-soluble vitamins by high-performance thin-layer chromatography using an aqueous micellar mobile phase. *J. Anal. Chem.* 62: 255–259.

Katay, G., Nemeth, Z., Szani, Sz., Kock, O., Albert, L., and E. Tyihak. 2004. Overpressured-layer chromatographic determination of ascorbigen (bound vitamin C) in *Brassica vegetables. J. Planar Chromatogr.—Mod. TLC* 17: 360–364.

Kawanabe, K. 1985. Application of thin-layer chromatography stick chromatographic identification test methods to drugs contained in preparations in the Pharmacopoeia of Japan. *J. Chromatogr. A* 333: 115–122.

Keefer, L.K. and D.E. Johnson. 1972. Magnesium hydroxide as a thin-layer chromatographic adsorbent. III. Application to separations of vitamin A and related carotenoids. *J. Chromatogr.* 69: 215–218.

Kivcak, B. and S. Akay. 2005. Quantitative determination of α-tocopherol in *Pistacia lentiscus, Pistacia lentiscus* var. *chia,* and *Pistacia terebinthus* by TLC-densitometry and colorimetry. *Fitoterapia* 76: 62–66.

Kivcak, B. and T. Mert. T. 2001. Quantitative determination of α-tocopherol in *Arbutus unedo* by TLC-densitometry and colorimetry. *Fitoterapia* 72: 656–661.

Kocjan, B. and J. Śliwiok. 1994. Chromatographic and spectroscopic comparison of the hydrophobicity of vitamins D_2 and D_3. *J. Planar Chromatogr.—Mod. TLC,* 7: 327–328.

Kolomoets, I.I. and Yu.I. Bidnichenko. 1992. Kilkiche viznaczennja retinolu acetatu ta stjefaglabrinu sulfatu densitomjetricznim metodom (in Russian). *Farm Zh (Kiev)* 3: 73–74.

Krzek, J. and A. Kwiecień. 1999. Densitometric determination of impurities in drugs Part IV. Determination of N-(4-aminobenzoyl)-L-glutamic acid in preparations of folic acid. *J. Pharm. Biomed. Anal.* 21: 451–457.

Kutsky, R.J. 1973. *Handbook of Vitamins and Hormones.* New York: Van Nostrand Reinhold Company.

Leray, C., Andriamampandry, M., Gutbier, G. et al. 1997. Quantitative analysis of vitamin E, cholesterol and phospolipid fatty acids in a single aliquot of human platelets and cultured endothelial cells. *J. Chromatogr. B* 696: 33–42.

Lichtenthaler, H.K. 1984. Prenyllipids including chlorophylls, carotenoids, prenylquinones, and fat-soluble vitamins. In *Handbook of Chromatography: Lipids,* Vol. II., H.K. Mangold, Ed., pp. 115–169. Boca Raton, FL: CRC Press.

Lichtenthaler, H.K., Börner, K., and C. Liljenberg. 1982. Separation of prenylquinones, prenylvitamins and prenols on thin-layer plates impregnated with silver nitrate. *J. Chromatogr.* 242: 196–201.

Lovelady, H.G. 1973. Separation of individual tocopherols from human plasma and red blood cells by thin-layer and gas—Liquid chromatography. *J. Chromatogr. A* 85: 81–92.

Madden, U.A. and H.M. Stahr. 1993. Reverse phase thin layer chromatography assay of vitamin K in bovine liver. *J. Liq. Chromatogr.* 16: 2825–2834.

Makowski, A., Adamek, E., and W. Baran. 2010. Use of photocatalytic reactions to visualize drugs in TLC. *J. Planar Chromatogr—Mod. TLC* 23: 84–86.

Manz, U., Struchen, E., and R. Zell. 1979. Quantitative Bestimmung von α-Tocopherol in Lebens- und Futtermitteln, Vitaminvormischugen und Konzentraten. *Mitt Gebiete Lebensm Hyd.* 70: 476–484.

Marciniec, B. and M. Stachowicz. 1989. Analiza produktów rozkładu leków. *Acta Pol. Pharm.* 46: 138–145.

Mazulin, O.V. and N.A. Kaloshina. 1997. The process of standardization of plants crude drugs of milfolil species determining the amount of vitamin K (in Russian) *Farm Zh (Kiev)* 5: 69–72.

Merck, E. 1980. *Dyeing Reagents for Thin Layer and Paper Chromatography.* Darmstadt, Germany: E. Merck.

Mitchell, K. and R.J. Fallon. 1990. The determination of ubiquinone profiles by reversed-phase high-performance thin-layer chromatography as an aid to the speciation of *Legionellaceae. J. Gen. Microbiol.* 136: 2035–2041.

Mohammad, A., Gupta, R., and S.A. Bhawani. 2009. Micelles activated planar chromatographic separation of hydrophilic vitamins. *Tenside Surfact. Deterg.* 46: 267–270.

Mohammad, A. and S. Laeeq. 2010. Application of water-in-oil microemulsion for chromatographic study of different groups of organic compounds. *Der Pharma Chemica* 2: 281–286.

Mohammad, A. and A. Zehra. 2008. Specific separation of thiamine from hydrophilic vitamins with aqueous dioxane on precoated silica TLC plates. *Acta Chromatogr.* 20: 637–642.

Mohammad, A. and A. Zehra. 2009. Simultaneous separation and identification of cyanocobalamin, thiamine, and ascorbic acid on polyoxyethylene sorbitan monoleate-impregnated silica layers with water as mobile phase. *J. Planar Chromatogr.—Mod. TLC* 22: 429–433.

Mohammad, A. and A. Zehra. 2010. Anionic–nonionic surfactants coupled micellar thin layer chromatography: Synergestic effect on simultaneous separation of hydrophilic vitamins, *J. Chromatogr. Sci.* 48: 145–149.

Nag, S.S. and S. Das. 1992. Identification and quantitation of panthenol and pantothenic acid in pharmaceutical preparations by thin-layer chromatography and densitometry. *J. AOAC Int.* 75: 898–901.

Nishioka, M., Kanosue, F., Miyamoto, E., Yabuta, Y., and F. Watanabe. 2009. TLC-bioautogram analysis of vitamin B_{12} compounds from boiled and dried Japanese anchovy (*Engraulis japonica*) products. *J. Liq. Chromatogr. Relat. Technol.* 32: 1175–1182.

Norman, A.W. and H.F. DeLuca. 1963. Chromatographic separation of mixtures of vitamin D_2, ergosterol, and tachysterol. *Anal. Chem.* 35: 1247–1250.

Olejnik, D., Gogolewski, M., and M. Nogala-Kalucka. 1997. Isolation and some properties of plastochromanol-8. *Nahrung* 41: 101–104.

Opong-Mensah, K. and W.R. Porter. 1988. Separation of some rodenticides and related compounds by thin layer chromatography. *J. Chromatogr.* 455: 439–443.

Orinak, A., Orinakova, R., and L. Turcaniova. 1998. Thin-layer chromatography of mixed-ligand zinc complexes coupled with atomic absorption spectrometric analysis of zinc. *J. Chromatogr. A* 825: 189–194.

Panahi, H.A., Kalal, H.S., and A. Rahimi. 2008. Separation of vitamin B_2 and B_{12} by impregnate HPTLC plates with boric acid. *World Acad. Sci. Eng. Technol.* 42: 193–195.

Panahi, H.A., Kalal, H.S., Rahimi, A., and E. Moniri. 2011. Isolation and quantitative analysis of B_1, B_2, B_6 and B_{12} vitamins using high-performance thin-layer chromatography. *Pharm. Chem. J.* 45: 125–129.

Parizkova, H. and J. Blattna. 1980. Preparative thin-layer chromatography of the oxidation products of retinyl acetate. *J. Chromatogr.* 191: 301–306.

Parys, W. and A. Pyka. 2010. Use of TLC and densitometry to evaluate the chemical stability of nicotinic acid and its esters on silica gel. *J. Liq. Chromatogr. Relat. Technol.* 33: 1038–1046.

Periquet, B., Bailly, A., Ghisolfi, J., and J.P. Thouvenot. 1985. Determination of retinyl palmitate in homogenates and subcellular fractions of rat liver by liquid chromatography. *Clin. Chim. Acta* 147: 41–49.

Perisic, N.U., Popovic, M.R., and T.Lj. Djaković. 1995. Quantitative determination of vitamin B-complex constituents by fluorescence quenching after TLC separation. *Acta Chromatogr.* 5: 144–150.

Pinelli, A., Witzke, F., and P.P. Nair. 1969. Separation of vitamin D from cholesterol by thin-layer chromatography. *J. Chromatogr. A* 42: 271–274.

Ponder, E.L., Fried, B., and J. Sherma. 2004. Thin-layer chromatographic analysis of hydrophilic vitamins in standards and from *Helisoma trivolvis* snails. *Acta Chromatogr.* 14: 70–81.

Postaire, E., Cisse, M., Le Hoang, M.D., and D. Pradeau. 1991. Simultaneous determination of water-soluble vitamins by over-pressure layer chromatography and photodensitometric detection. *J. Pharm. Sci.* 80: 368–370.

Proksa, B. and E. Slavikova. 1990. Chromatographic identification of yeast ubiquinines. *Pharmazie* 45: 936–937.

Pyka, A. 2003. Lipophilic vitamins. In *Handbook of Thin-Layer Chromatography*, J. Sherma and B. Fried, Eds., pp. 671–695. New York: Marcel Dekker, Inc.

Pyka, A. 2009a. Hydrophobic vitamins. In *Encyclopedia of Chromatography*, J. Cazes, Ed., pp. 1389–1399. Boca Raton, FL: Taylor & Francis Group.

Pyka, A. 2009b. Analytical evaluation of visualizing reagents used to detect tocopherol and tocopherol acetate on thin layer. *J. Liq. Chromatogr. Relat. Technol.* 32: 312–330.

Pyka, A. and D. Gurak. 2009. Use of RP-TLC and densitometry to analytical characteristic of vitamin K_1. *J. Liq. Chromatogr. Relat. Technol.* 32: 2097–2104.

Pyka, A. and W. Klimczok. 2007a. Application of densitometry for the evaluation of the separation effect of nicotinic acid derivatives. Part I. Nicotinic acid and its amides. *J. Liq. Chromatogr. Relat. Technol.* 30: 2317–2327.

Pyka, A. and W. Klimczok. 2007b. Application of densitometry for the evaluation of the separation effect of nicotinic acid derivatives. Part II. Nicotinic acid and its esters. *J. Liq. Chromatogr. Relat. Technol.* 30: 2419–2433.

Pyka, A. and W. Klimczok. 2007c. Application of densitometry for the evaluation of the separation effect of nicotinic acid derivatives. Part III. Nicotinic acid and its derivatives. *J. Liq. Chromatogr. Relat. Technol.* 30: 3107–3118.

Pyka, A. and W. Klimczok. 2008. Influence of impregnation of a mixture of silica gel and kieselguhr with cooper (II) sulphate (VI) on profile change of the spectrodensitograms and the R_F values of nicotinic acid and its derivatives. *J. Liq. Chromatogr. Relat. Technol.* 31: 526–542.

Pyka, A. and W. Klimczok. 2009. Use of thin layer chromatography to evaluate the stability of methyl nicotinate. *J. Liq. Chromatogr. Relat. Technol.* 32: 1299–1316.

Pyka, A., Nabiałkowska, D., Bober, K., and Dołowy, M. 2011. Comparison of NP-TLC and RP-TLC with densitometry to quantitative analysis of tocopherol acetate in pharmaceutical preparation. *J. Liq. Chromatogr. Relat. Technol.* 34: 2548–2564.

Pyka, A. and A. Niestrój. 2001. The application of topological indexes for prediction of the R_M values of tocopherols in RP-TLC. *J. Liq. Chromatogr. Relat. Technol.* 24: 2399–2413.

Pyka, A. and J. Sliwiok. 2001. Chromatographic separation of tocopherols. *J. Chromatogr. A* 935: 71–77.

Ramic, A., Medic-Saric, M., Turina, S., and I. Jasprica. 2006. TLC detection of chemical interactions of vitamins A and D with drugs. *J. Planar Chromatogr.—Mod. TLC* 19: 27–31.

Ranny, M. 1987. *Thin-Layer Chromatography with Flame Ionization Detection.* Prague, Czech Republic: Czechoslovak Academy of Sciences.

Rittich, B., Simek, M., and J. Coupek. 1977. Separation of naphthoquinones and lipophilic vitamins by gel and thin-layer chromatography. *J. Chromatogr. A* 133: 345–348.

Roomi, M.W. and C.S. Tsao. 1998. Thin-layer chromatographic separation of isomers of ascorbic acid and dehydroascorbic acid as sodium borate complexes on silica gel and cellulose plates. *J. Agric. Food Chem.* 46: 1406–1409.

Ropte, G. and K. Zieloft. 1985. Determination of nicotinamide in multivitamin preparations after thin layer chromatographic separation. *Pharmazie* 40: 793–794. (In Deutsche.)

Rutherford, A.J., Williams, D., and R.F. Bilton. 1991. Isolation of menaquinone 7 from Pseudomonas N.C.I.B. 10590: A natural electron acceptor for steroid A-ring dehydrogenations. *Biochem. Soc. Trans.* 19: 64S.

Rybakova, O.V., Safonova, E.F., and A.I. Slivkin. 2008. Determining tocopherols by thin-layer chromatography. *Pharm. Chem. J.* 42: 471–474.

Saari, J.C., Baker, E.M., and H.E. Sauberlich. 1967. Thin-layer chromatographic separation of the oxidative degradation products of ascorbic acid. *Anal. Biochem.* 18: 173–177.

Saden-Krehula, M. and M. Tajic. 1987. Vitamin D and its metabolites in the Pollen of pine. *Pharmazie* 42: 471–472.

Sakamoto, N., Kimura, M., Hiraike, H., and Y. Itokawa. 1996. Changes of phylloquinine and menaquinone-4 concentrations in rat liver after oral, intravenous and intraperitoneal administration. *Int. J. Vitam. Nutr. Res.* 66: 322–328.

Sakano, T., Nagaoka, T., Morimoto, A., and K. Hirauchi. 1986. Measurement of K vitamins in human and animals feces by high-performance liquid chromatography with fluorometric detection. *Chem. Pharm. Bull.* 34: 4322–4326.

Schlemmer, W. and E. Kammeri. 1973. Application of quantitative thin-layer chromatography in drug assay and stability testing: Determination of codeine phosphate, noscapine, diphenhydramine hydrochloride, phenylephrine hydrochloride, caffeine, etofyllin, Phenobarbital, and thiamine hydrochloride by in situ reflectance spectroscopy. *J. Chromatogr. A* 82: 143–149.

Schoepp-Cothenet, B., Lieutaund, C., Baymann, F. et al. 2009. Menaquinone as pool quinone in a purple bacterium. *Proc. Natl. Acad. Sci.* 106: 8549–8554.

Seher, A. 1959. Der analytische Nachweis synthetischer Antioxydantien in Speisefetten II: Trennung und Identificatiozierung synthetischer Antioxydantien durch Dünnschicht-Chromatographie. *Fette, Seifen, Anstrichm.* 61: 345–351.

Semenova, L.I. and D.I. Kuznetsov. 1984 Opredeljenije soctawa tokofjerolow fotodensitomjetriczeskim sposo-bom (in Russian). *Maslo. Zhir. Promst.* 5: 17–18. *Chem. Abstr.* 101: 71169b, 1984.

Shahidi, V., Amarowicz, R., Ablu-Ghrabia, H.A., Adel, A., and Y. Shehata. 1997. Endogenous antioxidants and stability of sesame oil as affected by processing and storage. *JAOCS* 74: 143–148.

Sherma, J. and M. Ervin. 1986. Quantification of niacin and niacinamide of vitamin preparations by densito-metric thin layer chromatography. *J. Liq. Chromatogr.* 9: 3423–3431.

Silva, D.H., Pereira, F.C., Zanoni, M.V., and M. Yoshida. 2001. Lipophylic antioxidants form *Iryanthera juru-ensis* fruits. *Phytochemistry* 57: 437–442.

Sirec, M., Miksa, Lj., Bican-Fister, T., Prosek, M., and E. Kucan. 1978. Quantitative in situ thin-layer chroma-tography of ergocalciferol in multivitamin tablets. *Chromatographia* 11: 217–219.

Skliar, M.I., Boland, R.L., Mourino, A., and G. Tojo. 1992. Isolation and identification of vitamin D_3, 25-hyroxyvitamin D_3, 1,25-dihydroxyvitamin D_3 and 1,24,25-trihydroxyvitamin D_3 in *Solanum Malacoxylon* incubated with ruminal fluid. *J. Steroid Biochem. Mol. Biol.* 43: 677–682.

Skorupa, A. and A. Gierak. 2011. Detection and visualization methods used in thin-layer chromatography. *J. Planar Chromatogr.—Mod. TLC* 24: 274–280.

Sliwiok, J. and B. Kocjan. 1992. Chromatographische Undersuchungen der hydrophoben Eigenschatfen von Tocopherolen. *Fat Sci. Technol.* 94: 157–159.

Sliwiok, J., Kocjan, B. Labe, B., Kozera, A., and J. Zalejska. 1993. Chromatographic studies of tocopherols. *J. Planar Chromatogr.—Mod. TLC* 6: 492–494.

Sliwiok, J., Podgorny, A., and A. Siwek. 1990. Chromatographic comparison of the hydrophobicity of vitamin A derivatives. *J. Planar Chromatogr.—Mod. TLC* 3: 429–30.

Stewart, B.A., Midland, S.L., and S.R. Byrn. 1984. Degradation of crystalline ergocalciferol [vitamin D_2, (3β,5Z,22E)-9,10-secoergosta-5,7,10(19),22-tetraen-3-ol]. *J. Pharm. Sci.* 73: 1322–1323.

Strohecker, R. and H.M. Henning. 1966. *Vitamin Assay, Tested Methods.* Weinheim, Germany: Verlag Chemie, GHBH.

Surai, P.F. 1988. An improved method of vitamin E estimation in foodstuffs. *Vopr. Pitan.* 3: 69–71. (In Russian.)

Takamura, K., Hoshino, H., Harima, N., Sugahara, T., and H. Amano. 1991. Identification of vitamin D_2 by thermospray-interface mass spectrometry. *J. Chromatogr.* 543: 241–243.

Tatsumi, S. and M. Izumitani. 1981. Determination of tocopherols by thin-layer chromatography and densitom-etry. *Eiyo Shokuryo* 34: 465–467.

Tazoe, M., Ichikawa, K., and T. Hoshino. 2000. Biosynthesis of vitamin B_6 in *Rhizobium. J. Biol. Chem.* 275: 11300–11305.

Thielemann, H. 1981a. Dünnschichtchromatografische Trennung und Identifizierung von Inhaltsstoffen des Mutivitaminpräparates Summavit®10 an Ferigfolien UV 254. *Pharmazie* 36: 574.

Thielemann, H. 1981b. Dünnschichtcgromatografische trennung and identufizierung der inhaltsstoffe des mul-tivitaminpräparates turigeran®. *Pharmazie* 36: 783.

Thierry-Palmer, M. and T.K. Gray. 1983. Separation of the hydroxylated metabolites of vitamin D_3 by high-performance thin-layer chromatography. *J. Chromatogr.* 262: 460–463.

Treadwell, G.E., Cairns, W.L., and D.E. Metzier, 1968. Photochemical degradation of flavins: V. Chromatographic studies of the products of photolysis of riboflavin. *J. Chromatogr. A* 35: 376–388.

Tyrpień, K., Schefer, R.R., Bachmann, S., and K. Albert. 2003. Development and application of new C_{30}—Modified TLC plates. *J. Planar Chromatogr.—Mod. TLC* 16: 256–262.

Ueta, K., Nishioka, M., Yabuta, Y., and F. Watanabe. 2010. TLC-bioautography analysis of vitamin B_{12} compound from the short-necked clam (*Ruditapes philipinarum*) extract used as a flavoring. *J. Liq. Chromatogr. Relat. Technol.* 33: 972–979.

Usui, Y., Nishimura, N., Kobayashi, N., Okanoue, T., Kimoto, M., and K. Ozawa. 1989. Measurement of vitamin K in human liver by gradient elution high-performance liquid chromatography using platinum-black catalyst reduction and fluorimetric detection. *J. Chromatogr. B* 81: 291–301.

Vostokov, V.M. and V.R. Kartaszov. 2006. Chromatograficzeskij kontrol biochimiczeskoi aktivnosti rzirorast-vorimych vitaminov (A,D,E) v piszczevoi I kormovoi prdykcii. *Chim. Chim. Technol.* 49: 115–118. (In Russian.)

Wardas, W. and A. Pyka. 1995. New visualizing agents for fatty vitamins in TLC. *Chem. Anal. (Warsaw)* 40: 67–72.

Watanabe, F., Abe, K., Fujita, T., Goto, M., Hiemori, M., and Y. Nakano. 1998. Effects of microwave heating on the los of vitamin B_{12} in foods. *J. Agric. Food Chem.* 46: 206–210.

Watanabe, F. and E. Miyamoto. 2003. Hydrophobic vitamins. In *Handbook of Thin-Layer Chromatography*, J. Sherma and B. Fried, Eds., pp. 589–605. New York: Marcel Dekker, Inc.

Watanabe, F. and E. Miyamoto. 2009. Hydrophobic vitamins. In *Encyclopedia of Chromatography*, J. Cazes, Ed., pp. 1157–1160. Boca Raton, FL: Taylor & Francis Group.

Watanabe, F. and E. Miyamoto. 2010. Vitamin B_{12} and related compound analysis by TLC. In *Encyclopedia of Chromatography*, J. Cazes, Ed., pp. 937–940. Boca Raton, FL: Taylor & Francis Group. DOI: 10.1081/ECHR3-120028859.

Watanabe, F., Takenaka, S., Katsura, H. et al. 2000. Characterization of a vitamin B_{12} compound in the edible purple laver, *Porphyra yezoensis*. *Biosci. Biotechnol. Biochem.* 64: 2712–2715.

Winkler, W. and U. Hachuła. 1997. Determination of vitamin B_{12} in pharmaceuticals after chromatographic separation. *J. Planar Chromatogr.—Mod. TLC* 10: 386–387.

Zeb, A. and M. Murkovic. 2010. Thin-layer chromatographic analysis of carotenoids in plant and animal samples. *J. Planar Chromatogr.—Mod. TLC* 23: 94–103.

Zempleni, J., McCormick, B., and D.M. Mock. 1997. Identification of biotin sulfone, bisnorbiotin methyl ketone, and tetra-norbiotin-*l*-sulfoxide in human urine. *Am. J. Clin. Nutr.* 65: 508–511.

Zempleni, J. and D.M. Mock. 1999. Advanced analysis of biotin metabolites in body fluids allows a more accurate measurement of biotin bioavailability and metabolism in humans. *J. Nutr.* 129: 494S–497S.

42 TLC of Antiseptics

Anna Apola, Mariusz Stolarczyk, and Jan Krzek

CONTENTS

42.1 INTRODUCTION

Antiseptics are a differentiated group of connections that are used for destroying microorganisms present on skin, mucosa, or in infected wounds. Disinfection means elimination of pathogenic microorganisms occurring apart from humans. Both antiseptics and disinfection means are characterized by an activity toward microorganisms such as fungi, bacteria, viruses, and protozoa. Antiseptics used for undamaged skin disinfection are iodophors, benzalkonium, nitrofuran, nitroprazine, or acryflavin; ethacridine, crystal violet, povidone-iodine, cetylpiridinium, and deqalinium are used on mucous membranes, while ethacridine or acryflavin are used for wound disinfection [1].

42.2 5-NITROFURAN DERIVATIVES

Antibacterial activity of 5-nitrofuran derivatives is conditioned by the presence of a nitro group in the 5th position of furan ring. Another common feature in the chemical structure of these substances is the presence of azomethylene group in the 2nd position substituted by a nitrogen atom with aliphatic or heterocyclic residue (e.g., hydantoin or oxazolidine ring) [2].

Nitrofurans in pharmaceutical preparations were determined using the TLC method [3] (Table 42.1).

Sample preparation: Powdered mass of the tablets was subjected to extraction with 2 mL of acetonitrile in a time of 0.5 h reaching 0.025 $g \cdot mL^{-1}$ concentration of the studied compound.

Stationary phase: Sorbfil PTSKh-AF-VUF plates (silica gel STKh-1 VE as a sorbent, dp = 8–12 μm, aluminum support, UV 254 nm) of size 10 × 10 cm, produced by Sorbpolimer Private Company (Krasnodar, Russia), were used in a traditional TLC method.

Capillaries made of melted quartz (without stationary phase on the internal surface), 0.53 mm i.d., manufactured by Phenomenex (USA), and sold by Akvilon (Moscow) were used in the capillary

TABLE 42.1
5-Nitrofuran Derivatives

Name of Preparation, Chemical Name of Substance	= R
Nitrofurazone, furacillin 5-Nitro-2-furaldehyde semicarbazone	
Nitrofurantoin, furantoin 1-[(5-Nitro-2-furyl)methylideneamino] imidazolidine-2,4-dione	
Furazolidone 3-{[(5-Nitro-2-furyl)methylene] amino}-1,3-oxazolidin-2-one	
Akritoin, furagin 1-[3-(5-Nitro-2-furyl)-2-propenylidenmino-2, 4-imidazolidindion	

Nifuratel, formirol, polmirol
5-[(Methylthio)methyl]-3-{[(1E)-(5-nitro-
  2-furyl)methylene]
  amino}-1,3-oxazolidin-2-one	

Nifuroxazide, ercefuryl 4-Hydroxy-N'-[(5-nitrofuran-2-yl)methylene] benzohydrazide	

TLC method. The external polymeric coating of capillaries was removed by keeping them in acetonitrile for 30 min. The sorbent coming from TLC plates was used for capillary packing.

Mobile phase: Acetonitrile–chloroform (1:1, v/v).

Detection: Both plates and capillaries were scanned using Epson Perfection 1260 flatbed scanner (China) (Table 42.2).

TABLE 42.2
R_f Values of 5-Nitrofuran Derivatives Obtained in Planar and Capillary TLC Analysis

Nitrofuran	Planar TLC R_f	Capillary TLC R_f
Nitrofural	0.30	0.32
Akritoine	0.46	0.40
Furazolidone	0.79	0.89
Furadonin	0.67	0.52
Ercefuryl	0.75	0.60

TABLE 42.3
R_f Values for 5-Nitrofuran
Derivatives Obtained in TLC Analysis

| | Development Systems | | |
| | I | II | III |
Nitrofurane	R_f	R_f	R_f
Furazolidone	0.40	0.21	0.30
Furaltadone	0.13	0.04	0.26

Nitrofurans are used for prevention purposes as an addition to feed mixtures for animals in amounts of 50–200 ppm. The method of these derivatives, determination in feed premixes was developed in the following conditions [4–6]:

Sample preparation: Suitable amount of fodder corresponding to 20 mg of nitrofurans was weighed, 10 mL of dimethylformamide was added, and the whole mixture was shaken for 30 min, and then centrifuged for 5 min. The supernatant was spread on the plates.

Stationary phase: Silica gel TLC plates (20 × 20 cm layer thickness 0.5 mm) with fluorescent indicator.

Mobile phase: I. Chloroform–acetone (70:30, v/v).
    II. 2-Butanol–diethyl ether–acetone (10:85:5, v/v).
    III. Dioxane–benzene (50:50, v/v).
    The chromatograms were developed without light access (Table 42.3).

Detection: Detection at $\lambda = 254$ nm.
Nitrofurans were also analyzed in other materials of natural origin, like eggs and milk, in the following conditions [7]:

Sample preparation: Furazolidone, nitrofurazone, furaltadone, and nitrofurantoin were analyzed after conduction of photochemical reaction with pyridine application.

Stationary phase: Silica gel HPTLC plates.

Mobile phase: I. Dichlorometane–acetonitrile–formic acid (87:10:3, v/v).
    II. Methanol–acetic acid (7:3, v/v).
    The plates were developed twice in eluent II.

Detection: Densitometric detection at $\lambda = 366$ nm.
Nitrofurans were examined in the presence of other substances in pork and beef using high-performance thin-layer chromatography [8]. Determination was performed for the following substances: chloramphenicol, nitrofurazone, nitrofurantoine, furaltadone, furazolidone, sulfamethazine, sulfadimethoxine, sulfadoxine, and sulfamethoxypyridazine.

Stationary phase: Silica gel HPTLC plates.

Mobile phase: Ethyl acetate–hexane (2:1, v/v).

Detection: Visualization of nitrofurans was conducted by sprinkling with pyridine and scanning of the stains at $\lambda = 366$ nm.
Visualization of sulfonamides and chloramphenicol was conducted in the following method: the plates were heated at a temperature of 110°C for 10 min, then they were sprinkled with tin chloride solution, left for 15 min in a dark place, heated for 15 min at a temperature of 110°C, sprinkled with sodium hydroxide solution, and sprinkled again with fluorescamine solution after drying.

42.3 QUINOLINE DERIVATIVES

Disinfection and antiseptic activity are demonstrated by 8-hydroxyquinoline derivatives, including clioquinol.

The structure of that compound contains a hydroxyl group in the 8th position, which conditions the possibility of complex formation with heavy metal ions, and this in turn enables penetration of 8-hydroxyquinoline derivative through microorganism cell membranes. Clioquinol demonstrates an antibacterial activity but also fights amoebas and protozoa, which makes this compound useful as intestinal antiseptic.

Thin-layer chromatography was used for clioquinol determination in pharmaceutical preparations [9–11] (Table 42.4).

Clioquinol is also present in creams and ointments. The analysis of clioquinol, except hydrocortisone and hydrocortisone acetate in creams and ointments, was conducted using high-performance thin-layer chromatography method [12].

Stationary phase: Silica gel HPTLC plates after washing the layers with dichloromethane–methanol (1:1) prior to sample application

Mobile phase: Hexane–ethyl acetate–acetic acid (20:30:1, v/v).

Detection: Densitometry.

Except betamethasone valerate, nipagin, and nipasol; also, clioquinol was determined in the cream [13].

Stationary phase: Silica gel TLC plates.

Mobile phase: Ethyl acetate–butanol–25% NH_3 (11:5:4, v/v).

Detection: Quantification by densitometry at 247 nm.

TABLE 42.4

Conditions of Clioquinol Determination in Pharmaceutical Preparations

System	Stationary Phase	Mobile Phase (v/v)	Detection
I	Machery Nagel precoated 20 × 20 cm polyamide II UV254 plates 0.2 mm thickness	Methanol–acetic acid (19:1)	Longwave UV
II	Silica Gel H (Merck) containing citric acid	Chloroform	Sprayed with ethanolic solution of 4-metyhylumbelliferone, exposed to ammonia vapor and observed under visible and longwave UV
III	Polyamide (WoelmP) powder with calcium sulfate coated on a glass plate	Methanol	1. UV at λ = 266 nm 2. Pauly reagent spray
IV [10]	Silica gel	Methanol–methoxyethanol–hydrochloric acid (88:10:2)	Information not available
V [11]	Silica Gel 60 HR containing fluorescence indicator F254 and pH = 5.7 phosphate buffer coated on a plate to 250 µm thickness	Triethylamine–dioxane–methylethyl ketone (80:15:5) Develop three times	1. Shortwave UV λ = 254 nm 2. Extraction of silica with acidified methanol and quantitation by spectrophotometry at 269 nm

42.4 PHENOLS AND DERIVATIVES

Phenol (carbolic acid) was introduced to medicine by Lister in the year 1865, and for many years it was the basic antiseptic used commonly. Although currently its application has been considerably limited, it deserves attention due to its application mainly in surgery, where it has played an important role. Strong bactericidal properties of phenol are attributed to its ability of bacterial protein denaturation. Phenol derivatives with strong activity and profitable parameters are currently used in medicine. Among the used phenol derivatives, the highest activity is demonstrated by the compounds containing a methylene bridge in *ortho* position to the phenol group, and their activity is additionally enhanced by the presence of chlorine atoms in the particle. Hexachlorophene is an example of such a substance.

Hexachlorophene was determined in tomatoes after previous leaves, exposition to that substance labeled with carbon ^{14}C in the following conditions [14]:

Sample preparation: Biological materials (leaves) were extracted with three amounts of diethyl ether (2 mL each).

Stationary phase: Preprepared 250 µm thick, silica gel G, 5 × 20 cm TLC plates.

Detection: The plates were scanned for radioactivity. Every plate was run at two different sensitivities, and at a voltage of 0.47 kV with Geiger gas flow rate of 0.75 L min^{-1} (Table 42.5).

Hexachlorophene was determined besides other bactericidal compounds such as 3,4,4′-trichlorocarbanilide (TCC), 4,4′-dichloro-3-trifluoromethylcarbanilide (Irgasan CF$_3$), 3,4′,5-tribromosalicylanilide (TBS), 3,5-dibromo-3′-trifluoromethylsalicylanilide (Flurophene), and zinc omadine in preparations for intimate hygiene in the following conditions [15]:

Sample preparation: Weighed amount of 10 g of soap was homogenized for 3 min with an addition of 100 mL dimethylformamide, and then filtrated.

Stationary phase: Silica Gel F$_{254}$ TLC plate.

Mobile phase: Benzene–ether (80:20, v/v).

Detection: Identification of substances is possible by detection at λ = 253 nm wavelength, as well as by dyeing with 4-aminoantipyrine and potassium ferricyanide (Table 42.6).

TABLE 42.5
Composition of Mobile Phase and R$_f$ Values for TLC Analysis of Hexachlorophene

Mobile Phase (v/v)	R$_f$ Value for Hexachlorophene
Heptane–acetic acid (9:1)	0.15
Hexane–diethyl ether–acetic acid (6:3:1)	0.63
Chloroform–methanol–ammonium hydroxide–water (70:30:2:2)	0.77

TABLE 42.6
R_f Values for TLC Analysis of Hexachlorophene in the Presence of Other Substances

Agent	R_f Value	Black Light	Color Reagent	Short Wave UV
Hexachlorophene	0.25–0.36	(−)	Red (+)	(+)
Zinc omadine	0.26–0.32	(+)	(−)	(+)
TCC	0.35–0.46	(−)	(−)	(+)
Irgasan	0.32–0.46	(−)	(−)	(+)
TBS	0.79–0.86	(+)	Orange (+)	(+)
Fluorophene	0.75–0.82	(+)	Orange (+)	(+)

42.5 AMMONIUM COMPOUNDS

Quaternary ammonium compounds demonstrate strong bactericidal activity toward Gram-positive bacteria and weaker activity toward Gram-negative ones. The activity of these connections is related to their ionization. An activity of these substances includes an increase in cell membrane permeability, and in higher concentrations on its destruction, inhibition of enzymatic processes, and formation of hardly soluble connections with proteins of microorganism cells.

That group includes dequalinium chloride.

Determination of dequalinium chloride was conducted in various forms of the drug such as water solution, ointment, and tablets using the method of high-performance thin-layer chromatography [16].

Sample preparation: Water solution used in the case of infection containing 1.5 mg of dequalinium chloride and 3.5 mg of benzalkonium chloride in 10 g of the solution was spread directly on the plate in an amount of 3.5 μL.

Ointment containing 4 mg of dequalinium chloride in 1 g of preparation was dissolved in methanol obtaining a solution of a concentration of about 4 mg mL^{-1} of dequalinium chloride, which was spread on the plates in an amount of 1 μL.

Two tablets containing 0.45 mg of dequalinium chloride each were dissolved in 2 mL of water at a temperature of 100°C in a flask of volume 10 mL in a water bath; after cooling to the temperature of 60°C, 7 mL of methanol was added to precipitate the sugars. It was filled up with methanol up to the volume of 10 mL at room temperature, centrifuged, and 5 μL of solution was spread on the plates.

Stationary phase: HPTLC silica gel Nano-TLC plate (10 × 10 cm with fluorescent dye).

Mobile phase: Methanol–1 M ammonium acetate in water (17:3, v/v).

Detection: In order to conduct the reaction with a fluorescence factor, the plate was submerged in a solution containing 20 g sodium tetraphenyl borate in 50 mL of water with an addition of 50 µL of concentrated hydrochloric acid for 2 s. Next, it was exposed for 10 min to UV radiation at 254 nm, and directly after that it was again exposed to UV radiation at 365 nm for 10 min. After these treatments, dequalinium chloride was transformed into fluorescence derivative. Next, the dry plate was submerged for 2 s in ethylene glycol–methanol (1:1) (v/v) mixture in order to strengthen the fluorescence. The plates were scanned in the range of 200–500 nm. The R_f value for dequalinium chloride is 0.58.

Dequalinium chloride may also be determined in pharmaceutical preparations in the following conditions [17]:

Stationary phase: Silica Gel F_{254} TLC plate.

Mobile phase: I. Methanol–8%$_{aq}$ ammonium acetate (19:1, v/v).
II. Ethyl acetate–methanol–anhydrous acetic acid (5:13:5, v/v).

Detection: Densitometrically at $\lambda = 240$ nm or after dyeing with Dragendorff reagent at $\lambda = 360$ nm.

42.6 ORGANIC ACID DERIVATIVES

Aseptins are solid substances, poorly soluble in water and easily soluble in organic solvents. Their antibacterial activity is independent of environmental pH. 4-Hydroxybenzoic acid with poor aseptic activity has not found an application, contrary to its esters, which are used mainly for preservation of pharmaceutical preparations.

Nipagins were examined in ointment except fusidic acid and butylhydroxyanisole [18] (Table 42.7).

Sample preparation: 10 mL of methanol was introduced to 2.5 g of an ointment and heated in a water bath at 70°C for 20 min. Next, the emulsion was cooled to 20°C, and then it was kept at 4°C for 30 min and filtered.

Stationary phase: TLC plates 10 × 10 cm (cut from 20 × 20 cm precoated silica gel 60 F_{254} on aluminum).

Mobile phase: Hexane–ethyl acetate–glacial acetic acid (6:3:1, v/v).

Detection: Densitometrically, R_f value for methyl hydroxybenzoate 0.64; for propyl hydroxybenzoate 0.72; and for butylhydroxyanisole 0.77.

TABLE 42.7
Esters of 4-Hydroxybenzoic Acid

Name of preparation	–R
Aseptinum M, Nipaginum M Ester metylowy kwasu 4-hydroksybenzoesowego	$-CH_3$
Aseptinum A, Nipaginum A Ester etylowy kwasu 4-hydroksybenzoesowego	$-C_2H_5$
Aseptinum P, Nipaginum P Ester propylowy kwasu 4-hydroksybenzoesowego	$-C_3H_7$

TLC method was used for aseptins, determination besides neomycin sulfate, polymixin B sulfate, and zinc bacytracin in ophthalmic ointment [19].

Sample preparation: 40 mL of methanol was introduced to 1.5 g of an ointment and heated in a water bath for 30 min. Next, the solution was cooled, filtered, and filled up with methanol up to a volume of 50 mL.

Stationary phase: TLC plates 10 × 10 cm (cut from 20 × 20 cm precoated silica gel aluminum TLC sheets).

Mobile phase: *n*-Pentane–glacial acetic acid (66:9, v/v).

Detection: Densitometrically at $\lambda = 260$ nm, R_f value for methyl hydroxybenzoate and propyl hydroxybenzoate were 0.11 and 0.20, respectively.

REFERENCES

1. Zejc, A. and M. Gorczyca. 2004. *Chemia Leków*. Warszawa, Poland: Wydawnictwo Lekarskie PZWL.
2. Kostowski, W. and Z. Herman. 2008. *Farmakologia. Podstawy farmakoterapii*. Warszawa, Poland: Wydawnictwo Lekarskie PZWL.
3. Berezkin, V.G., L.A. Onuchak, and E.N. Evtyugina. 2009. Capillary thin-layer chromatography of antibacterial nitrofuran derivatives. *Russ. J. Appl. Chem.* 82(2): 312–316.
4. Moretain, J.P., J. Boisseau, and G. Gayot. 1979. Thin-layer chromatographic analysis of nitrofurans in feed premixes. *J. Agric. Food Chem.* 27(2): 454–456.
5. Zoni, G. and E. Lauria. 1967. Separation and determination of various nitrofurans in combinations. *Boll. Chim. Farm.* 106(10): 706–709.
6. Bortoletti, B. and T. Perlotto. 1968. Separation, identification and quantitative determination of 5 nitrofurans with thin-layer chromatography. *Farmaco Prat.* 23(7): 371–376.
7. Echterhoff, M. and M. Petz. 1994. Quantitative HPTLC analysis of nitrofuran in egg and milk after prechromatographic derivatization. *Dtsch. Lebensm. Rdsch.* 90: 341–344.
8. Abjean, J.P. 1997. Planar chromatography for the multiclass, multiresidue screening of chloramphenicol, nitrofuran, and sulfonamide residues in pork and beef. *J. AOAC Int.* 80: 737–740.
9. Padmanabhan, G., I. Becue, and J.B. Smith. 1989. *Analytical Profiles of Drug Substances*, Vol. 18. New York: Academic Press, Inc.
10. Kubiak, E.J. and J.W. Munson. 1982. Analysis of iodochlorhydroxyquin in cream formulations and bulk drugs by high-performance liquid chromatography. *J. Pharm. Sci.* 71(8): 872–875.
11. Valle, R.O., D. Jimenez, G.S. Lopez, and I. Schroeder. 1978. Separation and quantification of iodochlorhydroxyquin and its homologues by thin layer chromatography. *J. Chromatogr. Sci.* 16(4): 162–165.
12. Sherma, J., B.P. Whitcomb, and K. Brubaker. 1990. Determination of clioquinol, hydrocortysone and hydrocortysone acetate in cream and ointment preparation by HPTLC-densitometry. *J. Planar Chromatogr.* 3: 189–190.
13. Indrayanto, G., I. Wahyuningsih, and R.J. Salim. 1997. Simultaneous densitometric determination of betamethasone valerate and clioquinol in cream, and its validation. *J. Planar Chromatogr.* 10: 204–207.
14. Van Auken, O.W. and M. Hulse. 1979. Translocation, distribution, and environmental degradation of hexachlorophene in tomatoes. *Arch. Environ. Contam. Toxicol.* 8: 213–230.
15. Graber, M.B., I.I. Domsky, and M.E. Ginn. 1969. A TLC method for identification of germicides in personal care products. *J. Am. Oil Chem. Soc.* 10: 529–531.
16. Hiegel, K. and B. Spangeberg. 2009. New method for the quantification of dequalinium cations in pharmaceutical samples by absorption and fluorescence diode array-layer chromatography. *J. Chromatogr. A* 1216: 5052–5056.
17. Nuti, V. and D. Bertini. 1987. Identification and quantitative determination of quaternary ammonium salts in medical aids. *Farmaco Ed. Prat.* 42: 335–343.
18. Krzek, J., U. Hubicka, J. Szczepańczyk, A. Kwiecień, and W. Rzeszutko. 2006. Simultaneous determination of fusidic acid, m- and p-hydroxybenzoates and butylhydroxyanisole by TLC with densitomatric detection in UV. *J. Liq. Chromatogr. Related Technol.* 29: 2129–2139.
19. Krzek, J., M. Starek, A. Kwiecień, and W. Rzeszutko. 2001. Simultaneous identification and quantitative determination of neomycin sulfate, polymixin B sulfate, zinc bacytracin and methyl and propyl hydroxybenzoates in ophthalmic ointment by TLC. *J. Pharm. Biomed. Anal.* 24: 629–636.

43 TLC of Sulfonamides

Irena Choma and Wioleta Jesionek

CONTENTS

43.1 INTRODUCTION

43.1.1 HISTORICAL INFORMATION

Sulfonamides (called also sulfa drugs) are bacteriostatic, synthetic antibiotics (chemotherapeutics) derived from the family of azo dyes containing the sulfanilamide group in their structure. They were the first systematically used antibacterial drugs discovered by Gerhard Domagk in 1935 [1], 6 years after the discovery of penicillin by Alexander Fleming. The clinical use of penicillin was not until 1944, while the first successful clinical trial of sulfonamides was carried out in 1936 at Queen Charlotte's Maternity Hospital in London [2,3].

The progenitor of sulfonamides, 4-aminobenzenesulfonamide (sulfanilamide), was synthesized in Australia by Gelmo, who had no knowledge about its antibacterial properties [4]. The history of sulfonamides as therapeutics started in the late 1920s, when Domagk began his investigations for anti-infective azo dyes [5]. He was then an employee of Bayer (IG Farben) among other chemists searching for drugs based on dyes (Figures 43.1 and 43.2). The group was very effective, giving, for instance, antimalarial chloroquine, discovered by Ehrlich's student Roehl [5–7]. After the 3-year study, in March 1935, Domagk published a paper on the dye KL730 (KL from Klarer), also called streptozon, prontosil, or prontosil rubrum because of its intensive red color. The drug was a derivative of the known azo dye chrysoidine, obtained by the addition of the sulfonamide group giving 4-[(2,4-diaminophenyl)azo]benzenesulfonamide. Two famous patients cured with prontosil rubrum from streptococcal septicemia were Domagk's daughter Hidlegard and Franklin Delano Roosevelt, Jr. In the laboratory tests, prontosil was effective in mice infected with streptococci but surprisingly was completely inactive *in vitro*. It was proved soon that only the sulfanilamide part of prontosil was an active antibacterial substance excreted in living organisms after cleavage of triaminobenzene [8,9] (Figure 43.3). Nota bene, Meitzsch and Klarer, the collaborators of Domagk and prontosil inventors, discovered at the end of 1931 that sulfanilamide itself (called later prontosil album or prontalbin) is active in vivo against β-hemolytic streptococcus. However, Domagk was

FIGURE 43.1   Domagk writing down microscopical findings in the laboratory diary (around 1934). (Photo by Bayer Archives, Leverkusen, Germany.)

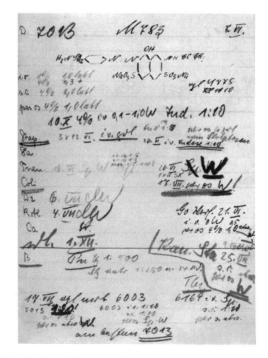

FIGURE 43.2   Domagk's handwritten laboratory diary about sulfonamide trials, June 1934. (Photo by Bayer archives, Leverkusen, Germany.)

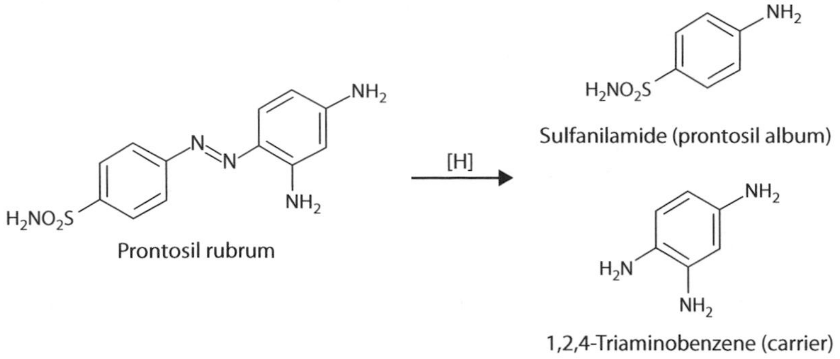

FIGURE 43.3 Metabolism of prontosil rubrum.

awarded alone the Nobel Prize in Physiology or Medicine in 1939. Hitler had forbidden Domagk to receive the award. Eventually, he got the Nobel diploma in 1947 but the money was reverted to the Nobel Foundation general pool. In his Nobel lecture, Domagk stressed the impact of Meitzsch and Klarer on sulfonamides discovery [10].

43.1.2 INFORMATION ABOUT THE DRUG GROUP

Thousands of sulfanilamide substitutions were obtained. They can be divided into two classes: amino group substituents and amide group substituents. Original prontosil rubrum belongs to the first class along with proseptasine, rubiazol, prontosil soluble, and soluseptasine (Figure 43.4). The activity of all of them was due to liberation of sulfanilamide after drug administration. It was proved later that only derivatives possessing substitution at the amide nitrogen are more active than sulfanilamide itself. The first amide-substitution sulfonamide was sulfapyridine, introduced in 1939 as M&B 693, then sulfathiazole (1940), and sulfadiazine (1941). Sulfapyridine was formerly very popular against pneumococcal and gonococcal infections—among others, it was used to treat Winston Churchill when he contracted pneumonia in December 1943. Sulfadiazine was modified to sulfamethazine (sulfadimethyldiazine), sulfamerazine (sulfamethyldiazine),

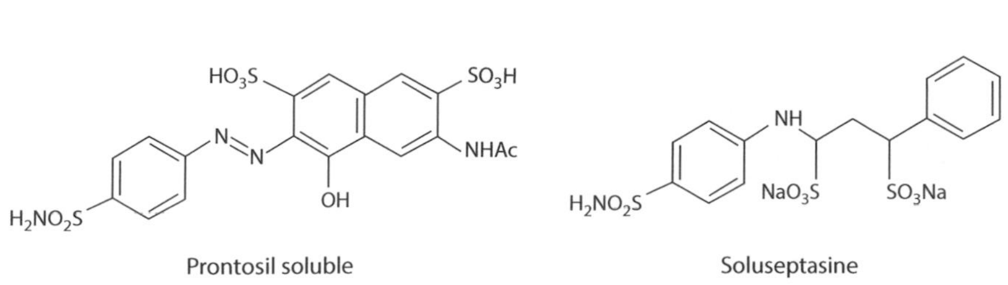

FIGURE 43.4 Sulfonamides: amino group substituents.

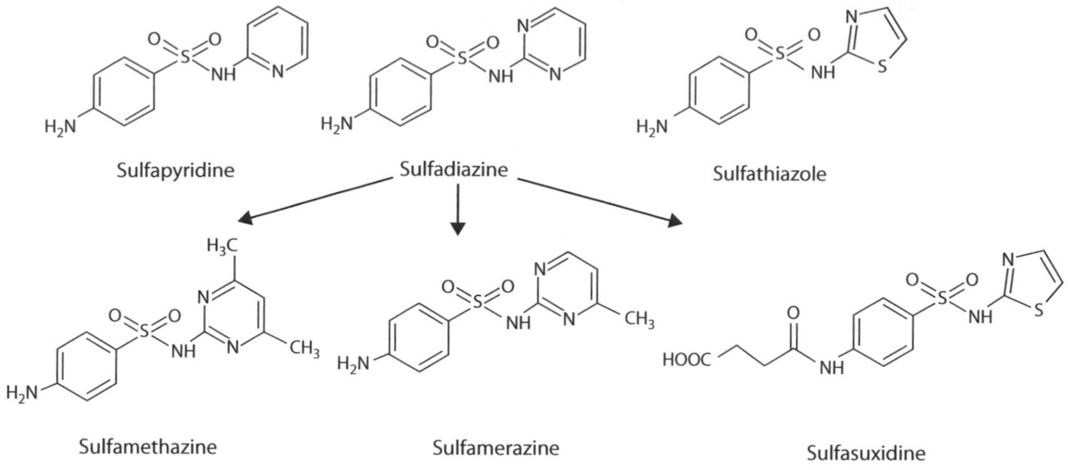

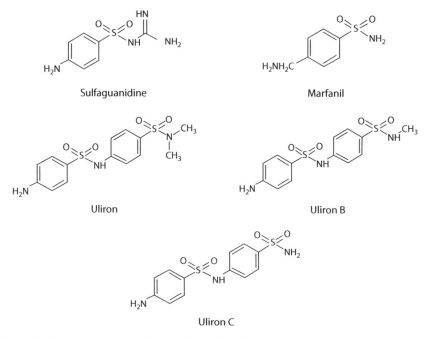

FIGURE 43.5 Sulfonamides: amide group substituents.

and sulfasuxidine (succinylsulfathiazole) (Figure 43.5). The most popular sulfa drugs were also sulfaguanidine, Uliron, Uliron B and C, as well as Marfanil, used extensively by Germans during World War II against gas gangrene. Marfanil differed from other sulfonamides—it was a derivative of sulfanilamide possessing the methylamino group instead of the amino one (Figure 43.6). Soon after the war, it was replaced by a more effective one against gangrene, penicillin. Sulfamethoxazole (Gantanol) usually used together with trimethoprim with the commercial names Bactrim, Septrin, Septra, or Biseptol are still in clinical use (Figure 43.7). About 30 sulfonamides have been used so far. They are used mostly in the treatment of urinary tract infections as well as of otitis, bronchitis, pneumonia, and sinusitis. Besides those mentioned earlier, the most popular are sulfalene, sulfadicramide, sulfamethoxypyridazine, sulfametrole, and sulfinpyrazone (Figure 43.8).

FIGURE 43.6 Sulfonamides: popular during World War II.

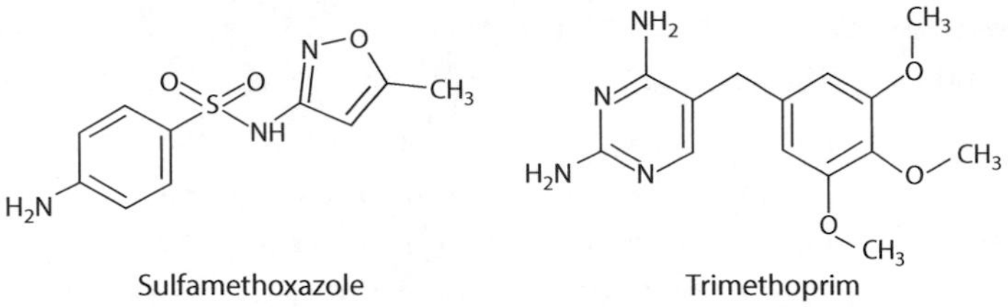

FIGURE 43.7 Sulfamethoxazole and trimethoprim (Bactrim).

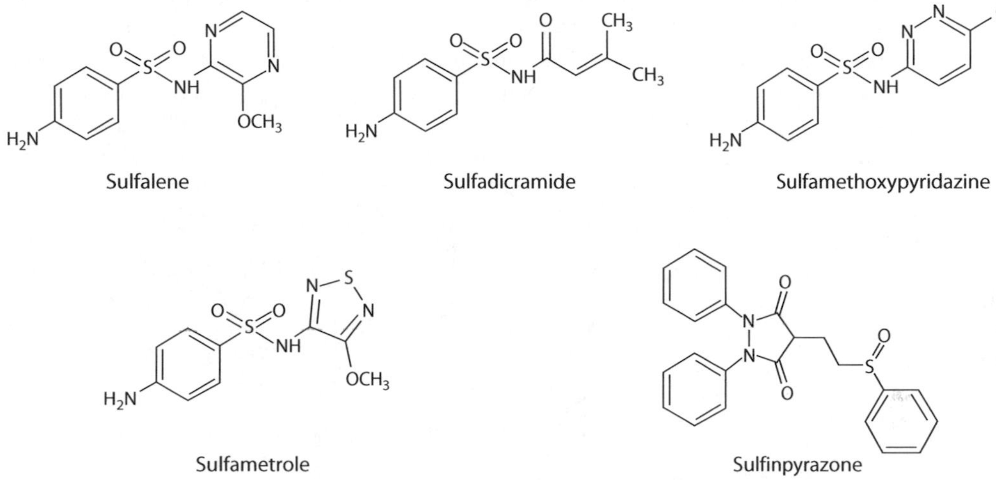

FIGURE 43.8 Sulfonamides: others popular nowadays.

43.2 CHROMATOGRAPHIC SYSTEMS USED IN THE DETERMINATION OF PARTICULAR DRUGS

43.2.1 VARIOUS ANALYTICAL TECHNIQUES

Because of their low cost and broad antibacterial spectrum, sulfonamide drugs are still widely used in human medicine as well as in veterinary, where they can be used both as therapeutic agents and as growth promoters in food-producing animals. Most of sulfonamides are toxic and allergenic; some of them could be even carcinogenic (sulfamethazine, for instance) [11,12]. Even trace residues may lead to the appearance of drug-resistant bacteria. Thus, controlling their level in the environment as well as in biological fluids and food of animal origin is an important analytical task [13–26]. The maximum residue levels (MRLs) established by the European Union as well as safe levels in the United States were set at 100 ppb for the total sulfonamide content in edible animal tissue, while at 10 ppb for milk in the United States (safe levels) [15,27,28].

Sulfonamides are detected mostly using microbial assays [29,30], receptor assays [31], or immunoassays [27,28,32–38], including enzyme-linked immunosorbent assay (ELISA) [28,34–36] and fluorescence polarization immunoassay (FPIA) [27,37,38]. Although they are excellent for fast screening and batch analysis, they are nonspecific ones. An alternative to them are instrumental methods like high-performance liquid chromatography (HPLC) [39–44], supercritical fluid chromatography (SFC) [45], gas chromatography (GC) [46–50], and capillary electrophoresis (CE) [51,52]. They usually require tedious sample preparation, expensive and sophisticated equipment, as well as skilled personnel. Extremely popular is the HPLC hyphenated with mass spectrometry due to its high sensitivity, broad linear range, accuracy, and precision [20,53–57].

43.2.2 TLC Systems Used in the Separation and/or Quantitation of Sulfonamide Drugs

Thin-layer chromatography (TLC) can be a method of choice because of its specific detection, low cost, simple equipment, and usually acceptable sensitivity, accuracy, and precision [26,39,58–60].

The TLC analysis of sulfonamides can be accomplished in both normal- and reversed-phase mode (NP- or RP-TLC). Silica gel, alumina, Florisil, and polyamide stationary phases are usually used in NP-TLC, while silanized silica, RP-2, RP-8, and RP-18 layers are used in RP-TLC. Nonaqueous and aqueous eluents are applied, respectively. Detection of sulfonamides is mostly performed on the fluorescence layers at 254 nm or after derivatization with the fluorescamine solution at 366 nm. The paper by Bieganowska et al. covers all the earlier mentioned variants of analysis [61]. The retention parameters of 15 sulfonamides for various chromatographic systems were compared by graphical correlations. Graphical illustrations of the influence of adsorbent on the R_m values are also presented (Figure 43.9). The tested drugs were sulfanilamide, sulfacarbamide, sulfaguanidine, sulfacetamide, sulfamethoxazole, sulfadicramide, sulfafurazole,

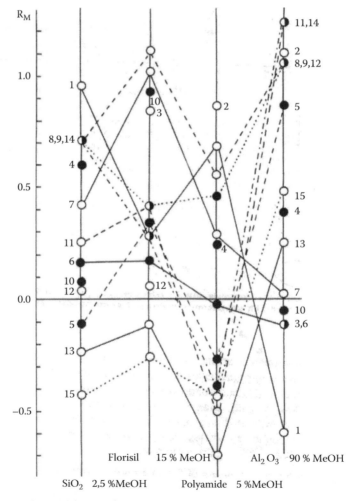

FIGURE 43.9 Graphical representation of dependence of sulfonamide R_m values on the nature of the adsorbent: mobile phase, methanol–chloroform. 1, SAN; 2, SAC; 3, SGUA; 4, SC; 5, SMX; 6, SDC; 7, SF; 8, ST; 9, SMTH; 10, SPX; 11, SDZ; 12, SMRZ; 13, SDD; 14, SSD; 15, SDMX. See List of Drug Abbreviations at the end of the chapter. (From Bieganowska, M.L. et al., *J. Planar Chromatogr. Mod. TLC*, 6, 121, 1993.)

sulfathiazole, sulfamethizole, sulfaproxyline, sulfadiazine, sulfamerazine, sulfadimidine, sulfisomidine, and sulfadimethoxine. The extended investigations of the retention behavior of the same drugs were reported in Part 2 of the paper [62]. Various sorbents were tested (silica gel, Florisil, polyamide, silanized silica) and eight various solvents mixed in the binary systems for both NP and RP chromatography. The graphical comparisons of R_m values obtained for various systems were presented. Florisil and polyamide seemed to be more selective adsorbents for sulfonamides than silica providing separation of all tested sulfonamides. Also in some RP systems, that is, on the silanized silica developed with water–dioxane or water–tetrahydrofuran (THF), all sulfonamides were separated.

Van Poucke tested various systems, both in the RP and NP modes, and found that the best for separation of 22 sulfonamides is 2D chromatography on cyanoplates developed successively with the following mobile phases: acetonitrile–THF–methanol–0.5 M aq. NaCl solution (16:3:11:70) and dichloromethane–methanol containing 2% ammonia (28%) (95:5), named by the authors as system 5 and system 6, respectively (Figure 43.10) [63].

Bičan-Fišter and Kajganovič used silica gel layers and ether or chloroform–methanol (10:1) as solvents for separation of 12 sulfonamides. The diazo reagent or p-dimethylaminobenzaldehyde (p-DAB) enabled detection of 0.25 µg of the drug. The technique was applied to the commercial preparations, for example, Sulfacombin (sulfadiazine, sulfadimidine, and sulfathiazole) and Trisulfon (sulfadiazine, sulfathiazole, and sulfamerazine) [64].

A similar system was used for separation of five sulfonamides (sulfanilic acid, sulfadiazine, Madribon [sulfadimethoxine], Gantrisin [sulfafurazole], and Gantrisin acetyl) by Wollish in his historical paper on TLC [65]. The drugs were separated on the silica gel G using chloroform–heptane–ethanol (1:1:1) and detected with p-DAB in ethanol.

Salting-out chromatography of chosen sulfonamides was performed on the silica gel developed with the aqueous solution of salts (kosmotropes, chaotropes, and neutral) [66]. The mechanism of hydrophobic interactions was confirmed by the quantitative structure–retention relationship

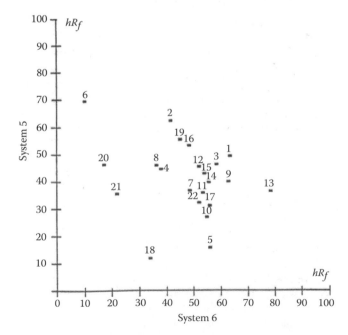

FIGURE 43.10 2D HPTLC of 22 sulfonamides on cyanoplates. 1, SDD; 2, SAN; 3, SDZ; 4, ST; 5, SQ; 6, SGUA; 7, STM; 8, FST; 9, SMDZ; 10, SMX; 11, SCPD; 12, SMPD; 13, SMP; 14, SMRZ; 15, SMTH; 16, SP; 17, SF; 18, SZP; 19, SC; 20, SCST; 21, PST; 22, SET. (From Van Poucke, L. et al., *J. Planar Chromatogr.*, 2, 395, 1989.)

(QSRR) analysis. The influence of salts on the decrease of retention was consistent with the salts order in the Hofmeister series [67].

Sulfonamides can be also separated on the silica gel impregnated with cobalt, copper, nickel, zinc, cadmium, and mercury salts [68]. There are also examples for polyamide impregnated with metal salts [69,70].

Walash used copper (II), cobalt (II), nickel (II), and cerium (IV) solutions for visualization of 15 sulfonamides. A solution of copper acetate in methanol produced spots of varying colors while acidic solution of ceric sulfate gave yellow or purple spots with all the sulfonamides studied [71].

The paper of Bieganowska and Petruczynik deals with the polar bonded phase, that is, aminopropyl silica gel using both aqueous and nonaqueous mobile phases containing organic modifiers and ion-pairing reagents. Various systems were used for determination of R_f and k values of 15 sulfonamides [72]. The spots were detected under ultraviolet (UV) light (254 nm).

There are also a few examples of RP chromatography of sulfonamides. One of the earliest is an article of Biagi [73]. The paper studies the relationship between the hydrophobicity constant π and the R_m values of sulfonamides obtained on homemade silanized silica plates developed with the aqueous buffer (sodium acetate–Veronal buffer 1:7 M) at pH 7.4. The stationary phase was obtained by impregnating the silica gel G layer with 5%, 10%, or 20% silicone DC 200 or 1-octanol solution in ether.

The RP-12 and RP-18 layers were used by Lepri et al. for the analysis of 20 sulfonamides and sulfanilic acid using both aqueous and nonaqueous eluents [74]. Additionally, 16 sulfonamides of the analyzed 19 were separated using 2D chromatography on the RP-18 layers. In the first direction, benzene–ethyl acetate–acetic acid (80:18:2) was used, and in the second, 1 M acetic acid in methanol.

RP-18, RP-12, and RP-2 modified silica were also applied by Okumura for separation of 10 sulfonamides. He used methanol–water or dioxane–water mixtures as mobile phases [75].

The retention behavior of 15 sulfonamides was investigated by thin-layer reversed-phase ion-pair chromatography. Retention and selectivity of the drugs were controlled by changes of pH, the kind and concentration of a counter ion, and the concentration of organic solvent in the mobile phase. The stepwise gradient improved the spot shape and the selectivity [76].

The silanized silica gel mixed with ethanol 95% and detergents (triethanolamine dodecyl-benzenesulfonate and N-dodecylpyridinium chloride) was used as the stationary phase in soap chromatography of 13 sulfonamides and sulfanilic acid by Lepri et al. [77]. The mobile phases were the mixtures of methanol in water of various acidity.

Srivastava et al. presented the method of separation and identification of 10 sulfonamides with p-DAB. He used the silica gel–calcium oxalate–impregnated plates developed with a mixture of ethyl acetate–chloroform–benzene (7:4:3). The separation was improved comparing with a similar system using plain silica gel G [78].

J. Kádár Pauncz used the ion-exchange resin precoated plates for qualitative and/or semiquantitative determination of sulfonamides and some other chemotherapeutics using as detection reagent DAB, p-dimethylaminocinnamaldehyde, and o-tolidine after chlorination. The mobile phases were the mixtures of buffered aqueous solutions and 2-methyl-2-propanol or ethanol [79].

Quantitative TLC was performed previously mainly using colorimetric or spectrophotometric determination from the developed spots [80–85]. The spots usually were scraped from the TLC plate after being located under UV light or by derivatization. Bićan-Fišter and Kajganović described the procedure for the sulfonamide determination in the mixtures and tablets [80]. After separation on the Kieselgel G plates with one of the three proposed solvents (depending on the mixture composition), each spot was extracted and determined colorimetrically by means of Bratton–Marshall reaction (with N-(1-naphthyl)ethylenediamine as a diazotization reagent). The typical examples of spectrophotometric determination are the papers by Cieri [81,82]. The experiments were performed on the silica gel H mixed with white phosphorus to facilitate detection of sulfonamides under UV light. The plates were developed with chloroform–methanol (88:12). The spots were located under UV light, scraped from the plate, and extracted with acidic alcohol or 0.1 N NaOH. The centrifuged extracts were read with a recording spectrophotometer. The recoveries of five standard sulfonamides (sulfathiazole, sulfadiazine, sulfamerazine, sulfamethazine, and sulfacetamide) varied from 86.4% to 99.7% [82], and after the improvement, the procedure was applied to commercial tablets giving recoveries close to 100% [81].

In the 1960s, the TLC quantitative analysis by direct spectrophotometry became popular [86]. The same phenomenon was related to the sulfonamide analysis. The most popular method of detection and quantitation became densitometry, which was applied for determination of sulfonamide pharmaceuticals in dosage forms [87–97], wastewater [98,99], body fluids, and food and feed [92,100–102].

43.3 SAMPLE PREPARATION IN CONTEXT OF MATRIX

43.3.1 Use of TLC in Pharmaceutical Analysis

As mentioned in Section 43.1, sulfonamides are broad-spectrum bacteriostatic antibiotics that act by interfering with bacterial synthesis of folic acid. Although parenteral preparations can be used, they are mostly available in the oral form. The preparations consist usually of two or three sulfonamides or sulfonamide(s) mixed with trimethoprim to increase the power of the drug. The two most frequently prescribed sulfonamides are sulfisoxazole (e.g., Gantrisin) and trimethoprim–sulfamethoxazole (e.g., Bactrim). TLC is a popular method in the analysis of dosage forms containing sulfonamides [80,82,87–97,103]. Several tablets are usually powdered and then the representative sample whose weight is equivalent to one tablet is taken. Then the drugs are extracted with solvent or solvent mixtures from the powdered tablets (or from the suspensions) and filtered; sometimes they need to be concentrated or diluted. The extracts are applied onto chromatographic plates, mostly silica gel, and developed with a proper mobile phase. The spots are quantitated usually with a densitometer in the reflectance mode at the valves related to the maximum absorbance of pure compounds. External calibration is usually performed using a set of standards. The most representative analytical procedures are described in Table 43.1.

43.3.2 Use of TLC in Water Samples

As it was mentioned in Section 43.1, the wide use of sulfonamides resulted in the appearance of their residues in the environment [13–15,104–106]. Most of them come from our households, hospitals, and from veterinary treatment. Significant amounts of unmetabolized sulfas can be released from the body with the urine or feces [14,15]. Because the drugs are not eliminated completely during sewage treatment, they will contaminate surface and ground water [13] and as a consequence also drinking water [105]. Micropollutants can come also from industry and landfills, from sewage treatment plants, as well as from aquaculture (e.g., from direct addition to the water, excessive feed, or excrements) and agriculture (e.g., from raw sewage and manure) [14]. The drugs entering the environment are detected at ppb or even ppm level. The paper of Babić et al. describes the method for separation and quantification of antibiotics (sulfadimidine, sulfadiazine, sulfaguanidine, and trimethoprim) extracted from spiked water samples. The method was validated for linearity, precision, limit of detection, and quantification. Precision (repeatability), expressed as relative standard deviation (RSD), was determined by repeated analysis (10 bends) of the mixed standard solutions at the same plate at concentrations of 0.5 µg per spot for SDD and SDZ and of 1 µg for SGUA and TMP (see Figure 43.11 and Tables 43.2 and 43.3 for details) [98]. The similar solid-phase extraction (SPE) procedure was applied for determination of seven antibacterials (enrofloxacin, oxytetracycline, penicillin G, sulfamethazine, sulfadiazine, sulfaguanidine, and TMP) in the production wastewater. The method of analysis was optimized and validated [99]. Linearity range was 100–400 ppb for SDZ and SMTZ and 300–1100 for SGUA; LOD and LOQ were 50 and 100 ppb, respectively, for SDZ and SMTZ and 200 and 300 ppb for SGUA (for other details see Table 43.4).

TABLE 43.1
TLC Determination of Sulfonamides in Dosage Forms

Tablets/Active Substances	Analytes	Sample Preparation	TLC/HPTLC Conditions	Validation	Ref.
Veltam F, Urimax F Tamsulosin (TAM) and finasteride (FINA) (0.4 mg + 5 mg)	TAM	30 tablets were powdered. A quantity of tablet equivalent to 1 mg of TAM and 12.5 mg of FINA was extracted (sonication) with MeOH, and filtered (nylon membrane filter) to get solution of TAM at 10 µg mL^{-1}	Plate: TLC silica gel F$_{254}$ Mobile phase: Toluene–MeOH–triethylamine (9:1.5:1) Detection: UV light at $\lambda = 270$ nm Densitometer, Scanner III Camag	Full validation Method compared to HPLC Recovery of TAM from tablets (two brands) 99.6% ± 1.2% or 100.6% ± 0.8% LOD 80 ng spot^{-1} LOQ 100 ng spot^{-1}	[87]
Trimosul, co-trimoxazole Sulfamethoxazole (SMX) 400 mg Trimethoprim (TMP)[a] 80 mg	SMX and impurities: SAN and SAA	Solution containing about 40 mg mL^{-1} of SMX in CHCl$_3$–MeOH (1:1) was prepared and applied directly on TLC plate for determination of SAA and SAN. The diluted solution (80 µg mL^{-1}) was used for direct determination of SMX and TMP content	Plate: TLC silica gel F$_{254}$ Mobile phase: CHCl$_3$–n-heptane–EtOH (1:1:1) Detection: UV light at $\lambda = 260$ nm Densitometer, Scanner II Camag	Full validation for SAN and SAA Recoveries: SMX: 102.95% ± 1.78% SAN: 104.5%, SAA: 97.6% LOD 4.5 and 4.1 ng µL^{-1} LOQ 13.5 and 12.4 ng µL^{-1} for SAN and SAA, respectively	[88]
Commercial tablets: I: SDZ + SMRZ + SMTZ II: ST + SDZ + SMRZ III: SC + SDZ + SMRZ IV: SC + SDZ + SMRZ + SMTZ	SC, ST, SDZ, SMRZ, SMTZ	About 20 tablets were weighed and ground to powder. Alcoholic solution of the portion of powder was spotted at the plate	Plate: silica gel H mixed with white phosphor Mobile phase: CHCl$_3$–MeOH (88:12) Detection: Spots were scraped from the plate, extracted with 0.1 NaOH, and centrifuged. The absorbance measured by spectrophotometer at 255 nm	No validation Recoveries of particular sulfonamides in various tablets composed of 3–4 components (tablets I–IV, column I) varied from 90.4% to 107.6%	[82]

Compound	Sample preparation	Method	Validation/Recovery	Reference
Commercial tablets: SM, SMX, SSX SM + SMX SM+SSX	One tablet was powdered and an amount of powder equivalent to ca. 4 mg of SM and 10 mg of SMX or 4 mg of SM and 16 mg of SSX was suspended in acetone–CHCl$_3$ (7:3) and sonicated. Then, the sample solution was filtered	Plate: HPTLC silica gel plate Mobile phase: CHCl$_3$–EtOAc (4:6) Detection: UV light at λ = 270 nm Densitometer, Scanner II Camag	Partial validation Recoveries: SM+SMX: 100.6 ± 0.89, 100.5 ± 1.69 respectively for SM and SMX SM+SSX: 100.1 ± 1.07, 101.2 ± 1.32 respectively for SM and SSX	[89]
Sulfamethoxazole (SMX) Trimethoprim(TMP) SMX Internal standard SAN	**Tablets** One tablet was powdered and an amount of powder equivalent to 25 mg of SMX and 5 mg of TMP was suspended in chloroform and sonicated. Then, internal standard was added and the mixture was diluted with acetone **Oral suspension** An aliquot of suspension containing 20 mg of SMX and 4 mg of TAM was diluted with distilled water. Then, internal standard was added and the solution was adjusted to volume with acetone	Plate: HPTLC silica gel plate Mobile phase: CHCl$_3$–EtOH (9:1) Detection: UV light at λ = 284 nm Densitometer, Zeiss PMQ II chromatogram spectrometer	Partial validation Recoveries of SMX Tablet: 99.6% ± 1.64% Suspension: 100% ± 1.30%	[90]
Sulfafurazole (SF) SF	One tablet was pulverized and an appropriate amount transferred to a volumetric flask to yield stock solution of about 20 mg/100 mL acetone. For TLC the stock solution was diluted 100 times	Plate: HPTLC silica gel plate with concentration zone Mobile phase: CHCl$_3$–EtOH (9:1) Detection: UV light at λ = 366 nm, fluorescence mode Densitometer, Scanner II Camag	Partial validation The method compared to HPLC Recoveries of SF: 95.65 ± 2.84 (TLC) 96.25 ± 3.41 (HPLC)	[91]
Commercial nimesulide tablets A and B NSD	One tablet (100 mg) was dissolved in MeOH and diluted to 10 and 20 μg mL^{-1}. Each solution was spotted to the plate	Plate: TLC silica gel F$_{254}$ plate Mobile phase: Toluene–acetone (10:1) Detection: UV light at λ = 310 nm Densitometer Scanner III Camag	No validation given Recovery: Product A: 99.16 ± 2.12 Product B: 99.66 ± 1.81	[92]

(continued)

TABLE 43.1 (continued)
TLC Determination of Sulfonamides in Dosage Forms

Tablets/Active Substances	Analytes	Sample Preparation	TLC/HPTLC Conditions	Validation	Ref.
Tablets, suspensions, suppositories composed of 3 or 4 components	SDZ, SMRZ, SMTZ, ST	Tablets: 20 tablets were weighed and powdered. A quantity equivalent to one sulfonamide tablet was dissolved in EtOH–ammonia solution and centrifuged	Plate: Silica gel G Mobile phases: CHCl$_3$–MeOH (90:10) CHCl$_3$–MeOH–ammonia solution (25%) (90:15:2.4) Ether 100% Detection: Spots localized with naphthylethylenediamine reagent were scraped, dissolved in 0.1 N HCL, and centrifuged. Sulfonamides determined colorimetrically by Bratton–Marshal reaction	Recoveries tablets: from 100% to 104.2% Suspensions: from 96.39% to 99.7% Suppositories: from 95.2% to 97.5%	[80]
Commercial creams, suppositories, tablets of SAN, SSX	SAN, SSX	A proper amount of a sample (cream containing SAN, suppositories containing SAN or SSX, powdered tablet containing SSX) was heated with acidic ethanol (EtOH–glacial acetic acid (99:1). Creams with SSX were dissolved in acidic EtOH. Tablets containing SSX were crushed and the powder was heated with acidic EtOH. Cooled solution was filtered	Plate: Whatman LKHPDF plates High-performance silica gel thin layers containing fluorescent phosphor Mobile phase: EtOAc–concentrated ammonium hydroxide (99:1) for SAN EtOAc–CHCl$_3$–MeOH (25:25:5) for SSX Detection: UV light at λ = 254 nm Kontes Chromaflex fiber-optic densitometer Identity of sulfas confirmed by reaction with Bratton–Marshall reagent or with fluorescamine	Recovery of sulfas Cream: 96%–1045% Suppository: 99.3%, 105% Tablet: 104%	[94]

a Trimethoprim (TMP), bacteriostatic antibiotic mainly used with sulfamethoxazole.

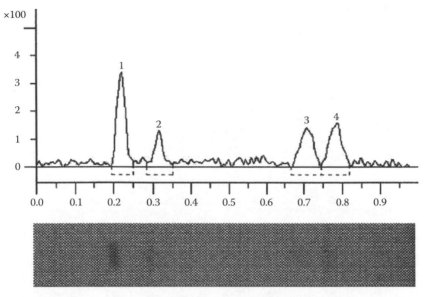

FIGURE 43.11 Chromatogram obtained from a standard mixture of the antibiotics: 1, SGUA; 2, TMP; 3, SDZ; 4, SDD. (From Babić, S. et al., *J. Planar Chromatogr. Mod. TLC*, 18, 423, 2005.)

TABLE 43.2

Regression Functions and Correlation Coefficients

Antibiotic	Regression Function	Correlation Coefficient (r^2)
Sulfadiazine	$y = 3659.2x + 311.67$	0.9979
Sulfadimidine	$y = 3840.3x + 271.05$	0.9904
Sulfaguanidine	$y = 2696.4x + 722.38$	0.9921
Trimethoprim	$y = 995.04x + 357.72$	0.9903

Source: Babić, S. et al., *J. Planar Chromatogr. Mod. TLC*, 18, 423, 2005.

Note: y, peak area; x, amount of sample (μg per spot).

TABLE 43.3

Results from Validation of Quantitative Determination of Antibiotics by TLC

Antibiotic	Linear Range (μg per Spot)	LOD (μg per Spot)	LOQ (μg per Spot)	RSD (%) ($n = 10$)
Sulfadiazine	0.1–1.0	0.05	0.1	5.11
Sulfadimidine	0.1–1.0	0.05	0.1	5.04
Sulfaguanidine	0.1–2.0	0.05	0.1	4.77
Trimethoprim	0.2–2.0	0.1	0.2	4.00

Source: Babić, S. et al., *J. Planar Chromatogr. Mod. TLC*, 18, 423, 2005.

TABLE 43.4
TLC for the Determination of Sulfonamides in Different Matrices

Matrix	Analyte	Sample Preparation	TLC/HPTLC Conditions	Recovery (%)/Other Information	Ref.
Water	SDZ, SDD, SGUA, TMP[a]	SPE, filtrated water samples (pH 4, 100 mL) were applied to preconditioned Oasis HLB cartridges, washed with 2 mL of 2% MeOH in water, and eluted with 2 × 5 mL organic solvent (EtOH, MeOH, ACN); filtrates were evaporated to dryness by rotary evaporation at 40°C; and residues dissolved in 1 mL MeOH	Plate: HPTLC silica gel F_{254} plate; Mobile phase: $CHCl_3$–MeOH (89:11); Detection: UV light at λ = 254 nm; 3CCD HV-C20, video camera Hitachi	**SDIAZ**: EtOH—79.5%, MeOH—89.4%, ACN—87.1%; **Sulfadimidine**: EtOH—84.2%, MeOH—93.3%, ACN—93.1%; **SGUA**: EtOH—9.4%, MeOH—12.7%, ACN—16.7%	[98]
Water	SGUA, SMTH, SDZ, TMP	SPE, filtrated water samples (pH 4, 100 mL) were applied to the preconditioned Oasis HLB cartridges, washed with 2 mL of 2% MeOH in water, and eluted with 2 × 5 mL MeOH; filtrates were evaporated to dryness by rotary evaporation at 40°C; and residues dissolved in 1 mL MeOH	Plate: HPTLC CN F_{254} plate; Mobile phase: 0.05 M oxalic acid–MeOH (1:19); Detection: UV light at λ = 254 and λ = 366 nm; 3CCD HV-C20, video camera Hitachi	**TMP**: wellspring water—99.3%, production water—103.2%; **SGUA**: wellspring water—6.2%, production water—no information; **SMETH**: wellspring water—92.8%, production water—102.2%; **SDIAZ**: wellspring water—92.6%, production water—92.7%	[99]
Salmon tissue	SDMX, SDZ, SMRZ, SMTZ, SP	MSPD, the sample was ground with C_{18} derivatized silica gel, washed with 10% toluene in hexane (discarded), and eluted with CH_2Cl_2. The CH_2Cl_2 extract was evaporated and reconstituted in MeOH	Plate: TLC silica gel plate; Mobile phase: EtOAc–n-butanol–MeOH–aqueous ammonia 30% (35:45:15:2); Detection: Plate sprayed with fluorescamine solution, scanned with Camag II densitometer, UV light at λ = 366 nm	Full validation; **SDMX**: 63%; **SDZ**: 61%; **SMRZ**: 63%; **SMTZ**: 60%; **SP**: 57%; LOD: 0.1 ppm SMRZ; LOD: 0.04 ppm for the others	[100]

Sample	Analytes	Sample preparation	Plate / Mobile phase / Detection	Results	Ref.
Swine plasma	SMTZ	5 mL of sample was pipetted into membrane cones, 20 mg of phenylbutazole was added, the cones were centrifuged at 2000 g for 30 min, and the supernatants were transferred to tubes. 2 μg SGUA was added as an internal standard and solution was extracted into 2.5 mL of EtOAc, then transferred to the tubes and evaporated to 0.2 mL. The condensed extract was spotted on TLC plate	Plate: TLC silica gel plate Mobile phase: EtOAc Detection: Plate dipped in fluorescamine solution; UV light at λ = 310 nm	**SMTZ:** LOD = 0.02–0.05 ppm	[111]
Swine liver	SMTZ, ST	Tissues were homogenized with $CHCl_3$–EtOAc (1:1). The analytes were extracted with carbonate buffer pH 10 and back-extracted into CH_2Cl_2 as ion pair with tetrabutylammonium hydroxide After evaporation of solvent the residue was dissolved in MeOH	Plate: TLC silica gel Derivatization with Bratton–Marshall reagent	Recoveries determined by HPLC: **SMTZ** 50.8% ± 4.2% **ST** 42.0% ± 4.7%	[120]
Pork muscle	SD, SDD, SDZ, SAN	The tissue (1 g) was extracted in ultrasonic bath with CH_2Cl_2 (1 mL), centrifuged; aqueous supernatant phase was separated from the organic phase. The latter spotted on TLC plate	Plate: TLC silica gel with concentration zone Mobile phase: $CHCl_3$–n-butanol (4:1) Detection: Plate treated with fluorescamine solution, UV light at λ = 365 nm	Screening: positive results for TLC confirmed by HPLC in the concentration range 34.4–190 ppb Sensitivity of the method 25 μg kg^{-1}	[121]
Milk	SMTZ	SPE, 10 mL milk sample with SBZ internal standard mixed with 10 mL of 0.2 M phosphate buffer (pH 5.7) passed through pretreated silica-C18 cartridge. The cartridge was washed with water and hexane. Sulfonamides were eluted with MeOH, purified over an acidic alumina column, and concentrated on an ion-exchange AG MP-1 resin	Plate: HPTLC silica gel plate Mobile phase: EtOAc–toluene (1:1) Detection: Plate dipped in fluorescamine solution, dried with N_2 and sprayed with 0.2 M H_3BO_3 UV light at λ = 366 nm densitometer Camag TLC Scanner II	Partial validation **SMTZ**: 88.3%–103.2% (average 96.07%) in fortified milk samples in 0.5–15 ppb range Amounts found in two incurred milk samples were 1.99 and 6.57 ppb, HPLC used for confirmation	[101]

(continued)

TABLE 43.4 (continued)
TLC for the Determination of Sulfonamides in Different Matrices

Matrix	Analyte	Sample Preparation	TLC/HPTLC Conditions	Recovery (%)/Other Information	Ref.
Milk	SDZ	MSPD, milk sample was mixed with pretreated C_{18} material; column bed was washed with 8 mL of hexane; and sulfonamides were eluted with 8 mL of CH_2Cl_2. The extract was dried and residues dissolved in 100 μL of MeOH	Plate: HPTLC silica gel plate F_{254} Mobile phase: Three multiple developments with MeOH, containing 2% ammonia (25%) and CH_2Cl_2 (30:70, 15:85, and 5.95), respectively for successive distances of 15, 30, 45 mm Detection: Plate sprayed with fluorescamine solution UV light at $\lambda = 366$ nm Densitometer, Camag scanner II	**SDZ:** 95% ± 2% for 2 g of C18 and 5 mL of milk LOD 10 ppb	[109]
Pork and beef muscle	ST, SDZ, SMTZ, SDMX, SD	SPE, the tissue samples were extracted with 2×10 mL EtOAc in ultrasonic bath for 15 min; after decantation, 30 mL of hexane was added to EtOAc extract (3:2), and filtered. The filtrate was passed onto Sep-Pak® silica cartridge and pretreated with 10 mL MeOH and 10 mL hexane–EtOAc (3:2). Sulfonamides were eluted with 3 mL of MeOH–ACN (1:9), evaporated to dryness, and dissolved in 100 μL MeOH		Partial validation No information on recovery LOD: 4 ppb	
Swine, cattle, and chicken tissue	SMTZ, SDZ, SQ, SAN, SD	SPE, 10 g of homogenized sample was extracted with 2×25 mL CH_2Cl_2. The extract was transferred into Sep-Pak® silica cartridge; the cartridge was rinsed with 5 mL of CH_2Cl_2, (discarded) and Sep-Pak® was dried in a stream of N_2 for 10 min; sulfonamides were eluted of 2.8 mL of phosphate buffer	Plate: HPTLC silica gel plate Mobile phase: $CHCl_3$–n-butanol (4:1) Detection: Plate sprayed with fluorescamine solution UV light at $\lambda = 366$ nm	No validation No information on recovery LOD = 0.1 mg kg^{-1}	[110]
Swine tissue	SMTZ, SP	Spiked liver sample was homogenized, with EtOAc, and centrifuged. EtOAc supernatant was partitioned with 10 mL 1 N HCl. The aqueous phase was adjusted to pH 6.5 and partitioned with CH_2Cl_2. The organic phase was evaporated to dryness and reconstituted in MeOH	Plate: Precoated TLC silica gel plate Mobile phase: $CHCl_3$–$tert$-butanol (80:20) Detection: Plate dipped in fluorescamine solution and dried for 15–30 min Spectrophotofluorometer with TLC scanning attachment	Validation **SMTZ:** 48.8% **SP:** 43.2%	[102]

Animal tissues	SDD, SAN, SDZ, ST, SMRZ, Na-SQ, SGUA, STM, SD, SPR, FST, SMDZ, SMX, Na-SCPD, SMMT, SPH, SMPD	SPE, sulfonamides extracted from homogenized tissue by $CHCl_3$–acetone (1:1), the acidified extract was concentrated and purified by means of a cation-exchange SPE column, ammonia vapor was passed through the column, after that the analytes were eluted with 3 mL MeOH, the solution was evaporated, and the residues were dissolved in 100 µL of acetone–MeOH (9:1)	Plate: HPTLC silica gel plate Mobile phase: Three solvent systems: v/v/v Solvent system A: EtOAc–MeOH–ammonia (28%) (30:15:1) Solvent system B: ACN–$CHCl_3$–ammonia (28%) (35:10:0.2) Solvent system C: $CHCl_3$–1-butanol–30°C–60°C petroleum ether (15:15:15) For 1D TLC system A For 2D TLC system A in one direction, system B or C in second direction Detection: Plates derivatized with fluorescamine solution	For all of sulfonamides LOD = 10 ng g^{-1}	[122]
Animal tissues: cattle and chicken	SAN, SDMX, SMTZ, SQ	Sulfonamides were extracted from tissues with $CHCl_3$–acetone (1:1), then with 1 N HCl–acetone–hexane. Organic phase was re-extracted with 1 N HCL and acidic extracts were combined. Then, pH was adjusted to basic one, sample extracted with $CHCl_3$ (rejected) and adjusted again to acidic pH	Plate: Silica gel G Mobile phase: Acetone–n-heptane–MeOH–ammonium hydroxide–n-butanol (36:10.5:4.5:5:5) Bratton–Marshall reaction was performed. Positive samples were partitioned to butanol and applied to TLC plate	Mean recoveries for liver, muscle, and kidney of cattle and chicken: At 0.1 ppm 90.6 ± 1.19 At 0.5 ppm 89.7 ± 0.32 At 1.0 ppm 90.8 ± 1.54 LOD: 0.1 ppm	[123]
Pork tissue	SMZ	Tissue samples were homogenized, twice sonicated with acetone and centrifuged. The combined supernatants were filtered and extracted with ether, and aqueous extract was diluted with water. Then it was twice extracted with methylene chloride and evaporated, and the sample was reconstituted in MeOH	TLC screening according to [80–82,127]. TLC was utilized to separate Bratton–Marshall positive reactants and used before GC analysis to separate and purify derivatized samples. Sulfa drug zones were identified with UV, scraped, eluted with acetone, and subjected to GC	Depletion of **SMZ** in swine fed at the rate 110 ppm in feed after 14, 21, and 28 days of withdrawal period. Tolerance level, 0.1 ppm, was achieved after 5 days for muscles, 7.5 days for kidney, and 9 days for liver LOD: 20 ppb in various tissues of swine	[124]

(continued)

TABLE 43.4 (continued)

TLC for the Determination of Sulfonamides in Different Matrices

Matrix	Analyte	Sample Preparation	TLC/HPTLC Conditions	Recovery (%)/Other Information	Ref.
Animal muscle, kidney, and serum	SMRZ, ST, SDZ, SP, SSD, SQ, SEP, SCST, SMX, SM,SPR, STM, SDMX, SL,SMP, FST, SCP, SCPD, SPH	The homogenized tissue extracted with water and EtOAc and filtered, serum centrifuged after adding sodium dodecylsulfate (SDS)	Plate: HPTLC silica gel plates Mobile phase: Various solvents in 1D or 2D mode Detection: Plates derivatized with fluorescamine solution	The lowest recoveries for **SAN**:8.72 ± 1.86, 12.44 ± 1.32, in kidney and muscle, respectively; the highest for **SL** in muscle 30.65 ± 4.52 In serum + SDS **SSD:**34.94 ± 3.04 **SAN:**31.28 ± 8.48 **SMX**:43.4 ± 3.36 LOD for the majority of 23 sulfonamides: 10 ppb	[125]
Animal urine: swine and cattle	SD, SQ, SMDX, SDD	Plates were developed in Soczewinski sandwich chamber. Samples of urine and standards were applied directly as a series of spots along the plate at about 2 cm distance (parallel to development direction), 5 spots each of 5 μL. Then they were concentrated and purified from the matrix ballast by developing with ACN under short cover plate; the samples concentrated on the upper edge of the plate were developed in a classical way	Plate: TLC Si 60 (without fluorescence indicator and concentration zone) Mobile phase: EtOAc–CH$_2$Cl$_2$ (4:1) Detection: Plate sprayed with fluorescamine UV light at λ = 366 nm	No information about recovery LOD: 0.1 μg mL^{-1} Procedure was tested for 243 swine and 184 cattle slaughtered for human consumption	[126]
Human plasma	NSD	The samples were extracted with CH$_2$Cl$_2$ (2 × 3 mL). The combined extract was evaporated to dryness and reconstituted in MeOH	Plate: TLC silica gel F$_{254}$ plate Mobile phase: Toluene–acetone (10:1) Detection: UV light at λ = 310 nm Densitometer Scanner III Camag	**NSD**: 97.10% ± 2.22%	[92]

[a] Trimethoprim (TMP) is a bacteriostatic antibiotic mainly used with sulfamethoxazole.

43.3.3 Use of TLC in Biological Samples

Quantitative analysis of sulfonamides in the biological samples is much more complicated than the analysis of pharmaceutical preparations. Biological matrices are extremely difficult (compared to water or tablets) because they can hinder determination of the analyte, being a source of interferences disturbing final results or even causing the damage of analytical equipment. Therefore, matrix effects are basic problems during sample preparation. The suitable pretreatment step is necessary to convert the original matrix to the one that enables determination of the analytes [17,107]. All the sample preparation steps have the same analytical purpose [108]:

- Analyte isolation from a matrix
- Reduction of interferences
- Analyte concentration
- Obtaining appropriate form of the sample

Sulfonamides are determined mostly in body fluids [92], animal tissues (e.g., muscles) [109–111], milk [101], eggs [112], and honey [113]. As described earlier, sulfonamide residues are also common in water [98,99]. Sulfonamides are well soluble in polar organic solvents, so the extraction process can be conducted using acetonitrile, acetone, chloroform, methylene chloride, or ethyl acetate [112,114,115]. Some of them are used for deproteinization, which facilitates further analysis [116]. The cleanup and enrichment procedures depend mainly on the type of matrix and solvents planned to be used for extraction. The typical techniques used in the preliminary steps of sample preparation are liquid–liquid extraction (LLE) [112,114], ultrafiltration, and membrane filtration [111]. The most effective techniques for cleanup and pre-concentration of the sample are SPE and matrix solid-phase dispersion (MSPD). The first one increasingly replaces the conventional LLE especially to analyze semi-volatile and nonvolatile liquid samples such as milk or water. Various sorbents were tested for sulfonamides sorption at the SPE cartridges. The most common is silica with the chemically bonded C18 group [93,94,96,110]. Before using, C18 cartridges are usually conditioned using methanol or deionized water [93,94,117]. Then the appropriate volume of water or other liquid phase containing drugs (typical volume is 100–500 mL) is filtered and percolated through the cartridge. After sorption, sulfonamides are eluted using methanol, ethanol, acetonitrile, or their mixtures as well as dichloromethane and phosphate buffer [93,95,97,110]. MSPD is considered to be the second important extraction technique, just after SPE. MSPD has a significant advantage over SPE for solid and semisolid matrices enabling extraction from inside of the cell of, for example, animal tissue [95,97,118,119]. The sample is blended with the sorbent (usually C18 silica), put into the cartridge (or sometimes the blending process is done directly inside the cartridge), and eluted with the solvent. Then, the solvent is evaporated to dryness and finally the residues are dissolved in the mobile phase or other solvents [95,97].

Table 43.4 presents chromatographic conditions as well as cleanup and concentration methods for various matrices containing sulfa drugs.

LIST OF DRUG ABBREVIATIONS

FINA	Finasteride
FST	Formosulfathiazole
Na-SCPD	Na-Sulfachlorpyridazine
Na-SQ	Na-Sulfaquinoxaline
NSD	Nimesulide
PST	Phthalylsulfathiazole
SZP	Salazopyrine
SCST	Succinylsulfathiazole

SBZ	Sulfabromomethazine
SAC	Sulfacarbamide
SC	Sulfacetamide
SCP	Sulfachlorpyrazine
SCPD	Sulfachlorpyridazine
SDZ	Sulfadiazine
SDC	Sulfadicramide
SDMX	Sulfadimethoxine
SDD	Sulfadimidine
SD	Sulfadoxine
SET	Sulfaethidole
SEP	Sulfaethoxypyridazine
SF	Sulfafurazole
SGUA	Sulfaguanidine
SL	Sulfalene
SMRZ	Sulfamerazine
SM	Sulfameter
SMTZ	Sulfamethazine
SMTH	Sulfamethizole
SMX	Sulfamethoxazole
SMDZ	Sulfamethoxidiazine
SMP	Sulfamethoxypyrazine
SMPD	Sulfamethoxypyridazine
SAN	Sulfanilamide
SAA	Sulfanilic acid
SPH	Sulfaphenazole
SP	Sulfapirydine
SPX	Sulfaproxyline
SPR	Sulfapyrazole
SQ	Sulfaquinoxaline
ST	Sulfathiazole
STM	Sulfatolamide
SSD	Sulfisomidine
SSX	Sulfisoxazole
SMMT	Sulfomonomethoxine
TAM	Tamsulosin
TMP	Trimethoprim

REFERENCES

1. Domagk, G. 1935. Ein Beitrag zur Chemotherapie der bacteriellen Infectionen. *Deut. Med. Woch.* 61:250–253.
2. Colebrook, L. and Kenny, M. 1936. Treatment of puerperal infections and experimental infection in mice with Prontosil. *Lancet* 227:1279–1281.
3. Mann, J. 2007. *Life Saving Drugs, The Elusive Magic Bullet.* London: The Royal Society of Chemistry.
4. Gelmo, P. 1908. Über sulfamide der p-amidobenzolsulfonsäure. *J. Prakt. Chem.* 77:369–382.
5. Wainwright, M. and Kristiansen, J. E. 2011. On the 75th anniversary of Prontosil. *Dyes Pigments* 88:231–234.
6. Wainwright, M. 2008. Dyes in the development of drugs and pharmaceuticals. *Dyes Pigments* 76:582–589.
7. Schulemann, W. 1932. Synthetic antimalarial preparations. *Proc. R. Soc. Med.* 25:897–905.
8. Tréfouël, J., Tréfouël, T., Nitti, F. et al. 1935. Activité du p-aminophénylsulfamide sur les infections expérimentales de la souris et du lapin. *Compt. Rend. Soc. Biol.* 20:756–758.

9. Fuller, A. T. 1937. Is p-aminobenzenesulfonamide the active agent in Prontosil therapy? *Lancet* 1:194–198.

10. Domagk, G. 1947. Further progress in chemotherapy of bacterial infections. *Nobel Lecture*, December 12.

11. Littlefield, N. 1988. Chronic toxicity and carcinogenicity of sulfamethazine I B_6CF_1 mice. *Fed. Reg.* 53:FR9492.

12. Poitier, L. A., Doerge, D. R., Gaylor, D. W. et al. 1999. An FDA review of sulfamethazine toxicity. *Regul. Toxicol. Pharmacol.* 30:217–222.

13. Lindsey, M. E., Meyer, M., and Thurman, E. M. 2001. Analysis of trace levels of sulfonamide and tetracycline antimicrobials in groundwater and surface water using solid-phase extraction and liquid chromatography/mass spectrometry. *Anal. Chem.* 73:4640–4646.

14. Hirsch, R., Ternes, T., Haberer, K. et al. 1999. Occurrence of antibiotics in the aquatic environment. *Sci. Total Environ.* 225:109–118.

15. Harting, C., Storm, T., and Jekel, M. 1999. Detection and identification of sulphonamide drugs in municipal waste water by liquid chromatography coupled with electrospray ionization tandem mass spectrometry. *J. Chromatogr. A* 854:163–173.

16. Comer, J. P. and Comer, I. 2006. Applications of thin-layer chromatography in pharmaceutical analyses. *J. Pharm. Sci.* 56:413–436.

17. Wang, S., Zang, H.-Y., Wang, L. et al. 2006. Analysis of sulphonamide residues in edible animal products: A review. *Food Addit. Contam.* 23:362–384.

18. Guggisberg, D., Mooser, A. E., and Koch, H. 1992. Methods for the determination of sulphonamides in meat. *J. Chromatogr.* 624:425–437.

19. Kennedy, D. G., McCracken, R. J., Cannavan, A. et al. 1998. Use of liquid chromatography—Mass spectrometry in the analysis of residues of antibiotics in meat and milk. *J. Chromatogr. A* 812:77–98.

20. Niessen, W. M. A. 1998. Analysis of antibiotics by liquid chromatography-mass spectrometry. *J. Chromatogr. A* 812:53–75.

21. Gentili, A., Perret, D., and Marchese, S. 2005. Liquid chromatography-tandem mass spectrometry for performing confirmatory analysis of veterinary drugs in animal-food products. *Trends Anal. Chem.* 24:704–733.

22. Horwitz, W. A. 1981. Analytical methods for sulfonamides in foods. Review of methodology. *J. Assoc. Off. Anal. Chem.* 64:104–130.

23. Agbaba, D. and Malenovic, A. 2011. New methods for drug analysis in biological samples and other matrixes. *J. Assoc. Off. Anal. Chem., Special Section* 94:665–785.

24. Nagaraja, P., Yathirajan, H. S., Sunitha, K. R. et al. 2002. A new, sensitive, and rapid spectrophotometric method for the determination of sulfa drugs. *J. Assoc. Off. Anal. Chem.* 85:869–874.

25. Choma, I. M. 2005. The use of thin-layer chromatography with direct bioautography for antimicrobial analysis. *LC-GC* 18:482–488.

26. Sherma, J. 2000. Thin-layer chromatography in food and agricultural analysis. *J. Chromatogr. A* 880:129–147.

27. Zang, S., Wang, Z., Nesterenko, I. et al. 2007. Fluorescence polarization immunoassay based on monoclonal antibody for the detection of sulphamethazine in chicken muscle. *Int. J. Food Sci. Technol.* 42:36–44.

28. Franek, M., Diblikova, I., Cernoch, I. et al. 2006. Broad-specificity immunoassays for sulfonamide detection: Immunochemical strategy for generic antibodies and competitors. *Anal. Chem.* 78:1559–1567.

29. Okerman, L., De Wasch, K., and Van Hoof, J. 1998. Detection of antibiotics in muscle tissue with microbiological inhibition tests: Effects of the matrix. *Analyst* 123:2361–2365.

30. Chung, H.-H., Lee, J.-B., Chung, Y.-H. et al. 2009. Analysis of sulfonamide and quinolone antibiotic residues in Korean milk using microbial assays and high performance liquid chromatography. *Food Chem.* 113:297–301.

31. Korsrud, G. O., Salisbury, C. D. C., Fesser, A. C. E. et al. 1994. Investigation of charm test II receptor assays for the detection of antimicrobial residues in suspect meat samples. *Analyst* 119:2737–2741.

32. Bjurling, P., Baxter, G. A., Caselunghe, M. et al. 2000. Biosensor of sulfadiazine and sulfamethazine residues in pork. *Analyst* 125:1771–1774.

33. Zang, H., Zang, Y., and Wang, S. 2008. Development of flow-through and dip-stick immunoassays for screening of sulfonamide residues. *J. Immunol. Methods* 337:1–6.

34. Zang, H., Wang, L., Zang, Y. et al. 2007. Development of an enzyme-linked immunosorbent assay for seven sulfonamide residues and investigation of matrix effects from different food samples. *J. Agric. Food Chem.* 55:2079–2084.

35. Adrian, J., Font, H., Diserents, J.-M. et al. Generation of broad specificity antibodies for sulfonamide antibiotics and development of an enzyme-linked immunosorbent assay (ELISA) for the analysis of milk samples. *J. Agric. Food Chem.* 57:385–394.

36. He, J., Shen, J., Suo, X. et al. 2005. Development of a monoclonal antibody-based ELISA for detection of sulfamethazine ang N^4-acetyl sulfamethazine in chicken breast muscle tissue. *Food. Chem. Toxicol.* 70:113–117.

37. Wang, Z., Zang, S., Ding, S. et al. 2005. Simultaneous determination of sulphamerazine, sulphamethazine and sulphadiazine in honey and chicken muscle by new monoclonal antibody-based fluorescence polarization immunoassay. *Food Addit. Contam.* 25:574–582.

38. Wang, Z., Zang, S., Nesterenko, I. et al. 2007. Monoclonal antibody-based fluorescence polarization immunoassay for sulfamethoxypyridazine and sulfachloropyridazine. *J. Agric. Food Chem.* 55:6871–6878.

39. Choma, I. M. 2003. Antibiotics. In *Handbook of Thin-Layer Chromatography*, 3rd edn., Revised and Expanded, J. Sherma and B. Fried, Eds., pp. 417–444. New York: Marcel Dekker.

40. Zang, Y., Xu, X., Qi, X. et al. 2012. Determination of sulfonamides in livers using matrix solid-phase dispersion extraction high-performance liquid chromatography. *J. Sep. Sci.* 35:45–52.

41. Arancibia, V., Valderrama, M., Rodriquez, P. et al. 2003. Quantitative extraction of sulfonamides in meats by supercritical methanol-modified carbon dioxide: A foray into real-world sampling. *J. Sep. Sci.* 26:1710–1716.

42. Agarwal, V. K. 1992. High-performance liquid chromatographic methods for the determination of sulfonamides in tissue, milk and eggs. *J. Chromatogr. A* 624:411–423.

43. Prez, N., Gutierrez, R., Noa, M. et al. 2002. Liquid chromatographic determination of multiple sulfonamides, nitrofurans, and chloramphenicols residues in pasteurized milk. *J. Assoc. Off. Anal. Chem.* 85:20–24.

44. Rambla-Alegre, H., Romero-Estere, J., and Cardla-Broch, S. 2011. Development and validation of micellar liquid chromatographic methods for the determination of antibiotics in different matrixes. *J. Assoc. Off. Anal. Chem.* 94:775–785.

45. Dost, K., Jones, D. J., and Davidson, G. 2000. Determination of sulfonamides by packed column supercritical fluid chromatography with atmospheric pressure chemical ionization mass spectrometric detection. *Analyst* 125:1243–1247.

46. Cannavan, A., Hewitt, S. A., Blanchflower, W. J. et al. 1996. Gas chromatography-mass spectrometric determination of sulfamethazine in animal tissue using a methyl/trimethylsilyl derivative. *Analyst* 121:1457–1461.

47. Stout, S. J., Steller, W. A., Manuel, A. J. et al. 1984. Confirmatory method for sulfamethazine residues in cattle and swine tissues, using gas chromatography-chemical ionization mass spectrometry. *J. Assoc. Off. Anal. Chem.* 67:142–144.

48. Simpson, R. M., Suhre, F. B. and Shafer, J. W. 1985. Quantitative gas chromatographic-mass spectrometric assay of five sulfonamide residues in animal tissue. *J. Assoc. Off. Anal. Chem.* 68:23–26.

49. Takatsuki, K. and Kikuchi, T. 1990. Gas chromatographic-mass spectrometric determination of six sulfonamide residues in egg and animal tissues. *J. Assoc. Off. Anal. Chem.* 73:886–892.

50. Carignan, G. and Carrier, K. 1991. Quantitation and confirmation of sulfamethazine residues in swine muscle and liver by LC and GC/MS. *J. Assoc. Off. Anal. Chem.* 74:379–382.

51. Chu, Q., Zhang, D., Wang, J. et al. 2009. Muli-residue analysis of sulfonamides in animal tissues by capillary zone electrophoresis with electrochemical detection. *J. Sci. Food Agric.* 89:2498–2504.

52. Hoff, R. and Kist, T. B. L. 2009. Analysis of sulfonamides by capillary electrophoresis. *J. Sep. Sci.* 32:854–866.

53. Porter, S. 1994. Confirmation of sulfonamide residues in kidney tissue by liquid chromatography-mass spectrometry. *Analyst* 119:2753–2756.

54. Pereira Lopes, R., Vasconcellos Augusti, D., De Souza, L. F. et al. 2011. Development and validation (according to the 2002/657/EC regulation) of a method to quantify sulfonamides in porcine liver by fast partition at very low temperature and LC-MS/MS. *Anal. Method* 3:606–613.

55. Won, S. Y., Lee, Ch. H., Chang, H. S. et al. 2011. Monitoring of 14 sulfonamide antibiotics residues in marine products using HPLC-PDA and LC-MS/MS. *Food Control* 22:1101–1107.

56. Zou, Q.-H., Wang, J., Wang, X.-F. et al. 2008. Application of matrix solid-phase dispersion and high-performance liquid chromatography for determination of sulfonamides on honey. *J. Assoc. Off. Anal. Chem.* 91:252–258.

57. Kaufmann, A., Sven, R., Bianca, R. et al. 2002. Quantitative LC/MS-MS determination of sulfonamides and some other antibiotics in honey. *J. Assoc. Off. Anal. Chem.* 85:853–860.

58. Choma, I. M. 2005. Antibiotics, analysis by TLC. In *Encyclopedia of Chomatography*, 2nd edn., J. Cazes, Ed., pp. 93–100. New York: Marcel Dekker.

59. Sherma, J. 1980. Paper and thin-layer chromatography. *Anal. Chem.* 52:276–289.

60. Poole, C. F. 2003. Thin-layer chromatography: Challenges and opportunities, *J. Chromatogr.* 1000:963–984.

61. Bieganowska, M. L., Doraczyńska-Szopa, A. D., and Petruczynik, A. 1993. The retention behavior of some sulfonamides on different thin layer plates. *J. Planar Chromatogr. Mod. TLC* 6:121–128.

62. Bieganowska, M. L., Doraczyńska-Szopa, A. D., and Petruczynik, A. 1995. The retention behavior of some sulfonamides on different TLC plates. 2. Comparison of the selectivity of the system and quantitative determination of hydrophobicity parameters. *J. Planar Chromatogr. Mod. TLC* 8:122–128.

63. Van Poucke, L., Rousseau, D., Van Peteghem, C., and De Spiegeleer, B. M. J. 1989. Two-dimensional HPTLC of sulphonamides on cyanoplates, using the straight-phase and reversed-phase character of the adsorbent. *J. Planar Chromatogr.* 2:395–397.

64. Bićan-Fišter, T. and Kajganović, V. 1963. Separation and identification of sulfonamides by thin-layer chromatography. *J. Chromatogr.* 11:492–495.

65. Wollish, E. G., Schmall, M., and Hawrylyshyn, M. 1961. Thin-layer chromatography. recent developments in equipment and applications. *Anal. Chem.* 33:1138–1142.

66. Flieger, J., Świeboda, R., and Tatarczak, M. 2007. Chemometric analysis of retention data from salting-out thin-layer chromatography in relations to structural parameters and biological activity oh chosen sulfonamides. *J. Chromatogr. B* 846:334–340.

67. Cacace, M. G., Landay, E. M., and Ramsden, J. J. 1997. The Hofmeister series: Salt and solvent effects on interfacial phenomena. *Q. Rev. Biophys.* 30:241–277.

68. Bhushan, R. and Ali, I. 1995. TLC separation of sulfonamides on impregnated silica gel layers, and their quantitative estimation by spectroscopy, *J. Planar Chromatogr. Mod. TLC* 8:245–247.

69. Szumiło, H. and Flieger, J. Chromatographic behavior of sulfonamides on polyamide impregnated with various metal salts. *J. Planar Chromatogr.* 9:462–466.

70. Flieger, J., Szumiło, H., and Giełzak-Koćwin, K. 1999. Investigation of conditions suitable for impregnation of polyamide of metal salts, and determination of the structures of the material obtained. *J. Planar Chromatogr.* 12:255–260.

71. Walash, M. I. and Agarwal, S. P. 1972. Characterization of sulfonamides by TLC using metal ions. *J. Pharm. Sci.* 61:277–278.

72. Bieganowska, M. L. and Petruczynik, A. 1996. Thin-layer and column chromatography of sulfonamides on aminopropyl silica gel. *Chromatographia* 43:654–658.

73. Biagi, G. L., Barbaro, A. M., Guerra, M. C. et al. 1974. Relationship between π and R_M values of sulfonamides. *J. Med. Chem.* 17:28–33.

74. Lepri, L., Desideri, P. G., and Coas, V. 1987. High-performance thin-layer chromatography of dinitropropyl-amino acids sulphonamides on silanized silica gel. *J. Chromatogr.* 405:394–400.

75. Okumura, T. and Nagaoka, T. 1980. Reversed-phase thin-layer chromatography of synthetic sulfonamide antibacterials by using alkylsilyl silica gels. *J. Liq. Chromatogr.* 3:1947–1960.

76. Bieganowska, M. L., Petruczynik, A., and Doraczyńska-Szopa, A. 1993. Thin-layer reversed-phase ion-pair chromatography of sulphonamides. *J. Pharm. Biomed. Anal.* 11:241–246.

77. Lepri, L., Desideri, P. G., and Heimler, D. 1978. Soap thin-layer chromatography of sulphonamides and aromatic amines. *J. Chromatogr.* 169:271–278.

78. Srivastava, S. P., Dua, V. K., and Saxena, R. C. 1979. Separation and identification of closely related sulpha drugs by TLC on silica gel-calcium oxalate impregnated plates. *Anal. Chem.* 299:207.

79. Kadar Pauncz, J. 2005. Application of ion-exchange resin coated plates for thin-layer chromatography of sulfonamides and some other chemotherapeutic compounds. *J. High Res. Chromatogr.* 4:287–291.

80. Bićan-Fišter, T. and Kajganović, V. 1964. Quantitative analysis of sulfonamide mixtures by thin-layer chromatography. *J. Chromatogr.* 16:503–509.

81. Cieri, U. R. 1969. Thin-layer chromatography and ultraviolet spectrophotometry of sulfonamide mixtures. *J. Chromatogr.* 45:421–431.

82. Cieri, U. R. 1970. Thin-layer chromatography and ultraviolet spectrophotometry of mixtures of sulfonamides. *J. Chromatogr.* 49:493–502.

83. Olivari, G. G. 1958. Direct spectrophotometric determination of a sulfonamide mixture on chromatograms after separation. *Boll. Chim. Farm.* 97:552–559.

84. Wagner, G. and Wandel, J. 1966. Experiments on the quantitative determination of sulfonamides separated by thin-layer chromatography. *Pharmazie* 21:105–109.

85. Papendick, V. E., Sutherland, J. W., and Williamson, D. E. 1969. Pharmaceuticals and related drugs. *Anal. Chem.* 41:190–215.

86. Spencer, R. D. and Beggs, B. H. 1965. Thin-layer chromatography in silica gel: Quantitative analysis by direct UV spectrophotometry. *J. Chromatogr.* 21:52–66.

87. Patel, D. B. and Patel, N. J. 2010. Validated RP-HPLC and TLC method for simultaneous estimation of Tamsulosin hydrochloride and Finasteride in combined dosage forms. *Acta Pharm.* 60:197–205.

88. Agbaba, D., Radovic, A., Vladimirov, S. et al. Simultaneous TLC determination of co-trimoxazole and impurities of sulfanilamide and sulfanilic acid in pharmaceuticals. *J. Chromatogr. Sci.* 34:460–464.

89. Salomies, H. 1993. Quantitative HPTLC of sulfonamides in pharmaceutical preparations. *J. Planar Chromatogr.* 6:337–340.

90. Tammiletho, S. A. 1985. High-performance thin-layer chromatographic determination of trimethoprim and sulphamethoxazole in pharmaceutical dosage forms. *J. Chromatogr.* 323:456–461.

91. Knupp, G., Pollmann, H., and Jonas, D. 1986. An improved HPTLC method for the rapid identification and quantification of sulfonamides. *Chromatographia* 22:210–224.

92. Pandya, K. K., Satia, M. C., Modi, I. A. et al. 1997. High-performance thin-layer chromatography for the determination of Nimesulide in human plasma, and its use in pharmacokinetic studies. *J. Pharm. Pharmacol.* 49:773–776.

93. Bertini, D. and Niti, V. 1990. Determination of sulfametrol in commercial dosage forms (Ladaprim) by TLC-densitometry. *Farmacology* 45:1129–1136.

94. Sherma, J. and Duncan, M. 1986. Determination of sulfanilamide and sulfasoxizole in drug preparation by quantitative HPTLC. *J. Liquid Chromatogr.* 9:1861–1868.

95. Georgirakis, M. 1980. Densitometric determination of sulfamethoxazole and trimethoprim in tablets. *Pharmazie* 35:804.

96. Li, T. 1988. Analysis of sulfamethoxazole pediatric tablets by TLC-densitometry. *Yaowu Fenxi Zazhi* 8:185–186.

97. Krzek, J. and Bielska, A. 1985. Densitometric determination of components in dispersed medicines. *Farm. Pol.* 41:534–537.

98. Babić, S., Ašperger, D., Mutavdžić, D. et al. 2005. Determination of sulfonamides and trimethoprim in spiked water samples by solid-phase extraction and thin-layer chromatography. *J. Planar Chromatogr. Mod. TLC* 18:423–426.

99. Babić, S., Ašperger, D., Mutavdžić, D. et al. 2007. Determination of veterinary pharmaceuticals in production wastewater by HPTLC-videodensitometry. *Chromatographia* 65:105–110.

100. Reimer, G. J. and Suarez, A. 1991. Development of a screening method for five sulfonamides in salmon muscle tissue using thin-layer chromatography. *J. Chromatogr.* 555:315–320

101. Unruh, J., Piotrowski, E., Schwartz, D. P. et al. 1990. Solid-phase extraction of sulphametazine in milk with quantitation at low ppb levels using thin-layer chromatography. *J. Chromatogr.* 519:179–187.

102. Thomas, M. H., Soroka, K. E., Simpson, M. R. et al. 1981. Determination of sulfamethazine in swine tissue by quantitative thin-layer chromatography. *J. Agric. Food Chem.* 29:621–624.

103. Klein, S. and Kho, B. T. 1962. Thin-layer chromatography in drug analysis I. Identification procedure for various sulfonamides in pharmaceutical combinations. *J. Pharm. Sci.* 51:966–970.

104. Göbel, A., McArdell, C. S., Suter, M. J., and Giger, W. 2004. Trace determination of macrolide and sulfonamide antimicrobials, a human sulfonamide metabolite, and trimethoprim in wastewater using liquid chromatography coupled to electrospray tandem mass spectrometry. *Anal. Chem.* 76:4756–4764.

105. Ye, Z. and Weinberg, H. S. 2007. Trace analysis of trimethoprim and sulfonamide, macrolide, quinolone, and tetracycline antibiotics in chlorinated drinking water using liquid chromatography electrospray tandem mass spectrometry. *Anal. Chem.* 79:1135–1144.

106. Renew, J. E. and Huang, Ch.-H. 2004. Simultaneous determination of fluoroquinolone, sulfonamide, and trimethoprim antibiotics in wastewater using tandem solid phase extraction and liquid chromatography-electrospray mass spectrometry. *J. Chromatogr. A* 1042:113–121.

107. Mutavdžić Pavlović, D., Babić, S., Horvat, A. J. M. et al. 2007. Sample preparation in analysis of pharmaceuticals. *Trends Anal. Chem.* 26:1062–1075.

108. Smith, R. M. 2003. Before the injection—Modern methods of sample preparation for separation techniques. *J. Chromatogr. A* 1000:3–27.

109. Van Poucke, L. S. G., Depoucq, G. C. I., and Van Peteghem, C. H. 1991. A quantitative method for the detection of sulfonamide residues in meat and milk samples with a high-performance thin-layer chromatographic method. *J. Chromatogr. Sci.* 29:423–426.

110. Haagsma, N., Dieleman, B., and Gortemaker, B. G. 1984. A rapid thin-layer chromatographic screening method for five sulfonamides in animal tissues. *Vet. Quart.* 6:8–12.

111. Bevill, R. F., Schemske, K. M., Luther, H. G. et al. 1978. Determination of sulfonamides in swine plasma. *J. Agric. Food Chem.* 26:1201–1203.

112. Roudaut, B. and Garnier, M. 2002. Sulphonamide residues in eggs following drug administration via the drinking water. *Food Addit. Contam.* 19:373–378.

113. Posyniak, A., Zmudzi, J., Niedzielska, J. et al. 2003. Sulfonamide residues in honey. Control and development of analytical procedure. *Apiacta* 38:249–256.

114. Papapanagiotou, E. P., Iossifidou, E. G., Psomas, I. E. et al. 2000. Simultaneous HPLC determination of sulphadiazine and trimethoprim in cultured gilthead SEA bream (*Sparus aurata* L.) tissues. *J. Liq. Chromatogr. Relat. Technol.* 23:2839–2849.

115. Stoev, G. and Michailova, A. 2000. Quantitative determination of sulfonamide residues in foods of animal origin by high-performance liquid chromatography with fluorescene detection. *J. Chromatogr. A* 871:37–42.

116. Barbieri, G., Bergamini, C., Ori, E. et al. 1995. Determination of sulfonamides in meat and meat products. *Ind. Aliment Italy* 34:1273–1276.

117. Koesukwiwat, U., Jayanta, S., and Leepipatpiboon, N. 2007. Validation of liquid chromatography-mass spectrometry multi-residue method for the simultaneous determination of sulfonamides, tetracyclines, and pyrimethamine in milk. *J. Chromatogr. A* 1140:147–156.

118. Barker, S. A. 2000. Applications of matrix solid-phase dispersion in food analysis. *J. Chromatogr. A* 880:63–68.

119. Barker, S. A. 2007. Matrix solid phase dispersion (MSPD). *J. Biochem. Biophys. Methods* 70:151–162.

120. Parks, O. W. 1982. Screening test for sulfamethazine and sulfathiazole in swine liver. *J. Assoc. Off. Anal. Chem.* 65:632–634.

121. Abjean, J. P. 1993. Screening of drug residues in food of animal origin by planar chromatography: Application to sulfonamides. *J. Planar Chromatogr.* 6:147–148.

122. Wyhowski de Bukanski, B., Degroodt, J.-M., and Beernaert, H. 1988. A two-dimensional high-performance thin-layer chromatographic screening method for sulphonamides in animal tissues. *Z. Lebens. Unters Forsch.* 187:242–245.

123. Özkazanç, N. and Kaya, S. 1983. Analysis of sulfonamides in uncooked edible tissues of animals. *A. U. Vet. Fak. Derg.* 30:624–638.

124. Saschenbrecker, P. W. and Fish, N. A. 1980. Sulfamethazine residues in uncooked edible tissues of pork following recommended oral administration and withdrawal. *Can. J. Comp. Med.* 44:338–345.

125. Schlatterer, B. 1983. Identification and determination of sulfonamides in tissues and serum of slaughter animals by high-performance thin layer and high-pressure liquid chromatography. *Z. Lebens. Unters Forsch.* 176:20–26.

126. Posyniak, A., Niedzielska, J., Semeniuk, S. et al. 1995. Screening for sulfonamide residues in urine by planar chromatography. *J. Planar Chromatogr.* 8:238–240.

127. Phillips, W. F. and Trafton, J. E. 1975. A screening method for sulfonamides extracted from animal tissues. *J. Assoc. Off. Anal. Chem.* 58:44–47.

44 TLC of Quinolones

Dorota Kowalczuk

CONTENTS

44.1 INTRODUCTION

Quinolones have been a dynamically developing class of antibacterials, inducing clinical and scientific interest since their discovery in the early 1960s, when the first representative of this group, nalidixic acid (patented in 1962), was introduced into medical practice. Quinolones are totally synthetic chemical compounds that exert antibacterial activity by inhibiting bacterial DNA gyrase-dependent processes (Kuhlmann et al. 1998) (Figure 44.1 shows the quinolones considered in this review).

The first-generation quinolones, systematically synthesized up to the end of the 1970s, were followed at the beginning of the 1980s by the highly active broad-spectrum antibacterial fluoroquinolones (Kuhlmann et al. 1998). Quinolones have been classified into four generations according to potency, clinical use, and pharmacokinetic and pharmacodynamic properties. The first generation of quinolones (e.g., nalidixic acid, oxolinic acid, piromidic acid, pipemidic acid, rosoxacin) has poor systemic distribution, minimal serum levels, and limited activity, and it is used primarily for gram-negative urinary tract infections. Fluoroquinolones (second, third, and fourth generations) have improved pharmacokinetic properties, increased activity against gram-negative bacteria, and a broader antimicrobial spectrum. Quinolones of the second generation (e.g., ciprofloxacin, ofloxacin, norfloxacin, lomefloxacin) are active against gram-negative bacteria (including *Pseudomonas* sp., not *Streptococcus pneumoniae*), some gram-positive bacteria (*Staphylococcus aureus*), and some atypical microbes. Third-generation drugs (e.g., levofloxacin, sparfloxacin, moxifloxacin) are effective against the same bacteria as the second generation with extended activity against gram-positive and atypical pathogens. Quinolones of the fourth generation (e.g., trovafloxacin) are effective against the same microbes as the third generation with enhanced significant activity against anaerobes (King et al. 2000).

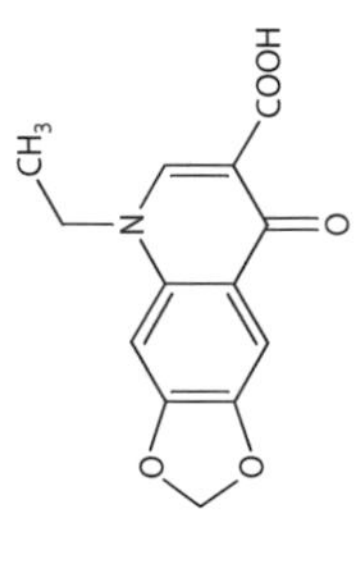

Flumequine (**FLU**)

Enoxacin (**ENO**)

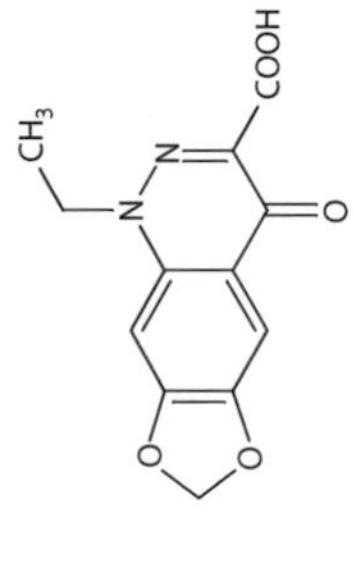

Pefloxacin (**PEF**)

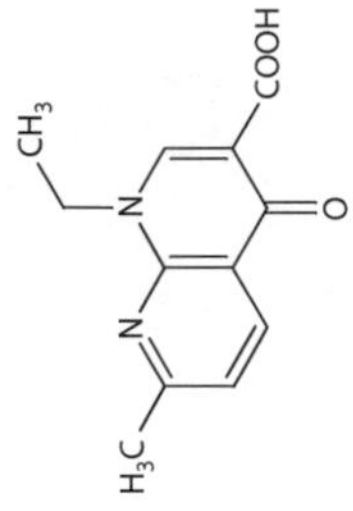

Danofloxacin (**DAN**)

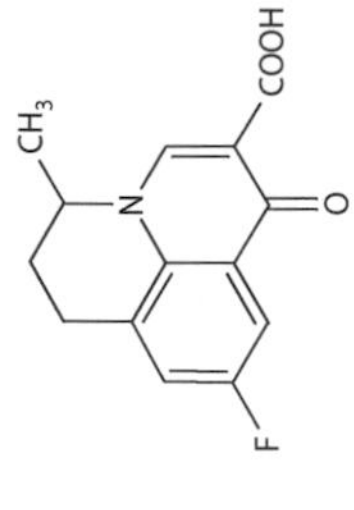

Nalidixic acid (**NAL**)

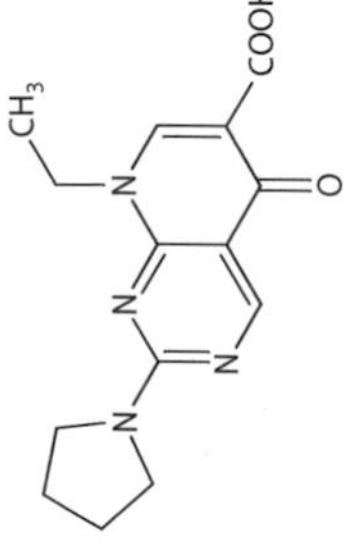

Rosoxacin (**ROS**)

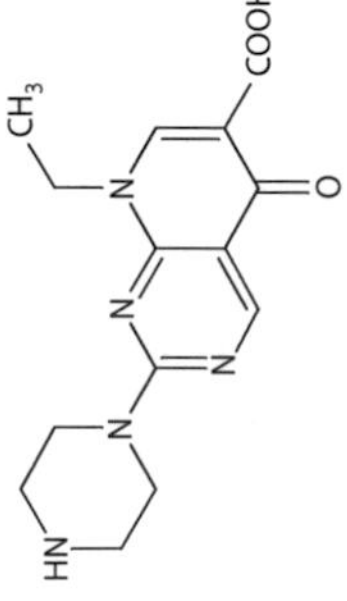

Norfloxacin (**NOR**)

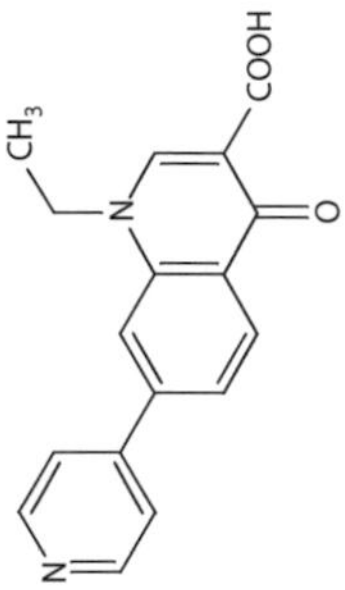

Fleroxacin (**FLE**)

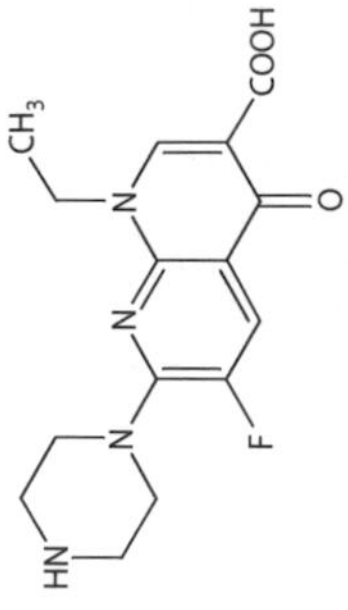

Cinoxacin (**CIN**)

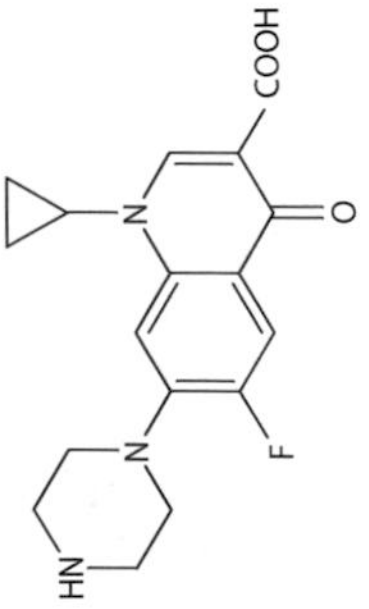

Pipemidic acid (**PIP**)

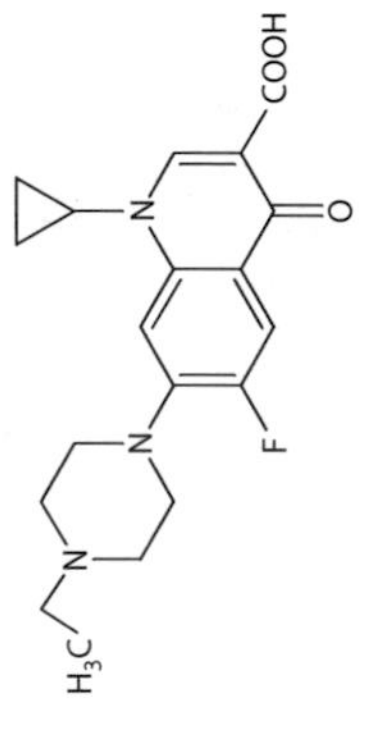

Enrofloxacin (**ENR**)

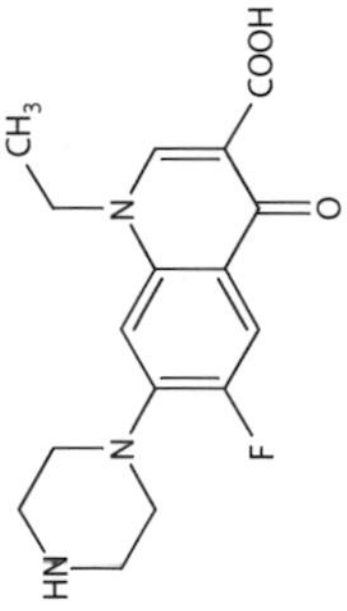

Sarafloxacin (**SAR**)

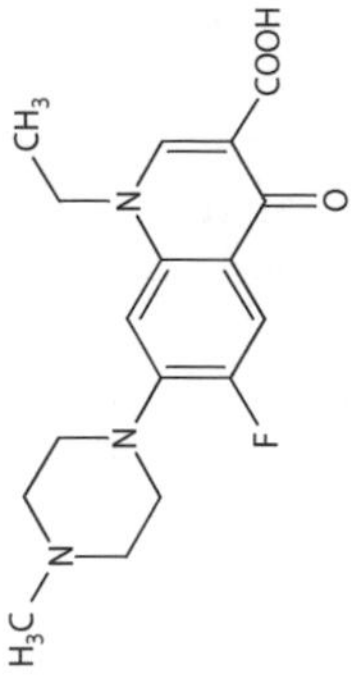

Oxolinic acid (**OXO**)

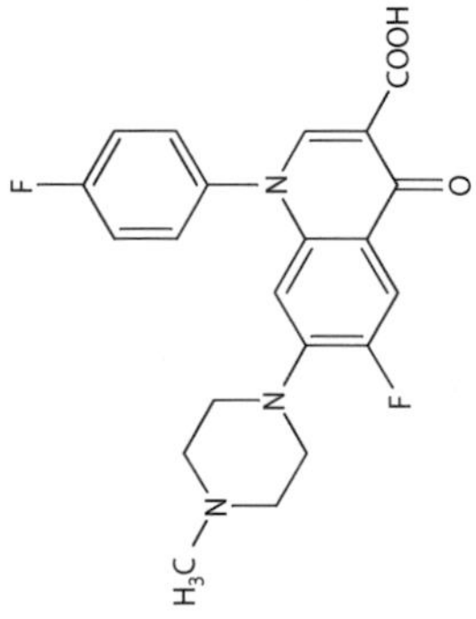

Piromidic acid (**PIR**)

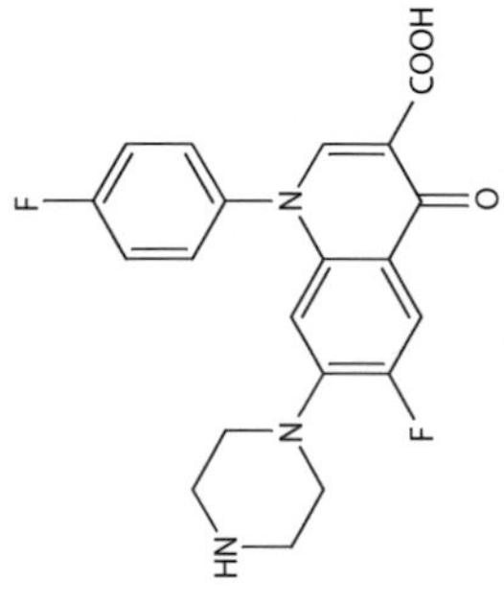

Ciprofloxacin (**CIP**)

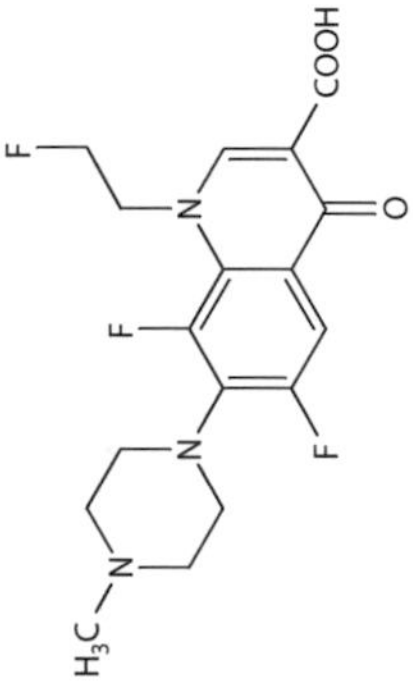

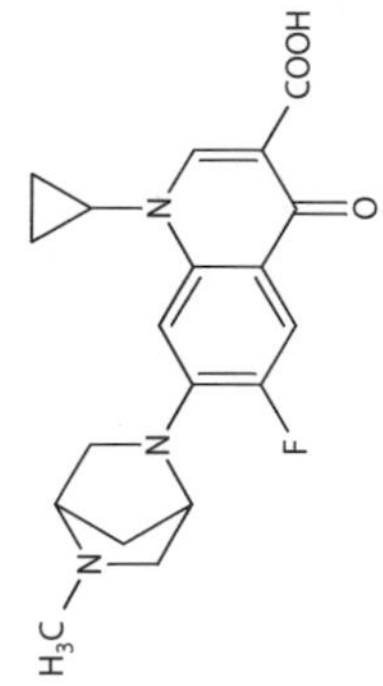

Difloxacin (DIF)

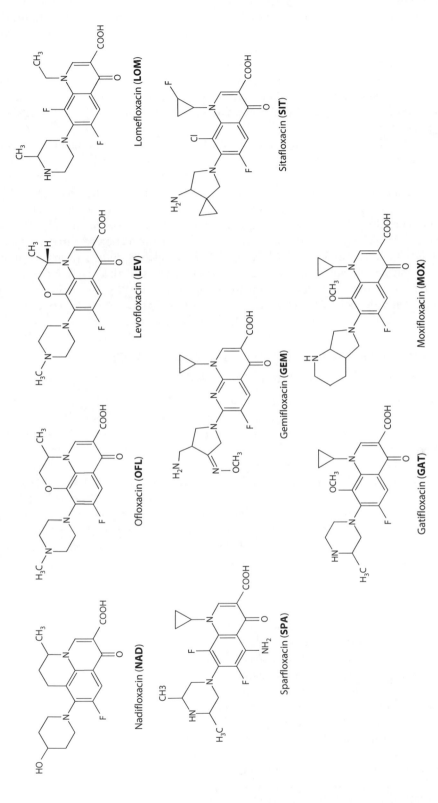

FIGURE 44.1 Structural formulae of the quinolones considered in this review.

The pharmacokinetic and in vitro potency profiles of fluoroquinolones determine the areas of clinical use, from the therapy of uncomplicated urinary tract infections, through complicated urinary tract infections, to more systemic use (sexually transmitted diseases, prostatitis, and skin and soft tissue infections) and to the treatment of respiratory tract infections (Andersson and MacGowan 2003; King et al. 2000).

The purpose of this chapter is to selectively review the latest achievements in thin-layer chromatography (TLC) and high-performance thin-layer chromatography (HPTLC) of 4-quinolone antibacterials and some related products published and/or abstracted since 1999. It also updates the previous reviews on this subject.

Belal et al. (1999) have reviewed most of the methods, including TLC, and reported the analysis of 4-quinolone drugs in pure forms, dosage forms, biological fluids, animal fish feed, etc. Another comprehensive review, dating from 2002, provided some information on the application of TLC for the analysis of quinolone residues in edible animal products (Hernández-Arteseros et al. 2002).

The present review provides more details on separation, detection, identification, and quantification of the quinolone drugs by TLC or HPTLC. These chromatographic methods were used for the determination of quinolone in pure form, monocomponent and combination preparations, biological material, fluids, and edible products. In addition, TLC was widely used in a variety of quinolone studies, such as analysis of quinolone photodegradation products, detection of quinolone antimicrobial activity (bioautography), control of radiochemical purity and stability of the labeled quinolones as radiotracers, determination of lipophilic character of quinolones, confirmation of homogeneity of new pharmaceuticals, and interaction of quinolones with cations.

44.2 TLC METHOD FOR SEPARATION, DETECTION, AND QUALITATIVE ANALYSIS OF QUINOLONES

Forgács and Csethati (1997) examined the interaction of 29 antibiotics, including NAL and NAL ethyl ester, with the anionic surfactant (sodium dodecyl sulfate [SDS]) by charge-transfer reversed-phase (RP)-TLC using silica gel $60F_{254}$ impregnated by the overnight predevelopment in n-hexane–paraffin oil (95:5, v/v) and water–methanol (0–85 vol.%) mixtures as mobile phases. Chromatograms were developed in the unsaturated sandwich chambers to the distance of about 150 mm. The drug spots were detected under ultraviolet (UV) light or by iodine vapor. The hydrophobicity of antibiotics and the relative strength of SDS–antibiotic interaction were calculated for each antibiotic–SDS pair. Relationship between the R_M values of antibiotics and the concentration of methanol (c_1) and SDS (c_2) in the mobile phase was calculated: $R_M = R_{M0} + b_1c_1 + b_2c_2$ (R_{M0}—hydrophobicity of antibiotic, b_1—specific hydrophobic surface of antibiotic, b_2—the relative strength of SDS–antibiotic interaction). The relative strength of interaction depended considerably on the molecular structure of the antibiotics. The interaction between SDS and antibiotics suggested that this interaction may have a marked influence on the biological efficiency of any pharmaceutical formulations simultaneously containing SDS and antibiotics.

Separation and identification of CIN, NAL, and OXO pure forms and pharmaceutical preparations were performed using normal-phase and RP-TLC (Hopkała and Kowalczuk 1999). In the normal-phase TLC, the best separation of all drugs was achieved on silica gel $60F_{254}$ plates with methanol–butanol–25% ammonia (5:5:2, v/v/v), ethyl acetate–dichloromethane–methanol–25% ammonia (3:3:5:2, v/v/v/v), and dichloromethane–triethylamine (10:1, v/v) as mobile phases, using the ascending technique in the saturated chamber or the horizontal technique in the unsaturated chamber. The separation in both the ascending and horizontal techniques was satisfactory, but the spots were more symmetrical in the ascending technique. The separation of drugs on RP-C18F_{254} plates in unsaturated horizontal chamber using tetrahydrofuran or acetonitrile–aqueous mobile phases (5:5 v/v) was also evaluated. The substances were identified by UV irradiation at $\lambda = 254$ nm and by the acidified 15% $FeCl_3$ solution or iodine reagent as the dyeing agents.

Wang et al. (1999) proposed the sensitive and highly selective fluorometric method for the detection of NAL and flumequine (FLU) on TLC plates by using sodium borohydride and hydrogen peroxide as fluorogenic spraying reagents. The samples (1 μL) of NAL, CIN, ENO, FLU, NOR, OFL, PIP, and PIR were spotted onto TLC silica gel 60 aluminum sheets, and the chromatograms were developed in a glass tank (an ascending mode) using chloroform–methanol–formic acid (10:2:0.03, v/v/v) as the mobile phase. The air-dried plates were sprayed successively with 0.75 M sodium borohydride and 15% hydrogen peroxide. The spots of the quinolones were visualized immediately under UV light at 366 nm. The very weak native fluorescence of NAL and FLU compared with the other quinolones after spraying with sodium borohydride and hydrogen peroxide strongly potentiated (the fluorescence intensity gradual decreased after heating). The detection limits (LODs; pmol) and fluorescence colors were found to be 30 for NAL (yellow-blue), 60 for FLU (orange), 10 for OFL (yellow), 80 for ENO (blue), 1250 for CIN (blue), 160 for NOR (blue), PIP (blue), and PIR (blue).

The other authors demonstrated that nonaqueous eluents of an ion-pairing reagent, di(2-ethylhexyl)orthophosphoric acid (HDEHP), in polar organic solvents are the effective mobile phases for the separation of quinolones, used in veterinary therapy, on diol silica gel F_{254} HPTLC plates (Soczewiński and Wójciak-Kosior 2001). Chromatograms with CIP, ENR, DIF, SAR, NOR, and FLU (0.2% solutions in methanol) were developed with mixtures of acetone–HDEHP (1%, 2.5%, 5%), ethyl acetate–HDEHP (5%, 10%, 15%), and ethyl acetate–HDEHP–methanol (9:1:1.25) as mobile phases in horizontal chambers over a path of 90 mm. Retention and selectivity in the adsorption-ion-association system was controlled by adjusting the concentration of HDEHP and changing the polar diluent. The observations of the slopes or R_M vs. log % HDEHP suggested the formation of 1:1 associates.

The same six veterinary fluoroquinolones—CIP, ENR, DIF, SAR, NOR, and FLU—were analyzed to search the optimal conditions for their separation (Choma 2003). Retention parameters (R_F, R_M, α) for the various chromatographic systems on the HPTLC Si60, HPTLC DIOL, HPTLC CN, and HPTLC NH_2 plates were compared and discussed. The best separations of the analyzed compounds were obtained using HPTLC silica gel plates and mobile phase consisted of dichloromethane, methanol, 2-propanol, and ammonia, however, without resolution of ENR and FLU. It was found that these drugs can be separated using 2D development. In the first direction, the chromatogram was developed with the dichloromethane–methanol–2-propanol–25% aqueous ammonia (3:3:5:2). In the second direction, perpendicular to the first, the chromatogram was developed with toluene–ethyl acetate–80% formic acid (60:30:10).

Similarly, the other two papers by Kamińska and Choma present the chromatographic behavior of the same veterinary fluoroquinolones. In the first paper (Kamińska and Choma 2009), authors discuss the various retention mechanisms of five amphoteric 7-piperazinylfluoroquinolones—SAR, DIF, NOR, ENR, and CIP (bases at low pH)—and FLU (neutral at low pH) on RP-C18W plates and the chaotropic effect on SiO_2-CN plates resulted from the use of acetonitrile–aqueous acidic (0.1 M citric acid) mobile phases with or without potassium perchlorate as chaotropic salt. On RP-C18W plates, all phases containing perchlorate anion gave lower retentions than those without perchlorate. In the mobile phases without an addition of perchlorate, retention of basic fluoroquinolones did not depend on the acetonitrile content. For all phases used (independently of the presence of perchlorate), retention of FLU decreased with the acetonitrile content. On cyanopropyl silica, the addition of perchlorate anion caused higher retention of basic fluoroquinolones, probably due to the chaotropic effect. This effect did not concern FLU, neutral at low pH.

In the second paper (Kamińska and Choma 2010), authors examined the influence of chaotropic perchlorate ion concentration on retention of six fluoroquinolones chromatographed on RP-C8 F_{254} plates using acetonitrile–aqueous acidic mobile phases containing constant concentration of citric acid (0.1 M) but various concentrations of potassium perchlorate (0.5–40 mM). Perchlorate, as chaotropic ion, making chaos in solvation shell of analytes, caused the increase in retention of basic fluoroquinolones. The retention increased with the increasing concentration of perchlorate ion in the mobile phase to achieve plateau for the mobile phases containing about 10–20 mM of chaotropic perchlorate.

Fluoroquinolones such as CIP, OFL, LEV, PEF, NOR, LOM, SPA, and MOX were also characterized with respect to chromatographic behavior on the HPTLC plates, coated with silica gel and covered with a phosphor film (Dorofeev et al. 2004). The chromatograms were developed in a glass TLC chamber (presaturated with the mobile phase vapor for 20–30 min), in the ascending mode, until the solvent front reached a level of about 85 mm. Finally, the spots were revealed by exposure to UV radiation at 254 nm. The mobile phase composition and the influence of polarity on the mobility of fluoroquinolones and the selectivity of the TLC system were studied. The optimum mobility and separation of the fluoroquinolone spots were observed in mobile phases of intermediate polarity containing diethyl ether, ethyl acetate, isoamyl alcohol, butanol, isopropyl alcohol, methanol, acetonitrile, and 25% ammonia. In these solvents and their mixtures, the R_F values were within 0.2–0.8. It was established that methanol is necessary to increase the TLC system selectivity. As a result, methanol–25% aqueous ammonia–ethyl acetate–acetonitrile (1:1:2:1) as the optimum mobile phase composition was selected. It was observed that isomers—OFL and LEV—were not distinguished by TLC on silica gel. Authors suggested that these drugs require stereoselective analytical technique. UV light at 254 and 365 nm as the detection method was employed. At a wavelength of 254 nm, most fluoroquinolones were detected as dark spots on the bright background, while OFL and MOX exhibited a bright violet emission and SPA produced a yellowish-green fluorescence. At a wavelength of 365 nm, most fluoroquinolones were detected as dark blue spots, while OFL and MOX exhibited bright blue emission, and the SPA spot was dark gray.

Separation of the fluoroquinolone, CIP, ENO, FLE, NOR, OFL, PEF, and SPA, was also tested by normal-phase TLC on silica gel F_{254} and cellulose F_{254} and by RP-TLC on chemically bonded silica gel (RP-C18) using numerous mobile phases and various methods of detection (Kowalczuk and Hopkala 2006). The separation of all drugs was achieved on silica gel with methanol–acetone–1 M citric acid–triethylamine (2.8:2:0.2:0.5, v/v/v/v) as mobile phase and using linear ascending 1D development over a path of 100 mm in a classical chamber (previously saturated for 30 min). Six of the seven compounds were separated on cellulose with dichloromethane–isopropanol–THF–25% ammonia (4:6:3:3, v/v/v/v) as mobile phase and using the horizontal development over a path of 100 mm in the unsaturated chamber. Under these conditions, the resolution of CIP from NOR was impossible. Five of the seven compounds were separated on silanized silica gel RP-C18 with mobile phase consisting of methanol–0.07 M phosphate buffer pH 6–10 mM benzyldimethyltetradecylammonium chloride (as ion-pairing reagent) (6:3:1, v/v/v) and using horizontal development over a path of 100 mm in the unsaturated chamber. The resolution of FLE, NOR, and OFL was not possible under these conditions. The separated compounds were detected under UV light or by the treatment of the plate surface with different dyeing agents. On silica gel plates, at a wavelength of 254 nm, SPA was detected as a dark spot on a green fluorescent background, while ENO and FLE as deep blue spots and CIP, NOR, OFL, and PEF as blue spots on a green fluorescent background. Using post-chromatographic derivatization, the separated drugs were detected as orange spots with Dragendorff's reagent, yellow-brown spots with Forrest's reagent, orange-brown spots with $FeCl_3$ in HCl, deep brown spots with iodic reagent, and blue spots with phosphomolybdic acid in H_2SO_4. On cellulose plates, at a wavelength of 254 nm, OFL appeared as a yellow-green spot on a blue fluorescent background and the other compounds appeared as deep blue spots.

Shu et al. (2007) performed interesting researches concerning the use of sigma- and pi-acceptors for the detection of CIP and OFL on TLC plates. The plates, coated with silica gel, were air-dried at room temperature, stored in an oven at 110°C for 30 min, and just before use allowed to cool to room temperature. The drug samples (reference CIP, Cipro®, ofloxacin) were spotted twice on a pair of coated plates, and the chromatograms were developed vertically in a tank saturated for 2 h, using a solvent system composed of chloroform, ethyl acetate, hexane, and water (1:3:0.5:1, v/v/v/v) until the solvent front reached 150 mm from the origin. The air-dried plates were sprayed with iodine or chloranilic acid in 1,4-dioxane as detecting reagents and examined for color formation and stability. The drugs (n-electron donors) reacted with iodine as sigma-acceptor and with chloranilic acid as pi-acceptor to form complexes. In another set of experiments, the plates were countersprayed with dimethylformamide (DMF) as stabilizing agent. Chloranilic acid gave a violet color with the drugs, which disappeared after 20 min. Iodine gave a reddish-brown color with the drugs, which changed to yellow after 6 min. On counterspraying with DMF, the reddish-brown color changed to yellow and there was no increase in intensity in the reference drug, but there was in the test drugs. However, the intensity of the violet color was generally increased. LODs were found to be 0.8, 1.2, 1.6 mg/mL for reference CIP, Cipro, and ofloxacin, respectively. Authors suggested that sigma- and pi-acceptors can be applied in qualitative detection of quinolones on thin-layer plates.

The selected quinolones (CIN, PIP) and fluoroquinolones (OFL, PEF) were also separated by TLC, and the results were estimated by calculating the most important chromatographic parameters (Bober 2008). Solutions of substances prepared in 0.2 M NaOH (CIN, PIP) and in 0.2 M HCl (OFL, PEF) were analyzed on TLC plates precoated with silica gel 60 F_{254} (activated at 120°C for 30 min) developed over a path of 75 mm in a classical chromatographic chamber (previously saturated for 30 min) using the mixtures of phosphate buffer pH 5.5–methanol (40:10, v/v) and acetonitrile–water–acetic acid (6:40:4, v/v/v) as mobile phases for quinolones and fluoroquinolones, respectively. The substances analyzed were visualized under UV light. The chromatographic parameters (R_F, ΔR_F, R_M, R_S) indicated good separation of CIN from PIP (ΔR_F 0.088) and OFL from PEF (ΔR_F 0.053). The obtained spots were not compact enough, and the fluoroquinolone spots were more broad (R_S 0.571) than quinolone spots (R_S 0.875).

The same quinolone antibiotics, in the same chromatographic conditions, but on three types of chromatographic plates—TLC silica gel 60F_{254}, TLC silica gel 60 (without fluorescein), and HPTLC silica gel 60 F_{254}—were analyzed in order to search the new methods of detection (Bober 2009). After the development, the plates were dipped in 0.05% solution of visualizing agents: Janus blue, methylene violet, gentian violet, methyl green, cresol red, malachite green, methylene blue, eosin yellowish, and metanil yellow. The visualizing effect was estimated: (1) directly after dipping in the solutions of visualizing agents, (2) after dipping and drying at a room temperature for 24 h, and (3) after dipping and drying at 120°C for 10 min. The LODs of the drugs were determined using the reagents that gave the best visualizing effects. The LODs were found to be 0.1 μg for CIN (on silica gel 60F_{254} plates using Janus blues directly after dipping in solution of visualizing agent and using methylene blue after drying at room temperature for 24 h), 0.1 μg for PIP (on silica gel 60F_{254} plates using Janus blues directly after dipping in solution of visualizing agent), 0.5 μg for OFL (on silica gel 60F_{254} plates using cresol red after drying at 120°C for 10 min), and 0.75 μg for PEF (on silica gel 60 plates using cresol red after drying at 120°C for 10 min).

44.3 TLC METHOD FOR DETERMINATION OF QUINOLONES IN PURE FORM AND FORMULATIONS

A simple, precise, and rapid HPTLC method was recommended for determination of CIP, ENR, LOM, NOR, OFL, and PEF in bulk drugs and the pharmaceutical preparations such as injectables/infusion and tablets (Argekar et al. 1996a). Silica gel 60F_{254} HPTLC aluminum plates as stationary phase and n-butanol–ethyl alcohol–6 M ammonia (4:1:2.2) as developing solvent system were employed. Linear response in the concentration range of 10–150 ng for all the six fluoroquinolones was obtained. Quantification was conducted by densitometric measurements at 280 nm for CIP, ENR, NOR, and PEF; 285 nm for LOM; and 295 nm for OFL.

Dhanesar (1999) proposed the application of the hydrocarbon-impregnated silica gel HPTLC plates for the direct quantification of many different classes of antibiotics including fluoroquinolone: CIP, NOR, and OFL. In order to facilitate direct quantification by densitometry, the standards and samples were dissolved in water, spotted on to the plate, and remained as single spots centered about the point of application. Detection levels were as low as 0.1 ng/spot. The relationship between peak area and quantity spotted was linear with correlation coefficients >0.95 for quantities <30 ng/spot and logarithmic for quantities >30 ng/spot. The direct sample determination resulted in rapid, high sample throughput with all samples being determined under the same conditions as the standards.

FLE, SPA, and CIN were determined in tablets by HPTLC videodensitometric method at $\lambda = 254$ nm, using silica gel 60F$_{254}$ plates (Kowalczuk and Hopkala 2001). Chromatograms were developed over a path of 50 mm in DS-horizontal chamber with a solvent system consisting of dichloromethane–isopropyl alcohol–25% ammonia (4:5:2, v/v/v). The separation of SPA, FLE, and CIN (R$_F$ 0.55, 0.46, and 0.40, respectively) was obtained when the solvent path was elongated to 80 mm. The calibration curves were achieved by plotting peak areas versus the drug concentrations in the range of 0.4–2.4 µg/band (r $\geq$ 0.997). The lowest detectable amounts of the drugs were tested using UV irradiation at 254 nm and various visualizing systems. The LODs were found to be 10 ng/spot for CIN (UV light) and 100 ng/spot for SPA (UV light, acidified phosphomolybdate reagent, and 15% FeCl$_3$ in 5% HCl) and FLE (UV light and 15% FeCl$_3$ in 5% HCl). The proposed method was precise (RSDs, from 2.60% to 5.27%) and accurate (recoveries, 96.10%–98.90% for SPA, 98.50%–103.20% for FLE, 99.30%–104.15% for CIN).

TLC–fluorescence scanning densitometry method was suggested also for the simultaneous determination of FLE and SPA (Feng 2001) using silica gel plates previously impregnated in sodium edetate solution of pH 7.0 (dried for at least 2 h at room temperature and then for 30 min at 110°C) with mixture of ethanol–ethyl acetate–1,2-dichloroethane–10% aqueous ammonia (40:30:20:10, v/v) as a mobile phase. FLE and SPA were separated from each other with the R$_F$ values of 0.4 and 0.59, respectively. The fluorescent spots were scanned in the fluorescent mode using 285 nm as the excitation wavelength and cutoff filter of 400 nm. Linear ranges were found to be 0.5–85 ng/spot (r = 0.9993) for FLE and 0.5–100 ng/spot (r = 0.9996) for SPA. The applicability of the method was confirmed by the recovery study. Recovery drugs from the spiked sample of serum and urine were 97.6%–102.0% with RSDs < 5.2%.

Novakovic et al. (2001) recommended the application of HPTLC method for the determination and the purity control of CIP-hydrochloride (CIP-HCl) in the coated tablets using silica gel 60F$_{254}$ plates. The HPTLC layer with the applied samples was exposed to ammonia vapor for 10 min and then developed over a path of 50 mm in a twin-trough developing chamber using a mixture of acetonitrile–ammonia 25%–methanol–methylene chloride (1:2:4:4, v/v/v/v). Chromatograms were scanned at 278 nm (CIP) or at 254 and 366 nm (the related substances). Under chromatographic conditions, the following compounds were detected: CIP (R$_F$ = 0.56) and ethylenediamine compound (R$_F$ = 0.41), desfluoro compound (R$_F$ = 0.45), by-compound A (R$_F$ = 0.48), and fluoroquinolonic acid (R$_F$ = 0.66) as the related substances and two unidentified impurities at R$_F$ = 0.18 and 0.26. LOD of the related substances was 0.04 µg/spot (S/N > 5), which corresponds to 0.1% of CIP in situ amount. The calibration function was polynomial from 6 to 116 ng/spot and linear from 23 to 70 ng/spot with a correlation coefficient of 0.9964 and RSD of 2.0%. The average found amount of CIP/tablet was 291.29 mg (from 100.10% to 104.16% of the declared amount) with RSD value of 1.02%. ANOVA and t-test revealed that the proposed HPTLC method is comparable to the official pharmacopoeia HPLC method and can be used for the determination of CIP in coated tablets. This method is suitable for the purity control of CIP-coated tablets when ethylenediamine compound, desfluoro compound, by-compound A, and fluoroquinolonic acid are considered as impurities with the limit of 0.2% each.

The alternative HPTLC-densitometric method was also developed for identification and quantification of CIP (R$_F$ = 0.61) and ethylenediamine compound (R$_F$ = 0.42), desfluoro compound (R$_F$ = 0.48), by-compound A (R$_F$ = 0.53), and fluoroquinolonic acid (R$_F$ = 0.68) as CIP degradation products in pharmaceutical preparations (Krzek et al. 2005). Silica gel 60 F$_{254}$ plates as the stationary phase and a mixture of chloroform–methanol–25% ammonia (43:43:14, v/v/v) as the mobile phase were used to separate individual constituents. UV-densitometric analysis was performed at 330 nm for fluoroquinolonic acid and at 277 nm for the other compounds. DMSO–methanol (1 + 1) was used to extract the drug constituents. The method showed high sensitivity (LOD 10–44 ng), a wide linear range (3–20 µg/mL), good precision (2.32%–6.46% RSD), and accuracy (recoveries, 98.62%–101.52%) for individual constituents.

TLC was also applied for the evaluation of the composition of various commercial preparations of enrofloxacin for veterinary use. Authors showed that only the one original brand product of enrofloxacin contained 5% of this drug, while the other four products contained 7.5% of ciprofloxacin (Sumano et al. 1994).

44.4 STABILITY-INDICATING HPTLC METHOD FOR QUINOLONE ANALYSIS

Motwani et al. (2006) developed stability-indicating HPTLC method for determination of GAT as a bulk drug and from polymeric nanoparticles using TLC aluminum plates precoated with silica gel $60F_{254}$ (prewashed by methanol and activated at 60°C for 5 min) as the stationary phase and a mixture of n-propanol–methanol–25% ammonia (5:1:0.9, v/v/v) as the mobile phase. This solvent system was found to give compact spots for GAT (R_F value of 0.60 ± 0.02). Densitometric analysis of GAT was carried out in the absorbance mode at 292 nm. The linear regression analysis data for the calibration plots showed good linear relationship with r = 0.9954 with respect to peak area in the concentration range of 400–1200 ng/spot. The method was validated for precision, accuracy, ruggedness, and recovery. The LOD and limit of quantitation (LOQ) were 2.73 and 8.27 ng/spot, respectively. GAT was subjected to acid and alkali hydrolysis, oxidation, photodegradation, and dry heat treatment. The drug exposed to photo- or UV degradation showed no additional peaks. The degraded products with R_F values of 0.01, 0.03, 0.51, and 0.75 for acid-induced degradation, 0.01 for base-induced degradation, and 0.54 and 0.52 for dry and wet heat degradation, respectively, were well separated from the pure drug (R_F = 0.60); consequently the method can be employed for the stability-indicating assay. Precision evaluation at three different concentration levels showed very low values of the RSD (<0.03%) for inter and intraday variation. Recovery of GAT from polymeric nanoparticles, after spiking with 50%, 100%, and 150% of additional drug, varied from 99.19% to 101.93%.

Sharma and Sharma (2011) suggested the alternative HPTLC-densitometric method for determination of GAT in bulk drug and tablets using precoated silica gel $60F_{254}$ aluminum sheets (pretreated with methanol and activated at 60°C for 5 min) as stationary phase and a mixture of toluene–acetic acid–triethylamine (4.0:2.5:0.5, v/v) as the mobile phase. Chromatograms were developed up to a distance of 50 mm in the chamber saturated for 30 min at 25°C. Densitometric measurements were made at 288 nm. Degradation tests such as acid and alkali hydrolysis, oxidation, photodegradation, and dry heat and wet heat treatment confirmed stability-indicating properties of the method. Chromatographic conditions were optimized to achieve good peak symmetry of GAT with R_F value of 0.46 and good separation of the active substance from degradation peaks. The calibration plot was linear over a concentration range of 200–400 ng/spot with a good correlation (r = 0.9989). LOD and LOQ values were found to be 40 and 88 ng/spot, respectively. The drug content and recovery from the spiked samples were found to be 99.90% and 99.98%–100.16%, respectively.

A stability-indicating HPTLC method, validated according to ICH, was also developed for the densitometric analysis of NAD in microemulsions using precoated silica gel aluminum plates $60F_{254}$ with a mixture of chloroform–methanol–formic acid (7.5:2.0:0.5, v/v) as the mobile phase (Kumar et al. 2010). This solvent system was found to give compact spot for NAD at R_F value of 0.39 ± 0.02, which was scanned at an absorption wavelength of 288 nm. The linear regression data for the calibration plots (r = 0.9981) were found with respect to peak area in the concentration range of 50–600 ng/spot. The LOD and LOQ were 9.4 and 20.5 ng, respectively. The drug was subjected to acid and alkaline hydrolysis, oxidation, photodegradation, and dry heat treatment. The peaks of degradation products were well resolved from the peak of the standard drug (R_F = 0.39) with significantly different R_F values: 0.64, 0.68, and 0.71 for acid hydrolysis; 0.19, 0.42, 0.47, and 0.50 for alkaline hydrolysis; 0.06, 0.43, 0.51, and 0.64 for hydrogen peroxide-induced oxidation; 0.03, 0.27, 0.49, and 0.63 for UV light degradation; 0.06, 0.54, and 0.58 for daylight degradation; and 0.43, 0.49, and 0.53 for dry heat degradation; consequently the method can be used for stability-indicating assay. Statistical analysis revealed that the method was precise (RSD, 0.06%–0.24%) and accurate (recovery, 99.1%–99.9%) for the determination of NAD in pharmaceutical dosage forms, especially microemulsions. The drug content was found to be 99.5% with RSD of 0.15%.

44.5 TLC METHOD FOR QUINOLONE DETERMINATION IN COMBINATION PREPARATIONS

4-Quinolone antibacterials are used in therapy with 5-nitroimidazole derivative agents. The application of the TLC method to the determination of this combination has been reported by many workers.

Argekar and Sawant (1999) proposed the application of HPTLC method for the simultaneous determination of CIP-HCl and tinidazole (TIN) in tablets. The drugs were separated on silica gel 60F$_{254}$ plates (prewashed by overnight dipping in the developing solvent and activated at 100°C for 30 min) with chloroform–methanol–toluene–triethylamine–water (2:2:1.6:1.5:0.4, v/v) as the mobile phase. The plates were developed to a distance of 60 mm in twin-trough chamber (R$_F$ values, 0.20 and 0.75 for CIP-HCl and TIN, respectively). Densitometric scanning was performed at 327 nm. The calibration curves were linear in the ranges of 100–350 (CIP-HCl) and 120–420 (TIN) ng/band (r = 0.999). LODs were found to be 1.9 and 5.05 ng/band for CIP-HCl and TIN, respectively. The recovery (standard addition method) varied between 99.22% and 103.11% (CIP-HCl) and between 96.28% and 101.17% (TIN). The RSD values obtained from tablet analysis were found to be 0.27%–1.03% for CIP-HCl and 1.09%–1.68% for TIN. The stability-indicating properties of the method were confirmed by degradation tests: daylight and acidic and alkaline and oxidizing conditions.

CIP-HCl was also determined in the presence of metronidazole (MET) in binary mixture using TLC method (Elkady and Mahrouse 2011). The separation was carried out on silica gel 60F$_{254}$ plates (prewashed by overnight dipping, in the developing solvent, and activated at 100°C for 30 min) using acetonitrile–ammonia–methanol–methylene chloride–hexane (1.3:1.1:2.0:3.0:1.0, v/v) as the developing system. The plates were developed by ascending development in the saturated chamber (the optimized chamber saturation time was 45 min) over a path of 80 mm. The bands were visualized under UV lamp at 254 nm. The bands developed were dense at R$_F$ of 0.63 ± 0.02 for CIP-HCl and 0.84 ± 0.02 for MET, respectively. Densitometric scanning was performed in the absorption mode at 280 nm. The drugs showed good linearity in concentration of 1.5–10 μg/band (r > 0.9991). LODs were found to be 0.37 and 0.32 μg/band for CIP-HCl and MET, respectively. The recovery experiments using standard addition method showed mean recoveries between 100.78% and 99.45% indicating good accuracy of the method. The values of precision (RSD) were 0.27%–0.89% for CIP-HCl and 1.31%–1.65% for MET. The proposed method was successfully applied to the determination of both drugs in commercial tablets. The mean content was found to be 99.40% with RSD = 0.74% for CIP-HCl and 99.38% with RSD = 0.85% for MET. The optimized TLC method proved to be specific, robust, and accurate for the quality control of the cited drugs.

A simple, rapid, and accurate HPTLC method was developed for the simultaneous determination of LEV and ornidazole (ORN) in tablet dosage form (Chepurwar et al. 2007). The method was based on the separation of the two drugs followed by densitometric measurements at 298 nm. The separation was carried out on TLC silica gel 60F$_{254}$ aluminum sheets using a mixture of n-butanol–methanol–ammonia (5:1:1.5, v/v/v) as the mobile phase. The drugs showed linearity in the concentration range of 50–250 and 100–500 ng/spot for LEV and ORN, respectively. The method was selective because no chromatographic interferences from the tablet excipients were found. The suitability of this HPTLC method for the quantitative determination of the compounds was proved by validation according to ICH guidelines.

The alternative HPTLC method for the simultaneous estimation of the LEV and ORN in the combined dosage forms was also suggested (Patel and Patel 2011). The chromatographic separation was performed on silica gel 60F$_{254}$ aluminum sheets (prewashed with methanol and activated in an oven at 50°C for 5 min) using the mobile phase comprising n-butanol–ethanol–8 M ammonia

(5:0.5:1.5, v/v/v). The ascending development to 80 mm was conducted in twin-trough glass chamber saturated for 30 min. The detection of spots was carried out densitometrically using a UV detector at 310 nm in absorbance mode. This system was found to give symmetrical peaks at R_F of 0.31 ± 0.003 for LEV and at R_F of 0.83 ± 0.008 for ORN. The calibration curve was found to be linear over the concentration range 40–140 and 80–280 ng/spot for LEV and ORN, respectively, with the correlation coefficient of 0.9992 (LEV) and 0.9989 (ORN). The LOD and LOQ were found to be 9.98 and 30.25 ng/spot, respectively, for LEV and 21.12 and 64.01 ng/spot, respectively, for ORN. The precision (RSD%) was 0.37–1.81 for LEV and 0.39–1.91 for ORN. The average recoveries obtained were $99.26\% \pm 0.88\%$ (LEV) and $100.4\% \pm 1.15\%$ (ORN). The proposed validated method was successfully applied for the determination of LEV and ORN in tablet dosage forms. The average assay was found to be $99.77\% \pm 0.99\%$ and $99.22\% \pm 1.14\%$ for LEV and ORN, respectively.

A simple and sensitive HPTLC method was also developed for the quantitative estimation of gatifloxacin (GAT) in the presence of ORN in its combined dosage forms (Suhagia et al. 2006). GAT and ORN were chromatographed on TLC silica gel 60 F_{254} aluminum sheets (prewashed with methanol and dried in air) using n-butanol–methanol–6 M ammonia (8:1:1.5, v/v) as the mobile phase in the twin-trough chamber previously saturated for 45 min. The plates were developed for up to 45 mm, dried in air, and scanned at 302 nm in the absorbance/reflectance mode. Chromatogram of the drug samples has shown resolution of GAT peak (R_F, 0.21 ± 0.02) and ORN peak (R_F, 0.76 ± 0.04). The linearity of GAT and ORN was in the range of 100–500 and 250–1250 ng/spot, respectively, with a correlation coefficient more than 0.9850. The LOD was found to be 40 ng/spot for GAT and 100 ng/spot for ORN. The intraday and interday coefficients of variation for both drugs were in the range of 0.68%–2.58% and 0.37%–3.62%, respectively. Recovery of GAT and ORN by the standard addition method at five levels of the calibration curve was found to be 98.03%–100.95% for GAT and 97.34%–101.88% for ORN. The proposed method was applied for the determination of both drugs in the combined dosage forms with amount found of 100.16–101.86 for GAT and 99.67–101.16 for ORN.

Similar to the previous papers, the HPTLC method was applied for simultaneous determination of NAL and MET in pharmaceutical preparations (Argekar et al. 1996b). NAL and MET were chromatographed on silica gel 60F_{254} HPTLC glass plates with ethyl acetate–chloroform–methanol–25% ammonia (2.5:2.5:1.5:0.5, v/v) as the mobile phase. In effect, a linear response was obtained in the concentration range of 20–120 ng for MET and 40–140 ng for NAL. The standard deviations for MET and NAL varied between 0.27 and 1.5; the recoveries varied between 98.48% and 101.14%.

44.6 TLC METHOD FOR ANALYSIS OF QUINOLONE PHOTODEGRADATION PRODUCTS

Hidalgo et al. (1993) determined the photodegradation kinetics of some quinolones using the HPTLC method. As a result of irradiation with UVA at 350 nm, photostability decreased in the following order: CIP > NOR > PIP > NAL > M-193324 > ROS > OXO. The photodegradation process in all cases followed first-order kinetics. The quantum yield of fluorescence for each quinolone was varied from 0.15E–2 to 5.15E–2. The results showed that quinolones that possess a piperazine ring at the C-7 position (CIP, NOR, PIP) are less photostable and more fluorescent.

HPTLC procedure combined with densitometry was also developed for qualitative and quantitative monitoring of the degradation of NAL in solutions irradiated with a high-pressure mercury lamp (Salomies and Koski 1996). The photodecomposed solutions chromatographed on silica gel 60 F_{254} plates using the mobile phase consisting methanol–water–ammonia (18:12:1) were scanned at 257 nm. The photodegradation rate increased with increasing temperature and with decreasing initial concentrations of the drug in the solution.

Torniainen et al. (1997) presented an interesting study concerning the TLC qualitative analysis of CIP decomposed photochemically in acidic solution to two major degradation products (artificial degradation under irradiation with high-pressure mercury lamp and daylight degradation). TLC analysis was performed on silica gel 60F$_{254}$ aluminum sheets using acetonitrile–10% ammonia with 0.3 M ammonium chloride (6.5:3.5, v/v) as the eluent system. The migration distance was 70 mm and the spots were detected under UV light at 254 and 366 nm. The main degradation product was 7-amino-1-cyclopropyl-6-fluoro-1,4-dihydro-4-oxo-3-quinolone carboxylic acid. The structure of the isolated compound was elucidated on the basis of information from chemical behavior in TLC and HPLC and from infrared, UV, mass, and nuclear magnetic resonance spectra.

44.7 TLC METHOD FOR DETERMINATION OF QUINOLONE RESIDUES AND TRACES

The traces or residues of quinolones were analyzed predominantly by TLC or HPTLC with fluorescence detection.

The application of TLC for the analysis of quinolone residues in edible animal products has been thoroughly reviewed by Hernández-Arteseros et al. (2002).

HPTLC–fluorescence (UV) method was proposed for qualitative and quantitative analysis of FLU and OXO. The method was proposed for quality control analysis of antibiotic added to fish feed and also for detection and quantification of residues in fish meat (Vega et al. 1995). Authors provided the information on the sample treatment (extraction, cleanup) and determination technique. Recovery efficiency from fish feed was over 97% with RSD of 1%–2%. For analysis of OXO residues in fish tissue, the plate was treated post chromatography with sulfuric acid–hydrochloric acid reagent and scanned in fluorescence mode (results in 10 ppb sensitivity).

TLC–fluorescence spectrodensitometry method was also reported for simultaneous determination of trace NOR, PEF, and CIP using silica gel plates previously treated with sodium edetate solution pH 7.0 and dried at room temperature and at 105°C for about 1 h, shortly before use (Pei-Lan Wang et al. 1997). Plates were developed with the mobile phase, chloroform–methanol–toluene–dichloromethane–aqueous ammonia (2.7:4.6:1.7:0.5:0.5, v/v), over a distance of 90 mm. The positions of fluorescence spots were observed in the UV light at 254 nm. Fluorescence scanning was performed using 278 nm as the excitation wavelength and the cutoff filter of 400 nm for detection. Under these conditions, NOR, PEF, and CIP were detected at R$_F$ values of 0.30, 0.42, and 0.54, respectively. The linear ranges were found to be 0–100 ng/spot (NOR, PEF) and 0–75 ng (CIP) with 0.998 < r < 0.9993. Determination of trace NOR, PEF, and CIP in serum and urine was successful. The RSD-expressed precision was <8.6%. The recovery of the three drugs from body fluids ranged from 96% to 108%.

Seven quinolones such as ENR, CIP, DAN, NOR, FLU, OXO, and NAL were determined in pig muscle by the HPTLC method based on solid-phase extraction (Juhel-Gaugain and Abjean 1998). The samples extracted and cleaned up by SP-extraction on RP-C8 SPE cartridges (conditioned with methanol, then KH$_2$PO$_4$ pH 7.4; washed with methanol–0.1 M ammonia solution, 75:25 v/v; eluted with methanol–1 M ammonia solution, 75:25, v/v) were chromatographed on silica gel 60 HPTLC plates with concentrating zone using methanol–ammonia (85:15, v/v) as the mobile phase. The chromatograms were developed to a distance of 40 mm above the concentrating zone in twin-trough chamber. The plates were first inspected under light at 312 nm, then sprayed with terbium chloride solution, heated for 5 min at 100°C, and again monitored under 312 nm UV. Under the used chromatographic system, DAN, NOR, CIP, OXO, NAL, FLU, and ENR were detected at the R$_F$ values of 0.34, 0.34, 0.43, 0.44, 0.58, 0.60, and 0.61, respectively (OXO, NAL, and FLU visible after terbium spraying). The method was validated to a level of 15 µg/kg for ENR, CIP, DAN, and NOR and 5 µg/kg for FLU, OXO, and NAL.

The other authors proposed micelle TLC–fluorometry for simultaneous determination of OFL, CIP, and SPA traces (Feng and Dong 2004). Authors found that the optimum molar ratio of SDS to EDTA in micelle solutions used as mobile phases was 0.01:0.1. OFL, CIP, and SPA are separated on the polyamide thin-layer sheet, with R$_F$ values of 0.72, 0.55, and 0.32, respectively. The fluorescence spots were scanned at the excitation wavelength of 282 nm and the cutoff filter of 400 nm. It was found that the linear ranges were in the concentration ranges of 1.0E–5 to 4E–4 mol/L for OFL, 1.0E–5 to 4.5E–4 mol/L for CIP, and 1.0E–5 to 4.2E–4 mol/L for SPA, and the LODs were 2.0E–6 mol/L for OFL,

1.5E–6 mol/L for CIP, and 1.6E–6 mol/L for SPA. For all drugs, the RSD values were in the range of 1.12%–5.82%, and the recoveries were found to be 96.7%–104.2% in urine and serum samples.

Lu and Feng (2005) proposed also the inclusion of TLC with fluorescence detection using β-cyclodextrin solutions with EDTA (molar ratio 0.005:0.1) as mobile phases for simultaneous determination of OFL, PEF, and SPA. The separation of the drugs was performed on the polyamide thin-layer sheet with R_F values of 0.74, 0.61, and 0.25, respectively, and chromatograms were scanned in the fluorescence mode using 278 nm for PEF, 280 nm for OFL, and 282 nm for SPA, as the excitation wavelengths and the cutoff filter of 400 nm. For the all drugs, the RSD values were in the range of 2.7%–6.9% and the recoveries were found to be 98.2%–104.6% in urine and serum samples.

For the determination of OFL residue in controlling pharmaceutical equipment cleanliness, an HPTLC method was developed using silica gel 60 plates and ethanol–conc. ammonia (4:1, v/v) as the mobile phase (Vovk and Simonovska 2011). The plates were developed to the middle of the plate (the migration distance of 40 mm) in a horizontal chamber (R_F of OFL 0.56), and the chromatograms were scanned densitometrically in the fluorescence mode at excitation wavelength of 313 nm. The simulated samples at a residue level of 1 mg/m^2 were prepared by spreading the OFL solution on 1, 5, and 10 dm^2 stainless steel surfaces, evaporation of the solvent, removal of the residue by two ethanol-wetted cotton swabs, and extraction with the mixture of ethanol and Na$_2$EDTA–water solution at pH 11 for 15 min with sonication. Authors found that the LOD of OFL, compared with the literature data, was highly satisfactory. The LOD and LOQ were 0.6 and 2 ng, respectively. Besides, it was shown that these limits can be improved by immersion of the developed plate into a solution of liquid paraffin–n-hexane (1:2, v/v) to approximately 0.25 and 0.9 ng, respectively. The LOD of the method using detection without paraffin–n-hexane was 3, 0.6, and 0.3 µg/m^2 by swabbing 1, 5, and 10 dm^2, respectively. The obtained RSD values (4.20% for 2 ng and 3.76% for 20 ng) responded to the acceptance criteria for system precision. The mean recovery (±RSD) at 1 mg/m^2 from 1, 5, and 10 dm^2 was 95.3% ± 3.78%, 88.6% ± 4.41%, and 89.7% ± 4.97%, respectively. The results confirmed that HPTLC with densitometry can also be used for quantitative analysis in the ng range.

44.8   TLC BIOAUTOGRAPHY OF QUINOLONE RESIDUES

TLC coupled with bioautography (contact, immersion, and direct bioautography (DB)) as a microbiological screening method is commonly used for the detection of antimicrobial activity. It is one of the screening techniques used in monitoring antibiotic residues in food and feed.

A series of publications involving the application of this method in the analysis of quinolone antibiotics was presented by Choma or Choma and coworkers.

The paper dated 2005 provides readers an overview of screening methods for the detection of antimicrobials, including dilution and diffusion methods and bioautography, mainly DB (TLC with microbiological detection) (Choma 2005).

TLC-DB using Chrom Biodip® Antibiotics Test Kit for semiquantitative determination of ENR and CIP standards was presented by Choma et al. (2004). TLC or HPTLC Si60 F$_{254}$ plates developed with dichloromethane–methanol–2-propanol–25% aqueous ammonia (3:3:5:2) as the mobile phase were immersed in inoculated nutrient medium and incubated to permit the growth of bacteria. After spraying with tetrazolium (MTT) salt, dehydrogenases of living bacteria convert MTT salt into purple formazan forming cream-white inhibition zones in the place of antibiotic spots. The inhibition zone areas depend on the volume spotted and the amount of antibiotic in the spot. Quantitative bioautographic analysis was performed by the regression analysis of the inhibition zones. The exponential relations (linear relation was possible only for a narrow range of concentrations) were proposed for the approximation of dependencies between areas of zone inhibitions and logarithm of the antibiotic amounts. The LOD for both antibiotics was 0.01 ppm, lower than maximum residue limits (MRLs) established for various food products.

TLC-DB, with matrix solid-phase dispersion (MSPD) as a pre-separation method, was used for the investigation of ENR and CIP residues in milk (Choma and Komaniecka 2005). Chromosorb W AW as a sorbent was used for MSPD of milk samples spiked with antibiotics. Two grams of Chromosorb W AW was mixed with 1 mL of the spiked milk sample and transferred to the syringe. The samples were defatted with hexane (aspirated by the water pump) and centrifuged, and the antibiotics were eluted with dichloromethane. The samples were chromatographed on silica gel 60 F_{254} TLC plate with dichloromethane–methanol–2-propanol–25% aqueous ammonia (3:3:5:2) as the mobile phase, dried, immersed briefly in the microorganism solution, incubated overnight at 28°C, and sprayed with the MTT salt solution, and the cream-white inhibition zones observed against a purple background were measured with a planimeter. The pre-separation method was combined with TLC bioautography (TLC-B) to obtain semiquantitative results. The best recoveries of the antibiotics from milk were obtained when milk was mixed not only with Chromosorb W AW but also with acetonitrile, which caused precipitation of milk proteins.

The similar procedure was applied for the determination of FLU residues in milk but with MSPD based on Chromaton N-AW as a pre-separation method for TLC-DB (Choma 2006a). The spiked milk samples were mixed with Chromaton N-AW and acetonitrile, added to precipitate of milk proteins (1 mL:2 g:1 mL), and transferred to a syringe-barrel cartridge. The samples were defatted with 10 mL of hexane (aspirated by the water pump) and centrifuged. The antibiotic was eluted with 10 mL of dichloromethane, the eluates were evaporated to dryness, and the residues were dissolved in the mobile phase. The samples were chromatographed on silica gel $60F_{254}$ TLC plate with dichloromethane–methanol–2-propanol–25% aqueous ammonia (3:3:5:2) as the mobile phase. The developed plate (to the top and continuously for 1 h) was dried, immersed briefly in the microorganism solution, incubated 20 h at 37°C, and sprayed with solution of salt. The cream-white inhibition zones observed against a purple background were measured. The established method was applied for semiquantitative determination of FLU residues in milk. The calibration curve as the function between the inhibition zone areas and logarithms of antibiotic amounts was plotted. The inhibition zone areas depend on the volume spotted and the amount of FLU in the spot.

The next paper of the same author (Choma 2006b) affects the screening of ENR and CIP residues in cows' milk using MSPD on a siliceous sorbent (Chromosorb, Chromaton) for the isolation and concentration of ENR and CIP residues. This pre-separation method, detailed previously (Choma and Komaniecka 2005, Choma 2006a), was combined with HPLC and with TLC-DB. Mean recoveries calculated for different levels of the antibiotics in milk were found to be 55.87%–77.05% for CIP and 71.39%–110.06% for ENR. The results revealed that MSPD-TLC-DB can be used for the screening of ENR and CIP at the maximum residue level stipulated for milk by the European Union.

TLC-B method (combination of TLC and microbiologic disk diffusion test) was also applied for the determination of various antibiotic residues including DAN and ENR in milk samples (Temamoğullari and Kaya 2010).

The two papers by Lewis et al. (2012a,b) demonstrate the application of HPTLC-BD as a screening technique for the detection of fluoroquinolones (CIP, ENR, NOR) and their bioactive environmental metabolites derived from photo- and microbial (aerobic) fermentation processes, as well as for the fractionation of fluoroquinolone environmental metabolites, using silica gel 60 F_{254} HPTLC aluminum sheets as the stationary phase and chloroform–methanol–ammonia (70:30:10, v/v/v) as the mobile phase for chromatographic analysis, using the low-intensity UV light or the iodine stain for visualization of the fractionated analytes, and using p-iodonitrotetrazolium violet for the detection of inhibition zones of bacterial growth.

44.9 TLC METHOD FOR QUINOLONE DETERMINATION IN BIOLOGICAL MATERIAL, FLUIDS, AND EDIBLE PRODUCTS

Hundt and Barlow (1981) presented the TLC method for determining plasma levels of NAL. The plasma samples were acidified with 1 M orthophosphoric acid and extracted with toluene. NAL on the plates was converted into a fluorescent compound by exposing to hydrogen chloride gas for 10 min, then to strong UV radiation from a mercury lamp for 10 min, and measured quantitatively using a spectrofluorometer.

TLC method coupled with densitometry was proposed for the determination of OFL in human plasma and pleural fluid (Warlich and Mutschler 1989). After extraction of OFL from biological samples with dichloromethane, chromatography was performed on TLC silica gel plate with a mobile phase consisting of ethanol and water in the tank saturated with vapors of concentrated ammonia. The authors found that the precision of the assay considerably increased when the fluorescence intensity of OFL was measured after spraying the plate with a citric acid solution and dipping the plate into paraffin or using a mixture of both components. The method showed a very low LOD (1 ng/mL) as well as good precision and linearity in the range 0.001–2.0 µg/mL for both plasma and pleural fluid.

For qualitative and quantitative determination of pyridonecarboxylic acids including NAL, OXO, and PIP in chicken plasma, the microbiological, spectrophotometric, TLC, and HPLC methods were applied (Kondo et al. 1994). TLC analysis was performed on a silica gel $60F_{254}$ plate using a solution of methanol–chloroform–acetic acid (3:1:1, v/v/v), which was found to be the most suitable solvent for separation. The LODs determined by this method were 0.5 µg/mL for NAL, 0.075 µg/mL for OXO, and 0.39 µg/mL for PIP.

TLC method was also applied for the detection of NAL and FLU in cultured fish and chicken using 0.1% metaphosphoric acid–methanol (3:7, v/v) as the extracting agent and RP-C18 SPE cleanup procedure (Wang et al. 1998). After chromatography, the sample on a silica gel plate was sprayed with sodium borohydride and hydrogen peroxide as fluorogenic reagents and visualized under longwave UV light. The LODs of the two drugs in 5 g samples of fish and chicken were 0.017–0.261 and 0.092–0.157 ppm, respectively.

HPTLC method was employed for the measurement of SPA in human plasma and for pharmacokinetic study (Mody et al. 1998). The plasma samples with SPA were mixed (using vortex mixer) with acetate buffer (pH 3.5) and extracted with 2 × 3 mL of dichloromethane on a vortex mixer and centrifuged. The combined dichloromethane extract was evaporated to dryness at 45°C, dissolved in dichloromethane, spotted on TLC silica gel $60F_{254}$ plate, and developed in a twin-trough chamber with a solvent system consisting of chloroform–methanol–formic acid–water (35:4:2:0.25, v/v/v/v), and the fluorescent spot of SPA (R_F value of SPA = 0.20) was scanned at 365 nm. The linear calibration curve was constructed over the range of 10–200 ng (r = 0.999). The recovery of SPA added to plasma at 0.1–0.8 µg/mL was 94.9%, the LOD in human plasma was 50 ng/mL, and the intra- and interday precision (RSD) was <1.3%. The method was also used for the determination of pharmacokinetic parameters of SPA after oral administration of two marketed preparations to healthy volunteers.

Choma et al. (1999) proposed the application of TLC screening method for determination of the chosen antibiotics from two classes of veterinary drugs, FLU (quinolones) and doxycycline (DOX; tetracycline), in milk using specially prepared silica gel plates with concentrating zone. The milk samples with antibiotics were injected on the TLC Si60 F_{254} plates into the middle of special regions of trapezoidal shape created by the incision into the concentrating zone of plate. The TLC plates were set into Soczewiński's sandwich chamber, the adsorbent layer (about 0.8 cm) was removed from the plates, the plates were predeveloped with hexane to remove a lipid fraction from the milk samples (cleanup procedure), and then they were developed to a distance of 15 cm with the solvent system to separate two antibiotics. The satisfactory separation was obtained with 0.1 M citric acid–MeOH (1:9). The spots were detected both by UV lamp at 254 nm (FLU) and 366 nm (DOX) and by densitometry (0.25 µg of the drugs was identified). Calibration curves were linear in the range of 0.1–0.5 µg.

GAT in human plasma was determined by HPTLC using paracetamol (PAR) as internal standard (Sowmiya et al. 2007). GAT in plasma was extracted with dichloromethane (after adding phosphate buffer pH 4.5) by vortex mixing and centrifuged at 35°C. The dichloromethane layer was evaporated to dryness with nitrogen gas, dissolved in ethanol, spotted on silica gel $60F_{254}$ plates, developed in the twin-trough chamber using chloroform–methanol–ethanol–ammonia (4:2:2:2% v/v/v/v) as the mobile phase, and scanned at 291 nm (R_F values: GAT = 0.70, PAR = 0.90). The calibration curve was plotted over the range 50–400 ng/spot (r = 0.990), the LOQ was 50 ng/spot, the intraday and interday RSD values were <7.1, and the recovery was in the range of 86.92%–87.72%.

Rote and Pingle (2009) determined GEM mesylate in human plasma by using HPTLC and HPLC. For HPTLC, the plasma samples with GEM mesylate and linezolid (LIN) as internal standard were extracted using chloroform–acetic acid (5.9:0.1, v/v) by vigorous vortex mixing and centrifuged. The extract was evaporated to dryness on hot plate, dissolved in methanol, spotted on TLC silica gel $60F_{254}$ plate (activated at 110°C for 20 min), developed in the twin-trough chamber (saturated for 20 min) with a mixture of ethyl acetate–methanol–ammonia (8.0:4.0:3.0 v/v/v) at a distance of 70 mm, and scanned at 254 nm (R_F values: SPA = 0.33, LIN = 0.69). The linearity was found over the range 50–600 ng/spot (r = 0.9952), the lower limit of quantitation (LLOQ) was 30 ng/mL, the recovery of GEM mesylate was found to be 80.01%–86.17%, and the intra-run precision and inter-run precision varied from 3.5% to 9.2% and from 1.75% to 4.5%, respectively. The stability of GEM mesylate in plasma was confirmed during three freeze–thaw cycles (−20°C), on bench during 12 h.

44.10 THIN-LAYER RADIOCHROMATOGRAPHY OF THE LABELED QUINOLONES

The labeled fluoroquinolones are potential radiopharmaceuticals for the detection of bacterial infections. Instant thin-layer chromatography (ITLC) method was often used for the determination of radiochemical purity and radiochemical stability of labeled fluoroquinolones.

The ITLC method was applied for the determination of radiochemical purity and stability in the serum of 99mTc-enrofloxacin and 99mTc-ciprofloxacin (Siaens et al. 2004) and 99mTc-difloxacin and 99mTc-pefloxacin (Motaleb 2010) using strips of silica gel–impregnated glass fiber sheets: one ITLC strip developed with acetone as the mobile phase for the pertechnetate content analysis (R_F of $99mTcO_4^-$ = 1; other species remained at the origin) and the second ITLC strip developed with ethanol–water–ammonium hydroxide (2:5:1, v/v/v) as the mobile phase for the technetium colloid analysis (R_F of $99mTcO_2$ = 0; other species migrate with the solvent front).

The radiochemical purity was also evaluated for the 99mTc-ciprofloxacin kit, suitable for scintigraphic imaging of infection (Kleisner et al. 2002), and 99mTc-labeled LEV (Naqvi et al. 2012), the subsequent radiotracer suitable for detecting sites of infection.

Shah et al. (2011) proposed the application of the radio-TLC method for the evaluation of the radiochemical stability of the 99mTc-labeled sitafloxacin dithiocarbamate in serum, as a potential radiotracer for *S. aureus* infection.

The other authors used ITLC method for quality control of 99mTc-ciprofloxacin preparation (Rodríguez-Puig et al. 2006). The percentages of the four species involved in the 99mTc-ciprofloxacin preparation were controlled using RP-C18 plates developed with saline solution–methanol–acetic acid (55:45:1, v/v/v) for technetium tartrate analysis (R_F of 99mTc-tartrate and $99mTcO_4^-$ = 0.9; R_F of 99mTc-ciprofloxacin = 0) and the TLC systems proposed previously (Siaens et al. 2004): ITLC-SG plates developed with acetone and ethanol–water–ammonium hydroxide (2:5:1) for analysis of pertechnetate and technetium colloid, respectively.

Quality control analysis with the help of ITLC was also performed for monitoring of 99mTc-moxifloxacin synthesis (Chattopadhyay et al. 2010).

44.11 TLC METHOD FOR DETERMINATION OF QUINOLONE LIPOPHILICITY

The R_M values as an expression of the lipophilic character of a series of antibacterial quinolones such as ENO, PIP, OFL, OXO, FLE, NAL, PIR, CIN, PEF, FLU, CIP, NOR, and MF 961 were measured at pH 9.0 and 1.2 using an RP-TLC system with silica gel GF_{254} plates impregnated with silicone DC 200 as a nonpolar stationary phase and aqueous buffer alone or mixed with various amounts of acetone, methanol, or acetonitrile as the mobile phase (Biagi et al. 1994).

The lipophilicity of some antibacterial active 4-imino-1,4-dihydrocinnoline-3-carboxylic acid and 4-oxo-1,4-dihydrocinnoline-3-carboxylic acid derivatives (isosteric analogues of quinolones) was also investigated by the experimental determination of log P and pK_a values and calculation using ACD/Labs system software (Lewgowd et al. 2007).

44.12 OTHER APPLICATIONS OF TLC IN QUINOLONE ANALYSIS

The TLC method was also used for the confirmation of homogeneity of erythromycin nalidixate, a new salt prepared from nalidixic acid and erythromycin base (Goswami et al. 1995), and for the confirmation of interaction between quinolones and divalent cations (Marshall and Piddock 1994).

REFERENCES

Andersson, M.I. and A.P. MacGowan. 2003. Development of the quinolones. *J Antimicrob Chemother* 51(Suppl. S1): 1–11.

Argekar, A.P., Kapadia, S.U., and S.V. Raj. 1996a. Determination of some fluoroquinolones in pharmaceutical preparations by HPTLC method. *Indian Drugs* 33(3): 107–111.

Argekar, A.P., Raj, S.V., and S.U. Kapadia. 1996b. Simultaneous determination of metronidazole and nalidixic acid in pharmaceutical dosage forms by HPTLC. *Indian Drugs* 33(4): 167–170.

Argekar, A.P. and J.G. Sawant. 1999. Simultaneous determination of ciprofloxacin hydrochloride and tinidazole in tablets by HPTLC. *J Planar Chromatogr Mod TLC* 12: 202–206.

Belal, F., Al-Majed, A.A., and A.M. Al-Obaid. 1999. Methods of analysis of 4-quinolone antibacterials. *Talanta* 50(4): 765–786.

Biagi, G.L., Barbaro, A.M., and M. Recanatini. 1994. Determination of lipophilicity by means of reversed-phase thin-layer chromatography. III. Study of the TLC equations for a series of ionizable quinolone derivatives. *J Chromatogr A* 678(1): 127–137.

Bober, K. 2008. Determination of selected quinolones and fluoroquinolones by use of TLC. *Anal Lett* 41(10): 1909–1913.

Bober, K. 2009. The visualizing agents for selected quinolones and fluoroquinolones. *J Liq Chromatogr Relat Technol* 32(20): 3049–3055.

Chattopadhyay, S., Saha Das, S., Chandra, S., De, K., Mishra, M., Ranjan Sarkar, B., Sinha, S., and Ganguly. 2010. Synthesis and evaluation of 99mTc-moxifloxacin, a potential infection specific imaging agent. *Appl Radiat Isot* 68(2): 314–316.

Chepurwar, S.B., Shirkhedkar, A.A., Bari, S.B., Fursule, R.A., and S.J. Surana. 2007. Validated HPTLC method for simultaneous estimation of levofloxacin hemihydrate and ornidazole in pharmaceutical dosage form. *J Chromatogr Sci* 45(8): 531–536.

Choma, I.M. 2003. TLC separation of fluoroquinolones: Searching for better selectivity. *J Liq Chromatogr Relat Technol* 26(16): 2673–2685.

Choma, I. 2005. The use of thin-layer chromatography with direct bioautography for antimicrobial analysis. *LC-GC Eur* 18(9): 482–488.

Choma, I.M. 2006a. Thin-layer chromatography-direct bioautography of flumequine residues in milk. *J Liq Chromatogr Relat Technol* 29(14): 2083–2093.

Choma, I.M. 2006b. Screening of enrofloxacin and ciprofloxacin residues in milk by HPLC and by TLC with direct bioautography. *J Planar Chromatogr Mod TLC* 19(108): 104–108.

Choma, I.M., Choma, A., Komaniecka, I., Pilorz, K., and K. Staszczuk. 2004. Semiquantitative estimation of enrofloxacin and ciprofloxacin by thin-layer chromatography—Direct bioautography. *J Liq Chromatogr Relat Technol* 27(13): 2071–2085.

Choma, I., Grenda, D., Malinowska, I., and Z.X. Suprynowicz. 1999. Determination of flumequine and doxycycline in milk by a simple thin-layer chromatographic method. *J Chromatogr B Biomed Sci Appl* 734(1): 7–14.

Choma, I.M. and I. Komaniecka. 2005. Matrix solid-phase dispersion combined with thin-layer chromatography—Direct bioautography for determination of enrofloxacin and ciprofloxacin residues in milk. *J Liq Chromatogr Relat Technol* 28(16): 2467–2478.

Dhanesar, S.C. 1999. Quantitation of antibiotics by densitometry on a hydrocarbon-impregnated silica gel HPTLC plate. Part V: Quantitation and evaluation of several classes of antibiotic. *J Planar Chromatogr Mod TLC*. 12(4): 280–287.

Dorofeev, V.L., Konovalov, A.A., Kochin, V. Yu., and A.P. Arzamastsev. 2004. TLC analysis of drugs of the fluoroquinolone group. *Pharm Chem J* 38(9): 510–512.

Elkady, E.F. and M.A. Mahrouse. 2011. Reversed-phase ion-pair HPLC and TLC-densitometric methods for the simultaneous determination of ciprofloxacin hydrochloride and metronidazole in tablets. *Chromatographia* 73(3–4): 297–305.

Feng, Y.-L. 2001. Determination of fleroxacin and sparfloxacin simultaneously by TLC-fluorescence scanning densitometry. *Anal Lett* 34(15): 2693–2700.

Feng, Y.-L. and C. Dong. 2004. Simultaneous determination of trace ofloxacin, ciprofloxacin, and sparfloxacin by micelle TLC-fluorimetry. *J Chromatogr Sci* 42(9): 474–477.

Forgács, E. and T. C. sethati.1997. Charge-transfer chromatographic study of the interaction of antibiotics with sodium dodecylsulfate. *J Pharm Biomed Anal* 15(9–10): 1295–1302.

Goswami, B.B., Manna, P.K., and S.K. Basu. 1995. Physico-chemical properties and biological activities of erythromycin nalidixate. *Arzneimittelforschung Drug Res* 45(7): 813–814.

Hernández-Arteseros, J.A., Barbosa, J., Compañó, R., and M.D. Prat. 2002. Analysis of quinolone residues in edible animal products. *J Chromatogr A* 945(1–2): 1–24.

Hidalgo, M.E., Pessoa, C., Fernández, E., and A.M. Cárdenas. 1993. Comparative determination of photodegradation kinetics of quinolones. *J Photochem Photobiol A Chem* 73(2): 135–138.

Hopkała, H. and D. Kowalczuk. 1999. Thin-layer chromatographic analysis of nalidixic acid, oxolinic acid and cinoxacin. *Acta Pol Pharm Drug Res* 56(1): 11–15.

Hundt, H.K.L. and E.C. Barlow. 1981. Thin layer chromatographic method for the quantitative analysis of nalidixic acid in human plasma. *J Chromatogr* 223(1): 165–172.

Juhel-Gaugain, M. and J.P. Abjean. 1998. Screening of quinolone residues in pig muscle by planar chromatography. *Chromatographia* 47(1–2): 101–104.

Kamińska, M. and I. Choma. 2009. Influence of perchlorate ion on the retention of fluoroquinolones in RP-TLC. *J Liq Chromatogr Relat Technol* 32(9): 1331–1341.

Kamińska, M. and I. Choma 2010. The influence of perchlorate ion concentration on the retention of fluoroquinolones in RP-TLC. *J Liq Chromatogr Relat Technol* 33(7–8): 894–902.

King, D.E., Malone, R., and S.H. Lilley. 2000. New classification and update on the quinolone antibiotics. *Am Fam Physician* 61(9): 2741–2748.

Kleisner, I., Komarek, P., Komarkova, I., and M. Konopkova. 2002. A new technique of 99mTc-ciprofloxacin kit preparation. *Nukl Med* 41(5): 224–229.

Kondo, F., Nagata, S., Tsai, C.E., and K. Saipanu. 1994. Determination of pyridonecarboxylic acids in plasma by reverse-phase high-performance liquid chromatography. *Microbios* 77(312): 181–189.

Kowalczuk, D. and H. Hopkala. 2001. Videodensitometric HPTLC determination of fleroxacin, sparfloxacin, and cinoxacin in tablets. *J Planar Chromatogr Mod TLC* 14(2): 126–129.

Kowalczuk, D. and H. Hopkala 2006. Separation of fluoroquinolone antibiotics by TLC on silica gel, cellulose and silanized layers. *J Planar Chromatogr Mod TLC* 19(109): 216–222.

Krzek, J., Hubicka, U., and J. Szczepańczyk. 2005. High-performance thin-layer chromatography with densitometry for the determination of ciprofloxacin and impurities in drugs. *J AOAC Int* 88(5): 1530–1536.

Kuhlmann, J., Dalhoff, A., and H.-J. Zeiler (Eds.). 1998. *Quinolone Antibacterials*. Berlin, Germany: Springer-Verlag.

Kumar, A., Sinha, S., Agarwal, S.P., Ali, J., Ahuja, A., and S. Baboota. 2010. Validated stability-indicating thin layer chromatographic determination of nadifloxacin in microemulsion and bulk drug formulations. *J Food Drug Anal* 18(5): 358–365.

Lewgowd, W., Stańczak, A., Ochocki, Z., and K. Rzeszowska-Modzelewska. 2007. Determination of lipophilicity and pKa measurement of some 4-imino-1,4-dihydrocinnoline-3-carboxylic acid and 4-oxo-1,4-dihydrocinnoline-3- carboxylic acid derivatives—Isosteric analogues of quinolones. *Acta Pol Pharm Drug Res* 64(3): 195–200.

Lewis, G., Juhasz, A., and E. Smith. 2012a. Detection of antibacterial-like activity on a silica surface: Fluoroquinolones and their environmental metabolites. *Environ Sci Pollut Res* 1–7. DOI: 10.1007/s11356-012-0781-8

Lewis, G., Juhasz, A., and E. Smith. 2012b. Mixture of environmental metabolites of fluoroquinolones: Synthesis, fractionation and toxicological assessment of some biologically active metabolites of ciprofloxacin. *Environ Sci Pollut Res* 1–11. DOI: 10.1007/s11356-012-0766-7

Lu, Z. and Y. Feng. 2005. Simultaneous determination of trace ofloxacin, pefloxacin and sparfloxacin by inclusion thin layer chromatography-fluorescence detection. *Fenxi Huaxue* 33(7): 999–1002.

Marshall, A.J.H. and L.J.V. Piddock. 1994. Interaction of divalent cations, quinolones and bacteria. *J Antimicrob Chemother* 34(4): 465–483.

Mody, V.D., Pandya, K.K., Satia, M.C., Modi, I.A., Modi, R.I., and T.P. Gandhi. 1998. High performance thin-layer chromatographic method for the determination of sparfloxacin in human plasma and its use in pharmacokinetic studies. *J Pharm Biomed Anal* 16: 1289–1294.

Motaleb, M.A. 2010. Radiochemical and biological characteristics of 99mTc-difloxacin and 99mTc-pefloxacin for detecting sites of infection. *J Label Comp Radiopharm* 53(3): 104–109.

Motwani, S.K., Khar, R.K., Ahmad, F.J., Chopra, S., Kohli, K., Talegaonkar, S. et al. 2006. Stability indicating high-performance thin-layer chromatographic determination of gatifloxacin as bulk drug and from polymeric nanoparticles. *Anal Chim Acta* 576(2): 253–260.

Naqvi, S.A.R., Ishfaq, M.M., Khan, Z.A., Nagra, S.A., Bukhari, I.H., Hussain, A.I. et al. 2012. 99mTc-labeled levofloxacin as an infection imaging agent: A novel method for labeling levofloxacin using cysteine · HCl as co-ligand and in vivo study. *Turkish J Chem* 36(2): 267–277.

Novakovic, J., Nesmerak, K., Nova, H., and K. Filka. 2001. An HPTLC method for the determination and the purity control of ciprofloxacin HCl in coated tablets. *J Pharm Biomed Anal* 25(5–6): 957–964

Patel, S.A. and N.J. Patel. 2011. Development and validation of HPTLC method for simultaneous estimation of levofloxacin and ornidazole in tablet dosage form. *J Pharm Educ Res* 2(2): 78–83.

Rodríguez-Puig, D., Piera, C., Fuster, D., Soriano, A., Sierra, J.M., Rubí, S., and J. Suades. 2006. A new method of [99mTc]-ciprofloxacin preparation and quality control. *J Label Compd Radiopharm* 49(13): 1171–1176.

Rote, A.R. and S.P. Pingle. 2009. Reverse phase-HPLC and HPTLC methods for determination of gemifloxacin mesylate in human plasma. *J Chromatogr B Anal Technol Biomed Life Sci* 877(29): 3719–3723.

Salomies, H. and Koski, S. 1996. Qualitative and quantitative determination of nalidixic acid in photodecomposed solutions by thin layer chromatography. *J Planar Chromatogr Mod TLC.* 9(2): 103–106.

Shah, S.Q., Khan, A.U., and M.R. Khan. 2011. Radiosynthesis and biological evaluation of 99mTcN-sitafloxacin dithiocarbamate as a potential radiotracer for *Staphylococcus aureus* infection. *J Radioanal Nucl Chem* 287(3): 827–832.

Sharma, M.C. and S. Sharma. 2011. Development and validation of TLC densitometric method for gatifloxacin in pharmaceutical formulations. *Int J PharmTech Res* 3(2): 1179–1185.

Shu, E.N., Muko, K.N., Ogbodo, S.O., Maduka, I.C., and M.N. Ezeunala. 2007. Detection of quinolones on thin layer chromatographic plates using sigma- and pi-acceptors in 1,4-dioxane. *Biomed Res* 18(2): 137–140.

Siaens, R.H., Rennen, H.J., Boerman, O.C., Dierckx, R., and G. Slegers. 2004. Synthesis and comparison of 99mTc-enrofloxacin and 99mTc-ciprofloxacin. *J Nucl Med* 45(12): 2088–2094.

Soczewiñski, E. and M. Wójciak-Kosior. 2001. Thin-layer chromatography of quinolones in ion-association systems with Di(2-ethylhexyl)orthophosphoric acid (HDEHP). *J Planar Chromatogr Mod TLC* 14(1): 28–33.

Sowmiya, G., Gandhimathi, M., Ravi, T., and K. Sireesaa. 2007. HPTLC method for the determination of gatifloxacin in human plasma. *Indian J Pharm Sci* 69(2): 301–302.

Suhagia, B.N., Shah, S.A., Rathod, I.S., Patel, H.M., Shah, D.R., and Marolia, B.P. 2006. Determination of gatifloxacin and ornidazole in tablet dosage forms by high-performance thin-layer chromatography. *Anal Sci* 22(5): 743–745.

Sumano, L.H., Gomez, R.B., Gracia, M.I., and L. Ruiz-Ramirez. 1994. The use of ciprofloxacin in veterinary proprietary products of enrofloxacin. *Vet Hum Toxicol* 36(5): 476–477.

Temamoğullari, F. and S. Kaya. 2010. Determination of various antibiotic residues in milk samples sold in ankara markets by thin layer chromotography and bioautographic method. *Kafkas Univ Vet Fak Derg* 16(2): 187–191.

Torniainen, K., Mattinen, J., Askolin, C.-P., and S. Tammilehto. 1997. Structure elucidation of a photodegradation product of ciprofloxacin. *J Pharm Biomed Anal* 15(7): 887–894.

Vega, M., Rios, G., Saelzer, R., and E. Herlitz. 1995. Analysis of quinolonic antibiotics by HPTLC. Oxolinic acid residue analysis in fish tissue. *J Planar Chromatogr Mod TLC* 8(5): 378–381.

Vovk, I. and B. Simonovska. 2011. Development and validation of a high-performance thin-layer chromatographic method for determination of ofloxacin residues on pharmaceutical equipment surfaces. *J AOAC Int* 94(3): 735–742.

Wang, M.-L., Chen, S.-C., and S.-C. Kuo. 1998. Thin-layer chromatographic detection of nalidixic acid and flumequine in cultured fish and chicken. *Chin Pharm J* 50(6): 313–317.

Wang, M.-L., Chen, S.-C., and S.-C. Kuo. 1999. Sodium borohydride and hydrogen peroxide as fluorogenic spray reagents for the detection of nalidixic acid and flumequine. *J Liq Chromatogr Relat Technol* 22(5): 771–775.

Wang, P.-L., Feng, Y.-L., and L. Chen. 1997. Simultaneous determination of trace norfloxacin, pefloxacin, and ciprofloxacin by TLC-fluorescence spectrodensitometry. *Microchem J* 56(2): 229–235.

Warlich, R. and E. Mutschler. 1989. Thin-layer chromatographic separation and in situ fluorimetric determination of ofloxacin in plasma and pleural fluid. *J Chromatogr B Biomed Sci Appl* 490(2): 395–403.

45 Thin-Layer Chromatography of Tuberculostatic Drugs

Anamaria Hosu and Claudia Cimpoiu

CONTENTS

45.1 INFORMATION ABOUT THE DRUG GROUP

Thin-layer chromatography (TLC) and also high-performance TLC (HPTLC) are important tools for screening of pharmaceuticals, giving reliable results with low costs, consuming less time, and more flexibility than other analytical method. They are used for the quality control of drug constituents, in the investigation of pharmaceutical active compounds and their degradation impurities, as well as in their analysis of biological samples.

Antituberculosis drugs are chemical compounds used to treat tuberculosis, an infection disease caused by *Mycobacterium tuberculosis* bacteria. Also, there are other types of bacteria from *Mycobacterium* family that can cause "atypical tuberculosis," infections that are sometimes clinically similar with typical tuberculosis. *M. tuberculosis* was first isolated in 1882 by Robert Koch who received the Nobel Prize for this discovery. Even if at the beginning of the twentieth century tuberculosis was the leading cause of mortality, nowadays it is considered to be a disease of the Third World countries and it can be prevented by the administration of vaccines or cured using the antituberculosis drugs.

Tuberculosis is a chronic disease and the tubercle bacilli are mostly located in intracellular inaccessible regions. They have periods of metabolic inactivity, resistance develops quickly, and sometimes they are present in atypical forms, less vulnerable to treatment. Therefore, treatment of tuberculosis is prolonged and requires the association of many antituberculosis drugs.

Antibiotics and antituberculosis chemotherapy includes a group of substances with different chemical structure, with bactericidal or bacteriostatic action against *M. tuberculosis* and atypical of bacteria. They can be divided into two groups:

1. Major (primary or first line) antituberculosis chemotherapeutic agents: ethambutol, isoniazid, rifampicin, pyrazinamide, rifabutin, streptomycin
2. Minor (second line) agents: ethionamide, prothionamide, cycloserine, p-aminosalicylic acid, dapsone, clofazimine, etc.

Determination of antituberculosis drugs is an important problem in pharmaceutical and clinical analysis. Different chromatographic methods, including TLC, HPTLC alone, or in combination with high-performance liquid chromatography (HPLC), have been used for the analysis of tuberculostatic drugs from different samples (Sorokoumova et al. 2008).

The aim of this chapter is to synthesize literature data regarding the use of TLC and HPTLC in the analysis of tuberculostatic drugs.

45.2 CHARACTERIZATION OF INDIVIDUAL TUBERCULOSTATIC DRUGS, METABOLITES, AND IMPURITIES FROM PHARMACEUTICAL FORMULATION

45.2.1 ETHAMBUTOL

Ethambutol ((2S,2′S)-2,2′-(ethane-1,2-diyldiimino)dibutan-1-ol, N,N′-ethylenebis-[(2S)-2-aminobutanol]-dihydrochloride) (Figure 45.1), commonly abbreviated EMB or simply E, has toxic action of *M. tuberculosis*, both acting on extracellular and intracellular bacilli. It has good bioavailability after oral administration, 75%–80% of a dose being absorbed from the gastrointestinal tract and distributed in body tissues and fluids. The metabolism of ethambutol involves an initial oxidation of the alcohol to an aldehydic intermediate, followed by conversion to a dicarboxylic acid (2,2′-(ethylenediimino)-di-butyric acid) corresponding to the terminal oxidation product of the parent compound (Peets and Buyske 1964). Approximately 50% of the initial dose is excreted unchanged in urine. Because of the rapid installation of drug resistance, ethambutol is usually administrated with isoniazid, rifampicin, and pyrazinamide.

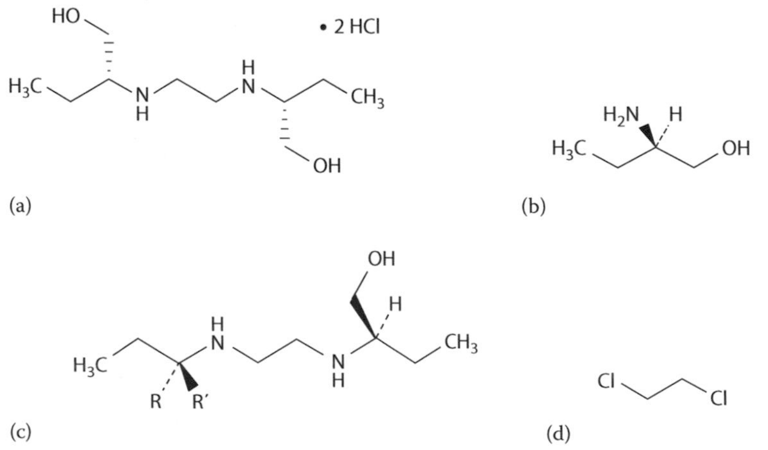

FIGURE 45.1 Chemical structures of ethambutol and their impurities that can occur in the pharmaceutical formulation. (a) Ethambutol dihydrochloride, (b) 2-Aminobutan-1-ol, (c) R = CH$_2$–OH, R′ = H (2R, 2′S)-2,2′-(ethylenediimino)dibutan-1-ol (meso-ethambutol), and (d) 1,2-dichloroethane (ethylene chloride).

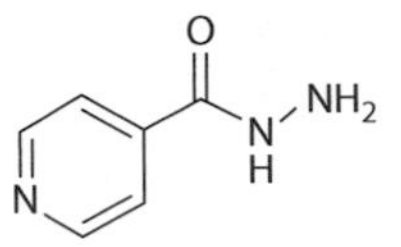

FIGURE 45.2 Chemical structure of isoniazid.

45.2.2 ISONIAZID

Isoniazid (pyridine-4-carbohydrazide, isonicotinylhydrazine (INH), Laniazid) (Figure 45.2) is a first-line tuberculostatic agent. At therapeutic levels, isoniazid has bactericidal action against actively growing intracellular and extracellular *M. tuberculosis* organisms and bacteriostatic action if the mycobacteria are slow growing. It has a good bioavailability after oral administration, but the absorption is reduced when isoniazid is administrated with food. The metabolism of isoniazid is achieved in the liver, by acetylation. Of the initial dose of isoniazid, 50%–70% is excreted in the urine within 24 h. Because there is a high frequency of resistance of bacilli being treated with isoniazid alone, it is administered in combination with other chemotherapeutic agents.

45.2.3 RIFAMPICIN

Rifampicin (R, RMP, RA, RF, or RIF) (Figure 45.3), also known as rifampin, is a semisynthetic compound derived from *Streptomyces mediterranei*. It is typically used in *Mycobacterium* infections

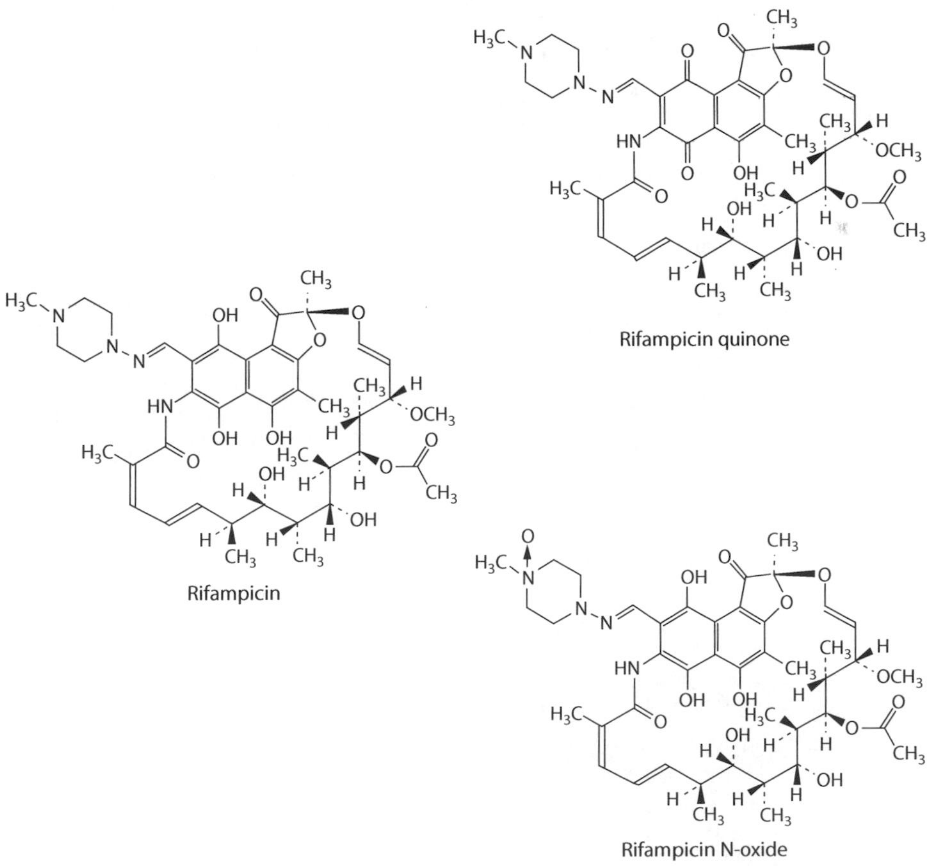

Rifampicin quinone

Rifampicin

Rifampicin N-oxide

FIGURE 45.3 Chemical structures of rifampicin and their impurities that can occur in the pharmaceutical formulation.

but also in the treatment of other infections caused by *Staphylococcus aureus*, *Listeria* species, *Legionella pneumophila*, *Haemophilus influenzae*, etc. Rifampicin is administered orally and is well absorbed from the gastrointestinal tract, being distributed in body tissues and fluids, including the cerebrospinal fluid. It is metabolized via deacetylation in the liver and eliminated in bile and, in a less quantity, in urine. A major disadvantage of rifampicin treatments is the fact that the resistance develops rapidly, so the association with other tuberculostatic drugs is necessary.

45.2.4 RIFABUTIN

Rifabutin (Rfb) (Figure 45.4), a semisynthetic derivative of rifamycin S, is a first-line antitubercular agent. It has a bactericidal action against *M. tuberculosis*, *Mycobacterium leprae*, and *Mycobacterium avium-intracellulare* and also against many gram-positive and gram-negative bacteria. Rifabutin is well absorbed from the gastrointestinal tract, and then distributed in body tissues and fluids, including the cerebrospinal fluid. Rifabutin is metabolized in liver, 25-*O*-desacetyl and

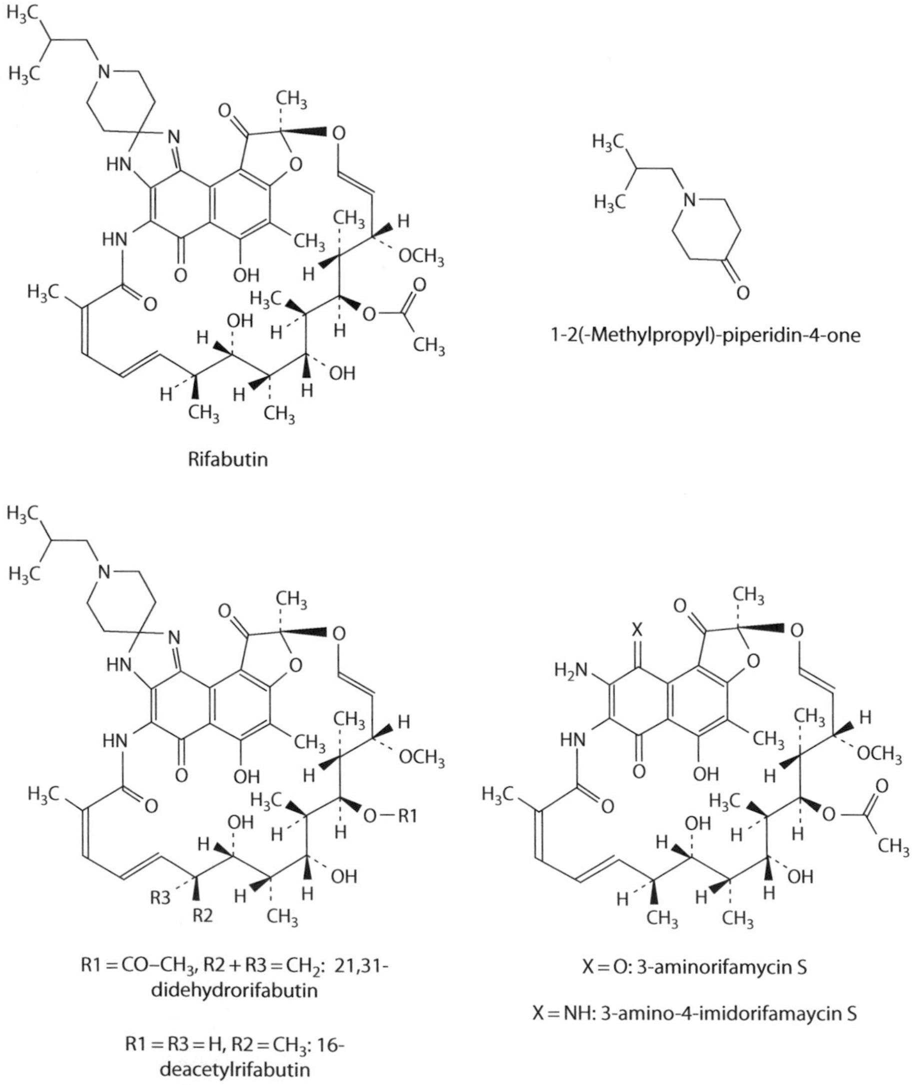

Rifabutin

1-2(-Methylpropyl)-piperidin-4-one

R1 = CO–CH₃, R2 + R3 = CH₂: 21,31-didehydrorifabutin

R1 = R3 = H, R2 = CH₃: 16-deacetylrifabutin

X = O: 3-aminorifamycin S

X = NH: 3-amino-4-imidorifamaycin S

FIGURE 45.4 Chemical structures of rifabutin and their impurities that can occur in pharmaceutical formulation.

31-hydroxy being the most important metabolites (Utkin et al. 1997). It is excreted in the urine, almost as metabolites. It is well tolerated by HIV-positive patients.

45.2.5 ETHIONAMIDE AND PROTHIONAMIDE

Ethionamide (2-ethylpyridine-4-carbothioamide) and prothionamide (2-propylpyridine-4-carbo-thioamide) (Figure 45.5) are second-line antitubercular agents, with similar properties, used in tuberculosis and leprosy treatment. They are nicotinic acid derivatives related to isoniazid. They can have both bactericidal and bacteriostatic action, depending on the concentration of the drug at the site of infection and the susceptibility of the infecting organism. They are almost completely absorbed after oral administration, being metabolized especially in the liver to an active sulfoxide metabolite and other inactive ones. A low quantity from the initial dose of ethionamide is excreted in urine, along with their metabolites. When the first-line antitubercular agents are ineffective or contraindicated, ethionamide and prothionamide are administrated with other tuberculostatic drugs.

45.2.6 CYCLOSERINE

Cycloserine ((R)-4-amino-1,2-oxazolidin-3-one) (Figure 45.6) is a second-line tuberculostatic agent. It can exert both bactericidal and bacteriostatic action, depending on the concentration of the drug at the site of infection and the susceptibility of the infecting organism. After oral administration, cycloserine is absorbed almost completely from gastrointestinal tract. It is used in tuberculosis treatment when one or more first-line tuberculostatic drugs cannot be used.

45.2.7 DAPSONE

Dapsone (4-[(4-aminobenzene)sulfonyl]aniline) (Figure 45.7) is a sulfone active against different bacteria, usually used in treatment of *M. leprae* infections. In this purpose, it is associated with rifampicin and clofazimine. Dapsone has a good bioavailability; it is metabolized in the liver and excreted in the urine.

45.2.8 CLOFAZIMINE

Clofazimine (*N*,5-bis(4-chlorophenyl)-3-(propan-2-ylimino)-3,5-dihydrophenazin-2-amine) (Figure 45.8) is a lipophilic compound that exerts a bactericidal effect against *M. leprae*, being used in combination

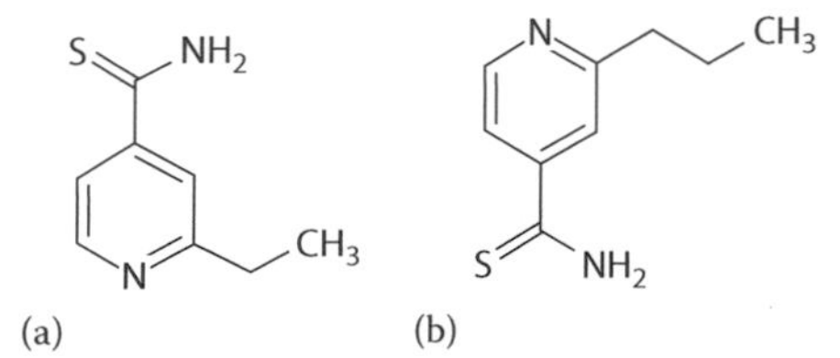

(a)   (b)

FIGURE 45.5 Chemical structures of ethionamide (a) and prothionamide (b).

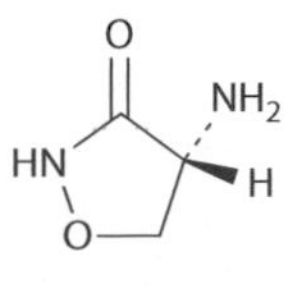

FIGURE 45.6 Chemical structure of cycloserine.

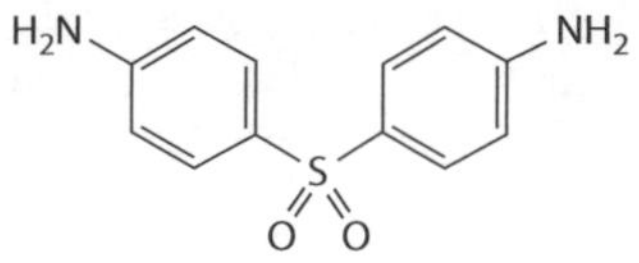

FIGURE 45.7 Chemical structure of dapsone.

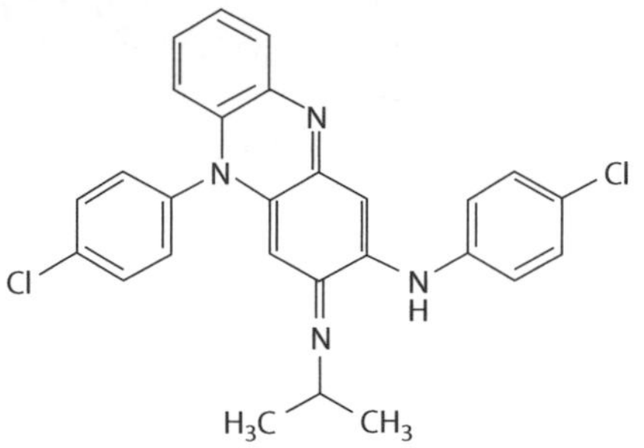

Clofazimine

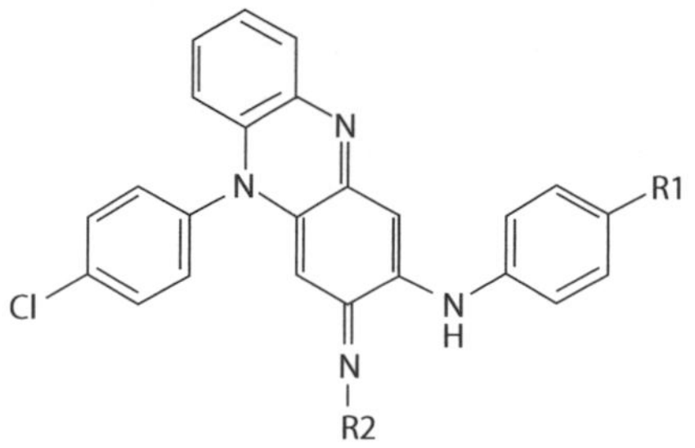

R1 = Cl, R2 = H: N,5-bis(4-chlorophenyl)-3-imino-3,5-
dihydrophenazin-2-amine

R1 = H, R2 = CH(CH$_3$)$_2$: 5-(4-chlorophenyl)-3-[(1-methyl
ethyl)imino]-3,5-dihydrophenazin-2-amine

FIGURE 45.8 Chemical structures of clofazimine and their impurities that can occur in pharmaceutical formulation.

with rifampicin and dapsone. The bioavailability and the rate of absorption of clofazimine are favored by food. It is metabolized in the liver, three metabolites being identified (Feng et al. 1981).

45.3 CHROMATOGRAPHIC SYSTEMS USED IN ANALYSIS OF TUBERCULOSTATIC DRUGS

The extraction of active compounds from the pharmaceutical formulation is usually simple, involving dissolving the substance to be examined in an appropriate solvent (European Pharmacopoeia 2005, Romanian Pharmacopoeia 2005, British Pharmacopoeia 2009, United States Pharmacopoeia 2007, Japanese Pharmacopoeia 2011), followed by centrifugation or filtration, in order to remove any solid material. Thus, ethambutol and cycloserine are usually dissolved in methanol, isoniazid is extracted in an acetone–water mixture (1:1, v/v), and rifampicin in chloroform. Rifabutin is soluble in methanol–methylene chloride mixture (1:1, v/v) and ethionamide and prothionamide in acetone. Some of tuberculostatic compounds are degraded at different pH values.

Different chromatographic systems were used for TLC and HPTLC analysis of tuberculostatic drugs, usually from pharmaceutical formulations. The most frequently used stationary phase in the analysis of these compounds is silica gel with florescence indicator (Kitamura et al. 1987, Izer et al. 1996, Le and Tao 1996, Kenyon et al. 2001). Mobile phases with different composition were tested for the separation of mixtures of tuberculostatic drugs with both similar chemical structures (Habel et al. 1997, Yun et al. 1997) and very different chemical structures (Avachat and Bhise 2010). Detection is usually done in UV light, but different derivatization reagents can be also used. Literature data are synthesized in Table 45.1.

Besides the analysis of the drugs, literature mentions the use of TLC and HPTLC as a quality control tool, in order to reveal the presence of impurities in pharmaceutical formulations (European Pharmacopoeia 2005, Romanian Pharmacopoeia 2005, British Pharmacopoeia 2009, Japanese Pharmacopoeia 2011). This is done following mainly the same procedure. Two solutions of the

TABLE 45.1
Chromatographic Conditions for the Separation of Tuberculostatic Drugs

Compounds	Sample	Stationary Phase	Mobile Phase	Detection	Reference
Ethambutol hydrochloride, isoniazid, pyrazinamide, rifampin, streptomycin sulfate	Tablets containing individual drug or drug combinations	TLC silica gel F_{254}	Methanol–conc. NH_4OH 25:0.38, v/v Methanol–acetone–conc. NH_4OH 13:17:1, v/v Ethyl acetate–acetic acid–conc. NH_4OH–water 3:3:1:1, v/v	UV light, VIS light after immersion in iodine–KI solution	Kenyon et al. (2001)
Ethambutol, isoniazid, pyrazinamide, aminosalicylic acid	Pharmaceutical formulation	RP-18	Acetonitrile–water 7:3, v/v modified by adding $CuCl_2$ 0.05 M	UV light, diode array detector (DAD)	Flieger et al. (2009)
Isoniazid	Tablets	TLC silica gel F_{254}	Methanol–chloroform 60:65, v/v	UV light	Kitamura et al. (1987)
Isoniazid, pyridoxine hydrochloride	Pharmaceutical formulation	TLC silica gel F_{254}	Acetone–carbon tetrachloride–6.5 M NH_4OH 21:7:2, v/v	UV light	Argekar and Kunjir (1996)
Isoniazid, salicylaldehyde, benzaldehyde derivatives of isoniazid	Serum	TLC silica gel F_{254}	Chloroform–methanol 9:1, v/v	UV light	Yun et al. (1997)
Isoniazid, acetylisoniazid	Serum	HPTLC silica gel F_{254}	Ethyl acetate–methanol 7:3, v/v	UV light	Habel et al. (1997)
Isoniazid, rifampicin	Pharmaceutical formulation	HPTLC silica gel F_{254} prewashed with methanol	Ethyl acetate–methanol–acetone–acetic acid 5:2:2:1, v/v	UV light	Tatarczak et al. (2005)
Isoniazid, rifabutin	Pharmaceutical formulation	HPTLC silica gel F_{254}	Dichloromethane–acetone–methanol 20:7:2, v/v	UV light	Avachat and Bhise (2010)
Rifampicin	Pharmaceutical formulation	HPTLC silica gel F_{254}	Chloroform–methanol–water 32:8:1, v/v	UV–VIS light	Jindal et al. (1994)
Rifampicin, rifampicin quinone	Eye drops	TLC silica gel	Chloroform–methanol 21:29, v/v	VIS light	Izer et al. (1996)
Rifampicin, troleandomycin, tylosin, rifamycin B, vancomycin, erythromycin	Standard compounds	TLC Kieselgel 60 WF_{254}, RP-18 W HPTLC	Methanol–water, 1-propanol–water in different proportions from 0% to 100%, v/v	Spraying with 20% perchloric acid	Nowakowska et al. (2002)
Rifampicin, troleandomycin, tylosin, rifamycin B, vancomycin, erythromycin	Standard compounds	LiChrospher silica gel	Alcohols–hexamethyldisiloxane, ketones–hexamethyldisiloxane in different proportions from 0% to 100%, v/v Alcohols–dimethyl sulfoxide, ketones–dimethyl sulfoxide in proportion from 0% to 50%, v/v	Spraying with conc. sulfuric acid–methanol, 1:4, v/v	Nowakowska (2004)

(continued)

TABLE 45.1 (continued)
Chromatographic Conditions for the Separation of Tuberculostatic Drugs

Compounds	Sample	Stationary Phase	Mobile Phase	Detection	Reference
Rifampicin, troleandomycin, tylosin, rifamycin B, erythromycin	Standard compounds	HPTLC silica gel RP-18	Ester or ketones in dimethyl sulfoxide or hexamethyldisiloxane in different portions from 0% to 100%, v/v	Spraying with conc. sulfuric acid–methanol, 1:4, v/v	Nowakowska (2005)
Rifampicin, troleandomycin, tylosin, rifamycin B, erythromycin	Standard compounds	RP-18	Alcohols–dimethyl sulfoxide, alcohols–hexamethyldisiloxane in different proportion from 0% to 100%, v/v	Spraying with conc. sulfuric acid–methanol, 1:4, v/v	Nowakowska (2006)
Rifampicin, rifamycin S, rifamycin SV, 3-formyl-rifamycin	Standard compounds	C-18, diphenyl	Hexane, cyclohexane, tetrahydrofuran, acetone, alcohols, different portion of hexane–chloroform, hexane–ethanol, methanol–water, acetonitrile–water	VIS light	Grassini-Strazza et al. (1986)
Prothionamide	Pharmaceutical formulation	TLC silica gel F_{254}	Chloroform–methanol 9:1, v/v	UV light	Le and Tao (1996)
Clofazimine	Plasma	TLC silica gel prewashed with chloroform–methanol 1:1, v/v	Toluene–acetic acid–water 50:50:4, v/v	VIS light	Lanyi and Dubois (1982)

pharmaceutical product with different concentrations are prepared and then applied on the chromatographic plate. On the same plate, solutions of possible impurities prepared using certified standard compounds are also applied. The evaluation is done after elution, by comparing the areas of spots. If the areas of spots corresponding to the impurities separated from sample are smaller than those of certified standard compounds, then the pharmaceutical product is considered pure.

According to Romanian Pharmacopoeia and Japanese Pharmacopoeia (Romanian Pharmacopoeia 2005, Japanese Pharmacopoeia 2011), ethambutol hydrochloride dissolved in methanol is determined on TLC silica gel G plates using as mobile phase a mixture of ethyl acetate–acetic acid–hydrochloric acid–water 11:7:1:1, v/v. After the elution to a distance of about 10–15 cm, the plate is air-dried and then heated at 105°C for 5 min. Detection can be done by spraying the cooled plate with cadmium acetate–ninhydrin solution and heating it at 90°C for 5 min (Romanian Pharmacopoeia 2005) or by using a ninhydrin–L-ascorbic acid solution and heating the plate at 105°C for 5 min (Japanese Pharmacopoeia 2011).

European Pharmacopoeia and British Pharmacopoeia proposed for the determination of ethambutol hydrochloride the use of the same stationary phase and a mobile phase consisting in conc. NH_4OH–water–methanol 10:15:75, v/v. The plate is developed, dried in air, and then heated at 110°C for 10 min. Detection is achieved after spraying the cooled plate with ninhydrin solution and heating it at 110°C for 10 min (European Pharmacopoeia 2005, British Pharmacopoeia 2009). The same compound analysis can be performed on TLC silica gel plates, using a mobile phase of methanol–NH_4OH 18:1, v/v. In this case, detection is done in VIS light, after spraying the developed plate with iodine solution (United States Pharmacopoeia 2007).

According to the European Pharmacopoeia and British Pharmacopoeia, isoniazid can be determined from injection and tablets, together with hydrazine sulfate, a possible impurity. Separation is achieved on TLC silica gel GF_{254} plates with an eluent mixture of ethyl acetate–acetone–methanol–water 50:20:20:10, v/v. Detection can be done in UV light or after spraying the developed plate with dimethylaminobenzaldehyde (European Pharmacopoeia 2005, British Pharmacopoeia 2009).

The monograph presented in the Romanian Pharmacopoeia presents the determination method of rifampicin together with 3-formyl-rifampicin, rifampin quinone, compounds that can occur in pharmaceutical formulation. After the elution with a chloroform–methanol 90:10, v/v mobile phase, the detection of compounds can be done in UV light, if TLC silica gel GF_{254} plates are used, or in VIS light, if TLC silica gel G prepared with phosphate buffer pH 7.4 plates are used as stationary phase (Romanian Pharmacopoeia 2005).

The United States Pharmacopeia (2007) indicates a method for determination of two mixtures of rifampicin and isoniazid and rifampicin, isoniazid, and pyrazinamide, from capsules. Silica gel F_{254} is used as stationary phase and the developing solvent is the mixture of acetone–glacial acetic acid 100:1, v/v. Detection is done in both cases in UV light.

European Pharmacopoeia (2005) and British Pharmacopoeia (2009) proposed a method for the determination of rifabutin on TLC silica gel GF_{254} plates, using a mixture of acetone–light petroleum 23:77, v/v mobile phase. After the elution of over 2/3 of the plate, detection is done in UV light or in VIS light, after the exposure of the plate to iodine vapor for about 5 min.

According to the European Pharmacopoeia (2005), British Pharmacopoeia (2009), and Japanese Pharmacopoeia (2011), ethionamide dissolved in acetone is determined on TLC silica gel GF_{254} plates and developed with methanol–chloroform 10:90, v/v mobile phase. Plates are dried in air and examined at 254 nm.

Dapsone dissolved in methanol is determined on silica gel G plates using a mobile phase consisting of concentrated. NH_4OH–methanol–ethyl acetate–heptane 1:6:20:20, v/v. After the elution in an unsaturated tank over a path of 15 cm, the plate is dried in air and sprayed with 0.1% w/v solution of 4-dimethylaminocinnamaldehyde in a mixture of HCl–ethanol 1:99, v/v, and visualized in VIS light (European Pharmacopoeia 2005, British Pharmacopoeia 2009).

The United States Pharmacopeia (2007) presents an HPTLC method for the determination of dapsone. The analysis is done on silica gel HPTLC plate, using a mobile phase of chloroform–acetone–n-butanol–formic acid 60:15:15:10, v/v. After the elution of 3/4 of the plate, it is dried in air and visualized after spraying with a 0.1% (w/v) solution of 4-dimethylaminocinnamaldehyde in a mixture of glacial acetic acid 1:1, v/v.

According to the European Pharmacopoeia, British Pharmacopoeia, and United States Pharmacopeia, clofazimine dissolved in methylene chloride is determined on TLC silica gel GF_{254}, eluted with propanol–methylene chloride 6:85, v/v (European Pharmacopoeia 2005, British Pharmacopoeia 2009) or with propanol–methylene chloride 10:1, v/v (United States Pharmacopoeia 2007). After the elution of 2/3 of the plate, it is dried horizontally in air for 5 min and then it is eluted again over 2/3 of the plate. Detection is done in UV light at 254 nm.

The literature presents a relatively small amount of information regarding TLC and HPTLC analysis of tuberculostatic drugs. Moreover, HPLC seems to be the most used technique for the determination of these compounds. Even so, TLC and HPTLC remain reliable analytical tools and good alternatives to HPLC techniques, used successfully in the quality control and in monitoring of tuberculostatic drugs.

REFERENCES

Argekar, A.P. and Kunjir, S.S. 1996. Simultaneous determination of isoniazid and pyridoxine hydrochloride in pharmaceutical preparations by high-performance thin-layer chromatography. *J. Planar Chromatogr.* 9:390–394.

Avachat, A.M. and Bhise, S.B. 2010. Stability-indicating validated HPTLC method for simultaneous analysis of rifabutin and isoniazid in pharmaceutical formulations. *J. Planar Chromatogr.* 23:123–128.

British Pharmacopoeia, 2009. The Stationery Office, London, U.K.*European Pharmacopoeia*, 5th edn. 2005. The European Pharmacopoeia Convention, Inc., Strasbourg, France.

Feng, P.C., Fenselau, C.C., and Jacobson, R.R. 1981. Metabolism of clofazimine in leprosy patients. *Drug Metab. Dispos.* 9:521–524.

Flieger, J., Paneth, P., Gielzak-Kocwin, K., and Tatarczak, M. 2009. Micropreparative isolation of Cu(II) complexes of isoniazid and ethambutol and determination of their structures. *J. Planar Chromatogr.* 22:83–88.

Grassini-Strazza, G., Nicoletti, I., Polcaro, C.M., Girelli, A.M., and Sanci, A. 1986. Effect of the mobile phase composition on the retention behaviour of diphenylsilica pre-coated plates. *J. Chromatogr.* 367:323–334.

Habel, D., Guermouche, S., and Guermouche, M.H. 1997. HPTLC determination of isoniazid and acetylisoniazid in serum. Comparison with HPLC. *J. Planar Chromatogr.* 10:453–456.

Izer, K., Török, I., Magyar-Pinter, G., Varsanyi, E., and Liptak, J. 1996. Study of stability of rifampicin eye drops. *Acta Pharm. Hung.* 66:157–163.

Japanese Pharmacopoeia XVI. 2011. The Minister of Health, Labour and Welfare, Tokyo, Japan.

Jindal, K.C., Chaudhary, R.S., Gangwal, S.S., Singla, A.K., and Khanna, A. 1994. High-performance thin-layer chromatographic method for monitoring degradation products of rifampicin in drug excipient interaction studies. *J. Chromatogr.* 685:195–199.

Kenyon, A.S., Layloff, T., and Sherma, J. 2001. Rapid screening of tuberculosis pharmaceuticals by thin layer chromatography. *J. Liq. Chromatogr. Relat. Technol.* 24:1479–1490.

Kitamura, K., Hatta, M., Fukuyama, S., Ito, M., Nakamura, Y., and Hozumi, K. 1987. Determination of isoniazid in tablets by second-derivative ultra-violet spectrophotometry of scraped-spot solution from thin-layer chromatography. *Anal. Chim. Acta* 201:357–361.

Lanyi, Z. and Dubois, J. 1982. Determination of clofazimine in human plasma by thin-layer chromatography. *J. Chromatogr.* 232:219–223.

Le, W. and Tao, H. 1996. Determination of prothionamide in its formulations by using an improved thin-layer chromatographic method. *Chin. J. Pharm. Anal.* 16:192–196.

Nowakowska, J. 2004. Analysis of selected macrocyclic antibiotics by HPTLC with non-aqueous mobile phases. *J. Planar Chromatogr.* 17:200–206.

Nowakowska, J. 2005. Normal and reversed-phase TLC separation of some macrocyclic antibiotics with non-aqueous mobile phases. *J. Planar Chromatogr.* 18:455–459.

Nowakowska, J. 2006. Effect of non-aqueous mobile phase composition on the retention of macrocyclic antibiotics in RP-TLC. *J. Planar Chromatogr.* 19:62–67.

Nowakowska, J., Halkiewicz, J., and Lukasiak, J. 2002. TLC determination of selected macrocyclic antibiotics using normal and reversed phases. *Chromatographia* 56:367–373.

Peets, E.A. and Buyske, D.A. 1964. Comparative metabolism of ethambutol and its L-isomer. *Biochem. Pharmacol.* 13:1403–1419.

Romanian Pharmacopoeia, 10th edn. 2005. Editura Medicală, Bucureşti, Romania.

Sorokoumova, G.M., Vostrikov, V.V., Selishcheva et al. 2008. Bacteriostatic activity and decomposition products of rifampicin in aqueous solution and liposomal composition. *Pharm. Chem. J.—USSR* 42:475–478.

Tatarczak, M., Flieger, J., and Szumilo, H. 2005. Simultaneous densitometric determination of rifampicin and isoniazid by high-performance thin-layer chromatography. *J. Planar Chromatogr.* 18:207–211.

United States Pharmacopoeia 30-NF 25 e-Book. 2007. United States Pharmacopoeia Convection, Rockville, MD.

Utkin, I., Koudriakova, T., Thompson, T. et al. 1997. Isolation and identification of major urinary metabolites of rifabutin in rats and humans. *Drug Metab. Dispos.* 25:963–969.

Yun, K., Wang, Y., Li, G., Meng, Q., and Wang, S. 1997. Rapid examination of isoniazid poisoning I. Determination of isoniazid in serum by thin-layer chromatography. *J. Pharm. Anal.* 17:116–118.

46 TLC of Antifungal and Antiprotozoal Drugs

Živoslav Lj. Tešić and Jelena Đ. Trifković

CONTENTS

46.1 TLC OF ANTIFUNGAL DRUGS

Fungi are microorganisms found in soil, water, and air and on plants, animals, and debris in general. Many fungi have pathogenic potential for humans. According to the tissues and organs affected, mycoses are classified into superficial mycoses (mycosis of the skin, nails, and hair) and mycoses subcutaneous, systemic, or deep. The vast majority of fungal infections are due to yeast of the genus *Candida* and fungi of the genus *Aspergillus*. However, infections by other fungi, rarer, are increasing in frequency.

Antifungal agents are a class of pharmaceuticals used in the treatment of fungal infection. Fungal infections are usually more difficult to treat than bacterial infections, because fungal organisms grow slowly and infections often occur in tissues that are poorly penetrated by antimicrobial agents (e.g., devitalized or avascular tissues). Superficial fungal infections involve cutaneous surfaces, such as skin, nails, and hair, and mucous membrane surfaces, such as oropharynx and vagina. Antifungals work by exploiting differences between mammalian and fungal cells to kill the fungal organism without dangerous effects on the host (Craig and Stitzel 1997). Unlike bacteria, both fungi and humans are eukaryotes, that is, fungal and human cells are similar at the molecular level. This makes it more difficult to find or design drugs that target fungi without affecting human cells.

A growing number of topical and systemic agents are available for the treatment of these infections. Table 46.1 summarizes antifungal agents according to the mechanism of their action and a class of chemical compounds to which they belong. Some of these compounds are primarily used topically to treat superficial dermatophytic, yeast, and mold infections, whereas others are administered orally for the treatment of systemic fungal infections (Table 46.1).

Antifungal agents can be analyzed by many different techniques, starting with classical ones (titrimetry), through optical and spectrophotometric, to electrochemical and separation methods. Reference pharmacopeia elaborations (Pharmacopoeia Internationalis 2006, European Pharmacopoeia 2008, United States Pharmacopeia 31 2008) require for antifungals, in most cases titrimetric method with potentiometrically determined endpoint. It is important to note that this titrimetric procedure is suitable for analyzing active substance individually but is not proper to be

TABLE 46.1
Antifungal Drugs

Mechanism of Action	Chemical Class	Generic Name	Chemical Name	Appliance
Polyene membrane disruptors	Polyene	Amphotericin B	(1R,3S,5R,6R,9R,11R,15S,16R,17R,18S,19E,21E,23E,25E,27E,29E,31E,33R,35S,36R,37S)-33-[(3-amino-3,6-dideoxy-β-D-mannopyranosyl)oxy]-1,3,5,6,9,11,17,37-octahydroxy-15,16,18-trimethyl-13-oxo-14,39-dioxabicyclo[33.3.1]nonatriaconta-19,21,23,25,27,29,31-heptaene-36-carboxylic acid	Systemic agent
		Nystatin	(1S,3R,4R,7R,9R,11R,15S,16R,17R,18S,19E,21E,25E,27E,29E,31E,33R,35S,36R,37S)-33-[(3-amino-3,6-dideoxy-β-D-mannopyranosyl)oxy]-1,3,4,7,9,11,17,37-octahydroxy-15,16,18-trimethyl-13-oxo-14,39-dioxabicyclo[33.3.1]nonatriaconta-19,21,25,27,29,31-hexaene-36-carboxylic acid	Topical agent
		Natamycin	(1R,3S,5R,7R,8E,12R,14E,16E,18E,20E,22R,24S,25R,26S)-22-[(3-amino-3,6-dideoxy-D-mannopyranosyl)oxy]-1,3,26-trihydroxy-12-methyl-10-oxo-6,11,28-trioxatricyclo[22.3.1.05,7]octacosa-8,14,16,18,20-pentaene-25-carboxylic acid	Topical agent
Ergosterol biosynthesis inhibitors	Imidazole	Clotrimazole	1-[(2-Chlorophenyl)(diphenyl)methyl]-1H-imidazole	Topical agent
		EZ	(RS)-1-{2-[(4-chlorophenyl)methoxy]-2-(2,4-dichlorophenyl)ethyl}-1H-imidazole	Topical agent
		Bifonazole	(RS)-1-[phenyl(4-phenylphenyl)methyl]-1H-imidazole	Topical agent
		Sertaconazole	1-{2-[(7-chloro-1-benzothiophen-3-yl)methoxy]-2-(2,4-dichlorophenyl)ethyl}-1H-imidazole	Topical agent
		Ketoconazole	cis-1-Acetyl-4-[4[[2-(2,4-dichlorophenyl)-2-(1H-imidazol-1-ylmethyl)-1,3-dioxolan-4-yl]methoxy]phenyl]piperazine	Topical and systemic agent
		MZ	(RS)-1-[2-(2,4-dichlorobenzyloxy)-2-(2,4-dichlorophenyl)ethyl]-1H-imidazole	Topical agent
		Oxiconazole	(E)-[1-(2,4-dichlorophenyl)-2-(1H-imidazol-1-yl)ethylidene][(2,4-dichlorophenyl)methoxy]amine	Topical agent
		Sulconazole	1-(2-[[(4-Chlorophenyl)methyl]sulfanyl]-2-(2,4-dichlorophenyl)ethyl)-1H-imidazole	Topical agent
		Tioconazole	(RS)-1-[2-[(2-chloro-3-thienyl)methoxy]-2-(2,4-dichlorophenyl)ethyl]-1H-imidazole	Topical agent
	Triazole	Terconazole	1-[4-[[(2S,4S)-2-(2,4-dichlorophenyl)-2-(1,2,4-triazol-1-ylmethyl)-1,3-dioxolan-4-yl]methoxy]phenyl]-4-propan-2-yl-piperazine	Topical agent
		Fluconazole	2-(2,4-Difluorophenyl)-1,3-bis(1H-1,2,4-triazol-1-yl)propan-2-ol	Systemic agent
		Voriconazole	(2R,3S)-2-(2,4-difluorophenyl)-3-(5-fluoropyrimidin-4-yl)-1-(1H-1,2,4-triazol-1-yl)butan-2-ol	Systemic agent
		Itraconazole	(2R,4S)-rel-1-(butan-2-yl)-4-{4-[4-[[(2R,4S)-2-(2,4-dichlorophenyl)-2-(1H-1,2,4-triazol-1-ylmethyl)-1,3-dioxolan-4-yl]methoxy}phenyl]piperazin-1-yl]phenyl]-4,5-dihydro-1H-1,2,4-triazol-5-one	Systemic agent
	Morpholine	Amorolfine	(±)-(2R*,6S*)-2,6-dimethyl-4-{2-methyl-3-[4-(2-methylbutan-2-yl)phenyl]propyl}morpholine	Topical agent

Allylamines and other squalene epoxidase inhibitors	Butenafine	[(4-*tert*-Butylphenyl)methyl](methyl)(naphthalen-1-ylmethyl)amine	Topical agent
	Naftifine	(2E)-N-methyl-N-(1-naphthylmethyl)-3-phenylprop-2-en-1-amine	Topical agent
	Terbinafine	[(2E)-6,6-dimethylhept-2-en-4-yn-1-yl](methyl)(naphthalen-1-ylmethyl)amine	Topical and systemic agent
	Tolnaftate (thiocarbamate)	O-2-naphthyl methyl(3-methylphenyl)thiocarbamate	Topical agent
Inhibitors of cell wall biosynthesis—echinocandins	Caspofungin	N-[(3S,6S,9S,11R,15S,18S,20R,21R,24S,25S)-3-[((1R)-3-amino-1-hydroxypropyl]-21-[(2-aminoethyl)amino]-6-[((1S,2S)-1,2-dihydroxy-2-(4-hydroxyphenyl)ethyl]- 11,20,25-trihydroxy-15-[((1R)-1-hydroxyethyl]-2,5,8,14,17,23-hexaoxo-1,4,7,13,16,22-hexaazatricyclo [22.3.0.09,13] heptacosan-18-yl]-10,12-dimethyltetradecanamide	Systemic agent
	Anidulafungin	N-[(3S,6S,9S,11R,15S,18S,20R,21R,24S,25S,26S)-6-[(1S,2R)-1,2-dihydroxy-2-(4-hydroxphenyl)ethyl]- 11,20,21,25-tetrahydroxy- 3,15-bis[((1R)-1-hydroxyethyl]- 26-methyl-2,5,8,14,17,23-hexaoxo-1,4,7,13,16,22-hexaazatricyclo [22.3.0.09,13] heptacosan-18-yl]- 4-[4-(pentyloxy)phenyl]phenyl}benzamide	Systemic agent
	Micafungin	{5-[((1S,2S)-2-((3S,6S,9S,11R,15S,18S,20R,21R,24S,25S,26S)-3-[((1R)-2-carbamoyl-1-hydroxyethyl]-11,20,21,25-tetrahydroxy-15-[((1R)-1-hydroxyethyl]-26-methyl-2,5,8,14,17,23-hexaoxo-18-[((4-{5-[4-(pentyloxy)phenyl]-1,2-oxazol-3-yl}benzene amido]-1,4,7,13,16,22-hexaazatricyclo[22.3.0.09,13]heptacosan-6-yl]-1,2-dihydroxyethyl]-2-hydroxyphenyl}oxidanesulfonic acid	Systemic agent
Drugs acting through other mechanisms	Flucytosine	4-Amino-5-fluoro-1,2-dihydropyrimidin-2-one	Systemic agent
	Griseofulvin	(2S,6′R)-7-chloro-2′,4,6-trimethoxy-6′-methyl-3H,4′H-spiro[1-benzofuran-2,1′-cyclohex[2]ene]-3,4′-dione	Systemic agent
	Haloprogin	1,2,4-Trichloro-5-[(3-iodoprop-2-yn-1-yl)oxy]benzene	Topical agent
	Ciclopirox	6-Cyclohexyl-1-hydroxy-4-methylpyridin-2(1H)-one	Topical agent
	Undecylenic acid	Undec-10-enoic acid	Topical agent

determined in complex matrices such as medicinal preparations. The same case occurs for spectrophotometric techniques. There is a necessity to analyze multi-ingredient samples by methods possessing separation potential.

Currently, the most extensively used techniques in pharmaceutical analysis are chromatographic methods. They enable separation, identification, and determination of huge amount of biologically active compounds. Among chromatographic techniques, the special focus should be given to liquid chromatography (LC), especially high-performance liquid chromatography (HPLC), thin-layer chromatography (TLC), and emerging ultra performance liquid chromatography (UPLC), and gas chromatography (GC) for the determination of volatile compounds (Ekiert et al. 2010).

Literature review revealed that HPLC methods are the most widely used analytical methods in pharmaceutical studies. However, they require expensive equipment, large amounts of mobile phases, high technology, and laborious sample preparation and are time consuming. Although there is a tendency in current pharmacopeias for favoring HPLC, TLC is still a very popular and frequently used analytical method in the pharmaceutical industry (Ferenczi-Fodor et al. 2011).

With its advantages of simplicity, economy, easy operation, and the need for only small amounts of solvent, TLC is used widely in various fields to separate or purify mixtures of chemical and biologically active compounds (Kaale et al. 2011). Because a new plate is used in each separation, the "memory effect" problems that occur with many other chromatographic techniques are not occurred in TLC separation, which makes TLC often used for direct analysis of crude samples with minimal purification procedures. Because it is performed under ambient conditions, TLC is one of the most suitable separation methods for high-throughput analysis. For example, many samples can be spotted on a TLC plate and separated simultaneously, or many plates can be analyzed in a tank containing the same mobile phase (Bele and Khale 2011).

Unlike HPLC, TLC is a simple and rapid chromatographic technique, but its separation efficiency is usually low. Many different TLC techniques have been developed to increase the separation, resolution, and reproducibility. High-performance thin-layer chromatography (HPTLC) uses gel particles having small diameters (4–6 µm) as the stationary phase, thereby increasing the number of interactions with the analyte molecules. HPTLC has a potential that meets the demands of a routine analytical technique due to its advantages of low operating cost, high sample throughput, and need for minimum sample cleanup. The major advantage is that several samples can be run simultaneously using a small quantity of mobile phase, unlike HPLC, thus lowering the analytical run times and cost per analysis. In pharmaceutical laboratories, there is always a need for faster, simpler, cheaper, and better-performing analytical methods. Further, TLC and HPTLC in instrumentalized mode using scanning after densitometry have been included as general methods in European Pharmacopoeia, permitting the use of planar chromatography for quantification at different stages of pharmaceutical research, development, and production (Marciniec et al. 2007). Further, one of the main advantages of planar chromatography is its ability to facilitate separations that can be successfully utilized to evaluate very different drug molecules, their impurities, and the metabolites in one run. Generally, the separations are discrete and very often complementary to other classified techniques such as HPLC and gas–liquid chromatography (GLC). Therefore, HPTLC can be a viable alternative for impurity profiling and characterization of newer drugs and unknown compounds.

TLC is the technique not used too extensively in the analysis of antifungal agents. There is a pity that researchers do not exploit it enough because it gives large analytical possibilities in conjunction with densitometry. The investigated literature, which comes from all relevant biomedical databases, covered TLC methods developed in the past. The presented methods are grouped according to the mechanism of action of the antifungal drugs. Chromatographic parameters that comprise the preparation of standard solution, selected stationary and mobile phase, development mode, detection, retention parameter, and appropriate reference are presented in Table 46.2.

Current analytical chemistry requires that all the methods should be validated. There is a necessity to prove that elaborated procedure is suitable for intended analytical purpose and leads repetitively to accurate results. There are many guidelines issued by medical authorities that advise

TABLE 46.2
Chromatographic Systems Used for TLC Analysis of Antifungal Drugs

Drug	Standard Solution	Plates	Stationary-Phase Pretreatment, Mobile-Phase Composition	Development Mode	Detection	Retention Factor	Reference
Amphotericin B	—	Silica gel	1. Activation at 110°C, 1 h; methanol–propan-2-ol–acetic acid (90 + 10 + 1) (v/v/v) 2. Activation at 110°C, 1 h; methanol–acetone–acetic acid (8 + 1 + 1) (v/v/v) 3. Activation at 110°C, 1 h; ethanol–ammonia–dioxane–water (8 + 1 + 1 + 1) (v/v/v) 4. Activation at 110°C, 1 h; butan-1-ol–pyridine–water (3 + 2 + 1) (v/v/v) 5. Butan-1-ol–ammonia–methanol–water (20 + 1 + 2 + 4) (v/v/v) 6. Chloroform–methanol–borate buffer (pH 8.3) (7 + 5 + 1) (v/v/v) 7. Butan-1-ol–ethanol–acetone–32% ammonia (2 + 5 + 1 + 3) (v/v/v)	—	1,2,5,6—0.2% p-dimethylaminobenzaldehyde in H_2SO_4, containing $FeCl_3$ 3,4—0.5% $KMnO_4$ 7—Spectrodensitometric	1. $R_F = 0.18$ 2. $R_F = 0.45$ 3. $R_F = 0.19$ 4. $R_F = 0.32$ 5. $R_F = 0.07$ 6. $R_F = 0.60$ 7. $R_F = 0.41$	Thomas (1976)
	—	Silica gel, poly(ethylene terephthalate) sheets	Methanol	—	UV lamp	$R_F = 0$ $R_F = 0.53$, main $R_F = 0.70$, main $R_F = 0.73$ $R_F = 0.76$	Kelly (1988)

(continued)

TABLE 46.2 (continued)
Chromatographic Systems Used for TLC Analysis of Antifungal Drugs

Drug	Standard Solution	Plates	Stationary-Phase Pretreatment, Mobile-Phase Composition	Development Mode	Detection	Retention Factor	Reference
	AmB (0.5 mg/mL) in the solvent mixture 30% dimethyl sulfoxide, 60% methanol, 10% distilled water	Silica gel	Methanol, activation at 120°C, 3 h; chloroform–methanol–borate buffer (pH 8.3) 4:5:1 (v/v/v)	Chamber saturation 20 min; ascending	UV lamp—366 nm Densitometry—385 nm Direct bioautography	$R_F = 0.46$, main $R_F = 0.31$, minor	Fittler et al. (2010)
Nystatin	—	Silica gel	1. Buffered with phosphate (pH 8); ethanol–ammonia–water (8 + 1 + 1) (v/v/v) 2. Buffered with phosphate (pH 8); butan-1-ol–acetic acid–water (3 + 1 + 1) (v/v/v) 3. Activation at 110°C, 1 h; methanol–propan-2-ol–acetic acid (90 + 10 + 1) (v/v/v) 4. Activation at 110°C, 1 h; methanol–acetone–acetic acid (8 + 1 + 1) (v/v/v) 5. Activation at 110°C, 1 h; ethanol–ammonia–dioxane–water (8 + 1 + 1 + 1) (v/v/v) 6. Activation at 110°C, 1 h; butan-1-ol–pyridine–water (3 + 2 + 1) (v/v/v)	—	1,2—10% KMnO$_4$ and 0.2% bromophenol blue 3,4—0.2% p-dimethylaminobenzaldehyde in H$_2$SO$_4$, containing FeCl$_3$ 5,6—0.5% KMnO$_4$	1. $R_F = 0.18$ 2. $R_F = 0.18$ 3. $R_F = 0.54$ 4. $R_F = 0.66$ 5. $R_F = 0.28$ 6. $R_F = 0.65$	Thomas (1976)

Compound	Solution	Stationary phase	Mobile phase	Conditions	Detection	Results	Reference
Natamycin	—	Silica gel	1. Buffered with phosphate (pH 8); ethanol–ammonia–water (8 + 1 + 1) (v/v/v) 2. Buffered with phosphate (pH 8); butan-1-ol-acetic acid–water (3 + 1 + 1) (v/v/v) 3. Activation at 110°C, 1 h; methanol–propan-2-ol-acetic acid (90 + 10 + 1) (v/v/v) 4. Activation at 110°C, 1 h; methanol–acetone–acetic acid (8 + 1 + 1) (v/v/v) 5. Activation at 110°C, 1 h; ethanol–ammonia–dioxane–water (8 + 1 + 1 + 1) (v/v/v) 6. Activation at 110°C, 1 h; butan-1-ol-pyridine–water (3 + 2 + 1) (v/v/v)	—	1,2—10% $KMnO_4$ and 0.2% bromophenol blue 3,4—0.2% p-dimethylaminobenzaldehyde in H_2SO_4, containing $FeCl_3$ 5,6—0.5% $KMnO_4$	1. $R_F = 0.34$ 2. $R_F = 0.34$ 3. $R_F = 0.40$ 4. $R_F = 0.54$ 5. $R_F = 0.18$ 6. $R_F = 0.55$	Thomas (1976)
Clotrimazole	Solution of clotrimazole (1 mg/mL) in chloroform	HPTLC silica gel	10%–90% of organic modifier 1. Acetone–n-hexane 2. Methanol–toluene 3. Methyl ethyl ketone–toluene	—	Exposure to iodine vapor	1. $R_M^0 = 2.65$, $C_0 = -0.52$ 2. $R_M^0 = 1.84$, $C_0 = -1.35$ 3. $R_M^0 = 3.78$, $C_0 = -1.57$	Aleksic et al. (2002)
Clotrimazole	Solution of clotrimazole (1 mg/mL) in methanol	HPTLC silica gel	Chloroform–acetone–ammonia (25%) (7 + 1 + 0.1) (v/v/v)	Chamber saturation 60 min; ascending	Scanned in reflectance/absorbance mode at 260 nm	Clotrimazole—$R_F = 0.75$ (2-chlorophenyl)-diphenyl methanol—$R_F = 0.94$ Imidazole—$R_F = 0.02$	Abdel-Moety et al. (2002)

(continued)

TABLE 46.2 (continued)
Chromatographic Systems Used for TLC Analysis of Antifungal Drugs

Drug	Standard Solution	Plates	Stationary-Phase Pretreatment, Mobile-Phase Composition	Development Mode	Detection	Retention Factor	Reference
	Solutions of clotrimazole (2.5 mg/mL) in 1:1 (v/v) chloroform–methanol	HPTLC silica gel	n-Butyl acetate–n-heptane–methanol–diethylamine (3 + 4.5 + 1 + 0.2) (v/v/v/v)	—	—	Clotrimazole—R_F = 0.28 Benzyl alcohol—R_F = 0.48	Čakar et al. (2005)
	2.5% solution of clotrimazole	HPTLC silica gel	Ethylene chloride–acetone (1 + 1) (v/v)—phase A Toluene–methanol (4 + 1) (v/v)—Phase B	—	254 nm	—	Marciniec et al. (2009)
	Solution of ketoconazole (2.5 mg/mL) in methanol	HPTLC silica gel	Methanol–distilled water–triethylamine (70 + 28 + 2) (v/v/v)	Chamber saturation 60 min; ascending	Scanned in reflectance/absorbance mode at 220 nm	Clotrimazole—R_F = 0.66 Degradation products—R_F = 0.28, 0.49	Mousa et al. (2008)
EZ	Solution of EZ nitrate in 1:1 (v/v) chloroform–methanol	HPTLC silica gel	n-Butyl acetate–carbon tetrachloride–methanol–diethylamine (3 + 6 + 2.5 + 0.5) (v/v/v/v)	—	Scanned in reflectance/absorbance mode at 230 nm	Peak 1—26.9 ± 0.3 mm Peak 2—62.7 ± 0.7 mm	Popović et al. (2004)
	Solution of EZ nitrate (5.0 mg/mL) in 1:1 (v/v) chloroform–methanol	HPTLC silica gel	Ethyl acetate–n-hexane–methanol–ammonia–diethylamine (0.5 + 4 + 0.8 + 0.4 + 2) (v/v/v/v/v)	—	Scanned in reflectance/absorbance mode at 230 nm	57.7 ± 0.7 mm	Čakar et al. (2004)
	Solution of EZ nitrate (2.5 mg/mL) in methanol	HPTLC silica gel	Hexane–isopropyl alcohol–triethylamine (80 + 17 + 3) (v/v/v)	Chamber saturation 60 min; ascending	Scanned in reflectance/absorbance mode at 225 nm	MZ nitrate—R_F = 0.68 Degradation products—R_F = 0.36, 0.53	Mousa et al. (2008)

Bifonazole	Solution of bifonazole (5.0 mg/mL) in 1:1 (v/v) chloroform–methanol	HPTLC silica gel	Ethyl acetate–n-hexane–methanol–ammonia–diethylamine (0.5 + 4 + 0.8 + 0.4 + 2) (v/v/v/v/v)	—	Scanned in reflectance/absorbance mode at 250 nm	53.2 ± 0.6 mm	Čakar et al. (2004)
	Solution of bifonazole (1 mg/mL) in methanol	HPTLC silica gel	Hexane–ethyl acetate–methanol–water–glacial acetic acid (42 + 40 + 15 + 2 + 1) (v/v/v/v/v)	—	260 nm	$R_F = 0.70$	Ekiert et al. (2008)
	Solution of bifonazole (1 mg/mL) in chloroform	HPTLC silica gel	10%–90% of organic modifier 1. Acetone–n-hexane 2. Methanol–toluene 3. Methyl ethyl ketone–toluene	—	Exposure to iodine vapor	1. $R_M^0 = 2.22$, $C_0 = -1.39$ 2. $R_M^0 = 2.26$, $C_0 = -1.51$ 3. $R_M^0 = 4.95$, $C_0 = -1.64$	Aleksic et al. (2002)
	Solution of clotrimazole (2.5 mg/mL) in 1:1 (v/v) chloroform–methanol	HPTLC silica gel	Ethyl acetate–n-heptane–methanol–diethylamine (3 + 4.5 + 1 + 0.2) (v/v/v/v)	—	Scanned in reflectance/absorbance mode at 230 nm	Bifonazole— $R_F = 0.32$ Benzyl alcohol— $R_F = 0.45$	Čakar et al. (2005)
	Solution of bifonazole (0.1 mg/mL) in ethanol	HPTLC silica gel	n-Heptane–ethyl acetate–acetone–diethylamine (4.5 + 4.5 + 1 + 0.4) (v/v/v/v)	—	Scanned in reflectance/absorbance mode at 260 nm	47.8 ± 0.8	Agbaba et al. (1991)
Ketoconazole	Solution of ketoconazole (1 mg/mL) in chloroform	HPTLC silica gel	10%–90% of organic modifier 1. Acetone–n-hexane 2. Methanol–toluene 3. Methyl ethyl ketone–toluene	—	Exposure to iodine vapor	1. $R_M^0 = 2.17$, $C_0 = -0.66$ 2. $R_M^0 = 5.61$, $C_0 = -2.25$ 3. $R_M^0 = 11.81$, $C_0 = -1.94$	Aleksic et al. (2002)

(continued)

TABLE 46.2 (continued)
Chromatographic Systems Used for TLC Analysis of Antifungal Drugs

Drug	Standard Solution	Plates	Stationary-Phase Pretreatment, Mobile-Phase Composition	Development Mode	Detection	Retention Factor	Reference
	Solution of ketoconazole (1 mg/mL) in methanol	HPTLC silica gel	Hexane–ethyl acetate–methanol–water–glacial acetic acid (42 + 40 + 15 + 2 + 1) (v/v/v/v/v)	—	230 nm	$R_F = 0.32$	Ekiert et al. (2008)
	Solution of ketoconazole (2.5 mg/mL) in methanol	HPTLC silica gel	Methanol–distilled water–triethylamine (70 + 28 + 2) (v/v/v)	Chamber saturation 60 min; ascending	Scanned in reflectance/absorbance mode at 243 nm	Ketoconazole—$R_F = 0.71$ Degradation products—$R_F =$ 0.40, 0.53, 0.81, 0.89	Mousa et al. (2008)
MZ	Solution of MZ (1 mg/mL) in ethanol	HPTLC silica gel	10%–90% of organic modifier 1. Acetone–n-hexane 2. Methanol–toluene 3. Methyl ethyl ketone–toluene	—	Exposure to iodine vapor	1. $R_M^0 = 2.42$, $C_0 = -1.32$ 2. $R_M^0 = 2.59$, $C_0 = -1.45$ 3. $R_M^0 = 4.87$, $C_0 = -1.60$	Aleksic et al. (2002)
	Solution of clotrimazole (2.5 mg/mL) in chloroform	HPTLC silica gel	n-Butyl acetate–carbon tetrachloride–methanol–diethylamine (3 + 6 + 2.5 + 0.5) (v/v/v/v)	—	Scanned in reflectance/absorbance mode at 230 nm	MZ—$R_F = 0.62$ Benzoic acid—$R_F = 0.33$	Čakar et al. (2005)
	Solution of MZ (2 mg/mL) in methanol	HPTLC silica gel	Toluene–chloroform–methanol (3.0 + 2.0 + 0.6) (v/v/v)	Chamber saturation 10 min, 25°C ± 2°C, relative humidity 60% ± 5%; ascending	Scanned in reflectance/absorbance mode at 240 nm	$R_F = 0.55$	Meshram et al. (2009)

Drug	Sample solution	Stationary phase	Mobile phase	Development	Detection	Results	Reference
	Solution of MZ nitrate (8 mg/mL) in 96% ethanol	HPTLC silica gel	Chloroform–acetone–glacial acetic acid (34 + 4 + 3) (v/v/v)	—	Scanned in reflectance/absorbance mode at 233 nm	MZ nitrate—$R_F = 0.26$ Betamethasone valerate—$R_F = 0.67$	Indrayanto et al. (1999)
	Solution of MZ nitrate (2.5 mg/mL) in methanol	HPTLC silica gel	Hexane–isopropyl alcohol–triethylamine (80 + 17 + 3) (v/v/v)	Chamber saturation 60 min; ascending	Scanned in reflectance/absorbance mode at 225 nm	MZ nitrate—$R_F = 0.90$ Degradation products—$R_F = 0.36, 0.49$	Mousa et al. (2008)
Tioconazole	Solution of tioconazole (1 mg/mL) in methanol	HPTLC silica gel	Organic modifier–methanol (30%–70% in 5% increments) Aqueous components 1. PB, 0.05 M, pH 7.4 2. PBS prepared from PB 0.05 M, pH 7.4 + 0.027 M KCl + 0.137 M NaCl 3. MOPS, 0.02 M, adjusted to pH 7.4 by addition of 1 M NaOH 4. MOPS 0.02 M + 0.2% n-decylamine adjusted to pH 7.4 by addition of 1 M NaOH 5. PB, 0.05 M, pH 11.0 10%–90% of organic modifier	Chamber saturation 60 min; ascending	254 nm	1. $R_M^0 = 5.49$ (± 0.26), $S = -7.26$ (± 0.44) 2. $R_M^0 = 4.85$ (± 0.13), $S = -6.41$ (± 0.22) 3. $R_M^0 = 4.79$ (± 0.05), $S = -5.05$ (± 0.07) 4. $R_M^0 = 7.47$ (± 0.06), $S = -4.90$ (± 0.08) 5. $R_M^0 = 5.78$ (± 0.22), $S = -7.17$ (± 0.35)	Giaginis et al. (2006)
Fluconazole	Solution of fluconazole (1 mg/mL) in chloroform	HPTLC silica gel	1. Acetone–n-hexane 2. Methanol–toluene 3. Methyl ethyl ketone–toluene	—	Exposure to iodine vapor	1. $R_M^0 = 2.75$, $C_0 = -1.51$ 2. $R_M^0 = 2.83$, $C_0 = -2.32$ 3. $R_M^0 = 9.25$, $C_0 = -1.83$	Aleksic et al. (2002)

(continued)

TABLE 46.2 (continued)
Chromatographic Systems Used for TLC Analysis of Antifungal Drugs

Drug	Standard Solution	Plates	Stationary-Phase Pretreatment, Mobile-Phase Composition	Development Mode	Detection	Retention Factor	Reference
	Solution of fluconazole (1 mg/mL) in methanol	HPTLC silica gel	Toluene–chloroform–methanol (1.2 + 3.0 + 0.4) (v/v/v)	Chamber saturation 15 min, 20°C ± 2°C, relative humidity 60% ± 5%; ascending	210 nm	R_F = 0.25 ± 0.024	Meshram et al. (2008b)
	Solution of fluconazole (1 mg/mL) in methanol	HPTLC silica gel	Hexane–ethyl acetate–methanol–water–glacial acetic acid (42 + 40 + 15 + 2 + 1) (v/v/v/v/v)	—	260 nm	RF = 0.43	Ekiert et al. (2008)
	Solution of fluconazole (1 mg/mL) in methanol	HPTLC silica gel	Ethyl acetate–methanol–ammonia–diaminoethane (85 + 10 + 5 + 0.5) (v/v/v/v)	—	216 nm	R_F = 0.4 ± 0.03	Shewiyo et al. (2011)
	1% solution of fluconazole in water	HPTLC silica gel	Chloroform–acetone–methanol–25% ammonia (4 + 4 + 1 + 0.1) (v/v/v/v)	—	254 nm	R_F = 0.53	Marciniec et al. (2007)
	Solution of fluconazole (10 mg/mL) in methanol	HPTLC silica gel	Butanol–water–acetic acid (8 + 2 + 1) (v/v)	Chamber saturation 20 min, 25°C ± 2°C, relative humidity 60% ± 5%; ascending	Scanned in reflectance/absorbance mode at 254 nm	Fluconazole— R_F = 0.67 ± 0.02 Impurity a— R_F = 0.49 ± 0.02 Impurity b— R_F = 0.79 ± 0.02	Ramesh et al. (2011)

Drug	Sample	Stationary phase	Mobile phase	Development conditions	Detection	Results	References
Voriconazole	Solution of voriconazole (1 mg/mL) in methanol	HPTLC silica gel	Toluene–methanol–triethylamine (6 + 4 + 0.1) (v/v/v)	Chamber saturation 15 min, 25°C; ascending	Scanned in reflectance/absorbance mode at 254 nm	$R_F = 0.72 \pm 0.03$	Dewani et al. (2011)
	Solution of voriconazole (1 mg/mL) in methanol	HPTLC silica gel	Methanol–toluene (3 + 7) (v/v)	Chamber saturation 15 min; ascending	Scanned in reflectance/absorbance mode at 255 nm	$R_F = 0.58 \pm 0.02$	Khetre et al. (2008)
Itraconazole	Solution of itraconazole (1 mg/mL) in toluene	HPTLC silica gel	10%–90% of organic modifier 1.Acetone–n-hexane 2. Methanol–toluene 3. Methyl ethyl ketone–toluene	—	Exposure to iodine vapor	1. $R_M^0 = 2.51$, $C_0 = -1.44$ 2. $R_M^0 = 4.83$, $C_0 = -1.64$ 3. $R_M^0 = 9.47$, $C_0 = -1.76$	Aleksic et al. (2002)
	Solution of itraconazole (1 mg/mL) in 3:1 (v/v) methanol–chloroform	HPTLC silica gel	Hexane–ethyl acetate–methanol–water–glacial acetic acid (42 + 40 + 15 + 2 + 1) (v/v/v/v/v)	—	260 nm	$R_F = 0.79$	Ekiert et al. (2008)
	Solution of itraconazole (1 mg/mL) in methanol	HPTLC silica gel	Toluene–chloroform–methanol (5 + 5 + 1.5) (v/v/v)	—	260 nm	$R_F = 0.52 \pm 0.02$	Parikh et al. (2011)
Terbinafine	Solution of terbinafine (1 mg/mL) in methanol	HPTLC silica gel	n-Hexane–acetone–glacial acetic acid (8 + 2 + 0.1) (v/v/v)	—	Scanned in reflectance/absorbance mode at 223 nm	$R_F = 0.43$	Suma et al. (2011)

(continued)

TABLE 46.2 (continued)
Chromatographic Systems Used for TLC Analysis of Antifungal Drugs

Drug	Standard Solution	Plates	Stationary-Phase Pretreatment, Mobile-Phase Composition	Development Mode	Detection	Retention Factor	Reference
	Solution of terbinafine (100 mg/mL) in methanol	HPTLC silica gel	Toluene–ethyl acetate–formic acid (4.5 + 5.5 + 0.1) (v/v/v)	Chamber saturation 15 min, 25°C ± 2°C, relative humidity 60% ± 5%; ascending	Scanned in reflectance/absorbance mode at 284 nm	$R_F = 0.31 \pm 0.02$	Ahmad et al. (2009)
Tolnaftate (thiocarbamate)	Tolnaftate topical solution diluted with methanol	HPTLC silica gel	Toluene–chloroform (4 + 1) (v/v)	Ascending	Scanned in reflectance/absorbance mode at 248 nm	$R_F = 0.59$	Meshram et al. (2008a)
Griseofulvin	Extracted from Penicillium urticae with chloroform	HPTLC silica gel	Chloroform–acetone (93 + 7) (v/v)	—	UV light, vis light after being sprayed with 50% H_2SO_4, and heated at 110°C for 30 min	$R_F = 0.65$	Cole et al. (1970)

how to perform validation. Criteria that need to be successfully fulfilled are selectivity, precision and linearity in defined range, and accuracy. Usually, there is also need to establish a limit of detection and quantification, robustness, and ruggedness and perform system suitability testing. All analytical procedures mentioned in this chapter were successfully validated. Analytical parameters of the stated methods are presented in Table 46.3.

46.1.1 POLYENE MEMBRANE DISRUPTORS: AMPHOTERICIN B, NYSTATIN, AND CONGENERS

Before mid-1950s, effective antifungal therapy was limited to topical applications of undecylenic acid derivatives, mixtures of benzoic acid and salicylic acid, and a few other agents of modest efficacy. No reliable treatments existed for the few cases of deep-seated systemic fungal infections that did occur. The discovery of the polyene antifungal agents, however, provided a breakthrough into both a new class of antifungal agents and the first drug to be effective against deep-seated fungal infections.

Polyenes are macrocyclic lactones with distinct hydrophilic and lipophilic regions. The hydrophilic region contains several alcohols, a carboxylic acid, and, usually, a sugar. The lipophilic region contains, in part, a chromophore of four to seven conjugated double bonds. The number of conjugated double bonds correlates directly with antifungal activity in vitro and, inversely, with the degree of toxicity to mammalian cells. The chemical structures of polyene antifungal drugs are presented in Figure 46.1.

In the review by Thomas (1976), the chemical and biological properties of amphotericin B, nystatin, and natamycin are examined in order to show how these properties can be used for the analysis and assay of the antibiotics. This review summarized all chemical analysis (paper, thin layer, column, and GC) of these compounds that have been carried out by a number of investigators in the period before 1976.

Thomas pointed at the considerable variation between the results of different groups as a reflection of the variation in the purity of the antibiotic and the experimental conditions used. Also, in all cited papers, chromatographic methods are rarely described in detail. Six different thin-layer chromatographic systems on silica gel G as stationary phase are presented for nystatin and natamycin (R_F values ranging from 0.18 to 0.66, i.e., 0.18 to 0.55, respectively) and seven chromatographic systems for amphotericin B (R_F values ranging from 0.07 to 0.60). Chromatographic conditions used for these separations are presented in Tables 46.2 and 46.3. All of the methods reviewed utilized tertiary and quaternary solvent systems.

Thomas emphasized on several factors negatively affecting the separation of polyenes, such as tailing, the heterogeneous nature of polyenes, degradation of substances during separation, unsuitability of the chromatographic methods for insoluble compounds, and possible complexation of polyenes with divalent metals present in the stationary phase.

Kelly performed a separation of amphotericin B components by combined TLC and HPLC. Contrary to chromatographic systems used in previously reported papers, in this study, amphotericin B was separated into five components by TLC using only methanol as a developing solvent. Namely, amphotericin B is not a homogeneous substance; it contains a main heptaene and several other heptaene and tetraene minor components. In order to ascertain the purity of the five amphotericin components, the TLC bands were subjected to further separation by HPLC. Besides the two major components found in other studies, present in the TLC bands 2 and 3, amphotericin B was also found to contain five other significant heptaenes, one tetraene, two fluorescent polyenes (pentaene and heptaenes), and at least five other significant components, which had appreciable absorbance at approximately 305 nm. It was concluded that all found compounds are similar in their hydrophilic characteristics but widely different in their hydrophobic properties.

The objective of the study of Fittler et al. (2010) was to evaluate formerly documented TLC methods for the analysis and separation of amphotericin B, its main and minor components, published by Thomas, and to develop a direct bioautographic method with the objective of measuring the antifungal activity of the components in the substances. It was concluded that by the use of optimized mobile phase, two components of amphotericin B samples can be separated on silica gel layers. Both the observed components have antifungal effect. According to these experiments, a mixture of

TABLE 46.3
Analytical Parameters of the Methods for Antifungal Drugs

Drug	Calibration Mode/Range	LOD (Unit/Spot, Band)	LOQ (Unit/Spot)	Precision (RSD%)	Recovery (%)	Matrix	Reference
Amphotericin B	Linear, 0.5–1000.0 ng/spot	Direct bioautography—0.8 ng / Densitometry—8 ng UV lamp—50 ng	—	—	—	Vetranal / Fungizone (fresh) / Fungizone (stored/degraded)	Fittler et al. (2010)
Clotrimazole	Second-degree polynomial, 5–25 µg/spot	2.5 µg	—	—	99.55% ± 0.74%	Cream / Vaginal tablets / Solution	Abdel-Moety et al. (2002)
	Linear, 1.0–2.0 µg/spot	0.03 µg	0.10 µg	2.60–3.51	100.7–101.2	Cream	Čakar et al. (2005)
	Linear	0.66 µg/20 µL	2.20 µg/20 µL	1.211	100.27 ± 1.114	Cream	Mousa et al. (2008)
EZ	Second-degree polynomial, 1–10 mg/mL			EZ base—1.78 / Nitrate—3.85	EZ base—100.8 / Nitrate—98.6	Vaginal pessaries / Spray solution	Popović et al. (2004)
	Second-degree polynomial, 500–1500 ng/spot	—	—	2.82	98.3	Cream	Čakar et al. (2004)
	Linear	1.39 µg/20 µL	4.63 µg/20 µL	1.478	100.21 ± 1.084	Cream	Moussa et al. (2008)
Bifonazole	Second-degree polynomial, 500–1500 ng/spot	—	—	2.29	100.2	Cream	Čakar et al. (2004)
	Linear, 0.5–10.0 µg/spot	0.03 µg	0.10 µg	1.31	—	—	Ekiert et al. (2008)
	Linear, 1.0–2.0 µg/spot	0.03 µg	0.10 µg	1.39–2.20	99.4–99.8	Cream	Čakar et al. (2005)
	Linear, 10.0–100.0 ng/spot	—	—	—	98–105	Lotion / Cream	Agbaba et al. (1991)

Drug	Linear range	LOD	LOQ	RSD	Recovery (%)	Sample	References
Ketoconazole	Linear, 0.5–10.0 µg/band	0.3 µg	0.5 µg	2.61	100.2–101.7	Tablets	Ekiert et al. (2008)
	Linear	1.3 µg/20 µL	4.37 µg/20 µL	1.638	99.89 ± 1.16	Tablets Cream	Mousa et al. (2008)
MZ	Linear, 2.0–3.0 µg/spot	0.1 µg	0.4 µg	0.72–3.08	100.3–100.9	Gel	Čakar et al. (2005)
	Linear, 600–1400 ng/spot	—	—	0.891–1.465	99.63 ± 1.46	Cream	Meshram et al. (2009)
	Linear, 5.3–16.0 µg/spot	0.68 µg	2.04 µg	0.89–1.67	100.49 ± 1.36	Cream	Indrayanto et al. (1999)
Fluconazole	Linear	1.34 µg/20 µL	4.47 µg/20 µL	2.050	100.00 ± 1.097	Cream	Mousa et al. (2008)
	Linear, 0.6–1.6 µg/spot	—	—	1.064–1.069	98.58–101.04	Tablets Capsule	Meshram et al. (2008b)
	Linear, 10.0–50.0 µg/band	5.0 µg	8.0 µg	3.45	99.7–102.4	Tablets	Ekiert et al. (2008)
	Linear	—	—	3.05	99.1 ± 2.1	Tablets	Shewiyo et al. (2011)
	Fluconazole—linear, 1–6 µg/spot	Fluconazole—0.031 µg	Fluconazole—0.098 µg	Fluconazole—0.36	Fluconazole—100.8 ± 2.93	Tablets	Ramesh et al. (2011)
	Impurity a—linear, 0.5–2.5 µg/spot	Impurity a—0.016 µg	Impurity a—0.045 µg	Impurity a—0.73	Impurity a—101.7 ± 3.00		
	Impurity b—linear, 0.5–2.5 µg/spot	Impurity b—0.014 µg	Impurity b—0.042 µg	Impurity b—0.77	Impurity b—101.5 ± 3.14		
Voriconazole	Linear, 50–400 ng/spot	—	50 ng	6.02	98.82	Human plasma	Dewani et al. (2011)
Itraconazole	Linear, 200–1000 ng/spot	12.05 ng	36.55 ng	<1.5	98–102	Tablets	Khetre et al. (2008)
	Linear, 0.5–10.0 µg/spot	0.10 µg	0.25 µg	2.19	—	—	Ekiert et al. (2008)
Terbinafine	Linear, 1000–6000 ng/spot	180.29 ng	546.34 ng	NMT 2	98–102	Capsule	Parikh et al. (2011)
	Linear, 200–1000 ng/spot	1.204 ng	3.648 ng	0.07320–0.6233	99.74 ± 0.2627	Tablets	Suma et al. (2011)
	Linear, 200–1000 ng/spot	10.5 ng	35.0 ng	1.08–2.12	97.8%–100.5%	Tablets Cream	Ahmad et al. (2009)
Tolnaftate (thiocarbamate)	Linear, 100–400 ng/spot	—	—	0.118	99.59 ± 0.18	Topical solution	Meshram et al. (2008a)

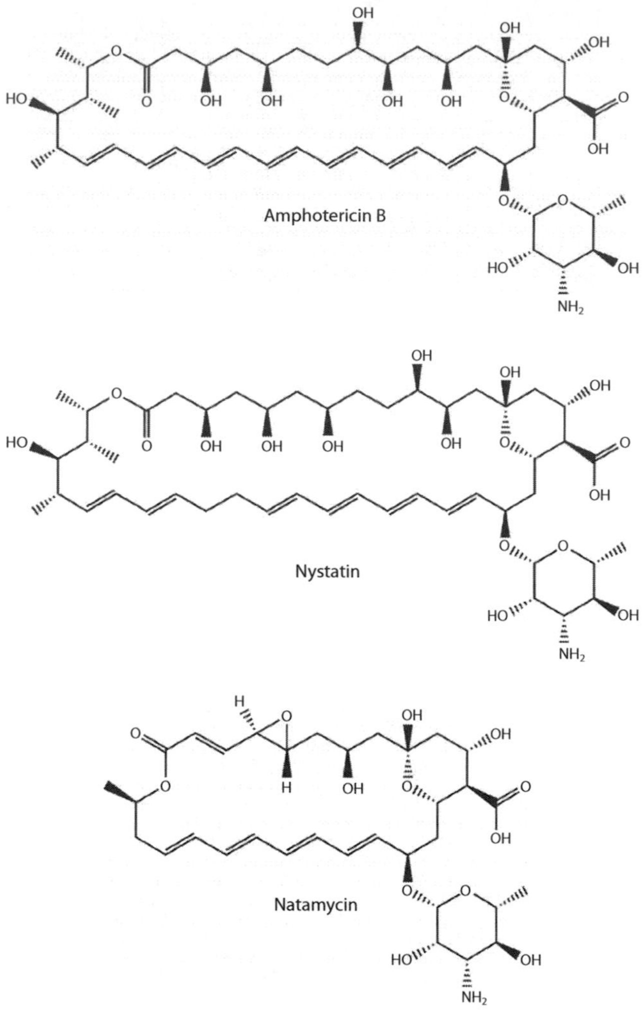

FIGURE 46.1 Structures of polyene antifungal drugs.

degradation products is produced during the storage of amphotericin B solutions. The degradation affects chromatographic behavior but does not alter antifungal effect strikingly. Also, among the observed detection methods (ultraviolet [UV] visualization, densitometry, and direct bioautography with *Candida albicans*), direct autobiography proved to be most sensitive.

46.1.2 ERGOSTEROL BIOSYNTHESIS INHIBITORS: AZOLES, IMIDAZOLES, AND TRIAZOLES

Fungicidal properties of azole agents were discovered in 1944 by D.W. Woolley. Chlormidazole is the first compound in this group that has been marketed and used in medical practice since 1958. It was recommended for topical application. Investigations led to the registration of a variety of azole drugs.

Azole antifungal drugs are synthetic compounds with broad-spectrum fungistatic activity. The characteristic chemical feature of azoles, from which their name is derived, is the presence of a five-membered aromatic ring containing either two or three nitrogen atoms. Imidazole rings have two nitrogens and three triazoles. In both cases, the azole ring is attached through N1 to a side chain containing at least one aromatic ring. Triazoles exhibit wider antifungal activity spectrum, better safety profile, and higher efficacy, compared with imidazole derivatives so far used. Imidazole- and triazole-containing agents are shown in Figures 46.2 and 46.3, respectively.

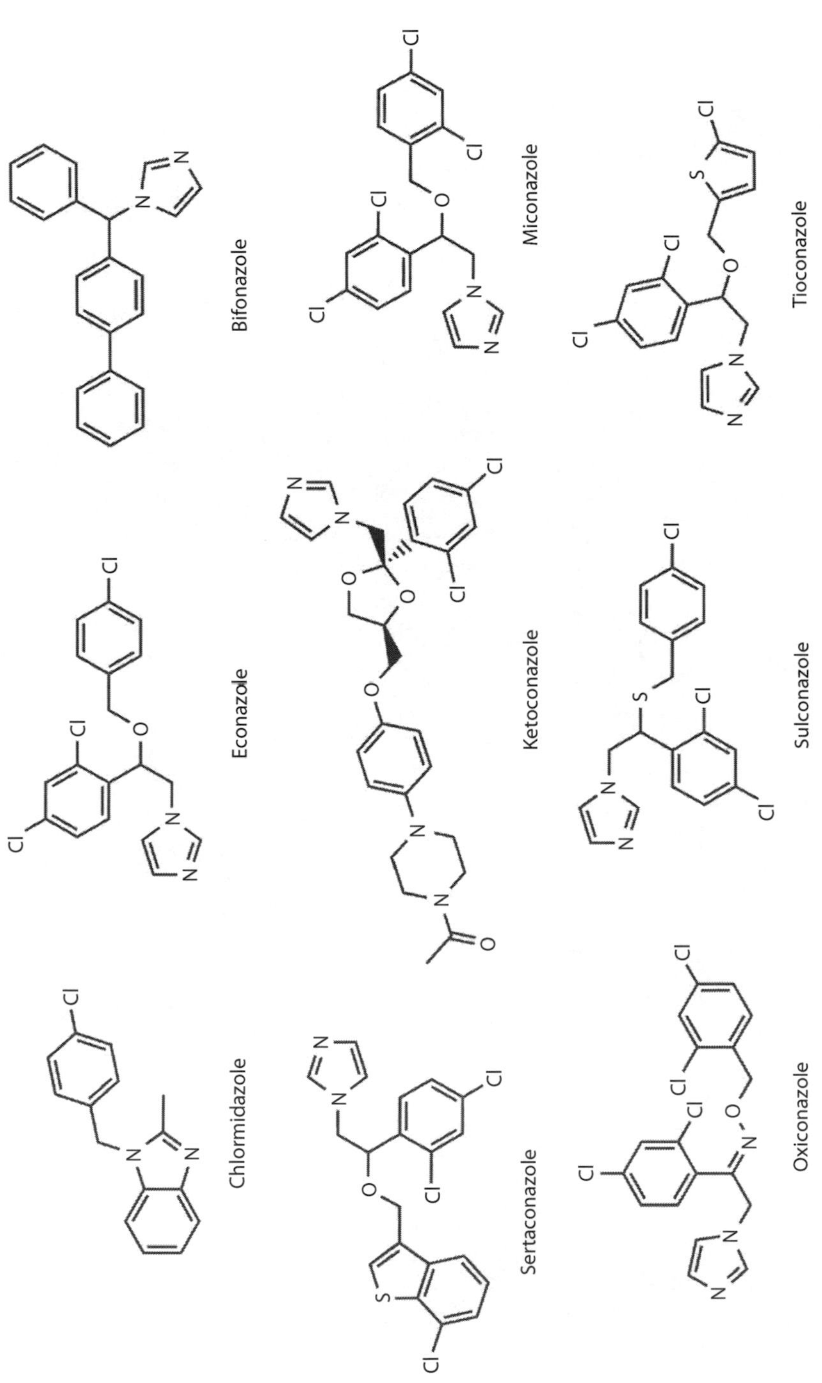

FIGURE 46.2 Structures of imidazole antifungal drugs.

(continued)

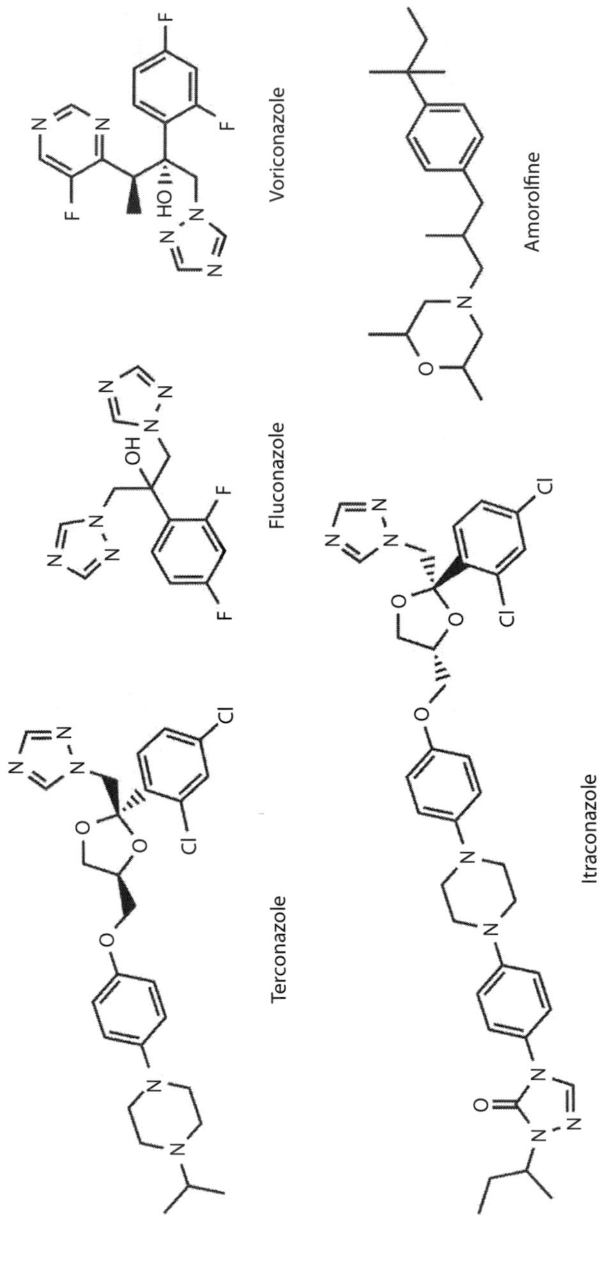

FIGURE 46.2 (continued)   Structures of imidazole antifungal drugs.

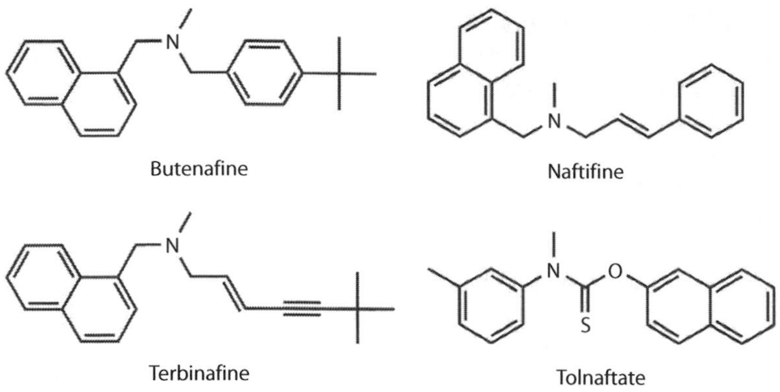

Butenafine

Naftifine

Terbinafine

Tolnaftate

FIGURE 46.3 Structures of triazole antifungal drugs.

The action of azole derivatives is based on the inhibition of biosynthesis of ergosterol in different stages. Ergosterol is the main component of fungal cell membrane. Its essential function is to be a bioregulator of the membrane fluidity, asymmetry, and integrity. This is the target of azole derivatives.

There are numerous papers describing the TLC determination of azole antifungal agents. The aim of these studies was to establish the identification and quantitative determination conditions for the analysis of the individual drugs and potentially their degradation products or identification, separation, and quantitative determination conditions for the simultaneous analysis of several antifungal compounds. Literature search revealed that the following azole antifungal agents are analyzed by TLC: imidazole drugs (clotrimazole, econazole [EZ], bifonazole, ketoconazole, miconazole [MZ], tioconazole), and triazole drugs (fluconazole, voriconazole, and itraconazole).

Several papers, dealing with the determination of individual azole antifungal agents, have been published. The first article on TLC of EZ nitrate was represented by Popović et al. (2004). They developed HPTLC and HPLC methods for the determination of EZ nitrate in pharmaceuticals and estimated their reliability by comparison of the results obtained by the two approaches. Chromatograms obtained by HPTLC indicate that the existence of two peaks corresponds to nitrate and EZ free base. Namely, diethylamine, used as a component of the mobile phase, converted the EZ salt into free form. HPLC chromatograms on polar amino column showed one peak ascribed to EZ, while nitrate peak was not detected. It was concluded that both methods can be successfully used for mentioned determination because there were no significant differences between results. Authors have, also, emphasized the advantages of HPTLC, that is, furnishing of two separate peaks for EZ base and nitrate, the possibility of high sample throughput, and simple and rapid sample preparation without tedious and time-consuming isolation procedures.

A simple and rapid densitometric method for routine quantitation of bifonazole in pharmaceutical formulations was presented by Agbaba et al. (1991). Its application to cream and lotion was described in detail. The sample pretreatment procedure used with established method eliminates the column chromatography extraction step, often an essential part of the analysis of creams and ointments.

A new simple, precise, accurate, and inexpensive HPTLC method was developed for the determination of ketoconazole (Saysin et al. 2010). The selected chromatographic system has been successfully applied to the determination of ketoconazole in various pharmaceutical dosage forms, shampoos, and creams. It was observed that common excipients in formulations do not interfere. The authors suggested that it should be used for routine analysis.

A new HPTLC method for the determination of fluconazole in pharmaceutical formulations (tablets and capsule), using an internal standard, was developed by Meshram et al. (2008b). The use of an internal standard helps eliminate analytical errors arising as a result of dilution of drugs. The proposed method, which was established and validated, using clotrimazole as internal standard, is based on HPTLC separation of the two compounds in normal-phase mode using three-component mobile phase, followed by densitometric measurement of the spots. The method was validated for accuracy, precision, specificity, and linearity. The results from tablet and capsule indicated satisfactory accuracy and precision. Excipients in both formulations did not interfere with the determination of the drug. The method can also be used for analysis of fluconazole in biological samples and in formulations containing two or more different antifungal agents.

The previously described method for the determination of fluconazole used mobile phase in which chloroform forms about 65%. Chloroform is considered an environmentally unfriendly chemical, and its use in routine applications may contribute to health and environmental hazards. Other published analytical methods dealing with this subject were developed for the simultaneous analysis of several azole drugs. Shewiyo et al. (2011) described the development and validation of an improved method for the analysis of fluconazole using HPTLC, required by the laboratory of Tanzania Food and Drugs Authority. The proposed method used four-component mobile phase consisting of no hazardous substances and specific only for fluconazole. It was successfully used to analyze the content of mentioned triazole antifungal drug in marketed tablet samples.

The first HPTLC method for the analysis of voriconazole was reported by Khetre et al. (2008). The method has been successfully applied in the analysis of marketed formulation. Voriconazole displays nonlinear pharmacokinetics in adults but has linear pharmacokinetics in children. Interindividual variability is generally high, both in children and adults, and diverse manifestations of toxicity are possibly attributed to high drug concentrations. This indicates the need to monitor voriconazole concentration in plasma after oral dose. A simple, selective, and sensitive HPLC for the determination of voriconazole in human plasma was developed and validated by Dewani et al. (2011). After the precipitation of plasma proteins with acetonitrile, the protein-free supernatant was chromatographed in normal-phase mode. Cephalexin was used as an internal standard. The results of validation proved that the developed method performs well with selectivity, precision, accuracy, stability, and linearity for the concentration range of voriconazole to be found in human plasma. The validated method covers a wide range of linearity and is therefore suitable for the determination of voriconazole in human plasma at different therapeutic dose levels. The present method involves minimal sample pretreatment, resulting in fast analysis. Also, it utilized protein precipitation as the sample preparation technique, which eliminates the drawbacks of less recovery due to liquid–liquid extraction or the use of solid-phase extraction cartridges, which is relatively costly. The HPTLC technique offers advantage of high throughput. As compared to Liquid chromatography-tandem mass spectrometry (LC-MSMS) and HPLC methods, the present method is economical, simple, and fast. The proposed method can be used for therapeutic drug monitoring in order to optimize drug dosage on an individual basis.

A simple, rapid, and sensitive HPTLC method for the quantification of itraconazole in raw material and pharmaceutical formulations has been reported by Parikh et al. (2011). Separation was achieved on silica gel using tricomponent mobile phase. The drug was not degraded under neutral and alkaline hydrolysis, UV and photolytic degradation, underelevated temperature, and humidity. Itraconazole is degraded under acidic hydrolysis and oxidative condition, but the degradation products were well resolved from individual bulk drug response. It was concluded that the developed method can effectively resolve drug from its excipients in capsule dosage form, and it can be used as a stability-indicating method. The specificity of the method was confirmed by peak purity of resolved peak.

The relative aqueous solubility of some azole antifungal agents allows it to be administered orally or intravenously. Therefore, it must be of appropriate chemical and microbiological purity as required by various pharmacopeias. These pharmacopeias accept the following methods of sterilization: saturated steam, dry heat, chemical sterilization (ethylene oxide), filtration, and ionizing radiation. In order to test whether a fluconazole can be sterilized by ionizing radiation, that is, could it retain its physicochemical and pharmacological properties, Marciniec et al. (2007) used a number of qualitative and quantitative methods such as scanning electron microscopy (SEM), nuclear magnetic resonance (NMR), UV and infrared (IR) spectroscopy, TLC and HPLC, and organoleptic analysis.

The TLC analysis conducted in four-component mobile phase reveals that one product of radiolysis was found, with R_F values of 0.80, for a sample irradiated to 100 kGy, and two, with R_F values of 0.77 and 0.82, for a sample irradiated to 200 kGy. Samples of fluconazole that received doses smaller than 100 kGy did not show any radiolysis product. Similar results were obtained with HPLC method. The remaining analytical methods (SEM, IR, and NMR) did not provide any conclusive information in respect of radiological stability of fluconazole. According to the obtained results, the authors concluded that fluconazole is a compound of low radiological stability and should not be sterilized using gamma, beta, or E-beam radiation. Similar results were previously published by Marciniec et al. (2004). They examined radiochemical stability of three fluorine-containing therapeutic substances: dexamethasone, fludrocortisone acetate (steroid derivatives), and fluconazole (azole derivative) using spectrophotometric (UV and IR) and chromatographic (TLC and HPLC) methods. It was found that fluconazole was too sensitive to electron beam irradiation and should be sterilized by other methods.

The effect of ionizing irradiation on clotrimazole has also been studied (Marciniec et al. 2009). The compound was subjected to ionization irradiation in the form of high-energy electron beam (25–800 kGy) from an accelerator. Before and after the irradiation, clotrimazole was analyzed by EPR, TLC, HPLC, and HPLC–MS. Chromatographic analysis by the TLC method performed with four-component mobile phases proved that before irradiation, clotrimazole was chromatographically pure, while after the irradiation with the doses 25–400 kGy, the results indicated the presence of one product of radiolysis of $R_F = 0.83$ (mobile phase A, presented in Table 46.2) or $R_F = 0.52$ (mobile phase B, presented in Table 46.2). The presence of the second product of radiolysis characterized by $R_F = 0.89$ (mobile phase A) was observed only after the irradiation with 800 kGy. On the basis of the HPLC–MS data, the main product of radiolysis is 1-(9-phenylfluoren-9-yl)-imidazole. Besides traces of (2-chlorophenyl)-diphenylmethanol, other impurities listed in the European Pharmacopoeia (European Pharmacopoeia 2008) have not been detected. Clotrimazole has been found to show relatively high resistance to ionizing irradiation (greater than fluconazole) and probably will be suitable for radiation sterilization but with doses lower than 25 kGy.

The imidazole antifungal derivatives are susceptible to degradation due to the effect of temperature, oxygen, light, and pH, and the degraded products may be pharmacologically inactive. Aqueous solutions of ketoconazole were investigated for stability (Enayatifard et al. 2005). It was refluxed at acidic and alkaline conditions. The degradation product(s) was separated by preparative TLC method. The product(s) structure was determined by Fourier transform infrared spectroscopy (FT-IR) and proton nuclear magnetic resonance (1HNMR). The results showed that after degradation of ketoconazole in acidic and alkaline conditions, only one compound was produced. The only degradation pathway of ketoconazole is deacetylation of molecule, and thus deacetyl ketoconazole was produced.

Clotrimazole is stable in alkaline medium but hydrolyzes in acid medium to (2-chlorophenyl)-diphenyl methanol and imidazole. A coupled TLC-densitometric method has been applied by Abdel-Moety et al. (2002) as a stability-indicating method for separation and quantification of clotrimazole alone or in the presence of by-product impurities and/or its acid degradation products, (2-chlorophenyl)-diphenyl methanol and imidazole. The results were evaluated and compared with those obtained by the official and reference methods. It has been seen that the proposed method is sufficiently accurate and quite reproducible for selective determination of the studied drug.

In addition, samples of fluconazole may contain some structurally related impurities derived from the manufacturing process, such as 2-(2-fluoro-4-(1H-1,2,4-triazol-1-yl) phenyl)-1,3-di (1H-1,2,4-triazol-1-yl)propan-2-ol (Impurity a), 1-(2,4-difluorophenyl)-2-(1H-1,2,4-Triazol-1-yl) ethanone (impurity b). The first HPTLC method for the simultaneous separation of fluconazole from its structurally related impurities has been reported by Ramesh et al. (2011). The selected chromatographic system offered compact spots for fluconazole from its impurity b, which is clearly visible under UV light, but impurity a is inactive under UV light, and it was visible when exposed to iodine vapor. The developed HPTLC method is suitable not only for separation and quantitative determination of active drug ingredient and the impurities to monitor the synthetic reactions but also for quality assurance of fluconazole in the presence of its structurally related impurities. The method was validated according to ICH guidelines and shown to be selective, repeatable, rapid, and accurate within the established ranges. Some antimycotic formulations, such as creams or gels, contain preservatives, as well. According to good manufacturing practice regulations, active substances and preservatives used for human drug formulations must be determined. Simultaneous determination of antimycotics and preservatives using TLC was described in only two papers. The first paper was presented by Čakar et al. (2004) and describes simple and specific HPTLC method for simultaneous determination of bifonazole and EZ and methyl- and propylparaben in medicinal creams using chromatographic system that consists of silica gel and five-component mobile phase. Based on the obtained results, it was concluded that direct densitometric determination is possible without previous isolation of the compounds of interest. Continuing their research, Čakar et al. (2005) also find optimal chromatographic conditions for simultaneous determination of the following pairs of antimycotics and preservatives: bifonazole and benzyl alcohol, clotrimazole and benzyl alcohol, and MZ and benzoic acid. The preparation of cream samples for the analysis was simple, enabling a direct densitometric determination without extraction. The determination of MZ and benzoic acid in a carbomer-based gel was performed after the extraction from an alkaline, that is, acidified medium, respectively. In all cases, a satisfactory separation of the two components was obtained.

Pharmacopeia methods do not report any method for the determination of these drugs either in the presence of their degradation products or in the presence of corticosteroid hormones. Mousa et al. (2008) developed a rapid, precise, and selective densitometric and HPLC method for the determination of some antifungal drugs—ketoconazole, clotrimazole, MZ nitrate, and EZ nitrate—in pure form or in the presence of their degradation products. A complete degradation of clotrimazole, MZ nitrate, and EZ nitrate was achieved by refluxing with hydrochloric acid for 5 h. On the other hand, the degradation products of ketoconazole were isolated using preparative TLC plates after refluxing for at least 5 h with hydrochloric acid. The method was based on different R_F values of the drug of choice and its products of acid degradation. The methods were successfully applied to determine these drugs in pharmaceutical dosage forms and in the presence of some corticosteroid hormones.

Several papers dealing with simultaneous analysis of several antifungal compounds have been published. TLC densitometry was used to separate, identify, and quantitate clotrimazole, MZ, and ketoconazole (alone or combined with other drugs) in various pharmacopeias or proprietary creams and ointments (Roychowdhury and Das 1996). Clotrimazole was extracted from the cream or ointment with ethyl alcohol, and MZ and ketoconazole were extracted from a mixture of equal volumes of chloroform and isopropyl alcohol. Active ingredients were separated from excipients and other drugs by TLC. The three azoles were well separated and easily identified in this chromatographic system. The separated azoles were visualized under short-wave UV light and quantitated by scanning densitometry at 220 nm by comparing the integrated areas of samples with those of standard (one azole was used as internal standard for the other). Recoveries from samples spiked with known amounts of azoles were excellent. The method was validated further by comparison with official liquid chromatographic methods.

The work performed by Ekiert et al. (2008) aimed to establish the identification, separation, and quantitative determination conditions for the simultaneous analysis of four antifungal compounds (ketoconazole, bifonazole, fluconazole, and itraconazole) for routine pharmaceutical monitoring using TLC with densitometric detection. All the papers where TLC has been used did not cover the combination of these drugs. An analysis under the established conditions for a mixture of the investigated compounds enabled both complete separation and identification based on R_F values. In previous analytical procedures, densitometric scanning was carried out at a wavelength of 220 nm (Roychowdhury and Das 1996), 230 nm (Čakar et al. 2005), or 250 nm (Čakar et al. 2004), but in this paper, 260 nm was considered to be the best one. At this wavelength, the risk of interferences is significantly lower, and it is the absorption maximum for three of the four investigated compounds (the exception being ketoconazole, 230 nm). The method parameters obtained from validation indicate its suitability for routine analysis of medicinal products.

In the publication of Selcuk et al. (1995), qualitative and quantitative studies have been performed on a variety of compound preparations that contain fluconazole and itraconazole. Fluconazole and itraconazole have been analyzed qualitatively along with other antifungal imidazole derivatives using TLC and HPLC. In TLC, 20 different solvent systems have been tried, and suitable RF values have been detected for active ingredients. In these studies, differentiating reagents have been used for active ingredients. In studies made by using HPLC method, fluconazole, itraconazole, clotrimazole, MZ nitrate, oxiconazole nitrate, ketoconazole, and tioconazole have been analyzed under the working conditions developed by us. Only the retention times of tioconazole and itraconazole have been found to be very close to each other. In quantitative studies, UV spectroscopy, HPLC, and densitometric methods have been used. In UV spectroscopy method, the absorbance value at 261 nm in methanol has been used for both of the active compounds. Quantitative determinations have been performed by using the regression equations calculated before. In quantitative determinations done with HPLC, ketoconazole has been used as the internal standard. Silica gel-coated plates have been applied in densitometric studies, and the regression equations calculated by using the peak areas of the active compound applied at varying concentrations, measured at 261 nm, have been used for quantitative calculations. The results obtained with three different methods have been compared using Student's t and Fisher's F tests.

Two simple, sensitive, and accurate methods were adopted for simultaneous determination of EZ nitrate–triamcinolone acetonide (TA) and MZ nitrate–hydrocortisone (HC) in binary drug mixtures (Marciniec et al. 2009). The first method depends on TLC densitometry via direct scanning of the analyzed drug mixtures at the corresponding wavelengths. The separation was carried out on precoated thin-layer chromatographic plates, silica gel 60 F254 with the developing mobile phase (chloroform–methanol, 10 + 0.5, v/v). The second method was based on the HPLC separation of the analytes on a column with a mobile phase consisting of methanol–water–acetonitrile (65 + 10 + 25, v/v/v), pH 7 followed by UV detection at 230 nm. The two proposed methods were successfully applied for the analysis of drugs under investigation in laboratory-prepared mixtures with corticosteroids and in their pharmaceutical dosage forms. In addition, these methods can be applied as stability-indicating techniques for the determination of EZ and MZ in the presence of their acid degradation products.

Indrayanto et al. (1999) published a simple densitometric method for routine analysis of the combination of betamethasone valerate and MZ nitrate, the active ingredients in cream pharmaceutical preparations. The TLC plate must be eluted two times due to the very low migration distance ($R_F = 0.06$) of MZ nitrate by the first elution. After second elution, the densitogram showed the spots of MZ nitrate and betamethasone valerate well separated from the spot of nipagin and nipasol. These preservatives could not be separated in selected system, so it appeared as one peak. All the relative standard deviation (RSD) of the repeatability and intermediate precession evaluations have values less than 2% and demonstrated that the accuracy and precision of the proposed method were satisfactory. Therefore, the proposed method is suitable for routine analysis of cream and products of similar composition in pharmaceutical industry's quality-control laboratories.

In the work of Meshram et al. (2008c), simultaneous analysis of clotrimazole and metronidazole in combined-dose tablets and cream was achieved for the first time by the use of densitometric TLC. The compounds were separated in normal-phase mode. Both peaks were symmetrical in nature, and no tailing was observed. Studies were also conducted to discover how accurately and specifically the analytes of interest were analyzed in the presence of other components (e.g., impurities and degradation products). The results from specificity studies showed that metronidazole undergoes marked degradation when exposed to alkaline, acidic, and oxidizing conditions for 24 h. Similarly, clotrimazole was degraded under acidic conditions. Metronidazole was found to be stable to heat and UV exposure. Although no extra peaks were observed in the densitograms, there were marked decreases in the peak areas of the drugs under some degradation conditions. It can therefore be inferred that the degradation products formed may be remaining at the origin or moving with the solvent front. Another possibility is that any degradation product is nonchromophoric or may not have substantial UV absorption at the wavelength used for analysis. Therefore, the proposed method could be successfully applied to pharmaceutical formulations.

Planar chromatography has been widely used to estimate the hydrophobicity of active drugs. Lipophilicity is one of the most commonly used physicochemical properties in quantitative structure-activity relationships (QSAR) and quantitative structure-retention relationships (QSRR) studies because it affects the permeation capacity of bioactive drugs through the hydrophobic membrane of the cells of the target organism. The property usually used for the characterization of hydrophobicity, from the date obtained by RP-TLC, is R_M^0, from the linear relationships between the solute R_M values and the concentration of organic modifier, and $C_0 = -R_M^0/m$, where m is a slope from previously mentioned relation.

Aleksic et al. (2002) examined the retention behavior of bifonazole, clotrimazole, fenticonazole, fluconazole, ketoconazole, MZ, metronidazole, and itraconazole, widely used antimycotic drugs, by TLC. The possible application of planar chromatography in QSAR studies of antimycotics was investigated by correlating lipophilicity parameters, R_M^0 and C_0, with calculated log P values. The obtained satisfying correlations indicate that TLC can be used to estimate the hydrophobicity of investigated drugs. Also, the linear correlation of C_0 with the antimicrobial activity of ketoconazole, MZ, fluconazole, clotrimazole, and itraconazole indicates the applicability of this measure of hydrophobicity in QSAR studies of antimycotic compounds. The best correlation equation was obtained for the binary mobile phase methanol–toluene.

The reversed-phase TLC retention behavior of a series of structurally diverse drugs, mostly basic compounds (including tioconazole), has been investigated with different aqueous mobile-phase components, by Giaginis et al. (2006). Phosphate buffer (PB), phosphate-buffered saline (PBS), and morpholinepropanesulfonic acid (MOPS), with or without an addition of n-decylamine, at pH 7.4, and PB at pH 11.0, were used with different proportions of methanol as the mobile phase. The effect of the buffer constituents on extrapolated R_M^0 values and the corresponding slopes, S, and their interrelationship was evaluated. The different sets of R_M^0 values were correlated with lipophilicity log P and log $D_{7.4}$. MOPS was found to be more suitable than PB for assessment of lipophilicity. The use of log D instead of log P, combined with the use of the ionization-correction term Q, led to improved correlation and revealed a reduced net effect of ionization on retention.

The development of resistance to multiple drugs is a major problem in the treatment of microbial infectious diseases. Multidrug-resistant *Staphylococcus aureus* (MRSA) and *Candida* sp., the major infectious agents, have been recently reported in quite a large number of studies. With more intensive studies for natural therapies, marine-derived products have been a promising source for the discovery of novel bioactive compounds. In this respect, a total of 45 marine fungi were isolated from the two sponges *Fasciospongia cavernosa* and *Dendrilla nigra* and screened for antimicrobial activity against pathogenic bacteria and fungi (Manilal et al. 2010). The novel basal media formulated in the present study resulted in increased frequency of fungal isolates when compared to all other media used in the present study. The cell-free supernatant of fungi exhibiting the broad spectrum of activity was subjected to chemical analysis using different chromatographic systems including TLC, column, and gas chromatography–mass spectrometry (GC–MS). Of the 15 fungal strains, 20% (3 strains) showed potential antagonistic activity against a panel of clinical pathogens used in the study. Based on the antimicrobial activity of the isolates, *Aspergillus clavatus* MFD15 was recorded as potent producer displaying 100% activity against the tested pathogenic organisms. The TLC of the crude ethyl acetate extract produced three spots with R_F values of 0.20, 0.79, and 0.95, respectively. The active TLC fraction was purified in column chromatography, which yielded 50 fractions. The active column fractions were combined and analyzed with FT-IR, UV–Vis, and GC–MS. The chemical analysis of the active compound envisaged the active compound to be a triazole, 1H-1,2,4-triazole-3-carboxaldehyde-5-methyl. The triazolic compound was bacteriostatic for *S. aureus* and bactericidal for *Escherichia coli*. The triazole-treated fabric showed 50% reduction in the growth of *E. coli*, *S. aureus*, and *Staphylococcus epidermidis*. Thus, the purified compound can find a place in the database for the development of fabrics with antimicrobial properties. This is the first report that envisaged the production of triazole antimicrobial compound from sponge associated marine fungi from the Indian coast.

46.1.3 ALLYLAMINES AND OTHER SQUALENE EPOXIDASE INHIBITORS

The group of agents generally known as allylamines strictly includes only naftifine and terbinafine, but because butenafine and tolnaftate function by the same mechanism of action, they are included in this class and are shown in Figure 46.4. These drugs have a more limited spectrum of activity than the azoles and are effective only against dermatophytes. Therefore, they are employed in the treatment of fungal infections of the skin and nails.

They act through the inhibition of the enzyme squalene epoxidase. The inhibition of this enzyme has two effects, a decrease in total sterol content of the fungal cell membrane and a buildup within the fungal cell of the hydrocarbon squalene, both of which appear to be involved in the fungi toxic mechanism of this class.

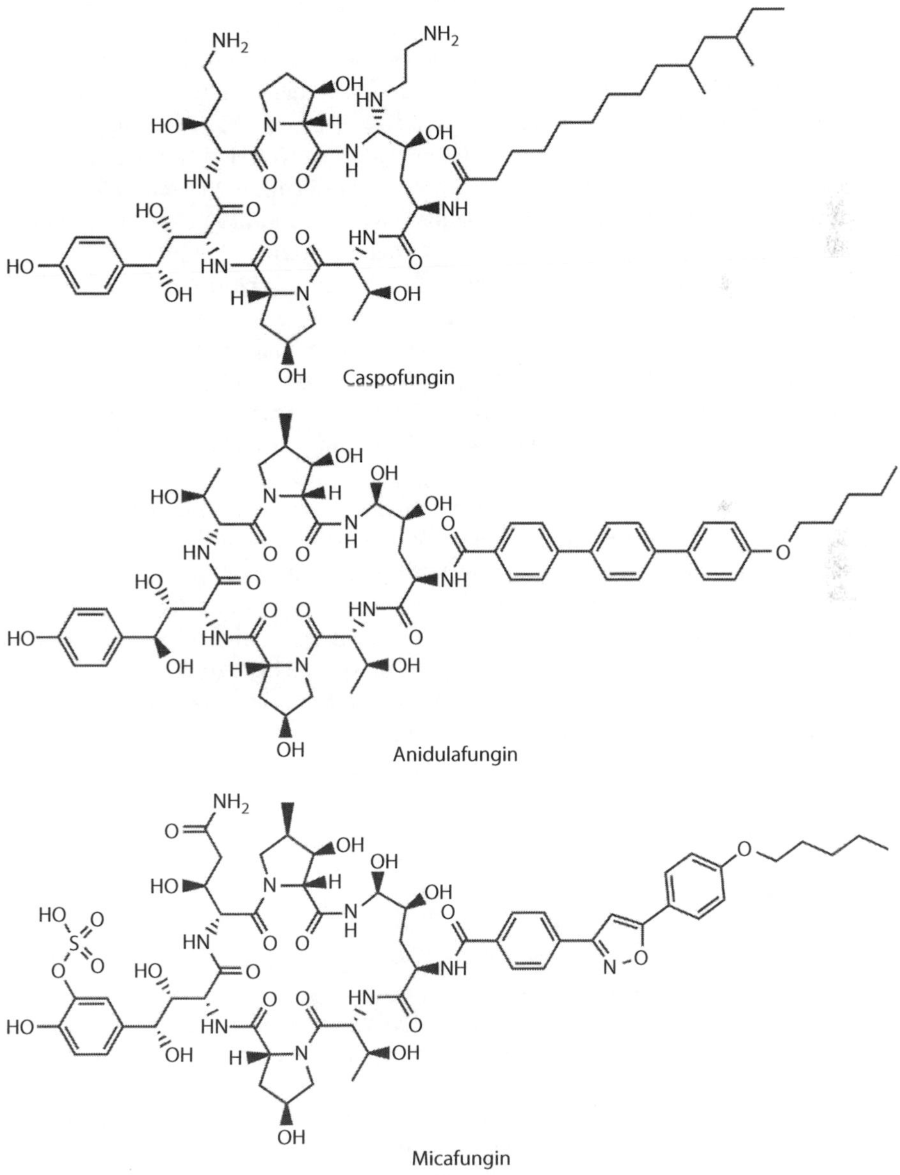

FIGURE 46.4 Structures of allylamines antifungal drugs.

Literature research reveals that TLC analyses of allylamines are limited to terbinafine and tolnaftate.

Terbinafine hydrochloride is an allylamine derivative with antifungal activity. The drug has been found to be a potent inhibitor of squalene epoxidase, which is an enzyme present in fungal and mammalian cell systems important in ergo sterol biosynthesis. The literature survey reveals that the drug has been determined in biological fluids (plasma, urine), tissues, nails, and hair by different analytical procedures, but no data were available about the HPTLC analysis until Ahmad et al. (2009) have not reported their study. The objective of that study was to develop and validate a rapid, simple, sensitive, and reproducible HPTLC method for the quantification of terbinafine as the bulk drug and in the tablet and cream formulations that would be applicable to stability studies. Chromatograms obtained from bulk terbinafine hydrochloride were compared with those obtained from formulations to assess the specificity and selectivity of the procedure. No peaks were observed at or near the R_F of terbinafine, indicating the high selectivity of the HPTLC method. It should be noted that when high concentrations of the drug were applied to TLC plates, one more band also appeared. The area of this band was much less than that of terbinafine. This band could be assumed to be of impurity, perhaps a residue from the manufacturing process. The chromatograms and drug recovery indicated that the degradation of terbinafine had not occurred in the tablet and cream formulations or in the bulk drug, after accelerated storage for 6 months at 40°C ± 1°C and 75% ± 5% RH. The drug content of terbinafine solution stored under these conditions was, however, significantly different ($P < 0.05$) from the initial drug content, indicating hydrolytic degradation of drug. As a result of hydrolytic decomposition of terbinafine, the chromatogram of its solution contained four additional peaks (degradation product peaks) with the terbinafine peak. The method also seems to be stability indicating, because the degradation product peaks were well resolved from the drug peak with significantly different R_F values. These results indicated that this HPTLC method is suitable for routine analysis and accelerated stability testing of terbinafine in pharmaceutical drug-delivery systems.

Later, Suma et al. (2011) developed a new HPTLC procedure that could be served as a stability-indicating assay method for terbinafine hydrochloride. Terbinafine was subjected to acid and alkaline hydrolysis, oxidative, photochemical, and thermal degradation in order to observe the behavior of its possible degradation products in selected chromatographic system. No significant degradation was observed under acid, base, and hydrogen peroxide exposure, that is, a sharp and symmetrical peak of terbinafine was only observed on chromatograms. However, terbinafine hydrochloride was found to be degraded when exposed to UV light at 36 h, that is, three additional peaks were found. The spot of the degradation product was well resolved from that of the drug. Statistical analysis proves that the method is suitable for the analysis of terbinafine hydrochloride as bulk drug and in pharmaceutical formulation without any interference from the excipients. Hence, this proposed method can be used for the routine analysis of terbinafine hydrochloride in pure, tablet form and in its degraded products.

Tolnaftate is used topically as a 1% solution, a powder, or a cream in the treatment or prophylaxis of superficial dermatophyte infections and of pityriasis versicolor. A literature survey revealed that the first report of the HPTLC analysis of mentioned antifungal drug was established and validated by Meshram et al. (2008a). The selected solvent mixture enabled good resolution and furnished dense and compact spot with minimum tailing. Specificity studies were also conducted to ascertain how accurately and specifically tolnaftate is analyzed in the presence of other components (impurities, degradation products, etc.). The analysis showed that degradation occurred when tolnaftate was subjected to acidic, alkaline, and oxidizing conditions and high temperature but not when it was exposed to UV. Major degradation of NaOH- and H_2O_2-treated samples resulted in marked reduction of both peak height and peak area. No extra peaks from degradation products were observed in extracts from any of the stressed solution; however, indicating the sample may have undergone degradation to a nonchromophoric product not detectable at the wavelength used for quantification. No peaks of by-product from synthesis were observed in the chromatograms studied. The results from validation showed that the method was fit for purpose and were comparable with those obtained by the use of other methods reported for the analysis of tolnaftate in pharmaceutical formulations.

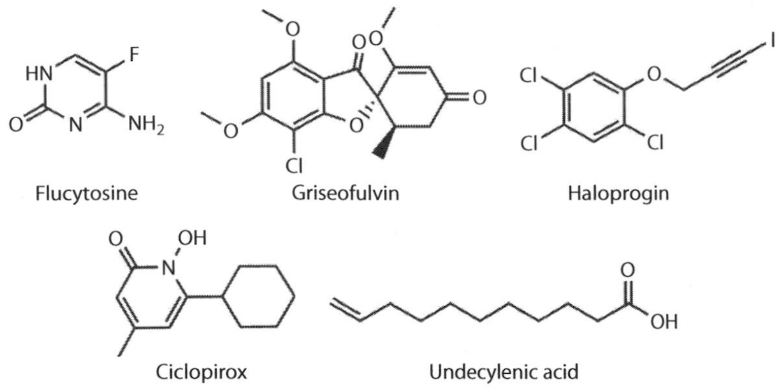

FIGURE 46.5 Structures of antifungal drugs acting through specific mechanism.

46.1.4 Drugs Acting through Other Mechanisms

Drugs that belong to this class of antifungal agents are presented in Figure 46.5.

Griseofulvin is an oral fungistatic agent used in the long-term treatment of dermatophyte infections caused by *Epidermophyton*, *Microsporum*, and *Trichophyton* spp. Produced by the mold *Penicillium griseofulvin*, this agent inhibits fungal growth by binding to the microtubules responsible for mitotic spindle formation, leading to defective cell wall development. Cole et al. (1970) described a rapid and accurate method for the determination of griseofulvin and dechlorogriseofulvin by TLC and GLC. Dechlorogriseofulvin is structurally related to griseofulvin, and can interfere with the analysis of griseofulvin. TLC was used for preliminary screening of griseofulvin or dechlorogriseofulvin, or both, since the difference in polarity between two compounds was insufficient for their complete separation (they have the same R_F value). Extracts that appear to contain griseofulvin or dechlorogriseofulvin, or both, are then simultaneously analyzed qualitatively and quantitatively by GLC.

46.2 TLC OF ANTIPROTOZOAL DRUGS

Antiprotozoal drugs are medicines that are used to treat a variety of diseases caused by protozoa. The protozoans, unlike bacteria and fungi, do not have a cell wall. They have a nucleus and a cytoplasm that are surrounded by a selectively permeable cell membrane. The cytoplasm contains organelles similar to those found in other animal and plant cells (e.g., mitochondria, Golgi apparatus, and endoplasmic reticulum). Thus, most of the antibiotics effective in inhibiting bacteria are not active against protozoans. There are numerous types of protozoal infections: amebiasis, giardiasis, babesiosis, Chagas' disease, leishmaniasis, malaria, sleeping sickness, toxoplasmosis, trichomoniasis, and pneumocystosis (also considered to be a fungal infection) (Lemke and Williams 2008).

Antiprotozoal drugs destroy protozoa or prevent their growth and ability to reproduce. Protozoans have little in common with each other, that is, agents effective against one pathogen may not be effective against another. The lack of effective antiprotozoal drugs has caused a renewed interest in the study of new chemotherapeutic compounds with better activities and fewer side effects. Although the mode of action of many antiprotozoals is not well understood, Table 46.4 summarizes agents according to the assumed or known modes of action and the chemical class to which they belong. Their structures are presented in Figure 46.6. Antiprotozoal agents are available in liquid, tablet, and injectable forms.

Investigated literature, which come from all relevant biomedical databases, covered TLC methods developed in the past. The presented methods are grouped according to the mechanism of action of the antifungal drugs. Chromatographic parameters, which comprise preparation of standard solution, selected stationary and mobile phase, development mode, detection, retention parameter, and appropriate reference, are presented in Table 46.5. Analytical parameters of the validated methods are presented in Table 46.6.

TABLE 46.4

Antiprotozoal Agents

Protozoal Disease	Drug Therapy	Chemical Class	Chemical Name
Amebiasis, giardiasis, and trichomoniasis	Metronidazole	Azole	2-(2-Methyl-5-nitro-1H-imidazol-1-yl) ethanol
	Tinidazole	Azole	1-(2-Ethylsulfonylethyl)-2-methyl-5-nitroimidazole
	Secnidazole	Azole	1-(2-Methyl-5-nitro-1H-imidazol-1-yl) propan-2-ol
	Ornidazole	Azole	1-Chloro-3-(2-methyl-5-nitro-1H-imidazol-1-yl)propan-2-ol
	Diloxanide furoate	Dichloroacetanide	4-[(Dichloroacetyl)(methyl)amino]phenyl furan-2-carboxylate
Pneumocystis	Pentamidine	Diamine	4,4′-[Pentane-1,5-diylbis(oxy)] dibenzenecarboximidamide
	Atovaquone	Naphthalene	*trans*-2-[4-(4-Chlorophenyl) cyclohexyl]-3-hydroxy-1,4-naphthalenedione
Trypanosomiasis	Pentamidine	Diamidine	4,4′-[Pentane-1,5-diylbis(oxy)] dibenzenecarboximidamide
	Eflornithine	Carboxylic acid	(RS)-2,5-diamino-2-(difluoromethyl) pentanoic acid
Malaria	Chloroquine	4-Aminoquinoline	(RS)-*N*′-(7-chloroquinolin-4-yl)-*N*,*N*-diethyl-pentane-1,4-diamine
	Hydroxychloroquine	4-Aminoquinoline	(RS)-2-[{4-[(7-chloroquinolin-4-yl)amino] pentyl}(ethyl)amino]ethanol
	Amodiaquine	4-Aminoquinoline	4-[(7-Chloroquinolin-4-yl)amino]-2-[(diethylamino)methyl]phenol
	Pamaquine	Diamidine	*N*,*N*-diethyl-*N*′-(6-methoxyquinolin-8-yl) pentane-1,4-diamine
	Primaquine	8-Aminoquinoline	(RS)-*N*-(6-methoxyquinolin-8-yl) pentane-1,4-diamine
	Mefloquine	4-Aminoquinoline	[(R*,S*)-2,8-bis(trifluoromethyl) quinolin-4-yl]-(2-piperidyl)methanol
	Pyrimethamine	2,4-Diaminopyrimidines	5-(4-Chlorophenyl)-6-ethyl-2,4-pyrimidinediamine
	Atovaquone	Naphthalene	*trans*-2-[4-(4-Chlorophenyl) cyclohexyl]-3-hydroxy-1,4-naphthalenedione
	Proguanil	Cycloguanil	1-(4-Chlorophenyl)-2-(*N*′-propan-2-ylcarbamimidoyl) guanidine
	Artesunate	Carboxylic acid	3R,5aS,6R,8aS,9R,10S,12R,12aR)-3,6,9-trimethyldecahydro-3,12-epoxypyrano[4,3-j]-1,2-benzodioxepin-10-yl hydrogen butanedioate

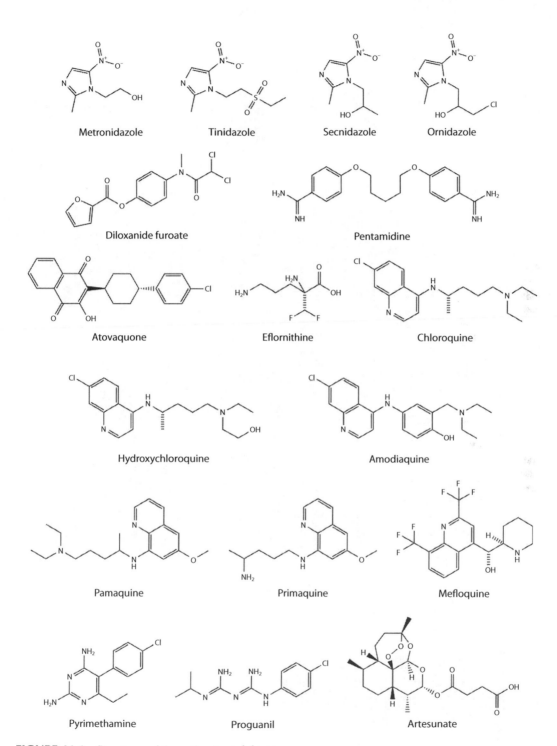

FIGURE 46.6 Structures of the antiprotozoal drugs.

TABLE 46.5
Chromatographic Systems Used for TLC Analysis of Antiprotozoal Drugs

Drug	Standard Solution	Plates	Stationary-Phase Pretreatment, Mobile-Phase Composition	Development Mode	Detection	Retention Factor	Reference
Metronidazole	Solution of metronidazole (10 mg/mL) in acetone	Silica gel	1. Acetone–ethyl acetate–ammonia (100 + 5 + 1) (v/v/v) 2. Acetone, chloroform and ammonia (100 + 15 + 1) (v/v/v)	—	Iodine vapor	1. $R_F = 0.467 \pm 0.006$ 2. $R_F = 0.657 \pm 0.021$	Awofisayo et al. (2010)
	Solution of metronidazole (1 mg/mL) in acetone	Silica gel	Aqueous PB (pH = 7.4) in methanol (50%–90%) (v/v)	—	Iodine vapor Scanned in reflectance/absorbance mode at 254 nm	$R_M^0 = -0.0149$	Dubey et al. (2009)
	Solution of metronidazole (1 mg/mL) in methanol	Silica gel	Acetonitrile–ammonia–methanol–methylene chloride–hexane (1.3 + 1.1 + 2.0 + 3.0 + 1.0) (v/v/v/v/v)	Chamber saturation 45 min, 25°C; ascending	Scanned in reflectance/absorbance mode at 280 nm	$R_F = 0.84 \pm 0.02$	Elkady and Mahrouse (2011)
	—	Silica gel	Benzene–ethyl acetate–toluene–methanol–glacial acetic acid (9.5 + 2.0 + 5.0 + 1.5 + 0.5) (v/v/v/v/v)	—	Scanned in reflectance/absorbance mode at 254 nm	Metronidazole— $R_F = 0.43 \pm 0.02$ Tetracycline hydrochloride— $R_F = 0.74 \pm 0.02$	Sharma and Sharma (2011)
	Solution of metronidazole (1 mg/mL) in water	HPTLC silica gel 60F$_{254}$	10%–90% of organic modifier 1. Acetone–n-hexane 2. Methanol–toluene 3. Methyl ethyl ketone–toluene	—	Exposure to iodine vapor	1. $R_M^0 = 2.88$, $C_0 = -1.55$ 2. $R_M^0 = 2.49$, $C_0 = -1.95$ 3. $R_M^0 = 3.18$, $C_0 = -2.06$	Aleksic et al. (2002)
	Solution of metronidazole (1 mg/mL) in methanol	HPTLC silica gel 60F$_{254}$	Acetone–chloroform–ethyl acetate (4 + 4 + 1) (v/v/v)	—	Scanned in reflectance/absorbance mode at 310 nm	Metronidazole—45.8 ± 0.3 mm 2-Methyl-5-nitroimidazole—66.6 ± 0.2 mm	Agbaba et al. (1998)

Drug	Sample	Stationary phase	Mobile phase	Conditions	Detection	R_F values	Reference
	Solution of metronidazole (1 mg/mL) in methanol	HPTLC silica gel 60F$_{254}$	Toluene–chloroform–methanol (3.0 + 2.0 + 0.6) (v/v/v)	Chamber saturation 10 min, 25°C ± 2°C, relative humidity 60% ± 5%; ascending	Scanned in reflectance/absorbance mode at 240 nm	$R_F = 0.34$	Meshram et al. (2009)
Tinidazole	Solution of tinidazole in ethanol	HPTLC silica gel	Methanol–diethyl ether–chloroform (1 + 9 + 3) (v/v/v)		Scanned in reflectance/absorbance mode at 314 nm	Tinidazole—$R_F = 0.24$ 2-Methyl-5-nitroimidazole— $R_F = 0.40$ 4-Nitroisomer— $R_F = 0.096$	Salo and Salomies (1996)
	Solution of tinidazole (10.0 µg/mL) in methanol	HPTLC silica gel	Toluene–ethyl acetate–methanol–glacial acetic acid (6.0 + 3.0 + 1.0 + 0.3) (v/v/v/v)	Chamber saturation 10 min; ascending	Scanned in reflectance/absorbance mode at 254 nm	$R_F = 0.61$	Vaidya et al. (2007)
	Solution of tinidazole (10 mg/mL) in acetone	Silica gel	1. Acetone–ethyl acetate–ammonia (100 + 5 + 1) (v/v/v) 2. Acetone–chloroform–ammonia (100 + 15 + 1) (v/v/v)	—	Iodine vapor	1. $R_F = 0.743 \pm 0.015$ 2. $R_F = 0.537 \pm 0.010$	Awofisayo et al. (2010)
	—	Silica gel	Chloroform–acetonitrile–acetic acid (60 + 40 + 2) (v/v/v)	—	Scanned in reflectance/absorbance mode at 320 nm	—	Guermouche et al. (1999)
	—	Silica gel	Chloroform–methanol–ammonia (9 + 1 + 0.1) (v/v/v)	—	Scanned in reflectance/absorbance mode at 335 nm	Tinidazole—$R_F = 0.79$ Furazolidone— $R_F = 0.63$	Tendolkar et al. (1995)
	—	Silica gel	Dichloromethane–methanol (9.6 + 0.25) (v/v)	—	Scanned in reflectance/absorbance mode at 280 nm	Tinidazole—$R_F = 0.28$ Diloxanide furoate— $R_F = 0.45$	Argekar and Powar (1999)
Ornidazole	Solution of ornidazole (100 ng/µL) in methanol	Silica gel	n-Butanol–methanol–toluene–ammonia (5 + 2 + 1 + 5) (v/v/v/v)	—	Scanned in reflectance/absorbance mode at 287 nm	Ornidazole— $R_F = 0.36 \pm 0.02$ Cefixime trihydrate— $R_F = 0.5 \pm 0.02$	Devika et al. (2010)

(continued)

TABLE 46.5 (continued)
Chromatographic Systems Used for TLC Analysis of Antiprotozoal Drugs

Drug	Standard Solution	Plates	Stationary-Phase Pretreatment, Mobile-Phase Composition	Development Mode	Detection	Retention Factor	Reference
	Solution of ornidazole in methanol	HPTLC silica gel	n-Butanol–methanol–ammonia (8 + 1 + 1.5) (v/v/v)	Chamber saturation 45 min, 25°C ± 2°C; ascending	Scanned in reflectance/absorbance mode at 302 nm	Ornidazole— $R_F = 0.76 \pm 0.04$ Gatifloxacin— $R_F = 0.21 \pm 0.02$	Suhagia et al. (2006)
	—	Silica gel	Toluene–n-butanol–triethylamine (8.5 + 2 + 0.5) (v/v/v)	—	Scanned in reflectance/absorbance mode at 285 nm	Ornidazole— $R_F = 0.51 \pm 0.007$ Cefuroxime axetil— $R_F = 0.67 \pm 0.009$	Ranjane et al. (2010)
Secnidazole	Solution of secnidazole (10 mg/mL) in acetone	Silica gel	1. Acetone–ethyl acetate–ammonia (100 + 5 + 1) (v/v/v) 2. Acetone, chloroform and ammonia (100 + 15 + 1) (v/v/v)	—	Iodine vapor	1. $R_F = 0.530 \pm 0.010$ 2. $R_F = 0.710 \pm 0.010$	Awofisayo et al. (2010)
Diloxanide furoate	Solution of diloxanide furoate (0.5 mg/mL) in ethanol	HPTLC silica gel	Chloroform–methanol (8 + 2) (v/v)	Chamber saturation 60 min; ascending	Scanned in reflectance/absorbance mode at 258 nm	—	Hasan et al. (2002)
	Solution of diloxanide furoate (200 µg/mL) in methanol	—	Cyclohexane–ethyl acetate (3 + 2) (v/v)	—	254 nm	Diloxanide furoate— $R_F = 0.60$ Furoic acid—$R_F = 0.01$ Diloxanide furoate— $R_F = 0.50$ Methylfuroate— $R_F = 0.75$	Gadkariem et al. (2004)

Drug	Sample	Stationary phase	Mobile phase	Conditions	Detection	Result	Reference
Chloroquine	Solution of chloroquine phosphate and potassium carbonate in methanol	HPTLC silica gel	Hexane–diethyl ether–methanol–diethylamine (37.5 + 37.5 + 25 + 0.5) (v/v/v/v)	—	254 nm	Peak—85 mm	Dwivedi et al. (2003)
	Solution of chloroquine in chloroform	Silica gel	Ethyl acetate–ethanol (absolute)–ammonia (25 + 2 + 2) (v/v/v)	—	UV light	Tablets—R_F = 0.23; 0.60 Syrups—R_F = 0.23; 0.38; 0.60	Abdelrahman et al. (1994)
Bulaquine	Solution of bulaquine in methanol containing 0.5% dimethyl octyl amine	HPTLC silica gel	Hexane–diethyl ether–methanol–diethylamine (37.5 + 37.5 + 25 + 0.5) (v/v/v/v)	—	254 nm	Peak—136 mm	Dwivedi et al. (2003)
Primaquine	Solution of primaquine and potassium carbonate in methanol	HPTLC silica gel	Hexane–diethyl ether–methanol–diethylamine (37.5 + 37.5 + 25 + 0.5) (v/v/v/v)	—	254 nm	Peak—72 mm	Dwivedi et al. (2003)
Artesunate	Solution of artesunate (1 mg/mL) in dimethyl sulfoxide	Silica gel	Paraffin–hexane (2 + 3) (v/v)	Chamber saturation 20 min; ascending	Iodine vapor	R_F = 0.04 ± 0.03	Argekar and Powar (1999)
Amodiaquine	Solution of amodiaquine (4 mg/mL) in dimethyl sulfoxide	Silica gel	Ethyl acetate–toluene (2.5 + 47.5) (v/v)	Chamber saturation 20 min; ascending	Iodine vapor	R_F = 0.06 ± 0.07	Argekar and Powar (1999)

TABLE 46.6
Analytical Parameters of the Methods for Antiprotozoal Drugs

Drug	Calibration Mode/Range	LOD (Unit/Spot, Band)	LOQ (Unit/Spot)	Precision (RSD%)	Recovery (%)	Matrix	Reference
Metronidazole	Linear, 300–700 ng/spot	—	—	0.426–0.768	98.92 ± 0.76	Cream	Meshram et al. (2009)
	Linear, 1.5–10 µg/band	0.32 µg	0.96 µg	1.24	—	Tablets	Elkady and Mahrouse (2011)
Tinidazole	Michaelis–Menten regression, 20–170 ng/spot	3 ng	>10 ng	2.4	—	Hydrolyzed solution	Salo and Salomies (1996)
	Linear, 300—800 µg/mL	120 µg	200 µg	—	100.02	Tablets	Vaidya et al. (2007)
	Linear, 1–10 ng/spot	—	—	6	96	Human serum	Guermouche et al. (1999)
Ornidazole	Linear, 900–2100 ng	20 ng	75 ng	0.2321	9.21–100.12	Tablets	Devika et al. (2010)
	Linear, 250–1250 ng/spot	100 ng	250 ng	0.82	97.34–101.88	Tablets	Suhagia et al. (2006)
	Linear, 100–500 ng/band	—	—	—	101.00 ± 1.192	Tablets	Ranjane et al. (2010)
Diloxanide furoate	Linear, 100–500 ng/spot	32.85 ng	99.54 ng	1.224–1.367	99.48 ± 0.18	Tablets Batch	Hasan et al. (2002)
Chloroquine	Linear, 4–33 µg/mL	0.59 µg	1.21 µg	0.625–3.24	—	Tablets Capsule	Dwivedi et al. (2003)
Bulaquine	Linear, 1.9–21.0 µg/mL	0.25 µg	1.52 µg	1.49–3.21	—	Tablets Capsule	Dwivedi et al. (2003)
Primaquine	Linear, 1–21 µg/mL	0.53 µg	1.07 µg	0.367–2.818	—	Tablets Capsule	Dwivedi et al. (2003)

46.2.1 AMEBIASIS, GIARDIASIS, AND TRICHOMONIASIS

Nitroimidazoles have been extensively evaluated in the treatment of trichomoniasis, giardiasis, liver, and intestinal amebiasis. The most widely used 5-nitroimidazoles are metronidazole, tinidazole, ornidazole, and secnidazole.

Metronidazole is the best documented antiprotozoal drug concerning TLC. It is a drug with antibacterial, antiprotozoal, and amebicidal activities. The major impurity in metronidazole drug substances and dosage formulations is 2-methyl-5-nitroimidazole, which originates from the synthesis. In accordance with international pharmacopeias, the 2-methyl-5-nitroimidazole content of metronidazole raw material and dosage formulations may vary over the range of 0.3%–0.5%. Metronidazole has been determined, as the main component of dosage formulations, by many analytical methods. Pharmacopeias describe TLC methods for semiquantitative determination of 2-methyl-5-nitroimidazole, but the first paper describes the quantification of this substance in metronidazole dosage forms using HPTLC that is presented by Agbaba et al. (1998). The selected chromatographic system enables the separation of two components with satisfactory accuracy and precision.

Metronidazole also has undesirable properties such as poor absorption, low aqueous solubility, and unwanted side effects, which have direct influence on its pharmacological and pharmacokinetic properties. Mahfouz et al. (1998) synthesized a series of identical twin esters of metronidazole and evaluated as potential prodrugs with improved physicochemical and pharmacokinetic properties. The synthesis of the twin esters was achieved by the interaction of metronidazole with the dicarboxylic acid anhydride or their dichloride. Their structures were verified by elemental and spectroscopic analyses. The lipophilicity of metronidazole and the prodrugs, expressed as R_M values, was determined using reversed-phase TLC and revealed enhanced lipophilic properties compared with metronidazole. Hydrolysis kinetics and antigiardial activity also suggest that the synthesized identical twin esters may be useful as a promising new prodrug form of metronidazole for oral drug delivery.

Later, the same group of authors (Aboul-Fadl and Mahfouz 1998) synthesized new nonidentical twin ester prodrugs by linking metronidazole and some antiprotozoal halogenated 8-hydroxyquinoline derivatives via dicarboxylic acid spacers with the aim of improving the therapeutic efficacy of both drugs. The synthesis necessitates the preparation of the precursor metronidazole hemisuccinate or hemiphthalate followed by the esterification with the respect to hydroxyquinoline derivative. To assess their suitability as prodrugs, the lipophilic properties, chemical stability, as well as in vitro and in vivo enzymatic hydrolysis were investigated. The lipophilic properties, expressed as R_M values, were determined using RP-TLC and showing enhanced lipophilicity as compared with the parent drugs.

One decade later, several novel aliphatic and aromatic ester prodrugs of metronidazole were synthesized, by utilizing the alcoholic functional group of metronidazole, with an aim to increase lipid solubility, so that it can be used to make injectable formulation (Dubey et al. 2009). The lipophilicity of the synthesized derivatives and standard metronidazole was determined experimentally using shake-flask method and by RP-TLC method and predicted *in silico* using software. The results showed that there is an increase in lipophilicity of the synthesized ester derivative with an increase in carbon chain length of acyl group when compared with metronidazole. This can lead to an increase in the transport across the biological membrane and into the cells by passive diffusion.

Several papers describe TLC methods for simultaneous determination of metronidazole and other antiprotozoal agents in their pure form and in pharmaceutical formulations. A ternary mixture containing metronidazole, diloxanide furoate, and mebeverine–HCl was determined by TLC and spectrophotometry (El-Ghobashy and Abo-Talib 2008). The normal-phase TLC with cyclohexane–ethyl acetate–methanol (12 + 7 + 2, v/v/v) mobile phase and spots scanned at 230, 257, and 221 nm, respectively, allowed good resolution and no significant interferences from excipients. Regression analysis showed good correlation in the selected ranges with excellent percentage recoveries.

Three methods are presented for simultaneous determination of diloxanide furoate and metronidazole, used in the presence of diloxanide furoate alkaline degradates and in pharmaceutical formulations, without previous separation (Abbas et al. 2011). The first method is chemometric-assisted spectrophotometry, in which principal component regression and partial least squares were applied. The second method is TLC densitometry, in which the binary mixture and degradates were separated on silica gel plates using a chloroform–acetone–glacial acetic acid (9.5 + 0.5 + 0.07, v/v/v) mobile phase and the bands were scanned at 254 nm. The last method is HPLC. The proposed methods were successfully applied for the analysis of diloxanide furoate and metronidazole in pharmaceutical formulations, and the results were statistically compared with a reported spectrophotometric method.

Several papers, dealing with simultaneous analysis of metronidazole and drugs with activity other than protozoal, have been published.

Metronidazole and MZ nitrate are used in association in the treatment as antiprotozoal and antibacterial medicine. The first TLC method for their simultaneous estimation has been reported by Meshram et al. (2009). The selected chromatographic system gave good resolution, and dense and compact spots along with minimum tailing and distance from each other. The whole procedure may be extended to pharmaceutical preparations and other applications on the same drug for routine screening. The method was further validated, and the results were found to be concurrent and comparable to the previously reported method.

According to their activity, metronidazole and ciprofloxacin hydrochloride are used in combination for the treatment of mixed aerobic/anaerobic infections. The first paper describing two methods (reversed-phase ion-pair HPLC and TLC-densitometric methods) for their simultaneous determination was presented by Elkady and Mahrouse (2011). Several developing systems and chromatographic conditions were tried in order to separate the mentioned drugs. The best chromatographic system was chosen, and the method has been validated. In order to confirm the reliability of TLC data, authors determined metronidazole and ciprofloxacin in three different pharmaceutical formulations, simultaneously by HPLC and TLC-densitometric methods. A paired t-test and F ratio were applied. The obtained values of t and F are lower than the tabulated ones, which leads to the conclusion that there is no significant difference between the suggested methods.

The determination of metronidazole and nalidixic was performed by TLC on silica gel with ethyl acetate–methanol–chloroform–25% ammonia (2.5 + 2.5 + 1.5 + 0.5, v/v/v/v) (Argekar et al. 1996); the determination of metronidazole and norfloxacin in tablets was based on the use of silica gel and chloroform–methanol–diethylamine–water (9 + 2 + 0.4 + 0.2, v/v/v/v) as the mobile phase (Zarapkar and Kanyawar 1999); metronidazole and spiramycin were separated using silica gel and methanol–chloroform (9 + 1, v/v) mobile phase (Maher and Youssef 2008); and the results of TLC separation of metronidazole and tetracycline hydrochloride (Sharma and Sharma 2011) are presented in Tables 46.5 and 46.6.

Two laboratories extensively investigated the use of HPTLC to perform assays on lamivudine–zidovudine, metronidazole, nevirapine, and quinine composite samples. To minimize the effects of differences in analysts' technique, the laboratories conducted the study with automatic sample application devices in conjunction with variable-wavelength scanning densitometers to evaluate the plates. The HPTLC procedures used relatively innocuous, inexpensive, and readily available chromatographic solvents used in the Minilab's TLC methods. The use of automatic sample applications in conjunction with variable-wavelength scanning densitometry demonstrated an average repeatability and an average reproducibility (Kaale et al. 2010).

Tinidazole is an antiparasitic drug belonging to the 5-nitroimidazole family. A few papers, dealing with the determination of individual tinidazole antiprotozoal agent, have been published (Patel et al. 1980, Guermouche et al. 1999).

The hydrolysis products of tinidazole have previously been identified as the 4-nitroisomer and 2-methyl-5-nitroimidazole, the latter being one of the starting materials for the synthesis of tinidazole. None of the earlier methods (Salomies 1992, Sanyal et al. 1992) could fully resolve the three compounds. Salo and Salomies (1996) presented a paper that describes the development and validation of an HPTLC method for the quantitation of tinidazole in hydrolyzed solutions. Methanol–diethyl ether–chloroform was used as the final mobile phase. The R_F values were strongly dependent on the amount of methanol in the mobile phase, since it has the largest solvent strength value comparing with two other solvents used. Closer studies revealed that diethyl ether is the major factor in creating the separation and is enhanced by the addition of chloroform, which also inhibits peak broadening. The results proved the method to be valid for monitoring the hydrolysis kinetics of tinidazole. It has been shown to be stability indicating. It is both precise and accurate, and the investigated substances could be quantified over a wide range of concentrations. The compounds were stable on the sorbent over a long period of time, a prerequisite for simultaneously analyzing the samples of a kinetic study. The method is also rugged with respect to mobile-phase composition, ambient temperature, and wavelength precision.

The subjects of other papers are simultaneous analysis of tinidazole and other drugs in combined dosage forms. Simple, precise, rapid, and selective HPTLC methods were developed for simultaneous determination of diloxanide furoate and tinidazole in tablets (chromatographic system presented in Table 46.5) (Argekar and Powar 1999), clotrimazole and tinidazole (chromatographic system presented in Table 46.5) (Vaidya et al. 2007), tinidazole and furazolidone in suspension (chromatographic system presented in Table 46.5) (Tendolkar et al. 1995), norfloxacin and tinidazole (silica gel/isopropanol–butanol–concentrated ammonia–water [25 + 50 + 5 + 25, v/v/v/v]; Argekar et al. 1996), ciprofloxacin and tinidazole in tablets (silica gel/chloroform–methanol–toluene–triethylamine–water [2 + 2 + 1.6 + 1.5 + 0.4, v/v/v/v/v]; Argekar and Sawant 1999), and norfloxacin and tinidazole (silica gel/n-butanol–ethanol–12.5% ammonia [4 + 1 + 2.2, v/v/v]; Mohammad et al. 2007).

A series of papers are dealing with the evaluation of TLC methods for determination of ornidazole in combined dosage forms. The new TLC methods for the simultaneous analysis of ornidazole and cefixime trihydrate (Devika et al. 2010), ornidazole and gatifloxacin (Suhagia et al. 2006), ornidazole and cefuroxime axetil (Ranjane et al. 2010), and ornidazole and levofloxacin hemihydrates (Surendra et al. 2007) were presented by Devika et al., Suhagia et al., Ranjane et al., and Surendra et al., respectively. The mobile-phase content, together with other chromatographic conditions like chamber saturation time, run length, sample application rate and volume, sample application positions, distance between tracks, and detection wavelength, was optimized to give reproducible R_F values, better resolution, and symmetrical peak shape for the two drugs. The methods were validated and found to be suitable for routine analysis of the selected drugs.

A simple, reliable, and rapid chemical identification system (RCIS) consisting of three color reactions based on the functional groups in the molecule (nitric acid reaction, sodium nitrite reaction, bismuth subnitrate reaction) and two TLC methods was developed for the preliminary detection of the 5-nitroimidazole drugs, metronidazole, tinidazole, and secnidazole (Awofisayo et al. 2010). Nitric acid test was not substituent specific as orange color was produced irrespective of the generic type of the nitroimidazole used. Sodium nitrite and bismuth subnitrate distinguished between metronidazole and the other two products as it gave a different color. The R_F values of the brands of tinidazole were higher than those of metronidazole and secnidazole in the solvent systems. This is probably due to its ethylsulfone substitution as against the primary and secondary alcohol functional group affiliations of metronidazole and secnidazole, respectively. In conclusion, the authors stated that by using the developed method, fraudulently labeled product 5-nitroimidazole antiprotozoal and antibacterial agents can be detected in approximately 40 min with limited reagents and a simple TLC technique. The method is rugged and simple and should be particularly handy for the use in detecting substandard products of the drug in the drug distribution chain where sophisticated equipment is often not available.

Diloxanide furoate is an amebicide that is effective against trophozoites in the intestinal tract. The structure of diloxanide furoate contains both an ester and a substituted amide linkage. Both linkages are liable to hydronium or hydroxide ion-catalyzed hydrolysis. However, the rate of alkaline-catalyzed hydrolysis of esters proceeds faster than that of the acid-catalyzed process. Also, amides are generally much more stable toward alkaline hydrolysis than esters. Accordingly, the ester group in the structure of diloxanide furoate is expected to undergo alkaline hydrolysis to yield diloxanide and furoic acid sodium salts, which on acidification give diloxanide and furoic acid.

Hasan et al. (2002) presented five new selective, precise, and accurate methods for the determination of diloxanide furoate in the presence of its degradation products, the first and second derivative spectrophotometry, RSD1 spectrophotometric method, pH-induced difference spectrophotometry, densitometry, and HPLC. The previously reported methods were used only for the determination of diloxanide furoate. The degradation products were prepared via alkaline hydrolysis of diloxanide furoate. TLC results indicated the presence of two degradation products, well separated from the main compound. Also, the validation parameters obtained by TLC and by nonaqueous titration reference method described in British Pharmacopoeia (BP) revealed no significant difference within a probability of 95%. However, the proposed method is far more sensitive and more selective than the BP method.

Later, Gadkariem et al. (2004) also performed TLC of the acidified basic hydrolysates of diloxanide furoate. Using the TLC method of the BP, the reaction mixture gave three spots. The spots were isolated from the TLC plates and further analyzed by UV spectrometry and HPLC. The results revealed that the spot coincided with that of the parent compound was actually a composite of diloxanide furoate and another alkaline degradation product of the methanolic diloxanide furoate solution. It was clear that the TLC system of the BP failed to resolve the latter alkaline degradate from the parent drug. In an attempt to separate this product from the drug, the authors developed a new TLC method and successfully separated the drug from its degradation products. The reaction mixture gave four spots. The third spot coincided with the parent compound, and the second spots corresponded to diloxanide, which was isolated and characterized as a hydrolytic product of diloxanide furoate. The degradation product with lowest R_F value corresponded to that of authentic furoic acid. The third degradate, with the highest R_F value, was methylfuroate formed through the alcoholysis of diloxanide furoate by the solvent methanol resulting in transesterification reaction.

46.2.2 Malaria

Malaria is a parasitic disease endemic in parts of the world where moisture and warmth permit the disease vector, mosquitoes of the genus Anopheles, to exist and multiply. The emergence of both drug-resistant strains of malarial parasites and insecticide-resistant strains of *Anopheles* has contributed significantly to the extensive reappearance of this infection. Malaria is one of the three major infectious diseases along with tuberculosis and AIDS. Patel and Patel (2010) presented a review of methods for determining antimalarial drugs in biological fluids. It has focused on the various analytical techniques for the assay of chloroquine, quinine, amodiaquine, mefloquine, proguanil, pyrimethamine, sulfadoxine, primaquine, and some of their metabolites. It was emphasized that the methods for determining antimalarial and their metabolites in biological samples have changed rapidly during the last 8–10 years with the increased use of chromatographic techniques. Chloroquine is still the most used antimalarial drug, and various methods of different complexity exist for the determination of chloroquine and its metabolites in biological fluids. The pharmacokinetics of chloroquine and other antimalarial drugs have been updated using these new methods. Various analytical techniques have been discussed, from simple colorimetric methods of intermediate selectivity and sensitivity to highly sophisticated, selective, and sensitive chromatographic methods applied in a modern analytical laboratory.

Chloroquine is by far the most extensively used chemotherapeutic and prophylactic drug for malaria. It is one of several 4-aminoquinoline derivatives that display antimalarial activity. Churchill presented a review with 23 references examining colorimetry, TLC, and ELISA methods used to detect chloroquine and its metabolites in human body fluids in the field in 1989 (Churchill 1989).

It has been noted that some patients do not apparently respond to treatment with chloroquine or responded differently to drugs from different suppliers. The investigation performed by Abdelrahman et al. (1994) involved screening various brands from a number of different dosage forms by TLC and HPLC to examine whether there were any detectable decomposition products.

The experiments were carried out directly and on the chloroformic extracts of the basified solutions of standard chloroquine, syrups, tablets, and chloroquine ampoules. The TLC screening of tablets and syrups showed two fluorescent spots for tablets, while for syrups, three spots were found. The chloroform extracts of the basified solutions of the two brands of dosage form gave similar results. It was concluded that for a proper evaluation of chloroquine in its dosage forms, it is necessary to develop a new method that can separate chloroquine from its degradation products.

The determination of chloroquine in human urine using two methods was compared (Estadieu et al. 1989). Dill–Glazko quality tests using eosin chloroformic solution gave a yellow to red-violet color of the organic phase after mixing with urine. The second method used chloroquine estimation from urine by ionex Whatman SA-2 paper, extraction of chloroquine from alkalized paper by diethyl ether, and TLC on silica gel plates of the organic phase residue. The chromatographic spots were visible in UV light or after detection with iodoplatinate–HCl reagent giving purplish color. This method was also suitable for monodesethylchloroquine, mefloquine, quinine, pyrimethamine, amodiaquine, and monodesethylamodiaquine, which gave the same color spots but had different R_F values. Sulfadoxine, a component of the pyrimethamine formulation Fansidar, was not detected by this method.

The HPTLC method for the determination of antimalarial drugs in body fluids was simplified to enable fast yet reliable identification under field conditions (Betschart et al. 1991). Urine is applied directly onto the silica gel plate without an extraction step. After drying, the plate is developed in an open horizontal development chamber, with the silica layer facing upward, using the solvent mixtures toluene–diethylamine–methanol (8 + 1 + 1, v/v/v). The plate can be read 20 min after application of the sample by means of a solar-powered UV lamp. Chloroquine, quinine, and their metabolites can be seen as fluorescent spots.

A simple, sensitive, and specific assay technique for the detection and semiquantification of chloroquine, amodiaquine, quinine, primaquine, sulfadoxine, and pyrimethamine in formulations and in human urine by using TLC was developed and tested in the laboratory (Lugimbana et al. 2006). The method involved developing test samples spotted on TLC chromatogram by diethylamine–toluene–isopropanol (1 + 4 + 5, v/v/v) as the eluting solvent. The selected solvent system enabled the elution and detection of all the tested antimalarial drugs in solution and those spiked in human urine. The detection limits for chloroquine, amodiaquine, quinine, and primaquine were the lowest at 0.00025 mg/mL. Sulfadoxine exhibited a detection limit of 0.0005 mg/mL, whereas that of pyrimethamine was 0.001 mg/mL. The results indicate the suitability of this technique in antimalarial drug quality and bioavailability studies. It is envisaged that this technique will adequately address the role of drug absorption and excretion in the chemotherapy of malaria as well as detect types of antimalarial drugs commonly used in the community.

The development of Aablaquine, a combination kit for antirelapse treatment of *Plasmodium vivax* malaria, consisting of chloroquine phosphate tablets and bulaquine capsules, arose the need for an analytical method for the simultaneous estimation of bulaquine, chloroquine, and primaquine (an impurity/degradation product of bulaquine). For that purpose, Dwivedi et al. (2003) developed a simple, sensitive, and reproducible HPLC and HPTLC methods. The HPTLC method provides a good separation of the mentioned antiprotozoal agents. The methods provided adequate sensitivity for the determination of bulaquine, primaquine, and chloroquine in bulk drug substance, in dosage forms, and to check primaquine, as an impurity in the bulaquine samples.

With a chemical structure significantly different from that of quinoline-based drugs, the natural product artemisinin and its derivatives have attracted the attention of many different groups regarding the mechanism of action of these potent antimalarials devoid of significant clinical resistance up to now in Africa. Artemisinin is a sesquiterpene lactone characterized by the presence of an endoperoxide that is associated with its potent antimalarial activity. Because artemisinin is chemically unstable and poorly soluble in water or oil, the carbonyl group at C-10 of the parent compound is often reduced to obtain dihydroartemisinin. Several derivatives have been developed by adding ether, ester, or other substituents to the hydroxyl group of dihydroartemisinin.

Adewuyi et al. (2011) developed and validated a TLC method for the determination of artemisinin-based derivatives—artesunate and amodiaquine—in tablet formulations. It is shown that our new TLC method achieved reproducibility, repeatability, linearity, and selectivity that compares favorably with those of GC, GC–MS, HPLC, HPTLC, spectrophotometry, and other methods reported regularly in literatures.

High efficacy of artemisinin-based combinations has been determined in areas where malaria is endemic. Also, sulfadoxine–pyrimethamine has been used extensively worldwide for malaria prophylaxis in areas endemic for chloroquine-resistant Plasmodium falciparum malaria. The article by Ayede et al. (2011) described a routine, simple, precise, economical, and TLC technique for the detection of sulfadoxine–pyrimethamine, artemether–lumefantrine in tablet dosage form, its metabolites, and artesunate + amodiaquine metabolites in human urine. Chromatography separations were performed on glass silica gel plates precoated with slurry of silica gel–water (1 + 3, v/v) mixture with ethyl acetate–toluene–glacial acetic acid (20 + 20 + 10, v/v/v) and ethyl acetate–toluene (1 + 2, v/v) as mobile phases for sulfadoxine–pyrimethamine, artemether–lumefantrine, and the metabolites, respectively. The drugs and metabolites were satisfactorily resolved with mean R_F values of 0.22 ± 0.12, 0.07 ± 0.04, and 0.15 ± 0.03, for sulfadoxine–pyrimethamine, artemether–lumefantrine, and the metabolites, respectively. Artemether–lumefantrine and sulfadoxine–pyrimethamine exhibited a detection limit of 0.05 mg/mL. The results indicate the suitability of this technique in antimalarial drug quality and bioavailability studies. The accuracy and reliability of the method were assessed by the evaluation of linearity, specificity, and precision. The method developed can be used for the analysis often or more formulation on a single plate and is a rapid and cost-effective quality-control tool for routine analysis of sulfadoxine–pyrimethamine, artemether–lumefantrine as the parent drug and in tablet formulations, as well as their metabolites in human biological fluids.

Proguanil, a biguanide derivative of pyrimidine, is the most active of a series of synthetic aryl biguanides compounds tested for antimalarial activity. Proguanil consists of a biguanide with two substituents p-chloro benzene and isopropyl groups at N^1 and N^5 positions. The basic structure of proguanil is responsible for its polar nature, which is similar for all biguanide drugs. The polar nature creates a problem of isolation and determination as is experienced for proguanil. However, the reaction of proguanil with a water-soluble fluorescent reagent like sodium benzoxazole-2-sulfonate is expected to produce a derivative that fluorescent intensely and can be determined easily. The observation of the derivatized solution under UV light gave an intense blue coloration, which was not observed with proguanil. This was an indication that fluorescence derivative resulted from the reaction. TLC was done to establish the chromatographic behavior. The derivative was detectable at low microgram level by TLC followed by the examination of the plate under UV light. The R_F of derivative was 0.48, while sodium benzoxazole-2-sulfonate gave R_F of 0.98. The derivative being a large molecule had a retention value of 0.48, which was expected and in agreement with the previous observation. This chromatographic behavior was noteworthy because it revealed that excess derivatizing reagent would not interfere with detection. The determination of proguanil was performed by HPLC (Adewuyi et al. 2010).

The world is confronted by the reality of large-scale counterfeit production of almost every conceivable drug. It is, therefore, desirable that deliberate and fraudulent labeling of drugs with respect to identity should be detected along the distribution chain as part of normal quality assurance processes. There is the need to make available useful methods for accurately validating the authenticity and identity of available medications. Portable labs that perform TLC provide a relatively inexpensive, versatile, and robust means of identifying substandard drugs at a fraction of the resources required for modern laboratory testing.

New technologies are making it easier to test the authenticity of drugs in field settings. A paper presented by Bate et al. (2009) compares two instruments that use the technologies of Raman spectrometry and near-IR spectrometry against TLC and disintegration testing to identify substandard drugs in the field. Researchers procured a range of antimalarial, antibiotic, and antimycobacterial drugs from cities in six countries: Ghana, India, Kenya, Nigeria, Tanzania, and Uganda. Semiquantitative TLC and disintegration tests, Raman spectrometry, and near-IR spectrometry were used to measure the concentration of active ingredients and excipients (spectrometry only) to determine whether the tested samples were of good quality. As expected, the spectrometers, which operate to more exacting standards, failed more samples than the less exacting methods of TLC and disintegration testing; nevertheless, even these methods failed a substantial minority of sampled drugs. Overall, the choice of technology will come down to a variety of factors: how quickly results are required (spectrometry is generally quicker; however, given that different but bioequivalent products produce different spectra, the methods must be established for all new brands, which means that the initial setup time can be longer for spectrometers), cost (TLC is less expensive), reliability of results to an uninitiated user (Raman spectrometry is generally more reliable), ease of transport (spectrometry is more easily transported), and transport across borders (near-IR spectrometry does not contain a laser and therefore involves the least bureaucracy). For most resource-constrained developing countries, the GPHF-Minilab® is the product of choice based on cost. However, as aid agencies import and purchase more drugs for developing countries (notably those to combat HIV/AIDS, tuberculosis, and malaria) and individuals and governments of the developing world become wealthier and purchase more pharmaceuticals, it might not be long before near-IR spectrometry and Raman spectrometers are deployed in even the poorest countries.

ACKNOWLEDGMENTS

The authors are grateful to the Ministry of Science of the Republic of Serbia (Grant No. 172017) for financial support and to Miss Mirjana Stanković for technical help.

REFERENCES

Abbas, S. S., N. E. Wagieh, M. Abdelkawy, and M. M. Abdelrahman. 2011. Simultaneous determination of diloxanide furoate and metronidazole in presence of diloxanide furoate degradation products. *J. AOAC Int.* 94(5): 1427–1439.

Abdel-Moety, E. M., F. I. Khattab, K. M. Kelani, and A. M. AbouAl-Alamein. 2002. Chromatographic determination of clotrimazole, ketoconazole and fluconazole in pharmaceutical formulations. *Il Farmaco* 57: 931–938.

Abdelrahman, A. N., E. I. A. Karim, K. E. E. Ibrahim. 1994. Determination of chloroquine and its decomposition products in various brands of different dosage forms by liquid chromatography. *J. Pharm. Biomed. Anal.* 12: 205–208.

Aboul-Fadl, T. and N. M. Mahfouz. 1998. Metronidazole twin ester prodrugs II. Non identical twin esters of metronidazole and some antiprotozoal halogenated hydroxyquinoline derivatives. *Sci. Pharm.* 66(4): 309–324.

Adewuyi, G. O., A. O. Abafe, and A. I. Ayede. 2011. Development and validation of a thin layer chromatographic method for the determination of artesunate and amodiaquine in tablet formulations. *Afr. J. Biotech.* 10(60): 13028–13038.

Adewuyi, G. O., O. Olubomehin, and A. W. Ayanniyi. 2010. High performance liquid chromatographic determination of proguanil after derivatisation with sodium benzoxazole-2-sulphonate. *Afr. J. Biotech.* 9(6): 900–905.

Agbaba, D., M. Đurković, J. Brborić, and D. Živanov-Stakić. 1998. Simultaneous HPTLC determination of metronidazole and its impurity 2-methyl-5-nitroimidazole in pharmaceuticals. *J. Planar Chromatogr.* 11: 447–449.

Agbaba, D., S. Vladimirov, and D. Živanov-Stakić. 1991. HPTLC determination of bifonazole in pharmaceutical formulation. *J. Planar Chromatogr.* 4: 164–165.

Ahmad, S., G. K. Jain, Md. Faiyazuddin et al. 2009. Stability-indicating high-performance thin-layer chromatographic method for analysis of terbinafine in pharmaceutical formulations. *Acta Chromatogr.* 21: 631–639.

Aleksic, M., S. Eric, D. Agbaba, J. Odovic, D. Milojkovic-Opsenica, and Ž. Tesic. 2002. Estimation of the hydrophobicity of antimycotic compounds by planar chromatography. *J. Planar Chromatogr.* 15: 414–417.

Argekar, A. P., S. U. Kapadia, and S. V. Raj. 1996. Simultaneous determination of norfloxacin and tinidazole in pharmaceutical preparations by high-performance thin-layer chromatography. *J. Planar Chromatogr. Mod. TLC* 9(3): 208–211.

Argekar, A. P. and S. G. Powar. 1999. Simultaneous determination of diloxanide furoate and tinidazole in tablets by high-performance thin-layer chromatography. *J. Planar Chromatogr. Mod. TLC* 12(6): 452–455.

Argekar, A. P., S. V. Raj, and S. U. Kapadia. 1996. Simultaneous determination of metronidazole and nalidixic acid in pharmaceutical dosage forms by HPTLC. *Indian Drugs* 33(4): 167–170.

Argekar, A. P. and J. G. Sawant. 1999. Simultaneous determination of ciprofloxacin hydrochloride and tinidazole in tablets by HPTLC. *J. Planar Chromatogr. Mod. TLC* 12(3): 202–206.

Awofisayo, O. S., O. A., Awofisayo, and N. Eyen. 2010. Development of a rapid chemical identification system (RCIS) for the detection of fraudulently labelled 5-nitroimidazole products. *Trop. J. Pharm. Res.* 9(2): 173–179.

Ayede, A. I., G. O. Adewuyi, and A. O. Abafe. 2011. Development and validation of a thin layer chromatographic method for the simultaneous determination of sulphadoxine–pyrimethamine, artemether–lumefantrine its metabolite and artesunate + amodiaquine metabolites in human urine. *Int. J. Chem.* 21(1): 41–53.

Bate, R., R. Tren, K. Hess, L. Mooney, and K. Porter. 2009. Pilot study comparing technologies to test for substandard drugs in field settings. *Afr. J. Pharm Pharmacol.* 3(4): 165–170.

Bele, A. A. and A. Khale. 2011. An overview on thin layer chromatography. *Int. J. Pharm. Sci. Res.* 2: 256–267.

Betschart, B., A. Sublet, and S. Steiger. 1991. Determination of antimalarial drugs under field conditions using thin-layer chromatography. *J. Planar Chromatogr. Mod. TLC* 4: 111–114.

Čakar, M., G. Popović, and D. Agbaba. 2005. High-performance thin-layer chromatography determination of some antimycotic imidazole derivatives and preservatives in medicinal creams and gel. *J. AOAC Int.* 88: 1544–1548.

Čakar, M., G. Popović, and S. Vladimirov. 2004. Simultaneous HPTLC determination of imidazole antimycotics and parabens in creams. *J. Planar Chromatogr.* 17: 177–180.

Churchill, F. C. 1989. Field-adapted assays for chloroquine and its metabolites in urine and blood. *Parasitol. Today* 5(4): 116, 121–126.

Cole, R. J., J. W. Kirksey, and C. E. Holaday. 1970. Detection of griseofulvin and dechlorogriseofulvin by thin-layer chromatography and gas–liquid chromatography. *Appl. Microbiol.* 19: 106–108.

Craig, C. R. and R. E. Stitzel. 1997. *Modern Pharmacology with Clinical Applications.* Baltimore, MD: Lippincott Williams & Wilkins.

Devika, G. S., M. Sudhakar, and J. V. Rao. 2010. Validated TLC densitometric method for the quantification of cefixime trihydrate and ornidazole in bulk drug and in tablet dosage form. *Der Pharm. Chem.* 2(6): 97–104.

Dewani, M. G., T. C. Borole, S. P. Gandhi, A. R. Madgulkar, and M. C. Damle. 2011. Development and validation of HPTLC method for determination of voriconazole in human plasma. *Der Pharm. Chem.* 3: 201–209.

Dubey, S., V. Jain, and G. B. Preethi. 2009. Evaluation of lipophilicity, antimicrobial activity and mutagenicity of some novel ester prodrugs of metronidazole. *Indian J. Chem.* 48B(11): 1571–1576.

Dwivedi, A. K., D. Saxena, and S. Singh. 2003. HPLC and HPTLC assays for the antimalarial agents chloroquine, primaquine and bulaquine. *J. Pharm. Biomed.* 33: 851–858.

Ekiert, R. J., J. Krzek, and W. Rzeszutko. 2008. Evaluation of a TLC densitometric method for analysis of azole antifungal agents. *Chromatographia* 67: 995–998.

Ekiert, R. J., J. Krzek, and P. Talik. 2010. Chromatographic and electrophoretic techniques used in the analysis of triazole antifungal agents. A review. *Talanta* 82: 1090–1100.

El-Ghobashy, M. R. and N. F. Abo-Talib. 2008. Application of derivative ratio and TLC-densitometric methods for determination of a ternary mixture containing metronidazole, diloxanide furoate and mebeverine hydrochloride. *Bull. Fac. Pharm.* 46(1): 75–86.

Elkady, E. F. and M. A. Mahrouse. 2011. Reversed-phase ion-pair HPLC and TLC-densitometric methods for the simultaneous determination of ciprofloxacin hydrochloride and metronidazole in tablets. *Chromatographia* 73: 297–305.

Enayatifard, R., N. Khalili, M. Rahimizadeh, and K. Morteza-Semnani. 2005. Investigation of degradation pathways of ketoconazole in aqueous media. *Chem. Indian J.* 2: 213–215.

Estadieu, M., J. Delmont, A. Durand, B. Ba, and A. Viala. 1989. Ion exchange paper adsorption and thin-layer chromatography for detection of chloroquine in urine for the study of chemosensitivity of Plasmodium falciparum. *Ann. Trop. Med. Parasitol.* 83(6): 577–581.

European Pharmacopoeia, 6th edn. Strasbourg, France: Council of Europe, 2008.

Ferenczi-Fodor, K., Z. Végh, and B. Renger. 2011. Impurity profiling of pharmaceuticals by thin-layer chromatography. *J. Chromatogr. A* 1218: 2722–2731.

Fittler, A., B. Kocsis, Z. Matus, and L. Botz. 2010. A Sensitive method for thin-layer chromatographic detection of amphotericin B. *J. Planar Chromatogr.* 23: 18–22.

Gadkariem, E. A., F. Belal, M. A. Abounassif, H. A. El-Obeid, and K. E. E. Ibrahim. 2004. Stability studies on diloxanide furoate: Effect of pH, temperature, gastric and intestinal fluids. *Il Farmaco* 59: 323–329.

Giaginis, C., D. Dellis, and A. Tsantili-Kakoulidou. 2006. Effect of the aqueous component of the mobile phase on RP-TLC retention and its implication in the determination of lipophilicity for a series of structurally diverse drugs. *J. Planar Chromatogr.* 19: 151–156.

Guermouche, M. H., D. Habel, and S. Guermouche. 1999. Assay of tinidazole in human serum by high-performance thin-layer chromatography-comparison with high-performance liquid chromatography. *J. AOAC Int.* 82(2): 244–247.

Hasan, N. Y., M. A. Elkawy, B. E. Elzeany, and N. E. Wagieh. 2002. Stability indicating methods for the determination of diloxanide furoate. *J. Pharm. Biomed.* 28: 187–197.

Indrayanto, G., S. Widjaja, and S. Sutiono. 1999. Simultaneous densitometric determination of betamethasone valerate and miconazole nitrate in cream, and its validation. *J. Liq. Chromatogr. Relat. Technol.* 22: 143–152.

Kaale, E., E. Reich, and T. P. Layloff. 2010. An interlaboratory investigation on the use of high-performance thin layer chromatography to perform assays of lamivudine-zidovudine, metronidazole, nevirapine, and quinine composite samples. *J. AOAC Int.* 93(6): 1836–1843.

Kaale, E., P. Risha, and T. Layloff. 2011. TLC for pharmaceutical analysis in resource limited countries. *J. Chromatogr. A* 1218: 2732–2736.

Kelly, P. K. 1988. Separation of amphotericin B components by combined thin-layer and high-performance liquid chromatography. *J. Chromatogr.* 437: 221–229.

Khetre, A. B., R. S. Darekar, P. K. Sinha, R. M. Jeswani, and M. C. Damle. 2008. Validated HPTLC method for determination of voriconazole in bulk and pharmaceutical dosage form. *Rasayan J. Chem.* 1: 542–547.

Lemke, T. L. and D. A. Williams. 2008. *Foye's Principles of Medicinal Chemistry*. Baltimore, MD: Lippincott Williams & Wilkins.

Lugimbana, L., H. M. Malebo, M. D. Segeja, J. A. Akida, L. N. Malle, and M. M. Lemnge. 2006. A simple technique for the detection of anti-malarial drug formulations and their presence in human urine. *Tanzan. Health Res. Bull.* 8(3): 149–154.

Maher, H. M. and R. M. Youssef. 2008. Development of validated chromatographic methods for the simultaneous determination of metronidazole and spiramycin in tablets. *Chromatographia* 69(3/4): 345–350.

Mahfouz, N. M., T. Aboul-Fadl, and A. K. Diab. 1998. Metronidazole twin ester prodrugs: Synthesis, physicochemical properties, hydrolysis kinetics and antigiardial activity. *Eur. J. Med. Chem.* 33(9): 675–683.

Manilal, A., B. Sabarathnam, G. S. Kiran, S. Sujith, C. Shakir, and J. Selvin. 2010. Antagonistic potentials of marine sponge associated fungi *Aspergillus clavatus* MFD15. *Asian J. Med. Sci.* 2: 195–200.

Marciniec B., K. Dettlaff, and J. Bafeltowska. 2004. The effect of fluorine substituent on radiochemical stability of some steroid and azole derivatives. *Ann. Acad. Med. Stetin.* 50: 77–82.

Marciniec, B., K. Dettlaff, E. Jaroszkiewicz, and J. Bafeltowska. 2007. Radiochemical stability of fluconazole in the solid state. *J. Pharm. Biomed.* 43: 1876–1880.

Marciniec, B., K. Dettlaff, and M. Naskrent. 2009. Influence of ionising irradiation on clotrimazole in the solid state. *J. Pharm. Biomed.* 50: 675–678.

Meshram, D. B., S. B. Bagade, and M. R. Tajne. 2008a. A simple HPTLC determination of tolnaftate in topical solution. *J. Planar Chromatogr.* 21: 283–287.

Meshram, D. B., S. B. Bagade, and M. R. Tajne. 2008b. A simple TLC method for analysis of fluconazole in pharmaceutical dosage forms. *J. Planar Chromatogr.* 21: 191–195.

Meshram, D. B., S. B. Bagade, and M. R. Tajne. 2008c. TLC-densitometric analysis of clotrimazole and metronidazole in combined dosage forms. *J. Planar Chromatogr.* 21: 277–282.

Meshram, D. B., S. B. Bagade, and M. R. Tajne. 2009. Simultaneous determination of metronidazole and miconazole nitrate in gel by HPTLC. *Pak. J. Pharm. Sci.* 22: 323–328.

Mohammad, M. A.-A., N. H. Zawilla, F. M. El-Anwar, and S. M. El-Moghazy Aly. 2007. Stability indicating methods for the determination of norfloxacin in mixture with tinidazole. *Chem. Pharm. Bull.* 55(1): 1–6.

Mousa, B. A., N. M. El-Kousy, R. I. El-Bagary, and N. G. Mohamed. 2008. Stability Indicating methods for the determination of some anti-fungal agents using densitometric and RP-HPLC methods. *Chem. Pharm. Bull.* 56: 143–149.

Parikh, S. K., J. B. Dave, C. N. Patel, and B. Ramalingan. 2011. Stability-indicating high-performance thin-layer chromatographic method for analysis of itraconazole in bulk drug and in pharmaceutical dosage form. *Pharm. Methods* 2: 88–94.

Patel, K. N. and J. K. Patel. 2010. Qualitative and quantitative analysis of antimalarial drugs and their metabolites in body fluids—A review. *J. Curr. Pharm. Res.* 2(1): 5–14.

Patel, R. B., A. A. Patel, T. P. Gandhi, P. R. Patel, V. C. Patel, and S. C. Manakiwala. 1980. New methods for the estimation of tinidazole. *Indian Drugs* 18(2): 76–78.

Pharmacopoeia Internationalis, 4th edn. Geneva, Switzerland: World Health Organization, 2006.

Popović, G., M. Čakar, K. Vučićević, S. Vladimirov, and D. Agbaba. 2004. Comparison of HPTLC and HPLC for determination of econazole nitrate in topical dosage forms. *J. Planar Chromatogr.* 17: 109–112.

Ramesh, B., P. S. Narayana, A. S. Reddy, and P. S. Devi. 2011. Spectrodensitometric evaluation and determination of fluconazole and its impurities in pharmaceutical formulations by high performance thin layer chromatography. *J. Pharm. Res.* 4: 1401–1404.

Ranjane, P. N., S. V. Gandhi, S. S. Kadukar, and K. G. Bothara. 2010. HPTLC determination of cefuroxime axetil and ornidazole in combined tablet dosage form. *J. Chromatogr. Sci.* 48(1): 26–28.

Roychowdhury, U. and S. K. Das. 1996. Rapid identification and quantitation of clotrimazole, miconazole, and ketokonazole in pharmaceutical creams and ointments by thin-layer chromatography-densitometry. *J. AOAC Int.* 79: 656–659.

Salo, J.-P. and H. Salomies. 1996. High performance thin layer chromatographic analysis of hydrolyzed tinidazole solutions. I. Development and validation method. *J. Pharm. Biomed.* 14: 1261–1266.

Salomies, H. 1992. Determination of tinidazole in dosage forms by quantitative HPTLC. *J. Planar Chromator. Mod. TLC* 5(4): 291–293.

Sanyal, S. N., A. K. Datta, and A. Chakrabarti. 1992. Stability indicating TLC method for the quantification of tinidazole in pharmaceutical dosage forms—Intravenous fluids. *Drug Dev. Ind. Pharm.* 18(19): 2095–2100.

Saysin S., B. Liawruangrath, and S. Liawruangrath. 2010. High-performance thin-layer chromatographic determination of ketoconazole in pharmaceutical formulations. *J. Cosm. Sci.* 61: 367–376.

Selcuk, F., O. Atay, and N. Noyanalpan. 1995. Qualitative and quantitative studies on the drugs containing N-substituted triazole structures. *FABAD Farmasotik Bilimler Dergisi* 20: 87–94.

Sharma, S. and M. C. Sharma. 2011. Development and validation of densitometric method for metronidazole and tetracycline hydrochloride in capsule dosage form. *Int. J. PharmTech. Res.* 3(2): 1169–1173.

Shewiyo, D. H., E. Kaale, P. G. Risha et al. 2011. Development and validation of a normal-phase HPTLC—Densitometric method for the quantitative analysis of fluconazole in tablets. *J. Planar Chromatogr.* 24: 529–533.

Suhagia, B. N., S. A. Shah, I. S. Rathod, H. M. Patel, D. R. Shah, and B. P. Marolia. 2006. Determination of gatifloxacin and ornidazole in tablet dosage forms by high-performance thin-layer chromatography. *Anal. Sci.* 22: 743–745.

Suma, B. V., K. Kannan, V. Madhavan, and R. N. Chandini. 2011. HPTLC method for determination of terbinafine in the bulk drug and tablet dosage form. *Int. J. Chem. Technol. Res.* 3: 742–748.

Surendra, K., S. D. Mageswari, R. Maheswari, N. Harikrishnan, C. Roosewelt, and V. Gunasekaran. 2007. Simultaneous estimation of levofloxacin hemihydrate and ornidazole in tablet dosage form by HPTLC. *Asian J. Chem.* 19(7): 5647–5651.

Tendolkar, N. M., B. S. Desai, J. S. Gaudh, and V. M. Shinde. 1995. Simultaneous determination of tinidazole and furazolidone in suspension by HPTLC and HPLC. *Anal. Lett.* 28(9): 1641–1653.

Thomas, A. H. 1976. Analysis and assay of polyene antifungal antibiotics. A review. *Analyst* 101: 321–340.

United States Pharmacopeia 31. The United States Pharmacopeial Convention, Rockville, MD, 2008.

Vaidya, V. V., S. N. Menon, G. R. Singh, M. B. Kekare, and M. P. Choukekar. 2007. Simultaneous HPTLC determination of clotrimazole and tinidazole in a pharmaceutical formulation. *J. Planar Chromatogr.* 20 (2): 145–147.

Zarapkar, S. S. and N. S. Kanyawar. 1999. Simultaneous determination of metronidazole and norfloxacin in tablets by high-performance thin-layer chromatography. *Indian Drugs* 36(5): 293–295.

47 TLC of β-Lactam Antibiotics

Monika Dąbrowska and Małgorzata Starek

CONTENTS

The history of antimicrobial preparations traces back in the beginning of the nineteenth century when Louis Pasteur observed that some microorganisms may suppress the growth and development of other microorganisms. They are the most commonly used and prescribed preparations among all antimicrobial medications. They are indicated in the treatment of infections caused by sensitive microbicidal as monotherapy or in combination with other medical preparations. In appropriate use, antibiotics are considered to be effective and safe remedies in the treatment of bacterial infections. Nowadays, many works have cited the thin-layer chromatography (TLC) coupled with various detections to study antibiotics alone and in the presence of impurities, degradation products, or metabolites.

47.1 PENICILLIN ANTIBIOTICS

Penicillin antibiotics are historically significant because they are the first drugs that were effective against many previously serious diseases, such as syphilis, and infections caused by *staphylococci* and *streptococci*. The penicillins, produced by various species of Penicillium, especially *Penicillium chrysogenum*, all possess the same bicyclic ring system, consisting of a β-lactam ring *cis*-fused to a thiazolidine ring. Over 100 penicillins, differing only in the nature of an *N*-acyl side chain, can be produced by fermentations to which an appropriate side-chain precursor has been added. Aliphatic or aryl-substituted aliphatic carboxylic acids or analogues, which can easily generate such acids in vivo, constitute acceptable side-chain precursors. In the absence of a suitable side-chain precursor, the fermentation of *P. chrysogenum* leads to the formation of a 6-aminopenicillanic acid (6-APA) and isopenicillin N. The former compound 6-APA is an important intermediate in the preparation of the so-called semisynthetic penicillins, produced by the acylation of the free amino group. All penicillins are β-lactam antibiotics and are used in the treatment of bacterial infections caused by susceptible, usually Gram-positive, organisms. Penicillins are still widely used today, though many types of bacteria are now resistant.

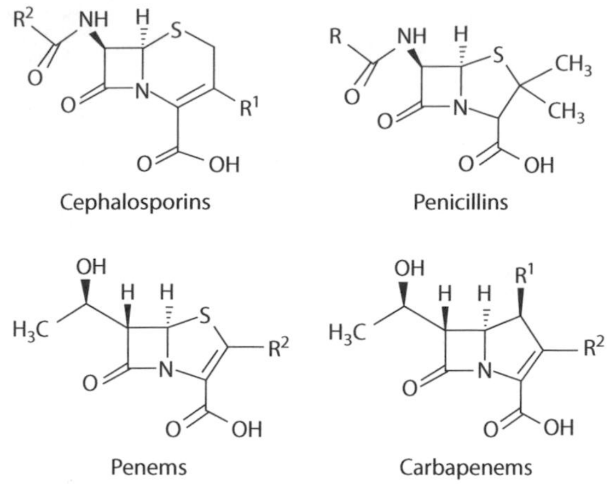

FIGURE 47.1 Chemical structure of β-lactam antibiotics.

47.1.1 SEPARATION AND DETERMINATION OF PENICILLINS

Nabi et al. described the TLC separation of penicillins on stannic arsenate–cellulose layers. Chromatography on thin layers of stannic arsenate mixed with cellulose layers resulted in differential migration of the studied penicillins. Several mobile phases were investigated to observe the behavior of the penicillins. Amoxicillin and ampicillin were selectively separated from binary and ternary synthetic mixtures and were also determined in commercially available drugs. The quantitative determination of the compounds, both singly and in combination, in commercial products, was performed using 2,3-dichloro-5,6-dicyano-*p*-benzoquinone as detection reagent. Granules of stannic arsenate were well powdered and mixed with cellulose powder in 1:4 ratio, and 10% CaSO$_4$ was added as binder. A slurry of this mixture in water was prepared and spread over glass plates, by means of applicator, to form uniform thin layers of 0.2 mm thick. The plates were then dried in an oven at 60°C. For qualitative analysis, solutions of the penicillins in methanol were applied to the plates by means of fine glass capillaries. After drying of the spots, the plates were developed with the desired mobile phase, to 15 cm from the point of application. After developing, they were air-dried and then placed in iodine chambers for the detection of the spots. The mixtures of penicillins were spotted on the plates, and the plates were developed. The adsorbent was scraped from the same regions of the sample plates, and the penicillins presented in the adsorbent were extracted with methanol. The extracts were determined spectrophotometrically, by measuring the absorbance at 460 nm using 2,3-dichloro-5,6-dicyano-*p*-benzoquinone as reagent. With ethyl methyl ketone:acetic acid:chloroform, R$_F$ values of penicillin and benzylpenicillin were high, whereas ampicillin and amoxicillin remained almost at the point of application. Penicillamine moved to some extent but with tailing. Clean separations of binary and ternary mixtures were achieved with acetone:acetic acid:chloroform and acetonitrile:acetic acid:chloroform. With these mobile phases, R$_F$ values of benzylpenicillin and penicillin were high, penicillamine migrated to the middle of the plate, and amoxicillin and ampicillin remained almost at the point of application. With acetonitrile:acetic acid:chloroform and acetone:acetic acid:chloroform mobile phases, R$_F$ values of amoxicillin and ampicillin were low compared with those of penicillamine, penicillin, and benzylpenicillin. For this reason, these mobile phases were chosen for selective separations of amoxicillin and ampicillin from synthetic mixtures of penicillins [1]. El Sadek developed a spectrodensitometric method for the determination of the penicillin V content in penicillin V benzathine after separation from closely related degradation products and benzathine on high-performance thin-layer chromatography (HPTLC) F$_{254}$ plates. Penicillin V potassium solution was prepared in distilled water with methanol, and penicillin V benzathine solution in pure and in syrup was prepared in methanol. Penicillin V was separated from benzathine, common additives, and related degradation products, on HPTLC F_{254} plates using acetone:chloroform:acetic acid (10:9:1, v/v/v). The spots were air-dried and measured at 230 nm. The R_F values for penicillin V were within the range of 0.25–0.75 [2]. Hendrickx et al. studied 18 penicillins on silica gel and silanized silica gel using 35 mobile phases. Silanized silica gel allows better separations than silica gel. Each penicillin can be separated from all others with an appropriate mobile phase. Any of the penicillins examined can be identified by combining the results obtained with a few mobile phases [3]. Gayen et al. reported the antibacterial and toxicological properties of enzymatically synthesized β-lactams by immobilized β-lactamase-free penicillin amidase produced by *Alcaligenes sp.* Methyl or ethyl esters of aromatic acids used as acyl donors were prepared, extracted with solvent, washed with water, dried over anhydrous sodium sulfate, and evaporated to dryness under reduced pressure. The homogeneity of the ester was tested by TLC on silica gel G plate using ethanol:chloroform:ammonium hydroxide (53:30:17, v/v/v) as solvent system. The spots were detected under (ultraviolet) UV light or in iodine vapor. The enzyme breads were separated by filtration. The filtrate was extracted with diethyl ether, and the extract was washed with sodium bicarbonate solution. The bicarbonate fraction was in turn extracted with ether, and the extract processed and evaporated to dryness. The homogeneity of the product was tested by TLC on silica gel G plate using butyl acetate:butanol:acetic acid:water (80:15:40:24, v/v/v/v) as solvent system. The spots were detected in iodine vapor [4].

Biagi et al. measured the R_M values and the influence of substituent groups on the R_M values of 11 penicillins by means of a reversed-phase (RP) TLC method using different concentrations of acetone in the mobile phase. Hydrophobicity was determined from the linear relationship between solute R_M values and the organic modifier content in the mobile phase [5].

Baltzer et al. studied the hydrolysis of mecillinam in aqueous solution (37°C) at pH 2–10. The degradation products observed by TLC and nuclear magnetic resonance (NMR) were identified and quantified. Mecillinam and the degradation product, (6R)-6-formamidopenicillanic acid, underwent reversible 6-epimerization in basic solution. Some of the thiazolidine derivatives formed epimerized at position 2. In contrast to penicillins, the degradation pattern of mecillinam becomes more complex with increasing pH. Rate constants for some processes were calculated [6]. Ellerbrok et al. studied penicillin-degrading activities of peptides from pneumococcal penicillin-binding proteins (PBPs). Trypsin treatment of native PBPs from *Streptococcus pneumoniae* resulted in the formation of stable peptides containing the β-lactam binding site with molecular masses ranging from 26 to 36 kDa. The concentration dependence of penicillin binding and the rate of release of the bound antibiotic were investigated. The nature of the released penicillin derivatives was analyzed, and two different pathways were described: hydrolysis of the penicilloyl–PBP ester linkage and fragmentation of the penicilloyl moiety. [³H]Benzylpenicillin, [³H]benzylpenicilloic acid, and [³H]phenylacetylglycine were used as standards. [3H]Benzylpenicilloic acid was generated by alkaline hydrolysis of [³H] benzylpenicillin in NaOH. For the identification of the products, [³H]benzylpenicillin was used for PBP labeling: in the case of the fragmentation pathway, radioactive phenylacetylglycine and non-radioactive N-formyl-D-penicillamine would be generated. Larger amounts of PBP 3 were used in order to compensate for the slower turnover reaction. After the microstep, one-half of the sample was digested with trypsin. The products formed during the subsequent 5 h incubation period were analyzed on TLC plates. Penicilloyl–PBP complexes, obtained after microsteps, were incubated in Tris/HCl pH 6.8, Triton X100. The supernatants, containing the turnover products, were lyophilized and dissolved in ethanol, spotted onto TLC silica gel G 60 with focusing zone plates, and developed in a mixture of n-butanol:water:acetic acid:ethanol (10:4:3:3, v/v) [7]. Płotkowiak et al. adapted UV and vis spectrophotometric methods after previous chromatographic separation (TLC/UV) and iodometric methods for the determination of azidocillin in the presence of its breakdown products. These methods were applied for kinetic measurement of changes in the azidocillin concentration with time in water solutions and solid state. Using TLC chromatography, several decomposition products of azidocillin were detected [8]. Qureshi et al. described the method for the analysis of degraded products of amoxicillin, ampicillin, and cloxacillin in drug formulations, obtained as a result of their base hydrolysis. Simultaneous spectrophotometric and volumetric determinations of the antibiotic were based on the neutralization of the degraded product by dilute hydrochloric acid to get a pH 2 to be conducive for redox titration using potassium iodate as titrant. The pathways of different degraded products and their oxidation mechanism were described on infrared (IR), TLC, and UV spectroscopic studies. The presence of 4-hydroxymethylene oxazolone was found on the basis of the results obtained by TLC (R_F 0.46) in 66% n-butanol:17% glacial acetic acid:17% water as a solvent system [9].

Omar et al. described a method for the determination of some penicillin derivatives and their additive and degradation products in the presence of each other. Penicillin derivatives were separated from their degradation and additive products on high-performance thin-layer silica gel G plates. The plates were developed in a linear chamber, air-dried, exposed to iodine vapor, and measured on a spectrodensitometer at 290 nm. Procaine and procaine penicillin were measured at 360 nm. For the reaction of penicillins with iodine, the formation of charge-transfer complexes was considered [10]. Ruzin et al. using a radioactive mannopeptimycin derivative bearing a photo-activation ligand have shown that antibiotics interact with the membrane-bound cell wall precursor lipid II [C55-MurNAc-(peptide)-GlcNAc], and this interaction is different from the binding of other lipid II-binding antibiotics such as vancomycin and mersacidin, and investigated the mechanism of methicillin and vancomycin. A TLC silica gel plates as stationary phase and isobutyric acid:1 M ammonia (5:3, v/v) as mobile phase were used. After separation, the plates were dried and exposed to x-ray film. Preparative quantities of both unlabeled and ^{14}C-labeled samples were applied to a preparative TLC plate and separated, and autoradiographs were detected by GS-710 densitometer [11]. Bijev and Hung prepared 12 penicillin derivatives for microbiological evaluation by N-acylation of ampicillin and amoxicillin with activated pyrrolecarboxylic acids via mixed anhydrides. An alternative synthetic approach via chloroanhydrides was checked and rejected because of the instability established for these intermediates. NMR and IR spectral data together with TLC confirmed the structure and purity of the new products. Antimicrobial tests in vitro indicated a reduction of the antibacterial activity compared with that of ampicillin and amoxicillin as reference antibiotics, but their minimal inhibitory concentrations were still in the range of 0.62–16 µg/mL against standard and clinical Gram-positive strains. Preliminary toxicological evaluations showed low toxicity [12].

Marciniec et al. reported the results of preliminary studies of the effect of β-irradiation on physical and chemical properties of salts of selected β-lactam antibiotics (sodium salt of ampicillin, azlocillin, benzylpenicillin, carbenicillin, and piperacillin and potassium salt of benzylpenicillin, ampicillin anhydricum, ampicillin trihydricum, amoxicillin trihydricum, and bacampicillin hydrochloride). The irradiation with β-rays was applied at a dose of 100 kGy. In the TLC method, aliquots of water solutions of the substances studied before and after irradiation were point-wise deposited on a foil covered with silica gel film 60 F_{254}. The process was conducted by the ascending technique in chambers previously saturated with mobile phase vapors. The substance studied before and after β-irradiation was subjected to TLC analysis on silica gel, with a mixture of octane 2-butyl:glacial acid (1.05 kg/L):buffer phosphoric solution (pH 6):2-butanol:methanol (150:75:45:27:10, v/v/v/v/v) as a mobile phase. The spots recorded were well developed and well visible, and no significant changes in the values of R_F were noted [13].

47.1.2 ANALYSIS OF PENICILLINS IN PHARMACEUTICAL FORMULATIONS

Torroba et al. selected a TLC method for determining the levels of penicillin, neomycin, and polymyxin in foot and mouth disease (FMD) vaccines. The working approach consisted of selecting the optimal working conditions for each of stages of the process: breaking the emulsion, extraction of the antibiotics, their purification, concentration, and identification and quantification by TLC. An FMD oil vaccine was used as a reference reagent. Several systems for purification of the aqueous extract of vaccines by partition with organic solvents and by development in opposite directions in TLC and by multiple development in the same phase were tested. The RP TLC plates were developed using methanol:0.1 M dipotassium phosphate (6:4, v/v). To visualize the localization of antibiotics on the TLC plates, methods based on the formation of colored compounds using J_2:starch, sodium azide:J_2, or ninhydrin:acetic acid, and second based on the formation of fluorescent derivatives with fluorescamine, were assayed. Developing with J_2:starch was selected for antibiotics of the penicillin family [14].

Gholipour et al. developed an HPTLC method for the simultaneous determination of ampicillin and amoxicillin. The method was applied for its quantitation in pharmaceutics, human blood plasma, and urine. Titanium(IV) silicate (a synthetic inorganic ion exchanger)-coated TLC plates were used. Titanium(IV) silicate was prepared by a dropwise addition of 0.25 M sodium silicate solution to 0.08 M titanic chloride solution in 0.2 M hydrochloric acid with constant stirring. The pH of the mixture was adjusted to 6.5 by the addition of 2 M sodium hydroxide solution. The gel formed was left to settle down overnight and then washed with distilled water until the supernatant was free from chloride, titanium, and silicate ions. The supernatant was removed completely, and a slurry was prepared by mixing the gel with gypsum as binder, in a conical flask with Teflon stopper, and shaking the flask vigorously. The slurry was poured immediately and used to coat glass plates with a 300 μm layer thickness. The plates were dried in an oven at 60°C for 2 h and then stored at room temperature inside a desiccator. Stock standard solutions of ampicillin and amoxicillin were prepared separately by dissolving compounds in water and methanol, respectively. Human plasma samples were stored under −20°C, and before use, the samples were thawed at room temperature. To plasma samples contained in a centrifuge tube, aqueous ampicillin and amoxicillin solution was added and mixed, and then acetonitrile was added. The mixture was centrifuged, and the supernatant was filled with deionized water till mark. Fresh urine samples were taken from healthy student volunteers, who abstained from any medications during the week preceding the study. To urine samples, aqueous ampicillin and amoxicillin solution was added and treated in methanol for the subsequent removal of proteins. The samples were agitated and placed in a centrifuge, and the upper liquid was removed. The plates were developed in a twin-trough TLC chamber, using a mixture of 0.1 M K_2HPO_4:0.1 M KH_2PO_4 (1:1, v/v) as mobile phase, dried at room temperature, and sprayed with fresh 1% solution of ninhydrin in ethanol after development to visualize the drugs. The spots have shown maximum absorbance at 546 nm. The results revealed R_F values for ampicillin and amoxicillin 0.62 and 0.92, respectively [15].

47.1.3 Penicillin's Analysis in Biological Material

Hurwitz and Carney identified and evaluated several ion-pair or adduct-forming additives that enhanced ampicillin partition behavior. At pH 3, picric acid and trichloroacetic acid increased the ampicillin aqueous–octanol partition coefficient 250 and 30 times, respectively. At pH 7, quaternary compounds gave the most significant increases in the partition coefficient. The values for an aqueous pH 7 chloroform system increased from zero in the absence of additives to 2.28, 1.86, 1.82, and 1.70 for equimolar amounts of benzalkonium, tetraheptylammonium, benzethonium, and cetalkonium chlorides, respectively. The extraction of ampicillin from aqueous solution was possible by adding a quaternary agent in an equimolar amount. The extraction of ampicillin from plasma required large molar excesses. Tetraheptylammonium chloride was added at a molar concentration 3 or 10 times greater than that of the ampicillin. The extracts were quantitated by TLC [16]. Graber et al. studied urinary excretion of ampicillin, amoxicillin, and oxacillin in five healthy volunteers. Determinations were carried out by chemical methods (PC, TLC, spectrophotometry). Besides the parent compounds and their penicilloic acids, an alpha-amino-substituted derivative was demonstrated. The total recovery of amoxicillin was nearly complete, and the recovery of ampicillin and oxacillin was about 50%. The combination of aminopenicillins with oxacillin did not alter significantly the excretion of the individual compounds [17]. Ulhnann and Wurst investigated four-hourly urine from volunteers and patients who had received penicillins orally or intravenously by means of

TLC and bioautography. The following penicillins were given orally: ampicillin, pivampicillin, bacampicillin, talampicillin, amoxicillin, azidocillin, indanyl carbenicillin, carfecillin, and epicillin. Penicillins given intravenously included carbenicillin, ticarcillin, azlocillin, and mezlocillin. Midstream urine was collected and centrifuged, and the supernatant was stored at −70°C until investigation. Urine from patients who were given penicillins was taken via an indwelling catheter after short infusion of the antibiotic. A TLC was performed using the ascending technique on prefabricated cellulose plates, and the solvent system consisted of n-propanol:ethanol:ethyl acetate:water (60:40:36:40, v/v/v/v). The plates were inserted and developed into a filter paper-lined chromatographic chamber, previously saturated with vapor of the solvent system. The plates were taken out of the tank and dried for 1 h at 40°C. For demonstration of antibacterially active spots, the plates were placed facedown on agar, which contained spores of *Bacillus subtilis*. The plates were incubated and the bioautogram was photographed and the diameters of the inhibition zones were noted. In order to identify the antibacterial activity of the various antibiotics, a standardized suspension of *Pseudomonas aeruginosa*, *Escherichia coli*, or *Staphylococcus aureus* was spread out on the surface of agar and used as detection media. The different spots on the TLC plates were detected by using a ninhydrin spray reagent. The plates were developed for color reaction for about 10 min at 110°C [18]. Okuno et al. analyzed the serum concentrations and urinary and biliary excretions of six penicillin derivatives including aspoxicillin in rats. The correlation between the values of pharmacokinetic parameters thus obtained and the R_M values measured by means of RP TLC was analyzed. Among the penicillins studied, hydrophilicity of amoxicillin was the highest, which was followed by aspoxicillin, ampicillin, *p*-hydroxypiperacillin, dehydroxyaspoxicillin, and piperacillin in descending order. The studies of correlation between the R_M values and the urinary or biliary excretion revealed that hydrophilic penicillins were almost excreted into urine, but more hydrophobic ones were mainly eliminated into the bile [19]. Ramirez et al. identified and quantified multiple antibiotic residues (ampicillin, benzylpenicillin, dicloxacillin, erythromycin, and chloramphenicol) in cow's milk by HPTLC combined with bioautography. *B. subtilis* ATCC 6633 as a test microorganism was used. All antibiotics were individually dissolved in methanol and kept at 4°C. Fresh milk from known origin, free from antimicrobial residues, was adjusted to pH 5–6 with hydrochloric acid. An aliquot milk sample was fortified with each antibiotic standard solution and extracted by mixing with acetonitrile. The liquid phase was decanted. Petroleum ether was added and shaken for 1 min, after which the upper phase (ether) was discarded. Sodium chloride was poured into the separating funnel and shaken. After that, dichloromethane was added to the acetonitrile phase and shaken again. The dichloromethane phase was drained into a round-bottomed flask and evaporated to dryness at 40°C in a rotary evaporator. The residue was reconstituted in methanol and used for spotting on HPTLC plates. The developing solvent used for HPTLC was a mixture of dichloromethane:acetone:methanol:glycerin (64:20:15:1, v/v/v/v). An HPTLC silica gel 60 precoated plates on prescored glass were supplied. After developing, the plate was placed facedown on a petri dish, contacting the silica layer with the inoculated media during 25 min. The petri dish was then inverted and pressed slightly to separate the HPTLC plate from the agar. Inverted petri dishes were then incubated for 18–24 h at 37°C. Inhibition zone diameters from fortified milk extracts were measured with vernier calipers and compared with antibiotic standard inhibition zones. The obtained R_F values were 0.11 for ampicillin, 0.18 for benzylpenicillin, and 0.19 for dicloxacillin [20]. Lin and Kondo accomplished the individual determination of residual drug concentrations in meat, using penicillin G in combination with aminoglycoside antibiotics, the differentiation with penicillin G, and the aminoglycoside groups by incubation of the individual drug with penicillinase. Each aminoglycoside treated with penicillinase produced a similar clear inhibition zone by the aminoglycoside alone, but penicillin G treated with penicillinase produced no inhibition zone even when a high concentration of penicillin G was used on the assay plates seeded with *B. subtilis* or *Micrococcus luteus*. Antibiotics were distinguished and identified according to their characteristic R_F values and colors by the TLC method using silica gel 60 F_{254} plates followed by spraying with various chemical reagents. A solution of n-butanol:acetic acid:water:*p*-toluenesulfonic acid (3:1:1:0.7, v/v/v/v) was a developing solvent for the separation [21]. Okerman et al. used a combination of three plates, seeded with strains of *M. luteus*, *Bacillus cereus*, or *E. coli*, for the detection of residues of β-lactam antibiotics, tetracyclines, and fluoroquinolones in poultry meat. β-Lactam antibiotics such as penicillin G, ampicillin, and amoxicillin give only inhibition zones on the plate with *M. luteus*. The limits of detection (LODs) of the antibiotics tested were penicillin G 0.9 ng, ampicillin 0.6 ng, and amoxicillin 1.0 ng on the plate with *M. luteus*. The three groups can also be detected on the plate seeded with *B. subtilis*; the LODs were 0.4 ng for penicillin G and 0.3 ng for ampicillin and amoxicillin. The test was applied to 228 broiler fillets and to 27 turkey thighs, originating from different poultry slaughterhouses. Nineteen broiler fillets contained inhibiting substances. The positive results of the inhibition tests were confirmed with a chromatographic technique. Amoxicillin residues were found in 16 samples [22]. Gafner described a method for detecting the following antimicrobial substances in feeds: penicillin, avilamycin, avoparcin, Zn bacitracin, erythromycin, flavomycin, furazolidone, lasalocid, monensin, narasin, salinomycin, spiramycin, tetracyclines, tylosin, and virginiamycin. Semiquantitative estimations of antibiotic content were compared with quantitative determinations of the previously mentioned substances in feeds. The method involves agar diffusion of buffered samples, a neutral extraction of polyether antibiotics followed by TLC, and an acid extraction for other antibiotics followed by TLC. After TLC development, an identification was achieved by bioautography with five test bacteria: *M. luteus* ATCC 9341, *S. aureus* ATCC 6538P, *Corynebacterium xerosis* NCTC 9755, *B. cereus* ATCC 11778, and *B. subtilis* ATCC 6633 [23]. Dolui et al. isolated and incubated different bacterial samples from the soil of different places of Dibrugarh in the presence of penicillin G for biotransformation to 6-amino penicillanic acid, which was detected by TLC. The bioconversion of penicillin G to 6-amino penicillanic acid was performed by using Gram-positive soil bacteria AKDD-7 and AKDD-9 both in free and immobilized states. Soil samples from different places of Dibrugarh were collected. Each soil sample was taken, sieved, and mixed with sterilized distilled water. Soil suspension was shaken, and supernatant was diluted. Diluted inoculum was aseptically transferred to agar petri plates, allowed to solidify, and incubated. The bacteria from each colony were transferred aseptically to sterilized agar slant and incubated. Each bacterial sample was grown in nutrient broth at 37°C for 48 h. Each broth culture was centrifuged in a cooling centrifuge. Cell pellets were collected by decanting off the supernatant. Free-cell suspension in saline was transferred to penicillin G solution. Supernatant was tested for the presence of penicillin G and 6-amino penicillanic acid by TLC using butanol:glacial acetic acid:water (3:1:1, v/v/v) as a solvent system and iodine vapor as detecting agent, by comparing with standard penicillin G and 6-amino penicillanic acid (R_F values 0.72 and 0.53, respectively). Two bacterial samples showed presence of 6-APA in the reaction mixture [24].

47.1.4 ANALYSIS OF PENICILLIN ANTIBIOTICS WITH BIOAUTOGRAPHY DETECTION

Kaya and Filazi used TLC/bioautography method for detecting antibiotic residues in animal tissue and analyzed residues of penicillin G, streptomycin, oxytetracycline, gentamicin, and neomycin by TLC/bioautography. For the extraction of antibiotic residues from milk samples, the proteins were precipitated by adding a mixture of acetonitrile:methanol:deionized water (40:20:20, v/v/v). It was centrifuged at 3000 rpm for 10 min, and the supernatant was used for analysis. Whatman cellulose TLC plates as stationary phase and a solvent system consisting of acetone:chloroform:*n*-propanol:impregnation liquid (16:20:27:16, v/v/v/v) as mobile phase were used. The TLC plates were placed in the development tanks prepared at least 1 h in advance, developed by applying sample extracts or antibiotics standards, and were dried and placed on the food-lot surface and allowed to contact agar for 20 min. At the end of contact period, the plates were removed and the bioplates were left for incubation for 16 h at 37°C. The inhibition zone diameters were measured using a caliper. The R_F value obtained for detecting penicillin G was 0.84 [25]. Qin et al. established an agar plate method to screen synergistic antibacterial agents other than β-lactamase inhibitors. By using this method, a strain *Aspergillus sp136* was selected for further studies.

From the metabolites of this strain, a synergistic antibacterial compound was isolated by bioautographic TLC assay-guided fractionation and identified as helvolic acid. The synergistic effect of helvolic acid to penicillin was about three times that of clavulanic acid to penicillin in agar diffusion assay on *B. cereus*. In checkerboard studies, helvolic acid exhibited synergistic effects with erythromycin on all tested multidrug-resistant *S. aureus* and with penicillin and tetracycline on some multidrug-resistant *S. aureus*. A pattern of enhanced killing was also found in time–kill studies on multidrug-resistant *S. aureus* [26].

47.2 CEPHALOSPORIN ANTIBIOTICS

Cephalosporins are a class of β-lactam antibiotics originally derived from the fungus *Acremonium*. Cephalosporin compounds were first isolated from cultures of *Cephalosporium acremonium* from a sewer in Sardinia in 1948 by Giuseppe Brotzu. Cephalosporins consist of a fused β-lactam-Δ^3-dihydrothiazine two-ring system and 7-aminocephalosporanic acid (7-ACA) and vary in side-chain substituents at C_3 (R_2) and C_7 (R_1). In the β-lactam ring, the substituents at C_3, C_4, and C_7 are important factors for their biological activity. The acylamido side chain at C_7 is the group governing largely the hydrophilic/hydrophobic character of these compounds. 7-ACA can be prepared chemically or biologically from cephalosporin C and serves as a precursor for the preparation of some valuable antibiotics. Based on the time of their discovery and microbiological properties, cephalosporins were divided into five generations. The progression in groups is associated with broadering of the Gram-negative antibacterial spectrum, reduction in activity against Gram-posiitve bacteria, and enhanced resistance to β-lactamase. Individual cephalosporins differ in their pharmacokinetic properties, plasma protein binding, and half-life.

47.2.1 SEPARATION AND DETERMINATION OF CEPHALOSPORINS

Misztal studied the retention behavior of five cephalosporins—cefsulodin, cephalothin, cefotaxime, cefoxitin, and cefamandole—by thin-layer reversed-phase ion-pair chromatography (RP IPC). The optimization of retention and selectivity of these compounds was carried out by changing the pH, the concentration of the ion-pairing counterion, and the concentration of the organic solvent in the aqueous mobile phase. The effects of various cationic and anionic pairing reagents in the mobile phase and the stationary phase were investigated. The experiments were carried out in a horizontal sandwich chamber using precoated TLC RP-18 $F_{254}S$ plates. The ion-pair-coated TLC RP-18 $F_{254}S$ plates were prepared by dipping the plate in a 3% ethanolic solution of the counterion. Sample solutes in phosphate buffer at pH 7.09 were applied on the plate and developed. Buffer solutions used as a mobile phase were prepared by dissolving 85% (w/v) orthophosphoric acid in water and adjusting the pH to the appropriate value with saturated sodium hydroxide solution. A stepwise gradient elution was carried out by introducing consecutively a series portion of the mobile phase, which contained decreasing concentrations of the ion-paring reagent. The separation was carried out in buffer solutions with pH 2.47 and 7.09 for the anionic and cationic ion-paring reagents. The spots of the compounds were detected under UV light at 254 nm. The plots show the effect of the amount of tetramethylammonium chloride, tetrabutylammonium hydroxide, and camphoric acid in the mobile phase or 3% camphoric acid in the stationary phase on retention of the compounds [27]. Dhanesar used scanning densitometry to evaluate a hydrocarbon-impregnated HPTLC plate for the direct quantitation of antibiotic without prior plate elution. Different volumes of deionized water and other aqueous solvents were spotted on to a hydrocarbon-impregnated HPTLC plate. The spots were evaluated by scanning densitometry at different wavelengths. A direct relation was found between the volume of solvent applied and the area and volume of plate damage. Different aqueous solvents created craters of different sizes; they all proved beneficial to the determination ceftriaxone dissolved in water and applied to the plate remained at the center of the spot, thereby facilitating direct quantitation by densitometry without the need for elution with a solvent [28,29]. Okumura performed silica gel thin-layer adsorption chromatography and RP TLC partition chromatography of cephalosporin antibiotics and steroidal hormones with chemically bonded dimethylsilyl silica gel in order to obtain suitable HPLC separation systems. A correlation of mobility between TLC and HPLC separation for cephalosporin antibiotics was obtained, and the possibility of direct transfer of chromatographic systems from TLC to HPLC for separation of these antibiotics was confirmed [30]. Eric et al. investigated the chromatographic behavior of cephalexin, cefaclor, ceftriaxone, cefixime, and cefotaxime on silica gel layers with binary mobile phases containing different amounts of organic modifier. The influence of type, chemical character, and position of substation of functional groups on the retention of analyzed compounds was of particular interest. Correlation of the hydrophobicity parameters with the lipophilicity parameter logP was established by the use of linear equations [31]. Dhanesar described the use of scanning densitometry for direct quantitation of 12 cephalosporins on a hydrocarbon-impregnated silica gel HPTLC plate without prior solvent elution. Cephalosporins were dissolved in water. The samples remained as a single spot centered around the point of application, thereby facilitating direct quantitation by densitometry at different wavelengths. Detection levels as low as 2 ng on the plate were measured. At low levels (<32 ng), antibiotic on the plate linear relationship was obtained, while at quantities >32 ng on a plate, a logarithmic relationship between peak area and quantity spotted was obtained [32]. Bhushan and Parshad applied EDTA as an impregnating reagent for the separation and identification of several cephalosporins, cefadroxil, cephalexin, cefazolin, cefotaxime, and cefaclor, on thin-layer silica gel plates. Cephalosporin samples were prepared from commercial capsules, tablets, or injection vials available for individual components. The standard solutions of each of the sample were prepared in 70% ethanol. Thin-layer plates were prepared by spreading a slurry of silica gel in water containing impregnating reagent EDTA disodium salt and then were dried in an oven at 60°C overnight. The three solvent systems were chosen, propionic acid:2-propanol:water (6:3:3, v/v/v), 2-propanol:water:ethyl acetate (5:3:3, v/v/v), and n-butanol:water:acetic acid (5:4:2, v/v/v). The chromatograms were developed, dried at 55°C, cooled to room temperature, and kept in iodine chamber for locating spots. Brownish spots were visualized [33]. Bhushan and Thiong'o tested two solvent systems for both normal-phase (NP) and RP TLC. Cephalosporin antibiotics, cefadroxil, cefazolin sodium, cephalexin, cefuroxime sodium, cefotaxime sodium, and cephradine, were separated on TLC plates impregnated with transition metal ions, Mn^{2+}, Fe^{2+}, Co^{2+}, Ni^{2+}, and Cu^{2+}, at different concentrations. TLC plates were prepared by spreading a slurry of silica gel G in distilled water in a ratio of 1:2 using the applicator. The plates were first dried at room temperature and then kept overnight in an oven at 60°C. For impregnated thin-layer plates, the slurry was prepared in aqueous solutions of different transition metal ions. The solution of each antibiotic was prepared in 70% ethanol. The impregnation was observed to have an effect on hR_F values, removed tailing of analytes, and improved the resolution. The results have been recorded for each metal ion and compared. The activation time of TLC plates impregnated with 0.1% $FeSO_4$ was found to affect both hR_F values and the resolution of cephalosporins. The solvent system, propanol:H_2O:butanol (15:3:1, v/v/v), resolved all cephalosporins under study by NP TLC [34]. Sharma et al. investigated a TLC determination of cephalexin and cefaclor. Chromatography of selected compounds was performed on precoated silica gel plates with concentrating zone by development in different mobile phases. Quantitative evaluation was performed by measuring the absorbance reflectance at 254 nm and by detecting reagents [35]. Tuzimski used 2D TLC to separate an eight-component mixture of cephalosporins. The separation was realized by development of the plate in two directions with different mobile phases in each one. Cefaclor, cefoperazone sodium, cefazolin sodium, cefotaxime sodium, cefoxitin sodium, cefuroxime sodium, and cephalothin sodium were dissolved in methanol; p-chlorophenacyl cephalothin ester was dissolved in acetone. A TLC was performed on glass-backed plates precoated with silica gel. The plate was developed in horizontal chambers in the first dimension with methanol:98% formic acid:ethyl acetate (29:1:70, v/v/v). After drying in a hood with an air circulating system, the plate was turned by 90° and developed in the second dimension with a methanol:toluene:ethyl acetate:98% formic acid (5:20:65:10, v/v/v/v) phase. Plates were scanned and video scanned. The spots were detected under UV illumination at 254 and 366 nm. Video scanning was performed to furnish a real picture of the plate with complete separation of the eight-component mixture of cephalosporins. The detection and quantitative determination of cephalothin sodium were performed at 240 nm. The obtained R_F values were in the range of 0.2–0.8 [36]. Misztal et al. separated cephalosporins belonging to three generations: cefoxitin sodium, cefsulodin sodium, cefalotin sodium, cefotaxime sodium, ceftriaxone sodium, cephalexin monohydrate, and cefamandole nafate. The TLC method with normal (NP TLC) and reversed (RP TLC) phases using a horizontal elution of chromatogram development was applied. Precoated TLC silica gel plates with the fluorescent indicator and silica gel RP C18 were used. Pure substances were diluted in phosphate buffer at pH 3.61. After developing, the plates were dried at room temperature, and the chromatograms were tested by UV irradiations or after spraying with the reagents. The effective mobile phases containing ethyl acetate:acetic acid:water (3:1:1, v/v/v) or n-butanol:acetic acid:water (20:1:2, v/v/v) and color-forming reagents, Dragendorff's reagent, iodine vapor, 0.2% ninhydrin solution, ethanol, 0.25% ninhydrin solution in 1% acetic acid, hexacyanoferrate(II) and ferric chloride(III), 2% p-dimethylamine, benzaldehyde in ethanol, potassium iodoplatinate (acidified with HCl), sulfuric acid–formaldehyde, and Marquis' reagent were tested. As concerns RP TLC analysis, a good separation was achieved using the mobile phases: phosphate buffer pH 2.65:tetrahydrofuran (9:1, v/v), (8:2, v/v), (7:3, v/v), and (6:4, v/v); phosphate buffer pH 2.65:methanol (8:2, v/v); as well as phosphate buffer pH 2.65:acetonitrile (8:2, v/v). Good separation was achieved using 15% ammonium acetate:acetonitrile:tetrahydrofuran (82:10:8, v/v/v), phosphate buffer pH 2.65:methanol (9:1, v/v) and (6:4, v/v), and phosphate buffer pH 2.65:acetonitrile (9:1, v/v) and (6:4, v/v). Greater selectivity was achieved by an increasing of tetrahydrofuran in the mobile phase up to 10%–40% vol. On TLC plates (NP), the best separation of the analyzed drugs was achieved using two mobile phases containing ethyl acetate:acetic acid:water (3:1:1, v/v/v) and n-butanol:acetic acid:water (20:1:2, v/v/v). The detection was carried out at 254 nm [37]. Choma established the optimal conditions for TLC analysis of eight cephalosporins, that is, cefaclor, cefoperazone, cefazolin, cefotaxime, cefoxitin, cefuroxime, cephalothin, and p-chlorophenacyl cephalothin ester. Retention parameters for various chromatographic systems were compared. Separation was achieved for several phases on plane silica gel and silica gel boned diol-, amino-, and cyanopropyl chains [38].

Fujisawa et al. reported the effectiveness of S-sulfocysteine on cephalosporin C production, using C. acremonium mutant that could not convert cysteine or inorganic sulfur to methionine. The proposed alternative pathway involving S-sulfocysteine is an important one for cysteine biosynthesis involved in cephalosporin C production by C. acremonium. The existence of S-sulfocysteine in cells of mutant 8650+/OAH−/SeMeR was subjected to TLC and paper electrophoresis. TLC chromatography was carried out on a cellulose plate with the following solvent systems: n-butanol:acetic acid:water (3:1:1, v/v/v), n-propanol:pyridine:acetic acid:water (15:10:3:10, v/v/v/v), methanol:NH_4OH:water (8:1:1, v/v/v), and n-propanol:water (8:2, v/v). The data presented in the work showed that S-sulfocysteine is superior to methionine as an effector of cephalosporin fragmentation by C. acremonium [39].

Nabi et al. described the separation and detection of different spontaneous, chemical, and enzymatic degradation products of cephalosporins. Suitable combination of mobile phase and spray/detection reagent enabled the identification of the products in aqueous preparations, biological fluids, and microbiological culture broths. The chromatographic behavior of some cephalosporins (cephalexin, cefadroxil, cefaclor, cefotaxime, ceftriaxone, cefoperazone, and ceftazidime) was studied on synthetic inorganic ion-exchanger (stannic oxide) layers. At least 20 different solvents were used as mobile phases, but the separation of antibiotics was achieved solely by the use of citrate and borate buffers of different pH. The TLC plates were prepared from a demineralized water slurry, dried in laboratory air overnight, and activated at 60°C for 1 h before spot application. The spots were detected by placing the TLC plates in an iodine vapor chamber. The iodine vapor dissolves in or forms weak charge-transfer complexes with organic compounds, and the cephalosporins show up as brown spots on a pale yellow background within a few minutes. After marking the zones for further reference, the exposure of the plates to air causes the iodine to sublime and the spots fade. The regions containing the cephalosporins were scraped from the plates, added to a demineralized water, filtered, and analyzed by a standard spectrophotometric method [40].

Dąbrowska and Krzek developed a method for the separation, identification, and quantitative determination of cefuroxime axetil diastereoisomers A and B in pharmaceutical formulations. Solutions for testing were prepared by dissolving the substance or powdered tablet in hot (60°C) aqueous $NH_4H_2PO_4$:methanol (31:19, v/v) mixture. By using HPTLC aluminum-backed cellulose plates activated by heating at 60°C as the stationary phase and 1% aqueous β-cyclodextrin:methanol (15:1, v/v) as a chiral mobile phase, separation of the constituents was achieved. The chromatogram spots differed significantly in retardation factors: R_F 0.87 for diastereoisomer A and 0.93 for diastereoisomer B. The individual diastereoisomers A and B exhibited maximum absorption at 285 nm, similar to the substance prior to separation. The identity of diastereoisomers was confirmed by ^{1}H NMR spectrometric analysis [41]. Dąbrowska and Krzek separated, identified, and quantitated the epimers of cefaclor by TLC. Solutions were prepared by dissolving the standard substance in 1.5% aqueous β-cyclodextrin solution:methanol (1:1, v/v) solution. For the analysis of tablets, the amount taken depended on the cefaclor content; this was then dissolved in 1:1 (v/v) 1.5% aqueous β-cyclodextrin solution:methanol, shaken for 10 min, and filtered. Chromatography was performed on aluminum foil-backed TLC plates coated with silica gel 60 F_{254}. The plates were impregnated with β-cyclodextrin, by development with 1:9 (v/v) 1.5% aqueous β-cyclodextrin solution:methanol, dried in open air, and then activated at the 60°C for 24 h. The plates were developed in a TLC chamber previously saturated with mobile phase vapor for 30 min at 5°C. Chromatograms were developed at 5°C with chloroform:methyl acetate:glacial acetic acid:water (4:4:4:1, v/vv/v) as mobile phase. Densitometric scanning at 274 nm was performed. Under these conditions, a good separation of the epimers was achieved with R_F values of 0.26 and 0.33. UV and ^{1}H NMR spectra were used to identify the epimers [42].

Yotsuji et al. used a series of monoanionic cephalosporins, cefoperazone, ceftezole, cefazolin, cefamandole, and cephalothin, which have different hydrophilicities to assay outer membrane permeation. The authors also examined the affinities of these drugs for PBPs, their outer membrane permeations, and stabilities to β-lactamase in *Bacteroides fragilis*. The used strains of *B. fragilis* were fully pathogenic and were recovered from human infections. The hydrophobic character of the drugs was expressed as the R_F value, which was measured by RP TLC. The polar mobile phase was composed of acetate:Veronal buffer (pH 7.0):methanol (4:1, v/v). A TLC silica gel 60 F_{254} siliconized precoated plates were used as the nonpolar stationary phase. A sample was dissolved in the acetate:Veronal buffer. The increased use of β-lactam antibiotics will result in new strategies by organisms to overcome the drug challenge [43]. Yotsuji et al. performed the analysis of the outer membrane permeation of *B. fragilis* by cephalosporins. The hydrophobic character of the antibiotics— cefamandole, cephradine, cefazolin, ceftezole, cefoperazone, cefsulodin, and cephalothin—was expressed as the R_F value, which was measured by RP TLC chromatography. The polar mobile phase consisted of acetate:Veronal buffer (pH 7.0):methanol (4:1, v/v) was used. TLC silica gel 60 F_{254} siliconized precoated plates were used as the nonpolar stationary phase. A sample was dissolved in the acetate:Veronal buffer. The hydrophilicity of cefoperazone was low; this was followed by the hydrophilicities of cephalothin, cefamandole, cefazolin, and ceftezole [44].

47.2.2 Analysis of Cephalosporins in Pharmaceutical Formulations

Darji et al. developed and validated an HPTLC method for the determination of cefpodoxime proxetil in dosage form. The substance was weighed and dissolved in and diluted up with methanol. The precoated silica gel 60 F_{254} aluminum sheets, prewashed with methanol, activated in an oven at 50°C for 30 min, as stationary phase, and chloroform:methanol:toluene (4:2:4, v/v/v) as mobile phase, were used. The chamber and plate saturation time was of 30 min. The detection of spot was carried out at 289 nm. The R_F value of 0.56 was obtained [45]. Date and Nagarsenke estimated cefpodoxime proxetil in self-nanoemulsifying formulations and conventional tablets. The antibiotic was chromatographed on a silica gel 60 F_{254} TLC plates using toluene:acetonitrile (6:4, v/v) as mobile phase. After chamber saturation (20 min), the plates were developed to a distance of 90 mm and then dried in hot air. Densitometric analysis was carried out in the absorbance mode of 234 nm. The solvent system offered optimum migration (R_F 0.29) and resolution of cefpodoxime proxetil from other excipients [46]. Coran et al. developed an HPTLC-densitometric method in order to obtain a procedure for routine analysis of cephalexin in pharmaceutical formulations. An HPTLC-precoated plate, silica gel 60 F_{254}, was prewashed by development with the mobile phase, air-dried, then oven conditioned at 120°C for 1 h, and cooled down in a desiccator. The solution was prepared by dissolving cephalexin in 0.01 N HCl and diluting with methanol. The standard and sample solutions were applied on the plate and developed in a horizontal developing chamber with ethyl acetate:acetic acid:water (7:2:l, v/v/v) as mobile phase. Chromatograms were scanned in reflectance mode at 263 nm [47]. Coran et al. carried out the assay on capsules containing cephalexin and excipients consisting in a 1:1 mixture of magnesium stearate and hydrogenated castor oil. The standard solution was prepared by dissolving cephalexin in 0.01 N HCl and diluting with methanol. HPTLC silica gel 60 F_{254} precoated plates were used. The plates were prewashed by development with the mobile phase, air-dried, then oven conditioned at 120°C for 1 h, and cooled down in a desiccator. The analysis was performed in a horizontal developing chamber with ethyl acetate:acetic acid:water (7:2:l, v/v/v) as mobile phase [47]. Jeswani et al. developed a stability-indicating HPTLC method for the analysis of cephalexin in bulk and dosage formulations. An HPTLC on silica gel plate precoated with silica gel 60 F_{254} with ethyl acetate:methanol:25% ammonia (6:4:1, v/v/v) was conducted. The hR_F value was 56. Densitometric quantification at 260 nm was performed. Cephalexin was subjected to forced degradation (acid, alkali, oxidation, thermal, photolytic). Dry heat studies were performed by keeping the drug sample in oven (80°C) for a period of 12 h. The photochemical stability of the drug was also studied by exposing the drug sample to UV light up to illumination of 200 W h/m^2 followed by fluorescent light up to illumination of 1200 lux h. On heating at 80°C in 0.1 N HCl (1 h), one peak of the degradation product was observed (R_F 0.76). On treatment with 0.01 N NaOH for 1/2 h, four new peaks of degradation products at R_F 0.15, 0.62, 0.78, and 0.82 were observed. On heating at 80°C in distilled water for 15 min, one new degradation peak at R_F 0.80 was detected [48]. Shah et al. developed and validated an HPTLC method for the determination of cefuroxime axetil in dosage form. The precoated silica gel 60 F_{254} plates as stationary phase and a mixture of chloroform:methanol:toluene (4:2:2, v/v/v) as mobile phase were used. TLC plates were prewashed with methanol. Activation of plates was done in an oven at 50°C for 5 min. The detection of spot was carried out at 290 nm. The mobile phase gave R_F values of 0.57 for active substance [49]. Shah et al. presented an HPTLC technique for the analysis of cefuroxime axetil in bulk drug and tablet formulations. Silica gel 60 F_{254} TLC plates prewashed with methanol were used as a stationary phase. The activation of plates was done in an oven at 50°C for 5 min. Chloroform:methanol:toluene (4:2:2, v/v/v) was used as mobile phase. Photometric measurements were performed at 290 nm in reflectance mode [49]. Sengar et al. reported an HPTLC method for the analysis of cefuroxime axetil and potassium clavulanate in a combined tablet dosage forms. The compounds were separated on aluminum foil plates precoated with silica gel F_{254}, with chloroform:methanol:toluene (4:3:3, v/v/v) as a mobile phase. Densitometric evaluation was performed at 225 nm. The R_F value was 0.77. The method was used for the analysis of the drugs in a pharmaceutical formulation with recovery of 100.05% ± 0.98% [50]. Sireesha et al. developed and validated an HPTLC method for the simultaneous determination of cefuroxime axetil and probenecid in combined dosage form. The precoated silica gel 60 F_{254} was used as the stationary phase. The mobile phase was a mixture of chloroform:acetonitrile:toluene:acetate buffer pH 6 (5:4:1:0.3 v/v/v/v). Detection of spots was carried out at 266 nm. The LOD and LOQ (limit of quantitation) for the drug combination were found to be 50 and 100 ng/spot, respectively [51]. Phattanawasin et al. established a TLC–image analysis method for stability-indicating studies and quantification of ceftriaxone sodium in bulk and pharmaceutical dosage forms. TLC was performed on aluminum foil-backed TLC plates coated with silica gel 60 RP 18 F_{254}S. The plate was developed in a TLC chamber previously saturated with a mixture of 15% w/v ammonium acetate buffer (pH 6.2):methanol:acetonitrile (12:0.5:0.25, v/v/v) for 20 min. After air-drying, an image of the developed plate under a UV lamp at 254 nm was taken with a digital camera set on a tripod 37 cm above the plate. The quantification of ceftriaxone sodium was performed by the use TLC videodensitometer. The mobile phase gave spot at R_F 0.58 for ceftriaxone. The ceftriaxone content of three commercially available products for injection was analyzed. Acid- and base-induced degradation was attempted by separately adding 0.1 M HCl and 0.02 M NaOH to stock solution, prepared in distilled water. For oxidative degradation 3% (v/v), hydrogen peroxide solution was added to stock solution. These mixtures were kept in the dark at room temperature for 1 h and analyzed immediately. A good separation was achieved between ceftriaxone and the degradation products obtained under different stress conditions, suggesting the method is stability indicating [52]. Sreekanth et al. developed and validated an HPTLC method for simultaneous estimation of ornidazole and cefixime in pure and pharmaceutical dosage form. It was performed on TLC plate precoated with silica gel 60 F_{254} as stationary phase using mobile phase consisting of methanol:water (60:40, v/v). The chamber was saturated for 10 min and maintained at 20°C ± 5°C temperature with 50%–60% humidity. The detection was carried out at 254 nm, showing R_F value 1.15 for cefixime. The percentage estimation of labeled claims of cefixime from market tablet was found to be 99.48 by height and 99.51 by area [53]. Krzek and Dąbrowska-Tylka developed a chromatographic and densitometric method for simultaneous identification and quantification of cefuroxime axetil and cefuroxime, which can be regarded as an impurity in oral forms of this ester. Solutions (substances and tablets) were prepared in methanol. The mobile phase containing of chloroform:ethyl acetate:glacial acetic acid:water (4:4:4:1, v/v/v/v) and TLC F_{254} plates as stationary phase enabled separation of the spots of both the ester of cefuroxime (R_F 0.93) and free cefuroxime (R_F 0.52) [54]. Ranjane et al. presented an HPTLC method for the determination of cefuroxime axetil and ornidazole in combined tablet dosage form. The separation was carried out on precoated silica gel 60 F_{254} aluminum plates using toluene:*n*-butanol:triethylamine (8.5:2:0.5, v/v/v) as mobile phase. The plates were prewashed with methanol and activated at 110°C for 5 min prior to chromatography. The optimized chamber saturation time for mobile phase was 15 min. The HPTLC plates were dried in a current of air. Densitometric scanning was performed at 285 nm. Standard stock solution was prepared by dissolving drug in methanol. The retention factor was found to be 0.67 for cefuroxime axetil. The method was successfully applied for the analysis of drugs in pharmaceutical formulation with the % assay 102.36 ± 0.775 for cefuroxime axetil [55]. Agbaba et al. used precoated silica gel HPTLC plates with concentration zones, without any pretreatment for an HPTLC assay of cephalexin and cefaclor in pharmaceuticals (capsules and syrup). The mobile phase containing methanol:ethyl acetate:acetone:water (5:2.5:2.5:1.5, v/v/v/v) was used. The stock standard methanol:aqueous solution of cephalexin monohydrate and cefaclor was freshly prepared in a mixture of methanol:water (9:1, v/v). The chromatogram was developed in a twin-trough chamber, previously saturated with the mobile phase. The measurement of each spot was carried out in situ at 265 nm using absorbance/reflectance mode [56]. Saleh et al. applied densitometric method at quality-control laboratories for the estimation of four α-aminocephalosprins, namely, cefaclor monohydrate, cefadroxil monohydrate, cefalexin anhydrous, and cephradine anhydrous, in pure form and different pharmaceutical dosage forms. Solutions of each compound were prepared in methanol. The analytical procedure was based on the separation of the studied cephalosporins on TLC aluminum sheets precoated with silica gel G F_{254} plates and quantitation of the separated spots with ninhydrin reagent. After developing in the tank presaturated with the mobile system vapor for at least 30 min before use, the plates were air-dried, viewed under UV lamp, sprayed with ninhydrin reagent, and then heated in air oven for 10 min at 110°C. The concentrations of 0.01 M NaOH and 0.5 M HCl were chosen to analyze the intact drug in the presence of its degradation products. The spots of degraded products were well resolved from the drug spot, and the additional spots were registered at different R_F values [57]. Eric-Jovanovic et al. used HPTLC plates with concentrating zone to develop an HPTLC method for the determination of ceftriaxone, cefixime, and cefotaxime in their dosage forms. Precoated silica gel plates with concentrating zones were used without any pretreatment. The mobile phase containing of ethyl acetate:acetone:methanol:water (5:2.5:2.5:1.5, v/v/v/v) was used. Stock standard solutions containing of ceftriaxone, cefixime, and cefotaxime were prepared in methanol. The chromatogram was allowed to develop in a twin-trough chamber previously saturated with the mobile phase. The measurement of each spot was carried out in situ at 270 nm using absorbance/reflectance mode. Migration distances were 34.7 for ceftriaxone, 66.9 for cefixime, and 85.19 for cefotaxime. The effect of the concentration of methanol in the mobile phase on the retention of compounds was investigated. The better separation of compounds was performed with content of methanol as organic modifier between 30% and 50% [58]. Singh and Maheshwari reported a TLC method with silica gel G–zinc ferrocyanide (SG–ZF) layers for the identification and separation of cephalosporins in pharmaceutical formulations. Pharmaceuticals containing cefadroxil, cephalexin, cefuroxime, cefixime, cefaclor, cefotaxime sodium, and ceftriaxone were tested. Standard solutions of each drug (tablets, capsules, and injection vial) were prepared by dissolving the drug in ethanol. Zinc ferrocyanide was prepared by mixing aqueous solutions of zinc chloride and potassium ferrocyanide in the volume ratio of 2:1. A gel thus obtained was allowed to settle for 24 h, filtered, washed with distilled water, and dried at 60°C in a hot air oven. The silica gel G and zinc ferrocyanide were taken in different ratios (1:0, 0:1, 0.5:1, 2:1, 4:1, w/w) and mixed with water. The mixture was stirred for 2 h using a magnetic stirrer. The resultant slurry was applied on the glass plates with the help of an applicator to give a 0.2 mm thick layer. The plates were dried at room temperature, activated at 100°C for 1 h, and stored in a closed chamber at room temperature until used. The level of mixing of zinc ferrocyanide on retention behavior and the hR_F values of eight cephalosporins were determined on thin layers with different SG–ZF ratios using acetone:acetic acid:chloroform (1:0.5:3.5, v/v/v) as a mobile phase. The hR_F values decrease with an increase in the amount of zinc ferrocyanide. This decrease in hR_F values seems due to zinc–amino complex formations. On pure silica gel TLC plates, the hR_F values were maximum. The best suitable mixing ratio of SG–ZF was 4:1 at which maximum separations were achieved. The amount of drug after separation was determined spectrophotometrically. The percentage recoveries in the range of 97.8%–100.3% indicated the success of separations in the mixtures [59]. Mohamed et al. developed a TLC method for the analysis of the cephalosporins, cefpodoxime proxetil, ceftriaxone sodium, ceftazidime pentahydrate, cefotaxime sodium, cefoperazone sodium, cefazolin sodium, and cefixime, in the bulk drug and pharmaceutical formulations. TLC was performed on aluminum sheets precoated with silica gel G 60 F_{254} as stationary phase. The mobile phase for the determined antibiotics composed of butanol:glacial acetic acid:hexane (4:4:7, v/v/v) for cefpodoxime proxetil, acetonitrile:water (15:7, v/v) for ceftriaxone sodium, ethyl acetate:methanol:water (1:9:2.5, v/v/v) for ceftazidime pentahydrate, butanol:ethanol:water (7:7:4, v/v/v) for cefotaxime sodium, ethyl acetate:methanol:water:acetone (3:1:1:2, v/v/v/v) for cefoperazone sodium, butanol:ethanol:water (4:7:4, v/v/v) for cefazolin sodium, and butanol:ethanol:water (9:7:4, v/v/v) for cefixime were used. The separated compounds were visualized as orange spots by spraying with Dragendorff's reagent. The mobile phases chosen for development gave compact spots for all the drugs with R_F values of 0.43–0.60 [60]. Dąbrowska et al. established the optimum chromatographic conditions for the separation of eight cephalosporins (cephalexin, cefadroxil, cefazolin, cefaclor, cefuroxime, cefuroxime axetil, ceftriaxone, and cefotaxime) in pharmaceutical formulations (capsules and powder for injection). A technique was used for the simultaneous analysis of these compounds without previous impregnation of the plates. Standard solutions and pharmaceuticals were prepared by dissolving in mixture of water:methanol (1:1, v/v). Chloroform:ethyl acetate:glacial acetic acid:water (4:4:4:1, v/v/v/v) was selected as mobile phase and precoated silica gel F_{254} plates as stationary phase. The chromatograms were dried in open air at room temperature, and the UV spectra were acquired in the range of 200–400 nm. The common analytical wavelength for all constituents was fixed at 275 nm, at which the absorbance of all analytes was >85% [61].

Dąbrowska et al. investigated the influence of acidic environment on the degradation processes of cefaclor and its inclusion complex of β-cyclodextrin and factors that determine the correlation between stability, described by kinetic and thermodynamic parameters. The experiments were carried out in acidic solutions with the presence of β-cyclodextrin, incubated for 3 h at appropriate temperatures. The cefaclor complexes with β-cyclodextrin were prepared in solutions and in tablets. The tablet was made by mixing the weighed amount of β-cyclodextrin with the amount of antibiotic in the following ratio: 1 portion of cefaclor:2 portions of β-cyclodextrin. Cefaclor was weighed, and the previously prepared β-cyclodextrin solution was added and shaken for 30 min. Afterward, the mixture was topped up to the specified volume with hydrochloric acid. On TLC chromatographic plates coated with silica gel 60 F_{254}, the relevant solutions were applied and then developed using the mixture chloroform:ethyl acetate:glacial acetic acid:water (4:4:4:1, v/v/v/v) as mobile phase. Densitometric measurement was developed at 274 nm. The hydrolysis product of R_F 0.68 was found to be 7–amino–3–chloro–8–oxo–5–tia–1–azabicyclo[4.2.0]oct–2–eno–2–carboxylic acid, while the second product R_F 0.87 was identified as phenylglycine [62]. Mohammad et al. described the stability-indicating TLC densitometry method for the determination of cefuroxime sodium in the presence of 5%–70% of its known hydrolytic degradation products: 3′hydroxy cephalosporin derivative, 3′acetoxy cephalosporin derivative, and cefuroxime lactone. Cefuroxime solution was prepared in methanol. Complete separation was achieved using silica gel F_{254} as the stationary phase and butanol:methanol:tetrahydrofuran:concentrated ammonium hydroxide (50:50:50:5, v/v/v/v) as the mobile phase. The spots were detected under UV lamp at 254 nm and scanned at 262 nm. Examination of impure cefuroxime sample gave four resolved spots with R_F 0.7 for cefuroxime and 0.62 and 0.83 for degradation products. The method was used for the analysis of the drugs in laboratory-prepared mixtures and dosage forms [63].

Saleh et al. described spectrophotometric methods for the determination of cephalothin sodium and cephradine. The methods were based on the charge-transfer complex formation between derivatives as n-donors and iodine, the o-acceptor, or 2,3-dichloro-5,6-dicyano-p-benzoquinone. In the spectrodensitometric method, cephalosporin derivatives were separated from their degradation or additive products on HPTLC silica gel G plates. The plates were developed, air-dried, exposed to iodine vapor, and measured on a spectrodensitometer at 290 and 360 nm for cephalothin and cephradine, respectively. The method was applied to the analysis of cephalothin and cephradine in dosage forms [64]. Palena and Mata selected cephalosporin as the substrate for thionation studies. Different conditions for thionation were attempted. The starting material was recovered unless high temperature (100°C), two equivalents of Lawesson's reagent, and 2 h of reaction were applied. After 40 min of heating monitored by TLC, the starting material was disappeared. The solvent was evaporated under reduced pressure. The residue was purified by flash column chromatography eluting with hexane:ethyl acetate. Analytical TLC was carried out with silica gel 60 F_{254} precoated aluminum sheets [65].

Fabre et al. investigated the feasibility of using fluorescamine detection of several cephalosporins (cephradine, cefaclor, cephalexin, and cefadroxil) after TLC development without the need of extraction before quantitation. The sensitivity of this detection was compared with those of UV absorption and fluorimetric o-phthalaldehyde detection on precoated silica gel 60 TLC plates. Some applications to the determination of antibiotics in pharmaceutical formulations (tablets and capsules diluted with distilled water) and biological fluids were studied. The amounts of each antibiotic were added to blank urine or plasma. For plasma samples, an additional treatment was previously applied. The mobile phase consisting of ethyl acetate:acetone:water:acetic acid (50:25:15:10, v/v/v/v) was used. For fluorimetric detection with fluorescamine, the plates were first sprayed with a solution of triethylamine in dichloromethane, dried under a stream of cold air, and then sprayed with an acetone solution of fluorescamine. For fluorimetric detection with o-phthalaldehyde, the plates were first sprayed with a borate buffer solution, dried under a stream of cold air, and then sprayed with an ethanolic solution of o-phthalaldehyde and mercaptoethanol. UV absorbance measurements were carried out at 262 nm. Fluorescence measurements were carried with an excitation wavelength of 365 nm and an emission filter with a cutoff at 460 nm. The R_F values of the cephalosporins in the used solvent system, measured after fluorescence quenching on a plate, were 0.24 for cephradine, 0.20 for cefaclor, 0.29 cephalexin, 0.21 cefadroxil, and 0.49 for cefotaxime [66].

47.2.3 Cephalosporin's Analysis in Biological Material

Kojima et al. studied the interaction of latamoxef with tobramycin in vitro. Solutions containing tobramycin alone, latamoxef alone, or both of these compounds in varying molar ratios were adjusting to pH 7.4 and incubated. Aliquot samples at a suitable time were subjected to paper electrophoresis and TLC. TLC analysis showed that the spot of latamoxef-radiated fluorescence with R_F value of 0.38. In the mixtures, there was a decrease in fluorescence of latamoxef compared with latamoxef alone. The UV spectrum of the mixture of tobramycin:latamoxef (1:2) showed a decrease in the intensity of absorption of latamoxef at 268 nm. These interactions between tobramycin and latamoxef were also observed in rat serum and its filtrate in vitro [67]. Blanchin et al. presented an alternative method to HPLC for monitoring the levels of cefotaxime, desacetylcefotaxime, cefmenoxime, and ceftizoxime in plasma and urine sample for patients under therapy with antibiotics. A loading of each cephalosporin was applied on precoated silica gel plates F_{254}. Solutions of cephalosporins were dissolved in a mixture of water:methanol (50:50, v/v). Recovery studies were carried out from spiked plasma samples treated for deproteination by an addition of ethanol per plasma. Spiked and blank urine samples were spotted without further treatment. A blank plasma and urine samples were spotted for comparison. The chromatograms were developed with the mobile phase, ethyl acetate:acetone:water:acetic acid (50:25:15:10, v/v/v/v), at different time intervals after the loading. The UV absorbance measurements were performed at 270 nm. The R_F values of the studied cephalosporins were 0.42 for cefotaxime, 0.48 cefmenoxime, and 0.49 for ceftizoxime. A noticeable fluorescence emission was observed in an alkaline medium for cefmenoxime, cefotaxime, and ceftizoxime on the TLC F_{254} plate. Fluorescence measurements were performed using an excitation wavelength of 313 nm and an emission at 390 nm [68].

Grzelak et al. developed a TLC screening method for the determination of cefacetrile and cefuroxime in milk. Two developments of TLC plates with concentrating zones were required: predevelopment with hexane, as a cleanup procedure to remove lipids from milk samples, and a proper development with methanol:toluene:ethyl acetate:98% formic acid (5:20:65:10, v/v/v/v) as mobile phase. The recoveries of both antibiotics were calculated over 5 days from the preparation of the samples. The obtained results on the second day of the experiment were 97.66% for cefacetrile and 86.13% for cefuroxime [69]. Choma et al. performed TLC/direct bioautography technique, which combines TLC with microbiological detection for the determination of cefacetrile residues in cow's milk. The excretion of cefacetrile in milk after the intramammary application at different times after injection was examined. The solution of cefacetrile was prepared in water. All solutions were stored at −18°C. Precoated silica gel Si60 F_{254} plates were used. After the application of antibiotic samples, bioautography was performed according to the Chrom Biodip® antibiotics test kit recipe. One bottle of nutrient medium was mixed with 0.5 M TRIS buffer, adjusted to pH 7.2 with 1 M HCl, and autoclaved. The sterile medium was then inoculated by pipetting in the *B. subtilis* spore suspension and incubated for 4 h at 37°C. The developed TLC plates were dried successively in air and a vacuum desiccator. They were immersed in the microorganism solution and incubated 20 h at 37°C. Next, the plates were sprayed with 3-[4,5-dimethylthiazol-2-yl]-2,5-diphenyltetrazolium bromide solution and left at room temperature for about 30 min. The cream-white inhibition zones were observed against a purple background. The plates were dried in air and scanned. The inhibition zone areas were measured with planimeter. The maximum concentration of cefacetrile in milk was observed 2 h after administration in cows [70].

47.3 PENEM ANTIBIOTICS

Among β-lactam antibiotics, penems show potent activity against a broad spectrum of bacteria, including β-lactamase-producing organisms. Although the penem skeleton was designed as a hybrid of penicillin and cephalosporin, the penem antibiotics show a more close structural relation to carbapenem. C6 alkyl substituents such as ethyl, hydroxymethyl, and hydroxyethyl moieties are characteristic substituents in penem antibiotics.

Long et al. synthesized new penems by structural modifications with ferrocenyl groups at the C2 position of the penem nucleus in order to enhance its antimicrobial activities and understand structure–activity relationships in these penem derivatives. The authors designed and synthesized a series of eight new penems having ferrocenyl group attached to the C2 position of the penem nucleus. The penem skeleton was chosen as a basic structure, because penems meet these criteria, having a broad spectrum from Gram-positive to Gram-negative bacteria including *P. aeruginosa* and an intrinsic bactericidal activity. The compounds showed satisfactory purity by TLC on 60 F_{254} plates visualized by UV light at 254 nm and/or by 6.3% w/v phosphomolybdic acid in ethanol [71].

Tanaka et al. estimated the enzyme-bound conformations of those C2 substituents by means of x-ray crystallography and molecular mechanics. The conformations of the substituents, in particular at C2 and C6 of the penem derivatives, would be useful for understanding structure–activity relationships in penem and related derivatives and also for the design of novel penem-related antibiotics. The compounds showed satisfactory purity by TLC on kieselgel 60 F_{254} plates, developed in dichloromethane and visualized by UV light at 254 nm and/or by 6.3% w/v phosphomolybdic acid in ethanol [72].

47.4 CARBAPENEM ANTIBIOTICS

Carbapenems are a class of β-lactam antibiotics with a broad spectrum of antibacterial activity. They have a structure that renders them highly resistant to most β-lactamases. Carbapenem antibiotics were originally developed from thienamycin, a naturally derived product of *Streptomyces cattleya*.

Jayaraman et al. described an application of (+)-(1S,2S)-2-amino-1-phenylpropan-1,3-diol in the formal total synthesis of carbapenems. The progress of the reactions was carefully monitored by TLC. The imines derived from (+)-(1S,2S)-2-amino-1-phenylpropan-1,3-diol furnished *cis*-β-lactams stereoselectively. The homochiral *cis*-β-lactams were converted into *cis*-β-lactams possessing aminol side chain at C4. The aminols were efficiently transformed into novel 4-cyano-*cis*-β-lactams, 4-acetoxy-β-lactam (which were well-proven starting materials in the synthesis of carbapenem antibiotics), and advanced starting materials for the synthesis of β-hydroxy aspartates [73].

Urbach et al. performed a complete comparative study of the chemical and biochemical reactivities of β-lactams. The authors explored a nontraditional approach for designing reactive β-lactams, and possibly new antibacterial agents, and illustrated the study of conformationally flexible 1,3-bridged β-lactams with the 13-membered-ring compounds derived from acetoxy azetidinone, commonly used as a precursor of thienamycin derivatives (carbapenem family). Manipulations were performed under an argon atmosphere in flame-dried glassware. The reaction process was monitored by TLC. TLC analyses were performed on aluminum plates coated with silica gel 60 F_{254} and visualized with UV at 254 nm and $KMnO_4$ solution [74].

Kapotas acclimatized a prototype sensitive strain of *P. aeruginosa* in vitro to imipenem by serial transfers in broth containing increasing concentrations of the antibiotic. Lipopolysaccharide chemical analysis revealed a marked, partially reversible increase in 2-keto-3-deoxyoctonate, total hexose, and heptose constituents; a readily extractable lipid chemical analysis and TLC revealed a marked, partially reversible increase in the phospholipid content of the outer membrane. Polar lipids in the fractions were separated by TLC in a multisolvent system and spotted on silica gel 60 plates. They were visualized by spraying the developer plates with 50% H_2SO_4 saturated with potassium dichromate and heating at 155°C for 45 min [75].

Mendez et al. carried out thermal and alkaline stability of meropenem (MEPM) in aqueous solution through the employment of stress conditions. The stability of MEPM was investigated in order to isolate and elucidate the mean degradation products involved in thermal and alkaline decomposition of MEPM in solution. The purification of thermal degradation product (45°C) involved a combination of preparative chromatographic techniques. The degradates were characterized by NMR and electrospray ionization mass spectrometry (ESI-MS). The thermal degradation product was a result of several chemical reactions, resulting in a pyrrolic derivative. Under alkaline conditions (0.1 N NaOH), MEPM was converted totally to the corresponding β-lactam ring-opened derivative, in sodium salt state. The degradation product was isolated by column chromatography and preparative TLC. The separation was carried out by the addition of mixtures of methylene chloride and methanol (increasing 4% of methanol in each portion until reaching the composition 40:60, v/v). Each fraction was analyzed by TLC applying methylene chloride:methanol (95:0.5, v/v) as mobile phase and revelation by UV 254 nm light. Those that showed to have degradation products were analyzed also by HPLC. The fractions containing the main degradation product were mixed and purified by preparative TLC. The spot corresponding to the degradation product (R_F 0.71) was removed and extracted with mixture of methylene chloride:methanol (50:50, v/v). The solvent of this final solution was distilled to dryness under reduced pressure, and the solid was stored. The thermal degradation product isolated by preparative TLC and column chromatography was the n-pyrrole derivative originated from the hydrolysis, decarboxylation, and alkyl chain elimination reactions [76].

Mendez et al. described an analytical study applied to the characterization of the antibiotic doripenem in bulk by different techniques: TLC, HPLC, UV spectrophotometry, IR spectroscopy, ^{1}H and ^{13}C NMR, and thermogravimetric analysis. The reference standard or bulk solution was prepared by dissolving the drug in methanol. Different chromatographic systems were tested, consisting of mixtures of ethyl acetate, chloroform, glacial acetic acid, and methanol in various proportions. Finally, the mobile phase consisting of butyl alcohol:glacial acetic acid:water (4:2:2, v/v/v) was used. The TLC plate was transferred to a glass chamber, previously lined with paper and saturated for 15 min with the mobile phase. After development, the plate was dried and examined under UV chamber (254 and 365 nm) and stained with iodine vapor. It was possible to observe the stain of doripenem at R_F value of 0.62, similar to the standard reference. The work was successfully applied to qualitative analysis of antibiotic [77]. Elragehy et al. described a TLC-densitometric method for the determination of MEPM singly in bulk form, pharmaceutical formulations, and/or in the presence of its major degradate. The degradation product has been isolated, via acid degradation, characterized, and confirmed. The selective quantification of MEPM was demonstrated. Different solvent systems were tried for the separation of MEPM and its degradate. Satisfactory results were obtained using a mobile phase consisting of n-butanol:acetone:water (4:3:3, v/v/v), with R_F 0.34 and 0.16 for MEPM and its degradate, respectively. The separation allows the determination of MEPM with no interference from its degradate. Scanning profile of different concentrations of MEPM at 298 nm was done [78]. Nakajima et al. clarified the mechanism of the drug–drug interaction using panipenem/betamipron (PAPM), MEPM, and doripenem (S-4661), a newly synthesized carbapenem. The antiepileptic valproic acid (VPA) ^{14}C-VPA-G was extracted with methanol from SEP-PAK C18 in which urine and bile, obtained after intravenous administration of ^{14}C-VPA to rats, were applied. The methanol extract was condensed and applied to a TLC plate. The TLC plate was developed with the solvent mixture of ethyl acetate:acetic acid:water (4:1:1, v/v/v). The ^{14}C-VPA-G fraction on the TLC plate, visualized by radioluminography using BAS 2000, was scrapped and extracted with methanol and then used as substrate. Under pentobarbital anesthesia, monkeys were killed by bleeding, and the liver and kidneys were removed and stored in chilled Krebs–Ringer bicarbonate buffer (KRB). After the connective tissues and fat had been removed, the liver and kidneys were cut into slices. One gram of liver and kidney slice was added to a flask containing KRB that had been bubbled with an $O_2:CO_2$ (95:5) gas mixture for more than 15 min. Next, ^{14}C-VPA and various concentrations of carbapenems (PAPM, MEPM, and S-4661) or flomoxef sodium (an analog compound of the cephem group) were added. The mixtures were finally bubbled again with the same gas mixture, incubated at 37°C, and homogenized. The portions of each homogenate were acidified by adding 1 N HCl and centrifuged, and the resultant supernatant was used for TLC analysis. The supernatant obtained from the incubation with ^{14}C-VPA was extracted with a solvent mixture of n-hexane:ethyl acetate (9:1, v/v) and then applied to DIAION HP-20. The column was washed with 0.1 N HCl and eluted with acetone. The solution was dried, dissolved into a small volume of 95% tetrahydrofuran aqueous solution, and subjected to TLC using silica gel 60 F$_{254}$ precoated plates with a developing solvent system of ethyl acetate:acetic acid:water (4:1:1, v/v/v). The TLC plate was brought into contact with x-ray film to obtain a radioautogram. Next, each fraction was scraped off, and its radioactivity was counted with a Tri-Carb 2200 CA liquid scintillation counter [79].

Okabe et al. devised a 1D silica gel TLC method for qualitative and quantitative analysis of carbapenem antibiotics in fermentation broths. The antibiotics were extracted from the broth filtrate with a solvent mixture of 1-butanol:chloroform (1:1, v/v) and developed on a silica gel thin-layer plate with a mobile phase of acetonitrile:0.75% acetic acid (9:2, v/v). The plates were visualized as reddish pink spots with the Ehrlich reagent and quantitated by densitometry [80].

Shibayama et al. investigated the distribution, metabolism, and excretion of CS-023, a new carbapenem, in rats and monkeys after a single intravenous administration of [^{14}C]CS-023. In addition, the drug's pharmacokinetics were examined in rats, dogs, and monkeys. Blood samples were collected with heparinized syringes from the jugular veins of rats and from the femoral veins of monkeys at predetermined times after drug administration. The blood samples were centrifuged at 4°C to obtain plasma samples. Aliquots were taken from each plasma sample for radioactivity measurement and mixed with the same volume of 0.2 M 3-morpholinopropanesulfonic acid (MOPS) pH 6.0. To the mixture, a fourfold volume of acetonitrile was added for rat plasma or a twofold volume of acetonitrile for monkey plasma. After centrifugation at 4°C, the supernatant was stored frozen at −20°C until analysis by TLC. Monkeys were housed individually in metabolic cages after dosing for separate collection of urine and feces over a 7-day period. A 0.5 M MOPS buffer (pH 6.0) was added to each of the urine receptacles cooled at around 4°C. Fecal samples were added to 0.2 M MOPS buffer (pH 6.0) and homogenized. Aliquots were taken from the urine samples and the fecal homogenates for measurement of radioactivity, and the remaining samples were stored frozen at −80°C until analysis by TLC. An aliquot of a plasma, urine, or fecal homogenate sample was spotted onto a TLC plate silica gel 60 F$_{254}$. The plate was developed with a mixture of a 3% NaCl solution:acetonitrile (2:1, v/v). After development, the plate was air-dried completely and put in contact with an imaging plate for about 24 h. The imaging plate was subjected to imaging analysis with a bioimaging analyzer. The amounts of CS-023 and its metabolites in each sample were calculated by multiplying the radioactive proportions of the respective spots in TLC with the total radioactivity in each sample [81].

47.5 SIMULTANEOUS STUDY OF SUBGROUPS OF β-LACTAM ANTIBIOTICS

Thangadurai et al. used unimpregnated TLC silica gel F$_{254}$ plates as an adsorbent for the separation of β-lactam and fluoroquinolone antibiotics. The plates were activated at 110°C for 30 min before use. Twelve types of β-lactam antibiotics—ampicillin, amoxicillin, benzylpenicillin, benzathine penicillin, bacampicillin, cloxacillin, cephalexin, cefotaxime, ceftriaxone, cefadroxil, cefazolin, and penicillin G procaine—were prepared by dissolving in methanol and analyzed. Nondestructive procedures, such as the use of UV light (254 and 356 nm), were used for the localization of separated spots. The plates were sprayed using two different methods with chromogenic reagents: iodine azide reagent and ninhydrin reagent heated at 110°C for the detection of β-lactams. The R$_F$ values in various solvent systems and the colors developed at each storage for β-lactam antibiotics were recorded [82]. Joshi et al. presented a method for the separation and detection of selected β-lactams. A variety of mobile phase were reported for use on ammonium chloride-impregnated silica gel G layers with fluorescent indicator. After several runs, the concentration of ammonium chloride was optimized as 0.2%. TLC was performed on silica gel G containing 13% calcium sulfate, iron, and 0.03% chloride and on readymade TLC silica gel 60 plates. Impregnated TLC plates were prepared by spreading a slurry of silica gel G in 0.2% ammonium chloride solution with a Stahl-type applicator and then drying overnight at 60°C in an oven. Ready-made silica gel 60 plates were impregnated by development with 0.2% ammonium chloride solution. These were dried overnight at 60°C. Detection was performed in UV at 365 nm. Propanol:acetic acid (4:1, v/v) enables successful separation of benzylpenicillin, ampicillin, and amoxicillin on silica gel G layers impregnated with 0.2% ammonium chloride, and butanol:acetic acid:water (4:1:2, v/v/v) enables the separation of five cephalosporins, cephalexin, cefoperazone, ceftriaxone, cefixime, and cefadroxil, on silica gel 60 plates impregnated with 0.2% ammonium chloride [83]. Lu and Long envisioned an alternative approach that could be used to generate chiral β-lactams to install a hydroxyl group at the C-3 methylene of L-cysteine-derived thiazolidines by aerobic photooxidation. The authors described the first asymmetric synthesis of N-protio monocyclic β-lactams from L-cysteine that is facilitated by the aerobic photooxidation of thiazolidine hydroxamate esters. The photooxidation of thiazolidines was examined using singlet O_2 generated from triplet O_2 via a 500 W halogen lamp in the presence of a photosensitizer, tetraphenylporphyrin. Various hydroxamate esters that would be useful in the synthesis of cephalosporins, monobactams, penems, oxamazins, and monosulfactams were evaluated in the reaction. The photooxidations were monitored by TLC; however, the hydroperoxide intermediates were often indiscernible on the plates. TLC was performed on silica gel layers with hexane:ethyl acetate (1:1, v/v). The study was resolved by co-spotting with Me_2S, which converted the hydroperoxide to thiohemiacetal on the TLC plate, and allowing the chemically stable product to be readily detected at a lower R_F value. The reaction scales ranged from 0.4 to 2.0 g with a slight to moderate decrease in the resulting yields when over 1.0 g of hydroxamate ester was used. In many instances, the yields after purification were ≥60%, with the highest consistently observed for thiazolidines bearing an N-benzoyl protecting group [84].

Sawai et al. described a detailed analysis of the outer membrane permeation of cephalosporins and penicillins. The analysis was made by using mutant strains of *E. coli*, *Proteus mirabilis*, and *Enterobacter cloacae*, which had lost their major outer membrane proteins. The analyzed antibiotics were kindly provided by the following pharmaceutical companies: benzylpenicillin, ampicillin, cloxacillin, carbenicillin, cefazolin, cephalothin, cephaloridine, piperacillin, and cefoxitin. The hydrophobic character of various antibiotics was expressed as the R_F value, which was measured by RP TLC. The polar mobile phase consisted of acetate:Veronal buffer (pH 7.0):methanol (4:1, v/v). TLC silica gel 60 F_{254} plates were used as the nonpolar stationary phase. A sample was dissolved in the acetate:Veronal buffer or methanol. After development at room temperature, the antibiotic was detected on the plate by using iodine vapor [85].

Husain et al. studied the chromatographic behavior of amoxicillin, ampicillin, cephalexin, cloxacillin, doxycycline, tetracycline, erythromycin, streptomycin, and co-trimoxazole that were studied on thin layers of titanic silicate inorganic ion exchanger, with organic, aqueous, and mixed aqueous–organic mobile phases. Capsules of amoxicillin, ampicillin, cephalexin, cloxacillin, streptomycin, doxycycline, and tetracycline were dissolved in ethanol. Titanium(IV) silicate ion-exchange plate was prepared by addition of 0.25 M sodium silicate solution to 0.08 M titanic chloride solution in 0.2 M HCl. The pH of the mixture was adjusted to 6.5 by 2 M NaOH. The gel formed was left to settle overnight and then washed with distilled water. The supernatant was removed completely, and a slurry was prepared by mixing the gel with gypsum. The slurry was poured into an automatic TLC plate coater and used to coat glass plates with a 300 μm layer. The plates were dried in an oven at 60°C for 2 h then stored at room temperature inside a desiccator. All used glassware was acid washed and light resistant. Antibiotic solutions were applied to the plates as circular spots. The plates were developed in ascending mode, without conditioning, in a twin-trough chamber. After development, the plates were dried in an air oven, and the antibiotics were detected with appropriate reagents (1%, w/v); ninhydrin in ethanol was used to locate amoxicillin, ampicillin, cephalexin, cloxacillin, gentamicin, and co-trimoxazole, and 5% (w/v) potassium dichromate in concentrated H_2SO_4 was used to locate streptomycin, erythromycin, tetracycline, and doxycycline. The effect of varying the volume ratio of the binary mobile phase methanol:0.1 M formic acid on the R_F values of the antibiotics was studied. A gradual increase in the concentration of ammonium sulfate results in a decrease in the R_F values of the antibiotics [86].

Chen et al. developed a fluorescent spot test method for specific detection of microbial β-lactamases. A β-lactam substrate solution consisting of a β-lactam antibiotic with an acyl side chain containing an α-amino group and an α-phenyl group, or its derivatives, was incubated with a β-lactamase-producing organism. The fluorescence developer consisted of 0.5 mM $HgCl_2$ in 0.5 M sodium citrate buffer pH 4.5, prepared in 0.5% formaldehyde aqueous solution, was used. It was added to the incubated β-lactam substrate solution, followed by heating the mixture at 45°C for 10 min. Production of fluorophore indicated β-lactamase activity. Each fluorophore was analyzed by TLC, and its chemical identity was determined. Using ampicillin as the penicillinase substrate and cephalexin as the cephalosporinase substrate, the method was carried out by using dropping bottles for storing and dispensing the substrate solutions and the fluorescence developer. The modified method provided more favorable conditions for the penicillinases to remain active during fluorescence development. Therefore, the sensitivity of the test was increased [87].

Shahverdi et al. screened various plants for their ability to decrease bacterial resistance to penicillin G and cephalexin, which are extensively used to treat infections caused by bacteria. An extract prepared from the *Ferula szowitsiana* roots and its active component (galbanic acid) on the antimicrobial activity of mentioned antibiotics were evaluated against *S. aureus*. Disk diffusion and broth dilution methods were used to determine the antibacterial activity of these drugs in the absence and presence of plant extract. Its various fractions were separated by TLC plate. The residue was fractionated by TLC on silica gel 60 F_{254} using petroleum ether:ethyl acetate (2:1, v/v) as the solvent system. TLC analysis of the acetone extract of *F. szowitsiana* roots showed at least six distinct fractions, which were visualized UV at 254 nm. The active component of plant extract involved in enhancement of penicillin G's and cephalexin's activities had R_F 0.336. In the presence of subinhibitory concentration of galbanic acid, the combination of galbanic acid and antibiotics of penicillin G for *S. aureus* decreased from 1 to 64 and for cephalexin a 128-fold decrease. The acetone extract of *F. szowitsiana* roots nor any of the fractions eluted from the preparative TLC plates showed any antimicrobial activity against test strains [88].

Kuroda et al. investigated farnesol's effects on antimicrobial susceptibilities of methicillin-susceptible and methicillin-resistant *S. aureus* to ampicillin, oxacillin, cefoxitin, bacitracin, teicoplanin, amikacin, ciprofloxacin, and clarithromycin by MIC (minimal inhibitory concentration) determination using the Etest, penicillin-binding protein PBP20 expression by Western blot analysis, β-lactamase secretion and activity by in vivo and in vitro farnesol inhibition assays, staphyloxanthin production by TLC, cell wall synthesis by [^{14}C]GlcNAc (where GlcNAc stands for *N*-acetylglucosamine) and [^{14}C]mevalonate incorporation assays, and TLC-based lipid extract profile analysis. Detection of staphyloxanthin production by *S. aureus* N315 was grown on Mueller-Hinton Tellurite (MHT) agar supplemented with sub-MIC concentrations of farnesol, oxacillin, bacitracin, teicoplanin, amikacin, or ciprofloxacin at 35°C for 20 h. The colonies were collected and suspended in distilled water. The cell pellet harvested after centrifugation was resuspended in phosphate-buffered saline containing lysostaphin and incubated at 37°C for 30 min. The cell lysate was mixed with ethyl acetate and centrifuged. An aliquot of the ethyl acetate phase was recovered and evaporated in the dark. The dried sample was dissolved in ethyl acetate, and the extract was separated on an HPTLC silica gel 60 plate with chloroform:methanol:water (65:25:4, v/v/v). The pigment spots were visually observed. For visualization of the whole lipid extracts, the plate was soaked in 10% sulfuric acid and baked at 150°C until the reaction was visible. The analysis of the C55 lipid carrier synthesis by HPTLC *S. aureus* N315 cells was labeled with [^{14}C]mevalonate, and during the labeling reaction, the cells were exposed to 1% Tween 80 as control, farnesol, bacitracin, or vancomycin. The cell pellets were suspended with chloroform:methanol (1:1, v/v), and the labeled lipids were extracted. An aliquot of the lipid extract was dried and resuspended in chloroform and then separated on an HPTLC silica gel 60 plate with chloroform:methanol:water:ammonia (88:48:10:1, v/v/v/v). The radioactive spots were detected after exposure to an imaging plate for 3 days and scanning [89].

REFERENCES

1. Nabi, S.A., Khan, M.A., Khowaja, S.N., and A. Alimuddin. 2006. Thin-layer chromatographic separation of penicillins on stannic arsenate–cellulose layers. *Acta Chromatogr.* 16:164–172.
2. El Sadek, M. 1986. Spectrodensitometric analysis of penicillin V benzathine oral suspension. *Analyst* 111:579–580.
3. Hendrickx, S., Roets, E., Hoogmartens, J., and H. Vanderhaeghe. 1984. Identification of penicillins by thin-layer chromatography. *J. Chromatogr. A* 291:211–218.
4. Gayen, J.R., Majee, S.B., Das, S., and T.B. Samanta. 2007. Antibacterial and toxicological evaluation of β-lactams synthesized by immobilized β-lactamase-free penicillin amidase produced by *Alicaligens sp. Indian J. Exp. Biol.* 45:1068–1072.
5. Biagi, G.L., Barbaro, A.M., Gamba, M.F., and M.C. Guerra. 1964. Partition data of penicillins determined by means of reversed-phase thin-layer chromatography. *J. Chromatogr. A* 41:371–379.
6. Baltzer, B., Lund, F., and N. Rastrup-Andersen. 1979. Degradation of mecillinam in aqueous solution. *J. Pharm. Sci.* 68:1207–1215.
7. Ellerbrok, H. and R. Hakenbeck. 1988. Penicillin-degrading activities of peptides from pneumococcal penicillin-binding proteins. *Eur. J. Biochem.* 171:219–224.
8. Płotkowiak, Z., Popielarz-Brzezińska, M., and M. Serafin. 2003. Methods for the assessment of quality and stability of azidocillin. *Acta Pol. Pharm.* 60:112–115.
9. Qureshi, S.Z., Qayoom, T., and M.I. Helalet. 1999. Simultaneous spectrophotometric and volumetric determinations of amoxycillin, ampicillin and cloxacillin in drug formulations: Reaction mechanism in the base catalysed hydrolysis followed by oxidation with iodate in dilute acid solution. *J. Pharm. Biomed. Anal.* 21:473–482.
10. Omar, N., Saleh, G., Neugebauer, M., and G. Rücker. 1988. Spectrodensitometric analysis of penicillins by detection with iodine. *Anal. Lett.* 21:1337–1345.
11. Ruzin, A., Singh, G., Severin, A. et al. 2004. Mechanism of action of the mannopeptimycins, a novel class of glycopeptide antibiotics active against vancomycin-resistant Gram-positive bacteria. *Antimicrob. Agents Chemother.* 48:728–738.
12. Bijev, A.T. and V. Hung. 2001. Synthesis and antimicrobial activity of new pyrrolecarboxylic acid derivatives of ampicillin and amoxicillin. *Arzneimittelforschung* 51:667–672.
13. Marciniec, B., Płotkowiak, Z., Wachowski, L., Kozak, M., and M. Popielarz-Brzezińska. 2002. Analytical study of β-irradiated antibiotics in the solid state. *J. Therm. Anal. Calorim.* 68:423–436.
14. Torroba, J., Varela-Diaz, V.M., Vivino, E.C., and J.A. Mesquita. 1993. Measurement of antibiotic levels in foot-and-mouth disease oil vaccines by chemical methods. *Biol. Centr. Panam. Fiebre Aftosa* 59:118–124.
15. Gholipour, V., Shokri, M., and S. Waqif-Husain. 2011. Determination of ampicillin and amoxicillin by high-performance thin-layer chromatography. *Acta Chromatogr.* 23:483–498.
16. Hurwitz, A.R. and C.F. Carney. 1978. Enhancement of ampicillin partition behavior. *J. Pharm. Sci.* 67:138–140.
17. Graber, H., Perenyi, T., Arr, M., and E. Ludwig. 1976. On human biotransformation of some penicillins. *Int. J. Clin. Pharmacol. Biopharmacol. Biopharm.* 14:284–289.
18. Ulhnann, U. and W. Wurst. 1979. Antibacterial active components in human urine after administration of penicillins. *Infection* 7:187–189.
19. Okuno, S., Maezawa, I., Sakuma, Y., Matsushita, T., and T. Yamaguchi. 1988. Effect of the hydroxyl group of the p-hydroxyphenyl moiety of aspoxicillin, a semisynthetic penicillin, on its pharmacokinetic property. *J. Antibiot.* 41:239–246.
20. Ramirez, A., Gutierrez, R., Diaz, G. et al. 2003. High-performance thin-layer chromatography-bioautography for multiple antibiotic residues in cow's milk. *J. Chromatogr. B* 784:315–322.
21. Lin, S.Y. and F. Kondo. 1994. Simple bacteriological and thin-layer chromatographic methods for determination of individual drug concentrations treated with penicillin-G in combination with one of the aminoglycosides. *Microbios* 77:223–229.
22. Okerman, L., Croubels, S., De Baere, S., Van Hoof, J., De Backer, P., and H. De Brabander. 2001. Inhibition tests for detection and presumptive identification of tetracyclines, beta-lactam antibiotics and quinolones in poultry meat. *Food Addit. Contam.* 18:385–393.
23. Gafner, J.L. 1999. Identification and semiquantitative estimation of antibiotics added to complete feeds, premixes, and concentrates. *J. AOAC Int.* 82:1–8.
24. Dolui, A.K., Sahana, S., and A. Kumar. 2012. Studies on production of 6-aminopenicillanic acid by free and κ-Carrageenan immobilized soil bacteria. *Indian J. Pharm. Edu. Res.* 46:70–74.

25. Kaya, S.E. and A. Filazi. 2010. Determination of antibiotic residues in milk samples. *Kafkas Univ. Vet. Fak. Deg.* 16:S31–S32.

26. Qin, L., Li, B., Guan, J., and G. Zhang. 2009. In vitro synergistic antibacterial activities of helvolic acid on multi-drug resistant *Staphylococcus aureus*. *Nat. Prod. Res: Form. Nat. Prod. Lett.* 23:309–318.

27. Misztal, G. 1999. Thin-layer reversed phase ion pair chromatography of some cephalosporins. *J. Liq. Chromatogr. Related Technol.* 22:1589–1598.

28. Dhanesar, S.C. 1998. Quantitation of antibiotics by densitometry on a hydrocarbon impregnated silica gel HPTLC plate. Part I: Evaluation of the HPTLC plate. *J. Planar Chromatogr.—Modern TLC* 11:195–200.

29. Dhanesar, S.C. 1998. Quantitation of antibiotics by densitometry on a hydrocarbon-impregnated silica gel HPTLC plate. Part II: Quantitation and evaluation of ceftriaxone. *J. Planar Chromatogr.—Modern TLC* 11:285–262.

30. Okumura, T. 1981. Application of thin-layer chromatography to high-performance liquid chromatographic separation of steroidal hormones and cephalosporin antibiotics. *J. Liq. Chromatogr.* 4:1053–1064.

31. Eric, S., Agbaba, D., Vladimirow, S., and D. Zivanov-Stakic. 2000. The retention behavior of several semisynthetic cephalosporins in planar chromatography. *J. Planar Chromatogr.—Modern TLC* 2:88–92.

32. Dhanesar, S.C. 1999. Quantitation of antibiotics by densitometry on a hydrocarbon-impregnated silica gel HPTLC plate. Part III: Quantitation and evaluation of cephalosporins. *J. Planar Chromatogr.—Modern TLC* 12:114–119.

33. Bhushan, R. and V. Parshad. 1996. Separation and identification of some cephalosporins on impregnated TLC plates. *Biomed. Chromatogr.* 10:258–260.

34. Bhushan, R. and G.T. Thiong'o. 2002. Separation of cephalosporins on thin silica layers impregnated with transition metal ions and by reversed-phase TLC. *Biomed. Chromatogr.* 16:165–174.

35. Sharma, S., Singh, S., and S. Baghel. 2003. Simple analysis of cephalexin and cefaclor in pharmaceutical preparation with UV detection. *J. Ultra Chem.* 1:13–18.

36. Tuzimski, T. 2004. Two-dimensional thin-layer chromatography of eight cephalosporins on silica gel layers. *J. Planar Chromatogr.* 17:46–50.

37. Misztal, G., Szalast, A., and H. Hopkała. 1998. Thin-layer chromatographic analysis of cephalosporins. *Chem. Anal.* 43:357–363.

38. Choma, I.M. 2007. TLC separation of cephalosporins: Searching for better selectivity. *J. Liq. Chromatogr. Related Technol.* 30:2231–2244.

39. Fujisawa, Y., Uchida, M., and M. Suzuki. 1982. S-sulfocysteine as a useful sulfur Skurce for cephalosporin C biosynthesis by *Cephalosporium acremonium*. *Agric. Biol. Chem.* 46:1519–1523.

40. Nabi, S.A., Laiq, E., and A. Islam. 2004. Selective separation and determination of cephalosporins by TLC on stannic oxide layers. *Acta Chromatogr.* 14:92–101.

41. Dąbrowska, M. and J. Krzek. 2010. Chiral separation of diastereoisomers of cefuroxime axetil by high-performance thin-layer chromatography. *J. AOAC Int.* 93:771–777.

42. Dąbrowska, M. and J. Krzek. 2010. Separation, identification, and quantitative analysis of the epimers of cefaclor by TLC–densitometry. *J. Planar Chromatogr.* 23:265–269.

43. Yotsuji, A., Mitsuyama, J., Hori, R., Yasuda, T., Saikawa, I., Inoue, M., and S. Mitsuhashi. 1988. Mechanism of action of cephalosporins and resistance caused by decreased affinity for penicillin-binding proteins in *Bacteroides fragilis*. *Antimicrob. Agents Chemother.* 32:1848–1853.

44. Yotsuji, A., Mitsuyama, J., Hori, R., Yasuda, T., Saikawa, I., Inoue, M., and S. Mitsuhashi. 1988. Outer membrane permeation of *Bacteroides fragilis* by cephalosporins. *Antimicrob. Agents Chemother.* 32:1097–1099.

45. Darji, B.H., Shah, N.J., Patel, A.T., and N.M. Patel. 2007. Development and validation of a HPTLC method for the estimation of cefpodoxime proxetil. *Indian J. Pharm. Sci.* 69:331–333.

46. Date, A.A. and M.S. Nagarsenke. 2007. HPTLC determination of cefpodoxime proxetil in formulations. *Chromatographia* 66:905–908.

47. Coran, S.A., Bambagiotti-Alberti, M., Giannellini, V., Baldi, A., Picchioni, G., and F. Paoli. 1998. Development of a densitometric method for the determination of cephalexin as an alternative to the standard HPLC procedure. *J. Pharm. Biomed. Anal.* 18:271–274.

48. Jeswani, R., Sinha, P., Topagi, K., and M. Damle. 2009. A validated stability indicating HPTLC method for determination of cephalexin in bulk and pharmaceutical formulation. *Int. J. PharmTech Res.* 3:527–538.

49. Shah, N.J., Shah, S.K., Patel, V.F., and N.M. Patel. 2007. Development and validation of a HPTLC method for the estimation of cefuroxime axetil. *Indian J. Pharm. Sci.* 69:140–142.

50. Sengar, M.R., Gandhi, S.V., Patil, U.P., Rajmane, V.S., and K.G. Bothara. 2001. A validated densitometric TLC method for analysis of cefuroxime axetil and potassium clavulanate in combined tablet dosage forms. *Acta Chromatogr.* 22:91–97.

51. Sireesha, K.R., Mhaske, D.V., Kadam, S.S., and S.R. Dhaneshwar. 2004. Development and validation of a HPTLC method for simultaneous estimation of cefuroxime axetil and probenecid. *Indian J. Pharm. Sci.* 66:278–282.

52. Phattanawasin, P., Sotanaphun, U., Sriphong, L., and I. Kanchanaphibool. 2011. Stability-indicating TLC-image analysis method for quantification of ceftriaxone sodium in pharmaceutical dosage forms. *J. Planar Chromatogr.* 24:30–34.

53. Sreekanth, N., Shivshanker, P., Shanmuga Pandiyan, P., and C. Rooswelt. 2007. Simultaneous determination and validation of ornidazole and cefixime by HPTLC in pure and pharmaceutical dosage forms. *Asian J. Chem.* 19:3621–3626.

54. Krzek, J. and M. Dąbrowska-Tylka. 2003. Simultaneous determination of cefuroxime axetil and cefuroxime in pharmaceutical preparations by thin-layer chromatography and densitometry. *Chromatographia* 58:231–234.

55. Ranjane, P.N., Gandhi, S.V., Kadukar, S.S., and K.G. Bothara. 2010. HPTLC determination of cefuroxime axetil and ornidazole in combined tablet dosage form. *J. Chromatogr. Sci.* 48:26–28.

56. Agbaba, D., Eric, S., Zivanov Stakic, D., and S. Vladimirov. 1998. HPTLC assay of cephalexin and cefaclor in pharmaceuticals. *Biomed. Chromatogr.* 12:133–135.

57. Saleh, G.A., Mohamed, F.A., El-Shaboury, S.R., and A.H. Rageh. 2010. Selective densitometric determination of four α-aminocephalosporins using ninhydrin reagent. *J. Chromatogr. Sci.* 48:68–75.

58. Eric-Jovanovic, S., Agbaba, D., Zivanov-Stakic, D., and S. Vladimirov. 1998. HPTLC determination of ceftriaxone, cefixime and cefotaxime in dosage forms. *J. Pharm. Biomed. Anal.* 18:893–898.

59. Singh, D.K. and G. Maheshwari. 2010. Chromatographic studies of some cephalosporins on thin layers of silica gel G–zinc ferrocyanide. *Biomed. Chromatogr.* 24:1084–1088.

60. Mohamed, F.A., Saleh, G.A., El-Shaboury, S.R., and A.H. Rageh. 2008. Selective densitometric analysis of cephalosporins using Dragendorff's reagent. *Chromatographia* 68:365–374.

61. Dąbrowska, M., Starek, M., and S. Pikulska. 2011. Simultaneous identification and quantitative analysis of eight cephalosporins in pharmaceutical formulations by TLC–densitometry. *J. Planar Chromatogr.* 24:23–29.

62. Dąbrowska, M., Krzek, J., and E. Miękina. 2012. Stability analysis of cefaclor and its inclusion complexes of β-cyclodextrin by thin-layer chromatography and densitometry. *J. Planar Chromatogr.* 25:127–132.

63. Mohammad, M.A., Zawilla, N.H., El-Anwar, F.M., and S.M. Aly. 2007. Column and thin-layer chromatographic methods for the simultaneous determination of acediasulfone in the presence of cinchocaine, and cefuroxime in the presence of its hydrolytic degradation products. *J. AOAC Int.* 90:405–413.

64. Saleh, G., Askal, H., and N. Omar. 1990. Use of charge-transfer complexation in the spectro-photometric and spectrodensitometric analysis of cephalotin sodium and cephradine. *Anal. Lett.* 23:833–841.

65. Palena, A.A.P. and E.G. Mata. 2005. Thionation of bicyclic β-lactam compounds by Lawesson's reagent. *ARKIVOC* 12:282–294.

66. Fabre, H., Blanchin, M.D., and B. Mandrou. 1985. Determination of cephalosporins utilising thin-layer chromatography with fluorescamine detection. *Analyst* 110:775–778.

67. Kojima, R., Ito, M., and Y. Suzuki. 1988. Studies on the nephrotoxicity of aminoglycoside antibiotics and protection from these effects. (5). Interaction of tobramycin with latamoxef in vitro. *J. Pharmacobiodyn.* 11:9–17.

68. Blanchin, M.D., Rondot-Dudragne, M.L., Fabre, H., and B. Mandrou. 1988. Determination of cefotaxime, desacetylcefotaxime, cefmenoxime and ceftizoxime in biological samples by fluorescence detection after separation by thin-layer chromatography. *Analyst* 113:899–902.

69. Grzelak, E., Malinowska, I., and I. Choma. 2009. Determination of cefacetrile and cefuroxime residues in milk by thin-layer chromatography. *J. Liq. Chromatogr. Related Technol.* 32:2043–2049.

70. Choma, I.M., Kowalski, C., Lodkowski, R., Burmańczuk, A., and I. Komaniecka. 2008. TLC-DB as an alternative to the HPLC method in the determination of cefacetril residues in cow's milk. *J. Liq. Chromatogr. Related Technol.* 31:1903–1912.

71. Long, B., He, Ch., Yang, Y., and J. Xiang. 2010. Synthesis, characterization and antibacterial activities of some new ferrocene-containing penems. *Eur. J. Med. Chem.* 45:1181–1188.

72. Tanaka, R., Oyama, Y., Imajo, S., Matsuki, S., and M. Ishiguro. 1997. Structure-activity relationships of penem antibiotics: Crystallographic structures and implications for their antimicrobial activities. *Bioorg. Med. Chem.* 5:1389–1399.

73. Jayaraman, M., Deshmukh, A.R., and B.M. Bhawal. 1996. Application of (+)-(1S,2S)-2-amino-1-phenylpropan-1,3-diol in the formal total synthesis of carbapenems, novel 4-cyano-β-lactams and β-hydroxy aspartates. *Tetrahedron* 52:8989–9004.

74. Urbach, A., Dive, G., Tinant, B., Duval, V., and J. Marchand-Brynaert. 2009. Large ring 1,3-bridged 2-azetidinones: Experimental and theoretical studies. *Eur. J. Med. Chem.* 44:2071–2080.

75. Kapotas, N.M. 1991. Acclimatization resistance of a *Pseudomonas aeruginosa* prototype strain to imipenem. *J. Antibiot.* 44:985–994.

76. Mendez, A., Chagastelles, P., Palma, E., Nardi, N., and E. Schapoval. 2008. Thermal and alkaline stability of meropenem: Degradation products and cytotoxicity. *Int. J. Pharm.* 350:95–102.

77. Mendez, A.S.L., Mantovani, L., Barbosa, F. et al. 2011. Characterization of the antibiotic doripenem using physicochemical methods—Chromatography, spectrophotometry, spectroscopy and thermal analysis. *Quim. Nova* 34:1634–1638.

78. Elragehy, N.A., Abdel-Moety, E.M., Hassan, N.Y., and M.R. Rezk. 2008. Stability-indicating determination of meropenem in presence of its degradation product. *Talanta* 77:28–36.

79. Nakajima, Y., Mizobuchi, M., Nakamura, M. et al. 2004. Mechanism of the drug interaction between valproic acid and carbapenem antibiotics in monkeys and rats. *DMD* 32:1383–1391.

80. Okabe, M., Kiyoshima, K., Kojima, I., Okamoto, R., Fukagawa, Y., and T. Ishikura. 1983. Thin-layer chromatographic analysis of carbapenem antibiotics in fermentation broths. *J. Chromatogr. A* 256:447–454.

81. Shibayama, T., Matsushita, Y., Kawai, K., Hirota, T., Ikeda, T., and S. Kuwahara. 2007. Pharmacokinetics and disposition of CS-023 (RO4908463), a novel parenteral carbapenem, in animals. *Antimicrob. Agents Chemother.* 51:257–263.

82. Thangadurai, S., Shukla, S.K., and Y. Anjaneyulu. 2002. Separation and detection of certain β-lactam and fluoroquinolone antibiotic drugs by thin layer chromatography. *Anal. Sci.* 18:97–100.

83. Joshi, S., Sharma, A., Rawat, M.S.M., and Ch. Dhiman. 2009. Development of conditions for rapid thin-layer chromatography of β-lactam antibiotics. *J. Planar Chromatogr.* 22:435–437.

84. Lu, X. and T.E. Long. 2011. Asymmetric synthesis of monocyclic β-lactams from L-cysteine using photochemistry. *Tetrahedron Lett.* 52:5051–5054.

85. Sawai, T., Hiruma, R., Kawana, N., Kaneko, M., Taniyasu, F., and A. Inami. 1982. Outer membrane permeation of β-lactam antibiotics in *Escherichia coli*, *Proteus mirabilis*, and *Enterobacter cloacae*. *Antimicrob. Agents Chemother.* 22:585–592.

86. Husain, S.W., Gholipour, V., and H. Sepahrian. 2004. Chromatographic behavior of antibiotics on thin layers of an inorganic ion-exchanger. *Acta Chromatogr.* 14:102–109.

87. Chen, K.C.S., Chen, L., and J.Y. Lin. 1994. Fluorescent spot test method for specific detection of β-lactamases. *Anal. Biochem.* 219:53–60.

88. Shahverdi, A.R., Fakhimi, A., Zarrini, G., Dehghan, G., and M. Iranshahi. 2007. Galbanic acid from *Ferula szowitsiana* enhanced the antibacterial activity of penicillin G and cephalexin against *Staphylococcus aureus*. *Biol. Pharm. Bull.* 30:1805–1807.

89. Kuroda, M., Nagasaki, S., and T. Ohta. 2007. Sesquiterpene farnesol inhibits recycling of the C55 lipid carrier of the murein monomer precursor contributing to increased susceptibility to β-lactams in methicillin-resistant *Staphylococcus aureus*. *JAC* 59:425–432.

48 TLC of Other Antibiotics

Małgorzata Starek and Monika Dąbrowska

CONTENTS

Antibiotics are different chemical compounds that kill (microbicidal action) or suppress the growth or replication of pathogenic or nonpathogenic microorganisms such as bacteria, fungi, and protozoans. The large number of antibiotics, besides the β-lactam group, currently available can be classified in a variety of ways, for example, by their chemical structure, microbial origin, or mode of action.

48.1   TETRACYCLINE ANTIBIOTICS

Tetracyclines are one of the first group of antibiotics. They are a group of broad-spectrum antibiotics whose general usefulness has been reduced with the onset of bacterial resistance. Despite this, they remain the treatment of choice for some specific indications. Tetracyclines exert a bacteriostatic effect on bacteria by binding reversible to the bacterial 30S ribosomal subunit and blocking incoming aminoacyl tRNA from binding to the ribosome acceptor site. They also bind to some extent to the bacterial 50S ribosomal subunit and may alter the cytoplasmic membrane causing intracellular components to leak from bacterial cells. Tetracyclines are generally used in the treatment of infections of the respiratory tract, sinuses, middle ear, urinary tract, and intestines, especially in patients allergic to β-lactams and macrolides. They remain the treatment of choice for infections caused by chlamydia (trachoma, psittacosis, salpingitis, urethritis, and Lymphogranuloma venereum infection), rickettsia (typhus, Rocky Mountain spotted fever), brucellosis, and spirochetal infections (borreliosis, syphilis, and Lyme disease). In addition, they may be used to treat anthrax, plague, tularemia, and Legionnaires' disease. They are also used in veterinary medicine on pigs and alike.

48.1.1   SEPARATION AND DETERMINATION OF TETRACYCLINES

Crecelius et al. analyzed tetracyclines using different particle suspensions by TLC–MALDI–TOF-MS. The majority of the results were obtained for a suspension of graphite in ethylene glycol that was found to yield better sensitivity in comparison to the other tested materials and dispersants. Extracted ion chromatograms have been constructed from the TLC–MALDI analysis of chlortetracycline, tetracycline, oxytetracycline, and minocycline. Calculation of the R_F value of the detected spots showed good agreement with the R_F values obtained by ultraviolet (UV) detection. Pretreatment of the aluminum-backed silica gel 60 F_{254} TLC plates was necessary in order to avoid the formation of metal–tetracycline complexes and hence to improve the separation. The plates were sprayed with an aqueous disodium EDTA solution (0.27 M, pH 8), dried for 30 min, and activated at 120°C for 30 min. The mobile phase dichloromethane–methanol–water (59:35:8, v/v/v) was saturated for 2 h prior to use. The plates were eluted, dried, and visualized under UV light at 254 nm [1]. Dong et al. described a TLC–fluorescence scanning densitometry for the simultaneous determination of tetracycline and 4-epimeric tetracycline, using a mobile phase system of chloroform–methanol–acetone–1% aqueous ammonia (10:22:53:15, v/v/v/v) and a silica gel thin layer, previously treated with 0.27 M sodium EDTA solution (pH 9). Compounds were separated with R_F values of 0.64 and 0.24 [2]. Meisen et al. reported the direct coupling of HPTLC with IR–MALDI–o-TOF-MS for the analysis of oxytetracycline, chlortetracycline, tetracycline, and doxycycline. The samples were applied on glass-backed silica gel 60 precoated HPTLC plates and glass-backed RP-18 W plates. The silica gel plates were predeveloped in saturated aqueous EDTA solution or dipped in 9% aqueous EDTA (pH 9.0), dried at room temperature for 30 min, and heated at 130°C for 2 h prior to use. The plates were developed in chloroform–methanol–5% aqueous EDTA (13:4:1 or 10:4:1, v/v/v). A solvent system containing methanol–acetonitrile–0.5 M oxalic acid pH 2.5 (1:1:4, v/v/v) was used for the development of RP-18 W plates. Tetracycline bands were visualized by dipping the plates for 10 s in a solution of 0.5% aqueous Fast Blue B. After treatment, the plates were heated at 110°C until the appearance of colored bands. Further investigations using UV spectroscopy and IR–MALDI–o-TOF-MS were performed from unstained plates [3]. Oka et al. presented a method for 8 tetracyclines using silica gel HPTLC, reversed-phase (RP)-TLC, and HPLC. TLC separation was carried out using methanol–acetonitrile–0.5 M oxalic acid solution pH 3 (1:1:4, v/v/v) on silica gel HPTLC and C_8 TLC plates. As a detection reagent, the diazonium salts including Fast Violet B was used. For the determination of residual rolitetracycline, it was effective to measure the amount of rolitetracycline as tetracycline by HPLC, HPTLC, and RP-TLC after its incubation for 5 min in methanol at 50°C [4]. Oka et al. developed a semiquantitative screening method for tetracyclines using detection on silica gel HPTLC and RP-TLC plates. Detection with 1% Fast Violet B Salt solution followed by heating on the silica gel HPTLC plate, and with 0.5% Fast Violet B Salt solution, and pyridine without heating on the RP-TLC plate given a good results [5]. Oka et al. determined tetracyclines using silica gel HPTLC. A predevelopment with saturated disodium ethylenediaminetetraacetate aqueous solution and the complementary use of three solvent systems enable a reliable identification. A change of measurement wavelength enabled the determination of overlapping spots [6]. Xie et al. described a TLC–fluorescence scanning densitometry method for the simultaneous determination of doxycycline, tetracycline, and oxytetracycline. With a mobile phase chloroform–methanol–acetone–1% aqueous ammonia (10:22:50:18, v/v/v/v) and a silica gel thin layer, previously treated with 0.27 M sodium EDTA solution (pH 9), all drugs were separated from each other. The R_F values were 0.52, 0.35, and 0.21. The fluorescent spots were scanned with a spectrodensitometer at 365 nm as the excitation wavelength, and the cutoff filter was set at 440 nm [7]. Fernandez et al. reported the separation and simultaneous determination of tetracycline, 4-epitetracycline, anhydrotetracycline, and 4-epianhydrotetracycline by TLC on kieselgel impregnated with EDTA at pH 7.5 with ethyl acetate–acetone–water (10:20:3, v/v/v) as solvent or on kieselgel with EDTA (pH 9) and water–acetone (1:10, v/v) as solvent [8]. Kapadia et al. presented a TLC method for the separation of hydrochlorides of tetracycline, oxytetracycline, and chlortetracycline employing three sequestering agents—disodium ethylenediaminetetraacetate, tartaric acid, and oxalic acid—and n-butanol saturated with water as the mobile phase [9]. Ragazzi et al. separated tetracyclines on a cellulose TLC layer by development with aqueous solutions of certain salts (magnesium, calcium, barium, and zinc chloride). The spots exhibit fluorescence in UV light [10]. Dijkhuis et al. determined a small amount of epitetracycline and chlortetracycline in tetracycline by TLC on kieselgel layers, impregnated with a citrate phosphate solution (pH 5.5), containing 10% glycerin. After elution with 0.1 M HCl, epitetracycline was measured at 356 nm, while for chlortetracycline, the fluorimetric method was used. The possible identities of three other impurities were discussed [11].

Oka et al. studied the combinations of a C_8 TLC plate with methanol–acetonitrile–0.5 M aqueous oxalic acid solution pH 2 (1:1:4, v/v/v) as the solvent system and C_{18} TLC plate with methanol–acetonitrile–0.5 M aqueous oxalic acid solution pH 2 (1:1:2, v/v/v) as the solvent system for parent tetracyclines and impurities in tetracycline, respectively [12]. Willekens described the TLC separation of impurities in tetracycline (anhydrotetracycline, epianhydrotetracycline, epitetracycline, and chlortetracycline), on a kieselgel layer impregnated with ethylene glycol–water–acetone–ethyl acetate (2:2:15:15, v/v/v/v) as a mobile phase. Spectrophotometry or direct TLC fluorimetry was used for detection [13]. Ascione et al. examined tetracyclines (demethylchlortetracycline, tetracycline, and chlortetracycline) and their derivatives in a wide variety of mixtures by TLC method. Acid-washed diatomaceous earth served as the support. It was treated with a buffer consisting of 0.1 M ethylenediaminetetraacetic acid (EDTA) disodium salt at pH 7, glycerin, and polyethylene glycol 400. The developing solvent was ethyl acetate saturated with 0.1 M EDTA at pH 7 [14]. Naidong et al. reported a TLC–densitometry method for the assay and purity control of oxytetracycline and doxycycline. With a mobile phase of dichloromethane–methanol–water (59:35:6, v/v/v) and a silica gel thin layer, previously sprayed with 10% sodium edentate solution (pH 9), the potential impurities were separated from the main components and from each other [15]. Naidong et al. described a TLC method with UV and fluorescence densitometry for the assay and purity control of metacycline. Dichloromethane–methanol–water (58:35:7, v/v/v) as a mobile phase and a silica gel thin layer as a stationary phase were used. The plates were previously sprayed with 10% sodium edentate solution (pH 9). The potential impurities of metacycline were separated [16]. Naidong et al. developed a TLC method with UV and fluorescence densitometry for the assay and purity control of minocycline. Before use, the silica gel plates were sprayed with a 10% solution of sodium edetate (EDTA; pH 9 with 10% solution of NaOH). The plates were dried 1 h at room temperature, and then at 110°C for 1 h, shortly before use. The chromatographic chamber was saturated with the mobile phase, dichloromethane–methanol–water (57:35:8, v/v/v), for 1 h prior to use. The plates were dried and the spots were measured with a CS-930 TLC scanner using absorption–reflection mode at 280 nm and in fluorescence mode with excitation at 400 nm. 4-Epiminocycline and 7-didemethylminocycline were separated from minocycline and from each other. 6-Deoxy-6-demethyltetracycline was selectively determined by fluorescence densitometry, while quantification of other impurities and the assay of minocycline were performed by UV densitometry [17]. Mowthorpe et al. presented the analysis of tetracycline and its impurities by TLC/matrix-assisted laser desorption/ionization mass spectrometry (TLC/MALDI/MS), with brushing a supersaturated matrix solution and deposition of matrix material by electrospraying. TLC silica gel $60F_{254}$ plates were employed. The following MALDI matrices were used: a-cyano-4-hydroxycinnamic acid (aCHCA), 2-(4-hydroxyphenylazo)benzoic acid, and 2,5-dihydroxybenzoic acid. On an aluminum plate, saturated solutions of aCHCA in acetone were deposited in order to create a homogenous crystal layer. Sample spots cut from the TLC plate were wetted with an extraction solution (ethanol or methanol–water, 1:1, v/v). The previously prepared matrix layer on the aluminum plate was pressed into the TLC plate. After transfer, the aluminum plate was removed, and the thus coated TLC spots were stuck to a conventional MALDI target using double-sided tape. Before use, the silica gel plates were sprayed with an aqueous solution of disodium EDTA (10%, pH 9) and dried at room temperature for 0.5 h, and at 120°C for 0.5 h. The TLC chamber was saturated with development solvent, dichloromethane–methanol–water (59:35:6, v/v/v), for 1 h prior to use. The plate was developed, dried, and visualized by UV light at 254 nm [18].

Jain et al. discussed subjecting minocycline to a variety of stress test conditions to establish inherent stability of the drug and to develop the stability-indicating HPTLC assay. The HPTLC method was used to investigate the kinetics of acidic and alkaline degradation processes by quantification of drug at different temperature, to calculate the activation energy and half-life for minocycline degradation. Precoated silica gel $60F_{254}$ on aluminum plates were used after spraying with a 10% EDTA solution (pH 9 with 10% NaOH solution). The plates were dried 1 h at room temperature, and at 110°C for 1 h, shortly before use. The mobile phase consisted of methanol–acetonitrile–isopropyl alcohol–water (5:4:0.5:0.5, v/v/v/v). Development was carried out in the chamber after saturation for 30 min with a mobile phase at 25°C and a relative humidity of 60%. Plates were dried and scanned densitometrically at 345 nm in the absorbance mode. The method was also employed for stability testing of commercial tablet and capsule formulation and for prepared polylactide co-glycolide–MC-nanoparticles. The drug undergoes acidic and basic degradation (2 N HCl or NaOH,), oxidation (6% H_2O_2, 2 h at 80°C), wet heat–induced degradation (2 h at boiling water bath), and photodegradation (sunlight for 48 h; UV irradiation at 254 nm for 48 h). The sample solutions showed five additional peaks at R_F values 0.02, 0.13, 0.19, 0.27, and 0.4 besides peak for minocycline (R_F 0.30) [19]. Liang et al. analyzed the stability of tetracycline in methanol solution by UV–VIS spectroscopy, HPLC, and TLC methods. After dissolution in methanol, the drug decomposed under the influence of light and atmospheric oxygen, forming more than 14 degradation products. HPTLC analysis was performed on the silica gel plates that were coated with saturated NaEDTA (pH 9) solution and heated at 105°C for 2 h. Methanol–dichloromethane–water (30:64:6, v/v/v) was used as the mobile phase. The plates were developed in a saturated chamber, dried, and submerged in a 30% solution of paraffin in n-hexane for 1 s. Fluorescence detection mode was used with a 360 nm excitation source and a 550 nm band-pass filter [20]. Makowski et al. described photocatalytic reactions performed in the presence of TiO_2 and anions of inorganic salts ($KMnO_4$, KBr, KI, KCl, $AgNO_3$) as a way of visualization of various types of therapeutic agents, that is, benzylpenicillin procaine, benzylpenicillin potassium, penicillic acid, tetracycline hydrochloride, oxytetracycline hydrochloride, and chlortetracycline hydrochloride, in TLC. Experiments were performed on aluminum foil silica gel 60 plates. Before use, plates were activated at 100°C for 30 min. The solutions were spotted on the plates and developed with butanol–anhydrous acetic acid–water (6:2:2, v/v/v) as a mobile phase. TLC plates were illuminated by use of UV lamp with the radiation at 366 nm. Plates were sprayed with four visualizing reagents: 0.1 M $KMnO_4$ with TiO_2, 1 M KI with TiO_2, 1 M KBr with TiO_2, and 1 M KCl with TiO_2. For the two last reagents, after spraying, the plates were illuminated for 10 min, sprayed with $AgNO_3$, and illuminated again for 3 min [21].

48.1.2 Analysis of Tetracyclines in Pharmaceutical Formulations

Izer et al. reported a TLC–densitometric method for the determination of the degradation products of oxytetracycline HCl formed in the "oxytetracycline-eyedrop." A chloroform–methanol–water (65:25:5, v/v/v) solvent system was applied on silica gel TLC plate, predeveloped with saturated disodium ethylenediaminetetraacetate aqueous solution. UV densitometry at 254 nm was used for quantitation. The chemical stability of oxytetracycline HCl in different eyedrops, the effects of pH, and the mode of the preparation and storage conditions on the decomposition process were discussed [22]. Liang et al. developed an HPTLC method with a scientifically operated charge-coupled device detector for the assay of tetracycline pharmaceutical products. Separation was provided on normal-phase (NP)- and RP-TLC plates with fluorescence detection. The existing impurities in tetracycline capsules were determined using HPLC and HPTLC techniques [23]. Simmons et al. described a 2D TLC procedure on microcrystalline cellulose for the determination of anhydrotetracycline and epianhydrotetracycline in degraded tetracycline tablets. Development was performed with 0.1 M EDTA–0.1% ammonium chloride solution and methanol-soluble excipients. Compounds were resolved by developing the chromatogram with chloroform saturated with EDTA–ammonium chloride solution [24].

48.1.3 Analysis of Tetracyclines in Biological Material

Jain et al. described the method for analysis of minocycline and its stability in human plasma. Chromatography was performed on aluminum silica gel $60F_{254}$ plates. Before use, they were sprayed with 10% aqueous disodium EDTA solution (pH 9.0 with 10% aqueous NaOH solution) and dried for 1 h at room temperature, and at 110°C for 1 h, shortly before use. Plasma samples were thawed at room temperature for 10 min and mixed with methanol. The solvent was evaporated at 37°C, and the residues were dissolved in methanol. Samples were applied to the plates. Development with methanol–acetonitrile–isopropanol–water (5:4:0.5:0.5, v/v/v/v) as a mobile phase was used in chamber, previously saturated with a mobile phase vapor for 30 min at 25°C and relative humidity 60%. Densitometric scanning was performed at 345 nm in absorbance mode. The stability of minocycline was investigated in human plasma samples, during storage at 20°C and 4°C. The first-order rate constant (k), half-life ($t_{1/2}$), and shelf life (t_{90}) of the drug in plasma were obtained from the slope of the straight lines at both temperatures. The half-life was 6.3 h at 20°C and 9.9 h at 4°C. Investigation of long-term freezer stability (−20°C) revealed that minocycline was stable in plasma for two months [25]. Jain et al. presented an HPTLC method for determination of minocycline in human plasma, saliva, and gingival fluid samples. The precoated silica gel aluminum $60F_{254}$ plates were used after spraying with a 10% EDTA solution (pH 9 with 10% NaOH). Methanol–acetonitrile–isopropyl alcohol–water (5:4:0.5:0.5, v/v/v/v) as a mobile phase was used. The chamber was saturated with a mobile phase for 30 min at 25°C at a relative humidity of 60%. Densitometric scanning was performed on a TLC scanner in absorbance mode at 345 nm after single-step extraction with methanol. The stability and degradation kinetics of minocycline in the studied biomatrices was studied. The solutions in methanol were stable at 20°C and for 6 days at 4°C. Frozen at −20°C, the drug was stable for at least 2 months and could tolerate two freeze–thaw cycles without losses higher than 10% [26]. Novakova developed a method for the identification of tetracycline in urine and gastric content, after extraction with ethyl acetate. TLC analysis on silufol layers impregnated with an aqueous solution of the disodium salt of EDTA (pH 5 or 7.4) was used. The mobile phase chloroform–methanol–Na_2EDTA (55:30:5, v/v/v) was suitable for the development. UV light of 254 or 366 nm and Fast Blue B reagent were used for the detection [27]. Oka et al. established a TLC/fast atom bombardment mass spectrometry (TLC/FABMS) method to identify residual antibiotics, oxytetracycline, tetracycline, chlortetracycline, and doxycycline in bovine tissues, including muscle, liver, and kidney. It consists of the following steps: extraction of drugs from tissues with McIlvaine buffer (pH 4) containing 0.1 M ethylenediaminetetraacetate (Na_2EDTA);

elution from prepacked Bond Elut CIS solid-phase extraction (SPE) cartridges, pretreated with Na_2EDTA, with ethyl acetate followed by methanol–ethyl acetate (5:95, v/v) as eluents; separation on a silica gel RP C_8 TLC plate with methanol–acetonitrile–0.5 M oxalic acid solution pH 2 (1:1:4, v/v) as a solvent system; and TLC/FABMS with a condensation technique employing thioglycerol as a matrix. Detection was performed by spraying with 2 M HCl, heating at 150°C for 1 min, and spraying with 28% aqueous ammonia, 0.5% aqueous Fast Blue BB Salt, and pyridine. The spots were located using a UV lamp at 254 nm. Next, the plate was placed on the TLC holder, a matrix was applied on the sample spot, and the TLC/FAB mass spectra were measured using a mass spectrometer with a TLC/FAB ion source [28].

Choma et al. described a method for determination of doxycycline and flumequine in milk. Two developments of TLC plate were needed: one as a cleanup procedure on precoated silica gel TLC Si 60 and Si 60F$_{254}$, with and without concentrating zone, the other as a proper analysis. The developing solvents were chloroform–methanol (1:1, v/v), acetonitrile–methanol (1:1, v/v), acetonitrile, methanol, 0.1 M citric acid–methanol (1:9; 1:4, v/v), 0.05 M citric acid–methanol (1:4, 1:3, 1:2, v/v), 0.01 M citric acid–methanol (1:4, 1:3, 1:2, v/v), 0.05 M citric acid–methanol–2-propanol (1:2:2, v/v/v), 0.05 M citric acid–methanol–1-propanol (1:2:2, v/v/v), 0.05 M citric acid–methanol–isopropyl ether (1:2:3, v/v/v), and 0.05 M citric acid–methanol–2-propanol (1:3:1, v/v/v). In proper analysis, solutions were injected on the TLC F$_{254}$ plates into the plate's concentrating zone. The plates were set into sandwich chamber, and predeveloped with hexane to remove a lipid fraction from the milk samples. Then, they were developed with a proper solvent system. The spots were detected by UV lamp at 254 and 366 nm and densitometrically [29]. Choma et al. reported a TLC method for detection of doxycycline and flumequine in milk. Samples were injected on TLC plates in the concentrating zone. The plates were predeveloped with hexane and acetone to remove lipid fractions from the milk samples. The plates were developed with 0.05 M citric acid–methanol–2-propanol (1:3:1, v/v/v) as a mobile phase. Plates without concentrating zones (not used for milk samples) were developed only in a mobile phase. After drying, spots were detected at 366 and 254 nm. Bioautography was performed according to Chrom Biodip Antibiotics Test Kit procedure. The TLC plates were immersed briefly in the microorganism solution, incubated overnight at 28°C, sprayed with 3-[4,5-dimethylthiazol-2-yl]-2,5-diphenyltetrazolium bromide solution, and incubated for 30 min. Yellow inhibition zones were visible against a purple background [30].

Ašperger et al. used the SPE and TLC for separation and determination of enrofloxacin, oxytetracycline, and trimethoprim, used in veterinary practice. SPE was performed on 47 mm poly(styrene–divinylbenzene [SDB]) Empore disks at pH 3 and 7.75. Chromatography was performed on HPTLC CN F$_{254}$s plates in a chamber saturated for 10 min at room temperature. Separation was conducted with 0.5 M oxalic acid solution–methanol (65:35, v/v) as a mobile phase. Spots were detected under UV light at 254 nm. Video densitometry was performed with 3CCD color video camera [31]. Naidong et al. presented a TLC method for the assay and purity control of tetracycline, chlortetracycline, and oxytetracycline in animal feeds and premixes. To premixes, methanol–1 M HCl (99:1, v/v) was added. The extracts were filtered and diluted with methanol. The sample solution was stable for at least 12 h at 4°C. Animal feeds were weighed. Two aliquots were fortified with laboratory-prepared TLC/HCl substandard in kieselgel. To each of the three aliquots, kieselgel was added and the samples were mixed. The powder was transferred into a glass column. Acidified methanol was added on the column and the extract was collected. The sample solution was filtered, kept at 4°C, and applied onto the silica gel TLC plate within 2 h. Before use, the plates were sprayed with a 10% solution of sodium edetate (pH 8 or 9) and dried for 1 h at room temperature and at 110°C for 1 h, shortly before use. The chamber was equilibrated with a mobile phase dichloromethane–methanol–water (59:35:6 for tetracycline, 60:35:5 for chlortetracycline, and 58:35:7, v/v/v, for oxytetracycline) for 1 h prior to use. The developed plate was dried at 105°C for 2 min and dipped vertically for 1 s in a tank containing liquid paraffin–hexane (30:70, v/v). The spots were detected by densitometer in fluorescence mode with 400 nm [32].

Mutavdžić et al. reported the study of different types of SPE material for preconcentration of enrofloxacin, norfloxacin, oxytetracycline, trimethoprim, sulfamethazine, sulfadiazine, sulfaguanidine, and penicillin G/procaine, from water samples. The compounds were extracted on different types of cartridge (Strata-X, Strata-X-C, Strata CN, Strata C18-E, Strata C8, Strata SCX, Strata SAX, Strata SDB-L, Strata SI-1, Strata FL-PR, and Strata NH2), with methanol as eluent. After elution, the solutions were evaporated to dryness and the residues were dissolved in methanol. The efficiency of extraction was determined by TLC on CN F_{254} HPTLC plates with 0.05 M oxalic acid solution–methanol (81:19, v/v) as a mobile phase. After development, the spots were visualized under UV light at 254 and 366 nm by means of a Hitachi HV-C20 highly sensitive 3CCD color video camera. For quantification, they were detected at 366 nm [33].

48.2 MACROLIDE ANTIBIOTICS

The term "macrolide" is used to describe drugs with a macrocyclic lactone ring of 12 or more elements. This class of compounds includes a variety of bioactive agents, such as antibiotics, antifungal drugs, prokinetics, and immunosuppressants. The 14-, 15-, and 16-membered macrolides are a widely used family of antibiotics. The macrolides are a group of antibiotics produced by various strains of *Streptomyces* (spore-forming bacteria that grow slowly in soil or water as a branching filamentous mycelium similar to that of fungi). They act by inhibiting protein synthesis, specifically by blocking the 50S ribosomal subunit. Macrolides inhibit the growth of bacteria and are often prescribed to treat rather common bacterial infections. The main clinical significance of macrolides consists in affecting gram-positive cocci and intracellular pathogens such as mycoplasma, chlamydia, campylobacter, and legionella.

48.2.1 SEPARATION AND DETERMINATION OF MACROLIDES

Vanderhaeghe et al. analyzed various macrolides and their esters in different chromatographic systems. The most useful system was ethyl acetate–ethanol (or isopropanol)–15% ammonium acetate pH 9.6 (9:4:8, v/v/v) on silica gel plates. For the separation of erythromycin esters, chloroform–ethanol–15% ammonium acetate pH 7 (or 3.5% ammonia) (85:15:1, v/v/v) gave the best results. The different components of erythromycin were separated on silanized silica gel plates with methanol–water–ammonium acetate pH 7 (50:20:10, v/v/v) [34]. Szulagyi et al. separated the natural complex of homologous components of primycin, in n-butanol or chloroform solvent systems containing a high percentage of organic acids. Detection by vanillin–sulfuric acid and by the Sakaguchi method (reaction with α-naphthol) was modified. The three main components of primycin (A_1, A_3, and B) as Sakaguchi positive and their proportions were in agreement with those of homogenous permethyl products prepared from primycin sulfate for structural studies [35]. Kondo reported the separation and quantitative analysis of propionyl derivatives of maridomycins. TLC was performed with addition reaction of gaseous iodine with 9-propionyl meridamycins on the plate, extraction of the 9-propionyl meridamycin–iodine complexes, and subsequent analysis of the amount of reacted iodine, using an automatic analysis system [36].

Tosti et al. analyzed the chromatographic behavior of roxithromycin, midecamycin, erythromycin, azithromycin, and erythromycin ethylsuccinate by salting out TLC with cellulose as an adsorbent. Ammonium sulfate was used to prepare the aqueous solutions of mobile phases. The plates were spotted with freshly prepared ethanol solutions of the antibiotics, except for erythromycin ethylsuccinate, which was dissolved in absolute acetone. Compounds were detected by exposing the plates to iodine vapor. Hydrophobicity was determined from the linear relationships between solute R_M values and the ammonium sulfate content in the mobile phase. Lipophilicity was correlated with calculated log P values [37].

Lazarevski et al. obtained the kinetic data, by quantitation of the hydrolytic degradation of erythromycin oxime and erythromycylamine, separated by TLC. The rate constants were determined at three temperatures (17°C, 26°C, and 36°C). The activation energy for the hydrolysis and methanolysis of the compounds was calculated [38]. Brisaert et al. used a combination of 2% erythromycin and 0.05% tretinoin in an alcohol/isopropanol lotion. Two parameters (pH and the concentration of butylhydroxytoluene as antioxidant) were investigated for their influence on the stability of erythromycin and/or tretinoin. Stability analysis was performed at 45°C in the dark. To quantitative drug analysis, a TLC method was used. Silica gel 60F$_{254}$ (with gypsum) plates was chosen. As a mobile phase, dichloromethane–methanol–ammonia (60:6:1, v/v/v) was used, and ethanol–chloroform–acetic acid–sulfuric acid–anisaldehyde (30:60:10:2:1, v/v/v/v/v) were the components of the dipping reagent. Spectrodensitometry was performed at 565 nm. The degradation of erythromycin was followed over 5 weeks. It seemed to be much faster than tretinoin degradation. Optimal stability was shown in the pH range of 8.2–8.6 for erythromycin and 7.2–8.2 for tretinoin, while the concentration of butylhydroxytoluene had no significant influence [39]. Gabriels et al. developed the analysis of tretinoin and erythromycin in a lotion in the presence of their degradation products and several excipients. Erythromycin was separated on a silica gel 60F$_{254}$ (with gypsum) plate and a mobile phase with dichloromethane–methanol–25% ammonia (60:6:1, v/v/v), tretinoin on a C$_{18}$ RP plate with acetonitrile–water (50:25, v/v) as a mobile phase, adding 1 mL acetic acid for the separation of the excipients and erythromycin. The plates were developed in a TLC tank, containing overnight saturated mobile phases at 25°C. After drying, they were dipped in a freshly prepared derivatization solution. The plate was heated 5 min at 110°C for qualitative, and longer for quantitative analysis. The spots were qualitatively observed in daylight or at 366 nm. For quantitative analysis, they were performed with scanner spectrophotometer in VIS in transmission mode (at 520 nm for tretinoin and 565 nm for erythromycin). Derivatization was done with a dipping reagent, consisting of anisaldehyde, sulfuric acid, and acetic acid (1%, 2%, and 10%) and dissolved in chloroform–94% alcohol (60:30, v/v) for erythromycin and 94% alcohol–water (50:40, v/v) for tretinoin [40].

Bhushan et al. reported the use of erythromycin as a chiral impregnating reagent while making thin silica gel plates to resolve dansyl-DL-amino acids (phenylalanine, valine, leucine, serine, glutamic acid, aspartic acid, norleucine, a-amino-n-butyric acid, methionine, and tryptophan) by TLC. Impregnated plates were prepared by spreading a slurry of silica gel in distilled water (containing of erythromycin). They were dried overnight at 60°C. Different combinations of 0.5 M aqueous NaCl–methyl cyanide–methanol were used as a developing system. The spots were revealed under UV radiation at 254 nm to greenish-yellow color. The effect of the concentration of erythromycin as impregnating reagent with silica gel was also studied. The best resolution was with 0.05% solution [41]. Bhushan et al. reported a direct resolution of racemic atenolol and propranolol into their enantiomers by TLC on silica gel G plates impregnated with optically pure L-tartaric acid, (R)-mandelic acid, and (–)-erythromycin as chiral selectors. The plates were activated for 8–10 h at 60°C. Chromatograms were developed in chambers preequilibrated with different solvent systems for 15 min. The plates were dried at 60°C for 10 min and spots were located in an iodine chamber. The influence of pH, temperature, and concentration of chiral selector was studied [42].

48.2.2 ANALYSIS OF MACROLIDES IN PHARMACEUTICAL FORMULATIONS

Khedr et al. developed a TLC for the analysis of azithromycin in bulk and capsule forms. Both potential impurity and degradation products can be estimated on a precoated silica gel TLC 60F$_{254}$ plate. The development system used was n-hexane–ethyl acetate–diethylamine (75:25:10, v/v/v) in the tank presaturated with the solvent vapor system for 30 min at room temperature. The separated bands were detected as brown to brownish-red spots after spraying with modified Dragendorff's solution. The R$_F$ values of azithromycin, azaerythromycin A, and the three degradation products were 0.54, 0.35, 0.40, 0.20, and 0.12. The forced degradation conditions include the effect of heat (in water, heated at 100°C for 30 min), moisture, light, acid–base hydrolysis (in dichloromethane with 0.05 N HCl or 1 N NaOH; heated at 60°C for 30 min), and oxidation (10% H$_2$O$_2$ solution; heated at 60°C for 30 min) [43]. Kwiecień et al. established a TLC method with densitometric detection for quantification of azithromycin in pharmaceutical preparations. Silica gel F_{254} plates were used with chloroform–ethanol–ammonia (6:14:0.2, v/v/v) as a mobile phase. The development was performed in a chamber after saturation with a mobile phase vapor for 10 min at 25°C. Plates were visualized by spraying with sulfuric acid–ethanol (1:4, v/v) and heating at 120°C for 5 min. Densitometric scanning was performed at 483 nm in reflectance mode. The R_F of azithromycin was 0.53 [44]. Liawruangrath et al. described an HPTLC method for the quantitation of erythromycin A, B, and C in pharmaceutical dosage forms. Separation was carried out on silica gel $60F_{254}$ plate using chloroform–methanol (95:5, v/v). It resulted in two spots with the R_F values of 0.08 and 0.14. The chloroform–methanol–acetic acid (90:10:1, v/v/v) as a solvent system could satisfactorily differentiate erythromycin derivatives (stearate, estolate, and ethylsuccinate). The plates were sprayed with 10% sulfuric acid solution and heated at 100°C for 10–15 min. The spots were quantified by a TLC scanner at 410 nm. It was found that ethyl acetate–ethanol–10% sodium acetate pH 9.5 (9:7:8, v/v/v) solvent system could be used for the differentiation of the erythromycin A, B, and C in standard erythromycin estolate by TLC with the R_F values of 0.57, 0.59, and 0.53, respectively [45]. Maher et al. presented HPTLC and LC methods for the simultaneous determination of metronidazole and spiramycin in their combined tablet formulation. For HPTLC, the samples were spotted on a precoated silica gel aluminum $60F_{254}$ plate. They were prewashed with methanol and activated at 100°C for 5 min prior to chromatography. The mobile phase consisted of methanol–chloroform (9:1 v/v). Development was carried out in a chamber saturated with the mobile phase for 30 min at 25°C. Densitometric scanning was performed at 240 nm in the reflectance/absorbance mode [46]. Hu et al. developed a Fast Chemical Identification System (FCIS) consisting of two-color reactions based on functional groups in molecules of antibiotics and two TLC methods, for screening of erythromycin, clarithromycin, roxithromycin, azithromycin, erythromycin ethylsuccinate, kitasamycin, leucomycin A3, acetylspiramycin, acetylkitasamycin, midecamycin, and meleumycin. The active ingredients were extracted from their oral preparations (tablets, capsules, granules, and suspensions) by absolute alcohol. Sulfuric acid reaction was first used to distinguish the macrolides from other types of drugs, and then 16-membered macrolides and 14-membered ones were distinguished by potassium permanganate reactions depending on the time of loss of color in the test solution. A TLC method was chosen for further identification. A TLC silica G F_{254} plate, the mobile phase A consisted of ethyl acetate–hexane–ammonia (100:15:15, v/v/v) for the identification of 14-membered macrolides, and the mobile phase B consisted of trichloromethane–methanol–ammonia (100:5:1, v/v/v) for the identification of 16-membered ones were used. After drying, plates were put in a closed tank with a quantity of iodine for several minutes until the spots appeared [47]. Richard et al. reported a method for the detection and quality control of erythromycin. TLC on sodium acetate-buffered silica gel plates enabled the differentiation of erythromycin, erythromycin estolate, and erythromycin ethylsuccinate from some degradation products and pharmaceutical excipients. The comparative behavior of 20 other antibiotics under these conditions was also presented [48].

48.2.3 ANALYSIS OF MACROLIDES IN BIOLOGICAL MATERIAL

Loya et al. presented an HPTLC method for the estimation of clarithromycin from human plasma. To a plasma sample containing clarithromycin, 0.1 N sodium carbonate was added. The drug was extracted with a mixture of n-hexane–ethyl acetate (1:1, v/v). After centrifugation at 7000 rpm for 20 min at 4°C, the organic layer was evaporated, and the residue was reconstituted with methanol and spotted on the plate. Separation was performed on precoated silica gel $60F_{254}$ plates, prewashed with methanol, and dried at 105°C–110°C for 30 min. The plates were developed in chamber previously saturated for 15 min with the solvent system, composed of ethyl acetate–methanol–15% ammonium acetate (pH 10.6) (9:4:8, v/v/v). The plates were dried, sprayed with derivatizing reagent comprising of xanthydrol–concentrated hydrochloric acid–acetone (18:5:90, v/v/v), and heated at 100°C–105°C for 90 s. The developed pink spots were detected by densitometer in absorbance mode at 506 nm. The R_F 0.62 was found. The method was applied to a pharmacokinetic study of clarithromycin 250 mg tablet in healthy human male volunteers. Stability was tested by subjecting the plasma controls to three freeze–thaw cycles, and for long-term stability studies, the concentration of clarithromycin was found out on 30th day, 60th day, and 90th day by storing samples at −20°C [49].

Petz et al. described the determination of erythromycin, tylosin, oleandomycin, and spiramycin in livestock products. The drugs were extracted from animal tissues, milk, and egg with acetonitrile at pH 8.5. Cleanup was done by adding sodium chloride and dichloromethane, evaporating the organic layer, and subsequent acid/base partitioning. The antibiotics were separated by TLC, reacted with xanthydrol, and detected as purple spots. Anisaldehyde–sulfuric acid, cerium sulfate–molybdic acid, phosphomolybdic acid, and Dragendorff's reagent proved to be less sensitive as visualizing agents. For quantitation, TLC plates were scanned at 525 nm. Bioautography with *Bacillus subtilis* was used to confirm results [50].

Shahverdi et al. analyzed the essential oil of *Cinnamomum zeylanicum* bark that enhanced the bactericidal activity of clindamycin and decreased the minimum inhibitory concentration of clindamycin required for atoxicogenic strain of *Clostridium difficile*. The mixture of water and essential oils, from plant material, flowed into a collection container to separate phases. The oil was fractionated by TLC on silica gel $60F_{254}$ plate using toluene–ethyl acetate (93:7, v/v) as the solvent system. The spots were visualized by 1% vanillin–sulfuric acid reagent and eluted using absolute ethanol. A TLC fraction with R_F 0.54 was found to enhance clindamycin activity. It was also subjected to GC–MS analysis for identification [51]. Sajewicz discussed establishing a SPE for isolation of josamycin (R_F 0.43), sulfamethoxazole (R_F 0.72), carbamazepine (R_F 0.56), diclofenac (R_F 0.19), and iopromide (R_F 0.93) from aqueous matrixes; quantifying these compounds by densitometric TLC; and testing the method by applying it to environmental samples (river water from South Poland). TLC was performed on precoated with octadecyl silica RP-18 $F_{254}S$ plates with acetonitrile–water–acetic acid (5:5:2, v/v/v) as a mobile phase, with detection at UV light from the deuterium lamp at 220 nm in reflectance mode [52].

Szabo et al. examined a TLC method for determination of bioactive guanidinium compounds, saturated antibiotics, that is, primycin, with UV absorbance. A separation system for the quality control of the end product and for monitoring the efficiency of isolation from the fermentation liquor was established. Fermentation broth was homogenized with methanol–n-butanol (1:1, v/v). After precipitation, the supernatant was spotted on precoated silica gel plates, washed with mixture of methanol–35% formic acid (1:1, v/v), in chamber, and dried. Plates were developed in chamber previously saturated with a mobile phase vapor. Two mobile phases were used: n-butanol–water–methanol–acetic acid (4:2:1:1, v/v/v/v) (mobile phase A) and chloroform–methanol–water–35% formic acid–n-butanol–formaldehyde (160:53:9:6:3:3, v/v/v/v/v/v) (mobile phase B). For mobile phase B, the development was repeated to obtain better resolution. The plates were dried in a vacuum chamber at 100°C and evaluated at 200 nm in reflectance mode. Among the spots obtained from primycin fermentation samples, guanidinium was visualized by means of a modified Sakaguchi-type reaction. The plates were placed in a tray containing a solution of 8-hydroxyquinoline in ethanol. After drying, they were immersed in a solution of N-bromosuccinimide for 1 min. The red spots obtained were then copied on a semitransparent paper [53]. Stassi et al. studied the erythromycin producer, *Saccharopolyspora erythraea* ER720, which was genetically engineered to produce 6,12-dideoxyerythromycin A, as the major macrolide in the fermentation broth. Antibiotic derivatives were extracted from supernatants of *S. erythraea* cultures with aqueous ammonia solution (pH 9) and ethyl acetate. The organic phase was removed, and the aqueous phase was reextracted with ethyl acetate. The organic phases were combined and dried. The residue was diluted in ethyl acetate and spotted onto silica gel TLC $60F_{254}$ plates. They were developed using isopropyl ether–methanol–ammonia (75:35:2, v/v/v). Compounds were visualized by spraying the plates with anisaldehyde–sulfuric acid–ethanol (1:1:9, v/v/v) and heating until color developed. 6,12-Dideoxyerythromycin A and 6-deoxyerythromycin D appear as blue spots [54]. Li et al. separated the metabolites of *Streptomyces hygroscopicus* 17997 by TLC. Compound from the TLC silica gel plates with anti-gram-positive bacteria activity and becoming red upon color reaction by 2 M NaOH was analyzed by HPLC. All procedures consisted of silica gel TLC, color reaction, HPLC, PCR detection, DNA sequence analysis of tgd gene, and LC-(+)–ESI–MS, were established for rapid identification of elaiophylin and its producer [55]. Szabo et al. presented a method for monitoring the level of primycin in the fermentation broth. Shaken fermentation broth was homogenized with a mixture of butanol–methanol (1:1, v/v). After sedimentation, the supernatant was applied on precoated polyester sheet. Plates were developed in a chamber previously saturated with the mobile phase consisting of chloroform–carbon tetrachloride–35% formic acid (40:10:7, v/v/v). After drying, the plate was immersed in sulfuric acid solution (sulfuric acid was added to refrigerated mixture of carbon tetrachloride–1-propanol [1:1, v/v]; heated at 108°C). The visualized chromatogram was evaluated by a TLC scanner in reflectance mode at 290 nm. The bioautographic detection of primycin was performed by the agar diffusion method with *B. subtilis* ATCC 6633. Samples were diluted with n-butanol–ethanol–distilled water (1:1:2, v/v/v), mixed, kept at room temperature for 1 h, diluted with 65 mM potassium phosphate buffer pH 8.0–ethanol (80:20, v/v), and pipetted into the holes of the agar plates. The plate was kept at 4°C for 2 h, incubated at 37°C for 16–18 h, and the inhibition zones were measured. For the bioautographic investigation of TLC, sheets were placed on the surface of the solidified agar plate. They were kept at 4°C for 1.5 h, and the strip was removed from the surface of the agar. The plate was incubated at 37°C for 16–18 h and evaluated [56]. Harindran et al. isolated from the biomass of *Streptomyces* CDRIL-312, a new antibiotic, HA-1-92 (oxohexaene macrolide), by extracting in butanol and further purified by silica gel column chromatography followed by preparative TLC. The crude powder of HA-1-92 was purified by silica gel column chromatography. The active fractions showing the desired UV absorption (at 404, 382, 361, 342 nm) were pooled and subjected to preparative TLC with precoated silica gel GF$_{254}$ plates and butanol–acetic acid–water (4:1:5, v/v/v) as a solvent system. The major active component was recovered with methanol, and the extract was concentrated, chilled at 4°C, filtered, and dried at 40°C. The purity of HA-1-92 was confirmed by TLC using different solvent systems showing single spots at R$_F$ values 0.32 (butanol–acetic acid–water, 4:1:5, v/v), 0.52 (butanol–methanol–water, 4:1:2, v/v), and 0.61 (pyridine–ethyl acetate–acetic acid–water, 5:5:1:3, v/v/v/v) [57]. El-Bondkly et al. discussed the mutation of the marine *Streptomyces* sp. AH2, bioactivity evaluation, fermentation, and isolation of the microbial metabolites (phenazine, 1-acetyl-β-carboline, perlolyrin, and erythromycin A). They described the ¹D and ²D NMR and ESI–MS data including ESI–MS² and MS³ patterns combined with HRESI–MS of erythromycin A. TLC was performed on Polygram SIL G UV$_{254}$. The fermentation medium was inoculated with 10% seeding of 10/14 mutant for 5 days at 28°C. The filtrate was extracted with ethyl acetate, and the organic extract was concentrated, washed with brine, dried over Na$_2$SO$_4$, and evaporated in vacuo to dryness. The dark-brown extract was flash chromatographed on silica gel of particle size eluting with an n-hexane–ethyl acetate gradient to deliver five fractions: (I) n-hexane, (II) n-hexane–ethyl acetate (9:1, v/v), (III) n-hexane–ethyl acetate (3:1, v/v), (IV) n-hexane–ethyl acetate (1:1, v/v), and (V) ethyl acetate. Purification of the fractions led to phenazine (I) as a greenish-yellow solid, perlolyrin (II) as two pale yellow solids, and erythromycin A as a white solid [58]. Ritzau et al. isolated elaiophylin and two methyl derivatives from the mycelium cake of *Streptomyces* strains HKI-0113 and HKI-0114. Purification was carried out by column chromatography on Sephadex LH-20. The fractions containing elaiophylins (monitored by TLC and antibacterial activity) were pooled and subjected to preparative TLC on silica gel 60 aluminum sheets with chloroform (run three times). The three zones staining blackish by 3% vanillin–concentrated H$_2$SO$_4$ (R$_F$ 0.3, 0.4, 0.5) were eluted by chloroform–methanol (1:1, v/v). Final purification was accomplished by preparative TLC on silica gel RP18 aluminum sheets with acetonitrile–water (83:17, v/v). The zones with R$_F$ 0.6, 0.7, and 0.8 were eluted by chloroform–methanol (1:1, v/v) and dried. The structures of derivatives were determined by MS and NMR investigations [59].

48.3   AMINOGLYCOSIDE ANTIBIOTICS

Aminoglycosides are considered to be the most toxic antibiotics. They possess potential nephro- and ototoxicity and may cause neuromuscular block. The main clinical value of aminoglycosides consists in management of nosocomial infections caused by aerobic gram-negative pathogens and also in the treatment of infective endocarditis. They can be used against certain gram-positive bacteria, but are not typically employed because other antibiotics are more effective and have fewer side effects. They are ineffective against anaerobic bacteria (bacteria that cannot grow in the presence of oxygen), viruses, and fungi. Only one aminoglycoside, paromomycin, is used against parasitic infection.

48.3.1 SEPARATION AND DETERMINATION OF AMINOGLYCOSIDES

Hubicka et al. developed a TLC–densitometric method for simultaneous identification and quantitation of amikacin, gentamicin, kanamycin, neomycin, netilmicin, and tobramycin. The separation was achieved on silica gel TLC plates without a fluorescent indicator, with methanol–25% ammonia–chloroform (3:2:1, v/v/v) as the mobile phase. The measurements were made at 500 nm after detection with a 0.2% ninhydrin solution in ethanol [60]. Bhushan et al. described the separation of aminoglycoside antibiotics by TLC on untreated silica gel layers using various combinations of solvent systems, dioxane–ammonia, methanol–ammonia, and propanol–ammonia. Antibiotics were visualized with the help of iodine vapors [61]. Roets et al. reported determination of the B and C components of neomycin sulfate by TLC using Whatman silica gel plates as the stationary phase. Samples were dissolved in water. The mobile phase consisted of methanol–20% NaCl solution (15:85, v/v). The chromatographic chamber was equilibrated with the mobile phase for 1 h prior to use. After development, the plate was dried at 105°C–110°C for 30 min. Detection of the spots was performed by filling with a solution of 4-chloro-7-nitrobenzo-2-oxa-l,3-diazole in methanol–acetone (1:l, v/v). The plate was heated at 80°C for 30 min to complete the derivatization reaction. The excess reagent was removed by developing the plate twice in methanol–acetone (1:1, v/v), and plate was dried at 80°C for 5 min. The sensitivity was enhanced by dipping the plate for 1 s in a 30% solution of liquid paraffin in n-hexane. The plate was dried at 80°C for 5 min; greenish-yellow fluorescent spots on a colorless background were obtained. The chromatograms were analyzed with a CS-990 TLC scanner. The R_F values obtained for the different antibiotics were 0.33 for neomycin A; 0.24 for neomycin B; 0.16 for neomycin C; 0.59 for paromamine; 0.51 for paromomycin I; 0.40 for paromomycin II; 0.47 for kanamycin A; 0.21 for kanamycin B; 0.68 for amikacin; 0.30 for tobramycin; 0.47 for apramycin; and 0.53, 0.57, 0.62 for gentamicins [62]. Funk et al. presented the determination of neomycin B and C and neamine (neomycin A) by TLC. Separation was achieved by ascending 1D development on silica gel $60F_{254}$ HPTLC or TLC plates at 50°C. Methanol–25% ammonia–acetone–chloroform (35:20:20:5, v/v/v/v) was used as a mobile phase. Postchromatographic derivatization was performed with fluorescamine. The fluorescence was stabilized and increased by dipping the developed chromatogram in a solution of dichloromethane–triethanolamine–light liquid paraffin (8:1:1, v/v/v) [63]. Wagman et al. analyzed the gentamicin antibiotic complex, consisting of three components, C_1, C_{1a}, and C_2, by means of paper and TLC using a solvent mixture consisting of chloroform–methanol–17% ammonium hydroxide (2:1:1, v/v/v). Methods were given for preparative separation of these antibiotics by use of cellulose and Chromosorb W chromatographic column procedures [64].

Hotta et al. studied a kanamycin group antibiotics, which were subjected to enzymatic acetylation by a cell-free extract containing an aminoglycoside 3-N-acetyltransferase, AAC(3)-X, derived from *S. griseus* SS-1198PR. Characterization of the incubated reaction mixtures by TLC and antibiotic assay revealed that a product-retaining activity was specifically formed from arbekacin, an anti-MRSA semisynthetic aminoglycoside [65]. Klimecka et al. presented a structural analysis of the BA2930 protein, a putative aminoglycoside acetyltransferase AcCoA, which may be a component of the bacterium's aminoglycoside resistance mechanism. They found that BA2930 deacetylates AcCoA in the presence of some aminoglycoside antibiotics and that some antibiotics were chemically modified by BA2930. To search for possible BA2930 substrates, they used TLC method. It showed that the enzyme was able to chemically modify several aminoglycoside antibiotics, that is, tobramycin, amikacin, and kanamycin. Antibiotic acetylation was carried out in reaction mixtures containing the protein, AcCoA, and antibiotic in an HEPES buffer, pH 7.5. Mixtures were spotted on silica gel 60F$_{254}$ plates and developed with 5% potassium phosphate. The antibiotics and their products were detected by spraying 0.5% ninhydrin reagent in acetone. The following compounds were used for the initial overnight test: kanamycin, paromomycin, neomycin, amikacin, streptomycin, dihydrostreptomycin, gentamicin, tobramycin, deferoxamine, moxalactam, and cefmetazole [66].

48.3.2 Analysis of Aminoglycosides in Pharmaceutical Formulations

Kunz et al. described a TLC method for the quantitative determination of netilmicin and gentamicins C$_t$, C$_1$, C$_2$, and C$_{2a}$ in pharmaceutical preparations. The components were separated on Ca or ClS RP layers (KC$_s$F or KC$_{18}$F) after its activation at 110°C for 30 min. Development was carried out with methanol–0.1 tool/L LiCl in 32% ammonia solution (5:25, v/v) as a mobile phase. The plates were dried and dipped (2 s) into a DOOB (2,2-diphenyl-l-oxa-3-oxonia-2-boratanaphthalene)-reagent solution. After this treatment, the plates were heated at 110°C–176°C for 10–20 min. After cooling, the layer was dipped in a solution of paraffin oil–n-hexane (1:6, v/v). Then a quantitative evaluation of the chromatogram was done [67]. Sekkat et al. reported a procedure for the determination of seven major aminoglycosides in commercial formulations (injections, capsules, eyedrops, solutions, and ointments). Separation was carried out on silica gel plates then located with ninhydrin and analyzed in situ a chromatogram spectrophotometer [68]. Decoster et al. assayed a neomycin sulfate powders, with D-(+)-α,α-trehalose on strongly alkaline ion-exchange resin (hydroxide form). The amount of neomycin B and C in commercial samples was determined, and the results were compared with the microbiological assay [69].

48.3.3 Analysis of Aminoglycosides in Biological Material

Bhogte et al. described an HPTLC analysis of gentamicin by in situ fluorodensitometric evaluation of its (4-chloro-7-nitrobenzo-2-oxa-1,3-diazole [NBD-Cl]) derivative. The drug was estimated as the bulk drug and from plasma and urine. The human plasma samples were supplemented with gentamicin sulfate and diluted in water. Trichloroacetic acid was added for the separation of plasma proteins. Blank human urine from a healthy volunteer was supplemented with gentamicin sulfate and diluted with water. The obtained solutions were spotted on precoated silica gel 60F$_{254}$ plates. The mobile phase chloroform–methanol–20% ammonium hydroxide (2.4:2.2:1.5, v/v/v) saturated with ammonia was used for development. The plate was dried at 120°C for 20 min, dipped in methanolic NBD-Cl for 2 s, and heated again at 120°C for 10 min. The cooled plate was rechromatographed in methanol in the same direction as the first development. Densitometric analysis was carried out in the fluorescence/reflectance mode at an excitation wavelength of 436 nm and a sharp cutoff K-500 filter. R$_F$ values of 0.34 for gentamicin C$_1$, 0.29 for gentamicin C$_2$, and 0.21 for gentamicin C$_{1a}$ were obtained [70]. Kunz et al. used a TLC, for the quantitative determination of netilmicin in human serum, and compared with fluorescence polarization immunoassay. A stock solution of netilmicin was made up of methanol–water (1:1, v/v). A portion of each serial dilution was mixed with pooled serum and applied to a washed (with methanol) and conditioned (with water) Bakerbond RP-18 cartridge. After applying the serum, the cartridges were washed with water, and netilmicin was eluted with ethanol–25% ammonia (4:1, v/v). RP LKC$_{18}$F was used as adsorbent layer (activated at 110°C for 30 min). The plates were developed in methanol–0.1 tool/L LiCl in 32% ammonia (5:25, v/v). The plate was dried, dipped into a 2,2-diphenyl-l-oxa-3-oxonia-2-boratanaphthalene reagent solution, and heated at 110°C–120°C for 10–20 min. After cooling, the layer was dipped into a solution of paraffin oil–n-hexane (1:6, v/v). The measurements were carried out with a KM 3 chromatogram spectrophotometer coupled with a D-2000 integrator or a TLC scanner or a FTR-20 scanner with MicroSys VMEbus. The excitation wavelength was set at 365 nm, and a F1 43 or K-400 filter was used for emission [71].

Campbell et al. presented a method for the purification of hygromycin B from biological fluids. Swine plasma, bovine serum, and bovine milk samples were spiked by adding hygromycin B to the fluid. Screen columns (copolymeric bonded silica with hydrophobic and cationic functions) were used for initial SPE. The columns were conditioned using of 5% diethylamine in methanol, water, and 2% phosphoric acid prior to sample application. The samples were acidified with 2% phosphoric acid and applied to the columns. The columns were washed using water followed by 5% diethylamine in methanol. The solvent was evaporated, reconstituted using water, and applied to the affinity columns followed by washing sequentially with deionized water and 10 mM sodium citrate buffer, pH 4 to elute the drug. Hygromycin B was detected using either fluorescamine derivatization or TLC. TLC analysis was performed on Whatman LHK-PD silica gel plate with preabsorbent zones. The plates were developed in acetone–ethanol–ammonium hydroxide (1:1:1, v/v/v), dried, and dipped for 4 s in 0.02% fluorescamine in acetone–hexane (1:15, v/v). After drying, they were sprayed with 0.2 M sodium citrate buffer, pH 3, and the fluorescent bands were visualized using UV light at 365 nm [72]. Medina et al. optimized the conditions for the isolation of hygromycin B from animal plasma or serum with a copolymeric bonded solid-phase silica column followed by TLC separation and detection of its fluorescence derivative. Bovine plasma fortified with hygromycin B was diluted with 2% phosphoric acid. The samples were applied to the Clean Screen column. Swine serum was diluted with distilled water, acidified with 20% trichloroacetic acid, mixed, centrifuged at 4°C for 30 min, and supernatant was applied to the column. Proteins in swine serum were precipitated with acetonitrile. The aqueous phase was acidified with 2% phosphoric acid prior to ion-exchange purification. Samples were fortified before and after acidification. Eluates were applied to Whatman LHK-PD silica gel TLC plates with preabsorbent zones. The plates were developed in acetone–ethanol–99.9% ammonia (1:1:1, v/v/v) and allowed to dry for 35 min at 80°C. Hygromycin B bands were visualized by dipping the plate into a 0.02% fluorescamine solution in acetone–hexane for 4 s. The plate was dried and sprayed with 0.2 M citrate buffer (pH 3). Fluorescent bands were detected at 366 nm. Fluorescamine derivatives of hygromycin B were sprayed with borate or citrate buffers to determine pH effects. The fluorescent bands were scanned at 395 nm excitation and 485 nm emission wavelengths. The TLC developing system separated gentamicin (R_F 0.64), neomycin (R_F 0.56), spectinomycin (R_F 0.52), and streptomycin (R_F 0.20) from hygromycin B (R_F 0.33). Results show that hygromycin B derivatives had a higher fluorescence in acidic than in basic solutions [73]. Kojima et al. described an interaction of tobramycin with latamoxef in rat serum and its filtrate in vitro. Solutions containing tobramycin alone, latamoxef alone, or both in varying molar ratios (1:1, 1:2, 1:4, 2:1) were incubated at 37°C for 0.5, 1, 3, and 5 h after adjusting to pH 7.4. Samples were subjected to paper electrophoresis and TLC. TLC analysis showed that the spot of latamoxef radiated fluorescence with R_F value 0.38. In the mixtures, there was a definite decrease in fluorescence of latamoxef, compared with that of latamoxef alone. The UV spectrum of the mixture of tobramycin–latamoxef (1:2) showed a decrease in the intensity of absorption of latamoxef at 268 nm [74]. Lin et al. determined residual drug concentrations in meat using penicillin G in combination with one of the aminoglycoside antibiotics. The differentiation of compounds was easily accomplished treated with PCase produced a similar clear inhibition zone by the aminoglycoside alone. These drugs could be identified according to their R_F values and colors by TLC method using silica gel 60F$_{254}$ plates followed by spraying with various chemical reagents. A solution of n-butanol–acetic acid–water–p-benzenesulfonic acid (3:1:1:0.7, v/v/v/v) was used for the separation [75]. Tsuda et al. presented a colorimetric bioassay of netilmicin, based on the discoloration of thymolphthalein by carbon dioxide produced by *B. subtilis*. Concentration of reagent was determined using a TLC scanner. The discoloration of blue color to white showed the proportionality between netilmicin concentrations and the degrees of discoloration of thymolphthalein [76].

48.4 PEPTIDE ANTIBIOTICS

Peptide antibiotics are composed of glycosylated cyclic or polycyclic nonribosomal peptides. The class is used in clinical practice since 1960 in the treatment of nosocomial infections caused by gram-positive microorganisms. They are also indicated for the treatment of MRSA (methicillin-resistant *Staphylococcus aureus*), MRSE (methicillin-resistant *S. epidermidis*), and enterococcal infections.

48.4.1 Separation and Determination of Peptide Antibiotics

Aszalos et al. described RP-TLC conditions for separation of 19 peptide-type antibiotics, with molecular weights between 102 and 25,000 and of different chemical characteristics. Twenty-seven different mobile phases were employed, representing three organic modifiers, three buffers, and three pH values. No empirical correlations between molecular weights, R_F, theoretical plate height, and plate number could be detected [77]. Stankovic et al. presented a purification of the naturally occurring ion channel-forming pentadecapeptide gramicidin A. It was isolated from the commercially available mixture of isomers after chromatography on HPTLC silica gel 60 plates. Plates were eluted with chloroform–methanol–water–acetic acid (250:30:4:1, v/v/v/v), visualized at 254 nm, and stained with p-dimethylaminobenzaldehyde followed by brief heating [78]. Pepinsky et al. purified gramicidin C from a mixture of gramicidins by preparative TLC. A maximum R_F difference (0.24) was attained on aluminum oxide plates. The antibiotic was identified by amino acid analysis, NMR, and UV absorbance spectroscopy [79]. Hou et al. reported radiochemical purity of ^{111}In-bleomycin complex (^{111}In-BLMC) by TLC (R_F 0.65). In 5% agarose gel electrophoresis in 0.02 M NaHCO$_3$, it migrated toward the anode. Autoradiographs of TLC and gel electrophoresis plates showed no change on storage for 3 weeks. They analyzed urine and plasma from untreated or glioma-bearing mice after injection of ^{111}In-BLMC by TLC and gel electrophoresis [80].

Pirett et al. investigated the functions of gramicidin S in its producer, *B. brevis Nagano*. After extraction, cultures were placed at room temperature overnight. After centrifugation, samples were applied to Bio-Dex silica gel TF-B flexible TLC strips. The chromatograms were developed in n-butanol–glacial acetic acid–water (60:15:20, v/v/v) [81]. Kitajima et al. reported the synthesis of TL-119, a peptide antibiotic isolated from a strain of *B. subtilis*, which is active against gram-positive bacteria. They described its physicochemical properties and biological activity and compared it with both natural TL-119 and natural A-3302-B. The synthetic compound three gave a single spot on HPTLC using ethyl acetate–ethanol (10:1, v/v) as a solvent system [82].

Armstrong et al. used a vancomycin as a chiral mobile phase additive for the TLC resolution of 6-aminoquinolyl-N-hydroxysuccinimidyl carbamate-derivatized amino acids, racemic drugs, and dansyl-amino acids. Separations were achieved for most of these compounds in the RP mode using diphenyl stationary phase and acetonitrile as the organic modifier [83]. Bhushan et al. described a TLC for monitoring of enantiomeric purity of atenolol, metoprolol, propranolol, and labetalol, using vancomycin as a chiral-impregnating reagent or as a mobile phase additive (0.56 mM aqueous vancomycin solution, pH 5.5). Solutions of the chiral selector were adjusted to pH 4.5, 5.5, 6.5 by addition of NH$_3$. TLC plates were prepared by spreading a slurry of silica gel G in these solutions and activation overnight at 60°C. Solutions of racemic β-blockers were prepared in methanol and applied to the plates. Chromatograms were developed in chambers previously equilibrated with the mobile phase at 16°C for 10–15 min. Plates were dried at 40°C, and spots were located in an iodine chamber. Mixtures of chloroform, acetic acid, ethanol, methanol, water, acetonitrile, and dichloromethane of different composition were evaluated to achieve separation of the enantiomers. The effect on resolution of variation of the amount of chiral selector used for impregnation was studied by using 0.28, 0.56, and 1.12 mM solutions of the chiral selector. The best resolution was obtained by use of the 0.56 mM solution [84]. Bhushan et al. used silica gel TLC plates impregnated with vancomycin, as chiral selector for the resolution of (±)-verapamil. Plates were prepared by spreading a slurry of silica gel G in distilled water, containing sterile vancomycin hydrochloride. TLC plates of pH 2 and 4 were prepared by adding a few drops of acetic acid, and the plates of pH 8 and 10 were prepared by adding 0.25% ammonia solution to the slurry. The plates were dried overnight at 60°C. Chromatograms were developed in chamber preequilibrated with the mobile phase, acetonitrile–methanol–water (15:2.5:2.5, v/v/v). The plates were dried and detected with iodine vapor. The effects of chiral selector, temperature, and pH on resolution were also studied [85]. Bhushan et al. presented a racemic resolution of dansyl-DL-amino acids by TLC on silica gel plates impregnated with 0.34 mM vancomycin, as chiral selector with the mobile phase containing acetonitrile–0.5 M

NaCl solution (10:4, 14:3, v/v). Spots were detected at 254 nm [86]. Taha et al. reported methods for the analysis and chiral discrimination of cetirizine. One of them based on the enantioseparation of cetirizine on silica gel TLC plates using different chiral selectors as mobile phase additives. The mobile phase enabling successful resolution was acetonitrile–water (17:3, v/v) containing 1 mM of chiral selector, namely, hydroxypropyl-β-cyclodextrin, chondroitin sulfate or vancomycin hydrochloride [87].

Ruzin et al. investigated the mechanism of methicillin and vancomycin. Samples were separated by TLC on silica gel plates in isobutyric acid–1 M ammonia (5:3; v/v). After separation, the plates were dried and exposed to x-ray film. Preparative quantities of both unlabeled and ^{14}C-labeled samples were applied to a preparative TLC plate and separated, and autoradiographs were detected by densitometer. By using a radioactive mannopeptimycin derivative bearing a photoactivation ligand, it was shown that antibiotics interact with the membrane-bound cell wall precursor lipid II [C55-MurNAc-(peptide)-GlcNAc], and this interaction is different from the binding of other lipid II-binding antibiotics such as vancomycin and mersacidin [88]. Schneider et al. studied the mechanism of action of friulimicin B, produced by the actinomycete *Actinoplanes friuliensis*. It formed a complex with bactoprenol phosphate without affecting membrane integrity. The products were extracted with butanol–pyridine acetate pH 4.2 (2:1, v/v) and analyzed by TLC. Radiolabeled spots were visualized by iodine vapor, excised from the silica plates, and quantified by scintillation counting [89].

48.4.2 ANALYSIS OF PEPTIDE ANTIBIOTICS IN PHARMACEUTICAL AND BIOLOGICAL SAMPLES

Datta et al. described a densitometric and spectrophotometric method for determination of components of bleomycin injections. They were separated by RP-TLC on silanized silica gel plates, using 2.5% aqueous ammonium nitrate–methanol (7:3, v/v) as a mobile phase. Detection was done at 291 nm by densitometry or spectroscopy after extracting the drugs from the adsorbent with the mobile phase [90]. Ikai et al. isolated a colistin-A and colistin-B from a commercial preparation by TLC and HPLC analysis. For TLC, a solvent system composed of n-butanol–0.04 M aqueous trifluoroacetic acid (1:1, v/v) was selected. FABMS was utilized to confirm the nature of the isolated components [91].

Jamre et al. developed an in vivo radionuclide generator by labeling bleomycin with ^{191}Os considering the beta emission (313 keV) and its suitable physical half-life (15.4 day). The complex can be used for both therapy and diagnosis based on the different natures of the parent and daughter characteristics. The complex of bleomycin with ^{191}Os-hexachloroosmate was obtained at the pH 2 in normal saline at 90°C in 48 h. Radio chromatography (RTLC) was performed using a TLC scanner Bioscan. Samples were spotted on Whatman no. 2 chromatography paper and developed in 10 mM pentetic acid or diethylene triamine pentaacetic acid solution as a mobile phase to discriminate free osmium from radiolabeled compound. A sample of ^{191}Os–bleomycin was kept at room temperature for 72 h while checked by RTLC every 2, 24, 48, and 72 h. The biodistribution study for ^{191}Os-hexachloroosmate and ^{191}Os–bleomycin were carried in wild-type mice up to 14 days. Lungs, liver, and spleen uptake increased 24–72 h after administration of ^{191}Os–bleomycin; 24 h after administration, radioactivity of the kidney increased and remained constant [92].

According to Selim et al. (2005), a *Paenibacillus* sp. strain B2, isolated from the mycorrhizosphere of sorghum colonized by *Glomus mosseae*, produces an antagonistic factor, which was isolated from the bacterial culture medium and purified by cation exchange, RP, and size exclusion chromatography. Solutions were spotted onto a TLC silica gel 60F$_{254}$ plate. Chromatograms were developed using methanol–n-butanol–25% ammonia–chloroform (14:4:9:12, v/v/v/v) as the mobile phase. The plates were dried at 100°C for 1.5 h, immersed in 0.2% ninhydrin ethanol solution for 15 min, and dried again at 100°C for 5 min. Visualized bands were excised by grating with a razor; the compounds were solubilized in methanol. The identity of active compounds was confirmed by comparing the inhibition zone with the bands on a TLC plate visualized by ninhydrin. The purified factor could be separated and analyzed by combined RP-LC–MS and MS–MS. After purification using the C_{18} Sep-Pak cartridge and elution in methanol, the solution was spotted onto a TLC plate and compared with polymyxin B and polymyxin E. Ninhydrin detection revealed bands with R_F values of 0.60, 0.73, and 0.79 for the active compounds; 0.68, 0.74, and 0.80 for polymyxin B; and 0.69 and 0.74 for polymyxin E, respectively [93]. Thomas et al. examined a sample of polymyxin B and E from different sources for its potency and composition by use of microbiological and chemical (TLC, GC, LC, HPLC) assays. Precoated silanized silica gel plates were used. They were placed in a tank after its saturation with the mobile phase, acetone–0.1 N HCl (25:75, v/v) containing 1% NaCl solution. The plates were dried at 110°C for 30 min. Spots were detected with ninhydrin reagent at 570 nm or by bioautography. The semiquantitative assessment of TLC by bioautography and densitometry was useful to compare the specific antibacterial activity of the major components. Polymyxins containing 6-methyloctanoic acid (B1, El) have been claimed to be more active than the corresponding polymyxin containing 6-methylheptanoic acid (B2, EJ), although when using *Escherichia coli*, no differences in antibacterial activity could be detected between the two major constituents of polymyxin B or polymyxin E [94].

48.5 CHLORAMPHENICOL, THIAMPHENICOL, AND FLORFENICOL

Chloramphenicol is a broad-spectrum antibiotic effective against rickettsiae, gram-positive and negative bacteria, and certain spirochetes. It is also used as a second-line agent in the treatment of tetracycline-resistant cholera. Thiamphenicol is the methylsulfonyl analogue of chloramphenicol and has a similar spectrum of activity, and florfenicol is its fluorinated synthetic analogue.

48.5.1 SEPARATION AND DETERMINATION OF CHLORAMPHENICOL, THIAMPHENICOL, AND FLORFENICOL

Ghulam et al. developed a TLC method for the determination of prednisolone acetate and chloramphenicol in presence of their degraded products. Conditions were maintained by refluxing reaction mixtures for 2 h at 80°C using parallel synthesizer including acidic (1 and 5 N HCl), alkaline (0.1, 1, and 5 N NaOH) and neutral (Milli-Q water) hydrolysis, oxidation (35% H_2O_2), and wet heat degradation. Dry heat degradation was conducted by heating in an oven at 90°C for 4 h. Separation was done on TLC glass plates, precoated with silica gel $60F_{254}$ using chloroform–methanol (14:1, v/v). Quantitative analysis was done by densitometric measurements at 243 nm for prednisolone acetate and 278 nm for chloramphenicol. Spots at R_F 0.21 and R_F 0.41 were recognized as chloramphenicol and prednisolone acetate, respectively [95]. Freimuller et al. analyzed different γ-irradiated chloramphenicol samples for impurities by recording melting point, solubility, pH, and occurrence of additional spots on thin-layer plates. Solution of chloramphenicol in acetone was applied to the plate with silica gel $60F_{254}$ without prior activation. Plates were prewashed with a mixture of methanol–methylene chloride (80:20, v/v) and dried. Development was carried out in saturated chamber using a mixture of water–methanol–chloroform (1:10:90, v/v/v) as mobile phase. Plates were dried and examined at 254 nm. The structure was confirmed by IR and ^{1}H or ^{13}C NMR and determined as the cyclic ketale condensation product of acetone and chloramphenicol [96].

Marciniec et al. reported the effects of irradiation that were tested by classical (organoleptic analysis), spectrophotometric (UV, IR, EPR), microscope observations (SEM), x-ray (x-ray diffraction [XRD]), chromatographic (TLC), and thermal (DSC) methods. Chloramphenicol was E-beam irradiated with doses in the range 25–400 kGy and the possible changes were detected. TLC silica gel kieselgel $60F_{254}$ plates were used. The mobile phase was ethyl acetate and ethyl acetate–methanol (99:1, 98:2, and 95:5, v/v). The traces were set with a quartz lamp working at 254 nm. The presence of radiolysis products was confirmed by TLC method in the compound irradiated by 100 kGy. The mobile phase used (ethyl acetate–methanol, 99:1, v/v) permitted detection of two new spots (R_F 0.82 and 0.36) on the chromatogram of CHF irradiated with 100 kGy and three new spots (R_F 0.82, 0.36, and 0.47) with 400 kGy [97]. Marciniec et al. described the effect of ionizing radiation of florfenicol to determine whether it can be sterilized using high-energy radiation. Florfenicol was irradiated by E-beam radiation to doses of 25–800 kGy, and changes in physicochemical properties were examined using chromatographic (TLC, HPLC), spectroscopic (NMR, MS), and hyphenated (HPLC-MS) methods. Silica gel $60F_{254}$ plates and n-hexane–ethyl acetate, as a mobile phase, were used. Traces were set with a quartz lamp working at 254 nm. After drying, the plates were sprayed with bromocresol green or ninhydrin reagent and dried for 10 min at 110°C. When a florfenicol was irradiated to 25 kGy, no radiolysis products were observed. Two additional spots (R_F 0.80 and 0.84) were observed for florfenicol irradiated to 100 kGy, three for 400 kGy (R_F 0.05, 0.80, and 0.84), and four for 800 kGy (R_F 0.05, 0.21, 0.80, and 0.84). It showed a reasonably good radiostability in the range of doses used for sterilization [98]. Marciniec et al. studied the effect of ionizing radiation on the physical and chemical properties of thiamphenicol in solid phase, by organoleptic analysis (form, color, smell, solubility, and clarity) and spectroscopic (UV, IR, EPR), chromatography (TLC), SEM observations, XRD, polarimetry, and thermal (DSC) methods. Samples were exposed to irradiation in a linear electron accelerator LAE 13/9 till they absorbed a dose of 25, 100, or 400 kGy. Silica gel $60F_{254}$ plates were used. The mobile phases were ethyl acetate and ethyl acetate–methanol (98:2, v/v). The traces were set with a quartz lamp at 254 nm. TLC revealed two additional spots (R_F 0.78 and 0.82) for the sample irradiated with 400 kGy, and one (R_F 0.77) with 100 kGy [99].

48.5.2 Analysis of Chloramphenicol in Pharmaceutical and Biological Samples

Vovk et al. developed a TLC–densitometry for the determination of chloramphenicol residues in controlling pharmaceutical equipment cleanliness. Methanol extract was applied on an HPTLC silica gel F_{254} plate. Plates were developed by using n-hexane–ethyl acetate (35:65, v/v) as developing solvent. Quantitation was performed densitometrically at 280 nm in the reflectance mode. Chloramphenicol was stable on the plate 2 h before and 24 h after development. It was stable during 7-day storage on the cotton swabs in the solvent at room temperature and in solution stored in darkness at 4°C [100].

Cravedi et al. discussed the possible metabolites following subcutaneous administration of [³H] chloramphenicol to duck; HPLC and TLC analyses showed that the two most important metabolites in 0 to 24 h excreta were chloramphenicol–oxamic acid and chloramphenicol–alcohol, which together accounted for about one-third of the radioactivity therein [101]. Vega et al. developed a TLC method for monitoring the dose of florfenicol in fish feed. Florfenicol was extracted with methanol. Chromatography was performed on HPTLC silica gel F_{254} plates with ethyl acetate–n-hexane (80:20, v/v) as a mobile phase. The plate was scanned at 223 nm with a spectrodensitometer. The optimum wavelengths were 223 nm for thiamphenicol and florfenicol and 282 nm for chloramphenicol [102].

El-Kersh et al. described the inactivation of chloramphenicol by the spores of a *Streptomyces* sp. isolated from Egyptian soil. Separation and purification of the chloramphenicol inactivation products was developed by TLC. The reaction mixture was centrifuged, and the supernatant was extracted with ethyl acetate. The extract was dried over anhydrous sodium sulfate and concentrated. The crude chloramphenicol transformation products in the extract were separated by preparative TLC on fluorescent silica gel G_{254} with acetic acid–chloroform mixture (1:4, v/v). The plates were examined under UV 254 light [103].

Hamburger et al. reported a direct bioautographic assay on TLC for various antibacterial compounds, that is, stigmasterol, sitosterol, and palmitic acid; quercetin and p-coumaric acid; tannic acid and sucrose; penicillin V; chloramphenicol; lignans; quassinoids; and camptothecins. A series of natural products and different stationary phases were tested in order to establish the isolation of these compounds from higher plants. Silica gel G $60F_{254}$ Al sheets were used. The plates were developed in chloroform–methanol (90:10, v/v). The chromatograms were dried and compounds were detected at 254 and 366 nm and marked on the TLC plate. The bacterial suspension of cryovial was diluted with nutrient broth. A sample of this suspension was dispersed evenly over a TLC plate. The plates were incubated overnight and sprayed with aqueous solution of 2,3,5-triphenyltetrazolium chloride, p-iodonitrotetrazolium violet, tetranitro blue tetrazolium, or methyl thiazolyl tetrazolium. The plates were again incubated at 37°C for 4 h and treated with 70% ethanol before evaluation of the bioautograms [104].

48.6   IONOPHORE ANTIBIOTICS

Ionophore antibiotics are a group of monocarboxylic polyether compounds. Their biological activity is due to their ability to disrupt the flow of ions (H^+, Li^+, Na^+, K^+, Ag^+) either into or out of cells (the formation of ion channels).

48.6.1   Separation and Determination of Ionophore Antibiotics

Aubiron et al. used a TLC coupled with flame ionization detection to separate and determine simultaneously abierixin, nigericin, and grisorixin, produced by *S. hygroscopicus* NRRL B 1865. Various proportions of chloroform, methanol, and formic or acetic acid were used in the developing solvent to determine changes in R_F values of the antibiotics and to allow conditions for maximum resolution to be obtained. Development on Chromarods SII with chloroform–methanol–formic acid (97:4:0.6, v/v/v) gave satisfactory separations [105]. Coutinho et al. studied the interaction between nystatin and small unilamellar vesicles of 1,2-dipalmitoyl-sn-glycero-3-phosphocholine, in gel (at 21°C) and liquid-crystalline (at 45°C) phases, by steady-state and time-resolved fluorescence measurements. The enhancement in the fluorescence intensity of the antibiotic was applied to study the membrane binding of nystatin, and it was shown that the antibiotic had an almost fivefold higher partition coefficient for the vesicles in a gel than in a liquid-crystalline phase. TLC of nystatin was performed on precoated silica gel 60 TLC plates with a mobile phase of chloroform–methanol–water (20:22:10, v/v/v). After development in the dark, the plates were dried and visualized with UV light or by exposure to iodine vapors. The detected spots were scraped and the compounds were eluted with methanol. After centrifugation, samples were subjected to gradient elution HPLC [106].

Khattab et al. described a study of increasing the productivity of nystatin and antibiotics produced by *S. noursei* NRRL 1714 through treatment with UV rays followed by intraspecific protoplast fusion. An experiment by TLC was carried out to detect the produced antibiotics by some selected superior mutants in comparison with the wild-type strain. To identify compounds in the fermentation broth, silica gel G 1500 plates were used. The solvent system used was n-butanol–acetic acid–water (4:1:1, v/v/v). For detection, plates were sprayed with 10% sulfuric acid in ethanol and heated at 100°C until the spots shared. For antibacterial separation, plates were sprayed with ninhydrin solution in n-butanol with acetic acid. The developed colored spots were recorded and photographed [107]. Mikhailova et al. determined a composition of sterol fractions of nystatin-resistant *Candida maltosa* strains using UV spectrometry, TLC, and GLC–MS. Resistance to nystatin was connected with the composition alterations of yeast cell sterols. TLC was developed on silufol UV_{254} plates, impregnated with 20% solution of $AgNO_3$, and activated at 110°C for 30 min with chloroform–acetone (10:1, v/v) solvent system. For detection, a solution of $FeCl_3$ in acetic acid with concentrated H_2SO_4 was used [108].

48.6.2   Analysis of Ionophore Antibiotics in Pharmaceutical and Biological Samples

Vanderkop et al. reported a TLC–bioautography method for monensin detection in poultry tissues (muscle, skeletal muscle, and liver and kidney). Quantitative analysis was not possible for samples from fatty tissue [109]. Weiss et al. described a method for the analysis of monensin and nigericin in animal tissues and fluids based on TLC–bioautography. For lasalocid, in addition to TLC for quantitation in chicken skin and fat, LC methods with fluorescence detection were developed [110].

Brooks et al. developed a spectrofluorometric determination of lasalocid in dog blood. The method was based on the intrinsic fluorescence of the compound in ethyl acetate. It can measure total levels of drug and any metabolites present. The specificity of the assay was verified by TLC separation of the dog blood extract, which indicated the presence of only intact drug [111].

Zotchev et al. initiated molecular genetic studies on *S. noursei* ATCC 11455, the producer of nystatin, and identified part of the gene cluster apparently governing the biosynthesis of an antibiotic. Culture supernatants were extracted with ethyl acetate and evaporated. The pellets were dissolved in ethyl acetate and subjected to analysis on TLC plates with ethyl acetate–methanol–water (100:2.5:1, v/v/v). For bioassay, TLC plates were placed for 30 min on agar plates inoculated with *Micrococcus luteus* ATCC 10240, which were incubated for 16 h at 35°C and screened visually for growth-inhibition zones [112].

48.7 RIFAMYCIN ANTIBIOTICS

Rifamycins are a family of antibiotics biosynthesized by a strain of *S. mediterranei*. There are various types of rifamycin forms, but the rifampicin form with a 4-methyl-1-piperazinaminyl group is by far the most clinically effective. Those antibiotics have a broad spectrum of bacteria, gram positive and gram negative, and *Mycobacterium tuberculosis*. They are used for the treatment of tuberculosis and prophylaxis of meningococcal infections.

48.7.1 SEPARATION AND DETERMINATION OF RIFAMYCINS

Zarzycki et al. studied the influence of temperature and mobile phase composition on retention of rifamycin B and rifampicin by RP-TLC, using wide-range (0%–100%) binary mixtures of methanol–water. Chromatography was performed on RP-18W HPTLC plates. The chambers were saturated with the vapor of the mobile phase under 1 atm pressure. Experiments were performed at temperatures of 5°C, 10°C, 20°C, 30°C, 40°C, 50°C, and 60°C. The plates were thermostated 25 min before development. When a new temperature was started, the chromatographic device was thermostated for at least 30 min. The cyclodextrins were visualized by spraying the plates with concentrated sulfuric acid–methanol (1:4, v/v) and heating at 140°C for 2–5 min. Compounds were visualized as gray and black spots on the white background. R_F values of the solute molecules were measured. Temperature changes of the chromatographic conditions produce significant differences in migration of the investigated compounds (differences in R_F values) [113].

48.7.2 ANALYSIS OF RIFAMYCINS IN PHARMACEUTICAL FORMULATIONS

Tatarczak et al. discussed the simultaneous determination of isoniazid and rifampicin in the pharmaceutical preparation by densitometric TLC. Silica gel 60F$_{254}$ plates and ethyl acetate–methanol–acetone–acetic acid (5:2:2:1, v/v/v/v) as mobile phases in horizontal Teflon DS chambers were used. Quantitative analysis was accomplished by UV–VIS densitometric scanning. The drugs were completely separated; retardation factors (hR$_F$) were obtained: 35.0 for isoniazid, 23.2 for rifampicin, 67.6 for pyrazinamide, 47.8 for p-aminosalicylic acid, and 16.9 for etambutol [114]. Kenyon et al. analyzed a convenience sample of 13 fixed-dose combination tuberculosis drugs from "The Fixed Dose Combination Project," using TLC as a screening method, and UV spectrophotometry or LC as confirmation. Each product was analyzed for isoniazid, rifampicin, and pyrazinamide content, using a TLC at 480 nm [115]. Izer et al. developed the best composition of rifampicin eyedrops for extemporaneous preparation in pharmacies. Stability of the eyedrops was studied. The preparation may be stored at −12°C for 1 month and may be subsequently used when stored at 2°C–8°C after the first opening for 5 days. Rifampicin and its degradation products were separated by TLC using silica gel layer, and chloroform–methanol (42:58, v/v) as eluent, and determined by densitometry at 475 nm (540 nm for rifampin quinone) [116]. Jindal et al. described the determination of rifampicin and its degradation components in drugs. The separation was performed on TLC silica gel plates with a mobile phase: chloroform–methanol–water (80:20:2.5, v/v/v). Spots were detected by densitometric evaluation of the chromatograms [117]. Avachat et al. determined the stability of rifabutin and isoniazid in a combination dosage form. Analysis was performed on aluminum foil HPTLC silica gel 60F$_{254}$ plates. A development with dichloromethane–acetone–methanol (20:7:2, v/v) as a mobile phase was performed in chamber, previously saturated with a mobile phase vapor for 25 min. The plates were dried at 40°C–50°C. Densitometric analysis was performed in absorbance mode, at 504 nm for rifabutin and 262 nm for isoniazid. R$_F$ value 0.84 for rifabutin and 0.48 for isoniazid were obtained. To verify the stability-indicating nature of the method, all the known forced degradation products of rifabutin were applied to a TLC plate [118].

48.8 OTHER SUBGROUPS OF ANTIBIOTICS

48.8.1 SEPARATION AND DETERMINATION OF OTHER ANTIBIOTICS

Pandey et al. presented an HPLC study of the daunorubicin complex from fermentation broth on an RP C$_8$ column with a solvent system composed of methanol–acidic water (pH 2 with phosphoric acid). Application of this system in conjunction with TLC in the identification of anthracycline antibiotics from the fermentation broth was discussed [119]. Joish et al. isolated several cultures of *Streptomyces* sp. (LC20, LC28, LC49, LC54, and WC241). One of them, LC28, produced a new antibiotic. Authors reported its isolation, purification, and characterization by TLC. The compound with peptide moiety was applied to silica gel Baker Si F$_{250}$ TLC plates and developed with chloroform–methanol (91, v/v). On the plate under UV light, the peptide appeared as a fluorescent band (R$_F$ 0.90). It was scraped off the plate and extracted with acetone. The homogeneity of the antibiotic was determined in different solvent systems by TLC. Spots were detected in iodine chamber [120].

48.8.2 ANALYSIS OF OTHER ANTIBIOTICS IN PHARMACEUTICAL AND BIOLOGICAL SAMPLES

Agbaba et al. reported an HPTLC method for the determination of metronidazole and its impurity 2-methyl-5-nitroimidazole in tablets, based on the use of silica gel 60F$_{254}$ with acetone–chloroform–ethyl acetate (4:4:1, v/v/v) as the mobile phase. The spots were detected at 254 nm [121].

Egorin et al. isolated and purified eleven metabolites of aclacinomycin A from the urine of four patients. First, they extracted the pooled urine, and the antibiotic and its metabolites were concentrated in n-butanol. Individual metabolites were subsequently isolated and purified by TLC. Purified urinary species were characterized and compared to standards of aclacinomycin and its known metabolites by TLC, HPLC, acid and enzymatic hydrolysis, and MS [122]. Colombo et al. analyzed the differential distribution of doxorubicin and daunorubicin within blood components, after an injection, from its metabolites and quantified by TLC scanning fluorescence technique. Doxorubicin accumulated in the plasma and red cells (RBC), white cells (WBC), and platelets (PT). In the presence of higher blood concentrations, the RBCs accumulated much more antibiotic than the plasma, WBC, and PT. It suggests that the RBC fraction has a greater capacity to concentrate the drug [123]. Shah et al. reported the radiolabeling of nitrofurantoin with technetium-99m to explain the bactericidal activity of this drug. Nitrofurantoin radiolabeling with technetium-99m (^{99m}Tc) was investigated using different concentrations of the drug, sodium pertechnetate (Na^{99m}TcO$_4$), and reducing agent (SnCl$_2$) at different pH ranges (5.1–6). The suitability of the ^{99m}Tc-nitrofurantoin was evaluated in terms of the radiochemical purity yield, in vitro stability in saline and serum, in vitro binding with *E. coli*, biodistribution in *E. coli*-infected model rat, and scintigraphic accuracy in *E. coli*-infected model rabbit. In serum, in vitro stability of the ^{99m}Tc-nitrofurantoin was evaluated by incubating the complex in saline with serum at 37°C for 120 min. Aliquots, at 1–120 min (in incubation period), were applied to the TLC strip. The strip was developed in acetone and ethanol–ammonia–water (2:1:5, v/v/v). The developed strips were divided at R_F 5 and counted for activity, using gamma rays well counter [124].

Yurt et al. investigated the ability of [^{131}I]linezolid to visualize soft-tissue infection of sterile inflammations in rats. Linezolid was labeled with ^{131}I, and potential of the radiolabeled antibiotic was investigated in inflamed rats with *S. aureus* and sterile-inflamed rats with turpentine oil. [^{131}I] linezolid showed good localization in bacterial inflamed tissue and can be used to detect inflammation in rats. Its efficiency was assessed by TLC–radiolabeling chromatography on silica gel 60F$_{254}$. Two solvent systems were used: n-butanol–ethanol–0.2 M ammonium hydroxide (5:2:1, v/v/v). R_F for Na^{131}I and [^{131}I]linezolid were 0.62 and 0.70. The sheets were scanned with Bioscan TLC scanner [125]. Valcarcel et al. tested a sample of lung of intoxicated pigs, by RP-HPLC in order to detect tiamulin residue. Samples were received labeled as growing and fattening (medicated and not medicated), extracted with ethyl acetate, and the solvent was evaporated. The residue was dissolved in ethyl acetate and separated on silica gel 60 plates with ethyl acetate as a mobile phase. The plate was sprayed with 97%–98% sulfuric acid and heated at 110°C until spots appear. Tiamulin gave a spot with R_F 0.85 [126].

Adrian et al. developed an immunochemical analytical method for sulfonamide antibiotic family. The approach consisted of raising polyclonal antibodies against an appropriately designed hapten, and their use on an indirect competitive enzyme-linked immunosorbent assay (ELISA) under heterologous conditions (different haptens as immunogen and as competitor). They demonstrated the excellent performance of ELISA to analyze these antibiotics in milk samples after dilution treatment. Purification of the reaction mixtures was accomplished by "flash" chromatography using silica gel as the stationary phase. The progress of the reaction was monitored by TLC using ethyl acetate as a mobile phase and precoated silica gel 60F$_{254}$ aluminum sheets as a stationary phase [127].

Naik et al. isolated and screened fluorescent pseudomonads from banana rhizospheric soil, for the production of enzymes and hormones and putative antibiotic-producing isolates. They were grown in the production media, and production of antibiotics was confirmed by TLC and HPLC. TLC was carried out on silica gel G 60 plate and activated at 110°C for 30 min. Ethanol solutions of antibiotic and the extract were spotted. Separation was performed using chloroform–methanol (9:1, v/v) for phenazine-1-carboxylic acid (PCA) and 2,4-diacetylphloroglucinol (DAPG) or chloroform–acetone (9:1, v/v) for pyoluteorin (PLT) and pyrrolnitrin (PRN). The PCA and DAPG were detected by UV irradiation at 254 nm. PLT spots were detected by spraying with an aqueous 0.5% Fast Blue RR Salt solution, and PRN—by spraying the plates with 2% p-dimethylaminobenzaldehyde dissolved in the ethanol–sulfuric acid (1:1, v/v). When the antibiotic isolates identified by PCR were subjected to fermentation, metabolites such as DAPG (yellowish white), PCA (greenish yellow), PRN (light yellow), and PLT (yellowish white) were detected in the production cultures. TLC and HPLC analyses confirmed the production of PCA (R_F 0.53), DAPG (R_F 0.77), PLT (R_F 0.50), and PRN (R_F 0.80) by the isolates [128]. Atta et al. reported the actinomycete culture AZ-H-A5, to produce a wide spectrum antimicrobial agent when cultivated on rice straw. The actinomycete AZ-H-A5 could be isolated from a soil sample collected from Helwan district, Egypt. The parameters controlling the biosynthetic process of its formation, including inoculum size, different pH values, different temperatures, different incubation period, and different carbon and nitrogen sources, potassium nitrate, K_2HPO_4, $MgSO_4\cdot7H_2O$, and KCl concentrations, were fully investigated. The active metabolite was extracted using ethyl acetate (pH 7). Its separation and purification was performed using TLC and column chromatography techniques. The purified compound was suggestive of being belonging to destomycin-A antibiotic produced by *S. rimosus* AZ-H-A5 [129]. Atta et al. described the isolation of a bacterial strain from Egyptian soil, which generates the antibiotic sparsomycin. The identification was based on the cultural, morphological, physiological, and biochemical characteristics. The parameters controlling the biosynthetic process include inoculum size, different pH values, different temperatures, different incubation period, different carbon and nitrogen sources, and different mineral salts concentrations (KNO_3, K_2HPO_4, $MgSO_4\cdot7H_2O$, and KCl) were investigated. The active metabolite was extracted using n-butanol at pH 7, isolated, and purified, and spectroscopic analysis and biological activities were determined using TLC and column chromatography techniques. Separation was done by TLC using chloroform–methanol (24:1, v/v) as the developing solvent. Obtained results revealed that the band at R_F 0.70 exhibited obvious inhibitory effects against the growth of both gram-positive and gram-negative bacteria and fungi [130]. Ilic et al. examined 20 different streptomycete (gram-positive filamentous bacteria) isolates, obtained from soils of southeast Serbia. The structure of the bioactive components was determined using UV/VIS, FTIR, and TLC. The extracts were concentrated in vacuo and adsorbed on a silica gel 100 column. After washing with chloroform and chloroform–methanol (3:1, v/v), active components were eluted with chloroform–methanol (1:1 and 1:3, v/v). The eluate was purified on a Sephadex LH-20. The active substances of new antibiotics were eluted with methanol and yielded of a yellow powder. Afterward, the TLC plates were air-dried. Bands were scraped under UV light, extracted with methanol, and filtered. Each band was bioassayed using *Botrytis cinerea*, and the active bands were purified again on TLC using the same solvent system and visualized using UV light or the iodine–sulfuric-acid color reaction as component A (R_F 0.70) and component B (R_F 0.78) [131].

Pope et al. developed a TLC, HPLC, and MS methods for the analysis of nybomycin, its derivatives deoxynybomycin, and nybomycin acetate, during fermentation and isolation. TLC and HPLC were used to confirm the presence of deoxynybomycin in the crude extracts of fermentation broths. TLC was carried out on silica gel $60F_{254}$ plates. Samples dissolved in ethanol, chloroform, or chloroform–methanol (2:1, v/v) were spotted and developed in the solvent system methylene chloride–methanol (2:1, v/v). The compounds were visualized as white fluorescent spots under UV at 366 nm or as dark spots at 254 nm. After TLC separation, the plate was tested by bioautography. Plates were dried, placed in a bioassay dish, overlayed with nutrient agar seeded with *B. subtilis*, and incubated overnight at 37°C. Zones of inhibition were visualized by flooding the plate with a solution of 2,3,5-triphenyl tetrazolium chloride in distilled water and incubating for 1 h at 37°C. Biological activity corresponding to deoxynybomycin was seen where a sample of the extract was spotted [132]. Itoh et al. described the identification of the producing organisms, fermentation, isolation, and physicochemical and biological properties of heptelidic acid. It was found in the culture filtrate of three different strains of fungi isolated from soil samples. This antibiotic was purified by column chromatography on silica gel or Sephadex LH-20, and finally by preparative TLC. The culture broth was filtered and extracted with ethyl acetate. The extract was concentrated and transferred into 2% $NaHCO_3$ solution (pH 3 with HCl). Antibiotic was reextracted twice with ethyl acetate, and the extract was concentrated, dissolved in chloroform, and applied onto a silica gel column. Eluate was collected and fractions were pooled and removed the solvent to yield of oil. The oily preparation was purified by TLC on silica gel F_{254}, using n-hexane–benzene–acetic acid (1:8:1, v/v/v) as a solvent system [133].

48.8.3 Analysis of Other Antibiotics with Bioautography Detection

Tamehiro et al. found a novel phospholipid antibiotic, bacilysocin, which accumulates within (or associates with) the cells of *B. subtilis* 168 and determined its structure by NMR and MS analyses. Bacilysocin was extracted with butanol from cells grown for 24 h. An extract was subjected to TLC with chloroform–methanol–water (60:25:4, v/v/v) as a development solvent. The bioassay was performed with *S. aureus* 209P as the test organism. TLC plates were placed in bioassay plates and overlaid with 0.5% Mueller–Hinton agar. The inhibitory zones were visible after 12 h of incubation at 37°C, followed by staining with GelCode Blue Stain reagent [134]. Fakhouri et al. studied a *Pseudomonas fluorescens* strain G308 isolated from barley leaves, which produces a novel antibiotic substance that was purified by preparative TLC and HPLC, and identified by LC–DAD, IR, LC–ESI(+)–MS, LC–ESI(–)–MS, GC–EIMS, LC–HRES(+)–MS, mass isotope ratios analysis, ^1H NMR, and ^{13}C NMR analysis. The TLC bioassay of the ethyl acetate extract of the culture filtrate exhibited two inhibition zones. The methanol extract was applied to TLC ALUGRAM-SIL-G UV_{254} plates and developed in *tert*-butylmethylether–hexane–methanol (50:40:3.5, v/v/v). The two substances had R_F values of 0.66 and 0.70. The properties of the bands were detected by the bioautographic method using *Cladosporium cucumerinum* and *Colletotrichum lagenarium* as test organisms. The procedure for detecting zones with antibiotic activity was repeated. The bands were scratched off from the preparative kieselgel 60F$_{254}$ TLC plates and developed in tert-butylmethylether–hexane–methanol (50:40:3.5, v/v/v). The separated compounds were eluted from silica gel with acetone [135]. Sadfi et al. reported the purification of active compounds isolated from growth media of the protagonists. Biologically active antibiotics were extracted with methanol from the interaction regions between the two protagonists on potato dextrose agar medium. Concentrated extracts were purified using SPE and TLC. The fraction was spotted onto silica gel 60F$_{254}$ TLC plates and developed using butanol–water–pyridine–toluene (10:6:6:1, v/v/v/v) as the solvent system. The spots were detected with a ninhydrin spray after drying the plate for 5–10 min at 110°C. All the purified fractions were tested by TLC bioassays on silica gel 60 TLC plates and dried 1 h in the laminar flow cabinet and one night at 4°C. Molten agar (7%) inoculated with a spore suspension of *Cladosporium* were overlayed on the TLC plate and incubated at 25°C in a humid chamber. After 48 h, halos of inhibition were measured. Results showed that *B. cereus* X16 produce more than one antibiotic. The partially purified compound was fractionated by preparative silica gel TLC. This assay showed that the antibiotic was highly mobile (R_F 0.7) on TLC plates developed in n-butanol–pyridine–toluene–water (10:6:6:1, v/v/v/v) and was almost immobile in methanol–water (2:1, v/v), ethanol–water (2 + 1, v/v), and n-butanol–acetic acid–water (3:1:1, v/v/v). The bands were visualized with a ninhydrin spray; however, the use of vanillin or CAM detectors failed to reveal the chromatogram. The chromatograms were red then yellow with three bands A, B, and C with R_F values being 0.7, 0.5, and 0.13 [136], respectively. Allen et al. found three transformants (*S. hygroscopicus* 3602 cloned in the plasmid vector pU61 transform *S. lividans* TK24) with the ability to produce an antibiotic lethal to a geldanamycin-sensitive strain of *B. subtilis*. Extracts of fermentation broth cultures were analyzed by TLC. The ethyl acetate extracts of all strains after growing were dried in vacuo, the residue resuspended in ethyl acetate, applied on precoated TLC plate, and developed in ethyl acetate–dichloromethane–n-hexane–methanol (9:6:6:1, v/v/v/v). For bioautography assays, the plates were overlaid with molten SNA seeded with *B. subtilis*, incubated for 24–48 h at 30°C, and scored for areas of antibiosis in the *B. subtilis* lawn. Antibiotic zones at R_F values of 0.04 and 0.08 were present in all samples. The results suggest that cloned inserts in three plasmids express functions when present in *S. lividans* TK24, which control the synthesis of compounds closely related, if not identical, to geldanamycin [137]. Gerber et al. reported the isolation, purification, and characterization of peniophorin A and B, produced by *Peniophora affinis* B 325. The separation was performed by TLC with precoated Polygram G/UV$_{254}$ plastic sheets, or Eastman Chromagram 13179 silica gel, and chloroform–acetone (90:10, v/v). Active spots were disclosed by bioautography with nutrient agar plates seeded with *Proteus vulgaris* or *P. aeruginosa*. They were also visualized with the fluorescent indicator and by sprays of dilute alkaline potassium permanganate (both), bromophenol blue (B only), saturated dinitrophenylhydrazine in 10% HCl (A only), and 10% HCl (A only). R_F values were for A 0.8 and for B 0.45. When *P. aeruginosa* was used as the test organism, only antibiotic B gave an active zone. Spots of the purified antibiotics on TLC plates became visible overnight in air. The used sprays showed that A was neutral; it possessed a reactive carbonyl group and gave a deep green color with 10% HCl. B was acidic and did not contain a reactive carbonyl group [138]. Tsuchida et al. described production, isolation, physicochemical, and biological properties of azicemicins A and B. Compounds were isolated from the culture broth of the strain MJ126-NF4 with 80% aqueous methanol. The active fractions were collected and concentrated. The residue was chromatographed on kieselgel 60 using mixtures of chloroform–methanol (20:1, 10:1, 7:1, 5:1, 4:1, 2:1, v/v). The crude antibiotic A was eluted with the mixture of chloroform–methanol (7:1, 5:1, v/v). The fractions that contained a UV-absorbing substance at R_F 0.26 on silica gel 60F$_{254}$ TLC plates with chloroform–methanol (10:1, v/v) were collected and concentrated to give a dark-yellow solid. The crude antibiotic B was obtained from the column chromatography eluting by the mixtures of chloroform–methanol

(4:1, 2:1, v/v). A compound was purified by silica gel TLC with chloroform–methanol–water (4:1:0.1, v/v/v) and was provided from giving a UV-absorbing spot at R_F 0.42 [139]. Mukhtar et al. studied streptogramin antibiotics that consist of a mixture of two components: cyclic polyunsaturated macrolactones (group A) and cyclic hexadepsipeptides (group B). They expressed recombinant Vgb (enzyme from *S. aureus*) in quantity and, using a combination of MS, NMR, and synthesis of model depsipeptides, show unequivocally that streptogramin B inactivation does not involve hydrolysis of the ester bond. Research expressed, purified, and characterized this enzyme and orthologues from *Bordetella pertussis* and *S. coelicolor* and demonstrated that elimination was the general mechanism of streptogramin B inactivation. TLC assay was performed on precoated silica gel $60F_{254}$ plates in the solvent systems: chloroform–methanol–acetic acid (95:5:3, v/v/v), ethyl acetate–hexane (1:1, v/v), and ethyl acetate–hexane (2:1, v/v) [140]. Pinchuk et al. determined the antagonistic activity of probiotic strain *B. subtilis* 3 against *Helicobacter pylori* in vitro and the cause of any inhibitory activity. Two antibiotics, detected by TLC and HPLC analysis, were found to be responsible for this activity. One of them was identified as amicoumacin A, an antibiotic with anti-inflammatory properties. TLC analysis was performed on precoated kieselgel 60 plates with chloroform–methanol–water (65:25:4, v/v/v) as a mobile phase. The bands were marked under UV light at 310 nm, with R_F values of 0.2, 0.43, 0.47, 0.8, and 0.85. These bands were submitted to an extraction procedure in water–isopropanol (30:70, v/v), and the activity of each band against *H. pylori* strains was tested by the agar well diffusion method [141]. Vivien et al. analyzed the chromatographic behaviors of albicidin, an antibiotic produced by *Xanthomonas albilineans*, by TLC. It was analyzed from the methanolic fractions from the cultures, which were pooled, vacuum dried, and separated with silica gel $60F_{254}$ TLC aluminum sheets using methanol as the mobile phase. Subsequently, strips of the TLC plate were placed for 2 h at room temperature on the surface of a top agar layer containing an *E. coli* indicator strain. Inhibition zones, observed after overnight incubation at 37°C, documented the positions of separated antibiotics. Two indicator strains sensitive to albicidin and two indicator strains resistant to albicidin were tested [142]. Hu et al. determined a 3,5-dihydroxy-4-isopropyl stilbene, an antibiotic produced by the bacterial symbiont *Photorhabdus luminescens* of the nematodes of the genus *Heterorhabditis* in nematode bacterium-infected insects, using HPLC or TLC for separation and UV for quantification. For the TLC analyses, kieselgel $60F_{254}$ TLC aluminum sheets, and a mixture of methanol–chloroform (1.5:98.5, v/v) as the developing solvent, were used. Spots were visualized with a UV lamp at 254 nm. The R_F value was 0.59 [143].

48.9 SIMULTANEOUS STUDY OF VARIOUS SUBGROUPS OF ANTIBIOTICS

48.9.1 SEPARATION AND DETERMINATION OF VARIOUS ANTIBIOTICS

Nowakowska et al. examined the chromatographic behavior of erythromycin, troleandomycin, tylosin, vancomycin, rifamycin B, and rifampicin on polyamide $11F_{254}$ TLC plates as stationary phase with five binary solvent mixtures (methanol–water, ethanol–water, propanol–water, acetonitrile–water, and tetrahydrofuran–water) as mobile phases in which the concentration of the organic modifier was from 0% to 100% (v/v) [144]. Nowakowska et al. separated erythromycin, tylosin, troleandomycin, vancomycin, rifamycin B, and rifampicin on NP- and RP-TLC plates using water, methanol, 1-propanol, and their binary solvent mixtures as mobile phases. The study of the influence of the number of carbon atoms of methanol and propanol, and sodium chloride on retention of these solutes, and the selection of antibiotics as chiral selector were discussed. Chromatography was performed on kieselgel 60W F_{254}s TLC and RP-18 W HPTLC plates. The plates were developed in chambers, previously saturated with a mobile phase vapor. The spots were visualized by spraying with 20% perchloric acid and heated for 10 min at 120°C. The R_F values were measured using a wide range (0%–100%, v/v) of methanol and 1-propanol in water, binary mobile phases [145]. Wayland et al. separated the antibiotics by TLC in combination with nitration, reduction, and diazotization to further distinguish colistin from polymyxin. Lincomycin, oleandomycin, and kanamycin were identified by TLC systems, and tetracyclines by paper chromatography. Under varying conditions of solvents and wavelengths, streptomycin, erythromycin, nystatin, vancomycin, novobiocin, cephalothin, and viomycin were identified spectrophotometrically. Color tests were devised for chloramphenicol, dihydrostreptomycin, and neomycin [146]. Bhushan et al. determined tetracycline antibiotics on a Co^{2+} (1%) impregnated silica gel layer using ethanol–acetic acid–water (10:6:6, v/v/v) as the mobile phase. Amino glycopeptide antibiotics are also separated on an untreated silica gel layer using the mobile phase n-butanol–formic acid–water (6:5:7, v/v/v). The spots were located by exposing the plates to iodine vapors [147]. Nowakowska reported an HPTLC separation of erythromycin, troleandomycin, tylosin, vancomycin, rifamycin B, and rifampicin by use of a variety of nonaqueous binary mobile phases. The effect on the retention of antibiotics of the nature of the mobile phase, especially the number of carbon atoms in the alcohols and ketones it contained, was studied. The use of these drugs as chiral selectors was also discussed. Chromatography was performed on aluminum-backed LiChrospher Si $60F_{254}s$ HPTLC plates in chambers previously saturated with a mobile phase vapor. A wide range of mixtures of alcohols and ketones with hexamethyldisiloxane in proportions from 0% to 100% (v/v) and with dimethyl sulfoxide in proportions from 0% to 50% (v/v) were used as mobile phases. Spots were visualized by spraying with a mixture of concentrated sulfuric acid–methanol (1:4, v/v) and heating for 10 min at 120°C. The compounds were visible as brown spots [148].

Nowakowska et al. described the retention behavior of aclarubicin and doxycycline on a variety stationary phases with modified n-alcohol mobile phases. Chromatography was performed on glass silica gel 60W $F_{254}s$, RP-18 $F_{254}s$, cellulose F, and polyamide 11 F_{254} TLC plates and on silica gel 60 F_{254} and RP-18 $WF_{254}s$ HPTLC plates. Chambers were previously saturated with a mobile phase vapor. Plates were developed with wide-range (0%–100%, v/v) mixtures of n-alcohols with dimethyl sulfoxide, hexamethyldisiloxane, acetonitrile, and water, as mobile phases. The antibiotics were detected by visualizing reagents: sulfuric acid–methanol (1:4, v/v), anisaldehyde–methanol–acetic acid–orthophosphoric acid–sulfuric acid (1:100:10:10:5, v/v/v/v/v), and 5% $AlCl_3$ in methanol. The compounds were visible in VIS and UV as multicolored spots [149]. Husain et al. discussed the retention behavior of amoxicillin, ampicillin, cephalexin, cloxacillin, doxycycline, tetracycline, erythromycin, gentamicin, streptomycin, and co-trimoxazole on thin layers of the inorganic ion-exchanger titanic silicate with organic, aqueous, and mixed aqueous–organic mobile phases. Titanium(IV) silicate was prepared by addition of 0.25 M sodium silicate solution to 0.08 M titanic chloride solution in 0.2 M HCl (pH 6.5 with 2 M NaOH). After mixing the gel with gypsum, the mixture was poured immediately into TLC plate coater and used to coat glass plates. The plates were dried at 60°C for 2 h. Various mobile phases were used: chloroform–methanol–ammonia (1:1:1, v/v/v), 1 M ammonium sulfate, 0.1 M formic acid, and 20% dipotassium hydrogen phosphate. Spots were detected with 1% ninhydrin in ethanol and 5% potassium dichromate in concentrated H_2SO_4. Salting out TLC using aqueous ammonium sulfate solutions revealed the dependence of R_F values on the concentration of salt in the mobile phase and the existence of a linear relationship between R_M and molarity of $(NH_4)_2SO_4$ for analyzed antibiotics. The effect of varying the volume ratio of the binary mobile phase methanol–0.1 M formic acid on the R_F values of the antibiotics was also studied [150].

Nowakowska reported an HPTLC separation of erythromycin, troleandomycin, tylosin, rifamycin B, and rifampicin, by use of a variety of nonaqueous binary mobile phases, the effect of the used mobile phases and of the number of carbon atoms of the alcohols in the mobile phases on the retention of the drugs, and the selection of antibiotics as chiral selectors. The retention behavior was examined on RP-18 $F_{254}s$ TLC plates as stationary phase with a wide range (0%–100%) of mixtures of different alcohols with dimethyl sulfoxide (DMSO) or hexamethyldisiloxane (HMDSO) as mobile phases. The substances were visualized by spraying the plates with a mixture of concentrated sulfuric acid–methanol (1:4, v/v) and heating at 120°C for 10 min, to obtain brown spots. The best separation was obtained with ethanol–HMDSO, propanol–HMDSO, and butanol–HMDSO mobile phases containing large amounts of alcohol, but it should be stressed that separation deteriorated with increasing carbon chain length. Use of HMDSO and DMSO in the mobile phases had very different effects on the migration of the investigated antibiotics. HMDSO strongly affected chromatographic retention, causing it to increase, whereas DMSO led to a substantial decrease in the retention of the drugs [151]. Nowakowska studied the retention and separation of erythromycin, tylosin, troleandomycin, rifamycin B, and rifampicin on NP- and RP-TLC plates with a variety of nonaqueous binary mobile phases; and the analysis of the effect of mobile phase composition and the number of carbon atoms of esters and ketones on the retention of the drugs, for potential use of the antibiotics as chiral selectors. Chromatography was performed on silica gel 60F$_{254}$s HPTLC and RP-18 TLC plates. The plates were developed at 20°C in chambers previously saturated with a mobile phase vapor. Binary mobile phases were prepared by mixing appropriate quantities of pure esters or ketones with hexamethyldisiloxane or dimethyl sulfoxide in proportions from 0% to 100% (v/v). The plates were sprayed with a mixture of concentrated sulfuric acid–methanol (1:4, v/v) and dried at 120°C for 10 min. The compounds were visible in visible light as brown spots [152].

48.9.2 Analysis of Various Antibiotics in Pharmaceutical Formulations

Sharaf et al. separated polymyxin B sulfate, neomycin sulfate, and bacitracin zinc in ointments. The ointments were dispersed in chloroform and the components were extracted with 0.1 N HCl. The stationary phase was silica gel G and the mobile phase consisted of a mixture of methanol–ethanol–methylene chloride–ammonium hydroxide–water (3:3:2:2:1.5, v/v/v/v/v) or methanol–isopropanol–methylene chloride–ammonium hydroxide–water (4:2:2:2:1.5, v/v/v/v/v). Detection was performed by spraying with a 0.2% solution of ninhydrin in 1-butanol [153]. Krzek et al. presented a simultaneous determination of neomycin sulfate, polymyxin B sulfate, and zinc bacitracin and auxiliary substances: methyl ester and propyl hydroxybenzoic acid used as preservative agents, by TLC–densitometric method. To separate, the silica gel TLC plates and two mobile phases were used: methanol–n-butanol–25% ammonia–chloroform (14:4:9:12, v/v/v/v) for antibiotics and n-pentane–glacial acetic acid (66:9, v/v) for hydroxybenzoates. Chromatograms were dried at 100°C for 1.5 h. The dried chromatograms were immersed in 0.2% ninhydrin ethanol solution for 15 min and dried at 100°C for 5 min. The peak areas were recorded densitometrically at 550 nm for antibiotics, and at 260 nm for hydroxybenzoates [154]. Krzek et al. determined oxytetracycline, tiamulin, lincomycin, and spectinomycin in veterinary drugs. TLC sheets precoated with silica gel and two mobile phases: n-butanol–ethanol–chloroform–25% ammonia (4:5:2:5, v/v) and 10% citric acid solution–n-hexane–ethanol (80:1:1, v/v), were used. Plates were dried at room temperature and detected at 350 nm for oxytetracycline; after spraying with 16% sulfuric acid and heating at 105°C for 10 min, at 430 nm for oxytetracycline, 450 nm for tiamulin, and 278 nm for lincomycin; and after spraying with Ehrlich's reagent (4-dimethylaminobenzaldehyde in 36% HCl–ethanol (25:75, v/v)), heating at 105°C for 10 min, and at 421 nm for spectinomycin [155]. Krzek et al. developed the determination of polymyxin B, framycetin, and dexamethasone in dental ointment. Silica gel 60F$_{254}$ plates and two mobile phases, methanol and methanol–n-butanol–25% ammonia–chloroform (14:4:9:12, v/v/v/v/), were used. Densitometric detection was made at 550 nm after spraying with 0.3% ninhydrin solution and recording at 245 nm [156]. Anokhina et al. used a silufol plates, unmodified and modified before use by the impregnation with 5% EDTA or phosphate buffer pH 5.8. Samples of drugs were applied to the plates and chromatographed in saturated TLC tanks with mobile phases: n-butanol–acetic acid–5% oxalic acid solution (10:5:4, v/v/v) for tetracycline, rifampicin, riboflavin, thiamine, and n-butanol–10% citric acid solution–water (2:2:2, v/v/v) for oxytetracycline, hexamethylenetetramine, and dibazol. The adsorption zones of rifampicin and riboflavin were colored. For the visualization of tetracycline or oxytetracycline, the plates were dried at 100°C for 10–15 min to obtain bright-yellow spots [157].

48.9.3 Analysis of Various Antibiotics in Biological Material

Baranowska et al. described NP-TLC and RP-TLC methods for analysis of L-arginine, its primary metabolites and selected drugs (dexamethasone, prednisolone, furosemide, vancomycin, amikacin, fluconazole, digoxin, captopril, dipyrone, metoprolol, and sildenafil), in model solutions, and spiked human urine. Chromatography was performed on glass TLC silica gel 60G F_{254} plates. They were developed at 22°C in glass chamber after saturation with a mobile phase vapor for 1 h. The drugs were separated with acetonitrile–water in different proportions as mobile phases. Drugs were visualized in an iodine-vapor chamber, as yellow–brown spots on a white background. The best resolution of all the drugs was obtained with acetonitrile–water (2:3, v/v) as a mobile phase. Human urine spiked with appropriate amounts of drug standards was diluted with water, applied directly to TLC plates, and developed in the same manner as for drugs solutions [158].

Tajick et al. detected routine poultry antibiotics residue in chicken meat tissues by TLC. Different tissues of chicken corpses were crumbled and extracted with 96% ethanol. The supernatant was evaporated and residues were resolved in methanol. The samples were spotted on silica F_{254} plates, activated in 120°C for 2 h, and developed in acetone–methanol (1:1, v/v) as a mobile phase. Spots were detected at 256 nm [159]. Salisbury et al. reported a TLC–bioautography method for determination of antibiotic residues in animal tissues: penicillin G, tetracycline, oxytetracycline, chlortetracycline, chloramphenicol, monensin, novobiocin, erythromycin, tylosin, oleandomycin, streptomycin, dihydrostreptomycin, gentamicin, and neomycin. TLC plates with preadsorbent spotting zones were warmed over a heating strip at 50°C during spotting. Whatman LKGD silica gel plate was developed in a saturated chamber containing chloroform–methanol–acetone–glycerin (49:30:20:1, v/v/v/v). Whatman LK2D cellulose plate was developed with acetone–chloroform–1-propanol–0.01 N phthalate buffer pH 3.75–glycerin (16:20:57:15.2:0.8, v/v/v/v/v). Whatman LKGD silica gel plate was developed with 1-butanol–methanol–acetic acid–water (15:30:9:36, v/v/v/v), dried, and redeveloped. For bioautography, medium was inoculated with *B. subtilis* ATCC 6633 spore suspension, mixed, poured into bioassay dish, and allowed to solidify. The TLC plate was placed face down on agar, left for 15 min to allow diffusion of antibiotics into the agar, and removed. The agar plates were incubated overnight at 37°C and examined for the presence, size, and location of zones of inhibition. Visualization was enhanced by spraying a 0.2% 2,3,5-triphenyltetrazolium chloride solution and incubating for 10 min [160].

48.9.4 ANALYSIS OF VARIOUS ANTIBIOTICS WITH BIOAUTOGRAPHY DETECTION

Ramirez et al. developed an HPTLC–bioautography method for quantitation of antibiotic residues (chloramphenicol, ampicillin, benzylpenicillin, dicloxacillin, and erythromycin) in cow's milk. Sample extracts with acetonitrile were spotted onto HPTLC silica gel 60 plates. They were developed in dichloromethane–acetone–methanol–glycerin (64:20:15:1, v/v/v/v) as the solvent system. Each plate was placed face down on a Petri dish, contacting the silica layer with the inoculated media (*B. subtilis* ATCC 6633) for 25 min. Inverted dishes were then incubated for 18–24 h at 37°C. Inhibition zone diameters were measured and compared with antibiotic standards inhibition zones [161]. Neidert et al. presented a method for the identification of the residues of 14 commonly used antibiotics in animal tissues. It is based on tissue extraction with methanol, and methanol–HCl (98:2, v/v), followed by TLC–bioautography. The extracts were spotted onto TLC plates and developed in suitable solvent systems. Obtained plates were placed on set medium seeded with *B. subtilis* and a bioautograph was produced. The location of zones of inhibition was used to identify antibiotic [162]. Kondo described a TLC–bioautographic procedure for 24 antibiotics. Seven TLC systems with an ammonium chloride solution in a graded concentration range were used. The antibiotics were divided into four groups showing the characteristic behavior of R_F values corresponding to similarities in chemical structure: β-lactam, aminoglycoside, macrolide, and tetracycline antibiotics. TLC–bioautography helped estimate the character of antibiotics and the characteristic change of R_F values, for classifying unknown residual antibiotics in animal samples [163]. Shareef et al. tested 75 samples of stored poultry products (liver, breast, and thigh muscle) for the presence of antibiotics residues (oxytetracycline, sulfadiazine, neomycin, and gentamicin) using TLC. Each sample was mixed with 96% ethanol. The supernatant was evaporated, dissolved in methanol, and spotted on TLC silica plates, which were activated in 120°C for 2 h before use. Plates were developed in acetone–methanol (1:1, v/v) as a mobile phase. Spots were observed on UV light at 256 nm. The results revealed 39 positive samples. No neomycin or gentamicin residues were detected in all tested samples [164]. Torroba et al. reported a TLC method for determining the levels of penicillin, neomycin, and polymyxin in foot-and-mouth disease (FMD) oil vaccines. Samples were applied on chromatofolio with silica gel and developed with methanol–acetone–chloroform–ammonia (3:2:2:2, v/v/v/v). Plates were sprayed with the ninhydrin solution and allowed to react for 5 min at 55°C, to obtain the purple spots. Results were discussed in terms of the application of TLC for quality control of immunogens and for studying postvaccinal reactions [165]. Ahmad et al. investigated the ability of alcoholic crude extracts and some fractions from 15 traditionally used Indian medicinal plants to inhibit the growth of extended spectrum β-lactamases (ESbLs)-producing multidrug-resistant enteric bacteria. The test bacteria *E. coli* and *Shigella* were resistant to 16–23 antibiotics. Interaction of crude extracts with tetracycline, ciprofloxacin, nalidixic acid, chloramphenicol, and streptomycin demonstrated synergistic interaction with tetracycline and ciprofloxacin. Phytochemical analysis and TLC–bioautography were carried out. Plant extract was prepared in 70% ethanol. To obtain various fractions of plants, dry extract was soaked in 97% acetone and refluxed at 50°C for 2 h. The filtrate was diluted in dimethyl sulfoxide. Other fractions were prepared in ethyl acetate and 97% methanol. Each solution was spotted on silica gel G F_{254} plates. Acetone–ethanol (1:1, v/v) was used as the solvent system. Developed plates were placed in petri plates. Culture was added to the agar medium and a thin layer was poured over chromatograms. Plates were incubated at 37°C for 24 h. Zone of inhibition of bacterial growth was seen around the active chromatogram spot. The spot was also confirmed by flooding the plates with a solution of p-iodonitrotetrazolium violet. Ten plant extracts showed synergistic interaction with tetracycline, while only three with tetracycline as well as ciprofloxacin. No interaction was detected with streptomycin, nalidixic acid, and chloramphenicol [166]. Gafner described semiquantitative estimations of avilamycin, avoparcin, Zn bacitracin, erythromycin, flavomycin, furazolidone, lasalocid, monensin, narasin, penicillin, salinomycin, spiramycin, tetracyclines, tylosin, and virginiamycin in feeds. The method involved agar diffusion of buffered samples, a neutral extraction of polyether antibiotics followed by TLC, and an acid extraction for other antibiotics followed by TLC. Identification after TLC was achieved by bioautography with the most sensitive microorganisms [167]. De Jong et al. developed a method for avoparcin, zinc bacitracin, spiramycin, tylosin, and virginiamycin, and the growth promoters carbadox and olaquindox. Methanol–water (1:1, v/v) and methanol–0.01 M HCl (9:1, v/v) were used for the extraction from feeds. The following techniques were applied: microbiological inhibition, high-voltage electrophoresis, TLC, HPLC, and LC–MS/MS. TLC detection was developed on silica gel plates by bioautography. For virginiamycin, antibiotic medium with pH 6.5 and, for tylosin and spiramycin, antibiotic medium with pH 7.6 were used. Both media inoculated with *M. luteus* ATCC 9341 and with addition of tylosin. Bacitracin analysis was carried out on cellulose with bioautography, with tryptic soy agar inoculated with *M. luteus* ATCC 10240, and with addition of neomycin [168]. Ergin et al. analyzed antibiotic residues in raw milk and pasteurized milk products sold in Ankara (Turkey), in terms of penicillin G, oxytetracycline, gentamicin, streptomycin, and neomycin by using TLC–bioautographic method. For extraction, acetonitrile–methanol–deionized water (40:20:20, v/v/v) mixture was added. The solutions were applied to the channels on cellulose plates, which were placed in the chambers after its saturation for 1 h with solvent system consisting of acetone–chloroform–n-propanol–impregnation liquid (16:20:27:16, v/v/v/v). The dried TLC plates were placed on the food-lot surface and allowed to contact agar with *B. subtilis* ATCC 6633 as test microorganism for 20 min. Then, plates were removed and bioplates were left for incubation for 16 h at 37°C. The inhibition zones diameters were detected [169]. Bossuyt et al. determined the different antibiotic residues (cloxacillin, dihydrostreptomycin, tetracycline, oxytetracycline, chlortetracycline, chloramphenicol, neomycin, novobiocin, bacitracin, erythromycin, oleandomycin, ampicillin, streptomycin, and oxacillin) in milk samples by TLC method [170]. Lata et al. discussed the characterization of endophytic bacteria associated with Echinacea, evaluated their resistance to antibiotic additives during propagation, and assessed their ability to produce secondary metabolite indole-3-acetic acid. TLC analysis was performed on silica gel Alugram Sil G UV$_{254}$ sheets using benzene–acetone–acetic acid (13:6:1, v/v/v), and RP-18 F$_{254}$S plates using methanol–water (65:35, v/v), with detection under UV at 254 nm. Antibiotic resistance was also assessed as a virulence factor. The majority of endophytic bacteria were resistant to the antibiotic kanamycin but susceptible to chloramphenicol. Recommendations for propagating Echinacea in vitro cultures involve the addition of chloramphenicol, tetracycline, and ampicillin, antibiotics that cause no side effects on these plant species [171]. Rusanova et al. isolated antibiotic substances produced by a transformed strain, and studied of physicochemical and biological properties of chromatographically pure components in order to identify them. The antibiotic complex of the transformed strain *S. werraensis* 1365T was extracted with methanol (1:2, w/v; pH 7) over 18 h. After evaporation, the components were purified and separated on a kieselgel 60 column with chloroform and chloroform–methanol mixture with increasing content of the latter (1%–50%). Additional preparative purification and separation of the components was performed on kieselgel 60 plates in a system of chloroform–benzene–methanol (30:20:7, v/v/v). Fractions were analyzed with TLC on kieselgel 60, in a mobile phase of chloroform–ethyl acetate (1:2, v/v) with subsequent bioautographic, chemical (with KMnO$_4$), or physical (fluorescence under UV excitation) detection. Results suggested that obtained substances were undecylprodigiosin, anisomycin, and copiamycin [172]. Kim et al. reported the biosynthetic pathway of 3-amino-5-hydroxybenzoic acid formation with cell-free extracts from the rifamycin B producer, *Amycolatopsis mediterranei* S699, and the ansatrienin A producer, *S. collinus* Tu1892. The results demonstrated the operation of a variant of the shikimate pathway in the formation of the mC7N units of ansamycin, and presumably also mitomycin. A solution of aminoSA and NADP+ in distilled water (pH 10 with 1 N NaOH) was prepared. The reaction was started by adding portions of cell-free extract of *E. coli* AB2834/pIA321 at room temperature. Since the pH of the reaction mixture decreased once the reaction started, it was repeatedly adjusted so as not to fall below 9.0. The reaction was monitored by TLC on silica gel with ethyl acetate–acetic acid–methanol–water (4:1:1:1, v/v/v/v) as a mobile phase [173]. Phillips et al. tried to evaluate the phenotypic diversity of streptomycete populations based on antibiotic resistance profiles. TLC was carried out on silica gel F$_{254}$ plates. The solvent systems used were ethyl acetate–n-hexane–dichloromethane–methanol (9:6:1:1, v/v/v/v) for ansamycins, geldanamycin, herbimycins A and C, and dichloromethane–methanol (9:1, v/v) for nigericin. Spots were observed under UV light or by spraying with 0.3% vanillin in ethanol plus 0.5% sulfuric acid and dried at 100°C for 10 min or by bioautography. Then, they were removed, a seeded overlay containing log phase *B. subtilis* cells was poured onto the base layer, dishes were incubated at 37°C overnight, and sprayed with *Aspergillus niger* spores in glycerol salts solution. Zones of inhibition were compared with spots on replicate TLC plates [174]. Gentile et al. used the antibiotic gradient-plate technique to generate producing *Actinomadura* strains that synthesize enhanced quantities of component 3A of a macrocyclic lactam compound Sch 38516. 79 colonies were selected from a mixture of rifampicin and spectinomycin gradient plates. The two described fermentation extracts produced an enhanced 3A component and was identified with silica gel TLC, HPLC, and bioautography [175].

REFERENCES

1. Crecelius, A., Clench, M.R., Richards, D.S., and V. Parr. 2002. Thin-layer chromatography–matrix-assisted laser desorption ionisation–time-of-flight mass spectrometry using particle suspension matrices. *J. Chromatogr. A* 958:249–260.
2. Dong, Ch., Xie, H., Shuang, S., and Ch. Liu. 1999. Determination of tetracycline and 4-epimeric tetracycline by TLC-fluorescence scanning densitometry. *Anal. Lett.* 32:1121–1130.
3. Meisen, I., Wisholzer, S., Soltwisch, J., et al. 2010. Normal silica gel and reversed phase thin-layer chromatography coupled with UV spectroscopy and IR-MALDI-o-TOF-MS for the detection of tetracycline antibiotics. *Anal. Bioanal. Chem.* 398:2821–2831.

4. Oka, H., Ikai, Y., Kawamura, N. et al. 1987. Improvement the chemical analysis of antibiotics X. Determination of eight tetracyclines using thin-layer and high-performance liquid chromatography. *J. Chromatogr. A* 393:285–296.

5. Oka, H., Uno, K., Harada, K.-I., Hayashi, M., and M. Suzuki. 1984. Improvement the chemical analysis of antibiotics VI. Detection reagents for tetracyclines in thin-layer chromatography. *J. Chromatogr. A* 295:129–139.

6. Oka, H., Uno, K., Harada, K.-I., Keneyama, Y., and M. Suzuki. 1983. Improvement of the chemical analysis of antibiotics I. Simple method for the analysis of tetracyclines on silica gel high-performance thin-layer plates. *J. Chromatogr. A* 260:457–462.

7. Xie, H.-Z., Dong, Ch., Fen, Y.-I., and Ch.-S. Liu. 1997. Determination of doxycycline, tetracycline and oxytetracycline simultaneously by TLC-fluorescence scanning densitometry. *Anal. Lett.* 30:79–90.

8. Fernandez, A.A., Noceda, V.T., and E.S. Carrera. 1969. Simultaneous separation and quantitative determination of tetracycline, anhydrotetracycline, 4-epitetracycline, and 4-epi-anhydrotetracycline in degraded tetracyclines by thin-layer chromatography. *J. Pharm. Sci.* 58:443–446.

9. Kapadia, G.J. and G.S. Rao. 1964. Circular thin-layer chromatography of tetracyclines. *J. Pharm. Sci.* 53:223–224.

10. Enrico, R. and V. Giovanni. 1977. Simple method for the quantitative analysis of tetracyclines by direct fluorimetry after thin-layer chromatography on cellulose plates. *J. Chromatogr. A* 132:105–114.

11. Dijkhuis, I.C. and M.R. Brommet. 1970. Determination of epitetracycline and chlortetracycline in tetracycline by quantitative thin-layer chromatography. *J. Pharm. Sci.* 59:558–560.

12. Oka, H., Uno, K., Harada, K.-I., and M. Suzuki. 1984. Improvement the chemical analysis of antibiotics III. Simple method for the analysis of tetracyclines on reversed-phase thin-layer plates. *J. Chromatogr. A* 284:227–234.

13. Willekens, G.J. 1975. Separation and quantitative determination of impurities in tetracycline. *J. Pharm. Sci.* 64:1681–1686.

14. Ascione, P.P., Zagar, J.B., and G.P. Chrekian. 1967. Tetracyclines I. Separation and examination by thin-layer chromatography. *J. Pharm. Sci.* 56:1393–1395.

15. Naidong, W., Geelen, S., Roets, E., and J. Hoogmartens. 1990. Assay and purity control of oxytetracycline and doxycycline by thin-layer chromatography—A comparison with liquid chromatography. *J. Pharm. Biomed. Anal.* 8:891–888.

16. Naidong, W., Hua, S., Verresen, K., Roets, E., and J. Hoogmartens. 1991. Assay and purity control of metacycline by thin-layer chromatography combined with UV and fluorescence densitometry—A comparison with liquid chromatography. *J. Pharm. Biomed. Anal.* 9:717–723.

17. Naidong, W., Hua, S., Roets, E., and J. Hoogmartens. 1995. Assay and purity control of minocycline by thin-layer chromatography using UV and fluorescence densitometry. A comparison with liquid chromatography. *J. Pharm. Biomed. Anal.* 13:905–910.

18. Mowthorpe, S., Clench, M.R., Cricelius, A., Richards, D.S., Parr, V., and L.W. Tetler. 1999. Matrix-assisted laser desorption/ionisation time-of-flight/thin layer chromatography/mass spectrometry-A rapid method for impurity testing. *Rapid Commun. Mass Spectrom.* 13:264–270.

19. Jain, N., Kumar, J.G., Jalees, A.F., and K.R. Krishen. 2007. Validated stability-indicating densitometric thin-layer chromatography: Application to stress degradation studies of minocycline. *Anal. Chim. Acta* 599:302–309.

20. Liang, Y., Denton, M.B., and R.B. Bates. 1998. Stability studies of tetracycline in methanol solution. *J. Chromatogr. A* 827:45–55.

21. Makowski, A., Adamek, E., and W. Baran. 2010. Use of photocatalytic reactions to visualize drugs in TLC. *J. Planar Chromatogr.* 23:84–86.

22. Izer, K., Torok, I., and G. Pinter-Magyar. 1994. Stability of oxytetracycline hydrochloride in eye-drops, prepared in pharmacies. *Acta Pharm. Hung.* 64:63–66.

23. Liang, Y., Simon, R.E., and M.B. Denton. 1999. Utilization of a scientifically operated charge-coupled device detector for high-performance thin-layer chromatographic analysis of tetracyclines. *Analyst* 124:1577–1582.

24. Simmons, D.L., Woo, H.S.L., Koorengevel, C.M., and P. Seers. 1966. Quantitative determination by thin-layer chromatography of anhydrotetracyclines in degraded tetracycline tablets. *J. Pharm. Sci.* 55:1313–1315.

25. Jain, G.K., Jain, N., Iqbal, Z., Talegaonkar, S., Ahmad, F.J., and R.K. Khar. 2007. Development and validation of an HPTLC method for determination of minocycline in human plasma. *Acta Chromatogr.* 19:197–205.

26. Jain, N., Jain, G.K., Iqbal, Z., Talegaonkar, S., Ahmad, F.J., and R.K. Khar. 2009. An HPTLC method for the determination of minocycline in human plasma, saliva, and gingival fluid after single step liquid extraction. *Anal. Sci.* 25:57–62.

27. Novakova, E. 1991. Detection of antibiotics in biological materials for purposes of toxicological analysis. I. Methodologic study for the detection of tetracycline. *Cesk. Farm.* 40:174–177.

28. Oka, H., Ikai, Y., Hayakawa, J. et al. 1993. Improvement of chemical analysis of antibiotics. 18. Identification of residual tetracyclines in bovine tissues by TLC/FABMS with a sample condensation technique. *J. Agric. Food Chem.* 41:410–415.

29. Choma, I., Grenda, D., Malinowska, I., and Z. Suprynowicz. 1999. Determination of flumequine and doxycycline in milk by a simple thin-layer chromatographic method. *J. Chromatogr. B* 734:7–14.

30. Choma, I., Choma, A., and K. Staszczuk. 2002. Direct bioautography–thin-layer chromatography of flumequine and doxycycline in milk. *J. Planar Chromatogr.* 15:187–191.

31. Ašperger, D., Mutavdžić, D., Babić, S., Horvat Alka, J.M., and M. Kaštelan-Macan. 2006. Solid-phase extraction and TLC quantification of enrofloxacin, oxytetracycline, and trimethoprim in wastewater. *J. Planar Chromatogr.* 19:129–134.

32. Naidong, W., Hua, S., Roets, E., and J. Hoogmartens. 2003. Assay and purity control of tetracycline, chlortetracycline and oxytetracycline in animal feeds and premixes by TLC densitometry with fluorescence detection. *J. Pharm. Biomed. Anal.* 33:85–93.

33. Mutavdžić, D., Babić, S., Ašperger, D., Horvat, A.J.M., and M. Kaštelan-Macan. 2006. Comparison of different solid-phase extraction materials for sample preparation in the analysis of veterinary drugs in water samples. *J. Planar Chromatogr.* 19:454–462.

34. Vanderhaeghe, H. and L. Kerremans. 1980. Thin-layer chromatography of macrolide antibiotics. *J. Chromatogr. A* 193:119–127.

35. Szulagyi, I., Mincsovics, E., and G. Kulcsar. 1984. Preparation of the homologous components (A_1, A_2 and B) of primycin by thin-layer chromatography. *J. Chromatogr. A* 295:141–151.

36. Kondo, K. 1979. Analytical studies of maridomycin II. Separation of 9-propionylmaridimycins by thin-layer chromatography. *J. Chromatogr. A* 169:337–342.

37. Tosti, T.B., Drljević, K., Milojković-Opsenica, D.M., and Ž.Lj. Tešić. 2005. Salting-out thin-layer chromatography of some macrolide antibiotics. *J. Planar Chromatogr.* 18:415–418.

38. Lazarevski, T., Radobolja, G., and S. Djokić. 1978. Erythromycin VI: Kinetics of acid-catalyzed hydrolysis of erythromycin oxime and erythromycylamine. *J. Pharm. Sci.* 67:1031–1033.

39. Brisaert, M., Gabriels, M., and J. Plaizier-Vercammen. 2000. Investigation of the chemical stability of an erythromycin–tretinoin lotion by the use of an optimization system. *Int. J. Pharm.* 197:153–160.

40. Gabriels, M., Brisaert, M., and J. Plaizier-Vercammen. 1999. Densitometric thin layer chromatographic analysis of tretinoin and erythromycin in lotions for topical use in acne treatment. *Eur. J. Pharm. Biopharm.* 48:53–58.

41. Bhushan, R. and V. Parshad. 1996. Thin-layer chromatographic separation of enantiomeric dansyl amino acids using a macrocyclic antibiotic as a chiral selector. *J. Chromatogr. A* 736:235–238.

42. Bhushan, R. and S. Tanwar. 2008. Direct TLC resolution of atenolol and propranolol into their enantiomers using three different chiral selectors as impregnating reagents. *Biomed. Chromatogr.* 22:1028–1034.

43. Khedr, A. and M. Sheha. 2003. Quantitative thin-layer chromatographic method of analysis of azithromycin in pure and capsule forms. *J. Chromatogr. Sci.* 41:10–16.

44. Kwiecień, A., Krzek, J., and Ł. Biniek. 2008. TLC–densitometric determination of azithromycin in pharmaceutical preparations. *J. Planar Chromatogr.* 21:177–181.

45. Liawruangrath, B. and S. Liawruangrath. 2001. High performance thin layer chromatographic determination of erythromycin in pharmaceutical preparations. *Chromatographia* 54:405–408.

46. Maher, H.M. and R.M. Youssef. 2009. Development of validated chromatographic methods for the simultaneous determination of metronidazole and spiramycin in tablets. *Chromatographia* 69:345–350.

47. Hu, Ch.-Q., Zou, W.-B., Hu, W.-S. et al. 2006. Establishment of a fast chemical identification system for screening of counterfeit drugs of macrolide antibiotics. *J. Pharm. Biomed. Anal.* 40:68–74.

48. Richard, G., Radecka, C., Hughes, D.W., and W.L. Wilson. 1972. Chromatographic differentiation of erythromycin and its esters. *J. Chromatogr. A* 67:69–73.

49. Loya, P. and P.D. Hamrapurkar. 2011. A simple, rapid, and sensitive HPTLC method for the estimation of clarithromycin: Application to single dose clinical study. *J. Planar Chromatogr.* 24:534–538.

50. Petz, M., Solly, R., Lymburn, M., and M.H. Clear. 1987. Thin-layer chromatographic determination of erythromycin and other macrolide antibiotics in livestock products. *J. AOAC* 70:691–697.

51. Shahverdi, A.R., Monsef-Esfahani, H.R., Tavasoli, F., Zaheri, A., and R. Mirjani. 2007. Trans-cinnamaldehyde from *Cinnamomum zeylanicum* bark essential oil reduces the clindamycin resistance of *Clostridium difficile* in vitro. *J. Food Sci.* 72:S55–S58.

52. Sajewicz, M. 2005. Use of densitometric TLC for detection of selected drugs present in river water in South Poland. *J. Planar Chromatogr.* 18:108–111.

53. Szabó, A., Erdélyi, B., Salát, J., and G. Máté. 2005. Densitometric determination of some bioactive guanidinium compounds without post-derivatization. *J. Planar Chromatogr.* 18:203–206.

54. Stassi, D., Post, D., Satter, M., Jackson, M., and G. Maine. 1998. A genetically engineered strain of *Saccharopolyspora erythraea* that produces 6,12-dideoxyerythromycin A as the major fermentation product. *Appl. Microbiol. Biotechnol.* 49:725–731.

55. Li, S., Wu, L., Chen, F., Wang, H., Sun, G., and Y. Wang. 2011. Rapid identification of elaiophylin from *Streptomyces hygroscopicus* 17997, a geldanamycin producer. *Sheng. Wu. Gong. Cheng. Xue. Bao.* 27:1109–1114.

56. Szabó, A., Kónya, A., Széll, V., Máté, G., and B. Erdélyi. 2007. Biological and chemical detections in adsorbent layer for monitoring microbial production of primycin. *J. Chromatogr. Sci.* 45:435–438.

57. Harindran, J., Gupte, T.E., and S.R. Naik. 1999. HA-1-92, a new antifungal antibiotic produced by *Streptomyces* CDRIL-312: Fermentation, isolation, purification and biological activity. *World J. Microbiol. Biotechnol.* 15:425–430.

58. El-Bondkly, A.M., Abd-Alla, H.I., Shaaban, M., and K.A. Shaaban. 2008. The electrospray ionization— Mass spectra of erythromycin a obtained from a marine *Streptomyces* sp. mutant. *Indian J. Pharm. Sci.* 70:310–319.

59. Ritzau, M., Heinze, S., Werner, F.F., Dahse, H.M., and U. Grafe. 1998. New macrodiolide antibiotics, 11-*O*-monomethyl- and 11,11′-*O*-dimethylelaiophylins, from *Streptomyces* sp. HKI-0113 and HKI-0114. *J. Nat. Prod.* 61:1337–1339.

60. Hubicka, U., Krzek, J., Woltyńska, H., and B. Stachacz. 2009. Simultaneous identification and quantitative determination of selected aminoglycoside antibiotics by thin-layer chromatography and densitometry. *J. AOAC Int.* 92:1068–1075.

61. Bhushan, R. and S. Joshi. 1994. TLC studies on certain aminoglycoside antibiotics. *Biomed. Chromatogr.* 8:315–316.

62. Roets, E., Adams, E., Muriithi, I.G., and J. Hoogmartens. 1995. Determination of the relative amounts of the B and C components of neomycin by thin-layer chromatography using fluorescence detection. *J. Chromatogr. A* 696:131–138.

63. Funk, W., Kuepper, T., Wirtz, A., and S. Netz. 1994. Quantitative TLC/HPTLC determination of neomycins A, B, and C. Part 1: Chromatographic separation and postchromatographic derivatization. *J. Planar Chromatogr.—Modern TLC* 7:10–13.

64. Wagman, G.H., Marquez, J.A., and M.J. Wienstein. 1968. Chromatographic separation of the components of the gentamicin complex. *J. Chromatogr. A* 34:210–215.

65. Hotta, K., Sunada, A., Ishikawa, J., Mizuno, S., Ikeda, Y., and S. Kondo. 1998. The novel enzymatic 3″′-N-acetylation of arbekacin by an amino glycoside 3-N-acetyltransferase of *Streptomyces* origin and there sulting activity. *J. Antibiot. (Tokyo)* 51:735–742.

66. Klimecka, M.M., Chruszcz, M., Font, J. et al. 2011. Structural analysis of a putative aminoglycoside N-acetyltransferase from *Bacillus anthracis*. *J. Mol. Biol.* 410:411–423.

67. Kunz, F.R. and H. Jork. 1988. Quantitative fluorometric in-situ determination of netilmicin and 4 gentamicins on TLC-RP-layers. *Fresenius Z. Anal. Chem.* 329:773–777.

68. Sekkat, M., Fabre, H., Simeon De Buochberg, M., and B. Mandrou. 1989. Determination of aminoglycosides in pharmaceutical formulations—I. Thin-layer chromatography. *J. Pharm. Biomed. Anal.* 7:883–892.

69. Decoster, W., Claes, P., and H. Vanderhaeghe. 1982. Chromatographic assay of neomycin B and C in neomycin sulfate powders. *J. Pharm. Sci.* 71:987–991.

70. Bhogte, Ch.P., Patravale, V.B., and V. Devarajan Padma. 1997. Fluorodensitometric evaluation of gentamicin from plasma and urine by high-performance thin-layer chromatography. *J. Chromatogr. B* 694:443–447.

71. Kunz, F.R., Jork, H., and H.E. Keller. 1993. Determination of netilmicin in serum by thin-layer densitometry and fluorescence polarization immunoassay. *Fresenius J. Anal. Chem.* 346:847–851.

72. Campbell, N.F., Hubbard, L.E., Mazenko, R.S., and M.B. Medina. 1997. Development of a chromatographic method for the isolation and detection of hygromycin B in biological fluids. *J. Chromatogr. B* 692:367–374.

73. Medina, M.B. and J.J. Unruh. 1995. Solid-phase clean-up and thin-layer chromatographic detection of veterinary aminoglycosides. *J. Chromatogr. B* 663:127–135.

74. Kojima, R., Ito, M., and Y. Suzuki. 1988. Studies on the nephrotoxicity of aminoglycoside antibiotics and protection from these effects. (5). Interaction of tobramycin with latamoxef in vitro. *J. Pharmacobiodyn.* 11:9–17.

75. Lin, S.Y. and F. Kondo. 1994. Simple bacteriological and thin-layer chromatographic methods for determination of individual drug concentrations treated with penicillin-G in combination with one of the aminoglycosides. *Microbios* 77:223–229.

76. Tsuda, Y., Wakamatsu, H., Yoshizawa, E., Kitagawa, E., and T. Fujimoto. 1993. Microbiological determination of netilmicin using a thin-layer chromatography scanner. *Jpn. J. Antibiot.* 46:1–7.

77. Aszalos, A. and A. Aquilar. 1984. Comparative study of peptide-type antibiotics in reversed-phase thin-layer chromatography and reversed-phase high-performance liquid chromatography. *J. Chromatogr. A* 290:83–96.

78. Stankovic, Ch.J., Delfino, J.M., and S.L. Schreiber. 1990. Purification of gramicidin A. *Anal. Biochem.* 184:100–103.

79. Pepinsky, R.B. and G.W. Feigenson. 1978. Purification of gramicidin C. *Anal. Biochem.* 86:512–518.

80. Hou, D.Y., Hoch, H., Johnston, G.S. et al. 1984. A new 111In-bleomycin complex for tumor imaging: Preparation, stability, and distribution in glioma-bearing mice. *J. Surg. Oncol.* 25:168–175.

81. Pirett, J.M. and A.L. Demain. 1983. Sporulation and spore properties of *Bacillus brevis* and its gramicidin S-negative mutant. *J. Gen. Microbiol.* 129:1309–1310.

82. Kitajima, Y., Waki, M., Shoji, J., and T.U.N. Izumiyal. 1990. Revised structure of the peptide lactone antibiotic, TL-119 and/or A-3302-B. *FEBS Lett.* 1:139–142.

83. Armstrong, D.W. and Y. Zhou. 1994. Use of a macrocyclic antibiotic as the chiral selector for enantiomeric separations by TLC. *J. Liq. Chromatogr.* 17:1695–1707.

84. Bhushan, R. and Ch. Agarwal. 2010. Resolution of beta blocker enantiomers by TLC with vancomycin as impregnating agent or as chiral mobile phase additive. *J. Planar Chromatogr.* 23:7–13.

85. Bhushan, R. and D. Gupta. 2005. Thin-layer chromatography separation of enantiomers of verapamil using macrocyclic antibiotic as a chiral selector. *Biomed. Chromatogr.* 19:474–478.

86. Bhushan, R. and G.T. Thiong'o. 2000. Separation of the enantiomers of dansyl-DL-amino acids by normal-phase TLC on plates impregnated with a macrocyclic antibiotic. *J. Planar Chromatogr.—Modern TLC* 13:33–36.

87. Taha, E.A., Salama, N.N., and S. Wang. 2009. Enantioseparation of cetirizine by chromatographic methods and discrimination by (1)H-NMR. *Drug Test Anal.* 1:118–124.

88. Ruzin, A., Singh, G., Severin, A. et al. 2004. Mechanism of action of the mannopeptimycins, a novel class of glycopeptide antibiotics active against vancomycin-resistant Gram-positive bacteria. *Antimicrob. Agents Chemother.* 48:728–738.

89. Schneider, T., Gries, K., Josten, M. et al. 2009. The lipopeptide antibiotic friulimicin B inhibits cell wall biosynthesis through complex formation with bactoprenol phosphate. *Antimicrob. Agents Chemother.* 53:1610–1618.

90. Datta, K., Das, S.K., and S.K. Roy. 1989. A rapid and inexpensive thin layer chromatographic method for quantitative analysis of bleomycin complex. *J. Liq. Chromatogr.* 12:949–956.

91. Ikai, Y., Oka, H., Hayakawa, J. et al. 1998. Isolation of colistin A and B using high-speed countercurrent chromatography. *J. Liq. Chromatogr. Relat. Technol.* 21:143–155.

92. Jamre, M., Salek, N., Jalilian, A.R. et al. 2011. Development of an in vivo radionuclide generator by labeling bleomycin with ^{191}Os. *J. Radioanal. Nucl. Chem.* 290:543–549.

93. Selim, S., Negrel, J., Govaerts, C., Gianinazzi, S., and D. Van Tuinen. 2005. Isolation and partial characterization of antagonistic peptides produced by *Paenibacillus* sp. strain B2 isolated from the sorghum mycorrhizosphere. *Appl. Environ. Microbiol.* 71:6501–6507.

94. Thomas, A.H., Thomas, J.M., and I. Holloway. 1980. Microbiological and chemical analysis of polymyxin B and polymyxin E (colistin) sulphates. *Analyst* 105:1068–1075.

95. Ghulam, M.S., Urooj, F., and S. Rahat. 2012. Stress degradation studies and development of stability-indicating TLC-densitometry method for determination of prednisolone acetate and chloramphenicol in their individual and combined pharmaceutical formulations. *Chem. Central J.* 6:7. doi: 10.1186/1752-153X-6-7.

96. Freimuller, S., Horsch, P., Andris, D., Zerbe, O., and H. Altorfer. 2001. Formation mechanism of solvent induced artifact arising from chromatographic purity testing of γ-irradiated chloramphenicol. *Chromatographia* 53:323–325.

97. Marciniec, B., Stawny, M., Kozak, M., and M. Naskrent. 2006. The effect of ionizing radiation on chloramphenicol. *J. Therm. Anal. Calorim.* 84:741–746.

98. Marciniec, B., Stawny, M., Kachlicki, P., Jaroszkiewicz, E., and M. Needham. 2009. Radiostability of florfenicol in the solid state. *Anal. Sci.* 25:1255–1260.

99. Marciniec, B., Stawny, M., Kozak, M., and M. Naskrent. 2008. The influence of radiation sterilization on thiamphenicol. *Spectrochim. Acta A* 69:865–870.

100. Vovk, I. and B. Simonovska. 2005. Development and validation of a thin-layer chromatographic method for determination of chloramphenicol residues on pharmaceutical equipment surfaces. *J. AOAC Int.* 88:1555–1561.

101. Cravedi, J.P., Baradat, M., Debrauwer, L., Alary, J., Tulliez, J., and G. Bories. 1994. Evidence for new metabolic pathways of chloramphenicol in the duck. *Drug Metab. Dispos.* 22:578–583.

102. Vega, M.H., Jara, E.T., and M.B. Aranda. 2006. Monitoring the dose of florfenicol in medicated salmon feed by planar chromatography (HPTLC). *J. Planar Chromatogr.* 19:204–207.

103. El-Kersh, T.A. and J.R. Plourde. 1976. Biotransformation of antibiotics I. Acylation of chloramphenicol by spores of *Streptomyces griseus* isolated from the Egyptian soli. *J. Antibiot.* 29:292–302.

104. Hamburger, M.O. and G.A. Cordel. 1987. A direct bioautographic TLC assay for compounds possessing antibacterial activity. *J. Nat. Prod.* 50:19–22.

105. Auboiron, S., Bauchart, D., and L. David. 1991. Separation and determination of polyether carboxylic antibiotics from *Streptomyces hygroscopicus NRRL B 1865* by thin-layer chromatography with flame ionization detection. *J. Chromatogr. A* 547:411–418.

106. Coutinho, A. and M. Prieto. 1995. Self-association of the polyene antibiotic nystatin in dipalmitoylphosphatidylcholine vesicles: A time-resolved fluorescence study. *Biophys. J.* 69:2541–2557.

107. Khattab, A.A. and A.M. El-Bondkly. 2006. Construction of superior *Streptomyces noursei* fusants for nystatin and antibacterial antibiotics production. *Arab. J. Biotech.* 9:95–106.

108. Mikhailova, N.P., Sorokoletova, E.F., Durasova, E.N., Vyunov, K.A., and O.I. Shapovalov. 1991. Sterol composition of nystatin-resistant *Candida maltosa* mutants. *Folia Microbiol.* 36:148–152.

109. Vanderkop, P.A. and J.D. MacNeil. 1989. Thin-layer chromatography/bioautography method for detection of monensin in poultry tissues. *J. AOAC* 72:735–738.

110. Weiss, G. and A. MacDonald. 1985. Methods for determination of ionophore-type antibiotic residues in animal tissues. *J. AOAC* 68:971–980.

111. Brooks, M.A., D'Arconte, L., DeSilva, J.A., Chen, G., and C. Crowley. 1975. Spectrofluorometric determination of the antibiotic lasalocid in blood. *J. Pharm. Sci.* 64:1874–1876.

112. Zotchev, S., Haugan, K., Sekurova, O., Sletta, H., Ellingsen, T.E., and S. Valla. 2000. Identification of a gene cluster for antibacterial polyketide-derived antibiotic biosynthesis in the nystatin producer *Streptomyces noursei* ATCC 11455. *Microbiology* 146:611–619.

113. Zarzycki, P.K., Nowakowska, J., Chmielewska, A., Wierzbowska, M., and H. Lamparczyk. 1997. Thermodynamic study of the retention behaviour of selected macrocycles using reversed-phase high-performance thin-layer chromatography plates and methanol-water mobile phases. *J. Chromatogr. A* 787:227–233.

114. Tatarczak, M., Flieger, J., and H. Szumiło. 2005. Simultaneous densitometric determination of rifampicin and isoniazid by high-performance thin-layer chromatography. *J. Planar Chromatogr.* 18:207–211.

115. Kenyon, T.A., Kenyon, A.S., Kgarebe, B.V., Mothibedi, D., Binkin, N.J., and T.P. Layloff. 1999. Detection of substandard fixed-dose combination tuberculosis drugs using thin-layer chromatography. *Int. J. Tuberc. Lung. Dis.* 3:S347–S350.

116. Izer, K., Torok, I., Magyarne Pinter, G., Varsanyi, L.E., and J. Liptak. 1996. Stability of rifampicin in eyedrops. *Acta Pharm. Hung.* 66:157–163.

117. Jindal, K.C., Chaudhary, R.S., Gangwal, S.S., Singla, A.K., and S. Khanna. 1994. High-performance thin-layer chromatographic method for monitoring degradation products of rifampicin in drug excipient interaction studies. *J. Chromatogr. A* 685:195–199.

118. Avachat, A.M. and S.B. Bhise. 2010. Stability-indicating validated HPTLC method for simultaneous analysis of rifabutin and isoniazid in pharmaceutical formulations. *J. Planar Chromatogr.* 23:123–128.

119. Pandey, R.C. and M.W. Toussaint. 1980. High-performance liquid chromatography and thin-layer chromatography of anthracycline antibiotics. Separation and identification of components of the daunorubicin complex from fermentation broth. *J. Chromatogr. A* 198:407–420.

120. Joish, R.Y., Sarakar, A., and S. Gurusiddaiah. 1986. Antifungal macrodiolide from *Streptomyces* sp. *Antimicrob. Agents Chemother.* 30:458–464.

121. Agbaba, D., Djurkovic, M., Brboric, J., and D. Zivanov-Stakic. 1998. Simultaneous HPTLC determination of metronidazole and its impurity 2-methyl-5-nitroimidazole in pharmaceuticals. *J. Planar Chromatogr.—Modern TLC* 11:447–449.

122. Egorin, M.J., Andrews, P.A., Nakazawa, H., and N.R. Bachur. 1983. Purification and characterization of aclacinomycin A and its metabolites from human urine. *Drug Metab. Dispos.* 11:167–171.

123. Colombo, T., Broggini, M., Garattini, S., and M.G. Donelli. 1981. Differential adriamycin distribution to blood components. *Eur. J. Drug Metab. Pharmacokinet.* 6:115–122.

124. Shah, S.Q., Khan, A.U., and M.R. Khan. 2011. Radiosynthesis of ^{99m}Tc-nitrofurantoin a novel radiotracer for in vivo imaging of *Escherichia coli* infection. *J. Radioanal. Nucl. Chem.* 287:417–422.

125. Yurt, L.F., Yilmaz, O., Durkan, K., Unak, P., and E. Bayrak. 2009. Preparation and biodistribution of [^{131}I] linezolid in animal model infection and inflammation. *J. Radioanal. Nucl. Chem.* 281:415–419.

126. Valcarcel, L. and M. Noa. 2000. Determination of tiamulin in lung and meal of intoxicated pigs after the consumption of medicated feed. *Rev. Salud Anim.* 22:59–60.
127. Adrian, J., Font, H., Diserens, J.-M., Sanchez-Baeza, F., and M.-Pp. Marco. 2009. Generation of broad specificity antibodies for sulfonamide antibiotics and development of an enzyme-linked immunosorbent assay (ELISA) for the analysis of milk samples. *J. Agric. Food Chem.* 57:385–394.
128. Naik, P.R., Sahoo, N., Goswami, D., Ayyadurai, N., and N. Sakthivel. 2008. Genetic and functional diversity among fluorescent *Pseudomonad*s isolated from the rhizosphere of banana. *Microb. Ecol.* 56:492–504.
129. Atta, H.M., Abul-Hamd, A.T., and H.G. Radwan. 2009. Production of destomycin-A antibiotic by *Streptomyces* sp. using rice straw as fermented substrate. *Commun. Agric. Appl. Biol. Sci.* 74:879–897.
130. Atta, H.M. and H.G. Radwan. 2012. Biochemical studies on the production of sparsomycin antibiotic by *Pseudomonas aeruginosa*, AZ-SH-B8 using plastic wastes as fermented substrate. *J. Saudi Chem. Soc.* 16:35–44.
131. Ilic, S.B., Konstantinovic, S.S., Todorovic, Z.B. et al. 2007. Characterization and antimicrobial activity of the bioactive metabolites in *Streptomycete* isolates. *Microbiology* 76:421–428.
132. Pope, J.A. Jr., Nelson, R.A., Schaffner, C.P., Rosen, R.T., and R.C. Pandey. 1990. Applications of thin layer chromatography, high performance liquid chromatography and mass spectrometry in the fermentation and isolation of the antibiotic nybomycin. *J. Ind. Microbiol.* 6:61–69.
133. Itoh, Y., Kodama, K., Furuya, K. et al. 1980. A new sesquiterpene antibiotic, heptelidic acid producing organisms, fermentation, isolation and characterization. *J. Antibiot.* 33:468–473.
134. Tamehiro, N., Okamoto-Hosoya, Y., Okamoto, S. et al. 2002. Bacilysocin, a novel phospholipid antibiotic produced by *Bacillus subtilis* 168. *Antimicrob. Agents Chemother.* 46:315–320.
135. Fakhouri, W., Walker, F., Vogler, B., Armbruster, W., and H. Buchenauer. 2001. Isolation and identification of N-mercapto-4-formylcarbostyril, an antibiotic produced by *Pseudomonas fluorescens*. *Phytochemistry* 58:1297–1303.
136. Sadfi, N., Cherif, M., Hajlaoui, M.R., Boudabbous, A., and R. Belanger. 2002. Isolation and partial purification of antifungal metabolites produced by *Bacillus cereus*. *Ann. Microbiol.* 52:323–337.
137. Allen, I.W. and D.A. Ritchie. 1994. Cloning and analysis of DNA sequences from *Streptomyces hygroscopicus* encoding geldanamycin biosynthesis. *Mol. Gen. Genet.* 243:593–599.
138. Gerber, N.N., Akram, S.S., and H.A. Lechevalier. 1980. Structures and antimicrobial activity of peniophorin A and B, two polyacetylenic antibiotics from *Peniophora affinis* Burt. *Antimicrobiol. Agents Chemother.* 17:636–641.
139. Tsuchida, T., Ilnuma, H., Kinoshita, N. et al. 1995. Azicemicins A and B, a new antimicrobial agent produced by *Amycolatopsis* I. Taxonomy, fermentation, isolation, characterization and biological activities. *J. Antibiot.* 48:217–221.
140. Mukhtar, T.A., Koteva, K.P., Hughes, D.W., and G.D. Wright. 2001. Vgb from *Staphylococcus aureus* inactivates streptogramin B antibiotics by an elimination mechanism not hydrolysis. *Biochemistry* 40:8877–8886.
141. Pinchuk, I.V., Bressollier, P., Verneuil, B. et al. 2001. In vitro anti-*Helicobacter pylori* activity of the probiotic strain *Bacillus subtilis* 3 is due to secretion of antibiotics. *Antimicrob. Agents Chemother.* 45:3156–3161.
142. Vivien, E., Pitorre, D., Cociancich, S. et al. 2007. Heterologous production of albicidin: A promising approach to overproducing and characterizing this potent inhibitor of DNA gyrase. *Antimicrob. Agents Chemother.* 51:1549–1552.
143. Hu, K., Li, J., and J.M. Webster. 1997. Quantitative analysis of a bacteria-derived antibiotic in nematode infected insects using HPLC–UV and TLC–UV methods. *J. Chromatogr. B* 703:177–183.
144. Nowakowska, J., Halkiewicz, J., and J.W. Łukasiak. 2001. The retention behavior of selected macrocyclic antibiotics on polyamide TLC plates. *J. Planar Chromatogr.* 14:350–354.
145. Nowakowska, J., Halkiewicz, J., and J.W. Lukasiak. 2002. TLC determination of selected macrocyclic antibiotics using normal and reversed phases. *Chromatographia* 56:367–373.
146. Wayland, S.Q. and P.J. Weiss. 1968. Identification of antibiotics in sensitivity disks. *J. Pharm. Sci.* 57:806–810.
147. Bhushan, R. and I. Ali. 1992. TLC separation of certain tetracycline and amino glycopeptide antibiotics. *Biomed. Chromatogr.* 6:196–197.
148. Nowakowska, J. 2004. Analysis of selected macrocyclic antibiotics by HPTLC with non-aqueous binary mobile phases. *J. Planar Chromatogr.* 17:200–206.

149. Nowakowska, J., Pikul, P., and P. Rogulski. 2010. TLC of aclarubicin and doxycycline with mixed *n*-alcohol mobile phases. *J. Planar Chromatogr.* 23:353–358.

150. Husain, S.W., Ghoulipour, V., and H. Sepahrian. 2004. Chromatographic behaviour of antibiotics on thin layers of an inorganic ion-exchanger. *Acta Chromatogr.* 14:102–109.

151. Nowakowska, J. 2006. Effect of non-aqueous mobile phase composition on the retention of macrocyclic antibiotics in RP-TLC. *J. Planar Chromatogr.* 19:64–67.

152. Nowakowska, J. 2005. Normal and reversed-phase TLC separations of some macrocyclic antibiotics with non-aqueous mobile phases. *J. Planar Chromatogr.* 18:455–459.

153. Sharaf, M.H.M., Sanchez, A.L., White, P.A., and R.G. Manning. 2002. TLC separation and identification of neomycin sulfate, polymyxin B sulfate, and bacitracin zinc in ointments. *J. Liq. Chromatogr. Relat. Technol.* 25:927–935.

154. Krzek, J., Starek, M., Kwiecień, A., and W. Rzeszutko. 2001. Simultaneous identification and quantitative determination of neomycin sulfate, polymixin B sulfate, zinc bacytracin and methyl and propyl hydroxybenzoates in ophthalmic ointment by TLC. *J. Pharm. Biomed. Anal.* 24:629–636.

155. Krzek, J., Kwiecień, A., Starek, M., Kierszniewska, A., and W. Rzeszutko. 2000. Identification and determination of oxytetracycline, tiamulin, lincomycin, and spectinomycin in veterinary preparations by thin-layer chromatography/densitometry. *J. AOAC Int.* 83:1502–1506.

156. Krzek, J., Maślanka, A., and P. Lipner. 2005. Identification and quantitation of polymyxin B, framycetin, and dexamethasone in an ointment by using thin-layer chromatography with densitometry. *J. AOAC Int.* 88:1549–1554.

157. Anokhina, T.A., Podlepich, L.V., Soin, A.L., Remezov, V.V., and O.A. Serdyukova. 1992. Qualitative estimation of the combined drugs rivicyclin, sulpenil, and geovet by TLC. *Pharm. Chem. J.* 26:919–920.

158. Baranowska, I., Markowski, P., Wilczek, A., Szostek, M., and M. Stadniczuk. 2009. Normal and reversed-phase thin-layer chromatography in the analysis of L-arginine, its metabolites, and selected drugs. *J. Planar Chromatogr.* 22:89–96.

159. Tajick, M.A. and B. Shohreh. 2006. Detection of antibiotics residue in chicken meat using TLC. *Int. J. Poult. Sci.* 5:611–612.

160. Salisbury, C.D.C., Rigby, Ch.E., and W. Chan. 1989. Determination of antibiotic residues in Canadian slaughter animals by thin-layer chromatography-bioautography. *J. Agric. Food Chem.* 37:105–108.

161. Ramirez, A., Gutierrez, R., Diaz, G. et al. 2003. High-performance thin-layer chromatography–bioautography for multiple antibiotic residues in cow's milk. *J. Chromatogr. B* 784:315–322.

162. Neidert, E., Saschenbrecker, P.W., and F. Tittiger. 1987. Thin layer chromatographic/bioautographic method for identification of antibiotic residues in animal tissues. *J. AOAC* 70:197–200.

163. Kondo, F. 1988. A simple method for the characteristic differentiation of antibiotics by TLC-bioautography in graded concentration of ammonium chloride. *J. Food Prot.* 51:786–789.

164. Shareef, A.M., Jamel, Z.T., and K.M. Yonis. 2009. Detection of antibiotic residues in stored poultry products. *Iraqi J. Vet. Sci.* 23:45–48.

165. Torroba, J., Ravela-Diaz, V.M., Vivino, E.C., and J.A. Mesquita. 1993. Measurement of antibiotic levels in foot-and-mouth disease oil vaccines by chemical methods. *Bol. Centr. Panam. Fiebre Aftosa* 59:118–124.

166. Ahmad, I. and F. Aqil. 2007. in vitro efficacy of bioactive extracts of 15 medicinal plants against ESbL-producing multidrug-resistant enteric bacteria. *Microbiol. Res.* 162:264–275.

167. Gafner, J.L. 1999. Identification and semiquantitative estimation of antibiotics added to complete feeds, premixes, and concentrates. *J. AOAC Int.* 82:1–8.

168. De Jong, J., Tomassen, M.J.H., Van Egmond, H.J. et al. Towards a control strategy for banned antibiotics and growth promoters in feed: The SIMBAG-FEED project. In: *Antimicrobial Growth Promoters: Where Do We Go from Here?* D. Barug, J. De Jong, A.K. Kies, and M.W.A. Verstegen (eds.), Vol. 978, pp. 211–234. ISBN: 987-90-76998-87-9.

169. Ergin, K.S. and A. Filazi. 2010. Determination of antibiotic residues in milk samples. *Kafkas Univ. Vet. Fak. Derg.* 16:S31–S35.

170. Bossuyt, R., Van Renterghem, R., and G. Waes. 1976. Identification of antibiotic residues in milk by thin-layer chromatography. *J. Chromatogr. A* 124:37–42.

171. Lata, H., Li, X.C., Silva, B., Moraes, R.M., and L. Halda-Alija. 2006. Identification of IAA-producing endophytic bacteria from micropropagated Echinacea plants using 16S rRNA sequencing. *Plant. Cell. Tiss. Organ Cult.* 85:353–359.

172. Rusanova, E.P., Alekhova, T.A., Fedorova, G.B., and G.S. Katrukha. 2000. An antibiotic complex produced by *Streptomyces werraensis* 1365T. *App. Biochem. Microbiol.* 36:486–490.

173. Kim, Ch.-G., Kirschning, A., Bergon, P. et al. 1996. Biosynthesis of 3-amino-5-hydroxybenzoic acid, the precursor of mC7N units in ansamycin antibiotics. *J. Am. Chem. Soc.* 118:7486–7491.
174. Phillips, L., Wellington, E.M.H., and S.B. Rees. 1994. The distribution of antibiotic resistance patterns within streptomycetes and their use in secondary metabolite screening. *J. Ind. Microbiol.* 13:53–62.
175. Gentile, F.A., Mayles, B.A., and P.L. Procopio. 1992. Enhancement of secondary metabolite production using the antibiotic gradient-plate technique. *Life Sci.* 50:287–293.

49 Thin-Layer Chromatography of Antiviral Drugs

Anna Gumieniczek and Anna Berecka

CONTENTS

49.1 TLC METHODS FOR THE ANALYSIS OF ANTIVIRAL DRUGS

Antiviral drugs are used to treat infections caused by viruses. Because viruses can only function within the cells of their hosts, it has been difficult to produce drugs that act specifically against viruses without damaging their host cells. The effectiveness of antiviral drugs is therefore limited. Fortunately, the majority of viral infections resolve spontaneously in most people and do not require specific medication. However, in individuals, whose ability to fight infection is impaired because of drug therapy (e.g., immunosuppressants) or disease (e.g., AIDS), antiviral treatment may be lifesaving.

It has been found that viruses use a number of virus-specific enzymes during replication. These enzymes and the processes they control are significantly different from those of the host cell to make them a useful target for the drugs. Consequently, antiviral drugs normally act by inhibiting viral nucleic acid synthesis, inhibiting attachment to and penetration of the host cell, or inhibiting viral protein synthesis.

In this chapter, a review of thin-layer chromatographic (TLC) methods elaborated for determination of antiviral drugs from different groups is presented. The chapter describes the methods applied for the analysis of these important drugs in pharmaceuticals and biological material as well as for their hydrophobicity/lipophilicity studies.

49.2 HOST CELL PENETRATION INHIBITORS: AMANTADINE AND OSELTAMIVIR

The mechanism by which amantadine (Figure 49.1) exerts its antiviral activity is not clearly understood. It appears to mainly prevent the release of infectious viral nucleic acid into the host cell by interfering with the function of the transmembrane domain of the viral M2 protein. Amantadine inhibits the replication of influenza A virus of each of the subtypes, but it has very little or no activity

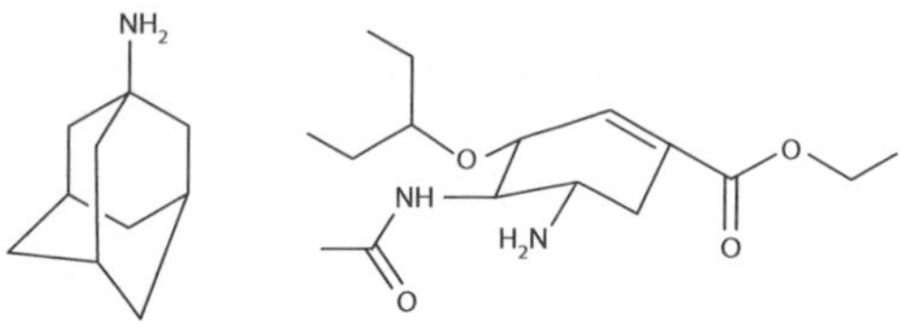

Amantadine   Oseltamivir

FIGURE 49.1   The chemical structures of the host cell penetration inhibitors, amantadine and oseltamivir.

against influenza B. Amantadine has good bioavailability on oral administration, being readily absorbed and distributed to most body fluids and tissues (Acosta and Flexner 2011).

A simple and accurate TLC method for quantitative determination of amantadine in pharmaceutical formulations was developed and validated. The method employed silica gel 60F$_{254}$ as stationary phase. The solvent system used for development consisted of n-hexane–methanol–diethylamine (80:40:5, v/v/v). The separated spots were visualized after spraying with modified Dragendorff's reagent. Amantadine was also subjected to accelerated stress conditions such as boiling, acid and alkaline hydrolysis, oxidation, and irradiation with UV light. The drug, however, was found to be stable under these stress conditions. The method was validated for linearity, limit of detection (LOD) and limit of quantitation (LOQ), precision, robustness, selectivity, and accuracy. The method was found to be linear in the range of 5–40 µg/spot with good correlation coefficient r = 0.9994. The LOD and LOQ values were 0.72 and 2.38 µg/spot, respectively. The method in terms of its sensitivity, accuracy, precision, and robustness met the ICH/FDA requirements (Askal et al. 2008).

Oseltamivir (Figure 49.1) is prescribed for the treatment of uncomplicated acute illness due to influenza infection that has been symptomatic for no more than 2 days. It is also used for the prophylaxis of influenza. The proposed mechanism of action of oseltamivir is the inhibition of influenza virus neuraminidase with the possibility of alteration of virus particle aggregation and release. The drug is readily absorbed from the gastrointestinal tract after oral administration with a bioavailability of 75% (Acosta and Flexner 2011).

High-performance TLC (HPTLC) method for analysis of oseltamivir in pharmaceutical dosage forms was developed and validated. It was performed on silica gel 60F$_{254}$ plates with ethyl acetate–acetic acid–water in the ratio of 7.5:1.5:1 (v/v/v) as mobile phase and densitometric detection at 265 nm. The system was found to give compact spots with R$_f$ value of ca. 0.57. The linear regression analysis data for the calibration plots showed good linear relationship with coefficient of regression value r^2 = 0.9989. The LOD and LOQ of oseltamivir were found to be 72 and 97 ng/spot, respectively (Sharma and Sharma 2010).

49.3 NUCLEIC ACID SYNTHESIS INHIBITORS: ACYCLOVIR AND RIBAVIRIN

These drugs act by inhibiting polymerases or reverse transcriptases required for viral nucleic acid chain formation (Figure 49.2).

Acyclovir is an antiviral drug effective against a number of herpes viruses, notably simplex, varicella zoster, and Epstein–Barr virus. The action of acyclovir is based on phosphorylation catalyzed by viral thymidine kinase and then by inhibition of viral DNA polymerase. The viral thymidine kinase is a more efficient catalyst for the phosphorylation of acyclovir then the thymidine kinase of the host cell. As a result, it preferentially competitively inhibits viral DNA polymerase and so prevents the virus from replication. It may be administered orally and by intravenous injection as well as topically (Acosta and Flexner 2011).

A simple and rapid TLC method was developed for the determination of acyclovir in pharmaceutical preparations, for example, tablets and creams (Sia et al. 2002). After extraction of the drug with a mixture of 96% alcohol and 0.05 M H$_2$SO$_4$ (9:1, v/v), the extracts were applied to silica gel plates that were developed with n-butanol–glacial acetic acid–water (15:9:6, v/v/v). Quantitative evaluation

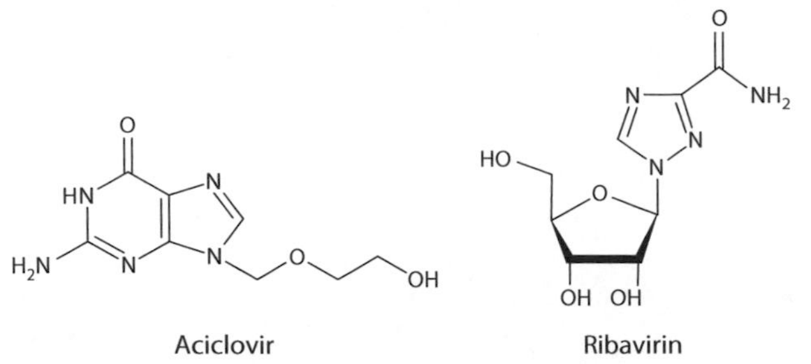

FIGURE 49.2   The chemical structures of the nucleic acid synthesis inhibitors, acyclovir and ribavirin.

was performed by measuring at the absorbance–reflectance mode at 277 nm. The results revealed that the accuracy and precision of the method were satisfactory in the range 80%–120% of the label claim. That work also showed that TLC plates could give results as good as HPTLC plates. The last ones were used by Dubhashi and Vavia (2000). This method is described later in the chapter concerning biological measurements.

The next drug ribavirin is active against a wide vanity of DNA and RNA viruses but the mechanism by which it acts is not fully understood. It is mainly used in aerosols to treat influenza and other respiratory viral infections. The mechanisms by which it acts may differ from one virus to another (Acosta and Flexner 2011).

A simple and accurate stability-indicating TLC method was developed and validated for quantitative determination of ribavirin in its bulk and capsule forms. The method employed silica gel $60F_{254}$ as stationary phase. The solvent system consisted of chloroform–methanol–acetic acid (60:15:15, v/v/v). The separated spots were visualized after spraying with anisaldehyde reagent. Ribavirin was also subjected to different stress conditions. The drug was found to undergo degradation under different stress conditions, but the degradation products were well resolved from the pure drug with significantly different R_f values. The method was found to be linear in the range of 5–40 μg/spot with a good correlation coefficient r = 0.9980. The LOD and LOQ values were 1.40 and 4.67 μg/spot, respectively. The label claim percentages were 98.8% ± 1.5% (Darwish et al. 2008).

49.4   TLC METHODS FOR DRUGS USED IN THERAPY OF HIV INFECTION

49.4.1   Nucleoside and Nucleotide Analogs and Non-Nucleoside Reverse-Transcriptase Inhibitors

The group of nucleoside analogs includes zidovudine, lamivudine, stavudine, and emtricitabine. This group contains tenofovir, while the group of non-nucleoside analogs includes efavirenz and nevirapine (Figure 49.3).

Deoxynucleotides are needed to synthesize the viral DNA and the respective analogs compete with them for incorporation into the growing viral DNA chain. Thus, viral DNA synthesis is halted in a process known as chain termination. In contrast, non-nucleoside inhibitors have a completely different mode of action. They block a reverse transcriptase by binding at a different site on the enzyme. They are not incorporated into the viral DNA but instead inhibit the movement of protein domains of reverse transcriptase that are needed to carry out the process of DNA synthesis (Flexner 2011).

A sensitive, selective, precise, and stability-indicating method was established and validated for analysis of zidovudine, both as bulk drug and in formulations. The method employed silica gel $60F_{254}$ HPTLC plates with toluene–carbon tetrachloride–methanol–acetone (3.5:3.5:2.0:1.0, v/v/v/v) as mobile phase. This system was found to give compact spots with R_f value 0.41 ± 0.02. Densitometric analysis was performed in the absorbance mode at 270 nm. The response was linear in the range 100–6000 ng/spot with a significantly high correlation coefficient $r^2 = 0.9980$.

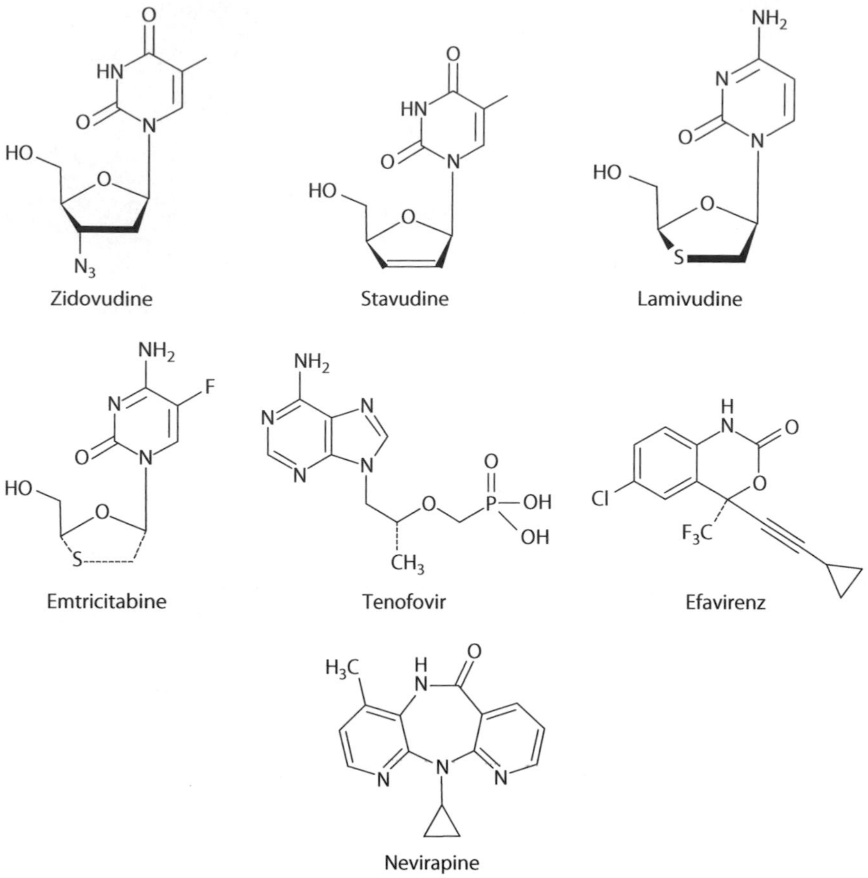

FIGURE 49.3 The chemical structures of the nucleoside and nucleotide analogs and nonnucleoside reverse-transcriptase inhibitors.

The LOD and LOQ were 20 and 40 ng/spot, respectively. Zidovudine was also subjected to acid and alkaline hydrolysis, oxidation, dry- and wet-heat treatment, and to photodegradation. It was found to undergo degradation under all these conditions except the dry-heat treatment. However, the degradation products were well separated from the pure drug with significantly different R_f values. The proposed HPTLC method was also used to investigate the kinetics of acid degradation. An Arrhenius plot was constructed and the activation energy was calculated (Kaul et al. 2004a).

Similar work was done for stavudine, both as bulk drug and in formulations. The solvent system consisted of toluene–methanol–chloroform–acetone (7.0:3.0:1.0:1.0, v/v/v/v). Analysis was carried out in the absorbance mode at 270 nm. This system was found to give compact spots with R_f value of 0.45 ± 0.05. Stavudine was subjected to acid and alkali hydrolysis, oxidation, dry-heat and wet-heat treatment, and photo- and UV degradation. The drug was found to be degraded under acidic and basic conditions, oxidation, and wet-heat degradation. Linear regression analysis data for the calibration plots showed a good linear relationship with $r^2 = 0.9997$ in the working concentration range of 300–1000 ng/spot. The LOD and LOQ were 10 and 30 ng/spot, respectively (Kaul et al. 2005).

The next work describes a validated HPTLC method for estimation of stavudine in capsules. Silica gel 60F$_{254}$ was used as stationary phase and toluene–methanol (7.5:2.5, v/v) as mobile phase. The wavelength selected for analysis was 270 nm. The linearity of detector response occurred in the range of 1.4–3.7 μg. The recovery was found to be 99.67% and 99.26% as per the peak height and peak area, respectively (Wadodkar et al. 2004).

A simple, accurate, precise, and rapid HPTLC method was developed and validated for efavirenz in bulk drug and tablets. The method employed TLC silica gel $60F_{254}$ as stationary phase. The mobile phase used was a mixture of toluene–ethyl acetate–formic acid (10:3:1, v/v/v). The detection was carried out at 254 nm. The calibration curve was found to be linear between 300 and 1800 ng/mL with regression coefficient of 0.9991. The accuracy of the proposed method was determined by recovery studies and was found to be 99.38%–99.68% (Kumar et al. 2011).

49.4.2 DRUGS FOR COMBINED ANTI-HIV THERAPY

Recommended first-line regimens in the treatment of HIV infection include the use of two nucleoside reverse-transcriptase inhibitors with either one or two protease inhibitors (described in Section 49.4.3) or with a nonnucleoside reverse-transcriptase inhibitor (Flexner 2011). Some of the recommended combinations are now available in the same pill or capsule. Therefore, rapid and simple TLC methods have been developed for quantitative determination of these antiviral agents in many combinations.

A simple, precise, rapid, selective, and cost-effective HPTLC method was developed for simultaneous estimation of lamivudine and zidovudine in bulk and their tablet dosage forms. Chromatography was performed on silica gel $60F_{254}$ plates with the mobile phase consisting of acetone–toluene–methanol (4:2:4, v/v/v). The plates were developed to a distance of 8.0 cm at ambient room temperature with earlier chamber saturation. The developed plates were scanned and the combined drugs were quantified at their wavelength of maximum absorption 272 and 267 nm for lamivudine and zidovudine, respectively. The R_f values of lamivudine and zidovudine were found to be ca. 0.34 and 0.74, respectively. The amount of the drugs spotted to the plate was in the range of 10–210 ng and 30–420 ng for lamivudine and zidovudine, respectively. The LOD and LOQ values for lamivudine and zidovudine were found to be 6.0 and 12.0 ng to 18 and 36 ng, respectively. The relative standard deviation (RSD) values for the precision study of lamivudine and zidovudine were obtained as 2.73% and 2.43%, respectively. In turn, percentage recoveries were obtained as 100.35 and 99.46 for lamivudine and zidovudine, respectively (Saini et al. 2010).

Next HPTLC method for simultaneous determination of lamivudine and zidovudine in a binary mixture was based on densitometric measurements of the spots at 276 and 271 nm for lamivudine and zidovudine, respectively. Separation was carried out on silica gel $60F_{254}$ plates using toluene–chloroform–methanol (1:6:3, v/v/v) as mobile phase. Second-order polynomial equations were obtained for the regression line in the ranges of 250–1400 and 250–1700 ng/spot for lamivudine and zidovudine, respectively. Correlation coefficient (r) values were 0.9998 for both drugs. In the method precision study, the coefficients of variation below 2% were obtained. The LOD and LOQ were 3.06 and 9.28 ng/spot for lamivudine and 3.34 and 10.13 ng/spot for zidovudine, respectively. Parameters such as mobile-phase composition, volume of mobile phase, time from spotting to development, and time from development to scanning were employed for the robustness study (Habte et al. 2009).

Similar work was done with silica gel $60F_{254}$ as stationary phase and a mixture of toluene–methanol–n-hexane (7:1.5:1.0, v/v/v) as mobile phase. Quantitation was carried out in the absorbance mode at 275 nm. The linearity ranges for lamivudine and zidovudine were found as 0.8–2.0 and 1.5–4.0 μg, respectively (Wankhede et al. 2006).

An interesting work was done where two laboratories extensively investigated the use of HPTLC to perform the assays of lamivudine–zidovudine, metronidazole, nevirapine, and quinine composite samples. The use of automatic sample applications in conjunction with variable-wavelength scanning densitometry allowed an average reproducibility among laboratory with RSD of 2.74% (Kaale et al. 2010).

An HPTLC method was also developed for simultaneous determination of lamivudine and tenofovir in bulk drugs and pharmaceutical dosage forms. Separation of the drugs was carried out on silica gel $60F_{254}$ using chloroform–methanol–toluene (8:2:2, v/v/v) as mobile phase. Densitometric measurement was carried out in the absorbance mode at 265 nm. The drugs were satisfactorily resolved with R_f values of ca. 0.27 and 0.51 for lamivudine and tenofovir, respectively. The linear regression analysis data for the calibration plots showed good linear relationship with $r^2 = 0.9999$

and 0.9996 in the concentration range of 60–210 ng/spot for each drug. The LOD and LOQ were 20 and 40 ng/spot, respectively, for lamivudine and 30 and 60 ng/spot, respectively, for tenofovir (Chandra et al. 2011).

A new simple, precise, accurate, and selective HPTLC method was developed for simultaneous analysis of tenofovir and emtricitabine in tablets. Chromatographic separation was achieved on silica gel $60F_{254}$ with toluene–methanol–ethyl acetate–acetic acid (4:2:5:0.1, v/v/v/v) as mobile phase. Detection was performed at 270 nm. The R_f values of tenofovir and emtricitabine were 0.52 and 0.40, respectively. The linearity of the method was proved in the range 120–600 ng/spot for tenofovir and 80–560 ng/spot for emtricitabine, while accuracy was 99.54% for tenofovir and 99.87% for emtricitabine (Rao et al. 2011). A similar two-component product was determined on silica gel $60F_{254}$. The mobile phase used was a mixture of chloroform–methanol (9:1, v/v). Detection of the spots was carried out at 265 nm. The calibration curve was found to be linear between 200 and 1000 ng with the regression coefficient of 0.9995 (Joshi et al. 2009).

An HPTLC method was developed for simultaneous determination of lamivudine, zidovudine, and nevirapine in pharmaceutical dosage forms. Chromatographic separation of the drugs was performed on silica gel $60F_{254}$ using n-hexane–chloroform–methanol (1:7:2, v/v/v) as mobile phase. A TLC scanner set at 275 nm was used in the reflectance–absorbance mode. These three drugs were satisfactorily resolved with R_f values of 0.22, 0.55, and 0.73 for lamivudine, zidovudine, and nevirapine, respectively. Calibration curves were polynomial in the range 100–1300 ng/band for both lamivudine and nevirapine and 100–1700 ng/band for zidovudine. Correlation coefficients (r) values were 0.9999 for lamivudine and 0.9998 for nevirapine and zidovudine. A low RSD (<2%) was found for both precision and robustness study showing that the proposed method was precise and robust. The method had sufficient accuracy of 98.93%, 99.45%, and 99.21% (Solomon et al. 2011).

The next chapter presents the development of simultaneous analysis of lamivudine, stavudine, and nevirapine using HPTLC silica gel $60F_{254}$ plates and the mobile phase comprising ethyl acetate, methanol, toluene, and ammonia 25% (38.7:19.4:38.7:3.2, v/v/v/v). The detection wavelength was 254 nm. The R_f values were 0.24 ± 0.03, 0.38 ± 0.04, and 0.69 ± 0.04. An F-test indicated that the calibration graphs were adequately linear at the evaluated concentration ranges. The% RSD for repeatability were found to be 0.62, 0.54, and 0.79, while the % RSD for intermediate precision were 1.66, 1.27, and 1.21. Most factors evaluated in the robustness test were found to have an insignificant effect on the selected responses at 95% confidence level. This method was successfully used to analyze the fixed-dose tablet samples of lamivudine, stavudine, and nevirapine (Shewiyo et al. 2011).

Simultaneous quantification of stavudine, lamivudine, and nevirapine by HPTLC method was also developed using a mobile phase of chloroform–methanol (9:1, v/v) on silica gel $60F_{254}$ with densitometry at 265 nm. The R_f values of stavudine, lamivudine, and nevirapine were 0.21–0.27, 0.62–0.72, and 0.82–0.93, respectively. Recovery values of 99.16%–101.89%, RSD of <0.7, and correlation coefficient of 0.9843–0.9999 were obtained. This TLC method was successfully compared with UV spectroscopy and reverse-phase high-performance liquid chromatography (RP-HPLC) methods (Anbazhagan et al. 2005).

49.4.3 HIV Protease Inhibitors

Currently, there are several HIV protease inhibitors approved for the treatment of HIV infection like indinavir, lopinavir, ritonavir, and nelfinavir (Figure 49.4). These medications work at the final stage of viral replication and attempt to prevent HIV from making new copies of itself by interfering with the HIV protease enzyme. As a result, the new copies of HIV are not able to infect new cells (Flexner 2011).

An HPTLC method for the analysis of indinavir both as bulk drug and in formulations was developed and validated. The method employed silica gel $60F_{254}$ plates and the solvent system consisted of carbon tetrachloride–chloroform–methanol–ammonia 10% (4:4.5:1.5:0.05, v/v/v/v).

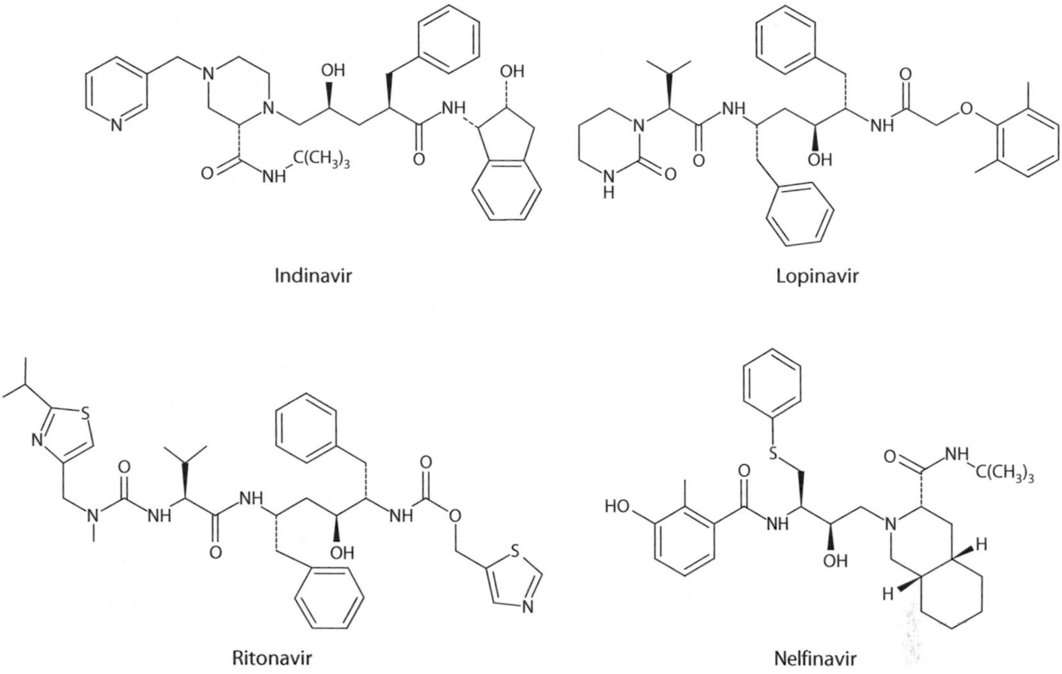

FIGURE 49.4 The chemical structures of HIV protease inhibitors.

Densitometric analysis was carried out in the absorbance mode at 260 nm. This system was found to give compact spots for indinavir with an R_f value of ca. 0.43. The drug was subjected to acid and alkali hydrolysis, oxidation, dry- and wet-heat treatment, and photodegradation. The drug was degraded under acidic and basic conditions, oxidation, dry- and wet-heat treatment, and photodegradation. However, the degradation products were well resolved from the pure drug with significantly different R_f values. Linearity of the proposed method was confirmed in the range of 100–6000 ng/spot with a significantly high value of correlation coefficient, $r^2 = 0.9970$. The LOD and LOQ were 40 and 120 ng/spot, respectively. Moreover, the proposed HPTLC method was utilized to investigate the kinetics of acid degradation process. Additionally, the Arrhenius plot was constructed and activation energy was calculated (Kaul et al. 2004b).

The next chapter describes a validated HPTLC method for the estimation of ritonavir in bulk and in pharmaceutical formulation. Determination was achieved on silica gel $60F_{254}$ using toluene–ethyl acetate–methanol–glacial acetic acid (7.0:2.0:0.5:0.5, v/v/v/v) as mobile phase. Quantification was done at 263 nm over the concentration range of 200–1000 ng/spot with recovery in the range of 98.00%–101.11%. The elaborated HPTLC method was compared with respective HPLC assay showing sufficient selectivity linearity, precision, and accuracy (Sudha et al. 2011).

In the next report, a method of analysis of nelfinavir both as bulk drug and in formulations was developed and validated. The method employed silica gel $60F_{254}$ as stationary phase. The solvent system consisted of toluene–methanol–acetone (7:1.5:1.5, v/v/v). Densitometric analysis was carried out in the absorbance mode at 250 nm. This system was found to give compact spots for nelfinavir with an R_f value of ca. 0.45. Nelfinavir was subjected to acid and alkali hydrolysis, oxidation, dry-heat treatment, and photodegradation. It was showed that the peaks of degraded products were well resolved from the pure drug with significantly different R_f values. The linear regression analysis data for the calibration plots showed good linear relationship with $r^2 = 0.999 \pm 0.002$ in the concentration range of 1000–6000 ng/spot. The method was validated for precision, robustness, and recovery. The LOD and LOQ were 60 and 140 ng/spot, respectively (Kaul et al. 2004c).

An HPTLC method was also established and validated for simultaneous determination of lopinavir and ritonavir in tablets. The method was based on densitometric measurements of their spots at 266 nm. The separation was carried out on silica gel $60F_{254}$ using ethyl acetate–ethanol–toluene–diethylamine (7:2.0:0.5:0.5, v/v/v/v) as mobile phase. Calibration curves were linear in the range of 8–20 μg/mL and 2–10 μg/mL for lopinavir and ritonavir, respectively (Patel et al. 2011).

A similar method with densitometry at 263 nm was developed and validated for simultaneous determination of these drugs in capsules. Separation was performed on silica gel $60F_{254}$ HPTLC plates using a mobile phase comprising toluene, ethyl acetate, methanol, and ammonia 25% (6.5:2.5:0.5:0.5, v/v/v/v), respectively. The detector response was linear in the range of 6.5–20.00 μg/spot and 1.5–5.00 μg/spot, for lopinavir and ritonavir, respectively. The LOD was found to be 1.5 and 4.6 ng/spot and LOQ was found to be 21.00 and 5.10 ng/spot, for lopinavir and ritonavir, respectively. The percentage recovery of lopinavir and ritonavir was found to be 99.90 ± 1.45 and 101.29 ± 1.95, respectively (Patel et al. 2009).

Also, a method with a densitometry at 263 nm, silica gel $60F_{254}$ HPTLC, and a mobile phase comprising toluene, ethyl acetate, methanol, and glacial acetic acid (7.0:2.0:0.5:0.5, v/v/v/v) was elaborated for such as combination. The detector response was linear in the range of 6.67–20.00 μg/spot and 1.67–5.00 μg/spot for lopinavir and ritonavir, respectively. The percentage recovery of lopinavir and ritonavir was found to be between 98.23–102.28 and 98.03–103.50, respectively (Sulebhavikar et al. 2008).

49.5　TLC ANALYSIS OF DIFFERENT ANTIVIRAL DRUGS IN BIOLOGICAL MATERIAL

An interesting work was done for determination of oseltamivir and its metabolite oseltamivir carboxylate in dried blood samples with a column-switching LC–MS/MS method, while online extraction was realized using the manual TLC–MS interface (Heinig et al. 2011).

A new simple, rapid, and selective HPTLC method was developed for quantitation of acyclovir during in vitro skin permeation studies. Separation of guinea pig skin proteins and acyclovir was achieved by employing the mobile phase consisting of chloroform–methanol–ammonia 25% (15:9:4, v/v/v) on silica gel $60F_{254}$ plates. Densitometric analysis was carried out at 255 nm. The LOD and LOQ were 30 and 50 ng, respectively. The calibration curve was linear in the range of 10–20 μg/mL (r = 0.9965). Intraday and interday variation studies gave an average 0.76% and 0.46% RSD for the three levels tested. Average recoveries of 101.8% and 100.1% were recorded for two marketed preparations studied. The method was employed to optimize a new topical liposomal gel formulation of acyclovir (Dubhashi and Vavia 2000).

A simple and inexpensive TLC assay for semiquantitative detection of saliva concentrations of nevirapine was elaborated by L'Homme et al. (2008). The method was validated in the study of an African target population. In this study, plasma and saliva nevirapine concentrations were assayed by HPLC method, while saliva concentrations of nevirapine were additively assayed by TLC. The TLC assay was found to be sensitive, specific, and robust in the detection of subtherapeutic concentration of nevirapine in saliva. It was concluded that the presented method was an attractive alternative to HPLC for therapeutic drug monitoring of nevirapine.

In the next chapters, simple and inexpensive TLC assays of nevirapine in blood or plasma are presented and compared with HPLC methods. The authors conclude that TLC was reasonably sensitive and specific for nevirapine detection (Chi et al. 2006, Dubuisson et al. 2004).

TLC method was also used for the study of radio-labeled conjugates of a novel thioether–phospholipid derivative of zalcitabine in cultured lymphocytes. The respective cell extracts were spotted on silica gel plates and developed with chloroform–methanol–ammonia 25% (50:30:10, v/v/v) and then with ethyl ether–hexane–formic acid 90% (60:90:4.5, v/v/v). The developed plates were exposed to a phosphor screen for 4 h and scanned using a phosphorimager (Kucera et al. 2001).

49.6 TLC FOR HYDROPHOBICITY/LIPOPHILICITY STUDIES OF DIFFERENT ANTIVIRAL AGENTS

The lipophilic character of a series of 5'-carbamates of zidovudine was studied by means of RP-TLC and RP-HPLC techniques giving the corresponding R_{MW} and log k'_w parameters. These values were compared with those obtained by the classical shake flask methodology. RP-TLC assays were performed on the basis of thermodynamically true R_M values using buffers of pH 7.4 and 12.03. In addition, the influence of the organic modifier was studied showing the superiority of methanol as compared with acetone as the organic modifier. Based on RP-TLC results, RP-HPLC studies were carried out using methanol and buffer pH 7.4 as mobile phase. It was concluded that chromatographic data (R_{MW} and log k'_w) proved to be reliable parameters for describing the lipophilic properties of the test compounds (Raviolo and Briñón 2005).

Similar work was done for studying the lipophilic properties of amino acid derivatives of zidovudine. These novel derivatives, obtained by association of zidovudine with the essential amino acids such as leucine, isoleucine, phenylalanine, valine, proline, and tryptophan, exhibited an increased log P as compared with the parent compound. All assays were performed using a buffer of pH 2 as the mobile phase. It was the pH at which the mentioned compounds were completely in their nonionized forms. In addition, good linear relationships were observed between log P values determined by the "shake flask" method and those obtained by chromatographic techniques and from theoretical calculations. These results demonstrated the applicability of the chromatographic methods to describe the lipophilic properties of this family of compounds (Moroni et al. 2002).

REFERENCES

Acosta, E. P. and C. Flexner. 2011. Antiviral agents (nonretroviral). In *Goodman & Gilman's the Pharmacological Basis of Therapeutics*. 12th Edn., eds., L.L. Brunton, B.A. Chabner, and B.C. Knollmann, McGrawHill Medical, New York, pp. 1593–1622.

Anbazhagan, S., N. Indumathy, P. Shanmugapandiyan, and S. K. Sridhar. 2005. Simultaneous quantification of stavudine, lamivudine and nevirapine by UV spectroscopy, reverse phase HPLC and HPTLC in tablets. *Journal of Pharmaceutical and Biomedical Analysis* 39: 801–804.

Askal, H. F., A. S. Khedr, I. A. Darwish, and R. M. Mahmoud. 2008. Quantitative thin-layer chromatographic method for determination of amantadine hydrochloride. *International Journal of Biomedical Science* 4: 155–160.

Chandra, P., A. S. Rathore, L. Sathiyanarayanan, and K. R. Mahadik. 2011. Application of high-performance thin-layer chromatographic method for the simultaneous determination of lamivudine and tenofovir disoproxil fumarate in pharmaceutical dosage form. *Journal of the Chilean Chemical Society* 56: 702–705.

Chi, B. H., A. Lee, E. P. Acosta, L. E. Westerman, M. Sinkala, and J. S. A. Stringer. 2006. Field performance of a thin-layer chromatography assay for detection of nevirapine in umbilical cord blood. *HIV Clinical Trials* 7: 263–269.

Darwish, I. A., H. F. Askal, A. S. Khedr, and R. M. Mahmoud. 2008. Stability-indicating thin-layer chromatographic method for quantitative determination of ribavirin. *Journal of Chromatographic Science* 46: 4–9.

Dubhashi, S. S. and P. R. Vavia. 2000. HPTLC method to study skin permeation of acyclovir. *Journal of Pharmaceutical and Biomedical Analysis* 23: 1017–1022.

Dubuisson, J. G., J. R. King, J. S. A. Stringer, M. L. Turner, C. Bennetto, and E. P. Acosta. 2004. Detection of nevirapine in plasma using thin-layer chromatography. *Journal of Acquired Immune Deficiency Syndromes* 35: 155–157.

Flexner, C. 2011. Antiretroviral agents and treatment of HIV infection. In *Goodman & Gilman's the Pharmacological Basis of Therapeutics*, 12th Edn., eds., L.L. Brunton, B.A. Chabner, and B.C. Knollmann, McGrawHill Medical, New York, pp. 1623–1663.

Habte, G., A. Hymete, and A.-M. I. Mohamed. 2009. Simultaneous separation and determination of lamivudine and zidovudine in pharmaceutical formulations using the HPTLC method. *Analytical Letters* 42: 1552–1570.

Heinig, K., T. Wirz, F. Bucheli, and A. Gajate-Perez. 2011. Determination of oseltamivir (Tamiflu®) and oseltamivir carboxylate in dried blood spots using off line or online extraction. *Bioanalysis* 3: 421–437.

Joshi, M., A. P. Nikalje, M. Shahed, and M. Dehghan. 2009. HPTLC method for the simultaneous estimation of emtricitabine and tenofovir in tablet dosage form. *Indian Journal of Pharmaceutical Sciences* 71: 95–97.

Kaale, E., P. Risha, E. Reich, and T. P. Layloff. 2010. An interlaboratory investigation on the use of high-performance thin layer chromatography to perform assays of lamivudine-zidovudine, metronidazole, nevirapine, and quinine composite samples. *Journal of AOAC International* 93: 1836–1843.

Kaul, N., H. Agrawal, A. R. Paradkar, and K. R. Mahadik. 2004a. Stability-indicating high-performance thin-layer chromatographic determination of zidovudine as the bulk drug and in pharmaceutical dosage forms. *Journal of Planar Chromatography-Modern TLC* 17: 264–274.

Kaul, N., H. Agrawal, A. R. Paradkar, and K. R. Mahadik. 2004b. Stability indicating high-performance thin-layer chromatographic determination of nelfinavir mesylate as bulk drug and in pharmaceutical dosage form. *Analytica Chimica Acta* 502: 31–38.

Kaul, N., H. Agrawal, A. R. Paradkar, and K. R. Mahadik. 2004c. The ICH guidance in practice: Stress degradation studies on indinavir sulphate and development of a validated specific stability-indicating HPTLC assay method. *Farmaco* 59: 729–738.

Kaul, N., H. Agrawal, A. R. Paradkar, and K. R. Mahadik. 2005. The ICH guidance in practice: Stress degradation studies on stavudine and development of a validated specific stability-indicating HPTLC assay method. *Journal of Chromatographic Science* 43: 406–415.

Kucera, G. L., C. L. Goff, N. Iyer et al. 2001. Cellular metabolism in lymphocytes of a novel thioether-phospholipid-AZT conjugate with anti-HIV-1 activity. *Antiviral Research* 50: 129–137.

Kumar, P., S. C. Dwivedi, and A. Kushnoor. 2011. Development and validation of HPTLC method for the determination of efavirenz as bulk drug and in tablet dosage form. *Research Journal of Pharmaceutical, Biological and Chemical Sciences* 2: 160–168.

L'Homme, R. F. A., E. P. Muro, J. A. H. Droste et al. 2008. Therapeutic drug monitoring of nevirapine in resource-limited settings. *Clinical Infectious Diseases* 47: 1339–1344.

Moroni, G. N., M. A. Quevedo, S. Ravetti, and M. C. Briñón. 2002. Lipophilic character of novel amino acid derivatives of zidovudine with anti HIV activity. *Journal of Liquid Chromatography and Related Technologies* 25: 1345–1365.

Patel, D. J., S. D. Desi, R. P. Savaliya, and D. Y. Gohil. 2011. Simultaneous HPTLC determination of lopinavir and ritonavir in combined dosage form. *Asian Journal of Pharmaceutical and Clinical Research* 4: 59–61.

Patel, G. F., N. R. Vekariya, and H. S. Bhatt. 2009. Application of TLC-densitometry method for simultaneous determination of lopinavir and ritonavir in capsule dosage form. *Oriental Journal of Chemistry* 25: 727–730.

Rao, J. R., S. A. Gondkar, and S. S. Yadav. 2011. Simultaneous HPTLC-densitometric analysis of tenofovir and emtricitabine in tablet dosage form. *International Journal of Pharm Tech Research* 3: 1430–1434.

Raviolo, M. A. and M. C. Briñón. 2005. Comparative study of hydrophobicity parameters of novel 5'-carbamates of zidovudine. *Journal of Liquid Chromatography and Related Technologies* 28: 2195–2209.

Saini, P. K., R. M. Singh, S. C. Mathur, G. N. Singh, S. Tuteja, and U. K. Singh. 2010. Simultaneous HPTLC method for estimation of lamivudine and zidovudine in bulk and in tablet dosage forms. *Indian Drugs* 47: 42–45.

Sharma, M. C. and S. Sharma. 2010. Development and validation of an HPTLC method for determination of oseltamivir phosphate in pharmaceutical dosage form. *Indian Drugs* 47: 68–72.

Shewiyo, D. H., E. Kaale, C. Ugullum et al. 2011. Development and validation of a normal-phase HPTLC method for the simultaneous analysis of lamivudine, stavudine and nevirapine in fixed-dose combination tablets. *Journal of Pharmaceutical and Biomedical Analysis* 54: 445–450.

Sia, T. K., L. Wulandari, and G. Indrayanto. 2002. TLC determination of acyclovir in pharmaceutical preparations, and validation of the method used. *Journal of Planar Chromatography-Modern TLC* 15: 42–45.

Solomon, G., A. Hymete, A.-M. I. Mohamed, and A. A. Bekhit. 2011. HPTLC-densitometric method development and validation for simultaneous determination of lamivudine, nevirapine and zidovudine in fixed dose combinations. *Thai Journal of Pharmaceutical Sciences* 35: 77–88.

Sudha, T., R. Vanitha, and V. Ganesan. 2011. Development and validation of RP-HPLC and HPTLC methods for estimation of ritonavir in bulk and in pharmaceutical formulation. *Der Pharma Chemica* 3: 127–134.

Sulebhavikar, A. V., U. D. Pawar, K. V. Mangoankar, and N. D. Prabhu-Navelkar. 2008. HPTLC method for simultaneous determination of lopinavir and ritonavir in capsule dosage form. *E-Journal of Chemistry* 5: 706–712.

Wadodkar, S. G., S. B. Wankhede, and K. R. Gupta. 2004. A validated HPTLC determination of stavudine in capsules. *Indian Drugs* 41: 300–302.

Wankhede, S. B., K. R. Gupta, M. R. Tajne, and S. G. Wadodkar. 2006. A validated HPTLC method for simultaneous estimation of lamivudine and zidovudine in tablets. *Asian Journal of Chemistry* 18: 2669–2672.

50 Thin-Layer Chromatography of Anticancer Drugs

Duygu Yeniceli

CONTENTS

50.1 INTRODUCTION: ANTICANCER DRUGS

Cancer, the uncontrolled growth of cells, is one of the leading causes of death throughout the world, for which the main treatments involve surgery, chemotherapy, and/or radiotherapy (Shewach and Kuchta 2009). Chemotherapy involves the use of low molecular weight drugs to selectively destroy tumor cells or at least limit their proliferation.

The main classes of chemotherapeutics are established according to their mechanism of action:

1. *Alkylating agents* alkylate DNA bases and block cell division.
2. *Antimetabolites* inhibit synthesis of precursor molecules essential for cell division.
3. *Antimitotics* block cell division by preventing microtubule function.
4. *Topoisomerase inhibitors* inhibit type I and type II topoisomerases that are essential enzymes that maintain the topology of DNA.
5. *Antibiotics* act by binding between base pairs with different mechanisms.
6. *Drugs with hormonal activity.*

This simple classification and Table 50.1 that includes the main classes and some examples are modified from a previous paper (Eckhardt 2002).

The first use of chemotherapy was in the early twentieth century, although it was not originally intended for that purpose. Mustard gas was used as a chemical warfare agent during World War I and was discovered to be a potent suppressor of blood production. A similar family of compounds known as nitrogen mustards was studied further during World War II at Yale University. It was reasoned that agent that damaged the rapidly growing white blood cells might have a similar effect on cancer. Therefore, in 1942, several patients with advanced lymphomas were given the drug intravenously, rather than by breathing the irritating gas (Papac 2001, Chabner and Roberts 2005).

TABLE 50.1
Anticancer Drugs

Alkylating agents	*Nitrogen mustards*: mechlorethamine, cyclophosphamide, ifosfamide, melphalan, chlorambucil
	Alkyl sulfonates: busulfan, treosulfan
	Nitrosourea analogs: lomustine, carmustine, fotemustine
	Triazenes: dacarbazine, temozolomide, procarbazine, altretamine
	Aziridines: mitomycin C
Antimetabolites	*Pyrimidine analogs*: 5-fluorouracil, capecitabine, tegafur, cytarabine, 5-azacytidine, gemcitabine
	Purine analogs: 6-mercaptopurine, 6-thioguanine, azathioprine, clofarabine, fludarabine
	Folate analogs: methotrexate, trimetrexate
	Adenosine analogs: cladribine, pentostatin
	Substituted ureas: hydroxyurea (hydroxycarbamide)
Platinum complexes	Cisplatin, carboplatin, oxaliplatin
Topoisomerase inhibitors	*Camptothecin analogs*: topotecan, irinotecan
	Podophyllotoxin analogs: etoposide, teniposide, amsacrine
Antimitotics (microtubule agents)	*Vinca alkaloids*: vincristine, vinblastine, vinorelbine, vindesine
	Taxanes: paclitaxel, docetaxel
Antibiotics	Doxorubicin, daunorubicin, bleomycin, actinomycin D, mithramycin
Hormonal agents	*SnRH analogs*: octreotide, lanreotide
	GnRH analogs: leuprolide, buserelin, goserelin, triptorelin
	Aromatase inhibitors: aminoglutethimide, fadrozole, vorozole, anastrozole, letrozole, exemestane, formestane
	Antiestrogens: toremifene, raloxifene, tamoxifen
	Antiandrogens: flutamide, bicalutamide
	Others: fluoxymesterone, diethylstilboestrol

Following the introduction of nitrogen mustard into clinical practice, more alkylating agents were developed and with more success. To this day, many of these agents are in clinical practice, most notably chlorambucil, melphalan, busulfan, and cyclophosphamide (Papac 2001).

Subsequently, antitumor antibiotics, platinum compounds, imidazole compounds, vinca alkaloids, taxols, camptothecin analogs, and biologic agents were shown to be effective for neoplastic diseases (Chabner and Roberts 2005).

50.2 THIN-LAYER CHROMATOGRAPHY: ANALYSIS OF ANTICANCER DRUGS

Chromatography is utilized extensively in the pharmaceutical industry for qualitative and quantitative analysis of drugs. Although thin-layer chromatography (TLC) is not as popular as high-performance liquid chromatography (HPLC) or gas chromatography (GC), it represents an important part of the market in chromatographic techniques because it has many advantages such as speed, versatility, flexibility, and low cost. Also, with the advances made in the technology, resolution, sensitivity, reproducibility, and performance have been further enhanced with high-performance TLC (HPTLC) (Linda 1991, Sherma 2010a).

A literature search of the last 20 years indicates that many TLC systems have been reported for the determination of anticancer drugs in pharmaceuticals, biological fluids, and plant materials. Also, three reviews including TLC of anticancer drugs to some extent have been published (Sherma 2010a,b; Nussbaumer et al. 2011).

50.2.1 Sample Preparation in Different Matrices

In sample preparation of pharmaceuticals, the most important parameter is the selection of a suitable solvent for the dissolution of the pharmaceutical preparations (Linda 1991). Pure methanol (MeOH) (Mhaske and Dhaneshwar 2007; Vadera et al. 2007; Akhtar et al. 2011) or a mixture of MeOH–water (1:1) (Perello et al. 2001; Paci et al. 2003; Bouligand et al. 2004, 2005; Gravel et al. 2005) has been used for this purpose.

Of course, the choice of optimal sample preparation techniques for biological samples is much more complex than for pharmaceuticals. Sample preparation procedures including protein precipitation with acetonitrile (ACN) for plasma samples and solid-phase extraction (SPE) for urine samples have been used for the determination of cyclophosphamide (Tasso et al. 1992, Yule et al. 1995) and ifosfamide (Boddy and Idle 1992; Boddy et al. 1993, 1995). Boddy and Idle (1992) used SPE for cerebrospinal fluid (CSF) samples, as well as urine samples.

For DNA studies, the sample preparation procedure requires a few steps. Adams et al. (1996) purified DNA with multiple ethanol (EtOH) precipitations and used Microcon 100 microconcentrators in some of the later experiments. In the final step, DNA was purified using QIAEX DNA binding resin.

Two steps of purification were also used for the analysis of cytarabine, 5-fluorouracil (5-FU), ftorafur, 6-azauridine (6-AZA), trifluorothymidine, uracil arabinoside, and uracil in the blood. This procedure included protein precipitation with MeOH and then SPE with C_{18} columns (Paw and Misztal 1997). In another study, blood and CSF samples were extracted with diethyl ether before spotting on a TLC sheet (Fung et al. 1998).

Sample extraction and purification methods are also very important for the successful analysis and isolation of natural compounds. Many extraction methods have been developed for the purification of yew extracts, including SPE on reversed-phase (RP) adsorbents (Glowniak et al. 1999) or on alumina (Hajnos et al. 2001a,b) or liquid–liquid extraction (LLE) with extraction solvents of different polarity. MeOH (Glowniak et al. 1999, Hajnos et al. 2002), hexane (followed by MeOH extraction) (Migas and Switka 2010), a mixture of chloroform and MeOH at different percentages (10:1, 9:1) (Wang et al. 2000, Sreekanth et al. 2009), or a buffer (Zocher et al. 1996) has been used. Glowniak and Mroczek (1999) used a procedure including different steps of extraction with MeOH, hexane, and dichloromethane (DCM), respectively, for the purification and isolation of taxoids. In some of these methods, SPE or LLE was followed by preparative TLC before the RP-HPLC analysis (Glowniak et al. 1996, 1999; Glowniak and Mroczek 1999; Hajnos et al. 2001b). Both LLE and SPE were also used (Hajnos et al. 2002). On the other hand, an HPTLC analysis of camptothecin and podophyllotoxin was performed with methanolic (Mishra et al. 2005, Dighe et al. 2007, Namdeo et al. 2010) or ethanolic extracts (Ahmad et al. 2007).

Wianowska et al. (2009) compared four types of solvent extraction methods (ultrasound- and microwave-assisted extraction, pressurized liquid extraction [PLE], and extraction in the Soxhlet apparatus) for paclitaxel, cephalomannine, and 10-deacetylbaccatin, taxoids recovered from common yew twigs. After the analysis of the extracts by HPLC, the greatest yields were obtained by multiple PLE.

50.2.2 Thin-Layer Chromatography of Anticancer Drugs
in Bulk Drug and Pharmaceuticals

TLC methods used for the determination of anticancer drugs in bulk drug and pharmaceuticals are summarized in Table 50.2. Some of these methods reported for quantitation of various anticancer drugs in capsules and infusion bags (Perello et al. 2001; Paci et al. 2003; Bouligand et al. 2004, 2005; Gravel et al. 2005) were part of a quality control program for pharmaceutical products prepared in the Department of Clinical Pharmacy at the Institut Gustave Roussy (Bourget et al. 2003).

TABLE 50.2
TLC of Anticancer Drugs in Bulk Drug and Pharmaceuticals

Compounds	Sample	Adsorbent	Solvent System	Detection	Reference
Fludarabine, cytarabine, gemcitabine, 5-FU	Infusion bags	Silica gel	EtOAc–MeOH–water (50:10:10)	UV 270 nm	Perello et al. (2001)
2,6-Disubstituted 7-methyl purines and 6-MP	Standard compounds	RP18 F_{254} plates	Acetone and buffer (sodium acetate–veronal, pH 7.0)	UV detection	Sochacka and Kowalska (2006)
Busulfan	Capsules, infusion bags	Silica gel 60 F_{254} plates	EtOAc–chloroform–MeOH (65:20:15)	4-Nitrobenzyl pyridine (NBP) in EtOH	Bouligand et al. (2004)
Cyclophosphamide	Capsules, infusion bags	Silica gel 60 F_{254} plates	DCM–MeOH–acetic acid (97:3:2)	1.25% phosphomolybdic acid in EtOH	Bouligand et al. (2005)
Irinotecan and topotecan	Infusion bags	Nano-SIL®–20 UV_{254} plates	Methylene chloride–MeOH–acetic acid–water (82:24:2:1)	Fluorescence ref. mode (exc. 366 nm, det. above 400 nm)	Gravel et al. (2005)
Vinca alkaloids	Infusion bags	Silica gel 60 F_{254} plates	DCM–MeOH (93:7)	Densitometry at 274 nm	Paci et al. (2003)
Irinotecan	Bulk drug, injectables	Silica gel 60F_{254} plates	Acetone–EtOAc–acetic acid (8.5:1.5:0.1)	UV 366 nm	Akhtar et al. (2011)
Tamoxifen citrate	Dissolution media	Silica gel 60 F_{254} HPTLC plates	Toluene–MeOH–glacial acetic acid (57:38:5)	Densitometry at 258 nm	Jamshidi and Sharifi (2009)
Camptothecin	Liposomes	Silica gel 60 HPTLC plates	Chloroform–MeOH–triethylamine–water (30:35:34:8)	–10% $CuSO_4$ acidified with 85% H_3PO_4—UV 254 and 366 nm	Saetern et al. (2005)
Dasatinib	Bulk drug, pharmaceuticals	Silica gel 60 F_{254} HPTLC plates	Toluene–chloroform (7:3)	Densitometry at 280 nm	Mhaske and Dhaneshwar (2007)
Imatinib mesylate	Bulk drug, pharmaceuticals	Silica gel 60 F_{254} HPTLC plates	Chloroform–MeOH (6:4)	Densitometry at 276 nm	Vadera et al. (2007)

Compound	Sample	Plate	Mobile phase	Detection	Reference
Leuprolide acetate	Bulk drug	Silica gel 60 F$_{254}$ HPTLC plates	EtOAc–MeOH–25% aqueous ammonia (60:30:10)	Densitometry at 280 nm	Jamshidi et al. (2006)
Leuprolide acetate	Implant polymeric matrix	Silica gel 60 F$_{254}$ HPTLC plates	Chloroform–MeOH (7:3)	Iodine vapor	Bahmanyar et al. (2008)
Anastrozole	Bulk drug, tablets	Silica gel TLC plates	Toluene–acetone–ammonia (6:4:0.3)	UV 200 nm	Bharati et al. (2010)
Bicalutamide	Bulk drug, liposomal formulation	Silica gel 60 F$_{254}$ HPTLC plates	Toluene–EtOAc (4.5:5.5)	Densitometry at 273 nm	Subramanian et al. (2009)
Aminoglutethimide, acetyl, and dansyl analogs	Standard compounds	Silica gel 60 F$_{254}$ TLC plates	30% Hydroxy trimethylpropylammonium–β-CD and MeOH (50:50)	UV 254 nm	Aboul-Enein et al. (2000)
Flutamide and its impurities	Standard compounds	Silica gel 60 F$_{254}$ TLC plates	Chloroform	Densitometry at 237 nm	Ferenczi-Fodor and Vegh (1993)
Oxaliplatin	Pharmaceutical preparation	Silica gel 60 F$_{254}$ TLC plates	MeOH–tetrahydrofuran–triethylamine–water (20:2:0.5:1.25)	UV 254 nm	Hernandez-Trejo et al. (2004)
Cisplatin and Pt compounds	Bulk drug, pharmaceuticals	Microcrystalline cellulose thin layers	0.1 M NaClO$_4$	I$_2$/4-nitroso-dimethylaniline (4-NDMA)	Lederer and Leipzig-Pagani (1998a)
Cisplatin and Pt compounds	Standard compounds	Silica gel 60 HPTLC plates	Acetone–water (95:5) or acetone–toluene–water (76:20:4)	4-NDMA	De Spiegeleer et al. (1984)
Carboplatin and Pt compounds	Bulk drug, pharmaceuticals	Microcrystalline cellulose thin layers	0.1 M NaClO$_4$	I$_2$/4-NDMA	Lederer and Leipzig-Pagani (1998b)
Mitoguazone dihydrochloride	Bulk drug	Silica gel TLC plates	Acetone–ammonium hydroxide–water (90:5:5)	Nitroprusside–ferricyanide reagent	Thomson et al. (1997)

Apart from these methods, Nowakowska et al. (2010) investigated the effect of mobile and stationary phases on the chromatographic behavior of the anticancer antibiotic aclarubicin and another antibiotic doxycycline. Various TLC and HPTLC plates were tried with solvent systems consisting of a wide range of mixtures (from 0% to 100%) of n-alcohols with dimethyl sulfoxide (DMSO), hexamethyldisiloxane, ACN, and water. Detection reagents were anisaldehyde–MeOH–acetic acid–orthophosphoric acid–H_2SO_4 (1:100:10:10:5), 5% $AlCl_3$ in MeOH, and H_2SO_4–MeOH (1:4).

Many papers have been published reporting enantiomer separations with TLC and in a few of them the separations of chiral anticancer drugs such as cyclophosphamide and aminoglutethimide were mentioned. According to these papers, the chiral separation of cyclophosphamide was achieved with a derivatization reagent, (–)-1-phenethyl alcohol, whereas the chromatography was performed on a chiral triacetylcellulose layer for aminoglutethimide (Lepri et al. 1994, Lepri 1997).

Aboul-Enein et al. (2000) reported the enantiomeric separation of aminoglutethimide, acetylaminoglutethimide, and dansyl aminoglutethimide by TLC with β-cyclodextrin (β-CD) and its derivatives as mobile-phase additives (described in Table 50.2). The effect of the type and amount of organic modifier as well as the concentration of β-CD derivative and the pH of the mobile phase on resolution were studied. Also, three types of maltodextrins were investigated for the resolution of the racemic compounds.

Dehydroepiandrosterone (DHEA) is the most abundant androgen that can be converted to other steroidal hormones and some of the DHEA derivatives show very high antiaromatase activity. The chromatographic behavior of DHEA derivatives was studied by HPTLC performed on a C_{18}-bonded phase with two aqueous eluents, acetone–water and dioxane–water. The correlation between the chromatographic lipophilic parameters $\left(R_M^0 \right)$ and the calculated log P values, as well as the biological activity, was investigated (Perisic-Janjic et al. 2004). In another study, quantitative relationships between normal-phase (NP) retention behavior and the structure of the DHEA derivatives were determined (Perisic-Janjic et al. 2005).

50.2.3 Thin-Layer Chromatography of Anticancer Drugs in Biological Fluids

The analysis of common anticancer drugs in biological fluids requires simple and cost-effective methods. Over the past 20 years, many TLC methods have been reported for this purpose as seen in Table 50.3.

Several papers were published reporting the determination of cyclophosphamide and ifosfamide in plasma and urine samples by the same research group. In these studies, the metabolites of cyclophosphamide and ifosfamide were determined as well as the main drugs (Boddy and Idle 1992; Boddy et al. 1993, 1995; Tasso et al. 1992; Yule et al. 1995).

Koskinen et al. (1997) used a "P-postlabeling method" for the analysis of DNA isolated from livers of rats receiving tamoxifen. The postlabeled DNA hydrolysis mixture was analyzed both by RP-HPLC with ^{32}P online detection and by TLC on polyethyleneimine plates followed by autoradiography.

50.2.4 Thin-Layer Chromatography of Anticancer Drugs in Plant Materials

Taxoids (taxanes) are diterpenes well known for their cytotoxic, antimitotic properties and they are used (especially paclitaxel—taxol) as antitumor drugs. Previously, TLC of taxanes as well as other diterpenes was reported in a chapter of "Thin Layer Chromatography in Phytochemistry" (Hajnos 2008).

TLC methods reported for purification, isolation, and/or quantitation of taxanes, vinca alkaloids, camptothecin, and podophyllotoxin from different plant parts are summarized in Table 50.4.

TLC was used in the analysis of taxoids for different purposes including quantitation (Matysik et al. 1995, Srinivasan et al. 1997), purification, and/or isolation as a sample preparation technique (Glowniak et al. 1996, Hajnos et al. 2001b, Sreekanth et al. 2009), and both purification and quantitation

TABLE 50.3
TLC of Anticancer Drugs in Biological Fluids

Compounds	Sample	Adsorbent	Solvent System	Detection	Reference
Cytarabine, ftorafur, 6-AZA, 5-FU	Human plasma	Silica gel 60 F$_{254}$ plates	BuOH–isopropyl alcohol–water (7:1:2)	UV 254 nm	Paw and Misztal (1997)
2-Thioguanine and 6-MP	Urine and serum samples	Silica gel 60 F$_{254}$ plates	MeOH	Iodine-azide reagent	Zakrzewski and Ciesielski (2006)
Cyclophosphamide and its metabolites	Plasma and urine samples	Silica gel HPTLC plates	BuOH–water (20:3) or chloroform–EtOH–glacial acetic acid (20:5:0.1) and DCM–MeOH–glacial acetic acid (18:12:0.1)	15% NBP in acetone and acetate buffer (pH 4; 8:2)	Tasso et al. (1992)
Cyclophosphamide and its metabolites	Plasma, urine samples	Silica gel HPTLC plates	BuOH–water (20:3)	5% NBP in acetone and acetate buffer (pH 4; 8:2)	Yule et al. (1995)
Ifosfamide and its metabolites	Plasma, urine samples	Silica gel TLC plates	DCM–dimethyl formamide–glacial acetic acid (90:8:1) and chloroform–MeOH–glacial acetic acid (90:60:1)	5% NBP in acetone–0.2 M acetate buffer (pH 4.6; 8:2)	Boddy et al. (1993, 1995)
Ifosfamide and its metabolites	Urine, plasma, and CSF samples	Silica gel 60 HPTLC plates	DCM–MeOH–glacial acetic acid (90:8:1) and chloroform–MeOH–glacial acetic acid (90:60:1)	5% NBP in acetone–0.2 M acetate buffer (pH 4.6; 8:2)	Boddy and Idle (1992)
Mitomycin C	DNA samples	Glass-backed KC$_{18}$ RP TLC plates	0.4 M ammonium formate, pH 6.2, and isopropanol–water (1:1)	Autoradiography	Reddy (1993)
Busulfan, chlorambucil, cyclophosphamide, melphalan	Calf thymus DNA	Polyethyleneimine (PEI)–cellulose TLC plates	2.3 M sodium phosphate, pH 4.6 or pH 5.8	Autoradiography	Adams et al. (1996)
^{14}C-taxol and its metabolites	Human liver microsomes	Silica gel 60 F$_{254}$ plates	Toluene–acetone–formic acid (60:39:1)	Bioimaging analyzing system	Fujino et al. (2001)
Carmustine, 4-hydroperoxy cyclophosphamide (4-HC), and paclitaxel	Monkey brain, blood, and CSF	Silica gel	Chloroform (carmustine), acetone–chloroform (1:1) (4-HC), acetone–chloroform (1:3) (paclitaxel)	Scintillation counting	Fung et al. (1998)

TABLE 50.4
TLC of Anticancer Drugs in Plant Materials

Compounds	Sample	Adsorbent	Solvent System	Detection	Reference
Vinca alkaloids	*C. roseus*	Amino-bonded silica gel HPTLC plates	Hexane–DCM–acetone–2–propanol (65:13:21:0.9)	Densitometry at 298 nm	Nagy-Turak and Vegh (1994)
Camptothecin	*N. foetida*—stem	Silica gel 60 F$_{254}$ plates	Toluene–ACN–glacial acetic acid (6.5:3.5:0.1)	Densitometry at 370 nm	Dighe et al. (2007)
Camptothecin	Callus and *N. foetida*—various parts	Silica gel 60 F$_{254}$ plates	Chloroform–EtOAc–MeOH (4:5:0.5)	UV 360 nm	Namdeo et al. (2010)
Podophyllotoxin	*P. hexandrum* Royle—tissue culture	RP18 F$_{254}$ TLC plates	ACN–water (50:50)	Densitometry at 217 nm	Mishra et al. (2005)
Podophyllotoxin	*P. hexandrum*—callus and roots	Silica gel GF$_{254}$ plates	ACN–water (4:6)	UV 210 nm	Ahmad et al. (2007)
Taxanes	*Taxus*—twigs and needles	Silica gel 60 HF$_{254}$ plates	Heptane–DCM–EtOAc (50:40:5)	UV 254 nm	Glowniak et al. (1996)
Taxanes	*Taxus*—needles	Silica Si 60 plates	Heptane–MeOH–chloroform (60:5:95, 70:5:95)	Densitometry at 243 nm	Matysik et al. (1995)
Taxanes	*Taxus baccata*—needles and stems	Si60 F$_{254S}$ TLC RP18W F$_{254S}$ and NH$_2$ F$_{254S}$ HPTLC plates	Heptane–EtOAc (5:5), MeOH–water (8:2), Chloroform–acetone (15:5)	UV 254, 366 nm	Migas and Switka (2010)
Taxol and 10-DABIII	*Gliocladium* sp. isolated from *T. baccata*	Silica gel G plates	Chloroform–MeOH (7:3), Chloroform–ACN (7:3) (Prep. TLC)	Anisaldehyde–sulfuric acid or vanillin–sulfuric acid	Sreekanth et al. (2009)
Taxanes	*Taxus* species—various parts	Silica gel GF plates	DCM–MeOH (95:5)	—	Vidensek et al. (1990)
Taxanes	*Taxus chinensis*, *T. baccata*—cell cult.	Silica gel GF$_{254}$ plates	Chloroform–ACN (4:1)	UV 254 nm, or vanillin–sulfuric acid	Srinivasan et al. (1997)
Taxanes	*Taxus* species—twigs, crude extracts, or CC isolated fractions	Silica gel 60 F$_{254}$ plates	Benzene–chloroform–acetone–MeOH (20:92.5:15:7.5), (8:37:6:3) DCM–dioxane–acetone–MeOH (84:10:5:1)	UV 254 nm, densitometry at 230 nm UV 366, 254, 230 nm	Glowniak and Mroczek (1999), Glowniak et al. (1999) Hajnos et al. (2001a,b, 2002)
10-deacetylbaccatin III, baccatin III, cephalomannine, paclitaxel					
Taxanes and taxine alkaloids	*T. baccata*—pollen	Silica gel	Chloroform–MeOH (24:1)	Dragendorff reagent Iodoplatinate reagent	Vanhaelen et al. (2002)

Taxanes	Taxus brevifolia—bark	Cyano-modified silica HPTLC plates	2D TLC: (1) DCM–hexane–acetic acid (9:10:1) (2) Water–ACN–MeOH–tetrahydrofuran (8:5:7:0.1)	–3% H_2SO_4 in EtOH + heating at 120°C for 5 min	Stasko et al. (1989)
		Diphenyl-modified silica plates	2D TLC: (1) hexane–isopropanol–acetone (15:2:3) (2) MeOH–water (7:3)	Anisaldehyde reagent + heating	
Taxanes as dansyl derivatives	Taxus species	Silica HPTLC plates	Cyclohexane–EtOAc–toluene–triethylamine–MeOH (9:6:4:3:1)	Densitometry at fluorescent mode (exc. at 254 nm, filter–440 nm)	He and He (1997)
Taxanes—yunnanxane	Taxus wallichiana—cell suspension cultures	Silica gel	Benzene–acetone (8:2)	UV 254 nm	Agrawal et al. (2000)
Taxanes—taxuyunnanines	Taxus yunnanensis	Silica gel	Isopropanol–chloroform (2:1)	10% H_2SO_4 in EtOH	Li et al. (2002)
Taxanes	T. chinensis—needles	Silica gel	Hexane–acetone, hexane–EtOAc, chloroform–MeOH (in different proportions)	10% H_2SO_4 in EtOH + heating	Shi et al. (1999)
Taxol	T. baccata—roots	Silica gel	Water-saturated EtOAc (I) Chloroform–MeOH (95:5) (II) EtOAc–MeOH–water (100:5:1) (III)	UV 230 nm	Zocher et al. (1996)
Taxol	Endophytic fungus (strain TF5)	Silica gel	Chloroform–MeOH (7:1), chloroform–ACN (7:3)	H_3PO_4–EtOH (2:8), H_2SO_4–MeOH (1:1) and others	Wang et al. (2000)

(Zocher et al. 1996, Wang et al. 2000, Migas and Switka 2010). It was also used for all of these processes: purification, separation, and isolation (Hajnos et al. 2001a). Apart from taxoids (*Taxus* species), the extracts of *Nothapodytes foetida* and *Podophyllum hexandrum* were analyzed by HPTLC for the content of camptothecin, lead structure of irinotecan topotecan, and podophyllotoxin, and lead structure of etoposide and teniposide (Mishra et al. 2005, Ahmad et al. 2007, Dighe et al. 2007, Namdeo et al. 2010).

Nagy-Turak and Vegh (1994) developed an overpressured layer chromatography (OPLC) method for the determination of bis-indole alkaloids (vinblastine and vincristine) from the extracts of *Catharanthus roseus*. Also, a test for the robustness of quantitative OPLC assay, using vinblastine as a practical example, was presented. This test was performed by saturated factorial experimental design, which enabled the number of experiments performed to be limited (Nagy-Turak et al. 1995).

50.3 STABILITY OF ANTICANCER DRUGS AND POSSIBLE DEGRADATION IMPURITIES

For bulk and pharmaceutical formulations, a good method for quality control should be able to simultaneously determine the parent drug, its impurities, and degradation products. In the review of Benizri et al. (2009), the stability of 34 cytotoxic drug preparations was evaluated, antineoplastic agents with sufficient chemical and physical stability were selected for home-based therapy, and a standardization of anticancer drug stability data was proposed.

Irinotecan (IRT), a camptothecin derivative, has a dynamic pH-dependent equilibrium between the closed-ring lactone and open-ring hydroxy acid form. The active and more hydrophobic lactone form is favored by acidic pH. So, Akhtar et al. (2011) developed a stability-indicating HPTLC method with an acidic mobile phase and IRT was subjected to acid and alkali hydrolysis, oxidation, thermal, and ultraviolet radiation treatments. IRT degraded under all these conditions with several degradation products. In another study following these treatments, anastrozole did not undergo degradation under acidic or oxidizing conditions, whereas it degraded under alkaline conditions (Bharati et al. 2010). Stability-indicating HPTLC methods were also reported for the determination of imatinib, dasatinib, and leuprolide acetate (LPA). All these drugs were subjected to acidic, basic, and oxidative degradation. Imatinib was also subjected to dry heat, whereas dasatinib was subjected to wet heat and photodegradation as well as the other conditions (Jamshidi et al. 2006, Mhaske and Dhaneshwar 2007, Vadera et al. 2007). Degradation products were well resolved from the main drugs in all these methods.

Structural stability of the entrapped LPA in poly(DL-lactide-co-glycolide) (PLGA) was studied after in vitro release of the drug using different spectroscopic methods and HPTLC. It was reported that the entrapped LPA suffered no structural changes during its stay in polymer formulation and remained stable during the first 24 h of release (Bahmanyar et al. 2008).

Hydrophobic environments like liposomes have been used for the stabilization of camptothecin lactone form. It has been reported that poor in vivo stability and low water solubility make liposomes an attractive formulation for camptothecin. Saetern et al. (2005) investigated the stability of camptothecin-containing liposomes by an HPTLC method. Although a pH around 6.0 was found to be more favorable for the stability of formulations, the storage stability of camptothecin-containing liposomes was inadequate even at pH 6, for a marketable pharmaceutical product.

Ferenczi-Fodor and Vegh (1993) investigated the effect of eluent flow rate on the reliability of quantitative measurements in an OPLC method. Validation parameters have been examined and OPLC data obtained at different flow rates compared with that of normal development in saturated and unsaturated chambers. This comparison was performed during the determination of flutamide and its three impurities.

5-FU is stable at a pH below 9 and is degraded to urea, fluoride, and an aldehyde derivative in alkaline solutions. 5-FU solution should be stored in the absence of light since UV irradiation results in changes in the UV spectrum. Also, 5-FU has been shown to be unstable in whole blood, primarily due to dihydropyrimidine dehydrogenase activity, and separating cells from plasma was found to be critical for obtaining accurate results (Breda and Baratte 2010).

In the European Pharmacopoeia (EP 2008), seven impurities of 5-FU including barbituric acid, isobarbituric acid (5-hydroxyuracil), uracil, 5-methoxyuracil, 5-chlorouracil, 2-ethoxy-5-fluoroura-cil (impurity F), and urea (impurity G) were specified. A TLC method was proposed for the determination of impurities F and G in which chromatography was performed on a TLC silica gel F_{254} plate with a mobile phase of MEOH–water–ethyl acetate (EtOAC) (15:15:70). For the detection of impurity F, UV light at 254 nm was used, whereas impurity G was detected after spraying with a mixture of dimethylaminobenzaldehyde in anhydrous EtOH and HCl. The limits for impurities F and G were reported as 0.25% and 0.2%, respectively.

Azathioprine (AZA) is metabolized to the active drug 6-mercaptopurine (6-MP) that is also commercially available (Nussbaumer et al. 2011). In addition, 6-MP is the principal in vitro degradation product of AZA, which is limited by the American Pharmacopoeia to less than 1% (w/w) by TLC (USP 2006). The TLC method reported in USP 2006 includes the application of the compounds on microcrystalline cellulose TLC plates and developing with butanol (BuOH), previously saturated with 6 N ammonium hydroxide (USP 2006).

Lederer and Leibzig-Pagani (1998a,b) investigated the aging of cisplatin and carboplatin solutions in water, 3.5 and 103 mM NaCl using TLC and thin-layer electrophoresis (TLE). Many platinum compounds including K_2PtCl_4, $K[Pt(NH_3)Cl_3]$, transplatin, and $[Pt(NH_3)_4Cl_2]$ as well as cisplatin and carboplatin were separated.

50.4 CONCLUDING COMMENTS

TLC is increasingly being used in industrial and clinical laboratories to determine anticancer drugs in pharmaceuticals, biological samples, and natural materials. Numerous examples, including important anticancer drugs such as busulfan, melphalan, cyclophosphamide, 5-fluorouracil, camptothecin, and podophyllotoxin analogs, have been presented in this chapter. Over the years, the instrumentation of TLC, especially HPTLC, has been improved and expanded resulting in a sensitivity comparable to that of HPLC and GC. Also, analysis with TLC is considerably faster, since multiple samples can be analyzed simultaneously. The speed, sensitivity, reproducibility, and automation achievable with improved TLC plates and densitometers suggest that TLC will continue to be a versatile and widely used analytical technique for quantitation of anticancer drugs.

ACKNOWLEDGMENTS

I am grateful to Prof. Dilek Dogrukol-Ak (Anadolu University, Turkey) for encouraging me to write this chapter and her invaluable feedback. And, many thanks to Prof. Ann Van Schepdael (KU Leuven, Belgium) for her support and kind editing of the manuscript. I also thank my colleagues Prof. Deirdre Cabooter (KU Leuven, Belgium) and Dr. Cigdem Belikusakli-Cardak (Anadolu University, Turkey) for their helpful suggestions. At last, thank you for reading this overview on TLC of anticancer drugs.

REFERENCES

Aboul-Enein, H. Y., M. I. El-Awady, and C. M. Heard. 2000. Enantiomeric separation of aminoglutethimide, acetylaminoglutethimide, and dansylaminoglutethimide by TLC with β-cyclodextrin and derivatives as mobile phase additives. *J. Liq. Chromatogr. Relat. Technol.* 23: 2715–2726.

Adams, S. P., G. M. Laws, R. D. Storer et al. 1996. Detection of DNA damage induced by human carcinogens in acellular assays: Potential application for determining genotoxic mechanisms. *Mutat. Res.* 368: 235–248.

Agrawal, S., S. Banerjee, and S. K. Chattopadhyay. 2000. Isolation of taxoids from cell suspension cultures of *Taxus wallichiana. Planta Med.* 66: 773–775.

Ahmad, R., V. K. Sharma, A. K. Rai et al. 2007. Production of lignans in callus culture of *Podophyllum hexandrum. Trop. J. Pharm. Res.* 6: 803–808.

Akhtar, N., S. Talegaonkar, R. K. Khar et al. 2011. A stability indicating HPTLC method for the analysis of irinotecan in bulk drug and marketed injectables. *J. Liq. Chromatogr. Relat. Technol.* 34: 1459–1472.

Bahmanyar, N., K. Haghbeen, A. Jamshidi et al. 2008. Studying the structural stability of leuprolide acetate after releasing from lactide-co-glycolide copolymer by different spectroscopic methods and HPTLC. *Iran. Polym. J.* 17: 345–352.

Benizri, F., B. Bonan, A. L. Ferrio et al. 2009. Stability of antineoplastic agents in use for home-based intravenous chemotherapy. *Pharm. World Sci.* 31: 1–13.

Bharati, P., A. Vinodini, A. S. Reddy et al. 2010. Development and validation of a planar chromatographic method with reflectance scanning densitometry for quantitative analysis of anastrozole in the bulk material and in tablet formulations. *J. Planar Chromatogr.* 23: 79–83.

Boddy, A. V. and J. Idle. 1992. Combined thin-layer chromatography-photography-densitometry for the quantification of ifosfamide and its principal metabolites in urine, cerebrospinal fluid and plasma. *J. Chromatogr. Biomed. Appl.* 575: 137–142.

Boddy, A. V., M. Proctor, D. Simmonds et al. 1995. Pharmacokinetics, metabolism and clinical effect of ifosfamide in breast cancer patients. *Eur. J. Cancer* 31: 69–76.

Boddy, A. V., S. M. Yule, R. Wyllie et al. 1993. Pharmacokinetics and metabolism of ifosfamide administered as a continuous infusion in children. *Cancer Res.* 53: 3758–3764.

Bouligand, J., A. Paci, L. Mercier et al. 2004. High-performance thin-layer chromatography with a derivatization procedure, a suitable method for the identification and quantitation of busulfan in various pharmaceutical products. *J. Pharm. Biomed. Anal.* 34: 525–530.

Bouligand, J., T. Storme, I. Laville et al. 2005. Quality control and stability study using HPTLC: Applications to cyclophosphamide in various pharmaceutical products. *J. Pharm. Biomed. Anal.* 38: 180–185.

Bourget, P., A. Paci, J. B. Rey et al. 2003. Contribution of high-performance thin-layer chromatography to a pharmaceutical quality assurance programme in a hospital chemotherapy manufacturing unit. *J. Pharm. Biomed. Anal.* 56: 445–451.

Breda, M. and S. Baratte. 2010. A review of analytical methods for the determination of 5-fluorouracil in biological matrices. *Anal. Bioanal. Chem.* 397: 1191–1201.

Chabner, B. A. and T. G. Roberts. 2005. Chemotherapy and the war on cancer. *Nat. Rev.* 5: 65–72.

De Spiegeleer, B., G. Slegers, W. Van Den Bossche et al. 1984. Quantitative analysis of cis-dichlorodiammineplatinum (II) by high-performance thin-layer chromatography. *J. Chromatogr.* 315: 481–487.

Dighe, V., R. T. Sane, G. Parekh et al. 2007. HPTLC quantitation of camptothecin in *Nothapodytes foetida* (Wight) Sleumer stem powder. *J. Planar Chromatogr.* 20: 131–133.

Eckhardt, S. 2002. History, present status and future prospects of anticancer drug therapy. *CME J. Gynecol. Oncol.* 3: 312–317.

Ferenczi-Fodor, K. and Z. Vegh. 1993. Validation of the quantitative planar chromatographic analysis of drug substances. 2: Comparison of some parameters in OPLC and TLC. *J. Planar Chromatogr.* 6: 256–258.

Fujino, H., I. Yamada, S. Shimada et al. 2001 Simultaneous determination of taxol and its metabolites in microsomal samples by a simple thin-layer chromatography radioactivity assay—Inhibitory effect of NK-104, a new inhibitor of HMG-CoA reductase, *J. Chromatogr. B* 757: 143–150.

Fung, L. K., M. G. Ewend, A. Sills et al. 1998. Pharmacokinetics of interstitial delivery of carmustine, 4-hydroperoxycyclophosphamide, and paclitaxel from a biodegradable polymer implant in the monkey brain. *Cancer Res.* 58: 672–684.

Glowniak, K. and T. Mroczek. 1999. Investigations on preparative thin-layer chromatographic separations of taxoids from *Taxus Baccata* L. *J. Liq. Chromatogr. Relat. Technol.* 22: 2483–2502.

Glowniak, K., T. Wawrynowicz, and M. Hajnos. 1999. The application of zonal thin-layer chromatography to the determination of paclitaxel and 10-deacetylbaccatin III in some *Taxus* species. *J. Planar Chromatogr.* 12: 328–335.

Glowniak, K., G. Zgorka, A. Jozefczyk et al. 1996. Sample preparation for taxol and cephalomannine determination in various organs of *Taxus* sp. *J. Pharm. Biomed. Anal.* 14: 1215–1220.

Gravel, E., P. Bourget, L. Mercier et al. 2005. Fluorescence detection combined with either HPLC or HPTLC for pharmaceutical quality control in a hospital chemotherapy production unit: Application to camptothecin derivatives. *J. Pharm. Biomed. Anal.* 39: 581–586.

Hajnos, M. L. 2008. TLC of Diterpenes. In *Thin Layer Chromatography in Phytochemistry*, eds., M. Waksmundzka-Hajnos, J. Sherma, and T. Kowalska, pp. 481–517. CRC Press: Boca Raton, FL.

Hajnos, M. L., K. Glowniak, and M. Waksmundzka-Hajnos. 2001a. Optimization of the isolation of some taxoids from yew tissues. *J. Planar Chromatogr.* 14: 119–125.

Hajnos, M. L., K. Glowniak, and M. Waksmundzka-Hajnos. 2002. Application of pseudo-reversed-phase systems to the purification and isolation of biologically active taxoids from plant material. *Chromatographia* 56: 91–94.

Hajnos, M. L., M. Waksmundzka-Hajnos, and J. Gawdzik. 2001b. Influence of the extraction mode on the yield of taxoids from yew tissues-preliminary experiments. *Chem. Anal.* 46: 831–838.

He, Y. B. and L. Y. He. 1997. Separation and determination of taxol and related compounds in *Taxus* by HPTLC and fluorescent derivatisation on the plate. In *Proceedings of the 9th International Symposium on Instrumental Chromatography*, Interlaken, Switzerland, pp. 171–177.

Hernandez-Trejo, N., A. Hampe, and R. H. Muller. 2004. A thin layer chromatography method to identify oxaliplatin in aqueous solution. *Pharm. Ind.* 66: 1545–1550.

Jamshidi, A., H. Mobedi, and F. Ahmad-Khanbeigi. 2006. Stability-indicating HPTLC assay for leuprolide acetate. *J. Planar Chromatogr.* 19: 223–227.

Jamshidi, A. and S. Sharifi. 2009. HPTLC analysis of tamoxifen citrate in drug-release media during development of an in-situ cross-linking delivery system. *J. Planar Chromatogr.* 22: 187–189.

Koskinen, M., H. Rajaniemi, and K. Hemminki. 1997. Analysis of tamoxifen-induced DNA adducts by 32P-postlabelling assay using different chromatographic techniques. *J. Chromatogr. B* 691: 155–160.

Lederer, M. and E. Leibzig-Pagani. 1998a. Studies of platinum (II) compounds in aqueous solution. Part 2. *cis*-Pt(NH$_3$)$_2$Cl$_2$. *Anal. Chim. Acta* 358: 61–68.

Lederer, M. and E. Leibzig-Pagani. 1998b. A note on the solution chemistry of carboplatin in aqueous solutions. *Int. J. Pharm.* 167: 223–228.

Lepri, L. 1997. Enantiomer separation by TLC. *J. Planar Chromatogr.* 10: 320–331.

Lepri, L., V. Coas, P. G. Desideri et al. 1994. Reversed phase planar chromatography of enantiomeric compounds on triacetylcellulose. *J. Planar Chromatogr.* 7: 376–381.

Li, S. H., H. J. Zhang, X. M. Niu et al. 2002. Taxuyunnanines S-V, new taxoids from *Taxus yunnanensis*. *Planta Med.* 68: 253–257.

Linda, L. 1991. Pharmaceuticals and drugs: Guidelines for analysis. In *Handbook of Thin-Layer Chromatography*, eds., J. Sherma and B. Fried, pp. 717–755. Marcel Dekker: New York.

Matysik, G., K. Glowniak, A. Jozefczyk et al. 1995. Stepwise gradient thin-layer chromatography and densitometric determination of taxol in extracts from various species of *Taxus*. *Chromatographia* 41: 485–487.

Mhaske, D. V. and S. R. Dhaneshwar. 2007. Stability indicating HPTLC and LC determination of dasatinib in pharmaceutical dosage form. *Chromatographia* 66: 95–102.

Migas, P. and M. Switka. 2010. TLC with an adsorbent gradient for the analysis of taxol in *Taxus baccata* L. *J. Planar Chromatogr.* 23: 286–288.

Mishra, N., R. Acharya, A. P. Gupta et al. 2005. A simple microanalytical technique for determination of podophyllotoxin in *Podophyllum hexandrum* roots by quantitative RP-HPLC and RP-HPTLC. *Curr. Sci.* 88: 1372–1373.

Nagy-Turak, A. and Z. Vegh. 1994. Extraction and in situ densitometric determination of alkaloids from *Catharanthus roseus* by means of overpressured layer chromatography on amino-bonded silica layers I. Optimization and validation of the separation system. *J. Chromatogr. A* 668: 501–507.

Nagy-Turak, A., Z. Vegh, and K. Ferenczi-Fodor. 1995. Validation of the quantitative planar chromatographic analysis of drug substances III. Robustness testing in OPLC. *J. Planar Chromatogr.* 8: 188–193.

Namdeo, A. G., A. Sharma, L. Sathiyanarayanan et al. 2010. HPTLC densitometric evaluation of tissue culture extracts of *Nothapodytes foetida* compared to conventional extracts for camptothecin content and antimicrobial activity. *Planta Med.* 76: 474–480.

Nowakowska, J., P. Pikul, and P. Rogulski. 2010. TLC of aclarubicin and doxycycline with mixed n-alcohol mobile phases. *J. Planar Chromatogr.* 23: 353–358.

Nussbaumer, S., P. Bonnabry, J. L. Veuthey et al. 2011. Analysis of anticancer drugs: A review. *Talanta* 85: 2265–2289.

Paci, A., L. Mercier, and P. Bourget. 2003. Identification and quantitation of antineoplastic compounds in chemotherapeutic infusion bags by use of HPTLC: Application to the vinca-alkaloids. *J. Pharm. Biomed. Anal.* 30: 1603–1610.

Papac, R. J. 2001. Origins of cancer therapy. *Yale J. Biol. Med.* 74: 391–398.

Paw, B. and G. Misztal. 1997. Thin-layer chromatographic analysis of antimetabolites of pyrimidine bases in human plasma. *Chem. Anal.* 42: 37–40.

Perello, L., S. Demirdjian, A. Dory et al. 2001. Application of high-performance, thin-layer chromatography to quality control of antimetabolite analogue infusion bags. *J. AOAC Int.* 84: 1296–1300.

Perisic-Janjic, N., T. Djakovic-Sekulic, S. Stojanovic et al. 2004. Evaluation of the lipophilicity of some dehydroepiandrosterone derivates using RP-18 HPTLC chromatography. *Chromatographia* 60: S201–S205.

Perisic-Janjic, N. U., T. Lj. Djakovic-Sekulic, S. Z. Stojanovic et al. 2005. HPTLC chromatography of androstene derivates application of normal phase thin-layer chromatographic retention data in QSAR studies.
Steroids 70: 137–144.

Reddy, M. V. 1993. C_{18} thin-layer chromatographic enhancement of the ^{32}P-postlabeling assay for aromatic or
bulky carcinogen-DNA adducts: Evaluation of adduct recoveries in comparison with nuclease P1 and
butanol methods. *J. Chromatogr.* 614: 245–251.

Saetern, A. M., M. Skar, A. Braaten et al. 2005. Camptothecin-catalyzed phospholipid hydrolysis in liposomes.
Int. J. Pharm. 288: 73–80.

Sherma, J. 2010a. Review of HPTLC in drug analysis: 1996–2009. *J. AOAC Int.* 93: 754–764.

Sherma, J. 2010b. Planar chromatography. *Anal. Chem.* 82: 4895–4910.

Shewach, D. S. and R. D. Kuchta. 2009. Introduction to cancer chemotherapeutics. *Chem. Rev.* 109: 2859–2861.

Shi, Q. W., T. Oritani, and T. Sugiyama. 1999. Three novel bicyclic taxane diterpenoids with verticillene skeleton from the needles of Chinese yew, *Taxus chinensis* var. *mairei. Planta Med.* 65: 356–359.

Sochacka, J. and A. Kowalska. 2006. Comparison of calculated values of the lipophilicity of 2,6-disubstituted
7-methylpurines with values determined by RPTLC. *J. Planar Chromatogr.* 19: 307–312.

Sreekanth, D., A. Syed, S. Sarkar et al. 2009. Production, purification, and characterization of taxol and
10-DABIII from a new endophytic fungus *Gliocladium* sp. Isolated from the Indian yew tree, *Taxus baccata. J. Microbiol. Biotechnol.* 19: 1342–1347.

Srinivasan, V., S. C. Roberts, and M. L. Shuler. 1997. Combined use of six-well polystyrene plates and thin
layer chromatography for rapid development of optimal plant cell culture processes: Application to taxane production by *Taxus* sp. *Plant Cell Rep.* 16: 600–604.

Stasko, M. W., K. M. Witherup, and T. J. Ghiorzi. 1989. Multimodal thin-layer chromatographic separation of
taxol and related compounds from *Taxus-brevifolia. J. Liq. Chromatogr.* 12: 2133–2143.

Subramanian, G. S., A. Karthik, A. Baliga et al. 2009. High-performance thin-layer chromatographic analysis
of bicalutamide in bulk drug and liposomes. *J. Planar Chromatogr.* 22: 273–276.

Tasso, M. J., A. V. Boddy, L. Price et al. 1992. Pharmacokinetics and metabolism of cyclophosphamide in
pediatric-patients. *Cancer Chemother. Pharmacol.* 30: 207–211.

The European Pharmacopoeia. 2008, 6th Ed., Council of Europe, pp. 1920–1922. Strasbourg, France.

The United States Pharmacopeia 29. 2006, 24th Ed., The United States Pharmacopeial Convention, pp. 225.
Rockville, MD.

Thomson, C. E., M. R. Gray, and M. P. Baxter. 1997. The use of capillary electrophoresis as part of a specificity testing strategy for mitoguazone dihydrochloride HPLC methods. *J. Pharm. Biomed. Anal.* 15:
1103–1111.

Vadera, N., G. Subramanian, and P. Musmade. 2007. Stability-indicating HPTLC determination of imatinib
mesylate in bulk drug and pharmaceutical dosage form. *J. Pharm. Biomed. Anal.* 43: 722–726.

Vanhaelen, M., J. Duchateau, and R. Vanhaelen-Fastre. 2002. Taxanes in *Taxus baccata* pollen: Cardiotoxicity
and/or allergenicity? *Planta Med.* 68: 36–40.

Vidensek, N., P. Lim, A. Campbell et al. 1990. Taxol content in bark, wood, root, leaf, twig and seedling from
several *Taxus* species. *J. Nat. Prod.* 53: 1609–1610.

Wang, J. F., G. L. Li, and H. Y. Lu. 2000. Taxol from *Tubercularia* sp strain TF5, an endophytic fungus of *Taxus
mairei. FEMS Microbiol. Lett.* 193: 249–253.

Wianowska, D., M. L. Hajnos, A. L. Dawidowicz et al. 2009. Extraction methods of 10-deacetylbaccatin III,
paclitaxel, and cephalomannine from *Taxus baccata* L. twigs: A comparison. *J. Liq. Chromatogr. Relat.
Technol.* 32: 589–601.

Yule, S. M., A. V. Boddy, M. Cole et al. 1995. Cyclophosphamide metabolism in children. *Cancer Res.* 55:
803–809.

Zakrzewski, R. and W. Ciesielski. 2006. Planar chromatography of heterocyclic thiols with detection by use of
the iodine-azide reaction. *J. Planar Chromatogr.* 19: 4–9.

Zocher, R., W. Weckwerth, C. Hacker et al. 1996. Biosynthesis of taxol: Enzymatic acetylation of 10-deacetyl-
baccatin-III to baccatin-III in crude extracts from roots of *Taxus baccata. Biochem. Biophys. Res.
Commun.* 229: 16–20.

51 Uncertainty Factors in the Enantioseparation of Chiral Drugs on Silica Gel Layers

Mieczysław Sajewicz and Teresa Kowalska

CONTENTS

51.1 MICROCRYSTALLINE CHIRALITY OF NATIVE SILICA GEL

Although silica gel has proved the best-performing thin-layer chromatographic (TLC) stationary phase, its application to direct enantioseparations can occasionally lead to certain striking effects, which will be discussed in this chapter. One source of these striking effects is the microcrystalline structure of silica gel.

Chemically, silica gel used as chromatographic stationary phase is identical with quartz (traditionally known as rock crystal), which is highly valued in optical technology for its ability to polarize (i.e., optically filter) the light beams of undefined or mixed polarization into a beam with well-defined polarization. The polarizing ability of quartz is due to its crystalline chirality, the phenomenon comprehensively discussed, for example, in papers [1,2]. Although it is widely claimed that silica gel employed in chromatography is amorphous, the technology of its precipitation cannot be so rigorous as to exclude randomly appearing microcrystalline (*ergo* chiral) enclaves. This presumption was verified in the experiment described in paper [3], which was carried out with use of the spectroscopy of circular dichroism (CD) on the silica gel samples for TLC, which were binder free and devoid of fluorizing agent (manufactured and generously donated by Merck KGaA, Darmstadt, Germany).

The obtained CD spectrum presented in Figure 51.1 [3] provides an unquestioned experimental proof of microcrystallinity and hence the chirality also with silica gel used for TLC. In the presented spectrum, two well-pronounced cotton bands are visible, one positive and one negative. As it will be shown in Section 51.3, microcrystalline chirality of silica gel exerts a considerable impact on the planar chromatographic enantioseparation results with chiral drugs (although our examples will focus on profen drugs only).

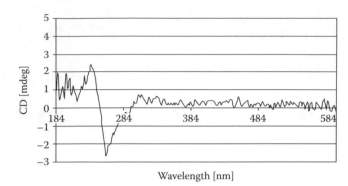

FIGURE 51.1   CD spectrum of binder-free silica gel for TLC, as a Nujol suspension. (From Sajewicz, M. et al., *J. Planar Chromatogr. Mod. TLC*, 19, 278, 2006.)

51.2   CONFIGURATIONAL LABILITY AND SPONTANEOUS OSCILLATORY CHIRAL CONVERSION OF PROFEN DRUGS (AND THEIR STRUCTURAL ANALOGS)

Configuration lability of profen drugs in physiological systems has been recognized ca. four decades ago and since that time, many reports have been published on the differentiated dynamics of chiral conversion, depending on individual profens and the physiological conditions considered. An informative coverage of this subject was provided in a review paper [4].

Symptomatically much later, the analogous phenomenon of the spontaneous chiral conversion was discovered with profen drugs (and with the chiral low molecular weight carboxylic acids in general), when dissolved and then stored in the abiotic aqueous and nonaqueous systems. The first paper on this subject was released in 2005 [5], and an exhaustive coverage of this phenomenon is provided, for example, in publications [6–9]. Right from the beginning [5], it became obvious that the spontaneous chiral conversion of profen drugs and their structural analogs is oscillatory in nature. With all these carboxylic acids dissolved in aqueous solvents, the mechanism of chiral conversion can be summarized by the following scheme [10]:

$$(+)\,\text{enantiomer} \quad \rightleftharpoons \quad \text{enolate ion} \quad \rightleftharpoons \quad (-)\,\text{enantiomer}$$

In anhydrous media and in the presence of trace amounts of water, the probable mechanism of chiral conversion is given as [11]

$$(+)\,\text{enantiomer} \quad \rightleftharpoons \quad \text{enol} \quad \rightleftharpoons \quad (-)\,\text{enantiomer}$$

A spontaneous and, hence, uncontrolled oscillatory chiral conversion of the low molecular weight carboxylic acids is a meaningful obstacle prohibiting an accurate quantification of the separated antimers not only by means of the chiral TLC but also using the chiral high-performance liquid chromatography. Due to the nonlinear nature of chiral conversion, concentrations of the two separated antimers alternately increase and decrease in the function of time, so that the strict quantification result can be viewed as unrepeatable and momentary only.

Another drawback of the attempts on an accurate chromatographic quantification of profen drugs (and the structurally related chiral carboxylic acids) is that they undergo not only nonlinear chiral conversion but at the same time nonlinear oligomerization also, as discussed in papers [8,9,12]. The parallel mechanism of the oscillatory chiral conversion and the oscillatory oligomerization of profen drugs can be illustrated by Scheme 51.1.

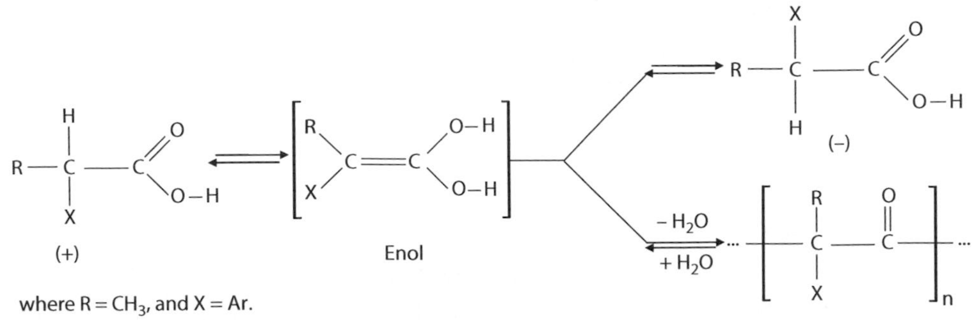

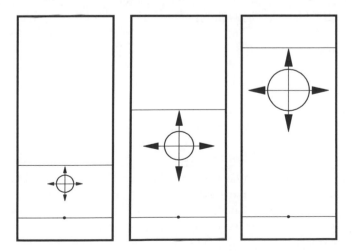

where R = CH₃, and X = Ar.

SCHEME 51.1 The parallel mechanism of the oscillatory chiral conversion and the oscillatory oligomerization of profen drugs. (From Sajewicz, M. et al., *J. Chromatogr. Sci.*, 50, 237, 2012.)

The oscillatory oligomerization of profen drugs is another spontaneous and hence uncontrolled factor, which along with the oscillatory chiral conversion makes quantification results of the enantioseparation lacking both repeatability and precision.

51.3 DRAWBACKS AND ADVANTAGES OF THIN-LAYER CHROMATOGRAPHIC ENANTIOSEPARATION OF CHIRAL DRUGS ON SILICA GEL LAYERS

Effective diffusion in planar chromatographic systems is 2D, occurring both in the direction of mobile phase flow and in the direction perpendicular to it. The 2D effective diffusion in planar chromatography is schematically illustrated by Figure 51.2 [13].

Effective diffusion is generally harmful to the separation process, because it results in band broadening and, therefore, reduction of the system's theoretical plate number and, hence, in overall deterioration of its resolving power. However, 2D effective diffusion can prove helpful in TLC enantioseparation, as its combination with the native microcrystalline chirality of silica gel can result in the 2D (vertical and horizontal) enantioseparations taking place in the 1D chromatographic development mode.

The phenomenon of the 2D enantioseparation in the 1D development mode can be illustrated by the example shown in Figure 51.3 [14].

FIGURE 51.2 Schematic illustration of the 2D effective diffusion in TLC by means of the three consecutive and time-dependent snapshots. (From Sajewicz, M. and Kowalska, T., *Acta Chromatogr.*, 22, 499, 2010.)

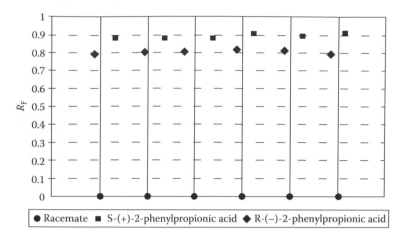

FIGURE 51.3 Schematic representation of the deviation from the vertical of the migration tracks of the enantiomers of 2-phenylpropionic acid on silica gel 60 F254 impregnated with *L*-arginine. The mobile phase was ACN–MeOH–H$_2$O, 5 + 1 + 0.75 (*v/v*). (From Sajewicz, M. et al., *J. Planar Chromatogr. Mod. TLC*, 19, 273, 2006.)

The 2D (vertical and horizontal) enantioseparation in the 1D development mode certainly enhances the resolution power of a given chromatographic system, yet the analyst has to be aware of this effect in advance. Consequently, the densitometric scanning of the respective chromatogram cannot be done with a narrow light beam along a linear migration track only, but instead it should embrace two broad enough margins (up to 10 mm from each side), so that the concentration profiles of the two enantioseparated antimers are never missed.

The two 10 mm scanning margins are important for one more reason. From the parallel and dense enough densitometric scans, one can reconstruct the 3D chromatographic bands and get a comprehensive information about the chromatographic behavior of a chiral analyte in the chromatographic system. As a matter of fact, not only the migration track of an analyte can undergo the deviation from linearity in the silica gel-based TLC systems, but the entire chromatographic band can assume a skewed shape, as it was probably first shown with the enantioseparated antimers of ibuprofen [15]. For the sake of example, the 2D and 3D reconstruction of the flurbiprofen antimers' skewed band shapes is shown in Figure 51.4 [16]. A novel insight in the issue of lateral relocations with chiral low molecular weight carboxylic acids in TLC is proposed in paper [17].

From the experiments on illumination of naproxen and ketoprofen (dissolved in 70% aqueous ethanol) with UV light ($\lambda = 254$ nm) followed by the spontaneous and long-lasting fluorescence effect in visible light, a conclusion was drawn as to relative stability of the enol form with these two compounds and maybe with profens in general [18]. An outcome of this experiment is schematically explained in Scheme 51.2 and it can be interpreted as an excitation of the conjugated aryl–enol π-electron system with UV light (hv_1), followed by the lower energy emission (hv_2) in the visible light range.

Upon the assumption regarding relative stability of the profen-derived enols, the chromatographic separation thereof can be anticipated from the (+) and (−) antimers. As enols lack the chirality center, no deviation of their chromatographic track from linearity can be expected. From the long-term experimental experience in the enantioseparation of profen drugs by means of the chiral TLC, it can be confirmed that the separation of certain profens (and of the structurally related compounds) sporadically yields in the three separate bands instead of two, with one species showing no lateral relocation (e.g., [19]).

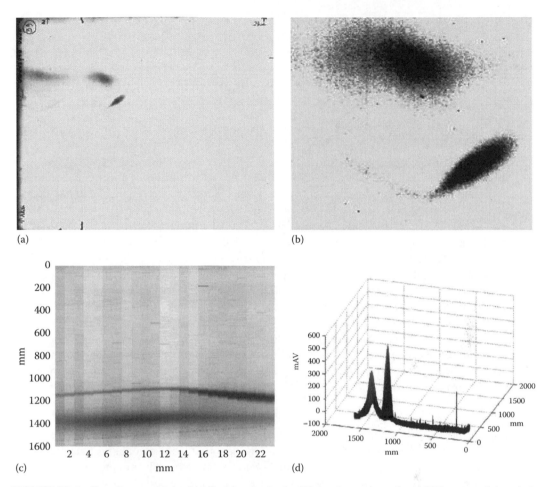

FIGURE 51.4 Enantioseparation of (±)flurbiprofen in the 2D development mode. (a) Videoscan of the whole plate; (b) the enlarged videoscan area showing the two skewed antimer bands; (c) 2D and (d) 3D reconstruction of the enantioseparated flurbiprofen antimer bands from the densitometric scans at the 1 mm intervals of the chromatogram track 30 mm wide. (From Sajewicz, M. et al., *Acta Chromatogr.*, 18, 226, 2007.)

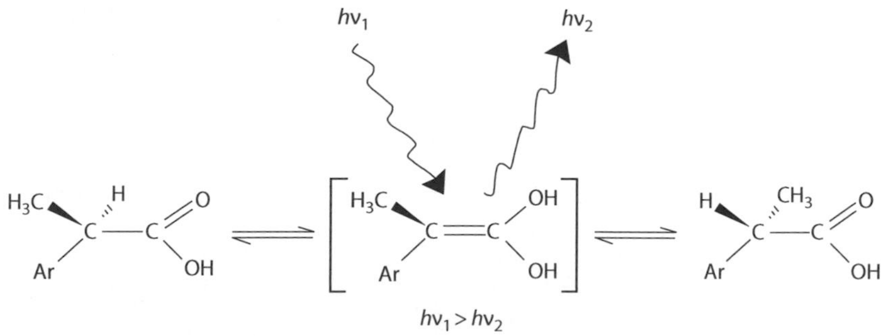

SCHEME 51.2 Schematic explanation of the pertinacious visible light blinking of the profen drug solution in 70% aqueous ethanol, when irradiated with UV light at λ = 254 nm. (From Sajewicz, M. and Kowalska, T., unpublished results.)

51.4 CONCLUSIONS

Based on the discussion given in this chapter, it has to be accepted that the TLC enantioseparation of the chiral low molecular weight carboxylic acid drugs (e.g., profens) on silica gel-based stationary phases is far from a simple and a straightforward task. In fact, successful enantioseparation of such antimer pairs needs an awareness of the following disturbing facts:

- Due to microcrystalline chirality of silica gel, the migration tracks of the separated antimers can be sidewise deviated from linearity for even up to 10 mm. In case of densitometric scanning of the chromatogram with a narrow light beam and along an expected linear migration track only, the analyst can miss his or her target, which is (an often skewed) chromatographic band of the separated chiral species. Therefore, it is highly recommended that the chromatograms be scanned not only along the expected linear migration track but also along the two 10 mm wide side margins, in parallel 1 mm intervals. Only in that way, the enantioseparated species cannot remain undetected or incompletely quantified.
- Due to the configurational lability of the chiral low molecular weight carboxylic acid antimers, an analyst should be aware that quantification of the enantiomer composition is handicapped by an acute lack of repeatability, due to the oscillatory chiral conversion (accompanied by the oscillatory oligomerization), uninterruptedly running in a given batch of the analyzed solution.
- From the results so far obtained, it seems that an additional phenomenon can also contribute to the overall analytical uncertainty. For example, it seems probable that the enolic form of profen drugs can be stable enough to occasionally appear on the chromatograms as the third separated species (which lacks the chirality center and, hence, does not show lateral relocation).

Summing up, the enantioseparation of profen drugs on the silica gel-based stationary phase demands a sufficient amount of experience and ingenuity from the analyst. On the other hand, the advantages of TLC enantioseparation can hardly be overestimated. The main benefits are preservation of the separation result, enabling an additional scrutiny of the separated species. Moreover, the oligomerization products, which cumulate in the stored and ageing profen solutions, are separated in the solid-phase extraction fashion and remain on the start of the chromatogram, in that way not interfering with (or superposing on) the enantioseparation result.

REFERENCES

1. H.D. Flack, Chiral and achiral crystal structures. *Helv. Chim. Acta*, **86**, 905–921 (2003).
2. I.-H. Suh, K.H. Park, W.P. Jensen, and D.E. Lewis, Molecules, crystals, and chirality. *J. Chem. Educ.*, 74, 800 (1997).
3. M. Sajewicz, H.-E. Hauck, G. Drabik, E. Namysło, B. Głód, and T. Kowalska, Tracing possible structural asymmetry of silica gel used for precoating thin-layer chromatographic plates. *J. Planar Chromatogr. Mod. TLC*, **19**, 278–281 (2006).
4. V. Wsol, L. Skalova, and B. Szotakova, Chiral inversion of drugs: Coincidence or principle? *Curr. Drug Metab.*, **5**, 517–533 (2004).
5. M. Sajewicz, R. Piętka, A. Pieniak, and T. Kowalska, Application of thin-layer chromatography (TLC) to investigating oscillatory instability of the selected profen enantiomers. *Acta Chromatogr.*, **15**, 131–149 (2005).
6. M. Sajewicz, R. Wrzalik, M. Gontarska, D. Kronenbach, M. Leda, I.R. Epstein, and T. Kowalska, In vitro chiral conversion, phase separation, and wave propagation in aged profen solutions. *J. Liq. Chromatogr. Relat. Technol.*, **32**, 1359–1372 (2009).
7. M. Sajewicz, M. Matlengiewicz, M. Leda, M. Gontarska, D. Kronenbach, T. Kowalska, and I.R. Epstein. Spontaneous oscillatory in vitro chiral conversion of simple carboxylic acids and its possible mechanism. *J. Phys. Org. Chem.*, **23**, 1066–1073 (2010).

8. M. Sajewicz, M. Gontarska, D. Kronenbach, M. Leda, T. Kowalska, and I.R. Epstein, Condensation oscillations in the peptidization of phenylglycine. *J. Syst. Chem.*, **1**, 7 (2010); DOI:10.1186/1759-2208-1-7.

9. M. Sajewicz, M. Dolnik, D. Kronenbach, M. Gontarska, T. Kowalska, and I.R. Epstein, Oligomerization oscillations of *L*-lactic acid in solutions. *J. Phys. Chem. A*, **115**, 14331–14339 (2011).

10. P. Belanger, J.G. Atkinson, and R.S. Stuart, Exchange reactions of carboxylic acid salts. Kinetics and mechanism. *Chem. J. Chem. Soc. D: Commun.* 1067–1068 (1969).

11. Y. Xie, H. Liu, and J. Chen, Kinetics of base catalyzed racemization of ibuprofen enantiomers. *Int. J. Pharm.*, **196**, 21–26 (2000).

12. M. Sajewicz, M. Gontarska, D. Kronenbach, E. Berry, and T. Kowalska, An HPLC-DAD and LC-MS study of condensation oscillations with *S*(+)-ketoprofen dissolved in acetonitrile. *J. Chromatogr. Sci.*, **50**, 237–244 (2012).

13. M. Sajewicz and T. Kowalska, On the mechanisms of enantiomer separations by chiral thin-layer chromatography on silica gel, and implications when densitometric detection is used. A mini review. *Acta Chromatogr.*, **22**, 499–513 (2010).

14. M. Sajewicz, R. Piętka, G. Drabik, E. Namysło, and T. Kowalska, On the stereochemically peculiar two-dimensional separation of 2-arylpropionic acids by chiral TLC. *J. Planar Chromatogr. Mod. TLC*, **19**, 273–277 (2006).

15. R. Bhushan and V. Parshad, Resolution of (±) ibuprofen using *L*-arginine-impregnated thin layer chromatography. *J. Chromatogr. A*, **721**, 369–372 (1996).

16. M. Sajewicz, M. Gontarska, D. Kronenbach, Ł. Wojtal, G. Grygierczyk, and T. Kowalska, Study of the oscillatory *in-vitro* transenantiomerization of the antimers of flurbiprofen and their enantioseparation by thin-layer chromatography (TLC). *Acta Chromatogr.*, **18**, 226–237 (2007).

17. J. Polanski, M. Sajewicz, M. Knas, M. Gontarska, and T. Kowalska, Lateral relocation in thin-layer chromatography. *J. Planar Chromatogr. Mod. TLC*, **25**, 208–213 (2012).

18. M. Sajewicz, M. Gontarska, and T. Kowalska, HPLC/DAD evidence of the oscillatory chiral conversion of phenylglycine. *J. Chromatogr. Sci.*, DOI:10.1093/chromsci/bmt033.

19. M. Sajewicz, M. Gontarska, M. Wróbel, and T. Kowalska, Enantioseparation and oscillatory transenantiomerization of S,R-(±)-ketoprofen, as investigated by means of thin layer chromatography with densitometric detection. *J. Liq. Chromatogr. Relat. Technol.*, **30**, 2193–2208 (2007).

Index

A

Absorption, distribution, metabolism and
extraction (ADME)
acetylation of acenocoumarol, 556
dermatan sulfate, 556
metabolism, warfarin, 552, 555
phase I reactions, phenprocoumon, 552, 555
phenprocoumon metabolism, 552
ticlopidine, rats, 556
TLC study, 552–554
ACE inhibitors, *see* Angiotensin converting enzyme
(ACE) inhibitors
Acetazolamide
HPTLC method, tablets and, 727
TLC method, thiazide diuretics and antihypertensive
agents, 725–727
Acethylcysteine
chemical name, 592
detection, 592
stationary and mobile phase, 592
Acetonitrile (ACN), 997
Acetylcholinesterase (AChE) inhibitors
donepezil, 450–451
galantamine, 451
rivastigmine, 451
tacrine, 451–452
ACN, *see* Acetonitrile (ACN)
Active pharmaceutical ingredient (API)
degradation products, 248
and HPLC, 247
TLC testing, 250
ADME, *see* Absorption, distribution, metabolism
and extraction (ADME)
Adsorption theory, *see* Polar adsorbents
AEDs, *see* Antiepileptic drugs (AEDs)
Aldosterone (ALD), 652, 670, 679, 681
Aldosterone antagonists, 717–718, 733–734
Allylamines and squalene epoxidase inhibitors
structures, 901
terbinafine hydrochloride, 902
tolnaftate, 902
Alzheimer's disease (AD)
AChE inhibitors, *see* Acetylcholinesterase (AChE)
inhibitors
chromatographic systems
compounds, 452, 454
dot-blot test, 452, 453
"pseudo-reversed systems", 453
TLC-AChE inhibitory assay, 455
TLC silica gel plate, 452, 453
description, 449
detection, 455–456
NMDA antagonist, 452
pharmacological treatment, 450
quantitative analysis, 456

Amantadine and oseltamivir, *see* Antiviral drugs, TLC
Ambroxol
chemical name, 592–593
detection, 593, 594
stationary and mobile phase, 593, 594
Amebiasis, giardiasis, and trichomoniasis, 911–914;
see also Nitroimidazoles
Amiloride, 731
Aminoglycoside antibiotics
in biological material
Clean Screen column, 960
hygromycin B, purification, 959–960
KM 3 chromatogram spectrophotometer, 959
netilmicin, 959
reagent concentration, 960
RP LKC$_{18}$F, 959
screen columns, 960
trichloroacetic acid, 959
nosocomial infections, 957–958
in pharmaceutical formulations
neomycin sulfate powder, assay, 959
quantitative determination, 959
separation and determination
antibiotic acetylation, 958
BA2930 protein, structural analysis, 958
cellulose usage, 958
Chromosorb W chromatographic column
procedures, 958
gentamicin antibiotic complex, 958
iodine vapors, 958
postchromatographic derivatization, 958
TLC-densitometric method, 958
Ammonium compounds
dequalinium chloride, 816
determination, 816–817
Amphetamine and derivatives
CAR, 424
chemical structure, 420, 421
extraction methods, 426
HPLC and TLC methods, 422
identification, toxic substances, 422
MDD, 424
MDEA, 424
3,4-methylenedioxymethamphetamine
(MDMA), 423
p-hydroxymethamphetamine (pOHMA), 423
screening procedure, 427
TOXI-LAB drug detection system, 425
Amphotericin B, *see* Polyene membrane disruptors
Anabolics
chromatographic parameters, determination, 650, 651
effects, 648
HPTLC, 649
MET, OXY, STA, metenolone, 646
NAN, 646, 648
NOR, 648

Printed and bound by CPI Group (UK) Ltd, Croydon, CR0 4YY

24/10/2024

01778310-0004